Earth, Moon, and Sun

Earth	
Mass	$M_E = 5.98 \times 10^{24}$ kg
Equatorial radius	$R_E = 6.378 \times 10^6$ m
Polar radius	$R'_E = 6.357 \times 10^6$ m
Mean density	5520 kg/m^3
Surface gravity	$g = 9.81$ m/s^2 = 32.2 ft/s^2
Period of rotation	1 sidereal day = 23 h 56 min 4 s = 8.616×10^4 s
Moment of inertia:	
about polar axis	$I = 0.331 M_E R_E^2$
about equatorial axis	$I' = 0.329 M_E R_E^2$
Mean distance from Sun	1.50×10^{11} m
Period of revolution (period of orbit)	1 year = 365 days 6 h = 3.16×10^7 s
Orbital speed	29.8 km/s
Moon	
Mass	7.35×10^{22} kg
Radius	1.74×10^6 m
Mean density	3340 kg/m^3
Surface gravity	1.62 m/s^2
Period of rotation	27.3 days
Mean distance from Earth	3.84×10^8 m
Period of revolution	1 sidereal month = 27.3 days
Sun	
Mass	$M_S = 1.99 \times 10^{30}$ kg
Radius	6.96×10^8 m
Mean density	1410 kg/m^3
Surface gravity	274 m/s^2
Period of rotation	~26 days
Luminosity	3.9×10^{26} W

Greek Alphabet

Α	α	alpha	Ν	ν	nu
Β	β	beta	Ξ	ξ	xi
Γ	γ	gamma	Ο	o	omicron
Δ	δ	delta	Π	π	pi
Ε	ε	epsilon	Ρ	ρ	rho
Ζ	ζ	zeta	Σ	σ	sigma
Η	η	eta	Τ	τ	tau
Θ	θ	theta	Υ	υ	upsilon
Ι	ι	iota	Φ	ϕ	phi
Κ	κ	kappa	Χ	χ	chi
Λ	λ	lambda	Ψ	ψ	psi
Μ	μ	mu	Ω	ω	omega

Physics

Hans C. Ohanian

UNION COLLEGE AND

RENSSELAER POLYTECHNIC INSTITUTE

W · W · NORTON & COMPANY

NEW YORK · LONDON

Published simultaneously in Canada by Penguin Books Canada Ltd.,
2801 John Street, Markham, Ontario L3R 1B4.
Printed in the United States of America.
First Edition

Book design by Antonina Krass
Makeup by Roberta Flechner
Production editor Frederick E. Bidgood
Picture research by Amy Boesky and Natalie Goldstein

Library of Congress Cataloging in Publication Data

Ohanian, Hans C.
Physics.

Includes index.
1. Physics. I. Title.
QC21.2.037 1985 530 84-25540
ISBN 0-393-95401-3

W. W. Norton & Company, Inc., 500 Fifth Avenue, New York, N.Y. 10110
W. W. Norton & Company Ltd., 37 Great Russell Street, London WC1B 3NU

1 2 3 4 5 6 7 8 9 0

To Susan Farnsworth Ohanian, writer,
who gently tried to teach me some of her craft.

Contents

PRELUDE: THE WORLD OF PHYSICS P1

Part I: The Large-Scale World P1
Part II: The Small-Scale World P11

1 | Measurement of Space, Time, and Mass 1

1.1 Space and Time 2
1.2 Frames of Reference 5
1.3 The Unit of Length 6
1.4 The Unit of Time 10
1.5 The Unit of Mass 12
1.6 Derived Units 14
Summary 19
Questions 19
Problems 20

2 | Kinematics in One Dimension 23

2.1 Average Speed 24
2.2 Average Velocity in One Dimension 25
2.3 Instantaneous Velocity 26
2.4 Acceleration 29
2.5 Motion with Constant Acceleration 31
2.6 The Acceleration of Gravity 33
Summary 37
Questions 38
Advice on Solving Problems 39
Problems 40

3 | Vectors 46

3.1 The Displacement Vector and the General Definition of a Vector 46
3.2 Vector Addition and Subtraction 48
3.3 The Position Vector; Components of Vectors 52
3.4 Vector Multiplication 55
Dot Product 56
Cross Product 57
3.5* Vectors and Coordinate Rotations 60
Summary 61
Questions 62
Problems 63

INTERLUDE A. THE ARCHITECTURE OF CRYSTALS

A.1 The Structure of Crystals A–1
A.2 Symmetry A–5
A.3 Close-Packed Structures A–8
A.4 Defects in Crystals A–12
Further Reading A–13
Questions A–14

4 | Kinematics in Three Dimensions 67

4.1 The Velocity and Acceleration Vectors 67
4.2 Motion with Constant Acceleration 70
4.3 The Motion of Projectiles 71
4.4 Uniform Circular Motion 76
4.5 The Relativity of Motion and the Galilean Transformations 79
Summary 82
Questions 83
Problems 84

5 | Dynamics — Newton's Laws 91

5.1 Newton's First Law 91
5.2 Newton's Second Law 94
5.3 The Superposition of Forces 99
5.4 Newton's Third Law 101
5.5 The Momentum of a Particle 104
5.6 The Angular Momentum of a Particle 106
5.7 Newtonian Relativity 109
Summary 111
Questions 111
Problems 114

* This section is optional.

6 | Dynamics — Forces and the Solution of the Equation of Motion 119

6.1 The Four Fundamental Forces 120
6.2 Weight 122
6.3 Motion with a Constant Force 125
6.4 Sliding Friction 129
6.5 Elastic Force of a Spring; Hooke's Law 135
6.6* Motion with a Variable Force 137
6.7 Dynamics of Uniform Circular Motion 141
Summary 144
Questions 145
Problems 147

7 | Work and Energy 153

7.1 Work in One Dimension 154
7.2 Work in Three Dimensions 158
7.3 Kinetic Energy 161
7.4 Gravitational Potential Energy 164
Summary 167
Questions 167
Problems 168

8 | Conservation of Energy 174

8.1 Conservative Forces 174
8.2 Potential Energy of a Conservative Force 176
8.3 Calculation of the Force from the Potential Energy 180
8.4 The Curve of Potential Energy 182
8.5 Other Forms of Energy 184
8.6 Mass and Energy 185
8.7 Power 187
Summary 190
Questions 191
Problems 191

9 | Systems of Particles 198

9.1 Momentum of a System of Particles 198
9.2 Center of Mass 202
9.3 The Motion of the Center of Mass 207
9.4 Energy of a System of Particles 209
9.5 Angular Momentum of a System of Particles 212
9.6* The Motion of a Rocket 213
Summary 215
Questions 216
Problems 217

* This section is optional.

10 | Collisions 223

10.1 Impulsive Forces 223
10.2 Collisions in One Dimension 227
10.3 Collisions in Two Dimensions 232
10.4* Collisions and Reactions of Nuclei and of Elementary Particles 236
Summary 239
Questions 239
Problems 240

INTERLUDE B. RADIATION AND LIFE

B.1 Penetrating Radiations B–1
B.2 Radioactive Decay B–3
B.3 Radiation Damage in Atoms B–5
B.4 Radiation Damage in Molecules and in Living Cells B–7
B.5 The Physiological Effects of Radiation B–8
B.6 Medical Applications of Radioisotopes B–10
Further Reading B–13
Questions B–13

11 | Kinematics of a Rigid Body 246

11.1 Motion of a Rigid Body 246
11.2 Rotation About a Fixed Axis 248
11.3 Motion with Constant Angular Acceleration 251
11.4 Kinetic Energy of Rotation; Moment of Inertia 253
11.5 Angular Momentum of a Rigid Body 259
Summary 263
Questions 264
Problems 265

12 | Dynamics and Statics of a Rigid Body 270

12.1 Torque 270
12.2 The Equation of Rotational Motion 272
12.3 Work, Energy, and Power in Rotational Motion 275
12.4 The Conservation of Angular Momentum 278
12.5 Rolling Motion 282
12.6 Precession of a Gyroscope 286
12.7 Statics of Rigid Bodies 288
12.8 Examples of Static Equilibrium 289
Summary 292
Questions 293
Problems 294

* This section is optional.

INTERLUDE C. ELEMENTARY PARTICLES

C.1 Exploring the Atom C–1
C.2 The Tools of High-Energy Physics C–5
C.3 The Multitude of Particles C–9
C.4 Conservation Laws C–11
C.5 Family Relationships C–12
C.6 Quarks C–13
C.7 Color and Charm C–15
Further Reading C–15
Questions C–16

13 | Gravitation 304

13.1 Newton's Law of Universal Gravitation 305
13.2 The Measurement of *G* 307
13.3 Circular Orbits 308
13.4 Elliptical Orbits; Kepler's Laws 311
13.5 Gravitational Potential Energy 316
13.6 The Gravitational Action of a Spherical Mass 322
13.7 Inertial and Gravitational Mass; the Principle of Equivalence 325
Summary 328
Questions 328
Problems 330

INTERLUDE D. THE BIG BANG AND THE EXPANSION OF THE UNIVERSE

D.1 The Expansion of the Universe D–1
D.2 The Age of the Universe and the Big Bang D–5
D.3 The Future of the Universe D–6
D.4 The Search for Numbers D–8
Further Reading D–9
Questions D–9

14 | Oscillations 337

14.1 Simple Harmonic Motion 337
14.2 The Simple Harmonic Oscillator 341
14.3 Kinetic Energy and Potential Energy 344
14.4 The Simple Pendulum 346
14.5 Other Oscillating Systems 350
The Physical Pendulum 351
The Torsional Pendulum 353
The Diatomic Molecule 354
14.6 Damped Oscillations and Forced Oscillations 356
Summary 359
Questions 359
Problems 361

15 | Waves 368

15.1 Wave Pulses 368
15.2 Traveling Waves 369
15.3 Speed of Waves on a String 372
15.4 Energy in a Wave; Power 374
15.5 The Superposition of Waves 377
15.6 Standing Waves 381
Summary 384
Questions 384
Problems 385

16 | Sound and Other Wave Phenomena 390

16.1 Sound Waves in Air 391
16.2 The Speed of Sound 395
16.3 The Doppler Effect 397
16.4 Water Waves 401
16.5 Seismic Waves 404
16.6 Diffraction 406
Summary 407
Questions 407
Problems 409

17 | The Theory of Special Relativity 416

17.1 The Speed of Light; the Ether 417
17.2 Einstein's Principle of Relativity 419
17.3 The Lorentz Transformations 423
17.4 The Time Dilation 429
17.5 The Length Contraction 431
17.6 The Combination of Velocities 433
17.7 Momentum and Energy 434
Summary 438
Questions 439
Problems 440

INTERLUDE E. GRAVITY AND GEOMETRY

E.1 The Principle of Equivalence E–1
E.2 The Deflection of Light and the Curvature of Space E–3
E.3 The Theory of General Relativity E–6
E.4 The Gravitational Time Dilation E–8
E.5 Black Holes E–10
Further Reading E–13
Questions E–14

18 | Fluid Mechanics 445

18.1 Density and Flow Velocity 446
18.2 Steady Incompressible Flow; Streamlines 447
18.3 Pressure 451
18.4 Pressure in a Static Fluid 452
18.5 Archimedes' Principle 456
18.6 Fluid Dynamics; Bernoulli's Equation 457
Summary 461
Questions 461
Problems 464

19 | Kinetic Theory of Gases 469

19.1 The Ideal-Gas Law 470
19.2 The Temperature Scale 473
19.3 Kinetic Pressure 475
19.4 The Energy of an Ideal Gas 477
Summary 479
Questions 480
Problems 480

20 | Heat 484

20.1 Heat as a Form of Energy 484
20.2 Thermal Expansion of Solids and Liquids 487
20.3 Conduction of Heat 490
20.4 Changes of State 492
20.5 The Specific Heat of a Gas 493
20.6 The Adiabatic Equation 496
Summary 499
Questions 499
Problems 501

21 | Thermodynamics 508

21.1 The First Law of Thermodynamics 509
21.2 The Carnot Engine 511
21.3 Entropy 516
21.4 The Second Law of Thermodynamics 520
Summary 525
Questions 525
Problems 527

INTERLUDE F. ENERGY, ENTROPY, AND ENVIRONMENT

F.1 The Conservation of Energy F–1
F.2 The Dissipation of Energy F–3
F.3 The Limits of Our Resources F–4
F.4 Alternative Energy Sources F–4
F.5 Pollution and the Environment F–10
F.6 Accidents and Safety F–11
Further Reading F–15
Questions F–15

22 | Electric Force and Electric Charge 531

22.1 The Electrostatic Force 532
22.2 Coulomb's Law 533
22.3 Charge Quantization and Charge Conservation 537
22.4 Conductors and Insulators; Frictional Electricity 539
Summary 542
Questions 542
Problems 543

23 | The Electric Field 546

23.1 The Superposition of Electric Forces 546
23.2 The Electric Field 549
23.3 Lines of Electric Field 553
23.4 Electric Dipole in an Electric Field 556
Summary 559
Questions 559
Problems 560

INTERLUDE G. FORCES, FIELDS AND QUANTA

G.1 Reaction Rates G-1
G.2 Conservation Laws and Symmetry G-4
G.3 Fields and Quanta G-4
G.4 The Unified Theory of Weak and Electromagnetic Forces G-7
G.5 Gluons and Chromodynamics G-8
Further Reading G-11
Questions G-12

24 | Gauss' Law 565

24.1 Electric Flux and the Number of Field Lines 565
24.2 Gauss' Law 567
24.3 Some Examples 569
24.4 Conductors in an Electric Field 572
Summary 575
Questions 575
Problems 576

25 | The Electrostatic Potential 580

25.1 The Electrostatic Potential 580
25.2 The Electrostatic Potential of a Point Charge 583
25.3 The Electrostatic Field as a Conservative Field 586
25.4 The Gradient of the Potential 586
25.5 The Potential and Field of a Dipole 590
25.6* The Mean-Value Theorem 591
Summary 595
Questions 595
Problems 596

26 | Electric Energy 601

26.1 Energy of a System of Point Charges 601
26.2 Energy of a System of Conductors 603
26.3 The Energy Density 604
Summary 607
Questions 608
Problems 608

INTERLUDE H. NUCLEAR FISSION

H.1 The Nucleus H-1
H.2 Fission H-4
H.3 Chain Reactions H-5
H.4 The Bomb H-7
H.5 The Effects of Nuclear Weapons H-9
H.6 Nuclear Reactors H-12
Further Reading H-15
Questions H-16

27 | Capacitors and Dielectrics 612

27.1 Capacitance 612
27.2 Capacitors in Combination 615
27.3 Dielectrics 617
27.4 Gauss' Law in Dielectrics 622
27.5 Energy in Capacitors 624
Summary 625
Questions 626
Problems 627

* This section is optional.

28 | Currents and Ohm's Law 631

28.1 Electric Current 631
28.2 Resistance and Ohm's Law 633
28.3* The Flow of Free Electrons 635
28.4 The Resistivity of Materials 637
28.5 Semiconductors 640
28.6 Resistances in Combination 641
Summary 643
Questions 643
Problems 644

29 | DC Circuits 649

29.1 Electromotive Force 649
29.2 Sources of Electromotive Force 651
Batteries 651
Electric Generators 652
Fuel Cells 652
Solar Cells 653
29.3 Single-Loop Circuits 654
29.4 Multiloop Circuits 656
29.5 Energy in Circuits; Joule Heat 658
29.6 The Hazards of Electric Currents 661
Summary 662
Questions 662
Problems 664

INTERLUDE I. ATMOSPHERIC ELECTRICITY

I.1 The Fair-Weather Electric Field I-1
I.2 Thunderstorms I-3
I.3 Generation of Electric Charge I-5
I.4 The Electric Field of a Thundercloud I-6
I.5 Lightning I-8
Further Reading I-11
Questions I-11

30 | The Magnetic Force and Field 668

30.1 The Magnetic Force 668
30.2 The Magnetic Field 672
30.3 The Biot–Savart Law 675
30.4* Relativity and the Magnetic Field 681
Summary 685
Questions 685
Problems 686

* This section is optional.

31 | Ampère's Law 692

31.1 Ampère's Law 692
31.2 Solenoids 694
31.3 Motion of Charges in Electric and Magnetic Fields 698
31.4 Force on a Wire 704
31.5 Torque on a Current Loop 705
Summary 707
Questions 708
Problems 709

32 | Electromagnetic Induction 714

32.1 Motional Electromotive Force 714
32.2 Faraday's Law 717
32.3 Some Examples; Lenz's Law 719
32.4 The Induced Electric Field 723
32.5 Inductance 726
32.6 Magnetic Energy 728
Summary 730
Questions 731
Problems 732

INTERLUDE J. PLASMA

J.1 The Fourth State of Matter J-1
J.2 Plasmas and the Magnetic Field J-3
J.3 Waves in Plasma J-6
J.4 Thermonuclear Fusion J-8
J.5 Plasma Confinement J-9
Further Reading J-12
Questions J-12

33 | Magnetic Materials 738

33.1 Atomic and Nuclear Magnetic Moments 739
33.2 Paramagnetism 741
33.3 Ferromagnetism 743
33.4 Diamagnetism 746
Summary 749
Questions 749
Problems 750

34 | AC Circuits 754

34.1 Simple AC Circuits with an External Electromotive Force 754
34.2 The Freely Oscillating LC Circuit 759
34.3 The LCR Circuit with an External Electromotive Force 763
34.4 The Transformer 770
Summary 771
Questions 772
Problems 773

INTERLUDE K. SUPERCONDUCTIVITY

K.1 Zero Resistance K-1
K.2 The Critical Magnetic Field K-2
K.3 The Meissner Effect K-3
K.4 Superconductors of the Second Kind K-4
K.5 The BCS Theory K-5
K.6 Technological Applications K-7
Further Reading K-11
Questions K-11

35 | The Displacement Current and Maxwell's Equations 778

35.1 The Displacement Current 778
35.2 Maxwell's Equations 782
35.3 Cavity Oscillations 783
35.4 The Electric Field of an Accelerated Charge 788
35.5* The Magnetic Field of an Accelerated Charge 792
Summary 795
Questions 795
Problems 797

36 | Light and Radio Waves 801

36.1 The Plane Wave 802
36.2 Plane Harmonic Waves; Polarization 806
36.3 The Generation of Electromagnetic Waves 810
36.4 Energy of a Wave 812
36.5 Momentum of a Wave 815
36.6 The Doppler Shift of Light 817
Summary 819
Questions 820
Problems 822

* This section is optional.

37 | Reflection and Refraction 830

37.1 Huygens' Construction 830
37.2 Reflection 831
37.3 Refraction 834
37.4 Spherical Mirrors 840
37.5 Thin Lenses 844
37.6 Optical Instruments 846
Summary 851
Questions 852
Problems 855

38 | Interference 861

38.1 The Standing Electromagnetic Wave 862
38.2 Thin Films 864
38.3 The Michelson Interferometer 866
38.4 Interference from Two Slits 867
38.5 Interference from Multiple Slits 872
Summary 876
Questions 876
Problems 878

39 | Diffraction 883

39.1 Diffraction by a Single Slit 883
39.2 Diffraction by a Circular Aperture; Rayleigh's Criterion 888
39.3* Babinet's Principle 892
Summary 894
Questions 895
Problems 896

40 | Quanta of Light 900

40.1 Blackbody Radiation 900
40.2 Energy Quanta 902
40.3 Photons and the Photoelectric Effect 907
40.4 The Compton Effect 910
40.5 Wave vs. Particle 912
Summary 915
Questions 915
Problems 916

* This section is optional.

41 | Atomic Structure and Spectral Lines 920

41.1 Spectral Lines 921
41.2 The Balmer Series and other Spectral Series 923
41.3 The Nuclear Atom 925
41.4 Bohr's Theory 928
41.5 The Correspondence Principle 933
41.6 Quantum Mechanics 934
Summary 935
Questions 936
Problems 937

INTERLUDE L. LASER LIGHT

L.1 Stimulated Emission L-1
L.2 Lasers L-2
L.3 Some Applications L-5
L.4 Holography L-9
Further Reading L-13
Questions L-13

Appendix 1: Index to Tables
Appendix 2: Mathematical Symbols and Formulas
Appendix 3: Perimeters, Areas, and Volumes
Appendix 4: Brief Review of Trigonometry
Appendix 5: Brief Review of Calculus
Appendix 6: The International System of Units (SI)
Appendix 7: Conversion Factors
Appendix 8: Best Values of Fundamental Constants
Appendix 9: The Chemical Elements and the Periodic Table
Appendix 10: Formula Sheets
Appendix 11: Answers to Even-Numbered Problems
Photo Credits
Index

Preface

This is a textbook for a two- or three-semester calculus-based physics course for science and engineering students. My main objectives in writing this book were to present a contemporary, modern view of classical mechanics and electromagnetism, and to offer the student a glimpse of what is going on in physics today. Thus, throughout the book, I encourage students to keep in mind the atomic structure of matter and to think of the material world as a multitude of restless electrons, protons, and neutrons. For instance, in the mechanics chapters, I emphasize that all macroscopic bodies are systems of particles; and in the electricity chapters, I introduce the concepts of positive and negative charge by referring to protons and electrons, not by referring to the antiquated procedure of rubbing glass rods with silk rags (which, according to experts on triboelectricity, can give the wrong sign if the silk has been thoroughly cleaned). I try to make sure that students are always aware of the limitations of the nineteenth-century fiction that matter and electric charge are continua. Blind reliance on this fiction has often been justified by the claim that engineering students need physics as a tool, and that the atomic structure of matter is of little concern to them. But if physics is a tool, it is also a work of art, and its style cannot be dissociated from its function. In this book I give a physicist's view of physics, because I believe that it is fitting that all students should gain some appreciation for the artistic style of the toolmaker.

The book contains two kinds of chapters: *core* chapters and *interlude* chapters. The 41 core chapters cover the essential topics of introductory physics: mechanics of particles, rigid bodies, and fluids; oscillations; wave motion; heat and thermodynamics; electricity and magnetism; optics; and quanta. They also include a chapter on special relativity, but this chapter is optional, as are a few sections of some other chapters (these are clearly marked in the table of contents and in the text). The 12 interludes present some of the fascinating discoveries and applications of physics today: crystal structure and symmetry, ionizing radiation, elementary particles, the expansion of the universe, general relativity, energy resources, fields and quanta, fission, atmospheric electricity, plasmas, superconductivity, and lasers. All of these interludes are optional — they rely on the core chapters, but the core chapters do not rely on them.

In order to accommodate students who are taking an introductory calculus course concurrently, derivatives are used slowly and hesitantly at first (Chapter 2), and routinely later on. Likewise, the use of integrals is postponed as far as possible (Chapter 7), and they come into heavy use only in the second volume (after Chapter 21). For students who need a review of calculus, Appendix 5 contains a concise primer on derivatives and integrals.

The organization of the core chapters is fairly traditional, with some innovations. I deviate from tradition by an early introduction of angular momentum, starting with the angular momentum of a particle instead of the angular momentum of a rigid body (Chapter 5). In my experience, for many students the first chapters are an expanded review of high-school physics, and students can profit from some new concepts at this stage (teachers who prefer to introduce angular momentum in conjunction with rigid bodies may postpone Section 5.6 until later). More notably, I start the study of magnetism (Chapter 30) with the law for the magnetic force between two moving point charges; this magnetic force is no more complicated than the force between two current elements, and the crucial advantage is that magnetism can be developed from the magnetic-force law in much the same way as electricity is developed from Coulomb's Law. This approach is consistent with the underlying philosophy of the book: particles are primary entities and should always be treated first, whereas macroscopic bodies and currents are composite entities which should be treated later. As another innovation, I include a simple derivation of the electric radiation field of an accelerated charge (Chapter 35); this calculation relies on Richtmeyer and Kennard's clever analysis of the kinks in the electric field lines of an accelerated charge (a set of computer-generated film loops available from the Educational Development Center shows how such kinks propagate along the field lines; these film loops tie in very well with the calculations of Chapter 35).

The chapters include generous collections of solved examples (about 275 altogether) and of problems (about 1700 altogether). Answers to the even-numbered problems are given in Appendix 11. The problems are grouped by sections, with the most difficult problems at the end of each section (exceptionally challenging problems are marked with an asterisk). I have tried to make the problems interesting to the student by drawing on realistic examples from technology, sports, and everyday life. Many of the problems are based on data extracted from engineering handbooks, car-repair manuals, *Jane's Book of Aircraft, The Guinness Book of World Records,* newspaper reports, etc. Many other problems deal with atoms and subatomic particles; these are intended to reinforce the atomistic view of the material world. In some cases, cognoscenti will perhaps consider the use of classical physics somewhat objectionable in a problem that really ought to be handled by quantum mechanics. But I believe that the advantages of familiarization with atomic quantities and magnitudes outweigh the disadvantages of a naïve use of classical mechanics.

Each chapter also includes a collection of qualitative questions intended to stimulate thought and to test the grasp of basic concepts (some of these questions are discussion questions that do not have a unique answer). Moreover, each chapter contains a brief summary of the main physical quantities and laws introduced in it. The virtue of these summaries lies in their brevity. They include essential definitions and equations, but no restatements of arguments or additional explanations, because the statements in the body of each chapter are adequate.

The inspiration for the 12 interlude chapters grew out of my unhappiness over a paradox afflicting the typical undergraduate physics curriculum: liberal-arts students in a nonmathematical physics course often get to see more of the beauty and excitement of the physics of today than science and engineering students in a calculus-based physics course. While liberal-arts students get a glimpse of quarks, black holes, or the Big Bang, science and engineering students are expected to calculate the motion of blocks on top of other blocks sliding down an inclined plane, or the motion of a baseball thrown in some direction or another by a man (or woman) riding in an elevator. To some extent this is unavoidable — science and engineering students need to learn and practice classical mechanics and electromagnetism, and they have little time left for dabbling in the arcane mysteries of contemporary physics. Nevertheless, most teachers will occasionally find an hour or two to tell their students a little of what is going on in physics today. I wrote the interludes to lend encouragement and support to such excursions to the frontiers of physics.

My choices of topics for the 12 interludes reflect the interests expressed by my students. Over the years, I have often been asked: When will we get to black holes? or Are you going to tell us about quarks? and I came to feel that such curiosity must not be allowed to whither away. Obviously, in the typical introductory course it will be impossible to cover all of the interludes (I have usually covered two per term), but the broad range of topics will permit teachers to select according to their own tastes. The interludes are mainly descriptive rather than analytic. In them, I try to avoid formulas and instead give the students a qualitative feeling for the underlying physics, keeping the discussion simple so that students can read them on their own. Thus, the interludes could be used for supplementary reading, not necessarily accompanied by lectures. For the inquisitive student, each interlude includes a collection of qualitative questions and an extensive annotated list of further readings.

The predominant system of units used in this text is SI. However, since American engineers continue to work with the British system, the text also includes examples and problems with these units. In the abbreviations for the units, I follow the dictates of the Conférence Générale des Poids et Mesures of 1971, although I deplore the majestic stupidity of the decision to replace the old, self-explanatory abbreviations amp, coul, nt, sec, °K by an alphabet soup of cryptic symbols A, C, N, s, K, etc. For the sake of clarity, I spell out the names of units in full whenever the abbreviations are likely to lead to ambiguity and confusion.

An excellent study guide for this book has been written by Professors Van E. Neie (Purdue University) and Peter J. Riley (University of Texas, Austin). This guide includes for every chapter a brief introduction laying out the objectives; a list of key terms for review; detailed commentaries on each of the main ideas; and a large collection of interesting sample problems, which alternate between worked problems (with full solutions) and guided problems (which provide step-by-step schemes that lead students to the solutions).

The book has been seven years in the making. The original manuscript went through several revisions, based both on my own experience with students at Union College and on extensive reviews and class testing at several other institutions. A preliminary edition of this book, reproduced photographically from typescript, was used in classes by Professors John R. Boccio (Swarthmore College), A. Douglas Davis (Eastern Illinois University), J. David Gavenda, (University of Texas, Austin), Frank Moscatelli (Swarthmore College), Harvey S. Picker (Trinity College), Peter J. Riley (University of Texas, Austin), Kenneth L. Schick (Union College), and Mark P. Silverman (Trinity College). The experience gained in these class tests permitted me to make many improvements. I am grateful to both teachers and students for sharing their reactions to the book with me.

I have greatly benefited from many comments by reviewers who carefully read my manuscript and suggested corrections and alterations. I am indebted to Professors John R. Boccio (Swarthmore College), Roger W. Clapp, Jr. (University of South Florida), A. Douglas Davis (Eastern Illinois University), Anthony P. French (Massachusetts Institute of Technology), J. David Gavenda (University of Texas, Austin), Roger D. Kirby (University of Nebraska), Roland M. Lichtenstein (Rensselaer Polytechnic Institute), Richard T. Mara (Gettysburg College), John T. Marshall (Louisiana State University), Harvey S. Picker (Trinity College), and Peter J. Riley (University of Texas, Austin) for very detailed, comprehensive reviews; and I am indebted to Professors John R. Albright (Florida State University), Frank A. Ferrone (Drexel University), James R. Gaines (Ohio State University), Michael A. Guillen (Harvard University), Walter Knight (University of California, Berkeley), Jean P. Krisch (University of Michigan), Hermann Nann (Indiana University), Norman Pearlman (Purdue University), P. Bruce Pipes (Dartmouth College), Jack Prince (Bronx Community College), Gerald A. Smith (Michigan State University), Julia A. Thompson and David Kraus (University of Pittsburgh), and Gary A. Williams (University of California, Los Angeles) for briefer reviews. I am also indebted to the experts who reviewed the interludes: Professors Edmond Brown (Rensselaer Polytechnic Institute; "The Architecture of Crystals"), Priscilla W. Laws (Dickinson College; "Radiation and Life"), Stephen Gasiorowicz (University of Minnesota; "Elementary Particles"), Alan H. Guth (Massachusetts Institute of Technology; "The Big Bang and the Expansion of the Universe"), Donald F. Kirwan (University of Rhode Island; "Energy, Entropy, and the Environment"), Malvin Ruderman (Columbia University; "Forces, Fields, and Quanta"), Irving Kaplan (Massachusetts Institute of Technology; "Nuclear Fission"), Bernard Vonnegut (State University of New York at Albany; "Atmospheric Electricity"), Sam Cohen (Princeton University Plasma Physics Laboratory; "Plasma"), and Margaret L. A. MacVicar (Massachusetts Institute of Technology; "Superconductivity").

I thank Dr. Richard D. Deslattes and Dr. Barry N. Taylor of the National Bureau of Standards for information on units, standards, and precision measurements. I thank some of my colleagues at Union College: Professor C. C. Jones prepared the beautiful photographs of interference and diffraction by light (Chapters 38 and 39); Professor Barbara C. Boyer gave valuable advice and assistance on photographs dealing with biological material; and the staff of the library helped me find many a tidbit of information needed for an example or a problem. Michael Rooks and Ben Hu independently checked the solutions to the problems and proposed corrections for some problems they found insoluble.

I thank the editorial staff of W. W. Norton & Co., who worked long and hard to bring all the pieces of this book together: Drake McFeely, editor, who gave the manuscript the most meticulous attention and with sharp eyes spotted many a sentence that required rephrasing and clarification; Christopher Lang, who guided the project through its early stages; Alicia Salomon, copy editor, who corrected subtle errors of grammar and enforced stylistic consistency; and Amy Boesky and Natalie Goldstein, who searched out many of the fascinating photographs that add interest to this book. And it is a pleasure to thank Terry Hynes for the typing of the first draft of the text from my handwritten manuscript; she was always eager to improve my style and skillfully supplied all those words and endings of words that my pen, somehow, left out when my thoughts raced ahead of my hand.

H. C. O.
February 1985

Physics

THE WORLD OF PHYSICS

Physics is the study of matter. In a very literal sense, physics is the greatest of all natural sciences: it encompasses the smallest particles, such as electrons and quarks; and it also encompasses the largest bodies, such as galaxies and the entire universe. The smallest particles and the largest bodies differ in size by a factor of more than 10^{40}! In the following pictures we will survey the world of physics and attempt to develop some rough feeling for the sizes of things in this world. This preliminary survey sets the stage for our explanations of the mechanisms that make things behave in the way they do. Such explanations are at the heart of physics, and they are the concern of the later chapters of this book.

The pictures fall into two sequences. In the first sequence we zoom out: we begin with a picture of a woman's face and proceed step by step to pictures of the entire Earth, the Solar System, the Galaxy, and the universe. This ascending sequence contains 27 pictures, with the scale decreasing in steps of factors of 10.

In the second sequence we zoom in: we again begin with a picture of the face and close in on the eye, the retina, the rod cells, the molecules, the atoms, and the subatomic particles. This descending sequence contains 15 pictures, with the scale increasing in steps of factors of 10.

Most of our pictures are photographs. Many of these have only become available in recent years; they were taken by high-flying U-2 aircraft, Landsat satellites, astronauts on the Moon, or sophisticated electron microscopes. For some of our pictures no photographs are available and we have to rely, instead, on carefully prepared drawings.

PART I: THE LARGE-SCALE WORLD

SCALE 1:1.5 This is Charlotte, an intelligent biped of the planet Earth, Solar System, Orion Arm, Milky Way Galaxy, Local Group, Local Supercluster. She is made of 5.1×10^{27} atoms, with 1.8×10^{28} electrons, the same number of protons, and 1.4×10^{28} neutrons.

0 $\quad 0.5 \times 10^0 \quad 10^0$ m

SCALE 1:1.5 × 10 Charlotte has a height of 1.7 meters and a mass of 55 kilograms. Her chemical composition (by mass) is 65% oxygen, 18.5% carbon, 9.5% hydrogen, 3.3% nitrogen, 1.5% calcium, 1% phosphorus, and 0.35% of other elements.

The matter in Charlotte's body and the matter in her immediate environment occur in three states of aggregation: *solid, liquid,* and *gas.* All these forms of matter are made of atoms and molecules, but solid, liquid, and gas are qualitatively different because the arrangements of the atomic and molecular building blocks are different.

In a solid each building block occupies a definite place. When a solid is assembled out of molecular or atomic building blocks, these blocks are locked in place once and for all and they cannot move or drift about except with great difficulty. This rigidity of the arrangement is what makes the aggregate hard — it makes a solid "solid." In a liquid the molecular or atomic building blocks are not rigidly connected. They are thrown together at random and they move about fairly freely, but there is enough adhesion between neighboring blocks to prevent the liquid from dispersing. Finally, in a gas the molecules or atoms are almost completely independent of one another. They are distributed at random over the volume of the gas and are separated by appreciable distances, coming in touch only occasionally during collisions. A gas will disperse spontaneously if it is not held in confinement by a container or by some restraining force.

The molecules of a gas are forever moving around at high speed. For example, at a temperature of 0°C the average speed of a molecule of nitrogen in air is about 450 meters/second, faster than the speed of sound. The speed of molecules is directly related to the temperature: the speed increases if the air is heated and decreases if the air is cooled. The molecules of a liquid also move; their speeds are not very different from those of the molecules of a gas. However, since there is little space between the molecules in a liquid, the motion is continually interrupted by collisions. Because of these frequent collisions, the path of a molecule consists of a series of random zigzags and the molecule takes a long time to wander from one part of the liquid to another. Even in a solid there is some motion of the building blocks. But the motion of each atom or molecule is merely a vibration around its assigned position — the atom behaves as if kept on a short leash and although it moves back and forth at a high speed, it never strays beyond some tight limits.

If one regards the motion of the atoms of a gas in a container as analogous to the bouncing of a few dice in a shaker, then the motion of atoms in a liquid is analogous to the random wandering of the dice in a shaker that has been loosely but completely filled with dice; and the motion of atoms in a solid is analogous to the impotent rattling of the dice in a shaker that has been tightly and regularly packed full of dice.

The eternal, dancing motion of the atoms is the key to the transformations of state from solid to liquid to gas and vice versa. If we heat a solid, the vibrational motion becomes more violent and the atoms or molecules finally shake themselves out of place — the solid softens and melts, turning into a liquid. If we heat this resulting liquid further, the motion ultimately becomes so violent that the adhesion between the atoms or molecules cannot prevent some of them from escaping from the surface of the liquid. Gradually more and more escape — the liquid evaporates, turning into a gas.

0 $\quad 0.5 \times 10^1 \quad 10^1$ m

SCALE 1:1.5 × 10² The building behind Charlotte is the New York Public Library, one of the largest libraries on Earth. This library holds 9,300,000 volumes, containing roughly 10% of the total accumulated knowledge of our terrestrial civilization.

SCALE 1:1.5 × 10³ The New York Public Library is located at the corner of Fifth Avenue and 42nd Street, in the middle of New York City.

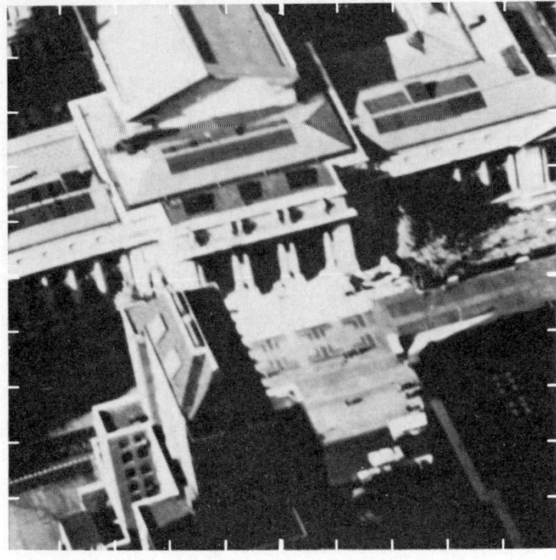

0 0.5×10^2 10^2 m

SCALE 1:1.5 × 10⁴ This aerial photograph shows an area of 1 kilometer × 1 kilometer in the vicinity of the New York Public Library. The streets in this part of the city are laid out in a regular rectangular pattern. The library is the building in the park in the upper middle of the picture. The Empire State Building shows up in the lower left of the picture. This skyscraper is 381 meters high and it casts a long shadow. Completed in 1930, it was for many years the tallest building in the world. Although the concrete and steel used in its construction are usually thought of as rigid materials, they have some elasticity — in a strong wind the top of the Empire State Building sways as much as 20 centimeters to a side.

This photograph was taken from an airplane flying at an altitude of 3658 meters. North is at the top of the photograph.

0 0.5×10^3 10^3 m

SCALE 1:1.5 × 10⁵ This photograph shows a large portion of New York City. We can recognize the library and its park as a small rectangular patch slightly above the center of the picture. The central mass of land is the island of Manhattan, with the Hudson River on the left and the East River on the right. Three bridges over the East River connect Manhattan with other parts of the city. The hundred-year-old Brooklyn Bridge is the southernmost of these bridges. It was the first steel-wire suspension bridge ever built and it stands as a spectacular and graceful achievement of nineteenth-century engineering.

This aerial photograph was taken by a U-2 aircraft flying at an altitude of about 20,000 meters.

0 0.5×10^4 10^4 m

SCALE 1:1.5 × 10^6 In this photograph, Manhattan is in the upper left quadrant. On this scale, we can no longer distinguish the pattern of streets in the city. However, we can clearly distinguish many highways, bridges, and causeways. For example, the thin white line crossing the dark water south of the tip of Manhattan is the Verrazano-Narrows Bridge. With a center span of 1300 m, it is one of the longest suspension bridges in the world. The vast expanse of water in the lower right of the picture is part of the Atlantic Ocean. The mass of land in the upper right is Long Island, with Long Island Sound to the north of it. Parallel to the south shore of Long Island we can see a string of very narrow islands; they almost look man-made. These are barrier islands; they are heaps of sand piled up by ocean waves in the course of thousands of years.

This photograph was taken by a Landsat satellite orbiting the Earth at an altitude of 920 kilometers.

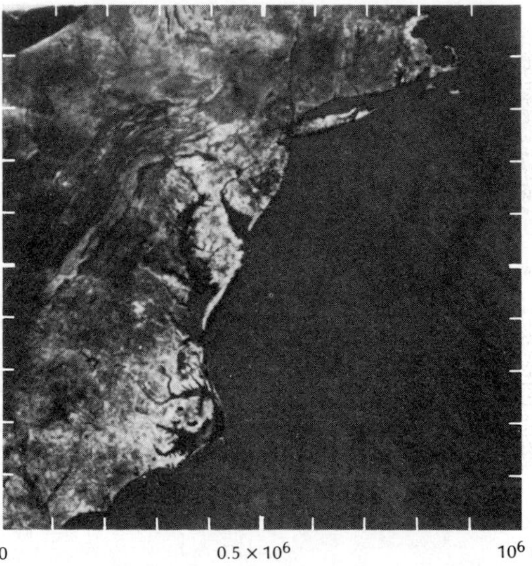

SCALE 1:1.5 × 10^7 Here we see the eastern coast of the United States, from Cape Cod to Cape Fear. Cape Cod is the hook near the northern end of the coastline, and Cape Fear is the promontory near the southern end of the coastline. If we move along the coast starting at the north, we first come to Long Island; then to Delaware Bay and Chesapeake Bay, two deep indentations in the coastline; and then to Cape Hatteras, at the extreme end of the large bulge of land thrusting eastward into the Atlantic. Several rivers show up as thin, dark lines: the Hudson running due south from the northern edge of the photograph, the Delaware flowing into Delaware Bay, and the Susquehanna and Potomac flowing into Chesapeake Bay. The wrinkles in the land west of Chesapeake Bay are the Appalachian Mountains. Note that on this scale no signs of human habitation are visible. However, at night the lights of large cities would stand out clearly.

This picture is a mosaic, assembled by joining together many Landsat photographs such as the one above.

SCALE 1:1.5 × 10^8 In this photograph, taken by the Apollo 16 astronauts during their trip to the Moon, we see a large part of the Earth. Through the gap in the clouds in the lower middle of the picture, we can see the coast of California and Mexico. We can recognize the peninsula of Baja California and the Gulf of California. In the middle right of the photograph we can recognize the Gulf of Mexico. Charlotte's location, the East Coast of the United States, is covered by a big system of swirling clouds in the upper right of the photograph.

Note that a large part of the area visible in this photograph is ocean. About 71% of the surface of the Earth is ocean; only 29% is land. The atmosphere covering this surface is about 100 kilometers thick; on the scale of this photograph, its thickness is about 0.7 millimeter. Seen from a large distance, the predominant colors of the planet Earth are blue (oceans) and white (clouds).

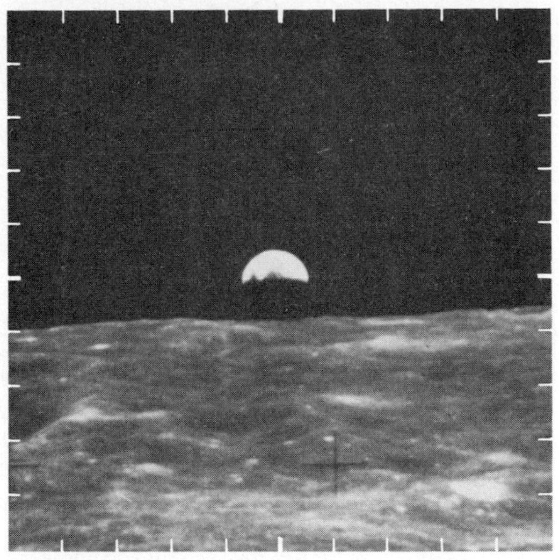

SCALE 1:1.5 × 10⁹ This photograph of the Earth was taken by the Apollo 16 astronauts standing on the surface of the Moon. Sunlight is striking the Earth from the top of the picture.

As is obvious from this and from the preceding photograph, the Earth is a sphere. Its radius is 6.38×10^6 meters and its mass is 5.98×10^{24} kilograms. Its chemical composition (by mass) is 38.6% iron, 28.6% oxygen, 14.4% silicon, 11.1% magnesium, 3.0% nickel, 1.6% sulfur, 1.3% aluminum, and 1.3% of other elements.

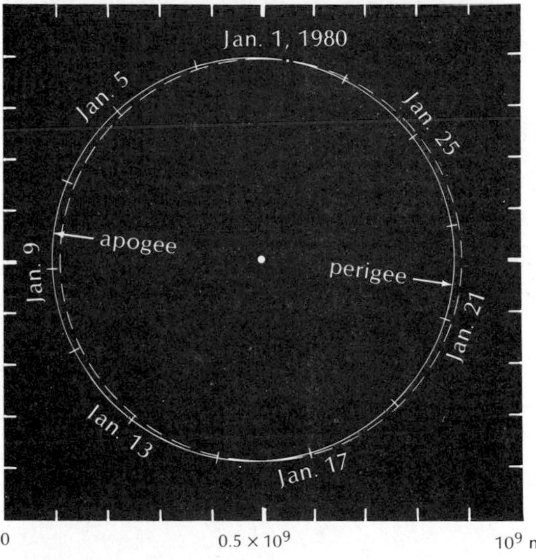

SCALE 1:1.5 × 10¹⁰ In this picture we see the Earth, the Moon, and its orbit. This picture is a drawing, not a photograph; none of our manned spacecraft has traveled sufficiently far away to take a photograph of such a panoramic view. (Many of the pictures on the following pages are also drawings.) As in the preceding picture, the Sun is far below the bottom of the picture. The position of the Moon is that of January 1, 1980. On this day, the Moon was almost full.

The orbit of the Moon around the Earth is an *ellipse*, but an ellipse that is very close to a circle. The solid curve in the picture is the orbit of the Moon and the dashed curve is a circle; by comparing these two curves we can see how much the ellipse deviates from a circle. The point on the ellipse closest to the Earth is called the *perigee* and the point farthest from the Earth is called the *apogee*. The distance between the Moon and the Earth is roughly 30 times the diameter of the Earth. The Moon takes 27.3 days to travel once around the Earth.

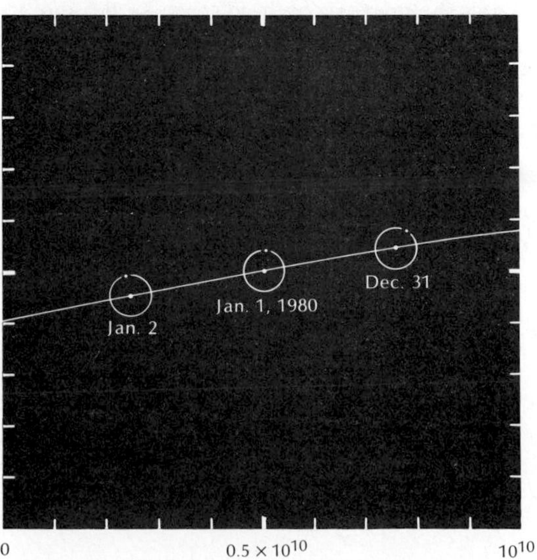

SCALE 1:1.5 × 10¹¹ This picture shows the Earth, the Moon, and portions of their orbits around the Sun. On this scale, both the Earth and the Moon look like small dots. Again, the Sun is far below the bottom of the picture. In the middle, we see the Earth and the Moon in their positions for January 1, 1980. On the right and on the left we see, respectively, the positions for 1 day before and 1 day after this date.

Note that the net motion of the Moon consists of the combination of two simultaneous motions: the Moon orbits around the Earth, which in turn orbits around the Sun. The net orbit of the Moon is only slightly more curved than the orbit of the Earth.

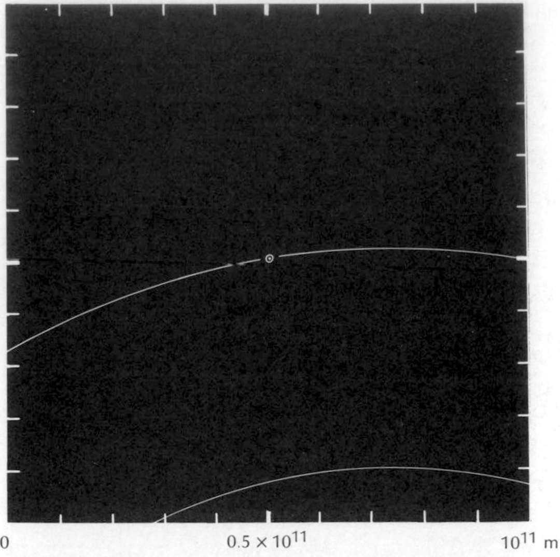

SCALE 1:1.5 × 10¹² Here we see the orbits of the Earth and of Venus. However, Venus itself is beyond the edge of the picture. The small circle is the orbit of the Moon. The dot representing the Earth is much larger than what it should be, although the draftsman has drawn it as minuscule as possible. On this scale, even the Sun is quite small; if it were included in this picture, it would be only 1 millimeter across.

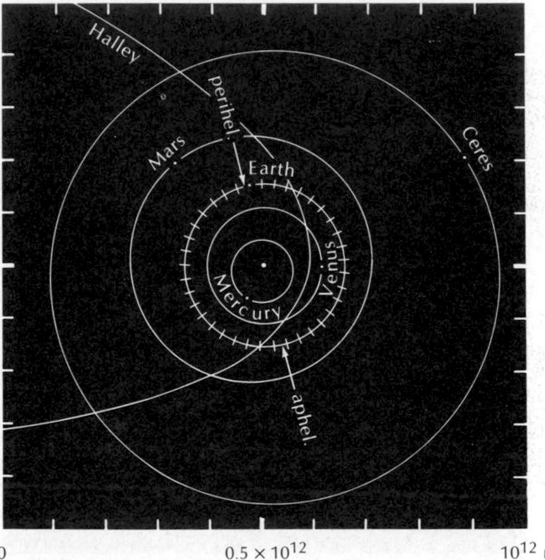

SCALE 1:1.5 × 10¹³ This picture shows the positions of the Sun and the inner planets: Mercury, Venus, Earth, and Mars. The positions of the planets are those of January 1, 1980. The orbits of all these planets are ellipses, but they are close to circles. The point of the orbit nearest to the Sun is called the *perihelion* and the point farthest from the Sun is called the *aphelion*. The Earth reaches perihelion about January 3 and aphelion about July 6 of each year.

All the planets travel around their orbits in the same direction: counterclockwise in our picture. The marks along the orbit of the Earth indicate the successive positions at intervals of 10 days. The orbits of the planets are not quite in the same plane. In the picture, we see the orbit of the Earth exactly face on; the orbits of the other planets are slightly tilted, but this tilt is not shown in the picture.

Beyond the orbit of Mars, a large number of asteroids orbit around the Sun. The four largest of these asteroids are Ceres, Pallas, Juno, and Vesta; the first of these has been included in our picture, but the others have been omitted to prevent excessive clutter. Furthermore, a large number of comets orbit around the Sun. Most of these have pronounced elliptical orbits. The comet Halley has been included in our picture.

The Sun is a sphere of radius 6.96×10^8 meters. On the scale of the picture, the Sun looks like a very small dot, even smaller than the dot drawn here. The mass of the Sun is 1.99×10^{30} kilograms. Its chemical composition (by mass) is 75% hydrogen, 23% helium, 0.8% oxygen, 0.4% carbon, 0.2% nitrogen, and about 0.6% of other elements. Since most of the mass of the universe is found inside of stars like the Sun, hydrogen and helium are the most abundant atoms in the universe. All the other atoms taken together account for less than 2% of the mass in the universe. Thus, the chemical elements found on the Earth and in our bodies must be regarded as mere traces, mere impurities. We are made of very rare stuff.

The matter in the Sun is in the *plasma* state, sometimes called the fourth state of matter. Plasma is a very hot gas in which violent collisions between the atoms in their random thermal motion have fragmented the atoms, ripping electrons off them. An atom that has lost one or more electrons is called an *ion*. Thus, plasma consists of a mixture of electrons and ions, all milling about at high speed and engaging in frequent collisions. These collisions are accompanied by the emission of light, making the plasma luminous.

In our universe plasma is by far the most abundant state of matter. Inside ordinary stars, such as the Sun, the temperature is so high that almost all atoms are ionized; only near the surface of the star can some atoms survive intact. Thus, all the stars are giant balls of plasma. Since most of the matter is found in stars, plasma is the predominant form of matter in the universe. Only a small fraction of matter is in the form of solids, liquids, and gases; these three states of matter are only found in planets, in neutron stars (pulsars), and in interstellar clouds of dust and gas.

SCALE 1:1.5 × 10¹⁴ This picture shows the positions of the outer planets of the Solar System: Jupiter, Saturn, Uranus, Neptune, and Pluto. On this scale, the orbits of the inner planets are barely visible. As in our other pictures, the positions of the planets are those of January 1, 1980.

The outer planets move slowly and their orbits are very large; thus they take a long time to go once around their orbit. The extreme case is that of Pluto, which takes 248 years to complete one orbit.

Uranus, Neptune, and Pluto are so far away and so faint that their discovery only became possible through the use of telescopes. Uranus was discovered in 1781, Neptune in 1846, and the tiny Pluto in 1930.

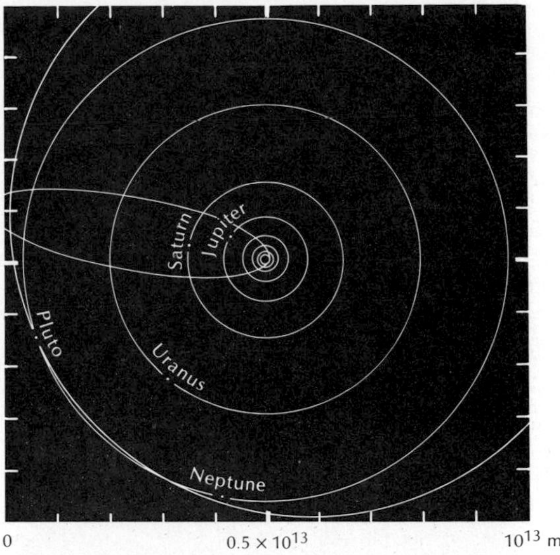

SCALE 1:1.5 × 10^{15} We see now that the Solar System is surrounded by a vast expanse of empty space. Actually, this space is not quite empty. The Solar System is encircled by a large cloud of millions of comets whose orbits crisscross the sky in all directions. Furthermore, the interstellar space in this picture and in the succeeding pictures contains traces of gas and of dust. The interstellar gas is mainly hydrogen; its density is typically 1 atom per cubic centimeter.

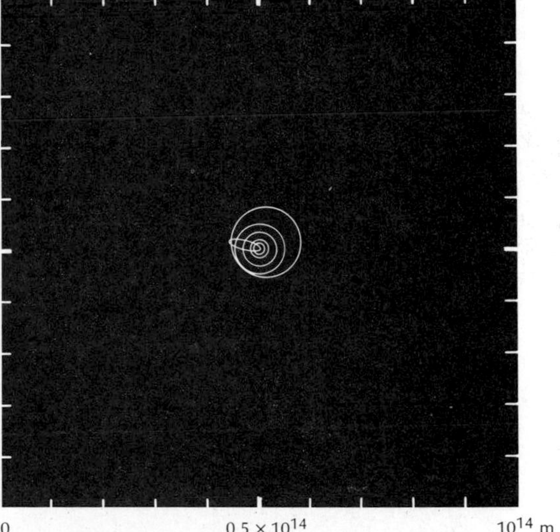

SCALE 1:1.5 × 10^{16} More empty space. The small circle is the orbit of Pluto.

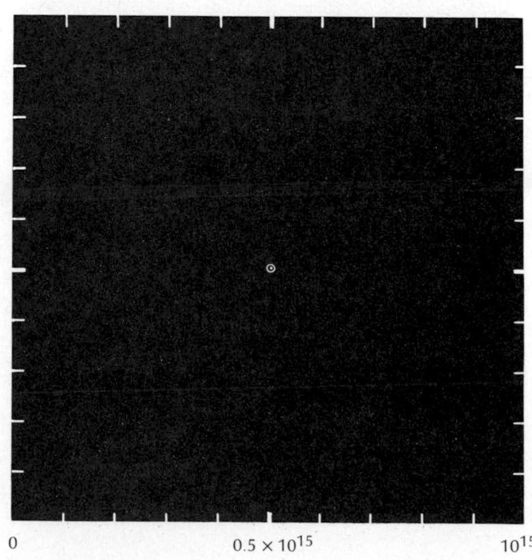

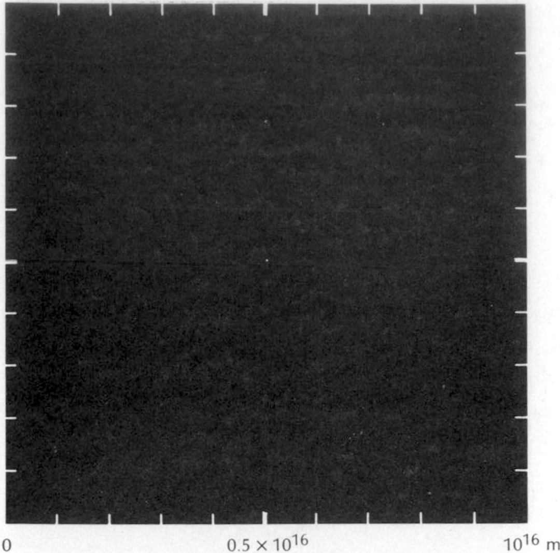

SCALE 1:1.5 × 10[17] And more empty space. On this scale, the Solar System looks like a minuscule dot, 0.1 millimeter across.

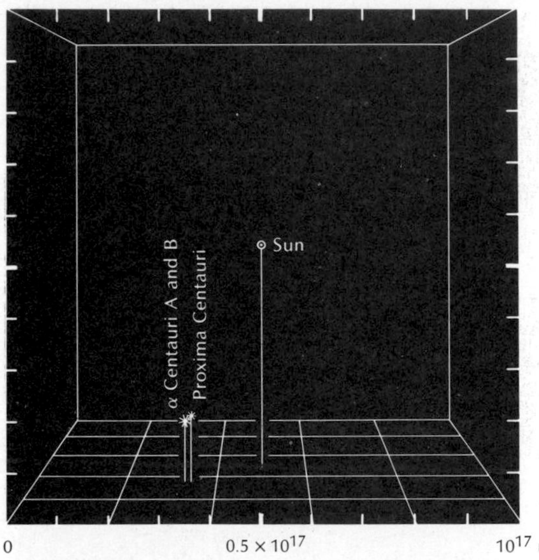

SCALE 1:1.5 × 10[18] Here, at last, we see the stars nearest to the Sun. The picture shows all the stars within a cubical box 10^{17} meters × 10^{17} meters × 10^{17} meters centered on the Sun: Alpha Centauri A, Alpha Centauri B, and Proxima Centauri. All three are in the constellation Centaurus, near the Southern Cross.

The star closest to the Sun is Proxima Centauri. This is a very faint, reddish star (a "red dwarf"), at a distance of 4.0×10^{16} meters from the Sun. Astronomers like to express stellar distances in light-years: Proxima Centauri is 4.2 light-years from the Sun, which means light takes 4.2 years to travel from this star to the Sun.

Proxima Centauri is too faint to be seen by the naked eye. The nearest stars that can be seen by the naked eye are Alpha Centauri A and Alpha Centauri B. The former is a bright star quite similar to our Sun; the latter is a fainter, orange star. These two stars are so close together that we need to use a telescope to distinguish between them. They form a double star, continually orbiting around each other. This double star is at a distance of 4.3 light-years from the Sun.

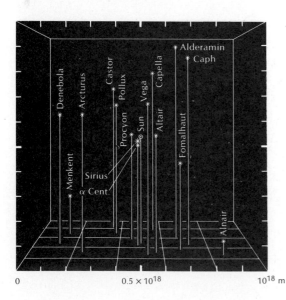

Denebola
Menkent
Arcturus
α Cent.
Sirius
Castor
Pollux
Procyon
Sun
Vega
Altair
Capella
Alderamin
Caph
Fomalhaut
Alnair
0
0.5 × 10^18
10^18 m

SCALE 1:1.5 × 10[19] This picture displays the brightest stars within a cubical box 10^{18} meters × 10^{18} meters × 10^{18} meters centered on the Sun. There are many more stars in this box besides those shown — the total number of stars in this box is close to 2000.

Sirius is the brightest of all the stars in the night sky. If it were at the same distance from the Earth as the Sun, it would be 28 times brighter than the Sun. Sirius has a much fainter companion very close to it.

SCALE 1:1.5 $\times$ 10^{20} Here we expand our box to 10^{19} meters $\times$ 10^{19} meters $\times$ 10^{19} meters, again showing only the brightest stars and omitting many others. The total number of stars within this box is close to 2 million. We recognize several clusters of stars in this picture: the Pleiades Cluster, the Hyades Cluster, the Coma Berenices Cluster, and the Perseus Cluster. Each of these has hundreds of stars crowded into a fairly small patch of sky.

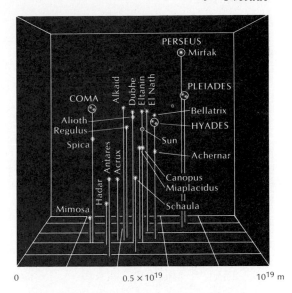

0 — 0.5×10^{19} — 10^{19} m

SCALE 1:1.5 $\times$ 10^{21} Now there are so many stars in our field of view that they appear to form clouds of stars. There are about a million stars in this photograph, and there are many more stars too faint to show up distinctly. Although this photograph is not centered on the Sun, it simulates what we would see if we could look toward the Solar System from very far away. The photograph shows a view of the Milky Way in the direction of the constellation Sagittarius. When we look from our Solar System in this direction, we see the clouds of stars in the neighboring spiral arm of our Galaxy (the next picture shows a galaxy and its spiral arms). This neighboring arm is the Sagittarius Spiral Arm; the cloud of stars in which our Sun is located belongs to the *Orion Spiral Arm*.

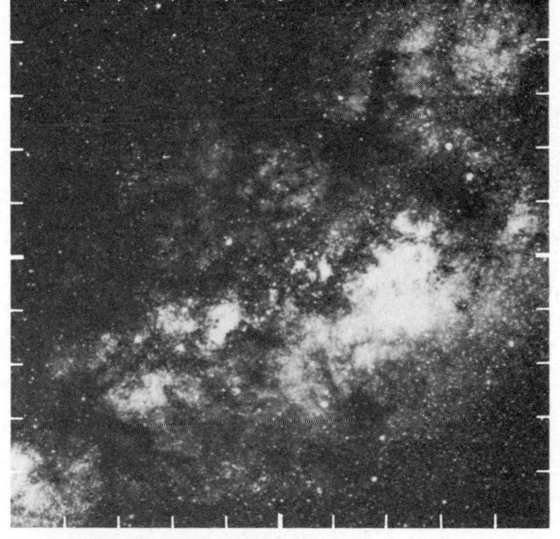

0 — 0.5×10^{20} — 10^{20} m

SCALE 1:1.5 $\times$ 10^{22} This is the galaxy NGC 5457 with its clouds of stars arranged in spiral arms wound around a central bulge. It is a spiral galaxy. The bright central bulge is the nucleus of the galaxy; it has a more or less spherical shape. The surrounding region, with the spiral arms, is the disk of the galaxy. This disk is quite thin; it has a thickness of only about 3% of its diameter. The stars making up the disk circle around the galactic center in a counterclockwise direction.

Our Sun is in a spiral galaxy of roughly similar shape and size: the *Galaxy of the Milky Way*. The total number of stars in this galaxy is about 10^{11}. The Sun is in one of the spiral arms, roughly one-third inward from the edge of the disk toward the center. The Sun takes about 250 million years to complete one circuit around the galactic center. Recent observations of the galactic center with radiotelescopes suggest that there is a large black hole at the center, a black hole several million times as massive as the Sun.

This photograph, like the remaining photographs of this section, was made with the great 5-meter telescope on Mt. Palomar.

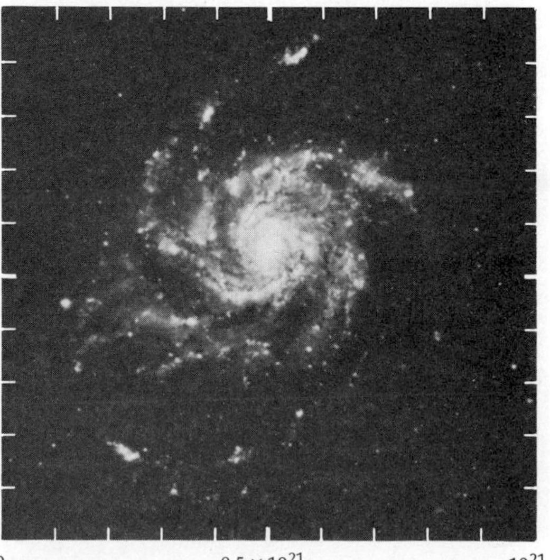

0 — 0.5×10^{21} — 10^{21} m

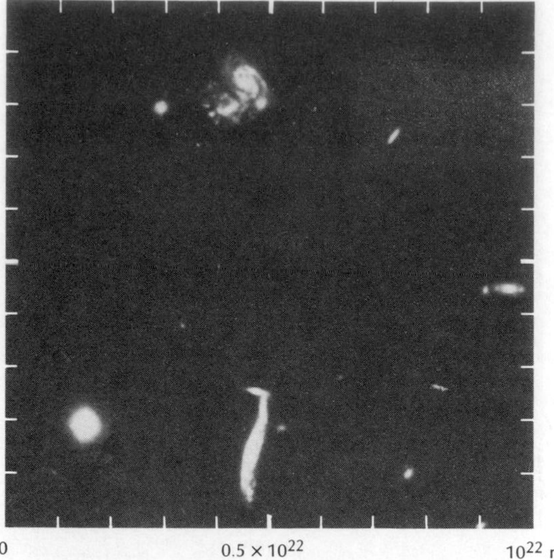

SCALE 1:1.5 × 10^{23} Galaxies are often found in clusters of several galaxies. Some of these clusters consist of just a few galaxies, others of hundreds or even thousands. The photograph shows a modest cluster, or group, of four galaxies beyond the constellation Hercules. This group lies within a much larger cluster of galaxies, called the Hercules Cluster. The group contains an elliptical galaxy like a luminous egg (lower left), two spiral galaxies close together (top), and a spiral with a bar (bottom).

Our Galaxy is part of another modest cluster, the *Local Group,* consisting of our own Galaxy, the great Andromeda galaxy, the Triangulum galaxy, the Large Magellanic Cloud, plus 16 other small galaxies. The smallest among these are "dwarf" galaxies with only 10^5 or so stars.

In this and the following photograph, the stars in the foreground have been erased to prevent their confusion with galaxies.

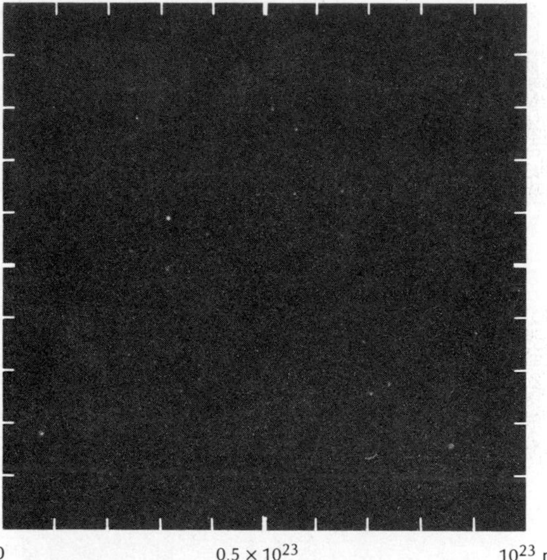

SCALE 1:1.5 × 10^{24} The Local Group lies on the fringes of a very large cluster of galaxies, called the *Local Supercluster.* This is a cluster of clusters of galaxies. At the center of the Local Supercluster is the Virgo Cluster with several thousand galaxies. Seen from a large distance, our supercluster would present a view comparable to this photograph, which shows a rich cluster beyond the constellation Hydra, a cluster that is at a large distance from us.

All the distant galaxies are moving away from us and away from each other. For instance, the Hydra Cluster is moving away from our Galaxy at the rate of 6030 kilometers per second. This motion of recession of the galaxies is analogous to the outward motion of, say, the fragments of a grenade after its explosion. The motion of the galaxies suggests that the universe began with a big explosion, the *Big Bang,* that launched the galaxies away from each other.

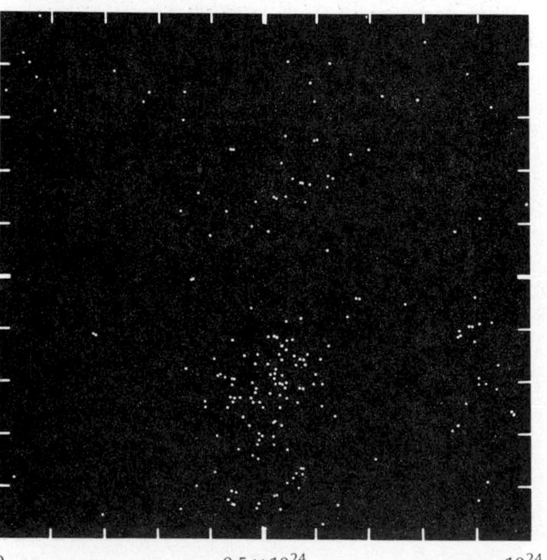

SCALE 1:1.5 × 10^{25} On this scale a galaxy equal in size to our own Galaxy would look like a fuzzy dot, 0.1 millimeter across. Thus, the galaxies are too small to show up clearly on a photograph. Instead we must rely on a plot of the positions of the galaxies. The plot shows the positions of about 200 of the brightest galaxies in an angular patch of the sky, 45° by 45°.

Since we are looking into a volume of space, some of the galaxies are in the foreground, some are in the background; but our plot takes no account of perspective. The distance scale has been computed for those galaxies that are at a middle distance, at about 1.4×10^8 light-years from the Earth; the scale is not valid for galaxies in the foreground or in the background.

The dense cluster of galaxies in the lower half of the picture is the Virgo Cluster. The loose cluster of galaxies in the upper half of the picture is the Coma Cluster. This cluster has roughly the same number of galaxies as the Virgo Cluster but is much farther away; hence only the very brightest galaxies of the Coma Cluster have been included in the picture.

SCALE 1:1.5 × 10^{26} Now there are so many galaxies in our field of view that it is impractical to plot the positions of individual galaxies. In this picture the sky has been divided into small squares, and the brightness of each square has been adjusted so that it is proportional to the number of galaxies in that square. The picture includes about 2.3 × 10^5 galaxies in an angular patch of the sky, 40° by 40°. The distance scale has been computed for those galaxies that are at a middle distance, at about 1.4 × 10^9 light-years from the Earth.

The galaxies in the picture are distributed more or less at random. If we look closely, we can recognize some chains or "filaments" of galaxies and some voids in between. But there are no large, distinctive features. On a large scale, the universe is quite uniform.

This is the last of our pictures in the ascending series. We have reached the limits of zoom-out. If we wanted to draw another diagram, 10 times larger than the above, we would need to know the shape and the size of the entire universe. We do not yet know that.

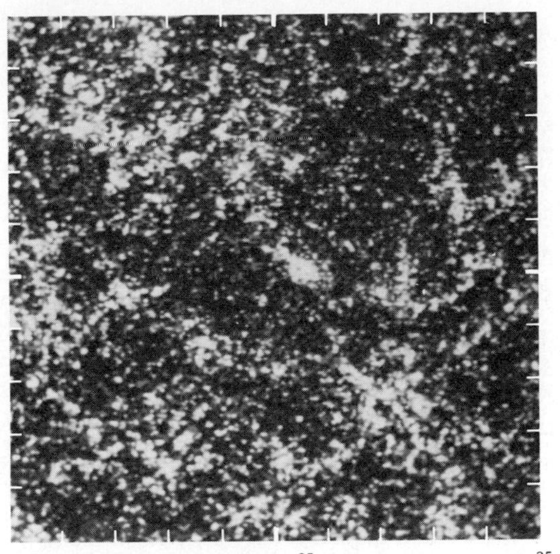

PART II: THE SMALL-SCALE WORLD

SCALE 1:1.5 We now return to Charlotte and zoom in on her eye. The surface of her skin appears smooth and firm. But this is an illusion. Matter appears continuous because the number of atoms in each cubic centimeter is extremely large. In a cubic centimeter of human tissue there are about 10^{23} atoms. This large number creates the illusion that matter is continuously distributed — we only see the forest and not the individual trees. The solidity of matter is also an illusion. The atoms in our bodies are mostly vacuum. As we will discover in the following pictures, within each atom the volume actually occupied by subatomic particles amounts to only about 1 part in 10^{13}.

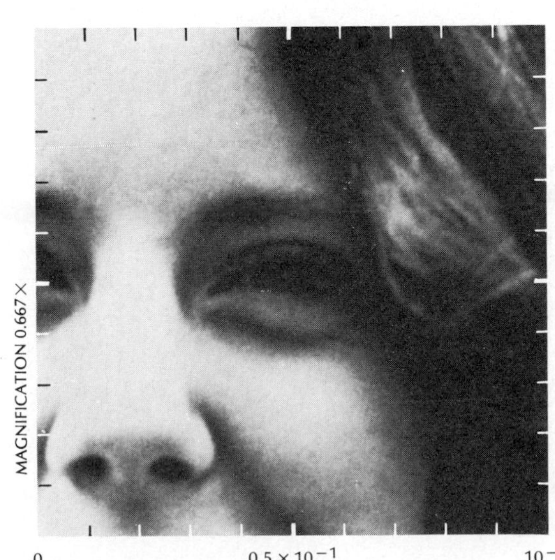

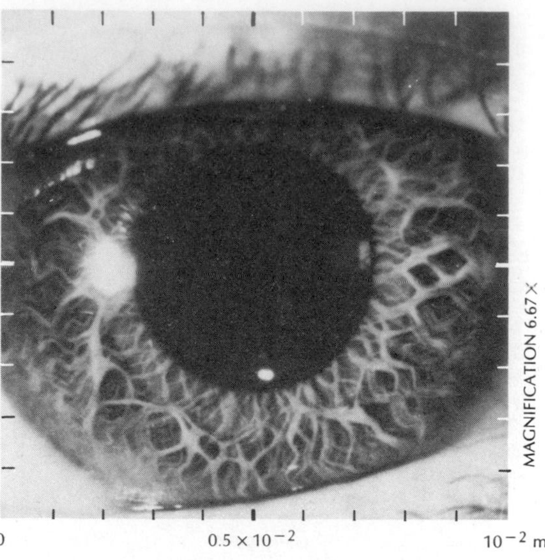

SCALE 1:1.5 × 10^{-1} Our eyes are very sophisticated sense organs; they collect more information than all our other sense organs taken together. The photograph shows the pupil and the iris of Charlotte's eye. Annular muscles in the iris change the size of the pupil and thereby control the amount of light that enters the eye. In strong light the pupil automatically shrinks to about 2 millimeters; in very weak light it expands to as much as 7 millimeters.

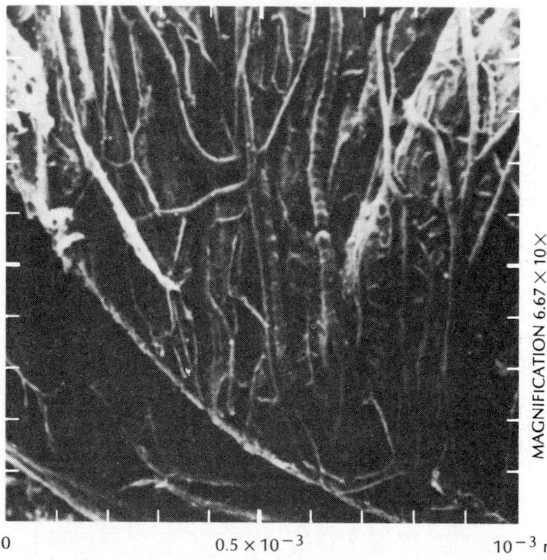

SCALE 1:1.5 × 10^{-2} This photograph shows the delicate network of blood vessels and nerve fiber bundles on the front surface of the retina, the light-sensitive membrane lining the rear interior of the eyeball. The rear surface of the retina is densely packed with two kinds of cells that sense light: cone cells and rod cells. In a human retina there are about 6 million cone cells and 120 million rod cells. The cone cells distinguish colors; the rod cells only distinguish brightness and darkness, but they are more sensitive than the cone cells and therefore give us vision in faint light ("night vision").

This and the following photographs were made with *electron microscopes.* An ordinary microscope uses a beam of light to illuminate the object; an electron microscope uses a beam of electrons. Electron microscopes can achieve much sharper contrast and much higher magnification than ordinary microscopes.

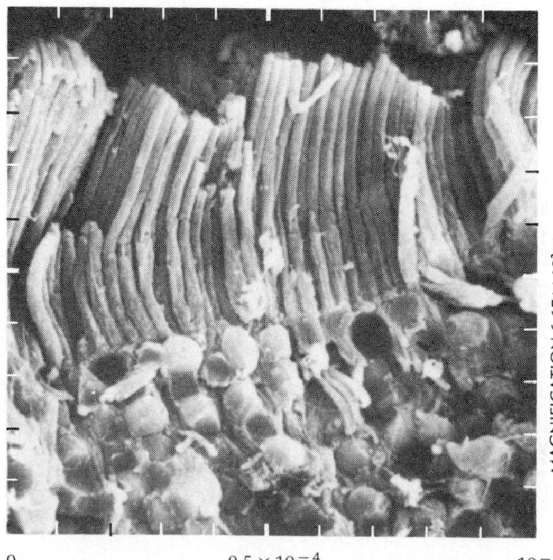

SCALE 1:1.5 × 10^{-3} Here we have a clear photograph of rod cells. To make this photograph, the retina was cut apart and the microscope was aimed at the edge of the cut. In the top half of the picture we see tightly packed rods. Each rod is connected to the main body of a cell. In the bottom part of the picture we can distinguish tightly packed cell bodies. The round balls lying about in the middle are the nuclei of some cells. The cutting of the retina has broken some cells apart and has exposed their nuclei.

SCALE 1:1.5 × 10^{-4} This is a close-up view of a few rods, showing the scarfed joints between the upper and the lower portions. The upper portions of the rods contain a special pigment — visual purple — that is very sensitive to light. The absorption of light by this pigment initiates a chain of chemical reactions that finally trigger nerve pulses from the eye to the brain.

MAGNIFICATION 6.67 × 10^3 ×

0 0.5 × 10^{-5} 10^{-5} m

SCALE 1:1.5 × 10^{-5} These are strands of DNA, or deoxyribonucleic acid, as seen with an electron microscope at very high magnification. DNA is found in the nuclei of cells. It is a long molecule made by stringing together a large number of nitrogenous base molecules on a backbone of sugar and phosphate molecules. The base molecules are of four kinds, the same in all living organisms. But the sequence in which they are strung together varies from one organism to another. This sequence spells out a message — the base molecules are the "letters" in the "words" of this message. The message contains all the genetic instructions governing the metabolism, growth, and reproduction of the cell.

The strands of DNA in the photograph are encrusted with a variety of small protein molecules. At intervals, the strands of DNA are wrapped around larger protein molecules that form lumps looking like the beads of a necklace.

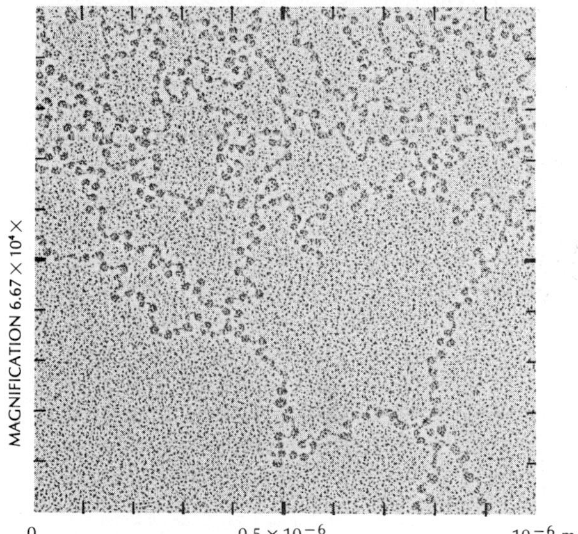

SCALE 1:1.5 × 10^{-6} The magnification in the preceding photograph is very difficult to exceed. Only in the case of a few special samples of materials and of atoms has it become possible to attain higher magnifications. Here we have a photograph of the tip of a fine platinum needle as viewed with an *ion microscope*. This kind of microscope uses helium ions instead of light to "illuminate" the object.

Each bright spot in the photograph is the image of a platinum atom. The needle is made of row upon row of atoms, arranged in a very beautiful symmetric pattern. Materials with a regular arrangement of atoms are called *crystals*. Here we have direct visual evidence that platinum is a crystal.

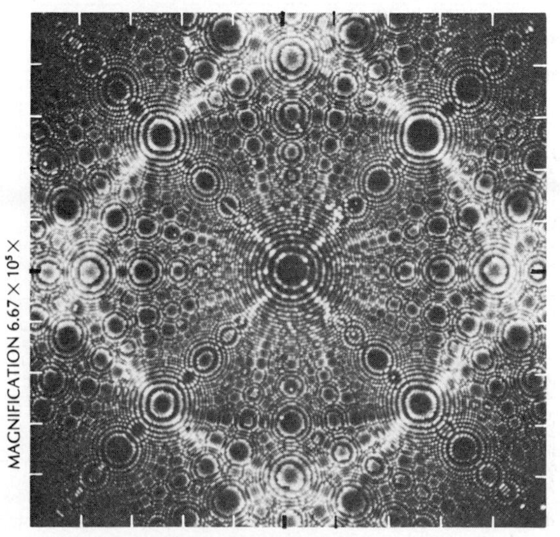

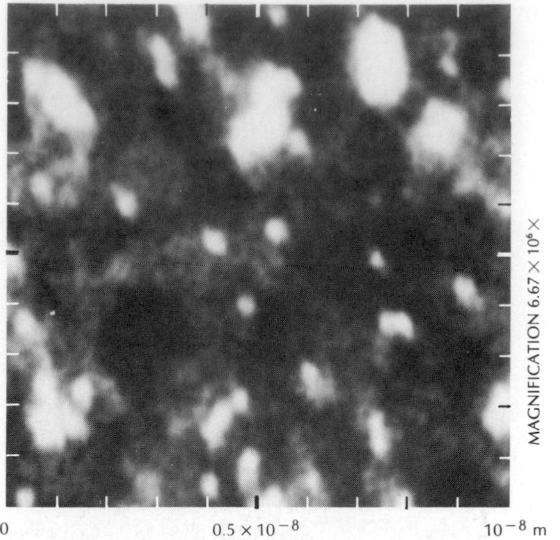

SCALE 1:1.5 × 10^{-7} This is a photograph of uranium atoms, taken with an electron microscope of extremely high power. The small white dots are individual uranium atoms; the large white blobs are clusters of uranium atoms. The atoms are stuck on a very thin film of carbon.

At present we know of more than 100 kinds of atoms or chemical elements. The *Periodic Table* P.1 is a list of these different kinds of atoms. The first entry in this table is the lightest atom — hydrogen (H) — with a mass of 1.67×10^{-27} kilogram; the last entry is one of the heaviest known atoms — hahnium (Ha) — with a mass 260 times as large. Some of the heavy atoms near the end of the table do not occur in nature; they can only be manufactured artificially in nuclear reactors or accelerators by "alchemy" or transmutation of other elements. Incidentally, nuclear physicists have claimed the discovery of two or three elements beyond hahnium. But these elements have not yet been officially baptized.

In Table P.1 two numbers are included with each kind of atom. The first number indicates the position of the atom in the table. This is called the *atomic number*. As we will see later, it represents the number of electrons belonging to the atom. The second number gives the *mass* of the atom in atomic mass units (1 atomic mass unit = 1 u = 1.66×10^{-27} kilogram). For example, the oxygen atom has numbers 8 and 15.9994, i.e., it is listed as the eighth entry in the table and it has a mass of 15.9994 u, or $15.9994 \times 1.66 \times 10^{-27}$ kilogram. With only a few exceptions, the masses of atoms in Table P.1 increase monotonically along the table.

The classical method for distinguishing between different atoms is, of course, chemical analysis — we identify the atoms by the reactions in which they engage. But a modern method that has become increasingly important is spectroscopic analysis — we identify the atoms by the light that they give off when stimulated by heat or by an electric current. In Table P.1, the atoms have been arranged in columns according to their chemical and spectroscopic properties. Different columns of this table contain atoms with similar properties; these columns are called *groups* and they are labeled IA, IIA, etc. For instance, the group IA contains hydrogen (H), lithium (Li), sodium (Na), potassium (K), etc.; these are the alkalis. The group 0 contains helium

Table P.1 THE PERIODIC TABLE OF CHEMICAL ELEMENTS[a]

IA	IIA	IIIB	IVB	VB	VIB	VIIB	VIII	VIII	VIII	IB	IIB	IIIA	IVA	VA	VIA	VIIA	0
1 H 1.0079																	2 He 4.00260
3 Li 6.941	4 Be 9.01218											5 B 10.81	6 C 12.011	7 N 14.0067	8 O 15.9994	9 F 18.99840	10 Ne 20.179
11 Na 22.98977	12 Mg 24.305											13 Al 26.98154	14 Si 28.0855	15 P 30.97376	16 S 32.06	17 Cl 35.453	18 Ar 39.948
19 K 39.098	20 Ca 40.08	21 Sc 44.9559	22 Ti 47.90	23 V 50.9414	24 Cr 51.996	25 Mn 54.9380	26 Fe 55.847	27 Co 58.9332	28 Ni 58.71	29 Cu 63.546	30 Zn 65.38	31 Ga 69.72	32 Ge 72.59	33 As 74.9216	34 Se 78.96	35 Br 79.904	36 Kr 83.80
37 Rb 85.4678	38 Sr 87.62	39 Y 88.9059	40 Zr 91.22	41 Nb 92.9064	42 Mo 95.94	43 Tc 98.9062	44 Ru 101.07	45 Rh 102.9055	46 Pd 106.4	47 Ag 107.868	48 Cd 112.40	49 In 114.82	50 Sn 118.69	51 Sb 121.75	52 Te 127.60	53 I 126.9045	54 Xe 131.30
55 Cs 132.9054	56 Ba 137.34	57–71 Rare Earths	72 Hf 178.49	73 Ta 180.947	74 W 183.85	75 Re 186.2	76 Os 190.2	77 Ir 192.22	78 Pt 195.09	79 Au 196.9665	80 Hg 200.59	81 Tl 204.37	82 Pb 207.2	83 Bi 208.9804	84 Po (210)	85 At (210)	86 Rn (222)
87 Fr (223)	88 Ra 226.0254	89-103 Acti-nides	104 Rf (257)	105 Ha (260)	106 (263)	107 (262)		109 (266)									

57 La 138.9055	58 Ce 140.12	59 Pr 140.9077	60 Nd 144.24	61 Pm (145)	62 Sm 150.4	63 Eu 151.96	64 Gd 157.25	65 Tb 158.9254	66 Dy 162.50	67 Ho 164.9304	68 Er 167.26	69 Tm 168.9342	70 Yb 173.04	71 Lu 174.97	Rare Earths (Lanthanides)
89 Ac (227)	90 Th 232.0381	91 Pa 231.0359	92 U 238.029	93 Np 237.0482	94 Pu (244)	95 Am (243)	96 Cm (247)	97 Bk (247)	98 Cf (251)	99 Es (254)	100 Fm (257)	101 Md (258)	102 No (259)	103 Lr (256)	Actinides

[a] In each box, the upper number is the *atomic number*. The lower number is the *atomic mass*. Numbers in parentheses denote the atomic mass of the most stable or best-known isotope of the element; all other numbers represent the average mass of a mixture of several isotopes as found in naturally occurring samples of the element.

(He), neon (Ne), argon (Ar), krypton (Kr), etc.; these are the noble gases. All atoms in a given group are chemically similar, i.e., they have the same valence and they engage in similar reactions. These atoms are also spectroscopically similar, i.e., they emit light with a similar pattern of colors; thus, the colors emitted by sodium display a pattern that has the same general features as the colors emitted by hydrogen.

Atoms are the building blocks of molecules and these, in turn, are the building blocks of everything surrounding us. Among the simplest, and most familiar, of all molecules are the water molecule and the oxygen molecule. The water molecule consists of two hydrogen atoms and one oxygen atom; that of oxygen of two oxygen atoms. But many other molecules are much more complex and much larger. For example, the molecules of DNA shown in the photograph on page 13 consist of about 100,000,000 atoms of carbon, hydrogen, oxygen, nitrogen, phosphorus, etc., assembled in long strands which are coiled into a helical structure. The molecules of synthetic polymers, such as nylon, polyethylene, synthetic rubber, etc., consist of an even larger number of atoms assembled in extremely long chains within which identical units are strung together one after another.

SCALE 1:1.5 × 10^{-8} This is a photograph of a neon atom. The atom is a sphere, with a somewhat fuzzy surface. The atoms of the noble-gas elements (helium, neon, argon, krypton, xenon, and radon) all have this same spherical shape, although their sizes are slightly different. The atoms of other elements have other shapes; for example, atoms of carbon have an approximately tetrahedral shape, with round edges.

This picture represents the highest magnification that has been attained to date. The picture was obtained with an *electron-holography microscope* by a two-stage process: first a photograph of the atom was taken with an electron microscope and then a photograph of this photograph was taken with an ordinary microscope. Normally, we would not expect that magnification of a photograph would yield any marked increase of visible detail. But the "photograph" prepared in the first stage was actually a holographic photograph, and this did contain extra information that could be brought out at the second stage by the special magnification technique.

All the pictures on the remaining pages of this section are drawings. At these higher magnifications no photographs are available.

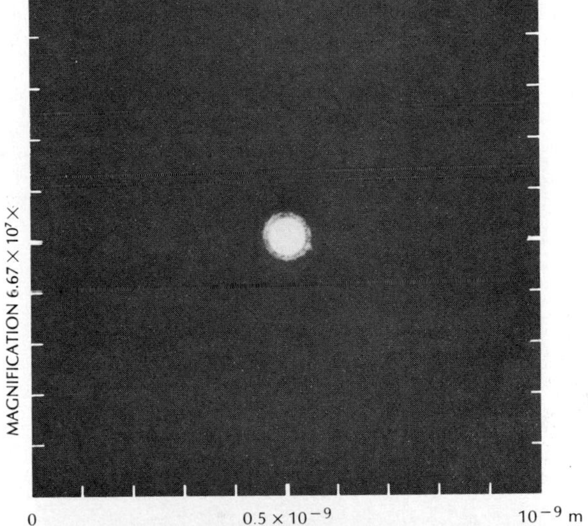

SCALE 1:1.5 × 10^{-9} The drawing shows the interior of an atom of neon. This atom consists of 10 electrons orbiting around a nucleus. In the drawing, the electrons have been indicated by small dots, and the nucleus by a slightly larger dot at the center of the picture. These dots have been drawn as small as possible, but even so the size of these dots does not give a correct impression of the actual size of the electrons and of the nucleus. The electron is smaller than any other particle we know; maybe the electron is truly pointlike and has no size at all. The nucleus has a finite size, but this size is much too small to show up on the drawing. Note that the electrons tend to cluster near the center of the atom. However, the overall size of the atom depends on the distance to the outermost electron; this electron defines the outer edge of the atom.

The mass of each electron is 9.1 × 10^{-31} kilogram, but most of the mass of the atom is in the nucleus; the 10 electrons of the neon atom have only 0.03% of the total mass of the atom.

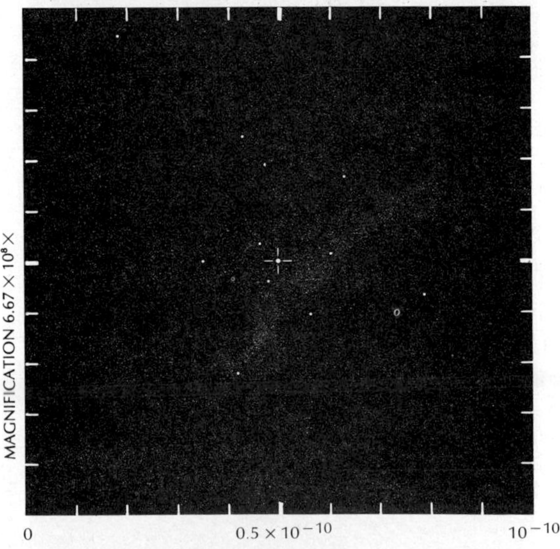

The electrons move around the nucleus in a very complicated motion. The drawing shows the electrons as they would be seen at one instant of time with a *Heisenberg microscope*. This is a hypothetical microscope that employs gamma rays instead of light rays to illuminate an object; no such microscope has yet been built.

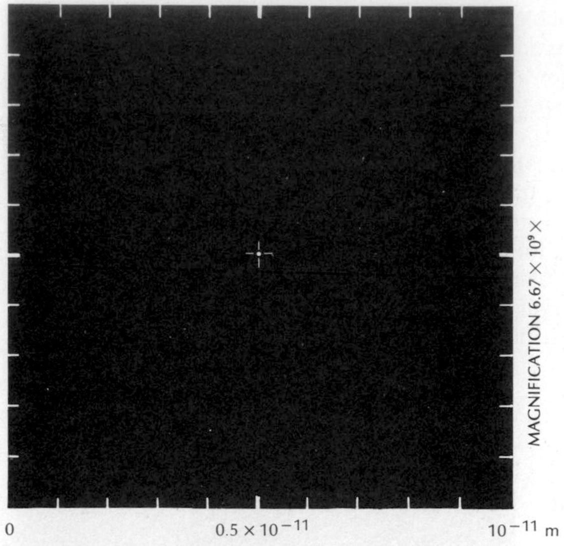

SCALE 1:1.5 × 10⁻¹⁰ Here we are closing in on the nucleus. We are seeing the central part of the atom. Only two electrons are in our field of view; the others are beyond the margin of the drawing. The size of the nucleus is still much smaller than the size of the dot at the center of the picture.

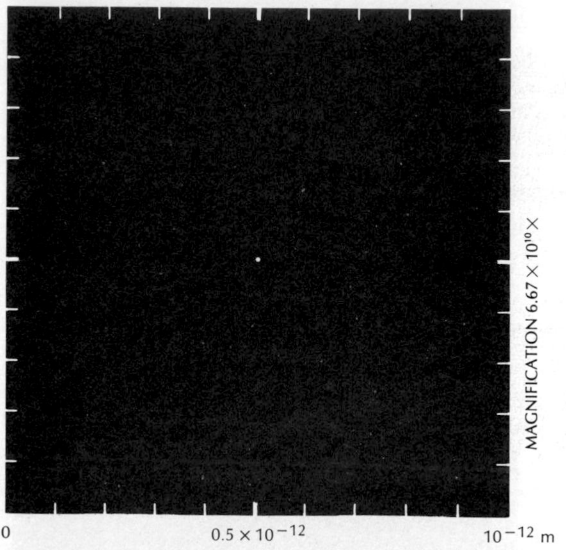

SCALE 1:1.5 × 10^{-11} In this drawing we finally see the nucleus in its true size. At this magnification, the nucleus of the neon atom looks like a small dot, 0.5 millimeter in diameter. Since the nucleus is extremely small and yet contains most of the mass of the atom, the density of the nuclear material is enormous. If we could assemble a drop of pure nuclear material of a volume of 1 cubic centimeter, it would have a mass of 2.3×10^{11} kilograms, or 230 million metric tons!

Our drawings show clearly that most of the volume within the atom is empty space. The nucleus only occupies a very small fraction of this volume.

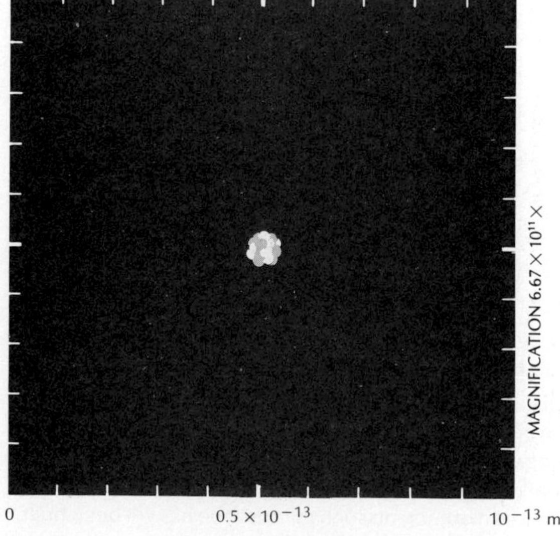

SCALE 1:1.5 × 10^{-12} We can now begin to distinguish the nuclear structure. The nucleus has a nearly spherical shape, but it is slightly lumpy.

SCALE 1:1.5 × 10^{-13} At this extreme magnification we can see the details of the nuclear structure. The nucleus of the neon atom is made up of 10 protons (white balls) and 10 neutrons (gray balls). Each proton and each neutron is a sphere with a diameter of about 2.1 × 10^{-15} meter, and a mass of 1.67 × 10^{-27} kilogram. In the nucleus, these protons and neutrons are tightly packed together, so tightly that they almost touch. The protons and neutrons move around the volume of the nucleus at high speed in a complicated motion.

Note that the number of protons matches the number of electrons, or the atomic number. Neon atoms always have 10 protons in their nucleus, but they do not always have 10 neutrons. In naturally occurring samples of neon gas, about 91% of the atoms have 10 neutrons, but 8.8% have 12 neutrons, and 0.2% have 11 neutrons. Atoms with the same numbers of protons but different numbers of neutrons are called *isotopes*. The naturally occurring isotopes of neon are designated ^{20}Ne, ^{22}Ne, and ^{21}Ne. The superscript on the chemical symbol indicates the sum of the number of protons and the number of neutrons; this superscript is called the *mass number*, since it is approximately proportional to the mass of the isotope.

Besides the isotopes ^{20}Ne, ^{21}Ne, and ^{22}Ne, neon also has the isotopes ^{17}Ne, ^{18}Ne, ^{19}Ne, ^{23}Ne, ^{24}Ne, and ^{25}Ne. However, the latter do not occur naturally; they can only be produced artificially by transmutation of elements in a nuclear reactor or accelerator. These artificial isotopes are extremely unstable, lasting for only a few minutes or

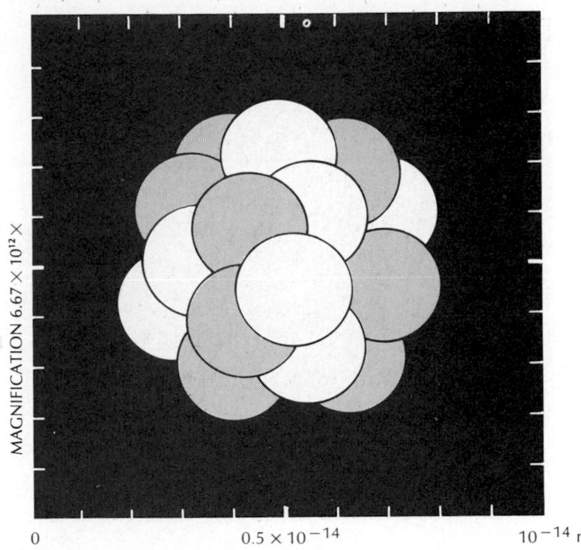

seconds before spontaneously decaying into fluorine or sodium.

All chemical elements have several isotopes. Hydrogen has three isotopes (^{1}H, or ordinary hydrogen; ^{2}H, or deuterium; ^{3}H, or tritium), helium has five isotopes (^{3}He, ^{4}He, ^{5}He, ^{6}He, ^{8}He), etc. Platinum has 34 isotopes, more than any other chemical element. Some of these isotopes occur in nature, others are unstable and can only be produced by artificial means. Table P.2 lists the isotopes of the first few elements.

Table P.2 EXCERPT FROM THE CHART OF ISOTOPES[a]

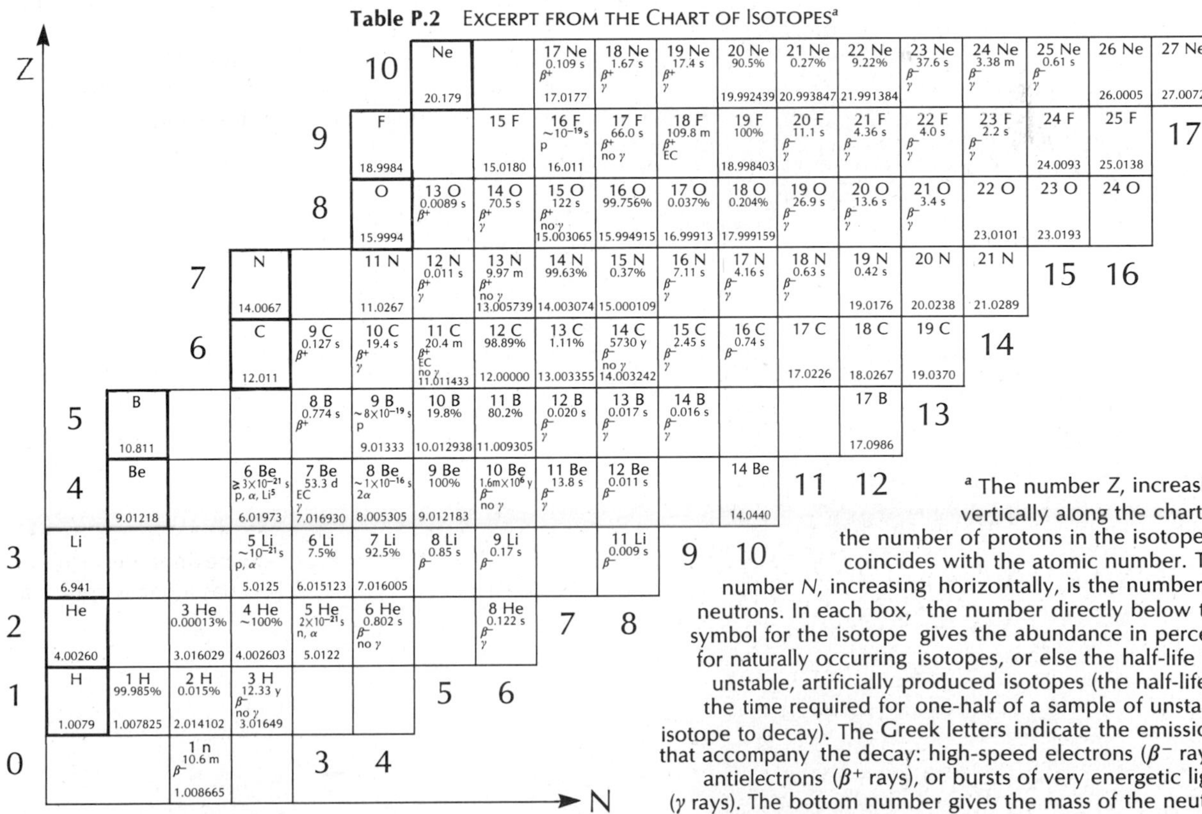

Z \ N		0	1	2	3	4	5	6	7	8	9	10	11	12	13	14	15	16	17
10							Ne 20.179		17 Ne 0.109 s β^+ 17.0177	18 Ne 1.67 s β^+ γ	19 Ne 17.4 s β^+ γ	20 Ne 90.5% 19.992439	21 Ne 0.27% 20.993847	22 Ne 9.22% 21.991384	23 Ne 37.6 s β^- γ	24 Ne 3.38 m β^- γ	25 Ne 0.61 s β^- γ	26 Ne 26.0005	27 Ne 27.0072
9						F 18.9984		15 F 15.0180	16 F ~10^{-19} s p 16.011	17 F 66.0 s β^+ no γ	18 F 109.8 m β^+ EC	19 F 100% 18.998403	20 F 11.1 s β^- γ	21 F 4.36 s β^- γ	22 F 4.0 s β^- γ	23 F 2.2 s β^- γ	24 F 24.0093	25 F 25.0138	
8						O 15.9994	13 O 0.0089 s β^+	14 O 70.5 s β^+ γ	15 O 122 s β^+ no γ 15.003065	16 O 99.756% 15.994915	17 O 0.037% 16.99913	18 O 0.204% 17.999159	19 O 26.9 s β^- γ	20 O 13.6 s β^- γ	21 O 3.4 s β^- γ	22 O 23.0101	23 O 23.0193	24 O	
7				N 14.0067		11 N 11.0267	12 N 0.011 s β^+ γ	13 N 9.97 m β^+ no γ 13.005739	14 N 99.63% 14.003074	15 N 0.37% 15.000109	16 N 7.11 s β^- γ	17 N 4.16 s β^- γ	18 N 0.63 s β^- γ	19 N 0.42 s 19.0176	20 N 20.0238	21 N 21.0289			
6				C 12.011	9 C 0.127 s β^+	10 C 19.4 s β^+ γ	11 C 20.4 m β^+ EC no γ 11.011433	12 C 98.89% 12.00000	13 C 1.11% 13.003355	14 C 5730 y β^- no γ 14.003242	15 C 2.45 s β^- γ	16 C 0.74 s β^-	17 C 17.0226	18 C 18.0267	19 C 19.0370				
5		B 10.811			8 B 0.774 s β^+	9 B ~8×10^{-19} s p 9.01333	10 B 19.8% 10.012938	11 B 80.2% 11.009305	12 B 0.020 s β^- γ	13 B 0.017 s β^- γ	14 B 0.016 s β^- γ			17 B 17.0986					
4		Be 9.01218		6 Be ≳3×10^{-21} s p, α, Li^5 6.01973	7 Be 53.3 d EC γ 7.016930	8 Be ~1×10^{-16} s 2α 8.005305	9 Be 100% 9.012183	10 Be 1.6m×10^6 y β^- no γ	11 Be 13.8 s β^- γ	12 Be 0.011 s β^-		14 Be 14.0440							
3	Li 6.941			5 Li ~10^{-21} s p, α 5.0125	6 Li 7.5% 6.015123	7 Li 92.5% 7.016005	8 Li 0.85 s β^-	9 Li 0.17 s β^-		11 Li 0.009 s β^-									
2	He 4.00260		3 He 0.00013% 3.016029	4 He ~100% 4.002603	5 He 2×10^{-21} s n, α 5.0122	6 He 0.802 s β^- no γ		8 He 0.122 s β^- γ											
1	H 1.0079	1 H 99.985% 1.007825	2 H 0.015% 2.014102	3 H 12.33 y β^- no γ 3.01649															
0			1 n 10.6 m β^- 1.008665																

[a] The number Z, increasing vertically along the chart, is the number of protons in the isotope; it coincides with the atomic number. The number N, increasing horizontally, is the number of neutrons. In each box, the number directly below the symbol for the isotope gives the abundance in percent for naturally occurring isotopes, or else the half-life for unstable, artificially produced isotopes (the half-life is the time required for one-half of a sample of unstable isotope to decay). The Greek letters indicate the emissions that accompany the decay: high-speed electrons (β^- rays), antielectrons (β^+ rays), or bursts of very energetic light (γ rays). The bottom number gives the mass of the neutral atom (nucleus plus Z electrons) in atomic mass units.

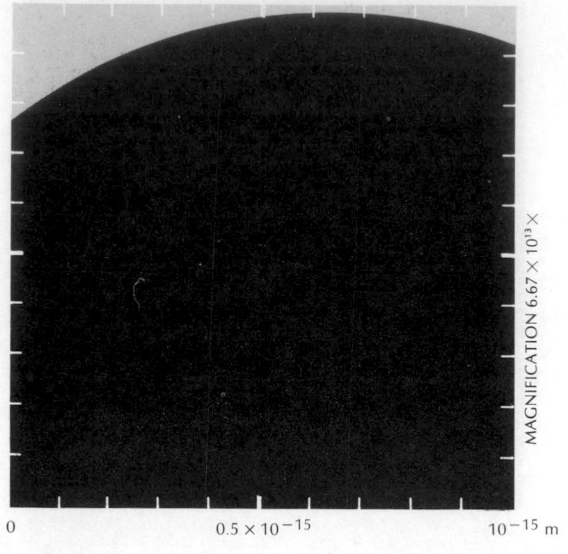

SCALE 1:1.5 × 10^{-14} This final picture shows three pointlike bodies within a proton. These pointlike bodies are *quarks* — each proton and each neutron is made of three quarks. Recent experiments have told us that the quarks are much smaller than protons, but we do not yet know their precise size. Hence the dots in the drawing probably do not give a fair description of the size of quarks. The quarks within protons and neutrons are of two kinds, called *up* and *down*. The proton consists of two *up* quarks and one *down* quark joined together; the neutron consists of one *up* quark and two *down* quarks joined together.

Besides these two kinds of quarks, physicists have discovered four other kinds, called by the quaint names *strange, charmed, top,* and *bottom*. These quaint quarks do not occur in pieces of ordinary matter; they can only be manufactured under exceptional circumstances in high-energy collisions between subatomic particles.

This final picture takes us to the limits of our knowledge of the subatomic world. As a next step we would like to zoom in on the quarks and show what they are made of. But we do not yet know whether they are made of anything else.

CHAPTER 1

Measurement of Space, Time, and Mass

The investigator of any phenomenon — an earthquake, a flash of lightning, a collision between two ships — must begin by asking the question: where and when did it happen? Phenomena happen at points in space and at points in time. A complicated phenomenon — such as the collision between two ships — is spread out over many points of space and time. But no matter how complicated, any phenomenon can be completely described by stating what happened at diverse points of space at successive instants of time. A happening at one point of space and one point of time is called an **event.** For example, the first contact between the bows of colliding ships is an event. The complete complicated phenomenon is a collection of such elementary events.

One of the simplest phenomena is the motion of a particle. In classical physics — the physics that reached perfection in the nineteenth century — an **ideal particle** is regarded as a pointlike object of no discernible size or internal structure. Such a particle may be thought of as an infinitesimal billiard ball, a ball so small that its size can be ignored for all practical purposes. At any given instant of time, this ideal particle occupies a single point of space. Furthermore, the particle has a mass. And that is all: if we know the position of the particle at each instant of time, and we know its mass, then we know everything that can be known about the particle. Position, time, and mass give a *complete* description of a classical particle.[1] Measurements of position, time, and mass are therefore of fundamental importance in physics.

Ideal particle

[1] We will ignore for now the possibility that the particle also has an electric charge. Electricity is the subject of Chapters 22–35.

1.1 Space and Time

To determine the position of an event, we first take some convenient point of space as **origin** and then measure the position of the event relative to this origin. To accomplish this, we imagine a grid of lines around the origin and check the location of the event within this grid; that is, we imagine that space is filled with three-dimensional "graph paper" and specify the position by means of coordinates read off this graph paper. The most common kinds of coordinates are **rectangular coordinates** x, y, z, which rely on a rectangular grid (Figure 1.1). The three mutually perpendicular lines through the origin are called the x, y, and z axes; the rectangular grid is erected on these axes. The coordinates x, y, z of a point simply indicate how far we must move parallel to the corresponding axis in order to go from the origin to the point. For example, the point P shown in Figure 1.1 has coordinates $x = 3$ units, $y = 5$ units, $z = 3$ units.

Systems of coordinates

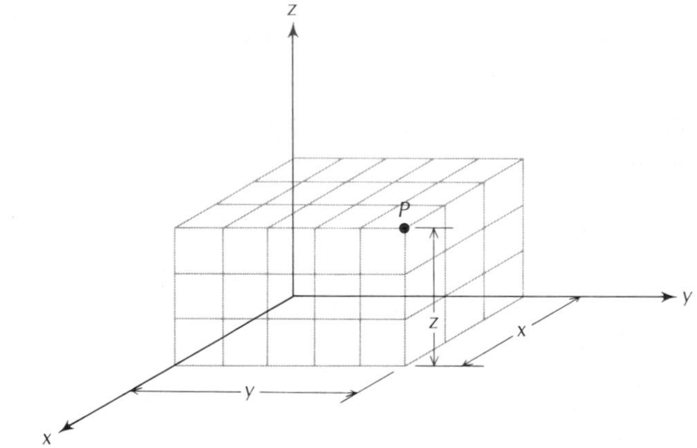

Fig. 1.1 Rectangular coordinates x, y, z.

Instead of rectangular coordinates, we can just as well use **spherical coordinates** or **cylindrical coordinates.** These rely on a spherical or cylindrical grid, respectively. The spherical coordinates are essentially the "latitude" angle, the "longitude" angle, and the radial distance from the origin (Figure 1.2); the cylindrical coordinates are the "longitude" angle (sometimes called the "bearing angle"), the height above the x–y plane, and the distance along the x–y plane (Figure 1.3). Which kind of coordinates is chosen is a matter of convenience. For example, an airport controller will usually describe the position of an approaching aircraft by means of a cylindrical grid with origin at the airport, i.e., he will give the bearing angle of the aircraft, its altitude above the horizontal plane, and its horizontal distance from the airport. On the other hand, the pilot of an aircraft on a transatlantic flight will find it more convenient to describe position by means of a spherical grid with origin at the center of the Earth, i.e., he will give the (geographic) latitude, longitude, and altitude (the latter is in obvious correspondence with the distance to the Earth's center).

Three dimensions of space

It is a fact of experience that, regardless of the kind of coordinate grid we choose, exactly *three* coordinates are necessary and sufficient to locate a point relative to an origin. This is the rigorous meaning of the assertion that the space in which we live is three dimensional. Using language adapted to a rectangular grid, we can say that space has length, width, and height (or depth).

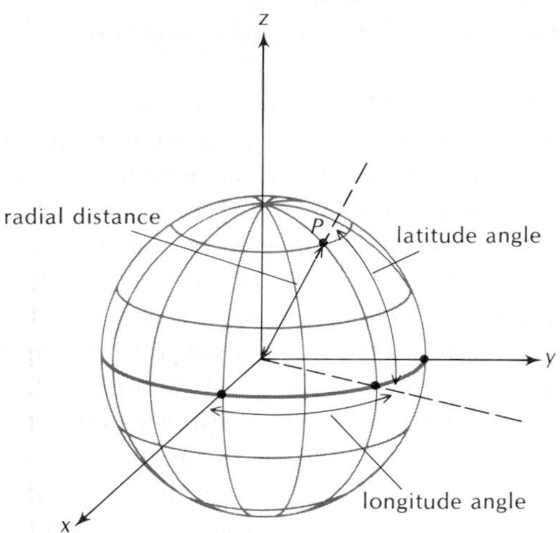

Fig. 1.2 Spherical coordinates: longitude, latitude, and radial distance.

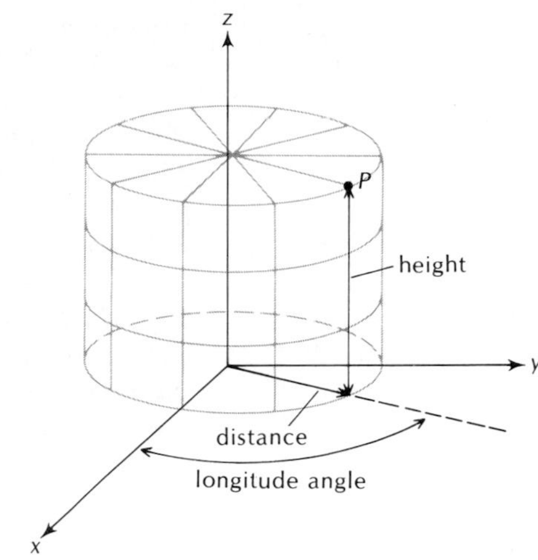

Fig. 1.3 Cylindrical coordinates: longitude, horizontal distance, and height.

That our space is three dimensional is its most elementary and important property. Next in importance is the property that the geometry of our space is **Euclidean,** at least to a very good approximation. This means that the postulates and theorems of Euclid are valid for the real world. For example, any arbitrary plane triangle laid out in space has 180° for the sum of its interior angles (Figure 1.4).[2]

Euclidean geometry

It must be emphasized that the three-dimensional character and the Euclidean geometry of our space rest on very direct **empirical evidence,** that is, evidence directly derived from observation and experimentation. We can perceive the three-dimensional character simply by feeling around space with our hands or looking about with our binocular vision. The Euclidean geometry requires observations of a more intricate kind. One way to check it is by precisely measuring the sum of the interior angles of a plane triangle. For a plane triangle defined by three points in the space in the vicinity of Earth, such an experiment can be done with surveying instruments; the measured result for the sum of interior angles agrees with 180° to within the experimental error, a few tenths of a second of arc. Thus, the direct experimental evidence shows that on or near the Earth, the Euclidean geometry is good to within better than a few parts in 10^6. (Some indirect evidence, based on the theory of relativity, indicates that the Euclidean geometry is probably good to within a few parts in 10^{10}.)

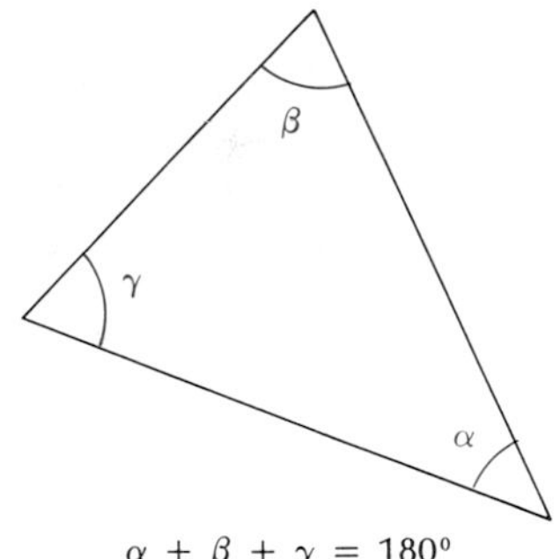

Fig. 1.4 A plane triangle with interior angles α, β, and γ.

Equal in importance to the property that space is Euclidean ranks the property that time is absolute. The meaning of this is as follows: Suppose we have two identical clocks, initially at rest side by side, synchronized and running at exactly the same rate. We put one of the clocks in motion and transport it along some path, perhaps with varying speed. Then we bring the traveling clock back to the location of its companion clock and compare them. If the two clocks are still synchronized, regardless of where the traveling clock was transported and with what velocity it was moved, then time is absolute, i.e., the

[2] Appendix 4 gives a review of trigonometry, including the definitions of degrees and radians. The Greek alphabet appears on the endpapers.

rate at which time elapses is independent of position and independent of velocity.[3]

Experiments indicate that time is absolute to a very high precision. Physicists take advantage of this fact in order to synchronize clocks in laboratories separated by some distance on Earth; they use a special portable atomic clock (see Section 1.4), which they first synchronize with the clock in one laboratory, and then transport to the other laboratory in order to transfer the synchronization.

The theory based on the assumption of an absolute time and a Euclidean geometry of space is called **classical** or **Newtonian physics.** The foundations of this theory were laid down by Isaac Newton in the seventeenth century. Newton asserted in his *Principia Mathematica* that

> absolute, true, and mathematical time, of itself, and from its own nature, flows equably without relation to anything external.

He took for granted that the geometry of space is Euclidean, and never even bothered to state this assumption explicitly. For more than 200 years Newton's theory stood unchallenged. But in the early part of this century, physicists found that the geometry of space is not exactly Euclidean and that time is not exactly absolute.

The breakdown of Euclidean geometry is most noticeable in the immediate vicinity of a very massive body, such as the Sun. Using experimental data concerning the propagation of light and radiowaves near the Sun ("bending" of light rays by the Sun), astronomers found that the sum of the interior angles of a triangle constructed with light rays deviates from 180° by 1 or 2 seconds of arc. (The exact value of the deviation depends on the size of the triangle, its orientation, and its distance from the Sun.[4]) This is a small, but significant, deviation from Euclidean geometry. It indicates that in the immediate vicinity of the Sun some of the postulates of Euclid fail; a new, non-Euclidean geometry is needed — the geometry of a curved space. The cause of this curvature of space is gravity. Physicists believe that curvature is generated not only by the gravity of the Sun, but also by the gravity of the Earth, and by the gravity of any other massive body. However, in the vicinity of the Earth, gravity is fairly weak, and therefore the curvature of space is so small as to be quite unnoticeable.

The failure of absolute time shows up in high-precision experiments with very accurate atomic and nuclear clocks. Physicists found that when these clocks are exposed to high velocities or to the effects of gravity at high altitude, they lose or gain time relative to clocks not so exposed. For instance, experiments with an extremely accurate atomic clock flying in an aircraft at 600 miles per hour at an altitude of 30,000 feet show that the combined effects of a velocity and gravity slow the rate of the clock down by about 1 part in 10^{12}, as compared to the rate of an identical clock that is kept stationary on the ground

[3] This, of course, relies on the tacit assumption that the clocks are designed in such a way that the shocks (accelerations) they receive during their motion do not directly (mechanically) affect their rate. For example, a good wristwatch is designed so that shocks do not affect it. In contrast, a pendulum clock is very sensitive to accelerations and would be quite unsuitable for this experiment.

[4] The question arises whether this deviation from Euclidean geometry is to be blamed on the light rays. But (indirect) experimental and theoretical evidence suggests that a deviation of about the same magnitude would persist even if the triangle were laid out with tightly stretched strings.

below.[5] If the clock were moving at much higher velocity, the effect would be larger — for a clock moving at 90% of the velocity of light, the rate at which time elapses would be 2.3 times slower than for a clock at rest. These "time-dilation" effects contradict the notion of an absolute time. However, as long as we are dealing with the velocities of everyday experience, which are very small compared to the velocity of light, the time-dilation effects are so small that they can be ignored.

The theories that account for these time-dilation effects and for the deviations from Euclidean geometry are the theory of Special Relativity and the theory of General Relativity. These theories were conceived by Albert Einstein in the early part of this century, and they brought about a revolutionary revision of our concepts of space and time.

Since the deviations from Euclidean geometry and from absolute time are so small as to be barely detectable with the most precise telescopes, radiotelescopes, and clocks, we will ignore Einstein's theories for the purposes of most of the following chapters and rely instead on Newton's theory with its Euclidean geometry and absolute time. Although Newton's theory is only an approximation to the real world, it is an eminently satisfactory one, quite adequate for the description of all the phenomena we encounter in everyday life and (almost) all the phenomena we encounter in the realm of engineering. Physics is full of such approximations made in the interest of simplicity. Nevertheless, physics is regarded as an "exact" science, because the physicist usually knows, more or less, what approximations can be made in the description of a given phenomenon without generating excessive errors.

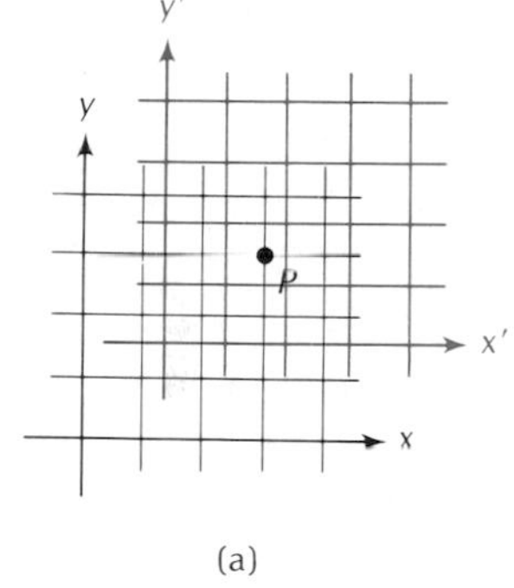

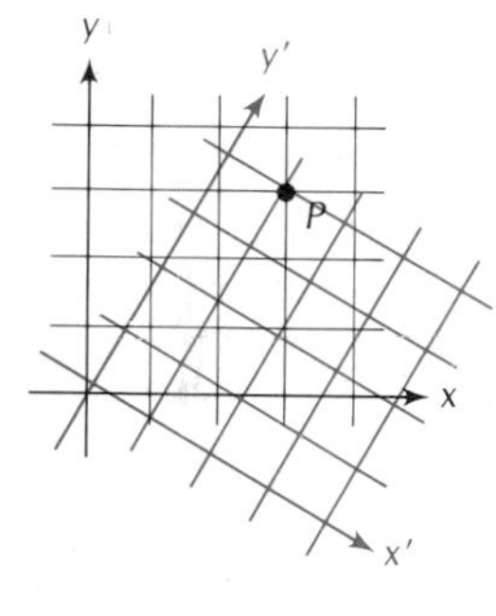

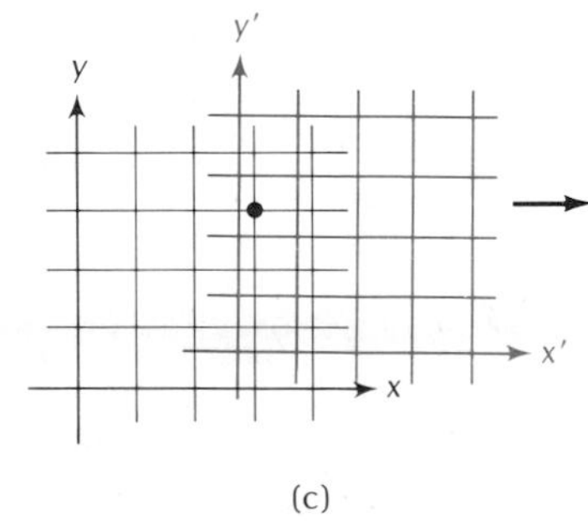

Fig. 1.5 Rectangular grids x–y and x'–y'. The rectangular grid x'–y' is displaced relative to the rectangular grid x–y.

1.2 Frames of Reference

To measure the position of an event, we take an origin, erect a coordinate grid, and measure the coordinates of the event with respect to this grid. This is a *relative measurement,* which is to say the coordinates we find depend on the choice of origin. Obviously, if two rectangular grids are based on origins that are displaced from one another, then the two sets of coordinates will have a corresponding difference (Figure 1.5a). It is also possible for two grids to differ in orientation (b), and this can lead to extra differences in the sets of coordinates. Furthermore, the origin of one grid may be in motion relative to another grid (c), in which case not only will the two sets of coordinates of a given point be different, but the difference will change in time. Position measurements are always relative; they only become meaningful when the coordinate grid used for the measurements is carefully specified.

The choice of origin of coordinates and the choice of its motion are matters of convenience. For example, the navigator of a ship often finds it convenient to take the midpoint of his ship as origin and to imagine a grid around this origin; the grid then moves with the ship.[6] If the navigator plots the track of a second ship on this grid, he can tell

[5] This number is based on the assumption that the aircraft is flying eastward above the equator; in this case the gravity effect tends to increase the clock rate, but the velocity effect decreases the rate and more than compensates for gravity. For details see Interlude E.

[6] In the U.S. Navy, the coordinates based on this moving grid are called "relative coordinates."

at a glance what the distance of closest approach will be, and whether the other ship is on a collision course (crosses the origin).

Reference frame

A coordinate grid together with a set of synchronized clocks is called a **reference frame.** The coordinate grid and the clocks are used to determine the space and time location of events. If the motion of a reference frame coincides with the motion of some given body — a particle, a ship, the Earth, or whatever — so that the body remains permanently at rest in this reference frame, then this reference frame is called the **rest frame** of that body. We can think of such a rest frame as rigidly attached to the body. A rest frame is often labeled with the name of the body to which it is attached; thus, we speak of the rest frame of the Earth (moving with the Earth), the rest frame of a ship (moving with the ship), etc.

The clocks associated with a given reference frame will be assumed to move with the reference frame. It is convenient to suppose that the clocks are arranged at regular intervals along the coordinate grid (Figure 1.6) so that whenever an event occurs anywhere, one of the clocks will be in the immediate vicinity and can be used to record the time.

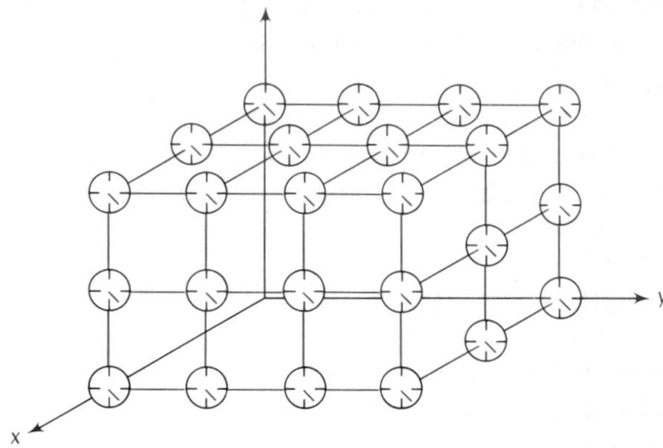

Fig. 1.6 A reference frame consisting of a coordinate grid and synchronized clocks.

As long as we deal with phenomena for which the notion of an absolute time is a satisfactory approximation, we do not really need to insist that each reference frame have its own clocks moving with it. If time is absolute, then all clocks in all reference frames agree and it is just as well to use the clocks of one chosen reference frame for all other reference frames. But when we wish to deal with relativistic phenomena involving velocities near the velocity of light (see Chapter 17), then the lack of an absolute time forces us to insist that each reference frame have its own set of synchronized clocks. In the relativistic domain both the rates of clocks and their synchronization are properties peculiar to a given reference frame, i.e., there is no absolute time, only a "relative" time associated with each reference frame. Although we will ignore this problem of the relativity of time until Chapter 17, for the sake of uniformity we will suppose that each reference frame always has its own clocks, whether the approximation of an absolute time applies or not.

1.3 The Unit of Length

In order to construct the coordinate grid to be used for a measurement of position, we need a **unit of length** that tells us where the points $x = 1$, $y = 1$, $z = 1$ are located along the respective axes. The unit of length is a basic amount of length against which any other amount of length can be compared.

Metric system

The most widely used system of units is the **metric system,** which is based on the **meter** as the unit of length, the **kilogram** as the unit of mass, and the **second** as the unit of time. As we will see in Section 1.6, units of length, time, and mass are necessary and sufficient for the measurement of any physical quantity.

meter, m

For many years, the standard of length that told us the amount of length in one meter (1 m) was the standard meter bar kept at the Bureau International des Poids et Mesures (BIPM, International Bureau of Weights and Measures) at Sèvres, France. This is a bar made of platinum–iridium alloy with a fine scratch mark near each end (Figure 1.7). By definition, the distance between these scratches was taken to be exactly one meter. The length of the meter was originally chosen so as to make the polar circumference of the Earth exactly 40 million meters (Figure 1.8); however, modern determinations of this circumference show it to be about 0.02% more than 40 million meters.

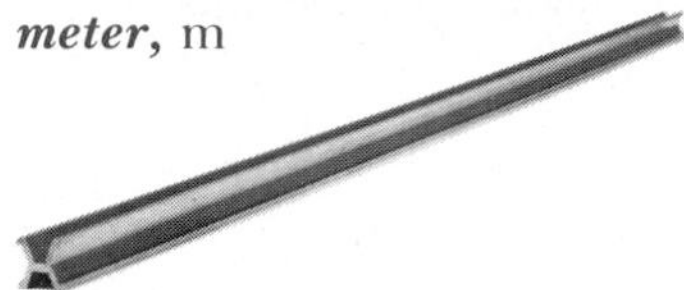

Fig. 1.7 International standard meter bar.

Copies of the prototype standard meter were constructed in France and distributed to other countries to serve as secondary standards. The length standards used in industry and engineering have been derived from these secondary standards. For example, Figure 1.9 shows a set of gauge blocks commonly used as length standards in machine shops.

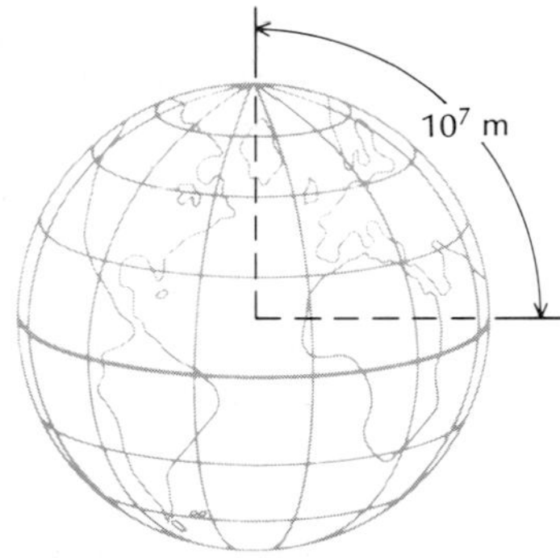

Fig. 1.8 One-quarter of the polar circumference of the Earth equals 10^7 m.

The precision of the standard meter is limited by the coarseness of the scratch marks at its ends. Although these scratches were made with great care, a microscope reveals them as irregular bands whose midpoints are somewhat uncertain. For the sake of higher precision, a new standard of length was adopted in 1960. This standard was the wavelength of the orange light emitted by krypton atoms. Just as a properly stimulated musical instrument, say, a flute, emits a sound wave consisting of a regular succession of wave crests and wave troughs, where the air is, respectively, compressed and rarified, a stimulated (excited) krypton atom emits a light wave consisting of a regular succession of wave crests and wave troughs of light. The distance from one wave crest to the next is called the **wavelength.** The wavelength of the light emitted by krypton atoms (under suitable conditions) serves as the standard of length. Since all krypton atoms are exactly alike, they all emit exactly the same wavelength, i.e., all krypton atoms are exactly in tune. This makes the atomic standard of length universally accessible to all laboratories — krypton is readily available everywhere and it is not necessary to keep a "prototype krypton atom" at the BIPM. The meter was defined as 1,650,763.73 times the wavelength of the light emitted by krypton. We can measure any unknown length by comparing it with the krypton wavelength. Figure 1.10 shows a high-precision interferometer used for such a comparison; essentially this interferometer counts the number of wavelengths that fit into the unknown length.

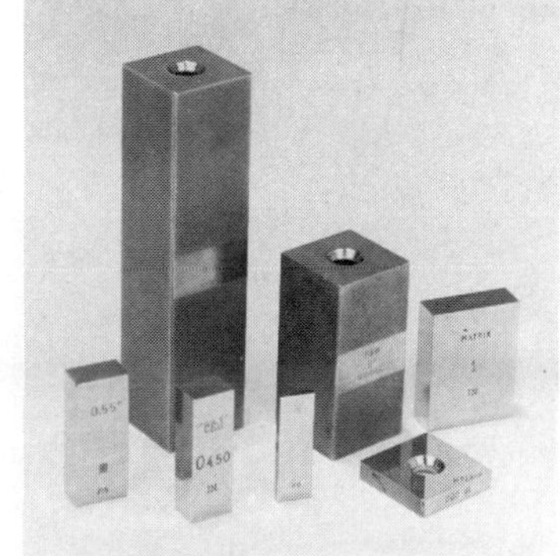

Fig. 1.9 Some gauge blocks. The numbers on the blocks give their length or thickness in inches.

The wavelength of the light emitted by a krypton atom suffers from a slight uncertainty that arises from processes occurring within the atom during the act of emission. This lack of sharpness of the wavelength limits the precision of the krypton standard. Recently, stabilized lasers have been developed which can attain much higher precision. The wavelength of the light generated by these lasers has an uncertainty of less than 2 parts in 10^{12}. This led to the adoption of a new

Fig. 1.10 Special interferometer used at the BIPM for the comparison of lengths. The interferometer automatically counts the wavelengths when the carriage C moves through a given length.

(and final) definition of the meter in 1983: the meter is the length of path traveled by a light wave in vacuum in a time interval of 1/299,792,458 second. This new definition hinges on the definition of the unit of time, which we will give in the next section. And it also hinges on the standard value of the speed of light — since the meter is adjusted so that light travels one meter in 1/299,792,458 second, the speed of light is (exactly)

$$c = 2.99792458 \times 10^8 \text{ meters per second}$$

The assignment of a fixed value to the speed of light makes good sense because, according to Einstein's theory, the speed of light is a fundamental, immutable constant, ideally suited to play the role of a standard of speed. Another advantage of the new definition of the meter is that any future improvement in the precision of the second will automatically lead to an improvement in the precision of the meter. Note that the definition of the meter does not mention stabilized lasers. However, the practical implementation of this definition relies on these lasers and on the uniformity of the light waves that they emit. Because of the uniformity of the light waves, the time interval between the emission of one wave crest and the next can be determined with extreme precision and, upon multiplying this time interval by the speed of light, we obtain the wavelength of the laser in meters. We can then use this laser wavelength as a standard for the measurement of an unknown length, in much the same way as we used the krypton wavelength.

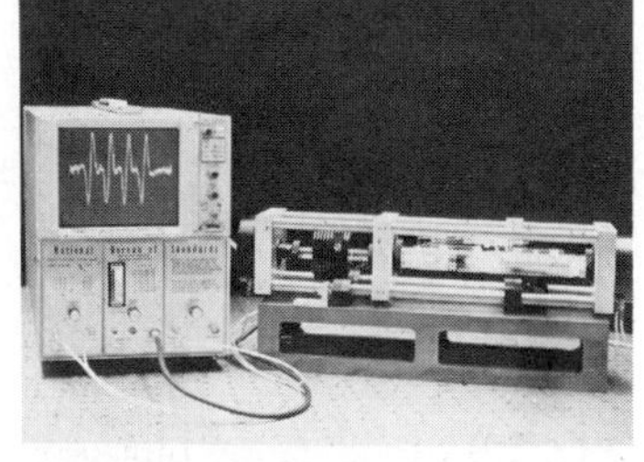

Fig. 1.11 Stabilized laser at the National Bureau of Standards.

By using a stabilized laser in conjunction with an interferometer (such as shown in Figure 1.11), we can measure a change of position (i.e., a displacement) of one or several millimeters with an error of only $\pm 10^{-12}$ m. For a change of position of the order of one or several meters, the errors tend to be somewhat larger, about $\pm 10^{-8}$ m. However, if we attempt to measure the length of a body (distance between the two ends), then the irregularity of the body's surface and the deformation of this surface by contact with the measuring device engender ambiguities in just what is meant by "length." These ambiguities prevent measurements of the length of a body to within better than $\pm 10^{-9}$ m.

Table 1.1 lists a few distances and sizes in meters, from the largest to the smallest (see also the Prelude). Some of these lengths are given with two significant figures, others with one significant figure. Of course, we can measure most of these lengths with much higher precision than that. For instance, we can measure the wavelength of light to within 9 or 10 significant figures, but to do this we would first have to decide just what kind of light to measure. Since Table 1.1 is only intended to present some typical examples of lengths, such high precision is not called for. Table 1.2 lists some multiples and submultiples of the meter and their abbreviations.

In the **British system of units**, the unit of length is the foot. By definition, one **foot** (1 ft) is exactly 0.3048 m. Table 1.3 gives multiples and submultiples of the foot; these multiples and submultiples are not decimal, a great computational nuisance which will lead to the complete abolition of this system of units in the near future. At present only a few dawdling countries, such as Brunei, Burma, Liberia, Yemen, and the United States, are still using the British system.

Table 1.1 SOME DISTANCES AND SIZES

Distance to boundary of observable universe	$\sim 1 \times 10^{26}$ m
Distance to Andromeda galaxy	2.1×10^{22} m
Diameter of our Galaxy	7.6×10^{20} m
Distance to nearest star (Proxima Centauri)	4.0×10^{16} m
Earth–Sun distance	1.5×10^{11} m
Radius of Earth	6.4×10^{6} m
Wavelength of radio wave (AM band)	$\sim 3 \times 10^{3}$ m
Length of ship *Queen Elizabeth*	3.1×10^{2} m
Height of man (average male)	1.8 m
Diameter of 5¢ coin	2.1×10^{-2} m
Diameter of red blood cell (human)	7.5×10^{-6} m
Wavelength of visible light	$\sim 5 \times 10^{-7}$ m
Diameter of smallest virus (potato spindle)	2×10^{-8} m
Diameter of atom	$\sim 1 \times 10^{-10}$ m
Diameter of atomic nucleus (iron)	8×10^{-15} m
Diameter of proton	2×10^{-15} m

Table 1.2 MULTIPLES AND SUBMULTIPLES OF THE METER

kilometer	1 km = 10^3 m
meter	1 m
centimeter	1 cm = 10^{-2} m
millimeter	1 mm = 10^{-3} m
micron	1 μm = 10^{-6} m
nanometer	1 nm = 10^{-9} m
Ångstrom	1 Å = 10^{-10} m
fermi	1 f = 10^{-15} m

Table 1.3 MULTIPLES AND SUBMULTIPLES OF THE FOOT

mile	1 mi = 5280 ft = 1609.38 m
yard	1 yd = 3 ft = 0.9144 m
foot	1 ft = 0.3048 m
inch	1 in. = $\frac{1}{12}$ ft = 2.540 cm

1.4 The Unit of Time

The unit of time in both the metric and British systems is the **second.** Originally one second (1 s) was defined as 1/(60 × 60 × 24) or 1/86,400 of a mean solar day. The solar day is the time interval between two successive passages of the Sun over a given meridian on the Earth, say, over the meridian of Greenwich. The length of the solar day depends on the rate of rotation of the Earth, which, unfortunately, is not quite uniform. There are not only periodic seasonal variations in the rotation rate (such variations can be averaged out by recourse to the *mean* solar day rather than an actual day), but also long-term variations.

The seasonal variations are caused by changes in the motions of air, water, and ice on the Earth. The long-term variations are mainly caused by friction that the ocean tides exert on the Earth; this leads to a gradual slowing down of the rotation rate. Thus, between 1900 and now, the time that the Earth takes to complete 365 rotations has increased by about 1 second.

second, s

To avoid such a lack of uniformity in the time scale, we now use an atomic standard of time. This standard is the period of one vibration of an atom of cesium. The second is defined as the time needed for 9,192,631,770 vibrations of a cesium atom. This coincides with the mean solar second as it was in the year 1900.

Figure 1.12 shows one of the atomic clocks at the National Bureau of Standards in Boulder, Colorado. The vibrations of the cesium atoms regulate the rate of this atomic clock just as the vibrations of a balance wheel regulate an ordinary wristwatch or the vibrations of a small tuning fork regulate an Accutron wristwatch. The roomful of complicated equipment shown in Figure 1.12 is needed to amplify the feeble vibrations of cesium atoms to a level that permits them to control the clock. The best cesium clocks are good to 1 part in 10^{13}, i.e., they lose or gain no more than 1 second in 300,000 years.

The time kept by the cesium clocks at the National Bureau of Standards is called **Universal Coordinated Time** (oddly enough, abbreviated UTC). Precise time signals keyed to the UTC time scale are continuously transmitted by the radio station WWV, Fort Collins, Colorado. These time signals can be picked up worldwide on shortwave receivers tuned to 2.5, 5, 10, 15, or 20 megahertz. The time an-

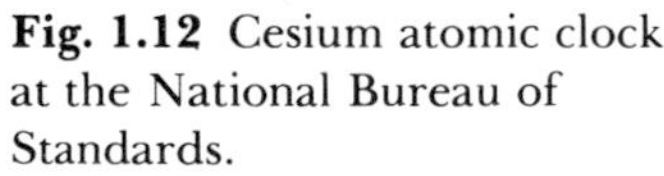

Fig. 1.12 Cesium atomic clock at the National Bureau of Standards.

nounced by WWV is Greenwich Mean Time, which is exactly 5 hours ahead of Eastern Standard Time.[7] Precise time signals are also announced continuously by telephone — the telephone number is (303) 499-7111. The radio signal is accurate to within 1×10^{-3} s,[8] and the telephone signal is accurate to within 30×10^{-3} s.

Incidentally, because the rotation of the Earth is slower than it used to be, Universal Coordinated Time gradually drifts ahead of mean solar time. To avoid any excessive divergence between these two time scales, the National Bureau of Standards inserts a leap second once a year, usually on December 31 — the last minute of the year has 61 seconds rather than 60 seconds.

For extremely high-precision comparisons between clocks located in different laboratories, radio signals are inadequate, because irregularities in their propagation through the atmosphere introduce uncertainties in their travel time from one laboratory to the other. To bypass this difficulty, portable atomic clocks have been built. These can be synchronized with a clock in one laboratory and then carried to another laboratory, transferring the synchronization. Figure 1.13 shows one such portable atomic clock; the clock and its batteries fit in a suitcase. This particular model makes use of the vibrations of rubidium atoms and, under typical transport conditions, it is accurate to 1 part in 10^{12}. The portable clock has been used for comparisons among the clocks of the National Bureau of Standards, the Naval Observatory (Washington, D.C.), and the Bureau International de l'Heure (Paris).

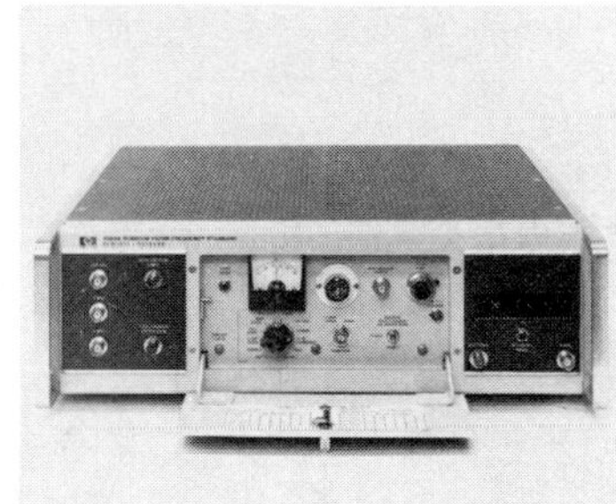

Fig. 1.13 Portable rubidium atomic clock built by Hewlett-Packard.

Table 1.4 lists some typical time intervals. There is a curious coincidence between the numbers appearing in Tables 1.1 and 1.4: the ratio of the longest and shortest lengths of objects in our universe is about 10^{40}, and the ratio of the longest and shortest times associated with objects in our universe is also about 10^{40}. As we will see in Section 1.5, the number 10^{40} recurs in Table 1.6, which lists typical masses: the ratio of the largest and smallest masses of objects in our universe is

Table 1.4 SOME TIME INTERVALS

Age of the universe	$\sim 4 \times 10^{17}$ s
Age of the Solar System	1.4×10^{17} s
Age of the oldest written records (Sumerian)	1.6×10^{11} s
Life-span of man (average)	2.2×10^{9} s
Travel time for light from nearest star	1.4×10^{8} s
Revolution of Earth (1 year)	3.2×10^{7} s
Rotation of Earth (1 day)	8.6×10^{4} s
Life-span of free neutron (average)	9.2×10^{2} s
Travel time for light from Sun	5×10^{2} s
Travel time for light from Moon	1.3 s
Period of heartbeat (human)	~ 0.9 s
Period of sound wave (middle C)	3.8×10^{-3} s
Period of radio wave (AM band)	$\sim 1 \times 10^{-5}$ s
Life-span of π^+ particle (average)	2.6×10^{-8} s
Period of light wave	$\sim 2 \times 10^{-15}$ s
Period of X ray	$\sim 3 \times 10^{-19}$ s
Life-span of shortest lived, unstable particle	$\sim 10^{-24}$ s

[7] Radio station CHU, Ottawa, Ontario, transmits similar time signals at 3.33, 7.33, and 14.67 megahertz.

[8] In order to take advantage of this high accuracy, the receiving station must make an allowance for the travel time of the radio signal. This travel time can amount to as much as $\frac{1}{10}$ s if the receiving station is at a remote location on the Earth's surface.

Table 1.5 MULTIPLES AND SUBMULTIPLES OF THE SECOND

century	1 century = 100 yr = 3.156×10^9 s
year	1 year = 3.156×10^7 s = 365.25 days
day	1 day = 86,400 s
hour	1 h = 3600 s
minute	1 min = 60 s
millisecond	1 ms = 10^{-3} s
microsecond	1 μs = 10^{-6} s
nanosecond	1 ns = 10^{-9} s
picosecond	1 ps = 10^{-12} s

Fig. 1.14 International standard kilogram.

about $(10^{40})^2$. Is this coincidence between these large numbers purely accidental or is it a clue pointing to a hidden connection between cosmology and elementary-particle physics? We do not know . . . yet.

Table 1.5 gives multiples and submultiples of the second.

1.5 The Unit of Mass

kilogram, kg

In the metric system, the unit of mass is the **kilogram.** The standard of mass is a cylinder of platinum–iridium alloy kept at the Bureau International des Poids et Mesures (Figure 1.14). By definition, the mass of this cylinder is exactly one kilogram (1 kg). Mass is the only fundamental unit for which we do not, as yet, have an atomic standard. It would obviously be logical to take the mass of, say, a hydrogen or krypton atom as the standard of mass. But it has so far been impossible to measure the platinum–iridium kilogram in terms of such an atomic standard with satisfactory accuracy. Comparison of an unknown mass with the platinum–iridium kilogram is much easier than comparison with the mass of a single atom.

Fig. 1.15 One-kilogram balance of the National Bureau of Standards.

Mass is measured with a balance, an instrument that compares the **weight** of an unknown mass with the weight of a standard mass. Weight is directly proportional to mass and hence equal weights imply equal masses (the precise distinction between mass and weight will be spelled out in Chapter 6). Figure 1.15 shows an extremely accurate balance especially designed by the National Bureau of Standards to handle the standard kilogram and similar masses. This balance has a pan at one end of the balance beam, and a permanently fixed counterweight at the other. To compare the weights of two masses, each of them in turn is placed in the pan and compared with the counterweight; this yields an indirect comparison of the two masses. This balance can compare two masses of about 1 kg to within a few parts in 10^9.

The masses of individual atoms cannot be measured with such a high precision. The most recent and best determination of the mass of an individual atom (in terms of kilograms) is accurate to within 1 part in 10^6. Essentially, such a determination is equivalent to a determination of **Avogadro's number** N_A, or the number of atoms per mole.[9] For in-

Mole

[9] One **mole** of any chemical element (or chemical compound) is defined as that amount of matter containing exactly as many atoms (or molecules) as there are atoms in 12 grams of carbon. The "atomic mass" of a chemical element (or the "molecular mass" of a compound) is the mass of one mole. Thus, according to the table of atomic masses (see Appendix 9), one mole of C has a mass of 12.0 grams, one mole of O_2 has a mass of 32.0 grams, one mole of H_2O has a mass of 18.0 grams, etc.

stance, since the mass of one mole of carbon is exactly 12 grams, the mass of one carbon atom is

$$m = \frac{12 \text{ grams}}{N_A} \tag{1}$$

To find N_A, it is necessary to count the number of atoms in one mole; this is most conveniently done in a crystal where the regular arrangement of the atoms makes the counting fairly straightforward. The best available data lead to

$$N_A = 6.0220 \times 10^{23} \tag{2}$$

with an uncertainty of about $\pm 3 \times 10^{18}$, or about 5 parts in 10^6.

Because of the uncertainty in N_A, an atomic standard of mass is at present not competitive with the standard kilogram. However, the masses of atoms can be measured *relative to one another* with considerably higher accuracy than relative to the kilogram. For this reason, the masses of atoms are often stated in terms of **atomic mass units.** In these units, the mass of the carbon atom is assigned the standard value

Atomic mass unit, u

$$[\text{mass of carbon atom}] = 12 \text{ atomic mass units} = 12 \text{ u}$$

and other masses of atoms are referred to this standard. In terms of kilograms, the value of 1 u is approximately 1.66057×10^{-27} kg.

Table 1.6 gives some examples of masses expressed in kilograms.

Table 1.7 is a list of multiples and submultiples of the kilogram.

Table 1.6 SOME MASSES

Observable universe	$\sim 10^{53}$ kg
Galaxy	4×10^{41} kg
Sun	2.0×10^{30} kg
Earth	6.0×10^{24} kg
Ship *Queen Elizabeth*	7.6×10^{7} kg
Jet airliner (Boeing 747, empty)	1.6×10^{5} kg
Skylab	7.0×10^{4} kg
Automobile	1.5×10^{3} kg
Man (average male)	73 kg
5¢ coin	5.2×10^{-3} kg
Raindrop	2×10^{-6} kg
Red blood cell	9×10^{-14} kg
Smallest virus (potato spindle)	4×10^{-21} kg
Atom (iron)	9.5×10^{-26} kg
Proton	1.7×10^{-27} kg
Electron	9.1×10^{-31} kg

Table 1.7 MULTIPLES AND SUBMULTIPLES OF THE KILOGRAM

metric ton (tonne)	$1 \text{ t} = 10^{3}$ kg
kilogram	1 kg
gram	$1 \text{ g} = 10^{-3}$ kg
milligram	$1 \text{ mg} = 10^{-6}$ kg
atomic mass unit	$1 \text{ u} = 1.66 \times 10^{-27}$ kg

The British engineering system uses *weight,* rather than mass, as a fundamental unit; the unit of weight is the **pound.** The unit of mass is then regarded as a derived unit; the name of this unit is **slug**:

$$1 \text{ slug} = 14.5939 \text{ kg} \tag{3}$$

However, in practice one often uses the pound-mass:

$$1 \text{ lb-mass} = 0.453592 \text{ kg} \tag{4}$$

1.6 Derived Units

Volume

Length, time, and mass are the fundamental units, or **base units** of the metric system. Any other physical quantity can be measured by introducing a **derived unit** constructed by some combination of the fundamental units. For example, **volume** can be measured with a derived unit that is the cube of the unit of length. Thus in the metric system the unit of volume is the cubic meter; Table 1.8 gives the submultiples of this unit.

Table 1.8 SUBMULTIPLES OF THE CUBIC METER

cubic meter	1 m^3
liter	1 liter = 10^{-3} m^3 = 10^3 cm^3
cubic centimeter	1 cm^3 = 10^{-6} m^3
cubic millimeter	1 mm^3 = 10^{-9} m^3
cubic foot	1 ft^3 = 0.02832 m^3
U.S. gallon	1 gal. = 0.003785 m^3

Density

Similarly, **density** can be measured with a derived unit that is the ratio of a unit of mass and a unit of volume. In the metric system, the unit of density is the kilogram per cubic meter (kg/m^3). We will see that velocity, acceleration, force, etc., are also measured with derived units. Can *all* physical quantities be given units that are some combinations of meters, seconds, and kilograms? Yes, they all can be given such derived units, but sometimes it is very awkward to do so. For example, electric charge could in principle be measured in $(kg)^{1/2} \cdot (m)^{3/2}/s$, but in practice this unit does not lend itself to precise experimental determination, and it is valuable to introduce an entirely different unit of charge, independent of meters, seconds, and kilograms.

In essence, three fundamental units are always sufficient because physics, in a way, reduces all the complicated properties of the world to particle properties. When faced with a complicated physical quantity, such as electric charge, the physicist always asks how this quantity would affect particles. And particles can be described totally by position as a function of time and by mass — requiring three, and only three, units. Note that there is no harm, and there may be some convenience, in using a system with more than three "fundamental" units; however, in such a system one or more of the "fundamental" units will in principle be redundant.

International System of Units (SI)

The **International System of Units** or **SI**,[10] used in this book, is the most widely accepted system of units for mechanical, thermodynamic, electric, and optical measurements. It is based on six "fundamental" units: the meter for length, the second for time, the kilogram for mass, the kelvin for temperature, the ampere for electric current, and the candela for luminosity of a light source.[11] The last three of these units are, strictly, redundant: temperature, electric current, and luminosity can be measured with suitable combinations of meters, seconds, and kilograms, but for practical reasons it happens to be convenient to make up independent units for these quantities.

The following examples give some hints for the solution of the problems at the end of the chapter.

EXAMPLE 1. The light-year is a unit of length often used by astronomers to measure the distance between stars. One light-year is the distance that light travels in one year. Given that the speed of light is 3.00×10^8 m/s, express one light-year in meters.

SOLUTION: According to Table 1.5 there are 3.156×10^7 s in one year. The distance that light travels in one year is therefore

$$[\text{distance}] = [\text{speed}] \times [\text{time}]$$

$$= 3.00 \times 10^8 \,\frac{\text{m}}{\text{s}} \times 3.156 \times 10^7 \text{ s}$$

$$= 9.47 \times 10^{15} \text{ m}$$

Note that in the second line of this calculation, the units of time cancel, and the final result has units of length, as it should. It is a general rule that in any calculation with the equations of physics, the units are to be multiplied and divided as though they were algebraic quantities, and this automatically yields the correct units for the final result — if it does not, there is some mistake in the calculation. Thus we will always find it worthwhile to keep track of the units in our calculations, because this provides us with more protection against costly mistakes. Any failure in the expected cancellations is a sure sign of trouble!

Furthermore, note that the final result has been rounded off to three significant figures, the same as the number of significant figures specified in the given data. Any extra significant figures in the final result would be unreliable and misleading. In fact, even the third significant figure in the answer is not quite reliable — a calculation starting with a more precise value of the speed of light shows that, to within three significant figures, one light-year equals 9.46×10^{15} m, and not 9.47×10^{15} m. It is always wise to question the accuracy of the last significant figure.

EXAMPLE 2. The density of water is 1.0×10^3 kg/m^3. Express this in g/cm^3 and in lb-mass/ft^3.

SOLUTION: Since 1 kg = 10^3 g and 1 m = 10^2 cm, we find

$$1.0 \times 10^3 \,\frac{\text{kg}}{\text{m}^3} = 1.0 \times 10^3 \times \frac{10^3 \text{ g}}{(10^2 \text{ cm})^3} = 1.0 \times 10^3 \times \frac{10^3 \text{ g}}{10^6 \text{ cm}^3}$$

$$= 1.0 \,\frac{\text{g}}{\text{cm}^3}$$

---

[10] Système International.

[11] Officially, the list of fundamental units of the SI system also includes the mole as a unit of "amount of substance."

Similarly, since 1 kg = 1/0.454 lb-mass and 1 m = 1/0.305 ft,

$$1.0 \times 10^3 \frac{\text{kg}}{\text{m}^3} = 1.0 \times 10^3 \times \frac{1/0.454 \text{ lb-mass}}{(1/0.305 \text{ ft})^3}$$

$$= 1.0 \times 10^3 \times \frac{(0.305)^3 \text{ lb-mass}}{0.454 \text{ ft}^3} = 62 \frac{\text{lb-mass}}{\text{ft}^3}$$

Conversion factors

An alternative method for the conversion of units from one system to another makes use of **conversion factors** and involves the following: Since 1 kg = 10^3 g, we have the identity

$$1 = \frac{10^3 \text{ g}}{1 \text{ kg}}$$

and, similarly,

$$1 = \frac{1 \text{ m}}{10^2 \text{ cm}}$$

This means that any quantity can be multiplied by 10^3 g/1 kg or 1 m/10^2 cm without changing its value. Thus, starting with 1.0×10^3 kg/m^3, we obtain

$$1.0 \times 10^3 \frac{\text{kg}}{\text{m}^3} = 1.0 \times 10^3 \frac{\text{kg}}{\text{m}^3} \times \frac{10^3 \text{ g}}{1 \text{ kg}} \times \left(\frac{1 \text{ m}}{10^2 \text{ cm}}\right)^3$$

$$= 1.0 \times 10^3 \times 10^{-3} \frac{\cancel{\text{kg}}}{\cancel{\text{m}^3}} \frac{\text{g}}{\cancel{\text{kg}}} \frac{\cancel{\text{m}^3}}{\text{cm}^3}$$

$$= 1.0 \frac{\text{g}}{\text{cm}^3}$$

Note that the kg and m cancel and only g/cm^3 is left, as desired.

Factors such as 10^3 g/kg and 1 m/10^2 cm, whose numerical values are 1, are called *conversion factors.* To change the units of a quantity, simply multiply the quantity by whatever conversion factors will bring about the cancellation of the old units.

EXAMPLE 3. We can obtain a rough estimate of the size of a molecule by means of the following simple experiment. Take a droplet of oil and let it spread out on a smooth surface of water. When the oil slick attains its maximum area, it consists of a monomolecular layer, i.e., it consists of a single layer of oil molecules which stand on the surface side by side. Given that an oil droplet of mass 8.4×10^{-7} kg and of density 920 kg/m^3 spreads out into an oil slick of maximum area 0.55 m^2, calculate the length of an oil molecule.

SOLUTION: The volume of the oil droplet is

$$[\text{volume}] = [\text{mass}]/[\text{density}]$$

$$= \frac{8.4 \times 10^{-7} \text{ kg}}{920 \text{ kg/m}^3} = 9.1 \times 10^{-10} \text{ m}^3$$

The volume of the oil slick must be exactly the same. This latter volume can be expressed in terms of the thickness and the area of the oil slick:

$$[\text{volume}] = [\text{thickness}] \times [\text{area}]$$

Consequently,

$$[\text{thickness}] = [\text{volume}]/[\text{area}]$$

$$= \frac{9.1 \times 10^{-10}\ \text{m}^3}{0.55\ \text{m}^2} = 1.7 \times 10^{-9}\ \text{m}$$

The length of a molecule is the same as this thickness, i.e., 1.7×10^{-9} m.

EXAMPLE 4. Some engineers have proposed that for long-distance travel between cities we should dig perfectly straight connecting tunnels through the Earth (Figure 1.16). A train running along such a tunnel would initially pick up speed in the first half of the tunnel as if running downhill; it would reach maximum speed at the midpoint of the tunnel; and it would gradually slow down in the second half of the tunnel, as if running uphill. Suppose such a tunnel were dug between San Francisco and Washington, D.C. The distance between these cities, measured along the Earth's surface, is 3900 km.
(a) What is the distance along the straight tunnel?
(b) What is the depth of the tunnel at its midpoint, somewhere below Kansas?
(c) What is the downward slope of the tunnel relative to the horizontal direction at San Francisco?

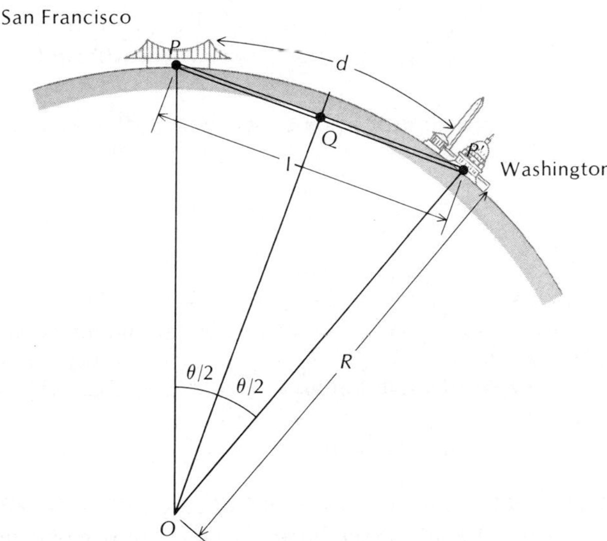

Fig. 1.16 Cross section through the Earth showing San Francisco and Washington and the straight tunnel connecting them. The distance d is measured along the surface of the Earth; the distance l is the length of the tunnel. The distance R is the radius of the Earth.

SOLUTION: (a) Figure 1.16 is a cross section through the Earth in the plane that contains the center of the Earth (O) and the two cities (P and P'). The circle represents the surface of the Earth. Figure 1.16 shows the tunnel connecting the two cities and the radial lines from the center of the Earth to the two cities. The angle, in radians, between these two radial lines is

$$\theta = \frac{[\text{distance along circle}]}{[\text{radius}]} = \frac{d}{R} \qquad (5)$$

With $d = 3.9 \times 10^3$ km and $R = 6.4 \times 10^3$ km, this yields

$$\theta = \frac{3.9 \times 10^3\ \text{km}}{6.4 \times 10^3\ \text{km}} = 0.609\ \text{radian}$$

To convert radians to degrees, we note that 2π radians $= 360°$, so the conversion factor is $360°/2\pi$ radians:

$$\theta = 0.609 \text{ radian} \times 360°/2\pi \text{ radian} = 34.9°$$

The distance QP is one side of the triangle OQP. Trigonometry tells us that in this triangle

$$QP = OP \sin(\theta/2)$$

or

$$QP = R \sin(\theta/2)$$

The length of the tunnel is twice QP:

$$l = 2QP = 2R \sin(\theta/2)$$

With $R = 6.4 \times 10^3$ km and $\theta = 34.9°$,

$$l = 2 \times 6.4 \times 10^3 \text{ km} \times \sin(34.9°/2)$$

$$= 3.8 \times 10^3 \text{ km}$$

(b) The distance OQ is another side of the right triangle OQP. Trigonometry tells us that

$$OQ = OP \cos(\theta/2) = R \cos(\theta/2)$$

The depth h of the tunnel at its midpoint equals the difference between R and OQ,

$$h = R - OQ = R - R \cos(\theta/2)$$

$$= R[1 - \cos(\theta/2)]$$

which gives

$$h = 6.4 \times 10^3 \text{ km} \times [1 - \cos(34.9°/2)]$$

$$= 2.9 \times 10^2 \text{ km}$$

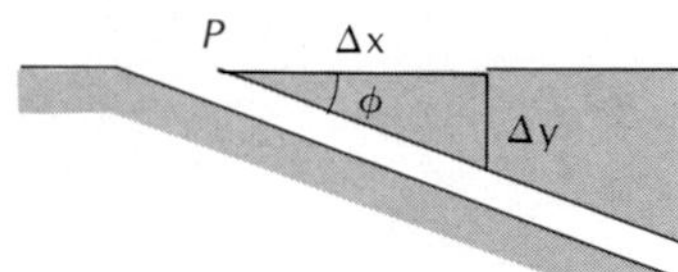

Fig. 1.17 At San Francisco, the tunnel goes down a vertical distance Δy while it advances a horizontal distance Δx. The ratio $\Delta y/\Delta x$ is the slope of the tunnel.

(c) The slope of a line relative to the horizontal direction is defined as the ratio of the vertical increment to the horizontal increment or decrement (Figure 1.17):

$$[\text{slope}] = \frac{\Delta y}{\Delta x} \tag{6}$$

By trigonometry, this is simply the tangent of the angle between the line and the horizontal,

$$[\text{slope}] = \tan \phi \tag{7}$$

Since the horizontal at San Francisco is perpendicular to OP, and the line of the tunnel is perpendicular to OQ, the angle ϕ equals the angle $\theta/2$. Thus

$$[\text{slope}] = \tan \phi = \tan(\theta/2) = \tan(34.9°/2) = 0.31$$

Engineers usually write slopes as ratios; in this notation, the slope is 31:100. Thus, at San Francisco the tunnel has to go down 31 m vertically for every 100 m it goes horizontally.

SUMMARY

Fundamental assumptions of classical physics:

The motion of a particle can be described by position as a function of time.
Space is three dimensional.
Space is Euclidean.
Time is absolute.

Reference frame: A coordinate grid with a set of synchronized clocks.

Units of length, time, and mass: meter, second, kilogram.

Standards of length, time, and mass: speed of light implemented by stabilized laser, cesium atomic clock, cylinder of platinum–iridium.

QUESTIONS

1. Try to estimate by eye the lengths, in centimeters or meters, of a few objects in your immediate environment. Then measure them with a ruler or meter stick. How good were your estimates?

2. How close is your watch to standard time right now? Roughly how many minutes does your watch gain or lose per month?

3. What is meant by the phrase *a point in time*?

4. Mechanical clocks (with pendulums) were not invented until the tenth century A.D. What clocks were used by the ancient Greeks and Romans?

5. By counting aloud "One, two, three, . . . , ten, one, . . ." at a reasonable rate you can measure seconds fairly accurately. Try to measure 30 seconds in this way. How good a timekeeper are you?

6. Pendulum clocks are affected by the temperature and pressure of air. Why?

7. During the debate over the "densepack" configuration of buried silos for American intercontinental ballistic missiles (ICBMs), the proponents argued that Soviet ICBMs would be ineffective against the "densepack" because the explosion of the first few incoming Soviet ICBMs would destroy most of the other incoming ICBMs ("fratricide"). A physicist was quick to point out (*The New York Times,* December 1, 1982) that the Soviets could avoid "fratricide" by making all their ICBMs explode at almost exactly the same instant. What method could be used to bring about simultaneous explosions within, say, 1 microsecond?

8. In 1761 an accurate chronometer built by John Harrison was tested aboard H.M.S. *Deptford* during a voyage at sea for 5 months. During this voyage, the chronometer accumulated an error of less than 2 minutes. For this achievement, Harrison was ultimately awarded a prize of £20,000 that the British government had offered for the discovery of an accurate method for the determination of geographical longitude at sea. Explain how the navigator of a ship uses a chronometer and observation of the position of the Sun in the sky to find longitude.

9. Lecky's *Wrinkles in Practical Navigation,* a famous nineteenth-century textbook of celestial navigation, recommends that each ship carry three chronometers for accurate timekeeping. What can the navigator do with three chronometers that he cannot do with two?

10. Suppose that by an "act of God" (or by the act of a thief) the standard kilogram at Sèvres were destroyed. Would this destroy the metric system?

11. Estimate the masses, in grams or kilograms, of a few bodies in your environment. Check the masses with a balance if you have one available.

12. Consider the piece of paper on which this sentence is printed. If you had available suitable instruments, what physical quantities could you measure about this piece of paper? Make the longest list you can and give the units. Are all these units derived from the meter, second, and kilogram?

13. Could we take length, time, and density as the three fundamental units? What could we use as a standard of density?

14. Could we take length, mass, and density as the three fundamental units? Length, mass, and speed?

PROBLEMS

Section 1.3

1. What is your height in feet? In meters?

2. With a ruler, measure the thickness of this book, excluding the cover. Deduce the thickness of each of the sheets of paper making up the book.

3. A football field measures 100 yd $\times$ $53\frac{1}{3}$ yd. Express this in meters.

4. Express the last four entries in Table 1.1 in inches.

5. Express the following fractions of an inch in millimeters: $\frac{1}{2}$, $\frac{1}{4}$, $\frac{1}{8}$, $\frac{1}{16}$, $\frac{1}{32}$, and $\frac{1}{64}$ in.

6. As seen from the Earth, the Sun has an angular diameter of 0.53°. The distance between the Earth and the Sun is 1.5×10^{11} m. From this, calculate the radius of the Sun.

7. Analogies can often help us to imagine the very large or very small distances that occur in astronomy or in atomic physics.
   (a) If the Sun were the size of a grapefruit, how large would the Earth be? How far away would the nearest star be?
   (b) If your head were the size of the Earth, how large would an atom be? How large would a red blood cell be?

8. The Earth is approximately a sphere of radius 6.37×10^6 m. Calculate the distance from the pole to the equator, measured along the surface of the Earth. Calculate the distance from the pole to the equator, measured along a straight line passing through the Earth.

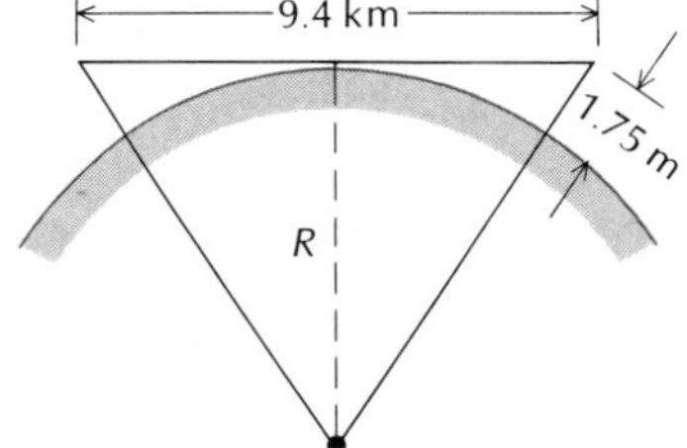

Fig. 1.18 The distance between the physicist and the pole is 9.4 km.

9. A nautical mile (nmi) equals 1.151 mi, or 1852 m. Show that a distance of 1 nmi along a meridian of the Earth corresponds to a change of latitude of 1 minute of arc.

10. A physicist plants a vertical pole at the waterline on the shore of a calm lake. When she stands next to the pole, its top is at eye level, 175 cm above the waterline. She then rows across the lake and walks along the waterline on the opposite shore until she is so far away from the pole that her entire view of it is blocked by the curvature of the surface of the lake, i.e., the entire pole is below the horizon (Figure 1.18). She finds that this happens when her distance from the pole is 9.4 km. From this information, deduce the radius of the Earth.

Section 1.4

11. What is your age in days? In seconds?

12. The age of the Earth is 4.5×10^9 years. Express this in seconds.

13. Each day at noon a mechanical wristwatch was compared with WWV time signals. The watch was not reset. It consistently ran late, as follows: June 24, late 4 s; June 25, late 20 s; June 26, late 34 s; June 27, late 51 s.
   (a) For each of the three 24-hour intervals, calculate the rate at which the wristwatch lost time. Express your answer in seconds lost per hour.
   (b) What is the average of the rates of loss found in part (a)?
   (c) When the wristwatch shows 10^h30^m on June 30, what is the correct WWV time? Do this calculation with the average rate of loss of part (b) and also with the largest of the rates of loss found in part (a). Estimate to within how many seconds the wristwatch can be trusted on June 30 after the correction for rate of loss has been made.

14. The navigator of a sailing ship seeks to determine his longitude by observing at what time (Greenwich Mean Time) the Sun reaches the zenith at his position (local noon). Suppose that the navigator's chronometer is in error and is late by 1 s compared to Greenwich Mean Time. What will be the consequent error of longitude (in minutes of arc)? What will be the error in position (in miles) if the ship is on the equator?

Section 1.5

15. What is your mass in pound-mass? In slugs? In kilograms? In atomic mass units?

16. What percentage of the mass of the Solar System is in the planets? What percentage is in the Sun? Use the data given in the table printed on the endpapers of the book.

17. The atomic mass of fissionable uranium is 235.0 g. What is the mass of a single uranium atom? Express your answer in kilograms and in atomic mass units.

18. The atom of uranium consists of 92 electrons, each of mass 9.1×10^{-31} kg, and a nucleus. What percentage of the total mass is in the electrons and what percentage is in the nucleus of the atom?

19. How many water molecules are there in 1 gallon (3.79 liters) of water? How many oxygen atoms? Hydrogen atoms?

20. (a) How many molecules of water are there in one cup of water? A cup is about 250 cm^3.
   (b) How many molecules of water are there in the ocean? The total volume of the ocean is 1.3×10^{18} m^3.
   (c) Suppose you pour a cup of water into the ocean, allow it to become thoroughly mixed, and then take a cup of water out of the ocean. On the average, how many of the molecules originally in the cup will again be in the cup?

21. How many molecules are there in one cubic centimeter of air? Assume that the density of air is 1.3 kg/m^3 and that it consists entirely of nitrogen molecules (N_2).

22. How many atoms are there in the Sun? The mass of the Sun is 1.99×10^{30} kg and its chemical composition (by mass) is approximately 70% hydrogen and 30% helium.

23. The chemical composition of air is (by mass) 75.5% N_2, 23.2% O_2, and 1.3% Ar. What is the average "molecular mass" of air; i.e., what is the mass of 6.02×10^{23} molecules of air?

24. How many atoms are there in a human body of 73 kg? The chemical composition (by mass) of a human body is 65% oxygen, 18.5% carbon, 9.5% hydrogen, 3.3% nitrogen, 1.5% calcium, 1% phosphorus, and 0.35% other elements (ignore the "other elements" in your calculation).

Section 1.6

25. The distance from our Galaxy to the Andromeda galaxy is 2.2×10^6 light-years. Express this distance in meters.

26. In analogy with the light-year, we can define the light-second as the distance light travels in one second and the light-minute as the distance light travels in one minute. Express the Earth–Sun distance in light-minutes. Express the Earth–Moon distance in light-seconds.

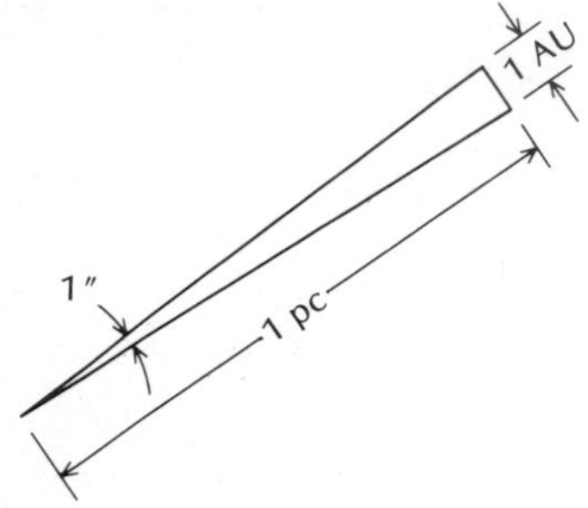

Fig. 1.19

27. Astronomers often use the astronomical unit (AU), the parsec (pc), and the light-year. The AU is the distance from the Earth to the Sun[12]; 1 AU = 1.496×10^{11} m. The pc is the distance at which 1 AU subtends an angle of exactly 1 second of arc (Figure 1.19). The light-year is the distance that light travels in 1 year.
   (a) Express the pc in AU.
   (b) Express the pc in light-years.
   (c) Express the pc and the light-year in meters.

28. The density of copper is 8.9 g/cm^3. Express this in kg/m^3, lb-mass/ft^3, lb-mass/in.^{3}, and slug/ft^3.

29. The federal highway speed limit is 55 mi/h. Express this in kilometers per hour, feet per second, and meters per second.

30. What is the volume of an average human body? (Hint: The density of the body is about the same as that of water.)

31. Meteorologists usually report the amount of rain in terms of the depth in inches to which the water would accumulate on a flat surface if it did not run off. Suppose that 1 in. of rain falls during a storm. Express this in cubic meters of water per square meter of surface. How many kilograms of water per square meter of surface does this amount to?

32. A fire hose delivers 600 gallons of water per minute. Express this in m^3/s. How many kilograms of water per second does this amount to? One gallon is 3.79 liters.

33. The total volume of the oceans of Earth is 1.3×10^{18} m^3. What percentage of the mass of the Earth is in the oceans?

34. The nucleus of an iron atom is spherical and has a radius of 4.6×10^{-15} m; the mass of the nucleus is 9.5×10^{-26} kg. What is the density of the nuclear material? Express your answer in metric tons per cubic centimeter.

35. Our Sun has a radius of 7.0×10^8 m and a mass of 2.0×10^{30} kg. What is the average density of the Sun? Express your answer in grams per cubic centimeter.

36. Pulsars, or neutron stars, typically have a radius of 20 km and a mass equal to that of our Sun (2.0×10^{30} kg). What is the average density of such a pulsar? Express your answer in metric tons per cubic centimeter.

37. The nuclei of all atoms have approximately the same density of mass. The nucleus of a copper atom has a mass of 1.06×10^{-25} kg and a radius of 4.8×10^{-15} m. The nucleus of a lead atom has a mass of 3.5×10^{-25} kg; what is its radius? The nucleus of an oxygen atom has a mass of 2.7×10^{-26} kg; what is its radius? Assume that the nuclei are spherical.

38. The table printed on the endpapers gives the masses and radii of the major planets. Calculate the average density of each planet and make a list of the planets in order of decreasing densities. Is there a correlation between the density of a planet and its distance from the Sun?

---

[12] Strictly, it is the semimajor axis of the Earth's orbit.

CHAPTER 2

Kinematics in One Dimension

Kinematics is the study of the geometry of motion: it deals with the mathematical description of motion in terms of position, velocity, and acceleration. Kinematics serves as a prelude to dynamics, which studies force as the cause of changes in motion. In this chapter, and in the next eight chapters, we will only be concerned with **translational motion** of a particle, which is defined as change of position of the particle as a function of time. In the case of an ideal particle — a body with no size and no internal structure — position as a function of time gives a complete description. In the case of a more complicated body — a man, a ship, or a planet — position as a function of time does not give a complete description. Such a complicated body has many internal parts which can rotate and move in relation to one another. Nevertheless, insofar as we are not interested in the size, orientation, and internal structure of a body, we may find it useful to concentrate on its translational motion and ignore all other motions. Under these circumstances we may pretend that the motion of the complicated body is particle motion. The position of the body is then to be specified by the position of some fiducial point[1] marked on it, such as its center, or its leading edge. For example, we can describe the motion of a ship steaming out of New York harbor as particle motion — for most purposes it will be sufficient to know the position of the center of the ship as a function of time. However, if the ship were to suffer a collision, we could not describe the motion of the ship during the collision as particle motion — in this case it would be essential to know the changes in internal structure and in orientation as a function of time.

Translational motion

[1] A fiducial point is a reference point, which we imagine painted on the body.

2.1 Average Speed

Average speed

Consider a particle in motion, i.e., a particle whose position changes as a function of time. The particle travels along a path, straight or curved. In some given time interval the particle will travel a certain distance measured along the path. The **average speed** of the particle is defined as the ratio of this distance to the magnitude of the time interval,

$$[\text{average speed}] = \frac{[\text{distance traveled}]}{[\text{time taken}]} \tag{1}$$

We see from this equation that the unit of speed is the unit of length divided by the unit of time. In the metric system, the unit of speed is the meter per second (m/s). In the British system, the unit of speed is the foot per second (ft/s). Table 2.1 gives some examples of typical speeds.

Table 2.1 SOME SPEEDS

Light	3.0×10^8 m/s
Recession of fastest known quasar	2.7×10^8 m/s
Electron around nucleus (hydrogen)	2.2×10^8 m/s
Earth around Sun	3.0×10^3 m/s
SST airliner (Tu-144, maximum airspeed)	7.1×10^2 m/s
Rifle bullet (muzzle velocity)	$\sim 7 \times 10^2$ m/s
Rotation of Earth (at equator)	4.6×10^2 m/s
Random motion of molecules in air (average)	4.5×10^2 m/s
Sound	3.3×10^2 m/s
Jet airliner (Boeing 747, maximum airspeed)	2.7×10^2 m/s
Cheetah (maximum)	28 m/s
Federal highway speed limit (55 mi/h)	25 m/s
Man (maximum)	12 m/s
Man (walking briskly)	1.3 m/s
Snail	$\sim 10^{-3}$ m/s
Glacier	$\sim 10^{-6}$ m/s
Rate of growth of hair (human)	3×10^{-9} m/s
Continental drift	$\sim 10^{-9}$ m/s
Rate of subsidence of Southeast England	$\sim 10^{-10}$ m/s

EXAMPLE 1. A runner takes 10 seconds to run a 100-meter dash. What is his average speed relative to the Earth?

SOLUTION: According to Eq. (1),

$$[\text{average speed}] = \frac{[\text{distance}]}{[\text{time taken}]} = \frac{100\ \text{m}}{10\ \text{s}} = 10\ \text{m/s}$$

EXAMPLE 2. The Earth moves around the Sun in a circular orbit of radius 1.50×10^8 km. What is the average speed of the Earth relative to the Sun?

SOLUTION: The Earth moves once around its orbit in 1 year, i.e., in 3.16×10^7 s. The distance covered in this time is $2\pi \times 1.50 \times 10^8$ km. Hence

$$[\text{average speed}] = \frac{2\pi \times 1.50 \times 10^8 \text{ km}}{3.16 \times 10^7 \text{ s}} = 29.8 \text{ km/s}$$

Motion and speed are *relative:* the value of the speed depends on the frame of reference with respect to which it is calculated. Example 1 gives the speed of the runner *relative to the surface of the Earth.* However, since the Earth is moving around the Sun, his speed *relative to the Sun* is approximately 29.8 km/s. Thus, questions regarding speed are meaningless unless the frame of reference is first specified. In everyday language "speed" usually means speed relative to the Earth's surface. The reference frame relative to which the speed is reckoned will usually be clear from the context. For example, in Table 2.1 the speed of the rifle bullet is reckoned relative to the rifle (which may be in motion relative to the Earth). We will be careful to specify the frame of reference whenever it is not clear from the context.

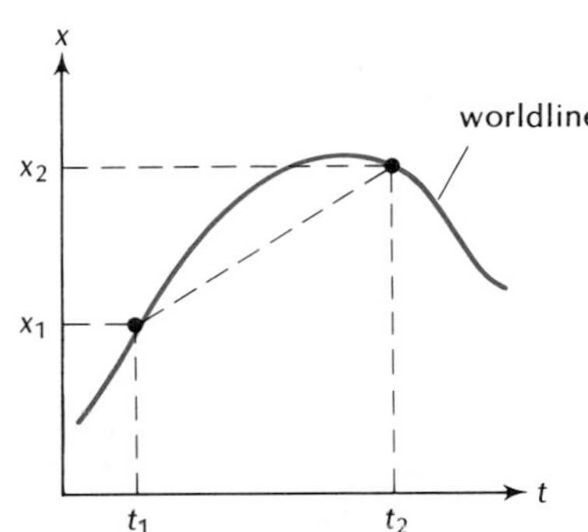

Fig. 2.1 Position as a function of time for a particle moving along the x axis. The curve in this plot is called the *worldline.* The straight line connecting the two points on the worldline has a slope $(x_2 - x_1)/(t_2 - t_1)$; this slope is the average velocity.

2.2 Average Velocity in One Dimension

For the rest of this chapter, we will consider the special case of motion in one dimension, that is, motion along a straight line. Without loss of generality we may assume that this straight line coincides with the x axis. The motion of the particle is then described by giving its x coordinate as a function of time. Graphically, the motion can be represented as a plot of the x coordinate vs. the time coordinate. For example, Figure 2.1 is such a plot of position vs. time for a particle moving along a straight line. The curve shown in Figure 2.1 gives a complete description of the (translational) motion. This curve is called the **worldline** of the particle.

Worldline

Suppose that at time t_1 the particle is at position x_1 and at a subsequent time t_2 the particle is at position x_2. The **average velocity** of the particle in this time interval is then defined as the change of position divided by the change of time,

$$\bar{v} = \frac{x_2 - x_1}{t_2 - t_1} \tag{2}$$

This expression may also be written as

$$\boxed{\bar{v} = \frac{\Delta x}{\Delta t}} \tag{3}$$

Average velocity

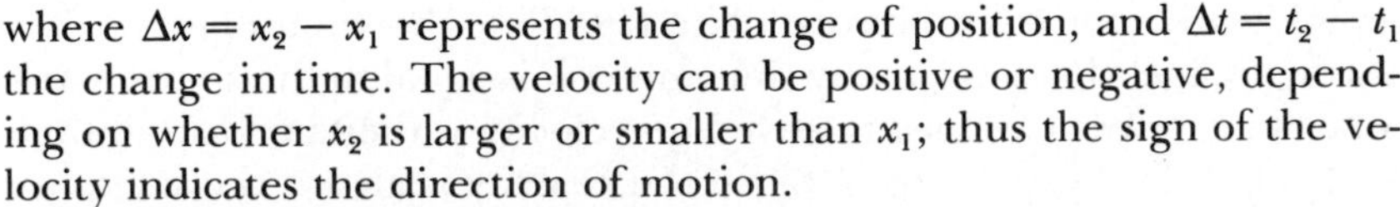

where $\Delta x = x_2 - x_1$ represents the change of position, and $\Delta t = t_2 - t_1$ the change in time. The velocity can be positive or negative, depending on whether x_2 is larger or smaller than x_1; thus the sign of the velocity indicates the direction of motion.

Graphically, the average velocity is the slope of the straight line connecting the points corresponding to $t = t_1$ and $t = t_2$ on the worldline (Figure 2.1).

The average velocity takes into account only the net change in position; it does not take into account how this change in position is accomplished in detail. For example, for any back and forth motion (round trip) in which the particle returns to its initial position, the average *velocity* [see Eq. (2)] is zero, just as though the particle had not moved at all. On the other hand, the average *speed* [see Eq. (1)] takes into account only the distance traveled; it does not take into account the direction of travel (speed is always a positive number) or whether the travel produced any net change of position. The following example will help to make this clear.

EXAMPLE 3. A runner runs 100 m on a straight track in 10 s and then walks back in 80 s. What are the average velocity and the average speed for each part of this motion and for the complete motion?

SOLUTION: The worldline of the runner is shown in Figure 2.2. The motion has two parts: the run ($0 \text{ s} \leq t \leq 10 \text{ s}$) and the walk ($10 \text{ s} \leq t \leq 90 \text{ s}$). The average velocity for the run is

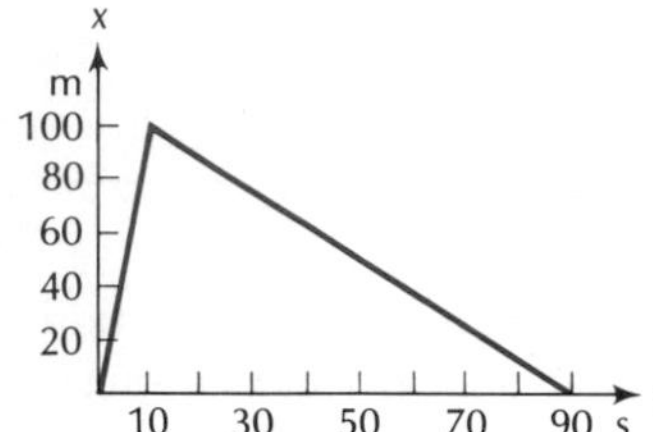

Fig. 2.2 Worldline of a runner.

$$\bar{v} = \frac{\Delta x}{\Delta t} = \frac{+100 \text{ m}}{10 \text{ s}} = +10 \text{ m/s}$$

The average velocity for the walk is

$$\bar{v} = \frac{-100 \text{ m}}{80 \text{ s}} = -1.2 \text{ m/s}$$

The average velocity for the entire motion is

$$\bar{v} = \frac{0 \text{ m}}{90 \text{ s}} = 0 \text{ m/s}$$

This is zero because the net change of position is zero.

The average speeds for the run and walk are, respectively, 10 m/s and 1.2 m/s. The average speed for the entire motion is

$$\frac{200 \text{ m}}{90 \text{ s}} = 2.2 \text{ m/s}$$

2.3 Instantaneous Velocity

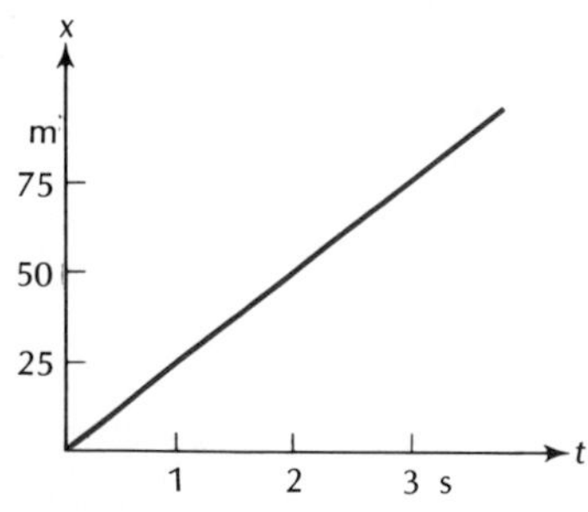

Fig. 2.3 Worldline of an automobile moving at constant velocity.

Figure 2.3 is a plot of the position of an automobile that is moving along a straight road at a constant velocity of 25 m/s (55 mi/h). The translational motion of the automobile can be regarded as particle motion. The position is measured from a fiducial point marked on the automobile to a point on the road. The worldline shown in Figure 2.3 is a straight line; the slope in any time interval is equal to the slope in any other time interval. Thus, the average velocity is the same for all time intervals; it is always 25 m/s. Since the velocity is always the same, we may regard the *instantaneous* velocity for this motion as identical to the average velocity.

Figure 2.4 is a plot of the position of an automobile that starts from rest, accelerates for 10 seconds, and then brakes and comes to a full stop 4.3 seconds later (the plot is based on data from a road test of a

Maserati sports car). Obviously, the velocity of this automobile is not constant — the worldline is not a straight line. How can we determine the velocity at each instant?

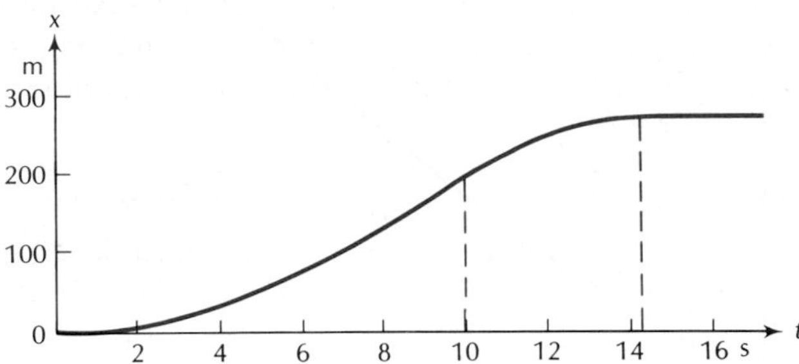

Fig. 2.4 Worldline of an automobile that accelerates (0 s $< t <$ 10 s), then brakes (10 s $< t <$ 14.3 s), and then stops ($t >$ 14.3 s) (based on data from a road test of a Maserati Bora by *Road & Track* magazine).

Consider the instant $t = 4$ s. In order to find an *approximate* value for the velocity at this instant, we take a small time interval of, say, 0.001 s centered on 4 s, that is, a time interval from 3.9995 to 4.0005 s. In this time interval the automobile covers some small distance Δx and we can approximate the actual (curved) worldline of the automobile by a straight worldline segment (Figure 2.5). According to the above paragraph, the instantaneous velocity associated with a straight worldline is simply the slope of the worldline; hence the instantaneous velocity at $t = 4$ s can be evaluated approximately as the slope of the short line segment shown in Figure 2.5. Whether this is a good approximation depends on how closely the straight line segment coincides with the actual curved worldline. Obviously, the approximation can be improved by taking a shorter time interval, 0.0001 s or even less. In the limiting case of extremely small time intervals (infinitesimal time intervals), the straight worldline segment has the direction of the

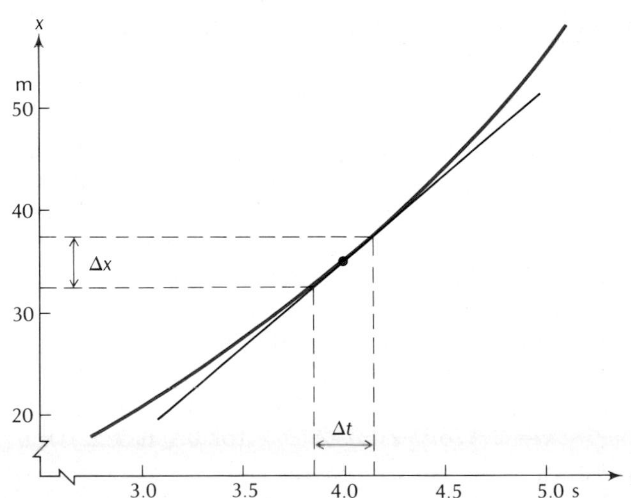

Fig. 2.5 Over a small time interval, we can approximate the worldline by a straight line.

Instantaneous velocity as slope

tangent to the worldline at point $t = 4$ s. Hence the instantaneous velocity at that time equals the *slope of the tangent to the worldline at that time.* For example, drawing the tangent at $t = 4$ s (Figure 2.6) and measuring its slope on the graph, we readily find that this slope is 17.0 m/s; hence the instantaneous velocity is $v = 17.0$ m/s at $t = 4$ s.

By drawing tangents at other points of the worldline and measuring their slope, we can obtain a complete table of instantaneous velocities

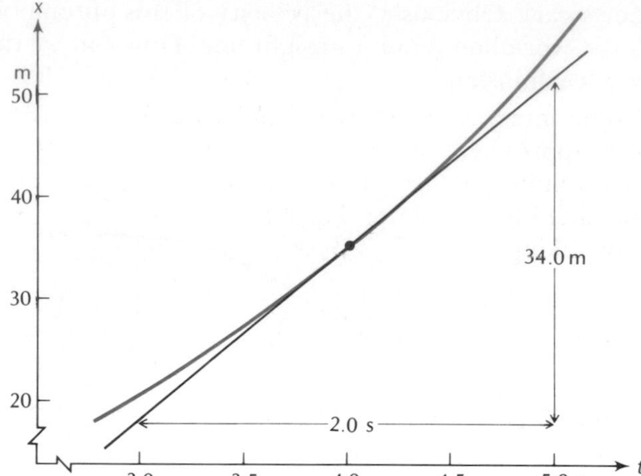

Fig. 2.6 The tangent to the worldline has a slope of 34.0 m/2.0 s = 17.0 m/s.

as a function of time. Figure 2.7 is a plot of the results of such a determination of the instantaneous velocities. The velocity is initially zero (zero slope in Figure 2.4), then increases, reaching a maximum of 34.9 m/s at $t = 10$ s (maximum slope in Figure 2.4), and finally decreases to zero at $t = 14.3$ s (zero slope in Figure 2.4).

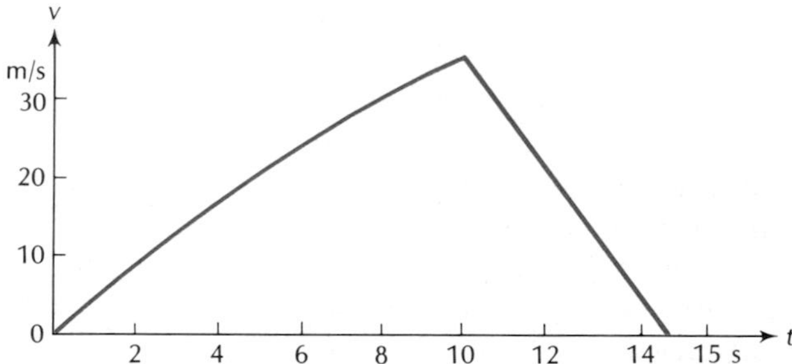

Fig. 2.7 Instantaneous velocity as a function of time.

The above is a graphical method for the determination of the instantaneous velocity; this method is convenient if the data have been given as a plot of position vs. time (Figure 2.4). However, if the data have been given analytically (as formulas), then an analytic method (based on formulas) is more convenient. Suppose that the worldline is described by an equation that gives x as a function of t. For example, the position of the automobile plotted in Figure 2.4 is described by the formulas

$$x = 2.376t^2 - 0.042t^3 \qquad \text{for } 0 \text{ s} \leq t \leq 10 \text{ s} \tag{4}$$

and

$$x = -555.8 + 115.4t - 4.022t^2 \qquad \text{for } 10 \text{ s} \leq t \leq 14.3 \text{ s} \tag{5}$$

where the distance is measured in meters and the time in seconds. As we will see in Eqs. (9) and (10), the instantaneous velocity can then be calculated directly from these expressions. We recall that as a first approximation to the instantaneous velocity, we took the average velocity

$$\bar{v} = \frac{\Delta x}{\Delta t} \tag{6}$$

for a small time interval $t = 0.001$ s centered around $t = 4$ s. To improve on this approximation, we took an even smaller time interval and finally an infinitesimal time interval. Thus the exact instantaneous velocity is obtained by evaluating Eq. (6) in the limiting case of vanishing Δt, i.e., by taking the limit as Δt tends to zero,

$$v = \lim_{\Delta t \to 0} \frac{\Delta x}{\Delta t} \tag{7}$$

In the notation of differential calculus this limit is written as

$$\boxed{v = \frac{dx}{dt}} \tag{8}$$

Instantaneous velocity as derivative

and is called the *derivative of x with respect to t.* Thus, the **instantaneous velocity** is the derivative of the position with respect to time.

According to the theorems of calculus, the derivative of the function ct^n with respect to time is cnt^{n-1} (where c is some arbitrary constant), that is,

$$\frac{d(ct^n)}{dt} = cnt^{n-1}$$

Furthermore, the derivative of the sum of two such functions is the sum of their derivatives.[2] Hence, the derivative of $2.376t^2 - 0.042t^3$ with respect to time is $2 \times 2.376t - 3 \times 0.042t^2$; and the derivative of $-555.8 + 115.4t - 4.022t^2$ with respect to time is $115.4 - 2 \times 4.022t$. The instantaneous velocities implied by the formulas (4) and (5) are then

$$v = \frac{dx}{dt} = 4.752t - 0.126t^2 \qquad \text{for } 0 \text{ s} \leq t \leq 10 \text{ s} \tag{9}$$

$$v = \frac{dx}{dt} = 115.4 - 8.044t \qquad \text{for } 10 \text{ s} \leq t \leq 14.3 \text{ s} \tag{10}$$

where velocity is measured in meters per second. At, say, $t = 4$ s, Eq. (9) gives $v = 4.752 \times 4 - 0.126 \times 4^2 = 17.0$ m/s, which agrees with the graphical determination.

2.4 Acceleration

Acceleration is the rate of change of velocity. If a particle has a velocity v_1 at time t_1 and a velocity v_2 at time t_2, then the **average accelera-**

[2] Appendix 5 contains a brief review of the rules for differentiation.

tion for this time interval is defined as the change of velocity divided by the change of time:

Average acceleration

$$\bar{a} = \frac{v_2 - v_1}{t_2 - t_1} = \frac{\Delta v}{\Delta t} \tag{11}$$

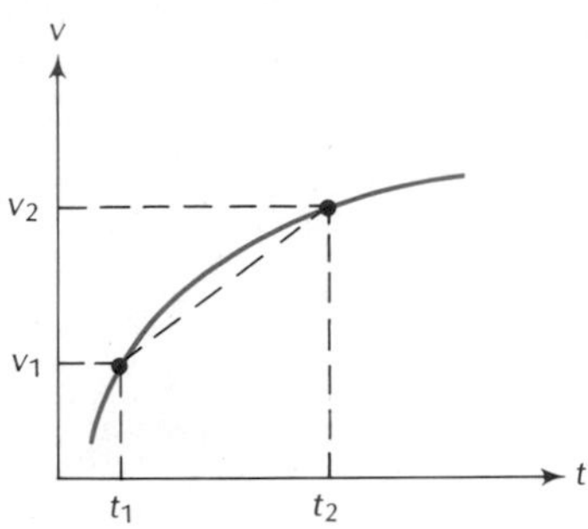

Fig. 2.8 The solid curve shows the instantaneous velocity as a function of time. The slope of the dashed line is the average acceleration.

On a plot of velocity vs. time, the average acceleration is the slope of the straight line connecting the points that correspond to $t = t_1$ and $t = t_2$ on the velocity curve (Figure 2.8).

The unit of acceleration is the unit of velocity divided by the unit of time. In the metric system the unit of acceleration is the (m/s)/s or m/s²; in the British system it is the ft/s². Table 2.2 gives some examples of accelerations.

Table 2.2 SOME ACCELERATIONS

Protons in Fermilab accelerator	9×10^{13} m/s²
Ultracentrifuge	3×10^{6} m/s²
Automobile crash (60 mi/h into fixed barrier)	1×10^{3} m/s²
Parachute opening (extreme)	3.2×10^{2} m/s²
Gravity on surface of Sun	2.7×10^{2} m/s²
Explosive seat ejection from aircraft (extreme)	1.5×10^{2} m/s²
Loss of consciousness in man ("blackout")	70 m/s²
Gravity on surface of Earth	9.8 m/s²
Braking of automobile	~8 m/s²
Gravity on surface of Moon	1.7 m/s²
Rotation of Earth (at equator)	3.4×10^{-2} m/s²

The **instantaneous acceleration** is the slope of the tangent to the velocity curve in a plot of velocity vs. time. For example, at $t = 4$ s, we can draw the tangent to the velocity curve of Figure 2.7 and find that the slope, or instantaneous acceleration, is 3.74 m/s². By drawing many such tangents at different times, we can prepare a table of instantaneous accelerations as a function of time for the automobile whose velocity is given by Figure 2.7. This acceleration is plotted in Figure 2.9. At the initial instant, $t = 0$, the acceleration is large (large slope in Figure 2.7); as the automobile gains velocity, the acceleration gradually drops (decreasing slope in Figure 2.7); at $t = 10$ s, the brakes are applied, leading to a very large negative value of the acceleration

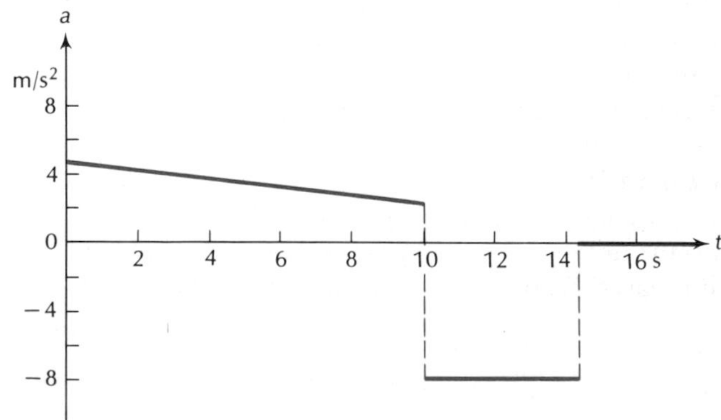

Fig. 2.9 Instantaneous acceleration as a function of time.

(negative slope in Figure 2.7); this negative acceleration, or **deceleration,** is maintained until the automobile comes to a halt. Note that the plot of Figure 2.9 has discontinuities at $t = 10$ s and $t = 14.3$ s; at $t = 10$ s the acceleration suddenly jumps from a positive value to a negative value and at $t = 14.3$ s it suddenly jumps to zero.

In the language of calculus, the instantaneous acceleration is the derivative of the instantaneous velocity with respect to time,

$$a = \lim_{\Delta t \to 0} \frac{\Delta v}{\Delta t} \qquad (12)$$

or

$$\boxed{a = \frac{dv}{dt}} \qquad (13)$$

Instantaneous acceleration

Equivalently, we can say that the acceleration is the second derivative of x with respect to time,

$$a = \frac{d}{dt}\left(\frac{dx}{dt}\right) \quad \text{or} \quad a = \frac{d^2x}{dt^2} \qquad (14)$$

For example, the accelerations calculated by differentiation of the formulas (9) and (10) are

$$a = 4.752 - 0.252t \qquad \text{for } 0 \text{ s} \leq t \leq 10 \text{ s} \qquad (15)$$

$$a = -8.044 \qquad \text{for } 10 \text{ s} \leq t \leq 14.3 \text{ s} \qquad (16)$$

where the acceleration is measured in m/s^2. According to Eq. (16), in the time interval between 10 and 14.3 s, the acceleration is constant; this agrees with the result obtained from the graphical method (see Figure 2.9).

2.5 Motion with Constant Acceleration

Constant acceleration implies a constant slope in the plot of velocity vs. time; thus the plot is a straight line. In this case the velocity increases (or decreases) by equal amounts in each 1-second interval. For example, in the interval between 10 and 14.3 s, the velocity plotted in Figure 2.7 decreases by 8.044 m/s in each second.

In the case of constant acceleration, there are some simple relations between acceleration, velocity, position, and time. Suppose that the initial velocity is v_0 and that the velocity increases at a constant rate given by the constant acceleration a. After a time t has elapsed, the velocity will have increased from the initial value v_0 to the value

$$\boxed{v = v_0 + at} \qquad (17)$$

Constant acceleration: v, a, and t

Suppose that the initial position is x_0. After a time t has elapsed, the position will have changed by an amount equal to the product of average velocity multiplied by time, i.e., the position will have changed from x_0 to

$$x = x_0 + \bar{v}t \tag{18}$$

Since the velocity increases uniformly with time, the average value of the velocity is simply the average of the initial value and the final value,

$$\bar{v} = \tfrac{1}{2}(v_0 + v) = \tfrac{1}{2}(v_0 + v_0 + at)$$

$$= v_0 + \tfrac{1}{2}at \tag{19}$$

Substituting this into Eq. (18), we find

$$x = x_0 + (v_0 + \tfrac{1}{2}at)t \tag{20}$$

i.e.,

$$x = x_0 + v_0 t + \tfrac{1}{2}at^2 \tag{21}$$

Thus, with constant acceleration, the change in position is

***Constant acceleration:** x, a, **and** t*

$$\boxed{x - x_0 = v_0 t + \tfrac{1}{2}at^2} \tag{22}$$

The right side consists of two terms: the term $v_0 t$ represents the change in position that the particle would suffer if moving at constant velocity, and the term $\frac{1}{2}at^2$ represents the effect of the acceleration.

Equations (17) and (22) express velocity and position in terms of time. By eliminating the time t between these two equations, we obtain a direct relation between position and velocity, which is sometimes useful. According to Eq. (17),

$$t = \frac{v - v_0}{a} \tag{23}$$

and if we substitute this into Eq. (22), we obtain

$$x - x_0 = v_0\left(\frac{v - v_0}{a}\right) + \tfrac{1}{2}a\left(\frac{v - v_0}{a}\right)^2 \tag{24}$$

which can be rearranged to give

***Constant acceleration:** x, a, **and** v*

$$\boxed{a(x - x_0) = \tfrac{1}{2}(v^2 - v_0^2)} \tag{25}$$

Figure 2.10 shows plots of position, velocity, and acceleration for motion with constant acceleration. In this figure the motion for negative values of t (instants before $t = 0$) has also been included with the assumption that a always has the same constant value. Note that the worldline in Figure 2.10a has the shape of a parabola; this shape of the worldline is a distinctive characteristic of motion with constant acceleration.

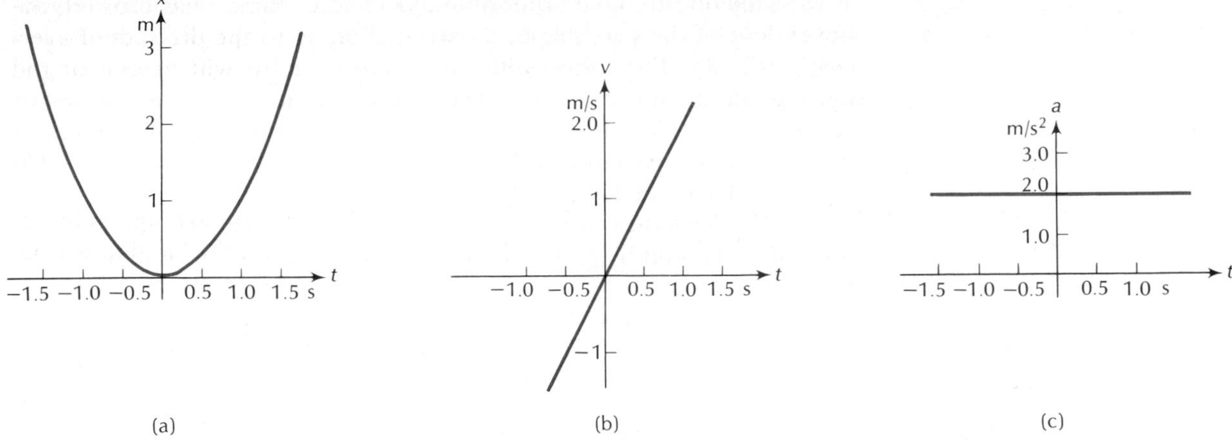

Fig. 2.10 (a) Position, (b) velocity, and (c) acceleration for an example of motion with constant acceleration.

EXAMPLE 4. An automobile initially traveling at 50 km/h crashes into a stationary, rigid barrier. The automobile comes to rest after its front end crumples by 0.40 m. Assuming constant deceleration during the collision, what is the value of the deceleration? How long does it take the automobile to stop?

SOLUTION: To solve this problem, we must first decide what equations to use. For motion with constant acceleration, the relevant equations are Eqs. (17), (22), and (25). The initial velocity, the final velocity, and the change in position are known quantities. To calculate the acceleration, it will be best to use Eq. (25), since there the acceleration appears as the *only* unknown quantity.

If we solve Eq. (25) for a, we find

$$a = \frac{v^2 - v_0^2}{2(x - x_0)}$$

The initial velocity is $v_0 = 50 \text{ km/h} = 50 \times 10^3 \text{ m}/3600 \text{ s} = 13.9 \text{ m/s}$. The final velocity is $v = 0$. The change in position corresponding to this change of velocity is $x - x_0 = 0.40$ m. Consequently,

$$a = \frac{v^2 - v_0^2}{2(x - x_0)} = \frac{-(13.9 \text{ m/s})^2}{2 \times 0.40 \text{ m}} = -2.4 \times 10^2 \text{ m/s}^2$$

We can then calculate the time the automobile takes to stop from Eq. (17):

$$t = \frac{v - v_0}{a} = \frac{-13.9 \text{ m/s}}{-2.4 \times 10^2 \text{ m/s}^2} = 5.8 \times 10^{-2} \text{ s}$$

Galileo Galilei, *1564–1642, Italian mathematician, astronomer, and physicist. He was professor at Pisa and at Padua, and became chief mathematician to Cosimo II de'Medici, grand duke of Tuscany. Galileo demonstrated experimentally that all bodies fall with the same acceleration, and he deduced that the trajectory of a projectile is a parabola. With a telescope of his own design, he discovered the satellites of Jupiter and sun spots. He vociferously defended the heliocentric system of Copernicus, for which he was sentenced by the Inquisition.*

2.6 The Acceleration of Gravity

A body released near the surface of the Earth will accelerate downward under the influence of the pull of gravity. If the frictional resistance of the air has been eliminated (by placing the body in an evacuated container), then the body is in **free fall** and the downward motion proceeds with constant acceleration. It is a remarkable fact that the value of this acceleration is exactly the same for all bodies released at the same location, i.e., the value of the acceleration is completely independent of the size, shape, chemical composition, etc., of the bodies (Figure 2.11). The universality of the rate of free fall is one of the most precisely and rigorously tested laws of nature; a long series of careful experiments have tested the equality of the rates of free fall of different bodies to within 1 part in 10^{10}, and in some special cases even to within 1 part in 10^{12}. [3]

The acceleration produced by gravity near the surface of the Earth is usually denoted by g. The numerical value of g is approximately

Acceleration of gravity, g

$$\boxed{g \cong 9.81 \text{ m/s}^2 \cong 32.2 \text{ ft/s}^2} \tag{26}$$

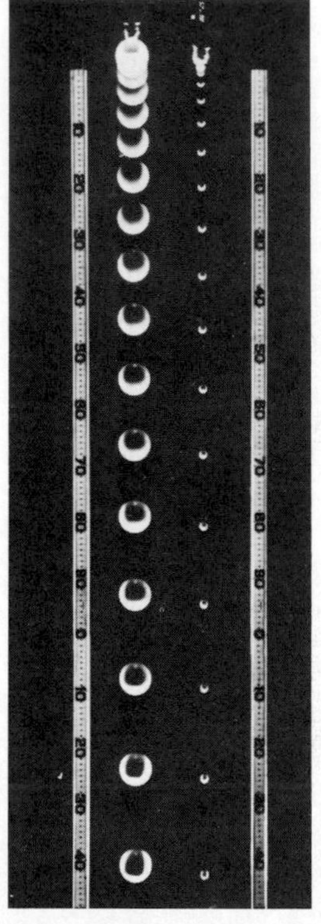

Fig. 2.11 Stroboscopic photograph of two free-falling balls of unequal mass released simultaneously. This photograph was made by leaving the shutter of the camera open, and triggering a flash at regular intervals.

The exact value of the acceleration varies from location to location on the Earth. For instance, the value of g increases with geographical latitude and decreases with altitude. Table 2.3 gives the values of g for a few locations. The variation with latitude shows up clearly in this table. However, 9.81 m/s² is accurate to within ± 0.02 m/s² for all of North America and for all altitudes up to about 6 km.

Table 2.3 VARIATION OF g WITH LATITUDE

Station	*Latitude*	g
Quito, Ecuador	0° N	9.780 m/s²
Madras, India	13° N	9.783
Hong Kong	22° N	9.788
Cairo, Egypt	30° N	9.793
New York, U.S.A.	41° N	9.803
London, England	51° N	9.811
Oslo, Norway	60° N	9.819
Murmansk, U.S.S.R.	69° N	9.825
Spitsbergen	80° N	9.831
North Pole	90° N	9.832

To describe vertical free-fall motion, take the x axis in the upward direction. Then $a = -g$ and Eqs. (17), (21), and (25) become

$$\boxed{v = v_0 - gt} \tag{27}$$

$$\boxed{x = x_0 + v_0 t - \tfrac{1}{2}gt^2} \tag{28}$$

Free-fall motion

$$\boxed{-g(x - x_0) = \tfrac{1}{2}(v^2 - v_0^2)} \tag{29}$$

Although these equations are strictly valid only for bodies falling in a vacuum, they are good approximations for dense bodies, such as pieces

[3] The tests to within 1 part in 10^{12} involved bodies falling toward the Sun rather than toward the Earth.

of metal or stone, released in air. For such bodies the frictional resistance is unimportant as long as the speed is low (the exact restriction to be imposed on the speed depends on the mass and shape of the body and on the desired accuracy). In the following calculations we will usually ignore the resistance of air, even when the speeds are not all that low.

EXAMPLE 5. At Acapulco, professional divers jump from a 36-m-high cliff into the sea. How long do they fall? What is their impact velocity?

SOLUTION: For this problem, the relevant equations are Eqs. (27), (28), and (29). The known quantities are the change of position and the initial velocity; the unknown quantities are the time of fall and the final velocity. To calculate the time from the known quantities, we will use Eq. (28), in which the time is the only unknown. With $x - x_0 = -36$ m and $v_0 = 0$, Eq. (28) yields

$$-36 \text{ m} = -\tfrac{1}{2}(9.8 \text{ m/s}^2)t^2$$

which can be solved for t,

$$t = \sqrt{\frac{2 \times 36}{9.8} \text{ s}^2} = 2.7 \text{ s}$$

The impact velocity is then

$$\begin{aligned} v = -gt &= -9.8 \text{ m/s}^2 \times 2.7 \text{ s} \\ &= -26 \text{ m/s} \end{aligned}$$

This is about 60 mi/h!

EXAMPLE 6. A powerful bow, such as used to establish world records in archery, can launch an arrow at a velocity of 90 m/s. How high will such an arrow rise if aimed vertically upward? How long will it take to return to the ground? How fast will it be going when it hits the ground? Ignore friction.

SOLUTION: At the highest point of the motion, the velocity is $v = 0$. Hence for the upward motion, we can regard the initial and final velocities as known. The height reached and the time are unknown. Equation (29) gives us the height in terms of the known quantities,

$$x - x_0 = \frac{-(v^2 - v_0^2)}{2g} = \frac{-0 - (90 \text{ m/s})^2}{2 \times 9.8 \text{ m/s}^2} = 4.1 \times 10^2 \text{ m}$$

Equatior (27) gives us the time for the upward motion:

$$t = \frac{v_0 - v}{g} = \frac{90 \text{ m/s} - 0}{9.8 \text{ m/s}^2} = 9.2 \text{ s}$$

The downward motion takes exactly as long as the upward motion; hence the time for the arrow to return to its starting point is 2×9.2 s $= 18.4$ s.

Since we are ignoring friction, the downward motion is simply the reverse of the upward motion. The velocity of the arrow when it hits the ground must therefore be the same as its initial velocity, except for sign; i.e., it must be -90 m/s. We can also see this from Eq. (29). With $x = x_0$, this equation yields $v^2 = v_0^2$, or $v = \pm v_0 = \pm 90$ m/s; the plus sign is for the upward motion, the minus sign for the downward motion.

EXAMPLE 7. Suppose you throw a stone straight up with an initial velocity of 15 m/s and, 2.0 s later, you throw a second stone straight up with the same initial velocity. The first stone going down will meet the second stone going up. At what height do the two stones meet?

SOLUTION: According to Eq. (28), the equation for the worldline of the first stone is

$$x_1 = x_0 + v_0 t - \tfrac{1}{2}gt^2 \tag{30}$$

The second stone is thrown 2.0 s later; hence the equation for its worldline is

$$x_2 = x_0 + v_0(t - 2.0\text{ s}) - \tfrac{1}{2}g(t - 2.0\text{ s})^2$$

We can write this as

$$x_2 = x_0 + v_0(t - t_2) - \tfrac{1}{2}g(t - t_2)^2 \tag{31}$$

where $t_2 = 2.0$ s. In these equations both the position and the time are unknown quantities — we have too many unknowns.

To proceed with the solution, we need an extra relationship between the unknowns. The extra relationship we have available is the condition that the two stones meet; expressed mathematically, $x_1 = x_2$ or

$$v_0 t - \tfrac{1}{2}gt^2 = v_0(t - t_2) - \tfrac{1}{2}g(t - t_2)^2 \tag{32}$$

This equation does not directly tell us the height of the meeting, but it does tell us the time of the meeting and, once we know this time, we can return to Eq. (30) to find the height. Simplifying Eq. (32), we obtain

$$v_0 t - \tfrac{1}{2}gt^2 = v_0 t - v_0 t_2 - \tfrac{1}{2}gt^2 + gtt_2 - \tfrac{1}{2}gt_2^2$$

or

$$0 = -v_0 t_2 + gtt_2 - \tfrac{1}{2}gt_2^2$$

Canceling a factor t_2 in this equation, we can solve to get t,

$$t = \frac{v_0}{g} + \tfrac{1}{2}t_2 = \frac{15\text{ m/s}}{9.8\text{ m/s}^2} + \tfrac{1}{2} \times 2.0\text{ s} = 2.53\text{ s}$$

According to Eq. (30), the height of the meeting is then

$$x_1 - x_0 = 15\text{ m/s} \times 2.53\text{ s} - \tfrac{1}{2} \times 9.8\text{ m/s}^2 \times (2.53\text{ s})^2$$

$$= 6.6\text{ m}$$

Acceleration is sometimes measured in multiples of a "standard" acceleration of gravity; we will call this "standard" acceleration **1 gee,** where

$$1\text{ gee} = 9.80665\text{ m/s}^2 \cong 9.81\text{ m/s}^2 \tag{33}$$

Note the distinction between g and gee: g is the acceleration of gravity on or near the surface of the Earth — its value is approximately 9.81 m/s^2, but its exact value depends on location (see Table 2.3). Gee is a unit of acceleration — its value is exactly 9.80665 m/s^2 by definition.

The acceleration of gravity at the Earth's surface is approximately 1 gee, the acceleration of gravity on the Moon's surface (see Table 2.2) is 0.17 gee, etc.

Measurements of g at different locations on the surface of the Earth are commonly performed by a pendulum method (we will describe this method in Chapter 14). However, in recent years the Bureau International des Poids et Mesures has developed a very sophisticated device which permits an extremely accurate determination of the value of g by directly observing the motion of a small projectile in free fall in an evacuated chamber. Figure 2.12 shows the projectile used in these measurements. This projectile is launched vertically upward by a rubber band and the time that it takes to rise a given distance and fall back is measured; the value of g can then be determined from this distance and time. For the sake of high precision, the distance the projectile travels is measured with an interferometer to within $\pm 10^{-9}$ m. This permits the determination of g to within 3 parts in 10^9, or $\pm 3 \times 10^{-8}$ m/s^2. One strange finding emerging from these high-precision experiments is that the value of g fluctuates slightly with time. From one year to the next g increases or decreases by as much as 4×10^{-7} m/s^2. The cause of these fluctuations is not known, but they are probably due to changes in the distribution of mass within the Earth.

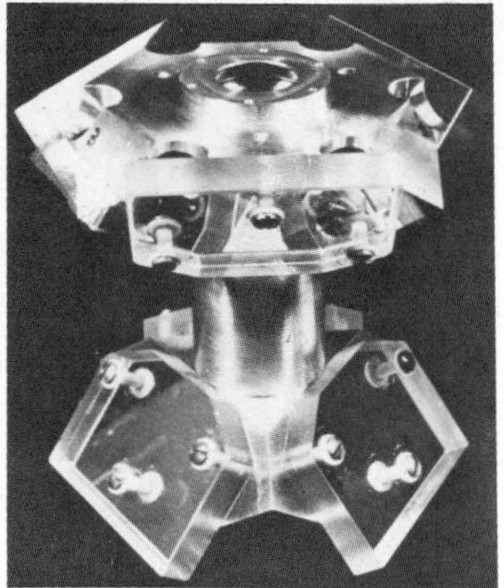

Fig. 2.12 Projectile used at BIPM for precise measurements of g.

Finally we will make some brief comments on the effects of the frictional resistance of air on bodies falling with high speeds. By holding a hand out of the window of a speeding automobile, you can readily feel that air offers a substantial frictional resistance to motion at speeds of a few tens of kilometers per hour; this frictional resistance increases with speed (roughly in proportion to the square of the speed). Hence a falling, accelerating body will experience a larger and larger frictional resistance as its speed increases. Ultimately, this resistance becomes so large that it counterbalances the pull of gravity — the body ceases to accelerate and attains a constant speed. This ultimate speed is called the **terminal speed.** The precise value of the terminal speed depends on the mass of the body, the size, and the shape; for instance, a sky diver with a closed parachute attains a terminal speed of about 200 km/h, whereas a sky diver with an open parachute attains a terminal speed of only about 18 km/h.

Terminal speed

SUMMARY

Average speed: $\dfrac{[\text{distance traveled}]}{[\text{time taken}]}$

Average velocity: $\bar{v} = \dfrac{\Delta x}{\Delta t}$

Instantaneous velocity: $v = \dfrac{dx}{dt}$

Average acceleration: $\bar{a} = \dfrac{\Delta v}{\Delta t}$

Instantaneous acceleration: $a = \dfrac{dv}{dt} = \dfrac{d^2x}{dt^2}$

Motion with constant acceleration: $v = v_0 + at$

$$x = x_0 + v_0 t + \tfrac{1}{2}at^2$$

$$a(x - x_0) = \tfrac{1}{2}(v^2 - v_0^2)$$

Acceleration of gravity: $g \cong 9.81\ \text{m/s}^2$

$$\cong 32.2\ \text{ft/s}^2$$

QUESTIONS

1. The motion of a runner can be regarded as particle motion, but the motion of a gymnast cannot. Explain.

2. According to newspaper reports, the world record for speed skiing is 126.238 mi/h. This speed was measured on a 100-meter "speed-trap." The skier took about 1.7 s to cross this trap. In order to calculate speed to six significant figures, we need to measure distance and time to six significant figures. What accuracy in distance and time does this require?

3. Do our sense organs permit us to feel velocity? Acceleration?

4. What is your velocity at this instant? Is this a well-defined question? What is your acceleration at this instant?

5. Does the speedometer of your car give speed or velocity? Does the speedometer care whether you drive eastward or westward along a road?

6. Suppose that at one instant of time the velocity of a body is zero. Can this body have a nonzero acceleration at this instant? Give an example.

7. Give an example of a body in motion with instantaneous velocity and acceleration of the same sign. Give an example of a body in motion with instantaneous velocity and acceleration of opposite signs.

8. Experienced drivers recommend that when driving in traffic you should stay at least 2 s behind the car in front of you. This is equivalent to a distance of about two car lengths for every 10 mi/h. Why is it necessary to leave a larger distance between the cars when the speed is larger?

9. In the seventeenth century Galileo Galilei measured the acceleration of gravity by rolling balls down an inclined plane. Why did he not measure the acceleration directly by dropping a stone from a tower?

10. Why did astronauts find it easy to jump on the Moon? If an astronaut can jump to a height of $\frac{1}{2}$ ft on the Earth (with his spacesuit), how high can he jump on the Moon?

11. According to Table 2.3, the value of g at the North Pole is 9.832 m/s². What do you expect for the value of g at the South Pole? Why?

12. Interpolate Table 2.3 to find the value of g at your latitude.

13. A particle is initially at rest at some height. If the particle is allowed to fall freely, what distance does it cover in the time from $t = 0$ s to $t = 1$ s? From $t = 1$ s to $t = 2$ s? From $t = 2$ s to $t = 3$ s? Show that these successive distances are in the ratios 1:3:5:7. . . .

14. An elevator is moving upward with a constant velocity of 5 m/s. If a passenger standing in this elevator drops an apple, what will be the acceleration of the apple relative to the elevator?

15. Some people are fond of firing guns into the air when under the influence of drink or patriotic fervor. What happens to the bullets? Is this practice dangerous?

16. A woman riding upward in an elevator drops a penny in the elevator shaft

when she is passing by the third floor. At the same instant, a man standing at the elevator door at the third floor also drops a penny in the elevator shaft. Which coin hits the bottom first? Which coin hits with the higher speed? Neglect friction.

17. If you take friction into account, how does this change your answers to Question 16?

18. Suppose that you drop a $\frac{1}{2}$-kg packet of sugar and a $\frac{1}{2}$-kg ball of lead from the top of a building. Taking air friction into account, which will take the shorter time to reach the ground? Suppose that you place the sugar and the lead in identical sealed glass jars before dropping them. Which will now take the shorter time?

19. Is air friction important in the falling motion of a raindrop? If a raindrop were to fall without friction from a height of 1000 ft and hit you, what would it do to you?

20. Galileo claimed in his *Dialogues* that "the variation of speed in air between balls of gold, lead, copper, porphyry, and other heavy materials is so slight that in a fall of 100 cubits a ball of gold would surely not outstrip one of copper by as much as four fingers. Having observed this, I came to the conclusion that in a medium totally void of resistance all bodies would fall with the same speed." One cubit is about 46 cm and four fingers are about 10 cm. According to Galileo's data, what is the maximum percent difference between the accelerations of the balls of gold and of copper?

Advice on Solving Problems

The solving of problems is an art; there is no simple recipe for obtaining the solutions. Most of the problems in this and the following chapters are applications of the formulas derived in the text. If you find it difficult to decide what formula to use, begin by looking at formulas that are valid under the given physical conditions of the problem and make a list of known and unknown quantities. Then try to spot a formula that expresses the unknowns in terms of the known quantities (see Examples 4, 5, and 6). Be discriminating in your selection of formulas — sometimes a formula will tempt you because it displays all the desired quantities, but it will be an invalid formula if the assumptions that went into its derivation are not satisfied in your problem. You will often find that you seem to have too many unknowns and too few equations. Then ask yourself the following questions: Are there any extra mathematical relationships that the conditions of the problem impose on the unknowns? Can you combine several equations to eliminate some of the unknowns? Are there any quantities that you can calculate from the known quantities? Do these calculated quantities bring you nearer to the answer? (See Example 7.)

It is good practice to solve all equations algebraically and only substitute numbers at the very end; this makes it easier to spot and correct mistakes.

When you have finished your calculations, always check whether your answer is plausible. For instance, if your calculation yields the result that a diver jumping off a cliff hits the water at 3000 km/h, then somebody has made a mistake somewhere!

And remember to round off your final answer to the same number of significant figures as given in the data for the problem.

PROBLEMS

Section 2.1

1. The speed of nerve impulses in mammals is typically 10^2 m/s. If a shark bites the tail of a 30-m-long whale, roughly how long will it take before the whale knows of this?

2. The world record for the 100-yard run is 9.0 seconds. What is the corresponding average speed in miles per hour?

3. In 1958 the nuclear-powered submarine *Nautilus* took 6 days and 12 hours to travel submerged 3150 miles across the Atlantic from Portland, England, to New York City. What was the average speed (in miles per hour) for this trip?

4. A galaxy beyond the constellation Corona Borealis is moving directly away from our Galaxy at the rate of 21,600 km/s. This galaxy is now at a distance of 1.4×10^9 light-years from our Galaxy. Assuming that the galaxy has always been moving at a constant speed, how many years ago was it right on top of our Galaxy?

5. On one occasion a tidal wave (tsunami) originating near Java was detected in the English Channel 32 h later. Roughly measure the distance from Java to England by sea (round the Cape of Good Hope) on a map of the world and calculate the average speed of the tidal wave.

6. A hunter shoots an arrow at a deer running directly away from him. When the arrow leaves the bow, the deer is at a distance of 40 m. When the arrow strikes, the deer is at a distance of 50 m. The speed of the arrow is 65 m/s. What must have been the speed of the deer? How long did the arrow take to travel to the deer?

7. In 1971, Francis Chichester, in the yacht *Gypsy Moth V*, attempted to sail the 4000 nautical miles (nmi) from Portuguese Guinea to Nicaragua in no more than 20 days.
   (a) What minimum average speed (in nautical miles per hour) does this require?
   (b) After sailing 13 days, he still had 1720 nmi to go. What minimum average speed did he require to reach his goal in the remaining 7 days? Knowing that his yacht could at best achieve a maximum speed of 10 nmi/h, what could he conclude at this point?

8. The fastest land animal is the cheetah, which runs at a speed of up to 101 km/h. The second fastest is the antelope, which runs at a speed of up to 88 km/h.
   (a) Suppose that a cheetah begins to chase an antelope. If the antelope has a head start of 50 m, how long does it take the cheetah to catch the antelope? How far will the cheetah have traveled by this time?
   (b) The cheetah can only maintain its top speed for about 20 s (and then has to rest), whereas the antelope can do it for a considerably longer time. What is the maximum head start the cheetah can allow the antelope?

9. The table printed on the endpapers gives the radii of the orbits of the planets ("mean distance from the Sun") and the times required for moving around the orbit ("period of revolution").
   (a) Calculate the speed of motion of each of the nine planets in its orbit around the Sun. Assume that the orbits are circular.
   (b) In a logarithmic graph of speed vs. radius, plot the logarithm of the speed of each planet and the logarithm of its radial distance from the Sun as a point. Draw a curve through the nine points. Can you represent this curve by a simple equation?

Section 2.2

10. Suppose you throw a baseball straight up so that it reaches a maximum height of 8.0 m and returns to you 2.5 s after you throw it. What is the average speed for this motion of the ball? What is the average velocity?

Sections 2.3 and 2.4

11. A Porsche racing car takes 2.2 s to accelerate from 0 to 96 km/h 60 mi/h). What is the average acceleration?

12. In an experiment with a water-braked rocket sled, an Air Force volunteer (?) was subjected to an acceleration of 82.6 gee for 0.04 s. What was his change of speed in this time interval?

13. (a) The solid line in Figure 2.13 is a plot of velocity vs. time for a Triumph sports car undergoing an acceleration test. By drawing tangents to the velocity curve, find the accelerations at time $t = 0$, 10, 20, 30, and 40 s; express these accelerations in gee.
(b) The dashed line in Figure 2.13 is a plot of velocity vs. time for the same car when coasting with its gears in neutral. Find the accelerations at time $t = 0$, 10, 20, 30, and 40 s.

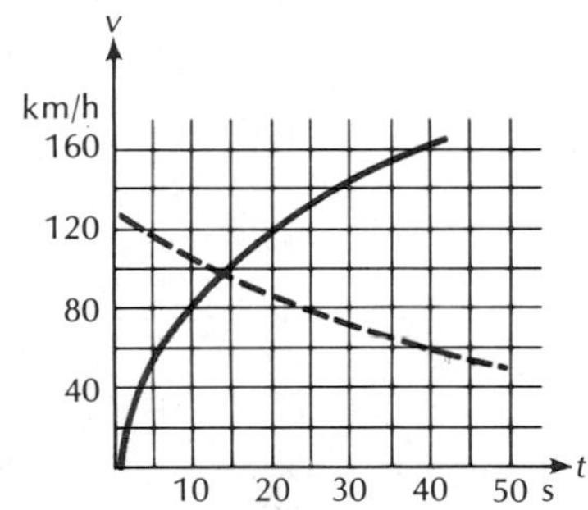

Fig. 2.13 Instantaneous velocity for a Triumph sports car undergoing an acceleration test.

14. A particle moves along the x axis with an equation for the worldline as follows:

$$x = 2.0 + 6.0t - 3.0t^2$$

where x is measured in meters and t is measured in seconds.
(a) What is the position of the particle at $t = 0.50$ s? What is the velocity? What is the acceleration?
(b) What is the position at $t = 2.0$ s? What is the velocity? What is the acceleration?
(c) What is the average velocity for the time interval $t = 0.50$ s to $t = 2.0$ s? What is the average acceleration?

15. Consider the solid curve showing velocity vs. time for the sports car described in Problem 13.
(a) From this curve, estimate the average velocity for the interval $t = 0$ s to $t = 5$ s. (Hint: The average velocity for a small time interval is roughly the average of the initial and final velocities; alternatively, it is roughly the velocity at the midpoint of the time interval.) Estimate how far the car travels in this time interval.
(b) Repeat the calculation of part (a) for every 5-s time interval between $t = 5$ s and $t = 45$ s.
(c) What is the *total* distance that the car will have traveled at the end of 45 s?

16. Table 2.4 gives the horizontal velocity as a function of time for a projectile of 100 lb-mass fired horizontally from a naval 6-inch gun. The velocity decreases with time because of the frictional drag of the air.

Table 2.4 EFFECT OF AIR RESISTANCE ON A PROJECTILE

Time	*Velocity*	*Time*	*Velocity*
0 s	2154 ft/s	1.80 s	1827 ft/s
0.30	2092	2.10	1779
0.60	2030	2.40	1733
0.90	1983	2.70	1686
1.20	1929	3.00	1646
1.50	1873		

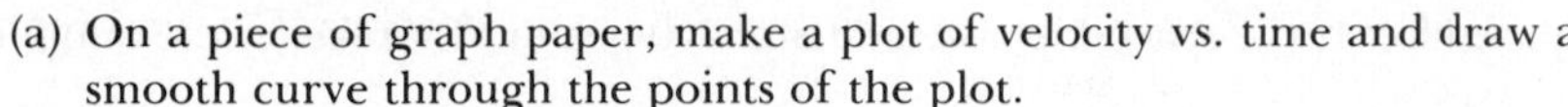

(a) On a piece of graph paper, make a plot of velocity vs. time and draw a smooth curve through the points of the plot.

(b) Estimate the average velocity for each time interval [the estimated average velocity for the time interval $t = 0$ s to $t = 0.30$ s is $\frac{1}{2}(2154 + 2092)$ ft/s, etc.]. From the average velocities calculate the distances that the projectile travels in each time interval. What is the total distance that the projectile travels in 3.0 s?

(c) By directly counting the squares on your graph paper, estimate the area (in units of s · ft/s) under the curve plotted in part (a) and compare this area with the result of part (b).

17. The instantaneous velocity of the projectile described in Problem 16 can be approximately represented by the formula (valid for $0 \text{ s} \leq t \leq 3 \text{ s}$)

$$v = 2152 - 200.6t + 10.7t^2$$

where v is measured in feet per second and t in seconds. Calculate the instantaneous accelerations of the projectile at $t = 0$ s, at $t = 1.50$ s, and at $t = 3.00$ s.

18. The equation of the worldline of a particle is given by

$$x = 2.0 \cos t$$

where x is measured in meters and t is measured in seconds.

(a) Roughly plot the worldline of this particle for the time interval $0 \text{ s} \leq t \leq 7.0 \text{ s}$.

(b) At what time does the particle pass the origin ($x = 0$)? What are its velocity and acceleration at this instant?

(c) At what time does the particle reach maximum distance from the origin? What are its velocity and acceleration at this instant?

19. The motion of a rocket burning its fuel at a constant rate while moving through empty interstellar space can be described by

$$x = u_{ex}t + u_{ex}(1/b - t) \ln(1 - bt)$$

where u_{ex} and b are constants (u_{ex} is the exhaust velocity of the gases at the tail of the rocket and b is proportional to the rate of fuel consumption).

(a) Find a formula for the instantaneous velocity of the rocket.

(b) Find a formula for the instantaneous acceleration.

(c) Suppose that a rocket with $u_{ex} = 3.0 \times 10^3$ m/s and $b = 7.5 \times 10^{-3}$/s takes 120 s to burn all its fuel. What is the instantaneous velocity at $t = 0$ s? At $t = 120$ s?

(d) What is the instantaneous acceleration at $t = 0$ s? At $t = 120$ s?

Section 2.5

20. The takeoff speed of a jetliner is 360 km/h. If the jetliner is to take off from a runway of length 2100 m, what must be its acceleration along the runway (assumed constant)?

21. A British 6-inch naval gun has a barrel 21.75 ft long. The muzzle speed of a projectile fired from this gun is 2154 ft/s. Assuming that upon detonation of the explosive charge the projectile moves along the barrel with constant acceleration, what is the magnitude of this acceleration? How long does it take the projectile to travel the full length of the barrel?

22. The nearest star is Proxima Centauri, at a distance of 4.2 light-years from the Sun. Suppose we wanted to send a spaceship to explore this star. To keep the astronauts comfortable, we want the spaceship to travel with a constant acceleration of 1.0 gee at all times (this will simulate ordinary gravity within the spaceship). If the spaceship accelerates at 1.0 gee until it reaches the midpoint

of its trip and then decelerates at 1.0 gee until it reaches Proxima Centauri, how long will the one-way trip take? What will be the speed of the spaceship at the midpoint? Do your calculations according to Newtonian physics (actually, the speed is so large that the calculation should be done according to relativistic physics; see Chapter 17).

23. In an accident on motorway M.1 in England, a Jaguar sports car made skid marks 276 m long while braking. Assuming that the deceleration was 1 gee during this skid (this is approximately the maxiumum deceleration that a car with rubber wheels can attain on ordinary pavements), calculate the initial speed of the car before braking.

24. The front end of an automobile has been designed so that upon impact it progressively crumples by as much as 0.7 m. Suppose that the automobile crashes into a solid brick wall at 80 km/h. During the collision the passenger compartment decelerates over a distance of 0.7 m. Assume that the deceleration is constant. What is the magnitude of the deceleration? If the passenger is held by a safety harness, is he likely to survive? (Hint: Compare the deceleration with the acceleration listed for a parachutist in Table 2.2.)

25. A jet-powered car racing on the Salt Flats in Utah went out of control and made skid marks 6 mi long. Assuming that the deceleration during the skid was about 1 gee, what must have been the initial speed of the car?

26. In a "drag" race a car starts at rest and attempts to cover 440 yd in the shortest possible time. The world record for a piston-engined car is 5.637 s; while setting this record, the car reached a final speed of 250.69 mi/h at the 440-yd mark.

(a) What was the average acceleration for the run?
(b) Prove that the car did not move with constant acceleration.
(c) What would have been the final speed if the car had moved with constant acceleration so as to reach 440 yd in 5.637 s?

27. In a "drag" race, a car with a rocket engine attained a final speed of 377.8 mi/h at the end of a 440-yd run, starting from rest. Assuming that the acceleration was constant during this run, find the value of the acceleration and find the time required to travel the given distance.

28. The operation manual of a passenger automobile states that the stopping distance is 50 m when the brakes are fully applied at 96 km/h. What is the deceleration?

29. In a collision, an automobile initially traveling at 50 km/h decelerates at a constant rate of 200 m/s^2. A passenger not wearing a seatbelt crashes against the dashboard. Before the collision, the distance between the passenger and the dashboard was 0.60 m. With what speed, relative to the automobile, does the passenger crash into the dashboard? Assume that the passenger has no deceleration before contact with the dashboard.

30. Differentiate Eq. (21) twice with respect to time and show that the result is consistent with the original assumption of constant acceleration.

31. In a large hotel, a fast elevator takes you from the ground floor to the 21st floor. The elevator takes 17 s for this trip: 5 s at constant acceleration, 7 s at constant velocity, and 5 s at constant deceleration. Each floor in the hotel has a height of 2.5 m. Calculate the values of the acceleration and deceleration (assume they are equal). Calculate the maximum speed of the elevator.

*32. The diagram of Figure 2.14 (copied from the operation manual of an automobile) describes the passing ability of the automobile at low speed. From the data supplied in this figure, calculate the acceleration of the automobile

---

* This symbol denotes especially challenging problems throughout the text.

during the pass and the time required for the pass. Assume constant acceleration.

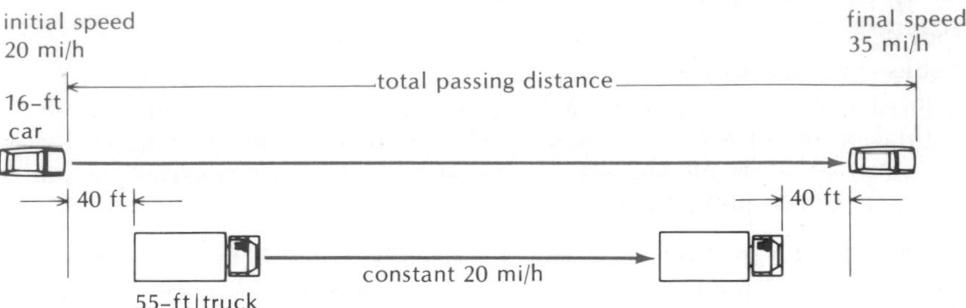

Fig. 2.14 Diagram from the operation manual of an automobile.

33. The driver of an automobile traveling at 60 mi/h perceives an obstacle on the road and slams on the brakes.
   (a) Calculate the total stopping distance (in feet). Assume that the reaction time of the driver is 0.75 s (so that there is a time interval of 0.75 s during which the automobile continues at constant speed while the driver gets ready to apply the brakes) and that the deceleration of the automobile is 0.80 gee when the brakes are applied.
   (b) Repeat the calculation of part (a) for initial speeds of 10, 20, 30, 40, and 50 mi/h. Make a plot of stopping distance vs. initial speed.

34. An automobile is traveling at 90 km/h on a country road when the driver suddenly notices a cow in the road 30 m ahead. The driver attempts to brake the automobile, but the distance is too short. With what velocity does the automobile hit the cow? Assume that, as in Problem 33, the reaction time of the driver is 0.75 s and that the deceleration of the automobile is 0.80 gee when the brakes are applied.

Section 2.6

35. An apple drops from the top of the Empire State Building, 380 m above street level. How long does the apple take to fall? What is its impact velocity on the street? Ignore air resistance.

36. Peregrine falcons dive on their prey with speeds of up to 130 km/h. From what height must a falcon fall freely to achieve this speed? Ignore air resistance.

37. The muzzle speed of a 22-caliber bullet fired from a rifle is 1200 ft/s. If there were no air resistance, how high would this bullet rise when fired straight up?

38. An engineer standing on a bridge drops a penny toward the water and sees the penny splashing into the water 3.0 s later. How high is the bridge?

Fig. 2.15 The volcano Loki.

39. The volcano Loki on Io, one of the moons of Jupiter, ejects debris to a height of 200 km (Figure 2.15). What must be the initial ejection velocity of the debris? The acceleration of gravity on Io is 1.80 m/s^2. There is no atmosphere on Io, hence no air resistance.

40. The nozzle of a fire hose discharges water at the rate of 280 liters/min at a speed of 26 m/s. How high will the stream of water rise if the nozzle is aimed straight up? How many liters of water will be in the air at any given instant?

41. According to an estimate, a man who survived a fall from a 185-ft cliff took 0.015 s to stop upon impact on the ground. What was his speed just before impact? What was his average deceleration during impact?

42. A golf ball released from a height of 1.5 m above a concrete floor bounces back to a height of 1.1 m. If the ball is in contact with the floor for 6.2×10^{-4} s, what is the average acceleration of the ball while in contact with the floor?

43. In 1978 the stuntman A. J. Bakunas died when he jumped from the 23rd floor of a skyscraper and hit the pavement. The air bag that was supposed to cushion his impact ripped.

(a) The height of his jump was 315 ft. What was his impact speed?
(b) The air bag was 12 ft thick. What would have been the man's deceleration had the air bag not ripped? Assume that his deceleration would have been uniform over the 12-ft interval.

44. At a height of 1500 m, a dive bomber in a vertical dive at 300 km/h shoots a cannon at a target on the ground. Relative to the bomber the initial speed of the projectile is 700 m/s. What will be the impact speed of the projectile on the ground? How long will it take to get there? Ignore air friction in your calculation.

*45. Suppose you throw a stone straight up with an initial speed of 15.0 m/s.

(a) If you throw a second stone straight up 1.00 s after the first, with what speed must you throw this second stone if it is to hit the first at a height of 11.0 m? (There are two answers. Are both plausible?)
(b) If you throw the second stone 1.30 s after the first, with what speed must you throw this second stone if it is to hit the first at a height of 11.0 m?

**46. Raindrops drip from a spout at the edge of a roof and fall to the ground. Assume that the drops drip at a steady rate of n drops per second (where n is large) and that the height of the roof is h.

(a) How many drops are in the air at one instant?
(b) What is the median height of these drops (i.e., the height above and below which an equal number of drops are found)?
(c) What is the average of the heights of these drops?

CHAPTER 3

Vectors

For the description of the position, velocity, and acceleration of a particle moving in two or three dimensions, the concept of a vector turns out to be very helpful. This concept is also useful for the description of many other physical quantities, such as force and momentum. The present chapter is an introduction to vectors and to the mathematics of vectors — their addition, subtraction, and multiplication. This chapter contains no physics; instead, it develops mathematical tools that we will need for handling the physics in subsequent chapters.

3.1 The Displacement Vector and the General Definition of a Vector

Displacement vector

We begin with the concepts of displacement and the displacement vector. The displacement of a particle is simply a change of its position. If a particle moves from a point P_1 to a point P_2, we can represent the change of position graphically by an arrow, or directed line segment, from P_1 to P_2. The directed line segment is the **displacement vector** of the particle. For example, if a ship moves from Liberty Island to the Battery in New York harbor, then the displacement vector is as shown in Figure 3.1. Note that the displacement vector only tells us where the final position (P_2) is in relation to the initial position (P_1); it does not tell us what path the ship followed between the two positions. Thus any of the paths shown by the dotted lines in Figure 3.2 results in the same final displacement vector. If we wanted to describe the complete path vectorially, then we would have to draw a sequence of vectors representing the displacement of the ship at different times (Figure 3.2).

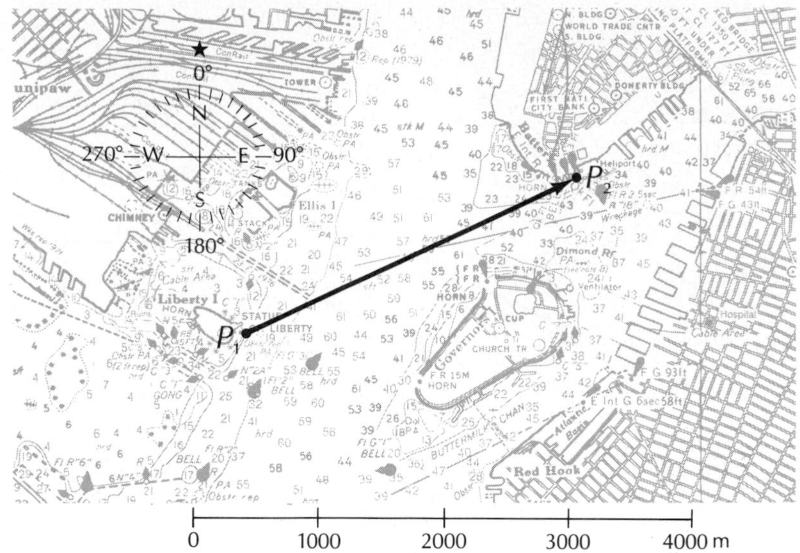

Fig. 3.1 Displacement vector for a ship moving from Liberty Island to the Battery in New York harbor. (Excerpt from National Ocean Survey Chart 12328.)

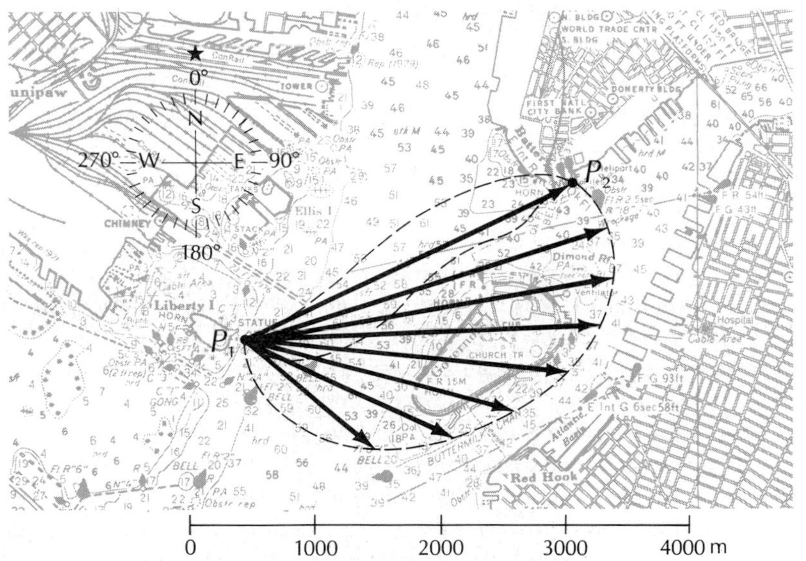

Fig. 3.2 Several alternative paths from Liberty Island to the Battery. All of these result in the same final displacement. Also shown are displacement vectors at successive times for a ship moving along one of these paths.

The displacement vector of Figure 3.1 has a *length* and a *direction.* Instead of describing the vector graphically by drawing a picture, we can describe it numerically by giving the numerical value of its length (in, say, meters) and the numerical value of the angle (in, say, degrees) it makes with some fiducial direction. For example, we can specify the displacement vector in Figure 3.1 by stating that it is 2890 m long and points at an angle of 65° east of north.

Since the displacement vector describes a *change* in position, any two line segments of identical length and direction represent equal vectors, regardless of whether the endpoints of the line segments are the same. Thus, the parallel directed line segments shown in Figure 3.3 do not represent different vectors; they all involve the same *change* of position

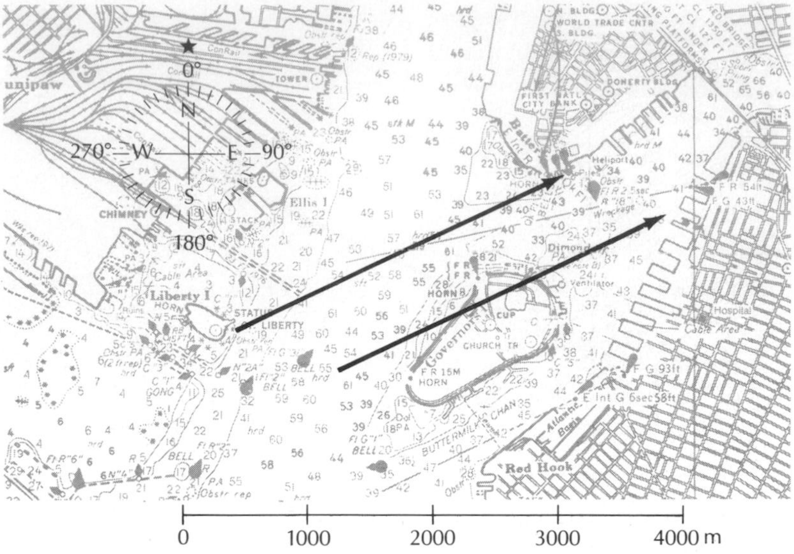

Fig. 3.3 These two displacement vectors are equal.

(same distance and same direction) and represent equal displacement vectors.

In printed books, vectors are usually indicated by boldface letters, such as **A,** and we will follow this convention. In handwritten calculations, an alternative notation consisting of either a small arrow, such as $\vec{A}$, or a wavy underline, such as $\underset{\sim}{A}$, is usually more convenient. We will denote the length or magnitude of the displacement vector **A** by $|\mathbf{A}|$ or, alternatively, by A.

The displacement vector serves as prototype for all other vectors. To decide whether some mathematical quantity is a vector, we compare its mathematical properties with those of the displacement vector. *Any quantity that has magnitude and direction and that behaves like*[1] *the displacement vector is a* **vector.** For example, velocity, acceleration, and force are vectors; they can be represented graphically by directed line segments of a length equal to the magnitude of the velocity, acceleration, or force (in some suitable units) and a corresponding direction.

Vector

By contrast, any quantity that has a magnitude but *no* direction is called a **scalar.** For example, length, time, mass, area, volume, density, and energy are scalars; they can be completely specified by their numerical magnitude. Note that the length of a displacement vector, such as the length 2890 m of the displacement vector in Figure 3.1, is a quantity that has magnitude but no direction, i.e., it is a scalar.

Scalar

3.2 Vector Addition and Subtraction

Since by definition all vectors have the mathematical properties of displacement vectors, we can investigate all the mathematical operations with vectors by looking at displacement vectors. The most important of these mathematical operations is **vector addition.**

Vector addition

Two displacements carried out in succession result in a net displace-

[1] The exact meaning of "behaves like" will be spelled out in Section 3.5.

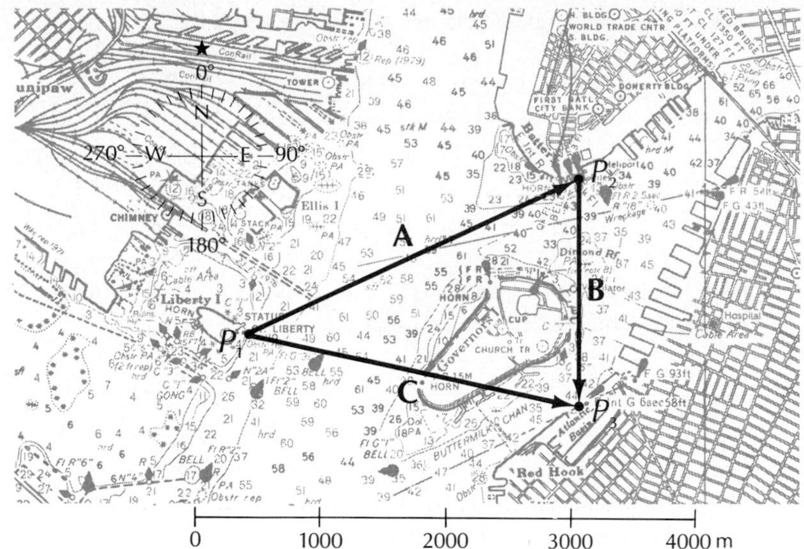

Fig. 3.4 The displacement **A** is followed by the displacement **B.** The net displacement is **C.**

ment, which can be regarded as the sum of the two individual displacements. For example, Figure 3.4 shows a displacement vector **A** (from P_1 to P_2) and a displacement vector **B** (from P_2 to P_3). The net displacement vector is the directed line segment from P_1 to P_3; this net displacement vector is denoted by **C.** This vector **C** can be regarded as the sum of the individual displacements,

Resultant

$$\mathbf{C} = \mathbf{A} + \mathbf{B} \tag{1}$$

The sum of two vectors is often called the **resultant** of these vectors. Thus **C** is called the resultant of **A** and **B.**

EXAMPLE 1. A ship moves from Liberty Island in New York harbor to the Battery and from there to the Atlantic Basin (Figure 3.4). The first displacement is 2890 m at 65° east of north; the second is 1830 m due south. What is the resultant?

SOLUTION: The resultant of the two displacement vectors **A** and **B** is the vector **C,** from the tail of **A** to the tip of **B.** For a graphical determination of **C,** we can measure the length of **C** directly on the chart using the scale of length marked on the chart, and we can measure the direction of **C** with a protractor (the way the navigator of the ship would solve the problem).

For a more precise numerical determination of **C,** we note that **A, B,** and **C** form a triangle. We can therefore find **C** by using the standard trigonometric methods.[2] The lengths of the known sides are $A = 2980$ m and $B = 1830$ m; the angle between these sides is 65° (Figure 3.5). By the law of cosines

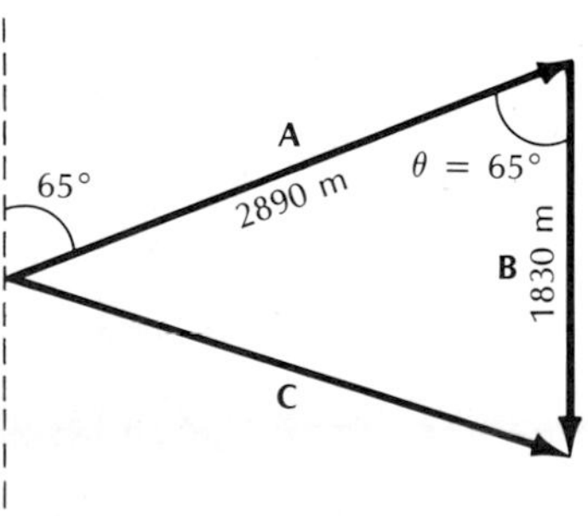

Fig. 3.5 Vector triangle.

$$C^2 = A^2 + B^2 - 2AB \cos 65° \tag{2}$$

from which

$$C = \sqrt{(2890)^2 + (1830)^2 - 2 \times 2890 \times 1830 \times \cos 65°} \text{ m}$$

$$= 2690 \text{ m}$$

---

[2] Appendix 4 gives a review of trigonometry.

By the law of sines the angle θ between C and A is given by

$$\frac{\sin\theta}{1830} = \frac{\sin 65^\circ}{2690}$$

and

$$\theta = 38.1^\circ$$

The angle between the northerly direction and **C** is then $38.1^\circ + 65^\circ = 103.1^\circ$. Thus the resultant displacement is 2690 m at 13.1° south of east.

The procedure for the addition of any arbitrary vectors mimics that for displacement vectors. If **A** and **B** are two arbitrary vectors (Figure 3.6a), then their resultant can be obtained by placing the tail of **B** on the head of **A;** the directed line segment connecting the tail of **A** to the head of **B** is the resultant (Figure 3.6b). Alternatively, the resultant can be obtained by placing the tail of **B** on the tail of **A** and drawing a parallelogram with **A** and **B** as two of the sides; the diagonal of the parallelogram is then the resultant (Figure 3.6c).

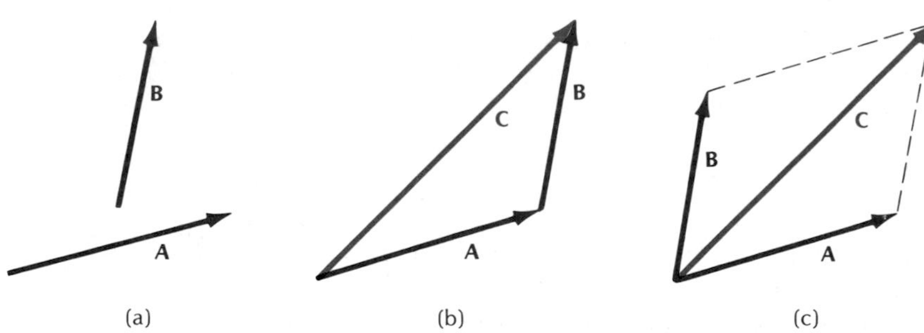

Fig. 3.6 The vector sum **A + B;** the resultant is **C.** (a) The two vectors **A** and **B.** (b) Addition of **A** and **B** by the tail-to-head method. (c) Addition of **A** and **B** by the parallelogram method.

Note that the order in which the two vectors are added makes no difference to the final result. Whether we place the tail of **A** on the head of **B** or the tail of **B** on the head of **A,** the resultant is the same (Figure 3.7). Hence

Commutative law

$$\boxed{\mathbf{A} + \mathbf{B} = \mathbf{B} + \mathbf{A}} \qquad (3)$$

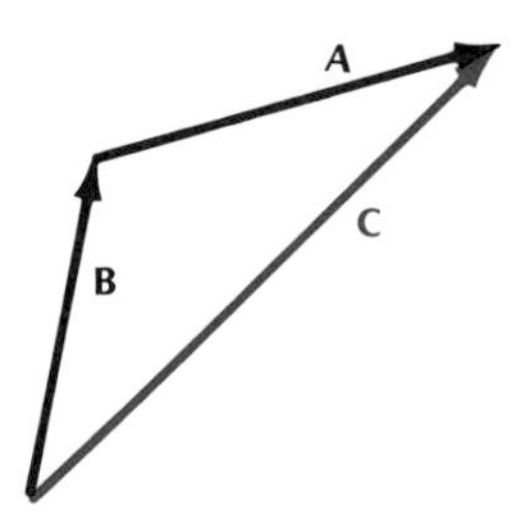

Fig. 3.7 The resultant **C** is the same as in Figure 3.6b.

This identity is called the **commutative law** for vector addition; it indicates that, just as in ordinary addition of numbers, the order of the terms is irrelevant.

Three or more vectors can be added in succession. For example, the resultant of three vectors **A, B,** and **D** can be obtained by first adding **A** and **B** and then adding **D** (Figure 3.8a). Alternatively, this resultant can be obtained by first adding **B** and **D** and then adding this to **A** (Figure 3.8b). Comparison of these figures shows that the grouping of terms makes no difference; in both cases we obtain the same final result. Hence

Associative law

$$\boxed{(\mathbf{A} + \mathbf{B}) + \mathbf{D} = \mathbf{A} + (\mathbf{B} + \mathbf{D})} \qquad (4)$$

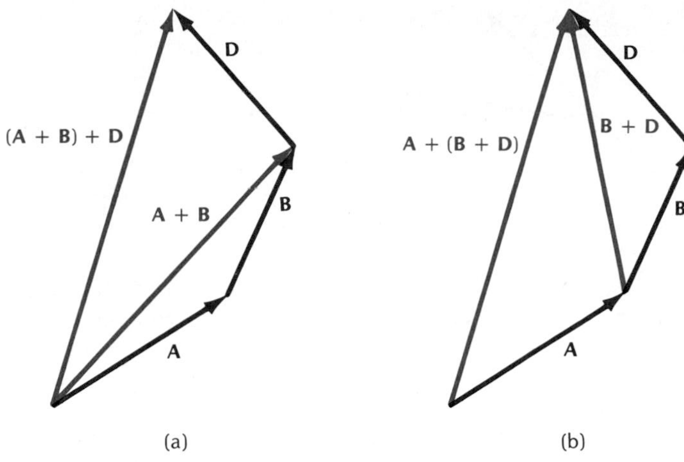

Fig. 3.8 (a) First add **A** and **B;** then add **D.** (b) First add **B** and **D;** then add this to **A.**

This is called the **associative law** of vector addition. Taken together, Eqs. (3) and (4) show that the resultant of a given number of vectors can be evaluated by summing the vectors in any convenient order.

The magnitude of the resultant of two (or more) vectors is usually less than the sum of the magnitudes of the vectors. Thus, if

$$\mathbf{C} = \mathbf{A} + \mathbf{B} \tag{5}$$

then

$$C \leq A + B \tag{6}$$

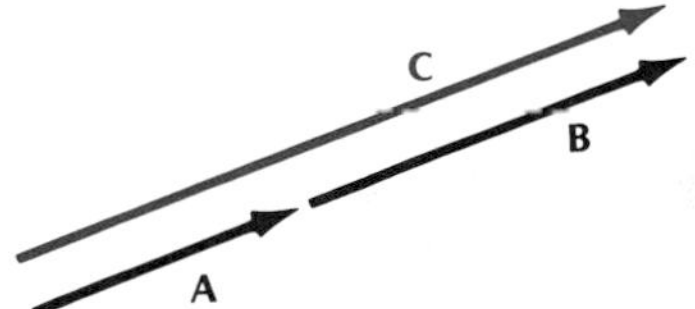

Fig. 3.9 Parallel vectors **A** and **B** and their resultant **C.**

This inequality simply expresses the fact that in a triangle (see Figure 3.6b) the length of any one side is less than the sum of the lengths of the other two sides. Only in the special case where **A** and **B** are parallel (Figure 3.9) will the magnitude of **C** equal the sum of the magnitudes of **A** and **B;** it can never exceed this sum.

The negative of a given vector **A** is a vector of the same magnitude, but opposite direction; this new vector is denoted by **—A** (Figure 3.10). Obviously,

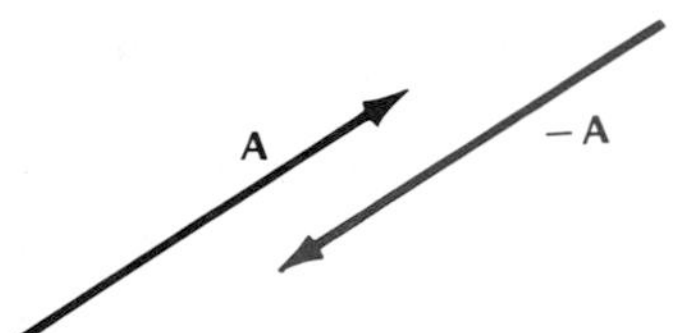

Fig. 3.10 The negative of the vector **A.**

$$\boxed{\mathbf{A} + (-\mathbf{A}) = \mathbf{0}} \tag{7}$$

which says that the sum of a vector and its negative gives a vector of zero magnitude.

The **subtraction of two vectors A** and **B** is defined as the sum of **A** and **—B,**

Vector subtraction

$$\boxed{\mathbf{A} - \mathbf{B} = \mathbf{A} + (-\mathbf{B})} \tag{8}$$

Figure 3.11a shows the vector sum of **A** and **—B.** Figure 3.11b shows that the same result can be obtained by placing the tail of **B** on the tail of **A** and drawing the same vector parallelogram as for addition: one of the diagonals of the parallelogram is then the sum **A + B,** and the other diagonal is the difference **A – B.**

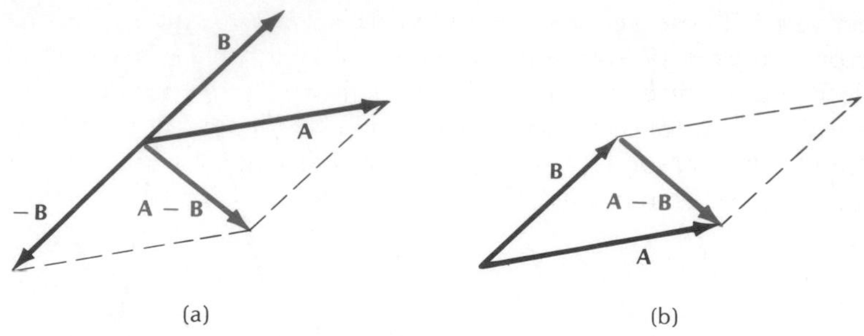

Fig. 3.11 The vector difference **A** − **B.** (a) Construction of **A** − **B** by means of the addition of **A** and −**B.** (b) Construction of **A** − **B** by means of the "other diagonal" of the vector parallelogram, from the head of **B** to the head of **A.**

A vector can be multiplied by a positive or negative number. For instance, if **A** is a given vector, then 3**A** is a vector of the same direction and of a magnitude three times as large (Figure 3.12a); and −3**A** is a vector of the opposite direction and, again, of a magnitude three times as large (Figure 3.12b). In particular, if we multiply a vector by −1, we will obtain the negative of that vector,

$$(-1)\mathbf{A} = -\mathbf{A}$$

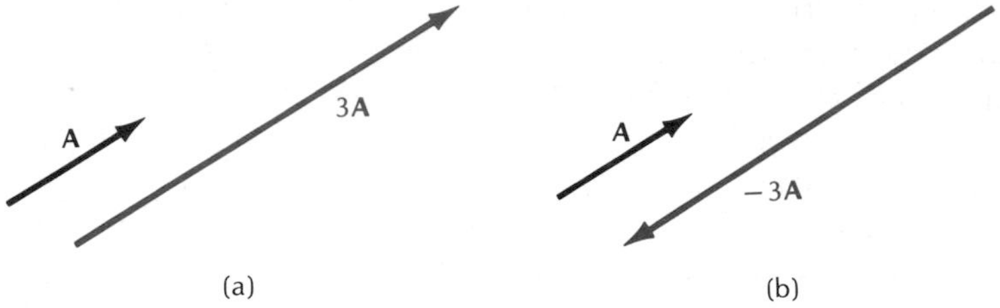

Fig. 3.12 (a) Vector **A** multiplied by 3. (b) Vector **A** multiplied by −3.

3.3 The Position Vector; Components of Vectors

To describe the position of a point P in three dimensions, we must choose an origin and construct a coordinate grid. If the grid is rectangular, then the position of a point will be given by the three rectangular coordinates, x, y, z. Alternatively, we can describe the position of a point by means of the displacement vector from the origin to the point (Figure 3.13). This displacement vector is called the **position vector** and is usually denoted by **r.**

Position vector

The explicit connection between **r** and x, y, z can be expressed mathematically as follows. We can define a vector $\hat{\mathbf{x}}$ (pronounced "x hat") that has a magnitude $|\hat{\mathbf{x}}| = 1$ and points in the positive x direction (Figure 3.14); likewise, we can define vectors $\hat{\mathbf{y}}$ and $\hat{\mathbf{z}}$ that have magnitudes $|\hat{\mathbf{y}}| = 1$ and $|\hat{\mathbf{z}}| = 1$ and point in the positive y and z directions, re-

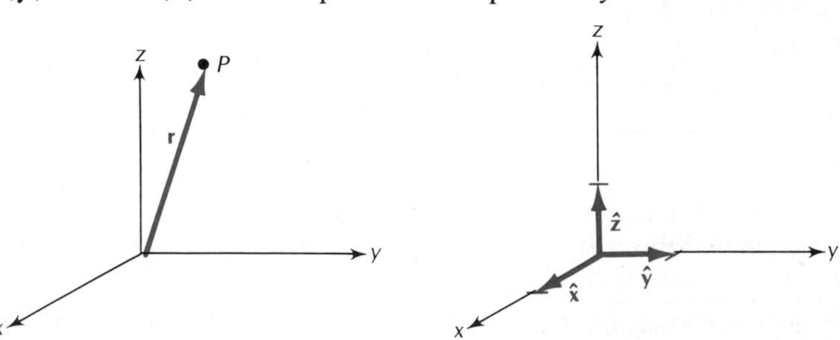

Fig. 3.13 The position vector **r** of the point P.

Fig. 3.14 The unit vectors $\hat{\mathbf{x}}$, $\hat{\mathbf{y}}$, $\hat{\mathbf{z}}$.

Unit vectors $\hat{\mathbf{x}}$, $\hat{\mathbf{y}}$, $\hat{\mathbf{z}}$

spectively.[3] These vectors are called **unit vectors.** Now consider the vector sum $x\hat{\mathbf{x}} + y\hat{\mathbf{y}} + z\hat{\mathbf{z}}$. This sum consists of a displacement of magnitude x in the x direction, followed by a displacement of magnitude y in the y direction, followed by a displacement of magnitude z in the z direction. This vector sum brings us from the origin to the point x, y, z (Figure 3.15). Hence this vector sum coincides with the position vector,

$$\boxed{\mathbf{r} = x\hat{\mathbf{x}} + y\hat{\mathbf{y}} + z\hat{\mathbf{z}}} \tag{9}$$

Components of vectors

The three numbers x, y, and z are called the **components** of the vector **r**. For example, if the point P has coordinates $x = 2$ m, $y = 3$ m, $z = 5$ m, then the position vector of P is

$$\mathbf{r} = (2\text{ m})\hat{\mathbf{x}} + (3\text{ m})\hat{\mathbf{y}} + (5\text{ m})\hat{\mathbf{z}} \tag{10}$$

and the x component of this vector is 2 m, the y component is 3 m, and the z component is 5 m.

Graphically, the components of **r** can be obtained by dropping perpendiculars from the point P to the three coordinate axes x, y, z. The intercepts of these perpendiculars with the axes give the components. Figure 3.16 shows the vector **r** of Eq. (10) and its components.

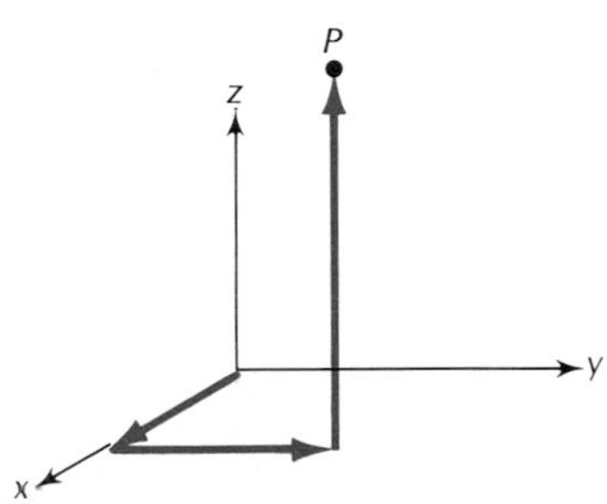

Fig. 3.15 The sum of three perpendicular displacements brings us from the origin to the point x, y, z.

For any arbitrary vector **A,** the x, y, and z components can be defined by analogy with those of the position vector. First the tail of the vector is placed at the origin of coordinates and perpendiculars are dropped from the tip of the vector to the coordinate axes. Once again the intercepts of these perpendiculars with the axes give the components (Figure 3.17). If we designate the numerical values of these components as A_x, A_y, and A_z, then the vector **A** can be expressed as

$$\boxed{\mathbf{A} = A_x\hat{\mathbf{x}} + A_y\hat{\mathbf{y}} + A_z\hat{\mathbf{z}}} \tag{11}$$

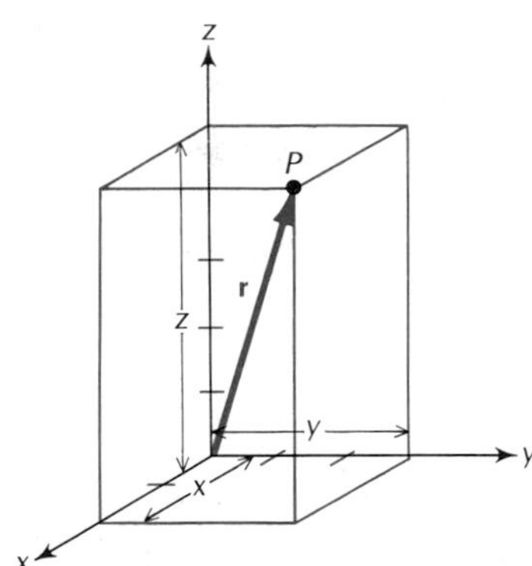

Fig. 3.16 The components x, y, z of the position vector **r**.

Figure 3.18 shows the special case of a vector **A** in two dimensions. We see from this figure that the x and y components of the vector can

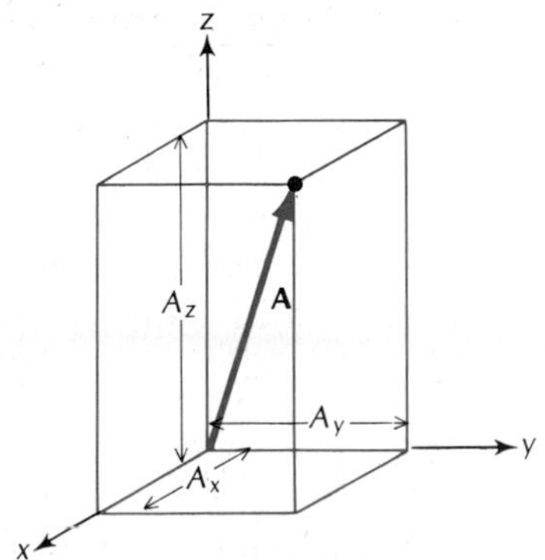

Fig. 3.17 The components A_x, A_y, A_z of an arbitrary vector **A**.

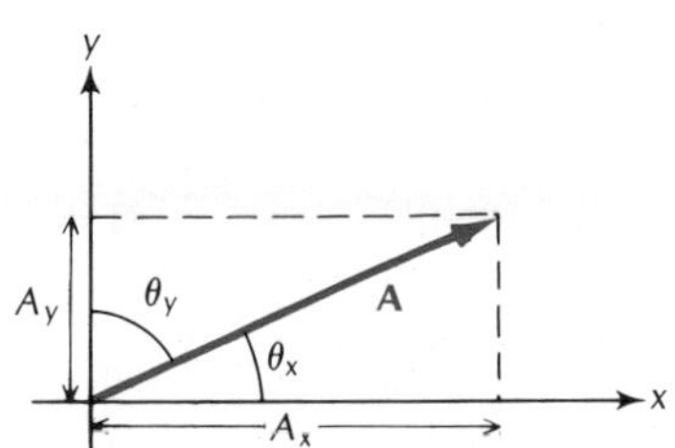

Fig. 3.18 Vector **A** in two dimensions and its components.

[3] Note that the magnitude of each unit vector is a pure number, without any meters, seconds, or kilograms. Nevertheless, when drawing the unit vectors (as in Figure 3.14), we will adopt the convention of showing these vectors as though they had a length of 1 meter; thus, $\hat{\mathbf{x}}$ starts at the point $x = 0$ meter and ends at $x = 1$ meter.

be expressed in terms of the magnitude of the vector and the angles that it makes with the x and y axes,

$$A_x = A \cos \theta_x \qquad A_y = A \cos \theta_y \tag{12}$$

Here θ_x is reckoned from the positive x axis toward the vector; values of θ_x larger than 90° give negative results for A_x. The same is true for θ_y.

Figure 3.18 also shows that, by the Pythagorean theorem, the magnitude of **A** can be expressed in terms of A_x and A_y,

$$A = \sqrt{A_x^2 + A_y^2} \tag{13}$$

Upon generalization to three dimensions, Eqs. (12) and (13) become

$$\boxed{A_x = A \cos \theta_x} \tag{14}$$

$$\boxed{A_y = A \cos \theta_y} \tag{15}$$

$$\boxed{A_z = A \cos \theta_z} \tag{16}$$

$$\boxed{A = \sqrt{A_x^2 + A_y^2 + A_z^2}} \tag{17}$$

Note that in the special case of the position vector, Eq. (17) takes the form

$$r = \sqrt{x^2 + y^2 + z^2} \tag{18}$$

which is a familiar expression for the distance between the point x, y, z and the origin of coordinates.

Expressed in terms of components, vector addition (or subtraction) turns out to be merely addition (or subtraction) of components. Thus, given two vectors

$$\mathbf{A} = A_x\hat{\mathbf{x}} + A_y\hat{\mathbf{y}} + A_z\hat{\mathbf{z}} \tag{19}$$

and

$$\mathbf{B} = B_x\hat{\mathbf{x}} + B_y\hat{\mathbf{y}} + B_z\hat{\mathbf{z}} \tag{20}$$

Their sum is

$$\begin{aligned} \mathbf{A} + \mathbf{B} &= A_x\hat{\mathbf{x}} + A_y\hat{\mathbf{y}} + A_z\hat{\mathbf{z}} + B_x\hat{\mathbf{x}} + B_y\hat{\mathbf{y}} + B_z\hat{\mathbf{z}} \\ &= (A_x + B_x)\hat{\mathbf{x}} + (A_y + B_y)\hat{\mathbf{y}} + (A_z + B_z)\hat{\mathbf{z}} \end{aligned} \tag{21}$$

Thus the x component of the resultant is the sum of the x components of **A** and **B**, etc.

EXAMPLE 2. The eye of the hurricane is 200 mi from Miami on a bearing of 30° south of east. A reconnaissance airplane is 100 mi due north of Miami. What displacement vector will bring the plane to the eye of the hurricane?

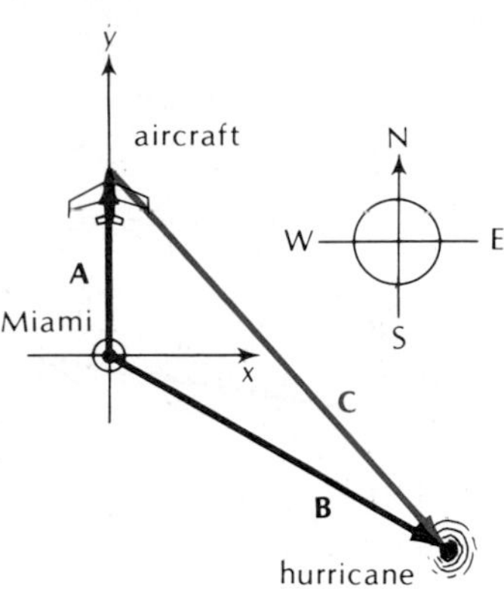

Fig. 3.19 Displacement vector **C** from the airplane to the eye of the hurricane.

SOLUTION: In Example 1 we saw how to use a graphical method and a trigonometric method for finding an unknown vector. Here we will see how to use the component method.

For this calculation we need to make a choice of coordinate system, i.e., a choice of origin and of axes. We can make this choice in any way that happens to be convenient. In Figure 3.19 we have placed the origin on Miami, with the x axis eastward and the y axis northward. In the coordinate system, the airplane has a position vector **A** with components

$$A_x = 0 \text{ mi} \qquad A_y = 100 \text{ mi}$$

The hurricane has a position vector **B** with components

$$B_x = 200 \text{ mi} \times \cos 30° = 173 \text{ mi}$$

$$B_y = 200 \text{ mi} \times \cos 120° = -100 \text{ mi}$$

The displacement **C** from the airplane to the hurricane is the *difference* between these position vectors, that is, $\mathbf{C} = \mathbf{B} - \mathbf{A}$. This vector **C** has components

$$C_x = B_x - A_x = 173 \text{ mi} - 0 \text{ mi} = 173 \text{ mi}$$

$$C_y = B_y - A_y = -100 \text{ mi} - 100 \text{ mi} = -200 \text{ mi}$$

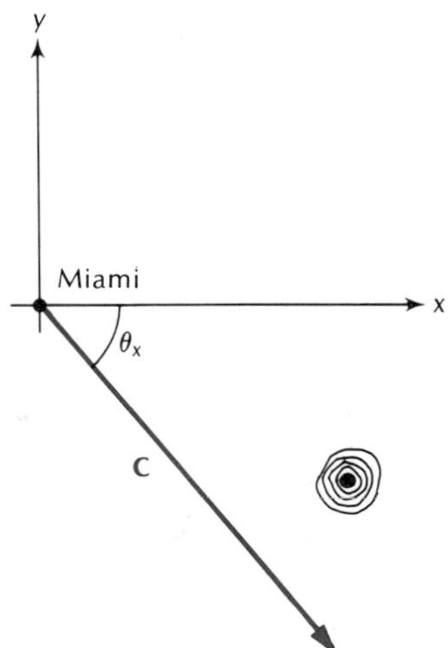

Fig. 3.20 The vector **C** has been parallel-transported so that its tail is at the origin.

Figure 3.20 shows the vector **C** with its tail placed at the origin. The magnitude of **C** is

$$C = \sqrt{C_x^2 + C_y^2} = \sqrt{(173)^2 + (200)^2} \text{ mi} = 264 \text{ mi}$$

and the angle between **C** and the x axis is given by

$$\tan \theta_x = \frac{-200}{173}$$

$$\theta_x = -49.1°$$

Hence a displacement of 264 mi at 49.1° south of east will bring the airplane to the hurricane.

3.4 Vector Multiplication[4]

There are several ways of multiplying vectors. The reason for this diversity is that in forming the "product" of two vectors, we must take into account both their magnitudes *and* their directions. Depending on how we combine these quantities, we obtain different kinds of vector products. The two most important vector products are the dot product and the cross product.

[4] This section may be postponed. The cross product will be used in Section 5.6 and the dot product will be used in Section 7.2.

Dot Product The **dot product** (also called the *scalar product* or the *inner product*) of two vectors **A** and **B** is denoted by $\mathbf{A} \cdot \mathbf{B}$. This quantity is simply the product of the magnitudes of the two vectors and the cosine of the angle ϕ between them (Figure 3.21),

Dot product

$$\boxed{\mathbf{A} \cdot \mathbf{B} = AB \cos \phi} \tag{22}$$

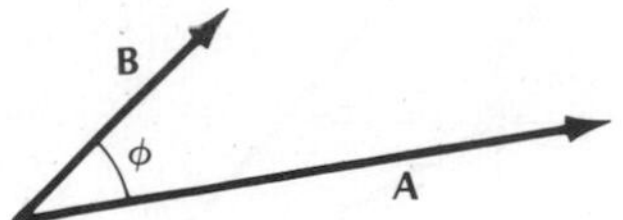
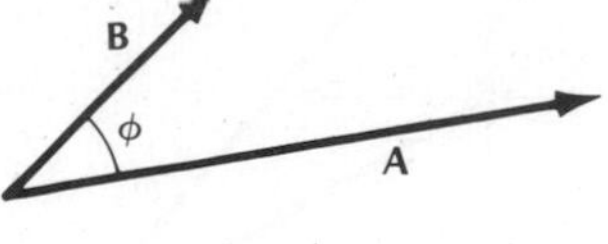

Fig. 3.21 The vectors **A** and **B** and the angle between them.

Thus, the dot product of two vectors simply gives a number, i.e., a scalar rather than a vector. The number will be positive if $\phi < 90°$ and negative if $\phi > 90°$. If the two vectors are perpendicular, then their dot product is zero. Note that the dot product is commutative; as in ordinary multiplication, the order of the factors is irrelevant:

$$\boxed{\mathbf{A} \cdot \mathbf{B} = \mathbf{B} \cdot \mathbf{A}}$$

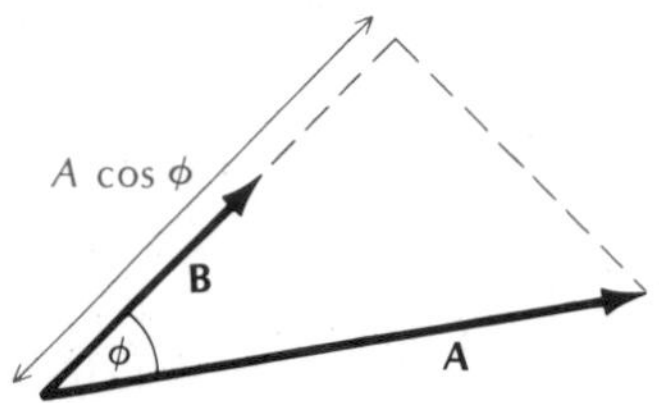

Fig. 3.22 The component of **A** along **B** is $A \cos \phi$; the component of **B** along **A** is $B \cos \phi$.

We see from Figure 3.22 that the dot product $\mathbf{A} \cdot \mathbf{B}$ can be regarded as B times the component of **A** along **B**, or as A times the component of **B** along **A**. The special case of the dot product of a vector with itself gives the square of the magnitude of the vector:

$$\boxed{\mathbf{A} \cdot \mathbf{A} = AA \cos 0 = A^2} \tag{23}$$

The dot products of the unit vectors with themselves and with each other are

$$\begin{array}{lll} \hat{\mathbf{x}} \cdot \hat{\mathbf{x}} = 1 & \hat{\mathbf{y}} \cdot \hat{\mathbf{y}} = 1 & \hat{\mathbf{z}} \cdot \hat{\mathbf{z}} = 1 \\ \hat{\mathbf{x}} \cdot \hat{\mathbf{y}} = \hat{\mathbf{y}} \cdot \hat{\mathbf{x}} = 0 & \hat{\mathbf{x}} \cdot \hat{\mathbf{z}} = \hat{\mathbf{z}} \cdot \hat{\mathbf{x}} = 0 & \hat{\mathbf{y}} \cdot \hat{\mathbf{z}} = \hat{\mathbf{z}} \cdot \hat{\mathbf{y}} = 0 \end{array} \tag{24}$$

By means of these equations it is possible to express the dot product of two arbitrary vectors in terms of their components[5]:

$$\begin{aligned} \mathbf{A} \cdot \mathbf{B} &= (A_x\hat{\mathbf{x}} + A_y\hat{\mathbf{y}} + A_z\hat{\mathbf{z}}) \cdot (B_x\hat{\mathbf{x}} + B_y\hat{\mathbf{y}} + B_z\hat{\mathbf{z}}) \\ &= A_xB_x\hat{\mathbf{x}} \cdot \hat{\mathbf{x}} + A_xB_y\hat{\mathbf{x}} \cdot \hat{\mathbf{y}} + A_xB_z\hat{\mathbf{x}} \cdot \hat{\mathbf{z}} \\ &\quad + A_yB_x\hat{\mathbf{y}} \cdot \hat{\mathbf{x}} + A_yB_y\hat{\mathbf{y}} \cdot \hat{\mathbf{y}} + A_yB_z\hat{\mathbf{y}} \cdot \hat{\mathbf{z}} \\ &\quad + A_zB_x\hat{\mathbf{z}} \cdot \hat{\mathbf{x}} + A_zB_y\hat{\mathbf{z}} \cdot \hat{\mathbf{y}} + A_zB_z\hat{\mathbf{z}} \cdot \hat{\mathbf{z}} \end{aligned} \tag{25}$$

With the values given by Eq. (24) this reduces to

Dot-product formula

$$\boxed{\mathbf{A} \cdot \mathbf{B} = A_xB_x + A_yB_y + A_zB_z} \tag{26}$$

[5] In working out this product, we make use of the fact that vector multiplication obeys the **distributive law:** $(\mathbf{C} + \mathbf{D}) \cdot \mathbf{E} = \mathbf{C} \cdot \mathbf{E} + \mathbf{D} \cdot \mathbf{E}$. To prove this law, we need only note that this product is the magnitude of **E** times the component of $\mathbf{C} + \mathbf{D}$ along **E**; and that the component of $\mathbf{C} + \mathbf{D}$ along any direction is equal to the sum of the component of **C** plus the component of **D**.

Thus the dot product is simply the sum of the products of the x, y, and z components of the two vectors.

Finally, note that the components of a vector are equal to the dot product of the vector and the corresponding unit vectors. For example,

$$\hat{\mathbf{x}} \cdot \mathbf{A} = \hat{\mathbf{x}} \cdot (A_x\hat{\mathbf{x}} + A_y\hat{\mathbf{y}} + A_z\hat{\mathbf{z}})$$

$$= A_x\hat{\mathbf{x}} \cdot \hat{\mathbf{x}} + A_y\hat{\mathbf{x}} \cdot \hat{\mathbf{y}} + A_z\hat{\mathbf{x}} \cdot \hat{\mathbf{z}} = A_x \tag{27}$$

EXAMPLE 3. Find the dot product of the vectors **A** and **B** of Example 2.

SOLUTION: The vector **A** has a magnitude $A = 100$ mi and **B** has a magnitude $B = 200$ mi; the angle between the vectors is $\phi = 120°$. Hence

$$\mathbf{A} \cdot \mathbf{B} = AB \cos \phi = 100 \text{ mi} \times 200 \text{ mi} \times \cos 120°$$

$$= -10{,}000 \text{ mi}^2 \tag{28}$$

Alternatively, the calculation can be done by components:

$$\mathbf{A} \cdot \mathbf{B} = A_xB_x + A_yB_y + A_zB_z$$

$$= 0 \text{ mi} \times 173 \text{ mi} + 100 \text{ mi} \times (-100 \text{ mi}) + 0 \text{ mi} \times 0 \text{ mi}$$

$$= -10{,}000 \text{ mi}^2 \tag{29}$$

This agrees with Eq. (28).

EXAMPLE 4. Find the angle between the vectors $\mathbf{A} = 2\hat{\mathbf{x}} + \hat{\mathbf{y}}$ and $\mathbf{B} = -\hat{\mathbf{x}} + 3\hat{\mathbf{y}} + 2\hat{\mathbf{z}}$.

SOLUTION: Since

$$\mathbf{A} \cdot \mathbf{B} = AB \cos \phi$$

the cosine of the angle between the vectors is

$$\cos \phi = \frac{\mathbf{A} \cdot \mathbf{B}}{AB} = \frac{A_xB_x + A_yB_y + A_zB_z}{\sqrt{A_x^2 + A_y^2 + A_z^2}\,\sqrt{B_x^2 + B_y^2 + B_z^2}}$$

$$= \frac{(2)(-1) + (1)(3) + (0)(2)}{\sqrt{5}\,\sqrt{14}} = 0.120$$

This implies an angle of 83.1°.

CROSS PRODUCT The **cross product** (also called the *vector product*) of two vectors **A** and **B** is denoted by **A** × **B**. This quantity is a *vector* with a magnitude equal to the product of the magnitude of the two vectors and the sine of the angle between them. Thus if we write the resulting vector as

$$\mathbf{C} = \mathbf{A} \times \mathbf{B} \tag{30}$$

then the magnitude of this vector is

$$\boxed{C = AB \sin \phi} \tag{31}$$

Cross product

The direction of **C** is defined as being perpendicular to the plane formed by **A** and **B** (Figure 3.23). The direction of **C** along the perpendicular is given by the **right-hand rule:** *put the fingers of your right hand along* **A** *(Figure 3.24a) and curl them toward* **B** *in the direction of the smaller angle from* **A** *to* **B** *(Figure 3.24b); the thumb then points along* **C.** Note that the fingers must be curled from the first vector in the product toward the second. Thus **A** × **B** is not the same as **B** × **A.** For the latter product the fingers must be curled from **B** toward **A** (rather than vice versa); hence, the direction of the vector **B** × **A** is opposite to that of **A** × **B,**

Right-hand rule

$$\mathbf{B}\times\mathbf{A}=-\mathbf{A}\times\mathbf{B} \tag{32}$$

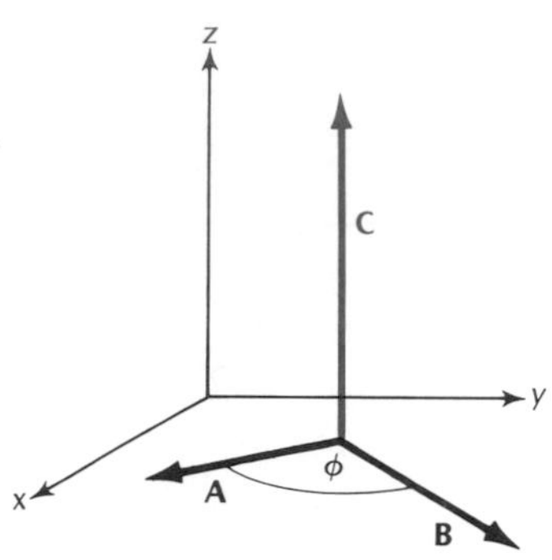

Fig. 3.23 The vectors **A** and **B** and their cross product **C** = **A** × **B.**

Thus, the cross product of two vectors is *not* commutative; in contrast to ordinary multiplication, the result does depend on the order of the factors.

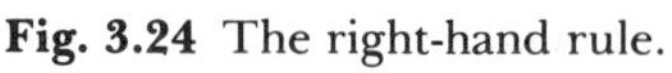
Fig. 3.24 The right-hand rule.

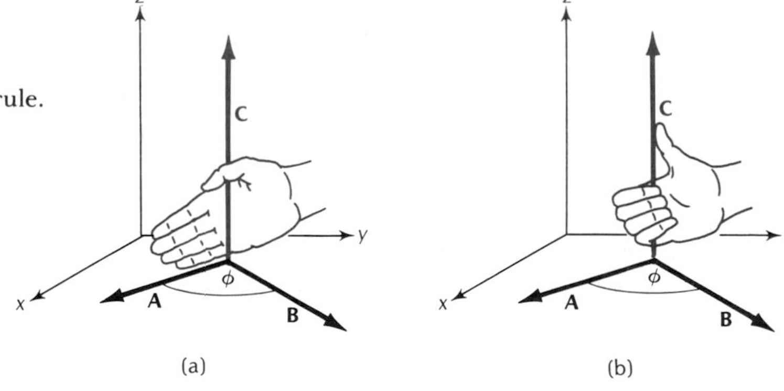

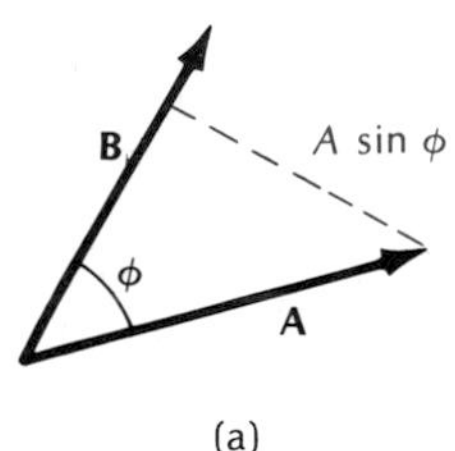

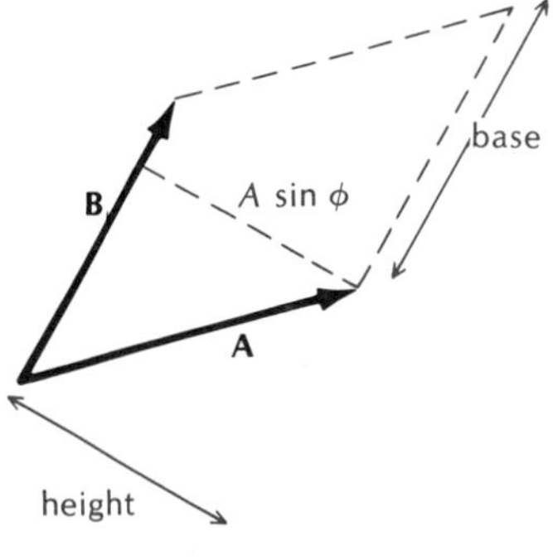

Fig. 3.25 (a) $A \sin \phi$ is the component of **A** perpendicular to **B.** (b) $AB \sin \phi$ is the area of the parallelogram.

As we see from Figure 3.25a, the magnitude of **A** × **B** is equal to the product of the magnitude of **B** times the component of **A** perpendicular to **B** (or the magnitude of **A** times the component of **B** perpendicular to **A**). Furthermore, from Figure 3.25b, we see that the magnitude of **A** × **B** is equal to the area of the parallelogram formed out of the vectors **A** and **B.**

If the vectors **A** and **B** are parallel, then their cross product is zero; in particular, the cross product of any vector with itself is zero,

$$\mathbf{A}\times\mathbf{A}=0 \tag{33}$$

The cross products of the unit vectors are

$$\begin{aligned}
&\hat{\mathbf{x}}\times\hat{\mathbf{x}}=0 &&\hat{\mathbf{y}}\times\hat{\mathbf{y}}=0 &&\hat{\mathbf{z}}\times\hat{\mathbf{z}}=0\\
&\hat{\mathbf{x}}\times\hat{\mathbf{y}}=-\hat{\mathbf{y}}\times\hat{\mathbf{x}}=\hat{\mathbf{z}} &&\hat{\mathbf{z}}\times\hat{\mathbf{x}}=-\hat{\mathbf{x}}\times\hat{\mathbf{z}}=\hat{\mathbf{y}} &&\hat{\mathbf{y}}\times\hat{\mathbf{z}}=-\hat{\mathbf{z}}\times\hat{\mathbf{y}}=\hat{\mathbf{x}}
\end{aligned} \tag{34}$$

Using these equations, we can express the cross product of two arbitrary vectors in terms of their components,

$$\begin{aligned}
\mathbf{A} \times \mathbf{B} &= (A_x\hat{\mathbf{x}} + A_y\hat{\mathbf{y}} + A_z\hat{\mathbf{z}}) \times (B_x\hat{\mathbf{x}} + B_y\hat{\mathbf{y}} + B_z\hat{\mathbf{z}}) \\
&= A_xB_x\hat{\mathbf{x}} \times \hat{\mathbf{x}} + A_xB_y\hat{\mathbf{x}} \times \hat{\mathbf{y}} + A_xB_z\hat{\mathbf{x}} \times \hat{\mathbf{z}} \\
&\quad + A_yB_x\hat{\mathbf{y}} \times \hat{\mathbf{x}} + A_yB_y\hat{\mathbf{y}} \times \hat{\mathbf{y}} + A_yB_z\hat{\mathbf{y}} \times \hat{\mathbf{z}} \\
&\quad + A_zB_x\hat{\mathbf{z}} \times \hat{\mathbf{x}} + A_zB_y\hat{\mathbf{z}} \times \hat{\mathbf{y}} + A_zB_z\hat{\mathbf{z}} \times \hat{\mathbf{z}} \\
&= 0 + A_xB_y\hat{\mathbf{z}} - A_xB_z\hat{\mathbf{x}} \\
&\quad - A_yB_x\hat{\mathbf{z}} + 0 + A_yB_z\hat{\mathbf{x}} \\
&\quad + A_zB_x\hat{\mathbf{y}} - A_zB_y\hat{\mathbf{x}} + 0 \qquad (35)
\end{aligned}$$

and, collecting terms, we have

$$\boxed{\begin{aligned}\mathbf{A} \times \mathbf{B} &= (A_yB_z - A_zB_y)\hat{\mathbf{x}} + (A_zB_x - A_xB_z)\hat{\mathbf{y}} \\ &\quad + (A_xB_y - A_yB_x)\hat{\mathbf{z}}\end{aligned}} \qquad (36)$$

Cross-product formula

[This messy result can be compactly written as a determinant,

$$\mathbf{A} \times \mathbf{B} = \begin{vmatrix} \hat{\mathbf{x}} & \hat{\mathbf{y}} & \hat{\mathbf{z}} \\ A_x & A_y & A_z \\ B_x & B_y & B_z \end{vmatrix} \qquad (37)$$

By multiplying out this determinant according to the rules for 3×3 determinants, one can check that this is the same as Eq. (36).]

EXAMPLE 5. What is the cross product of the vectors **A** and **B** of Example 2?

SOLUTION: Figure 3.26 shows the vectors **A** and **B**. The magnitudes of the vectors are 100 mi and 200 mi, respectively; the angle between them is 120°. Hence the magnitude of the cross product is

$$\begin{aligned}
C = |\mathbf{A} \times \mathbf{B}| &= AB \sin \phi \\
&= 100 \text{ mi} \times 200 \text{ mi} \times \sin 120° \\
&= 17{,}300 \text{ mi}^2
\end{aligned}$$

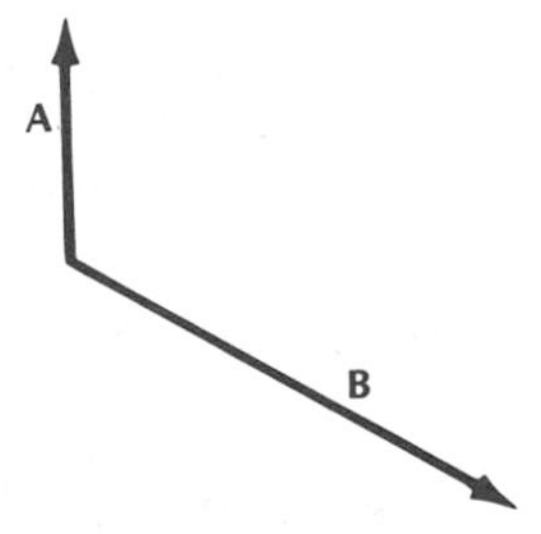

Fig. 3.26 The vectors **A** and **B** of Example 2.

The direction of **C** is perpendicular and into the plane of Figure 3.26.

We can also do this calculation by means of the cross-product formula of Eq. (36). With $A_x = 0$ mi, $A_y = 100$ mi, $A_z = 0$ mi, $B_x = 173$ mi, $B_y = -100$ mi, $B_z = 0$ mi, the first two terms in Eq. (36) drop out, since they involve factors of A_z or B_z which are zero. The last term gives

$$\begin{aligned}
\mathbf{A} \times \mathbf{B} &= 0 + 0 + (A_xB_y - A_yB_x)\hat{\mathbf{z}} \\
&= [0 \text{ mi} \times (-100 \text{ mi}) - 100 \text{ mi} \times 173 \text{ mi}]\hat{\mathbf{z}} \\
&= -17{,}300\hat{\mathbf{z}} \text{ mi}^2
\end{aligned}$$

This vector points along the negative z axis, i.e., again into the plane of Figure 3.26.

3.5 Vectors and Coordinate Rotations[6]

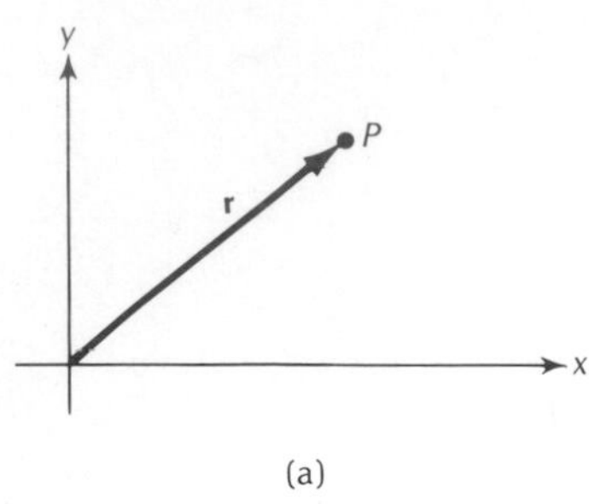

(a)

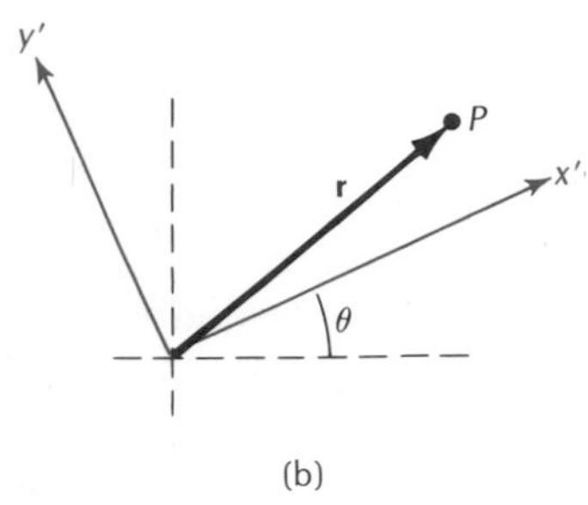

(b)

Fig. 3.27 Two sets of rectangular coordinate axes: (a) x, y and (b) x', y'.

According to the definition we gave in Section 3.1, a vector is a mathematical quantity that behaves like a displacement vector. This definition was somewhat vague, since we did not spell out precisely what is meant by the "behavior" of a displacement vector. A branch of advanced mathematics called tensor analysis tells us that the crucial behavior is that under coordinate rotations. In this section we will briefly discuss coordinate rotations. For the sake of simplicity we will focus on rotations in two dimensions; of course, a similar discussion can be given in three dimensions. Figure 3.27 shows two sets of rectangular coordinate axes; the axes x', y' are rotated by an angle θ relative to the axes x and y. Figure 3.27 also shows a point P and its position vector. The position vector **r** in Figures 3.27a and 3.27b is exactly the same (the displacement from the origin is the same), but the values of the components of the position vector depend on the choice of coordinates, and these values are different. In the first coordinate system we have

$$\mathbf{r} = x\hat{\mathbf{x}} + y\hat{\mathbf{y}} \tag{38}$$

and in the second coordinate system

$$\mathbf{r} = x'\hat{\mathbf{x}}' + y'\hat{\mathbf{y}}' \tag{39}$$

The relationship between the quantities x, y and x', y' is

Coordinate rotation

$$\boxed{x' = x \cos\theta + y \sin\theta} \tag{40}$$

$$\boxed{y' = -x \sin\theta + y \cos\theta} \tag{41}$$

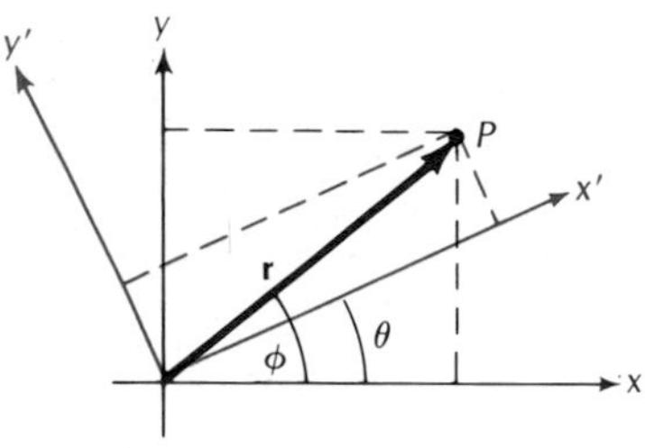

Fig. 3.28 The angles θ and ϕ.

To derive these two equations, we use Figure 3.28. In terms of the angles θ and ϕ shown in this figure, we have

$$x = r\cos\phi \qquad y = r\sin\phi \tag{42}$$

$$x' = r\cos(\phi - \theta) \qquad y' = r\sin(\phi - \theta) \tag{43}$$

Using the trigonometric identity $\cos(\phi - \theta) = \cos\phi\cos\theta + \sin\phi\sin\theta$, we obtain

$$\begin{aligned} x' &= r\cos\phi\cos\theta + r\sin\phi\sin\theta \\ &= x\cos\theta + y\sin\theta \end{aligned}$$

which is Eq. (40). The derivation of Eq. (41) is similar.

Taken together, Eqs. (40) and (41) are **transformation equations**; they show how the rotation from one coordinate system to another changes the components of the position vector, i.e., the coordinates of a point.

[6] This section is optional. It provides some useful background for the Galilean transformations and the Lorentz transformations.

Rigorous vector definition

We are now ready for a rigorous definition of a vector: *in two dimensions, a **vector** is an object with two components* A_x *and* A_y *that transform under a coordinate rotation in exactly the same way as the components of the position vector.* Thus, if the vector **A** has the form

$$\mathbf{A} = A_x\hat{\mathbf{x}} + A_y\hat{\mathbf{y}} \tag{44}$$

in the first coordinate system and the form

$$\mathbf{A} = A'_x\hat{\mathbf{x}}' + A'_y\hat{\mathbf{y}}' \tag{45}$$

in the second coordinate system, then the respective components are related as follows:

$$A'_x = A_x \cos\theta + A_y \sin\theta \tag{46}$$

$$A'_y = -A_x \sin\theta + A_y \cos\theta \tag{47}$$

According to the rigorous statement given above, vectors are defined by their transformation properties. In three dimensions the situation is similar, although the equations for three-dimensional rotations of coordinates are rather more messy; the three components of the vector must obey transformation equations identical to those for the three components of the position vector.

The magnitude of any vector is unchanged by a coordinate rotation. This is obvious from the fact that the rotation does not change the vector; it only changes its components. Mathematically, the preservation of the magnitude can be expressed as

$$\sqrt{A'^2_x + A'^2_y} = \sqrt{A^2_x + A^2_y} \tag{48}$$

Alternatively, this identity can be verified by explicit calculation: simply substitute the expressions (46) and (47) into the left side of Eq. (48); the sines and cosines will cancel.

The dot product of any two vectors is also unchanged by a coordinate rotation. This is a simple consequence of the fact that a rotation changes neither the magnitudes of the two vectors, nor the angle between them. The preservation of the dot product can be expressed as

$$A'_xB'_x + A'_yB'_y + A'_zB'_z = A_xB_x + A_yB_y + A_zB_z \tag{49}$$

Again, this identity can be verified by explicit calculation.

According to the rigorous definitions, a scalar is a numerical quantity that is unchanged by coordinate rotations. Thus, the length of any vector is a scalar and the dot product of any two vectors is a scalar. Any numerical constant such as π, or $\sqrt{2}$, or 666, or whatever, is of course also a scalar.

SUMMARY

Vector: Quantity with magnitude and direction; it behaves like a displacement vector.

Addition of vectors: Use the parallelogram method or the tail-to-head method; alternatively, add the components.

Unit vectors: $\hat{\mathbf{x}}$, $\hat{\mathbf{y}}$, $\hat{\mathbf{z}}$

Position vector: $\mathbf{r} = x\hat{\mathbf{x}} + y\hat{\mathbf{y}} + z\hat{\mathbf{z}}$

Components of a vector:

$$\mathbf{A} = A_x\hat{\mathbf{x}} + A_y\hat{\mathbf{y}} + A_z\hat{\mathbf{z}}$$

$$A_x = A\cos\theta_x$$

$$A_y = A\cos\theta_y$$

$$A_z = A\cos\theta_z$$

$$A = \sqrt{A_x^2 + A_y^2 + A_z^2}$$

Dot product: $\mathbf{A}\cdot\mathbf{B} = AB\cos\phi$

$$= A_xB_x + A_yB_y + A_zB_z$$

Cross product: magnitude: $|\mathbf{A}\times\mathbf{B}| = AB\sin\phi$
direction: Right-hand rule

Behavior of a vector under coordinate rotation:

$$A'_x = A_x\cos\theta + A_y\sin\theta$$

$$A'_y = -A_x\sin\theta + A_y\cos\theta$$

QUESTIONS

1. A large oil tanker proceeds from Kharg Island (Persian Gulf) to Rotterdam via the Cape of Good Hope. A small oil tanker proceeds from Kharg Island to Rotterdam via the Suez Canal. Are the displacement vectors of the two tankers equal? Are the distances covered equal?

2. An airplane flies from Boston to Houston and back to Boston. Is the displacement zero in the reference frame of the Earth? In the reference frame of the Sun?

3. Does a vector of zero magnitude have a direction? Does it matter?

4. Can the magnitude of a vector be negative? Zero?

5. Two vectors have nonzero magnitude. Under what conditions will their sum be zero? Their difference?

6. Is it posible for the sum of two vectors to have the same magnitude as the difference of the two vectors?

7. Three vectors have the same magnitude. Under what conditions will their sum be zero?

8. The magnitude of a vector is never smaller than the magnitude of any one component of the vector. Explain.

9. Two vectors have nonzero magnitude. Under what conditions will their dot product be zero? Their cross product?

10. Suppose $\mathbf{A}\cdot\mathbf{B} > 0$. What can you conclude about the angle between **A** and **B**?

11. If **A** and **B** are any arbitrary vectors, then $\mathbf{A}\cdot(\mathbf{B}\times\mathbf{A}) = 0$. Explain.

12. Why is there no vector division? (Hint: If **A** and **B** are given and if $A = \mathbf{B}\cdot\mathbf{C}$, then there exist several vectors **C** that satisfy this equation, and similarly for $\mathbf{A} = \mathbf{B}\times\mathbf{C}$.)

13. Assume that **A** is some nonzero vector. If $\mathbf{A}\cdot\mathbf{B} = \mathbf{A}\cdot\mathbf{C}$, can we conclude that $\mathbf{B} = \mathbf{C}$? If $\mathbf{A}\times\mathbf{B} = \mathbf{A}\times\mathbf{C}$, can we conclude that $\mathbf{B} = \mathbf{C}$? What if *both* $\mathbf{A}\cdot\mathbf{B} = \mathbf{A}\cdot\mathbf{C}$ and $\mathbf{A}\times\mathbf{B} = \mathbf{A}\times\mathbf{C}$?

PROBLEMS

Section 3.2

1. The displacement vector **A** has a length of 350 m in the direction 45° west of north; the displacement vector **B** has a length of 120 m in the direction 20° east of north. Find the magnitude and direction of the resultant of these vectors.

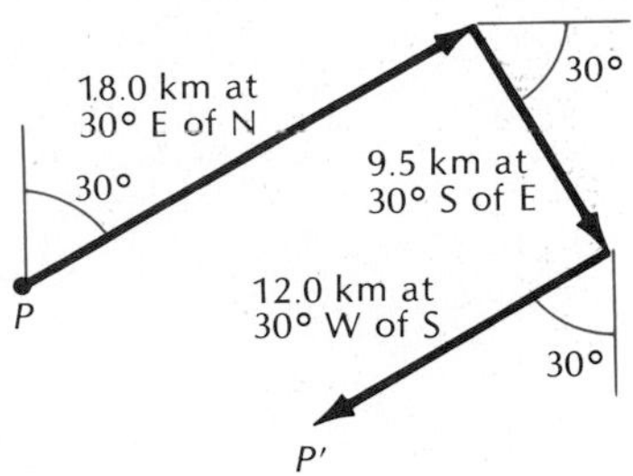

Fig. 3.29 Successive displacement vectors of an aircraft.

2. Figure 3.29 shows the successive displacements of an aircraft flying a search pattern. The initial position of the aircraft is P and the final position is P'. What is the net displacement (magnitude and direction) between P and P'? Find the answer both graphically (by carefully drawing a page-size diagram with protractor and ruler and measuring the resultant) and trigonometrically (by solving triangles).

3. A sailboat tacking against the wind moves as follows: 3.2 km at 45° east of north, 4.5 km at 50° west of north, 2.6 km at 45° east of north. What is the net displacement for the entire motion?

4. Three displacement vectors **A, B,** and **C** are, respectively, 4 cm at 30° west of north, 8 cm at 30° east of north, and 3 cm due north. Carefully draw these vectors on a sheet of paper. Find **A + B + C** graphically. Find **A + B − C** graphically.

5. During the maneuvers preceding the Battle of Jutland, the British battle cruiser *Lion* moved as follows (distances are in nautical miles): 1.2 mi due north, 6.1 mi at 38° east of south, 2.9 mi at 59° east of south, 4.0 mi at 89° east of north, and 6.5 mi at 31° east of north.
   (a) Draw each of these displacement vectors and draw the net displacement vector.
   (b) Graphically or algebraically find the distance between the initial position and the final position.

6. The Earth moves around the Sun in a circle of radius 1.50×10^{11} m at (approximately) constant speed.
   (a) Taking today's position of the Earth as origin, draw a diagram showing the position vector 3 months, 6 months, 9 months, and 12 months later.
   (b) Draw the displacement vector between the 0-month and the 3-month position; the 3-month and the 6-month position, etc. Calculate the magnitude of the displacement vector for one of these 3-month intervals.

7. Both Singapore and Quito are (nearly) on the Earth's equator; the longitude of Singapore is 104° East and that of Quito is 78° West. What is the magnitude of the displacement vector between these cities? What is the distance between them measured along the equator?

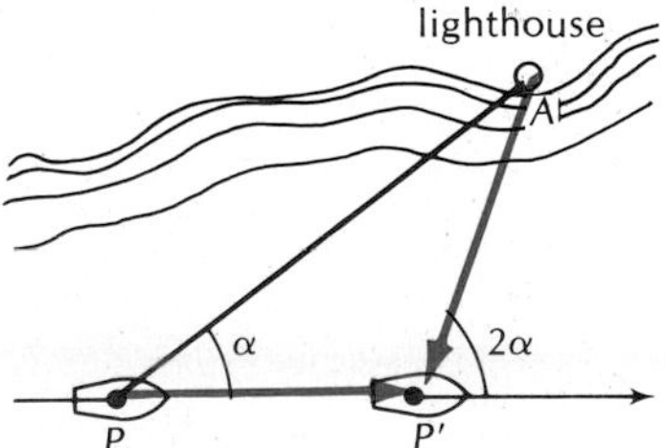

Fig. 3.30

8. By a method known as "doubling the angle on the bow," the navigator of a ship can determine his position relative to a fixed point, such as a lighthouse. Figure 3.30 shows the (straight) track of a ship passing by a lighthouse. At the point P, the navigator measures the angle α between the line of sight to the lighthouse and the direction of motion of the ship. He then measures how far the ship advances through the water until the angle between the line of sight and the direction of motion is twice as large as it was initially. Prove that the magnitude of the displacement vector PP' equals the magnitude of the position vector AP' of the ship relative to the lighthouse.

9. The radar operator of a stationary Coast Guard cutter observes that at 10^h30^m an unidentified ship is at a distance of 9.5 mi on a bearing of 60° east of north and at 11^h10^m the unidentified ship is at a distance of 4.2 mi on a bearing of 33° east of north. What is the displacement vector of the unidentified ship at 11^h10^m? Assuming that the unidentified ship continues on the same course at the same speed, what will be its displacement vector at 11^h30^m? What will be its distance and bearing from the cutter?

*10. The fastest crossing of the Atlantic by sail was achieved in 1916 by the four-masted ship *Lancing,* which sailed from New York (latitude 40°48′ North, longitude 73°58′ West) to Cape Wrath, Ireland (latitude 38°36′ North, longitude 5°1′ West), in $6\frac{3}{4}$ days. What was the magnitude of the displacement vector for this trip?

Section 3.3

11. A vector length of 5.0 m is in the x–y plane at an angle of 30° with the x axis. What is the x component of this vector? The y component?

12. A displacement vector has a magnitude of 12.0 km in the direction 40° west of north. What is the north component of this vector? The west component?

13. Air traffic controllers usually describe the position of an aircraft relative to the airport by altitude, horizontal distance, and bearing. Suppose an aircraft is at altitude 500 m, distance 15 km, and bearing 35° east of north. What are the x, y, and z components (in meters) of the position vector? The x axis is east, the y axis is north, and the z axis is vertical.

14. The displacement vectors **A** and **B** are in the x–y plane. Their components are $A_x = 3$ cm, $A_y = 2$ cm, $B_x = -1$ cm, $B_y = 3$ cm.
   (a) Draw a diagram showing these vectors.
   (b) Calculate the resultant of **A** and **B**. Draw the resultant in your diagram.

15. A vector in the x–y plane has a magnitude of 8.0 units. The angle between the vector and the x axis is 52°. What are the x, y, and z components of this vector?

16. Given that a vector has a magnitude of 6.0 units and makes angles of 45° and 85° with the x and y axes, respectively, find the x and y components of this vector. Does the given information determine the z component? What can you say about the z component?

17. An air traffic controller notices that one aircraft approaching the airport is at altitude 2500 ft, (horizontal) distance 12.0 mi, and bearing 20° south of east. A second aircraft is at altitude 3500 ft, distance 11.0 mi, and bearing 25° south of east. What is the displacement vector from the first aircraft to the second? Express your answer in terms of altitude, (horizontal) distance, and bearing.

18. What is the magnitude of the vector $3\hat{\mathbf{x}} + 2\hat{\mathbf{y}} - \hat{\mathbf{z}}$?

19. Find a vector that has the same direction as $3\hat{\mathbf{x}} - 6\hat{\mathbf{y}} + 2\hat{\mathbf{z}}$ but a magnitude of 2 units.

20. A vector has components $A_x = 5.0$, $A_y = -3.0$, $A_z = 1.0$. What is the magnitude of this vector? What is the angle between this vector and the x axis? The y axis? The z axis?

21. Suppose that $\mathbf{A} = -5\hat{\mathbf{x}} - 3\hat{\mathbf{y}} + \hat{\mathbf{z}}$ and $\mathbf{B} = 2\hat{\mathbf{x}} + \hat{\mathbf{y}} - 3\hat{\mathbf{z}}$. Calculate the following.
   (a) $\mathbf{A} + \mathbf{B}$
   (b) $\mathbf{A} - \mathbf{B}$
   (c) $2\mathbf{A} - 3\mathbf{B}$

22. Given that $\mathbf{A} = 6\hat{\mathbf{x}} - 2\hat{\mathbf{y}}$ and $\mathbf{B} = -4\hat{\mathbf{x}} - 3\hat{\mathbf{y}} + 8\hat{\mathbf{z}}$, find a vector **C** such that $3\mathbf{A} - 2\mathbf{C} = 4\mathbf{B}$.

Section 3.4

23. Calculate the dot product of the vectors $5\hat{\mathbf{x}} - 2\hat{\mathbf{y}} + \hat{\mathbf{z}}$ and $2\hat{\mathbf{x}} - \hat{\mathbf{z}}$.

24. Calculate the dot product of the vectors **A** and **B** described in Example 1.

25. Find the magnitude of the vector $-2\hat{\mathbf{x}} + \hat{\mathbf{y}} + 2\hat{\mathbf{z}}$. Find the magnitude of the vector $3\hat{\mathbf{x}} - 6\hat{\mathbf{y}} + 2\hat{\mathbf{z}}$. Find the angle between these two vectors.

26. A vector **A** has components $A_x = 2$, $A_y = -1$, $A_z = -4$. Find a vector (give its components) that has the same direction as **A** but a magnitude of 1 unit.

27. Find a unit vector which points exactly halfway between the directions of the two vectors $4\hat{\mathbf{x}} + 2\hat{\mathbf{y}}$ and $-\hat{\mathbf{x}} + 3\hat{\mathbf{y}} + 2\hat{\mathbf{z}}$.

28. Suppose that $\mathbf{A} = 3\hat{\mathbf{x}} + 4\hat{\mathbf{y}}$ and $\mathbf{B} = -\hat{\mathbf{x}} + 3\hat{\mathbf{y}} - 2\hat{\mathbf{z}}$. Find the component of **A** along the direction of **B**. Find the component of **B** along the direction of **A**.

29. Find the angle between the diagonal of a cube and one of its edges. (Hint: Suppose that the edges of the cube are parallel to the vectors $\hat{\mathbf{x}}$, $\hat{\mathbf{y}}$, and $\hat{\mathbf{z}}$. What vector is then parallel to the diagonal?)

30. The displacement vector **A** has a length of 6.0 m in the direction 30° east of north; the displacement vector **B** has a length of 8.0 m in the direction 40° south of east. Find the magnitude and the direction of $\mathbf{A} \times \mathbf{B}$.

31. Calculate the cross product of the vectors **A** and **B** described in Example 1.

32. Given that $\mathbf{A} = 2\hat{\mathbf{x}} - 3\hat{\mathbf{y}} + 2\hat{\mathbf{z}}$ and $\mathbf{B} = -3\hat{\mathbf{x}} + 4\hat{\mathbf{z}}$, calculate the cross product $\mathbf{A} \times \mathbf{B}$.

33. The vectors **A**, **B**, and **C** have components $A_x = 3$, $A_y = -2$, $A_z = 2$, $B_x = 0$, $B_y = 0$, $B_z = 4$, $C_x = 2$, $C_y = -3$, $C_z = 0$. Calculate the following.
   (a) $\mathbf{A} \cdot (\mathbf{B} + \mathbf{C})$
   (b) $\mathbf{A} \times (\mathbf{B} + \mathbf{C})$
   (c) $\mathbf{A} \cdot (\mathbf{B} \times \mathbf{C})$
   (d) $\mathbf{A} \times (\mathbf{B} \times \mathbf{C})$

34. Find a unit vector perpendicular to both $4\hat{\mathbf{x}} + 3\hat{\mathbf{y}}$ and $-\hat{\mathbf{x}} - 3\hat{\mathbf{y}} + 2\hat{\mathbf{z}}$.

*35. Show that the magnitude of $\mathbf{A} \cdot (\mathbf{B} \times \mathbf{C})$ is the volume of the parallelepiped determined by **A**, **B**, and **C**.

*36. Show that $\mathbf{A} \times (\mathbf{B} \times \mathbf{C}) = \mathbf{B}(\mathbf{A} \cdot \mathbf{C}) - \mathbf{C}(\mathbf{A} \cdot \mathbf{B})$. (Hint: Choose the orientation of your coordinate axes in such a way that **B** is along the x axis and that **C** is in the x–y plane.)

Section 3.5

37. With respect to a given coordinate system, a vector has components $A_x = 5$, $A_y = -3$, $A_z = 0$.
   (a) What are the components A'_x and A'_y of this vector in a new coordinate system whose x' and y' axes make angles of 30° with the old x and y axes?
   (b) Calculate the length of the vector from its A_x and A_y components. Calculate the length of the vector from its A'_x and A'_y components.

38. In the vicinity of New York City, the direction of magnetic north is 11°55′ west of true north (i.e., a magnetic compass needle points 11°55′ west of north). Suppose that an aircraft flies 5.0 mi on a bearing of 56° east of magnetic north.
   (a) What are the north and east components of this displacement in a coordinate system based on the direction of magnetic north?
   (b) What are the north and east components of the displacement in a coordinate system based on the direction of true north?

39. Verify Eq. (48) by explicit substitution of Eqs. (46) and (47).

40. Verify Eq. (49) by explicit substitution of Eqs. (46) and (47).

41. In one rectangular coordinate system, a vector has components $A_x = 5.00$, $A_y = 3.00$, $A_z = 0$. In another coordinate system, it has components $A'_x = 0.10$,

$A'_y = -5.83$, $A'_z = 0$. Show that the two coordinate systems are related by a rotation. What is the angle of rotation?

42. A vector has components $A_x = 6$, $A_y = -3$, $A_z = 0$ in a given rectangular coordinate system. Find a new coordinate system such that the only nonzero component of the vector is A'_x.

43. Show that if the axes x', y' are rotated by an angle θ relative to the axes x, y, then the corresponding unit vectors are related as follows:

$$\hat{\mathbf{x}}' = \hat{\mathbf{x}} \cos \theta + \hat{\mathbf{y}} \sin \theta$$

$$\hat{\mathbf{y}}' = \hat{\mathbf{y}} \cos \theta - \hat{\mathbf{x}} \sin \theta$$

44. Suppose that the coordinates x', y', z' are related to the coordinates x, y, z by a rotation through an angle θ about the z axis [as in Eqs. (40) and (41)]. Suppose that the coodinates x'', y'', z'' are related to x', y', z' by a rotation through an angle ϕ about the x' axis.

(a) What is the equation that relates the x'', y'', z'' coordinates to the x', y', z' coordinates?

(b) What is the equation that relates the x'', y'', z'' coordinates to the x, y, z, coordinates?

THE ARCHITECTURE OF CRYSTALS[1]

Atoms and molecules are the building blocks within all the pieces of matter in our immediate environment — they are the building blocks within sticks, stones, water, air, our own bodies, and the entire Earth. The number of atoms in ordinary pieces of matter is extremely large. For example, the number of copper atoms in a penny coin is about 3×10^{22}. This large number of atoms creates the illusion that matter is continuously distributed. Yet the physical and chemical properties of a substance, such as copper, are entirely determined by the properties of its atoms. The density of copper, its elasticity and strength, its color, its melting point and boiling point, its thermal and electrical characteristics, etc., all hinge on the properties of the atomic building blocks and on the manner in which these building blocks are assembled into a large-scale structure. Unfortunately, the calculation of the macroscopic properties of a piece of matter from the microscopic properties of its constituent atoms is a formidable mathematical task because the number of atoms in even a small piece of matter is so extremely large. Physicists have devoted much effort to the study of metals and minerals, solids with a regular structure, which makes them more amenable to mathematical analysis. The building blocks in metals and minerals are arranged in an orderly, repetitive pattern, reminiscent of the orderly pattern of soldiers standing on parade. Solids with such an orderly, repetitive arrangement of atoms or molecules are called **crystals.** The study of all the conceivable geometrical arrangements of the atomic or molecular building blocks in a crystal is called **crystallography;** alternatively it might be called the architecture of crystals.

Regularity in the arrangement of the building blocks implies **symmetry.** In its precise mathematical meaning, a symmetry of a body is any geometric operation that leaves the body unaltered. For instance, the human body has bilateral symmetry, or reflection symmetry: we can exchange the right side of the body point by point with the left side, leaving the body unaltered (Figure A.1). In this chapter we will become acquainted with a variety of other geometric symmetries and we will uncover some of the relationships between these symmetries and the physical properties of crystals. But the significance of the concept of symmetry in physics runs much deeper than is apparent from the concrete geometric examples in this chapter. For instance, the repetitive pattern of behavior of the atoms listed in the periodic table (see Table P.1) can be traced to an underlying symmetry of the equation governing the motion of the electrons in the atoms. And the regular pattern of behavior of the elementary particles in "families" of particles (see Interlude C) can be traced to a symmetry of the parameters describing the internal structure of these particles. Such abstract mathematical symmetries play a key role in physics. In fact, much of modern theoretical physics can be described as a search for symmetry.

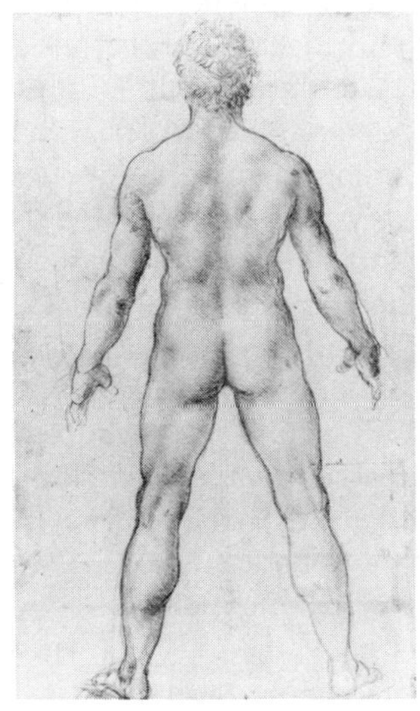

Fig. A.1 *Male nude.* Red chalk drawing by Leonardo da Vinci. *(Courtesy the Royal Library, Windsor Castle.)*

A.1 THE STRUCTURE OF CRYSTALS

In all solids the building blocks, whether molecules or atoms, are permanently locked into their positions. What distinguishes a crystalline solid, such as a metal, from a noncrystalline solid, such as wood or glass, is the regularity of the arrangement of the building blocks — they form an orderly, repetitive lattice, somewhat like a three-dimensional, rectangular coor-

[1] This chapter is optional.

dinate grid.[2] Thus, metals and other crystalline solids can be regarded as giant supermolecules in which the atoms have a carefully organized arrangement, just as they have in an ordinary molecule. A crystalline solid can be regarded as a three-dimensional polymer rather than a stringlike, one-dimensional polymer.

The regularity of the arrangement of the building blocks of crystals can be seen very clearly in Figures A.2–A.5. The first of these shows a "crystal" made of tobacco necrosis virus particles. Each individual virus particle is a giant molecule, nearly spherical in shape with a diameter of about 250 Å; because of the relatively large size of the molecule, the structure of this crystal can be seen with an electron microscope. Figure A.3 shows the regular arrangement of the molecules in a barium titanate crystal; this picture was taken with a new, very powerful electron microscope. Higher magnifications and more complicated devices are needed to see the structure of other crystals. Figure A.4 shows the arrangement of iron disulfide molecules in a marcasite crystal. This picture was produced with a **two-wavelength microscope,** which employs both X rays and visible light. Finally, Figure A.5 shows the atoms on the surface of the spherical tip of a needle of platinum. The orderly arrangement of the layers of platinum atoms manifests itself in the beautiful symmetry of this picture; the large-scale pattern of circles and lines is due to the protruding ridges of the atomic layers, ridges that correspond to the intersection of the curved surface of the needle with different atomic layers.

[2] If a small sample of molten metal is very suddenly cooled (quenched), its atoms will sometimes freeze in a disordered (amorphous) configuration, rather like a glass. In the following we will ignore such exceptional, freakish states of metals.

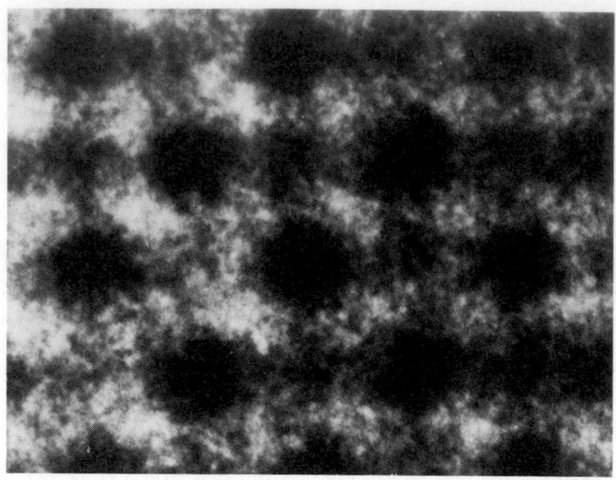

Fig. A.3 This electron-microscope photograph shows the arrangement of $BaTiO_3$ molecules in a crystal of barium titanate. The dark spheres are titanate ions and the lighter spheres are barium ions; the distance between them is about 2 Å. *(Courtesy R. Gronsky, Lawrence Berkeley Laboratories.)*

Figure A.5 was prepared with an **ion microscope,** which "illuminates" the needle with ions rather than with ordinary light. Briefly, the mechanism of this microscope is as follows: A very sharp platinum needle, ending in a hemisphere a few hundred angstroms in radius, is placed in a glass vessel containing rarified he-

Fig. A.2 Electron microscope photograph of a crystal of tobacco necrosis virus particles. The magnification is about 50,000×.

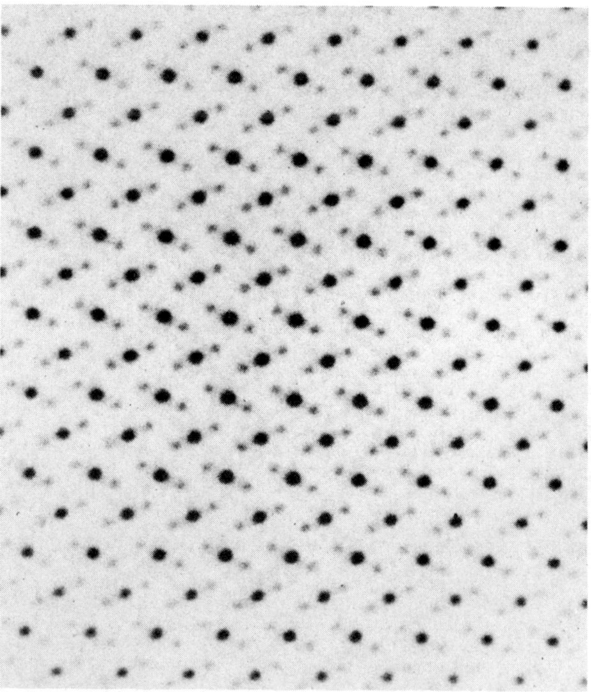

Fig. A.4 Arrangement of FeS_2 molecules in a crystal of marcasite. The large dots are iron atoms and the smaller dots are sulfur atoms; the distance between an iron atom and the nearest sulfur atom is about 1 Å.

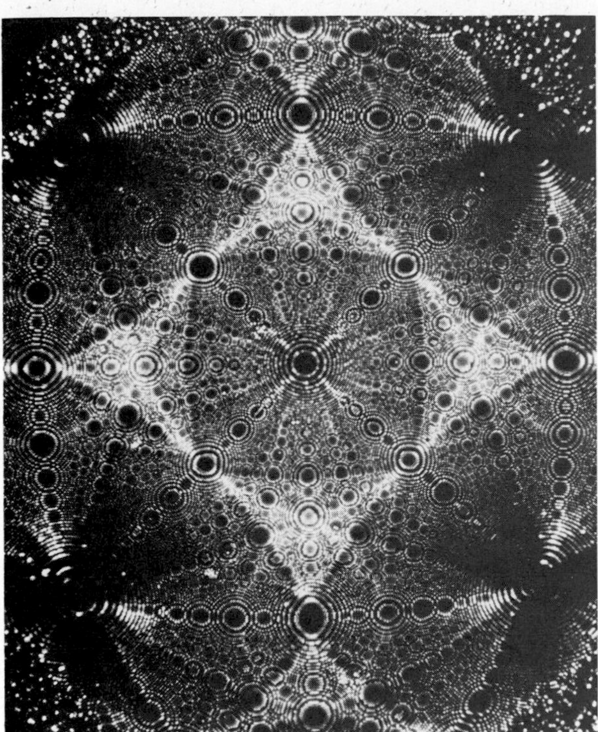

Fig. A.5 Image of the tip of a platinum needle viewed with an ion microscope. The radius of the tip is approximately 1000 Å and the magnification is approximately 5×10^5X. The positions of the dots show the positions of individual atoms, but the size of the dots is much larger than the size of the atoms. *(Courtesy T. T. Tsong, Pennsylvania State University.)*

lium gas. A very high voltage is then applied to the needle. This generates strong electric forces near the tip of the needle and any helium atom that ventures near the surface of the needle will be torn apart — it will be transformed into an ion. The electric forces then push the helium ion away from the needle at high speed. The ion travels in a straight path from the needle to a fluorescent screen, similar to a TV screen. The photograph in Figure A.5 shows the pattern of the impacts of ions on this screen. This pattern is a highly magnified representation of this pattern of ridges and indentations of the surface of the needle.

Ion microscopy has only been successfully applied to a very few crystals. A more common method used to discover the internal regularity of crystals is **X-ray diffraction**; this method does not produce spectacular visual displays such as in the above figures, but it does give us the most detailed information about the structure of crystals. The method is based on the following: if a crystal is illuminated with a beam of X rays, these rays do not all pass through in a straight line; rather, some are deflected and emerge at different angles. This deflection of X rays is called diffraction. The deflection of balls in a pinball machine is a crude analog of the diffraction of X rays by a crystal. The atoms are represented by the pins, and the X rays are the balls that make their way along the "alleys" between the pins. The incident balls are deflected by the pins one way or another and they finally emerge in certain directions; similarly, the X rays are scattered by the atoms and they finally emerge from the crystals in a few select directions. A photographic plate is used to register the emerging X rays; Figures A.6 and A.7 show some of the patterns of impact points of X rays on a photographic plate placed behind a crystal exposed to an X-ray beam. The dark spot in the center of these pictures is the impact point of the undeflected beam; the other spots, or **Laue spots,** were made by de-

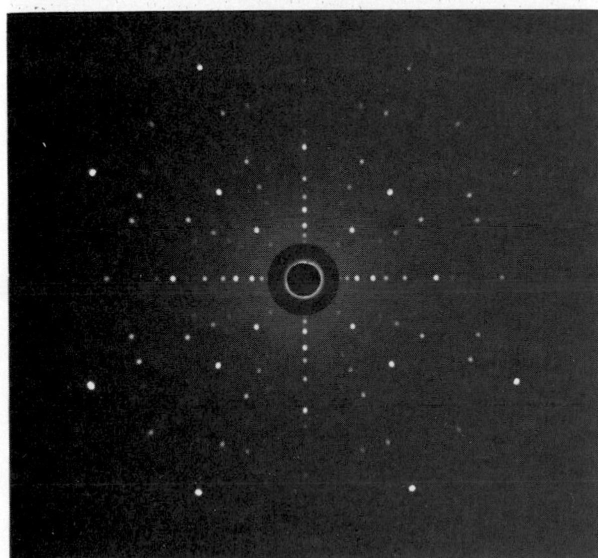

Fig. A.6 Pattern of Laue spots produced by a silicon crystal. *(Courtesy R. P. Goehner, General Electric.)*

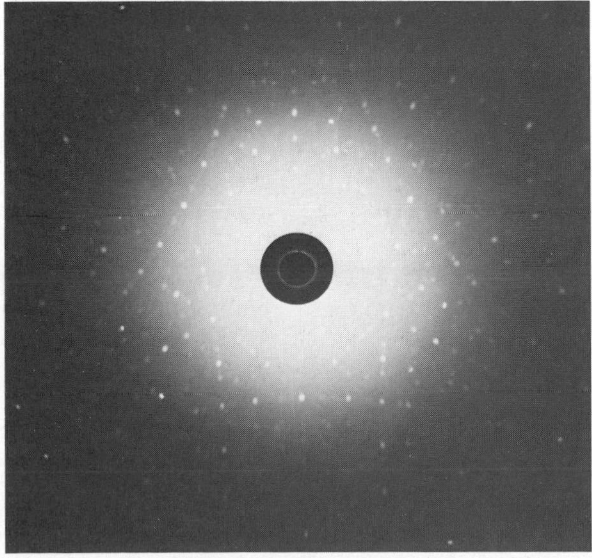

Fig. A.7 Pattern of Laue spots produced by an Al_2O_3 crystal. *(Courtesy R. P. Goehner, General Electric.)*

flected rays. The regularity of the pattern of Laue spots corresponds to the regularity of the arrangement of atoms within the crystal. From a careful study of the pattern of spots, crystallographers can deduce the arrangement of the atoms within the crystal.

Occasionally, the underlying regularity of the microscopic structure manifests itself in the gross macroscopic shape of the crystal. For example, the rectangular shape of grains of ordinary salt (NaCl) is a direct manifestation of the rectangular (or, more precisely, cubic) arrangement of the atoms in this crystal. Figure A.8 shows grains of salt magnified a hundred times; if one crushes these small rectangular crystallites, the resulting fragments will again be rectangular. Figure A.9 shows a single, large crystal of topaz; the regularity of the angles between the faces of this crystal is a manifestation of the regularity of the arrangement of the atoms in this crystal. But perhaps the most beautiful example is found in the shape of ice crystals in snowflakes (Figure A.10); the hexagonal pattern in these crystals is a direct manifestation of the underlying hexagonal arrangement of the water molecules. Note that although the center of the ice crystal is a very precise hexagon, the six "trees," or dendrites, that grow outward from the corners of the hexagon are not identical. These ice trees are about as similar as one might expect of, say, neighboring pine trees growing in some particular spot of the woods. The ice trees on a given crystal tend to be similar for the same reason that the pine trees are similar — they grew up together in the same environment.

Metals, such as iron, tin, or brass, usually do not display any obvious macroscopic features that hint at their underlying regularity. These metals are **polycrystalline,** that is, they are not made of a single crystal but, rather, of many small crystallites, or microcrystals, jumbled together. Such crystallites form the angular, flaky pattern that can be seen on the surface of the zinc plating commonly used on "galvanized" sheet metal (Figure A.11). In other metals the edges of the crystallites can be made visible by etching the surface of the metal with acid. The acid penetrates more deeply along the edges of the crystallites, where the atoms do not fit together very well; hence the acid outlines these edges in relief (Figure A.12).

Fig. A.8 Grains of salt; magnification 100×.

Fig. A.9 Crystal of topaz.

Fig. A.10 Snowflakes. All these snowflakes have a hexagonal pattern.

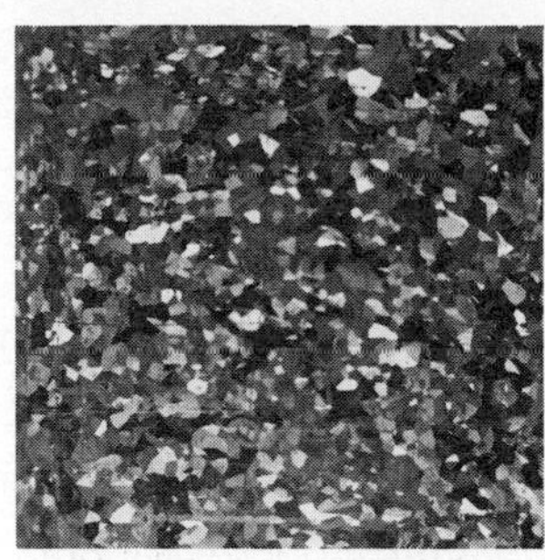

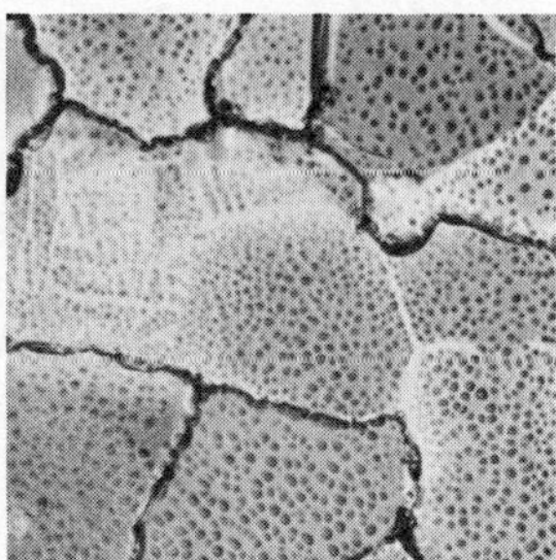

Fig. A.11 (left) Crystals on the surface of galvanized sheet metal. Approximately natural size.

Fig. A.12 (right) Boundaries of crystals in a lead–tin alloy, made visible by etching the surface with acid. Magnification 250×. *(Courtesy H. B. Huntington, Rensselaer Polytechnic Institute.)*

A.2 SYMMETRY

The essential features in the structure of a crystal are *regularity* and *repetition*. The fundamental building blocks — whether atoms or molecules — are stacked together in a three-dimensional array of regular geometric shapes that repeat and repeat in all directions almost forever. Of course, any real crystal has a beginning and an end; however, the number of repeated building blocks is many thousands or many millions in any one direction and, for the purposes of mathematical description, it is useful to pretend that the crystal is infinite.

To study the geometric properties of the arrangement of building blocks in a crystal, we need to become familiar with different kinds of symmetries. It will be best to begin with some simple examples of symmetries in two dimensions.

Figure A.13a shows a drawing by the artist M. C. Escher. The pattern of this drawing relies on regularity and repetition; for instance, the design of each of the four angels is essentially the same — the pattern has symmetry. To arrive at a precise description of the symmetries, let us ask what operations we can perform with this pattern that leave it in a condition indistinguishable from its initial condition. First of all, we can imagine exchanging each point above the horizontal midline with a corresponding point below the midline (Figure A.13b). In this operation the midline acts like a mirror — the operation is a **reflection** about the midline. The reflected pattern is indistinguishable from the original pattern. Hence reflection about the horizontal midline is symmetry of the pattern. Obviously, reflection about the vertical midline is also a symmetry.

The pattern has some other symmetries. For instance, we can perform a **rotation** of this pattern by 180° about an axis through its center perpendicular to its face (Figure A.13b). The rotated pattern is indistinguishable from the original pattern. The axis of this rotation is a symmetry axis; it is called a twofold symmetry axis because it involves two symmetric orientations (0° and 180°).

Finally, we can imagine exchanging each point with a corresponding point on the opposite side of the center (Figure A.13b). This operation can be described more precisely as a replacement of the point of posi-

(a)

(b)

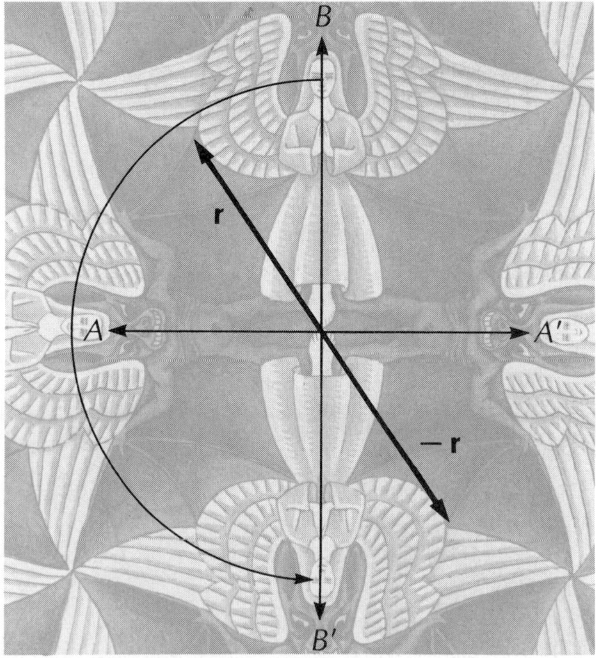

Fig. A.13 (a) A drawing by M. C. Escher. (b) The symmetries of the drawing are reflection about the line *AA'*, reflection about the line *BB'*, rotation by 180° about the center, and inversion about the center.

(a)

(b)
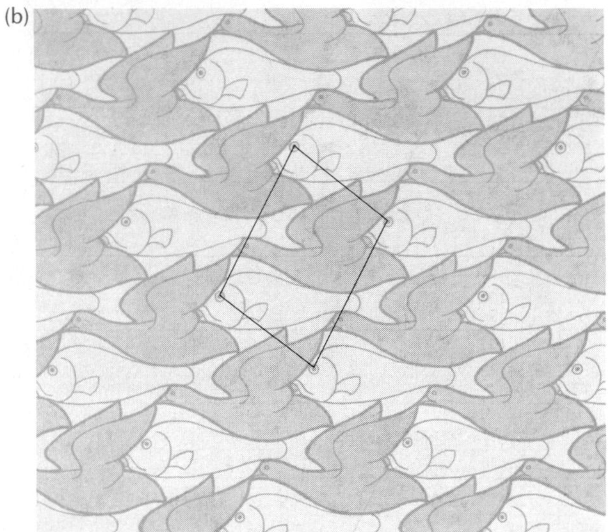

(c)
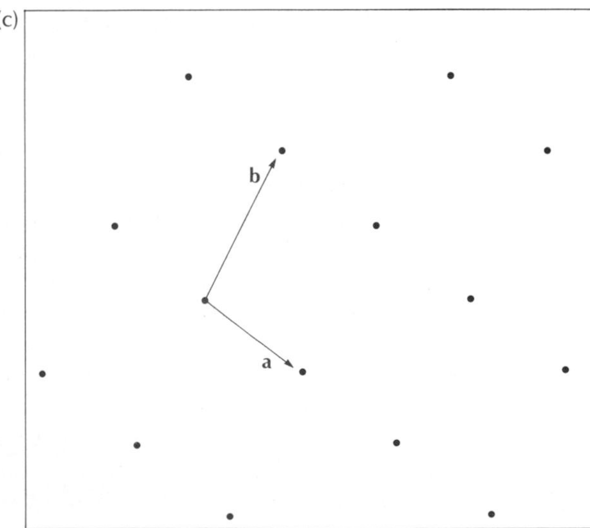

Fig. A.14 (a) Another drawing by M. C. Escher. (b) The primitive cell. (c) The lattice points and the primitive vectors.

tion vector **r** (relative to the center) by the point of position vector $-\mathbf{r}$. This operation is called an **inversion** through the center. Although it is another symmetry of the pattern, it is not an independent symmetry: under inversion each point suffers exactly the same net displacement as under a rotation of 180°. However, this identity between inversion and rotation only holds true for a two-dimensional object. For a three-dimensional object, inversion through a point and rotation are independent operations.

Figure A.14a shows another drawing by Escher. This pattern relies on repetition in an obvious way. Let us pretend that the pattern repeats forever in both directions. Then the operation of translation by one step in an oblique downward direction or by one step in an oblique upward direction leaves the pattern unchanged. Hence **translation** is a symmetry of this pattern.

The pattern of Figure A.14a can be regarded as a two-dimensional "crystal" and it is therefore worthwhile to investigate its geometric properties in some detail. The entire pattern consists of identical basic building blocks arranged in a regular lattice. Figure A.14b shows one such basic building block. In the language of crystallography, the basic building block is called a **primitive cell;** we will adopt this terminology. To describe the position of a primitive cell, we take one corner of the cell as a fiducial point. The positions of the cells within the complete pattern can then be indicated by the lattice points shown in Figure A.14c. The displacement vectors from one lattice point to the two nearest lattice points are called the **primitive vectors** of the lattice. We will designate these vectors by **a** and **b** (Figure A.14c). Any displacement vector from one lattice point to another lattice point is then a (positive or negative) multiple of one of the primitive vectors or a sum of such multiples, i.e., any such displacement vector is of the form

$$\Delta\mathbf{r} = m\mathbf{a} + n\mathbf{b}$$

where m and n are positive or negative integers or zero. The primitive vectors completely characterize the translation symmetry of the lattice.

A two-dimensional pattern can simultaneously have translation, rotation, and reflection symmetries. Figure A.15a shows yet another drawing by Escher, a pattern with several simultaneous symmetries. Figure A.15b shows the basic building block, or primitive cell. The translation symmetry is characterized by primitive vectors of equal length. Obviously, rotation is an extra symmetry of the pattern. A rotation by 180° about a perpendicular axis through the point P is a symmetry; this axis is a twofold symmetry axis since it involves two symmetric orientations of the pattern (0° and

(a)

(b)

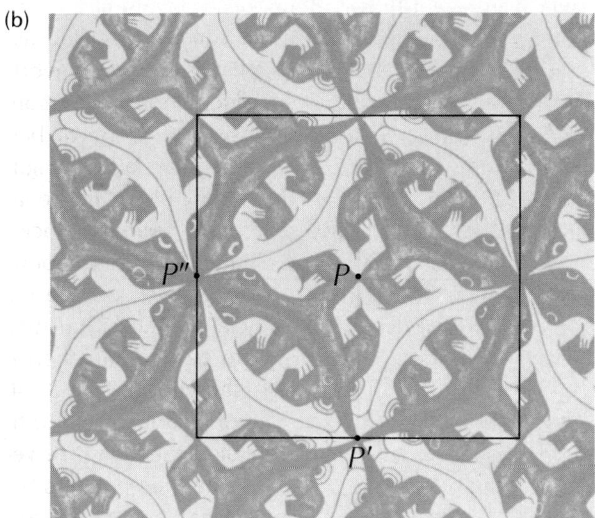

Fig. A.15 (a) Another drawing by M. C. Escher. (b) The primitive cell.

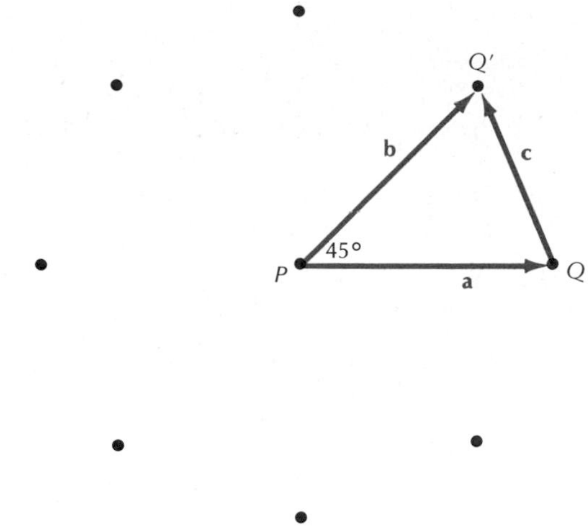

Fig. A.16 Two primitive vectors **a** and **b** connect P to two lattice points Q and Q'.

180°). A rotation by 90°, or by a multiple of 90°, about a perpendicular axis through either of the points P' or P'' is also a symmetry; each of these axes is a fourfold symmetry axis since they all involve four symmetric orientations (0°, 90°, 180°, and 270°).

In general, the coexistence of translation symmetry and rotation symmetry imposes restrictions on the primitive vectors and on the axes that characterize these symmetries. The analysis of these restrictions shows that the only allowed rotation axes are either twofold, threefold, fourfold, or sixfold, i.e., any rotation symmetry must involve an angle of 180°, 120°, 90°, or 60°. To see how such restrictions come about, let us examine in detail why a rotation symmetry with an angle of less than 60° is incompatible with translation symmetry.

Figure A.16 shows a lattice point P. Suppose that the perpendicular through P is a symmetry axis, say, an eightfold symmetry axis. Then there must exist eight lattice points that are nearest neighbors of P; these lattice points must be uniformly spaced with angles of 45° between them (Figure A.16). The two primitive vectors **a** and **b** connect P to two of these lattice points, say, Q and Q' (Figure A.16). By definition, the primitive vectors are the *shortest* vectors joining any adjacent lattice points. But this immediately leads to a contradiction, because the vector **c** connecting the lattice points Q and Q' is shorter than **a** and **b**. Hence an eightfold symmetry axis is impossible. Obviously, a similar argument can be used to rule out any rotational symmetry for which the angle between **a** and **b** is less than 60° — if the angle is exactly 60°, then the length of **c** is the same as that of **a** and **b**, which is (barely) acceptable. Incidentally, a slight modification of this argument can be used to establish that rotation symmetry with an angle of 72° (a fivefold axis) is also incompatible with translation symmetry.

Besides translation, rotation, and reflection symmetries, a two-dimensional pattern can have yet another symmetry that involves a joint reflection and translation, neither of which is a symmetry by itself. Figure A.17, a woodcut by Escher, illustrates this other symmetry. If we reflect the pattern about the vertical line AA' and then shift the pattern through the displacement indicated by the vector **d**, the pattern is left unchanged. The line AA' is called a **glide line.**

The most general symmetry of a two-dimensional "crystal" is some combination of translation symmetry, rotation symmetry, reflection symmetry, and glide-line symmetry. By taking all these symmetries into account and by investigating all the requirements for their compatibility, one can show that a two-dimensional

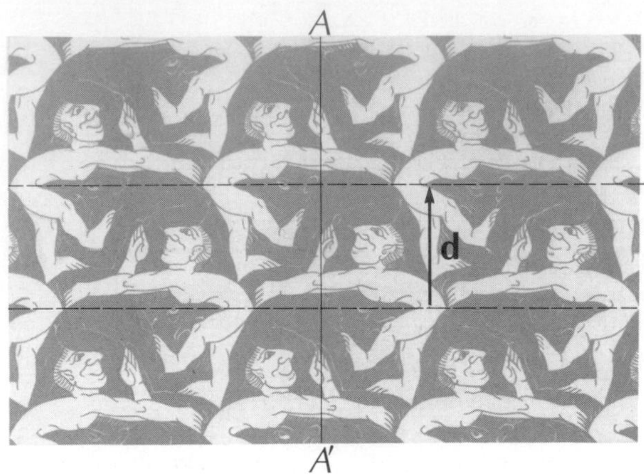

Fig. A.17 A woodcut by M. C. Escher. The line *AA'* is a glide line.

"crystal" structure must have one or another of 17 distinct kinds of symmetry combinations.

In three dimensions symmetry combinations are more complicated. The basic building blocks are three dimensional and there are three independent primitive vectors. Furthermore, while in the two-dimensional case the inversion through a point could be ignored (it coincides with at 180° rotation), in the three-dimensional case this symmetry must be taken into consideration separately; and there is yet another possible symmetry that involves a joint rotation and translation, neither of which is a symmetry by itself (the corresponding rotation and translation axis is called a **screw axis**). Thus, the most general symmetry of a three-dimensional crystal is some combination of translation symmetry, rotation symmetry, reflection symmetry, inversion symmetry, glide-plane symmetry, and screw-axis symmetry. It turns out that a three-dimensional crystal must have one or another of 230 distinct kinds of symmetry combinations.[3]

A.3 CLOSE-PACKED STRUCTURES

The crystals formed by the atoms or molecules of a given substance may involve cubic, rectangular, hexagonal, rhombohedral, or other lattices. Just what configuration the building blocks will adopt depends on the details of the interatomic forces. These forces are essentially electric attractions and repulsions between the electrons and nucleus of one atom and those of another. But differences in the arrangement of electrons within the atoms of different elements lead to differences in the corresponding interatomic forces. Crudely, we might say that the atoms of different chemical elements have different shapes and hence fit together in different ways when packed together in a crystal.

Some atoms have a perfect spherical shape — the force that such an atom exerts on another atom is completely independent of the relative orientation of the atoms. Helium, neon, argon, and krypton are atoms of this kind. At normal temperatures they form gases, but at very low temperatures they will aggregate into crystals. The atoms of, say, helium attract one another and so they will settle into a configuration in which each is as close to its neighbors as possible, a configuration called **close packed.**

The geometric problem of close-packing such atoms is similar to the problem of close-packing oranges into a crate or stacking cannonballs in a mound. The recipe for close packing is simple: begin by assembling a first layer (Figure A.18); the atoms of each row fit into the notches of the adjacent rows. Note that the examination of Figure A.18 reveals a hexagonal pattern — each atom is surrounded by six neighboring atoms. Next, construct a second layer on top of the first; to achieve close packing, it is obviously necessary to place each atom of this second layer in a hollow of the first (Figure A.19). Likewise, place another layer on top of the second. There are two alternative ways of placing this third layer. Either the atoms of the third layer are placed directly above those of the first layer (the third layer then duplicates the first layer; Figure A.20a), or else the atoms of the third layer are placed with a lateral offset relative to those of the first layer (the third layer is then shifted relative to the first; Figure A.20b), but when a fourth layer is next added, it dupli-

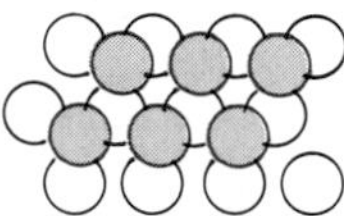

Fig. A.18 (left) A layer of close-packed spherical atoms.

Fig. A.19 (right) Second layer of close-packed atoms.

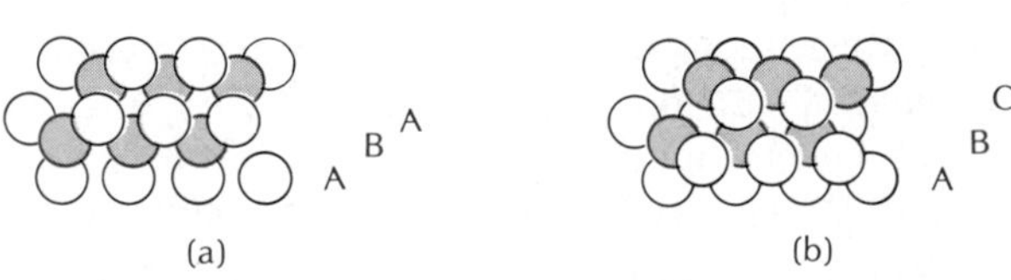

Fig. A.20 Third layer of close-packed atoms: (a) configuration *ABA* and (b) configuration *ABC*.

[3]These kinds of symmetry combinations are called **space groups.** Symmetry combinations involving all the above operations *except* translation are called **point groups.**

cates either the first or the second layer. Thus there are at most three distinct layers, which we can designate by the letters *A, B,* and *C.* Any general close-packed configuration can then be described by a sequence of these letters, such as *ABCBA* . . . , in which adjacent letters are different.

In crystals, close packing is usually accomplished by regular repetition of two distinct layers *AB,* or else by regular repetition of three distinct layers *ABC.* The configuration is then *ABAB* . . . or else *ABCABC* . . . ; in the former the third layer is a duplicate of the first, and in the latter the fourth layer is a duplicate of the first. Both of these configurations are equally close packed. And both have a high degree of symmetry.

Figure A.21 shows atoms in the *ABAB* . . . configuration; this figure gives an "exploded" view in which the interatomic distances have been exaggerated for the sake of clarity. Obviously, there is a threefold rotation axis passing vertically through the center of Figure A.21. There are also three twofold rotation axes passing horizontally through the center. Besides, there are several vertical and horizontal planes of reflection.

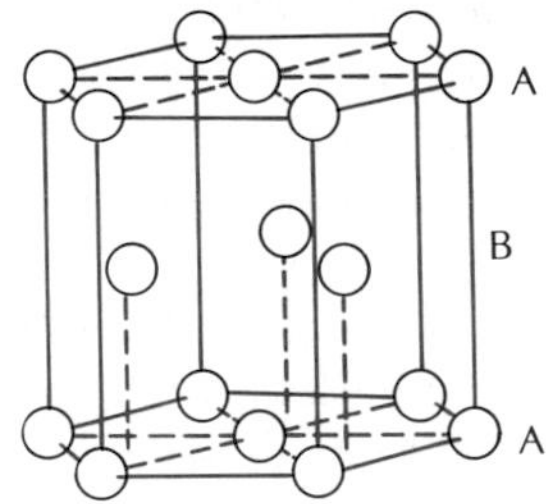

Fig. A.21 Close-packed configuration with layers *ABA.*

Figure A.22a shows atoms in the *ABCABC* . . . configuration. To discover the symmetries of this configuration, it is best to begin with the observation that the atoms are actually arranged in a cubic structure (Figure A.22b). This kind of cubic structure is called a **face-centered cube** because there is an atom at the center of each face. The *ABCABC* . . . configuration therefore has all the symmetries of a cube. There is a fourfold rotation axis passing through the center of each face, a threefold rotation axis through each vertex, a twofold rotation axis through the middle of each edge, and there are several planes of reflection (to visualize these symmetries, it is best to manipulate a cube).

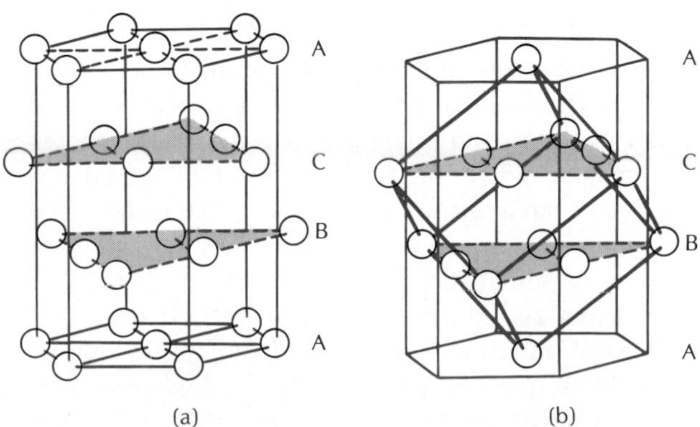

Fig. A.22 (a) Close-packed configuration with layers *ABCA.* (b) The atoms of this close-packed configuration are arranged in a face-centered cube.

Which of the two close-packed configurations will be adopted by the atoms of a particular element depends on what forces an atom in one layer exerts on an atom two or more layers away — these forces between remote atoms are weak but they decide the issue. Helium crystallizes in the *ABAB* . . . configuration; neon, argon, and krypton crystallize in the *ABCABC* . . . configuration.

Many metals, such as magnesium, zinc, aluminum, and copper, also adopt one or the other of these simple close-packed structures. This is perhaps somewhat unexpected. The atoms of metals do not have the spherical symmetry of the atoms of the noble gases; so why do they adopt close packing, which, as we saw, is characteristic of spheres? The answer to this puzzle is found in the electronic structure of these atoms. The atoms of metal are nonspherical because the outermost one or two electrons of each atom stick out in asymmetric ways; but except for these outermost electrons, the atoms are spherical — the atomic cores are spherical. When such atoms aggregate in a crystal, they release the outermost electrons and the residual spherical cores then settle very nicely in a close-packed arrangement.

What happens to the released electrons? They remain free to wander all over the crystal, forming a gas of electrons. Thus, a metal consists of two interpenetrating components: a lattice made of atomic cores and a gas of electrons. Note that the atomic cores are ions, i.e., atoms with missing electrons. This reminds one of a plasma where ions and electrons coexist in an intimate mixture. The difference between a metal and a plasma lies in the arrangement of the ions: in a metal the ions form a lattice, while in a plasma the ions form a gas.

The presence of a gas of free electrons inside a metal accounts for many of the characteristic properties of metals. For instance, metals have a remarkable ability to conduct heat and electricity — the electron gas carries heat and electricity very efficiently from one end of the metal to the other.

Besides the close-packed crystal lattice described above, there are many other kinds of crystal lattices.

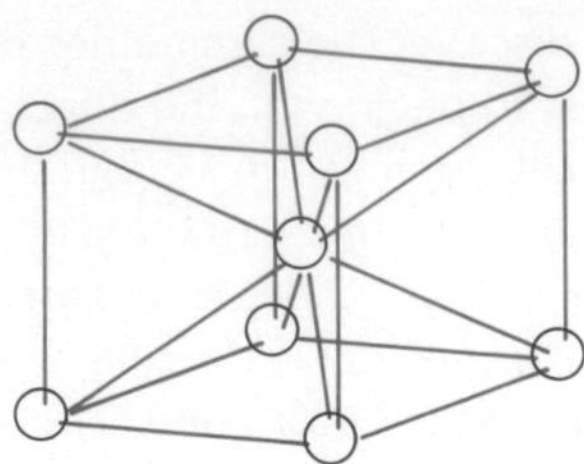

Fig. A.23 Structure of sodium crystal.

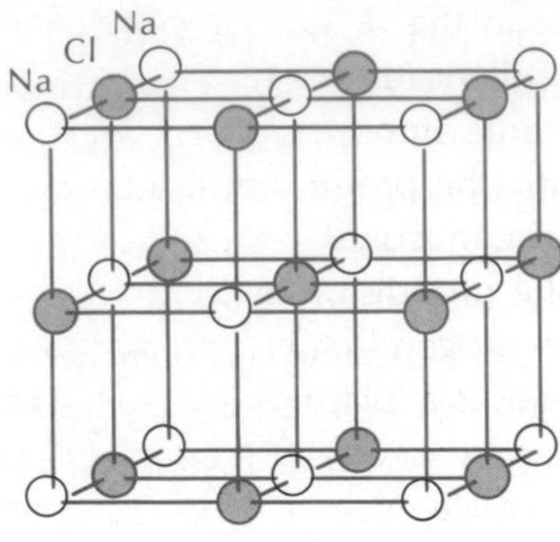

Fig. A.26 Structure of NaCl crystal. The white balls are sodium atoms and the colored balls are chlorine atoms.

For example, Figure A.23 shows the crystal structure of sodium; this structure is called a **body-centered cube** because there is an atom at the center of each cube. Figures A.24–A.26 show the crystal structures of diamond, graphite, and sodium chloride (ordinary salt). These figures give a glimpse of the variety of beautiful symmetric arrangements found in crystals.

Many of the physical properties of solids can be readily explained in terms of their crystal structure. Diamond is hard and graphite is soft, yet both consist of pure carbon (diamond burns like coal if ignited at high temperature). The qualitative explanation of this difference is as follows. In diamond, each carbon atom is surrounded by four other carbon atoms; each atom is at the center of a tetrahedron with four surrounding atoms at the vertices of this tetrahedron (Figure A.24). The central carbon atom forms bonds with these surrounding atoms (it is known from chemistry that carbon atoms tend to form four tetrahedral bonds). These strong interlocking bonds give the diamond crystal its extreme hardness and high melting point. In graphite the carbon atoms are arranged in parallel sheets; within each sheet the atoms form adjoining hexagonal rings (Figure A.25). The bonds within each sheet are fairly strong, but the bonds between one sheet and the next are fairly weak. Hence the sheets easily slide over one another. This sliding motion gives graphite its softness and its lubricating quality, properties that find practical application in graphite ("lead") pencils and graphite grease.

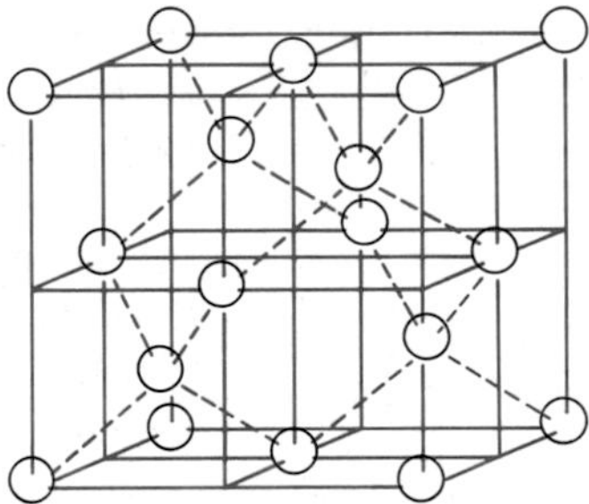

Fig. A.24 Structure of diamond crystal.

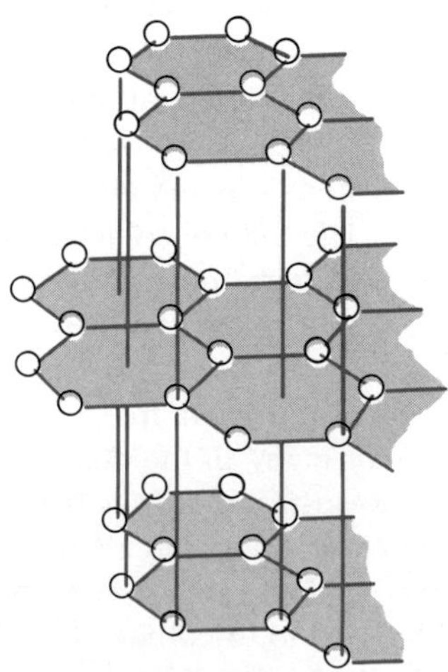

Fig. A.25 Structure of graphite crystal.

The ability of many metals to form alloys with hydrogen, carbon, nitrogen, and boron can also be readily explained in terms of their crystal structure. These alloys are solid solutions; the atoms of solute penetrate into the lattice of the metal and occupy the remaining empty spaces, or interstices, between the atoms of metal. Surprisingly, the close-packed structure of metals is more favorable to the formation of such a solid solution than other structures. How can a close-packed lattice accommodate extra atoms? The answer to this puzzle is that although the net volume of the interstitial spaces in a crystal of close-packed spheres is a small fraction of the total volume of the crystal, the shape of each interstitial space is such that it can easily accommodate an extra sphere, provided that the extra sphere is somewhat smaller. A simple calculation shows that an extra sphere will fit into an interstitial space if its radius is smaller than 0.59 times the radius of the close-packed spheres. Atoms of hydrogen, carbon, nitrogen, and boron have sizes that are not far from this limit and they therefore fit quite nicely into the interstitial spaces of metals.

Incidentally, alloys of hydrogen with metals are of

great interest in the technology of fuels. Pure hydrogen is an excellent fuel — when burned with oxygen, it releases a large amount of energy without producing any pollutants (the product of the combustion is pure water). Hydrogen could be used as a substitute for gasoline in the propulsion of automobiles. However, a tank full of hydrogen in the form of a liquid or a compressed gas poses a frightful explosive hazard. It would be much safer to store the hydrogen within a chunk of metal. For instance, magnesium readily absorbs hydrogen, forming an alloy or solid solution. Magnesium can store up to 67 kg of hydrogen per cubic meter of (porous) magnesium, so the storage of an amount of hydrogen equivalent to 20 gal. of gasoline would require about 80 gal. of hydrogen–magnesium alloy. Although this is an inconveniently large volume (and weight), it does eliminate the explosion hazard. To supply the engine, the hydrogen can be gradually released from the magnesium by means of a small amount of heat applied to the alloy.

A.4 DEFECTS IN CRYSTALS

Crystals found in nature are never perfect. A large crystal is likely to contain within it a number of microcrystals jumbled together at different angles (Figure A.27); the mismatch at the boundaries of the microcrystals is a conspicuous defect of the crystal. But even within a single microcrystal, one often finds a variety of defects. Sometimes atoms are missing from the lattice, leaving empty sites (Figure A.28). Sometimes extra atoms are squeezed between the regular rows and columns of the lattice (Figure A.29). Furthermore, sometimes a few lattice sites are occupied by atoms of a different element; the presence of such a different atom tends to distort the crystal lattice in the immediate vicinity (Figure A.30). These three kinds of isolated defects are called, respectively, **vacancies, interstitials,** and **substitutional impurities.**

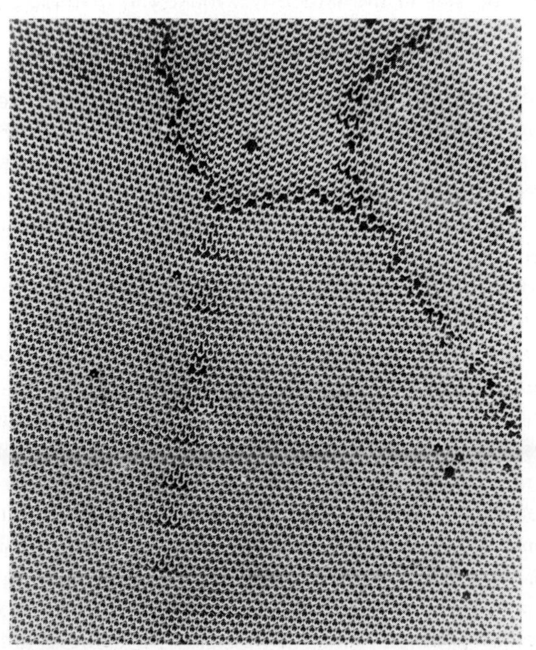

Fig. A.27 A raft of small soap bubbles floating on water. The raft consists of several "crystals" with different orientations. The boundaries between these crystals are very conspicuous. *(Courtesy C. S. Smith, MIT.)*

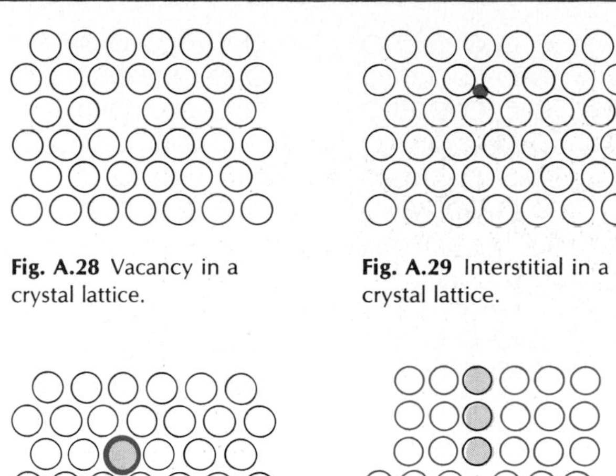

Fig. A.28 Vacancy in a crystal lattice.

Fig. A.29 Interstitial in a crystal lattice.

Fig. A.30 Substitutional impurity in a crystal lattice.

Fig. A.31 Dislocation in a crystal lattice.

Figure A.31 shows a lattice with a large defect: the crystal has an extra atomic plane slipped between the regular layers. This kind of defect is called a **dislocation.** The crystals in ordinary pieces of cast metal have an abundance of such defects — anywhere between 10^5 and 10^9 dislocations per square centimeter.

Dislocations have a crucial effect on the strength of the metal. If a bar of metal is subjected to a large bending force, it will become permanently deformed, i.e., its crystal planes will suffer a permanent displacement. This displacement does not occur by the simultaneous slippage of an entire atomic layer over the adjacent layer — such a motion would involve the simultaneous disruption of all the atomic bonds joining one layer to the next and that would require a prohibitively large force. Rather, the layer moves step by step, by means of propagating dislocation. Figure A.32 shows how the dislocation propagates. Note that only one atom moves at a time and hence no very large force is required to generate this motion. The displacement of crystal layers by the propagation of a dislocation is analogous to the displacement of a rug by the passage of a wrinkle: to move a heavy rug a few inches, it is much easier to run a wrinkle down the rug than to pull the whole rug along all at once by brute force.

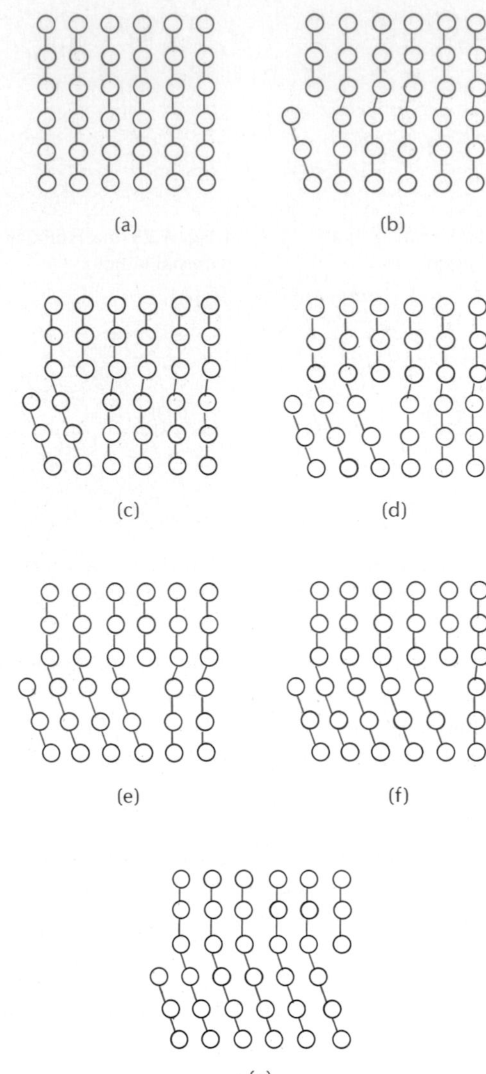

Fig. A.32 Propagation of a dislocation. The dislocation moves from left to right.

Dislocations move through a crystal so easily that metals would be rather soft were it not for a snag: when a moving dislocation collides with another dislocation or when it reaches the boundary between one microcrystal and the adjacent one, it finds it hard to keep going — the sudden change in the orientation of the crystal planes at such places acts as an obstacle. Hence a crystal with many defects will be hard. The hardening of iron and steel by cold-working takes advantage of this effect. The treatment generates dislocations and breaks up the microcrystals into smaller microcrystals; the corresponding increase in the number of obstacles increases the resistance that the metal offers to deformation.

The presence of substitutional atoms also tends to harden a metal. The distortion of the lattice in the vicinity of the substitutional atoms acts as an obstacle to any moving dislocation. This effect is largely responsible for the fact that the alloy of two metals (e.g., copper and tin) is harder than either of the pure metals.

The industrial production of steel is a prominent practical example of how the mechanical properties of a material can be controlled by deliberate modification of its crystal structure. Ordinary steel is an alloy of iron and carbon (up to 1.7% carbon). At high temperature (above 1000°C), the atoms in iron crystals adopt a close-packed structure, a face-centered cube (see Figure A.22). As we remarked in the preceding section, the carbon atoms fit into the interstitial spaces of the close-packed iron lattice. However, at room temperature, the atoms in iron crystals adopt a new cubic structure, a body-centered cube (see Figure A.23). The transformation from the face-centered to the body-centered structure occurs at a temperature between 700°C and 900°C (depending on the amount of carbon). Thus, drastic changes occur in a crystal of steel as it cools from an initially high temperature.

What happens to the carbon atoms during this drastic change of the structure of the iron lattice depends crucially on the rate of cooling. If the steel is subjected to very sudden cooling (quenching) by immersion in water or oil, the carbon atoms do not have time to move from their sites; they remain fixed in their positions as the iron lattice changes. But since the carbon atoms do not fit into the interstitial spaces of the body-centered lattice, this lattice will be strongly distorted in the vicinity of each carbon atom. As we have seen, such defects in the lattice will make the material hard — the quenched steel is very hard. Unfortunately, this steel is also quite brittle — a sudden blow can crack and shatter it. If the steel is cooled slowly, the carbon atoms have time to segregate and the steel develops crystallites of two different kinds: one kind consists of almost pure iron and the other consists of molecules of Fe_3C (iron carbide). These two kinds of crystallites form alternating plates (or lamellae) of irregular size and shape. The resulting steel is fairly soft.

The two varieties of steel described above are extremes. For most applications the steel should be neither excessively hard (and brittle) nor excessively soft. One method for producing steel of intermediate qualities relies on careful heat treatment. By slowly heating (tempering) hard steel, one can encourage its carbon atoms to segregate and to form crystallites of iron compounds. Skillful control of the temperature and the duration of this heat treatment allows metallurgists to achieve a wide range of grades of hardness and shock resistance. It is even possible to achieve different mechanical qualities in different parts of one single piece of steel. For instance, in the manufacture of armor plate, the outer face of the plate is made very hard and the inner part somewhat softer; when struck

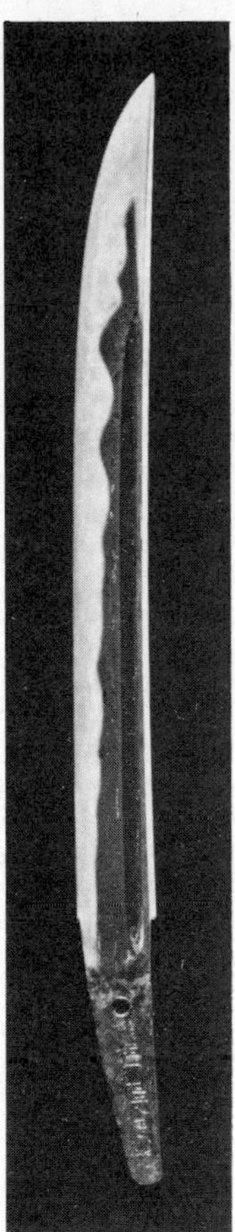

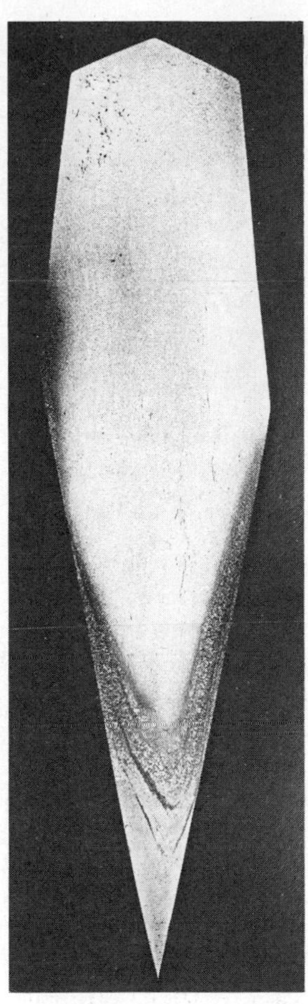

Fig. A.33 (left) Blade by the sixteenth-century sword smith Hiromitsu. Note the fine texture in the bright zone along the cutting edge.

Fig. A.34 (right) Cross section of a blade attributed to the thirteenth-century sword smith Nagamitsu. The core is soft steel, and the cutting edge hard steel. The zone in between consists of different kinds of steel in layers produced by the forging process and heat treatment. *(Courtesy C. S. Smith, MIT.)*

by a projectile, the outer part resists penetration while the inner part absorbs the shock and prevents the outer part from shattering.

The sword blades forged by the old Japanese sword smiths achieve their remarkable strength by a similar combination of a very hard cutting edge backed by a soft, elastic core. The intermingling of the two kinds of steel gives the surface of these blades their rich, distinctive texture (Figures A.33 and A.34).

Vacancies and impurities in a crystal lattice also have drastic effects on the optical and electric properties of the crystal. For instance, the deep red color of ruby is due to chromium impurities and the deep blue color of sapphire is due to chromium, iron, and titanium impurities; in both of these precious stones, the bulk of the crystal is aluminum oxide, but the impurities determine the color.

All of solid-state electronics depends on impurities. The diodes and transistors in solid-state circuitry and the solar cells that generate electric power from sunlight are made of semiconductors such as silicon or germanium. A pure semiconductor is a fairly poor conductor of electricity — it has some electrons within it, but not enough to carry a large electric current. However, selected impurities introduced into the crystal of semiconductor — such as indium impurities in a crystal of silicon — will yield up some of their electrons to the crystal, i.e., these impurities act as sources of free electrons. This dramatically improves the semiconductor's ability to conduct electricity. The manufacture of solid-state devices relies on such manipulation of the electronic properties of materials by artificially "doping" the crystals with impurities. As in other practical applications of crystals, we again see that the imperfections put the crystal at our service. Perfect crystals are beautiful, but imperfect crystals are useful.

Further Reading

Snow Crystals by W. A. Bentley and W. J. Humphreys (Dover, New York, 1962) contains a spectacular collection of more than 2000 photographs of snowflakes. *On Growth and Form* by D.W. Thompson (Cambridge University Press, Cambridge, 1961) gives many beautiful examples of symmetry and regularity in plants and animals. P. W. Medawar described this book as "beyond comparison the finest work of literature in all the annals of science that have been recorded in the English tongue."

Symmetry by H. Weyl (Princeton University Press, Princeton, N.J., 1952) is a classic study by a renowned mathematical physicist of the symmetries found in plants, animals, art, and architecture. *Symmetry Discovered* by J. Rosen (Cambridge University Press, Cambridge, 1975) gives a careful elementary introduction to the concept of symmetry and includes some applications of symmetry in the solution of problems in physics. *The Ambidextrous Universe* by M. Gardner (Basic Books, New York, 1964) is a charming exploration of the role of reflection symmetry in biology, chemistry, and physics.

Symmetry Aspects of M. C. Escher's Periodic Drawings by C. H. MacGillavry (Abrams, New York, 1976) presents a mathematical analysis of the symmetries of Escher's drawings. *The Mathematics of Islamic Art* (Metropolitan Museum of Art, New York, 1979) is a teaching packet with 20 beautiful slides that provide an introduction to symmetry in Islamic art.

A Search for Structure by C. S. Smith (MIT Press, Cambridge, 1982) is a collection of elegant and perceptive essays by an expert metallurgist that describe how the microscopic structure of materials has influenced their use in art and technology. In his investigations of history and art, Smith establishes that the first discovery of useful materials and processes almost always occurred in the decorative arts, and the practical applications were only recognized later.

Crystals and Crystal Growing by A. H. Holden and P. Singer (Doubleday, New York, 1960) provides a nice, elementary introduction to the general properties of crystals. This book includes recipes for growing crystals. It also includes a list of further books and articles. *Crystals and Light* by E. A. Wood (Van Nostrand, New York, 1964) is a concise introduction to the symmetry of crystals, with a minimum of mathematics. It contains a thorough discussion of how the internal structure affects the external shape. The second half of the book deals with the propagation of light in crystals and with the relationship between optical properties and symmetry. *Crystals and X Rays* by H. S. Lipson (Wykeham, London, 1970) describes the diffraction of X rays by crystals and how we can extract information about the structure of crystals and molecules from X-ray diffraction experiments. The book includes an account of the first discovery of a crystal structure (NaCl) by W. H. Bragg and W. L. Bragg.

The following is a list of some recent articles dealing with crystals and symmetry:

"Snow Crystals," C. Knight and N. Knight, *Scientific American,* January 1973

"Crystallography," A. Guinier, *Physics Today,* February 1975

"Geometry of Surface Layers," P. J. Estrup, *Physics Today,* April 1975

"Disclinations," W. F. Harris, *Scientific American,* December 1977

"Electron Microscopy of Atoms in Crystals," J. M. Cowley and S. Iijima, *Physics Today,* March 1977

"Hydrogen Storage in Metal Hydrides," J. J. Reilly and G. D. Sandrock, *Scientific American,* February 1980

Questions

1. Can you think of any animals whose external shape has a symmetry other than bilateral symmetry?

2. List the symmetries of (a) a circle, (b) a triangle, (c) a rectangle, and (d) a square. In what sense does a circle have more symmetry than a square, and a square more symmetry than a rectangle?

3. List the symmetries of (a) a sphere, (b) a cylinder, (c) a cube, (d) a parallelepiped, and (e) an ellipsoid of revolution.

4. What are the symmetries of this book? Assume that the book is closed and ignore the print on the pages and on the cover.

Fig. A.35 Glazed tile panel; Nishapur, thirteenth to fourteenth century.

5. What are the rotation and the reflection symmetries of the pattern in Figure A.5? In Figure A.6?

6. Try to make a drawing of interlocking figures, in the style of Escher. Why is this difficult?

7. Identify the primitive cell and the primitive vectors for the sodium chloride crystal of Figure A.26.

8. Find a primitive cell and the primitive vectors for one of the "crystals" shown in Figure A.27.

9. Figure A.35 shows an old Iranian wall panel. Identify the primitive cell and the primitive vectors. List the rotation and the reflection symmetries of this pattern.

10. For shipment to stores, round cans of canned goods are packed in cardboard boxes, usually 24 cans to a box. Are these cans close packed? Why does the manufacturer not take advantage of close packing?

11. If you buy apples by the bushel (a unit of volume), do you get more weight if you select small apples or large apples? Assume that the apples are close packed.

12. Consider a plane array of close-packed circles of equal size. What fraction of the area is between the circles?

13. What defects can you see in the "crystals" shown in Figure A.27?

CHAPTER 4

Kinematics in Three Dimensions

We now take up the description of particle motion in three dimensions. This is a straightforward generalization of the one-dimensional case we studied in Chapter 2; essentially, three-dimensional motion consists of three one-dimensional motions occurring simultaneously. We will see that the vector formalism developed in the preceding chapter permits us to describe the motion in a very concise manner.

4.1 The Velocity and Acceleration Vectors

In the case of one-dimensional motion along a straight line we defined the average velocity in Eq. (2.3):

$$\bar{v} = \frac{\Delta x}{\Delta t} \tag{1}$$

where Δx is the change in position in the time interval Δt.

In three-dimensional motion, the change in position, or the displacement, will involve simultaneous changes in x, y, and z; correspondingly, the **average velocity** has three components:

$$\bar{v}_x = \frac{\Delta x}{\Delta t} \qquad \bar{v}_y = \frac{\Delta y}{\Delta t} \qquad \bar{v}_z = \frac{\Delta z}{\Delta t} \tag{2}$$

In vector notation this can be written compactly as

$$\boxed{\bar{\mathbf{v}} = \frac{\Delta \mathbf{r}}{\Delta t}} \tag{3}$$

Average velocity

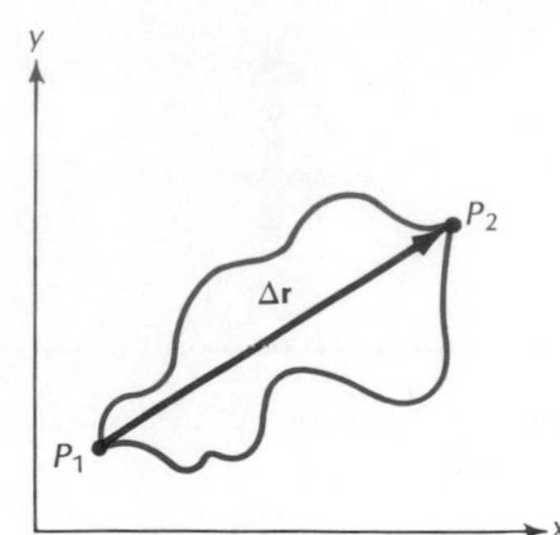

Fig. 4.1 Two alternative paths connecting the points P_1 and P_2. If the motions along these paths take the same time, then they have the same average velocity, regardless of the difference in the lengths of the paths.

where **r** is the position vector [Eq. (3.9)] and $\Delta\mathbf{r}$ is the change in position, i.e, it is the displacement

$$\Delta\mathbf{r} = \Delta x\ \hat{\mathbf{x}} + \Delta y\ \hat{\mathbf{y}} + \Delta z\ \hat{\mathbf{z}} \tag{4}$$

The average velocity given by Eq. (3) is the displacement (a vector) multiplied by $1/\Delta t$ (a number or scalar); thus, velocity is a *vector* quantity.

As in the case of one-dimensional motion (see Section 2.2), the average velocity takes into account only the net displacement in the time interval Δt. Hence the average velocity ignores the details of the motion; it gives no credits for back and forth motion or for the length of the path, so that two motions between points P_1 and P_2 that take the same time have the same average velocity, regardless of the exact path taken (Figure 4.1).

The **instantaneous velocity** gives a more precise description of the motion. As in the one-dimensional case, the instantaneous velocity is obtained by evaluating the average velocity in the limit of vanishing Δt,

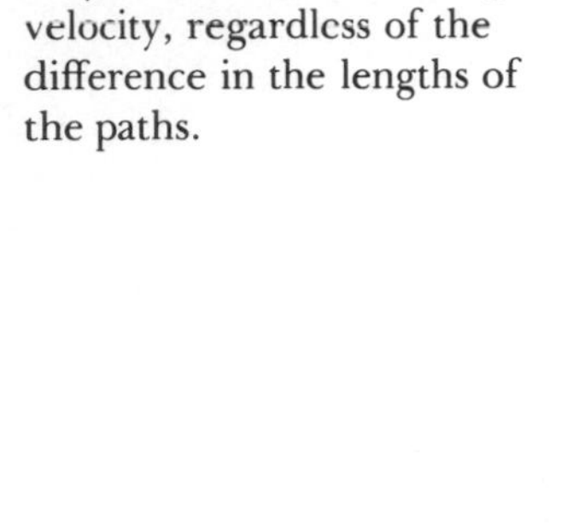

$$v_x = \frac{dx}{dt} \qquad v_y = \frac{dy}{dt} \qquad v_z = \frac{dz}{dt} \tag{5}$$

In vector notation these equations become

$$\mathbf{v} = v_x\hat{\mathbf{x}} + v_y\hat{\mathbf{y}} + v_z\hat{\mathbf{z}} \tag{6}$$

or

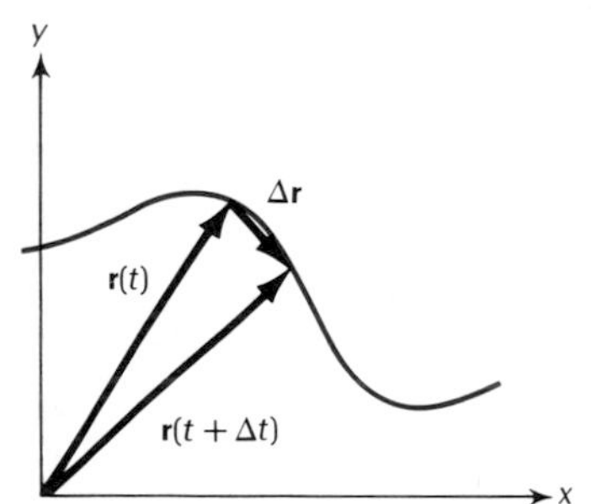

Fig. 4.2 Position vectors of a particle at times t and $t + \Delta t$. Their difference is $\Delta\mathbf{r}$.

$$\mathbf{v} = \frac{dx}{dt}\hat{\mathbf{x}} + \frac{dy}{dt}\hat{\mathbf{y}} + \frac{dz}{dt}\hat{\mathbf{z}} \tag{7}$$

Since the unit vectors $\hat{\mathbf{x}}$, $\hat{\mathbf{y}}$, and $\hat{\mathbf{z}}$ are time independent (constant), they have zero time derivative; hence $\hat{\mathbf{x}}\ dx/dt = d(\hat{\mathbf{x}}x)/dt$, etc., and Eq. (7) may also be written as

$$\mathbf{v} = \frac{d}{dt}(\hat{\mathbf{x}}x + \hat{\mathbf{y}}y + \hat{\mathbf{z}}z) \tag{8}$$

or

Instantaneous velocity

$$\boxed{\mathbf{v} = \frac{d\mathbf{r}}{dt}} \tag{9}$$

Thus, the **instantaneous velocity vector** is the time derivative of the position vector.

The velocity vector of Eq. (9) has the direction of the tangent to the path of the particle. This can be seen from Figure 4.2, which shows the position vectors of a particle at the two times t and $t + \Delta t$. In the limiting case $\Delta t \to 0$, the vector $\Delta\mathbf{r}$ will be tangent to the path and hence the velocity vector will also be tangent to the path.

EXAMPLE 1. A gun aimed southward fires a heavy projectile with a muzzle velocity of 90.0 m/s at an angle of 34° with the horizontal. What are the components of the instantaneous velocity of the projectile in the reference frame of the ground? The x, y, and z axes point south, east, and up, respectively.

SOLUTION: According to Figure 4.3, the velocity vector is in the x–z plane; it makes an angle of 34° with the x axis, an angle of 56° with the z axis, and an angle of 90° with the y axis. Hence

$$v_x = v \cos 34° = 90.0 \text{ m/s} \times \cos 34° = 74.6 \text{ m/s}$$

$$v_y = v \cos 90° = 0 \text{ m/s}$$

$$v_z = v \cos 56° = 90.0 \text{ m/s} \times \cos 56° = 50.3 \text{ m/s}$$

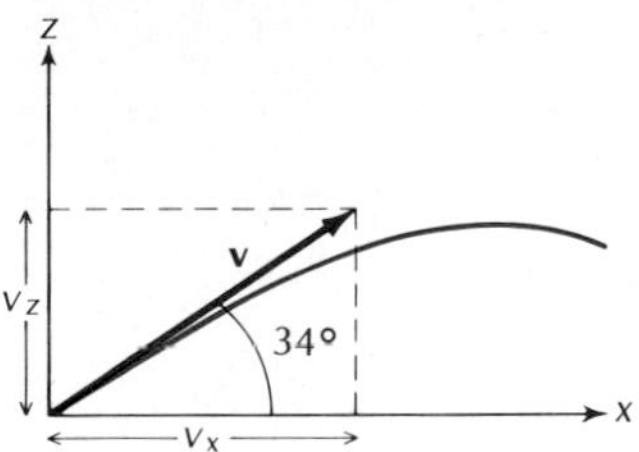

Fig. 4.3 Velocity vector of a projectile.

In everyday language, the words *velocity* and *speed* are synonymous. However, we will adopt the convention that the *speed* is the magnitude of the velocity vector:

$$[\text{speed}] = v = \left|\frac{d\mathbf{r}}{dt}\right| \tag{10}$$

or, in rectangular coordinates,

$$[\text{speed}] = v = \sqrt{v_x^2 + v_y^2 + v_z^2} \tag{11}$$

According to this precise definition, velocity is a vector and speed a scalar; velocity has magnitude and direction, while speed only has magnitude. Note that Eqs. (10) and (11) give the *instantaneous* speed; in contrast, the *average* speed is the time average of this instantaneous speed, i.e., it is the total path length divided by the time taken [see Eq. (2.1)].

The definitions of average and instantaneous acceleration in three dimensions are straightforward generalizations of those in one dimension. For the **average acceleration** we have

$$\bar{a}_x = \frac{\Delta v_x}{\Delta t} \qquad \bar{a}_y = \frac{\Delta v_y}{\Delta t} \qquad \bar{a}_z = \frac{\Delta v_z}{\Delta t} \tag{12}$$

or

$$\boxed{\bar{\mathbf{a}} = \frac{\Delta \mathbf{v}}{\Delta t}} \tag{13}$$

Average acceleration

and for the **instantaneous acceleration** we have

$$a_x = \frac{dv_x}{dt} = \frac{d^2x}{dt^2} \qquad a_y = \frac{dv_y}{dt} = \frac{d^2y}{dt^2} \qquad a_z = \frac{dv_z}{dt} = \frac{d^2z}{dt^2} \tag{14}$$

or

$$\boxed{\mathbf{a} = \frac{d\mathbf{v}}{dt} = \frac{d^2\mathbf{r}}{dt^2}} \tag{15}$$

Instantaneous acceleration

It is an important consequence of this definition that there is an acceleration whenever the velocity changes either in magnitude or in *direction*. Thus, a motion at constant speed is an accelerated motion whenever its direction changes.

EXAMPLE 2. An automobile enters a 90° curve at a constant speed of 25 m/s (or 55 mi/h) and emerges from this curve 6.0 s later. What is the average acceleration for this time interval?

SOLUTION: Figure 4.4a shows the path of the automobile with the initial velocity $\mathbf{v}_1$ and the final velocity $\mathbf{v}_2$. Figure 4.4b shows the change $\Delta\mathbf{v} = \mathbf{v}_2 - \mathbf{v}_1$. The vector triangle is a right isosceles triangle; its hypotenuse is

$$|\Delta\mathbf{v}| = |v_1|/\cos 45^\circ$$

$$= (25 \text{ m/s})/\cos 45^\circ$$

$$= 35.4 \text{ m/s}$$

Hence the average acceleration has a magnitude

$$|\bar{\mathbf{a}}| = \left|\frac{\Delta\mathbf{v}}{\Delta t}\right| = \frac{35.4 \text{ m/s}}{6.0 \text{ s}} = 5.9 \text{ m/s}^2$$

The direction of this acceleration is along the vector $\mathbf{v}_2 - \mathbf{v}_1$.

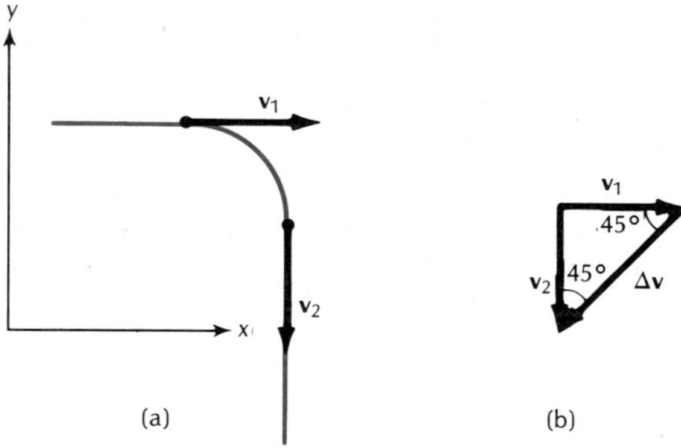

Fig. 4.4 (a) Path of an automobile through a 90° curve. (b) The initial and the final velocity vectors $\mathbf{v}_1$ and $\mathbf{v}_2$.

4.2 Motion with Constant Acceleration

For a particle moving in three dimensions with constant acceleration, we can derive some equations relating acceleration, velocity, position, and time analogous to the equations that hold for motion in one dimension (see Section 2.5). We begin with Eq. (13), which tells us that a constant acceleration **a** produces a velocity change $\mathbf{a}\,\Delta t$ in a time interval Δt. Suppose that the initial velocity vector is $\mathbf{v}_0$; then the velocity vector after a time t has elapsed will be

Constant acceleration:* v, a, *and t

$$\boxed{\mathbf{v} = \mathbf{v}_0 + \mathbf{a}t} \tag{16}$$

Furthermore, a mathematical argument similar to that used in the one-dimensional case [see Eqs. (2.18)–(2.22)] leads us to the following expression for the change of the position vector:

$$\boxed{\mathbf{r} - \mathbf{r}_0 = \mathbf{v}_0 t + \tfrac{1}{2}\mathbf{a}t^2} \tag{17}$$

Constant acceleration:* r, a, *and t

The x, y, and z components of Eqs. (16) and (17) are, respectively,

$$v_x = v_{0x} + a_x t \tag{18}$$

$$v_y = v_{0y} + a_y t \tag{19}$$

$$v_z = v_{0z} + a_z t \tag{20}$$

and

$$x - x_0 = v_{0x} t + \tfrac{1}{2} a_x t^2 \tag{21}$$

$$y - y_0 = v_{0y} t + \tfrac{1}{2} a_y t^2 \tag{22}$$

$$z - z_0 = v_{0z} t + \tfrac{1}{2} a_z t^2 \tag{23}$$

Equations (18)–(23) state that the x, y, and z components of the motion proceed completely independently of one another. Thus, the x acceleration only affects the x velocity, and the change in the x position is entirely determined by the x acceleration and the initial x velocity. Figure 4.5 shows an experimental demonstration of this independence between different components of the motion. Two balls were released simultaneously from a platform above the surface of the Earth; one was merely dropped from rest, the other was launched with an initial horizontal velocity. Since the vertical accelerations are the same, Eqs. (20) and (23) predict that the vertical components of the motions are exactly the same for both balls; only the horizontal components of the motions should differ. The stroboscopic images of the two balls at successive instants of time show that they do in fact fall downward in unison, reaching equal heights at the same time. Thus, the vertical components of the position vectors are the same for both balls; only the horizontal components differ.

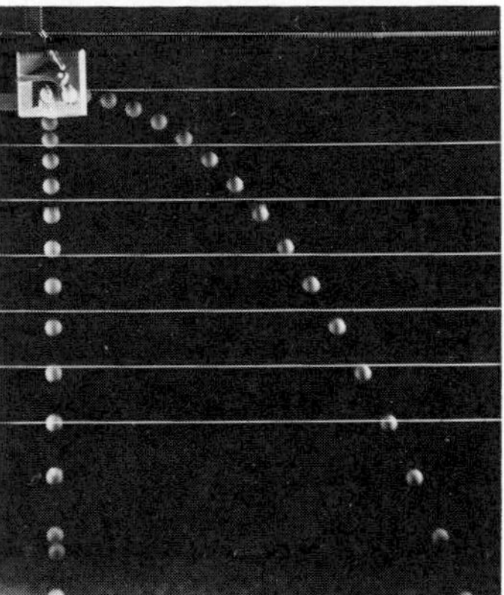

Fig. 4.5 Stroboscopic photograph of two balls that have been released simultaneously from a platform; one ball has horizontal velocity, the other does not.

4.3 The Motion of Projectiles

Near the surface of the Earth the pull of gravity gives a freely falling body a downward acceleration of about 9.81 m/s². If we ignore friction with the air, this is the only acceleration that the body experiences when thrown from some initial position with some initial velocity. The motion of such a body is therefore motion with constant acceleration.

The initial velocity of the thrown body, or projectile, will in general have both a vertical and a horizontal component. If we take the z axis in the direction of the upward vertical and the x axis in the direction of the initial horizontal velocity, we have $a_x = 0$, $a_y = 0$, $a_z = -g = -9.81$ m/s², and $v_{0y} = 0$. (Note that a_z is *negative* because the z axis points upward and the acceleration of gravity is downward.) Furthermore, let us take $x_0 = 0$, $y_0 = 0$, and $z_0 = 0$ for the sake of simplicity. The components of the velocity and position will then be, according to Eqs. (18)–(23),

Equations for projectile motion

$$v_x = v_{0x} \tag{24}$$

$$v_y = 0 \tag{25}$$

$$v_z = v_{0z} - gt \tag{26}$$

$$x = v_{0x}t \tag{27}$$

$$y = 0 \tag{28}$$

$$z = v_{0z}t - \tfrac{1}{2}gt^2 \tag{29}$$

With our choice of coordinates, the value of y is initially zero and always remains zero. Thus the motion of the projectile is confined to the x–z plane — the motion is *two-dimensional.* We will from now on ignore the y coordinate entirely in our study of projectile motion.

EXAMPLE 3. Consider the projectile fired from the gun described in Example 1. What is the maximum height reached by the projectile? What is the horizontal distance at this instant?

SOLUTION: The relevant equations are Eqs. (24)–(29). The known quantities are the components of the initial velocity; according to Example 1, these components are $v_{0x} = 74.6$ m/s and $v_{0z} = 50.3$ m/s. Furthermore, we know that the projectile reaches maximum height when $v_z = 0$. In terms of these values of v_{0z} and v_z, Eq. (26) determines the time of maximum height,

$$0 = -gt + 50.3 \text{ m/s}$$

or

$$t = \frac{50.3 \text{ m/s}}{g} = \frac{50.3 \text{ m/s}}{9.81 \text{ m/s}^2} = 5.13 \text{ s}$$

With this value of the time, we can calculate the maximum height,

$$\begin{aligned} z &= v_{0z}t - \tfrac{1}{2}gt^2 \\ &= 50.3 \text{ m/s} \times 5.13 \text{ s} - \tfrac{1}{2} \times 9.81 \text{ m/s}^2 \times (5.13 \text{ s})^2 \\ &= 129 \text{ m} \end{aligned}$$

and the horizontal distance,

$$x = v_{0x}t = 74.6 \text{ m/s} \times 5.13 \text{ s} = 383 \text{ m}$$

In our calculation we have ignored the effects of air resistance. For a heavy projectile with such a fairly low velocity, this is a good approximation.

EXAMPLE 4. In low-level bombing (at "smokestack level") a World War II bomber releases a bomb at a height of 150 ft above the surface of the sea while in horizontal flight at a constant speed of 200 mi/h. How long does the bomb take to fall to the surface? How far ahead (horizontally) of the point of release is the point of impact?

SOLUTION: It is convenient to place the origin of coordinates 150 ft above the level of the sea, with the x axis along the horizontal path of the bomber (Figure 4.6). The initial velocity of the bomb is the same as that of the bomber:

$$v_{0x} = 200 \text{ mi/h} = 293 \text{ ft/s}$$

$$v_{0z} = 0$$

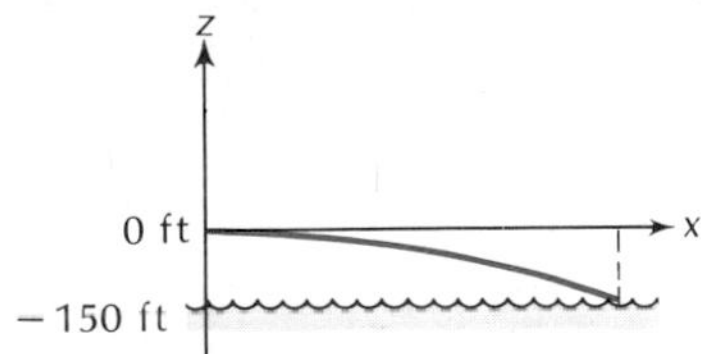

Fig. 4.6 Trajectory of a bomb.

When the bomb reaches the level of the sea, its vertical position is $z = -150$ ft (relative to the origin!). With these known quantities, Eq. (29) gives us the time of impact:

$$-150 \text{ ft} = -\tfrac{1}{2}gt^2$$

or

$$t = \sqrt{2 \times 150 \text{ ft}/g} = \sqrt{2 \times 150 \text{ ft}/(32.2 \text{ ft/s}^2)} = 3.05 \text{ s}$$

At this time, the horizontal position of the bomb is

$$x = v_{0x}t = 293 \text{ ft/s} \times 3.05 \text{ s} = 894 \text{ ft}$$

Fig. 4.7 A "string" of bombs released from a bomber. The bombs continue to move forward with the same horizontal velocity as that of the bomber.

Note that the bomber moves exactly the same *horizontal* distance in this time; that is, the bomb always remains directly below the bomber because both have exactly the same horizontal velocity $v_{0x} = 293$ ft/s. Figure 4.7 shows bombs released from a bomber at successive instants of time.

In our calculation we have again ignored air resistance. For low-level bombing, this is an acceptable approximation, but in high-level bombing the resistive corrections are very important.

The path of a bomb, or any other projectile, is a parabola. The mathematical proof of this statement rests on Eqs. (27) and (29). Suppose that time is reckoned from the instant at which the projectile reaches its maximum height (so that $t < 0$ before the instant of maximum height, and $t > 0$ after). Then $v_{0z} = 0$ and

$$x = v_{0x}t \tag{30}$$

$$z = -\tfrac{1}{2}gt^2 \tag{31}$$

Combining these equations, one finds

$$z = -\tfrac{1}{2}g(x/v_{0x})^2 \tag{32}$$

This equation has the form $z = [\text{constant}] \times x^2$, which is the equation of a **parabola** in the x–z plane (Figure 4.8).

If air resistance is negligible, the path of a projectile launched with some horizontal and vertical velocity is always some portion of a parabola. In the case of a bomb dropped by an airplane, the relevant portion starts at the apex and goes down to the ground. In the case of a shot or bullet fired from a gun, the relevant portion begins at ground level, rises, and comes down to the target. In the latter case, it is often important to calculate the **maximum height** reached, the **time of flight** (time between the instants of launch and impact), and the **range** (distance between the points of launch and impact). For the following calculation we will assume that the muzzle of the gun and the target are at the same height.

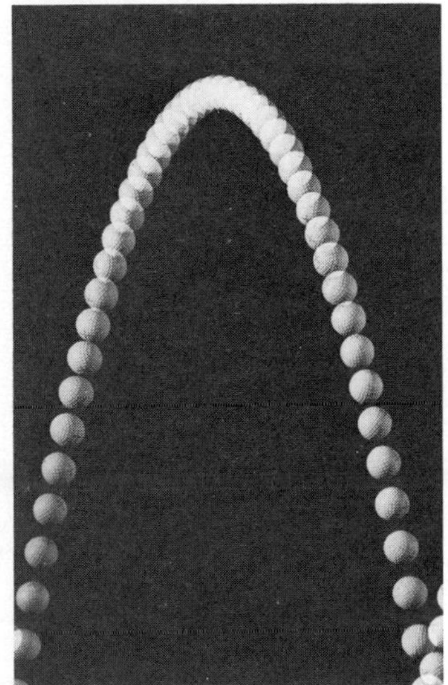

Fig. 4.8 The trajectory of a projectile (golf ball) in free fall is a parabola. *(Courtesy H. E. Edgerton, MIT.)*

We take the origin of coordinates at the launch point. The projectile is launched with some initial horizontal (v_{0x}) and vertical (v_{0z}) velocity. It reaches maximum height at the time t_{max}, when

$$0 = v_{0z} - gt_{max}$$

Consequently,

$$t_{max} = v_{0z}/g \tag{33}$$

At this time the height reached is

$$z_{max} = v_{0z}t_{max} - \tfrac{1}{2}gt_{max}^2 = v_{0z}(v_{0z}/g) - \tfrac{1}{2}g(v_{0z}/g)^2 \tag{34}$$

that is,

$$z_{max} = \tfrac{1}{2}v_{0z}^2/g \tag{35}$$

To find the time of impact, we must evaluate Eq. (29) with $z = 0$,

$$0 = v_{0z}t - \tfrac{1}{2}gt^2 \tag{36}$$

This equation has two solutions: $t = 0$ and $t = 2v_{0z}/g$. The latter solution gives the time of flight,

$$t_{flight} = 2v_{0z}/g \tag{37}$$

Comparison of Eqs. (33) and (37) shows that the time of flight is exactly twice the time required to reach maximum height; this is so because the motion is symmetric about the apex of the parabola.

The range is simply the horizontal distance attained at the time $t = t_{flight}$,

$$x_{max} = v_{0x}t_{flight} = 2v_{0x}v_{0z}/g \tag{38}$$

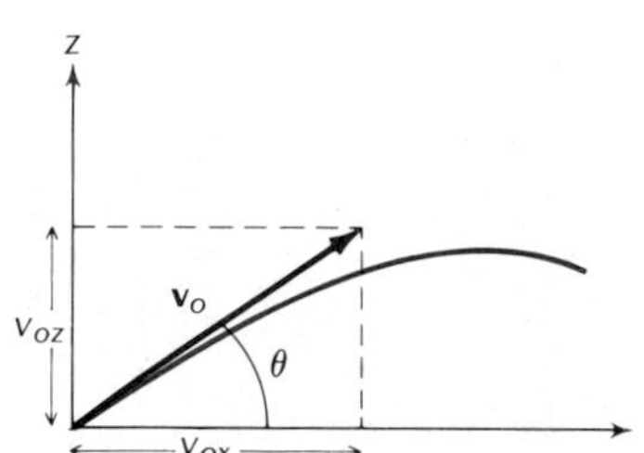

Fig. 4.9 The elevation angle θ gives the initial direction of motion of the projectile.

The formulas for z_{max}, t_{flight}, and x_{max} can be expressed in terms of the magnitude of the initial velocity (the "muzzle velocity" or, more precisely, the muzzle speed) and the angle between the horizontal and the direction of the initial velocity (the "elevation angle"). Figure 4.9 shows the initial velocity vector $\mathbf{v}_0$ and the elevation angle θ; we have

$$v_{0x} = v_0 \cos \theta \tag{39}$$

$$v_{0z} = v_0 \sin \theta \tag{40}$$

Hence

Maximum height

$$z_{max} = \frac{v_0^2 \sin^2 \theta}{2g} \tag{41}$$

Time of flight

$$t_{flight} = \frac{2v_0 \sin \theta}{g} \tag{42}$$

$$x_{max} = \frac{2v_0^2 \sin\theta \cos\theta}{g}$$

We can also write this last formula as

$$\boxed{x_{max} = \frac{v_0^2 \sin 2\theta}{g}} \qquad (43)$$

Range

Keep in mind that the calculation leading to Eqs. (41)–(43) assumed that the launch point and the impact point are at the same height ($z = 0$). If the ground is not flat, the projectile may reach impact before or after the time given by Eq. (42) and the range will be shorter or longer than that given by Eq. (43). Furthermore, keep in mind that although Eqs. (41)–(43) are often handy for the solution of problems of projectile motion, they do not give us a complete description of the motion. The complete description of the motion is contained in Eqs. (24)–(29), and we can always extract anything we want to know about the motion from these equations.

Figure 4.10 shows the trajectories of projectiles as a function of elevation angle. Note that the range is maximum for $\theta = 45°$ and that angles that are equal amounts above or below 45° yield the same range; for instance, 30° and 60° yield the same range.

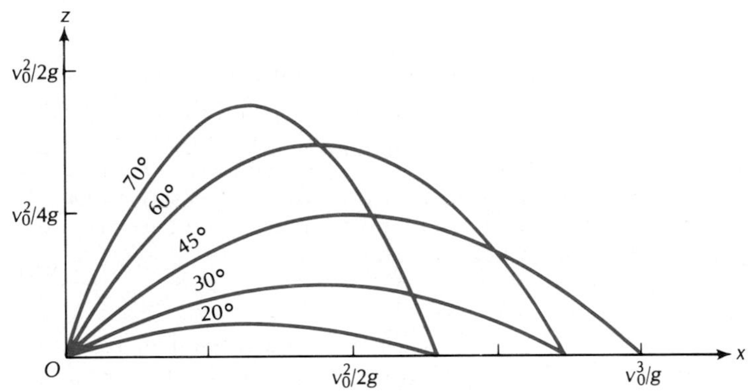

Fig. 4.10 Trajectories of projectiles of the same muzzle speed but different elevation angles.

EXAMPLE 5. The world record for the discus throw set by M. Wilkins in 1976 was 232.5 ft. What is the least initial speed required to achieve this range?

SOLUTION: To achieve the largest range with the least initial speed, the thrower must launch the discus with an elevation angle of 45°. Equation (43) then gives

$$v_0^2 = gx_{max}/\sin 2\theta = 32.2 \text{ ft/s}^2 \times 232.5 \text{ ft}/\sin 90°$$

$$= 7.49 \times 10^3 \text{ ft}^2/\text{s}^2$$

and

$$v_0 = 86.5 \text{ ft/s}$$

The curve labeled 45° in Figure 4.10 shows the trajectory of the discus.

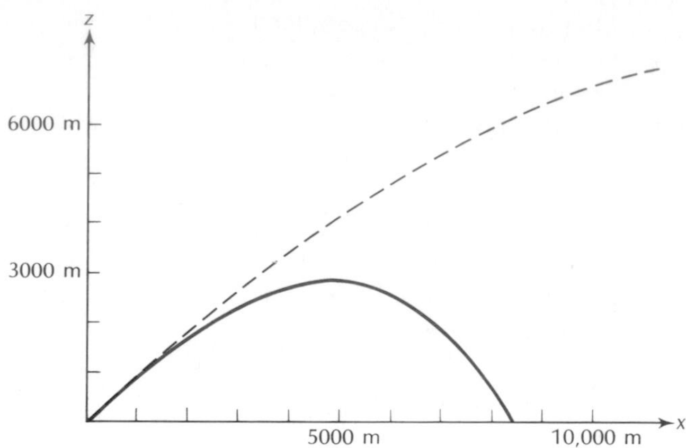

Fig. 4.11 Trajectory of a projectile taking into account air resistance. The projectile is a pointed cylinder of diameter 7.7 cm, mass 6.9 kg, launched with $v_0 = 550$ m/s, $\theta = 45°$. The dashed line shows the theoretical trajectory without air resistance. (Based on C. Cranz and K. Becker, *Exterior Ballistics.*)

Ballistic curve

In Figure 4.10, as in all calculations of this section, air resistance has been neglected. Figure 4.11 shows the trajectory of a high-speed projectile when air resistance is included. The trajectory is a **ballistic curve**; it cannot be described by a simple mathematical formula, but it can be calculated numerically by taking into account the empirically measured magnitude of the air resistance. Obviously, the trajectory for such a projectile is very different from the trajectory without air resistance predicted by the simple formulas of this section. If we use these simple formulas in the calculation of the motion of a high-speed projectile, our results will bear only a very vague resemblance to reality.

(a)

(b)

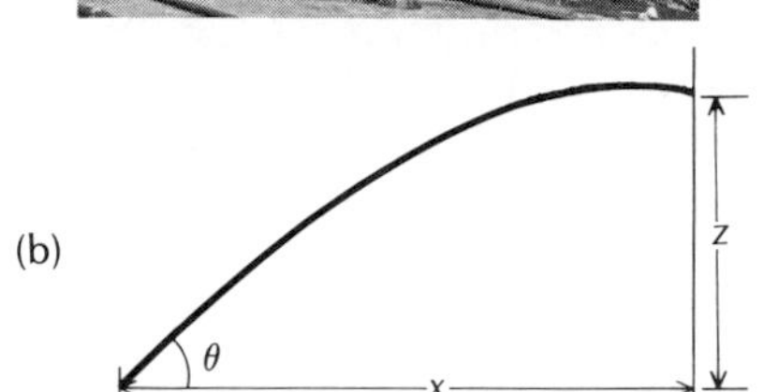

Fig. 4.12 (a) Stream of water from a fire hose follows the trajectory of a projectile. (b) Coordinates of the launch point and of the impact point.

EXAMPLE 6. A fireman standing at a horizontal distance x from a burning building aims the stream of water from a fire hose up at an angle θ (Figure 4.12). The water leaves the nozzle of the fire hose with a speed v_0. At what height z will the stream of water strike the building?

SOLUTION: Each water particle moves like a projectile. However, we cannot use Eqs. (41)–(43) because the launch point and the impact point are *not* at the same height. We must begin with Eqs. (27) and (29):

$$x = v_{0x}t$$

$$z = v_{0z}t - \tfrac{1}{2}gt^2$$

From the first of these equations we obtain $t = x/v_{0x}$, which, when substituted into the second equation, gives us the vertical height of the impact point,

$$z = \frac{v_{0z}}{v_{0x}}x - \frac{1}{2}\frac{gx^2}{{v_{0x}}^2}$$

With $v_{0x} = v_0 \cos\theta$ and $v_{0z} = v_0 \sin\theta$, this becomes

$$z = x \tan\theta - \frac{1}{2}\frac{gx^2}{v_0^2 \cos^2\theta} \tag{44}$$

4.4 Uniform Circular Motion

Uniform circular motion

Uniform circular motion is motion with constant speed along a circular path. Figure 4.13 shows the position at different times of a particle in uniform circular motion. All the velocity vectors shown have the same magnitude (same speed), but they differ in direction. Because of this change of direction, uniform circular motion is *accelerated motion.*

Consider a particle moving with constant speed v on a circle of radius r. To find the value of the instantaneous acceleration, we must look at the velocity change in a very short time interval Δt. Figure 4.14a shows the particle at two positions $\mathbf{r}_1$ and $\mathbf{r}_2$ a short time apart. The figure also shows the two velocity vectors $\mathbf{v}_1$ and $\mathbf{v}_2$; both velocity vectors have the same magnitude $|\mathbf{v}_1| = |\mathbf{v}_2| = v$. The position vectors make a small angle $\Delta\theta$ with one another, and so do the velocity vectors. Figure 4.14b shows the two velocity vectors tail to tail and their difference $\Delta\mathbf{v} = \mathbf{v}_2 - \mathbf{v}_1$. If Δt is very small, then $\Delta\theta$ (measured in radians) will also be small and the magnitude of $\Delta\mathbf{v}$ will be approximately

$$|\Delta\mathbf{v}| \cong v\,\Delta\theta \tag{45}$$

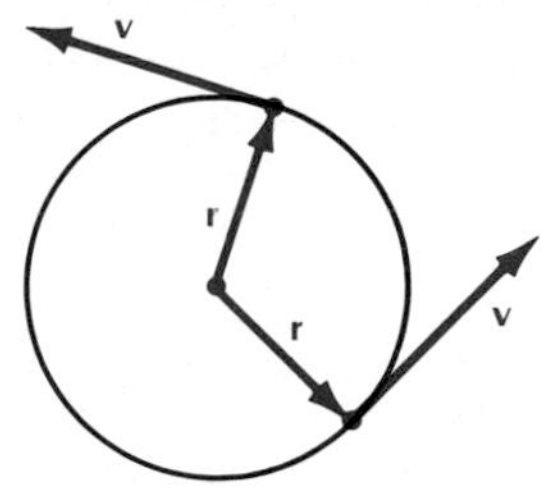

Fig. 4.13 Position and velocity vectors for a particle in uniform circular motion.

To justify this approximation, note that in Figure 4.14c the length of the circular arc of radius v connecting the tips of $\mathbf{v}_1$ and $\mathbf{v}_2$ is nearly equal to the length of the straight line segment connecting these tips, an approximation which becomes exact in the limit $\Delta\theta \to 0$. Since the straight line segment has a length $|\Delta\mathbf{v}|$ and the circular arc has a length [radius] × [angle in radians] $= v \times \Delta\theta$, their approximate equality implies Eq. (45).

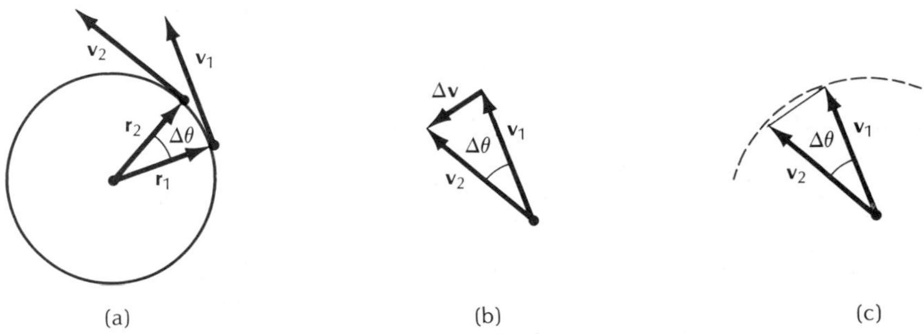

Fig. 4.14 (a) The position vectors at times t and $t + \Delta t$. (b) The velocity vectors at times t and $t + \Delta t$. The difference between the velocity vectors is $\Delta\mathbf{v}$. (c) The circular arc nearly coincides with the straight line segment connecting the tips of the vectors $\mathbf{v}_1$ and $\mathbf{v}_2$.

The magnitude of the acceleration is therefore

$$a = \frac{|\Delta\mathbf{v}|}{\Delta t} \cong v\frac{\Delta\theta}{\Delta t} \tag{46}$$

To evaluate the ratio $\Delta\theta/\Delta t$, we note that the time interval between the two positions shown in Figure 4.14a is

$$\Delta t = \frac{[\text{path length}]}{[\text{speed}]} = \frac{[\text{radius}] \times [\text{angle in radians}]}{[\text{speed}]}$$

$$= r\,\Delta\theta/v$$

Hence Eq. (46) becomes

$$a \cong v\frac{\Delta\theta}{r\,\Delta\theta/v} = \frac{v^2}{r} \tag{47}$$

In the limit $\Delta t \to 0$, this relation becomes exact, so that the acceleration for uniform circular motion is

Centripetal acceleration

$$a = v^2/r \qquad (48)$$

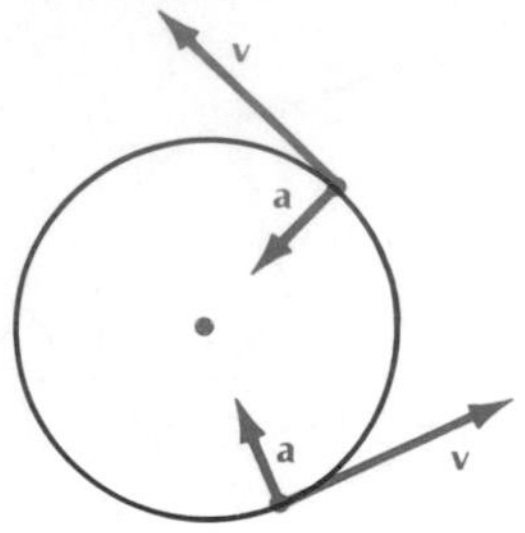

Fig. 4.15 Acceleration vectors for a particle in uniform circular motion.

The *direction* of this acceleration remains to be determined. From Figure 4.14b it is clear that in the limit $\Delta\theta \to 0$, the direction of $\Delta\mathbf{v}$ will be perpendicular to the velocity vectors $\mathbf{v}_1$ and $\mathbf{v}_2$ (which will be parallel in this limit); hence the acceleration is perpendicular to the velocity. Since the velocity vector corresponding to circular motion is tangential to the circle, the acceleration points along the radius, toward the center of the circle. Figure 4.15 shows the velocity and acceleration vectors at several positions along the circular path. The acceleration of uniform circular motion, with a magnitude given by Eq. (48) and a direction toward the center of the circle, is called **centripetal acceleration.**

EXAMPLE 7. In tests of the effects of high acceleration on the human body, astronauts at the National Aeronautics and Space Administration (NASA) Manned Spacecraft Center in Houston are placed in a capsule that is whirled around a path of radius 15 m at the end of a revolving girder (Figure 4.16). If the girder makes 24 revolutions per minute, what is the acceleration of the capsule? How many gee does this amount to?

Fig. 4.16 Centrifuge at the Manned Spacecraft Center in Houston.

SOLUTION: The circumference of the circular path is $2\pi \times [\text{radius}] = 2\pi \times 15$ m. Since the capsule takes (1/24) min, or (60/24) s, to go around this circumference, the speed is

$$v = \frac{2\pi \times 15 \text{ m}}{60/24 \text{ s}} = 38 \text{ m/s}$$

From Eq. (48), the centripetal acceleration is then

$$a = \frac{v^2}{r} = \frac{(38 \text{ m/s})^2}{15 \text{ m}} = 95 \text{ m/s}^2$$

This acceleration is 9.7 gee.

EXAMPLE 8. An automobile enters a semicircular curve of radius 100 ft. If the wheels of the automobile can tolerate a maximum acceleration of $0.82g$ without skidding, what is the maximum permissible speed?

SOLUTION: Equation (48) gives

$$v^2 = ra = 100 \text{ ft} \times 0.82g = 100 \text{ ft} \times 0.82 \times 32 \text{ ft/s}^2$$

$$= 2.6 \times 10^3 \text{ ft}^2/\text{s}^2$$

or

$$v = 51 \text{ ft/s} = 35 \text{ m/h}$$

Note that the acceleration of a particle, or an automobile, in uniform circular motion is always centripetal (toward the center) and never **centrifugal** (away from the center). However, with respect to the reference frame of the automobile, any free body — such as an apple that the driver of the automobile drops over the side — accelerates away from the center with a centrifugal acceleration of magnitude equal to the centripetal acceleration of the automobile with respect to the ground (Figure 4.17). This centrifugal acceleration of the apple only exists in the reference frame of the automobile; it does not exist in the reference frame of the ground, where the apple merely continues its horizontal motion with constant velocity while the automobile accelerates away from it.

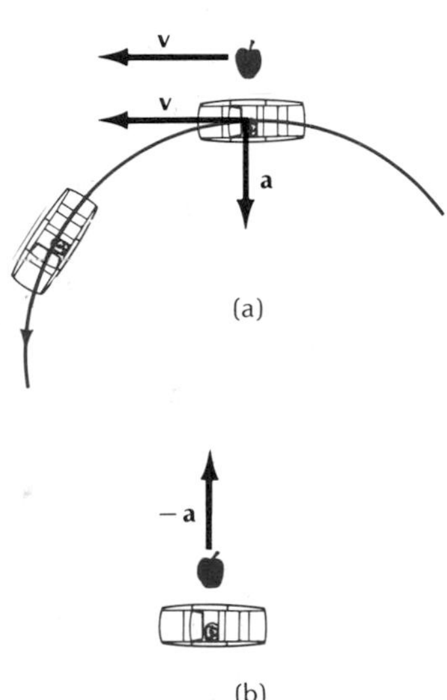

Fig. 4.17 (a) Automobile in uniform circular motion. The driver has just released an apple, which now moves with constant velocity **v** relative to the ground. The automobile accelerates away from the apple with the centripetal acceleration **a.** (b) Relative to the automobile, the apple has a centrifugal acceleration **−a.**

4.5 The Relativity of Motion and the Galilean Transformations

Motion is relative — the values of the velocity and acceleration of a particle depend on the frame of reference in which these quantities are evaluated. For example, consider one reference frame attached to the ground and a second reference frame attached to a ship moving due east at a constant speed of 10 m/s. Suppose that observers in both reference frames measure and plot the coordinates of a sailboat passing by (Figure 4.18). The coordinates and the time measured in the first reference frame will be denoted by x, y, z, and t; those measured in the second reference frame will be denoted by x', y', z', and t'. The values of these two sets of coordinates measured in the two reference frames will then be different. However, the coordinates obtained in one reference frame are related to those obtained in the other reference frame by a simple transformation formula which can be deduced as follows.

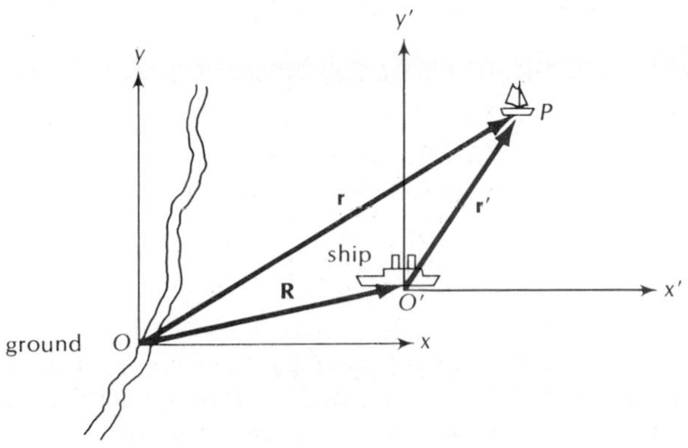

Fig. 4.18 The coordinate grid x'–y' of the ship moves relative to the coordinate grid x–y of the ground.

We begin with the time coordinate. Since time is absolute, at least to a very good approximation (see Section 1.1), we can assume without loss of generality that the clocks in both reference frames are permanently synchronized; hence the measured values of the time coincide,

$$\boxed{t' = t} \tag{49}$$

Next we consider the space coordinates. Figure 4.18 shows the coordinate grids associated with both the ground and the ship; these coordinate grids have the same orientation (the axes are parallel), but they move past one another at a speed of 10 m/s. The position vector of the sailboat is designated by $\mathbf{r}$ in the coordinate grid of the ground, and by $\mathbf{r}'$ in the coordinate grid of the ship. To establish the mathematical relationship between $\mathbf{r}$ and $\mathbf{r}'$, we will make the simplifying (but unnecessary) assumption that the origins O and O' of the two coordinate grids coincide at the time $t = 0$. The displacement $\mathbf{R}$ from the origin O to the origin O' at any later time is then

$$\mathbf{R} = \mathbf{V}t \tag{50}$$

where $\mathbf{V}$ is the velocity of the second coordinate grid with respect to the first, i.e., $\mathbf{V}$ is the velocity of the ship with respect to the ground. Inspection of Figure 4.18 then indicates that the displacement $\mathbf{r}$ is simply the sum of the displacements $\mathbf{R}$ and $\mathbf{r}'$,

$$\mathbf{r} = \mathbf{r}' + \mathbf{R} \tag{51}$$

or

$$\mathbf{r} = \mathbf{r}' + \mathbf{V}t \tag{52}$$

However, there is a catch in this argument: $\mathbf{r}'$ is a vector defined in the reference frame of the ship, while $\mathbf{R}$ is a vector defined in the reference frame of the ground; and before we add these vectors according to the usual rules of vector addition, we ought to ask whether these rules remain valid for vectors defined in *different* reference frames. The justification for the vector addition in Eq. (51) rests on the fact that length is absolute — at least to a very good approximation (see Section 1.1). As a consequence, the observers in the reference frame of the ground and those in the reference frame of the ship agree on the measured length (and direction) of the line segment from the ship to the sailboat (from O' to P), i.e., the value of the displacement vector $\mathbf{r}'$ is the same regardless of whether it is measured in the reference frame of the ship or in that of the ground. It is therefore quite legitimate to add this vector $\mathbf{r}'$ to the vector $\mathbf{R}$, as in Eq. (51), since we may pretend that both vectors have been measured in the reference frame of the ground.

Equation (52) can be rewritten as

$$\boxed{\mathbf{r}' = \mathbf{r} - \mathbf{V}t} \tag{53}$$

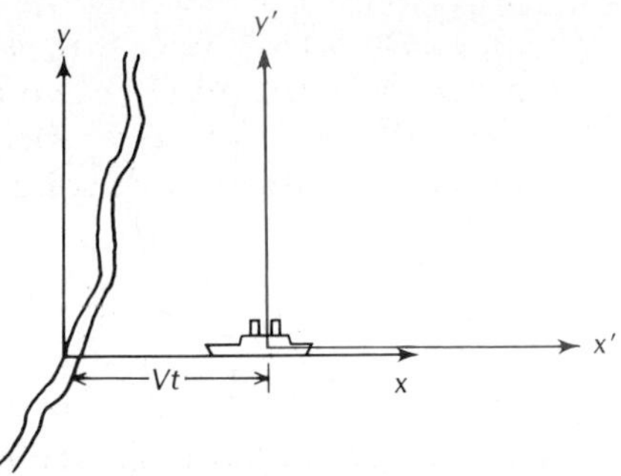

Fig. 4.19 The coordinate grid x'–y' moves along the x axis of the coordinate grid x–y.

If the velocity **V** is along the x axis, as in Figure 4.19, then the components of this equation are

$$x' = x - Vt \tag{54}$$

$$y' = y \tag{55}$$

$$z' = z \tag{56}$$

Taken together, Eqs. (49), (54), (55), and (56) constitute the transformation equations which change the time and the space coordinates of one reference frame into those of another. These transformation equations for reference frames in uniform motion relative to one another are called the **Galilean transformations.** It is worth emphasizing that these equations hinge on the assumption that length and time are absolute. As we have seen in Chapter 1, these assumptions lie at the foundations of Newton's theory of space and time. The Galilean transformation equations are therefore valid as long as Newton's theory is valid, that is, as long as we are dealing with the velocities of everyday experience, which are small compared to the velocity of light. New transformation equations are required when the velocities are compa rable to the velocity of light, and they will be the subject of Chapter 17.

Galilean transformations

The transformation equations for velocity and acceleration can be obtained directly from Eq. (53) by differentiation. Thus, the velocity of the sailboat relative to the ship of Figure 4.18 is

$$\mathbf{v}' = \frac{d\mathbf{r}'}{dt'} = \frac{d\mathbf{r}'}{dt} = \frac{d}{dt}(\mathbf{r} - \mathbf{V}t) = \frac{d\mathbf{r}}{dt} - \mathbf{V} \tag{57}$$

i.e.,

$$\boxed{\mathbf{v}' = \mathbf{v} - \mathbf{V}} \tag{58}$$

This shows that the velocity of the sailboat relative to the ship is merely the difference between the velocities of sailboat and ship relative to the ground. If this relation seems intuitively obvious, it is only because the absolute character of time and length are so deeply engrained in our intuition.

Likewise,

$$\mathbf{a} = \frac{d\mathbf{v}'}{dt'} = \frac{d\mathbf{v}'}{dt} = \frac{d}{dt}(\mathbf{v} - \mathbf{V}) = \frac{d\mathbf{v}}{dt} - 0 \tag{59}$$

i.e.,

$$\mathbf{a}' = \mathbf{a} \tag{60}$$

Hence the accelerations of the sailboat measured in the two reference frames are exactly the same; this, of course, depends on the assumption that $\mathbf{V}$ is constant — if not, then the two accelerations $\mathbf{a}'$ and $\mathbf{a}$ would differ by $d\mathbf{V}/dt$.

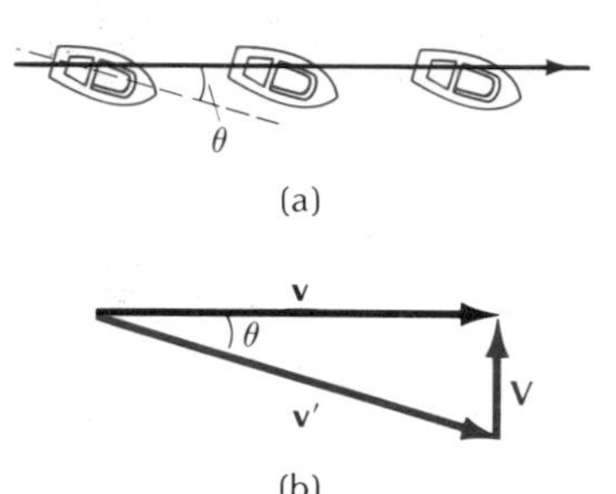

Fig. 4.20 (a) The path of the boat is toward the east, but the heading of the boat is south of east. (b) The velocity vector $\mathbf{V}$ of the water relative to the ground, and the velocity vector $\mathbf{v}'$ of the boat relative to the water. The sum of these vectors equals the velocity $\mathbf{v}$ of the boat relative to the ground.

EXAMPLE 9. Off the coast of Miami, the Gulf Stream current has a velocity of 4.8 km/h in a direction due north. The captain of a motorboat wants to travel on a straight course from Miami to Bimini Island, due east of Miami. His boat has a speed of 18 km/h relative to the water. In what direction must he head his boat? What will be its speed relative to the ground?

SOLUTION: Obviously, the captain must head his boat somewhat to the south to compensate for the Gulf Stream current. Figure 4.20a shows the path of the boat relative to the ground, and the heading of the boat. Figure 4.20b shows the velocity vector $\mathbf{V}$ of the water relative to the ground and the velocity vector $\mathbf{v}'$ of the boat relative to the water. According to Eq. (58), the velocity vector $\mathbf{v}$ of the boat relative to the ground is the sum of $\mathbf{v}'$ and $\mathbf{V}$,

$$\mathbf{v} = \mathbf{v}' + \mathbf{V} \tag{61}$$

By hypothesis, this vector $\mathbf{v}$ points due east (Figure 4.20b). Since the vector triangle is a right triangle, the angle between $\mathbf{v}'$ and the eastward direction is given by

$$\sin\theta = V/v'$$

so that

$$\theta = \sin^{-1}\left(\frac{V}{v'}\right) = \sin^{-1}\left(\frac{4.8 \text{ km/h}}{18 \text{ km/h}}\right) = 15^\circ$$

Thus, the boat must head 15° south of east. The speed of the boat relative to the ground is

$$v = v' \cos\theta = (18 \text{ km/h}) \times \cos 15^\circ = 17 \text{ km/h}$$

SUMMARY

Average velocity: $\bar{\mathbf{v}} = \dfrac{\Delta\mathbf{r}}{\Delta t}$

Instantaneous velocity: $\mathbf{v} = \dfrac{d\mathbf{r}}{dt}$

Average acceleration: $\bar{\mathbf{a}} = \dfrac{\Delta\mathbf{v}}{\Delta t}$

Instantaneous acceleration: $\mathbf{a} = \frac{d\mathbf{v}}{dt} = \frac{d^2\mathbf{r}}{dt^2}$

Motion with constant acceleration: $\mathbf{v} = \mathbf{v}_0 + \mathbf{a}t$

$$\mathbf{r} - \mathbf{r}_0 = \mathbf{v}_0 t + \tfrac{1}{2}\mathbf{a}t^2$$

Motion of a projectile: $v_x = v_{0x} = v_0 \cos\theta$

$$v_z = v_{0z} - gt = v_0 \sin\theta - gt$$

$$x = v_{0x}t$$

$$z = v_{0z}t - \tfrac{1}{2}gt^2$$

Range, maximum height, and time of flight (over flat ground): $x_{max} = \frac{v_0^2 \sin 2\theta}{g}$

$$z_{max} = \frac{v_0^2 \sin^2\theta}{2g}$$

$$t_{flight} = \frac{2v_0 \sin\theta}{g}$$

Centripetal acceleration in uniform circular motion: $a = v^2/r$

Galilean transformations: $t' = t$

$$\mathbf{r}' = \mathbf{r} - \mathbf{V}t$$

$$\mathbf{v}' = \mathbf{v} - \mathbf{V}$$

QUESTIONS

1. The speedometer of your automobile shows that you are proceeding at a steady 55 mi/h. Is it nevertheless possible that your automobile is in accelerated motion?

2. Can an automobile have eastward instantaneous velocity and northward instantaneous acceleration? Give an example.

3. Consider an automobile that is rounding a curve and braking at the same time. Draw a diagram showing the relative directions of the instantaneous velocity and acceleration.

4. A projectile is launched over level ground. Its initial velocity has a horizontal component v_{0x} and a vertical component v_{0y}. What is the average velocity of the projectile between the instants of launch and of impact?

5. What is the acceleration of a projectile when it reaches the top of its trajectory?

6. Are the velocity and the acceleration of a projectile ever perpendicular? Parallel?

7. If you throw a crumpled piece of paper, its trajectory is not a parabola. How does it differ from a parabola and why?

8. If a projectile is subject to air resistance, then the elevation angle for maximum range is not 45°. Do you expect the angle to be larger or smaller than 45°?

9. Figure 4.11 shows the trajectory of a high-speed projectile subject to strong air friction. The trajectory is not symmetric about its highest point. Why not? Would you expect the ascending or the descending portion of the trajectory to take more time?

10. Pitchers are fond of throwing curve balls. How does the trajectory of such a ball differ from the simple parabolic trajectory we studied in this chapter? What accounts for the difference?

11. A particle travels once around a circle with uniform circular motion. What is the average velocity and the average acceleration?

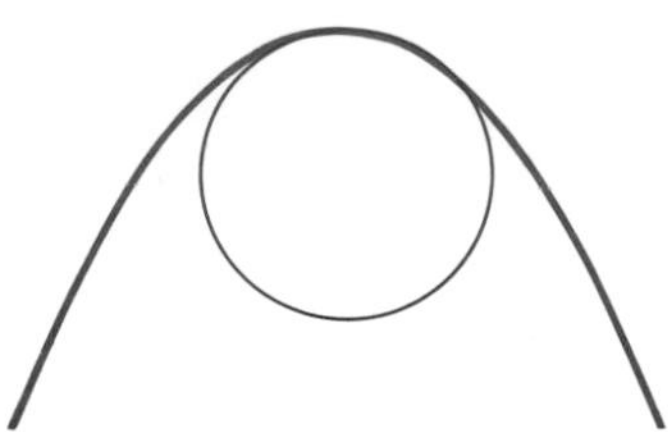

Fig. 4.21 The osculating circle.

12. A pendulum is swinging back and forth. Is this uniform circular motion? Draw a diagram showing the directions of the velocity and the acceleration at the top of the swing. Draw a similar diagram at the bottom of the swing.

13. Suppose that at the top of its parabolic trajectory a projectile has a horizontal speed v_{0x}. The segment at the top of the parabola can be approximated by a circle, called the osculating circle (Figure 4.21). What is the radius of this circle? (Hint: The projectile is instantaneously in uniform circular motion at the top of the parabola.)

14. Why do raindrops fall down at a pronounced angle with the vertical when seen from the window of a speeding train? Is this angle necessarily the same as that of the path of a water drop sliding down along the outside surface of the window?

15. When a sailboat is sailing to windward ("beating"), the wind feels much stronger than when the sailboat is sailing downwind ("running"). Why?

16. Rain is falling vertically. If you run through the rain, at what angle should you hold your umbrella? If you don't have an umbrella, should you bend forward while running?

17. In the reference frame of the ground, the path of a sailboat beating to windward makes an angle of 45° with the direction of the wind. In the reference frame of the sailboat, the angle is somewhat smaller. Explain.

18. Suggest a method for measuring the speed of falling raindrops.

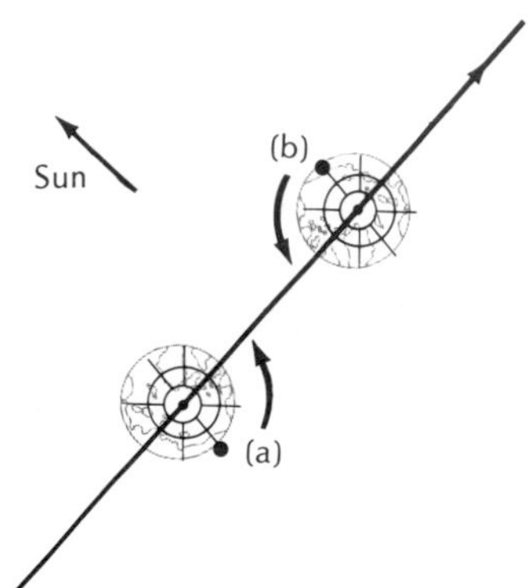

Fig. 4.22 Rotational and translational motions of the Earth: (a) at midnight and (b) at noon.

19. According to a theory proposed by Galileo, the tides on the oceans are caused by the Earth's rotational motion about its axis combined with its translational motion around the Sun. At midnight these motions are in the same direction; at noon they are in opposite directions (Figure 4.22). Thus, any point on the Earth alternately speeds up and slows down. Galileo was of the opinion that the speeding up and slowing down of the ocean basins would make the water slosh back and forth, thus giving rise to tides. What is wrong with this theory? (Hint: What is the acceleration of a point on the Earth? Does this acceleration depend on the translational motion?)

PROBLEMS

Section 4.1

1. A sailboat tacking against the wind moves as follows: 3.2 km at 45° east of north, 4.5 km at 50° west of north, and 2.6 km at 45° east of north. The entire motion takes 1 h 15 min.
   (a) What is the total displacement for this motion?
   (b) What is the average velocity for this motion?
   (c) What is the speed if it is assumed to be constant?

2. In one-half year, the Earth moves halfway around its orbit, a circle of radius 1.50×10^{11} m centered on the Sun. What is the average speed and what is the magnitude of the average velocity for this time interval?

3. The fastest bird is the spine-tailed swift, which reaches speeds of 106 mi/h. Suppose that you wish to shoot such a bird with a 22-caliber rifle that fires a bullet with a speed of 1200 ft/s. If you fire at the instant when the bird is 100 ft directly overhead, how many feet ahead of the bird must you aim the rifle? Ignore gravity in this problem.

4. The position vector of a particle is given by the following function of time:

$$\mathbf{r} = (6 + 2t^2)\hat{\mathbf{x}} + (3 - 2t + 3t^2)\hat{\mathbf{y}}$$

where distance is measured in meters and time in seconds.

(a) What is the instantaneous velocity vector at $t = 2$ s? What is the magnitude of this vector?

(b) What is the instantaneous acceleration vector at $t = 2$ s? What is the magnitude of this vector?

5. A particle is moving in the x–y plane; the components of its position vector are

$$x = A \cos bt \qquad y = A \sin bt$$

where A and b are constants.

(a) What are the components of the instantaneous velocity vector? The instantaneous acceleration vector?

(b) What is the magnitude of the instantaneous velocity? The instantaneous acceleration?

Sections 4.2 and 4.3

6. The fastest recorded speed of a baseball thrown by a pitcher is 100.9 mi/h, achieved by Nolan Ryan in 1974 at Anaheim Stadium. If the baseball leaves the pitcher's hand with a horizontal velocity of this magnitude, how far will the ball have fallen vertically by the time it has traveled 20 yd horizontally?

7. At Acapulco, professional divers jump from a 36-m-high cliff into the sea (compare Example 2.5). At the base of the cliff, a rocky ledge sticks out for a horizontal distance of 6.4 m. With what minimum horizontal velocity must the divers jump off if they are to clear this ledge?

8. A stunt driver wants to make his car jump over 10 cars parked side by side below a horizontal ramp (Figure 4.23). With what minimum speed must he drive off the ramp? The vertical height of the ramp is 6.0 ft and the horizontal distance he must clear is 80 ft.

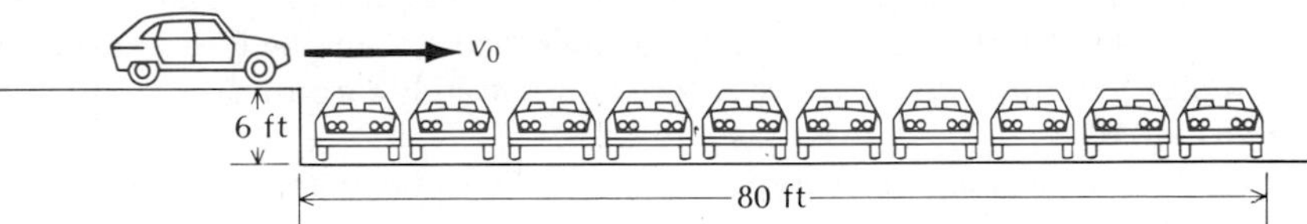

Fig. 4.23

9. A particle has an initial position vector $\mathbf{r} = 0$ and an initial velocity $\mathbf{v}_0 = 3\hat{\mathbf{x}} + 2\hat{\mathbf{y}}$ (where distance is measured in meters and velocity in meters per second). The particle moves with a constant acceleration $\mathbf{a} = \hat{\mathbf{x}} - 4\hat{\mathbf{y}}$ (measured in m/s^2). At what time does the particle reach a maximum y coordinate? What is the position vector of the particle at that time?

10. Volcanoes on the Earth eject rocks at speeds of up to 700 m/s. Assume that the rocks are ejected in all directions; ignore the height of the volcano and ignore air friction.

(a) What is the maximum height reached by rocks?

(b) What is the maximum horizontal distance reached by rocks?

(c) Is it reasonable to ignore air friction in these calculations?

11. In a circus act at the Ringling Bros. and Barnum & Bailey Circus, a "human cannonball" was fired from a large cannon with a muzzle velocity of 54 mi/h. Assume that the firing angle was 45° from the horizontal. How many seconds did the human cannonball take to reach maximum height? How high did he rise? How far from the cannon did he land?

12. According to a reliable report, in 1795 a member of the Turkish embassy in England shot an arrow to a distance of 482 yd. According to a less reliable report, a few years later the Turkish Sultan Selim shot an arrow to 972 yd. In each of these cases calculate what must have been the minimum initial speed of the arrow.

13. A golfer claims that a golf ball launched with an elevation angle of 12° can reach a horizontal range of 250 m. Ignoring air friction, what would the initial speed of such a golf ball have to be? What maximum height would it reach?

14. A gunner wants to fire a gun at a target at a horizontal distance of 12,500 m from his position.

(a) If his gun fires with a muzzle velocity of 700 m/s and if $g = 9.81\ \text{m/s}^2$, what is the correct elevation angle? Pretend that there is no air resistance.

(b) If the gunner mistakenly assumes $g = 9.80\ \text{m/s}^2$, by how many meters will he miss the target?

15. According to the *Guinness Book of World Records*, during a catastrophic explosion at Halifax on December 6, 1917, William Becker was thrown through the air for some 1600 yards and was found, still alive, in a tree. Assume that Becker left the ground and returned to the ground (ignore the height of the tree) at an angle of 45°. With what speed did he leave the ground? How high did he rise? How long did he stay in flight?

16. A hay-baling machine throws each finished bundle of hay 8 ft up in the air so it can land on a trailer waiting 16 ft behind the machine. What must be the speed with which the bundles are launched? What must be the angle of launch?

17. The motion of a ballistic missile can be regarded as the motion of a projectile, because along the greatest part of its trajectory the missile is in free fall. Suppose that a missile is to strike a target 3000 km away. What minimum speed must the missile have at the beginning of its trajectory? What maximum height does it reach when launched with this minimum speed? How long does it take to reach its target? For these calculations assume that $g = 9.8\ \text{m/s}^2$ everywhere along the trajectory.

18. The natives of the South American Andes throw stones by means of slings which they whirl over their heads. They can accurately throw a $\frac{1}{5}$-kg stone to a distance of 50 m.

(a) What is the minimum speed with which the stone must leave the sling to reach this distance?

(b) Just before release, the stone is being whirled around a circle of radius 1.0 m with the speed calculated in part (a). How many revolutions per second does the stone make?

19. A large stone-throwing engine designed by Archimedes could throw a 77-kg stone over a range of 180 m. What must have been the initial speed of the stone if thrown at an initial angle of 45° with the horizontal?

20. The nozzle of a fire hose ejects 150 gal. of water per minute at a speed of 86 ft/s. How far away will the stream of water land if the nozzle is aimed at an angle of 35° with the horizontal? How many gallons of water are in the air at any given instant?

21. According to an ancient Greek source, a stone-throwing machine on one occasion achieved a range of 730 m. If this is true, what must have been the minimum initial speed of the stone as it was ejected from the engine? When thrown with this speed, how long would the stone have taken to reach its target?

22. The muzzle velocity for a Lee-Enfield rifle is 630 m/s. Suppose you fire this rifle at a target 700 m away and at the same level as the rifle.

(a) In order to hit the target, you must aim the barrel at a point above the target. How many meters above the target must you aim? Pretend that there is no air resistance.
(b) What will be the maximum height that the bullet reaches along its trajectory?
(c) How long does the bullet take to reach the target?

23. In artillery, it is standard practice to fire a sequence of trial shots at a target before commencing to fire "for effect." The artillerist first fires a shot short of the target, then a shot beyond the target, and then makes the necessary adjustment in elevation so that the third shot is exactly on target. Suppose that the first shot fired from a gun aimed with an elevation angle of 7°20′ lands 180 m short of the target; the second shot fired with an elevation of 7°35′ lands 120 m beyond the target. What is the correct elevation angle to hit the target?

24. The world record for the javelin throw by a woman, established in 1976 by Ruth Fuchs in Berlin, was 69.11 m (226 ft 9 in.). If Fuchs had thrown her javelin with the same initial velocity in Buenos Aires rather than in Berlin, how much farther would it have gone? The acceleration of gravity is 9.8128 m/s^2 in Berlin and 9.7967 m/s^2 in Buenos Aires. Pretend that air resistance plays no role in this problem.

25. At a location where $g = 9.81$ m/s^2, the range of a gun is 950 m when fired at an elevation of angle 40°. If we move this gun to a new location where $g = 9.78$ m/s^2, what will be its range when fired at the same elevation angle? Ignore air resistance.

26. (a) A golfer wants to drive a ball to a distance of 260 yd. If he launches the ball with an elevation angle of 14°, what is the appropriate initial speed? Ignore air resistance.
(b) If the speed is too great by 2 ft/s, how much farther will the ball travel when launched at the same angle?
(c) If the elevation angle is 0.5° larger than 14°, how much farther will the ball travel if launched with the speed calculated in part (a)?

27. Show that for a projectile launched with an elevation angle of 45°, the maximum height reached is one-quarter of the range.

28. During a famous jump in Richmond, Virginia, in 1903, the horse Heatherbloom with its rider jumped over an obstacle 8 ft 8 in. high while covering a horizontal distance of 37 ft. At what angle and with what speed did the horse leave the ground? Make the (somewhat doubtful) assumption that the motion of the horse is particle motion.

29. With what elevation angle must you launch a projectile if its range is to equal twice its maximum height?

30. A gun fires a shot with a muzzle velocity of 160 m/s at a target at a horizontal distance of 1500 m.
(a) What is the appropriate elevation angle?
(b) If the adjustment of elevation angle is in error by 5 minutes of arc, by how many meters will the shot miss the target?

31. The gun of a coastal battery is emplaced on a hill 150 ft above the water level. It fires a shot with a muzzle velocity of 2000 ft/s at a ship at a horizontal distance of 13,000 yd. What elevation angle must the gun have if the shot is to hit the ship? Pretend that there is no air resistance.

32. In a flying ski jump, the skier acquires a speed of 110 km/h by racing down a steep hill and then lifts off into the air from a horizontal ramp. Beyond this ramp, the ground slopes downward at an angle of 45°.
(a) Assuming that the skier is in free-fall motion after he leaves the ramp, at what distance down the slope will he land?
(b) In actual jumps, skiers reach distances of up to 165 m. Why does this not agree with the result you obtained in part (a)?

33. A battleship steaming at 45 km/h fires a gun at right angles to the longitudinal axis of the ship. The elevation angle of the gun is 30° and the muzzle velocity of the shot is 720 m/s; the gravitational acceleration is 9.80 m/s². What is the range of this shot in the reference frame of the ground? Pretend that there is no air resistance.

*34. The maximum speed with which you can throw a stone is about 25 m/s (a professional baseball pitcher can do much better than this). Can you hit a window 50 m away and 13 m up from the point where the stone leaves your hand? What is the maximum height of a window you can hit at this distance?

*35. A gun on the coast (at sea level) fires a shot at a ship which is heading directly toward the gun at a speed of 40 km/h. At the instant of firing, the distance to the ship is 15,000 m. The muzzle velocity of the shot is 700 m/s. Pretend that there is no air resistance.

(a) What is the required elevation angle for the gun? Assume $g = 9.80$ m/s².

(b) What is the time interval between firing and impact?

*36. A ship is steaming at 30 km/h on a course parallel to a straight shore at a distance of 17,000 m. A gun emplaced on the shore (at the water level) fires a shot with a muzzle velocity of 700 m/s when the ship is at the point of closest approach. If the shot is to hit the ship, what must be the elevation angle of the gun? How far ahead of the ship must the gun be aimed? Give the answer to the latter question both in meters and in minutes of arc. Pretend that there is no air resistance. (Hint: Solve this problem by the following method of successive approximations. First calculate the time of flight of the shot, neglecting the motion of the ship; then calculate how far the ship moves in this time; and then calculate the elevation angle and the aiming angle required to hit the ship at this new position.)

Section 4.4

37. An ultracentrifuge spins a small test tube in a circle of radius 10 cm at 1000 revolutions per second. What is the centripetal acceleration of the test tube? How many gee does this amount to?

38. The blade of a circular saw has a diameter of 8 in. If this blade rotates at 7000 revolutions per minute (its maximum safe speed), what is the speed and what is the centripetal acceleration of a point on the rim?

39. At the Fermilab accelerator (one of the world's largest atom smashers), protons are forced to travel in an evacuated tube in a circular orbit of diameter 2.0 km. The protons have a speed nearly equal to the speed of light (99.99995% of the speed of light). What is the centripetal acceleration of these protons? Express your answer in m/s² and in gee.

40. An automobile travels at a steady 90 km/h along a road leading over a small hill. The top of the hill is rounded so that, in the vertical plane, the road approximately follows an arc of circle of radius 70 m. What is the centripetal acceleration of the automobile at the top of the hill?

41. The Earth moves around the Sun in a circular path of radius 1.50×10^{11} m at uniform speed. What is the magnitude of the centripetal acceleration of the Earth toward the Sun?

42. An automobile has wheels of diameter 25 in. What is the centripetal acceleration of a point on the rim of this wheel when the automobile is traveling at 60 mi/h?

*43. The table printed on the endpapers lists the radii of the orbits of the planets ("mean distance from the Sun") and the times required for moving around the orbit ("period of revolution"). For the following problem assume that the planets move along circular orbits at constant speed.

(a) Calculate the magnitude of the centripetal acceleration of each planet.

(b) In a logarithmic graph of centripetal acceleration vs. radius, plot the log of the acceleration vs. the log of the radius and draw a smooth curve through the points. Can you discover a simple equation that describes the acceleration as a function of radius?

Section 4.5

44. On a rainy day, a steady wind is blowing at 30 km/h. In the reference frame of the *air,* the raindrops are falling vertically with a speed of 10 m/s. What is the magnitude and direction of the velocity of the raindrops in the reference frame of the ground?

45. A lump of concrete falls off a crumbling overpass and strikes an automobile traveling on a highway below. The lump of concrete falls 5 m before impact and the automobile has a speed of 90 km/h.

(a) What is the speed of impact of the lump in the reference frame of the automobile?

(b) What is the angle of impact?

46. On a rainy day, raindrops are falling with a vertical velocity of 10 m/s. If an automobile drives through the rain at 25 m/s (55 mi/h), what is the velocity (magnitude and direction) of the raindrops relative to the automobile?

47. A battleship steaming at 13 m/s fires a shot in the forward direction. The elevation angle of the gun is 20° and the muzzle speed of the shot is 660 m/s. What is the velocity vector of the shot relative to the ground?

48. A wind of 30 m/s is blowing from the west. What will be the speed, relative to the ground, of a sound signal traveling due north? The speed of sound, relative to air, is 330 m/s.

49. On a windy day, a hot-air balloon is ascending at a rate of 0.5 m/s relative to the air. Simultaneously, the air is moving with a horizontal velocity of 12 m/s. What is the velocity (magnitude and direction) of the balloon relative to the ground?

50. A blimp is motoring at constant altitude. The airspeed indicator on the blimp shows that its speed relative to the air is 20 km/h and the compass shows that the heading of the blimp is 10° east of north. If the air is moving over the ground with a velocity of 15 km/h due east, what is the velocity (magnitude and direction) of the blimp relative to the ground? For an observer on the ground, what is the angle between the longitudinal axis of the blimp and the direction of motion?

51. A sailboat is moving in a direction 50° east of north at a speed of 8.5 mi/h. The wind measured by an instrument aboard the sailboat has an apparent (relative to the sailboat) speed of 20 mi/h coming from an apparent direction of 10° east of north. Find the true (relative to the ground) speed and direction of the wind.

52. Raindrops, driven by the wind, are falling with a speed of 7.0 m/s at an angle of 10° with the vertical. What is the angle with the vertical that the trajectories of these raindrops make in the reference frame of an airplane traveling at 300 km/h in the same direction as the wind?

53. A wind is blowing at 50 km/h from a direction 45° west of north. The pilot of an airplane wishes to fly on a route due north from an airport. The airspeed of the airplane is 250 km/h.

(a) In what direction must the pilot point the nose of the airplane?

(b) What will be the airplane's speed relative to the ground?

54. At the entrance of Ambrose Channel at New York Harbor, the tidal current at one time of the day has a velocity of 4.2 km/h in a direction 20° south of east. Consider a ship in this current; suppose that the ship has a speed of 16 km/h relative to the water. If the helmsman keeps the bow of the ship

aimed due north, what will be the actual velocity (magnitude and direction) of the ship relative to the ground?

55. A white automobile is traveling at a constant speed of 90 km/h on a highway. The driver notices a red automobile 1.0 km behind, traveling in the same direction. Two minutes later, the red automobile passes the white automobile.

(a) What is the average speed of the red automobile relative to the white?

(b) What is the speed of the red automobile relative to the ground?

56. Two automobiles travel at equal speeds in opposite directions on two separate lanes of a highway. The automobiles move at constant speed v_0 on straight parallel tracks separated by a distance h. Find a formula for the rate of change of the distance between the automobiles as a function of time; take the instant of closest approach as $t = 0$. Plot v vs. t for $v_0 = 60$ km/h, $h = 50$ m.

57. A ferryboat on a river has a speed v relative to the water. The water of the river flows with a speed V relative to the ground. The width of the river is d.

(a) Show that the ferryboat takes a time $2d/\sqrt{v^2 - V^2}$ to travel across the river and back.

(b) Show that the ferryboat takes a time $2dv/(v^2 - V^2)$ to travel a distance d up the river and back. Which trip takes a shorter time?

58. An AWACS aircraft is flying at high altitude in a wind of 150 km/h from due west. Relative to the air, the heading of the aircraft is due north and its speed is 750 km/h. A radar operator on the aircraft spots an unidentified target approaching from northeast; relative to the AWACS aircraft, the bearing of the target is 45° east of north, and its speed is 950 km/h. What is the speed of the unidentified target relative to the ground?

CHAPTER 5

Dynamics — Newton's Laws

Dynamics is the study of forces and their effects on the motion of bodies. So far we have dealt only with the mathematical description of motion — the definitions of position, velocity, and acceleration, and the relationships among these quantities. We did not inquire what causes a body to accelerate. In this chapter we will see that the cause of acceleration is force. The fundamental properties of force and the relationship between force and acceleration are given by Newton's three laws. The first of these laws describes the natural state of motion of a free body on which no forces are acting, whereas the other two laws deal with the behavior of bodies under the influence of forces. The first law was actually discovered by Galileo Galilei early in the seventeenth century, but it remained for Isaac Newton, in the second half of the seventeenth century, to formulate a coherent theory of forces and to lay down a complete set of equations from which the motion of bodies under the influence of arbitrary forces can be calculated.

5.1 Newton's First Law

Everyday experience suggests that a force — a push or pull — is needed to keep an object moving at constant velocity. For example, if the wind pushing a sailboat suddenly dies out, the boat will coast along for some distance, but it will gradually slow down, stop, and remain stopped until a new gust of wind comes along. However, what actually slows down the sailboat is not the *absence* of a propulsive force but, rather, the *presence* of friction forces. Under ideal frictionless conditions a body in motion would continue to move forever. Experiments with pucks or gliders riding on a cushion of air on a low-friction air table or air track give some hint of the persistence of motion; but in order to eliminate friction entirely, it is best to use bodies moving in a

Sir Isaac Newton, *1642–1727, English mathematician and physicist, widely regarded as the greatest scientist of all time. He was professor at Cambridge, president of the Royal Society, and later became master of the Mint. His brilliant discoveries in mechanics were published in 1687 in his book* Principia Mathematica, *one of the glories of the Age of Reason. In this book, Newton laid down the laws of motion and the law of universal gravitation, and he demonstrated that planets in the sky as well as bodies on the Earth obey the same mathematical equations. For over 200 years, Newton's laws stood as the unchallenged basis of all our attempts at a scientific explanation of the physical world; even today, these laws still remain as the essential ingredient of engineering physics. Newton was also the foremost mathematician of his time, and he shares with the philosopher Gottfried Leibniz the credit for discovering the calculus.*

vacuum, without even air against which to rub. Observations of particles moving in evacuated tubes as well as observations of celestial bodies — planets, satellites, and comets — moving in the emptiness of interplanetary space show that a body left to itself persists indefinitely in its state of uniform motion. A body on which no force is acting is called a **free body** and the experimental observations concerning such bodies are summarized in **Newton's First Law:**

Newton's First Law

> *A body at rest remains at rest and a body in motion continues to move at constant velocity unless acted upon by an external force.*

This law is also called the **law of inertia,** since it expresses the tendency of bodies to maintain their original state of motion.

One crucial restriction on Newton's First Law concerns the choice of reference frame: the law is not valid in all reference frames, but only in certain special frames. It is obvious that if this law is valid in one given reference frame, then it cannot be valid in a second reference frame that has an accelerated motion relative to the first. For example, in the reference frame of the ground, a ball at rest on the floor of a train station remains at rest, but in the reference frame of an accelerating train leaving the station, the same ball has a "spontaneous" acceleration toward the rear of the train, in contradiction to Newton's First Law. Those special reference frames in which the law is valid are called **inertial reference frames.** Thus the reference frame of the ground is an inertial reference frame, but that of the accelerating train is not.

Inertial reference frame

How do we know whether or not a given reference frame is inertial? We can only tell by making a test: take a free body, i.e., a body isolated from all external forces, and observe its motion; if the body persists in its state of uniform motion, then the reference frame is inertial. This test procedure suggests that Newton's First Law is nothing but a rule for finding the "right" reference frames. But the First Law is much more than that. The decisive fact is that after *one* body has been used to test that the reference frame is inertial, all *other* free bodies will obey the First Law in that reference frame without any further quibble. With this in mind, we can reformulate the essence of the First Law as follows: *if one free body remains at rest or in uniform motion in a given reference frame, then so will all other free bodies.* This is a law of nature, i.e., it is a statement about the behavior of the physical world that can be verified by experiments. Thus, the First Law plays a dual role: it is a law of nature as well as a definition of what is meant by an inertial reference frame. Several other laws of physics, including Newton's Second Law (see below), also play such dual roles.

Note that if some given reference frame is inertial, any other reference frame in uniform translational motion relative to the first will also be inertial, and any other reference frame in accelerated motion relative to the first will not be inertial. Thus, any two inertial reference frames can differ only by some constant relative velocity[1]; they cannot differ by an acceleration. This implies that, as measured with respect to inertial reference frames, *acceleration is absolute:* when a particle has some acceleration in one inertial reference frame, then it will have exactly the same acceleration in any other. The velocity of the particle is relative; the coordinates of one inertial reference frame are related to those of another by the Galilean transformations of Eq. (4.53) and the velocities are related by the simple additive law of Eq. (4.58).

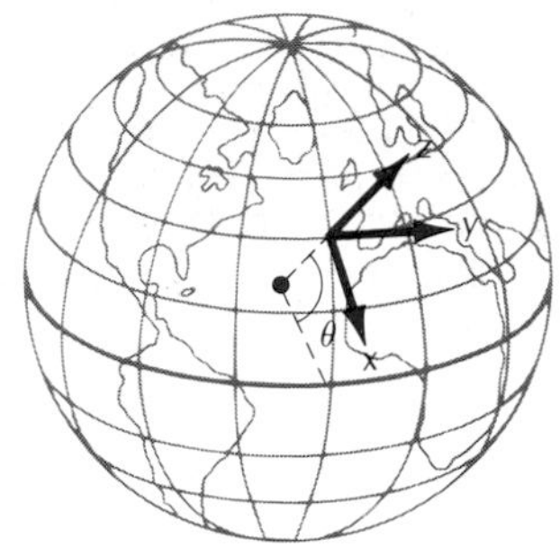

Fig. 5.1 A reference frame with the origin of coordinates fixed at the surface of the Earth.

Finally, we must answer a very important question: which of the reference frames in practical use for everyday measurements are inertial? For the description of everyday phenomena, the most commonly used reference frame is one attached to the ground, with the origin of coordinates fixed at some point on the surface of the Earth (Figure 5.1). Although crude experiments indicate that this reference frame is inertial (for example, an isolated railroad car at rest on a level track remains at rest), more precise experiments show that this reference frame is not inertial. A famous experiment of this kind is Foucault's pendulum experiment (Figure 5.2). When a long pendulum on a twist-free suspension is allowed to oscillate back and forth, observation shows that the pendulum gradually drifts sideways, i.e., the orientation of the plane of oscillation gradually rotates relative to the Earth. Of course, the pendulum bob is not a free body, since gravity and the pendulum string both tug on it; but neither of these forces can cause the transverse rotational motion that is observed. Hence this motion indicates that Newton's First Law is not valid in the reference frame of the Earth. The explanation is that the Earth rotates about its axis — it rotates relative to an inertial reference frame. Since rotational motion has a centripetal acceleration, a reference frame attached to the Earth has an acceleration and is not an inertial reference frame. Thus, the rotational motion of the plane of oscillation of the Foucault pendulum arises from the rotation of the Earth: the pendulum tends to swing in a fixed plane, but the Earth rotates under it.

Fig. 5.2 Foucault pendulum at the Smithsonian Institution, Washington, D. C.

Although a reference frame attached to the ground is not exactly inertial, the numerical value of the centripetal acceleration of points on the surface of the Earth is fairly small. Figure 5.1 shows the origin of coordinates at a point on the surface of the Earth; this point is in uniform circular motion around the Earth's axis. If the origin is at a geographical latitude θ, then the radius of the circular motion is $r = R_E \cos\theta$ (where R_E is the radius of the Earth; Figure 5.3) and the speed of the motion is $v = 2\pi r/T$ (where $T =$ 23 h 56 min, the time for one rotation of the Earth[2]). The centripetal acceleration is then

[1] They can also differ in the orientation of the axes. But, in the present context, such a difference is of no interest.

[2] The period of rotation of the Earth with respect to an inertial reference frame is 23 h 56 min rather than 24 h. The reason is that, as seen from the Earth, the Sun moves with respect to the background of "fixed" stars. Consequently, in order to catch up with the Sun, the Greenwich meridian must turn through a bit more than one revolution; this makes the time between successive meridian passages of the Sun (solar day, 24 h) slightly longer than the time between successive meridian passages of a fixed star (sidereal day, 23 h 56 min). The difference is small and can often be neglected.

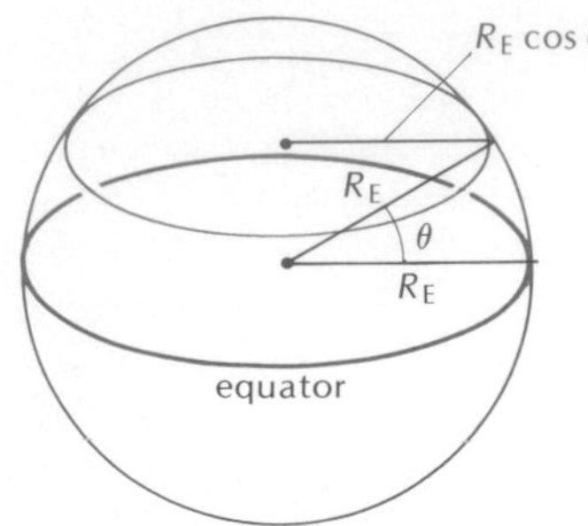

Fig. 5.3 A point on the surface of the Earth at latitude θ moves along a circle of radius $R_E \cos\theta$ as the Earth rotates.

$$a = \frac{v^2}{r} = \left(\frac{2\pi r}{T}\right)^2 \frac{1}{r} = 4\pi^2 \frac{R_E \cos\theta}{T^2} \tag{1}$$

At the equator ($\cos\theta = 1$), this gives the result $a = 0.034$ m/s^2; at other latitudes, this result is reduced by a factor of $\cos\theta$. For most practical purposes, we can neglect this small centripetal acceleration and we can regard a reference frame attached to the ground as inertial to a good approximation. What contributes to the success of this approximation is that, to some extent, the effects of the noninertial acceleration of the reference frame can be hidden among the effects of gravity. For example, at the equator (say, in Quito, Ecuador), the centripetal acceleration of the reference frame gives a free body a relative centrifugal acceleration of 0.034 m/s^2 upward; simultaneously, gravity gives it an acceleration of 9.814 m/s^2 downward; the result is a net acceleration of 9.780 m/s^2 downward. Tables of values of g (e.g., Table 2.3) always list this net acceleration, i.e., these tables already take into account the acceleration of the reference frame. When using such tables one can therefore pretend that the reference frame attached to the ground is inertial. However, there are limits to this pretense — the Foucault experiment shows that not all the effects of the acceleration of the reference frame can be hidden in the value of g. Incidentally, most of the variation of g with geographical latitude (see Table 2.3) arises from the variation of the centripetal acceleration with latitude [see Eq. (1)]. Some extra variation arises from the ellipsoidal shape of the Earth, which makes the pull of gravity somewhat stronger at the poles than at the equator.

Hereafter, unless otherwise stated, we will take it for granted that the reference frames in which we express the laws of physics are inertial reference frames, either exactly inertial or at least so nearly inertial that no appreciable deviation from Newton's First Law occurs within the region of space and time on which we wish to concentrate our attention.

Ernst Mach (makh), *1838–1916, Austrian philosopher and physicist, professor at Prague and at Vienna. He performed a profound analysis of the logical foundations of physics. His book* The Science of Mechanics *is a brilliant critical examination of the historical development of Newtonian mechanics.*

5.2 Newton's Second Law

Newton's Second Law establishes the relationship between force, acceleration, and mass. Before we can deal with this law, we need a precise definition of mass. The standard of mass (standard kilogram) has already been described in Section 1.5. It remains to set up a procedure for comparing an unknown mass with this standard. The common procedure makes use of a balance (see Figure 1.15). But this is not quite satisfactory — a balance will not function in interstellar space, far from the pull of gravity of the Earth or any other celestial body. It is therefore desirable to contrive a measurement procedure that can be used anywhere in the universe. This is a question of principle, but it also is a practical question. During the Skylab mission, three astronauts were kept for about 2 months under zero-gravity conditions in a capsule orbiting the Earth, and scientists who wanted a daily record of the astronauts' masses had to invent some measurement device not subject to the limitations of an ordinary balance.

A perfectly general procedure, originally proposed by Ernst Mach, for the measurement of an unknown mass is the following. Take the body of unknown mass and take the standard of mass and let them interact, i.e., let them exert forces on each other. The mechanism involved is irrelevant. For example, the bodies may exert forces on each other by means of springs, rubber bands, strings, etc. The only restriction is that the device used to communicate the forces from one body to the other must itself have a mass that is negligible compared to both the unknown mass and the standard mass. Under the influence of their mutual push or pull, the body of unknown mass and the standard of mass will accelerate away from or toward one another. Designate the magnitudes of the accelerations of these bodies (in an inertial reference frame) by a and a_s, respectively, and denote the values of their masses by m and m_s, respectively. The unknown mass is then defined by the relation

$$\boxed{\frac{m}{m_s} = \frac{a_s}{a}} \tag{2}$$

Definition of mass

According to this relation, the ratio of the masses is the inverse ratio of the accelerations, i.e., the unknown mass is large if its acceleration is small. This is of course quite reasonable — a large mass has large inertia and is therefore hard to accelerate. The precise definition given by Eq. (2) expresses the intuitive notion that mass is a measure of the resistance that a body offers to changes in its velocity.

Since $m_s = 1$ kg, we can also put Eq. (2) in the form

$$m = \frac{a_s}{a} \times 1 \text{ kg} \tag{3}$$

EXAMPLE 1. An unknown mass is connected to the standard mass by a (nearly) massless spring. The pull of the spring gives the masses (instantaneous) accelerations $a = 0.5$ m/s^2 and $a_s = 2.50$ m/s^2 in opposite directions. What is the value of the unknown mass?

SOLUTION: By Eq. (3)

$$m = \frac{a_s}{a} \times 1 \text{ kg} = \frac{2.50 \text{ m/s}^2}{0.50 \text{ m/s}^2} \times 1 \text{ kg} = 5.0 \text{ kg}$$

Now that we have given a precise definition of mass, we can state Newton's Second Law, which tells us what acceleration a force produces when acting on a body of a certain mass. Qualitatively, a force is any push or pull exerted on the body. It is intuitively obvious that such a push or pull has a direction as well as a magnitude — in fact, force is a vector quantity. The precise quantitative definition of force is included in the Second Law and we will discuss this definition below. For the sake of simplicity, we assume that only one force is acting on the body, but we will remove this restriction in the next section. **Newton's Second Law** says:

> *A force acting on a body gives it an acceleration which is in the direction of the force and has a magnitude inversely proportional to the mass of the body,*

Newton's Second Law

$$\mathbf{a} = \mathbf{F}/m \tag{4}$$

or

$$\boxed{m\mathbf{a} = \mathbf{F}} \tag{5}$$

The Second Law is subject to the same restrictions as the First Law: it is valid only in inertial reference frames. Furthermore, the Second Law, just like the First Law, plays a dual role: it is a law of nature and it also serves as a precise definition of force. To measure a given force — say, the force generated by a spring that has been stretched a certain amount — one applies this force to the standard mass. If the resulting acceleration of the standard mass is a_s, then the force has a magnitude

$$F = m_s a_s = 1\ \text{kg} \times a_s \tag{6}$$

After the standard mass has been used to measure the force, any other masses to which this same force is applied will be found to obey the Second Law. In regard to these other masses, the Second Law is a law of nature — it is an assertion about the physical world that can be verified by experiments.

Units of force

The metric unit of force is the **newton** (1 N); this is the force that will give the standard mass an acceleration of 1 m/s^2,

$$1\ \text{newton} = 1\ \text{N} = 1\ \text{kg} \cdot \text{m/s}^2 \tag{7}$$

Table 5.1 lists the magnitudes of some typical forces.

In the European countries, the newton is the only officially recognized unit of force. However, in practical usage an alternative unit of force has gained wide acceptance: the **kilopond** (kp), where

$$1\ \text{kilopond} = 1\ \text{kp} = 9.80665\ \text{N} \tag{8}$$

The conversion factor from newtons to kiloponds has the same numerical value as the "standard" acceleration of gravity, 1 gee = 9.80665 m/s^2; the reason for this coincidence is that the weight of 1 kg is 1 kp, i.e., the downward gravitational pull that the Earth exerts on 1 kg is 1 kp (the precise meaning of weight will be further discussed in Section 6.2).

The British unit of force is the **pound** (lb); this unit can be defined in terms of newtons:

$$1\ \text{pound} = 1\ \text{lb} = 4.44822\ \text{N} \tag{9}$$

The British unit of mass is derived from the unit of force. Thus, in the British system, the units of length, time, and force are regarded as

Table 5.1 SOME FORCES

Gravitational pull of Sun on Earth	3.5×10^{22} N
Gravitational pull of Earth on Moon	2.0×10^{20} N
Thrust of Saturn V rocket engines	3.3×10^{7} N
Pull of large tugboat	1×10^{6} N
Thrust of jet engines (Boeing 747)	7.7×10^{5} N
Pull of large locomotive	5×10^{5} N
Decelerating force on automobile during braking	1×10^{4} N
Force between two protons in a nucleus	$\sim 10^{4}$ N
Accelerating force on automobile	7×10^{3} N
Gravitational pull of Earth on man	7.3×10^{2} N
Maximum upward force exerted by forearm (isometric)	2.7×10^{2} N
Gravitational pull of Earth on 5¢ coin	5.1×10^{-2} N
Force between electron and nucleus of atom (hydrogen)	8×10^{-8} N

primary and all other units are regarded as derived. The unit of mass is called a **slug;** this is the mass that will acquire an acceleration of 1 ft/s^2 when subjected to a force of 1 lb,

$$1 \text{ slug} = \frac{1 \text{ lb}}{1 \text{ ft/s}^2} = 1 \text{ lb} \cdot \text{s}^2/\text{ft}$$

In terms of kilograms,

$$\begin{aligned} 1 \text{ slug} &= 1 \text{ lb} \cdot \text{s}^2/\text{ft} \times 4.44822 \text{ N/lb} \times 1 \text{ ft}/0.3048 \text{ m} \\ &= 14.5939 \text{ N} \cdot \text{s}^2/\text{m} \\ &= 14.5939 \text{ kg} \end{aligned}$$

Although the pound is strictly a unit of force, in commercial and industrial usage it is regarded as a unit of mass:

$$1 \text{ lb-mass} = 0.453592 \text{ kg} = 1/32.1740 \text{ slug} \qquad (10)$$

Note that the conversion factor from pound-mass to slug has the same numerical value as the "standard" acceleration of gravity, 1 gee = 32.1740 ft/s^2; this is so because the weight of 1 lb-mass is 1 lb (see Section 6.2).

EXAMPLE 2. The jet-engined car *Spirit of America* (Figure 5.4), which set a speed record on the Salt Flats of Utah, had a mass of 9000 lb-mass and its engine could develop up to 15,000 lb of thrust. What acceleration could this car achieve?

SOLUTION: Expressed in slugs, the mass of the car is

$$\text{m} = 9000 \text{ lb-mass} \times \frac{1}{32.2} \frac{\text{slug}}{\text{lb-mass}} = 280 \text{ slugs}$$

If the force on this mass is 15,000 lb, Newton's Second Law tells us that the acceleration is

$$a = \frac{F}{m} = \frac{15{,}000 \text{ lb}}{280 \text{ slugs}} = 53.6 \text{ ft/s}^2$$

which is 1.67 gee.

Fig. 5.4 *Spirit of America.*

This brings us to the question of the practical measurement of mass and force. In laboratories on the Earth, the most common and most precise mass measurements are carried out with beam balances that compare the weights of the masses (Figure 5.5). The precise meaning of "weight" will be given in Section 6.2; as we will see, measurements of mass via weight give results compatible with those obtained by the primary procedure based on Eq. (3). Likewise, measurements of force are often carried out by matching the unknown force with a known weight. Alternatively, forces can be measured with a spring balance (Figure 5.6) by matching the unknown force with a known force supplied by a stretched spring; the spring can be calibrated by hanging known weights on it and checking how far it stretches.

For the measurement of the masses of the astronauts during the Skylab mission, neither beam balances nor conventional spring bal-

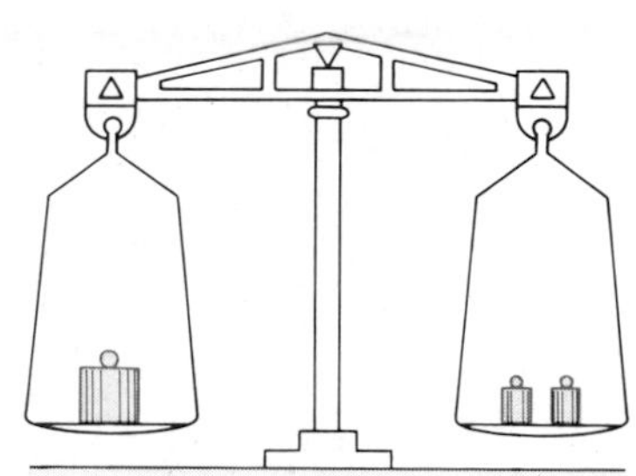

Fig. 5.5 Beam balance.

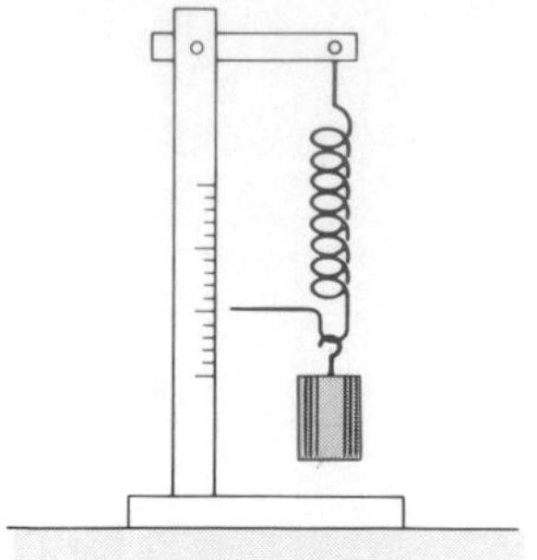

Fig. 5.6 Spring balance.

ances would have been of any use. The design of the device used aboard Skylab was based on Eq. (5). In essence, the device applied a known force to the astronaut's body and measured the resulting acceleration.[3] Equation (5) then permitted the calculation of the mass of the astronaut. The known force was supplied by a calibrated torsional spring attached to a small chair into which the astronaut was strapped (Figure 5.7).

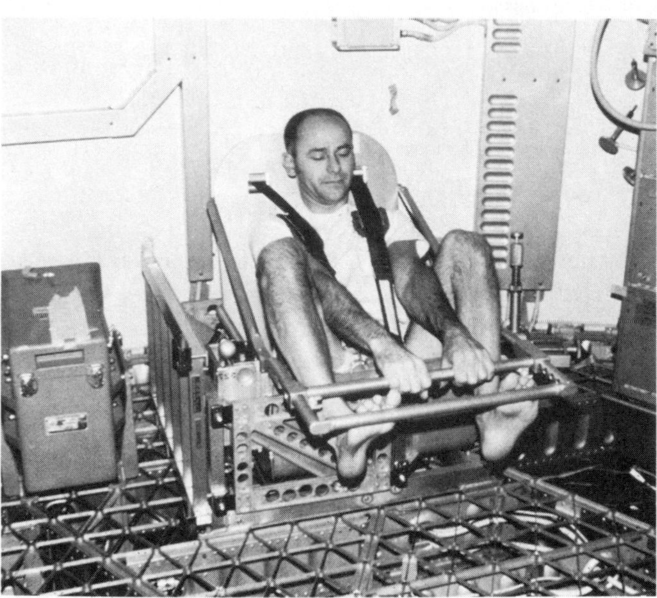

Fig. 5.7 Body-mass measurement device on Skylab.

EXAMPLE 3. The chair of the mass measurement device aboard Skylab had a mass of 15.3 kg. Suppose that with astronaut J. R. Lousma sitting in this chair, a force of 2.067 N applied to the chair produced an acceleration of 2.040×10^{-2} m/s^2. What was the mass of Lousma?

SOLUTION: If we designate the total mass of Lousma plus the chair by m, then Newton's Second Law yields

$$m = \frac{F}{a} = \frac{2.067 \text{ N}}{2.040 \times 10^{-2} \text{ m/s}^2} = 101.3 \text{ kg}$$

Of this total mass, 15.3 kg belong to the chair. Hence, 101.3 kg − 15.3 kg = 86.0 kg must belong to Lousma.

The masses of electrons and protons and the masses of ions (atoms with missing electrons or with excess electrons) are also measured with a device based on Eq. (5). In this device, called a mass spectrometer, a steady succession of particles are shot into an evacuated vessel where, in the absence of forces, they would form a straight beam. The mass spectrometer then applies known electric and magnetic forces to the particles, causing a deflection of the beam. The measured value of the deflection indicates the acceleration, and Eq. (5) then permits the calculation of the masses of the particles. Table 5.2 lists the masses of the electron, the proton, and the neutron (the latter particle does not re-

[3] In practice, the device actually measured the period of back-and-forth oscillations of the chair; this amounts to an indirect measurement of the acceleration.

spond to electric and magnetic forces, and hence its mass cannot be determined directly by the mass-spectrometer method; but its mass can be deduced from the masses of ions containing known numbers of neutrons in their atomic nuclei).

Table 5.2 THE MASSES OF THE ELECTRON, PROTON, AND NEUTRON

Masses of electron, proton, and neutron

Particle	*Mass*[a]
Electron	9.110×10^{-31} kg
Proton	1.673×10^{-27}
Neutron	1.675×10^{-27}

[a] More precise values of these masses will be found in Appendix 8.

5.3 The Superposition of Forces

More often than not, a body will be subjected to the simultaneous action of several forces. For example, Figure 5.8 shows a barge under tow by two tugboats. The forces acting on the barge are the pull of the first tow rope, the pull of the second tow rope, and the frictional resistance of the water[4]; these forces are indicated by the arrows in Figure 5.8. Newton's Second Law tells us what each of these forces would do

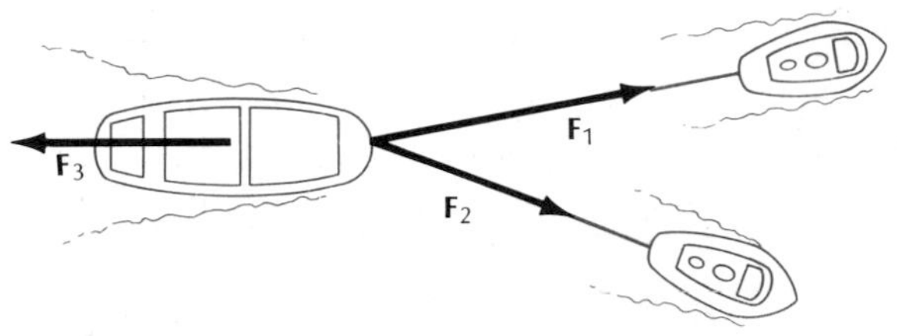

Fig. 5.8 A barge under tow by two tugboats.

if acting by itself. The question now is: How can we calculate the simultaneous effect of several forces? The answer is supplied by the **principle of superposition** for forces:

If several forces $\mathbf{F}_1$, $\mathbf{F}_2$, $\mathbf{F}_3$, . . . , *act simultaneously on a body, then the acceleration is the same as that produced by a single force given by*

$$\boxed{\mathbf{F}_{\text{net}} = \mathbf{F}_1 + \mathbf{F}_2 + \mathbf{F}_3 + \cdots} \qquad (11)$$

Principle of superposition

The single force that has the same effect as the combination of the individual forces is called the **net force** or the **resultant force.** According to Eq. (11), the net force is merely the vector sum of the individual forces. In terms of the net force, Newton's Second Law becomes

Net force

$$m\mathbf{a} = \mathbf{F}_{\text{net}} \qquad (12)$$

[4] These are the horizontal forces. There are also vertical forces: the downward pull of gravity (weight) and the upward pressure of the water (buoyancy); these vertical forces will be ignored (they cancel each other).

Equation (11) can be interpreted as follows: each force ($\mathbf{F}_1$, $\mathbf{F}_2$, $\mathbf{F}_3, \ldots$) acting by itself produces its own acceleration ($\mathbf{a}_1 = \mathbf{F}_1/m$, $\mathbf{a}_2 = \mathbf{F}_2/m$, $\mathbf{a}_3 = \mathbf{F}_3/m, \ldots$); all the forces acting together produce a net acceleration **a**, which is simply the sum of the individual accelerations, that is,

$$\mathbf{a} = \mathbf{a}_1 + \mathbf{a}_2 + \mathbf{a}_3 + \cdots \tag{13}$$

$$\mathbf{a} = \frac{1}{m}(\mathbf{F}_1 + \mathbf{F}_2 + \mathbf{F}_3 + \cdots) \tag{14}$$

Equation (14) is equivalent to Eq. (12). Thus, the principle of superposition of forces is equivalent to the assertion that each force produces an acceleration independently of the presence or absence of other forces.

We must emphasize that this principle is a law of nature which has the same status as Newton's Laws. The most precise empirical test of this principle emerges from the study of planetary motion. There, one finds that the net force on a planet is indeed the sum of all the gravitational pulls exerted by the Sun and by the other planets. Somewhat less precise tests of this principle can be performed in a laboratory experiment by pulling on a body with known forces in known directions.

EXAMPLE 4. Suppose that the two towropes in Figure 5.8 pull with forces of 2×10^5 and 1.5×10^5 lb, and that these forces make angles of 10° and 20° with the long axis of the barge (Figure 5.9). Suppose that the friction force is zero. What is the net force on the barge?

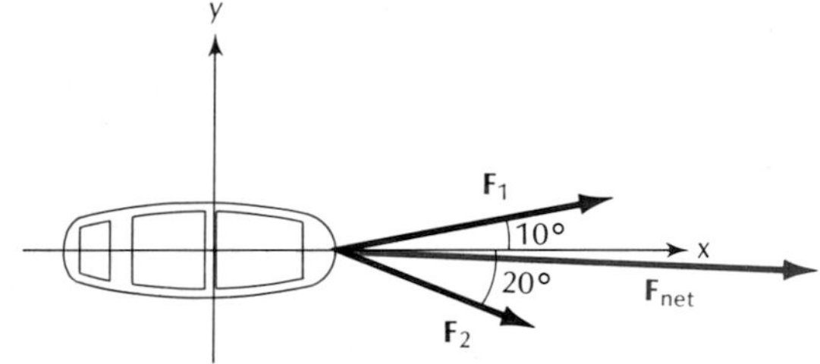

Fig. 5.9 The forces $\mathbf{F}_1$ and $\mathbf{F}_2$ pull on the barge. The friction force is assumed to be zero. The net force is the vector sum of $\mathbf{F}_1$ and $\mathbf{F}_2$.

SOLUTION: The net force is

$$\mathbf{F}_{\text{net}} = \mathbf{F}_1 + \mathbf{F}_2$$

where $\mathbf{F}_1$ is the force of the first towrope and $\mathbf{F}_2$ that of the second. With the x and y axes as shown in Figure 5.9, the net force has components

$$F_{\text{net},x} = 2.0 \times 10^5 \text{ lb} \times \cos 10° + 1.5 \times 10^5 \text{ lb} \times \cos 20°$$

$$= 3.38 \times 10^5 \text{ lb}$$

$$F_{\text{net},y} = 2.0 \times 10^5 \text{ lb} \times \sin 10° - 1.5 \times 10^5 \text{ lb} \times \sin 20°$$

$$= -1.66 \times 10^4 \text{ lb}$$

This net force is shown in Figure 5.9.

5.4 Newton's Third Law

Consider a tugboat pushing on a barge (Figure 5.10a) with a force of, say, 2×10^5 lb. This force of 2×10^5 lb describes the action of the tugboat on the barge. However, there is also a reciprocal action (or reaction) of the barge on the tugboat; the presence of the barge slows down the motion of the tugboat — the barge pushes back on the tugboat. Thus the mutual interaction of the tugboat and the barge involves two forces: the "action" force of the tugboat on the barge and the "reaction" force of the barge on the tugboat. These forces are said to form an **action–reaction pair.** Which of the forces is regarded as "action" and which as "reaction" is irrelevant. It may seem reasonable to regard the push of the tugboat as an action; then the push of the barge back on the tugboat is a reaction. However, it is equally valid to regard the push the barge exerts on the tugboat as an action and then the push of the tugboat is a reaction. The important point is that forces always occur in pairs; each of them cannot exist without the other.

Action and reaction

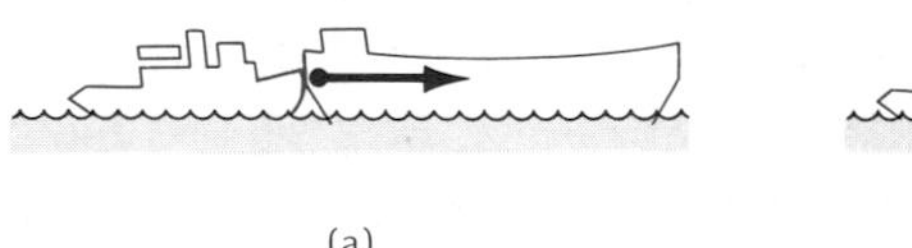

(a)

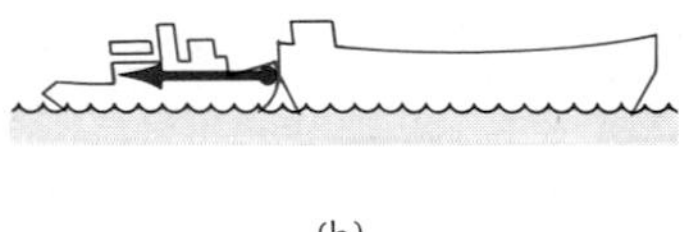

(b)

Fig. 5.10 (a) Tugboat pushes on barge (action). (b) Barge pushes on tugboat (reaction).

Newton's Third Law gives the quantitative relationship between the action force and the corresponding reaction force:

> *Whenever a body exerts a force on another body, the latter exerts a force of equal magnitude and opposite direction on the former.*

Newton's Third Law

To return to the example given above, the tugboat exerts a force of 2×10^5 lb on the barge; hence by the Third Law the barge exerts an opposite force of 2×10^5 on the tugboat (Figure 5.10b). Note that although the forces are of equal magnitude, they act on different bodies and their effects are quite different: the first force gives an acceleration to the barge (if there is no other force acting on the barge), whereas the second force merely slows the tugboat and prevents it from accelerating as much as it would if the barge were not there. Thus, although action and reaction are forces of equal magnitude and of opposite direction, their effects do not cancel because *they are applied to different bodies.*

Incidentally, the force that propels the tugboat forward is another example of a reaction force: the propeller of the tugboat presses against the water (action) and the water consequently presses on the propeller (reaction) and pushes the tugboat along. Figure 5.11 shows the two forces[5] exerted on the tugboat; the reaction of the barge is in the backward direction and the reaction of the water is in the forward direction.

Fig. 5.11 Forces acting on tugboat.

[5] Only the horizontal forces are shown. There are also vertical forces (weight and buoyant force), but they cancel each other.

Fig. 5.12 Automobile pushes on ground; ground pushes on automobile.

Reaction forces play a crucial role in all machines that produce locomotion by pushing against the ground, water, or air. For example, an automobile moves by pushing backward on the ground with its wheels; the reaction of the ground then pushes the automobile forward (Figure 5.12). A man walks by pushing backward on the ground; the reaction of the ground then pushes the man forward (Figure 5.13), etc.

Note that reaction forces exist even if the two interacting bodies are not in direct contact, so that the forces between them must bridge the intervening empty space. For example, consider an apple in free fall at some height above the ground. The Earth pulls on the apple by means of gravity. If this pull has a magnitude of, say, 2.5 N, then the Third Law requires that the apple pull on the Earth with an opposite force of 2.5 N (Figure 5.14). This reaction force is also a form of gravity — it is the gravity that the apple exerts on the Earth. However, the effect of the apple on the motion of the Earth is insignificant because the mass of the Earth is so large (about 6×10^{24} kg) that a force of only 2.5 N can hardly move it.

Fig. 5.13 Man pushes on ground; ground pushes on man.

Fig. 5.14 Earth pulls on apple; apple pulls on Earth.

EXAMPLE 5. A tugboat tows an empty barge of mass 25,000 kg by means of a strong steel cable of mass 200 kg (Figure 5.15). If the tugboat exerts a pull of 3000 N on the cable, what is the acceleration of the barge? What is the tension in the cable at its forward end? At its rearward end? At its midpoint? Assume that the cable is horizontal and does not sag, and ignore friction on the barge.

Fig. 5.15 A tugboat tows a barge by means of a steel cable.

SOLUTION: The force of 3000 N must accelerate both the cable and the barge; i.e., it must accelerate a total mass of $m_{\text{barge}} + m_{\text{cable}}$. Hence the resulting acceleration is

$$a = \frac{F}{m_{\text{barge}} + m_{\text{cable}}} = \frac{3000\ \text{N}}{25{,}000\ \text{kg} + 200\ \text{kg}} = 0.119\ \text{m/s}^2$$

By **tension** at the end of the cable is meant the force with which the cable pulls on what is attached to it. Since the tugboat pulls on the cable with a force of 3000 N, Newton's Third Law requires that the cable pull on the tugboat with an equally large force, i.e., the tension at the forward end is

$$T_1 = 3000 \text{ N}$$

To find the tension at the rearward end, we note that in order to accelerate the barge at the rate of 0.119 m/s², the cable must exert a pull of $m_{\text{barge}}a$. Hence the tension at the rearward end is

$$T_2 = m_{\text{barge}}a = 25{,}000 \text{ kg} \times 0.119 \text{ m/s}^2 = 2976 \text{ N}$$

By tension at the midpoint of the cable is meant the force with which the forward half of the cable pulls on the rearward half (or vice versa; Figure 5.16). To find this force, we note that it must accelerate both the rearward half of the cable and the barge, i.e., it must accelerate a total mass of

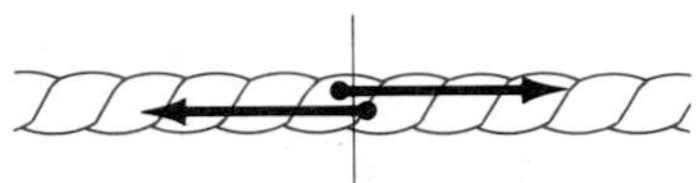

Fig. 5.16 Forward portion of cable pulls on rearward portion, and vice versa.

$$m_{\text{barge}} + \tfrac{1}{2}m_{\text{cable}}$$

Hence the required tension is

$$T_3 = (m_{\text{barge}} + \tfrac{1}{2}m_{\text{cable}})a$$

$$= (25{,}000 \text{ kg} + 100 \text{ kg}) \times 0.119 \text{ m/s}^2 = 2988 \text{ N}$$

From this calculation it is evident that the tension in the cable steadily decreases along its length from a value of 3000 N at the forward end to a value of 2976 N at the rearward end. The difference between the tensions at the ends is 24 N; this force is of course exactly what is required to accelerate the cable at the rate of 0.119 m/s².

In the preceding example, the difference of 24 N between the tensions at the two ends of the cable is only a small fraction of the tension — the difference is about 1% of the tension. For practical purposes one can usually pretend that a freely hanging cable, rope, string, or chain has the same tension everywhere along its entire length, i.e., the cable transmits the tension without change. This assumption is justified whenever the mass of the cable is negligible compared to the mass pulled by the cable. In subsequent problems we will always neglect the mass of the cable unless it is explicitly stated otherwise.

Incidentally, note that the definition of mass by means of Eq. (2) hinges on Newton's Third Law and on the assumption that the device that transmits the force from the unknown mass to the standard mass is itself nearly massless. If so, the device will transmit a push or a pull from the unknown mass to the standard mass without change — the two bodies will experience forces of equal magnitude and their accelerations will be inversely proportional to their masses.

EXAMPLE 6. In order to pull an automobile out of the mud in which it is stuck, a man stretches a rope tautly from the front end of the automobile to a stout tree. He then pushes sideways against the rope at the midpoint (Figure 5.17). When he pushes with a force of 900 N, the angle between the two

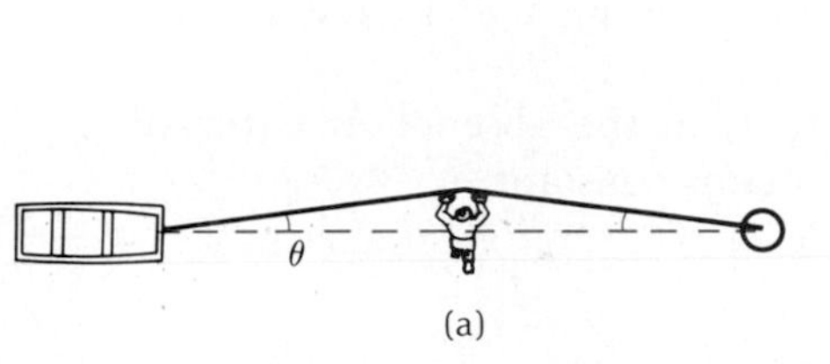

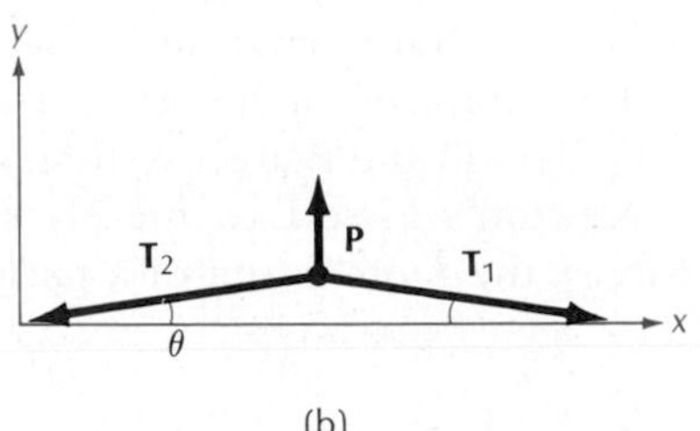

Fig. 5.17 (a) The rope is stretched between the automobile and the tree. The man pushes at the midpoint.
(b) Forces acting on the point of the rope where the man pushes. The angle θ is 5°.

halves of the rope on his right and left is 170°. What is the tension in the rope under these conditions?

SOLUTION: Figure 5.17b shows the three forces acting on the rope at the point where the man pushes. These forces are the push **P** and the tensions $\mathbf{T}_1$ and $\mathbf{T}_2$ toward the right and left; the magnitudes of these tensions are equal, $|\mathbf{T}_1| = |\mathbf{T}_2| = T$. Since this point of the rope has no acceleration, the three forces must cancel. With the x and y axes as shown in Figure 5.17b, the components of the forces are

$$P_x = 0 \qquad P_y = P$$

$$T_{1,x} = T\cos\theta \qquad T_{1,y} = -T\sin\theta$$

$$T_{2,x} = -T\cos\theta \qquad T_{2,y} = -T\sin\theta$$

The net force has an x component

$$F_{\text{net},x} = 0 + T\cos\theta - T\cos\theta$$

and a y component

$$F_{\text{net},y} = P - T\sin\theta - T\sin\theta$$

The x component is identically zero, regardless of the value of T. To make the y component zero, we must satisfy the equation

$$P - 2T\sin\theta = 0$$

This tells us that the tension is

$$T = \frac{P}{2\sin\theta} = \frac{900\text{ N}}{2\sin 5°} = 5.2 \times 10^3\text{ N}$$

Thus, with this rope trick the push of 900 N generates a tension that is almost six times as large. Of course, once the automobile moves forward, the angle θ will increase and the tension will decrease. To take full advantage of the rope trick the man must then shorten the rope before pushing again.

5.5 The Momentum of a Particle

Newton's laws can be expressed very neatly in terms of momentum, a new quantity of great importance in physics. The **momentum** of a particle is defined as the product of its mass and velocity:

Momentum

$$\boxed{\mathbf{p} = m\mathbf{v}} \qquad (15)$$

The momentum $\mathbf{p}$ is a vector that has the same direction as the velocity vector, but a magnitude that is m times as large.

The units of momentum are kg · m/s in the metric system and slug · ft/s in the British system.

Newton's First Law merely states that, in the absence of external forces, the momentum of a particle remains constant,

$$\boxed{\mathbf{p} = [\text{constant}]} \qquad (16)$$

Newton's First Law in terms of momentum

It is customary to say that a quantity that remains constant during motion is **conserved.** Thus Eq. (16) states that the momentum of a free particle is conserved. Of course, we could equally well say that the velocity of a free particle is conserved; but the great importance of momentum will emerge when we study the motion of a system of many particles exerting forces on one another (see Chapter 9). We will find that the total momentum of such a system is conserved — any momentum lost by one particle is compensated by a momentum gain of some other particle or particles. Conserved quantities are useful in physics because they permit us to make some predictions about the motion of a particle or a system of particles without any need for a tedious calculation of the full details of the motion. In later sections we will become acquainted with other quantities that are conserved under suitable conditions, such as angular momentum and energy.

To express the Second Law in terms of momentum, we make use of the fact that the mass of the particle is constant, and we transform $m\mathbf{a}$ into the following:

$$m\mathbf{a} = m\frac{d\mathbf{v}}{dt} = \frac{d}{dt}(m\mathbf{v}) = \frac{d}{dt}(\mathbf{p})$$

The Second Law can then be written

$$\boxed{\frac{d\mathbf{p}}{dt} = \mathbf{F}} \qquad (17)$$

Newton's Second Law in terms of momentum

This says that the rate of change of momentum equals the force.

Finally, note that we can also express Newton's Third Law in terms of momentum. Since the action force is exactly opposite to the reaction force, the rate of change of momentum generated by the action force on one body is exactly opposite to the rate of change of momentum generated by the reaction force on the other body. Hence whenever two bodies exert forces on one another, the resulting changes of momentum are equal and opposite. This fact will lead us to a general law of conservation of momentum (see Section 9.1).

EXAMPLE 7. A tennis player smashes a ball of mass 60 g at a vertical wall. The ball hits the wall at right angles with a speed of 40 m/s and bounces straight back with the same speed. What is the change of the momentum of the ball during the impact?

SOLUTION: Take the positive x axis along the direction of the initial motion of the ball. The momentum of the ball before impact is then in the positive direction:

$$p_x = mv_x = 0.060 \text{ kg} \times 40 \text{ m/s} = 2.4 \text{ kg} \cdot \text{m/s}$$

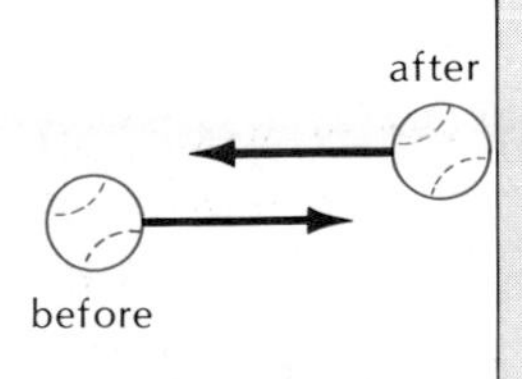

Fig. 5.18 A tennis ball bounces off a wall.

The momentum of the ball after impact has the same magnitude but opposite direction (Figure 5.18):

$$p'_x = -2.4 \text{ kg} \cdot \text{m/s}$$

The change of momentum is

$$\Delta p_x = p'_x - p_x = -2.4 \text{ kg} \cdot \text{m/s} - 2.4 \text{ kg} \cdot \text{m/s}$$
$$= -4.8 \text{ kg} \cdot \text{m/s}$$

5.6 The Angular Momentum of a Particle[6]

For a particle moving under the influence of some given force, the momentum is not a conserved quantity. However, if the force satisfies some special conditions, then there exists a conserved quantity closely related to the momentum. This conserved quantity is the **angular momentum,** defined as the cross product of the position vector and the momentum,

Angular momentum of a particle

$$\boxed{\mathbf{L} = \mathbf{r} \times \mathbf{p}} \tag{18}$$

We recall that according to the definition of the cross product given in Section 3.4, the magnitude of the vector $\mathbf{r} \times \mathbf{p}$ is

$$L = rp \sin\theta$$

where θ is the angle between **r** and **p** (Figure 5.19). The vector $\mathbf{r} \times \mathbf{p}$ is in the direction specified by the right-hand rule.

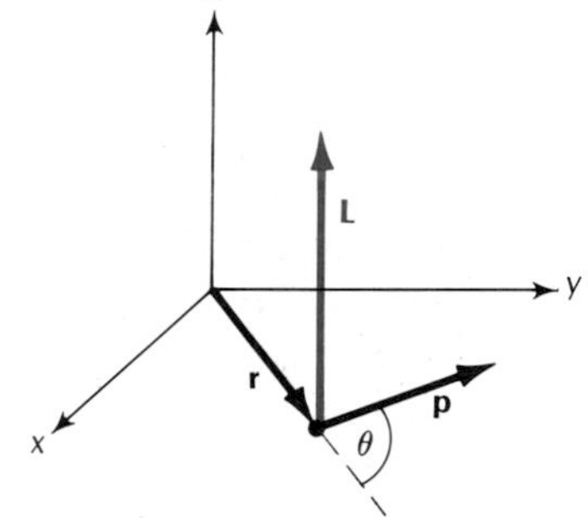

Fig. 5.19 Position, momentum, and angular momentum vectors of a particle. The vector **L** is perpendicular to both **r** and **p.**

The unit of angular momentum is the product of the units of length and momentum. In the metric system the unit of angular momentum is therefore m · kg · m/s, or kg · m²/s. In the British system the unit is slug · ft²/s.

The angular momentum of a particle moving with constant velocity in the absence of forces is constant. Figure 5.20 makes this clear. It shows the line of motion of the particle and the origin of coordinates. Obviously, the direction of **L** is constant (out of the plane of Figure 5.20). The magnitude of **L** is $rp \sin\theta$. In this expression the quantity p is constant because the motion proceeds at uniform velocity; and the quantity $r \sin\theta$ is also constant because it merely represents the short-

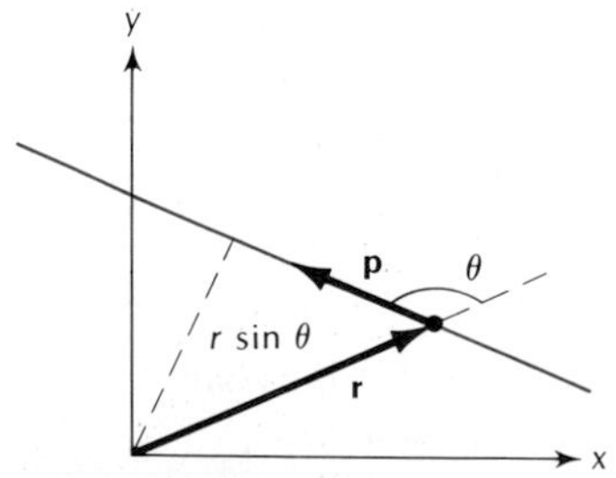

Fig. 5.20 The distance between the origin and the line of motion is $r \sin\theta$.

[6] This section can be postponed until Section 11.5 is reached.

est distance between the origin and the line of motion, and is therefore not affected by the motion along the line. Hence both the magnitude and the direction of **L** remain constant.

This conservation of angular momentum of a free particle is of no great interest — it tells us nothing new about the motion of the particle. For a somewhat more interesting case of conservation of angular momentum, consider a particle in uniform circular motion, such as a stone being whirled along a circle at the end of a string, or the Earth moving along its circular orbit around the Sun. Figure 5.21 shows the origin at the center of the circle; since the position vector is always perpendicular to the velocity vector, the magnitude of the angular-momentum vector is

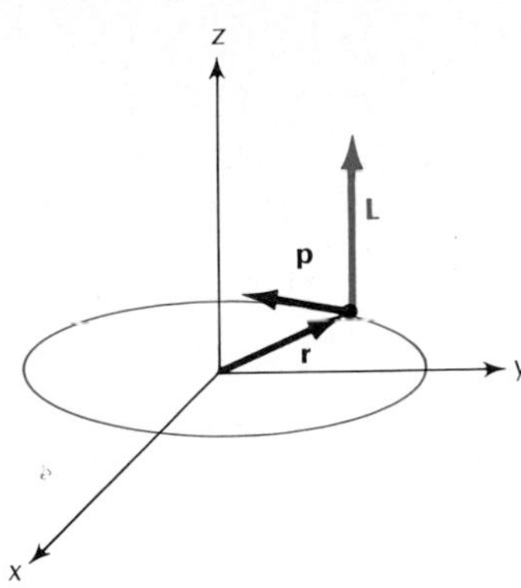

Fig. 5.21 A particle in uniform circular motion. The vector **L** is perpendicular to the plane of the circle.

$$\boxed{L = rp = rmv} \tag{19}$$

The direction of the angular-momentum vector is perpendicular to the plane of the circle (Figure 5.21). As the particle moves around the circle, **L** remains constant in magnitude and direction.

EXAMPLE 8. The Earth moves around the Sun with (approximately) uniform circular motion. If the origin of coordinates is placed on the Sun, what is the angular momentum of the Earth?

SOLUTION: The radius of the circle is 1.5×10^{11} m and the speed of the Earth is 3.0×10^4 m/s (see Example 2.2); the mass of the Earth is 6.0×10^{24} kg. Hence

$$L = rmv = 1.5 \times 10^{11}\ \text{m} \times 6.0 \times 10^{24}\ \text{kg} \times 3.0 \times 10^4\ \text{m/s}$$

$$= 2.7 \times 10^{40}\ \text{kg} \cdot \text{m}^2/\text{s}$$

We must remember that the angular momentum in Eqs. (18) and (19) is reckoned relative to a specific origin of coordinates. Obviously, any shift of the origin of coordinates alters the position vector **r** and hence the angular momentum. It is therefore meaningless to speak of the angular momentum without first specifying the origin of the coordinates. Thus, the angular momentum of a particle in uniform circular motion is constant if the origin of coordinates is placed at the center of the circle, as in Figure 5.21, but not if the origin is placed elsewhere.

Now that we have looked at some special cases of conservation of angular momentum, let us inquire into the general conditions that must be imposed on the force to guarantee the conservation of angular momentum. The answer to this question is contained in the following theorem:

Angular momentum and central forces

The angular momentum of a particle is conserved whenever the force on the particle is a central force, i.e., a force that points toward or away from a fixed central point.

The pull of the string on a stone being whirled along a circle and the gravitational pull of the Sun on the Earth moving along its orbit are examples of such central forces.

To establish the theorem, we begin with a calculation of the rate of change of the angular momentum. For this, we differentiate Eq. (18) with respect to time:

$$\frac{d\mathbf{L}}{dt} = \frac{d}{dt}(\mathbf{r} \times \mathbf{p}) \tag{20}$$

We can evaluate the derivative of the vector product on the right side of this equation by the usual rules for differentiating a product, but we must take care not to change the order of the factors. Hence

$$\frac{d\mathbf{L}}{dt} = \frac{d\mathbf{r}}{dt} \times \mathbf{p} + \mathbf{r} \times \frac{d\mathbf{p}}{dt} \tag{21}$$

The first term on the right side is

$$\frac{d\mathbf{r}}{dt} \times \mathbf{p} = \mathbf{v} \times (m\mathbf{v}) = m(\mathbf{v} \times \mathbf{v})$$

This is zero, since the cross product of a vector with itself is always zero. According to Eq. (17), the second term on the right side of Eq. (21) is

$$\mathbf{r} \times \frac{d\mathbf{p}}{dt} = \mathbf{r} \times \mathbf{F}$$

Therefore Eq. (21) becomes

Rate of change of angular momentum

$$\boxed{\frac{d\mathbf{L}}{dt} = \mathbf{r} \times \mathbf{F}} \tag{22}$$

So if we want $d\mathbf{L}/dt = 0$, we need $\mathbf{r} \times \mathbf{F} = 0$. Thus, **F** must be parallel to **r,** that is, **F** must point toward or away from the origin. This establishes our theorem.

The statement

$$\frac{d\mathbf{L}}{dt} = 0$$

or

Conservation of angular momentum

$$\mathbf{L} = [\text{constant}] \tag{23}$$

is called the **law of conservation of angular momentum.** Although this law is of limited validity, it is of great importance in physics because many forces are central. Thus, Eq. (23) is valid for planets, satellites, and comets moving under the influence of central gravitational forces (see Chapter 13), and it is valid for electrons moving under the influence of electric central forces around the nucleus of an atom (see Chapter 22). The law is especially useful in calculations involving the elliptical orbits of such bodies.

EXAMPLE 9. Vanguard I, the second American artificial satellite, moved in an elliptical orbit around the Earth with a perigee distance of 7.02×10^6 m and an apogee distance of 10.3×10^6 m (Figure 5.22). At perigee, the speed of this satellite was 8.22×10^3 m/s. What was the speed at apogee?

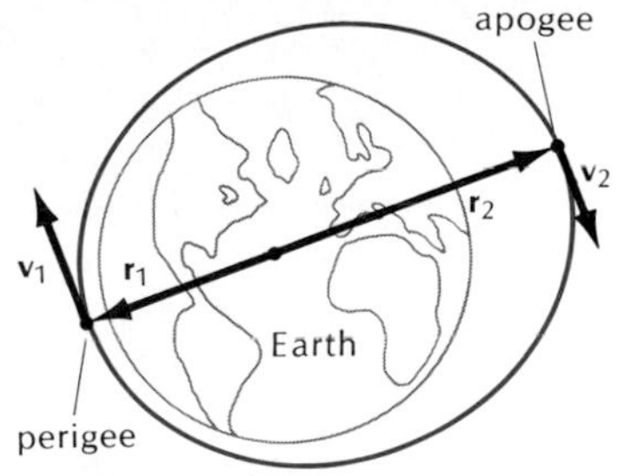

Fig. 5.22 Orbit of the Vanguard I satellite.

SOLUTION: The force holding the satellite in its orbit is the force of gravity; this is a central force, directed toward the center of the Earth (we will study this force in detail in Chapter 13). According to the above argument, the angular momentum of the satellite will be conserved. At perigee, the radius vector $\mathbf{r}_1$ is perpendicular to the velocity vector $\mathbf{v}_1$. Hence, the magnitude of the angular momentum is

$$L = mv_1 r_1$$

Likewise, at apogee,

$$L = mv_2 r_2$$

Conservation of angular momentum then gives us

$$mv_2 r_2 = mv_1 r_1$$

or

$$v_2 = \frac{r_1}{r_2} v_1 = \frac{7.02 \times 10^6 \text{ m}}{10.3 \times 10^6 \text{ m}} \times 8.22 \times 10^3 \text{ m/s}$$

$$= 5.60 \times 10^3 \text{ m/s}$$

If the force is *not* central, then the angular momentum is *not* conserved. The rate of change of the angular momentum is then given by Eq. (22). The quantity $\mathbf{r} \times \mathbf{F}$ is called the **torque** of the force **F** on the particle:

Torque

$$\boldsymbol{\tau} = \mathbf{r} \times \mathbf{F} \tag{24}$$

In terms of the torque, Eq. (22) becomes

$$\frac{d\mathbf{L}}{dt} = \boldsymbol{\tau} \tag{25}$$

This equation says that the rate of change of angular momentum equals the torque. Obviously, Eq. (25) is analogous to Eq. (17), which says that the rate of change of (linear) momentum equals the force. We will leave the deeper exploration of the concept of torque to Chapter 12; for now, we regard torque as merely a convenient shorthand for $\mathbf{r} \times \mathbf{F}$.

5.7 Newtonian Relativity

As has been emphasized in the early part of this chapter, Newton's First and Second Laws are valid only in inertial reference frames. These laws therefore distinguish from the set of all conceivable reference frames a subset of preferred, inertial reference frames. Different inertial reference frames are in uniform translational motion relative to one another. If a given event is observed in two such reference frames, then the measured coordinates are related by the Galilean transformations [Eqs. (4.54)–(4.56)].

Newton's laws do not make any intrinsic distinction between two inertial reference frames. For instance, Newton's laws in the reference frame of a ship steaming away from the shore at constant velocity are exactly the same as in the reference frame of the shore[7]; these laws depend only on the *acceleration,* which is the same in both reference frames. Hence the equations governing the motion of billiard balls on a pool table aboard the ship are exactly the same as on shore. No experiment with billiard balls, or any other mechanical experiment, aboard the ship will reveal its uniform motion with respect to the shore. Only *changes* in the uniform motion of the ship can be detected. For example, if the ship rolls, or pitches, or strikes an iceberg and suddenly decelerates, then the billiard balls will misbehave, developing "spontaneous" accelerations relative to the ship.

Thus, uniform motion is relative, whereas accelerated motion is absolute; i.e., the uniform motion of a reference frame can only be detected relative to another reference frame, whereas the accelerated motion of a reference frame can be detected by experiments within that reference frame. The inertial guidance systems used aboard ships, aircraft, and missiles take advantage of the absolute character of acceleration to keep track of the motion of the reference frame. These guidance systems rely on **accelerometers** to measure absolute accelerations. Essentially, an accelerometer consists of a spring with one end attached to the reference frame and the other end attached to a mass (a spring balance; Figure 5.23). When the reference frame accelerates, the inertia of the mass stretches the spring; the amount of stretch is directly proportional to the acceleration. High-precision accelerometers used in inertial guidance systems are capable of measuring accelerations to within better than $\pm 10^{-6}$ gee. A complete inertial guidance system contains three accelerometers to measure the acceleration along three perpendicular axes; the system also contains gyroscopes which maintain the axes in a fixed orientation (see Chapter 12). From a knowledge of the acceleration (both the magnitude and direction) as a function of time and a knowledge of the initial position and velocity, a computer can automatically calculate the position and velocity of the ship at any later time.

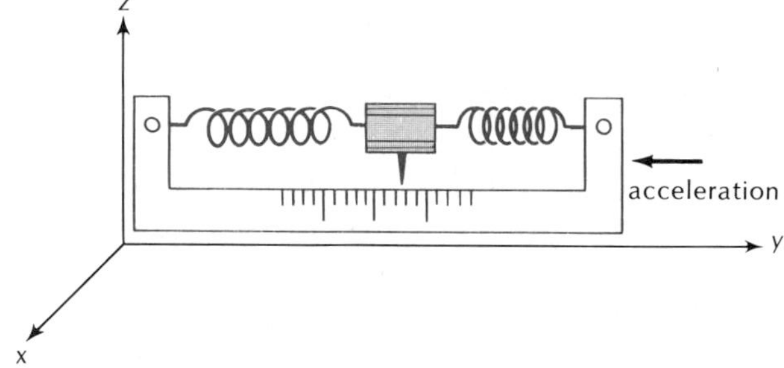

Fig. 5.23 Accelerometer.

The impossibility of detecting uniform translational motion by experiments within an inertial reference frame was recognized by Galileo, who argued that it is impossible to distinguish between a

[7] For the purposes of this example, we will pretend that the reference frame of the Earth is inertial.

"stationary" Earth and a "moving" Earth by mechanical experiments performed on the Earth's surface.[8] The assertion that the laws of mechanics are the same in all inertial reference frames is called the Principle of **Galilean** or **Newtonian Relativity.**

Principle of Newtonian relativity

SUMMARY

Newton's First Law: In an inertial reference frame, a body at rest remains at rest and a body in motion continues to move at constant velocity unless acted upon by an external force.

Definition of mass: $m/m_s = a_s/a$

Newton's Second Law: $m\mathbf{a} = \mathbf{F}$

Superposition of forces: $\mathbf{F}_{\text{net}} = \mathbf{F}_1 + \mathbf{F}_2 + \mathbf{F}_3 + \cdots$

Newton's Third Law: Whenever a body exerts a force on another body, the latter exerts a force of equal magnitude and opposite direction on the former.

Momentum: $\mathbf{p} = m\mathbf{v}$

$$\frac{d\mathbf{p}}{dt} = \mathbf{F}$$

Angular momentum: $\mathbf{L} = \mathbf{r} \times \mathbf{p}$

$$\frac{d\mathbf{L}}{dt} = \mathbf{r} \times \mathbf{F}$$

Angular momentum for uniform circular motion: $L = rmv$

Conservation of angular momentum (for central force):

$$\mathbf{L} = [\text{constant}]$$

QUESTIONS

1. If a glass stands on a table on top of a sheet of paper, you can remove the paper without touching the glass by jerking the paper away very sharply. Explain why the glass more or less stays put.

2. Tribes of natives in the Amazon jungle use extremely long and heavy arrows (3 m or more). Why? (Hint: What is likely to happen to an arrow flying through dense jungle?)

3. Make a critical assessment of the following statement: An automobile is a device for pushing the air out of the way of the passenger so that his body can continue to its destination in its natural state of motion at uniform velocity.

4. When rounding a curve in your automobile, you get the impression that a force tries to pull you toward the outside of the curve. Is there such a force?

[8] However, Galileo suffered from some confusion: he thought (erroneously) that the ocean tides result from a combination of the Earth's translational motion and its rotational motion about its axis. If this were so, it would make an absolute distinction between a stationary and a moving Earth.

5. If the Earth were to stop spinning (other things remaining equal), the value of g at all points of the surface except the poles would become slightly larger. Why?

6. Does the mass of a body depend on the frame of reference from which we observe the body? Answer this by appealing to the definition of mass.

7. Suppose that a (strange) body has negative mass. Suppose you tie this body to a body of positive mass of the same magnitude by means of a stretched rubber band. Describe the motion of the two bodies.

8. Does the magnitude or the direction of a force depend on the frame of reference?

9. A fisherman wants to reel in a large dead shark hooked on a thin fishing line. If he jerks the line, it will break; but if he reels it in very gradually and smoothly, it will hold. Explain.

10. The following statements appeared in *Tennis Trade* magazine:

> The racquet is an instrument of work and a transmitter of energy. The amount of energy you transfer to the ball can best be defined as a Force (F). The amount of this Force can then be equated to its mass times its acceleration.
>
> $$F = ma$$
>
> F = Force is the amount of energy you transfer to the ball by the racquet.
>
> m = Mass is the weight of the racquet divided by gravity, or $m = w/g$; gravity is generally 32 ft/sec and for all practical purposes, simply refer to it as weight.
>
> a = Acceleration is the speed with which you swing the racquet.
>
> From the above equation ($F = ma$), you can see that the faster one swings the racquet the greater the force will be transmitted to the ball.

Which of these statements are false? (If a statement makes no sense, regard it as false.) Would any of these statements help you to become a better tennis player?

11. If a body crashes into a water surface at high speed, the impact is almost as hard as on a solid surface. Explain.

12. Moving downwind, a sailboat can go no faster than the wind. Moving across the wind, a sailboat can go faster than the wind. How is this possible? (Hint: What are the horizontal forces on the sail and on the keel of a sailboat?)

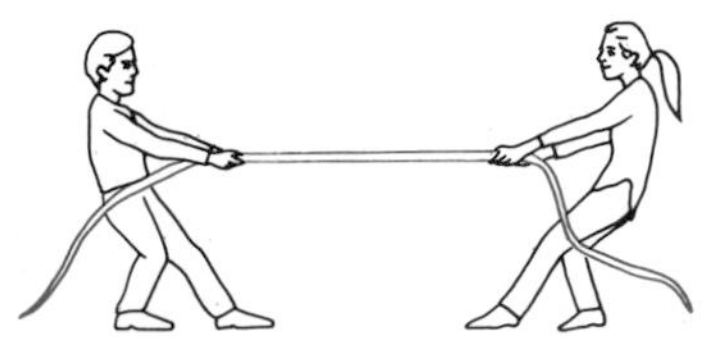

Fig. 5.24 A boy and a girl in a tug-of-war.

13. In the situation described in Example 6, why is it best to push at the midpoint of the rope?

14. A boy and girl are engaged in a tug-of-war (Figure 5.24). (a) Draw a diagram showing the horizontal forces on the boy, (b) on the girl, and (c) on the rope. Which of these forces are action–reaction pairs?

15. In an experiment performed in 1654, Otto von Guericke, mayor of Magdeburg and inventor of the air pump, gave a demonstration of air pressure before Emperor Ferdinand. He had two teams of 15 horses each pull in opposite directions on two evacuated hemispheres held together by nothing but air pressure. The horses failed to pull these hemispheres apart (Figure 5.25). If each horse exerted a pull of 300 kp, what was the tension in the harness attached to each hemisphere? If the harness attached to one of the hemispheres had simply been tied to a stout tree, what would have been the tension exerted by a single team of horses hitched to the other harness? What would have been the tension exerted by the two teams of horses hitched in series to the other

Fig. 5.25 The Magdeburg hemispheres.

harness? Can you guess why von Guericke hitched up his horses in the way he did?

16. In a tug-of-war, two teams of children pull on a rope (Figure 5.26). Is the tension constant along the entire length of the rope? Along what portion of the rope is it constant?

Fig. 5.26 Two teams of children in a tug-of-war.

17. When you are standing on the Earth, your feet exert a force (push) against the surface. Why does the Earth not accelerate away from you?

18. When an automobile accelerates on a level road, the force that produces this acceleration is the push of the road on the wheels. If so, why does the automobile need an engine?

19. You are in a small boat in the middle of a calm lake. You have no oars, and you cannot put your hands in the water because the lake is full of piranhas. The boat carries a large load of coconuts. How can you get to the shore?

20. A tennis ball in horizontal flight bounces off a vertical wall. Is the momentum conserved?

21. An automobile travels at constant speed along a road consisting of two straight segments connected by a curve in the form of an arc or circle. Taking the center of the circle as origin, what is the direction of the angular momentum of the automobile? Is the angular momentum constant as the automobile travels along this road?

22. Is the angular momentum of the orbital motion of a planet constant if we choose an origin of coordinates *not* centered on the Sun?

23. A pendulum is swinging back and forth. Is the angular momentum of the pendulum bob constant?

24. You are inside a ship that is trying to make headway against the strong current of a river. Without looking at the shore or other outside markers, is there any way you can tell whether the ship is making any progress?

25. A submarine uses its inertial guidance system to determine position and velocity at any instant without the need of visual observation of shore points, stars, or other outside markers. Does this conflict with the Newtonian Principle of Relativity?

PROBLEMS

Section 5.2

1. On a flat road, a Maserati sports car can accelerate from 0 to 80 km/h (0 to 50 mi/h) in 5.8 s. The mass of the car is 1620 kg. What are the average acceleration and the average force on the car?

2. The Grumman F-14B fighter plane has a mass of 36,000 lb-mass and its engines develop a thrust of 60,000 lb when at full power. What is the maximum horizontal acceleration that this plane can achieve? Ignore friction.

3. Pushing with both hands, a sailor standing on a pier exerts a horizontal force of 60 lb on a destroyer of 3750 short tons.[9] Assuming that the mooring ropes do not interfere and that the water offers no resistance, what is the acceleration of the ship? How far does the ship move in 60 s?

4. A woman of 57 kg is held firmly in the seat of her automobile by a lap-and-shoulder seat belt. During a collision, the automobile decelerates from 50 to 0 km/h in 0.12 s. What is the average force that the seat belt exerts on the woman? Express the answer in kiloponds.

5. A heavy freight train has a total mass of 16,000 metric tons. The locomotive exerts a pull of 670,000 N on this train. What is the acceleration? How long does it take to increase the speed from 0 to 50 km/h?

6. With brakes fully applied, a 1500-kg automobile decelerates at the rate of 8.0 m/s^2 on a flat road. What is the braking force acting on the automobile? Draw a diagram showing the direction of motion of the automobile and the direction of the braking force.

7. Consider the impact of the automobile on a barrier described in Example 2.4. If the mass of the automobile is 1400 kg, what is the average force acting on the automobile during the deceleration?

8. The projectile fired by the gun described in Problem 2.21 has a mass of 100 lb-mass. What is the average force on this projectile as it moves along the barrel?

9. Figure 2.13 shows the plot of velocity vs. time for a Triumph sports car coasting along with its gears in neutral. From the values of the deceleration at the times $t = 0$, 10, 20, 30, and 40 s (see Problem 2.13b), calculate the friction force that the car experiences at these times. The mass of the car is 1160 kg. Make a plot of friction force vs. velocity.

10. Table 2.4 gives the velocity of a projectile as a function of time. The projectile slows down because of the friction force exerted by the air. For the first 0.30-s time interval and for the last 0.30-s time interval, calculate the average friction force.

11. A proton moving in an electric field has an equation of motion

$$\mathbf{r} = (5.0 \times 10^4 t)\hat{\mathbf{x}} + (2.0 \times 10^4 t - 2.0 \times 10^5 t^2)\hat{\mathbf{y}} - (4.0 \times 10^5 t^2)\hat{\mathbf{z}}$$

where distance is measured in meters and time in seconds. The proton has a mass of 1.7×10^{-27} kg. What are the components of the force acting on this proton? What is the magnitude of the force?

12. The speed of a projectile traveling horizontally and slowing down under the influence of air friction can be approximately represented by

$$v = 2152 - 200.6t + 10.7t^2$$

[9] The short ton is 2000 lb. In the context of this problem it is 2000 lb-mass.

where v is measured in feet per second and t in seconds; the mass of the projectile is 100 lb-mass (see Problems 2.16 and 2.17). Find a formula for the force of air friction as a function of time.

Section 5.3

13. While braking, an automobile of mass 1200 kg decelerates along a level road at 0.80 gee. What is the horizontal force that the road exerts on the automobile? The vertical force? Ignore the friction of the air.

14. In 1978, in an accident at a school in Harrisburg, Pennsylvania, several children lost parts of their fingers when a nylon rope suddenly snapped during a giant tug-of-war among 2300 children. The rope was known to have a breaking tension of 13,000 lb. Each child can exert a pull of approximately 30 lb. Was it safe to employ this rope in this tug-of-war?

15. At lift-off, the Saturn V rocket used for the Apollo missions has a mass of 2.45×10^6 kg.

(a) What is the minimum thrust that the rocket engines must develop to achieve lift-off?
(b) The actual thrust that the engines develop is 2.98×10^7 N. What is the vertical acceleration of the rocket at lift-off?
(c) At burnout, the rocket has spent its fuel and its remaining mass is 0.75×10^6 kg. What is the acceleration just before burnout? Assume that the motion is still vertical and that the strength of gravity is the same as when the rocket is on the ground.

16. The Earth exerts a gravitational pull of 2.0×10^{20} N on the Moon; the Sun exerts a gravitational pull of 4.3×10^{20} N on the Moon. What is the net force on the Moon when the angular separation between the Earth and the Sun is 90° as seen from the Moon?

17. A sailboat is propelled through the water by the combined action of two forces: the push ("lift") of the wind on the sail and the push of the water on the keel. Figure 5.27 shows the magnitudes and the directions of these forces acting on a medium-sized sailboat (this oversimplified diagram does not include the drag of wind and water). What is the resultant of the forces in Figure 5.27?

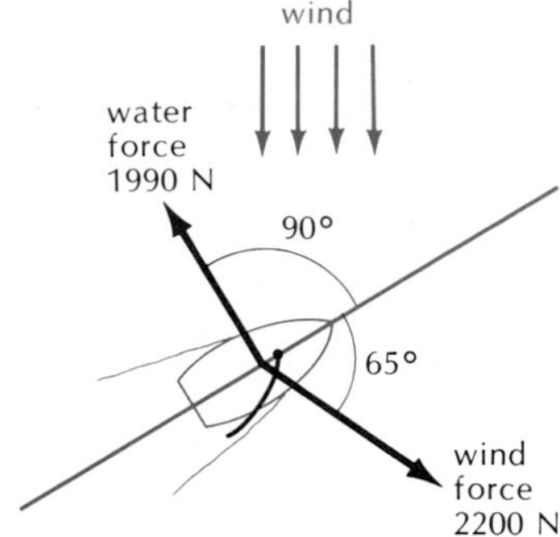

Fig. 5.27 Forces on a sailboat. The angles are measured relative to the line of motion.

18. A boat is tied to a dock by four (horizontal) ropes. Two ropes, with a tension of 60 lb each, are at right angles to the dock. Two other ropes, with a tension of 80 lb each, are at an angle of 20° with the dock (Figure 5.28). What is the resultant of these forces?

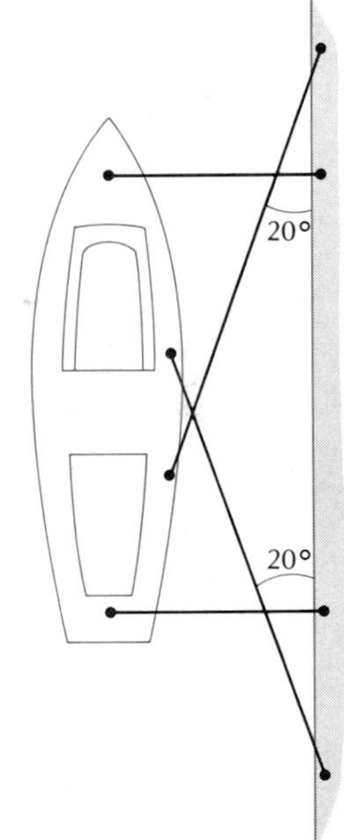

Fig. 5.28 Ropes holding a boat at a dock.

19. In a tug-of-war, a jeep of mass 1400 kg and a tractor of mass 2000 kg pull on a horizontal rope in opposite directions. At one instant, the tractor pulls on the rope with a force of 1.50×10^4 N while its wheels push horizontally against the ground with a force of 1.60×10^4 N. Calculate the instantaneous accelerations of the tractor and of the jeep; calculate the horizontal push of the wheels of the jeep. Assume the rope does not stretch or break.

Section 5.4

20. A woman stands on a chair. Her mass is 60 kg and the mass of the chair is 20 kg. What is the force that the chair exerts on the woman? What is the force that the floor exerts on the chair?

21. A diver of mass 75 kg is in free fall after jumping off a high platform.

(a) What is the force that the Earth exerts on the diver? What is the force that the diver exerts on the Earth?
(b) What is the acceleration of the diver? What is the acceleration of the Earth?

22. A long freight train consists of 250 cars each of mass 64 metric tons. The pull of the locomotive accelerates this train at the rate of 0.043 m/s² along a level track. What is the tension in the coupling that holds the first car to the locomotive? What is the tension in the coupling that holds the last car to the next to last car? Ignore friction.

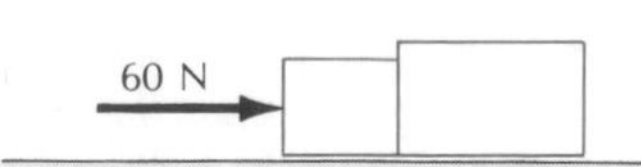

Fig. 5.29 Two boxes in contact.

23. Two heavy boxes of masses 20 kg and 30 kg sit on a smooth, frictionless surface. The boxes are in contact and a horizontal force of 60 N pushes horizontally against the smaller box (Figure 5.29). What is the acceleration of the two boxes? What is the force that the smaller box exerts on the larger box? What is the force that the larger box exerts on the smaller box?

24. An archer pulls the string of her bow back with her hand with a force of 40 lb. If the two halves of the string above and below her hand make an angle of 120° with each other, what is the tension in each half of the string?

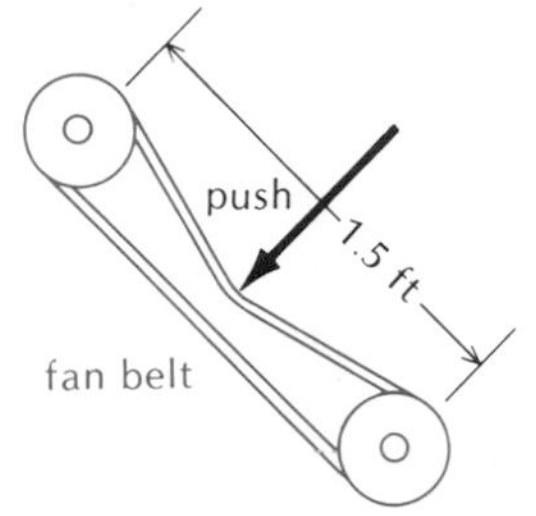

Fig. 5.30 Push on a fan belt.

25. A mechanic tests the tension in a fan belt by pushing against it with his thumb (Figure 5.30). The force of the push is 30 lb and it is applied to the midpoint of a segment of belt 1.5 ft long. The lateral displacement of the belt is 1.0 in. What is the tension in the belt (while the mechanic is pushing)?

26. On a sailboat, a rope holding the foresail passes through a block (a pulley) and is made fast on the other side to a cleat (Figure 5.31). The two parts of the rope make an angle of 140° with each other. The sail pulls on the rope with a force of 1200 kp. What is the force that the rope exerts on the block?

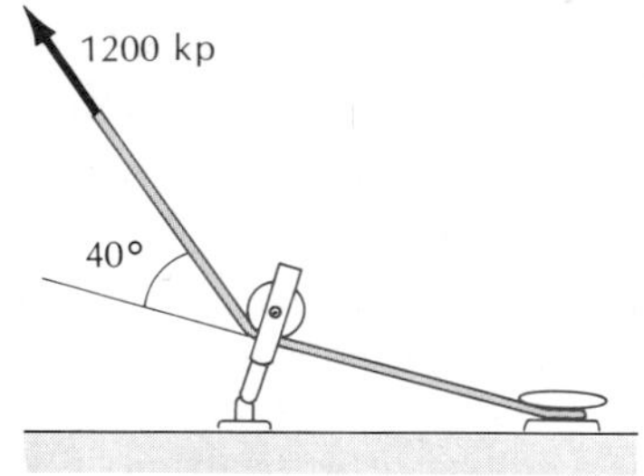

Fig. 5.31 The left end of the rope is attached to the sail, the right end is attached to a cleat.

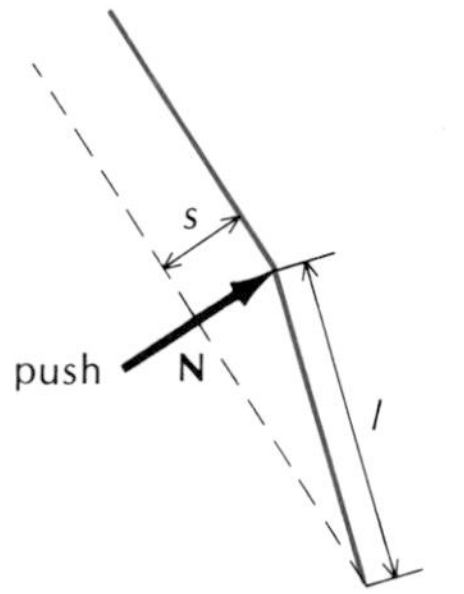

Fig. 5.32 Push on a wire rope.

27. A sailor tests the tension in a wire rope holding up a mast by pushing against the rope with his hand at a distance l from the lower end of the rope. When he exerts a transverse push N, the wire rope suffers a transverse displacement s (Figure 5.32).

(a) Show that for $s \ll l$ the tension in the wire rope is given approximately by the formula

$$T = Nl/s$$

In your calculation, assume that the distance to the upper end of the rope is effectively infinite, i.e., the total length of the rope is much larger than l.

(b) What is the tension in a rope that suffers a transverse displacement of 2.0 cm under a force of 150 N applied at a distance of 1.5 m from the lower end?

28. On a windy day, a small tethered balloon is held by a long string making an angle of 70° with the ground. The vertical buoyant force on the balloon (exerted by the air) is 15 lb. During a sudden gust of wind, the (horizontal) force of the wind is 45 lb; the tension in the string is 29.5 lb. What are the magnitude and direction of the force on the balloon?

29. A crate weighing 2000 kg is hanging from a crane at the end of a cable 12 m long. If we attach a horizontal rope to this crate and gradually apply a pull of 1800 N, what angle will the cable finally make with the vertical?

30. A horse, walking along the bank of a canal, pulls a barge. The horse exerts a pull of 300 N on the barge at an angle of 30° (Figure 5.33). The bargeman relies on the rudder to steer the barge on a straight course parallel to the bank. What transverse force (perpendicular to the bank) must the rudder exert on the barge?

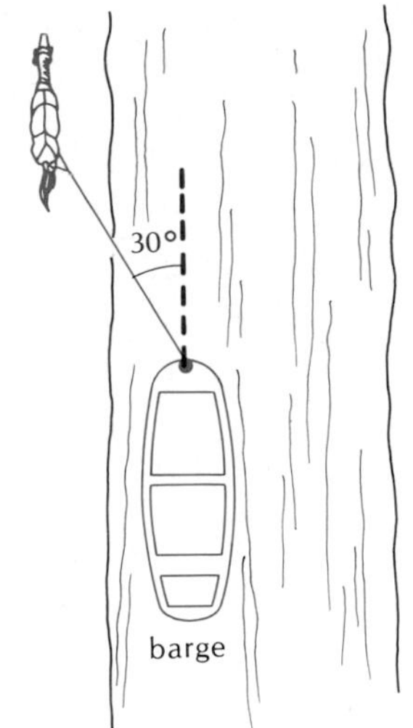

Fig. 5.33 Horse pulling barge.

Section 5.5

31. Using the entries listed in Tables 1.6 and 2.1, find the magnitude of the momentum for each of the following: Earth moving around Sun, jet airliner at maximum airspeed, automobile at federal speed limit, man walking, electron moving around a nucleus.

32. What is the momentum of an automobile of 2000 lb-mass moving at 40 mi/h? If a truck of 16,000 lb-mass is to have the same momentum as the automobile, what must be its speed?

33. Consider the proton with the equation of motion given in Problem 11. What are the components of the momentum of this proton at time $t = 0$?

34. The push that a bullet exerts during impact on a target depends on the momentum of the bullet. A Remington .244 rifle, used for hunting deer, fires a bullet of 90 grains (1 grain is $\frac{1}{7000}$ lb) with a speed of 3200 ft/s. A Remington .35 rifle fires a bullet of 200 grains with a speed of 2210 ft/sec. What is the momentum of each bullet?

35. An electron of mass 9.1×10^{-31} kg is moving in the x–y plane; its speed is 2.0×10^5 m/s and its direction of motion makes an angle of 25° with the x axis. What are the components of the momentum of the electron?

36. A sky diver of mass 75 kg is in free fall. What is the rate of change of his momentum?

*37. The Earth moves around the Sun in a circle of radius 1.5×10^{11} m at a speed of 3.0×10^4 m/s. The mass of the Earth is 6.0×10^{24} kg. Calculate the magnitude of the rate of change of the momentum of the Earth from these data. (Hint: The magnitude of the momentum does not change, but the direction does.)

Section 5.6

38. At the Fermilab accelerator, protons of momentum 5.3×10^{-16} kg · m/s travel around a circular path of diameter 2.0 km. What is the orbital angular momentum of one of these protons? Assume that the origin is at the center of the circle. Draw a diagram showing the direction of the angular momentum in relation to the path of the proton.

39. Prior to launching a stone from a sling, a Bolivian native whirls the stone at 3.0 revolutions per second around a circle of radius 0.75 m. The mass of the stone is 0.15 kg. What is the angular momentum of the stone relative to the center of the circle? Draw a diagram showing the direction of motion and the direction of the angular momentum.

40. A communications satellite of mass 100 kg is in a circular orbit of radius 4.22×10^7 m around the Earth. The orbit is in the equatorial plane of the Earth and the satellite moves along it from west to east with a speed of 4.90×10^2 m/s. What are the magnitude and the direction of the angular momentum of this satellite?

41. According to Bohr's (oversimplified) theory, the electron in the hydrogen atom moves in one or another of several possible circular orbits around the nucleus. The radii and the orbital velocities of the three smallest orbits are, respectively, 0.529×10^{-10} m, 2.18×10^6 m/s; 2.12×10^{-10} m, 1.09×10^6 m/s; and 4.76×10^{-10} m, 7.27×10^5 m/s. For each of these orbits calculate the or-

bital angular momentum of the electron, with the origin at the center. How do these angular momenta compare?

42. A high-speed meteorite moves past the Earth along an (almost) straight line. The mass of the meteorite is 150 kg, its speed relative to the Earth is 60 km/s, and its distance of closest approach to the center of the Earth is 1.2×10^4 km.

(a) What is the angular momentum of the meteorite in the reference frame of the Earth (origin at the center of the Earth)?
(b) What is the angular momentum of the Earth in the reference frame of the meteorite (origin at the center of the meteorite)?

43. A train of mass 1500 metric tons runs along a straight track at 85 km/h. What is the angular momentum (magnitude and direction) of the train about a point 50 m to the side of the track, left of the train? About a point on the track?

44. Consider the motion of the Earth around the Sun as described in Example 8. Take as origin the point at which the Earth is today and treat the Earth as a particle.

(a) What is the angular momentum of the Earth about this origin today?
(b) What will be the angular momentum of the Earth about the same origin three months later? Six months later? Nine months later? Is the angular momentum conserved?

45. Explorer I, the first American artificial satellite, had an elliptical orbit around the Earth with a perigee distance of 6.74×10^6 m and an apogee distance of 8.91×10^6 m. The speed of this satellite was 6.21×10^3 m/s at apogee. Calculate the speed at perigee.

46. Halley's comet orbits the Sun in an elliptical orbit. When the comet is closest to the Sun (perihelion), its distance is 8.75×10^{10} m and its speed is 5.46×10^4 m/s. When the comet is farthest from the Sun (aphelion), its speed is 9.08×10^2 m/s. Calculate the aphelion distance of Halley's comet.

47. The orbit of the Earth around the Sun is not quite a circle but, rather, an ellipse of very small elongation (small eccentricity). At the point of closest approach (perihelion), the Earth–Sun distance is 1.47×10^{11} m; at the point of farthest recession (aphelion) the Earth–Sun distance is 1.52×10^{11} m. By what factor is the speed of the Earth at perihelion greater than the speed at aphelion?

48. The electron in a hydrogen atom moves around the nucleus under the influence of the electric force of attraction, a central force pulling the electron toward the nucleus. According to the Bohr theory, one of the possible orbits of the electron is an ellipse of angular momentum $2\hbar$ with a distance of closest approach $(1 - 2\sqrt{2}/3)a_0$ and a distance of farthest recession $(1 + 2\sqrt{2}/3)a_0$, where $\hbar$ and a_0 are two atomic constants with the numerical values 1.05×10^{-34} kg · m²/s ("Planck's constant") and 5.3×10^{-11} m ("Bohr radius"), respectively. In terms of $\hbar$ and a_0, find the speed of the electron at the points of closest approach and farthest recession; then evaluate numerically.

CHAPTER 6

Dynamics — Forces and the Solution of the Equation of Motion

Newton's Second Law is usually called the **equation of motion.** If the force on a particle is known, then the Second Law determines the acceleration, and from this the complete motion of the particle can be calculated. Thus, in principle, the motion of the particle is completely predictable. As the great French mathematician and astronomer Pierre Simon de Laplace expressed it,

> If an intellect were to know, for a given instant, all the forces that animate nature and the condition of all the objects that compose her, and were also capable of subjecting these data to analysis, then this intellect would encompass in a single formula the motions of the largest bodies in the universe as well as those of the smallest atom; nothing would be uncertain for this intellect, and the future as well as the past would be present before its eyes.

Pierre Simon, marquis de Laplace, *1749–1827, professor at the École Militaire, Paris. He investigated the theory of the mutual disturbances that the planets exert on one another, and used Newton's equations to establish that the Solar System is stable — in the long run the mutual disturbances average to zero. For his great contributions to celestial mechanics he was called the "Newton of France."*

To find a solution of the equation of motion means to find a force $\mathbf{F}$ and a position vector $\mathbf{r}(t)$ which is a function of time such that the equation $m\mathbf{a} = \mathbf{F}$ is satisfied. For a physicist, the typical problem involves a known force and an unknown motion; for example, the physicist knows the force between the planets and the Sun and he seeks to calculate the motion of these bodies. But for an engineer, the reverse problem with a known motion and an unknown force is often of practical importance; for example, the engineer knows that a train is to round a given curve at 60 mi/h and he seeks to calculate the forces that the track and the wheels must withstand. A special problem with known motion is the problem of statics; here we know that the body is at rest ($\mathbf{v} = 0$ and $\mathbf{a} = 0$) and we wish to compute the forces that will maintain this condition of equilibrium. Thus, depending on the circumstances, we can regard either the right side or the left side of the equation $m\mathbf{a} = \mathbf{F}$ as an unknown that is to be calculated from what we know about the other side.

Before we look at some examples of solutions of the equation of motion, we will briefly discuss some general properties of forces.

6.1 The Four Fundamental Forces

In everyday experience we encounter an enormous variety of forces: the gravity of the Earth which pulls all bodies downward, contact forces between rigid bodies that resist their interpenetration, elastic forces that oppose the deformation of springs and beams, pressure forces exerted by air or water on bodies immersed in them, adhesive forces exerted by a layer of glue bonding two surfaces, friction forces that resist the motion of a surface sliding over another, electrostatic forces between two electrified bodies, magnetic forces between the poles of magnets, and so on.

Besides these forces that act in the macroscopic world of everyday experience, there are many others that act in the microscopic world of atomic and nuclear physics. There are intermolecular forces that attract or repel molecules to or from each other, interatomic forces that bind atoms into molecules or repel them if they come too close together, atomic forces within the atom that hold its parts together, nuclear forces that act on the parts of the nucleus, and even more esoteric forces which only act for a brief instant when subnuclear particles are made to suffer violent collisions in high-energy experiments performed in accelerator laboratories.

The four fundamental forces

Yet, at a fundamental level, this bewildering variety of forces involves only four different kinds of force. The four fundamental forces are the gravitational force, the electromagnetic force, the "strong" force, and the "weak" force.

The **gravitational force** is a mutual attraction between all masses. If we reckon the strength of the forces in terms of their effect on elementary particles, then gravitation is the weakest of the four forces. The gravitational attraction between two neighboring protons in a nucleus is only about 10^{-34} N, which is completely insignificant. On the surface of the Earth, we feel the force of gravity only because the mass of the Earth is so large; although the force between the individual particles in our bodies and in the Earth is insignificant, our bodies and the Earth contain a very large number of particles and their forces add up.

The **electromagnetic force** is an attraction or repulsion between electric charges. The electric and the magnetic forces, once considered to be separate, are now grouped together because they are closely related: the magnetic force is nothing but an extra electric force that acts whenever charges are in motion. The electric force is of medium strength; between two neighboring protons, it is about 90 N. Of all the forces, the electric force plays the largest role in our lives. With the exception of the Earth's gravity, every force in our immediate macroscopic environment is electric. Contact forces between rigid bodies, elastic forces, pressure forces, adhesive forces, friction forces, etc., are nothing but electric forces between the charged particles in the atoms of one body and those in the atoms of another.

The **"strong" force** acts mainly within the nuclei of atoms. It plays the role of the nuclear glue that prevents the pieces of the nucleus from flying apart. This nuclear force is called "strong" because it is the strongest of the four forces. Between two neighboring protons in a nucleus, this force is typically 10,000 N. It can be either attractive or repulsive: the strong force will push the protons apart if they come too near to each other, and it will pull them together if they begin to drift too far apart.

Finally, the **"weak" force** only manifests itself in certain reactions among elementary particles. Most of the reactions caused by the weak force are radioactive decay reactions; they involve the spontaneous breakup of a particle into several other particles (see Section C.3). This force is called "weak" because it is very weak; between two neighboring protons in a nucleus, its strength is estimated at only about 10^{-2} N.

Table 6.1 lists the four fundamental forces in order of increasing strength.

Table 6.1 THE FOUR FUNDAMENTAL FORCES

Force	*Acts on*	*Strength*[a]	*Range*
Gravitational	All masses	10^{-34} N	Infinite
Weak	Most elementary particles	10^{-2}	Less than 10^{-17} m
Electromagnetic	Electric charges	10^{2}	Infinite
Strong	Nuclear particles	10^{4}	10^{-15} m

[a]The strength listed here is the force (in newtons) between two protons separated by a distance equal to their diameter, 2×10^{-15} m.

Gravitation and electromagnetism both bridge empty space and reach from one particle to another, even if the intervening distance is very large. A spectacular instance of such **action-at-a-distance** is the gravitation of the Sun: the Earth feels the gravitational pull of the Sun even though there is a gap of 100 million miles between these bodies. Another instance is the magnetism of the Earth: a compass needle feels the pull of the magnetic north pole even when it is thousands of miles away. In Chapter 23 we will inquire into the mechanism involved in the transmission of force through empty space; we will see that this involves a **field,** or a disturbance that spreads through space and carries the interaction from one body to another. But for now, we will ignore the mechanism and simply think of forces as bridging space and acting at a distance. Note that even so-called contact forces, such as friction or the force that opposes the interpenetration of solid bodies, rely on action at some finite distance. When your finger pushes against a page of this book, the subatomic particles making up two neighboring atoms in "contact" are not really touching, and the forces must bridge the gap of empty space between these particles.

Action-at-a-distance

The strong force also bridges space, reaching from one particle to another. However, its reach is very short, not much more than 10^{-15} m. The reach of the weak force is even shorter; it only acts if the particles are just about interpenetrating one another. The distance over which a force acts is called its **range.** Table 6.1 lists the ranges of the four forces.

In the next few sections we will look at the practical aspects of a few forces of great importance in the macroscopic world: the gravity of the Earth (or weight), the friction force, and the elastic force of a spring.

6.2 Weight

Weight is the pull of the Earth's gravity, that is, weight is a *force*. Consequently, weight is a vector quantity — it has a direction (downward) as well as a magnitude. The units of weight are the units of force, that is, newtons or pounds.

Consider a body of mass m in free fall near the surface of the Earth. Under the influence of gravity, this body will accelerate downward with an acceleration g. According to Newton's Second Law, the magnitude of the gravitational force that causes this acceleration must be

$$F = ma = mg \tag{1}$$

This gravitational force is what we call the weight. Usually we will denote the weight by the vector symbol **w**. The magnitude of **w** is

Weight

$$\boxed{w = mg} \tag{2}$$

If the body is not in free fall but is held in a stationary position by some supports, then the weight is of course still the same as in Eq. (2); however, the supports prevent this weight from producing the downward motion.

EXAMPLE 1. What is the weight of a 74-kg man? Assume $g = 9.81\ \mathrm{m/s^2}$.

SOLUTION: By Eq. (2)

$$w = mg = 74\ \mathrm{kg} \times 9.81\ \mathrm{m/s^2}$$
$$= 726\ \mathrm{N}$$

Alternatively, in British units the mass is 74 kg × 1 slug/14.6 kg = 5.07 slugs, and

$$w = mg = 5.07\ \mathrm{slugs} \times 32.2\ \mathrm{ft/s^2}$$
$$= 163\ \mathrm{lb}$$

Since the value of g depends on location, the weight of a body also depends on its location. For example, if a 74-kg man travels from London ($g = 9.81\ \mathrm{m/s^2}$) to Hong Kong ($g = 9.79\ \mathrm{m/s^2}$), his weight will decrease from 726 N to about 724 N, a difference of nearly 2 N. And if this man were to travel to a point 6400 km above the surface of the Earth ($g = 2.45\ \mathrm{m/s^2}$), his weight would decrease to 181 N! This example illustrates an essential distinction between mass and weight: mass is an *intrinsic* property of a body, measuring the resistance (inertia) with which the body opposes changes in its motion; the definition of mass is designed in such a way that a given body has the same mass regardless of its position in the universe. Weight is an *extrinsic* property of a body,

Mass vs. weight

measuring the pull of gravity on the body; it depends on the (gravitational) environment in which the body is located and is therefore a function of position.

In practice, the mass of a body is usually measured by "weighing" the body on a balance, either a beam balance or a spring balance. A beam balance with equal arms will be in equilibrium if the weight (force) on each balance pan is the same (Figure 6.1a). Since, to a very high accuracy, the value of g is the same at both balance pans, the equality of weights implies the equality of masses; thus, a beam balance can be used to compare masses via their weight. A spring balance can likewise be used to compare masses via their weight (Figure 6.1b). The spring balance can be directly calibrated to read in mass units by suspending known masses from it (of course, this calibration will only remain accurate as long as the balance is kept in a fixed location where g has some fixed value).

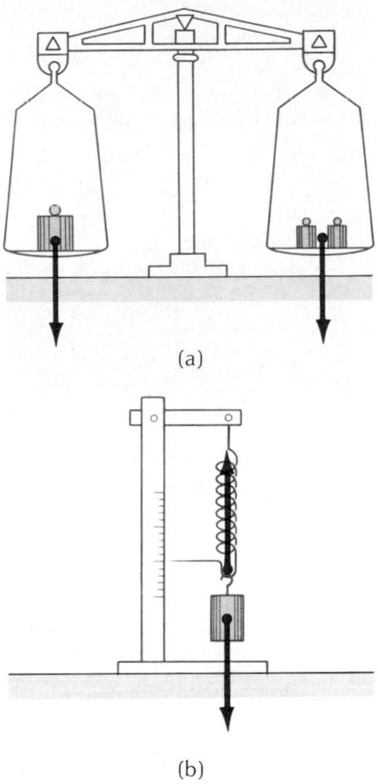

(a)

(b)

Fig. 6.1 (a) A beam balance is in equilibrium when the weights in the two pans are equal. (b) A spring balance is in equilibrium when the weight matches the force exerted by the spring.

As was pointed out in Section 5.2, in commercial and industrial usage the pound is regarded as a unit of mass, with 1 lb-mass = 0.453592 kg, or 1/32.1740 slug. By definition, 1 lb-mass has a weight of exactly 1 lb at a location where g has the "standard" value 1 gee = 9.80665 m/s², or 32.1740 ft/s²:

$$\begin{aligned}\begin{bmatrix}\text{weight of 1 lb-mass under} \\ \text{``standard'' conditions}\end{bmatrix} &= \frac{1}{32.1740}\ \text{slug} \times 32.1740\ \text{ft/s}^2 \\ &= 1\ \text{lb}\end{aligned}$$

Whether 1 lb-mass has a weight of 1 lb at some other location depends on the value of g. The weight is directly proportional to g, so that

$$\begin{aligned}[\text{weight of 1 lb-mass}] &= \frac{1}{32.1740}\ \text{slug} \times g \\ &= 1\ \text{lb} \times \left(\frac{g}{32.1740\ \text{ft/s}^2}\right)\end{aligned}$$

In everyday usage, *mass* and *weight* are often confused. The typical labels on packages of commercial goods display this confusion. Such labels state, "This package has a weight of 1 lb," whereas they should state, "This package has a mass of 1 lb-mass." The balances used to measure commercial goods compare an unknown mass with a set of known, standard masses; thus these balances are calibrated in pounds-mass, not in pounds-weight. Because of the common misuse of the word *weight,* it is often necessary to guess its intended meaning from the context in which it appears.[1]

A body deep in intergalactic space, far from any star or planet, will not experience any gravitational pull at all — the weight of the body will be exactly zero. This condition of weightlessness can be simulated in the vicinity of the Earth by means of a freely falling reference frame. Consider an observer in free fall, say, a diver who has jumped off a springboard and is accelerating downward with the acceleration

[1] The Soviet Union seems to be the only country free of this confusion. Russian labels on packages give the mass (MACCA).

Fig. 6.2 Diver and apple in free fall. *(Courtesy Helmut Ohanian.)*

of gravity. If the diver releases an apple from his hand, the apple accelerates downward at the same rate as the diver; that is, the apple remains at rest relative to the diver (Figure 6.2).

Thus, in the rest frame of the diver (a freely falling reference frame accelerating downward with the acceleration of gravity!) a body at rest remains at rest and, more generally, a body in motion continues to move at constant velocity. This means that in such a reference frame the gravitational pull is *apparently* zero; the weight is *apparently* zero. Of course, this simulated weightlessness arises from the accelerated motion of the reference frame — in the unaccelerated, inertial reference frame of the ground, the weight of the apple is certainly not zero. Nevertheless, if the diver insists on looking at things from his own reference frame, he will judge the weight of the apple, and also the weight of his own body, as zero. This sensation of weightlessness is also simulated within an airplane flying along a parabola, imitating the motion of a (frictionless) projectile (Figure 6.3); and it is also simulated in a spacecraft orbiting the Earth (Figure 6.4). Both of these motions are free-fall motions.

Fig. 6.3 Astronauts training in an airplane.

Fig. 6.4 Astronauts floating in the cargo bay of the Space Shuttle.

EXAMPLE 2. In preparation for walking in low-gravity conditions on the surface of the Moon, astronauts at NASA trained on a Moon simulator consisting of an inclined plane across which they could walk while suspended in a harness (Figure 6.5a). Suppose that the pull of the harness is parallel to the inclined plane. What must be the angle of inclination if the forces perpendicular to the plane are to simulate the conditions on the Moon? The acceleration of gravity on the surface of the Moon is $g_{\text{Moon}} = 1.6\ \text{m/s}^2$.

SOLUTION: Suppose that the astronaut takes a small jump so that his feet momentarily lose contact with the plane. Then the only forces acting on the astronaut are his weight **w** and the tension **T** supplied by the harness. Figure

6.5b shows these forces (we will pretend that the astronaut can be regarded as a particle so that all the forces act at one point). Taking the x axis parallel to the inclined plane and the y axis perpendicular, we can resolve the weight **w** into two components: a component w_x parallel to the plane and a component w_y perpendicular to the plane. These components have the values

$$w_x = -mg \sin \theta \qquad w_y = -mg \cos \theta$$

Since the harness restrains the astronaut from moving in the x direction, the component w_x is of no interest in the present problem (the force w_x is actually compensated by the tension **T**). The component w_y produces an acceleration in the y direction,

$$a_y = \frac{w_y}{m} = -g \cos \theta$$

In order to simulate Moon conditions, we want this acceleration to match the acceleration of gravity on the Moon, so that

$$-g \cos \theta = -g_{\text{Moon}}$$

From this we find

$$\cos \theta = \frac{g_{\text{Moon}}}{g} = \frac{1.6 \text{ m/s}^2}{9.8 \text{ m/s}^2} = 0.16$$

and $\theta = \cos^{-1}(0.16) = 81°$.

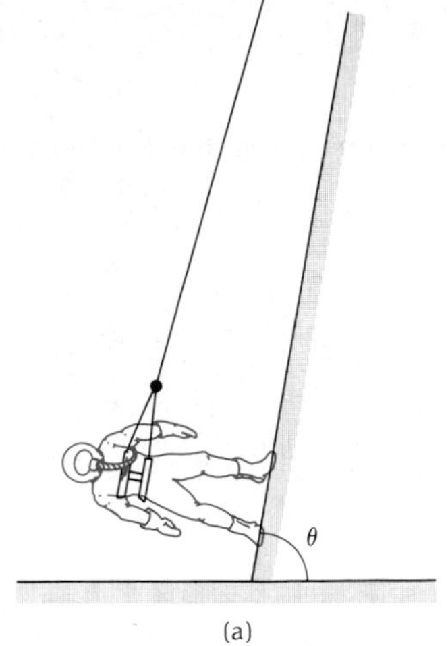

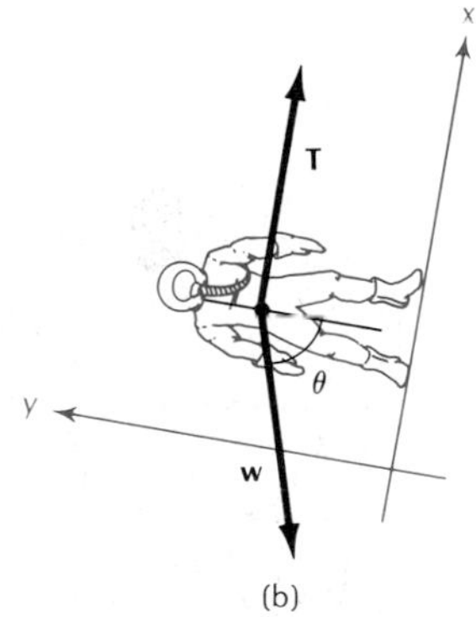

Fig. 6.5 (a) Astronaut on inclined plane. The astronaut can walk forward and backward, and he can jump perpendicularly to the plane. (b) Forces acting on the astronaut during a small jump.

6.3 Motion with a Constant Force

If the force acting on a particle is constant, then the acceleration is also constant. The motion is then given by Eqs. (4.18)–(4.23), with $a_x = F_x/m$, $a_y = F_y/m$, and $a_z = F_z/m$.

The most obvious example of motion with constant force is the motion of projectiles. If we neglect air resistance, the force on the projectile is the weight, $F = w = mg$. The weight is constant provided that g is constant along the path of the projectile, a provision that is well satisfied for most projectiles except ballistic missiles, which reach very large heights and cover very great distances. With the z axis in the vertical upward direction, the acceleration of the projectile is $a_x = 0$, $a_y = 0$, and $a_z = -g$. We have already discussed the resulting motion in Section 4.3.

Another example of motion with constant force is the motion of a body sliding on an inclined plane. Figure 6.6a shows a block sliding down a smooth, frictionless plane inclined at an angle θ to the horizontal. There are two forces acting on the block: the weight **w** pointing vertically downward and the **normal force N** pointing in a direction perpendicular to the inclined plane. The force **N** is a contact force that arises from the repulsion between the atoms of the block and the atoms of the plane. This repulsion prevents the block from penetrating the plane. The contact force acts uniformly all over the bottom surface of the block, but in Figure 6.6a the force is shown as though acting only at the center of the surface. The contact force is perpendicular to

Normal force

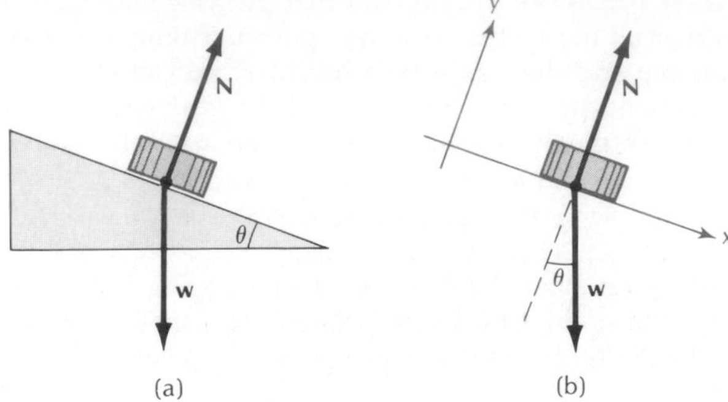

Fig. 6.6 (a) Block sliding down a plane. (b) "Free-body" diagram showing all the forces on the block.

the plane because we are assuming that there is no friction. If there were friction, then the force would have both a component perpendicular to the plane and a component parallel to the plane; the latter would be a friction force in the opposite direction of the sliding motion of the block along the plane.

Figure 6.6b shows the block (regarded as a particle) and all the forces acting on it: these forces are the weight **w** (vertically downward) and the normal force **N** (perpendicular to the plane). The inclined plane is not shown in this diagram — the effect of the plane is entirely contained in the force **N.** Such a diagram, showing the body and all the forces acting on it, but not showing the surrounding bodies that exert these forces, is called a **"free-body" diagram.** (In this context *free* does not mean free of force; it means that the surroundings are represented by the exerted forces.) Looking at Figure 6.6b, we see that the net force on our block is the vector sum of **N** and **w.** Taking the x axis parallel to the plane and the y axis perpendicular, we find that the components of these two forces are

"Free-body" diagram

$$N_x = 0 \qquad N_y = N$$

$$w_x = mg \sin \theta \qquad w_y = -mg \cos \theta$$

and the components of the net force are

$$F_x = mg \sin \theta \qquad F_y = N - mg \cos \theta \tag{3}$$

Newton's equation of motion then gives the corresponding components of the acceleration,

$$a_x = F_x / m = g \sin \theta \tag{4}$$

$$a_y = F_y / m = N/m - g \cos \theta \tag{5}$$

Equation (4) indicates that the acceleration of the block along the plane is $g \sin \theta$. In the case of a flat surface ($\theta = 0$), there is no acceleration; and in the case of a vertical surface ($\theta = 90°$), the acceleration is that of free fall. Both of these extreme cases are as expected.

Equation (5) can be used to evaluate N. Since the motion is necessarily parallel to the plane, the acceleration a_y perpendicular to the plane is identically zero; hence

$$0 = N/m - g \cos \theta \tag{6}$$

or

$$N = mg \cos \theta \qquad (7)$$

In the case of a horizontal surface ($\theta = 0$), the normal force has a magnitude mg, i.e., the magnitude matches the weight; and in the case of a vertical surface ($\theta = 90°$), the normal force vanishes.

EXAMPLE 3. A very steep portion of a railroad track in Guatemala has a slope of 1:11 (Figure 6.7a). What force is required to move a boxcar of 20 metric tons at constant speed along this track? Ignore friction and treat the boxcar as a particle.

SOLUTION: The forces on the boxcar are the weight **w**, the normal force **N**, and the force **T** that pulls the boxcar along the track. Figure 6.7b shows these forces in a "free-body" diagram. The net force **F** is the vector sum **w** + **N** + **T**. The x and y components of the net force are

$$F_x = mg \sin \theta - T \qquad (8)$$

$$F_y = N - mg \cos \theta \qquad (9)$$

Since the acceleration is supposed to vanish, $F_x = 0$ and therefore

$$T = mg \sin \theta$$

For a slope of 1:11, the angle θ is $\theta = \tan^{-1}(1/11) = 5.19°$. Hence

$$T = 20 \times 10^3 \text{ kg} \times 9.81 \text{ m/s}^2 \times \sin 5.19°$$

$$= 1.78 \times 10^4 \text{ N}$$

Note that the force **T** has this value whenever the acceleration is zero — it does not matter whether the boxcar is moving up or down or is at rest.

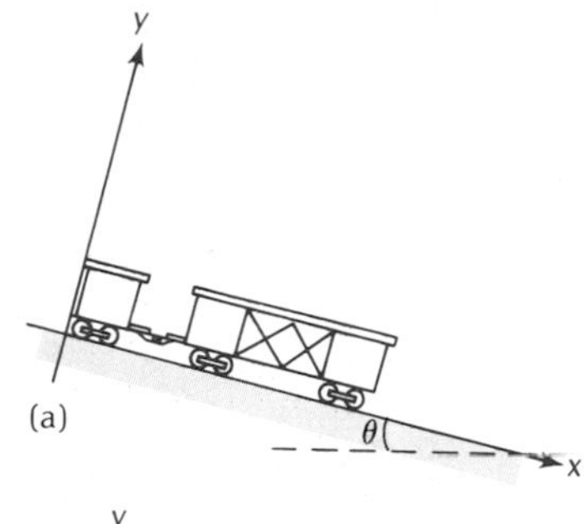

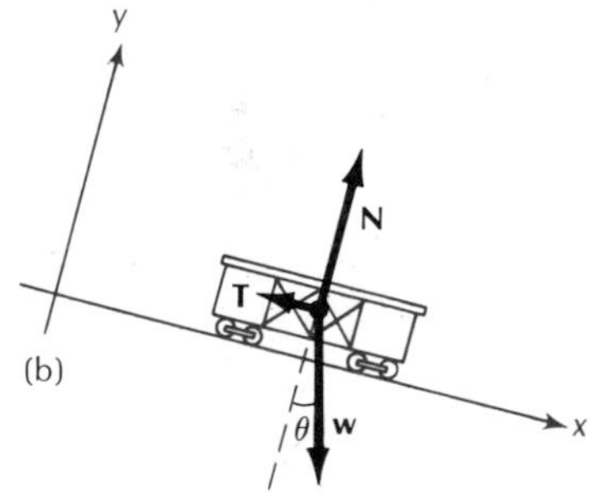

Fig. 6.7 (a) Boxcar on a steep railroad track. (b) "Free-body" diagram for the boxcar.

Yet another example of motion with constant force is illustrated in Figure 6.8a. Two masses, m_1 and m_2, hang on the two ends of a string which runs over a frictionless pulley. Since the two masses are linked by the string, it is necessary to solve their equations of motion simultaneously. We will assume that the string and the pulley are massless and that the pulley runs freely, without offering any resistance to the motion of the string. Under these conditions, the string merely transmits the tension from one mass to the other; consequently, the upward

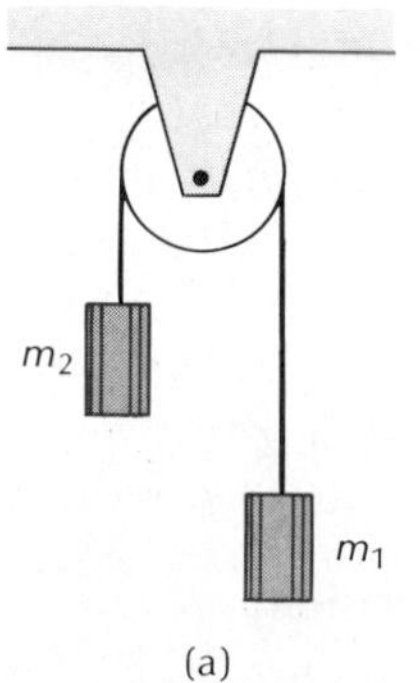

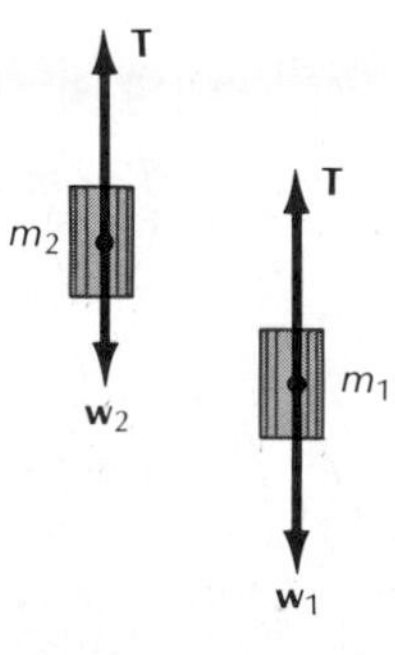

Fig. 6.8 (a) Two bodies connected by a string. (b) "Free-body" diagrams for the masses m_1 and m_2.

forces exerted by the string on each mass are exactly equal. (Keep in mind that this equality of tensions is a consequence of the special conditions of this example — if the pulley had a mass or if a brake or a motor were coupled to the shaft of the pulley, preventing it from running freely, then the tensions would *not* be equal.)

Figure 6.8b shows "free-body" diagrams for the masses m_1 and m_2; $\mathbf{T}$ represents the tension and $\mathbf{w}_1$ and $\mathbf{w}_2$ represent the weights. The vertical force on m_1 is $T - w_1$; the vertical force on m_2 is $T - w_2$. Consequently, the equation for the vertical motion of each mass is

$$T - w_1 = m_1 a_1 \tag{10}$$

$$T - w_2 = m_2 a_2 \tag{11}$$

where force and acceleration are regarded as positive when directed upward. Since the two masses are tied together by a fixed length of string, their accelerations are always of the same magnitude and in opposite directions, i.e.,

$$a_1 = -a_2 \tag{12}$$

With this equation and with $w_1 = m_1 g$ and $w_2 = m_2 g$, we obtain

$$T - m_1 g = m_1 a_1 \tag{13}$$

$$T - m_2 g = -m_2 a_1 \tag{14}$$

These are two simultaneous equations for the two unknowns T and a_1. To solve these equations, we simply subtract each side of the second equation from each side of the first equation; this gives

$$T - m_1 g - (T - m_2 g) = m_1 a_1 - (-m_2 a_1)$$

Here the unknown T cancels out and leaves us with a simple equation for a_1,

$$-m_1 g + m_2 g = m_1 a_1 + m_2 a_1$$

We can then immediately solve for a_1:

$$a_1 = \frac{m_2 - m_1}{m_1 + m_2} g \tag{15}$$

Substituting this result into Eq. (13), we obtain a simple equation for T:

$$T - m_1 g = m_1 \frac{m_2 - m_1}{m_1 + m_2} g$$

which gives

$$T = m_1 \frac{m_2 - m_1}{m_1 + m_2} g + m_1 g$$

or, combining the two terms on the right side,

$$T = \frac{2gm_1m_2}{m_1 + m_2} \tag{16}$$

As we might have expected, Eq. (15) shows that the acceleration is zero if the masses are equal — the two masses are then in equilibrium.

EXAMPLE 4. A passenger elevator consists of an elevator cage of 2000 lb (empty) and a counterweight of 2200 lb connected by a cable running over a pair of pulleys (Figure 6.9). Neglect the masses of the cable and of the pulleys. (a) What is the upward acceleration of the elevator cage if the pulleys are permitted to run freely? What is the tension in the cable? (b) What are the tensions in the cable if the pulleys are locked (by means of a brake) so that the elevator remains stationary?

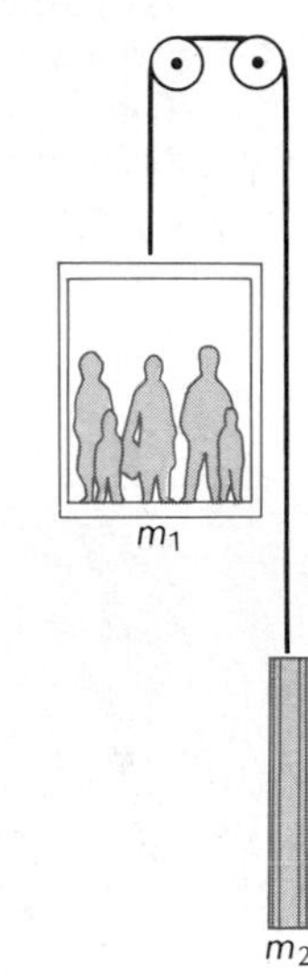

Fig. 6.9 Elevator with counterweight.

SOLUTION: (a) With $m_1 = 2000$ lb/g and $m_2 = 2200$ lb/g, Eq. (15) gives

$$a_1 = \frac{2200 \text{ lb}/g - 2000 \text{ lb}/g}{2200 \text{ lb}/g + 2000 \text{ lb}/g}\, g = \frac{200}{4200}\, g$$

$$= 0.048g = 1.53 \text{ ft/s}^2$$

and Eq. (16) gives

$$T = \frac{2g \times 2000 \text{ lb}/g \times 2200 \text{ lb}/g}{2000 \text{ lb}/g + 2200 \text{ lb}/g} = 2095 \text{ lb}$$

(b) If the pulleys are locked, the tension in the cable on either side of the pulleys must match the weight hanging from that side. Thus

$$T_1 = m_1 g = 2000 \text{ lb} \quad \text{and} \quad T_2 = m_2 g = 2200 \text{ lb}$$

These tensions are *not* equal.

6.4 Sliding Friction

Friction forces, which we have ignored up to now, play an important role in our environment and provide us with many interesting examples of motion with constant force. Suppose that a block of steel, in the shape of a brick, slides on a tabletop of steel. If the block has some initial velocity, friction will gradually slow it down and ultimately stop it. If the steel surfaces are clean and dry, the block will decelerate at the rate of about 6 m/s², or 0.6 gee. Figure 6.10 shows the forces acting on the block. The weight **w** acts downward with a magnitude mg. The normal force **N** exerted by the table on the block acts upward. This normal force is a contact force that keeps the block from penetrating the table. The magnitude of the normal force must be mg so that it exactly balances the weight. The friction force **f** acts horizontally, in a direction opposite to the motion. This force, just like the normal forcc, is a contact force which acts over the entire bottom surface of the block; however, for the sake of simplicity, in Figure 6.10 it is shown as though acting at the center of the surface.

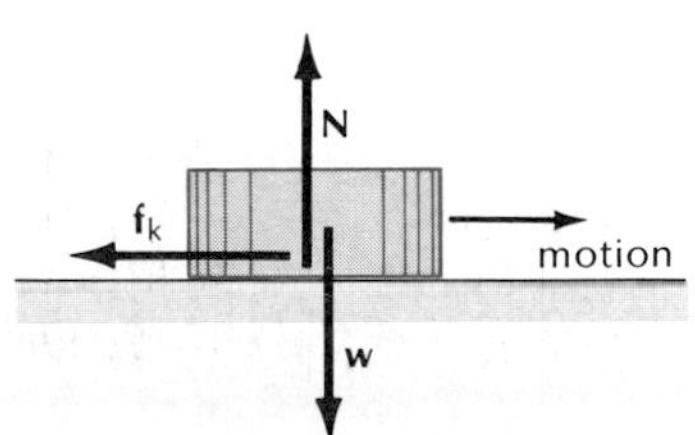

Fig. 6.10 Forces on a block sliding on a plate.

The friction force arises from adhesion between the two metals: the atoms in the block form bonds with the atoms in the tabletop. The

Fig. 6.11 Fragment of soft steel adhering to the surface of hard steel of a ball bearing. The fragment was plucked out of a slider of soft steel that was pushed strongly against the surface. Magnification 1600×. *(Courtesy S. J. Calabrese, Rensselaer Polytechnic Institute.)*

Fig. 6.12 Surface of a finely polished, mirror-smooth ball bearing of steel, magnified 3300×. The ridges were produced by the machining of the ball bearing. *(Courtesy S. J. Calabrese, Rensselaer Polytechnic Institute.)*

bonds between the atoms in the two metals can be so strong that occasionally small fragments of steel are plucked out from their surface and they adhere to the opposing surface (Figure 6.11). However, the adhesion does not occur uniformly over the entire bottom surface of the block. On a microscopic scale, the apparently smooth surface of a machined and polished metal contains a great many irregular protuberances (Figure 6.12). When two such surfaces are placed one over the other, the microscopic area of contact is much smaller than the apparent macroscopic area. It is rather like turning Switzerland upside down and placing it on top of Austria — only the tips of the mountains will touch. Thus, for two metal surfaces, intimate contact will occur only at isolated spots at the tips of the protuberances — and only at these spots will the atoms of one surface adhere to those of the other. During sliding, the surfaces are instantaneously "spot-welded" together; then the welds rupture and new welds form, etc.

Although at the microscopic level the phenomenon of friction is very complicated, at the macroscopic level the resulting friction force can often be described by a simple empirical law, first enunciated by Leonardo da Vinci: *The magnitude of the force of friction between dry, unlubricated surfaces sliding one over the other is proportional to the normal force pressing the surfaces together and is independent of the (macroscopic) area of contact and of the relative speed.*

Friction involving surfaces in relative motion is called **kinetic friction** (or sliding friction). According to the above law, the force of kinetic friction can be written mathematically as

Kinetic friction

$$\boxed{f_k = \mu_k N} \tag{17}$$

where μ_k is the **coefficient of kinetic friction,** a characteristic constant of the materials involved.

Note that Eq. (17) states that the *magnitudes* of $\mathbf{f}_k$ and $\mathbf{N}$ are proportional; the *directions* of these forces are, however, very different: $\mathbf{N}$ is normal to the surface of contact, and $\mathbf{f}_k$ is parallel to this surface, in a direction opposite to that of the motion.

The above simple "law" of friction lacks the general validity of, say, Newton's laws. It is only approximately valid and it is phenomenological, i.e., it is merely a descriptive summary of empirical observations which does not rest on any detailed theoretical understanding of the mechanism that causes friction. Deviations from this simple law occur at extremely high and extremely low speeds; in the former case the friction tends to be smaller and in the latter case larger. However, we can ignore these deviations in many everyday engineering problems in which the speeds are not at these extremes. The simple law is then quite a good approximation for a wide range of materials, and is at its best for metals.

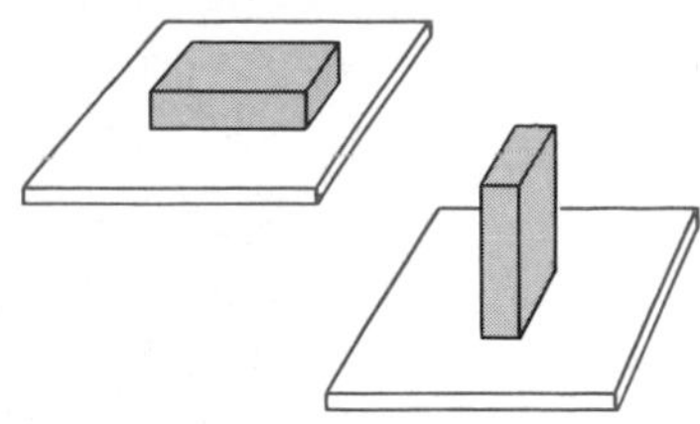

Fig. 6.13 Steel block on a steel plate.

The fact that the friction force is independent of the (macroscopic) area of contact means that the friction force of the block sliding on the steel tabletop is the same whether the block slides on a large face or on one of the small faces (Figure 6.13). This may seem surprising at first: since friction results from adhesion, we might expect the friction force to be larger when the block slides on its large face, because the contact area is larger. However, what determines the amount of adhesion is not the macroscopic contact area, but the microscopic contact area, and the latter is pretty much independent of whether the block rests on a large face or on a small face. The steel of the block only makes contact with the steel of the tabletop at the tips of the irregular protuberances projecting from its surface; the normal force squeezes the protuberances of one surface against those of the other and causes these protuberances to deform to some extent so that they mate more closely with each other. When the block rests on a large face, the number of contacting protuberances is larger than when it rests on a small face, but since the normal force is distributed over a larger number of such protuberances, the deformation of each is less than when the block rests on a small face. Thus, in one case there is a large number of contacting protuberances, each involving a small area; and in the other case there is a somewhat smaller number of contacting protuberances, each involving a somewhat larger area. The net result is that in both cases the sum of all the microscopic contact areas is the same and, consequently, the friction force is the same.

Leonardo da Vinci, *1452–1519, Italian artist, engineer, and scientist. Famous for his brilliant achievements in painting, sculpture, and architecture, Leonardo also made pioneering contributions to science. His insatiable curiosity led him to investigate problems in mathematics, astronomy, mechanics, hydraulics, geology, botany, and anatomy.*

Table 6.2 lists typical friction coefficients μ_k for a variety of materials. The values given are only approximate. The friction depends to some extent on the condition of the surfaces and also on the temperature. For example, if two surfaces of the same metal are cleaned while in a vacuum so as to prevent any contamination, they will then adhere very strongly and the friction will be very large — the two surfaces "weld" together, forming a single chunk of metal. The coefficients

Table 6.2 KINETIC AND STATIC FRICTION COEFFICIENTS[a]

Materials	μ_k	μ_s
Steel on steel	0.6	0.7
Steel on lead	0.9	0.9
Steel on copper	0.4	0.5
Copper on cast iron	0.3	1.1
Copper on glass	0.5	0.7
Waxed ski on snow		
at −10°C	0.2	—
at 0°C	0.05	—
Rubber on concrete	~1	~1

[a] The friction coefficient depends on the condition of the surfaces. The values in this table are typical for dry surfaces but not entirely reliable.

listed in Table 6.2 involve surfaces exposed to air; consequently the metal surfaces are actually covered by films of oxide and there will be very few contacts between the pure metals.

The low friction of ice and snow is due to the formation of a lubricating film of water between the ice surface and the other surface. The sliding process heats the top layer of the ice and melts it; the body skidding on the ice actually floats on this film of water. High speeds and a temperature near the melting point of ice favor this lubrication mechanism.

Incidentally, although the independence of the friction force of the area of (macroscopic) contact is an excellent approximation for metals, it fails for some other materials, e.g., plastics and rubber. For these materials, the coefficient of friction depends on the shape of the sliding surfaces.

To the extent that the friction force is constant, it is easy to calculate the motion of objects with friction.

EXAMPLE 5. A ship is launched toward the water on a slipway making an angle of 5° with the horizontal (see Figure 6.14a). The coefficient of kinetic friction between the bottom of the ship and the slipway is $\mu_k = 0.08$. What is the acceleration of the ship along the slipway?

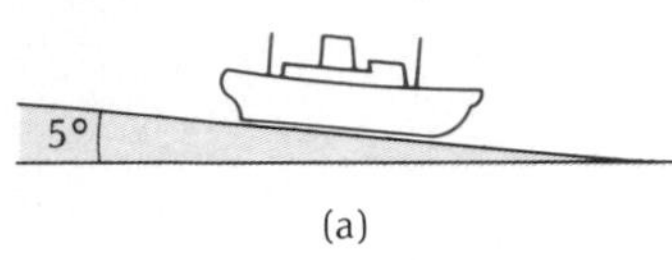

(a)

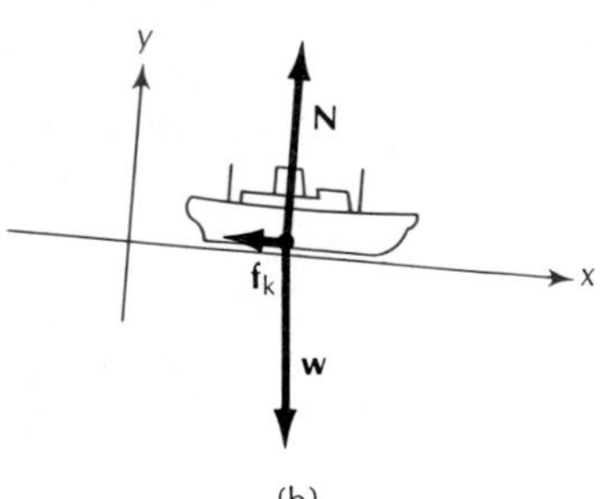

(b)

Fig. 6.14 (a) Ship on slipway. (b) "Free-body" diagram for the ship.

SOLUTION: Figure 6.14b is a "free-body" diagram for the ship. The forces shown are the weight ($w = mg$), the normal force ($N = mg \cos \theta$), and the friction force ($f_k = \mu_k N$). The magnitude of the latter is

$$f_k = \mu_k N = \mu_k mg \cos \theta \tag{18}$$

With the x axis parallel to the plane of the slipway, the net force has an x component $F_x = w_x - f_k$, or $F_x = mg \sin \theta - \mu_k mg \cos \theta$. Hence,

$$ma_x = mg \sin \theta - \mu_k mg \cos \theta$$

or

$$\begin{aligned} a_x &= (\sin \theta - \mu_k \cos \theta)g \\ &= (\sin 5° - 0.08 \cos 5°)g \\ &= 0.0075g = 0.073 \text{ m/s}^2 \end{aligned}$$

EXAMPLE 6. A man pushes a heavy crate over a smooth floor. The man pushes downward and forward, so that his push makes an angle of 30° with the horizontal (Figure 6.15a). The mass of the crate is 60 kg and the coefficient of kinetic friction is $\mu_k = 0.50$. What force must the man exert to keep the crate moving at uniform velocity?

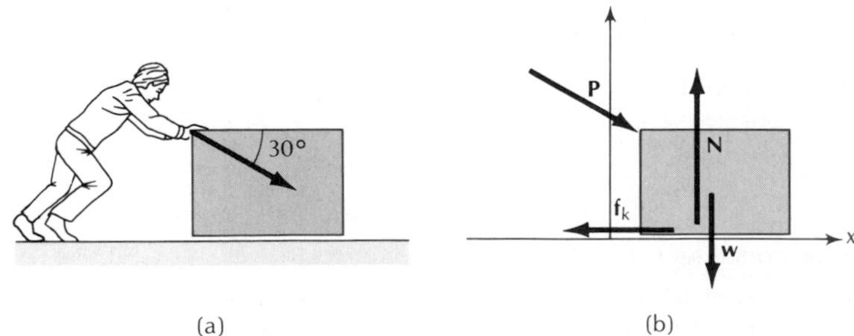

Fig. 6.15 (a) Man pushing a crate. (b) "Free-body" diagram of the crate.

SOLUTION: Figure 6.15b shows a "free-body" diagram for the crate. The forces acting on the crate are the push **P** of the man, the weight **w,** the normal force **N,** and the friction force $\mathbf{f}_k$. Note that because the man pushes the crate down against the floor, the magnitude of the normal force is not equal to *mg;* we will have to treat the magnitude of the normal force as an unknown. The horizontal and vertical components of the forces are, respectively,

$$P_x = P\cos 30° \qquad P_y = -P\cos 60°$$

$$w_x = 0 \qquad w_y = -mg$$

$$N_x = 0 \qquad N_y = N$$

$$f_{k,x} = -\mu_k N \qquad f_{k,y} = 0$$

Since the acceleration of the crate is zero in both the x and y directions, the net force in each of these directions must be zero,

$$P\cos 30° + 0 + 0 - \mu_k N = 0$$

$$-P\cos 60° - mg + N + 0 = 0$$

These are two equations for the two unknowns P and N. By adding μ_k times the second of these equations to the first, we find

$$P(\cos 30° - \mu_k \cos 60°) - \mu_k mg = 0$$

from which

$$P = \frac{\mu_k mg}{\cos 30° - \mu_k \cos 60°}$$

$$= \frac{0.50 \times 60\text{ kg} \times 9.8\text{ m/s}^2}{\cos 30° - 0.50 \times \cos 60°} = 4.8 \times 10^2\text{ N}$$

Friction forces also act between two surfaces at rest. If a force is exerted against the side of, say, a steel block initially at rest on a steel tabletop, the block will not move unless the force is sufficiently large to overcome the friction that holds it in place. For example, on a 1-kg steel block resting on steel, a lateral push of about 7 N must be applied to start the motion. Friction between surfaces at rest is called **static friction.** The maximum magnitude of the static friction force, that is, the magnitude that this force attains when the lateral push is just about to start the motion, can be described by an empirical law quite similar to that for the kinetic friction force: *The magnitude of the maximum force of friction between dry, unlubricated surfaces at rest with respect to each other is proportional to the normal force and independent of the (macroscopic) area of contact.* Mathematically,

$$\boxed{f_{s,max} = \mu_s N} \qquad (19)$$

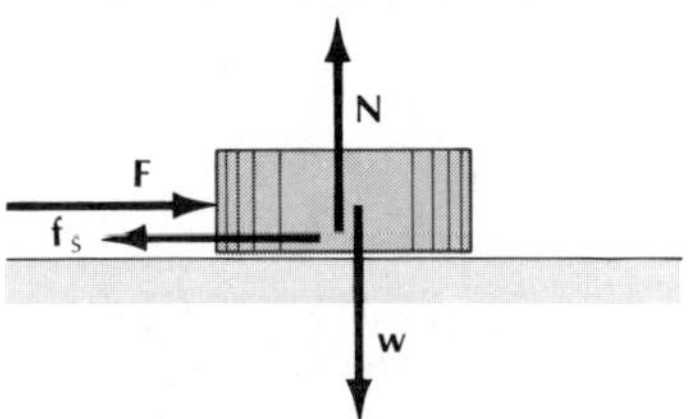

Fig. 6.16 Forces on a steel block at rest on a steel plate.

where μ_s is the **coefficient of static friction.** The direction of $f_{s,\,max}$ is such as to oppose the lateral push that tries to move the body (Figure 6.16).

The force in Eq. (19) carries the subscript *max* because it represents the largest friction force that the surfaces can support without beginning to slide, i.e., $f_{s,max}$ is the friction force at the "breakaway" point, when the lateral push is just about to start the motion. Of course, if the lateral push is below this critical value, then the friction force f_s is also below $f_{s,max}$ and exactly matches the magnitude of the lateral push. Thus, in general,

Static friction

$$\boxed{f_s \leq \mu_s N} \tag{20}$$

For most materials $\mu_s > \mu_k$, i.e., the extreme static friction force is larger than the kinetic friction force. This implies that if the lateral push applied to the block is large enough to start it moving, it will more than compensate for the subsequent kinetic friction and it will therefore accelerate the block continuously.

Table 6.2 includes some typical values of the coefficient of friction μ_s.

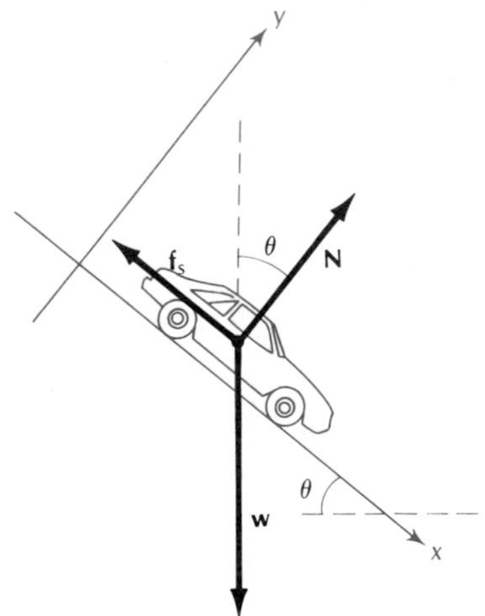

Fig. 6.17 "Free-body" diagram for an automobile parked on a very steep street.

EXAMPLE 7. The coefficient of static friction of hard rubber on a street surface is $\mu_s = 0.8$. What is the steepest slope of a street on which an automobile with rubber tires (and locked wheels) can rest without slipping?

SOLUTION: The "free-body" diagram is shown in Figure 6.17. The angle θ is assumed to be at its maximum value, i.e., the friction force has its maximum value $f_{s,max} = \mu_s N = \mu_s mg \cos\theta$. The component of the net force along the inclined plane is

$$\begin{aligned} F_x &= w_x - f_{s,max} \\ &= mg \sin\theta - \mu_s mg \cos\theta \end{aligned} \tag{21}$$

If the automobile is to remain stationary, $F_x = 0$ and

$$0 = mg \sin\theta - \mu_s mg \cos\theta$$

or

$$\mu_s = \tan\theta \tag{22}$$

With $\mu_s = 0.8$, this gives $\theta = 39°$, or a slope of 4:5.

EXAMPLE 8. An automobile is braking on a level road. What is the maximum deceleration that the automobile can achieve without skidding? As in the preceding example, assume that the tires of the automobile have a coefficient of static friction $\mu_s = 0.8$.

SOLUTION: If the wheels are rolling without skidding, their rubber surface does *not* slide on the street surface (the point of contact between the wheel and the street is instantaneously at rest on the street — we will further explore this feature of rolling motion in Section 12.5). Hence, the relevant friction force is the *static* friction force. The maximum value of this force is

$$f_{s,\,max} = \mu_s N$$

where the normal force N is simply the weight mg of the automobile. The deceleration is then given by

$$ma = -f_{s,\,max} = -\mu_s mg$$

which yields

$$a = -\mu_s g = -0.8 \times 9.8\ m/s^2 = -7.8\ m/s^2$$

6.5 Elastic Force of a Spring; Hooke's Law

A body is said to be **elastic** if it suffers a deformation when a stretching or compressing force is applied to it. For example, suitable forces can stretch a coil spring or a rubber band and they can bend a flexible rod or a beam of metal or wood. Even bodies normally regarded as rigid, such as the balls of a ball bearing made of hardened steel, are somewhat elastic — they will deform if a sufficiently large force is applied to them.

Robert Hooke, *1635–1703, English experimental physicist, curator and secretary of the Royal Society. He was regarded as the foremost mechanic of his time, making many improvements in telescopes and other astronomical instruments, clocks, and watches. Hooke recognized that the motion of the planets must be considered as a mechanical problem, and he proposed the inverse-square law for the gravitational force.*

The force with which a body resists deformation is called its **restoring force.** If we stretch a spring by pulling with one hand on each end, we can feel the restoring force opposing our pull. Under static conditions the restoring force and the force that produces the deformation are an action–reaction pair.

Hooke's law

The restoring force with which an elastic body opposes whatever pulls on it often obeys a simple empirical law known as **Hooke's law:** *the magnitude of the restoring force is directly proportional to the deformation.* This is not a general law of physics — the *exact* restoring force produced by the deformation of an elastic body depends in a complicated way on the shape of the body and on the detailed properties of the material of the body. Hooke's law is only an approximate, phenomenological description of the restoring force. However, it is often a very good approximation, provided that the deformation is small.

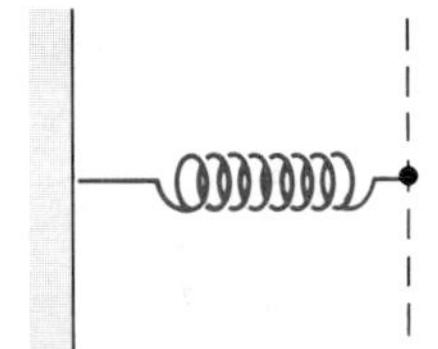

Fig. 6.18 Spring, relaxed.

As a special case, consider a coil spring. Figure 6.18 shows such a spring in its relaxed state; it is loosely coiled so that it can be compressed as well as stretched. Suppose that we attach the left end of the spring to a rigid support (wall) and we apply a stretching or compressing force to the right end. Under the influence of this force, the spring will settle in a new equilibrium configuration such that the restoring force exactly balances the externally applied force. We can measure the deformation of the spring by the displacement that the right end undergoes relative to its initial position. In Figure 6.19 this displacement is denoted by x. A positive value of x corresponds to an elongation of the spring and a negative value corresponds to a compression. Clearly, x is nothing but the change in the length of the spring.

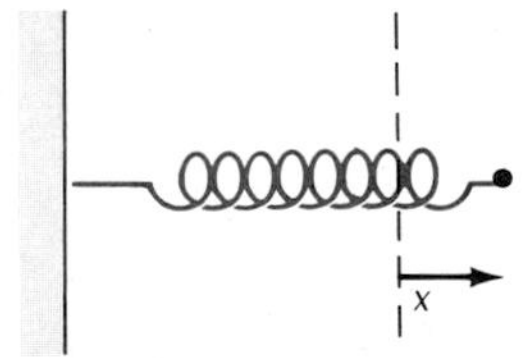

Fig. 6.19 Spring, stretched by a length x.

Hooke's law for the value of the restoring force, then, is that the force is directly proportional to x,

Restoring force of a spring

$$\boxed{F = -kx} \tag{23}$$

Spring constant

The constant k is the **spring constant**; it is a positive number. The spring constant is a measure of the stiffness of the spring — a stiff spring has a high value of k, and a soft spring a low value of k. The unit of the spring constant is the newton per meter (N/m) or pound per foot (lb/ft). The negative sign in Eq. (23) indicates that the restoring force opposes the deformation: if the spring is elongated (positive x), then the restoring force is negative and opposes the external stretching force (Figure 6.20a); if the spring is compressed (negative x), then the restoring force is positive and opposes the external compressing force (Figure 6.20b).

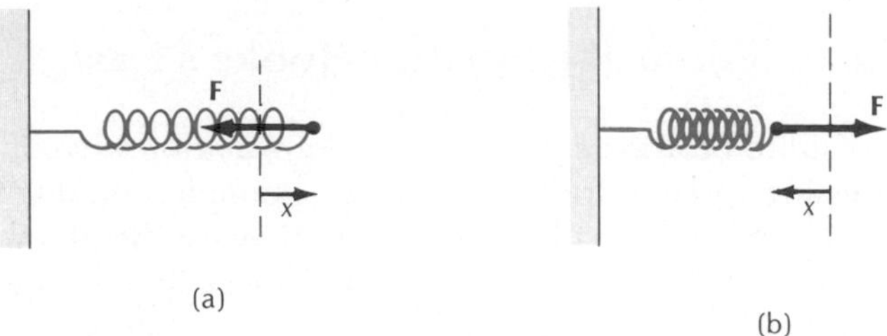

Fig. 6.20 (a) If x is positive, then F is negative. (b) If x is negative, then F is positive.

Coil springs usually obey Hooke's law quite closely unless the deformation is excessively large. If a spring is stretched beyond its elastic limit, it will suffer a permanent deformation and not snap back to its original shape when released. Such damage to the spring destroys the simple proportionality given by Eq. (23). Furthermore, if a spring is stretched much beyond its elastic limit, it will break and then, of course, the restoring force disappears altogether. Figure 6.21 is a plot of the restoring force versus length for a small steel spring stretched beyond its elastic limit.

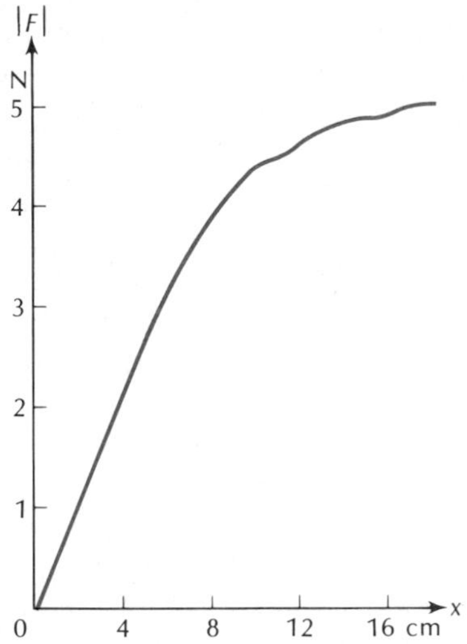

Fig. 6.21 Restoring force vs. length for a small steel spring. When stretched beyond 7 cm, the spring suffered a permanent deformation.

EXAMPLE 9. The manufacturer's specifications for the coil spring for the front suspension of a Triumph sports car call for a spring of 10 coils with a relaxed length of 0.316 m, and a length of 0.205 m when under a load of 399 kg. What is the spring constant?

SOLUTION: The restoring force that will balance the weight of 399 kg is $F = 399\text{ kg} \times 9.81\text{ m/s}^2 = 3.91 \times 10^3\text{ N}$. The corresponding change of length is $x = 0.205\text{ m} - 0.316\text{ m} = -0.111\text{ m}$. Hence,

$$k = -\frac{F}{x} = -\frac{3.91 \times 10^3\text{ N}}{-0.111\text{ m}} = 3.52 \times 10^4\text{ N/m}$$

EXAMPLE 10. If the spring described in the preceding example is cut into two equal pieces, what will be the spring constant of each piece?

SOLUTION: If the spring is cut into two equal pieces, the spring constant of each piece will be twice as large, i.e., $k' = 7.04 \times 10^4$ N/m. This can be best understood by noting that if a 5-coil spring is to be compressed by, say, 10 cm, each coil must be deformed by 2 cm; whereas if a 10-coil spring is to be compressed by 10 cm, each coil must be deformed by only 1 cm. The force required to deform one coil by 2 cm is twice as large as the force required to deform one coil by 1 cm. Consequently, the spring constant of the 5-coil spring is twice as large as the spring constant of the 10-coil spring.

This example leads to a general rule: other things being equal, the spring constant is inversely proportional to the number of coils. With this rule we can understand how the suspension springs of an automobile can be stiffened by means of clamps that rigidly hold two adjacent coils so as to prevent their relative motion. Such a clamp makes the two coils act as one, i.e., it effectively reduces the total number of coils.

6.6 Motion with a Variable Force[2]

If the force acting on a body is a known function of position or of time, then Newton's Second Law determines the acceleration of the body at each instant, and from this acceleration the complete motion can be calculated. The examples of the preceding sections illustrate how such calculations are done in the simple case of a constant force. However, this case is an exception — most forces that occur in nature are *not* constant. In this section we will take a brief look at an example involving a motion with a variable force. Mathematically, the solution of the equation of motion with a variable force is often quite difficult. Consider one-dimensional motion (along the x axis) with a force that is a known function of position, $F = F(x)$. The equation of motion is

$$m\frac{d^2x}{dt^2} = F(x) \qquad (24)$$

Differential equation

This is a **differential equation** for the function $x(t)$, i.e., an equation that contains the function $x(t)$ and its derivatives. The function $x(t)$ is regarded as the unknown; finding a solution of the differential equation means finding a function $x(t)$ such that Eq. (24) is satisfied. The solution will be mathematically simple only if $F(x)$ is some very special function. For example, if the force is that of a spring obeying Hooke's law, $F(x) = -kx$, then Eq. (24) can be solved fairly easily in terms of simple mathematical functions (see Section 14.2). But, in general, the solution of Eq. (24) cannot be expressed in "closed" form, i.e., it cannot be expressed in terms of well-known, standard mathematical functions such as sines, cosines, or exponentials.

If no convenient exact solution is available, one may have recourse to approximate solutions obtained by numerical methods. To illustrate such a numerical solution, we will look at the example of an inverse-square force,

[2] This section is optional.

$$F(x) = -\frac{A}{x^2} \tag{25}$$

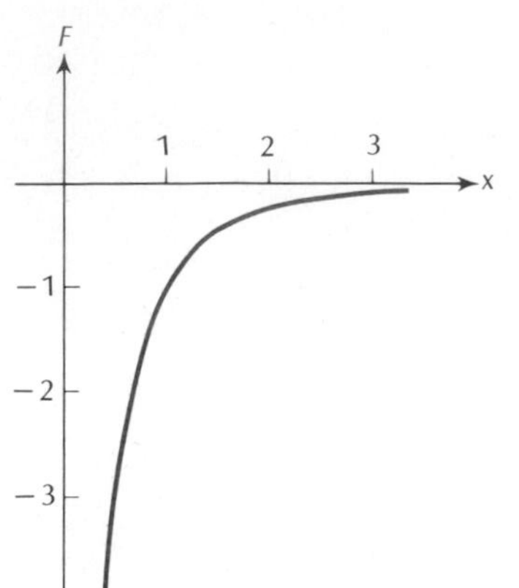

Fig. 6.22 The inverse-square force as a function of x, with $A = 1$.

where A is a constant. At points close to $x = 0$, this force has a very large value; the point $x = 0$ is called the **center of force.** The direction of the force is toward this center[3] and the magnitude decreases as the distance x from this center increases. According to Eq. (25), whenever x increases by a factor of two, the force decreases by a factor of four, etc. Figure 6.22 is a plot of $F(x)$.

The equation of motion of the particle exposed to this force is then

$$m\frac{d^2x}{dt^2} = -\frac{A}{x^2}$$

or

$$a = -\frac{1}{x^2}\frac{A}{m} \tag{26}$$

This equation describes, for example, the motion of a high-altitude rocket, such as a ballistic missile, that has been fired straight up and that is in free fall after its engines have been cut off. The only force on this rocket is then the gradually decreasing pull of the Earth's gravity — this force varies in proportion to the inverse square of the distance from the Earth's center (see Chapter 13). For now we do not want to worry about the physical situation described by Eq. (26); instead, we want to concentrate on the mathematical solution of this equation.

To solve the equation numerically, we regard the time variable as made up of discrete intervals, each of which lasts a time Δt. The instants of time that characterize the beginning and the end of each of these small intervals are

$$\begin{aligned} t_0 &= 0 \\ t_1 &= \Delta t \\ t_2 &= 2\ \Delta t \\ t_3 &= 3\ \Delta t, \qquad \text{etc.} \end{aligned} \tag{27}$$

and the values of x at these times are

$$\begin{aligned} &x_0, \quad \text{position at } t = t_0 \\ &x_1, \quad \text{position at } t = t_1 \\ &x_2, \quad \text{position at } t = t_2 \\ &x_3, \quad \text{position at } t = t_3, \qquad \text{etc.} \end{aligned} \tag{28}$$

We now approximate the exact worldline of the particle by a sequence

[3] The minus sign in Eq. (25) indicates that the force is in the negative direction; i.e., it is toward the origin whenever x is on the positive x axis. Throughout the following calculation we will assume $x > 0$.

of straight worldline segments (Figure 6.23). This means that the velocities within each time interval are regarded as constant,

$$\begin{aligned} &v_0, \quad \text{velocity for } t_0 \le t < t_1 \\ &v_1, \quad \text{velocity for } t_1 \le t < t_2 \\ &v_2, \quad \text{velocity for } t_2 \le t < t_3, \quad \text{etc.} \end{aligned} \tag{29}$$

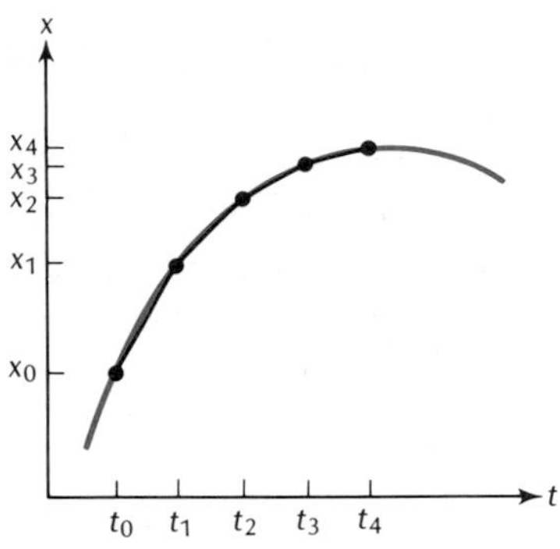

Fig. 6.23 The worldline of a particle can be approximated by short straight segments.

Hence, the values of x at succeeding times t_i and t_{i+1} are related by

$$x_{i+1} = x_i + v_i\,\Delta t \tag{30}$$

where $i = 0, 1, 2, 3,$ etc.

The acceleration [Eq. (26)] will gradually change the velocity of the particle. In the context of our approximation, all of this change is supposed to occur at the ends of the time intervals (Figure 6.23). The values of the velocities at succeeding times t_i and t_{i+1} differ by about $a\,\Delta t$, where we take the acceleration a at x_i, that is, $a = -(1/x_i^2)\,(A/m)$. Thus

$$v_{i+1} = v_i - \frac{1}{x_i^2}\frac{A}{m}\,\Delta t \tag{31}$$

The two sets of equations for position and velocity [Eqs. (30) and (31), respectively] can be solved recursively, beginning with the initial values x_0 and v_0, which are assumed known. The segmented worldline obtained in this way is an approximation to the precise curved worldline; the approximation becomes exact in the limit $\Delta t \to 0$. Of course, we cannot achieve this limit in our numerical work — we will have to be satisfied with some small, but finite, value at Δt.

For the following calculation we will assume that $A = 1.0$ and that the mass of the particle is $m = 1.0$ (in some suitable units). Equations (30) and (31) then become

$$x_{i+1} = x_i + v_i\,\Delta t \tag{32}$$

$$v_{i+1} = v_i - \Delta t/x_i^2 \tag{33}$$

For the initial values we take $x_0 = 1.0$ and $v_0 = 1.0$. For the time interval we take $\Delta t = 0.1$; this choice of Δt is small enough to give reasonable accuracy without excessive labor. The calculation then proceeds step by step:

$t_0 = 0$

$x_0 = 1.0$

$v_0 = 1.0$

$t_1 = 0.1$

$$x_1 = x_0 + v_0\,\Delta t = 1.0 + 1.0 \times 0.1 = 1.100$$

$$v_1 = v_0 - \frac{\Delta t}{x_0^2} = 1.0 - \frac{0.1}{(1.0)^2} = 0.900$$

$$t_2 = 0.2$$

$$x_2 = x_1 + v_1\,\Delta t = 1.100 + 0.90 \times 0.1 = 1.190$$

$$v_2 = v_1 - \frac{\Delta t}{x_1^2} = 0.900 - \frac{0.1}{(1.10)^2} = 0.817$$

and so on.

Table 6.3 gives the result of the complete calculation (done on a pocket calculator) and Figure 6.24 gives a plot of position as a function of time. At the time $t = 2.5$, the particle reaches a maximum distance $x \cong 1.99$ from the center of force; at this time the velocity changes from positive to negative and the particle begins to move back. Clearly, what has happened is that the attractive force has halted the outward motion of the particle and begun to pull it back in. The subsequent inward motion is nothing but the outward motion in reverse.

Table 6.3 NUMERICAL SOLUTION OF EQS. (32) AND (33)

t_i	x_i	v_i
0.0	1.000	1.000
0.1	1.100	0.900
0.2	1.190	0.817
0.3	1.272	0.747
0.4	1.346	0.685
0.5	1.415	0.630
0.6	1.478	0.580
0.7	1.536	0.534
0.8	1.589	0.492
0.9	1.638	0.452
1.0	1.684	0.415
1.1	1.726	0.380
1.2	1.763	0.346
1.3	1.798	0.314
1.4	1.829	0.283
1.5	1.857	0.253
1.6	1.883	0.224
1.7	1.905	0.196
1.8	1.924	0.168
1.9	1.941	0.141
2.0	1.956	0.115
2.1	1.967	0.088
2.2	1.976	0.063
2.3	1.982	0.037
2.4	1.986	0.012
2.5	1.987	−0.014

The dots in Figure 6.24 show the approximate results of Table 6.3 and the smooth curve shows the exact solution of the equation of motion. Our approximation is quite good, and it can be made even better by repeating the calculation using a shorter time interval, say, $\Delta t = 0.05$. Such a reduction of the time interval by, say, a factor of 2 can also serve as a self-consistent test of the approximation: if the numerical calculations with $\Delta t = 0.1$ and $\Delta t = 0.05$ give just about the same results, then one can be reasonably confident that the results are pretty close to the exact solution.

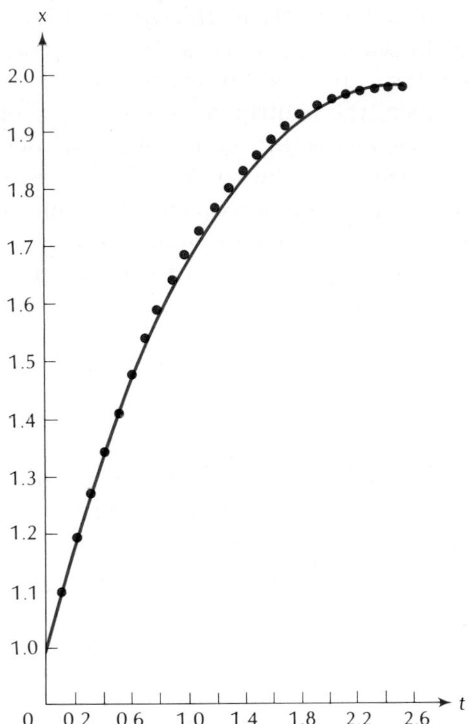

Fig. 6.24 The dots represent the results listed in Table 6.3. The smooth curve is the exact worldline.

Incidentally, the method of numerical solution adopted in the above example is not the most efficient. For example, one can improve the precision of the calculation by inserting in place of the velocity v_i in Eq. (32) the average velocity $\frac{1}{2}(v_i + v_{i+1})$. When programming numerical calculations on an electronic digital computer, it usually pays to take advantage of such tricks. But the main reason for the high precision attainable with modern electronic computers is their high speed of computation, which makes it feasible to proceed by a very large number of steps, each involving only a very small value of Δt. For example, for the accurate numerical solution of the equation of motion of a planet, astronomers trace out the orbit by taking several hundred thousand steps around the Sun.

However, these refinements must not be allowed to obscure the important principle involved in Eq. (30) and (31). These equations very explicitly display the essential *deterministic* feature of the equation of motion: given the initial values of position and velocity, a step-by-step calculation leads us inevitably to the values of position and velocity at any later time.

6.7 Dynamics of Uniform Circular Motion

As we saw in Section 4.4, uniform circular motion is accelerated motion with a centripetal acceleration. If the motion proceeds with speed

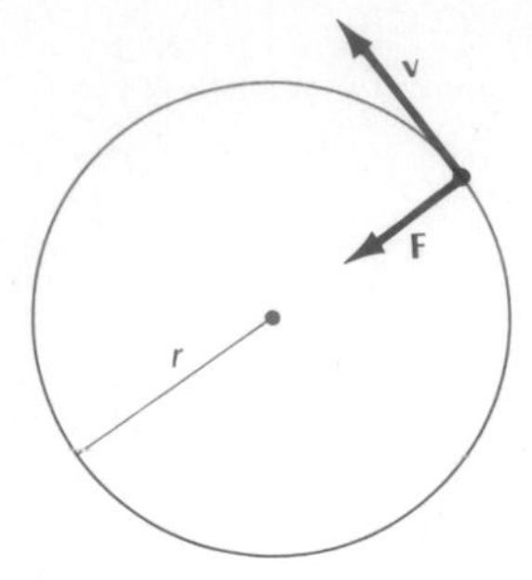

Fig. 6.25 Centripetal force for a particle in uniform circular motion.

v along a circle of radius r, the magnitude of the centripetal acceleration is

$$a = v^2/r \tag{34}$$

According to Newton's Second Law, this acceleration must be caused by a force having the same direction as that of the acceleration; i.e., the direction of the force must be toward the center of the circle (Figure 6.25). For instance, the centripetal acceleration of a stone being whirled around a circle at the end of a string is caused by the pull of the string toward the center of the circle, and the centripetal acceleration of the Earth moving in its (nearly) circular orbit around the Sun is caused by the gravitational pull of the Sun.

The magnitude of the force required to maintain uniform circular motion is

Force for uniform circular motion

$$\boxed{ma = mv^2/r} \tag{35}$$

This equation can be used to calculate the magnitude of the force if the speed of the motion is known, or it can be used to calculate the speed if the force is known. The following are several examples of such calculations involving different kinds of forces.

EXAMPLE 11. The Earth moves around the Sun in a circular orbit of radius 1.50×10^{11} m. The speed of the Earth is 2.98×10^4 m/s (see Example 2.2) and its mass is 5.98×10^{24} kg. Calculate the magnitude of the force required to maintain this motion.

SOLUTION: From Eq. (35)

$$ma = \frac{mv^2}{r} = \frac{5.98 \times 10^{24}\text{ kg} \times (2.98 \times 10^4\text{ m/s})^2}{1.50 \times 10^{11}\text{ m}}$$

$$= 3.54 \times 10^{22}\text{ N}$$

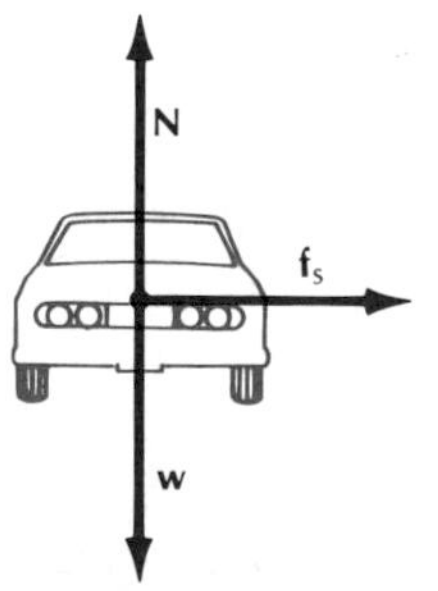

Fig. 6.26 "Free-body" diagram for an automobile rounding a curve.

EXAMPLE 12. What is the maximum speed with which an automobile can round a curve of radius 100 m? Assume that the road is flat and that the coefficient of static friction between the tires and the road surface is $\mu_s = 0.8$.

SOLUTION: The "free-body" diagram of the automobile is given in Figure 6.26. The weight balances the normal force, i.e., $N = mg$. The horizontal friction force must provide the centripetal force; hence the magnitude of the friction force must be

$$f_s = mv^2/r$$

The friction is *static* because, by assumption, there is no lateral slippage. At the maximum speed the friction force has its maximum value $f_s = \mu_s mg$ so that

$$\mu_s mg = mv^2/r$$

This yields

$$v = \sqrt{\mu_s g r}$$

$$= \sqrt{0.8 \times 9.81\ \text{m/s}^2 \times 100\ \text{m}}$$

$$= 28.0\ \text{m/s} \qquad (36)$$

EXAMPLE 13. At a speedway in Texas, a curve of radius 500 m is banked at an angle of 22° (Figure 6.27a). If the driver of a racing car does not wish to rely on friction, at what speed should he take this curve?

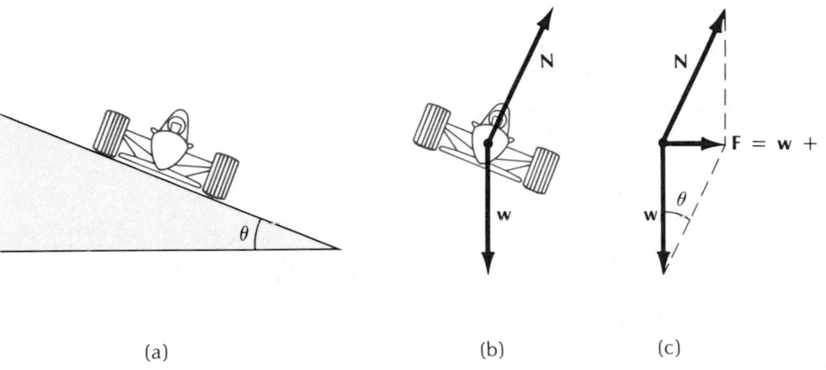

Fig. 6.27 (a) Car on a banked curve. (b) "Free-body" diagram for a car rounding a banked curve. (c) Resultant of the forces **N** and **w**.

SOLUTION: The "free-body" diagram of the car is shown in Figure 6.27b. Friction is assumed to be absent and hence the normal force **N** and the weight **w** are the only forces acting on the racing car.[4] The resultant of these forces must provide the centripetal force. Hence the resultant must be horizontal, as in Figure 6.27c. From this figure we see that the magnitude of the resultant is $F = w \tan\theta$, which must coincide with the magnitude of the centripetal force,

$$w \tan\theta = mv^2/r$$

or

$$mg \tan\theta = mv^2/r$$

which yields

$$v = \sqrt{rg \tan\theta}$$

$$= \sqrt{500\ \text{m} \times 9.81\ \text{m/s}^2 \times \tan 22^\circ}$$

$$= 44.5\ \text{m/s} \qquad (37)$$

This is 99.6 mi/h. If the car goes faster than this, it will tend to skid up the embankment; if it goes slower it will tend to skid down the embankment unless friction holds it there.

EXAMPLE 14. A pilot in a fast jet aircraft loops the loop (Figure 6.28). Assume that the aircraft maintains a constant speed of 200 m/s and that the radius of the loop is 1.5 km. (a) At the bottom of the loop, what is the apparent weight that the pilot feels, i.e., the force with which the seat of his pants presses against his chair? Express the answer as a multiple of his normal weight. (b) What is his apparent weight at the top of the loop?

Fig. 6.28 Jet looping the loop.

[4] Air resistance also acts on the car, but is compensated by the propulsive force that the wheels produce by reaction on the ground.

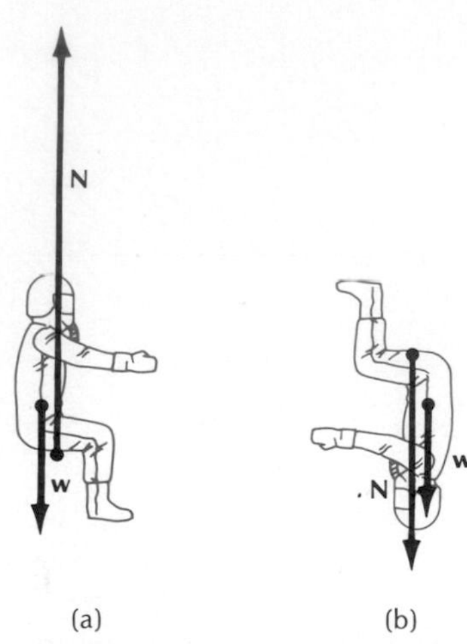

Fig. 6.29 "Free-body" diagrams of jet pilot at (a) the bottom and (b) the top of the loop.

SOLUTION: (a) Figure 6.29a shows a "free-body" diagram for the pilot at the bottom of the loop. The forces acting on him are the (true) weight **w** and the normal force **N** exerted by the chair. The net vertical upward force is $N - mg$ and this must provide the centripetal acceleration:

$$N - mg = \frac{mv^2}{r}$$

Hence

$$N = mg + \frac{mv^2}{r} = mg\left(1 + \frac{v^2}{gr}\right)$$

With $v = 200$ m/s and $r = 1.5 \times 10^3$ m, this gives

$$N = mg\left(1 + \frac{(200\ \text{m/s})^2}{9.8\ \text{m/s}^2 \times 1.5 \times 10^3\ \text{m}}\right)$$

$$= mg \times 3.7$$

Thus, the apparent weight is larger than the normal weight by a factor of 3.7. In these circumstances, the pilot would say that he is experiencing 3.7 gee (or "pulling 3.7 gee") because he feels as though gravity had increased by a factor of 3.7.

(b) Figure 6.29b shows the "free-body" diagram at the top of the loop. The net vertical downward force is $N + mg$ and

$$N + mg = \frac{mv^2}{r}$$

Hence

$$N = -mg + \frac{mv^2}{r} = mg\left(-1 + \frac{v_2}{gr}\right)$$

This gives

$$N = mg\left(-1 + \frac{(200\ \text{m/s})^2}{9.8\ \text{m/s}^2 \times 1.5 \times 10^3\ \text{m}}\right)$$

$$= mg \times 1.7$$

The apparent weight is larger than the normal weight by a factor of 1.7. Note that the pilot presses *up* against the chair. Thus, he will have the impression that the direction of gravity is reversed — the sky is below his feet and the Earth is above his head. This sensation can be very disorienting for inexperienced pilots.

SUMMARY

The four fundamental forces: gravitational, "weak," electromagnetic, "strong"

Weight: magnitude: $w = mg$
direction: Downward

Kinetic friction: $f_k = \mu_k N$

Static friction: $f_s \leq \mu_s N$

Restoring force of a spring (Hooke's law): $F = -kx$

Force required for uniform circular motion: magnitude: mv^2/r
direction: Centripetal

QUESTIONS

1. According to the adherents of parapsychology, some people are endowed with the supernormal power of psychokinesis, i.e., spoon-bending-at-a-distance via mysterious psychic forces emanating from the brain. Physicists are confident that the only forces acting between pieces of matter are those listed in Table 6.1, none of which are implicated in psychokinesis. Given that the brain is nothing but a (very complicated) piece of matter, what conclusions can a physicist draw about psychokinesis?

2. If you carry a spring balance from London to Hong Kong, do you have to recalibrate it? If you carry a beam balance?

3. When you stretch a rope horizontally between two fixed points, it always sags a little, no matter how great the tension. Why?

4. What are the forces on a soaring bird? How can the bird gain altitude without flapping its wings?

5. An automobile is parked on a street. (a) Draw a "free-body" diagram showing the forces acting on the automobile. What is the net force? (b) Draw a "free-body" diagram showing the forces that the automobile exerts on the Earth. Which of the forces in diagrams (a) and (b) are action–reaction pairs?

6. A ship sits in calm water. What are the forces acting on the ship? Draw a "free-body" diagram for the ship.

7. Some old-time roofers claim that when walking on a rotten roof, it is important to "walk with a light step so that your full weight doesn't rest on the roof." Can you walk on a roof with less than your full weight? What is the advantage of a light step?

8. In a tug-of-war on sloping ground the party on the low side has the advantage. Why?

9. The label on a package of sugar claims that the contents are "1 lb or 454 g." What is wrong with this statement?

10. A physicist stands on a bathroom scale in an elevator. When the elevator is stationary, the scale reads 160 lb. Describe qualitatively how the reading of the scale will fluctuate when the elevator makes a trip to a higher floor.

11. How could you use a pendulum suspended from the roof of your automobile to measure its acceleration?

12. When an airplane flies along a parabolic path similar to that of a projectile, the passengers experience a sensation of weightlessness. How would the airplane have to fly to give the passengers a sensation of enhanced weight?

13. A frictionless chain hangs over two adjoining inclined planes (Figure 6.30a). Prove the chain is in equilibrium, i.e., the chain will not slip to the left or to the right. [Hint: One method of proof, due to the seventeenth-century engineer and mathematician Simon Stevin, asks you to pretend that an extra piece of chain is hung from the ends of the original chain (Figure 6.30b). This makes it possible to conclude that the original chain cannot slip.]

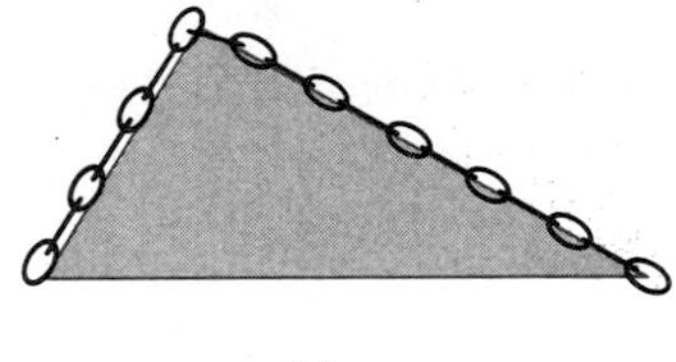

(a)

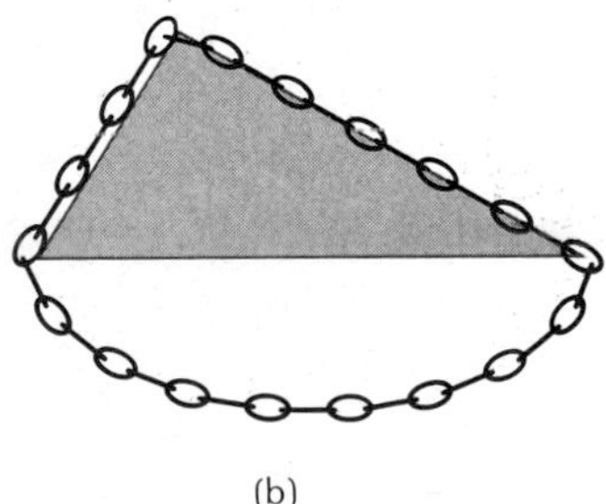

(b)

Fig. 6.30

14. Seen from a reference frame moving with the wave, the motion of a surfer is analogous to the motion of a skier down a mountain.[5] If the wave were to last forever, could the surfer ride it forever? In order to stay on the wave as long as possible, in what direction should the surfer ski the wave?

15. Excessive polishing of the surfaces of a block of metal increases its friction. Explain.

16. Some drivers like to spin the wheels of their automobiles for a quick start. Does this give them greater acceleration? (Hint: $\mu_s > \mu_k$.)

17. Cross-country skiers like to use a ski wax that gives their skis a large coefficient of static friction, but a low coefficient of kinetic friction. Why is this useful? How do "waxless" skis achieve the same effect?

18. Designers of locomotives usually reckon that the maximum force available for moving the train ("tractive force") is $\frac{1}{4}$ or $\frac{1}{5}$ of the weight resting on the drive wheels of the locomotive. What value of the friction coefficient between the wheels and the track does this implicitly assume?

19. When an automobile with rear-wheel drive accelerates from rest, the maximum acceleration that it can attain is less than the maximum deceleration that it can attain while braking. Why? (Hint: Which wheels of the automobile are involved in acceleration? In braking?)

20. Can you think of some materials with $\mu_k > 1$?

21. For a given initial speed, the stopping distance of a train is much longer than that of a truck. Why?

22. Why does the traction on snow or ice of an automobile with rear-wheel drive improve when you place extra weight over the rear wheels?

Fig. 6.31 Drag racer at the start of a race.

23. Why are wet streets slippery?

24. In order to stop an automobile on a slippery street in the shortest distance, it is best to brake as hard as possible without initiating a skid. Why does skidding lengthen the stopping distance? (Hint: $\mu_s > \mu_k$.)

25. Suppose that in a panic stop, a driver locks the wheels of his automobile and leaves skid marks on the pavement. How can you deduce his initial speed from the length of the skid marks?

26. Hot-rod drivers in drag races find it advantageous to spin their wheels very fast at the start so as to burn and melt the rubber of their tires (Figure 6.31). How does this help them to attain a larger acceleration than expected from the static coefficient of friction?

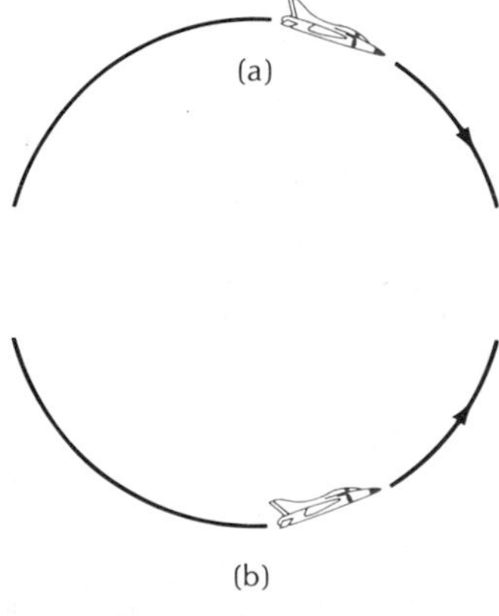

Fig. 6.32 Aircraft (a) beginning a dive and (b) pulling out of a dive.

27. A curve on a highway consists of a quarter of a circle connecting two straight segments. If this curve is banked perfectly for motion at some given speed, can it be joined to the straight segments without a bump? How could you design a curve that is banked perfectly along its entire length and merges smoothly into straight segments without any bump?

28. Automobiles with rear engines (such as the old VW "Beetle") tend to oversteer, i.e., in a curve their rear end tends to swing toward the outside of the curve, turning the car excessively into the curve. Explain.

29. (a) If a pilot in a fast aircraft very suddenly pulls out of a dive (Figure 6.32b), he will suffer "blackout" caused by loss of blood pressure in the brain. If he suddenly begins a dive while climbing (Figure 6.32a), he will suffer "redout" caused by excessive blood pressure in the brain. Explain.

(b) A pilot wearing a G suit — a tightly fitting garment that squeezes the tissues of the legs and abdomen — can tolerate 5 gee while pulling out of a dive.

[5] There is, however, one complication: surf waves grow higher as they approach the beach. Ignore this complication.

How does the G suit prevent blackout? A pilot can tolerate no more than −2 gee while beginning a dive. Why does the G suit not help against redout?

30. While rounding a curve at high speed, a motorcycle rider leans the motorcycle toward the center of the curve. Why?

PROBLEMS

Section 6.2

1. What is the actual weight (in pounds) of a "1-lb" bag of sugar in New York? In Hong Kong? In Quito? Use the data given in Table 2.3.

2. A bar of gold of mass 500 g is transported from Paris ($g = 9.8094$ m/s^2) to San Francisco ($g = 9.7996$ m/s^2).
   (a) What is the decrease of the weight of the gold? Express your answer in ounces.
   (b) The price of gold is $400 an ounce. Does this mean that the bar of gold is worth less in San Francisco?

3. At launch, the engines of a Saturn V rocket develop a (vertical) thrust of 3.4×10^6 kp. The rocket has a mass of 2.4×10^6 kg. What is its upward acceleration?

4. An elevator accelerates upward at 1.8 m/s^2. What is the normal force on the feet of an 80-kg passenger standing in the elevator? By how much does this force exceed his weight?

Section 6.3

5. A boy on a skateboard rolls down a hill of slope 1:5. What is his acceleration? What speed will he reach after rolling for 50 m? Ignore friction.

6. A skier of mass 75 kg is sliding down a frictionless hillside inclined at 35° to the horizontal.
   (a) Draw a "free-body" diagram showing all the forces acting on the skier (regarded as a particle); draw a separate diagram showing the resultant of these forces.
   (b) What is the magnitude of each force? What is the magnitude of the resultant?
   (c) What is the acceleration of the skier?

7. Suppose that the last car of a train becomes uncoupled while the train is moving upward on a slope of 1:6 at a speed of 30 mi/h. How far will the car coast up the slope before it stops? Ignore friction.

8. A bobsled slides down an icy track making an angle of 30° with the horizontal. How far must the bobsled slide in order to attain a speed of 90 km/h if initially at rest? When will it attain this speed? Assume that the motion is frictionless.

9. Figure 6.33 shows a spherical ball hanging on a string on a smooth frictionless wall. The mass of the ball is m, its radius is R, and the length of the string is l. Find the normal force between the ball and the wall. Show that $N \to 0$ as $l \to \infty$.

10. Figure 6.34 shows two masses hanging from a string running over a pulley (see also Figure 6.8a). Such a device can be used to measure the acceleration of gravity; it is then called **Atwood's machine**. If the masses are nearly equal, then the acceleration a of the masses will be much smaller than g; this makes it convenient to measure a and then to calculate g by means of Eq. (15). Suppose that an experimenter using masses $m_1 = 400.0$ g and $m_2 = 402.0$ g finds that the masses move a distance of 0.50 m in 6.4 s starting from rest. What value of g does this imply? Assume the pulley is massless.

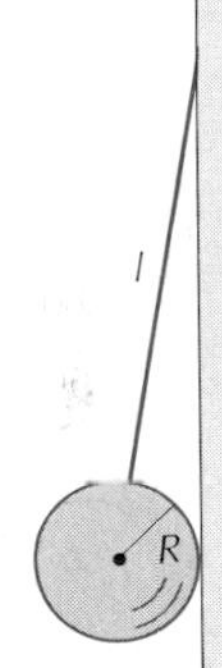

Fig. 6.33

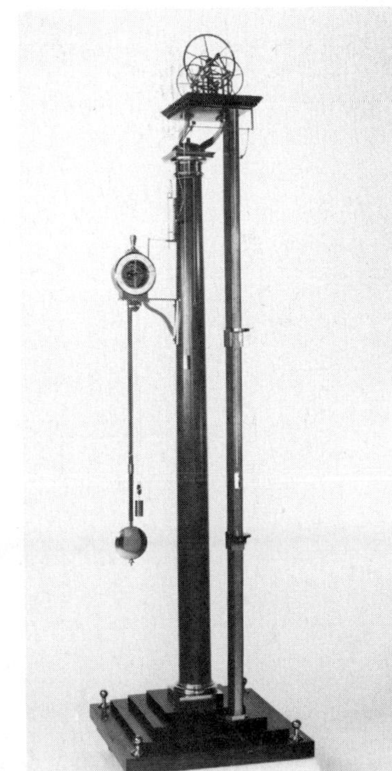

Fig. 6.34 Atwood's machine.

11. If the elevator described in Example 4 carries four passengers weighing 150 lb each, what speed will the elevator attain running down freely from a height of 30 ft, starting from rest?

12. During takeoff, a jetliner is accelerating along the runway at 1.2 m/s^2. In the cabin, a passenger holds a pocket watch by a chain (a plumb). What angle will the chain make with the vertical during this acceleration?

13. In a closed subway car, a girl holds a helium-filled balloon by a string. While the car is traveling at constant velocity, the string of the balloon is exactly vertical.

(a) While the subway car is braking, will the string be inclined forward or backward relative to the car?

(b) Suppose that the string is inclined at an angle of 20° with the vertical and remains there. What is the acceleration of the car?

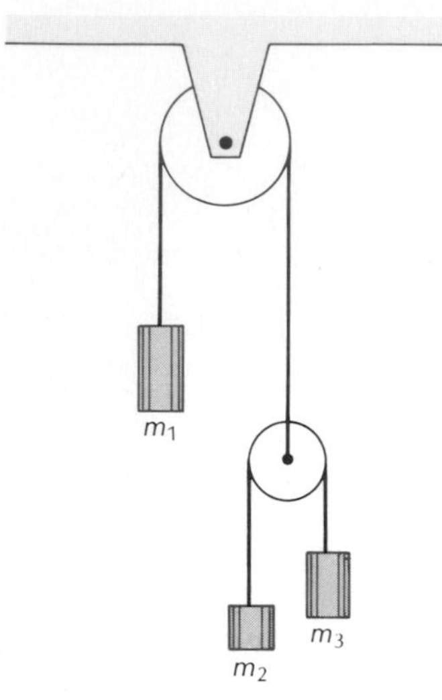

Fig. 6.35

*14. A mass m_1 hangs from one end of a string passing over a frictionless, massless pulley. A second frictionless, massless pulley hangs from the other end of the string (Figure 6.35). Masses m_2 and m_3 hang from a second string passing over this second pulley. Find the acceleration of the three masses, and find the tensions in the two strings.

Section 6.4

15. The ancient Egyptians moved large stones by dragging them across the sand in sleds. How many Egyptians were needed to drag an obelisk of 700 short tons? (A short ton is 2000 lb.) Assume that $\mu_k = 0.3$ for the sled on sand and that each Egyptian exerted a horizontal force of 80 lb.

16. The base of a winch is bolted to a mounting plate with four bolts. The base and the mounting plate are flat surfaces made of steel; the friction coefficient of these surfaces in contact is $\mu_s = 0.4$. The bolts provide a normal force of 600 lb each. What maximum static friction force will act between the steel surfaces and help to oppose lateral slippage of the winch on its base?

17. According to tests performed by the manufacturer, an automobile with an initial speed of 65 km/h has a stopping distance of 20 m on a level road. Assuming that no skidding occurs during braking, what is the value of μ_s between the wheels and the road required to achieve this stopping distance?

18. A crate sits on the load platform of a truck. The coefficient of friction between the crate and the platform is $\mu_s = 0.4$. If the truck stops suddenly, the crate will slide forward and crash into the cab of the truck. What is the maximum braking deceleration that the truck may have if the crate is to stay put?

19. When braking (without skidding) on a dry road, the stopping distance of a sports car with a high initial speed is 38 m. What would have been the stopping distance of the same car with the same initial speed on an icy road? Assume that $\mu_s = 0.95$ for the dry road and $\mu_s = 0.20$ for the icy road.

20. If the coefficient of static friction between the tires of an automobile and the road is $\mu_s = 0.8$, what is the minimum distance the automobile needs in order to stop without skidding from an initial speed of 55 mi/h?

21. In a remarkable accident on motorway M.1 (in England), a Jaguar car initially speeding at 100 mi/h skidded 950 ft before coming to rest. Assuming that the wheels were completely locked during the skid, find the value of the coefficient of kinetic friction between the wheels and the road.

22. Because of a failure of its landing gear, an airplane has to make a belly landing on the runway of an airport. The landing speed of the airplane is 90 km/h and the coefficient of kinetic friction between the belly of the airplane and the runway is $\mu_k = 0.6$. How far will the airplane slide along the runway?

23. During braking, a truck has a steady deceleration of 7.0 m/s². A box sits on the platform of this truck. The box begins to slide when the braking begins and, after sliding a distance of 2.0 m (relative to the truck), it hits the cab of the truck. With what speed (relative to the truck) does the box hit? The coefficient of kinetic friction for the box is $\mu_k = 0.50$.

24. The "Texas" locomotives of the old T&P railway had a weight of 440,000 lb, of which 300,000 lb rested on the driving wheels. What maximum acceleration could such a locomotive attain (without slipping) when pulling a train of 100 boxcars weighing 40,000 lb each on a level track? Assume that the coefficient of static friction between the driving wheels and the track is 0.25.

25. The friction force (including air friction and rolling friction) acting on an automobile traveling at 65 km/h amounts to 500 N. What slope must a road have if the automobile is to roll down this road at a constant speed of 65 km/h (with its gears in neutral)? The mass of the automobile is 1.5×10^3 kg.

26. In a downhill race, a skier slides down a 40° slope. Starting from rest, how far must he slide down the slope in order to reach a speed of 80 mi/h? How many seconds does it take him to reach this speed? The friction coefficient between his skis and the snow is $\mu_k = 0.1$. Ignore the resistance offered by the air.

27. To measure the coefficient of static friction of a block of plastic on a plate of steel, an experimenter places the block on the plate and then gradually tilts the plate. The block suddenly begins to slide when the plate makes an angle of 38° with the horizontal. What is the value of μ_s?

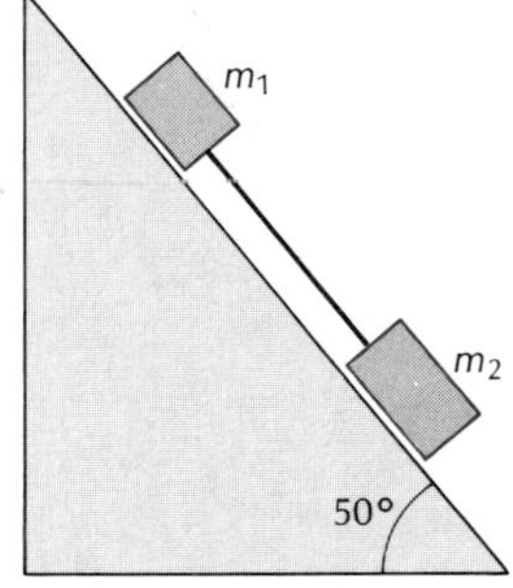

Fig. 6.36 Two masses connected by a string sliding down a ramp.

28. A block of wood rests on a sheet of paper lying on a table. The coefficient of static friction between the block and the paper is $\mu_s = 0.7$ and that between the paper and the table is $\mu_s = 0.5$. If you tilt the table, at what angle will the block begin to move?

29. On a level road, the stopping distance for an automobile is 35 m for an initial speed of 90 km/h. What is the stopping distance of the same automobile on a similar road with a downward slope of 1:10?

30. Two masses, of 2.0 kg each, connected by a string slide down a ramp making an angle of 50° with the horizontal (Figure 6.36). The mass m_1 has a coefficient of kinetic friction 0.60 and the mass m_2 has a coefficient of kinetic friction 0.40. Find the acceleration of the masses and the tension in the string.

31. A girl pulls a sled along a level dirt road by means of a rope attached to the front of the sled (Figure 6.37). The mass of the sled is 40 kg, the coefficient of kinetic friction is $\mu_k = 0.60$, and the angle between the rope and the road is 30°. What pull must the girl exert to move the sled at constant velocity?

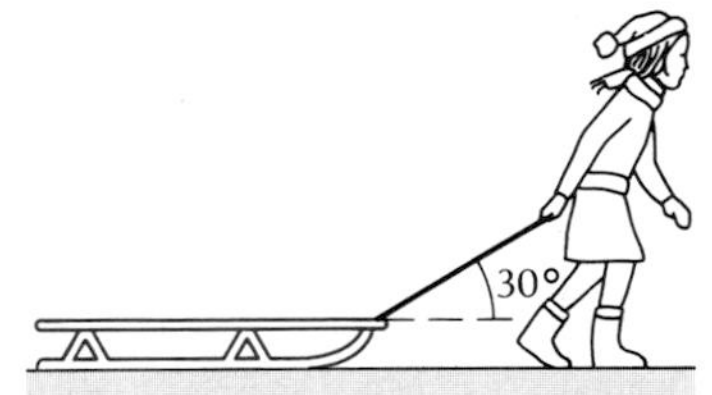

Fig. 6.37 Girl pulling a sled.

32. A box is being pulled along a level floor at constant velocity by means of a rope attached to the front end of the box. The rope makes an angle θ with the horizontal. Show that for a given mass m of the box and a given coefficient of kinetic friction μ_k, the tension required in the rope is minimum if $\tan\theta = \mu_k$. What is the tension in the rope when at this optimum angle?

33. Consider the man pushing the crate described in Example 6. Assume that instead of pushing down at an angle of 30°, he pushes down at an angle θ. Show that he will not be able to keep the crate moving if θ is larger than $\tan^{-1}(1/\mu_k)$.

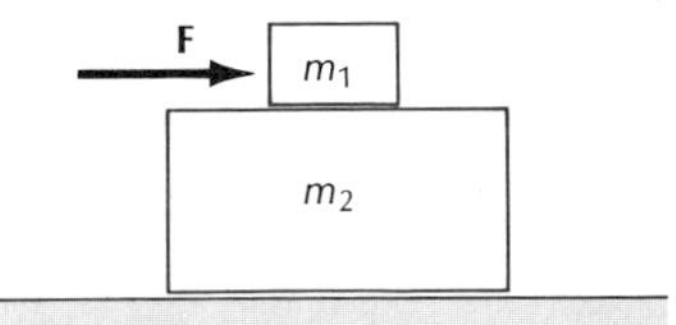

Fig. 6.38

34. A block of mass m_1 sits on top of a larger block of mass m_2 which sits on a flat surface (Figure 6.38). The coefficient of kinetic friction between the upper and lower blocks is μ_1, and that between the lower block and the flat surface is μ_2. A horizontal force **F** pushes against the upper block, causing it to slide; the friction force between the blocks then causes the lower block to slide

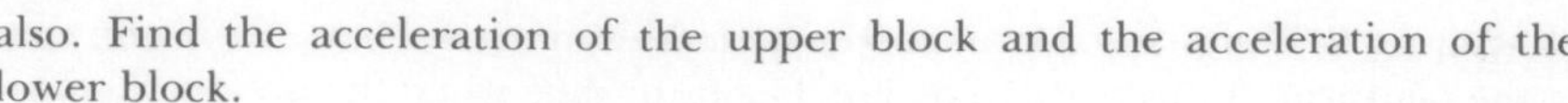

also. Find the acceleration of the upper block and the acceleration of the lower block.

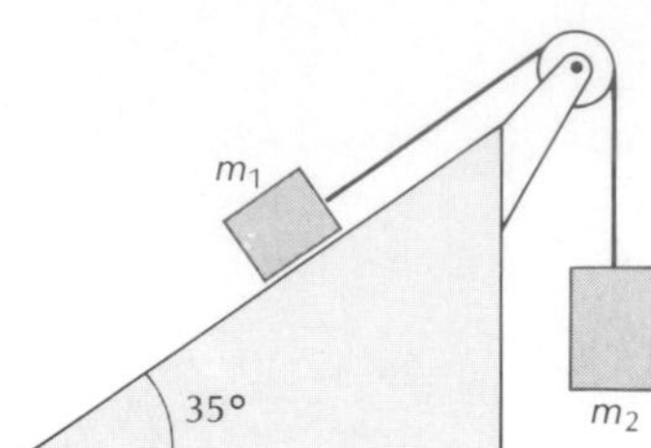

Fig. 6.39

35. Two masses $m_1 = 1.5$ kg and $m_2 = 3.0$ kg are connected by a thin string running over a massless pulley. One of the masses hangs from the string; the other mass slides on a 35° ramp with a coefficient of kinetic friction $\mu_k = 0.40$ (Figure 6.39). What is the acceleration of the masses?

36. A man pulls a sled up a ramp by means of a rope attached to the front of the sled (Figure 6.40). The mass of the sled is 80 kg, the coefficient of kinetic friction between the sled and the ramp is $\mu_k = 0.70$, the angle between the ramp and the horizontal is 25°, and the angle between the rope and the ramp is 35°. What pull must the man exert to keep the sled moving at constant velocity?

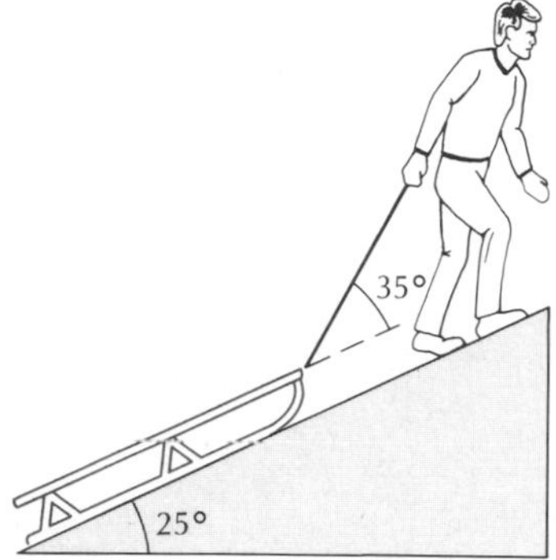

Fig. 6.40

*37. Two blocks are sliding down an inclined plane making an angle θ with the horizontal. The leading block has a coefficient of kinetic friction μ_k; the trailing block has a coefficient of kinetic friction $2\mu_k$. A string connects the two blocks; this string makes an angle ϕ with the ramp (Figure 6.41). Find the acceleration of the blocks and the tension in the string.

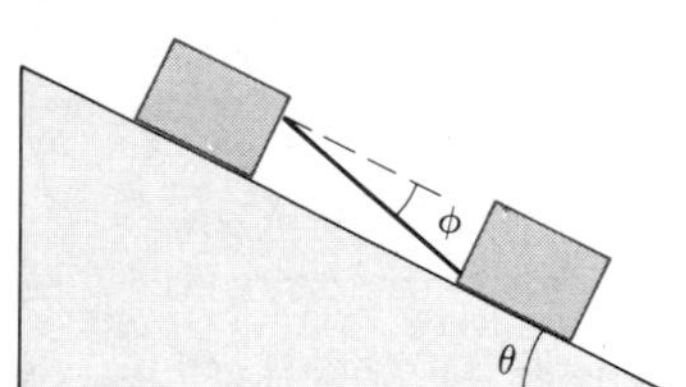

Fig. 6.41 Two blocks connected by a string sliding down an inclined plane.

Section 6.5

38. Attempting to measure the force constant of a spring, an experimenter clamps the upper end of the spring in a vise and suspends a mass of 1.5 kg from the lower end. This stretches the spring by 0.20 m. What is the force constant of the spring?

39. A spring with a force constant $k = 150$ N/m has a relaxed length of 0.15 m. What force must you exert to stretch this spring to twice its length? What force must you exert to compress this spring to one-half its length?

40. The body of an automobile is held above the axles of the wheels by means of four springs, one near each wheel. Assume that the springs are vertical and that the forces on all the springs are the same. The mass of the automobile is 1200 kg and the spring constant of each spring is 2.0×10^4 N/m. When the automobile is stationary on a level road, how far are the springs compressed from their relaxed length?

41. A rubber band of relaxed length 2.5 in. stretches to 4.0 in. under a force of 0.22 lb, and to 6.5 in. under a force of 0.44 lb. Does this rubber band obey Hooke's law?

42. Suppose that a uniform spring with a constant $k = 120$ N/m is cut into two pieces, one twice as long as the other. What are the spring constants of the two pieces?

43. Show that if two springs, of constants k_1 and k_2, are connected in series (Figure 6.42), the net spring constant k of the combination is given by

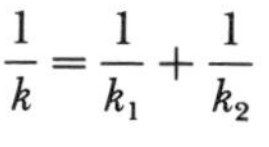

$$\frac{1}{k} = \frac{1}{k_1} + \frac{1}{k_2}$$

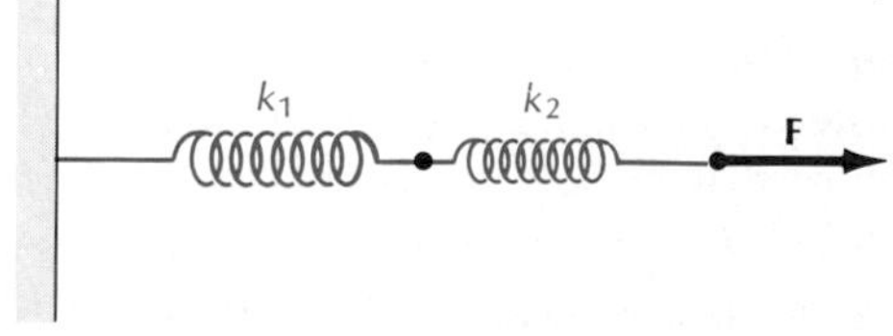

Fig. 6.42 Springs in series.

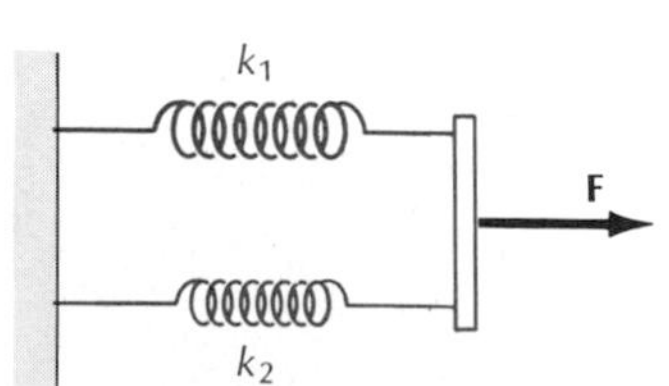

Fig. 6.43 Springs in parallel.

44. Show that if two springs, of constants k_1 and k_2, are connected in parallel (Figure 6.43), the net spring constant k of the combination is given by

$$k = k_1 + k_2$$

Section 6.6

45. A particle of mass 2.0 kg constrained to move along the x axis experiences a force $F = -40x$, where F is measured in newtons and x in meters. The initial position of the particle is $x = 0.2$ m and the initial velocity is zero. By numerical solution of the equation of motion, find how long it takes the particle to reach the origin $x = 0$. (Hint: Take time intervals of about 0.05 s.)

46. Under the influence of a friction force proportional to the velocity, a particle has an acceleration $a = -0.2v$, where a is measured in m/s^2 and v in m/s. The initial velocity of the particle is 10 m/s. By numerical solution of the above equation, find how long it takes for the velocity to decrease by a factor of 2. (Hint: Use time intervals of 0.20 s.)

*47. Equation (26) describes the motion of a ballistic missile when its engines have cut off and the missile is coasting upward under the influence of the gradually decreasing gravity of the Earth. For a missile of m = 1000 kg, the value of the force constant is $A = 3.99 \times 10^{17}$ N · m^2. Suppose that the initial conditions are $x_0 = 6.5 \times 10^6$ m and $v_0 = 6.0 \times 10^3$ m/s. Solve the equation of motion numerically and find the maximum value of x that the missile reaches.

Section 6.7

48. A swing consists of a seat supported by a pair of ropes 5 m long. A 60-kg woman sits in the swing. Suppose that the speed of the woman is 5.0 m/s at the instant the swing goes through its lowest point. What is the tension in each of the two ropes? Ignore the masses of the seat and of the ropes.

49. The Moon moves around the Earth in a circular orbit of radius 3.8×10^8 m in 27 days. The mass of the Moon is 7.3×10^{22} kg. From these data, calculate the magnitude of the force required to keep the Moon in its orbit.

50. A man of 80 kg is standing in the cabin of a Ferris wheel of radius 30 m rotating at 1 rev/min. What is the force that the feet of the man exert on the floor of the cabin when he reaches the highest point? The lowest point?

51. The highest part of a road over the top of a hill follows an arc of a vertical circle of radius 50 m. With what minimum speed must you drive an automobile along this road if its wheels are to lose contact with the road at the top of the hill?

52. A man holds a pail full of water by the handle and whirls it around a vertical circle at constant speed. The radius of this circle is 0.9 m. What is the minimum speed that the pail must have at the top of its circular motion if the water is not to spill out of the upside-down pail?

53. A stone of 2 lb attached to a string is being whirled around a vertical circle of radius 3.0 ft. Assume that during this motion the speed of the stone is constant. If at the top of the circle, the tension in the string is (just about) zero, what is the tension in the string at the bottom of the circle?

54. In ice speedway races, motorcycles run at a high speed on an ice-covered track and are kept from skidding by long spikes on their wheels. Suppose that a motorcycle runs around a curve of radius 100 ft at a speed of 60 mi/h. What is the angle of inclination of the force exerted by the track on the wheels?

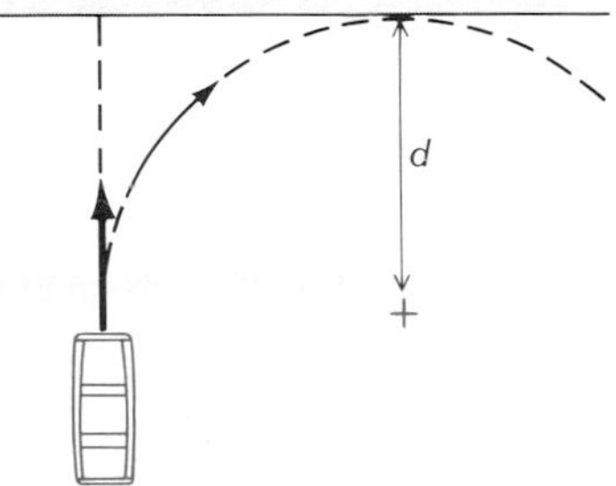

Fig. 6.44 Automobile approaching a brick wall.

55. An automobile traveling at speed v on a level surface approaches a brick wall (Figure 6.44). When the automobile is at a distance d from the wall, the driver suddenly realizes that he must either brake or turn. If the coefficient of static friction between the tires and the surface is μ_s, what is the minimum distance that the driver needs to stop (without turning)? What is the minimum distance that the driver needs to complete a 90° turn (without braking)? What is the safest tactic for the driver?

56. A few copper coins are lying on the (flat) dashboard of an automobile. The coefficient of static friction between the copper and the dashboard is 0.5.

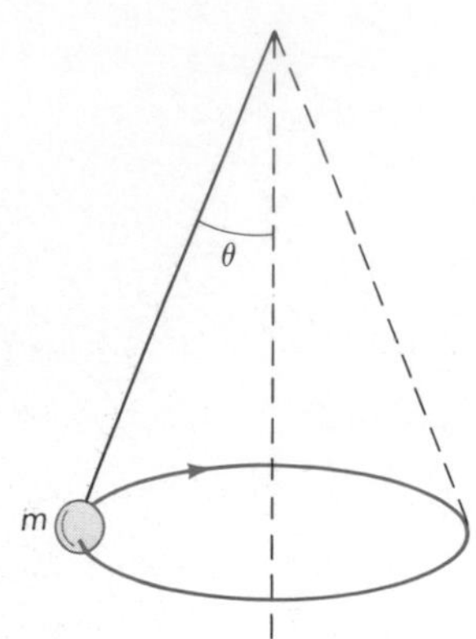

Fig. 6.45 Mass suspended from a string swinging around in a circle (conical pendulum).

Suppose the automobile rounds a curve of radius 300 ft. At what speed of the automobile will the coins begin to slide? The curve is *not* banked.

57. What is the maximum speed with which an automobile can round a curve of radius 120 m? Assume that the road is flat (no banking) and the coefficient of static friction between the tires and the road is $\mu_s = 0.8$.

58. A curve of radius 1200 ft has been designed with a banking angle such that an automobile moving at 45 mi/h does not have to rely on friction to stay in the curve. What is the banking angle?

59. An airplane flies in a horizontal circular path of radius 5000 ft at 200 mi/h. At what angle should the wings be banked? [Hint: The force exerted by the air on the wings (lift) is perpendicular to the wings.]

60. A mass is attached to the lower end of a string of length l; the upper end of the string is held fixed. Suppose that the string initially makes an angle θ with the vertical. With what horizontal velocity must we launch the mass so that it continues to travel at constant speed along a horizontal circular path under the influence of the combined forces of the tension of the string and gravity? This device is called a **conical pendulum** (Figure 6.45).

61. An automobile of mass 1200 kg rounds a curve at a speed of 25 m/s. The radius of the curve is 400 m and its banking angle is 6°. What is the magnitude of the normal force on the automobile? The friction force?

*62. A circle of rope of mass m and radius r is spinning about its center so that each point of the rope has a speed v. Calculate the tension in the rope.

*63. The rotor of a helicopter consists of two blades 180° apart. Each blade has a weight of 300 lb and a length of 12 ft. What is the tension in each blade at the hub when rotating at 320 rev/min? Pretend that each blade is a uniform thin rod.

*64. Assume that the Earth is a sphere and that the force of gravity (mg) points precisely toward the center of the Earth. Taking into account the rotation of the Earth about its axis, calculate the angle between the direction of a plumb line and the direction of the Earth's radius as a function of latitude. What is this deviation angle at a latitude of 45°?

*65. A curve of radius 120 m is banked at an angle of 10°. If an automobile with wheels with $\mu_s = 0.9$ is to round this curve without skidding, what is the maximum permissible speed?

CHAPTER 7

Work and Energy

Conservation laws

Conservation laws play an important role in the world of matter. Such laws assert that some quantity is conserved, i.e., some quantity remains constant even when matter suffers drastic changes involving motions, collisions, and reactions. One familiar example of a conservation law is the conservation of mass. Expressed in its simplest form, this law asserts that the mass of a given atom, or a given subatomic particle, remains constant. In the two preceding chapters we took this conservation law for granted and we treated the particle mass appearing in Newton's Second Law as a time-independent, constant quantity (in $m\mathbf{a} = \mathbf{F}$ the particle mass appears as a time-independent constant of proportionality). In everyday life we always rely implicitly on the conservation of mass. For instance, in the chemical plants that reprocess the uranium fuel for nuclear reactors, the batches of uranium compounds are carefully weighed at several checkpoints to make sure that no uranium is diverted for nefarious purposes. Obviously, this procedure would make no sense if the particle masses were changeable, so that the net mass of a batch could increase or decrease spontaneously. Other conservation laws involving mechanical quantities are the conservation of energy, momentum, and angular momentum. Furthermore, elementary-particle physicists have discovered several other conservation laws involving esoteric quantities such as baryon number, lepton number, etc.

Physicists take advantage of conservation laws in several ways. In mechanics, they use the conservation laws for energy, momentum, and angular momentum to make predictions about some aspect of the motion of a particle or of a system of particles when it is undesirable or impossible to calculate the full details of the motion from Newton's Second Law. This is especially helpful in those cases where the force law is not known exactly (some examples of this kind will be treated in Chapter 9). In atomic, nuclear, and elementary-particle physics, physi-

James Prescott Joule (jool), *1818–1889, English physicist. He established experimentally that heat is a form of mechanical energy, and he made the first direct measurement of the mechanical equivalent of heat. By a series of meticulous mechanical, thermal, and electrical experiments, Joule provided empirical proof of the general law of conservation of energy.*

cists use the conservation laws to decide what reactions are possible — if a hypothetical reaction violates a conservation law, then it is impossible. But, besides these more or less practical aspects, conservation laws also have a very deep theoretical significance. There is an intimate connection between conservation laws and symmetry: the existence of a quantity that remains constant during the motions, collisons, and reactions of matter indicates the presence of a mathematical symmetry in the equations for the basic forces governing these phenomena. Such a mathematical symmetry means that certain terms in the equation can be interchanged without affecting its validity. Although the details of this connection are beyond the scope of our discussion, it is obvious that the laws of force must have some special features if they are to avoid violations of the conservation laws. In the search for the basic laws of force, the conservation laws therefore give the crucial clues.

Both the present chapter and the next deal with the conservation of energy. This conservation law is one of the most fundamental laws of nature. Although we will derive this law from Newton's laws, it is actually much more general and remains valid even when Newton's laws fail. No violation of the law of conservation of energy has ever been discovered.

7.1 Work in One Dimension

In spelling out the definition of work, we will begin with the simple case of one-dimensional motion and then proceed to the general case of three-dimensional motion. Consider a particle moving in one dimension, say, along the x axis. If a constant force F_x, also along the x axis, acts on the particle, then the **work** done by this force on the particle as it moves some given distance is defined as the product of the force and the displacement,

Work done by constant force

$$\boxed{W = F_x\,\Delta x} \tag{1}$$

This rigorous definition of work is motivated by our intuitive notion of what constitutes "work" — if we push a heavy crate over a floor, the work that we perform is proportional to the magnitude of the force we have to exert and to the distance we move the crate. But the ultimate justification of our definition lies in the consequences of this definition, to be discussed in the following sections.

Note that F_x is positive if the force is in the positive x direction and negative if in the negative x direction, i.e., F_x is the component of the force along the x axis. According to Eq. (1), the work is positive if the force and the displacement are in the same direction, and negative if they are in opposite directions. Note that if the work is negative, then the particle (by its reaction force) does positive work on the agent that exerts the force — negative work *on* the particle means positive work *by* the particle, and positive work *on* the particle means negative work *by* the particle.

Equation (1) gives the work done by one of the forces acting on the particle. If several forces act, then Eq. (1) can be used to calculate the work done by each force.

In the metric system the unit of work is the **joule** (J), which is the work done by a force of 1 N during a displacement of 1 m:

$$1 \text{ joule} = 1 \text{ J} = 1 \text{ N} \cdot \text{m} = 1 \text{ kg} \cdot \text{m}^2/\text{s}^2 \qquad (2)$$

joule, J

In the British system the corresponding unit is the **foot-pound** (ft · lb), which is the work done by a force of 1 lb during a displacement of 1 ft:

$$1 \text{ ft} \cdot \text{lb} = 1 \text{ ft} \times 1 \text{ lb}$$

$$= 1.356 \text{ J} \qquad (3)$$

EXAMPLE 1. A 2000-lb elevator cage descends 1300 ft within a skyscraper. What is the work done by gravity on the elevator cage during this displacement?

SOLUTION: With the x axis arranged vertically upward (Figure 7.1), the displacement is negative, $\Delta x = -1300$ ft, and the weight is also negative, $F_x = -mg = -2000$ lb. Hence the work is

$$W = F_x \, \Delta x = (-2000 \text{ lb}) \times (-1300 \text{ ft})$$

$$= 2.60 \times 10^6 \text{ ft} \cdot \text{lb} \qquad (4)$$

Note that the work done by gravity is completely independent of the details of the motion; the work only depends on the total vertical displacement and not on the velocity or acceleration of the motion.

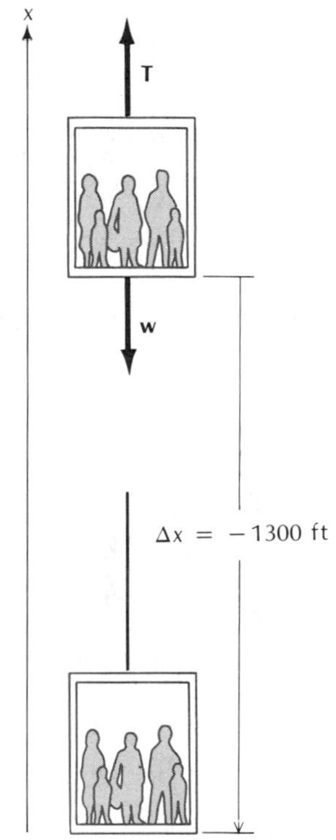

Fig. 7.1 A descending elevator.

EXAMPLE 2. Assuming that the elevator cage of Example 1 descends at constant velocity, what is the work done by the tension of the suspension cable?

SOLUTION: For motion at constant velocity the tension force must exactly balance the weight. Therefore the tension force has the same magnitude as the weight but the opposite direction, $F_x = T = +2000$ lb. The work done by this force is then

$$W = F_x \, \Delta x = (2000 \text{ lb}) \times (-1300 \text{ ft})$$

$$= -2.60 \times 10^6 \text{ ft} \cdot \text{lb} \qquad (5)$$

Note that the work done by the tension is exactly the opposite of the work done by gravity — the total work done by both forces together is zero. However, the result (5) depends implicitly on the assumptions made about the motion. Only for unaccelerated motion does the tension remain constant at 2000 lb. For example, if the elevator cage were allowed to move downward with the acceleration g, then the tension would be zero; the work done by the tension would then also be zero, while the work done by gravity would still be 2.60×10^6 ft · lb.

Although the rigorous definition of work contained in Eq. (1) agrees to some extent with our intuition of what constitutes "work," the rigorous definition clashes with our intuition in some instances. For example, consider a man holding a bowling ball in a fixed position in his outstretched hand (Figure 7.2). Our intuition suggests that the man does work — yet Eq. (1) indicates that no work is done on the ball,

Fig. 7.2 Man holding a bowling ball.

since $\Delta x = 0$! The resolution of this conflict hinges on the observation that, although the man does no work *on the ball,* he does work *within his own muscles* and, consequently, grows tired of holding the ball. A contracted muscle is never in a state of complete rest: within it, atoms, cells, and muscle fibers engage in complicated chemical and mechanical processes which involve motion and work. These internal motions go on continuously while the external shape of the muscle remains rigid. This means that work is done, and wasted, internally within the muscle while no work is done externally on the bone to which the muscle is attached or on the bowling ball supported by the bone. The man does work, but no useful, external work. Obviously, it is very inefficient to hold up a weight by means of rigidly contracted muscles. The human machine would be much more efficient if it had some mechanism for locking the joints (with a pin or a clamp) whenever rigidity was desired; then no muscular effort would be required to hold a weight and no internal work would be wasted.

Another conflict between our intuition and the rigorous definition of work arises when we contemplate a body in motion. Suppose that the man of Figure 7.2 with the bowling ball in his hand rides in an elevator moving steadily upward. In this case, the displacement is not zero and the force (push) exerted by the hand on the ball does work — the displacement and the force are both in the same direction and consequently the hand continuously does positive work on the ball. Nevertheless, to the man the ball feels no different when riding in the elevator than when standing on the ground. The man plays a passive role: the elevator floor does work on his feet and the man transmits a suitable fraction of this work to the bowling ball. This example illustrates that the amount of work depends on the frame of reference. In the reference frame of the ground, the ball is moving upward and work is done on it; in the reference frame of the elevator, the ball is at rest and no work is done on it. The lesson we learn from this is that before proceeding with a calculation of work, we must be careful to specify the frame of reference.

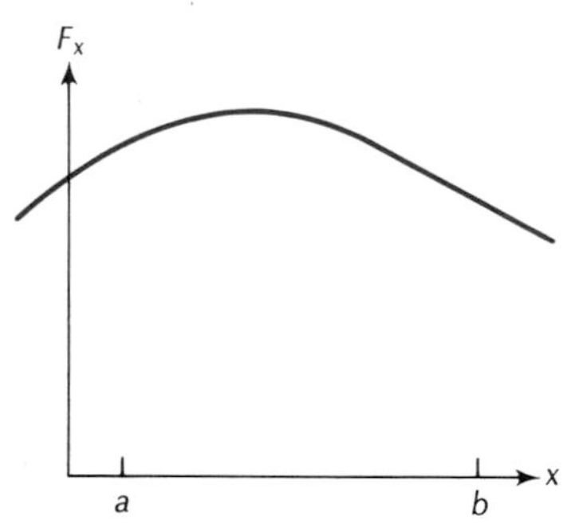

Fig. 7.3 Plot of F_x vs. x.

Next we will have to consider the case of a force that is not constant. Suppose that the force is some function of position,

$$F_x = F_x(x) \tag{6}$$

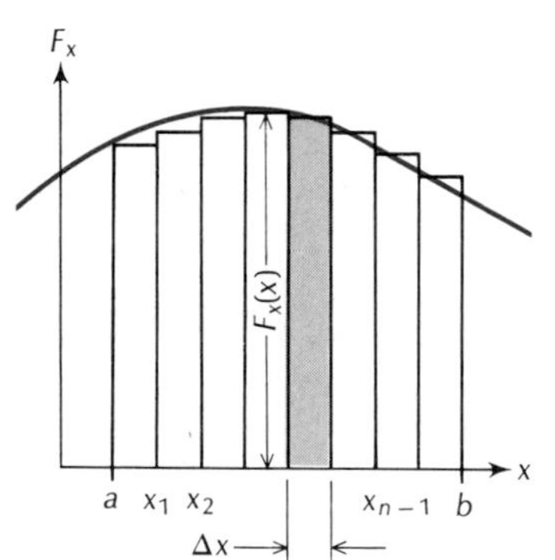

Fig. 7.4 The total displacement from a to b has been divided into small intervals of length Δx.

Figure 7.3 shows a plot of typical function $F_x(x)$. To evaluate the work done by this force on a particle during a displacement from $x = a$ to $x = b$, we divide the total displacement into a large number of small intervals, each of length Δx (Figure 7.4). The beginnings and ends of these intervals are located at $x_0, x_1, x_2, \ldots, x_n$, where $x_0 = a$ and $x_n = b$. Within each interval the force can be regarded as approximately constant — within the interval x_i to x_{i+1} (where $i = 0, 1, 2, 3, \ldots$), the force is approximately $F_x(x_i)$. This approximation is obviously at its best if Δx is very small. The work done by this force as the particle moves from x_i to x_{i+1} is then

$$\Delta W_i = F_x(x_i)\,\Delta x \tag{7}$$

and the total work done as the particle moves from a to b is simply the sum of all the small amounts of work associated with the small intervals:

$$W = \sum_{i=0}^{i=n-1} \Delta W_i = \sum_{i=0}^{i=n-1} F_x(x_i)\, \Delta x \tag{8}$$

Note that $F_x(x_i)\, \Delta x$ is the area of the rectangle of height $F_x(x_i)$ and base Δx (see Figure 7.4). Thus, Eq. (8) gives the sum of all the rectangular areas shown in Figure 7.4.

Equation (8) is of course only an approximation for the work. In order to improve this approximation, we must use a smaller interval Δx. In the limiting case $\Delta x \to 0$ (and $n \to \infty$), we obtain an exact expression for the work. Thus, the exact definition for the work done by a variable force is

$$W = \lim_{\Delta x \to 0} \sum_i F_x(x_i)\, \Delta x \tag{9}$$

This expression is called the **integral** of the function $F_x(x)$ between the limits a and b. The usual notation for this integral is

$$\boxed{W = \int_a^b F_x(x)\, dx} \tag{10}$$

Work done by variable force

where $\int$ is called the integral sign and $F_x(x)\, dx$ is called the integrand. The quantity in Eq. (9) or (10) is exactly equal to the area bounded by the curve representing $F_x(x)$, the x axis, and the vertical lines $x = a$ and $x = b$ in Figure 7.3; areas above the x axis are reckoned positive and areas below the x axis are reckoned negative.[1]

EXAMPLE 3. A spring exerts a restoring force of $F_x(x) = -kx$ on a particle attached to it (compare Section 6.5). What is the work done by the spring on the particle when it moves from $x = x_1$ to $x = x_2$?

SOLUTION: By Eq. (10),

$$W = \int_{x_1}^{x_2} F_x(x)\, dx = \int_{x_1}^{x_2} (-kx)\, dx \tag{11}$$

To evaluate this integral we rely on a theorem of calculus which states that the (definite) integral of the function x^n is

$$\int_a^b x^n\, dx = \left[\frac{x^{n+1}}{n+1}\right]_a^b = \frac{b^{n+1}}{n+1} - \frac{a^{n+1}}{n+1}$$

Therefore

$$W = \int_{x_1}^{x_2} (-\, kx)\, dx = -k\left[\tfrac{1}{2}x^2\right]_{x_1}^{x_2} = -\tfrac{1}{2}k(x_2^2 - x_1^2) \tag{12}$$

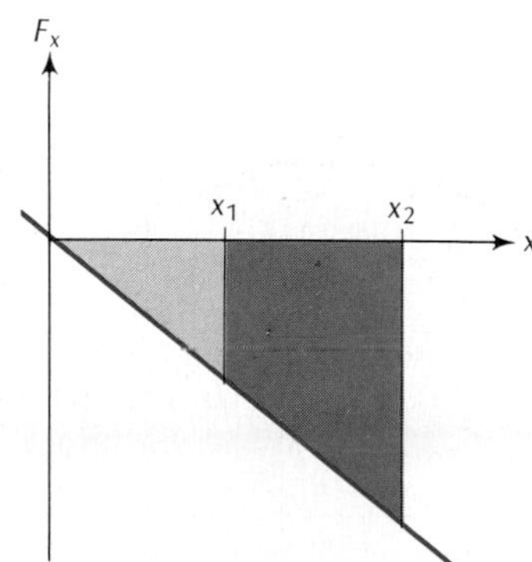

Fig. 7.5 The straight line is a plot of the function $F_x(x) = -kx$. The quadrilateral area to the right of x_1 and the left of x_2 represents the work.

This result can also be obtained by calculating the area in a plot of force v. position. Figure 7.5 shows the force $F_x(x) = -kx$ as a function of x. The trian-

---

[1] A brief review of integrals is given in Appendix 5.

gular area between the origin and $x = x_2$ is $\frac{1}{2}$[base] × [height] $= \frac{1}{2}x_2 \times kx_2$, or $\frac{1}{2}kx_2^2$. Likewise, the triangular area between the origin and $x = x_1$ is $\frac{1}{2}kx_1^2$. The quadrilateral area that represents the work is the difference between the two triangular areas, i.e., $\frac{1}{2}kx_2^2 - \frac{1}{2}kx_1^2$. Taking into account that areas below the x axis must be reckoned as negative, we see that this area calculation agrees with Eq. (12).

7.2 Work in Three Dimensions

In the case of three-dimensional motion, the definition of work must take into account the different directions of force and displacement. If a constant force **F** acts on a particle, then the **work** done by this force on the particle while it undergoes a displacement $\Delta\mathbf{r}$ is defined as the dot product of force and displacement,

$$\boxed{W = \mathbf{F} \cdot \Delta\mathbf{r}} \tag{13}$$

According to Eq. (3.22), this dot product is equal to

$$W = F\,\Delta r \cos\theta \tag{14}$$

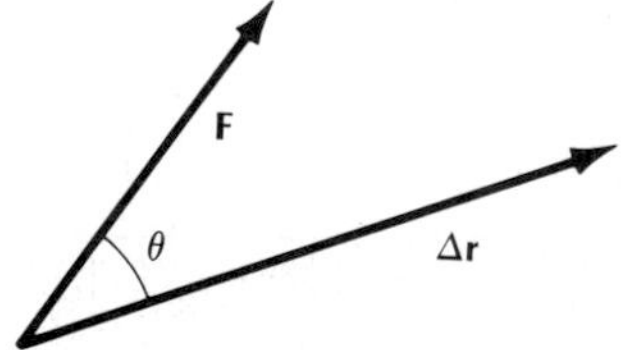

Fig. 7.6 A constant force **F** acts during a displacement $\Delta\mathbf{r}$.

where θ is the angle between the force and the displacement (Figure 7.6). This expression can be regarded as either the magnitude of **F** multiplied by the component of $\Delta\mathbf{r}$ along **F,** or else the magnitude of $\Delta\mathbf{r}$ multiplied by the component of **F** along $\Delta\mathbf{r}$. The work is positive or negative, depending on the value of angle θ. If the force is parallel to the direction of motion ($\theta = 0$) or antiparallel to the direction of motion ($\theta = 180°$), then the work is simply $F\,\Delta r$ or $-F\,\Delta r$, respectively; this is just as in the one-dimensional case. If the force is perpendicular to the direction of motion ($\theta = 90°$), then the work vanishes; for instance, this means that neither the normal force **N** on a solid body sliding on a surface nor the centripetal force on a body in circular motion does any work — both of these forces are always perpendicular to the displacement.

Note that, in view of Eq. (3.26) for the dot product, Eq. (13) can also be written in terms of components,

Work done by constant force (three dimensions)

$$\boxed{W = F_x\,\Delta x + F_y\,\Delta y + F_z\,\Delta z} \tag{15}$$

which shows that each separate component of the force does work as though the motion were one-dimensional [compare Eq. (1)].

EXAMPLE 4. What is the work done by gravity and by the normal force on a block that moves up an inclined plane to a height Δz?

SOLUTION: Figure 7.7a shows the inclined plane; the block moves up the full length of this plane so that the displacement has a magnitude

$$\Delta r = \Delta z / \sin\phi$$

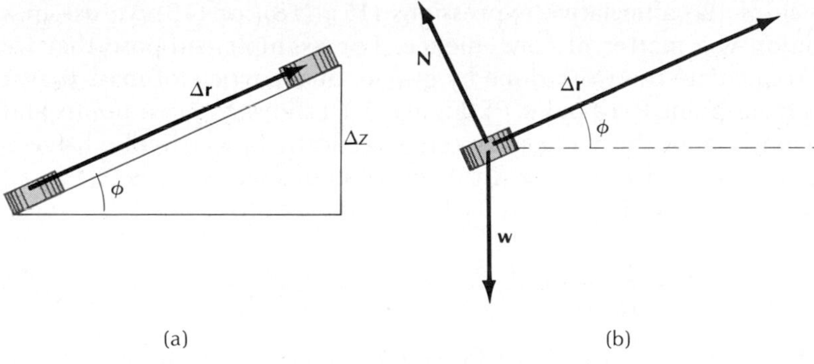

Fig. 7.7 (a) A block undergoing a displacement along an inclined plane. (b) "Free-body" diagram showing the weight, the normal force, and the displacement of the block.

Figure 7.7b is a "free-body" diagram of the block. The work done by gravity is

$$W = mg\ \Delta r \cos\left(\frac{\pi}{2} + \phi\right) = mg\frac{\Delta z}{\sin\phi}(-\sin\phi) = -mg\Delta z$$

The work done by the normal force is zero, since this force makes an angle of 90° with the displacement.

The definition of work in the case of a three-dimensional force that is not constant involves an integral, just as in the analogous case of a one dimensional force. Suppose that a particle moves along some given path from point P_1 to point P_2 (Figure 7.8). We can then divide the path into short segments, each of length $\Delta\mathbf{r}$. Within each such segment the force and the direction of motion are approximately constant and the work done on the particle is

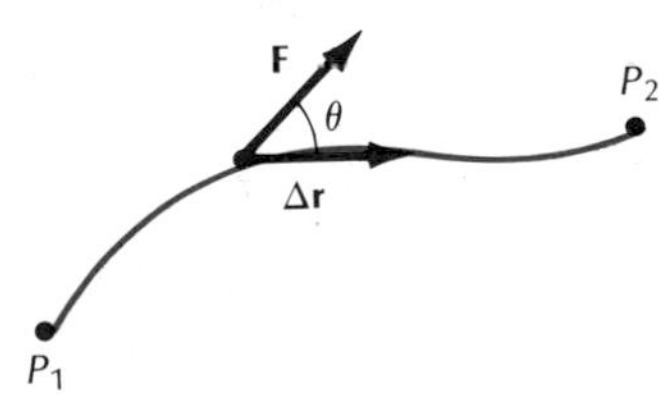

Fig. 7.8 A force **F** acts during a small displacement $\Delta\mathbf{r}$.

$$\Delta W = F\ \Delta r \cos\theta \tag{16}$$

The total work is approximately the sum of all such small amounts of work. In the limit $\Delta r \to 0$, this sum becomes the integral

$$W = \int_{P_1}^{P_2} F \cos\theta\ dr \tag{17}$$

where dr is the element of length along the path followed by the particle. Equation (17) can also be written more compactly as

$$W = \int_{P_1}^{P_2} \mathbf{F} \cdot d\mathbf{r} \tag{18}$$

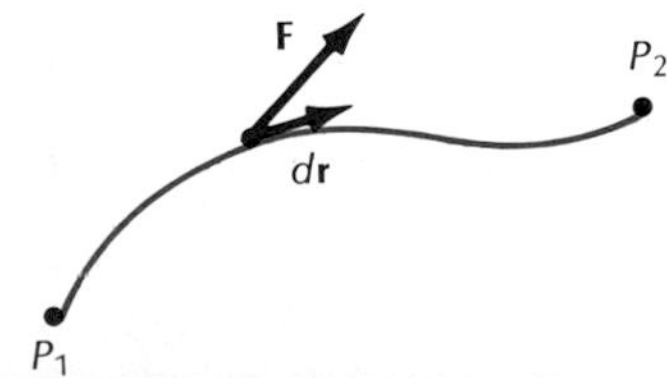

Fig. 7.9 The infinitesimal displacement $d\mathbf{r}$ is tangent to the path.

where $d\mathbf{r}$ is a vector of magnitude dr and of direction tangent to the path (Figure 7.9). Alternatively, this can be written

$$W = \int_{P_1}^{P_2} (F_x\ dx + F_y\ dy + F_z\ dz)$$

or

$$\boxed{W = \int_{P_1}^{P_2} F_x\ dx + \int_{P_1}^{P_2} F_y\ dy + \int_{P_1}^{P_2} F_z\ dz} \tag{19}$$

Work done by variable force (three dimensions)

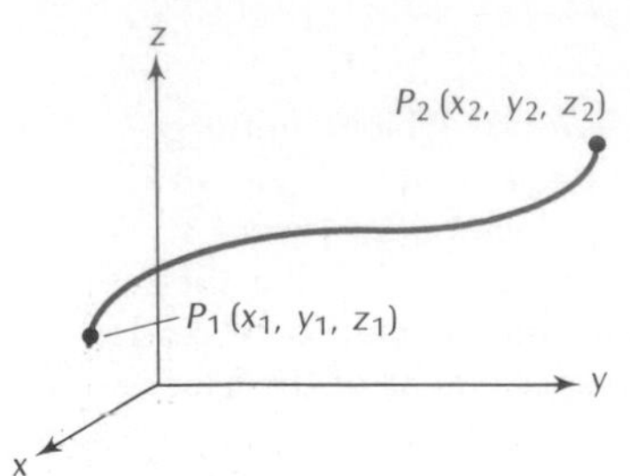

Fig. 7.10 Path of a particle. The z axis is directed vertically upward.

Which of the alternative expressions (17), (18), or (19) we use in a calculation is a matter of convenience. For example, suppose that we want to calculate the work done by gravity on a particle of mass m as it moves from point P_1 to point P_2. Figure 7.10 shows the two points and their coordinates; the z axis is directed vertically upward. The force is $F_x = 0$, $F_y = 0$, and $F_z = -mg$. In terms of coordinates, P_1 is (x_1, y_1, z_1) and P_2 is (x_2, y_2, z_2). Hence, Eq. (19) immediately gives

$$W = \int_{z_1}^{z_2} (-mg)\, dz = -mg\Big[z\Big]_{z_1}^{z_2} = -mg(z_2 - z_1) \tag{20}$$

This shows that the work done by gravity depends *only* on the vertical separation between P_1 and P_2 — neither the horizontal separation nor the exact shape of the path between P_1 and P_2 is of any relevance. The result (20) is usually written

Work done by gravity

$$\boxed{W = -mg\, \Delta z} \tag{21}$$

where $\Delta z = z_2 - z_1$ is the change in height between P_1 and P_2; this agrees with Example 4.

EXAMPLE 5. At a construction site a laborer pushes horizontally against a large bucket full of concrete of total mass 600 kg suspended from a crane by a 20-m cable (Figure 7.11a). How much work must the laborer do on the bucket to push it slowly 2 m away from the vertical? How much work does gravity do on the bucket?

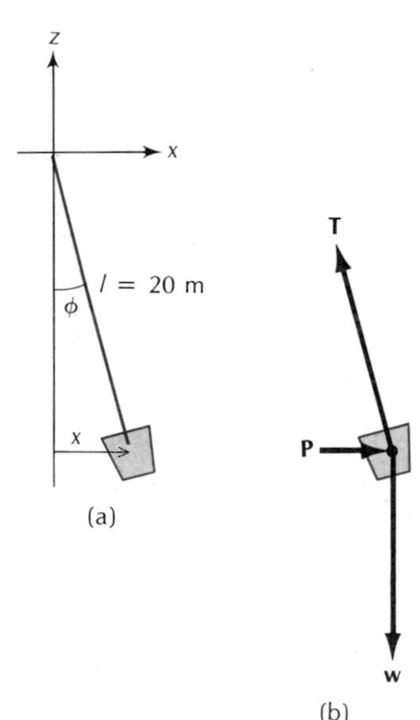

Fig. 7.11 (a) Bucket hanging on a cable. (b) "Free-body" diagram showing the forces on the bucket.

SOLUTION: Figure 7.11b is a "free-body" diagram of the bucket, regarded as a particle, displaced sideways by an angle ϕ. The forces on the bucket are the weight **w**, the tension **T** of the cable, and the push **P** of the laborer. The motion is slow, i.e., the acceleration is just about zero, and therefore the forces must balance. For the x components and the z components of the forces, this implies

$$-T \sin \phi + P = 0$$

$$T \cos \phi - mg = 0$$

Eliminating T between these two equations, we find

$$P = mg \tan \phi$$

Since the push is entirely in the x direction, we can use Eq. (19) with $F_x = P$, $F_y = 0$, and $F_z = 0$. This gives

$$W = \int_0^{2\text{ m}} mg \tan \phi\, dx$$

Here x is the variable of integration and we therefore ought to express $\tan \phi$ as a function of x. Instead of doing this, let us adopt ϕ as the variable of integration and express dx in terms of $d\phi$. From Figure 7.11a,

$$\sin \phi = x/l$$

where l is the length of the cable. Hence

$$dx = l \cos \phi\, d\phi$$

Furthermore, when $x = 2$ m, $\sin\phi = \frac{2}{20}$ and $\phi = 0.10$ radian $= 5.74°$. Our expression for the work then becomes

$$W = \int_0^{0.10} mg \tan\phi \, (l \cos\phi \, d\phi) = \int_0^{0.10} mgl \sin\phi \, d\phi$$

$$= mgl\Big[-\cos\phi\Big]_0^{0.10} = mgl(1 - \cos 0.10) = mgl(1 - \cos 5.74°)$$

$$= 600 \text{ kg} \times 9.8 \text{ m/s}^2 \times 20 \text{ m} \times 0.0050 = 590 \text{ J} \qquad (22)$$

The work done by gravity can be calculated directly from Eq. (20). The initial value of z is $z_1 = -l$ and the final value is $z_2 = -l(\cos 5.74°)$. Hence the work is

$$W = -mg(-l \cos 5.74° + l)$$

$$= -mgl(1 - \cos 5.74°) \qquad (23)$$

Comparing this with Eq. (22), we see that the work done by gravity is exactly the opposite of the work done by the push, i.e., it is -590 J. In the next section we will find the explanation of this coincidence.

7.3 Kinetic Energy

We will now derive an important identity between the work done on a particle and the change of speed of the particle. Suppose that the particle moves from a point P_1 to a point P_2 and that during this motion the *net* force acting on it is **F**. This net force may be a function of position and of time, but in any case the work done by **F** on the particle is

$$W = \int_{P_1}^{P_2} \mathbf{F} \cdot d\mathbf{r} \qquad (24)$$

or

$$W = \int_{P_1}^{P_2} (F_x \, dx + F_y \, dy + F_z \, dz) \qquad (25)$$

Consider the first term in the integrand. Since

$$F_x = m\frac{dv_x}{dt}$$

the integral of the first term is

$$\int_{P_1}^{P_2} F_x \, dx = \int_{P_1}^{P_2} m\frac{dv_x}{dt} \, dx \qquad (26)$$

The velocity v_x is a function of time; but for the purpose of the integration in Eq. (26), it is better to regard the velocity as a function of position (along the worldline of the particle, time and position are in one-to-one correspondence and therefore it is always possible to change from one of these variables to the other). Then

$$\frac{dv_x}{dt} = \frac{dv_x}{dx}\frac{dx}{dt} = \frac{dv_x}{dx}v_x = v_x\frac{dv_x}{dx} \tag{27}$$

and, consequently,

$$\int_{P_1}^{P_2} F_x\,dx = \int_{P_1}^{P_2} m\frac{dv_x}{dt}\,dx = \int_{P_1}^{P_2} mv_x\frac{dv_x}{dx}\,dx$$

$$= \int_{v_{1,x}}^{v_{2,x}} mv_x\,dv_x = m\left[\tfrac{1}{2}v_x^2\right]_{v_{1,x}}^{v_{2,x}}$$

$$= \tfrac{1}{2}mv_{2,x}^2 - \tfrac{1}{2}mv_{1,x}^2 \tag{28}$$

where $v_{1,x}$ is the x velocity at P_1 and $v_{2,x}$ is the x velocity at P_2.

The other two terms in Eq. (25) can be transformed similarly, with the final result

$$W = \int_{P_1}^{P_2} (F_x\,dx + F_y\,dy + F_z\,dz)$$

$$= \tfrac{1}{2}mv_{2,x}^2 - \tfrac{1}{2}mv_{1,x}^2 + \tfrac{1}{2}mv_{2,y}^2 - \tfrac{1}{2}mv_{1,y}^2$$

$$+ \tfrac{1}{2}mv_{2,z}^2 - \tfrac{1}{2}mv_{1,z}^2$$

or

$$W = \tfrac{1}{2}mv_2^2 - \tfrac{1}{2}mv_1^2 \tag{29}$$

where, according to the usual notation, $v^2 = v_x^2 + v_y^2 + v_z^2$.

Equation (29) shows that the work done by the net force between P_1 and P_2 is proportional to the change in the square of the velocity. In Examples 2 and 5 we looked at motions in which the speed was constant; i.e., the square of the velocity was constant. According to Eq. (29), the net work should then be zero and, of course, this is exactly what we found by the explicit and lengthy calculations contained in these examples.

The quantity

Kinetic energy

$$\boxed{K = \tfrac{1}{2}mv^2} \tag{30}$$

is called the **kinetic energy** of the particle. With this terminology, Eq. (29) says that *the work done on the particle by the net force equals the change in the kinetic energy,*

$$W = K_2 - K_1 \tag{31}$$

or

Work–energy theorem

$$\boxed{W = \Delta K} \tag{32}$$

This result is called the **work–energy theorem.** It is important to keep in mind that the work in Eqs. (29), (31), and (32) must be evaluated with the *net* force, i.e., all the forces that do work on the particle must be included.

When positive work is done on a particle initially at rest, its kinetic energy increases. The acquired kinetic energy gives the particle a capacity to do work: if the moving particle is allowed to push against some obstacle, then this obstacle does negative work on the particle and simultaneously the particle does positive work on the obstacle (the mutual forces are opposite and they are an action–reaction pair). When the particle does work, its kinetic energy decreases — the total amount of work the particle can deliver on the obstacle is equal to its kinetic energy. Thus the kinetic energy represents accumulated work; it represents latent work which under suitable conditions can become active work.

The acquisition of kinetic energy through work and the subsequent production of work by this kinetic energy is neatly illustrated in the operation of a waterwheel driven by falling water. In a flour mill of an old Spanish Colonial design, the water runs down from a reservoir in a steep, open channel (Figure 7.12). The motion of a water particle is essentially that of a particle sliding down an inclined plane. If we ignore friction, then the only force that does work on the water particle is gravity. This work is positive; consequently, the kinetic energy of the water increases and attains a maximum value at the lower end of the channel. The stream of water emerges from this channel with high kinetic energy and hits the blades of the waterwheel; the water pushes on the wheel, turns it, and gives up its kinetic energy while doing work — and the wheel turns the millstones and does useful work on them. Thus, the work that gravity does on the descending water is ultimately converted into useful work, with the kinetic energy playing an intermediate role in this process. The operation of turbines in a hydroelectric power station involves the same principle; however, the situation is somewhat more complicated because the water flows in a closed pipe rather than in an open channel and pressure plays an important role.

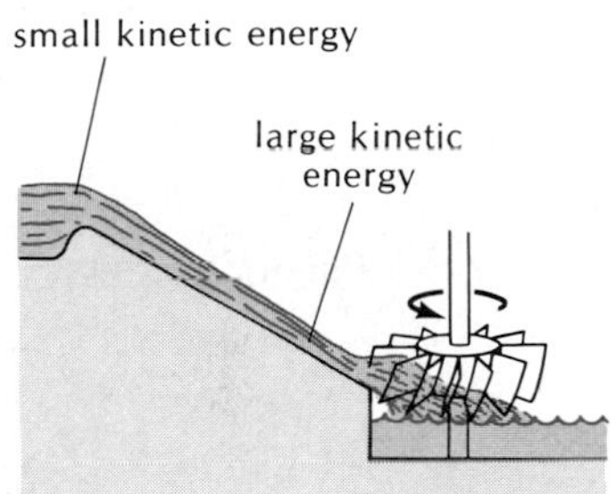

Fig. 7.12 Horizontal waterwheel.

The units of kinetic energy are the same as the units of work: the joule and the foot-pound. Table 7.1 lists some typical kinetic energies.

Table 7.1 Some Kinetic Energies

Orbital motion of Earth	2.6×10^{33} J
Rotational motion of Earth	2.1×10^{29} J
Ship *Queen Elizabeth* (at cruising speed)	9×10^{9} J
Jet airliner (Boeing 747 at maximum speed)	7×10^{9} J
Automobile (at 55 mi/h)	5×10^{5} J
Rifle bullet	4×10^{3} J
Man walking	60 J
Highest energy found in a single cosmic ray	50 J
Falling raindrop	4×10^{-5} J
Proton from large accelerator (Fermilab)	1.6×10^{-7} J
Fission fragments of one uranium-235 nucleus	2.6×10^{-11} J
Proton in nucleus	3×10^{-12} J
Electron in atom (hydrogen)	2.2×10^{-18} J
Air molecule (at room temperature)	6.2×10^{-21} J

It is sometimes convenient to express the kinetic energy directly in terms of momentum by means of the formula

Kinetic energy as a function of momentum

$$K = \frac{p^2}{2m} \tag{33}$$

By substituting $p = mv$ into this, we immediately recognize that Eq. (33) agrees with Eq. (30).

7.4 Gravitational Potential Energy

Potential energy

As we have seen, the kinetic energy represents the capacity of a particle to do work by virtue of its velocity. We will now become acquainted with another form of energy that represents the capacity of the particle to do work by virtue of its position in space. This is the **potential energy.** In this section, we will examine the special case of gravitational potential energy for a particle moving under the influence of the constant gravitational force and we will formulate a law of conservation of energy for such a particle. In the next chapter we will examine other cases of potential energy and formulate the general law of conservation of energy.

When the constant force of gravity, $F_z = -mg$, acts on a particle undergoing a displacement from (x_1, y_1, z_1) to (x_2, y_2, z_2), it does an amount of work [see Eq. (20)]

$$W = -mg(z_2 - z_1)$$

We can write this as

$$W = -U(z_2) + U(z_1) \tag{34}$$

where the function

Gravitational potential energy

$$U(z) = mgz \tag{35}$$

is called the **gravitational potential energy** of the particle. According to Eq. (34), the change in the potential energy between the points z_1 and z_2 equals the negative of the work done by gravity on the particle moving between these points.

The gravitational potential energy represents the capacity of the particle to do work by virtue of its height above the surface of the Earth. A particle high above the surface of the Earth is endowed with a large amount of latent work which can be exploited and converted into actual work by allowing the particle to push against some obstacle as it descends; the total amount of work that can be extracted during the descent is equal to the change in potential energy. Of course, the work extracted in this way really arises from the Earth's gravity — the particle can do work on the obstacle because gravity is doing work on the particle. Hence the gravitational potential energy is really a joint property of both the particle and the Earth; it is a property of the configuration of the particle–Earth system.

If the only force acting on a particle is gravity, then by combining Eqs. (31) and (34) we can obtain a relation between potential energy and kinetic energy,

$$-U(z_2) + U(z_1) = K_2 - K_1 \tag{36}$$

which we can rewrite as follows:

$$K_1 + U(z_1) = K_2 + U(z_2) \tag{37}$$

This equality indicates that the quantity $K + U(z)$ is a constant of the motion, i.e., it has the same value at the point P_2 as at the point P_1. Thus

$$K + U(z) = [\text{constant}] \tag{38}$$

The sum of the kinetic and potential energies is called the **mechanical energy** of the particle. It is usually designated by the symbol E:

$$\boxed{E = K + U(z)} \tag{39}$$

Mechanical energy

This energy represents the total capacity of the particle to do work by virtue of both its velocity and its position.

Equation (38) shows that if the only force acting on the particle is gravity, then the mechanical energy remains constant:

$$\boxed{E = [\text{constant}]} \tag{40}$$

Law of conservation of mechanical energy

This is the **law of conservation of mechanical energy.**

Apart from its practical significance in terms of work, mechanical energy is very useful in the study of the motion of the particle. Since the sum of the potential and kinetic energies must remain constant during the motion, an increase in one must be compensated by a decrease in the other; this means that during the motion, kinetic energy is converted into potential energy and vice versa. If we make use of the formulas for K and U, Eq. (40) becomes

$$\tfrac{1}{2}mv^2 + mgz = E = [\text{constant}] \tag{41}$$

This shows explicitly how the particle trades speed for height during the motion: whenever z increases, v must decrease (and conversely) so as to keep the sum of the two terms on the left side of Eq. (41) constant. Incidentally, Eq. (2.29) is essentially Eq. (41) written in a slightly different form.

An important facet of Eq. (41) is that it is valid not only for a particle in free fall (a projectile), but also for a particle sliding on a surface or a track of arbitrary shape, provided that there is no friction. Of course, under these conditions besides the gravitational force there also acts the normal force; but this force does no work and hence does not affect the left side of Eq. (36) or any of the subsequent Eqs. (37)–(41).

Christiaan Huygens (hoigens), *1629–1695, Dutch mathematician and physicist. He invented the pendulum clock, made improvements in the manufacture of telescope lenses, and discovered the rings of Saturn. Huygens investigated the theory of collisions of elastic bodies and the theory of oscillations of the pendulum, and he stated the law of conservation of mechanical energy for motion under the influence of gravity.*

This gives us a glimpse of the elegance and power of the principle of conservation of mechanical energy. Whereas the derivation of Eq. (2.29) is valid only for a freely falling particle, the derivation of Eq. (41) is valid under much more general conditions.

EXAMPLE 6. A bobsled run leading down a hill at Lake Placid (the site of the 1980 Winter Olympics) descends 148 m from its highest point to its lowest. Suppose that a bobsled, initially at rest at the highest point, slides down this run without friction. What speed will the bobsled attain at the lowest point?

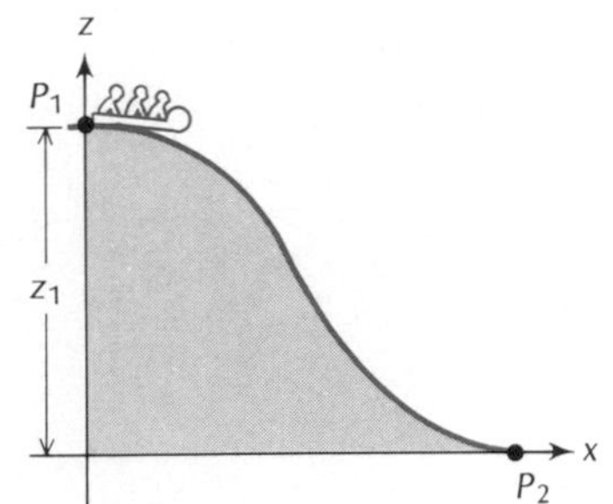

Fig. 7.13 A bobsled run.

SOLUTION: The coordinates of the highest and the lowest points are $z_1 = 148$ m and $z_2 = 0$ m, respectively (Figure 7.13). According to Eq. (39), the energy at the start of the motion is

$$E = \tfrac{1}{2}mv_1^2 + mgz_1 = 0 + mgz_1 \tag{42}$$

and the energy at the end of the motion is

$$E = \tfrac{1}{2}mv_2^2 + mgz_2 = \tfrac{1}{2}mv_2^2 + 0 \tag{43}$$

The conservation of energy implies that Eqs. (42) and (43) are equal,

$$\tfrac{1}{2}mv_2^2 = mgz_1$$

or

$$v_2 = \sqrt{2gz_1} = \sqrt{2 \times 9.8 \text{ m/s}^2 \times 148 \text{ m}} = 54 \text{ m/s}$$

Note that to obtain this answer by direct computation of forces and accelerations would have been extremely difficult — it would have required detailed knowledge of the shape of the run on the hill. With the law of energy conservation we can bypass these complexities.

EXAMPLE 7. A pendulum consists of a mass m tied to one end of a string of length l. The other end of the string is attached to a fixed point of support (Figure 7.14a). Suppose that the pendulum is initially held at an angle of 45° with the vertical. If the pendulum is released from this position, what will be the tension in the string at the instant the mass passes through its lowest position?

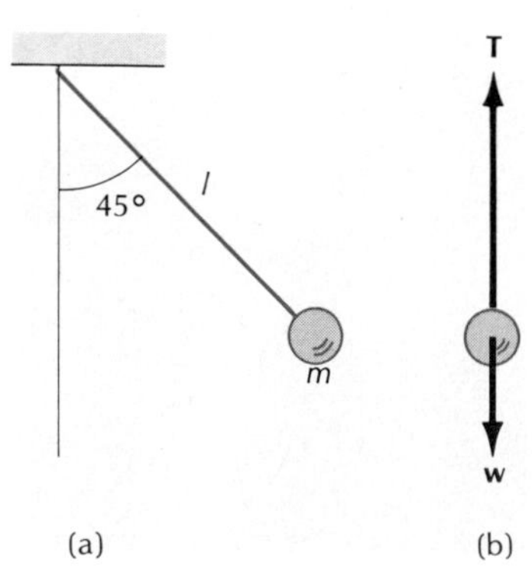

Fig. 7.14 (a) A pendulum. Initially the pendulum makes an angle of 45° with the vertical. (b) "Free-body" diagram for the mass m at the instant it reaches the lowest point.

SOLUTION: First, we must calculate the speed attained by the mass. If the z coordinate is measured upward from the point of support, the initial z coordinate is $z_1 = -l \cos \theta = -l \cos 45°$ and the lowest z coordinate is $z_2 = -l$. The energy at the initial position is

$$E = \tfrac{1}{2}mv_1^2 + mgz_1 = 0 - mgl \cos 45° \tag{44}$$

and the energy at the lowest position is

$$E = \tfrac{1}{2}mv_2^2 + mgz_2 = \tfrac{1}{2}mv_2^2 - mgl \tag{45}$$

The conservation of energy then implies

$$\tfrac{1}{2}mv_2^2 - mgl = -mgl \cos 45°$$

from which

$$v_2^2 = 2(gl - gl \cos 45°) = 2gl(1 - \cos 45°)$$

From this, we find that the centripetal acceleration is

$$v_2^2/l = 2g(1 - \cos 45°) \tag{46}$$

The mass multiplied by the centripetal acceleration gives us the magnitude of the centripetal force. This force has two contributions: a positive (upward) contribution due to the tension T in the string and a negative (downward) contribution due to the weight mg (Figure 7.14b). Thus, the net centripetal force is $T - mg$ so that

$$mv_2^2/l = T - mg$$

or

$$2mg(1 - \cos 45°) = T - mg$$

Consequently,

$$\begin{aligned} T &= mg + 2mg(1 - \cos 45°) \\ &= (3 - \sqrt{2}/2)mg \end{aligned} \tag{47}$$

SUMMARY

Work done by a constant force: $W = \mathbf{F} \cdot \Delta\mathbf{r} = F_x\,\Delta x + F_y\,\Delta y + F_z\,\Delta z$

Work done by a variable force:

$$W = \int_{P_1}^{P_2} \mathbf{F} \cdot d\mathbf{r} = \int_{P_1}^{P_2} F_x\,dx + \int_{P_1}^{P_2} F_y\,dy + \int_{P_1}^{P_2} F_z\,dz$$

Work done by gravity: $W = -mg\,\Delta z$

Kinetic energy: $K = \frac{1}{2}mv^2$

$$= \frac{p^2}{2m}$$

Work–energy theorem: $W = \Delta K$

Gravitational potential energy: $U(z) = mgz$

Mechanical energy: $E = K + U(z)$

Conservation of mechanical energy: $E = [\text{constant}]$

QUESTIONS

1. Does the work of a force on a body depend on the frame of reference in which it is calculated? Give some examples.

2. Does your body do work (external or internal) when standing at rest? When walking steadily along a level road?

3. Consider a pendulum swinging back and forth. During what part of the motion does the weight do positive work? Negative work?

4. According to Eq. (30), $K = \frac{1}{2}mv_x^2 + \frac{1}{2}mv_y^2 + \frac{1}{2}mv_z^2$. Does this mean that the kinetic energy has x, y, and z components?

5. Consider a woman steadily climbing a flight of stairs. The external forces on the woman are her weight and the normal force of the stairs against her feet. During the climb, the weight does negative work, while the normal force does no work. Under these conditions how can the kinetic energy of the woman remain constant? (Hint: The entire woman cannot be regarded as a particle, since her legs are not rigid; but the upper part of her body can be regarded as a particle, since it is rigid. What is the force of her legs against the upper part of her body? Does this force do work?)

6. An automobile increases its speed from 50 to 60 mi/h. What is the percent increase of kinetic energy? What is the percent reduction of travel time for a given distance?

7. Two blocks in contact slide past one another and exert friction forces on one another. Can the friction force *increase* the kinetic energy of one block? Of both? Does there exist a reference frame in which the friction force decreases the kinetic energy of both blocks?

8. When an automobile with rear-wheel drive is accelerating on, say, a level road, the horizontal force of the road on the rear wheels gives the automobile momentum, but it does not give the automobile any energy because the point of application of this force (point of contact of wheel on ground) is instantaneously at rest if the wheel is not slipping. What force gives the body of the automobile energy? Where does the energy come from? (Hint: Consider the force that the rear axle exerts against its bearings.)

9. A dictionary gives the following definition of *zoom:* "to climb in an airplane suddenly and sharply at an angle greater than normal, using the energy of momentum." Does this make any sense?

10. Why do elevators have counterweights? (See Figure 6.9.)

11. A parachutist jumps out of an airplane, opens a parachute, and lands safely on the ground. Is the mechanical energy for this motion conserved?

12. If you release a tennis ball at some height above a hard floor, it will bounce up and down several times, with a gradually decreasing amplitude. Where does the ball suffer a loss of mechanical energy?

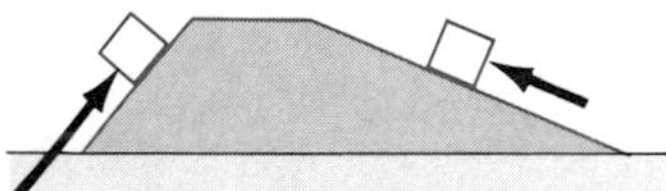

Fig. 7.15 Two ramps of different steepness.

13. Two ramps, one steeper than the other, lead from the floor to a loading platform (Figure 7.15). It takes more force to push a (frictionless) box up the steeper ramp. Does this mean it takes more work to raise the box from the floor to the platform?

14. Consider the two ramps described in the preceding question. Taking friction into account, which ramp requires less work for raising a box from the floor to the platform?

15. A stone is tied to a string. Can you whirl this stone in a vertical circle with constant speed? Can you whirl this stone with constant energy? For each of these two cases, describe how you must move your hand.

PROBLEMS

Section 7.1

1. If it takes a horizontal force of 300 N to push a stalled automobile along a level road, how much work must you do to push this automobile a distance of 5.0 m?

2. In an overhead lift, a champion weight lifter raises 560 lb from the floor to a height of 6.5 ft. How much work does he do?

3. Suppose that the force required to push a saw back and forth through a piece of wood is 8 lb. If you push this saw back and forth 30 times, moving it forward 5 in. and back 5 in. each time, how much work do you do?

4. A man pushes a crate along a flat concrete floor. The mass of the crate is 120 kg and the coefficient of friction between the crate and the floor is $\mu_k = 0.5$. How much work does the man do if, pushing horizontally, he moves the crate 15 m?

5. By means of a towrope, a girl pulls a sled loaded with firewood along a level, icy road. The coefficient of friction between the sled and the road is $\mu_k = 0.10$ and the mass of the sled plus its load is 150 kg. The towrope is attached to the front of the sled and makes an angle of 30° with the horizontal. How much work must the girl do on the sled to pull it 1.0 km?

6. A man pushes a heavy box up an inclined ramp making an angle of 30° with the horizontal. The mass of the box is 60 kg and the coefficient of kinetic friction between the box and the ramp is 0.45. How much work must the man do to push the box to a height of 2.5 m? Assume that the man pushes on the box in a direction parallel to the surface of the ramp.

7. An elevator consists of an elevator cage and a counterweight attached to the ends of a cable which runs over a pulley (Figure 7.16). The mass of the cage (with its load) is 1200 kg and the mass of the counterweight is 1000 kg. The elevator is driven by an electric motor attached to the pulley. Suppose that the elevator is initially at rest at the first floor of the building, and the motor makes the elevator accelerate upward at the rate of 1.5 m/s^2.

(a) What is the tension in the part of the cable attached to the elevator cage? What is the tension in the part of the cable attached to the counterweight?

(b) The acceleration lasts exactly 1.0 s. How much work has the electric motor done in this interval? Ignore friction forces and ignore the mass of the pulley.

(c) After the acceleration interval of 1.0 s, the motor pulls the elevator upward at constant speed until it reaches the third floor, exactly 10.0 m above the first floor. What is the total amount of work that the motor has done up to this point?

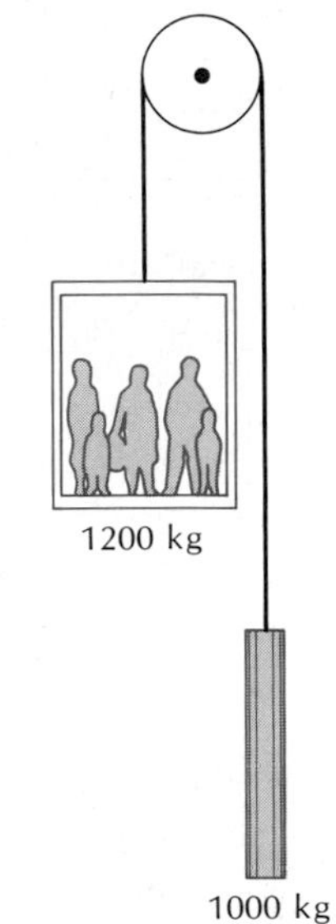

Fig. 7.16 Elevator cage and counterweight.

8. A man pulls a cart along a level road by means of a short rope stretched over his shoulder and attached to the front end of the cart. The friction force that opposes the motion of the cart is 250 N.

(a) If the rope is attached to the cart at shoulder height, how much work must the man do to pull the cart 50 m at constant speed?

(b) If the rope is attached to the cart below shoulder height so that it makes an angle of 30° with the horizontal, what is the tension in the rope? How much work must the man now do to pull the cart 50 m? Assume that the friction force is unchanged.

9. A spring has a force constant of 3.5×10^4 N/m. The spring is initially relaxed. How much work must you do to compress the spring by 0.10 m? How much work must you do to compress the spring a further 0.10 m?

Section 7.2

10. A particle moves along the x axis from $x = 0$ to $x = 2$. A force of magnitude $F(x) = 2x^3 + 8x$ acts on the particle (the distance x is measured in meters, and the force in newtons). Calculate the work done by the force $F(x)$ during this motion.

11. A particle moves in the x–y plane from the origin $x = 0$, $y = 0$ to the point $x = 2$, $y = -1$ while under the influence of a force $\mathbf{F} = 3\hat{\mathbf{x}} + 2\hat{\mathbf{y}}$. How much work does this force do on the particle during this motion? The distances are measured in meters and the force in newtons.

12. During a storm a sailboat is anchored in a 10-m-deep harbor. The wind pushes against the boat with a steady horizontal force of 7000 N.

(a) The anchor rope that holds the boat in place is 50 m long and is

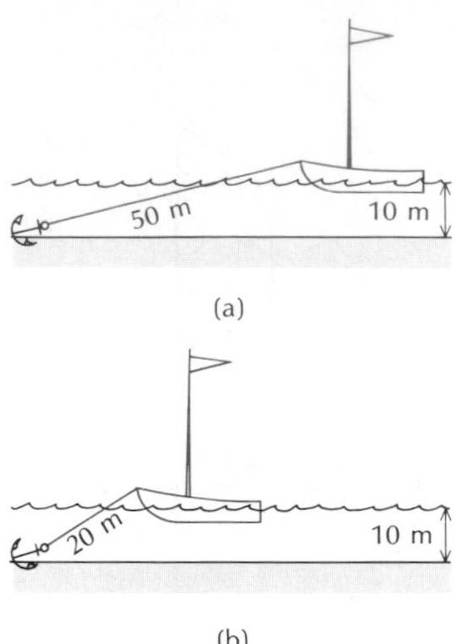

Fig. 7.17 A sailboat at anchor.

stretched straight between the anchor and the boat (Figure 7.17a). What is the tension in the rope?

(b) How much work must the crew of the sailboat do to pull in 30 m of the anchor rope, bringing the boat nearer to the anchor (Figure 7.17b)? What is the tension in the rope when the boat is in this new position?

13. Suppose that the force acting on a particle is a function of position; the force has components $F_x = 4x^2 + 1$, $F_y = 2x$, $F_z = 0$, where force is measured in newtons and distance in meters. What is the work done by this force if the particle moves on a straight line from $x = 0$, $y = 0$, $z = 0$ to $x = 2$, $y = 2$, $z = 0$?

14. A horse pulls a sled along a snow-covered curved ramp. Seen from the side, the surface of the ramp follows an arc of a circle of radius R (Figure 7.18). The pull of the horse is always parallel to this surface. The mass of the sled is m and the coefficient of sliding friction between the sled and the surface is μ_k. How much work must the horse do on the sled to pull it to a height $(1 - \sqrt{2}/2)R$, corresponding to an angle of 45° along the circle (Figure 7.18)? How does this compare with the amount of work required to pull the sled to the same height along a straight ramp of length R?

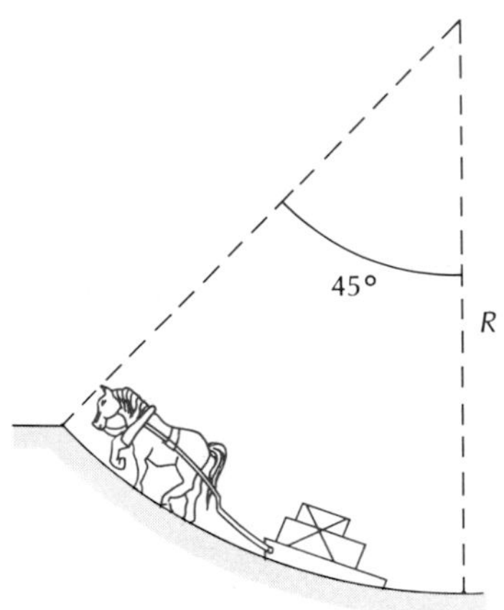

Fig. 7.18 A horse pulling a sled along a curved ramp.

Section 7.3

15. Calculate the kinetic energy that the Earth has owing to its motion around the Sun.

16. The electron in a hydrogen atom has a speed of 2.2×10^6 m/s. What is the kinetic energy of this electron?

17. The Skylab satellite disintegrated when it reentered the atmosphere. Among the pieces that crashed down on the surface of the Earth, one of the heaviest was a lead-lined film vault of 1770 kg which had an estimated impact speed of 120 m/s on the surface. What was its kinetic energy? How many kilograms of TNT would we have to explode to release the same amount of energy? (One kilogram of TNT releases 4.6×10^6 J.)

18. An automobile of mass 1600 kg is traveling along a straight road at 80 km/h.

(a) What is the kinetic energy of this automobile in the reference frame of the ground?

(b) What is the kinetic energy in the reference frame of a motorcycle traveling in the same direction at 60 km/h?

(c) What is the kinetic energy in the reference frame of a truck traveling in the opposite direction at 60 km/h?

19. According to statistical data, the probability that an occupant of an automobile suffers a lethal injury when involved in a crash is proportional to the square of the speed of the automobile.

(a) At a speed of 50 mi/h, the probability is approximately 3%. What are the probabilities at 60 mi/h, 70 mi/h, and 80 mi/h?

(b) Do the data suggest that the injuries are related to the kinetic energy of the automobile or to the momentum?

20. For the projectile described in Problem 2.16, calculate the initial kinetic energy ($t = 0$) and calculate the final kinetic energy ($t = 3.0$ s). How much energy does the projectile lose to friction in 3.0 s?

21. The mass of the bullet and the muzzle velocity for several rifles manufactured around 1900 are as follows: Lee-Enfield, 13.9 g, 628 m/s; Lebel, 15.0 g, 632 m/s; Mauser, 14.7 g, 638 m/s; Springfield, 9.7 g, 792 m/s.

(a) For each of these rifles compute the momentum and the kinetic energy of the bullet.

(b) Roughly, the momentum determines the push that the bullet can exert upon impact on the target and the kinetic energy determines the damage (breakage) that the bullet can do within the target. Which of these bullets can exert the largest push? Which can do the most damage?

22. The fastest skier is Franz Weber, who attained 126.24 mi/h on a steep slope at Velocity Peak, Colorado. The fastest runner is Robert Hayes, who briefly attained 27.89 mi/h on a level track. Assume that the skier and the runner each have a mass of 162 lb-mass. What is the kinetic energy of each? By what factor is the kinetic energy of the skier larger than that of the runner?

23. A large stone-throwing engine designed by Archimedes could throw a 77-kg stone over a range of 180 m. Assume the stone is thrown at an initial angle of 45° with the vertical.

(a) Calculate the initial kinetic energy of this stone.

(b) Calculate the kinetic energy of the stone at the highest point of the trajectory.

24. With the brakes fully applied, a 1500-kg automobile decelerates at the rate of 8.0 m/s^2.

(a) What is the braking force acting on the automobile?

(b) If the initial speed is 90 km/h, what is the stopping distance?

(c) What is the work done by the braking force in bringing the automobile to a stop from 90 km/h?

(d) What is the change in the kinetic energy of the automobile?

25. The velocity of small bullets can be roughly measured with ballistic putty. When the bullet strikes a slab of putty, it penetrates a distance that is roughly proportional to the kinetic energy. Suppose that a bullet of velocity 160 m/s penetrates 0.8 m into the putty and a second, identical bullet fired from a more powerful gun penetrates 1.2 cm. What is the velocity of the second bullet?

Section 7.4

26. It has been reported that at Cherbourg, France, waves smashing on the coast lifted a boulder of 7000 lb over a 20-ft wall. What minimum energy must the waves have given to the boulder?

27. A 75-kg man walks up the stairs from the first to the third floor of a building, a height of 10 m. How much work does he do against gravity? Compare your answer with the food energy he acquires by eating an apple (see Table 8.1).

28. What is the kinetic energy and what is the gravitational potential energy (relative to the ground) of a jetliner of weight 160,000 lb cruising at 550 mi/h at an altitude of 30,000 ft?

29. A golf ball of mass 50 g released from a height of 1.5 m above a concrete floor bounces back to a height of 1.0 m. What is the amount of energy lost during the impact?

30. Surplus energy from an electric power plant can be temporarily stored as gravitational energy by using this surplus energy to pump water from a river into a reservoir at some altitude above the level of the river. If the reservoir is 250 m above the level of the river, how much water (in cubic meters) must we pump in order to store 2×10^{13} J?

31. The track of a cable car on Telegraph Hill in San Francisco rises more than 200 ft from its lowest point. Suppose that a car is ascending at 8 mi/h along the track when it breaks away from its cable at a height of exactly 200 ft. It will then coast up the hill some extra distance, stop, and begin to race down the hill. What speed does the car attain at the lowest point of the track? Ignore friction.

32. In pole vaulting, the jumper achieves great height by converting his kinetic energy of running into gravitational potential energy. The pole plays an intermediate role in this process. When the jumper leaves the ground, part of his translational kinetic energy has been converted into kinetic energy of rotation (with the foot of the pole as the center of rotation) and part has been con-

verted into elastic potential energy of deformation of the pole. When the jumper reaches his highest point, all of this energy has been converted into gravitational potential energy. Suppose that a jumper runs at a speed of 10 m/s. If the jumper converts all of the corresponding kinetic energy into gravitational potential energy, how high will his center of mass rise? The actual height reached by pole vaulters is 5.7 m (measured from the ground). Is this consistent with your calculation?

33. Because of a brake failure, a bicycle with its rider careens down a steep hill 150 ft high. If the bicycle starts from rest and if there is no friction, what is the final speed attained at the bottom of the hill?

34. Under favorable conditions, an avalanche can reach extremely great speeds because the snow rides down the mountain on a cushion of trapped air that makes the sliding motion nearly frictionless. Suppose that a mass of 2×10^7 kg of snow breaks loose from a mountain and slides down into a valley 500 m below the starting point. What is the speed of the snow when it hits the valley? What is its kinetic energy? The explosion of 1 short ton (2000 lb) of TNT releases 4.2×10^9 J. How many tons of TNT release the same energy as the avalanche?

35. A roller coaster near St. Louis is 110 ft high at its highest point.
(a) What is the maximum speed that a car can attain by rolling down from the highest point if initially at rest? Ignore friction.
(b) Some people claim that cars reach a maximum speed of 62 mi/h. If this is true, what must be the initial speed of a car at the highest point?

36. In some barge canals built in the nineteenth century, barges were lifted from a low level of the canal to a higher level by means of wheeled carriages. In a French canal, barges of 70 metric tons were placed on a carriage of 35 tons which was pulled, by a wire rope, to a height of 12 m along an inclined track 500 m long.
(a) What was the tension in the wire rope?
(b) How much work was done to lift the barge and carriage?
(c) If the cable had broken just as the carriage reached the top, what would have been the final speed of the carriage when it crashed at the bottom?

37. A parachutist of mass 60 kg jumps out of an airplane at an altitude of 800 m. Her parachute opens and she lands on the ground with a speed of 5.0 m/s. How much energy has been lost to air friction in this jump?

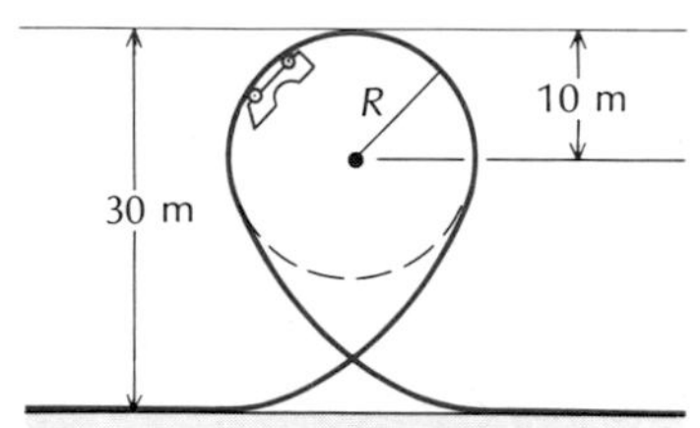

Fig. 7.19 A roller coaster with a full loop.

38. A center fielder throws a baseball of mass 0.17 kg with an initial speed of 28 m/s and elevation angle of 30°. What is the kinetic energy and what is the potential energy of the baseball when it reaches the highest point of its trajectory? Ignore friction.

39. A stone is tied to a string of length R. A man whirls this stone in a vertical circle. Assume that the energy of the stone remains constant as it moves around the circle. Show that if the string is to remain taut at the top of the circle, the speed of the stone at the bottom of the circle must be at least $\sqrt{5gR}$.

40. In a loop coaster at an amusement park, cars roll along a track that is bent in a full vertical loop (Figure 7.19). If the upper portion of the track is an arc of a circle of radius $R = 10$ m, what is the minimum speed that a car must have at the top of the loop if it is not to fall off? If the highest point of the loop has a height $h = 30$ m, what is the minimum speed with which the car must enter the loop at its bottom?

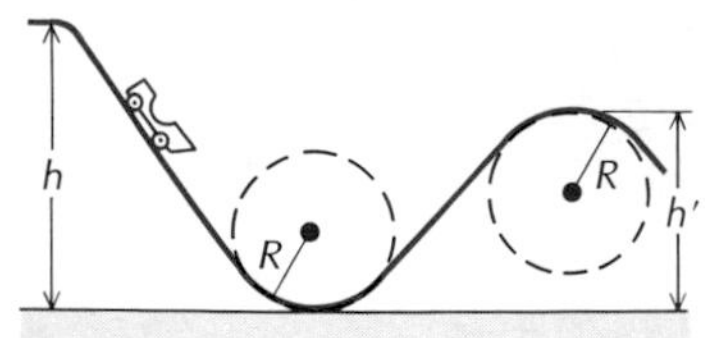

Fig. 7.20 A roller coaster.

41. You are to design a roller coaster in which cars start from rest at a height $h = 100$ ft, roll down into a valley, and then up a mountain (Figure 7.20).
(a) What is the speed of the cars at the bottom of the valley?
(b) If the passengers are to feel 8 gee at the bottom of the valley, what must be the radius R of the arc of circle that fits the bottom of the valley?

(c) The top of the next mountain is an arc of circle of the same radius R. If the passengers are to feel 0 gee at the top of this mountain, what must be its height h'?

*42. One portion of the track of a toy roller coaster is bent into a full vertical circle of radius R. A small cart rolling on the track enters the bottom of the circle with a speed $2\sqrt{gR}$. Show that this cart will fall off the track before it reaches the top of the circle and find the (angular) position at which the cart loses contact with the track.

*43. A particle initially sits on top of a large smooth sphere of radius R (Figure 7.21). The particle begins to slide down the sphere, without friction. At what angular position will the particle lose contact with the surface of the sphere? Where will the particle land on the ground?

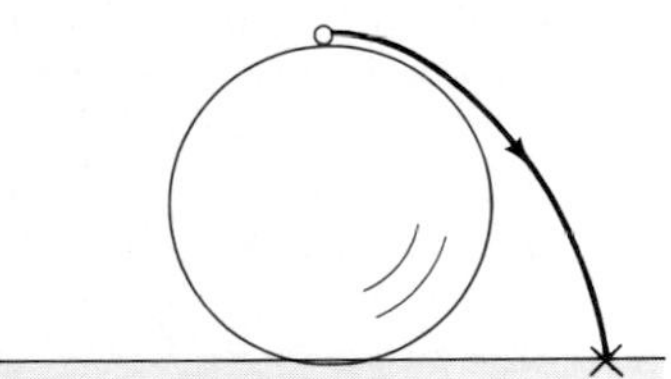

Fig. 7.21 Particle sliding down a sphere.

CHAPTER 8

Conservation of Energy

In the preceding chapter we found how to formulate a law of conservation of mechanical energy for a particle moving under the influence of the gravitational force. Now we will seek to formulate a law of conservation of mechanical energy when other forces act on the particle, and we will state the general law of conservation of energy.

8.1 Conservative Forces

The example of a particle moving under the influence of the gravitational force showed us how to obtain a constant energy by taking the sum of the kinetic and potential energies. In order to formulate a similar conservation law for a particle moving under the influence of some other force, we have to begin by asking what conditions must be met by the force if a potential energy is to exist. The answer to this question is that the force must be conservative.

Conservative force

The definition of a **conservative force** is as follows. Consider a particle that moves from a point P_1 to a point P_2 along some path. Suppose that the force **F** that acts on the particle is a function of position only, so that the force does not depend on the velocity of the particle and does not depend (explicitly[1]) on time. The work done by the force on the particle can be calculated from Eq. (7.18),

$$W = \int_{P_1}^{P_2} \mathbf{F} \cdot d\mathbf{r} \tag{1}$$

[1] Since the position of the particle is a function of time, the force at the position of the particle is also a function of time. However, we will assume that the force at any given position is the same regardless of the time at which the particle arrives at that position.

The force **F** is *conservative if this work depends only on the position of the points P_1 and P_2 and not on the shape of the path between P_1 and P_2.* Thus, all the paths shown in Figure 8.1 give exactly the same result for the integral in Eq. (1).

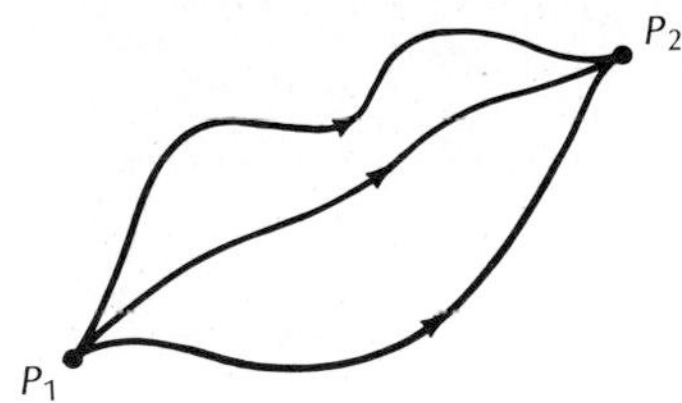

Fig. 8.1 Several alternative paths for a particle moving from P_1 to P_2.

The characteristic property of a conservative force can also be expressed another way: the force is conservative if the work is exactly zero for any round trip along a closed path. Figure 8.2 shows such a closed path which starts and ends at the point P_1. If the work for this complete path is zero, then the work done between P_1 and some intermediate pont P_2 must exactly cancel the work done between P_2 and P_1, i.e.,

$$0 = \underset{\text{(path I)}}{\int_{P_1}^{P_2} \mathbf{F} \cdot d\mathbf{r}} + \underset{\text{(path II)}}{\int_{P_2}^{P_1} \mathbf{F} \cdot d\mathbf{r}} \tag{2}$$

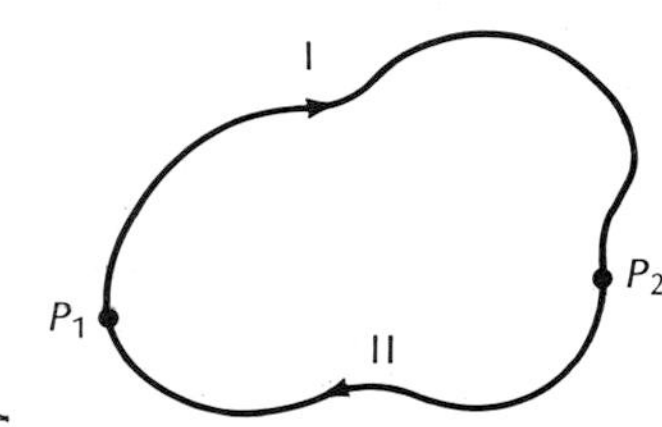

Fig. 8.2 Round trip along a closed path.

If we exchange the upper and lower limits on the second integral, we obtain

$$0 = \underset{\text{(path I)}}{\int_{P_1}^{P_2} \mathbf{F} \cdot d\mathbf{r}} - \underset{\text{(path II)}}{\int_{P_1}^{P_2} \mathbf{F} \cdot d\mathbf{r}} \tag{3}$$

But this simply states that the work done in going from P_1 to P_2 along path I is the same as the work done along path II, i.e., the work is path independent. Thus the criterion of path independence is equivalent to the criterion of zero work for any round trip.

Gravity is an obvious example of a conservative force. According to Eq. (7.20), the work done by gravity between the points P_1 and P_2 with coordinates (x_1, y_1, z_1) and (x_2, y_2, z_2) is a function of only z_1 and z_2, i.e., the work depends only on the location of the points, and not on the path between them.

The force exerted by a spring in one-dimensional motion is another example of a conservative force. According to Eq. (7.12), the work is a function of only x_1 and x_2, i.e., it again depends only on the location of the points and not on how much the particle zigzagged back and forth while getting from one point to the other.

The force of kinetic friction is an example of a force that is *not* conservative. Obviously, the longer the path, the greater the work done by friction. Thus the criterion of path independence fails — the work depends on the length of the path and not just on the location of the endpoints. The criterion of zero work for a round trip also fails — the work done by friction is negative for all portions of the path and no cancellation is possible.

However, the nonconservative character of friction is to some extent an illusion which arises from our failure to keep careful track of the motion of all the particles involved. According to the criterion of zero work for any round trip, we ought to examine motions in which all the particles return to their starting points. If we slide a metal block back and forth on a tabletop, this condition is never quite met because the atoms in the block and in the tabletop are disturbed in an uncontrollable manner. When we reverse the motion of the block, the motion of the atoms will not be reversed: during the return trip the atoms do not return to their original place; instead they are even more disturbed. Hence, from a microscopic point of view, we recognize that the work done by the friction force on the block during a round trip is not a fair test of the conservative character of the adhesion forces involved in the friction mechanism. If all the atomic motions are taken into account, then it actually turns out that these adhesion forces *are* conservative. Of course, for practical purposes, we adopt a macroscopic point of view and ignore the atomic motions — friction forces must then be regarded as nonconservative.

8.2 Potential Energy of a Conservative Force

Whenever the force acting on a particle is conservative, it is possible to construct a corresponding potential energy by the following recipe: Take a reference point P_0; this point may be the origin of coordinates, or some point at a large distance from the region where the force acts, or any other convenient point. At the point P_0, assign to the potential energy some value $U(P_0)$; this value may be any convenient number, for instance, $U(P_0) = 0$. At any other point P, assign to the potential energy the value

Potential energy

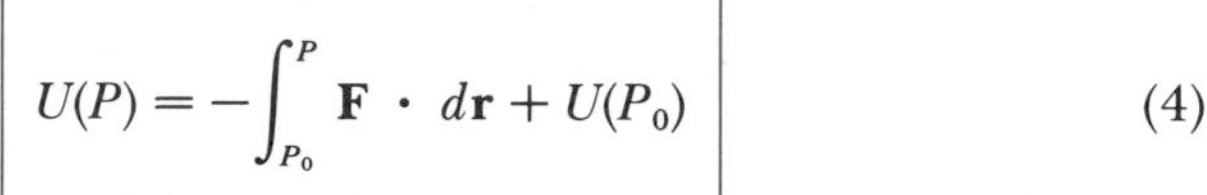

$$U(P) = -\int_{P_0}^{P} \mathbf{F} \cdot d\mathbf{r} + U(P_0) \qquad (4)$$

Joseph Louis, Comte Lagrange (lagranj), *1736–1813, French mathematician and theoretical astronomer, director of the Berlin Academy and professor at the École Polytechnique, Paris. He was the most eminent mathematician of the eighteenth century, barring only the great Leonhard Euler. In the elegant mathematical treatise* Analytical Mechanics *— which is probably the only book on physics completely devoid of diagrams — Lagrange formulated Newtonian mechanics in the language of advanced calculus and introduced the general definition of the potential-energy function. Lagrange is also known for his calculations of the motion of planets and for his influential role in securing the adoption of the metric system of units.*

The integral appearing here is to be evaluated along any path connecting P_0 and P — in view of the conservative character of **F**, any choice of path will give the same result. The integral may therefore be regarded as a function of the position of P (it is also a function of the position of P_0, but since the latter point is kept fixed, this dependence can be ignored). Thus, $U(P)$ is some well-defined function of position.

We can now readily verify that the potential-energy function constructed according to the recipe of the preceding paragraph has all the properties we wish. First of all, the change in the potential energy between two points P_1 and P_2 equals the negative of the work done by the force between these two points, as shown by the following equalities:

$$\begin{aligned} U(P_2) - U(P_1) &= -\int_{P_0}^{P_2} \mathbf{F} \cdot d\mathbf{r} + U(P_0) + \int_{P_0}^{P_1} \mathbf{F} \cdot d\mathbf{r} - U(P_0) \\ &= -\int_{P_0}^{P_2} \mathbf{F} \cdot d\mathbf{r} - \int_{P_1}^{P_0} \mathbf{F} \cdot d\mathbf{r} \\ &= -\int_{P_1}^{P_2} \mathbf{F} \cdot d\mathbf{r} \qquad (5) \end{aligned}$$

Note that $U(P_0)$ cancels in this relation. Thus, the choice of P_0 and the choice of $U(P_0)$ in no way affects the calculation of the work from the potential energy. This justifies the somewhat cavalier attitude adopted in the preceding paragraph: the constant $U(P_0)$ can be given any arbitrary value because the quantity of physical significance is the *change* in the potential energy rather than the potential energy itself.

Next, we can show that the total mechanical energy is conserved. Equation (7.31) states that the change in kinetic energy equals the work, i.e.,

$$K_2 - K_1 = W = \int_{P_1}^{P_2} \mathbf{F} \cdot d\mathbf{r} \tag{6}$$

But, according to Eq. (5), the integral appearing in Eq. (6) is simply $U(P_1) - U(P_2)$. Hence

$$K_2 - K_1 = U(P_1) - U(P_2)$$

or

$$K_2 + U(P_2) = K_1 + U(P_1)$$

This proves that $K + U$ is a constant of the motion,

$$K + U = [\text{constant}] \tag{7}$$

As in the gravitational case, the **mechanical energy** is defined as the sum of the kinetic and potential energies,

$$\boxed{E = K + U} \tag{8}$$

Mechanical energy

and Eq. (7) shows that this energy is conserved,

$$\boxed{E = [\text{constant}]} \tag{9}$$

Law of conservation of mechanical energy

This is the **law of conservation of mechanical energy.**

As a concrete illustration of these general mathematical results, consider the motion of a particle under the influence of the force of gravity, $F = -mg$. For the reference point P_0, take the origin $x = 0$, $y = 0$, $z = 0$, and for the potential energy at this point, take $U(P_0) = 0$. Then the general recipe (4) gives[2]

$$U(P) = -\int_0^x F_x\,dx' + \int_0^y F_y\,dy' + \int_0^z F_z\,dz'$$

$$= -\int_0^z F_z\,dz' = -\int_0^z (-mg)\,dz' = mgz$$

Hence the potential energy is a function of only z,

$$U(P) = U(z) = mgz$$

and the conserved mechanical energy is

$$E = K + U = \tfrac{1}{2}mv^2 + mgz$$

[2] The variables of integration in this integral have been written as x', y', z' in order to distinguish them from the limits of integration x, y, z.

This, of course, agrees with Eq. (7.41). Thus, our general recipe for the potential energy reproduces the formula for gravitational potential energy that we derived in Section 7.4.

As another illustration, consider the motion of a particle in one dimension (along the x axis) under the influence of a force $F_x(x) = -kx$ produced by an elastic spring. The corresponding potential energy can be constructed as follows. For the reference point P_0 take $x = 0$, and for the potential energy at this point take $U(P_0) = 0$. Then[3]

$$U(x) = -\int_0^x (-kx')\, dx' = k\left[\tfrac{1}{2}x'^2\right]_0^x$$

or

Potential energy of a spring

$$\boxed{U(x) = \tfrac{1}{2}kx^2} \tag{10}$$

Hence the conserved mechanical energy is

$$E = K + U = \tfrac{1}{2}mv^2 + \tfrac{1}{2}kx^2 \tag{11}$$

This expression for the energy gives us some information about the general features of the motion; it shows how, say, an increase of x requires a decrease of v so as to keep E constant.

Note that if instead of the arbitrary choice $U(P_0) = 0$ we had made the equally good (and equally arbitrary) choice $U(P_0) = 3.0$ J, then the energy in Eqs. (10) and (11) would have had an extra 3.0 J added to it. This only changes the base value from which the energy is reckoned; it does not alter the fact that the energy is conserved — and that is what is really important.

EXAMPLE 1. A child's toy gun shoots a dart by means of a compressed spring. The constant of the spring is 3.2×10^2 N/m and the mass of the dart is 8.0 g. Before shooting, the spring is compressed by 6.0 cm and the dart is placed in contact with the spring (Figure 8.3); the spring is then released. What will be the speed of the dart when the spring reaches its equilibrium length?

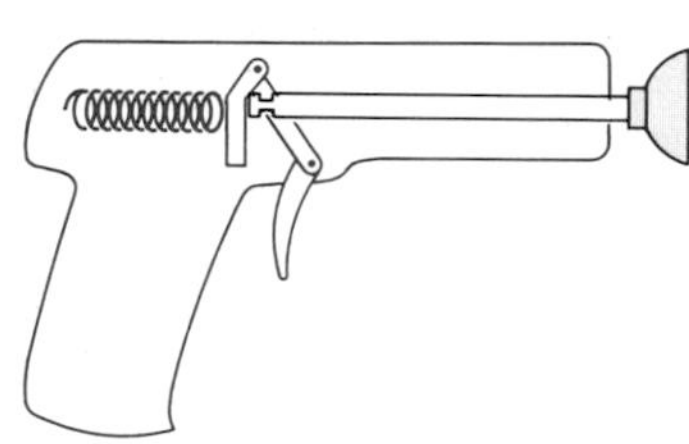

Fig. 8.3 A toy gun. The spring is compressed 6.0 cm.

SOLUTION: The dart can be regarded as a particle moving under the influence of a force $F = -kx$. According to Eq. (11), the initial energy is

$$E = \tfrac{1}{2}mv_1^2 + \tfrac{1}{2}kx_1^2 = 0 + \tfrac{1}{2}kx_1^2 \tag{12}$$

When the spring reaches its equilibrium length ($x = 0$), the energy will be

$$E = \tfrac{1}{2}mv_2^2 + \tfrac{1}{2}kx_2^2 = \tfrac{1}{2}mv_2^2 + 0 \tag{13}$$

Conservation of energy demands that Eqs. (12) and (13) be equal,

$$\tfrac{1}{2}mv_2^2 = \tfrac{1}{2}kx_1^2$$

[3] The variable of integration in this integral has been written as x' in order to distinguish it from the limit of integration x.

Hence the speed of the dart will be

$$v_2 = \sqrt{k/m}\, x_1$$

$$= \sqrt{(3.2 \times 10^2\ \text{N/m})/8.0 \times 10^{-3}\ \text{kg}} \times 6.0 \times 10^{-2}\ \text{m}$$

$$= 12\ \text{m/s}$$

As another illustration, consider the motion of a particle in one dimension with the force $F(x) = -A/x^2$, i.e., the inverse-square force mentioned in Section 6.6. For the reference point P_0 take $x = \infty$, and for the potential energy at this point take $U(P_0) = 0$. Then

$$U(x) = -\int_{\infty}^{x} \frac{-A}{x'^2}\, dx' = A\left[-\frac{1}{x'}\right]_{\infty}^{x} = -\frac{A}{x} \tag{14}$$

The corresponding energy is

$$E = K + U = \tfrac{1}{2}mv^2 - \frac{A}{x} \tag{15}$$

This expression again shows how changes in x and v must be related.

EXAMPLE 2. A particle of mass m is moving along the x axis under the influence of the force $F(x) = -A/x^2$; assume that (in some suitable units) $m = 1.0$ and $A = 1.0$. The initial position and velocity of the particle are $x = 1.0$ and $v = 1.0$. What is the energy of the particle? What is the maximum distance that the particle can reach along the (positive) x axis?

SOLUTION: We have already analyzed this motion in detail by means of numerical approximation methods in Section 6.6. It will be instructive to compare our earlier approximate result for the maximum distance with the exact result obtained from energy conservation.

The energy is given by Eq. (15) with $A = 1.0$, $m = 1.0$, $x = 1.0$, and $v = 1.0$:

$$E = \tfrac{1}{2}mv^2 - \frac{A}{x} = \tfrac{1}{2} \times 1.0 \times (1.0)^2 - \frac{1.0}{1.0} = -\tfrac{1}{2}$$

The units here are the same as in Section 6.6 and will not be further specified.

The energy at any later time must be the same:

$$-\tfrac{1}{2} = \tfrac{1}{2}mv^2 - \frac{A}{x}$$

When the particle reaches maximum distance along the x axis, the velocity will be zero, i.e.,

$$-\tfrac{1}{2} = -\frac{A}{x}$$

which gives $x = 2.0$. This exact result agrees fairly well with the approximate result, $x \cong 1.986$, that we obtained by numerical methods (see Table 6.3).

If several conservative forces simultaneously act on a particle, then the net potential energy is the sum of all the potential energies of all these forces. The total mechanical energy is the sum of the kinetic energy and the net potential energy; the total mechanical energy is, of course, conserved.

Finally, what if both a conservative and a nonconservative force, such as friction, act on a particle? Then the change in the kinetic energy plus the change in the potential energy of the conservative force will be equal to the work done by the friction force.

$$\Delta K + \Delta U = W_{\text{friction}}$$

that is,

Loss of mechanical energy by friction

$$\boxed{\Delta E = W_{\text{friction}}} \tag{16}$$

This equation for the change of energy in the presence of a friction force is not nearly as useful in the investigation of the motion of a particle as is the equation of energy conservation that would hold if the friction force were absent. The latter equation tells us how arbitrary changes in position and in speed are related (see Examples 1 and 2). The former equation does no such thing because the value of W_{friction} depends not only on the net change of position, but also on the details of the motion between the initial and the final positions; thus, W_{friction} cannot be evaluated explicitly until the details of the motion are known. The main use of Eq. (16) is in the evaluation of W_{friction} *after* position and speed have been calculated by other means.

8.3 Calculation of the Force from the Potential Energy

It is sometimes desirable to calculate the force from the potential energy function. Such a calculation may seem irrelevant, since, according to Eq. (4), the potential energy is defined in terms of the force. However, in modern theoretical physics potential energies play a very important role — in many instances interactions are primarily described in terms of the corresponding potential energy and the force is only calculated afterward.

Suppose that the points P and P_0 in Eq. (4) are separated only by an infinitesimal displacement $d\mathbf{r}$; then $U(P)$ will differ from $U(P_0)$ only by an infinitesimal quanity

$$dU = U(P) - U(P_0) = -\mathbf{F} \cdot d\mathbf{r}$$

We can also write this as

$$dU = -F_x\,dx - F_y\,dy - F_z\,dz \tag{17}$$

Now assume that the infinitesimal displacement is entirely in the x direction. Then $dy = 0$ and $dz = 0$, so that

$$dU = -F_x\,dx \tag{18}$$

or

$$F_x = -\frac{dU}{dx} \tag{19}$$

Thus, the x component of the force is the derivative of U with respect to x, *with y and z being held constant;* i.e., when we differentiate the function U with respect to x in Eq. (19), both y and z are to be treated just as any other constants. Since the differentiation only applies to one of the variables, the derivative appearing in Eq. (19) is called a **partial derivative** and is usually written with the following special notation:

$$\boxed{F_x = -\frac{\partial U}{\partial x}} \tag{20}$$

Force as the derivative of potential energy

Likewise,

$$F_y = -\frac{\partial U}{\partial y} \tag{21}$$

$$F_z = -\frac{\partial U}{\partial z} \tag{22}$$

For instance, if U is the potential energy of an elastic spring, $U(x) = \frac{1}{2}kx^2$, then Eq. (20) yields a force

$$F_x = -\frac{\partial U}{\partial x} = -\frac{\partial}{\partial x}\left(\tfrac{1}{2}kx^2\right) = -kx$$

which is as expected. Note that in this instance U does not depend on y or z and therefore Eqs. (21) and (22) yield $F_y = 0$ and $F_z = 0$ — there is no force in the x and y directions.

EXAMPLE 3. In a diatomic molecule the atoms can move relative to one another within certain limits. According to a simple theory of interatomic forces, the potential energy for the motion of each atom is

$$U(x) = U_0\left(e^{-2(x-x_0)/b} - 2e^{-(x-x_0)/b}\right)$$

Here e = 2.718 . . . ; U_0, x_0, and b are constants; and x is the distance from the one atom to the midpoint of the molecule (Figure 8.4). What is the force on the atom? For what value of x is this force zero?

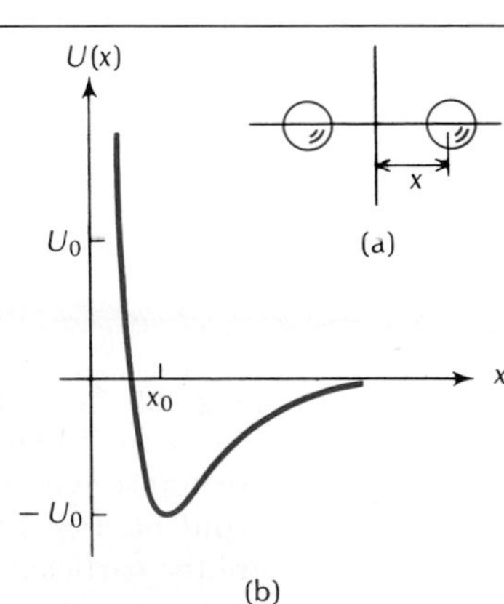

Fig. 8.4 (a) A molecule consisting of two atoms. (b) Plot of the potential energy as a function of x.

SOLUTION: By Eq. (20), the force is the negative of the derivative of U with respect to x:

$$F_x = -\frac{\partial U}{\partial x} = -U_0\left(-\frac{2}{b}e^{-2(x-x_0)/b} + \frac{2}{b}e^{-(x-x_0)/b}\right)$$

The force is zero if

$$-\frac{2}{b}e^{-2(x-x_0)/b}+\frac{2}{b}e^{-(x-x_0)/b}=0$$

which happens when $x = x_0$.

8.4 The Curve of Potential Energy

If a particle of some given energy is moving in one dimension under the influence of a conservative force, then Eq. (8) permits us to calculate the speed of the particle as a function of position. Suppose that the potential energy is $U = U(x)$; then Eq. (8) states

$$E = \tfrac{1}{2}mv^2 + U(x)$$

or

$$v^2 = \frac{2}{m}[E - U(x)] \tag{23}$$

Since the left side of this equation is never negative, we can immediately conclude that the particle must always remain within a range of values of x for which $U(x) \leq E$. If the particle reaches a point at which $U(x) = E$, then $v = 0$, i.e., the particle will stop at this point and its motion will reverse. Such a point is called a **turning point** of the motion.

Turning point

According to Eq. (23), v^2 is directly proportional to $E - U(x)$, i.e., v^2 is large wherever the difference between E and $U(x)$ is large. We can therefore gain some insights into the qualitative features of the motion by drawing a graph of potential energy on which it is possible to display the difference between E and $U(x)$. For example, Figure 8.5 shows the curve of potential energy for an atom in a diatomic molecule (compare Example 3). The equation of this curve is

$$U(x) = U_0(e^{-2(x-x_0)/b} - 2e^{-(x-x_0)/b}) \tag{24}$$

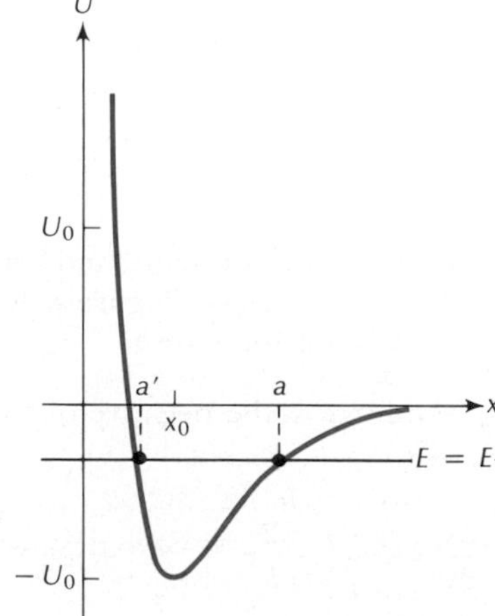

Fig. 8.5 Potential-energy curve for an atom in a diatomic molecule. The horizontal line is the energy level of the particle. The turning points are at $x = a$ and at $x = a'$.

On this graph, we indicate the value of the energy of the particle by a horizontal line. We will call this horizontal line the **energy level** of the particle. At any point x, we can then see the difference between E and $U(x)$ at a glance; according to Eq. (23), this tells us v^2. For instance, suppose that a particle has an energy $E = E_1$. Figure 8.5 shows this energy level. Obviously, the particle has maximum speed at the point $x = x_0$, where the separation between the energy level and the potential-energy curve is maximum. The speed gradually decreases as the particle moves toward the right. The potential-energy curve intersects the energy level at $x = a$; at this point the speed of the particle will reach zero, i.e., this point is a turning point of the motion. The particle then moves toward the left, again attaining its greatest speed at $x = x_0$. The speed gradually decreases as the particle continues to move toward the left, and the speed reaches zero at $x = a'$, the second turning point of the motion. Here the particle begins to move toward the right, etc. Thus the particle continues to move back and forth between the two turning points — the particle is confined between the two turning points. The regions $x > a$ and $x < a'$ are forbidden regions; the region $a' \leq x \leq a$ is permitted. The particle is said to be in a **bound orbit.** The motion is periodic, i.e., it repeats again and again whenever the particle returns to its starting point.

Energy level

Bound orbit

The location of the turning points depends on the energy. For a particle with a lower energy level the turning points are closer together. The lowest possible energy level intersects the potential-energy curve at its minimum (see $E = -U_0$ in Figure 8.6); the two turning points then merge into the single point $x = x_0$. A particle with this lowest possible energy cannot move at all — it remains stationary at $x = x_0$. The point $x = x_0$ is an **equilibrium point.** Note that the potential-energy curve has zero slope at $x = x_0$; this corresponds to zero force, $F_x = -\partial U/\partial x = 0$ (see Example 3).

Equilibrium point

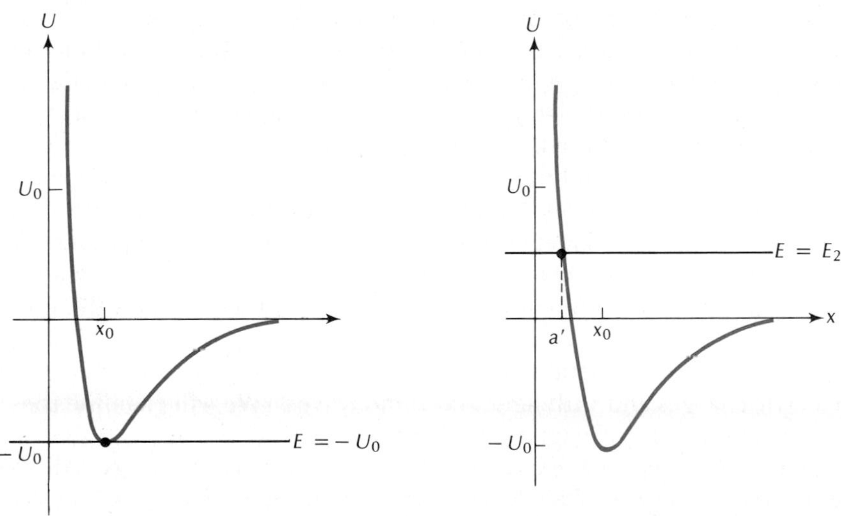

Fig. 8.6 (left) Energy level of a stationary particle.

Fig. 8.7 (right) Energy level of a particle in an unbound orbit. There is a single turning point at $x = a'$.

In Figure 8.5, the right side of the potential-energy curve never rises above $U = 0$. Consequently, if the energy level is above this value (for instance $E = E_2$; see Figure 8.7), then there is only one single turning point on the left, and no turning point on the right. A particle with energy E_2 will continue to move toward the right forever; it is not confined. Such a particle is said to be in an **unbound orbit.**

Unbound orbit

Of course, the above qualitative analysis based on the curve of potential energy cannot tell us the details of the motion such as, say, the travel time from one point to another. But the qualitative analysis is useful because it gives us a quick survey of the types of orbits that are possible for different values of the energy. If we need more details, we must solve Eq. (23). In view of $v = dx/dt$, Eq. (23) is a differential equation for x; this differential equation can be solved by integration, but the integrals are usually rather difficult to evaluate, especially if the potential is a complicated function such as $U(x)$ of Eq. (24).

8.5 Other Forms of Energy

As we have seen in Section 8.2, if the forces acting on a particle are conservative, then the mechanical energy of the particle is conserved. But if some of the forces acting on the particle are not conservative, then the mechanical energy of the particle will not always remain constant. For instance, if friction forces are acting on a body, they do negative work on it and decrease its mechanical energy.

However, it is a remarkable fact about our physical universe that whenever mechanical energy is lost by a body, this energy never disappears — it is merely transmuted into other forms of energy. Thus, in the case of friction, the mechanical energy lost by the body is transformed into kinetic and potential energy of the atoms in the body and in the surface against which it is rubbing. The energy that the atoms acquire in the rubbing process is disorderly kinetic and potential energy — it is spread among the atoms in an irregular, random fashion. The disorderly kinetic and potential energy of the atoms of a body is what is meant by **heat.** Hence friction produces heat.

Heat

Heat is a form of energy, but whether it is to be regarded as a new form of energy or not depends on the point of view one wishes to adopt. Taking a macroscopic point of view, we ignore the atomic motions; then heat is to be regarded as distinct from mechanical energy. Taking a microscopic point of view, we recognize heat as kinetic and potential energy of the atoms; then heat is to be regarded as mechanical energy. (We will further discuss heat in Chapter 20.)

Chemical energy and nuclear energy are two other forms of energy. The former is kinetic and potential energy of the electrons within the atoms; the latter is kinetic and potential energy of the protons and neutrons within the nuclei of atoms. As in the case of heat, whether these are to be regarded as new forms of energy depends on the point of view.

Electric and magnetic energy are forms of energy associated with electric charges and with light and radio waves. (We will examine these forms of energy in Chapters 26 and 32.)

Table 8.1 lists some examples of different forms of energy. All the energies in Table 8.1 have been expressed in joules. However, for reasons of tradition and convenience, some other energy units are often used in specialized areas of physics and engineering. The energy of atomic and subatomic particles is usually measured in **electron-volts** (eV), where

Alternative energy units

$$1 \text{ electron-volt} = 1 \text{ eV} = 1.602 \times 10^{-19} \text{ J}$$

The multiples of this unit are 1 keV = 10^3 eV, 1 MeV = 10^6 eV, and 1 GeV = 10^9 eV.

Table 8.1 SOME ENERGIES

Nuclear fuel in Sun	1×10^{45} J
Explosion of a supernova	1×10^{44} J
Fossil fuel available on Earth	2.0×10^{23} J
Yearly energy expenditure of the United States	8×10^{19} J
Volcanic explosion (Krakatoa)	6×10^{18} J
Annihilation of 1 kg of matter–antimatter	9.0×10^{16} J
Nuclear fuel in nuclear reactor	1×10^{16} J
Explosion of thermonuclear bomb (1 megaton)	4.2×10^{15} J
Fission of 1 kg of uranium	8.2×10^{13} J
Gravitational potential energy of jet airliner (Boeing 747 at 30,000 ft)	2×10^{10} J
Lightning flash	1×10^{9} J
Combustion of 1 gal. of gasoline	1.3×10^{8} J
Daily food intake of man (3000 kcal)	1.3×10^{7} J
Explosion of 1 kg of TNT	4.6×10^{6} J
Metabolization of one apple (110 kcal)	4.6×10^{5} J
Kinetic energy of running man	4×10^{3} J
One push-up	3×10^{2} J
Fission of one uranium nucleus	3.2×10^{-11} J
Annihilation of electron–positron pair	1.6×10^{-13} J
Energy of ionization of hydrogen atom	2.2×10^{-18} J

The energy supplied by electric power plants is usually expressed in **kilowatt-hours** (kW · h):

$$1 \text{ kilowatt-hour} = 1 \text{ kW} \cdot \text{h} = 3.600 \times 10^6 \text{ J}$$

And the thermal energy supplied by the combustion of fuels is expressed in **kilocalories** (kcal):

$$1 \text{ kilocalorie} = 1 \text{ kcal} = 4.187 \times 10^3 \text{ J}$$

or in British **thermal units** (Btu):

$$1 \text{ Btu} = 1.055 \times 10^3 \text{ J}$$

We will learn more about the definitions of these units in later chapters.

Hermann von Helmholtz, *1821–1894, Prussian surgeon, biologist, mathematician, and physicist, professor at Berlin. His scientific contributions ranged from the invention of the ophthalmoscope and studies of the physiology and physics of vision and hearing, to the measurement of the speed of light and studies in theoretical mechanics. Helmholtz formulated the general law of conservation of energy, treating it as a consequence of the basic laws of mechanics and electricity.*

All these forms of energy can be transformed into one another. For example, in an internal combustion engine, chemical energy of the fuel is transformed into heat and kinetic energy; in a hydroelectric power station, gravitational potential energy of the water is transformed into electric energy; in a nuclear reactor, nuclear energy is transformed into heat, light, kinetic energy, etc. However, in any such transformation process the sum of all the energies of all the pieces of matter involved in the process remains constant: *the form of the energy changes, but the total amount of energy does not.* This is the **general law of conservation of energy.**

General law of conservation of energy

8.6 Mass and Energy

One of the great discoveries made by Einstein early in this century is that energy can be transformed into mass and mass can be transformed into energy. Thus, *mass is a form of energy.* The amount of energy contained in an amount m of mass is

Mass is energy

$$\boxed{E = mc^2} \tag{25}$$

where c is the speed of light ($c = 3.0 \times 10^8$ m/s). The most spectacular demonstration of this relationship is found in the annihilation of matter and antimatter (see Interlude C). If a proton collides with an antiproton, or an electron with an antielectron, they react violently and annihilate each other in an explosion that generates an intense flash of very energetic light. In this reaction the mass of the particles is entirely converted into energy of light. According to Eq. (25), the annihilation of just 1 metric ton of matter and antimatter (500 kg of each) would release an amount of energy

$$E = mc^2 = 10^3 \text{ kg} \times (3 \times 10^8 \text{ m/s})^2 = 9 \times 10^{19} \text{ J} \tag{26}$$

This is enough energy to satisfy the needs of the United States for a full year. Unfortunately, antimatter is not readily available in large amounts. On Earth, antiparticles can only be obtained from reactions induced by the impact of beams of high-energy particles on a target. These collisions occasionally result in the creation of a particle–antiparticle pair. Such pair creation is the reverse of pair annihilation: the creation process transforms some of the kinetic energy of the collision into mass and a subsequent anihilation merely gives back the original energy.

But the relationship between energy and mass of Eq. (25) also has another aspect. *Energy has mass.* Whenever the energy of a body is changed, its mass (and weight) is changed. The change in mass that corresponds to a given change of energy is

Energy has mass

$$\boxed{\Delta m = \Delta E/c^2} \tag{27}$$

For instance, if the kinetic energy of a body increases, its mass (and weight) increases. At speeds small compared to the speed of light, the mass increment is not noticeable. But, as a body approaches the speed of light, the mass increment becomes very large. The electrons produced by the Stanford Linear Accelerator provide an extreme example of this effect: these electrons have a speed of 99.99999997% of the speed of light and their mass is 44,000 times the mass of electrons at rest!

Rest mass

The mass that a particle has when at rest is sometimes called its **rest mass** and the corresponding energy (Eq. (25)) is called the **rest-mass energy.** The masses listed in tables of particles (see, e.g., Table C.1) are always the rest masses. In Chapter 17 we will study the theory of Special Relativity and obtain a formula for the increase of mass with velocity. However, in all other chapters we will neglect the dependence of mass on velocity because the effect is insignificant at the velocities that we encounter in everyday experience.

The fact that energy has mass indicates that energy is a form of mass. Conversely, as we have seen above, mass is a form of energy. Hence mass and energy must be regarded as essentially the same thing.

The laws of conservation of mass and conservation of energy are therefore not two independent laws — each implies the other. For example, consider the fission reaction of uranium inside the reactor vessel of a nuclear power plant. The complete fission of 1 kg of uranium yields an energy of 8.2×10^{13} J. The reaction conserves energy — it merely transforms nuclear energy into heat, light, and kinetic energy, but does not change the total amount of energy. The reaction also conserves mass — if the reactor vessel is hermetically sealed and thermally insulated from its environment, then the reaction does not change the mass of the contents of the vessel. However, if we open the vessel during or after the reaction, and let some of the heat and light escape, then the mass of the residues will not match the mass of the original amount of uranium. The mass of the residues will be about 0.1% smaller than the mass of the uranium. This mass defect represents the mass carried away by the energy that has escaped. Thus the often repeated statement that nuclear reactions convert mass into energy is misleading. True, the mass of the residues is less than the mass of the original uranium, but the escaped energy carries with it just the right amount of mass to balance the accounts: the net mass remains constant. A nuclear reaction merely transforms energy into new forms of energy and mass into new forms of mass. In this regard a nuclear reaction is not essentially different from a chemical reaction. The net mass remains constant in any chemical reaction, but the mass of the residues of an exothermic chemical reaction is slightly less than the original mass. The heat released in a chemical reaction carries away mass, but, in contrast to a nuclear reaction, this amount of mass is so small as to be quite immeasurable.

James Watt, *1736–1819, Scottish inventor and engineer. He modified and improved an earlier steam engine designed by Thomas Newcomen, adding a separate condenser. Watt introduced the* horsepower *as a unit of mechanical power.*

8.7 Power

The power delivered by a force to a body is the rate at which the force does work on that body. If the force does an amount of work ΔW in an interval of time Δt, then the **average power** is

$$\boxed{\bar{P} = \frac{\Delta W}{\Delta t}} \tag{28}$$

Average power

The **instantaneous power** is the time derivative of the work,

$$\boxed{P = \frac{dW}{dt}} \tag{29}$$

Instantaneous power

In the metric system, the unit of power is the **watt** (W), where

$$1 \text{ watt} = 1 \text{ W} = 1 \text{ J/s} \tag{30}$$

watt, W

In the British system, the unit of power is the ft · lb/s. In practice, engineers often prefer to measure power in **horsepower** (hp) units, where

***horsepower,* hp**

$$1 \text{ horsepower} = 1 \text{ hp} = 550 \text{ ft} \cdot \text{lb/s} = 745.7 \text{ W} \tag{31}$$

This is roughly the rate at which a (very strong) horse can do work.

EXAMPLE 4. An elevator has a weight of 2000 lb. How many horsepower must the motor deliver to the elevator if it is to raise the elevator at the rate of 6 ft/s? The elevator has no counterweight.

SOLUTION: By means of the elevator cable, the motor must exert an upward force of 2000 lb to raise the elevator. If the elevator is raised a distance Δz, the work done by this force is

$$\Delta W = F\,\Delta z$$

and the rate at which work is done is

$$P = \frac{\Delta W}{\Delta t} = F\frac{\Delta z}{\Delta t} = Fv \tag{32}$$

This gives

$$\begin{aligned} P &= 2000 \text{ lb} \times 6 \text{ ft/s} \\ &= 1.2 \times 10^4 \text{ ft} \cdot \text{lb/s} \\ &= 1.2 \times 10^4 \text{ ft} \cdot \text{lb/s} \times \frac{1 \text{ hp}}{550 \text{ ft} \cdot \text{lb/s}} = 22 \text{ hp} \end{aligned}$$

Equation (32) is a special instance of the following formula, which expresses the instantaneous power in terms of force and velocity. During an infinitesimal displacement $d\mathbf{r}$, a force $\mathbf{F}$ will perform an amount of work

$$dW = \mathbf{F} \cdot d\mathbf{r}$$

The instantaneous power delivered by this force is then

$$P = \frac{dW}{dt} = \mathbf{F} \cdot \frac{d\mathbf{r}}{dt} \tag{33}$$

which leads to the formula

Power delivered by a force

$$\boxed{P = \mathbf{F} \cdot \mathbf{v}} \tag{34}$$

EXAMPLE 5. A horse drags a sled up a steep snow-covered street of slope 1:7 (Figure 8.8a). The sled has a mass of 300 kg and the coefficient of sliding friction between the sled and the snow is 0.12. If the horse pulls parallel to the surface of the street and delivers a power of 1.0 hp, what is the maximum (constant) speed with which the horse can drag the sled? What fraction of the horse's power is expended against friction?

SOLUTION: Figure 8.8b is a "free-body" diagram for the sled, showing the weight ($w = mg$), the normal force ($N = mg\cos\theta$), the friction force ($f_k = \mu_k N$), and the pull of the horse (T). Since the acceleration along the street is zero,

$$0 = T + w_x - f_k = T - mg \sin \theta - f_k$$

so that

$$T = mg \sin \theta + f_k = mg \sin \theta + \mu_k mg \cos \theta$$

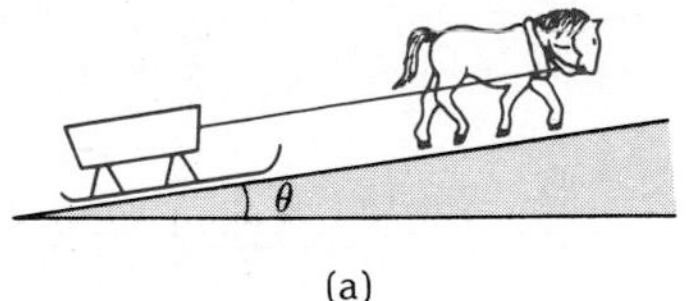

(a)

The power delivered by the horse is

$$P = \mathbf{T} \cdot \mathbf{v} = (mg \sin \theta + \mu_k mg \cos \theta)v$$

from which

$$v = \frac{P}{mg(\sin \theta + \mu_k \cos \theta)}$$

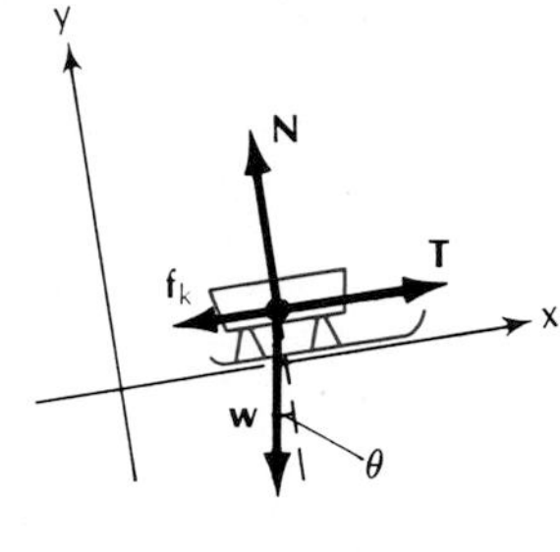

(b)

Fig. 8.8 (a) Horse dragging a sled and (b) the "free-body" diagram for the sled.

With $\theta = \tan^{-1} \frac{1}{7} = 8.13°$, this yields

$$v = \frac{746 \text{ W}}{300 \text{ kg} \times 9.8 \text{ m/s}^2 \times (\sin 8.13° + 0.12 \cos 8.13°)}$$

$$= 0.98 \text{ m/s}$$

The power exerted by the friction force is

$$P_{\text{friction}} = \mathbf{f}_k \cdot \mathbf{v} = -f_k v = -\mu_k mg \cos \theta \times v$$

$$= -0.12 \times 300 \text{ kg} \times 9.8 \text{ m/s}^2 \times \cos 8.13° \times 0.98 \text{ m/s}$$

$$= -3.4 \times 10^2 \text{ W} = -0.46 \text{ hp}$$

and the power exerted by gravity is

$$P_{\text{gravity}} = \mathbf{w} \cdot \mathbf{v} = w_x v = -mg \sin \theta \times v$$

$$= -300 \text{ kg} \times 9.8 \text{ m/s}^2 \times \sin 8.13° \times 0.98 \text{ m/s}$$

$$= -4.1 \times 10^2 \text{ W} = -0.54 \text{ hp}$$

Thus 46% of the horse's power is expended against friction and 54% against gravity.

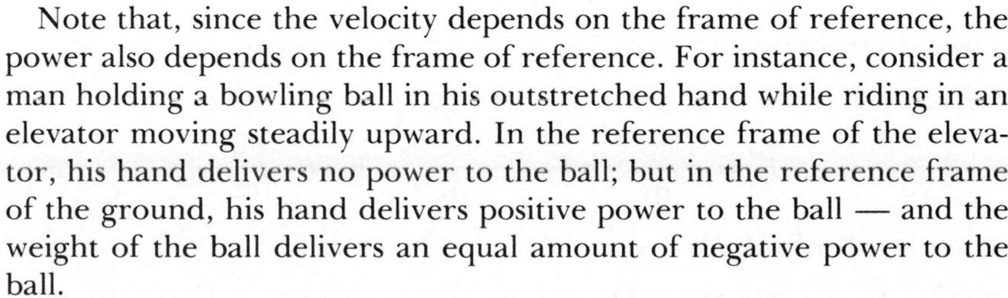

Note that, since the velocity depends on the frame of reference, the power also depends on the frame of reference. For instance, consider a man holding a bowling ball in his outstretched hand while riding in an elevator moving steadily upward. In the reference frame of the elevator, his hand delivers no power to the ball; but in the reference frame of the ground, his hand delivers positive power to the ball — and the weight of the ball delivers an equal amount of negative power to the ball.

The above equations all refer to *mechanical* power. In general, power is the rate at which energy is transformed from one form to another or transported from one place to another. Table 8.2 gives some examples of different kinds of power.

Incidentally, multiplication of a unit of power by a unit of time gives a unit of energy. An example of this is the kilowatt-hour (kW · h), already mentioned in Section 8.5:

$$1 \text{ kilowatt-hour} = 1 \text{ kW} \cdot \text{h} = 10^3 \text{ W} \cdot 3600 \text{ s}$$

$$= 3.6 \times 10^6 \text{ J}$$

This is the unit commonly used to measure electric energy delivered by power plants.

Table 8.2 SOME POWERS

Light and heat emitted by the Sun	3.9×10^{26} W
Solar light and heat incident on the Earth	1.7×10^{17} W
Mechanical power generated by hurricane	2×10^{13} W
Total power used in United States (average)	2×10^{12} W
Large electric power plant	$\sim 10^9$ W
Jet airliner engines (Boeing 747)	2.1×10^8 W
Automobile engine	1.5×10^5 W
Radio emission by large radio transmitter	1×10^5 W
Solar light and heat per square meter at Earth	1.4×10^3 W
Electricity used by toaster	1×10^3 W
Work output of man (athlete at maximum)	2×10^2 W
Electricity used by light bulb	1×10^2 W
Heat output of man (average)	1×10^2 W
Heat and work output of bumblebee (in flight)	2×10^{-2} W
Atom radiating light	$\sim 10^{-10}$ W

SUMMARY

Conservative force: The work done by the force does not depend on the path; it depends only on the positions of the endpoints of the path.

Potential energy of a conservative force:

$$U(P) = -\int_{P_0}^{P} \mathbf{F} \cdot d\mathbf{r} + U(P_0)$$

Mechanical energy: $E = K + U$

Conservation of energy: $E = [\text{constant}]$

Potential energy of a spring: $U(x) = \frac{1}{2}kx^2$

Loss of mechanical energy by friction: $\Delta E = W_{\text{friction}}$

Force as derivative of potential energy:

$$F_x = -\frac{\partial U}{\partial x} \qquad F_y = -\frac{\partial U}{\partial y} \qquad F_z = -\frac{\partial U}{\partial z}$$

Mass is a form of energy: $E = mc^2$

Energy has mass: $\Delta m = \Delta E / c^2$

Average power: $\overline{P} = \dfrac{\Delta W}{\Delta t}$

Instantaneous power: $P = \dfrac{dW}{dt}$

Mechanical power delivered by a force: $P = \mathbf{F} \cdot \mathbf{v}$

QUESTIONS

1. A body slides on a smooth horizontal plane. Is the normal force of the plane on the body a conservative force? Can we define a potential energy for this force according to the recipe in Section 8.2?

2. If you stretch a spring so far that it suffers a permanent deformation, is the force exerted by the spring during this operation conservative?

3. Is there any frictional dissipation of mechanical energy in the motion of the planets of the Solar System or in the motion of their satellites? (Hint: Consider the tides.)

4. What happens to the kinetic energy of an automobile during braking without skidding? With skidding?

5. Consider a stone thrown vertically upward. If we take friction into account, we see that $\frac{1}{2}mv^2 + mgz$ must *decrease* as a function of time. From this, prove that the stone will take longer for the downward motion than for the upward motion.

6. An automobile travels down a road leading from a mountain peak to a valley. What happens to the gravitational potential energy of the automobile? How is it dissipated?

7. Suppose you wind up a watch and then place it into a beaker full of nitric acid and let it dissolve. What happens to the potential energy stored in the spring of the watch?

8. News reporters commonly speak of "energy consumption." Is it accurate to say that energy is *consumed*? Would it be more accurate to say that energy is *dissipated*?

9. The explosive yield of thermonuclear bombs (Figure 8.9) is usually reported in kilotons or megatons of TNT. Would the explosion of a 1-megaton hydrogen bomb really produce the same effects as the explosion of 1 megaton of TNT (a mountain of TNT several hundred feet high)?

Fig. 8.9 Thermonuclear explosion.

10. When you heat a potful of water, does its mass increase?

11. Since mass is a form of energy, why don't we measure mass in the same units as energy? How could we do this?

12. In the annihilation of matter and antimatter, a particle and an antiparticle — such as a proton and an antiproton, or an electron and an antielectron — disappear explosively upon contact, giving rise to an intense flash of light. Is energy conserved in this reaction? Is mass conserved?

13. It takes about 5000 hp to keep an 86-ft yacht moving at its top speed of 55 mi/h (50 knots). What happens to this power?

14. In order to travel at 80 mi/h, an automobile of average size needs an engine delivering about 40 hp to overcome the effects of air friction, road friction, and internal friction (in the transmission and drive train). Why do most drivers think they need an engine of 150 or 200 hp?

PROBLEMS

Section 8.1

1. A particle moves along the x axis under the influence of a variable force $F_x = 2x^3 + 1$ (where force is measured in newtons and distance in meters).
   (a) Show that this force is conservative; i.e., show that for any back-and-forth motion that starts and ends at the same place (round trip), the work done by the force is zero.

(b) Show that the same is true for any force $F_x = F_x(x)$ that is an arbitrary function of position.

2. Show that if a force is a function of the velocity of the particle, then it cannot be a conservative force.

3. A particle moves along a circle $x^2 + y^2 = R^2$ in the x–y plane. Suppose that a force with components $F_x = -y$ and $F_y = x$ (where force is measured in newtons and distance in meters) acts on the particle. Show that this force is not conservative.

Section 8.2

4. The force acting on a particle moving along the x axis is given by the formula $F_x = K/x^4$, where K is a constant. Find the corresponding potential energy. Assume that $U(x) = 0$ for $x = \infty$.

5. A particle moves along the x axis under the influence of a variable force $F_x = 5x^2 + 3x$ (where force is measured in newtons and distance in meters). What is the potential energy associated with this force? Assume that $U(x) = 0$ at $x = 0$.

6. The spring from an automobile suspension has a spring constant 3.53×10^4 N/m (see Example 6.9). How much work must you do to compress this spring from its relaxed length of 0.316 m to 0.205 m?

7. A mass m hangs on a vertical spring of spring constant k.
(a) How far will this hanging mass have stretched the spring from its relaxed length?
(b) If you now push up on the mass and lift it until the spring reaches its relaxed length, how much work will you have done against gravity? Against the spring?

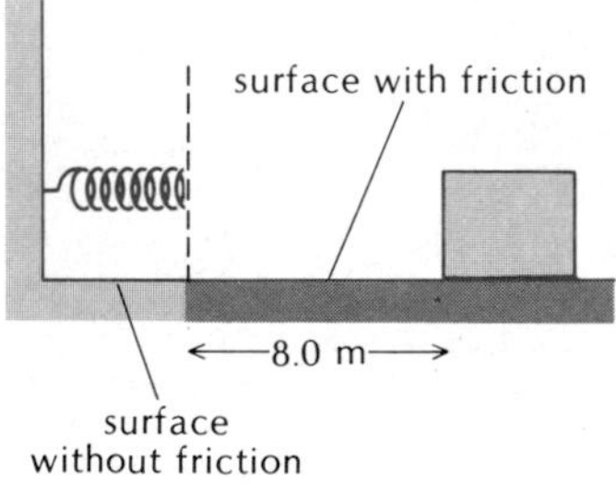

Fig. 8.10 Block released from a spring.

8. A 3.0-kg block is released from a compressed spring whose force constant is 120 N/m. After leaving the spring, it travels over a horizontal surface, with coefficient of friction 0.20, for a distance of 8.0 m before coming to rest (Figure 8.10).
(a) What was its maximum kinetic energy?
(b) How far was the spring compressed before being released?

9. A bow may be regarded mathematically as a spring. The archer stretches this "spring" and then suddenly releases it so that the bowstring pushes against the arrow. Suppose that when the archer stretches the "spring" 1.7 ft, he must exert a force of 35 lb to hold the arrow in this position. If he now releases the arrow, what will be the speed of the arrow when the "spring" reaches its equilibrium position? The weight of the arrow is 0.045 lb. Pretend that the "spring" is massless.

10. The four wheels of an automobile of mass 1200 kg are suspended below the body by vertical springs of constant $k = 7.0 \times 10^4$ N/m. If the forces on all wheels are the same, what will be the maximum instantaneous deformation of the springs if the automobile is lifted by a crane and dropped on the street from a height of 0.8 m?

11. A rope can be regarded as a long spring; when under tension, it stretches and stores elastic potential energy. Consider a nylon rope similar to that which snapped during a giant tug-of-war at a school in Harrisburg (see Problem 5.14). Under a tension of 13,000 lb (applied at its ends), the rope of initial length 1000 ft stretches to 1300 ft. What is the elastic energy stored in the rope at this tension? What happens to this energy when the rope breaks?

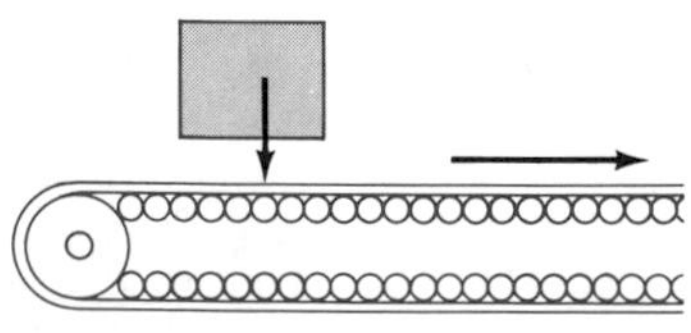

Fig. 8.11 Package dropped onto a conveyer belt.

12. A package is dropped on a horizontal conveyor belt (Figure 8.11). The mass of the package is m, the speed of the conveyor belt is v, and the coefficient of kinetic friction for the package on the belt is μ_k. For what length of time will the package slide on the belt? How far will it move in this time? How much energy is dissipated by friction?

Section 8.3

13. The potential energy of a particle moving in the x–y plane is $U = a/(x^2 + y^2)^{1/2}$, where a is a constant. What is the force on the particle? Draw a diagram showing the particle at the position x,y and the force vector.

14. The potential energy of a particle moving along the x axis is $U(x) = K/x^2$, where K is a constant. What is the corresponding force acting on the particle?

15. According to theoretical calculations, the potential energy of two quarks (see Section C.6) separated by a distance r is $U = \eta r$, where $\eta = 1.18 \times 10^{24}$ eV/m. What is the force between the two quarks? Express your answer in newtons.

Section 8.4

16. The potential energy of one of the atoms in the hydrogen molecule may be taken to be (see Example 3)

$$U(x) = U_0(e^{-2(x - x_0)/b} - 2e^{-(x - x_0)/b})$$

with $U_0 = 2.36$ eV, $x_0 = 0.37$ Å, and $b = 0.34$ Å.[4] Under the influence of the force corresponding to this potential, the atom moves back and forth along the x axis within certain limits. If the energy of the atom is $E = -1.15$ eV, what will be the turning points of the motion, i.e., at what positions x will the kinetic energy be zero? [Hint: Solve this problem graphically by making a careful plot of $U(x)$; from your plot find the values of x that yield $U(x) = -1.15$ eV.]

17. Suppose that the potential energy of a particle moving along the x axis is

$$U(x) = \frac{b}{x^2} - \frac{2c}{x}$$

where b and c are positive constants.

(a) Plot $U(x)$ as a function of x; assume $b = c = 1$ for this purpose. Where is the equilibrium point?
(b) Suppose the energy of the particle is $E = -\frac{1}{2}c^2/b$. Find the turning points of the motion.
(c) Suppose that the energy of the particle is $E = \frac{1}{2}c^2b$. Find the turning points of the motion. How many turning points are there in this case?

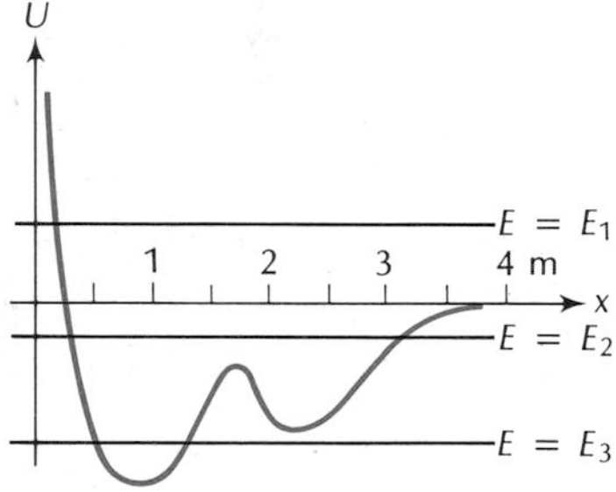

Fig. 8.12 Plot of $U(x)$ vs. x.

18. A particle moves along the x axis under the influence of a conservative force with a potential energy $U(x)$. Figure 8.12 shows the plot of $U(x)$ vs. x. Figure 8.12 shows several alternative energy levels for the particles: $E = E_1$, $E = E_2$, and $E = E_3$. Assume that the particle is initially at $x = x_0$. For each of the three alternative energies describe the motion qualitatively, answering the following questions.

(a) Roughly, where are the turning points (right and left)?
(b) Where is the speed of the particle maximum? Where is the speed minimum?
(c) Is the orbit bound or unbound?

Section 8.5

19. Express the last three entries in Table 8.1 in electron-volts.

20. The chemical formula for TNT is $CH_3C_6H_2(NO_2)_3$. The explosion of 1 kg of TNT releases 4.6×10^6 J. Calculate the energy released per molecule of TNT. Express your answer in electron-volts.

21. Using the data of Table 8.1, calculate the amount of gasoline that would be required if all the energy requirements of the United States were to be met

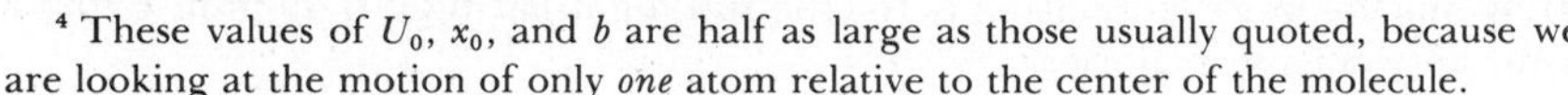

[4] These values of U_0, x_0, and b are half as large as those usually quoted, because we are looking at the motion of only *one* atom relative to the center of the molecule.

by the consumption of gasoline. How many gallons per day would have to be consumed?

22. The following table lists the fuel consumption and the passenger capacity of several vehicles. Assume that the energy content of the fuel is that of gasoline (see Table 8.1). Calculate the amount of energy used by each vehicle per passenger per mile. Which is the most energy-efficient vehicle? The least energy efficient?

Vehicle	*Passenger capacity*	*Fuel consumption*
Motorcycle	1	60 mi/gal.
Snowmobile	1	12
Automobile	4	12
Intercity bus	45	5
Jetliner	360	0.1
Concorde SST	110	0.12

Section 8.6

23. The atomic bomb dropped on Hiroshima had an explosive energy equivalent to that of 20,000 tons of TNT, or 8.4×10^{23} J. How many kilograms of rest mass must have been converted into energy in this explosion?

24. How much energy will be released by the annihilation of one electron and one antielectron (both initially at rest)? Express your answer in electron-volts.

25. The mass of the Sun is 2×10^{30} kg. The thermal energy in the Sun is about 2×10^{41} J. How much does the thermal energy contribute to the mass of the Sun?

26. In a collision between an electron and an antielectron, the two particles can annihilate and create a proton and an antiproton. The reaction

$$e + \bar{e} \rightarrow p + \bar{p}$$

converts the rest-mass energy and kinetic energy of the electron and antielectron into the rest-mass energy of the proton and antiproton. Assume that the electron and the antielectron collide head on with opposite velocities of equal magnitudes and that the proton and antiproton are at rest immediately after the reaction. Calculate the kinetic energy of the electron required for this reaction; express your answer in electron-volts.

Section 8.7

27. For an automobile traveling at a steady speed of 65 km/h, the friction of the air and the rolling friction of the ground on the wheels provide a total external friction force of 500 N.

(a) At what rate does this force remove momentum from the automobile? At what rate does it remove energy?

(b) To keep the automobile going at constant velocity, the momentum loss and the energy loss must be compensated for. What body supplies the necessary momentum? What body supplies the necessary energy?

28. In 1979, B. Allen flew a very lightweight propeller airplane across the English Channel. His legs, pushing bicycle pedals, supplied the power to turn the propeller. To keep the airplane flying, he had to supply about 0.30 hp. How much energy did he supply for the full flight lasting 2 h 49 min? Express your answer in kilocalories.

29. The ancient Egyptians and Romans relied on slaves as a source of mechanical power. One slave, working desperately by turning a crank, can deliver about 200 W of mechanical power (at this power the slave would not last long). How many slaves would be needed to match the output of a modern automobile engine (150 hp)? How many slaves would an ancient Egyptian have to own in order to command the same amount of power as the average per capita power used by residents of the United States (14 kW)?

30. An electric clock uses 2 W of electric power. How much electric energy (in kilowatt-hours) does this clock use in 1 year? What happens to this electric energy?

31. Nineteenth-century engineers reckoned that a laborer turning a crank can do steady work at the rate of 5000 ft · lb/min. Suppose that four laborers working a manual crane attempt to lift a load of 9 short tons (1 short ton = 2000 lb). If there is no friction, what is the rate at which they can lift this load? How long will it take them to lift the load 15 ft?

32. The driver of an automobile traveling on a straight road at 80 km/h pushes forward with his hands on the steering wheel with a force of 50 N. What is the rate at which his hands do work on the steering wheel in the reference frame of the ground? In the reference frame of the automobile?

33. A crane is powered by an electric motor delivering 60 hp. What is the maximum speed with which this crane can raise a load of 10 short tons (1 short ton = 2000 lb)? Assume that 28% of the power of the motor is lost to friction within the crane.

34. A horse walks along the bank of a canal and pulls a barge by means of a long horizontal towrope making an angle of 35° with the bank. The horse walks at the rate of 5 km/h and the tension in the rope is 400 N. What horsepower does the horse deliver?

35. A 2000-lb automobile accelerates from 0 to 50 mi/h in 7.5 s. What are the initial and the final translational kinetic energies of the automobile? What is the average power delivered by the engine in this time interval? Express your answer in horsepower.

36. A six-cylinder internal combustion engine, such as used in an automobile, delivers an average power of 150 hp while running at 3000 rev/min. Each of the cylinders fires once every two revolutions. How much energy does each cylinder deliver each time it fires?

37. The ancient Egyptians moved large stones by dragging them across the sand in sleds. Suppose that 6000 Egyptians are dragging a sled with a coefficient of sliding friction $\mu_k = 0.3$ along a level surface of sand.

(a) If each Egyptian exerts a force of 80 lb, what is the maximum weight they can move at constant speed?

(b) If each Egyptian delivers a mechanical power of 0.20 hp, what is the maximum speed with which they can move this weight?

38. An automobile engine typically has an efficiency of about 25%, i.e., it converts about 25% of the chemical energy available in gasoline into mechanical energy. Suppose that an automobile engine has a mechanical output of 110 hp. At what rate (in gallons per hour) will this engine consume gasoline? See Table 8.1 for the energy content of gasoline.

39. In a braking test, a 990-kg automobile takes 2.1 s to come to a full stop from an initial speed of 60 km/h. What is the amount of energy dissipated in the brakes? What is the average power dissipated in the brakes? Ignore external friction in your calculation and express the power in horsepower.

40. The takeoff speed of a DC-3 airplane is 100 km/h. Starting from rest, the airplane takes 10 s to reach this speed. The mass of the (loaded) airplane is 11,000 kg. What is the average power delivered by the engines to the airplane during takeoff?

41. Equations (2.9) and (2.15) give the velocity and the acceleration of an accelerating Maserati sports car as a function of time. The mass of this automobile is 1770 kg. What is the instantaneous power delivered by the engine to the automobile? Plot the instantaneous power as a function of time in the time interval from 0 to 10 s. At what time is the power maximum?

42. The ship *Globtik Tokyo,* a supertanker, has a weight of 720,000 short tons (1 short ton = 2000 lb) when fully loaded.
   (a) What is the kinetic energy of the ship when her speed is 16 mi/h?
   (b) The engines of the ship deliver a power of 44,000 hp. According to the energy requirements, how long a time does it take the ship to reach a speed of 16 mi/h, starting from rest? Make the assumption that 50% of the engine power goes into friction or into stirring up the water and 50% remains available for the translational motion of the ship.
   (c) How long a time does it take the ship to stop from an initial speed of 16 mi/h if her engines are put in reverse? Estimate roughly how far she will travel during this time.

43. At Niagara Falls, 6200 m^3 per second of water fall down a height of 49 m.
   (a) What is the rate (in watts) at which gravitational potential energy is wasted by the falling water?
   (b) What is the amount of energy (in kilowatt-hours) wasted in 1 year?
   (c) Power companies get paid about 5 cents per kilowatt-hour of electric energy. If all of the gravitational potential energy wasted at Niagara Falls could be converted into electric energy, how much money would this be worth?

44. The movement of a grandfather clock is driven by a 5-kg weight which drops a distance of 1.5 m in the course of a week. What is the power delivered by the weight to the movement?

45. In a waterfall on the Alto Paraná river (between Brazil and Paraguay), the height of fall is 110 ft and the average rate of flow is 470,000 ft^3 of water per second. What is the power wasted by this waterfall?

46. A 60,000-lb truck has a 550-hp engine. What is the maximum speed with which this truck can move up a 10° slope?

47. In order to overcome air friction and other mechanical friction, an automobile of weight 3200 lb requires a power of 20 hp from its engine to travel at 40 mi/h on a level road. What power does the same automobile require to travel uphill on an incline of slope 1:10 at the same speed? Downhill on the same incline at the same speed?

48. With the gears in neutral, an automobile rolling down a long incline of slope 1:10 reaches a terminal speed of 60 mi/h. At this speed the rate of decrease of the gravitational potential energy matches the power required to overcome air friction and other mechanical friction. What power (in horsepower) must the engine of this automobile deliver to drive it at 60 mi/h on a level road? The weight of the automobile is 3300 lb.

49. When jogging at 12 km/h on a level road, a 70-kg man uses 750 kcal/h. How many kilocalories per hour does he require when jogging up a 1:10 incline at the same speed? Assume that the frictional losses are independent of the value of the slope.

50. Each of the two Wright "Cyclone" engines on a DC-3 airplane generates a power of 850 hp. The mass of the loaded plane is 10,900 kg. The plane can climb at the rate of 260 m/min. When the plane is climbing at this rate, what percentage of the engine power is used to do work against gravity?

51. A fountain sends a stream of water 10 m up in the air. The base of the stream is 10 cm across. What power is expended to send the water to this height?

52. The record of 203.1 km/h (126.2 mi/h) for speed skiing set by Franz Weber at Velocity Peak, Colorado, was achieved on a mountain slope inclined downward at 51°. At this speed, the force of friction (air and sliding friction) balances the pull of gravity along the slope, so that the motion proceeds at constant velocity.
   (a) What is the rate at which gravity does work on the skier? Assume that the mass of the skier is 75 kg.
   (b) What is the rate at which sliding friction does work? Assume that the coefficient of friction is $\mu_k = 0.03$.
   (c) What is the rate at which air friction does work?

53. A windmill for the generation of electric power has a propeller of diameter 1.8 m. In a wind of 40 km/h, this windmill delivers 200 W of electric power.
   (a) At this wind speed, what is the rate at which the air carries kinetic energy through the circular area swept out by the propeller? The density of air is 1.29 kg/m^3.
   (b) What percentage of the kinetic energy of the air passing through this area is converted into electric energy?

54. The energy of sunlight arriving at the surface of the Earth amounts to about 1 kW per square meter of surface (facing the Sun). If all of the energy incident on a collector of sunlight could be converted into useful energy, how many square meters of collector area would we need to satisfy all of the energy demands in the United States? See Table 8.1 for the energy expenditure of the United States.

55. A small electric kitchen fan blows 8.5 m^3/min of air at a speed of 5.0 m/s out of the kitchen. The density of air is 1.3 kg/m^3. What electric power must the fan consume to give the ejected air the required kinetic energy?

56. The final portion of the Tennessee River has a downward slope of 0.39 ft to the mile. The rate of flow of water in the river is 10,000 ft^3/s. Assume that the speed of the water is constant along the river. How much power is wasted by friction of the water against the riverbed per mile?

57. Off the coast of Florida, the Gulf Stream has a speed of 4.6 km/h and a rate of flow of 2.2×10^3 km^3/day. At what rate is kinetic energy flowing past the coast? If all this kinetic energy could be converted into electric power, how many kilowatts would it amount to?

58. The Sun emits energy in the form of radiant heat and light at the rate of 3.9×10^{26} W. At what rate does this energy carry away mass from the Sun? How much mass does this amount to in 1 year?

59. The reaction that supplies the Sun with energy is

$$\text{H} + \text{H} + \text{H} + \text{H} \rightarrow \text{He} + [\text{energy}]$$

(The reaction involves several intermediate steps, but this need not concern us now.) The mass of the hydrogen (H) atom is 1.00813 u and that of the helium (He) atom is 4.00388 u.
   (a) How much energy is released in the reaction of four hydrogen atoms (by the conversion of rest mass into energy)?
   (b) How much energy is released in the reaction of 1 kg of hydrogen atoms?
   (c) The Sun releases energy at the rate of 3.9×10^{26} W. At what rate (in kg/s) does the Sun consume hydrogen?
   (d) The Sun contains about 1.5×10^{30} kg of hydrogen. If it continues to consume hydrogen at the same rate, how long will the hydrogen last?

CHAPTER 9

Systems of Particles

So far we have dealt almost exclusively with the motion of a single particle. Now we will begin to study systems of several particles interacting with each other via some forces. Since chunks of ordinary matter are made of particles (electrons, protons, and neutrons), all the macroscopic bodies that we encounter in our everyday environment are in fact many-particle systems containing a very large number of particles. However, for most practical purposes it is not desirable to adopt such an extreme microscopic point of view. For example, in dealing with a collision among several automobiles we may find it convenient to pretend that each of the automobiles is a particle and not inquire into their internal structure — we then regard the colliding automobiles as a system of particles which exert forces on each other when in contact. Likewise, in dealing with the Solar System, we may find it convenient to pretend that each planet and each satellite is a particle — we then regard the Solar System as a system of such planet and satellite particles loosely held together by gravitation and orbiting around the Sun and around each other.

The equations of motion of a system of many particles are often hard, and sometimes impossible, to solve. It is therefore necessary to make the most of any information that can be extracted from the general conservation laws. In the following sections we will see how the laws of conservation of momentum, energy, and angular momentum apply to a system of particles.

9.1 Momentum of a System of Particles

The momentum of a single particle was defined in Section 5.5; it is the product of the mass and velocity of the particle:

$$\mathbf{p} = m\mathbf{v}$$

Now consider a system of n particles. The total momentum of the system of particles is simply the sum of the individual momenta of all the particles. Thus, if $\mathbf{p}_1 = m_1\mathbf{v}_1$, $\mathbf{p}_2 = m_2\mathbf{v}_2$, . . . , $\mathbf{p}_n = m_n\mathbf{v}_n$ are the momenta of the particles, then the total momentum is

Momentum of a system of particles

$$\boxed{\mathbf{P} = \mathbf{p}_1 + \mathbf{p}_2 + \cdots + \mathbf{p}_n} \tag{1}$$

or

$$\mathbf{P} = \sum_{i=1}^{n} \mathbf{p}_i$$

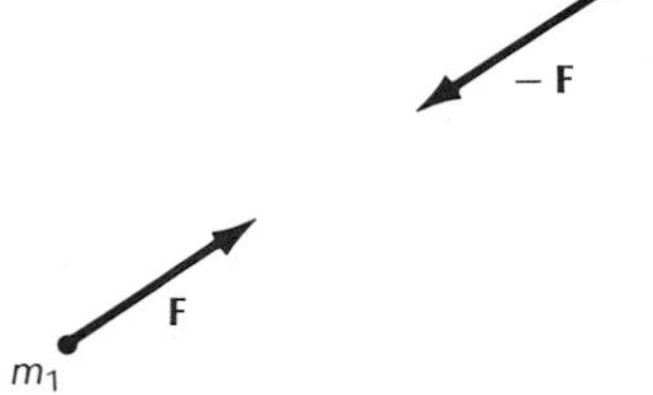

Fig. 9.1 Two particles exerting forces on each other.

The simplest of all many-particle systems consists of just two particles exerting some mutual force on one another (Figure 9.1). Let us assume that the two particles are isolated from the rest of the universe so that, besides their mutual forces, they experience no extra forces of any kind. If the force exerted by particle 2 on particle 1 is **F**, then, by Newton's Third Law, the force exerted by particle 1 on particle 2 is $-\mathbf{F}$ and hence the equation of motion of each particle is

$$\frac{d\mathbf{p}_1}{dt} = \mathbf{F} \tag{2}$$

$$\frac{d\mathbf{p}_2}{dt} = -\mathbf{F} \tag{3}$$

If we add these two equations together, we obtain

$$\frac{d\mathbf{p}_1}{dt} + \frac{d\mathbf{p}_2}{dt} = \mathbf{F} + (-\mathbf{F}) \tag{4}$$

that is,

$$\frac{d}{dt}(\mathbf{p}_1 + \mathbf{p}_2) = 0 \tag{5}$$

This shows that the total momentum of the two-particle system is a constant of the motion:

Law of conservation of momentum

$$\boxed{\mathbf{P} = \mathbf{p}_1 + \mathbf{p}_2 = [\text{constant}]} \tag{6}$$

This is the **law of conservation of momentum.** Note that this law is a direct consequence of Newton's Third Law: the total momentum is a constant because the equality of action and reaction keeps the momentum changes of the two particles exactly equal in magnitude but opposite in direction — the particles merely exchange some momentum by means of their mutual forces.

Conservation of momentum is a powerful tool which permits us to calculate some general features of the motion even when we are ignorant of the detailed features of the interparticle forces.

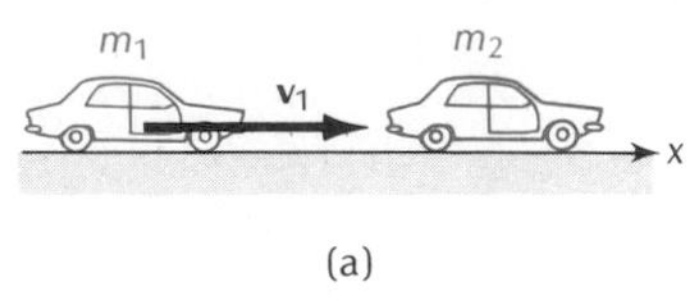

(a)

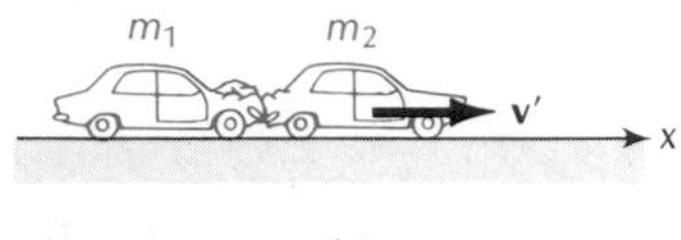

(b)

Fig. 9.2 (a) A speeding automobile crashes into a parked automobile. (b) After the collision, the two automobiles remain joined.

EXAMPLE 1. An automobile of mass 1500 kg traveling at 25 m/s (55 mi/h) crashes into a similar parked automobile. The two automobiles remain joined after the collision. What is the velocity of the wreck immediately after the collision? Neglect friction against the road.

SOLUTION: Under the assumptions of the problem, the only horizontal forces are the mutual forces of one automobile on the other. Thus, momentum conservation applies to the horizontal component of the momentum — the value of this component must be the same before and after the collision. Before the collision the (horizontal) velocity of one car is $v_1 = 25$ m/s and that of the other is $v_2 = 0$. With the x axis along the direction of motion (Figure 9.2), the total momentum is $P_x = m_1v_1 + m_2v_2 = m_1v_1$, where $m_1 = m_2 = 1500$ kg. After the collision both automobiles have the same velocity, $v_1' = v_2' = v'$, and the total momentum is

$$P_x' = m_1v_1' + m_2v_2' = (m_1 + m_2)v'$$

Since the momenta P_x and P_x' are equal,

$$m_1v_1 = (m_1 + m_2)v'$$

or

$$v' = \frac{m_1v_1}{m_1 + m_2} = \frac{1500 \text{ kg} \times 25 \text{ m/s}}{1500 \text{ kg} + 1500 \text{ kg}} = 12.5 \text{ m/s}$$

Obviously, the forces acting during the collision are extremely complicated. Momentum conservation allows us to bypass these complications and obtain the answer directly. Incidentally, it is easy to check that kinetic energy is *not* conserved in this collision; some of this energy is used up to produce structural changes in the automobiles.

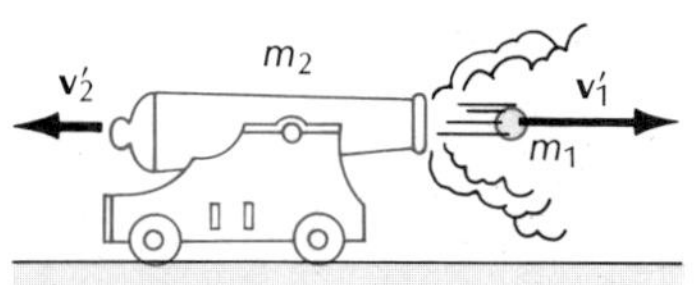

Fig. 9.3 Recoil of a gun.

EXAMPLE 2. A gun used on board an eighteenth-century warship is mounted on a carriage which allows the gun to roll back each time it is fired (Figure 9.3). The weight of the gun (including the carriage) is 4500 lb. The gun fires a 13-lb shot horizontally with a velocity of 1600 ft/s. What is the recoil velocity of the gun?

SOLUTION: The total momentum of the shot plus gun must be the same before the firing and just after the firing. Before, the total momentum is zero. After, the (horizontal) velocity of the shot is $v_1' = 1600$ ft/s and that of the gun is v_2'; hence the total momentum is $P_x' = m_1v_1' + m_2v_2'$, where $m_1 = 13/32.2$ slug is the mass of the shot and $m_2 = 4500/32.2$ slug is the mass of the gun (including the carriage). Thus,

$$0 = m_1v_1' + m_2v_2'$$

or

$$v_2' = -\frac{m_1}{m_2}v_1'$$

$$= -\frac{13/32.2}{4500/32.2} \times 1600 \text{ ft/s} = -4.6 \text{ ft/s}$$

The negative sign indicates that the recoil velocity is opposite to the velocity of the shot.

The conservation law of Eq. (6) depends on the absence of "extra" forces. If the particles are not isolated from the rest of the universe, then besides the mutual forces exerted by one particle on the other, there are also forces exerted by other bodies not belonging to the particle system; the former forces are called **internal forces** and the latter **external forces.** For example, if the two particles are near the Earth, then gravity will act on them and play the role of an external force. To take such external forces into account, we must modify Eqs. (2) and (3). If the external force on particle 1 is $\mathbf{F}_{1,\text{ext}}$, then the total force on this particle is $\mathbf{F} + \mathbf{F}_{1,\text{ext}}$ and the equation of motion will be

Internal and external forces

$$\frac{d\mathbf{p}_1}{dt} = \mathbf{F} + \mathbf{F}_{1,\text{ext}} \tag{7}$$

Likewise,

$$\frac{d\mathbf{p}_2}{dt} = -\mathbf{F} + \mathbf{F}_{2,\text{ext}} \tag{8}$$

These equations lead to

$$\frac{d}{dt}(\mathbf{p}_1 + \mathbf{p}_2) = \mathbf{F}_{1,\text{ext}} + \mathbf{F}_{2,\text{ext}} \tag{9}$$

The sum $\mathbf{F}_{1,\text{ext}} + \mathbf{F}_{2,\text{ext}}$ is simply the total external force on the particle system. Thus, Eq. (9) states that the rate of change of the total momentum of the two-particle system equals the total *external* force on the system.

For a system containing more than two particles, we can obtain similar results. If the system is isolated so that there are no external forces, then the mutual interparticle forces acting between pairs of particles merely transfer momentum from one particle of the pair to the other, just as in Eq. (5). Since all the internal forces necessarily arise from such forces between pairs of particles, these internal forces cannot change the total momentum. For example, Figure 9.4 shows three particles exerting forces on one another. Consider particle 1; the mutual forces between particles 1 and 2 exchange momentum between these two, while the mutual forces between particles 1 and 3 exchange momentum between those two. But none of these momentum transfers will change the total momentum. The same holds for particles 2 and 3. Consequently, the total momentum of an isolated system obeys the conservation law

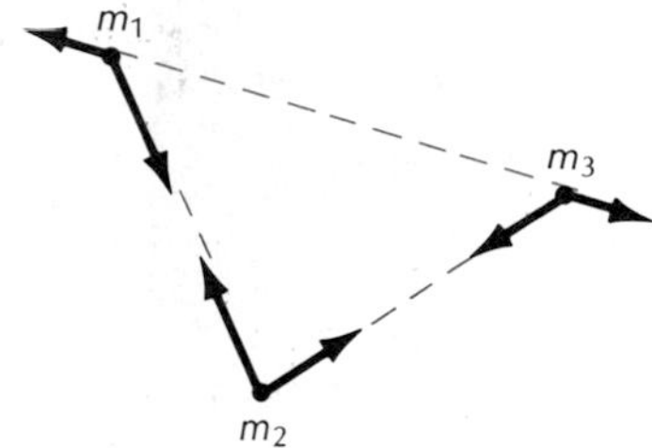

Fig. 9.4 Three particles exerting forces on each other.

$$\boxed{\mathbf{P} = [\text{constant}]} \tag{10}$$

If besides the internal forces, there are external forces, then the latter will change the momentum. The rate of change can be calculated in essentially the same way as for the two-particle system and, again, the rate of change of the total momentum is equal to the total external force. We can write this as

Rate of change of momentum

$$\boxed{\frac{d\mathbf{P}}{dt} = \mathbf{F}_{\text{ext}}} \tag{11}$$

where $\mathbf{F}_{\text{ext}}$ is the external force on the system.

Equations (10) and (11) have exactly the same mathematical form as Eqs. (5.16) and (5.17) and may be regarded as the generalizations for a system of particles of Newton's First and Second Laws. As we will see in Section 9.3, Eq. (11) is an equation of motion for the system — it determines the overall translational motion of the system.

EXAMPLE 3. During a rainstorm the volume of rain falling on 1 m^2 of ground in one hour amounts to 0.1 m^3. The raindrops hit the ground with a vertical velocity of 10 m/s. What is the average force per unit area (force per square meter) that the impact of the raindrops exerts on the ground?

SOLUTION: We can calculate the force from the rate at which the rain transfers momentum to the ground. The mass of water that falls on 1 m^2 in one hour is $m = 0.1 \text{ m}^3 \times 10^3 \text{ kg/m}^3 = 1 \times 10^2$ kg, and the momentum contributed by this mass is $\Delta P = mv = 1 \times 10^2 \text{ kg} \times 10 \text{ m/s} = 1 \times 10^3 \text{ kg} \cdot \text{m/s}$. The average rate of momentum transfer to the ground, or the average force, is then

$$F = \frac{\Delta P}{\Delta t} = \frac{1 \times 10^3 \text{ kg} \cdot \text{m/s}}{3600 \text{ s}} = 0.3 \text{ N}$$

9.2 Center of Mass

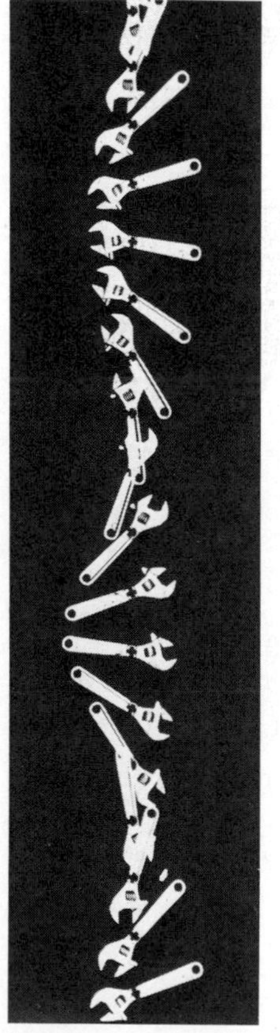

Fig. 9.5 A wrench moving freely in the absence of external forces. The center of mass, marked with a cross, moves with uniform velocity, just like a free particle.

In our study of kinematics and dynamics in the preceding chapters, we have always ignored the sizes of the bodies; even when analyzing the motion of the large body — a railroad car or a ship — we pretended that the motion could be treated as particle motion, position being described by reference to some fiducial point marked on the body. In reality, large bodies are systems of particles and their motion obeys Eq. (11) for a system of particles. This equation can be converted into an equation of motion, containing just one acceleration rather than the rate of change of momentum of the entire system, by making reference to one special point in the body: the **center of mass.** The equation that describes the motion of this special point has the same mathematical form as the equation of motion of a particle, i.e., the motion of the center of mass mimics particle motion (Figure 9.5).

The position of the center of mass of a system is merely the average position of the mass of the system. For example, if the system consists of two particles each of a mass of 1 kg, then the center of mass is halfway between them (Figure 9.6). In any system consisting of n particles of equal mass — such as a piece of pure metal with atoms of only one kind — the position vector of the center of mass is simply the average of the position vectors of all the particles,

$$\mathbf{r}_{\text{CM}} = \frac{\mathbf{r}_1 + \mathbf{r}_2 + \cdots + \mathbf{r}_n}{n} \tag{12}$$

If the system consists of particles of unequal mass, then the position of the center of mass can be calculated by first subdividing the particles into fragments of equal mass. This of course means that in the average over the positions of the fragments, the positions of particles of large mass will have to be included more often than the positions of particles of small mass — the number of times the position of a particle must be included in the average is in direct proportion to its mass. This leads to the following general expression for the position of the center of mass of the system of particles:

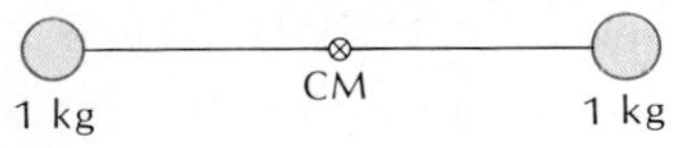

CM

1 kg 1 kg

Fig. 9.6 Two particles of equal mass and their center of mass.

$$\mathbf{r}_{\text{CM}} = \frac{m_1\mathbf{r}_1 + m_2\mathbf{r}_2 + \cdots + m_n\mathbf{r}_n}{m_1 + m_2 + \cdots + m_n} \tag{13}$$

or

$$\boxed{\mathbf{r}_{\text{CM}} = \frac{m_1\mathbf{r}_1 + m_2\mathbf{r}_2 + \cdots + m_n\mathbf{r}_n}{M}} \tag{14}$$

Center of mass of a system of particles

where $M = m_1 + m_2 + \cdots + m_n$ is the total mass.

For numerical calculations it is usually best to treat the x, y, and z components of Eq. (14) separately:

$$x_{\text{CM}} = \frac{1}{M}(m_1x_1 + m_2x_2 + \cdots + m_nx_n) \tag{15}$$

$$y_{\text{CM}} = \frac{1}{M}(m_1y_1 + m_2y_2 + \cdots + m_ny_n) \tag{16}$$

$$z_{\text{CM}} = \frac{1}{M}(m_1z_1 + m_2z_2 + \cdots + m_nz_n) \tag{17}$$

EXAMPLE 4. The distance between the centers of the atoms of potassium and bromine in a potassium bromide (KBr) molecule is 2.82 Å. Treating the atoms as particles, find the center of mass.

SOLUTION: In Figure 9.7 the atoms of bromine and potassium are shown as particles. The bromine atom is at $x_1 = 0$ and the potassium atom is at $x_2 =$ 2.82 Å. The center of mass is at

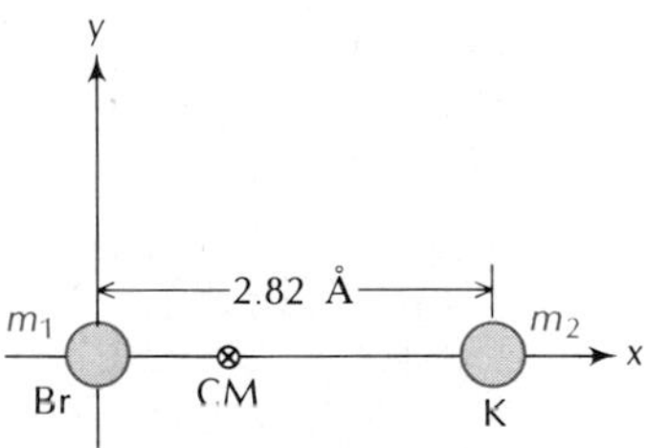

Fig. 9.7 Atoms of bromine (Br) and potassium (K), regarded as particles.

$$x_{\text{CM}} = \frac{m_1x_1 + m_2x_2}{m_1 + m_2} \tag{18}$$

or, since $x_1 = 0$,

$$x_{\text{CM}} = \frac{m_2}{m_1 + m_2}x_2 \tag{19}$$

The masses of atoms of bromine and potassium are $m_1 = 79.9$ u and $m_2 = 39.1$ u, respectively (see Appendix 9). This gives

$$x_{\text{CM}} = \frac{39.1\ \text{u}}{79.9\ \text{u} + 39.1\ \text{u}} \times 2.82\ \text{Å} = 0.93\ \text{Å}$$

This center of mass is shown in Figure 9.7. Note that, according to Eq. (18), the position of the center of mass divides the line segment connecting the atoms in the ratio $m_1{:}m_2$.

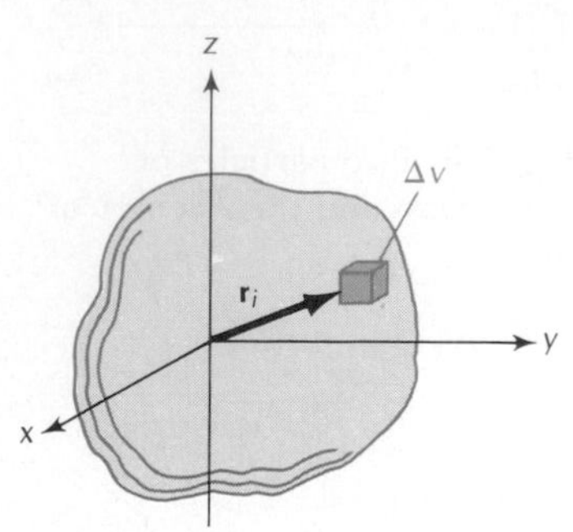

Fig. 9.8 Volume element Δv in a continuous mass distribution.

The position of the center of mass of a solid body can, in principle, be calculated from Eq. (13) — a solid is a collection of atoms, each of which can be regarded as a particle. However, it would be awkward to deal with the 10^{23} or so atoms that make up a chunk of matter the size of a coin. It is more convenient to pretend that matter in bulk has a smooth and continuous distribution of mass over its entire volume. The amount of mass in each region is then described by the *density* or the mass per unit volume.

To find the center of mass of such a continuous mass distribution, imagine that the volume is divided into small volume elements, each of size Δv (Figure 9.8). Then the mass in one of these volume elements is

$$\Delta m_i = \rho\, \Delta v \tag{20}$$

where ρ is the mass density at the position of the volume element (this density may be a function of position). According to Eq. (13), the position of the center of mass is then approximately

$$\mathbf{r}_{\mathrm{CM}} = \frac{1}{M}\sum_{i=1}^{n} \mathbf{r}_i\, \Delta m_i = \frac{1}{M}\sum_{i=1}^{n} \mathbf{r}_i \rho\, \Delta v \tag{21}$$

This approximation becomes exact in the limit $\Delta v \to 0$:

$$\mathbf{r}_{\mathrm{CM}} = \frac{1}{M}\lim_{\Delta v \to 0} \sum_i \mathbf{r}_i \rho\, \Delta v \tag{22}$$

The limit is of course the same as the integral

Center of mass of a continuous distribution of mass

$$\boxed{\mathbf{r}_{\mathrm{CM}} = \frac{1}{M}\int \mathbf{r}\rho\, dv} \tag{23}$$

In terms of components, Eq. (23) is

$$x_{\mathrm{CM}} = \frac{1}{M}\int x\rho\, dv \tag{24}$$

$$y_{\mathrm{CM}} = \frac{1}{M}\int y\rho\, dv \tag{25}$$

$$z_{\mathrm{CM}} = \frac{1}{M}\int z\rho\, dv \tag{26}$$

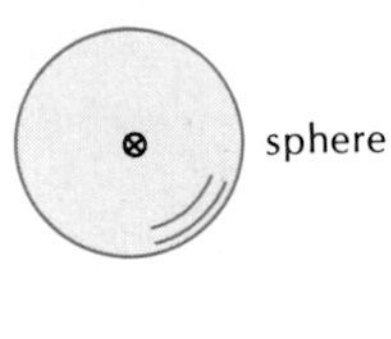

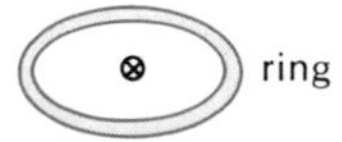

Fig. 9.9 Several bodies for which the center of mass coincides with the geometric center.

If the density of the body is uniform, then the position of the center of mass is simply the average position of all the points making up its volume (this average is usually called the *centroid* in mathematics). If the body has a symmetric shape, this average position will often be obvious by inspection. For instance, a homogeneous sphere, or a ring, or a circular plate, or a parallelepiped will have its center of mass at the geometrical center (Figure 9.9).

EXAMPLE 5. A thin rod of cross-sectional area A is bent in the shape of a semicircle of radius R (Figure 9.10). Where is the center of mass of the rod?

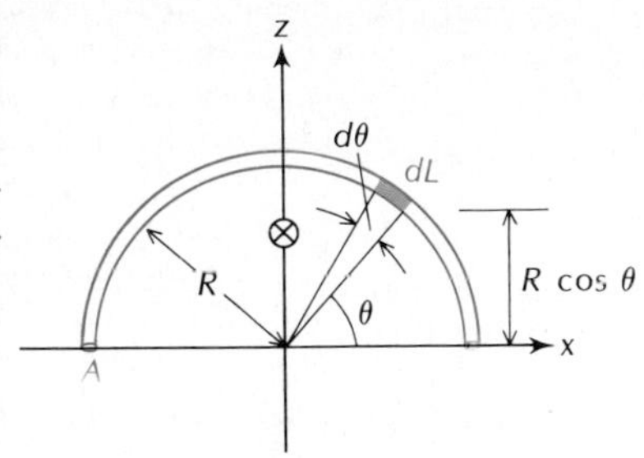

Fig. 9.10 Rod in the shape of a semicircle.

SOLUTION: Assume that the rod is in the z–y plane, and that the center of the semicircle is at the origin. In view of the symmetry of the rod, the center of mass will be somewhere on the z axis. To find the z coordinate of the center of mass, we use Eq. (26),

$$z_{\text{CM}} = \frac{1}{M}\int z\rho\, dv$$

Consider a small segment dl of the rod (Figure 9.10); this segment subtends a small angle $d\theta$ so that

$$dl = R\, d\theta$$

The volume of rod within this small angle is the volume of a small cylinder of base A and height dl:

$$dv = A\, dl = AR\, d\theta$$

The z coordinate of this small volume is $z = R \sin\theta$. Hence

$$\begin{aligned} z_{\text{CM}} &= \frac{1}{M}\int z\rho\, dv \\ &= \frac{1}{M}\int (R\sin\theta)\rho AR\, d\theta \\ &= \frac{1}{M}\rho AR^2\int \sin\theta\, d\theta \end{aligned}$$

Taking into account that the limits of integration for the angle θ are 0° and 180°, we obtain

$$\begin{aligned} z_{\text{CM}} &= \frac{1}{M}\rho AR^2\int_{0^\circ}^{180^\circ} \sin\theta\, d\theta \\ &= \frac{1}{M}\rho AR^2\Big[-\cos\theta\Big]_{0^\circ}^{180^\circ} \\ &= \frac{2}{M}\rho AR^2 \end{aligned}$$

The total volume of the rod is that of a cylinder of base A and height πR; therefore the density of the rod is $\rho = M/v = M/(A\pi R)$. Substituting this into our equation for z_{CM}, we find the final result

$$z_{\text{CM}} = \frac{2}{M}\frac{M}{A\pi R}AR^2 = \frac{2}{\pi}R$$

Thus, the distance of the center of mass from the center of the circle is $(2/\pi)R$, or $0.637R$.

EXAMPLE 6. The Great Pyramid at Gizeh (Figure 9.11) has a height of 481 ft. Assuming that the entire volume is completely filled with stone of uniform density, find the center of mass.

Fig. 9.11 The Great Pyramid.

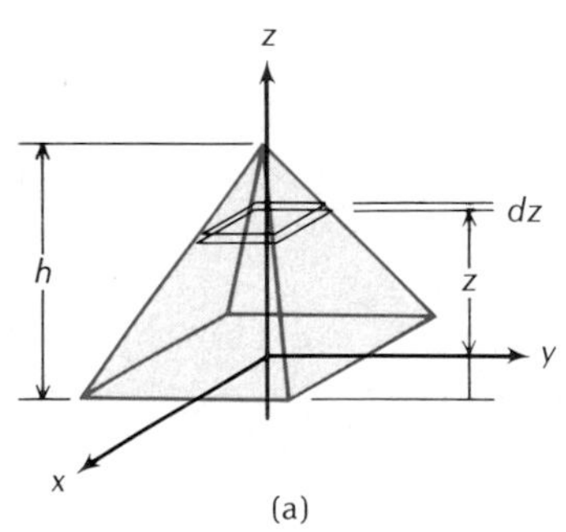

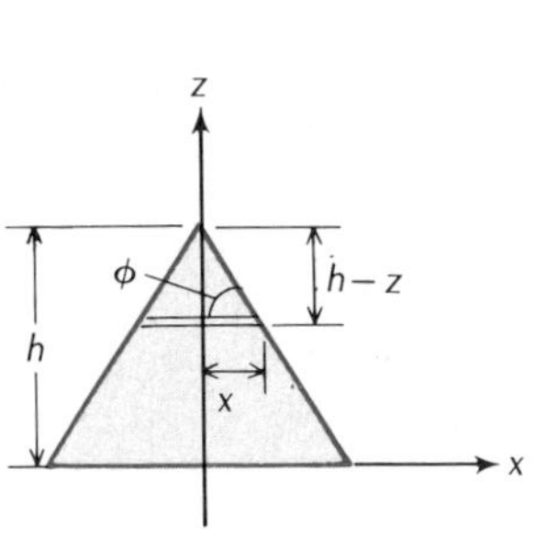

Fig. 9.12 (a) The pyramid and a thin horizontal slab within the pyramid. (b) Side view ("elevation") of the pyramid.

SOLUTION: Because of symmetry, the center of mass will be on the vertical line through the apex; in Figure 9.12a this line coincides with the z axis. To find the z coordinate of the center of mass, consider a thin horizontal slab of thickness dz at a height z above the ground. The entire pyramid can be regarded as built up of such slabs. The horizontal face of the slab is a square measuring $2x$ by $2x$; thus the volume of the slab is approximately $(2x)^2\,dz$, i.e.,

$$dv = (2x)^2\,dz$$

From Figure 9.12b we obtain $x = (h - z)/\tan\phi$, where h is the height of the pyramid and ϕ the angle between one side and the ground. Hence

$$dv = 4\frac{(h-z)^2}{\tan^2\phi}\,dz$$

Equation (26) then gives

$$z_{\text{CM}} = \frac{1}{M}\int z\rho\,dv = \frac{1}{M}\int_0^h 4z\rho\,\frac{(h-z)^2}{\tan^2\phi}\,dz$$

$$= \frac{4\rho}{M\tan^2\phi}\int_0^h z(h-z)^2\,dz$$

$$= \frac{4\rho}{M\tan^2\phi}\int_0^h (zh^2 - 2z^2h + z^3)\,dz$$

$$= \frac{4\rho}{M\tan^2\phi}\left[\tfrac{1}{2}z^2h^2 - \tfrac{2}{3}z^3h + \tfrac{1}{4}z^4\right]_0^h$$

$$= \frac{4\rho}{M\tan^2\phi}\,\frac{h^4}{12} \tag{27}$$

The density of the pyramid is given by $\rho = \dfrac{M}{v}$, where v is the total volume,

$$v = \int dv = \int_0^h 4\frac{(h-z)^2}{\tan^2 \phi}\,dz = \frac{4}{\tan^2 \phi}\frac{h^3}{3} \tag{28}$$

The combination of Eqs. (27) and (28) yields

$$z_{CM} = \tfrac{1}{4}h = \tfrac{1}{4} \times 481 \text{ ft} = 120 \text{ ft} \tag{29}$$

Note that the final result is independent of the angle ϕ, i.e., for any pyramid the center of mass is one-quarter of the height above the base.

Incidentally, the gravitational potential energy of an extended body near the surface of the Earth can be expressed as a function of the position of the center of mass. According to Eq. (7.35), the potential energy of a single particle at a height z above the ground is mgz. For a system of particles, the total potential energy is then[1]

$$V = (m_1 z_1 + m_2 z_2 + \cdots + m_n z_n)g \tag{30}$$

Comparison with Eq. (17) shows that the quantity in parentheses is Mz_{CM}. Hence

$$\boxed{V = Mgz_{CM}} \tag{31}$$

This expression for the gravitational potential energy of the system has the same mathematical form as for a single particle — it is as though the entire mass of the system were located at the center of mass.

9.3 The Motion of the Center of Mass

As a system of particles moves, so does the center of mass. We will now derive the equation of motion of the center of mass. We begin with an expression for the **velocity of the center of mass:**

Velocity of the center of mass

$$\mathbf{v}_{CM} = \frac{d\mathbf{r}_{CM}}{dt} = \frac{1}{M}\left(m_1\frac{d\mathbf{r}_1}{dt} + m_2\frac{d\mathbf{r}_2}{dt} + \cdots + m_n\frac{d\mathbf{r}_n}{dt}\right) \tag{32}$$

$$= \frac{1}{M}(m_1\mathbf{v}_1 + m_2\mathbf{v}_2 + \cdots + m_n\mathbf{v}_n) \tag{33}$$

Hence

$$m_1\mathbf{v}_1 + m_2\mathbf{v}_2 + \cdots + m_n\mathbf{v}_n = M\mathbf{v}_{CM} \tag{34}$$

---

[1] Do not confuse the potential energy V in this equation with the volume v in Eqs. (20)–(28).

But the left side of this equation is the total momentum of the system. Consequently,

Momentum of a system of particles

$$\boxed{\mathbf{P} = M\mathbf{v}_{\text{CM}}} \tag{35}$$

that is, the total momentum of the system equals the total mass times the velocity of the center of mass. This equation is the analog of the familiar equation ($\mathbf{p} = m\mathbf{v}$) for the momentum of a single particle.

The equation of motion now follows from Eq. (11),

$$\mathbf{F}_{\text{ext}} = \frac{d\mathbf{P}}{dt} = \frac{d}{dt}(M\mathbf{v}_{\text{CM}}) = M\frac{d\mathbf{v}_{\text{CM}}}{dt} \tag{36}$$

This can also be written

Acceleration of the center of mass

$$\boxed{M\mathbf{a}_{\text{CM}} = \mathbf{F}_{\text{ext}}} \tag{37}$$

where $\mathbf{a}_{\text{CM}} = d\mathbf{v}_{\text{CM}}/dt$ is the **acceleration of the center of mass.** Equation (37) for a system of particles is obviously the analog of Newton's equation of motion for a single particle. The center of mass moves as though it were a particle of mass M under the influence of a force $\mathbf{F}_{\text{ext}}$.

This result justifies some of the approximations we made in previous chapters. For instance, in Example 6.5 we treated a ship sliding down a slipway as a particle. Equation (37) shows that this treatment is legitimate: the center of mass of the ship under the influence of the external forces (gravity and friction) moves parallel to the slipway, just as though it were a particle on an inclined plane with the external forces acting directly on it.

If the net external force vanishes, then the acceleration of the center of mass also vanishes; hence the center of mass remains at rest or moves with uniform velocity.

EXAMPLE 7. During a space "walk" an astronaut floats in space 8.0 m from his Gemini spacecraft orbiting the Earth. He is tethered to the spacecraft by a long umbilical cord; to return, he pulls himself in by this cord. How far does the spacecraft move toward him? The mass of the spacecraft is 3500 kg and the mass of the astronaut, including his space suit, is 110 kg.

SOLUTION: Astronaut and spacecraft exert equal and opposite forces on one another (via the cord); the astronaut is pulled toward the spacecraft and the spacecraft is pulled toward the astronaut. In the absence of external forces, the center of mass of the astronaut–spacecraft system remains at rest. Thus, the spacecraft and the astronaut both move toward the center of mass and there they meet.

With the x axis as in Figure 9.13b, the x coordinate of the center of mass is

$$x_{\text{CM}} = \frac{m_1 x_1 + m_2 x_2}{m_1 + m_2}$$

where $m_1 = 3500$ kg is the mass of the spacecraft and $m_2 = 110$ kg is the mass of the astronaut. Strictly, the coordinates x_1 and x_2 of the spacecraft and the as-

(a)

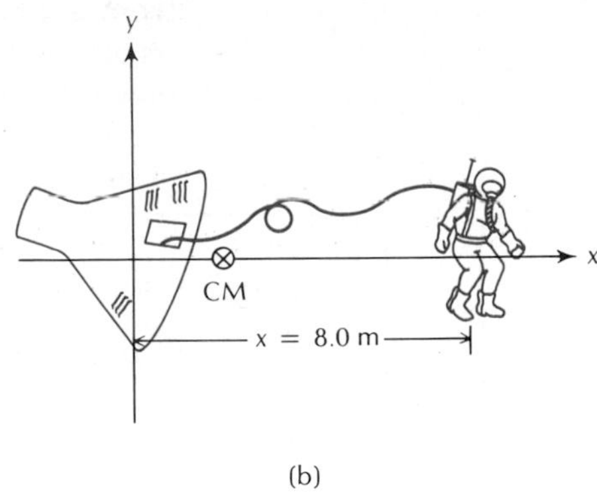

(b)

Fig. 9.13 (a) Astronaut on a space "walk" during the Gemini 4 mission. (b) Astronaut and spacecraft.

tronaut should correspond to the centers of mass[2] of these bodies, but, for the sake of simplicity, we neglect their size and treat both as particles. The initial values of the coordinates are $x_1 = 0$ and $x_2 = 8.0$ m; hence

$$x_{CM} = \frac{m_2}{m_1 + m_2} \times 8.0 \text{ m} = \frac{110 \text{ kg}}{3500 \text{ kg} + 110 \text{ kg}} \times 8.0 \text{ m}$$

$$= 0.24 \text{ m}$$

During the pulling in, the spacecraft will move from $x_1 = 0$ to $x_1 = 0.24$ m; simultaneously the astronaut will move from $x_2 = 8.0$ m to $x_2 = 0.24$ m.

9.4 Energy of a System of Particles

Since Eq. (35) for the momentum of a system of particles resembles the expression for the momentum of a single particle, one is tempted to guess that the expression for the kinetic energy for a system of particles also resembles that for a single particle and has the form $K = \frac{1}{2}Mv_{CM}^2$. But this is wrong! The total kinetic energy of a system is usually greater than $\frac{1}{2}Mv_{CM}^2$. In what follows, we will see why this is so.

The total kinetic energy of a system of particles is simply the sum of the individual kinetic energies of all the particles,

$$K = \tfrac{1}{2}m_1v_1^2 + \tfrac{1}{2}m_2v_2^2 + \cdots + \tfrac{1}{2}m_nv_n^2 \tag{38}$$

In order to rewrite this in a form involving $\mathbf{v}_{CM}$, we introduce the velocities relative to the center of mass. The particle velocities $\mathbf{v}_1, \mathbf{v}_2, \ldots,$ $\mathbf{v}_n$ are reckoned in some given reference frame, say, the reference

[2] The center of mass of a man standing erect is within his body at about the height of the navel.

frame of the ground. We now want to look at these particle velocities from a reference frame moving with the center of mass — the rest frame of the center of mass, or the "CM frame." The velocity of this new reference frame with respect to the old is $\mathbf{v}_{CM}$ and for the particle velocities in the new reference frame the Galilean transformation [Eq. (4.58)] gives the values

$$\mathbf{u}_1 = \mathbf{v}_1 - \mathbf{v}_{CM} \qquad \mathbf{u}_2 = \mathbf{v}_2 - \mathbf{v}_{CM}, \text{ etc.} \tag{39}$$

This yields

$$\mathbf{v}_1 = \mathbf{u}_1 + \mathbf{v}_{CM} \qquad \mathbf{v}_2 = \mathbf{u}_2 + \mathbf{v}_{CM}, \text{ etc.} \tag{40}$$

Inserting Eqs. (40) into Eq. (38), we obtain

$$\begin{aligned} K &= \tfrac{1}{2}m_1(\mathbf{u}_1 + \mathbf{v}_{CM})^2 + \tfrac{1}{2}m_2(\mathbf{u}_2 + \mathbf{v}_{CM})^2 + \cdots \\ &= \tfrac{1}{2}m_1(\mathbf{u}_1^2 + 2\mathbf{u}_1 \cdot \mathbf{v}_{CM} + \mathbf{v}_{CM}^2) + \tfrac{1}{2}m_2(\mathbf{u}_2^2 + 2\mathbf{u}_2 \cdot \mathbf{v}_{CM} + \mathbf{v}_{CM}^2) + \cdots \\ &= [\tfrac{1}{2}m_1\mathbf{u}_1^2 + \tfrac{1}{2}m_2\mathbf{u}_2^2 + \cdots] + [m_1\mathbf{u}_1 + m_2\mathbf{u}_2 + \cdots] \cdot \mathbf{v}_{CM} \\ &\qquad + \tfrac{1}{2}[m_1 + m_2 + \cdots]\mathbf{v}_{CM}^2 \end{aligned} \tag{41}$$

The first bracket on the right side of Eq. (41) is nothing but the kinetic energy as reckoned in the CM frame; we will call this the **internal kinetic energy** (it is also often called the CM energy):

$$\boxed{K_{int} = \tfrac{1}{2}m_1u_1^2 + \tfrac{1}{2}m_2u_2^2 + \cdots + \tfrac{1}{2}m_nu_n^2} \tag{42}$$

The quantity inside the second bracket is zero, since it equals

$$\begin{aligned} m_1(\mathbf{v}_1 - \mathbf{v}_{CM}) + m_2(\mathbf{v}_2 - \mathbf{v}_{CM}) + \cdots &= (m_1\mathbf{v}_1 + m_2\mathbf{v}_2 + \cdots) \\ &\quad - (m_1 + m_2 + \cdots)\mathbf{v}_{CM} \\ &= (m_1\mathbf{v}_1 + m_2\mathbf{v}_2 + \cdots) - M\mathbf{v}_{CM} \end{aligned} \tag{43}$$

which, indeed, is zero by Eq. (34). The quantity within the last bracket is simply the total mass.

Taking all of this into account, we can reduce Eq. (41) to

Kinetic energy of a system of particles

$$\boxed{K = K_{int} + \tfrac{1}{2}Mv_{CM}^2} \tag{44}$$

Thus, the total kinetic energy of a system of particles contains two terms: the translational kinetic energy $\frac{1}{2}Mv_{CM}^2$ of the center of mass, calculated just as though the center of mass were a particle of mass M and velocity $\mathbf{v}_{CM}$; and the internal kinetic energy K_{int}, which is the energy of motion as reckoned in the CM frame.

If the system of particles is a solid body whose particles are not moving relative to the center of mass, then $K_{int} = 0$. But such absence of "internal" motion requires that the body have no rotational motion about the center of mass (no change of orientation) and also that the particles be rigidly connected to one another (no change of shape, no vibration). The second requirement is very unrealistic; the atoms of a solid body, e.g., a chunk of metal, are never at rest — they vibrate back and forth about their equilibrium positions at high speed (typically ~400 m/s). Thus, the internal kinetic energy of a solid body is quite large. The kinetic energy associated with the disorganized, random motions of atoms within a body is **heat energy.** If we are interested only in the macroscopic translational and rotational motion of a solid body, we can often ignore the energy of these internal microscopic motions, because it usually remains constant; i.e., the heat energy remains constant. However, if friction forces act on the solid body, they will generate heat in the body, and we cannot ignore this energy transfer when we seek to balance the energy accounts.

EXAMPLE 8. Two automobiles, each of mass 1500 kg, travel in the same direction along a straight road. The speed of the leading automobile is 25 m/s and the speed of the trailing automobile is 15 m/s. If we regard these automobiles as a system of two particles, what is the translational kinetic energy of the center of mass? What is the internal kinetic energy?

SOLUTION: With the x axis along the direction of motion, the velocity of the center of mass along this axis is [see Eq. (33)]

$$v_{\text{CM}} = \frac{m_1 v_1 + m_2 v_2}{m_1 + m_2}$$

$$= \frac{1500\text{ kg} \times 25\text{ m/s} + 1500\text{ kg} \times 15\text{m/s}}{3000\text{ kg}}$$

$$= 20\text{ m/s}$$

Hence the translational kinetic energy of the center of mass is

$$\tfrac{1}{2}(m_1 + m_2)v_{\text{CM}}^2 = \tfrac{1}{2}(3000\text{ kg}) \times (20\text{ m/s})^2$$

$$= 6.0 \times 10^5\text{ J}$$

The velocities of the automobiles relative to the center of mass are $u_1 = 25\text{m/s} - 20\text{ m/s} = 5\text{ m/s}$ and $u_2 = 15\text{ m/s} - 20\text{ m/s} = -5\text{ m/s}$. The internal kinetic energy is then

$$K_{\text{int}} = \tfrac{1}{2}m_1 u_1^2 + \tfrac{1}{2}m_2 u_2^2$$

$$= \tfrac{1}{2} \times 1500\text{ kg} \times (5\text{ m/s})^2 + \tfrac{1}{2} \times 1500\text{ kg} \times (-5\text{ m/s})^2$$

$$= 3.7 \times 10^4\text{ J}$$

It is easy to check that the sum of these two kinetic energies has the same value as $\tfrac{1}{2}m_1 v_1^2 + \tfrac{1}{2}m_2 v_2^2$.

If the internal and external forces acting on a system of particles are conservative, then the system will have a potential energy. Unless we specify the forces, we cannot write down an explicit formula for the potential energy; but in any case, this potential energy will be some function of the positions of all the particles. The total energy is then the sum of the total kinetic energy [Eq. (44)] and the potential energy. This total energy will be conserved during the motion of the system of particles.

9.5 Angular Momentum of a System of Particles

The angular momentum of a single particle with momentum **p** and position vector **r** has been defined in Section 5.6:

$$\mathbf{L} = \mathbf{r} \times \mathbf{p}$$

The total angular momentum of a system of particles is obtained by summing the individual angular momenta of all the particles:

Angular momentum of a system of particles

$$\boxed{\mathbf{L} = \sum_{i=1}^{n} \mathbf{r}_i \times \mathbf{p}_i} \tag{45}$$

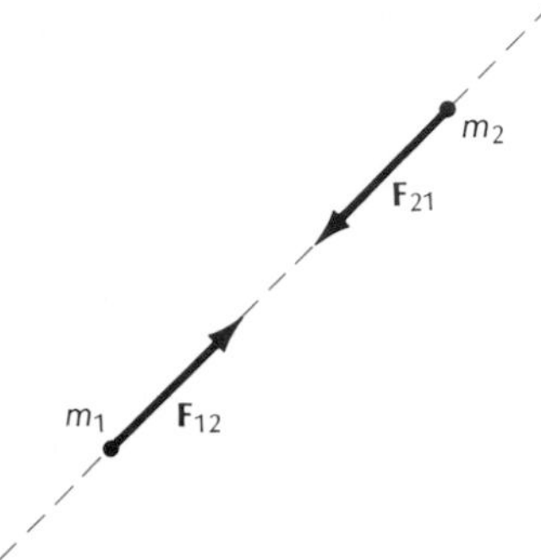

Fig. 9.14 The forces that two particles exert on one another are of equal magnitude and of opposite direction. Furthermore, we assume that the forces act along the line joining the particles.

Let us examine the rate of change of this total angular momentum. By a calculation similar to that of Section 5.6 [see Eqs. (5.20)–(5.22)], we find

$$\frac{d\mathbf{L}}{dt} = \sum_{i=1}^{n} \mathbf{r}_i \times \mathbf{F}_i \tag{46}$$

where $\mathbf{F}_i$ is the force on particle i. According to Newton's Third Law, the mutual forces of the particles of the system occur in action–reaction pairs, so that the mutual forces of a pair of particles are equal in magnitude and opposite in direction. Let us now introduce the extra assumption that the two forces in an action–reaction pair are not only equal in magnitude and opposite in direction, but also lie along the line joining the particles (Figure 9.14). Then the corresponding values of $\mathbf{r} \times \mathbf{F}$ are also equal in magnitude and opposite in direction. This can be seen from Figure 9.15, which shows one pair of particles: the magnitudes of the vectors $\mathbf{r}_1 \times \mathbf{F}_{12}$ and $\mathbf{r}_2 \times \mathbf{F}_{21}$ are $F_{12} r_1 \sin\theta_1$ and $F_{21} r_2 \sin\theta_2$, respectively; these magnitudes are equal because the magnitudes of the forces F_{12} and F_{21} are equal and the factors $r_1 \sin\theta_1$ and $r_2 \sin\theta_2$ are also equal (each of these factors represents the perpendicular distance from the origin to the line joining the particles). The directions of the vectors $\mathbf{r}_1 \times \mathbf{F}_{12}$ and $\mathbf{r}_2 \times \mathbf{F}_{21}$ are opposite because the forces are opposite. Hence the mutual forces between pairs of particles contribute equal and opposite terms to the right side of Eq. (46). In the net sum, all these terms cancel and only the terms involving external forces remain:

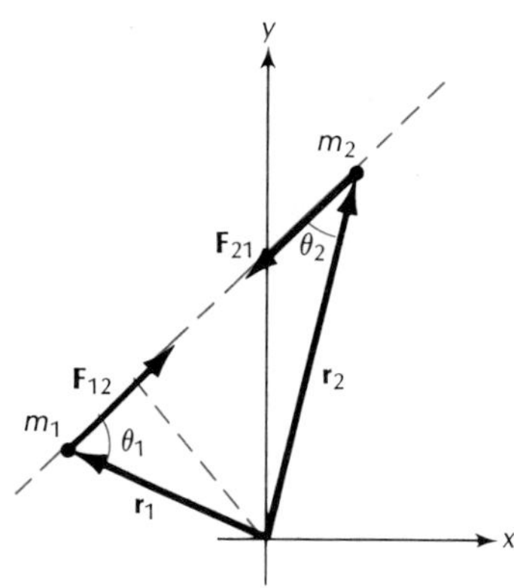

Fig. 9.15 The perpendicular distance from the origin to the line joining the particles is shown in color.

$$\frac{d\mathbf{L}}{dt} = \sum_{i=1}^{n} \mathbf{r}_i \times \mathbf{F}_{i,\text{ext}} \tag{47}$$

As we mentioned in Section 5.6, the quantity $\mathbf{r}_i \times \mathbf{F}_i$ is the **torque** of the force $\mathbf{F}_i$ on particle i. Hence the terms appearing on the right side of Eq. (47) are the torques caused by the external forces and the net sum of all these terms is the total external torque. If we write this total external torque as τ_{ext}, then Eq. (47) takes the form

Rate of change of angular momentum

$$\boxed{\frac{d\mathbf{L}}{dt} = \tau_{\text{ext}}} \tag{48}$$

This equation for the rate of change of the angular momentum of a system is analogous to Eq. (11) for the rate of change of the (linear) momentum of a system.

If the external forces are such that the total external torque is zero, then the angular momentum of the system is conserved,

Law of conservation of angular momentum

$$\boxed{\mathbf{L} = [\text{constant}]} \tag{49}$$

This is the **law of conservation of angular momentum.** We will use this law in Chapter 12 in our study of the motion of a rigid body.

9.6 The Motion of a Rocket[3]

The propulsion of any kind of vehicle — automobile, ship, aircraft — depends on reaction forces: the machinery exerts a backward push against its environment — road, water, air — and the reaction of the environment pushes the vehicle forward (see Section 5.4). The propulsion of a spacecraft in empty space is more difficult; since there is nothing in the environment to push against, it is necessary for the machinery to supply its own medium on which to push. In the operation of a rocket engine, this medium consists of the exhaust gas. The rocket engine produces a large quantity of hot, high-pressure gas from the combustion of liquid or solid fuel in a combustion chamber and then ejects this gas at high speed at the tail end of the rocket. The rocket pushes on the gas and the reaction force of the gas propels the rocket forward (Figure 9.16). This can be regarded as a recoil mechanism — just as a gun recoils when it ejects a projectile (see Example 2), the rocket recoils when it ejects the particles of gas.

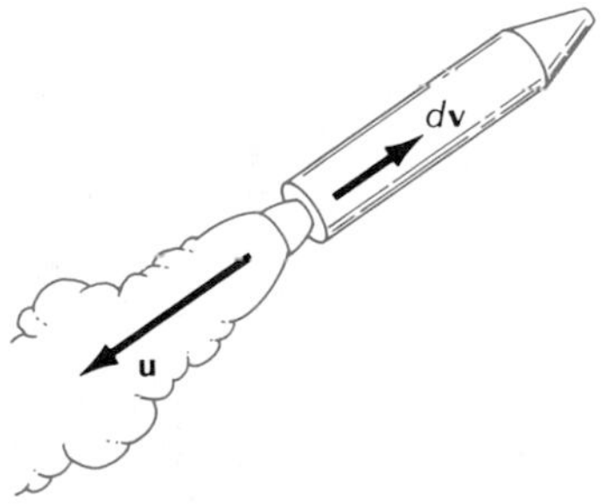

Fig. 9.16 In the chosen reference frame, the velocity of the rocket is zero at time t. A short time dt later, the velocity of the rocket is $d\mathbf{v}$. The velocity of the exhaust is $\mathbf{u}$.

To obtain a simple equation of motion for a rocket, let us assume that the gas particles ejected by the rocket engine all have the same exhaust velocity u (relative to the rocket) and move in an exactly backward direction. This assumption is somewhat unrealistic, since the exhaust actually contains particles with a distribution of speeds and, moreover, the exhaust will tend to spread out laterally, giving the particles a distribution of directions; however, the assumption is not a bad approximation if the numerical value of u is taken to be some average exhaust velocity. Let us further assume that there are no external forces, i.e., the rocket is in deep interstellar space with no nearby gravitating bodies. We can then use the conservation of momentum to obtain the equation of motion.

As gas is ejected from the tail end of the rocket, the mass remaining in the rocket decreases. Suppose that the mass at time t is $M(t)$. To find the acceleration at this time, it is best to examine the conservation of momentum in an inertial reference frame which is instantaneously at rest relative to the rocket, but which does not participate in its acceleration. In this reference frame, the velocity of the rocket at time t is zero [$v(t) = 0$], but the acceleration is not zero ($dv/dt \neq 0$).

When making use of the conservation of total momentum for a system, we must make sure that the system contains a well-defined number of particles. In the present case, we will take as our system the

[3] This section is optional.

rocket (with its fuel inside) at the time t; the total momentum of this system is zero in the chosen reference frame. A short interval dt later, an amount $-dM$ of fuel has been converted into exhaust gases.[4] Our system now consists of the rocket plus these exhaust gases; the total momentum of this system must still be zero. The new velocity of the rocket is dv and hence its momentum is $M\,dv$; the velocity of the exhaust gases is $-u$ and their momentum is $-u(-dM)$, or $u\,dM$. The sum of these momenta must then be zero:

$$M\,dv + u\,dM = 0 \tag{50}$$

Expressing this in terms of time derivatives, we obtain

$$M\frac{dv}{dt} + u\frac{dM}{dt} = 0 \tag{51}$$

or

$$M\frac{dv}{dt} = -u\frac{dM}{dt} \tag{52}$$

This is the equation of motion of the rocket — it gives the acceleration of the rocket in the absence of external forces. Although we have derived this equation by means of a rather special reference frame, it is valid in any inertial reference frame because the value of the acceleration is the same in all such reference frames. The quantity $-u\,dM/dt$ appearing on the right side of Eq. (52) plays the role of propulsive force (it equals mass times acceleration); this quantity is called the **thrust** of the rocket.

If $M(t)$ is a known function of time, i.e., if the rocket has a known rate of consumption of fuel, then we can solve Eq. (52) to find the motion. However, even without detailed knowledge of the rate of fuel consumption, we can find a general relation between the change of velocity and the amount of fuel that must be consumed to achieve this change of velocity. According to Eq. (50),

$$dv = -u\frac{dM}{M} \tag{53}$$

from which

$$\int dv = -u\int\frac{dM}{M} \tag{54}$$

If the initial velocity, initial mass, final velocity, and final mass are denoted by v_0, M_0, v, and M, respectively, then the limits of integration in Eq. (54) are as follows:

$$\int_{v_0}^{v} dv = -u\int_{M_0}^{M}\frac{dM'}{M'} \tag{55}$$

[4] Since the mass of the rocket is a decreasing function of time, dM is negative and $-dM$ is positive.

This gives

$$v - v_0 = -u \ln\left(\frac{M}{M_0}\right)$$

or

$$\boxed{v - v_0 = u \ln\left(\frac{M_0}{M}\right)} \qquad (56)$$

Rocket equation

where ln stands for the natural logarithm. The difference between M_0 and M represents the fuel consumed. Thus, Eq. (56) permits the direct calculation of the terminal velocity of a rocket from its terminal mass, amount of fuel, and exhaust velocity.

EXAMPLE 9. The rocket engines of the Saturn V rocket, used for the Apollo and Skylab missions, burn a mixture of kerosene and liquid oxygen (Figure 9.17). Under ideal conditions the exhaust gases from the combustion of this fuel have an exhaust velocity of 3.1×10^3 m/s. The mass of the rocket at liftoff is 2.45×10^6 kg, of which 1.70×10^6 kg are kerosene and liquid oxygen. In the absence of gravity, what would be the terminal velocity of the rocket at burnout?

Fig. 9.17 *Apollo 11* liftoff.

SOLUTION: At burnout the remaining mass is 2.45×10^6 kg $- 1.70 \times 10^6$ kg $= 0.75 \times 10^6$ kg. Equation (56) then gives, with $v_0 = 0$,

$$v = u \ln\left(\frac{M_0}{M}\right) = (3.1 \times 10^3 \text{ m/s})\left[\ln\left(\frac{2.45 \times 10^6}{0.75 \times 10^6}\right)\right]$$

$$= 3.7 \times 10^3 \text{ m/s}$$

SUMMARY

Momentum of a system of particles: $\mathbf{P} = \mathbf{p}_1 + \mathbf{p}_2 + \cdots + \mathbf{p}_n$

Rate of change of momentum:

$$\frac{d\mathbf{P}}{dt} = \mathbf{F}_{\text{ext}}$$

Conservation of momentum (in the absence of external forces):

$$\mathbf{P} = [\text{constant}]$$

Center of mass: $\mathbf{r}_{\text{CM}} = \dfrac{m_1\mathbf{r}_1 + m_2\mathbf{r}_2 + \cdots + m_n\mathbf{r}_n}{M}$

$$\mathbf{r}_{\text{CM}} = \frac{1}{M}\int \mathbf{r}\rho\, dv$$

Momentum of a system of particles: $\mathbf{P} = M\mathbf{v}_{\text{CM}}$

Motion of the center of mass: $M\mathbf{a}_{\text{CM}} = \mathbf{F}_{\text{ext}}$

Kinetic energy of a system of particles:

$$K = K_{\text{int}} + \tfrac{1}{2}Mv_{\text{CM}}^2$$

Angular momentum of a system of particles:

$$\mathbf{L} = \mathbf{r}_1 \times \mathbf{p}_1 + \mathbf{r}_2 \times \mathbf{p}_2 + \cdots + \mathbf{r}_n \times \mathbf{p}_n$$

Rate of change of angular momentum:

$$\frac{d\mathbf{L}}{dt} = \tau_{\text{ext}}$$

Conservation of angular momentum (in absence of external torque):

$$\mathbf{L} = [\text{constant}]$$

Rocket equation: $v - v_0 = u \ln\left(\dfrac{M_0}{M}\right)$

QUESTIONS

1. When the nozzle of a fire hose discharges a large amount of water at high speed, several strong firemen are needed to hold the nozzle steady. Explain.

2. When firing a shotgun, a hunter always presses it tightly against his shoulder. Why?

3. As described in Example 2, guns on board eighteenth-century warships were often mounted on carriages (see Figure 9.3). What was the advantage of this arrangement?

4. Hollywood movies often show a man being knocked over by the impact of a bullet while the man who shot the bullet remains standing, quite undisturbed. Is this reasonable?

5. Where is the center of mass of this book when it is closed? Mark the center of mass with a cross.

6. Roughly, where is the center of mass of this book when it is open, as it is at this moment?

7. In a high jump (Figure 9.18), is it possible for the body of the jumper to pass over the bar while his center of mass passes under? What would he gain by this?

8. A fountain shoots a stream of water vertically into the air. Roughly, where is the center of mass of the water that is in the air at one instant? Is the center of mass higher or lower than the middle height?

9. Consider the moving wrench shown in Figure 9.5. If the center of mass on this wrench had not been marked, how could you have found it by inspection of this photograph?

10. Is it possible to propel a sailboat by mounting a fan on the deck and blowing air on the sail? Is it better to mount the fan on the stern and blow air toward the rear?

11. Cyrano de Bergerac's sixth method for propelling himself to the Moon was as follows: "Seated on an iron plate, to hurl a magnet in the air — the iron follows — I catch the magnet — throw again — and so proceed indefinitely." What is wrong with this method (other than the magnet being too weak)?

12. Within the Mexican jumping bean, a small insect larva jumps up and down. How does this lift the bean off the table?

Fig. 9.18 High jumper passing over the bar.

13. An elephant jumps off a cliff. Does the Earth move upward while the elephant falls?

14. A juggler stands on a balance, juggling five balls (Figure 9.19). On the average, will the balance register the weight of the juggler plus the weight of the five balls? More than that? Less?

Fig. 9.19 Juggler on a balance.

15. Suppose you fill a rubber balloon with air and then release it so that the air spurts out of the nozzle. The balloon will fly across the room. Explain.

16. The combustion chamber of a rocket engine is closed at the front and at the sides, but it is open at the rear (Figure 9.20). Explain how the pressure of the gas on the walls of this combustion chamber gives a net forward force that propels the rocket.

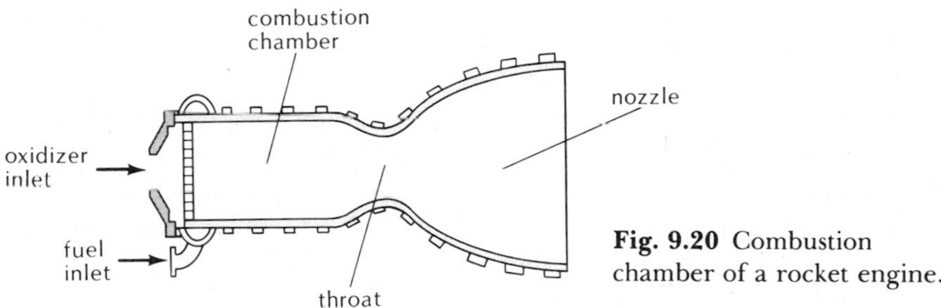

Fig. 9.20 Combustion chamber of a rocket engine.

17. Can the terminal velocity of a rocket [see Eq. (56)] ever exceed the exhaust velocity u of the gas? What mass ratio M_0/M would this require?

18. What is the advantage of multiple-stage rockets over single-stage rockets?

PROBLEMS

Section 9.1

1. Calculate the change of kinetic energy in the collision between the two automobiles described in Example 1.

2. Find the recoil velocity for the gun described in Example 2 if the gun is fired with an elevation angle of 20°.

3. A typical warship built around 1800 (such as the U.S.S. *Constitution*) carried 15 long guns on each side. These guns fired a shot weighing 24 lb with a muzzle velocity of about 1600 ft/s. The weight of the ship was about 4000 short tons (1 short ton = 2000 lb). Suppose that all of the 15 guns on one side of the ship are fired (almost) simultaneously in a horizontal direction at right angles to the ship. What is the recoil velocity of the ship? Ignore the resistance offered by the water.

4. Two automobiles, moving at 40 mi/h in opposite directions, collide head on. One automobile has a weight of 1500 lb; the other a weight of 3200 lb. After the collision both remain joined together. What is the velocity of the wreck? What is the change of the velocity of each automobile during the collision?

5. The nucleus of an atom of radium (mass 3.77×10^{-25} kg) suddenly ejects an alpha particle (mass 6.68×10^{-27} kg) of an energy of 7.26×10^{-16} J. What is the velocity of recoil of the nucleus? What is the kinetic energy of the recoil?

6. A lion of mass 120 kg leaps at a hunter with a horizontal velocity of 12 m/s. The hunter has an automatic rifle firing bullets of mass 15 g with a muzzle velocity of 630 m/s and he attempts to stop the lion in midair. How many bullets

would the hunter have to fire into the lion to stop its horizontal motion? Assume the bullets stick inside the lion.

7. A Maxim machine gun fires 450 bullets per minute. Each bullet has a mass of 14 g and a velocity of 630 m/s.
   (a) What is the average force that the impact of these bullets exerts on a target? Assume the bullets penetrate the target and remain embedded in it.
   (b) What is the average rate at which the bullets deliver their kinetic energy to the target?

8. A vase falls off a table and hits a smooth floor, shattering into three fragments of equal mass. Two of the fragments leave the point of impact with velocities of equal magnitude v at right angles. What is the magnitude and direction of the velocity of the third fragment?

9. The nucleus of an atom of radioactive copper undergoing beta decay simultaneously emits an electron and a neutrino. The momentum of the electron is 2.64×10^{-22} kg · m/s, that of the neutrino is 1.97×10^{-22} kg · m/s, and the angle between their directions of motion is 30°. The mass of the residual nucleus is 63.9 u. What is the recoil velocity of the nucleus?

10. An automobile of mass 1500 kg and a truck of 3500 kg collide at an intersection. Just before the collision the automobile was traveling north at 80 km/h and the truck was traveling east at 50 km/h. After the collision both vehicles remain joined together.
   (a) What is the velocity (magnitude and direction) of the vehicles immediately after collision?
   (b) How much kinetic energy is lost during the collision?

11. The solar wind sweeping past the Earth consists of a stream of particles, mainly hydrogen ions of mass 1.7×10^{-27} kg. There are about 10^7 ions per cubic meter and their speed is 4×10^5 m/s. What force does the impact of the solar wind exert on an artificial Earth satellite that has an area of 1.0 m^2 facing the wind? Assume that upon impact the ions at first stick to the surface of the satellite.

12. The nozzle of a fire hose ejects 200 gal./min of water at a speed of 86 ft/s. Estimate the recoil force on the nozzle. By yourself, can you hold this nozzle steady in your hands?

13. A spaceship of frontal area 25 m^2 passes through a cloud of interstellar dust at a speed of 1.0×10^6 m/s. The density of dust is 2.0×10^{-18} kg/m^3. If all the particles of dust that impact on the spaceship stick to it, find the average decelerating force that the impact of the dust exerts on the spaceship.

14. An automobile is traveling at a speed of 80 km/h through heavy rain. The raindrops are falling vertically at 10 m/s and there are 7.0×10^{-4} kg of raindrops in each cubic meter of air. For the following calculation assume that the automobile has the shape of a rectangular box 2 m wide, 1.5 m high, and 4 m long.
   (a) At what rate (in kg/s) do raindrops strike the front and top of the automobile?
   (b) Assume that when a raindrop hits, it initially sticks to the automobile, although it falls off later. At what rate does the automobile give momentum to the raindrops? What is the horizontal drag force that the impact of the raindrops exerts on the automobile?

15. The record for the heaviest rainfall is held by Unionville, Maryland, where 1.23 in. of rain (3.12 cm) fell in an interval of 1 min. Assuming that the impact velocity of the raindrops on the ground was 10 m/s, what must have been the average impact force on each square meter of ground during this rainfall?

Section 9.2

16. A 130-lb woman and a 160-lb man sit on a seesaw, 12 ft. long. Where is their center of mass? Neglect the mass of the seesaw.

17. Consider the system Earth–Moon; use the data in the table printed on the endpapers. How far from the center of the Earth is the center of mass of this system?

18. In order to balance the wheel of an automobile, a mechanic attaches a piece of lead alloy to the rim of the wheel. The mechanic finds that if he attaches a piece of 40 g at a distance of 20 cm from the center of a wheel of 30 kg, the wheel is perfectly balanced, i.e., the center of the wheel coincides with the center of mass. How far from the center of the wheel was the center of mass before the mechanic balanced the wheel?

19. The distance between the oxygen and each of the hydrogen atoms in a water (H_2O) molecule is 0.958 Å; the angle between the two oxygen–hydrogen bonds is 105° (Figure 9.21). Treating the atoms as particles, find the center of mass.

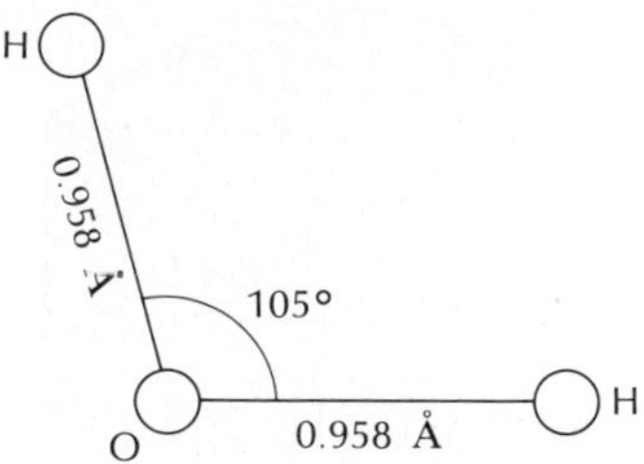

Fig. 9.21 Water molecule.

20. Figure 9.22 shows the shape of a nitric acid (HNO_3) molecule and its dimensions. Treating the atoms as particles, find the center of mass of this molecule.

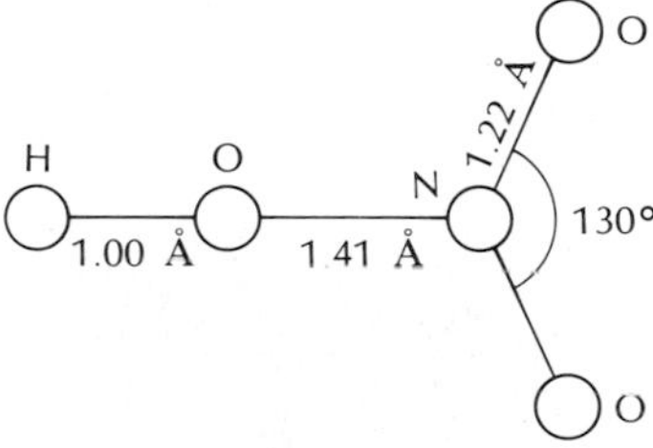

Fig. 9.22 Nitric acid molecule.

21. Figure 13.8a shows the positions of the three inner planets (Mercury, Venus, and Earth) on January 1, 1980. Measure angles and distances off this figure and find the center of mass of the system of these planets (ignore the Sun). The masses of the planets are listed in Table 13.1.

22. The Local Group of galaxies consists of our Galaxy and its nearest neighbors. The masses of the most important members of the Local Group are as follows (in multiples of the mass of the Sun): our Galaxy, 2×10^{11}; the Andromeda galaxy, 3×10^{11}; the Large Magellanic cloud, 2.5×10^{10}; and NGC598, 8×10^{9}. The x, y, z coordinates of these galaxies are, respectively, as follows (in thousands of light-years): (0, 0, 0); (1640, 290, 1440); (8.5, 56.7, −149); and (1830, 766, 1170). Find the coordinates of the center of mass of the Local Group. Treat all the galaxies as point masses.

23. Suppose we take the semicircular rod described in Example 5 and we add to it a straight rod of length $2R$ fitted between the ends of the semicircular rod (Figure 9.23). Where is the center of mass of this system?

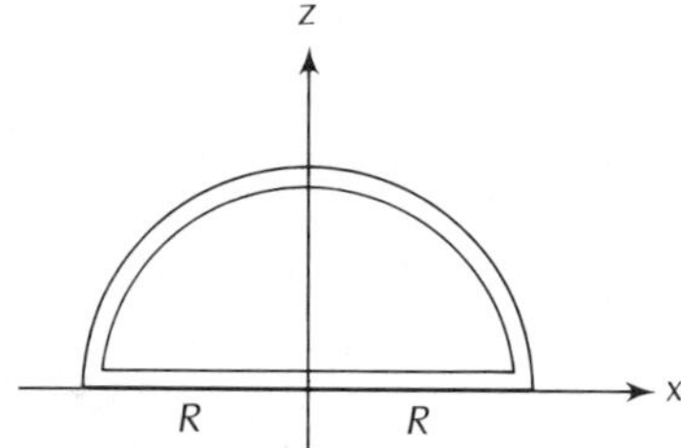

Fig. 9.23 A semicircular rod and a straight rod joined together.

24. Three uniform square pieces of sheet metal are joined along their edges so as to form three of the sides of a cube (Figure 9.24). The dimensions of the squares are $L \times L$. Where is the center of mass of the joined squares?

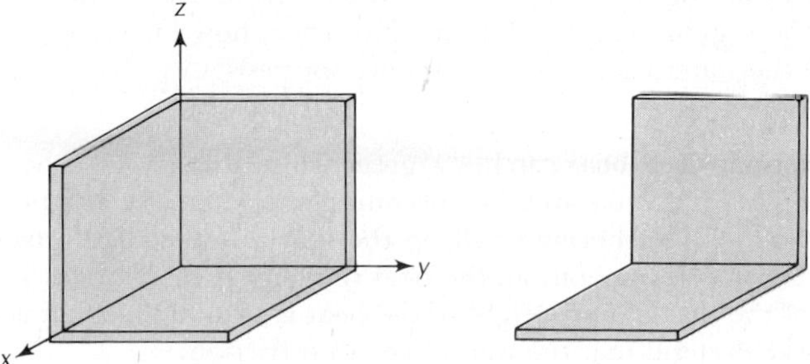

Fig. 9.24 (left) Three square pieces of sheet metal joined together along their edges.

Fig. 9.25 (right) Two uniform squares of sheet metal joined along one edge.

25. Two uniform squares of sheet metal of dimension $L \times L$ are joined at a right angle along one edge (Figure 9.25). One of the squares has twice the mass of the other. Find the center of mass of the combined squares.

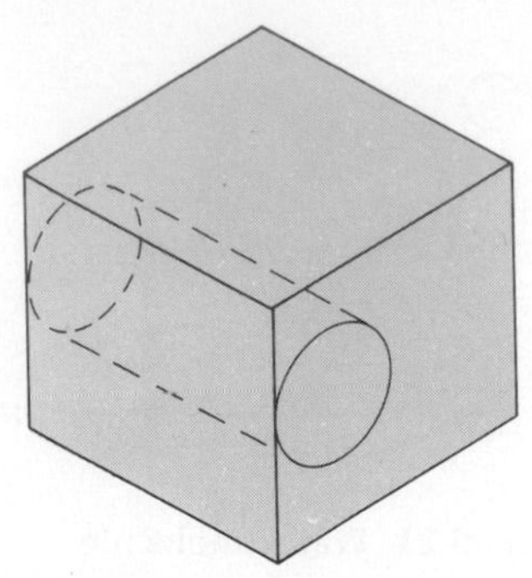

Fig. 9.26 Iron cube with a hole.

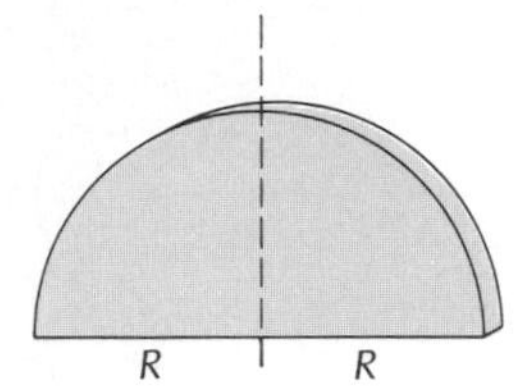

Fig. 9.27 Semicircle of sheet metal.

26. A cube of iron has dimensions $L \times L \times L$. A hole of radius $\frac{1}{4}L$ has been drilled all the way through the cube, so that one side of the hole is tangent to one face along its entire length (Figure 9.26). Where is the center of mass of the drilled cube?

27. A semicircle of uniform sheet metal has radius R (Figure 9.27). Find the center of mass. (Hint: Regard the semicircle as assembled of many thin, concentric semicircular rods; use the result of Example 5.)

*28. Suppose that water drops are released from a point at the edge of a roof with a constant time interval Δt between one water drop and the next. The drops fall a distance l to the ground. If Δt is very short (so that the number of water drops falling through the air at any given instant is very large), show that the center of mass of the falling drops is at a height of $\frac{2}{3}l$ above the ground. From this, deduce that the time-average height of a projectile launched from the ground and returning to the ground is $\frac{2}{3}$ of its maximum height. (This theorem is useful in the calculation of the average air pressure and air resistance encountered by the projectile.)

29. The great pyramid at Gizeh has a mass of 6.6×10^6 metric tons and a height of 147 m (see Example 6). Assume that the mass is uniformly distributed over the volume of the pyramid.
(a) How much work must the ancient Egyptian laborers have done against gravity to pile up the stones in the pyramid?
(b) If each laborer delivered work at an average rate of 4×10^5 J/h, how many man-hours of work have been stored in this pyramid?

30. Mount Fuji has approximately the shape of a cone. The half-angle at the apex of this cone is 65° and the height of the apex is 3800 m. At what height is the center of mass? Assume that the material in Mount Fuji has uniform density.

31. Show that the center of mass of a uniform flat triangular plate is at the point of intersection of the lines drawn from the vertices to the midpoints of the opposite sides.

32. A lock on the Champlain canal is 240 ft long and 30 ft wide; the lock has a lift of 12 ft, i.e., the difference between the water levels of the canal on one side of the lock and on the other side is 12 ft. How much gravitational potential energy is wasted each time the lock goes through one cycle (involving the filling of the lock with water from the high level and then the spilling of this water to the low level)?

Section 9.3

33. A tugboat of mass 4000 metric tons and a ship of mass 28,000 metric tons are joined by a long towrope of 400 m. Both vessels are initially at rest in the water. If the tugboat reels in 200 m of towrope, how far does the ship move relative to the water? The tugboat? Ignore the resistance that the water offers to the motion.

34. A fisherman in a boat catches a great white shark with a harpoon. The shark struggles for a while and then becomes limp when at a distance of 1000 ft from the boat. The fisherman pulls in the shark by the rope attached to the harpoon. During this operation, the boat (initially at rest) moves 150 ft in the direction of the shark. The weight of the boat is 12,000 lb. What is the weight of the shark? Pretend that the water exerts no friction.

35. A 75-kg man climbs the stairs from the ground to the fourth floor of a building, a height of 15 m. How far does the Earth recoil in the opposite direction as the man climbs?

36. A 6000-kg truck stands on the deck of an 80,000-kg ferryboat. Initially the ferry is at rest and the truck is located at its front end. If the truck now drives 15 m along the deck toward the rear of the ferry, how far will the ferry

move forward relative to the water? Pretend that the water has no effect on the motion.

37. In a molecule, such as the potassium bromide (KBr) molecule of Example 4, the atoms usually execute a rapid vibrational motion about their equilibrium positions. Suppose that in an isolated KBr molecule the speed of the potassium atom is 5.0×10^3 m/s at one instant (relative to the center of mass). What is the speed of the bromine atom at the same instant?

38. While moving horizontally at 5.0×10^3 m/s at an altitude of 2.5×10^4 m, a ballistic missile explodes and breaks apart into two fragments of equal mass which fall freely. One of the fragments has zero speed immediately after the collision and lands on the ground directly below the point of the explosion. Where does the other fragment land? Ignore the friction of air.

Section 9.4

39. Repeat the calculation of Example 8 if the two automobiles travel in the *same* direction.

40. A projectile of 100 lb fired from a gun has a speed of 2100 ft/s. The projectile explodes in flight, breaking apart into a fragment of 70 lb and a fragment of 30 lb (we assume that no mass is dispersed in the explosion). Both fragments move along the original direction of motion. The speed of the first fragment is 1500 ft/s and that of the second is 3500 ft/s.

(a) Calculate the translational kinetic energy of the center of mass before the explosion. Calculate the internal kinetic energy before the explosion.

(b) Calculate these quantities after the explosion. Where does the extra internal kinetic energy come from?

41. Consider the automobile collision described in Problem 4. What is the internal kinetic energy before the collision? After the collision?

42. Regard the automobile and the truck described in Problem 10 as a system of two particles.

(a) What is the translational kinetic energy of the center of mass before the collision? What is the internal kinetic energy?

(b) What is the translational kinetic energy of the center of mass after the collision? What is the internal kinetic energy?

43. The typical speed of the vibrational motion of the iron atoms in a piece of iron at room temperature is 360 m/s. What is the total internal kinetic energy of a 1-kg chunk of iron?

Section 9.5

44. The Earth and the Moon move in circular orbits around their common center of mass. The center-to-center distance between the Earth and the Moon is 3.84×10^8 m and the time required for one orbit is 27.3 days. What is the orbital angular momentum of the Moon? Of the Earth? What is the total orbital angular momentum? Treat both bodies as particles.

Section 9.6

45. In order to achieve a thrust of 3.3×10^7 N, at what rate (in metric tons per second) must the engines of the Saturn V rocket consume their fuel? Assume that the exhaust velocity of the hot gas from the engines is 2900 m/s.

46. If a rocket, initially at rest, is to attain a terminal velocity of a magnitude equal to the exhaust velocity, what fraction of the initial mass must be fuel?

47. A rocket burns a kerosene–oxygen mixture. The complete burning of 1.0 kg of kerosene requires 3.4 kg of oxygen; this burning releases about 4.2×10^7 J of thermal energy. Suppose that all of this thermal energy is con-

verted into kinetic energy of the reaction products (4.4 kg). What will be the exhaust velocity of the reaction products?

*48. Rockets used for space exploration usually consist of several stages sitting on top of one another and fired in sequence. When the first stage has burned out, it is jettisoned and the second stage is ignited, etc. Show that the terminal velocity of the last stage of such a multiple-stage rocket is never more than $v = u \ln(M_0/M)$, where M_0 is the initial mass of the multiple-stage rocket and M is the terminal mass of the last stage at burnout. Assume that all stages have the same exhaust velocity and ignore gravity. (Hint: Compare the multiple-stage rocket with an ideal rocket consisting entirely of fuel except for the terminal mass M.)

CHAPTER 10

Collisions

The collision between two bodies — an automobile and a solid wall (Figure 10.1), a ship and an iceberg, a molecule of oxygen and a molecule of nitrogen, an alpha particle and a nucleus of a gold atom — involves a violent change of the motion, a change brought about by very strong forces that begin to act suddenly when the bodies come into contact, last a short time, and then cease just as suddenly when the bodies separate. The forces that act during a collision are usually rather complicated so that their complete theoretical description is impossible (e.g., an automobile collision) or at least very difficult (e.g., a nuclear collision). However, even without exact knowledge of the force law, we can make some predictions about the collision by taking advantage of the general laws of conservation of momentum and energy. In the following sections we will see what constraints these laws impose on the motion of the colliding bodies.

The study of collisions is an important tool in the experimental investigation of atoms, nuclei, and elementary particles. All subatomic bodies are too small to be made visible with any kind of microscope. Just as a surgeon who cannot see the interior of the wound uses probes to feel the condition of the tissues, a physicist who cannot see the interior of an atom uses probes to "feel" for subatomic structures. The probe used by physicists in the exploration of subatomic structures is simply a stream of fast-moving particles — electrons, protons, neutrons, alpha particles, or others. These projectiles are aimed at a target containing a sample of the atoms, nuclei, or elementary particles under investigation. From the manner in which the projectiles collide and react with the target, physicists can deduce some of the properties of the subatomic structures in the target.

10.1 Impulsive Forces

The force that two colliding bodies exert on one another acts only for a short time, giving a brief but strong push. This force is called an **im-**

Impulsive force

pulsive force. During the collision, the impulsive force is much stronger than any other forces that may be present; consequently the impulsive force produces a large change in the motion while the other forces produce only small and insignificant changes. For example, during the automobile collision shown in Figure 10.1, the only important force is the push of the wall on the front end of the automobile; the effects produced by gravity and by the normal force of the road during the collision are insignificant.

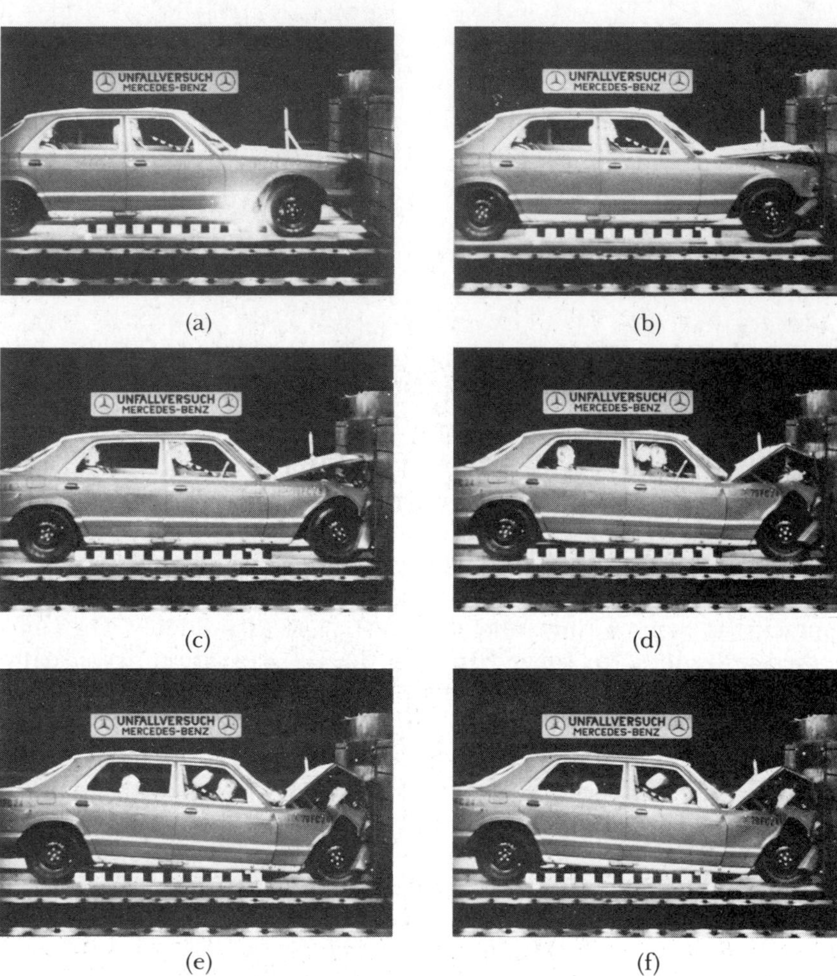

Fig. 10.1 Crash test of a Mercedes-Benz automobile. The photographs show an impact at 49 km/h on a rigid barrier. The first photograph was taken 5×10^{-3} s after the initial contact; the others were taken at intervals of 20×10^{-3} s thereafter. The automobile remains in contact with the barrier for 0.120 s; it then recoils from the barrier with a speed of 4.7 km/h. The checkered bar on the ground has a length of 2 m.

Suppose that a collision lasts a short time Δt, say, from $t = 0$ to $t = \Delta t$, and that during this time an impulsive force **F** acts on one of the colliding bodies. The force is zero before $t = 0$ and it is zero after $t = \Delta t$, but is large between these times. For example, Figure 10.2 shows a plot of the force experienced by an automobile in a collision with a solid wall lasting 0.120 s. The force is zero before $t = 0$ and after $t = 0.120$ s, and varies in a complicated way between these times. The **impulse** delivered by such a force **F** to the body is defined as the integral of the force over time,

Impulse

$$\boxed{\mathbf{I} = \int_0^{\Delta t} \mathbf{F}\, dt} \tag{1}$$

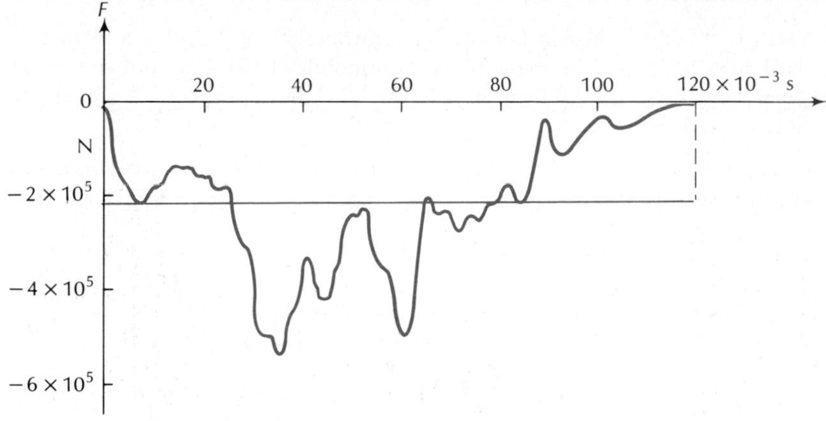

Fig. 10.2 Force on the automobile as a function of time during the impact shown in Figure 10.1. The colored horizontal line gives the time-average force. (From data supplied by Mercedes-Benz of North America, Inc.)

According to this equation, the x component of the impulse for the force shown in Figure 10.2 is the area between the curve $F_x(t)$ and the t axis, and similarly for the y component and the z component.

The units of impulse are N · s or kg · m/s in the metric system and lb · s in the British system; these units are the same as those of momentum.

By means of the equation of motion

$$\mathbf{F} = \frac{d\mathbf{p}}{dt} \tag{2}$$

we can transform Eq. (1) into

$$\mathbf{I} = \int_0^{\Delta t} \mathbf{F}\, dt = \int_0^{\Delta t} \frac{d\mathbf{p}}{dt}\, dt = \int d\mathbf{p} = \mathbf{p}' - \mathbf{p} \tag{3}$$

where $\mathbf{p}$ is the momentum before the collision (at time $t = 0$) and $\mathbf{p}'$ is the momentum after the collision (at time $t = \Delta t$). Thus, the impulse of a force is simply equal to the momentum change produced by this force. However, since the force acting during a collision is usually not known in detail, Eq. (3) is not very helpful for calculating momentum changes. It is often best to apply Eq. (3) in reverse for calculating the average force from the known momentum change. The time-average force is defined by

$$\overline{\mathbf{F}} = \frac{1}{\Delta t} \int_0^{\Delta t} \mathbf{F}\, dt \tag{4}$$

In a plot of force as a function of time, such as shown in Figure 10.2, the time-average force simply represents the mean height of the function above the t axis; this mean height is shown by the red horizontal line in Figure 10.2. By means of Eq. (3) we can write the time-average force as

$$\overline{\mathbf{F}} = \frac{1}{\Delta t} (\mathbf{p}' - \mathbf{p}) \tag{5}$$

This relation gives a quick estimate of the average impulsive force if the collision time and the momentum change are known.

EXAMPLE 1. The collision between the automobile and wall shown in Figure 10.1 lasts 0.120 s. The mass of the automobile is 1700 kg and the initial and final speeds are $v = 13.6$ m/s and $v' = -1.3$ m/s, respectively. Evaluate the impulse and the time-average force from these data.

SOLUTION: With the x axis along the direction of the initial motion, the change in momentum is

$$p'_x - p_x = mv' - mv$$

$$= 1700 \text{ kg} \times (-1.3 \text{ m/s}) - 1700 \text{ kg} \times 13.6 \text{ m/s}$$

$$= -2.53 \times 10^4 \text{ kg} \cdot \text{m/s}$$

Hence the impulse is $I_x = -2.53 \times 10^4$ kg · m/s and the time-average force is

$$\overline{F}_x = \frac{1}{\Delta t}(p'_x - p_x) = \frac{-2.53 \times 10^4 \text{ kg} \cdot \text{m/s}}{0.120 \text{ s}}$$

$$= -2.11 \times 10^5 \text{ N}$$

Note that since the mutual forces on two bodies engaged in a collision are an action–reaction pair of equal magnitudes and of opposite directions, the corresponding impulses are equal and opposite. For instance, in Example 1 the impulse on the automobile is $I_x = -2.53 \times 10^4$ kg · m/s and the impulse on the wall is

$$I_{x,\text{ wall}} = 2.53 \times 10^4 \text{ kg} \cdot \text{m/s}$$

This, of course, expresses momentum conservation: the momentum changes of the two colliding bodies are of equal magnitudes and opposite directions.

It is often not possible to calculate the motion of the colliding bodies by direct solution of Newton's equation of motion because the impulsive forces that act during the collision are not known in sufficient detail. Consequently, we must glean whatever information we can from the general laws of conservation of momentum and energy, which do not depend on details of the forces. In some simple instances these general laws permit the deduction of the motion after the collision from what is known about the motion before the collision.

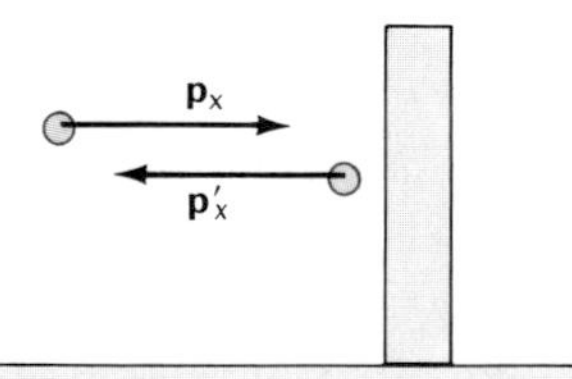

Fig. 10.3 Recoil of a ball thrown against a wall.

EXAMPLE 2. A "superball" made of rubberlike plastic is thrown against a hard, smooth wall. The ball strikes the wall from a perpendicular direction with speed v. Assuming energy conservation, find the speed of the ball after the collision.

SOLUTION: The only force on the ball is a normal force exerted by the wall; this force reverses the motion of the ball (Figure 10.3). Since the wall is very massive, the reaction force of the ball will not give it any appreciable velocity. Hence the kinetic energy of the system, both before and after the collision, is merely the kinetic energy of the ball. Conservation of this kinetic energy then requires that the ball rebound with a speed v equal to the incident speed.

Note that although the kinetic energy of the ball remains constant, the momentum does not remain constant (also see Example 5.7). If the x axis is in the direction of the initial motion, then the momentum of the ball before the collision is $p_x = mv$ and after the collision it is $p'_x = -mv$; hence the change of momentum is $p'_x - p_x = -2mv$. The wall suffers an equal and opposite momentum change of $+2mv$ so that the total momentum of the system is conserved. The wall can acquire a momentum $2mv$ *without* acquiring any appreciable velocity because its mass is large. For instance, if $m = 0.1$ kg and $v = 10$ m/s, then $2mv = 2 \times 0.1 \text{ kg} \times 10 \text{ m/s} = 2.0 \text{ kg} \cdot \text{m/s}$; and if the mass of the wall is 1000 kg, then its recoil velocity is given by $1000 \text{ kg} \times v_{\text{wall}} = 2.0$ kg m/s, which yields $v_{\text{wall}} = 2 \times 10^{-3}$ m/s. Note that the corresponding kinetic energy is also quite small, $\frac{1}{2} \times 1000 \text{ kg} \times (2 \times 10^{-3} \text{ m/s})^2 = 2 \times 10^{-3}$ J.[1] As a general rule, a body of very large mass can absorb momentum without any appreciable change in its velocity or kinetic energy.

Elastic collision

A collision in which the kinetic energy is conserved is called **elastic.** Collisions between macroscopic bodies are usually not elastic — during the collision some of the kinetic energy is transformed into heat by the internal friction forces and some is used up in doing work to change the internal configuration of the bodies. For example, the automobile collision shown in Figure 10.1 is highly inelastic; almost the entire kinetic energy is used up in doing work on the automobile parts, changing their shape. On the other hand, the collision of a "superball" and a hard wall or the collision of two billiard balls comes pretty close to being elastic (the noise made by these collisions indicates that some sound energy is generated from kinetic energy; hence these collisions cannot be *exactly* elastic).

Collisions between "elementary" particles — such as electrons, protons, and neutrons — are often elastic. These particles have no internal friction forces which could dissipate kinetic energy. A collision between such particles can only be inelastic if it involves the creation of new particles; such new particles may arise either by conversion of some of the available kinetic energy into mass or else by transmutation of the old particles by means of a change of their internal structure. Collisions of this kind will be discussed in Section 10.4.

10.2 Collisions in One Dimension

Elastic collision in one dimension

The collision of two boxcars on a straight railroad track is an example of a one-dimensional collision. More generally, the collision of any two bodies that approach head-on and recoil along their original line of motion is one dimensional. Obviously, such collisions will occur only under rather exceptional circumstances; nevertheless, we find it worthwhile to study one-dimensional collisions because they display in a simple way some of the features that reappear in two- and three-dimensional collisions.

In an **elastic** one-dimensional collision between two particles, the laws of conservation of momentum and of energy completely determine the final velocities in terms of the initial velocities. In the following calculations we will assume that one particle (the "target") is initially at rest, and the other (the "projectile") is initially in motion.

Figure 10.4a shows the particles before the collision and Figure 10.4b shows them after; the x axis is along the direction of motion. We will designate the x components of the velocity of particle 1 ("projec-

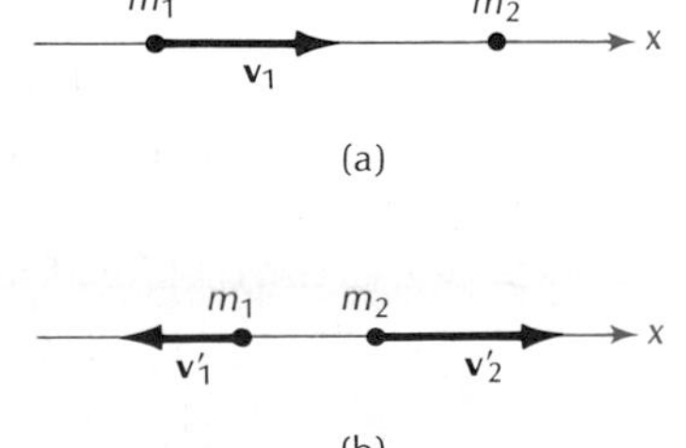

Fig. 10.4 (a) Before the collision, particle 2 is at rest and particle 1 has velocity $\mathbf{v}_1$. (b) After the collision, particle 1 has velocity $\mathbf{v}'_1$ and particle 2 has velocity $\mathbf{v}'_2$.

[1] This calculation is not quite accurate. The wall is attached to the Earth, and it therefore cannot move freely — it can at most rock back and forth. Our crude calculation merely provides an upper limit on v_{wall}.

tile") and particle 2 ("target") by v_1 and v_2 before the collision, and by v_1' and v_2' after the collision, respectively. Since particle 2 is initially at rest, $v_2 = 0$.

Conservation of momentum states that

$$m_1 v_1 = m_1 v_1' + m_2 v_2' \tag{6}$$

and conservation of energy states that

$$\tfrac{1}{2} m_1 v_1^2 = \tfrac{1}{2} m_1 v_1'^2 + \tfrac{1}{2} m_2 v_2'^2 \tag{7}$$

These two equations can be rearranged as follows:

$$m_1(v_1 - v_1') = m_2 v_2' \tag{8}$$

$$\tfrac{1}{2} m_1 (v_1 - v_1')(v_1 + v_1') = \tfrac{1}{2} m_2 v_2'^2 \tag{9}$$

After dividing Eq. (9) by Eq. (8) we obtain

$$\tfrac{1}{2}(v_1 + v_1') = \tfrac{1}{2} v_2' \tag{10}$$

The advantage of this equation is that it does not contain any squares of the velocities.

We regard the initial velocities v_1, v_2 as known and the final velocities v_1', v_2' as unknown. Equations (10) and (6) taken together are then a simple (linear) system of two equations for the two unknowns v_1', v_2'. The simultaneous solution of these equations leads to

$$\boxed{v_1' = \frac{m_1 - m_2}{m_1 + m_2} v_1} \tag{11}$$

Speeds after a one-dimensional elastic collision

and

$$\boxed{v_2' = \frac{2m_1}{m_1 + m_2} v_1} \tag{12}$$

EXAMPLE 3. An empty boxcar of mass $m_1 = 20$ metric tons rolling on a straight track at 5 m/s collides with a loaded stationary boxcar of mass $m_2 = 65$ metric tons. Assuming that the cars bounce off one another elastically, find the velocities after the collisions.

SOLUTION: With $m_1 = 20$ tons and $m_2 = 65$ tons, Eqs. (11) and (12) yield

$$v_1' = \frac{20\text{ tons} - 65\text{ tons}}{20\text{ tons} + 65\text{ tons}} \times 5\text{ m/s} = -2.6\text{ m/s}$$

$$v_2' = \frac{2 \times 20\text{ tons}}{20\text{ tons} + 65\text{ tons}} \times 5\text{ m/s} = 2.4\text{ m/s}$$

Note that if the mass of the target is much larger than the mass of the projectile ($m_2 \gg m_1$), then Eqs. (11) and (12) give $v_1' \cong -v_1$ and $v_2' \cong 0$, which means that the projectile bounces off with a reversed velocity and the target remains stationary. Conversely, if the mass of the projectile is much larger than the mass of the target ($m_1 \gg m_2$), then $v_1' = v_1$ and $v_2' = 2v_1$, which means that the projectile plows right on and the target bounces off with *twice* the speed of the projectile. This second case can be understood in terms of the preceding case: in the rest frame of m_1, m_2 is approaching with velocity $-v_1$ and bounces off with velocity $+v_1$, that is, the velocity of the particle of small mass changes by $2v_1$.

EXAMPLE 4. Inside a nuclear reactor containing uranium fuel, the fission reactions produce an abundant flux of fast neutrons of a speed of about 2×10^7 m/s. These neutrons are used to trigger more fission reactions. But before they can be so used, they must be slowed down to a much lower speed. The slowing down is accomplished by collisions: the uranium fuel in the reactor is surrounded by water or by graphite and the neutrons lose their kinetic energy in collisions with the nuclei of these materials. By what factor is the speed of a neutron reduced in a head-on collision with a stationary carbon nucleus? With a hydrogen nucleus? The masses of a neutron, a carbon nucleus, and a hydrogen nucleus are 1.0087 u, 11.9934 u, and 1.0073 u, respectively.

SOLUTION: The final speed of the neutron is given by Eq. (11):

$$v_1' = \frac{m_1 - m_2}{m_1 + m_2} v_1$$

For a collision with carbon, we substitute the approximate values $m_1 = 1.01$ u for the mass of the neutron and $m_2 = 12.00$ u for the mass of the carbon nucleus:

$$v_1' = \frac{1.01 - 12.00}{1.01 + 12.00} v_1 = -0.84 v_1$$

For a collision with a proton, we need to substitute more precise values of the masses; with $m_1 = 1.0087$ u for the mass of the neutron and $m_2 = 1.0073$ u for the mass of the proton,

$$v_1' = \frac{1.0087 - 1.0073}{1.0087 + 1.0073} v_1 = 6.9 \times 10^{-4} v_1$$

This shows that a single head-on collision with a proton will just about stop a neutron, but a single head-on collision with a carbon nucleus will only reduce the speed of the neutron by 16%. Hence several collisions with carbon nuclei are needed to reduce the speed of a neutron to a small value.

Although Eqs. (11) and (12) are based on the assumption of a stationary target ($v_2 = 0$), they can also be useful for solving problems with a moving target ($v_2 \neq 0$). The trick is first to solve the problem in the initial reference frame of the target [where Eqs. (11) and (12) are valid] and then transform the velocities to any other reference frame by means of the Galilean transformation equations.

One interesting feature of an elastic collision is that the *relative velocity* $v_2 - v_1$ reverses during the collision, i.e., the relative velocity changes its sign but not its magnitude. This can be readily seen from Eq. (10), which, upon rearrangement, reads

$$v_1' - v_2' = -v_1 \qquad (13)$$

The quantity on the left side of this equation is the relative velocity after the collision and the quantity on the right side is the negative of the relative velocity before the collision (remember $v_2 = 0$).

We can gain some insight into this preservation of the relative velocity by looking at the expression for the kinetic energy of the system in terms of the center-of-mass velocity, that is, Eq. (9.44):

$$K = \tfrac{1}{2}Mv_{CM}^2 + \tfrac{1}{2}m_1u_1^2 + \tfrac{1}{2}m_2u_2^2 \tag{14}$$

where u_1 and u_2 are the velocities relative to the center of mass:

$$u_1 = v_1 - v_{CM} = v_1 - \frac{m_1v_1 + m_2v_2}{m_1 + m_2} = \frac{m_2(v_1 - v_2)}{m_1 + m_2} \tag{15}$$

$$u_2 = v_2 - v_{CM} = v_2 - \frac{m_1v_1 + m_2v_2}{m_1 + m_2} = \frac{m_1(v_2 - v_1)}{m_1 + m_2} \tag{16}$$

In these equations the velocities may be evaluated either before, during, or after the collision. If we square Eqs. (15) and (16) and substitute them into Eq. (14), we readily find the result

$$K = \tfrac{1}{2}Mv_{CM}^2 + \frac{1}{2}\frac{m_1m_2}{m_1 + m_2}(v_1 - v_2)^2 \tag{17}$$

The first term on the right side of this equation represents the kinetic energy associated with the translational motion of the center of mass; the second term represents the "internal" kinetic energy associated with motion relative to the center of mass. The first term is necessarily conserved because v_{CM} is constant in the absence of external forces. The second term must then also be constant, since otherwise the total kinetic energy would *not* be conserved. Thus, the preservation of the magnitude of the relative velocity $v_2 - v_1$ in an elastic collision expresses the conservation of the "internal" kinetic energy.

Inelastic collision

If the collision is **inelastic,** then the only conservation law that is applicable is the conservation of momentum. This, by itself, is insufficient to calculate the velocities of both particles after the collision. However, if the collision is **totally inelastic** so that a maximum amount of kinetic energy is lost, then the velocities after the collision can be calculated. If there is a maximum loss of kinetic energy, then the two particles will have zero relative velocity after the collision, i.e., they will stick together. This can be understood from Eq. (17): the first term on the right side of this equation remains constant whether the collision is elastic or inelastic; only the second term can change — in a totally inelastic collision this term will become zero and consequently the relative velocity $v_1' - v_2'$ after the collision becomes zero. If both particles have the same velocity and so remain together, then their velocity must necessarily coincide with the velocity of the center of mass.

EXAMPLE 5. Suppose that the two boxcars of Example 3 couple during the collision and remain locked together. What is the velocity of the combination after the collision? How much kinetic energy is dissipated during the collision?

SOLUTION: With $v_2 = 0$, conservation of momentum [see Eq. (6)] states that

$$m_1v_1 = m_1v_1' + m_2v_2'$$

After the collision the velocities of both cars are equal, $v_1' = v_2'$, so that

$$m_1 v_1 = m_1 v_1' + m_2 v_1'$$

Hence the final velocity of both cars is

$$v_1' = \frac{m_1 v_1}{m_1 + m_2} = \frac{20 \text{ tons}}{20 \text{ tons} + 65 \text{ tons}} \times 5 \text{ m/s} = 1.2 \text{ m/s}$$

(Note that the velocity of the center of mass is

$$v_{\text{CM}} = \frac{m_1 v_1 + m_2 v_2}{m_1 + m_2}$$

or, since $v_2 = 0$,

$$v_{\text{CM}} = \frac{m_1 v_1}{m_1 + m_2}$$

The final velocity therefore coincides with the velocity of the center of mass, just as it should.)

The lost kinetic energy corresponds to the second term on the right side of Eq. (17), evaluated with the initial velocities $v_1 = 15$ m/s and $v_2 = 0$:

$$[\text{loss of kinetic energy}] = \frac{1}{2}\frac{m_1 m_2}{m_1 + m_2}(v_1 - v_2)^2$$

$$= \frac{1}{2}\frac{20 \text{ tons} \times 65 \text{ tons}}{20 \text{ tons} + 65 \text{ tons}}(5 \text{ m/s})^2$$

$$= 1.9 \times 10^2 \text{ tons} \cdot (\text{m/s})^2$$

This loss of kinetic energy can of course also be calculated by taking the difference between the initial and final total kinetic energies:

$$\tfrac{1}{2}m_1 v_1^2 + \tfrac{1}{2}m_2 v_2^2 - \tfrac{1}{2}(m_1 + m_2)v_{\text{CM}}^2$$

$$= \tfrac{1}{2} \times 20 \text{ tons} \times (5 \text{ m/s})^2 + 0 - \tfrac{1}{2}(20 \text{ tons} + 65 \text{ tons}) \times (1.2 \text{ m/s})^2$$

$$= 1.9 \times 10^2 \text{ tons} \cdot (\text{m/s})^2$$

EXAMPLE 6. Figure 10.5a shows a **ballistic pendulum,** a device used in the past century to measure the speed of bullets. The pendulum consists of a large block of wood of mass m_2 suspended from long wires. Initially, the pendulum is at rest. The bullet of mass m_1 strikes the block horizontally and remains stuck in it. The impact of the bullet puts the block in motion, causing it to swing upward to a height h (Figure 10.5b). In a test of a Springfield rifle firing

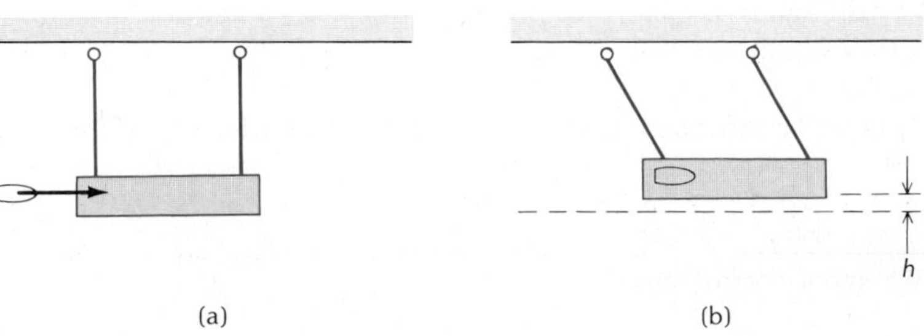

Fig. 10.5 (a) Impact of a bullet on a ballistic pendulum. (b) After the impact, the pendulum swings to a height h.

a bullet of 9.7 g, a ballistic pendulum of 4.0 kg swings up to a height of 19 cm. What was the speed of the bullet before impact?

SOLUTION: The collision of the bullet with the block of wood is totally inelastic. Since the collision takes only a very short time, the external forces (gravity, pull of wires) can be neglected and the momentum is conserved during the collision. Immediately after the collision, bullet and block move with the velocity of the center of mass (compare Example 5),

$$v_{CM} = \frac{m_1 v_1}{m_1 + m_2}$$

During the subsequent swinging motion of the pendulum, the total energy (kinetic plus potential) is conserved. At the bottom of the swing, the energy is kinetic, $\frac{1}{2}(m_1 + m_2)v_{CM}^2$, and at the top of the swing it is potential, $(m_1 + m_2)gh$. Hence

$$\tfrac{1}{2}(m_1 + m_2)v_{CM}^2 = (m_1 + m_2)gh$$

from which $v_{CM} = \sqrt{2gh}$ and therefore

$$v_1 = \frac{m_1 + m_2}{m_1}\sqrt{2gh} \qquad (18)$$

$$= \frac{0.0097\text{ kg} + 4.0\text{ kg}}{0.0097\text{ kg}}\sqrt{2 \times 9.8\text{ m/s}^2 \times 0.19\text{ m}}$$

$$= 8.0 \times 10^2\text{ m/s}$$

Note that during the collision, momentum is conserved but not energy (the collision is inelastic); and that during the swinging motion, energy is conserved but not momentum (the swinging motion proceeds under the influence of the "external" forces of gravity and the tension of the wires).

10.3 Collisions in Two Dimensions

In an *elastic* two-dimensional collision between two particles, the laws of conservation of momentum and of energy do not suffice to determine the final velocities in terms of the initial velocities. Conservation gives us three equations: two equations from the conservation of the two separate components of momentum and one equation from the conservation of energy. But there are four unknowns: two components of velocity for each of the two particles. Thus, we cannot calculate the final motion completely. However, the conservation laws provide some useful information by placing severe restrictions on the possible final motions.

We will again assume that one of the particles is initially at rest and the other is initially in motion. Figure 10.6a shows the particles before the collision and Figure 10.6b shows them after. Particle 1 ("projectile") initially moves along a straight line[2]; but when it comes close to

[2] We ignore external forces.

particle 2 ("target"), it feels a force and begins to deflect. Note that the collision shown in Figure 10.6 is not head-on — the initial line of motion of particle 1 passes to one side of particle 2. The particles never quite come into contact, but we will assume that the force reaches from one to the other, bridging the empty space between. The region over which the force acts (shown in red in Figure 10.6) is called the **interaction region.** What processes occur within this region need not concern us — we only need to know that these processes conserve energy and momentum. All the changes of motion occur within the interaction region; outside of this region the particles move with uniform velocity along straight paths. For example, if the projectile is an electron and the target an atom, then the interaction region is about equal to the volume of the atom — nothing happens to the electron until it penetrates the atom and begins to feel the presence of the atomic electrons and the atomic nucleus.

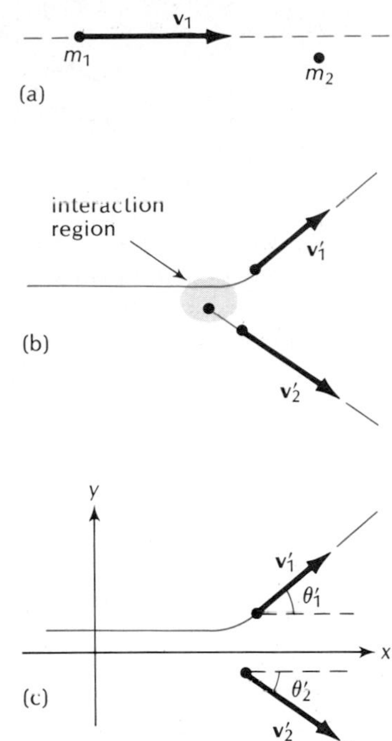

Fig. 10.6 (a) Before the collision, particle 2 is at rest and particle 1 has velocity $\mathbf{v}_1$. (b) After the collision, particle 1 has velocity $\mathbf{v}'_1$ and particle 2 has velocity $\mathbf{v}'_2$. (c) The velocities $\mathbf{v}'_1$ and $\mathbf{v}'_2$ make angles θ'_1 and θ'_2 with the x axis, respectively.

The initial line of motion of particle 1 and the initial point of particle 2 define a plane (the plane of the page in Figure 10.6). If the motion is to be two dimensional, the particles must always remain in this plane. Obviously this will be the case if the force acting between them is in the plane. Among the forces that play an important role in the physical world, both the electric force and the gravitational force between particles (see Chapters 13 and 22) lie along the line joining the particles; thus, these forces are in the plane of the motion and collisions involving these forces may be treated as two dimensional.

We will designate the velocity of particle 1 before the collision by $\mathbf{v}_1$, and the velocities of particles 1 and 2 after the collision by $\mathbf{v}'_1$ and $\mathbf{v}'_2$. The velocity before the collision is along the x axis and the velocities after the collision make angles θ'_1 and θ'_2 with the x axis (Figure 10.6c).

Conservation of the x component of momentum and of the y component of momentum implies, respectively,

Conservation of momentum and energy in a two-dimensional elastic collision

$$m_1 v_1 = m_1 v'_1 \cos\theta'_1 + m_2 v'_2 \cos\theta'_2 \tag{19}$$

$$0 = m_1 v'_1 \sin\theta'_1 - m_2 v'_2 \sin\theta'_2 \tag{20}$$

Conservation of energy gives

$$\tfrac{1}{2} m_1 v_1^2 = \tfrac{1}{2} m_1 v_1'^2 + \tfrac{1}{2} m_2 v_2'^2 \tag{21}$$

Note that, according to the present notation, the quantities v_1, v'_1, and v'_2 are speeds or magnitudes of velocities; that is, they are all positive (this is in contrast to the notation of Section 10.2, where these quantities were positive or negative, depending on direction).

Equations (19), (20), and (21) constitute three restrictions on the four quantities v'_1, v'_2, θ'_1, and θ'_2; if, besides the initial speed v_1, one of these four quantities is known, then the other three can be calculated.

EXAMPLE 7. A neutron with an initial speed of 3.0×10^7 m/s collides elastically with a helium nucleus. The neutron is deflected through an angle of 60° (Figure 10.7). What is its final speed?

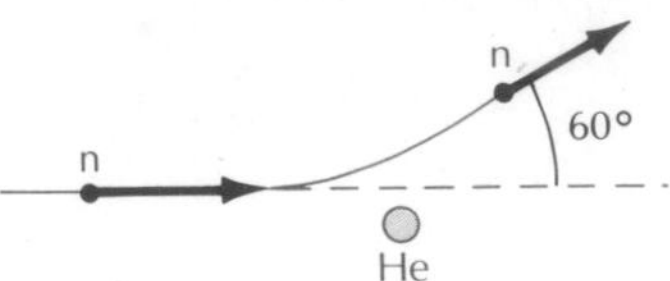

Fig. 10.7 Collision of a neutron with a helium nucleus. The neutron is deflected through an angle of 60°.

SOLUTION: We must eliminate v_2' and θ_2' between Eqs. (19), (20), and (21). We begin with Eqs. (19) and (20):

$$m_2 v_2' \cos \theta_2' = m_1 v_1 - m_1 v_1' \cos \theta_1' \tag{22}$$

$$m_2 v_2' \sin \theta_2' = m_1 v_1' \sin \theta_1' \tag{23}$$

We square these equations and then add them together side to side, making use of the identity $\sin^2 \theta_1' + \cos^2 \theta_1' = 1$:

$$m_2^2 v_2'^2 = m_1^2 v_1^2 - 2m_1^2 v_1 v_1' \cos \theta_1' + m_1^2 v_1'^2$$

This yields an expression for $v_2'^2$ which can be substituted into Eq. (21):

$$\tfrac{1}{2} m_1 v_1^2 = \tfrac{1}{2} m_1 v_1'^2 + \frac{1}{2}\left(\frac{m_1^2}{m_2} v_1^2 - 2\frac{m_1^2}{m_2} v_1 v_1' \cos \theta_1' + \frac{m_1^2}{m_2} v_1'^2\right)$$

Upon rearrangement this becomes

$$\left(\frac{m_1}{m_2} + 1\right) v_1'^2 - 2\frac{m_1}{m_2} v_1 v_1' \cos \theta_1' + \left(\frac{m_1}{m_2} - 1\right) v_1^2 = 0 \tag{24}$$

which is a quadratic equation for the unknown v_1'.

Before proceeding, it is convenient to substitute numerical values into Eq. (24), since otherwise the solution becomes very messy. The neutron mass is $m_1 = 1.0$ u and the helium mass is $m_2 = 4.0$ u. Hence

$$(\tfrac{1}{4} + 1) v_1'^2 - \tfrac{1}{2} v_1 v_1' \cos 60° + (\tfrac{1}{4} - 1) v_1^2 = 0$$

or

$$1.25 v_1'^2 - 0.25 v_1 v_1' - 0.75 v_1^2 = 0$$

Using the standard formula for the solution of a quadratic equation with v_1' as the unknown, we find

$$v_1' = \frac{0.25 v_1 \pm \sqrt{(0.25)^2 v_1^2 + 4 \times 1.25 \times 0.75 v_1^2}}{2 \times 1.25}$$

Obviously, only the positive sign makes sense. This yields

$$v_1' = 0.88 v_1 = 0.88 \times 3.0 \times 10^7 \text{ m/s}$$

$$= 2.6 \times 10^7 \text{ m/s}$$

If the collision is *inelastic,* then the conservation law of Eq. (21) for the energy is not applicable. The conservation laws of Eqs. (19) and (20) for the x and y components of the momentum remain applicable, and they place restrictions on the possible final motions. Since the number of available equations in the inelastic case is smaller than in the elastic case, we are now even less able to calculate the final motion completely. However, there is an exception: if the collision is *totally inelastic,* then the laws of conservation of momentum by themselves are sufficient to calculate the velocities after the collision. As we saw in Section 10.2, in a totally inelastic collision, the two particles will stick together and, therefore, their final velocity coincides with the velocity of the center of mass. There are then only two unknowns: the two components of the velocity of the center of mass.

Although in the above discussion we emphasized the calculation of the final motion from the known initial motion, in practical applications of momentum conservation we often need to calculate the initial motion (or some aspect of the initial motion) from the known final motion. As the following example shows, this involves the same set of equations; it is merely a matter of keeping in mind which quantities are known, and which unknown.

EXAMPLE 8. A white automobile of mass 1100 kg and a black automobile of mass 1300 kg collide at an intersection. The investigation of this collision discloses that just before the collision the white automobile was traveling due east, and the black automobile was traveling due north (Figure 10.8). After the collision, the wrecked automobiles remained joined together and their tires made skid marks 18.7 m long in a direction 30° north of east before coming to rest. What was the speed of each automobile before the collision? Was one of them exceeding the legal speed limit of 55 mi/h (25 m/s)? Assume that the wheels of both automobiles remained locked after the collision and that the coefficient of kinetic friction between the locked wheels and the pavement is $\mu_k = 0.80$.

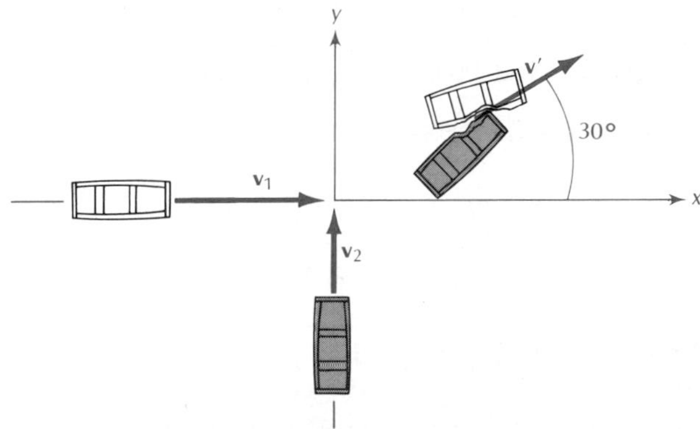

Fig. 10.8 An automobile collision. Before the collision the velocities of the automobiles were $\mathbf{v}_1$ and $\mathbf{v}_2$; after the collision both velocities were $\mathbf{v}'$.

SOLUTION: We designate the velocities of the white and the black automobile before the collision by $\mathbf{v}_1$ and $\mathbf{v}_2$, respectively. Their velocity after the collision is $\mathbf{v}'$, the same for both. The velocity $\mathbf{v}_1$ is along the x axis; $\mathbf{v}_2$ is along the y axis; and $\mathbf{v}'$ is at an angle $\theta' = 30°$ with the x axis (Figure 10.8). Conservation of the x component and of the y component of momentum gives

$$m_1 v_1 = (m_1 + m_2)v' \cos\theta' \tag{25}$$

$$m_2 v_2 = (m_1 + m_2)v' \sin\theta' \tag{26}$$

In order to calculate the speeds v_1 and v_2 from these equations, we need to know v', the speed immediately after the collision. This speed is determined by the length of the skid marks. The deceleration of the skidding automobiles is

$$a = \frac{f_k}{m_1 + m_2} = \frac{\mu_k(m_1 + m_2)g}{m_1 + m_2} = \mu_k g$$

According to Eq. (2.25), the speed v' is related to the length l of the skid marks:

$$al = \tfrac{1}{2}v'^2$$

Hence

$$v' = \sqrt{2al} = \sqrt{2\mu_k gl} = \sqrt{2 \times 0.80 \times (9.8 \text{ m/s}^2) \times 18.7 \text{ m}}$$

$$= 17 \text{ m/s}$$

Equations (25) and (26) then yield

$$v_1 = \frac{m_1 + m_2}{m_1} v' \cos\theta' = \frac{1100 \text{ kg} + 1300 \text{ kg}}{1100 \text{ kg}} \times 17 \text{m/s} \times \cos 30°$$

$$= 32 \text{ m/s}$$

and

$$v_2 = \frac{m_1 + m_2}{m_2} v' \sin\theta' = \frac{1100 \text{ kg} + 1300 \text{ kg}}{1300 \text{ kg}} \times 17 \text{m/s} \times \sin 30°$$

$$= 16 \text{ m/s}$$

The white automobile *was* speeding.

10.4 Collisions and Reactions of Nuclei and of Elementary Particles[3]

To explore the structure of atoms, nuclei, and elementary particles, physicists use beams of high-energy projectiles as probes. These beams, usually consisting of a stream of electrons or protons produced by a particle accelerator, are made to impact on a target containing the atoms, nuclei, etc., to be explored. The projectiles penetrate deep into the subatomic structures and there they engage in collisions. The manner in which the projectiles bounce off or react gives physicists some clues about the structures responsible for the collisions (see Interlude C).

In elastic collisions between nuclei or elementary particles, only the direction of motion changes; the particles themselves do not change. The process of deflection of such particles in a collision is called **scattering.** In inelastic collisions between nuclei or elementary particles, kinetic energy is lost (or gained) by the alteration of the internal structure of the particles and by the conversion of energy into rest mass (or vice versa); this is a process of destruction of old particles and creation of new particles.

Scattering

In some high-energy collisions the number of particles created may be extremely large. Figure 10.9 shows a spectacular inelastic collision between a very energetic cosmic-ray particle and a nucleus. The collision took place within the emulsion of a photographic film; in this medium the particles left visible tracks because their passage damaged (exposed) the film. The tracks show that a great many new particles were created in the collision and that they sped away from the scene of the accident; obviously a large amount of kinetic energy must have been converted into rest mass.

[3] This section is optional.

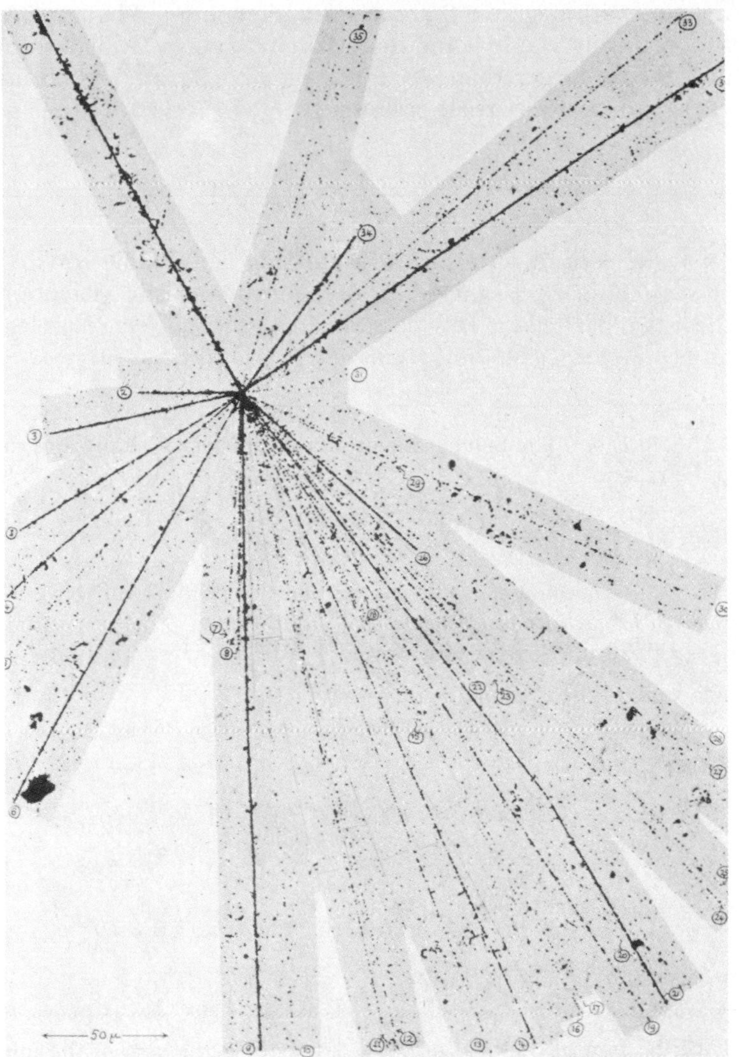

Fig. 10.9 Tracks of particles in a photographic emulsion (enlarged about 330×). The heavy track on the upper left was made by a cosmic ray. This cosmic ray suffered a collision with the nucleus of an atom. The tracks of many new particles emerge from the scene of the accident.

In order to write down the energy balance for an inelastic collision involving the destruction and creation of particles, we find it useful to introduce a quantity Q which is the net change in the rest-mass energy. If the particles present before the reaction have masses $m_1, m_2, \ldots, m_n$ and those present after the collision have masses $m_1', m_2', \ldots, m_r'$, then

Rest-mass energy change in an inelastic collision

$$\boxed{Q = (m_1 + m_2 + \cdots + m_n)c^2 - (m_1' + m_2' + \cdots + m_r')c^2} \qquad (27)$$

Thus, Q is positive if the net rest-mass energy decreases in the collision and Q is negative if it increases. The conservation of total energy (kinetic and rest-mass energy) then reads

$$\boxed{K + Q = K'} \qquad (28)$$

where K and K' represent the kinetic energies before and after the collision.

Obviously, a reaction with a positive Q releases kinetic energy and a reaction with negative Q absorbs kinetic energy. The latter kind of reaction is only viable if the initial kinetic energy available for inelastic processes is larger than $-Q$. According to Eq. (17), the initial kinetic energy for a two-particle collision is

$$K = \tfrac{1}{2}Mv_{CM}^2 + \frac{1}{2}\frac{m_1 m_2}{m_1 + m_2}(v_1 - v_2)^2 \qquad (29)$$

Condition for inelastic reaction

Only the second term on the right side is available for inelastic processes (compare Example 5). Hence an inelastic reaction involving these two particles is only possible if this second term is at least as large as $-Q$, i.e., *if the internal kinetic energy is at least as large as* $-Q$.

EXAMPLE 9. The bombardment of a ^{7}Li nucleus with protons can give rise to the reaction

$$p + {}^7\text{Li} \rightarrow {}^7\text{Be} + n$$

What is the minimum kinetic energy that the proton must have if it is to induce this reaction on a stationary ^{7}Li nucleus? The relevant masses are $m_p = 1.0073$ u, $m_{Li} = 7.0144$ u, $m_{Be} = 7.0147$ u, and $m_n = 1.0087$ u.

SOLUTION: The Q value for the reaction is

$$\begin{aligned} Q &= (m_p + m_{Li})c^2 - (m_{Be} + m_n)c^2 \\ &= (1.0073 + 7.0144)\ \text{u} \times c^2 - (7.0147 + 1.0087)\ \text{u} \times c^2 \\ &= -\ 0.0017\ \text{u} \times c^2 \\ &= -\ 0.0017\ \text{u} \times 1.66 \times 10^{-27}\ \text{kg/u} \times (3.0 \times 10^8\ \text{m/s})^2 \\ &= -2.6 \times 10^{-13}\ \text{J} \end{aligned}$$

Since the sum of the final masses is larger than the sum of the initial masses, the reaction can only proceed by the conversion of some kinetic energy into rest mass: 2.6×10^{-13} J of kinetic energy must be converted into rest mass. With $v_{Li} = 0$, the condition on the second term on the right side of Eq. (29) is

$$\frac{1}{2}\frac{m_p m_{Li}}{m_p + m_{Li}} v_p^2 \geq -Q$$

or

$$\begin{aligned} \tfrac{1}{2}m_p v_p^2 &\geq \frac{m_p + m_{Li}}{m_{Li}} \times (-Q) \qquad (30) \\ &\geq \frac{1.0073 + 7.0144}{7.0144} \times 2.6 \times 10^{-13}\ \text{J} \\ &\geq 3.0 \times 10^{-13}\ \text{J} \end{aligned}$$

Note that the total initial kinetic energy is simply the kinetic energy of the proton. Equation (30) therefore shows that it is *not sufficient* for the total initial kinetic energy to match $-Q$. The total kinetic energy must be larger than $-Q$ by a factor of $(m_p + m_{Li})/m_{Li}$ because part of this kinetic energy is tied up in the motion of the center of mass and is therefore not available for the inelastic process.

The minimum initial kinetic energy that makes a reaction viable is called the **threshold energy.** For instance, in Example 9 the threshold energy is 3.0×10^{-13} J. If the reaction happens at exactly the threshold energy, then the reaction products have minimum energy, i.e., they have zero relative velocity and remain together at the center of mass.

SUMMARY

Impulse: $\mathbf{I} = \int_0^{\Delta t} \mathbf{F}\, dt$

$$= \mathbf{p}' - \mathbf{p}$$

Elastic collision: Kinetic energy is conserved.

Totally inelastic collision: Maximum amount of kinetic energy is lost; particles remain joined together.

Speeds in one-dimensional elastic collision:

before: $v_1 \neq 0, \quad v_2 = 0$

after: $v_1' = \dfrac{m_1 - m_2}{m_1 + m_2} v_1, \quad v_2' = \dfrac{2m_1}{m_1 + m_2} v_1$

Change of rest-mass energy in inelastic collision:

$$Q = (m_1 + m_2 + \cdots + m_n)c^2 - (m_1' + m_2' + \cdots + m_r')c^2$$

Condition for reaction with negative Q**:** $K_{\text{int}} > |Q|$

QUESTIONS

1. In experimental simulations of automobile collisions, baboons were used as passengers. When wearing only lap belts, 50% of the animals suffered fatal injuries in collisions at 32 gee. When wearing lap belts and shoulder harnesses, 50% of the animals suffered fatal injuries in collisions at 100 gee. When protected by air bags, the animals suffered only minor injuries in collisions of up to 120 gee. (Accelerations of 32 and 100 gee are typical of automobile impacts into a solid wall at 35 and 60 mi/h, respectively.) Why are air bags better than seat belts (in head-on collisions)? What are some of the advantages and disadvantages of airbags and seat belts?

2. A (foolish) stuntman wants to jump out of an airplane at high altitude without a parachute. He plans to jump while tightly encased in a strong safe which can withstand the impact on the ground. How would you convince the stuntman to abandon this project?

3. Each year 52,000 people die in automobile accidents in the United States; even more suffer severe injuries sometimes resulting in amputation, blindness, or paralysis. Why have automobile manufacturers failed to design an automobile that would permit its occupants to walk away from a 60-mi/h crash? Is this technologically impossible?

4. Is the collision shown in Figure 10.1 totally inelastic? For the sake of safety, would it be desirable to design automobiles so that their collisions are elastic or inelastic?

5. Statistics show that, on the average, the occupants of a heavy ("full-size") automobile are more likely to survive a crash than those of a light ("compact") automobile. Why would you expect this to be true?

6. In Joseph Conrad's tale *Gaspar Ruiz,* the hero ties a cannon to his back and, hugging the ground on all fours, fires several shots at the gate of a fort. How does the momentum absorbed by Ruiz compare with that absorbed by the gate? How does the energy absorbed by Ruiz compare with that absorbed by the gate?

Fig. 10.10 A five-pendulum toy.

7. Give an example of a collision between two bodies in which *all* of the kinetic energy is lost to inelastic processes.

8. Explain the operation of the five-pendulum toy shown in Figure 10.10.

9. In order to split a log with a small ax, you need a greater impact speed than you would need with a large ax. Why? If the energy required to split the log is the same in both cases, why is it more tiring to use a small ax? (Hint: Think about the kinetic energy of your arms.)

10. If you throw an (elastic) baseball at an approaching train, the ball will bounce back at you with an increased speed. Explain.

11. You are investigating the collision of two automobiles at an intersection. The automobiles remained joined together after this collision and their wheels made measurable skid marks on the pavement before they came to rest. Assume that during skidding all the wheels remained locked so that the deceleration was entirely due to kinetic friction. You know the direction of motion of the automobiles before the collision (drivers are likely to be honest about this), but you do not know the speeds (drivers are likely to be dishonest about this). What do you have to measure at the scene of the accident to calculate the speeds of both automobiles before the collision?

PROBLEMS

Section 10.1

1. A stunt man of mass 77 kg "belly-flops" on a shallow pool of water from a height of 11 m. When he hits the pool, he comes to rest in about 0.05 s. What is the average braking force that the water and the bottom of the pool exert on his body during this time interval?

2. A large ship of 700,000 metric tons steaming at 20 km/h runs aground on a reef, which brings it to a halt in 5.0 s. What is the average force on the ship? What is the average deceleration?

3. The photographs of Figure 10.1 show the impact of an automobile on a rigid wall.
   (a) Measure the positions of the automobile on these photographs and calculate the average velocity for each of the 20×10^{-3} –s intervals between one photograph and the next; calculate the average acceleration for each time interval from the change between one average velocity and the next.
   (b) The mass of this automobile is 1700 kg. Calculate the average force for each time interval.
   (c) Make a plot of this force as a function of time and find the impulse by estimating the area under this curve.

4. The "land divers" of Pentecost Island (New Hebrides) jump from platforms 70 ft high. Long liana vines tied to their ankles jerk them to a halt just short of the ground. If the pull of the liana takes 0.02 s to halt the diver, what is the average acceleration of the diver during this time interval? If the weight of the diver is 140 lb, what is the corresponding average force on his ankles?

Section 10.2

5. Meteor Crater in Arizona (Figure 10.11), a hole 180 m deep and 1300 m across, was gouged in the surface of the Earth by the impact of a large meteorite. The mass and speed of this meteorite have been estimated at 2×10^9 kg and 10 km/s, respectively, before impact.
   (a) What recoil velocity did the Earth acquire during this (inelastic) collision?
   (b) How much kinetic energy was released for inelastic processes during the collision? Express this energy in the equivalent of tons of TNT; 1 ton of TNT releases 4.2×10^9 J upon explosion.

Fig. 10.11 Meteor crater in Arizona.

6. It has been reported (fallaciously) that the deer botfly can attain a maximum airspeed of 818 mi/h, i.e., 366 m/s. Suppose that such a fly, buzzing along at this speed, strikes a stationary hummingbird and remains stuck in it. What will be the recoil velocity of the hummingbird? The mass of the fly is 2 g; the mass of the hummingbird is 50 g.

7. A projectile of 45 kg has a muzzle velocity of 656.6 m/s when fired horizontally from a gun held in a rigid support (no recoil). What will be the muzzle velocity (relative to the ground) of the same projectile when fired from a gun that is free to recoil? The mass of the gun is 6.6×10^3 kg. (Hint: The kinetic energy of the gun–projectile system is the same in both cases.)

8. An automobile approaching an intersection at 10 km/h bumps into the rear of another automobile standing at the intersection with its brakes off and its gears in neutral. The mass of the moving automobile is 1200 kg and that of the stationary automobile is 700 kg. If the collision is elastic, find the velocities of both automobiles after the collision.

9. Two automobiles weighing 1200 and 3000 lb collide head-on while moving at 50 mi/h in opposite directions. After the collision the automobiles remain locked together.
   (a) Find the velocity of the wreck immediately after the collision.
   (b) Find the kinetic energy of the two-automobile system before and after the collision.
   (c) The front end of each automobile crumples by 2.0 ft during the collision. Find the acceleration (relative to the ground) of the passenger compartments of each automobile; make the assumption that these accelerations are constant during the collision.

10. A speeding automobile impacts on the rear of a parked automobile. After the impact the two automobiles remain locked together and they skid along the pavement with all their wheels locked. An investigation of this accident establishes that the length of the skid marks made by the automobiles after the impact was 18 m; the mass of the moving automobile was 2200 kg and that of the parked automobile was 1400 kg, and the coefficient of friction between the wheels and the pavement was 0.95.
   (a) What was the speed of the two automobiles immediately after impact?
   (b) What was the speed of the moving automobile before impact?

11. A ship of 3.0×10^4 metric tons steaming at 40 km/h strikes an iceberg of 8.0×10^5 metric tons. If the collision is totally inelastic, what fraction of the initial kinetic energy of the ship is converted into inelastic energy? What fraction remains as kinetic energy of the ship–iceberg system?

12. The impact of the head of a golf club on a golf ball can be approximately regarded as an elastic collision. The mass of the head of the golf club is 0.15 kg and that of the ball is 0.045 kg. If the ball is to acquire a speed of 60 m/s in the collision, what must be the speed of the club before impact?

13. In karate, the fighter makes his hand collide at high speed with the target; this collision is inelastic and a large portion of the kinetic energy of the hand becomes available to do damage in the target. According to a crude estimate, the energy required to break a concrete block (28 cm $\times$ 15 cm $\times$ 1.9 cm sup-

ported only at its short edges) is of the order of 10 J. Suppose the fighter delivers a downward hammer-fist strike with a speed of 12 m/s to such a concrete block. In principle, is there enough energy to break the block? Assume that the fist has a mass of 0.4 kg.

14. The impact of a hammer on a nail can be regarded as an elastic collision between the head of the hammer and the nail. Suppose that the mass of the head of the hammer is 0.50 kg and it strikes a nail of mass 12 g with an impact speed of 5 m/s. How much energy does the nail acquire in this collision?

15. A proton of energy 1.6×10^{-13} J is moving toward a proton at rest. What is the velocity of the center of mass of the system?

16. Figure 13.8a shows the positions of the three inner planets (Mercury, Venus, Earth) on January 1, 1980. Measuring angles off this figure and using the data on masses, orbital radii, and periods given in Table 13.1, find the velocity of the center of mass of this system of three planets.

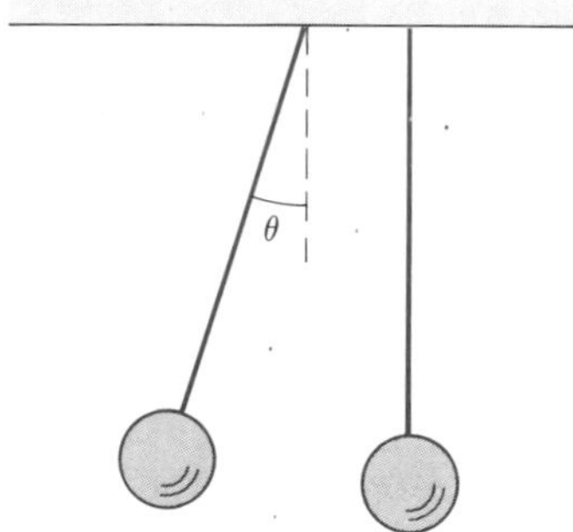

Fig. 10.12

17. Suppose that a neutron in a nuclear reactor initially has an energy of 4.8×10^{-13} J. How many head-on collisions with carbon nuclei must this neutron make before its energy is reduced to 1.6×10^{-19} J?

18. A proton of energy 8.0×10^{-13} J collides head-on with a proton at rest. How much energy is available for inelastic reactions between these protons?

19. (a) Two identical steel balls are suspended from strings of length l so that they touch when in their equilibrium position (Figure 10.12). If we pull one of the balls back until its string makes an angle θ with the vertical and then let go, it will collide elastically with the other ball. How high will the other ball rise?
(b) Suppose that instead of steel balls we use putty balls. They will then collide inelastically and remain stuck together. How high will the balls rise?

20. Two small balls are suspended side by side from two strings of length l so that they touch when in their equilibrium position. Their masses are m and $2m$, respectively. If the left ball (of mass m) is pulled aside and released from a height h, it will swing down and collide with the right ball (of mass $2m$) at the lowest point. Assume the collision is elastic.
(a) How high will each ball swing after the collision?
(b) Both balls again swing down and they collide once more at the lowest point. How high will each swing after this second collision?

Fig. 10.13

21. On a smooth, frictionless table, a billiard ball of velocity v is moving toward two other aligned billiard balls in contact (Figure 10.13). What will be the velocity of each ball after impact? Assume that all balls have the same mass and that the collisions are elastic. (Hint: Treat this as two successive collisions.)

22. Repeat Problem 21 but assume that the middle ball has twice the mass of each of the others.

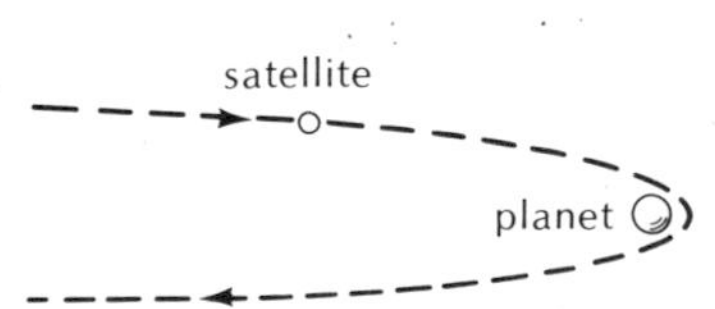

Fig. 10.14 Satellite "colliding" with a planet.

23. If an artificial satellite, or some other body, approaches a planet at fairly high speed at a suitable angle, it will whip around the planet and recede in a direction almost opposite to the initial direction of motion (Figure 10.14). This can be regarded approximately as a one-dimensional "collision" between the satellite and the planet; the collision is elastic. In such a collision the satellite will gain kinetic energy from the planet, provided that it approaches the planet along a direction opposite to the direction of the planet's motion. This slingshot effect has been used to boost the speed of both Voyager spacecraft as they passed near Jupiter. Consider the head-on "collision" of a satellite of initial speed 10 km/s with the planet Jupiter, which has a speed of 13 km/s. What is the maximum gain of speed that the satellite can achieve?

*24. A turbine wheel with curved blades is driven by a high-velocity stream of water that impinges on the blades and bounces off (Figure 10.15). Under ideal conditions the velocity of the water particles after collision with the blade is exactly zero so that all of the kinetic energy of the water is transferred to the turbine wheel. If the speed of the water particles is 90 ft/s, what is the ideal speed of the turbine blade? (Hint: Treat the collision of a water particle and the blade as a one-dimensional elastic collision.)

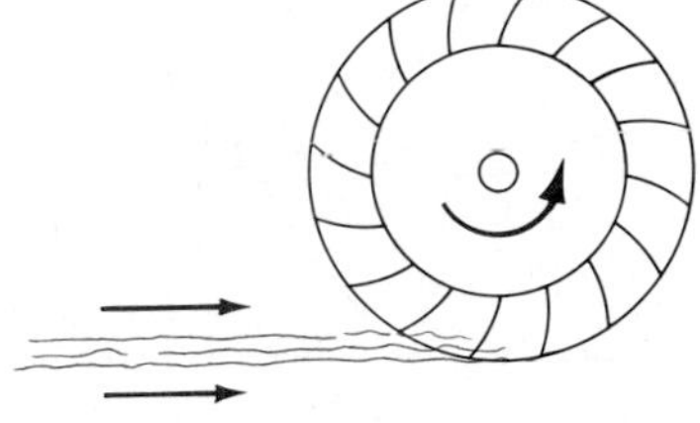

Fig. 10.15 An undershot turbine wheel.

*25. In Section 10.2 we showed that for an elastic one-dimensional collision the relative velocity reverses during the collision. If v_2 is zero, we can express this condition as $v_1' - v_2' = -v_1$ [see Eq. (13)]; if v_2 is not zero, we can express this condition as $v_1' - v_2' = -(v_1 - v_2)$. For a partially inelastic collision the relative velocity after the collision will have a smaller magnitude than the relative velocity before the collision. We can express this mathematically as $v_1' - v_2' = -e(v_1 - v_2)$, where $e < 1$ is called the **coefficient of restitution.** For some kinds of bodies, the coefficient e is a constant, independent of v_1 and v_2.

(a) Show that in this case the final internal energy is less than the initial internal kinetic energy by a factor of e^2, that is, $K_{\text{int}} = e^2 K_{\text{int}}$.

(b) Derive formulas analogous to Eqs. (11) and (12) for the velocities v_1' and v_2' in terms of v_1 and v_2.

Section 10.3

26. Two hockey players of mass 80 kg collide while skating at 7.0 m/s. The angle between their initial directions of motion is 130°.

(a) Suppose that the collision is totally inelastic. What is their velocity immediately after collision?

(b) Suppose that the collision lasts 0.030 s. What is the magnitude of the average acceleration of each player during the collision?

27. Two automobiles of equal mass collide at an intersection. After the collision, they remain joined together and skid, with locked wheels, before coming to rest. The length of the skid marks is 18 m and the coefficient of friction between the locked wheels and the pavement is 0.80. Each driver claims his speed was less than 14 m/s (30 mi/h) before the collision. Prove that at least one driver is lying.

28. Your automobile of mass $m_1 = 900$ kg collides at a traffic circle with another automobile of mass $m_2 = 1200$ kg. Just before the collision your automobile was moving due east and the other automobile was moving 40° south of east. After the collision the two automobiles remain entangled while they skid, with locked wheels, until coming to rest. Your speed before the collision was 14 m/s. The length of the skid marks is 17.4 m and the coefficient of kinetic friction between the tires and the pavement is 0.85. Calculate the speed of the other automobile before the collision.

29. On July 27, 1956, the ships *Andrea Doria* (40,000 metric tons) and *Stockholm* (20,000 metric tons) collided in the fog south of Nantucket Island and remained locked together (for a while). Immediately before the collision the velocity of the *Andrea Doria* was 22 knots at 15° east of south and that of the *Stockholm* was 19 knots at 48° east of south (1 knot = 1 nmi/h = 1.15 mi/h).

(a) Calculate the velocity (magnitude and direction) of the combined wreck immediately after the collision.

(b) Find the amount of kinetic energy that was converted into other forms of energy by inelastic processes during the collision.

(c) The large amount of energy absorbed by inelastic processes accounts for the heavy damage to both ships. How many pounds of TNT would have to be exploded to obtain the same amount of energy as was absorbed by inelastic processes in the collision? The explosion of 1 lb of TNT releases 2.1×10^6 J.

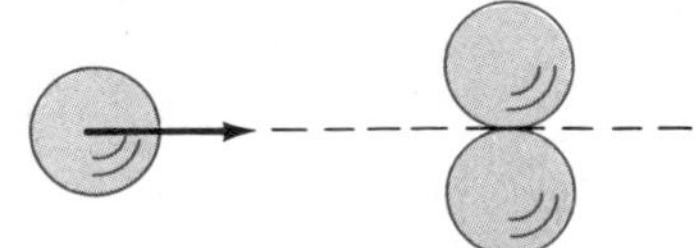

Fig. 10.16

30. Two billiard balls are placed in contact on a smooth, frictionless table. A third ball moves toward this pair with velocity v in the direction shown in Figure 10.16. What will be the velocity (magnitude and direction) of the three balls after the collision? The balls are identical and the collisions are elastic.

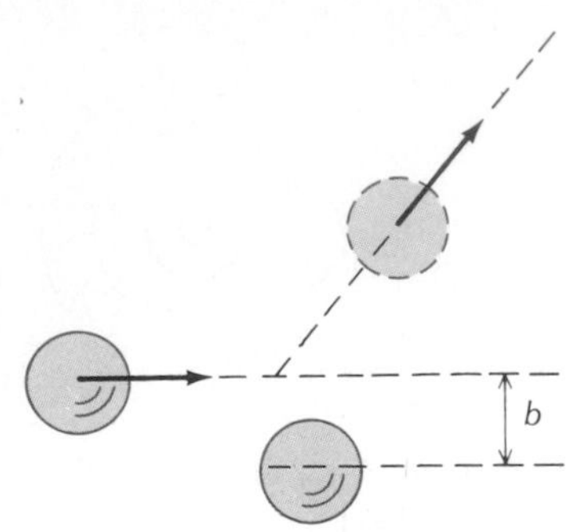

Fig. 10.17

31. A billiard ball of mass m and radius R moving with speed v on a smooth frictionless table collides elastically with an identical stationary billiard ball glued firmly to the surface of the table.

(a) Find a formula for the angular deflection suffered by the moving billiard ball as a function of the impact parameter b (defined in Figure 10.17). Assume the billiard balls are very smooth so that the force during contact is entirely along the center-to-center line of the balls.

(b) Find a formula for the magnitude of the momentum change suffered by the billiard ball.

32. For the collision of the neutron and the helium nucleus described in Example 7, calculate the velocity of recoil of the helium nucleus (magnitude and direction).

33. A nuclear reactor designed and built in Canada (CANDU) contains heavy water (D_2O). In this reactor, the fast neutrons are slowed down by collisions with the deuterium nuclei of the heavy-water molecule.

(a) By what factor will the speed of a neutron be reduced in a head-on collision with a deuterium nucleus? The mass of this nucleus is 2.01 u.

(b) After how many head-on collisions with deuterium nuclei will the speed be reduced by the same factor as in a single head-on collision with a proton?

Fig. 10.18

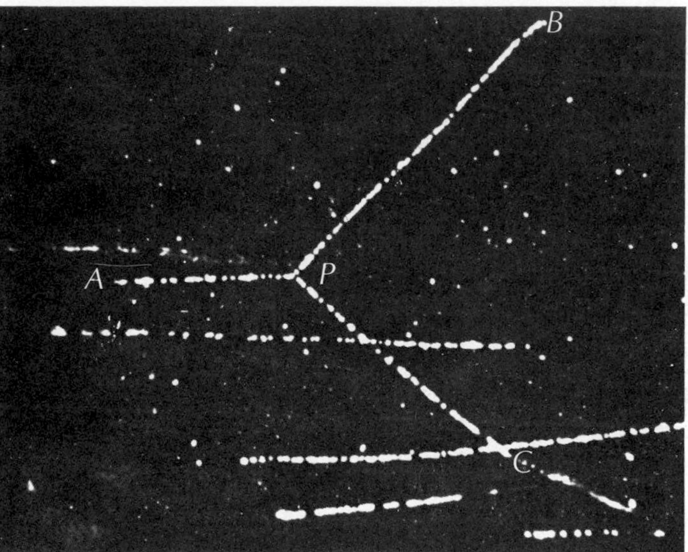

34. The photograph of Figure 10.18 shows an elastic collision between two protons in a bubble chamber. Initially, one of the protons (track AP) has an energy of 8.0×10^{-13} J and the other is at rest (at P). After the collision both protons are in motion (tracks PB and PC). Measure the angles between the initial direction of motion and the final directions of motion and then calculate the final energy of each proton.

*35. Repeat the calculations of Problem 31 if the second billiard ball is initially stationary but not glued to the table. (Hint: First solve this problem in the center-of-mass frame; there, the motion of the balls is symmetric. Then transform the angle of the deflected motion into the reference frame of the table.)

*36. A proton with kinetic energy 5.0×10^6 eV collides elastically with a lithium nucleus and is deflected through an angle of 30°. What energy does the lithium nucleus acquire in this collision and what will be its direction of motion? The mass of the proton is 1.0 u and the mass of the lithium nucleus is 7.0 u.

*37. A moving proton collides elastically with another proton that is initially at rest. Show that the angle between the tracks of the two protons after the collision is 90°; assume that the collision is not head-on.

Section 10.4

38. What is the Q value for the reaction

$$\nu + \mathrm{n} \rightarrow \mathrm{p} + \mathrm{e}?$$

The neutrino (ν) has zero mass.

39. If struck by a neutron of sufficient energy, the nucleus of helium can split into two nuclei of isotopes of hydrogen according to the reaction

$$\mathrm{n} + {}^{4}\mathrm{He} \rightarrow {}^{3}\mathrm{H} + {}^{2}\mathrm{H}$$

The Q value for this reaction is -2.80×10^{-12} J. What minimum kinetic energy must the neutron have if it is to initiate this reaction by impact on a (initially) stationary helium nucleus?

40. Consider the creation of a pion in a collision between two protons. The reaction is described by the equation

$$\mathrm{p} + \mathrm{p} \rightarrow \mathrm{p} + \mathrm{p} + \pi^{0}$$

If one of the protons is initially stationary, what minimum kinetic energy must the other proton have to make this reaction possible? The mass of the pion is 2.4×10^{-28} kg. Pretend that the Newtonian formulas for kinetic energy remain valid (even though the speed of the proton is near the speed of light).

RADIATION AND LIFE[1]

Ordinary light is the most familiar of all forms of radiation. It consists of waves of electric and magnetic energy streaming out from a source — the flame of a candle, the hot filament of a light bulb, the Sun, or whatever. A pulse of light, or light ray, propagates outward from its source in a straight line. Light rays can propagate through thousands of kilometers of air, and hundreds of meters of water, or glass, or other transparent materials. Yet light rays are easily stopped by a thin layer of opaque material such as a sheet of paper, aluminum foil, or human skin. Light has only a small penetrating power.

Several forms of invisible radiation — X rays, alpha rays, beta rays, gamma rays, cosmic rays — have a very large penetrating power. These rays can easily pass through thick layers of material opaque to light. For example, X rays pass through human bodies, gamma rays pass through massive walls of concrete, and cosmic rays occasionally pass through entire mountains. Some of these penetrating radiations are closely related to ordinary light — they consist of extremely energetic waves of electric and magnetic energy; others are corpuscular — they consist of a stream of fast-moving particles. When these radiations hit a human body, their impact is painless yet their effect can be lethal.

B.1 PENETRATING RADIATIONS

X rays[2] are generated in collisions between electrons and atoms. In an X-ray tube (Figure B.1), a stream of energetic electrons, moving at high speed in a vacuum, strikes a block or target of metal; the X rays are then emitted by the incident electrons when these are suddenly decelerated and stopped by collisions with the target atoms, and they are also emitted by the atomic electrons within the target atoms when these are disturbed (excited) by the impact of the incident electrons. X rays are waves of electric and magnetic energy. These waves are qualitatively similar to light waves, but the wavelength of X-ray waves is typically 10,000 times shorter than the wavelength of light waves.

Alpha rays (α), **beta rays** (β), and **gamma rays** (γ) are nuclear radiations — these rays are emitted by spontaneous processes that take place inside the nuclei of atoms. For example, the nuclei of atoms of radiocarbon and of radiostrontium emit beta rays, nuclei of radiocobalt emit beta rays and gamma rays, nuclei of radium and of uranium emit alpha rays and gamma rays, etc. (Figures B.2 and B.3). Atoms that emit alpha, beta, or gamma rays are said to be **radioactive.** Among these rays, the most penetrating are the gamma rays — some of them will penetrate one meter of concrete.

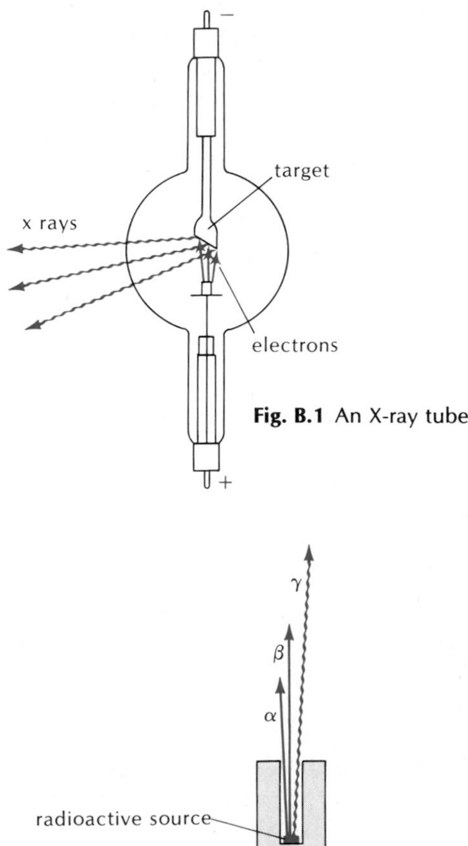

Fig. B.1 An X-ray tube.

Fig. B.2 Some radioactive materials, such as radium, emit alpha, beta, and gamma rays.

[1] This chapter is optional.

[2] Discovered by the German physicist Wilhelm Konrad Röntgen, 1845–1943. For this discovery, Röntgen was awarded the first Nobel Prize in 1901.

Fig. B.3 Tracks of alpha rays in a cloud chamber containing air with an excess of water vapor. The alpha rays trigger the formation of fine trails of water droplets, which make them visible.

The least penetrating are the alpha rays — most of them will be stopped by a sheet of heavy paper.

The investigation of gamma rays has established that they are essentially high-energy X rays (megavolt X rays). The only difference between the former and the latter is their place of origin: X rays are produced by the motion of electrons in the volume of the atom, while gamma rays are produced by the motion of protons and neutrons inside the nucleus.

Beta rays are particles — they are particles of either negative or positive electric charge that are created by a spontaneous reaction inside nuclei and then ejected at high speed. Experiments with beta rays show that the negative beta particles (β^-) are nothing but high-speed electrons, and the positive beta particles (β^+) are nothing but high-speed antielectrons, or positrons. When a beta-ray antielectron is stopped by some absorbing material, it sooner or later collides with an electron and both annihilate; this mutual annihilation of matter and antimatter generates a flash of gamma rays.

Alpha rays are also particles — they have a positive electric charge and a relatively large mass (about 7300 times the mass of a beta particle). Experiments reveal that the alpha particle has exactly the same structure as the nucleus of a helium atom. To produce an alpha ray, the radioactive nucleus somehow assembles part of its material into a heliumlike structure and ejects this fragment at high speed.

Cosmic rays are another kind of penetrating radiation. These rays come from outer space and they continually bombard the Earth from all directions. Cosmic rays are particles, mainly high-energy protons. Some of these particles originate on the Sun; others originate somewhere in the depths of interstellar space. The solar contribution is highly variable — during solar flares it can become extremely large and cause disturbances in the upper atmosphere ("magnetic storms" or aurora borealis). The interstellar contribution is constant in time and it reaches the Earth from all directions with equal strength; it includes particles of extremely high energies — some of these cosmic rays have an energy much in excess of any energy ever achieved by our most potent accelerating machines. We do not know exactly where the interstellar cosmic rays come from; possibly they are generated by supernova explosions of stars. When cosmic-ray particles hit the top of our atmosphere, large numbers of a variety of new particles are produced in the violent collisions between the incident cosmic rays and the nuclei of air atoms (Figure B.4). The particles produced in such collisions are called **secondary cosmic rays** to distinguish them from the incident particles, or **primary cosmic rays.** Most of the cosmic rays that transverse the atmosphere and reach sea level are secondary particles — about 75% of these are muons, particles similar to electrons, but heavier (see Interlude C).

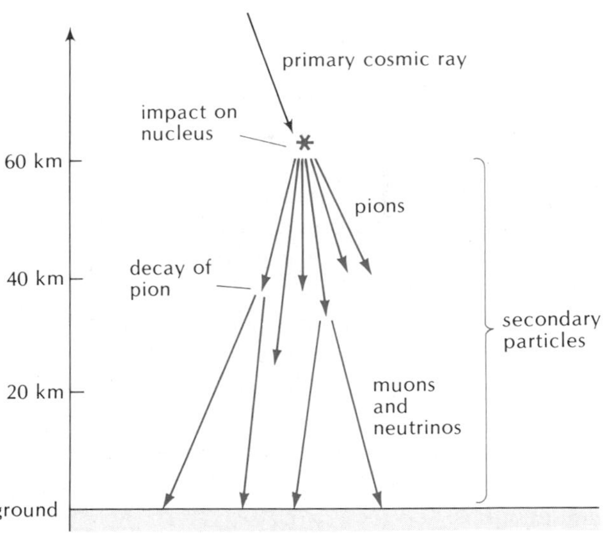

Fig. B.4 Primary cosmic rays (black) striking the upper atmosphere generate showers of secondary cosmic rays (color), which reach the ground.

Because of their exceedingly high energy, cosmic rays are by far the most penetrating kind of radiation. They not only transverse the thickness of the atmosphere, but sometimes penetrate thousands of meters into the ground — cosmic-ray muons have been known to penetrate more than a kilometer of rock and to reach some of the deepest mine pits.

Table B.1 summarizes some of the properties of

Table B.1 PENETRATING RADIATIONS

Radiation	Mass	Electric charge[b]	Typical penetration depth[a] Air	Water	Lead	Description
X rays	0	0	100 m	10 cm	1 mm	Very energetic form of light
Alpha rays	4.0 u	+2e	10 cm	0.1 mm	0.01 mm	Same structure as helium nucleus
Beta rays	0.00055 u	±e	1 m	1 mm	0.1 mm	High-energy electrons or positrons
Gamma rays	0	0	1 km	1 m	10 cm	Extremely energetic form of light
Cosmic rays, primary	Mixed	Mixed		Very large		Mostly high-energy protons
Cosmic rays, secondary	Mixed	Mixed		Very large		Mostly high-energy muons
Neutrons	1.0 u	0	1 km	1 m	10 cm	—

[a] X rays, gamma rays, and neutrons do not have a sharply defined penetration depth — they are attenuated gradually; the numbers in the table give the thickness of material that attenuates beams of these rays by a factor of 1000. All these numbers are rough approximations; exact values depend on the energy of the radiation.

[b] The quantity e appearing in this column represents the electric charge of the proton. Thus, an alpha particle has twice the electric charge of the proton, etc.

penetrating radiations. Neutrons have been included in this table even though they are not emitted spontaneously by radioactive materials. However, neutrons are emitted in great abundance during the fission reactions (see Interlude H) that take place in a nuclear reactor or in a nuclear bomb; hence these artificial devices are intense sources of neutron "rays." Neutrons are also emitted by some nuclei when suitably stimulated with other kinds of radiation; for example, beryllium releases neutrons when exposed to bombardment by alpha rays. Small neutron sources ("neutron howitzers") can be constructed by simply mixing beryllium with radium or some other alpha emitter; within such a mixture the beryllium is then under continuous alpha bombardment, which provokes it to emit neutrons.

B.2 RADIOACTIVE DECAY

Although all atoms of a given chemical element have exactly the same chemical properties, they do not all have exactly the same physical properties: some atoms have more mass than others. The differences in mass among the atoms of a given chemical element hinge on the numbers of neutrons in the nuclei — all the atoms have the same number of protons in their nuclei, but they have different numbers of neutrons. Atoms with the same number of protons but with different numbers of neutrons are called **isotopes.** Thus, the isotope ^{14}C of carbon has six protons and eight neutrons in its nucleus, ^{13}C has six protons and seven neutrons, and ^{12}C (which is the most abundant isotope in naturally occurring samples of carbon) has six protons and six neutrons. The superscript on the chemical symbol is the sum of the number of protons and neutrons, called the **mass number.** Carbon has several other isotopes; Table P.2 lists all the isotopes of carbon, as well as the isotopes of a few other elements.

Most isotopes are unstable, i.e., they decay by a spontaneous nuclear reaction and transmute themselves into another, more stable element. The unstable isotopes are **radioactive** — their spontaneous nuclear reactions involve the emission of alpha rays, beta rays, or gamma rays. For example, the isotope ^{14}C is unstable and decays into ^{14}N with the emission of a beta ray,

$$^{14}C \rightarrow {}^{14}N + \beta^- \tag{1}$$

The decays of the following isotopes of sodium, cobalt, strontium, radium, and uranium are further examples of radioactive decays with emission of beta rays and alpha rays:

$$^{22}Na \rightarrow {}^{22}Ne + \beta^+ \tag{2}$$

$$^{60}Co \rightarrow {}^{60}Ni + \beta^- \tag{3}$$

$$^{90}Sr \rightarrow {}^{90}Y + \beta^- \tag{4}$$

$$^{226}Ra \rightarrow {}^{222}Ru + \alpha \tag{5}$$

$$^{238}U \rightarrow {}^{234}Th + \alpha \tag{6}$$

Note that all of the reactions (1)–(6) involve **transmutation of elements:** carbon into nitrogen, sodium into neon, cobalt into nickel, etc. Such transmutations are a characteristic feature of alpha and beta decays.

In many cases of alpha and beta decays, the nucleus emerges from the reaction with an excess of internal energy; the nucleus will then eliminate this energy by emitting a gamma ray. For example, the ^{60}Ni nucleus formed by the beta decay of ^{60}Co contains extra internal energy and emits a gamma ray,

$$^{60}Ni \rightarrow {}^{60}Ni + \gamma \tag{7}$$

Thus emission of γ rays by a sample of radioactive ^{60}Co is a two-step process:

$$^{60}Co \rightarrow {}^{60}Ni + \beta^- \tag{8}$$
$$\quad\quad \hookrightarrow {}^{60}Ni + \gamma \tag{9}$$

Note that although a transmutation of elements [Eq. (8)] precedes the emission of the gamma ray, the emission itself [Eq. (9)] does not involve transmutation. Gamma-ray emission from a nucleus is analogous to light emission from an atom; in both cases energy is released in consequence of a sudden change of motion of the constituents of a system (protons and neutrons of a nucleus, or electrons of an atom), but the constituents themselves do not change.

If we initially have a given amount of some radioactive element, then this amount will gradually decrease in time as more and more of the atoms decay. Measurements show that the quantitative law that describes this decay process is very simple: if a certain fraction of the initial amount of radioactive material decays in a certain time interval, then the same fraction of the remainder decays in the next (equal) time interval, and the same fraction of the new remainder decays in the next (equal) time interval, etc. For example, suppose we initially have 1 g of radioactive strontium. This decays by beta decay,

$$^{90}Sr \rightarrow {}^{90}Y + \beta^- \tag{10}$$

In this reaction, strontium is called the **parent** material and yttrium is the **daughter** material. Measurements show that it takes 29 years for one-half of the initial amount of parent to decay. The law of radioactive decay then asserts that during the next 29 years, one-half of the remainder will decay, etc. Hence the amounts of parent material left at times $t = 0$, 29, 58, and 87 years will be 1, $\frac{1}{2}$, $\frac{1}{4}$, and $\frac{1}{8}$ g, respectively (Figure B.5). This means that the amounts left after the lapse of equal time intervals form a simple geometric progression. The time required for the decay of one-half of the amount of parent material is called the **half-life.** For radioactive strontium, the half-life is 29 years.

In a sample of radioactive material, a certain number of decays will occur during each second. This decay rate is called the **activity** of the sample. Usually, the activity is expressed in **curies:**[3]

$$\begin{aligned} 1 \text{ curie} &= 1 \text{ Ci} \\ &= 3.7 \times 10^{10} \text{ disintegrations per second} \end{aligned} \tag{11}$$

For example, the activity of 1 g of pure radium is just about 1 Ci. Commercially available sources of penetrating radiation carry a label indicating their strength in curies and a label with a warning symbol (Figure B.6). Any source of a strength more than about 10^{-6} Ci is regarded as potentially dangerous. Health author-

[3] After the French scientists Pierre Curie, 1859–1906, and Marie Sklowdowska Curie, 1867–1934. Both are famous for their work on radioactivity, and their discovery of radium and polonium. They shared a Nobel Prize with A. H. Becquerel in 1903. Marie Curie was also awarded a Nobel Prize in Chemistry in 1911.

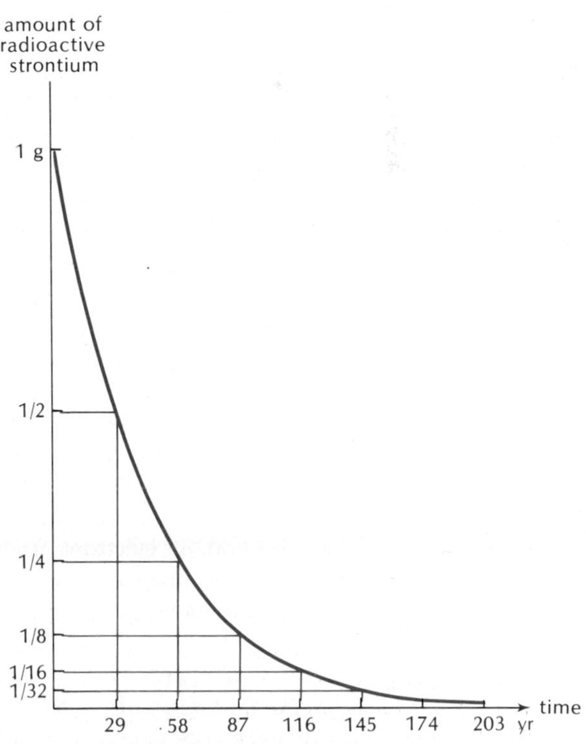

Fig. B.5 Decay of radioactive strontium. At $t = 0$ the initial amount of strontium is 1 g; the curve shows the amount at later times. The equation of the curve is $n = n_0 \exp[-t (\ln 2)/t_{1/2}]$, where $t_{1/2}$ is the half-life.

Fig. B.6 This warning sign is usually magenta on a yellow background.

ities require that such sources be handled only by qualified persons. Table B.2 lists the strengths of some typical radioactive sources used in diverse applications.

Table B.2 STRENGTHS OF RADIOACTIVE SOURCES

Source	*Radiation*	*Strength*
Fission products released by 20-kiloton atomic bomb	α, β, γ	6×10^{11} Ci
Nuclear power reactor	α, β, γ	10^{10}
^{144}Ce as heat source for thermoelectric generator	β, γ	10^{6}
^{60}Co in high-level industrial irradiation cell	β, γ	10^{6}
^{60}Co in teletherapy unit	β, γ	$\sim 10^{3}$
^{131}I for eradication of thyroid	β, γ	0.1
^{131}I for thyroid scan	β, γ	$5\text{–}50 \times 10^{-6}$
^{226}Ra in fluorescent paint in wristwatch	β, γ	$\sim 10^{-6}$
Natural ^{40}K in human body	β, γ	10^{-7}

B.3 RADIATION DAMAGE IN ATOMS

When radiation penetrates a material, it disrupts the atoms, i.e., it disturbs the motion of the electrons in the atoms and sometimes it even ejects electrons entirely from the atoms. The latter kind of disruption is called **ionization** — the ejection of electrons from their atoms leaves behind charged atoms or ions. Because of this, alpha, beta, and gamma rays are often called **ionizing radiation.**

As an alpha, beta, or gamma ray penetrates through a material, it gradually spends its energy in the disruption of the atoms along its track. The larger the initial energy of the ray, the farther it will penetrate before all its energy is exhausted. The details of the mechanism of energy loss depend on the type of radiation.

Alpha rays or other electrically charged particles zipping through matter exert electric forces on the atomic electrons and the latter exert electric forces on the former. Since the alpha ray passes by the electron very quickly, the mutual interaction can be regarded as a collision. An alpha ray is so massive that such collisions with atomic electrons will not deflect it very much — the alpha ray violently pushes the electrons aside and continues on its straight path (Figure B.7).[4]

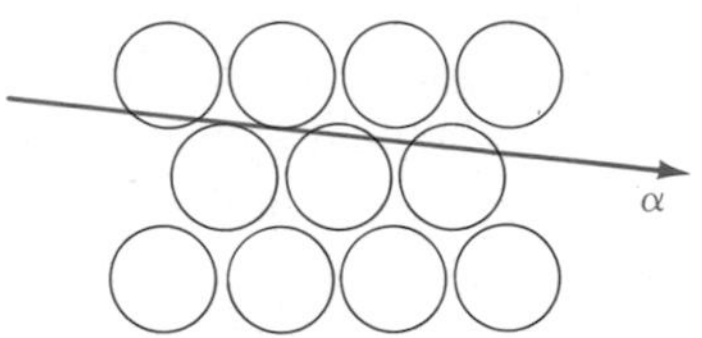

Fig. B.7 An alpha ray passing through atoms.

Although each collision with an electron only slows the alpha ray down a very little bit and produces only a small loss of kinetic energy, the number of electrons that the alpha ray encounters along its path is very large and hence the cumulative loss of kinetic energy is rapid. Alpha rays produce intense ionization along their path (Figure B.8); thus, they lose their energy quickly and are stopped by a thin layer of material.

Beta rays penetrating matter also suffer collisions with atomic electrons. However, since their mass is equal to the mass of atomic electrons, they cannot shoulder these electrons aside without suffering large

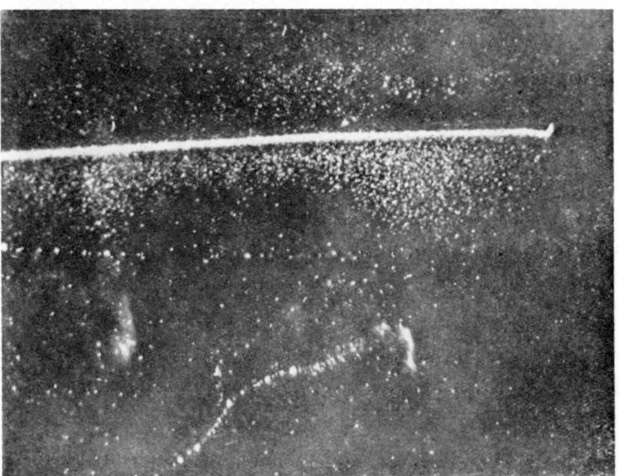

Fig. B.8 Tracks of an alpha and beta ray in a cloud chamber. The track of the alpha ray is heavy and straight. The track of the beta ray is faint and tortuous.

[4] Collisions with the atomic nuclei of the matter are so rare as to be insignificant.

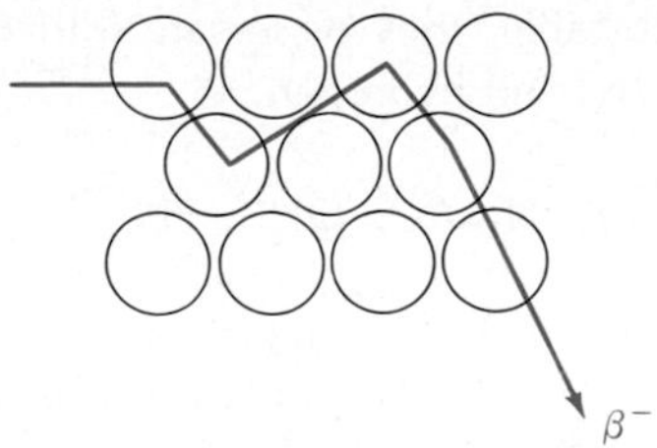

Fig. B.9 A beta ray passing through atoms.

deflections themselves (Figure B.9). Thus the path of a beta ray is not straight, but tortuous, full of erratic twists and turns (see Figure B.8). Besides, since the electric charge of a beta ray is of lesser magnitude than that of an alpha ray, the electric forces and disturbances that it exerts on nearby atomic electrons are of a correspondingly lesser magnitude. The beta ray generates less intense ionization than an alpha ray, dissipates its energy more slowly, and penetrates a deeper layer of material.

Gamma rays, in passing through matter, behave quite differently from alpha and beta rays. Whereas the latter engage in a great many collisions with atomic electrons and thereby dissipate their energy more or less continuously along their path, a gamma ray will sometimes penetrate a thick layer of material without any energy loss. What happens to an individual gamma ray is a matter of chance: as a gamma ray attempts to pass through a layer of material, there is some chance that the gamma ray will collide with an electron and lose its energy and there is some chance that it will pass through without collision and without energy loss. When a gamma ray finally does interact with an atomic electron, it loses all — or a large part — of its energy and the electron takes off at high speed, leaving the atom (Figure B.10a and b). If the gamma ray has a sufficiently high energy, an energy of at least twice the electron rest-mass energy (twice m_ec^2), then it can also lose energy by an alternative mechanism: when passing near a nucleus, such a gamma ray can spend its energy in the creation of an electron–positron pair (Figure B.10c). Note that neither of these primary energy transfer mechanisms causes much direct damage in the material — only one atom is involved in each case. However, the high-energy electrons (and positrons) that inherit the original gamma-ray energy behave exactly as beta rays and they will damage a large number of atoms along their path.

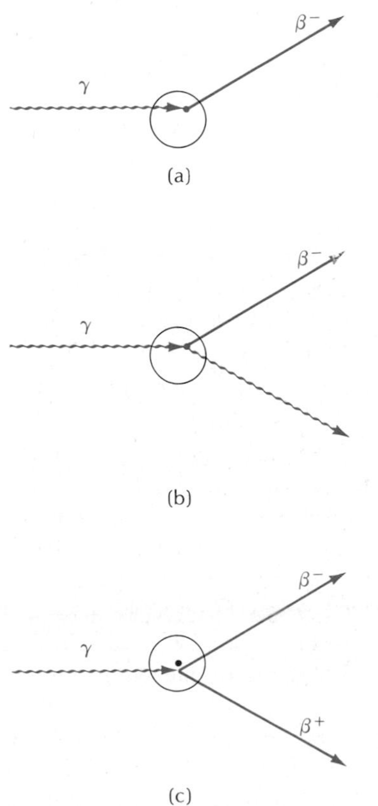

Fig. B.10 A gamma ray passing through an atom may (a, b) lose all or some of its energy to an electron or (c) create an electron–positron pair.

Neutrons passing through matter do not interact with the atomic electrons — the neutrons have no electric charge and electrons scarcely feel them.[5] The attenuation of a neutron beam by a layer of material is due to direct contact collisions between neutrons and atomic nuclei. Since the nuclei are small, such collisions are not very frequent and therefore neutrons, just like gamma rays, can penetrate very thick layers of material without stopping. The impact of a neutron on a nucleus is a matter of chance; when such impact finally occurs, the neutron may either bounce off the nucleus, losing some energy, or else be absorbed by the nucleus. Under the impact the nucleus recoils and, if the recoil energy is large, forms a "nuclear ray," which behaves similarly to an alpha ray but causes much heavier ionization. If the nucleus absorbs the neutron, it becomes radioactive: the capture of a neutron gives the nucleus extra internal energy, which it will eliminate by undergoing one or more radioactive transformations. The secondary alpha, beta, or gamma rays emitted in this process will then damage the material in the usual way.

When a material is exposed to alpha, beta, or gamma radiation for some length of time, it will absorb a certain amount of radiation energy. What alterations

[5] Even in a head-on collision, an electron and a neutron will usually just pass through each other with no effect on either.

this absorbed energy will produce in the properties of the material depends on how much energy is absorbed per unit mass of the material. The amount of absorbed radiation energy per kilogram of material is called the **absorbed dose;** it is measured in **rad,**

$$1 \text{ rad} = 0.01 \text{ J/kg} \tag{12}$$

For example, if a human body of 75 kg receives a dose of 600 rad over the entire volume of the body (a "whole-body exposure"), then the total amount of absorbed energy is

$$\begin{aligned}[\text{absorbed energy}] &= [\text{absorbed dose}] \times [\text{mass of body}] \\ &= 600 \text{ rad} \times 75 \text{ kg} \\ &= 600 \times 0.01 \text{ J/kg} \times 75 \text{ kg} \\ &= 450 \text{ J}\end{aligned}$$

Absorption of such a dose of radiation is very likely to be lethal.

B.4 RADIATION DAMAGE IN MOLECULES AND IN LIVING CELLS

Radiation damages atoms by the forcible ejection of electrons from the atom (ionization). In biological tissues the damaged atoms or molecules will then engage in chemical reactions that disrupt the normal chemical processes. Such an interference with the normal processes of living tissues can have grave consequences.

Since biological tissues are composed of 70% to 90% water, most of the primary impacts of radiation on molecules will be impacts on water molecules. These impacts lead to the decomposition of water and the subsequent formation of a variety of extremely corrosive molecules (radiolysis of water).

The first step in the decomposition of water is the ionization of a water molecule by the impact of radiation; this results in a positive water ion and a free electron,

$$[\text{radiation}] + H_2O \rightarrow H_2O^+ + e^- \tag{13}$$

The electron travels some distance and is then captured by another water molecule;[6] this results in a negative water ion,

$$e^- + H_2O \rightarrow H_2O^- \tag{14}$$

Both the positive and the negative ions (H_2O^+ and H_2O^-) are unstable; they dissociate into radicals and ions of hydroxyl and hydrogen,

$$H_2O^+ \rightarrow OH^\cdot + H^+ \tag{15}$$

$$H_2O^- \rightarrow OH^- + H^\cdot \tag{16}$$

These radicals $OH^\cdot$ and $H^\cdot$ are electrically neutral but they have "loose ends," i.e., they have unsaturated chemical bonds; these radicals are therefore very eager to attach themselves to something — they are very reactive.

The reactive hydroxyl and hydrogen radicals attack the complicated organic molecules — proteins, enzymes, nucleic acids — that are the basic building blocks of living cells. Furthermore, by reacting with one another and with molecules of their environment, the hydroxyl and hydrogen radicals produce some other corrosive compounds that inflict additional damage on the organic molecules. The damage done to the organic molecules can take several forms: breakage of bonds, leading to fragmentation of the long molecular chains that serve as backbones for the molecules; rearrangement of bonds, leading to changes in the chemical behavior of molecules; and formation of extra bonds, leading to cross-linking either within one molecule or between neighboring molecules. Such damage to the organic molecules in a living cell disrupts the normal chemical processes that sustain life. If the disruption is severe, the life of the cell ends.

Massive doses of radiation (many thousands of rad) can cause an almost immediate cessation of cellular metabolism followed by cellular disintegration. However, even a much smaller dose of radiation (a few hundred rad or less) can cause "reproductive death" of cells by destroying their ability to undergo cell division and to reproduce. Since the life of any complex organism depends on the continual replacement of old cells by new ones, the "reproductive death" of cells will ultimately bring about the physiological death of the organism even though the damaged cells continue to metabolize and, except for their inability to reproduce, may seem chemically and physically quite normal.

There exists considerable evidence that reproductive cell death is due to damage to the chromosomes of the cell — specifically, damage to the DNA molecules within the chromosomes. DNA is the most crucial molecule in the cell; it is the molecule that contains all the information regarding the construction and operation of the cell, and it therefore directs all the processes within the cell. The DNA molecules and the chromosomes that contain them are both quite ra-

[6] If the electron has sufficiently high energy, it can, of course, produce some ionization before it is captured.

diosensitive — both DNA and chromosomes are subject to breakage when exposed to radiation. Furthermore, experiments show that even very low levels of radiation, which produce no overt morphological changes in the cell, interfere with DNA synthesis and consequently inhibit cell division.

The sensitivity of a cell to radiation depends on the phase of the life of the cell. The cell is most susceptible to suffer lethal damage if the radiation strikes it during the period of cell division (mitosis) just when the DNA and the chromosomes are attempting duplication. Obviously, cells that divide frequently are more likely to get caught in this delicate stage than cells that divide rarely or not at all. This leads to a general rule: the cells with the highest radiosensitivity are those that have the highest rate of cell division (law of Bergonié and Tribondeau). For example, in mammals the cells in the bone marrow, lymphatic tissues, lining of the intestine, ovaries and testes, and the cells in the embryo undergo very frequent cell divisions — all these cells are very easily damaged by radiation. In contrast, the cells in the brain, muscles, bones, liver, and kidneys undergo little or no cell division — these cells are relatively resistant to damage by radiation.

The damage that a given absorbed dose of radiation inflicts on a tissue also depends on the kind of radiation. Some kinds of radiation, e.g., alpha rays, are more efficient than others in damaging and killing cells. The kill efficiency of alpha rays is high because they produce very intense ionization along their path. When an alpha ray passes through tissues, it will damage many molecules in each cell; this concentrated damage is likely to have a lethal effect. The kill efficiency of gamma rays is much lower. When a gamma ray passes through tissues, it will damage only a few molecules in each cell; hence the damage will be spread out over a larger number of cells and the injuries to individual cells are not as likely to be lethal. Injured cells have a chance to repair themselves gradually and to recover.

Because of such differences in kill efficiency, the absorbed dose (in rad) is not a good indicator of the permanent biological damage. To obtain a measure of the biological damage, the absorbed dose must be multiplied by a correction factor that accounts for the kill efficiency of the radiation. The correction factor is called the RBE (*R*elative *B*iological *E*ffectiveness). Table B.3 gives the RBE for diverse kinds of radiation; the values in this table are approximate because the exact RBE depends on the energy of the rays. The RBE of X rays is unity by definition; according to Table B.3, alpha rays are 10 to 20 times more damaging than X rays, fast neutrons are 10 times more damaging than X rays, etc.

Table B.3 RBE VALUES

Radiation	*RBE*
Alpha rays	10–20
Beta rays	1
Gamma rays	1
X rays	1
Neutrons (fast)	10
Neutrons (slow)	5

A reasonably good indicator of the biological damage in human tissues is the **"equivalent" absorbed dose,** which is the absorbed dose multiplied by the RBE:

$$[\text{"equivalent" absorbed dose}] = \text{RBE} \times [\text{absorbed dose}]$$

Since RBE is a pure number, the unit of "equivalent" absorbed dose is the same as that of absorbed dose (rad); however, in order to keep track of these two kinds of doses, the unit of "equivalent" absorbed dose is not written as rad, but as **rem** (*r*ad *e*quivalent *m*an). For example, if a human body receives a 100-rad absorbed dose of fast neutrons, the "equivalent" absorbed dose is 10×100 rems $= 1000$ rems. Incidentally, this is an absolutely lethal dose. Note that doses of 100 rad of fast neutrons, 1000 rad of X rays, 1000 rad of gamma rays, etc., all yield the same "equivalent" dose.

B.5 THE PHYSIOLOGICAL EFFECTS OF RADIATION

The effect of radiation on man depends not only on the total "equivalent" dose, but also on how it is delivered. A dose that is delivered all at once (acute) is more dangerous than a dose spread out over a long period of time (chronic); a dose delivered to the entire volume of the body (whole-body exposure) is more dangerous than a dose delivered to only some part of the body; a dose delivered to a radiosensitive part of the body (bone marrow, lymphatic tissue, gastrointestinal tract) is more dangerous than a dose delivered to a radioresistant part (brain, muscles, bones, etc.).

Table B.4 summarizes the effects of absorption of a whole-body dose delivered all at once. The information on radiation sickness and radiation death contained in this table is based on observations of the victims of the bombings of Hiroshima, Nagasaki, and the Bikini atoll; observations of the victims of nuclear reactor accidents; and extensive studies of laboratory animals subjected to irradiation.

Doses below 100 rems do not produce radiation

Table B.4 EFFECTS OF ACUTE RADIATION DOSES

Dose (whole body)	*Critical organ*	*Effect*	*Incidence of death*
0–100 rems	Blood	Some blood-cell destruction	None
100–200	Blood-forming tissue (bone marrow, lymphatic tissue)	Decrease in white blood-cell count	None
200–600	Blood-forming tissue	Severe decrease of white blood-cell count, internal hemorrhage, infection, loss of hair	0–80% within 2 months
600–1000	Blood-forming tissue	Same	80–100% within 2 months
1000–5000	Gastrointestinal tract	Diarrhea, fever, electrolyte imbalance	Nearly 100% within 2 weeks
5000 and above	Central nervous system	Convulsions, tremor, lack of coordination, lethargy	100% within 2 days

sickness, although there may be some temporary changes in the blood-cell count.

Doses between 100 and 200 rems produce radiation sickness with nausea and vomiting. There is a reduction of the number of white blood cells due to damage to the bone marrow, which generates these cells. If no secondary complications arise, the victim recovers in a few weeks.

Doses from 200 to 600 rems lead to severe radiation sickness with, possibly, a lethal outcome. A dose of 300 to 350 rems is sufficient to kill 50% of the exposed population if no medical treatment is available. With an exposure in this range there is severe damage to the bone marrow and a concomitant drastic reduction of the white blood-cell count; this renders the body very susceptible to infections.

Doses between 600 and 1000 rems involve qualitatively similar, but more massive, damage. Survival is unlikely, but not impossible.

Doses above 1000 rems are almost invariably fatal. Exposure at this level destroys the lining of the intestine; this eliminates the control of the body over its fluid balance and also lays the interior of the body open to massive bacterial and viral invasion.

Doses of more than 5000 rems result in cerebral death by direct damage to the brain; at extreme doses, death can be nearly instantaneous.

Although low doses of radiation do not cause radiation sickness, they can cause other damage within the body. Irradiation of a pregnant woman, even at a low level, is likely to interfere with the normal development of the embryo. The embryonic tissues are extremely radiosensitive and a dose of 15 rems or even less during the first 2 months of pregnancy may induce monstrous deformations of the embryonic organs.

Furthermore, radiation exposure produces delayed effects such as cancers, cataracts, and genetic defects, which only make their appearance years, or generations, later. For example, among the survivors of the bombings in Japan, leukemia was two or three times as frequent as normally expected.

Even very small doses of radiation carry with them a (small) risk of cancer. It has been estimated that a yearly dose of 200 millirems delivered to every inhabitant of the United States would lead to an average of 7100 extra cancer deaths each year. Besides, such doses of radiation increase the incidence of mutations in cells and therefore carry with them an increased risk of genetic defects. A yearly dose of 200 millirems delivered to every inhabitant of the United States would lead to an average of 6000 extra genetic defects in newborn children each year.

These numbers are directly relevant because they correspond to the average amount of radiation that inhabitants of the United States actually receive from diverse sources (see Table B.5).

Table B.5 AVERAGE ANNUAL RADIATION DOSES PER INHABITANT OF THE UNITED STATES[a]

Source	*Dose (per year)*
Medical and dental X rays	70 millirems
Cosmic rays	50
Radioactivity of ground and buildings	50
Natural ^{40}K in human body	20
Other natural radioactivity in human body	4
Radioactivity of air	5
Total	~200 millirems

[a] From B. L. Cohen, *Nuclear Science and Society*.

Note that one-third of the yearly dose is due to medical practice; this contribution could be substantially reduced (by a factor of about two) by eliminating unnecessary diagnostic X-ray procedures and by restricting the size of the X-ray beam to the size of the film that is being exposed. Table B.6 gives the doses received from various medical X-ray procedures.

The maximum permissible dose for persons occupationally exposed to radiation — radiologists, nuclear reactor operators, experimental physicists — has been arbitrarily set at 5 rems/year.

Table B.6 RADIATION DOSES FROM DIAGNOSTIC X RAYS[a]

Diagnostic procedure	*Doses*[b]
Mammography (breast)	250–300 millirem
Upper gastrointestinal series	150–400
Middle spine	150–400
Lower gastrointestinal series	90–250
Lower back and spine	70–250
Upper spine	40–80
Gallbladder	25–60
Skull	20–50
Chest	5–35
Dental (whole mouth)	10–30

[a] From P. W. Laws and The Public Citizen Health Research Group, *The X-Ray Information Book*.

[b] This is the effective dose of whole-body exposure that produces the same average damage as the actual localized exposure of the target organ and the surrounding organs.

B.6 MEDICAL APPLICATIONS OF RADIOISOTOPES

In medical practice radioactive isotopes, or **radioisotopes,** are used both for diagnosis and for therapy — they are used both as tracers to investigate bodily organs that are suspected of disease and as sources of intense radiation to destroy diseased tissues of, say, a malignant tumor.

As tracers, radioisotopes are injected into the body and their subsequent flow through the bloodstream and their accumulation in different organs, or in different parts of one organ, gives an indication of some of the physical and chemical processes that take place deep within the body.

Tracers are routinely used to measure blood flow, total blood volume, metabolic rates of diverse organs, etc. One of the most interesting and informative tracer techniques is **scanning**; this technique produces a picture of the distribution of radioisotope within the body. The radioisotopes suited to this purpose are gamma emitters; since gamma rays have great penetrating power, the gamma emissions of a radioisotope in an organ deep within the body can easily reach a detector placed outside of the body — the detector can "see" the organ by the gamma-ray "light" that it emits. Unfortunately, gamma rays cannot be focused on a screen or photographic plate in the same way that light rays can be focused. When a gamma ray encounters a lens made of glass (or any other material), it is likely to pass straight through (without any refraction); and if it is deflected, then the deflection is random. However, the gamma-ray detector can be given directional sensitivity by means of a **collimator,** a lead block with many finely drilled holes, all pointed at one spot (Figure B.11). Rays from the aiming spot pass through the collimator and reach the detector; rays from elsewhere strike the lead and are lost. To build up a picture of the radioisotope distribution, the detector and its collimator must be moved from spot to spot and the intensity at each place must be recorded; this procedure is called scanning. Alternatively, a picture can be made by means of a **pinhole camera** consisting of a plate of lead with a fine hole, behind which is placed a sheet of photographic film. The gamma rays that pass through the hole form an image on the film (Figures B.12 and B.13).

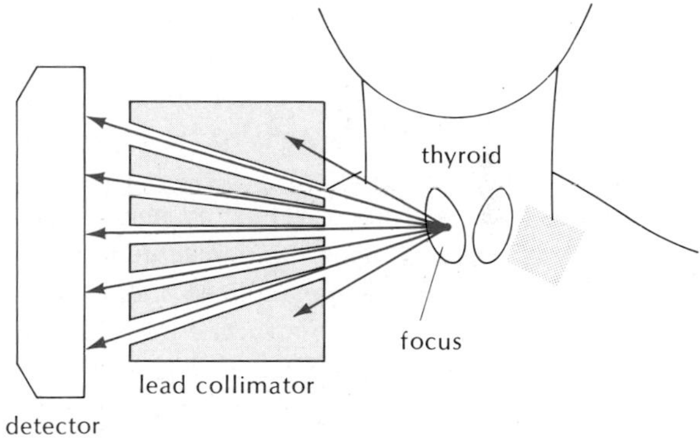

Fig. B.11 A collimator.

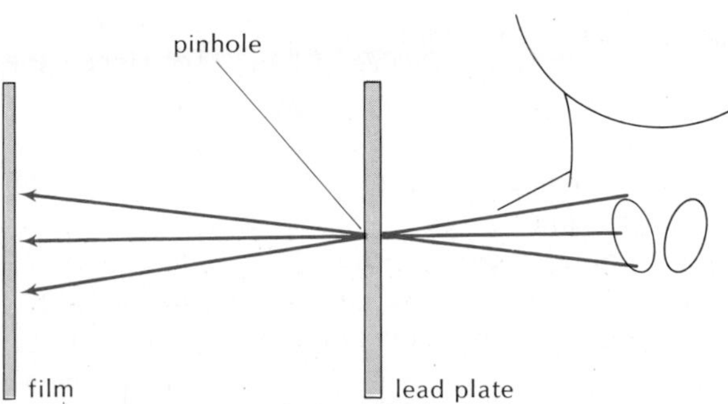

Fig. B.12 A pinhole camera.

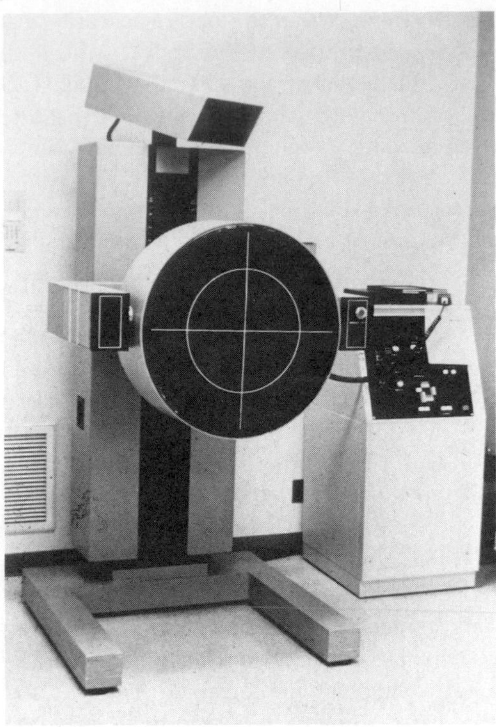

Fig. B.13 Siemens gamma camera. *(Courtesy Nuclear Medicine Section, Hospital of the University of Pennsylvania.)*

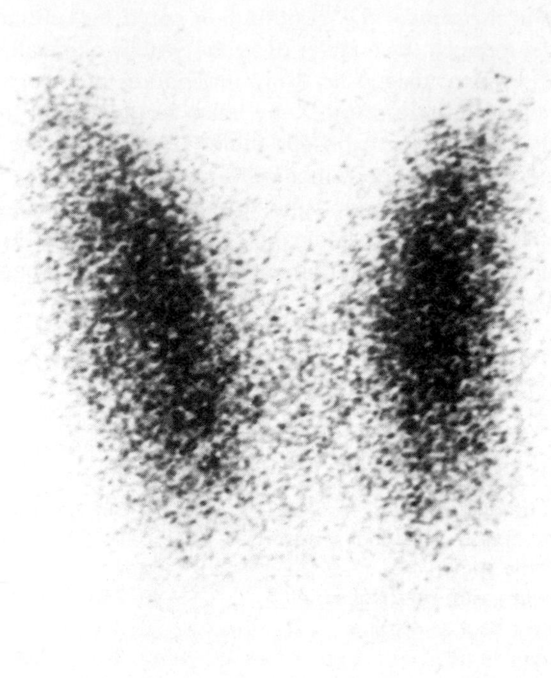

Fig. B.14 Pinhole-camera scan of normal thyroid showing uniform concentration of ^{123}I. *(Courtesy Michael G. Velchik, M.D., Nuclear Medicine Section, Hospital of the University of Pennsylvania.)*

Different radioisotopes in conjunction with different chemical carriers are used to take pictures of different parts of the body. The choice of radioisotope and the choice of carrier substance depends on the organ that one wishes to "see." For example, Figure B.14 shows a scan of the thyroid gland of a healthy person; this scan was made with a ^{123}I compound injected into the bloodstream. The thyroid metabolizes iodine and hence this element tends to concentrate there — the thyroid then glows in its gamma ray "light." Figure B.15 shows a scan of an abnormal thyroid of a patient suffering from cancer of the thyroid.

Brain scans can be made either with ^{123}I or with ^{99}Tc.[7] These radioisotopes are incorporated in certain protein molecules (albumin) that are not absorbed by normal brain tissue; however, the tissue of a brain tumor is abnormal and absorbs these protein molecules with their radioactive ^{123}I or ^{99}Tc. Hence the tumor begins to emit gamma "light" and becomes visible on a scan. Figures B.16 and B.17 are, respectively, scans of a normal brain and of an abnormal brain with malignant tumors.

Scans of other organs — lungs, heart, liver, kidneys, bones — can be made with appropriate radioisotopes. For example, cancer of the bone can be detected with

[7] The ^{99}Tc used for this purpose is the daughter of ^{99}Mo. This daughter is a gamma-ray emitter.

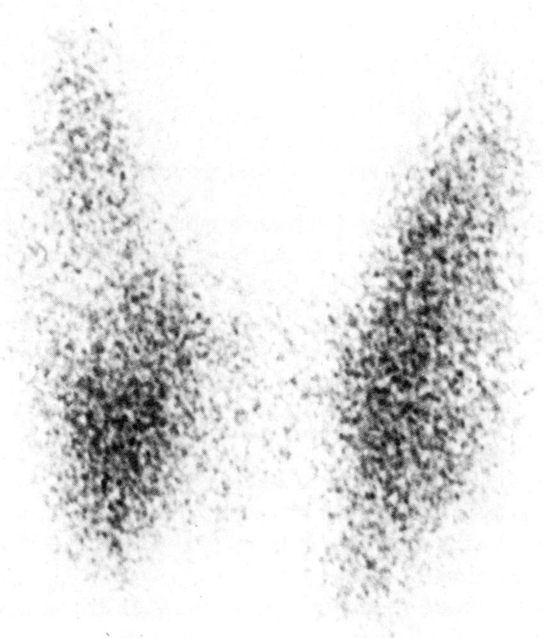

Fig. B.15 Pinhole-camera scan of abnormal thyroid. The left lobe displays a focal area of low concentration of ^{123}I, indicating the presence of a tumor. *(Courtesy Michael G. Velchik, M.D., Nuclear Medicine Section, Hospital of the University of Pennsylvania.)*

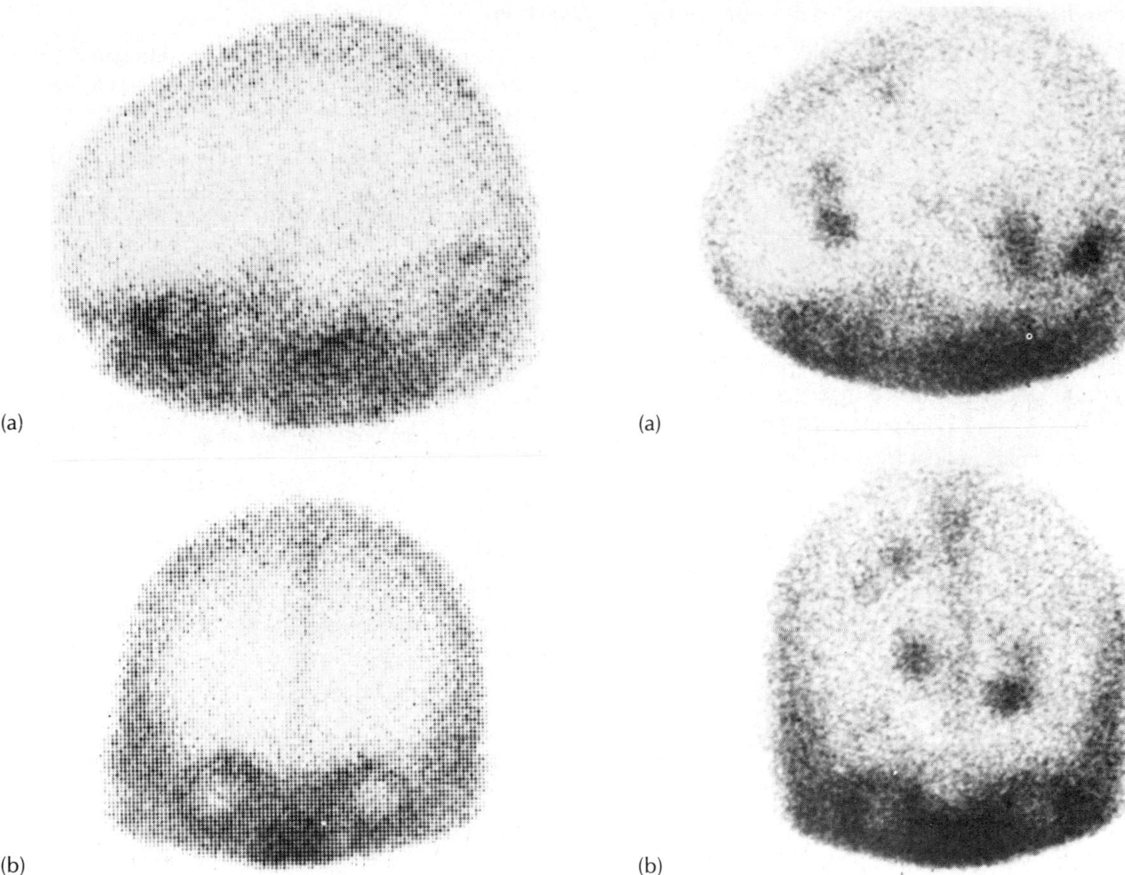

Fig. B.16 Collimator scan of normal human brain showing normal distribution of ^{99}Tc: (a) lateral view; (b) anterior view. *(Courtesy Michael G. Velchik, M.D., Nuclear Medicine Section, Hospital of the University of Pennsylvania.)*

Fig. B.17 Pinhole-camera scan of abnormal brain with several metastatic tumors caused by breast cancer: (a) lateral view; (b) anterior view. *(Courtesy Michael G. Velchik, M.D., Nuclear Medicine Section, Hospital of the University of Pennsylvania.)*

^{90}Sr; this radioisotope is chemically similar to Ca and therefore tends to concentrate in the bones. Cancerous tissues are regions of high metabolic activity and they absorb the ^{90}Sr more greedily than adjacent normal tissues; the consequent high concentration of radioisotope shows up in the scan picture.

When radioisotopes are used as sources of intense radiation for the destruction of diseased tissue, they may be placed either outside or inside the body. *External* sources of gamma rays are used for radiation therapy in much the same way as intense X rays; the advantage of gamma rays over megavolt X rays lies in the energy distribution of the rays — all gamma rays emitted from a particular radioisotope source have the same energy, while the X rays generated in X-ray tubes always have mixed energies. This uniformity of the gamma-ray energy permits a more precise delivery into the deep layers of the body. Figure B.18 shows a ^{60}Co teletherapy unit; this machine contains a pellet of ^{60}Co (several hundred or a thousand curies) in a heavy lead container. The gamma rays emerge through a small channel in the lead, forming a beam that can be

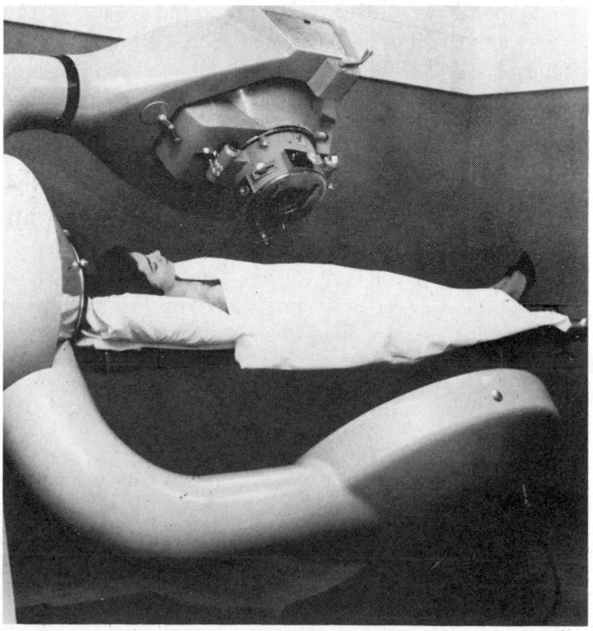

Fig. B.18 A ^{60}Co radiation therapy unit.

aimed at the diseased tissue. Some of these machines are designed so that during operation they can be continuously rotated around the patient; the gamma-ray beam then swings around the target spot but always remains centered on it. With this arrangement the gamma rays are highly concentrated at the "focal point" and they achieve a maximum amount of damage at this point; at the same time the gamma rays are spread out over the adjacent layers of the body and they do less damage to healthy tissues. *Internal* sources of beta and gamma rays can be used to deliver radiation to specific locations at short range. Needles made of radium or cobalt can be inserted directly into a tumor to deliver very concentrated doses of radiation; ribbons or small pellets ("seeds") of some other radioisotopes can be implanted for the same purpose. Also, one can take advantage of body chemistry to concentrate a radioisotope compound in a specific organ. For example, the partial destruction of an overactive thyroid can be accomplished with a radioiodine compound similar to that used in thyroid scanning, but administered in a much larger dose.

Further Reading

Radiation and Life by E. J. Hall (Pergamon Press, Oxford, 1976) and, at a somewhat more advanced level, *Ionizing Radiation and Life* by V. Arena (Mosby, Saint Louis, 1971) are introductions to radiation, its effects on living tissues, and its application in medicine. *Biological Effects of Radiation* by J. Coggle (Wykeham, London, 1973) provides a very clear and concise survey of the action of radiation at the cellular level, with emphasis on mammals. *The X-Ray Information Book* by P. W. Laws and The Public Citizen Health Research Group (Farrar, Straus, and Giroux, New York, 1983) gives very sensible advice on how to keep your exposure to medical X rays to a minimum.

The Effects of Nuclear Weapons, edited by S. Glasstone (U.S. Government Printing Office, Washington, D.C., 1964) includes a description of severe radiation injury (radiation sickness). *Energy, Ecology, and the Environment* by R. Wilson and W. J. Jones (Academic Press, New York, 1974) contains an informative chapter on low-level radiation injuries and radiation hazards to man.

The following are some recent articles dealing with radiation:

Sources and Effects of Ionizing Radiation, Scientific Committee on the Effects of Atomic Radiation, United Nations, 1977

"The Biological Effects of Low-Level Ionizing Radiation," A. C. Upton, *Scientific American,* February 1982

"Neutrons in Science and Technology," D. A. Bromley, *Physics Today,* December 1983

Questions

1. The penetration depth of alpha, beta, and gamma rays, and neutrons in water is about 10 times as large as in lead. To what is this difference due?

2. In 1968 a team of physicists under the direction of Luis Alvarez used cosmic rays to "X-ray" the pyramid of Kephren. The aim of this project was to detect secret chambers hidden in the volume of stone. Explain how a cosmic-ray detector placed under the pyramid in a subterranean vault, discovered long ago, can be used to detect empty cavities in the stone above. (No cavities were found.)

3. Use Table P.2 to make a list of all the isotopes of oxygen. What are the number of protons and the number of neutrons in the nucleus of each of these isotopes?

4. According to Table P.2, what ionizing radiations are emitted by the isotopes of carbon?

5. Why do alpha and beta emissions involve transmutation of elements, but gamma emission does not?

6. The isotope ^{24}Na decays by emission of a negative beta ray. What is the daughter isotope produced in this decay?

7. Nuclear fission bombs produce a long-lived radioactive isotope of strontium (^{90}Sr), an element that is chemically similar to calcium. Explain why radioactive strontium in the environment poses a severe hazard to man.

8. Suppose that a Geiger counter placed next to a sample of radioactive isotope initially registers 3620 counts per second, and 30 minutes later it registers 450 counts per second. What is the half-life of the radioactive sample?

9. The isotope ^{14}C is used in the radioactive dating of samples of material of biological origin, such as wood. The half-life of this isotope is 5570 years. As long as a tree is alive and breathes, the concentration of ^{14}C relative to ^{12}C in the wood will be the same as in the atmosphere, about 1 part in 10^{12}. But when the tree dies, the ^{14}C is not replenished anymore and its decay leads to a gradual decrease of its concentration relative to that of the stable isotope ^{12}C. Suppose that in a sample of wood found in an ancient tomb, 75% of the original amount of ^{14}C has decayed. How old is the tomb?

10. Tritium (^{3}H) is a radioactive isotope of hydrogen which occurs naturally in small concentrations in ordinary water in the environment. The half-life of this isotope is 12 years. Describe how you could take advantage of this isotope to determine the age of a bottle full of wine that your wine merchant claims is 25 years old.

11. If you irradiate a sample of material with alpha, beta, or gamma rays, is the sample likely to become radioactive? What if you irradiate it with neutrons?

12. Some artificial satellites carry small nuclear reactors as power sources. What danger does this pose for people on Earth?

13. Can you guess why fast neutrons have a higher RBE than slow neutrons?

14. Insects are extremely resistant to radiation — most insect species can tolerate a radiation dose of up to 2000 or even 100,000 rad. Explain this radiation resistance. (Hint: The cells of mature insects divide only rarely.)

15. How many dental X rays were you subjected to during the course of the year? According to the numbers given in Table B.6, how many millirems did you receive from these X rays?

16. The stone used in the construction of Grand Central Station, New York, is slightly radioactive. The radiation dose to people in the station is between 90 and 550 millirems per year. How does this compare with the average dose for inhabitants of the United States (see Table B.5)?

17. According to Table B.5, the natural ^{40}K in your body gives you a radiation dose of 20 millirems per year. Estimate what extra dose you receive if your body is in close contact with other human bodies (in a crowded room) for 24 hours.

18. Astronauts in a spacecraft above the atmosphere receive a radiation dose of 100 millirems per day from cosmic rays. How many days can an astronaut stay in such a spacecraft before he attains the maximum permissible occupational dose of 5 rems (within any given year)?

19. Some new "antipersonnel" bombs discharge a large number of plastic pellets upon explosion. The material in these pellets has been chosen so that an X ray of a wound will not reveal the pellets. Discuss this technological achievement from the point of view of physics, medicine, and morals.

CHAPTER 11

Kinematics of a Rigid Body

Rigid body

A body is **rigid** if its particles do not move relative to one another. Thus, the body has a fixed shape and all its parts have a fixed position relative to one another. No real body is absolutely rigid. For instance, consider a ball of hardened steel, such as found in a ball bearing. On a microscopic scale this body is certainly not rigid — the atoms in the steel are in a state of continuous thermal agitation; they randomly move back and forth about their equilibrium positions. On a macroscopic scale we can ignore these atomic motions because, on the average, they do not change the shape of the body. But even so, the ball is not quite rigid — if subjected to a very heavy load, the ball will suffer a noticeable deformation. In this chapter and in the following chapter, we will ignore such deformations produced by the forces acting on bodies. We will study the motion of bodies under the assumption that rigidity is a good approximation.

11.1 Motion of a Rigid Body

Rotational motion

A rigid body can simultaneously have two kinds of motion: it can change its position in space and it can change its orientation in space. Change in position is translational motion; as we saw in Chapter 9, this motion can be conveniently described by the motion of the center of mass. Change in orientation is **rotational motion;** it can be proved mathematically that any change of orientation of a rigid body can be regarded as a rotation about some axis.[1]

[1] This is **Euler's theorem.**

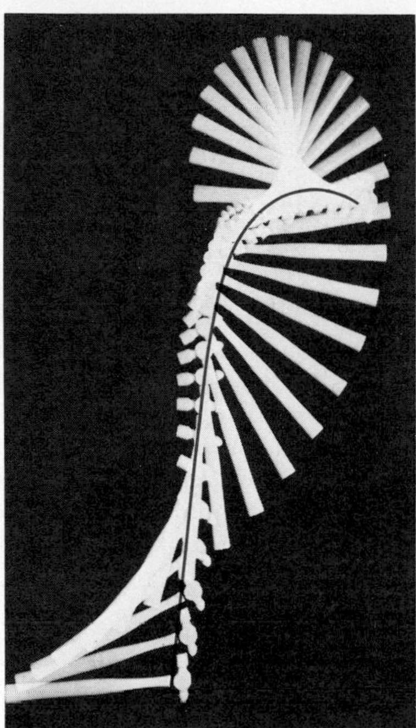

Fig. 11.1 A hammer in free fall under the influence of gravity. The center of mass of the hammer, marked by the colored line, moves with constant vertical acceleration g, just like a particle in free fall. (Courtesy Harold E. Edgerton, MIT.)

As an example consider the motion of a hammer thrown up in the air (Figure 11.1). The orientation of the hammer changes relative to fixed coordinates attached to the ground. Instantaneously, the body rotates about a horizontal axis, say, a horizontal axis that passes through the center of mass (in Figure 11.1 this axis sticks out of the plane of the page and is marked with a dot). The complete motion can be described as a simultaneous rotation of the hammer about this axis and a translation of the axis along a parabolic path. It is worth keeping in mind that the choice of axis is somewhat arbitrary: instead of using a horizontal axis passing through the center of mass of the hammer, we could just as well use a horizontal axis passing through any other point, say, through the end of the handle. The hammer rotates in relation to both of these axes; the only difference lies in their translational motion — the axis through the end of the hammer undergoes a much more complicated translational motion than the axis through the center of mass. This emphasizes that there are two ways in which rotation is relative: first, it is necessary to specify the fixed reference frame with respect to which the motion occurs; and second, it is necessary to specify some point of the body through which the axis of rotation is supposed to pass (by hypothesis, this point then does *not* rotate). The freedom in the choice of axis of rotation will come in handy in the discussion of the motion of a rolling wheel in Section 12.5. However, it is often most convenient to suppose that the axis passes either through the center of mass or through some point which is held fixed by rigid mechanical constraints (e.g., a point on a fixed pivot or an axle).

In the above example of the thrown hammer the axis of rotation always remains horizontal. In the general motion of a rigid body, the axis of rotation can have any direction and can also change its direction. To describe such a complicated motion, it is convenient to separate the rotation into three components along three perpendicular axes. According to nautical and aeronautical terminology, the three components of the rotational motion are called **roll, pitch,** and **yaw** (Figure 11.2). However, in the following sections we will not deal with

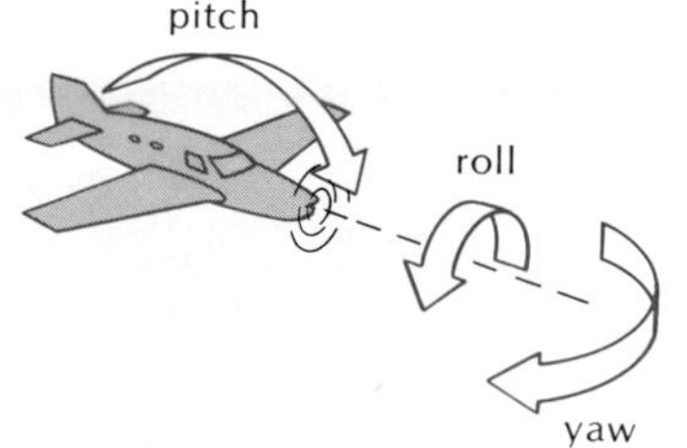

Fig. 11.2 The three independent rotational motions of an aircraft.

this general complicated three-component rotation; we will only deal with rotation about a single, fixed axis.

EXAMPLE 1. One interesting feature of rotations about different axes is that they *do not commute,* i.e., the net result of two successive rotations depends on which rotation is done first, which second. Show that this is so by rotating a book about two different axes.

SOLUTION: Figure 11.3a shows a book lying on a table and facing you. Perform two rotations: first, a clockwise rotation by 90° about the vertical axis; and second, a clockwise rotation by 180° about the horizontal axis pointing away from you (Figures 11.3b and c).

Next, perform the same rotations on a similar book, but in the opposite order: first, a clockwise rotation by 180° about the horizontal axis pointing away from you; and second, a clockwise rotation by 90° about the vertical axis (Figures 11.4b and c). Comparison of Figures 11.3c and 11.4c shows that the final results are different, i.e., the results depend on the order in which the rotations were carried out.

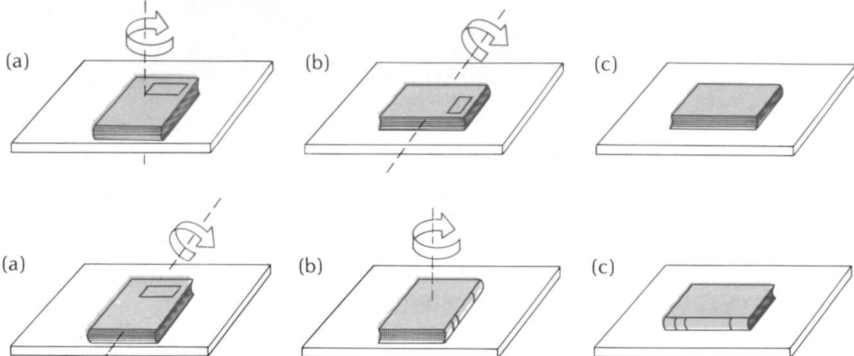

Fig. 11.3 Rotation through 90° about a vertical axis followed by a rotation through 180° about a horizontal axis.

Fig. 11.4 Rotation through 180° about a horizontal axis followed by a rotation through 90° about a vertical axis.

11.2 Rotation About a Fixed Axis

Figure 11.5 shows a rigid body rotating about a fixed axis. During the rotational motion, each point of the body remains at a given distance from this axis and moves along a circle centered on the axis. To describe the orientation of the body at any instant, we select one particle in the body and use it as a reference point; any particle will do as reference point, provided that it is not on the axis of rotation. The circular motion of this fiducial particle (labeled P in Figure 11.5) is then representative of the rotational motion of the entire rigid body and the angular position of this particle is representative of the angular position of the entire rigid body.

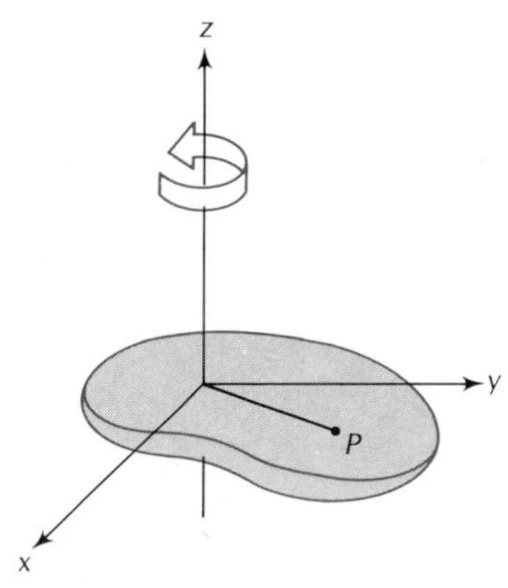

Fig. 11.5 Rotation of a rigid body about a fixed axis (z axis).

Figure 11.6 shows the rigid body as seen from along the axis of rotation. The coordinates in Figure 11.6 have been chosen so that the z axis is along the direction of rotation while the y and x axes are in the plane of the circle traced out by the motion of the fiducial particle.

The angular position of the fiducial particle — and hence the angular orientation of the entire rigid body — can be described by the position angle ϕ between the position vector **r** and the x axis.[2] Conventionally, the angle ϕ is reckoned as positive when measured in a counterclockwise direction (as in Figure 11.6). We will always measure the position angle ϕ in radians. Consequently, the length of the path between the x axis and the point P is

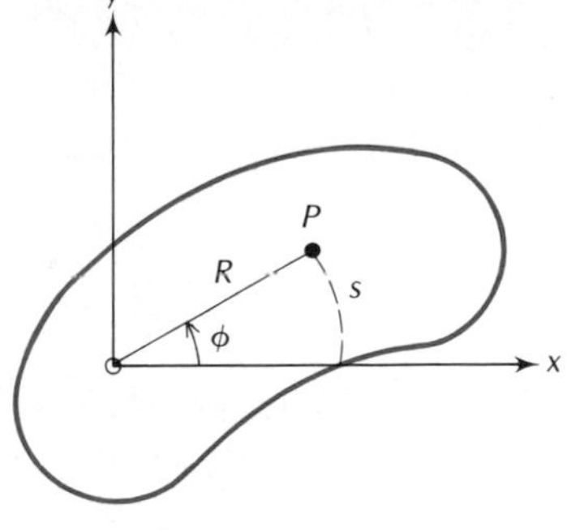

Fig. 11.6 Fiducial particle P on a rigid body. The radius of the circle traced out by the motion is the same as the length of the position vector, $R = \sqrt{x^2 + y^2}$.

$$s = \phi R \tag{1}$$

where R is the radius of the circle traced out by the motion. For a rotating body the position angle ϕ is some function of time, $\phi = \phi(t)$. The definition of the angular velocity for rotational motion is mathematically analogous to the definition of velocity for one-dimensional motion (see Sections 2.2 and 2.3). The **average angular velocity** is defined as

Average angular velocity

$$\boxed{\overline{\omega} = \frac{\Delta\phi}{\Delta t}} \tag{2}$$

where $\Delta\phi$ is the change in the angle and Δt the corresponding change in the time. The **instantaneous angular velocity** is defined as

Instantaneous angular velocity

$$\boxed{\omega = \frac{d\phi}{dt}} \tag{3}$$

The units of angular velocity are radians per second (radian/s). The radian is the ratio of two lengths [compare Eq. (1)] and hence is a pure number, with no dimensions of length, time, or mass; hence, dimensionally, 1 radian/s is the same thing as 1/s. However, to prevent confusion in calculations, it is often useful to retain the dimensionally vacuous label *radian.*

If the body rotates with a constant angular velocity, then one can also measure the rate of rotation in terms of the **frequency,** or number of revolutions per unit time. Since each complete revolution involves a change of ϕ by 2π radians, the time per revolution, or the **period,** is

Period

$$T = 2\pi/\omega \tag{4}$$

and the frequency of revolution is

Frequency

$$\nu = 1/T \tag{5}$$

or

$$\nu = \omega/2\pi \tag{6}$$

This expresses the frequency in terms of the angular velocity. The units of frequency are revolutions per second (rev/s). Dimensionally,

[2] A list of Greek letters and their names appears on the endpapers.

1 rev/s is the same thing as 1/s; but, as in the case of radians, to prevent confusion it is useful to retain the label *rev.* If the rate of rotation of the body is not uniform, then it is best to avoid the use of the word *frequency;* this word is usually reserved for repetitive motions in which each revolution takes the same time as the preceding one.

The definition of the **average angular acceleration** is

Average angular acceleration

$$\overline{\alpha} = \frac{\Delta\omega}{\Delta t} \tag{7}$$

and that of the **instantaneous angular acceleration** is

Instantaneous angular acceleration

$$\alpha = \frac{d\omega}{dt} \tag{8}$$

or

$$\alpha = \frac{d^2\phi}{dt^2} \tag{9}$$

The units of angular acceleration are radian/s^2.

Equations (3) and (8) give the angular velocity and angular acceleration of the rigid body; i.e., they give the angular velocity and acceleration of every particle in the body. It is interesting to focus on one of these particles and evaluate its *linear* velocity and acceleration in terms of the corresponding angular quantities. If the particle is at a distance R from the axis (Figure 11.7), then the length along the circular path of the particle is [compare Eq. (1)]

$$s = \phi R \tag{10}$$

Since R is a constant, differentiation of this equation gives

$$\frac{ds}{dt} = R\frac{d\phi}{dt} \tag{11}$$

Here ds/dt is the speed at which the particle moves along its circular path and $d\phi/dt$ is the angular velocity; hence Eq. (11) is equivalent to

$$v = R\omega \tag{12}$$

This shows that the (linear) speed of the particle around the axis increases in direct proportion to the radius: the farther a particle of the rigid body is from the axis, the faster it moves.

Differentiation of Eq. (12) yields

$$\frac{dv}{dt} = R\frac{d\omega}{dt} \tag{13}$$

Here dv/dt is the rate of change of the speed of the circular motion, i.e., it is the acceleration *along* the circular path. Since this acceleration represents a change in the length of the velocity vector, it is in the same direction as the velocity vector, i.e., it is tangent to the circle (Figure 11.7). If we write this tangential acceleration as a_{tan} and take into account that $d\omega/dt$ is the angular acceleration, Eq. (13) becomes

$$a_{tan} = R\alpha \tag{14}$$

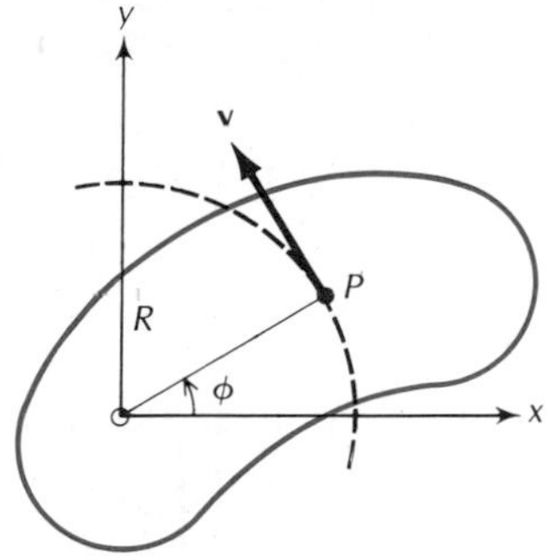

Fig. 11.7 Instantaneous velocity of a particle in a rotating rigid body.

Note that the tangential acceleration of Eq. (14) is not the complete acceleration of the particle. As we saw in Section 4.4, a particle in circular motion has a centripetal acceleration of magnitude

$$a_{cent} = \frac{v^2}{R} = R\omega^2 \tag{15}$$

The net acceleration is the vector sum of a tangential vector of magnitude a_{tan} and a radial vector of magnitude a_{cent}; this is illustrated in Figure 11.8 (and will be further explored in Example 3). The tangential acceleration is due to the change of *magnitude* of the velocity and the centripetal acceleration is due to the change of *direction* of the velocity.

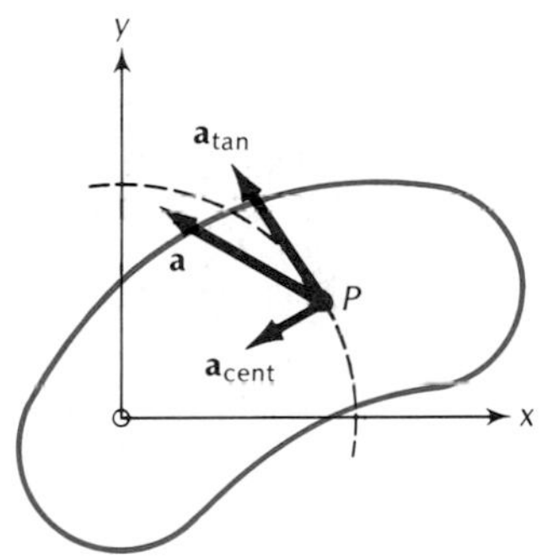

Fig. 11.8 Instantaneous acceleration of a particle in a rotating rigid body.

11.3 Motion with Constant Angular Acceleration

If the rigid body rotates with a constant angular acceleration, then the angular velocity increases at a constant rate and, after a time t has elapsed, the velocity will reach the value

Equations for constant angular acceleration

$$\omega = \omega_0 + \alpha t \tag{16}$$

where ω_0 is the initial value at $t = 0$.

The angular position ϕ can be calculated from this angular velocity by exactly the same methods as were used in Section 2.5 to calculate x from v [see Eqs. (2.17)–(2.21)]. The result is

$$\phi = \phi_0 + \omega_0 t + \tfrac{1}{2}\alpha t^2 \tag{17}$$

Furthermore, the methods of Section 2.5 lead to an identity between acceleration, position, and velocity [see Eqs. (2.23)–(2.25)]:

$$\alpha(\phi - \phi_0) = \tfrac{1}{2}(\omega^2 - \omega_0^2) \tag{18}$$

Equations (16), (17), and (18) permit a complete description of the motion of wheels or other bodies rotating with constant angular acceleration. These equations for angular motion are mathematically analogous to Eqs. (2.17), (2.21), and (2.25) for translational motion with constant acceleration.[3]

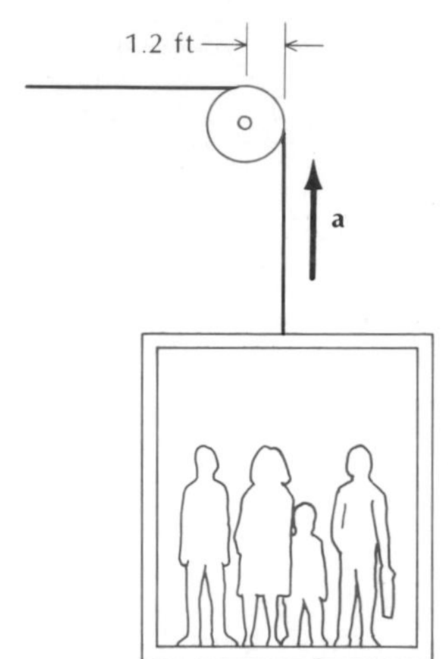

Fig. 11.9 The cable supporting an elevator runs over a rotating wheel.

EXAMPLE 2. The cable supporting an elevator runs over a wheel of radius 1.2 ft (Figure 11.9). If the elevator ascends with an upward acceleration of 1.53 ft/s² (as in Example 6.4), what is the angular acceleration of the wheel? How many turns does the wheel make if this accelerated motion lasts 5.0 s starting from rest? Assume that the cable runs over the wheel without slipping.

SOLUTION: If there is no slipping, the speed of the cable must always coincide with the speed of a point on the rim of the wheel. The acceleration $a = 1.53\ \text{ft/s}^2$ of the cable must then coincide with the tangential acceleration of a point on the rim of the wheel,

$$a = a_{\text{tan}} = R\alpha$$

where $R = 1.2$ ft is the radius of the wheel; hence

$$\alpha = a/R = (1.53\ \text{ft/s}^2)/1.2\ \text{ft} = 1.28/\text{s}^2$$

$$= 1.28\ \text{radian/s}^2$$

According to Eq. (17), the angular displacement in 5.0 s is

$$\phi - \phi_0 = \omega_0 t + \tfrac{1}{2}\alpha t^2$$

$$= 0 + \tfrac{1}{2}(1.28\ \text{radian/s}^2) \times (5.0\ \text{s})^2$$

$$= 15.9\ \text{radians}$$

Each revolution comprises 2π radians; thus 15.9 radians is the same as $15.9/2\pi$ revolutions, or 2.54 revolutions.

EXAMPLE 3. Find the net instantaneous acceleration (tangential and centripetal) of a point on the rim of the wheel described in the preceding example at the instant $t = 1.00$ s.

SOLUTION: We already know the tangential acceleration,

$$a_{\text{tan}} = 1.53\ \text{ft/s}^2$$

To find the centripetal acceleration, we need the angular velocity. According to Eq. (16),

$$\omega = \omega_0 + \alpha t = 0 + (1.28\ \text{radian/s}^2) \times 1.00\ \text{s}$$

$$= 1.28\ \text{radian/s}$$

[3] The analogy between rotational and translational quantities can serve as a useful mnemonic for remembering the equations for rotational motion. The following is a list of analogous equations:

$v = dx/dt$	→	$\omega = d\phi/dt$
$a = d^2x/dt^2$	→	$\alpha = d^2\phi/dt^2$
$v = v_0 + at$	→	$\omega = \omega_0 + \alpha t$
$x = x_0 + v_0 t + \frac{1}{2}at^2$	→	$\phi = \phi_0 + \omega_0 t + \frac{1}{2}\alpha t^2$
$a(x - x_0) = \frac{1}{2}(v^2 - v_0^2)$	→	$\alpha(\phi - \phi_0) = \frac{1}{2}(\omega^2 - \omega_0^2)$

and, according to Eq. (15), the centripetal acceleration is

$$a_{\text{cent}} = \omega^2 R = (1.28\ \text{radian/s})^2 \times 1.2\ \text{ft}$$

$$= 1.95\ \text{ft/s}^2$$

The net acceleration of the point is the vector sum of $a_{\tan}$ and a_{cent} (Figure 11.10); this net acceleration has a magnitude

$$a = \sqrt{a_{\tan}^2 + a_{\text{cent}}^2} = 2.48\ \text{ft/s}^2$$

and makes an angle of 38° with the radius.

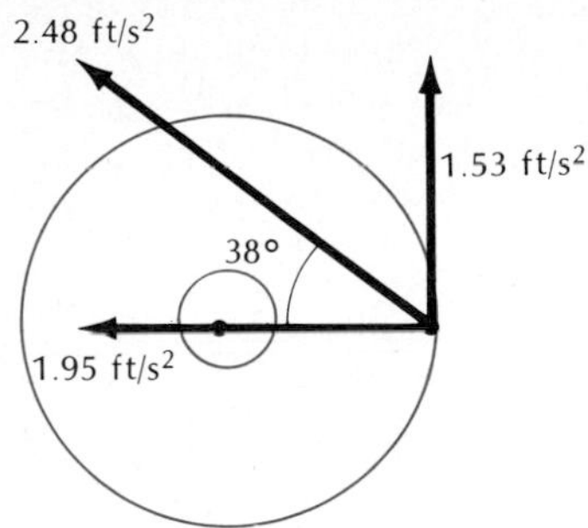

Fig. 11.10 Acceleration of a point on the rim of the wheel.

11.4 Kinetic Energy of Rotation; Moment of Inertia

As for any system of particles, the total kinetic energy K of a rotating rigid body is simply the sum of the individual kinetic energies of all the particles. If the particles have masses m_i and velocities $\mathbf{v}_i$ (where $i = 1, 2, \ldots, n$), then

$$K = \sum_{i=1}^{n} \tfrac{1}{2} m_i v_i^2$$

In a rigid body rotating about a given axis, all the particles move with the same angular velocity ω along circular paths. The speeds of the particles along their paths are given by Eq. (12):

$$v_i = R_i\omega \qquad (i = 1, 2, 3, \ldots) \tag{19}$$

and hence the total kinetic energy is

$$K = \sum_{i=1}^{n} \tfrac{1}{2} m_i R_i^2 \omega^2 \tag{20}$$

We will write this as

$$\boxed{K = \tfrac{1}{2} I \omega^2} \tag{21}$$

Kinetic energy of rotation

where

$$\boxed{I = \sum_{i=1}^{n} m_i R_i^2} \tag{22}$$

Moment of inertia of a system of particles

is the **moment of inertia** of the rotating body about the given axis. The units of moment of inertia are kg · m² or slug · ft².

Note that Eq. (21) has a mathematical form reminiscent of the familiar expression $\frac{1}{2}mv^2$ for the kinetic energy of a single particle — the moment of inertia replaces the mass and the angular velocity replaces the linear velocity. As we will see in the next chapter, the moment of inertia is a measure of the resistance that a body offers to changes in its rotational motion, just as mass is a measure of the resistance that a body offers to changes in its translational motion. Such analogies between rotational and translational quantities make it easy to remember the equations for rotational motion.

Equation (22) shows that the moment of inertia—and consequently the kinetic energy for a given value of ω—is large if most of the mass of the body is at a large distance from the axis of rotation. This is very reasonable: for a given value of ω, particles at a large distance from the axis move with high speed and therefore have a large kinetic energy.

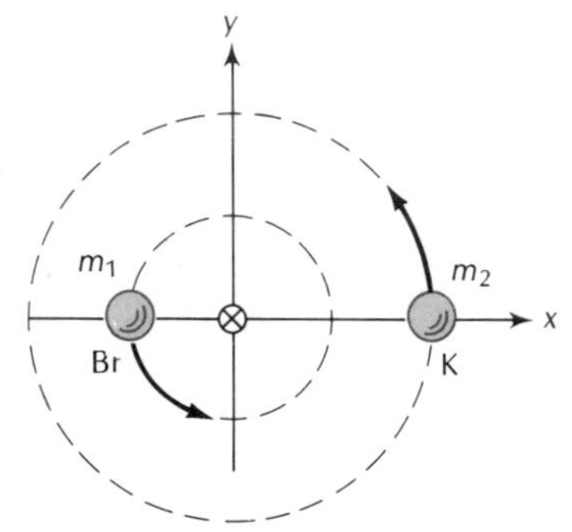

Fig. 11.11 The bromine and potassium atoms revolve about their common center of mass.

EXAMPLE 4. Suppose that the molecule of potassium bromide (KBr) described in Example 9.4 rotates about its center of mass (Figure 11.11). What is the moment of inertia of the molecule about this axis? Suppose that the molecule rotates with an angular velocity of 1.0×10^{12} radian/s. What is the rotational kinetic energy?

SOLUTION: It is best to place the origin of coordinates at the center of mass (see Figure 11.11). According to the results obtained in Example 9.4, the coordinate of the Br atom is then $x_1 = -0.93$ Å, and that of the K atom is $x_2 = +1.89$ Å. The corresponding masses are $m_1 = 79.9$ u and $m_2 = 39.1$ u. The moment of inertia is

$$I = m_1 x_1^2 + m_2 x_2^2$$

$$= 79.9 \text{ u} \times (0.93 \text{ Å})^2 + 39.1 \text{ u} \times (1.89 \text{ Å})^2$$

$$= 2.09 \times 10^2 \text{ u} \cdot \text{Å}^2$$

$$= 2.09 \times 10^2 \text{ u} \cdot \text{Å}^2 \times \frac{1.66 \times 10^{-27} \text{ kg}}{1 \text{ u}} \times \left(\frac{10^{-10} \text{ m}}{1 \text{ Å}}\right)^2$$

$$= 3.47 \times 10^{-45} \text{ kg} \cdot \text{m}^2$$

The kinetic energy is

$$K = \tfrac{1}{2} I \omega^2$$

$$= \tfrac{1}{2} \times 3.47 \times 10^{-45} \text{ kg} \cdot \text{m}^2 \times (1.0 \times 10^{12} \text{ radian/s})^2$$

$$= 1.73 \times 10^{-21} \text{ J}$$

This kinetic energy could equally well have been obtained by first calculating the individual velocities of the atoms ($v_1 = R_1\omega$, $v_2 = R_2\omega$) and then adding the corresponding individual kinetic energies.

If we regard the mass of a solid body as continuously distributed over its volume, then we can calculate the moment of inertia by converting Eq. (22) into an integral. As in the calculation of the position of the center of mass [see Eqs. (9.21)–(9.26)], we imagine that the volume is divided into small volume elements with masses

$$\Delta m_i = \rho \, \Delta V \tag{23}$$

where ρ is the density. Then

$$I = \sum_{i=1}^{n} m_i R_i^2 = \sum_{i=1}^{n} \rho \, R_i^2 \, \Delta V \qquad (24)$$

In the limit $\Delta V \to 0$, this becomes the integral

$$\boxed{I = \int \rho R^2 \, dV} \qquad (25)$$

Moment of inertia of a continuous mass distribution

In some simple cases, it is possible to find the moment of inertia without having recourse to an explicit calculation of the integral in Eq. (25). For example, if the rigid body is a thin hoop (Figure 11.12) or a thin cylindrical shell of radius R_0 (Figure 11.13) rotating about its axis of symmetry, then *all* of the mass of the body is at the same distance from the axis of rotation — the moment of inertia is then simply the total mass M of the hoop or shell multiplied by its radius R_0 squared,

$$I = MR_0^2 \qquad (26)$$

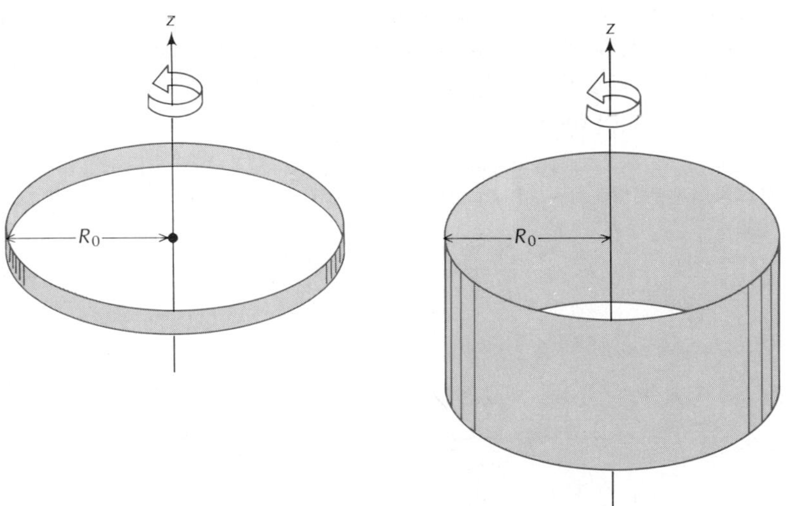

Fig. 11.12 (left) A thin hoop rotating about its axis of symmetry.

Fig. 11.13 (right) A thin cylindrical shell rotating about its axis of symmetry.

The following are some examples of explicit evaluations of the integral of Eq. (25).

EXAMPLE 5. Find the moment of inertia of a thin disk of uniform density of mass M and radius R_0 rotating about its axis.

SOLUTION: Figure 11.14 shows the disk with the z axis along the axis of rotation. Suppose the disk has a thickness l. The disk can be regarded as made up of a large number of thin concentric hoops fitting one around another. Figure 11.14 shows one such hoop of radius R, width dR, and thickness l. The volume of the hoop is $dV = 2\pi Rl \, dR$; with this, Eq. (25) gives

$$I = \int_0^{R_0} \rho R^2 2\pi Rl \, dR = 2\pi \rho l \int_0^{R_0} R^3 \, dR = 2\pi \rho l \, \frac{R_0^4}{4} \qquad (27)$$

The total mass of the disk is the volume $\pi R_0^2 l$ multiplied by the density ρ,

$$M = \pi R_0^2 l \rho$$

Hence $\rho l = M/\pi R_0^2$, which, when substituted into Eq. (27), yields

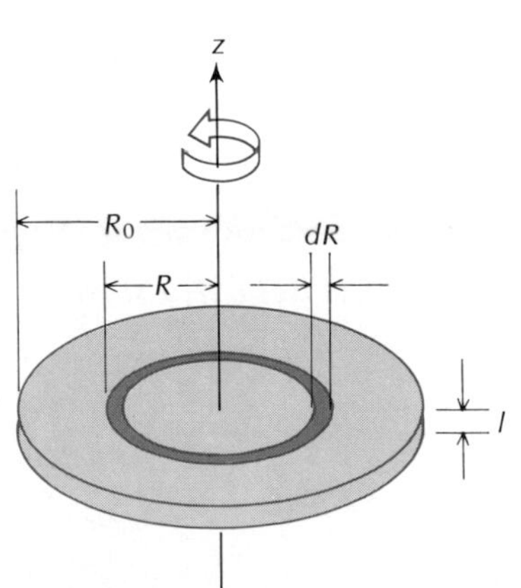

Fig. 11.14 A disk rotating about its axis of symmetry.

$$I = \tfrac{1}{2}MR_0^2 \tag{28}$$

for the moment of inertia of the disk. Incidentally, this formula also gives the moment of inertia of a cylinder, since such a body is simply a very thick disk. The above calculation is equally valid for thin and for thick disks; the final formula is independent of l.

EXAMPLE 6. Find the moment of inertia of a uniform thin rod of length l and mass M rotating about an axis perpendicular to the rod and through its center.

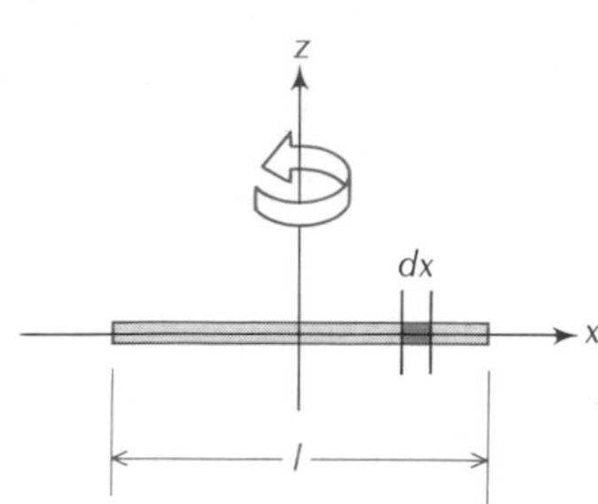

Fig. 11.15 A thin rod rotating about an axis through its midpoint.

SOLUTION: Figure 11.15 shows the rod lying along the x axis; the axis of rotation is the z axis. Consider a small slice dx of the rod. If the cross-sectional area of the rod is A, then $dV = A\,dx$ and Eq. (25) becomes

$$I = \int_{-l/2}^{l/2} \rho x^2 A\,dx = \rho A\left[\frac{x^3}{3}\right]_{-l/2}^{l/2} = \tfrac{1}{12}\rho A l^3 \tag{29}$$

Since Al is the volume of the rod, ρAl is its total mass and therefore the moment of inertia can be written

$$I = \tfrac{1}{12}Ml^2 \tag{30}$$

EXAMPLE 7. Repeat the calculation of the preceding example for an axis through one end of the rod.

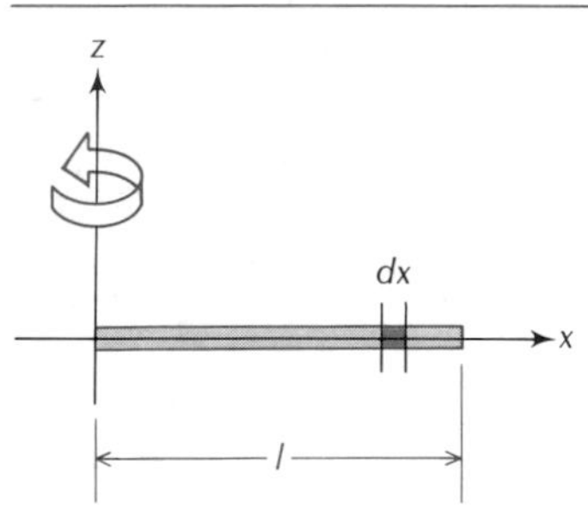

Fig. 11.16 A thin rod rotating about an axis through one end.

SOLUTION: Figure 11.16 shows the rod and the axis of rotation. The rod extends from $x = 0$ to $x = l$; hence instead of Eq. (29) we obtain

$$I = \int_0^l \rho x^2 A\,dx = \rho A\left[\frac{x^3}{3}\right]_0^l = \tfrac{1}{3}\rho A l^3 \tag{31}$$

and instead of Eq. (30), we obtain

$$I = \tfrac{1}{3}Ml^2 \tag{32}$$

Comparison of Eqs. (30) and (32) makes it very clear that the value of the moment of inertia depends on the location of the axis of rotation. The moment of inertia is small if the axis passes through the center of mass, and large if it passes through the end of the rod.

We can prove a general theorem, the **parallel-axis theorem,** that relates the moment of inertia I_{CM} about an axis through the center of mass to the moment of inertia I about a parallel axis through some other point. The theorem asserts that

Parallel-axis theorem

$$\boxed{I = I_{CM} + Md^2} \tag{33}$$

where M is the total mass of the body and d the distance between the two axes. It is a corollary of Eq. (33) that the moment of inertia about an axis through the center of mass is always less than that about any other parallel axis.

A simple and instructive proof of the theorem is as follows: Figure 11.17 shows the body rotating about a fixed axis and also shows an alternative, parallel axis through the center of mass. The kinetic energy of the body rotating about the fixed axis is

$$K = \tfrac{1}{2} I \omega^2 \tag{34}$$

However, we can express the kinetic energy in an alternative form. According to Eq. (9.44) the kinetic energy of an arbitrary system is the sum of the translational energy of motion of the center of mass and the "internal" energy of motion relative to the center of mass,

$$K = \tfrac{1}{2} M v_{CM}^2 + K_{int} \tag{35}$$

In Figure 11.17 the center of mass moves along a circle of radius d around the axis. Hence

$$v_{CM} = \omega d \tag{36}$$

and the first term on the right side of Eq. (35) is

$$\tfrac{1}{2} M v_{CM}^2 = \tfrac{1}{2} M \omega^2 d^2 \tag{37}$$

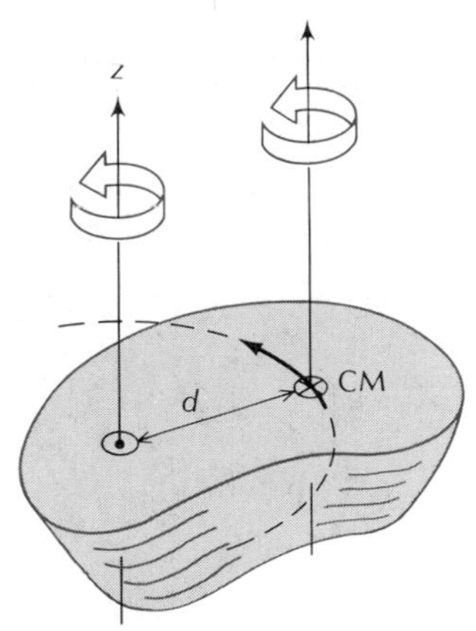

Fig. 11.17 Two alternative parallel axes of rotation of a rigid body. The z axis is fixed. The center-of-mass axis moves along a circle of radius d around the z axis. The body is in rotational motion relative to each of these axes.

Let us next look at the second term on the right side of Eq. (35). Although the rotational motion of the body in Figure 11.17 is most simply described as a rotation about the fixed axis, it can also be described as a rotation about the moving axis through the center of mass. Each time the body completes one revolution about the fixed axis, it also completes one revolution about this center-of-mass axis; consequently, the angular velocity of the rotation around the center-of-mass axis is also ω. The kinetic energy of the rotational motion relative to the center of mass is then

$$K_{int} = \tfrac{1}{2} I_{CM} \omega^2 \tag{38}$$

Thus, the total kinetic energy of Eq. (35) is

$$K = \tfrac{1}{2} M \omega^2 d^2 + \tfrac{1}{2} I_{CM} \omega^2 = \tfrac{1}{2}(Md^2 + I_{CM})\omega^2 \tag{39}$$

The values given by Eqs. (34) and (39) must be equal; this implies that

$$I = I_{CM} + Md^2$$

which is the result we wanted to obtain.

We can also prove another theorem, the **perpendicular-axis theorem,** which relates the moments of inertia of a thin flat plate (e.g., a sheet of metal) about three mutually perpendicular axes. Figure 11.18 shows a flat plate of arbitrary shape in the x–y plane. The plate may rotate either about an axis perpendicular to the plate (the z axis), or about an axis in the plane of the plate (the x axis or the y axis). We will call the moments of inertia for rotation about these alternative axes I_z, I_x, and I_y, respectively. Then the perpendicular–axis theorem asserts that

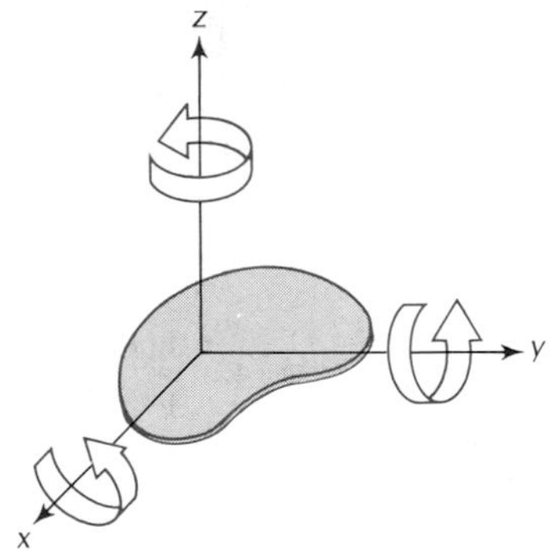

Fig. 11.18 A thin, flat plate. The plate may rotate about either the x axis, the y axis, or the z axis.

Perpendicular-axis theorem

$$\boxed{I_z = I_x + I_y} \tag{40}$$

The proof of the theorem is very simple: For rotation about the z axis, the distance from the axis of rotation to a particle in the plate is

$$R = \sqrt{x^2 + y^2}$$

hence

$$I_z = \int \rho R^2 \, dV = \int \rho(x^2 + y^2) \, dV$$

For rotation about the x axis, $R = y$ and

$$I_x = \int \rho y^2 \, dV$$

Likewise, for rotation about the y axis, $R = x$ and

$$I_y = \int \rho x^2 \, dV$$

Comparison of these three formulas makes it obvious that the sum of I_x and I_y equals I_z.

Table 11.1 gives the moments of inertia of a variety of rigid bodies about axes through their centers of mass; all the bodies are assumed to have uniform density. The parallel-axis theorem can be used to find the moment of inertia about any axis parallel to the center-of-mass axis. Furthermore, by means of combinations of the formulas of Table 11.1, it is possible to find the moments of inertia of some other bodies.

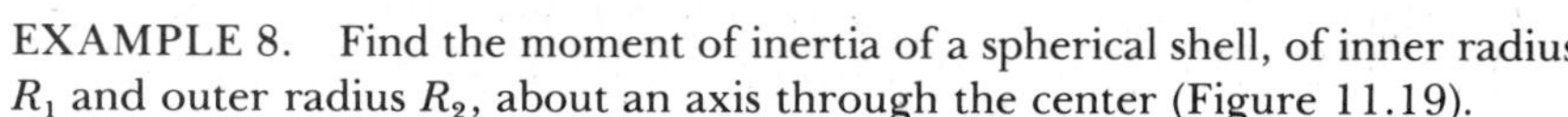

EXAMPLE 8. Find the moment of inertia of a spherical shell, of inner radius R_1 and outer radius R_2, about an axis through the center (Figure 11.19).

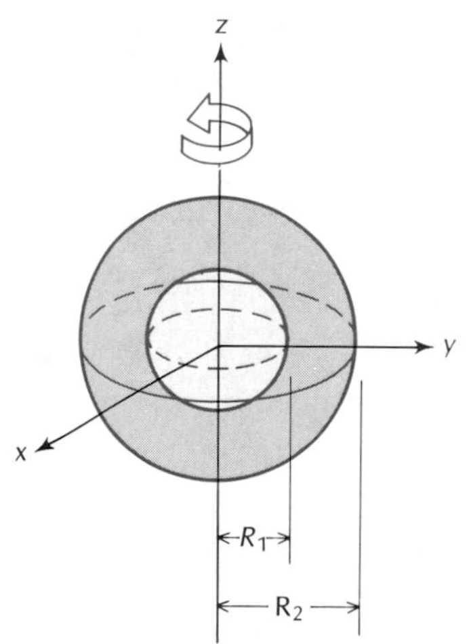

Fig. 11.19 A thick spherical shell of inner radius R_1 and outer radius R_2.

SOLUTION: According to Table 11.1, the moment of inertia of a solid sphere of radius R_2 is

$$I = \frac{2}{5} M R_2^2$$

Since $M = \rho V = \rho \times \frac{4}{3}\pi R_2^3$, this is the same as

$$I = \frac{2}{5} \frac{4\pi}{3} \rho R_2^5$$

The shell of Figure 11.19 is a solid sphere with a concentric spherical hole. Correspondingly, the inertia is that of a solid sphere of radius R_2 minus that of a concentric sphere of radius R_1:

$$I = \frac{2}{5} \frac{4\pi}{3} \rho (R_2^5 - R_1^5)$$

The volume of the shell is

$$\frac{4\pi(R_2^3 - R_1^3)}{3}$$

and hence the density is

$$\rho = M/[4\pi(R_2^3 - R_1^3)/3]$$

where M is now the mass of the shell. This leads to

$$I = \frac{2}{5} \frac{R_2^5 - R_1^5}{R_2^3 - R_1^3} M$$

For the inertia of the spherical shell.

Table 11.1 Some Moments of Inertia

Body	Moment of inertia
Thin hoop about symmetry axis	MR^2
Thin hoop about diameter	$\frac{1}{2}MR^2$
Disk or cylinder about symmetry axis	$\frac{1}{2}MR^2$
Cylinder about diameter through center	$\frac{1}{4}MR^2 + \frac{1}{12}Ml^2$
Thin rod about perpendicular axis through center	$\frac{1}{12}Ml^2$
Thin rod about perpendicular axis through end	$\frac{1}{3}Ml^2$
Sphere about diameter	$\frac{2}{5}MR^2$
Thin spherical shell about diameter	$\frac{2}{3}MR^2$

11.5 Angular Momentum of a Rigid Body

In Section 5.6 we defined the angular momentum of a particle as

$$\mathbf{L} = \mathbf{r} \times \mathbf{p} = m\mathbf{r} \times \mathbf{v}$$

where $\mathbf{r}$ is the position vector of the particle (relative to a given origin) and $\mathbf{p}$ is the momentum. The total **angular momentum of a rigid body** is the sum of the angular momenta of all the particles in the body. If these particles have masses m_i, velocities $\mathbf{v}_i$, and position vectors $\mathbf{r}_i$ (relative to a given origin of coordinates), then the total angular momentum is

$$\mathbf{L} = \sum_{i=1}^{n} m_i \mathbf{r}_i \times \mathbf{v}_i \qquad (41)$$

Angular momentum of a rigid body

As we pointed out in Section 5.6, the value of the angular momentum obtained from this formula depends on the choice of origin of coordinates. For the calculation of the angular momentum of a rigid body rotating about a fixed axis, it is usually convenient to choose an origin on the axis of rotation or at the center of mass. Since the angular momentum is a vector, it has three separate components; in general, these components are complicated functions of the geometry, orientation, and angular velocity of the body.

Incidentally, making use of the metric unit of energy, we see that the unit of angular momentum can be conveniently written as joule-second. Since $1\ \text{J} = 1\ \text{kg}\cdot\text{m}^2/\text{s}^2$, this unit agrees with the unit $\text{kg}\cdot\text{m}^2/\text{s}$ of angular momentum that we introduced in Section 5.6. Table 11.2 gives the values of some typical angular momenta.

Table 11.2 SOME ANGULAR MOMENTA

Orbital motion of all planets in Solar System	$3.2 \times 10^{43}\ \text{J}\cdot\text{s}$
Orbital motion of Earth	$2.7 \times 10^{40}\ \text{J}\cdot\text{s}$
Rotation of Earth	$5.8 \times 10^{33}\ \text{J}\cdot\text{s}$
Helicopter rotor (320 rev/min)	$5 \times 10^{4}\ \text{J}\cdot\text{s}$
Automobile wheel (55 mi/h)	$1 \times 10^{2}\ \text{J}\cdot\text{s}$
Electric fan	$1\ \text{J}\cdot\text{s}$
Frisbee	$1 \times 10^{-1}\ \text{J}\cdot\text{s}$
Toy gyroscope	$1 \times 10^{-1}\ \text{J}\cdot\text{s}$
Phonograph record ($33\frac{1}{3}$ rev/min)	$6 \times 10^{-3}\ \text{J}\cdot\text{s}$
Bullet fired from rifle	$2 \times 10^{-3}\ \text{J}\cdot\text{s}$
Orbital motion of electron in atom	$1.05 \times 10^{-34}\ \text{J}\cdot\text{s}$
Spin of electron	$0.53 \times 10^{-34}\ \text{J}\cdot\text{s}$

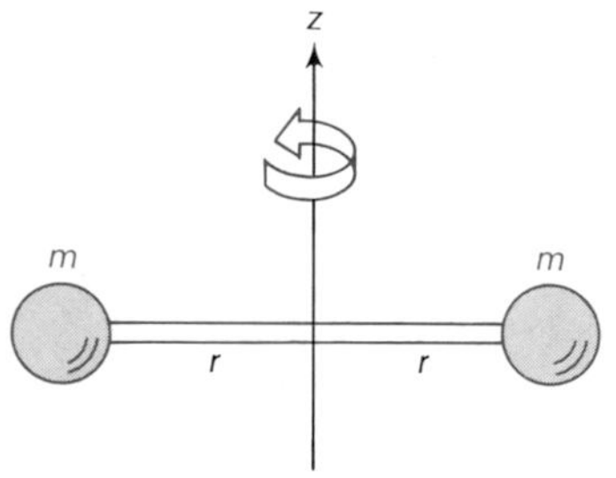

Fig. 11.20 Two particles of mass m at the ends of a massless rod.

EXAMPLE 9. Figure 11.20 shows a rigid body consisting of two particles of mass m attached to the ends of a (nearly) massless rigid rod of length $2r$. The body rotates with angular velocity ω about a perpendicular axis through the center of the rod. Find the angular momentum.

SOLUTION: Since the instantaneous position vector of each particle is perpendicular to the instantaneous velocity vector, the magnitude of the angular momentum vector $m\mathbf{r} \times \mathbf{v}$ of each particle is simply mrv. The direction of the angular momentum vector of each particle is parallel to the axis of rotation. The total angular momentum is

$$L = mrv + mrv = 2mrv = 2mr^2\omega$$

where we have taken into account that $v = r\omega$.

EXAMPLE 10. Suppose that the body described in the preceding example rotates with angular velocity ω about an axis inclined at an angle θ with respect to the rod (Figure 11.21). Find the angular momentum.

SOLUTION: The distance between each mass and the axis of rotation is

$$R = r \sin\theta \tag{42}$$

and the velocity of each mass has a magnitude

$$v = \omega R = \omega r \sin\theta \tag{43}$$

The direction of the velocity is perpendicular to the position vector. Hence the angular momentum of each mass has a magnitude

$$m\,|\mathbf{r} \times \mathbf{v}| = mrv = m\omega r^2 \sin\theta \tag{44}$$

The direction of the angular momentum of each mass is perpendicular to both the velocity and the position vectors, as specified by the right-hand rule. The angular momentum vector of each mass is shown in Figure 11.21; these vectors are parallel to each other, they are in the plane of the axis and rod, and they make an angle of $90° - \theta$ with the axis. The total angular momentum is then the vector sum of these individual angular momenta. This vector is in the same direction as the individual angular momentum vectors and has a magnitude twice as large as Eq. (44):

$$L = 2m\omega r^2 \sin\theta \tag{45}$$

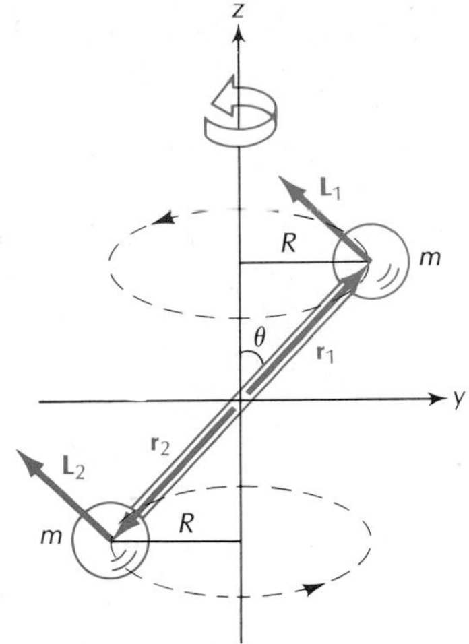

Fig. 11.21 Two particles of mass m at the ends of a rod that makes an angle θ with the axis of rotation. At the instant shown here the angular momentum vectors $\mathbf{L}_1$ and $\mathbf{L}_2$ of the two particles are both in the y–z plane.

At the instant shown in Figure 11.21, the rod is in the z–y plane. Hence the instantaneous angular momentum is also in this plane and has components

$$L_z = L\cos(90° - \theta) = 2m\omega r^2 \sin^2\theta \tag{46}$$

$$L_y = -L\cos\theta = -2m\omega r^2 \sin\theta\cos\theta \tag{47}$$

As the body rotates, so does the angular momentum vector; i.e., the components of this vector change but its magnitude remains fixed at the value given by Eq. (45).

Note that Eq. (46) can also be written

$$L_z = 2m\omega R^2 \tag{48}$$

where $R = r\sin\theta$ is the distance between each mass and the axis of rotation. Since $2mR^2$ is simply the moment of inertia of our bodies about the z axis, Eq. (48) is the same as

$$L_z = I\omega \tag{49}$$

As we will see below, this formula is of general validity.

The preceding example shows that the angular momentum vector of a rotating body need not always lie along the axis of rotation. However, if the axis of rotation is also an axis of symmetry of the body, then the angular momentum vector will lie along this axis (Figure 11.22). For such a symmetric body, each particle on one side of the axis has a counterpart on the other side of the axis. The angular momenta of these two particles have the same component along the axis (z components in Figure 11.22), but opposite components transverse to the axis (y components in Figure 11.22). Thus, the net angular momentum contributed by this pair of particles — and any other pair of particles — has only a component along the axis.

In any case, for the remainder of this section, we will focus our attention exclusively on the component of the angular momentum *along the axis of rotation.* Figure 11.23 shows an arbitrary rigid body rotating about a fixed axis, which coincides with the z axis. The relevant component of the angular momentum is then the z component. To evaluate this component, we consider one of the particles of the body (see

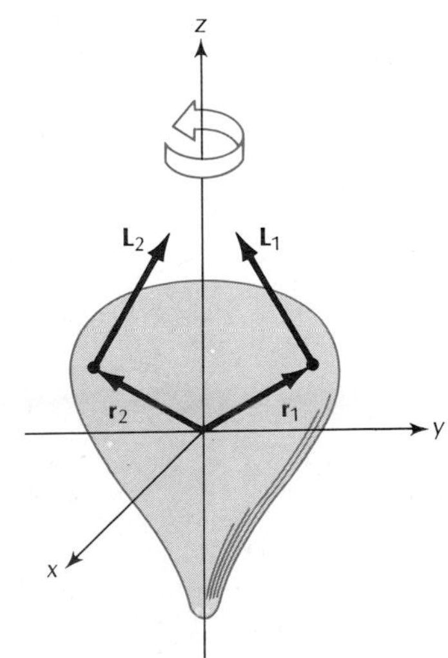

Fig. 11.22 A rigid body rotating about its axis of symmetry.

Figure 11.23). The position vector **r** of the particle makes an angle θ with the z axis. By a calculation similar to that which led to Eqs. (45) and (48), we find that the magnitude of the angular momentum of this particle is $m\omega r^2 \sin\theta$ and the z component of the angular momentum is $m\omega R^2$, where, as in Eq. (48), R is the distance between the particle and the axis of rotation. Hence the z component of the total angular momentum of all the particles in the rotating body is

$$L_z = \sum_{i=1}^{n} m_i R_i^2 \omega \tag{50}$$

or

Component of angular momentum along the axis of rotation

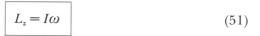

$$\boxed{L_z = I\omega} \tag{51}$$

where $I = \Sigma m_i R_i^2$ is the usual expression [Eq. (22)] for the total moment of inertia of the body (about the z axis). This establishes the general validity of the simple formula [Eq. (49)] that we found in Example 10: the component of the angular momentum along the axis of rotation is the product of moment of inertia and angular velocity. This result will be very useful in the next chapter when we study the dynamics of rotational motion. Equation (51) is reminiscent of the familiar expression $p = mv$ for the (linear) momentum of a particle — the moment of inertia again replaces the mass and the angular velocity replaces the linear velocity.

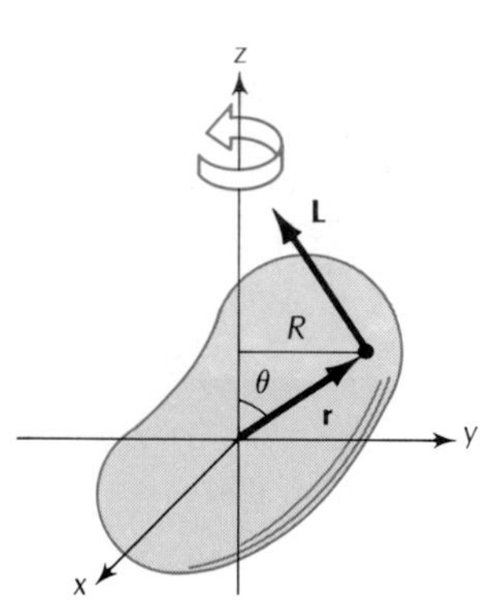

Fig. 11.23 An arbitrary rigid body rotating about a fixed axis.

If we solve Eq. (51) for ω,

$$\omega = L_z/I \tag{52}$$

and substitute this into the formula for the rotational kinetic energy, we obtain

$$K = \tfrac{1}{2}I\omega^2 = \tfrac{1}{2}I(L_z/I)^2 \tag{53}$$

or

$$K = \tfrac{1}{2}L_z^2/I \tag{54}$$

This is the analog of Eq. (7.33).

Spin

The angular momentum associated with the rotational motion of a rigid body is often called the **spin** to distinguish it from the angular momentum corresponding to the translational motion. For example, the Earth has spin angular momentum because of its rotation about its axis and it has orbital momentum because of its motion around the Sun.

EXAMPLE 11. The moment of inertia of the Earth about an axis through the poles is 8.1×10^{37} kg·m². What is the spin angular momentum? What is the rotational kinetic energy?

SOLUTION: The frequency of rotation is one revolution per day.[4]

[4] Strictly, the frequency is one revolution per sidereal day (23 h 56 min). But within the accuracy of our present calculation, we can ignore this small correction.

$$\nu = 1/\text{day} = 1/8.6 \times 10^4 \text{ s} = 1.2 \times 10^{-5}/\text{s} \tag{55}$$

and

$$\omega = 2\pi\nu = 7.3 \times 10^{-5} \text{ radian/s} \tag{56}$$

The angular momentum is then

$$L_z = I\omega$$

$$= 8.1 \times 10^{37} \text{ kg} \cdot \text{m}^2 \times 7.3 \times 10^{-5} \text{ radian/s}$$

$$= 5.9 \times 10^{33} \text{ kg} \cdot \text{m}^2/\text{s}$$

and the kinetic energy

$$K = \tfrac{1}{2}L_z^2/I$$

$$= \tfrac{1}{2}(5.9 \times 10^{33} \text{ kg} \cdot \text{m}^2/\text{s})^2/(8.1 \times 10^{37} \text{ kg} \cdot \text{m}^2)$$

$$= 2.1 \times 10^{29} \text{ J}$$

The subatomic particles — electrons, protons, and neutrons — all have intrinsic spin angular momenta. The magnitudes of the spin angular momenta of electrons, protons, and neutrons are equal: they all are 0.53×10^{-34} J · s. This magnitude of angular momentum is usually expressed as $\frac{1}{2}\hbar$, where $\hbar = 1.05 \times 10^{-34}$ J · s is a fundamental unit of angular momentum which plays a pervasive role in modern quantum physics.[5] (The constant $\hbar$, pronounced *h bar,* is Planck's constant h divided by 2π, $\hbar = h/2\pi$). The spin angular momentum of a subatomic particle is analogous to the spin angular momentum that the Earth possesses because of its rotation about its axis. However, this analogy between the spin of a subatomic particle and the classical angular momentum of rotation of a rigid body must not be taken too seriously — the spin of subatomic particles involves quantum mechanics in a very essential way. In contrast to a classical angular momentum, the spin of a subatomic particle has an absolutely fixed and immutable magnitude. According to modern quantum physics, external forces cannot change the magnitude of the spin of such a particle; they can only change the direction of this spin.

SUMMARY

Average angular velocity: $\bar{\omega} = \dfrac{\Delta\phi}{\Delta t}$

Instantaneous angular velocity: $\omega = \dfrac{d\phi}{dt}$

Average angular acceleration: $\bar{\alpha} = \dfrac{\Delta\omega}{\Delta t}$

Instantaneous angular acceleration: $\alpha = \dfrac{d\omega}{dt}$

Speed of particle on rotating body: $v = R\omega$

[5] See Appendix 8 for a more precise value of $\hbar$.

Acceleration of particle on rotating body:

$$a_{\text{tan}} = R\alpha$$

$$a_{\text{cent}} = R\omega^2$$

Motion with constant angular acceleration:

$$\omega = \omega_0 + \alpha t$$

$$\phi - \phi_0 = \omega_0 t + \tfrac{1}{2}\alpha t^2$$

$$\alpha(\phi - \phi_0) = \tfrac{1}{2}(\omega^2 - \omega_0^2)$$

Moment of inertia: $I = \sum_{i=1}^{n} m_i R_i^2$

$$I = \int \rho R^2 \, dV$$

Kinetic energy of rotation: $K = \tfrac{1}{2}I\omega^2$

Parallel-axis theorem: $I = I_{\text{CM}} + Md^2$

Perpendicular-axis theorem (for a flat plate in the *x–y* plane):

$$I_z = I_x + I_y$$

Component of angular momentum along axis of rotation: $L_z = I\omega$

QUESTIONS

1. A spinning flywheel in the shape of a disk suddenly shatters into many small fragments. Draw the trajectories of a few of these small fragments; assume that the fragments do not interfere with each other.

2. You may have noticed that in some old movies the wheels of carriages or stagecoaches seem to rotate backward. How does this come about?

3. Relative to an inertial reference frame, what is your angular velocity right now about an axis passing through your center of mass?

4. Consider the wheel of an accelerating automobile. Draw the instantaneous acceleration vectors for a few points on the rim of the wheel.

5. When engineers design the teeth for gears, they usually make the sides of the teeth round (Figure 11.24). Can you guess why?

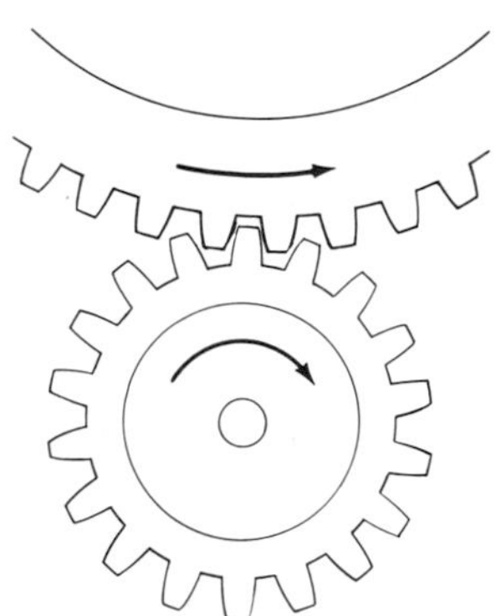

Fig. 11.24 Gears.

6. The hands of a watch are small rectangles with a common axis passing through one end. The minute hand is long and thin; the hour hand is short and thicker. Both hands have the same mass. Which has the larger moment of inertia? Which has the greater kinetic energy and angular momentum?

7. What configuration and what axis would you choose to give your body the smallest possible moment of inertia? The greatest?

8. About what axis through the center of mass is the amount of inertia of this book largest? Smallest? Assume the book is closed.

9. A circular hoop made of thin wire has a radius R and mass M. About what axis must you rotate this hoop to obtain the minimum moment of inertia? What is the value of this minimum?

10. Automobile engines and other internal combustion engines have flywheels attached to their crankshafts. What is the purpose of these flywheels? (Hint: Each explosive combustion is one of the cylinders of such an engine gives a sudden push to the crankshaft. How would the crankshaft respond to this push if it had no flywheel?)

11. What is the direction of the angular momentum of rotation of the Earth?

12. A bicycle is traveling east along a level road. What is the direction of the angular momentum of its wheels?

PROBLEMS

Section 11.2

1. The minute hand of a wall clock has a length of 20 cm. What is the speed of the tip of this hand?

2. Quito is on the Earth's equator; New York is at latitude 41° North. What is the angular velocity of each city about the Earth's axis of rotation? What is the linear speed of each?

3. An aircraft passes directly over you with a speed of 900 km/h at an altitude of 10,000 m. What is the angular velocity of the aircraft (relative to you) when directly overhead? Three minutes later?

4. The outer edge of the grooved area of a long-playing record is at a radial distance of 5.75 in. from the center; the inner edge is at a radial distance of 2.50 in. The record rotates at $33\frac{1}{3}$ rev/min. The needle of the pick-up arm takes 25 min to play the record and in that time interval it moves uniformly and radially from the outer edge to the inner edge. What is the radial speed of the needle? What is the speed of the outer edge relative to the needle? What is the speed of the inner edge relative to the needle?

5. Consider the phonograph record described in Problem 4. What is the total length of the groove in which the needle travels?

6. An automobile has wheels with a radius of 1.0 ft. What is the angular velocity (in radians per second) and the frequency (in revolutions per second) of the wheels when the automobile is traveling at 55 mi/h?

7. An automobile has wheels of diameter 0.63 m. If the automobile is traveling at 80 km/h, what is the instantaneous velocity vector (relative to the ground) of a point at the top of a wheel? At the bottom? At the front?

8. Because of the rotation of the Earth about its axis, a stone resting on a tower at, say, the equator has an eastward velocity in an inertial reference frame centered on the Earth but not rotating with it. In this reference frame, the foot of the tower also has an eastward velocity, but slightly smaller than that of the stone (Figure 11.25).

(a) Show that if the height of the tower is h, the difference in the magnitude of these velocities is ωh, where ω is the angular velocity of rotation of the Earth.

(b) If the stone drops from the tower, its excess eastward velocity will cause it to land at some distance from the foot of the tower. Show that it will land at a distance of about $\omega h\sqrt{2h/g}$ eastward of the foot of the tower. Evaluate for $h = 30$ m.

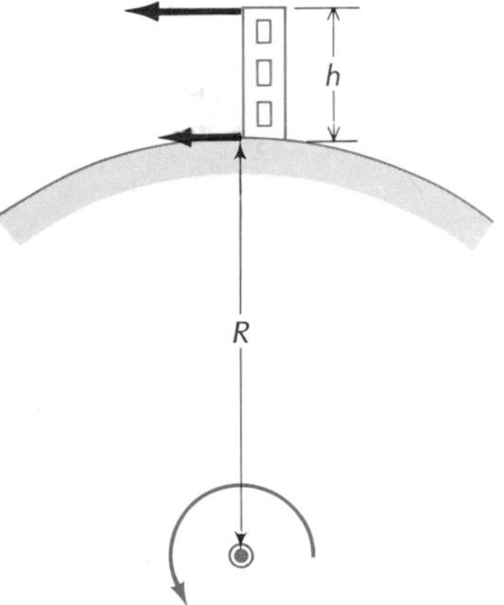

Fig. 11.25

Section 11.3

9. The blade of a circular saw of diameter 8 in. accelerates uniformly from rest to 7000 rev/min in 1.2 s. What is the angular acceleration? How many revolutions will the blade have made by the time it reaches full speed?

10. An automobile accelerates uniformly from 0 to 50 mi/h in 6.0 s. The automobile has wheels of radius 12 in. What is the angular acceleration of the wheels?

11. When you turn off the motor, a phonograph turntable initially rotating at $33\frac{1}{3}$ rev/min makes 25 revolutions before it stops. Calculate the angular deceleration of this turntable; assume it is constant.

12. The rotation of the Earth is slowing down. In 1977, the Earth took 1.01 s longer to complete 365 rotations than in 1900. What was the average angular deceleration of the Earth in the time interval from 1900 to 1977?

13. An automobile engine accelerates at a constant rate from 200 rev/min to 3000 rev/min in 7.0 s and then runs at constant speed.
(a) Find the angular velocity and the angular acceleration at $t = 0$ (just after acceleration begins) and at $t = 7.0$ s (just before acceleration ends).
(b) A flywheel with a radius of 0.6 ft is attached to the shaft of the engine. Calculate the tangential and the centripetal acceleration of a point on the rim of the flywheel at the times given above.
(c) What angle does the net acceleration vector make with the radius at $t = 0$ and at $t = 7.0$ s? Draw diagrams showing the wheel and the acceleration vector at these times.

Section 11.4

14. According to spectroscopic measurements, the moment of inertia of an oxygen molecule about an axis through the center of mass and perpendicular to the line joining the atoms is 1.95×10^{-46} kg · m². The mass of an oxygen atom is 2.66×10^{-26} kg. What is the distance between the atoms? Treat the atoms as pointlike particles.

15. The moment of inertia of the Earth about its polar axis is $0.331M_{\text{E}}R_{\text{E}}^2$, where M_{E} is the mass and R_{E} the equatorial radius. Why is this moment of inertia smaller than that of a sphere of uniform density? What would the radius of a sphere of uniform density have to be if its mass and moment of inertia are to coincide with those of the Earth?

16. Problem 9.20 gives the dimensions of a molecule of nitric acid (HNO_3). What is the moment of inertia of this molecule when rotating about the symmetry axis passing through the H, O, and N atoms? Treat the atoms as pointlike particles.

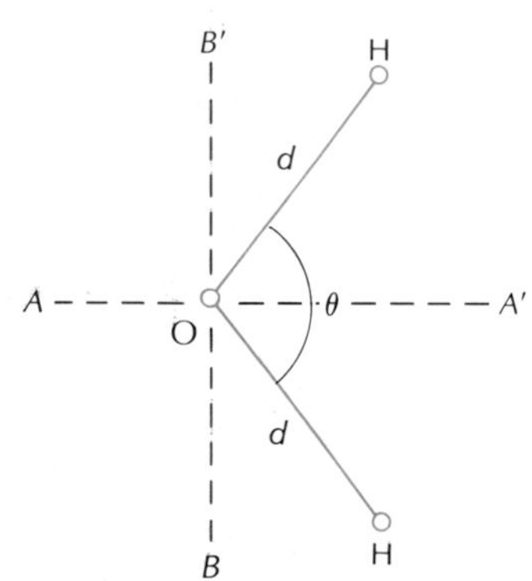

Fig. 11.26 Water molecule.

17. The water molecule has the shape shown in Figure 11.26. The distance between the oxygen and the hydrogen atoms is d and the angle between the hydrogen atoms is θ. From spectroscopic investigations it is known that the moment of inertia of the molecule is 1.93×10^{-47} kg · m² for rotation about the axis AA' and 1.14×10^{-47} kg · m² for rotation about the axis BB'. From this information and the known values of the masses of the atoms, determine the values of d and θ. Treat the atoms as pointlike.

18. What is the moment of inertia (about the axis of symmetry) of a piece of pipe of inner radius R_1, outer radius R_2, and mass M?

19. An empty beer can has a mass of 50 g, a length of 12 cm, and a radius of 3.3 cm. Find the moment of inertia of the can about its axis of symmetry. Assume that the can is a perfect cylinder of sheet metal with no ridges, indentations, or holes.

20. Suppose that a supertanker of mass 6.6×10^8 kg moves from Venezuela (latitude 10° North) to Holland (latitude 53° North). What is the change of the moment of inertia of the Earth–ship system?

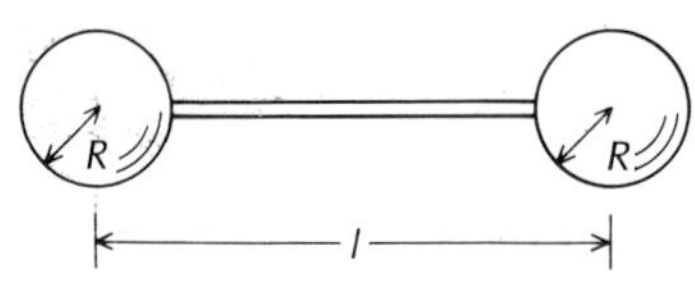

Fig. 11.27 A dumbbell.

21. A dumbbell consists of two uniform spheres of mass M and radius R joined by a thin rod of mass m and length l (Figure 11.27). What is the moment of inertia of this device about an axis through the center of the rod perpendicular to the rod? About an axis along the rod?

22. Suppose that the Earth consists of a spherical core of mass $0.22M_{\text{E}}$ and radius $0.54R_{\text{E}}$ and a surrounding mantle (a spherical shell) of mass $0.78M_{\text{E}}$ and outer radius R_{E}. Suppose that the core is of uniform density and the mantle is also of uniform density. According to this model, what is the moment of inertia of the Earth? Express your answer as a multiple of $M_{\text{E}}R_{\text{E}}^2$.

23. In order to increase her moment of inertia about a vertical axis, a spinning figure skater stretches out her arms horizontally; in order to reduce her moment of inertia, she brings her arms down vertically along her sides. Calculate the change of moment of inertia between these two configurations of the arms. Assume that each arm is a thin, uniform rod of length 0.60 m and mass 2.8 kg hinged at the shoulder at a distance of 0.20 m from the axis of rotation.

24. Find the moment of inertia of a thin rod of mass M and length L about an axis through the center inclined at an angle θ with respect to the rod.

25. Use the parallel-axis theorem to derive Eq. (32) from Eq. (30).

26. Given that the moment of inertia of a sphere about a diameter is $\frac{2}{5}MR^2$, show that the moment of inertia about an axis tangent to the surface is $\frac{7}{5}MR^2$.

27. Find a formula for the moment of inertia of a uniform, thin, square plate (mass m, dimension $l \times l$) rotating about an axis that coincides with one of its edges.

28. What is the moment of inertia of a thin, flat plate in the shape of a semicircle rotating about the straight side (Figure 11.28)? The mass of the plate is M and the radius is R.

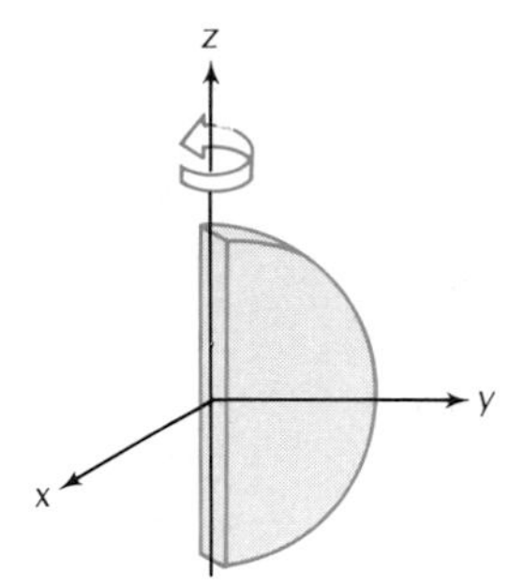

Fig. 11.28

29. A solid cylinder capped with two solid hemispheres rotates about its axis of symmetry (Figure 11.29). The radius of the cylinder is R, its height is h, and the total mass (hemispheres included) is M. What is the moment of inertia?

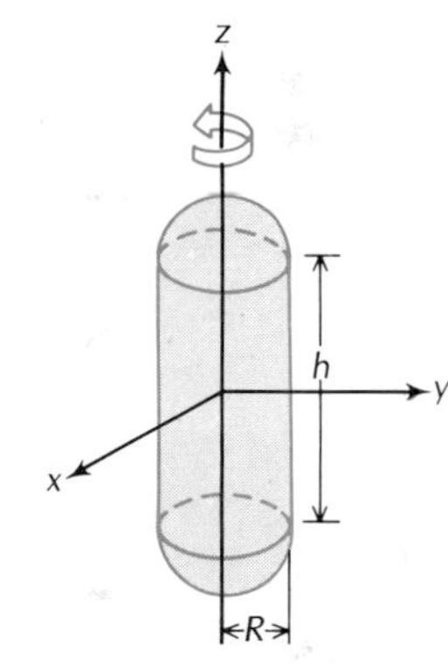

Fig. 11.29

30. A hole of radius r has been drilled in a circular, flat plate of radius R (Figure 11.30). The center of the hole is at a distance d from the center of the circle. The mass of this body is M. Find the moment of inertia for rotation about an axis through the center of the circle, perpendicular to the plate.

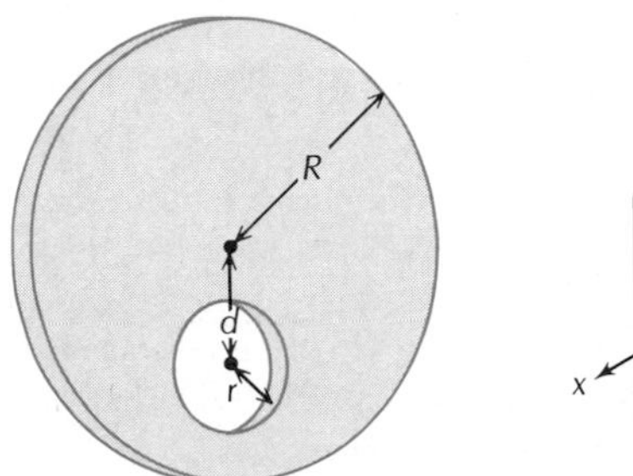

Fig. 11.30

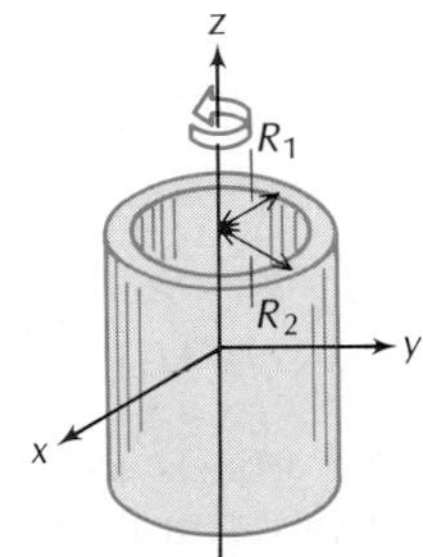

Fig. 11.31

31. A piece of steel pipe (Figure 11.31) has an inner radius R_1, an outer radius R_2, and a mass M. Find the moment of inertia about the axis of the pipe.

32. Find the moment of inertia of a flywheel of mass M made by cutting four large holes of radius r out of a uniform disk of radius R (Figure 11.32). The holes are centered at a distance $R/2$ from the center of the flywheel.

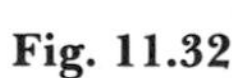

Fig. 11.32

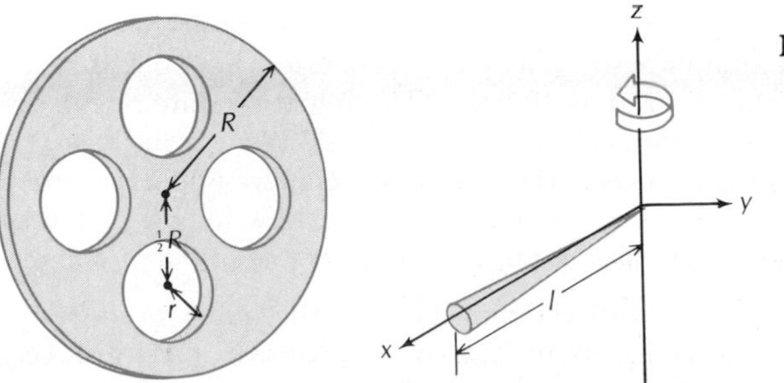

Fig. 11.33

33. Show that the moment of inertia of a long, very thin cone (Figure 11.33) about an axis through the apex and perpendicular to the centerline is $\frac{3}{5}Ml^2$, where M is the mass and l the height of the cone.

34. The mass distribution within the Earth can be roughly approximated by several concentric spherical shells, each of constant density. The following table gives the outer and the inner radius of each shell and its mass (expressed as a fraction of the Earth's mass):

Shell	*Outer radius*	*Inner radius*	*Fraction of mass*
1	6400 km	5400 km	0.28
2	5400	4400	0.25
3	4400	3400	0.16
4	3400	2400	0.20
5	2400	0	0.11

Use these data to calculate the moment of inertia of the Earth about its axis.

35. An airplane propeller consists of three radial blades, each of length 1.8 m and mass 20 kg. What is the kinetic energy of this propeller when rotating at 2500 rev/min? Assume that each blade is (approximately) a uniform rod.

36. The drilling pipe of an oil rig is 2 km long, 15 cm in diameter, and has a mass of 20 kg per meter of length. Assume that the wall of the pipe is very thin.

(a) What is the moment of inertia of this pipe rotating about its longitudinal axis?

(b) What is the kinetic energy when rotating at 1 rev/s?

37. Engineers have proposed that large flywheels be used for the temporary storage of surplus energy generated by electric power plants. A suitable flywheel would be 12 ft in diameter, weigh 300 short tons (1 short ton = 2000 lb), and spin at 3000 revolutions per minute. What is the kinetic energy of rotation of this flywheel? Give the answer in both foot-pounds and kilowatt-hours. Assume that the moment of inertia of the flywheel is that of a uniform disk.

38. An automobile weighing 3000 lb has wheels 30.0 in. in diameter weighing 60.0 lb each. Taking into account the rotational kinetic energy of the wheels about their axles, what is the total kinetic energy of the automobile when traveling at 50.0 mi/h? What percentage of the kinetic energy belongs to the rotational motion of the wheels about their axles? Pretend that each wheel has a mass distribution equivalent to that of a uniform disk.

39. Derive the formula for the moment of inertia of a thin disk of mass M and radius R rotating about a diameter. (Hint: Use the perpendicular-axis theorem.)

40. Derive the formula for the moment of inertia of a thin hoop of mass M and radius R rotating about a diameter.

41. Pulsars are rotating stars made almost entirely of neutrons closely packed together. The rate of rotation of most pulsars gradually decreases because rotational kinetic energy is gradually converted into other forms of energy by a variety of complicated "frictional" processes. Suppose that a pulsar of mass 1.5×10^{30} kg and radius 20 km is spinning at the rate of 2.1 rev/s and is slowing down at the rate of 1.0×10^{-15} rev/s^2. What is the rate (in joules per second or watts) at which the rotational energy is decreasing? If this rate of decrease of the energy remains constant, how long will it take the pulsar to come to a stop? Treat the pulsar as a sphere of uniform density.

42. For the sake of directional stability, the bullet fired by a rifle is given a spin angular velocity about its axis by means of spiral grooves ("rifling") cut into the barrel. The bullet fired by a Lee-Enfield rifle is (approximately) a uniform cylinder of length 3.18 cm, diameter 0.790 cm, and mass 13.9 g. The bullet emerges from the muzzle with a translational velocity of 628 m/s and a spin angular velocity of 2.47×10^3 rev/s. What is the translational kinetic en-

ergy of the bullet? What is the rotational kinetic energy? What fraction of the total kinetic energy is rotational?

43. Find a formula for the moment of inertia of a uniform, thin, square plate (mass M, dimension $l \times l$) rotating about an axis through the center and perpendicular to the plate.

*44. Find the moment of inertia of a uniform cube of mass M and edge l. Assume the axis of rotation passes through the center of the cube and is perpendicular to two of the faces.

*45. A cone of mass M has a height h and a base diameter R. Find its moment of inertia about its axis of symmetry.

*46. Derive the formula given in Table 11.1 for the moment of inertia of a sphere.

Section 11.5

47. According to a simple (but erroneous) model, the proton is a uniform rigid sphere of mass 1.67×10^{-27} kg and radius 1.0×10^{-15} m. The spin angular momentum of the proton is 5.3×10^{-35} J · s. According to this model, what is the angular velocity of rotation of the proton? What is the linear velocity of a point on its equator? What is the rotational kinetic energy? How does this rotational energy compare with the rest-mass energy mc^2?

48. A phonograph turntable is a uniform disk of radius 6.0 in. and mass 3.0 lb-mass. If this turntable accelerates from 0 rev/min to 78 rev/min in 2.5 s, what is the average rate of change of the angular momentum in this time interval?

49. The propeller shaft of a cargo ship has a diameter of 0.29 ft, a length of 87 ft, and a weight of 2600 lb. What is the rotational kinetic energy of this propeller shaft when it is rotating at 200 rev/min? What is the angular momentum?

50. The Sun rotates about its axis with a period of about 25 days. Its moment of inertia is $0.20M_SR_S^2$, where M_S is its mass and R_S its radius. Calculate the angular momentum of rotation of the Sun. Calculate the total orbital angular momentum of all the planets; make the assumption that each planet moves in a circular orbit of radius equal to its mean distance from the Sun listed in Table 13.1. What percentage of the angular momentum of the Solar System is in the rotational motion of the Sun?

51. The friction of the tides on the coastal shallows and the ocean floors gradually slows down the rotation of the Earth. The period of rotation (length of a sidereal day) is gradually increasing by 0.0016 s per century. What is the angular deceleration (in radians/s^2) of the Earth? What is the rate of decrease of the rotational angular momentum? What is the rate of decrease of the rotational kinetic energy? The moment of inertia of the Earth about its axis is $0.331M_ER_E^2$, where M_E is the mass of the Earth and R_E its equatorial radius.

*52. The spin angular momentum of the Earth has a magnitude of 5.9×10^{33} kg · m^2/s. Because of forces exerted by the Sun and the Moon, the spin angular momentum gradually changes direction, describing a cone of half angle 23.5° (Figure 11.34). The angular momentum vector takes 26,000 years to swing once around this cone. What is the magnitude of the rate of change of the angular momentum vector, i.e., what is the value of $|d\mathbf{L}/dt|$?

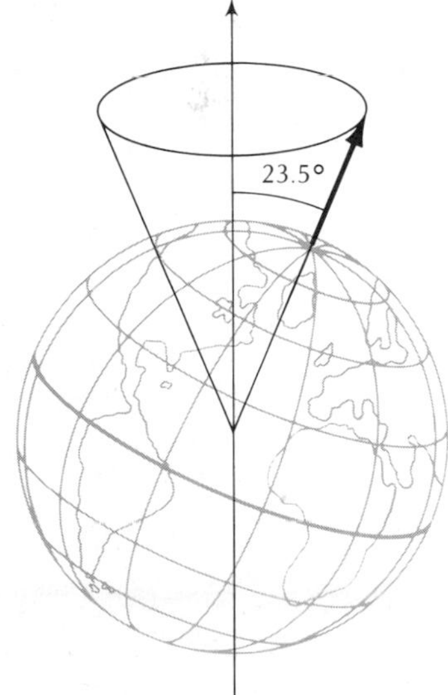

Fig. 11.34

CHAPTER 12

Dynamics and Statics of a Rigid Body

As we saw in Chapter 6, Newton's Second Law is the equation that determines the translational motion of a body. In this chapter we will derive an equation that determines the rotational motion of a rigid body. Just as Newton's equation of motion gives us the translational acceleration and permits us to calculate the change in velocity and position, the analogous equation for rotational motion gives us the angular acceleration and permits us to calculate the change in angular velocity and angular position. The equation for rotational motion is not a new law of physics, distinct from Newton's three laws. Rather, it is a consequence of these laws.

12.1 Torque

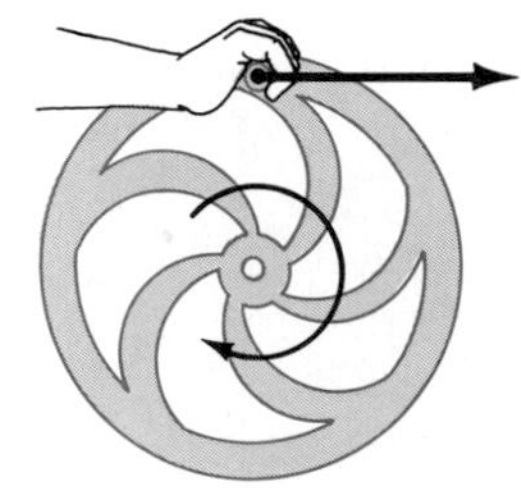

Fig. 12.1 Push of hand against crank of a wheel.

Before we lay down the laws for the rotational motion of a rigid body, we must discuss the concept of **torque** or **moment of a force.** As we will see in the next section, this quantity plays a role in rotational motion analogous to that played by the force in translational motion — it produces angular acceleration, just as force produces linear acceleration. For instance, the push of your hand against a crank on a wheel (Figure 12.1) exerts a torque or "twist" that produces angular acceleration of the wheel. The concept of torque was briefly mentioned in Sections 5.6 and 9.5; we will now spell out the definition of torque in detail. Suppose that a force **F** acts on a particle having a position vector **r** with respect to a given origin of coordinates; then the torque of this force with respect to the origin is the vector quantity

Torque of a force

$$\boxed{\boldsymbol{\tau} = \mathbf{r} \times \mathbf{F}} \qquad (1)$$

In terms of the angle θ between the position vector and the force (see Figure 12.2), the magnitude of the torque is

$$\tau = rF \sin \theta \qquad (2)$$

and the direction of the torque is perpendicular to both the position vector and the force, as indicated by the right-hand rule (Figure 12.2).

The quantity $r \sin \theta$ appearing in Eq. (2) has a simple geometric interpretation: it is the perpendicular distance between the line of action of the force and the origin of coordinates (Figure 12.3a); this perpendicular distance is called the **moment arm** of the force. Hence Eq. (2) states that the magnitude of the torque equals the magnitude of the force multiplied by the moment arm.

Alternatively, the quantity $F \sin \theta$ can be given an interpretation: it is the component of the force along a direction perpendicular to the position vector, i.e., it is the transverse component of the force (Figure 12.3b). Hence Eq. (3) states the magnitude of the torque equals the radial distance multiplied by the transverse component of the force.

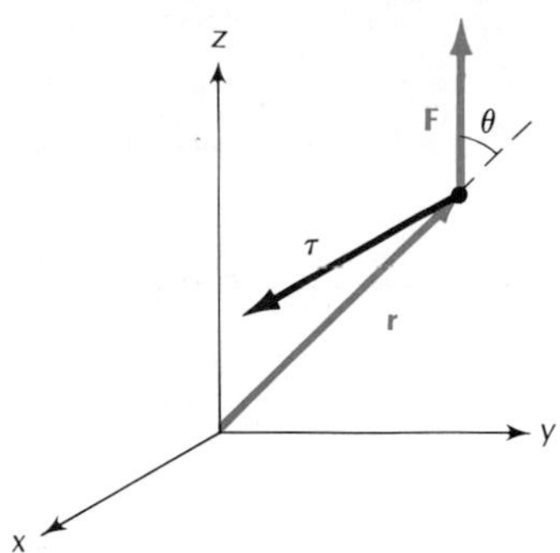

Fig. 12.2 Position vector, force vector, and torque vector. Here $\boldsymbol{\tau}$ is parallel to the x axis and $\mathbf{r}$ and $\mathbf{F}$ are in the y–z plane.

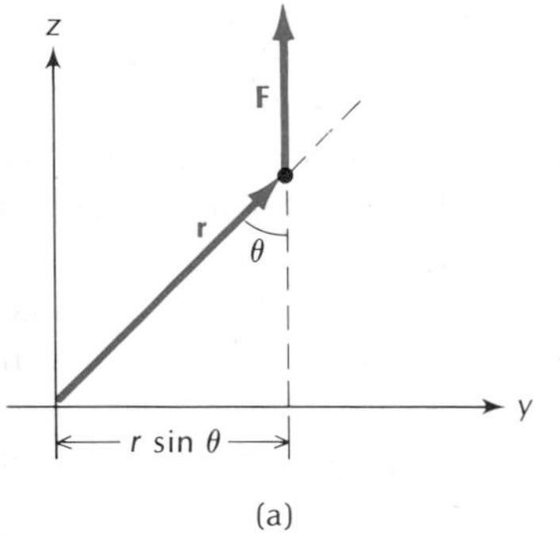

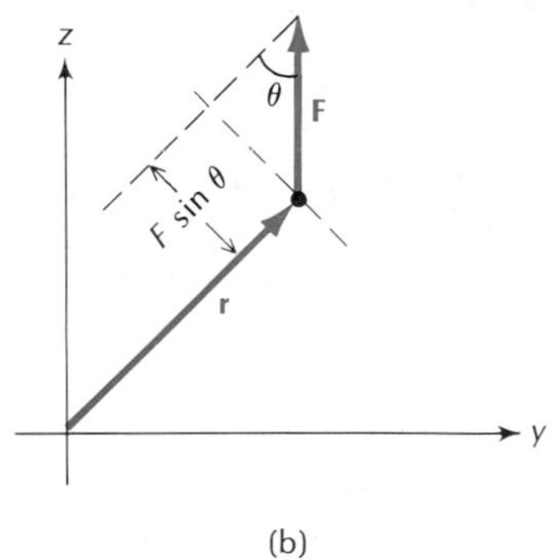

Fig. 12.3 (a) Here $r \sin \theta$ is the perpendicular distance between the line of action of the force and the origin of coordinates. (b) Here $F \sin \theta$ is the component of **F** perpendicular to **r**.

If the force is parallel to the position vector, then $\tau = 0$; if the force is perpendicular to the position vector, then $\tau = rF$ (Figure 12.4). Thus, as a function of the angle θ, a force of given magnitude generates a maximum torque if it is perpendicular to the position vector. If the particle on which the force acts belongs to a rigid body, then we can readily recognize that this dependence of the torque on the angle θ agrees with our intuition about the rotational effect of a force. For instance, if we want to make a wheel rotate about an origin at the center of the wheel, we must push on some point of the wheel transversely to the radial line; if we push along the radial line, we will merely displace the wheel in the direction of this push, without rotation.

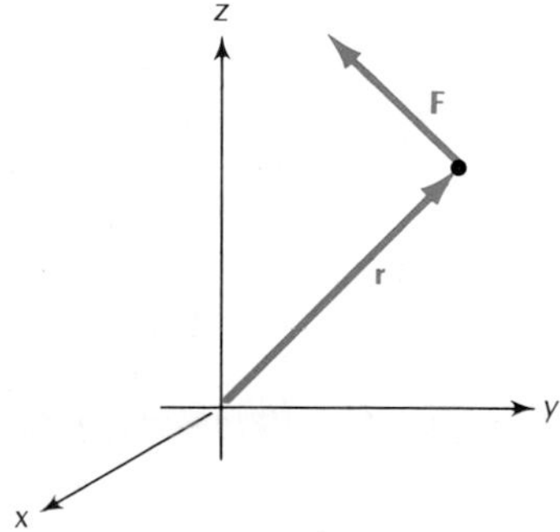

Fig. 12.4 An example of a force is perpendicular to the position vector.

It is important to keep in mind that the torque of a given force depends on the choice of origin; for instance, if we were to change the origin of coordinates of Figure 12.2 so that it sits on the line of action of the force, then the torque would be zero. This is why we call the expansion (1) the torque *with respect to the origin* or the torque *about the origin.*

In the metric system the unit of torque is the newton-meter. This means that the unit of torque is the same as the unit of work. Since $1\text{N} \cdot \text{m} = 1\text{ J}$, we could in principle measure torque in joules; however, in order to maintain at least some distinction between the units of torque and work, we will insist on the exclusive use of newton-meters for the former and joules for the latter.

In the British system the unit of torque is the pound-foot.

EXAMPLE 1. During a sudden stop, the horizontal braking force exerted by the road on the front wheel of a bicycle with a front-wheel brake is 600 N. What is the torque of this force about the center of mass of the bicycle and its rider? The center of mass is 95 cm above the road and 70 cm behind the point of contact of the front wheel with the ground.

SOLUTION: In Figure 12.5 the origin of coordinates is at the center of mass. The distance between the origin and the point of application of the force is

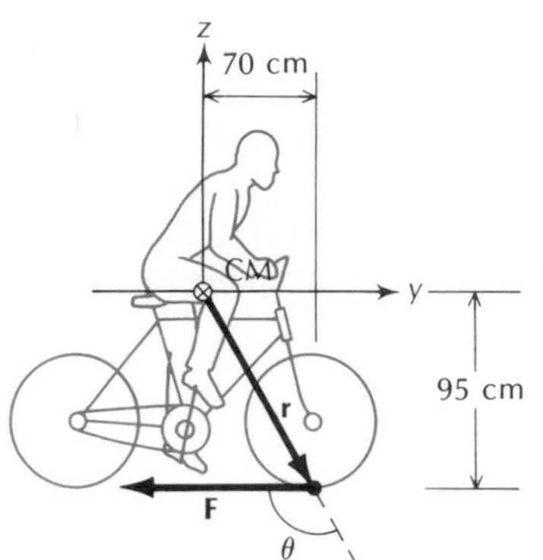

Fig. 12.5 Bicycle with front-wheel brake. The braking force **F** produces a torque about the center of mass (CM).

$$r = \sqrt{(0.70 \text{ m})^2 + (0.95 \text{ m})^2} = 1.18 \text{ m}$$

The sine of the angle between the position vector and the force is $\sin\theta = 0.95 \text{ m}/1.18 \text{ m} = 0.81$. Hence the magnitude of the torque is

$$\tau = rF\sin\theta = 1.18 \text{ m} \times 600 \text{ N} \times 0.81 = 570 \text{ N}\cdot\text{m}$$

To obtain the direction of $\boldsymbol{\tau}$ from the right-hand rule, we must place our fingers along the angle from **r** to **F** (see Figure 12.5); the direction of $\boldsymbol{\tau}$ is along the thumb, into the plane of the page. It is intuitively obvious that the torque exerted by the braking force will tend to flip the bicycle over its front wheel, i.e., this torque tends to produce clockwise rotation. Sudden braking with a front-wheel brake is hazardous!

Alternatively, we can calculate the components of $\boldsymbol{\tau}$ by means of Eq. (3.36). With the coordinates shown in Figure 12.5:

$$\mathbf{r} = 0\hat{\mathbf{x}} + (0.70 \text{ m})\hat{\mathbf{y}} - (0.95 \text{ m})\hat{\mathbf{z}}$$

$$\mathbf{F} = 0\hat{\mathbf{x}} - (600 \text{ N})\hat{\mathbf{y}} + 0\hat{\mathbf{z}}$$

so that Eq. (3.36) gives

$$\mathbf{r}\times\mathbf{F} = [0.70\times 0 - (0.95)\times(-600)]\ (\text{N}\cdot\text{m})\hat{\mathbf{x}} + 0\hat{\mathbf{y}} + 0\hat{\mathbf{z}}$$

$$= -(570 \text{ N}\cdot\text{m})\hat{\mathbf{x}} \tag{3}$$

This agrees with the preceding calculation in both magnitude and direction.

12.2 The Equation of Rotational Motion

Suppose that a system of particles is subjected to some external forces. According to the general result of Section 9.5, valid in any inertial reference frame, the rate of change of the total angular momentum of the system is then

$$\boxed{\frac{d\mathbf{L}}{dt} = \boldsymbol{\tau}_{\text{ext}}} \tag{4}$$

where $\boldsymbol{\tau}_{\text{ext}}$ is the total external torque, i.e., it is the sum of the external torques acting on all the particles in the system.

If the system of particles we are dealing with is a rigid body, then Eq. (4) is the dynamical equation that determines the rotational motion of this rigid body. Since this equation is a vector equation, it has three components:

$$\frac{dL_x}{dt} = \tau_{\text{ext},x} \tag{5}$$

$$\frac{dL_y}{dt} = \tau_{\text{ext},y} \tag{6}$$

$$\frac{dL_z}{dt} = \tau_{\text{ext},z} \tag{7}$$

These three equations are exactly what is necessary and sufficient to determine all three components of the rotational motion of a rigid body; in the terminology introduced in Section 11.1, the three equations given above determine the angular accelerations of the roll, pitch, and yaw motions of the rigid body.

Equation (4) for rotational motion is the analog of the equation $d\mathbf{P}/dt = \mathbf{F}_{\text{ext}}$ for translational motion. Together these equations entirely determine the rotational motion of a rigid body. It is worth emphasizing that our equations for the motion of a rigid body are not new laws, distinct from Newton's laws; our equations are nothing but consequences of Newton's laws as applied to a system of many particles.

In most of the following examples, we will be dealing with a rigid body that rotates about a fixed axis. We then only need one of Eqs. (5), (6), and (7). If the axis of rotation is along the z axis, the relevant equation is

$$\frac{dL_z}{dt} = \tau_z \tag{8}$$

Here we have omitted the label *ext* for the sake of brevity; this will cause no confusion, because in the following examples we will never concern ourselves with anything but external torques. According to Eq. (11.51),

$$L_z = I\omega \tag{9}$$

and hence

$$I\frac{d\omega}{dt} = \tau_z \tag{10}$$

or

$$\boxed{I\alpha = \tau_z} \tag{11}$$

Equation of rotational motion

This equation relates the angular acceleration about the axis of rotation to the torque about this axis. Obviously, this is the analog of Newton's equation $ma = F$.[1]

EXAMPLE 2. A phonograph turntable driven by an electric motor accelerates at a constant rate from 0 to $33\frac{1}{3}$ rev/min in a time of 2.0 s. The turntable is a uniform disk of metal, of a mass 1.5 kg and a radius 13 cm. What torque about the axis is required to drive this turntable? If the driving wheel makes contact with the turntable at its outer rim (Figure 12.6), what is the force that it must exert?

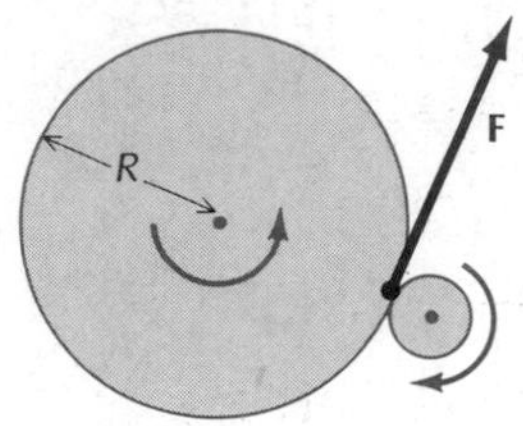

Fig. 12.6 The driving wheel exerts a force on the turntable.

SOLUTION: The final angular velocity is $\omega = 2\pi \times 33.3$ radian/60 s = 3.49 radian/s and the angular acceleration is

$$\alpha = \Delta\omega/\Delta t = (3.49 \text{ radian/s})/2.0 \text{ s} = 1.75 \text{ radian/s}^2$$

The moment of inertia of the turntable is that of a disk,

$$I = \tfrac{1}{2}MR^2 = \tfrac{1}{2} \times 1.5 \text{ kg} \times (0.13 \text{ m})^2$$

$$= 1.27 \times 10^{-2} \text{ kg} \cdot \text{m}^2$$

Hence the required torque is

$$\tau_z = I\alpha = 1.27 \times 10^{-2} \text{ kg} \cdot \text{m}^2 \times 1.75 \text{ radian/s}^2$$

$$= 2.22 \times 10^{-2} \text{ N} \cdot \text{m}$$

The driving force is perpendicular to the radius vector so that $\tau_z = RF$ and

$$F = \tau_z/R = 2.22 \times 10^{-2} \text{ N} \cdot \text{m}/0.15 \text{ m} = 0.147 \text{ N}$$

EXAMPLE 3. Two masses m_1 and m_2 are suspended from a string which runs, without slipping, over a pulley (Figure 12.7a). The pulley has a radius R and a moment of inertia I about its axle. Find the acceleration of the masses.

SOLUTION: We have already found the motion of this system in Section 6.3, where we neglected the inertia of the pulley. Now we will take this inertia into account.

Suppose that the tensions in the two parts of the string attached to the masses are T_1 and T_2. These tensions are shown in Figure 12.7b. (Note that T_1 and T_2 are not equal. If the moment of inertia of the pulley were zero, then these tensions would be equal. But if the moment of inertia is not zero, then a difference between T_1 and T_2 is required for acceleration.) The equations of motion of the two masses m_1 and m_2 are

$$T_1 - m_1g = m_1a$$

[1] The analogy between rotational and translational quantities mentioned in Section 11.3 can be extended as follows:

$$ma = F \quad \rightarrow \quad I\alpha = \tau_z$$
$$p = mv \quad \rightarrow \quad L_z = I\omega$$
$$K = \tfrac{1}{2}mv^2 \quad \rightarrow \quad K = \tfrac{1}{2}I\omega^2$$

As we will see in the next section, this analogy also encompasses the equations for work and power:

$$W = \int F\,dx \quad \rightarrow \quad W = \int \tau_z\,d\phi$$
$$P = Fv \quad \rightarrow \quad P = \tau_z\omega$$

and

$$T_2 - m_2 g = -m_2 a$$

where the acceleration a is reckoned as positive if the mass m_1 moves upward.

Figure 12.7c shows the pulley and the forces acting on it. (Since the string does not slip, it behaves as though it were attached to the pulley at the points of first contact; this is why the tension forces are shown acting at the points of first contact.) The upward supporting force at the axle generates no torque about the axle. The tension forces T_1 and T_2 generate torques $-RT_1$ and RT_2 about the axle, the torque being reckoned as positive if it tends to produce counterclockwise acceleration. The equation of rotational motion of the pulley is then

$$I\alpha = \tau = -RT_1 + RT_2$$

The angular and linear accelerations are related by $\alpha = a/R$. Hence

$$Ia/R = -R(m_1 g + m_1 a) + R(m_2 g - m_2 a)$$

from which

$$a = \frac{m_2 - m_1}{m_1 + m_2 + I/R^2} g \tag{12}$$

Note that if the mass of the pulley is small, then I/R^2 can be neglected; with this approximation Eq. (12) does reduce to Eq. (6.15), which was obtained without taking into account the inertia of the pulley.

A device of this kind, called **Atwood's machine,** can be used to determine the value of g (see Figure 6.34). For this purpose it is best to use masses m_1 and m_2 which are nearly equal. Then a is much smaller than g and easier to measure; the value of g can be calculated from the measured value of a according to Eq. (12).

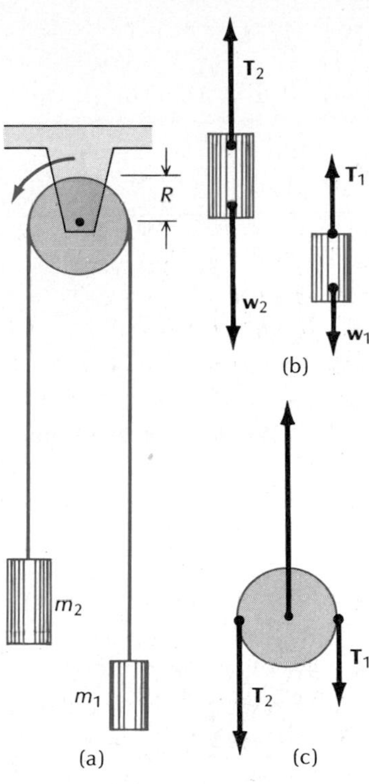

Fig. 12.7 (a) Masses m_1 and m_2 hanging from a string passing over a pulley of radius R and moment of inertia I. (b) "Free-body" diagrams for the masses m_1 and m_2. (c) "Free-body" diagram for the pulley.

12.3 Work, Energy, and Power in Rotational Motion

We will now calculate the work done by an external force on a rigid body rotating about a fixed axis. Figure 12.8 shows the body as seen from along the axis; the force is applied at some point of the body at a distance R from the axis of rotation. For a start we will assume that the force has no component parallel to the axis; in Figure 12.8 the force is therefore entirely in the plane of the page. The work done by the force is

$$dW = \mathbf{F} \cdot d\mathbf{r} \tag{13}$$

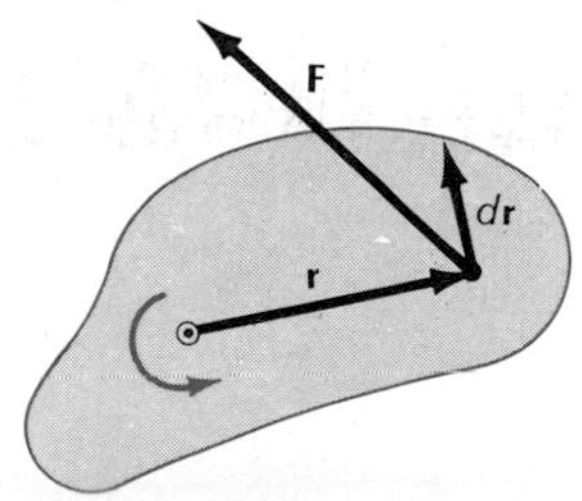

Fig. 12.8 Force applied to a rigid body rotating about a fixed axis.

where $d\mathbf{r}$ is the displacement of the particle on which the force acts, i.e., it is the displacement of the point on the rigid body at which the force is applied. The dot product $\mathbf{F} \cdot d\mathbf{r}$ can also be expressed as the magnitude of the displacement multiplied by the component of the force along the direction of the displacement. When the body rotates through an angle $d\phi$, the displacement has a magnitude $|d\mathbf{r}| = R\,d\phi$ and a tangential direction. Hence we can express Eq. (13) in terms of the tangential component of the force,

$$dW = F_{\tan} R\,d\phi \tag{14}$$

According to the definition of torque [see the discussion following Eq. (2)], the quantity $F_{tan}R$ is the torque of the force about the axis of rotation. If, as always, we take the z axis along the axis of rotation, then $F_{tan}R = \tau_z$ and

$$dW = \tau_z \, d\phi \tag{15}$$

This is the rotational analog of our familiar equation $dW = F_x \, dx$ for translational motion. The work done during a finite angular displacement is obtained by integrating Eq. (15):

Work done by torque

$$\boxed{W = \int \tau_z \, d\phi} \tag{16}$$

Although Eqs. (15) and (16) were obtained under the assumption that the force has no component parallel to the axis of rotation, this restriction is actually superfluous: any component of the force along the direction of the axis neither does work (the displacement in this direction is zero) nor generates any torque about the axis; consequently, such a component can be ignored in the above calculation and Eqs. (15) and (16) are of general validity.

The work done by the torque changes the rotational kinetic energy of the body. If the initial and final angular velocities are ω_1 and ω_2, respectively, then the change in the kinetic energy of the body is

$$W = \tfrac{1}{2}I\omega_2^2 - \tfrac{1}{2}I\omega_1^2 \tag{17}$$

This is the work–energy theorem for rotational motion. The proof of Eq. (17) is entirely analogous to the proof of the corresponding equation for translational motion [Eq. (7.32)].

If the force acting on the body is conservative — such as the force of gravity or the force of a spring — then the work equals the negative of the change of the potential energy, so that Eq. (17) becomes

$$-U_2 + U_1 = \tfrac{1}{2}I\omega_2^2 - \tfrac{1}{2}I\omega_1^2$$

or

$$\tfrac{1}{2}I\omega_1^2 + U_1 = \tfrac{1}{2}I\omega_2^2 + U_2$$

This equation expresses the **conservation of mechanical energy in rotational motion,**

Conservation of mechanical energy

$$\boxed{E = \tfrac{1}{2}I\omega^2 + U = [\text{constant}]} \tag{18}$$

From Eq. (15) we find that the **power** delivered by a torque is

$$P = \frac{dW}{dt} = \tau_z \frac{d\phi}{dt}$$

or

Power delivered by torque

$$\boxed{P = \tau_z \omega} \tag{19}$$

Thus, in rotational motion the power is torque times angular velocity, just as in translational motion it is force times velocity.

EXAMPLE 4. Consider the motion of the turntable of Example 2. Calculate the work done by the torque during the acceleration. Calculate the average power.

SOLUTION: The total angular displacement during the 2.0-s interval is

$$\Delta\phi = \tfrac{1}{2}\alpha t^2 = \tfrac{1}{2} \times (1.75 \text{ radian/s}^2) \times (2.0 \text{ s})^2$$

$$= 3.49 \text{ radians}$$

Since the torque is constant, the work is

$$W = \tau_z \, \Delta\phi = 2.22 \times 10^{-2} \text{ N} \cdot \text{m} \times 3.49 \text{ radians}$$

$$= 7.72 \times 10^{-2} \text{ J}$$

The average power is

$$\overline{P} = W/t = 7.72 \times 10^{-2} \text{ J}/2.0 \text{ s} = 3.86 \times 10^{-2} \text{ W}$$

(Alternatively, we can find the work done from the change in the rotational kinetic energy. The initial kinetic energy is zero and the final kinetic energy is $\frac{1}{2}I\omega^2$ so that

$$W = \tfrac{1}{2}I\omega^2 - 0$$

$$= \tfrac{1}{2} \times 1.27 \times 10^{-2} \text{ kg} \cdot \text{m}^2 \times (3.49 \text{ radian/s})^2$$

$$= 7.72 \times 10^{-2} \text{ J})$$

EXAMPLE 5. A meter stick is initially standing vertically on the floor. If the meter stick falls over, with what angular velocity will it hit the floor? Assume that the end in contact with the floor does not slip.

SOLUTION: The motion of the meter stick is rotation about a fixed axis passing through the point of contact with the floor (Figure 12.9). The stick is a uniform rod of mass M and length l. Its moment of inertia about the point of contact is $Ml^2/3$ (see Table 11.1) and the rotational kinetic energy is therefore $Ml^2\omega^2/6$. The gravitational potential energy is Mgz_{CM}. When the meter stick is standing vertically, the energy is

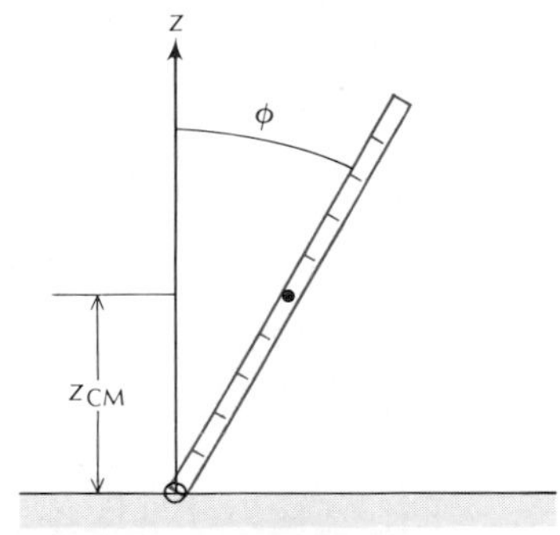

Fig. 12.9 Meter stick rotating about its lower end.

$$E = \tfrac{1}{6}Ml^2\omega_1^2 + Mgz_{\text{CM}} = 0 + Mgl/2$$

When the meter hits the floor,

$$E = \tfrac{1}{6}Ml^2\omega_2^2 + Mgz_{\text{CM}} = \tfrac{1}{6}Ml^2\omega_2^2 + 0$$

Conservation of energy therefore implies

$$\tfrac{1}{6}Ml^2\omega_2^2 = Mgl/2$$

from which

$$\omega_2^2 = 3g/l$$

or

$$\omega_2 = \sqrt{3g/l} = \sqrt{(3 \times 9.8 \text{ m/s}^2)/1.0 \text{ m}} = 5.4 \text{ radian/s}$$

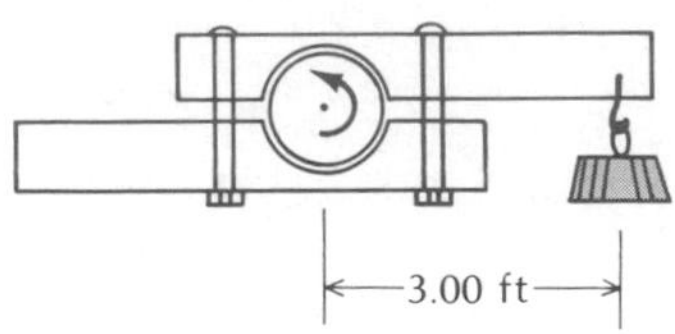

Fig. 12.10 Prony dynamometer.

EXAMPLE 6. A Prony **dynamometer,** used for the measurement of the power output of an engine, consists of two equal beams clamped loosely around a flywheel mounted to the shaft of the engine (Figure 12.10). The frictional drag between the spinning flywheel and the clamp tends to turn the beams counterclockwise; a weight attached to the right beam tends to turn them clockwise. The weight must be adjusted so that the opposing torques of weight and friction are in balance and the beams remain in a static, horizontal position. In a test of an automobile engine driving a flywheel of 1.0-ft radius at 2500 rev/min, a weight of 73.0 lb balances the beams when hung at a distance of 3.00 ft from the axis of the flywheel. What is the power output of the engine?

SOLUTION: The torque of the weight w acting at a distance l from the axis of the rotation is wl. In equilibrium this torque on the beams must have the same magnitude as the frictional torque of the flywheel on the beams. By Newton's Third Law, the frictional torque of the beams *on the flywheel* must then also have the same magnitude. The power of this latter frictional torque is then

$$P = \tau_z \omega = wl\omega$$

where ω is the angular velocity of the flywheel. The mechanical power removed by friction from the flywheel is equal to the mechanical power delivered by the engine to the flywheel. Hence the above equation gives the power output of the engine:

$$\begin{aligned} P &= wl\omega \\ &= 73.0 \text{ lb} \times 3.00 \text{ ft} \times \left(2500 \frac{\text{rev}}{\text{min}} \times \frac{2\pi \text{ radians}}{1 \text{ rev}} \times \frac{1 \text{ min}}{60 \text{ s}}\right) \\ &= 5.73 \times 10^4 \text{ ft} \cdot \text{lb/s} = 104 \text{ hp} \end{aligned}$$

This quantity is usually called the **brake horsepower** of the engine.

12.4 The Conservation of Angular Momentum

If the external torque on a system of particles is zero, then Eq. (4) asserts that the total angular momentum of the system is conserved,

$$\mathbf{L} = [\text{constant}] \tag{20}$$

This conservation law can be used in much the same way as the conservation of total (linear) momentum to calculate those features of the motion of the system that do not depend on the details of the internal forces.

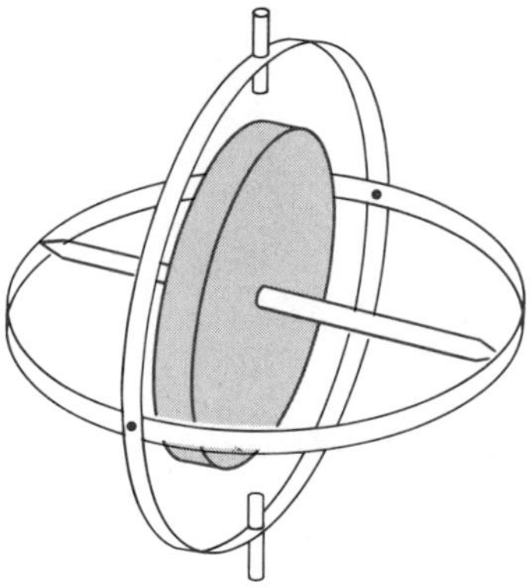

Fig. 12.11 Gyroscope in gimbals.

A gyroscope, consisting of a flywheel mounted in gimbals (Figure 12.11), provides us with a nice illustration of the conservation law. The angular momentum of the flywheel lies along its axis of rotation. (As we saw in Section 11.5, the angular momentum of a symmetric body rotating about its axis of symmetry always lies along this axis.) Since there are no torques on the flywheel, except for the very small and negligible frictional torques in the pivots of the gimbals, the angular momentum remains constant both in magnitude and direction.

Hence the angular velocity remains constant and the orientation of the axis remains fixed in space — the gyroscope can be carried about, its base twisted and turned in any way, and yet the axis always continues to point in its original direction. High-precision gyroscopes are used in inertial guidance systems for ships, aircraft, missiles, and spacecraft (Figure 12.12); they provide an absolute reference direction relative to which the direction of travel of the vehicle can be reckoned. In such applications, three gyroscopes aimed along mutually perpendicular axes establish the orientation of an absolute x, y, z coordinate grid.

The rotation of the Earth about its axis also illustrates the conservation of angular momentum. The axis of rotation makes an angle of 23.5° with the plane of the Earth's orbit around the Sun (Figure 12.13). There is no external torque on the Earth, and its angular momentum remains constant.[2] Hence, as the Earth revolves in its orbit, the orientation of the axis of rotation remains fixed; the axis always moves parallel to itself.

For a rigid body rotating about a fixed axis (the z axis) with $\tau_z = 0$, the **conservation law for angular momentum** reduces to $L_z = 0$, or

$$\boxed{I\omega = [\text{constant}]} \tag{21}$$

The following example involves the application of this conservation law.

Fig. 12.12 A Sperry C-12 directional gyroscope, such as currently used aboard many commercial aircraft.

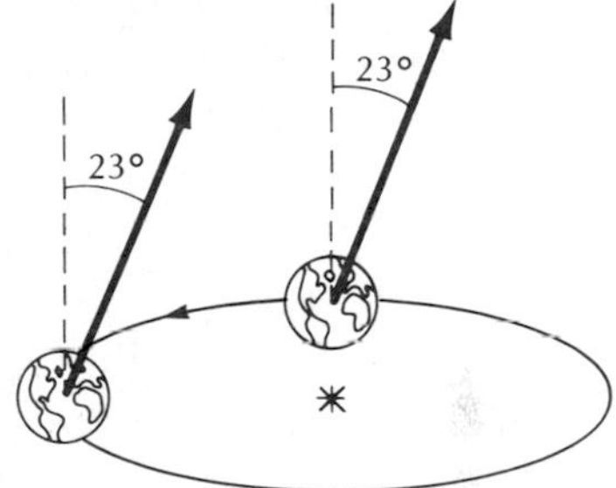

Fig. 12.13 The axis of rotation of the Earth maintains a fixed direction as the Earth moves around the Sun.

EXAMPLE 7. Suppose that the phonograph turntable described in Example 2 is coasting (with the motor disengaged) at $33\frac{1}{3}$ rev/min when a stack of 10 phonograph records suddenly drops down on it. What is the angular velocity immediately after the drop? What is the kinetic energy before and after the drop? Each record has a mass of 0.17 kg and a radius of 15.2 cm.

SOLUTION: Since the records drop suddenly, the external (frictional) torque can be ignored and the angular momentum of the complete system (turntable plus records) is conserved. The angular momentum before the drop is $I\omega$ (where ω is the initial angular velocity and I moment of inertia of the turntable) and the angular momentum after the drop is $I'\omega'$ (where ω' is the final angular velocity and I' the moment of inertia of turntable and records combined). Hence,

$$I\omega = I'\omega'$$

and

$$\omega' = \frac{I}{I'}\omega \tag{22}$$

The moment of inertia of the turntable is (see Example 2)

$$I = 1.27 \times 10^{-2}\ \text{kg}\cdot\text{m}^2$$

[2] Actually, both the Moon and the Sun exert some small torque on the Earth by their gravitational attraction on the equatorial bulge of the Earth. This torque gradually changes the direction of the angular momentum of the Earth and the orientation of the Earth. This is what astronomers call the **precession of the equinoxes.**

and the moment of inertia of the 10 records is

$$I_r = 10 \times \tfrac{1}{2} M_r R_r^2 = 10 \times \tfrac{1}{2} \times 0.17 \text{ kg} \times (0.152 \text{ m})^2$$

$$= 1.96 \times 10^{-2} \text{ kg} \cdot \text{m}^2$$

Accordingly,

$$\omega' = \frac{I}{I'}\,\omega = \frac{I}{I + I_r}\,\omega$$

$$= \frac{1.27 \times 10^{-2}}{1.27 \times 10^{-2} + 1.96 \times 10^{-2}} \times 3.49 \text{ radian/s}$$

$$= 1.37 \text{ radian/s}$$

The initial kinetic energy is simply that of the turntable,

$$K = \tfrac{1}{2} I \omega^2$$

$$= \tfrac{1}{2} \times 1.27 \times 10^{-2} \text{ kg} \cdot \text{m}^2 \times (3.49 \text{ radian/s})^2$$

$$= 7.73 \times 10^{-2} \text{ J}$$

The final kinetic energy is that of turntable and records together,

$$K' = \tfrac{1}{2} I' \omega'^2$$

$$= \tfrac{1}{2} \times 3.23 \times 10^{-2} \text{ kg} \cdot \text{m}^2 \times (1.37 \text{ radian/s})^2$$

$$= 3.03 \times 10^{-2} \text{ J}$$

Note that some kinetic energy is lost. When the records drop on the turntable, they slide for a few instants until friction (within the system) brings their speed up to that of the turntable; during this process friction dissipates some of the kinetic energy of the turntable.

EXAMPLE 8. Water flowing along an open channel drives an undershot waterwheel of radius 2.2 m (Figure 12.14). The water approaches the wheel with a speed of 5.0 m/s and leaves with a speed of 2.5 m/s; the amount of water passing by is 300 kg per second. At what rate does the water deliver angular momentum to the wheel? What is the torque that the water exerts on the wheel? If the speed of the rim of the wheel is 3 m/s, what is the power delivered to the wheel?

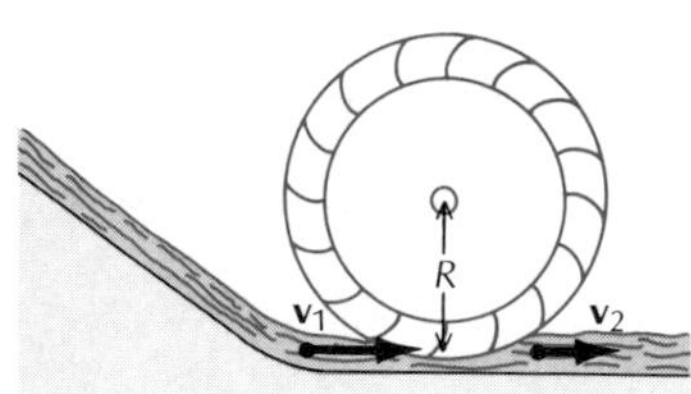

Fig. 12.14 Undershot waterwheel.

SOLUTION: Consider a small mass dm of water which approaches the wheel with a speed v_1 and leaves with a speed v_2. The angular momentum of this mass, reckoned about an origin at the center of the wheel at a radial distance R, decreases from $(dm)v_1R$ to $(dm)v_2R$ as it passes through the wheel. Hence the angular momentum lost by the water is $(dm)(v_1 - v_2)R$ and, by conservation, this must be the angular momentum gained by the wheel (we ignore friction of the water against the channel). The rate at which the wheel acquires angular momentum from the water is then

$$\frac{dL}{dt} = \frac{dm}{dt}(v_1 - v_2)R \tag{23}$$

where dm/dt is the rate at which water flows through the channel (in kilograms of water per second). Inserting the relevant numbers, we obtain

$$\frac{dL}{dt} = 300 \text{ kg/s} \times (5.0 \text{ m/s} - 2.5 \text{ m/s}) \times 2.2 \text{ m}$$

$$= 1.7 \times 10^3 \text{ kg} \cdot \text{m}^2/\text{s}^2 = 1.7 \times 10^3 \text{ J}$$

The torque contributed by the water equals this rate of transfer of angular momentum,

$$\tau_z = dL/dt = 1.7 \times 10^3 \text{ N} \cdot \text{m}$$

The angular velocity of the wheel is $\omega = (3 \text{ m/s})/2.2 \text{ m} = 1.36 \text{ radian/s}$ and the power is therefore

$$P = \tau_z \omega = 1.7 \times 10^3 \text{ N} \cdot \text{m} \times 1.36 \text{ radian/s}$$

$$= 2.3 \times 10^3 \text{ J/s} = 2.3 \text{ kW}$$

EXAMPLE 9. In an experiment first performed during the Skylab mission, an astronaut floating freely in the gravity-free environment of his orbiting spacecraft demonstrates how he can change his rotational velocity by changing his moment of inertia. Initially the astronaut holds his body erect while rotating about his center of mass at a rate of 0.17 rev/s (Figure 12.15a). He then contracts his body to a fetal position (Figure 12.15b). What is his new rate of rotation? What is the change in rotational kinetic energy? Assume that the moment of inertia is 18 kg · m² in the erect position and 5.5 kg · m² in the fetal position.

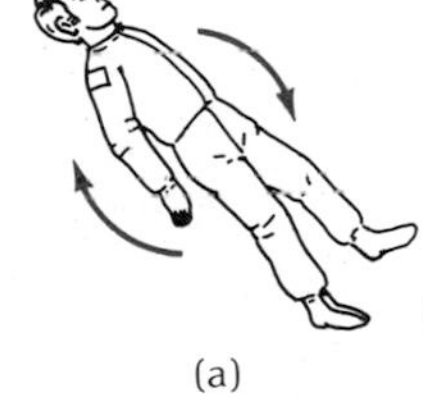

(a)

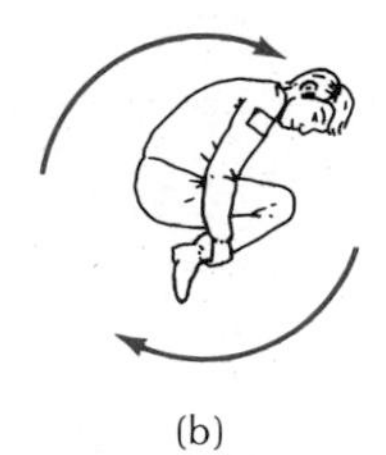

(b)

Fig. 12.15 Astronaut in Skylab.

SOLUTION: The initial angular momentum is $I\omega$ and the final angular momentum is $I'\omega'$. Since there are no external torques, the angular momentum is conserved:

$$I\omega = I'\omega'$$

With $\nu = \omega/2\pi$, this yields

$$\nu' = \frac{I}{I'}\nu$$

$$= \frac{18 \text{ kg} \cdot \text{m}^2}{5.5 \text{ kg} \cdot \text{m}^2} \times 0.17 \text{ rev/s} = 0.56 \text{ rev/s}$$

The change in kinetic energy is

$$\tfrac{1}{2}I'\omega'^2 - \tfrac{1}{2}I\omega^2 = \tfrac{1}{2}I'(2\pi\nu')^2 - \tfrac{1}{2}I(2\pi\nu)^2$$

$$= \tfrac{1}{2} \times 5.5 \text{ kg} \cdot \text{m}^2 \times (2\pi \times 0.56/\text{s})^2$$

$$-\tfrac{1}{2} \times 18 \text{ kg} \cdot \text{m}^2 \times (2\pi \times 0.17/\text{s})^2$$

$$= 33.6 \text{ J} - 10.3 \text{ J} = 23.3 \text{ J}$$

This increase of kinetic energy comes from the work the astronaut does while contracting his body.

Divers in free fall after jumping off a diving board and figures skaters pirouetting on (nearly) frictionless ice also use this method of changing their rotational velocity by changing their moment of inertia.

EXAMPLE 10. In another experiment the astronaut demonstrates how he can change the orientation of his body witout having recourse to external forces. How can the astronaut turn his body about-face? How can the astronaut turn his body upside down?

Fig. 12.16 Astronaut in Skylab.

SOLUTION: To set his body rotating about a longitudinal axis, the astronaut swings one of his feet (or both) in a circle (Figure 12.16a). His foot then has an angular momentum and, by conservation, the rest of his body must have an opposite angular momentum, i.e., it must turn in the opposite direction. This turning motion continues as long as, and only as long as, the astronaut keeps his foot swinging.

To set his body rotating about a transverse axis, the astronaut swings his arms in a lateral circle (Figure 12.16b).

(Incidentally, these experiments explain how a cat always manages to land on its feet. When dropped toward the floor from an upside-down position, the cat turns its body by means of very quick swinging motions of his hind legs and tail.)

12.5 Rolling Motion

In our study of the rotational motion of a rigid body in the preceding sections, we dealt only with rotation about a fixed axis. This restriction made it easy to convert the equation for the rate of change of angular momentum ($d\mathbf{L}/dt = \boldsymbol{\tau}$) into an equation for the rate of change of the angular velocity ($I\ d\omega/dt = \tau_z$), from which the motion can be calculated. Of course, we can relax this restriction somewhat and deal with rotation about an axis that has a uniform translational motion; such a motion of the axis makes no essential difference, since it can be eliminated by a simple change of reference frame.

Now we will deal with a somewhat more complicated case of combined translational and rotational motion which is of considerable practical importance: the motion of a body rolling on some surface without slipping. A typical example is the motion of a railroad wheel rolling along its track. The motion can be described as a rotation of the wheel about an axis through its center plus a translation of the axis along the track (Figure 12.17a). If there is no slipping, then these two motions are coupled: the translational speed v and the rotational velocity ω are related by

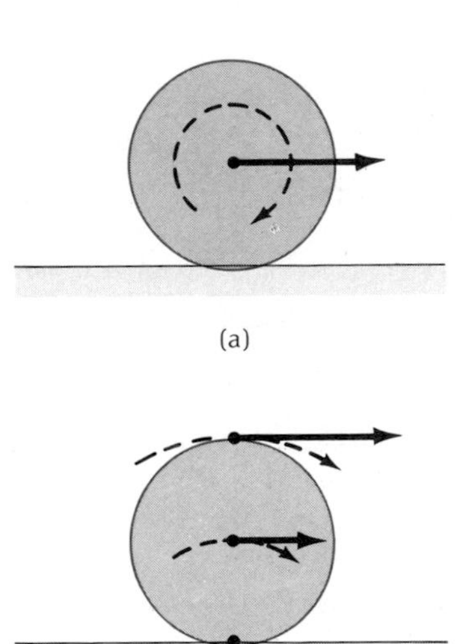

Fig. 12.17 The motion of a rolling wheel can be regarded either as (a) rotation about an axis through the center plus a translation of this center or else as (b) rotation about an axis through the instantaneous point of contact.

$$v = \omega R \tag{24}$$

where R is the radius of the wheel.

Alternatively, the motion of the wheel can be described as pure rotation about an instantaneous axis that passes through the instantaneous point of contact (Figure 12.17b). As we have seen in Section 11.1, there is always some freedom in the choice of the axis of rotation — if the wheel rotates about an axis through its center, it also rotates about any parallel axis through any other point; furthermore, the angular velocities about all these alternative axes are the same. Of all these alternative axes, the one passing through the point of contact is the most interesting: if there is no slipping, then the point of contact of the wheel with the track is instantaneously at rest on the track; hence the axis passing through this point is an **instantaneously fixed axis.**

Instantaneously fixed axis for rolling motion

The instantaneous velocities of the points shown in Figure 12.17b, as well as the instantaneous velocities of any other points on the wheel, can be obtained by means of the usual addition procedure for velocities. For example, the uppermost point of the wheel has a forward velocity ωR relative to the center, which itself has a velocity v relative to the track; hence the net forward velocity of the uppermost point is

$$v + \omega R = v + v = 2v \tag{25}$$

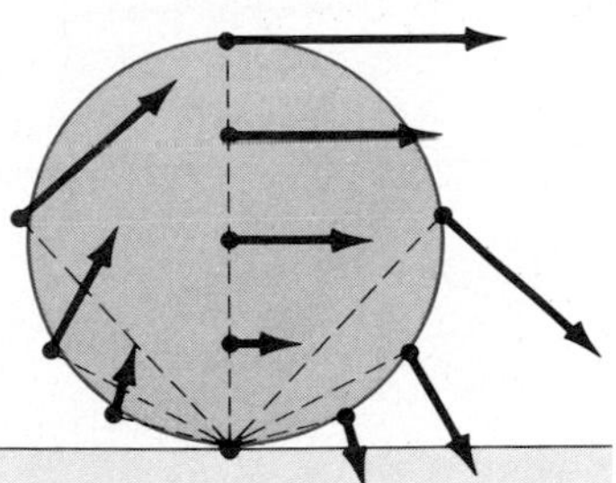

Fig. 12.18 Instantaneous velocity vectors of points on a rolling wheel.

Likewise, the net forward velocity of the lowermost point of the wheel is

$$v - \omega R = v - v = 0 \tag{26}$$

This, of course, means that the lowermost point is at rest, just as required by the condition of no slipping. Figure 12.18 shows the instantaneous velocities of selected points on the wheel. The magnitudes of the velocities increase in direct proportion to their distance from the point of contact; such an increase is a characteristic feature of rotation about a point.

Figure 12.19 is a time-exposure photograph showing the path traced out by a small light bulb attached to the rim of a rolling wheel. Note that whenever the light bulb touches the ground, the horizontal component of the motion vanishes; furthermore, the vertical component of the motion reverses direction, i.e., this component also vanishes instantaneously. The curve traced out in Figure 12.19 is called a **cycloid.**

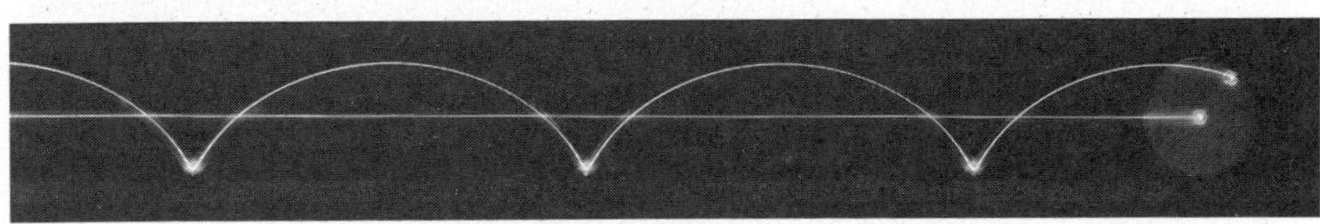

Fig. 12.19 Path of a point on the rim of a rolling wheel.

The existence of an instantaneous fixed axis in rolling motion (without slipping) enables us to deal with this motion by the methods developed in the preceding sections. The rotation about the instantaneous fixed axis obeys our old equation $I\,d\omega/dt = \tau_z$. From this, we can calculate the motion of a rolling body on which external forces and torques act.

EXAMPLE 11. A piece of steel pipe weighing 800 lb rolls down a ramp inclined at 30° to the horizontal. What is the acceleration if the pipe rolls without slipping? What is the magnitude of the friction force that acts at the point of contact between the pipe and the ramp?

SOLUTION: Figure 12.20a shows the pipe on the ramp; Figure 12.20b shows the forces in a "free-body" diagram. The only force that generates a torque about the instantaneous axis through the point of contact is the weight. The weight Mg effectively acts at the center of mass of the pipe,[3] at a distance R from the point of contact; since the angle between the weight and the radial line is θ, the torque of the weight about the point of contact is

$$\tau_z = MgR \sin\theta$$

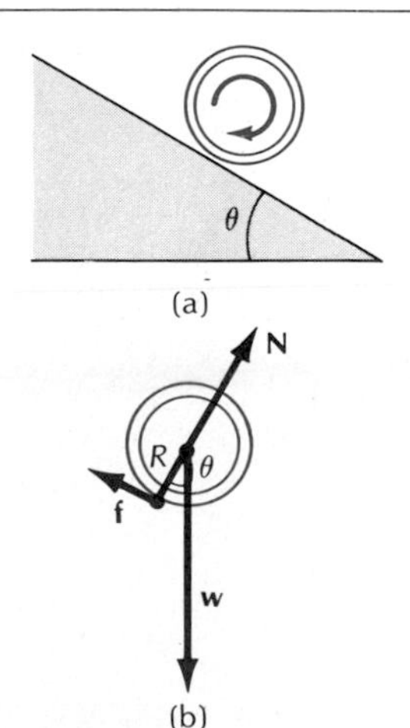

Fig. 12.20 (a) A pipe rolling down a ramp. (b) "Free-body" diagram for the pipe.

The moment of inertia of the pipe about the center of mass is $I_{CM} = MR^2$ and, by the parallel-axis theorem, the moment of inertia about the point of contact

[3] We will prove this in the next section.

is $I = I_{CM} + MR^2 = MR^2 + MR^2 = 2MR^2$. Consequently the equation of motion

$$I\alpha = \tau_z$$

becomes

$$2MR^2\alpha = MgR \sin\theta$$

or

$$\alpha = \frac{1}{2R} g \sin\theta$$

The linear acceleration down the ramp is

$$a = \alpha R = \tfrac{1}{2} g \sin\theta \tag{27}$$

$$= \tfrac{1}{2} \times 32.2 \text{ ft/s}^2 \times \sin 30° = 8.05 \text{ ft/s}^2$$

Note that Eq. (27) shows that the acceleration of the rolling pipe is exactly half as large as that of a particle sliding down a frictionless ramp of the same angle.

The friction force f can be obtained from the equation for the translational motion along the ramp:

$$Ma = Mg \sin\theta - f \tag{28}$$

from which we have

$$f = M(g \sin\theta - a)$$

$$= M(g \sin\theta - \tfrac{1}{2} g \sin\theta) = \tfrac{1}{2} Mg \sin\theta \tag{29}$$

$$= \tfrac{1}{2} \times (800 \text{ lb}) \times \sin 30° = 200 \text{ lb}$$

EXAMPLE 12. Suppose that the pipe of the preceding example starts from rest and rolls a distance of 10 ft along the ramp. What is the total kinetic energy at this instant? What is the translational kinetic energy? What is the kinetic energy of rotation about the center of mass?

SOLUTION: The total kinetic energy is simply the rotational kinetic energy about the instantaneous axis through the point of contact,

$$K = \tfrac{1}{2} I\omega^2 = \tfrac{1}{2}(2MR^2)(v/R)^2 = Mv^2$$

In terms of the distance and acceleration, the change of velocity is given by

$$v^2 - v_0^2 = 2ax$$

or

$$v = \sqrt{2ax} = \sqrt{2 \times 8.05 \text{ ft/s}^2 \times 10 \text{ ft}} = 12.7 \text{ ft/s}$$

so that

$$K = Mv^2 = \frac{800}{32.2} \text{ slug} \times (12.7 \text{ ft/s})^2$$

$$= 4.00 \times 10^3 \text{ ft} \cdot \text{lb}$$

The translational kinetic energy is

$$\tfrac{1}{2}Mv_{CM}^2 = \tfrac{1}{2}Mv^2 = 2.00 \times 10^3 \text{ ft} \cdot \text{lb}$$

The rotational kinetic energy about the center of mass is

$$K_{int} = \tfrac{1}{2}I_{CM}\omega^2 = \tfrac{1}{2}(MR^2)(v/R)^2 = \tfrac{1}{2}Mv^2$$

$$= 2.00 \times 10^3 \text{ ft} \cdot \text{lb}$$

Note that the translational and rotational kinetic energies add up to the total kinetic energy, as they should according to Eq. (11.35). Note, furthermore, that the total kinetic energy of 4.0×10^3 ft · lb is exactly equal to the change in gravitational potential energy,

$$Mgh = Mg \times 10 \text{ ft} \times \sin 30°$$

$$= 800 \text{ lb} \times 10 \text{ ft} \times \sin 30°$$

$$= 4.0 \times 10^3 \text{ ft} \cdot \text{lb}$$

This indicates that although the force of friction has a magnitude of 200 lb, it *does no work.* The reason why friction does no work in the rolling motion is that the point of the wheel at which the friction acts is the point of contact — and that point is always instantaneously at rest, i.e., it has no displacement in the direction of the friction force.

The solution of the problem of the rolling pipe in Example 11 relied on a very special trick: we reckoned the angular momentum and the torque about an instantaneous fixed axis. But this trick only applies to rolling without slipping. The general translational and rotational motion of a rigid body subjected to arbitrary forces has neither an axis that is fixed nor an axis that is in uniform translational motion. For example, the axis of rotation of the hammer shown in Figure 11.1 has an *accelerated* translational motion. Since our derivation of the equation $d\mathbf{L}/dt = \boldsymbol{\tau}$ in Section 9.5 assumed an inertial reference frame, the equation is not necessarily valid for rotational motion about an axis that has an accelerated translational motion. To deal with this general motion, we can begin by calculating the acceleration of the center of mass according to $M\mathbf{a}_{CM} = \mathbf{F}_{ext}$. Then we can calculate the rotational motion about an axis through the center of mass according to the following theorem: *If both the angular momentum* $\mathbf{L}$ *and the torque* $\boldsymbol{\tau}$ *are reckoned with respect to the center of mass, then the equation* $d\mathbf{L}/dt = \boldsymbol{\tau}$ *remains valid even when the center of mass has some arbitrary acceleration.* We will not attempt the proof of this theorem but only remark that the motion of the thrown hammer shown in Figure 11.1 agrees with the consequences of the theorem: since gravity exerts no torque about the center of mass of the hammer,[4] the angular momentum about the center of mass ought to remain constant; inspection of Figure 11.1 indicates that the angular velocity of the hammer, and consequently also its angular momentum about the center of mass, does indeed remain constant.

Rate of change of angular momentum relative to center of mass

If we apply the theorem quoted in the preceding paragraph to the special case of rolling motion, the final results will be in agreement

[4] We will prove this in the next section.

with those obtained by our method of the instantaneous fixed axis. The following example illustrates this agreement.

EXAMPLE 13. Recalculate the acceleration of the pipe described in Example 11 by starting with the torque about the center of mass.

SOLUTION: The torque about the center of mass of the pipe is entirely due to the friction force. From Figure 12.20b we see that this torque is

$$\tau_{z,\mathrm{CM}} = Rf$$

The moment of inertia of the pipe about the center of mass is $I_{\mathrm{CM}} = MR^2$. By the theorem quoted above, the torque about the center of mass equals the rate of change of the angular momentum about the center of mass, i.e.,

$$I_{\mathrm{CM}}\alpha = \tau_{z,\mathrm{CM}}$$

or

$$MR^2\alpha = Rf$$

With $\alpha = a/R$, this becomes

$$Ma = f \tag{30}$$

The equation for the translational motion is of course still Eq. (28),

$$Ma = Mg \sin\theta - f \tag{31}$$

Equations (30) and (31) are a simultaneous system of equations for the unknowns a and f. The solution of these equations gives

$$f = \tfrac{1}{2}Mg \sin\theta$$

and

$$a = \tfrac{1}{2} \sin\theta$$

These results agree with the results in Eqs. (27) and (29) that we obtained in our previous calculation.

12.6 Precession of a Gyroscope

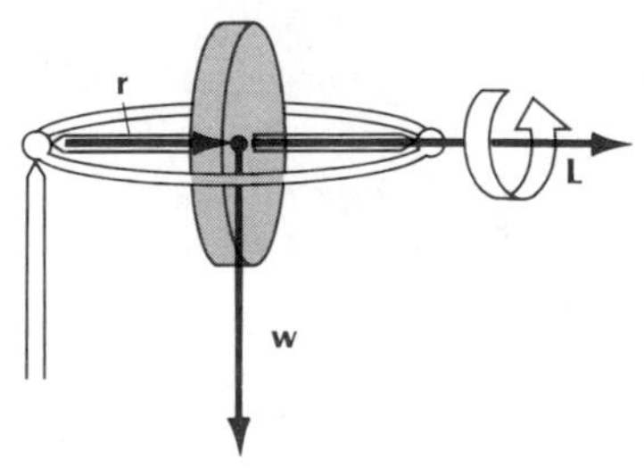

Fig. 12.21 Spinning gyroscope supported at one end.

In Section 12.2 we asserted that Eq. (4) is sufficient to determine the rotational motion of a rigid body. Let us apply this equation to a gyroscope or symmetric top which consists of a symmetric flywheel mounted on an axle. Figure 12.21 shows such a flywheel within a rigid framework which helps in the manipulation of the flywheel; the framework is nearly massless and plays no role in the following calculations. The flywheel spins at high speed in nearly frictionless bearings and is supported at its left end by a pivot. For a symmetric rigid body, such as the flywheel, the angular momentum is along the axis of rotation and has the directon shown in the figure.

Suppose now that after placing the left end of the axis on the pivot, we release the flywheel. Intuitively we expect that the force of gravity pulling on the center of mass will cause the right end of the axis to swing downward. But in this our intuition misleads us — what actually happens is quite different. To find out, we must use Eq. (4). We can write this equation as

$$d\mathbf{L} = \boldsymbol{\tau}\, dt \tag{32}$$

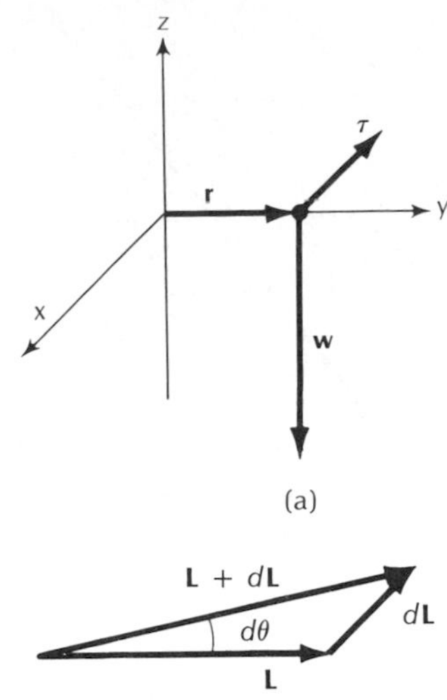

Fig. 12.22 (a) The torque $\boldsymbol{\tau}$ is perpendicular to **r** and **w**. (b) Hence $d\mathbf{L}$ is perpendicular to **L**.

The torque is caused by the weight **w** acting at the center of mass. For the calculation of this torque, we will take the pivot as origin. According to the right-hand rule, the direction of the torque $\mathbf{r} \times \mathbf{w}$ is then into the plane of Figure 12.22a. Hence the change $d\mathbf{L}$ of the angular momentum is also into the plane of the figure. This means that the weight does not dip the axis downward; rather it deflects the axis horizontally in a direction perpendicular to the applied force. More generally, whenever we exert a lateral push or pull against the axis of a gyroscope, the resulting motion of the axis will be perpendicular to the push or pull — a vertical push will deflect the axis horizontally and a horizontal push will deflect the axis vertically.

From Eq. (32) we can calculate the rate at which gravity causes the gyroscope to move horizontally. If the mass of the flywheel is M and its distance from the pivot is r, then

$$\tau = rMg \tag{33}$$

and

$$dL = rMg\, dt \tag{34}$$

From Figure 12.22b we see that the angle $d\theta$ through which the axis swings in a time dt is

$$d\theta = \frac{dL}{L} = \frac{rMg\, dt}{L}$$

or

$$\frac{d\theta}{dt} = \frac{rMg}{L} \tag{35}$$

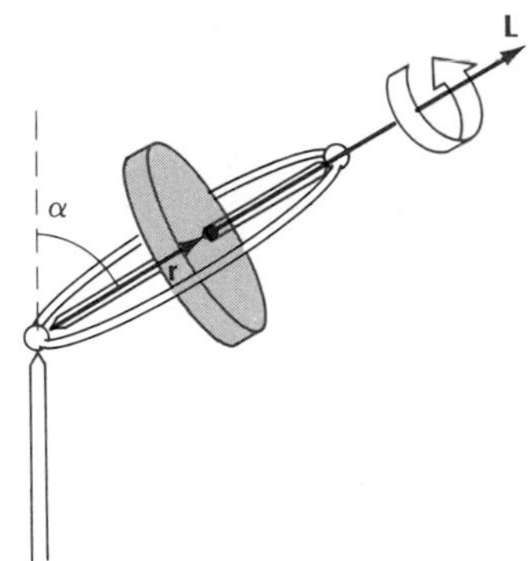

Fig. 12.23 Spinning gyroscope with axis inclined at an angle α.

This is the angular velocity with which the axis swings around in the horizontal plane. This motion of the axis is called **precession** and the angular velocity of Eq. (35) is called the **precession frequency:**

Precession frequency

$$\boxed{\omega_p = \frac{rMg}{L}} \tag{36}$$

Although we have obtained this result under the assumption that the axis is initially horizontal, it is easy to show that the precession proceeds with the same frequency if the axis is inclined at some arbitrary angle with the vertical (Figure 12.23). The result can therefore be ap-

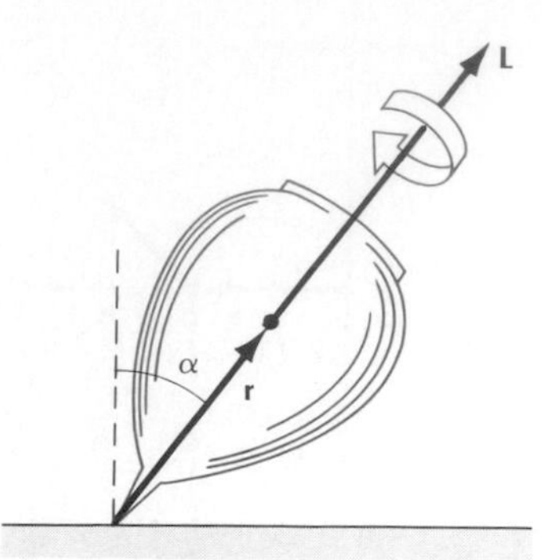

Fig. 12.24 Spinning top.

plied to an ordinary top, whose angle of inclination can of course never reach 90° (Figure 12.24).

One crucial ingredient in the above calculations is the assumption that the gyroscope spins at high speed. If the gyroscope does not spin at all or spins only slowly, then it will flop down when released from the horizontal position shown in Figure 12.21. Even when the gyroscope spins at high speed, it will at first swing downward a little bit and then swing back up; and it will continue to repeat this down-and-up motion as it precesses in a horizontal direction. For a fast-spinning gyroscope this oscillatory vertical motion is so small that it is hardly noticeable. Our calculation gave no indication of this vertical motion because we assumed that the angular momentum is entirely associated with the spin of the gyroscope on its axis. But this is not quite true — the precession involves the gradual motion of the mass around a horizontal circle and there is some small amount of extra angular momentum associated with this motion. Because of this extra angular momentum, the total angular momentum vector does not lie exactly along the axis of the gyroscope. A precise calculation which takes this small deviation into account can explain the small up-and-down motions of the gyroscope, but this is a rather complicated calculation and we will not attempt it here.

12.7 Statics of Rigid Bodies

The basic problem of statics is the calculation of the forces necessary to hold a body in equilibrium. If a rigid body is to remain at rest, its translational and rotational accelerations must be zero. Hence the condition for the equilibrium of a rigid body is that *the sum of external forces and the sum of external torques on the body be zero.* The torques can be reckoned about any axis that happens to be convenient. With a bit of practice one learns to recognize which choice of axis will most quickly lead to the solution of the problem.

Static equilibrium

The force of gravity plays an important role in many problems of statics. The force of gravity on a body is distributed over all parts of the body, each part being subjected to a force proportional to its mass. However, for the purpose of calculating the torque exerted by gravity on a *rigid* body, *the entire gravitational force may be regarded as acting at the center of mass.* The proof of this rule is easy: suppose that we release the body and permit it to fall freely from an initial condition of rest. Since all particles in the body fall at the same rate, the body will not change its orientation as it falls. This absence of angular acceleration implies that gravity does not generate any torque about the center of mass. Hence if we want to simulate gravity by a single force acting at one point of the rigid body, that point will have to be the center of mass so that this single force does not generate any torque either. Both gravity and the single force that replaces it then produce exactly the same rotational motion about the center of mass (namely, no motion) and they are therefore equivalent in the equations of motion of the body. Note, however, that this equivalence only holds for a rigid body — for an elastic body or fluid the distribution of gravity over all parts of the material plays an essential role in the equations of motion of all these parts.

12.8 Examples of Static Equilibrium

The following are some examples of solutions of problems in statics. Note that the first step of the solution is always a careful enumeration of all the forces acting on the body. To keep track of these forces, it helps to draw them on a "free-body" diagram.

EXAMPLE 14. A locomotive weighing 90 tons (180,000 lb) is one-third of the way across a bridge 300 ft long. The bridge is made of a uniform iron girder, weighing 900 tons, which rests on two piers (Figure 12.25a). What is the load on each pier?

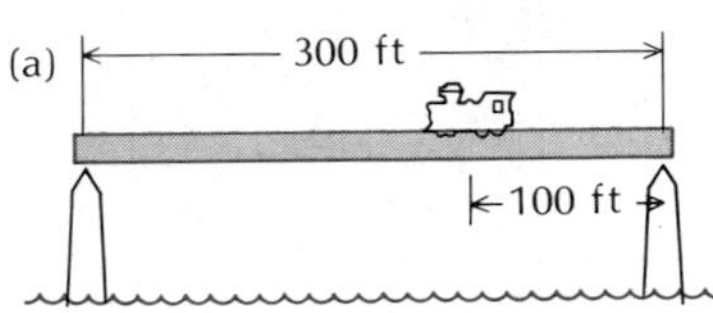

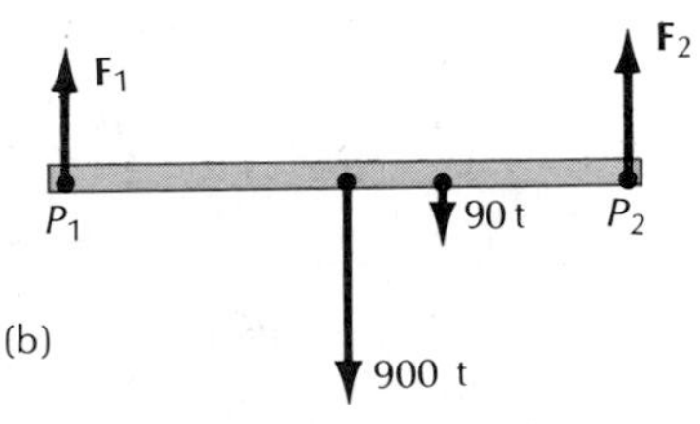

Fig. 12.25 (a) Bridge with a locomotive on it. (b) "Free-body" diagram for the bridge.

SOLUTION: Figure 12.25b is a "free-body" diagram for the bridge showing all the forces on it; the weight of the bridge is shown acting at the center of the mass. The bridge is static and hence the net torque on the bridge reckoned about any point must be zero. Setting the torque about the point P_2 equal to zero, we have

$$150 \text{ ft} \times 900 \text{ tons} + 100 \text{ ft} \times 90 \text{ tons} - 300 \text{ ft} \times F_1 = 0$$

or

$$F_1 = 480 \text{ tons}$$

Likewise, setting the torque about the point P_1 equal to zero, we have

$$-150 \text{ ft} \times 900 \text{ tons} - 200 \text{ ft} \times 90 \text{ tons} + 300 \text{ ft} \times F_2 = 0$$

or

$$F_2 = 510 \text{ tons}$$

EXAMPLE 15. The mast of a sailboat is held by three steel cables attached as shown in Figure 12.26a. Each of the front cables has a tension of 5.0×10^3 N. What is the tension in the rear cable? What force does the foot of the mast exert on the sailboat? Neglect the weight of the mast and assume that the foot of the mast is hinged.

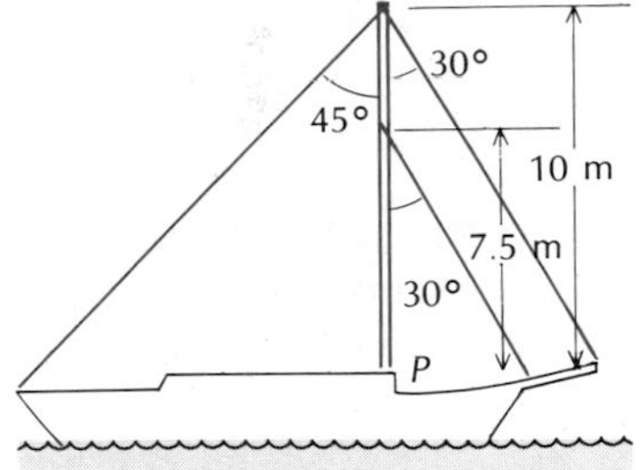

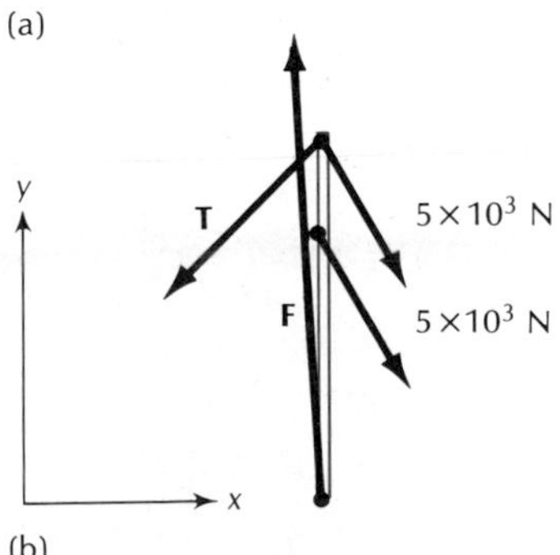

Fig. 12.26 (a) Steel cables staying a mast. (b) "Free-body" diagram for the mast.

SOLUTION: Figure 12.26b is a "free-body" diagram displaying the forces on the mast. To find the tension T in the rear cable it is convenient to reckon the net torque on the mast about the point P. This torque must be zero:

$$10 \text{ m} \times T \times \sin 45° - 7.5 \text{ m} \times 5.0 \times 10^3 \text{ N} \times \sin 30°$$
$$- 10 \text{ m} \times 5.0 \times 10^3 \text{ N} \times \sin 30° = 0$$

and

$$T = 6.2 \times 10^3 \text{ N}$$

The net force on the mast must also be zero. Figure 12.26b shows the forces on the mast; the force **F** exerted by the boat against the mast has horizontal and vertical components F_x and F_y. The net horizontal force is

$$F_x + 5.0 \times 10^3 \cos 60° + 5.0 \times 10^3 \cos 60° - 6.2 \times 10^3 \cos 45° = 0$$

and the net vertical force is

$$F_y - 5.0 \times 10^3 \cos 30° - 5.0 \times 10^3 \cos 30° - 6.2 \times 10^3 \cos 45° = 0$$

which gives

$$F_x = -0.62 \times 10^3 \text{ N}$$

$$F_y = 13.0 \times 10^3 \text{ N}$$

The force exerted by the mast on the boat is opposite to this.

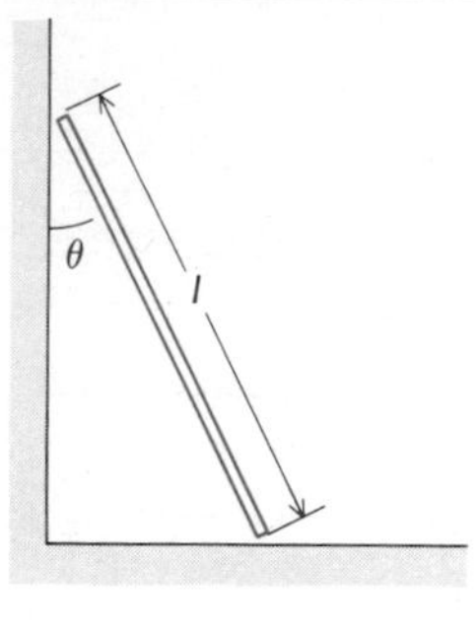

(a)

EXAMPLE 16. The bottom end of a meter stick rests on the floor and the top end rests against a wall (Figure 12.27a). If the coefficient of static friction between the stick and the floor and wall is $\mu_s = 0.4$, what is the maximum angle that the stick can make with the wall without slipping?

SOLUTION: Figure 12.27b shows a "free-body" diagram. The forces acting on the ends of the meter stick are shown resolved into horizontal (x) and vertical (y) components. If the stick is about to slip, the friction forces have their maximum values

$$f_1 = \mu_s N_1 \qquad f_2 = \mu_s N_2$$

The weight of the stick acts vertically downward at the center of mass.

The sum of the torques of these forces must be zero. If we reckon the torques about the point of contact with the floor, this condition becomes

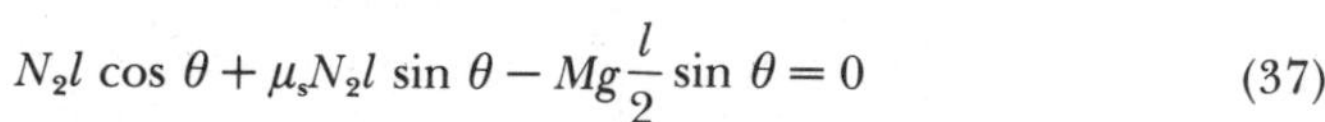

$$N_2 l \cos\theta + \mu_s N_2 l \sin\theta - Mg\frac{l}{2}\sin\theta = 0 \tag{37}$$

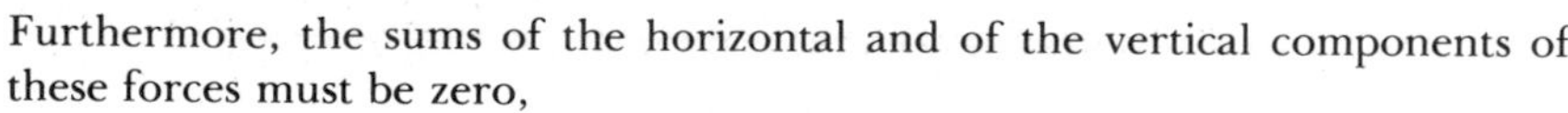

Furthermore, the sums of the horizontal and of the vertical components of these forces must be zero,

$$-\mu_s N_1 + N_2 = 0$$

$$\mu_s N_2 + N_1 - Mg = 0$$

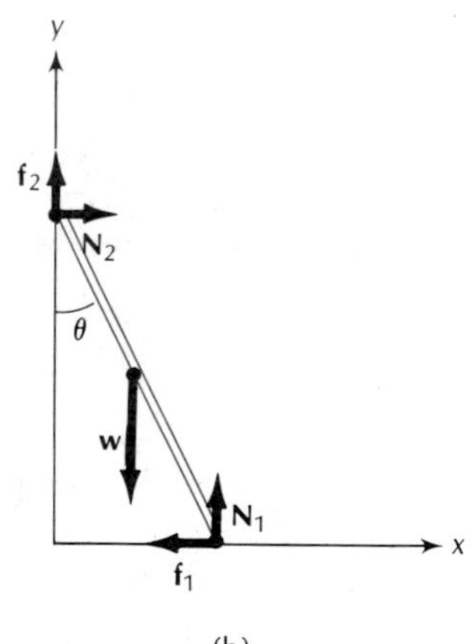

(b)

Fig. 12.27 (a) Meter stick leaning against a wall. (b) "Free-body" diagram for the meter stick.

Solving the last two equations for N_1 an N_2, we obtain

$$N_1 = Mg/(\mu_s^2 + 1)$$

$$N_2 = \mu_s Mg/(\mu_s^2 + 1)$$

and substituting these results into Eq. (37), we obtain

$$\frac{\mu_s Mgl}{\mu_s^2 + 1}\cos\theta + \frac{\mu_s^2 Mgl}{\mu_s^2 + 1}\sin\theta - \frac{Mgl}{2}\sin\theta = 0$$

Here, we can cancel Mgl and then solve for $\sin\theta/\cos\theta$:

$$\frac{\sin\theta}{\cos\theta} = \tan\theta = \frac{2\mu_s}{1 - \mu_s^2} \tag{38}$$

With $\mu_s = 0.4$ this gives

$$\tan\theta = \frac{2 \times 0.4}{1 - (0.4)^2} = 0.95$$

or $\theta = 44°$.

EXAMPLE 17. In order to hold a mooring rope against the strong pull of a ship, a sailor wraps the rope several times around a cylindrical bollard (Figure 12.28). He then finds that by pulling on the tail end with a fairly small tension

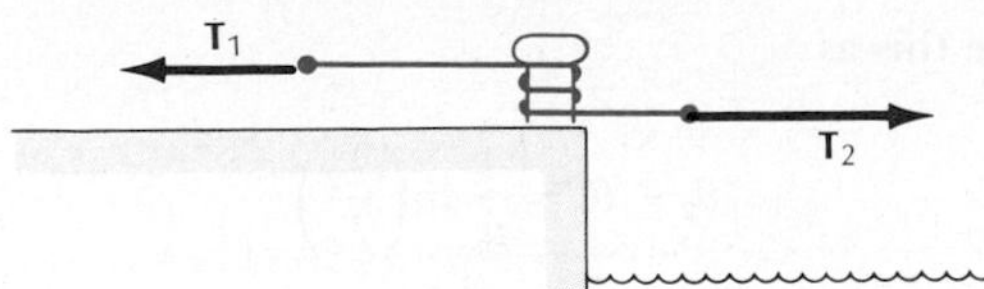

Fig. 12.28 Rope wrapped around a bollard.

$\mathbf{T}_1$, he can hold the rope steady against the much larger tension $\mathbf{T}_2$ on the other end. Assume that the coefficient of friction of the rope against the bollard is $\mu_s = 0.5$. Assume that $T_2 = 3.0 \times 10^4$ N and assume that the sailor can exert at most a pull $T_1 = 500$ N. How many turns must the sailor wrap around the bollard?

SOLUTION: Figure 12.29 shows a cross section through the bollard, with the rope wrapped around. Consider a small segment of rope subtending a small angle $d\theta$. Figure 12.30 shows a "free-body" diagram of this small segment of rope. The forces on this small segment are the tensions **T** and **T′** at the ends (which are of approximately equal magnitude), the normal force **N**, and the friction force $\mathbf{f}_s$. Under the conditions of the problem, the rope is about to slip, so that the friction force has its maximum value, $f_s = \mu_s N$. The vertical component of each of the tension forces in Figure 12.30 is $T \sin(d\theta/2)$; since $d\theta$ is small, this equals $T(d\theta/2)$.[5] The equilibrium of the radial forces (vertical forces in Figure 12.30) requires

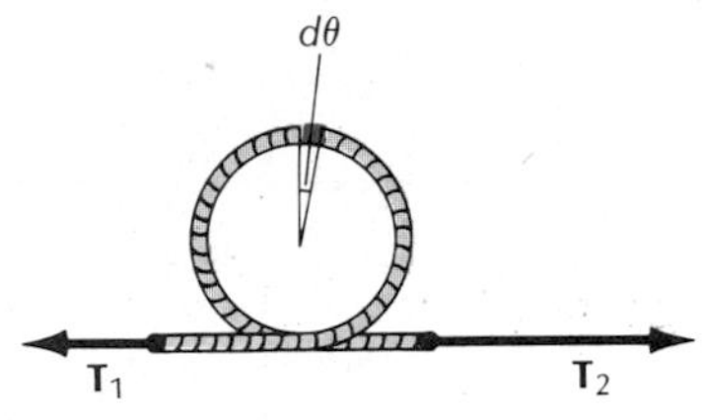

Fig. 12.29 Cross section through the bollard, with the rope wrapped around.

$$N - T(d\theta/2) - T(d\theta/2) = 0$$

from which

$$N = T\,d\theta$$

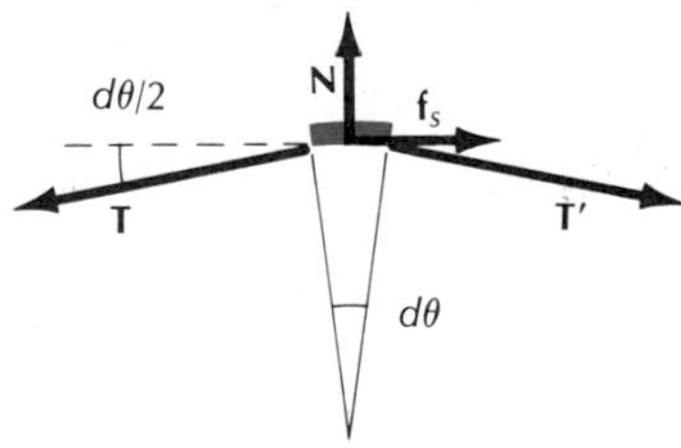

Fig. 12.30 "Free-body" diagram of the rope segment.

The friction force is then

$$f_s = \mu_s N = \mu_s T\,d\theta \tag{39}$$

The equilibrium of the tangential forces (horizontal forces in Figure 12.30) then requires that the tension on the right of the segment of rope be slightly larger than the tension on the left so that the increment dT of tension matches the friction force,

$$dT = \mu_s T\,d\theta$$

or

$$\frac{dT}{T} = \mu_s\,d\theta \tag{40}$$

To find the total change of tension along the entire wrapped portion of the rope, we must integrate Eq. (40):

$$\int_{T_1}^{T_2} \frac{dT}{T} = \mu_s \int_{\theta_1}^{\theta_2} d\theta$$

This yields

$$\Big[\ln T\Big]_{T_1}^{T_2} = \mu_s\Big[\theta\Big]_{\theta_1}^{\theta_2}$$

or

$$\ln T_2 - \ln T_1 = \mu_s(\theta_2 - \theta_1) \tag{41}$$

[5] $\sin\alpha \cong \alpha$ if α is small.

We can also write this as

$$\theta_2 - \theta_1 = \frac{1}{\mu_s} \ln\left(\frac{T_2}{T_1}\right) \tag{42}$$

With the numbers specified in our problem,

$$\theta_2 - \theta_1 = \frac{1}{0.5} \ln\left(\frac{3.0 \times 10^4}{500}\right) = 8.2 \text{ radian}$$

or

$$\theta_2 - \theta_1 = 8.2/2\pi = 1.3 \text{ turns}$$

EXAMPLE 18. A rectangular box 2 m high, 1 m wide, and 1 m deep stands on the platform of a truck (Figure 12.31a). What is the maximum forward acceleration of the truck that the box can withstand without toppling over? Assume that the box is loaded with a material of uniform density and that it does not slide.

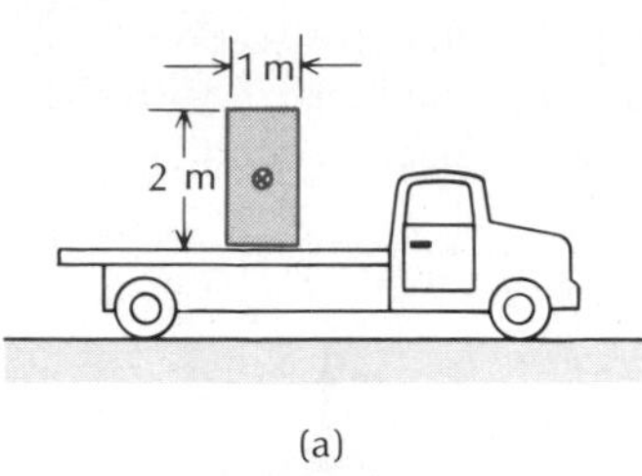

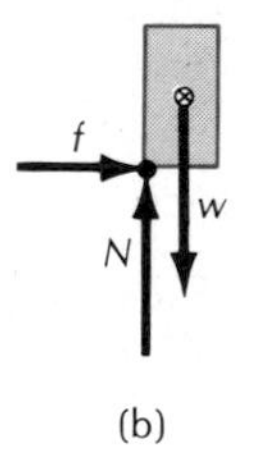

Fig. 12.31 (a) Box on a truck. (b) "Free-body" diagram for the box.

SOLUTION: Strictly, this is not a problem of statics, since the translational motion is accelerated; however, the rotational motion involves a question of equilibrium and can be treated by the methods of this section. Under the condition of the problem, the forces on the box are as shown in Figure 12.31b. Both the normal force **N** and the friction force **f** act at the corner. (When the box is about to topple, it only makes contact with the platform at the rear corner.) Gravity acts at the center of mass, which is 1 m above and 0.5 m in front of the corner. If the box is to remain upright, the net torque about the center of mass must be zero:

$$1.0 \text{ m} \times f - 0.5 \text{ m} \times N = 0 \tag{43}$$

Expressions for f and N can be obtained from the equations for the horizontal and vertical translational motions; the horizontal acceleration is a and the vertical acceleration is zero:

$$f = Ma$$

$$N - Mg = 0$$

Thus, Eq. (43) becomes

$$1.0 \text{ m} \times Ma - 0.5 \text{ m} \times Mg = 0$$

or

$$a = 0.5 \text{ g} = 4.91 \text{ m/s}^2$$

If the acceleration exceeds this value, rotational equilibrium fails and the box topples.

SUMMARY

Torque: $\boldsymbol{\tau} = \mathbf{r} \times \mathbf{F}$

Equation of rotational motion (fixed axis or axis through the center of mass):

$$I\alpha = \tau_z$$

Work done by torque: $W = \int \tau_z \, d\phi$

Conservation of energy: $E = \frac{1}{2}I\omega^2 + U = [\text{constant}]$

Power delivered by torque: $P = \tau_z \omega$

Conservation of angular momentum: $I\omega = [\text{constant}]$

Rolling motion (without slipping): The axis through the point of contact is instantaneously a fixed axis.

Precession of gyroscope: $\omega_p = rMg/L$

Static equilibrium: The sums of external forces and of external torques on a rigid body are zero. Gravity effectively acts at the center of mass.

QUESTIONS

1. Suppose you push down on the rim of a stationary phonograph turntable. What is the direction of the torque you exert about the center of the turntable?

2. Many farmers have been injured when their tractors suddenly flipped over backward while pulling a heavy piece of farm equipment. Can you explain how this happens?

3. According to popular belief, a falling piece of bread always lands on its buttered side. Why or why not?

4. Rifle bullets are given a spin about their axis by spiral grooves ("rifling") in the barrel of the gun. What is the advantage of this?

5. You are standing on a frictionless turntable (like a phonograph turntable, but sturdier). How can you turn 180° without leaving the turntable or pushing against any exterior body?

6. If you give a hard-boiled egg resting on a table a twist with your fingers, it will continue to spin. If you try doing the same with a raw egg, it will not. Why?

7. A tightrope walker uses a balancing pole to keep steady (Figure 12.32). How does this help?

Fig. 12.32 A tightrope walker.

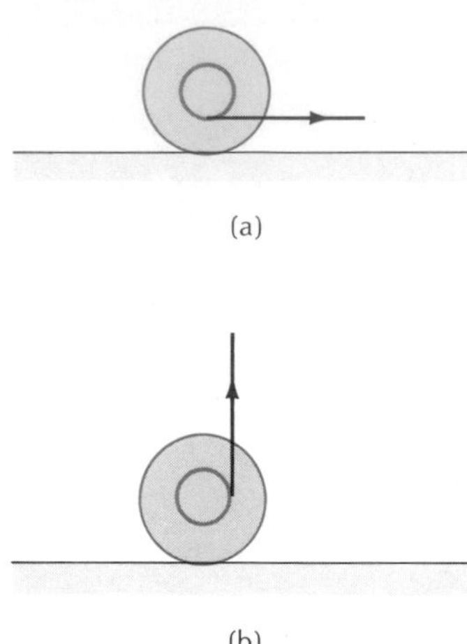

Fig. 12.33 Yo-yo resting on a table: (a) string pulls horizontally; (b) string pulls vertically.

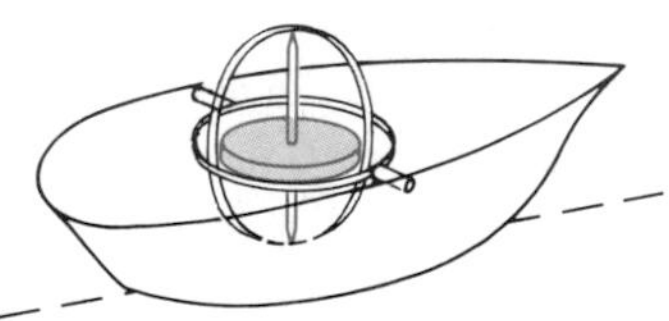

Fig. 12.34 Schick stabilizer.

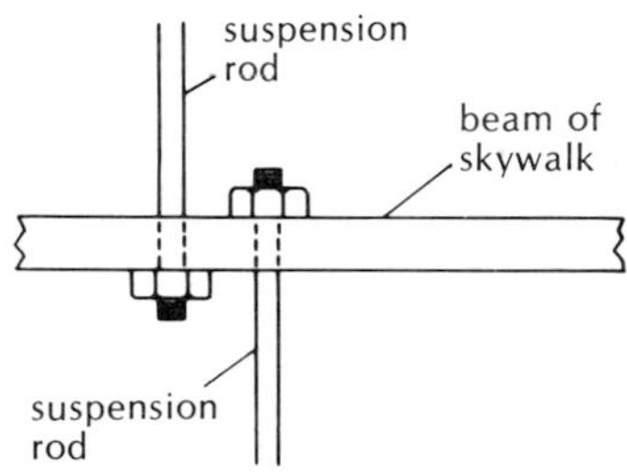

Fig. 12.35 Beams of skywalk.

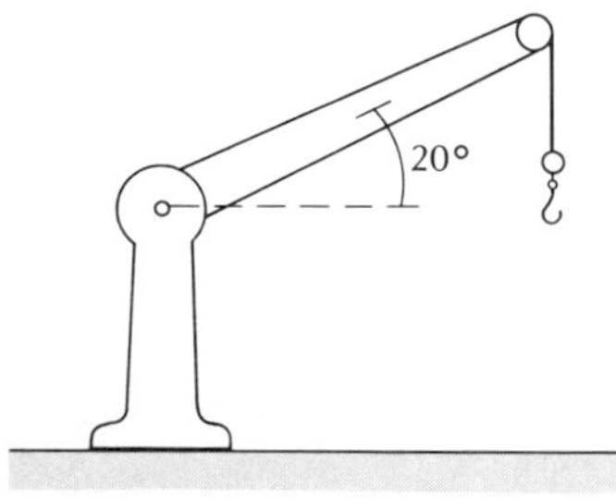

Fig. 12.36 Small crane.

8. Why do helicopters need a small vertical propeller on their tail?

9. The rotation of the Earth is subject to small seasonal variations (see Section 1.4). Does this mean that angular momentum is not conserved?

10. Why does the front end of an automobile dip down when the automobile is braking sharply?

11. The friction of the tides against the ocean coasts and the ocean shallows is gradually slowing down the rotation of the Earth. What happens to the lost angular momentum?

12. An automobile is traveling on a straight road at 55 mi/h. What is the speed, relative to the ground, of the lowermost point on one of its wheels? The topmost point? The midpoint?

13. A sphere and a hoop of equal masses roll down an inclined plane without slipping. Which will get to the bottom first? Will they have equal kinetic energies when they reach the bottom?

14. A yo-yo rests on a table (Figure 12.33). If you pull the string horizontally, which way will it move? If you pull vertically?

15. Stand a pencil vertically on its point on a table and let go. The pencil will topple over.
   (a) If the table is very smooth, the point of the pencil will slip in the direction opposite to that of the toppling. Why?
   (b) If the table is somewhat rough, or covered with a piece of paper, the point of the pencil will jump in the direction of the toppling. Why? (Hint: During the early stages of the toppling, friction holds the point of the pencil fixed; thus the pencil acquires a horizontal momentum.)

16. The Schick stabilizer, which was installed on a luxury yacht, consists of a large flywheel mounted in a frame within the ship (Figure 12.34). The frame permits the gyroscope to rotate freely about the horizontal axis and about the vertical axis. Describe what happens to the flywheel and to the ship if a wave tries to heel the ship to the left or to the right.

17. The collapse of several "skywalks" at the Hyatt Regency hotel in Kansas City on July 17, 1981, with the loss of 114 lives, was due to a defective design of the suspension system. Instead of suspending the beams of the skywalks directly from single, long steel rods anchored at the top of the building, some incompetent engineers decided to use several short steel rods joining the beams of each skywalk to those of the skywalk above (Figure 12.35). Criticize this design, keeping in mind that the beams are made of a much weaker material than the rods.

18. If the legs of a table are of exactly the same length and if the floor is exactly flat, then the weight of the table will be equally distributed over all four legs. But if there are small deviations from exactness, then the weight will not be equally distributed. Is it possible for all of the weight to rest on three legs? On two?

PROBLEMS

Section 12.1

1. The operating instructions for a small crane specify that when the boom is at an angle of 20° above the horizontal (Figure 12.36), the maximum safe load for the crane is 500 lb. Assuming that this maximum load is determined by the maximum torque that the pivot can withstand, what is the maximum safe load for 40°? For 60°?

2. A simple manual winch consists of a drum of radius 4.0 cm to which is at-

tached a handle of radius 25 cm (Figure 12.37). When you turn the handle, the rope winds up on the drum and pulls the load. Suppose that the load carried by the rope is 2500 N. What force must you exert on the handle to hold this load?

3. The repair manual for an automobile specifies that the cylinder-head bolts are to be tightened to a torque of 46 lb · ft. If a mechanic uses a wrench of length 8.0 in. on such a bolt, what perpendicular force must he exert on the end of this wrench to achieve the correct torque?

4. While braking, a 1500-kg automobile decelerates at the rate of 8.0 m/s^2. What is the magnitude of the braking force that the road exerts on the automobile? What torque does this force generate about the center of mass of the automobile? Will this torque tend to lift the front end of the automobile or tend to depress it? Assume that the center of mass of the automobile is 60 cm above the surface of the road.

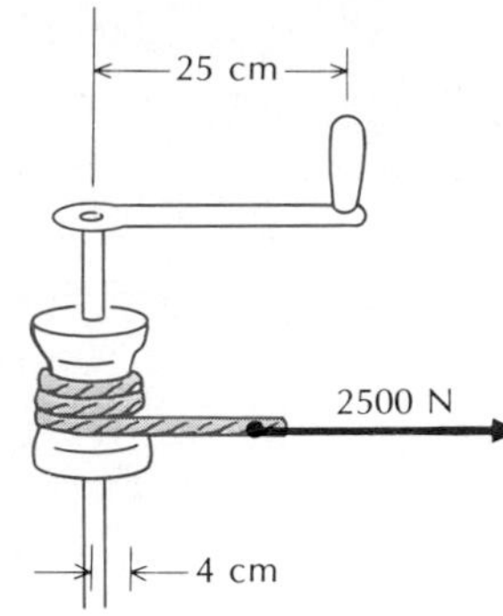

Fig. 12.37 Manual winch.

Section 12.2

5. Because of friction, the turntable of a phonograph gradually slows down when the motor is disengaged and finally stops. The initial angular speed of the turntable is $33\frac{1}{3}$ rev/min and the time it takes to stop is 1.2 min. The turntable is a disk of mass 2.0 kg and radius 15 cm. What is the average frictional torque on the turntable?

6. The center span of a revolving drawbridge consists of a uniform steel girder of mass 300 metric tons and length 25 m. This girder can be regarded as a uniform thin rod. The bridge opens by rotating about a vertical axis through its center. What torque is required to open this bridge in 60 s? Assume that the bridge first accelerates uniformly through an angular interval of 45° and then decelerates uniformly through an angular interval of 45° so as to come to rest after rotating by 90°.

7. An automobile of mass 1200 kg has four brake drums of diameter 25 cm. The brake drums are rigidly attached to the wheels of diameter 60 cm. The braking mechanism presses brake pads against the rim of each drum and the friction between the pad and the rim generates a torque that slows the rotation of the wheel. Assume that all four wheels contribute equally to the braking. What torque must the brake pads exert on each drum in order to decelerate the automobile at 0.8 gee? If the coefficient of friction between the pad and the drum is $\mu_k = 0.6$, what normal force must the brake pad exert on the rim of the drum? Ignore the masses of the wheels.

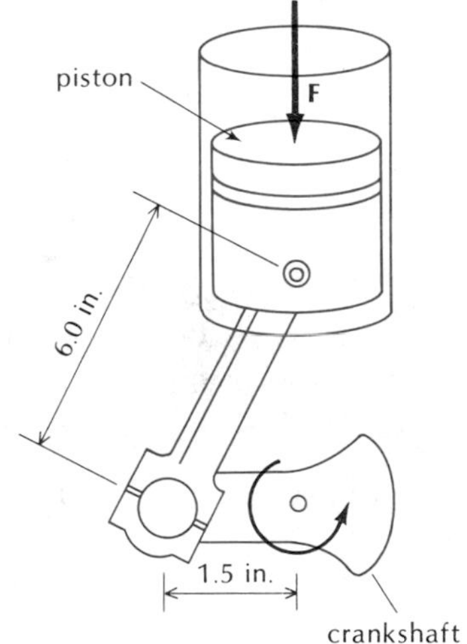

Fig. 12.38

8. In one of the cylinders of an automobile engine, the gas released by internal combustion pushes on the piston, which, in turn, pushes on the crankshaft by means of a piston rod (Figure 12.38). If the crankshaft experiences a torque of 23 lb · ft and if the dimensions of the crankshaft and piston rod are as in Figure 12.38, what must be the force of the gas on the piston when the crankshaft is in the horizontal position as in Figure 12.38? Ignore friction, and ignore the masses of the piston and rod.

9. The original Ferris wheel, built by George Ferris, had a radius of 38 m and a mass of 9.7×10^5 kg. Assume that all of its mass was uniformly distributed along the rim of the wheel. If the wheel was initially rotating at 0.5 rev/min, what constant torque had to be applied to bring it to a full stop in 30 s? What force exerted on the rim of the wheel would have given such a torque?

10. The pulley of an Atwood machine is a brass disk of mass 120 g. When using masses $m_1 = 450.0$ g and $m_2 = 455.0$ g, an experimenter finds that the larger mass descends 1.6 m in 8.0 s, starting from rest. What is the value of g?

11. A disk of mass M is free to rotate about a fixed horizontal axis. A string is wrapped around the rim of this disk and a mass m is attached to this string (Figure 12.39). What is the downward acceleration of the mass?

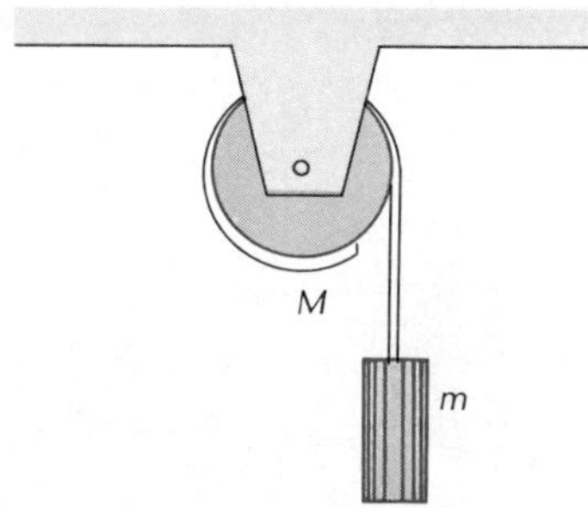

Fig. 12.39

12. The wheel of an automobile has a mass of 25 kg and a diameter of 70 cm. Assume that the wheel can be regarded as a uniform disk.
   (a) What is the angular momentum of the wheel when the automobile is traveling at 25 m/s (55 mi/h) on a straight road?
   (b) What is the rate of change of the angular momentum of the wheel when the automobile is traveling at the same speed along a curve of radius 80 m?
   (c) For this rate of change of angular momentum, what must be the torque on the wheel? Draw a diagram showing the path of the automobile, the angular momentum vector of the wheel, and the torque vector.

13. Consider the airplane propeller described in Problem 11.35. If the airplane is flying around a curve of radius 500 m at a speed of 360 km/h, what is the rate of change of the angular momentum of the propeller? What torque is required to change the angular momentum at this rate? Draw a diagram showing $\mathbf{L}$, $d\mathbf{L}/dt$, and $\boldsymbol{\tau}$.

14. A heavy hatch on a ship is made of a uniform plate of steel that measures 1.2 m × 1.2 m and has a mass of 400 kg. The hatch is hinged along one side; it is horizontal when closed and opens upward. A torsional spring assists in the opening of the hatch. The spring exerts a torque of 2.0×10^3 N · m when the hatch is horizontal and a torque of 0.3×10^3 N · m when the hatch is vertical; in the range of angles between horizontal and vertical, the torque decreases linearly with the angle (e.g., the torque is 1.15×10^3 N · m when the hatch is at 45°).
   (a) At what angle will the hatch be in equilibrium so that the spring exactly compensates the torque due to the weight?
   (b) What minimum push must a sailor exert on the hatch to open it from the closed position? To close it from the open position? Assume the sailor pushes perpendicularly on the hatch at the edge that is farthest from the hinge.

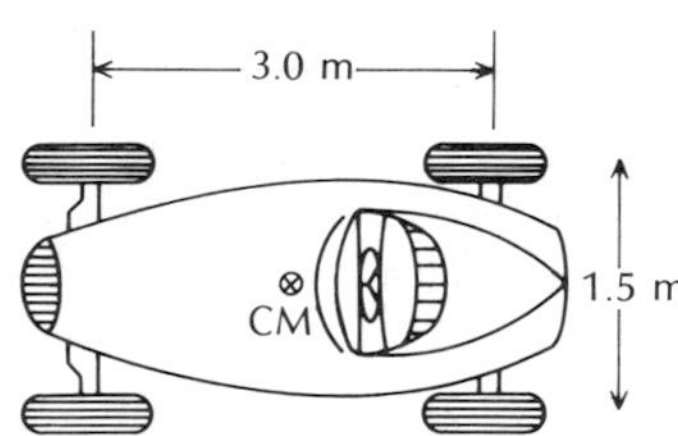

Fig. 12.40

15. An automobile has the arrangement of wheels shown in Figure 12.40. The mass of this automobile is 1800 kg, the center of mass is at the midpoint of the rectangle formed by the wheels, and the moment of inertia about a vertical axis through the center of mass is 2200 kg · m². Suppose that during braking in an emergency, the left front and rear wheels lock and begin to skid while the right wheels continue to rotate just short of skidding. The coefficient of static friction between the wheels and the road is $\mu_s = 0.90$ and the coefficient of kinetic friction is $\mu_k = 0.50$. Calculate the instantaneous angular acceleration of the automobile about the vertical axis.

16. A ship of mass 7.0×10^8 kg is steaming due north at a speed of 33 km/h. The ship is at latitude 45° North. Taking your origin at the center of the Earth, calculate the angular momentum of the ship due to the eastward rotation of the Earth. What is the rate of change of this angular momentum? What is the torque that the ship exerts on the Earth?

Section 12.3

17. The engine of an automobile delivers a maximum torque of 150 lb · ft when running at 4600 rev/min and it delivers a maximum power of 142 hp when running at 5750 rev/min. What power does the engine deliver when running at maximum torque? What torque does it deliver when running at maximum power?

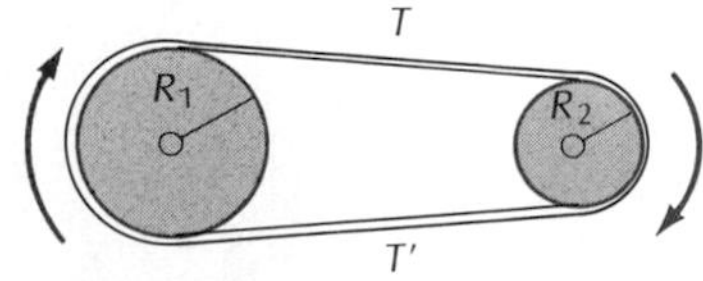

Fig. 12.41

18. The flywheel of a motor is connected to the flywheel of a pump by a drive belt (Figure 12.41). The first flywheel has a radius R_1, and the second a radius R_2. While the motor wheel is rotating at a constant angular velocity ω_1, the tensions in the upper and the lower portions of the drive belt are T and T', respectively. Assume that the drive belt is massless.
   (a) What is the angular velocity of the pump wheel?
   (b) What is the torque of the drive belt on each wheel?

(c) By taking the product of torque and angular velocity, calculate the power delivered by the motor to the drive belt, and the power removed by the pump from the drive belt. Are these powers equal?

19. The Wright Cyclone engine on a DC-3 airplane delivers a power of 850 hp with the propeller revolving steadily at 2100 rev/min. What is the torque exerted by air resistance on the propeller?

20. A bicycle and its rider have a mass of 90 kg. While accelerating from rest to 12 km/h, the rider turns the pedals through three full revolutions. What torque must the rider exert on the pedals? Assume that the torque is constant during the acceleration and ignore friction within the mechanism of the bicycle.

21. A meter stick is initially standing vertically on the floor. If the meter stick falls over, with what angular velocity will it hit the floor? Assume that the end in contact with the floor experiences no friction and slips freely.

22. A meter stick is held to a wall by a nail passing through the 60-cm mark (Figure 12.42). The meter stick is free to swing about this nail. If the meter stick is released from an initial horizontal position, what angular velocity will it attain when it swings through the vertical position?

Fig. 12.42

23. A uniform solid sphere of mass M and radius R hangs from a string of length $R/2$. Suppose the sphere is released from an initial position making an angle of 45° with the vertical (Figure 12.43).

(a) Calculate the angular velocity of the sphere when it swings through the vertical position.

(b) Calculate the tension in the string at this instant.

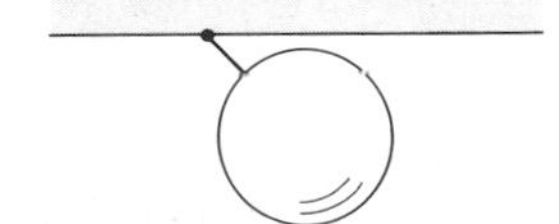

Fig. 12.43

24. The maximum (positive) acceleration an automobile can achieve on a level road depends on the maximum torque the engine can deliver to the wheels.

(a) The engine of a Maserati sports car delivers a maximum torque of 325 lb · ft to the gearbox. The gearbox steps down the rate of revolution by a factor of 2.58; i.e., whenever the engine makes 2.58 revolutions, the wheels make 1 revolution. What is the torque delivered to the wheels? Ignore frictional losses in the gearbox.

(b) The weight of the car (including fuel, driver, etc.) is 3900 lb and the radius of its wheels is 1.0 ft. What is the maximum acceleration? Ignore the moment of inertia of the wheels and frictional losses.

Section 12.4

25. A very heavy freight train made up of 250 cars has a total weight of 17,000 short tons (1 short ton = 2000 lb). Suppose that such a train accelerates from 0 to 40 mi/h on a track running exactly east from Quito, Ecuador (on the equator). The force that the engine exerts on the Earth will slow down the rotational motion of the Earth. By how much will the angular velocity of the Earth have decreased when the train reaches its final speed? Express your answer in revolutions per day. The moment of inertia of the Earth is $0.33M_ER_E^2$.

26. There are 1.1×10^8 automobiles in the United States, each of an average mass of 2000 kg. Suppose that one morning all these automobiles simultaneously start to move in an eastward direction and accelerate to a speed of 80 km/h.

(a) What total angular momentum about the axis of the Earth do all these automobiles contribute together? Assume that the automobiles travel at an average latitude of 40°.

(b) How much will the rate of rotation of the Earth change because of the action of these automobiles? Assume that the axis of rotation of the Earth remains fixed. The moment of inertia of the Earth is 8.1×10^{37} kg · m^2.

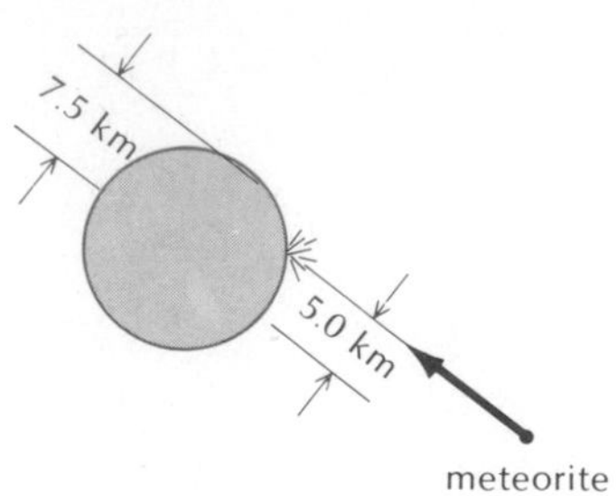

Fig. 12.44

27. Phobos is a small moon of Mars. For the purposes of the following problem, assume that Phobos has a mass of 5.8×10^{15} kg and that it has the shape of a uniform sphere of radius 7.5×10^{3} m. Suppose that a meteorite strikes Phobos 5.0×10^{3} m off center (Figure 12.44) and remains stuck. If the momentum of the meteorite was 3×10^{13} kg · m/s before impact and the mass of the meteorite is negligible compared to the mass of Phobos, what is the change in the rotational angular velocity of Phobos?

28. A woman stands in the middle of a small rowboat. The rowboat is floating freely and experiences no friction against the water. The woman is initially facing east. If she turns around 180° so that she faces west, through what angle will the rowboat turn? Assume that the woman performs her turning movement at constant angular velocity and that her moment of inertia remains constant during this movement. The moment of inertia of the rowboat about the vertical axis is 20 kg · m² and that of the woman is 0.80 kg · m².

29. Two automobiles both of 1200 kg and both traveling at 30 km/h collide on a frictionless icy road. They were initially moving on parallel paths in opposite directions, with a center-to-center distance of 1.0 m (Figure 12.45). In the collision, the automobiles lock together, forming a single body of wreckage; the moment of inertia of this body about its center of mass is 2.5×10^{3} kg · m².

Fig. 12.45

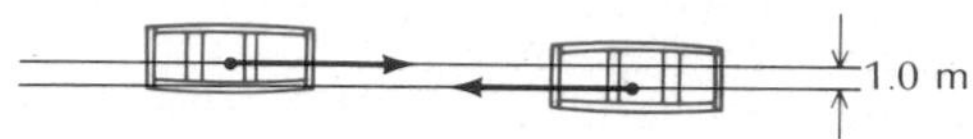

(a) Calculate the angular velocity of the wreck.
(b) Calculate the kinetic energy before the collision and after the collision. What is the change of kinetic energy?

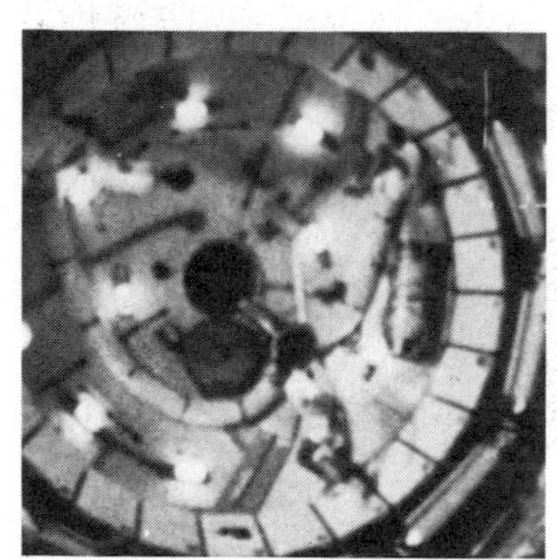

Fig. 12.46

30. In one experiment performed under weightless conditions in Skylab, the three astronauts ran around the inside wall of the spacecraft so as to generate artificial gravity for their bodies (Figure 12.46). Assume that the center of mass of each astronaut moves around a circle of radius 2.5 m; treat the astronauts as particles.
(a) With what speed must each astronaut run if the average normal force on his feet is to equal his normal weight (mg)?
(b) Suppose that before the astronauts begin to run, Skylab is floating in its orbit without rotating. When the astronauts begin to run clockwise, Skylab will begin to rotate counterclockwise. What will be the angular velocity of Skylab when the astronauts are running steadily with the speed calculated above? Assume that the mass of each astronaut is 70 kg and that the moment of inertia of Skylab about its longitudinal axis is 3×10^{5} kg · m².
(c) How often must the astronauts run around the inside if they want Skylab to rotate through an angle of 30°?

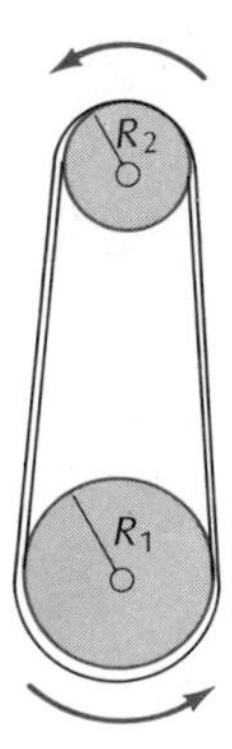

Fig. 12.47

31. A flywheel rotating freely on a shaft is suddenly coupled by means of a drive belt to a second flywheel sitting on a parallel shaft (Figure 12.47). The initial angular velocity of the first flywheel is ω; that of the second is zero. The flywheels are uniform disks of masses M_1, M_2, and of radii R_1, R_2, respectively. The drive belt is massless and the shafts are frictionless.
(a) Calculate the final angular velocity of each flywheel.
(b) Calculate the kinetic energy lost during the coupling process. What happens to this energy?

32. A thin rod of mass M and length l hangs from a pivot at its upper end. A ball of clay of mass m and of horizontal velocity v strikes the lower end at right angles and remains stuck (a totally inelastic collision). How high will the rod swing after this collision?

33. If the melting of the polar ice caps were to raise the water level on the Earth by 10 m, by how much would the day be lengthened? Assume that the

moment of inertia of the ice in the polar ice caps is negligible (they are very near the axis), and assume that the extra water spreads out uniformly over the entire surface of the Earth (i.e., neglect the area of the continents compared with the area of the oceans). The moment of inertia of the Earth (now) is $8.1 \times 10^{37}\ \text{kg} \cdot \text{m}^2$.

*34. A rod of mass M and length l is lying on a flat, frictionless surface. A ball of putty of mass m and initial velocity v at right angles to the rod impacts on the rod at a distance $l/4$ from the center (Figure 12.48). The collision is inelastic and the putty adheres to the rod. Find the translational and rotational motions of the rod after this collision.

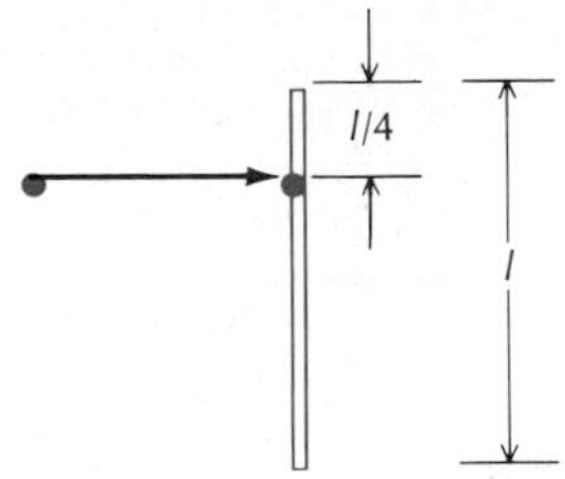

Fig. 12.48 A ball of putty impacts on a rod.

Section 12.5

35. A hula-hoop rolls down a slope of 1:10 without slipping. What is the (linear) acceleration of the hoop?

36. A uniform cylinder rolls down a plane inclined at an angle θ with the horizontal. Show that if the cylinder rolls without slipping, the acceleration is $a = \frac{2}{3}g \sin \theta$.

37. The spare wheel of a truck, accidentally released on a straight road leading down a steep hill, rolls down the hill without slipping. The mass of the wheel is 60 kg and its radius is 0.40 m; the mass distribution of the wheel is approximately that of a uniform disk. At the bottom of the hill, at a vertical distance of 120 m below the point of release, the wheel slams into a telephone booth. What is the total kinetic energy of the wheel just before impact? How much of this kinetic energy is translational energy of the center of mass of the wheel? How much is rotational kinetic energy about the center of mass? What is the speed of the wheel?

38. Galileo measured the acceleration of gravity by rolling a sphere down an inclined plane. Suppose that, starting from rest, a sphere takes 1.6 s to roll a distance of 3.00 m down a 20° inclined plane. What value of g can you deduce from this?

39. A yo-yo consists of a uniform disk with a string wound around the rim. The upper end of the string is held fixed. The yo-yo unwinds as it drops. What is its downward acceleration?

40. An automobile with rear-wheel drive is accelerating at 8.0 m/s². The wheels of the automobile are uniform disks of mass 25 kg and radius 0.38 m. What *horizontal* force does the front wheel exert on the road? What *horizontal* force does the axle exert on the wheel?

41. A barrel of mass 200 kg and radius 0.5 m rolls down a 40° ramp without slipping. What is the value of the friction force acting at the point of contact between barrel and ramp? Treat the barrel as a cylinder with uniform mass density.

42. A bowling ball sits on the smooth floor of a subway car. If the car has a horizontal acceleration a, what is the acceleration of the ball? Assume that the ball rolls without slipping.

43. A hoop rolls down an inclined ramp. The coefficient of static friction between the hoop and the ramp is μ_s. If the ramp is very steep, the hoop will slip while rolling. Show that the critical angle of inclination at which the hoop begins to slip is given by $\tan \theta = 2\mu_s$.

44. A solid cylinder rolls down an inclined plane. The angle of inclination θ of the plane is large so that the cylinder slips while rolling. The coefficient of kinetic friction between the cylinder and the plane is μ_k. Find the rotational and the translational acceleration of the cylinder. Show that the translational acceleration is the same as that of a block sliding down the plane.

45. Suppose that a tow truck applies a horizontal force of 1000 lb to the front end of an automobile similar to that described in Problem 11.38. Taking into account the rotational inertia of the wheels and ignoring frictional losses, what is the acceleration of the automobile? What is the percentage difference between this value of the acceleration and the value calculated by neglecting the rotational inertia of the wheels?

*46. A cart consists of a body and four wheels on frictionless axles. The body has a mass m. The wheels are uniform disks of mass M and radius R. Taking into account the moment of inertia of the wheels, find the acceleration of this cart if it rolls without slipping down an inclined plane making an angle θ with the horizontal.

Section 12.6

47. In order to stabilize a ship against rolling, an inventor proposes that a large flywheel spinning at high speed should be mounted within the ship (Figure 12.49). The axis of the flywheel lies across the ship and the bearings are rigidly attached to the side of the ship.

Fig. 12.49 Ship with a flywheel.

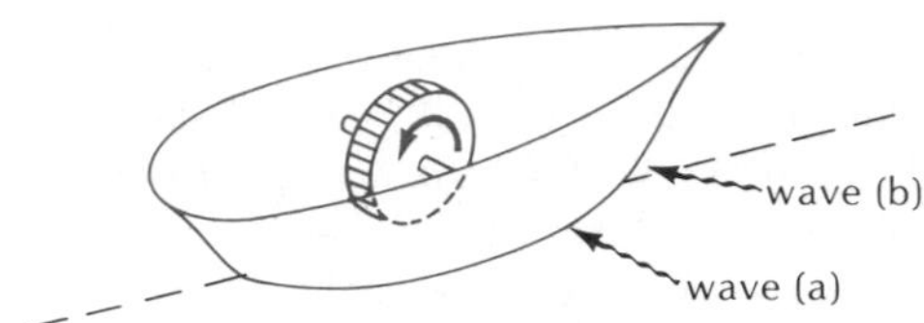

(a) If a wave hits the ship broadside and attempts to roll (capsize) the ship to the left, what will be the response of the ship? In answering this question, assume that the response of the ship is dominated by the effect of the flywheel. With words or with a diagram, carefully describe how the orientation of the ship will change.
(b) If a wave hits the bow of the ship and attempts to push the bow to the left, what will be the response of the ship?

48. Suppose that the flywheel of the gyroscope shown in Figure 12.23 is a uniform disk of mass 250 g and radius 3.5 cm. The distance of this flywheel from the point of support is 4.0 cm. What is the precession frequency if the flywheel is spinning at 120 rev/s?

49. The angular momentum of rotation (spin) of a neutron is

$$\tfrac{1}{2}\hbar = 5.3 \times 10^{-35}\ \text{J} \cdot \text{s}$$

Suppose that, when placed in a certain magnetic field, the spin of the neutron precesses at the rate of 2.4×10^7 radian/s about a line perpendicular to the spin, just like the gyroscope of Figure 12.21. What is the magnitude of the torque on the neutron?

Section 12.7

50. A door made of a uniform piece of wood measures 3 ft by 7 ft and weighs 40 lb. The door is entirely supported by two hinges, one at the bottom corner and one at the top corner. Find the force (magnitude and direction) that the door exerts on each hinge. Assume that the *vertical* force on each hinge is the same.

51. Figure 12.50 shows the arrangement of wheels on a passenger engine of the Caledonian Railway. The numbers give the distances between the wheels in feet and the downward forces that each wheel exerts on the track in short tons (1 short ton = 2000 lb; the numbers for the forces include both the right

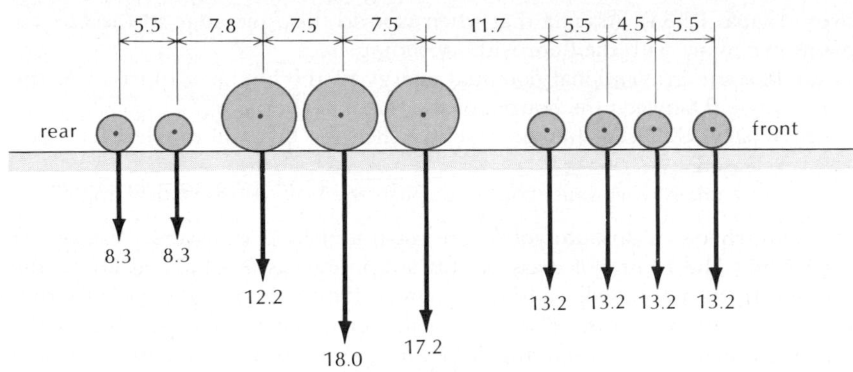

Fig. 12.50

and the left wheels). From the information given, find how far the center of mass of the engine is behind the front wheel.

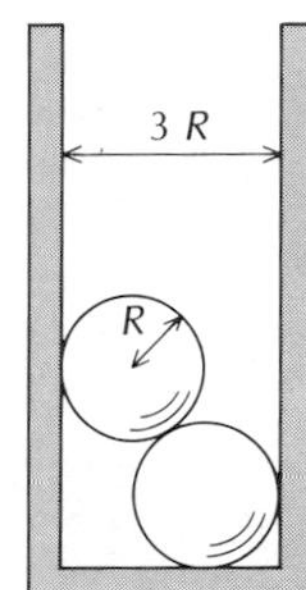

Fig. 12.51

52. Two smooth balls of steel of mass m and radius R are sitting inside a tube of radius $1.5R$. The balls are in contact with the bottom of the tube and with the wall (at two points; see Figure 12.51). Find the contact force at the bottom and at the two points on the wall.

53. An automobile with a wheelbase (distance from the front wheels to the rear wheels) of 3.0 m has its center of mass at a point midway between the wheels at a height of 0.65 m above the road. When the automobile is on a level road, the force with which each wheel presses on the road is 3100 N. What is the normal force with which each wheel presses on the road when the automobile is standing on a steep road of slope 3:10 with all the wheels locked?

54. The human forearm (including the hand) can be regarded as a lever pivoted at the joint of the elbow and pulled upward by the tendon of the biceps (Figure 12.52a). The dimensions of this lever are given in Figure 12.52b. Suppose that a load of 25 kg rests in the hand. What upward force must the biceps exert to keep the forearm horizontal? What is the downward force at the elbow joint? Neglect the weight of the forearm.

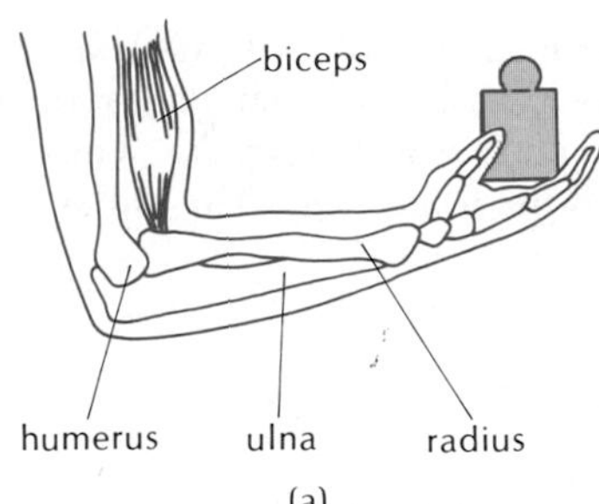

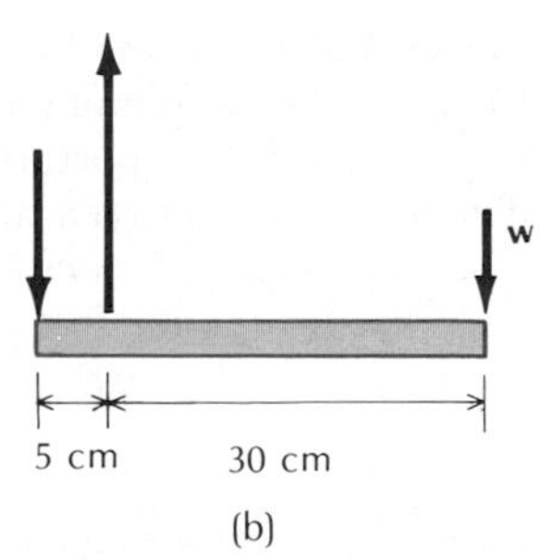

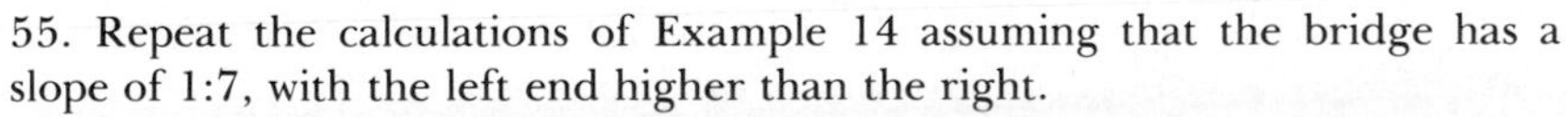

Fig. 12.52

55. Repeat the calculations of Example 14 assuming that the bridge has a slope of 1:7, with the left end higher than the right.

56. Consider the meter stick leaning against a wall described in Example 16. If the stick makes an angle of 40° with the wall, how hard can you press down vertically on the top of the stick with your hand before slipping begins?

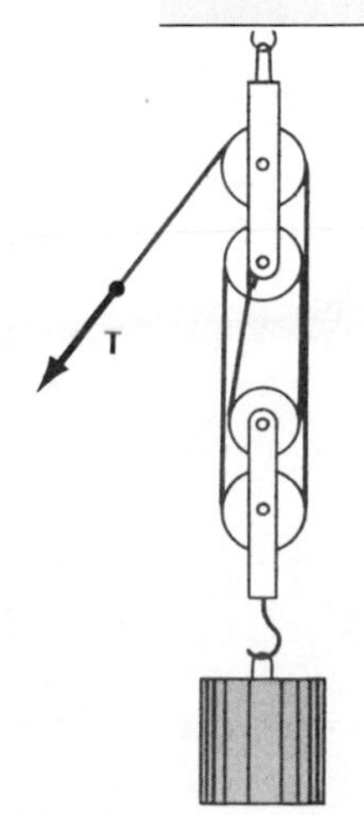

Fig. 12.53

57. A rope hoist consists of four pulleys assembled in two pairs with rigid straps, with a rope wrapped around as shown in Figures 12.53. A load of 300 kg hangs from the lower pair of pulleys. What tension must you apply to the rope to hold the load steady? Treat the pulleys and the rope as massless, and ignore any friction in the pulleys.

58. A wooden box is filled with material of uniform density. The box (with its contents) has a mass of 80 kg; it is 0.6 m wide, 0.6 m deep, and 1.2 m high.

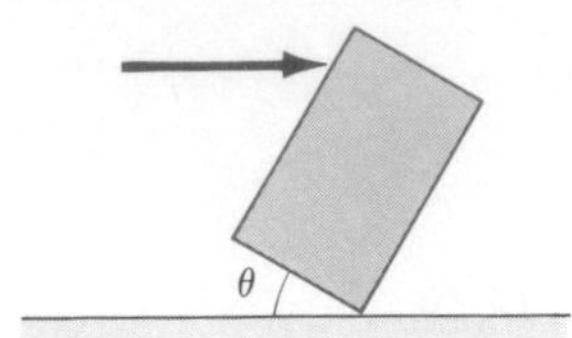

Fig. 12.54

The box stands on a level floor. By pushing against the box, you can tilt it over (Figure 12.54). Assume that when you do this, one edge of the box remains in contact with the floor without sliding.

(a) Plot the gravitational potential energy of the box as a function of the angle θ between the bottom of the box and the floor.

(b) What is the critical angle beyond which the box will topple over when released?

(c) How much work must you do to push the box to this critical angle?

59. The wheels of an automobile are separated by a transverse distance of $l = 1.5$ m. The center of mass of this automobile is $h = 0.60$ m above the ground. If the automobile is driven around a flat (no banking) curve of radius $R = 25$ m with an excessive speed, it will topple over sideways. What is the speed at which it will begin to topple? Express your answer in terms of l, h, and R; then evaluate numerically. Assume that the wheels do not skid.

60. The front and rear wheels of an automobile are separated by a distance of 3.0 m (wheelbase). The center of mass of this automobile is at a height of 0.60 m above the ground. Suppose that this automobile has rear-wheel drive and that it is accelerating along a level road at 6 m/s². When the automobile is parked, 50% of its weight rests on the front wheels and 50% on the rear wheels. What is the weight distribution when it is accelerating? Pretend that the body of the automobile remains parallel to the road at all times.

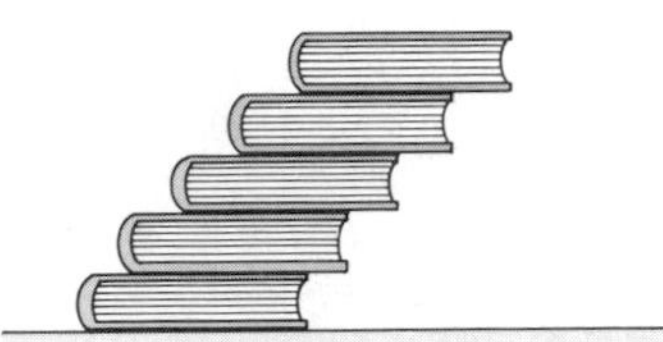

Fig. 12.55

61. By stacking books (or bricks), you can build a leaning tower (Figure 12.55). Suppose that the height of the tower is 1 m and it is made of many very thin books measuring 20 cm × 20 cm. What is the maximum angle of inclination for which the tower will be stable?

62. A wooden box, filled with a material of uniform density, stands on a concrete floor. The box has a mass of 75 kg and is 0.5 m wide, 0.5 m long, and 1.5 m high. The coefficient of friction between the box and the floor is $\mu_s = 0.80$. If you exert a (sufficiently strong) horizontal push against the side of the box, it will either topple over or start sliding without toppling over, depending on how high above the level of the floor you push. What is the maximum height at which you can push if you want the box to slide? What is the magnitude of the force you must exert to start the sliding?

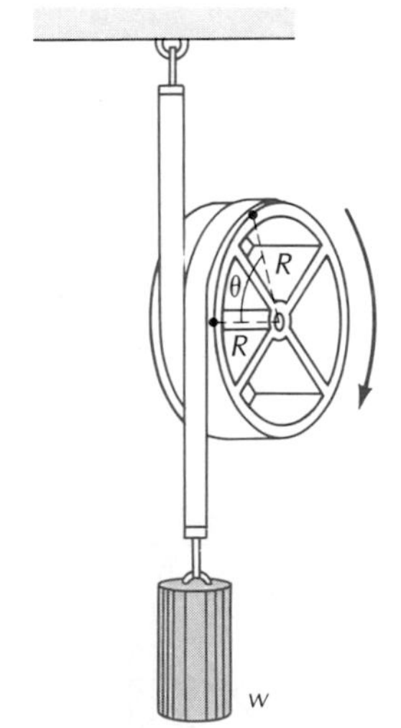

Fig. 12.56

*63. A power brake invented by Lord Kelvin consists of a strong flexible belt wrapped once around a spinning flywheel (Figure 12.56). One end of the belt is fixed to an overhead support; the other end carries a weight w. The coefficient of kinetic friction between the belt and the wheel is μ_k. The radius of the wheel is R and its angular velocity is ω.

(a) Show that the tension in the belt is

$$T = we^{-\mu_k \theta}$$

as a function of the angle of contact (Figure 12.56).

(b) Show that the net frictional torque the belt exerts on the flywheel is

$$\tau = wR(1 - e^{-2\pi\mu_k})$$

(c) Show that the power dissipated by friction is

$$P = wR\omega(1 - e^{-2\pi\mu_k})$$

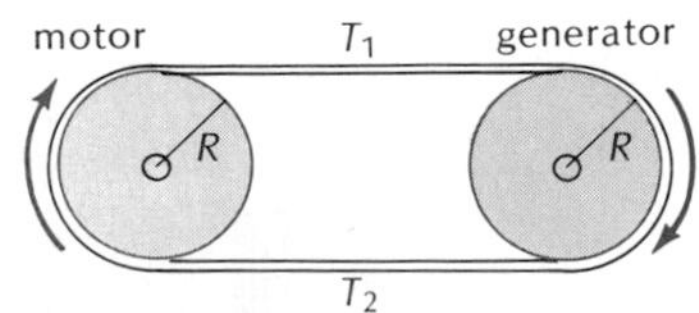

Fig. 12.57

*64. The flywheel of a motor is connected to the flywheel of an electric generator by a drive belt (Figure 12.57). The flywheels are of equal size, each of radius R. While the flywheels are rotating, the tensions in the upper and the lower portions of the drive belt are T_1 and T_2, respectively, so that the drive belt exerts a torque $\tau = (T_2 - T_1)R$ on the generator. The coefficient of static friction between the flywheel and the drive belt of μ_s. Assume that the tension in the drive belt is as low as possible with no slipping and that the drive belt is massless. Show that under these conditions

$$T_1 = \frac{\tau}{R} \frac{1}{e^{\mu_s \pi} - 1}$$

$$T_2 = \frac{\tau}{R} \frac{1}{1 - e^{-\mu_s \pi}}$$

*65. Two playing cards stand on a table leaning against each other so as to form an "A-frame" roof. The frictional coefficient between the bottoms of the cards and the table is μ_s. What is the maximum angle that the cards can make with the vertical without slipping?

ELEMENTARY PARTICLES[1]

The search for the elementary, indivisible building blocks of matter is the most fundamental problem in physics. Early in this century physicists discovered that the atom is not an elementary, indivisible unit — each atom consists of electrons orbiting around a nucleus. And then they discovered that the nucleus is not an elementary, indivisible unit — each nucleus consists of protons and neutrons tightly packed together. In the 1930s physicists began to build accelerating machines producing beams of energetic protons or electrons that could serve as projectiles; with these "atom smashers" physicists could split the nucleus. In the 1950s they built much larger and more powerful accelerating machines; with these new machines they attempted to split the proton and the neutron. But the result of these attempts was chaos: under the impact of the very energetic projectiles, the proton and the neutron do not split into any simple "subprotonic" pieces. Instead, the violent collisions of protons and neutrons generate a variety of new, exotic particles by the conversion of kinetic energy into mass. For want of a better name, these new particles were called "elementary particles." However, most of these elementary particles are more massive and more complicated than protons and neutrons — they are obviously not elementary, indivisible units. Only in recent years has some order emerged from this chaos. Physicists have found convincing circumstantial evidence that protons, neutrons, and other elementary particles are made of very small, compact subunits. The subprotonic building blocks are called *quarks*.

C.1 EXPLORING THE ATOM

We do not have available any microscope sufficiently powerful to look inside an atom. But we can use a rather crude technique to discover what is inside: smash the atom to pieces, identify the pieces, and then try to figure out how these pieces ought to fit together when assembled in an atom.

It is easy to remove small pieces from the atoms of a gas by means of an electric discharge. For this purpose we place the gas in a glass tube with two electrodes connected to a source of high voltage (Figure C.1). If the gas is at atmospheric pressure, nothing will happen — the gas will not conduct electricity. But if we partially evacuate the tube with a vacuum pump so that the gas is at very low pressure, an electric discharge will start inside the tube and an electric current will pass through the gas from one electrode to the other. The character of this electric discharge varies with the pressure. If the pressure is about 10^{-3} atmosphere, the discharge is luminous — the gas emits a bright glow of light with a characteristic color. The familiar red-pink glow of neon signs is due to this kind of electric discharge in a tube containing neon gas. If the pressure is even lower, the luminosity disappears even though the electric current continues to flow through the gas. If the pressure is below 10^{-6} atmosphere, a new phenomenon occurs in the tube: the negative electrode (cathode) emits invisible rays that propagate in straight lines through the nearly empty space within the tube. Although the rays by themselves are invisible, their presence can easily be perceived when they strike the atoms of residual gas in the tube; the gas then phosphoresces in a brilliant greenish or bluish color (Figure C.2).

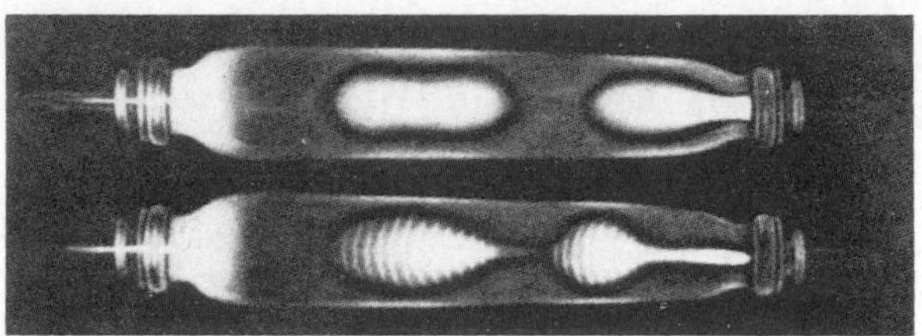

Fig. C.1 Electric discharge in a gas. The electrodes at the ends of the tubes are connected to a source of high voltage.

Experimental investigations by J. J. Thomson[2] established that these cathode rays are beams of very small particles with a negative electric charge. Figure C.3 shows the apparatus used by Thomson for the meas-

[1] This chapter is optional.

[2] **Sir Joseph Thomson,** 1856–1940, English experimental physicist, director of the Cavendish Laboratory at Cambridge, and president of the Royal Society. His discovery of the electron marks the beginning of modern atomic physics. Thomson received the Nobel Prize in 1906 for his investigations on electric discharges in gases. His experiments with beams of positive ions led to the first separation of the isotopes of a chemical element.

Fig. C.2 A beam of electrons in a cathode-ray tube. The electrons follow a clockwise circular path because they are being deflected by magnetic forces.

urement of lateral deflections of cathode rays by electric and magnetic forces. He showed that the cathode rays could be attracted or repelled by electric charges stored on the two plates in the tube, and that they could also be pushed one way or another by a small electromagnet installed just outside the tube. By balancing the electric and magnetic deflections, he managed to measure the ratio [electric charge]/[mass] of the particles making up the cathode rays. Under the reasonable assumption — later verified by additional measurements — that the charge has the same magnitude as that of a hydrogen ion, he concluded that the mass of the particle is about 1800 times smaller than that of a hydrogen atom. This particle was given the name of **electron.** J. J. Thomson's discovery of the electron, just before the end of the nineteenth century, was the first discovery of a subatomic elementary particle.

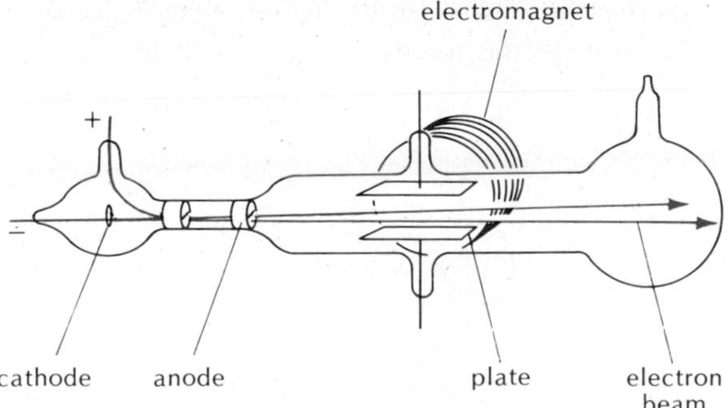

Fig. C.3 (top) Cathode-ray tube of J. J. Thomson. One pole of an electromagnet is behind the tube; the other pole has been omitted for the sake of clarity. (bottom) Schematic diagram.

Since electrons can be knocked out of all kinds of materials, it is obvious that they must be a universal ingredient in the makeup of atoms. To find out more about the structure of the interior of the atom, we must use a probe to "feel" the interior. A beam of projectiles aimed at the atom can serve as a probe — from the manner in which the incident projectiles are deflected, or **scattered,** by the structures in the atom's interior, we can deduce the details of this structure. This procedure is analogous to probing the interior of, say, a peach by means of a stream of shotgun pellets; if we fire pellets at the peach, some of them will pass almost straight through the flesh of the peach, but others will bounce off the hard kernel in the middle; and from a careful count of how many pellets bounce which way, we can in principle deduce the shape and position of the kernel.

The first such experiments probing the interior of atoms were performed around 1910 under the direction of Ernest Rutherford. In these experiments the alpha particles emitted by radioactive radium were used as projectiles.[3] When a beam of these projectiles was aimed at a thin foil of gold (Figure C.4), most of

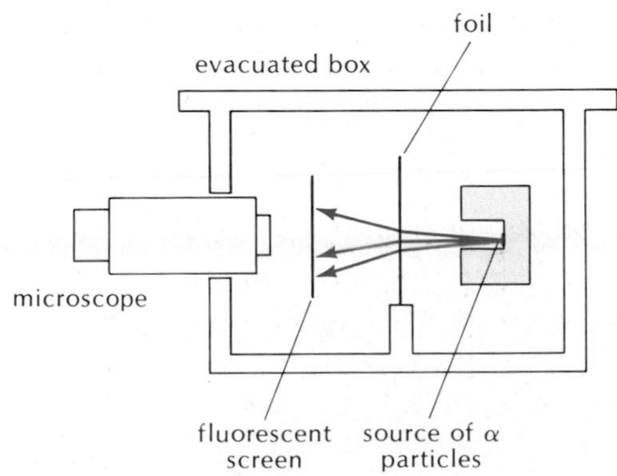

Fig. C.4 Rutherford's apparatus.

[3]Alpha particles have exactly the same structure as nuclei of helium. These particles are emitted spontaneously by many radioactive substances. Rutherford and his experiments with alpha particles will be further discussed in Chapter 41.

them passed through the atoms of gold with very little deflection. But once in a while some of the projectiles were deflected through large angles, bouncing off in a backward direction. Rutherford immediately recognized that these bouncing projectiles must have struck some hard and dense obstacle within the atom. The scattering of the projectiles revealed the presence of a very small, very massive kernel in the middle of the atom. This kernel is loaded with positive electric charge. The size of this kernel, or **nucleus,** is much smaller than the size of the atom — typically the diameter of the kernel is about 10,000 times smaller than the diameter of the atom. Yet it contains almost the entire mass of the atom — typically about 99.98% of the total mass. The remainder of the mass of the atom is accounted for by the cortege of electrons orbiting around the nucleus.

Crudely, the atom resembles a miniature solar system with the nucleus in the role of Sun and the electrons orbiting around in the role of planets (Figure C.5). But this similarity is only superficial: the electrons move around the nucleus in all directions, forming a three-dimensional cloud rather than a flat planetary system. Furthermore, the forces are very different: what holds the electrons in their orbits around the nucleus is the electric force of attraction between the negative electric charge of the electrons and the positive charge of the nucleus; what holds the planets in their orbits is gravity.

The number of electrons in an atom coincides with the atomic number listed in Table P.1. The hydrogen atom has one electron, helium has two, lithium three, etc. As a whole, the atom is electrically neutral because the negative electric charge of electrons is balanced by a corresponding positive electric charge of the nucleus. Thus, the hydrogen nucleus has an electric charge exactly opposite to that of an electron, the helium nucleus has an electric charge twice as large as that of the hydrogen nucleus, etc.

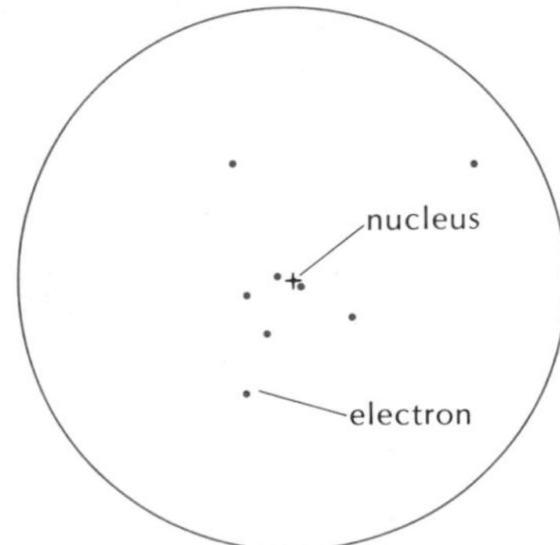

Fig. C.5 Planetary model of an oxygen atom. The position of the nucleus is marked by a cross. The atom has eight electrons, represented by the dots.

In later experiments Rutherford proceeded to explore the interior of the nucleus, again using a beam of projectiles as a probe. He found that if the projectiles were energetic enough to penetrate the nucleus, they would often split it into two pieces, two smaller nuclei. Thus, in 1919 Rutherford achieved the first nuclear transmutation, converting the nucleus of one chemical element into nuclei of other chemical elements. In these experiments Rutherford found that the smallest piece that he could split off a nucleus was a **proton** — a particle with a mass nearly equal to the mass of a hydrogen atom, and with a positive electric charge exactly the opposite of the charge of the electron. The close coincidence between the mass of a proton and the mass of a hydrogen atom has a simple explanation: the nucleus of a hydrogen atom is one single proton, so that the mass of the hydrogen atom is the sum of the mass of a proton and the mass of an electron.

The nuclei of other atoms all contain protons — the number of protons in the nucleus of the atom exactly matches the number of electrons orbiting around the nucleus. But these nuclei also contain particles of another kind. What other kind became clear in 1932 when J. Chadwick discovered that a **neutron** would sometimes emerge from a nucleus when under bombardment by alpha particles. The neutron has a mass nearly equal to that of the proton, but it has no electric charge. The study of nuclear reactions, induced by bombardment with a variety of projectiles, established that each nucleus consists of protons and neutrons tightly stuck together (Figure C.6). Each proton or neutron is a small ball, about 2×10^{-15} m across; in the nucleus these small balls are so closely packed together that they touch, or nearly touch, much like the water molecules in a droplet of water.

In the experimental investigations of the composition and structure of nuclei, the accelerators invented in the early 1930s played a crucial role. These machines, commonly called atom smashers, produce intense beams of electrically charged particles and accelerate these particles to high energies by pushing them with electric forces. The first such accelerators

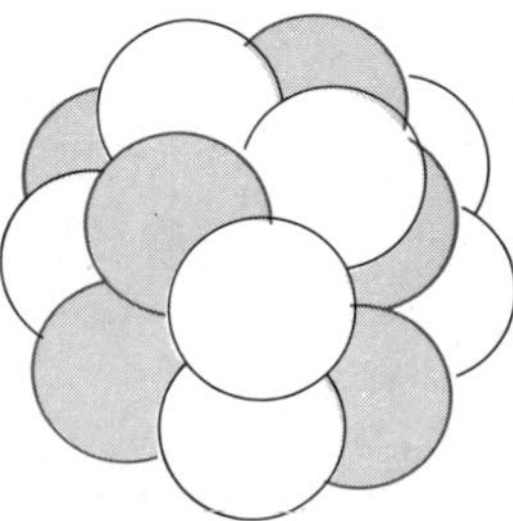

Fig. C.6 The nucleus of an atom. The colored balls are protons; the white balls are neutrons.

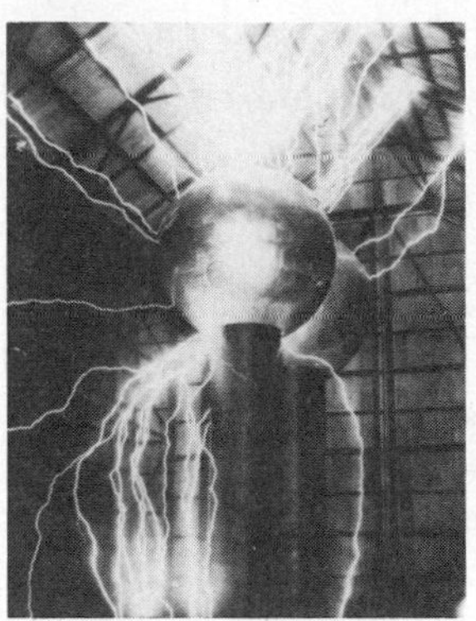

Fig. C.7 The accelerators of (left) Van de Graaff and (right) Cockroft and Walton.

were built in 1932 by R. J. Van de Graaff and by J. D. Cockcroft and E. T. S. Walton (Figure C.7); both of these accelerators produced beams of energetic protons. By bombarding nuclei with the protons from their machine, Cockcroft and Walton succeeded in inducing the first nuclear transmutation with artificially accelerated particles. Soon thereafter, E. O. Lawrence invented the cyclotron — a very efficient accelerator in which the particles travel round and round a circular (or spiral) path under the influence of magnetic forces, while at the same time they acquire higher and higher energies from the carefully timed push of an electric force (Figure C.8). The gigantic accelerators used in particle physics today are direct descendants of the cyclotron.

Table C.1 lists the main characteristics of the particles that serve as building blocks of all atoms and nuclei. The electric charge in this table is expressed in multiples of the electric charge of the proton; thus, the value of +1 for the proton is a matter of definition and the value of −1 for the electron simply indicates that the electron has an electric charge exactly opposite to that of the proton. The last column of the table lists the spin, or intrinsic angular momentum, of the particle, expressed in multiples of the basic value $\hbar = 1.05 \times 10^{-34}$ J · s; according to our table, all the constituents of the atom have a spin of $\frac{1}{2}\hbar$.

Table C.1 SUBATOMIC PARTICLES

Particle	*Mass*	*Electric charge*	*Diameter*	*Spin*
Electron, e	9.110×10^{-31} kg	−1	0 (?)	$\frac{1}{2}$
Proton, p	1.673×10^{-27}	+1	2×10^{-15} m	$\frac{1}{2}$
Neutron, n	1.675×10^{-27}	0	2×10^{-15} m	$\frac{1}{2}$

What are protons and neutrons made of? In order to smash protons and neutrons by bombardment, we need projectiles of much higher energy than that required to smash an atom or a nucleus. Atomic, nuclear, and elementary-particle physicists like to measure the energy of their projectiles in million electron-volts (MeV) or in billion electron-volts (GeV). Expressed in joules, these units of energy are

$$1 \text{ MeV} = 10^6 \text{ eV} = 1.6 \times 10^{-13} \text{ J}$$

$$1 \text{ GeV} = 10^9 \text{ eV} = 1.6 \times 10^{-10} \text{ J}$$

Although these amounts of energy are small by the standards of the macroscopic world of everyday experience, they are fairly large by the standards of the microscopic world of atoms and nuclei. A projectile of a bombarding energy of a few electron-volts (eV) can shatter an atom, and a projectile of a bombarding energy of a few MeV can shatter a nucleus.

Protons and neutrons are much "harder" than a nucleus — to make a dent in a proton, we need a bombarding energy of a few hundred or a thousand MeV. The acceleration of a projectile to such a high energy requires big, sophisticated accelerator machines. High energy — one or several GeV — characterizes the realm of particle physics. Collisions between particles at these high energies are capable of creating a large

(a)

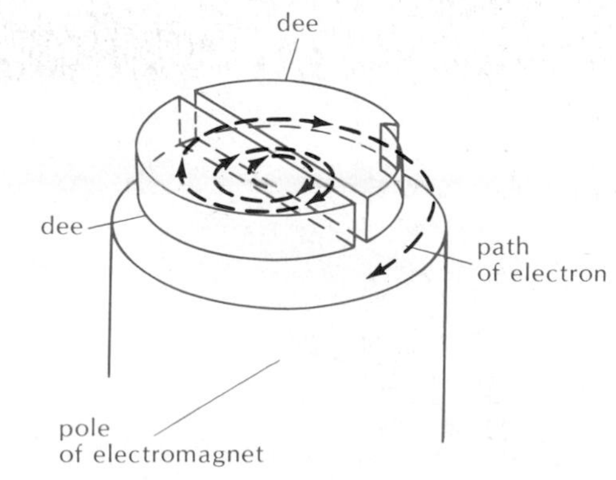

(b)

Fig. C.8 (a) Lawrence's cyclotron. (b) Schematic diagram. Each time the electron crosses from one half circle ("dee") into the other half circle, it receives an extra push.

variety of new particles by a process of conversion of energy into mass. The detection of these new particles and the measurement of their properties involves some very special equipment.

C.2 THE TOOLS OF HIGH-ENERGY PHYSICS

The two largest accelerating machines are the proton synchrotrons at the Fermi National Accelerator Laboratory (Fermilab, near Chicago) and at the Centre Européen de la Recherche Nucléaire (CERN, on the Swiss–French border near Geneva). Figure C.9 shows an overall view of Fermilab; the accelerator is buried underground in a circular tunnel 6 km in circumference. The CERN accelerator is slightly larger, about 7 km in circumference. Their gigantic size makes these accelerators two of the largest pieces of machinery ever built by man. The Fermilab accelerator produces beams of protons with an energy of 1000 GeV and a speed of 99.99995% of the speed of light. The protons travel in an evacuated circular steel pipe (Figure C.10). Large magnets placed along this pipe exert forces on the protons, preventing their escape — the protons move as if on a circular racetrack. At regular intervals the pipe is joined to cavities connected to high-voltage generators; in each of these cavities, electric forces impel the protons to higher energy. After several hundred thousand circuits around the racetrack, the protons reach their final energy of 1000 GeV.

Incidentally, before the protons are allowed to enter the giant circular racetrack, they must pass through several smaller preliminary accelerators. At Fermilab there are three preliminary accelerators: the protons generated by a proton gun are first given an energy of about 1 MeV by an electrostatic generator; then their energy is raised to 200 MeV by a linear accelerator; and then it is raised further to 8 GeV by a "small" circular accelerator (Figure C.11). Only after the protons have passed through these stages do they enter the main ring.

Fig. C.9 Panoramic view of Fermilab.

Fig. C.10 View of the underground tunnel housing the main accelerator at Fermilab. The long row of magnets encases the beam pipe.

Fig. C.11 The linear accelerator and the small circular accelerator at Fermilab. The large main accelerator is in the background.

The next largest accelerating machine is the Stanford Linear Accelerator (SLAC), 2 miles long. Figure C.12 shows the gallery sitting above the tunnel in which the machine is buried. This accelerator produces a beam of electrons with an energy of 22 GeV. The speed of electrons of this energy is within 8 cm/s of the speed of light; thus, the Stanford machine holds the speed record for artificially accelerated particles. Table C.2 lists some other large accelerators.

Once the particles have been given their maximum energy, they are guided out of the accelerator by steering magnets and made to crash against a target consisting of a block of metal or a tankful of liquid. Within the material of the target, the high-energy particles collide violently with the protons or neutrons of the nuclei. The reactions that take place in these collisions create an abundance of new particles by conversion of energy into mass. Unfortunately, not all of the kinetic energy of the incident particles can participate

Fig. C.12 Stanford Linear Accelerator gallery. The klystrons standing in a row in this gallery provide electric power to accelerate the electrons in the tube buried underneath.

Table C.2 Some Large Accelerators

Accelerator	*Start of operation*	*Projectiles*	*Kinetic energy*
Brookhaven Alternating Gradient Synchrotron (AGS), New York	1961	Protons	33 GeV
Stanford Linear Accelerator (SLAC), California	1961	Electrons	22
Cornell Electron Synchrotron, New York	1967	Electrons	12
Serpukhov Proton Synchrotron, U.S.S.R.	1967	Protons	76
Fermilab Main Ring, Illinois	1972	Protons	500
Deutsches Elektronen Synchrotron (DESY), Germany	1974	Electrons	22[a]
CERN Super Proton Synchrotron (SPS), Switzerland–France	1976	Protons	500[b]
Fermilab Tevatron, Illinois	1984	Protons	1000
CERN Large Electron–Positron Storage Ring (LEP)	1988?	Electrons	50[c]
Serpukhov UNK, U.S.S.R.	1990?	Protons	3000

[a]Used for 2×22-GeV head-on collisions.
[b]Used for 2×270-GeV head-on collisions.
[c]For 2×50-GeV head-on collisions.

in these reactions. As we saw in Section 10.4, the kinetic energy associated with the motion of the center of mass remains constant during collisions and is therefore not available for inelastic reactions; only the "internal" kinetic energy associated with motion relative to the center of mass is available for inelastic reactions and can be used up in such reactions. According to Eq. (10.27), this available energy for a proton of kinetic energy 500 GeV colliding with a stationary proton is 250 GeV. However, this calculation ignores relativistic effects; and since the speed of the incident proton is very close to the speed of light, these effects are actually very important. A correct relativistic calculation indicates the available energy is much less — it is only 31 GeV. Thus, collisions between a high-energy

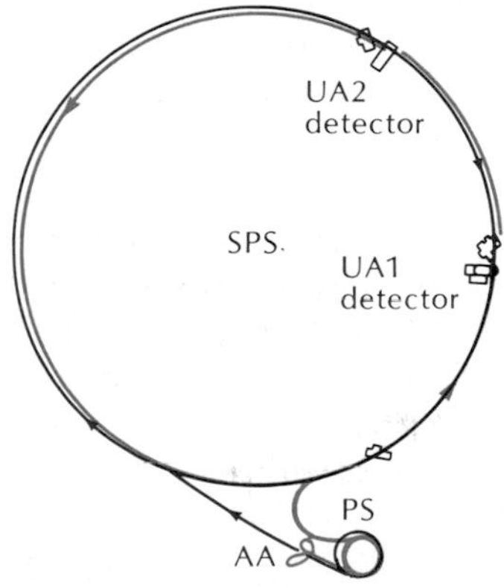

Fig. C.13 In the SPS accelerator at CERN, high-energy protons and antiprotons travel in opposite directions around a ring of circumference 7 km. The protons and antiprotons are produced by a medium-energy accelerator (PS, lower right). The antiprotons are temporarily stored in an accumulator ring (AA), while the protons are sent directly into the large ring. When enough antiprotons have accumulated, they are also sent into the large ring, and both kinds of particles are accelerated to their final, high energy. The orbits of the protons (black) and the antiprotons (color) are slightly separated, but they intersect at two points, where the detectors UA 1 and UA 2 are located.

Fig. C.14 The UA 2 detector at CERN. This detector surrounds the point of intersection of the proton and antiproton orbits.

proton and a stationary proton are quite inefficient.

The efficiency improves drastically if two high-energy protons are made to collide head on. The available energy is then the sum of the energies of the two protons. At CERN, a beam of protons and a beam of antiprotons of 270 GeV each are made to travel in opposite directions around the accelerator in slightly different orbits that cross at two places (Figures C.13 and C.14). Where the two orbits cross, the two beams meet nearly head on. Several other accelerator laboratories have recently installed similar storage rings with which they can bring about head-on collisions between particles.

When a beam of high-energy particles crashes into a stationary target or into an oncoming beam, the violent collisions send pieces flying off in all directions. Exotic reactions generate a wide variety of new particles. In order to observe the particles that emerge from the scene of these collisions, physicists use several kinds of particle detectors. Some of these — scintillation counters and Čerenkov counters — signal the passage of each electrically charged particle by giving off brief (and weak) flashes of light. These flashes of light are detected by sensitive photomultiplier tubes ("electric eyes") that convert the light into a pulse of electric current; electronic devices automatically count the number of these pulses and display them as particle counts on a digital readout, or scaler (Figure C.15). Other detectors — bubble chambers and spark chambers — render the track of an electrically charged particle visible and record these tracks on photographs.

A **bubble chamber** is a tank filled with a superheated liquid, usually liquid hydrogen. A liquid is said to be superheated when its temperature is slightly above the boiling point. Such a liquid is unstable — it is about to start boiling but it will usually not start until some disturbance triggers the formation of the first few bubbles. A charged particle zipping through the chamber provides just the kind of disturbance the liquid is waiting for — a fine stream of bubbles forms in the wake of the particle's passage. High-speed cameras can take a picture of these bubble tracks before they disperse and disappear in the turmoil of subsequent widespread bubbling and boiling of the liquid.

Fig. C.15 Scalers at an experimental station at Fermilab. These scalers store and count electric pulses arriving from particle detectors.

Fig. C.16 The CERN bubble chamber, surrounded by the large magnet which almost completely hides it from view.

Fig. C.17 The CERN bubble chamber before its installation in the magnet.

Figure C.16 shows a bubble chamber at CERN. The bubble chamber is surrounded by a large electromagnet, which aids in the identification of the particles passing through the bubble chamber. The magnetic force generated by this magnet pushes the particles into curved orbits as they pass through the chamber. The direction of the curvature depends on the sign of the electric charge and the magnitude of the curvature depends on the momentum of the particles; thus, measurement of the curvature of a bubble track tells us the sign of the electric charge of the particle and the momentum of the particle. Figure C.17 shows the exterior of the bubble chamber during construction. The portholes are for cameras and for flash lamps. Photographs are taken simultaneously with several cameras at different angles so as to obtain a stereoscopic view of the particle tracks. A chamber of the kind shown in Figure C.16 can go through about 10 complete boiling cycles per second; each time a completely different set of tracks will be recorded on the photograph. Sometimes experimenters must take several hundred thousand photographs before finding one with the special kind of collision they are looking for.

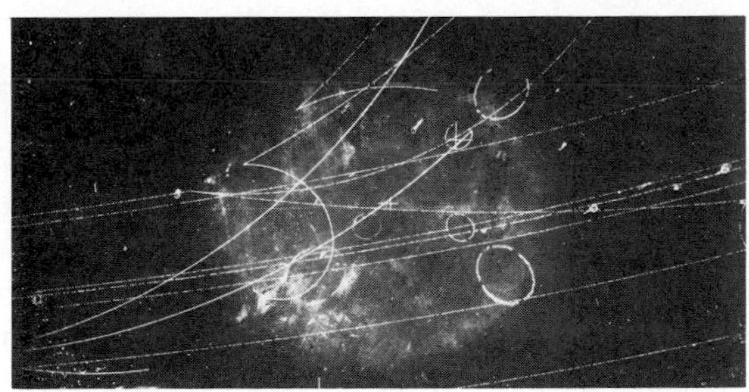

(a)

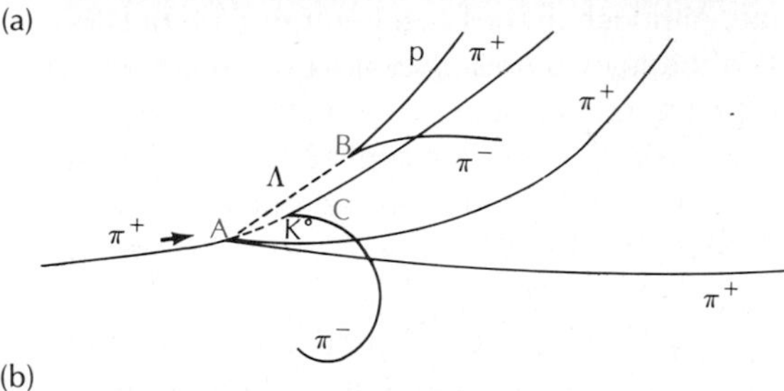

(b)

Fig. C.18 (a) Tracks of pions passing through bubble chamber. (b) Collision of a pion and a proton in the bubble chamber. Tracing based on (a).

The photograph in Figure C.18 shows a high-energy beam of positive pions (π^+) entering the chamber. The tracks of the pions, entering the field of view from the left, are slightly curved upward by the magnetic force exerted by the electromagnet. Pions are particles heavier than electrons but lighter than protons; they are generated in great abundance whenever protons impact on nuclei. A beam of pions can be produced by simply aiming a beam of protons from the accelerator at a target (a thick chunk of metal); many pions then come flying out of the far side of the target and they can be assembled in a beam and guided to the bubble chamber by an array of electromagnets.

At point *A* in Figure C.18, one of the pions hits one of the protons sitting at rest at the center of a hydrogen atom. Both the pion and the proton are destroyed by this violent collision — they cease to exist. In their place two new pions, a kaon (K^0), and a lambda particle (Λ) emerge from the scene of the accident. Both the kaon and the lambda particle are electrically neutral and hence leave no visible tracks in the bubble chamber. Nevertheless, it is possible to reconstruct their trajectories because, after a short while, both of these particles decay spontaneously into new particles that do leave tracks — at point *B* the lambda decays into a proton and a negative pion (π^-) and at point *C* the kaon decays into two pions. These reactions can be summarized as follows:

$$\pi^+ + p \rightarrow \Lambda + K^0 + \pi^+ + \pi^+ \qquad (1)$$

$$K^0 \rightarrow \pi^+ + \pi^- \qquad (2)$$

$$\Lambda \rightarrow p + \pi^- \qquad (3)$$

The net result of these reactions is the conversion of one proton and one pion into one proton and five pions. Thus, the mass after the reactions is much larger than the mass before, i.e., the *Q* value of the reaction is negative [see Eq. (10.27)]. The excess mass comes from the conversion of energy into mass — some of the kinetic energy of the incident pion has been converted into mass.

This conversion of kinetic energy into mass plays a crucial role in the discovery of new particles. Almost all the new particles discovered during the last 20 years are considerably heavier than protons and neutrons. Physicists need powerful accelerators to produce the large kinetic energies that must be supplied for the manufacture of these new heavy particles.

Although bubble chambers yield the best pictures of particle tracks, they are very complex, very large, and very expensive machines. **Spark chambers** yield somewhat cruder pictures of particle tracks, but they are much simpler. A spark chamber consists of several parallel screens or thin plates of metal, each separated from the next by a gap of a few centimeters (Figure C.19). The plates are connected to a high-voltage supply. Any charged particle passing through the chamber ionizes the air and thereby triggers the formation of electric sparks between the screens. A succession of sparks marks the passage of the particle through the chamber (Figure C.20). Cameras record these sparks photographically; usually mirrors are placed around the spark chamber so that a single photograph can simultaneously record several views of the track of sparks seen from several angles.

Fig. C.19 Spark chamber.

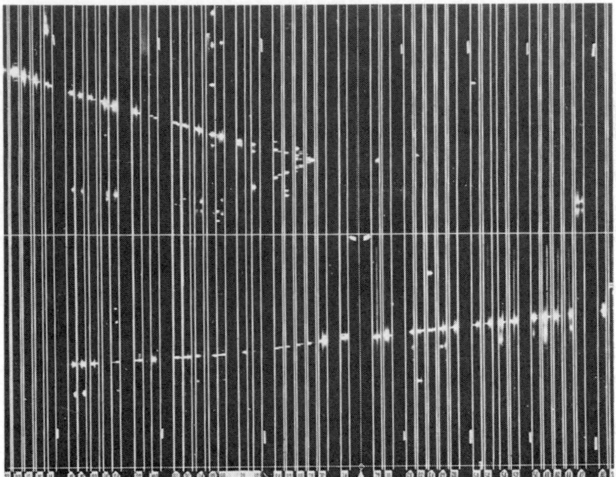

Fig. C.20 Sparks in spark chamber.

Fig. C.21 Physicists of the team responsible for the discovery of the upsilon particle pose with their equipment at Fermilab.

Experiments in high-energy physics involve such a large array of equipment that they are run by teams — sometimes several dozen physicists in a team working in relays around the clock so as to make the most efficient use of their expensive machinery (Figure C.21).

C.3 THE MULTITUDE OF PARTICLES

As physicists built accelerators of higher and higher energy, they discovered more and more new particles. By now, close to 300 kinds of particles are known and a few extra particles are found every year. It seems that as more energy is put into a collision, more particles and heavier particles will be produced — there seems to be no end in sight. Whenever a new, more powerful accelerator comes into operation, it is sure to discover a few additional particles.

The known particles fall into three groups: leptons, baryons, and mesons. The last two groups are collectively designated as **hadrons,** or **strongly interacting particles,** because they exert "strong" forces on one another.

There are six different **leptons** (see Table C.3). Among these, the electron (e) is the most familiar. The muon (μ) is very similar to an electron; it has the same electric charge as an electron but its mass is about 200 times as large. The tau (τ) is an esoteric particle, discovered only recently in high-energy experiments at several accelerator laboratories. The neutrinos (ν_e, ν_μ, and ν_τ) are particles of zero mass and zero electric charge. Although muons and neutrinos are not constituents of ordinary atoms, they are both quite abundant in nature. Large numbers of muons are generated at the top of the Earth's atmosphere by the impact of cosmic rays; these muons are secondary cosmic rays that penetrate to sea level and hit our bodies, doing some radiation damage (see Interlude B). Large numbers of neutrinos are released by the nuclear reactions in the core of the Sun; these neutrinos penetrate through the Earth and through our bodies, but they do very little damage — our bodies and even the entire bulk of the Earth are nearly transparent to neutrinos.

In Table C.3 the masses of the particles are expressed in MeV energy units (the "mass" listed in this table is really mass $\times$ c^2), and the electric charge and

Table C.3 THE LEPTONS

Particle	*Mass*	*Electric charge*	*Spin*
e	0.511 MeV	−1	$\frac{1}{2}$
ν_e	0	0	$\frac{1}{2}$
μ	105.7	−1	$\frac{1}{2}$
ν_μ	0	0	$\frac{1}{2}$
τ	1780	−1	$\frac{1}{2}$
ν_τ	0	0	$\frac{1}{2}$

Table C.4 THE BARYONS[a]

Particle (and mass)[b]	Electric charge	Spin	Particle (and mass)[b]	Electric charge	Spin
N(939)	0, 1	$\frac{1}{2}$	Λ(1600)	0	$\frac{1}{2}$
N(1440)	0, 1	$\frac{1}{2}$	Λ(1670)	0	$\frac{1}{2}$
N(1520)	0, 1	$\frac{3}{2}$	Λ(1690)	0	$\frac{3}{2}$
N(1535)	0, 1	$\frac{1}{2}$	Λ(1800)	0	$\frac{1}{2}$
N(1650)	0, 1	$\frac{1}{2}$	Λ(1800)	0	$\frac{1}{2}$
N(1675)	0, 1	$\frac{5}{2}$	Λ(1820)	0	$\frac{5}{2}$
N(1680)	0, 1	$\frac{5}{2}$	Λ(1830)	0	$\frac{5}{2}$
N(1700)	0, 1	$\frac{1}{2}$	Λ(1890)	0	$\frac{3}{2}$
N(1710)	0, 1	$\frac{1}{2}$	Λ(2100)	0	$\frac{7}{2}$
N(1720)	0, 1	$\frac{3}{2}$	Λ(2110)	0	$\frac{5}{2}$
N(2190)	0, 1	$\frac{7}{2}$	Λ(2350)	0	$\frac{9}{2}$
N(2220)	0, 1	$\frac{9}{2}$	Σ(1193)	1, 0, −1	$\frac{1}{2}$
N(2250)	0, 1	$\frac{9}{2}$	Σ(1385)	1, 0, −1	$\frac{3}{2}$
N(2600)	0, 1	$\frac{11}{2}$	Σ(1660)	1, 0, −1	$\frac{1}{2}$
Δ(1232)	2, 1, 0, −1	$\frac{3}{2}$	Σ(1670)	1, 0, −1	$\frac{3}{2}$
Δ(1620)	2, 1, 0, −1	$\frac{1}{2}$	Σ(1750)	1, 0, −1	$\frac{1}{2}$
Δ(1700)	2, 1, 0, −1	$\frac{3}{2}$	Σ(1775)	1, 0, −1	$\frac{5}{2}$
Δ(1900)	2, 1, 0, −1	$\frac{1}{2}$	Σ(1915)	1, 0, −1	$\frac{5}{2}$
Δ(1905)	2, 1, 0, −1	$\frac{5}{2}$	Σ(1940)	1, 0, −1	$\frac{3}{2}$
Δ(1910)	2, 1, 0, −1	$\frac{1}{2}$	Σ(2030)	1, 0, −1	$\frac{7}{2}$
Δ(1920)	2, 1, 0, −1	$\frac{3}{2}$	Σ(2250)	1, 0, −1	?
Δ(1930)	2, 1, 0, −1	$\frac{5}{2}$	Ξ(1318)	0, −1	$\frac{1}{2}$
Δ(1950)	2, 1, 0, −1	$\frac{7}{2}$	Ξ(1530)	0, −1	$\frac{3}{2}$
Δ(2420)	2, 1, 0, −1	$\frac{11}{2}$	Ξ(1820)	0, −1	$\frac{3}{2}$
Λ(1116)	0	$\frac{1}{2}$	Ξ(2030)	0, −1	?
Λ(1405)	0	$\frac{1}{2}$	Ω^-(1672)	−1	$\frac{3}{2}$
Λ(1520)	0	$\frac{3}{2}$	Λ_c^+(2282)	1	$\frac{1}{2}$

[a] From *Review of Particle Properties,* by The Berkeley Particle Data Group, April 1984.
[b] The mass is expressed in MeV.

Table C.5 THE MESONS[a]

Particle (and mass)[b]	Electric charge	Spin	Particle (and mass)[b]	Electric charge	Spin
π(140)	1, 0, −1	0	ψ(3685)	0	1
η(549)	0	0	ψ(3770)	0	1
ρ(770)	1, 0, −1	1	ψ(4030)	0	1
ω(783)	0	1	ψ(4160)	0	1
η'(958)	0	0	ψ(4415)	0	1
S*(975)	0	0	Υ(9460)	0	1
δ(980)	1, 0, −1	0	χ_b(9875)	0	?
ϕ(1020)	0	1	χ_b(9895)	0	?
B(1235)	1, 0, −1	1	χ_b(9915)	0	?
f(1270)	0	2	Υ(10025)	0	1
A_1(1270)	1, 0, −1	1	χ_b(10255)	0	?
D(1285)	0	1	χ_b(10270)	0	?
ε(1300)	0	0	Υ(10355)	0	1
A_2(1320)	1, 0, −1	2	Υ(10575)	0	1
E(1420)	0	1	K±(494)	1, −1	0
ι(1440)	0	0	K^0(498)	0, 0	0
f′(1525)	0	2	K*(892)	1, 0, −1	1
ρ'(1600)	1, 0, −1	1	Q_1(1280)	1, 0, −1	1
ω(1670)	0	3	κ(1350)	1, 0, −1	0
A_3(1680)	1, 0, −1	2	Q_2(1400)	1, 0, −1	1
ϕ'(1680)	0	1	K*(1430)	1, 0, −1	2
g(1690)	1, 0, −1	3	L(1770)	1, 0, −1	2
θ(1690)	0	2	K*(1780)	1, 0, −1	3
ϕ(1850)	0	3	D^0(1865)	0, 0	0
h(2030)	0	4	D±(1869)	1, −1	0
η_c(2980)	0	0	D*±(2010)	1, −1	1
J/ψ(3100)	0	1	D^{*0}(2010)	0, 0	1
χ(3415)	0	0	F±(1971)	1, −1	0
χ(3510)	0	1	B±(5271)	1, −1	0
χ(3555)	0	2	B^0(5274)	0, 0	0

[a] From *Review of Particle Properties,* by The Berkeley Particle Data Group, April 1984.
[b] The mass is expressed in MeV.

the spin are expressed, respectively, in multiples of the proton charge and in multiples of the basic value $\hbar = 1.05 \times 10^{-34}$ J · s, just as in Table C.1.

Besides the leptons of Table C.3, there are six antileptons: the antielectron, the antimuon, the antitau, and the three antineutrinos. These antiparticles have an opposite electric charge and some other opposite properties, but they have exactly the same mass and spin as the corresponding particles.

The **baryons** are the most numerous group of particles (see Table C.4). The proton and the neutron are baryons; in Table C.4 they have been listed as a pair under the single entry N(939) because of close similarities (the proton and the neutron are much more similar to each other than to any other particle). For the same reason, other baryons have also been listed as pairs, triplets, or quadruplets under a single entry. The first column of the table gives the baryon name consisting of a letter and a number in parentheses; the number is (approximately) the baryon mass expressed in MeV energy units. The second column gives the alternative electric charges of the several baryons that share a common name, and the third column gives the spin.

For every baryon in Table C.4 there exists an antibaryon. As in the case of leptons, these antiparticles have an opposite charge but the same mass and spin as the corresponding particles.

Finally, the **mesons** are another numerous group of particles. They are listed in Table C.5. Mesons with close similarities are listed under a single entry; for instance, π(140) includes the positive pion π^+, the negative pion π^-, and the neutral pion π^0. Note that all mesons have spin 0, or 1, or 2, etc.; while all baryons have spin $\frac{1}{2}$, or $\frac{3}{2}$, or $\frac{5}{2}$, etc. This is the distinctive difference between mesons and baryons. For every meson there exists an antimeson; however, these antiparticles have already been included in Table C.5. For example, the antiparticle to the π^+ is the π^- (and vice versa). The antiparticle to the π^0 is the π^0; this means that when two π^0 mesons meet, they can annihilate each other.

We know much more about these particles than their mass, electric charge, and spin. Almost all of the known particles are unstable; they decay, spontaneously falling apart into several other particles. The only absolutely stable particles are the electron, the proton, and the neutrinos. Not even the neutron is stable; it decays into a proton, an electron, and an antineutrino,[4]

$$n \rightarrow p + e + \bar{\nu} \tag{4}$$

This reaction is called **beta decay,** since it is the source of the beta particles (energetic electrons) emitted by some radioactive nuclei (see Interlude B). An isolated, free neutron typically lasts for only 1000 s before decaying. However, inside a nucleus the beta decay is inhibited by the presence of the nuclear protons; the survival of the neutrons inside the nuclei of ordinary atoms hinges on this inhibition mechanism. Thus, the stability of nuclei depends on a delicate symbiotic relationship between neutrons and protons — the presence of protons stabilizes the neutrons against decay and the presence of neutrons stabilizes the protons against the disruptive electric forces that repel one proton from another.

Figure C.18 contains two examples of decays: the decay of a kaon and that of a lambda particle. Each of these particles lives for about 10^{-10} s from the instant of production to the instant of decay. By the standards of high-energy physics a time interval of 10^{-10} s is rather long. If a particle lives that long, physicists can do experiments on it and with it; for example, kaons live long enough to be assembled in a beam and shot at a target. In the terminology of high-energy physics, any particle that lives 10^{-10} s or even 10^{-14} s is regarded as **stable.** If this is stable, what is unstable? Most of the particles listed in Tables C.4 and C.5 have lifetimes of about 10^{-23} s. Such particles are regarded as unstable.

A particle that lives only 10^{-23} s is incapable of making a visible track in a bubble chamber. Thus, such an unstable particle cannot be detected directly, but its existence can be inferred from a careful study of the rates of reactions of stable particles engaged in collisions. The unstable particle participates in these reactions as an intermediate, ephemeral state, which causes a characteristic increase of the reaction rate whenever the energy of the stable particles coincides with the energy required for the production of the unstable particle. Because of their effects on reaction rates, the unstable particles are often called **resonances.**

[4] A bar over a letter indicates an antiparticle; thus, the letter ν stands for neutrino, and $\bar{\nu}$ stands for antineutrino.

C.4 CONSERVATION LAWS

To bring some order into the chaotic multitude of the 300 or so known baryons and mesons, we would like to group these particles into separate families consisting of particles of closely related behavior patterns. We can learn about the behavior patterns of particles by studying the reactions in which they engage (or do not engage), just as we can learn about the behavior patterns of atoms by studying the chemical reactions in which they engage. From the study of reactions of atoms, chemists have learned how to classify atoms into families according to their valence — atoms in a given valence family engage in similar reactions. In the periodic table (see Table P.1), all the atoms listed in the same column have the same valence and belong to the same family. Physicists have searched for analogous classification schemes for particles, but the search has been hindered by the greater complexity of the reactions. When atoms react chemically, they combine in diverse ways but they do not disappear — in chemical reactions atoms are permanent. When particles react, they transmute into other particles — in high-energy reactions particles are not permanent. To find a classification scheme, physicists must first look for something that remains permanent in these reactions. They must look for **conserved quantities.**

In earlier chapters we have already come across some conserved quantities: energy (or mass), momentum, and angular momentum. In a later chapter we will study the conservation of electric charge. But besides these more or less familiar conserved quantities, several esoteric conserved quantities have been discovered in experiments with high-energy particles. Among these new conserved quantities, the most important are **baryon number, lepton number, isospin,** and **strangeness.**

The concept of **baryon number** is easy to grasp. Each baryon has a baryon number +1, each antibaryon −1, and all other particles have 0. The conservation law for baryon number then simply states that the net baryon number remains constant in any reaction. For example, consider the reaction

$$\pi^- + p \rightarrow \Lambda + K^0 \tag{5}$$

which occurs when high-energy pions collide with protons. The baryon numbers for π^-, p, Λ, and K^0 are, respectively, 0, 1, 1, and 0; hence in the above reaction the net baryon number is 0 + 1 before and 1 + 0 after, i.e., there is no change in the net baryon number.

The **lepton number** works in a similar way. Each lepton has a lepton number +1, each antilepton −1, and all other particles have 0. The net lepton number remains constant in any reaction.

Isospin is somewhat more complicated because it is a quantity with several components; that is, isospin is a vector quantity. The conservation law for isospin states that the net sum of the isospin vectors of all the particles involved in a reaction remains constant. Obviously, this is somewhat similar to the conservation law for the momentum vector or for the angular momentum vector. However, the isospin is not a vector in our ordinary three-dimensional space but, rather, a vector in an abstract mathematical space that theoretical physicists invented as a computational tool for the description of the "strong" force.

The **strangeness** number is similar to the baryon and lepton numbers. Each hadron has a strangeness number: the proton has strangeness 0, the kaon has +1, the lambda has −1, the pion has 0, etc. The conservation law for strangeness states that the net strangeness number remains constant in any reaction. For example, in reaction (5) the net strangeness is $0+0$ before and it is $+1-1$ after; i.e., there is no change.

The conservation laws for energy, momentum, angular momentum, electric charge, baryon number, and lepton number are absolute — no violation of any of them has ever been discovered.[5] By contrast, both the conservation laws for isospin and for strangeness are *approximate* conservation laws — they are valid for some reactions, but fail in some others. This, of course, raises the question of what possible meaning can be attached to a "law" that works sometimes and fails sometimes. The answer is that we usually know beforehand whether the conservation law applies or not. The rule is that reactions caused by the strong force will obey all the conservation laws, but reactions caused by the electromagnetic or weak forces will not. It is usually easy to tell what force is involved in a reaction: reactions involving the strong force tend to be fast; reactions involving the other forces tend to be (relatively) slow. For example, reaction (5) is a fast reaction brought about by the strong force, whereas reaction (4) is a slow reaction brought about by the weak force. The relationship between different kinds of forces and conservation laws will be further discussed in Interlude G.

[5] According to a new unified theory of the strong, electromagnetic, and weak forces, the conservation law for baryon number is only approximate. If so, the proton would be unstable — once in a rare while, a proton would decay into leptons. Several experiments are in progress that attempt to detect this hypothetical rare decay of the proton.

C.5 FAMILY RELATIONSHIPS

Baryon number, lepton number, spin, isospin, strangeness, and other conserved quantities can be regarded as labels attached to the particles. The particles are not permanent — during high-energy reactions particles are freely created and destroyed. But the labels are permanent — during reactions the conserved quantities do not change and the labels are merely reshuffled. This persistence of the labels suggests that they should be of primary importance in any attempt at classification of the particles. Modern classification schemes group the particles into families according to the values of their labels. The most successful of these classification schemes is SU(3) or the **Eightfold Way** proposed by M. Gell-Mann and Y. Ne'eman in 1961. According to this scheme, each family consists of particles of the same spin, the same baryon number, and nearly the same mass; within the family the isospin and the strangeness vary in a regular way. Figure C.22 shows one of these families of particles: the **baryon**

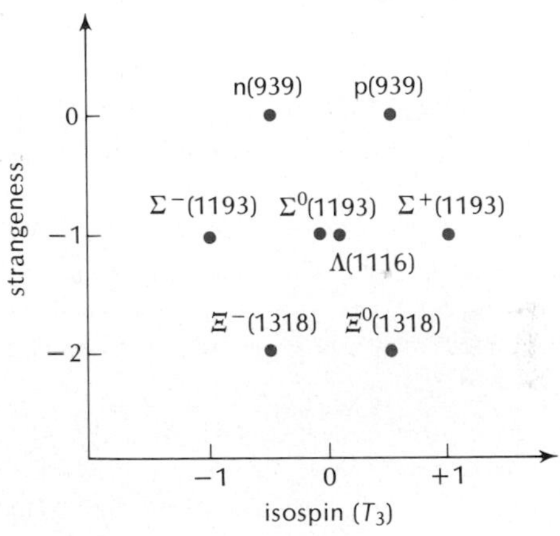

Fig. C.22 The baryon octet arranged according to strangeness and isospin.

octet family. This family contains the proton and the neutron plus six other particles. All particles in this family have spin $\frac{1}{2}$. The position of a particle within the family is determined by the isospin (marked along the horizontal axis) and the strangeness (marked along the vertical axis). The regular arrangement of the particles within the family and the hexagonal symmetry of Figure C.22 reflect an internal symmetry of the particles. This symmetry is not geometric (like the symmetry of a crystal); instead it involves mathematical transformations in an abstract space of several dimensions.

Figure C.23 shows another family of particles: the **baryon decuplet** family. This family contains 10 particles, all with spin $\frac{3}{2}$.

The Eightfold Way accommodates all known baryons and mesons into families, or **multiplets,** of the

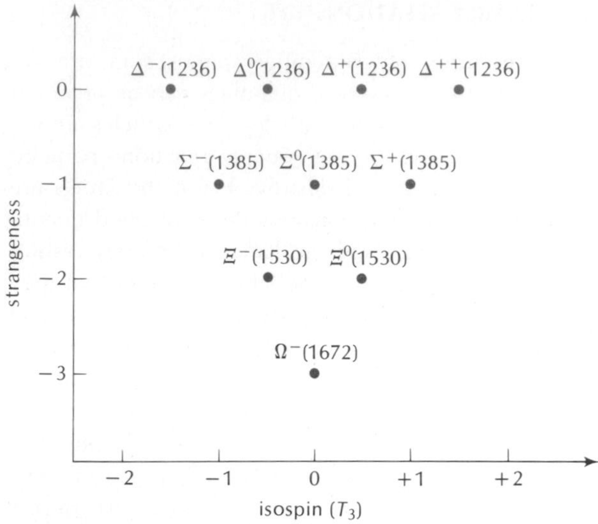

Fig. C.23 The baryon decuplet arranged according to strangeness and isospin.

kind shown in Figures C.22 and C.23. However, some of the families are incomplete, i.e., some positions in the family are left open. Presumably this means that some family members yet remain to be discovered.

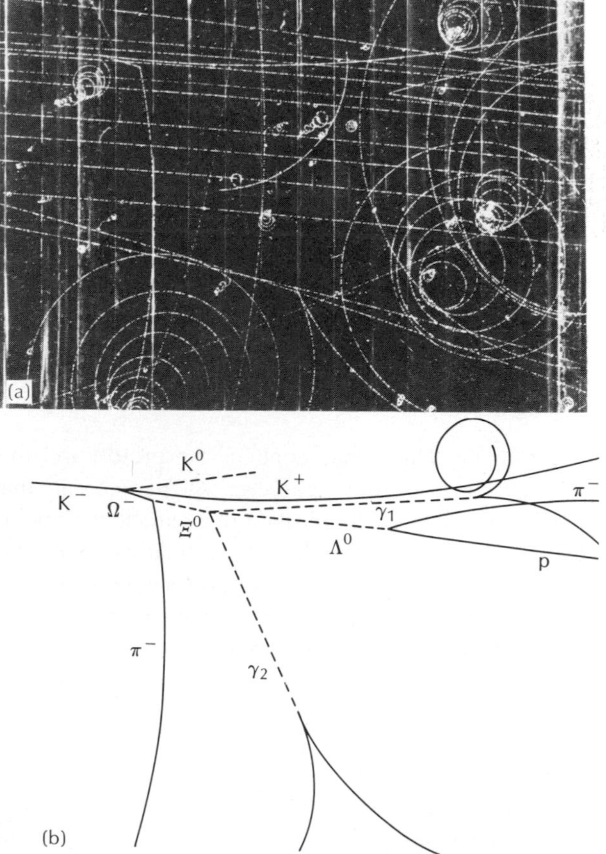

Fig. C.24 (a) First photograph of the track of an Ω^- particle in a bubble chamber. (b) Tracing based on (a).

Hence the Eightfold Way predicts the existence of new particles. The most famous instance of such a prediction was the case of the omega particle. When theoreticians first assembled the baryon decuplet family, the position in the lowest corner of Figure C.23 was still open — the omega had not yet been discovered. Theoreticians predicted all of the properties of the omega by virtue of its membership and position in the decuplet family. Experimenters saw this as a challenge and immediately began to search for the missing particle — and within a few short and exciting months they found it, in February of 1964. Figure C.24 is a bubble-chamber photograph of a track made by the first omega ever detected.

C.6 QUARKS

Let us now return to our initial question. What are the ultimate, indivisible building blocks of matter? We know of nearly 300 particles. It is inconceivable that all these hundreds of particles are *elementary* particles. It is likely that most of them, or maybe all of them, are composite particles made of just a few truly elementary building blocks.

To discover these building blocks, physicists have tried to break protons into pieces by bombarding them with projectiles of very high energy. Unfortunately, if the energy of the projectile is large enough to make a dent in a proton, then it is also large enough to create new particles during the collision. This abundant creation of particles confuses the issue — we can never be quite sure which of the pieces that come flying out of the scene of the collision are newly created particles and which are fragments of the original proton. In fact, none of the pieces ever found in such collision experiments seem likely candidates for an elementary building block. Typically, the particles that emerge from a collision between a high-energy projectile and a proton are pions, kaons, lambdas, deltas, etc., all of which seem to be even less elementary than a proton.

Although brute-force collision experiments have failed to fragment protons into elementary building blocks, somewhat more subtle experiments have provided us with some evidence that distinct building blocks do indeed exist inside protons. At the Stanford Linear Accelerator, very high-energy electrons were shot at protons; these electrons served as probes to "feel" the interior of the protons. The experiments showed that occasionally the bombarding electrons were deflected through large angles, bouncing off sharply from the interior of a proton. These deflections indicate the presence of some lumps or hard kernels in the interior of the proton, just as the deflections of alpha particles by atoms indicate the presence of a

Table C.6 THREE QUARKS

Quark	*Mass*[a]	*Electric charge*	*Spin*
u	336 MeV	$\frac{2}{3}$	$\frac{1}{2}$
d	338	$-\frac{1}{3}$	$\frac{1}{2}$
s	540	$-\frac{1}{3}$	$\frac{1}{2}$

[a] Based on theoretical estimates

hard kernel (nucleus) in the interior of the atom (see Section C.1). Protons and all the other baryons and mesons seem to be composite bodies made of several distinct pieces. In contrast, electrons and the other leptons seem to be indivisible bodies with no internal structure. In recent experiments the electron has been probed with beams of extremely energetic particles to within 10^{-18} m of its center. Even at these extremely short distances, no substructures of any kind were found. Thus, the electron appears to be a pointlike body with no size at all, a truly elementary particle.

Even before the experimental evidence for lumps inside protons became available, theoretical physicists had proposed models in which all the known baryons and mesons are regarded as constructed out of a few fundamental building blocks. According to these theories, the similarity between particles in a given family or multiplet reflects the similarity of their internal construction, just as the similarities between atoms in a group of the periodic table reflect the similarity of their internal construction.

The most successful of these theories is the quark model proposed by Gell-Mann and by G. Zweig. In this model, all particles are constructed of three kinds of fundamental building blocks called **quarks.** Gell-Mann took the word *quark* from *Finnegan's Wake,* a book by James Joyce quopping with weird words. The three quarks are usually labeled **up, down,** and **strange,** or simply **u, d,** and **s.** They all have spin $\frac{1}{2}$ and electric charges of $\frac{2}{3}$, $-\frac{1}{3}$, and $-\frac{1}{3}$, respectively (see Table C.6). In the Eightfold Way classification scheme, the quarks belong to a family of their own: the triplet family (Figure C.25). Of course, each quark — like any other particle — has an antiparticle. These antiparticles belong to another family, the antitriplet family (Figure C.26).

To make the ordinary particles out of the quarks, the latter must be glued together in diverse ways. For example, a proton is made of two u quarks and one d quark (Figure C.27). A neutron is made of two d quarks and one u quark (see Figure C.28). A positive pion is made of one u quark and one d antiquark (Figure C.29), etc. By exercising some care in what quarks are glued together to make what particle, one can build up all the known baryons and mesons and explain their family relationships. Besides, one can predict some properties, such as reaction rates, of the composite particles from the (assumed) properties of the quarks. The experimental confirmation of these predictions is strong circumstantial evidence for the quark model.

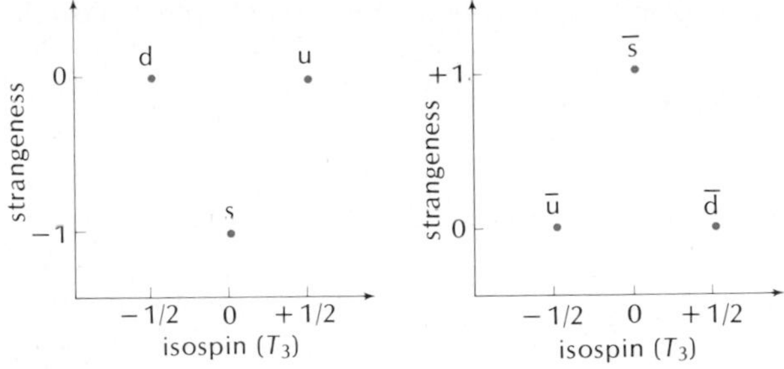

Fig. C.25 (left) The quark triplet arranged according to strangeness and isospin.

Fig. C.26 (right) The antiquark antitriplet arranged according to strangeness and isospin.

There is only one snag: in spite of prodigious efforts, experimental searches for quarks have been unsuccessful. Experimenters have looked for quarks in the debris of the very energetic collisions of particles from accelerators. They have looked for quarks in the debris of the even more energetic collisions caused by the impact of cosmic rays on the atmosphere of the Earth; they have looked for quarks in samples of water, air, dust, meteorites, lava, limestone, iron, graphite, nio-

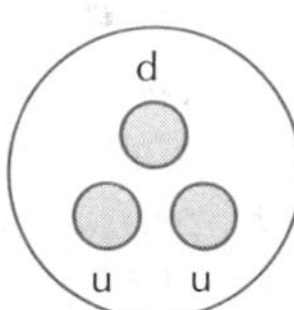

Fig. C.27 Structure of the proton: two u quarks and one d quark. The sizes of the quarks have not been drawn to scale. Their sizes are probably much smaller than the size of the proton.

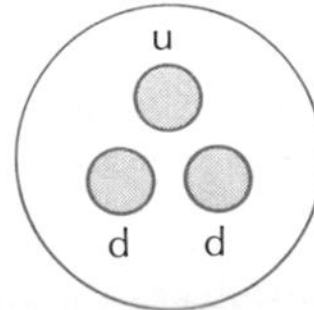

Fig. C.28 Structure of the neutron: two d quarks and one u quark.

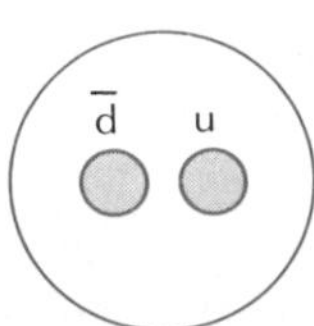

Fig. C.29 Structure of a positive pion (π^+): one u quark and one d antiquark.

bium, plankton, seaweed, mineral oil, vegetable oil, moon rocks, etc. In all these searches the distinctive feature that would permit the unambiguous identification of a quark is its *fractional charge* — in contrast to normal particles, which all have an electric charge of 0, ± 1, ± 2, etc., the quarks have charges of $\pm\frac{1}{3}$ or $\pm\frac{2}{3}$.

The searches have failed to turn up any quarks. Only in a very few instances was anything found that could *possibly* have been a quark, and even in these exceptional cases the evidence was not sufficiently convincing to establish the existence of the quark beyond reasonable doubt.

Physicists now suspect that quarks are permanently confined inside the ordinary particles so that there is no way to break a quark out of, say, a proton. Although the details of the confinement mechanism are not known, it seems that the quarks are held in place by an exceptionally strong force, which prevents their escape. This new force is called the "color" force; we will discuss its properties in the next section.

C.7 COLOR AND CHARM

The simple quark model with three quarks (and three antiquarks) described in the preceding section suffers from a few deficiencies. These deficiencies became apparent as soon as theoretical physicists began to study the precise arrangement of quarks within normal particles and the forces that hold these quarks together. First of all, for certain technical reasons having to do with the laws of quantum mechanics, physicists found it necessary to postulate that each of the three quarks exists in three different varieties. These varieties differ by a new label that has been called **color.** Of course, this "color" has nothing to do with real color; it is merely a (somewhat unimaginative) name for a new property of matter. The different quark colors are **red, green,** and **blue.** Thus, there is a *red* u quark, a *green* u quark, and a *blue* u quark, etc.

Color is a very subtle property of matter; it usually remains hidden inside the ordinary particles. All the normal particles are "colorless" — they consist of several quarks with an equal mixture of all three colors. For instance, one of the three quarks inside the proton is *red,* one is *green,* and one is *blue*. Nevertheless, color plays a very important role in the theory of the force that holds the quarks together. As we mentioned in the preceding section, the attractive force between the quarks is believed to be extremely strong. Between the quarks in, say, a proton, the force is about 10,000 lb. This extremely strong force is a **color force,** i.e., this force acts between any two "colored" particles. Thus, the source of this force is color, in much the same way as the source of gravity is mass and the source of electric force is electric charge.

A further modification of the quark model emerged from the study of weak and electromagnetic forces. Physicists have long had the ambition to find a unified theory of forces, i.e., a theory in which all forces are regarded as different aspects of a single fundamental force. A few years ago they took a first step in this direction by producing a unified theory of weak and electromagnetic forces (see Interlude G). In order to make this theory fit the experimental observations, they had to postulate the existence of a fourth quark, different from the u, d, and s quarks. This new hypothetical quark was labeled **charmed,** or c.

The hypothesis of the c quark soon received firm experimental support. In 1974 a team of experimenters under the direction of S. Ting at the Brookhaven accelerator and an independent team under the direction of B. Richter at the Stanford accelerator discovered the J/ψ meson and some other related ψ mesons (see Table C.5), all of which have exceptionally long lifetimes. It seems that these particles contain one c quark with an electric charge of $\frac{2}{3}$ and an (estimated) mass of 1500 MeV.

Unfortunately, the proliferation of quarks did not stop with four quarks. In 1977 a team of experimenters at Fermilab discovered the Υ mesons. These are by far the most massive hadrons known (see Table C.5). It seems that each of these contains a new and very massive quark. This fifth kind of quark has been labeled **bottom.** Theoretical considerations suggest that there should also exist a sixth quark, labeled **top.** In 1984, experimenters at CERN confirmed the existence of this sixth quark.

Each of the six quarks (up, down, strange, charmed, top, bottom) comes in three varieties of color (*red, green, blue*); furthermore, for each quark there exists an antiquark, which comes in three varieties of "anticolor" (anti*red,* anti*blue,* and anti*green*). Altogether, this amounts to 36 quarks — and the proliferation of quarks will apparently not end with this. Some new theories postulate the existence of several dozen quarks (plus their color varieties plus their antiquarks). This continuing proliferation raises the question of whether matter really has such a large number of elementary building blocks. Pushing forward our search for the ultimate building blocks, we have uncovered layers of structures within layers of structure — nuclei within atoms, protons and neutrons within nuclei, quarks within protons and neutrons. Is there another layer within quarks?

Further Reading

Elementary Particles by C. N. Yang (Princeton University Press, Princeton, 1962), *The Discovery of Subatomic Particles* by S. Weinberg (Freeman, San Francisco, 1983), and *From X-*

Rays to Quarks by E. Segré (Freeman, San Francisco, 1980) give accounts of the early history of the discovery of subatomic particles. The exciting story of the recent discoveries of quarks and other exotic particles is told with journalistic verve in *The Key to the Universe* by N. Calder (Viking, New York, 1977).

The Stuff of Matter by H. Fritzsch (Basic Books, New York, 1983) and *From Atoms to Quarks* by J. S. Trefil (Scribner, New York, 1980) are excellent, up-to-date introductions to the world of particle physics. *The Cosmic Code* by H. R. Pagels (Simon and Schuster, New York, 1982) explores the mysteries of particle physics and of quantum physics; it is beautifully written. Other worthwhile elementary introductions to particle physics are *The Particle Play* by P. C. Polkinghorne (Freeman, San Francisco, 1979) and *What Is the World Made Of?* by G. Feinberg (Doubleday, Garden City, 1977).

Many articles dealing with particle physics have been published in *Scientific America* and in *Physics Today:*

"Quarks with Color and Flavor," S. L. Glashow, *Scientific American,* October 1975

"The Search for New Families of Elementary Particles," D. B. Kline, A. K. Mann, and C. Rubbia, *Scientific American,* January 1976

"The Confinement of Quarks," Y. Nambu, *Scientific American,* November 1976

"Fundamental Particles with Charm," R. F. Schwitters, *Scientific American,* October 1977

"The Tevatron," R. R. Wilson, *Physics Today,* October 1977

"Heavy Leptons," M. L. Perl and W. T. Kirk, *Scientific American,* March 1978

"When Is a Particle?" S. D. Drell, *Physics Today,* June 1978

"The Upsilon Particle," L. M. Lederman, *Scientific American,* October 1978

"The Spin of the Proton," A. D. Krisch, *Scientific American,* May 1979

"The Bag Model of Quark Confinement," K. A. Johnson, *Scientific American,* July 1979

"The Next Generation of Particle Accelerators," R. R. Wilson, *Scientific American,* January 1980

"The Decay of the Proton," S. Weinberg, *Scientific American,* June 1981

"U.S. Particle Accelerators at Age 50," R. R. Wilson, *Physics Today,* November 1981

"The Search for the Intermediate Vector Bosons," D. B. Cline, C. Rubbia, and S. van der Meer, *Scientific American,* March 1982

"Quarkonium," E. D. Bloom and G. J. Feldmann, *Scientific American,* May 1982

"The Structure of Quarks and Leptons," H. Harari, *Scientific American,* April 1983

"The Inner Structure of the Proton," M. Jacob and P. Landshoff, *Scientific American,* March 1980

"Particles with Naked Beauty," N. B. Mistry, R. A. Poling, and E. H. Thorndike, *Scientific American,* July 1983

Several of the references listed in Interlude G are closely related to particle physics. The *Resource Letter NP-1* in the *American Journal of Physics,* February 1980, gives a long list of further references at a more advanced level.

Questions

1. Why are high-energy accelerators necessary for the production and discovery of new, massive particles?

2. Physicists are contemplating the construction of a 20,000-GeV (or 20-TeV) accelerator with a circumference of 200 km. This monster would cost more than a billion dollars. Can such an expenditure be justified?

3. Neutrons do not make tracks in a bubble chamber. Why not?

4. If neutrons had a somewhat smaller mass, then the (slow) electrons in atoms could combine with the protons in the nucleus according to reaction

$$e + p \rightarrow n + \nu$$

What consequences would this have for the existence of atoms and for the existence of life?

5. The names *baryon, meson,* and *lepton* come from the Greek *barys* (heavy), *mesos* (middle), and *leptos* (thin, slender). These names were originally intended to indicate the masses of the particles. According to the lists of particles and masses given in this chapter, is it true that the baryons have the largest masses and the leptons the smallest?

6. Count the number of particles in the tables of leptons, baryons, and mesons.

7. How does the antiproton differ from the proton? The antineutron from the neutron?

8. Why does a particle that lives only 10^{-23} s not make a track in a bubble chamber? (Hint: Suppose the particle moves at the maximum conceivable speed; how far will it travel in 10^{-23} s?)

9. Which of the following reactions are forbidden by an absolute conservation law?

$$\pi^+ + p \rightarrow \Lambda + K^0$$

$$K^- + p \rightarrow K^- + p + \pi^0$$

$$\pi^- + n \rightarrow \pi^- + \pi^0 + \pi^0$$

$$K^- + n \rightarrow \Sigma^- + \pi^0$$

$$e + \nu \rightarrow \pi^- + \pi^0$$

10. Is strangeness conserved in the following reactions?

$$\pi^+ + n \rightarrow \Sigma^+ + K^0$$

$$\Lambda \rightarrow p + \pi^-$$

$$K^0 \rightarrow \pi^+ + \pi^-$$

11. How many quarks are there in a hydrogen atom? In a water molecule? (The oxygen nucleus contains eight protons and eight neutrons.)

12. How many quarks are created in the net reaction (1)–(3)?

13. A particle is made of one d quark and one u antiquark. What is the electric charge of this particle? What is this particle?

14. According to some recent speculations, quarks are made of smaller constituents variously called prequarks, preons, or rishons (from the Hebrew word for first or primary). Can you think of a better name for the constituents of the quarks?

CHAPTER 13

Gravitation

The high accuracy of celestial mechanics is legendary. Calculations based on Newton's laws of motion and on Newton's law of gravitation yield long-range predictions for the motions of the planets, satellites, and comets; these predicted motions agree very precisely with astronomical observations. For example, the predicted planetary angular positions agree with the observed positions to within a few seconds of arc, even after a lapse of tens of years. Most of the limitations in the accuracy of celestial mechanics arise not from any defect in Newton's theory, but from the approximations that must be made to simplify the lengthy computations.

By the nineteenth century Newton's theory of gravitation had proved itself so trustworthy that when astronomers noticed an irregularity in the motion of Uranus, they could not bring themselves to believe that the theory was at fault. Instead, they suspected that a new unknown planet caused these irregularities by its gravitational pull on Uranus. J. C. Adams and U. J. J. Leverrier proceeded to calculate the expected position of this hypothetical planet — and the new planet, later named Neptune, was immediately found at just about the expected position. This discovery of Neptune was a spectacular success of the theory of gravitation.

Newton's theory of gravitation has had many other great successes, but it has also had a (minor) failure. Leverrier discovered an enigmatic discrepancy in the motion of Mercury; roughly, Mercury wanders ahead of its predicted position by 43 seconds of arc in each century. This (small) deviation cannot be explained by Newton's theory. It was finally explained by Einstein's theory of gravitation (General Relativity), a new theory based on a radical revision of our fundamental concepts of space and time (see Interlude E).

Apart from the barely noticeable defect in the motion of Mercury,

Newton's theory of celestial mechanics has stood up remarkably well to the test of centuries; and it remains one of the most accurate and successful theories in all of physics, that is to say, in all of science.

13.1 Newton's Law of Universal Gravitation

Within the Solar System, planets orbit around the Sun and satellites orbit around planets. These circular, or nearly circular, motions require a centripetal force pulling the planets toward the Sun and the satellites toward the planets. It was Newton's great discovery that this interplanetary force holding celestial bodies in their orbits is of the same kind as the force of gravity that causes apples, and other things, to fall downward near the surface of the Earth. The **law of universal gravitation** formulated by Newton states:

> *Every particle attracts every other particle with a force directly proportional to the product of their masses and inversely proportional to the square of the distance between them.*

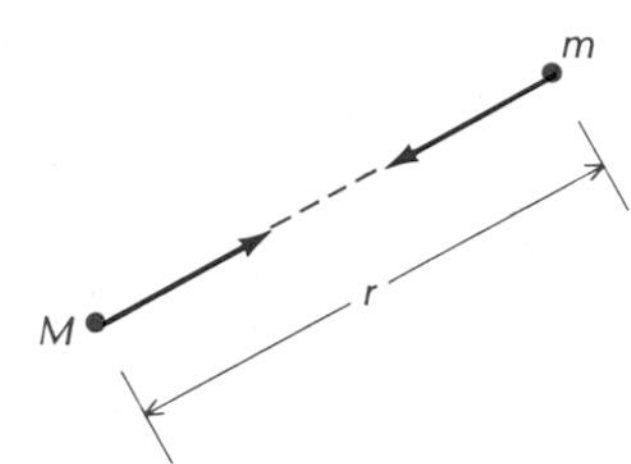

Fig. 13.1 Two particles attract each other gravitationally. The forces are of equal magnitude and opposite direction.

Expressed mathematically, the magnitude of the gravitational force that two particles of masses M and m separated by a distance r exert on each other is

$$\boxed{F = \frac{GMm}{r^2}} \qquad (1)$$

Law of universal gravitation

where G is a universal constant. The direction of the force on each particle is toward the other particle.

Figure 13.1 shows the direction of the force on each particle. Note that the two forces are of equal magnitude and opposite direction; they form an *action–reaction pair.*

The constant G is known as the **gravitational constant.** In metric units its value is

$$\boxed{G = 6.67 \times 10^{-11}\ \mathrm{N \cdot m^2/kg^2}} \qquad (2)$$

Gravitational constant, G

EXAMPLE 1. What is the gravitational force between a 70-kg man and a 70-kg woman separated by a distance of 10 m? Treat both masses as particles.

SOLUTION: From Eq. (1),

$$F = \frac{GMm}{r^2}$$

$$= \frac{6.67 \times 10^{-11}\ \mathrm{N \cdot m^2/kg^2} \times 70\ \mathrm{kg} \times 70\ \mathrm{kg}}{(10\ \mathrm{m})^2}$$

$$= 3.3 \times 10^{-9}\ \mathrm{N}$$

This is a very small force, but as we will see in Section 13.2, the measurement of such small forces is not beyond the reach of very sensitive instruments.

The gravitational force of Eq. (1) is an inverse-square force: it decreases by a factor of 4 when the distance increases by a factor of 2; it decreases by a factor of 9 when the distance increases by a factor of 3; etc. Figure 13.2 is a plot of the magnitude of the force **F** as a function of the distance. Although **F** decreases with distance, it never quite reaches zero. Thus every particle in the universe continually attracts every other particle at least a little bit, even if the distance between them is very, very large.

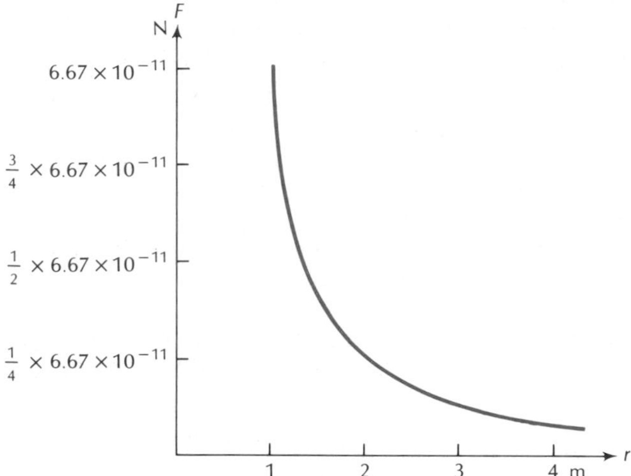

Fig. 13.2 Magnitude of the gravitational force exerted by one particle of 1 kg on another particle of 1 kg.

The gravitational force does not require any contact between the interacting particles. In reaching from one remote particle to another, the gravitational force somehow bridges the empty space between the particles. This is called *action-at-a-distance*. For now we will not inquire into the mechanism that transmits the force over the empty space, but we will explore this question later in Chapter 23 and Interlude G.

It is also quite remarkable that the gravitational force between two particles is unaffected by the presence of intervening masses. For example, a particle in Washington attracts a particle in Peking with exactly the force given by Eq. (1), even though all of the bulk of the Earth lies between Washington and Peking. This means that it is impossible to shield a particle from the gravitational attraction of another particle.

Since the gravitational attraction between two particles is completely independent of the presence of other particles, it follows that the net gravitational force between two bodies (e.g., the Earth and the Moon, or the Earth and an apple) is merely the vector sum of the individual forces between all the particles making up the bodies, i.e., the gravitational force obeys the principle of linear superposition. We will prove in Section 13.6 that this implies that the net gravitational force between spherical masses acts just as though each mass were concentrated at the center of its respective sphere. Since the Sun, the planets, and their satellites are almost exactly spherical, this important theorem permits us to treat all these celestial bodies as pointlike particles in all calculations concerning their gravitational attractions. For instance, the gravitational force exerted by the Earth on a particle is

$$F = \frac{GM_E m}{r^2} \tag{3}$$

where r is the distance from the *center* of the Earth (Figure 13.3). This expression, however, is only valid if the particle is outside the Earth. If the particle is inside the Earth (e.g., in a mine shaft), then the force is smaller (see Section 13.6).

If the particle is at the surface of the Earth, at a radius $r = R_E$, then Eq. (3) gives a force

$$F = \frac{GM_E m}{R_E^2} \tag{4}$$

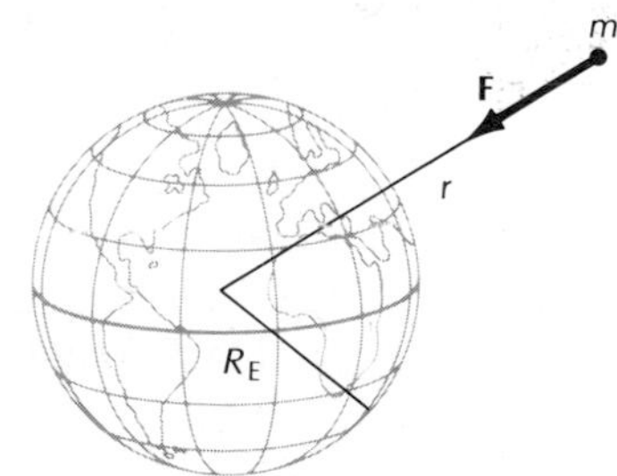

Fig. 13.3 The gravitational force exerted by the Earth on a particle is directed toward the center of the Earth.

The corresponding acceleration of the mass m is

$$a = \frac{F}{m} = \frac{GM_E}{R_E^2} \tag{5}$$

But this acceleration is what we usually call the acceleration of gravity:

Acceleration of gravity, g

$$\boxed{g = \frac{GM_E}{R_E^2}} \tag{6}$$

If the particle is at high altitude above the surface of the Earth, then the acceleration of gravity is less than that given by Eq. (6). At a distance r from the center, Eq. (3) gives an acceleration

$$a = \frac{GM_E}{r^2} \tag{7}$$

which can also be written

$$a = \frac{R_E^2}{r^2} g \tag{8}$$

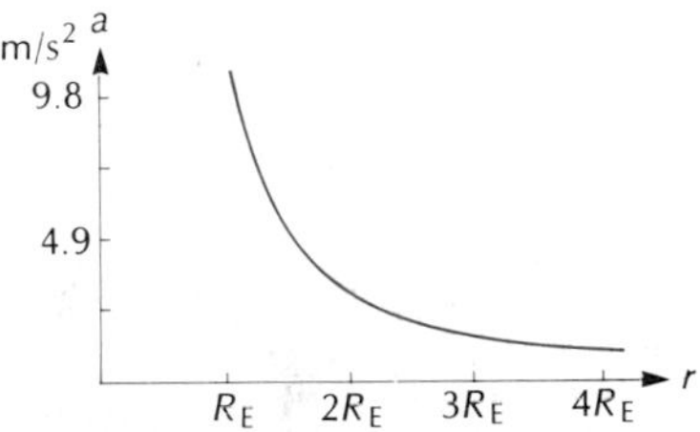

Fig. 13.4 The acceleration of gravity as a function of distance from the center of the Earth.

where, as in Eq. (6), g is the acceleration at the surface. Figure 13.4 is a plot of this acceleration as a function of radial distance.

13.2 The Measurement of *G*

The gravitational constant G is rather difficult to measure with precision. The trouble is that the gravitational forces between masses of laboratory size are extremely small, and thus a very delicate apparatus is needed to detect these forces. Measurements of G are usually done with a Cavendish **torsion balance** (Figure 13.5). Two equal, small spherical masses m, m' are attached to a lightweight horizontal beam which is suspended at its middle by a thin vertical fiber. When the beam is left undisturbed, it will settle into an equilibrium position such that the fiber is completely untwisted. If two equal large masses M, M' are brought near the masses m, m', the gravitational attraction between each small mass and its neighboring large mass tends to rotate the beam counterclockwise (as seen from above). The twist of the fiber opposes this rotation and the net result is that the beam settles into a new equilibrium position in which the torque on the beam generated by the gravitational attraction between the masses is exactly balanced by the

Fig. 13.5 The torsion balance used by Cavendish.

torque generated by the twisted fiber. The gravitational constant can then be calculated from the measured values of the angular displacement between the two equilibrium positions, the values of the masses and their distances, and the value of the force constant[1] of the fiber. [A preliminary measurement of the force constant (torque constant) of the fiber is required; in Example 14.6 we will describe a simple procedure for doing this.]

Several modern methods for the measurement of G rely on clever modifications of the basic Cavendish balance described above. The best available data lead to $G = 6.673 \times 10^{-11}\ \text{N} \cdot \text{m}^2/\text{kg}^2$ with an uncertainty of about 0.01%. Thus G is only known to within four significant figures, while most other fundamental constants of physics are known to within six or eight significant figures. Fortunately, the lack of precision in the measurements of G does not affect the high accuracy of celestial mechanics: the gravitational force exerted by, say, the Sun only depends on the product $G \times$ [mass of the Sun] and this *product* can be determined very precisely from planetary observations even though the individual factors cannot be determined very precisely.

Henry Cavendish, *1731–1810, English experimental physicist and chemist. Cavendish's main work was on the chemistry of gases; he isolated hydrogen and identified it as a separate chemical element. His torsion balance for the absolute measurement of the gravitational force was based on an earlier design used by Coulomb for the measurement of the electric force.*

Incidentally, the mass of the Earth can be calculated from Eq. (6) using the known values of G, R_E, and g:

$$M_E = \frac{R_E^2 g}{G} = \frac{(6.38 \times 10^6\ \text{m})^2 \times 9.81\ \text{m/s}^2}{6.67 \times 10^{-11}\ \text{N} \cdot \text{m}^2/\text{kg}}$$

$$= 5.98 \times 10^{24}\ \text{kg} \qquad (9)$$

For a more precise calculation it is necessary to take into account that the Earth is an ellipsoid rather than a sphere.

The above calculation would seem to be a rather roundabout way to arrive at the mass of the Earth, but there is no direct route, since we cannot place the Earth on a balance. Because the calculation requires a prior measurement of the value of G, the Cavendish experiment has often been described figuratively as "weighing the Earth."

13.3 Circular Orbits

Although the mutual gravitational forces of the Sun on a planet and of a planet on the Sun are equal in magnitude, the mass of the Sun is

[1] Just as a spring has a spring constant, a fiber has a force constant that characterizes its torque.

much larger than the mass of a planet and hence its acceleration is much smaller. It is therefore an excellent approximation to regard the Sun as fixed and immovable. It then only remains to investigate the motion of the planet. If we designate the masses of the Sun and planet by M_S and m, and their separation by r, then the magnitude of the gravitational force on the planet is

$$F = \frac{GM_S m}{r^2} \tag{10}$$

This force points toward the center of the Sun, i.e., the center of the Sun is the center of force (Figure 13.6).

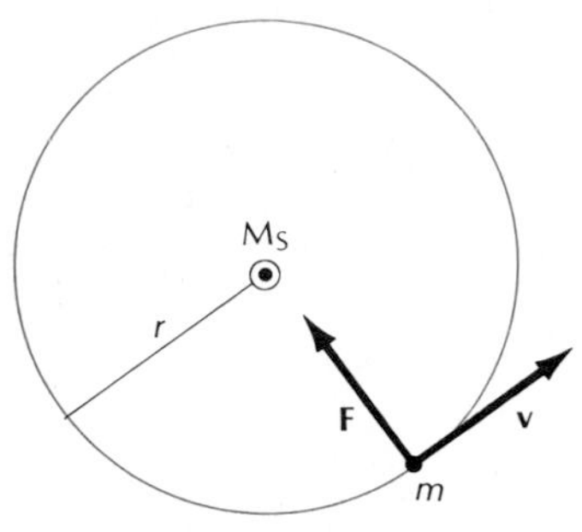

Fig. 13.6 Circular orbit of a planet around the Sun.

The simplest conceivable motion of a planet is uniform circular motion. The gravitational force acts as centripetal force. If the speed of the planet is v, the centripetal acceleration is v^2/r and the equation of motion is

$$\frac{GM_S m}{r^2} = \frac{mv^2}{r} \tag{11}$$

or

$$\boxed{\frac{GM_S}{r} = v^2} \tag{12}$$

The speed of the planet is related to the circumference of the orbit and to the time T for one revolution:

$$v = \frac{2\pi r}{T} \tag{13}$$

Nicolas Copernicus, *1473–1543, Polish astronomer. In his book* De Revolutionibus Orbium Coelestium *he formulated the heliocentric system for the description of the motion of the planets, according to which the Sun is immovable and the planets orbit around it. This new system — ardently defended by Galileo — gradually supplanted the old Ptolemaic system, according to which the Earth is immobile at the center of the universe.*

The time T is called the **period** of the orbit. With this, Eq. (12) becomes

$$\frac{GM_S}{r} = \frac{4\pi^2 r^2}{T^2} \tag{14}$$

or

Period for circular orbit

$$\boxed{T^2 = \frac{4\pi^2}{GM_S} r^3} \tag{15}$$

This says that the square of the period is proportional to the cube of the radius of the orbit, with a constant of proportionality depending on the mass of the central body.

EXAMPLE 2. Both Venus and the Earth have approximately circular orbits around the Sun. The period of Venus is 0.615 year and the period of the Earth is 1 year. According to Eq. (15), by what factor do the sizes of the two orbits differ?

SOLUTION: From Eq. (15) the orbital radius is proportional to the $\frac{2}{3}$ power of the period. Thus

$$\frac{r_E}{r_V} = \frac{T_E^{2/3}}{T_V^{2/3}} = \frac{(1 \text{ year})^{2/3}}{(0.615 \text{ year})^{2/3}} = 1.38$$

Of course, an equation similar to Eq. (15) also applies to the circular motion of a moon or artificial satellite around a planet. In this case the planet plays the role of central body and in Eq. (15) its mass appears in place of the mass of the Sun.

EXAMPLE 3. Equation (15) can be used to find the mass of the Sun from the observed values of the orbital radius and period of a planet. The (mean) radius of the orbit of the Earth is 1.496×10^{11} m. Find the mass of the Sun.

SOLUTION: The period of the orbital motion of the Earth is 1 year $= 3.156 \times 10^7$ s. Consequently, the mass of the Sun must be

$$M_S = \frac{4\pi^2 r^3}{GT^2}$$

$$= \frac{4\pi^2 \times (1.496 \times 10^{11} \text{ m})^3}{6.673 \times 10^{-11} \text{ N} \cdot \text{m}^2/\text{kg}^2 \times (3.156 \times 10^7 \text{ s})^2}$$

$$= 1.989 \times 10^{30} \text{ kg}$$

A similar method can be used to find the mass of a planet from the observed values of the orbital radius and the period of one of its satellites.

EXAMPLE 4. The *Early Bird* communications satellite is in a circular equatorial orbit around the Earth. The period of the orbit is exactly 1 day so that the satellite always holds a fixed station relative to the rotating Earth. What must be the radius of such a "synchronous" or "geostationary" orbit?

SOLUTION: Since the central body is the Earth, the equation analogous to Eq. (15) is

$$T^2 = \frac{4\pi^2}{GM_E} r^3 \tag{16}$$

and

$$r = \left(\frac{GM_E T^2}{4\pi^2}\right)^{1/3}$$

$$= \left(\frac{6.67 \times 10^{-11} \text{ N} \cdot \text{m}^2/\text{kg}^2 \times 5.98 \times 10^{24} \text{ kg} \times (24 \times 60 \times 60 \text{ s})^2}{4\pi^2}\right)^{1/3}$$

$$= 4.23 \times 10^7 \text{ m}$$

Fig. 13.7 Orbit of a "geostationary" satellite around Earth.

The orbit is shown in Figure 13.7, which is drawn to scale. A number of communication satellites have been placed in this geostationary orbit. These satellites routinely relay radio and TV signals from one continent to another.

13.4 Elliptical Orbits; Kepler's Laws

Although the orbits of the planets around the Sun are approximately circular, none of these orbits are *exactly* circular. Figure 13.8 shows the orbits of the planets Mercury, Venus, Earth, Mars, Jupiter, and Saturn (the orbits of Uranus, Neptune, and Pluto are considerably larger and would have to be plotted on a separate diagram; see the Prelude). The orbits of Mercury and Mars (and also of Pluto) are noticeably different from circles. We will not attempt the general solution of the equation of motion for such noncircular orbits. A complete calculation shows that with the inverse-square force of Eq. (10), the planetary orbits are ellipses. This is **Kepler's First Law:**

Kepler's First, Second, and Third Laws

The orbits of the planets are ellipses with the Sun at one focus.

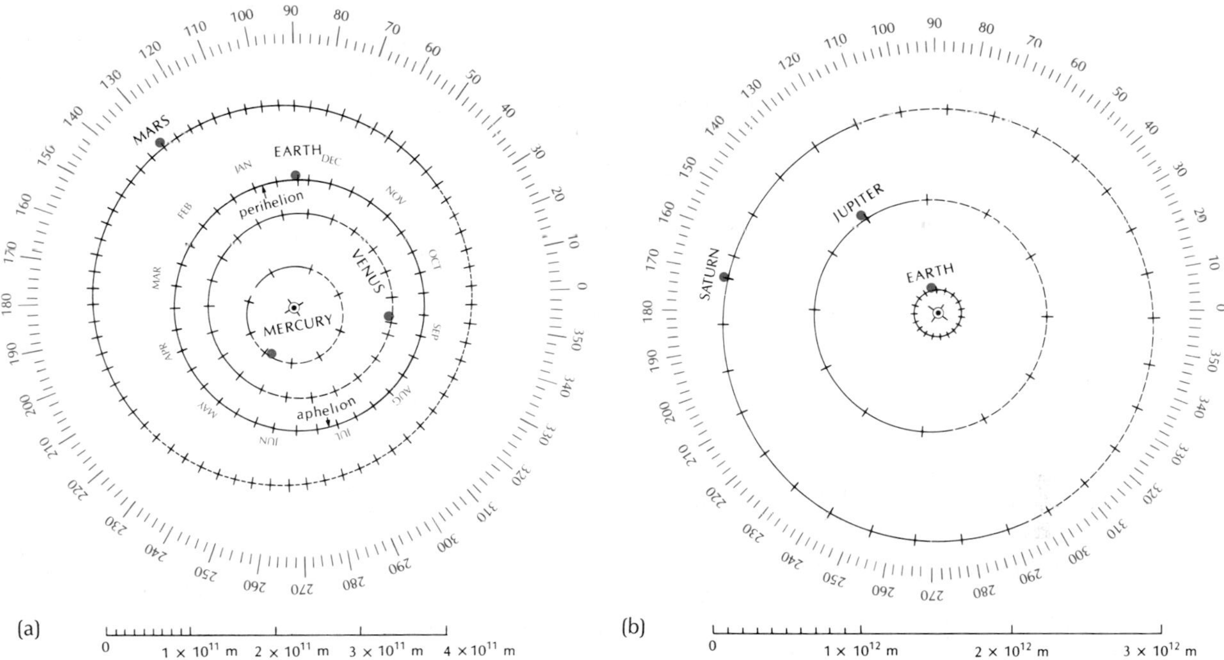

Fig. 13.8 (a) Orbits of Mercury, Venus, Earth, and Mars. The colored dots indicate the positions of the planets on January 1, 1980. The tick marks indicate the positions at intervals of 10 days. The orbit of the Earth is shown in the plane of the page. The orbits of the other planets are slightly tilted relative to this plane; the portions of the orbits above or below this plane are represented by solid or dashed lines, respectively. (b) Orbits of Jupiter and Saturn. The tick marks indicate the positions at intervals of 1 year.

Figure 13.9 shows an elliptical planetary orbit (for the sake of clarity, the elongation of this ellipse has been exaggerated; actual planetary orbits have only very small elongations). The point closest to the Sun is called the **perihelion;** the point farthest from the Sun is called the **aphelion.** The sum of the perihelion and the aphelion distances is the major axis of the ellipse. The distance from the center of the ellipse to the perihelion (or aphelion) is the semimajor axis; this distance equals the average of the perihelion and aphelion distances.

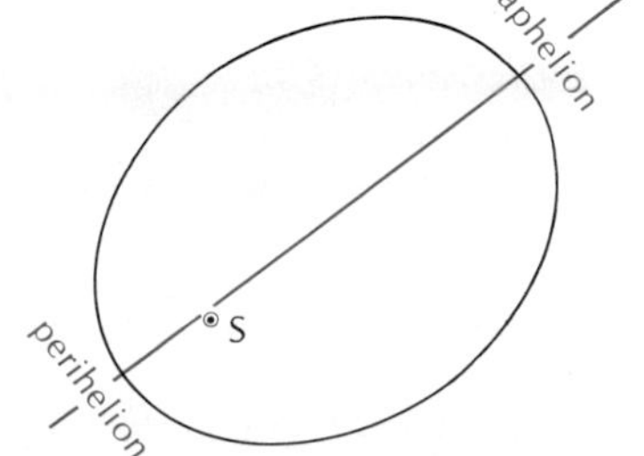

Fig. 13.9 Elliptical orbit. The Sun is at one focus of the ellipse.

Kepler originally discovered this law and his other two laws (see below) early in the seventeenth century, by direct analysis of the avail-

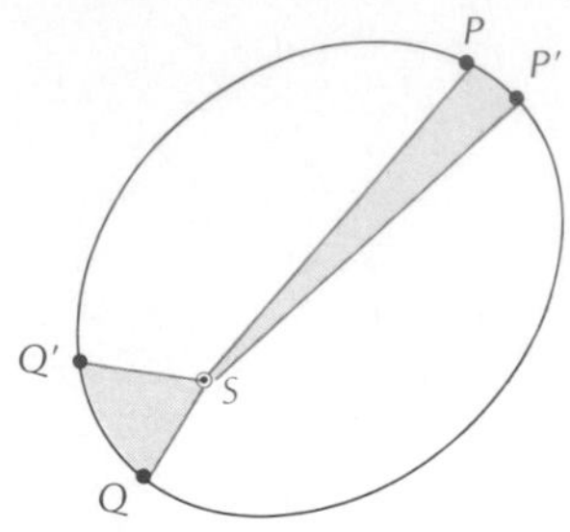

Fig. 13.10 For equal time intervals, the areas $QQ'S$ and $PP'S$ are equal.

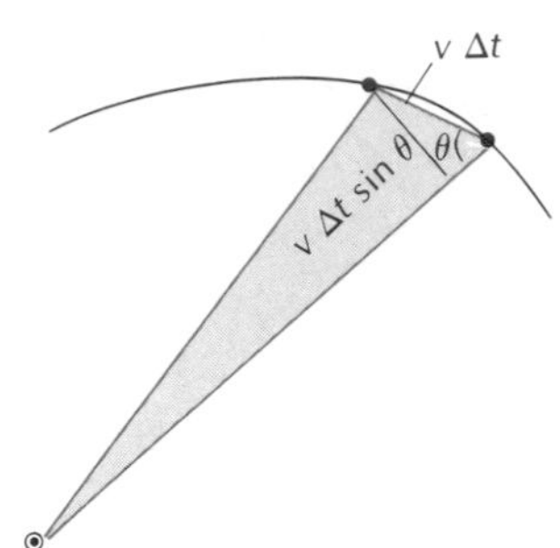

Fig. 13.11 The base of the triangle is r, and the height of the triangle is $v\ \Delta t \sin\theta$.

Johannes Kepler, *1571–1630, German astronomer, professor at Graz and at Linz, and court mathematician to the Emperor Rudolph II. Kepler relied on the theoretical framework of the Copernican system and he extracted his three laws by a meticulous analysis of the observational data on planetary motions collected by the great Danish astronomer Tycho Brahe.*

able observational data on planetary motions. Hence Kepler's laws were originally purely phenomenological statements, i.e., they described the phenomenon of planetary motion but did not explain its causes. This explanation only came later, when Newton laid down his laws of motion and his law of universal gravitation and deduced the features of planetary motion from these fundamental laws.

Kepler's Second Law describes the variation in the speed of the motion:

> *The radial line segment from the Sun to the planet sweeps out equal areas in equal times.*

Figure 13.10 illustrates this law. The two shaded areas are equal and the planet takes equal times to move from P to P' and from Q to Q'. According to Figure 13.10, the speed of the planet is obviously larger when it is near the Sun (at Q) than when it is far from the Sun (at P).

This Second Law, also called the law of areas, expresses the conservation of angular momentum. The gravitational force is a *central* force, i.e., it always is along the direction of the radial line from the center of the Sun to the planet. We know from Section 5.6 that for such a force the angular momentum is a constant of the motion,

$$\mathbf{L} = m\mathbf{r} \times \mathbf{v} = [\text{constant}] \tag{17}$$

The magnitude of this angular momentum is

$$\boxed{L = mrv \sin\theta = [\text{constant}]} \tag{18}$$

We can relate this magnitude of the angular momentum to the rate at which the radial line segment sweeps out area. Figure 13.11 shows the planet at two positions separated by a short time interval Δt. The shaded triangle represents approximately the area swept out in the time Δt. This area has a magnitude $\frac{1}{2}rv\ \Delta t \sin\theta$. Hence the rate at which area is swept out is

$$\frac{\Delta A}{\Delta t} = \tfrac{1}{2}rv \sin\theta \tag{19}$$

Comparing this with the magnitude of the angular momentum of Eq. (18), we see that $\Delta A/\Delta t$ is proportional to L. Since the latter quantity is a constant of the motion, the former quantity must also be constant.

Kepler's Third Law relates the period of the orbit to the size of the orbit:

> *The square of the period is proportional to the cube of the semimajor axis of the planetary orbit.*

This Third Law, or law of periods, is nothing but the generalization of Eq. (15) to elliptical orbits.

Table 13.1 gives the orbital data on the planets of the Solar System. The mean distance is the average of the perihelion and aphelion distances, i.e., it is the semimajor axis. The difference between the peri-

Table 13.1 THE PLANETS

Planet	Mass	Mean distance from Sun (semimajor axis)	Perihelion distance	Aphelion distance	Period
Mercury	3.30×10^{23} kg	57.9×10^6 km	45.9×10^6 km	69.8×10^6 km	0.241 year
Venus	4.87×10^{24}	108	107	109	0.615
Earth	5.98×10^{24}	150	147	152	1.00
Mars	6.42×10^{23}	228	207	249	1.88
Jupiter	1.90×10^{27}	778	740	816	11.9
Saturn	5.67×10^{26}	1430	1350	1510	29.5
Uranus	8.70×10^{25}	2870	2730	3010	84.0
Neptune	1.03×10^{26}	4500	4460	4540	165
Pluto	6.6×10^{23}	5910	4420	7380	248

Table 13.2 THE MAIN MOONS OF JUPITER

Moon	Mass	Mean distance from Jupiter (semimajor axis)	Perijove distance	Apjove distance	Period
Io	8.9×10^{22} kg	422×10^3 km	422×10^3 km	422×10^3 km	1.77 days
Europa	4.9×10^{22}	671	671	671	3.55
Ganymede	1.5×10^{23}	1070	1068	1071	7.16
Callysto	1.1×10^{23}	1883	1870	1896	16.69

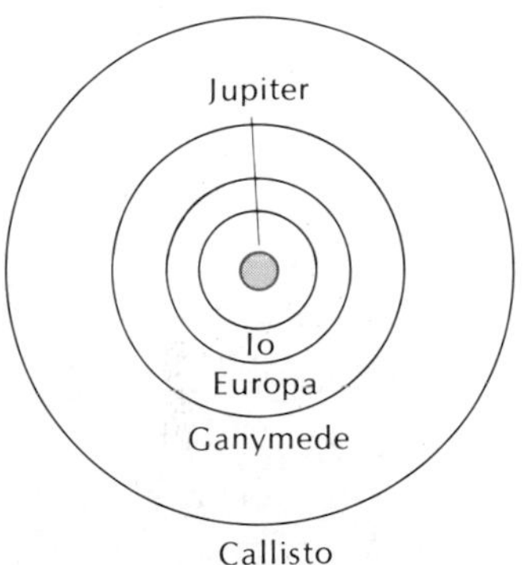

Fig. 13.12 Orbits of the main moons of Jupiter.

Table 13.3 THE FIRST ARTIFICIAL SATELLITES OF THE EARTH

Satellite	Mass	Mean distance from center of Earth (semimajor axis)	Perigee distance	Apogee distance	Period
Sputnik I	83 kg	6.97×10^3 km	6.60×10^3 km	7.33×10^3 km	96.2 min
Sputnik II	3000	7.33	6.61	8.05	104
Explorer I	14	7.83	6.74	8.91	115
Vanguard I	1.5	8.68	7.02	10.3	134
Explorer III	14	7.91	6.65	9.17	116
Sputnik III	1320	7.42	6.59	8.25	106

helion and aphelion distances gives an indication of the elongation of the ellipse.

Kepler's three laws apply not only to planets but also to satellites and to comets.[2] For example, Figure 13.12 shows the orbits of the main moons of Jupiter and Figure 13.13 shows the orbits of a few of the many artificial satellites of the Earth. All these orbits are ellipses. Table 13.2 gives the orbital data for the main moons of Jupiter; besides these four moons, Jupiter has another eight, but these other moons are much smaller. Table 13.3 gives the orbital data for the first

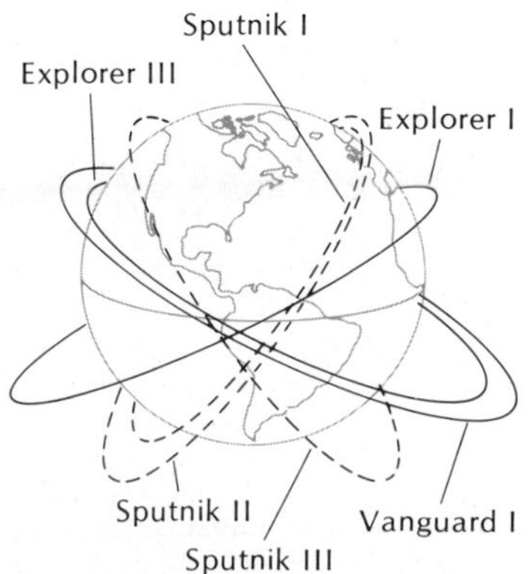

Fig. 13.13 Orbits of the first artificial satellites of the Earth.

[2] Kepler's laws only apply to *periodic* comets, i.e., those that return at regular intervals. Some comets only make a single pass by the Sun and never return; their orbits are parabolas or hyperbolas and neither Kepler's First nor Third Law applies to them.

six artificial satellites of the Earth (Figures 13.14 and 13.15); these satellites were launched in 1957 and 1958 and, except for Vanguard I, they all burned up in the atmosphere after a few months or a few years because they were not sufficiently far from the Earth to avoid the effects of residual atmospheric friction.

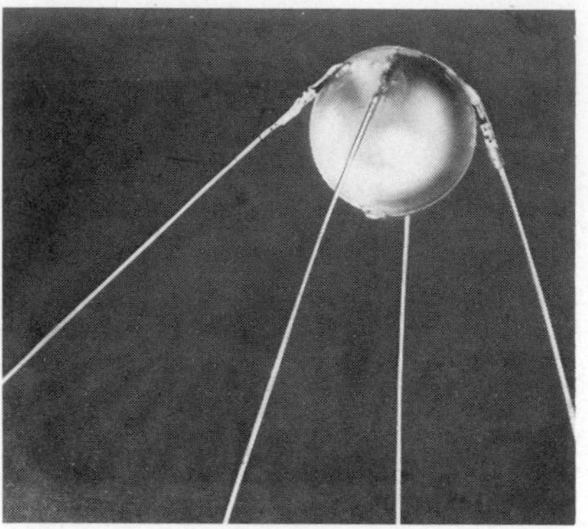

Fig. 13.14 Sputnik I, the first artificial satellite of the Earth.

Fig. 13.15 Vanguard I.

EXAMPLE 5. An astronaut in a spacecraft is in a circular orbit of radius 9.6×10^3 km around the Earth. At one point of the orbit, he briefly fires the thrusters of his spacecraft in the forward direction so as to reduce his speed. This places him in a new elliptical orbit with apogee equal to the radius of the old orbit, but with a smaller perigee (Figure 13.16). Suppose that the perigee of the new orbit is 7.0×10^3 km. Compare the periods of the old and new orbits.

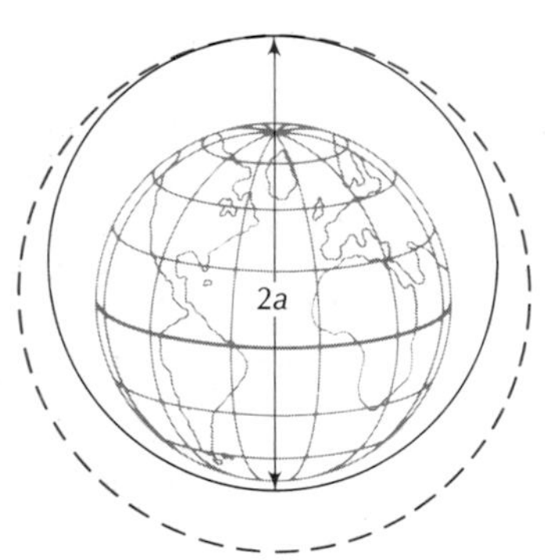

Fig. 13.16 Old and new orbits of the spacecraft. The old orbit is a circle, the new an ellipse.

SOLUTION: According to Eq. (16), the period of the old, circular orbit is

$$T_{\text{old}} = \sqrt{\frac{4\pi^2}{GM_E} r^3}$$

$$= \sqrt{\frac{4\pi^2 \times (9.6 \times 10^6 \text{ m})^3}{6.67 \times 10^{-11} \text{ N} \cdot \text{m}^2/\text{kg}^2 \times 5.98 \times 10^{24} \text{ kg}}} = 9.4 \times 10^3 \text{ s}$$

According to Kepler's Third Law, the period of the new, elliptical orbit is given by an equation similar to Eq. (16), but with r replaced by the semimajor axis a of the ellipse:

$$T_{\text{new}} = \sqrt{\frac{4\pi^2}{GM_E} a^3}$$

With $a = \frac{1}{2}(9.6 \times 10^3 \text{ km} + 7.0 \times 10^3 \text{ km}) = 8.3 \times 10^3$ km,

$$T_{\text{new}} = \sqrt{\frac{4\pi^2 \times (8.3 \times 10^6 \text{ m})^3}{6.67 \times 10^{-11} \text{ N} \cdot \text{m}^2/\text{kg}^2 \times 5.98 \times 10^{24} \text{ kg}}} = 7.5 \times 10^3 \text{ s}$$

Thus, the period of the new orbit is about 20% shorter than the period of the old orbit — even though the astronaut's maneuver has *reduced* his speed at apogee, he takes a shorter time to complete the orbit! The explanation is, of course, that the maneuver has *increased* his speed at perigee and also has shortened the distance around the orbit.

In a slightly generalized form, Kepler's laws also apply to stars orbiting about each other. For example, Figure 13.17 shows the orbits of the two stars in the binary system Krüger 60. The two stars have comparable masses and hence both have comparable accelerations — neither can be regarded as remaining fixed. However, the center of mass of the system can be regarded as remaining fixed. Under the influence of their mutual gravitational attraction, the stars orbit about each other, each star describing an ellipse with the center of mass at the focus.

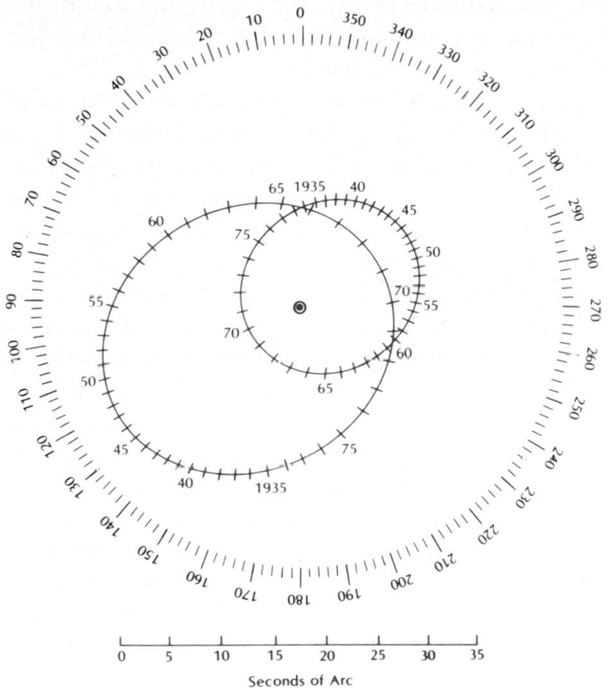

Fig. 13.17 The orbits of the two stars of the binary system Krüger 60. The center of mass is at the focus of each ellipse.

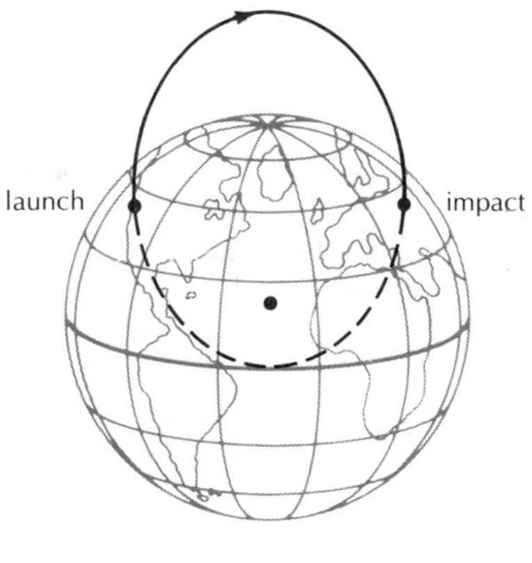

launch

impact

Fig. 13.18 Orbit of an intercontinental ballistic missile.

Of course, Kepler's Laws also apply to the motion of a projectile near the Earth. For instance, Figure 13.18 shows the trajectory of an intercontinental ballistic missile (ICBM). During most of this trajectory, the only force acting on the missile is the gravity of the Earth; the thrust of the engines and the friction of the atmosphere act only during relatively short initial and final segments of the trajectory (on the scale of Figure 13.18 these initial and final segments are too small to show). The trajectory is a portion of an elliptical orbit cut short by impact on the Earth. Similarly, the motion of an ordinary low-altitude projectile, such as a cannon ball, is also a portion of an elliptical orbit (if we ignore atmospheric friction). In Chapter 4 we pretended that gravity was constant in magnitude and in direction; with these approximations, we found that the orbit of a projectile was a parabola. Although the exact orbit of a projectile is an ellipse, the parabola approximates this ellipse quite well over the relatively short distance involved in ordinary projectile motion (Figure 13.19).

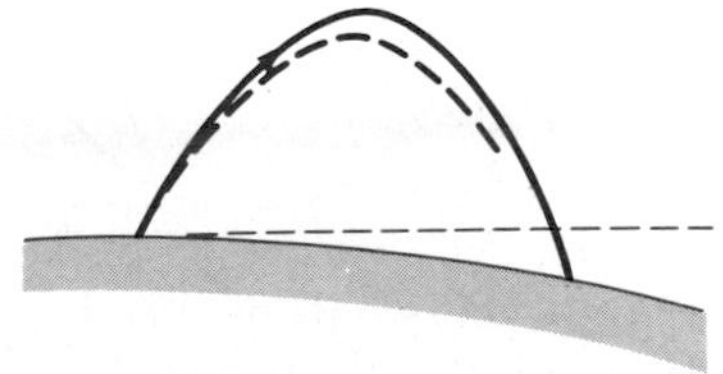

Fig. 13.19 The parabola (dashed curve) approximates the ellipse (solid curve) for small altitudes.

The connection between projectile motion and orbital motion is neatly illustrated by a *Gedankenexperiment*[3] due to Newton: Imagine

[3] *Gedankenexperiment* is German for "thought experiment," an experiment that can be done in principle but that has never been done in practice.

that we fire a projectile horizontally from a gun emplaced on a high mountain (Figure 13.20). If the muzzle velocity is fairly low, the projectile will arc toward the Earth and strike near the base of the mountain. The trajectory is a segment of a parabola, or, more precisely, a segment of an ellipse. If we increase the muzzle velocity, the projectile will describe larger and larger arcs. Finally, if the muzzle velocity is just large enough, the rate at which the trajectory curves downward is precisely matched by the curvature of the surface of the Earth — the projectile never hits the Earth and keeps on falling forever while moving in a circular orbit. This example makes it very clear that orbital motion is free-fall motion.

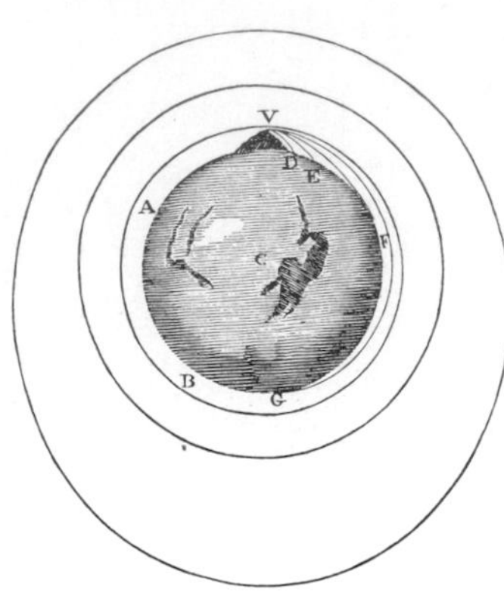

Fig. 13.20 A *Gedankenexperiment* by Newton. The trajectory of a fast projectile is a circular orbit.

Finally, we note that in our mathematical description of planetary motion we have neglected the gravitational forces that the planets exert on one another. These forces are much smaller than the force exerted by the Sun, but in a precise calculation they must be taken into account. The net force on any one planet is then a function of the position of all the other planets. The solution of the equation of motion involves a **many-body problem:** the motions of all the planets are coupled together and the calculation of the motion of one planet requires the simultaneous calculation of the motion of all the other planets. No exact solution of this many-body problem exists — it is necessary to have recourse to numerical methods. Thus, Kepler's simple laws only describe the planetary motions in first approximation, and the interplanetary forces generate diverse perturbations in Kepler's laws. The most common such perturbation is a gradual drift of the orientation of the Keplerian elliptical orbit. For example, because of the action of the interplanetary forces, the orientation of the major axis of the elliptical orbit of the Earth drifts 104 seconds of arc in each century (Figure 13.21); most of this perturbation is due to the gravitational force that Jupiter exerts on the Earth.

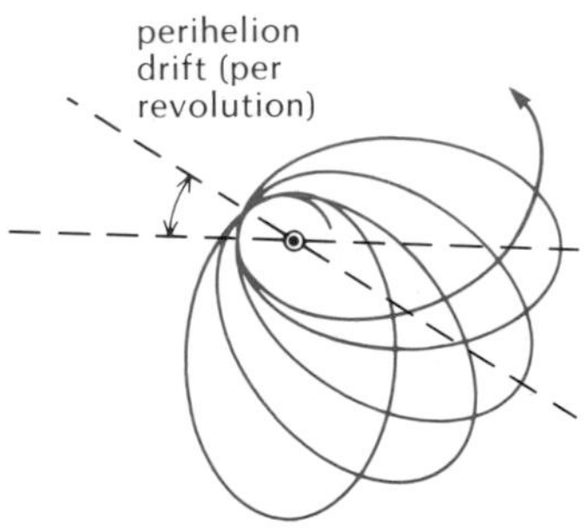

Fig. 13.21 Precession of the perihelion of a planetary orbit (exaggerated).

13.5 Gravitational Potential Energy

The gravitational force is a *conservative* force, i.e., the work done by this force on a particle moving from a point P_1 to a point P_2 depends on the position of these points but not on the shape of the path connecting them. To prove this, consider the gravitational force GMm/r^2 that acts between two particles of masses M and m. Suppose that M remains stationary at the origin and that m moves. Figure 13.22 shows two alternative paths connecting the initial and final positions of m. The work along a path is $\int \mathbf{F} \cdot d\mathbf{r}$. To evaluate this, we approximate each path by a series of steplike segments, each step being either a radial line or an arc of a circle (Figure 13.22). The circular segments do not contribute anything to the work; along these segments the work is zero because the force is perpendicular to the displacement. The radial segments do contribute to the work; these segments contribute *equally* for each of the two alternative paths because for every radial segment belonging to the first path there is an equal radial segment belonging to the second path, and the magnitude of the force is exactly the same at equal radial distances from the central mass M. Hence the work is completely independent of the shape of the path.

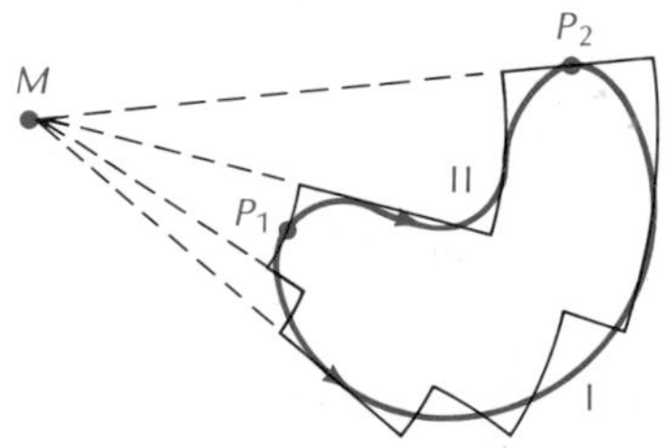

Fig. 13.22 Two paths (I and II) connecting the points P_1 and P_2. The paths can be approximated by radial segments and by arcs of circles.

We can then construct the gravitational potential energy by following the recipe of Section 8.2. For the reference point P_0 required in this recipe, we take a point at infinite distance[4] from the central mass M; for the value of the potential at this point, we take $U(P_0) = 0$. Then

$$U(r) = -\int_{\infty}^{r} \mathbf{F} \cdot d\mathbf{r} \tag{20}$$

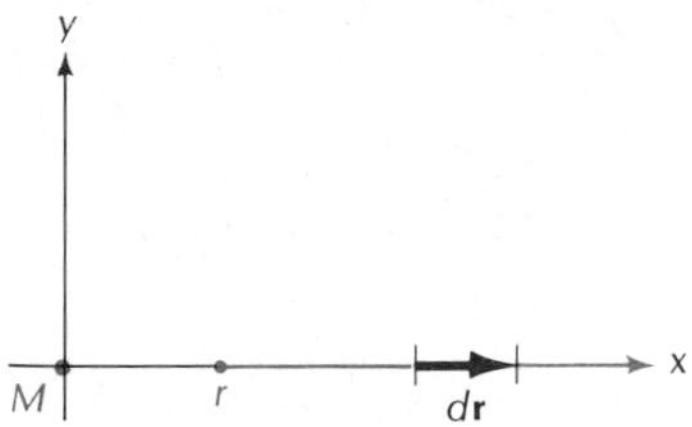

Fig. 13.23 Path from the point r to ∞.

Any path connecting ∞ and r may be used in the evaluation of this integral. The most convenient choice is a straight radial path (Figure 13.23). If we place the x axis along this path, then $\mathbf{F} = -(GMm/x^2)\hat{\mathbf{x}}$ and $d\mathbf{r} = \hat{\mathbf{x}}\, dx$, so that $\mathbf{F} \cdot d\mathbf{r} = -(GMm/x^2)$ and

$$U(r) = -\int_{\infty}^{r} -\left(\frac{GMm}{x^2}\right) dx = -\left[\frac{GMm}{x}\right]_{\infty}^{r} \tag{21}$$

or

Gravitational potential energy

$$\boxed{U(r) = -\frac{GMm}{r}} \tag{22}$$

Figure 13.24 is a plot of this potential energy as a function of distance. The potential energy *increases* with distance (it increases from a large negative value toward zero). Such an increase of potential energy with distance is of course characteristic of an attractive force. Note that $U(r)$ is really the *mutual* potential energy of both particles M and m. However, if M is a very large mass which does not move (e.g., the Sun), then it is convenient to regard $U(r)$ as the potential energy of the mass m which does move.

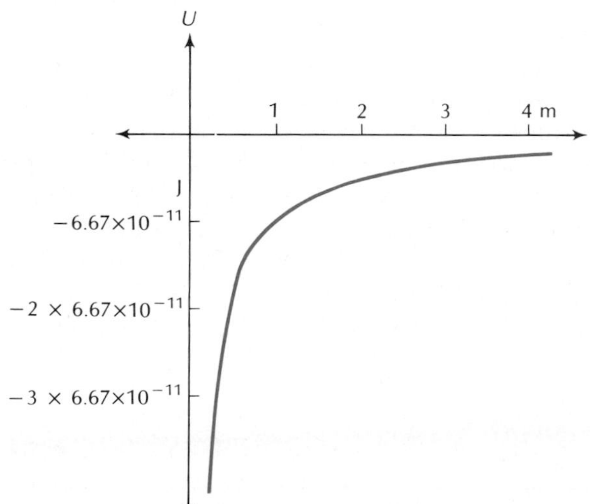

Fig. 13.24 Potential energy for a particle of 1 kg attracted gravitationally by another particle of 1 kg.

The total energy is the sum of the potential energy and the kinetic energy. If the mass M is stationary, then the kinetic energy is entirely due to the motion of the mass m and

[4] The *direction* in which this point P_0 lies is irrelevant. All points at infinite distance have the same potential.

Conservation of energy

$$E = K + U = \tfrac{1}{2}mv^2 - \frac{GMm}{r} = [\text{constant}] \tag{23}$$

This total energy remains constant during the motion. As in Chapter 8, by examination of the energy, we can draw some general conclusions concerning the motion. Obviously, Eq. (23) implies that if r increases, v must decrease, and conversely.

Let us now investigate the possible orbits around, say, the Sun from the point of view of their energy. For a circular orbit, the orbital velocity is [see Eq. (12)]

$$v = \sqrt{GM_S/r} \tag{24}$$

and the kinetic energy is

$$K = \tfrac{1}{2}mv^2 = \frac{GM_Sm}{2r} \tag{25}$$

Hence the total energy is

$$E = \tfrac{1}{2}mv^2 - \frac{GM_Sm}{r} = \frac{GM_Sm}{2r} - \frac{GM_Sm}{r} \tag{26}$$

or

Energy for circular orbit

$$E = -\frac{GM_Sm}{2r} \tag{27}$$

Consequently, the total energy is negative and is exactly one-half the potential energy.

For an elliptical orbit the total energy is also negative. It can be shown that the energy can still be written in the form of Eq. (27), but the quantity r must be taken equal to the semimajor axis of the ellipse. The total energy does not depend on the shape of the ellipse, but only on its overall size. Figure 13.25 shows several orbits with exactly the same total energy. Incidentally, the angular momenta of these orbits differ; the circular orbit has the highest angular momentum and the very elongated ellipse has the lowest.

If the energy is nearly zero, then the size of the orbit is very large.

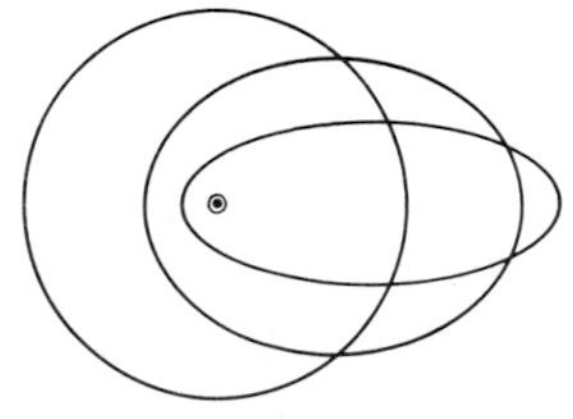

Fig. 13.25 Orbits of the same total energy. All these orbits have the same semimajor axis.

Table 13.4 SOME COMETS

Comet	Perihelion distance	Aphelion distance	Period
Encke	50.5×10^6 m	612×10^6 m	3.3 years
Pons-Winnecke	173	828	6.12
Biela	129	926	6.62
D'Arrest	206	857	6.70
Finlay	158	920	6.84
Faye	249	891	7.44
Halley	87.8	5280	76.0

Such orbits are characteristic of comets, many of which have elliptical orbits that extend far beyond the edge of the Solar System (Figure 13.26 and Table 13.4). If the energy is exacty zero, then the "ellipse" extends all the way to infinity and never closes; such an "open ellipse" is actually a parabola (Figure 13.27). Equation (23) indicates that if the energy is zero, the comet will reach infinite distance with zero velocity (if $r = \infty$, then $v = 0$). By considering the reverse of this motion, we recognize that a comet that is initially at rest at a very large distance from the Sun will fall along this type of parabolic orbit.

If the energy is positive, then the orbit again extends all the way to infinity and again fails to close; such an open orbit is a hyperbola. The comet will then reach infinite distance with some nonzero velocity and continue moving along a straight line (Figure 13.28).

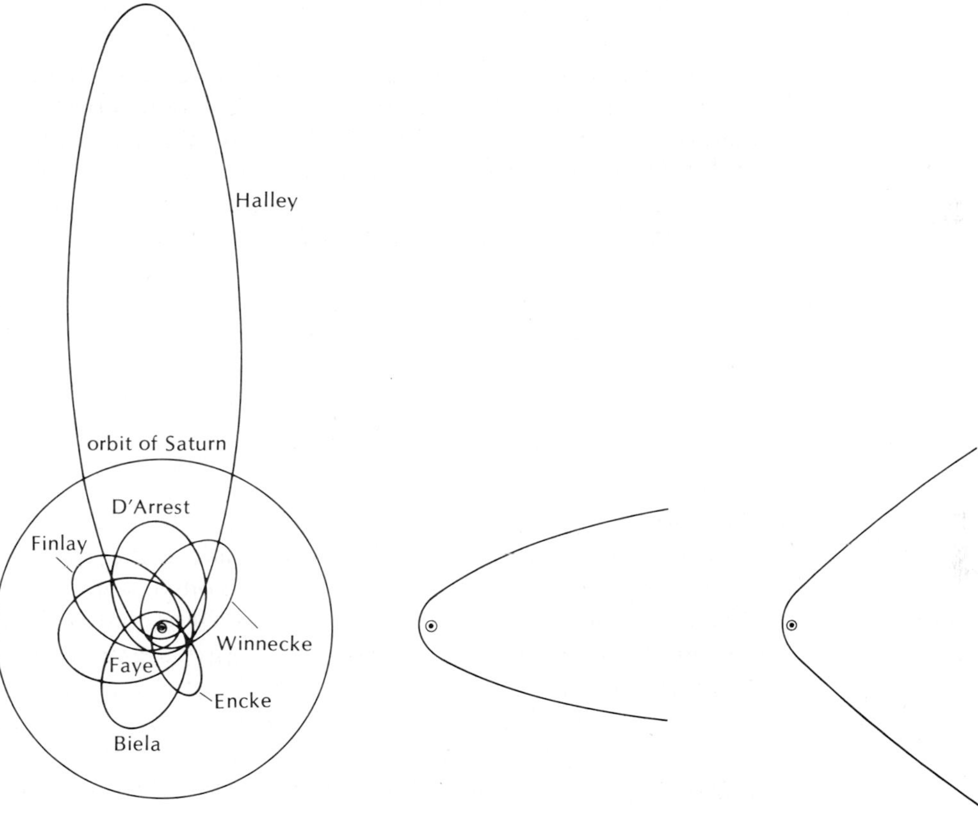

Fig. 13.26 Orbits of some periodic comets.

Fig. 13.27 Orbit of zero energy — a parabola.

Fig. 13.28 Orbit of positive energy — a hyperbola.

EXAMPLE 6. A meteroid (a chunk of rock) is initially at rest in interplanetary space at a large distance from the Sun. Under the influence of gravity, the meteroid begins to fall toward the Sun along a straight radial line. With what speed does it strike the Sun?

SOLUTION: The energy of the meteroid is

$$E = \tfrac{1}{2}mv^2 - \frac{GM_S m}{r} = [\text{constant}]$$

Initially, both the kinetic and potential energies are zero ($v = 0$ and $r \cong \infty$). Hence at any later time

$$\tfrac{1}{2}mv^2 - \frac{GM_Sm}{r} = 0$$

or

Escape velocity

$$\boxed{v = \sqrt{2GM_S/r}} \tag{28}$$

which is called the escape velocity (see below). At the moment of impact $r = R_S = 6.96 \times 10^8$ m and therefore

$$v = \sqrt{2GM_S/R_S} \tag{29}$$

$$= \sqrt{2 \times 6.67 \times 10^{-11}\ \text{N}\cdot\text{m}^2/\text{kg}^2 \times 1.99 \times 10^{30}\ \text{kg}/6.96 \times 10^8\ \text{m}}$$

$$= 6.18 \times 10^5\ \text{m/s} = 618\ \text{km/s}$$

The quantity given by Eq. (29) is called the **escape velocity** because it is the minimum initial velocity with which a body must be launched upward from the surface of the Sun if it is to escape and never fall back. We can recognize this by looking at the motion of the meteroid in reverse: it starts with a velocity of 618 m/s at the surface of the Sun and gradually slows down as it rises, but it never quite stops until it reaches a very large distance ($r \cong \infty$).

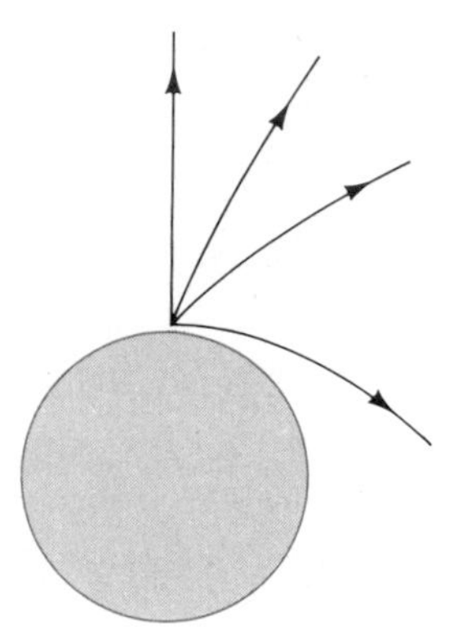

Fig. 13.29 Different parabolic orbits with the same starting point and initial speed.

Note that the direction in which a body is launched is immaterial — the body will succeed in its escape whenever the direction of launch is above the horizontal. Of course, the escape route that the body takes will depend on the direction of launch (Figure 13.29).

The escape velocity for a body launched from the surface of the Earth can be calculated from a formula analogous to Eq. (28), provided that we ignore atmospheric friction and the pull of the Sun on the body. Atmospheric friction will be absent if we launch the body from just above the atmosphere, and the pull of the Sun has only a small effect on the escape velocity if we contemplate a body that "escapes" to a distance of, say, $r = 100R_E$ or $200R_E$ rather than $r = \infty$. For such a body the escape velocity is approximately $\sqrt{2GM_E/R_E} =$ 11.2 km/s.

EXAMPLE 7. From the data given in Table 13.1, find the speed of Mercury at perihelion and at aphelion.

SOLUTION: We designate the distances and speeds at perihelion and at aphelion by r_1, v_1, r_2, and v_2, respectively. At both of these points, the velocity is perpendicular to the radius; hence the corresponding angular momenta simply have magnitudes mv_1r_1 and mv_2r_2. Conservation of angular momentum then tells us

$$mv_1r_1 = mv_2r_2 \tag{30}$$

Conservation of energy tells us

$$\tfrac{1}{2}mv_1^2 - \frac{GM_Sm}{r_1} = \tfrac{1}{2}mv_2^2 - \frac{GM_Sm}{r_2} \tag{31}$$

Obviously, the mass m cancels in both of these equations. From Eq. (30), $v_2 = v_1r_1/r_2$, which when substituted into Eq. (31) yields

$$\tfrac{1}{2}v_1^2 - \frac{GM_S}{r_1} = \frac{1}{2}\left(\frac{v_1 r_1}{r_2}\right)^2 - \frac{GM_S}{r_2}$$

or

$$v_1^2 = \frac{2GM_S(1/r_1 - 1/r_2)}{1 - r_1^2/r_2^2} \tag{32}$$

This can be simplified to read

$$v_1^2 = 2GM_S \frac{r_2}{r_1(r_1 + r_2)} \tag{33}$$

so that the perihelion speed is

$$v_1 = \sqrt{2 \times 6.67 \times 10^{-11}\ \text{N} \cdot \text{m}^2/\text{kg}^2 \times 1.99 \times 10^{30}\ \text{kg} \times \frac{69.8}{45.9 \times (69.8 + 45.9) \times 10^9\ \text{m}}}$$

$$= 5.91 \times 10^4\ \text{m/s}$$

The aphelion speed is then

$$v_2 = v_1 \frac{r_1}{r_2} = 5.91 \times 10^4\ \text{m/s} \times \frac{45.9}{69.8}$$

$$= 3.88 \times 10^4\ \text{m/s}$$

Finally, let us examine the gravitational potential energy of a particle in the vicinity of the Earth. According to Eq. (22),

$$U(r) = -\frac{GM_E m}{r} \tag{34}$$

The *change* in the potential energy between the point r and a point on the surface of the Earth is then

$$\Delta U = U(r) - U(R_E) = -\frac{GM_E m}{r} + \frac{GM_E m}{R_E}$$

$$= GM_E m \frac{r - R_E}{rR_E} \tag{35}$$

If the point r is near the surface of the Earth so that $r \cong R_E$, then we can approximate the product rR_E by R_E^2. Furthermore, the difference $r - R_E$ is simply the height z above the surface so that

$$\Delta U = \frac{GM_E m}{R_E^2} z \tag{36}$$

But by Eq. (6), $GM_E/R_E^2 = g$ and hence

$$\Delta U = mgz \tag{37}$$

This, of course, is our old expression for the potential energy of gravity near the surface of the Earth (see Section 7.4). Consequently, the old expression is an approximation to the exact potential energy; this approximation is valid if the height z is much smaller than the radius of the Earth ($z \ll R_E$).

13.6 The Gravitational Action of a Spherical Mass

In Section 13.1 we claimed that the gravitational attraction exerted by a spherical mass distribution is the same as if all the mass were concentrated at the center. This famous theorem is due to Newton. The theorem is valid as long as the attracted body is *outside* of the spherical mass distribution. In what follows, we will prove this theorem and also investigate what happens if the attracted body is inside the mass distribution.

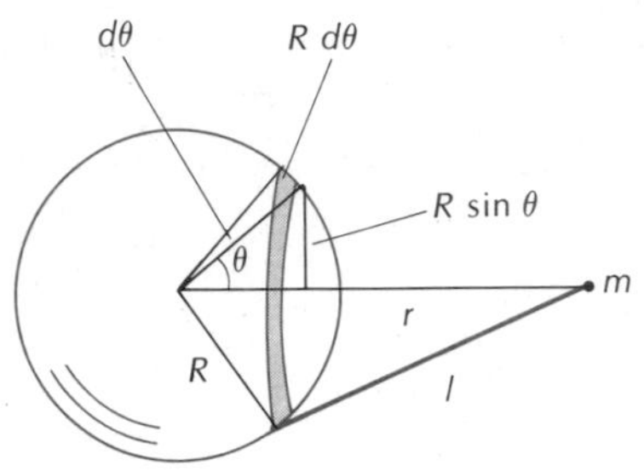

Fig. 13.30 A thin spherical shell. Particle m is outside the shell.

A spherical mass distribution can be regarded as constructed of a collection of thin spherical shells, each one nested inside the other. We will begin with a calculation of the gravitational attraction that such a shell exerts on a particle outside it. Figure 13.30 shows the shell, of mass M and radius R, and a particle of mass m at a distance r from the center of the shell. The net gravitational force exerted by the shell on the particle m is the vector sum of all the forces exerted by all the particles of the shell. This vector sum is somewhat messy to evaluate directly. It is easier first to evaluate the net potential energy; the force can be calculated afterward from the potential energy by taking derivatives [see Eqs. (8.20)–(8.22)].

The net potential energy between the shell and the particle m is the sum of all the individual potential energies between the particles in the shell and the particle m. If we divide the shell into small mass elements ΔM_i, then this net potential energy is approximately

$$U = \sum_i \left(-\frac{Gm\,\Delta M_i}{r_i} \right) \tag{38}$$

where r_i is the distance between ΔM_i and m. To obtain the potential energy exactly, it is necessary to proceed to the limit $\Delta M_i \to 0$ so that the sum becomes an integral. For the mass element ΔM_i it is convenient to take a ring cut from the shell as in Figure 13.30; obviously, the entire shell can be constructed out of such rings. The ring shown in Figure 13.30 is at a distance $r_i = l$ from the particle m. The ring has a width $R\,d\theta$, a radius $R \sin\theta$, and a circumference $2\pi R \sin\theta$; hence the surface area of the ring is $2\pi R^2 \sin\theta\, d\theta$. The mass of the ring is proportional to this surface area. Since the total mass M is uniformly distributed over the total area $4\pi R^2$ of the shell, the proportionality between mass and area yields

$$\Delta M_i = M \times \frac{2\pi R^2 \sin\theta\, d\theta}{4\pi R^2} = \tfrac{1}{2} M \sin\theta\, d\theta$$

for the mass of the ring. Inserting this in Eq. (38) and proceeding to the limit $\Delta M_i \to 0$, we obtain the integral

$$U = -\int \frac{GmM}{2} \frac{\sin\theta\, d\theta}{l} \tag{39}$$

If we apply the law of cosines to the triangle with sides R, r, and l, we find

$$l^2 = R^2 + r^2 - 2rR \cos \theta \tag{40}$$

Let us take the differential of this, remembering that r and R are constants:

$$2l\, dl = 2rR \sin \theta\, d\theta$$

or

$$\frac{\sin \theta\, d\theta}{l} = \frac{dl}{rR}$$

We can now substitute this expression into Eq. (39) and obtain

$$U = -\int \frac{GmM}{2} \frac{1}{rR}\, dl = -\frac{GmM}{2rR} \int dl$$

The largest value of l is $r + R$ and the smallest is $r - R$. Hence

$$U = -\frac{GmM}{2rR} \int_{r-R}^{r+R} dl = -\frac{GmM}{2rR} \left[l \right]_{r-R}^{r+R} \tag{41}$$

$$= -\frac{GmM}{2rR}(2R)$$

or

$$U = -\frac{GmM}{r} \tag{42}$$

This result shows that the potential energy behaves exactly as though all of the mass of the shell were at its center. Thus, the force between the shell and the particle must also behave exactly as though all the mass were at its center.

A spherical mass distribution is nothing but a collection of shells. Since each shell acts as though its mass were at its center, the complete spherical mass distribution will also act as though its mass were at its center. Note that this conclusion remains unchanged if the density of the mass distribution is some function of radius (as, e.g., it is in the case of the Earth, where the inner layers are much denser than the outer).

Furthermore, the spherical mass distribution acts as though its mass were at its center, not only in regard to the force it exerts on a particle, but also in regard to the force it feels from this particle. We can deduce that this is so from the equality of action and reaction (Newton's Third Law): since the force felt by the particle acts as though all the mass of the spherical distribution were at its center, the equal and opposite force felt by the spherical mass distribution must act likewise. Thus, it follows that the mutual gravitational forces exerted and experienced by the (nearly) spherical planets are as though the mass of each planet were concentrated in a point at its center.

In the above we have assumed that the particle that experiences the force is outside of the spherical mass distribution. To see what happens if the particle is inside, we begin by calculating the gravitational attraction that a shell exerts on a particle inside it (Figure 13.31). The calculation proceeds exactly as above. The only change is in Eq. (41): the largest value of l is now $R + r$ and the smallest is $R - r$. With these new limits of integration, Eq. (41) becomes

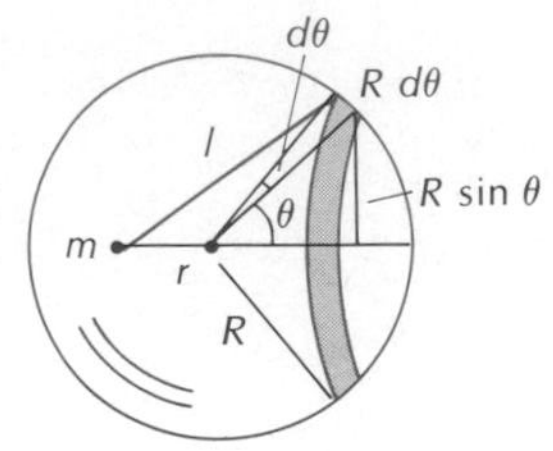

Fig. 13.31 A thin spherical shell. Particle m is inside the shell.

$$U = -\frac{GmM}{2rR}\left[l\right]_{R-r}^{R+r} = -\frac{GmM}{2rR}(2r)$$

or

$$U = -\frac{GMm}{R} \tag{43}$$

This means that inside the shell, the potential energy of the particle m is *independent* of the position r. But a constant potential energy implies the absence of any force. We have therefore obtained the surprising result that a spherical shell exerts *no gravitational force* on a particle inside it.

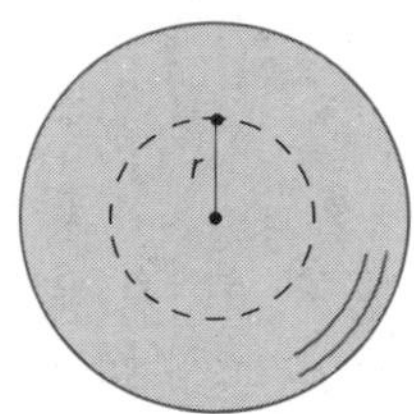

Fig. 13.32 A particle inside the Earth.

For a spherical mass distribution this has the following consequence: consider a particle within the mass distribution; for example, consider a particle inside a mine shaft dug in the Earth (Figure 13.32). All the spherical shells of radius larger than the radius at which the particle is located do not exert any force. Thus the force is entirely due to the mass contained in a radius smaller than the radius at which the particle is located; this mass exerts a force just as though it were concentrated at the center. The force may then be written as

Gravitational force on a particle inside a spherical mass

$$\boxed{F = \frac{GmM(r)}{r^2}} \tag{44}$$

where $M(r)$ is the amount of mass contained within the radius r.

EXAMPLE 8. A uniform sphere has mass M and radius R. Find the gravitational force on a particle of mass m at radius $r < R$.

SOLUTION: The mass contained within the radius r is directly proportional to the volume $4\pi r^3/3$. The total mass M is distributed over a volume $4\pi R^3/3$. Hence the proportionality between mass and volume implies

$$M(r) = M\frac{4\pi r^3/3}{4\pi R^3/3} = \frac{Mr^3}{R^3}$$

and

$$F = \frac{Gm}{r^2}\frac{Mr^3}{R^3} = \frac{GmM}{R^3}r \tag{45}$$

Hence the force *increases* in direct proportion to the radius. Of course, at $r = R$, the force ceases to increase and begins to decrease as $1/r^2$ (Figure 13.33).

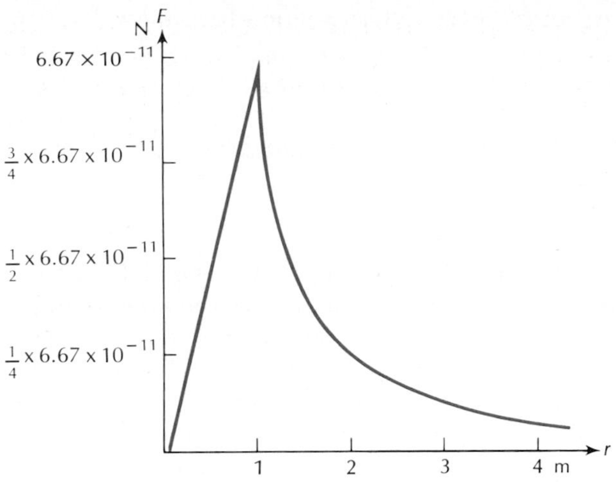

Fig. 13.33 Magnitude of the gravitational force exerted by a uniform sphere of mass 1 kg and radius 1 m on a particle of mass 1 kg.

13.7 Inertial and Gravitational Mass; the Principle of Equivalence

The definition of mass given in Section 5.2 relies on a comparison of the accelerations of two interacting masses, the unknown mass and the standard mass. Briefly explained, the unknown mass and the standard mass are made to exert forces on each other, and the ratio of the consequent accelerations then gives the inverse ratio of the masses [see Eqs. (5.2) and (5.3)]. According to this definition, mass is a measure of inertia, i.e., it is a measure of the opposition that a body offers to any attempts at changing its state of motion. Hence the mass measured in this way is often called the **inertial mass.**

Inertial mass

However, in everyday practice mass is measured by means of a balance which compares the weights of two masses rather than their inertia. Figure 5.5 shows an analytical balance with two equal arms. The unknown mass is placed in one pan of the balance and the standard mass, or a suitable multiple or submultiple of the standard mass, is placed in the other pan. The balance will be in equilibrium if the weights on the two pans are equal. Thus, the balance compares the gravitational force that the Earth exerts on the masses. The mass measured in this way is called the **gravitational mass.**

Gravitational mass

It is of fundamental importance to verify that the two methods for the measurement of mass agree, i.e., that the inertial mass and the gravitational mass of any body are equal. For the standard mass (standard kilogram) this equality of inertial and gravitational masses holds true by hypothesis, but for any other body the equality must be tested by experiment.

To see what this involves, suppose that two bodies have inertial masses m and m' and weights w and w', respectively. Suppose further that, as tested by a balance, the weights are exactly the same,

$$w = w' \tag{46}$$

With $w = mg$ and $w' = m'g$, this equation becomes

$$mg = m'g \tag{47}$$

Upon cancellation of the factor of g on both sides, we then find that

$$m = m' \tag{48}$$

i.e., the inertial masses are also exactly the same. Thus the weights, or gravitational masses, of the two bodies are the same if and only if their inertial masses are the same. If one of the bodies is the standard of mass or a multiple or submultiple of the standard, which has equal gravitational and inertial masses by hypothesis, then it follows that the other body also has equal inertial and gravitational masses.

However, this argument hinges on an implicit assumption: the factors of g on both sides of Eq. (47) can be canceled if and only if these factors are the same, i.e., if and only if both masses have exactly the same acceleration of free fall. Hence the equality of inertial and gravitational masses will hold true for a given body if and only if this body has the same acceleration of free fall as the standard mass. The experimental question that must be answered is then this: are the rates of free fall of different kinds of bodies really exactly the same?

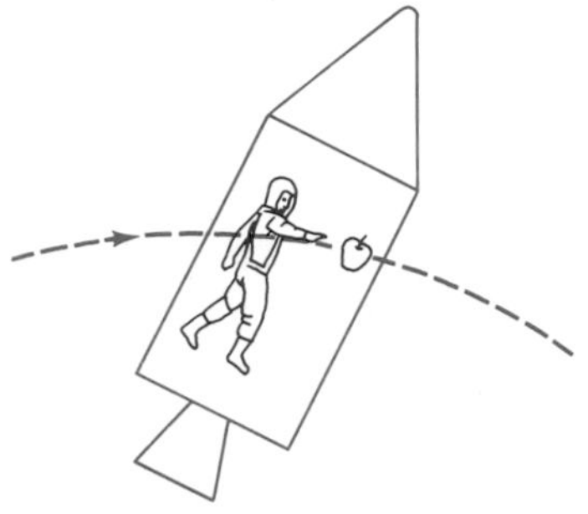

Fig. 13.34 Astronaut and apple in orbit.

The first experiments that sought to answer this question were performed by Galileo and by Newton. A series of much more precise and much more complete experiments were performed early in this century by Lorand von Eötvös, who compared the rates of free fall of samples of platinum, copper, water, copper sulfate, asbestos, etc., and found that all these bodies fall at the same rate. The most recent and most precise versions of these experiments have verified that the rates of free fall of bodies made of different substances are equal to within better than 1 part in 10^{12} (see Section E.1 for some details on these experiments). Hence the inertial and gravitational masses of a body are equal to within better than 1 part in 10^{12} and they are probably exactly equal.

Fig. 13.35 Astronaut in Skylab trying to capture floating, weightless trash bags.

As we have already noted in Section 6.2, the universality of the rate of free fall leads to the phenomenon of weightlessness for a freely falling observer. Consider, for example, an astronaut orbiting the Earth in a spacecraft, such as Skylab. The astronaut, and the spacecraft, and any other free body within or near the spacecraft are all in free fall — they accelerate at exactly the same rate and have no acceleration relative to one another. Any body the astronaut releases from his fingers will simply float in midair (Figure 13.34). Hence, in the accelerated reference frame of the astronaut, a condition of apparent weightlessness prevails: bodies behave as though they were at some remote place of the universe, far from the gravity of Earth, Sun, or other stars (Figure 13.35). This shows that a suitably accelerated motion of the reference frame can cancel the effects of gravity.

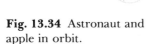

More generally, an accelerated motion of the reference frame can either decrease or increase the apparent gravity (Figure 13.36). For example, if a spacecraft far from the Earth, Sun, or other stars accelerates in some direction with a rate of, say, 9.8 m/s^2 and an astronaut within this spacecraft releases an apple from his fingers, then the apple will accelerate in the backward direction relative to the astronaut. Hence in the reference frame of the astronaut, the apple behaves exactly as though it were under the influence of ordinary gravity — the accelerated motion of the reference frame simulates the effects of gravity.

The similarity between the effects of gravity and the effects of a suitably accelerated motion of the frame of reference is called the **Principle of Equivalence.** Einstein took this principle as a starting point for his theory of General Relativity (see Interlude E).

Principle of Equivalence

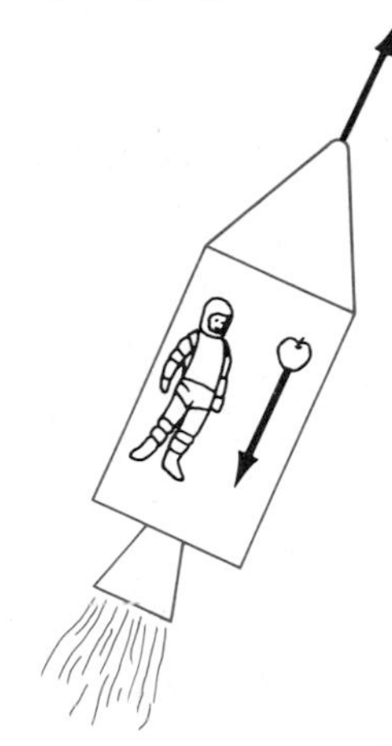

Fig. 13.36 Astronaut and apple in accelerated spacecraft.

Incidentally, the cancellation of gravity by means of accelerated motion plays an important role in our lives. The Earth, and everything on it, is subjected to the gravitational pull of the Sun, Moon, and other celestial bodies. Yet we do not feel this pull — the Earth is in an accelerated free-fall motion toward these bodies, and objects on the Earth behave as weightless in regard to the gravitational pull of the celestial bodies.

However, this cancellation of gravitational effects by the free-fall motion is not quite perfect. The **tides** observed on the oceans of the Earth constitute a small residual effect which the free-fall motion fails to cancel (Figure 13.37). For example, the lunar tides arise because the water nearest the Moon experiences a gravitational pull which is a bit too strong, and the water farthest from the Moon a gravitational pull which is a bit too weak, to match the free-fall motion of the Earth's center. Thus the water on the near side bulges toward the Moon and the water on the far side bulges away from the Moon. If we neglect the orbital motion of the Moon, we see that the locations of the tidal bulges remain more or less fixed in space while the Earth, and its oceans, turns relative to them once per day. This gives rise to two high lunar tides and two low lunar tides per day (Figure 13.38). The solar tides involve the same mechanism, but with the Sun as the attracting body.

Moon

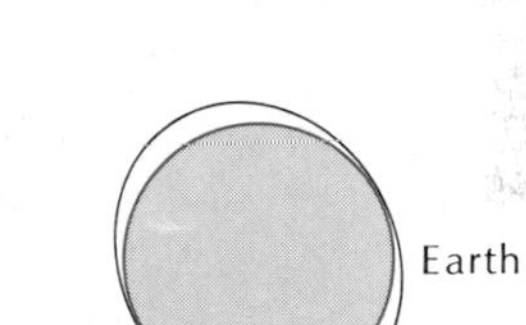

Fig. 13.37 Tidal bulges on the Earth, generated by the pull of the Moon.

(a)

(b)

Fig. 13.38 (a) High and (b) low tide in the Bay of Fundy.

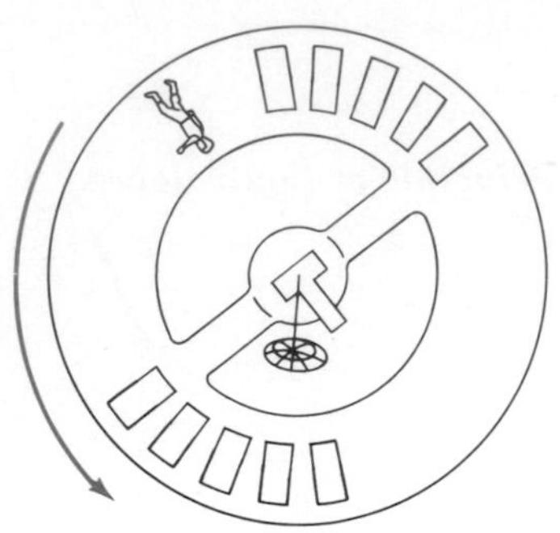

Fig. 13.39 A rotating space station.

The generation of gravity, or pseudo-gravity, by means of acceleration will play an important role in the design of the space stations of the future. For example, Figure 13.39 shows a space station in the shape of a large spinning wheel. Each point on the rim of the wheel has a centripetal acceleration and hence at each such point the apparent gravity is along the radial direction. For the inhabitants of the space station, "up" is toward the center of the wheel and "down" is radially outward. By adjusting the rate of spin of the wheel, the strength of the apparent gravity can be made equal to that of the normal gravity found on Earth.

SUMMARY

Law of universal gravitation: $F = \dfrac{GMm}{r^2}$

$$G = 6.67 \times 10^{-11}\ \text{N}\cdot\text{m}^2/\text{kg}^2$$

Acceleration of gravity on Earth: $g = \dfrac{GM_E}{R_E^2}$

Circular orbit around Sun: $v^2 = \dfrac{GM_S}{r}$

Kepler's First Law: The orbits of the planets are ellipses with the Sun at one focus.

Kepler's Second Law: The radial line segment from the Sun to the planet sweeps out equal areas in equal times.

Kepler's Third Law: The square of the period is proportional to the cube of the semimajor axis of the planetary orbit.

Gravitational potential energy: $U = -\dfrac{GMm}{r}$

Energy for circular orbit around the Sun: $E = -\dfrac{GM_S m}{2r}$

Escape velocity for Earth: $v = \sqrt{2GM_E/R_E}$

QUESTIONS

1. Can you directly feel the gravitational pull of the Earth with your sense organs? (Hint: Would you feel anything if you were in free fall?)

2. According to a tale told by Professor R. Lichtenstein, some apple trees growing in the mountains of Tibet produce apples of negative mass. In what direction would such an apple fall if it fell off its tree? How would such an apple hang on the tree?

3. Eclipses of the Moon can occur only at full Moon. Eclipses of the Sun can occur only at new Moon. Why?

4. Explain why the sidereal day (the time of rotation of the Earth relative to the stars, or 23 h 56 min 4 s) is shorter than the mean solar day (the time between successive passages of the Sun over a given meridian, or 24 h). (Hint:

The rotation of the Earth around its axis and the revolution of the Earth around the Sun are in the same direction.)

5. Communications satellites are usually placed in an orbit or radius of 4.2×10^7 m so that they remain stationary above a point on the Earth's equator (see Example 4). Can we place a satellite in an orbit so that it remains stationary above the North Pole?

6. When an artificial satellite — such as the ill-fated Skylab — experiences friction against the residual atmosphere of the Earth, the radius of the orbit decreases while at the same time the speed of the satellite *increases.* Explain.

7. Suppose that an airplane flies around the Earth along the equator. If this airplane flies *very* fast, it would not need wings to support itself. Why not?

8. The mass of Pluto was not known until 1978 when a moon of Pluto was finally discovered. How did the discovery of this moon help?

9. It is easier to launch an Earth satellite into an eastward orbit than into a westward orbit. Why?

10. Would it be advantageous to launch rockets into space from a pad at very high altitude on a mountain? Why has this not been done?

11. Describe how you could play squash on a small, round asteroid (with no front wall). What rules of the game would you want to lay down?

12. According to an NBC news report of April 5, 1983, a communications satellite launched from the space shuttle went into an orbit as shown in Figure 13.40. Is this believable?

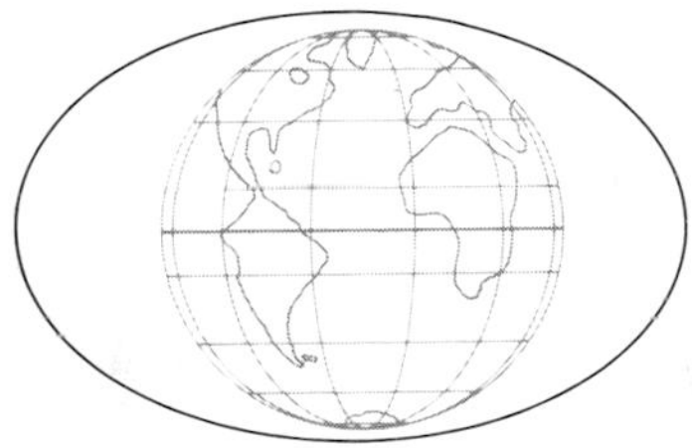

Fig. 13.40 Communications satellite orbit.

13. If we take into account both the gravitational pull exerted by the Sun and that exerted by other planets, is the orbital angular momentum of the Earth constant? Is the total angular momentum of the Solar System constant?

14. Why were the Apollo astronauts able to jump much higher on the Moon than on Earth (Figure 13.41)? If they had landed on a small asteroid, could they have launched themselves into a parabolic or hyperbolic orbit by a jump?

Fig. 13.41 The jump of the astronaut.

15. The Earth reaches perihelion on January 3 and aphelion on July 6. Why is it not warmer in January than in July?

16. When the Apollo astronauts were orbiting around the Moon at low altitude, they detected several mass concentrations ("mascons") below the lunar surface. What is the effect of a mascon on the orbital motion?

17. An astronaut in a circular orbit above the Earth wants to take his spacecraft into a new circular orbit of larger radius. Give him instructions on how to do this.

18. A Japanese and an American astronaut are in two separate spacecrafts in the same circular orbit around the Earth. The Japanese is slightly behind the American and he wants to overtake him. The Japanese fires his thrusters in the *forward* direction, braking for a brief instant. This changes his orbit into an ellipse. One orbital period later, the astronauts return to the vicinity of their initial positions, but the Japanese is now ahead of the American. He then fires his thrusters in the *backward* direction. This restores his orbit into the original circle. Carefully explain the steps of this maneuver, drawing diagrams of the orbits.

19. The gravitational force that a hollow spherical shell of mass exerts on a particle in its interior is zero. Does this mean that such a shell acts as a gravity shield?

20. Consider an astronaut launched in a rocket from the surface of the Earth and then placed in a circular orbit around the Earth. Describe the astronaut's weight (measured in an inertial reference frame) at different times during this trip. Describe the astronaut's *apparent* weight (measured in his own reference frame) at different times.

21. Several of our astronauts suffered severe motion sickness while under conditions of apparent weightlessness aboard Skylab. Since the astronauts were not being tossed about (as in an airplane or a ship in a storm), what caused this motion sickness? What other difficulties does an astronaut face in daily life under conditions of weightlessness?

22. An astronaut on Skylab lights a candle. Will the candle burn like a candle on Earth?

23. Astrology is an ancient superstition according to which the planets influence phenomena on the Earth. The only force that can reach over the large distances between the planets and act on pieces of matter on the Earth is gravitation (planets do not have electric charge and they therefore do not exert electric forces; some planets do have magnetism, but their magnetic forces are too weak to reach the Earth). Given that the Earth is in free fall under the action of the net gravitational force of the planets and the Sun, is there any way that the gravitational forces of the planets can affect what happens on the Earth?

24. The Sun has a much larger mass than the Moon, and yet the tides generated by the Sun are only about half as large as the tides generated by the Moon. Does this make sense?

25. The two daily lunar tides are usually not of equal height. Why not?

26. The Moon always shows the same face to the Earth, i.e., its period of rotation and its period of revolution about the Earth coincide. Can you guess the reason for this coincidence?

27. Extremely high tides ("springs") occur when the Moon is full and when the Moon is new. Why?

PROBLEMS

Section 13.1

1. Two supertankers, each with a mass of 700,000 metric tons, are separated by a distance of 2.0 km. What is the gravitational force that each exerts on the other? Treat them as particles.

2. What is the gravitational force between two protons separated by a distance equal to their diameter, 2.0×10^{-15} m?

3. Somewhere between the Earth and the Moon there is a point where the gravitational pull of the Earth on a particle exactly balances that of the Moon. At what distance from the Earth is this point?

4. Calculate the value of the acceleration of gravity at the surfaces of Venus, Mercury, and Mars. Use the data on planetary masses and radii given in the table printed on the endpapers.

5. The asteroid Ceres has a diameter of 1100 km and a mass of (approximately) 7×10^{20} kg. What is the value of the acceleration of gravity at its surface? What would be your weight (in pounds) if you were to stand on this asteroid?

6. Mimas, a small moon of Saturn, has a mass of 3.8×10^{19} kg and a diameter of 500 km. What is the maximum angular velocity with which we can make this moon rotate about its axis if pieces of loose rock sitting on its surface at its equator are not to fly off?

Section 13.3

7. The Midas II spy satellite was launched into a circular orbit at a height of 500 km above the surface of the Earth. Calculate the orbital period and the orbital speed of this satellite.

8. The Sun is moving in a circular orbit around the center of our Galaxy. The radius of this orbit is 3×10^4 light-years. Calculate the period of the orbital motion and calculate the orbital speed of the Sun. The mass of our Galaxy is 4×10^{41} kg and all of this mass can be regarded as concentrated at the center of the Galaxy.

9. Table 13.5 lists some of the moons of Saturn. Their orbits are circular.
   (a) From the information given, calculate the periods and orbital speeds of all these moons.
   (b) Calculate the mass of Saturn.

10. The Discoverer II satellite had an approximately circular orbit passing over both poles of the Earth. The radius of the orbit was about 6.67×10^3 km. Taking the rotation of the Earth into account, if the satellite passed over New York City at one instant, over what point of the United States would it pass after completing one more orbit?

Table 13.5 SOME MOONS OF SATURN

Moon	*Distance from Saturn*	*Period*	*Orbital speed*
Tethys	2.95×10^5 km	1.89 days	—
Dione	3.77	—	—
Rhea	5.27	—	—
Titan	12.22	—	—
Iapetus	35.60	—	—

11. The binary star system PSR 1913 + 16 consists of two neutron stars orbiting about their common center of mass with a period of 7.75 h. Assume that the stars have equal masses and that their orbits are circular with a radius of 8.67×10^8 m.
   (a) What are the masses of the stars?
   (b) What are their speeds?

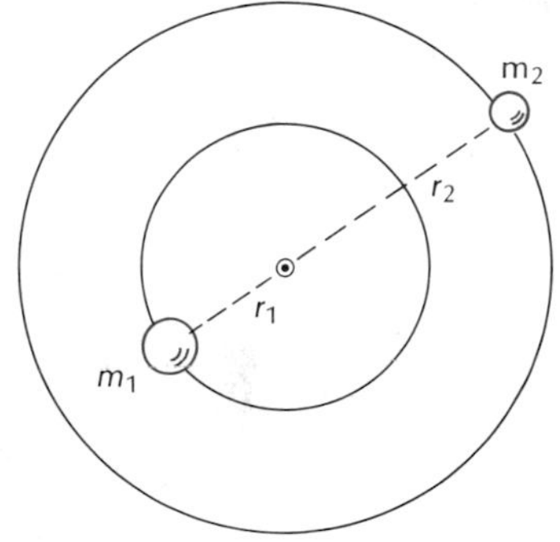

Fig. 13.42 Binary star system.

12. Figure 13.17 shows two stars orbiting about their common center of mass in the binary system Krüger 60. From the relative size of their orbits, determine the ratio of their masses.

13. A binary star system consists of two stars of masses m_1 and m_2 orbiting about each other. Suppose that the orbits of the stars are circles of radii r_1 and r_2 centered on the center of mass (Figure 13.42). Show that the period of the orbital motion is given by

$$T^2 = \frac{4\pi^2}{G(m_1 + m_2)} (r_1 + r_2)^3$$

14. The binary system Cygnus X-1 consists of two stars orbiting about their common center of mass under the influence of their mutual gravitational forces. The orbital period of the motion is 5.6 days. One of the stars is a supergiant with a mass 25 times the mass of the Sun. The other star is believed to be a black hole with a mass of about 10 times the mass of the Sun. From the information given, determine the distance between these stars; assume that the orbits of both stars are circular.

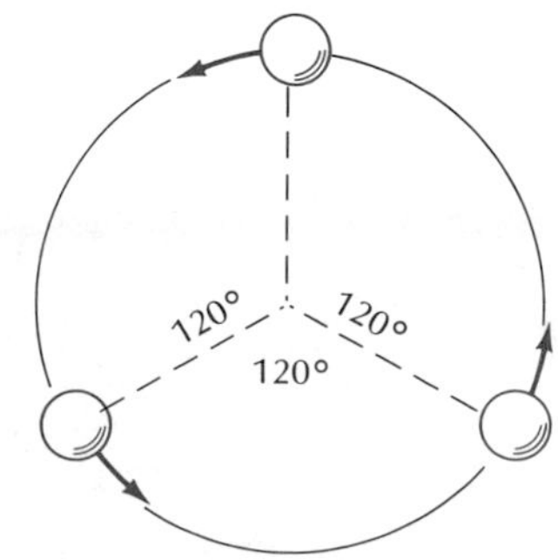

Fig. 13.43 Three identical stars orbiting around their common center of mass.

15. A hypothetical triple-star system consists of three stars orbiting about each other. For the sake of simplicity, assume that all three stars have equal masses and that they move along a common circular orbit maintaining an angular separation of 120° (Figure 13.43). In terms of the mass M of each star and the orbital radius R, what is the period of the motion?

16. Take into account the rotation of the Earth in the following problem:
   (a) Cape Canaveral is at a latitude of 28° North. What eastward speed (relative to the ground) must a satellite be given if it is to achieve a low-alti-

tude circular orbit (Figure 13.44)? What westward speed must the satellite be given if it is to travel along the same orbit in the opposite direction? For the purpose of this problem, pretend that "low altitude" means essentially "zero altitude."

(b) Suppose that the satellite has a mass of 14.0 kg. What kinetic energy must the launch vehicle give to the satellite for an eastward orbit? For a westward orbit?

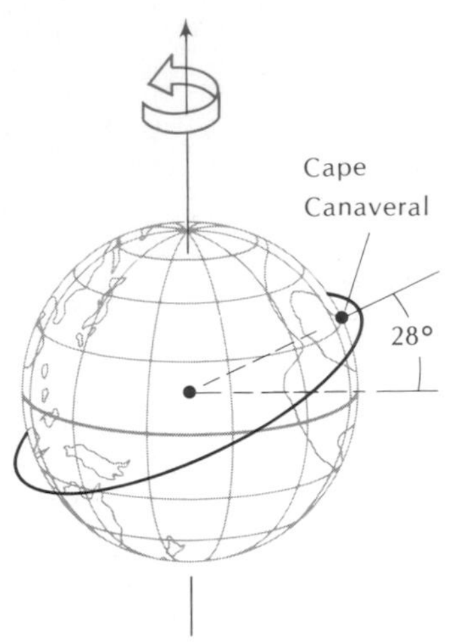

Fig. 13.44 Orbit of satellite launched from Cape Canaveral.

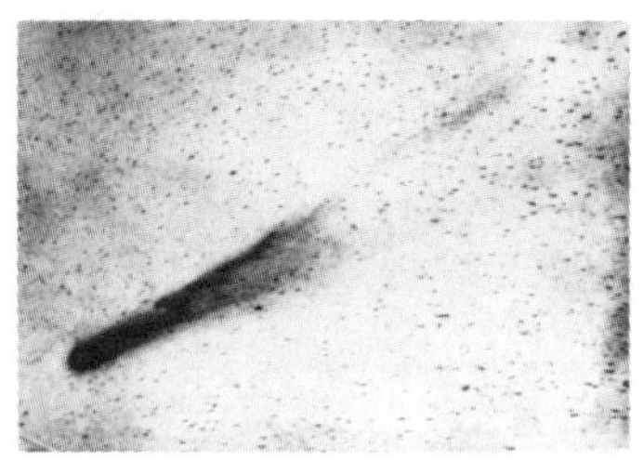

Fig. 13.45 Halley's comet, photographed in 1910.

Section 13.4

17. Halley's comet (Figure 13.45) orbits the Sun in an elliptical orbit (the comet is due to reach perihelion in 1986). When the comet is at perihelion, its distance from the Sun is 8.75×10^{10} m, and its speed is 5.46×10^{4} m/s. When the comet is at aphelion, its distance is 5.26×10^{12} m. What is the speed at aphelion?

18. The Explorer X satellite had an orbit with perigee 175 km and apogee 181,200 km above the surface of the Earth. What was the period of this satellite?

19. Calculate the orbital periods of Sputnik I and Explorer I from their apogee and perigee distances given in Table 13.3.

Section 13.5

20. What is the kinetic energy and what is the gravitational potential energy for the orbital motion of the Earth around the Sun? What is the total energy?

21. The Voskhod I satellite, which carried Yuri Gagarin into space in 1961, had a mass of 4.7×10^{3} kg. The radius of the orbit was (approximately) 6.6×10^{3} km. What were the orbital speed, the orbital angular momentum, and the orbital energy of this satellite?

22. The Andromeda galaxy is at a distance of 2.1×10^{22} m from our Galaxy. The mass of Andromeda is 6×10^{41} kg and the mass of our Galaxy is 4×10^{41} kg.

(a) Gravity accelerates the galaxies toward each other. As reckoned in an inertial reference frame, what is the acceleration of Andromeda? What is the acceleration of our Galaxy? Treat both galaxies as point particles.

(b) The speed of Andromeda *relative to our Galaxy* is 266 km/s.[5] What is the speed of Andromeda and what is the speed of our Galaxy *relative to the center of mass* of the two galaxies?

(c) What is the kinetic energy of each galaxy relative to the center of mass? What is the total energy (kinetic and potential) of the system of the two galaxies? Will the two galaxies eventually escape from each other?

23. The motor of a Scout rocket uses up all the fuel and stops when the rocket is at an altitude of 200 km above the surface of the Earth and is moving vertically at 8.50 km/s. How high will this rocket rise? Ignore any residual atmospheric friction.

24. Neglect the gravity of the Moon, neglect atmospheric friction, and neglect the rotational velocity of the Earth in the following problem: A long time ago, Jules Verne, in his book *From Earth to the Moon* (1865) suggested sending an expedition to the Moon by means of a projectile fired from a gigantic gun.

(a) With what muzzle velocity must a projectile be fired vertically from a gun on the surface of the Earth if it is to (barely) reach the distance of the Moon?

(b) Suppose that the projectile has a mass of 2000 kg. What energy must the gun deliver to the projectile? The explosion of 1 short ton (2000 lb)

[5] This is actually the component of the velocity along the line of sight, i.e., it is the radial component of the velocity. However, for the purpose of this problem, assume that it is the total velocity.

of TNT releases 4.2×10^9 J. How many tons of TNT are required for firing this gun?
(c) If the gun barrel is 500 m long, what must be the average acceleration of the projectile during firing?

25. What is the escape velocity for a projectile launched from the surface of our Moon?

26. An artificial satellite of 3500 kg made of aluminum is in a circular orbit at a height of 100 km above the surface of the Earth. Atmospheric friction removes energy from the satellite and causes it to spiral downward so that it ultimately crashes into the ground.
(a) What is the initial orbital energy (gravitational plus kinetic) of the satellite? What is the final energy when the satellite comes to rest on the ground? What is the energy change?
(b) Suppose that all of this energy is absorbed in the form of heat by the material of the satellite. Is this enough heat to melt the material of the satellite? To vaporize it? The heats of fusion and of vaporization of aluminum are given in Table 20.4.

27. According to one theory, glassy meteorites (tektites) found on the surface of the Earth originate in volcanic eruptions on the Moon. With what minimum speed must a volcano on the Moon eject a stone if it is to reach the Earth? With what speed will this stone impact on the surface of the Earth? In this problem ignore the orbital motion of the Moon around the Earth; use the data for the Earth–Moon system listed in the tables printed on the endpapers. (Hint: When the rock reaches the intermediate point where the gravitational pulls of the Moon and the Earth cancel, it must have zero velocity.)

28. A spacecraft is launched with some initial velocity toward the Moon from 300 km above the surface of the Earth.
(a) What is the minimum initial speed required if the spacecraft is to coast all the way to the Moon without using its rocket motors? For this problem pretend that the Moon does not move relative to the Earth. The masses and radii of the Earth and the Moon and their distance are listed in the tables printed on the endpapers. (Hint: When the spacecraft reaches the point in space where the gravitational pulls of the Earth and the Moon cancel, it must have zero velocity).
(b) With what speed will the spacecraft impact on the Moon?

29. The Pons–Brooks comet had a speed of 47.30 km/s when it reached its perihelion point, 1.160×10^8 km from the Sun. Is the orbit of this comet elliptic, parabolic, or hyperbolic?

30. At a radial distance of 2.00×10^7 m from the center of the Earth, three artificial satellites (I, II, III) are ejected from a rocket. The three satellites I, II, III are given initial speeds of 5.47 km/s, 4.47 km/s, and 3.47 km/s, respectively; the initial velocities are all in the tangential direction.
(a) Which of the satellites I, II, III will have a circular orbit? Which will have elliptical orbits? Explain your answer.
(b) Draw the circular orbit. Also, on the top of the same diagram draw the elliptical orbits of the other satellites; label the orbits with the names of the satellites. (Note: You need not calculate the exact sizes of the ellipses, but your diagram should show whether the ellipses are larger or smaller than the circle.)

31. (a) Since the Moon (*our* moon) has no atmosphere, it is possible to place an artificial satellite in a circular orbit that skims along the surface of the Moon (provided that the satellite does not hit any mountains!). Suppose that such a satellite is to be launched from the *surface* of the Moon by means of a gun that shoots the satellite in a horizontal direction. With what velocity must the satellite be shot out from the gun? How long does the satellite take to go once around the Moon?
(b) Suppose that a satellite is shot from the gun with a horizontal velocity

of 2.00 km/s. Make a rough sketch showing the Moon and the shape of the satellite's orbit; indicate the position of the gun on your sketch.
(c) Suppose that a satellite is shot from the gun with a horizontal velocity of 3.00 km/s. Make a rough sketch showing the Moon and the shape of the satellite's orbit. Is this a closed orbit?

32. Suppose that a projectile is fired horizontally from the surface of the Moon with an initial speed of 2.0 km/s. Roughly sketch the orbit of the projectile. What maximum height will this projectile reach? What will be its speed when it reaches maximum height?

33. Sputnik I, the first Russian satellite (1957), had a mass of 83.5 kg; its orbit reached perigee at a height of 225 km and apogee at 959 km. Explorer I, the first American satellite (1958), had a mass of 14.1 kg; its orbit reached perigee at a height of 368 km and apogee at 2540 km. What was the orbital energy of each of these satellites?

34. The orbits of most meteoroids around the Sun are nearly parabolic.
(a) With what speed will a meteoroid reach a distance from the Sun equal to the distance of the Earth from the Sun? (Hint: In a parabolic orbit the speed at any radius equals the escape velocity at that radius. Why?)
(b) Taking into account the Earth's orbital speed, what will be the speed of the meteoroid *relative to the Earth* in a head-on collision with the Earth? In an overtaking collision? Ignore the effect of the gravitational pull of the Earth on the meteoroid.

35. Calculate the perihelion and the aphelion speed of Encke's comet. The perihelion and aphelion distances of this comet are 5.06×10^7 km and 61.25×10^7 km.

36. The Explorer XII satellite was given a tangential velocity of 10.39 km/s when at perigee at a height of 457 km above the Earth. Calculate the height of apogee.

37. Prove that the orbital energy of a planet or a comet in an elliptical orbit around the Sun can be expressed as

$$E = -\frac{GM_S m}{r_1 + r_2}$$

where r_1 and r_2 are, respectively, the perihelion and aphelion distances. [Hint: Start with the sum of the kinetic and potential energies at perihelion and use Eq. (33).]

38. Suppose that a comet is originally at rest at a distance r_1 from the Sun. Under the influence of the gravitational pull, the comet falls radially toward the Sun. Show that the time it takes to reach a radius r_2 is

$$t = \int_{r_2}^{r_1} \frac{dr}{\sqrt{2GM_S/r - 2GM_S/r_1}}$$

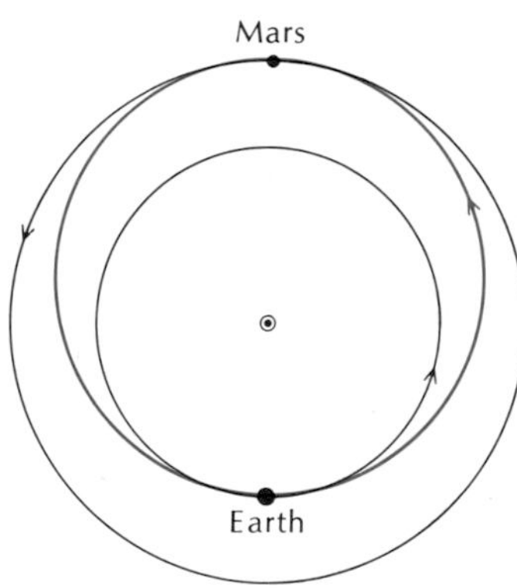

Fig. 13.46 Orbit for a space probe on trip to Mars.

*39. The Earth has an orbit of radius 1.50×10^8 km around the Sun; Mars has an orbit of radius 2.28×10^8 km. In order to send a spacecraft from the Earth to Mars, it is convenient to launch the spacecraft into an elliptical orbit whose perihelion coincides with the orbit of the Earth and whose aphelion coincides with the orbit of Mars (Figure 13.46).
(a) To achieve such an orbit, with what speed (relative to the Earth) must the spacecraft be launched? Ignore the pull of the gravity of the Earth and Mars on the spacecraft.
(b) With what speed (relative to Mars) does the spacecraft approach Mars at the aphelion point? Assume that Mars actually is at the aphelion point when the spacecraft arrives.
(c) How long does the trip from Earth to Mars take?

(d) Where must Mars be (in relation to the Earth) at the instant the spacecraft is launched? Where will the Earth be when the spacecraft arrives at its destination? Draw a diagram showing the relative positions of Earth and Mars at these two times.

*40. Repeat the calculations of Problem 39 for the case of a spacecraft launched on a trip to Venus. The orbit of Venus has a radius of 1.08×10^8 km.

*41. If an artificial satellite, or some other body, approaches a moving planet on a hyperbolic orbit, it can gain some energy from the motion of the planet and emerge with a larger speed than it had initially. This is the slingshot effect already discussed in a special case in Problem 10.23. Suppose that the line of approach of the satellite makes an angle θ with the line of motion of the planet and the line of recession of the satellite is parallel to the line of motion of the planet (Figure 13.47; the planet can be regarded as moving on a straight line during the time interval in question). The speed of the planet is u and the initial speed of the satellite is v.

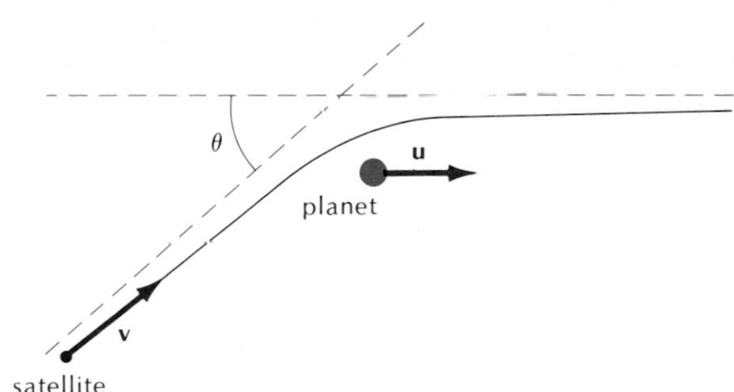

Fig. 13.47 Trajectory of artificial satellite passing by planet.

(a) Show that the final speed of the satellite is

$$v' = u + \sqrt{v^2 + u^2 - 2uv \cos \theta}$$

(Hint: In the reference frame of the planet, the initial speed of the satellite is the same as the final speed. Why?)

(b) Show that the satellite will not gain any speed in this encounter if $\theta = 0$ and show that the satellite will gain maximum speed if $\theta = 180°$.

(c) If a satellite with $v = 3$ km/s approaches Jupiter at an angle of $\theta = 20°$, what will be its final speed?

Section 13.6

42. The following table lists radii measured from the center of the Earth and the fraction of the Earth's mass that lies within this radius:

r	Fraction of mass
1400 km	0.024
2400	0.11
3400	0.31
4400	0.47
5400	0.72
6400	1.00

Calculate the value of the acceleration of gravity at each radius. Plot this value as a function of radius.

43. Calculate the change of the acceleration of gravity between a point on the surface of the sea and a point at a depth of 4000 m below the surface of the sea. For the purposes of this problem, assume that the Earth is a sphere and that it is covered with water to a depth of more than 4000 m. Is the acceleration of gravity at this depth larger or smaller than at the surface?

*44. The gravitational potential energy of a small mass m placed on the surface of a large spherical body of radius R and mass M is $U = -GMm/R$. Find an expression for the gravitational energy if the mass m is placed inside the spherical body, at some radius $r < R$. Assume that the mass of the body is uniformly distributed over the sphere of radius R.

Section 13.7

45. According to one design studied by NASA, a large space colony in orbit around the Earth would consist of a torus of diameter 1.8 km, looking somewhat like a gigantic bicycle tire. In order to generate artificial gravity of 1 gee, how fast must this space colony rotate about its axis?

THE BIG BANG AND THE EXPANSION OF THE UNIVERSE[1]

Cosmology studies the universe at large, its size, its shape, and its evolution. Seeking to grasp the universe, the mind of the cosmologist has to wander over distances as great as 10 billion light-years and over times as long as 10 billion years or longer.

Until the early part of this century, astronomers thought the universe to be much smaller. They thought that the farthest stars at the edge of our Galaxy were about 30,000 light-years away and that there was nothing but dark, empty space beyond that distance. But in the 1920s, Edwin Hubble used the 100-inch telescope on Mt. Wilson to establish that the faint, whispy "nebulae" found in all parts of the sky were actually gigantic conglomerations of stars, similar to our own Galaxy but located at a very great distance from us. He developed methods for measuring these enormous cosmic distances and discovered that some of the remote galaxies were more than half a billion light-years away from us.

To begin a theoretical analysis of the universe, cosmologists must make a basic hypothesis: the laws of physics that hold in other parts of the universe are the same as those that hold in our part of the universe. We have some observational evidence in favor of this hypothesis. For instance, the light and the radiowaves reaching us from remote galaxies have the same basic features as light and radiowaves produced in our laboratories on the Earth. But, of course, there is no direct way of verifying the hypothesis in detail. In practice, the hypothesis of the universal applicability of the laws of physics seems to work quite well and it enables us to make sense out of the observational data that astronomers and radio astronomers collect with their telescopes and radiotelescopes.

Of all the forces of nature, gravity plays the dominant role in shaping the large-scale features and the evolution of the universe. Hence Newton's law of gravitation will be the main tool for the theoretical investigations of this chapter. Strictly speaking, Newton's theory is not entirely adequate for a description of the universe because at the large cosmic distances, relativistic effects become important; thus, we really ought to use Einstein's theory of General Relativity (see Interlude E) in our investigations. However, it turns out that Newton's theory is surprisingly successful in the study of some of the questions concerning the dynamics of the universe, and in this chapter we will be able to get along quite well without relativity. Of course, there are many questions concerning the geometry of the universe, its size, and its shape that Newton's theory cannot answer. But even if we avoid all the items that require relativity, there are still many interesting items left.

D.1 THE EXPANSION OF THE UNIVERSE

In the night sky we can see about 2000 stars with the naked eye. A telescope reveals many more stars and shows that they are part of a large conglomerate or cloud of stars. This cloud is called the **Galaxy,** or the **Milky Way.** It contains about 10^{11} stars arranged in an irregular disklike region some 10^5 light-years in diameter. The disk has a central bulge and it has spiral arms along which stars are concentrated (Figure D.1).

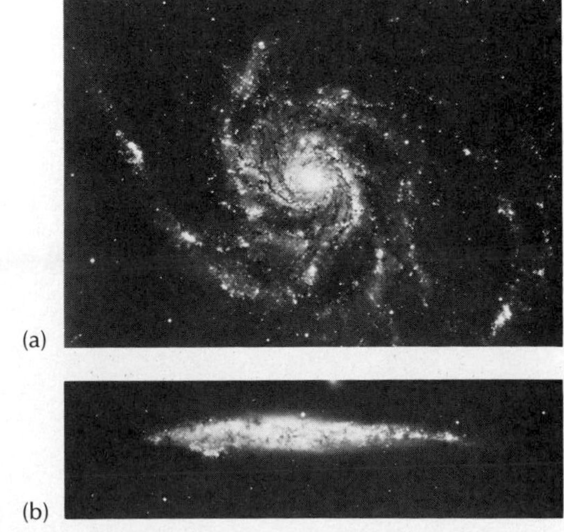

Fig. D.1 These pictures give an impression of what our Galaxy looks like viewed (a) face on and (b) edge on. The pictures are actually photographs of two distant galaxies (NGC 5457 and NGC 4631) similar to our Galaxy.

[1] This chapter is optional.

There are many external galaxies beyond our Galaxy. With our large telescopes we can see altogether about 10^{11} galaxies. There are supergiant galaxies with 10^{13} stars each, and there are dwarf galaxies with "only" 10^6 stars. There are spherical galaxies and elliptical galaxies, like luminous globes and eggs; there are spiral galaxies and barred-spiral galaxies, like whirling pinwheels; and there are irregular galaxies with the weirdest shapes (Figures D.2–D.8).

All the galaxies are in motion. Many of them congregate in clusters, orbiting about each other, sometimes accidentally colliding. But let us ignore the details of the motion of the galaxies and concentrate on the large-scale features of the motion. We then find that on a large scale, the galaxies have a motion of recession — all the distant galaxies are moving away from us. For example, Figure D.9 shows several distant galaxies, all of which are speeding away from us at up to 6×10^7 m/s. There are some objects that are racing away from us even faster than that — the extreme case is that of the quasar QSO (OH471), which has a speed of about 90% of the speed of light.

The recession velocities of galaxies can be determined by the **red-shift** method. This method hinges on the fact that the light from a receding source will look redder to us than the light from a stationary source (conversely, the light from an approaching source will look bluer). The shift of color, or, more precisely, the shift of wavelength of the light, is directly related to the velocity. We will learn more about this in the discussion of the Doppler effect in Chapter 16. To find the recession velocity, astronomers need only measure how much the color of the light received from atoms in some distant galaxy is shifted relative to the color emitted by similar atoms in our laboratories on Earth.

Fig. D.2 Spiral galaxy (NGC 3031) in Ursa Major. The plane of this galaxy is inclined to our line of sight; face on, this galaxy would look circular.

Fig. D.3 Spiral galaxy (NGC 4565) in Coma Berenices, seen edge on. Note how thin this galaxy is.

Fig. D.4 Spiral galaxy (NGC 5194) in Canes Venatici. This beautiful spiral is called the "Whirlpool" galaxy.

Fig. D.5 Unusual galaxy (NGC 5128) in Centaurus. Note the thick lane of dust surrounding this galaxy.

Fig. D.6 Spiral galaxy (NGC 7217) in Pegasus.

Fig. D.7 Elliptical galaxy (NGC 4486) in Virgo. This is one of the brightest galaxies known.

Fig. D.8 Peculiar spiral galaxy (NGC 2623) in Cancer.

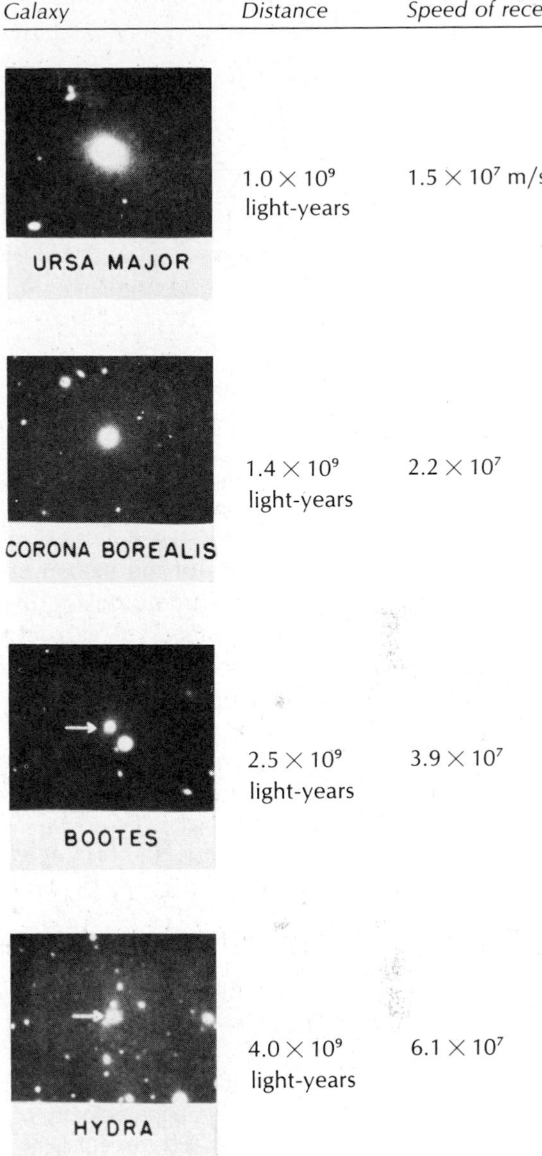

Galaxy	*Distance*	*Speed of recession*
URSA MAJOR	1.0×10^9 light-years	1.5×10^7 m/s
CORONA BOREALIS	1.4×10^9 light-years	2.2×10^7
BOOTES	2.5×10^9 light-years	3.9×10^7
HYDRA	4.0×10^9 light-years	6.1×10^7

Fig. D.9 Several distant galaxies.

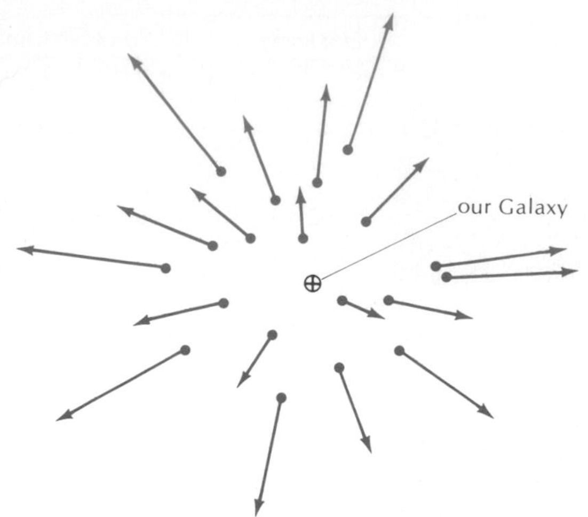

Fig. D.10 The motion of recession of distant galaxies (expansion of the universe).

Figure D.10 shows the velocities of the motion of recession of distant galaxies relative to our Galaxy. By measuring the velocities and distances of galaxies, Hubble discovered that the motion of recession obeys a very simple rule: the velocity of each galaxy is directly proportional to its distance, that is, nearby galaxies move slowly and distant galaxies move fast. This proportionality is called **Hubble's Law.** Mathematically, it can be expressed as

$$v = H_0 r \tag{1}$$

Where H_0 is Hubble's constant. If r is expressed in light-years, the numerical value of Hubble's constant is

$$H_0 = 1.7 \times 10^4 \text{ (m/s)/(million light-years)} \tag{2}$$

For example, the galaxy beyond the constellation Hydra shown in Figure D.9 has a recession velocity of 6.1×10^7 m/s and it is at a distance of 3.5×10^9 light-years in good agreement with Eq. (1).

Incidentally, astronomers determine the enormous cosmic distance from us to other galaxies by the brightness or "headlight" method. This method relies essentially on the following: faraway galaxies look faint to us and nearby galaxies look bright — just as, on a dark road, the headlights of a faraway automobile look faint and the headlights of a nearby automobile look bright. If all galaxies generated precisely the same quantity of light, then differences in their apparent brightness as observed by our telescopes would be entirely due to differences in their distances — there would then be a simple mathematical relationship between apparent brightness and distance. However, in practice, there are some complications; two galaxies can be at the same distance yet differ in apparent brightness because one has more stars and generates more light than the other. Such intrinsic differences between galaxies must be taken into account when using the brightness method. Astronomers have developed clever techniques for selecting galaxies or portions of galaxies of standard brightness, but rather large uncertainties still remain in the distance determinations. Correspondingly, there are uncertainties in the value of the Hubble constant. The uncertainty in Eq. (2) is at least 25% and possibly more.

Although Figure D.10 gives the misleading impression that our Galaxy is at the center of the universe and that all other galaxies are fleeing away from us, our Galaxy does not occupy any special spot in the universe. The other galaxies are not just fleeing away from us; they are fleeing away from each other. All galaxies are receding from all other galaxies — the universe is expanding. An extraterrestrial astronomer sitting on that galaxy beyond Hydra (Figure D.9) would see our Galaxy and all other galaxies fleeing away from him. Hence our spot in the universe is pretty much the same as every other spot. Cosmologists believe that this overall uniformity holds not only in regard to the expansion, but also in regard to all other general features of the universe. For instance, the numbers and types of galaxies that the extraterrestrial astronomer finds in his neighborhood will, on the average, be the same as we find in our neighborhood — the universe is pretty much the same everywhere. This assertion of large-scale uniformity of the universe is called the **Cosmological Principle.**

The motion of recession of the galaxies can be described by a simple analogy. When a grenade explodes in midair, the fragments of shrapnel spurt out in all directions. Different fragments may have different velocities and, in a given amount of time, they will reach different distances. After a time t, the position of a fragment having a velocity v will be

$$r = vt \tag{3}$$

If we rewrite this as

$$v = r/t \tag{4}$$

we see that at any given time the fragments that are at the greatest distances are those with the highest velocities. This proportionality of velocity and distance has the same form as Hubble's Law [see Eq. (1)]. Thus, Hubble's Law suggests that the galaxies were set in motion by a primordial cosmic explosion many years ago and have been more or less coasting along ever since. The only fault with the analogy between the motions of shrapnel fragments and galaxies is that in the case of the former there is a clearly defined center of explosion, while in the case of the latter the explosion occurred simultaneously everywhere and there is no special center of burst. Incidentally, in the expansion of the universe, only the distances between the galaxies increase; the galaxies themselves do not expand. This is also in agreement with the grenade analogy, where, of course, only the distances between the shrapnel fragments increase while the fragments themselves remain of constant size.

D.2 THE AGE OF THE UNIVERSE AND THE BIG BANG

The explosion that started the expansion of the universe is called the **Big Bang.** We can reckon how long ago this happened by comparing Eqs. (1) and (4). Clearly, the inverse of the Hubble constant must coincide with the expansion time,

$$t = 1/H_0 \tag{5}$$

or

$$t = \frac{1}{1.7 \times 10^4} \times \frac{\text{million light-years}}{\text{m/s}}$$

$$= \frac{1}{1.7 \times 10^4} \times \frac{9.5 \times 10^{21}\ \text{m}}{\text{m/s}}$$

$$= 5.6 \times 10^{17}\ \text{s} = 1.8 \times 10^{10}\ \text{years} \tag{6}$$

However, in this calculation of the age of the universe, we have ignored the possibility that the velocity of galaxies may change with time. For instance, since gravity pulls the galaxies toward each other, we might expect that gravity tends to inhibit the motion of recession and tends to slow down the expansion of the universe. This would imply that the velocities of all galaxies were somewhat larger in the past and, consequently, the true age of the universe ought to be somewhat smaller than the 18 billion years indicated by our naïve calculation. But in any case, this number gives us a rough estimate of the age.

Incidentally, because the universe started some finite time ago, only the light from those parts of it that are sufficiently near can have reached us. The speed of light is $c = 3.00 \times 10^8$ m/s = 1 light-year/year, and in a time t it travels a distance

$$ct = (1\ \text{light-year/year}) \times (1.8 \times 10^{10}\ \text{year})$$

$$= 1.8 \times 10^{10}\ \text{light-year}$$

This distance is the radius of the **observable universe.**[2] Everything within this radius we can see (given sufficiently powerful telescopes); anything beyond we cannot see because the light has not yet had enough time to reach us. Note that as time increases, the radius ct increases; i.e., the observable universe includes more and more of the total universe.

The idea of the Big Bang can be put to a direct test: if the universe originated 18 billion or so years ago, then nothing in the universe can be older than that. The Earth and the Sun have an age of only 4.5 billion years — for a serious test of the Big Bang hypothesis, we need to look for much older objects. The oldest objects that we can reliably date are the globular clusters of stars found near our galaxy. Each of these is an aggregate of 10^5 or 10^6 stars, looking rather like a great swarm of bees in the sky (Figure D.11). The theory of stellar evolution permits one to calculate the age of such a cluster from the observed color and luminosity of its stars. As stars reach old age, they enter a red-giant stage, turning a reddish color and swelling to several hundred times normal size. A count of the number of such old red giants in a cluster permits one to deduce the age of the cluster — a young cluster will contain few red giants and an old cluster will contain many. Careful calculations indicate that the oldest globular clusters have ages of about 10 billion to 16 billion years, in good agreement with the expansion time given by Eq. (6).

The age of the oldest globular clusters suggests the date at which the first stars were born in our Galaxy. There is another method by which we can date the birth of stars: by the age of chemical elements. All the elements, with the exception of hydrogen and helium, were synthesized by nuclear reactions in the interior of very massive stars born soon after our Galaxy came into being. These stars survived only a short time and then exploded as supernovas, spurting their chemical elements all over the place, some of these elements

Fig. D.11 The globular cluster (NGC 5272) in Canes Venatici.

[2] This value of the radius of the observable universe is only an approximation. For an exact calculation we need to use the theory of General Relativity, taking into account that space and time are curved.

eventually winding up in the cloud of gas and dust that was destined to become the Solar System. The ages or the atoms of these elements — e.g., uranium and thorium — can be determined by a method that relies on their radioactivity. These atoms suffer radioactive decay as they grow old and the amount of decay gives an indication of their age. Such radioactive-decay measurements tell us that the atoms are somewhere between 7 billion and 15 billion years old. Thus, the age of the elements agrees with the expansion time to within the experimental uncertainties.

A discordance in the ages deduced from the globular clusters, the chemical elements, and the recessional motion of the galaxies would have proved the Big Bang hypothesis wrong. The concordance of these ages does not prove the hypothesis right, but it encourages us to look for further evidence.

The most decisive item of evidence for the Big Bang is that some of the radiant heat given off by the primordial explosion can still be found in the sky today. At the initial instant of the Big Bang, the universe must have been densely filled with compacted matter, at very high temperature and pressure. The matter must have been expanding at a colossal rate so that even today it continues to coast along because of the momentum it had in the beginning. The matter must initially have been hotter than white hot, emitting intense light. The whole universe must have been filled with a fireball of a brightness much greater than that of the Sun or of a thermonuclear explosion.

Originally, the radiant heat emitted by this primordial fireball was in the form of very penetrating gamma rays and X rays. These rays are essentially light waves of very short wavelength. But as the universe expanded, these light waves expanded with it. Thus, the wavelengths of the fireball light still remaining in the sky at present are much longer than the wavelengths of the original light — they are larger in the same proportion as the present intergalactic distances are larger than the original distances. Theoretical calculations suggest that at present the wavelength should be about 1 mm. This is much longer than the wavelength of visible light — it is the wavelength of radiowaves (microwaves).

This kind of radiation was discovered in 1964 by A. A. Penzias and R. W. Wilson, two scientists working at Bell Laboratories with very sensitive microwave communication equipment (Figure D.12). They found that the entire sky is noisy; there is radiation coming at the Earth from all directions. Scientists at Princeton immediately recognized the cosmological significance of this discovery. Subsequently, many experimenters took careful measurements of this **cosmic background radiation** and established that the radiation is strongest at a wavelength of about 2 mm; above or below this wavelength, the amount of radiation gradually decreases. The radiation has all the characteristics of radiant heat — it is the same kind of radiant heat as emitted by a "hot" body of a temperature of 3°C above absolute zero. What has happened here is that the extremely hot radiant heat from the primordial fireball has gradually cooled down as the universe expanded and by now its temperature has come pretty close to absolute zero; as the universe continues to expand, the temperature will continue to drop. The cosmic background radiation is residual radiant heat left in the sky by the Big Bang. It is direct material evidence for the Big Bang.

Fig. D.12 Horn antenna at Bell Lab, Holmdel, N.J. This antenna was designed for microwave communication experiments with the Echo and Telstar satellites.

Further evidence for the Big Bang is supplied by the study of the abundance of helium in the universe. The chemical composition of our universe is about 74% hydrogen (by mass) and 24% helium, with only traces of other elements. All this hydrogen and almost all this helium are primordial, i.e., they were formed in the hot primordial fireball, soon after the beginning of the universe.[3] The temperature of the fireball was 10^9 °C when the universe was 3 minutes old. At this temperature, hydrogen is subject to nuclear fusion leading to the formation of helium. Theoretical calculations show that the fusion reactions lead to an abundance of about 75% hydrogen and 25% helium, in remarkable agreement with the observed abundance. This confirms the picture of a hot Big Bang.

D.3 THE FUTURE OF THE UNIVERSE

If the galaxies were to continue their present motion with constant velocity, they would gradually recede into the distance, looking smaller and smaller, dimmer

[3] Stars obtain their energy from the nuclear fusion of hydrogen, which produces helium. However, the total mass of helium produced by all of the stars since the beginning amounts to only a few percent.

and dimmer; ultimately they would fade from our sight — and our Galaxy would be left alone.

The motion of recession, however, does not proceed with constant velocity. The force of gravitation attracts every galaxy to every other galaxy. This mutual attraction tends to decelerate the expansion. Of course, the pull of gravity is very weak because the intergalactic distances are so enormously large. But even a very small deceleration can be important in the long run. If the motion ever stops, then the galaxies will begin to fall back toward each other. The universe will then contract. As the galaxies come closer together, the gravitational pull will become stronger and the velocities of contraction will become larger. Finally, the galaxies will collide with one another and the universe will collapse in a terminal cosmic implosion.

The big question in cosmology is this: Will the universe expand forever? Or will it come to a stop and contract? To find the answer to this question, we must calculate the deceleration of the motion of recession.

Since our only concern is the average motion of the galaxies, we will pretend that the galaxies are uniformly distributed throughout the universe, forming a gas of galaxies. The expansion of the universe is then equivalent to the expansion of this gas. To find the law of expansion of the gas, consider a spherical region centered on, say, our Galaxy; the spherical region is supposed to be small compared to the size of the universe, but to contain very many galaxies. Imagine that, by magic, we remove all of the galaxies from this sphere, leaving an empty hole. The rest of the universe has spherical symmetry about this hole. Hence, by the theorem of Section 13.6, gravity in the hole will be exactly zero. It then follows that if we put the galaxies back into the spherical region, their motion will be completely unaffected by the rest of the universe — only the galaxies in the spherical region exert gravitational forces within the spherical region. Consider now one of the galaxies at the surface of this region at a radial distance r from us (Figure D.13). The gravitational force that the mass in the spherical region exerts on that one galaxy is as though all of the mass were concentrated at the center (see Section 13.6). Consequently, the acceleration of that one galaxy is

$$\frac{dv}{dt} = -\frac{GM}{r^2} \tag{7}$$

where M is the mass in the spherical region. This equation determines how the size of the spherical region varies with time.

The equation of motion (7) is of the same form as for a ballistic missile coasting away from the Earth along a radial line. The radial motion continually decelerates, but whether it ever stops and reverses depends on the initial velocity: if the initial velocity is larger than the escape velocity [see Eq. (13.29)], the radial distance continues to increase forever; if the initial velocity is smaller than the escape velocity, the radial motion comes to a halt and then reverses. If we regard the present instant as the initial instant and the present radius as the starting radius, the initial velocity is [see Eq. (1)]

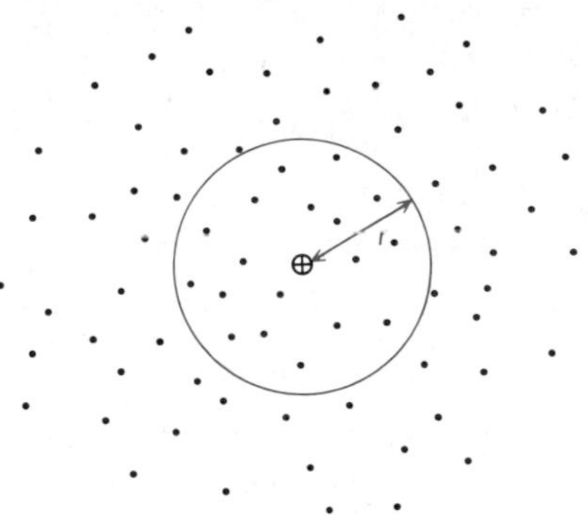

Fig. D.13 A spherical region of our universe.

$$v_0 = H_0 r_0 \tag{8}$$

and the escape velocity is

$$v_{crit} = \sqrt{2GM/r_0} \tag{9}$$

In these equations the subscript 0 has been inserted to emphasize that the quantities are evaluated at the present instant. The mass M can be expressed in terms of the average density of mass in the universe:

$$M = \frac{4\pi}{3} r_0^3 \rho_0 \tag{10}$$

so that

$$v_{crit} = \sqrt{\frac{8\pi}{3} G r_0^2 \rho_0} \tag{11}$$

The condition for a permanently expanding universe is then

$$v_0 \geq v_{crit} \tag{12}$$

which is equivalent to

$$H_0 r_0 \geq \sqrt{\frac{8\pi}{3} G r_0^2 \rho_0} \tag{13}$$

or

$$\rho_0 \leq \frac{3}{8\pi G} H_0^2 \tag{14}$$

Likewise, the condition for an ultimately contracting universe is

$$\rho_0 > \frac{3}{8\pi G} H_0^2 \tag{15}$$

If we insert the numerical value $H_0 = 1/5.6 \times 10^{17}$ s [see Eq. (6)] and the known numerical value for the gravitational constant on the right sides of Eqs. (14) and (15), we obtain the conditions

$$\rho_0 \leq 5.7 \times 10^{-27}\ \text{kg/m}^3 \quad \text{(permanent expansion)}$$
$$\rho_0 > 5.7 \times 10^{-27}\ \text{kg/m}^3 \quad \text{(ultimate contraction)} \tag{16}$$

In principle, this makes it very simple to predict the future evolution of the universe: we only need to measure the average density of mass and check whether it is larger or smaller than the critical value 5.7×10^{-27} kg/m³. In practice, we are handicapped by the uncertainties in the mass density and, to a lesser extent, uncertainties in the value of the right side of Eq. (16).

D.4 THE SEARCH FOR NUMBERS

The basic cosmological parameters that describe the large-scale behavior of our universe are the velocity of expansion [or the Hubble constant; see Eq. (1)], the rate of change of this velocity, the age of the universe, and the mass of the universe. If we knew the values of these parameters, we could predict the future evolution of the universe. For instance, Eq. (16) permits us to make such a prediction from the observed values of the mass density and the velocity of expansion (or the value of the Hubble constant). Alternative equations involving the age of the universe or the deceleration of the velocity can be used similarly.

The search for numbers has been the great problem of observational cosmology. Although much progress has been made in recent years, the values of the cosmological parameters remain rather uncertain. This makes it hard to decide on a final answer to the big question of the future evolution of the universe.

What do we know about the average mass density? When reckoning the mass of the universe, we must take into account the mass belonging to galaxies and also whatever mass is to be found in the intergalactic space between the galaxies. The mass of the galaxies is rather uncertain — different methods of mass determination give values that differ by as much as a factor of 10. The source of this difficulty is that besides visible stars, a galaxy may contain a large amount of invisible hidden mass in the form of small, very faint stars, or black holes. According to the best available estimates, the total visible mass associated with galaxies has a density of 2×10^{-28} kg/m³. This is below the critical value given of Eq. (16). Hence, unless there is some extra, hidden mass somewhere in the galaxies or between the galaxies, the universe will continue to expand forever.

The intergalactic space may or may not contain extra hidden mass in the form of black holes, hydrogen gas at very high temperature (plasma), or neutrinos. Mass in such forms would be very hard to detect and we therefore cannot place any very tight limits on the amounts of such extra mass. There is, however, some circumstantial evidence suggesting that the extra mass cannot be very large. This evidence hinges on the abundance of deuterium in the universe. Deuterium is an isotope of hydrogen — the nucleus of hydrogen consists of one proton, while the nucleus of deuterium consists of one proton plus one neutron. Deuterium is found wherever hydrogen is found; for example, in water on the Earth, one molecule in 6000 is a molecule of "heavy water" in which one of the hydrogen atoms in H_2O is replaced by a deuterium atom. The Sun contains deuterium mixed with its hydrogen and so do all other stars. Astrophysicists believe that this deuterium was created in nuclear reactions during the Big Bang, long before galaxies and stars were born. From theoretical calculations of deuterium production, we learn that a large mass density of the universe interferes with the formation of deuterium because it favors some nuclear reactions that destroy deuterium. Since the universe now contains a fair amount of deuterium, the mass density during the Big Bang must have been below a certain limit; consequently, the mass density now must also be below a corresponding limit — at most 6×10^{-28} kg/m³. This value is considerably below the critical value of Eq. (16) and indicates a forever-expanding universe. Unfortunately, the argument concerning deuterium is not completely rigorous and we cannot rule out the possibility that a somewhat larger amount of mass may be hidden somewhere.

Will the universe continue to expand or will it ultimately contract? The available data are not yet quite precise enough for a firm prognosis. A forever-expanding universe gives the best fit to all the facts as we know them. But there are enough uncertainties in our measurements and enough loopholes in our arguments that the possibility of a contracting universe cannot be dismissed. It will be a while before we know the ultimate fate of the universe.

Further Reading

Galaxies by T. Ferris (Sierra Book Club, San Francisco, 1980; reprinted in paperback by Stewart, Tabori, and Chang, New York, 1982) is a splendid pictorial survey of the universe with awesome photographs of distant galaxies and with informative, if somewhat fragmented, text.

Galaxies by H. Shepley (Harvard University Press, Cambridge, 1972) is a classic introduction to the study of galaxies, including some description of the tools used by astronomers. *Galaxies: Structure and Evolution* by R. J. Tayler (Wykeham, London, 1978) gives a short course in galactic structure at a slightly more advanced level.

The Realm of the Nebulae by E. Hubble (Dover, New York, 1958) is a reprint of Hubble's own account of the first measurements of extragalactic distances and of the discovery of his law. First published in 1936, the details in this account are out of date, but it retains its historical value.

Cosmology by E. R. Harrison (Cambridge University Press, Cambridge, 1981) is an excellent introduction to modern cosmological theory at an elementary level; it is very clearly written and each chapter contains an exhaustive list of further references. *The Big Bang* by J. Silk (Freeman, San Francisco, 1980) and *Black Holes, Quasars, and the Universe* by H. L. Shipman (Houghton Mifflin, Boston, 1980) give good, concise introductions to cosmology with a somewhat more observational orientation.

The Structure of the Universe by E. L. Schatzmann (McGraw-Hill, New York, 1968) and *The Nature of the Universe* by C. Kilmister (Dutton, New York, 1971) are nice, but very elementary, introductions.

Modern Cosmology by D. W. Sciama (Cambridge University Press, Cambridge, 1971) and *Principles of Modern Cosmology* by M. Berry (Cambridge University Press, Cambridge, 1976) are two good books at an intermediate level. Both use some calculus, but avoid the full machinery of Einstein's equations.

The First Three Minutes by S. Weinberg (Basic Books, New York, 1977) gives a brilliant description of the early stages of the expansion of the universe.

The Red Limit by T. Ferris (Morrow, New York, 1977) and *Violent Universe* by N. Calder (Viking, New York, 1969) tell of the recent discoveries and of the astronomers, astrophysicists, and cosmologists who made these discoveries.

The following are some recent articles dealing with cosmology:

"Will the Universe Expand Forever?" J. R. Gott, J. E. Gunn, D. N. Schramm, and B. M. Tinsley, *Scientific American,* March 1976

"The Curvature of Space in a Finite Universe," J. J. Callahan, *Scientific American,* August 1976

"The Clustering of Galaxies," E. J. Groth, P. J. E. Peebles, M. Seldner, and R. M. Soneira, *Scientific American,* November 1977

"The Cosmic Background Radiation and the New Ether Drift," R. A. Muller, *Scientific American,* May 1978

"The New Inflationary Universe," M. M. Waldrop, *Science,* January 1983

"The Future of the Universe," C. Teplitz and V. L. Teplitz, *Scientific American,* March 1983

"The Early Universe and High-Energy Physics," D. N. Schramm, *Physics Today,* April 1983

"The Structure of the Early Universe," J. D. Barrow and J. Silk, *Scientific American,* April 1980

The *Resource Letter RC-1* in the *American Journal of Physics,* March 1976, lists some further references.

Questions

1. Why do astronomers not rely on triangulation to measure the distances of galaxies?

2. If you look at the night sky, you see that the distribution of stars is not uniform. What main deviation from uniformity do you see, and to what is this deviation due?

3. The last figure of Part I of the Prelude shows a plot of the angular distribution of galaxies in a region of the sky. According to this plot, the angular distribution is pretty much uniform (isotropic). However, this plot gives us no direct evidence that the distribution is also uniform in depth (homogeneous). If you assume that the Earth does not occupy a preferred, central spot in the universe, can you conclude that isotropy implies homogeneity?

4. Figure D.14 shows the positions of four galaxies, with our own Galaxy at the center. Draw a figure showing the positions of these four galaxies at a later time, when the universe is twice as large.

Fig. D.14

5. Consider the galaxies shown in Figure D.15; the arrows are the velocity vectors of these galaxies in the reference frame of our Galaxy. According to the Galilean addition law for velocities, find the velocity vectors of these galaxies (including our own Galaxy) in the reference frame of the galaxy *P*; do this by graphically subtracting the velocity vectors.

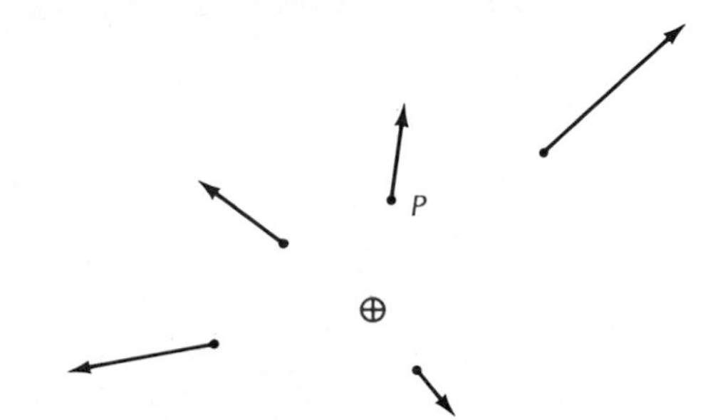

Fig. D.15

6. If the velocity of recession of the galaxies were not proportional to the distance ($v = H_0 r$) but, rather, proportional to the distance squared ($v = H_0 r^2$) or proportional to some other power of the distance, then our Galaxy would occupy a preferred, central spot in the universe. Explain.

7. The value of the Hubble constant that Hubble had deduced from the available data in 1936 was 1.6×10^5 (m/s)/(million light-years). The corresponding expansion time [see Eq. (5)] is 1.9×10^9 years. How does this compare with the age of the Earth and with the age of globular clusters? With what problem was Hubble faced?

8. If the expansion of the universe is slowing down, can the Hubble constant be truly constant?

9. Imagine a galaxy very close to the edge of the observable universe. This galaxy is moving away from us at a speed nearly equal to the speed of light. Does this mean that this galaxy will reach the edge of the observable universe and disappear from sight?

10. Assume that the universe stops expanding and begins to contract. Would the light from distant galaxies continue to exhibit a red shift?

11. Describe the difference between the final states of the universe for the cases $\rho_0 < 3H_0^2/(8\pi G)$ and $\rho_0 = 3H_0^2/(8\pi G)$.

12. How will life ultimately end if the universe continues to expand forever? If it contracts?

13. Why are black holes, neutrons, or gravitational waves in intergalactic space hard to detect?

14. Where were the atoms in your body made?

CHAPTER 14

Oscillations

The motion of a particle or of a system of particles is **periodic** if it repeats again and again at regular intervals of time. The orbital motion of a planet around the Sun, the uniform rotational motion of a phonograph turntable, the back-and-forth motion of a piston in an automobile engine, the swinging motion of a pendulum in a grandfather clock, and the vibration of a guitar string are examples of periodic motions. A periodic back-and-forth or swinging motion of a body is called an **oscillation.** Thus, the motion of the piston is an oscillation, and so is the motion of the pendulum, and the motion of the individual particles of the guitar string.

Oscillation

In this chapter we will examine in great detail the motion of a mass oscillating back and forth under the push and pull exerted by an ideal spring. The mathematical equations that we will develop for the description of this mass–spring system are of great importance because analogous equations recur in the description of all other oscillating systems.

14.1 Simple Harmonic Motion

In this section we will consider the special case of one-dimensional periodic motion involving a particle that oscillates back and forth along some path. For example, Figure 14.1 shows the worldline of a particle oscillating back and forth along the x axis between $x = -0.3$ m and $x = +0.3$ m. The worldline in this special example has a very simple mathematical form:

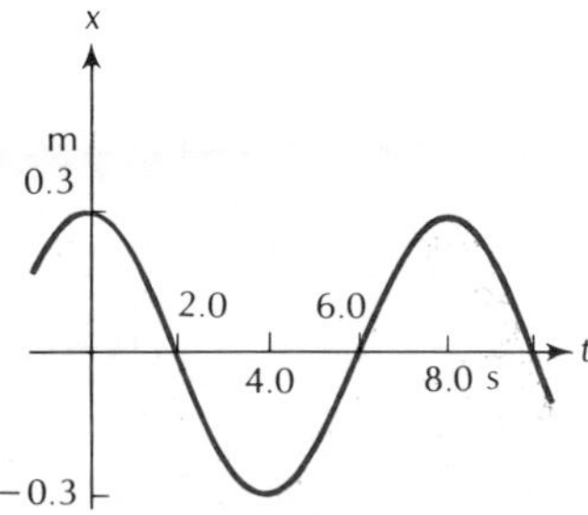

Fig. 14.1 Worldline of a particle oscillating along the x axis with simple harmonic motion.

$$x = 0.3 \cos \frac{\pi}{4} t \qquad (1)$$

where distance is measured in meters and time in seconds. Cosines and sines are called **harmonic functions;** accordingly, the position of the particle given by Eq. (1) is a harmonic function of time. At $t = 0$ the particle is at maximum distance from the origin ($x = 0.3$ m) and is just starting to move; at $t = 2.0$ s it passes through the origin ($x = 0$); at $t = 4.0$ s it reaches maximum distance, but on the negative x axis ($x = -0.3$ m); at $t = 6.0$ s it again passes through the origin. Finally, at $t = 8.0$ s it returns to maximum distance on the positive x axis, exactly as at $t = 0$ — it has completed one cycle of the motion and is ready to begin the next cycle. Thus the period of the motion is

$$T = 8.0 \text{ s} \tag{2}$$

and the frequency of the motion, or the rate of repetition of the motion, is

$$\nu = \frac{1}{T} = \frac{1}{8.0 \text{ s}} = 0.125/\text{s} \tag{3}$$

The points $x = 0.3$ m and $x = -0.3$ m, at which the x coordinate attains its maximum and minimum are the turning points of the motion and the point $x = 0$ is the midpoint or the equilibrium point.

Equation (1) is an example of **simple harmonic motion.** More generally, a motion is simple harmonic if the position as a function of time has the form

Simple harmonic motion

$$\boxed{x = A\cos(\omega t + \delta)} \tag{4}$$

Amplitude of the motion

Angular frequency

The quantities A, ω, and δ are constants. The quantity A is called the **amplitude of the motion**; it is simply the distance between the midpoint ($x = 0$) and either turning point ($x = +A$ or $x = -A$). The quantity ω is called the **angular frequency**; it is directly related to the period and to the frequency of the motion. To establish this relation, note that an increase of t by $2\pi/\omega$ changes the function of Eq. (4) into

$$\begin{aligned} x &= A\cos[\omega(t + 2\pi/\omega) + \delta] \\ &= A\cos(\omega t + \delta + 2\pi) = A\cos(\omega t + \delta) \end{aligned} \tag{5}$$

i.e., such an increase of t does not change the function at all because the function merely repeats itself. The **period** of the motion is the time interval for one such repetition:

Period

$$\boxed{T = \frac{2\pi}{\omega}} \tag{6}$$

and **frequency** of the motion is

Frequency

$$\boxed{\nu = \frac{1}{T} = \frac{\omega}{2\pi}} \tag{7}$$

The units of angular frequency are radians per second. The units of frequency are cycles per second. Dimensionally, both of these units are equivalent to 1/s; but, to prevent confusion between them, it is useful to retain the labels *radian* and *cycle.* The unit of frequency is often called a **hertz** (Hz):

***hertz,* Hz**

$$1 \text{ hertz} = 1 \text{ Hz} = 1 \text{ cycle per second}$$

Note that the above equations connecting *angular frequency,* period, and frequency are formally the same as the equations connecting *angular velocity,* period, and frequency of uniform rotational motion [see Eqs. (11.4) and (11.6)]. As we will see later in this section, this coincidence arises from a special geometrical relationship between simple harmonic motion and uniform circular motion. Aside from this special relationship, angular frequency has nothing to do with angular velocity, even though both quantities are labeled by the same letter ω and are measured in the same units.

In terms of the period or the frequency, Eq. (4) can be written as

$$x = A\cos\left(\frac{2\pi}{T}t + \delta\right) \tag{8}$$

or as

$$x = A\cos(2\pi\nu t + \delta) \tag{9}$$

Phase and phase constant

The argument of the cosine function is called the **phase** of the oscillation and the quantity δ is called the **phase constant.** This constant determines at what time the particle reaches the point of maximum displacement. At this instant

$$\omega t_{max} + \delta = 0 \tag{10}$$

i.e.,

$$t_{max} = -\delta/\omega \tag{11}$$

Hence the particle reaches the point of maximum displacement at a time δ/ω *before* $t = 0$ (Figure 14.2). Of course the particle also passes through this point at periodic intervals before and after this time.

Note that since

$$A\cos(\omega t + \delta) = A\sin(\omega t + \delta + \pi/2)$$
$$= A\sin[\omega t + (\delta + \pi/2)] \tag{12}$$

the cosine function of Eq. (4) can be replaced by a sine function by merely changing the phase constant by $\pi/2$ radians, or 90°. Thus, simple harmonic motion can be equally well expressed in terms of the sine function. Alternatively, the trigonometric identity for the cosine of the sum of two angles gives

$$A\cos(\omega t + \delta) = A\cos\delta\cos\omega t - A\sin\delta\sin\omega t$$
$$= (A\cos\delta)\cos\omega t - (A\sin\delta)\sin\omega t \tag{13}$$

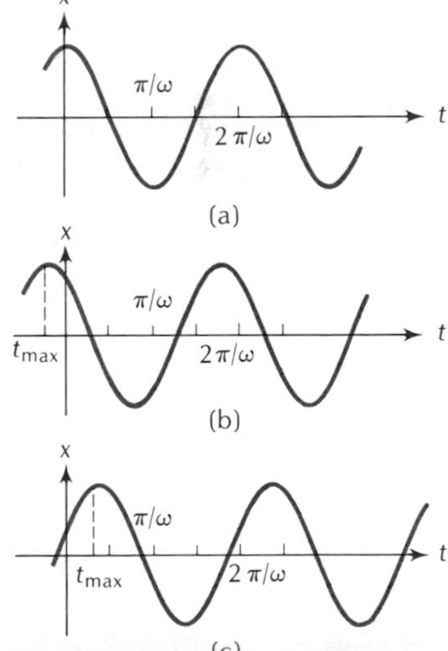

Fig. 14.2 Examples of worldlines of particles with simple harmonic motion with different phase constants. (a) $\delta = 0$. The particle has maximum displacement at $t = 0$. (b) $\delta = \pi/4$ (or 45°). The particle has maximum displacement before $t = 0$. (c) $\delta = -\pi/3$ (or −60°). The particle has maximum displacement after $t = 0$.

This expresses simple harmonic motion as a superposition of sine and cosine functions, each with zero phase constant and with amplitudes $A \cos \delta$ and $-A \sin \delta$.

There is a simple geometric relationship between simple harmonic motion and uniform circular motion. Consider a particle moving with angular velocity ω along a circle of radius A (Figure 14.3). If at $t = 0$ the angular position of the particle is $\theta = \delta$, then the position angle at any later time is

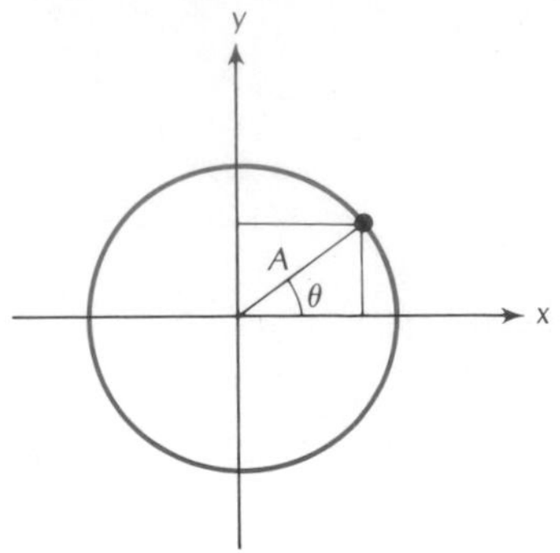

Fig. 14.3 Particle with uniform circular motion on a circle of radius A.

$$\theta = \omega t + \delta \tag{14}$$

and the x and y coordinates of the particle are (see Figure 14.3)

$$x = A \cos(\omega t + \delta)$$

$$y = A \sin(\omega t + \delta) = A \cos(\omega t + \delta - \pi/2) \tag{15}$$

Hence the x coordinate of a particle in uniform circular motion has simple harmonic motion, and the angular velocity of the circular motion coincides with the angular frequency of the harmonic motion. Likewise, the y coordinate of this particle has simple harmonic motion, but the phase differs by $\pi/2$ (or 90°) from the motion of the x coordinate.

This geometric relationship is often used mechanically to generate simple harmonic motion from circular motion. It is only necessary to place a slotted arm over a peg which is attached to a wheel in uniform circular motion (Figure 14.4). The slot is vertical and the arm is constrained to move horizontally. The peg then drags the arm back and forth and makes it move with simple harmonic motion.

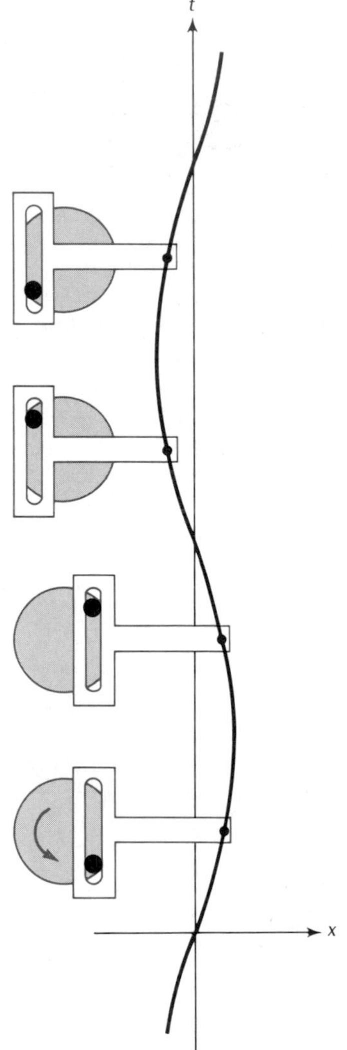

Fig. 14.4 Rotating wheel with a peg driving a slotted arm back and forth. The drawing shows several "snapshots" of the wheel at different times.

Finally, let us compute the velocity and acceleration corresponding to simple harmonic motion. If the displacement is

$$x = A \cos(\omega t + \delta) \tag{16}$$

then the velocity is

$$\frac{dx}{dt} = -A\omega \sin(\omega t + \delta) \tag{17}$$

and the acceleration is

$$\frac{d^2x}{dt^2} = -A\omega^2 \cos(\omega t + \delta) \tag{18}$$

Comparison of Eqs. (16) and (18) shows that

$$\boxed{\frac{d^2x}{dt^2} = -\omega^2 x} \tag{19}$$

i.e., the acceleration is always proportional to the displacement, but oppositely directed. This is a characteristic feature of simple harmonic motion, a fact that will be useful in the next section.

14.2 The Simple Harmonic Oscillator

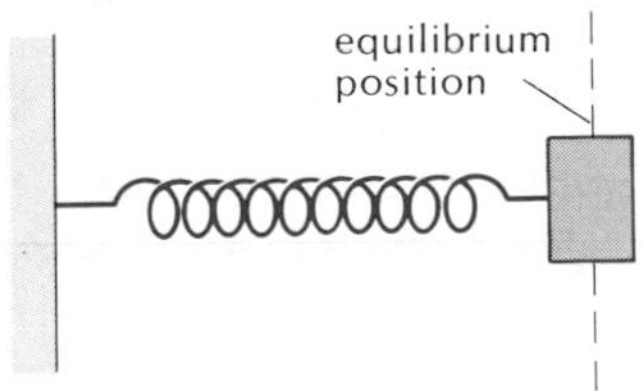

Fig. 14.5 Mass attached to a spring.

The **simple harmonic oscillator** consists of a mass coupled to an ideal, massless spring which obeys Hooke's law. One end of the spring is attached to the mass and the other is held fixed (Figure 14.5). We will ignore gravity and friction so that the spring force is the only force acting on the mass. This system has an equilibrium position corresponding to the relaxed length of the spring. If the system is initially not in this equilibrium position, then the spring supplies a restoring force which impels the mass toward the equilibrium position. But inertia causes the mass to overshoot this equilibrium position and the mass then oscillates back and forth — forever if there is no friction.

A practical example of such a mass–spring system is the device used aboard Skylab for the daily measurement of the masses of the astronauts (Figure 14.6); here, the astronaut and the chair into which he is strapped play the role of the mass of Figure 14.5, and a torsional spring plays the role of the coil spring of Figure 14.5. Aside from such direct practical applications, the great importance of the simple harmonic oscillator is that many physical systems are mathematically equivalent to simple harmonic oscillators, i.e., these systems have an equation of motion of the same mathematical form as the simple harmonic oscillator. A pendulum, the balance wheel of a watch, a tuning fork, the air in an organ pipe, and the atoms in a molecule are systems of this kind; the restoring force and the inertia are of the same mathematical form in these systems as in the simple harmonic oscillator and one can transcribe the general mathematical results from the latter to the former.

Fig. 14.6 Body-mass measurement device used on Skylab. The torsional spring is attached to the corners of the triangular frame on the left.

To derive the equation of motion of the simple harmonic oscillator, we begin with Hooke's law for the force exerted by the spring on the mass (compare Section 6.5):[1]

$$F = -kx \tag{20}$$

Here the displacement is measured from the equilibrium position. The constant k is the spring constant. The force is negative if x is positive (stretched spring; see Figure 14.7a) and the force is positive if x is negative (compressed spring; see Figure 14.7b). Of course, Eq. (20) is only valid as long as the spring is not stretched or compressed too much. Excessive stretch would ultimately break the spring, and excessive compression would bring the coils of the spring into contact — in either case Eq. (20) would fail. We will assume that the displacement remains within limits acceptable to the spring.

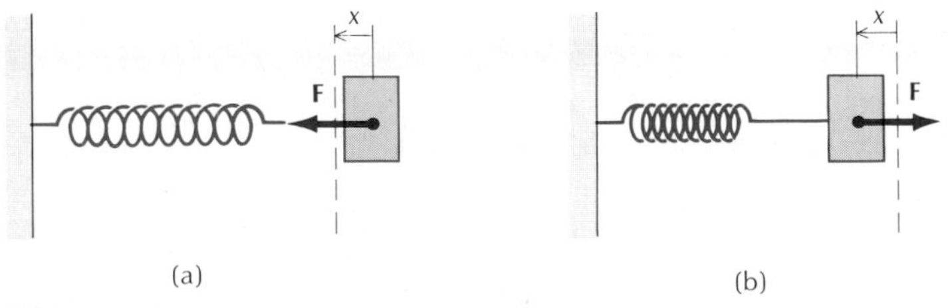

Fig. 14.7 (a) Positive displacement of the mass; the force is negative. (b) Negative displacement of the mass; the force is positive.

[1] Note that in Section 6.5 we were dealing with static conditions (x was constant or varied only very slowly), whereas now we are dealing with dynamic conditions (x varies with time). We will make the assumption that even when x varies with time, the force law of Eq. (20) remains applicable. This is equivalent to the assumption that the spring is massless, because, if so, no force is required to accelerate the spring and all of the restoring force remains available to accelerate the mass attached to the spring.

With the force as given by Eq. (20), the equation of motion of the mass is

Equation of motion of a simple harmonic oscillator

$$m\frac{d^2x}{dt^2} = -kx \tag{21}$$

Rather than attempt to solve this equation by the standard mathematical techniques for the solution of differential equations, let us make use of our knowledge of simple harmonic motion. First we rewrite Eq. (21) as

$$\frac{d^2x}{dt^2} = -\frac{k}{m}x \tag{22}$$

This says that the acceleration is directly proportional to the displacement. As we saw in the preceding section, such a proportionality is a characteristic feature of simple harmonic motion. In fact, comparison of Eqs. (19) and (22) shows that they are identical, provided that

$$\omega^2 = \frac{k}{m} \tag{23}$$

We know from Chapter 6 that, for given initial conditions, the equation of motion completely determines the motion. Since Eqs. (19) and (22) are identical, we can conclude that the motion of a mass on a spring is simple harmonic motion with an angular frequency

Angular frequency of a simple harmonic oscillator

$$\omega = \sqrt{k/m} \tag{24}$$

The position as a function ot time is then

$$x = A\cos(\sqrt{k/m}t + \delta) \tag{25}$$

As in Eq. (13), we can also write this in the form

$$x = (A\cos\delta)\cos(\sqrt{k/m}t) - (A\sin\delta)\sin(\sqrt{k/m}t) \tag{26}$$

The constants A and δ remain to be determined. These constants can be expressed in terms of the initial conditions of the motion, i.e., the initial position x_0 and velocity v_0 at $t = 0$. According to Eqs. (17) and (25),

$$x_0 = A\cos(0 + \delta) = A\cos\delta \tag{27}$$

$$v_0 = -\sqrt{k/m}A\sin(0 + \delta) = \sqrt{k/m}A\sin\delta \tag{28}$$

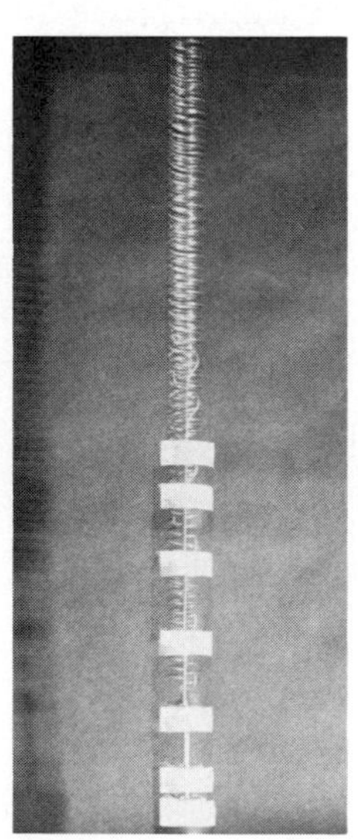

Fig. 14.8 Stroboscopic photograph of an oscillating mass on a spring. Note that the mass moves slowly at the extremes of its motion.

If we take the values of $A\cos\delta$ and $A\sin\delta$ from Eqs. (27) and (28) and insert them into Eq. (26), we obtain

$$x = x_0\cos(\sqrt{k/m}t) + v_0\sqrt{m/k}\sin(\sqrt{k/m}t) \tag{29}$$

which expresses the motion in terms of the initial conditions.

Note that the frequency of the motion of the simple harmonic oscillator depends *only* on the spring constant and the mass. The frequency of the oscillator will always be the same, regardless of the amplitude with which the oscillator has been set swinging; this property of the oscillator is called **isochronism.**

Isochronism

Figure 14.8 shows a multiple-exposure photograph of the oscillations of a mass on a spring. The motion of the mass is simple harmonic.

EXAMPLE 1. A mass of 400 kg is moving along the x axis under the influence of the force of a spring with $k = 3.5 \times 10^4$ N/m. There are no other forces acting on the mass. The equilibrium point is at $x = 0$. Suppose that at $t = 0$, the mass is at $x = 0$ and has a velocity of 2.4 m/s in the positive direction. What is the frequency of oscillation? What is the amplitude? Where will the mass be at $t = 0.60$ s?

SOLUTION: The angular frequency of oscillation is

$$\omega = \sqrt{\frac{k}{m}} = \sqrt{\frac{3.5 \times 10^4 \text{ N/m}}{4.0 \times 10^2 \text{ kg}}} = 9.4 \text{ radian/s}$$

and the frequency is

$$\nu = \frac{\omega}{2\pi} = 1.5/\text{s} = 1.5 \text{ Hz}$$

Since $x_0 = 0$ and $v_0 = 2.4$ m/s, Eq. (29) gives

$$x = 2.4 \text{ m/s} \times \sqrt{\frac{m}{k}} \sin \sqrt{\frac{k}{m}}\, t$$

$$= 0.26 \text{ m} \times \sin (9.4t/\text{s})$$

The amplitude of oscillation is 0.26 m and at $t = 0.60$ s the position of the mass will be

$$x = 0.26 \text{ m} \times \sin(9.4 \times 0.60 \text{ radian}) = -0.16 \text{ m}$$

EXAMPLE 2. A mass m hangs vertically from a spring of spring constant k (Figure 14.9). Find the motion of this system, taking gravity into account.

SOLUTION: We arrange the x axis downward with origin at the point corresponding to the relaxed length of the spring. The spring force is then $-kx$. Besides this force, there is also the force mg due to gravity. Hence the equation of motion is

$$m \frac{d^2x}{dt^2} = -kx + mg \tag{30}$$

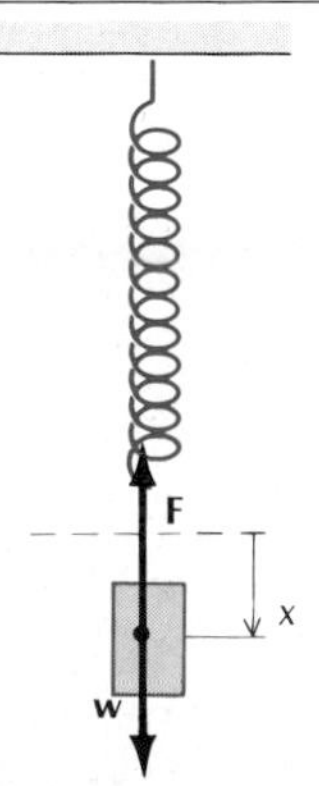

Fig. 14.9 Mass hanging from a spring. At the equilibrium point, the spring force must match the weight.

It is convenient to express this equation in terms of a new variable x' which represents the displacement from the new equilibrium position. When the mass is in equilibrium, the spring must be stretched just enough so that the spring force matches the force of gravity, i.e.,

$$k\,\Delta x = mg \tag{31}$$

Hence at equilibrium the spring is stretched an amount

$$\Delta x = mg/k \tag{32}$$

If x' represents the displacement measured from this equilibrium position, then

$$x' = x - \Delta x = x - mg/k \tag{33}$$

or

$$x = x' + mg/k \tag{34}$$

Inserting this expression for x in Eq. (30), we obtain

$$m\frac{d^2}{dt^2}\left(x' + \frac{mg}{k}\right) = -k\left(x' + \frac{mg}{k}\right) + mg \tag{35}$$

which simplifies to

$$m\frac{d^2}{dt^2}x' = -kx' \tag{36}$$

This is the usual equation of motion for the simple harmonic oscillator [see Eq. (21)]. Hence the solution for the coordinate x' is the usual harmonic function characteristic of simple harmonic motion,

$$x' = A\cos(\omega t + \delta) \tag{37}$$

with

$$\omega = \sqrt{k/m} \tag{38}$$

The solution for the coordinate x is then

$$x = x' + mg/k = A\cos(\omega t + \delta) + mg/k \tag{39}$$

Thus the mass executes simple harmonic motion around the new equilibrium point. Note that gravity does not affect the frequency of oscillation of this motion; it only affects the location of the equilibrium point.

14.3 Kinetic Energy and Potential Energy

The kinetic energy of a mass in simple harmonic motion is

$$K = \tfrac{1}{2}mv^2 = \tfrac{1}{2}m[-A\omega\sin(\omega t + \delta)]^2 \tag{40}$$

$$= \tfrac{1}{2}m\omega^2A^2\sin^2(\omega t + \delta) \tag{41}$$

This can also be written

$$K = \tfrac{1}{2}kA^2\sin^2(\omega t + \delta) \tag{42}$$

The potential energy associated with the force $F = -kx$ is [see Eq. (8.10)]

$$U = \tfrac{1}{2}kx^2 \tag{43}$$

For simple harmonic motion this becomes

$$U = \tfrac{1}{2}k[A\cos(\omega t + \delta)]^2$$

$$= \tfrac{1}{2}kA^2\cos^2(\omega t + \delta) \tag{44}$$

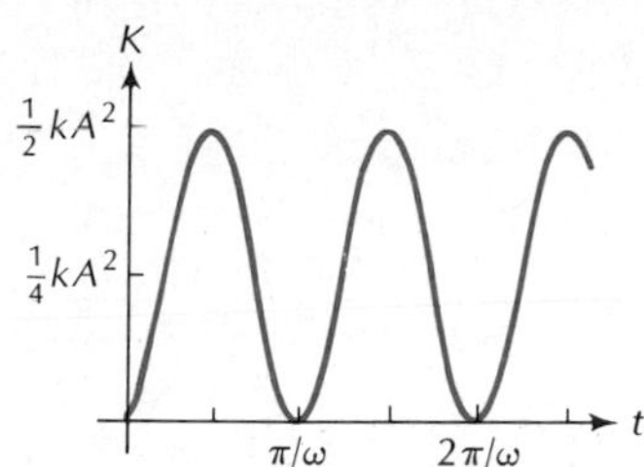

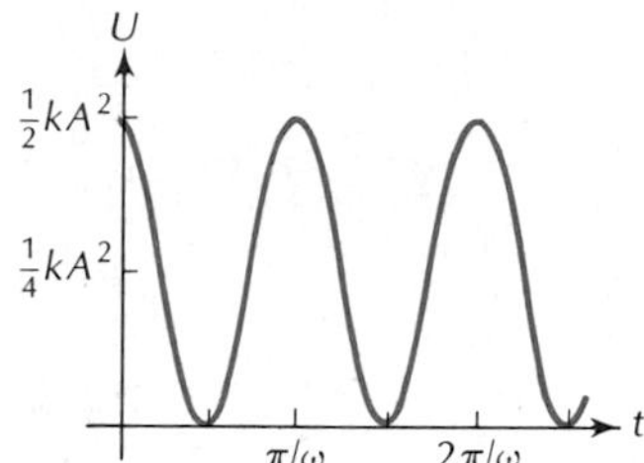

Fig. 14.10 Kinetic energy and potential energy of a simple harmonic oscillator as a function of time (for $\delta = 0$).

The kinetic energy and the potential energy are both functions of time. Each oscillates between a minimum value of 0 and a maximum value of $\frac{1}{2}kA^2$ (Figure 14.10). When the mass passes the equilibrium position ($x = 0$), the kinetic energy is maximum (maximum speed) and the potential energy is zero; when the mass reaches the turning point, the kinetic energy is zero and the potential energy is maximum (maximum displacement).

Since the force $F = -kx$ is conservative, the total energy $E = K + U$ is a constant of the motion. This conservation law for the energy can be verified explicitly by taking the sum of Eqs. (42) and (44),

$$E = K + U$$

$$= \tfrac{1}{2}kA^2\sin^2(\omega t + \delta) + \tfrac{1}{2}kA^2\cos^2(\omega t + \delta)$$

$$= \tfrac{1}{2}kA^2[\sin^2(\omega t + \delta) + \cos^2(\omega t + \delta)]$$

or

$$\boxed{E = \tfrac{1}{2}kA^2} \tag{45}$$

Energy of a simple harmonic oscillator

Note that by means of Eq. (45), the maximum displacement can be expressed in terms of the energy:

$$x_{\text{max}} = A = \sqrt{2E/k} \tag{46}$$

Likewise, the maximum velocity can be expressed in terms of the energy:

$$\boxed{E = \tfrac{1}{2}mv_{\text{max}}^2}$$

or

$$v_{\text{max}} = \sqrt{2E/m} \tag{47}$$

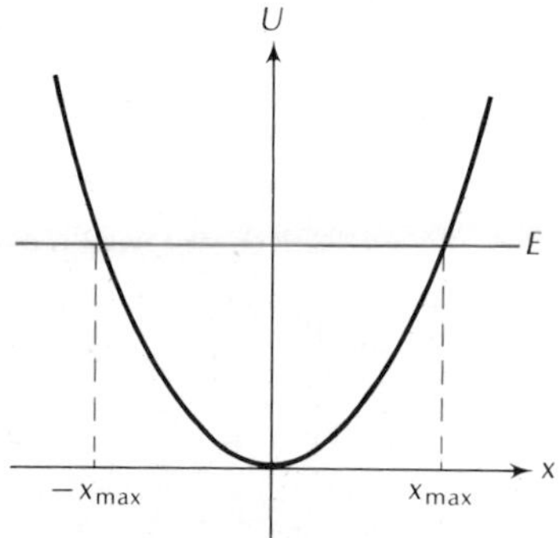

Fig. 14.11 Potential energy of simple harmonic oscillator as a function of x. The horizontal line shows the energy level E.

According to Eq. (46), the amplitude of oscillation is large whenever the energy is large. This increase of the amplitude of oscillation with energy can be readily understood by examining the curve of potential energy. For the simple harmonic oscillator, the potential energy is $U = \frac{1}{2}kx^2$; Figure 14.11 shows a plot of this potential energy as a function of x. The horizontal line shows an energy level E. The turning points of the motion are at $+x_{max}$ and $-x_{max}$. If we were to increase the height of the energy level, we would spread these turning points farther apart, and we would increase the amplitude of oscillation.

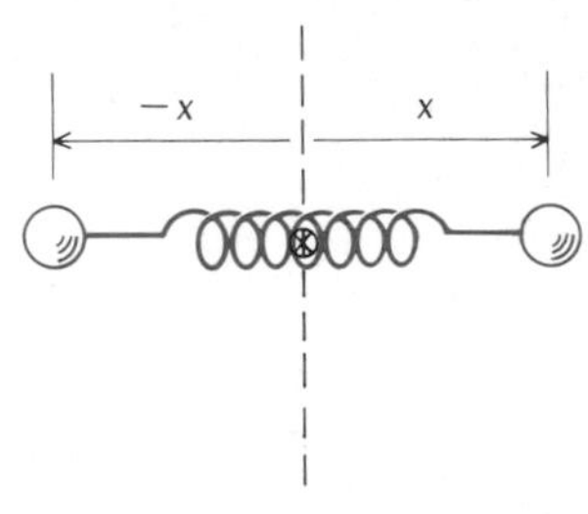

Fig. 14.12 A hydrogen molecule, represented as two masses joined by a spring. The masses move symmetrically relative to the center of mass.

EXAMPLE 3. The hydrogen molecule (H_2) may be regarded as two masses joined by a spring (Figure 14.12). The center of the spring (center of mass) can be regarded as fixed so that the molecule consists of two identical simple harmonic oscillators vibrating in opposite directions. The spring constant is 1.13×10^3 N/m and the mass of each hydrogen atom is 1.67×10^{-27} kg. Suppose that the vibrational energy of the molecule is 1.3×10^{-19} J. Find the corresponding amplitude of oscillation and the maximum velocity.

SOLUTION: Each atom has half the energy of the molecule; thus the energy per atom is

$$E = 6.5 \times 10^{-20}\ \text{J}$$

and the amplitude and the velocity for each atom is

$$x_{max} = \left(\frac{2E}{k}\right)^{1/2} = \left(\frac{2 \times 6.5 \times 10^{-20}\ \text{J}}{1.13 \times 10^3\ \text{N/m}}\right)^{1/2}$$

$$= 1.1 \times 10^{-11}\ \text{m}$$

$$v_{max} = \left(\frac{2E}{m}\right)^{1/2} = \left(\frac{2 \times 6.5 \times 10^{-20}\ \text{J}}{1.67 \times 10^{-27}\ \text{kg}}\right)^{1/2}$$

$$= 8.8 \times 10^3\ \text{m/s}$$

14.4 The Simple Pendulum

A **simple pendulum** consists of a bob suspended by a string or a rod (Figure 14.13). The bob is assumed to behave like a particle of mass m and the string is assumed massless. Gravity acting on the particle provides the restoring force. When in equilibrium, the pendulum hangs vertically, just like a plumb line. When released at some angle with the vertical, the pendulum will swing back and forth in a fixed vertical plane containing the equilibrium position and the initial position of the string (Figure 14.14).[2] The motion is two-dimensional; however, the position of the pendulum can be completely described by a single variable: the angle between the string and the vertical (Figure 14.13). This angle will be reckoned as positive on the right side of the vertical, and negative on the left side.

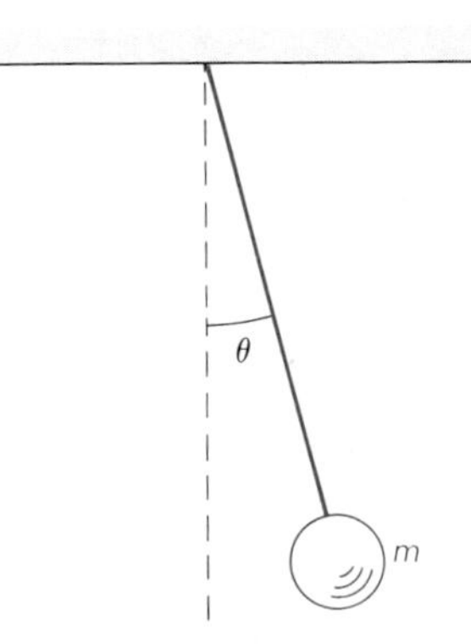

Fig. 14.13 A pendulum. The angle θ is reckoned as positive if the deflection of the pendulum is toward the right, as in this figure.

Since the particle and the string swing as a rigid unit, the motion can be regarded as rotation about a horizontal axis through the point of suspension, and the equation of motion is that of a rigid body [see Eq. (12.11)]:

$$I\alpha = \tau \tag{48}$$

[2] We ignore the Foucault effect, that is, the slow rotation of the plane of oscillation of the pendulum caused by the rotation of the Earth (see Section 5.1).

Here the moment of inertia and the torque are reckoned about the horizontal axis through the point of suspension.

Figure 14.15 shows the external forces acting on the string–particle system. The external forces are the weight **w** acting on the mass m and the suspension force **S** acting on the string at the point of support. The suspension force exerts no torque, since its point of application is on the axis of rotation. The weight exerts a torque

$$\tau = -mgl \sin \theta \tag{49}$$

where l is the length of the pendulum. The minus sign in Eq. (49) indicates that this is a restoring torque which tends to pull the pendulum to its equilibrium position.

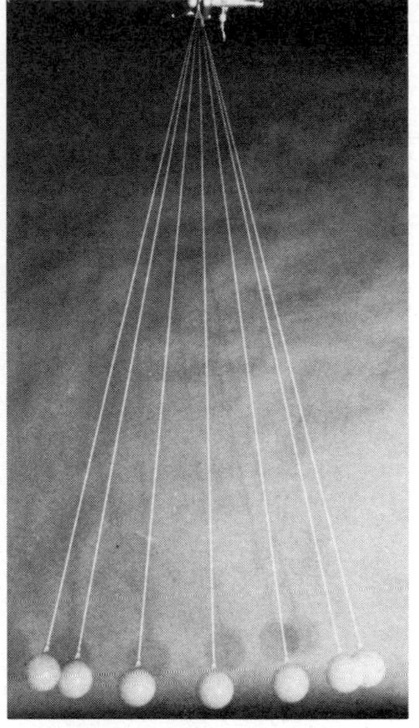

Fig. 14.14 Stroboscopic photograph of swinging pendulum.

The moment of inertia is simply that of the particle of mass m at a distance l from the axis,

$$I = ml^2 \tag{50}$$

Hence the equation of motion is

$$ml^2 \frac{d^2\theta}{dt^2} = -mgl \sin \theta \tag{51}$$

or

$$l \frac{d^2\theta}{dt^2} = -g \sin \theta \tag{52}$$

We will only solve this equation of motion in the special case of small oscillations of the pendulum. If θ is small, then we can make the approximation[3]

$$\sin \theta \cong \theta \tag{53}$$

where the angle is measured in radians. The equation of motion then becomes

$$l \frac{d^2\theta}{dt^2} = -g\theta \tag{54}$$

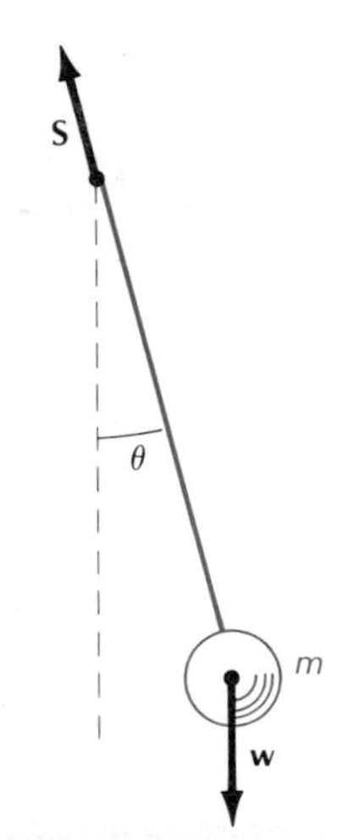

Fig. 14.15 "Free-body" diagram for the spring–particle system.

This differential equation has the same form as Eq. (21); θ replaces x, l replaces m, and g replaces k. Hence the angular motion is simple harmonic,

$$\theta = A \cos(\omega t + \delta) \tag{55}$$

with a frequency

Angular frequency of a simple pendulum

$$\boxed{\omega = \sqrt{g/l}} \tag{56}$$

The period of the pendulum is then

[3] See Appendix 5.

Period of a simple pendulum

$$\boxed{T = 2\pi/\omega = 2\pi\sqrt{l/g}} \tag{57}$$

Note that this period depends only on the length of the pendulum and on the acceleration of gravity; it does not depend on the mass of the pendulum bob or on the amplitude of oscillation.

EXAMPLE 4. What is the length of the "seconds" pendulum at a place where $g = 9.81\ \text{m/s}^2$? The "seconds" pendulum has a period of exactly 2.0 s so that each one-way swing takes exactly 1.0 s.

SOLUTION: From Eq. (57)

$$l = \left(\frac{T}{2\pi}\right)^2 g = \left(\frac{2.0\ \text{s}}{2\pi}\right)^2 \times 9.81\ \text{m/s}^2 = 0.994\ \text{m}$$

The kinetic energy of the pendulum can be calculated from Eq. (11.21)[4]:

$$K = \tfrac{1}{2}I\left(\frac{d\theta}{dt}\right)^2 = \tfrac{1}{2}ml^2[\omega A \sin(\omega t + \delta)]^2$$

$$= \tfrac{1}{2}mglA^2 \sin^2(\omega t + \delta) \tag{58}$$

The potential energy is simply the gravitational potential energy mgh, where h represents the height of the mass above the equilibrium point. From Figure 14.16 we see that

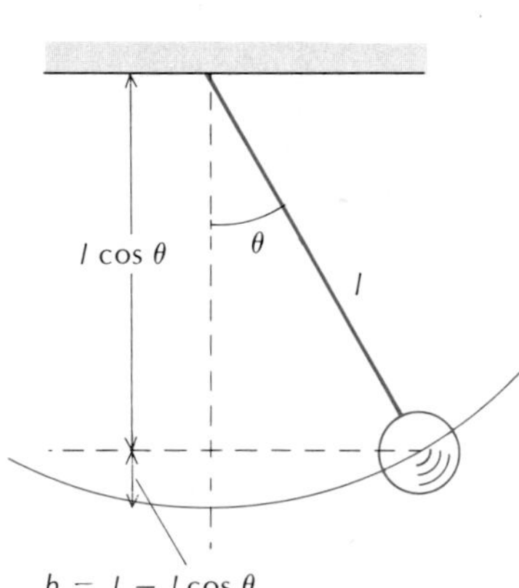

Fig. 14.16 The height h of the mass above its equilibrium point is the difference between l and $l \cos \theta$.

$$h = l - l\cos\theta \tag{59}$$

so that

$$U = mgh = mgl(1 - \cos\theta) \tag{60}$$

If the angle θ is small, then we can take advantage of the approximation[5]

$$\cos\theta \cong 1 - \tfrac{1}{2}\theta^2$$

This leads to the expression

$$U = mgl(1 - 1 + \tfrac{1}{2}\theta^2) = \tfrac{1}{2}mgl\theta^2 \tag{61}$$

$$= \tfrac{1}{2}mglA^2 \cos^2(\omega t + \delta) \tag{62}$$

The total energy $E = K + U$ is of course constant, as the following formulas show:

[4] Keep in mind that the angular velocity $\omega = d\theta/dt$ of Chapter 11 is not the same thing as the angular frequency ω of the present chapter. Only for uniform circular motion are these two ω's the same.

[5] See Appendix 5.

$$\begin{aligned} E &= K + U \\ &= \tfrac{1}{2}mglA^2 \sin^2(\omega t + \delta) + \tfrac{1}{2}mglA^2 \cos^2(\omega t + \delta) \\ &= \tfrac{1}{2}mglA^2 \end{aligned} \tag{63}$$

Fig. 14.17 This electromechanical clock designed by W. H. Shortt served as the U.S. frequency standard from 1924 to 1929. The clock has two pendulums: a master pendulum (in the evacuated cannister at left) and a slave pendulum (at right) that drives the clock mechanism. The master controls the slave by means of periodic electric signals.

The most familiar application of pendulums is the construction of pendulum clocks. Up to about 1950 the most accurate clocks were pendulum clocks of a special design, which were kept inside airtight flasks placed in deep cellars to protect them from disturbances due to the variation of the atmospheric pressure and temperature (Figure 14.17). The best of these high-precision pendulum clocks were accurate to within a few thousandths of a second per day.

Another important application of pendulums is the measurement of the acceleration of gravity g. For this purpose it is only necessary to time the swings of a pendulum of known length; the value of g can then be calculated from Eq. (57). For precise determinations of g, instead of a string and bob one uses a rigid pendulum consisting of a solid bar swinging about one end (we will discuss this kind of pendulum in the next section). This pendulum method is very convenient and very precise, but modern electronic instrumentation permits one to achieve even higher precision by direct timing of the free-fall motion of a small projectile in an evacuated chamber (see Section 2.6).

Pendulums can also be used to test whether the acceleration of gravity is the same for masses made of different materials. If two pendulum bobs made of different materials experience slightly different values of g, then their periods of oscillation would be slightly different, even if the lengths of the strings suspending them were exactly the same; if the two pendulums are released simultaneously, they would gradually get out of step. This method was used by Galileo and by Newton in the early experiments testing the universality of the rate of free fall.

Finally, we emphasize that the approximation of Eq. (53) is only valid for small angles. If the amplitude of oscillation is more than a few degrees, then Eq. (53) fails and the motion of the pendulum deviates from simple harmonic motion. For instance, at large amplitudes the period of the pendulum depends on the amplitude — the larger the amplitude, the larger the period. This lack of isochronism indicates that the pendulum does not behave any more as a simple harmonic oscillator. Table 14.1 shows how the period of a pendulum increases with amplitude.

Table 14.1 INCREASE OF PERIOD OF SIMPLE PENDULUM

Amplitude	*Increase of period*
0°	0.0%
15°	0.4
30°	1.7
45°	4.0
60°	7.3
90°	18.0
135°	52.8
175°	187.7
180°	∞

14.5 Other Oscillating Systems

As we saw in the preceding section, the simple pendulum behaves approximately as a simple harmonic oscillator. The crucial requirement for this approximation is that the amplitude of the oscillations around the equilibrium position be small; if so, then the effective restoring force is directly proportional to the (angular) displacement, and the equation of motion has the same mathematical form as that of the simple harmonic oscillator [see Eqs. (52)–(54)]. Many other physical systems behave approximately as simple harmonic oscillators when oscillating with small amplitude around an equilibrium position. The reason is that near the equilibrium position, the effective force is usually directly proportional to the displacement. This can easily be seen by writing a Taylor-series[6] expansion for the force as a function of displacement. Suppose that the position of the system (a particle, rigid body, atom, or whatever) is described by the variable x. If we place the origin at the equilibrium point, then the Taylor series for the force as a function of x is of the form

$$F(x) = F(0) + \left(\frac{dF}{dx}\bigg|_{x=0}\right)x + \frac{1}{2}\left(\frac{d^2F}{dx^2}\bigg|_{x=0}\right)x^2 + \cdots \tag{64}$$

We are assuming here that the motion is in one dimension. It is also possible to deal with motion in three dimensions — then *each* component of the force has a series expansion and, furthermore, each of the series must include x, y, and z terms; but we will not deal with these complications. Incidentally, if the motion is described by an angular variable, then it is best to write a Taylor series involving this angular variable rather than the x variable [see Eq. (53)].

By hypothesis, the point $x = 0$ is an equilibrium point, i.e., a point at which the force vanishes,

$$F(0) = 0 \tag{65}$$

Furthermore, if the displacement is small, then the second-order and higher-order terms in Eq. (64) can be neglected compared to the first-order term. Consequently, Eq. (64) leads to the approximation

$$F(x) \cong \left(\frac{dF}{dx}\bigg|_{x=0}\right)x \tag{66}$$

which says that the force is directly proportional to the displacement. Let us write this equation as

$$F(x) = -kx \tag{67}$$

with

$$k = -\frac{dF}{dx}\bigg|_{x=0} \tag{68}$$

[6] See Appendix 5 for some examples of Taylor series.

Equation (67) has the familiar form of Hooke's law. We therefore see that this law is a general approximation formula which describes forces near equilibrium points.

If the derivative of the force is negative, then $k > 0$ and Eq. (67) shows that the force is a *restoring* force analogous to the familiar spring force. Under these conditions the equilibrium is **stable:** if the system is displaced slightly from $x = 0$, the force tends to pull it back. If the derivative of the force is positive, then $k < 0$ and then Eq. (67) shows that the force is *repulsive.* This means that the equilibrium is **unstable:** if the system is displaced slightly from $x = 0$, the force tends to push it away even farther.

Stable and unstable equilibria

Figure 14.18 gives some examples of equilibria. The automobile shown in Figure 14.18a rests at the bottom of a valley; this is stable equilibrium: if we push the automobile forward or backward, it tends to return to its original position.[7] The automobile shown in Figure 14.18b rests on the top of a hill; this is unstable equilibrium: if we push the automobile ever so slightly, it will roll down the hill. Finally, the automobile shown in Figure 14.18b rests on a level road; this is neutral equilibrium.

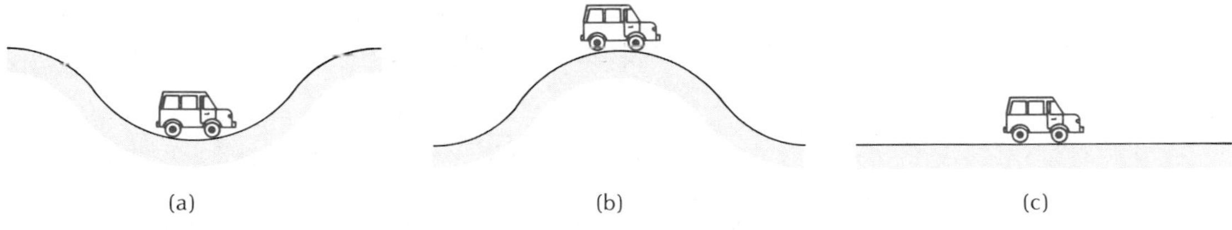

Fig. 14.18 Stationary automobile in (a) stable, (b) unstable, and (c) neutral equilibrium.

We may then conclude that systems in motion near a stable equilibrium point will usually behave as simple harmonic oscillators — such systems will oscillate back and forth through the equilibrium point.[8] In the rest of this section we will look at several examples of such systems.

THE PHYSICAL PENDULUM The **physical pendulum** consists of a solid body which is suspended from a horizontal axis (Figure 14.19). Under the influence of gravity, the body will swing back and forth. The position of the pendulum can be described by the angle θ. The theory of the physical pendulum is much the same as that of the simple pendulum. The equation of motion is that of a rigid body,

$$I\frac{d^2\theta}{dt^2} = \tau \tag{69}$$

where the torque and the moment of inertia are reckoned about the horizontal axis. The only external torque on the system is that exerted by the force of gravity. This torque can be evaluated by pretending that the point of action of the force is the center of mass. If the dis-

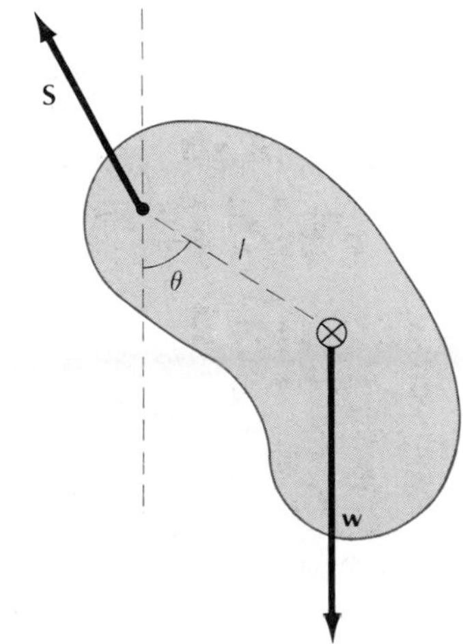

Fig. 14.19 "Free-body" diagram for a physical pendulum.

[7] The motion is two dimensional, instead of one dimensional as in the discussion of Eqs. (64)–(68). However, the motion can be described by a single variable — for instance, by the angle measured along the circular arc of the bottom of the valley.

[8] However, this conclusion fails if $dF/dx = 0$ at the equilibrium point. Then the higher order terms in Eq. (64) must be taken into account.

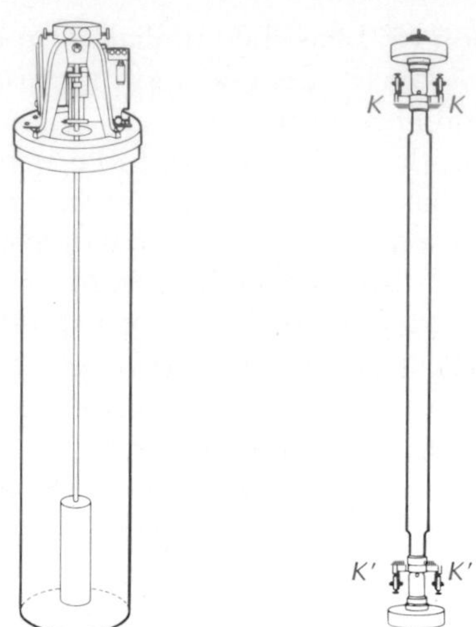

Fig. 14.20 (left) The pendulum of the Shortt clock (see Figure 14.17) in its cannister. As in most high-precision pendulum clocks, the bob is a cylinder attached to a rod.

Fig. 14.21 (right) Reversible pendulum used for the very accurate determination of the acceleration of gravity at Potsdam, Germany, in 1905, which for many years served as the basis of comparison for determinations of the acceleration of gravity at other stations. When in use, the pendulum swings on the knife edges *KK* resting on a support (not shown). The acceleration of gravity is calculated from the measured period of swing. The pendulum can be reversed, so that it swings on the knife edges *K′K′*. This reversal permits elimination of air friction from the calculation.

tance between the axis and the center of mass is l, then (see Figure 14.19)

$$\tau = -Mgl \sin\theta \tag{70}$$

where M is the mass of the body. Hence

$$I\frac{d^2\theta}{dt^2} = -Mgl \sin\theta$$

For small oscillations about the vertical, we can make the same approximation as in the case of the simple pendulum. We then obtain the approximate equation

$$I\frac{d^2\theta}{dt^2} = -Mgl\theta \tag{71}$$

This equation is, again, analogous to Eq. (21). The moment of inertia I plays the role of mass and Mgl plays the role of force constant. Consequently, the motion is simple harmonic with a frequency

Angular frequency of a physical pendulum

$$\boxed{\omega = \sqrt{Mgl/I}} \tag{72}$$

Comparing this to Eq. (56), we see that the simple pendulum is a special case of a physical pendulum with $I = Ml^2$. Figure 14.20 shows the physical pendulum used in an accurate pendulum clock.

As we mentioned in the preceding section, measurements of the acceleration of gravity are commonly done with a physical pendulum. Figure 14.21 shows a pendulum designed for such measurements.

EXAMPLE 5. A physical pendulum consists of a uniform spherical ball of mass M and radius R suspended from a massless string of length L. Taking into account the size of the ball, what is the period of small oscillations of this pendulum?

SOLUTION: The moment of inertia of a sphere about its center of mass is $\frac{2}{5}MR^2$. In Figure 14.22, the sphere is rotating about the point of suspension, at a distance $R + L$ from the center of mass. According to the parallel-axis theorem [see Eq. (11.33)], the moment of inertia of the sphere about this point is

$$I = I_{CM} + M(R + L)^2 = \tfrac{2}{5}MR^2 + M(R + L)^2$$

$$= M[\tfrac{2}{5}R^2 + (R + L)^2]$$

The pull of gravity acts at the center of mass of the sphere. Hence the length l in Eq. (72) is $l = R + L$ and the frequency of oscillation is

$$\omega = \sqrt{\frac{Mgl}{I}} = \sqrt{\frac{Mg(R + L)}{M[\tfrac{2}{5}R^2 + (R + L)^2]}}$$

$$= \sqrt{\frac{g(R + L)}{\tfrac{2}{5}R^2 + (R + L)^2}} \qquad (73)$$

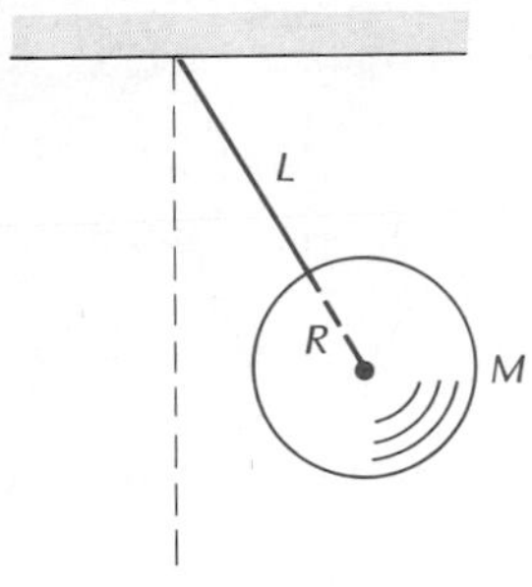

Fig. 14.22 Physical pendulum consisting of a sphere suspended by a massless string.

THE TORSIONAL PENDULUM The **torsional pendulum** consists of a rigid body swinging back and forth about a fixed axis — in this regard the torsional pendulum is quite similar to the physical pendulum; however, the restoring force is supplied by a spring rather than by gravity. Figure 14.23 shows a torsional pendulum with a spiral spring; one end of the spring is attached to the axis; the other end is held fixed by a support. Such a spiral spring has an equilibrium configuration and exerts a restoring torque if disturbed from this equilibrium configuration. Figure 14.24 shows an example of a torsional pendulum with a torsional suspension fiber. The pendulum has the shape of a symmetric dumbbell, and the fiber on which it hangs serves both as axis of rotation and as "spring." If the pendulum is turned through some arc about the vertical axis, the fiber is twisted and exerts a restoring torque as it tries to untwist.

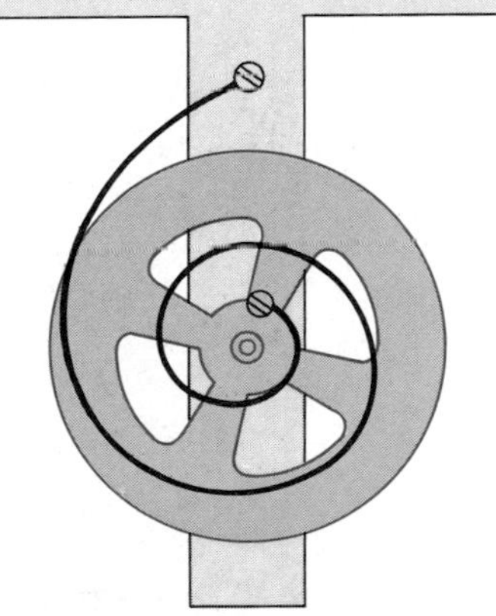

Fig. 14.23 Torsional pendulum consisting of a wheel with a spiral spring.

Under the assumption that the angular displacement of the torsional pendulum from its equilibrium position is small,[9] the restoring torque is proportional to this angular displacement,

$$\tau = -\kappa\theta \qquad (74)$$

The quantity κ is the **torsional constant** of the spring or fiber; its units are N · m/radian. Equation (74) is the angular analog of Eq. (20). The equation of motion of the rigid body on which the torque acts is

$$I\frac{d^2\theta}{dt^2} = -\kappa\theta$$

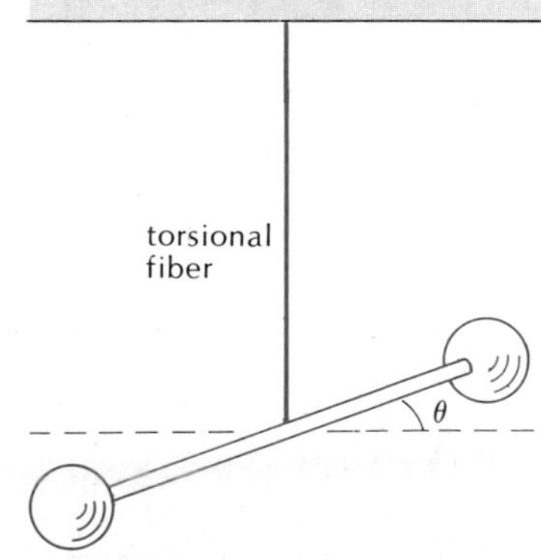

Fig. 14.24 Torsional pendulum consisting of a dumbbell suspended from a fiber.

This is, again, an equation of a single harmonic oscillator. The angular motion is then simple harmonic with a frequency

Angular frequency of a torsional pendulum

$$\boxed{\omega = \sqrt{\kappa/I}} \qquad (75)$$

[9] Just how small is small enough depends on the design of the spring. Some torsional pendulums (with long spiral springs or long, thin suspension fibers) can be twisted through several turns without introducing appreciable deviations into the proportionality of Eq. (74).

Fig. 14.25 Balance wheel of a watch.

Torsional pendulums with spiral springs ("hairsprings") find an important application in mechanical watches and chronometers. The rate of such a watch is controlled by the oscillations of the balance wheel, a torsional pendulum. Figure 14.25 shows the balance wheel of a wristwatch and its hairspring; the screws on the rim of the wheel are used for fine adjustments of the moment of inertia.

The Cavendish balance of Section 13.2 is essentially a torsional pendulum with a suspension fiber. In this balance the suspension fiber is usually made of a very thin strand of quartz that has a very low restoring torque (small value of κ); correspondingly, the period of the rotational oscillations is very long.

EXAMPLE 6. The dumbbell of a Cavendish balance consists of two equal masses of 0.025 kg connected by a nearly massless rod 0.40 m long. When set in motion, the balance rotates back and forth with a period of 3.8 min. Find the value of the torsional constant from these data.

SOLUTION: The angular frequency of the oscillations is

$$\omega = \frac{2\pi}{T} = \frac{2\pi}{3.8 \times 60 \text{ s}} = 2.76 \times 10^{-2} \text{ radian/s}$$

and the moment of inertia of the dumbbell is

$$I = mR^2 + mR^2 = 2mR^2 = 2 \times 0.025 \text{ kg} \times (0.20 \text{ m})^2 = 2.0 \times 10^{-3} \text{ kg} \cdot \text{m}^2$$

Hence, from Eq. (75),

$$\kappa = I\omega^2 = 2.00 \times 10^{-3} \text{ kg} \cdot \text{m}^2 \times (2.76 \times 10^{-2}/\text{s})^2 = 1.52 \times 10^{-6} \text{ N} \cdot \text{m/radian}$$

This shows how the torsional constant can be determined from a simple measurement of the period of oscillation. Since κ is very small, it would be extremely difficult to determine κ by a direct measurement of torque and angular displacement.

THE DIATOMIC MOLECULE The atoms of a **diatomic molecule,** such as the hydrogen atoms in H_2, are bound to each other; but they are not rigidly bound. The interatomic force that holds the atoms in the molecule permits some freedom of movement (this force arises from the attractions and repulsions between the positive and negative electric charges within one atom and those within the other). If the atoms are widely separated, the net force between them is attractive; but if they are very close together, the force is repulsive and opposes their interpenetration. At one particular distance the force is zero — the atoms are then in equilibrium.

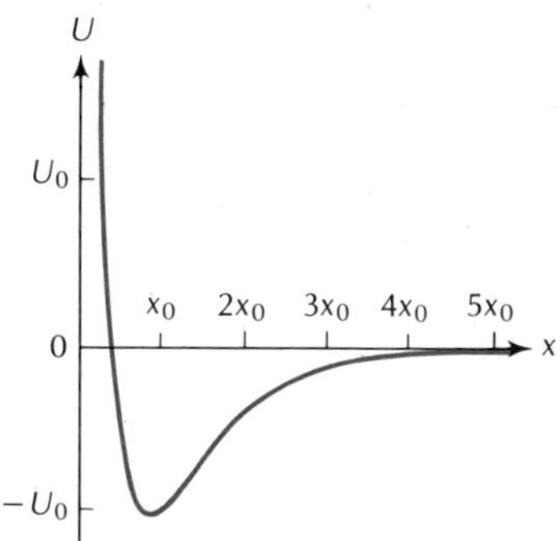

Fig. 14.26 Potential energy of one H atom in an H_2 molecule. Here x is the distance of the atom from the center of mass.

Figure 14.26 shows a plot of the potential energy of a hydrogen atom in an H_2 molecule. The x coordinate represents the position of the atom relative to the center of mass. Since the two atoms have equal masses, the center of mass is always midway between the two atoms (Figure 14.27). If the coordinate of one atom is x, that of the other must be $-x$; the distance between the atoms is therefore $2x$. The atoms overlap completely when $x = 0$; as Figure 14.26 shows, the potential energy is then very large and positive, i.e., it takes a very large amount of work to squeeze the atoms together. The potential energy has a minimum $U = -U_0$ at $x = x_0$, and the potential energy tends to zero as

to the square of the amplitude and as the latter decreases, so does the former. It is instructive to calculate the fractional energy loss in a time equal to one period. According to Eq. (45),

$$E = [\text{constant}] \times A^2 = [\text{constant}] \times e^{-\gamma t} \tag{82}$$

Hence

$$\ln E = -\gamma t + \ln[\text{constant}] \tag{83}$$

The time derivative of this equation gives us

$$\frac{1}{E}\frac{dE}{dt} = -\gamma \tag{84}$$

For a motion such as shown in Figure 14.29, the damping is not excessively strong and the energy loss per period is small; then the energy loss ΔE per period divided by the period is approximately equal to the derivative dE/dt,

$$\frac{\Delta E}{T} \cong \frac{dE}{dt} \tag{85}$$

or

$$\frac{\Delta E}{T} \cong -\gamma E \tag{86}$$

Accordingly, the fractional energy loss per period is

$$\frac{\Delta E}{E} = -\gamma T \tag{87}$$

This relation is usually written in terms of the angular frequency:

$$\boxed{\frac{\Delta E}{E} = -2\pi\frac{\gamma}{\omega}} \tag{88}$$

Energy loss of a damped oscillator

The quantity $2\pi|E/\Delta E|$ is called the "Q" of the damped oscillator:

$$Q = 2\pi|E/\Delta E|$$

or

$$\boxed{Q = \omega/\gamma} \tag{89}$$

Q

In an oscillator of high Q, the oscillations continue for a long time with only a small loss of amplitude. In an oscillator of low Q, the amplitude is quickly damped by friction. Essentially, Q is the **ringing time,** or the time it takes for the oscillations to dampen appreciably, expressed in units of one period. To see this, note that the amplitude given by Eq.

(81) only approaches zero asymptotically — the oscillations never stop completely. However, after a characteristic time

$$t^* = 2/\gamma \tag{90}$$

the amplitude will have decreased to

$$A = A_0 e^{-\gamma t^*/2} = A_0 e^{-\gamma(2/\gamma)/2} = A_0 e^{-1} \tag{91}$$

i.e., the amplitude will have decreased by a factor of e. We can therefore regard the characteristic time given by Eq. (90) as the ringing time of the oscillator. If we express this time in multiples of the period, we obtain

$$\frac{t^*}{T} = \frac{2}{\gamma T} = \frac{2}{\gamma(2\pi/\omega)}$$

According to Eq. (85), this is

$$\frac{t^*}{T} = \frac{2}{2\pi(1/Q)} = \frac{Q}{\pi}$$

Thus, except for a factor of π, Q is the number of periods contained in t^*.

Mechanical oscillators of low friction, such as tuning forks or piano strings, have Q's of a few thousand, i.e., they "ring" for a few thousand periods before their oscillations fade so much that they become hardly noticeable.

If the oscillations of a damped oscillator are to be maintained at a constant level, it is necessary to exert some extra force on the oscillator so that the energy fed into the oscillator by this new force compensates for the energy lost to friction. An extra force is also needed to start the oscillations of any oscillator, damped or not, by supplying the initial energy for the motion. Any such extra force exerted on an oscillator is called a **driving force.** A familiar example is the "pumping" force that must be exerted on a swing (a pendulum) to start it moving and to keep it going at a constant amplitude. This is an example of a *periodic* driving force. If the period of the driving force coincides with the period of the natural oscillations of the oscillator, then even a quite small driving force can gradually build up large amplitudes. Essentially, what happens is that under these conditions the driving force steadily feeds energy into the oscillations and the amplitude of the latter grows until the friction becomes so large that it inhibits further growth. Thus, the ultimate amplitude that is reached depends on friction; in an oscillator of low friction, or high Q, this ultimate amplitude can be extremely large. The buildup of a large amplitude by the action of a driving force in tune with the natural frequency of an oscillator is called **resonance** or **sympathetic oscillation.**

Driving force

Resonance

The phenomenon of resonance plays a crucial role in many pieces of mechanical machinery — if one vibrating part of a machine is driven at resonance by a perturbing force originating from some other part, then the amplitude of oscillation can build up to a violent level and shake the machine apart. Such dangerous resonant effects can occur not only in moving pieces of machinery, but also in structures which are normally regarded as static. In a famous accident that took place in 1850 in Angers (France), the stomping of 487 soldiers marching over a suspension bridge excited a resonant swinging motion of the bridge; the motion quickly rose to a disastrous level and broke the bridge apart, causing the death of 226 of the soldiers.

SUMMARY

Simple harmonic motion: $x = A\cos(\omega t + \delta)$

Period: $T = 2\pi/\omega$

Frequency: $\nu = 1/T = \omega/2\pi$

Equation of motion of simple harmonic oscillator:

$$m\frac{d^2x}{dt^2} = -kx$$

Angular frequency of simple harmonic oscillator: $\omega = \sqrt{k/m}$

Energy of simple harmonic oscillator: $E = \frac{1}{2}kA^2$

$$= \frac{1}{2}mv_{max}^2$$

Angular frequency and period of simple pendulum:

$$\omega = \sqrt{g/l} \qquad T = 2\pi\sqrt{l/g}$$

Angular frequency of physical pendulum: $\omega = \sqrt{Mgl/I}$

Angular frequency of torsional pendulum: $\omega = \sqrt{\kappa/I}$

Fractional energy loss per period of damped oscillator:

$$\frac{\Delta E}{E} = -2\pi\frac{\gamma}{\omega}$$

***Q* of damped oscillator:** $Q = \omega/\gamma$

QUESTIONS

1. Is the motion of the piston of an automobile engine simple harmonic motion? How does it differ from simple harmonic motion?

2. According to Section 14.1, the superposition of two simple harmonic motions along the x and y axes gives uniform circular motion if their amplitudes are equal and their phases differ by $\pi/2$.
   (a) Draw a picture of the orbit if the amplitudes of the x and y motions are not equal.
   (b) Draw a picture of the orbit if the amplitudes are equal but the frequency of the x motion is twice as large as the frequency of the y motion (this orbit is a Lissajous figure).

3. In our calculation of the frequency of the simple harmonic oscillator, we ignored the mass of the spring. Qualitatively, how does the mass of the spring affect the frequency?

4. A simple harmonic oscillator has a frequency of 1.5 Hz. What will happen to the frequency if we cut the spring in half and attach both halves to the mass so that both springs push jointly?

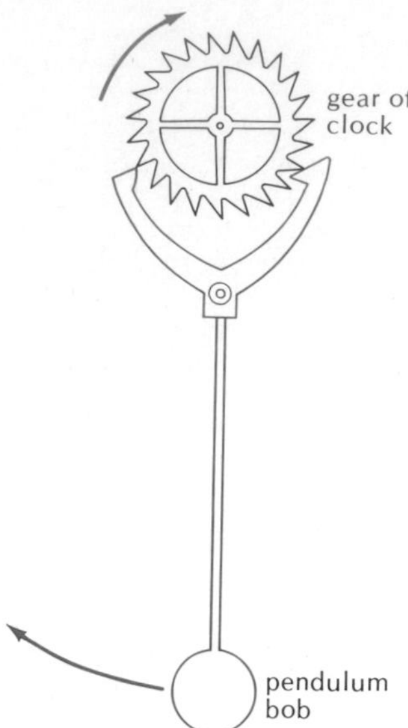

Fig. 14.30 Escapement of a pendulum clock.

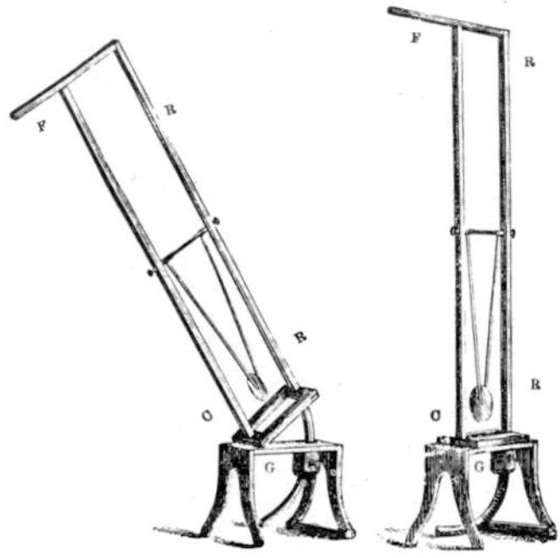

Fig. 14.31 Huygens' tilted pendulum.

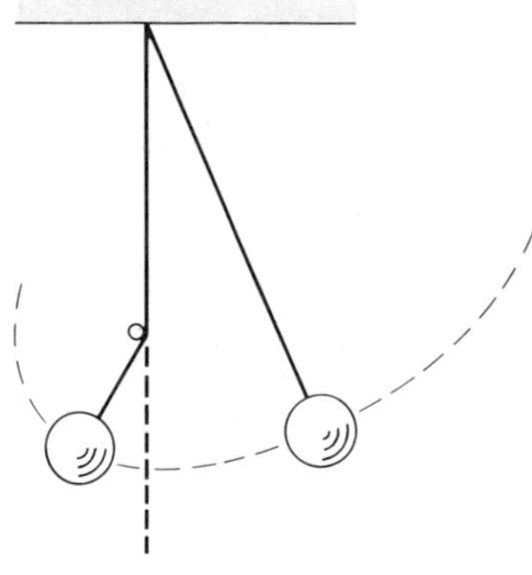

Fig. 14.32 An "interrupted" pendulum.

5. A grandfather clock is regulated by a pendulum. If the clock is running late, how must we readjust the length of the pendulum?

6. Figure 14.30 shows the escapement of a pendulum clock, i.e., the linkage that permits the pendulum to control the rotation of the wheels of the clock. Explain how the wheel turns as the pendulum swings.

7. Would a pendulum clock keep good time on a ship?

8. Galileo claimed that the oscillations of a pendulum are isochronous, even for an amplitude of oscillation as large as 30°. What is your opinion of this claim?

9. Why would you expect a pendulum oscillating with an amplitude of nearly (but not quite) 180° to have a very long period?

10. Can a pendulum oscillate with an amplitude of more than 180°?

11. Figure 14.31 shows a "tilted pendulum" designed by Christian Huygens in the seventeenth century. When the pendulum is tilted, its period is longer than when the pendulum is vertical. Explain.

12. An "interrupted" pendulum consists of a simple pendulum of length l with a nail placed at a distance $\frac{3}{4}l$ below the point of support. If this pendulum is released from one side, it will begin to wrap around the nail as soon as it passes through the vertical position (Figure 14.32). What is the period of this pendulum?

13. Most grandfather clocks have a lenticular pendulum bob which supposedly minimizes friction by "slicing" through the air. However, experience has shown that a cylindrical pendulum bob experiences less air friction. Can you suggest an explanation?

14. Galileo described an experiment to compare the acceleration of gravity of lead and of cork:

> I took two balls, one of lead and one of cork, the former being more than a hundred times as heavy as the latter, and suspended them from two equal thin strings, each four or five bracchia long. Pulling each ball aside from the vertical, I released them at the same instant, and they, falling along the circumferences of the circles having the strings as radii, passed thru the vertical and returned along the same path. This free oscillation, repeated more than a hundred times, showed clearly that the heavy body kept time with the light body so well that neither in a hundred oscillations, nor in a thousand, will the former anticipate the latter by even an instant, so perfectly do they keep step.

Since air friction affects the cork ball much more than the lead ball, do you think Galileo's results are credible?

Newton reported a more careful experiment that avoided the inequality of friction:

> I tried the thing in gold, silver, lead, glass, sand, common salt, wood, water, and wheat. I provided two equal wooden boxes. I filled the one with wood, and suspended an equal weight of gold (as exactly as I could) in the centre of oscillation of the other. The boxes, hung by equal threads of 11 feet, made a couple of pendulums perfectly equal in weight and figure . . . and, placing the one by the other, I observed them to play together forwards and backwards for a long while, with equal vibrations. . . . And by these experiments, in bodies of the same weight, one could have discovered a difference of matter less than the thousandth part of the whole.

Explain how Newton's experiment was better than Galileo's.

15. A simple pendulum hangs below a table, with its string passing through a small hole in the tabletop. Suppose you gradually pull the string while the pen-

dulum is swinging. What happens to the frequency of oscillation? To the (angular) amplitude?

16. Consider a cone (a) lying on its side, (b) standing on its base, and (c) standing on its apex on a flat table (Figure 14.33). For which of these positions is the equilibrium stable, unstable, or neutral?

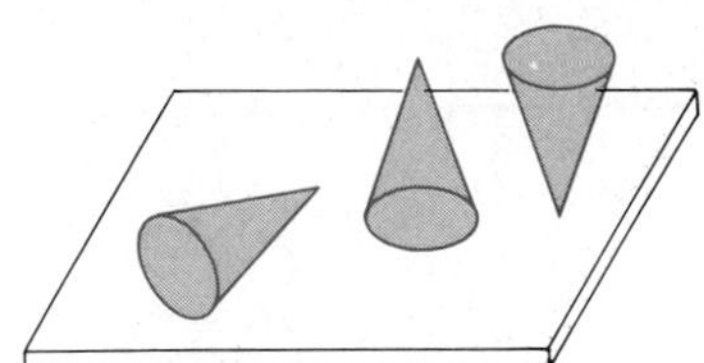

Fig. 14.33 A cone lying on its side, standing on its base, and standing on its apex.

17. Shorter people have a shorter length of stride, but a higher rate of step when walking "naturally." Explain.

18. A girl sits on a swing whose ropes are 1.5 m long. Is this a simple pendulum or a physical pendulum?

19. A simple pendulum consists of a particle of mass m attached to a string of length l. A physical pendulum consists of a body of mass m attached to a string in such a way that the center of mass is at a distance l from the point of support. Which pendulum has the shorter period?

20. Is Eq. (73) valid for a negative value of L? What is the meaning of negative L?

21. All metals expand slightly with increasing temperature. The balance wheel of a watch is made of metal. What will happen to the rate of the watch when the temperature increases? Can you think of some way to compensate for such temperature errors?

22. Suppose that the spring in the front-wheel suspension of an automobile has a natural frequency of oscillation equal to the frequency of rotation of the wheel at, say, 50 mi/h. Why is this bad?

23. When marching soldiers are about to cross a bridge, they break step. Why?

PROBLEMS

Section 14.1

1. A particle moves back and forth along the x axis between the points $x = 0.20$ m and $x = -0.20$ m. The period of the motion is 1.2 s, and it is simple harmonic. At the time $t = 0$, the particle is at $x = 0$ and its velocity is positive.
   (a) What is the frequency of the motion? The angular frequency?
   (b) What is the amplitude of the motion?
   (c) What is the phase constant?
   (d) At what time will the particle reach the point $x = 0.20$ m? At what time will it reach the point $x = -0.10$ m?
   (e) What is the speed of the particle when it is at $x = 0$? What is the speed of the particle when it reaches the point $x = -0.10$ m?

2. The motion of the piston in an automobile engine is approximately simple harmonic. Suppose that the piston travels back and forth over a distance of 8.50 cm and that the piston has a mass of 1.2 kg. What are its maximum acceleration and maximum speed if the engine is turning over at its highest safe rate of 6000 rev/min? What is the maximum force on the piston?

3. A given point on a guitar string (say, the midpoint of the string) executes simple harmonic motion with a frequency of 440 Hz and an amplitude of 1.2 mm. What is the maximum speed of this motion? The maximum acceleration?

4. A particle moves as follows as a function of time:

$$x = 3.0\cos(2.0t + \pi/3)$$

where distance is measured in meters and time in seconds.

(a) What is the amplitude of this simple harmonic motion? The frequency? The angular frequency? The period?
(b) At what time does the particle reach the equilibrium point? The turning point?

5. Experience shows that from one-third to one-half of the passengers in an airliner can be expected to suffer motion sickness if the airliner bounces up and down with a peak acceleration of 0.4 gee and a frequency of about 0.3 Hz. Assume that this up-and-down motion is simple harmonic. What is the amplitude of the motion?

6. The frequency of oscillation of a mass attached to a spring is 3.0 Hz. At time $t = 0$, the mass has an initial displacement of 0.20 m and an initial velocity of 4.0 m/s.
(a) What is the position of the mass as a function of time?
(b) When will the mass first reach a turning point? What will be its acceleration at that time?

Section 14.2

7. The body-mass measurement device used aboard Skylab consisted of a chair supported by a spring. The device was calibrated before the space flight by placing a standard mass of 66.91 kg in the chair; with this mass the period of oscillation of the chair was 2.088 s. During the space flight astronaut Lousma sat in the chair; the period of oscillation was then 2.299 s. What was the mass of the astronaut? Ignore the mass of the chair.

8. The body of an automobile of mass 1100 kg is supported by four vertical springs attached to the axles of the wheels. In order to test the suspension, a man pushes down on the body of the automobile and then suddenly releases it. The body rocks up and down with a period of 0.75 s. What is the spring constant of each of the springs? Assume that all the springs are identical and that the compressional force on each spring is the same; also assume that the shock absorbers of the automobile are completely worn out so that they do not affect the oscillation frequency.

9. Deuterium is an isotope of hydrogen. The mass of the deuterium atom is 1.998 times larger than the mass of the hydrogen atom. Given that the frequency of vibration of the H_2 molecule is 1.31×10^{14} Hz (see Example 7), calculate the frequency of vibration of the D_2 molecule. Assume the "spring" connecting the atoms is the same in H_2 and D_2.

10. Calculate the frequency of vibration of the HD molecule consisting of one atom of hydrogen and one of deuterium. See Problem 9 for necessary data.

11. A mass $m = 2.5$ kg hangs from the ceiling by a spring with $k = 90$ N/m. Initially, the spring is in its unstretched configuration and the mass is held at rest by your hand. If, at time $t = 0$, you release the mass, what will be its position as a function of time?

12. The wheel of a sports car is suspended below the body of the car by a vertical spring with a spring constant 1.1×10^4 N/m. The mass of the wheel is 14 kg and the diameter of the wheel is 61 cm.
(a) What is the frequency of up-and-down oscillations of the wheel? Regard the wheel as a mass on one end of a spring and regard the body of the car as a fixed support for the other end of the spring.
(b) Suppose that the wheel is slightly out of round, having a bump on one side. As the wheel rolls on the street it receives a periodic push each time the bump comes in contact with the street. At what speed of the translational motion of the car will the frequency of this push coincide with the natural frequency of the up-and-down oscillations of the wheel? What will happen to the car at this speed? (Note: This problem is not quite realistic because the elasticity of the tire also contributes a restoring force to the up-and-down motion of the wheel.)

13. A mass m slides on a frictionless plane inclined at an angle θ with the horizontal. The mass is attached to a spring, parallel to the plane (Figure 14.34); the spring constant is k. How much is the spring stretched at equilibrium? What is the frequency of the oscillations of this mass up and down the plane?

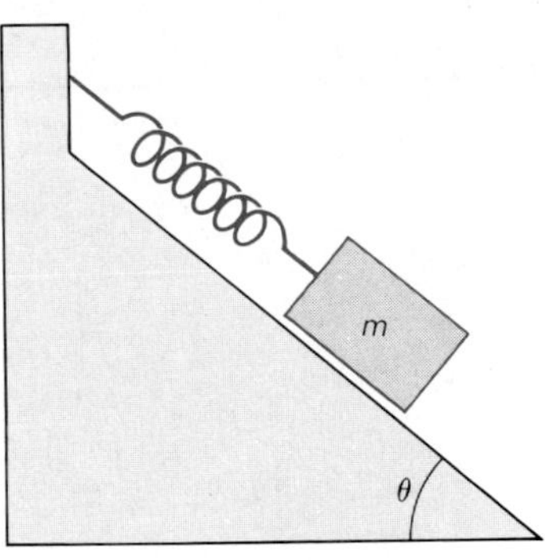

Fig. 14.34 Mass sliding on an inclined frictionless plane.

14. Two identical masses slide with one-dimensional motion on a frictionless plane under the influence of three identical springs attached as shown in Figure 14.35. The magnitude of each mass is m and the spring constant of each spring is k.

(a) Suppose that at time $t = 0$, the masses are at their equilibrium positions and their instantaneous velocities are $v_1 = -v_2$. Find the position of each mass as a function of time. What is the frequency of the motion?

(b) Suppose that at time $t = 0$, the masses are at their equilibrium positions and their instantaneous velocities are $v_1 = v_2$. Find the position of each mass as a function of time. What is the frequency of the motion?

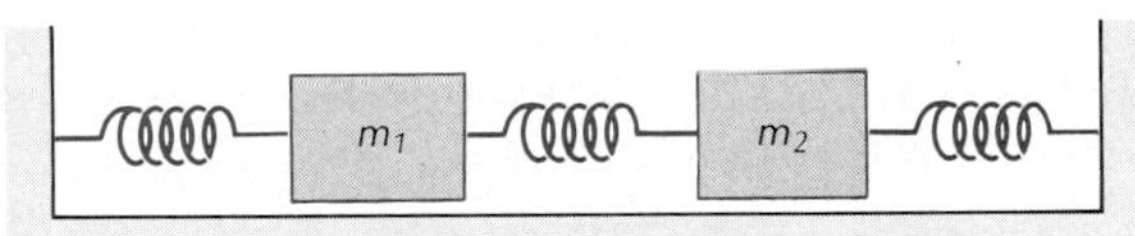

Fig. 14.35 Two masses sliding on a frictionless plane.

*15. A cart consists of a body and four wheels on frictionless axles. The body has a mass m. The wheels are uniform disks of mass M and radius R. The cart rolls, without slipping, back and forth on a horizontal plane under the influence of a spring attached to one end of the cart (Figure 14.36). The spring constant is k. Taking into account the moment of inertia of the wheels, find a formula for the frequency of the back-and-forth motion of the cart.

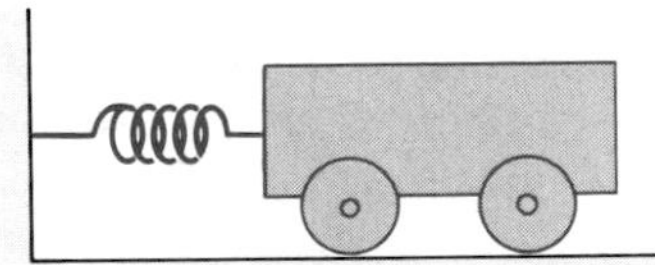

Fig. 14.36

Section 14.3

16. Suppose that a particle of mass 0.24 kg acted upon by a spring undergoes simple harmonic motion with the parameters given in Problem 1.

(a) What is the total energy of this motion?

(b) At what time is the kinetic energy zero? At what time is the potential energy zero?

(c) At what time is the kinetic energy equal to the potential energy?

17. A mass of 8.0 kg is attached to a spring and oscillates with an amplitude of 0.25 m and a frequency of 0.60 Hz. What is the energy of the motion?

18. The separation between the equilibrium positions of the two atoms of a hydrogen molecule is 1.0 Å. Using the data given in Example 7, calculate the value of the vibrational energy that corresponds to an amplitude of vibration of 0.5 Å for each atom. Is it valid to treat the motion as a small oscillation if the energy has this value?

19. Although it is usually a good approximation to neglect the mass of a spring, sometimes this mass must be taken into account. Suppose that a uniform spring has a relaxed length l and a mass m'; a mass m is attached to the end of the spring. The mass m' is uniformly distributed along the spring. Suppose that if the moving end of the spring has a speed v, all other points of the spring have speeds directly proportional to their distance from the fixed end, e.g., a point midway between the moving and the fixed end has a speed $\frac{1}{2}v$.

(a) Show that the kinetic energy in the spring is $\frac{1}{6}m'v^2$ and that the kinetic energy of the mass m and the spring is

$$K = \tfrac{1}{2}mv^2 + \tfrac{1}{6}m'v^2 = \tfrac{1}{2}(m + \tfrac{1}{3}m')v^2$$

Consequently, the effective mass of the combination is $m + \frac{1}{3}m'$.

(b) Show that the frequency of oscillation is $\omega = \sqrt{k/(m + \frac{1}{3}m')}$.

(c) Suppose that the spring described in Example 1 has a mass of 5 kg. The frequency of oscillation of the 400-kg mass attached to this spring will then be somewhat smaller than calculated in Example 1. How much smaller? Express your answer as a percentage of the answer of Example 1.

20. Two masses m_1 and m_2 are joined by a spring of spring constant k. Show that the frequency of vibration of these masses along the line connecting them is

$$\omega = \sqrt{\frac{k(m_1 + m_2)}{m_1 m_2}}$$

(Hint: The center of mass remains at rest.)

Section 14.4

21. A "seconds" pendulum is a pendulum that has a period of exactly 2.0 s: — each one-way swing of the pendulum therefore takes exactly 1.0 s. What is the length of the seconds pendulum in Paris ($g = 9.809$ m/s^2), Buenos Aires ($g = 9.797$ m/s^2), and Washington, D.C. ($g = 9.801$ m/s^2)?

22. A grandfather clock controlled by a pendulum of length 0.9932 m keeps good time in New York ($g = 9.803$ m/s^2).

(a) If we take this clock to Austin ($g = 9.793$ m/s^2), how many minutes per day will it fall behind?

(b) In order to adjust the clock, by how many millimeters must we shorten the pendulum?

23. The pendulum of a grandfather clock has a length of 0.994 m. If the clock runs late by 1 minute per day, how much must you shorten the pendulum to make it run on time?

24. A mass suspended from a parachute descending at constant velocity can be regarded as a pendulum. What is the frequency of pendulum oscillations of a human body suspended 7 m below a parachute?

25. The pendulum of a pendulum clock consists of a rod of length 0.99 m with a bob of mass 0.40 kg. The pendulum bob swings back and forth along an arc of length 20 cm.

(a) What are the maximum velocity and the maximum acceleration of the pendulum bob along the arc?

(b) What is the force that the pendulum exerts on its support when it is at the midpoint of its swing? At the endpoint? Neglect the mass of the rod in your calculations.

26. The pendulum of a regulator clock consists of a mass of 120 g at the end of a (massless) wooden stick of length 44 cm.

(a) What is the total energy (kinetic plus potential) of this pendulum when oscillating with an amplitude of 4°?

(b) What is the speed of the mass when at its lowest point?

27. Galileo claimed to have verified experimentally that a pendulum oscillating with an amplitude as large as 30° has the same period as a pendulum of identical length oscillating with a much smaller amplitude. Suppose that you let two pendulums of length 1.5 m oscillate for 10 min. Initially, the pendulums oscillate in step. If the amplitude of one of them is 30° and the amplitude of the other is 5°, by what fraction of a (one-way) swing will the pendulums be out of step at the end of the 10-min interval? What can you conclude about Galileo's claim?

Section 14.5

28. In windup clocks a strong torsional spring is used to store mechanical energy. Suppose that each week a clock requires four full turns of the winding

key to keep running. The initial turn requires a torque of 0.30 N · m and the final turn a torque of 0.45 N · m.

(a) What amount of mechanical energy do you store in the spring when winding the clock?

(b) What is the consumption of mechanical power by the clock?

(c) What is the torsional spring constant?

29. The balance wheel of a watch, such as that shown in Figure 14.25, can be approximately described as a hoop of diameter 1.0 cm and mass 0.60 g. Each of the screws, whose masses are included in the mass given for the hoop, has a mass of 0.020 g. Suppose the watch runs fast by 1.2 minutes per day. To adjust the watch so that it keeps perfect time, by how much must we increase the moment of inertia of the balance wheel? If we want to achieve this increment by moving one of the screws outward in a radial direction, how far must we move the screw?

30. Show that the potential energy of a torsional pendulum is $U = \frac{1}{2}\kappa\theta^2$. [Hint: Begin with Eq. (12.16) for the work done by the torque.]

31. To test that the acceleration of gravity is the same for a piece of iron and a piece of brass, an experimenter takes a pendulum of length 1.800 m with an iron bob and another pendulum of the same length with a brass bob and starts them swinging in unison. After swinging for 12 min, the two pendulums are no more than one-quarter of a (one-way) swing out of step. What is the largest difference between the values of g for iron and for brass consistent with these data? Express your answer as a fractional difference.

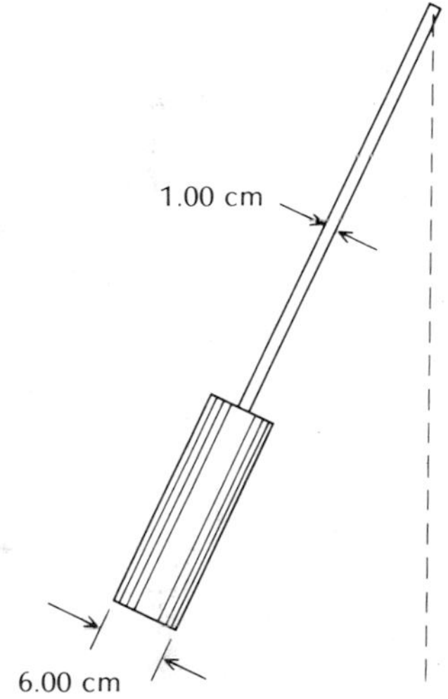

Fig. 14.37 A physical pendulum.

32. A pendulum consists of a brass rod with a brass cylinder attached to the end (Figure 14.37). The diameter of the rod is 1.00 cm and its length is 90.00 cm; the diameter of the cylinder is 6.00 cm and its length is 20.00 cm. What is the period of this pendulum?

33. At the National Bureau of Standards in Washington, D.C., the value of the acceleration of gravity is 9.80095 m/s². Suppose that at this location a very precise physical pendulum, designed for measurements of the acceleration of gravity, has a period of 2.10356 s. If we take this pendulum to a new location at the U.S. Coast and Geodetic Survey, also in Washington, D.C., it has a period of 2.10354 s. What is the value of the acceleration of gravity at this new location? What is the percentage change of the acceleration between the two locations?

34. Calculate the natural period of the swinging motion of a human leg. Treat the leg as a rigid physical pendulum with axis at the hip joint. Pretend that the mass distribution of the leg can be approximated as two rods joined rigidly end to end. The upper rod (thigh) has a weight of 15 lb and a length of 17 in.; the lower rod (shin plus foot) has a weight of 9 lb and a length of 18 in. Using a watch, measure the period of the natural swinging motion of *your* leg when you walk at a normal rate. Compare with the calculated number.

35. A hole has been drilled through a meter stick at the 30-cm mark and the meter stick has been hung on a wall by a nail passing through this hole. If the meter stick is given a push so that it swings about the nail, what is the period of the motion?

36. A physical pendulum has the shape of a disk of radius R. The pendulum swings about an axis perpendicular to the plane of the disk and at distance l from the center of the disk.

(a) Show that the frequency of the oscillations of this pendulum is

$$\omega = \sqrt{\frac{gl}{\frac{1}{2}R^2 + l^2}}$$

(b) For what value of l is this frequency at a maximum?

37. Suppose that the physical pendulum in Figure 14.19 is a thin rigid rod of

mass m suspended at one end. Suppose that this rod has an initial position $\theta = 20°$ and an initial angular velocity $\omega = 0$. Calculate the force **F** that the support exerts on the pendulum at this initial instant (give horizontal and vertical components).

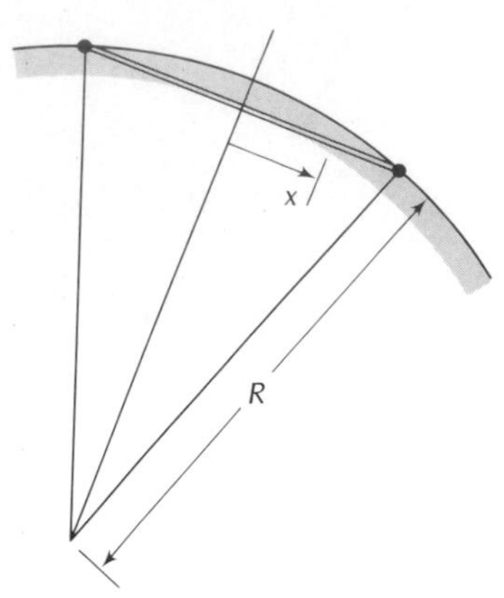

Fig. 14.38

38. The door of a house is made of wood of uniform thickness. The door weighs 60 lb and measures 6 ft 3 in. × 3 ft. The door is held shut by a torsional spring with $\kappa = 22$ lb · ft/radian arranged so that it exerts a torque of 40 lb · ft when the door is fully open (at right angles to the wall of the house). What angular speed does the door attain if it slams shut from the fully open position? What linear speed does the edge of the door attain?

39. According to a proposal described in Example 1.4, very fast trains could travel from one city to another in straight subterranean tunnels (see Figure 14.38). For the following calculations, assume that the density of the Earth is constant so that, according to Eq. (13.45), the acceleration of gravity as a function of the radial distance r from the center of the Earth is $g = (GM/R^3)r$.

(a) Show that the component of the acceleration of gravity along the track of the train is

$$g_x = -(GM/R^3)x$$

where x is measured from the midpoint of the track (see Figure 14.38).

(b) Ignoring friction, show that the motion of the train along the track is simple harmonic motion with a period independent of the length of the track,

$$T = 2\pi\sqrt{\frac{R^3}{GM}}$$

(c) Starting from rest, how long would a train take to roll freely along its track from San Francisco to Washington, D.C.? What would be its maximum speed (at the midpoint)? Use the numbers calculated in Example 1.4 for the length and depth of the track.

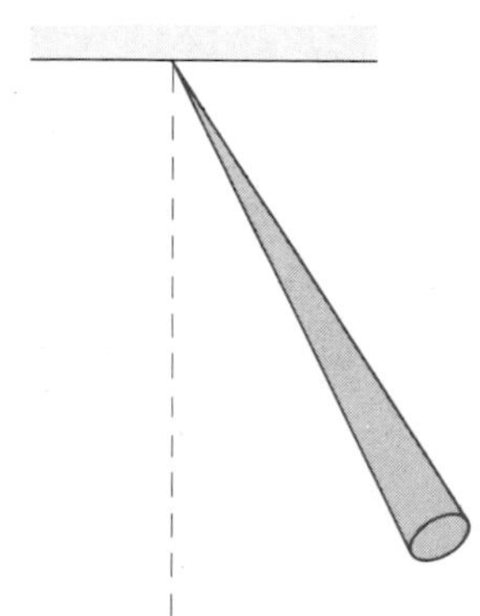

Fig. 14.39

*40. A physical pendulum consists of a long, thin cone suspended at its apex (Figure 14.39). The height of the cone is l. What is the period of this pendulum?

41. The net gravitational force on a particle placed midway between two equal spherical bodies is zero. However, if the particle is placed some distance away from this equilibrium point, then the gravitational force is not zero.

(a) Show that if the particle is at a distance x from the equilibrium point in a direction toward one of the bodies, then the force is approximately $4GMmx/r^3$, where M is the mass of each spherical body, m is the mass of the particle, and $2r$ is the distance between the spherical bodies. Assume $x \ll r$.

(b) Show that if the particle is at a distance x from the equilibrium point in a direction perpendicular to the line connecting the bodies, then the force is approximately $-2GMmx/r^3$, where the negative sign indicates that the direction of the force is toward the equilibrium point.

(c) What is the frequency of small oscillations of the mass m about the equilibrium point when moving in a direction perpendicular to the line connecting the bodies? Assume that the bodies remain stationary.

*42. The motion of a simple pendulum is given by

$$\theta = A\cos(\sqrt{g/l}\,t)$$

(a) Find the tension in the string of this pendulum; assume that $\theta \ll 1$.

(b) The tension is a function of time. At what time is the tension maximum? What is the value of this maximum tension?

*43. The total energy of a body of mass m orbiting the Sun is

$$E = \left[\tfrac{1}{2}m\left(\frac{dr}{dt}\right)^2 + \tfrac{1}{2}mr^2\left(\frac{d\theta}{dt}\right)^2\right] - \frac{GmM_S}{r}$$

where the quantity in brackets represents the kinetic energy written in terms of the radial and the tangential component of the velocity.

(a) Show that in terms of the (constant) angular moment L, the energy can be written

$$E = \tfrac{1}{2}m\left(\frac{dr}{dt}\right)^2 + \frac{1}{2}\frac{L^2}{mr^2} - \frac{GmM_S}{r}$$

(b) For a circular orbit the radius is constant: $r = r_0$. For a nearly circular orbit, the radius differs from r_0 only by a small quantity: $r = r_0 + x$. Show that in terms of the small quantity x, the energy for a nearly circular orbit is approximately

$$E = \tfrac{1}{2}m\left(\frac{dx}{dt}\right)^2 + \frac{3}{2}\frac{L^2}{mr_0^4}x^2 - \frac{GmM_S}{r_0^3}x^2 - \frac{GmM_S}{2r_0}$$

The last term in this expression is a constant (independent of x).

(c) Show that, except for an additive constant, this expression for the energy coincides with the equation for the energy of a harmonic oscillator, provided that we identify

$$k = 3L^2/mr_0^4 - 2GmM_S/r_0^3$$

Show that this equals $k = GmM_S/r_0^3$.

(d) What is the frequency of small radial oscillations about the circular orbit? Show that this frequency equals the frequency of the circular orbit. Make a sketch of the shape of the orbit that results from the combination of revolution around the circle and small oscillations along the radius.

Section 14.6

44. A pendulum of length 1.50 m is set swinging with an initial amplitude of 10°. After 12 min, friction has reduced the amplitude to 4°. What is the value of γ for this pendulum?

45. The pendulum of a grandfather clock has a length of 0.994 m and a mass of 1.2 kg.

(a) If the pendulum is set swinging, the friction of the air reduces its amplitude of oscillation by a factor of 2 in 13.0 min. What is the value of γ for this pendulum?

(b) If we want to keep this pendulum swinging at a constant amplitude of 8°, we must supply mechanical energy to it at a rate sufficient to make up for the frictional loss. What is the required mechanical power?

CHAPTER 15

Waves

In Chapters 11 and 12 we studied the rotational motion of a rigid body. This is a collective motion of all the particles in the body. Although the number of particles may be very large, the equations for the rotational motion are fairly simple because the particles are rigidly connected so that they do not move relative to one another. We will now study the **wave motion** of a deformable body — water, air, strings, elastic solids, etc. This is a collective motion of the particles in the body, but here the particles do move relative to one another and they exert time-dependent forces on one another. Nevertheless, the equations of motion remain fairly simple because each particle only interacts with its nearest neighbors. Wave motion is a disturbance propagating from one particle to the next in a regular manner.

Wave motion

For the sake of simplicity, in this chapter we will concentrate on the mathematical analysis of wave motion in a string. However, most of our results also apply to wave motion in other elastic bodies. In the next chapter we will examine some details of wave motion in water (sea waves and tidal waves), in air (sound waves), and in the crust of the Earth (seismic waves).

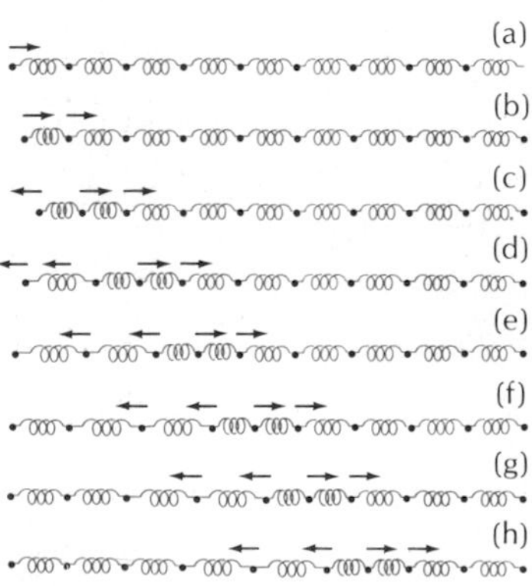

Fig. 15.1 Particles joined by springs. A transverse disturbance propagates from left to right. The particles move up and down.

15.1 Wave Pulses

Consider a tightly stretched elastic string, such as a long rubber cord. If we snap one end of the string back and forth with a flick of the wrist, a disturbance travels along the string. Figure 15.1 shows in detail how such a traveling disturbance comes about. The string may be regarded as a row of particles joined by small, massless springs. When we jerk the first particle to one side, it will pull the second particle to the same side, and this will pull the third, etc. If we then jerk the first particle back to its original position, it will pull the second particle back, and this will likewise pull the third, etc. As the motion is transmitted from one particle to the next particle, the disturbance propagates along the row of particles. Such a disturbance is called a **wave pulse.**

Alternatively, we can generate a disturbance by suddenly pushing the first particle toward the second. Figure 15.2 shows such a compressional disturbance propagating along the row of particles. The wave of Figure 15.1 is called a **transverse wave**, and that of Figure 15.2 a **longitudinal wave.**

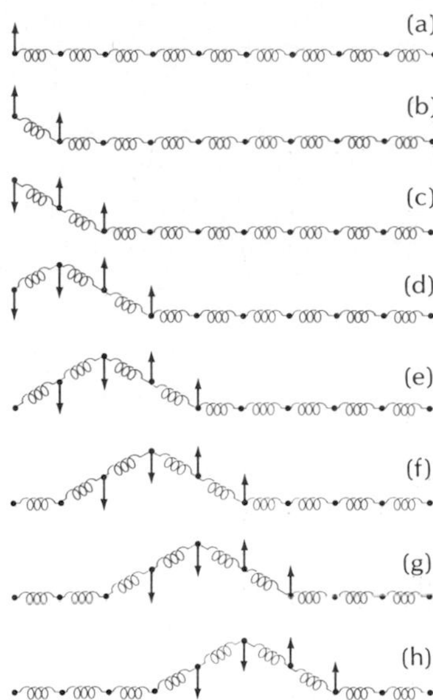

Fig. 15.2 Particles joined by springs. A longitudinal disturbance propagates from left to right. The particles move back and forth.

Note that although the wave pulse travels along the string, the particles do not — they merely move back and forth around their equilibrium positions. Also note that in the region of the wave pulse, the string has kinetic energy (due to the back-and-forth motion of particles) and potential energy (due to the stretching of the springs between the particles). Hence a wave pulse traveling along the string carries energy with it — the wave transports energy from one end of the string to the other.

Wave motion in water, air, or any other elastic body displays the same general features. The wave is a propagating deformation or compression of the body communicated by pushes and pulls from one particle to the next. The wave transports energy without transporting particles.

15.2 Traveling Waves

Consider a transverse wave pulse traveling along a string with some speed v. Let us make the assumption that the shape of the wave pulse remains constant as it travels (we will justify this assumption in the next section). To describe this propagation mathematically, we must begin with the shape of the wave pulse at the initial time, $t = 0$. If the string is stretched in the x direction and its deformation is in the y direction, then its initial shape can be represented by some function of x:

$$y = f(x) \qquad \text{at initial time } t = 0 \tag{1}$$

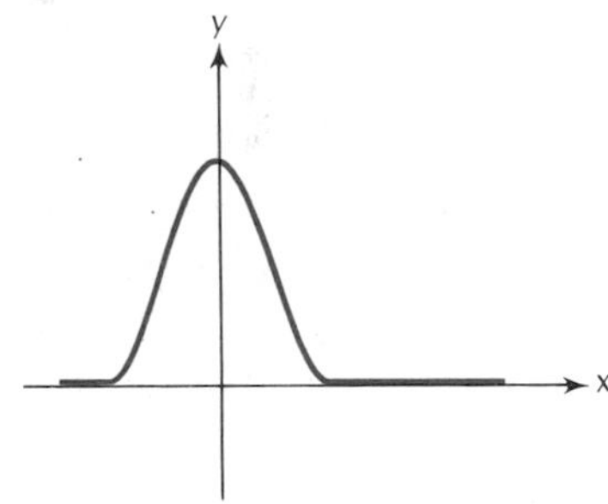

Fig. 15.3 Wave pulse at $t = 0$.

For example, Figure 15.3 shows a function with a peak at $x = 0$. At some later time t, the wave pulse will have shifted a distance to the right. The peak $f(0)$ will have moved from $x = 0$ to $x = vt$. The function $f(x)$ must then be replaced by a new function in which the value of the argument is shifted by a distance vt. Hence the new function describing the wave is of the form

General wave function

$$y = f(x - vt) \qquad \text{at later time } t \tag{2}$$

Note that according to Eq. (1) the peak $f(0)$ is at $x = 0$, whereas according to Eq. (2) the peak $f(0)$ is at $x = vt$; this is exactly as it should be (Figure 15.4). Equation (2) is a general expression for a wave traveling in the positive x direction.

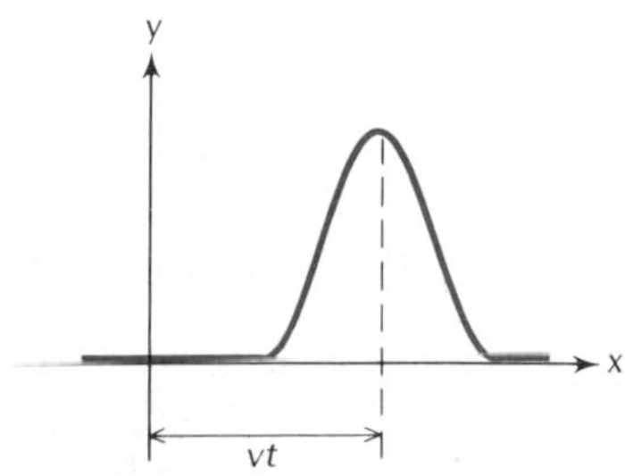

Fig. 15.4 Wave pulse at $t > 0$. The peak has traveled a distance vt.

Similar reasoning shows that a wave traveling in the negative x direction is described by a function of the form

$$y = f(x + vt) \qquad \text{at later time } t \tag{3}$$

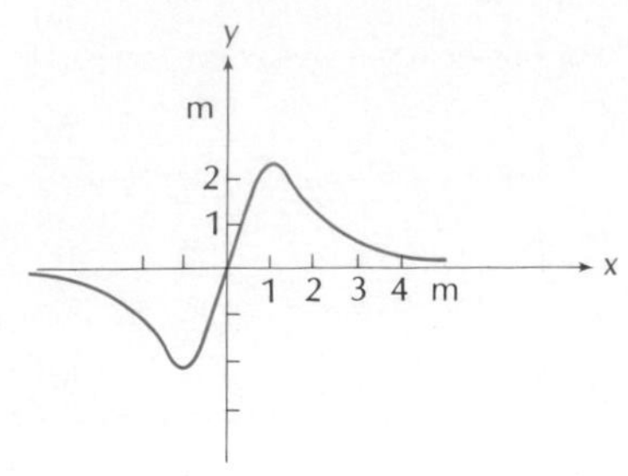

Fig. 15.5 Plot of wave pulse at $t = 0$.

EXAMPLE 1. Suppose that at an initial time $t = 0$, the shape of a wave pulse on a string is represented by the function

$$y = f(x) = \frac{3x}{1 + x^4} \qquad \text{at initial time } t = 0$$

where y and x are in meters. Suppose that this wave pulse has a velocity $v = 2$ m/s toward the positive x direction. What function represents the wave pulse at time t? Plot this function when $t = 1$ s.

SOLUTION: Figure 15.5 shows the plot of the function specified above. This function has a crest at $x = 1$ and a trough at $x = -1$. The plot gives the shape of the wave pulse at $t = 0$. To find the shape at $t > 0$, we must shift the wave pulse a distance vt toward the right. According to our general argument, we accomplish this by replacing x in $f(x)$ by $x - vt$:

$$y = f(x - vt) = \frac{3(x - vt)}{1 + (x - vt)^4} \qquad \text{at later time } t$$

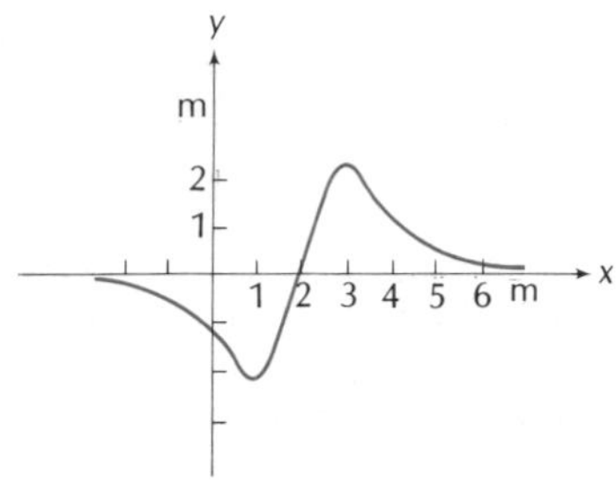

Fig. 15.6 Plot of wave pulse at $t = 1$s.

If $v = 2$ m/s and $t = 1$ s, this becomes

$$y = \frac{3(x - 2)}{1 + (x - 2)^4}$$

Figure 15.6 shows the plot of this new function. As expected, this new plot has the same shape as the old plot of Figure 15.5, but it is shifted toward the right by 2 meters.

Let us now look at the important special case of **harmonic waves.** These waves are sine or cosine functions of position and time. At the initial time $t = 0$, the shape of the wave is

$$y = A \cos kx \tag{4}$$

Amplitude and wave number

The constant A is called the **amplitude** of the wave and k is called the **wave number.**[1]

Figure 15.7 is a plot of Eq. (4). Obviously, the harmonic wave can be regarded as a regular succession of wave pulses. The maxima of the wave (largest positive deformations) are called **wave crests;** they occur at

$$kx = 0,\ 2\pi,\ 4\pi,\ 6\pi, \text{ etc.} \tag{5}$$

Fig. 15.7 A harmonic wave.

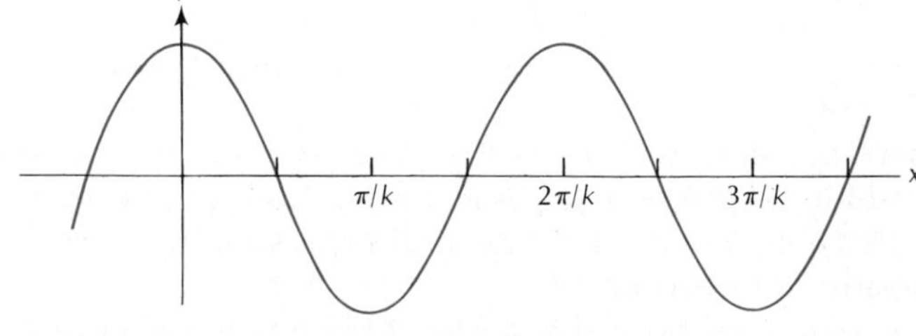

[1] Do not confuse the wave number k with the spring constant of the preceding chapter. The former has nothing to do with the latter!

The minima of the wave (largest negative deformations) are called **wave troughs;** they occur at

$$kx = \pi,\ 3\pi,\ 5\pi, \text{ etc.} \tag{6}$$

The distance from one crest to the next (or one trough to the next) is called the **wavelength,** usually designated by λ. From Eq. (5) we find that

$$\boxed{\lambda = \frac{2\pi}{k}} \tag{7}$$

Wavelength

At any later time the wave of Eq. (4) will have traveled some distance to the right or to the left. The new equation describing the wave is then

$$\boxed{y = A \cos k(x - vt)} \quad \text{for a wave traveling in positive } x \text{ direction}$$

or elsc

$$\boxed{y = A \cos k(x + vt)} \quad \text{for a wave traveling in negative } x \text{ direction} \tag{8}$$

Harmonic waves

The shape of the shifted wave coincides with the shape of the initial wave when the distance traveled is exactly one, or two, or three, etc., wavelengths. Thus, the shape of the wave is a periodic function of time. The **period** of the wave is the time in which the wave travels a distance equal to the wavelength:

$$\boxed{T = \frac{\lambda}{v}} \tag{9}$$

Period

The **frequency** of the wave is

$$\boxed{\nu = \frac{1}{T} = \frac{v}{\lambda}} \tag{10}$$

Frequency

which can be written compactly as

$$\lambda\nu = v \tag{11}$$

We can also introduce the **angular frequency**

$$\boxed{\omega = 2\pi\nu = kv} \tag{12}$$

Angular frequency

In terms of these quantities, we can express the wave function of Eq. (8) in the alternative forms

$$y = A\cos\left(\frac{2\pi}{\lambda}x - 2\pi\nu t\right) \tag{13}$$

and

$$y = \text{A}\cos(kx - \omega t) \tag{14}$$

Note that a given particle, at the position $x = x_0$ on the string, has a sideways displacement

$$y = A\cos(kx_0 - \omega t) \tag{15}$$

If we compare this with Eq. (14.4), we recognize that this particle executes harmonic motion with amplitude A and angular frequency ω.

Although we have derived the preceding formulas in the context of waves on a string, we can also apply these formulas to other kinds of waves, provided that we give the "deformation" y a suitable interpretation. For instance, to describe waves on the surface of the sea, we must regard y as the height of the water above the mean sea level; to describe seismic waves, we must regard y as the displacement of a lump of earth from its original position, etc.

EXAMPLE 2. Ocean waves of a period of 10 s have a speed of 16 m/s. What is the wavelength of these waves? What is the (horizontal) distance between a wave crest and a wave trough?

SOLUTION: From Eq. (9) the wavelength is

$$\lambda = Tv = 10\text{ s} \times 16\text{ m/s} = 160\text{ m}$$

The distance between a wave crest and a wave trough is one-half a wavelength, that is, 80 m.

15.3 Speed of Waves on a String

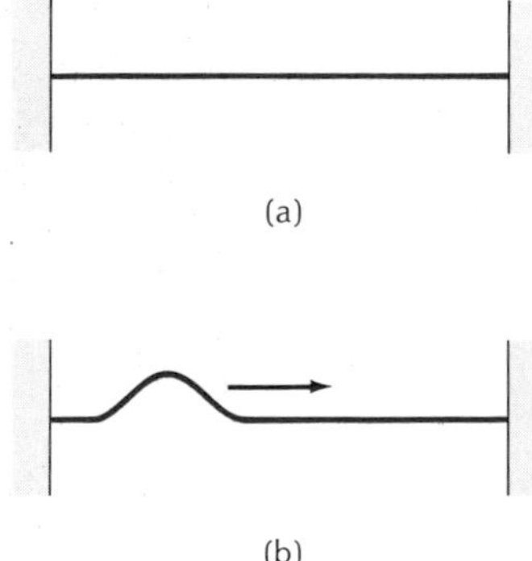

Fig. 15.8 (a) A tightly stretched string; the tension is F. (b) A wave pulse on the string; the speed of this wave pulse is v.

The speed of waves depends on the characteristics of the medium. In some cases the speed also depends on the wavelength; for instance, ocean waves of long wavelength — such as tidal waves — have a larger speed than waves of short wavelength. In this section we will derive an expression for the speed of waves on a string. These waves are simple to treat because their speed does not depend on their wavelength.

Figure 15.8a shows an elastic string tightly stretched between two end points. The tension in the string is F and the density of the string is μ kilograms per meter of length. Of course, the string is a system of a large number of particles, but for our present purposes it is convenient to regard the mass as continuously distributed along the string. Figure 15.8b shows a wave pulse propagating along the string. We will assume that the amplitude of the wave pulse is very small (that is, very small compared to the length of the string); then the wave pulse only produces a small perturbation in the tension and, to a good approximation, the tension everywhere along the string is constant, as in the static case.

It is convenient to analyze the motion of the string in a reference frame moving to the right with a velocity equal to that of the wave pulse. In this reference frame the wave pulse is at rest and the entire string travels to the left. Each segment of string travels along a curved path having the shape of the wave pulse. Figure 15.9a shows one short segment Δl rounding this curved path at speed v. Over the short length Δl, the path can be approximated as an arc of a circle of radius R; the segment Δl subtends a small angle $\Delta\theta$ of this circle. Instantaneously, the segment Δl is in uniform circular motion. The mass of the segment is $\mu\,\Delta l$ and its centripetal acceleration is v^2/R. The forces on the segment are the tensions acting at its ends (Figure 15.9a); the vector sum of these forces is a centripetal force of magnitude $F\,\Delta\theta$ (Figure 15.9b). Hence the equation of motion is

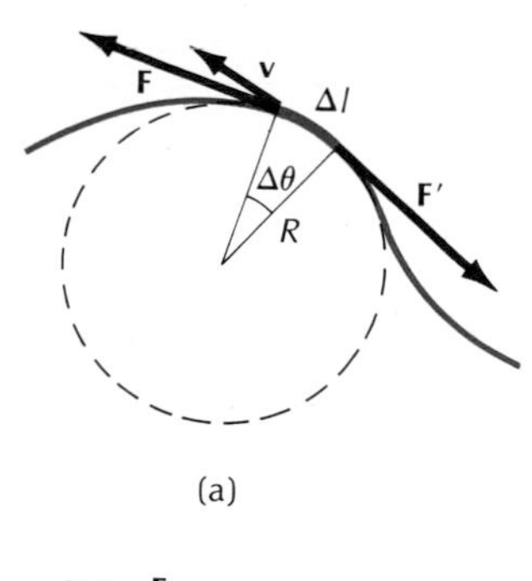

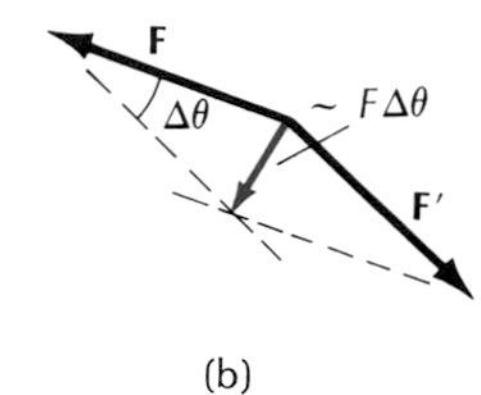

Fig. 15.9 (a) Forces on a short segment Δl of the string.
(b) Resultant force.

$$\mu\,\Delta l\,\frac{v^2}{R} = F\,\Delta\theta \tag{16}$$

or

$$\mu\,\Delta l v^2 = FR\,\Delta\theta \tag{17}$$

Since $R\,\Delta\theta$ equals the length Δl of the segment, Eq. (17) can be simplified to

$$\mu v^2 = F \tag{18}$$

This yields a speed

$$\boxed{v = \sqrt{F/\mu}} \tag{19}$$

Speed of a wave on a string

This equation expresses the speed in terms of the tension and the density of the string. Note that the speed is large if the tension is large and the mass density small. This is intuitively reasonable — a large tension can move a small mass very quickly. Of course, Eq. (19) applies not only to strings but also to tightly stretched wires and cables.

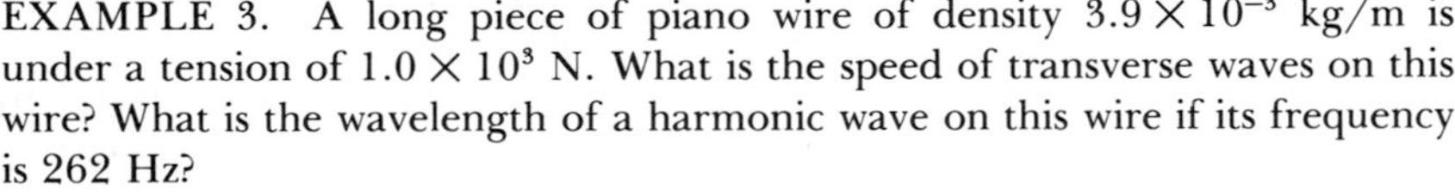

EXAMPLE 3. A long piece of piano wire of density 3.9×10^{-3} kg/m is under a tension of 1.0×10^3 N. What is the speed of transverse waves on this wire? What is the wavelength of a harmonic wave on this wire if its frequency is 262 Hz?

SOLUTION: From Eq. (19) the velocity is

$$v = \sqrt{\frac{F}{\mu}} = \sqrt{\frac{1.0 \times 10^3\ \text{N}}{3.9 \times 10^{-3}\ \text{kg/m}}} = 5.1 \times 10^2\ \text{m/s}$$

Consequently, the wavelength is

$$\lambda = v/\nu = (5.1 \times 10^2\ \text{m/s})/(262/\text{s}) = 1.9\ \text{m}$$

Since in our derivation of Eq. (19) we made no special assumption about the shape of the wave, it is clear that the wave speed is independent of the shape. Consequently, all the different portions of a wave pulse (different shapes) propagate at the same speed, i.e., the wave pulse propagates as though it were in rigid translation along the string.

A harmonic wave can be regarded as a succession of alternating positive and negative pulses. Since all such pulses have the same speed, all harmonic waves on a string have the same speed, independent of wavelength.

Although a wave on a string is a rather special and simple case of wave motion, Eq. (19) illustrates a general feature of wave motion. In broad terms, this equation states that the speed of the wave depends on the restoring force and on the inertia of the medium. This is true for all kinds of wave motion. In all cases some force within the medium opposes its deformation — tension tends to keep the string straight, gravity tends to keep the surface of the sea smooth, elastic forces tend to keep the crust of the Earth as it is. But if there is an initial disturbance, then the restoring force will cause it to propagate (compare Figure 15.1) with a speed depending on the magnitude of the restoring force and on the magnitude of the inertia or, equivalently, the density of mass. In general, the speed will be large if the restoring force is large and the density of mass is small.

The lack of dependence of speed on shape is a rather special feature of waves on a string. For many other kinds of waves, the speed does depend on the shape so that different parts of a wave pulse travel at different speeds. Consequently, the wave pulse changes shape as some of its parts get ahead, or fall behind, other parts. Because of this, the wave pulse usually tends to spread out, becoming more and more shallow. A medium that gives rise to such behavior is called a **dispersive medium.**

Dispersive medium

In contrast to the case of a harmonic wave on a string, a harmonic wave in a dispersive medium cannot be regarded as simply a succession of wave pulses, because the pulses change their shape, whereas a harmonic wave does not. There is then no simple connection between the speed of a wave pulse and the speed of a harmonic wave. To distinguish between these speeds, we call the speed of the peak of a wave pulse the **group velocity** and the speed of a harmonic wave the **phase velocity.**

Group velocity and phase velocity

The group velocity is also sometimes called the signal velocity because it describes the propagation of signals in the medium. If we want to send a signal by means of a wave, we must use a wave pulse rather than a harmonic wave; the latter has no beginning and no end — it lasts forever and is therefore useless as a signal.

The phase velocity of harmonic waves in a dispersive medium depends on wavelength. For instance, ocean waves of long wavelength have a higher velocity than those of short wavelength. We will discuss this characteristic of ocean waves in the next chapter.

15.4 Energy in a Wave; Power

A transverse wave on a string has kinetic energy because the particles are in motion, and it has potential energy because work is required to stretch the string. Consider a small interval dx. The mass of string within this interval is $\mu\ dx$ and the velocity of the piece of string is dy/dt; hence the kinetic energy associated with this piece of string is

$$dK = \tfrac{1}{2}\mu\ dx\left(\frac{dy}{dt}\right)^2 \qquad (20)$$

To find the potential energy, we note that the piece of string is slightly elongated while the wave passes. The wave stretches the string from its original length dx to a new length $\sqrt{dx^2 + dy^2}$ (Figure 15.10). Hence the change of length is

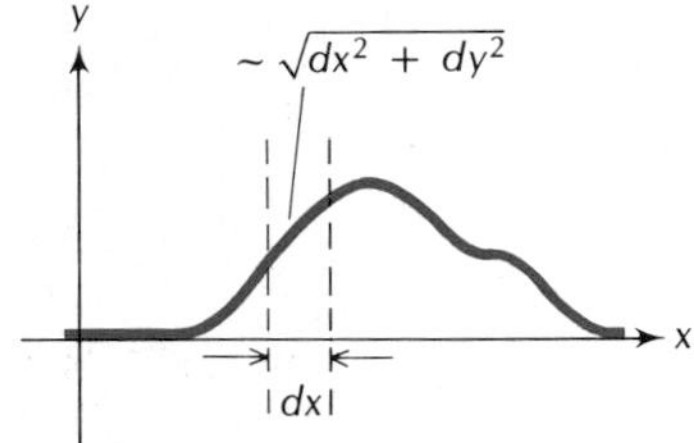

Fig. 15.10 A short segment of string included between x and $x + dx$.

$$\delta l = \sqrt{dx^2 + dy^2} - dx \tag{21}$$

or

$$\delta l = dx\left[\sqrt{1 + \left(\frac{dy}{dx}\right)^2} - 1\right] \tag{22}$$

We will assume that dy/dx is small (so that the string only makes a small angle with its original, horizontal direction). Then we can take advantage of approximation

$$\sqrt{1 + \left(\frac{dy}{dx}\right)^2} \cong 1 + \frac{1}{2}\left(\frac{dy}{dx}\right)^2 \tag{23}$$

to obtain

$$\delta l = \frac{1}{2}\left(\frac{dy}{dx}\right)^2 dx \tag{24}$$

The potential energy associated with the interval dx is simply the work that must be done against the tension F to stretch the string by the amount δl, i.e.,

$$dU = F\,\delta l = \tfrac{1}{2}F\left(\frac{dy}{dx}\right)^2 dx \tag{25}$$

The total energy associated with the interval dx is the sum of Eqs. (20) and (25),

$$dE = dK + dU = \tfrac{1}{2}\mu\left(\frac{\partial y}{\partial t}\right)^2 dx + \tfrac{1}{2}F\left(\frac{\partial y}{\partial x}\right)^2 dx \tag{26}$$

Here we have introduced the standard notation for partial derivatives to indicate that in the first term on the right side of Eq. (26) the function y must be differentiated with respect to t only, and in the second term with respect to x only.

If we divide Eq. (26) by dx, we obtain the amount of energy per unit length, or the **energy density,** of the wave:

Energy density of a wave

$$\frac{dE}{dx} = \tfrac{1}{2}\mu\left(\frac{\partial y}{\partial t}\right)^2 + \tfrac{1}{2}F\left(\frac{\partial y}{\partial x}\right)^2 \tag{27}$$

If we are dealing with a harmonic wave, then the required derivatives are

$$\frac{\partial y}{\partial t} = \frac{\partial}{\partial t}[A\cos(kx - \omega t)] = \omega A\sin(kx - \omega t) \tag{28}$$

$$\frac{\partial y}{\partial x} = \frac{\partial}{\partial x}[A\cos(kx - \omega t)] = -kA\sin(kx - \omega t) \tag{29}$$

so that

$$\frac{dE}{dx} = \tfrac{1}{2}(\mu\omega^2 + Fk^2)A^2 \sin^2(kx - \omega t) \tag{30}$$

In view of Eqs. (12) and (19),

$$Fk^2 = F\omega^2/v^2 = \mu\omega^2 \tag{31}$$

and therefore

$$\frac{dE}{dx} = \mu\omega^2 A^2 \sin^2(kx - \omega t) \tag{32}$$

Figure 15.11 is a plot of this energy density at one instant of time. Note that there is no energy at the wave crests — the velocity of the particles is zero at the crests (no kinetic energy) and the string is unstretched at the crests (no potential energy).

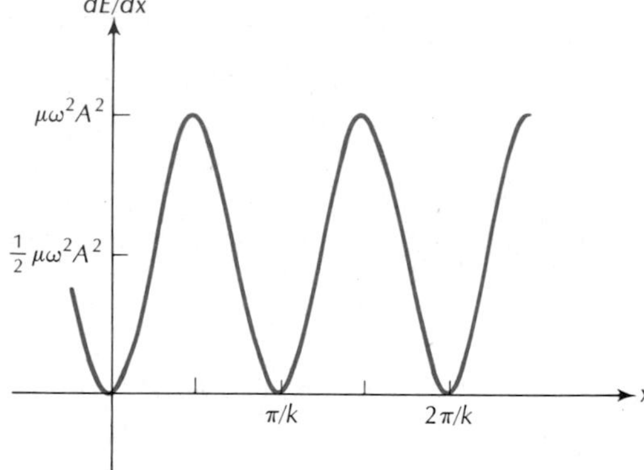

Fig. 15.11 Energy density in a harmonic wave at one instant of time.

The power transported by the wave is simply the amount of energy that moves past a given point on the string in one unit of time. The energy must travel with the wave at the wave speed v. Hence the energy in the interval dx takes a time

$$dt = \frac{dx}{v} \tag{33}$$

to move out of this interval (of course, as fast as the energy moves out of one end of the segment, it is replaced with energy moving in at the other end of the segment). Combining Eqs. (32) and (33), we obtain an expression for the (instantaneous) **power** transported by the harmonic wave:

Power of a wave

$$P = \frac{dE}{dt} = v\frac{dE}{dx} = v\mu\omega^2 A^2 \sin^2(kx - \omega t) \tag{34}$$

The power arriving at any given position oscillates as a function of time between a maximum value $v\mu\omega^2 A^2$ and a minimum value zero.

Note that the power is proportional to the wave velocity, the square of the frequency, and the square of the amplitude. This proportionality turns out to be true not only for waves on a string, but also for other kinds of waves.

15.5 The Superposition of Waves

Superposition principle

Most kinds of waves obey a **superposition principle:** when two or more waves arrive at any given point, the *resultant instantaneous deformation is the sum of the individual instantaneous deformations.* Such a superposition means that the waves do not interact; they have no effect on one another. Each wave propagates as though the other wave were not present. For instance, if the sound waves from a violin and a flute reach our ear simultaneously, then each of these waves produces a displacement of the air molecules just as though it were acting alone and the net displacement of the air molecules is the (vector) sum of these individual displacements.

For sound waves of ordinary intensity, the superposition principle is very well satisfied. However, if the sound wave is extremely strong, as in a shock wave, then superposition fails. What happens here is that the passage of an extremely strong sound wave can heat the air and this will affect the propagation of a second sound wave arriving at the same place. The same is true for other kinds of waves: when the wave is very intense, superposition fails. In this section we will not worry about these extreme conditions and we will assume that superposition is a good approximation.

As a first example of superposition, let us consider two waves propagating in the same direction with the same frequency and amplitude, but different phases. These waves might be waves on a string, in air, on the surface of water, or whatever. The wave functions describing the individual waves are

$$y_1 = A\cos(kx - \omega t) \tag{35}$$

$$y_2 = A\cos(kx - \omega t + \delta) \tag{36}$$

Here δ is the phase difference. According to the superposition principle, the wave function for the resultant wave is

$$y = y_1 + y_2 = A\cos(kx - \omega t) + A\cos(kx - \omega t + \delta) \tag{37}$$

With the trigonometric identity

$$\cos\alpha + \cos\beta = 2\cos\tfrac{1}{2}(\alpha + \beta)\cos\tfrac{1}{2}(\alpha - \beta) \tag{38}$$

we can transform Eq. (37) into

$$y = 2A\cos(kx - \omega t + \tfrac{1}{2}\delta)\cos\tfrac{1}{2}\delta \tag{39}$$

The resultant wave has the same frequency as the original waves and its amplitude [the factor multiplying $\cos(kx - \omega t + \frac{1}{2}\delta)$] is $2A\cos\frac{1}{2}\delta$. If δ is zero, the two waves are in phase; they meet crest to crest and trough

Constructive and destructive interference

to trough, reinforcing each other (Figure 15.12). This is **constructive interference.** If $\delta = \pi$, the two waves are exactly out of phase; they meet crest to trough, canceling each other. This is **destructive interference;** the resultant wave is zero (Figure 15.13).

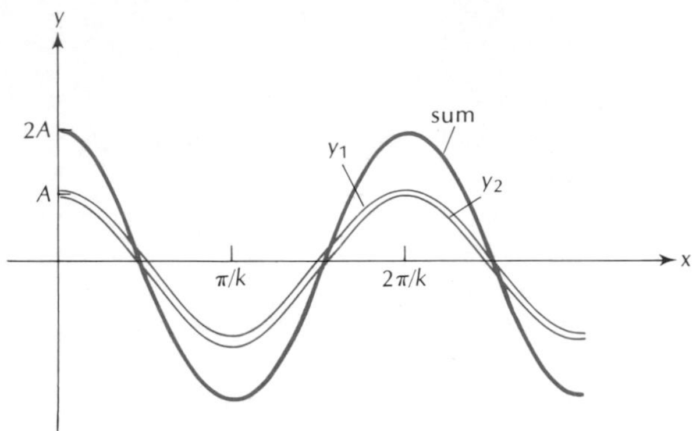

Fig. 15.12 Constructive interference of two waves.

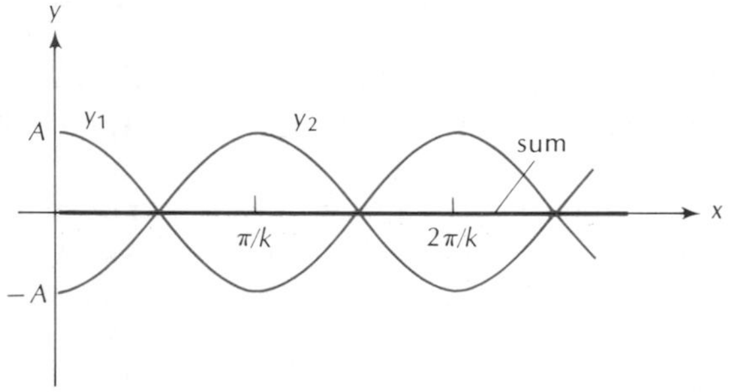

Fig. 15.13 Destructive interference of two waves; the waves cancel everywhere.

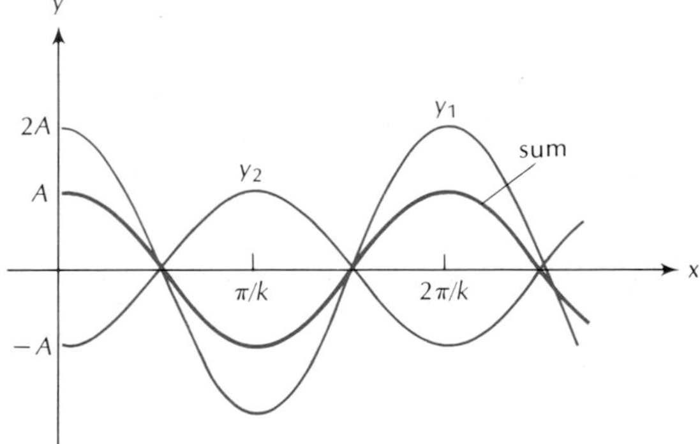

Fig. 15.14 Destructive interference of two waves of different amplitude. The sum is small, but not zero.

If the two waves have different amplitudes, then their destructive interference will not lead to a total cancellation of the waves; some portion of whatever wave has the larger amplitude will be left over (Figure 15.14).

As another example of superposition, let us consider two waves of slightly different frequencies:

$$y_1 = A\cos(k_1x - \omega_1 t) \tag{40}$$

$$y_2 = A\cos(k_2x - \omega_2 t) \tag{41}$$

Using the same trigonometric identity as before, we obtain

$$\begin{aligned} y &= y_1 + y_2 \\ &= 2A\cos[\tfrac{1}{2}(k_1 + k_2)x - \tfrac{1}{2}(\omega_1 + \omega_2)t]\cos[\tfrac{1}{2}(k_1 - k_2)x - \tfrac{1}{2}(\omega_1 - \omega_2)t] \end{aligned} \tag{42}$$

Let us examine this at an initial time $t = 0$. We can then write Eq. (42) as

$$y = 2A\cos[\tfrac{1}{2}(\Delta k)x]\cos(\bar{k}x) \tag{43}$$

where Δk is the difference in wave numbers,

$$\Delta k = k_1 - k_2 \tag{44}$$

and $\bar{k}$ is the average wave number

$$\bar{k} = \tfrac{1}{2}(k_1 + k_2) \tag{45}$$

If Δk is small compared to $\bar{k}$, the expression on the right side of Eq. (43) can be interpreted as a cosine wave of wave number $\bar{k}$ with an amplitude $2A\cos[\frac{1}{2}(\Delta k)x]$, i.e., the amplitude of the wave is a slowly varying function of position.

Figure 15.15 shows a plot of the wave given by Eq. (43). The amplitude of this wave is said to be **modulated:** the amplitude is large at $x = 0$ where the two superposed waves interfere constructively; it then gradually decreases and becomes zero at $x = \pi/\Delta k$, where the two waves interfere destructively. Then again it increases and becomes large at $x = 2\pi/\Delta k$, etc.

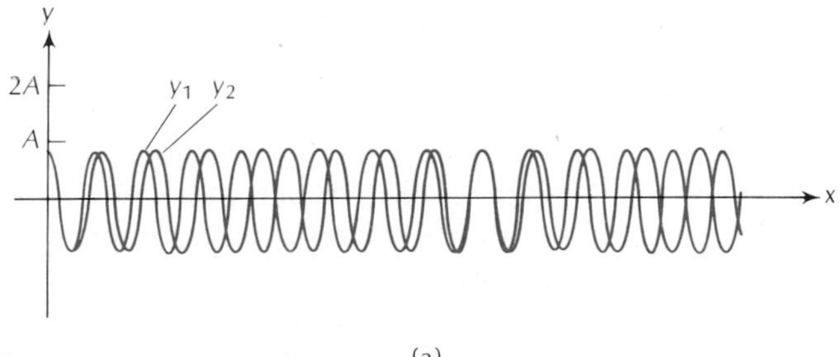

(a)

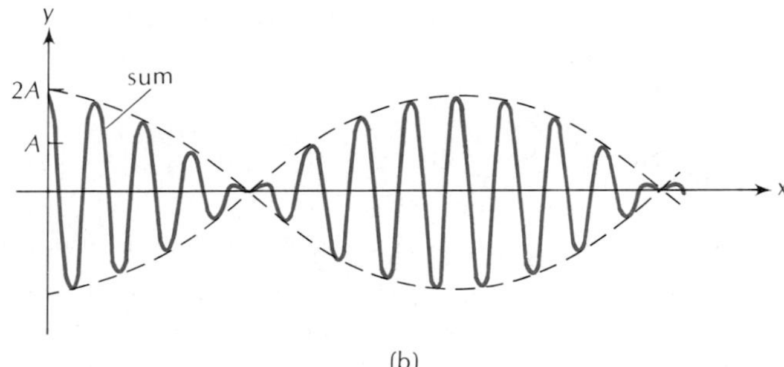

(b)

Fig. 15.15 Superposition of two waves of slightly different wavelengths and frequencies. The dashed line shows the wave envelope or the average amplitude.

With the passing of time, the entire pattern of Figure 15.15 moves to the right with the wave velocity. This gives rise to the phenomenon of **beats.** At any given position the amplitude of the wave pulsates — first the amplitude of the oscillations is large, then it becomes small, then large again, etc. The frequency with with the wave pulsates is called the **beat frequency.** Since the time interval between one amplitude maximum and the next in Figure 15.15 is $\Delta t = \Delta x/v = 2\pi/(\Delta k v)$, the beat frequency is

$$\nu_{\text{beat}} = \frac{1}{\Delta t} = \frac{v\,\Delta k}{2\pi} = \frac{vk_1}{2\pi} - \frac{vk_2}{2\pi}$$

or

Beat frequency

$$\boxed{\nu_{\text{beat}} = \nu_1 - \nu_2} \tag{46}$$

Thus the beat frequency is simply the difference between the two wave frequencies.

EXAMPLE 4. Suppose that a pair of flutes generates sound waves of frequency 264 Hz and 262 Hz, respectively. What is the beat frequency?

SOLUTION: According to Eq. (46),

$$\nu_{\text{beat}} = 264\ \text{Hz} - 262\ \text{Hz} = 2\ \text{Hz}$$

Hence a listener will hear a tone of average frequency 263 Hz but with an amplitude pulsating 2 times per second.

Beats are a sensitive indication of small frequency differences and they are very useful in the tuning of musical instruments. For example, to bring the two flutes of Example 4 in tune, the musicians listen to the beats and, by trial, adjust one of the flutes so as to reduce the beat frequency; when the beat disappears entirely (zero beat frequency), the two flutes will be generating waves of exactly equal frequency.

Fourier's theorem and Fourier series

By the superposition of harmonic waves of different amplitude and frequency, we can construct some rather complicated wave shapes. In fact, it can be shown that any arbitrary periodic wave can be constructed by the superposition of a sufficiently large number of sinusoidal and cosinusoidal harmonic waves. This is **Fourier's theorem.** For example, Figure 15.16a shows a square wave (plotted at $t = 0$) that alternates periodically between positive and negative values. The wavelength of this wave is L. To construct this wave by the superposition of harmonic waves we must take the following combination, which is called a **Fourier series:**

$$y = \frac{4A}{\pi} \sin \frac{2\pi x}{L} + \frac{4A}{3\pi} \sin \frac{6\pi x}{L} + \frac{4A}{5\pi} \sin \frac{10\pi x}{L} + \cdots \tag{47}$$

These are harmonic waves of wavelength L, $L/3$, $L/5$, etc. The three dots on the right side stand for extra harmonic waves of even shorter wavelength; for an accurate representation of the square wave we need an infinite series of extra harmonic waves of smaller and smaller wavelength, but, as Figure 15.16b shows, the three waves on the right side of Eq. (47) already give a rough approximation to the square wave.

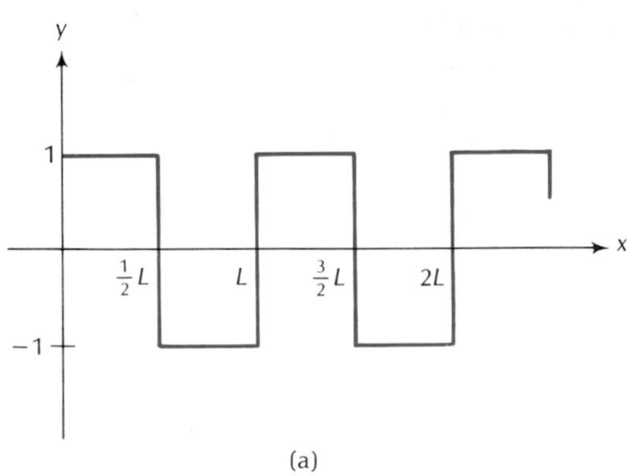

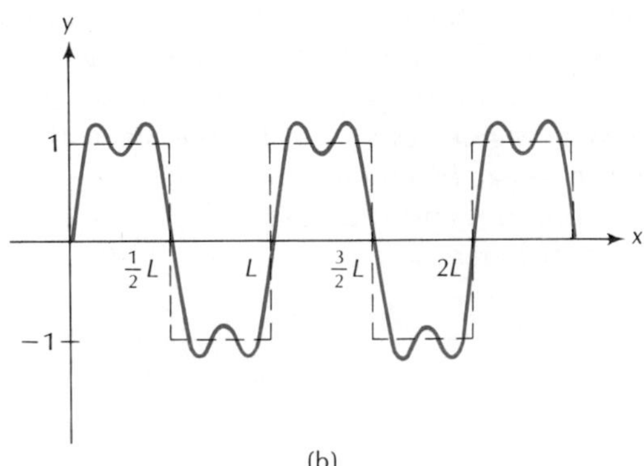

Fig. 15.16 (a) A square wave. (b) Superposition of the first three terms of the Fourier series of Eq. (47) for the square wave.

15.6 Standing Waves

Let us now consider the superposition of two waves of the same amplitude and frequency but of opposite directions of propagation. If the individual wave functions are

$$y_1 = A\cos(kx - \omega t) \tag{48}$$

$$y_2 = A\cos(kx + \omega t) \tag{49}$$

then the resultant wave function is

$$\boxed{y = y_1 + y_2 = 2A\cos kx\cos \omega t} \tag{50}$$

Standing wave

where again we have used the trigonometric identity of Eq. (38).

The expression on the right side of Eq. (50) describes a **standing wave.** This wave travels neither right nor left; its peaks remain at fixed positions while the entire wave increases and decreases in unison. The entire wave pulsates with a frequency ω, as indicated by the overall factor $\cos \omega t$. Figure 15.17 shows the standing wave at successive instants of time.

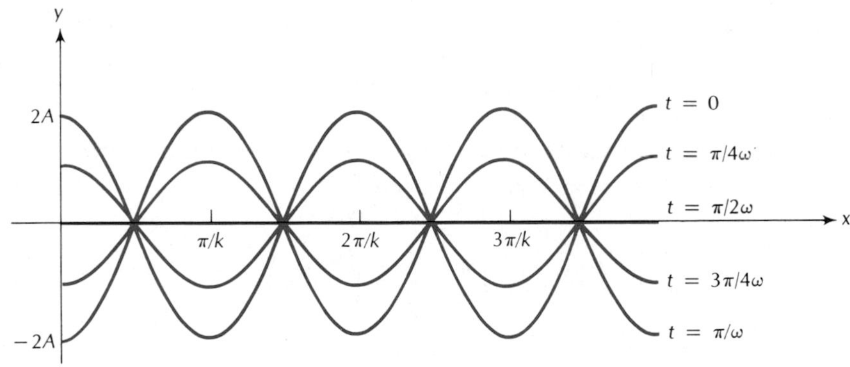

Fig. 15.17 A standing wave at successive instants of time.

If the wave function of Eq. (50) represents the motion of a string, then each particle of this string executes simple harmonic motion. However, in contrast to the case of a traveling wave, where the amplitude of oscillation of each particle is the same, the amplitude of oscillation now depends on position: it has the value $A\cos kx_0$ at the position x_0.

The positions at which the amplitude of oscillation is maximum are given by

$$kx = 0,\ \pi,\ 2\pi,\ \text{etc.} \tag{51}$$

With $k = 2\pi/\lambda$ this becomes

$$x = 0,\ \frac{\lambda}{2},\ \lambda,\ \frac{3\lambda}{2},\ \text{etc.} \tag{52}$$

The maxima are due to constructive interference between the two waves of Eqs. (48) and (49).

Likewise, the positions at which the amplitude is zero are given by

$$kx = \frac{\pi}{2}, \frac{3\pi}{2}, \frac{5\pi}{2}, \text{ etc.} \tag{53}$$

or

$$x = \tfrac{1}{4}\lambda, \tfrac{3}{4}\lambda, \tfrac{5}{4}\lambda, \text{ etc.} \tag{54}$$

Nodes and antinodes

The minima are due to destructive interference between the two waves. The minima of the standing waves are called **nodes** and the maxima are called **antinodes.**

Boundary condition

So far, in our discussion of the waves on a string, we have assumed that the string is very long and we have ignored the endpoints of the string. Obviously, something rather drastic must happen to a wave arriving at an endpoint. Figure 15.18 shows a string of length L with both ends fixed on rigid supports. Thus, there is a **boundary condition** at the endpoints: the deformation y must be zero at these points at all times. This imposes a serious restriction on the possible waves permitted on the string. To find the possible waves, we note that a standing wave with nodes at the endpoints satisfies our boundary condition. Figures 15.19–15.21 show several standing waves of this kind (plotted at the time $t = 0$). If one end of the string is at $x = 0$ and the other at $x = L$, the wave functions corresponding to Figures 15.19–15.21 are, respectively,

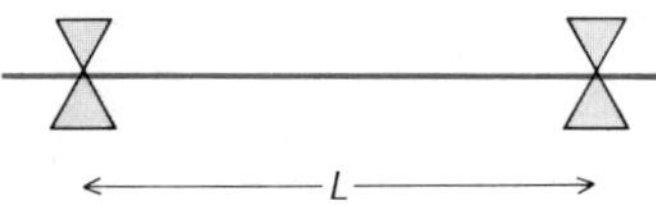

Fig. 15.18 A tightly stretched string with fixed ends. The next figures shows possible standing waves on this string.

$$y_1 = A \sin\left(\frac{\pi}{L}x\right)\cos\left(\frac{\pi}{L}vt\right) \tag{55}$$

$$y_2 = A \sin\left(\frac{2\pi}{L}x\right)\cos\left(\frac{2\pi}{L}vt\right) \tag{56}$$

$$y_3 = A \sin\left(\frac{3\pi}{L}x\right)\cos\left(\frac{3\pi}{L}vt\right) \tag{57}$$

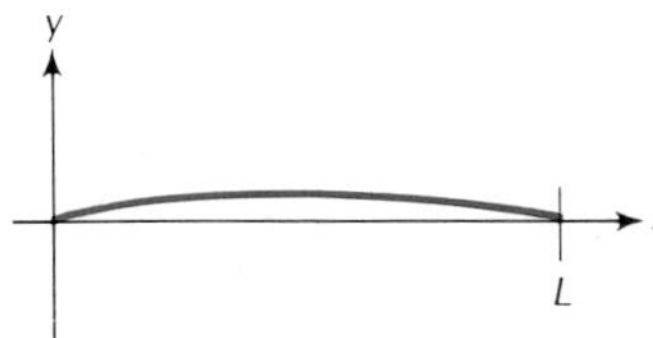

Fig. 15.19 The fundamental mode.

Fig. 15.20 The first harmonic mode.

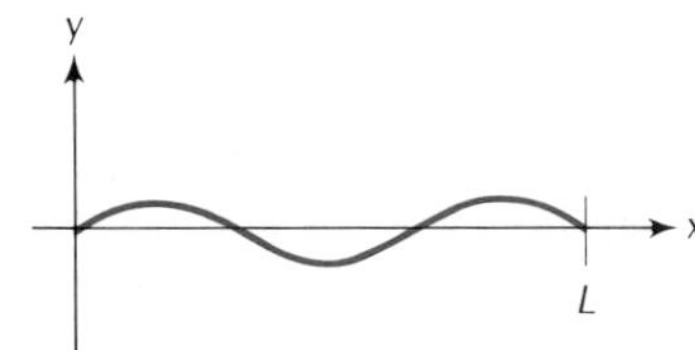

Fig. 15.21 The second harmonic mode.

Normal mode

These possible motions of the string are called the **normal modes.** Equation (55) gives the **fundamental** mode, Eq. (56) the **first harmonic** mode, Eq. (57) the **second harmonic** mode, and similar equations give higher harmonic modes. The wavelengths of these modes are

Wavelengths of normal modes

$$\boxed{\lambda_1 = 2L, \qquad \lambda_2 = L, \qquad \lambda_3 = \tfrac{2}{3}L, \text{ etc.}} \tag{58}$$

and the frequencies are

$$\nu_1 = \frac{v}{2L}, \qquad \nu_2 = \frac{v}{L}, \qquad \nu_3 = \frac{3v}{2L}, \text{ etc.}$$

The frequencies of these modes are often called the *normal frequencies, proper frequencies,* or **eigenfrequencies** of the string. We can write the following general formula for these frequencies:

$$\boxed{\nu_n = \frac{nv}{2L} \qquad n = 1, 2, 3, \ldots} \tag{59}$$

Eigenfrequencies

This formula clearly shows that all the eigenfrequencies are multiples of the fundamental frequency $v/2L$.

In general, any arbitrary motion of the string will be some superposition of several of the above normal modes. Which modes will be present depends on how the motion is started. For instance, when a guitar player plucks a string on his guitar near the middle, he will excite the fundamental mode and also the second harmonic mode and (to a lesser extent) some of the higher modes.

EXAMPLE 5. The middle C string of a piano vibrates with a frequency of 261.6 Hz when excited in its fundamental mode. What are the frequencies of the first, second, and third harmonic modes of this string?

SOLUTION: According to Eq. (59), ν_2 equals $2\nu_1$, ν_3 equals $3\nu_1$, and ν_4 equals $4\nu_1$. Hence the frequencies of the first, second, and third harmonics are, respectively, 2×261.6 Hz, 3×261.6 Hz, and 4×261.6 Hz.

The normal modes of a long, thin elastic rod or beam fixed at both ends are mathematically similar to the normal modes of a string. However, such an elastic body can experience both transverse deformations (like those of a string) and rotational, or "torsional," deformations. Figure 15.22 shows a spectacular example of a torsional standing wave

Fig. 15.22 Standing wave on the Tacoma Narrows bridge, July 1, 1940.

in the span of a bridge at Tacoma, Washington. This standing wave was excited by a wind blowing across the bridge, which generated vortices in resonance with one of the normal modes of the span. The bridge oscillated for several hours, with increasing amplitude, and then broke apart.

SUMMARY

General wave function: $y = f(x \mp vt)$

Wavelength: $\lambda = 2\pi/k$

Harmonic wave: $y = A \cos k(x \mp vt)$

$$= A \cos(kx \mp \omega t)$$

Frequency: $\nu = v/\lambda$

Angular frequency: $\omega = 2\pi\nu$

Speed of wave on a string: $v = \sqrt{F/\mu}$

Power of wave on a string: $P \propto v\omega^2 A^2$

Superposition principle for two or more waves: The net instantaneous deformation is the sum of the individual instantaneous deformations.

Constructive interference: Waves meet crest to crest

Destructive interference: Waves meet crest to trough

Beat frequency: $\nu_{\text{beat}} = \nu_1 - \nu_2$

Standing harmonic wave: $y = A \cos kx \cos \omega t$

Node: Point of zero oscillation

Antinode: Point of maximum oscillation

Wavelengths of normal modes of string: $\lambda = 2L, L, \frac{2}{3}L, \ldots$

Eigenfrequencies: $\nu = \frac{v}{2L}, \frac{v}{L}, \frac{3v}{2L}, \ldots$

QUESTIONS

1. You have a long, thin steel rod and a hammer. How must you hit the end of the rod to generate a longitudinal wave? A transverse wave?

2. A wave pulse on a string transports energy. Does it also transport momentum? To answer this question, imagine a washer loosely encircling the string at some place; what happens to the washer when the wave pulse strikes it?

3. The strings of a guitar are made of wires of different thicknesses (the thickest wires are manufactured by wrapping copper or brass wire around a strand

of steel). Why is it impractical to use wire of the same thickness for all the strings?

4. According to Eq. (19), the speed of a wave on a string increases by a factor of 2 if we increase the tension by a factor of 4. However, in the case of a rubber string, the speed increases by more than a factor of 2 if we increase the tension by a factor of 4. Why are rubber strings different?

5. A harmonic wave is traveling along a string. Where in this wave is the kinetic energy at maximum? The potential energy? The total energy?

6. Suppose that two strings of different densities are knotted together to make a single long string. If a wave pulse travels along the first string, what will happen to the wave pulse when it reaches the junction? (Hint: If the second string had the same density as the first string, the wave pulse would proceed without interruption; if the second string were much denser than the first, the wave pulse would be totally reflected.)

7. Figure 15.17 shows a standing wave on a string. At time $t = \pi/2\omega$, the amplitude of the wave is everywhere zero. Does this mean the wave has zero energy at this instant?

8. After an arrow has been shot from a bow, the bowstring will oscillate back and forth, forming a standing wave. Which of the eigenmodes shown in Figures 15.19–15.21 do you expect to be present?

9. In tuning a guitar or violin, by what means do you change the frequency of a string?

10. A mechanic can make a rough test of the tension in the spokes of a wire wheel by striking the spokes with a wrench or a small hammer. A spoke under tension will ring, but a loose spoke will not. Explain.

11. What is the purpose of the frets on the neck of a guitar or a mandolin?

PROBLEMS

Section 15.2

1. Suppose that at time $t = 0$ a wave pulse on a string has a shape described by the function

$$y = \frac{9 \times 10^{-2}}{9 + x^2}$$

where y and x are measured in meters.
   (a) If this wave travels in the positive x direction at a speed of 2 m/s, what is the function that describes the wave at a time t?
   (b) Make a rough plot of the shape of the wave at $t = 0$ and at $t = 3$ s.

2. An ocean wave has a wavelength of 120 m and a period of 8.77 s. Calculate the frequency, angular frequency, wave number, and speed of this wave.

3. The speed of tidal waves in the Pacific is about 740 km/h.
   (a) How long does a tidal wave take to travel from Japan to San Francisco, a distance of 8000 km?
   (b) If the wavelength of the wave is 300 km, what is its frequency?

4. Suppose that the function $y = 6.0 \times 10^{-3} \cos(20x + 4.0t + \pi/3)$ describes a wave on a long string (distance is measured in meters and time in seconds).
   (a) What are the amplitude, wavelength, wave number, frequency, angular frequency, direction of propagation, and speed of this wave?
   (b) At what time does this wave have a maximum at $x = 0$?

5. A harmonic wave on a string has an amplitude of 2.0 cm, a wavelength of 1.2 m, and a velocity of 6.0 m/s in the positive x direction. At time $t = 0$, this wave has a crest at $x = 0$.

(a) What are the period, frequency, angular frequency, and wave number of this wave?

(b) What is the mathematical equation describing this wave as a function of x and t?

6. Ocean waves smash into a breakwater at the rate of 12 per minute. The wavelength of these waves is 39 m. What is their speed?

7. The velocity of sound in fresh water at 15°C is 1440 m/s and at 30°C it is 1530 m/s. Suppose that a sound wave of frequency 440 Hz penetrates from a layer of water at 30°C into a layer of water at 15°C. What will be the change in the wavelength? Assume that the frequency remains unchanged.

8. A light wave of frequency 5.5×10^{14} Hz penetrates from air into water. What is its wavelength in air? In water? The speed of light is 3.0×10^8 m/s in air and 2.3×10^8 m/s in water; assume that the frequency remains the same.

9. Ocean waves of wavelength 100 m have a speed of 6.2 m/s; ocean waves of wavelength 20 m have a speed of 2.8 m/s.[2] Suppose that a sudden storm at sea generates waves of all wavelengths. The long-wavelength waves travel fastest and reach the coast first. A fisherman standing on the coast first notices the arrival of 100-m waves; 10 hours later he notices the arrival of 20-m waves. How far is the storm from the coast?

Section 15.3

10. A clothesline of length 10 m is stretched between a house and a tree. The clothesline is under a tension of 50 N and it has a density of 6.0×10^{-2} kg/m. How long does a wave pulse take to travel from the house to the tree and back?

11. A wire rope used to support a radio mast has a length of 20 m and a density of 0.8 kg/m. When you give the wire rope a sharp blow at the lower end and generate a wave pulse, it takes 1 s for this wave pulse to travel to the upper end and to return. What is the tension in the wire rope?

12. A nylon rope of length 80 ft is under a tension of 3000 lb. The total weight of this rope is 6.0 lb. If a wave pulse starts at one end of this rope, how long does it take to reach the other end?

13. A string of density μ is tied to a second string of density μ'. A harmonic wave of speed v traveling along the first string reaches the junction and enters the second string. What will be the speed v' of this wave in the second string? Your answer should be a formula involving μ, μ', and v.

*14. The end of a long string of mass density μ is knotted to the beginning of another long string of mass density μ' (the tensions in these strings are equal). A harmonic wave travels along the first string toward the knot. This incident wave will be partially transmitted into the second string, and partially reflected. The frequencies of all these waves are the same. With the knot at $x = 0$, we can write the following expressions for the incident, reflected, and transmitted waves:

$$y_1 = A_{\text{in}} \cos(kx - \omega t)$$

$$y_2 = A_{\text{ref}} \cos(kx + \omega t)$$

$$y_3 = A_{\text{trans}} \cos(k'x - \omega t)$$

[2] These values are group velocities, or signal velocities.

Show that

$$A_{\text{ref}} = \frac{k - k'}{k + k'} A_{\text{in}} = \frac{\sqrt{\mu} - \sqrt{\mu'}}{\sqrt{\mu} + \sqrt{\mu'}} A_{\text{in}}$$

$$A_{\text{trans}} = \frac{2k}{k + k'} A_{\text{in}} = \frac{2\sqrt{\mu}}{\sqrt{\mu} + \sqrt{\mu'}} A_{\text{in}}$$

[Hint: At $x = 0$, the displacement of the string must be continuous, $y_1 + y_2 = y_3$; if not, the string would break at the knot. Furthermore, the slope of the string must be continuous, $dy_1/dx + dy_2/dx = dy_3/dx$; if not, the string would have a kink and the (massless) knot would receive an infinite acceleration.]

Section 15.4

15. A harmonic wave travels along a string of mass density 5.0×10^{-3} kg/m. The amplitude of the wave is 2.0×10^{-2} m, its frequency is 60 Hz, and the tension in the string is 20 N. What is the average power transported by this wave?

16. The average waves generated by a wind of 30 knots blowing on the open sea for a long time (steady-state conditions) have a height of 11 ft, a period of 11 s, and a speed of 57 ft/s. The average waves generated by a wind of 40 knots have a height of 20 ft, a period of 15 s, and a speed of 78 ft/s. What is the ratio of the powers carried by these two kinds of waves?

17. Consider the wave on a string described in Problem 4. What is the energy density as a function of time at $x = 0.5$ m? The mass per unit length of the string is 6.0×10^{-3} kg/m.

Section 15.5

18. Consider the wave function $y = 3.0 \cos(5.0x - 8.0t) + 4.0 \sin(5.0x - 8.0t)$. Show that this wave function can be written in the form $y = A \cos(5.0x - 8.0t + \delta)$. What are the values of A and δ?

19. A thin wire of length 1.0 m vibrates in a superposition of the fundamental mode and the second harmonic. The wave function is

$$y = 0.006 \sin \pi x \cos 400\pi t - 0.004 \sin 3\pi x \cos 1200\pi t$$

where y and x are measured in meters and t in seconds.

(a) What is the deformation at $x = 0.5$ m as a function of time?

(b) Plot this deformation as a function of time in the interval $0 \text{ s} \leq t \leq 0.005$ s.

20. Two ocean waves with $\lambda = 100$ m, $\nu = 0.125$ Hz and $\lambda = 90$ m, $\nu = 0.132$ Hz arrive at a seawall simultaneously. What is the beat frequency of these waves?

21. A guitar player attempts to tune his instrument perfectly with the help of a tuning fork. If the guitar player sounds the tuning fork and a string on his guitar simultaneously, he perceives beats at a frequency of 4 per second. The tuning fork is known to have a frequency of 294 Hz. What fractional increase (or decrease) of the tension of the guitar string is required to bring the guitar in tune with the tuning fork? From the available information, can you tell whether an increase or decrease of tension is required?

22. Figure 15.23 shows the height of tide at Pakhoi. These tides can be regarded as a wave. The shape of the curve in Figure 15.23 indicates that the wave consists of two harmonic waves of slightly different frequencies beating against each other. What are the frequencies of the two harmonic waves? What are the periods? Which is caused by the Moon, which by the Sun?

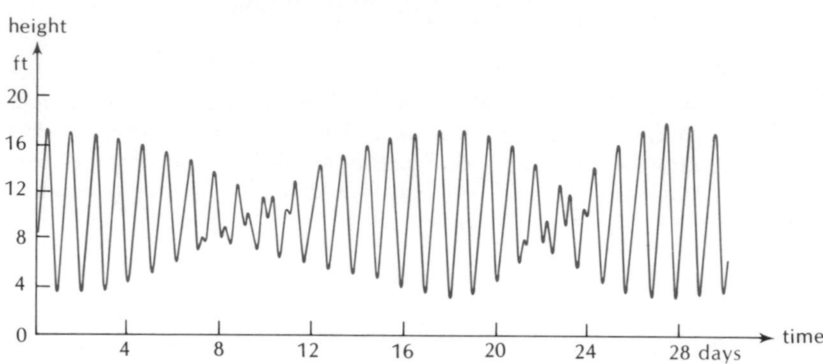

Fig. 15.23 Height of the tide at Pakhoi as a function of time.

23. A piano wire of length 1.5 m fixed at its end vibrates in its second harmonic mode. The frequency of vibration is 440 Hz and the amplitude at the midpoint of the wire is 0.4 mm. Express this standing wave as a superposition of traveling waves. What are the amplitudes and speeds of the traveling waves?

Section 15.6

24. The fundamental mode of the G string of a violin has a frequency of 196 Hz. What are the frequencies of the first, second, third, and fourth harmonics?

25. Suppose that a vibrating mandolin string of length 0.34 m vibrates in a mode with five nodes (including the nodes at the ends) and four antinodes. What harmonic is this? What is its wavelength?

26. A telegraph wire made of copper is stretched tightly between two telephone poles 50 m apart. The tension in the wire is 500 N and the mass per unit length is 2×10^{-2} kg/m. What is the frequency of the fundamental mode? The first harmonic?

27. A violin has four strings; all the strings have (approximately) equal tensions and lengths but they have different densities (kg/m) so that when excited in their fundamental modes they vibrate at different frequencies. The fundamental frequencies of the four strings are 196, 294, 440, and 659 Hz. What must be the ratios of the densities of the strings?

28. The fundamental mode of the G string in a mandolin has a frequency of 196 Hz. The length of this string is 0.34 m and its density is 4.0×10^{-3} kg/m. What is the tension on this string?

29. The middle C string of a piano is supposed to vibrate at 261.6 Hz when exited in its fundamental mode. A piano tuner finds that in a piano that has tension of 900 N on this string, the frequency of vibration is too low (flat) by 15 Hz. How much must he increase the tension of the string to achieve the correct frequency?

30. The D string of a violin vibrates in its fundamental mode with a frequency of 294 Hz and an amplitude of 2.0 mm. What are the maximum velocity and the maximum acceleration of the midpoint of the string?

31. Many men enjoy singing in shower stalls because their voice resonates in the cavity of the shower stall. Consider a shower stall measuring 1 m $\times$ 1 m $\times$ 2.5 m. What are the four lowest resonant frequencies of standing sound waves in such a shower stall?

32. The wire rope supporting the mast of a sailboat from the rear is under a large tension. The rope has a length of 9.0 m and a mass per unit length of 0.22 kg/m.
   (a) If a sailor pushes on the rope sideways at its midpoint with a force of 150 N, he can deflect it by 7.0 cm. What is the tension in the rope?
   (b) If the sailor now plucks the rope near its midpoint, the rope will vibrate back and forth like a guitar string. What is the frequency of the fundamental mode?

33. Some automobiles are equipped with wire wheels. The spokes of these wheels are made of short segments of thick wire installed under large tension. Suppose that one of these wires is 9.0 cm long, 0.40 cm in diameter, and under a tension of 2200 N. The wire is made of steel; the density of steel (mass per unit volume) is 7.8 g/cm^3. To check the tension, a mechanic gives the spoke a light blow with a wrench near its middle. With what frequency will the spoke ring? Assume that the frequency is that of the fundamental mode.

34. A light wave of wavelength 5.0×10^{-7} m strikes a mirror perpendicularly. The reflection of the wave by the mirror makes a standing wave with a node at the mirror. At what distance from the mirror is the nearest antinode? The nearest node?

35. A wave on the surface of the sea with a wavelength of 3.0 m and a period of 4.4 s strikes a seawall oriented perpendicularly to its path. The reflection of the wave by the seawall sets up a standing wave. For such a wave, there is an antinode at the seawall. How far from the seawall will there be nodes?

36. A mandolin has a string of length 0.34 m. Suppose that the string vibrates in its second harmonic mode at a frequency of 588 Hz and an amplitude of 1.2 mm. The mass per unit length of the string is 1.8×10^{-3} kg/m. What is the energy density along the string? Where is the energy density maximum? Where minimum? What is the value of the energy density at its maximum?

*37. A piano wire of length 1.8 m vibrates in its fundamental mode. The frequency of vibration is 494 Hz; the amplitude is 3.0×10^{-3} m. The mass per unit length of the wire is 2.2×10^{-3} kg/m. What is the energy of vibration of the entire wire?

CHAPTER 16

Sound and Other Wave Phenomena

In this chapter we will study some important examples of wave propagation in two and in three dimensions: water waves on the surface of water, sound waves in the volume of air, and seismic waves in the volume of the Earth. In a later chapter we will study the propagation of light waves in a transparent medium and in vacuum. All these waves can be described graphically by their **wave fronts,** that is, the locations of the wave crests at a given instant of time. For example, Figures 16.1 and 16.2 show wave fronts of a water wave and of a sound wave.

Wave front

As time passes, the wave fronts spread outward, expanding as they move away from their source. This spreading of the waves is a characteristic feature of wave propagation in two or three dimensions. It implies that the intensity at a given wave front decreases as the wave front increases in size. Consider the case of circular water waves spreading out from a point of disturbance caused by, say, the impact of

Fig. 16.1 (left) Water wave spreading out on the surface of a pond. The wave fronts are concentric circles.

Fig. 16.2 (right) Sound wave spreading out in air. The wave fronts are concentric spheres. The sound wave has been made visible by means of a small electric light bulb attached to a microphone that controlled the brightness of the bulb. The bulb and microphone were swept through the space in front of the telephone earpiece along arcs, as indicated by the fine pattern of ridges.

a pebble on the water's surface. At some time a wave front has a radius r_1 and its energy is distributed along its circumference $2\pi r_1$. At a later time the wavefront has a larger radius r_2 and the same amount of energy is distributed along a larger circumference $2\pi r_2$. The intensity of the wave is proportional to the energy delivered by the wave per unit length along the circumference; hence the intensity decreases in inverse proportion to the radial distance from the source. By a similar argument we will find that the intensity of a sound wave, or some other three-dimensional wave, decreases as the inverse square of the radius.

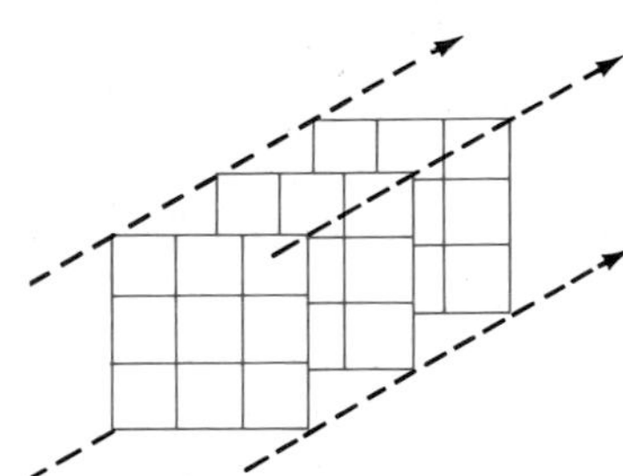

Fig. 16.3 Parallel wave fronts.

At a very large distance from a pointlike source, the circular or spherical wave fronts of a two- or three-dimensional wave can be regarded as nearly straight or flat, provided that we concentrate our attention on a small region (Figure 16.3). Such waves with parallel wave fronts are called **plane waves.**

Plane wave

16.1 Sound Waves in Air

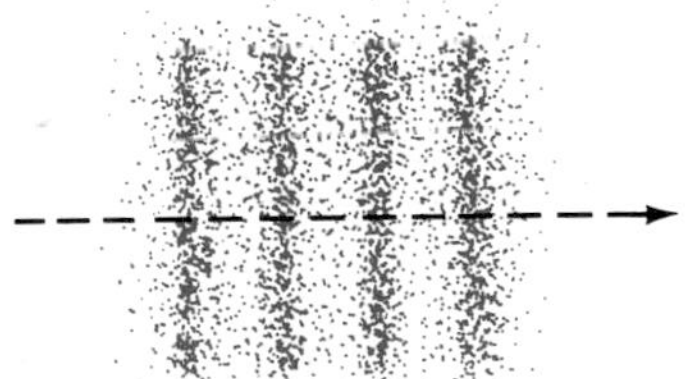

Fig. 16.4 Density changes in air in a sound wave.

A sound wave in air is a longitudinal wave in which pushes are communicated from molecule to molecule. The restoring force for such a wave is due to the pressure of air — wherever the density of molecules is higher than normal, the pressure also is higher than normal and pushes the molecules apart. Figure 16.4 shows a sound wave in air. The wave consists of alternating zones of low and high density. At the zones of low density the molecules have suffered maximum displacement — they have been displaced away from the zone. At the zones of high density, the molecules have suffered minimum displacement — they have remained stationary while molecules from the adjacent zones of maximum displacement have converged upon them. In Figure 16.4 the displacements and the density enhancements are shown much exaggerated. Even an extremely intense sound wave, such as that produced by a jet airliner at takeoff, involves displacements of only about 10^{-4} m and density changes of only about 1%.

The range of frequencies audible to the human ear extends from 20 to 20,000 Hz. These limits are somewhat variable; for instance, the ears of older people are less sensitive to high frequencies. Frequencies above 20,000 Hz are called **ultrasound;** some animals — dogs, cats, and bats — can hear these frequencies.

Ultrasound

The **intensity** of a sound wave is the power transported by this wave per square meter of wave front; the units of intensity are W/m^2. At a frequency of 10^3 Hz, the minimum intensity audible to the human ear is 1.2×10^{-12} W/m^2. This intensity is called the **threshold of hearing.** There is no upper limit for the intensity of sound; however, an intensity above 1 W/m^2 produces a painful sensation in the ear.

Intensity of sound

The intensity of sound is usually expressed on a logarithmic scale called the **intensity level.** The unit of intensity level is the **decibel** (dB); this is a dimensionless unit. The definition of intensity level is as follows: we take an intensity of 0.468×10^{-12} W/m^2 as our standard of intensity which corresponds to 0 dB; an intensity 10 times as large corresponds to 10 dB; an intensity 100 times as large corresponds to 20 dB; an intensity 1000 times as large corresponds to 30 dB, etc. (This logarithmic scale is intended to agree with our subjective impression of the intensity of sounds — we tend to underestimate sounds of great

Intensity level of sound
Decibel

intensity.) Mathematically, the relationship between intensities in watts per square meter and intensity levels in decibels is

$$[\text{intensity level in dB}] = 10 \log\left(\frac{[\text{intensity in W/m}^2]}{0.468 \times 10^{-12}\ \text{W/m}^2}\right) \tag{1}$$

Thus, the threshold of hearing (1.2×10^{-12} W/m^2) corresponds to 4 dB and the threshold of pain (1 W/m^2) corresponds to 120 dB. Note that in consequence of the mathematical properties of logarithms, a multiplicative increase of intensity by a factor of 2 yields an additive increase in intensity level of 10 log 2 dB, or 3.01 dB. Table 16.1 lists some examples of sounds of different intensities.

Table 16.1 SOME SOUND INTENSITIES

Sound	*Intensity level*
Rupture of eardrum	160 dB
Jet engine (at 30 m)	130
Threshold of pain	120
Rock music	115
Thunder (loud)	110
Subway train (New York City)	100
Heavy street traffic	70
Normal conversation	60
Whisper	20
Normal breathing	10
Threshold of hearing	4

According to Fourier's theorem (see Section 15.5), a sound wave of arbitrary shape can be regarded as a superposition of harmonic waves. The relative intensity of the harmonic waves in this superposition determines the timbre (or quality) of the sound. Pure noise, or **white noise,** consists of a mixture of harmonic waves of all frequencies with equal intensities. The musical notes emitted by a piano or a violin consist of a mixture of just a few harmonic waves: the fundamental and its first few harmonics. Figure 16.5 shows the wave forms emitted by an oboe and a violin when the musical note A is played on these instru-

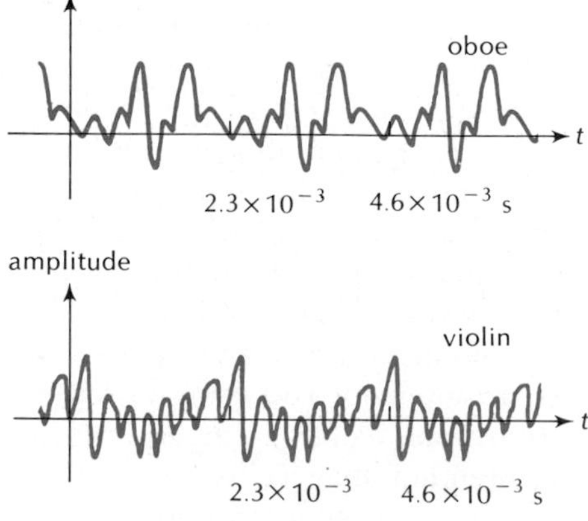

Fig. 16.5 Wave forms emitted by an oboe and by a violin playing the note A. The wave pattern repeats 440 times per second.

ments. In both cases the wave is periodic, repeating at the rate of 440 cycles per second; but the shape of the wave and the intensity of the harmonics is quite different in each case. It is because of this difference in the shapes of the waves that the ear can distinguish between diverse musical instruments.

Table 16.2 THE CHROMATIC MUSICAL SCALE

Note	Frequency[a]
C	261.7 Hz
C#	277.2
D	293.7
D#	311.2
E	329.7
F	349.2
F#	370.0
G	392.0
G#	415.3
A	440.0
A#	466.2
B	493.9

[a] Based on a frequency of 440 Hz for A.

Table 16.2 gives the frequencies of the 12 notes of the chromatic musical scale. These frequencies are based on the system of equal temperament; successive frequencies in this scale differ by a factor of $(2)^{1/12}$. The first entry in this table is middle C, with a frequency of 261.7 Hz (Figure 16.6). Any musical note not listed in Table 16.2 can be obtained by multiplying (or dividing) the given frequencies by a factor of 2, or 4, or 8, etc. Notes that differ by a factor of 2 in frequency are said to be separated by an **octave.** For example, C one octave above middle C has a frequency of 523.3 Hz, C two octaves above middle C has a frequency of 1046.6 Hz, etc.

Fig. 16.6 Middle C.

Octave

Incidentally, for a musician the absolute values of these frequencies are not as important as the ratios of the frequencies. If an orchestra tunes its instruments so that their middle C has a frequency of, say, 255 Hz, this will not do any noticeable harm to their music, provided that the frequencies of all the other notes are also decreased in proportion.

As a sound wave spreads out from its source, its intensity falls off because the area of the wave front grows larger and therefore the energy per unit area grows smaller. In a homogeneous medium the intensity of the wave decreases with the inverse square of the distance from the source. Figure 16.7 helps to make this clear; it shows a spherical wave front at successive instants of time. The wave front grows from an old radius r_1 to a new radius r_2; correspondingly, its area grows from $4\pi r_1^2$ to $4\pi r_2^2$. The total power carried by this wave front remains the same; hence the intensity, or power per unit area, varies as the inverse square of the distance:

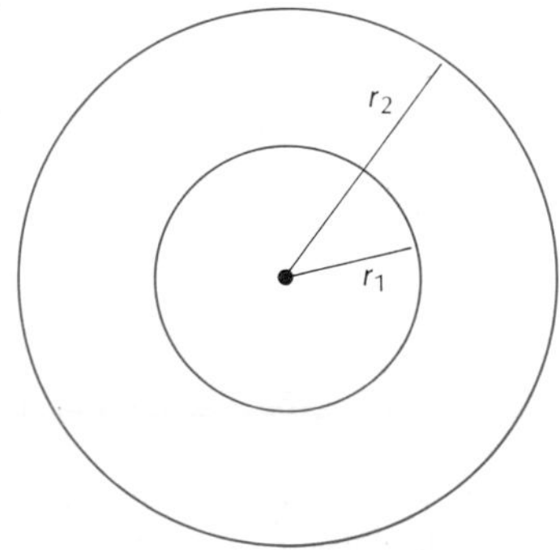

Fig. 16.7 Concentric spherical wave fronts of a sound wave in air.

$$I_2 = \frac{r_1^2}{r_2^2} I_1 \qquad (2)$$

Decrease of intensity of a sound wave

In the derivation of this simple result, we have taken for granted that the medium is homogeneous so that the wave velocity does not depend on position. We know from Section 15.4 that the power carried by the wave is proportional to the velocity; hence, if the velocity were to vary, we would have to take this extra complication into account. Furthermore, we have taken for granted that the medium is perfectly elastic so that the wave does not dissipate any of its energy by friction.

Although Figure 16.7 shows the wave fronts spreading out uniformly in all directions, Eq. (2) remains true if the wave is a beam, such as the beam of a loudspeaker with a horn, aimed in some preferred direction. In this case the intensity falls off as $1/r^2$ along the direction of the beam.

EXAMPLE 1. At a distance of 60 m from a jet airliner engaged in takeoff, the intensity level of sound is 120 dB. What is the intensity (in W/m^2) and the intensity level (in decibels) at a distance of 180 m?

SOLUTION: According to Eq. (1), an intensity level of 120 dB corresponds to

$$I_1 = 4.68 \times 10^{-12}\ W/m^2 \times 10^{[\text{intensity level in dB}]/10}$$

$$= 0.468 \times 10^{-12}\ W/m^2 \times 10^{120/10}$$

$$= 0.468\ W/m^2$$

In Eq. (2) the intensities are measured in W/m^2. At a distance of 180 m the intensity is

$$I_2 = (r_1^2/r_2^2)I_1 = (60\ m/180\ m)^2 \times 0.468\ W/m^2$$

$$= 5.2 \times 10^{-2}\ W/m^2$$

When converted to decibels, this gives an intensity level of

$$10 \log\left(\frac{5.2 \times 10^{-2}\ W/m^2}{0.468 \times 10^{-12}\ W/m^2}\right) = 110\ dB$$

Fig. 16.8 Acoustic micrograph of reverse surface of a penny coin. This picture was made with sound waves of 50 megahertz, which were sent through the coin (the obverse surface of the coin was machined smooth). Note the mottling; this is due to crystal grains in the copper. (Courtesy R. S. Gilmore, General Electric Research and Development Center.)

Ultrasonic waves of very high frequency do not propagate very well through air — they are rapidly dissipated by air molecules. However, these waves propagate readily through liquids and solids and, in recent years, this has led to the development of some interesting practical applications of ultrasonic waves. For instance, such waves are now used in place of X rays to take pictures of the interior of the human body; this permits the examination of the fetus in the body of a pregnant woman and avoids the damage that X rays might do to the very sensitive tissues of the fetus. The ultrasonic "cameras" that take such pictures employ sound waves of a frequency of about 10^6 Hz. Further development of this technique has led to the construction of acoustic microscopes. The most powerful of these devices employ ultrasound waves of a frequency in excess of 10^9 Hz to make highly magnified pictures of small samples of materials. The wavelength of sound waves of such extremely high frequency is about 10^{-6} m, roughly the same as the wavelength of ordinary light waves. The micrographs made by experimental acoustic microscopes compare favorably with micrographs made by ordinary optical microscopes (Figures 16.8 and 16.9).

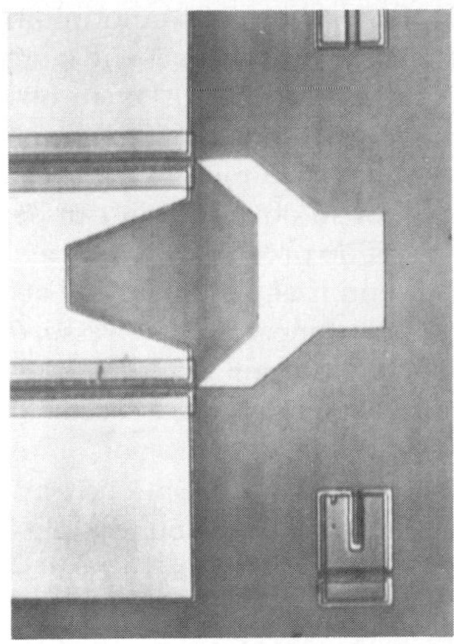

Fig. 16.9 (left) Acoustic micrograph of a portion of a transistor (250×) and (right) detail (1000×). This picture was made with sound waves of 2.7 gigahertz. Note that these sound waves give a clear view through several layers of material. (Courtesy, C. F. Quate and L. Lam, Stanford University.)

16.2 The Speed of Sound

As in the case of a wave on a string, the speed of sound in air depends on the restoring force and on the density. Since the restoring force is due to the pressure of air, we expect the speed of sound to be a function of the air pressure and the air density. A somewhat involved calculation shows that the theoretical formula for the speed of sound is

$$v = \sqrt{1.40\frac{p_0}{\rho_0}} \tag{3}$$

where p_0 and ρ_0 designate the unperturbed pressure and density, respectively. Note that this formula does display the expected increase of speed with restoring force and decrease with density. Under standard conditions (temperature 0°C, $p_0 = 1$ atm $= 1.01 \times 10^5$ N/m², $\rho_0 =$ 1.29 kg/m³), the value of the **speed of sound** is 331 m/s.

Speed of sound in air

The speed of sound in liquids and in metals is higher than that in air because the restoring force is much larger — liquids and solids offer much more opposition to compression than do gases. Table 16.3 gives the values of the speed of sound in some materials.

Table 16.3 The Speed of Sound in Some Materials

Material	v
Air	
0°C, 1 atm	331 m/s
20°C, 1 atm	344
100°C, 1 atm	386
Helium, 0°C, 1 atm	965
Water (distilled)	1497
Water (sea)	1531
Aluminum	5104
Iron	5130
Glass	5000–6000
Granite	6000

A simple method for the measurement of the speed of sound in air takes advantage of standing waves in a tube open at one end and closed at the other (Figure 16.10). The standing wave must then obviously have a displacement node at the closed end, since the motion of the air is restricted by the wall at this end. The wave must have a displacement antinode at the open end; this is not so obvious, but can be understood by first considering the pressure. The pressure at the open end must remain constant because the open end is accessible to the atmosphere and hence any incipient decrease or increase of pressure would immediately lead to an inflow or outflow of air from the surrounding atmosphere, canceling the pressure change. Thus, the atmosphere behaves as a reservoir of constant pressure. This is not quite exact because some sound waves are radiated from the open end and produce a periodic oscillation of the pressure of the surrounding atmosphere, but it is a good approximation because the pressure oscillations of the surrounding atmosphere are much smaller than the oscillations within the tube. The open end is therefore a pressure node; as we saw in the preceding section, this corresponds to a displacement antinode.

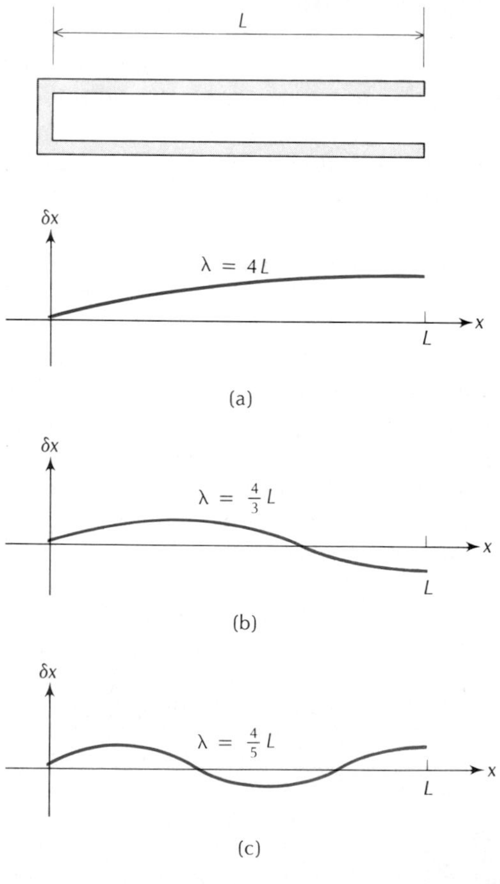

Fig. 16.10 Possible standing waves in a tube open at one end: (a) the fundamental mode, (b) the first harmonic, and (c) the second harmonic.

Wavelengths and eigenfrequencies for standing waves in a tube open at one end

With these boundary conditions at the ends of the tube, the possible standing waves are those shown in Figure 16.10. If the length of the tube is L, then the wavelengths of the normal modes are

$$\boxed{\lambda_1 = 4L, \qquad \lambda_2 = \tfrac{4}{3}L, \qquad \lambda_3 = \tfrac{4}{5}L, \ \ldots} \tag{4}$$

and the corresponding eigenfrequencies are

$$\nu_1 = \frac{v}{4L}, \qquad \nu_2 = \frac{3v}{4L}, \qquad \nu_3 = \frac{5v}{4L}, \ldots \tag{5}$$

or, in general,

$$\boxed{\nu_n = n\frac{v}{4L} \qquad n = 1, 3, 5, \ldots} \tag{6}$$

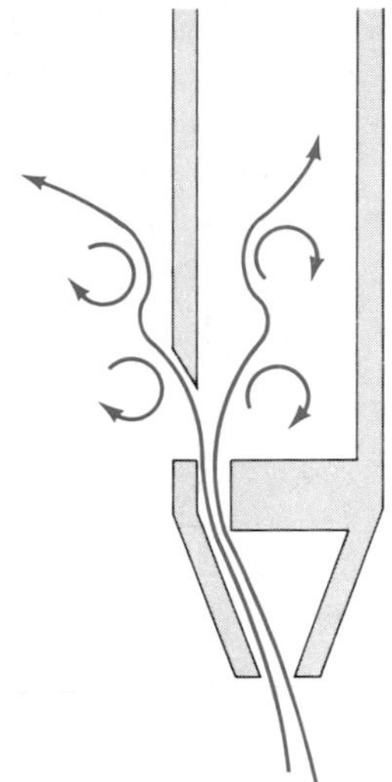

Fig. 16.11 Vortices at the blowhole of an organ pipe.

Note that the expressions (4) for the wavelengths of the normal modes of a tube differ from the expressions (15.58) for the wavelengths of the normal modes of a string. This is due to the difference in boundary conditions: the tube has a node at one end and an antinode at the other end, whereas the string has nodes at both ends.

The speed of sound can be determined by measuring the resonant frequency of a tube of known length. This method is convenient but not very accurate, because the walls of the pipe exert a certain amount of friction on the air, which slightly affects the velocity of sound waves.

Most musical instruments involve standing waves in cavities of diverse shapes. Organs, flutes, trumpets, and other wind instruments are essentially tubes within which standing waves are excited by a stream of air blown across a blowhole. That a steady stream of air can excite oscillations depends on a subtle phenomenon occurring at the edges of the blowhole. As the air streams past the edge, it forms a vortex (Figure 16.11). This vortex soon breaks away from the edge and is replaced by another vortex, and another, etc. The regular succession of vortices constitutes an oscillation of the stream of air and this can excite the oscillation of a standing wave by resonance.

The resonant frequency of the tube is determined by its length. In many wind instruments — flutes, trumpets, bassoons — the effective length of the tube can be varied by opening or closing valves.

Stringed instruments — violins, guitars, mandolins — use a resonant cavity to amplify and modify the sound of the string. These cavities are mechanically coupled to the string, and the vibrations of the latter excite resonant vibrations of the former. The resonant vibrations not only involve standing waves in the air in the cavity, but also standing waves in the solid material (wood) of the walls. Because the area of the body of, say, a violin is much larger than the area of its strings, the body pushes against much more air and radiates sound more effectively than the strings. Hence most of the sound from a violin emerges from its body.

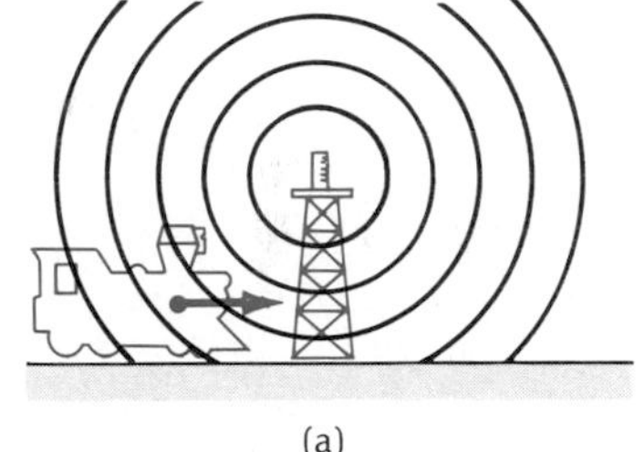

(a)

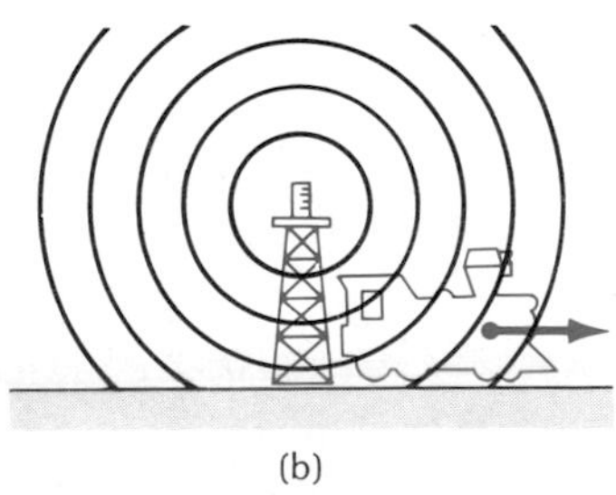

(b)

Fig. 16.12 (a) Train approaching a siren. The train encounters *more* wave fronts per unit time than when stationary. (b) Train receding from a siren. The train encounters *fewer* wave fronts per unit time than when stationary.

16.3 The Doppler Effect

Under standard conditions, the speed of a sound wave is 331 m/s when measured in the rest frame of the air. But when measured in a reference frame moving through the air, the speed of the sound wave will be larger or smaller, depending on the direction of motion of the reference frame. For example, if a train moving at 30 m/s approaches a stationary siren (Figure 16.12a), then the speed of the sound waves relative to the train will be 361 m/s; if the train moves away from the siren (Figure 16.12b), the speed will be 301 m/s. .

The motion of the train affects not only the speed of the sound waves, but also their frequency. For instance, if the train approaches the siren, it runs head on into the sound waves (Figure 16.12a) and hence encounters more wave fronts per unit time than if it were stationary; if the train recedes from the siren, it runs with the sound waves (Figure 16.12b) and hence encounters fewer wave fronts. Consequently, a receiver on the train will detect a higher frequency when approaching the siren, and a lower frequency when receding. To calculate the frequency shift, we note that in the reference frame of the ground we have the usual relation between the frequency, speed, and wavelength of sound,

$$\nu = v/\lambda \tag{7}$$

and in the reference frame of the train we have a corresponding relation

$$\nu' = v'/\lambda \tag{8}$$

The wavelengths in Eqs. (7) and (8) are exactly the same because the distance between wave crests does not depend on the reference frame. Dividing Eq. (8) by Eq. (7), we obtain

$$\frac{\nu'}{\nu} = \frac{v'}{v} \tag{9}$$

We will designate the speed of the train acting as receiver of sound waves by V_R. In the reference frame of the train, the speed of sound is $v' = v \pm V_R$, where the positive sign corresponds to motion of the train toward the source of sound and the negative sign to motion away from the source of sound. Consequently, Eq. (9) yields

$$\nu' = \nu\left(\frac{v \pm V_R}{v}\right)$$

or

Frequency at moving receiver

$$\boxed{\nu' = \nu\left(1 \pm \frac{V_R}{v}\right)} \qquad \begin{array}{l} + \text{ for approaching receiver} \\ - \text{ for receding receiver} \end{array} \tag{10}$$

The fractional difference between the frequencies in the two reference frames is then

Doppler shift for moving receiver

$$\boxed{\frac{\Delta\nu}{\nu} = \frac{\nu' - \nu}{\nu} = \pm\frac{V_R}{v}} \tag{11}$$

This frequency change due to motion of the receiver is called the **Doppler shift.**

EXAMPLE 2. Suppose that the stationary siren emits a tone of a frequency 440 Hz as the train approaches it at 30 m/s. What is the frequency received on the train?

SOLUTION: From Eq. (10)

$$\nu' = \nu\left(1 + \frac{V_R}{v}\right) = 440\text{ Hz}\left(1 + \frac{30\text{ m/s}}{331\text{ m/s}}\right) = 480\text{ Hz}$$

Of course, Eq. (10) applies not only to sound waves, but also to water waves and other waves. Note that if the receiver is moving away from the source at a velocity $V_R = v$, then the frequency ν' is zero; this simply means that the receiver is moving exactly with the waves so that no wave fronts catch up with it. If the receiver is moving away at a velocity greater than v, then Eq. (10) gives a negative frequency; this means that the receiver overruns the waves from behind.

EXAMPLE 3. A motorboat speeding at 15 m/s is moving in the same direction as a group of water waves of frequency 0.17/s and velocity 9.3 m/s (relative to the water). What is the frequency with which wave crests pound on the motorboat?

SOLUTION: We again use Eq. (10):

$$\nu' = \nu\left(1 - \frac{V_R}{v}\right) = \left(0.17/\text{s}\right)\left(1 - \frac{15\text{ m/s}}{9.3\text{ m/s}}\right)$$
$$= -0.10/\text{s}$$

The negative sign indicates that the motorboat overtakes the waves at the rate of 0.10/s, i.e., one wave every 10 seconds.

A shift between the frequency emitted by a source of sound waves (or other waves) and the frequency detected by a receiver will also occur if the source is in motion and the receiver is at rest. For example, if a train approaching a station blows its whistle, the successive wave fronts will be crowded together in the forward direction and spaced apart in the rearward direction (Figure 16.13). Consequently, a stationary receiver will detect a higher frequency when in front of the train, and a lower frequency when in rear. As the train rushes by the stationary receiver, the detected frequency suddenly changes from high to low. This explains the sudden drop in pitch that you hear when standing next to the railroad track as a whistling train passes by you.

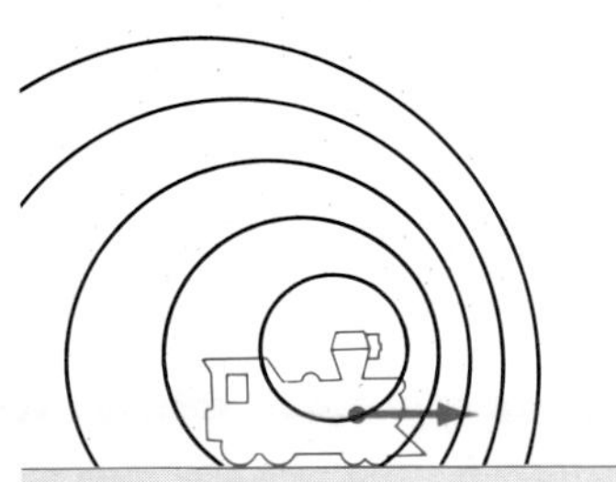

Fig. 16.13 Train emitting sound waves while in motion. The wavelength ahead of the train is shorter, and that behind the train is longer than when the train is stationary.

We will designate the speed of the train when it acts as an emitter of sound waves by V_E. To calculate the frequency change produced by motion of the emitter, we begin by noting that in the time $1/\nu$ corresponding to one period, the train travels a distance $(1/\nu)V_E$ and hence the wavelength is shortened from its normal value λ to a new value $\lambda' = \lambda \mp V_E/\nu$. The new frequency is therefore

$$\nu' = \frac{v}{\lambda'} = \frac{v}{\lambda \mp V_E/\nu} = \frac{v}{v/\nu \mp V_E/\nu} \tag{12}$$

or

Frequency at receiver for moving emitter

$$\boxed{\nu' = \nu\left(\frac{1}{1 \mp V_E/v}\right)} \quad \begin{array}{l} - \text{ for approaching emitter} \\ + \text{ for receding emitter} \end{array} \tag{13}$$

This gives the Doppler shift for a moving emitter.

EXAMPLE 4. Suppose that the whistle of a train emits a tone of a frequency 440 Hz as the train approaches a stationary observer at 30 m/s. What frequency does the observer hear?

SOLUTION: According to Eq. (13),

$$\nu' = \nu\frac{1}{1 - V_E/v} = \frac{440 \text{ Hz}}{1 - (30 \text{ m/s})/(331 \text{ m/s})}$$

$$= 484 \text{ Hz}$$

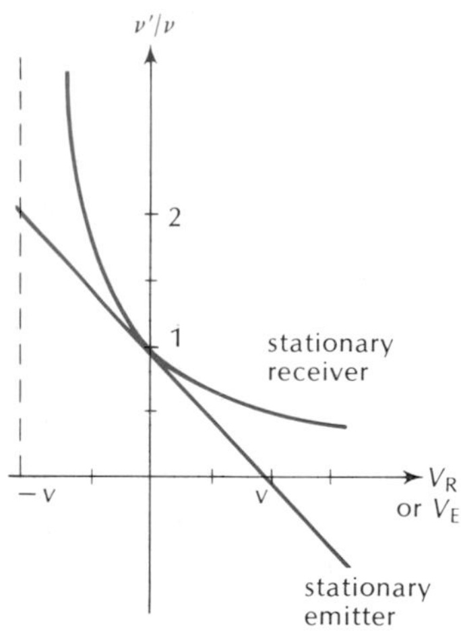

Fig. 16.14 The ratio ν'/ν as a function of velocity.

If we compare the results of Examples 2 and 4, we see that motion of the source and motion of the receiver have nearly the same effect on the frequency — in both examples the frequency is increased by about 10%. This symmetry of the Doppler shifts has to do with the low speed of the motion. Whenever V_E is small compared to v, we can use the approximation

$$\frac{1}{1 - x} = 1 + x + \cdots \tag{14}$$

in Eq. (13) to obtain the approximation

$$\frac{1}{1 \mp V_E/v} \cong 1 \pm V_E/v + \cdots \tag{15}$$

As a consequence

$$\nu' \cong \nu(1 \pm V_E/v) \tag{16}$$

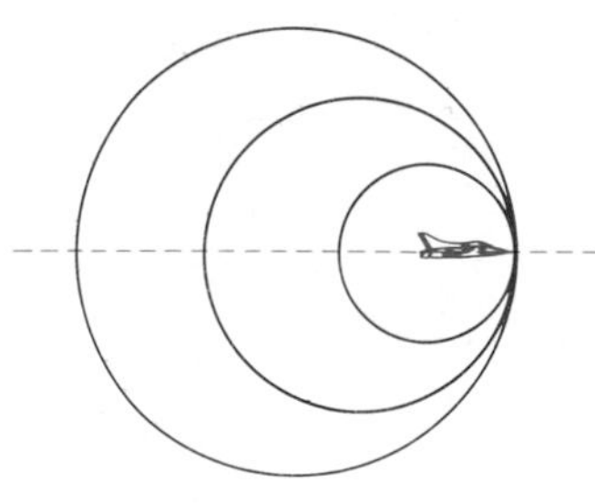

Fig. 16.15 A subsonic aircraft of a speed very close to the speed of sound emitting sound waves.

This equation has the same form as Eq. (10) and shows that, at low velocities, the frequency shift is the same whether the emitter or the receiver is moving. Figure 16.14 is a plot of ν'/ν as a function of V_R or V_E.

Finally, let us consider the case of an emitter, such as a fast aircraft, moving with a velocity nearly equal to the velocity of sound. If the aircraft emits sound of some frequency ν, then Eq. (13) indicates that the frequency received at points just ahead of the airplane is very large — in the limiting case $V_E \to v$, the frequency becomes infinite. This is so because all the wave fronts are infinitely bunched together and they all arrive at almost the same instant as the aircraft (Figure 16.15). If the velocity of the aircraft exceeds the velocity of sound, then the aircraft will overtake the wave fronts (Figure 16.16). In this case the sound is always confined to a conical region that has the aircraft as its apex and moves with the aircraft at the speed V_E; ahead of this region the air has not yet been disturbed, although it will be disturbed when the aircraft moves sufficiently far to the right. The cone is called the **Mach cone.**

Mach cone

The half angle of the apex of the Mach cone is given by

$$\boxed{\sin\theta = v/V_E} \tag{17}$$

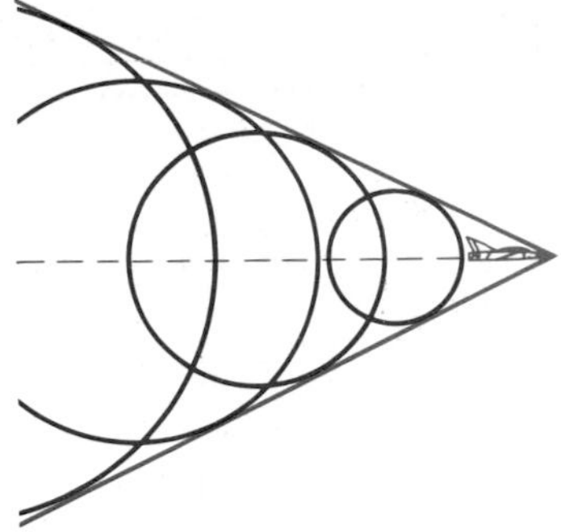

Fig. 16.16 A supersonic aircraft emitting sound waves.

This can readily be seen from Figure 16.17, which shows the aircraft at a time t and the wave front that was emitted by the aircraft at time zero. The radius of the wave front is vt. This radius is the opposite side of a right triangle of angle θ and of hypotenuse V_Et. Consequently,

$$\sin\theta = vt/V_Et \tag{18}$$

which is equivalent to Eq. (17).

Any supersonic aircraft (or other body) will generate a Mach cone, regardless of whether or not it carries an artificial source of sound aboard (Figure 16.18). The motion of the aircraft through the air creates a pressure disturbance which spreads outward with the speed of sound and forms the cone. The cone trails behind the aircraft much as the wake trails behind a ship. The sharp pressure disturbance at the surface of the cone is heard as a loud bang whenever the cone sweeps over the ear. This is the **sonic boom.** For a large aircraft, such as the Concorde SST, the noise level of the sonic boom reaches the pain threshold even if the aircraft is 20 km away.

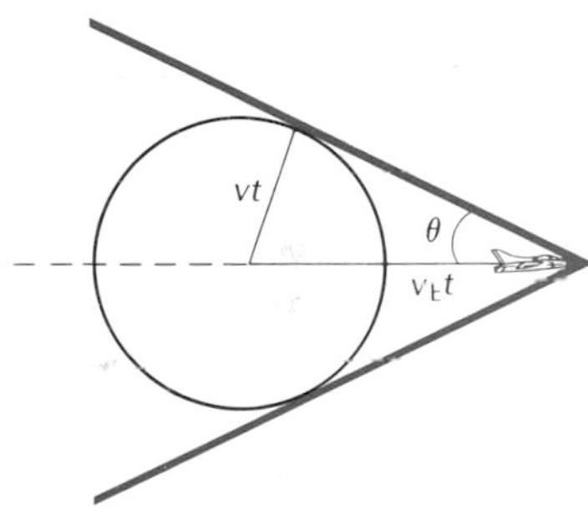

Fig. 16.17 Mach cone.

Fig. 16.18 A .30-caliber bullet and its Mach cone. The bullet has just smashed through a plate of Plexiglas. Note the faint Mach cones of the many Plexiglas fragments. (Courtesy H. E. Edgerton, MIT.)

16.4 Water Waves

Capillary waves and gravity waves

Waves on the surface of water are of two kinds: **capillary waves** and **gravity waves.** The former are ripples of fairly short wavelength — no more than a few centimeters — and the restoring force that produces them is the surface tension of water. This is nothing but the attraction between neighboring water molecules; surface tension tends to keep the surface of, say, a pond flat just as though a (weak) rubber sheet were stretched along the surface.

Gravity waves have wavelengths above half a meter; typically, their wavelengths range from several meters to several hundred meters. The restoring force that produces these waves is the pull of gravity, which tends to keep the water surface at its lowest level. In the following discussion we will deal exclusively with gravity waves.

The speed of these waves depends on their wavelength and also on the depth h of the water. We will consider two extreme cases: deep water ($h \gg \lambda$) and shallow water ($h \ll \lambda$). If the water is *deep,* the speed is independent of the depth, but is dependent on the wavelength. It can be shown that the formula for the speed is[1]

Speed of wave in deep water

$$\boxed{v = \sqrt{g\lambda/2\pi}} \tag{19}$$

This formula is exact if the depth is infinite, and it is a good approximation whenever the depth is more than three wavelengths. The frequency of the wave is

$$\nu = v/\lambda = \sqrt{g/2\pi\lambda} \tag{20}$$

and the period is

$$T = \lambda/v = \sqrt{2\pi\lambda/g} \tag{21}$$

Note that this expression for the period resembles the expression for the period of a simple pendulum [$T = 2\pi\sqrt{l/g}$; see Eq. (14.57)]; the resemblance is no coincidence, since the motion of the particles in a water wave is essentially a swinging motion under the influence of gravity. Figure 16.19 is a plot of the speed of waves in deep water as a function of their wavelength.

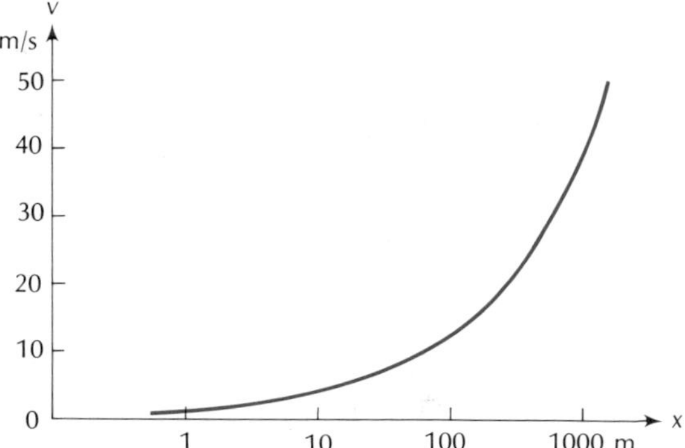

Fig. 16.19 Speed of ocean waves in deep water.

If the water is *shallow,* the speed is dependent on the depth, but is independent of the wavelength. The formula for the speed is[2]

Speed of wave in shallow water

$$\boxed{v = \sqrt{gh}} \tag{22}$$

[1] Actually, this is the phase velocity (see the discussion in Section 15.3). The value of the group velocity of these waves is half as large.

[2] This is the phase velocity and it is also the group velocity. For these waves, the two velocities coincide.

This formula is a good approximation whenever the depth is less than $\frac{1}{10}$ of the wavelength.

In a water wave, whether in deep or in shallow water, the motion of the water particles is not only up and down but also back and forth. Figure 16.20 shows the orbits of a few water particles in a wave in deep water. The particles on the surface move in circles of a radius

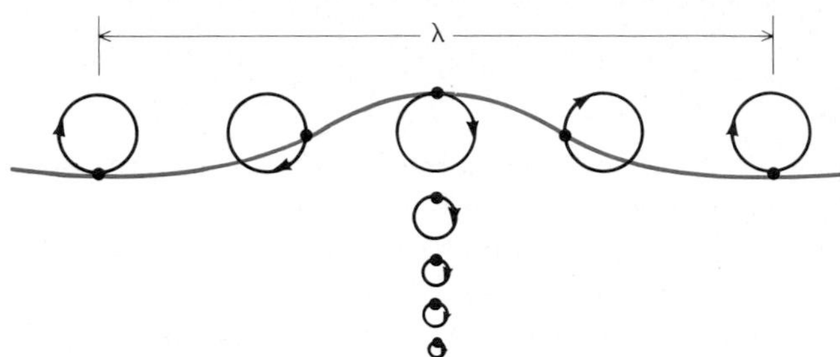

Fig. 16.20 Motion of water particles in an ocean wave in deep water.

equal to the amplitude of the wave. The particles below the surface also move in circles, but of smaller radius. Note that the shape of the surface is sinusoidal. This is a good approximation for waves of small amplitude (compared to their wavelength). Waves of larger amplitude tend to form sharper peaks and flatter troughs (Figure 16.21). If the amplitude is excessively large, then the crest tends to break (Figure 16.22); empirically, one finds that a wave breaks when its height (measured from trough to crest) is more than one-seventh of its length.

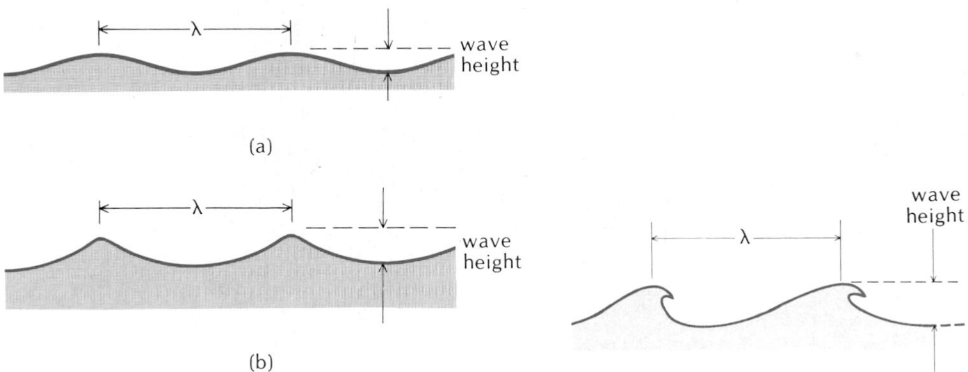

Fig. 16.21 (a) Ocean wave of small amplitude. (b) Ocean wave of large amplitude.

Fig. 16.22 Breaking wave crests.

The waves on the ocean are generated by the wind. Turbulence and eddies in the wind play a large role in this, giving the water repeated pushes that build up waves. The typical amplitudes and wavelengths that will be generated depend on the speed of the wind, how long it blows, and over how large an area. Oceanographers can forecast the sea conditions from the wind conditions. Strong winds blowing for a long time over a large area will generate gigantic waves. In one recorded incident, the U.S.S. *Ramapo* encountered a 34-m wave (from trough to crest) during a 7-day storm in the North Pacific.

Waves of extremely long wavelength are produced by earthquakes of the ocean floor. The sudden motion of the ocean floor disturbs the water and generates waves on the surface, called **tidal waves** or **tsu-**

namis. The wavelengths of these waves are typically 100–400 km. This means that the wavelength is much larger than the depth of the water (the mean depth of the Pacific Ocean is 4.3 km). Under these conditions the water of the ocean must be regarded as shallow; the formula of Eq. (22) for the speed is then applicable. For tsunamis in the Pacific Ocean, this gives a speed of

Fig. 16.23 Tsunami striking the beach near the Putumaile Hospital at Hilo, Hawaii, on April 1, 1946.

$$v = \sqrt{9.8 \text{ m/s}^2 \times 4.3 \times 10^3 \text{ m}} = 205 \text{ m/s}$$

$$= 740 \text{ km/h!}$$

In the open sea, tsunamis have a height of only a meter or less; they pass under ships without being noticed. But when they run into the shallow water of the continental shelf, the leading waves slow down [see Eq. (22)] — this causes the waves to pile up and their height then reaches tens of meters. After the great Alaskan earthquake of 1964, a tidal wave 70 m high appeared at Valdez on the Gulf of Alaska. Figure 16.23 shows a tsunami, about 5 m high, reaching the shore at Hilo, Hawaii. Tsunamis easily travel from one end of the Pacific basin to the other. For example, in 1960 a tsunami originating in southern Chile damaged harbors as far away as Mexico and California (Figure 16.24).

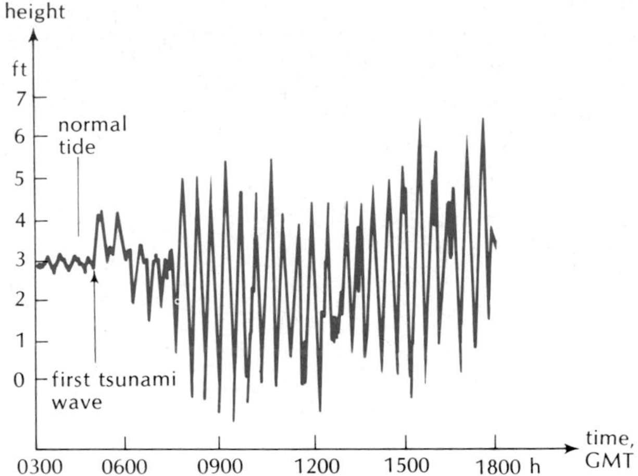

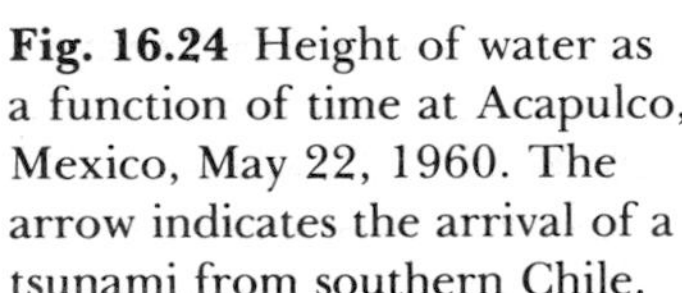

Fig. 16.24 Height of water as a function of time at Acapulco, Mexico, May 22, 1960. The arrow indicates the arrival of a tsunami from southern Chile.

16.5 Seismic Waves

Seismic waves are propagating vibrations in the body of the Earth. Where these vibrations reach the surface of the Earth, they are felt as a trembling of the ground, or an earthquake (Figure 16.25). The propagation of seismic waves in earth involves a mechanism similar to the propagation of sound waves in the air — the restoring force is due to the elasticity of earth. However, the periods of seismic waves are usually longer than $\frac{1}{10}$ s, i.e., the periods are longer than those of audible sound. The amplitudes of seismic waves range from as small as 10^{-6} m to as large as several meters (Figure 16.26).

The initial disturbance that generates the seismic wave is a sudden rupture within the solid crust of the Earth. The place of rupture may be several kilometers or several hundred kilometers below the surface. The point on the surface directly above the rupture is called the **epi-**

Fig. 16.25 Cracks in the pavement on Union Street in San Francisco after the great earthquake of 1906.

center of the earthquake. The waves that spread outward from the initial disturbance are of two kinds: transverse waves and longitudinal waves. Seismologists designate these as **S waves** and **P waves.** The former (*S*hear waves) can only propagate through the solid crust and mantle of the Earth, whereas the latter (*P*ressure waves) can also penetrate through the liquid core. When these waves reach the surface of the Earth or the core of the Earth, they are reflected, i.e., they bounce off at an angle (Figure 16.27). During this reflection the S and P waves can generate yet another kind of wave: a **surface wave.** This wave propagates along the surface of the Earth or along the surface of the core of the Earth; it is similar to a water wave propagating along the surface of the ocean.

S and P waves

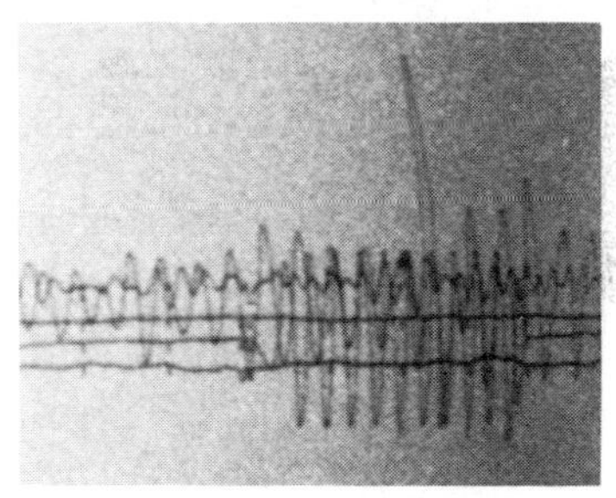

Fig. 16.26 Seismic wave recorded by a seismometer. The sudden increase in the amplitude of the wriggles indicates the arrival of the wave.

The total energy of the seismic waves generated during a large earthquake may be as much as 10^{17} or 10^{18} J. The energy is usually expressed by the **Richter magnitude scale,** according to which the magnitude M of the earthquake is

Richter magnitude scale

$$\boxed{M = 0.67 \log E - 2.9} \qquad (23)$$

where E is the energy in joules. For example, an earthquake with an energy of 10^{17} J has a Richter magnitude $M = 8.5$ — about the magnitude of the great San Francisco earthquake of 1906.

The speed of seismic waves depends on the characteristics of the material within the Earth. P waves have speeds of about 5 km/s in the crust and speeds as great as 14 km/s in the deepest part of the mantle; S waves have speeds of 3–8 km/s. The difference in speed between P and S waves implies that the P waves reach the surface ahead of the S waves. The time lag between these waves can be used to calculate the distance they have traveled.

Careful analysis of the propagation of seismic waves reveals the characteristics of the material through which they have traveled. What we know about the interior of the Earth is largely due to the analysis of seismic waves. For instance, we know that the Earth has a liquid core, because S waves (transverse waves) do not propagate through this core (see Figure 16.27). In recent years the study of seismic waves has acquired military significance. Underground explosions of nuclear bombs generate seismic waves that can be detected by sensitive seismometers over large distances. This permits us to monitor underground nuclear explosions all around the globe.

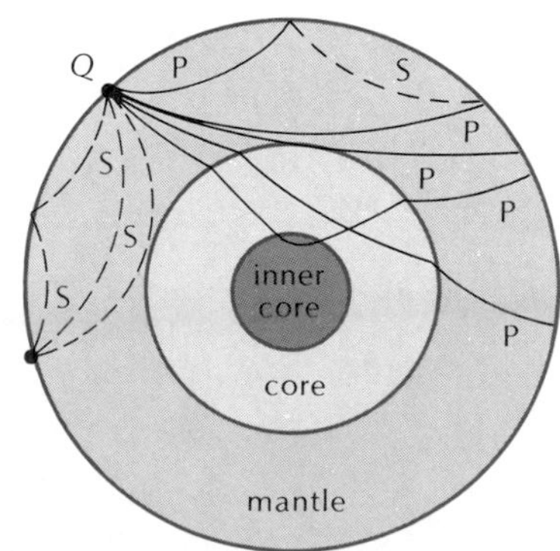

Fig. 16.27 Propagation of S and P waves within the Earth. The waves originate at point Q near the surface of the Earth. The S waves cannot propagate in the (liquid) core.

16.6 Diffraction

Fig. 16.28 Ocean waves incident on breakwater.

It is a characteristic feature of waves that they will deflect around the edges of obstacles placed in their path and penetrate into the shadow zone behind the obstacle. For example, Figure 16.28 shows water waves striking a breakwater at the entrance to a harbor. The region directly behind the breakwater is outside of the direct path of the waves, but nevertheless waves reach this region because each wave front spreads sideways once it has passed the entrance to the harbor. This lateral spreading of the wave fronts can be easily understood: the breakwater cuts a segment out of each wave front. The segments of wave front cannot just keep moving straight on as though nothing had happened — the end of each segment is a vertical wall of water where the breakwater has chopped off the wave front. The water at the ends will immediately begin to run out sideways, producing a disturbance at the edge of the segment. This disturbance continues to spread and gradually forms the curved wave fronts to the left and to the right of the main beam of the wave.

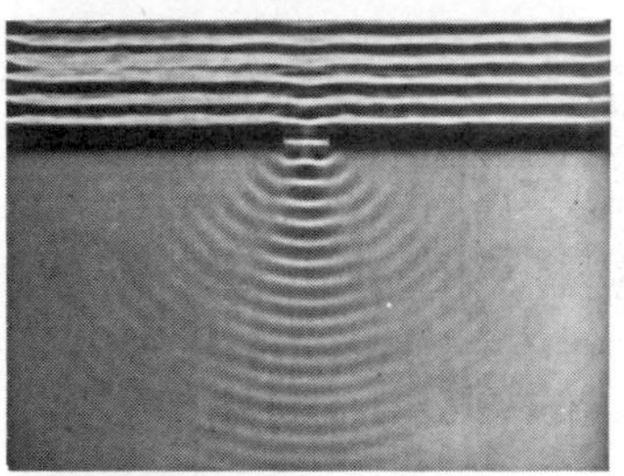

Fig. 16.29 Water waves of fairly long wavelength in a ripple tank exhibit strong diffraction when passing through a gap.

The deflection of waves by the edge of an obstacle is called **diffraction.** It is a general rule that the amount of diffraction suffered by a wave passing through a gap depends on the ratio of wavelength to the size of the gap. An increase of the wavelength (or a decrease of the gap size) makes the diffraction effects more pronounced; a decrease of the wavelength (or an increase of the gap size) makes the diffraction effects less pronounced. For instance, Figure 16.29 shows diffraction of waves of relatively long wavelength by a small gap; the waves spread out very strongly, forming divergent, fanlike beams of concentric wave fronts. Figure 16.30 shows diffraction of waves of shorter wavelength; here the wave spreads out only slightly; most of the wave remains within a straight beam of nearly parallel wave fronts. Note that in Figure 16.30 the beam has a fairly well-defined edge — the region in the "shadow" of the breakwater remains nearly undisturbed while the region in front of the gap receives the full impact of the waves.

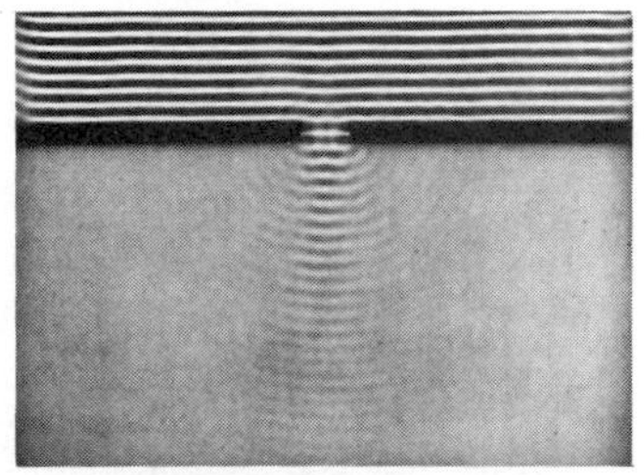

Fig. 16.30 Water waves of shorter wavelength exhibit less diffraction.

The fanlike beams of waves originating at the gap constitute a **diffraction pattern.** In Figure 16.29 there is a central beam and two clearly recognizable secondary beams on each side. The beams are separated by nodal lines along which the wave amplitude is zero. Figures 16.31 and 16.32 show the diffraction patterns generated by a small island. Note that if the wavelength is large compared to the size of the island, then there exists no shadow zone; instead, the island merely produces some distortion of the waves.

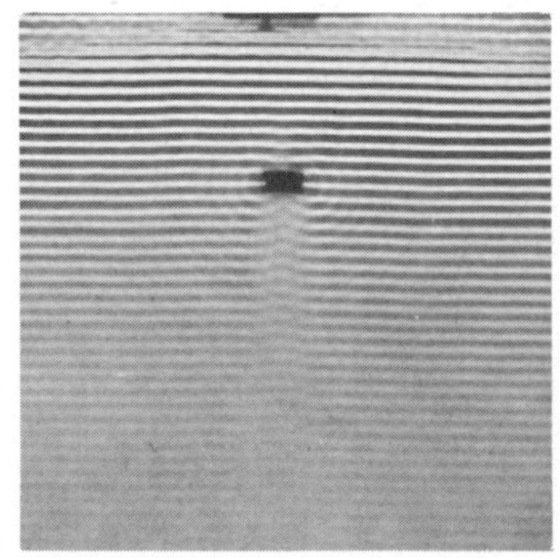

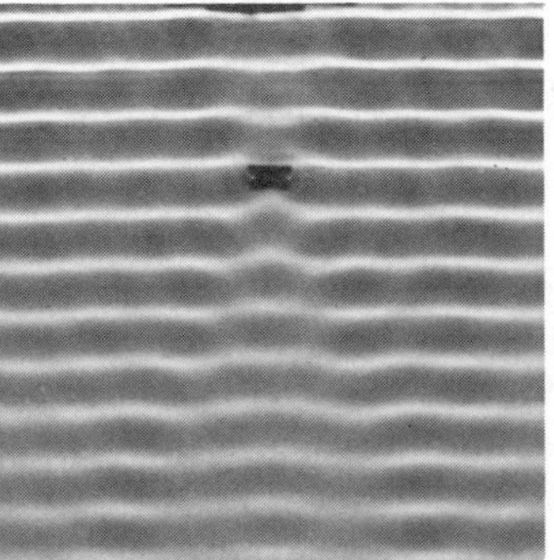

Fig. 16.31 (left) Diffraction pattern generated by a small island; short wavelength.

Fig. 16.32 (right) Diffraction pattern generated by a small island; long wavelength.

The diffraction pattern generated by some obstacle depends in a complicated way on the shape of the obstacle. In Chapter 39 we will develop some mathematical formulas for the intensity of the beams of diffracted waves. Although the main concern of Chapter 39 is the behavior of light waves, the same mathematical treatment applies to other kinds of waves. Incidentally, one of the crucial experiments that convinced physicists early in the nineteenth century that light is a wave was a diffraction experiment. Augustin Fresnel, following up the earlier investigations of Thomas Young, showed that light exhibits diffraction effects quite similar to those of any other kind of wave.

Diffraction also plays an important role in the propagation of sound. We can listen to, and talk to, a person standing out of our line of sight beyond the corner of, say, a house because the sound waves diffract around the corner.

SUMMARY

Decrease of intensity of sound wave with distance:

$$I_2 = I_1(r_1^2/r_2^2)$$

Speed of sound in air: 331 m/s

Standing waves in pipe open at one end:

$$\lambda_1 = 4L, \qquad \tfrac{4}{3}L, \qquad \tfrac{4}{5}L, \ldots$$

Doppler shift:

stationary emitter: $\nu' = \nu(1 \pm V_R/v)$

stationary receiver: $\nu' = \dfrac{\nu}{1 \mp V_E/v}$

Doppler shift for small speed: $\dfrac{\Delta\nu}{\nu} = \pm\dfrac{V_R}{v}$ or $\pm\dfrac{V_E}{v}$

Mach cone: $\sin\theta = v/V_E$

Speed of water waves:

deep water: $v = \sqrt{g\lambda/2\pi}$

shallow water: $v = \sqrt{gh}$

QUESTIONS

1. Could an astronaut be heard playing the violin while standing on the surface of the moon?

2. You can estimate your distance from a bolt of lightning by counting the seconds between seeing the flash and hearing the thunder, and then dividing by three to obtain the distance in kilometers. Explain this rule.

3. A hobo can hear a very distant train by placing an ear against the rail. How does this help? (Hint: Ignoring frictional losses, how does the intensity of sound decrease with distance in air? In the rail?)

4. If you speak while standing in a corner with your face toward the wall, you will sometimes notice that your voice sounds unusually loud. Explain.

5. Why does the wind whistle in the rigging of a ship or in the branches of a tree?

6. How does a flutist play different musical notes on a flute?

7. When inside a boat, you can often hear the engine noises of another boat much more loudly than when on deck. Can you guess why?

8. Many men like singing in shower stalls because the stall somehow enhances their voice. How does this happen? Would the effect be different for men and women?

9. The intensity level of sound near a rock band is 115 dB. What is the intensity level of sound near two such rock bands playing together?

10. The pitch of the vowels produced by the human voice is determined by the frequency of standing waves in several resonant cavities (larynx, pharynx, mouth, and nose). In an amusing demonstration experiment, a volunteer inhales helium gas and then speaks a few words. As long as his resonant cavities are filled with helium gas, the pitch of his voice will be much higher than normal. Given that the speed of sound in helium is about three times as large as in air, calculate the factor by which the eigenfrequencies of his resonant cavities will be higher than normal.

11. The pipes that produce the lowest frequencies in a great organ are very long, usually 16 ft. Why must they be long? These pipes are also very thick. Why would a thin pipe give a poor performance?

12. Some of the old European opera houses and concert halls renowned for their acoustic excellence have very irregular walls, heavily encrusted with an abundance of stucco ornamentation which reflects sound waves in almost all directions. How does the sound reaching a listener in such a hall differ from the sound reaching a listener in a modern concert hall with four flat, plain walls?

13. Electric guitars amplify the sound of the strings electronically. Do such guitars need a body?

14. Does the temperature of the air affect the pitch of a flute? A guitar?

15. The human auditory system is very sensitive to small differences between the arrival times of a sound signal at the right and left ears. Explain how this permits us to perceive the direction from which a sound signal arrives.

16. The depth finder (or "fish finder") on a boat sends a pulse of sound toward the bottom and measures the time an echo takes to return. The dial of the depth finder displays this echo time on a scale directly calibrated in distance units. Experienced operators can tell whether the bottom is clean rock, or rock covered by a layer of mud, or whether a school of fish is swimming somewhere above the bottom. What echo times would you expect to see displayed on the dial of the depth finder in each of these instances?

17. The helmsman of a fast motorboat heading toward a cliff sounds his horn. A woman stands on the top of the cliff and listens. Compare the frequency of the horn, the frequency heard by the woman, and the frequency heard by the helmsman in the echo from the cliff. Which of these three is the highest frequency? Which the lowest?

18. Two automobiles are speeding in opposite directions while sounding their horns. Describe the changes of pitch that each driver hears as they pass by one another.

19. A man is standing north of a woman while a strong wind is blowing from the south. If the man and the woman yell at each other, how does the wind affect the pitch of the voice of each as heard by the other?

20. A Concorde SST passing overhead at an altitude of 20 km produces a

sonic boom with an intensity level of 120 dB lasting about half a second. How does this compare with some other loud noises? Would it be acceptable to let this aircraft make regular flights over populated areas?

21. Many people have reported seeing UFOs traveling through air noiselessly at speeds much greater than the speed of sound. If the UFO consisted of a solid impenetrable body, would you expect its motion to produce a sonic boom? What can you conclude from the absence of sonic booms?

22. When an ocean wave approaches a beach, its height increases. Why?

23. Occasionally ocean waves passing by a harbor entrance will excite very high standing waves ("seiche") within the harbor. Under what conditions will this happen?

24. Seismic waves of the S and P types have different speeds. Explain how a scientist at a seismometer station can take advantage of this difference in speed to determine the distance between his station and the point of origin of the waves.

25. The amplitude of an ocean wave initially decreases as the wave travels outward from its point of origin; but when the wave has traveled a quarter of the distance around the Earth, its amplitude *increases.* Explain how this comes about. (Hint: If the wave were to travel half the distance around the Earth, it would converge on a point, if no continents block its progress.)

26. Underground nuclear explosions generate seismic waves. How could you discriminate between the seismic waves received from such an explosion and the seismic waves from an earthquake? (Hint: Would you expect an explosion to produce mainly S waves or mainly P waves?)

27. You are standing on the surface of the Earth; a seismic wave approaches from below. If the wave is an S wave, how will it shake you? If the wave is a P wave?

28. Why do the Japanese build their houses out of wood slats and paper?

29. Figure 16.33 shows a seismometer, an instrument used to detect and measure seismic waves. A vertical post is firmly set in the ground and a large mass is suspended from it by a rigid horizontal beam and a diagonal wire. The beam ends in a sharp point that rests against the post; the beam is therefore free to swing in the horizontal plane. Describe how the beam will swing if the ground moves and tilts the post. For what direction of incidence of a seismic wave is this seismometer most sensitive?

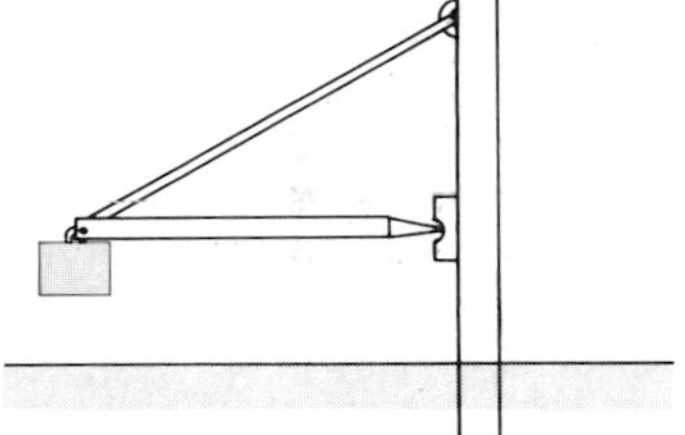

Fig. 16.33 Seismometer.

30. If you are standing on the south side of a house, you can speak to a friend standing on the east side, out of sight around the corner. How do your sound waves reach into the shadow zone?

PROBLEMS[3]

Section 16.1

1. The range of frequencies audible to the human ear extends from 20 to 20,000 Hz. What is the corresponding range of wavelengths?

2. The lowest musical note available on a piano is A, three octaves below that listed in Table 16.2; and the highest note available is C, four octaves above that listed in Table 16.2. What are the frequencies of these notes?

3. A violin has four strings, each of them 0.326 m long. When vibrating in

[3] In all the problems assume that the speed of sound in air is 331 m/s, unless otherwise stated.

their fundamental modes, the four strings have frequencies of 196, 294, 440, and 659 Hz, respectively.

(a) What is the wavelength of the standing wave on each string? What is the wavelength of the sound wave generated by the string?
(b) What is the frequency and the wavelength of the first harmonic mode on each string? What is the corresponding wavelength of the sound wave generated by each string?
(c) According to Table 16.2, to what musical tones do the frequencies calculated above correspond?

4. A mandolin has strings 34.0 cm long fixed at their ends. When the mandolin player plucks one of these strings, exciting its fundamental mode, this string produces the musical note D (293.7 Hz; see Table 16.2). In order to produce other notes of the musical scale, the player shortens the string by holding a portion of the string against one or another of several frets (small transverse metal bars) placed underneath the string. The player shortens the string by one fret to produce the note D#, by two frets to produce the note E, by three frets to produce the note F, etc. Calculate the correct spacing between the successive frets of the mandolin for one complete octave. Assume that the string always vibrates in its fundamental mode and assume that the tension in the string is always the same.

5. For each of the entries in Table 16.1, calculate the intensity in W/m^2.

6. Suppose that a whisper has an intensity level of 20 dB at a distance of 0.5 m from the speaker's mouth. At what distance will this whisper be below your threshold of hearing?

7. A loudspeaker receives 8 W of electric power from an audio amplifier and converts 3% of this power into sound waves. Assuming that the loudspeaker radiates the sound uniformly over a hemisphere (a vertical and horizontal angular spread of 180°), what will be the intensity and the intensity level at a distance of 10 m in front of the loudspeaker?

8. An old-fashioned hearing trumpet has the shape of a flared funnel, with a diameter of 8 cm at its wide end and a diameter of 0.7 cm at its narrow end. Suppose that all of the sound energy that reaches the wide end is funneled into the narrow end. By what factor does this hearing trumpet increase the intensity of sound (measured in W/m^2)? By how many decibels does it increase the intensity level of sound?

Section 16.2

9. Spectators at soccer matches often notice that they hear the sound of the impact of the ball on the player's foot (or head) sometime after seeing this impact. If a spectator notices that the delay time is about 0.5 s, how far is he from the player?

10. In the past century a signal gun was fired at noon at most harbors so that the navigators of the ships at anchor could set their chronometers. This method is somewhat inaccurate, because the sound signal takes some time to travel the distance from gun to ship. If this distance is 2.0 mi, how long does the signal take to reach the ship? Can you suggest a better method for signaling noon?

11. In fresh water, sound travels at a speed of 1460 m/s. In air, sound travels at a speed of 331 m/s. Suppose that an explosive charge explodes on the surface of a lake. A woman with her head in the water hears the bang of the explosion and, lifting her head out of the water, she hears the bang again 5 s later. How far is she from the site of the explosion?

12. In order to measure the depth of a ravine, a physicist standing on a bridge drops a stone and counts the seconds between the instant he releases the stone

and the instant he hears it strike some rocks at the bottom. If this time interval is 6.0 s, how deep is the ravine? Take into account the travel time of the sound signal, but ignore air friction.

13. A bat can sense its distance from the wall of a cave (or whatever) by emitting a sharp ultrasonic pulse that is reflected by the wall. The bat can tell the distance from the time the echo takes to return.

(a) If a bat is to determine the distance of a wall 10 m away with an error of less than ± 0.5 m, how accurately must it sense the time interval between emission and return of the pulse?

(b) Suppose that a bat flies into a cave filled with methane (swamp gas). By what factor will this gas distort the bat's perception of distances? The speed of sound in methane is 432 m/s.

14. The ultrasonic range finder on a new automatic camera sends a pulse of sound to the target and determines the distance by the time an echo takes to return.

(a) If the range finder is to determine a distance of 50 cm with an error no larger than ± 2 cm, how accurately (in seconds) must it measure the travel time?

(b) If you aim this camera at an object placed beyond a sheet of glass (a window or a glass door), on what will the camera focus?

15. Because the human auditory system is very sensitive to small differences between the arrival times of a sound signal at the right and the left ear, we can perceive the direction from which a sound signal arrives to within about 5°. Suppose that a source of sound (a ringing bell) is 10 m in front of and 5° to the left of a listener. What is the difference in the arrival times of sound signals at the left and right ears? The separation between the ears is about 15 cm.

16. Consider a tube of length L open at both ends. Show that the eigenfrequencies of standing sound waves in this tube are

$$\nu_n = n\frac{v}{2L} \qquad n = 1, 2, 3, \ldots$$

Draw diagrams similar to those of Figure 16.10 showing the displacement amplitude for each of the first four standing waves.

17. The largest pipes in a great organ usually have a length of about 16 ft. These pipes are open at both ends so that a standing sound wave will have a displacement antinode at each end. What is the frequency of the fundamental mode of such a pipe?

18. A flute can be regarded as a tube open at both ends. It will emit a musical note if the flutist excites a standing wave in the air column in the tube.

(a) The lowest musical note that can be played on a flute is C (261.7 Hz; see Table 16.2). What must be the length of the tube? Assume that the air column is vibrating in its fundamental mode (see Problem 16).

(b) In order to produce higher musical notes, the flutist opens valves arranged along the side of the tube. Since the holes in these valves are large, an open valve has the same effect as shortening the tube. The flutist opens one valve to play C#, two valves to play D, etc. Calculate the successive spacings between the valves of a flute for one complete octave. (The actual spacings used on flutes differ slightly from the results of this simple theoretical evaluation because the mouth cavity of the flutist also resonates and affects the frequency.)

19. The human ear canal is approximately 2.7 cm long. The canal can be regarded as a tube open at one end and closed at the other. What are the eigenfrequencies of standing waves in this tube? The ear is most sensitive at a frequency of about 3000 Hz. Would you expect that resonance plays a role in this?

20. Consider a tube of length L closed at both ends. Show that the eigenfrequencies of standing sound waves in this tube are

$$\nu_n = n\frac{v}{2L} \qquad n = 1, 2, 3, \ldots$$

Draw diagrams similar to those of Figures 16.10 showing the displacement amplitude for each of the first four standing waves.

Section 16.3

21. In an experiment performed shortly after the discovery of the Doppler effect by the Austrian physicist Christian Doppler in 1842, several trumpeters were placed on a train and told to play a steady musical tone. As the train sped by a listener standing on the side of the track, the pitch perceived by the listener suddenly dropped by a noticeable amount. Suppose that the train had a speed of 60 km/h and that the trumpets on the train sounded the note of E (329.7 Hz; see Table 16.2). What was the frequency of the note perceived by the listener on the ground when the train was approaching? When the train was receding? Approximately to what musical notes do these Doppler-shifted frequencies correspond?

22. Ocean waves with a wavelength of 330 ft have a period of 8 s. A motorboat, with a speed of 30 ft/s, heads directly into the waves. What is the speed of the waves relative to the motorboat? With what frequency do wave crests hit the front of the motorboat?

23. The horn of an automobile emits a tone of frequency 520 Hz. What frequency will a pedestrian hear when the automobile is approaching at a speed of 85 km/h? Receding at the same speed?

24. Two automobiles are driving on the same road in opposite directions. The speed of the first automobile is 90.0 km/h and that of the second is 60.0 km/h. The horns of both automobiles emit tones of frequency 524 Hz. Calculate the frequency that the driver of each automobile hears coming from the other automobile. Assume that there is no wind blowing along the road.

25. Repeat Problem 24 under the assumption that a wind of 40 km/h blows along the road in the same direction as that of the faster automobile.

26. A train approaches a mountain at a speed of 75 km/h. The train's engineer sounds a whistle that emits a frequency of 420 Hz. What will be the frequency of the echo that the engineer hears reflected off the mountain?

27. Suppose that a moving train carries a source of sound and also a receiver of sound so that both have the same velocity relative to the air. Show that in this case the Doppler shift due to motion of the source cancels the Doppler shift due to motion of the receiver — the frequency detected by the receiver is the same as the frequency generated by the source.

28. The whistle on a train generates a tone of 440 Hz as the train approaches a station at 30 m/s. A wind blows at 20 m/s in the same direction as the motion of the train. What is the frequency that an observer standing at the station will hear?

29. Figure 16.18 shows the shock wave of a bullet speeding through air. Measure the angle of the Mach cone and calculate the speed of the bullet.

30. The Concorde SST has a cruising speed of 2160 km/h.
 (a) What is the half angle of the Mach cone generated by this aircraft?
 (b) If the aircraft passes directly over your head at an altitude of 12,000 m, how long after this instant will the shock wave strike you?

31. You may have noticed that at the instant a fast (but subsonic) jet aircraft

passes directly over your head, the sound it makes seems to come from a point behind the aircraft.

(a) Show that the direction from which the sound seems to come makes an angle θ with the vertical such that $\sin\theta = V_E/v$, where V_E is the speed of the aircraft and v the speed of sound.

(b) If you hear the sound from an angle of 30° behind the aircraft, what is the speed of the aircraft?

Section 16.4

32. A giant, freak wave encountered by a weather ship in the North Atlantic was 77.0 ft high from trough to crest; its wavelength was 1150 ft and its period 15.0 s. Calculate the maximum vertical acceleration of the ship as the wave passed underneath; calculate the maximum vertical velocity. Assume that the motion of the ship was purely vertical.

33. A rule of thumb known to naval architects is that the speed of a ship through the water cannot exceed the speed of an ocean wave of wavelength equal to the length of the hull of the ship.[4] This limiting speed is called the **hull speed** of the ship. The reason for this rule is that a ship passing through the water makes waves; when the speed of the ship reaches the hull speed, the crest of one of these waves will be at the bow of the ship, and the crest of the next at the stern. The ship is then permanently trapped in a wave trough — any increase of engine power (or sail power) will merely push the ship against the wave crest at its bow; this will feed more energy into the wave, making it higher, but produce only an insignificant increase in the speed of the ship.

(a) Show that the hull speed (in meters per second) is given by the formula $v = 1.2\sqrt{L}$, where L is the length of the ship (in meters). Naval architects prefer to measure speed in knots (1 knot = 1 nautical mi/h = 1.15 mi/h) and length in feet. Show that in these units the formula becomes $v = 1.3\sqrt{L}$.

(b) What is the hull speed for a sailboat of length 30 ft? A clipper ship of length 200 ft?

34. The passenger of an airplane flying over a ship notices that ocean waves are smashing into the ship regularly at the rate of 10 per minute. He also notices that the length of the ship is about the same as three wavelengths. Deduce the length of the ship from this information; assume the ship is at rest.

35. The National Ocean Survey has deployed buoys off the Atlantic Coast to measure ocean waves. Such a buoy detects waves by the vertical acceleration that it experiences as it is lifted and lowered by the waves. In order to calibrate the device that measures the acceleration, scientists placed the buoy on a Ferris wheel at an amusement park. If the *vertical* acceleration (as a function of time) of a buoy riding on a Ferris wheel of a radius 20 ft rotating at 6 rev/min is to simulate the vertical acceleration of a buoy riding a wave, what would be the amplitude and wavelength of this wave? Assume that the waves are in deep water and that the buoy always rides on the surface of the water.

36. Figure 16.24 is a record of a tsunami striking the coast of Mexico.

(a) Approximately, what was the frequency of this wave?

(b) What was its wavelength and its speed while still in the open sea? Assume that the depth of the open sea (Pacific) is 4.3 km.

37. In the open sea, a tsunami usually has an amplitude less than 30 cm and a wavelength longer than 80 km. Assume that the speed of the tsunami is 740 km/h. What is the maximum vertical velocity and acceleration that such a tsunami will give to a ship floating on the water? Will the crew of the ship notice the passing of the tsunami?

---

[4] This rule does not apply to boats that skim along the surface of the water, relying on hydrodynamic lift to keep most of their hull out of the water.

38. The Bay of Fundy (Nova Scotia) is about 250 km long and about 100 m deep.
   (a) What is the speed of waves of long wavelength in the bay?
   (b) What are the frequency and the period of the fundamental mode of oscillation of the bay? Use the speed calculated in part (a) and treat the bay as a long, narrow tube open at one end and closed at the other.
   (c) The period of the tidal pull exerted by the Moon is about 12 h. Would you expect that the very large tidal oscillations (with a height of up to 50 ft) observed in the Bay of Fundy are due to resonance?

39. Suppose that a tsunami spreads out along the surface of the Pacific Ocean in concentric circles from its source. If the distance from the source is less than 1000 or 2000 km, the surface of the Earth can be (approximately) regarded as flat and the intensity of the wave therefore decreases as the inverse of the distance (see the argument given in the introduction to this chapter). However, if the distance is several thousand kilometers, then the curvature of the surface of the Earth must be taken into account in the calculation of the intensity.
   (a) Show that the intensity of the wave varies as $1/\sin(r/R)$, where r is the distance measured along the surface of the Earth and R is the radius of the Earth.
   (b) Show that if $r \ll R$, this variation of intensity is proportional to $1/r$, as expected.
   (c) Show that for $r > 10{,}000$ km, the intensity *increases* with increasing distance.
   (d) Suppose that the tsunami has an amplitude of 30 cm when at a distance of 4000 km from its source. What will be the amplitude at 8000 km? At 10,000 km? At 12,000 km? (Hint: The intensity is proportional to the square of the amplitude.)

40. In the open sea, where the depth h of the water is large, the amplitude of a tsunami is small; but when the tsunami enters the coastal shallows, where h is small, the amplitude becomes large.
   (a) Show that the amplitude varies as $A \propto 1/h^{1/4}$. (Hint: As in the case of a wave on a string, the power carried by a tsunami is proportional to $v\omega^2A^2$. Assume that the power in the wave remains constant as it enters the coastal shallows; also note that the frequency remains constant.)
   (b) Suppose that a tsunami has an amplitude of 35 cm when in the open sea, where $h = 4.3$ km. If this wave reaches a coastal region where $h = 10$ m, what will be its amplitude? Its height from trough to crest?

41. Experienced sailors know that when waves traveling in one direction encounter a tidal stream flowing in the opposite direction, the waves become dangerously steep and high. Show that if a wave of speed v in still water runs into a stream of water flowing in the opposite direction at speed V, then the wavelength will decrease by a factor $1 - V/v$. Qualitatively, explain why the amplitude of the wave will increase.

Section 16.5

42. In the crust of the Earth, seismic waves of the P type have a speed of about 5 km/s; waves of the S type have a speed of about 3 km/s. Suppose that after an earthquake, a seismometer placed at some distance first registers the arrival of P waves and 9 min later the arrival of S waves. What is the distance between the seismometer and the source of the waves?

43. By what factor is the energy in the seismic waves of an earthquake of magnitude 8.0 on the Richter scale larger than in those of an earthquake of magnitude 4.0?

44. The shock of an underground nuclear explosion generates seismic waves that can be detected at a large distance with sensitive seismometers. What is the magnitude, on the Richter scale, of the earthquake generated by an under-

ground explosion equivalent to 1 megaton of TNT? The explosion of 1 ton of TNT releases an energy of 4.2×10^9 J. Assume that 30% of the energy of the explosion goes into seismic waves.

45. The great earthquake that struck Lisbon, Portugal, in 1755 had an (estimated) magnitude of 9 on the Richter scale. What was the energy of this earthquake? Express your answer in the equivalent of megatons of TNT; the explosion of 1 ton of TNT releases 4.2×10^9 J.

46. Many inhabitants of Tangshan, China, reported that during the catastrophic earthquake of July 28, 1976, they were thrown 2 m into the air as if by a "huge jolt from below."

(a) With what speed must a body be thrown upward to reach a height of 2 m?

(b) Assume that the vertical motion of the ground was simple harmonic with a frequency of 1 Hz. What amplitude of the vertical motion is required to generate a speed equal to that calculated in part (a)?

CHAPTER 17

The Theory of Special Relativity

As we saw in Chapter 5, Newton's laws of motion are equally valid in every inertial reference frame. Consequently, no mechanical experiment can detect any intrinsic difference between two inertial reference frames. This is the Newtonian principle of relativity. The example already mentioned in Section 5.7 illustrates this principle very concretely: the behavior of billiard balls on a pool table on a ship steaming away from shore at constant velocity is not different from the behavior of similar billiard balls on the shore. Experiments with such billiard balls aboard the ship will not reveal the uniform motion of the ship. To detect this motion, the crew of the ship must take sightings of points on the shore or use some other navigational technique that fixes the position and velocity of the ship in relation to the shore. Hence, in regard to mechanical experiments, uniform translational motion of our inertial reference frame is always relative motion — it can only be detected as motion of our reference frame with respect to another reference frame. There is no such thing as *absolute* motion through space.

The question naturally arises whether the relativity of mechanical experiments also applies to electric, magnetic, optical, and other experiments, some of which we will encounter in later chapters. Do any of these experiments permit us to detect an absolute motion of our reference frame through space? Albert Einstein answered this question in the negative. He laid down a principle of relativity for *all* laws of physics. Before we deal with the details of Einstein's theory of relativity, we will briefly describe why nonmechanical experiments — and, in particular, experiments with light — might be expected to detect absolute motion which mechanical experiments cannot detect.

17.1 The Speed of Light; the Ether

Since the laws of mechanics are the same in all inertial reference frames, it seems quite natural to assume that the laws of electricity, magnetism, and optics are also the same in all inertial reference frames. But this assumption immediately leads to a paradox concerning the velocity of light. As we will see in Chapter 36, light is an oscillating electric and magnetic disturbance propagating through space. The laws of electricity and magnetism permit us to deduce that the speed of propagation of this disturbance must always be 3.00×10^8 m/s, in all reference frames. The trouble with this deduction is that according to the Galilean addition law for velocity [Eq. (4.58)], the speed of light ought *not* to be the same in all reference frames. For instance, imagine that an alien spaceship approaching the Earth with a velocity of, say, 1.00×10^8 m/s flashes a light signal toward the Earth; if this light signal has a velocity of 3.00×10^8 m/s in the reference frame of the spaceship, then the Galilean addition law tells us that it ought to have a velocity of 4.00×10^8 m/s in the reference frame of the Earth.

To resolve this paradox, we must either give up the notion that the laws of electricity and magnetism (and the values of the speed of light) are the same in all inertial reference frames or else we must give up the Galilean law for the addition of velocities. Both alternatives are unpleasant: the former means that we must abandon all hope for a principle of relativity embracing electricity and magnetism, and the latter means that together with the Galilean velocity transformation [Eq. (4.58)], we must throw out the Galilean coordinate transformations [Eqs. (4.49)–(4.56)] as well as the intuitively "obvious" notions of absolute time and absolute length from which these transformation equations were derived.

Since the failure of a relativity principle embracing electricity and magnetism seems to be the lesser of two evils, let us first explore this alternative. Let us assume that there exists a preferred inertial reference frame in which the laws of electricity and magnetism take their simplest form. In this reference frame the speed of light has its standard value 3.00×10^8 m/s (which we will designate by c), whereas in any other reference frame it is larger or smaller according to the Galilean addition law. The propagation of light is then analogous to the propagation of sound. There exists a preferred reference frame in which the equations for the propagation of sound waves in, say, air take their simplest form: the reference frame in which the air is at rest. In this reference frame, sound has its standard speed of 331 m/s. In any other reference frame, the equations for the propagation are more complicated, but the velocity of propagation can always be obtained very directly from the Galilean addition law. For instance, if a wind of 40 m/s (a hurricane) is blowing over the surface of the Earth, then sound waves have a speed of 331 m/s relative to the air, but their speed relative to the ground depends on direction — downwind the speed is 371 m/s, whereas upwind it is 291 m/s.

This analogy between the propagation of sound and of light suggests that there exists some pervasive medium that serves as the propagator of light. Presumably this ghostly medium fills all of space, even the interplanetary and interstellar space which is normally regarded as a vacuum. The physicists of the nineteenth century called this hypothetical medium the **ether** and they attempted to describe light waves as oscillations of the ether. The preferred reference frame in which light has its standard velocity is then the reference frame in which the ether is at rest. The existence of such a preferred reference frame would imply that velocity is absolute — the ether frame sets an absolute standard of rest and the velocity of any body could always be referred to this frame. For instance, instead of describing the velocity of the Earth relative to a material object, such as the Sun, we could always describe its velocity relative to the ether.

Ether

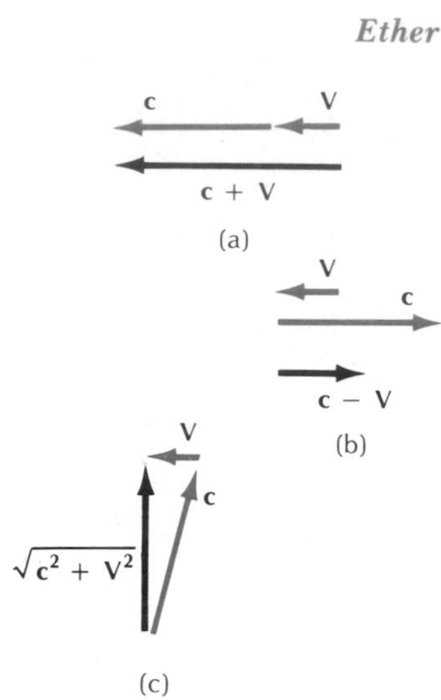

Fig. 17.1 The velocity of light is **c** relative to the ether and the velocity of the ether is **V** relative to the laboratory. The velocity of light relative to the laboratory is then the vector sum **c + V.** (a) If **c** and **V** are parallel, the magnitude of the vector sum is $c + V$. (b) If **c** and **V** are antiparallel, the magnitude of the vector sum is $c - V$. (c) If **c** and **V** are perpendicular, the magnitude of the vector sum is $\sqrt{c^2 - V^2}$.

Presumably the Earth has some nonzero velocity relative to the ether. Even if the Earth were at rest in the ether at one instant, this condition could not last, since the Earth continually changes its motion as it orbits around the Sun. The motion of the ether past the Earth was called the *ether wind* by the nineteenth-century physicists. If the Sun is at rest in the ether, then the ether wind would have a velocity opposite to that of the Earth around the Sun — about 30 km/s; if the Sun is in motion, then the ether wind would vary with the seasons — smaller than 30 km/s during one-half of the year and greater than 30 km/s during the other half.

Experimenters attempted to detect this ether wind by its effect on the propagation of light. A light wave on the surface of the Earth would have a greater speed when moving downwind and a reduced speed when moving upwind or across the wind. If the speed of the ether wind "blowing" through a laboratory is V, then the speed of light in this laboratory is $c + V$ for a light signal with downwind motion, $c - V$ for upwind motion, and $\sqrt{c^2 - V^2}$ for motion perpendicular across the wind (Figure 17.1) If V has a value of about 30 km/s, then the increase or decrease of the speed of light amounts to only about 1 part in 10^4 — a very sensitive apparatus is required for the detection of this small change.[1]

Michelson–Morley experiment

In a famous experiment first performed in 1881 and often repeated thereafter, A. A. Michelson and E. W. Morley attempted to detect small changes in the speed of light by means of an interferometer. In this instrument (Figure 17.2) a light beam is split into two separate beams which are made to travel along mutually perpendicular paths of equal lengths. One beam travels along one arm of the instrument parallel to the direction of the (hypothetical) ether wind and the other beam travels along the other arm perpendicular to the direction of the ether wind. At the ends of the arms, the light beams are reflected by mirrors and they return to the point at which they separated. The waves will meet crest to crest (constructive interference) or crest to trough (destructive interference), depending on the difference, if any, in travel time. This interference phenomenon serves as a sensitive indicator of any difference of travel time (further details will be given in Section 38.3).

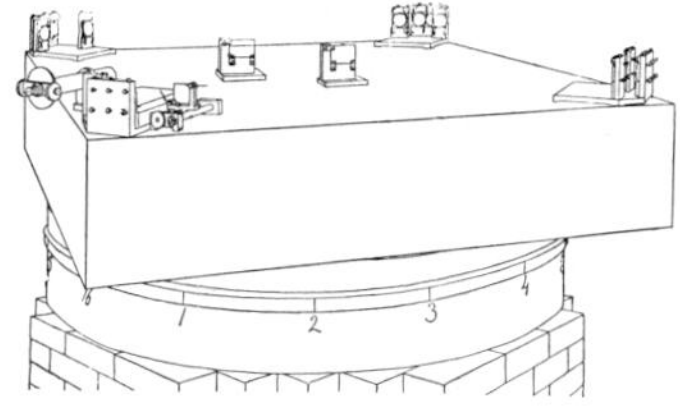

Fig. 17.2 The apparatus of Michelson and Morley.

Michelson and Morley failed to detect any ether wind. The sensitivity of their original experiment was such that a wind of 5 km/s would have produced a noticeable effect. Since the expected wind is about 30 km/s, the experimental result contradicts the ether theory of the propagation of light. Later, more refined versions of the experiment

[1] In practice, the experiments make a comparison between the average speeds for *round trips,* upwind and downwind vs. across the wind. The average speeds for these round trips differ by only 1 part in 10^8 — this imposes *extreme* demands on the sensitivity of the apparatus.

established that if there were an ether wind, its speed would certainly be less than 1.5 km/s. What is more, some new experiments with gamma rays[2] have shown that if there is any ether wind, its speed is less than about 3 m/s. The experimental evidence therefore establishes beyond reasonable doubt that the motion of the Earth has no effect on the propagation of light. The propagation of light is *not* analogous to the propagation of sound — there is no preferred reference frame. As the Earth moves around the Sun, its velocity changes continuously, and the Earth shifts from one inertial reference frame to another. But all these inertial reference frames appear to be completely equivalent in regard to the propagation of light.

Albert Einstein, *1879–1955, German (and Swiss, and American) theoretical physicist, professor at Zurich, at Berlin, and at the Institute for Advanced Study at Princeton. Einstein was the most celebrated physicist of this century. He formulated the theory of Special Relativity in 1905 and the theory of General Relativity in 1916. In the judgment of Bertrand Russell "relativity is probably the greatest synthetic achievement of the human intellect up to the present time." Einstein also made incisive contributions to modern quantum theory, for which he received the Nobel Prize in 1921. Einstein spent the last years of his life in an unsuccessful quest for a unified theory of forces that was supposed to incorporate gravity and electricity in a single set of equations.*

17.2 Einstein's Principle of Relativity

Since both the laws of mechanics and the laws for the propagation of light fail to reveal any intrinsic distinction between different inertial reference frames, in 1905 Einstein took a bold step forward and laid down a general hypothesis concerning *all* the laws of physics. This hypothesis is the **Principle of Relativity:**

All the laws of physics are the same in all inertial reference frames.

This deceptively simple principle forms the foundation of the theory of Special Relativity. Since the laws for the propagation of light are included in the laws of physics, one immediate corollary of the Principle of Relativity is

The speed of light (in vacuum) is the same in all inertial reference frames; it always has the value $c = 3.0 \times 10^8$ *m/s.*

As we pointed out in the preceding section, this invariance of the speed of light conflicts with the Galilean law for the addition of velocities. We will therefore have to throw out this law and we will also have to throw out the Galilean coordinate transformation on which it is based. In the next section we will see what new coordinate transformation replaces the Galilean transformation.

The invariance of the speed of light also requires that we give up some of our intuitive, everyday notions of space and time. Obviously, the fact that a light signal should always have a speed of 3.0×10^8 m/s, regardless of how hard we try to move toward it or away from it in a fast spaceship, does violence to our intuition. This strange behavior of light is only possible because of a strange behavior of length and time. As we will see later in this section, neither length nor time is absolute — they both depend on the reference frame in which they are measured and suffer contraction or dilation when the reference frame changes.

Before we can inquire into the consequences of the Principle of Relativity, we must carefully describe the construction of reference frames and the synchronization of clocks. A reference frame is a coordinate grid and a set of clocks (Figure 17.3) which can be used to determine the space and time coordinates of any event, i.e., any point in space and time. The grid intersections correspond directly to the space coor-

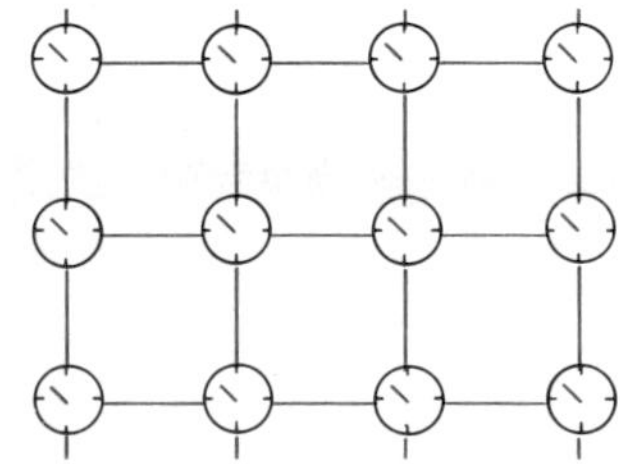

Fig. 17.3 A reference frame consisting of a coordinate grid and synchronized clocks.

[2] Gamma rays are a form of very energetic light (see Interlude B).

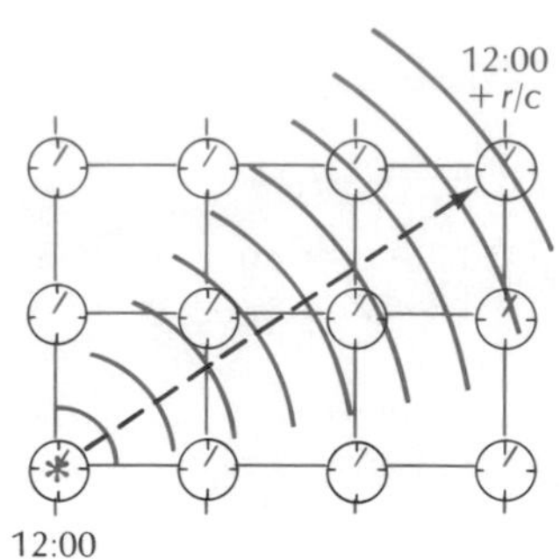

Fig. 17.4 Synchronization procedure for two clocks separated by a distance r. A flash of light is sent from one clock to the other.

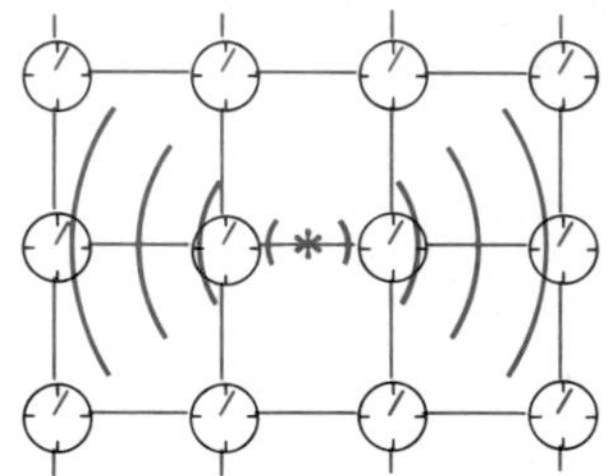

Fig. 17.5 Alternative synchronization procedure. A flash of light is sent from the midpoint toward each clock.

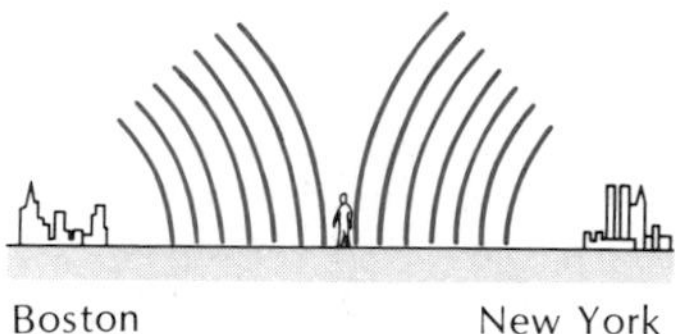

Fig. 17.6 Observer on the ground at the midpoint between Boston and New York.

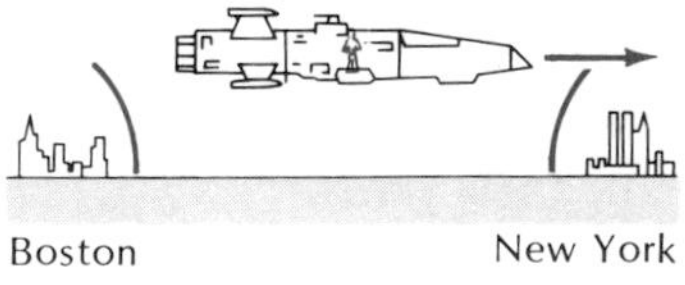

Fig. 17.7 Observer in a fast spaceship, instantaneously at the midpoint between Boston and New York. Light flashes have just been emitted.

dinates (with respect to some fiducial point chosen as origin). The time registered by the clock nearest the event is the time coordinate. Of course, the clocks must all be synchronized with the master clock sitting at the origin of coordinates. This can be accomplished by sending out a flash of light or a radio signal from the origin when the clock at this point indicates, say, 12:00 noon. In our synchronization procedure we must take into account that light requires a finite time to reach a clock at some distance from the origin. The flash of light will take a time r/c to reach a clock at a distance r from the origin. The clock at this distance is therefore synchronized with the clock at the origin if it indicates a time 12:00 noon $+ r/c$ at the instant when the light reaches it (Figure 17.4). Alternatively, the synchronization can be accomplished by sending out a flash of light from a point exactly midway between the clock at the origin and the other clock. The two clocks are synchronized if both show exactly the same time when the light from the midpoint reaches them (Figure 17.5). Note that both of these synchronization procedures depend on the universality of the speed of light. If the speed of light were not constant, but were dependent on the reference frame and on direction (as in the ether theory), then we could not achieve synchronization by the simple procedures outlined above.

One immediate consequence of our synchronization procedures is that *simultaneity is relative,* i.e., the simultaneity of two events depends on the reference frame. The following is a concrete example: suppose that two streetlamps in Boston and New York City are briefly turned on and off at exactly 6:00 P.M. Eastern Standard Time. The emissions of the brief flashes of light from the two streetlamps are two events. These two events are simultaneous in the reference frame of the Earth. However, in the reference frame of a fast spaceship passing by the Earth in the direction from Boston to New York, these two events are *not simultaneous* — as judged by the clocks on board the spacecraft, the streetlamp in Boston is turned on slightly later than the street lamp in New York.

To see how this difference between the two reference frames comes about, let us examine the procedure for testing simultaneity. In the reference frame of the Earth, an observer can test for simultaneity by placing himself exactly at the midpoint between Boston and New York (Figure 17.6); this observer will receive the flashes of light from the streetlamps in the two cities at the same instant. Thus, the observer will confirm that in the reference frame of the Earth, the lamps were turned on simultaneously.

In the reference frame of the spaceship, an observer can likewise test for simultaneity by placing himself exactly at the midpoint between Boston and New York at 6:00 P.M. (Figure 17.7 shows the spaceship observer as seen from the Earth at the instant when he is at the midpoint.) Since this observer moves with the spaceship, he will not be at this midpoint at a later (or earlier) time; but for the test of simultaneity, what matters is that the flashes of light from the two streetlamps *originate* at equal distances from the observer. The spaceship observer moves toward the flash of light that originated at New York and away from the flash of light that originated at Boston; obviously, the New York flash will encounter the observer first, and the Boston flash will catch up later (Figure 17.8 shows the spaceship observer as seen from the Earth at the instant when the New York flash encounters him). Hence the spaceship observer will conclude that the flashes were not emitted simultaneously — according to his reckoning, the lamp in Boston was turned on late!

Although this qualitative argument shows that simultaneity depends on the reference frame, it does not tell us by how much. A careful quantitative calculation shows that for a spaceship traveling at, say, 90% of the speed of light, the lamp in Boston is turned on late by about 10^{-3} s.

If simultaneity is relative, then the synchronization of clocks is also relative. In the reference frame of the Earth, clocks in Boston and New York are synchronized, i.e., the hands of these clocks reach the 6:00 P.M. position simultaneously. But in the reference frame of the spaceship, the clock in Boston is judged to be late — in the same way as the lamp is turned on later, the pointers of the clock reach the 6:00 P.M. position later than in New York. Figure 17.9 shows the clocks belonging to the reference frame of the Earth as observed at one instant from the spaceship.

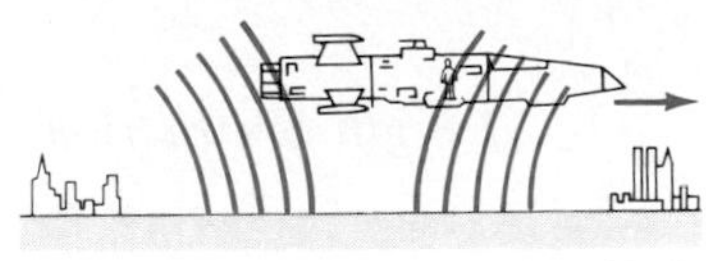

Fig. 17.8 The observer on the spaceship encounters the light flash emanating from New York.

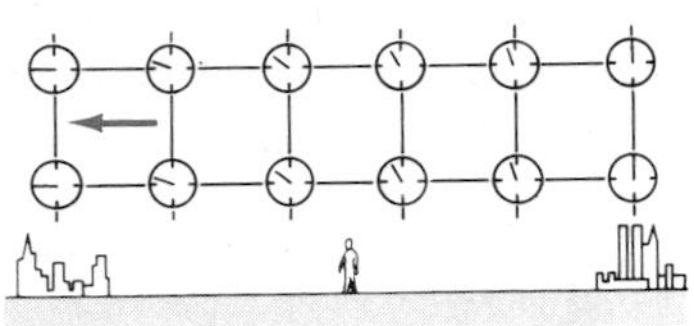

Fig. 17.9 Clocks of the reference frame of the Earth as observed at one instant of spaceship time (see Fig. 17.15 for a more accurate picture).

The effect is symmetric. In the reference frame of the spaceship, all the clocks on board are synchronized. But, as observed in the reference frame of the Earth, the clocks on the front part of the spaceship are late. Figure 17.10 shows the clocks belonging to the reference frame of the spaceship as observed at one instant from the Earth. Note that in Figure 17.9 we are viewing the reference frame of the Earth moving past the spaceship and in Figure 17.10 we are viewing the reference frame of the spaceship moving past the Earth. In either case the clocks on the *leading edge* of the moving reference frame are *late.*

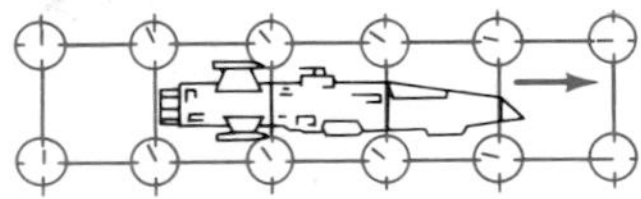

Fig. 17.10 Clocks of the reference frame of the spaceship as observed at one instant of Earth time (see Fig. 17.14 for a more accurate picture).

The relativity of synchronization is a direct consequence of the invariance of the speed of light (our procedures for testing simultaneity depend crucially on the speed of light). The breakdown of absolute simultaneity implies that time is relative. There exists no absolute time coordinate; instead, there only exist relative time coordinates associated with particular reference frames.

Time dilation

The relativity of time shows up not only in the synchronization of clocks, but also in the rate of clocks. A clock on board the spaceship suffers **time dilation:** the rate of the moving clock is slow compared to the rate of identically manufactured clocks at rest on the Earth. To see how this comes about, imagine that the experimenters in the spaceship set up a 300-m track perpendicular to the direction of motion of the spaceship (Figure 17.11).[3] If the experimenters use one of their clocks to measure the time of flight of a light signal that goes from one end of the track to the other and returns, they will find that the light signal takes a time of $600 \text{ m}/(3.00 \times 10^8 \text{ m/s}) = 2.0 \times 10^{-6}$ s to complete the trip. But experimenters on the Earth see that the light signal has simultaneous vertical and horizontal motions (Figure 17.12). For the experimenters on the Earth, the light signal travels a total distance *larger* than 600 m. Since the speed must still be 3.00×10^8 m/s, they will find that according to their clocks the light signal now takes a time *longer* than 2.0×10^{-6} s to complete the trip. Thus, a given time interval registered by a clock on the spaceship is registered as a longer time interval by the clocks on the Earth. The clock on the spaceship runs slow when judged by the clocks on Earth. Note that the experiment involves *one* clock of the spaceship, but several (synchronized) clocks on the Earth, because the light signal does not return to the point at which it started on Earth.

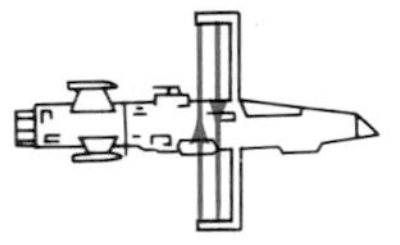

Fig. 17.11 Spaceship with a "racetrack" for a light pulse.

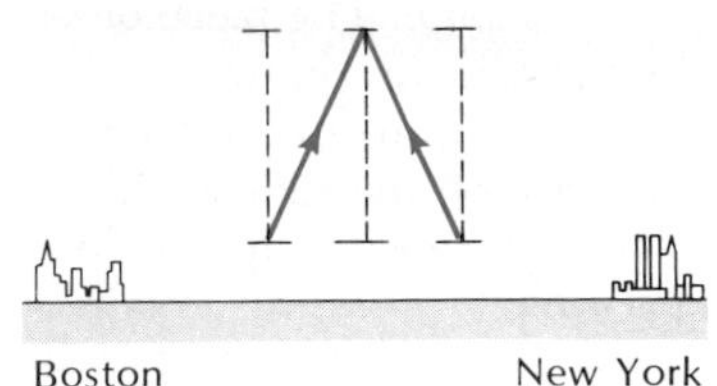

Fig. 17.12 The trajectory of the light pulse as observed from the Earth.

[3] The track does not have to be perpendicular, but the analysis is simplest if it is.

Length contraction

It turns out that because synchronization is relative, length is also relative. A measuring rod, or any other object, on board the spaceship suffers **length contraction** along the direction of motion — the length of the moving measuring rod will be short when compared with the length of identically manufactured measuring rods at rest on the Earth. The reason for this is that the length measurement of a moving object depends on simultaneity, and since simultaneity is relative, so is length. Suppose that the spaceship carries a measuring rod that has a length of, say, 300 km in the reference frame of the spaceship (Figure 17.13). To measure the length of this rod in the reference frame of the Earth, we station observers in the vicinity of New York and Boston with instructions to ascertain the positions of the front and rear ends of the measuring rod at one instant of time, say, 6:00 P.M. Eastern Standard Time. But when the observers on the Earth do this, the observers on the spaceship will claim that the position measurement was not done simultaneously and that the observers in Boston measured the position of the rear end at a later time. In the extra time, the rear end moves an extra distance to the right and hence the distance between the positions measured for the rear and front ends will be reduced. From the point of view of the observers on the spaceship, it is therefore immediately obvious that the length measured by the observers on the Earth will be *short.* Figures 17.14 and 17.15 show the reference frame of the spaceship moving past the Earth and the reference frame of the Earth moving past the spaceship; respectively. In these figures the length contraction has been included (it was left out in Figures 17.9 and 17.10).

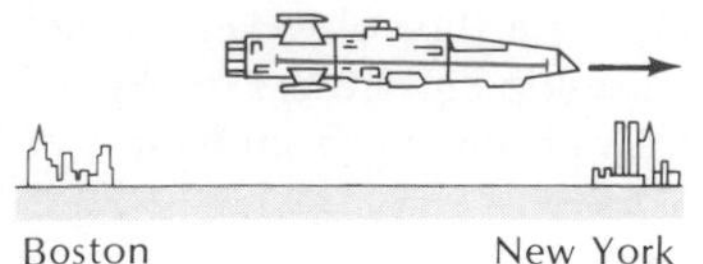

Fig. 17.13 Spaceship with a measuring rod.

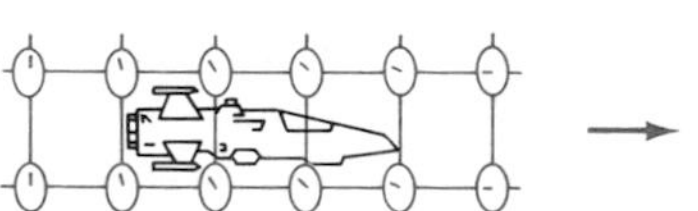

Fig. 17.14 Reference frame of the spaceship as observed at one instant of Earth time (including length contraction).

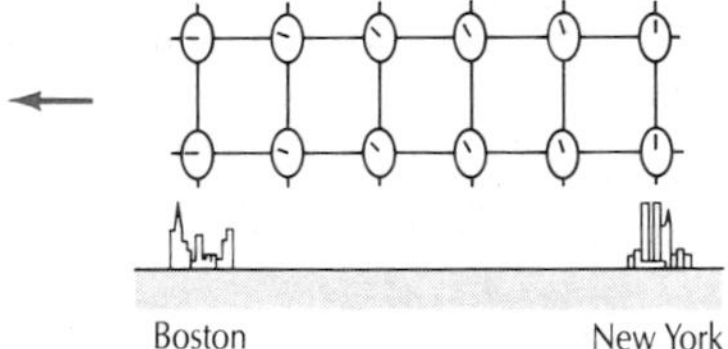

Fig. 17.15 Reference frame of the Earth as observed at one instant of spaceship time (including length contraction).

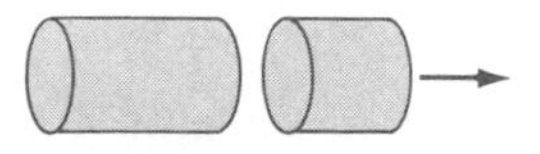

Fig. 17.16 Two identical pieces of pipe. The piece on the left is at rest in the reference frame of the Earth; the piece on the right is at rest in the reference frame of the spaceship.

Incidentally, the contraction effect only applies to lengths along the direction of motion of the spaceship. Lengths perpendicular to the direction of motion are not affected. The proof of this is by contradiction: Imagine that we have two identically manufactured pieces of pipe, one at rest on Earth and one at rest on the spaceship (Figure 17.16). If the motion of the spaceship relative to the Earth were to bring about a transverse contraction of the spaceship pipe, then, by the Principle of Relativity, the motion of the Earth relative to the spaceship would have to bring about a contraction of the Earth pipe. These contraction effects are contradictory, since in one case the spaceship pipe would fit inside the Earth pipe and in the other case it would fit outside.

Just like the relativity of time, the relativity of length is a direct consequence of the invariance of the speed of light. The dependence of simultaneity, time intervals, and lengths on the reference frame is a drastic departure from the physics of Newton. It was Einstein's great discovery that Newton's universal, absolute time and absolute length do not exist. This destroys much of the conceptual basis of Newtonian physics and compels us to adopt new definitions of momentum, energy, angular momentum, etc. In Section 17.7 we will introduce some of these new relativistic formulas. As we shall see, the differences between the old nonrelativistic formulas and the new relativistic formulas are very small if the speeds are small compared to the speed of light. This means that for the motion of airplanes, ships, automobiles, etc., Newtonian physics, although not exact, is sufficiently accurate for all practical purposes.

17.3 The Lorentz Transformations

In Newtonian physics the space and time coordinates in one inertial reference frame are related to those in another by the Galilean transformations (see Section 4.5), which rely on absolute time and length. In relativistic physics, we must construct a new set of transformation equations which take into account the relativity of time and length. The new transformation equations are called the **Lorentz transformations.** These transformation equations express the fundamental characteristics of relativistic space and time — they express the geometry of **spacetime.** In the following discussion we will emphasize these geometric properties of the Lorentz transformations by drawing a great many diagrams.

Spacetime

For the sake of simplicity, we will first deal with one-dimensional motion, so that the position x as a function of t gives a complete description of the motion. We recall that in Chapter 2 we introduced the wordline of a particle as the plot of position versus time. In relativistic physics it is customary to arrange the time axis vertically and the space axis horizontally. For example, Figure 17.17 shows the x and t axes associated with the reference frame of the Earth[4] and it shows the worldline of an automobile traveling in the positive x direction. Before time zero, the automobile was at rest at $x = 0$ and its worldline coincided with the t axis; it then accelerated as indicated by the changing slope of the worldline; finally, it reached a uniform velocity as indicated by the constant slope of the worldline. The worldline consists of individual events, or spacetime points; for instance, the departure of the automobile at $x = 0$, $t = 0$ is an event. In relativistic physics, the complete diagram is called a **spacetime diagram,** or a **Minkowski diagram.**

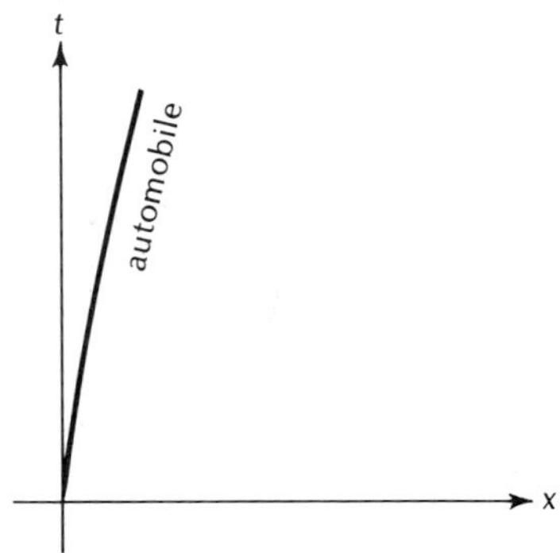

Fig. 17.17 Worldline of an automobile.

Spacetime diagram

Now suppose that we wish to describe the motion of the automobile from the point of view of a spaceship zipping past the Earth in the positive x direction at high speed. This spaceship carries its own coordinate grid and clocks, i.e., it carries its own inertial reference frame. The motion of the automobile can then be measured with respect to the new x' and t' axes associated with this new reference frame. It is then easy to make a new plot of the worldline of the automobile with respect to the new axes. But instead of replotting all the points of the worldline, let us keep the worldline of Figure 17.17 and rearrange the coordinate axes. The change of the coordinate axes will then graphically represent the change of reference frame.

[4] We will pretend that this is an inertial reference frame.

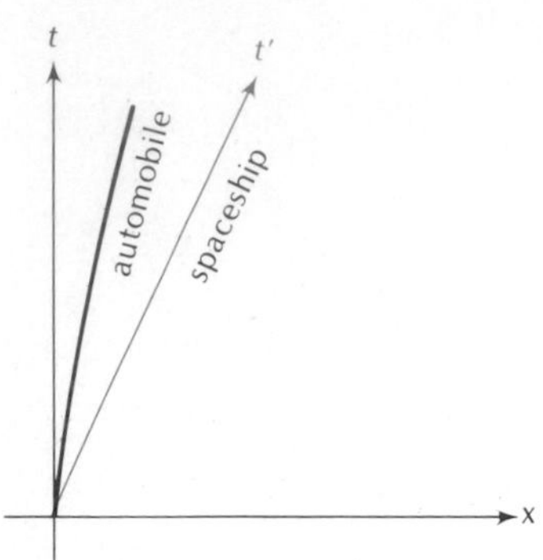

Fig. 17.18 Worldline of a spaceship. This worldline coincides with the t' axis.

How can we draw new x' and t' axes on top of Figure 17.17 so that the same worldline represents the motion of the automobile with respect to the spaceship? For the sake of practice, we will do this first under the pretense that Newton's notion of absolute space and absolute time is correct.

Figure 17.18 shows the worldline of the midpoint of the spaceship moving with uniform velocity V past the Earth. We will regard this midpoint as the origin of the x' coordinates; thus, the slanted line in Figure 17.18 is the worldline of the origin of the spaceship coordinates. Along this line, x' remains permanently zero — only t' increases. Thus, the slanted line is the t' axis of the reference frame of the spaceship. Where is the x' axis? This axis is the set of all those events (or spacetime points) that occur at $t' = 0$. But since in Newtonian physics time is absolute, the clocks of the spaceship and those of the Earth can be synchronized so that $t' = 0$ is the same as $t = 0$. Consequently, the x' axis coincides with the x axis. Figure 17.19 shows the grid of the new x' and t' coordinates; this figure also shows the grid of the old x and t coordinates. The new coordinate grid is slanted with respect to the old one.

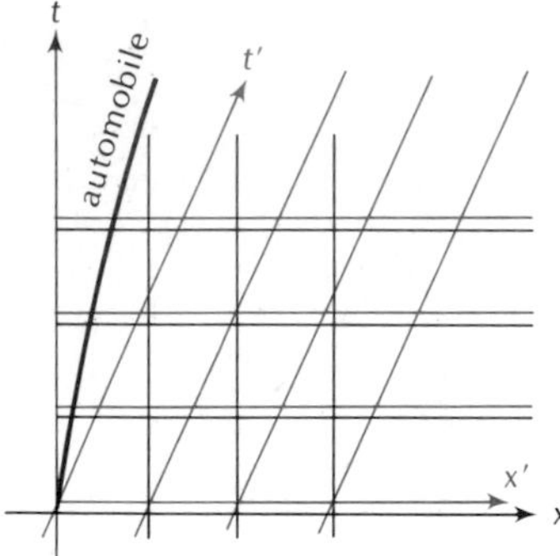

Fig. 17.19 Coordinate grids for the reference frame of the Earth (black) and for the reference frame of the spaceship (colored).

Figure 17.19 is a graphical representation of the transformation of coordinates from x, t to x', t'. The old and the new coordinates of any event can be read off by drawing appropriate lines intercepting the axes. From Figure 17.20 we readily see that the mathematical relationship between the values of the old and the new coordinates of an event is

$$t' = t \tag{1}$$

$$x' = x - Vt \tag{2}$$

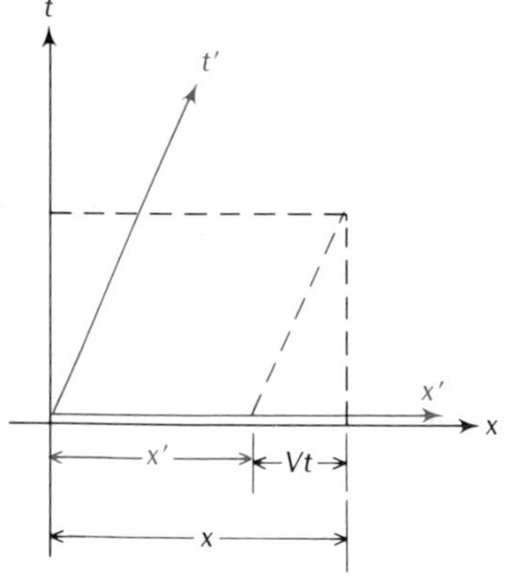

Fig. 17.20 Graphical method of deriving the Galilean equations.

where V is the velocity of the spaceship relative to the Earth. These equations are the Galilean equations [Eqs. (4.49) and (4.54)]. Although these equations are not new to us, we have succeeded here in deriving them by a graphical method relying on spacetime diagrams. We will now use a similar method to derive the relativistic transformation equations.

The crucial difference between the Galilean and the relativistic transformation equations is that the latter are supposed to leave the speed of a light signal invariant. We begin by plotting the worldline of a light signal. Since the speed of light is inconveniently large, it will be better to introduce new units so that the speed of light has the numerical value $c = 1$. We can do this by taking as our unit of length the distance that light travels in one second. We will call this unit the **light-second**[5]:

$$1 \text{ light-second} = 3.00 \times 10^8 \text{ m/s} \times 1 \text{ s}$$

$$= 3.0 \times 10^8 \text{ m}$$

In terms of this new unit of length, the speed of light is 1 light-second per second:

light-second

$$c = 1 \text{ light-second/s} \tag{3}$$

[5] This unit is analogous to the light-year, the distance that light travels in 1 year.

Note that when we express any speed in terms of these units, we are actually expressing it as a multiple of the speed of light. For example, an electron moving with a speed of 5.0×10^7 m/s has

$$v = 5.0 \times 10^7 \text{ m/s} \times \frac{1 \text{ light-second}}{3.00 \times 10^8 \text{ m}}$$

$$= 0.166 \frac{\text{light-second}}{\text{s}}$$

or

$$v = 0.166c \tag{4}$$

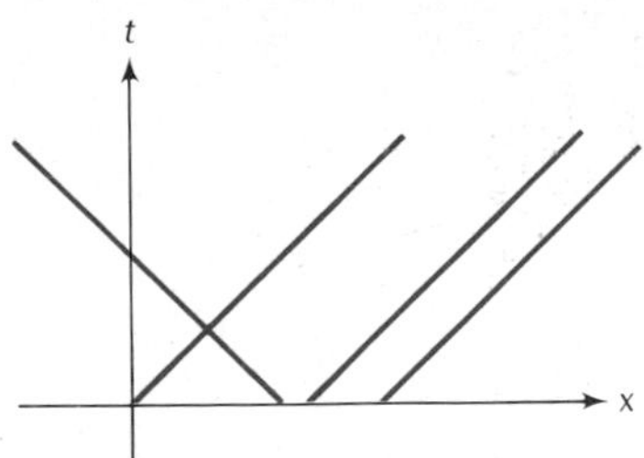

Fig. 17.21 Worldlines of several light signals. Three of these signals travel in the positive x direction; one travels in the negative x direction.

Since the numerical value of the speed of light in our new units is $c = 1$, we will *omit factors of c or c^2* in all of the equations of this section and Sections 17.4–17.6. For instance, we will write Eq. (4) as

$$v = 0.166$$

Note that this means that our equations are applicable only when the distances and times appearing in them are evaluated in light-seconds and seconds, respectively; we will have to remember this whenever we carry out numerical calculations with these equations (as in Example 1 below).

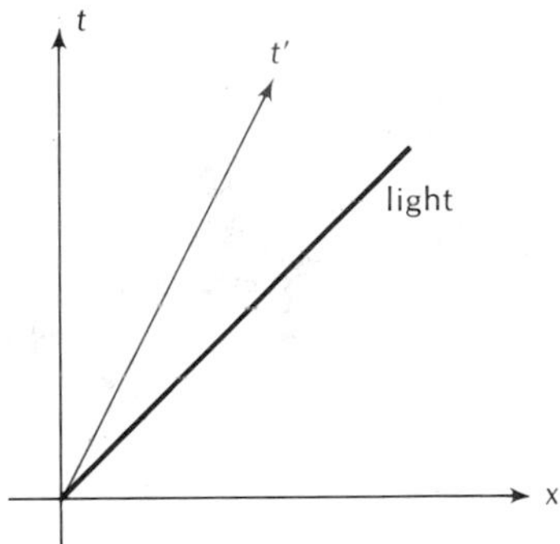

Fig. 17.22 The worldline of the spaceship coincides with the t' axis. The worldline of a light signal that starts at the origin lies halfway between the x and the t axes.

If the units along the x and t axes of the spacetime diagram are light-seconds and seconds, then the worldline of a light signal has a slope of 1. Figure 17.21 shows the worldlines of several light signals which start at different points along the x axis.

How can we draw new x' and t' axes on top of Figure 17.21 so that the worldline of a light signal has the same speed with respect to both the old and new axes? Figure 17.22 shows the worldline of a light signal that starts at the origin. The worldline lies exactly halfway between the x and t axes. This makes the x and t coordinates of any event of this worldline equal, just as required by the condition that the speed be 1 length unit per 1 time unit. Let us examine the behavior of this light signal from the point of view of a spaceship moving at high speed with respect to the Earth. Figure 17.22 shows the worldline of the midpoint of a spaceship, which coincides with the t' axis. The t' axis is a slanted axis just as in Figure 17.18. What about the x' axis? It is here that the invariance of the speed of light enters our calculations. The worldline of the light signal must lie halfway between the x' and t' axes so that the x' and t' coordinates of any event on this worldline will be equal, as required by the condition that the speed of light be 1 length unit per 1 time unit. Hence these axes must be symmetrically placed with respect to the worldline light signal — both the x' and t' axes are slanted (Figure 17.23).

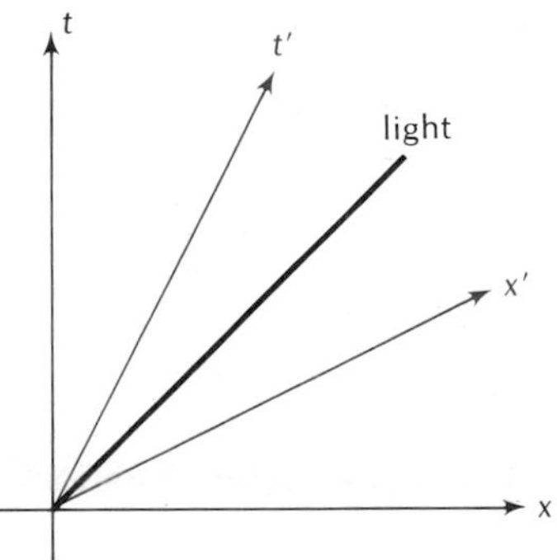

Fig. 17.23 The t' and x' axes are symmetrically placed with respect to the worldline of the light signal.

The symmetry of the arrangement of the x' and t' axes in Figure 17.23 suggests that the equations relating the old and the new coordinates are

$$x' = x - Vt \tag{5}$$

$$t' = t - Vx \tag{6}$$

The first of these equations is the same as Eq. (2); it expresses the slant of the x' axis. The second equation expresses the slant of the t' axis. Note the symmetry between these two equations — the space and time coordinates are simply interchanged between Eq. (5) and (6).

However, the above equations are not yet quite correct. Our choice of units ($c = 1$) assures that the scales along the x and t axes are the same. But this does not mean that the scales along the new axes are the same as those along the old axes. In the case of the Galilean transformations, Eqs. (1) and (2), this problem of scale did not arise; there, distance intervals and time intervals were absolute and the scale of the new axes could be directly obtained from the old. In the case of the relativistic transformations, we will have to figure out the proper scale by other means.

Since the scale of the new coordinates differs by some unknown factor from that of the old, we will include an extra factor γ in our transformation equations:

$$x' = \gamma(x - Vt) \tag{7}$$

$$t' = \gamma(t - Vx) \tag{8}$$

Note that the *same* scale factor γ appears in each equation; this is necessary so that the x' and t' coordinates of any event along the worldline of our light signal remain equal.

To find an explicit expression for the scale factor γ we can proceed as follows. First solve Eqs. (7) and (8) for x and t. Regarding x and t as "unknowns" in these equations, we find

$$x = \frac{1}{\gamma(1 - V^2)}(x' + Vt') \tag{9}$$

$$t = \frac{1}{\gamma(1 - V^2)}(t' + Vx') \tag{10}$$

These equations are the inverses of Eqs. (7) and (8); the former express the old coordinates in terms of the new and the latter express the new in terms of the old. But there is an alternative way to express the old coordinates in terms of the new: instead of beginning with rectangular axes for the Earth's reference frame and then constructing slanted axes for the spaceship's reference frame, we could have begun with rectangular axes for the spaceship and then constructed slanted axes for the Earth. This would have led us to a pair of equations similar to Eqs. (7) and (8), with the new and old coordinates exchanged and with one further alteration: the velocity V must be replaced by $-V$. This change of sign of the velocity merely reflects the fact that the velocity of the Earth relative to the spaceship is opposite to the velocity of the spaceship relative to the Earth. Thus the equations would have been

$$x = \gamma(x' + Vt') \tag{11}$$

$$t = \gamma(t' + Vx') \tag{12}$$

By comparing these two equations with Eqs. (9) and (10), we see that they coincide, provided that

$$\gamma = \frac{1}{\gamma}\frac{1}{1 - V^2} \tag{13}$$

or

$$\gamma = \frac{1}{\sqrt{1 - V^2}} \tag{14}$$

Consequently, our transformation equations become

Lorentz transformations

$$x' = \frac{1}{\sqrt{1 - V^2}}(x - Vt) \tag{15}$$

$$t' = \frac{1}{\sqrt{1 - V^2}}(t - Vx) \tag{16}$$

and the inverse transformation equations become

$$x = \frac{1}{\sqrt{1 - V^2}}(x' + Vt') \tag{17}$$

$$t = \frac{1}{\sqrt{1 - V^2}}(t' + Vx') \tag{18}$$

These equations are the **Lorentz transformations**.[6] They replace the Galilean transformations and accomplish the remarkable feat of making the speed of light the same in all reference frames. Although in the above discussion we labeled the two inertial reference frames as "Earth" and "spaceship," Eqs. (15)–(18) are quite general and give the coordinate transformation between any two inertial reference frames.

We can now put the correct scale on the diagram of Figure 17.23. According to Eq. (15), the line $x' = 1$ intersects the x axis (that is, $t = 0$) at the point

$$x = \sqrt{1 - V^2}$$

For instance, if $V = \frac{1}{2}$, then the intersection point is

$$x = \sqrt{1 - \tfrac{1}{4}} = 0.866$$

This determines the location of the unit point on the x' axis and immediately permits us to construct the complete coordinate grid (Figure

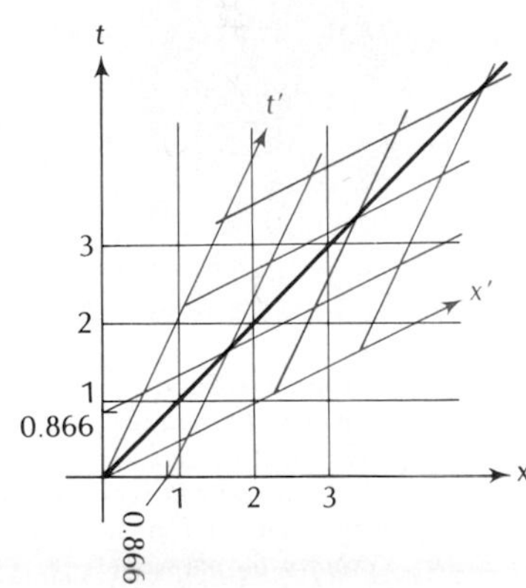

Fig. 17.24 Coordinate grids for the reference frame of the Earth (black) and for the reference frame of the spaceship (color).

[6] In conventional units (x in meters and t in seconds) the Lorentz transformation equations are

$$x' = \frac{1}{\sqrt{1 - V^2/c^2}}(x - Vt)$$

$$t' = \frac{1}{\sqrt{1 - V^2/c^2}}\left(t - \frac{Vx}{c^2}\right)$$

17.24). The new coordinates of any event can then be read off this grid. Alternatively, the coordinates can be calculated from Eqs. (15) and (16).

The Lorentz transformations treat space and time symmetrically. A change of reference frame intricately mixes the space and time coordinates of an event. What is a pure space interval or a pure time interval in one reference frame becomes a mixture of both space and time intervals in another reference frame. As H. Minkowski wrote, "henceforth space by itself, and time by itself, are doomed to fade away into mere shadows, and only a kind of union of the two will preserve an independent reality."

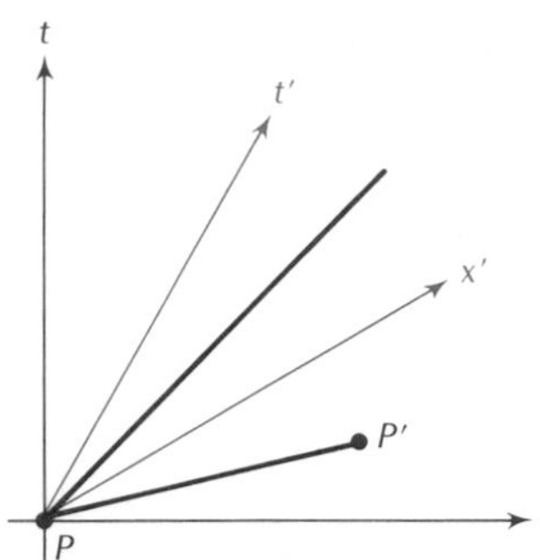

Fig. 17.25 Worldline of a hypothetical signal of speed greater than that of light.

One interesting conclusion we can draw from the Lorentz transformations is that no signal can propagate with a speed greater than the speed of light. The proof is by contradiction: Suppose that in the reference frame of the Earth we send out a signal of speed greater than that of light. The worldline of this signal is shown in Figure 17.25; the signal is emitted at the point P and is received at the point P' at a later time. But in the reference frame of the spaceship, the point P' has an earlier time coordinate than the point P — hence in this reference frame the signal is received *before* it is emitted. This is absurd and rules out the existence of such a signal.

If the relative velocity between the two reference frames is small compared to the velocity of light, then the Lorentz transformations approximately coincide with the Galilean transformations. In our units, small velocity simply means $V \ll 1$. Hence

$$\sqrt{1 - V^2} \cong 1$$

and Eqs. (15) and (16) reduce to

$$x' \cong x - Vt \tag{19}$$

$$t' \cong t - Vx \tag{20}$$

Hendrik Antoon Lorentz, *1853–1928, Dutch theoretical physicist, professor at Leiden. He investigated the relationship between electricity, magnetism, and mechanics. In order to explain the observed effect of magnetic fields on emitters of light (Zeeman effect) he postulated the existence of electrons, for which he was awarded the Nobel Prize in 1902. He derived the Lorentz-transformation equations by some tangled mathematical arguments, but he was not aware that these equations hinge on a new concept of space and time.*

Furthermore, the term Vx in the second equation can be neglected if V and x are velocities and distances of the magnitudes typically encountered on the Earth. For instance, suppose that the "spaceship" reference frame is actually that of a supersonic airplane moving at 600 m/s. Expressed as a multiple of the velocity of light, this gives $V = 2.0 \times 10^{-6}$, a very small number. Hence even if x has a large value, say, $x = 10^7$ m $= 3.3 \times 10^{-2}$ light-seconds, the term Vx has a magnitude of only $2.0 \times 10^{-6} \times 3.3 \times 10^{-2}$ s $= 6.6 \times 10^{-8}$ s — this is such a small time interval that it usually can be ignored. We are then entitled to throw out the term Vx in Eq. (20) and obtain

$$x' \cong x - Vt \tag{21}$$

$$t' \cong t \tag{22}$$

These, of course, are the Galilean transformations. The differences between relativistic physics and Newtonian physics are therefore quite insignificant at low velocities.

Finally, for the sake of completeness we must write down the equations for the transformation of the transverse coordinates. Since we have assumed that the relative motion of the reference frames is along the direction of the x axis, the transverse directions are those of the y and z axes. In Section 17.2 we found that transverse lengths do not suffer any contraction. Hence the coordinate transformations for y and z must simply be

$$y' = y \tag{23}$$

$$z' = z \tag{24}$$

Lorentz transformations for t and z

EXAMPLE 1. Imagine that, as in the example discussed qualitatively in Section 17.2, a spaceship travels in the Boston–New York direction at a speed $V = 0.90$. At one instant of spaceship time, the clock in New York shows 6:00 P.M. Eastern Standard Time. What time does the clock in Boston show at this instant? The distance between New York and Boston is 290 km.

SOLUTION: For convenience we take the origin of the Earth's reference frame at New York ($x = 0$) and the origin of Earth time at 6:00 P.M. ($t = 0$). According to Eq. (16), the spaceship time corresponding to the event $x = 0$, $t = 0$ in New York is

$$t' = \frac{0 - V \cdot 0}{\sqrt{1 - V^2}} = 0 \tag{25}$$

With the x axis in the Boston–New York direction (see Figure 17.6), the x coordinate of Boston is $x = -290\ \text{km} = -2.9 \times 10^5\ \text{m} = -9.7 \times 10^{-4}$ light-seconds. The event in Boston has the same t' coordinate as that in New York, i.e., $t' = 0$. Hence Eq. (16) gives

$$0 = \frac{t - Vx}{\sqrt{1 - V^2}} = \frac{t + 0.90 \times 9.7 \times 10^{-4}\ \text{s}}{\sqrt{1 - V^2}}$$

This leads to

$$t = -0.90 \times 9.7 \times 10^{-4}\ \text{s} = -8.7 \times 10^{-4}\ \text{s}$$

that is, the clocks in Boston show a time 8.7×10^{-4} s *before* 6:00 P.M.

17.4 The Time Dilation

We have already discovered in Section 17.2 that a clock in motion relative to some inertial reference frame will run slow as compared to clocks at rest in that reference frame. From the Lorentz transformation equations, we can derive a quantitative formula for this effect. Suppose the clock is at rest in a spaceship that has velocity V relative to the Earth. Consider two consecutive ticks of the clock; these ticks are two events between which there are certain coordinate differences. According to Eq. (18), the coordinate differences obey the relation

$$\Delta t = \frac{1}{\sqrt{1 - V^2}} (\Delta t' + V\, \Delta x') \tag{26}$$

Since the clock is at rest in the spaceship, its x' coordinate does not change, i.e., $\Delta x' = 0$. Hence

Time dilation (clock in spaceship)

$$\boxed{\Delta t = \frac{1}{\sqrt{1 - V^2}}\,\Delta t'} \qquad \text{clock at rest in spaceship} \tag{27}$$

This is the **time-dilation** formula. It shows that the time between consecutive ticks measured in the reference frame of the Earth is greater than the time between consecutive ticks measured in the reference frame of the spaceship. As measured by the clocks on the Earth, the clock on the spaceship runs slow by a factor of $1/\sqrt{1 - V^2}$. The time-dilation effect is symmetric: as measured by the clocks on the spaceship, a clock on the Earth runs slow by the same factor,

Time dilation (clock on Earth)

$$\boxed{\Delta t' = \frac{1}{\sqrt{1 - V^2}}\,\Delta t} \qquad \text{clock at rest on Earth} \tag{28}$$

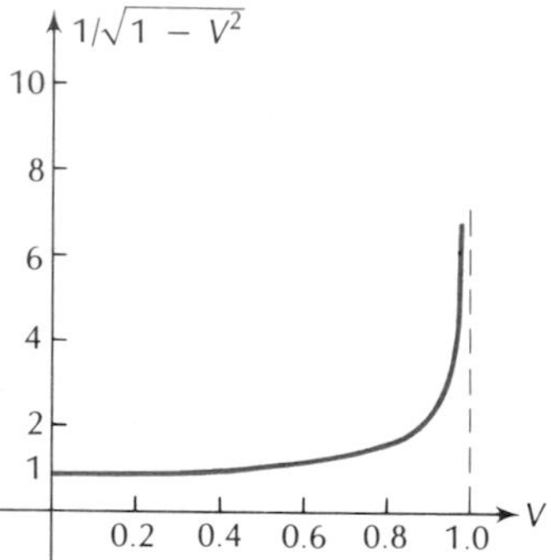

Fig. 17.26 Time-dilation factor as a function of V.

The derivation of Eq. (28) can be based on Eq. (16) with $\Delta x = 0$.

Figure 17.26 is a plot of the time-dilation factor $1/\sqrt{1 - V^2}$ as a function of V.

The slowing down of the rate of lapse of time applies to all physical processes — atomic, nuclear, biological, etc. At low speeds the time-dilation effect is insignificant but at high speeds it can become quite appreciable. Very drastic time-dilation effects have been observed in the decay of short-lived elementary particles. For instance, a muon (see Interlude C) usually decays in about 2.2×10^{-6} s; but if it is moving at high speed through our laboratory, then the internal processes that produce the decay will slow down and, as reckoned by the clocks in our laboratory, the muon lives a longer time. In an accurate experiment performed at the Centre Européen de Recherches Nucléaires (CERN) laboratory near Geneva, muons of a velocity $V = 0.9994$ were found to have a lifetime 29 times as long as muons at rest — in excellent agreement with the prediction based on Eq. (27).

At everyday speeds the time-dilation effect is extremely small. For example, consider a clock aboard an airplane traveling at 300 m/s over the ground. In our units of light-seconds and seconds this corresponds to $V = 1.0 \times 10^{-6}$. To evaluate the time-dilation factor, it is convenient to take advantage of the Taylor-series expansion

$$\frac{1}{\sqrt{1 - V^2}} = 1 + \frac{V^2}{2} + \cdots \tag{29}$$

which gives

$$\frac{1}{\sqrt{1 - V^2}} \cong 1 + 5 \times 10^{-13}$$

that is, the clock in the airplane will only slow down by 5 parts in 10^{13}. However, such a small change is not beyond the reach of modern atomic clocks. In a recent experiment, scientists placed portable atomic clocks on board commercial airliners and kept them flying for several days, making a complete trip around the world. Before and after the trip, the clocks were compared with an identical clock that was kept on the ground. The flying clocks were found to have lost time — in one instance the total time lost because of the motion of the clock was about 10^{-7} s.

The time-dilation effects leads to the famous **twin "paradox,"** which we can state as follows: A pair of identical twins, Terra and Stella, celebrate their twentieth birthday on Earth. Then Stella boards a spaceship that carries her at a speed of $V = 0.99$ to Proxima Centauri, 4 light-years away; the spaceship immediately turns around and brings Stella back to the Earth. According to the clocks on the Earth, this trip takes about 8 years, so that Terra's age will be 28 years when the twins meet again. But Stella has benefited from time dilation — from the point of view of the reference frame of the Earth, the spaceship clocks run slow by a factor

$$\sqrt{1 - V^2} = \sqrt{1 - (0.99)^2} = 0.14$$

Hence the 8 years of travel registered by the Earth clocks amount to only $8 \times 0.14 = 1.1$ years according to the spaceship clocks, so that Stella's biological age on return will only be 21.1 years. Stella will be younger than Terra. The paradox arises when we examine the elapsed times from the point of view of the reference frame of the spaceship. In this reference frame the Earth is moving; hence the Earth clocks run slow — and Terra should be younger than Stella.

The resolution of this paradox hinges on the fact that our time-dilation formula is only valid if the time of a moving clock is measured from the point of view of an *inertial* reference frame. The reference frame of the Earth is (approximately) inertial, and therefore our calculation of the time dilation of the spaceship clocks is valid. But the reference frame of the spaceship is not inertial — the spaceship must decelerate when it reaches Proxima Centauri, stop, and then accelerate toward the Earth. Therefore, we cannot use the simple time-dilation formula to find the time dilation of the Earth clocks from the point of view of the spaceship reference frame. The "paradox" results from the misuse of this formula.

A detailed analysis of the behavior of the Earth clocks from the point of view of the spaceship reference frame shows that the Earth clocks run slow as long as the spaceship is moving with uniform velocity, but that the Earth clocks run fast when the spaceship is undergoing its acceleration at Proxima Centauri. The time that the Earth clocks gain during the accelerated portion of the trip more than compensates for the time they lose during the other portions of the trip. This confirms that Stella will be younger than Terra, even from the point of view of the spaceship reference frame.

17.5 The Length Contraction

Suppose that a rigid body, such as a meter stick, is at rest in a spaceship moving relative to the Earth. The length of this body along the direction of motion is $\Delta x'$ in the reference frame of the spaceship. As pointed out in Section 17.2, to find the length in the reference frame of the Earth, we must measure the position of the forward end and the rear end of the meter stick at the same time t. The measurements at the two ends can be regarded as two events separated by some coordinate difference. According to Eq. (15), coordinate differences obey the transformation law

$$\Delta x' = \frac{1}{\sqrt{1 - V^2}} (\Delta x - V\, \Delta t) \tag{30}$$

and since $\Delta t = 0$,

$$\Delta x' = \frac{1}{\sqrt{1 - V^2}} \Delta x \tag{31}$$

or

Length contraction

$$\boxed{\Delta x = \sqrt{1 - V^2}\Delta x'} \quad \text{body at rest in spaceship} \tag{32}$$

This is the formula for **length contraction.** According to this formula the length of the body measured in the reference frame of the Earth is shorter than the length measured in the reference frame of the spaceship by a factor of $\sqrt{1 - V^2}$. This effect is again symmetric: a body at rest on the Earth will suffer from contraction when measured by instruments on board the spaceship.

The length contraction has not been tested directly by experiment. There is no practical method for a high-precision measurement of the length of a fast-moving body. Our best bet might be high-speed photography, but even this is nowhere near accurate enough, since the contraction is extremely small even at the highest speeds that we can impart to a macroscopic body.

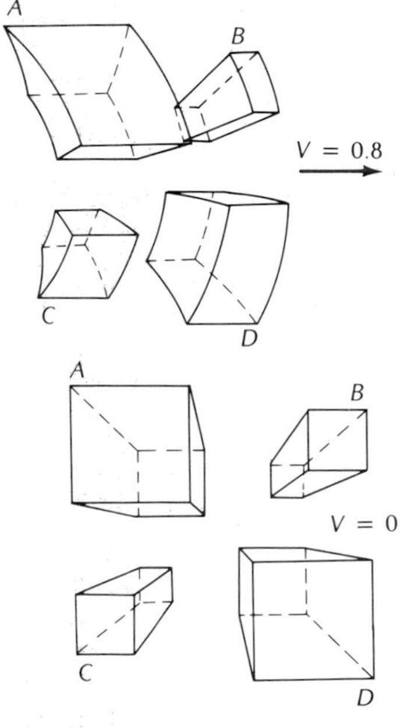

Fig. 17.27 Computer simulation of a photograph of boxes moving at very high speed. [Based on *Am. J. Phys., 33,* 534 (1965).]

Incidentally, if we could take a sharp photograph of a macroscopic body zipping by at a speed of, say, $V = 0.8$, the photographic image would not only show the contraction, but also a strong distortion of the shape of the body. A photograph never shows the surface of the body as it is, but rather as it was at the instant the light from it began to travel to the camera. The light that makes the photographic image has to reach the camera at the instant its shutter is open. If light from near and far parts of the body is to arrive at the camera at the same instant, the light from the far parts must start out earlier than that from the near parts. Thus the image shows different parts of the body at different times — far parts at an early time and near parts at a late time. The image is therefore a distorted picture of the body. Figure 17.27 is a computer simulation of the photographic image that a camera would record when aimed at some rectangular boxes traveling past the camera at $V = 0.8$. The length contraction is clearly noticeable along the midline, but everywhere else the image is severely distorted. Figure 17.28 is a similar computer simulation of the photographic images of trains of boxcars traveling at $V = 0.5$ and at $V = 0.9$. Again, the length contraction shows up clearly near the center of the pictures, but elsewhere the image is distorted. Note that the boxcars seem to turn their rear ends toward the camera.

17.6 The Combination of Velocities

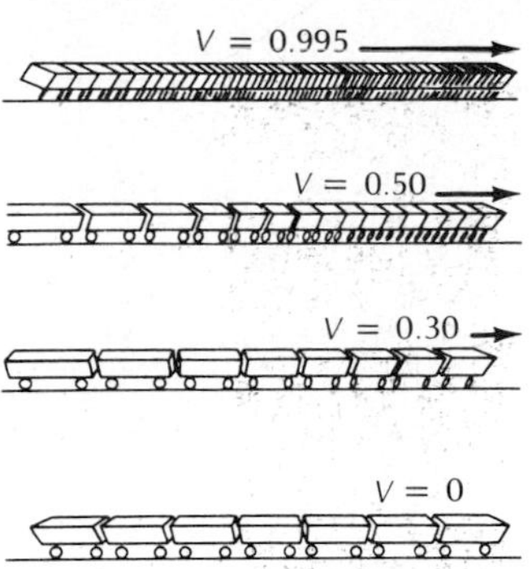

Fig 17.28 Computer simulation of a photograph of a train of boxcars moving at very high speed. [Based on *Am. J. Phys.*, *38*, 971 (1970).]

Our search for a new theory of space and time was in part motivated by the conflict between the Galilean addition law for velocities and the invariance of the speed of light. Since the Lorentz transformations were specifically designed to incorporate the invariance of the speed of light, we certainly expect that the relativistic "addition" law, or combination law, for velocities will avoid the above conflict.

We can derive the relativistic combination law for the x component of velocity by simply taking differentials of the Lorentz transformation equations:

$$dx' = \gamma\,(dx - V\,dt) \tag{33}$$

$$dt' = \gamma\,(dt - V\,dx) \tag{34}$$

Hence

$$\frac{dx'}{dt'} = \frac{dx - V\,dt}{dt - V\,dx} = \frac{dx/dt - V}{1 - V\,dx/dt} \tag{35}$$

But dx/dt is the velocity v_x of the particle, light signal, or whatever, with respect to the reference frame of the Earth; dx'/dt' is likewise the velocity v_x' with respect to the reference frame of the spaceship. Hence Eq. (35) may be written

$$\boxed{v_x' = \frac{v_x - V}{1 - v_x V}} \tag{36}$$

Combination of velocities

This is to be compared with the Galilean equation

$$v_x' = v_x - V \tag{37}$$

It is the denominator in Eq. (36) that makes all the difference. For instance, suppose that v_x is the velocity of a light signal along the x axis in the reference frame of the Earth. Then $v_x = 1$ and Eq. (36) gives

$$v_x' = \frac{1 - V}{1 - V} = 1 \tag{38}$$

Thus, as required, the velocity of the light signal has exactly the same magnitude in the reference frame of the spaceship.

The inverse combination law that expresses the velocity v_x in terms of v_x' is

$$v_x = \frac{v_x' + V}{1 + v_x' V} \tag{39}$$

The combination law for light velocities has been explicitly tested in an experiment at CERN involving a beam of very fast pions. These particles decay spontaneously by a reaction that emits a flash of very intense, very energetic light (gamma rays, discussed in Interlude B). Hence such a beam of pions can be regarded as a high-speed light source. In the experiment the velocity of the pions relative to the laboratory was $V = 0.99975$. The Galilean combination law would then predict laboratory velocities of $v_x = 1.99975$ for light emitted in the forward direction and of $v_x = 0.00025$ for light emitted in the backward direction. But the experiment confirmed the relativistic combination law — the laboratory velocity of the light was the same in all directions.

EXAMPLE 2. A spaceship approaching the Earth at $V = 0.40$ fires a rocket at the Earth. If the velocity of the rocket is $v'_x = 0.80$ in the reference frame of the spaceship, what is its velocity in the reference frame of the Earth?

SOLUTION: By Eq. (39)

$$v_x = \frac{v'_x + V}{1 + v'_x V} = \frac{0.80 + 0.40}{1 + 0.80 \times 0.40} = 0.91$$

In conventional units this is $0.91c$, or $0.91 \times 3.0 \times 10^8$ m/s $= 2.7 \times 10^8$ m/s.

Equation (36) only gives the transformation law for the component of velocity parallel to the motion of the reference frame (x component). The components of the velocity perpendicular to the motion of the reference frame (y and z components) also have a transformation law different from the Galilean transformation law. Without giving the derivations, we state the results:

Combination of velocities for y and z components

$$v'_y = \frac{v_y}{(1 - Vv_x)\sqrt{1 - V^2}}$$
$$v'_z = \frac{v_z}{(1 - Vv_x)\sqrt{1 - V^2}} \qquad (40)$$

17.7 Momentum and Energy

The drastic revision that the theory of Special Relativity imposed on the Newtonian concepts of space and time implies a corresponding revision of the concepts of momentum and energy. The formulas for momentum and energy and the equations expressing their conservation are intimately tied to the transformation equations of the space and time coordinates. To see that this is so, we briefly examine the Newtonian (nonrelativistic) case.

The Newtonian momentum of a particle of mass m and velocity $\mathbf{v}$ is

$$\mathbf{p} = m\mathbf{v} \qquad (41)$$

and the Galilean transformation for velocity is

$$\mathbf{v}' = \mathbf{v} - \mathbf{V} \qquad (42)$$

Accordingly, the transformation equation for momentum is

$$\mathbf{p}' = m\mathbf{v}' = m\mathbf{v} - m\mathbf{V} = \mathbf{p} - m\mathbf{V} \tag{43}$$

i.e., the momentum $\mathbf{p}'$ in the new reference frame differs from the momentum $\mathbf{p}$ in the old reference frame only by a constant quantity (a quantity independent of the velocity $\mathbf{v}$ of the particle). Hence if the total momentum of a system of colliding particles is conserved in one reference frame, it will also be conserved in another frame — the law of conservation of momentum obeys the Principle of Relativity. This shows that the nonrelativistic formula for momentum and the nonrelativistic addition law for velocities match in just the right way.

What happens when we replace the Galilean combination law for velocity [Eq. (42)] by the relativistic law [Eqs. (36) and (40)]? If the law of conservation of momentum is to obey the Principle of Relativity, we must design a new relativistic formula for momentum that matches the new relativistic combination law for velocities. It turns out that the correct relativistic formula for momentum is

$$\boxed{\mathbf{p} = \frac{m\mathbf{v}}{\sqrt{1 - v^2/c^2}}} \tag{44}$$

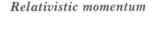

Relativistic momentum

Note that in this formula we have included the proper factors of c, instead of omitting them as in the preceding sections. Thus, the formula is equally valid in units of light-seconds and seconds and in units of meters and seconds; this will make it easier to compare the relativistic and nonrelativistic formulas for momentum. We will forgo the derivation of Eq. (44), but we will verify in Example 5 that this equation does indeed agree with the Principle of Relativity.

If the velocity of the particle is small compared to the velocity of light, then

$$\sqrt{1 - v^2/c^2} \cong 1$$

and Eq. (44) becomes approximately

$$\mathbf{p} \cong m\mathbf{v} \tag{45}$$

This shows that for low velocities the relativistic and nonrelativistic formulas coincide. At high velocities the formulas differ drastically — the relativistic momentum becomes infinite as the velocity of the particle approaches the velocity of light. Figure 17.29 is a plot of the magnitude of $\mathbf{p}$ as a function of v.

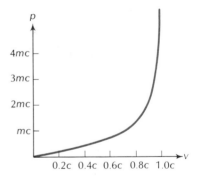

Fig 17.29 Momentum of a particle as a function of speed.

EXAMPLE 3. An electron in the beam of a TV tube has a speed of 1.0×10^8 m/s. What is the momentum of this electron?

SOLUTION: For this electron, $v/c = (1.0 \times 10^8 \text{ m/s})/(3.0 \times 10^8 \text{ m/s}) = 0.33$. According to Eq. (44), the momentum is then

$$p = \frac{mv}{\sqrt{1 - v^2/c^2}}$$

$$= \frac{9.1 \times 10^{-31} \text{ kg} \times 1.0 \times 10^8 \text{ m/s}}{\sqrt{1 - (0.33)^2}}$$

$$= 9.7 \times 10^{-23} \text{ kg} \cdot \text{m/s}$$

Note that if we had calculated the momentum according to the nonrelativistic equation $p = mv$, we would have obtained 9.1×10^{-23} kg · m/s, and we would have been in error by about 6%.

It turns out that we also need a new formula for kinetic energy. Again, without derivation, we state that the relativistic formula for kinetic energy is

Relativistic kinetic energy

$$K = \frac{mc^2}{\sqrt{1 - v^2/c^2}} - mc^2 \tag{46}$$

To compare this with our old, nonrelativistic formula, we can make use of the Taylor-series expansion:

$$\frac{1}{\sqrt{1 - x^2}} = 1 + x^2/2 + \cdots \tag{47}$$

Thus, if the velocity of the particle is small compared to the velocity of light, Eq. (46) becomes

$$K \cong mc^2\left(1 + \frac{v^2}{2c^2}\right) - mc^2 = \tfrac{1}{2}mv^2 \tag{48}$$

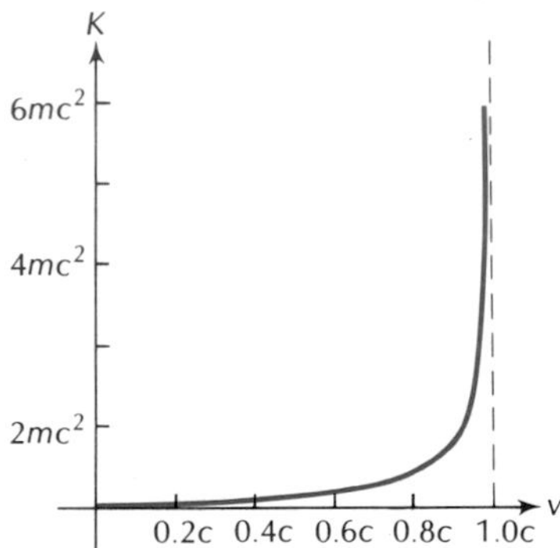

Fig 17.30 Kinetic energy of a particle as a function of speed.

Again, for low velocities the relativistic and nonrelativistic formulas coincide.

The relativistic kinetic energy becomes infinite as the speed of the particle approaches the speed of light. This indicates that, for any particle, the speed of light is unattainable, since it is impossible to supply a particle with an infinite amount of energy. Figure 17.30 is a plot of the kinetic energy as a function of v.

EXAMPLE 4. The maximum speed that electrons achieve in the Stanford Linear Accelerator is $0.99999999967c$. What is the kinetic energy of an electron moving with this speed?

SOLUTION: For v/c near 1, it is convenient to write

$$\sqrt{1 - v^2/c^2} = \sqrt{1 + v/c}\,\sqrt{1 - v/c} \cong \sqrt{2}\,\sqrt{1 - v/c} \tag{49}$$

In our case the value of $1 - v/c$ is 3.3×10^{-10}. Hence

$$K = mc^2\left(\frac{1}{\sqrt{1 - v^2/c^2}} - 1\right)$$

$$= 9.1 \times 10^{-31}\ \text{kg} \times (3.0 \times 10^8\ \text{m/s})^2\left(\frac{1}{\sqrt{2}\,\sqrt{3.3 \times 10^{-10}}} - 1\right)$$

$$= 3.2 \times 10^{-9}\ \text{J}$$

As we already mentioned in Section 8.6, mass is a form of energy. According to Einstein's mass–energy relation, a particle of mass m at rest has an energy mc^2 [see Eq. (8.25)]; this relation can be derived from the theory of Special Relativity but, as with other fundamental equations in this section, we will not give a proof. The total energy of a free particle in motion is the sum of its rest-mass energy mc^2 and its kinetic energy K:

$$E = mc^2 + K = mc^2 + \frac{mc^2}{\sqrt{1 - v^2/c^2}} - mc^2 \tag{50}$$

This leads to a simple formula for the total relativistic energy of a free particle:

$$\boxed{E = \frac{mc^2}{\sqrt{1 - v^2/c^2}}} \tag{51}$$

Relativistic total energy

It is easy to verify (see Problem 24) that this relativistic energy can be expressed as follows in terms of the relativistic momentum:

$$E = \sqrt{c^2p^2 + m^2c^4} \tag{52}$$

For a particle moving at a speed close to the speed of light, the first term within the square root in Eq. (52) is much larger than the second term. Hence, for such an ultrarelativistic particle we can neglect m^2c^4 and obtain

$$E \cong cp \tag{53}$$

Thus, the momentum and the energy of an ultrarelativistic particle are directly proportional.

We also mentioned in Section 8.6 that Einstein's mass–energy relation means not only that mass is a form of energy, but also that energy has mass. For instance, the mass associated with the energy E of Eq. (51) is

$$m' = \frac{E}{c^2} = \frac{m}{\sqrt{1 - v^2/c^2}} \tag{54}$$

The effective mass of a body in motion is larger by a factor of $1/\sqrt{1 - v^2/c^2}$ than the mass of the body when at rest — a body in motion offers more resistance to acceleration and has more weight than the body at rest.

EXAMPLE 5. Show that, with the relativistic formulas for momentum and energy, the laws of conservation of momentum and energy obey the Principle of Relativity, i.e., show that if the total momentum and energy of a system of colliding particles are conserved in one reference frame, then they will also be conserved in any other reference frame.

SOLUTION: For the sake of simplicity, we will only deal with one-dimensional motion, along the x axis. The momentum and energy of a particle are then

$$p_x = \frac{mv_x}{\sqrt{1 - v_x^2/c^2}} \quad \text{and} \quad E = \frac{mc^2}{\sqrt{1 - v_x^2/c^2}}$$

The relativistic combination law for velocities tells us that the velocity of the particle in a new reference frame moving with velocity V along the x axis of the old reference frame is $v_x' = (v_x - V)/(1 - v_xV/c^2)$. Hence the momentum of the particle in the new reference frame is

$$p_x' = \frac{mv_x'}{\sqrt{1 - v_x'^2/c^2}} = \frac{m(v_x - V)}{1 - v_xV/c^2} \frac{1}{\sqrt{1 - (1/c^2)[(v_x - V)/(1 - v_xV/c^2)]^2}}$$

$$= \frac{m(v_x - V)}{\sqrt{(1 - v_xV/c^2)^2 - (v_x - V)^2/c^2}}$$

Since $(1 - v_xV/c^2)^2 - (v_x - V)^2/c^2 = (1 - V^2/c^2)\,(1 - v_x^2/c^2)$, this becomes

$$p_x' = \frac{1}{\sqrt{1 - V^2/c^2}} \frac{mv_x}{\sqrt{1 - v_x^2/c^2}} - \frac{V}{\sqrt{1 - V^2/c^2}} \frac{m}{\sqrt{1 - v_x^2/c^2}}$$

In terms of p_x and E we can write this as

$$p_x' = \frac{1}{\sqrt{1 - V^2/c^2}} p_x - \frac{V/c^2}{\sqrt{1 - V^2/c^2}} E \qquad (55)$$

This equation shows that the momentum p_x' in the new reference frame is a linear superposition of the momentum and energy in the old reference frame, with constant coefficients (the coefficients are independent of the velocity v_x of the particle). The same will therefore hold true for the total momentum of the system of all the colliding particles. Consequently, if the momentum and energy of this system are conserved in the old reference frame, the momentum will necessarily be conserved in the new reference frame.

Note that the conservation of momentum in the new reference frame depends on conservation of both momentum and energy in the old reference frame. However, this does not mean that the momentum in the new reference frame is only conserved if the collisions are elastic. The total relativistic energy is conserved even if the collisions are inelastic; such collisions merely convert kinetic energy into rest-mass energy (internal energy of the particles) without changing the total energy.

Equation (55) is a Lorentz transformation equation for momentum. There is also a Lorentz transformation equation for energy:

$$E' = \frac{1}{\sqrt{1 - V^2/c^2}} E - \frac{V}{\sqrt{1 - V^2/c^2}} p_x \qquad (56)$$

(we will leave the proof of this to Problem 39). The transformation equations of Eqs. (55) and (56) for momentum and energy are completely analogous to the transformation equations of Eqs. (15) and (16) for position and time. From Eq. (56) we deduce that if the momentum and energy of a system of particles are conserved in the old reference frame, then the energy will necessarily be conserved in the new reference frame.

SUMMARY

Principle of Relativity: All the laws of physics are the same in all inertial reference frames.

Lorentz transformations:[7] $x' = \dfrac{x - Vt}{\sqrt{1 - V^2}}$

$$t' = \frac{t - Vx}{\sqrt{1 - V^2}}$$

[7]The units of distance and time are light-seconds and seconds, respectively.

Time dilation[7]**:** $\Delta t = \dfrac{1}{\sqrt{1-V^2}}\,\Delta t'$ (clock at rest in spaceship)

Length contraction[7]**:** $\Delta x = \sqrt{1-V^2}\,\Delta x'$ (body at rest in spaceship)

Combination of velocities[7]**:** $v'_x = \dfrac{v_x - V}{1 - v_x V}$

Relativistic momentum: $\mathbf{p} = \dfrac{m\mathbf{v}}{\sqrt{1-v^2/c^2}}$

Relativistic kinetic energy and total energy:

$$K = \frac{mc^2}{\sqrt{1-v^2/c^2}} - mc^2$$

$$E = \frac{mc^2}{\sqrt{1-v^2/c^2}}$$

QUESTIONS

1. An astronaut is inside a closed space capsule coasting through interstellar space. Is there any way that the astronaut can measure the speed of the capsule without looking outside?

2. Why did Michelson and Morley use *two* light beams, rather than a single light beam, in their experiment?

3. According to a reliable source,[8] when Einstein was a boy he wondered about the following question: a runner holds a mirror at arm's length in front of his face. Can he see himself in the mirror if he runs at (almost) the speed of light? Answer this question both according to the ether theory and according to the theory of Special Relativity.

4. Consider the piece of paper on which one page of this book is printed. Which of the following properties of the piece of paper are absolute, i.e., which are independent of whether the paper is at rest or in motion relative to you? (a) The thickness of the paper, (b) the mass of the paper, (c) the volume of the paper, (d) the number of atoms in the paper (e) the chemical composition of the paper, (f) the speed of the light reflected by the paper, and (g) the color of the colored print on the paper.

5. Two streetlamps, one in Boston and the other in New York City, are turned on at exactly 6:00 P.M. Eastern Standard Time. Find a reference frame in which the streetlamp in New York was turned on late.

6. According to the theory of Special Relativity, the time order of events can be reversed under certain conditions. Does this mean that a sparrow might fall from the sky before it leaves the nest?

7. Because of the rotational motion of the Earth about its axis, a point on the equator moves with a speed of 460 m/s relative to a point on the North Pole. Does this mean that a clock placed on the equator runs more slowly than a similar clock placed on the pole?

8. According to Jacob Bronowski, author of the *Ascent of Man,* the explanation of time dilation is as follows: If you are moving away from a clock tower at a speed nearly equal to the speed of light, you keep pace with the light that the face of the clock sent out at, say, 11 o'clock. Hence if you look toward the

[8] E. F. Taylor and J. A. Wheeler, *Spacetime Physics.*

clock tower, you always see its hands at the 11 o'clock position. Is this explanation correct? What is wrong with it?

9. Suppose you wanted to travel into the future and see what the twenty-fifth century is like. In principle, how could you do this? Could you ever return to the twentieth century?

10. According to the qualitative arguments of Section 17.2, a light signal traveling along a track placed perpendicular to the direction of motion of the spaceship (see Figure 17.11) takes a longer time to complete a round trip when measured by the clocks on the Earth than when measured by the clocks on the spaceship. Would the same be true for a light signal traveling along a track placed parallel to the direction of motion? Explain qualitatively.

11. A cannonball is perfectly round in its own reference frame. Describe the shape of this cannonball in a reference frame relative to which it has a speed of $0.95c$. Is the volume of the cannonball the same in both reference frames?

12. Could we use the argument based on the two identical pieces of pipe (see Figure 17.16) to prove that lengths *along* the direction of motion are not affected? Why not?

13. A rod at rest on the ground makes an angle of 30° with the x axis in the reference frame of the Earth. Will the angle be larger or smaller in the reference frame of a spaceship moving along the x axis?

14. In the charming tale "City Speed Limit" by George Gamow,[9] the protagonist, Mr. Tompkins, finds himself riding a bicycle in a city where the speed of light is very low, roughly 30 km/h. What weird effects must Mr. Tompkins have noticed under these circumstances?

15. Suppose that a *very* fast runner holding a long horizontal pole runs through a barn open at both ends. The length of the pole (in its rest frame) is 6 m and the length of the barn (in *its* rest frame) is 5 m. In the reference frame of the barn the pole will suffer length contraction and, at one instant of time, all of the pole will be inside the barn. However, in the reference frame of the runner, the barn will suffer length contraction and all of the pole will never be inside the barn at one instant of time. Is this a contradiction?

16. If the beam from a revolving searchlight is intercepted by a distant cloud, the bright spot will move across the surface of the cloud very quickly, with a speed that can easily exceed the speed of light. Does this conflict with our conclusion of Section 17.3, that nothing can move faster than the speed of light?

17. Figure 17.24 shows the x'–t' coordinate grid for a spaceship moving in the positive x direction. Draw a corresponding diagram showing the coordinate grid for a spaceship moving in the negative x direction.

18. Why can a spaceship not travel as fast as or faster than the speed of light?

PROBLEMS

Section 17.3

1. (a) In a spacetime diagram, plot the worldline of an electron moving in the positive x direction at a speed of 2.0×10^8 m/s. In your diagram use units of seconds for t and light-seconds for x. What is the slope of this worldline?
   (b) In the same diagram plot the worldline of a supersonic aircraft moving in the positive x direction at 500 m/s. What is the slope of this worldline?

2. Suppose that a spaceship moves relative to the Earth at a constant velocity of $V = 0.33c$.

[9]George Gamow, *Mr. Tompkins in Wonderland.*

(a) Carefully draw a diagram such as that shown in Figure 17.24. Draw the coordinate grids for x, t and x', t' in different colors and make sure that you have the correct angles between the axes and the correct scales along the axes.

(b) On your diagram draw the worldline of an electron of velocity $v_x = 0.66c$. What is the slope of this worldline relative to the x, t axes? Measure the slope of the worldline relative to the x', t' axes and thereby find v'_x. Does your result agree with Eq. (36)?

3. A spaceship moves relative to the Earth at a speed $V = 0.5c$. Assume that the spaceship moves along the *negative* x axis.

(a) Draw a diagram analogous to Figure 17.24. Draw the coordinate grids for x, t and x', t' in different colors and make sure that you have the correct angles between the axes and the correct scales along the axes.

(b) Plot the spacetime point $x = 2$, $t = 3$ in your diagram. Read the values of the x' and t' coordinates of this event directly off your diagram. Check that your graphical determination of the values of x' and t' agrees with the computation of these values from Eqs. (15) and (16).

4. The captain of a spaceship traveling away from Earth in the x direction at $V = 0.80c$ observes that a nova explosion occurs at a point with spacetime coordinates $t' = -6.0 \times 10^8$ s, $x' = 6.2 \times 10^8$ light-seconds, $y' = 4.0 \times 10^8$ light-seconds, $z' = 0$ as measured in the reference frame of the spaceship. He reports this event to Earth via radio.

(a) What are the spacetime coordinates of the explosion in the reference frame of the Earth? Assume that the master clock of the spaceship coincides with the master clock of the Earth at the instant the spaceship passes by the Earth and that the origin of the spaceship x', y', and z' coordinates is at the midpoint of the spaceship.

(b) Will the Earth receive the report of the captain before or after astronomers on the Earth see the nova explosion in their telescopes? No calculation is required for this question.

5. Consider the situation described in Problem 4. Since light takes some time to travel from the nova to the spaceship, the spacetime coordinates that the captain reports are not directly measured, but rather deduced from the time of arrival and the direction of the nova light reaching the spaceship.

(a) At what time (t' time) did the nova light reach the spaceship?

(b) If the captain sends a report to Earth via radio as soon as he sees the supernova, at what time (t time) does the Earth receive the report?

(c) At what time do Earth astronomers see the nova?

6. At $11^h0^m0^s$ A.M. a boiler explodes in the basement of the Museum of Modern Art in New York City. At $11^h0^m0.0003^s$ A.M. a similar boiler explodes in the basement of a soup factory in Camden, New Jersey, at a distance of 150 km from the first explosion. Show that in the reference frame of a spaceship moving at a speed greater than $V = 0.60c$ from New York toward Camden, the first explosion occurs *after* the second.

7. Show that the Lorentz transformations [Eqs. (15) and (16)] can be written in the form

$$x' = x \cosh \theta - t \sinh \theta$$

$$t' = -x \sinh \theta + t \cosh \theta$$

where $\tanh \theta = V$. In this form the Lorentz transformation is reminiscent of a rotation (see Section 3.5).

Section 17.4

8. In 1961 the cosmonaut G. S. Titov circled the Earth for 25 h at a speed of 7.8 km/s. According to Eq. (27), what was the time-dilation factor of his body clock relative to the clocks on Earth? By how many seconds did his body clock fall behind during the entire trip?

9. Muons are unstable particles which — if at rest in a laboratory — decay after a time of only 2.0×10^{-6} s. Suppose that a muon is created in a collision between a cosmic ray and an oxygen nucleus at the top of the Earth's atmosphere, at an altitude of 20 km above sea level.

(a) If the muon has a downward speed of 2.97×10^8 m/s relative to the Earth, at what altitude will it decay? Ignore gravity in this calculation.

(b) Without time dilation, at what altitude would the muon have decayed?

10. In a test of the relativistic time-dilation effect, physicists compared the rate of vibration of nuclei of iron moving at different speeds. One sample of iron nuclei was placed on the rim of a high-speed rotor; another sample of similar nuclei was placed at the center. The radius of the rotor was 10 cm and it rotated at 35,000 rev/min. Under these conditions, what was the speed of the rim of the rotor relative to the center? According to Eq. (27), what was the time-dilation factor of the sample at the rim compared to the sample at the center?

11. Suppose that a special breed of cat (*Felix Einsteiniensis*) lives for exactly 7 years according to its own body clock. When such a cat is born, we put it aboard a spaceship and send it off at $V = 0.8c$ toward the star Alpha Centauri. How far from the Earth (reckoned in the reference frame of the Earth) will the cat be when it dies? How long after the departure of the spaceship will a radio signal announcing the death of the cat reach us? The radio signal is sent out from the spaceship at the instant the cat dies.

12. If cosmonauts from the Earth wanted to travel to the Andromeda galaxy in a time of no more than 10 years as reckoned by clocks aboard their spaceship, at what (constant) speed would they have to travel? How much time would have elapsed on Earth after 10 years of time on the spaceship? The distance to the Andromeda galaxy is 2.2×10^6 light-years.

13. Because of the rotation of the Earth, a point on the equator has a speed of 460 m/s relative to a point at the North Pole. According to the time-dilation effect described in Section 17.4, by what factor do the rates of two clocks differ if one is located on the equator and the other at the North Pole? After 1 year has elapsed, by how many seconds will the clocks differ? Which clock will be ahead?[10]

*14. The star Alpha Centauri is 4.4 light-years away from us. Suppose that we send a spaceship on an expedition to this star. Relative to the Earth the spaceship accelerates at a constant rate of 0.1 gee until it reaches the midpoint 2.2 light-years from Earth. The spaceship then decelerates at a constant rate of 0.1 gee until it reaches Alpha Centauri. The spaceship performs the return trip in the same manner.

(a) What is the time required for the complete trip according to the clocks on the Earth? Ignore the time that the spaceship spends at its destination.

(b) What is the time required for the complete trip according to the clocks on the spaceship? Assume that the *instantaneous* time-dilation factor is still $\sqrt{1 - V^2}$, even though the velocity V is a function of time.

Section 17.5

15. According to the manufacturer's specifications, a spaceship has a length of 200 m. At what speed (relative to the Earth) will this spaceship have a length of 100 m in the reference frame of the Earth?

16. What is the percent length contraction of an automobile traveling at 60 mi/h?

17. Suppose that a proton speeds by the Earth at $V = 0.8c$ along a line parallel to the axis of the Earth.

[10] The results of this problem cannot be compared directly with experiments because the clocks also suffer a gravitational time-dilation effect which must be calculated from the theory of General Relativity.

(a) In the reference frame of the proton, what is the polar diameter of the Earth? The equatorial diameter?
(b) In the reference frame of the proton, how long does the proton take to travel from the point of closest approach to the North Pole to the point of closest approach to the South Pole? In the reference frame of the Earth, how long does this take?

Section 17.6

18. A collision between two gamma rays creates an electron and an anitelectron that travel away from the point of creation in opposite directions, each with a speed of $0.95c$ in the laboratory. What is the speed of the antielectron in the rest frame of the electron, and vice versa?

19. Find the inverses of Eq. (40), i.e., express v_y and v_z in terms of v'_x, v'_y, and v'_z.

20. In the reference frame of the Earth, a projectile is fired in the x–y plane with a speed of $0.8c$ at an angle of 30° with the x axis. What will be the speed and the angle with the x' axis in the reference frame of a spaceship passing by the Earth in the x direction at a speed of $0.5c$? [Hint: Use Eqs. (36) and (40).]

21. Derive Eq. (40) for v'_y and v'_z.

Section 17.7

22. A particle has a kinetic energy equal to its rest-mass energy. What is the speed of this particle?

23. The yearly energy expenditure of the United States is about 8×10^{19} J. Suppose that all of this energy could be converted into kinetic energy of an automobile of mass 1000 kg. What would be the speed of this automobile?

24. Derive Eq. (52).

25. What is the percent difference between the Newtonian and the relativistic value for the momentum of a meteoroid reaching the Earth at a speed of 72 km/s?

26. The speed of an electron in a hydrogen atom is 2.2×10^6 m/s. What is the percent deviation between the Newtonian and the relativistic value of the kinetic energy of an electron at this speed? [Hint: Take the Taylor-series expansion of Eq. (47) one term further.]

27. What is the kinetic energy of a spaceship of rest mass 50 metric tons moving at a speed of $0.5c$? How many metric tons of a matter–antimatter mixture would have to be consumed to make this much energy available?

28. At the Fermilab accelerator, protons are given kinetic energies of 1.6×10^{-7} J. By how many meters per second does the speed of such a proton differ from the speed of light? What is the momentum of such a proton?

29. Consider the electrons of a speed $0.99999999967c$ produced by the Stanford Linear Accelerator. According to Eq. (54), what is the mass of one of these electrons? By what factor is this mass greater than the mass of an electron at rest?

30. The most energetic cosmic rays have energies of about 50 J. Assume that such a cosmic ray consists of a proton. By how much does the speed of such a proton differ from the speed of light? Express your answer in meters per second.

31. Table C.2 lists several large accelerators that produce high-energy protons or electrons. In each case, calculate the difference $c - v$ between the speed of light and the speed of the particle. [Hint: $1 - v^2/c^2 = (1 + v/c) \times (1 - v/c) \cong 2(1 - v/c)$ if $v \cong c$.]

32. At the Fermilab accelerator protons of kinetic energy 1.6×10^{-7} J are made to move along a circle of diameter 2.0 km.
(a) What is the centripetal acceleration of these protons?

(b) According to Eq. (54), what is their mass?
(c) What is the centripetal force on the protons?

33. Suppose that a spaceship traveling at $0.8c$ through our solar system suffers an inelastic collision with a small meteoroid of mass 2.0 kg.
(a) What is the kinetic energy of the meteoroid in the reference frame of the spaceship?
(b) In the collision all of this kinetic energy suddenly becomes available for inelastic processes that damage the spaceship. The effect on the spaceship is similar to an explosion. How many tons of TNT will release the same explosive energy? One ton of TNT releases 4.2×10^9 J.

34. A K^0 particle at rest decays spontaneously into a π^+ particle and a π^- particle. What will be the speed of each of the latter? The mass of the K^0 is 8.87×10^{-28} kg and the masses of the π^+ and π^- particles are 2.49×10^{-28} kg.

35. Free neutrons decay spontaneously into a proton, an electron, and an antineutrino:

$$\mathrm{n} \rightarrow \mathrm{p} + \mathrm{e} + \bar{\nu}$$

The neutron has a rest mass 1.6749×10^{-27} kg; the proton, 1.6726×10^{-27} kg; the electron, 9.11×10^{-31} kg; and the antineutrino zero (or nearly zero). Assume that the neutron is at rest.
(a) What is the energy released in this decay?
(b) Suppose that in one such decay, this energy is shared between the proton and the antineutrino so that these particles emerge from the point of decay with equal and opposite momenta while the electron remains at rest. What is the momentum of the proton and of the antineutrino?
(c) What is the energy of the proton? What is the energy of the antineutrino? [Hint: The antineutrino is always ultrarelativistic; use Eq. (53).]

36. Derive Eq. (56).

37. A K^0 particle moving at a speed of $0.60c$ through the laboratory decays into a muon and an antimuon.
(a) In the rest frame of the K^0, what is the speed of each muon? The mass of the K^0 is 8.87×10^{-28} kg and the masses of the muon and the antimuon are 1.88×10^{-28} kg each.
(b) Assume that the muon moves in a direction parallel to the original direction of motion of the K^0 and that the antimuon moves in the opposite direction. What are the speeds of the muon and the antimuon with respect to the laboratory?

*38. At the Brookhaven AGS accelerator, protons of kinetic energy 5.3×10^{-9} J are made to collide with protons at rest.
(a) What is the speed of a moving proton in the laboratory reference frame?
(b) What is the speed of a reference frame in which the two colliding protons have the same speed (and are moving in opposite directions)?
(c) What is the energy of each proton in the latter reference frame?

*39. In a given reference frame, a particle of mass m has a momentum with components p_x, p_y, p_z and an energy E. Show that the Lorentz transformation equations for the momentum and the energy are

$$p'_x = \frac{p_x - VE/c^2}{\sqrt{1 - V^2/c^2}}$$

$$p'_y = p_y$$

$$p'_z = p_z$$

$$E' = \frac{E - Vp_x}{\sqrt{1 - V^2/c^2}}$$

where, as in Example 5, the velocity of the new reference frame is along the x axis of the old.

GRAVITY AND GEOMETRY[1]

Few theories can compare in the accuracy of their predictions with Newton's theory of universal gravitation. But in spite of its spectacular successes, Newton's theory is not perfect. As we mentioned in Chapter 13, the planet Mercury wanders slightly ahead of its predicted position. This deviation, called the perihelion precession of Mercury, amounts to 43 seconds of arc per century. The failure of all attempts to explain this small deviation led astronomers to question the validity of Newton's theory. Besides, from the point of view of relativity there is a serious defect in this theory. According to Newton's law of universal gravitation, any gravitational disturbance propagates instantaneously from one mass to another. For instance, if the Sun suddenly were to grow a large bulge, the gravitational effect of this bulge should be felt at the Earth instantaneously. But this would amount to propagation of a signal with infinite speed, which is impossible according to the theory of Special Relativity.

After the discovery of Special Relativity, Einstein set to work on a new theory of gravitation and in 1915 he formulated his famous theory of General Relativity. In the preceding chapter we saw that for reference frames in motion, space and time are not absolute; in this chapter we will see that in the vicinity of gravitating bodies, space and time are distorted. In essence, the content of Einstein's new theory of gravitation is that the orbital motion of planets does not arise from a gravitational force, but from the distortion of spacetime. Gravity is due to curved spacetime.

E.1 THE PRINCIPLE OF EQUIVALENCE

The most important fact about gravity is that, at any given place, all bodies fall at the same rate. This fact was firmly established by the experiments of Galileo Galilei late in the sixteenth century. Galileo not only checked that bodies of unequal weight fall at the same rate, but he also checked that bodies made of different substances fall at the same rate. In his most sensitive experiments, he compared the periods of oscillation of pendulums of equal length but with bobs made of different substances; any inequality of the free-fall accelerations of the bobs would then show up as an inequality of their periods (see Section 14.4). Galileo compared pendulum bobs of cork, lead, and other substances and found them equal. Such pendulum experiments were repeated by Newton with some refinements to compensate for the friction between the pendulum bobs and air.

Experiments testing the universality of free-fall rates for different substances have come to be called **Eötvös experiments,** after Lorand von Eötvös, a Hungarian physicist who, around 1920, did very extensive tests of very high precision (Figure E.1). Modified versions of these experiments have been used to test the free-fall rates toward the Earth and toward the Sun. The basic method underlying these experiments can most easily be described in the case of free fall toward the Sun: we attach the two masses that are to be compared to the beam of a Cavendish balance. The masses, the Cavendish balance, our laboratory, and the entire Earth are in free fall toward the Sun. If one of the masses is falling faster than the other, then the beam of the balance will tend to turn toward the Sun, i.e., the beam will experience a torque. The Eötvös experiment attempts to measure this torque. The torque is exactly zero if, and only if, the free-fall rates of the two masses are exactly equal.

In the hope of discovering some small deviation between free-fall rates, Eötvös and his successors tested many common metals and also a wide variety of other substances — snakewood, tallow, asbestos, radioac-

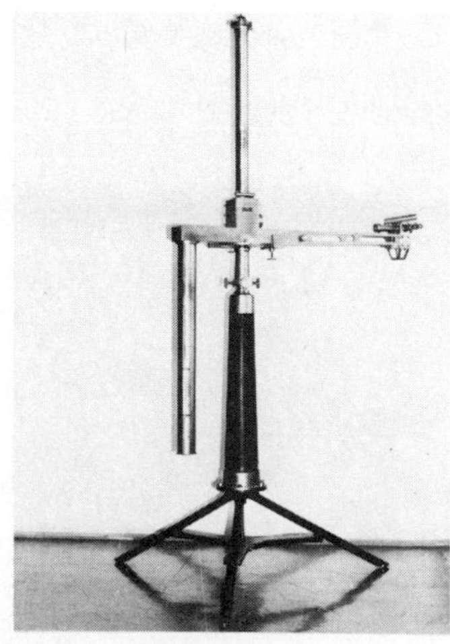

Fig. E.1 One of the Cavendish torsional balances used by Eötvös. The balance beam is inside the horizontal box. The suspension fiber is inside the upper vertical tube.

[1] This chapter is optional.

tive strontium, etc. — but they found no evidence for any deviation. The latest and most precise experiments by R. H. Dicke (at Princeton University) and V. B. Braginsky (at Moscow University) were done on gold, platinum, and aluminum. The atoms of gold and aluminum are about as different as can be — one is very heavy and the other very light. Nevertheless, the experiments show that these atoms fall at exactly the same rate to within the instrumental error of the apparatus, about 1 part in 10^{12}.

Figure E.2 shows the apparatus used by Braginsky. The balance beam is star shaped, with eight prongs carrying small masses of aluminum and platinum. The reason for this curious configuration of the beam is that local gravitational disturbances, such as trucks or streetcars driving on nearby streets, do not disturb it as easily as they would an ordinary single beam. Of course, the beam is placed in a good vacuum (10^{-8} mmHg) and the entire apparatus is carefully kept in thermal isolation. Because there is no air rubbing on the beam, the balance has next to no friction; if set in motion, it swings for several years.

The delicate experiments of Dicke and Braginsky confirm with extreme precision what Galileo already recognized four centuries ago: under the influence of gravity all bodies fall at exactly the same rate. As we already mentioned in earlier chapters, one consequence of this universality of the rate of free fall is that free fall simulates the condition of weightlessness, or absence of gravity. For example, Figure E.3 shows a photograph of the interior of the Skylab satellite in orbit around the Earth. Skylab and the astronauts within it are in free fall. The astronauts feel weightless; they are in a 0-g environment. The reason for this apparent absence of gravitational forces is that any object released by one of the astronauts falls at exactly the same rate as the astronaut, i.e., relative to the astronaut it seems not to fall at all.

The freely falling reference frame of Skylab simulates an inertial reference frame: a body at rest remains at rest and a body in motion continues to move with constant velocity along a straight line (Figure E.4). A reference frame in free fall under the influence of gravity is called a **local inertial reference frame.** This kind of reference frame must be regarded as having a small size — it can simulate a true inertial reference frame only over a small range of distances. For instance, if we attempt to extend the reference frame of Skylab over an appreciable part of the orbit, then the decrease of the gravitational acceleration with distance from the Earth would cause a free body in the upper part of the reference frame to accelerate less than a body in the lower part — both of these bodies would then have an acceleration relative to the midpoint of the reference frame (Figure E.5). Although these deviations between the behaviors of bodies in a local inertial reference frame and in a true inertial reference frame are most noticeable if the distances between the

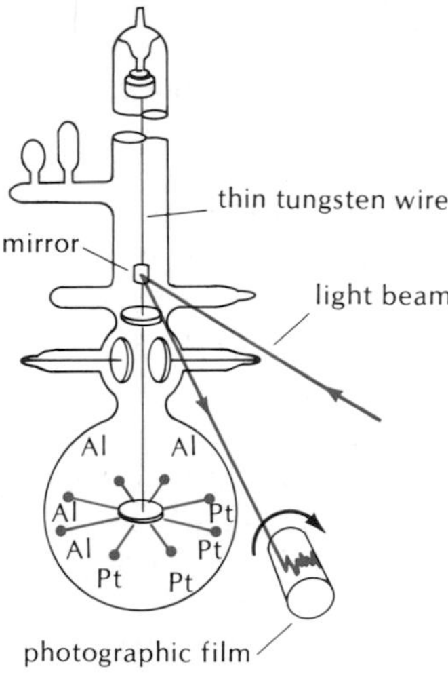

Fig. E.2 The apparatus of Braginsky.

Fig. E.3 Astronaut floating freely in Skylab.

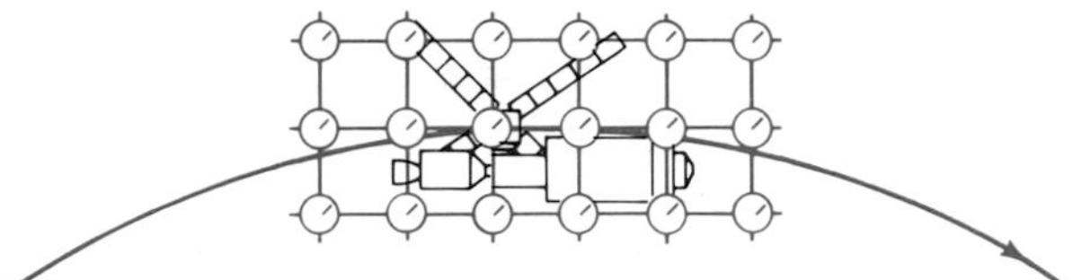

Fig. E.4 The reference frame of Skylab is in free fall.

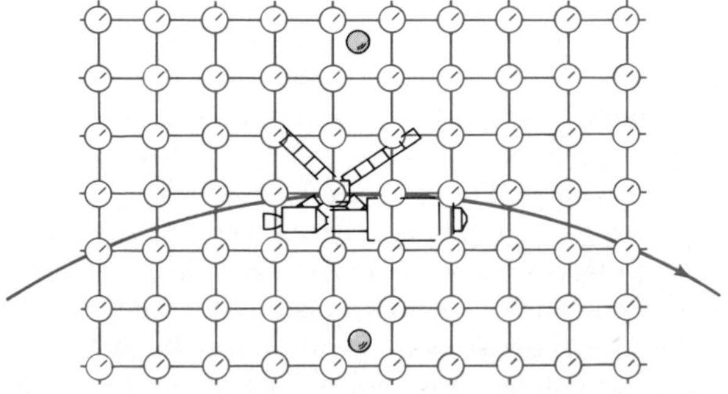

Fig. E.5 A large reference frame in free fall near the Earth. Bodies (balls) that are above or below the midpoint of this reference frame accelerate relative to the reference frame.

bodies are large, a very sensitive instrument can detect the deviations even when the distances are small. For instance, a pair of modern, sensitive accelerometers placed at opposite ends of the Skylab satellite could easily have detected the difference in the gravitational accelerations at the two ends. Hence the elimination of gravity by free fall is not quite perfect. But the residual effects are so small that we will ignore them in the following.

The extremely precise experiments on the equality of the rates of free fall establish that, as far as mechanical experiments are concerned, there is no intrinsic distinction between a local inertial reference frame and a true inertial reference frame. This indicates that there is not only a relativity of uniform translational motion but, to some extent, also a relativity of accelerated motion — the astronauts in a freely falling spacecraft cannot detect their acceleration by any mechanical experiments performed within their spacecraft. In 1911, some years before he formulated his new theory of gravitation, Einstein conjectured that this general relativity of motion applies not only to mechanical experiments, but also to any other experiments. He formulated this conjecture as his famous **Principle of Equivalence,** a hypothesis about the character of gravitation:

> *In a local inertial reference frame, the effects of gravity are absent; in such a reference frame all the laws of physics are the same as in a true inertial reference frame in interstellar space, far from any gravitating masses. Conversely, in an accelerating reference frame in interstellar space there is artificial gravity; in such a reference frame the laws of physics are just as though the reference frame were at rest on or near a gravitating body.*

This principle lies at the basis of the theory of General Relativity. In the following sections we will explore a few consequences of this principle.

E.2 THE DEFLECTION OF LIGHT AND THE CURVATURE OF SPACE

One of the most obvious conclusions we can draw from the Principle of Equivalence is that the speed of light ought to be the same everywhere in a gravitational field. Imagine that everywhere around the Sun there are small spaceships (Skylabs), instantaneously at rest but in free fall (i.e., these spaceships have just been released and they are about to fall toward the Sun, but they have not yet acquired any appreciable velocity). In each of them gravity is effectively absent; and when experimenters aboard them measure the speed of light, the Principle of Equivalence demands that they all obtain the same result, the same as in interstellar space, 3.00×10^8 m/s.

Another conclusion we can draw from the Principle of Equivalence is that light will be deflected by the Sun's gravity. And this leads to a paradox: if light is deflected by the Sun, then light must be slowed down by the Sun — and that contradicts the previous conclusion that the speed of light is constant. Before we explore the consequences of this paradox, let us spell out why we expect that light will be deflected by the Sun and also slowed down by the Sun.

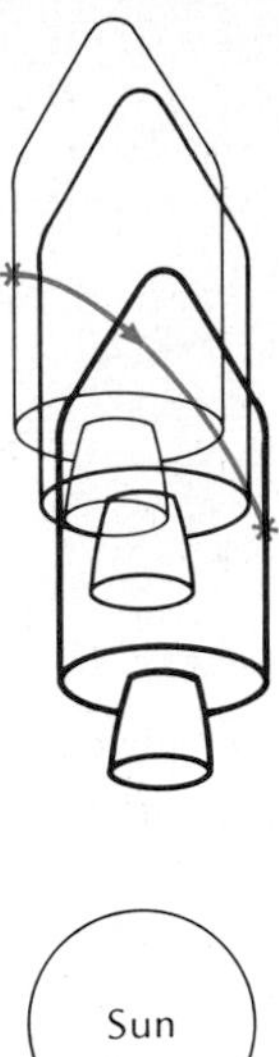

Fig. E.6 A spaceship in free fall. A light ray propagates from the left side of the spaceship to the right.

Consider one of the small spaceships in free fall toward the Sun. Since in this spaceship gravity is virtually absent, a light ray will propagate across this spaceship in a straight line, just as in interstellar space. But if the light ray follows a straight path relative to the spaceship, then it must follow a curved path relative to the Sun. Figure E.6 shows a light ray propagating from one side of the spaceship to the other; the light ray starts at the midpoint of the left side and ends at the midpoint of the right side; since the spaceship accelerates downward, the light ray must also accelerate downward at the same rate if it is to follow a straight (horizontal) path relative to the spaceship. Thus, the Principle of Equivalence leads to the conclusion that in the vicinity of the Sun a light ray must follow a curved path — the Sun "bends" rays of light (Figure E.7). If

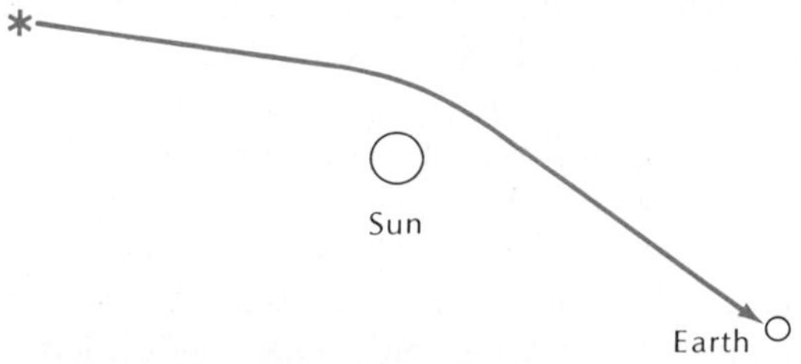

Fig. E.7 Curved trajectory of light ray passing near the Sun. The bending of this trajectory has been wildly exaggerated in this diagram.

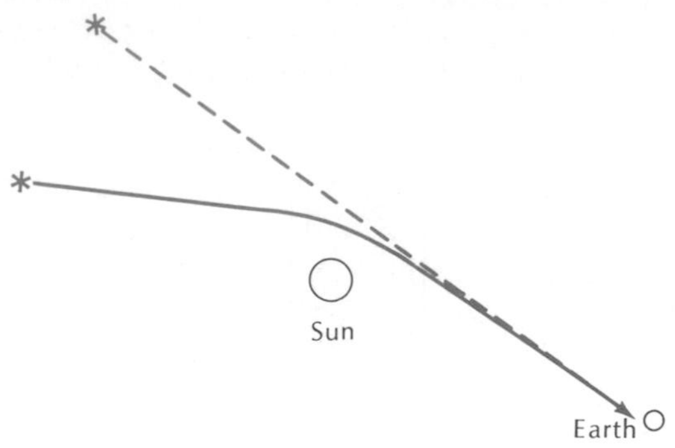

Fig. E.8 An astronomer on the Earth sees the star at a virtual position different from its real position.

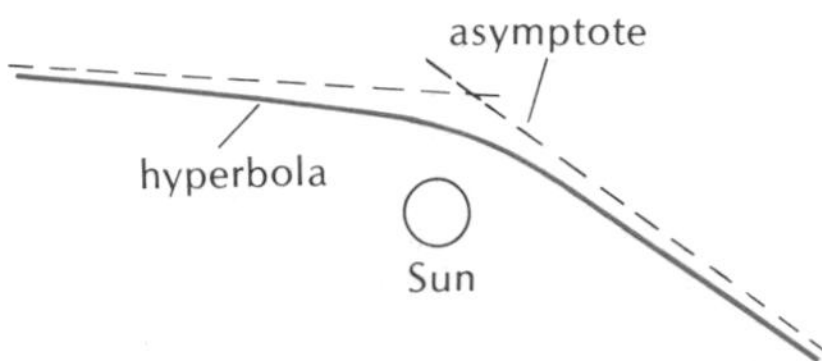

Fig. E.9 According to Laplace, light should follow a hyperbolic orbit.

the light originates from a star, then an astronomer on the Earth will see this star displaced from its usual position whenever the Sun happens to be near the line of sight (Figure E.8).

The deflection of light by the gravity of the Sun need not surprise us. Light is energy, hence light has mass. We therefore expect that the Sun will attract light and we expect that a ray of light passing near the Sun has an hyperbolic orbit (Figure E.9) just like the orbit of a fast comet. The possibility of this effect was already recognized by the French mathematician and astronomer Laplace more than a century before Einstein. But the effect is very small: a ray passing very close to the Sun is only deflected by a few seconds of arc. It was not until 1919 that astronomers made an attempt to measure this tiny deflection. At that time Einstein had calculated the exact value of the expected deflection from his theory of General Relativity and this stimulated astronomers to make the attempt.

It is a very difficult measurement to make. One needs to look at a ray of light that comes from a star, passes near the Sun, and reaches the Earth. That means that one must observe the star when it is near the Sun in the sky; it is of course only possible to do this during a total eclipse of the Sun. Unfortunately, eclipses usually happen at inconvenient places such as the jungles of Brazil or the deserts of the Sudan and the primitive conditions at the observation site are not conducive to extremely high precision. All the telescopes, mirrors, cameras, etc., must be carried to these locations and carefully protected from disturbances while the eclipse photographs are being taken (Figure E.10). Since 1919, astronomers have gone to

Fig. E.10 Eclipse instruments at Sobral, Brazil, for the observation of the eclipse of 1919.

great pains with these eclipse observations and they have sent out a dozen expeditions to remote places, but they have not succeeded in reducing the experimental uncertainties to a satisfactory level. Although all the measurements found a deflection somewhere between 1.5 and 2.0 seconds of arc for a light ray that just grazes the Sun, there are substantial and unexplained differences between the individual measurements.

Rather more reliable measurements of the deflection have recently been made with radio waves. Many objects in the sky are sources of radio waves. Among these, the best suited for the deflection measurement are the *quasars*; they are radio sources of very small angular size, which act as pinpoints of radio waves just as ordinary stars act as pinpoints of light. To measure the deflection of the radio waves, one must wait for some day of the year when the Sun and a quasar are nearly lined up. For the quasar 3C279 this happens every 8th of October and radio astronomers have taken advantage of this alignment to measure the deflection on several occasions. The best of these measurements indicate that a radio wave just grazing the Sun is deflected by about 1.7 or 1.8 seconds of arc.

The deflection of light or radio waves implies that the speed of these waves is reduced when they go near the Sun. To understand this, look at Figure E.11, which shows wave fronts marching along. How do these wave fronts deflect? If every portion of the wave front had the same speed as every other portion, there would be no deflection. If there is to be a deflection,

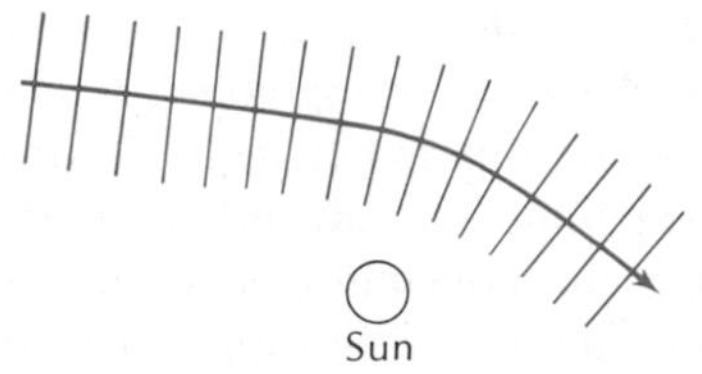

Fig. E.11 Wave fronts of a light wave passing by the Sun.

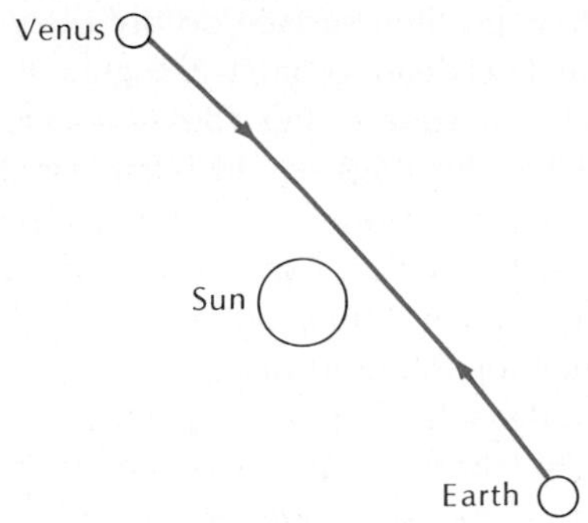

Fig. E.12 A radar pulse emitted from the Earth reaches Venus and returns to the Earth.

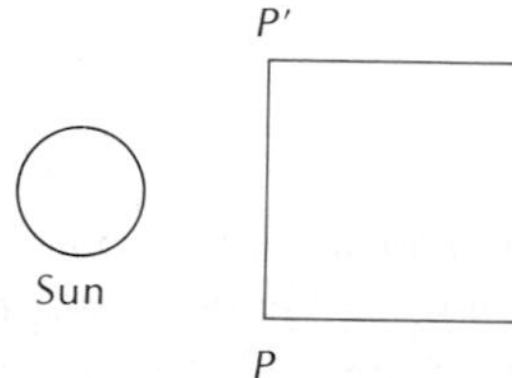

Fig. E.13 A large square laid out with tightly stretched strings near the Sun. The side *PP'* nearest the Sun is longer than the other sides, although in this diagram drawn on a flat two-dimensional surface all sides look to be the same length.

the portion of wave front nearest the Sun must slow down so that this portion begins to lag behind — the wave front gradually turns like a row of soldiers swinging to a new direction of advance.

The slowing down of radio waves by the Sun has been confirmed very directly by means of radar signals sent past the Sun. Radio astronomers have used their giant antennas as transmitters to send radar signals to Mercury and to Venus; the signals then bounce back to Earth. If the target planet happens to be near occultation (i.e., nearly eclipsed by the Sun), then these radio waves must pass close to the Sun both on the outbound and inbound portions of the trip (Figure E.12). The experiments show that the radar echo from Venus is delayed by about 200 microseconds as compared to the echo obtained when the radar pulse does not pass near the Sun. This small time delay can be readily detected by accurate atomic clocks. The delay shows very directly that the radio signal is slowed down by gravity. A similar slowing down has also been detected in radio signals sent to the *Mariner* 6 and 7 spacecraft and retransmitted back to Earth.

So, all our experiments show that light slows down near the Sun — yet according to the Principle of Equivalence, it should not. What is the answer to this paradox? The answer must be that there is more distance from here to Venus than we suppose there is. Light does not move slower, it merely goes farther and therefore takes longer. In the immediate vicinity of the Sun, distances are longer than expected on the basis of Euclidean geometry. For example, consider a "square" formed by laying out sides at right angles; the segment nearest the Sun is longer than the segment farthest from the Sun (Figure E.13). Ordinary Euclidean geometry fails — we have a space that is not flat, but a space that is curved and distorted.

To shape a mental image or, more precisely, an analogy of this distortion, it is helpful to think of a hypothetical two-dimensional space rather than our real three-dimensional space. Figure E.14 shows a flat two-dimensional surface; this is the analog of our real three-dimensional space with the Sun absent. Figure E.15 shows what happens when a "sun" is placed in this two-dimensional space: the space acquires a dip just as the surface of a waterbed acquires a dip when you sit in the center. The distance between the points *P* and *P'* measured along the curved surface is longer than the distance measured along a straight line, and light traveling from *P* to *P'* along the curved surface will take longer than what we would expect if we ignored the dip. This resolves the paradox: the speed of light is the same everywhere, but light that passes near the Sun must go an extra distance and hence is delayed. How much extra distance? For a light ray grazing the Sun, the extra distance from Earth to Venus is about 30 km.

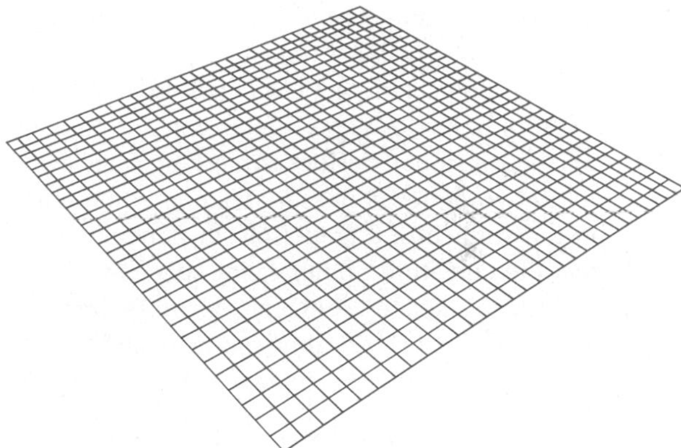

Fig. E.14 A flat two-dimensional surface.

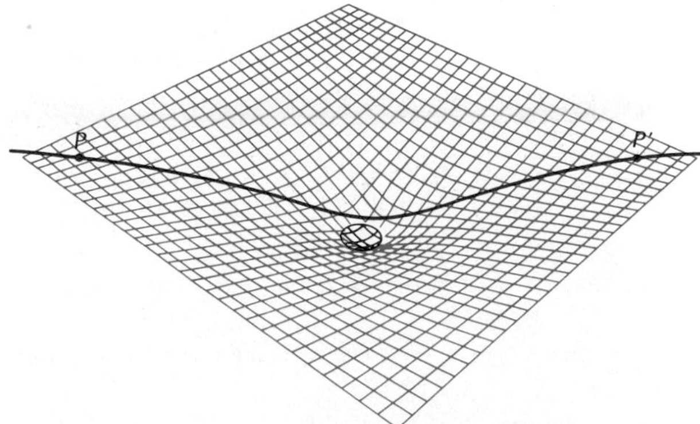

Fig. E.15 If a "sun" is placed on this surface, the surface becomes curved. The shortest line between two points *P* and *P'* is longer than on the flat surface.

This compelling argument shows how the Principle of Equivalence leads us to the conclusion that the space near the Sun is curved. Curved space is the starting point of Einstein's theory of General Relativity.

E.3 THE THEORY OF GENERAL RELATIVITY

The fundamental tenet of Einstein's theory of General Relativity is that gravity is caused by curvature. The Sun, or any other massive body, produces curvature in the space surrounding it and this curvature affects the motion of light and of planets, making them move in the way they do. The Sun does not exert any force on the light or the planets — it distorts space and also distorts time, and the light or planet merely moves on the "straightest" possible line in this curved spacetime. Thus, Einstein's theory of General Relativity is a geometrical theory of gravitation. John Wheeler has coined the name *geometrodynamics* for Einstein's theory. This name draws attention to the essential feature of the theory: the dynamical, curved spacetime geometry which acts on and reacts to matter.

It is beyond our powers of imagination to visualize a curved three-dimensional space. To visualize a curved three-dimensional space we would have to visualize a fourth or even a fifth dimension into which the three-dimensional space can curve — and we just cannot visualize any four-dimensional object.

Fortunately, one can detect the curvature of a space without stepping into a higher dimension and looking at it from the outside. Consider some little bugs living on a curved two-dimensional surface, e.g., some mites living on the surface of a grapefruit. In principle, they can detect the curvature of their living space by experiments within this space. They can do measurements on geometrical figures on their surface and check whether the geometry is Euclidean or something different. If the living space of the bugs is the surface of a sphere, then they will find that the circles and triangles that they draw on their surface do not obey the usual theorems of Euclidean geometry. Figure E.16 shows the surface of a sphere and shows a circle and its radius (or what the bugs would regard as its radius); obviously the circumference of the circle is shorter than 2π times the radius. Figure E.16 also shows a triangle on the surface of the sphere; obviously the sum of the interior angles of this triangle is larger than 180°. Incidentally, the sides of this triangle are drawn so that they are the shortest lines connecting the vertex points, that is, the sides are drawn by stretching a string between the vertex points.

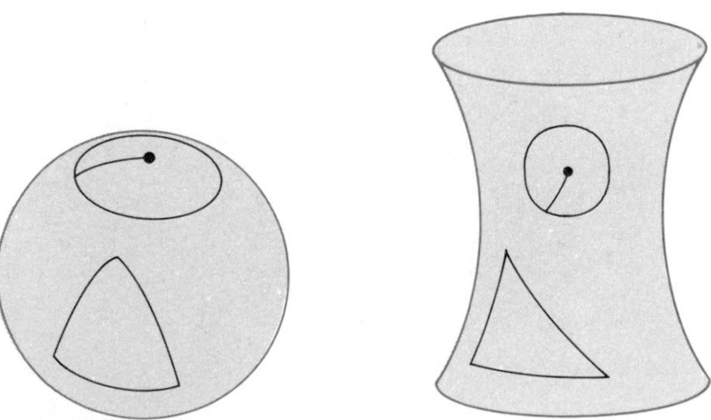

Fig. E.16 (left) The surface of a sphere is a curved two-dimensional surface.

Fig. E.17 (right) The surface of a hyperboloid is another curved two-dimensional surface.

Circles and triangles can also be used to detect the curvature of other kinds of curved surfaces. Figure E.17 shows the surface of a hyperboloid. On this surface the circumference of a circle is longer than 2π times the radius, and the sum of the interior angles of a triangle is smaller than 180°.

According to General Relativity, the space inside the Sun or any gravitating body has characteristics analogous to those of a spherical surface: circumferences are too small and interior angles of triangles too large. The space outside and near the Sun has characteristics analogous to those of a hyperboloidal surface: circumferences are too large and interior angles of triangles too small. These deviations from Euclidean geometry are implicitly contained in Einstein's equations that relate the curvature of the space to the mass of the body. Einstein's equations are rather complicated, but for a spherical gravitating body of uniform density the deviations from Euclidean geometry can be expressed quite simply as follows. A circumference that is too small is equivalent to a radius that is too large. Let us define the radius excess ΔR as the amount by which the radius of the body is too large when compared with its circumference C,

$$\Delta R = R - C/2\pi \tag{1}$$

Then the Einstein equations state that the radius excess is directly proportional to the mass M of the body:

$$\Delta R \cong \frac{4GM}{3c^2} \tag{2}$$

[Alternatively, we can define the radius excess as the amount by which the radius is too large when compared with the surface area of the body:

$$\Delta R = R - \sqrt{A/4\pi} \tag{3}$$

It turns out that this radius excess has the same value as in Eq. (2).] The Einstein equations also yield a formula for the excess angle of a triangle immersed in the body (such as shown in Figure E.18):

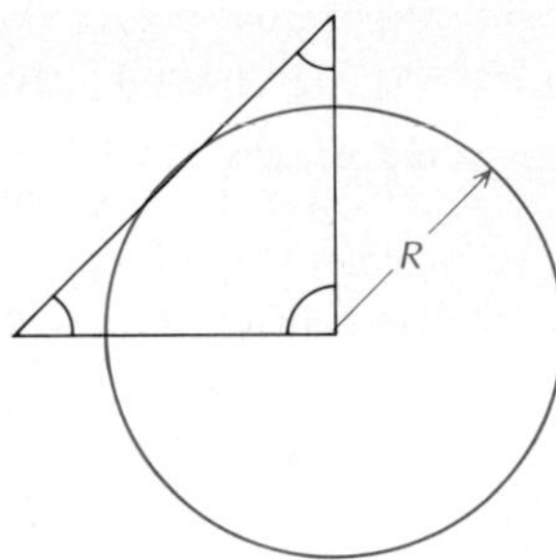

Fig. E.18 A triangle laid out with tightly stretched strings immersed in a spherical body.

$$\Delta\theta \cong \frac{\sqrt{2}GM}{Rc^2} \tag{4}$$

When we apply these formulas to the Earth, pretending that the density is constant, we find the following: the circumference of the Earth is smaller than $2\pi R$ by 5.9 mm, the area of the surface of the Earth is smaller than $4\pi R^2$ by 0.95 km^2, and the sum of interior angles of a triangle immersed in the Earth is larger than 180° by 2.0×10^{-4} seconds of arc. The corresponding numbers for the Sun cannot be obtained directly from our formulas because the density of the Sun is far from constant. A calculation taking into account the variation of density leads to the following: the circumference of the Sun is smaller than $2\pi R$ by 23 km, the area of the surface of the Sun is smaller than $4\pi R^2$ by 5.4×10^7 km^2, and the sum of interior angles of a triangle differs from 180° by 0.62 seconds of arc.

For the Sun and the Earth, these deviations from the expected Euclidean values are so small that it is hopeless to attempt to measure them directly. The curvature of space only shows up indirectly in its effects on the propagation of light mentioned above. It also shows up in an effect on planetary motion: the **perihelion precession.** In flat space Newton's theory tells us that planets move in ellipses around the Sun. But when one attempts to fit such an elliptical motion into the curved, bowl-like space surrounding the Sun, the orbit precesses, i.e., it changes its orientation a little bit on each revolution. This precession effect can be understood in terms of the curvature of space. The portion of bowl along which a planet moves can be approximated by a conical surface, tangent to the bowl at about the radius of the planetary motion. Figure E.19 shows an ellipse drawn on a flat plane; to make this flat plane into a cone that approximates the bowl-like curved space, we must cut a wedge out of the plane and join the edges together. But this introduces a kink in the orbit (Figure E.19). When the planet reaches this kink, it will overshoot the old elliptical orbit and go into a new elliptical orbit whose major axis has a slightly different orientation. Since this happens whenever the planet comes near perihelion, the net effect is a gradual change of orientation of the elliptical orbit (Figure E.20); this is the perihelion precession. For Mercury the precession calculated from General Relativity is 43 seconds of arc per century, in excellent agreement with the observed value.

However, this is not the whole story. The motion of planets, and the motion of light, is not only affected by the curvature of space, but also by the curvature of time. In fact, what we have is not separate space and time but, rather, a single four-dimensional **curved spacetime.** What is meant by curvature of spacetime is even harder to visualize than curvature of space. In the case of space, we can always have recourse to the analogy of a two-dimensional curved space; but in the case of spacetime, there is no such analogy. To get at least some feeling for curved spacetime, let us go back to Figure E.13 showing a "square." The side near the Sun is longer than the side far from the Sun, i.e., cur-

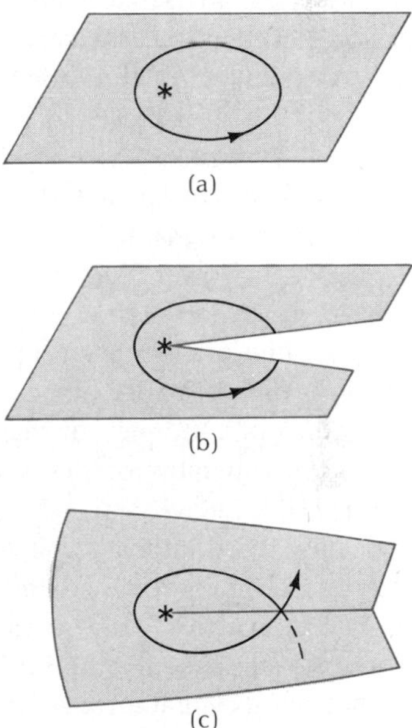

Fig. E.19 (a) Elliptical orbit of a planet on a flat plane. If we deform the flat plane into a cone (b), the orbit acquires a kink (c).

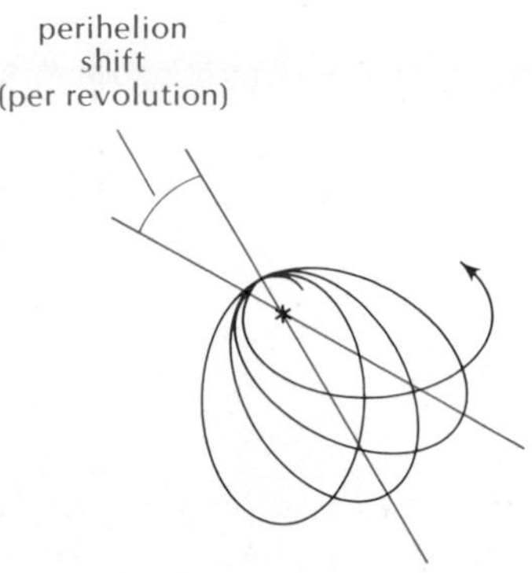

Fig. E.20 Perihelion precession of the orbit of a planet. The amount of precession has been wildly exaggerated.

vature of space means that distances near the Sun are longer. Likewise, curvature of spacetime means that time near the Sun is longer. Expressed another way, clocks near the Sun run slower than clocks far from the Sun. And in general, a clock in a strong gravitational field near a large mass runs slower than a clock far from a mass. This effect is called **gravitational time dilation** or **gravitational red shift.** As we will see in the next section, this effect is a direct consequence of the Principle of Equivalence.

E.4 THE GRAVITATIONAL TIME DILATION

We know from the theory of Special Relativity that when a clock is in motion relative to another clock, it suffers time dilation. We will now see that the theory of General Relativity predicts an extra kind of time dilation undergone by clocks at rest in a gravitational field.

Consider a stationary laboratory near a gravitating body, say, a laboratory on the surface of the Earth. Suppose that a pair of identical clocks are installed on the floor and on the ceiling of this laboratory (Figure E.21); the lower clock is then closer to the Earth. We want to compare the frequencies of ticking of these clocks. Since the clocks are separated by some distance, we need to send a signal from one clock to the other. Let us couple the lower clock to a radio transmitter that emits a wave of the same frequency as that of the clock; the radio wave travels to the second clock and we can compare the frequency of the arriving radio wave with the frequency of that clock. This permits us to check the rates of the clocks against each other.

To figure out how these intervals will compare, we appeal to the Principle of Equivalence and replace the stationary laboratory near the Earth by an accelerated laboratory in interstellar space (Figure E.22). If the latter laboratory has an acceleration $g = 9.8\ \text{m/s}^2$, then the behavior of clocks within this accelerated laboratory is the same as within a stationary laboratory on Earth. It is convenient to analyze the motion of the radio waves from the point of view of an inertial reference frame in which the laboratory is initially at rest when the clock on the floor sends out its first radio waves. If the distance between the clocks is h, then the waves take a time h/c to reach the clock on the ceiling. But in this time interval the acceleration g gives the latter clock a velocity gh/c relative to our inertial reference frame. Thus, when the radio waves arrive, the receiver of these waves has a recessional velocity gh/c and therefore the frequency detected by the receiver is subject to a Doppler shift. This Doppler shift reduces the frequency from a value ν at the emitter to a smaller value ν' at the receiver. Thus, as compared to the clock on the ceiling, the clock on the floor runs slow.

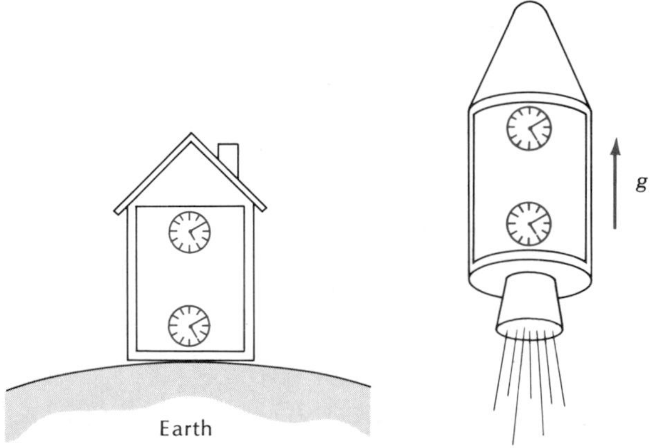

Fig. E.21 (left) Stationary laboratory with clocks on the Earth.

Fig. E.22 (right) Accelerated laboratory in interstellar space.

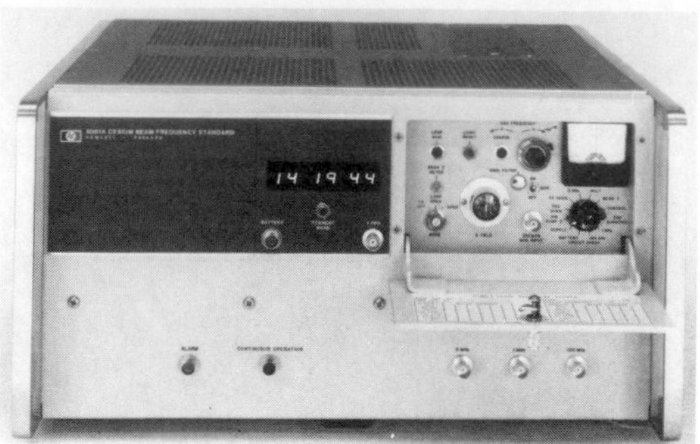

Fig. E.23 Hewlett-Packard portable atomic clock.

The time dilation produced by the gravity of the Earth is very small: a clock sitting on the surface of the Earth runs slow by only 1 part in 10^9 compared to a clock far above the surface. However, even these small effects are within reach of extremely precise atomic clocks. In 1972 scientists demonstrated the gravitational time dilation by carrying portable atomic clocks to an altitude of about 30,000 ft and comparing them with identical clocks kept on the ground. The clocks were carried to this altitude by commercial jets — the experimenters simply bought tickets for a round-the-world trip on Pan Am, TWA, and American Airlines. They took one trip around the Earth in an eastward direction, and one westward, each trip taking about three days, of which about two days were spent airborne. Four portable Hewlett-Packard cesium beam clocks were carried on these trips (Figure E.23); continual comparisons of these traveling clocks served as protection against occasional sudden, small jumps in rate to which the cesium clocks are prone. When the traveling clocks were brought back down and compared with the clock of the U.S. Naval Observatory in Washington, D.C., it was found that they had gained about 1.5×10^{-7} s.[2]

[2] To obtain this result, the experimenters had to take into account the special-relativistic time dilation produced by the motion of the clocks (see Chapter 17).

Although this experiment with traveling clocks is the most direct test of gravitational time dilation, it is not the first test, or the latest, or the most accurate. The time dilation has also been tested by red-shift measurements. Atoms sitting on the surface of the Sun or on the surface of some other star can be regarded as clocks. These atoms tick slower than atoms far from the Sun's surface and consequently the frequency of the light they emit will be low compared to the frequency of faraway atoms. Hence light emitted by, say, potassium atoms on the Sun will exhibit a red shift when it is received on Earth and is compared to light emitted by potassium atoms on Earth. According to the theory of General Relativity, the frequency should be low by 2 parts in 10^6. The measured values agree with this theoretical value.

Because of the large uncertainties inherent in these measurements, the red-shift observations on sunlight and starlight are not a very precise test of General Relativity. A somewhat better red-shift observation was made on the Earth using a nucleus, rather than an atom, as a clock. A vibrating nucleus emits gamma rays (just as a vibrating atom emits light). The gamma ray emitted by, say, an iron nucleus at the surface of the Earth will exhibit a red-shift when it ascends to some height above the surface. Although the red shift is only 2 parts in 10^{15} for a height increase of 20 m, it can be detected by allowing the gamma ray to hit another iron nucleus, which will try to absorb it. The absorption depends very critically on frequency — even a tiny reduction of frequency drastically diminishes the chances of absorption. Physicists at Harvard University performed an experiment of this kind in a 20-m-high tower (Figure E.24). They placed some radioactive iron (which emits gamma rays) at the bottom of the tower and some extra iron (to serve as detector) at the top. With this they managed to test the red shift to 1%.

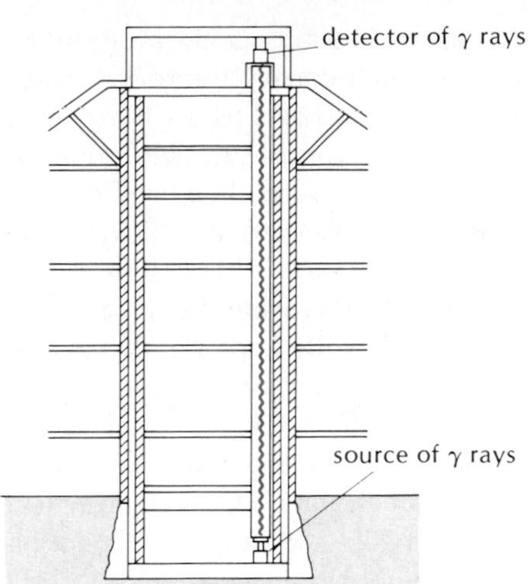

Fig. E.24 Emitter (bottom) and absorber (top) of gamma rays in the Jefferson Physical Laboratory at Harvard University.

Finally, we must mention the most precise red-shift experiment done to date. To get a large red shift, it is desirable to send a clock to high altitude above the Earth. So, in 1976 physicists from the Smithsonian Astrophysical Observatory sent a Scout rocket with an atomic clock to a height of 10,000 km above the Atlantic Ocean (Figure E.25). At this height the clock is

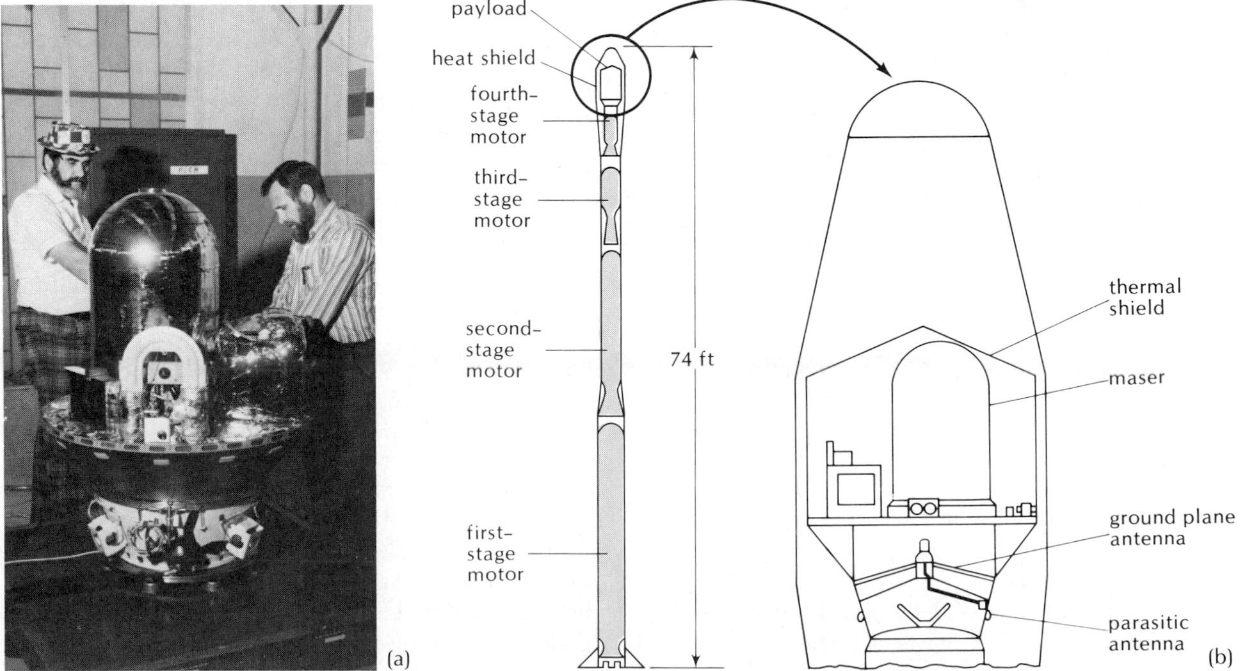

Fig. E.25 (a) Maser clock (within dome) before installation in the rocket. (b) Position of the maser clock in the bulbous nose cone of the Scout rocket. (Courtesy R. F. C. Vessot and M. L. Levine, Smithsonian Astrophysical Observatory.)

expected to run faster than a clock on the ground by 4.5 parts in 10^{10}. Of course, the rocket clock shatters when the rocket crashes and so the comparison of clock rates has to be made in flight, during the two hours or so that the payload spends in free fall. The rocket clock was linked to a radio transmitter that transmitted the "ticks" of the clock down to the ground. There the ticks were compared with ticks of another, identical clock kept on the ground. Apart from the tricky technical problems of protecting the rocket clock from the traumatic shock of the takeoff acceleration, the extreme temperature variations encountered in spaceflight, and the changing strength of the Earth's magnetic field, this experiment had to resolve another problem: since the rocket is in continuous motion with a fairly large velocity relative to the ground, the radio signals received on the ground have a Doppler shift. So, it is necessary to sort out this Doppler shift from the gravitational red shift (or, rather, blue shift, since the rocket clock will be fast). The Doppler shift was detemined by much the same method as used by the police in their Doppler radar systems. When a beam of radar waves is sent to an oncoming automobile, the waves are received by the surface of the automobile with some Doppler shift since the automobile is moving toward the waves. When the waves bounce off the automobile and travel back to the policeman's radar receptor, they acquire an extra Doppler shift because the automobile acts as a moving source while reflecting the waves. In the rocket experiment, a radio signal from the ground was aimed at the rocket and retransmitted by the rocket back to the ground by means of a retransmitter, or transponder. This radio signal suffers a Doppler shift both when going out to the rocket and when returning to the ground, i.e., it returns with a net Doppler shift of twice the one-way Doppler shift. The signal also suffers a gravitational frequency shift — but this is a red shift on the outgoing part of the trip and a blue shift on the return part; hence the gravitational time dilation does not affect the net frequency of a round-trip signal. The round-trip signal can therefore be used to make the necessary Doppler correction to the clock signal. The result of this elaborate experiment confirmed the theoretical value of the gravitational time dilation to within 0.01%.

E.5 BLACK HOLES

Within the solar system the effects of Einstein's theory of General Relativity are minuscule — we can only detect the deviations between Newton's theory and Einstein's theory by means of delicate instruments of very high precision. To find drastic deviations, we must seek out places in the sky where there are gravitational fields much stronger than anywhere in the Solar System. The most spectacular relativistic effects are displayed by **black holes.** These are the disembodied remains of stars that have collapsed under their own weight, crushing their material to a singular state of infinite density. After the collapse nothing is left of the star except an extremely intense gravitational field. There is good circumstantial evidence that several such collapsed stars exist in the sky in orbits around ordinary stars belonging to our Galaxy.

A black hole is black because neither light nor anything else can escape the overwhelming grip of its gravitational field. Particles can enter the black hole, but no particle can ever emerge. Electrons, protons, atoms, spaceships, astronauts, or whatever that enter a black hole must abandon all hope of ever escaping again. Once they enter the interior of the black hole, they are trapped; they are inexorably pulled deeper and deeper toward the center. Gravity in a black hole is so strong that even light, which moves at higher speed than anything else, is pulled down.

In terms of escape velocity, we can get some rough understanding of how gravity traps particles in a black hole. Recall that at the surface of the Earth, the escape velocity is 11 km/s (see Section 13.5) — a projectile launched with this velocity, or with more than this velocity, shrugs off the gravitational pull of the Earth and keeps on rising forever; a projectile launched with less than this velocity is gradually brought to a stop by the gravitational pull and then falls back to the Earth. For a body that is more compact, or more massive, than the Earth, the associated escape velocity is more than 11 km/s. For a starlike spherical body of mass M and radius R, the escape velocity will be [see Eq. (13.29)]

$$v = \sqrt{2GM/R} \tag{5}$$

If the mass of the body is so large (or the radius so small) that this velocity exceeds the speed of light, then light will not be able to escape — the body will appear black. This was already deduced by Laplace some 200 years ago:

> A luminous star, of the same density as the Earth, and whose diameter should be two hundred and fifty times larger than that of the Sun, would not, in consequence of its attraction, allow any of its rays to arrive at us; it is therefore possible that the largest luminous bodies in the universe may, through this cause, be invisible.

What Laplace did not know is something that we learned from the theory of relativity: if light cannot escape, nothing else can either, because nothing can move faster than light. Thus Laplace discovered that the strong gravitational pull might make some "stars" black, but he did not discover that such "stars" had to be holes.

If we take $v = c$ in Eq. (5), we find that the radius within which the mass M must be contained for escape to be impossible is

$$R = 2\frac{GM}{c^2} \tag{6}$$

This critical radius is called the **Schwarzschild radius.** For a body of mass equal to that of the Sun, the value of the critical radius is

$$R = \frac{2 \times 6.67 \times 10^{-11}\ \mathrm{N/m^2 \cdot kg^2} \times 2.0 \times 10^{30}\ \mathrm{kg}}{(3.0 \times 10^8\ \mathrm{m/s})^2}$$

$$= 3.0\ \mathrm{km}$$

If the radius of the body is smaller than that, any particle approaching closer than 3.0 km will be trapped.

The boundary of the region of no escape that surrounds the black hole on all sides like a ghostly iron curtain, permitting entrance but preventing exit, is called the **one-way membrane** or the **horizon.** This membrane demarcating the edge of the black hole is not made of anything; it is not a substance, but merely a place, the place of no return. According to Eq. (6) a black hole of mass equal to that of the Sun has a one-way membrane of radius 3.0 km (Figure E.26); a black hole of a mass equal to that of the Earth would have a one-way membrane of a radius of just 9 mm (Figure E.27).

Although to talk of escape velocity is helpful, it is also somewhat misleading. According to Newton's theory, an object launched in the upward direction from a point inside the black hole first moves up for some distance, then stops, and then falls back. What actually happens in a black hole is quite different: the object never even begins to move upward; in a black hole all motion is toward the center. Only Einstein's theory of General Relativity can give an adequate picture of how objects move in the interior of a black hole. It turns out that, by a happy coincidence, a careful calculation based on the latter theory yields exactly the same formula (6) for the no-escape radius as our naïve calculation based on Newton's theory.

Fig. E.26 A black hole of radius 3 km on top of New York City.

Fig. E.27 A black hole of mass equal to that of the Earth would have the size of this dot.

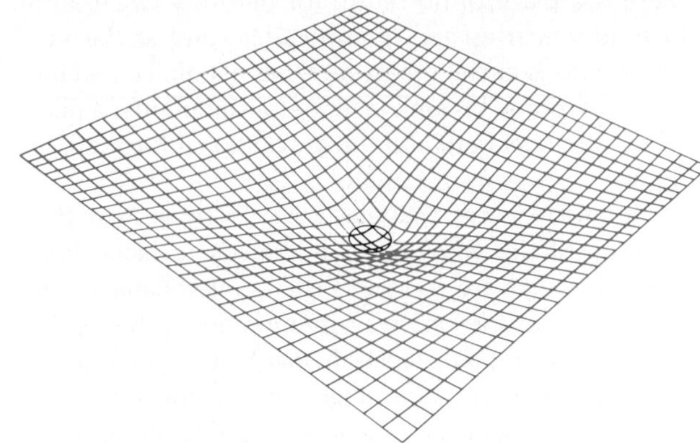

Fig. E.28 Two-dimensional surface with a "sun" at the center.

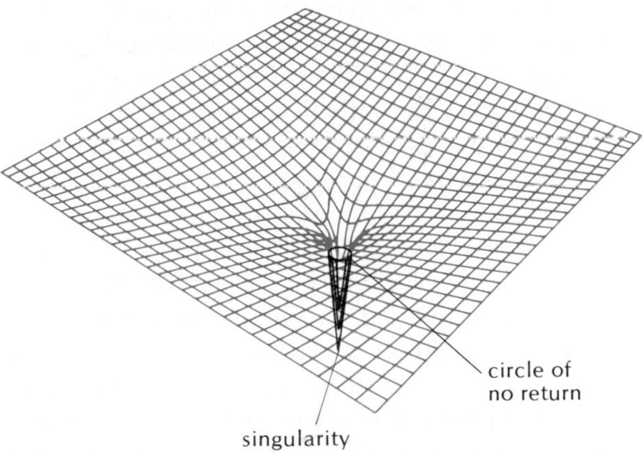

Fig. E.29 Two-dimensional surface with a black hole at the center. The circle marks the place of no return. At the kink at the bottom, the curvature of the surface is infinite.

According to General Relativity, the space and time near any gravitating body are curved; near and inside a black hole, the curvature is extreme — space and time are greatly distorted. In terms of the two-dimensional analog described in Section E.3, a gravitating body or "sun" produces a dip in the surrounding space (Figure E.28). If we replace the "sun" by a black hole, the dip becomes much deeper (Figure E.29). Within the dip there is a place of no return (or one-way circle) from beyond which escape is impossible; and at the very center of this circle, the surface has a kink, i.e., an infinite distortion. This kink indicates a **singularity,** a place where the curvature and the gravitational force are infinite.

Although Figures E.28 and E.29 are suggestive, they do not tell the whole tale. For inside a black hole, space and time are — in a sense — interchanged. The distance from the one-way membrane to the central singularity is not a distance in space but, rather, an interval of time, i.e., the singularity is not a point in space, but a point in time. Correspondingly, inside a black hole the gravitational fields are not static; rather, they are time dependent and they evolve, becoming stronger and stronger until finally they become infinite everywhere in the interior of the black hole. This violent evolution is very quick; in a black hole of one solar mass, it only takes 10^{-5} s. Paradoxically, seen from the outside, the black hole does not appear to evolve at all; it just sits there for ever and ever. The resolution of this paradox hinges on one of the most spectacular features of a black hole: its gravitational time dilation.

As we saw in Section E.4, a clock placed near any gravitating body will run slow. This gravitational time dilation is one aspect of the distortion of space and time in the vicinity of a gravitating body. Near the Earth the distortion is small, and the time dilation is small. But near a black hole the distortion and the time dilation are very large. At the one-way membrane, the time dilation becomes infinite — a clock placed next to the membrane will be slowed down so much that, compared to a clock far away from the black hole, it will appear to have stopped. If we imagine an astronaut in a space capsule parked right at the edge of the black hole, all his life functions (pulse rate, alpha rhythm, voluntary and involuntary muscular contractions) will almost be at a standstill — to the scientists at "mission control" on a mother spaceship stationed far away, outside of the strong gravitational field, he will appear to have frozen. Yet the astronaut would not perceive himself as having slowed down; he would merely perceive the surrounding world as having speeded up.

This time dilation resolves the paradox of how the inside of the black hole can change while the outside does not. As judged from outside, any changes at or inside the one-way membrane take an infinite amount of time. Hence, from the outside point of view, these changes never happen and the black hole seems completely static.

There are several objects in the sky that astrophysicists suspect contain black holes. The best known of these is Cygnus X-1, a strong source of X rays in the constellation Cygnus. Figure E.30 shows a picture of this object taken with an X-ray telescope installed on the *Einstein* satellite launched in 1978. Figure E.31 is a picture of the same object taken with an optical telescope. It seems that Cygnus X-1 is a binary system consisting of two components: one is an ordinary star of about 30 solar masses; this star is the source of the ordinary light visible in Figure E.31. The other is a very compact body of about 10 solar masses; this is the source of the X rays shown in Figure E.30. These two components orbit about each other with a period of 5.6 days; the distance between the two components is quite small.

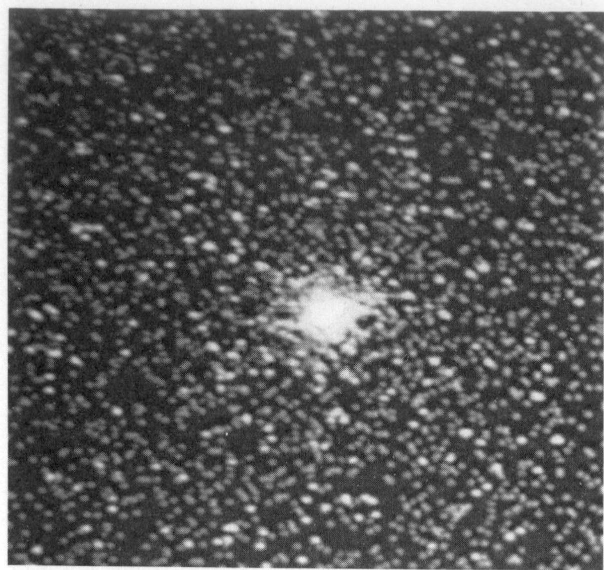

Fig. E.30 X-ray picture of Cygnus X-1.

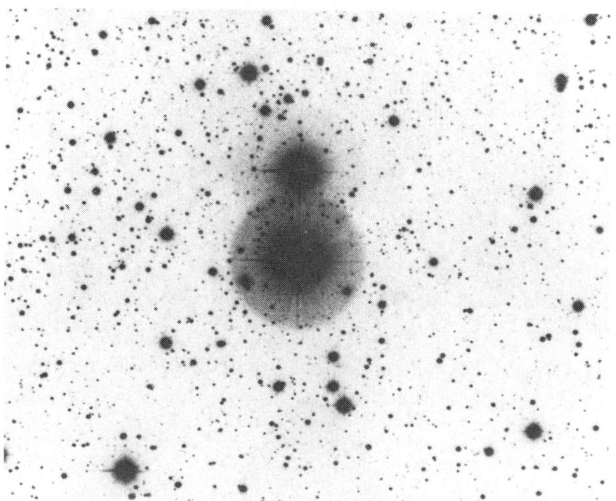

Fig. E.31 Optical photograph of Cygnus X-1. The star of interest is in the center of this photograph. (Hale Observatories photograph courtesy J. Kristian.)

The compact body is probably a black hole. It produces the X rays by an **accretion** mechanism. The black hole is so near the surface of the star that it raises a large tidal bulge and pulls a steady stream of material off that surface, material that spirals around the black hole, forming a disk (called the accretion disk; Figure E.32). As the material spirals and gradually falls toward the black hole, it is accelerated by the strong gravitational fields and heated by compression. The material reaches temperatures of up to 100 million degrees Celsius. At such temperatures the violent thermal collisions between the particles of the material release a copious flux of X rays. The key to this mechanism for the production of X rays is the small size or, equivalently, the high density of the black hole. When a mass of 10 suns is contained in a small region, the gravitational fields in the immediate neighborhood will be very intense; the enormous heating of the material becomes possible only by the action of these intense gravitational fields.

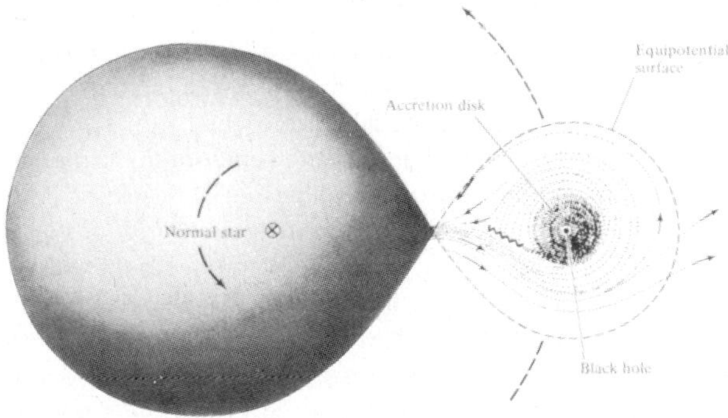

Fig. E.32 The normal star and the black hole orbit about their common center of mass. The black hole raises a large tidal bulge on the star. Material streaming from the star toward the black hole forms an accretion disk.

From the characteristics of the observed X-ray flux, astrophysicists have deduced that the size of the compact object is less than 300 km across. To the best of our knowledge, there is no way in which a mass of 10 suns can be confined in a region 300 km across and remain in equilibrium — at such densities gravity would overwhelm all other forces and would immediately bring about the total collapse of the mass. The compact object in Cygnus X-1 probably collapsed a long time ago and is now a black hole.

There are several other objects similar to Cygnus X-1 in the sky. All these objects are powerful sources of X rays, and astrophysicists believe that they consist of a more or less normal star and a black hole orbiting about each other. The black hole steadily pulls materials off the star, swallows it, and emits the X rays. Note that the X rays do not come from the interior of the black hole; they come from the accretion disk just outside of the black hole. After the in-falling material enters the one-way membrane, any X rays that it emits will remain trapped. The black hole is black, but the material outside it glows with X-ray light.

Some astronomers speculate that there may be a very large number of black holes in our Galaxy and in other galaxies. It may be that most of the mass of some galaxies is in the form of black holes. But the only black holes we can hope to detect with our Earth-bound instruments are those that are swallowing gas or dust from their environment and are producing X rays or other radiation. An isolated black hole is very effectively camouflaged against the black background of outer space. Isolated black holes are the hidden reefs of outer space. They are invisible. They can be detected only by their gravitational pull.

Further Reading

Albert Einstein: Creator and Rebel by B. Hoffmann and H. Dukas (Viking, New York, 1972) is a charming biography that traces how Einstein conceived his theory of General Relativity.

The Riddle of Gravitation by P. G. Bergmann, one of Einstein's co-workers (Scribner, New York, 1968); *Space, Time, and Gravity* by R. M. Wald (University of Chicago Press, Chicago, 1977); and *Cosmology* by E. R. Harrison (Cambridge University Press, Cambridge, 1981) are clear and simple introductions to the theory of General Relativity, curved space, black holes, and the expanding universe.

Space, Time and Gravitation by A. Eddington (Cambridge University Press, Cambridge, 1968) is an old classic that explains the basic concepts of the theory.

Chapters 3–5 of *Principles of Gravitation and Cosmology* by M. Berry (Cambridge University Press, Cambridge, 1976) present some of the mathematics of curved space and General Relativity, concentrating on those aspects that can be handled by simple calculus.

General Relativity by R. Geroch (University of Chicago Press, Chicago, 1978) and *Talking About Relativity* by J. L. Synge (North-Holland, Amsterdam, 1970) are nice introductions to the geometry of four-dimensional spacetime; the former is filled with abundant examples of spacetime diagrams.

Flatland by E. A. Abott (Dover, New York, 1952) is an enchanting tale of the inhabitants of an imagined two-dimensional world and of their difficulties in visualizing the third dimension. *Geometry, Relativity and the Fourth Dimension* by R. v. b. Rucker (Dover, New York, 1977) is a worthy sequel to *Flatland* that provides a gentle introduction to curved space and to four-dimensional spacetime.

The bizarre properties of black holes and the observational evidence for their existence are described in *Black Holes, Quasars, and the Universe* by H. L. Shipman (Houghton Mifflin, Boston, 1980), *Black Holes* by W. Sullivan (Doubleday, Garden City, N.J., 1979), *Black Holes in Space* by P. Moore and I. Nicolson (Norton, New York, 1974), and *The Cosmic Frontiers of General Relativity* by W. J. Kaufmann (Little, Brown, New York, 1977); the last of these books includes a chapter on their weird spacetime geometry and the singularities of rotating black holes. More on the singularities in black holes and in the Big Bang will be found in two books by P. C. W. Davies, *Space and Time in the Modern Universe* (Cambridge University Press, Cambridge, 1977) and *The Edge of the Universe* (Cambridge University Press, Cambridge, 1981).

The Search for Gravity Waves by P. C. W. Davies (Cambridge University Press, Cambridge, 1980) is a concise summary of the properties of gravitational waves and of the recent experiments and observations.

The following articles deal with black holes and with observational evidence for them:

"Introducing the Black Hole," R. Ruffini and J. A. Wheeler, *Physics Today,* January 1971

"Black Holes," R. Penrose, *Scientific American,* May 1972

"The Search for Black Holes," K. S. Thorne, *Scientific American,* December 1974

"X-Ray Emitting Double Stars," H. Gursky and E. P. J. van den Heuvel, *Scientific American,* March 1975

"The Black Hole in Astrophysics: The Origin of the Concept and Its Role," S. Chandrasekhar, *Contemporary Physics,* Vol. 14, 1974

The *Resource Letters GR-1 and GI-1* in the *American Journal of Physics,* February 1968 and June 1982, lists many books and articles on General Relativity at an introductory level and at more advanced levels.

Questions

1. If you were to discover a body that did not accelerate in a gravitational field at the same rate as all other bodies, how would this affect Einstein's theory of General Relativity?

2. The Earth is in free fall in the gravitational field of the Sun, the Moon, and the planets. Does this eliminate all external gravitational effects in the reference frame of the Earth?

3. According to both Newton's and Einstein's theories of gravitation, a ray of light will be deflected by the Sun. However, each of these theories assigns a different cause to this deflection. Explain.

4. Why do astronomers find it more convenient to measure the deflection of rays by the Sun with radio waves rather than with light waves?

5. Suppose that you measure the speed of light in a laboratory on the Earth and suppose that you repeat this measurement in a laboratory in interstellar space, far from any gravitational influences. Will you get the same result in both cases?

6. One of Euclid's postulates states that given a straight line and a separate point, there exists one and only one straight line that passes through the point and does not intersect the given straight line (parallel postulate). What happens to this postulate in the two-dimensional curved spaces shown in Figures E.16 and E.17?

7. Is the space in this room flat or curved? If you had available measurement instruments of extreme precision, how could you check your answer?

8. Suppose that you stretch a long massless string so that it passes by the Sun. Would you expect this string to coincide with the path of a light ray?

9. Why did Einstein name his theory of gravitation *General Relativity?*

10. There are two possible circumstances under which an astronaut exploring interstellar space will suffer time dilation relative to his comrades on Earth. What are they?

11. Because of the rotational motion of the Earth, a point on the equator has a speed of 460 m/s relative to a point on the pole. Because of the equatorial bulge of the Earth, a point on the equator is 21 km farther from the center of the Earth than a point on the pole. According to the special relativistic time-dilation effect, will a clock on the equator tick faster or slower than an identical clock on the pole? According to the gravitational time-dilation effect? (It turns out that these two time-dilation effects cancel each other if both clocks are at sea level.)

12. Suppose that you dangle a long rope into a black hole from a safe distance. If you attempt to pull the rope out, what will happen?

13. A suicidal scientist jumps off a spaceship into a black hole. What effects does he perceive as he crosses the one-way membrane? What effects do his colleagues watching from the spaceship perceive?

14. According to recent astronomical observations, the X-ray source LMC X-1 is a binary star system one of whose members has a mass about nine times the mass of the Sun and a radius of only a few hundred kilometers. What can you conclude about this star?

CHAPTER 18

Fluid Mechanics

A fluid is a system of particles loosely held together by their own cohesive forces or by the restraining forces exerted by the walls of a container. In contrast to the particles in a rigid body, which are permanently locked into fixed positions, the particles in a fluid body are more or less free to wander about within the volume of the body. A fluid will flow, i.e., it will change its shape in response to external forces. Both liquids and gases are fluids — a body of water or a body of air will change its shape in response to the forces exerted by the container. The difference between these two kinds of fluids is that liquids are incompressible, or nearly so, whereas gases are compressible. A body of water has a constant volume independent of the shape and size of the container, whereas a body of air has a variable volume — the air always spreads so as to entirely fill the container, and it can be made to expand or contract by increasing or decreasing the size of the container.

Fig. 18.1 Folded layers of rock at Scapegoat Mountain, Montana.

Solids can sometimes behave as fluids. For example, in the manufacture or "drawing" of wire, a metal is forced to flow plastically through a small nozzle; such plastic flow of a metal requires a very strong pull on the wire as it emerges through the nozzle. The photograph in Figure 18.1 shows another example of plastic flow: here layers of rock gradually changed their shape under the influence of forces acting over a few million years.

What distinguishes the flow of a solid from that of a liquid or gas is the large pressure and large stress required to initiate the flow. The solid will not flow at all unless the applied forces are in excess of some threshold value, whereas the liquid or gas will flow even if the forces are very small. This is true even of liquids of high viscosity, such as molasses or pitch; even if the force on them is small, they will flow — albeit slowly.

In this chapter we will concentrate on the motion of **perfect fluids,** i.e., those that do not offer any resistance to flow except through their inertia.

18.1 Density and Flow Velocity

Although a fluid is a system of particles, the number of particles in, say, a cubic centimeter of water is so large that it is not feasible to describe the state of the fluid in terms of the masses, positions, velocities, and forces of all the individual particles. Instead, we will describe the state of the fluid in terms of its **density, velocity of flow,** and **pressure.** These quantities give us a macroscopic description of the fluid — they tell us the *average* behavior of the particles in small regions within the fluid. For example, if the velocity of flow of water in a firehose is 4 m/s, this does not mean that all the water molecules have this velocity. The water molecules have a high-speed thermal motion of about 900 m/s; they move in short zigzags because they frequently collide with one another. This thermal motion of a water molecule is random; the motion is as likely to be in a direction opposite to the flow as along the flow (Figure 18.2). The flow of the water molecules along the fire hose at 4 m/s represents a slow drift superimposed on the much faster random zigzag motion. However, on a macroscopic scale we notice only the drift and not the random small-scale motion — we notice only the *average* motion of the water molecules.

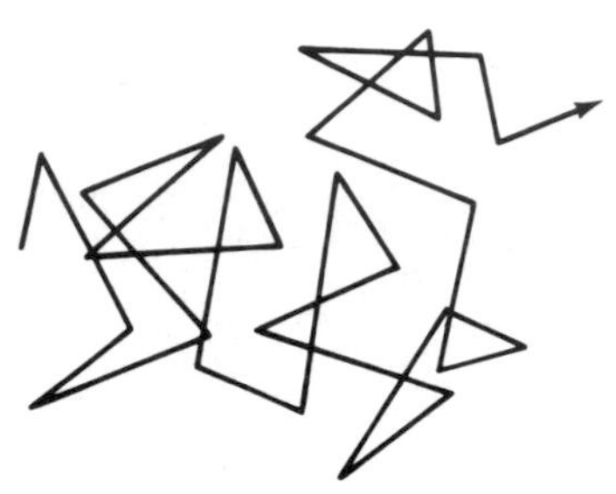

Fig. 18.2 Motion of a water molecule in water. The straight segments are typically about 10^{-10} m long.

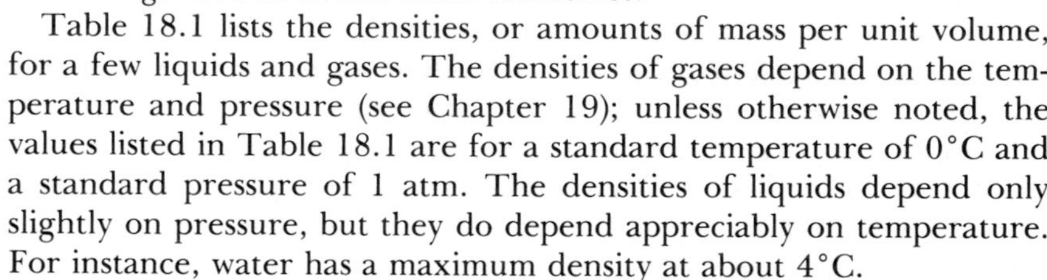

Table 18.1 lists the densities, or amounts of mass per unit volume, for a few liquids and gases. The densities of gases depend on the temperature and pressure (see Chapter 19); unless otherwise noted, the values listed in Table 18.1 are for a standard temperature of 0°C and a standard pressure of 1 atm. The densities of liquids depend only slightly on pressure, but they do depend appreciably on temperature. For instance, water has a maximum density at about 4°C.

In general, both the density ρ and the flow velocity $\mathbf{v}$ in a fluid will be functions of position and of time,

$$\rho = \rho(x,y,z,t) \tag{1}$$

$$\mathbf{v} = \mathbf{v}(x,y,z,t) \tag{2}$$

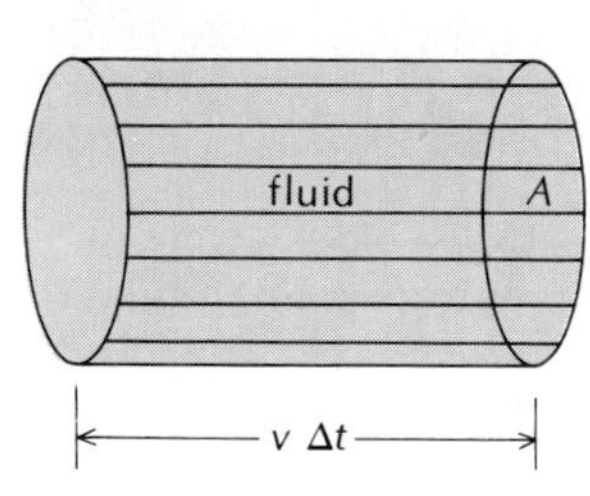

Fig. 18.3 Flow of fluid across an area A.

The magnitude of $\mathbf{v}$ equals the volume of fluid that flows across a unit area perpendicular to $\mathbf{v}$ in a unit time. Figure 18.3 helps to make this clear; it shows a (fixed) area A and a volume of fluid about to cross this area. The fluid that crosses the area in a time Δt is initially in a cylinder of base A and length $v\ \Delta t$ (Figure 18.3). The amount of fluid volume that crosses the area is therefore

$$\Delta V = Av\ \Delta t \tag{3}$$

and the amount that crosses per unit area and per unit time is

$$\frac{1}{A}\frac{\Delta V}{\Delta t} = v \tag{4}$$

Table 18.1 DENSITIES OF SOME FLUIDS[a]

Fluid	ρ
Water	
0°C	999.8 kg/m³
4°C	1,000.0
20°C	998.2
100°C	958.4
Sea water, 15°C	1,025
Mercury	13,600
Sodium, liquid at 98°C	929
Texas crude oil, 15°C	875
Gasoline, 15°C	739
Olive oil, 15°C	920
Human blood, 25°C	1,060
Air	
0°C	1.29
20°C	1.20
Water vapor, 100°C	0.598
Hydrogen	0.0899
Helium	0.178
Nitrogen	1.25
Oxygen	1.43
Carbon dioxide	1.98
Propane	2.01

[a] At 0°C and 1 atm, unless otherwise noted.

EXAMPLE 1. The water in a fire hose of diameter 6.4 cm ($2\frac{1}{2}$ in.) has a flow velocity of 4.0 m/s. At what rate does this hose deliver water? Give the answer both in m³/s and in kg/s.

SOLUTION: The cross-sectional area of the hose is $A = \pi(3.2 \times 10^{-2}\ \text{m})^2 = 3.2 \times 10^{-3}\ \text{m}^2$. Hence the rate of delivery is

$$\frac{\Delta V}{\Delta t} = Av = 3.2 \times 10^{-3}\ \text{m}^2 \times 4.0\ \text{m/s} \tag{5}$$

$$= 1.3 \times 10^{-2}\ \text{m}^3/\text{s}$$

To find the rate of delivery in terms of the mass of water, we must multiply $\Delta V/\Delta t$ by ρ:

$$\frac{\Delta m}{\Delta t} = \rho\frac{\Delta V}{\Delta t} = 1.0 \times 10^3\ \text{kg/m}^3 \times 1.3 \times 10^{-2}\ \text{m}^3/\text{s} \tag{6}$$

$$= 13\ \text{kg/s}$$

18.2 Steady Incompressible Flow; Streamlines

Graphically we can represent the velocity of flow in a fluid by drawing the velocity vector at each point of space; for example, Figure 18.4 shows the velocity vectors for gas streaming out of a star immediately after a supernova explosion. The picture of all these velocity vectors is often called the **velocity field** of the fluid. If the velocity field is a function of time, then we must draw separate diagrams for different times.

Velocity field

Steady flow

Most of the examples in this chapter will deal with **steady flow,** for which the velocity at each point of space is independent of time. Obviously, steady flow requires a source that can continuously supply fluid and a sink that can continuously withdraw an equal amount of fluid. The supernova explosion does not yield steady flow — the supply of gas in such an explosive event is clearly time dependent. Figure 18.5 shows an example of steady flow of water around a cylindrical obstacle. The water enters in a broad stream from a source on the left and disappears in a similar broad stream toward a sink on the right.[1]

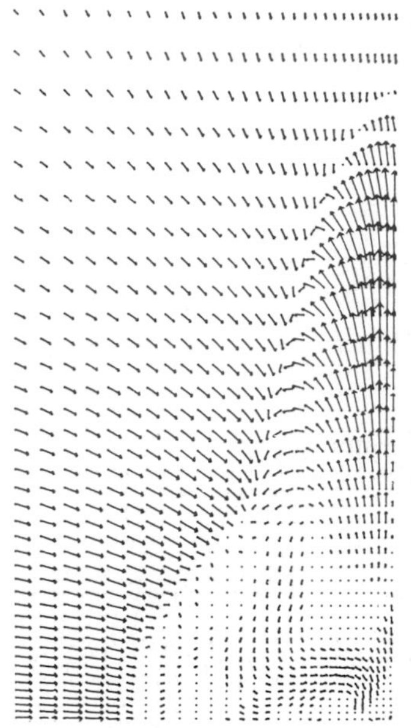

Fig. 18.4 The velocity vectors show the flow of a jet of gas expelled by a very massive, magnetized star, which is collapsing gravitationally after a supernova explosion has ripped off its outer layers. The largest velocity vectors have a magnitude of 4.4×10^7 m/s. (Courtesy J. M. LeBlanc and J. R. Wilson, Lawrence Livermore Laboratory.)

For the flow of an incompressible fluid, such as water, the picture of velocity vectors can be replaced by an alternative graphical representation. Suppose we focus our attention on a small volume element of water, say, 1 mm^3 of water, and we take note of the motion of this 1 mm^3 from the source to the sink. The small volume element will trace out a **streamline**. Neighboring volume elements will trace out neighboring streamlines. In Figure 18.6 we show the pattern of streamlines for the same steady flow of water that we already represented in Figure 18.5 by means of velocity vectors. The streamlines on the far left (and far right) of Figure 18.6 are evenly spaced to indicate the uniform and parallel flow in this region.

The information contained in the pattern of streamlines is equivalent to that in the velocity field. If we know the velocity field, we can trace out the motion of a small volume element and therefore construct the streamlines. But the converse is also true — if we know the streamlines, we can reconstruct the velocity field. We can do this by means of the following rule: the direction of the velocity at any one point is tangent to the streamline and the magnitude is proportional to

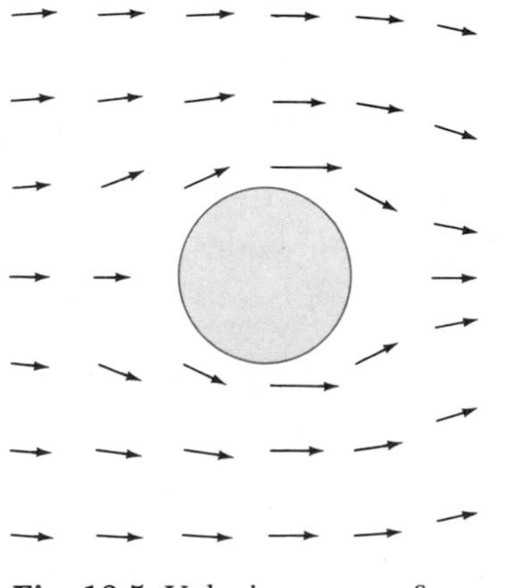

Fig. 18.5 Velocity vectors for water flowing around a cylinder.

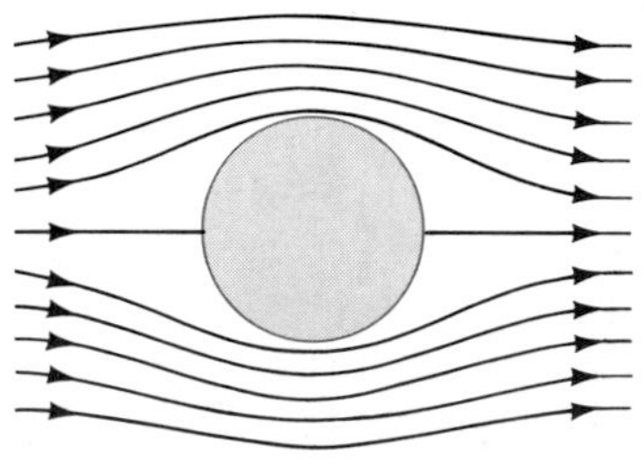

Fig. 18.6 Streamlines for water flowing around a cylinder.

[1] Incidentally, Figure 18.5 is based on the pretense that water is free of friction (free of viscosity). This is a reasonable approximation, except at the surface of the obstacle. There, the friction between the water and the surface slows the water to a standstill and produces a thin layer in which the flow of the water is very complicated (turbulent). Figure 18.5 does not show this layer.

the density of the streamlines. The first part of the rule is self-evident. To establish the second part, consider a bundle of streamlines forming a pipelike region called a **stream tube.** Any fluid inside the stream tube will have to move along the tube; it cannot cross the surface of the tube because streamlines never cross. The tube therefore plays the same role as a pipe made of some impermeable material — it serves as a conduit for the fluid. If we assume that the tube is very narrow, so that its cross-sectional area is infinitesimal, the velocity of flow will vary along the length of the tube but it will be nearly the same at all points on a given cross-sectional area. For instance, on the area A_1 (Figure 18.7) the velocity is v_1, and on the area A_2 the velocity is v_2. In a time Δt the fluid volume that enters across the area A_1 is $v_1 A_1 \, \Delta t$ and the fluid volume that leaves across the area A_2 is $v_2 A_2 \, \Delta t$. The amount of fluid that enters must match the amount that leaves, since, under steady conditions, fluid cannot accumulate in the segment of tube between A_1 and A_2; hence

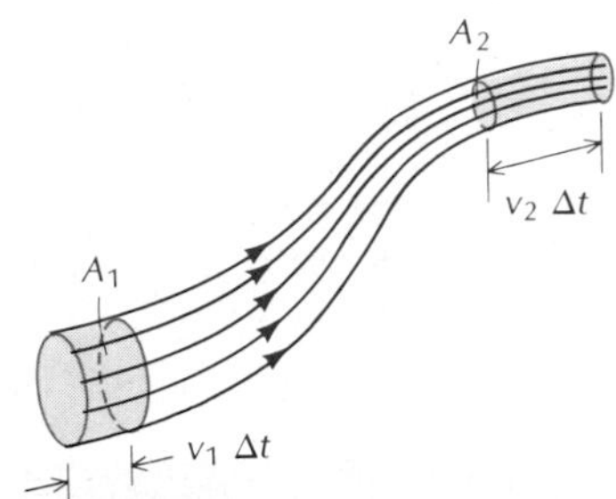

Fig. 18.7 A stream tube.

$$v_1 A_1 \, \Delta t = v_2 A_2 \, \Delta t \tag{7}$$

or

$$\boxed{v_1 A_1 = v_2 A_2} \tag{8}$$

Continuity equation

This relation is called the **continuity equation.** It shows that along any thin stream tube, the speed of flow is inversely proportional to the cross-sectional area of the stream tube. The density of streamlines inside the tube is the number of such lines divided by the cross-sectional area; since the number of stream lines entering A_1 is necessarily the same as that leaving A_2, the density of streamlines is inversely proportional to the cross-sectional area. This implies that the speed at any point in a fluid is directly proportional to the density of streamlines at that point. For example, in Figure 18.6 the speed of the water is large at the top and bottom of the obstacle (large density of streamlines) and smaller to the left and right (smaller density of streamlines).

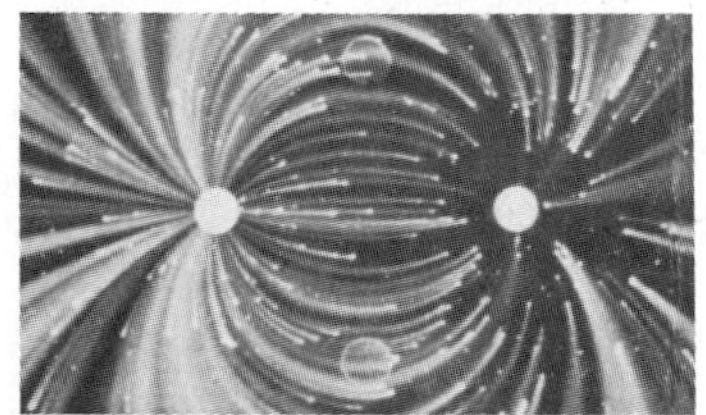

Fig. 18.8 Streamers of dye indicate the streamlines in water flowing from a source (right) to a sink (left).

In experiments on fluid flow, the streamlines of a fluid can be made directly visible by several clever techniques. If the fluid is water, one can place dye particles at diverse points within the volume of water; the dye will then be carried along by the flow and it will mark the streamlines. The photograph of Figure 18.8 shows a pattern of streamlines made visible with this technique. The water emerges from a pointlike source on the left and disappears into a pointlike sink on the right. Small grains of potassium permanganate create the streamers of dye while dissolving in the water.

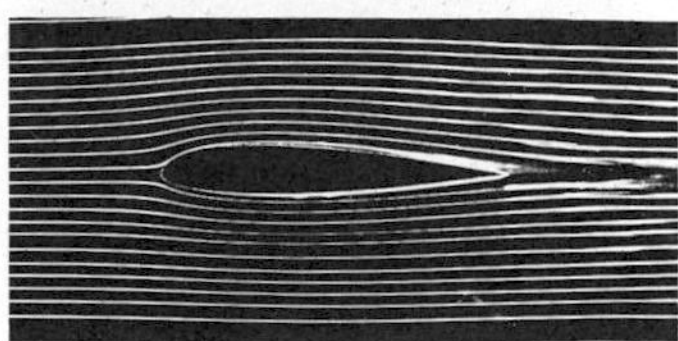

Fig. 18.9 Fine trails of smoke indicate the steamlines in air flowing around the wing of an airplane.

If the fluid is air, one can make the streamlines visible by releasing smoke from small jets at diverse points within the flow of air. The photograph of Figure 18.9 shows trails of smoke marking the streamlines in air flowing past a scale model of the wing of an airplane in a wind tunnel. The experimental investigation of such streamline patterns plays an important role in airplane design. Incidentally, the flow of air can be regarded as nearly incompressible provided that the velocity of flow is well below the velocity of sound (331 m/s). Although the air will suffer some changes of density in its flow around obstacles, the changes are usually small enough to be neglected.

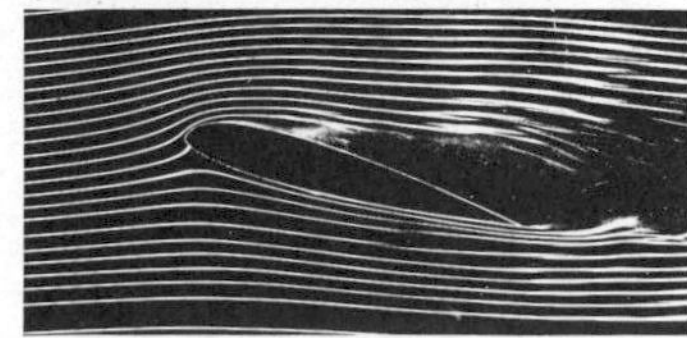

Fig. 18.10 Here the wing is in a partial stall, and the flow behind the wing has become turbulent.

Finally, Figure 18.10 shows an example of turbulent flow. In the region behind the wing, the streamers of smoke become twisted and chaotic. This is due to the generation of vortices in this region. As the vortices form, grow, break away, and disappear in quick succession, the velocity of flow fluctuates violently. The flow of the fluid becomes very unsteady and very irregular. The formation of vortices and the onset of turbulence have to do with the viscosity of the fluid. It is a general rule that vortices and turbulence will develop in a fluid of given viscosity whenever the velocity of flow, the size of the obstacle, or both exceed a certain limit.

Fig. 18.11 The world's tallest fountain.

EXAMPLE 2. The world's tallest fountain (at Fountain Hills, Arizona; Figure 18.11) shoots water to a height of 170 m at the rate of 26,000 liters/min. What is the flow velocity at the base of the fountain? At a height of 100 m? What is the diameter of the water column at the base? At a height of 100 m? Assume that the flow is steady and incompressible, and that the falling water does not interfere with the rising water. Ignore friction.

SOLUTION: The initial speed required to attain a height of 170 m is determined by Eq. (2.29),

$$\tfrac{1}{2}v_0^2 = g(x - x_0)$$

with $x - x_0 = 170$ m, this gives

$$v_0 = \sqrt{2g(x - x_0)} = \sqrt{2 \times 9.8 \text{ m/s}^2 \times 170 \text{ m}}$$

$$= 58 \text{ m/s}$$

The speed at a height of 100 m is also determined by Eq. (2.29),

$$\tfrac{1}{2}v^2 = \tfrac{1}{2}v_0^2 - g(x - x_0)$$

with $x - x_0 = 100$ m and $v_0 = 58$ m/s this gives

$$v = \sqrt{v_0^2 - 2g(x - x_0)} = \sqrt{(58 \text{ m/s})^2 - 2 \times 9.8 \text{ m/s}^2 \times 100 \text{ m}}$$

$$= 37 \text{ m/s}$$

To calculate the cross-sectional area at the base, we use Eq. (4),

$$A_0 = \frac{1}{v_0}\frac{\Delta V}{\Delta t}$$

The rate of delivery of the fountain is

$$\frac{\Delta V}{\Delta t} = \frac{2.6 \times 10^4 \text{ liters}}{\text{min}} = \frac{26 \text{ m}^3}{60 \text{ s}}$$

and therefore

$$A_0 = \frac{1}{58 \text{ m/s}} \times \frac{26 \text{ m}^3}{60 \text{ s}} = 7.5 \times 10^{-3} \text{ m}^2$$

The diameter of a circle of this area is $2\sqrt{A_0/\pi} = 9.8 \times 10^{-2}$ m.

To calculate the cross-sectional area at a height of 100 m, we use the continuity equation,

$$vA = v_0 A_0$$

or

$$A = \frac{v_0}{v}A_0 = \frac{58\ \text{m/s}}{37\ \text{m/s}} \times 7.5 \times 10^{-3}\ \text{m}^2 = 1.2 \times 10^{-2}\ \text{m}^2$$

The diameter of a circle of this area is $2\sqrt{A_0/\pi} = 1.2 \times 10^{-1}$ m. Note that the water column is narrow at the base and widens at the top. The photograph in Figure 18.11 does not show this widening because the column of rising water is hidden in a curtain of falling water; besides, the emerging water column does not have a uniform velocity across the width of the nozzle, and it is affected by friction and by entrainment of air.

18.3 Pressure

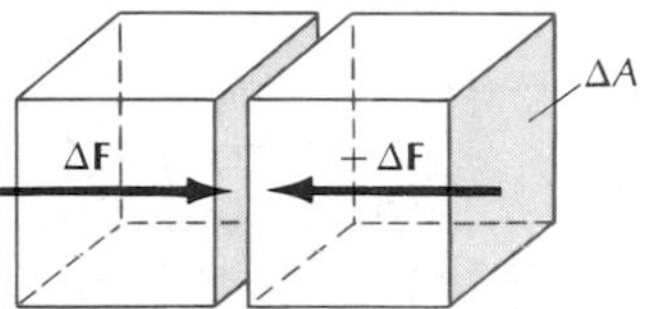

Fig. 18.12 Adjacent small cubes of fluid exerting forces on each other.

The **pressure** within a fluid is defined in terms of the force that a small volume of fluid exerts on an adjacent volume or on the adjacent wall of a container. Figure 18.12 shows two small adjacent volumes of fluid of cubical shape. The cube of fluid on the left presses against the cube on the right, and vice versa. Suppose that the magnitude of the perpendicular force between the two cubes is ΔF and that the area of one face of one of the cubes is ΔA; then the pressure is defined by

$$\boxed{p = \frac{\Delta F}{\Delta A}} \qquad (9)$$

Pressure

where the limit $\Delta A \to 0$ is to be taken. According to this definition, pressure is simply the force per unit area. Note that the pressure p is a scalar quantity, without direction. We cannot associate a direction with the pressure, because each small volume exerts pressure forces in all directions and the pressure forces in all directions are equal (in the limit $\Delta A \to 0$).

The metric unit of pressure is the N/m², which has recently been baptized with the name **pascal** (Pa):

$$1\ \text{Pa} = 1\ \text{N/m}^2 \qquad (10)$$

The British unit of pressure is the lb/in.². Other units in common use are the **atmosphere** (atm),

$$1\ \text{atm} = 1.013 \times 10^5\ \text{N/m}^2 = 14.7\ \text{lb/in.}^2 \qquad (11)$$

the millimeter of mercury (mmHg), or **torr,**

$$1\ \text{mmHg} = 1\ \text{torr} = \tfrac{1}{760}\ \text{atm} \qquad (12)$$

and the **millibar** (mbar),

$$1\ \text{mbar} = 10^2\ \text{N/m}^2 = 0.750\ \text{mmHg} \qquad (13)$$

Table 18.2 gives diverse examples of values of pressure.

Blaise Pascal, *1623–1662, French scientist. He made important contributions to mathematics and is regarded as the founder of modern probability theory. In physics, he performed experiments on atmospheric pressure and on the equilibrium of fluids.*

Table 18.2 SOME PRESSURES

Core of neutron star	1×10^{38} N/m²
Center of Sun	2×10^{16} N/m²
Highest sustained pressure achieved in laboratory	5×10^{11} N/m²
Center of Earth	4×10^{11} N/m²
Bottom of Pacific Ocean (5.5-km depth)	6×10^{7} N/m²
Water in core of nuclear reactor	1.6×10^{7} N/m²
Overpressure[a] in automobile tire	2×10^{5} N/m²
Air at sea level	1.0×10^{5} N/m²
Overpressure at 7 km from 1-megaton explosion	3×10^{4} N/m²
Air in funnel of tornado	2×10^{4} N/m²
Overpressure in human heart	
systolic	1.6×10^{4} N/m²
diastolic	1.1×10^{4} N/m²
Lowest vacuum achieved in laboratory	10^{-12} N/m²

[a] The *overpressure* is the amount of pressure in excess of normal atmospheric pressure.

EXAMPLE 3. In 1934 C. W. Beebe and O. Barton descended to a depth of 923 m below the surface of the ocean in a steel bathysphere of diameter 1.45 m (Figure 18.13). The pressure at this depth is 9.3×10^6 N/m². Under these conditions, what is the force pressing one-half of the steel sphere toward the opposite half?

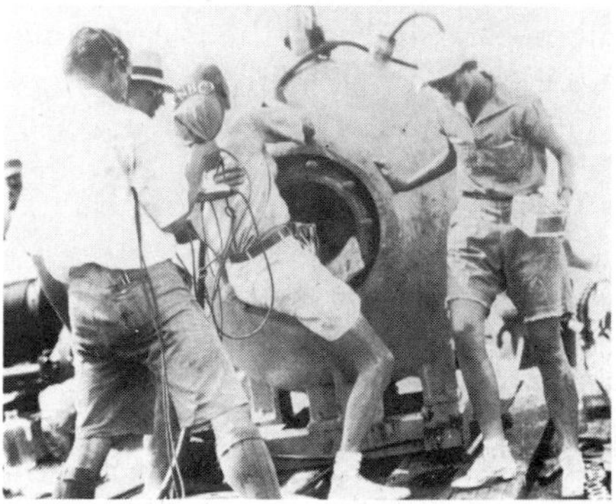

Fig. 18.13 Explorer C. W. Beebe entering the bathysphere.

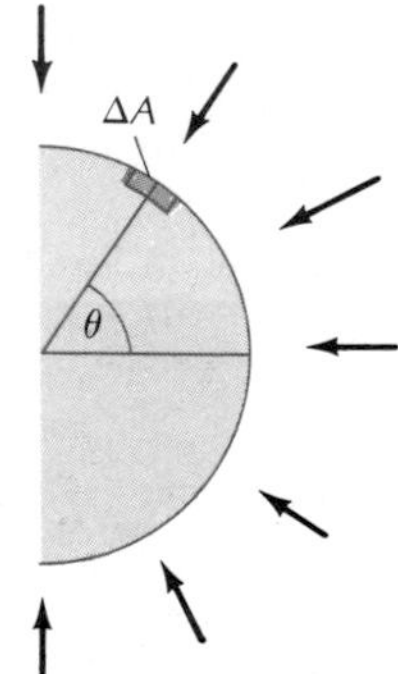

Fig. 18.14 Pressure forces on the sphere.

SOLUTION: Figure 18.14 is a diagram of the right half of the steel sphere. The arrows show the pressure force exerted by the water. On a small area element dA of the sphere, the pressure force is $p\ dA$ and the horizontal component of this force is $p\ dA \cos\theta$. But $dA \cos\theta$ is the projection of the area dA on a vertical plane, i.e., it is the fraction of the area facing to the right, in silhouette. For the complete half sphere, the net area facing to the right is πR^2 and the net force toward the left is therefore

$$p\pi R^2 = 9.3 \times 10^6\ \text{N/m}^2 \times \pi \times (0.72\ \text{m})^2$$

$$= 1.5 \times 10^7\ \text{N}$$

18.4 Pressure in a Static Fluid

We will now consider a fluid in static equilibrium so that the velocity of flow is everywhere zero. An example of static fluid is the air in a closed room with no air currents, or the water in a bottle. At first we will ignore gravity and pretend that the only forces acting on the fluid are those exerted by the walls of the container. Under these conditions the pressure at all points within the fluid must be the same. To see that this is so, take two points P_1 and P_2 in the fluid and consider a long, thin parallelepiped with bases at these two points (Figure 18.15). The fluid outside the parallelepiped exerts pressure forces on the fluid inside the parallelepiped. The compoment of these forces along the long direction of the parallelepiped is entirely due to the forces on the bases at P_1 and P_2. If the parallelepiped of fluid is to remain static, these forces on the opposite bases must be equal in magnitude. Hence the pressures at P_1 and P_2 must be equal.

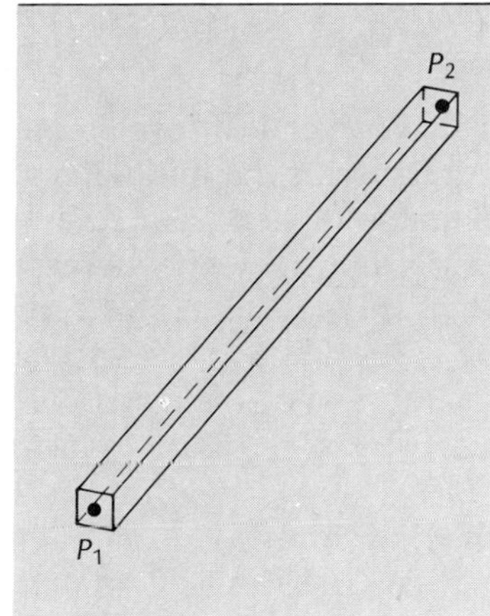

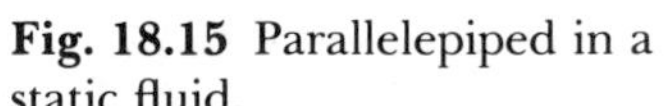

Fig. 18.15 Parallelepiped in a static fluid.

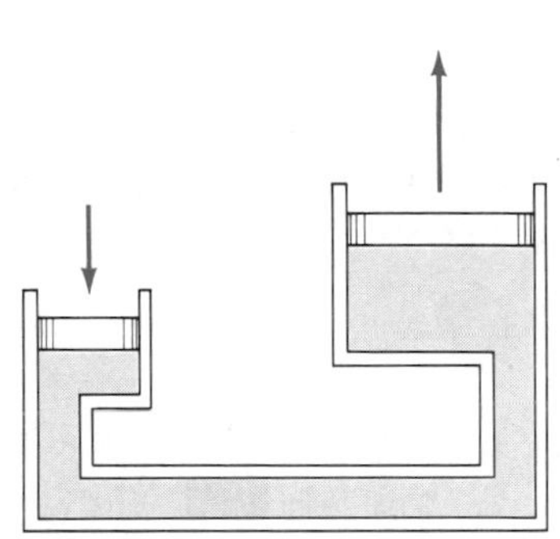

Fig. 18.16 Hydraulic press (schematic).

Pascal's Principle

The rule that the pressure is the same at all points in a static fluid is called **Pascal's Principle.** For example, the pressure of the air is the same at all points of a room — if the pressure is 1 atm in one corner of the room, it will be the same at any other point of the room. Pascal's Principle finds widespread application in the design of hydraulic presses, jacks, and remote controls. Figure 18.16 is a schematic diagram of a hydraulic press consisting of two cylinders with pistons, one small and one large. The cylinders are filled with an incompressible fluid and they are connected by a pipe. By pushing down on the small piston, we increase the pressure in the fluid; this increases the force on the large piston. Since the pressure is the same on each piston, the forces on the pistons are in the same ratio as the areas of their faces; thus, a small force on the small piston will generate a large force on the large piston. The brake systems and other control systems on automobiles, trucks, and aircraft employ similar arrangements. Figure 18.17 shows the hydraulic brake system of an automobile. The brake pedal pushes on the small master piston, and the resultant pressure of the brake fluid pushes the large slave piston and activates the brakes.

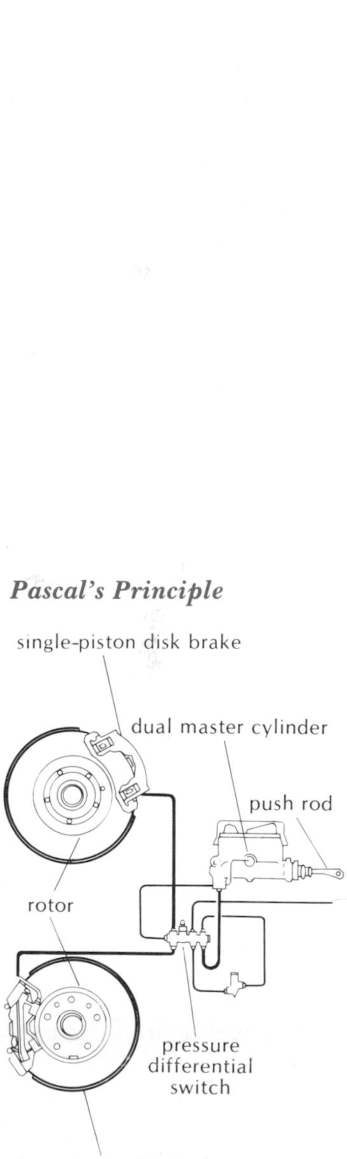

Fig. 18.17 Hydraulic brake system of an automobile.

EXAMPLE 4. The diameter of the small piston in Figure 18.16 is 0.75 in. and that of the large piston is 3.5 in. If you exert a force of 30 lb on the master piston, what force will this generate on the slave piston?

SOLUTION: The forces are in the ratio of the areas of the pistons:

$$F_2 = F_1 \frac{A_2}{A_1} = 30 \text{ lb} \times \frac{(3.5)^2}{(0.75)^2} = 650 \text{ lb}$$

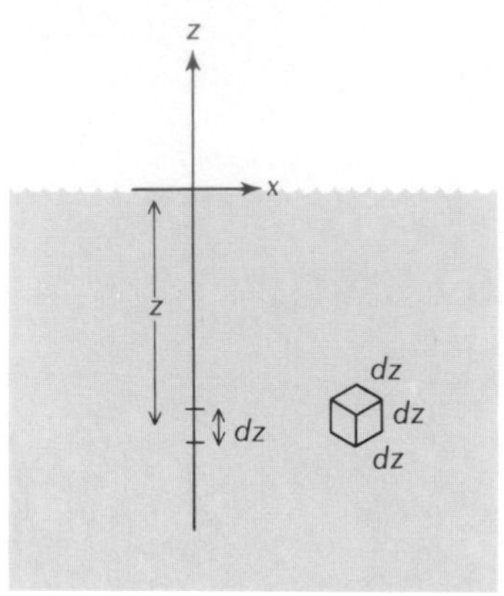

Fig. 18.18 A small cube of fluid at a depth z below the surface.

Next, we want to take into account the effect of gravity. For a static fluid in the gravitational field of the Earth, such as the water of a lake, the pressure force at any given depth must support the weight of all the overlaying mass of fluid; consequently, the pressure must increase as a function of depth. To derive the dependence of pressure on depth, we must consider the condition for the equilibrium of a small volume of fluid. Figure 18.18 shows a small cube at some depth below the surface of the fluid. As always, the z coordinate is reckoned as positive in the upward direction; hence the depth of the cube corresponds to some negative value of z. The dimension of the cube is $dz \times dz \times dz$ and hence its weight is

$$g\,dm = g\rho\,dz \times dz \times dz \tag{14}$$

The weight of the cube must be balanced by the vertical forces contributed by the pressure. Suppose that the pressure at the midpoint of the cube is p; the pressure at the top of the cube is then $p - \frac{1}{2}dp$ and the pressure at the bottom is $p + \frac{1}{2}dp$, where dp represents the increase of pressure in the interval dz. Since the area of the faces is $dz \times dz$, the vertical pressure force on the cube will be

$$dz \times dz(p + \tfrac{1}{2}dp) - dz \times dz(p - \tfrac{1}{2}dp) = dz \times dz \times dp \tag{15}$$

Setting Eqs. (14) and (15) equal to each other, we find

$$dz \times dz \times dp = -g\rho\,dz \times dz \times dz \tag{16}$$

or

$$dp = -\,g\rho\,dz \tag{17}$$

This formula gives the increase in pressure for a small increase in depth. For a compressible fluid the density ρ will be a function of p and, consequently, a function of z, and we cannot proceed further without knowledge of the explicit form of this function. However, for an incompressible fluid the density is constant and therefore Eq. (17) shows that the change in pressure is directly proportional to the depth,

Hydrostatic pressure for incompressible fluid

$$\boxed{p - p_0 = -g\rho z} \tag{18}$$

Here p_0 is the pressure at the surface of the fluid.

EXAMPLE 5. What is the pressure at a depth of 10 m below the surface of a lake? The pressure of air at the surface of the lake is 1.0 atm.

SOLUTION: With $p_0 = 1$ atm $= 1.01 \times 10^5$ N/m^2, $\rho = 1.00 \times 10^3$ kg/m^3, and $z = -\,10$ m, Eq. (18) gives

$$\begin{aligned} p &= p_0 - g\rho z \\ &= 1.01 \times 10^5 \text{ N/m}^2 + 9.81 \text{ m/s}^2 \times 1.00 \times 10^3 \text{ kg/m}^3 \times 10 \text{ m} \\ &= 1.99 \times 10^5 \text{ N/m}^2 \end{aligned}$$

Thus, the pressure is 1 atm at the surface and about 2 atm at a depth of 10 m, i.e., the pressure increases by about 1 atm in 10 m.

EXAMPLE 6. What is the change in atmospheric pressure between the basement of a house and the attic, at a height of 10 m above the basement? The density of air is 1.29 kg/m³.

SOLUTION: Although air is compressible fluid, the change in its density is small if the change in altitude is small, as it is in the present example. Therefore Eq. (18) is a good approximation:

$$p - p_0 = g\rho z = -9.8\ \text{m/s}^2 \times 1.29\ \text{kg/m}^3 \times 10\ \text{m}$$

$$= -\ 1.3 \times 10^2\ \text{N/m}^2 = -0.95\ \text{mmHg}$$

Hence the pressure decreases by about 1 mmHg in 10 m. This decrease of pressure can be readily detected by carrying an ordinary barometer from the basement to the attic of the house.

Several simple instruments for the measurement of pressure make use of a column of liquid. Figure 18.19 shows a **mercury barometer** consisting of a tube of glass, about 1 m long, closed at the upper end and open at the lower end. The tube is filled with mercury, except for a small evacuated space at the top. The bottom of the tube is immersed in an open bowl filled with mercury. The atmospheric pressure acting on the exposed surface of mercury in the bowl prevents the mercury from flowing out of the tube. At the level of the exposed surface, the pressure exerted by the column of mercury is ρgh, where $\rho = 1.360 \times 10^4$ kg/m³ is the density of mercury and h the height of the mercury column. For equilibrium, this pressure must match the atmospheric pressure:

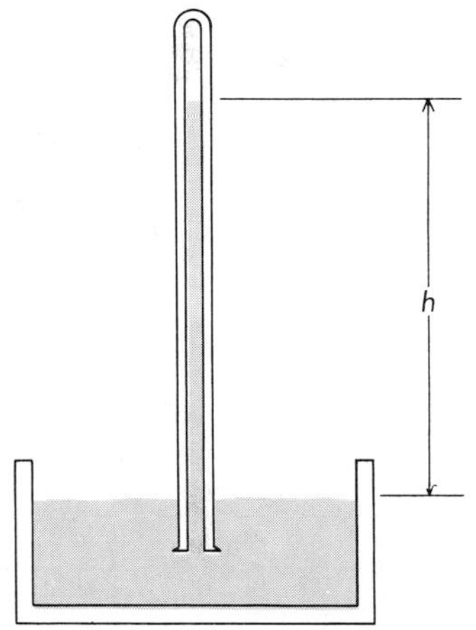

Fig. 18.19 Mercury barometer.

$$p_{\text{at}} = \rho gh \tag{19}$$

This permits a simple determination of the atmospheric pressure from a measurement of the height of the mercury column.

In view of the direct correspondence between the atmospheric pressure and the height of the mercury column, the pressure is often quoted in terms of this height, usually expressed in millimeters [see Eq. (12)]. However, for accurate work we must take into account that the acceleration of gravity depends on location and that the density of mercury depends on temperature. The convention that we will follow is that the height of the mercury column to be quoted as an indicator of pressure is not the actual height of the column but, rather, the height that the column would have if g had the standard value 9.8066 m/s² and ρ had the standard value 1.3593×10^4 kg/m³. The average value of the **atmospheric pressure** at sea level is 760 mmHg; by definition, this is 1 atm. Hence

Atmospheric pressure

$$1\ \text{atm} = 760\ \text{mmHg} = 0.760\ \text{m} \times \rho \times g$$

$$= 0.760\ \text{m} \times 1.3595 \times 10^4\ \text{kg/m}^3 \times 9.8066\ \text{m/s}^2$$

$$= 1.013 \times 10^4\ \text{N/m}^2 \tag{20}$$

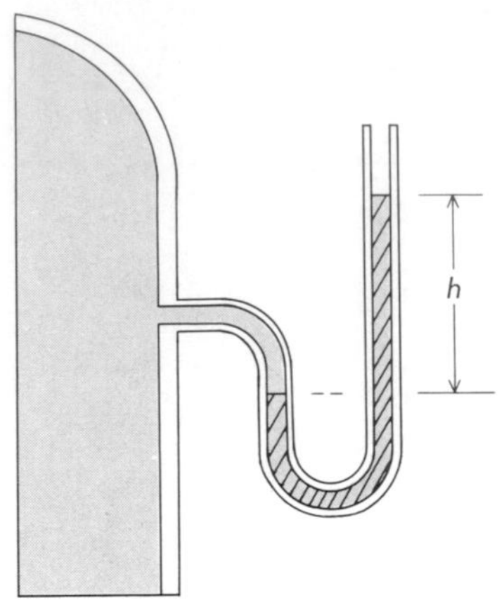

Fig. 18.20 Open-tube manometer.

Figure 18.20 shows an open-tube **manometer,** a device for the measurement of the pressure of a fluid contained in a tank. The tube contains mercury, or water, or oil. One side of the tube is in contact with the fluid in the tank; the other is in contact with the air. The fluid in the tank therefore presses down on one end of the mercury column and the air presses down on the other end. The difference h in the heights of the levels of mercury at the two ends gives the difference in the pressure at the two ends,

$$p - p_{\text{at}} = \rho g h \tag{21}$$

Hence this kind of manometer indicates the amount of pressure in the tank in excess of the atmospheric pressure. This excess is called the **overpressure** or **gauge pressure.** It is good to keep in mind that many gauges used in engineering practice are calibrated in terms of overpressure rather than absolute pressure. For instance, the pressure gauges used for automobile tires read overpressure.

18.5 Archimedes' Principle

Buoyant force

If a body is partially or totally immersed in a static fluid, the pressure of the fluid exerts a force on the body that is directed vertically upward. This force is called the **buoyant force.** The magnitude of this force is given by **Archimedes' Principle:**

Archimedes' Principle

The buoyant force on a body has the same magnitude as the weight of the fluid displaced by the body.

The proof of this famous principle is very simple. Imagine that we replace the immersed volume of a body by an equal volume of fluid (Figure 18.21). The volume of fluid will then be in static equilibrium. Obviously, this requires a balance between the resultant of all the pressure forces acting on the surface enclosing this volume of fluid and the weight of the fluid. But the pressure forces on the surface of the original immersed body are exactly the same as the pressure forces on the surface of the volume of fluid by which we have replaced it. Hence the magnitude of the resultant of the pressure forces acting on the original body must equal the weight of the displaced fluid.

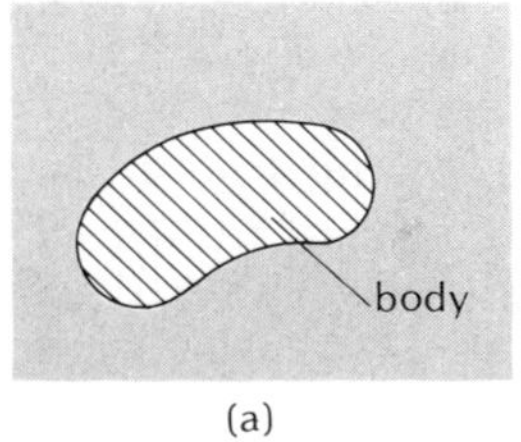

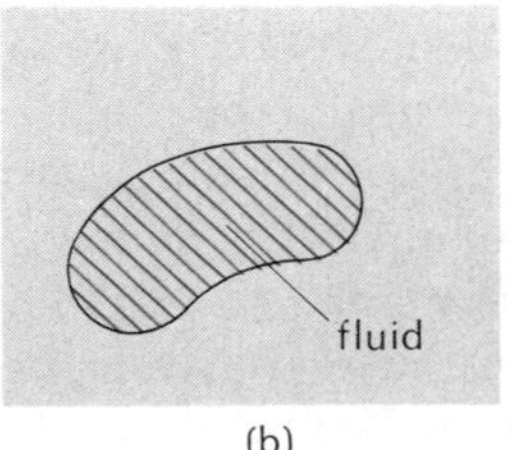

Fig. 18.21 (a) Submerged body. (b) Volume of fluid of the same shape as the body.

EXAMPLE 7. A hot-air balloon has a volume of 2.20×10^3 m^3; it is filled with hot air of density 0.96 kg/m^3. What maximum load can this balloon lift when surrounded by cold air of density 1.29 kg/m^3?

SOLUTION: The mass of cold air displaced by the balloon is 1.29 kg/m^3 $\times$ 2.20 $\times$ 10^3 m^3 = 2.84 $\times$ 10^3 kg. The weight of this cold air is g $\times$ 2.84 $\times$ 10^3 kg and this is therefore the buoyant force on the balloon. The buoyant force must support both the weight of the hot air and the load. The weight of the hot air is $g \times 0.96$ kg/m^3 $\times$ 2.20 $\times$ 10^3 m^3 = $g \times 2.11 \times 10^3$ kg. Hence the weight of the load can be at most

$$g \times 2.84 \times 10^3 \text{ kg} - g \times 2.11 \times 10^3 \text{ kg} = g \times 730 \text{ kg}$$

i.e., the mass of the load can be at most 730 kg. Note that in this context, the "load" includes the hull of the balloon, the gondola, and everything attached to it.

EXAMPLE 8. A chunk of ice floats in water (Figure 18.22). What percentage of the volume of ice will be above the level of the water? The density of ice is 917 kg/m^3.

SOLUTION: If the mass of the chunk of ice is, say, 1000 kg, it must displace 1 m^3 of water. The volume of ice below the water level is then 1 m^3 and the total volume is 1000 kg/(917 kg/m^3) = 1.091 m^3. The fraction of ice above the water level is therefore 0.091 m^3/1.091 m^3 = 0.083, or 8.3%.

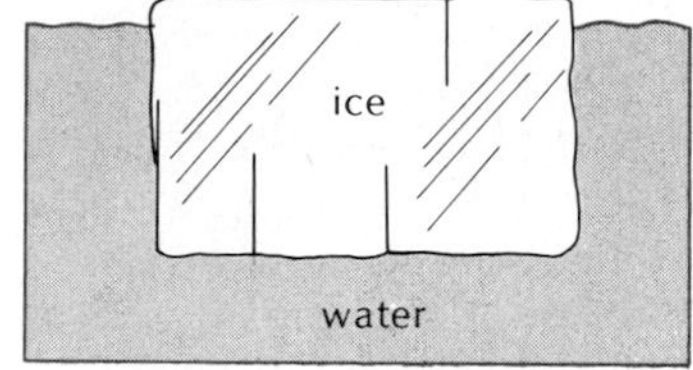

Fig. 18.22 Ice floating in water.

18.6 Fluid Dynamics; Bernoulli's Equation

Although it is not difficult to derive an equation of motion for a fluid corresponding to Newton's Second Law for a particle, the solution of this equation is usually quite laborious. One of the troubles is that the pressure affects the motion of the fluid and in turn the motion affects the pressure. Hence we must deal with simultaneous equations for motion and for pressure. When faced with such an awkward set of equations, we find it very profitable to make use of conservation theorems. In the following discussion we will formulate the conservation theorem for energy in the special case of steady flow of an incompressible fluid with no friction.

We know from Section 18.2 that steady incompressible flow can be described by streamlines. As in the derivation of the equation of continuity, we consider a bundle of streamlines forming a thin stream tube. The fluid flows inside this tube as though the surface of the tube were an impermeable pipe. Figure 18.23a shows a segment of this "pipe"; this segment contains some mass of fluid. Figure 18.23b shows the same mass of fluid at a slightly later time — the fluid has moved toward the right. During this movement the pressure at the left and at the right end of the segment does some work on the mass of fluid. By energy conservation, this work must equal the change of kinetic and potential energy. To express this mathematically, we begin by calculating the work done by the pressure. As the left end of the mass of fluid moves through a distance Δl_1, the work done by the pressure is

$$\Delta W_1 = p_1 A_1 \, \Delta l_1 \tag{22}$$

Since the product $A_1 \, \Delta l_1$ is the volume ΔV vacated by the retreat of the fluid on the left end, we can also write this as

$$\Delta W_1 = p_1 \, \Delta V \tag{23}$$

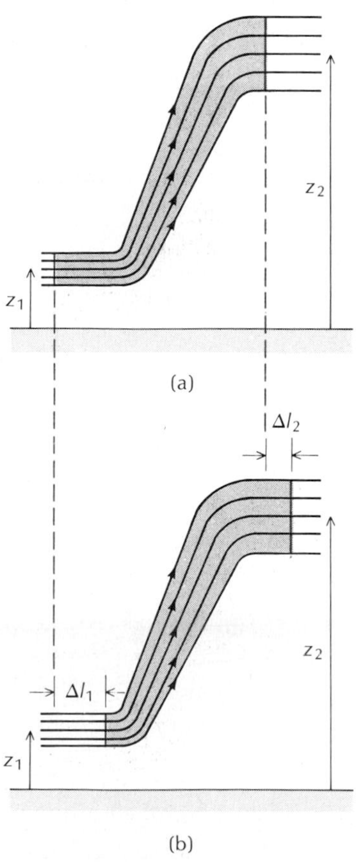

Fig. 18.23 (a) A thin stream tube. (b) Motion of fluid along the stream tube.

Likewise, the work done by the pressure at the right end is

$$\Delta W_2 = -p_2\,\Delta V \tag{24}$$

Daniel Bernoulli (bernooyee), *1700–1782, Swiss physician, physicist, and mathematician. He was professor of anatomy and natural philosophy at Basel, where he wrote his great treatise* Hydrodynamica, *which included the equation named after him. Several other members of the Bernoulli family made memorable contributions to mathematics and physics.*

Note that the *same* volume ΔV appears in Eqs. (23) and (24) — since the fluid is incompressible, the volume vacated by the retreat of the fluid at one end must equal the volume occupied by the advance of the fluid at the other end. The net work is then

$$\Delta W = \Delta W_1 + \Delta W_2 = p_1\,\Delta V - p_2\,\Delta V \tag{25}$$

The change of kinetic and potential energy is entirely due to the changes at the ends of the mass of fluid; everywhere else the shift of the fluid merely replaces fluid of some kinetic and potential energy with fluid of exactly the same kinetic and potential energy. The change at the ends involves replacing a mass Δm of fluid, of velocity v_1 at height z_1, by an equal mass, of velocity v_2 at height z_2. The corresponding change of kinetic and potential energy is

$$\Delta K + \Delta U = \tfrac{1}{2}\Delta m v_2^2 - \tfrac{1}{2}\Delta m v_1^2 + \Delta m g z_2 - \Delta m g z_1 \tag{26}$$

Setting the expressions on the right sides of Eqs. (25) and (26) equal to each other, we find

$$p_1\,\Delta V - p_2\,\Delta V = \tfrac{1}{2}\Delta m v_2^2 - \tfrac{1}{2}\Delta m v_1^2 + \Delta m g z_2 - \Delta m g z_1 \tag{27}$$

Upon rearrangement this becomes

$$\frac{1}{2}\frac{\Delta m}{\Delta V}v_1^2 + \frac{\Delta m}{\Delta V}gz_1 + p_1 = \frac{1}{2}\frac{\Delta m}{\Delta V}v_2^2 + \frac{\Delta m}{\Delta V}gz_2 + p_2 \tag{28}$$

Since $\Delta m/\Delta V$ is the density ρ of the fluid, we can also express this as

$$\tfrac{1}{2}\rho v_1^2 + \rho g z_1 + p_1 = \tfrac{1}{2}\rho v_2^2 + \rho g z_2 + p_2 \tag{29}$$

or as

Bernoulli's equation

$$\boxed{\tfrac{1}{2}\rho v^2 + \rho g z + p = [\text{constant}]} \tag{30}$$

This is **Bernoulli's equation.** It states that along any streamline, the sum of the density of kinetic energy, density of potential energy, and pressure is a constant. If the fluid is static, so that $v = 0$, Bernoulli's equation reduces to

$$p - [\text{constant}] = -\rho g z \tag{31}$$

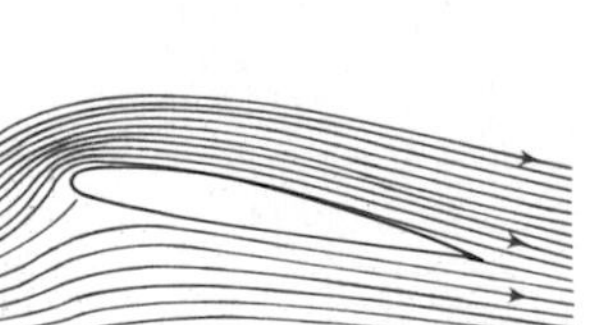

Fig. 18.24 Flow of air around an airfoil.

This coincides with Eq. (18) for the pressure in a static fluid.

Note that according to Bernoulli's equation, the velocity along any given streamline must *decrease* wherever the pressure *increases.* Intuitively, we might have expected that where the pressure is high, the velocity is large; but energy conservation demands exactly the opposite. The relation between pressure and velocity plays an important role in the design of wings for airplanes. Figure 18.24 shows an airfoil and the streamlines of air flowing around it. The shape of the airfoil has been designed so that along its upper part the velocity is large (high density of streamlines) and along its lower part the velocity is small (low density of streamlines). Consider now one streamline passing just over the airfoil and one just under the airfoil. At a large distance to the right, the fluid on all streamlines has the same pressure and the same velocity. Bernoulli's equation applied to each of the two streamlines therefore tells us that in the region just above the airfoil, the pressure is low and in the region just below the airfoil, the pressure is high. This leads to a net upward force, or **lift,** on the airfoil — this lift supports the airplane in flight. As Figure 18.24 shows, at a large distance to the left, the streamlines have a slight downward trend; what has happened is that the air gave some upward momentum to the airfoil and, in return, acquired an equal amount of downward momentum. Ultimately, the downward flow of air presses against the ground and transmits the weight of the airplane to the ground.

To conclude this chapter we will work out some examples involving Bernoulli's equation.

EXAMPLE 9. A water tank has a (small) hole near its bottom at a depth h from the top surface (Figure 18.25). What is the speed of the stream of water emerging from the hole?

SOLUTION: Qualitatively, one of the streamlines for the water flowing out of this tank will look as shown in Figure 18.25. Since the hole is small, the water level at the top of the tank drops only very slowly. We can therefore take $v_1 = 0$ in Eq. (29). Furthermore, the pressure at the top and in the emerging stream of water is the same: $p_1 = p_2 = p_{at}$. This gives

$$\rho g z_1 + p_{at} = \tfrac{1}{2}\rho v_2^2 + \rho g z_2 + p_{at} \tag{32}$$

which yields

$$v_2 = \sqrt{2g(z_1 - z_2)} = \sqrt{2gh} \tag{33}$$

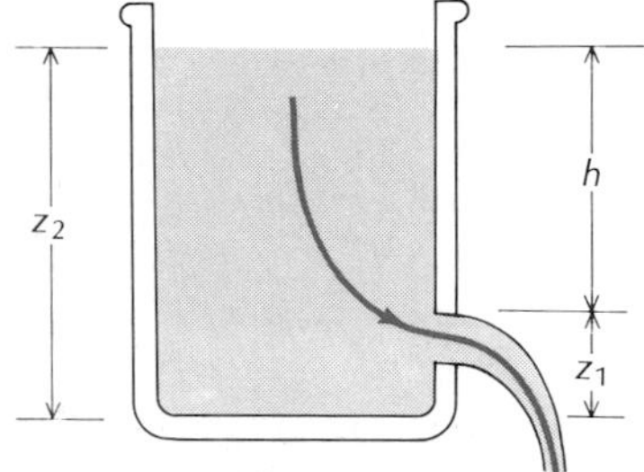

Fig. 18.25 Streamline for water flowing out of a tank.

Thus, the speed of the emerging water is exactly what it would be if the water were to fall freely from a height h. This, of course, expresses conservation of energy: when a drop of water flows out at the bottom, the loss of potential energy of the water in the tank is equivalent to the removal of a drop of water from the top; the conversion of this potential energy into kinetic energy will give the drop the speed of free fall.

EXAMPLE 10. The **Venturi flowmeter** is a simple device that measures the velocity of a fluid flowing in a pipe. It consists of a constriction in the pipe with a cross-sectional area A_2 that is smaller than the cross-sectional area A_1 of the pipe itself (Figure 18.26). Small holes in the constriction and in the pipe permit the measurement of the pressures at these points by means of a manometer. Express the velocity of flow in terms of the pressure difference.

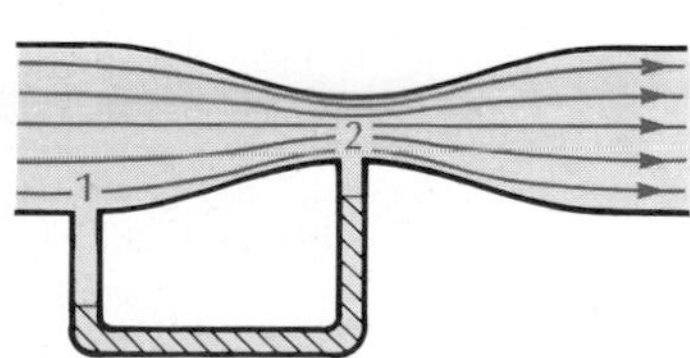

Fig. 18.26 Venturi flowmeter. The manometer measures the pressure difference between points 1 and 2.

SOLUTION: The velocities at points 1 and 2 are related to the cross-sectional areas by the continuity equation [Eq. (8)].

$$v_2 = v_1 A_1 / A_2$$

With $z_1 = z_2 = 0$, Bernoulli's equation then gives us the following for the pressure difference:

$$p_1 - p_2 = \tfrac{1}{2}\rho v_2^2 - \tfrac{1}{2}\rho v_1^2 = \tfrac{1}{2}\rho v_1^2[(A_1/A_2)^2 - 1]$$

so that

$$v_1 = \sqrt{\frac{2(p_1 - p_2)}{\rho[(A_1/A_2)^2 - 1]}}$$

EXAMPLE 11. The overpressure in a fire hose of diameter 6.4 cm is 3.5×10^5 N/m^2 and the speed of flow is 4.0 m/s. The fire hose ends in a metal tip of diameter 2.5 cm (Figure 18.27). What are the pressure and velocity of the water in the tip?

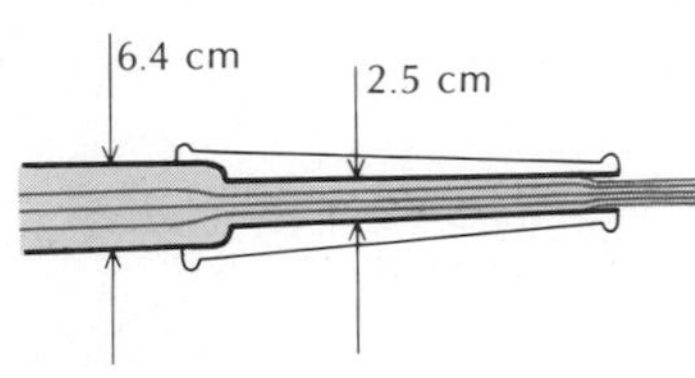

Fig. 18.27 A fire-hose tip.

SOLUTION: As in the preceding example, the velocity of the water in the tip is determined by the equation of continuity:

$$v_2 = v_1 \frac{A_1}{A_2} = 4.0 \text{ m/s} \times \frac{(6.4 \text{ cm})^2}{(2.5 \text{ cm})^2} = 26.2 \text{ m/s}$$

The pressure of water in the tip can then be calculated from Bernoulli's equation with $z_1 = z_2 = 0$:

$$p_2 = p_1 + \tfrac{1}{2}\rho v_1^2 - \tfrac{1}{2}\rho v_2^2$$

To convert this into an equation involving overpressures, we subtract p_{at} from both sides:

$$p_2 - p_{at} = p_1 - p_{at} + \tfrac{1}{2}\rho v_1^2 - \tfrac{1}{2}\rho v_2^2$$

With $p_1 - p_{at} = 3.5 \times 10^5$ N/m^2, this yields

$$\begin{aligned} p_2 - p_{at} &= 3.5 \times 10^5 \text{ N/m}^2 + \tfrac{1}{2} \times 1.0 \times 10^3 \text{ kg/m}^3 \times (4.0 \text{ m/s})^2 \\ &\quad - \tfrac{1}{2} \times 1.0 \times 10^3 \text{ kg/m}^3 \times (26.2 \text{ m/s})^2 \\ &= 1.4 \times 10^4 \text{ N/m}^2 \end{aligned}$$

EXAMPLE 12. What is the speed of the water just outside of the tip of the fire hose of Example 11?

SOLUTION: The velocity of the water just outside the tip can be directly calculated from Bernoulli's equation. We again set $z_1 = z_3 = 0$ and we set the pressure p_3 equal to the atmospheric pressure p_{at}:

$$\begin{aligned} v_3^2 &= v_1^2 + \frac{2p_1}{\rho} - \frac{2p_{at}}{\rho} = v_1^2 + \frac{2}{\rho}(p_1 - p_{at}) \\ &= (4.0 \text{ m/s})^2 + \frac{2 \times 3.5 \times 10^5 \text{ N/m}^2}{1.0 \times 10^3 \text{ kg/m}^3} \\ &= 716 \text{ (m/s)}^2 \end{aligned}$$

and

$$v_3 = 26.8 \text{ m/s}$$

Note that the diameter of the stream of water just outside of the tip is slightly smaller than the diameter in the tip. This follows from the equation of continuity,

$$A_3 = A_2 \frac{v_2}{v_3} = A_2 \frac{26.2 \text{ m/s}}{26.8 \text{ m/s}} = A_2 \times 0.98$$

Since the cross-sectional area A_2 has a diameter of 2.5 cm, the cross-sectional area A_3 will have a diameter $2.5\ \text{cm} \times \sqrt{0.98} = 2.5\ \text{cm} \times 0.99$. Thus, the stream of water contracts by about 1% as it leaves the nozzle (Figure 18.27). Intuitively, we might have expected that the stream expands as it leaves the nozzle, but the opposite is the case.

SUMMARY

Continuity equation (for incompressible fluid): $vA = [\text{constant}]$

Definition of pressure: $p = \dfrac{\Delta F}{\Delta A}$

Atmospheric pressure (average): $1\ \text{atm} = 760\ \text{mmHg}$

$$= 1.0 \times 10^5\ \text{N/m}^2$$

Hydrostatic pressure (for incompressible fluid): $p - p_0 = -g\rho z$

Archimedes' Principle: The buoyant force has the same magnitude as the weight of displaced fluid.

Bernoulli's equation: $\frac{1}{2}\rho v^2 + \rho g z + p = [\text{constant}]$

QUESTIONS

1. According to popular belief, blood is denser than water. Is this true?

2. The sheet of water of a waterfall is thick at the top and thin at the bottom. Explain.

3. If you place a block of wood on the bottom of a swimming pool, why does the pressure of the water not keep it there?

4. Explain what holds a suction cup on a smooth surface.

5. Newton gave the following description of his classic experiment with a rotating bucket:

> If a vessel, hung by a long cord, is so often turned about that the cord is strongly twisted, then filled with water, and held at rest together with the water; thereupon, by the sudden action of another force, it is whirled about the contrary way, and while the cord is untwisting itself, the vessel continues for some time in this motion; the surface of the water will at first be plain, as before the vessel began to move; but after that, the vessel, by gradually communicating its motion to the water, will make it begin sensibly to revolve, and recede by little and little from the middle, and ascend to the sides of the vessel, forming itself into a concave figure (as I have experienced), and the swifter the motion becomes, the higher will the water rise, till at last, performing its revolutions in the same times with the vessel, it becomes relatively at rest in it.

Explain why hydrostatic equilibrium requires that the rotating water be higher at the rim of the bucket than at the center.

6. When a doctor measures the blood pressure of a patient he places the cuff on the arm, at the same vertical level as the heart. What would happen if he were to place the cuff around the leg?

7. Scuba divers have survived short intervals of free swimming at depths of 1400 ft. Why does the pressure of the water at this depth not crush them?

8. Figure 18.28 shows a glass vessel with vertical tubes of different shapes. Explain why the water level in all the tubes is the same.

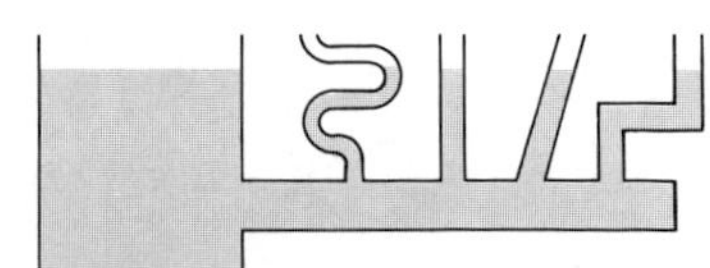

Fig. 18.28 Glass vessel with vertical tubes.

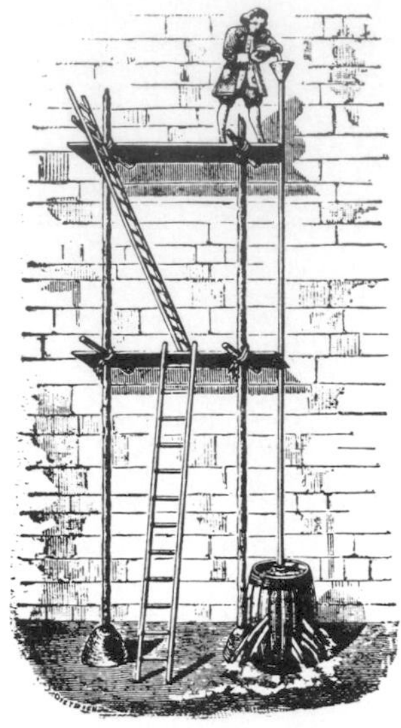

Fig. 18.29 Blaise Pascal's experiment.

Fig. 18.30 Grain silos.

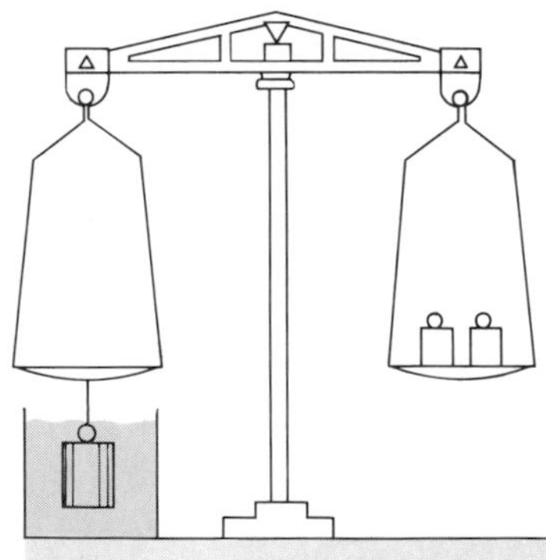

Fig. 18.31 (above) Weighing a body underwater.

9. In a celebrated experiment, Blaise Pascal attached a long metal funnel to a tight cask (Figure 18.29). When he filled the cask by pouring water into this funnel, the cask burst. Explain.

10. Face masks for scuba divers have two indentations at the bottom into which the diver can stick thumb and forefinger to pinch his nose shut. What is the purpose of this arrangement?

11. Figure 18.30 shows grain silos held together by circumferential steel bands. Why has the farmer placed more bands near the bottom than near the top?

12. The level of a large oil slick on the sea is slightly higher than the level of the surrounding water. Explain.

13. Flooding on the low coasts of England and the Netherlands is most severe when the following three conditions are in coincidence: onshore wind, full or new moon, and low barometric pressure. Can you explain this?

14. Why do some men or women float better than others? Why do they all float better in salt water than in plain water?

15. Would you float if your lungs were full of water?

16. Will a stone float in a tub full of mercury?

17. An ice cube floats in a glass full of water. Will the water level rise or fall when the ice melts?

18. The density of a solid body can be determined by first weighing the body in air and then weighing it again when it is immersed in water (Figure 18.31). How can you deduce the density from these two measurements?

19. A buoy floats on the water. Will the flotation level of the buoy change when the atmospheric pressure changes?

20. Will the water level in a canal lock rise or fall if a ship made of steel sinks in the lock? What if the ship is made of wood?

21. While training for the conditions of weightlessness they would encounter in an orbiting spacecraft, NASA astronauts were made to float submerged in a large water tank (Figure 18.32). Small weights attached to their spacesuits gave the astronauts neutral buoyancy. To what extent does such simulated weightlessness imitate true weightlessness?

22. A girl standing in a subway car holds a helium balloon on a string. Which way will the balloon move when the car accelerates?

23. A slurry of wood chips is used in some tanning operations. What would happen to a man were he to fall into a vat filled with such slurry?

Fig. 18.32 (right) Astronaut floating underwater.

24. Figure 18.33 shows a wide pipe with a constriction, and a thin tube that connects the constricted part of the pipe with the wide part. Suppose that the fluid in the thin tube is initially at rest, and the fluid in the wide pipe is flowing from left to right. What will happen to the fluid in the tube? What if the fluid in the pipe were flowing from right to left?

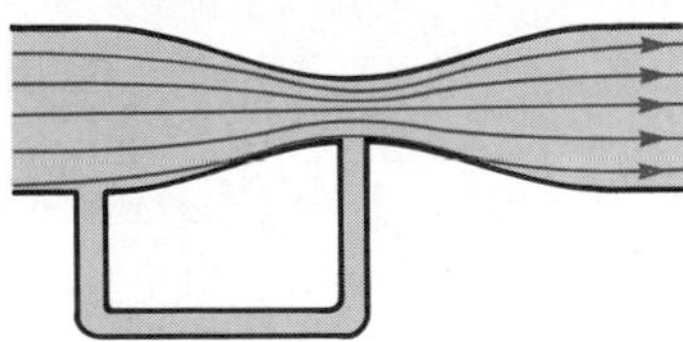

Fig. 18.33

25. The bathyscaphe *Trieste,* which set a record of 35,817 ft in a deep dive in the Marianas Trench in 1960, consists of a large tank filled with gasoline below which hangs a steel sphere for carrying the crew (Figure 18.34). Can you guess the purpose of the tank of gasoline?

Fig. 18.34 The bathyscaphe *Trieste.*

26. Gasoline vapor from, say, a small leak in the fuel system poses a very serious hazard in a motorboat, but only a minor hazard in an automobile. Explain. (Hint: Gasoline vapor is denser than air.)

27. How does a submarine dive? Ascend?

28. How does a balloonist control the ascent and descent of a hot-air balloon?

29. When you release a bubble of air while underwater, the bubble grows in size as it ascends. Explain.

30. A "Cartesian diver" consists of a small inverted bottle floating inside a larger bottle whose mouth is covered by a rubber membrane (Figure 18.35). By depressing the membrane, you can increase the water pressure in the large bottle. How does this affect the buoyancy of the small bottle?

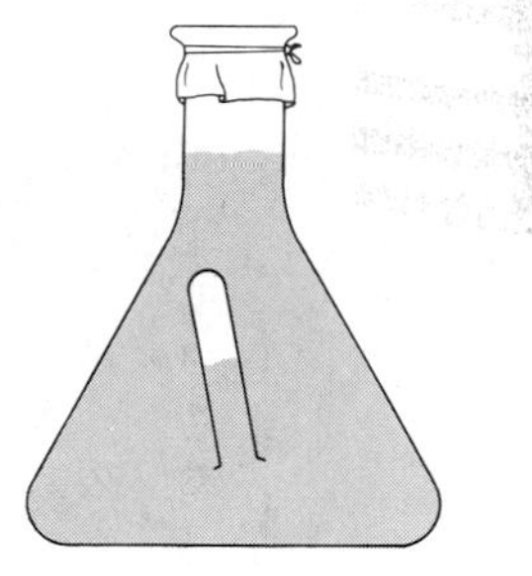

Fig. 18.35 A Cartesian diver.

31. Why are the continuity equation and Bernoulli's equation only *approximately* valid for the flow of air?

32. To throw a curve ball, the baseball pitcher gives the ball a spinning motion about a vertical axis. The air on the left and right sides of the ball will then have slightly different speeds. Using Bernoulli's equation, explain how this creates a lateral deflecting force on the ball.

33. Hurricanes with the lowest central barometric pressures have the highest wind speeds. Observations show that the square of the wind speed in a hurricane is roughly proportional to the difference between the barometric pressure outside the hurricane and the barometric pressure in the hurricane. Show that this proportionality is expected from Bernoulli's equation. (Hint: The streamlines of air start outside the hurricane and gradually spiral in toward the "eye.")

34. If you place a ping-pong ball in the jet of air from a vacuum cleaner hose aimed vertically upward, the ping-pong ball will be held in stable equilibrium within this jet. Explain this by means of Bernoulli's equation. (Hint: The speed of air is maximum at the center of the jet.)

PROBLEMS

Section 18.1

1. The following table, taken from a firefighter's manual, lists the rate of flow of water (in gallons per minute) through a fire hose of diameter 1.5 in. connected to a nozzle of given diameter; the listed rate of flow will maintain a pressure of 50 lb/in.^{2} in the nozzle:

RATE OF FLOW FOR A 1.5-IN. HOSE

Rate of flow	*Nozzle diameter*	*Nozzle pressure*
25 gal./min	$\frac{3}{8}$ in.	50 lb/in.^{2}
50	$\frac{1}{2}$	50
75	$\frac{5}{8}$	50

For each case calculate the velocity of flow (in feet per second) in the hose and in the nozzle (1 gal. = 0.134 ft^3).

2. A fountain shoots a stream of water vertically upward. Assume that the stream is inclined very slightly to one side so that the descending water does not interfere with the ascending water. The upward velocity at the base of the column of water is 15 m/s.
   (a) How high will the water rise?
   (b) The diameter of the column of water is 7.0 cm at the base. What is the diameter at the height of 5 m? At the height of 10 m?

Section 18.3

3. Within the funnel of a tornado, the air pressure is much lower than normal — about 200 mbar as compared to the normal value of 1010 mbar. Suppose that such a tornado suddenly envelops a house; the air pressure inside the house is 1010 mbar and the pressure outside suddenly drops to 200 mbar. This will cause the house to burst explosively. What is the net outward pressure force on a 12 m $\times$ 3 m wall of this house? Is the house likely to suffer less damage if all the windows and doors are open?

4. What is the downward force that air pressure (1 atm) exerts on the upper surface of a sheet of paper ($8\frac{1}{2}$ in. $\times$ 11 in.) lying on a table? Why does this force not squash the paper against the table?

5. At a distance of 7 km from a 1-megaton nuclear explosion, the blast wave has an overpressure of 3×10^4 N/m^2. Calculate the force that this blast wave exerts on the front of a standing man; the frontal area of the man is 0.7 m^2. (The actual force on a man exposed to the blast wave is larger than the result of this simple calculation because the blast wave will be reflected by the man and this leads to a substantial increase of pressure.)

6. The overpressure in the tires of a 2800-lb automobile is 35 lb/in.^{2}. If each tire supports one-fourth the weight of the automobile, what must be the area of each tire in contact with the ground? Pretend the tires are completely flexible.

7. The shape of the wing of an airplane is carefully designed so that, when the wing moves through the air, a pressure difference develops between the bottom surface of the wing and the top surface; this supports the weight of the airplane. A fully loaded DC-3 airplane has a mass of 10,900 kg. The (bottom) surface area of its wings is 92 m^2. What is the average pressure difference between the top and bottom surfaces when the airplane is in flight?

8. Pressure gauges used on automobile tires read the overpressure, i.e., the amount of pressure in excess of atmospheric pressure. If a tire has an over-

pressure of 35.0 lb/in.^{2} on a day when the barometric pressure is 28.5 in.Hg, what will be the overpressure when the barometric pressure increases to 30.3 in.Hg? Assume that the volume and the temperature of the tire remain constant.

9. Commercial jetliners have pressurized cabins enabling them to carry passengers at a cruising altitude of 10,000 m. The air pressure at this altitude is 210 mmHg. If the air pressure inside the jetliner is 760 mmHg, what is the net outward force on a 1 m × 2 m door in the wall of the cabin?

10. In 1654 Otto von Guericke, the inventor of the air pump, gave a public demonstration of air pressure (see Figure 5.25). He took two hollow hemispheres of copper, whose rims fitted tightly together, and evacuated them with his air pump. Two teams of 15 horses each, pulling in opposite directions, were unable to separate these hemispheres. If the evacuated sphere had a radius of 40 cm and the pressure inside it was nearly zero, what force would each team of horses have to exert to pull the hemispheres apart?

11. The baggage compartment of the DC-10 airliner is under the floor of the passenger compartment. Both compartments are pressurized at a normal pressure of 1 atm. In a disastrous accident near Orly, France, in 1974, a faulty lock permitted the baggage compartment door to pop open in flight, depressurizing this compartment. The normal pressure in the passenger compartment then caused the floor to collapse, jamming the control cables. At the time the airliner was flying at an altitude of 12,500 ft, where the air pressure is 0.64 atm. What was the net pressure force on a 1 ft × 1 ft square of the floor?

Section 18.4

12. Porpoises dive to a depth of 1700 ft. What is the water pressure at this depth?

13. A tanker is full of oil of density 880 kg/m^3. The flat bottom of the hull is at a depth of 26 m below the surface of the surrounding water. Inside the hull, oil is stored with a depth of 30 m (Figure 18.36). What is the pressure of the water on the bottom of the hull? The pressure of the oil? What is the net vertical pressure force on 1 m^2 of bottom?

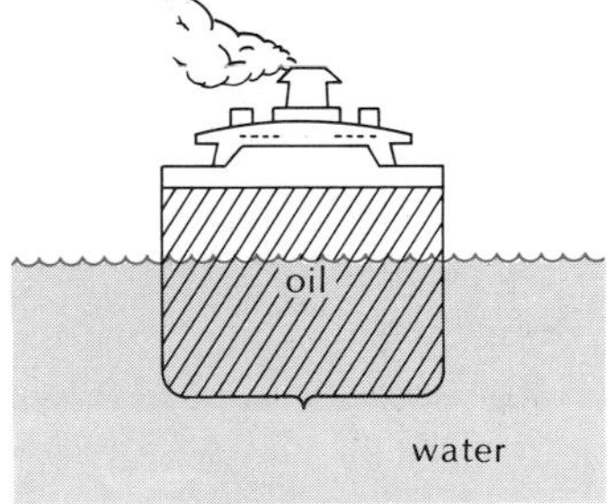

Fig. 18.36 Cross section of a tanker.

14. A pencil sharpener is held to the surface of a desk by means of a rubber "suction" cup measuring 6 cm × 6 cm. The air pressure under the suction cup is zero and the air pressure above the suction cup is 1 atm.
   (a) What is the magnitude of the pressure force pushing the cup against the table?
   (b) If the coefficient of static friction between the rubber and the table is 0.9, what is the maximum transverse force the suction cup can withstand?

15. (a) Calculate the mass of air in a column of base 1 m^2 extending from sea level to the top of the atmosphere. Assume that the pressure at sea level is 760 mmHg and that the value of the acceleration of gravity is 9.81 m/s^2, independent of height.
   (b) Multiply your result by the surface area of the Earth to find the total mass of the entire atmosphere.

16. Suppose that a zone of low atmospheric pressure (a "low") is at some place on the surface of the sea. The pressure at the center of the "low" is 2.5 in.Hg less than the pressure at a large distance from the center. By how much will this cause the water level to rise at the center?

17. (a) Under normal conditions the human heart exerts a pressure of 110 mmHg on the arterial blood. What is the arterial blood pressure in the feet of a man standing upright? What is the blood pressure in the brain? The feet are 140 cm below the level of the heart; the brain is 40 cm above the level of the heart; the density of human blood is 1055 kg/m^3.

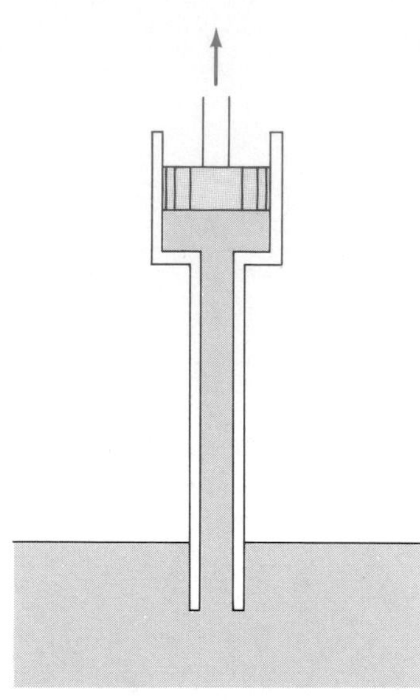

Fig. 18.37 Suction pump.

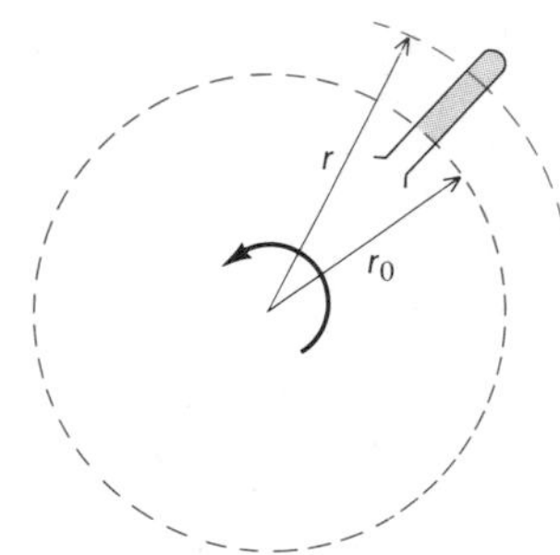

Fig. 18.38 Test tube in an ultracentrifuge.

Fig. 18.39 Fully loaded and empty supertankers.

(b) Under conditions of stress the human heart can exert a presure of up to 190 mmHg. Suppose that an astronaut were to land on the surface of a large planet where the acceleration of gravity is 61 m/s^2. Could the astronaut's heart maintain a positive blood pressure in his brain while he is standing upright? Could the astronaut survive?

18. A **"suction" pump** consists of a piston in a cylinder with a long pipe leading down into a well (Figure 18.37). What is the maximum height to which such a pump can "suck" water?

19. A man whirls a bucket full of water around a vertical circle at the rate of 0.70 rev/s. The surface of the water is at a radial distance of 1.0 m from the center.

(a) What is the pressure difference between the surface of the water and a point 1.0 cm below the surface when the bucket is at the lowest point of the circle? You may assume that a point on the surface and a point 1.0 cm below have practically the same centripetal acceleration.

(b) What if the bucket is at the highest point of the circle?

20. A test tube filled with water is being spun around in an ultracentrifuge with angular velocity ω. The test tube is lying along a radius and the free surface of the water is at radius r_0 (Figure 18.38).

(a) Show that the pressure at radius r within the test tube is

$$p = \tfrac{1}{2}\rho\omega^2(r^2 - r_0^2)$$

where ρ is the density of the water. Ignore gravity and ignore atmospheric pressure.

(b) Suppose that $\omega = 3.8 \times 10^5$ radian/s and $r_0 = 10$ cm. What is the pressure at $r = 13$ cm?

Section 18.5

21. What is the buoyant force on a human body of volume 7.4×10^{-2} m^3 when totally immersed in air? In water?

22. Icebergs commonly found floating in the North Atlantic are 30 m high (above the water) and 400 m × 400 m across. The density of ice is 920 kg/m^3.

(a) What is the total volume of such an iceberg (including the volume below the water)?

(b) What is the total mass?

23. You can walk on water if you wear very large shoes shaped like boats. Calculate the length of the shoes that will support you; assume that each shoe is 1 ft × 1 ft in cross section.

24. A gasoline barrel, made of steel, has a mass of 20 kg when empty. The barrel is filled with 0.12 m^3 of gasoline with a density of 730 kg/m^3. Will the full barrel float in water? Neglect the volume of the steel.

25. A round log of wood of density 600 kg/m^3 floats in water. The diameter of the log is 30 cm. How high does the upper surface of the log protrude from the water?

26. The supertanker *Globtik London* has a mass of 240,000 tons (2.2×10^8 kg) when empty and it can carry up to 480,000 tons (4.4×10^8 kg) of oil when fully loaded. Assume that the shape of its hull is approximately that of a rectangular parallelepiped 380 m long, 60 m wide, and 40 m high (Figure 18.39).

(a) What is the draft of the empty tanker, that is, how deep is the hull submerged in the water? Assume that the density of (sea) water is 1.02×10^3 kg/m^3.

(b) What is the draft of the fully loaded tanker?

27. A supertanker has a draft (submerged depth) of 30 m when in sea water (density $\rho = 1.02 \times 10^3$ kg/m^3). What will be the draft of this tanker when it

enters a river estuary with fresh water (density $\rho = 1.00 \times 10^3$ kg/m^3)? Assume that the sides of the ship are vertical.

28. When a body is weighed in air on an analytical balance, a correction must be made for the buoyancy contributed by the air. Suppose that the weights used in the balance are made of brass, of density 8.7×10^3 kg/m^3, and that the density of air is 1.3 kg/m^3. By how many percent must you increase (or decrease) the mass indicated by the balance in order to obtain the true mass of a body of density 1.0×10^3 kg/m^3? 5.0×10^3 kg/m^3? 10.0×10^3 kg/m^3?

29. The bottom half of a tank is filled with water ($\rho = 1.0 \times 10^3$ kg/m^3) and the top half is filled with oil ($\rho = 8.5 \times 10^2$ kg/m^3). Suppose that a rectangular block of wood of mass 5.5 kg, 30 cm long, 20 cm wide, and 10 cm high, is placed in this tank. How deep will the bottom of the block be submerged in the water?

*30. A **hydrometer,** the device used to determine the density of battery acid, wine, or other fluids, consists of a bulb with a long vertical stem (Figure 18.40). The device floats in the liquid with the bulb submerged and the stem protruding above the surface. The density of the liquid is directly related to the length h of the protruding portion of the stem. Suppose that the bulb is a sphere of radius R and that the stem is a cylinder of radius R' and length l; the mass of both together is M. Derive a formula for the density of the liquid in terms of h, R, R', l, and M.

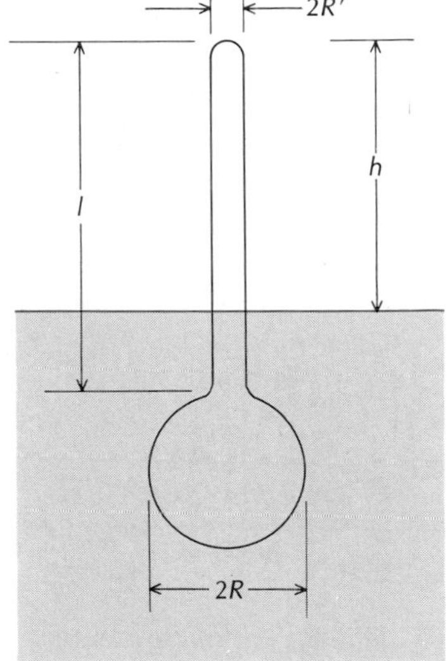

Fig. 18.40 Hydrometer.

*31. A ship that displaces a weight of water equal to its own weight is in equilibrium with regard to vertical motion. If the ship were placed lower in the water, the buoyant force would exceed gravity and the ship would surge upward. If the ship were placed higher in the water, the buoyant force would be less than gravity and the ship would sink downward. Suppose that the sides of a ship are vertical above and below its normal waterline.

(a) Show that the frequency of small up-and-down oscillations of the ship is roughly $\omega = \sqrt{A\rho g/M}$, where A is the horizontal area bounded by the waterline, M is the mass of the ship, and ρ is the density of water. (This formula is only a rough aproximation because, as the ship moves up and down, the water also has to move; hence the effective inertia is greater than M.)

(b) What is the frequency of such up-and-down oscillations for the fully loaded supertanker described in Problem 26?

*32. You drop a pencil, point down, into the water. If you release the pencil from a height of 0.040 m (measured from the point to the water level), how far will it dive? Treat the pencil as a cylinder of radius 0.40 cm, length 19 cm, and mass 4.0 g. Ignore the frictional and inertial resistance of the water, but take into account the buoyant force.

Section 18.6

33. A thin stream of water emerges vertically from a small hole on the side of a water pipe and ascends to a height of 1.2 m. What is the pressure inside the pipe? Assume the water inside the pipe is nearly static.

34. If you blow a stream of air with a speed of 7 m/s out of your mouth, what must be the overpressure in your mouth?

35. A pump has a horizontal intake pipe at a depth h below the surface of a lake. What is the maximum speed for steady flow of water into this pipe? (Hint: Inside the pipe the maximum speed of flow corresponds to zero pressure.)

36. To fight a fire on the fourth floor of a building, firemen want to use a hose of diameter $2\frac{1}{2}$ in. to shoot 250 gal./min of water to a height of 40 ft.

(a) With what minimum speed must the water leave the nozzle of the fire hose if it is to ascend 40 ft?

(b) What pressure must the water have inside the fire hose? Ignore friction.

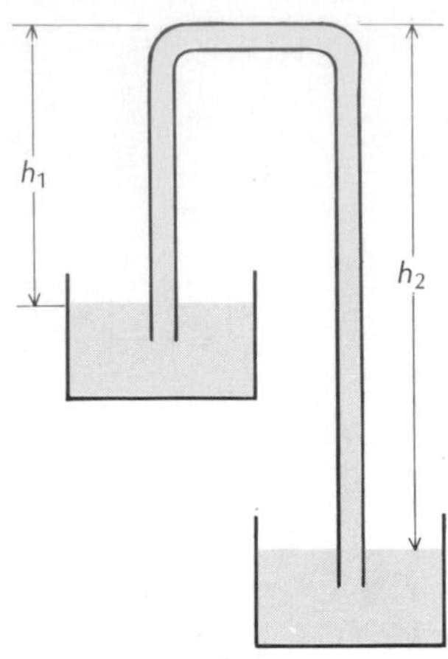

Fig. 18.41 Siphon.

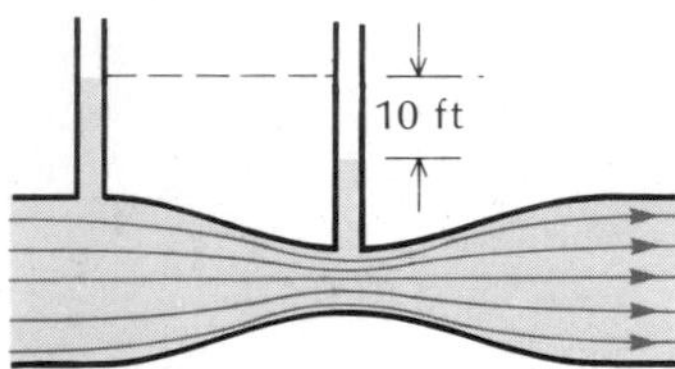

Fig. 18.42 Venturi flowmeter.

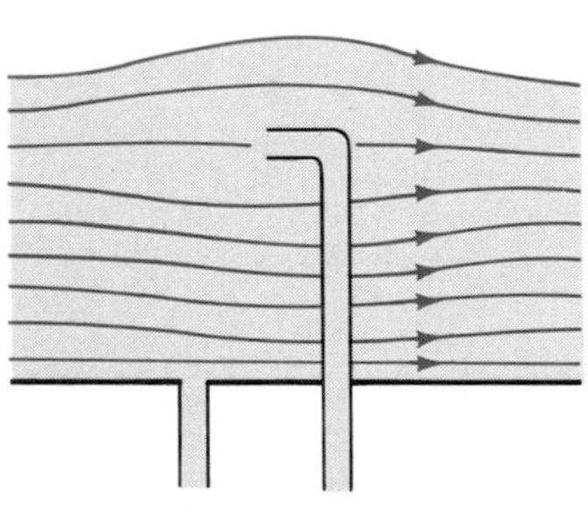

Fig. 18.43 Pitot tube.

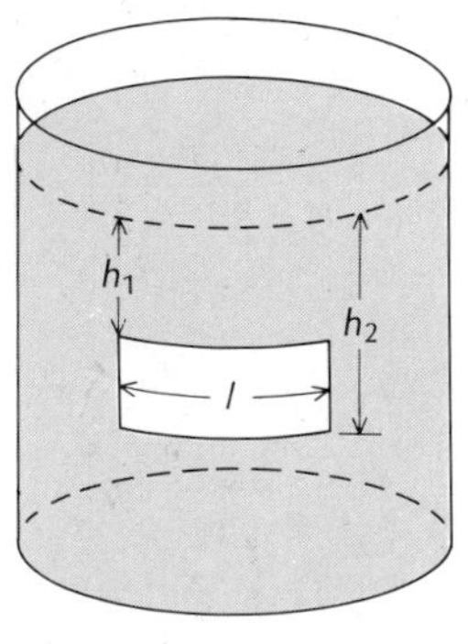

Fig. 18.44 Tank with a rectangular opening.

37. Streams of water from fire hoses are sometimes used to disperse crowds. Suppose that the stream of water emerging from the fire hose described in Example 11 impinges horizontally on a man. The collision of the water with the man is totally inelastic.

(a) What is the force that the stream of water exerts on the man, i.e., what is the rate at which the water delivers momentum to the man?
(b) What is the rate at which the water delivers energy?

38. A **siphon** is an inverted U-shaped tube that is used to transfer liquid from a container at a high level to a container at a low level (Figure 18.41).

(a) Using the lengths shown in Figure 18.41 and the density ρ of the liquid, find a formula for the speed with which the liquid emerges from the lower end of the siphon.
(b) Find the pressure at the highest point of the siphon.
(c) By setting this pressure equal to zero, find the maximum height h_1 with which the siphon can operate.

39. A pump of a fire engine draws 300 gal. of water per minute from a well with a water level 15 ft below the pump and discharges this water into a fire hose of diameter $2\frac{1}{2}$ in. at a pressure of 80 lb/in.^{2}. In the absence of friction, what power (in hp) does this pump require?

40. In the calculation of pressures in fire hoses, it is often necessary to take into account the frictional losses suffered by the water as it flows along the hose. Consider a horizontal hose of diameter $1\frac{1}{2}$ in. and length 100 ft carrying 100 gal./min of water. According to tables used by firemen, a pressure of 90 lb/in.^{2} at the upstream end of this hose will result in a pressure of only 50 lb/in.^{2} at the downstream end; the loss of 40 lb/in.^{2} is attributed to friction. At what rate (in hp) does friction remove energy from the water?

41. A Venturi flowmeter in a water main of diameter 12 in. has a constriction of diameter 4 in. Vertical pipes are connected to the water main and to the constriction (Figure 18.42); these pipes are open at their upper ends and the water level within them indicates the pressure at their lower ends. Suppose that the difference in the water levels in these two pipes is 10 ft. What is the velocity of flow in the water main? What is the rate (in gallons per second) at which water is delivered?

42. The **Pitot tube** is used for the measurement of the flow speeds of fluids, such as the flow speed of air past the fuselage of an airplane. It consists of a bent tube protruding into the airstream (Figure 18.43) and another tube opening flush with the fuselage. The pressure difference between the air in the two tubes can be measured with a manometer. Show that in terms of the pressures p_1 and p_2 in the two tubes, the speed of airflow is

$$v = \sqrt{\frac{2(p_2 - p_1)}{\rho}}$$

where ρ is the density of air. (Hint: p_1 is simply the static air pressure. To find p_2, consider a streamline reaching the opening of the bent tube; at this point, the velocity of the air is zero.)

*43. A rectangular opening on the side of a tank has a width l. The top of the opening is at a depth h_1 below the surface of the water and the bottom is at a depth h_2 (Figure 18.44). Show that the volume of water that emerges from the opening in a unit time is

$$\tfrac{2}{3}l\sqrt{2g}(h_2^{3/2} - h_1^{3/2})$$

*44. A cylindrical tank has a base area A and a height l; it is initially full of water. The tank has a small hole of area A' at its bottom. Calculate how long it will take all the water to flow out of this hole.

CHAPTER 19

Kinetic Theory of Gases

One cubic centimeter of air contains about 2.7×10^{19} molecules. The enormous magnitude of this number makes it hopeless to calculate the motion of the molecules in detail. We have no way of ascertaining the initial position and velocity of each molecule; and even if we had, the calculation of the simultaneous motions of such an enormous number of molecules is far beyond the capabilities of even the fastest conceivable computer.

In lack of a microscopic description involving the individual positions and velocities of the molecules of the gas, we must be content with a macroscopic description involving just a few variables that characterize the *average* conditions in the volume of gas. For example, we will regard the mass as a macroscopic variable. The assertion that the mass of a cubic centimeter of air is 1.29×10^{-6} kg does not mean that every cubic centimeter of air has exactly the same mass as every other cubic centimeter. The air molecules wander about and sometimes a few more, or a few less, molecules will be within any given cubic centimeter of space. We can only say what the mass will be on the average. In practice the fluctuations of the mass in a cubic centimeter of air are quite small — typically about $10^{-8}\%$.

Other macroscopic parameters characterizing the average condition of a gas that can be measured with large-scale laboratory instruments are the number of moles, the volume, the density, the pressure, and the temperature.

Kinetic theory attempts to derive relations between the macroscopic parameters by investigating the average behavior of the microscopic parameters that characterize the individual molecules. Thermodynamics also attempts to derive relations between macroscopic parameters, but without making any use of microscopic properties. Instead, thermodynamics relies only on a few general principles — such as conservation of energy — that do not depend on the details of the structure of matter and do not even depend on the existence of atoms. In the present chapter we will study the kinetic theory of gases and in Chapter 21 we will study the laws of thermodynamics and their consequences.

19.1 The Ideal-Gas Law

The pressure, volume, and temperature of a gas obey some simple laws. Before we state these laws, let us recall the definition of pressure (see Section 18.3). Imagine that the gas is divided into small adjacent cubical volumes. The pressure is the force that one of these cubes exerts on an adjacent cube (or on an adjacent wall) divided by the area of one face of the cube; i.e., the pressure is the force per unit area. As we know from Section 18.4, the pressure is the same throughout the entire volume of a container of gas (within a container of gas, gravity causes a small decrease of pressure from the bottom to the top, but this decrease of pressure can usually be ignored).

Consider now a given amount of gas, say, n moles of gas.[1] Experiments show that — to a good approximation — the pressure p, the volume V, and the temperature T of the gas are related by the **ideal-gas law**

Ideal-gas law

$$\boxed{pV = nRT} \tag{1}$$

Universal gas constant R

Here R is the **universal gas constant** with the value

$$\boxed{R = 8.314 \text{ J/K}} \tag{2}$$

William Thomson, Lord Kelvin, *1824–1907, British physicist and engineer, professor at the University of Glasgow. Besides inventing the absolute temperature scale, he made many other contributions to the theory of heat. He was first to state the principle of dissipation of energy incorporated in the Second Law of Thermodynamics. In engineering, Kelvin made crucial improvements in the design of cables for submarine telegraphy and in the design of the mariner's compass.*

The temperature in Eq. (1) is measured on the **absolute temperature scale** and the unit of temperature is the **kelvin** or the **degree absolute,** abbreviated K. We have not previously given the definition of temperature and of the unit of temperature because Eq. (1) plays a dual role: it is a law of physics and also serves for the definition of temperature. This is by now a familiar story — in Chapter 5 we already came across laws that play such a dual role.

We will give the details concerning the definition of temperature in the next section. For now it will suffice to note that the zero of temperature is the absolute zero, $T = 0$ K. According to Eq. (1), the pressure of the gas vanishes at this point. Actually, the gas will liquefy or even solidify before the absolute zero of temperature can be reached; when this happens, Eq. (1) becomes inapplicable. We also note that the freezing point of water corresponds to a temperature of 273.15 K and the boiling point of water corresponds to 373.15 K. Hence there is an interval of exactly 100 K between the freezing and boiling points.

[1] As we saw in Chapter 1, 1 mole of any chemical element (or chemical compound) is that amount of matter that contains exactly as many atoms (or molecules) as there are atoms in 12 g of carbon. The "atomic mass" of a chemical element (or the "molecular mass" of a compound) is the mass of 1 mole. Thus, according to the table of atomic masses (Appendix 9), 1 mole of C has a mass of 12.0 g, 1 mole of O_2 has a mass of 32.0 g, 1 mole of N_2 has a mass of 28.0 g, etc.

The ideal-gas law is a simple relation between the macroscopic parameters that characterize a gas. At normal densities and pressures, real gases obey this law quite well; but if a gas is compressed to an excessively high density, then its behavior will deviate from this law. For instance, at 1-atm pressure the volume of 1 mole of oxygen gas is only about 0.13% smaller than predicted by the ideal-gas law, but at 20-atm pressure the volume is about 2.3% smaller than predicted. In the following we will always take it for granted that the pressures and densities are low enough so that deviations from Eq. (1) can be neglected. An **ideal gas** is a gas that obeys Eq. (1) *exactly*. The ideal gas is a limiting case of a real gas when the density and the pressure of the latter tend to zero. The ideal gas may be thought of as consisting of atoms of infinitesimal size which only exert forces over an infinitesimal range; the atoms then exert no force at all, except for the instantaneous impulsive forces during collisions with the walls of the container holding the gas.

Ideal gas

The ideal-gas law incorporates two laws: **Boyle's law** and **Gay-Lussac's law** (or **Charles' law**). Boyle's law asserts that if the temperature is constant, then the product of pressure and volume must remain constant as a given amount of gas is compressed or expanded,

Boyle's law and Gay-Lussac's law

$$pV = [\text{constant}] \qquad \text{for } T = [\text{constant}] \tag{3}$$

Gay-Lussac's law asserts that if the pressure is constant, the ratio of volume to temperature remains constant as a given amount of gas is heated or cooled,

$$V/T = [\text{constant}] \qquad \text{for } p = [\text{constant}] \tag{4}$$

The relation given by Eq. (3) can be tested experimentally by placing a sample of gas in a cylinder with a movable piston surrounded by some substance at a constant temperature (say, a water bath; see Figure 19.1). The relation of Eq. (4) can be tested with a similar cylinder–piston arrangement with a fixed weight mounted on the piston and a heat source below the piston (Figure 19.2). Together, the experimental tests of Eqs. (3) and (4) amount to a test of the ideal-gas law.

Robert Boyle, *1627–1691, English experimental physicist. He was one of the younger sons of the first earl of Cork, and he used the financial means at his disposal to set up laboratories of his own at Oxford and at London. Boyle invented a new air pump, with which he performed the experiments on gases that led to discovery of the law named after him.*

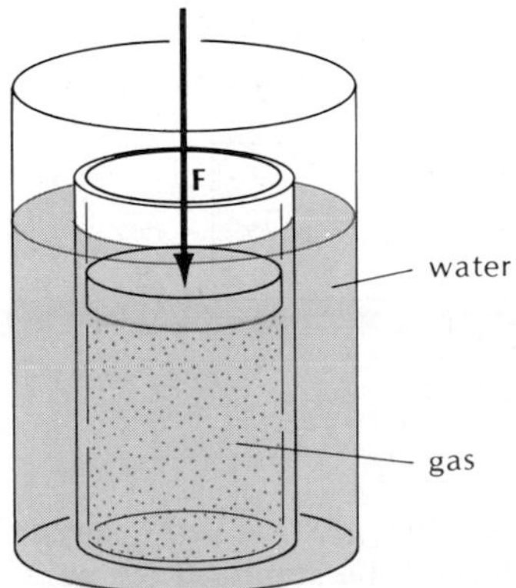

Fig. 19.1 Compression of a gas by a force applied to the piston. The container with the gas is surrounded by a water bath that keeps the temperature constant.

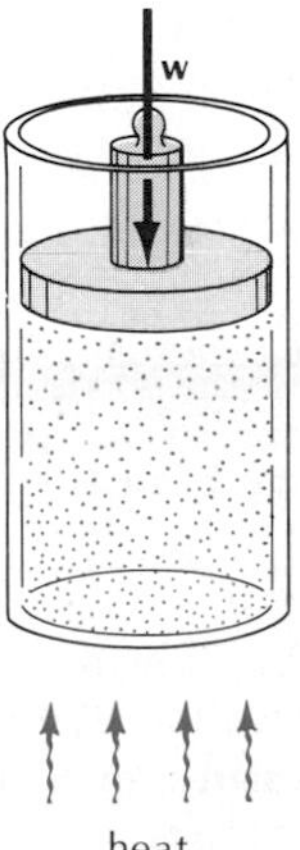

Fig. 19.2 Expansion of a gas kept at constant pressure by a weight on the piston. The gas is being heated.

EXAMPLE 1. Early in the morning at the beginning of a trip, the tires of an automobile are cold (280 K) and their air is at a pressure of 3.0 atm. Later in the day after a long trip on hot pavements, the tires are hot (330 K). What is the pressure? Assume that the volume of the tires remains constant.

SOLUTION: At constant volume the pressure is proportional to the temperature [see Eq. (1)]. Hence

$$p_2 = p_1 \times \frac{T_2}{T_1} = 3.0 \text{ atm} \times \frac{330 \text{ K}}{280 \text{ K}} = 3.5 \text{ atm}$$

Note that pressure gauges for automobile tires are commonly calibrated to read **overpressure,** i.e., the excess above atmospheric pressure. Thus, the pressure gauge would read 2.0 atm, or 29 lb/in.^{2}, in the morning; and 2.5 atm, or 37 lb/in.^{2}, later in the day (if one assumes that the atmospheric pressure is constant at 1 atm).

Equation (1) can also be written in terms of the number of molecules. If the number of moles is n, then the number of molecules is

$$N = N_A n \tag{5}$$

where N_A is **Avogadro's number,**[2]

Avogadro's number, N_A

$$\boxed{N_A = 6.0220 \times 10^{23} \text{ molecules/mole}} \tag{6}$$

With this, Eq. (1) becomes

$$pV = \frac{N}{N_A} RT$$

or

Ideal-gas law

$$\boxed{pV = NkT} \tag{7}$$

where

Boltzmann's constant, k

$$\boxed{k = \frac{R}{N_A} = \frac{8.314 \text{ J/K}}{6.022 \times 10^{23}} = 1.381 \times 10^{-23} \text{ J/K}} \tag{8}$$

The quantity k is called **Boltzmann's constant.**[3]

Standard temperature and pressure (STP)

EXAMPLE 2. The mean molecular mass of air[4] is 29.0 g. What is the density of air (in kg/m^3) at 273 K and 1 atm pressure, called the **standard temperature and pressure, or STP**?

[2] See Appendix 8 for a more precise value of this constant.

[3] See Appendix 8 for a more precise value of this constant.

[4] Air consists, on the average, of 75.54% nitrogen, 23.1% oxygen, and 1.3% argon by mass.

SOLUTION: In metric units, 1 atm is 1.01×10^5 N/m². Hence the number of molecules in 1 m³ is

$$N = \frac{pV}{kT} = \frac{1.01 \times 10^5 \text{ N/m}^2 \times 1 \text{ m}^3}{1.38 \times 10^{-23} \text{ J/K} \times 273 \text{ K}} = 2.68 \times 10^{25}$$

This is $2.68 \times 10^{25}/6.02 \times 10^{23}$ moles, or 44.5 moles. Since the mass per mole is 29.0 g, the density is

$$\rho = 44.5 \times 29.0 \times 10^{-3} \text{ kg/m}^3 = 1.29 \text{ kg/m}^3$$

19.2 The Temperature Scale

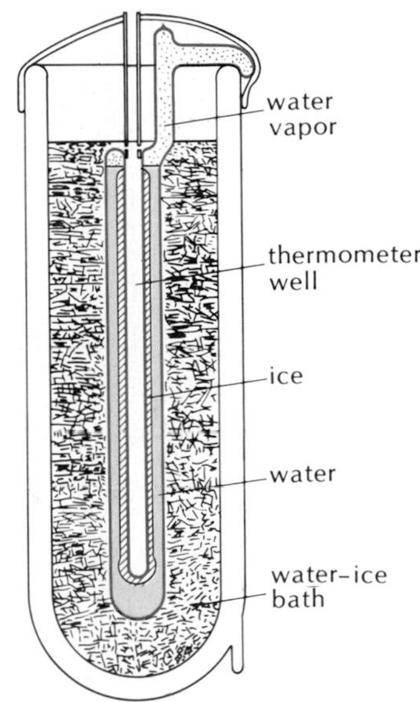

Fig. 19.3 Triple-point cell of the National Bureau of Standards. The inner tube (red) contains water, water vapor, and ice in equilibrium.

To use Eq. (1) for the definition of temperature, we take a fixed amount of some gas (say, helium) and place it in an airtight, nonexpanding container (say, a Pyrex glass bulb). According to Eq. (1), the pressure for a gas kept in such a constant volume is directly proportional to the temperature. Thus, a simple measurement of pressure gives us the temperature.

To calibrate the scale of this thermometer, we must choose a standard reference temperature. The internationally accepted standard is the temperature of the **triple point of water,** i.e., the temperature at which water, ice, and water vapor coexist in a closed vessel. Figure 19.3 shows a triple-point cell used to achieve the standard temperature. This standard temperature has been assigned the value 273.16 kelvin, or 273.16 K. If the bulb of the gas thermometer is placed in thermal contact with this cell so that it attains a temperature of 273.16 K, it will read some pressure p_{tri}. At any other temperature T, it will read a pressure p which is greater or smaller by a factor p/p_{tri}. Hence the temperature T may be expressed as

Absolute temperature scale

$$T = 273.16 \text{ K} \times \frac{p}{p_{\text{tri}}} \tag{9}$$

This equation calibrates our thermometer. The temperature scale defined in this way is called the **ideal-gas temperature scale** or the **absolute temperature scale.**

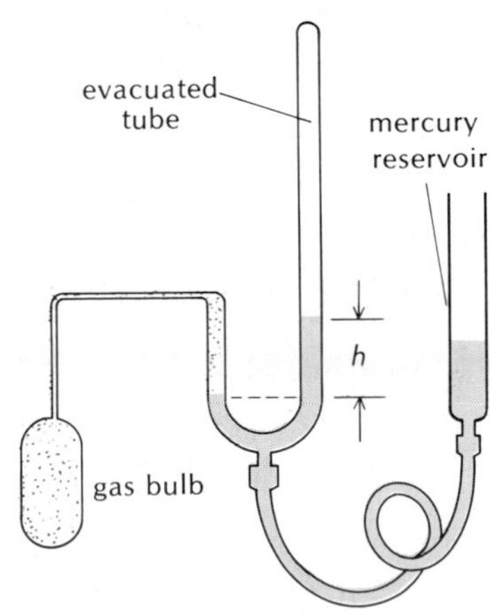

Fig. 19.4 Constant-volume gas thermometer.

When connecting a pressure gauge to the bulb of gas, we must take special precautions to ensure that the operation of the pressure gauge does not alter the volume available to the gas. Figure 19.4 shows an arrangement that will serve our purposes; this device is called a **constant-volume gas thermometer.** The pressure gauge used in this thermometer consists of a closed-tube manometer; one branch of the manometer is connected to the bulb of gas and the other branch consists of a closed, evacuated tube. The difference h in the heights of the levels of mercury in these two branches is proportional to the pressure of the gas. The manometer is also connected to a mercury reservoir. During the operation of the thermometer, this reservoir must be raised or lowered so that the level of mercury in the left branch of the manometer tube always remains at a constant height; this keeps the gas in the bulb at a constant volume. The bulb of this thermometer may be put in contact with any substance whose temperature we wish to measure and the pressure registered by the manometer then gives us the absolute temperature.

Table 19.1 lists some examples of temperatures of diverse bodies.

For everyday and industrial use, the ideal-gas thermometer is somewhat inconvenient and is often replaced by mercury-bulb thermometers, bimetallic strips, resistance thermometers, or thermocouples (Figures 19.5–19.8). These must be calibrated in terms of the ideal-gas thermometer if they are to read absolute temperature.

Although the absolute temperature scale is the only scale of fundamental significance, several other temperature scales are in practical use. The **Celsius scale** (formerly known as the *centigrade scale*) is shifted by 273.15 degrees relative to the absolute scale,

$$\boxed{T_C = T - 273.15} \tag{10}$$

Note that on the Celsius scale, absolute zero is at $-273.15°C$. The triple point of water, is then at 0.01 °C, and the boiling point at 100 °C.

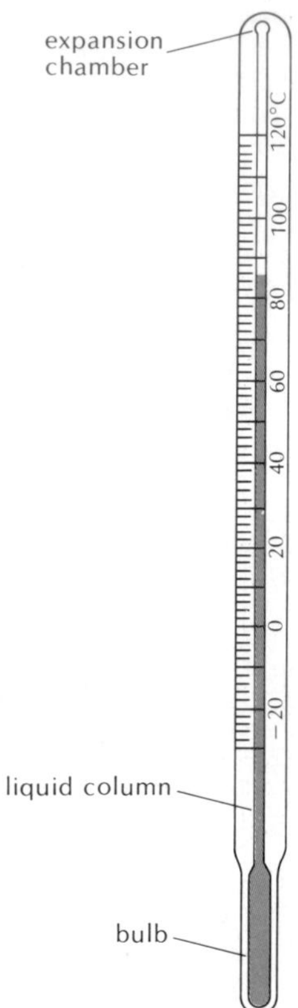

Fig. 19.5 Mercury-bulb thermometer. The thermal expansion of the mercury in the bulb causes it to rise in the capillary tube, indicating the temperature.

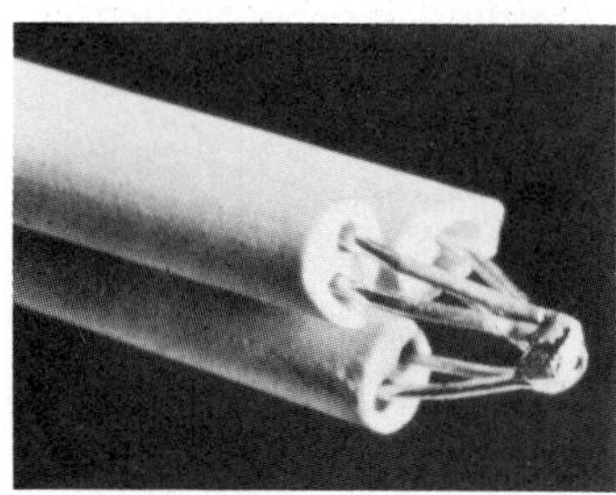

Fig. 19.8 Thermocouple thermometer. Each of the three thermocouples shown here consists of one wire of platinum and one of platinum–rhodium alloy joined at their ends. If the other ends of the wires are connected to an external electric circuit, they will generate a (weak) electric current, depending on the temperature.

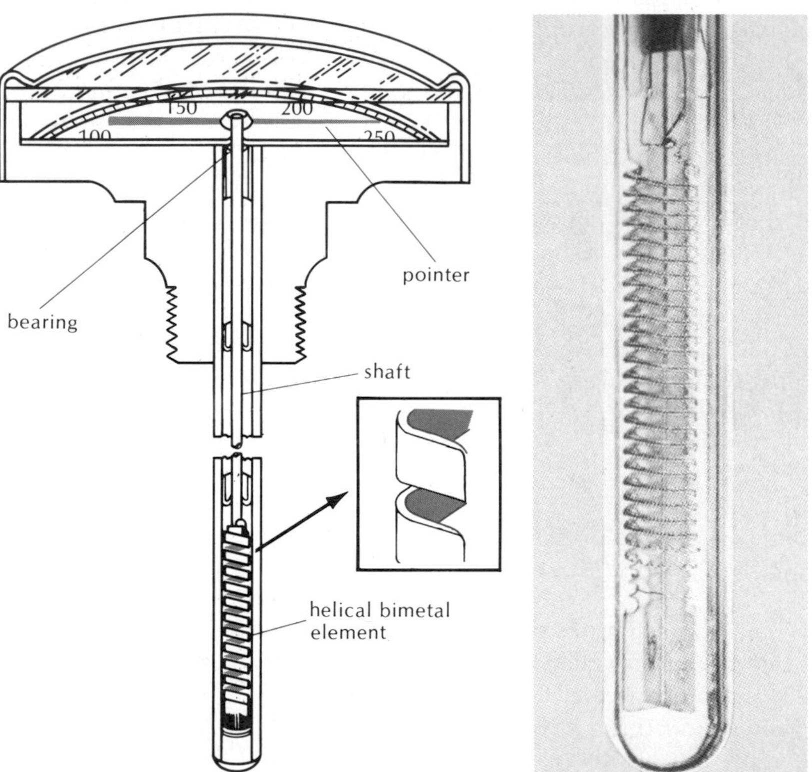

Fig. 19.6 Bimetallic-strip thermometer. The helix consists of two joined bands of different metals. With a change of temperature, the bands expand or contract by different amounts, which tends to coil or uncoil the helix, and turns the pointer.

Fig. 19.7 Platinum-resistance thermometer. The resistance that the fine coil of platinum wire offers to an electric current serves as indicator of temperature.

The **Fahrenheit scale** is shifted relative to the Celsius scale and, furthermore, uses degrees of smaller size, each degree Farenheit corresponding to $\frac{5}{9}$ degree Celsius:

$$\boxed{T_F = \tfrac{9}{5}T_C + 32} \tag{11}$$

On this scale the freezing point of water is at 32°F and the boiling point at 212°F. Figure 19.9 can be used for rough conversions between the Fahrenheit and Celsius scales.

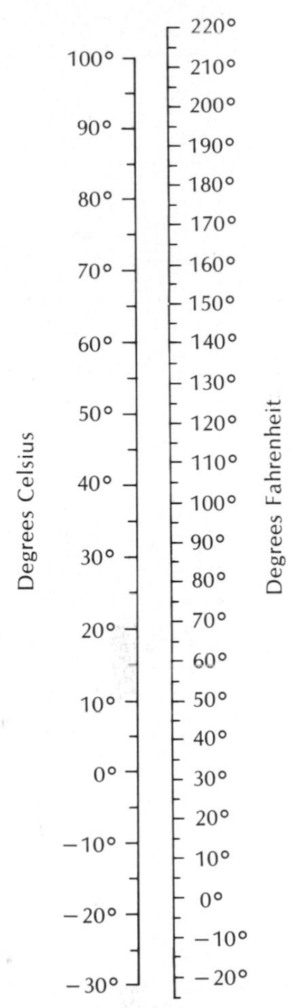

Fig. 19.9 The correspondence between the Fahrenheit scale and the Celsius scale.

Table 19.1 SOME TEMPERATURES

	Absolute temperature	*Celsius temperature*
Interior of hottest stars	10^9 K	10^9 °C
Center of H-bomb explosion	10^8 K	10^8 °C
Highest temperature attained in laboratory (plasma)	6×10^7 K	6×10^7 °C
Center of Sun	1.5×10^7 K	1.2×10^7 °C
Surface of Sun	4.5×10^3 K	4.2×10^3 °C
Center of Earth	4×10^3 K	3.7×10^3 °C
Acetylene flame	2.9×10^3 K	2.6×10^3 °C
Melting of iron	1.8×10^3 K	1.5×10^3 °C
Melting of lead	6.0×10^2 K	3.3×10^2 °C
Boiling of water	373 K	100°C
Human body	310 K	37°C
Surface of Earth (average)	287 K	14°C
Freezing of water	273 K	0°C
Liquefaction of nitrogen	77 K	−196°C
Liquefaction of hydrogen	20 K	−253°C
Liquefaction of helium	4.2 K	−269°C
Interstellar space	3 K	−270°C
Lowest temperature attained in laboratory	5×10^{-8} K	≅−273.15°C

19.3 Kinetic Pressure

The pressure of a gas against the walls of its container is due to the impact of the molecules on the walls. We will now calculate this pressure by means of kinetic theory by considering the average motion of the molecules of gas. We will assume that the container is a cube of side L, that the gas molecules only collide with the walls but not with each other, and that these collisions are elastic. These assumptions are not necessary, but they simplify the calculations.

Figure 19.10 shows the container filled with gas molecules. The motion of each molecule can be resolved into x, y, and z components. Consider one molecule and its motion in the x direction. The component of velocity in this direction is v_x and the magnitude of this velocity remains constant, since the collisions with the wall are elastic. The time that the molecule takes to move from the face at $x = 0$ to $x = L$ and back to $x = 0$ is

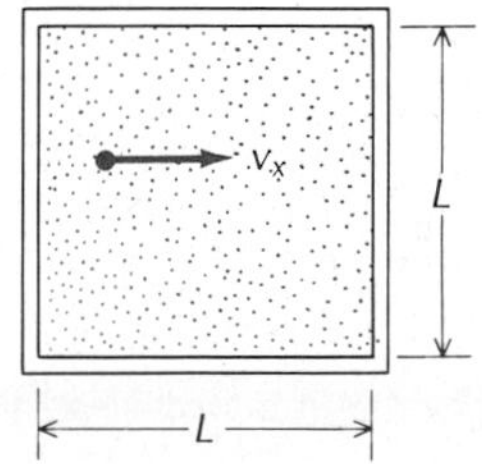

Fig. 19.10 Gas molecules in a container.

$$\Delta t = 2L/v_x$$

This is therefore the time between one collision with the face at $x = 0$ and the next collision with the same face. When the molecule strikes the face, its x velocity is reversed from $-v_x$ to $+v_x$ (Figure 19.10). Hence during each collision at $x = 0$, the x momentum changes by

$$\Delta p_x = 2mv_x \tag{12}$$

where m is the mass of the molecule. The average rate at which the molecule transfers momentum to the face at $x = 0$ is then

$$\frac{\Delta p_x}{\Delta t} = \frac{2mv_x}{2L/v_x} = \frac{mv_x^2}{L} \tag{13}$$

This is the average force that the impacts of this one molecule exert on the wall. To find the total force exerted by the impacts of all the N molecules, we must multiply the force given in Eq. (13) by N; and to find the pressure, we must divide by the area L^2 of the wall. This gives a pressure

$$p = \frac{N}{L^2}\frac{mv_x^2}{L} \tag{14}$$

or, in terms of the volume $V = L^3$,

$$p = \frac{Nmv_x^2}{V} \tag{15}$$

In this calculation we have made the implicit assumption that all the molecules have the same velocity. This is, of course, not true — the molecules of the gas have a distribution of velocities. To account for this, we must replace the force in Eq. (13) due to one given molecule by the average over all the molecules. Consequently, we must replace v_x^2 by an average over all the molecules in the container. We will designate this average by $\overline{v_x^2}$. Equation (15) then becomes

$$p = \frac{Nm\overline{v_x^2}}{V} \tag{16}$$

To proceed further, we note that on the average, molecules are just as likely to move in the x, y, or z direction. Hence the average values of $\overline{v_x^2}$, $\overline{v_y^2}$, and $\overline{v_z^2}$ are equal,

$$\overline{v_x^2} = \overline{v_y^2} = \overline{v_z^2} \tag{17}$$

The sum of the squares of the components is the square of the magnitude of the velocity,

$$\overline{v_x^2} + \overline{v_y^2} + \overline{v_z^2} = \overline{v^2} \tag{18}$$

and therefore each of the terms on the left side of Eq. (18) must equal $\frac{1}{3}\overline{v^2}$. We can then write Eq. (16) as

$$pV = \frac{Nm\overline{v^2}}{3} \tag{19}$$

This expresses the pressure in terms of the average square of the speed of the molecules.

Let us now compare this result with the ideal-gas law [Eq. (7)],

$$pV = NkT \tag{20}$$

Obviously, the right sides of Eqs. (19) and (20) must be equal,

$$\frac{m\overline{v^2}}{3} = kT \tag{21}$$

This shows that the average speed is a simple function of the temperature. The square root of $\overline{v^2}$ is called the **root-mean-square speed,** or rms speed, and it is usually designated by v_{rms}. If we take the square root of both sides of Eq. (21), we find

$$v_{\mathrm{rms}} = \sqrt{\overline{v^2}} = \sqrt{\frac{3kT}{m}} \tag{22}$$

Root-mean-square speed

This root-mean-square speed may be regarded as the typical speed of the molecules of the gas. Incidentally, there are other ways of calculating a typical speed; for example, we may want to know the average of all the molecular speeds or the most probable of all the molecular speeds. These other typical speeds turn out to be approximately the same as v_{rms}, but their calculation requires some further knowledge of the distribution of molecular speeds.

EXAMPLE 3. What is the root-mean-square speed of nitrogen molecules in air at 0°C? Of oxygen molecules?

SOLUTION: The molecular mass of N_2 molecules is 28.0 g. Hence the mass of one molecule is

$$m = \frac{28.0\ \mathrm{g}}{N_A} = \frac{28.0\ \mathrm{g}}{6.02 \times 10^{23}} = 4.65 \times 10^{-26}\ \mathrm{kg}$$

and Eq. (22) yields

$$v_{\mathrm{rms}} = \sqrt{\frac{3kT}{m}} = \sqrt{\frac{3 \times 1.38 \times 10^{-23}\ \mathrm{J/K} \times 273\ \mathrm{K}}{4.65 \times 10^{-26}\ \mathrm{kg}}}$$

$$= 493\ \mathrm{m/s}$$

Likewise, for O_2 molecules the molecular mass is 32.0 g, the mass of one molecule is 5.32×10^{-26} kg, and v_{rms} is 461 m/s.

19.4 The Energy of an Ideal Gas

From Eq. (21) we find that the average kinetic energy of a molecule of ideal gas is

$$\tfrac{1}{2}m\overline{v^2} = \tfrac{3}{2}kT \tag{23}$$

If we multiply this by the total number of molecules, we will obtain the total translational kinetic energy of all the molecules. In an ideal gas, the molecules exert no forces on one another (they do not collide) and hence there is no intermolecular potential energy. The kinetic energy is then the total energy,

$$E = K = \tfrac{3}{2}NkT \tag{24}$$

Kinetic energy of an ideal monatomic gas

This formula tells us how much energy is stored in the thermal motion of the gas. Note that we have assumed that the molecules behave like pointlike particles — each molecule has translational kinetic energy, but no internal energy. Monatomic gases, such as helium, argon, and krypton behave in this way.

EXAMPLE 4. What is the thermal kinetic energy in 1 kg of He gas at 0°C? How much extra energy must be supplied to this gas to increase its temperature to 60°C (at a constant volume)?

SOLUTION: Since $Nk = nR$ [see Eqs. (5) and (8)], we can also write Eq. (24) as

$$E = \tfrac{3}{2}nRT \qquad (25)$$

The number of moles in 1 kg of He is

$$n = \frac{1 \text{ kg}}{4.0 \text{ g/mole}} = 250 \text{ moles}$$

Hence

$$E = \tfrac{3}{2}nRT = \tfrac{3}{2} \times 250 \times 8.3 \text{ J/K} \times 273 \text{ K} = 8.5 \times 10^5 \text{ J}$$

The extra energy needed to increase the temperature to 60°C is

$$\Delta E = \tfrac{3}{2}nR\,\Delta T = \tfrac{3}{2} \times 250 \times 8.3 \text{ J/K} \times 60 \text{ K} = 1.9 \times 10^5 \text{ J}$$

Diatomic gases, such as N_2 and O_2, store additional amounts of energy in the internal motions of the atoms within each molecule. The molecules of these gases may be regarded as two pointlike particles rigidly connected together (a dumbbell; see Figure 19.11). If such a molecule collides with another molecule or with the wall of the container, it will usually start rotating about its center of mass. We therefore expect that, on the average, an appreciable fraction of the energy of the gas will be in the form of this kind of rotational kinetic energy.

Fig. 19.11 A diatomic molecule represented as two pointlike particles joined by a rod.

The molecule may rotate about either of the two axes through the center of mass perpendicular to the line joining the atoms (Figure 19.11). If the moments of inertia about these axes are I_1 and I_2, and the corresponding angular velocities are ω_1 and ω_2, then the kinetic energy for these two rotations is

$$\tfrac{1}{2}I_1\omega_1^2 + \tfrac{1}{2}I_2\omega_2^2 \qquad (26)$$

The average kinetic energy is

$$\tfrac{1}{2}I_1\overline{\omega_1^2} + \tfrac{1}{2}I_2\overline{\omega_2^2} \qquad (27)$$

where, as above, the bars denote the average over all the molecules of the gas.

To find out the value of these average rotational energies, let us return to Eq. (23) and write it in terms of the *x*, *y*, and *z* components of velocity:

$$\tfrac{1}{2}m\overline{v_x^2} + \tfrac{1}{2}m\overline{v_y^2} + \tfrac{1}{2}m\overline{v_z^2} = \tfrac{3}{2}kT \qquad (28)$$

We know that the average *x*, *y*, and *z* speeds are equal. Hence Eq. (28) asserts that the kinetic energy for each component of the motion has the value $\tfrac{1}{2}kT$,

$$\tfrac{1}{2}m\overline{v_x^2} = \tfrac{1}{2}kT, \qquad \tfrac{1}{2}m\overline{v_y^2} = \tfrac{1}{2}kT, \qquad \tfrac{1}{2}m\overline{v_z^2} = \tfrac{1}{2}kT, \qquad (29)$$

It turns out that this is true not only for the components of translational motion, but also for rotational motion. The general result is known as the *equipartition theorem:*

Equipartition theorem

> *Each translational or rotational component of the random thermal motion of a molecule has an average kinetic energy of* $\frac{1}{2}kT$.

We will not prove this theorem, but we will make use of it.[5]

According to this theorem, each of the terms in Eq. (27) has a value $\frac{1}{2}kT$. The total average rotational kinetic energy is then kT and when we add this to the translational kinetic energy $\frac{3}{2}kT$ we obtain a total kinetic energy

$$kT + \tfrac{3}{2}kT = \tfrac{5}{2}kT \tag{30}$$

for one molecule. The energy of all the molecules taken together is then

$$E = \tfrac{5}{2}NkT \tag{31}$$

Ludwig Boltzmann, *1844–1906, Austrian theoretical physicist. He held positions as professor of physics at Munich, Vienna, and Leipzig, and made crucial contributions in the kinetic theory of gases and in statistical mechanics.*

Note that in this calculation we have ignored the possibility of rotation about the longitudinal axis of the molecule. This means we have ignored the rotation of the atoms about an axis through them, just as we have ignored this kind of rotation of the atoms of a monatomic gas. Furthermore, we have ignored the vibrational motion of the atoms of the diatomic molecule. The interatomic forces do not really hold these atoms rigidly; rather, the forces act somewhat like springs (see Example 14.3) and permit a restricted back-and-forth vibration. The reason why we ignored these motions in our calculation lies beyond the realm of classical physics; it lies in the realm of quantum physics. There we find that the rotation of atoms about their own axis and the vibration of atoms in a molecule do not occur unless the temperature is rather high, 400°C or more. As we will see in Section 20.5, the results of Eqs. (24) and (31) actually agree pretty well with experiments, provided that we do not exceed this temperature limit.

SUMMARY

Ideal-gas law: $pV = nRT$ $\quad R = 8.31$ J/K
$pV = NkT$ $\quad k = 1.38 \times 10^{-23}$ J/K

Temperature scales: Absolute: T
Celsius: $T_C = T - 273.15$
Fahrenheit: $T_F = \frac{9}{5}T_C + 32$

Root-mean-square speed: $v_{rms} = \sqrt{3kT/m}$

Kinetic energy of ideal monatomic gas: $K = \frac{3}{2}NkT$

Equipartition theorem: Each translational or rotational component of the random thermal motion of a molecule has an average kinetic energy of $\frac{1}{2}kT$.

[5] Each translational or rotational component of the motion is called a **degree of freedom.** With this terminology, the equipartition theorem can be stated as: Each degree of freedom has an average energy of $\frac{1}{2}kT$.

QUESTIONS

1. Why do meteorologists usually measure the temperature in the shade rather than in the sun?

2. Why are there no negative temperatures on the absolute temperature scale?

3. The temperature of the ionized gas in the ionosphere of the Earth is about 2000 K, but the density of this gas is extremely low, only about 10^5 gas particles per cubic centimeter. If you were to place an ordinary mercury thermometer in the ionosphere, would it register 2000 K? Would it melt?

4. The temperature of the intergalactic space is 3 K. How can empty space have a temperature?

5. At the airport of La Paz, Bolivia, one of the highest in the world, pilots find it preferable to take off early in the morning or late at night, when the air is very cold. Why?

6. If you release a rubber balloon filled with helium, it will rise to a height of a few thousand meters and then remain stationary. What determines the height reached? Is there an optimum pressure to which you should inflate the balloon to reach greatest height?

7. How can you use a barometer as an altimeter?

8. Explain why a real gas behaves like an ideal gas at low densities but not at high densities.

9. Helium and neon approach the behavior of an ideal gas more closely than do any other gases. Why would you expect this?

10. If you open a bottle of perfume in one corner of a room, it takes a rather long time for the smell to reach the opposite corner (assuming that there are no air currents in the room). Explain why the smell spreads slowly, even though the typical speeds of perfume molecules are 300–400 m/s.

11. Ultrasound waves of extremely short wavelength cannot propagate in air. Why not?

12. In our calculation of the pressure on the walls of a box (see Section 19.3) we have ignored gravity. If we take gravity into account, the pressure on the bottom of the box will be greater than that at the top. Show that the pressure difference is $p - p_0 = (N/V)\,mgL$. (Hint: When a molecule falls from the top to the bottom, its speed increases according to $v_y^2 - v_{0_y}^2 = 2gL$.)

13. Prove that it is impossible for all of the molecules in a gas to have the same speeds and to keep these speeds forever. (Hint: Consider an elastic collision between two molecules with the same speed. Will the speeds remain constant if the initial lines of motion are not parallel?)

14. If you increase the absolute temperature by a factor of 2, by what factor will you increase the average speed of the molecules of gas?

15. Air consists of a mixture of nitrogen (N_2), oxygen (O_2), and argon (A). Which of these molecules has the highest average speed? The lowest?

16. Equipartition of energy applies not only to atoms and molecules, but also to macroscopic "particles" such as, e.g., golf balls. If so, why do golf balls remain at rest on the ground instead of flying through the air like molecules?

PROBLEMS

Sections 19.1 and 19.2

1. Express the last six temperatures listed in Table 19.1 in terms of degrees Fahrenheit.

2. The hottest place on Earth is Al' Aziziyah, Libya, where the temperature has soared to 136.4°F. The coldest place is Vostok, Antarctica, where the temperature has plunged to −126.9°F. Express these temperatures in degrees Celsius and in kelvin.

3. In summer when the temperature is 30°C, the pressure within an automobile tire is 30 lb/in.^{2}. What will be the pressure within this tire in winter when the temperature is 0°C? Assume that no air is added to the tire and that no air leaks from the tire; assume that the volume of the tire remains constant.

4. What is the number of molecules in 1 cm^3 of ideal gas at 273 K and 1 atm?

5. What is the volume of 1 mole of gas at 273 K and 1-atm pressure?

6. Repeat the calculation of Example 1 assuming that, because of the increase of pressure, the volume of the tire increases by 5%.

7. The lowest pressure attained in a "vacuum" in a laboratory on the Earth is 1×10^{-16} atm. Assuming a temperature of 20°C, what is the number of molecules per cubic centimeter in this vacuum?

8. Clouds of interstellar hydrogen gas have densities of up to 10^{10} atoms/m^3 and temperatures of up to 10^4 K. What is the pressure in such a cloud?

9. The following table gives the pressure and density of the Earth's upper atmosphere as a function of altitude:

Altitude	*Pressure*	*Density*
20,000 m	5.5×10 mbar	8.8×10^{-2} kg/m^3
32,000	8.7	1.2×10^{-2}
53,000	5.8×10^{-1}	7.1×10^{-4}
90,000	1.8×10^{-3}	3.2×10^{-6}

Calculate the temperature at each altitude. The mean molecular mass for air is 29.0 g.

10. What is the density (in kilograms per cubic meter) of helium gas at 1 atm at the temperature of boiling helium liquid (see Table 19.1)?

11. How much does the frequency of middle C (see Table 16.2) played on a flute change when the air temperature drops by 30°C? [Hint: The speed of sound in air is given by Eq. (16.3).]

12. Suppose your pour 10 g of water into a 1-liter jar and seal it tightly. You then place the jar into an oven and heat it to 500°C (a dangerous thing to do!). What will be the pressure of the vaporized water?

13. A scuba diver releases an air bubble of diameter 1.0 cm at a depth of 15 m below the surface of a lake. What will be the diameter of this bubble when it reaches the surface? Assume that the temperature of the bubble remains constant.

14. Consider the automobile tire described in Example 1. If the volume of this tire is 2.5×10^{-2}m^3, what is the mass of air inside it? The mean molecular mass of air is 29.0 g.

15. The helium atom has a volume of about 3×10^{-30} m^3. What fraction of a volume of helium gas at STP is actually occupied by atoms?

16. A carbon dioxide (CO_2) fire extinguisher has an interior volume of 0.10 ft^3. The extinguisher has a weight of 13 lb when empty and a weight of 18 lb when fully loaded with CO_2. At a temperature of 20°C, what is the pressure of CO_2 in the extinguisher?

17. (a) When you heat the air in a house, some air escapes because the pressure inside the house must remain the same as the pressure outside.

Suppose you heat the air from 10°C to 30°C. What fraction of the mass of air originally inside will escape?

(b) If the house were completely airtight, the pressure would have to increase as you heat the house. Suppose that the initial pressure inside the house is 1.0 atm. What is the final pressure? What force does the excess inside pressure exert on a window 1.0 m high and 1.0 m wide? Do you think the window can withstand this force?

18. During the volcanic eruption of Mt. Pelée on the island of Martinique in 1902, a *nuée ardente* (burning cloud) of very hot gas rolled down the side of the volcano and killed the 30,000 inhabitants of Saint-Pierre. The temperature in the cloud has been estimated at 700°C. Assume that this cloud consisted of a gas of high molecular mass. What must have been this molecular mass to make the cloud as dense as, or denser than, the surrounding air (at 20°C)?

19. A typical hot-air balloon has a volume of 77,500 ft^3 and a weight of 1600 lb (including balloon, gondola, four passengers, and a propane tank). If the temperature of the external air is 20°C, what must be the minimum temperature of the internal air in the balloon to achieve lift-off? The density of the external air is 1.20 kg/m^3.

20. A research balloon floats in the upper atmosphere where the pressure (outside and also inside the balloon) is $2.2 \times 10^2 N/m^2$ and the temperature is 0°C. The volume of the balloon is 8.5×10^5 m^3 and it is filled with helium. What payload (including the mass of the fabric but excluding the helium) can this balloon carry?

21. A sunken ship of steel is to be raised by making the upper part of the hull airtight and then pumping compressed air into it while letting the water escape through holes in the bottom. The mass of the ship is 50,000 metric tons and it is at a depth of 60 m. How much compressed air (in kilograms) must be pumped into the ship? The temperature of the air and the water is 15°C.

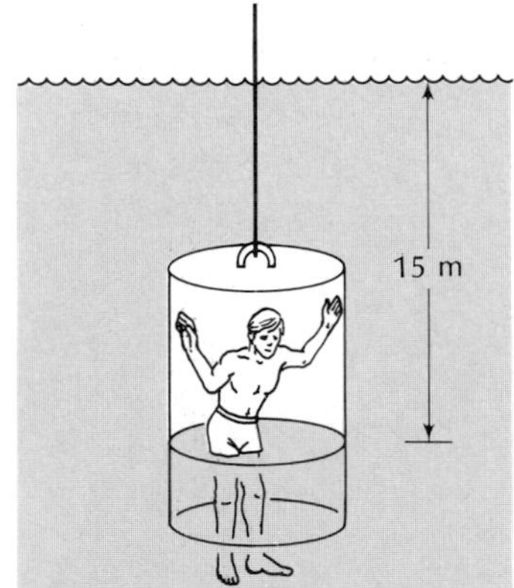

Fig.19.12 Submerged diving bell.

22. A **diving bell** is a cylinder closed at the top and open at the bottom; when it is immersed in the water, any air initially in the cylinder remains trapped in the cylinder. Suppose that such a diving bell, 2 m high and 1.5 m across, is immersed to a depth of 15 m measured from water level to water level (see Figure 19.12).

(a) How high will the water rise within the diving bell?

(b) If compressed air is pumped into the bell, water will be expelled from the bell. How much air (in kilograms) must be pumped into the bell, and at what pressure, to get rid of all the water? Assume that the temperature of the air is 15°C.

23. At high altitudes, pilots and mountain climbers must breathe an enriched mixture containing more oxygen than the standard concentration of 21% found in ordinary air at sea level. At an altitude of 11,000 m, the atmospheric pressure is 0.22 atm. What oxygen concentration is required at this altitude if with each breath the same number of oxygen molecules is to enter the lungs as for ordinary air at sea level?

24. Air is 75.54% nitrogen (N_2), 23.1% oxygen (O_2), and 1.3% argon (A) by mass. From this information and from the molecular masses of N_2, O_2, and A, deduce the mean molecular mass of air.

*25. (a) The gas at the center of the Sun is 38% hydrogen and 62% helium at a temperature of 15.0×10^6 K and a density of 1.48×10^5 kg/m^3. What is the pressure?

(b) The gas at a distance of 20% of the solar radius from the center of the Sun is 71% hydrogen and 29% helium at a temperature of 9.0×10^6 K and a density of 3.6×10^4 kg/m^3. What is the pressure?

*26. Show that if the temperature in the atmosphere is independent of altitude, then the pressure as a function of altitude is

$$p = p_0 e^{-mgh/kT}$$

where m is the average mass per molecule of air. (This formula is applicable only for altitudes less than about 2 km; higher up, the temperature depends on the altitude.)

Section 19.3

27. According to Eq. (16.3) the speed of sound in air is $\sqrt{1.4p/\rho}$.
   (a) Show by means of the ideal-gas law that this expression equals $\sqrt{1.4kT/m}$, where m is the average mass per molecule of air.
   (b) Show that, in terms of the rms speed, the latter expression equals $\sqrt{1.4/3}v_{rms}$ or $0.68v_{rms}$.
   (c) Calculate the speed of sound in air at temperatures of 0°C, 10°C, 20°C, and 30°C.

28. What is the average kinetic energy of an oxygen molecule in air at STP? A nitrogen molecule?

29. What is the rms speed of a helium atom at 0°C? At −269°C?

30. A wind is blowing at 60 km/h. Consider the motion of a molecule of nitrogen in this wind. The random thermal motion has a speed of about 500 m/s; this motion consists of zigzags, each leg of which typically takes 1×10^{-10} s. How many zigzags does the molecule make as it moves 1 cm downwind? What total distance does the molecule travel as it moves 1 cm downwind?

31. The rms speed of nitrogen molecules in air at some temperature is 493 m/s. What is the rms speed of hydrogen molecules in air at the same temperature? The atomic mass of nitrogen is 14.007 g and that of hydrogen is 1.008 g.

32. One method for the separation of the rare isotope ^{235}U (used in nuclear bombs and reactors) from the abundant isotope ^{238}U relies on diffusion through porous membranes. Both isotopes are first made into a gas of uranium hexafluoride (UF_6). The molecules of $^{235}UF_6$ have a higher rms speed and they will diffuse faster through a porous membrane than the molecules of $^{238}UF_6$. The molecular masses of these two molecules are 349 and 352 g, respectively. What is the percent difference between their rms speeds at a given temperature?

*33. An open window in a room measures 1.0 m × 1.5 m. In their random motion, molecules of air move out through the window and in through the window; on the average, the number moving in matches the number moving out.
   (a) *Estimate* the number moving out in 1 second; assume STP conditions.
   (b) If the room measures 5 m × 5 m × 2.5 m, how long does it take for a number of molecules equal to the number in the room to move out through the window?

Section 19.4

34. What is the thermal kinetic energy in 1 kg of oxygen gas at a temperature of 20°C? What fraction of this energy is translational? What fraction is rotational?

35. Assume that air consists of the diatomic gases O_2 and N_2. How much must we increase the thermal energy of 1 kg of air in order to increase its temperature by 1°C?

36. A container is divided into two equal compartments by a partition. One compartment is initially filled with helium at a temperature of 250 K; the other is filled with oxygen at a temperature of 310 K. Both gases are at the same pressure. If we remove the partition and allow the gases to mix, what will be their final temperature?

CHAPTER 20

Heat

Heat is a form of energy: it is the kinetic and potential energy of the random microscopic motion of molecules, atoms, electrons, and other particles. Today, heat is often called **thermal energy.** But until well into the nineteenth century, scientists did not have a clear understanding of the concept of energy and they thought that heat was an invisible, imponderable fluid called the "caloric." The first experiments to give conclusive evidence on the nature of heat were performed by Count Rumford, who showed that the mechanical energy lost in friction is converted into heat. The practical development of steam engines for the industrial generation of mechanical energy from heat motivated a careful examination of the theoretical principles underlying the operation of such engines and led to the discovery of the law of conservation of energy and to the recognition that heat is a form of energy. Steam engines and other heat engines do not create energy; they merely convert thermal energy into mechanical energy.

20.1 Heat as a Form of Energy

Long before physicists recognized that heat is the kinetic and potential energy of the random microscopic motion of atoms, they had defined heat in terms of the temperature changes it produces in a body. The traditional unit of heat is the **calorie** (cal), which is the amount of heat needed to raise the temperature of 1 g of water by 1°C. The kilocalorie is 1000 cal:

calorie

$$1 \text{ kcal} = 10^3 \text{ cal}$$

In the British system the unit of heat is the **British thermal unit** (Btu), which is the heat needed to raise the temperature of 1 lb-mass of water by 1°F. The relationship between these units is

British thermal unit

$$1 \text{ Btu} = 0.252 \text{ kcal}$$

Incidentally, the "calories" marked on packages of food in grocery stores are actually kilocalories, sometimes also called large calories.

The heat necessary to raise the temperature of 1 kg of a material by 1°C is called the **specific heat capacity** or the **specific heat**, usually designated by the symbol c. Thus, water has a specific heat capacity

Specific heat capacity

$$\boxed{c = 1 \text{ kcal/kg} \cdot {}^\circ\text{C}}$$

by definition. The specific heat varies from substance to substance; Table 20.1 lists the specific heats of some common materials. The specific heat also varies with the temperature. For example, the specific heat of water varies by about 1% between 0°C and 100°C, reaching a minimum at 35°C. (This variation must be taken into account for a precise definition of the calorie: a calorie is the heat needed to raise the temperature of 1 g of water from 14.5°C to 15.5°C.) Finally, the specific heat depends on the pressure to which the material is subjected during the heating. All the values in Table 20.1 were obtained at room temperature (20°C) and at a constant pressure of 1.0 atm.

Benjamin Thompson, Count Rumford, *1753–1814, American-British scientist, minister of war and of police in Bavaria. On the basis of experimental observations that he collected while supervising the boring of cannon, Rumford argued against the prevailing view that heat is a substance, and he proposed that heat is nothing but the random microscopic motion of the particles within a body. Robert von Mayer, 1814–1878, German physician and physicist, calculated the mechanical equivalent of heat by comparing the work done on a gas during adiabatic compression with the consequent increase of temperature. Finally, J. P. Joule measured this quantity directly by means of his famous experiment.*

Table 20.1 SOME SPECIFIC HEATS[a]

Substance	c
Aluminum	0.214 kcal/kg · °C
Brass	0.092
Copper	0.092
Iron, steel	0.11
Lead	0.031
Tin	0.054
Silver	0.056
Invar	0.120
Mercury	0.033
Water	1.00
Ice (− 10°C)	0.530
Ethyl alcohol	0.581
Glycol	0.571
Mineral oil	0.5
Glass, thermometer	0.20
Marble	0.21
Granite	0.19
Sea water	0.93

[a]At room temperature and 1 atm, unless otherwise noted.

The values in Table 20.1 give the heat required to increase the temperature of 1 kg of a given substance by 1°C; for a mass m of this substance, the heat ΔQ and the increase of temperature ΔT are related by

Relation between heat and temperature changes

$$\Delta Q = mc\,\Delta T \tag{1}$$

Since heat is a form of energy, it can be transformed into mechanical energy, and vice versa. The transformation of heat into work is accomplished by a steam engine, a steam turbine, or a similar device; we will examine the theory of such heat engines in the next chapter. The transformation of work into heat requires no complicated machinery — any kind of friction will convert work into heat. Because heat is a form of energy, the kilocalorie is a unit of energy and it must be possible to express it in joules. The conversion factor between these units is called the **mechanical equivalent of heat.**

The traditional method for the measurement of this quantity is **Joule's experiment.** A set of falling weights drive a paddle wheel which churns the water in a thermally insulated bucket (Figure 20.1). The churning raises the temperature of the water by a measurable amount, converting a known amount of potential energy into a known amount of heat. The best available experimental results give

Mechanical equivalent of heat

$$1 \text{ cal} = 4.186 \text{ J}$$

for the mechanical equivalent of heat.[1]

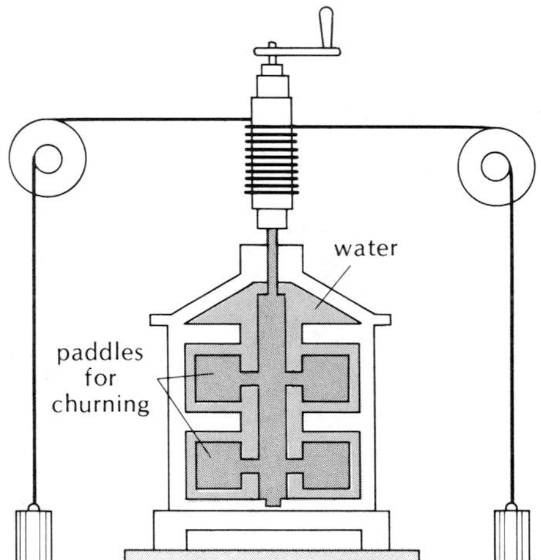

Fig. 20.1 Joule's apparatus.

EXAMPLE 1. When an automobile is braking, the friction between the brake drums and the brake shoes converts translational kinetic energy into heat. If a 2000-kg automobile brakes from 25 m/s (55 mi/h) to 0 m/s, how much heat is generated by the brakes? If each of the four brake drums has a mass of 9.0 kg of iron of specific heat 0.11 kcal/kg · °C, how much does the temperature of the brake drums rise? Assume that all the heat accumulates in the brake drums (there is not enough time for the heat to leak away into the air) and that the heat in all brake drums is the same.

[1] This is the conventional calorie. Several other calories, defined in slightly different ways, are in use, for instance the **"International Table" calorie** (4.1868 J) and the **thermochemical calorie** (4.1840 J), often used by chemists. When reading chemistry textbooks, always check what calorie is being used.

SOLUTION: The initial kinetic energy of the automobile is

$$K = \tfrac{1}{2}mv^2 = \tfrac{1}{2} \times 2000 \text{ kg} \times (25 \text{ m/s})^2 = 6.5 \times 10^5 \text{ J}$$

Expressed in kilocalories, this gives the amount of heat

$$6.2 \times 10^5 \text{ J} \times \frac{1 \text{ kcal}}{4.19 \times 10^3 \text{ J}} = 1.5 \times 10^2 \text{ kcal}$$

The total mass of iron to be heated is 4×9.0 kg. Hence the corresponding temperature increase is [see Eq. (1)]

$$\Delta T = \frac{\Delta Q}{mc} = \frac{1.5 \times 10^2 \text{ kcal}}{4 \times 9.0 \text{ kg} \times 0.11 \text{ kcal/kg} \cdot {}^\circ\text{C}} = 38^\circ\text{C}$$

20.2 Thermal Expansion of Solids and Liquids

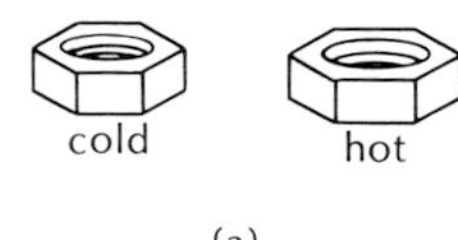

(a)

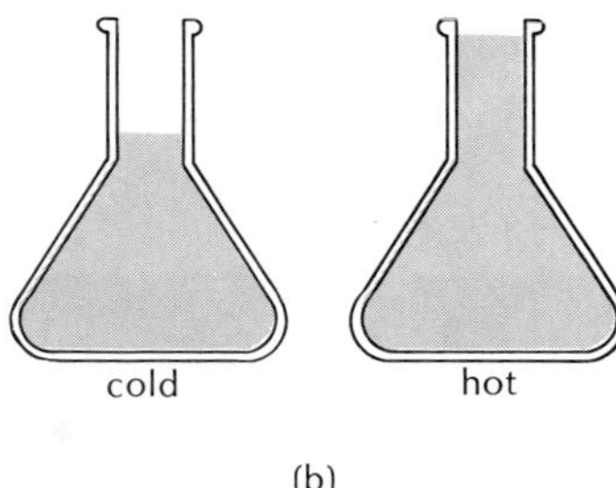

(b)

Fig. 20.2 (a) Thermal expansion of a solid. (b) Thermal expansion of a liquid. The expansion of the flask has been neglected.

As we saw in the preceding chapter, the volume of a given amount of gas will increase with the temperature [if the pressure is held constant; see Eq. (19.4)]. Such an increase in volume with temperature also occurs for solids and liquids; this phenomenon is called **thermal expansion.** However, the thermal expansion of solids and liquids is much less than that of gases. For example, if we raise the temperature of a piece of iron by 100°C, we will only increase its volume by 0.36%. During the thermal expansion the solid retains its shape, but all its dimensions increase in proportion. Figure 20.2a shows the thermal expansion of a piece of metal; for the sake of clarity, the expansion has been exaggerated. An expanding liquid does not, of course, retain its shape; the liquid will merely fill more of the container that holds it. Figure 20.2b shows the thermal expansion of a liquid.

The thermal expansion of a solid can be best described by the increase in the linear dimensions of the solid. The increment in the length is directly proportional to the increment of temperature and to the original length,

$$\boxed{\Delta L = \alpha L \, \Delta T} \qquad (2)$$

Coefficient of linear expansion

The constant of proportionality α is called the **coefficient of linear expansion.** Table 20.2 lists the values of this coefficient for a few materials.

The increment in the volume of the solid is directly proportional to the temperature and to the original volume,

$$\boxed{\Delta V = \beta V \, \Delta T} \qquad (3)$$

Coefficient of cubical expansion

Here the constant of proportionality β is called the **coefficient of cubical expansion.** This coefficient is three times the coefficient of linear expansion,

$$\beta = 3\alpha$$

Table 20.2 COEFFICIENTS OF EXPANSION[a]

Solids	α	*Liquids*	β
Aluminum	$24 \times 10^{-6}/°C$	Alcohol, ethyl (99%)	$1.01 \times 10^{-3}/°C$
Brass	19	Carbon tetrachloride	1.18
Concrete	~12	Ether	1.51
Copper	17	Gasoline	0.95
Glass		Glycerine	0.49
commercial	11	Olive oil	0.68
Pyrex	3.3	Mercury	0.18
Invar	0.9		
Iron, steel	~12		
Lead	29		
Quartz, fused	0.50		

[a]At room temperature.

To see how this relationship comes about, consider a solid in the shape of a cube of edge L and volume $V = L^3$. A small increment ΔL in the length can be treated as a differential and consequently $\Delta V = 3L^2\,\Delta L$, which gives

$$\Delta V = 3L^3(\Delta L/L) = 3V(\alpha\,\Delta T)$$

Comparing this with Eq. (3), we see that, indeed, $\beta = 3\alpha$.

The increment in the volume of a liquid can be described in terms of the same equation [Eq. (3)] as the increment in the volume of a solid. Table 20.2 also gives some values of coefficients of cubical expansion for some liquids.

Water has not been included in this table because its behavior is rather unusual: from 0°C to 3.98°C, the volume *decreases* with temperature, but not uniformly; above 3.98°C, the volume increases with temperature. Figure 20.3 shows the volume of 1 kg of water as a function of the temperature. The strange behavior of the density of water at low temperatures can be traced to the crystal structure of ice. Water molecules have a rather angular shape that prevents a tight fit of these molecules; when they assemble in a solid, they adopt a very complicated crystal structure with large gaps. As a result, ice has a lower density than water — the density of ice is 917 kg/m^3 and the volume of 1 kg of ice is 1091 cm^3. At a temperature slightly above the freezing point, water is liquid, but some of the water molecules already have assembled themselves into microscopic (and ephemeral) ice crystals; these microscopic crystals give the cold water an excess volume.

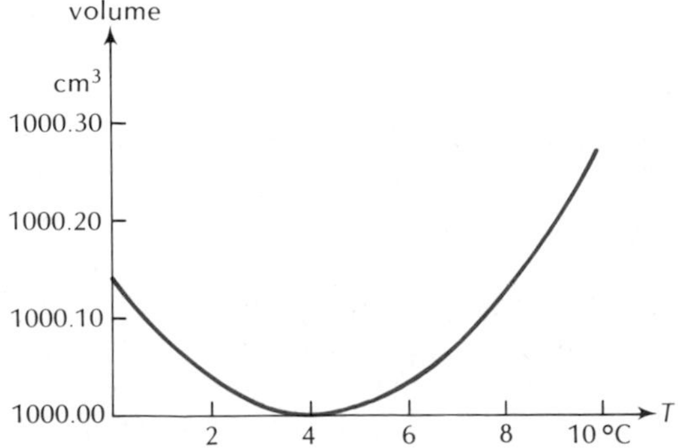

Fig. 20.3 Volume of 1 kg of water as a function of temperature.

The maximum in the density of water at about 4°C has an important consequence for the ecology of lakes. In winter the layer of water on the surface of the lake cools, becomes denser than the lower layer, and sinks to the bottom. This process continues until the temperature of the entire volume of the lake reaches 4°C. Beyond this point, the cooling of the surface layer will make it *less dense* than the lower layers; thus the surface layer stays in place, floating on top of the lake. Ultimately, this surface layer freezes, becoming a sheet of ice while the volume of the lake remains at 4°C. The sheet of ice inhibits the heat loss from the lake, especially if covered with a blanket of snow. Besides, any further heat loss merely causes some thickening of the sheet of ice, without disturbing the deeper layers of water, which remain at a stable temperature of 4°C — fish and other aquatic life can survive the winter in reasonable comfort.

EXAMPLE 2. A glass vessel of volume 200 cm^3 is filled to the rim with mercury. How much of the mercury will overflow the vessel if we raise the temperature by 30°C?

SOLUTION: The volume of mercury will increase by

$$\Delta V_{\mathrm{Hg}} = \beta_{\mathrm{Hg}} V\,\Delta T = 0.18 \times 10^{-3}/°\mathrm{C} \times 200\ \mathrm{cm}^3 \times 30°\mathrm{C} = 1.08\ \mathrm{cm}^3$$

The volume of the glass vessel will increase just as though all of the vessel were filled with glass (solid glass):

$$\Delta V_{\mathrm{glass}} = \beta_{\mathrm{glass}} V\,\Delta T = 3\alpha_{\mathrm{glass}} V\,\Delta T$$

$$= 3 \times 11 \times 10^{-6}/°\mathrm{C} \times 200\ \mathrm{cm}^3 \times 30°\mathrm{C} = 0.20\ \mathrm{cm}^3$$

The difference 1.08 cm^3 − 0.20 cm^3 = 0.88 cm^3 is the volume of mercury that will overflow.

Ordinary thermometers and thermostats make use of thermal expansion to sense changes of temperature. The mercury-bulb thermometer (Figure 19.5) consists of a glass bulb filled with mercury connected to a capillary tube. Thermal expansion makes the mercury overflow into the capillary tube and increases the length of the mercury column; this length indicates the temperature. The bimetallic-strip thermometer (Figure 19.6) consists of two parallel strips of different metals (e.g., aluminum and iron) welded together and curled into a spiral. The differential thermal expansion increases the length of one side of the strip more than that of the other side; this causes the strip to curl up more tightly and rotates the upper end of the spiral relative to the lower end; a pointer attached to the upper end indicates the temperature.

Incidentally, our ability to erect large buildings and other structures out of reinforced concrete hinges on the fortuitous coincidence of the coefficients of expansion of iron and concrete (see Table 20.2). Reinforced concrete consists of iron rods in a concrete matrix. If the coefficients of expansion for these two materials were appreciably different, then the daily and seasonal temperature changes would cause the iron rods to move relative to the concrete — ultimately, the iron rods would work loose and the reinforcement would come to an end.

20.3 Conduction of Heat

If you put one end of an iron poker or rod into the fire and hold the other end in your hand, you will feel the end in your hand gradually become warmer. This is an example of heat transfer by **conduction.** The atoms and electrons in the hot end of the rod have greater kinetic and potential energies than those in other parts of the rod. In random collisions these energetic atoms and electrons share some of their energy with their less energetic neighbors; these in turn share their energy with their neighbors, etc. The result is a gradual diffusion of thermal energy from the hot end of the rod to the cold end.

Conduction

Metals are excellent conductors of heat and also excellent conductors of electricity. The high thermal and electric conductivities of a metal are due to an abundance of "free" electrons within the volume of the metal; these are electrons that have become detached from their atoms — they wander all over the volume of the metal with little hindrance, but they are held back by the surface of the metal. The free electrons behave like the particles of a gas and the metal acts like a bottle holding this gas. Typically, a free electron will move past a few hundred atoms before it suffers a collision. Because the electrons move such fairly large distances between collisions, they can travel very quickly from one end of a metallic rod to the other. Thus the motion of the free electrons transports the thermal energy much more efficiently than does the vibrational motion of the atoms.

Heat flow

We can describe the transfer of heat quantitatively by the **heat flow**, or the **heat current**; this is the amount of heat that passes by some given place on the rod per unit time. We will use the symbol $\Delta Q/\Delta t$ for heat flow. The metric unit of heat flow is the joule per second; however, in practice the preferred unit is the calorie per second or, in the British system, the British thermal unit per second.

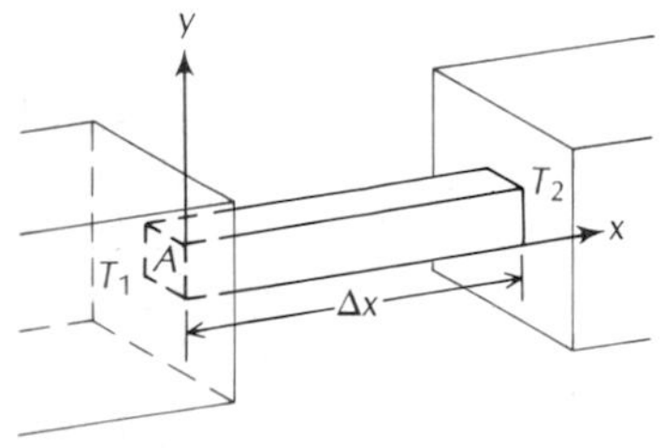

Fig. 20.4 A rod of cross-sectional area A conducting heat from a high-temperature reservoir (T_2) to a low-temperature reservoir (T_1).

Consider a rod of cross-sectional area A and length Δx (Figure 20.4). Assume that the cold end of the rod is kept at a constant temperature T_1 and the hot end at a constant temperature T_2 so that the difference of temperature between the ends is $\Delta T = T_2 - T_1$. If the ends are kept at these constant temperatures for a while, the temperatures at all other points of the rod will settle to final steady values. Under these steady-state conditions, the heat flow along the rod is directly proportional to the temperature difference and to the cross-sectional area, and inversely proportional to the length,

$$\boxed{\frac{\Delta Q}{\Delta t} = -kA\frac{\Delta T}{\Delta x}} \qquad (4)$$

The direction of the heat flow is, of course, from the hot end of the rod toward the cold end; the negative sign in our equation indicates this direction of flow: if the rod lies along the x axis and T increases in the positive x direction ($\Delta T/\Delta x > 0$), then the heat flows in the negative x direction ($\Delta Q/\Delta t < 0$). The constant of proportionality k is called the **thermal conductivity.** Table 20.3 lists values of k for some materials.[2]

Thermal conductivity

[2] The thermal conductivity constant must not be confused with the Boltzmann constant; both are designated with the same letter k, but they are not related.

Table 20.3 SOME THERMAL CONDUCTIVITIES[a]

Substance	*k*
Aluminum	49 cal/(s · m · °C)
Copper	92
Iron, cast	11
Steel	11
Lead	8.3
Silver	97
Ice, 0°C	0.3
Snow, 0°C, compact	0.05
Glass, crown	0.25
Porcelain	0.25
Concrete	0.2
Styrofoam	0.002
Wood, pine	0.03
Down	0.0046

[a] At room temperature, unless otherwise noted.

Equation (4) can be regarded as an empirical law that has been verified by many experiments. Alternatively, this equation can be derived by a detailed study of the process of diffusion of thermal energy along the rod. For the sake of brevity, we will accept the equation as an empirical law and not attempt any derivation. Equation (4) also applies to heat conduction through a wide slab or a plate; such a piece of material can be regarded as a rod of very short length and very large cross-sectional area.

For a rod or a slab, the temperature decreases linearly from the hot end to the cold end; hence the ratio $\Delta T/\Delta x$ equals the temperature gradient dT/dx and we can write our equation for heat flow as

$$\boxed{\frac{\Delta Q}{\Delta t} = -kA\frac{dT}{dx}} \tag{5}$$

Equation of heat conduction

In this form the equation is valid for a rod or a slab with a cross section or a conductivity that depends on position — the equation gives us the heat flow at a given position x along the rod, in terms of the values of k, A, and dT/dx at that position.

EXAMPLE 3. The bottom of a tea kettle consists of a layer of stainless steel 0.050 cm thick welded to a layer of copper 0.030 cm thick. The area of the bottom of the kettle is 300 cm². The copper sits in contact with a hot plate at a temperature of 101.2°C and the steel is covered with boiling water at 100.0°C. What is the rate of heat transfer through the bottom of the kettle from the hot plate to the water?

SOLUTION: The heat flow in the copper and in the steel must be the same, i.e.,

$$k_{\text{Cu}}A\left.\frac{dT}{dx}\right|_{\text{Cu}} = k_{\text{Fe}}\,A\left.\frac{dT}{dx}\right|_{\text{Fe}}$$

Furthermore, the sum of the temperature changes across the copper and the steel must equal the net temperature change of $\Delta T = 1.2°\text{C}$,

$$\left.\frac{dT}{dx}\right|_{\mathrm{Cu}} \Delta x_{\mathrm{Cu}} + \left.\frac{dT}{dx}\right|_{\mathrm{Fe}} \Delta x_{\mathrm{Fe}} = \Delta T$$

By combining these equations, we can solve for $dT/dx|_{\mathrm{Cu}}$:

$$\left.\frac{dT}{dx}\right|_{\mathrm{Cu}} = \frac{\Delta T}{\Delta x_{\mathrm{Cu}} + \Delta x_{\mathrm{Fe}} k_{\mathrm{Cu}}/k_{\mathrm{Fe}}}$$

This gives a heat flow

$$\frac{\Delta Q}{\Delta t} = -k_{\mathrm{Cu}} A \left.\frac{dT}{dx}\right|_{\mathrm{Cu}}$$

$$= \frac{-k_{\mathrm{Cu}} A\, \Delta T}{\Delta x_{\mathrm{Cu}} + \Delta x_{\mathrm{Fe}} k_{\mathrm{Cu}}/k_{\mathrm{Fe}}} = \frac{-A\, \Delta T}{\Delta x_{\mathrm{Cu}}/k_{\mathrm{Cu}} + \Delta x_{\mathrm{Fe}}/k_{\mathrm{Fe}}}$$

$$= \frac{-3.0 \times 10^{-2}\ \mathrm{m}^2 \times 1.2^\circ\mathrm{C}}{3.0 \times 10^{-4}\ \mathrm{m}/92\ \mathrm{cal}/(\mathrm{s}\cdot\mathrm{m}\cdot{}^\circ\mathrm{C}) + 5.0 \times 10^{-4}\ \mathrm{m}/11\ \mathrm{cal}/(\mathrm{s}\cdot\mathrm{m}\cdot{}^\circ\mathrm{C})}$$

$$= -7.4 \times 10^2\ \mathrm{cal/s}$$

Convection and radiation

Besides conduction, there are two other mechanisms of heat transfer: convection and radiation. In **convection** the heat is stored in a moving fluid and it is carried from one place to another by the motion of this fluid. In **radiation** the heat is carried from one place to another by electromagnetic waves (light waves, infrared waves, radio waves, etc.). All three mechanisms of heat transfer are neatly illustrated by the operation of a hot-water heating system in a house. In this system the heat is carried from the boiler to the radiators in the rooms by means of water flowing in pipes (convection); the heat then diffuses through the metallic walls of the radiators (conduction); and finally it spreads from the surface of the radiator into the volume of the room (radiation, supplemented by some convection of air heated by direct contact with the radiator).

Radiation is the only mechanism of heat transfer that can carry heat through a vacuum; for instance, the heat of the Sun reaches the Earth by radiation. We will study thermal radiation in Chapter 40.

20.4 Changes of State

Heat of fusion and of vaporization

Heat absorbed by a body will not only increase the temperature, but it will also bring about a change of state from solid to liquid or from liquid to gas when the body reaches its melting point or its boiling point. While the body is melting or boiling, it absorbs some amount of heat without any increase of temperature. This heat is required to loosen or break the bonds that hold the atoms inside the solid or liquid. The heat absorbed during the transformation of state is called **heat of fusion** or **heat of vaporization,** as the case may be. Table 20.4 lists the heats of fusion and vaporization for a few substances (at a pressure of 1.0 atm).

The values listed in Table 20.4 depend on the pressure. The decrease of the boiling point of water with a decrease of pressure is a phenomenon familiar to people living at high altitude; for instance, in Denver, Colorado, at an altitude of 1600 m, the mean pressure is 0.96 atm and the boiling point of water is 90°C.

Table 20.4 HEATS OF FUSION AND VAPORIZATION[a]

Substance	*Melting point*	*Heat of fusion*	*Boiling point*	*Heat of vaporization*
Water	0°C	79.7 kcal/kg	100°C	539 kcal/kg
Nitrogen	−210	6.2	−196	47.8
Oxygen	−218	3.3	−183	51
Helium	—	—	−269	5.97
Hydrogen	−259	15.0	−253	107
Aluminum	660	95.3	2467	2520
Copper	1083	48.9	2567	1240
Iron	1535	65	2750	1620
Lead	328	6.8	1740	203
Tin	232	14.2	2270	463
Silver	962	23.7	2212	563
Tungsten	3410	44	5660	1180
Mercury	−39	2.7	357	69.7
Carbon dioxide[b]	−79	—	—	138

[a]At a pressure of 1 atm.
[b]Undergoes direct vaporization (sublimation) from solid to gas.

EXAMPLE 4. How many ice cubes must be added to a bowl containing 1 liter of boiling water at 100°C so that the resulting mixture reaches a temperature of 40°C? Assume that each ice cube has a mass of 20 g and that the bowl and the environment do not exchange any heat with the water.

SOLUTION: The heat released by the hot water during cooling from 100°C to 40°C is

$$\Delta Q = 1\ \text{kg} \times 1\ \text{kcal/(kg}\cdot{}^\circ\text{C)} \times 60^\circ\text{C} = 60\ \text{kcal}$$

If the total mass of ice is m, then the heat absorbed by this mass during fusion and subsequent heating from 0°C to 40°C is

$$\Delta Q = m \times 79.7\ \text{kcal/kg} + m \times 1\ \text{kcal/(kg}\cdot{}^\circ\text{C)} \times 40^\circ\text{C}$$

These amounts of heat must be equal,

$$m \times 79.7\ \text{kcal/kg} + m \times 40\ \text{kcal/kg} = 60\ \text{kcal}$$

which yields

$$m = 0.50\ \text{kg}$$

This is 500/20 or 25 ice cubes.

20.5 The Specific Heat of a Gas

If we heat a gas, the increase of temperature causes an increase of the pressure and this tends to bring about an expansion of the gas. The

Specific heat at constant pressure

value of the specific heat of the gas depends on whether the container permits this expansion. If the container is perfectly rigid, the heating proceeds at constant volume (Figure 20.5). For gases it is customary to reckon the specific heat per mole (rather than per kilogram). The **specific heat at constant volume** is designated by C_V; it is the heat needed to raise the temperature of 1 mole of gas by 1°C. If we are dealing with n moles of gas, the heat absorption and the temperature increase at constant volume are related by

$$\Delta Q = nC_V \, \Delta T \tag{6}$$

Specific heat at constant volume

If the container is fitted with a vertical piston, the heating proceeds at a constant pressure determined by the weight of the piston (Figure 20.6). The **specific heat at constant pressure** is designated by C_p. For n moles of gas, the heat absorption and the temperature increase at constant pressure are related by

$$\Delta Q = nC_p \, \Delta T \tag{7}$$

We expect C_p to be larger than C_V because if we supply some amount of heat to the container of Figure 20.6, only part of this heat will go into a temperature increase of the gas; the rest will be converted into work as the gas, expanding, lifts the piston. Let us calculate the difference between the two heat capacities.

At constant volume the gas does no work. Hence all the heat absorbed will go into the energy of the gas,

$$\Delta Q = \Delta E$$

or, according to Eq. (6),

$$nC_V \, \Delta T = \Delta E \tag{8}$$

At constant pressure the gas does work against the piston. Suppose that the piston has an area A and is displaced by a small distance Δx (Figure 20.7). The force of the gas on the piston is pA and the work done by the gas is $pA \, \Delta x$,

Fig. 20.5 A gas kept at constant volume while being heated.

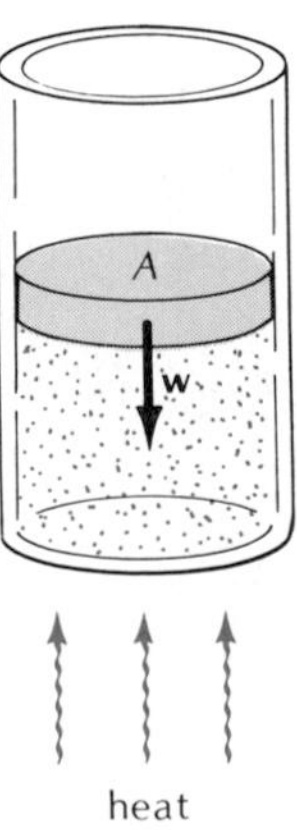

Fig. 20.6 A gas kept at constant pressure while being heated.

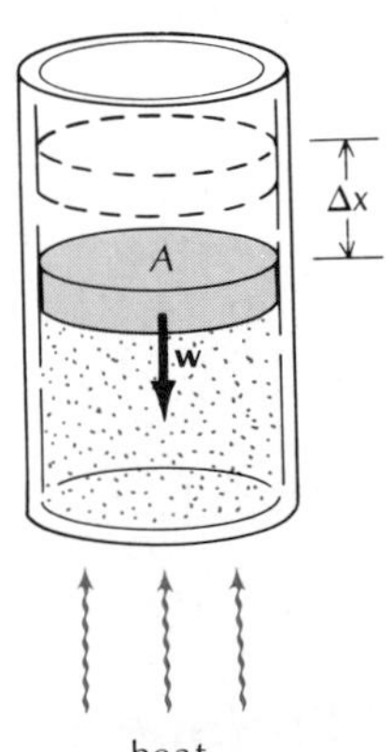

Fig. 20.7 Displacement of the piston by the expanding gas.

$$\Delta W = pA\ \Delta x \tag{9}$$

The product $A\ \Delta x$ is simply the change of volume ΔV. Hence

$$\Delta W = p\ \Delta V \tag{10}$$

The heat absorbed must provide both the energy of the gas and the work done by the gas,

$$\Delta Q = \Delta E + \Delta W = \Delta E + p\ \Delta V \tag{11}$$

so that, according to Eq. (7),

$$nC_p\ \Delta T = \Delta E + p\ \Delta V \tag{12}$$

In an ideal gas the energy E is a function of only the temperature [see Eqs. (19.24) and (19.31)]; consequently, if the temperature increment at constant pressure has the same value as the temperature increment at constant volume, the increase ΔE in the energy must be the same. We can therefore insert Eq. (8) into Eq. (12) and obtain

$$nC_p\ \Delta T = nC_V\ \Delta T + p\ \Delta V \tag{13}$$

Hence

$$C_p - C_V = \frac{p\ \Delta V}{n\ \Delta T} \tag{14}$$

The quantity on the right side can be evaluated by appealing to the ideal-gas law. At constant pressure, Eq. (19.1) gives

$$p\ \Delta V = nR\ \Delta T \tag{15}$$

so that

$$\boxed{C_p - C_V = R} \tag{16}$$

Relation between specific heats

The numerical value of R is 8.31 J/K · mole, or 1.99 cal/K · mole. Hence Eq. (16) shows that C_p is larger than C_V by about 2 cal/K · mole.

Note that although the above general argument does permit us to evaluate the difference $C_p - C_V$, it does not permit us to find the individual values of C_p and C_V. For this we must turn to kinetic theory. According to Eq. (19.24), we then find, for a monatomic gas,

$$\Delta E = \tfrac{3}{2}nR\ \Delta T \tag{17}$$

By the definition of C_V [Eq. (8)], this yields

$$C_V = \tfrac{3}{2}R \tag{18}$$

and by Eq. (16)

$$C_p = C_V + R = \tfrac{5}{2}R \tag{19}$$

Likewise, for a diatomic gas

$$C_V = \tfrac{5}{2}R \tag{20}$$

and

$$C_p = \tfrac{7}{2}R \tag{21}$$

Table 20.5 lists the specific heats of some gases. In the cases of monatomic and diatomic gases, the agreement with Eqs. (18)–(21) is reasonable, but there are some appreciable deviations due to our oversimplifications in kinetic theory. Note that in all cases the difference $C_p - C_V$ agrees quite precisely with Eq. (16).

Table 20.5 SPECIFIC HEATS OF SOME GASES[a]

Gas	C_V	C_p	$C_p - C_V$
Helium (He)	3.00 cal/K · mole	4.98 cal/K · mole	1.98 cal/K · mole
Argon (A)	3.00	5.00	2.00
Nitrogen (N_2)	4.96	6.95	1.99
Oxygen (O_2)	4.96	6.95	1.99
Carbon monoxide (CO)	4.93	6.95	2.02
Carbon dioxide (CO_2)	6.74	8.75	2.01
Methane (CH_4)	6.48	8.49	2.01

[a] At STP.

20.6 The Adiabatic Equation

If an amount of gas at high pressure and temperature is placed in a container fitted with a piston, the gas will push the piston outward and do work on it. As the gas expands, its temperature decreases — the expansion process converts thermal energy into useful mechanical energy. This process is at the core of steam engines, automobile engines, and other heat engines. In this section we will investigate the equations for the expansion of a gas. We will assume that the gas is thermally insulated so that it neither receives heat from its environment nor loses any. The temperature changes of the gas are then entirely due to the work that the gas does on its environment. Such a process occurring without exchange of heat is called **adiabatic.**

Adiabatic process

If the volume of the gas increases by a small amount dV, the work done by the gas on this piston is [see Eq. (10)]

$$dW = p\, dV \tag{22}$$

The heat absorbed is zero; hence the change of energy of the gas is

$$dE = -dW = -p\, dV \tag{23}$$

The change of energy can also be expressed in terms of the temperature change [see Eq. (8)]:

$$dE = nC_V\, dT \tag{24}$$

Combining Eqs. (23) and (24), we have

$$nC_V\, dT = -p\, dV \tag{25}$$

By differentiating the ideal-gas law $nRT = pV$, we find

$$nR\, dT = p\, dV + V\, dp \tag{26}$$

If we eliminate dT between Eqs. (25) and (26), we obtain

$$-Rp\, dV = C_V p\, dV + C_V V\, dp \tag{27}$$

or

$$-\frac{R + C_V}{C_V}\frac{dV}{V} = \frac{dp}{p} \tag{28}$$

Let us write this as

$$-\gamma\frac{dV}{V} = \frac{dp}{p} \tag{29}$$

where

$$\boxed{\gamma = \frac{R + C_V}{C_V} = \frac{C_p}{C_V}} \tag{30}$$

γ

The quantity γ is the ratio of the two kinds of heat capacities; for instance, for an ideal monatomic gas, $\gamma = C_p/C_V = \frac{5}{3}$. To obtain an equation linking p and V, we must integrate Eq. (29). The integrals of each side are natural logarithms with an additive constant of integration:

$$-\gamma \ln V = \ln p + [\text{constant}] \tag{31}$$

Using the properties of logarithms, we can rewrite this as

$$-\ln(pV^\gamma) = [\text{constant}] \tag{32}$$

that is,

$$\boxed{pV^\gamma = [\text{constant}]} \tag{33}$$

Adiabatic equation

This is the equation for the adiabatic expansion of a gas. With this we can readily calculate the drop of temperature that the gas suffers as it expands. Figure 20.8 shows the adiabatic curves $pV^\gamma = [\text{constant}]$ in a p–V diagram for an ideal monatomic gas, with $\gamma = \frac{5}{3}$. For comparison the figure also shows the isothermal curves $T = [\text{constant}]$, or $pV = [\text{constant}]$.

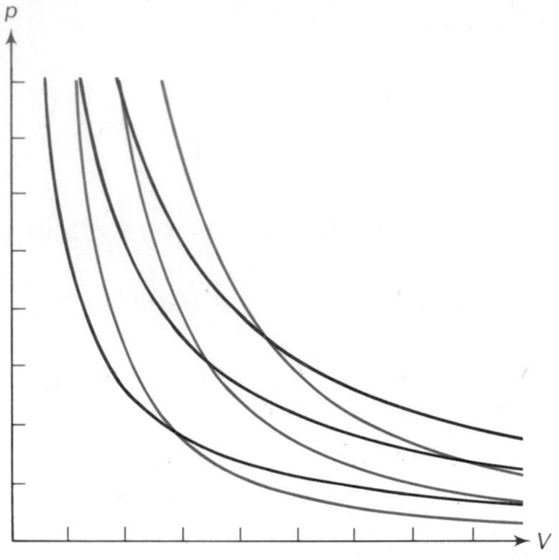

Fig. 20.8 Adiabatic curves (colored) and isothermal curves (black) for an ideal monotomic gas. Each curve is characterized by a different value of the constant pV^γ or T (the numerical values of these constants have not been listed; they depend on the number of moles of the gas). Note that the adiabatic curves are steeper than the isothermal curves. Thus, when a gas expands adiabatically and evolves downward along one of the adiabatic curves, it crosses into regions of lower temperature.

EXAMPLE 5. In one of the cylinders of an automobile engine, after the explosive combustion of the fuel, the gas has an initial pressure of 3.4×10^6 N/m² and an initial volume of 50 cm³. As the piston moves outward, the gas expands nearly adiabatically to a final volume of 250 cm³. What is the final pressure? Assume that $\gamma = 1.40$.

SOLUTION: By Eq. (33),

$$p_1 V_1^\gamma = p_2 V_2^\gamma \tag{34}$$

so that

$$p_2 = p_1\left(\frac{V_1}{V_2}\right)^\gamma = 3.4 \times 10^6 \text{ N/m}^2 \times \left(\frac{50 \text{ cm}^3}{250 \text{ cm}^3}\right)^{1.4}$$

$$= 3.6 \times 10^5 \text{ N/m}^2$$

EXAMPLE 6. How much work does the gas described in the preceding example do during its adiabatic expansion?

SOLUTION: The work done during a small increase of volume is $p\,dV$ and the work done during the entire expansion is

$$W = \int_{V_1}^{V_2} p\,dV \tag{35}$$

According to Eq. (33),

$$p = p_1 V_1^\gamma / V^\gamma \tag{36}$$

so that

$$W = \int_{V_1}^{V_2} \frac{p_1 V_1^\gamma}{V^\gamma} dV = p_1 V_1^\gamma - \left(\frac{1}{\gamma - 1}\frac{1}{V_2^{\gamma-1}} + \frac{1}{\gamma - 1}\frac{1}{V_1^{\gamma-1}}\right)$$

$$= \frac{1}{\gamma - 1} p_1 V_1 \left[1 - \left(\frac{V_1}{V_2}\right)^{\gamma - 1}\right] \tag{37}$$

With the numbers of Example 4, this gives

$$W = \frac{1}{1.40 - 1} \times 3.4 \times 10^6 \text{ N/m}^2 \times 5.0 \times 10^{-5} \text{ m}^3 \left[1 - \left(\frac{50 \text{ cm}^3}{250 \text{ cm}^3} \right)^{1.40-1} \right]$$

$$= 2.0 \times 10^2 \text{ J}$$

SUMMARY

Specific heat of water: $c = 1$ kcal/kg · °C

Mechanical equivalent of heat: 1 cal = 4.186 J

Thermal expansion: $\Delta L = \alpha L / \Delta T$

$$\Delta V = \beta V \, \Delta T$$

Conduction of heat: $\dfrac{\Delta Q}{\Delta t} = -kA \dfrac{dT}{dx}$

Relation between specific heats of a mole of gas: $C_p - C_V = R$

Adiabatic equation for gas: $pV^\gamma - [\text{constant}]$

$$\gamma = C_p / C_V$$

QUESTIONS

1. Can the body heat from a crowd of people produce a significant temperature increase in a room?

2. The expression "cold enough to freeze the balls off a brass monkey" originated aboard ships of the British Navy where cannonballs of lead were kept in brass racks ("monkeys"). Can you guess how the balls might fall off a "monkey" on a very cold day?

3. If the metal lid of a glass jar is stuck, it can usually be loosened by running hot water over the lid. Explain.

4. On hot days, bridges expand. How do bridge designers prevent this expansion from buckling the road?

5. At regular intervals, oil pipelines have lateral loops (shaped like a **U**; see Figure 20.9). What is the purpose of these loops?

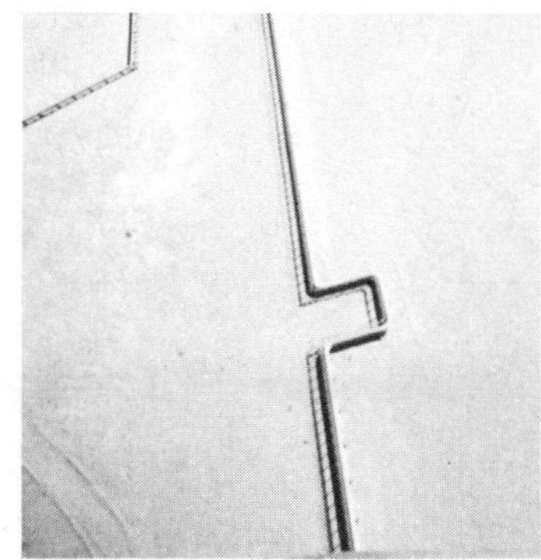

Fig. 20.9 Oil pipeline.

6. When you heat soup in a metal pot, the soup rises at the rim of the pot and falls at the center. Explain.

7. A sheet of glass will crack if heated in one spot. Why?

8. When aluminum wiring is used in electrical circuits, special terminal connectors are required to hold the ends of the wires securely. If an ordinary brass screw were used to hold the end of an aluminum wire against a brass plate, what is likely to happen during repeated heating and cooling of the circuit?

9. The frequency of the ticking of a mechanical wristwatch depends on the moment of inertia of its balance wheel. If the watch is subjected to an increase of temperature, the balance wheel expands and its moment of inertia increases, which makes the watch run slow. Good wristwatches have a built-in temperature compensation in their balance wheels, so that the moment of inertia stays constant. Design such a compensated balance wheel using two metals of different coefficients of thermal expansion.

10. Suppose that a piece of metal and a piece of wood are at the same temperature. Why does the metal feel colder to the touch than the wood?

11. In lack of better, some nineteenth-century explorers in Africa measured altitude by sticking a thermometer into a pot of boiling water. Explain.

12. Can you guess why an alloy of two metals usually has a lower melting point than either pure metal?

13. In the cooling system of an automobile, how is the heat transferred from the combustion cylinder to the cooling water (conduction, convection, or radiation)? How is the heat transferred from the water in the engine to the water in the radiator? How is the heat transferred from the radiator to the air?

14. A fan installed near the ceiling of a room blows air down toward the floor. How does such a fan help to keep you cool in summer and warm in winter?

15. It is often said that an open fireplace sends more heat up the chimney than it delivers to the room. What is the mechanism for heat transfer to the room? For heat transfer up the chimney?

16. A large fraction of the heat lost from a house escapes through the windows (Figure 20.10). This heat is carried to the windowpane by convection — hot air at the top of the room descends along the windowpane, giving up its heat. Suppose that the windows are equipped with venetian blinds. In order to minimize the heat loss, should you close the blinds so that the slats are oriented down and away from the window or up and away?

Fig. 20.10 Thermal photograph of a house.

17. Fiber glass insulation used in the walls of houses has a shiny layer of aluminum foil on one side. What is the purpose of this layer?

18. Is it possible to add heat to a system without changing its temperature? Give an example.

19. Why is boiling oil much more likely to cause severe burns on skin than boiling water?

20. Would you expect the melting point of ice to increase or decrease with an increase of pressure?

21. A very cold ice cube, fresh out of the freezer, tends to stick to the skin of your fingers. Why?

22. What is likely to happen to the engine of an automobile if there is no antifreeze in the cooling system and the water freezes?

23. If an evacuated glass vessel, such as a TV tube, fractures and implodes, the fragments fly about with great violence. From where does the kinetic energy of these fragments come?

24. When you boil water and convert it into water vapor, is the heat you supply equal to the change of the internal energy of the water?

25. A gas is in a cylinder fitted with a piston. Does it take more work to compress the gas isothermally or adiabatically?

26. If you let some air out of the valve of an automobile tire, it feels cold. Why?

27. According to the result of Section 10.2, when a particle of small mass collides elastically with a body of very large mass, the particle gains kinetic energy if the body of large mass was approaching the particle before the collision. Using this result, explain how the collisions between the particles of gas and the moving piston lead to an increase of temperature during an adiabatic compression.

28. The air near the top of a mountain is usually cooler than that near the bottom. Explain this by considering the adiabatic expansion of a parcel of air carried from the bottom to the top by an air current.

29. When a gas expands adiabatically, its temperature decreases. How could you take advantage of this effect to design a refrigerator?

30. A sample of ideal gas is initially confined in a bottle at some given temperature. If we break the bottle and let the gas expand freely into an evacuated chamber of larger volume, will the temperature of the gas change?

PROBLEMS

Section 20.1

1. The immersible electric heating element in a coffee maker converts 620 W of electric power into heat. How long does this coffee maker take to heat 1.0 liter of water from 20°C to 100°C? Assume that no heat is lost to the environment.

2. The body heat released by children in a school makes a contribution toward heating the building. How many kilowatts of heat do 1000 children release? Assume that the daily food intake of each child has a chemical energy of 2000 kcal and that this food is burned at a steady rate throughout the day.

3. In 1847 Joule attempted to measure the frictional heating of water in a waterfall near Chamonix in the French Alps. If the water falls 120 m and all of its gravitational energy is converted into thermal energy, how much does the temperature of the water increase? Actually, Joule found no increase of temperature because the falling water cools by evaporation.

4. A nuclear power plant takes in 5×10^6 m^3 of cooling water per day from a river and exhausts 1200 megawatts of waste heat into this water. If the temperature of the inflowing water is 20°C, what is the temperature of the outflowing water?

5. Your metabolism extracts about 100 kcal of chemical energy from one apple. If you want to get rid of all this energy by jogging, how far must you jog? At a speed of 12 km/h, jogging requires about 750 kcal/h.

6. For basic subsistence a human body requires a diet with about 2000 kcal/day. Express this power in watts.

7. By turning a crank, you can do mechanical work at the steady rate of

0.15 hp. If the crank is connected to paddles churning 1 gal. of water, how long must you churn the water to raise its temperature by 5°C?

8. You can warm the surfaces of your hands by rubbing one against the other. If the coefficient of friction between your hands is 0.6 and if you press your hands together with a force of 60 N while rubbing them back and forth at an average speed of 0.5 m/s, at what rate (in calories per second) do you generate heat on the surfaces of your hands?

9. Problem 8.56 gives the relevant numbers for frictional losses in the Tennessee river. If all the frictional heat were absorbed by the water and if there were no heat loss by evaporation, how much would the water temperature rise per mile?

10. A simple gadget for heating water for showers consists of a black plastic bag holding 10 liters of water. When hung in the sun, the bag absorbs heat. On a clear, sunny day, the power delivered by sunlight per unit area facing the Sun is 1.0×10^3 W/m^2. The bag has an area of 0.10 m^2 facing the sun. How long does it take for the water to warm from 20°C to 50°C? Assume that the bag loses no heat.

11. The beam dump at the Stanford Linear Accelerator consists of a large tank with 12 m^3 of water into which the accelerated electrons can be aimed when they are not wanted elsewhere. The beam carries 3.0×10^{14} electrons/s; the kinetic energy per electron is 3.2×10^{-9} J. In the beam dump this energy is converted into heat.

(a) What is the rate of production of heat?

(b) If the water in the tank does not lose any heat to the environment, what is the rate of increase of temperature of the water?

12. A solar collector consists of a flat plate that absorbs the heat of sunlight. A water pipe attached to the back of the plate carries away the absorbed heat (Figure 20.11). Assume that the solar collector has an area of 4.0 m^2 facing the sun and that the power per unit area delivered by sunlight is 1.0×10^3 W/m^2. What is the rate at which water must circulate through the pipe if the temperature of the water is to increase by 40°C as it passes through the collector?

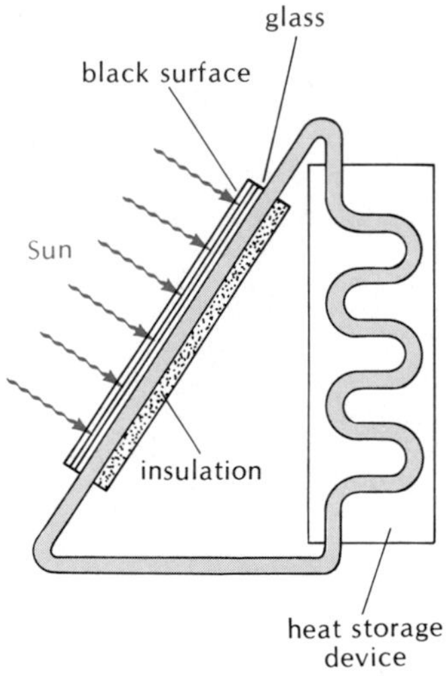

Fig. 20.11 Collector of solar heat.

Section 20.2

13. Consider the waterwheel described in Example 12.8. How much does the temperature of the water increase as it passes through the wheel?

14. The tallest building in the world is the Sears Tower in Chicago, which is 1454 ft high. It is made of concrete and steel. How much does its height change between a day when the temperature is 95°F and a day when the temperature is −20°F?

15. The height of the Eiffel Tower is 321 m. How much does its height change when the temperature changes from −20°C to 35°C?

16. Machinists use gauge blocks of steel as standards of length. A one-inch gauge block is supposed to have a length of 1 in., to within $\pm 10^{-6}$ in. In order to keep the length of the block within this tolerance, how precisely must the machinist control the temperature of the block?

17. A mechanic wants to place a sleeve (pipe) of copper around a rod of steel. At a temperature of 18°C, the sleeve of copper has an inner diameter of 0.998 cm and the rod of steel has a diameter of 1.000 cm. To what temperature must the mechanic heat the copper to make it fit round the steel?

18. (a) Segments of steel railroad rails are laid end to end. In an old railroad, each segment is 60 ft long. If they are originally laid at a temperature of 20°F, how much of a gap must be left between adjacent segments if they are to just barely touch at a temperature of 110°F?

(b) In a modern railroad, each segment is 2600 ft long, with a special expansion joint at each end. How much of a gap must be left between adjacent segments in this case?

19. A spring made of steel has a relaxed length of 0.316 m at a temperature of 20°C. By how much will the length of this spring increase if we heat it to 150°C? What compressional force must we apply to the hot spring to bring it back to its original length? The spring constant is 3.5×10^4 N/m.

20. A wheel of metal has a moment of inertia I at some given temperature. Show that if the temperature increases by ΔT, the moment of inertia will increase by approximately $\Delta I = 2\alpha I \, \Delta T$.

21. The pendulum (rod and bob) of a pendulum clock is made of brass.

(a) What will be the fractional increment in the length of this pendulum if the temperature increases by 20°C? What will be the fractional increase in the period of the pendulum?

(b) The pendulum clock keeps good time when its temperature is 15°C. How much time (in seconds per day) will the clock lose when its temperature is 35°C?

22. (a) The density of gasoline is 730 kg/m³ when the temperature is 0°C. What will be the density of gasoline when the temperature is 30°C?

(b) The price of gasoline is 120 cents per gallon. What is the price per kilogram at 0°C? What is the price per kilogram at 30°C? (Note that 1 gal. = 3.80×10^{-3} m³.) Is it better to buy cold gasoline or warm gasoline?

23. An ordinary mercury thermometer consists of a glass bulb to which is attached a fine capillary tube. As the mercury expands, it rises up the capillary tube. Given that the bulb has a volume of 0.20 cm³ and that the capillary tube has a diameter of 7.0×10^{-3} cm, how far will the mercury column rise up the capillary tube for a temperature increase of 10°C? Ignore the expansion of the glass and ignore the expansion of the mercury in the capillary tube.

*24. In order to compensate for deviations caused by temperature changes, a pendulum clock built during the last century for an astronomical observatory uses a large cylindrical glass tube filled with mercury as a pendulum bob. This tube is held by a brass rod and bracket (Figure 20.12); the combined length of the rod and bracket is l (measured from the point of suspension of the pendulum). Neglecting the mass of the brass and the glass and neglecting the expansion of the glass, show that the height of the mercury in the glass tube must be

$$(2\alpha_{\text{brass}}/\beta_{\text{mercury}})l$$

if the center of mass of the mercury is to remain at a fixed distance from the point of suspension, regardless of temperature.

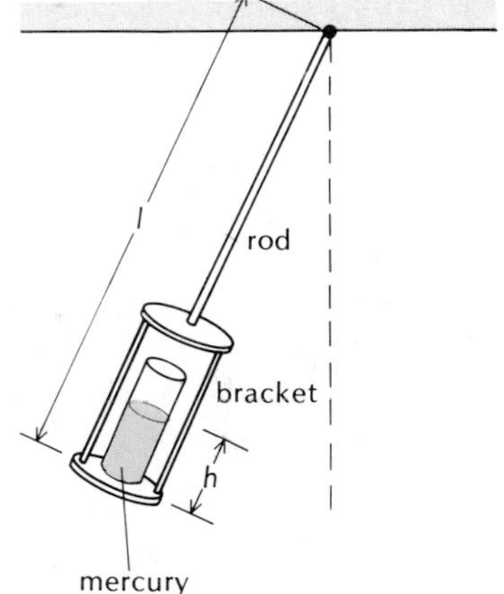

Fig. 20.12 Pendulum with a temperature compensator.

Section 20.3

25. The walls of an igloo are made of compacted snow, 30 cm thick. What thickness of styrofoam would provide the same insulation as the snow?

26. A pan of aluminum, filled with boiling water, sits on a hot plate. The bottom area of the pan is 300 cm² and the thickness of the aluminum is 0.10 cm. If the hot plate supplies 2000 W of heat to the bottom of the pan, what must be the temperature of the hot plate?

27. A rod of steel 0.70 cm in diameter is surrounded by a tight copper sleeve of inner diameter 0.70 cm and outer diameter 1.00 cm. What will be the heat flow along this compound rod if the temperature gradient along the rod is 50°C/cm? What fraction of the heat flows in the copper? What fraction in the steel?

28. A styrofoam box, used for the transportation of medical supplies, is filled with dry ice (carbon dioxide) at a temperature of −79°C. The box measures 30 cm × 30 cm × 40 cm and its walls are 4 cm thick. If the outside surface of the box is at a temperature of 20°C, what is the rate of loss of dry ice by vaporization?

29. The icebox on a sailboat measures 60 cm × 60 cm × 60 cm. The contents of this icebox are to be kept at a temperature of 0°C for 4 days by the gradual melting of a block of ice of 20 kg, while the temperature of the outside of the box is 30°C. What minimum thickness of styrofoam insulation is required for the walls of the icebox?

30. A man has a skin area of 1.8 m^2; his skin temperature is 34°C. On a cold winter day, the man wears a whole-body suit insulated with down. The temperature of the outside surface of his suit is −25°C. If the man can stand a heat loss of no more than 100 kcal/h, what is the minimum thickness of down required for his suit?

31. The end of a rod of copper 0.50 cm in diameter is welded to a rod of silver of the same diameter. Each rod is 6.0 cm long. What is the heat flow along these rods if the free end of the copper rod is in contact with boiling water and the free end of the silver rod is in contact with ice? What is the temperature of the junction? Assume that there is no heat loss through the lateral surfaces of the rods.

32. On a cold winter day, the water of a shallow pond is covered with a layer of ice 6.0 cm thick. The temperature of the air is −20°C and the temperature of the water is 0°C. What is the (instantaneous) rate of growth of the thickness of the ice (in centimeters per hour)? Assume that wind chill keeps the top surface of the ice at exactly the temperature of the air, and assume that there is no heat transfer through the bottom of the pond.

33. Suppose that the pond described in the preceding problem has a layer of compacted snow 3.0 cm thick on top of the ice. What is the rate of growth of the thickness of the ice?

34. Several slabs of different materials are piled one on top of another. All slabs have the same face area A, but their thickness and conductivities are Δx_i and k_i, respectively. Show that the heat flow through the pile of slabs is

$$\frac{\Delta Q}{\Delta t} = \frac{-A\,\Delta T}{\sum_{i=1}^{n} \Delta x_i/k_i}$$

where ΔT is the temperature difference between the bottom of the first slab and the top of the last slab.

*35. A cable used to carry electric power consists of a metallic conductor of radius r_1 encased in an insulator of inner radius r_1 and outer radius r_2. The metallic conductor is at a temperature T_1 and the outer surface of the insulator is at a temperature T_2. Show that the radial heat flow across a length l of the insulator is given by

$$\frac{\Delta Q}{\Delta t} = \frac{2\pi k(T_1 - T_2)l}{\ln(r_2/r_1)}$$

where k is the conductivity of the insulator.

Section 20.4

36. Thunderstorms obtain their energy by condensing the water vapor contained in humid air. Suppose that a thunderstorm succeeds in condensing *all* the water vapor in 10 km^3 of air.

(a) How much heat does this release? Assume the air is initially at 100% humidity and that each cubic meter of air at 100% humidity (at 20°C and

1 atm) contains 1.74×10^{-2} kg of water vapor. The heat of vaporization of water is 585 kcal/kg at 20°C.

(b) The explosion of an A bomb releases an energy of 2×10^{10} kcal. How many A bombs does it take to make up the energy of one thunderstorm?

37. During a rainstorm lasting 2 days, 3 in. of rain fell over an area of 10^3 mi^2.
 (a) What is the total mass of the rain (in kilograms)?
 (b) Suppose that the heat of vaporization of water in the rain clouds is 580 kcal/kg. How many calories of heat are released during the formation of the total mass of rain by condensation of the water vapor in the rain clouds?
 (c) Suppose that the rain clouds are at a height of 1500 m above the ground. What is the gravitational potential energy of the total mass of rain before it falls? Express your answer in calories.
 (d) Suppose that the raindrops hit the ground with a speed of 10 m/s. What is the total kinetic energy of all the raindrops taken together? Express your answer in calories. Why does your answer to part (c) not agree with this?

38. If you pour 0.50 kg of molten lead at 328°C into 2.5 liters of water at 20°C, what will be the final temperatures of the water and the lead? The specific heat of (solid) lead has an average value of 3.4×10^{-2} kcal/kg · °C over the relevant temperature range.

39. Suppose you drop a cube of titanium of mass 0.25 kg into a Dewar (a thermos bottle) full of liquid nitrogen at −196°C. The initial temperature of the titanium is 20°C. How many kilograms of nitrogen will boil off as the titanium cools from 20°C to −196°C? The specific heat of titanium is 8.2×10^{-2} kcal/kg · °C.

40. While jogging on a level road, your body generates heat at the rate of 750 kcal/h. Assume that evaporation of sweat removes 50% of this heat and convection and radiation the remainder. The evaporation of 1 kg (or 1 liter) of sweat requires 580 kcal. How many kilograms of sweat do you evaporate per hour?

41. The Mediterranean loses a large volume of water by evaporation. This loss is made good, in part, by currents flowing into the Mediterranean through the straits joining it to the Atlantic and the Black Sea. Calculate the rate of evaporation (in km^3/h) of the Mediterranean on a clear summer day from the following data: the area of the Mediterranean is 2.9×10^6 km^2, the power per unit area supplied by sunlight is 1×10^3 W/m^2, and the heat of vaporization of water is 580 kcal/kg (at a temperature of 21°C). Assume that all the heat of sunlight is used for evaporation.

Section 20.5

42. A TV tube of glass with zero pressure inside and atmospheric pressure outside suddenly cracks and implodes. The volume of the tube is 2.5×10^{-2} m^3. During the implosion, the atmosphere does work on the fragments of the tube and on the layer of air immediately adjacent to the tube. This amount of work represents the energy released in the implosion. Calculate this energy. If all of this energy is acquired by the fragments of glass, what will be the mean speed of the fragments? The total mass of the glass is 2.0 kg.

43. The rear end of an air conditioner dumps 3000 kcal/h of waste heat into the air outside a building. A fan assists in the removal of this heat. The fan draws in 15 m^3/min of air at a temperature of 30°C and ejects this air after it has absorbed the waste heat. With what temperature does the air emerge?

44. If we heat 1.00 kg of hydrogen gas from 0°C to 50.0°C in a cylinder with a piston keeping the gas at a constant pressure of 1.0 atm, we must supply

1.69×10^2 kcal of heat. How may kilocalories of heat must we supply to heat the same amount of gas from 0°C to 50.0°C in a container of constant volume?

45. On a winter day you inhale cold air at a temperature of −30°C and at 0% humidity. The amount of air you inhale is 0.45 kg per hour. Inside your body you warm and humidify the air; you then exhale the air at a temperature of 37°C and 100% relative humidity. At a temperature of 37°C, each kilogram of air at 100% relative humidity contains 0.041 kg of water vapor. How many calories are carried out of your body by the air that passes through your lungs in one hour? Take into account both the heat needed to warm the air at constant pressure and the heat needed to vaporize the moisture that the exhaled air carries out of your body. The specific heat of air at constant pressure is 0.25 kcal/°C · kg; the heat of vaporization of water at 37°C is 576 kcal/kg.

46. A helium balloon consists of a large bag loosely filled with 600 kg of helium at an initial temperature of 10°C. While exposed to the heat of the Sun, the helium gradually warms to a temperature of 30°C. The heating proceeds at a constant pressure of 1.0 atm. How much heat does the helium absorb during this temperature change?

47. The theoretical expression for the speed of sound in a gas is $\sqrt{\gamma p/\rho}$ [compare Eq. (16.3)]. Calculate the speed of sound in carbon dioxide at STP.

*48. An air conditioner removes heat from the air of a room at the rate of 2000 kcal/h. The room measures 5.0 m × 5.0 m × 2.5 m and the pressure is constant at 1.0 atm.

(a) If the initial temperature of the air in the room is 30.0°C, how long does it take the air conditioner to reduce the temperature of the air by 5.0°C? Pretend that the mass of air in the room is constant.

(b) As the air in the room cools, it contracts slightly and draws in some extra air from the outside; hence the mass of air is not exactly constant. Repeat your calculation taking into account this increase of the mass of air. Assume that the extra air enters with an initial temperature of 30°C. Does the result of your second calculation differ appreciably from that of your first calculation?

Section 20.6

49. In one of the combustion chambers of a diesel engine, the piston compresses the air–fuel mixture from an initial volume of 630 cm^3 to a final volume of 30 cm^3. The initial temperature of the mixture is 40°C. Assuming that the compression is adiabatic with $\gamma = 1.4$, what is the final temperature?

50. Suppose that a submarine suddenly breaks up at a depth of 300 m below sea level. The air in the submarine will then be suddenly (adiabatically) compressed. If the initial pressure of the air is 1.0 atm and the initial temperature is 20°C, what will be the final pressure and temperature? For air, $\gamma = 1.4$.

51. The air in an automobile tire is at an overpressure of 35 lb/in.^{2} and at a temperature of 20°C. The atmospheric pressure outside the tire is 14.7 lb/in.^{2}. If you let some air escape through the valve, what will be the temperature of the emerging air? Assume that the escaping air expands adiabatically with $\gamma = 1.4$.

52. Suppose that you have a sample of oxygen gas at a pressure of 300 atm and a temperature of −29°C. Assume that oxygen behaves as an ideal gas with $\gamma = 1.4$. If you suddenly (adiabatically) let this gas expand to a final pressure of 1 atm, what will be the final temperature? This method of cooling by adiabatic expansion was used by Cailletet in 1877 to liquify oxygen (the liquefaction point of oxygen at 1 atm is −183°C).

53. A fire extinguisher is filled with 1.0 kg of compressed nitrogen gas at a pressure of 1.2×10^6 N/m^2 and at a temperature of 20°C.

(a) What is the volume of this gas?

(b) If you open the nozzle of the fire extinguisher, the gas will escape expanding adiabatically to atmospheric pressure (1.01×10^5 N/m^2). What will be the volume and the temperature of the expanded gas? The value of γ for nitrogen is 1.4.

*54. By means of a hand pump, you inflate an automobile tire from 0 lb/in.^{2} to 35 lb/in.^{2} (overpressure). The volume of the tire remains constant at 3.5 ft^3 during this operation. How much work must you do on the air with the pump? Assume that each stroke of the pump is an adiabatic process and that the air is initially at STP.

*55. In an air pressure gun (BB gun), the projectile is driven along the barrel by the pressure of air that is suddenly released into the barrel. Initially, the air is held at high pressure in a small reservoir; when this air is suddenly released into the barrel, it expands adiabatically and does work on the projectile, giving it kinetic energy. You are told to design a gun that shoots a projectile of 1.0 g with a muzzle velocity of 100 m/s from a barrel of length 25 cm and diameter 0.5 cm. The volume of the reservoir is 1.0×10^{-5} m^3. What must be the minimum initial pressure in the reservoir? How much air (in kilograms, at 20°C) must this reservoir contain initially? For air, $\gamma = 1.4$. Ignore friction in the barrel.

CHAPTER 21

Thermodynamics

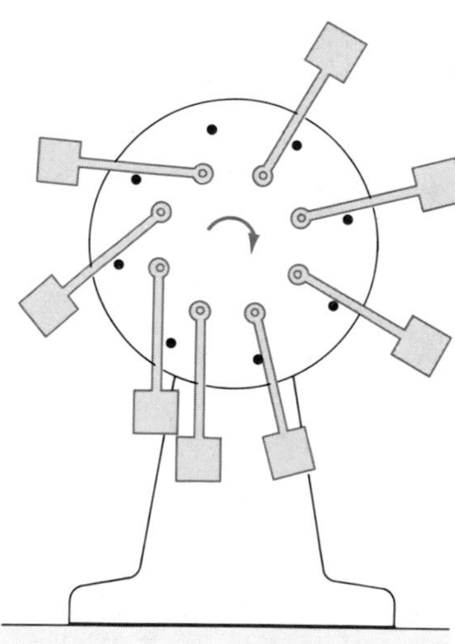

Fig. 21.1 A hypothetical perpetual motion machine.

Thermodynamics is the description of the behavior of physical systems in terms of purely macroscopic parameters. Such a macroscopic, large-scale description is necessarily somewhat crude, since it overlooks all of the small-scale, microscopic details. However, in practice these microscopic details are often irrelevant. For instance, an engineer investigating the behavior of the combustion gases in the cylinder of an automobile engine can get by reasonably well with such macroscopic quantities as temperature, pressure, density, and heat capacity.

The most important application of thermodynamics concerns the conversion of one form of energy into another, especially the conversion of heat into other forms of energy. These conversions are governed by the two fundamental laws of thermodynamics. The first of these is essentially a general statement of the law of conservation of energy and the second is a statement about the maximum efficiency attainable in the conversion of heat into work.

The study of thermodynamics was inaugurated by nineteenth-century engineers who wanted to know what ultimate limitations the laws of physics impose on the operation of steam engines and other machines that generate mechanical energy. They soon recognized that perpetual motion machines are impossible. A **perpetual motion machine of the first kind** is a (hypothetical) device that supplies an endless output of work without any input of fuel or any other input of energy. Figure 21.1 shows a proposed design for such a machine. Weights are attached to the rim of a wheel by short pivoted rods resting against pegs. With the rods in the position shown, there is an imbalance in the weight distribution causing a clockwise torque on the wheel; as the wheel turns, the rod coming to the top presumably flips over, maintaining the imbalance. This perpetual torque would not only keep the wheel turning, but would also continually deliver energy to the axle of the wheel. Of course, a detailed analysis shows that the machine will not perform as intended — the wheel actually settles in a static equilibrium configuration such that the top rod just barely fails to flip over. The First Law of Thermodynamics, or the law of conservation of energy, directly tells us of the failure of this machine: after one revolution of the wheel, the masses all return to their initial positions, their potential energy returns to its initial value, and they will not have delivered net energy to the motion of the wheel.

A **perpetual motion machine of the second kind** is a device that extracts thermal energy from air or from the water of the oceans and converts it into mechanical energy. Such a device is obviously not forbidden by conservation laws. The oceans are an enormous reservoir of thermal energy; if we could extract this thermal energy, a temperature drop of just 1 °C of the oceans would supply the energy needs of the United States for the next 50 years. But, as we will see, the Second Law of Thermodynamics tells us that conversion of heat into work requires not only a heat source, but also a heat sink. Heat flows out of a warm body only if there is a colder body that can absorb it. If we want heat to flow from the ocean into our machine, we must provide a low-temperature heat sink toward which the heat will tend to flow spontaneously. Since no convenient low-temperature sink is available, the extraction of heat from the oceans is impossible. We cannot build a perpetual motion engine of the second kind.

21.1 The First Law of Thermodynamics

Consider some amount of gas with a given initial volume V_1, pressure p_1, and temperature T_1. The gas is in a container fitted with a piston (Figure 21.2). Suppose we compress the gas to some smaller volume V_2 and also lower its temperature to some smaller value T_2; the pressure will then reach some new value p_2. (Such a compression and cooling process is of practical importance in the liquifaction of gases, say, oxygen or nitrogen — before the gas can be liquefied it must be compressed and cooled.) Obviously we can reach the new state V_2, p_2, T_2 from the old state V_1, p_1, T_1 in a variety of ways. For instance, we may first compress the gas and then cool it. Or else we may first cool it and then compress it. Or we may go through small alternating steps of compressing and cooling. In order to compress the gas, we must do work on it [see Eq. (20.10)]; and in order to cool the gas, we must remove heat from it. If we extract an amount of work ΔW and we transfer an amount of heat ΔQ, we will change the internal energy of the gas by an amount

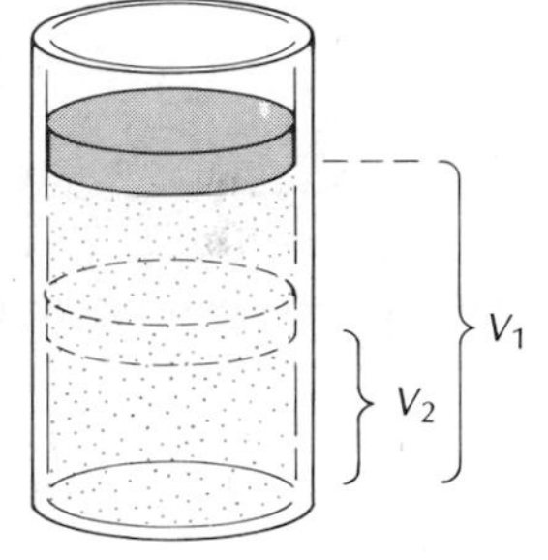

Fig. 21.2 Compression of a gas.

$$\Delta E = \Delta Q - \Delta W \qquad (1)$$

Note the sign conventions in this equation: ΔQ is positive if we add heat to the gas and negative if we remove heat; ΔW is positive if the gas does work on us and negative if we do work on the gas.

The values of ΔQ and ΔW depend on the process. If we first compress the gas and subsequently cool it, then during the first stage ΔW is negative and ΔQ zero; and during the second stage ΔW is zero and ΔQ is negative. If we first cool the gas and then compress it, the values of ΔW and ΔQ will be quite different. Yet, it turns out that regardless of what sequence of operations we use to transform the gas from its initial state p_1, V_1, T_1 to its final state p_2, V_2, T_2, the net change ΔE in the internal energy is always the same: ΔQ and ΔW vary, but the sum of ΔQ and $-\Delta W$ is a constant. This is the **First Law of Thermodynamics:**

Whenever we employ some process involving heat and work to change a system from an initial state characterized by certain values of the macroscopic parameters to a final state characterized by new values of the macroscopic parameters, the change in the internal energy of the system

First Law of Thermodynamics

$$\boxed{\Delta E = \Delta Q - \Delta W} \tag{2}$$

has a fixed value which does not depend on the details of the process.

Note that the First Law tells us that energy is conserved — the change of the internal energy equals the input of heat and work. But the First Law tells us more than that. If we describe a system in terms of the detailed microscopic positions and velocities of all of its constituent particles, then energy conservation is a theorem of mechanics. But if we describe a system in terms of nothing but macroscopic parameters, then it is not at all obvious that we have available enough information to determine the energy and to formulate a conservation law. The First Law of Thermodynamics tells us that a knowledge of the macroscopic parameters is indeed sufficient to determine the internal energy of the system.

As an example of how the First Law permits us to draw some conclusions about a physical system, consider the following experiment. Take a thermally insulated bottle with ideal gas at some temperature T_1 and, by means of a pipe with a stopcock, connect this to another insulated bottle which is evacuated (Figure 21.3). If we suddenly open the stopcock, the gas will rush from the first bottle into the second until the pressures are equalized. Experimentally, we find that this process of free expansion does not change the temperature of the gas — when the gas attains equilibrium and stops flowing, the final temperature of both bottles are equal to the initial temperature T_1. What can we deduce from this experimental observation? Since the bottles are thermally insulated from their environment, the expansion process neither adds nor removes heat from the gas, that is, $\Delta Q = 0$. Furthermore, the expansion process involves no work (except for an insignificant amount required for turning the stopcock), that is $\Delta W = 0$. Consequently, the energy of the gas does not change,

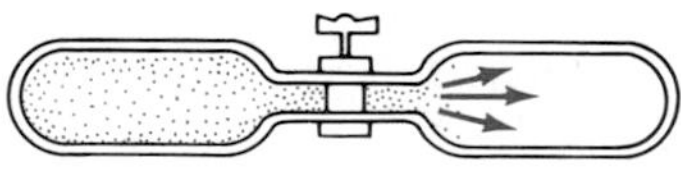

Fig. 21.3 Free expansion of a gas.

$$\Delta E = \Delta Q - \Delta W = 0 \tag{3}$$

This shows that a change of volume does not change the energy; i.e., the energy of the ideal gas is *not* a function of the volume. According to the First Law, the energy of the gas is supposed to be some function of the macroscopic parameters V, p, and T. Since the ideal-gas law allows us to express p in terms of V and T, the energy may be regarded as a function of the two variables V and T. But Eq. (3) shows that a change of volume does not affect the energy; consequently, the energy of the ideal gas is a function of the *temperature* alone.

Note that this conclusion is consistent with Eqs. (19.24) and (19.31), obtained from kinetic theory, which tell us explicitly what function of the temperature is involved. However, the calculations of kinetic theory depend on some questionable assumptions concerning the ideal behavior of a gas, whereas the arguments of thermodynamics are free of these. The conclusions of thermodynamics are therefore of very general validity, but the price we pay for this is that the information we obtain from thermodynamics is not as detailed as that which we obtain from kinetic theory.

The conclusions of thermodynamics apply only to the equilibrium states of a system, i.e., those static states into which the system settles when mass transfer, heat transfer, and all chemical and other reactions come to an end. For instance, for gas in the two bottles shown in Figure 21.3, the initial state (gas confined to one bottle) is an equilibrium state, and the final state (gas evenly distributed over both bottles) is also an equilibrium state; however, the intermediate state, when gas is rushing from the full bottle into the empty bottle immediately after we open the stopcock, is not an equilibrium state — thermodynamics does not tell us anything about this intermediate, time-dependent state of the gas. In the following discussion we will always suppose that the system under consideration is in an equilibrium state.

21.2 The Carnot Engine

If an engine is continually to convert thermal energy into mechanical energy, it must operate cyclically. At one end of each cycle it must return to its initial configuration so that it can repeat the process of conversion of heat into work over and over again. Steam engines and automobile engines are obviously cyclic — after one (or sometimes two) revolutions, they return to their initial configuration. These engines are not 100% efficient. The condenser of a steam engine and the exhaust of an automobile engine both eject a substantial amount of heat into the environment; this waste heat represents lost energy.[1]

In this section we will calculate the efficiency of an ideal heat engine which converts heat into work with a minimum of waste. The engine absorbs heat from a heat reservoir at high temperature, converts this heat partially into work, and rejects the remainder as waste heat into a reservoir at low temperature. In this context, a **heat reservoir** is simply a body that remains at constant temperature even when heat is removed from or added to it. In practice, the high-temperature heat reservoir is often a boiler whose temperature is kept constant by the controlled combustion of some fuel and the low-temperature reservoir is usually the atmosphere of the Earth, whose large volume permits it to absorb the waste heat without appreciable change in temperature.

Heat reservoir

Figure 21.4 is a flow chart for the energy, showing the heat Q_1 flowing into the engine from the high-temperature reservoir, the heat Q_2 (waste heat) flowing out of the engine into the low-temperature reservoir, and the work generated. The work generated is the difference between Q_1 and Q_2:

$$W = Q_1 - Q_2 \tag{4}$$

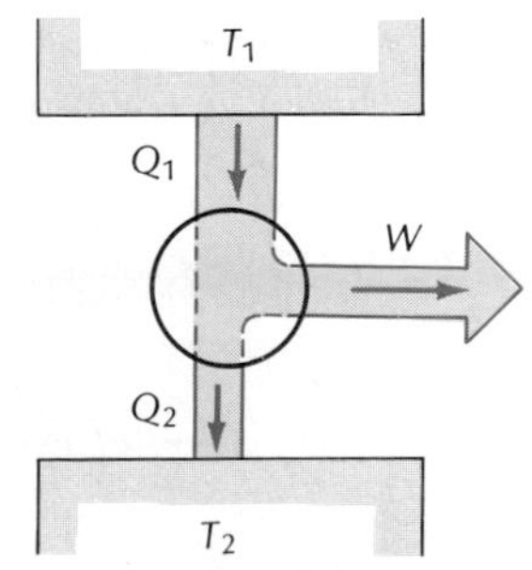

Fig. 21.4 Flow chart for an ideal heat engine.

[1] Besides, there are frictional losses.

The **efficiency** of the engine is defined as the ratio of this work to the heat absorbed from the high-temperature reservoir,

Efficiency

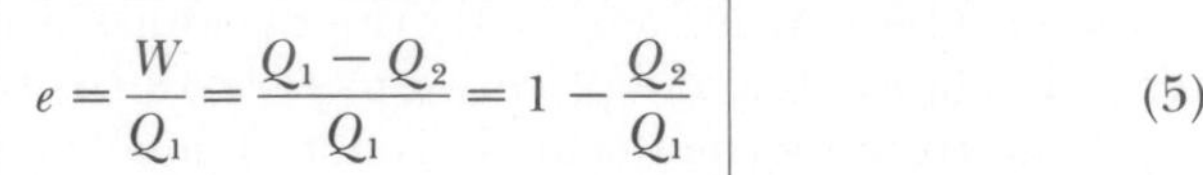

$$e = \frac{W}{Q_1} = \frac{Q_1 - Q_2}{Q_1} = 1 - \frac{Q_2}{Q_1} \qquad (5)$$

Sadi Carnot (karno), *1796–1832, French engineer and physicist. In his book* On the Motive Power of Heat *he formulated the theory of the conversion of heat into work.*

This says that if $Q_2 = 0$ (no waste heat), then the efficiency would be $e = 1$, or 100%. If so, the engine would convert the high-temperature heat *totally* into work. As we will see later, this extreme efficiency is unattainable. Even under ideal conditions the engine will produce some waste heat. It turns out that the efficiency of an ideal engine depends only on the temperatures of the heat reservoirs.

We will now calculate the efficiency of an ideal heat engine which converts heat into work by means of the push of a gas against a piston. As we will show in the next section, for maximum efficiency the thermodynamic processes within the engine should be **reversible.** This means that the engine can, in principle, be operated in reverse and it then converts work into heat at the same rate as it converts heat into work when operating in the forward direction. To achieve reversibility, the motion of the piston must be sufficiently slow so that the gas is always in an equilibrium configuration. If the piston were to have a sudden motion, a pressure wave (shock wave) would travel through the gas and the motion of this pressure wave could not be reversed by giving the piston a sudden motion in the opposite direction — this would merely create a second pressure wave. Furthermore, to achieve reversibility, the temperature of the gas must coincide with the temperature of the heat reservoir during contact. If the gas were to have, say, a lower temperature than that of the heat reservoir with which it is in contact, heat would rush from the reservoir into the gas and this flow of heat could not be reversed by any manipulation of the piston. For reversibility the gas must remain in mechanical equilibrium with the piston and in thermal equilibrium with the heat reservoir at all times.

Fig. 21.5 Carnot engine: a gas-filled cylinder with a piston.

An engine that operates with reversible processes is called a **Carnot engine.** Figure 21.5 shows the engine, a gas-filled cylinder fitted with a piston. The cylinder can be put in thermal contact with either of the heat reservoirs of temperatures T_1 or T_2, so that heat can flow from the reservoir into the cylinder, or vice versa. To describe the operation of this engine, it is best to use a p–V diagram (Figure 21.6). Each point in this diagram represents an equilibrium configuration of the gas; p and V can be read directly from the diagram and T can then be calculated from the ideal-gas equation. The initial volume and pressure of the gas are V_1 and p_1. The operation of the Carnot engine takes the gas through a sequence of four steps with varying volume and pressure, but at the end of the last step the gas returns to its initial volume and pressure. This sequence of four steps is called the **Carnot cycle.**

Carnot cycle

a. We begin the cycle by placing the cylinder in contact with the high-temperature heat reservoir. This maintains the temperature of the gas at the constant value T_1. The gas is now allowed to expand from the initial volume V_1 to a new volume V_2 (see Figure 21.6). During this expansion the gas does work on the piston, i.e., the engine absorbs heat from the reservoir and converts it into work.

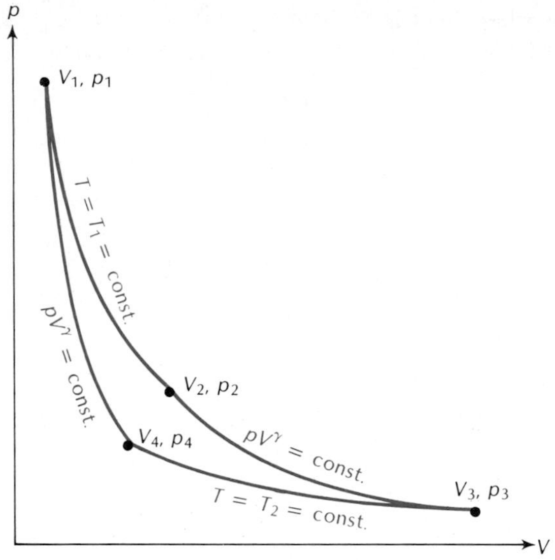

Fig. 21.6 The Carnot cycle shown on a p–V diagram.

b. When the gas has reached volume V_2 and pressure p_2, we remove it from the heat reservoir and allow it to continue the expansion adiabatically; during this expansion the temperature falls.
c. When the temperature has fallen to T_2, we stop the piston and place the gas in contact with the low-temperature heat reservoir. The volume at this instant is V_3 and the pressure is p_3. We now begin to push the piston back toward its starting position, i.e., we compress the gas. This means that the engine converts work into heat and ejects this heat into the low-temperature reservoir.
d. When the gas has reached volume V_4 and pressure p_4, we remove it from the reservoir and continue to compress it adiabatically until the volume and the pressure return to their initial values V_1 and p_1.

To find the efficiency, we need to calculate the heat Q_1 that the engine absorbs from the high-temperature reservoir in step (a) and the heat Q_2 that it ejects into the low-temperature reservoir in step (c). During the heat-absorption process, the temperature is constant; since the energy E of the ideal gas is a function of only the temperature, it follows that $\Delta E = 0$ and, by Eq. (2), the absorbed heat Q_1 equals the work done by the gas,

$$Q_1 = \Delta W \tag{6}$$

During the expansion from V_1 to V_2, the pressure is

$$p = nRT_1/V \tag{7}$$

and the work done by the gas during a small displacement is

$$p\,dV = nRT_1\,dV/V \tag{8}$$

By integrating this, we find

$$\Delta W = \int p\,dV = \int_{V_1}^{V_2} nRT_1\frac{dV}{V} = nRT_1 \ln\left(\frac{V_2}{V_1}\right) \tag{9}$$

Thus, the absorbed heat is

$$Q_1 = nRT_1 \ln\left(\frac{V_2}{V_1}\right) \tag{10}$$

Likewise, we readily find that the ejected heat is

$$Q_2 = nRT_2 \ln\left(\frac{V_3}{V_4}\right) \tag{11}$$

The ratio of these is

$$\frac{Q_2}{Q_1} = \frac{T_2}{T_1} \frac{\ln(V_3/V_4)}{\ln(V_2/V_1)} \tag{12}$$

During the adiabatic expansion and the adiabatic compression of the gas, the pressure and volume obey the relation given in Eq. (20.33):

$$pV^\gamma = [\text{constant}] \tag{13}$$

By the ideal-gas law this is equivalent to

$$TV^{\gamma-1} = [\text{constant}] \tag{14}$$

Applying this to the adiabatic expansion and compression, we obtain, respectively,

$$T_1 V_2^{\gamma-1} = T_2 V_3^{\gamma-1} \tag{15}$$

and

$$T_1 V_1^{\gamma-1} = T_2 V_4^{\gamma-1} \tag{16}$$

Dividing these equations into each other, we find

$$\left(\frac{V_2}{V_1}\right)^{\gamma-1} = \left(\frac{V_3}{V_4}\right)^{\gamma-1} \tag{17}$$

or

$$\frac{V_2}{V_1} = \frac{V_3}{V_4} \tag{18}$$

It follows from this that the two logarithms in Eq. (12) are equal and they cancel. This leaves us with

$$\frac{Q_2}{Q_1} = \frac{T_2}{T_1} \tag{19}$$

The efficiency of the Carnot engine is therefore

Efficiency of a Carnot engine

$$\boxed{e = 1 - \frac{Q_2}{Q_1} = 1 - \frac{T_2}{T_1}} \tag{20}$$

This expresses the efficiency in terms of the temperatures of the heat reservoirs. Note that an efficiency of $e = 1$ (or 100%) can only be achieved if $T_2 = 0$, that is, if the low-temperature reservoir is at the absolute zero of temperature. Unfortunately, we have no such absolutely cold reservoir available; even if we could devise some clever method for using interstellar space as a heat dump, the temperature would still be $T_2 = 3$ K and not zero.

EXAMPLE 1. The boiler of a steam engine produces steam at a temperature of 500°C. The engine exhausts its waste heat into the atmosphere where the temperature is 20°C. Theoretically, what is the limiting efficiency of this engine?

SOLUTION: As we pointed out above, the efficiency of an engine can never be greater than that of a Carnot engine. Thus, the efficiency cannot be greater than that given by Eq. (20) with $T_1 = 773$ K and $T_2 = 293$ K:

$$e = 1 - \frac{T_2}{T_1} = 1 - \frac{293}{773} = 0.62 \tag{21}$$

Hence only 62% of the heat can be converted into work. For every 62 J of work produced by the engine, at least 100 J must be supplied in the form of energy of fuel.

The actual performance of steam engines falls considerably below the limit given in Eq. (20). Typical efficiencies of the steam engines used in electric power plants are near 40%.

If the Carnot engine is operated in reverse, it uses up work to transfer heat from the low-temperature reservoir to the high-temperature reservoir. This is the principle involved in refrigerators, air conditioners, and "heat pumps." The minimum amount of work required to operate these devices can be calculated by means of Eq. (20).

EXAMPLE 2. On a cold day a homeowner uses a **heat pump** (essentially, an air conditioner operating in reverse) to extract heat from the outside air and inject it into his home. If the outside temperature is −10°C and the inside temperature is 20°C, what is the minimum amount of work that the heat pump must do to pump 1 kcal of heat from the outside to the inside?

SOLUTION: At the least, the heat pump will require as much work as a Carnot engine. For such an engine the ratio of the heats is given by Eq. (19):

$$\frac{Q_2}{Q_1} = \frac{T_2}{T_1} = \frac{263}{293} = 0.90 \tag{22}$$

If $Q_2 = 1$ kcal, then $Q_1 = Q_2/0.90 = 1.11$ kcal. The difference $Q_1 - Q_2$ represents the work that must be done; hence the work is 0.11 kcal.

Note that by the expenditure of 0.11 kcal of work, the heat pump delivers a total of 1.11 kcal of heat into the house. This is obviously a much cheaper heating method than the expenditure of 1.11 kcal of fuel or of electric energy in a conventional furnace or electric heater.

Incidentally, we can use the relation given by Eq. (20) between the efficiency of a Carnot engine and the temperatures of heat reservoirs for a measurement of temperature. We only need to run the Carnot engine between a heat reservoir at the unknown temperature and a heat reservoir at a known temperature (say, the temperature of the tri-

Thermodynamic temperature scale

ple point of water); a direct measurement of the efficiency then determines the unknown temperature. The temperature scale defined by this procedure is called the **thermodynamic temperature scale.** It coincides with the ideal-gas temperature scale, since, as is clear from the derivation leading to Eq. (20), the variables T_1 and T_2 are ideal-gas temperatures.

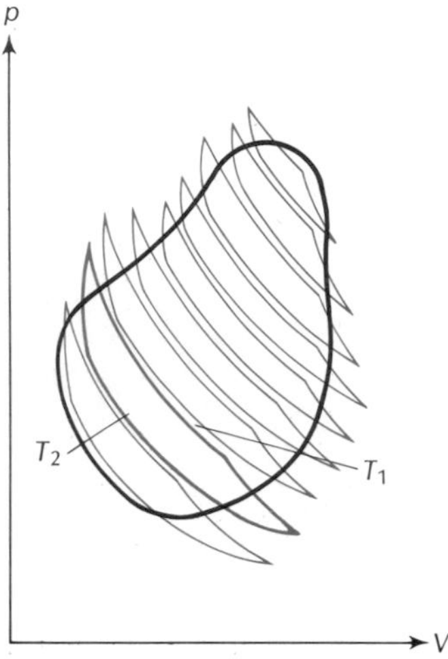

Fig. 21.7 A reversible cycle shown on a p–V diagram. This cycle can be approximated as a collection of Carnot cycles.

21.3 Entropy

The Carnot cycle described in the preceding section involves several steps of expansion and compression of an ideal gas. During each step the gas is kept either at constant temperature (isothermal process) or at constant energy (adiabatic process). It is easy to see that any reversible cycle involving any processes whatsoever can be regarded as a collection of Carnot cycles. In a reversible cycle, the gas goes through a sequence of equilibrium configurations and the state of the gas at each instant can be represented by a point in a p–V diagram. The complete cycle is then represented by a closed curve in the p–V diagram. Figure 21.7 shows how this cycle can be approximated by a collection of Carnot cycles. Obviously, the approximation becomes exact in the limit of very many, very small Carnot cycles. The sum of the amounts of heat and work associated with the small cycles equals the amount of heat and work associated with the original cycle.

This equivalence between an arbitrary reversible cycle and a collection of Carnot cycles leads to the following **theorem of Clausius:** *the integral of dQ/T around any reversible cycle is zero,*

Theorem of Clausius (reversible process)

$$\int \frac{dQ}{T} = 0 \tag{23}$$

Rudolph Clausius, *1822–1888, German mathematical physicist, professor at Zurich and at Bonn. He was one of the creators of the science of thermodynamics. He contributed the concept of entropy, as well as the restatement of the Second Law of Thermodynamics.*

In this equation the change of heat is reckoned as positive if heat flows into the system and negative if it flows out. The proof of the theorem is simple: the cycle can be regarded as a collection of small Carnot cycles. Hence Eq. (23) will be true, provided that it is true for each Carnot cycle. Consider one of the Carnot cycles; it involves the absorption of heat Q_1 at temperature T_1 and the rejection of heat Q_2 at temperature T_2. Thus, the integral $\int dQ/T$ for this cycle may be expressed as

$$\int \frac{dQ}{T} = \frac{Q_1}{T_1} - \frac{Q_2}{T_2} \tag{24}$$

where the negative sign has been inserted in front of Q_2 because this is heat that *leaves* the system. According to Eq. (19), the right side of Eq. (24) is zero — this confirms Eq. (23) for each of the small Carnot cycles and hence establishes the Clausius theorem.

In the p–V diagram a reversible cycle is represented by a closed curve. The Clausius theorem therefore asserts that the integral of dQ/T around any closed curve is zero. From this we can immediately deduce the following: if two points A_1 and A_2 in the p–V diagram are connected by a curve (representing a reversible process), then the integral

$$\int_{A_1}^{A_2} \frac{dQ}{T} \tag{25}$$

only depends on the points A_1 and A_2, but not on the shape of the curve connecting these points. For a proof, consider two curves I and II connecting A_1 and A_2 (Figure 21.8). By the Clausius theorem the integral around the closed curve that starts at A_1, goes to A_2, and returns to A_1 must be zero:

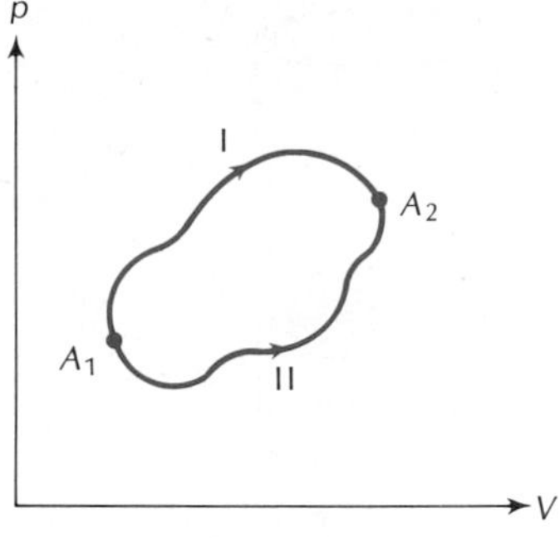

Fig. 21.8 Two paths connecting the points A_1 and A_2 in a p–V diagram.

$$0 = \underset{\text{(curve I)}}{\int_{A_1}^{A_2} \frac{dQ}{T}} + \underset{\text{(curve II)}}{\int_{A_2}^{A_1} \frac{dQ}{T}} \tag{26}$$

i.e.,

$$\underset{\text{(curve I)}}{\int_{A_1}^{A_2} \frac{dQ}{T}} = -\underset{\text{(curve II)}}{\int_{A_2}^{A_1} \frac{dQ}{T}} = \underset{\text{(curve II)}}{\int_{A_1}^{A_2} \frac{dQ}{T}} \tag{27}$$

Hence the integrals along curves I and II are indeed equal.

Obviously, this mathematical argument is quite analogous to that given in Section 8.1 in the study of conservative forces. There the path independence of the integral of $\mathbf{F} \cdot d\mathbf{r}$ permitted us to define a potential-energy function [see Eq. (8.4)]. Likewise, the curve independence of the integral of dQ/T permits us to define a new function

$$\boxed{S(A) = \int_{A_0}^{A} \frac{dQ}{T} + S(A_0)} \tag{28}$$

Entropy (reversible process)

Here A is an arbitrary point in the p–V plane and A_0 is a fixed reference point at which the function S has some prescribed standard value. The new function defined by Eq. (28) is called the **entropy.** The units of entropy are calories per kelvin (cal/K) or joules per kelvin (J/K). The entropy in thermodynamics plays a role somewhat analogous to that of the potential energy in mechanics. Just as the potential energy allows us to make some predictions about the possible motions of a mechanical system (see Section 8.4), the entropy allows us to make some predictions about the possible behavior of a thermodynamic system. And just as we can calculate the force in a mechanical system by differentiating the potential (see Section 8.3), we can calculate certain forcelike quantities in a thermodynamic system by differentiating the entropy.

For an ideal gas, we can express the entropy as an explicit function of p and V [see Eq. (33) below]. Before we give a derivation of this function, we will look at two special examples of how to use Eq. (28) for the calculation of changes of entropy.

EXAMPLE 3. One mole of ideal gas in a cylinder fitted with a piston is made to expand slowly (reversibly) from an initial volume $V_1 = 10^3\ \text{cm}^3$ to a final volume $V_2 = 2 \times 10^3$ cm^3 (Figure 21.9). The cylinder is in contact with a heat reservoir so that, throughout the expansion process, the gas is held at constant temperature. What is the change of entropy of the gas?

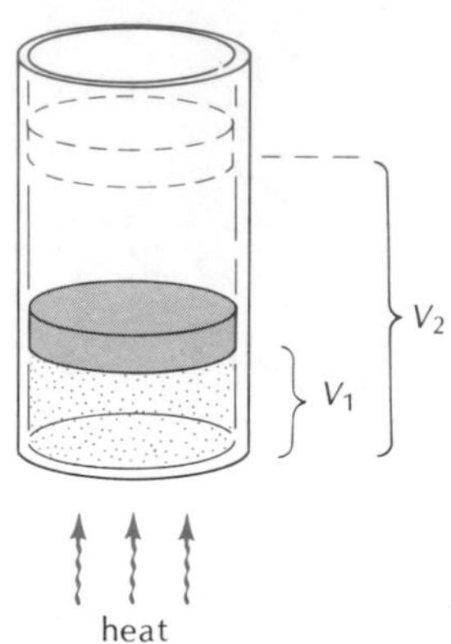

Fig. 21.9 Expansion of a gas at constant temperature. The gas absorbs heat from the heat reservoir.

SOLUTION: The internal energy of an ideal gas is a function of the temperature alone. Consequently, during an isothermal expansion the internal energy remains constant, i.e., $dE = 0$. By Eq. (2), the heat absorbed from the reservoir will be

$$dQ = dW = p\,dV$$

and

$$S_2 - S_1 = \int \frac{dQ}{T} = \int_{V_1}^{V_2} \frac{p}{T} dV$$

From the ideal-gas law, $p/T = nR/V = R/V$ and hence

$$S_2 - S_1 = \int_{V_1}^{V_2} R\frac{dV}{V} = R \ln\left(\frac{V_2}{V_1}\right)$$

Numerically, this gives

$$S_2 - S_1 = 1.99 \text{ cal/K} \times \ln\left(\frac{2 \times 10^3 \text{cm}^3}{10^3 \text{cm}^3}\right)$$

$$= 1.38 \text{ cal/K}$$

EXAMPLE 4. One mole of ideal gas is initially contained in a thermally insulated bottle of volume $V_1 = 10^3$ cm^3. A pipe connects this bottle to a second evacuated bottle; the combined volume of both bottles is $V_2 = 2 \times 10^3$ cm^3 (Figure 21.10). If we suddenly open the stopcock and let the gas rush (irreversibly) into the second bottle so that it finally fills the volume of both bottles, what is the change of entropy of the gas?

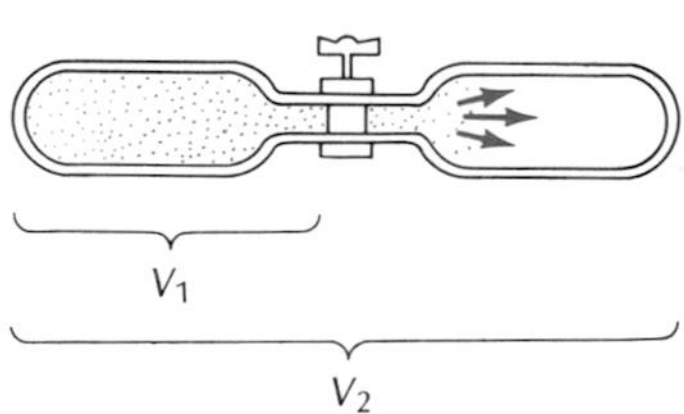

Fig. 21.10 Free expansion of a gas.

SOLUTION: During the free expansion of an ideal gas, the internal energy remains constant and hence the temperature remains constant. Therefore the initial state and the final state of the freely expanding gas are exactly the same as for the isothermally expanding gas of Example 3. Since the entropy is only a function of the initial and final states, the entropy change of the freely expanding gas must be the same as in Example 3,

$$S_2 - S_1 = 1.38 \text{ cal/K}$$

Note that during a free expansion, the gas is not in equilibrium and the pressure and volume are not well defined. Hence this process cannot be represented by a curve in the p–V plane; thus, we cannot directly apply Eq. (28) to this process. To calculate the entropy change, we must first find some reversible process that transforms the gas from the same initial state to the same final state as the free-expansion process. Such a reversible process can be represented by a curve in the p–V diagram and the integral of Eq. (28) can then be evaluated along this curve. In Example 4, isothermal expansion provided us with a suitable reversible process so that, by making a comparison, we could calculate the change of the entropy.

Let us now evaluate the entropy change of an ideal gas that expands from an arbitrary pressure and volume p_1, V_1 to p_2, V_2. We choose some curve in the p–V plane connecting the initial point p_1, V_1 with the final point p_2, V_2. Then

$$S_1 - S_2 = \int_{p_1,V_1}^{p_2,V_2} \frac{dQ}{T} \tag{29}$$

By Eqs. (2) and (20.8),

$$dQ = dE + p\,dV = nC_V\,dT + p\,dV \tag{30}$$

where n is the number of moles of gas. Hence

$$S_2 - S_1 = \int nC_V \frac{dT}{T} + \int \frac{p}{T}\,dV \tag{31}$$

Substituting $p/T = nR/V$ into this, we obtain

$$S_2 - S_1 = \int nC_V \frac{dT}{T} + \int nR\,\frac{dV}{V} \tag{32}$$

Provided that C_V is constant, both of these integrals are easy to do:

$$S_2 - S_1 = nC_V \ln\left(\frac{T_2}{T_1}\right) + nR \ln\left(\frac{V_2}{V_1}\right) \tag{33}$$

Entropy of ideal gas

The temperatures T_2 and T_1 can be expressed in terms of the pressures and volumes ($T = pV/nR$), but it is often convenient to leave Eq. (33) as it is.

Note that Eq. (33) shows very explicitly that the change of entropy does not depend on the detailed steps by which the gas changes from its initial state to its final state. It only depends on the initial and final values of the macroscopic parameters T and V (or p and V).

We have so far discussed only the entropy of an ideal gas. But the Clausius theorem and the definition Eq. (28) of the entropy are valid in general, for any thermodynamic system consisting of any combination of solids, liquids, and gases. The state of any such system can be described by some macroscopic parameters analogous to p and V, and reversible processes in the system can be described by curves in the "space" of these macroscopic parameters. The integral in Eq. (28) can then be evaluated along one of the curves connecting the initial and final states. Any curve or any process will give us the same value for the integral; we must only make sure that the process is reversible.

EXAMPLE 5. A heat reservoir at a temperature of $T_1 = 400$ K is briefly put in thermal contact with a reservoir at $T_2 = 300$ K. If 1 cal of heat flows from the hot reservoir to the cold reservoir, what is the change of the entropy of the system consisting of both reservoirs?

SOLUTION: Obviously, the heat flow from the hot to the cold reservoir is irreversible (more about this in the next section). To evaluate the entropy change by means of Eq. (28), we must invent some process that reversibly takes the system from the same initial state to the same final state. We can imagine that 1 cal flows from the hot reservoir into an auxiliary reservoir of a temperature just barely below 400 K and that simultaneously 1 cal flows from another auxiliary reservoir of a temperature just barely above 300 K into the cold reservoir. Then all processes are reversible and Eq. (28) gives us the change of entropy

$$S(A) - S(A_0) = -\frac{\Delta Q}{T_1} + \frac{\Delta Q}{T_2} = \Delta Q\left(\frac{1}{T_2} - \frac{1}{T_1}\right) \tag{34}$$

$$= 1\ \text{cal}\left(\frac{1}{300\ \text{K}} - \frac{1}{400\ \text{K}}\right) = 8.3 \times 10^{-4}\ \text{cal/K}$$

Thus, the entropy *increases* when heat flows from a hot reservoir to a cold reservoir. As we will see in the next section, all natural processes have a tendency to increase the entropy.

21.4 The Second Law of Thermodynamics

Second Law of Thermodynamics (Kelvin–Planck)

As we saw in Section 21.2, a heat engine operating with an ideal gas as working substance has a limited efficiency — it fails to convert all the heat into work and instead produces waste heat. The **Second Law of Thermodynamics** asserts that this is a limitation from which all heat engines suffer. As formulated by Kelvin and Max Planck,[2] this law simply states: *An engine operating in a cycle cannot transform heat into work without some other effect on its environment.*

Carnot's theorem

It is an immediate corollary of this law that the efficiency of any heat engine operating between two heat reservoirs of high and low temperature is never greater than the efficiency of a Carnot engine. Furthermore, the efficiency of any reversible engine equals the efficiency of a Carnot engine. The proof of these statements, known as **Carnot's theorem,** is by contradiction. Imagine a heat engine of very high efficiency which converts heat from a reservoir at a high temperature into work and only ejects a small amount of waste heat into a reservoir at a low temperature. We can then use the work output of this engine to drive a Carnot engine in reverse, pumping the waste heat from the low-temperature reservoir to the high-temperature reservoir. By hypothesis the given engine is more efficient than a Carnot engine; hence only part of its work output will be needed to drive the reversed Carnot engine and return all of the waste heat to the high-temperature reservoir. The remainder of the output constitutes available work (see Figure 21.11 for a flow chart). The net effect of the joint operation of both engines is then the complete conversion of heat into work, without waste heat, in contradiction with the Second Law. We can only avoid this contradiction if the efficiency of any engine is never greater than that of a Carnot engine.

To prove that the efficiency of any reversible engine equals that of a Carnot engine, we again consider the net effect of the joint operation of the two engines with the Carnot engine running in the forward direction and its work output driving the other reversible engine in the backward direction (Figure 21.12). By an argument similar to that given above, it now follows that the efficiency of the Carnot engine cannot be greater than that of the other engine. Thus, the efficiency of each engine can be no greater than that of the other, i.e., they both must have exactly the same efficiency.

[2] See Section 40.2 for a short biography of Max Planck.

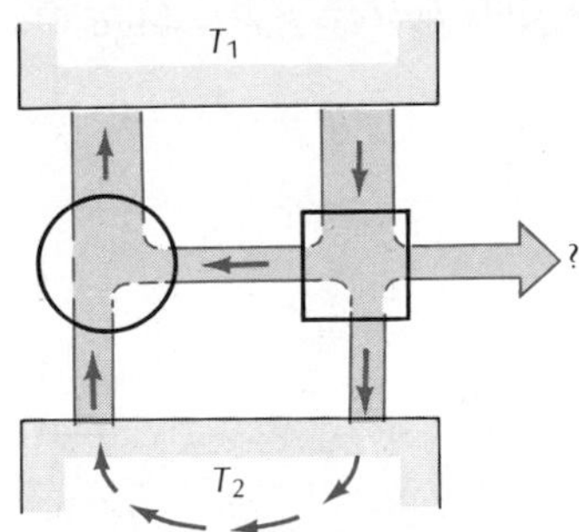

Fig. 21.11 Flow chart for an arbitrary reversible engine (box) and a Carnot engine (circle) connected together. The arbitrary engine drives the Carnot engine in reverse.

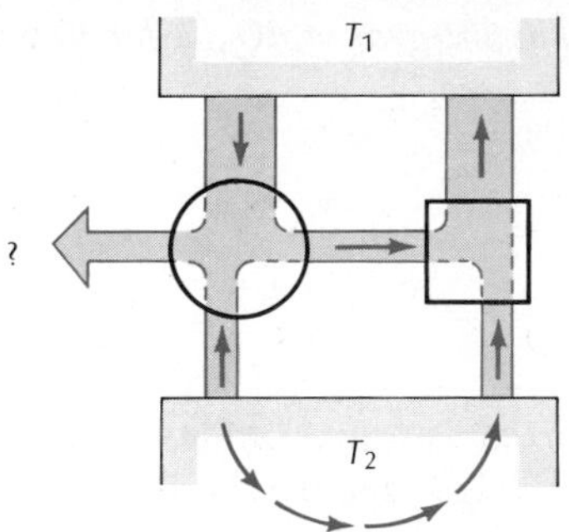

Fig. 21.12 Flow chart for an arbitrary reversible engine (box) and a Carnot engine (circle) connected together. The Carnot engine drives the arbitrary engine in reverse.

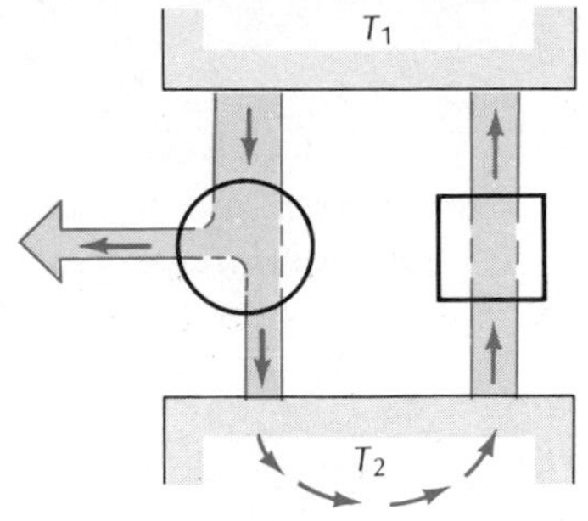

Fig. 21.13 Flow chart for a refrigerator (box) and a Carnot engine (circle). The refrigerator pumps heat from the cold to the hot reservoir without work input.

We recall that a perpetual motion machine of the second kind is an engine that takes heat energy from a reservoir and completely converts it into work. Thus, the Second Law asserts that no perpetual motion machines of the second kind exist. Essentially, the operation of any heat engine hinges on the temperature difference between two heat reservoirs. The heat in the high-temperature reservoir has a high "potential" for work and that in the low-temperature reservoir has a low "potential." If we were to place the two reservoirs in thermal contact, heat would rush from the high-temperature to the low-temperature reservoir. By interposing a heat engine in the path of this rush of heat, we can force the heat to do useful work. Thus, heat engines depend on the tendency of heat to flow from a hot reservoir to a cold reservoir. The Second Law can be expressed in an alternative form that is based on this characteristic of the flow of heat. This formulation, due to Rudolph Clausius, states: *An engine operating in a cycle cannot transfer heat from a cold reservoir to a hot reservoir without some other effect on its environment.* This means that a refrigerator cannot be operated without some work input.

Second Law of Thermodynamics (Clausius)

The Clausius and Kelvin–Planck formulations of the Second Law are equivalent — each implies the other. To see this, suppose that the Clausius statement is false and that there exists a refrigerator that can make heat flow from a cold to a hot reservoir without any work input. If such a device is operated jointly with an ordinary heat engine, it can be used to remove the waste heat from the cold reservoir and return it to the hot reservoir. The net effect would be the complete conversion of heat from the hot reservoir into work (Figure 21.13), in contradiction to the Kelvin–Planck statement. Conversely, suppose the Kelvin–Planck statement is false and a heat engine exists that converts heat from a hot reservoir completely into work. Then the work output of this engine can be used to operate an ordinary refrigerator and the net effect would be the transfer of heat from a cold reservoir to the hot reservoir (Figure 21.14), in contradiction to the Clausius statement.

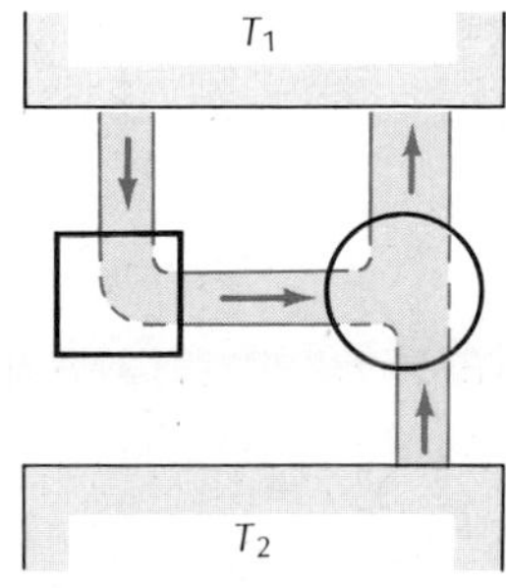

Fig. 21.14 Flow chart for an engine (box) driving a Carnot engine (circle) in reverse. The engine converts heat completely into work, and the Carnot engine acts as a refrigerator.

As we saw in Example 5, a flow of heat from a hot reservoir to a cold reservoir leads to an increase of entropy. This suggests that we express the Second Law in terms of changes of the entropy. To do that, we need the following theorem, a generalization of the theorem of Clausius: *The integral of dQ/T for any cyclic irreversible process is less than or equal to zero,*

Theorem of Clausius (irreversible process)

$$\boxed{\int \frac{dQ}{T} \leq 0} \tag{35}$$

The proof is quite similar to the proof for reversible cycles given in Section 21.3. Figure 21.15 shows an arbitrary irreversible cycle; the axes in this figure have been labeled with p and V, but they could equally well represent some other macroscopic parameters describing an arbitrary thermodynamic system. The cycle consists of a reversible portion indicated by a solid line and an irreversible portion indicated symbolically by the dotted line; this latter portion cannot be represented as a curve in the p–V plane. The cycle can be regarded as a collection of small irreversible cycles consisting of adiabatic and isothermal processes (Figure 21.16). Consider one of these cycles involving the absorption of heat Q_1 at temperature T_1 and the rejection of heat Q_2 at temperature T_2. As in Eq. (24), we can express the integral $\int dQ/T$ for this cycle as

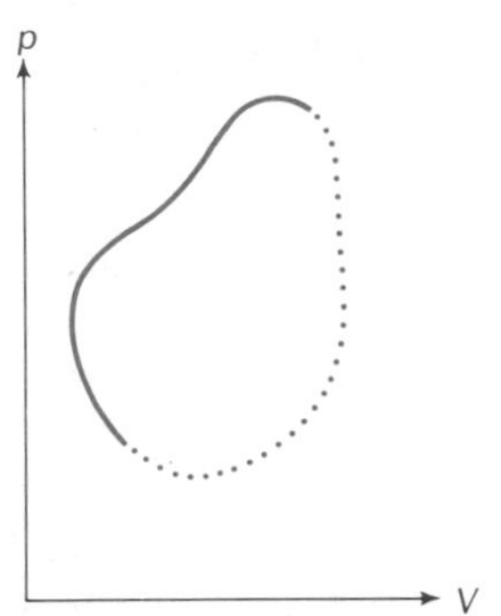

Fig. 21.15 An irreversible cycle. The dotted portion of the path cannot be represented as a curve in the p–V diagram.

$$\int \frac{dQ}{T} = \frac{Q_1}{T_1} - \frac{Q_2}{T_2} \tag{36}$$

By the Second Law of Thermodynamics, the efficiency of this cycle can at most equal that of a Carnot cycle operating between the same temperatures. For the latter cycle $Q_1/T_1 - Q_2/T_2 = 0$; hence for our cycle $Q_1/T_1 - Q_2/T_2 \leq 0$ and consequently the right side of Eq. (36) is less than or equal to zero. This establishes the inequality of Eq. (35) for each small cycle and hence it also establishes this inequality for the entire irreversible cycle.

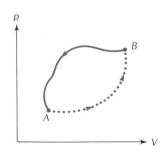

Fig. 21.16 The irreversible cycle of Figure 21.15 can be approximated by a collection of small irreversible cycles.

Let us now see what Eq. (35) says about the entropy change associated with an irreversible process. Suppose that a system in a state A suffers some irreversible process that brings it to state B (Figure 21.17). We can then imagine some reversible process that takes the system back to state A; in Figure 21.17 this process is represented by a solid line. For the complete cycle $A \to B \to A$, we have

$$\int \frac{dQ}{T} \leq 0 \tag{37}$$

or

$$\underset{\text{(irreversible)}}{\int_A^B \frac{dQ}{T}} + \underset{\text{(reversible)}}{\int_B^A \frac{dQ}{T}} \leq 0 \tag{38}$$

Fig. 21.17 An irreversible process takes the system from the point A to the point B. A reversible process returns the system from B to A.

But, by the definition of entropy,

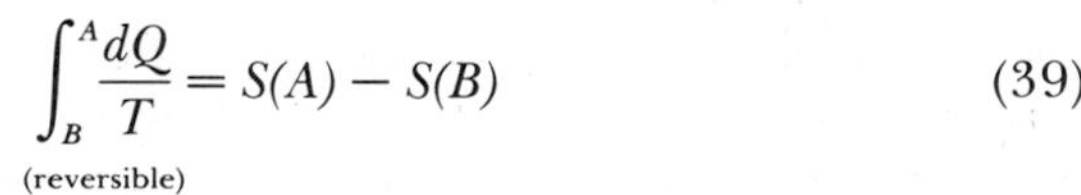

$$\underset{\text{(reversible)}}{\int_B^A \frac{dQ}{T}} = S(A) - S(B) \tag{39}$$

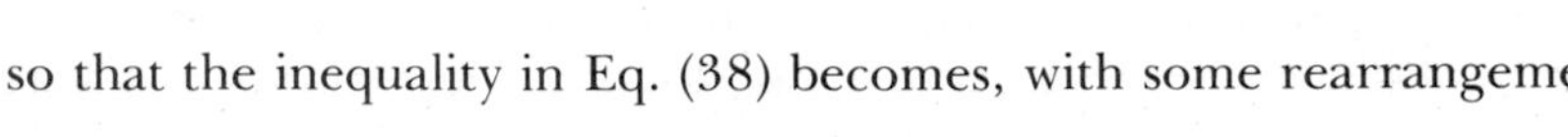

so that the inequality in Eq. (38) becomes, with some rearrangement,

$$S(B) - S(A) \geq \int_A^B \frac{dQ}{T} \quad \text{(irreversible)} \tag{40}$$

Entropy (irreversible process)

This inequality may be regarded as a reformulation of the Second Law of Thermodynamics. In the particular case of an isolated system which exchanges no heat with its surroundings ($dQ = 0$), the inequality becomes

$$S(B) - S(A) \geq 0 \tag{41}$$

This says that *the entropy of an isolated system never decreases* — it either increases or stays constant.

We have already come across two examples of such an increase of entropy in irreversible processes in isolated systems. The first was the spontaneous expansion of a gas released from a bottle (see Example 4) and the second was the spontaneous flow of heat from a hot reservoir to a cold reservoir (see Example 5). In both of these cases, we found by explicit calculation that the entropy of the final configuration was larger than that of the initial configuration.

It is easy to think of other examples. For instance, consider the effect of friction forces on a mechanical system, such as a stone sliding down a hill (in this case the system consists of the stone and the environment). Friction converts mechanical energy into heat; this is obviously an irreversible process and it makes a positive contribution to the right side of Eq. (40). Thus, it brings about an increase of the entropy of the system.

EXAMPLE 6. A large stone, of mass 80 kg, slides down a hill of a vertical height of 100 m and stops at the bottom. What is the increase of entropy of the stone plus the environment? Assume that the temperature of the environment (hill and air) is 270 K.

SOLUTION: All of the initial mechanical energy of the stone is converted into heat:

$$\Delta Q = mgz = 80\text{ kg} \times 9.8\text{ m/s}^2 \times 100\text{ m}$$

$$= 7.8 \times 10^4\text{ J} = 1.87 \times 10^4\text{ cal} \tag{42}$$

This heat is delivered to the environment at a temperature of 270 K.

Obviously, the process is irreversible. To calculate the entropy, we must imagine a reversible process that brings the stone down the hill and delivers the heat into the environment. In principle, we can use an elevator that lets the stone down slowly without friction and extracts work while removing the potential energy; and afterward we can use a heat reservoir of a temperature barely above 270 K to reversibly supply the correct amount of heat [Eq. (42)] to the stone's environment. The first of these processes makes no contribution to the entropy and the second makes a contribution

$$S(B) - S(A) = \frac{\Delta Q}{T} = \frac{1.87 \times 10^4\text{ cal}}{270\text{ K}} = 69\text{ cal/K}$$

From a microscopic point of view, the increase of the entropy of a system is an increase of disorder. This is very obvious in our example of the conversion of mechanical energy into heat by friction. The translational kinetic energy of a macroscopic body is ordered energy — all the particles in the body move in the same direction with the same speed. Heat is disordered energy — the particles move in random directions with a mixture of speeds.

The increase of disorder also holds true in our other examples of spontaneous expansion of a gas and spontaneous flow of heat. When a gas is confined to a small volume, it has more order than when confined to a larger volume; we can best recognize this if we imagine that the small volume is *very* small — if all the particles of gas are at just about one point, the system clearly has a very high degree of order. When a given amount of heat is added to a reservoir of low temperature, it causes more disorder than when the same amount of heat is added to a reservoir of high temperature; in both reservoirs the added heat generates extra random motion and extra disorder, but in the cold reservoir the percent increment of the random motions is larger than in the hot reservoir and consequently the extra disorder is larger.

Walther Hermann Nernst, *1864–1941, German physicist and chemist. He was a pioneer in physical chemistry and received the Nobel Prize in chemistry in 1920 for his discovery of the Third Law of Thermodynamics.*

The connection between entropy and disorder can be given a precise mathematical meaning, but the details would require a discussion of statistical mechanics and of information theory — and here we cannot go into this. The Second Law can be reformulated to say that the processes in an isolated system always tend to increase the disorder.

The human activities on the Earth, like any other process in nature, cause an increase of disorder. Of course, some of our activities result in an increase of order in some portion of the system. For instance, when we extract dispersed bits of metal from ores and assemble them into a watch, we are obviously increasing the order of the bits of metal. However, we can only do this by simultaneously generating disorder somewhere else — the smelting of ores and the machining of metals demands an input of energy which is converted into waste heat and increases the disorder of our environment. The net result is always an increase of disorder. All our activities depend on a supply of highly ordered energy in the form of chemical or nuclear fuels, or light from the Sun that can "soak up" disorder while becoming degraded into waste heat. We continually convert useful, ordered energy into useless, disordered energy.

Third Law of Thermodynamics

We end this chapter with a brief statement of the **Third Law of Thermodynamics.** This law, as formulated by Walther Nernst, asserts that *the entropy of a system at absolute zero is a universal constant (independent of all the macroscopic parameters describing the system) which may be set equal to zero.* We can understand this law in terms of the connection between entropy and disorder, mentioned above. As we lower the temperature of a system, we decrease the random thermal motions and we decrease the disorder. According to classical kinetic theory, the random thermal motions cease completely as the temperature of the system approaches absolute zero.[3] The system then tends to settle into a state of minimum disorder, i.e., a state of minimum entropy.

Incidentally, a physical system can never quite reach the temperature of absolute zero; it can only approach this temperature asymptoti-

[3] Actually, owing to quantum effects, the motion never ceases completely — at absolute zero there remains a zero-point motion. But this zero-point motion is not a disordered motion, and it does not contribute to the entropy.

cally. This unattainability of the absolute zero can be shown to be mathematically equivalent to the vanishing of the entropy at absolute zero. Hence, in an alternative formulation, the Third Law of Thermodynamics asserts that the absolute zero of temperature is unattainable. The lowest temperature achieved in laboratory experiments to date is 5×10^{-8} K.

SUMMARY

First Law of Thermodynamics: $\Delta E = \Delta Q - \Delta W$

Efficiency of heat engine: $e = 1 - \dfrac{Q_2}{Q_1}$

Efficiency of Carnot engine: $e = 1 - \dfrac{T_2}{T_1}$

Carnot's theorem: The efficiency of any engine cannot exceed that of a Carnot engine; the efficiency of any reversible engine equals that of a Carnot engine.

Clausius theorem (reversible cycle): $\displaystyle\int \frac{dQ}{T} = 0$

Definition of entropy: $S(A) = \underset{\text{(reversible)}}{\displaystyle\int_{A_0}^{A} \frac{dQ}{T}} + S(A_0)$

Second Law of Thermodynamics: An engine operating in a cycle cannot transform heat into work without some other effect on its environment *or* an engine operating in a cycle cannot transfer heat from a cold reservoir to a hot reservoir without some other effect on its environment.

Clausius theorem (irreversible cycle): $\displaystyle\int \frac{dQ}{T} \leq 0$

Change of entropy in irreversible process: $S(B) - S(A) \geq \underset{\text{(irreversible)}}{\displaystyle\int_{A}^{B} \frac{dQ}{T}}$

Third Law of Thermodynamics: At absolute zero the entropy is zero.

QUESTIONS

1. Figure 21.18 shows a perpetual motion machine designed by M. C. Escher. Exactly at what point is there a defect in this design?

Fig. 21.18 Waterfall. Lithograph by M. C. Escher.

2. An inventor proposes the following scheme for the propulsion of ships on the ocean without an input of energy: cover the hull of the ship with copper sheets and suspend an electrode of zinc in the water at some distance from the hull; since sea water is an electrolyte, the hull of the ship and the electrode will then act as the terminals of a battery which can deliver energy to an electric motor propelling the ship. Will this scheme work? Does it violate the First Law of Thermodynamics?

3. The **Zeroth Law of Thermodynamics,** found in some textbooks, states that *whenever two bodies are individually in thermal equilibrium with a third body, then they are also in thermal equilibrium with each other.* If we take as the third body

the ideal-gas thermometer described in Section 19.2, then the Zeroth Law states that two bodies at the same ideal-gas temperature are in thermal equilibrium with each other. Is this assertion self-evident? Is it necessary to perform an experiment to check that two bodies at the same temperature are in thermal equilibrium, i.e., that there is no heat flow between them? If the Zeroth Law were not true, would the concept of temperature make any sense?

4. An electric power plant consists of a coal-fired boiler that makes steam, a turbine, and an electric generator. The boiler delivers 90% of the heat of combustion of the coal to the steam; the turbine converts 50% of the heat of the steam into mechanical energy, and the electric generator converts 99% of this mechanical energy into electric energy. What is the overall efficiency of generation of electric power?

5. The lowest-temperature heat sink available in nature is interstellar and intergalactic space with a temperature of 3 K. Why don't we use this heat sink in the operation of a Carnot engine?

6. In some showrooms, salesmen demonstrate air conditioners by simply plugging them into an outlet without bothering to install them in a window or a wall. Does such an air conditioner cool the showroom or heat it?

7. Can mechanical energy be converted completely into heat? Give some examples.

8. If you leave the door of a refrigerator open, will it cool the kitchen?

9. Which of the following processes are irreversible, which are reversible? (a) Burning of a piece of paper. (b) Slow descent of an elevator attached to a perfectly balanced counterweight. (c) Breaking of a windowpane. (d) Explosion of a stick of dynamite. (e) Electric pump lifting water from a low-level reservoir to a high-level reservoir. (f) Block being slowly pushed up a frictionless inclined plane.

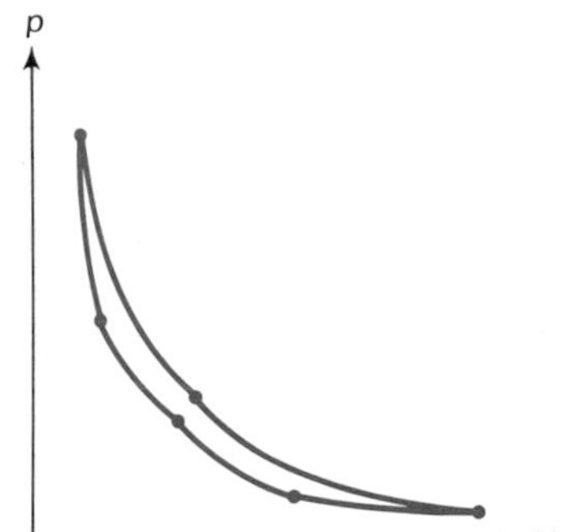

Fig. 21.19 A modified Carnot cycle.

10. Consider a modified Carnot cycle consisting of an isothermal expansion, an adiabatic expansion, an isothermal compression, an adiabatic compression, another isothermal compression, and another adiabatic compression (see Figure 21.19). Does the efficiency of this cycle differ from that of the standard Carnot cycle?

11. If we measure the efficiency of a Carnot engine directly, we can use Eq. (20) to calculate the temperature of one of the heat reservoirs. What are the advantages and the disadvantages of such a thermodynamic determination of temperature?

12. In lack of a heat sink, energy cannot be extracted from an ocean at uniform temperature. However, within the oceans there are small temperature differences between the warm water on the surface and the cooler water in the depths. Design a Carnot engine that exploits this temperature difference.

13. Why are heat pumps used for heating houses in mild climates but not in very cold climates?

14. Does the Second Law of Thermodynamics forbid the spontaneous flow of heat between two bodies of equal temperature?

15. The inside of an automobile parked in the sun becomes much hotter than the surrounding air. Does this contradict the Second Law of Thermodynamics?

16. According to a story by George Gamow,[4] on one occasion Mr. Tompkins was drinking a highball when all of a sudden one small part at the surface of the liquid became hot and boiled with violence, releasing a cloud of steam, while the remainder of the liquid became cooler. Is this consistent with the First Law of Thermodynamics? With the Second Law?

[4] G. Gamow, *Mr. Tompkins in Paperback.*

17. A vessel is divided into two equal volumes by a partition. One of these volumes contains helium gas, and the other contains argon at the same temperature and pressure. If we remove the partition and allow the gases to mix, does the entropy of the system increase?

18. Does the motion of the planets around the Sun generate entropy?

19. Does static friction generate entropy?

20. Consider the process of emission of light by the surface of the Sun followed by absorption of this light by the surface of the Earth. Does this entail an increase of entropy?

21. Suppose that a box contains gas of extremely low density, say, only 50 molecules of gas altogether. The molecules move at random and it is possible that once in a while all of the 50 molecules are simultaneously in the left half of the box, leaving the right half empty. At this instant the entropy of the gas is less than when the gas is more or less uniformly distributed throughout the box. Is this a violation of the Second Law of Thermodynamics? What are the implications for the range of validity of this law?

22. **Maxwell's demon** is a tiny hypothetical creature that can see individual molecules. The demon can make heat flow from a cold body to a hot body as follows: Suppose that a box initially filled with gas at uniform temperature and pressure is divided into two equal volumes by a partition, equipped with a small door which is closed but can be opened by the demon (Fig. 21.20). Whenever a molecule of above-average speed approaches the door from the left, the demon quickly opens the door and lets it through. Whenever a molecule of below-average speed approaches from the right, the demon also lets it through. This selective action of the demon accumulates hot gas in the right volume, and cool gas in the left volume. Does this violate the Second Law of Thermodynamics? Do any of the activities of the demon involve an *increase* of entropy?

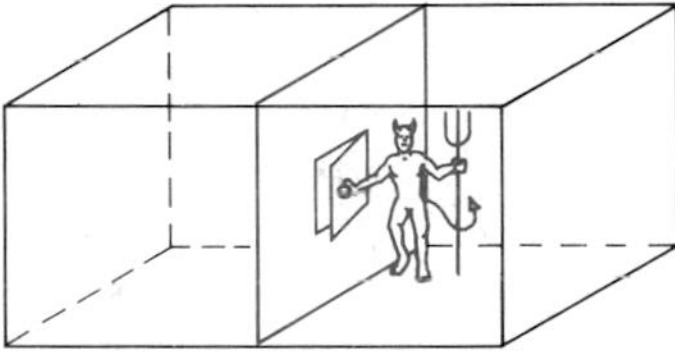

Fig. 21.20 Maxwell's demon.

23. According to one cosmological model based on Einstein's theory of General Relativity, the universe oscillates — it expands, then contracts, then expands, and then contracts, etc., etc. If the Second Law of Thermodynamics is valid, can each cycle of oscillation be the same as the preceding cycle?

24. **Negentropy** is defined as the negative of the entropy ([negentropy] $= -S$). Explain the following statement: "In our everyday activities on the Earth, we do not consume energy, but we consume negentropy."

25. The amount of energy dissipated in the United States per year is 8×10^{19} J. Roughly, what is the entropy increase that results from this dissipation?

26. If you tidy up a messy room you are producing a decrease of disorder. Does this violate the Second Law?

PROBLEMS

Section 21.1

1. Consider 1 kg of water at 100°C surrounded by air at a constant pressure of 1 atm. If we vaporize all of this water and convert it into steam at 100°C and 1 atm pressure, what is the heat absorbed by the water? The work done by the water vapor against atmospheric pressure? The change of the internal energy of the water? Assume that steam obeys the ideal-gas law.

2. A ball of lead of mass 0.25 kg drops from a height of 0.80 m, hits the floor, and remains there at rest. Assume that all the heat generated during the impact remains within the lead. What are the values of ΔQ, ΔW, and ΔE for the lead during this process? What is the increase of temperature of the lead?

3. At a pressure of 1 atm, the heat of vaporization of water is 539 kcal/kg; this is the heat required to convert 1 kg of water at 100°C into water vapor at

the same temperature. Given that the specific heat of water is $c = 1$ kcal/kg · °C and that of water vapor is (approximately[5]) $c_p = 0.48$ kcal/kg · °C, use the First Law of Thermodynamics to calculate the heat of vaporization of water at 20°C. [Hint: Instead of directly converting water at 20°C into water vapor, we can first heat the water to 100°C, then vaporize it, and then cool the vapor to 20°C (without condensation). The net work done against atmospheric pressure is the same during this indirect process as during direct vaporization (why?); hence, according to the First Law, the heat absorbed during each process is also the same.]

4. Calculate the heat of vaporization of water at a temperature of 140°C, at a pressure of 1 atm. Use the data given in Problem 3.

5. Calculate the heat of vaporization of water at a temperature of 100°C, at a pressure of 2 atm. Assume that water vapor behaves as an ideal gas.

Section 21.2

6. Each of the two engines of a DC-3 airplane produces 1100 hp. The engines consume gasoline; the combustion of 1 kg of gasoline yields 44×10^6 J. If the efficiency of the engines is 20%, at what rate do the two engines consume gasoline?

7. A nuclear power plant generates 1000 megawatts of electric (or mechanical) power. If the efficiency of this plant is 33%, at what rate does the plant generate waste heat? If this waste heat is to be removed by passing water from a river through the plant, and if the water is to suffer a temperature increase of at most 8°C, how many cubic meters of water per second are required?

8. Equation 20.37 gives the work done during the adiabatic expansion (or contraction) of a gas. Use this equation to calculate the amount of work done by the gas during the adiabatic expansion and the contraction in the Carnot cycle described in Figure 21.6. Verify explicitly that the amount of the work for the complete Carnot cycle coincides with $Q_1 - Q_2$.

9. A gun may be regarded as a thermodynamic engine. The barrel of the gun is a cylinder; the shot moving along the barrel is a piston; and the hot gases released by the combustion of the explosive charge do work on this piston, giving it a large kinetic energy. During the first few instants after the detonation of the charge, there is simultaneous combustion and expansion; but after the combustion is complete, the expansion proceeds adiabatically. Experiments with a gun with a barrel of length 12 ft and diameter 6 in. firing a shot of weight 100 lb have shown that when the combustion ends, the shot is 3.2 ft along the barrel and has a speed of 1020 ft/s; the pressure of the gas at this instant is 16 tons/in.^{2}.

(a) How much work does the adiabatically expanding gas do on the shot as it travels the remaining 8.8 ft along the barrel? Assume that $\gamma = 1.2$.

(b) What will be the muzzle velocity of the shot? Ignore friction.

10. A Carnot engine has 40 g of helium gas as its working substance. It operates between two reservoirs of temperatures $T_1 = 400$°C and $T_2 = 60$°C. The initial volume of the gas is $V_1 = 2.0 \times 10^{-2}$ m^3, the volume after isothermal expansion is $V_2 = 4.0 \times 10^{-2}$ m^3, and the volume after the subsequent adiabatic expansion is $V_3 = 5.0 \times 10^{-2}$ m^3.

(a) Calculate the work generated by this engine during one cycle.

(b) Calculate the heat absorbed and the waste heat ejected during one cycle.

11. A Carnot engine operates between two heat reservoirs of temperature 500°C and 30°C, respectively.

(a) What is the efficiency of this engine?

---

[5] The specific heat of water is somewhat dependent on temperature.

(b) If the engine generates 1.5×10^3 J of work, how many calories of heat does it absorb from the hot reservoir? Eject into the cold reservoir?

12. In principle, nuclear reactions can achieve temperatures of the order of 10^{11} K. What is the efficiency of a Carnot engine taking in heat from such a nuclear reaction and exhausting waste heat at 300 K?

13. An automobile engine takes heat from the combustion of gasoline, converts part of this heat into mechanical work, and ejects the remainder into the atmosphere. The temperature attainable by the combustion of gasoline is about 2100°C and the temperature of the atmosphere is 20°C. What is the maximum theoretical efficiency of an engine operating between these temperatures? (The actual efficiency attained by an automobile engine is typically 0.2, much lower than the theoretical maximum.)

14. On a hot day a house is kept cool by an air conditioner. The outside temperature is 32°C and the inside temperature is 21°C. Heat leaks into the house at the rate of 9000 kcal/h. What is the minimum mechanical power that the air conditioner requires to hold the inside temperature constant?

15. A scheme for the extraction of energy from the oceans attempts to take advantage of the temperature difference between the upper and lower layers of ocean water. The temperature at the surface in tropical regions is about 25°C; the temperature at a depth of 1000 ft is about 5°C.
(a) What is the efficiency of a Carnot engine operating between these temperatures?
(b) If a power plant operating at the maximum theoretical efficiency generates 1 megawatt of mechanical power, at what rate does this power plant release waste heat?
(c) The power plant obtains the mechanical power and the waste heat from the surface water by cooling this water from 25°C to 5°C. At what rate must the power plant take in surface water?

16. In a nuclear power plant, the reactor produces steam at 520°C and the cooling tower eliminates waste heat into the atmosphere at 30°C. The power plant generates 500 megawatts of electric (or mechanical) power.
(a) If the efficiency is that of a Carnot engine, what is the rate of release of waste heat (in megawatts)?
(b) Actual efficiencies of nuclear power plants are about 33%. For this efficiency, what is the rate of release of heat?

17. An air conditioner removes 8000 Btu/h of heat from a room at a temperature of 70°F and ejects this heat into the ambient air at a temperature of 80°F. This air conditioner requires 950 W of electric power.
(a) How much mechanical power would a Carnot engine, operating in reverse, require to remove this heat at the same rate?
(b) By what factor is the power required by the air conditioner larger than that required by the Carnot engine?

18. An ice-making plant consists of a reversed Carnot engine extracting heat from a well-insulated ice box. The temperature in the icebox is −5°C and the temperature of the ambient air is 30°C. Water, of an initial temperature of 30°C, is placed in the icebox and allowed to freeze and to cool to −5°C. If the ice-making plant is to produce 10,000 kg of ice per day, what mechanical power is required by the Carnot engine?

19. The boiler of a power plant supplies steam at 1000°F to a turbine which generates mechanical power. The steam emerges from the turbine at 500°F and enters a steam engine that generates extra mechanical power. The steam is finally released into the atmosphere at a temperature of 100°F. Assume that the conversion of heat into work proceeds with the maximum efficiency permitted by Carnot's theorem.
(a) What is the efficiency of the turbine? Of the steam engine?
(b) What is the net efficiency of both engines acting together? How does it

compare with the efficiency of a single engine operating between 1000°F and 100°F?

*20. One liter of water is initially at the same temperature as the surrounding air, 30°C. You wish to cool this water to 5°C by transferring heat from the water into the air. What is the minimum amount of work that you must supply in order to accomplish this? (Hint: Take a sequence of Carnot engines operating in reverse; use each engine to reduce the temperature by an infinitesimal amount.)

Sections 21.3 and 21.4

21. Using Eq. (33), prove that in the adiabatic expansion of an ideal gas, the final entropy equals the initial entropy.

22. On a winter day heat leaks out of a house at the rate of 2.5×10^4 kcal/h. The temperature inside the house is 21°C and the temperature outside is −5°C. At what rate does this process produce entropy?

23. Consider the air conditioner described in Problem 17. Calculate the rate of increase of entropy contributed by the operation of this air conditioner.

24. Your body generates about 2000 kcal of heat per day. Estimate how much entropy you generate per day. Neglect the (small) amount of entropy that enters your body in the food you consume.

25. A steam engine operating between reservoirs at temperatures of 900°F and 80°F has an efficiency of 40%. The engine delivers 2000 hp of mechanical power. At what rate does this engine generate entropy?

26. Suppose that 1.0 kg of water freezes while at 0°C. What is the change of entropy of the water during this freezing process?

27. (a) Consider a material with a constant specific heat capacity c. Show that if we gradually supply heat to a mass m of this material, increasing its temperature from T_1 to T_2, the increase of entropy of the mass m is

$$S_2 - S_1 = mc \ln(T_2/T_1)$$

(b) What is the increase of entropy in 1.0 kg of water that is heated from 20°C to 80°C?

28. Consider the process of dissolving ice cubes in water as described in Example 20.4. What is the change of entropy of the system? (Hint: Use the formula derived in Problem 27.)

29. For an automobile moving at a constant speed of 65 km/h on a level road, rolling friction, air friction, and friction in the drive train absorb a mechanical power of 12 kW. At what rate do these processes generate entropy? The temperature of the environment is 20°C.

30. An automobile of 2100 kg moving at 80 km/h brakes to a stop. In this process the kinetic energy of the automobile is first converted into thermal energy of the brake drums; this thermal energy later leaks away into the ambient air. Suppose that the temperature of the brake drums is 60°C when the automobile stops, and that the temperature of the air, and the final temperature of the brake drums, is 20°C.

(a) How much entropy is generated by the conversion of mechanical energy into thermal energy of the brake drums?

(b) How much extra entropy is generated as the heat leaks away into the air?

31. At Niagara Falls, 5700 m^3/s of water fall through a vertical distance of 50 m, dissipating all of their gravitational energy. Calculate the rate of increase of entropy contributed by this falling water. The temperature of the environment is 20°C.

ENERGY, ENTROPY, AND ENVIRONMENT[1]

Ours is an age of mechanization. We use electric motors and internal combustion engines to run gadgetry and machinery ranging from electric toothbrushes to supertankers. The intensive use of power from purely electric and mechanical sources is a fairly recent development in the history of our civilization. Up to the middle of the nineteenth century, human and animal muscles were the main source of power. The men of ancient Egypt, ancient Rome, and medieval Europe tilled and harvested their fields and erected their pyramids, aqueducts, and cathedrals by hand; oxen, horses, and mules pulled plows or carts and carried loads. The only purely mechanical sources of power were windmills, waterwheels, and sailing ships. The power developed by windmills and waterwheels was modest — about 2 kW for a waterwheel and about 10 kW for a windmill, although the large water machine constructed at Marly in 1684 to pump water for the fountains of Versailles developed a power of about 100 kW. Sailing ships were the most powerful machines prior to the mechanical age — in a gale of wind, the sails of a clipper ship running at its maximum speed of 20 knots developed a propulsive power of about 15,000 kW (or about 20,000 hp).

Around the middle of the nineteenth century, steam engines became the principal sources of power in the industrial world. Later on, turbines and internal combustion engines came into use and these remain at present our principal sources of power. All these engines consume prodigious amounts of fossil fuel in the form of oil, coal, and natural gas — in the next 30 years the United States will consume about as much fuel as it has in the 200 years since its birth. This voracious consumption raises the question of how much longer our reserves of fuel will last and what will replace them when they finally do run out.

F.1 THE CONVERSION OF ENERGY

Energy is always conserved; it cannot be consumed, but only converted from one form to another. Our civilization does not face a problem of energy consumption, but one of the energy dissipation. We are converting useful forms of enegy into useless forms of energy.

Consider for instance, the sequence of energy conversions that take place in an automobile propelled by a gasoline engine. The gasoline contains chemical energy (gasoline releases 3.4×10^7 J/liter when burned with oxygen at atmospheric pressure) and the engine partially converts this chemical energy into mechanical energy. Figure F.1 is a flow chart showing the successive energy conversions for a typical automobile traveling at 65 km/h on a level road. The engine delivers the mechanical energy to the transmission and axle of the automobile. From there, one part of the energy goes into the body of the automobile, compensating for the losses that the body suffers because of friction against the air, and another part goes into the wheels, compensating for the losses due to rolling friction against the ground. The engine consumes fuel at the rate of 7.6 liter/h, corresponding to an input of 72 kW of chemical energy, but only 9 kW of this emerges as propulsive power doing useful mechanical work against the external friction forces of air and road. Most of the energy is dissipated in the conversion process. Note that the largest waste occurs in the engine itself — in a gasoline engine most of the chemical energy of the fuel is ejected into the exhaust gases. Diesel engines achieve a somewhat higher efficiency because they burn their fuel more completely.

However, in part the inefficiency of the engine is a

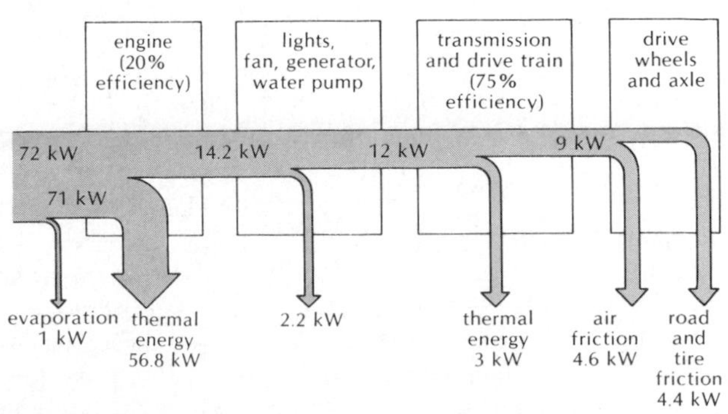

Fig. F.1 Flow chart summarizing energy conversions in automobile. (Based on R. H. Romer, *Energy: An Introduction to Physics.*)

[1] This chapter is optional.

consequence of the basic laws of thermodynamics. The gasoline engine is a heat engine — it converts the heat released by the burning fuel into mechanical energy. The efficiency of an ideal heat engine is [see Eq. (21.20)]

$$e = 1 - \frac{T_2}{T_1} \tag{1}$$

where T_1 is the temperature of the heat source and T_2 that of the heat sink. In the gasoline engine the heat source is the hot gas produced by combustion and the heat sink is the air surrounding the engine. The complete combustion of hydrocarbon fuel gives the residual gas a temperature of about 2400 K. The air surrounding the engine will typically have a temperature of 300 K. Hence the maximum attainable efficiency is

$$e = 1 - \frac{300\text{ K}}{2400\text{ K}} = 0.88 \tag{2}$$

Gasoline engines, and other engines that burn hydrocarbon fuels, do not come anywhere near this maximum efficiency, because they do not burn the fuel completely, they eject exhaust gases at temperatures much higher than 300 K, and they suffer from internal friction.

In any case, Eq. (2) tells us there is a limit to the conversion of chemical energy of fuel into mechanical energy. An appreciable part of the chemical energy of fuel is disordered energy. If we want to convert disordered energy into ordered energy, we have to pay a price. In general, the disorder inherent in a given kind of energy can be measured by the corresponding entropy. Table F.1 lists some kinds of energy and the corresponding entropy per joule. The kinetic and potential energies of macroscopic bodies are the most ordered kinds of energy (least entropy) and the thermal fireball radiation is the most disordered kind (most entropy). Energy of low entropy can be totally converted into energy of higher entropy; for instance, the gravitational potential energy of the water stored behind a dam can, in principle, be converted into an equal amount of mechanical or electric energy (in practice, hydroelectric power stations attain an efficiency of 85%). Energy of high entropy can only be

Table F.1 ENERGY AND ENTROPY

Kind of energy	*Entropy per joule*
Kinetic (macroscopic)	0 J/K
Gravitational potential	0
Electric potential	0
Thermal, from nuclear reactions	10^{-11}
Thermal, within Sun	10^{-7}
Sunlight	10^{-5}
Thermal, from chemical reactions	$\sim 10^{-4}$
Thermal, in Earth's environment	3×10^{-3}
Thermal, cosmic background radiation	3×10^{-1}

Table F.2 PRACTICAL ENERGY CONVERSION EFFICIENCIES

Conversion from	*Conversion to:* *Mechanical*	*Gravitational potential*	*Electric*	*Radiant (light)*	*Chemical*	*Thermal*
Mechanical	—	—	99% (electric generator)	—	—	100% (brake drum)
Gravitational potential	86% (water turbine)	—	85% (hydroelectric plant)	—	—	—
Electric	93% (electric motor)	80% (pumped storage)	—	40% (gas laser)	72% (storage battery)	100% (heating coil)
Radiant (light)	—	—	12% (solar cell)	—	0.6% (photosynthesis)	100% (solar furnace)
Chemical	~30% (animal muscle)	—	91% (dry cell battery)	15% (chemical laser)	—	88% (furnace of steam boiler)
Thermal	47% (steam turbine, rocket engine)	—	7% (thermocouple)	3% (light bulb)	—	—

partially converted into energy of low entropy; some of the energy must be converted into energy of even higher entropy so as to guarantee the net increase of entropy required by the Second Law of Thermodynamics.

In practice, the efficiency of energy conversion is often much below the theoretical limit set by thermodynamics. Table F.2 lists the maximum efficiencies attainable with present-day devices.

F.2 THE DISSIPATION OF ENERGY

Most of the mechanical and electric energy supplied to our machinery as well as the heat used in our homes and factories is generated from fossil fuels — oil, coal, and natural gas. A small part is contributed by hydroelectric dams and nuclear reactors. Figure F.2 summarizes the rate of energy dissipation in the United States since 1850; the figure includes an extrapolation up to the year 2000.

Essentially all of this energy is dissipated into our environment, ultimately becoming degraded heat at a temperature of about 300 K. The example of the automobile mentioned in the preceding section illustrates this dissipation process. The engine of the automobile releases thermal energy with the exhaust gases; as these mix with the air, their temperature gradually drops to 300 K, and the thermal energy becomes degraded. Simultaneously, the engine supplies mechanical energy to the body and wheels of the car; air and road friction convert this energy into heat and this also becomes degraded as it spreads into the environment. The net result is the conversion of chemical energy into degraded thermal energy. The unburned hydrocarbons in the exhaust represent some small amounts of chemical energy that is not degraded directly. But natural reactions in the atmosphere gradually decompose these hydrocarbons, so that ultimately their energy will also be degraded. In Figure F.2 no distinction has been made between such delayed dissipation and immediate dissipation, i.e., Figure F.2 is based on the yearly rate of consumption of fuel.

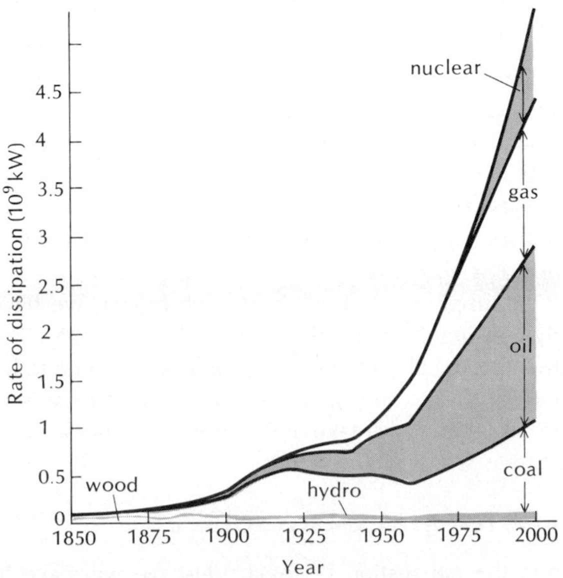

Fig. F.2 Rate of energy dissipation in the United States. (Based on *Scientific American,* September 1971.)

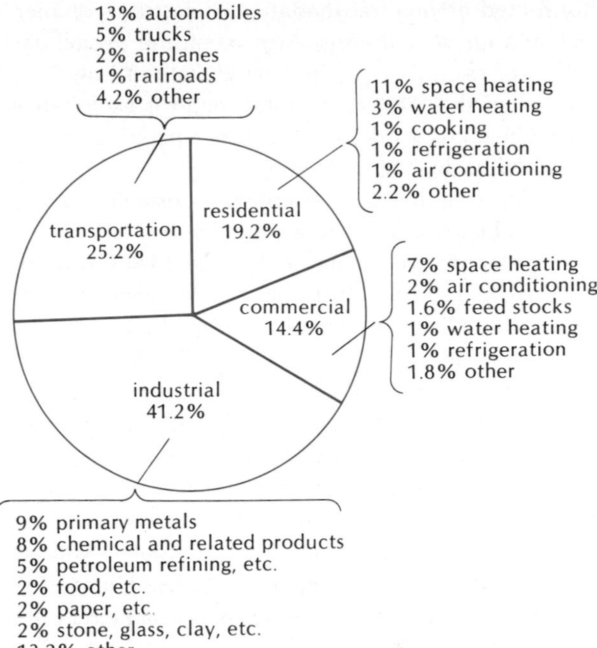

Fig. F.3 Distribution of energy dissipation in the United States. (Based on R. L. Loftness, *Energy Handbook.*)

At present (1980) the rate of dissipation of energy in the United States is 2.2×10^6 MW.[2] Worldwide, the rate of dissipation is about four times as large. The United States has by far the highest per capita rate of energy dissipation — about 14 kW per capita. Although this is in part due to the very high standard of living enjoyed by Americans and to the prodigious expenditure of energy in heavy industry and in the manufacture of military hardware, it is also in part due to the careless use of energy — inefficient trucks and automobiles, wasteful industrial installations, poor thermal insulation in homes. The standard of living enjoyed by Swedes and by Germans is also very high, but their per capita rate of dissipation of energy is much lower — about 6 kW per capita.

Figure F.3 shows how the dissipation of energy is

[2] Large amounts of power are usually measured in megawatts,

$$1 \text{ MW} = 10^6 \text{ W}$$

Large amounts of energy are often measured in Quad,

$$1 \text{ Quad} = 10^{15} \text{ Btu} = 1.05 \times 10^{18} \text{ J}$$

or in Q,

$$1 \text{ Q} = 10^{18} \text{ Btu} = 1.05 \times 10^{21} \text{ J}$$

distributed among transportation, industrial, commercial, and residential users. At present, the overall efficiency of energy use in the United States is only 0.41; this means that 41% of the total energy is exploited as useful heat and work before it is dissipated, and 59% is wasted or dissipated directly.

In almost all the uses of energy, substantial savings are possible. According to a study by the National Bureau of Standards, savings as high as 30% should be possible in industry. We could also significantly increase the efficiency of transportation by heavier use of trains. Trains are more energy efficient than airplanes, trucks, or automobiles — a train can transport a given amount of freight over a given distance with only $\frac{1}{4}$ of the energy expenditure required by a truck, or $\frac{1}{60}$ the energy expenditure required by an airplane. With improved insulation in residential homes, we could cut the consumption of heating fuel — with insulation a typical home requires 45% less heating fuel than a home without any insulation. The added insulation can pay for itself out of savings on the fuel bill in just a few years. Furthermore, the simple expedient of lowering the thermostat in winter can save 15 to 20% of the heating fuel for a 5°F reduction of indoor temperature.[3]

F.3 THE LIMITS OF OUR RESOURCES

Worldwide, the rate of dissipation of energy is increasing by a factor of 2 every 10 years. Most of this energy is being generated from fossil fuels. The total amount of such fuels stored in natural deposits under the surface of the Earth is limited. Hence, within the foreseeable future the rising demand will exhaust our supplies of fossil fuels. How much longer will these fuels last?

The supply of oil and natural gas will be exhausted fairly soon; we have already consumed almost a fourth of the original supply of these fuels (see Table F.3). Only a small fraction of the oil will be left 50 years from now; next to none will be left 100 years from now (Figure F.4). The exploitation of oil shale and of tar sand could provide extra oil — up to 10 times more oil than the total obtained from liquid petroleum deposits. However, it is not yet clear to what extent we can economically extract this oil from the stone in which it is absorbed.

Our supply of coal will last much longer; we have only consumed about 2% of the original supply. The rate of production of coal will probably continue to increase well into the next century; at current rates of consumption, it will take 300 or 400 years before coal becomes scarce (Figure F.5).

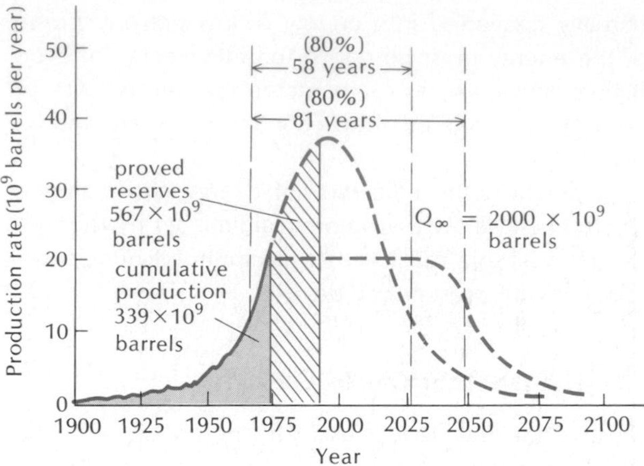

Fig. F.4 Estimated world production of oil. Beyond the year 2000, production is expected to decline because oil wells will run dry and few new wells will be discovered. The total supply is $Q_\infty = 2000 \times 10^9$ barrels. The upper curve assumes that 80% of this total is consumed in 58 years; the lower curve assumes 81 years. (Based on M. K. Hubbert, *American Journal of Physics,* November 1981.)

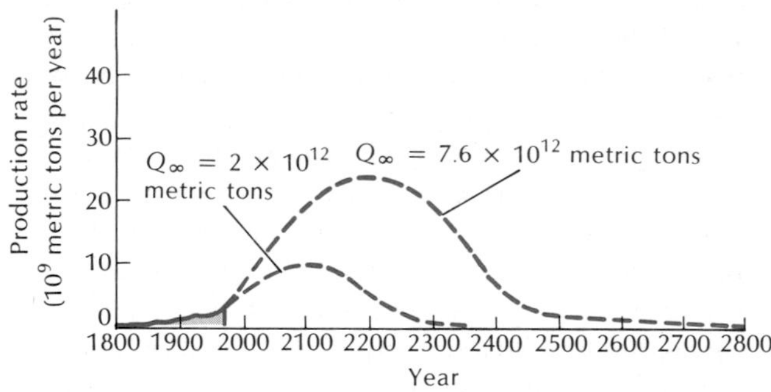

Fig. F.5 Estimated world production of coal. The upper curve assumes a total supply $Q_\infty = 7.6 \times 10^{12}$ metric tons; the lower curve assumes $Q_\infty = 2 \times 10^{12}$ metric tons. (Based on M. K. Hubbert, *American Journal of Physics,* November 1981.)

Table F.3 THE WORLDWIDE SUPPLY OF FOSSIL FUELS

Fuel	*Original supply*[a]	*Remaining supply (1980)*
Oil	12×10^{21} J	9.4×10^{21} J
Natural gas	11×10^{21}	9.8×10^{21}
Coal	188×10^{21}	184×10^{21}
Oil shales (rich deposits)	90×10^{21} (?)	90×10^{21}

[a] From M. K. Hubbert, *American Journal of Physics,* November 1981; data on shales from *The Global 2000 Report to the President* by the Council on Environmental Quality and the Department of State.

F.4 ALTERNATIVE ENERGY SOURCES

Upon the exhaustion of fossil fuels, we will have to rely on alternative sources of energy to satisfy the in-

[3] Incidentally, such a reduction of indoor temperature is good for the health of most people.

Fig. F.6 Nuclear power station at Three Mile Island.

creasing demand. Some of the most interesting alternatives are nuclear energy, solar energy, geothermal energy, and tidal energy.

Nuclear Energy from Fission

At present, nuclear reactors supply about 12% of the worldwide electric power (Figure F.6). In these nuclear reactors, energy is generated by the **fission** of uranium — the nuclei of uranium are made to split into two or more pieces, a reaction that releases a large amount of energy in the form of heat. The same kind of reaction also generates the explosive energy of an atomic bomb; but while in an atomic bomb the energy is released in a sudden burst, in a nuclear reactor the energy is released gradually, under controlled conditions (see Interlude H for details on nuclear fission).

The total fission of 1 kg of uranium releases as much heat as the combustion of 3×10^6 kg of coal. However, commercial reactors only "burn" a few percent of the fresh uranium fuel placed in them; the rest must be reprocessed chemically before it can be used again. Furthermore, the efficiency of nuclear power plants is lower than that of conventional power plants — in a nuclear power plant only about 30% of the thermal energy released by the uranium is converted into electric energy, while in a modern conventional power plant about 40% of the thermal energy of coal or oil is converted into electric energy.

Naturally occurring uranium is a mixture of two isotopes: 99.3% of ^{238}U and 0.7% of ^{235}U. Only the rare isotope ^{235}U will burn in an ordinary nuclear reactor. Since it is rather difficult to separate the isotopes completely, most nuclear reactors have been designed to operate with "enriched" uranium (several percent ^{235}U) rather than pure uranium ^{235}U; in this mixture, only the ^{235}U burns, while the ^{238}U remains behind. The worldwide supply of high-grade ores suitable for the (partial) extraction of ^{235}U is rather limited (Table F.4); and this supply is likely to become exhausted before the end of the century.

Fortunately, the abundant isotope ^{238}U can be converted, by nuclear alchemy, into ^{239}Pu, an artificial isotope of plutonium which will burn in a nuclear reactor. The conversion is accomplished in a breeder reactor. In such a reactor an initial supply of ^{239}Pu is surrounded by a blanket of ^{238}U. As the plutonium fuel burns, it not only releases heat, but it also releases an intense stream of fast neutrons which impinge on the uranium and convert it into plutonium. Thus, the reactor not only produces energy, but also new fuel — in an efficient breeder, the amount of new fuel generated by conversion of uranium exceeds the amount of initial fuel consumed. By taking advantage of this breeding process we gain access to the large amount of energy locked in the worldwide supply of ^{238}U (see Table F.4). Our supply of fuel for nuclear reactors would then last more than a hundred years. Because of the high efficiency of breeders, we can even exploit low-grade uranium ores that are good enough for ^{238}U extraction but not good enough for ^{235}U extraction. Our supply of fuel for nuclear reactors would then last several hundred years. Furthermore, breeding is also possible with ^{232}Th, an isotope of thorium. There are extensive deposits of low-grade thorium ores in granite, such as the Conway granite of New Hampshire. In principle there is enough fuel for breeder reactors to last us several thousand years.

For the immediate future, nuclear fission appears to be the most practical source of energy to replace oil and coal. According to estimates based on recent trends, nuclear fission is expected to contribute about 50% of electric power requirements of the United States in the year 2000. Unfortunately, nuclear fission yields rather dirty energy — the fission reactions generate dangerous radioactive residues. Nuclear power plants must be carefully designed to hold these radio-

Table F.4 THE WORLDWIDE SUPPLY OF NUCLEAR FUELS

Fuel	*Original supply*[a]	*Remaining supply (1980)*
Uranium-235 (high-grade ores)	20.7×10^{21} J	20.6×10^{21} J
Uranium-238 (high-grade ores)	1×10^{24}	1×10^{24}
Uranium-238 and thorium-232 (low-grade ores)	3×10^{27}	3×10^{27}
Deuterium	7×10^{30}	7×10^{30}

[a] Uranium data from *The Global 2000 Report to the President;* other data from article by J. L. Tuck in *Cosmology, Fusion & Other Matters.*

active residues in confinement. The cumbersome safety features that must be incorporated in the design make the construction and maintenance of nuclear power plants extremely expensive. Furthermore, when the load of fuel of a nuclear reactor has been spent, the residual radioactive wastes must be removed to a safe place to be held in storage for hundreds of years until their radioactivity has died away.

Nuclear Energy from Fusion

Nuclear fusion is an attractive source of energy because it bypasses some of the safety problems associated with nuclear fission. In fusion reactions energy is generated by merging nuclei of hydrogen, deuterium, or tritium;[4] these nuclei combine to form helium. For instance, the heat emitted by the Sun is generated by a fusion reaction that burns hydrogen into helium; and the explosive energy of a hydrogen bomb is generated by a fusion reaction that burns deuterium or tritium into helium. Note that the process of fusion is the reverse of fission: light nuclei (e.g., hydrogen) release energy when they merge; heavy nuclei (e.g., uranium) release energy when they split.

In order to initiate a fusion reaction, the nuclei must be subjected to extremely high temperatures and pressures. For instance, the fusion reactions at the center of the Sun involve a temperature of 15×10^6 K and a pressure of 10^9 atm. Nuclear reactions that require such extreme temperatures are called **thermonuclear.** Schemes for the development of fusion power on Earth intend to use deuterium or tritium because their reaction rates are much faster than that of hydrogen; however, the temperature required for the fusion of these nuclei is about 10^8 K — even higher than in the Sun! Under these conditions the deuterium or tritium will be in the form of plasma and it cannot be contained by a conventional reactor vessel of steel or ceramic material. One scheme for fusion attempts to suspend the plasma in the middle of an evacuated vessel by means of magnetic fields; the development of this scheme is the objective of most of our modern plasma research (discussed in Interlude J). Another scheme for fusion attempts to extract the energy by exploding small pellets of a deuterium–tritium mixture in a combustion chamber by hitting them with an intense laser beam or an intense electron beam (Figure F.7); the beam suddenly heats the pellet to such a high temperature that fusion begins — the pellet explodes like a miniature hydrogen bomb. After the thermal energy from the exploded pellet has been extracted from the combustion chamber, the next pellet is dropped in, etc. (see Interlude L).

One advantage of fusion over fission is that we have available an enormous supply of deuterium. In the oceans of the Earth, 1 water molecule in 6000 is heavy water, with one deuterium atom replacing one of the hydrogen atoms in the molecule. The total supply of deuterium is about 2×10^{16} kg — enough fusion fuel for millions of years! Furthermore, the fusion reaction does not produce any appreciable amounts of radioactive residues; in contrast to fission, fusion yields very clean energy.

(a)

(b)

Fig. F.7 (a) Small pellets consisting of glass microballoons filled with deuterium–tritium mixture and (b) the target chamber within which the pellets are exploded at the Lawrence Livermore Laboratory.

Solar Energy, Indirect

Solar energy is an inexhaustible source of energy; it will remain available as long as the Sun lasts — 5 billion years or longer. The total solar energy intercepted by the Earth is 2.8×10^{10} MW, but so far we are using only a small fraction of this. Aside from agriculture and forestry, our only large-scale exploitation of solar power is in hydroelectric power stations. This is an indirect exploitation of solar energy. The oceans may be regarded as the boilers of hydroelectric plants and the Sun is the source of heat. The water evaporated by the

[4] Deuterium and tritium are isotopes of hydrogen: ^{2}H and ^{3}H.

Fig. F.8 Hydroelectric power station at the Fontana dam on the Little Tennessee River.

Fig. F.9 Large windmill generator on the island of Culebra. This generator supplies about 20% of the electric power used on the island.

Sun falls as rain on mountains, collects behind a dam, and drives the turbines of the power stations (Figure F.8). At present, hydroelectric power contributes about 25% to the worldwide generation of electric power. The maximum conceivable contribution from hydroelectric plants is limited by suitable sites for dams and could reach at most 3×10^6 MW sometime in the future. Since this represents only a small fraction of the solar energy reaching the Earth, it will pay to develop other means of capturing this energy.

Windmills are another indirect exploitation of solar energy. Winds obtain their energy from the solar heat absorbed by the surface of the Earth — the atmosphere acts as a giant heat engine converting this heat into kinetic energy of motion of the air. A crude estimate suggests that the total available wind power is about 10^5 MW. Unfortunately, the power of wind is rather dilute. In a wind of 10 m/s (35 km/h), a vertical area of 1 m^2 intercepts kinetic energy at the rate of 1.3 kW. Even if it were possible to extract all of this energy,[5] the generation of large amounts of power would require a windmill, or an array of windmills, presenting a very large frontal area to the wind. Figure F.9 shows a windmill on the island of Culebra, Puerto Rico, which generates up to 200 kW. Small windmills for pumping water and grinding grain were in common use in the United States during the last century and the early part of this century; around 1900 these windmills supplied about 25% of the total power generated for purposes other than transportation. The availability of cheap electric power led to the neglect of these very useful machines.

Solar Energy, Direct

When concentrated by mirrors or lenses, sunlight can

[5] Careful analysis shows that at best 59% can be extracted.

Fig. F.10 Solar "furnace" at Odeillo.

be used directly to heat water or other materials. Figure F.10 is a view of the large parabolic mirror of the solar "furnace" at Odeillo, Font-Romeu, France. Sixty-three flat mirrors throw the sunlight at the parabolic mirror which concentrates about 600 kW of solar energy into a focal spot 30 cm across. The intense solar heat — up to 4000°C — has been used for high-temperature research, but this solar furnace has not been designed to generate power.

A similar arrangement of mirrors can be used to collect solar heat for a power plant; the heat can be focused on a boiler that generates steam for a steam turbine. Figure F.11 shows an experimental solar power station in California's Mojave Desert.

Instead of concentrating the solar energy with large mirrors, it is possible to collect the energy by circulating water through heat pipes in a large array of absorbers. Small solar collectors circulating water through heat pipes are now available for installation on

Fig. F.11 The solar power station "Solar One" in California's Mojave Desert consists of collecting mirrors spread over an area of about 0.5 km^2. The boiler on the central tower generates enough steam for a turbine delivering 10^4 kW of electric power.

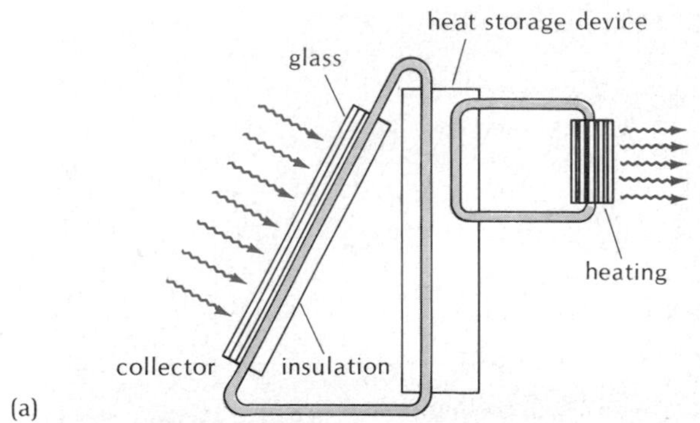

Fig. F.12 Solar heating system for a home: (a) schematic diagram and (b) collectors on roof. (Courtesy Advance Energy Technologies, Inc.)

the roofs of private homes; these solar collectors can supply hot water for kitchens and bathrooms and also for radiators to warm the house (Figure F.12).

In principle, the simplest method for obtaining electric power from sunlight is the direct conversion of the energy of light into electric energy. This can be done with a **solar cell** (see Section 29.2), a thin wafer of two types of silicon that generates an electric current when exposed to light. Large numbers of such solar cells assembled in panels have been used to supply electricity for artificial satellites. Figure F.13 shows the solar panels of the Skylab spacecraft; these supplied most of the electric power required by the astronauts living aboard the spacecraft. Figure F.14 shows a small solar panel designed for charging a 12-volt battery on a sailing yacht.

Solar cells are rather inefficient; they convert at best about 15% of the energy of sunlight into electric energy. They are also very expensive and, until we develop some cheap technique of mass production, we will not be able to build large-scale power plants with solar cells. However, some experimental power plants of modest size are already in operation. Figure F.15 shows the array of solar cells of a power plant at the Papago Reservation in Arizona. This array delivers 3.5 kW of electric power.

Fig. F.13 Solar panels on Skylab.

Fig. F.14 Small solar panel on sailing yacht.

Fig. F.15 Arrays of solar cells mounted on panels on the Carriza Plains, California. These solar cells now deliver 6000 kW of electric power, but have a capacity of 160,000 kW.

The suggestion has been made that a large power plant with solar panels 12 km long might be placed in orbit high above the Earth. It would then receive direct sunlight 24 hours a day and it would not be affected by clouds. The electric power could be converted into microwaves and beamed to a large receiving antenna on the surface of the Earth (Figure F.16). The output of such a power station would be about 4000 MW at the Earth terminal.

Geothermal Energy

The interior of the Earth is hot. The temperature in the Earth increases gradually with depth and reaches 4500°C at the center of the Earth. At some exceptional locations, hot rocks at a temperature of several-hundred degrees are found at a depth of just a few kilometers. Hot springs, steam vents, and geysers bring some of the heat from the interior of the Earth to the surface. These natural sources of geothermal energy are being exploited in a few places such as Larderello (near Pisa, Italy), the Geysers (California), Cerro Prieto (Mexico), and Wairakei (New Zealand). Figure F.17 shows one of the geothermal power plants at the Geysers. These power plants operate with steam blowing out from wells drilled down to depths of as much as 2100 m. The power plant shown in Figure F.17 delivers 110 MW. The total power generated by the Geysers is 700 MW, which makes it the largest of all the geothermal sources in operation.

Geothermal sources near populated areas can be used directly to provide heat and steam for homes and for industrial processes. In Iceland nearly half the population lives in houses heated by geothermal sources; almost all the houses in Reykjavík are heated in this manner.

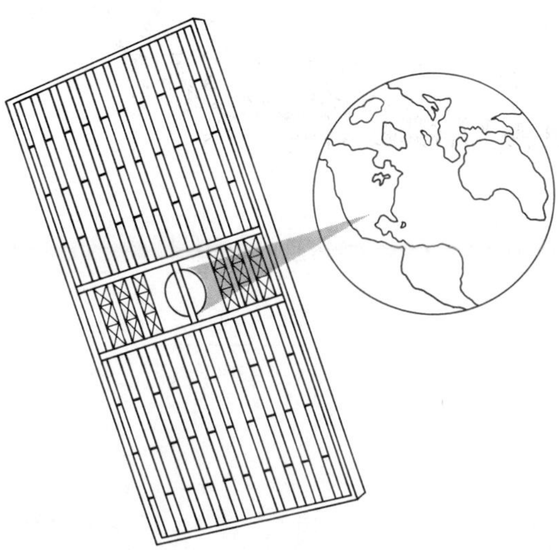

Fig. F.16 Proposed design for a satellite solar power station. The giant array of solar cells measures 4 km × 12 km.

Fig. F.17 Unit 18 of Pacific Gas and Electric Company's geothermal power station at the Geysers.

It may be possible to supplement the natural supply of geothermal steam by the artificial exploitation of the thermal energy of hot rocks. In some sites, rocks at a temperature of several hundred degrees are near enough to the surface of the Earth to be reached by drilling. If the rock can be fractured with high-pressure water or with explosives, then water can be circulated through it. The water would absorb the heat of the rock and make steam. Intensive development of such projects could contribute perhaps 10% of the electric power that will be required in the United States 20 or so years from now.

Tidal Energy

Along the coasts of the oceans, the tides caused by the Moon (and, to a lesser extent, by the Sun) regularly raise and lower the level of the water. In some special

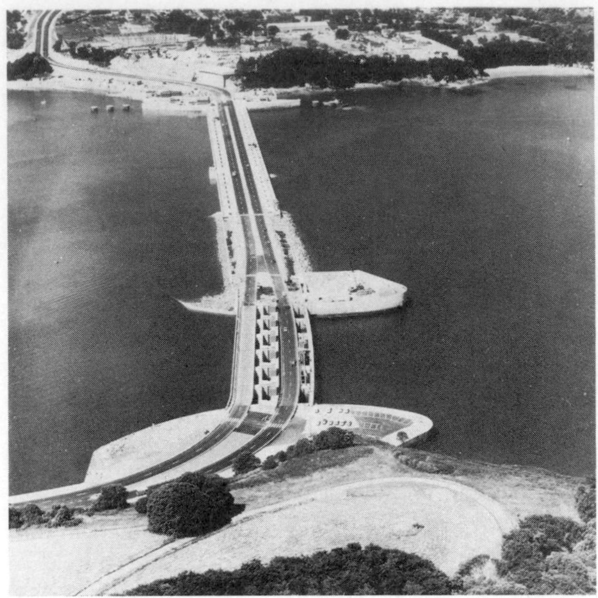

Fig. F.18 Tidal power plant at La Rance.

locations such as the Bay of Fundy (Nova Scotia) and the estuary of the La Rance river (France), the lift of the water is more than 10 m. We can generate hydroelectric power by capturing the water in a reservoir behind a dam at high tide and letting it run out through turbines at low tide. This method has been employed with great success at La Rance (Figure F.18), where special low-speed turbines generate an average power of 250 MW; these turbines are reversible — they generate power both when the water rushes in to fill the reservoir and when it rushes out.

Like windmills or solar cells, tidal power is completely free of pollution. Of course, the dam has some impact on aquatic life, but much less than the very high dam used for a conventional hydroelectric power station. Unfortunately, there are only a few potential sites for the development of tidal power — we need high tides and we need a bay or estuary that can be closed off by a dam. One suitable site that has been discussed sporadically for many years is Passamaquody Bay on the border between Maine and Canada. This site could generate as much as 1800 MW. Another suitable site is the Minas-Cobequid Basin in the Bay of Fundy; this could generate about 10 times as much. But altogether, tidal power could only supply a few percent of our total energy requirements.

Incidentally, the tidal energy derives from the rotational energy of the Earth. The friction of the tides against the shores and the bottoms of the oceans gradually slows the rate of rotation of the Earth. A tidal power station increases the effective "friction" of the tide and therefore slows the rotation even more. But the effect of a tidal power station on the rate of rotation of the Earth is insignificant compared to the effect of natural friction.

F.5 POLLUTION AND THE ENVIRONMENT

Pollution is a by-product of the operation of our power plants. Coal-fired power plants spew out ash and stack gases, internal combustion engines emit exhaust gases, and nuclear reactors leak small amounts of radioactive materials. Besides, all power plants release large amounts of waste heat, which constitutes thermal pollution. What is most noticeable about all this pollution are the immediate, local effects. Sometimes these effects reach catastrophic levels, as in the great London "killer smog" of December 1952, when nearly 4000 people died from heart trouble, pneumonia, and bronchial diseases triggered by pollutants, from coal-burning furnaces, that accumulated in the air held stagnant in the city by exceptional atmospheric conditions. However, although the dispersed effects of pollution are less noticeable, they may be even more serious in the long run.

The combustion of fossil fuels releases millions of tons of pollutants into the air each year (see Table F.5). Of these pollutants, the oxides of sulfur (mainly SO_2) pose the most serious threat to our health. For example, it has been estimated that in the year 1968 more than 50,000 deaths in the United States could be attributed to air pollution. The oxides of sulfur are released by power plants burning coal containing sulfur. Much of the American coal has a high sulfur content and it is rather expensive to remove this sulfur from the fuel before combustion or to remove it from the stack gases after combustion.

Both nitrogen oxides and hydrocarbons are released by internal combustion engines. Reactions between these pollutants and the oxygen in air produce the brownish photochemical smog for which Los Angeles

Table F.5 ATMOSPHERIC POLLUTANTS FROM FOSSIL FUELS IN THE UNITED STATES (1980)[a]

Pollutant	*Amount per year*	*Major source*
Carbon monoxide	85×10^6 metric tons	Automobiles
Hydrocarbons	22×10^6	Automobiles
Fly ash	8×10^6	Industrial plants
Sulfur oxides	24×10^6	Power plants
Nitrogen oxides	21×10^6	Automobiles, power plants

[a] From *Environmental Quality 1981*, 12th Annual Report of the Council on Environmental Quality.

is famous. Incidentally, although this smog is very lethal for plant life, it is only an irritant for animal life; the Los Angeles smog does not kill directly, at least not in the short run.

An important worldwide effect of the combustion of fossil fuels is the increase of the carbon dioxide content of the atmosphere. The combustion of hydrocarbons releases mainly carbon monoxide, but this combines with oxygen in the atmosphere to form the dioxide. The concentration of carbon dioxide in the atmosphere has increased by about 10% in the last 50 years. The increase would have been even larger than this were it not for the absorption of carbon dioxide by the water of the oceans. The oceans act as a reservoir of carbon dioxide, removing a large fraction of the carbon dioxide injected into the air.

Some climatologists worry that the continuing combustion of hydrocarbon fuels might lead to a drastic increase of the temperature of the Earth's surface. This is the "greenhouse effect." Like the glass on a greenhouse, carbon dioxide is transparent to sunlight, but absorbs heat radiation (infrared); thus, it permits sunlight to enter the atmosphere, but prevents the escape of heat. The accumulation of carbon dioxide in the atmosphere therefore causes an increase of temperature. Even a small increase of the Earth's temperature could have a drastic impact on the climate. It could possibly lead to a catastrophic "runaway greenhouse effect" by triggering the release of extra carbon dioxide from the oceans, which would lead to extra heating, and further release of carbon dioxide, etc. However, the physics of the atmosphere is rather complicated and a compensation mechanism might come into play: a slight increase of temperature would encourage the evaporation of water and the formation of clouds; since clouds reflect sunlight, they tend to keep the temperature down. According to a recent (1983) report of the Environmental Protection Agency, the average global temperature is likely to increase by about 5°C by the year 2100. This will lead to major disruptions in the climate patterns, inflicting severe droughts on some regions and severe floods on others. The temperature increase over the polar regions will be even larger, with substantial melting of polar ice and snow. This will raise the sea level by about 2 meters and cause the flooding of many low-lying coastal areas.

The thermal pollution from our power plants is of little consequence on a global scale — the total input of waste heat into our environment is small compared to the input of solar heat. A simple estimate indicates that the total heat from all our dissipated energy has only increased the temperature of the Earth by about 0.01°C. However, on a local scale the disposal of the waste heat poses serious problems. A nuclear or a coal-burning power plant generating 500 MW of electricity will also generate 1200 MW of waste heat. This heat must be dumped into the environment. The most convenient heat sink is flowing water. The amount of cooling water required by a large power plant is enormous — about 10^7 m^3/day; for this reason, many power plants are located on river banks, lake shores, or ocean coasts. Figure F.19 is an infrared photograph displaying the heat released into a river. The temperature of the plume of warm water emerging from the outlet is several degrees higher than the normal temperature; this makes the water unsuitable for some species of fish. Besides, the water intake often sucks in and kills huge quantities of fish and fish larvae.

In an effort to avoid the thermal pollution of rivers, lakes, and estuaries, some new power plants make use of cooling towers in which the air is the heat sink. These cooling towers act much like the radiator in an automobile: the heat is brought out of the power plant by water circulating in pipes; in the cooling tower a draft of air passes over these pipes, taking away their heat. Sometimes the cooling is assisted by water sprayed over the pipes. Figure F.20 shows cooling towers of different types.

F.6 ACCIDENTS AND SAFETY

The generation of electric power — whether by means of coal-burning, oil-burning, hydroelectric, or

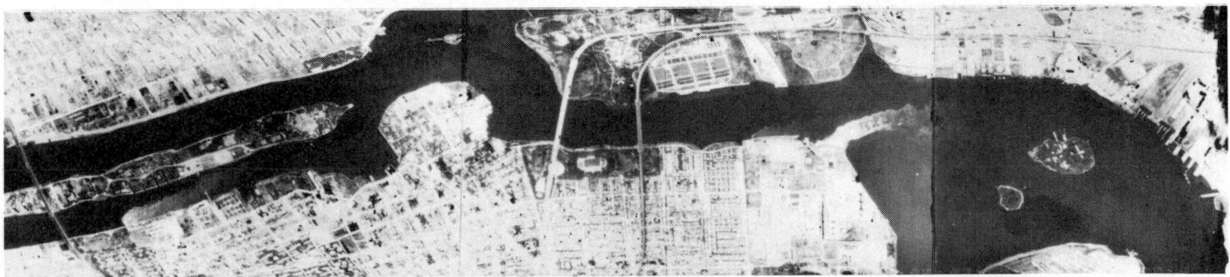

Fig. F.19 Infrared photograph showing plume of hot water released into a river by a power plant.

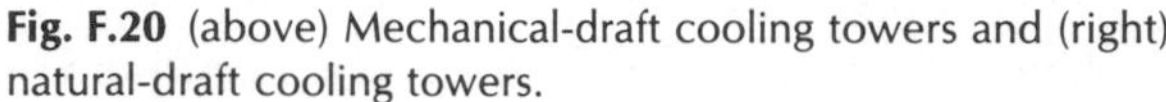

Fig. F.20 (above) Mechanical-draft cooling towers and (right) natural-draft cooling towers.

nuclear power plants — always involves the manipulation of potentially dangerous concentrations of energy: chemical energy in coal or oil, gravitational energy in water held behind a high dam, or nuclear energy in fissionable and radioactive materials. If this energy overwhelms our controls, the result is disaster: fires and explosions in coal mines, oil tankers, and oil storage depots; dam failures and flash floods; and reactor fires, meltdowns, and explosions.

Of late, the safety of nuclear power plants has been the subject of much public discussion because an accident in such a power plant could have terrifying consequences. The worst reactor accidents to date are listed in Table F.6; none of these accidents led to a large loss of life.

Although ordinary nuclear reactors obtain their energy from the same physical process as an atomic bomb, they will not explode like a bomb. In the nuclear reactors commonly employed in the United States for the generation of electric power, the core of the reactor consists of rods of uranium fuel surrounded by water (Figure F.21). The water carries away the heat and it also plays the role of a catalyst — it enhances the rate of the fission reactions (see Interlude H). If the core is accidentally subjected to violent overheating, the fission reactions will tend to slow down and come to a halt. The reason is twofold. First, the reaction is intrinsically less efficient at high temperature. And second, the overheating will drive water out of the core; without a catalyst, the nuclear reaction will slow down. In contrast to a conventional fire, an overheated nuclear "fire" simply dies out.

If the overheating of the core is extreme, so that the fuel rods melt and form a puddle in the bottom of the reactor vessel, then the concentration of uranium could cause an atomic explosion, but it would only be a very minor atomic explosion. Incidentally, a reactor with plutonium fuel (breeder reactor) is much more dangerous, because plutonium is much more explo-

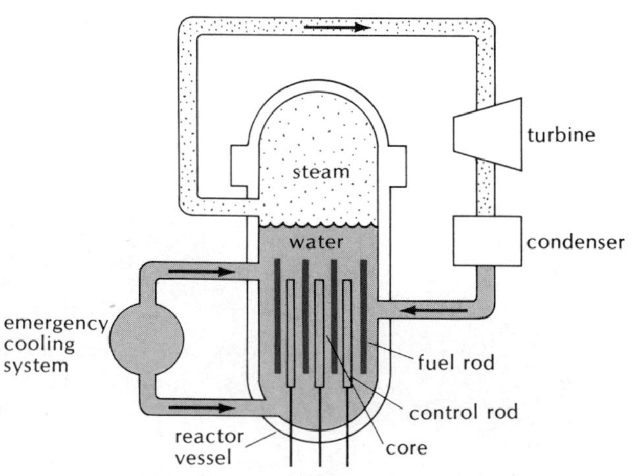

Fig. F.21 Schematic diagram of a nuclear reactor.

Table F.6 REACTOR ACCIDENTS

Location and date	*Reactor type*	*Accident*
Chalk River, Canada, December 12, 1952	Heavy water, experimental	Control rods jammed. Reactor overheated, with fuel meltdown, hydrogen explosion, and bursting of reactor vessel. Some radioactive contamination escaped into atmosphere; no casualties.
Windscale, England, October 7, 1957	Gas cooled, converter	Intentional overheating (for maintenance) ignited a fire in the uranium–graphite core. Radioactive iodine escaped and contaminated neighboring cattle pastures; no casualties.
Idaho Falls, Idaho, January 3, 1961	Light water, experimental	Control rod was mishandled by operators, and reactor went momentarily supercritical. All three operators were killed by massive radiation overexposure.
Lagoona Beach, Michigan, October 5, 1966	Breeder, experimental	A loose piece of metal blocked flow of coolant (liquid sodium), leading to overheating and partial meltdown of fuel. Core was damaged beyond repair. No radioactive contamination; no casualties.
Brown's Ferry, Alabama, March 22, 1975	Boiling water, power	Fire destroyed the electrical cables connecting reactor to control room; emergency cooling system went out of control and shut itself off; only a makeshift arrangement of pumps succeeded in maintaining sufficient cooling. No radioactive contamination; no casualties.
Three-Mile Island, Pennsylvania, March 28, 1979	Pressurized water, power	Failure in pump system triggered shutdown of supply of cooling water to the core. Reactor overheated; pressure relief valve opened, then failed to close, permitting water in core to boil; contaminated water flooded reactor housing; hydrogen bubble formed in reactor vessel, creating an explosion hazard and interfering with cooling for several days. Lining of some fuel rods melted; core was severely damaged. Some radioactive steam vented into atmosphere; no casualties.

sive than uranium. In the United States plans for the construction of plutonium reactors have been temporarily abandoned, in part because of the explosion hazard.

It is a great comfort to know that a nuclear reactor will not imitate a nuclear bomb, but the large quantity of radioactive fission residues within the reactor poses a great hazard. Radioactivity is extremely efficient at destroying life (see Interlude B). If a failure of the reactor vessel were to permit some of the radioactive material to blow out of the reactor and scatter over the surrounding countryside, damage from this radioactive fallout could be disastrous. To prevent the leakage of radioactive gas and dust, most reactors are housed in a containment shell — a gas-tight enclosure made of steel or concrete (Figure F.22).

The amount of radioactive material within a reactor is so large that even when the fission reactions stop,

Fig. F.22 Construction of the containment shell for the nuclear reactor at the Beaver Valley 2 power station, Duquesne.

the reactor continues to generate heat because of the energy released in the delayed radioactive decay of the fission residues. While the reactor is in operation the energy released by this delayed radioactivity is about 7% of the total fission energy. Once the reactor is shut down, this radioactivity gradually declines. If the cooling system fails, the heat released by radioactivity will melt down the core of the reactor — and it might even melt a hole in the reactor vessel and in the containment shell.

Scenarios for conceivable reactor accidents have been analyzed in great detail. For an ordinary nuclear reactor, the worst that might happen is a loss-of-coolant accident, i.e., the loss of the water that normally circulates through the core. Such an accident could occur if there were a sudden break in the pipe feeding the water into the core; high-pressure water mixed with steam would then blow out of the break, leaving the core with insufficient cooling. As we have seen above, both the increase of temperature and the loss of water tend to cut off the fission chain. However, the heat flow from the decay of radioactive fission residues does not cut off; and as this heat accumulates in the core, the temperature escalates catastrophically.

To prevent this, reactors are equipped with emergency cooling systems. As the primary coolant escapes, the emergency cooling system pumps water into the core. This must be done very fast, within seconds — otherwise the temperature of the fuel rods might reach 1500°C, at which point a violent reaction begins between steam and the zirconium metal used as a lining on the fuel rods; this not only releases extra heat, but also hydrogen gas, which can cause a (chemical) hydrogen explosion. If the emergency cooling system succeeds in bringing the temperature under control, there still remains the problem of the radioactivity that escapes with the water and steam from the reactor vessel. If all goes well, this radioactive mixture will be held in check by the containment shell.

However, if by mischance the emergency cooling system were to fail, the fuel rods would melt and their radioactive material would drip down into the bottom of the reactor vessel. The hot, molten mass of radioactive material endowed with its own internal source of heat would then gradually melt through the bottom of the reactor and, in a few hours, spill on the floor below. And then it would melt its way through the floor, through the containment shell, and into the earth. The motion of the molten mass of radioactive material digging its way through the earth as though it were heading to the antipode has been called the "China syndrome." Some recent calculations indicate that it will run out of heat and solidify after penetrating no more than 15 m into the ground.

The breach of containment would release some radioactive contamination into the atmosphere, mainly in the form of radioactive iodine, krypton, and xenon. In the worst conceivable case — involving substantial radioactive contamination caused by a meltdown of a reactor located near a densely populated area — several thousand people might die from acute radiation sickness, and those that survive a heavy exposure of radiation are susceptible to develop fatal cancer at a later time (see Interlude B). According to a reactor safety study of the Atomic Energy Commission [the WASH-740 Report (updated), completed in 1965 but kept secret for many years], a major reactor accident near a city could conceivably kill up to 45,000 people and injure 100,000. According to a more recent study (the WASH-1400 Report, or the Rasmussen Report, released in 1975), the casualty figures are reduced to 3300 killed and 45,000 severely injured. The difference between these casualty estimates reflects the difference in the assumed fallout: the earlier study assumed that secondary explosions (e.g., hydrogen explosions) would blow a substantial part of the radioactive material out of the containment; while the later, more optimistic study supposes it much more likely that most radioactive materials will remain trapped in the molten mass at the reactor site and only some iodine, krypton, and xenon gas will escape. In any case, regardless of the precise numbers of casualties, the possible consequences of a major reactor accident are appalling.

Promoters of the nuclear energy program argue that a major disaster could only happen through an extremely improbable concatenation of mischances. They estimate that the probability for a major disaster is extremely small. For example, according to one calculation, the chance of death for any one individual in a given year through a nuclear reactor accident is only 1 in 5×10^9 — much smaller than the chance of death through automobile accident (1 in 4000) or airplane accident (1 in 100,000). Opponents of the nuclear energy program have disputed this calculation by pointing out that nuclear power plants have only been in operation for a few years and we have not yet accumulated enough statistical data for any accurate evaluation of chances of accident. Although the exact numbers are in dispute, it is undoubtedly true that the risk of death from a nuclear reactor accident is less than that from many other kinds of accidents. What conclusion are we to draw from such a comparison of risks of accidents? The answer depends on the point of view that we adopt. From the point of view of an administrator in charge of national energy policies, the nuclear energy program makes sense; the cost–benefit ratio for a nuclear power plant is favorable, or at least more favorable than for airplanes or automobiles. But from the point of view of a citizen behind whose backyard the power company is erecting a nuclear reactor, a cost–benefit analysis is of little consolation. The citizen knows he can avoid airplane and automobile accidents by avoiding airplanes and automobiles, but there is nothing he can do to avoid a nuclear-reactor accident except to insist that the greatest and most conscientious safety precautions be taken in the design, construction, and operation of the reactor.

Further Reading

Energy, Ecology, and the Environment by R. Wilson and W. J. Jones (Academic, New York, 1974) deals with the energy problem from the point of view of physics; *Energy: An Introduction to Physics* by R. H. Romer (Freeman, San Francisco, 1976) deals with physics from the point of view of energy. Both books contain a wealth of information on energy resources, conversion, dissipation, and costs.

The article "The World's Evolving Energy System" by M. K. Hubbert in the *American Journal of Physics,* November 1981, gives an authoritative, up-to-date review of our fossil fuel resources. *Ecoscience* by P. R. Ehrlich, A. H. Ehrlich, and J. H. Holdren (Freeman, San Francisco, 1977) is an encyclopedic textbook on population, resources, and environment; it includes excellent annotated lists of further books and articles. *Energy Resources* by J. T. McMullan, R. Morgan, and R. B. Murray (Wiley, New York, 1977) is a short and neat textbook which emphasizes energy technology. *Man, Energy, Society* by E. Cook (Freeman, San Francisco, 1976) is a broad survey of global energy resources, as well as a brilliant outline of the history of man's use of energy and the consequent effects on social fabric and lifestyle; it includes extensive lists of further books and articles.

Energy from Heaven and Earth by E. Teller (Freeman, San Francisco, 1979) is an informal discussion of the energy problem and its possible solutions, enlivened by many a shrewd remark; it is noteworthy that although Teller believes in the safety of existing reactors, he would prefer to bury reactors deep in the ground. *The Global 2000 Report to the President* by the Council of Environmental Quality and the Department of State (Government Printing Office, Washington, D.C., 1981) is a study of the probable changes in the world's population, resources, and environment in the next 20 years.

The entire issue of *Scientific American,* September 1971, is devoted to diverse aspects of the energy problem. The reprint volume *Energy* by S. F. Singer is a selection of more recent articles from *Scientific American,* among them an incisive article by H. Bethe with forceful arguments for the necessity of fission power. Another useful reprint volume is *Energy and Man: Technical and Social Aspects of Energy,* edited by M. G. Morgan (IEEE Press, New York, 1975), with a wide selection of articles on energy technology and on social issues. *Nuclear Science and Society* by B. L. Cohen (Doubleday, Garden City, N.Y., 1974) is a nicely written introduction to nuclear energy and its benefits and risks. *Cosmology, Fusion, and Other Matters,* edited by F. R. Reines (University of Colorado Press, Boulder, 1972), contains an excellent chapter by J. L. Tuck on fusion. *The Accident Hazards of Nuclear Power Plants* by R. E. Webb (University of Massachusetts Press, Amherst, 1976) and *The Menace of Atomic Energy* by R. Nader and J. Abbots (Norton, New York, 1977) give warning of the dangers of nuclear reactors.

The following are some recent articles not included in the above reprint volumes:

"The Reprocessing of Nuclear Fuels," W. P. Bebbington, *Scientific American,* December 1976

"World Coal Production," E. D. Griffith and A. W. Clarke, *Scientific American,* January 1979

"Progress Towards a Tokamak Fusion Reactor," H. P. Furth, *Scientific American,* August 1979

"Acid Rain," G. E. Likens, R. F. Wright, J. N. Galloway, and T. J. Butler, *Scientific American,* October 1979

"Energy-Storage Systems," F. R. Kalhammer, *Scientific American,* December 1979

"World Uranium Resources," K. S. Deffeyes and I. D. MacGregor, *Scientific American,* January 1980

"The Safety of Fission Reactors," H. W. Lewis, *Scientific American,* March 1980

"Energy," W. Sassin, *Scientific American,* September 1980

"Oil Mining," R. A. Dick and S. P. Wimpfen, *Scientific American,* October 1980

"Toward a Rational Strategy for Oil Exploration," H. W. Menard, *Scientific American,* January 1981

"Catastrophic Releases of Radioactivity," S. A. Fetter and K. Tsipis, *Scientific American,* April 1981

"The Fuel Economy of Light Vehicles," C. L. Gray and F. von Hippel, *Scientific American,* May 1981

"Gas-Cooled Nuclear Reactors," H. M. Agnew, *Scientific American,* June 1981

"Carbon Dioxide and World Climate," R. Revelle, *Scientific American,* August 1982

"A Debate on Radioactive-Waste Disposal," F. A. Donath and R. O. Pohl, *Physics Today,* December 1982

"The Engineering of Magnetic Fusion Reactors," R. W. Conn, *Scientific American,* October 1983

Questions

1. According to the flow chart for the energy conversions in an automobile (see Figure F.1), where is the greatest potential for possible increases in efficiency?

2. Figure F.1 shows the flow chart for the energy conversions in an automobile traveling on a level road. Qualitatively, what changes would you expect in the flow chart if the automobile were traveling uphill? Downhill?

3. Give an example of conversion of mechanical kinetic energy into gravitational potential energy. How efficient is this conversion in your example?

4. Can you think of any device that converts chemical energy directly into mechanical energy without an intermediate stage of thermal energy?

5. If you convert solar energy into electric energy (by means of a solar cell) and then convert this into mechanical energy (by means of an electric motor), what is the overall conversion efficiency? Use the numbers given in Table F.1.

6. Throughout the nineteenth century most of the power required for agriculture was supplied by horses. What are the advantages and disadvantages of operating a farm on horsepower?

7. Sailing ships obtain their power from the wind, which is free. Nevertheless, early in the twentieth century shipowners found that sailing ships could not compete economically with steamships. What are the economic disadvantages of sailing ships?

8. According to Figure F.2, roughly what percentage of our energy supply will be due to nuclear power by the year 2000?

9. When Dmitrii Mendeleev made his first acquaintance with petroleum, he declared, "This material is too valuable to burn," a sentiment echoed in more recent times by the late Shah of Iran. Explain.

10. Why is it difficult to separate the isotopes ^{238}U and ^{235}U?

11. A fusion reactor would produce a large amount of tritium (^{3}H), a radioactive isotope of hydrogen. If the tritium were accidentally released into the environment, it is likely to contaminate the water. Explain.

12. A temperature of 60°F indoors feels quite chilly. But the same temperature outdoors feels comfortable. Why?

13. In terms of the law of heat conduction [Eq. (20.5)] explain why less heat is lost through the walls of a house if the internal temperature is lowered.

13. According to a simple rule, the amount of heat required to keep a house at a comfortable temperature over a number of days is proportional to the number of days multiplied by the deviation of the average outside temperature from 65°F. This product is called the number of **degree-days,**

$$[\text{number of degree-days}] = [\text{number of days}] \times (65°\text{F} - \overline{T})$$

For instance, if the average temperature is 45°F over a period of 3 days, the number of degree-days is 3 × 20, or 60. What would the average temperature have to be to give 60 degree-days in a single day?

15. Some people believe that they can save energy by lowering the thermostat in their home at night, while others argue that the extra heating required to rewarm the home in the morning more than balances what is saved at night. Who is right?

16. According to recommendations by experts, houses in the colder parts of the United States would derive maximum benefits from triple-glazed windows on the north side, but only double-glazed windows on the south side. Explain this, taking into account that a sheet of glass partially reflects sunlight.

17. The dissipation of the power obtained from burning fossil fuels or nuclear fuels delivers *extra* heat to the surface of the Earth, but the dissipation of the power obtained from hydroelectric generators or from windmills does not. Explain.

18. In what kind of orbit would we have to place the satellite power station pictured in Figure F.16?

19. Acid rain contains nitric and sulfuric acids. From what do these acids come?

20. The fly ash spewed out by smokestacks in the United States was 35 million metric tons in 1968, but only 8 million tons in 1980. What brought about this decrease?

21. The largest man-made disaster (apart from bombings and other acts of war) was the bursting of the South Fork dam in Johnstown, Pennsylvania, on May 31, 1889; this caused 2209 deaths. How does this compare with the estimated number of deaths expected in a nuclear reactor accident?

22. Edward Teller has recommended that all nuclear reactors be buried 200 feet under the ground. What would be the advantages?

CHAPTER 22

Electric Force and Electric Charge

Ordinary matter — solids, liquids, and gases — consists of atoms, each with a nucleus surrounded by a swarm of electrons. For example, Figure 22.1 shows the structure of an atom of neon. At the center of this atom there is a nucleus made of ten protons and ten neutrons packed very tightly together — the diameter of the nucleus is only about 6×10^{-15} m. Moving around this nucleus there are ten electrons; these electrons are confined to a roughly spherical region about 1×10^{-10} m across. Figure 22.1 shows the electrons at one instant of time. Since the electrons move around the nucleus very quickly, for most purposes the average distribution of electrons is more relevant than the instantaneous distribution. Figure 22.2 is a picture of the average distribution of electrons in a neon atom. This picture gives a good impression of the spherical shape of this atom.

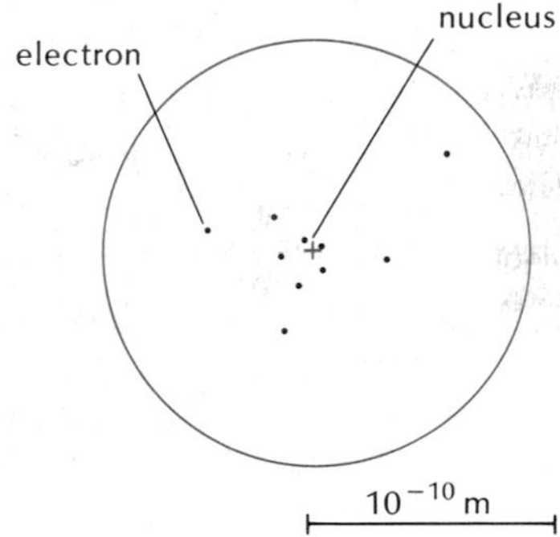

Fig. 22.1 Neon atom as seen at one instant of time with a (hypothetical) gamma-ray microscope.

The atom somewhat resembles the Solar System, with the nucleus as Sun and the electrons as planets. In the Solar System the force that holds a planet near the Sun is the gravitational force. What is the force that holds an electron near the nucleus? The force is the **electric force** of attraction between the electron and the protons in the nucleus. The

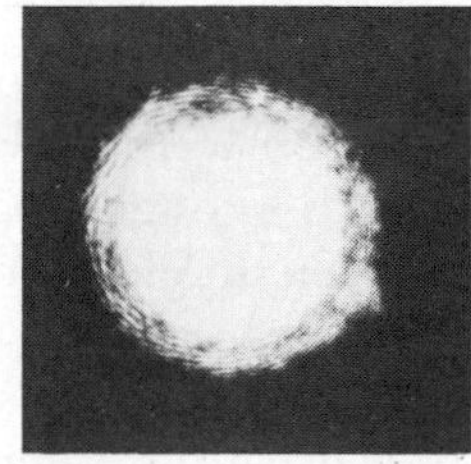

Fig 22.2 Neon atom as seen with an electron-holography microscope. The atom looks like a spherical cloud. The density, or brightness, at any point of this cloud is roughly proportional to the probability for finding an electron at this point. The magnification is $2 \times 10^8 \times$. (Courtesy L. S. Bartell, University of Michigan.)

Electric force

electric force not only holds electrons inside atoms, but also holds atoms together in molecules or crystals, and holds these building blocks together in large-scale macroscopic structures — a rock, a tree, a human body, a skyscraper, or a supertanker. All the mechanical "contact" forces of everyday experience — the push of a hand against a door, the pull of an elevator cable, the pressure of water against the hull of a ship — are nothing but the combined electric forces of many atoms. Thus our immediate environment is dominated by electric forces.

In the following chapters we will study electric forces and their effects. For a start (Chapters 22–29) we will assume that the particles exerting these forces are at rest or moving only very slowly (quasistatic).

Electrostatic force

The electric forces are then called **electrostatic forces.** Later on (Chapters 30–34) we will consider the forces when the particles are moving with uniform velocity or nearly uniform velocity. Under these conditions the electric forces are modified — besides the electrostatic force there arises a **magnetic force.** The magnetic force may be regarded as a supplement to the electric force, a supplement depending on the velocity of the particles. The combined electrostatic and magnetic forces are called **electromagnetic forces.** Finally, we will consider the forces when the particles are moving with accelerated motion (Chapters 35 and 36). The electromagnetic forces are then further modified with a drastic consequence, that is, the emission of electromagnetic waves.

Electricity was first discovered through friction. The ancient Greeks noticed that, when rubbed, rods of amber gave off sparks and attracted small bits of straw or feathers. In the nineteenth century technical applications of electricity were gradually developed, but it was only in the twentieth century that the pervasive presence of electric forces holding together all the matter of our environment was recognized.

We will begin our study of electricity with the fundamental electric force between charged particles rather than following the historical route, because the origin of frictional electricity remains rather mysterious. Even now, physicists have no precise understanding of the detailed mechanism that generates electric charge on rubbed objects or why some rubbed objects acquire positive charge and some negative charge.

22.1 The Electrostatic Force

The electric force between, say, an electron and a proton resembles gravitation in that it decreases in proportion to the inverse square of the distance. But the electric force is much stronger than the gravitational force. The electric attraction between an electron and a proton (at any given distance) is about 2×10^{39} times as strong as the gravitational attraction. Thus the electric force is by far the strongest force felt by an electron in an atom. The other great difference between the gravitational and the electric force is that gravitation is always attractive, whereas electric forces can be attractive or repulsive. The electric forces between electron and proton, electron and electron, and proton and proton all have the same magnitude (for the same distance); the electron–proton force is attractive, but the electron–electron and proton–proton forces are repulsive. Table 23.1 gives a qualitative summary of the electric forces between the particles in an atom.

Table 22.1 ELECTRIC FORCES (QUALITATIVE)

Particles	*Force*
Electron and proton	Attractive
Electron and electron	Repulsive
Proton and proton	Repulsive
Neutron and anything	Zero

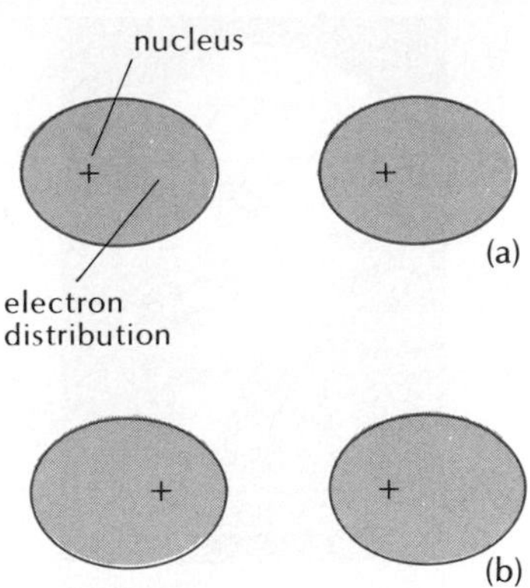

Fig. 22.3 (a) Two neighboring distorted atoms. The colored regions represent the average distributions of the electrons. The electrons of the left atom are closer to the nucleus of the right atom than to its electrons. (b) The nucleus of the left atom is closer to the nucleus of the right atom than to its electrons.

The "contact" force between two atoms close together actually arises from a counterplay of the attractive and repulsive electric forces between the electrons of the atoms and the protons in their nuclei. The force between two neighboring atoms depends on the exact location of the electrons and the nuclei. If the distribution of the electrons around the nucleus is somewhat distorted so that the electrons in one atom are closer to the nucleus of the neighboring atom than to its electrons, then the force between these atoms will be attractive. Figure 22.3a shows a distortion that leads to an attractive force; the distortion may either be intrinsic to the structure of the atom or induced by the presence of the neighboring atom. Figure 22.3b shows a distortion that leads to a repulsive force.

If two macroscopic bodies are separated by some distance, the electric attractions and repulsions between them tend to cancel, provided the bodies are neutral, i.e., provided their number of protons matches their number of electrons. For example, if the macroscopic bodies are a man and a woman separated by a distance of 10 m, then each electron in the man will be strongly attracted by the protons in the woman, but simultaneously it will be strongly repelled by the electrons in the woman; these forces cancel each other. Only when the surfaces of the two macroscopic bodies are very near one another ("touching") will the atoms in one surface exert a net force on those in the other surface.

Benjamin Franklin, *1706–1790, American scientist, statesman, and inventor. Although he is most often remembered for his hazardous experiments with a kite in a thunderstorm, which demonstrated that lightning is an electric phenomenon, and for his invention of lightning rods, Franklin also made other significant contributions to the experimental and theoretical studies of electricity, and he was admired and honored by the leading scientific associations in Europe. Among these contributions was his formulation of the law of conservation of electric charge and his introduction of the modern notation for plus and minus charges, which he regarded as an excess or deficiency of "electric fluid."*

22.2 Coulomb's Law

As stated in the preceding section, the electric forces between electron and proton, electron and electron, and proton and proton all have the same magnitude. Obviously, this implies that the force does not depend on the mass of these particles. Instead, it depends on a new quantity: the **electric charge.** For the mathematical formulation of the law of electric force between electrons, protons, and other particles we assign to each particle an electric charge which is either positive, negative, or zero. The charge on the proton is taken as positive and that on the electron negative. Experiments show that the absolute values of these charges are exactly the same. We will designate the charges of the proton and electron by $+e$ and $-e$, respectively. The charge on the neutron is zero. Table 22.2 summarizes these values of the charges.

Charles Augustin de Coulomb (koolom), *1736–1806, French physicist. A military engineer by profession, he retired at age 53 from his post as superintendent of waters and fountains so that he could fully pursue scientific research. With the torsion balance, which he invented, he established that the electric force between small charged balls obeys an inverse-square law.*

Table 22.2 ELECTRIC CHARGES OF PROTONS, ELECTRONS, AND NEUTRONS

Particle	*Charge*
Proton, p	$+e$
Electron, e	$-e$
Neutron, n	0

In terms of these electric charges, we can then state that the electric force between charges of like sign is repulsive and that the electric force between charges of unlike sign is attractive.

The precise value of the electric force that one charged particle exerts on another is given by **Coulomb's Law:**

> *The magnitude of the electric force that a particle exerts on another particle is directly proportional to the product of their charges and inversely proportional to the square of the distance between them. The direction of the force is along the line joining the particles.*

Thus, the magnitude of the force that a particle of charge q' exerts on a particle of charge q at a distance r is

$$F = [\text{constant}] \times \frac{|q'q|}{r^2} \tag{1}$$

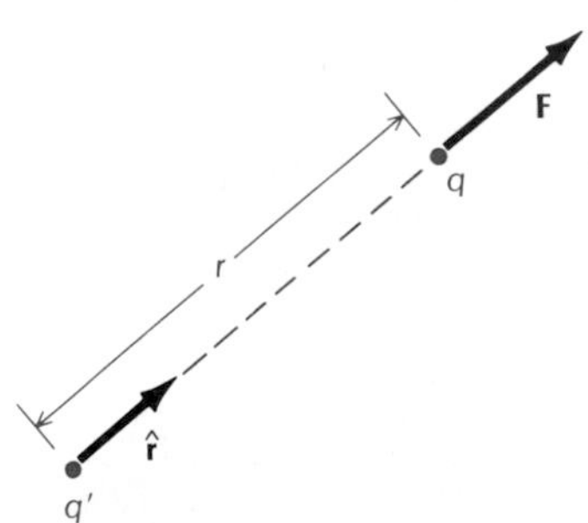

Fig. 22.4 Two point charges q' and q. The unit vector $\hat{\mathbf{r}}$ points from q' to q.

To express the direction of the force vectorially, we will use the unit vector $\hat{\mathbf{r}}$ pointing from the charge q' to the charge q (Figure 22.4). In terms of this unit vector, we can write the following equation for the force that q' exerts on q:

$$\mathbf{F} = [\text{constant}] \times \frac{q'q}{r^2}\,\hat{\mathbf{r}} \tag{2}$$

This vector equation gives us both the magnitude and the direction of **F**. If the charges are of like sign, then the product $q'q$ is positive and **F** is parallel to $\hat{\mathbf{r}}$ (repulsive force). If the charges are of unlike sign, then the product $q'q$ is negative and **F** is antiparallel to $\hat{\mathbf{r}}$ (attractive force).

The numerical value of the charge of the proton and the numerical value of the constant in Eqs. (1) and (2) depend on the system of units.

coulomb, C

In the SI system of units the electric charge is measured in **coulombs** (C) and the corresponding numerical values are

Charge of proton, e

$$\boxed{e = 1.60 \times 10^{-19}\ \text{C}} \tag{3}$$

and

$$[\text{constant}] = 8.99 \times 10^{9}\ \text{N}\cdot\text{m}^2/\text{C}^2 \tag{4}$$

For obscure historical reasons this constant is traditionally written in the complicated form

$$[\text{constant}] = \frac{1}{4\pi\varepsilon_0} \tag{5}$$

with

$$\boxed{\varepsilon_0 = 8.85 \times 10^{-12}\ \text{C}^2/(\text{N} \cdot \text{m}^2)} \tag{6}$$

Permittivity constant

The quantity ε_0 is called the **permittivity constant.**[1]

Using the permittivity constant, Coulomb's Law for the force that a particle of charge q' exerts on a particle of charge q then becomes

$$\boxed{\mathbf{F} = \frac{1}{4\pi\varepsilon_0}\frac{q'q}{r^2}\,\hat{\mathbf{r}}} \tag{7}$$

Coulomb's Law in vector notation

This equation applies to particles — electrons and protons — and also to any small charged bodies, provided that the sizes of these bodies are much less than the distance between them; such bodies are called **point charges.** Equation (7) obviously resembles Newton's Law for the gravitational force (see Section 13.1); the constant $1/4\pi\varepsilon_0$ is analogous to the gravitational constant G and the charge is analogous to the mass.

Point charge

In the SI system the coulomb is defined in terms of a standard electric current: one coulomb is the amount of electric charge that a current of one ampere delivers in one second. Unfortunately the definition of the standard current involves magnetic fields and we will therefore have to postpone the question of the precise definition of ampere and coulomb to a later chapter. For the time being, we may define the coulomb by the electric repulsive force between two equal charges at a standard distance from one another: the charge on each of two equally charged small bodies is one coulomb if they repel with a force of 8.99×10^9 N when their distance is one meter. (Incidentally, this shows that charge, like everything else, can be defined in terms of the fundamental units of mass, length, and time.)

EXAMPLE 1. Compare the magnitudes of the gravitational force of attraction and of the electric force of attraction between the electron and the proton in a hydrogen atom. According to Newtonian mechanics, what is the acceleration of the electron? Assume that the distance between the two particles is 0.53×10^{-10} m.

SOLUTION: The gravitational force is

$$\frac{GmM}{r^2} = \frac{6.67 \times 10^{-11}\ \text{N}\cdot\text{m}^2 \cdot \text{kg}^{-2} \times 9.11 \times 10^{-31}\ \text{kg} \times 1.67 \times 10^{-27}\ \text{kg}}{(0.53 \times 10^{-10}\ \text{m})^2}$$

$$= 3.6 \times 10^{-47}\ \text{N}$$

The electric force is

[1] The values of e and ε_0 in Eqs. (3) and (6) have been rounded off to three significant figures. The best available values of these constants are listed in Appendix 8.

$$\frac{1}{4\pi\varepsilon_0}\frac{e^2}{r} = 8.99 \times 10^9\ \text{N} \cdot \text{m}^2/\text{C}^2 \times \frac{(1.60 \times 10^{-19}\ \text{C})^2}{(0.53 \times 10^{-10}\ \text{m})^2} = 8.2 \times 10^{-8}\ \text{N}$$

The ratio of these forces is $8.2 \times 10^{-8}\ \text{N}/3.6 \times 10^{-47}\ \text{N} = 2.3 \times 10^{39}$.

Since the gravitational force is very small compared to the electric force, it can be neglected. The acceleration of the electron is then

$$a = \frac{F}{m} = \frac{8.2 \times 10^{-8}\ \text{N}}{9.1 \times 10^{-31}\ \text{kg}} = 9.0 \times 10^{22}\ \text{m/s}^2$$

EXAMPLE 2. How much negative and how much positive charge is there on the electrons and protons of a cup of water (0.25 kg)?

SOLUTION: The molecular mass of water is 18 g; hence 250 g of water amounts to 250/18 moles. Each mole has 6.0×10^{23} molecules, giving $6.0 \times 10^{23} \times (250/18)$ molecules in a cup. Each molecule consists of two hydrogen atoms (one electron apiece) and one oxygen atom (8 electrons). Thus, there are 10 electrons in each molecule and the total negative charge on all the electrons is $-6.0 \times 10^{23} \times 250/18 \times 10 \times 1.6 \times 10^{-19}\ \text{C} = -1.3 \times 10^7\ \text{C}$. The positive charge on the protons is the opposite of this.

EXAMPLE 3. What is the magnitude of the attractive force exerted by the electrons in a cup of water on the protons in a second cup of water at a distance of 10 m?

SOLUTION: According to the preceding example, the charge on the electrons is -1.3×10^7 C, and the charge on the protons is $+1.3 \times 10^7$ C. If we treat both of these charges (approximately) as point charges, the force is

$$F = \frac{1}{4\pi\varepsilon_0}\frac{qq'}{r^2}$$

$$= 9.0 \times 10^9\ \text{N} \cdot \text{m}^2 \cdot \text{C}^{-2} \times \frac{(-1.3 \times 10^7\ \text{C}) \times (+1.3 \times 10^7\ \text{C})}{(10\ \text{m})^2}$$

$$= -1.5 \times 10^{22}\ \text{N}$$

This is the weight of 10^{18} tons! This enormous attractive force is precisely canceled by an equally large repulsive force exerted by the protons in one cup on the protons in the other cup. Thus, the cups exert no net force on each other.

Coulomb's Law of Force is an empirical fact about the mutual interactions of charges. This law was first discovered in the eighteenth century by simple and direct experiments with small balls and globes on which electric charges were placed. Nowadays, the strongest observational evidence for Coulomb's Law comes from atomic mechanics, just as the strongest observational evidence for Newton's law comes from celestial mechanics. Of course, the motion of an electron inside an atom must be described by quantum mechanics rather than classical mechanics. The electron has no well-defined orbit and it is not possible to check on the force law by direct measurement of orbital positions; however, orbital energies can be measured and compared with theoretical values calculated from Coulomb's Law. In the best of these calculations, which take into account not only quantum theory but also relativity theory and many subtle effects arising from the interplay of these theories, the orbital energies of the hydrogen atom have been calculated to *nine significant figures*. To within this precision the theoretical and experimental values agree. This represents one of the greatest triumphs of modern theoretical physics and of Coulomb's Law — even the legendary precision of celestial mechanics and of Newton's Law of Gravitation pale by comparison. There is no question that our understanding of the electric force surpasses our understanding of any other force in nature.

22.3 Charge Quantization and Charge Conservation

Not only electrons and protons exert electric forces on one another, but so do many other particles. The magnitudes of these electric forces are given by Eq. (7) with the appropriate values of the electric charges. Table 22.3 lists the electric charges of some particles; a more complete list will be found in Section C.3. The charges of antiparticles are always opposite to those of the corresponding particles; e.g., the antielectron (or positron) has charge $+e$, the antiproton has charge $-e$, the antineutron has charge 0, etc.

All the particles that have been found in nature have charges that are some integer multiple of the fundamental charge e, i.e., the charges are always 0, $\pm e$, $\pm 2e$, $\pm 3e$, etc. Why no other charges exist is a mystery for which classical physics offers no explanation. (Recent investigations suggest that the unified theory of fields may supply an explanation; see Interlude G.) Much effort has been expended on experimental searches for charges of $\frac{1}{3}e$ and $\frac{2}{3}e$ (these are the charges of the quarks, the elementary constituents supposedly found inside protons and neutrons; see Interlude C), but neither such fractional charges nor any other fractional charges have ever been found.

Table 22.3 Electric Charges of Some Particles

Particle	Charge
Photon, γ	0
Neutrino, ν	0
Electron, e	$-e$
Muon, μ	$-e$
Pion, π^0	0
Pion, π^+	$+e$
Pion, π^-	$-e$
Proton, p	$+e$
Neutron, n	0
Delta, Δ^{++}	$+2e$
Delta, Δ^+	$+e$
Delta, Δ^0	0
Delta, Δ^-	$-e$

Quantization of charge

Since charges only exist in discrete packets, we say that **charge is quantized** — the fundamental charge e is called the quantum of charge. However, in a description of the charge distribution on macroscopic bodies, the discrete nature of charge can often be ignored and it is usually sufficient to treat the charge as a continuous "fluid" with a charge density (C/m^3) that varies more or less smoothly as a function of position. This is analogous to describing the mass distribution of a solid, liquid, or gas by a mass density (kg/m^3) which ignores the fact that on a microscopic scale the mass is concentrated in atoms. A solid, liquid, or gas seems smooth because there are very many atoms in each cubic millimeter (about 10^{16} atoms/mm^3 in a gas under standard conditions) and the distances between atoms are very small. Likewise, charge distributions placed on wires or other conductors will seem smooth because they are due to very many electrons (or protons) in each cubic millimeter.

Conservation of charge

The electric charge is a **conserved quantity:** in any reaction involving charged particles, the total charges before and after the reaction are always the same. Here are some examples of reactions in which particles are destroyed, yet the charge remains constant:

matter–antimatter annihilation,

$$\begin{array}{lccccc} & [\text{electron}] & + & [\text{antielectron}] & \rightarrow & 2[\text{photons}] \\ \text{charges:} & -e & + & e & \rightarrow & 0 \end{array} \tag{8}$$

radioactive disintegration of a neutron,

$$\begin{array}{lccccccc} & [\text{neutron}] & \rightarrow & [\text{proton}] & + & [\text{electron}] & + & [\text{antineutrino}] \\ \text{charges:} & 0 & \rightarrow & e & + & (-e) & + & 0 \end{array} \tag{9}$$

pion creation in a high-energy proton–proton collision,

$$\begin{array}{lccccccccc} & [\text{proton}] & + & [\text{proton}] & \rightarrow & [\text{neutron}] & + & [\text{proton}] & + & [\text{pion}, \pi^+] \\ \text{charges:} & e & + & e & \rightarrow & 0 & + & e & + & e \end{array} \tag{10}$$

No reaction that creates or destroys electric charge has ever been found in nature. In an effort to test the law of charge conservation, physicists have looked for hypothetical reactions such as

$$[\text{electron}] \rightarrow [\text{neutrino}] + [\text{photon}] \tag{11}$$

that would destroy negative charge. There is strong experimental evidence that such reactions never happen.

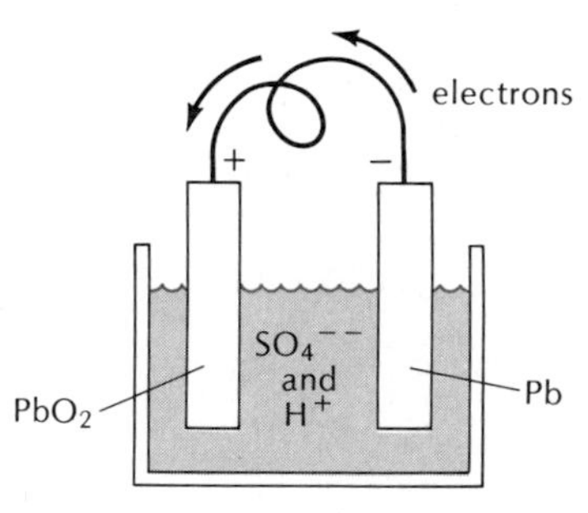

Fig. 22.5 Lead–acid battery.

Charge is of course also conserved in chemical reactions. For instance, in a lead–acid battery (automobile battery), plates of lead and of lead dioxide are immersed in an electrolytic solution of sulfuric acid (Figure 22.5). The reactions that take place on these plates involve sulfate ions (SO_4^{--}) and hydrogen ions (H^+); the reactions release electrons at the lead plate and absorb electrons at the lead-dioxide plate:

$$\begin{array}{lccccccc} & Pb & + & SO_4^{--} & \rightarrow & PbSO_4 & + & 2[\text{electrons}] \\ \text{charges:} & 0 & + & (-2e) & \rightarrow & 0 & + & (-2e) \end{array} \tag{12}$$

$$\begin{array}{lccccccccccc} & PbO_2 & + & 4H^+ & + & SO_4^{--} & + & 2[\text{electrons}] & \rightarrow & PbSO_4 & + & 2H_2O \\ \text{charges:} & 0 & + & 4e & + & (-2e) & + & (-2e) & \rightarrow & 0 & + & 0 \end{array} \tag{13}$$

The plates of such a battery are connected by an external circuit (e.g., a wire) and the electrons released by the reaction (12) travel from one plate to the other via this external circuit, forming an electric current (Figure 22.5).

EXAMPLE 4. A fully "charged" battery contains a large amount of sulfuric acid in the electrolytic solution (H_2SO_4 in the form of SO_4^{--} ions and H^+ ions). As the battery delivers charge to the external circuit connecting its poles, the amount of sulfuric acid in solution gradually decreases. Suppose that while discharging completely, the positive pole of an automobile battery delivers a charge of 1.8×10^5 C through the external circuit. How many grams of sulfuric acid will be used up in this process?

SOLUTION: Since each electron has a charge of -1.6×10^{-19} C, the number of electrons in -1.8×10^5 C is $1.8 \times 10^5/1.6 \times 10^{-19}$, or 1.1×10^{24}. According to the reactions (12) and (13), whenever two electrons are transferred from the lead to the lead-dioxide plate, two sulfate ions are absorbed (one at each plate). Thus 1.1×10^{24} sulfate ions will be used up, i.e., 1.1×10^{24} molecules of sulfuric acid will be used up. The required number of moles of sulfuric acid is therefore $1.1 \times 10^{24}/6.02 \times 10^{23} = 1.9$ and, since the molecular mass of sulfuric acid is 98 g per mole, the required mass of sulfuric acid is $1.9 \times 98 \text{ g} = 183$ g.

The conservation of charge in chemical reactions is a trivial consequence of the conservation of electrons and protons. All such reactions involve nothing but a rearrangement of electrons and protons; during this, the numbers of electrons and protons remain constant. Obviously, the net charge must then also remain constant.

The same argument applies to all macroscopic electric processes, such as the operation of electrostatic machines and generators, the flow of currents on wires, the storage of charge in capacitors, the electric discharge of thunderclouds, etc. All such processes involve nothing but a rearrangement of electrons and protons. Consequently, the net charge must remain constant.

22.4 Conductors and Insulators; Frictional Electricity

Conductors and insulators

A conductor — such as copper, aluminum, or iron — is a material that permits the motion of electric charge through its volume. An insulator — such as glass, porcelain, rubber, or nylon — is a material that does not permit the motion of electric charge. Thus, when we place some electric charge on one end of a conductor, it immediately spreads out over the entire conductor until it finds an equilibrium distribution[2]; when we place some charge on one end of an insulator, it stays in place.

Free electrons

All metals are good conductors. The motion of charge in metals is due to the motion of electrons. In a metal, some of the electrons of each atom are **free,** that is, they are not bound to any particular atom although they are bound to the metal as a whole. The free electrons come from the outer parts of the atoms. The outer electrons of the atom are not very strongly attached and readily come loose; the inner electrons are firmly bound to the nucleus of the atom and are likely to

[2] We will study the conditions for the equilibrium of electric charge on a conductor in Chapter 24. It turns out that when the charges finally reach equilibrium, they will all sit on the surface of the conductor (provided the conductor is homogeneous).

stay put. The free electrons wander through the entire volume of the metal, suffering occasional collisions, but they only experience a restraining force when they hit the surface of the metal. The electrons are held inside the metal in much the same way as the particles of a gas are held inside a container — the particles of gas can wander through the volume of the container, but they are restrained by the walls. In view of this analogy, electrons in a metallic container are often said to form a **free-electron gas.** If one end of a metallic conductor has an excess or deficit of electrons, the motion of the free-electron gas will quickly distribute this excess or deficit to other parts of the metallic conductor.

Free-electron gas

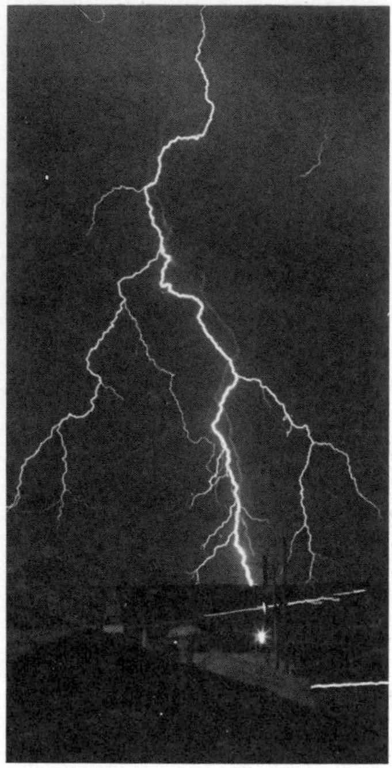

Fig. 22.6 Lightning.

The charging of a body of metal is usually accomplished by the removal or addition of electrons. A body will acquire a net positive charge if electrons are removed and a net negative charge if electrons are added. Thus, positive charge on a body of metal is simply a deficit of electrons, and negative charge an excess of electrons.

Liquids containing ions (atoms or molecules with missing electrons or with excess electrons) are also good conductors. For instance, a solution of common salt in water contains ions of Na^+ and Cl^-. The motion of charge through the liquid is due to the motion of these ions. Liquid conductors with an abundance of ions are called **electrolytes.**

Incidentally, very pure, distilled water is a poor conductor, but ordinary water is a good conductor because it contains some ions contributed by dissolved impurities. The ubiquitous water in our environment makes many substances into conductors. For example, earth (ground) is a reasonably good conductor, mainly because of the presence of water. Furthermore, on a humid day many insulators acquire a microscopic surface film of water and this permits electric charge to leak away along the insulator; thus, on humid days it is difficult to store electric charge on bodies supported by insulators.

Ordinary gases are insulators, but ionized gases are good conductors. For example, air is an insulator, but the ionized air found in the path of a lightning bolt is a good conductor (Figure 22.6). This ionized air is a mixture of positive ions and free electrons; the motion of charge in such a mixture is due mainly to the motion of the electrons. An ionized gas is called a **plasma.**

Plasma

We will end this chapter with a few brief comments on frictional electricity. One can accumulate electric charge on a glass rod merely by rubbing it with a piece of silk. The silk becomes negatively charged and the glass positively charged — the rubbing motion between the surfaces of the silk and the glass rips charges off one of these surfaces and makes them stick to the other, but the detailed mechanism is not well understood. It is believed that what is usually involved is a transfer of ions from one surface to the other. Contaminants residing on the rubbed surfaces play a crucial role in frictional electricity. If glass is rubbed with an absolutely clean piece of silk or other textile material, the glass becomes negatively charged rather than positively charged. Ordinarily pieces of silk apparently have such a large amount of dirt on their surfaces that the charging process is dominated by the dirt rather than by the silk. Even air can act as contaminant for some surfaces; for instance, careful experiments on the rubbing of platinum with silk show that in vacuum the platinum becomes negatively charged, but in air it becomes positively charged.

The electric charge that can be accumulated on the surface of a body of ordinary size (a centimeter or more) by rubbing may be as much as 10^{-9} to 10^{-8} coulomb per square centimeter. If the charge concentration on a body is higher than that, it will cause an electrical discharge into the surrounding air (a corona discharge; see Section I.4); a higher charge concentration can subsist only on a small body or on a small spot on a large body.

Induction

Once we have accumulated some charge on, say, a rod of glass, we can produce charges on other bodies by a process of **induction,** as follows: First we bring the glass rod near a metallic body supported on an insulating stand. The positive charge on the rod will then attract free electrons to the near side of the body and leave a deficit of free electrons on the far side (Figure 22.7a); thus, the near side will acquire negative charge and the far side positive charge. If we next momentarily connect the far side to the ground, the positive charge will be neutralized by an influx of electrons from the ground (virtually, the positive charge leaks away; Figure 22.7b). This leaves the metallic body with a net negative charge. When we finally withdraw the glass rod, this charge will remain on the metallic body, distributing itself over its entire surface (Figure 22.7c).

Some old electrostatic machines accumulate large amounts of charge by a repetitive operation of induction. Figure 22.8 illustrates the principle. Two metallic bodies on insulating stands (A and A') initially carry some small positive and negative charges. In order to increase these charges, we place two small metallic bodies B and B' nearby. Charges will then be induced on B and B' as shown in Figure 22.8a. If we now momentarily connect the far sides of B and B' with a wire, the charges on these sides will cancel, leaving B with a negative charge and B' with a positive charge (Figure 22.8b). Next, we put B' in contact with A and B in contact with A' (Figure 22.8c); this transfers the charges of the small bodies almost entirely to the larger bodies. The small bodies are

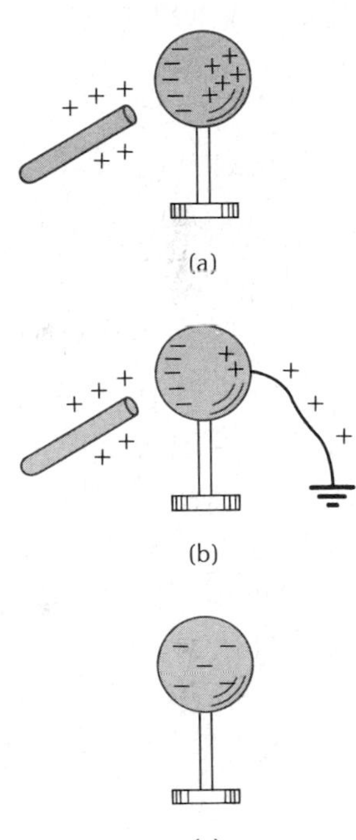

Fig. 22.7 (a) The positively charged glass rod induces a charge distribution on the metallic sphere. (b) When the far side of the sphere is connected to the ground by means of a wire, the positive charge leaks away. (c) Finally, only negative charge remains on the sphere.

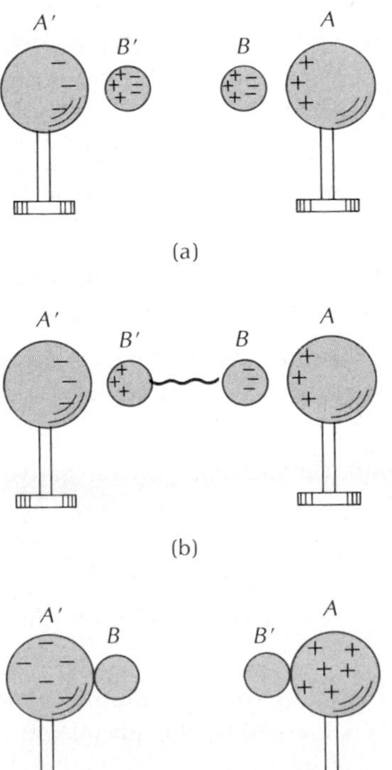

Fig. 22.8 (a) The large spheres induce charge distributions on the small spheres. (b) When the neighboring sides of the small spheres are connected by means of a wire, the charges of these sides cancel. (c) The large spheres absorb the charge of the small spheres.

Fig. 22.9 A Wimshurst machine.

then left uncharged and we can again repeat the sequence of operations. The net result is a stepwise increase of the electric charges on A and A'. In electrostatic machines — such as Wimshurst's machine (Figure 22.9) — the entire sequence of manipulations of the bodies B and B' is carried out automatically when the experimenter turns a crank.

SUMMARY

Electric charge may be positive, negative, or zero: like charges repel, unlike charges attract.

Coulomb's Law: $F = \dfrac{1}{4\pi\varepsilon_0}\dfrac{qq'}{r^2}$

Permittivity constant: $\varepsilon_0 = 8.85 \times 10^{-12}\ \mathrm{C^2/N \cdot m^2}$

$$\frac{1}{4\pi\varepsilon_0} = 8.99 \times 10^9\ \mathrm{N \cdot m^2/C^2}$$

Charge of proton: $e = 1.60 \times 10^{-19}$ C

Charge of electron: $-e = -1.60 \times 10^{-19}$ C

Charge conservation: In any reaction the net electric charge remains constant.

Conductor: Permits motion of charge.

Insulator: Does not permit motion of charge.

QUESTIONS

1. Suppose that the Sun has a positive electric charge and each of the planets a negative electric charge, and suppose that there is no gravitational force. In what way would the motions of the planets predicted by this "electric" model of the Solar System differ from the observed motions?

2. The protons in the nucleus of an atom repel each other electrically. What holds the protons (and the neutrons) together, and prevents the nucleus from bursting apart?

3. Describe how you would set up an experiment to determine whether the electric charges on an electron and a proton are exactly the same.

4. Assume that neutrons have a small amount of electric charge, say, positive charge. Discuss some of the consequences of this assumption for the behavior of matter.

5. If we were to assign a positive charge to the electron and a negative charge to the proton, would this affect the mathematical statement [Eq. (7)] of Coulomb's Law?

6. In the cgs, or Gaussian, system of units, Coulomb's Law is written as $F = qq'/r^2$. In terms of grams, centimeters, and seconds, what are the units of electric charge in this system?

7. Could we use the electric charge of an electron as an atomic standard of electric charge to define the coulomb? What would be the advantages and disadvantages of such a standard?

8. Besides electric charge, what other physical quantities are conserved in reactions among particles? Which of these quantities are quantized?

9. Since the free electrons in a piece of metal are free to move any which way, why don't they all fall to the bottom of the piece of metal under the influence of the pull of gravity?

10. If the surface of a piece of metal acts like a container in confining the free electrons, why can't we cause these electrons to spill out by drilling hole in the surface?

11. If you rub a plastic comb, it will attract hairs or bits of paper, even though they have no net electric charge. Explain.

12. Some old-fashioned physics textbooks define positive electric charge as the kind of charge that accumulates on a glass rod when rubbed with silk. What is wrong with this definition?

13. When you rub your shoes on a carpet, you sometimes pick up enough electric charge to feel an electric shock if you subsequently touch a radiator or some other metallic body connected to the ground. Why is this more likely to happen in winter than in summer?

14. Some automobile operators hang a conducting strap on the underside of their automobile, so that this strap drags on the street. What is the purpose of this arrangement?

15. Some electric charge has been deposited on a ping-pong ball. How could you find out whether the charge is positive or negative?

16. Two aluminum spheres of equal radii hang from the ceiling on insulating threads. You have a glass rod and a piece of silk. How can you give these two spheres exactly equal amounts of electric charge?

PROBLEMS

Section 22.2

1. Within a typical thundercloud there are electric charges of -40 C and $+40$ C separated by a vertical distance of 5 km. Treating these charges as pointlike, find the magnitude of the electric force of attraction between them.

2. A crystal of NaCl (common salt) consists of a regular arrangement of ions of Na^+ and Cl^-. The distance from one ion to its neighbor is 2.82×10^{-10} m. What is the magnitude of the electric force of attraction between the two ions? Treat the ions as point charges.

3. The two protons in the nucleus of a helium atom are at a distance of 2×10^{-15} m from each other. What is the magnitude of the electric force of repulsion that they exert on each other? What would be the acceleration of each if this were the only force acting on them? Treat the protons as point charges.

4. An alpha particle (charge $+2e$) is launched at high speed toward a nucleus of uranium (charge $+92e$). What is the magnitude of the electric force on the alpha particle when it is at a distance of 5×10^{-14} m from the nucleus? What is the corresponding instantaneous acceleration of the alpha particle?

5. According to recent theoretical and experimental investigations, the subnuclear particles are made of quarks and of antiquarks (see Interlude C). For example, a positive pion is made of a u quark and a d antiquark. The electric charge on the u quark is $\frac{2}{3}e$ and that on the d antiquark is $\frac{1}{3}e$. Treating the quarks as classical particles, calculate the electric force of attraction between the quarks in the pion if the distance between them is 1.0×10^{-15} m.

6. A small charge of 2×10^{-6} C is at the point $x = 2$ m, $y = 3$ m in the x–y plane. A second small charge of -3×10^{-6} is at the point $x = 4$ m, $y = -2$ m. What is the electric force that the first charge exerts on the second? What is

the force that the second charge exerts on the first? Express your answers as vectors, with x and y components.

7. Two tiny chips of plastic of mass 5×10^{-5} g are separated by a distance of 1 mm. Suppose that they carry equal and opposite electrostatic charges. What must the magnitude of the charge be if the electric attraction between them is to equal their weight?

8. Although the best available experimental data are consistent with Coulomb's Law, they are also consistent with a modified Coulomb's Law of the form

$$F = \frac{1}{4\pi\varepsilon_0} \frac{q_1 q_2}{r^2} e^{-r/r_0}$$

where r_0 is a constant with the dimensions of a length and a numerical value which is known to be no less than 10^9 m and is probably much larger. Assuming that $r_0 = 10^9$ m, what is the fractional deviation between Coulomb's Law and the modified Coulomb's Law for $r = 1$ m? For $r = 10^4$ m? (Hint: Use the approximation $e^x \cong 1 + x$ for small x.)

9. A proton is at the origin of coordinates. An electron is at the point $x = 0.40$ Å, $y = 0.20$ Å, $z = 0.15$ Å. What are the x, y, and z components of the electric force that the proton exerts on the electron? That the electron exerts on the proton?

10. Precise experiments have established that the magnitudes of the electric charges of an electron and a proton are equal to within an experimental error of $\pm 10^{-21}e$ and that the electric charge of neutron is zero to within $\pm 10^{-21}e$. Making the worst possible assumption about the combination of errors, what is the largest conceivable electric charge of an oxygen atom consisting of 8 electrons, 8 protons, and 8 neutrons? Treating the atoms as point particles, compare the electric force between two such oxygen atoms with the gravitational force between these atoms. Is the net force attractive or repulsive?

11. Suppose that under the influence of the electric force of attraction the electron in a hydrogen atom orbits around the proton on a circle of radius 0.53×10^{-10} m. What is the orbital speed? What is the orbital period?

12. According to Bohr's theory of the atom, the electron in a hydrogen atom orbits around the nucleus in a circular orbit. The force that holds the electron in this orbit is the Coulomb force. The size of the orbit depends on the angular momentum — the smallest possible orbit has an angular momentum $\hbar = 1.05 \times 10^{-34}$ J · s; the next possible orbit has angular momentum $2\hbar$; the next, $3\hbar$, etc.

(a) Calculate the radius of each of these three possible circular orbits.

(b) In general, show that if a circular orbit has an angular momentum $n\hbar$ (where $n = 1, 2, 3, \ldots$), then its radius is

$$r = \frac{4\pi\varepsilon_0}{m_e e^2} n^2 \hbar^2$$

(c) Evaluate this radius for $n = 1$.

13. A maximum electric charge of 7.5×10^{-6} C can be placed on a metallic sphere of radius 15 cm before the surrounding air suffers electric breakdown (sparks). How many excess electrons (or missing electrons) does the sphere have when breakdown is about to occur?

14. How many electrons are in a paper clip of iron of mass 0.3 g?

15. The electric charge in one mole of protons is called **Faraday's constant.** What is its numerical value?

16. What is the number of electrons and of protons in a human body of mass 73 kg? The chemical composition of the body is given in Problem 1.24.

17. Suppose that you remove all the electrons in a copper penny of mass 2.7 g and place them at a distance of 2.0 m from the remaining copper ions. What would be the electric force of attraction on the electrons?

18. How many extra electrons would we have to place on the Earth and on the Moon so that the electric repulsion between these bodies cancels their gravitational attraction? Assume that the numbers of extra electrons on the Earth and on the Moon are in the same proportion as the radial dimensions of these bodies (6.38:1.74).

19. At a place directly below a thundercloud, the induced electric charge on the surface of the Earth is $+10^{-7}$ coulomb per square meter of surface. How many singly charged positive ions per square meter does this represent? The number of atoms on the surface of a solid is typically 3×10^{20} per square meter. What fraction of these atoms must be ions to account for the above electric charge?

20. The electric charge flowing through an ordinary 110-volt, 150-watt light bulb is 1.5 C/s. How many electrons per second does this amount to?

Section 22.3

21. Consider the following hypothetical reactions involving the collision between a high-energy proton (from an accelerator) and a stationary proton (in the nucleus of a hydrogen atom serving as target):

$$p + p \rightarrow n + n + \pi^+$$

$$p + p \rightarrow n + p + \pi^0$$

$$p + p \rightarrow n + p + \pi^+$$

$$p + p \rightarrow p + p + \pi^0 + \pi^0$$

$$p + p \rightarrow n + p + \pi^0 + \pi^-$$

where the symbols p, n, π^+, π^-, and π^0 stand for proton, neutron, positively charged pion, negatively charged pion, and neutral pion. Which of these reactions are impossible?

22. Consider the reaction

$$Ni^{++} + 4H_2O \rightarrow NiO_4^{--} + 8H^+ + [\text{electrons}]$$

How many electrons does this reaction release?

23. We can silver-plate a metallic object, such as a spoon, by immersing the spoon and a bar of silver in a solution of silver nitrate ($AgNO_3$). If we then connect the spoon and the silver bar to an electric generator and make a current flow from one to the other, the following reactions will occur at the immersed surfaces (Figure 22.10):

$$Ag^+ + [\text{electron}] \rightarrow Ag_{(\text{metal})}$$

$$Ag_{(\text{metal})} \rightarrow Ag^+ + [\text{electron}]$$

The first reaction deposits silver on the spoon and the second removes silver from the silver bar. How many electrons must we make flow from the silver bar to the spoon in order to deposit 1 g of silver on the spoon?

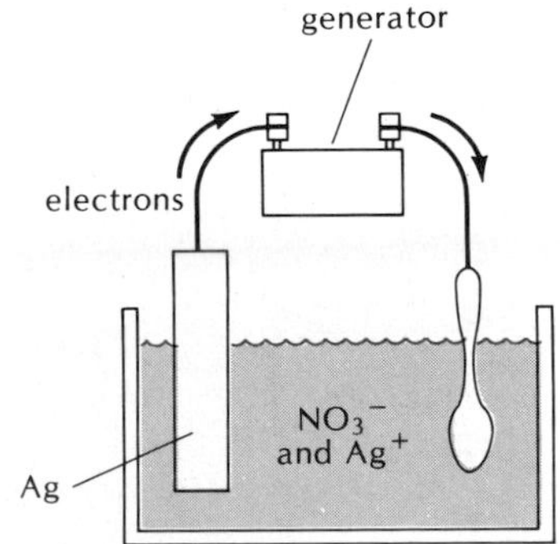

Fig. 22.10

CHAPTER 23

The Electric Field

Up to this point we have taken the view that the gravitational forces and the electric forces between particles are **action-at-a-distance**, i.e., a particle exerts a direct gravitational or electric force on another particle even though these particles are not touching. Although such an interpretation of gravitational and electric forces as a ghostly tug-of-war between distant particles is suggested by Newton's Law of gravitation and by Coulomb's Law of electric force, Newton himself expressed considerable misgivings about this interpretation. In his own words:

> It is inconceivable, that inanimate brute matter, should, without the mediation of something else, which is not material, operate upon and affect other matter without mutual contact. That Gravity should be innate, inherent and essential to Matter so that one Body may act upon another at a Distance thro' a *Vacuum* without the Mediation of anything else, by and through which their Action and Force may be conveyed from one to another, is to me so great an Absurdity that I believe no Man who has in philosophical Matters a competent Faculty of thinking can ever fall into it.

Field

According to the modern view, there is indeed an entity that conveys the force from one particle to another by contact. This entity is the **field**. In the present chapter we will become acquainted with the electric field that conveys the electric force from one charge to another distant charge. But we must first take a look at what happens to electric forces when many electric charges interact simultaneously.

23.1 The Superposition of Electric Forces

According to Eq. (22.7), the electric force exerted by a point charge q' on a point charge q is

$$\mathbf{F} = \frac{1}{4\pi\varepsilon_0} \frac{qq'}{r^2} \hat{\mathbf{r}} \tag{1}$$

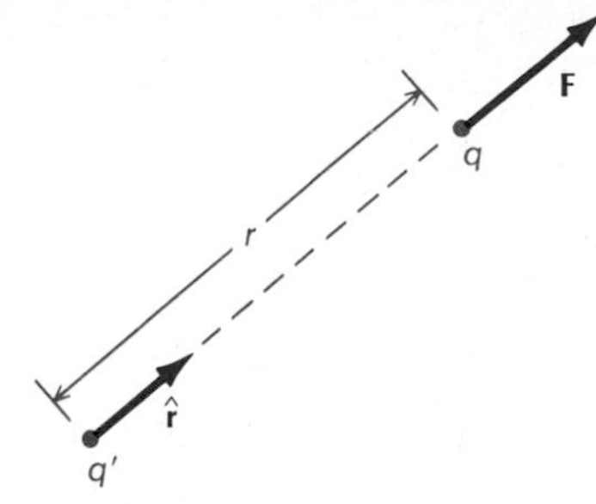

Fig. 23.1 A charge q' exerts an electric force **F** on the charge q.

where $\hat{\mathbf{r}}$ is a unit vector pointing from the charge q' toward the charge q (Figure 23.1).

If several point charges $q_1, q_2, q_3, \ldots$, simultaneously exert electric forces on a charge q, then the collective force on q is obtained by taking the vector sum of the individual forces. Thus, the force on q is

$$\mathbf{F} = \mathbf{F}_1 + \mathbf{F}_2 + \mathbf{F}_3 + \cdots$$

$$= \frac{1}{4\pi\varepsilon_0} \frac{qq_1}{r_1^2} \hat{\mathbf{r}}_1 + \frac{1}{4\pi\varepsilon_0} \frac{qq_2}{r_2^2} \hat{\mathbf{r}}_2 + \frac{1}{4\pi\varepsilon_0} \frac{qq_3}{r_3^2} \hat{\mathbf{r}}_3 + \cdots \tag{2}$$

where $r_1, r_2, r_3, \ldots$, and $\hat{\mathbf{r}}_1, \hat{\mathbf{r}}_2, \hat{\mathbf{r}}_3, \ldots$, are the distances and unit vectors, respectively (Figure 23.2).

Principle of linear superposition

Equation (2) expresses the **principle of linear superposition** of electric forces. According to Eq. (2), the force contributed by each charge is independent of the presence of the other charges; for instance, charge q_2 does not affect the interaction of q_1 with q, it merely adds its own interaction with q. This simple combination law is an important *empirical* fact about electric forces. Since the contact forces (pushes and pulls) of everyday experience are electric forces, they will likewise obey the superposition principle and they can be combined by simple vector addition. Incidentally, the gravitational forces on the Earth and within the Solar System also obey the linear superposition principle. Thus, all the forces in our immediate environment obey this principle (see Section 5.3).

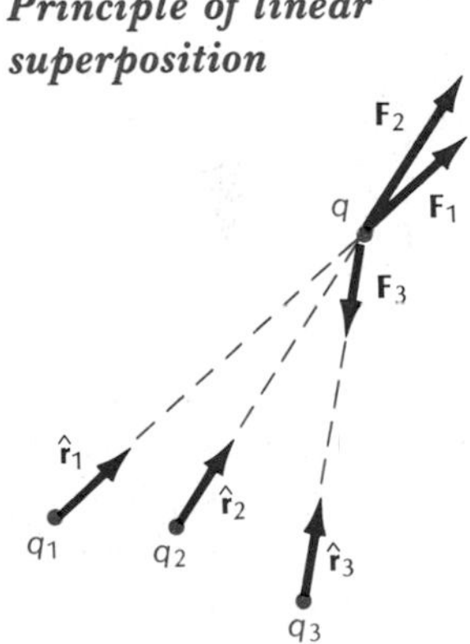

Fig. 23.2 Charges q_1, q_2, q_3 exert electric forces $\mathbf{F}_1, \mathbf{F}_2, \mathbf{F}_3$ on the charge q.

EXAMPLE 1. Point charges Q and $-Q$ are separated by a distance d. A point charge q is equidistant from these charges, at a distance x from their midpoint (Figure 23.3). What is the electric force on q?

SOLUTION: With the choice of axes shown in Figure 23.3, the distance between the charges $\pm Q$ and q is $\sqrt{x^2 + d^2/4}$. Hence the magnitudes of the Coulomb forces exerted by $+Q$ and $-Q$ are

$$F_1 = F_2 = \frac{1}{4\pi\varepsilon_0} \frac{qQ}{x^2 + d^2/4} \tag{3}$$

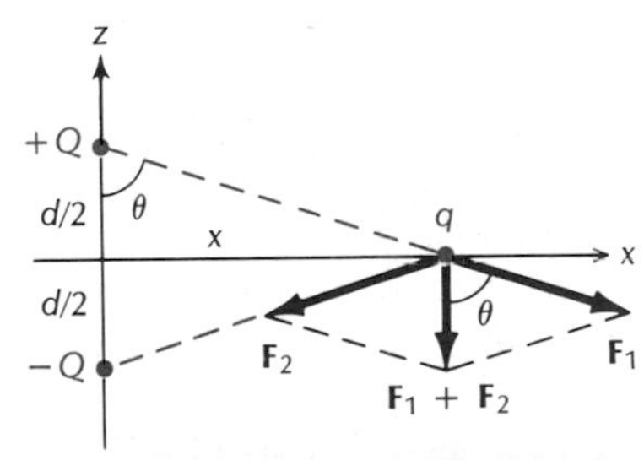

Fig. 23.3 The charges $+Q$ and $-Q$ exert forces $\mathbf{F}_1$ and $\mathbf{F}_2$ on the charge q.

The vector $\mathbf{F}_1$ points away from Q and $\mathbf{F}_2$ points toward $-Q$. In the vector sum $\mathbf{F}_1 + \mathbf{F}_2$, the x components obviously cancel and only the z component survives. The latter component has a magnitude

$$F_z = -\frac{1}{4\pi\varepsilon_0} \frac{qQ}{x^2 + d^2/4} \cos\theta - \frac{1}{4\pi\varepsilon_0} \frac{qQ}{x^2 + d^2/4} \cos\theta \tag{4}$$

With $\cos\theta = \frac{1}{2}d/(x^2 + d^2/4)^{1/2}$, this gives

$$F_z = -\frac{1}{4\pi\varepsilon_0} \frac{qQd}{(x^2 + d^2/4)^{3/2}} \tag{5}$$

Note that at large distance from the two charges, d^2 can be neglected compared to x^2 so that $(x^2+d^2/4)^{3/2} \cong x^3$. The force is then proportional to $F \propto 1/x^3$, i.e., the force decreases in inverse proportion to the cube of the distance. Thus, although the force generated by each point charge is an inverse-square force, the net force has a quite different behavior because at large distance the force generated by one charge tends to cancel the force generated by the other charge.

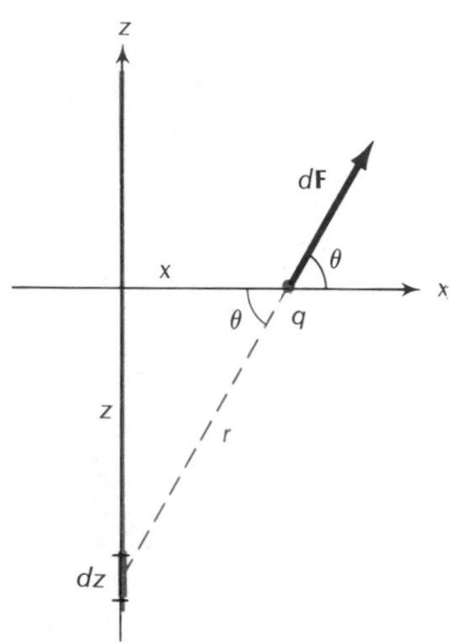

Fig. 23.4 A small segment dz of a line of charge exerts a force on the charge q. The angle θ is reckoned as positive if z is positive, and negative if z is negative.

EXAMPLE 2. Charge is distributed uniformly along a very long, thin line (say, along a string of silk). If the amount of the charge is λ coulomb per meter of line, what is the electric force on a charge q placed near the line?

SOLUTION We will pretend that the line of charge is of infinite length; this simplifies the calculation and introduces no appreciable error since, in any event, most of the force on q is contributed by those portions of the line which are nearest to q.

Figure 23.4 shows the line of charge along the z axis and the charge q on the x axis, at a distance x from the origin. To find the force, we must perform an integration, regarding the line of charge as made up of infinitesimal line elements, each of which can be treated as a point charge (Figure 23.4). Before proceeding with such an integration it is always best to determine the *direction* of the force by a preliminary, qualitative argument. The force generated by the line element dz shown in Figure 23.4 has both x and z components. Upon integration, the z component will cancel against the z component contributed by a line element lying at the same distance above the origin. Therefore the net force will only have an x component.

The line element dz carries a charge $dQ=\lambda\, dz$. Since this line element can be treated as a point charge, it generates a force

$$dF = \frac{1}{4\pi\varepsilon_0}\frac{q\,dQ}{r^2} = \frac{1}{4\pi\varepsilon_0}\frac{q\lambda\,dz}{r^2} \tag{6}$$

This force has an x component:

$$dF_x = \frac{1}{4\pi\varepsilon_0}\frac{q\lambda\cos\theta\,dz}{r^2} \tag{7}$$

The net force is then

$$F_x = \int_{-\infty}^{+\infty} \frac{1}{4\pi\varepsilon_0}\frac{q\lambda\cos\theta}{r^2}\,dz \tag{8}$$

To evaluate this, it is convenient to express all variables in terms of θ. From Figure 23.4,

$$z = x\tan\theta \tag{9}$$

i.e.,

$$dz = x\sec^2\theta\,d\theta \tag{10}$$

Furthermore,

$$r = x\sec\theta \tag{11}$$

With these substitutions we obtain an integral over the angle θ; in terms of this angle, the limits of integration are $\theta=-90°$ and $\theta=90°$, or, in radians, $\theta=-\pi/2$ and $\theta=\pi/2$:

$$F_x = \frac{1}{4\pi\varepsilon_0}\frac{q\lambda}{x}\int_{-\pi/2}^{\pi/2}\cos\theta\,d\theta \tag{12}$$

Since

$$\int_{-\pi/2}^{\pi/2} \cos\theta \, d\theta = \left[\sin\theta\right]_{-\pi/2}^{\pi/2} = 2$$

we find

$$F_x = \frac{1}{2\pi\varepsilon_0} \frac{q\lambda}{x} \qquad (13)$$

Note that this force decreases in inverse proportion to the distance (not the square of the distance). The direction of the force is radially away from the line of charge.

23.2 The Electric Field

As we remarked in the introduction to this chapter, the simplest interpretation of Coulomb's Law is that the electric force between charges is action-at-a-distance, i.e., a charge q' exerts a direct force on a charge q even though these charges are separated by a large distance and are not touching. Such an interpretation of electric force leads to serious difficulties in the case of moving charges. Suppose that we suddenly move the charge q' somewhat nearer to the charge q; then the electric force has to increase. But the required increase cannot occur instantaneously — the increase can be regarded as a signal from q' to q and it is a fundamental rule of Special Relativity that no signal can propagate faster than the speed of light (as discussed in Section 17.3). This suggests that, when we suddenly move the charge q', some kind of disturbance travels through empty space from q' to q at the speed of light and, when this disturbance reaches q, it adjusts the electric force to the new increased value (Figure 23.5). Thus charges exert forces on one another by means of disturbances that they generate in the space surrounding them. These disturbances are called **electric fields.**

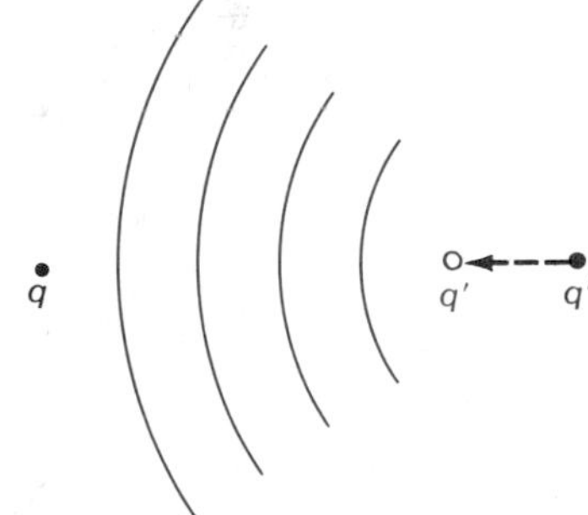

Fig. 23.5 A disturbance emanates from q' and reaches q.

Electric field

Fields are a form of matter — they are endowed with energy and with momentum and they therefore exist in a material sense. In the context of the above example it is easy to see why the disturbance, or field, generated by the sudden displacement of q' must carry momentum and energy: as we suddenly move q' toward q, the force on q' immediately increases according to Coulomb's Law, but the increase in the force on q will be delayed until a signal has had time to propagate from q' to q carrying the information regarding the changed position of q'. Thus action and reaction will be temporarily out of balance, i.e., the momentum of the two particles will not be conserved. In order to maintain an overall momentum balance, the momentum missing from the particles must be transferred to the field, i.e., the field must acquire momentum. This special case of the two charged particles is representative of the general case — it can be demonstrated rigorously that, when relativity is taken into account, the momentum and energy of a general system of interacting particles cannot be conserved by itself. An extra entity, such as the field, is needed to take up the momentum and energy missing from the particles.

Although the above arguments for the existence of fields arose from the problem of charges in motion, we will now adopt the very natural view that the forces on charges at rest involve the same mechanism. We suppose that each charge generates a permanent, static disturbance in the space surrounding it and that this disturbance exerts forces on other charges. Thus we take the view that the electric interaction between charges is **action-by-contact:** a charge q' generates a field which fills the surrounding space and exerts forces on any other charges that it touches. The field serves as the mediator of the force according to the scheme

Action-by-contact

$$\text{charge } q' \rightarrow \text{field of charge } q' \rightarrow \text{force on charge } q \qquad (14)$$

The formal definition of the electric field is as follows: To find the field at a given position, take a point charge q (a "test charge") and place it at that position. The charge will then feel an electric force **F**; the electric field is defined as the force **F** divided by the charge q,

Relationship between electric force and field

$$\boxed{\mathbf{E} = \mathbf{F}/q} \qquad (15)$$

Thus, the electric field is simply the force per unit positive charge.[1]

The unit of electric field is the newton/coulomb.[2] Table 23.1 gives the magnitudes of some typical electric fields.

Table 23.1. SOME ELECTRIC FIELDS

At surface of uranium nucleus	2×10^{21} N/C
At surface of pulsar	$\sim 10^{14}$ N/C
At orbit of electron in hydrogen atom	6×10^{11} N/C
In X-ray tube	5×10^{6} N/C
Electrical breakdown of air	3×10^{6} N/C
In Van de Graaff accelerator	2×10^{6} N/C
Within lightning bolt	10^{4} N/C
Under thundercloud	1×10^{4} N/C
Near radar transmitter (FPS-6)	7×10^{3} N/C
In sunlight (rms)	1×10^{3} N/C
In atmosphere (fair weather)	1×10^{2} N/C
In beam of small laser (rms)	1×10^{2} N/C
In fluorescent lighting tube	10 N/C
In radio wave	$\sim 10^{-1}$ N/C
Within household wiring	$\sim 3 \times 10^{-2}$ N/C
In thermal radiation in intergalactic space (rms)	3×10^{-6} N/C

The electric field surrounding a charge or a distribution of charges is a function of position. For example, according to Coulomb's law, a point charge q' exerts a force $(1/4\pi\varepsilon_0)q'q\hat{\mathbf{r}}/r^2$ on a charge q; hence the electric field generated by q' at a distance r is

[1] The procedure involved in this definition of the electric field implicitly assumes that all the charges that generate the field **E** remain fixed in their positions while the test charge is brought up. To avoid disturbances to these charges, it is usually convenient to take a very small charge q.

[2] Newton/coulomb is the same thing as volt/meter (see Section 25.1).

$$\mathbf{E} = \frac{1}{4\pi\varepsilon_0} \frac{q'}{r^2} \hat{\mathbf{r}} \qquad (16)$$

Electric field of point charge

Note that the electric field is a vector. Its direction must be specified either by unit vectors [as in Eq. (16)] or else by components.

The net electric field generated by any collection of charges can be calculated by linear superposition of the individual fields, in much the same way as is done for electric forces.

EXAMPLE 3. A charge Q is uniformly distributed along the circumference of a thin ring of radius R. What is the electric field on the axis of the ring?

SOLUTION: We regard the ring as made up of infinitesimal line elements ds (Figure 23.6), each of which can be treated as a point charge. Before proceeding with the integration of the contributions of all such line elements, let us determine the direction of the net electric field. The field generated by the line element ds shown in Figure 23.6 has both a horizontal and a vertical component. Obviously, for any given line element there is an equal line element on the opposite side of the ring's center that contributes an electric field of opposite horizontal component, i.e., the horizontal components cancel pairwise. The net electric field is therefore vertical.

The charge per unit length along the circumference is $Q/2\pi R$; hence, the charge in the line element ds is $dQ = ds(Q/2\pi R)$. At a height z above the plane of the ring (Figure 23.6), the electric field contributed by the charge element dQ has a magnitude

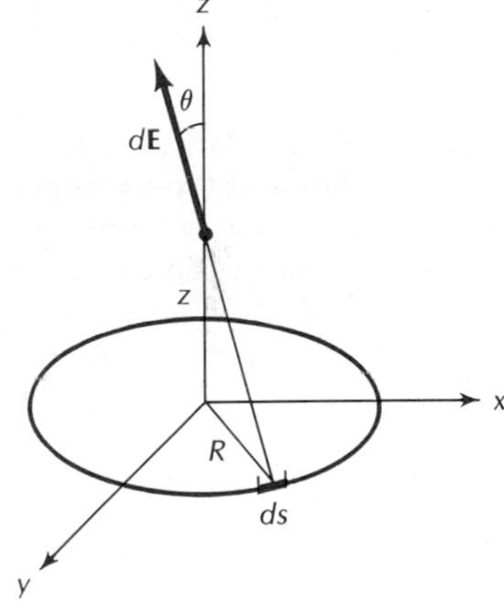

Fig. 23.6 A small segment ds of a ring of charge contributes an electric field $d\mathbf{E}$.

$$dE = \frac{1}{4\pi\varepsilon_0} \frac{dQ}{r^2} = \frac{1}{4\pi\varepsilon_0} \frac{Q\,ds}{2\pi R} \frac{1}{z^2 + R^2} \qquad (17)$$

This field has a vertical component

$$dE_z = \frac{1}{4\pi\varepsilon_0} \frac{Q\,ds}{2\pi R} \frac{\cos\theta}{z^2 + R^2} \qquad (18)$$

Consequently the net electric field is

$$E_z = \int \frac{1}{4\pi\varepsilon_0} \frac{Q}{2\pi R} \frac{\cos\theta}{z^2 + R^2}\,ds \qquad (19)$$

Since the integrand has the same value at all points of the circumference, the integral is of the form [constant] $\times \int ds$; this equals [constant] $\times\, 2\pi R$, since the length of the circumference is $2\pi R$. Hence,

$$E_z = \frac{1}{4\pi\varepsilon_0} \frac{Q\cos\theta}{z^2 + R^2} = \frac{1}{4\pi\varepsilon_0} \frac{Qz}{(z^2 + R^2)^{3/2}} \qquad (20)$$

where $\cos\theta = z/\sqrt{z^2 + R^2}$ has been inserted. In vector notation,

$$\mathbf{E} = \frac{1}{4\pi\varepsilon_0} \frac{Qz}{(z^2 + R^2)^{3/2}} \hat{\mathbf{z}} \qquad (21)$$

Note that for $z = 0$, $\mathbf{E} = 0$; and for $z \gg R$, $\mathbf{E} \cong Q\hat{\mathbf{z}}/(4\pi\varepsilon_0 z^2)$, which is the electric field of a point charge.

EXAMPLE 4. What is the electric field generated by a large sheet (sheet of paper) carrying a uniform charge density of σ coulomb per square meter?

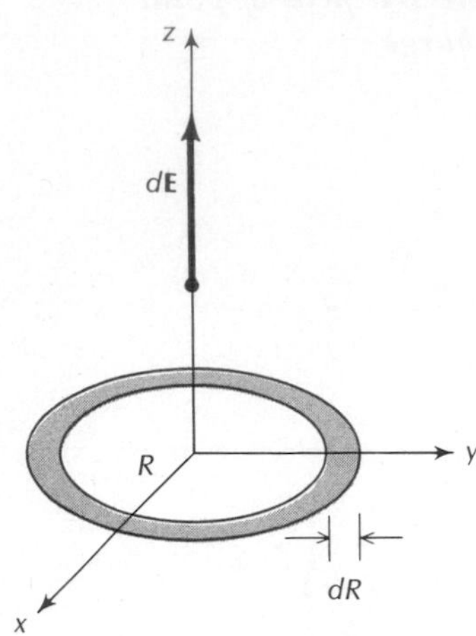

Fig. 23.7 A very large sheet of charge lies in the x–y plane. A thin ring of charge within this sheet produces an electric field $d\mathbf{E}$.

SOLUTION: We will pretend that the sheet is infinitely large. The sheet can be regarded as made up of a collection of rings. Figure 23.7 shows one of these rings with radius R, width dR; this ring has an area $2\pi R dR$ and a charge $dQ = (2\pi R\, dR) \times \sigma$. This produces a vertical electric field [see Eq. (21)],

$$dE = \frac{1}{4\pi\varepsilon_0} \frac{2\pi R\sigma z\, dR}{(z^2 + R^2)^{3/2}} \tag{22}$$

The net electric field is therefore

$$E = \frac{2\pi\sigma z}{4\pi\varepsilon_0} \int_0^\infty \frac{R\, dR}{(z^2 + R^2)^{3/2}} \tag{23}$$

where $R = 0$ is the radius of the smallest ring and $R = \infty$ the radius of the largest.

With the substitution of $u = R^2$, the integral becomes

$$\int_0^\infty \frac{R\, dR}{(z^2 + R^2)^{3/2}} = \int_0^\infty \frac{(1/2)\, du}{(z^2 + u)^{3/2}} = \left[\frac{-1}{(z^2 + u)^{1/2}}\right]_0^\infty = \frac{1}{z}$$

and therefore

$$E = \frac{2\pi\sigma z}{4\pi\varepsilon_0} \frac{1}{z} = \frac{\sigma}{2\varepsilon_0} \tag{24}$$

In vector notation,

Electric field of flat sheet

$$\boxed{\mathbf{E} = \frac{\sigma}{2\varepsilon_0}\hat{\mathbf{z}}} \tag{25}$$

This is a constant electric field, i.e., the electric field of a uniform sheet of charge is the same at points near and at points far from the plane of the sheet. (Of course, if the sheet of charge is of finite size, then this constancy of the electric field is only true in the vicinity of the sheet; at a very large distance from a *finite* sheet of charge the electric field will resemble that of a point charge.)

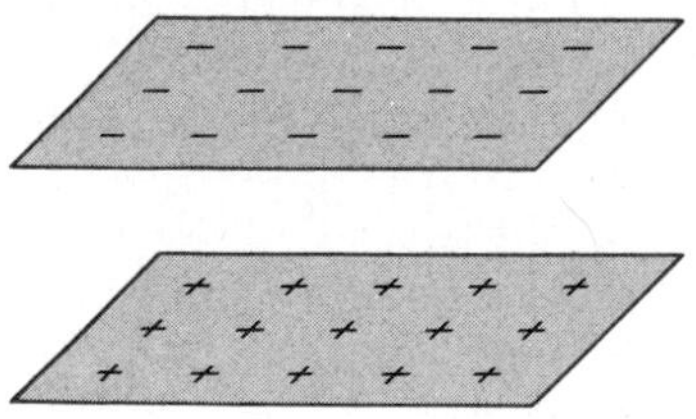

Fig. 23.8 Two very large sheets of charge.

The electric field of two parallel uniform sheets with opposite charges (Figure 23.8) can be obtained by superposing the fields generated by each sheet. In the space between the sheets, the two fields add together, giving a net field

$$E = \sigma/\varepsilon_0 \tag{26}$$

where σ is the magnitude of the charge per unit area on either sheet. Above and below the pair of sheets, the two fields cancel.

EXAMPLE 5. An electron is placed in the uniform electric field between the two charged sheets of Figure 23.8. (a) If the magnitude of the field is 3.0×10^4 N/C, what is the acceleration of the electron? (b) Suppose the electron is initially at rest on the negative sheet and then moves toward the positive sheet under the influence of the electric force. With what speed will it reach the positive sheet? The distance between the sheets is 1.0 cm.

SOLUTION: The force on the electron has a magnitude $F = eE$ and the acceleration has a magnitude

$$a = \frac{eE}{m_e} = \frac{(1.6 \times 10^{-19}\ \text{C}) \times (3 \times 10^4\ \text{N/C})}{0.91 \times 10^{-30}\ \text{kg}} = 5.3 \times 10^{15}\ \text{m/s}^2$$

For constant acceleration, $v^2 - v_0^2 = 2a(x - x_0)$ [see Eq. (2.25)]. With $v_0 = 0$, $a = 5.3 \times 10^{15}\ \text{m/s}^2$, and $x - x_0 = 0.01$ m this gives

$$v = \sqrt{2a(x - x_0)} = 1.0 \times 10^7\ \text{m/s}$$

23.3 Lines of Electric Field

The electric field can be represented graphically by drawing, at any given point of space, a vector whose magnitude and direction are those of the electric field at that point. Figure 23.9 shows the electric field vectors in the space surrounding a positive point charge.

Alternatively, the electric field can be represented graphically by **field lines.** These lines are drawn in such a way that, at any given point, the tangent to the line has the direction of the electric field. Furthermore, the density of lines is directly proportional to the magnitude of the electric field; that is, where the lines are closely spaced the electric field is strong and where the lines are far apart the field is weak. Figure 23.10 shows the electric field lines of a positive point charge and Figure 23.11 those of a negative point charge. The arrows on these lines indicate the direction of the electric field along each line.

Field lines

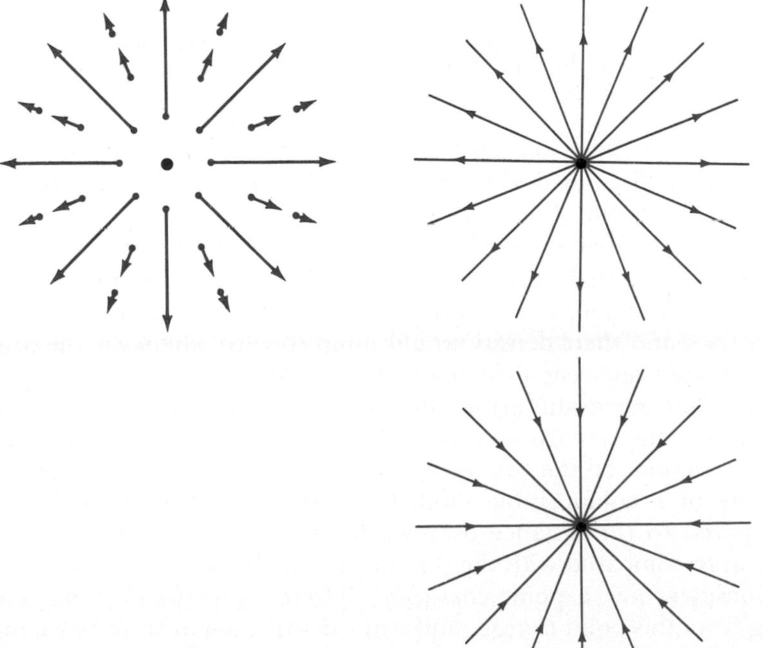

Fig. 23.9 (top left) Electric field vectors surrounding a positive point charge.

Fig. 23.10 (top right) Electric field lines of a positive point charge. Note that in three dimensions the lines spread out in all three directions of space, whereas the diagram shows the lines spreading out only in the two directions within the page. This gives a misleading impression of the density of field lines as a function of distance. This limitation of two-dimensional diagrams should always be kept in mind when looking at pictures of field lines.

Fig. 23.11 (bottom) Electric field lines of a negative point charge.

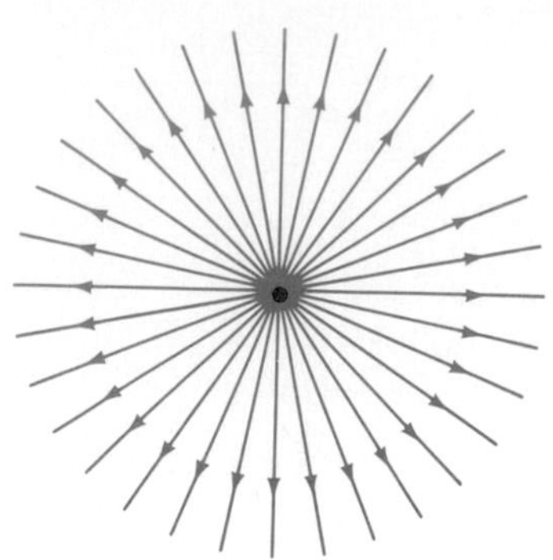

Fig. 23.12 Electric field lines of a positive point charge, twice as large as in Figure 23.10.

When we draw a pattern of field lines, we must begin each line on a positive point charge and end on a negative point charge. Since the magnitude of the electric field is directly proportional to the amount of electric charge, the number of field lines that we draw emerging from a positive charge must be proportional to the charge. Figure 23.12 shows the electric field lines of a positive charge twice as large as that of Figure 23.10. We will adopt the convention that the number of field lines emerging from a charge Q is Q/ε_0; hence the number of lines emerging from one coulomb of charge is $1/\varepsilon_0 = 1.13 \times 10^{11}$. This normalization is very convenient for making computations with field lines, but it is not always quite practical for making drawings — depending on the magnitude of the charge, it sometimes yields an enormous number of field lines so that drawing them becomes an unbearable chore, sometimes a fractional number, so that drawing them becomes altogether meaningless. For instance, according to our normalization, a proton with $e = 1.6 \times 10^{-19}$ C, has $e/\varepsilon_0 = 1.8 \times 10^{-8}$ field line. Such a number makes good sense in a computation, but no sense at all in a drawing. If we want a draftsman to prepare a drawing of the field lines of a proton, we will first have to alter our normalization. From an artistic point of view, it is desirable that the spacing between the field lines be small compared to the distance from the charge or any other relevant distance; this produces a clear picture of the spatial dependence of the electric field. In case of need, we can alter the normalization — but we must be careful to maintain a fixed normalization throughout any given computation or series of drawings.

The inverse-square law for the electric field of a point charge can be "derived" from the picture of field lines — it is easy to show that the density of lines necessarily obeys an inverse-square law. Before we do this, we must give a precise definition of density of field lines. Density of lines is the number of lines per unit area, i.e., the number of lines intercepted by a small area A erected perpendicularly to the lines (Figure 23.13) divided by the magnitude of this area,

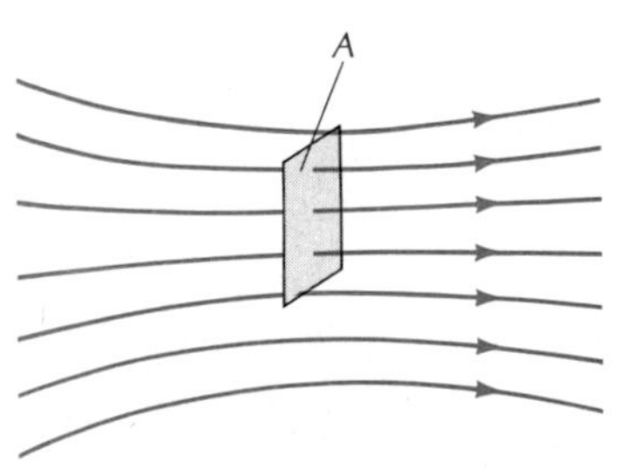

Fig. 23.13 A small area A intercepts some field lines.

$$[\text{density of lines}] = \frac{[\text{number of intercepted lines}]}{A}$$

The area A used in this equation must be small, but not too small: it must be small compared to the distance over which the electric field varies appreciably, but large compared to the spacing between the field lines, so that it intercepts a large number of field lines. If we choose A too small, the density of field lines would not be a well-defined, continuous function of position — the number of intercepted field lines and their density would jump to zero whenever the area A fits between adjacent field lines. By keeping A sufficiently large, we smooth out irrelevant, erratic fluctuations in the density of field lines. Obviously, the restrictions on the choice of the area A are analogous to the restrictions on the volume element used for the computation of the density of a compressible fluid: the volume element must be small compared to the distance over which the density varies appreciably, but large compared with the distance between the molecules.

Consider now a point charge q'. There will be q'/ε_0 lines emerging from this point charge, uniformly distributed over all radial directions. At a distance r from the point charge, these lines are uniformly distributed over the area $4\pi r^2$ of a concentric sphere, i.e., there are

$(q'/\varepsilon_0)/4\pi r^2$ lines per unit area. Hence the density of lines decreases in proportion to the inverse square of the distance. In fact, with our normalization regarding the number of lines per coulomb of charge, the density of lines is not only proportional to, but exactly *equal* to the magnitude of the electric field [see Eq. (16)]. This "derivation" of the Coulomb Law is really only a consistency check — it is because the Coulomb Law is an inverse-square law that the field can be represented by field lines; any other dependence on distance would make it impossible to draw continuous field lines that start and end only on charges.

Figure 23.14 shows the field lines generated jointly by a positive and a negative charge of equal magnitudes; in the midplane between the charges the electric force has the magnitude given by Eq. (5). Figure 23.15 shows the field lines of a pair of equal positive charges and Figure 23.16 those of a pair of unequal positive and negative charges. Figure 23.17 shows the field lines of a large, uniformly charged sheet with a positive density of charge.

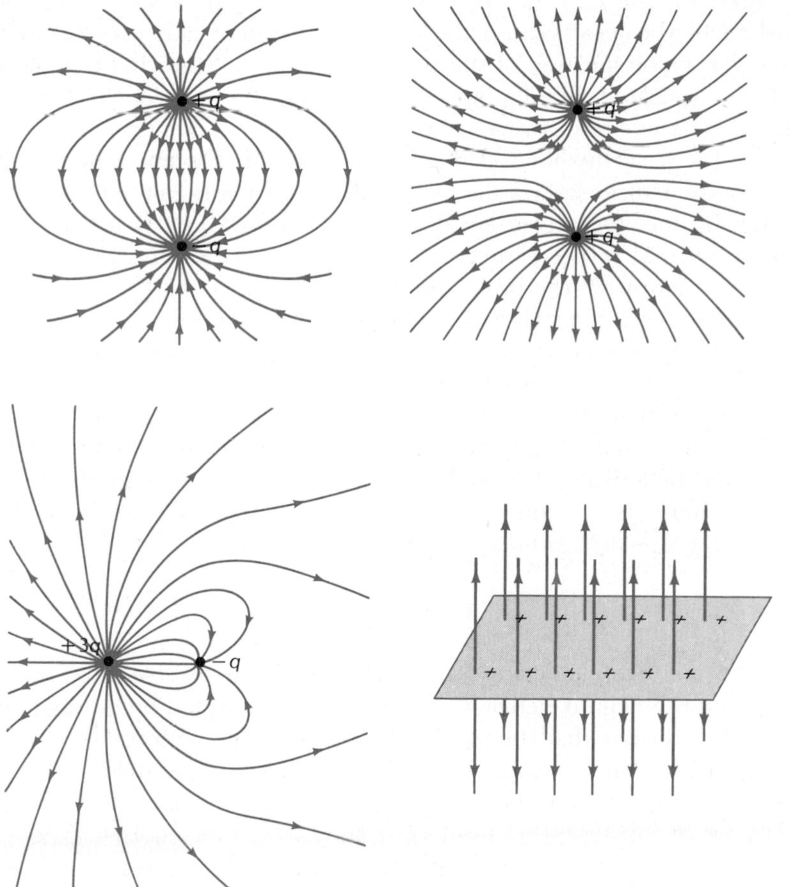

Fig. 23.14 (left) Field lines generated by positive and negative charges of equal magnitudes.

Fig. 23.15 (right) Field lines generated by two positive charges of equal magnitudes.

Fig. 23.16 (left) Field lines generated by positive and negative charge of unequal magnitudes. The positive charge has three times the magnitude of the negative charge.

Fig. 23.17 (right) Field lines of a very large sheet of charge.

Note that in all cases, the field lines start on positive charges and end on negative charges — the positive charges are **sources** of field lines and the negative charges are **sinks.** Also, note that field lines never intersect (except where they start or end at point charges). If the lines ever were to intersect, the electric field would have *two* directions at the point of intersection; this is impossible.

The above pictures of field lines help us to develop some intuitive feeling for the spatial dependence of the electric fields surrounding diverse arrangements of electric charges. But we must not fall into the trap of thinking of the field lines as physical objects. The field lines are merely mathematical crutches to aid our imagination.

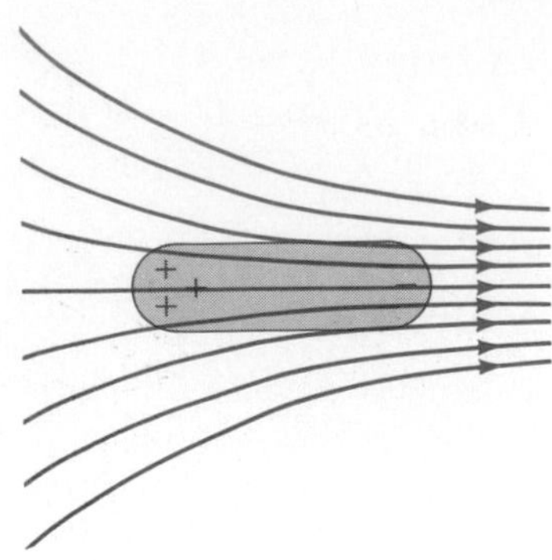

Fig. 23.18 An elongated body with positive charge at one end, negative charge on the other end. The body is placed in a nonuniform electric field.

23.4 Electric Dipole in an Electric Field

If a neutral body is placed in a given electric field, we might expect that the body experiences no force. However, this expectation is not always realized. A neutral body may contain within it separate positive and negative charges (of equal magnitudes) and it is possible that the electric force on one of these charges is larger than on the other; the body then experiences a net force. Such an imbalance of the forces on the negative and positive charges will happen if the electric field is stronger at the location of one kind of charge than at the location of the other. For example, the body shown in Figure 23.18 with positive charges on one end and negative charges on the other end will be pushed to the left because the electric field that acts on the body is stronger at the location of the negative charges. Note that this electric field — indicated by field lines in Figure 23.18 — is not the field generated by the body; rather it is an electric field generated by some other charges (not shown in Figure 23.18). The electric field that acts on a body is often called the **external field;** in contrast, the field generated by the body itself is called the **self-field.** The latter field only exerts internal forces within the body and does not contribute to the net force acting on the body from the outside.

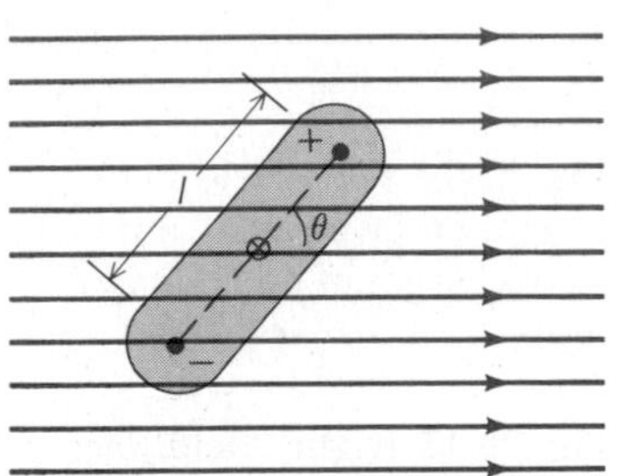

Fig. 23.19 Electric dipole in a uniform electric field. The dots indicate the average positions of the positive and the negative charges, respectively.

If the external electric field is uniform, then the forces on the positive and negative charges in a neutral body cancel and there is no net force. However, there may still be a torque. Figure 23.19 shows a neutral body in a uniform electric field. The body carries equal positive and negative charges $\pm Q$, with the average positions of these charges separated by a distance l. Such a body is called an **electric dipole.** Obviously there is a torque on the body. The torque of each force about the center is $-\frac{1}{2}lQE \sin\theta$ and the total torque is

Electric dipole

$$\tau = -lQE \sin\theta \tag{27}$$

where θ is the angle between the direction of the electric field and the line from the negative charge to the positive charge. The minus sign in Eq. (27) indicates that the torque is clockwise, in the sense of negative θ; the torque tends to align the body with the electric field.

One can write Eq. (27) as

$$\tau = -pE \sin\theta \tag{28}$$

where

$$\boxed{p = lQ} \tag{29}$$

Dipole moment

The quantity p is called the **dipole moment** of the body; it is simply the charge multiplied by the separation between the charges. The units of dipole moment are meter · coulomb.

In vector notation, we can write the torque as

$$\boxed{\boldsymbol{\tau} = \mathbf{p} \times \mathbf{E}} \qquad (30)$$

Torque on dipole

where vector **p** is directed from the negative charge toward the positive charge.

Corresponding to the torque (28) there exists a potential energy that equals the amount of work that *you* must do against the electric forces to twist the dipole through some angle. If you want to twist the dipole through some angle, you must supply a torque $-\tau = +pE \sin\theta$ [opposite to the torque (28)] and do work[3]

$$U = \int_{\theta_0}^{\theta} -\tau \, d\theta' = \int_{\theta_0}^{\theta} pE \sin\theta' \, d\theta'$$

$$= pE\Big[-\cos\theta'\Big]_{\theta_0}^{\theta} = -pE\cos\theta + pE\cos\theta_0$$

It is customary to take the starting angle as $\theta_0 = 90°$. Then

$$U = -pE\cos\theta \qquad (31)$$

or

$$\boxed{U = -\mathbf{p} \cdot \mathbf{E}} \qquad (32)$$

Potential energy of dipole

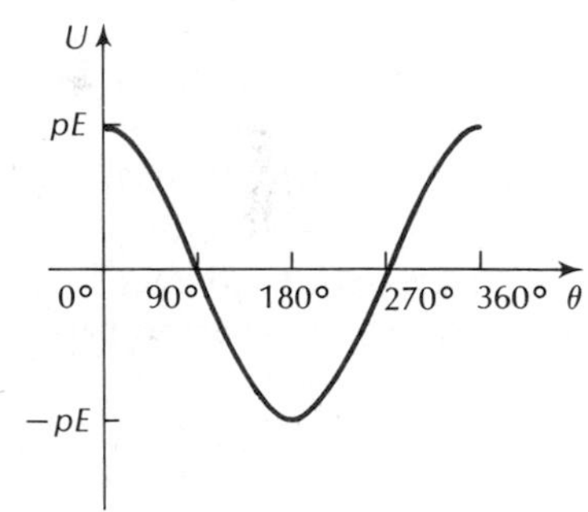

Fig. 23.20 Potential energy of an electric dipole as a function of angle.

This potential energy has a minimum when **p** and **E** are parallel, a maximum when they are antiparallel. Figure 23.20 is a plot of the potential energy U vs. the angle θ.

Many asymmetric molecules have **permanent dipole moments** due to an excess of electrons on one end of the molecule and a corresponding deficit on the other. This means that the molecule has a negative charge on one end and positive charge on the other. For example, Figure 23.21 shows a water molecule. In this molecule the electrons tend to concentrate on the oxygen atom; in Figure 23.21, the left side of the molecule is negatively charged and the right side positively charged. Since the average positions of the positive and negative charges do not coincide, the water molecule has a dipole moment. For a water molecule in water vapor, $p = 6.1 \times 10^{-30}$ C · m.

Molecules of atoms that do not have a permanent dipole moment may acquire a temporary dipole moment when placed in an electric field. The opposite electric forces on the negative and positive charges can distort the molecule and produce a charge separation. Such a dipole moment, which only lasts as long as the molecule is immersed in the electric field, is called an **induced dipole moment.** The magnitude of the induced dipole moment is directly proportional to the magnitude of the electric field. (This proportionality holds for electric fields of the strengths we will be concerned with in this text; however, it fails for electric fields of extreme strength.)

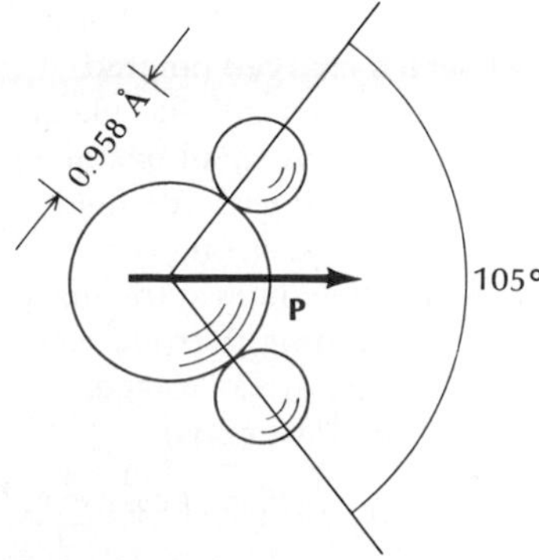

Fig. 23.21 A water molecule.

[3] The variable of integration in this integral has been written θ' in order to distinguish it from the limit of integration θ.

EXAMPLE 6. A molecule of water with a dipole moment of 6.1×10^{-30} C · m is placed in an electric field of 2.0×10^5 N/C. What is the difference between the potential energies for parallel and antiparallel orientations?

SOLUTION: When the dipole moment is parallel to the electric field, the potential energy is

$$U = -pE \cos 0° = -6.1 \times 10^{-30}\ \text{C} \cdot \text{m} \times 2.0 \times 10^5\ \text{N/C}$$

$$= -1.2 \times 10^{-24}\ \text{J}$$

When the dipole moment is antiparallel,

$$U = -pE \cos 180° = +6.1 \times 10^{-30}\ \text{C} \cdot \text{m} \times 2.0 \times 10^5\ \text{N/C}$$

$$= 1.2 \times 10^{-24}\ \text{J}$$

Hence the energy difference between the parallel and antiparallel orientations is 2.4×10^{-24} J.

The tendency for alignment of a dipole with an electric field can be exploited to make the field lines visible. For this purpose, small bits of thread are suspended in oil in a container placed in the electric field; alternatively, small grass seeds are scattered on a sheet of paper placed in the electric field. The electric field induces a dipole moment along the long axis of the bit of thread or the grass seed, and the torque on the dipole then aligns the bit of thread or the grass seed with the electric field. Figures 23.22 and 23.23 show two photographs of field lines made visible by bits of thread (Figures 28.1 and 28.2 show photographs of field lines made visible with grass seeds).

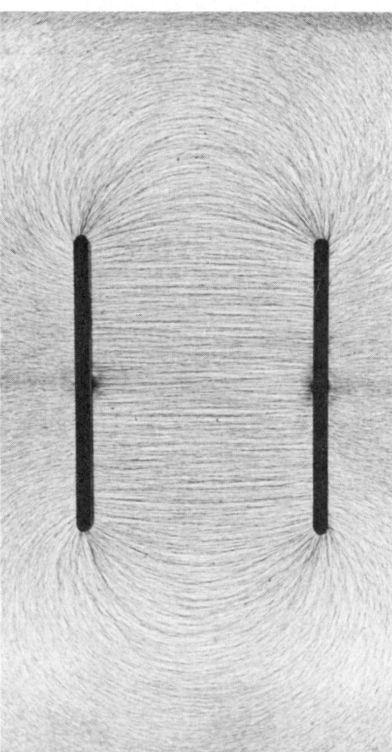

Fig. 23.22 (left) Field lines between a pair of charged parallel plates, made visible by small bits of thread aligned with the field lines. Note the fringing, or spreading, of the field lines near the edges of the plates. (Courtesy H. Waage, Princeton University.)

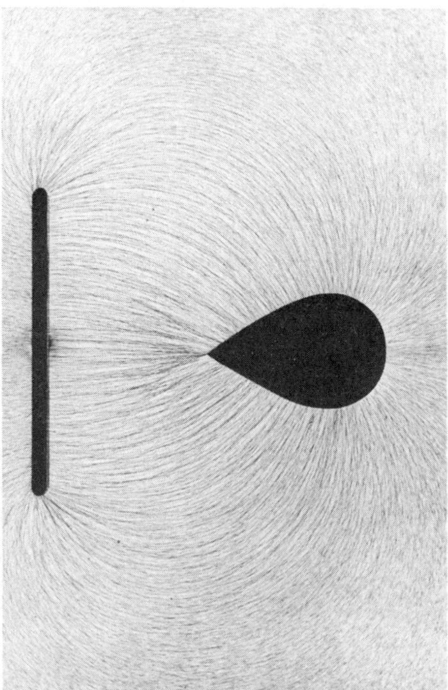

Fig. 23.23 (right) Field lines between a charged pointed body and a charged flat plate, made visible by small bits of thread aligned with the field lines. Note the strong concentration of field lines at the sharp point, indicating a strong electric field. (Courtesy H. Waage, Princeton University.)

SUMMARY

Superposition principle: Electric fields combine additively (as vectors).

Definition of electric field: $\mathbf{E} = \mathbf{F}/q$

Electric field of point charge: $\mathbf{E} = \dfrac{1}{4\pi\varepsilon_0}\dfrac{q'}{r^2}\,\hat{\mathbf{r}}$

Electric field of uniformly charged plane: $E = \sigma/2\varepsilon_0$

Electric dipole moment: $p = lQ$

Torque on dipole: $\tau = \mathbf{p} \times \mathbf{E}$

Potential energy of dipole: $U = -\mathbf{p} \cdot \mathbf{E}$

QUESTIONS

1. Does it make any difference whether the value of the charge q in the equation defining the electric field [Eq. (15)] is positive or negative?

2. In the cgs, or Gaussian, system of units, Coulomb's Law is written as $F = qq'/r^2$. In terms of grams, centimeters, and seconds, what are the units of the electric field in this system?

3. How would you formally define a gravitational field vector? Is the unit of gravitational field the same as the unit of electric field? According to your definition, what are the magnitude and the direction of the gravitational field at the surface of the Earth?

4. During days of fair weather, the Earth has an atmospheric electric field that points vertically down. This electric field is due to charges on the surface of the Earth. What must be the sign of these charges?

5. A large flat sheet measures $L \times L$; the sheet carries a uniform distribution of charge. Roughly how far from the center of the sheet would you expect the electric field to be markedly different from the uniform electric field of an infinitely large sheet?

6. Figure 23.24 shows diagrams of hypothetical field lines corresponding to some static charge distributions, which are beyond the edge of the diagram. What is wrong with these field lines?

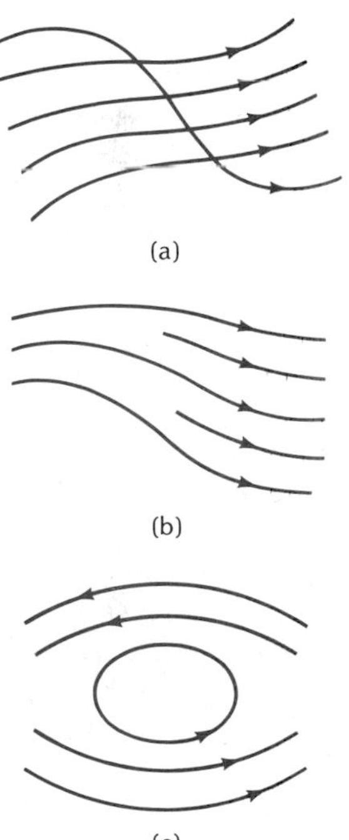

Fig. 23.24

7. If a positive point charge is released from rest in an electric field, will its orbit coincide with a field line? What if the point charge has zero mass?

8. A **tube of force** is the volume enclosed between a bundle of adjacent field lines (Figure 23.25; such a tube of force is analogous to a flow tube in hydrodynamics, and the field lines are analogous to stream lines). Along such a tube of force, the magnitude of the electric field varies in inverse proportion to the cross-sectional area of the tube. Explain.

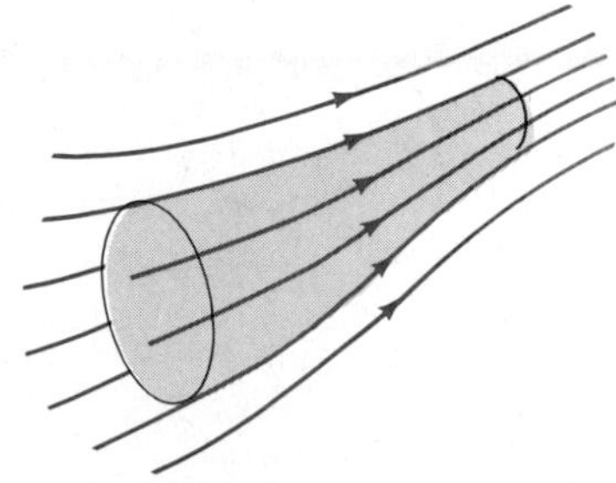
Fig. 23.25

9. A negative point charge $-q$ sits in front of a very large flat sheet with a uniform distribution of positive charge. Make a rough sketch of the pattern of field lines. Is there any point where the electric field is zero?

10. A very long straight line of positive charge lies along the z axis. A very large flat sheet of positive charge lies in the x–y plane. Sketch a few of the field lines of the net electric field produced by both of these charge distributions acting together.

11. A large, flat, thick slab of insulator has positive charge uniformly distrib-

uted over its volume. Sketch the field lines on both sides and inside the slab; pay careful attention to the starting points of the field lines.

12. If our universe is topologically closed, so that a straight line drawn in any direction returns on itself from the opposite direction, can the net charge in the universe be different from zero?

13. How could you build a "compass" that indicates the direction of the electric field?

14. When a neutral metallic body, insulated from the ground, is placed in an electric field, it develops a charge separation, acquiring positive charge on one end and negative charge on the other. This means the body acquires an induced dipole moment. How is the direction of this dipole moment related to the direction of the electric field?

15. By inspection of Figure 23.22, make a rough, qualitative plot of the electric field strength as a function of position along a line parallel to the plates, midway between the plates.

16. One electric dipole is at the origin, oriented parallel to the z axis. Another electric dipole is at some distance on the x axis. The electric field of the first dipole then exerts a torque on the second dipole. For what orientation of the second dipole is the potential energy minimum?

PROBLEMS

Sections 23.1 and 23.2

1. Electric fields as large as 3.4×10^5 N/C have been measured by airplanes flying through thunderclouds. What is the force on an electron exposed to such a field? What is its acceleration?

2. The Earth has not only a magnetic field, but also an atmospheric electric field. During days of fair weather (no thunderclouds), this atmospheric electric field has a strength of about 100 N/C and it points down. Taking into account this electric field and also gravity, what will be the acceleration (magnitude and direction) of a grain of dust of mass 1.0×10^{-18} kg carrying a single *electron* charge?

3. **Millikan's experiment** measures the elementary charge e by the observation of the motion of small oil droplets in an electric field. The oil droplets are charged with one or several elementary charges and, if the (vertical) electric field has the right magnitude, the electric force on the droplet will balance its weight, holding the drop suspended in midair. Suppose that an oil droplet of radius 1.0×10^{-4} cm carries a single elementary charge. What electric field is required to balance the weight? The density of oil is 0.80 g/cm^3.

4. In an X-ray tube (see Figure B.1), electrons are exposed to an electric field of 8×10^5 N/C. What is the force on an electron? What is its acceleration?

5. According to a theoretical estimate, at the surface of a neutron star of mass 1.4×10^{30} kg and radius 1.0×10^4 m there is an electric field of magnitude 6×10^3 N/C pointing vertically up. Show that the corresponding electric force on a proton more than balances the gravitational force on the proton.

6. A long hair, taken from a girl's braid, has a mass of 1.2×10^{-3} g. The hair carries a charge of 1.3×10^{-9} C distributed along its length. If we want to suspend this hair in midair, what (uniform) electric field do we need?

7. The electric field in the electron gun of a TV tube is supposed to accelerate electrons uniformly from 0 to 3.3×10^7 m/s within a distance of 1.0 cm. What electric field is required?

8. Electric breakdown (sparks) occurs in air if the electric field reaches

3×10^6 N/C. At this field strength, free electrons present in the atmosphere are quickly accelerated to such large velocities that upon impact on atoms they knock electrons off the atom and thereby generate an avalanche of electrons. How far must a free electron move under the influence of the above electric field if it is to attain a kinetic energy of 3×10^{-19} J (that is sufficient to produce ionization)?

9. The nuclei of the atoms in a chunk of metal lying on the surface of the Earth would fall to the bottom of the metal if their weight were the only force acting on them. Actually, within the interior of any metal exposed to gravity there exists a very small electric field that points vertically up. The corresponding electric force on a nucleus just balances the weight of the nucleus. Show that for a nucleus of atomic number Z, mass m, the required field has a magnitude $mg/(Ze)$. What is the numerical value of this electric field in a chunk of iron?

10. The hydrogen atom has a radius of 0.53×10^{-10} m. What is the magnitude of the electric field that the nucleus of the atom (a proton) produces at this radius?

11. What is the strength of the electric field at the surface of a uranium nucleus? The radius of the nucleus is 7.4×10^{-15} m and the electric charge is $92e$. For the purposes of this problem the electric charge may be regarded as concentrated at the center.

12. Figure 23.26 shows the arrangement of nuclear charges (positive charges) of a KBr molecule. Find the electric field that these charges produce at the center of mass at a distance of 0.93×10^{-10} m from the Br atom.

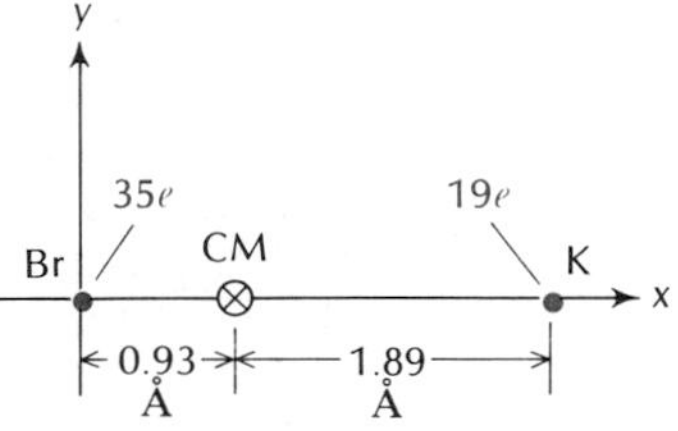

Fig. 23.26 The positive (nuclear) charges in a KBr molecule.

13. The distance between the oxygen nucleus and each of the hydrogen nuclei in an H_2O molecule is 0.958 Å; the angle between the two hydrogen atoms is 105° (Figure 23.27). Find the electric field produced by the nuclear charges (positive charges) at the point P at a distance of 1.2 Å to the right of the oxygen nucleus.

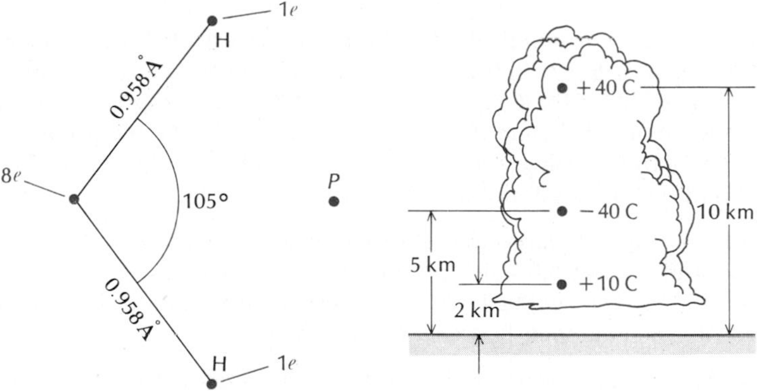

Fig. 23.27 (left) The positive (nuclear) charges in a water molecule.

Fig. 23.28 (right) Charges in a thundercloud.

14. Figure 23.28 shows the charge distribution within a thundercloud. There is a charge of 40 C at a height of 10 km, −40 C at 5 km, and 10 C at 2 km. Treating these charges as pointlike, find the electric field (magnitude and direction) that they produce at a height of 8 km and a horizontal distance of 3 km.

15. Suppose that an airplane flies through the thundercloud described in Problem 14 at the 8-km level. Plot the magnitude of the electric field as a function of position along the path of the airplane; start with the airplane 10 km away from the thundercloud.

16. Suppose that the charge distribution of a thundercloud can be approxi-

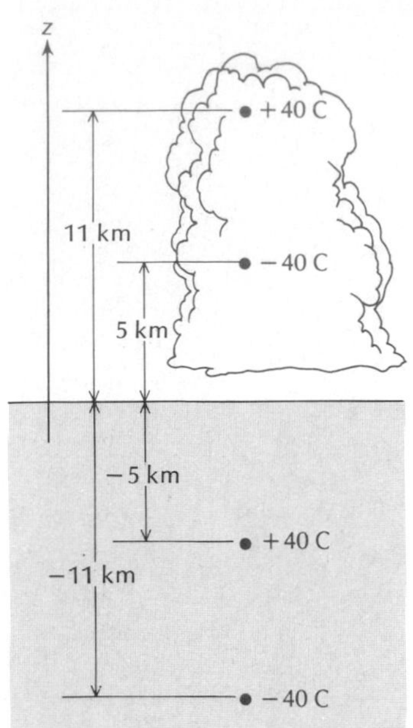

Fig. 23.29 Charges in a thundercloud and their images in the ground.

mated by two point charges, a negative charge of −40 C at a height of 5 km (above ground) and a positive charge of 40 C at a height of 11 km. To find the electric field strength at the ground we must take into account that the ground is a conductor and that the charge of the thundercloud induces charges on the ground. It can be shown that the effect of the induced charges can be simulated by a point charge of 40 C at 5 km *below ground* and a point charge of −40 C at 11 km below ground (Figure 23.29); these fictitious charges are called *image charges* (described in Section I.4). By adding the electric field of the image charges to that of the two real charges in the thundercloud, calculate the magnitude of the electric field at a point on the ground directly below the thundercloud charges. Similarly, calculate the magnitude of the electric field at horizontal distances of 2, 4, 6, 8, and 10 km. Plot the field vs. the distance.

17. Consider eight of the ions of Cl^- and Na^+ in a crystal lattice of common salt. The ions are located at the vertices of a cube measuring 2.82×10^{-10} m on an edge. Calculate the magnitude of the electric force that seven of these ions exert on the eight.

18. Each of two very long, straight, parallel lines carries a positive charge of λ coulomb per meter. The distance between the lines is d. Find the electric field at a point equidistant from the lines, with a distance $2d$ from each line. Draw a diagram showing the direction of the electric field.

19. Two infinite lines of silk with uniform charge distributions of λ coulomb per meter lie along the x and the y axes, respectively. Find the electric field at a point with coordinates x, y, z; assume that $x > 0, y > 0, z > 0$.

20. A semi-infinite line carrying a uniform charge distribution of λ coulomb lies along the positive x axis from $x = 0$ to $x = \infty$. Find the components of the electric field at the point with coordinates x, y, with $z = 0$; assume $x > 0, y > 0$.

21. A semi-infinite line with a uniform charge distribution of $+\lambda$ coulomb per meter lies along the positive x axis from $x = 0$ to $x = \infty$. Another semi-infinite line with a charge distribution of $-\lambda$ coulomb per meter lies along the negative x axis from $x = 0$ to $x = -\infty$. Find the electric field at a point on the y axis.

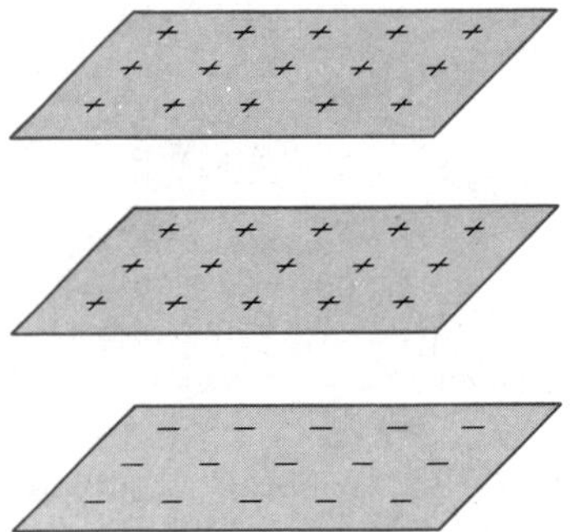

Fig. 23.30 Three parallel sheets.

22. Electric charge is uniformly distributed over each of three large, parallel sheets of paper (Figure 23.30). The charges per unit area on the sheets are 2×10^{-6} C/m², 2×10^{-6} C/m², and -2×10^{-6} C/m², respectively. The distance between one sheet and the next is 1.0 cm. Find the strength of the electric field **E** above the sheets, below the sheets, and in the space between the sheets. Find the direction of **E** at each place.

23. Each of two very large, flat sheets of paper carries a uniform positive charge distribution of 3.0×10^{-4} C/m². The two sheets of paper intersect at an angle of 45° (Figure 23.31). What are the magnitude and the direction of the electric field at a point between the two sheets?

24. Two large sheets of paper intersect at right angles. Each sheet carries a uniform distribution of positive charge (Figure 23.32). The charge per unit

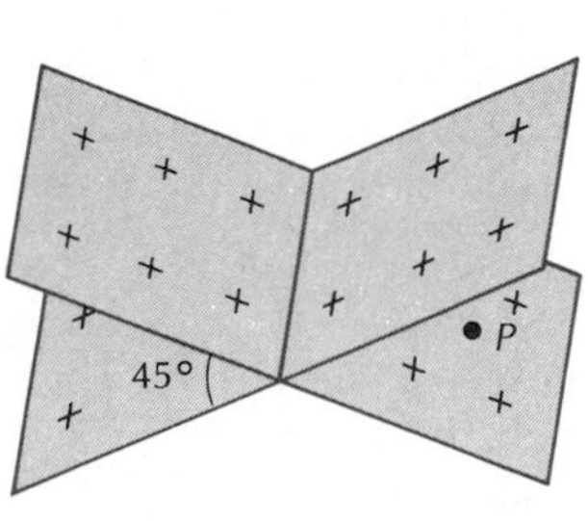

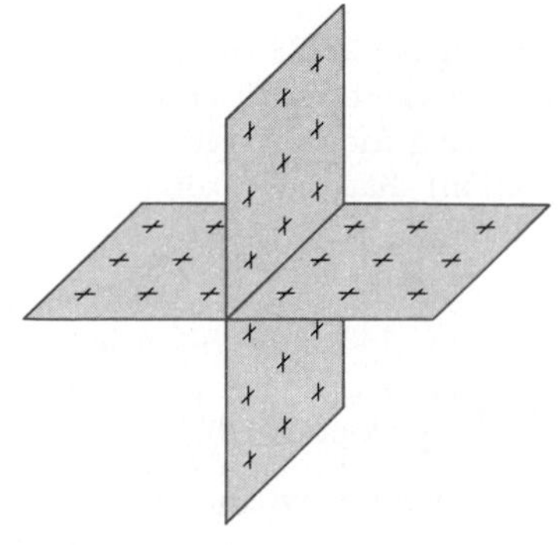

Fig. 23.31 (left) Two sheets intersecting at 45°.

Fig. 23.32 (right) Two sheets intersecting at right angles.

area on the sheets is 3×10^{-6} C/m². Find the magnitude of the electric field in each of the four quadrants. Draw the field lines in each quadrant.

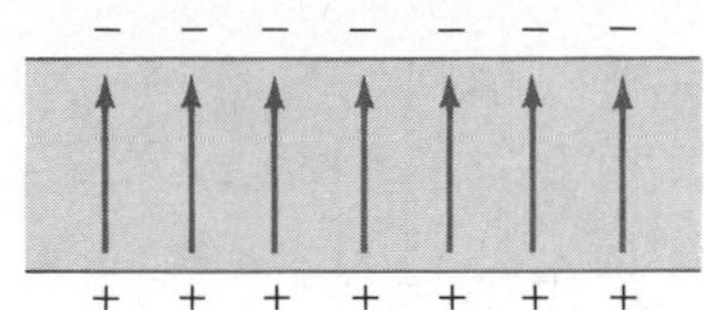

Fig. 23.33 A horizontal slab of iron.

25. The electric field within a chunk of metal exposed to the Earth's gravity (see Problem 9) is due to a distribution of surface charge. Suppose that we have a slab of iron oriented horizontally (Figure 23.33). What must be the surface charge densities on the upper and lower surfaces?

26. A total amount of charge Q is uniformly distributed along a thin, straight plastic rod of length l.
   (a) Find the electric field at the point P, at a distance x from one end of the rod (Figure 23.34).
   (b) Find the electric field at point P', at a distance y from the midpoint of the rod (see Figure 23.34).

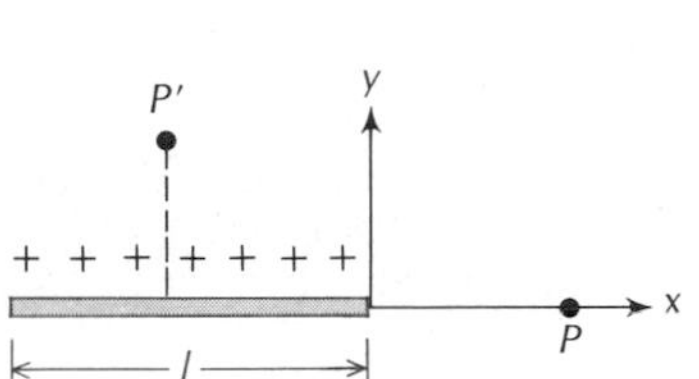

Fig. 23.34 A thin rod, with a uniform distribution of charge.

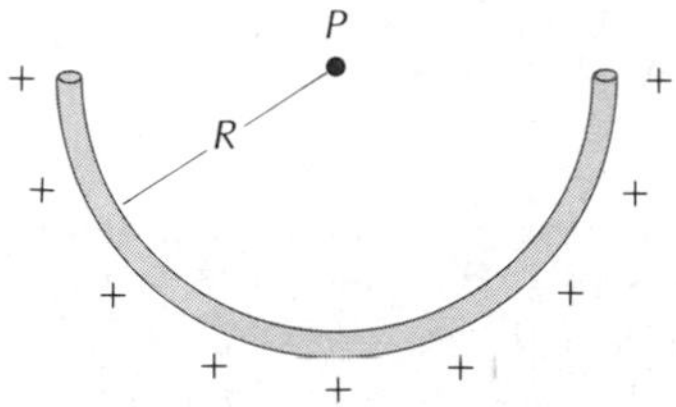

Fig. 23.35 A semicircular rod, with a uniform distribution of charge.

27. A thin, plastic rod is bent so that it has the shape of a semicircle of radius R (Figure 23.35). An amount of charge Q is uniformly distributed along the rod. What is the electric field at the center of the circle?

28. A Plexiglas square of dimension $l \times l$ has a uniform charge density of magnitude λ coulomb per meter along its edges. Two of the edges are positive and two are negative (Figure 23.36). Find the electric field at the center of the square.

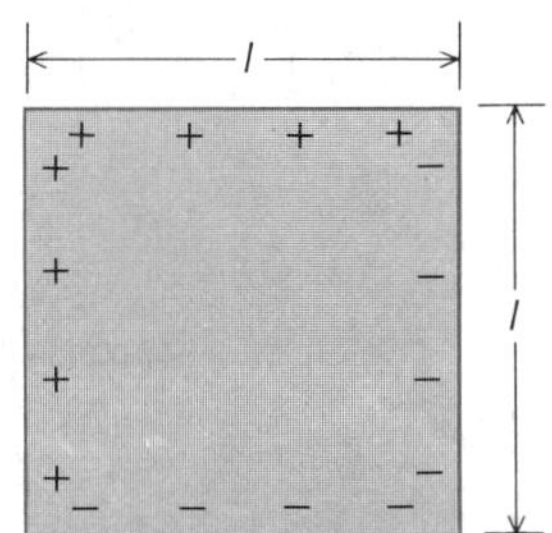

Fig. 23.36 A square, with charge along its edges.

29. Three thin glass rods carry charges Q, Q, and $-Q$, respectively. The length of each rod is l and the charge is uniformly distributed along each rod. The rods form an equilateral triangle. Calculate the electric field at the center of the triangle.

30. A disk of radius R carries an amount of charge Q uniformly distributed over its surface. Find the electric field at a point on the axis of the disk at a distance z from the center. Show that in the limiting case $z \gg R$, the result reduces to that for a point charge.

*31. Two thin, semi-infinite rods lie in the same plane. They make an angle of 45° with each other and they are joined by another thin rod bent along an arc of circle of radius R, with center at P (see Figure 23.37). All the rods carry a uniform charge distribution of λ coulomb per meter. Find the electric field at the point P.

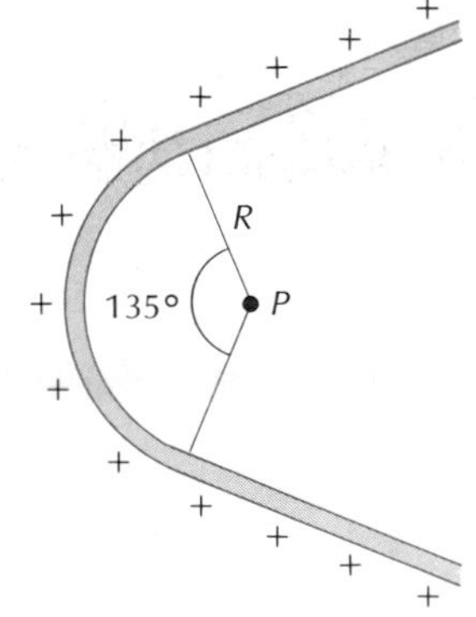

Fig. 23.37 Rods with uniform distributions of charge.

*32. A thin, semi-infinite rod with a uniform charge distribution of λ coulomb per meter lies along the positive x axis from $x = 0$ to $x = \infty$; a similar rod lies along the positive y axis from $y = 0$ to $y = \infty$. Calculate the electric field at a point in the x–y plane in the first quadrant.

*33. A cylindrical Plexiglas tube of length l, radius R carries a charge Q uniformly distributed over its surface. Find the electric field on the axis of the tube at one of its ends.

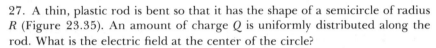

*34. Two thin rods of length L carry equal charges Q uniformly distributed over their lengths. The rods are aligned, and their nearest ends are separated by a distance x (Figure 23.38). What is the electric force of repulsion between these rods?

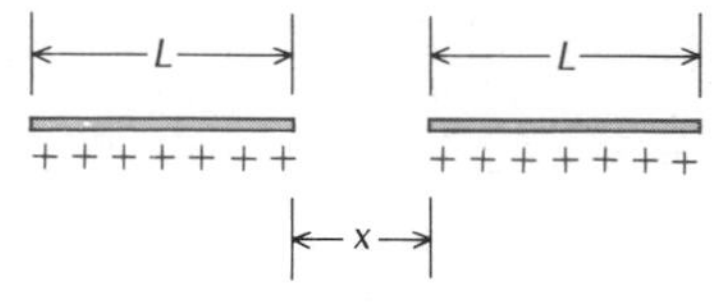

Fig. 23.38

Section 23.4

35. The two charges of ± 40 C in the thundercloud of Figure 23.28 form a dipole. What is the dipole moment?

36. In a hydrogen atom, the electron is at a distance of 0.53×10^{-10} m from a proton.
   (a) What is the instantaneous dipole moment of this system?
   (b) Taking into account that the electron moves around the proton on a circular orbit, what is the time-average dipole moment of this system?

37. (a) Pretend that the HCl molecule consists of (pointlike) ions of H^+ and Cl^- separated by a distance of 1.0×10^{-10} m. If so, what would be the dipole moment of this system?
   (b) The observed dipole moment is 3.4×10^{-30} C · m. Can you suggest a reason for this discrepancy?

38. The dipole moment of a HCl molecule is 3.4×10^{-30} C · m. Calculate the magnitude of the torque that an electric field of 2.0×10^6 N/C exerts on this molecule when the angle between the electric field and the longitudinal axis of the molecule is 45°.

39. The dipole moment of the water molecule is 6.1×10^{-30} C · m. In an electric field, a molecule with a dipole moment will tend to settle into an equilibrium orientation such that the dipole moment is parallel to the electric field. If disturbed from this equilibrium orientation, the molecule will oscillate like a torsional pendulum. Calculate the frequency of small oscillations of this kind for a water molecule about an axis through the center of mass (and perpendicular to the plane of the three atoms) when the molecule is in an electric field of 2.0×10^6 N/C. The moment of inertia of the molecule about this axis is 1.93×10^{-47} kg · m^2.

FORCES, FIELDS, AND QUANTA[1]

As noted in Chapter 6, the bewildering varieties of forces that we find in nature — gravitational attractions, elastic forces, viscous forces, friction forces, electric forces, intermolecular and interatomic forces, nuclear forces, etc. — all result from just four fundamental forces: the gravitational force, the electromagnetic force, the "strong" force, and the "weak" force.

The gravitational force acts on everything — it acts on all forms of matter. One usually says it acts on the masses of particles, but it would be more accurate to say that it acts on both mass and energy, i.e., not only are particles under the sway of gravitational attractions, but so is energy. For example, the thermal energy of the gas in the Sun attracts the Earth — about 1 part in 10^6 of the Sun's pull is due to its thermal energy. Since energy has mass (see Chapter 17), it should come as no surprise that energy, and hence any form of matter, is subject to gravitational interactions.

The electric force acts between electric charges, and the magnetic force acts between magnets and between electric currents. However, a closer look reveals that the force between, say, two magnets is really nothing but a kind of extra electric force between the moving charges inside the magnets. Thus, the magnetic force is not fundamentally different from the electric force; these two forces are merely two aspects of a single force called the electromagnetic force, just as space and time are two aspects of a single entity called spacetime. This unification of electricity and magnetism is a consequence of the theory of relativity, just as is the unification of space and time. When Einstein formulated the theory of relativity and proved that space and time transform into one another under Lorentz transformations, he also proved that electric and magnetic fields transform into one another. In Section 30.4 we will examine a special instance of such a transformation; we will see that an electric field transforms into a combination of electric and magnetic fields when we go from one inertial reference frame to another. Electromagnetism is the best-known example of a **unified field theory,** that is, a theory that treats two fields as two aspects of a single underlying field.

The "strong" force acts on the protons and neutrons in nuclei and it acts on all the baryons and mesons. The particles that interact via the strong force are often called **hadrons;** thus, baryons and mesons are hadrons, but leptons are not. (Interlude C contains tables of baryons, mesons, and leptons.)

The "weak" force is important for neutrinos: it is the only force that neutrinos ever experience (apart from the gravitational force, which is insignificant for them under laboratory conditions); hence this force is crucial in the collisions between neutrinos and other matter. This force also acts on all other leptons, and on baryons and mesons, but in these cases the effects of the weak force are often hidden behind the much larger effects produced by the strong or the electromagnetic force. The weak force is deeply implicated in many reactions that bring about the disruption of the unstable particles. For example, the weak force is responsible for the disruption, or decay, of the neutron.

Table G.1 lists the strength of each of the fundamental forces and also the range, or the maximum distance over which this force can reach from one particle to another. This table contains the same information as Table 6.1, but in the new table we have expressed the strengths of the forces relative to that of the strong force to which we arbitrarily have assigned a strength of 1.[2]

In this chapter we will explore the mechanism that generates these forces. To attain a full understanding of this mechanism, we would have to begin with a study of quantum theory, which we cannot do here. We will therefore have to leave out many details and deal only with the broad qualitative features of the mechanism.

G.1 REACTION RATES

A basic assumption of Newtonian physics is that the motion of a particle can be described by position as a function of time. But on a microscopic level, this assumption breaks down: in consequence of their quantum-mechanical properties, particles acquire uncertainties in position and in velocity, and their tra-

[1] This chapter is optional.

[2] The strengths of the forces depend on the energies of the particles. The values in the table are appropriate for low energies.

Table G.1 THE FOUR FUNDAMENTAL FORCES

Force	*Acts on*	*Relative strength*	*Range*
Gravitational	All forms of matter (all forms of energy)	10^{-38}	Infinite
Weak	Leptons, baryons, and mesons	10^{-6}	Less than 10^{-17} m
Electromagnetic	Charged particles	10^{-2}	Infinite
Strong	Baryons and mesons (hadrons)	1	$\sim 10^{-15}$ m

jectories are not well defined. Worse yet, in high-energy reactions among particles, the number of particles is not well defined. In such reactions particles are created, and destroyed, and created again. Thus, the particles are evanescent objects with uncertain trajectories. Under these circumstances, the concept of force ceases to have an exact meaning. Hence particle physicists prefer to speak of **interactions,** that is, the action of particles on other particles; they speak of "strong" interactions, electromagnetic interactions, "weak" interactions, and gravitational interactions. Mathematically, these interactions can be described by formulas that specify the amount of energy for each interaction — energies remain meaningful even at a microscopic level, and therefore energies are more relevant than forces. Nevertheless, for the purpose of the following qualitative discussion, let us continue to speak of forces even though this intuitive concept is ambiguous and ought to be replaced by a more precise and sophisticated mathematical concept of interaction energies.

For the experimental particle physicist the phenomena most readily accessible to observation are collisions, either collisions in which new particles are created by the destruction of some old particles or by the conversion of energy into mass (inelastic reactions), or collisions in which particles merely bounce off one another (elastic reactions). The rates of these reactions depend on the energy, momentum, and angular momentum of the colliding particles. And they also depend on the strength of the force between the particles — the stronger the force, the faster the reaction will proceed. Figure G.1 shows an extremely fast reaction recorded in a photographic emulsion. A very high-energy cosmic ray — probably a proton — smashes into the nucleus of an atom and produces a "shower" of particles. Among the particles created in such collisions are pions, kaons, protons, electrons, etc. These particles are created by the strong force between the incident proton and the protons and neutrons in the nucleus. The time scale of the reaction can be estimated as follows: the incident proton travels at nearly the speed of light and hence passes through the nucleus very quickly; since the dimension of the nucleus is of the order of 10^{-15} m, the time available for the interaction can be calculated by dividing this distance by the speed of light; this gives about 10^{-23} s. Thus, the typical time scale for such reactions is extremely short and the reactions are extremely fast.

The strong force can bring about not only particle

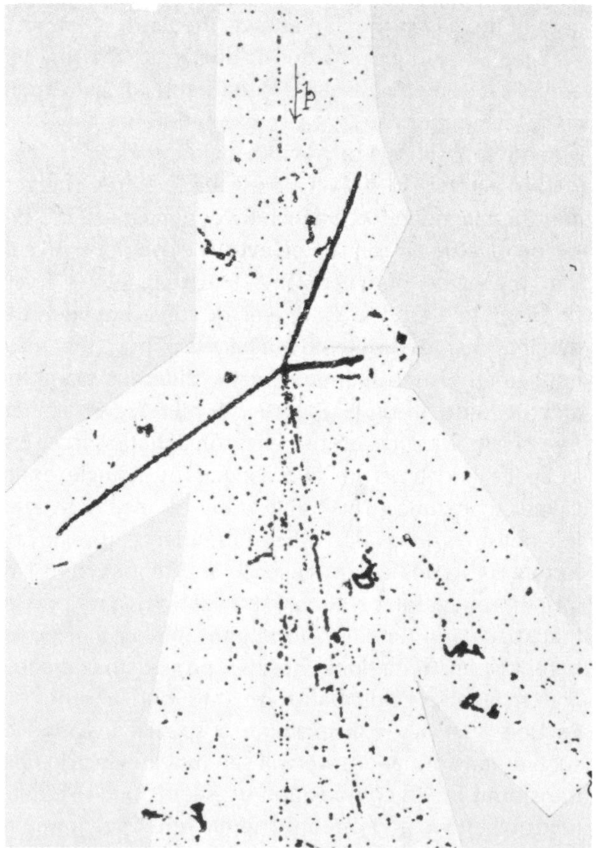

Fig. G.1 Tracks of particles in a photographic emulsion. The faint track at the top was made by a high-energy cosmic ray. This cosmic ray collided with a nucleus of an atom in the photographic emulsion. In this collision many new particles were created. Their tracks diverge from the point of collision.

creation but also particle decay. The time scale for such decays is about the same as the time scale for creation: about 10^{-23} s. Hence those particles that decay strongly lead a very ephemeral existence; they are very unstable and burst apart within an instant of their creation. For example, the ρ meson decays into two pions (π) by the strong reaction

$$\rho^- \rightarrow \pi^- + \pi^0$$

within 10^{-23} s. An object that only lasts such a short time cannot be detected directly; its existence can only be inferred from the correlations observed in the motion of the decay products (see Interlude C).

Reactions caused by electromagnetic interactions take longer. Figure G.2 shows the decay of a pion into an electron (e), antielectron ($\overline{\mathrm{e}}$), and gamma ray (γ) by the electromagnetic reaction

$$\pi^0 \rightarrow \mathrm{e} + \overline{\mathrm{e}} + \gamma$$

which takes about 10^{-14} s (this meson can also decay into two gamma rays, which is a somewhat faster reaction). Although 10^{-14} s is a short time by ordinary standards, it is a long time by the standards of high-energy physicists; a particle that takes this "long" to decay is regarded by them as stable!

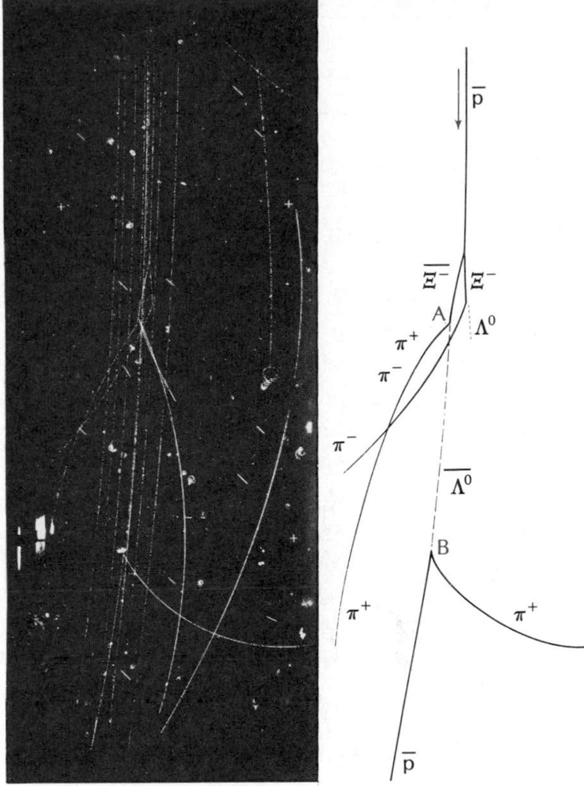

Fig. G.3 (left) Photograph of tracks of particles in a bubble chamber and (right) tracing based on the photograph. The dashed line indicates the trajectory of a $\overline{\Lambda}$ particle; it is created at A, travels a distance of about 20 cm, and then decays into a $\overline{\mathrm{p}}$ and a π^+ at B.

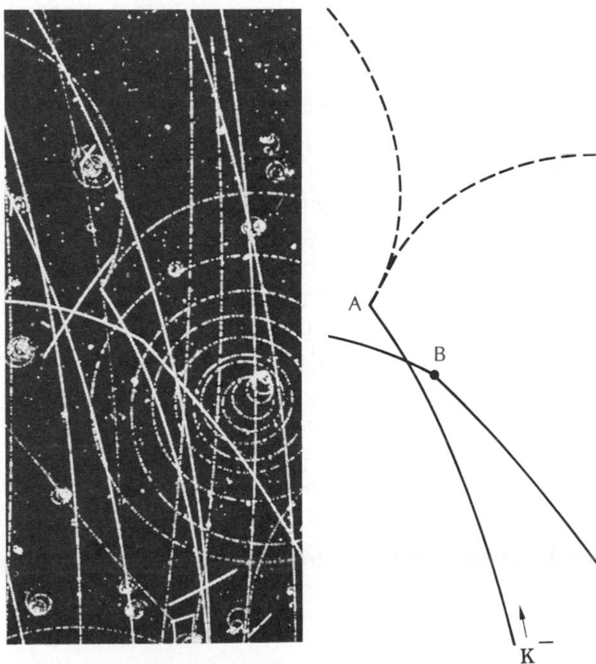

Fig. G.2 (left) Photograph of tracks of particles in a bubble chamber and (right) tracing based on the photograph. Neutral particles do not make visible tracks in the bubble chamber, but their extrapolated trajectories have been indicated by dashed lines in the tracing. A Λ baryon and a π^0 were created in a collision at A. The π^0 decayed immediately into an e, $\overline{\mathrm{e}}$, and γ. The tracks of the e and $\overline{\mathrm{e}}$ can be seen diverging from the point of decay, but the γ made no track in the photograph.

The reactions involving the weak force take much longer. Figure G.3 shows the decay of an antilambda baryon ($\overline{\Lambda}$) into an antiproton ($\overline{\mathrm{p}}$) and a pion (π^+),

$$\overline{\Lambda} \rightarrow \overline{\mathrm{p}} + \pi^+$$

This reaction is weak and takes about 10^{-10} s. By the standards of high-energy physicists, this is a *very* long time. Moving at close to the speed of light, the $\overline{\Lambda}$ travels a distance of several centimeters between its creation and its decay.

Note that the above reactions, with time scales of 10^{-23}, 10^{-14}, and 10^{-10} s for strong, electromagnetic, and weak forces, are intended only as more or less typical examples. Rates of other reactions involving these forces can differ by several, or even by many, powers of 10 from the above. Thus, the decay of the neutron into a proton, an electron, and an antineutrino,

$$\mathrm{n} \rightarrow \mathrm{p} + \mathrm{e} + \overline{\nu}$$

Table G.2 FORCES AND CONSERVED QUANTITIES

Force	*Energy, momentum, and angular momentum*	*Charge*	*Baryon number*	*Lepton number*	*Strangeness*	*Parity*	*Isospin*
Gravitational	✓	✓	✓	✓			
Weak	✓	✓	✓	✓			
Electromagnetic	✓	✓	✓	✓	✓	✓	
Strong	✓	✓	✓	✓	✓	✓	✓

is a weak reaction that is extremely slow — it takes about 15 min. This delay in the reaction is to be blamed on the small amount of energy available for the decay process. Although there is some overlap among the rates of the reactions generated by the different kinds of forces, as a rough rule we can state that the strong force generates the fastest reactions, the electromagnetic force slower reactions, and the weak force the slowest reactions.[3] The gravitational force between individual particles is so feeble that it is incapable of generating reactions that can be observed in our laboratories.

G.2 CONSERVATION LAWS AND SYMMETRY

All reactions conserve energy, momentum, angular momentum, and electric charge, i.e., the energy, momentum, angular momentum, and electric charge before and after any reaction among particles are exactly the same. Furthermore, all reactions ever observed conserve baryon number and lepton number (as discussed in Section C.4).

Besides these exactly conserved quantities there are some others that are only approximately conserved; these other quantities — isospin, parity, and strangeness — are conserved in some reactions but not in all (see Section C.4). Which reactions conserve what quantities depends on the type of force involved in the reaction; Table G.2 lists the four forces and some of the quantities conserved by each.

The strong force obeys all conservation laws; the electromagnetic interaction violates the conservation of isospin; the weak force violates the conservation of isospin, of parity, and of strangeness; and the gravitational force violates the same laws as the weak force. The violation of conservation laws in gravitation is an indirect effect. All forms of energy gravitate, and when the weak and electromagnetic energies gravitate, they can produce reactions that violate some of the conservation laws — the weak and electromagnetic forces contaminate gravitation with their violations.

There is a profound connection between conservation laws governing a physical system and the symmetries of that system. This connection is contained in the following general theorem: *to every symmetry there corresponds a conservation law.* A symmetry of the system is any operation that leaves the system invariant, i.e., unchanged. To understand the meaning of the general theorem, it will be best to examine a particular case, say, the conservation of momentum. This conservation law can be regarded as a consequence of the invariance of a system under translations in space. Consider a system of particles, such as a water molecule; if it makes no difference whether we place this system at one point of space or at another, then the system has invariance under translations in space. Obviously, if an external force were acting on the system, it would make a difference where we placed the system — at some points of space the potential energy associated with the external force would be high, at others it would be low. Hence invariance under translations in space demands the absence of external forces and this, in turn, implies the conservation of the momentum of the system.

By similar arguments, one can show that the conservation of angular momentum is related to invariance under rotations in space; conservation of energy is related to invariance under translations in time; and the conservation of other quantities listed in Table G.2 is related to other symmetries of a somewhat more abstruse kind, which we cannot discuss in detail here.

The exploitation of symmetries has proved very profitable in the study of the strong and the weak forces. Both of these forces are extremely complicated; the known symmetries help to place restrictions on the mathematical formulas describing these forces. High-energy physicists are engaged in a continuing search for further symmetries (and further conservation laws).

G.3 FIELDS AND QUANTA

According to classical theory, forces are mediated by fields. Distant particles do not act on one another directly; rather, each particle generates a field of force and this field acts on the other particles. As we saw in Section 23.2, the existence of fields is required by

[3] The reaction rates depend on energy. This comparison of the reaction rates is valid for the typical energies attained in laboratories.

conservation of energy and momentum. Fields play the role of storehouses of energy and momentum; the energy and momentum stored in the fields balance any excess or deficit in the energy and momentum of the interacting particles engaged in (nonuniform) motion. For example, radio waves and light waves are electromagnetic fields that store the energy and momentum lost by charges moving on radio antennas and in atoms. The most spectacular example of the conversion of particle energy into field energy occurs in the annihilation of matter with antimatter: if an electron collides with an antielectron, the two particles annihilate one another, giving off a burst of very energetic light, or gamma rays. In this annihilation the energy of the particles — including their rest-mass energy — is completely converted into field energy. The reverse reaction is also possible: if a gamma ray collides with a charged particle, it can create an electron–antielectron pair. In such a pair creation, the energy of the gamma ray is converted into the energy of the pair of particles (Figure G.4).

Each of the four fundamental forces is mediated by fields of its own. Hence there are gravitational fields, electromagnetic fields, strong fields, and weak fields. On a macroscopic scale, the gravitational and electric fields of everyday experience are smooth functions of space and time; these functions can be calculated by classical field theory. The electric-field problems that we solved in Chapter 23 are examples of such classical calculations. However, on a microscopic scale, physics is ruled by quantum theory and not by classical theory. Quantum theory introduces a new fundamental feature: the quantization of energy. It turns out that the energy stored in fields is not smoothly distributed; rather the energy is found in **quanta,** that is, small packets or lumps of energy.

Fig. G.4 The two spiraling tracks in this bubble-chamber photograph were made by an electron and an antielectron. These particles were created by a high-energy gamma ray in a collision with the electron of a hydrogen atom in the bubble chamber. The long, slightly curved downward track was made by the recoiling electron.

The quanta of electric and magnetic energy in a radio wave or a light wave are called **photons.** In an ordinary radio or light wave, the photons are not directly noticeable, just as the molecules in air are not directly noticeable. There are so many molecules in each cubic centimeter of air — about 10^{19} molecules per cubic centimeter under standard conditions — that for most practical purposes air can be described as a continuous fluid with a smoothly varying pressure and density. Likewise, there are so many photons in each cubic centimeter of a light wave — about 10^7 photons per cubic centimeter in bright sunlight — that for most purposes the wave can be described as a smoothly varying electric and magnetic field. However, the "lumps" in light can be detected by placing electrons in their path and watching for collisions; the recoil of an electron hit by a photon announces the presence of the latter. In their collisions with electrons, the photons display particle properties: they have energy, momentum, and spin-angular momentum. Their mass is zero, as is required by the theory of relativity, according to which only a zero-mass particle can move at the speed of light yet have finite energy [see Eq. (17.51)].

Table G.3 lists the quanta corresponding to the fields that mediate the four kinds of forces. Our belief in gravitons rests largely on theoretical considerations; these particles have not yet been detected and, given the extremely small strengths of gravitational forces, it is unlikely that they will be directly observed soon.

Table G.3 FIELDS AND THEIR QUANTA

Field	*Quanta*	*Mass*[a]
Gravitational	Gravitons	0 MeV
Weak	W particles	81,000
	Z particles	93,000
Electromagnetic	Photons	0
Strong	Pions and some other mesons	140

[a] The masses are expressed in energy units, as in Tables C.3–C.5.

Pions are easy to observe; they are bona fide particles, even more so than photons. Pions have energy, momentum, electric charge (0, +e, or −e, depending on the type of pion; see Table C.5). Their spin is zero, but in contrast to photons their mass is not zero. The notion that a "solid" particle such as a pion should play the role of quantum of a field seems a bit bizarre. But quantum theory teaches us that pions, like all other particles, have a dual character: they are both waves and particles (see Section 41.6). In their character of waves, pions are described by a wave field; taking into account that the energy of this field must be quantized, it turns out that the corresponding quanta are nothing but the pions themselves. This identification of particles and quanta holds in general: *all particles are quanta of fields.* Pions are quanta of the **pion field,** electrons are quanta of the **electron field,** protons are quanta of the **proton field,** etc. Unfortunately, this means that there is one kind of field for each kind of particle — there are very many kinds of fields. Since it is unlikely that the many, many particles we know of are all elementary, it is unlikely that the many, many fields are fundamental. Thus, if the proton is a composite structure made of three quarks (see Section C.6), then the proton field is also a composite structure made of a combination of three quark fields.

In view of this multitude of new fields, the question arises whether each of the new fields will act as mediator and carry some kind of force between particles. The answer is yes: all fields can play an intermediate role and mediate forces between particles. For example, the electron field can act between two photons, permitting them to exert forces on one another. Diverse meson fields (ρ-meson field, ω-meson field, etc.) can act between two protons, permitting them to exert strong forces in addition to the forces generated via the pion field; in fact, the diversity of these additional corrections is one of the reasons why the theoretical analysis of the strong force is so dreadfully difficult.

At the quantum level, we can picture the field of force generated by a particle as a swarm of quanta buzzing around the particle. For example, we can picture the electric field surrounding an electron or any other charged particle as a swarm of photons. The swarm is in a state of everlasting activity — the charged particle continually emits and reabsorbs the photons of the swarm. Emission is creation of a photon; absorption is annihilation of a photon. Hence we can say that the electric field arises from the continual interplay of three fundamental processes: creation, propagation, and annihilation of photons. The action of one charged particle on another involves a sequence of these three fundamental processes: a photon is emitted by one particle, propagates through the intervening distance, and is absorbed by the other particle. This exchange process can be represented graphically by a diagram showing the worldlines of the particles (Figure G.5); such a diagram is called a **Feynman diagram.** The photon exchanged between the two electrons is called a **virtual photon** because it lasts only a very short time and, being reabsorbed by an electron, is undetectable by any direct experiment. The steady attractive or repulsive force between two charged particles is generated by continual repetition of this exchange process. This is action-by-contact with a vengeance — at a fundamental level, all forces reduce to local acts of creation and destruction involving particles in direct contact.

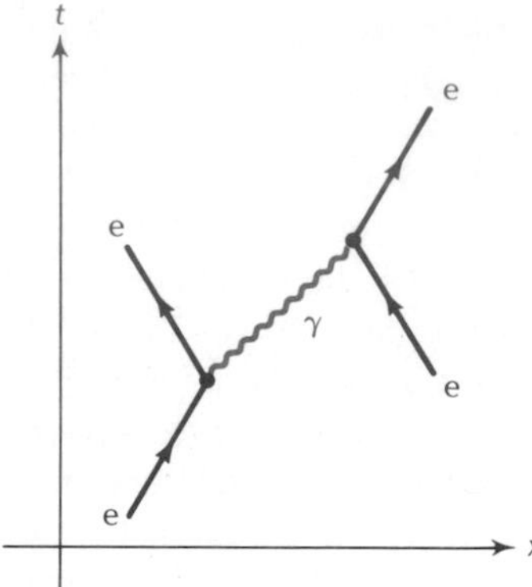

Fig. G.5 Feynman diagram representing the exchange of a virtual photon between two electrons. The solid black lines indicate the worldlines of the two electrons, with the t axis plotted vertically and the x axis plotted horizontally. The wavy colored line indicates the worldline of the photon. The electron on the left emits this photon and the electron on the right absorbs it.

In terms of a simple analogy, we can easily understand how the exchange of particles brings about forces. Imagine two boys tossing a ball back and forth between them (Figure G.6); it is intuitively obvious that this produces a net repulsive force between the boys due to the recoil they suffer when throwing or catching the ball. Our intuition suggests that no such exchange process can ever produce attraction. However, imagine two Australian boys tossing a boomerang back and forth between them (Figure G.7); it is then obvious that this produces an attractive force between the boys.[4] Whether a photon exchanged between two charges behaves like a ball or like a boomerang depends on the signs of the charges. Quantum calculations, which take into account the wave nature of all the particles involved, show that the net force is attractive for unlike charges and repulsive for like charges, as it should be.

This theory of electric forces is called **quantum electrodynamics** (or QED); it was developed by P. A. M. Dirac, R. P. Feynman, J. Schwinger, S. Tomonaga,[5] and

[4] One of the flaws in this analogy is that the momentum of the boomerang changes during its flight through the air. But this is not a fatal flaw because we can assume that the initial and the final direction of motion are the same (a 180° turn), so that the boomerang merely borrows momentum from the air, without any *net* change.

[5] Paul Adrien Maurice Dirac, 1902–1984, English physicist; he received the Nobel Prize in 1933 for his contributions to quantum theory. Richard P. Feynman, 1918–, American physicist, Julian Schwinger 1918–, American physicist, and Sin-Itiro Tomonaga, 1906–1979, Japanese physicist, shared the Nobel Prize in 1965 for their work on quantum electrodynamics.

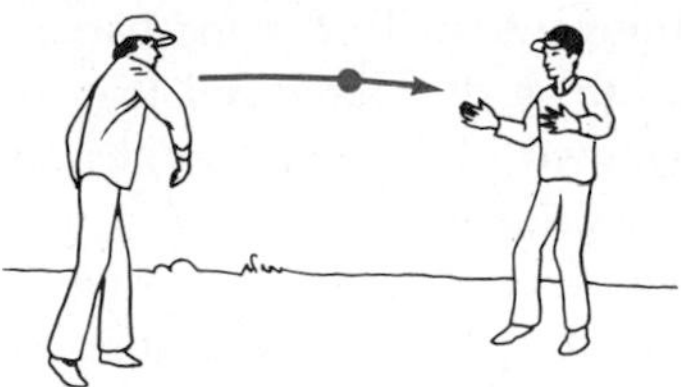

Fig. G.6 Two boys throw a ball back and forth.

Fig. G.7 Two boys throw a boomerang back and forth.

other physicists between 1930 and 1950. Quantum electrodynamics is the most accurate theory in all of physics, and in all of science. Some theoretical calculations in quantum electrodynamics have been carried out to nine significant figures and delicate experimental measurements have confirmed these calculations exactly.

The gravitational, weak, and strong forces are also generated by the exchange of virtual particles: gravitons, W and Z particles, and pions (and some other mesons). It is a general rule of quantum theory that the range of the force is inversely related to the mass of the particle that serves as the carrier of the force. Thus, the photons and the gravitons that are the carriers of the electromagnetic and the gravitational force have zero mass — the ranges of these forces are infinite. The pions that are the carriers of the strong force have a fairly large mass — the range of this force is short. The W and Z particles that are the carriers of the weak force have a very large mass — the range of this force is very short.

G.4 THE UNIFIED THEORY OF WEAK AND ELECTROMAGNETIC FORCES

It has long been one of the aspirations of theoretical physicists to achieve the unification of the forces of nature. The theory of electromagnetic forces is an example of such a unification — the electric and the magnetic forces are nothing but different aspects of a single underlying force. Theoretical physicists tend to regard this example as rather trite — they are much more interested in a grandiose scheme for the unification of all four forces. Einstein devoted much effort to an attempt at joining the gravitational and the electromagnetic forces, but he failed to reach this goal.

Some years ago, S. Weinberg, A. Salam, and S. Glashow[6] achieved the unification of weak and electromagnetic forces. This new theory has shown that, in spite of the seemingly drastic differences between their characters, the weak and electromagnetic forces are basically the same — they are merely two aspects of a single **electroweak force.**

If we seek to unify the weak and electromagnetic forces, we must regard the carriers of these forces — the quanta whose exchange generates the forces — as closely related. Thus, the proton and the W and Z particles ought to be very similar particles that differ only in small details; all of these particles ought to belong to a family, just like, say, the proton and neutron belong to a family (see Section C.5). The trouble is that the photon and the W and Z particles do not look at all similar: the photon has zero mass while the W and Z particles have very large masses, masses much larger than that of any other particle discovered so far. The unified theory predicts the masses of the W and Z particles by the following argument: If the weak and electromagnetic forces are essentially the same, then they must also have the same strength. The fact that the experimentally observed strengths seem quite different (see Table G.1) is attributed to the masses of the W and Z particles — under certain conditions a strong force can have the appearance of a weak force if the particle that carries the force is very massive. A theoretical calculation shows that at a fundamental level the weak and the electromagnetic forces can be regarded as having the same strength provided that the W and the Z particles have masses of about 80 times the mass of a proton. These large masses explain both the strength of the weak force and its short range.

But here we run into an obstacle that would seem to defeat our efforts to construct a unified theory: on the one hand, we need very massive W and Z particles to account for the relative strengths of the weak and electromagnetic forces, and on the other hand, if these particles are very massive, then they are very different from the photon, which is massless. To overcome this obstacle we need to find a clever explanation of how particles can be related — can be members of the same family — and yet have very different masses.

[6] Stephen Weinberg, 1933–, American physicist, Abdus Salam, 1926–, Indian physicist, and Sheldon Lee Glashow, 1932–, American physicist, shared the Nobel Prize in 1979 for their unified theory of weak and electromagnetic forces.

Such an explanation is supplied by the scheme of **"broken symmetry,"** which could also be called "hidden symmetry." According to this scheme, the photon and the W and Z particles are indeed closely related, but the relationship is hidden from direct view. A discussion of "broken symmetry" is beyond the scope of this textbook, but the following is a simple example that illustrates the meaning of such a symmetry. Consider a pencil standing vertically on its tip on a table. This equilibrium configuration has rotational symmetry: the configuration is invariant under rotations about a vertical axis passing through the pencil. However, the equilibrium is unstable, and sooner or later the pencil will fall flat on the table and settle into a new, stable equilibrium configuraton. When this happens, the rotational symmetry will be destroyed because the pencil has to fall in one definite direction. Such a breakdown of symmetry is called **spontaneous** because it is not attributable to the equation of motion. In fact, the equation of motion is rotationally symmetric, i.e., it has the same form for all the directions in which the pencil can fall, and it therefore does not select any preferred directon. Generalizing from this simple example, we can say that a spontaneously broken symmetry is a symmetry that is present in the equation of motion but absent from the actual equilibrium configuration into which the system settles.

Photons and W and Z particles can be regarded as some kind of equilibrium configurations of fields. Their equations of motion (field equations) are rotationally symmetric; here, the relevant rotation is not an ordinary rotation, but rather a rotation in an abstract mathematical space, a rotation that transforms photons and W and Z particles into each other, and thereby defines their family relationships. In the abstract mathematical space, the photons and the W and Z particles can assume a symmetric equilibrium configuration, in which they all have the same mass. But this equilibrium configuration is unstable and, under normal laboratory conditions, photons and W and Z particles settle into a different, stable equilibrium configuration, with a spontaneous breakdown of the rotational symmetry. The observed mass differences between these particles are a by-product of this breakdown of symmetry. Thus, photons and W and Z particles are closely related, but under ordinary laboratory conditions, their relationship is hidden from view. The symmetry between photons and W and Z particles can be restored by giving all these particles very high energies, in excess of 100 GeV; at such high energies, the particles can attain their symmetric equilibrium configuration, and the symmetry between them would then become manifest. Such high energies are difficult to achieve in our laboratories, but they were readily available during the very early stages of the Big Bang, when the universe was younger than 10^{-10} s and had a temperature in excess of 10^{15} K. It is believed that at these early times the symmetry between photons and W and Z particles was unbroken.

One notable consequence of the unified theory of weak and electromagnetic interactions is that it provides a natural explanation of the quantization of electric charge. As we saw in Section 22.3, the electric charge on all known particles is some integral multiple of the basic unit of charge, $e = 1.60 \times 10^{-19}$ C. It can be shown that this quantization rule emerges as an immediate consequence of the rotational symmetry of the unified theory.[7]

From the experimental point of view, an early success of the unified theory was its prediction of certain hitherto unknown forces between baryons and neutrinos. The weak forces between electrons and neutrinos can be accounted for by exchange of two kinds of W particles: a positively charged particle W^+ and a negatively charged particle W^-. However, the photon and these two charged W particles do not make a complete family by themselves. The scheme of broken symmetry demands that the family contain one more particle: the electrically neutral Z, also called Z^0. The force generated by the exchange of these neutral Z particles is called the neutral-current force. The existence of this new force was soon confirmed by experimenters on high-energy baryon–neutrino collisions at CERN.

The most recent and most impressive success of the unified theory was its prediction of the masses of the W and Z particles. These particles were detected in 1982 in experiments at the proton–antiproton collider at CERN. The experiments involved the observation of about a billion head-on collisions between protons and antiprotons of the same energy, 270 GeV. A few dozen W and Z particles were produced in these collisions (Figure G.8). The measured masses of the W and the Z particles are, respectively, 81 GeV and 93 GeV (expressed in energy units). These measured values are within 2 GeV of the theoretically predicted values. This excellent agreement constitutes a brilliant confirmation of the unified theory of weak and electromagnetic interactions.

G.5 GLUONS AND QUANTUM CHROMODYNAMICS

As described in Interlude C, protons, neutrons, and other hadrons are made of small fundamental particles

[7] The unified theory of weak and electromagnetic forces explains the quantization of electric charge only for particles within given families of particles. In order to explain the quantization of electric charge for all particles in all families, it is necessary to go one step further and unify the weak, electromagnetic, and *strong* forces (see next section).

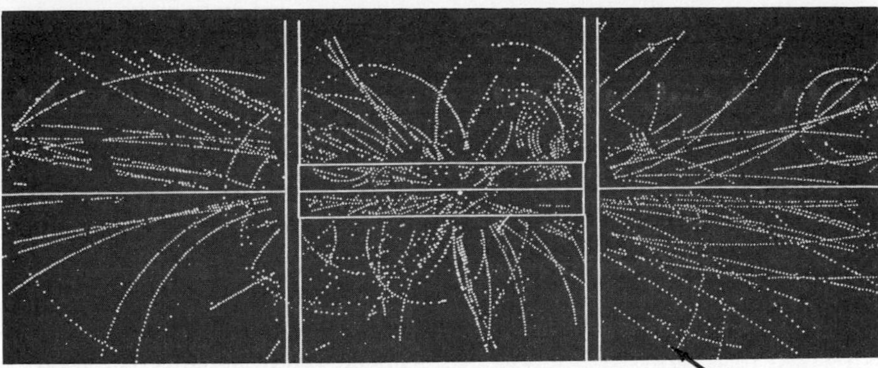

Fig. G.8 Tracks of particles produced in a very energetic head-on collision between a proton of 270 GeV that entered from the right and an antiproton of 270 GeV that entered from the left. In this collision a W^- particle was created. It immediately decayed into an electron and a neutrino; the track of the former (indicated by the arrow) can be seen emerging toward the lower right.

called **quarks.** According to recent theoretical and experimental investigations, there exist six kinds, or "flavors," of quarks: **up, down, strange, charmed, top,** and **bottom;** and each of these can have any of three "colors": **red, green,** and **blue.** For example, a proton is made of two u quarks and one d quark; one of these (but not always the same one) is *red,* one *green,* and one *blue.* Furthermore, for each of the quarks there exists an antiquark. An antiproton is made of two u antiquarks and one d antiquark; one of these is anti*red,* one anti*green,* and one anti*blue.*

The quarks are confined inside the proton by very strong mutual attractive forces. These forces between quarks are color forces — the source of these forces is color just as the source of electric forces is electric charge. Each of the three varieties of color (*red, green,* and *blue*) is analogous to a kind of positive electric charge and each of the "anticolors" is analogous to a kind of negative electric charge. A body is color neutral if it contains equal amounts of all three colors or if it contains equal amounts of color and anticolor, just as a body is electrically neutral if it contains equal amounts of positive and negative charge. Thus, protons, as well as other ordinary particles, are color neutral.

The color force is a fundamental force that should be included in our table of fundamental forces instead of the strong force (Table G.1). The color force is closely related to the strong force — the latter is actually a special instance of the former. The relationship between the color force and the strong force is analogous to the relationship between the electric force and the intermolecular force. As we saw in Section 22.1, the force between two electrically neutral atoms or molecules is a residual electric force resulting from an imperfect cancellation among the attractions and repulsions of the charges in the two atoms or molecules. Likewise, the strong force between, say, two "colorless" protons is a residual color force resulting from an imperfect cancellation among the attractions and repulsions of the quarks in the two protons. Thus, the "strong" force between protons is no more than a pale reflection of the much stronger color forces acting within each proton.

The color force has some very remarkable properties. All the forces listed in Table G.1 *decrease* with distance, but the color force remains constant as the distance between the quarks increases. This persistence of the color force brings about the confinement of quarks. If one of the quarks in, say, a proton is somewhat separated from its companions by the violent impact of a collision, the color attraction pulls it back into its original position as soon as the collision ends. The quarks behave as though linked by rubber strings. During a collision, the rubber strings stretch, but after the collision they again contract (Figure G.9). This analogy is perhaps fairly close to the truth; according to one theory, the color force is communicated from one quark to another along tightly bundled field lines concentrated in thin channels, similar to the strings shown in Figure G.9.

Although the color force keeps the quarks on a short leash and pulls them back sharply whenever they wander too far apart, the force allows the quarks to move rather freely within certain limits. This behavior is in accord with the string analogy — if the quarks stretch the strings, they experience strong restraining forces, but if they stay within the limits set by the relaxed lengths of the strings, they feel almost no force. This situation has been described by the picturesque phrases "infrared slavery" and "ultraviolet freedom" (in this context *infrared* refers to long distances and *ultraviolet* refers to short distances).

In an extremely violent collision, one of the strings

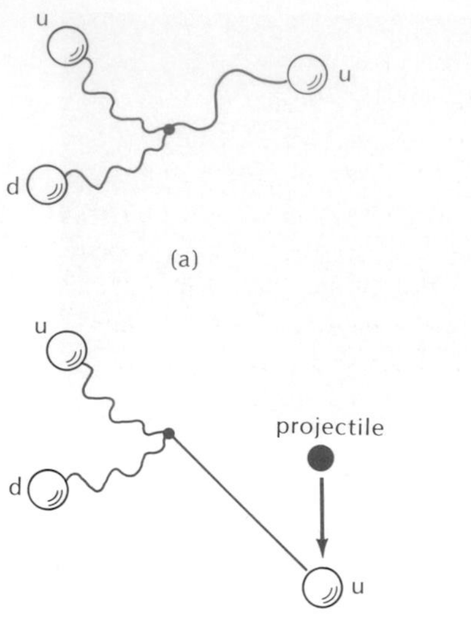

Fig. G.9 (a) The three quarks in a proton held together by "strings." (b) When a projectile collides with one of the quarks, the "string" stretches.

might break. But such a break does not result in a free quark with a dangling piece of string hanging out. Rather, the energy stored in the stretched string creates a quark–antiquark pair at the site of the break (Figure G.10). What happens in the breaking of a string is analogous to what happens in the breaking of a bar magnet. Such a magnet has a north pole and a south pole and we might hope to obtain a free north pole and south pole by simply breaking the magnet at the middle. But experience with magnets shows that this does not give us two magnets, each with a single pole. Rather, the act of breaking creates a new south pole and a new north pole, and we are left with two complete magnets (Figure G.11). The creation of new quarks during the breaking of strings makes it impossible to obtain isolated quarks; instead, the breaking of

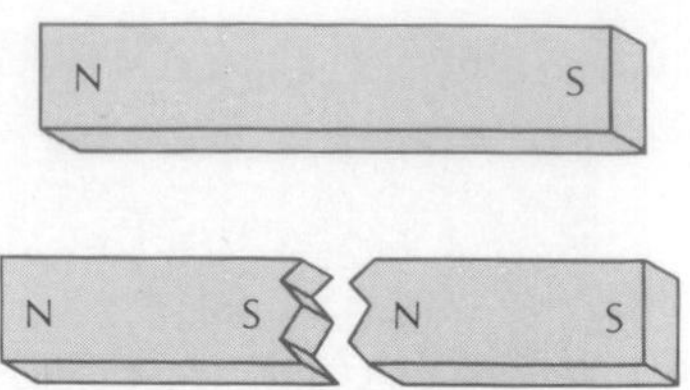

Fig. G.11 A magnet and a broken magnet.

the string results in the creation of an extra, normal particle. For example, Figure G.10 shows how the breaking of a string inside the proton leads to the creation of a pion.

At a fundamental level, the color force between quarks is due to an exchange of virtual particles between the quarks. The particle that acts as the carrier of the color force is the **gluon.** Figure G.12 shows a Feynman diagram representing the exchange of a gluon between two quarks. Such an exchange of a gluon between two colored quarks is analogous to the exchange of a photon between two charged particles (see Figure G.5). However, the gluon exchange is a rather more complicated process than photon exchange — the gluons themselves have color whereas the photons do not have any electric charge. Whenever a quark emits a gluon, this gluon takes color away from the quark and thereby changes the color of the quark. For this reason, the quark entering a vertex in Figure G.12 has a different color than the quark leaving the vertex. The color of the gluon is the difference between the colors of these quarks; for example, the gluon emerging from the left vertex in Figure G.12 has colors *green* and anti*blue*, i.e., it has two colors. Thus, the exchange of a gluon between two quarks entails an exchange of colors. The force between, say, the three quarks within a proton is due to continual, repet-

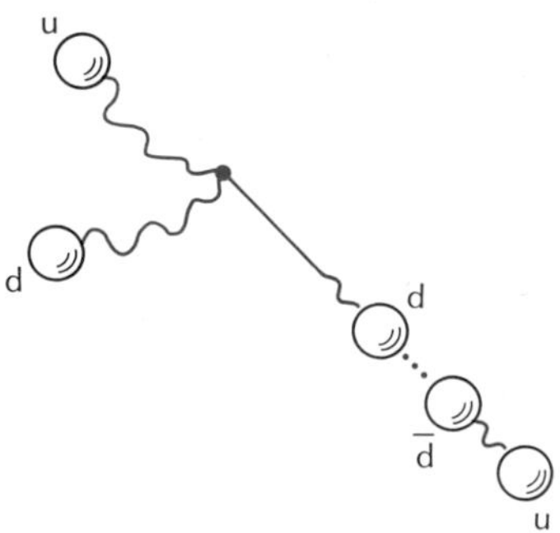

Fig. G.10 In a very violent collision, the string breaks. A quark–antiquark pair (d and $\bar{d}$) is created at the site of the break. The $\bar{d}$ antiquark remains tied to the u quark; this combination is a pion, π^+.

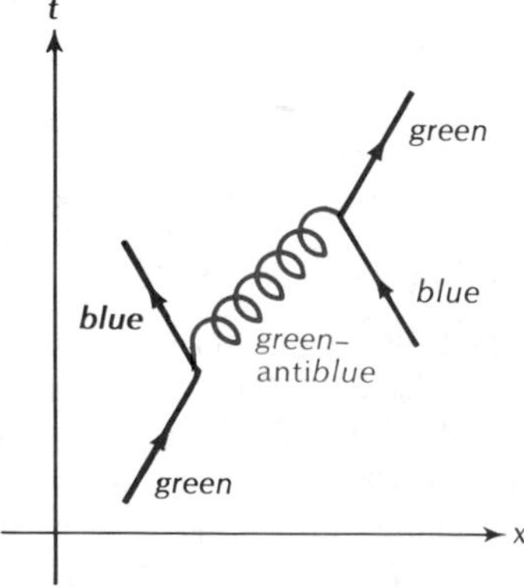

Fig. G.12 Exchange of a gluon between two quarks. The quark on the left suffers a color change from *green* to *blue;* that on the right a color change from *blue* to *green.*

itive exchanges of gluons between the quarks. During this process, the color of each quark changes again and again; for instance, the u quark within the proton is sometimes *red,* sometimes *green,* and sometimes *blue.*

The theory of the color force is called **quantum chromodynamics** (or QCD). The equations of this theory are much more complicated than those of quantum electrodynamics. Because the calculations in chromodynamics are exceedingly difficult and tedious, most of the theoretical predictions that have been obtained so far are qualitative rather than quantitative. Nevertheless, experiments performed at the DESY accelerator in 1978 did reveal some circumstantial evidence confirming the existence of the gluons. In these experiments, physicists attempted to manufacture gluons in electron–antielectron collisions of extremely high energy. Since the gluons only last for a short instant of time, there is no hope of seeing them directly. But chromodynamics predicts that in some violent electron–antielectron collisions, a quark, an antiquark, and a gluon will be produced. Each of these three particles decays into several pions; when emerging from the scene of the collision, these pions should then be arranged in three distinct jets spurting out at distinct angles. The experiments confirmed the existence of such triple jets. Figure G.13 shows a picture of such jets produced at the proton–antiproton collider at CERN.

As a next step, some theoretical physicists are speculating on a possible connection between gluons, photons, and W and Z particles. As we saw in Section G.4, photons and W and Z particles are part of a single family of particles and, consequently, the electromagnetic and weak forces are merely different aspects of a single electroweak force. If it should turn out that gluons are also part of the same family of particles, then the electromagnetic, the weak, and the strong force would all be different aspects of just one single, basic kind of force. If so, we would have achieved a deep unification of three of the fundamental forces of physics. Then there would only remain the question of how to incorporate gravitation into this grand scheme. Some clever suggestions on how to approach this question have been made, but the answer still remains hidden.

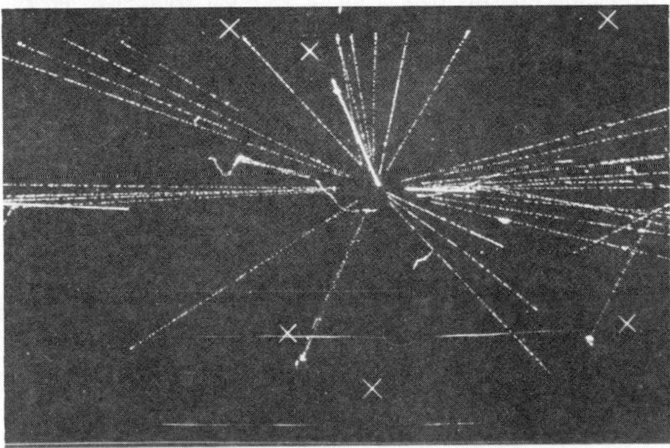

Fig. G.13 Tracks of particles produced in a very energetic head-on collision between a proton of 270 GeV that entered from the left and an antiproton of 270 GeV that entered from the right. Among the multitude of particles created in this collision, we can distinguish three jets: one jet of three particles in the lower right, one of five particles in the upper center, and one of four particles in the upper left.

Further Reading

The World of Elementary Particles by K. W. Ford (Blaisdell, New York, 1965) gives a simple, nonmathematical introduction to fields, quanta, and the generation of force by exchange of quanta. Unfortunately, this book is by now somewhat outdated, as it was written before the discoveries of the unified theory of weak and electromagnetic forces, and of quantum chromodynamics. These recent discoveries are described in *The Moment of Creation* by J. S. Trefil (Scribner's, New York, 1983), an exceptionally lucid and readable book, which includes a discussion of the cosmological implications of the unified theory. The recent discoveries are also briefly described in several of the books already mentioned in Interlude C:

Quarks: The Stuff of Matter by H. Fritzsch (Basic Books, New York, 1983)

From Atoms to Quarks by J. S. Trefil (Scribner's, New York, 1980)

The Cosmic Code by H. R. Pagels (Simon and Schuster, New York, 1982)

What Is the World Made Of? by G. Feinberg (Doubleday, Garden City, 1977)

The following articles deal with diverse aspects of the fundamental forces:

"Field Theory," F. J. Dyson, *Scientific American,* April 1953

"Unified Theories of Elementary Particle Interactions," S. Weinberg, *Scientific American,* July 1974

"The Detection of Neutral Weak Currents," A. K. Mann and C. Rubbia, *Scientific American,* December 1974

"Light as a Fundamental Particle," S. Weinberg, *Physics Today,* June 1975

"The Confinement of Quarks," Y. Nambu, *Scientific American,* November 1976

"Supergravity and the Unification of the Laws of Physics," D. Z. Freedman and P. van Nieuwenhuizwen, *Scientific American,* February 1978

"Gauge Theories of the Forces between Elementary Particles," G. 'tHooft, *Scientific American,* June 1980

"Antiproton–Proton Colliders, Intermediate Bosons," D. Cline and C. Rubbia, *Physics Today,* August 1980

"Unified Theory of Elementary-Particle Forces," H. Georgi and S. L. Glashow, *Physics Today,* September 1980
"A Unified Theory of Elementary Particles and Forces," H. Georgi, *Scientific American,* April 1981
"The Development of Field Theory in the Last 50 Years," W. F. Weisskopf, *Physics Today,* November 1981
"The Search for Intermediate Vector Bosons," D. B. Cline, C. Rubbia, and S. van der Meer, *Scientific American,* March 1982
"Glueballs," K. Ishikawa, *Scientific American,* November 1982
"The Lattice Theory of Quark Confinement," C. Rebbi, *Scientific American,* February 1983
"The Structure of Quarks and Leptons," H. Harari, *Scientific American,* April 1983

Questions

1. The strengths of the fundamental forces depend on the energies of the particles. In the case of the gravitational force, the strength increases with the energy. Why would you expect this to be true?

2. The boomerang analogy described in Figure G.7 is defective in that the boomerang requires the presence of air. What would be the motion of a boomerang in vacuum?

3. Show that if classical physics is valid, the emission of a photon by a free electron conflicts with energy conservation. (Hint: Consider the emission process in the reference frame in which the electron is initially at rest.)

4. The W particle can have either a positive charge (W^+) or a negative charge (W^-). Figure G.14 shows the Feynman diagram for the decay of the neutron (n) via exchange of a W^-; the end products are a proton (p), an electron (e), and an antineutrino ($\bar{\nu}$). Can you guess the Feynman diagram for the decay of the antineutron?

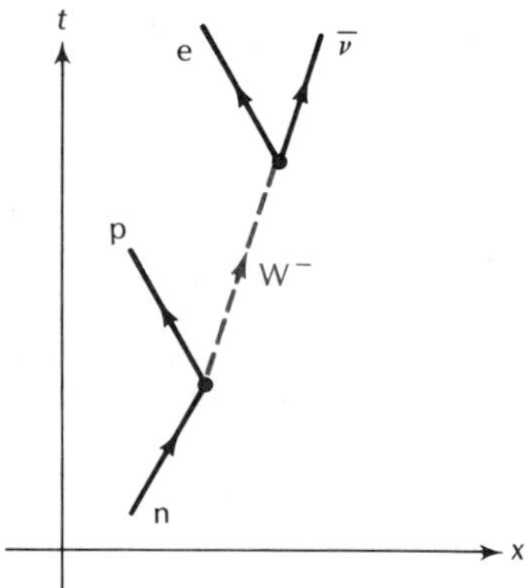

Fig. G.14

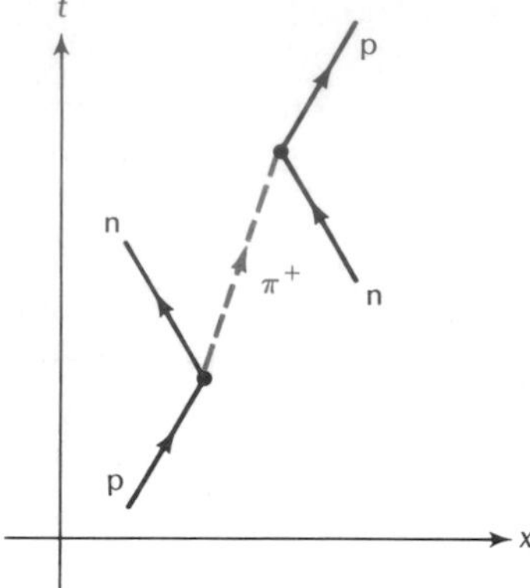

Fig. G.15

5. Figure G.15 shows the Feynman diagram for the exchange of a π^+ between a proton and a neutron; note that the proton changes into a neutron, and vice versa. Draw corresponding diagrams for the exchange of a π^- and of a π^0.

6. Give another simple example of spontaneous breaking of symmetry similar to the example of the pencil in Section G.4.

7. In Figure G.13, a large number of particles emerge in the longitudinal direction (toward the right and the left). Why is this expected, whereas the emergence of particles in the transverse direction (upward and downward) is surprising? (Hint: Consider a head-on collision between two aircraft; which way do you expect most fragments to spurt out?)

8. Figure G.10 shows the creation of a π^+ pion in a very violent collision. What is the particle (consisting of three quarks) left behind in this collision?

9. Suppose that in Figure G.10 the quark–antiquark pair created at the site of the break in the string is a u–$\bar{u}$ pair instead of a d–$\bar{d}$ pair. What kind of particle (consisting of two quarks) will then be created?

10. A pion consists of a u quark and a $\bar{u}$ antiquark tied together by a string. What kinds of particles (consisting of two quarks) can be created if the string breaks?

11. Given that each gluon has a color and an anticolor, how many different kinds of gluons are there?

12. A glueball consists of two or more gluons bound together in a composite structure (without quarks). The net color of glueballs is neutral; they always contain equal net amounts of a color and its anticolor. Suppose that a glueball consists of two gluons, one of which is *green*–anti*blue*. What must be the colors of the other gluon? Suppose that a glueball consists of three gluons, one *green*–anti*blue,* one *red*–anti*green*. What must be the colors of the third gluon?

CHAPTER 24

Gauss' Law

Although the electric field of any given charge distribution can be calculated by means of Coulomb's Law, as in Examples 1–4 of the preceding chapter, this method often involves the evaluation of tedious integrals. Fortunately, there exists another method for calculating the electric field. This method relies on a theorem called Gauss' Law. That law is a consequence of Coulomb's Law and it therefore contains no new physics. It does, however, contain some new mathematics which supplies an elegant shortcut for calculating the electric field of a given charge distribution provided that this charge distribution has a certain amount of symmetry. This means that Gauss' Law does not help in every problem, but when it does help it works wonders. Before we present Gauss' Law and some examples of its use, we need to introduce the concept of electric flux.

24.1 Electric Flux and the Number of Field Lines

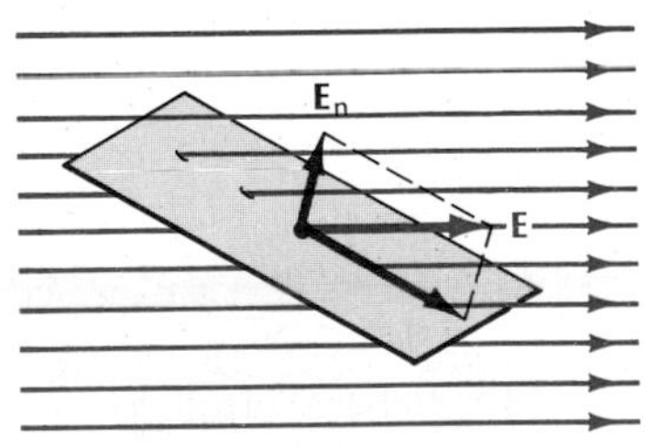

Fig. 24.1 Flat rectangular surface immersed in a uniform electric field.

Consider a mathematical (i.e., imagined) surface in the shape of a rectangle of area A. Suppose that this surface is immersed in a constant electric field E (Figure 24.1). This electric field makes an angle with the surface; the electric-field vector has a component tangential to the surface and a component normal (i.e., perpendicular) to the surface. The **electric flux** Φ through the surface is defined as the product of the area A by the magnitude of the normal component of the electric field,

$$\Phi = E_n A \quad (1)$$

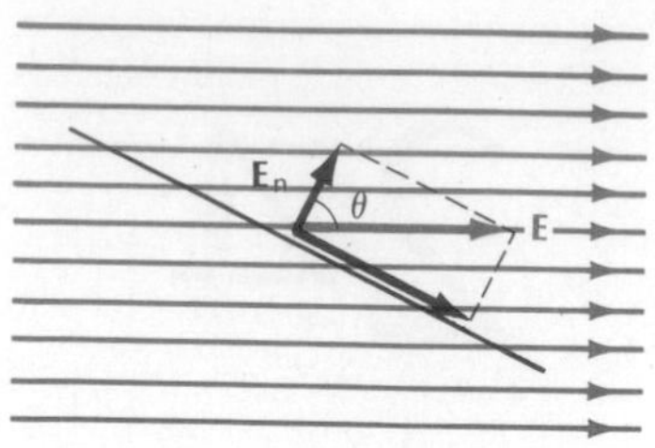

Fig. 24.2 The perpendicular to the surface makes an angle θ with the field lines. Note that the surface intercepts six field lines, i.e., the electric flux through the surface is $\Phi = 6$.

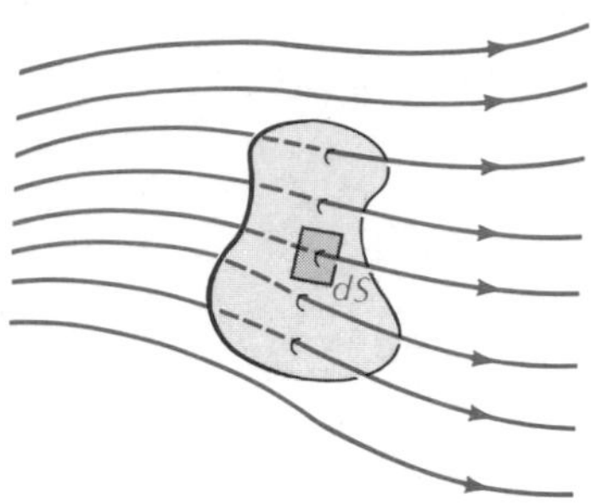

Fig. 24.3 An arbitrary surface immersed in an arbitrary electric field. The flux is $\Phi = 5$.

The normal component E_n can also be written as $E \cos \theta$, where θ is the angle between $\mathbf{E}$ and the perpendicular erected on the surface (Figure 24.2). Hence

$$\Phi = EA \cos \theta \tag{2}$$

The quantity $A \cos \theta$ can be interpreted as the projection of the area A onto a plane perpendicular to the electric field, that is, $A \cos \theta$ can be regarded as that part of the area A that faces the electric field. According to Section 23.3, the magnitude E of the electric field is numerically equal to the number of field lines intercepted by a unit area facing the electric field. Hence $EA \cos \theta$ must be numerically equal to the number of field lines intercepted by the area A: *the electric flux* Φ *through an area is equal to the number of field lines intercepted by the area.* Note that this equality hinges on the normalization adopted in Section 23.3 — flux and number of lines are equal if and only if electric field and number of lines per unit area are equal, and the latter is true if and only if we adopt the normalization that Q/ε_0 lines emerge from each charge Q.

More generally, consider a mathematical surface of arbitrary shape immersed in an electric field of arbitrary strength (Figure 24.3). Then we can define the electric flux by subdividing the surface into infinitesimal plane areas dS within each of which the electric field is nearly constant; the flux through one such infinitesimal area is $d\Phi = E \cos \theta \, dS$ and the total flux through the surface is the integral obtained by summing all these infinitesimal contributions,

Electric flux

$$\boxed{\Phi = \int E \cos \theta \, dS} \tag{3}$$

According to the above arguments, this electric flux is again numerically equal to the number of field lines intercepted by the surface. Note that lines going through the surface in one direction give a positive contribution to the flux, lines going in the opposite direction give a negative contribution (Figure 24.4). This means that the perpendiculars to the surface, with respect to which the angle θ is reckoned, must all be erected on the same side of the surface (e.g., on the right side of the surface of Figure 24.4).

Finally, consider a *closed* surface immersed in an electric field (Figure 24.5). The electric flux through this surface is given by Eq. (3) with the integration extending over all the area of the closed surface. This flux is equal to the number of field lines intercepted by the surface. The number can be positive or negative. For any closed surface we adopt the convention that the angle θ is reckoned with respect to

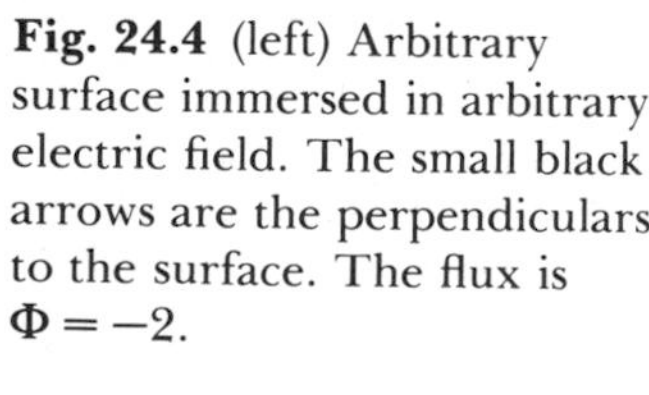

Fig. 24.4 (left) Arbitrary surface immersed in arbitrary electric field. The small black arrows are the perpendiculars to the surface. The flux is $\Phi = -2$.

Fig. 24.5 (right) A closed surface immersed in an electric field.

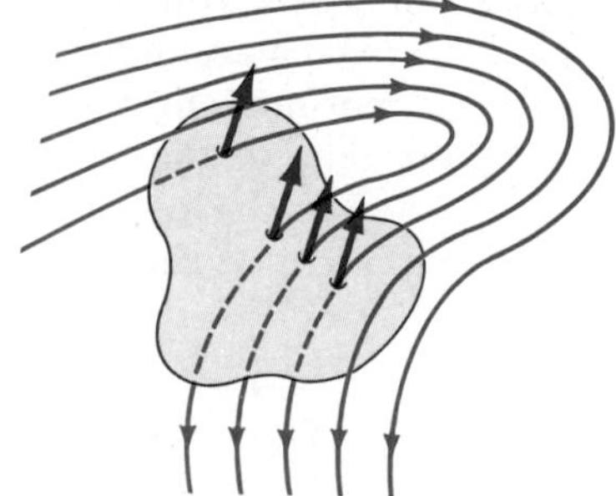

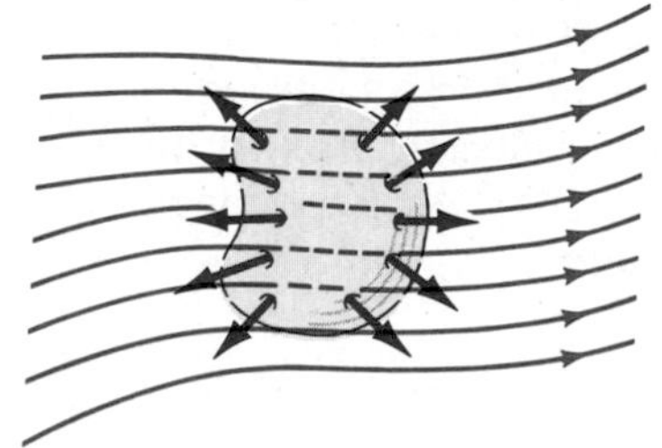

perpendiculars erected on the *outside* of the closed surface; then field lines leaving the volume enclosed by the surface make a positive contribution to the flux, and lines entering the volume make a negative contribution.

EXAMPLE 1. A point charge q is inside a closed spherical surface (Figure 24.6). What is the flux of its electric field through this surface?

SOLUTION: A positive charge q is the starting point of q/ε_0 outward field lines and all of these will pierce the surface. Hence the flux must be q/ε_0. Note that by using the concept of field lines, we are able to get the answer without explicit calculation of any integral; the explicit integration of the flux integral $\int E \cos\theta \, dS$ for the configuration shown in Figure 24.6 is rather messy (unless the charge is at the exact center of the surface).

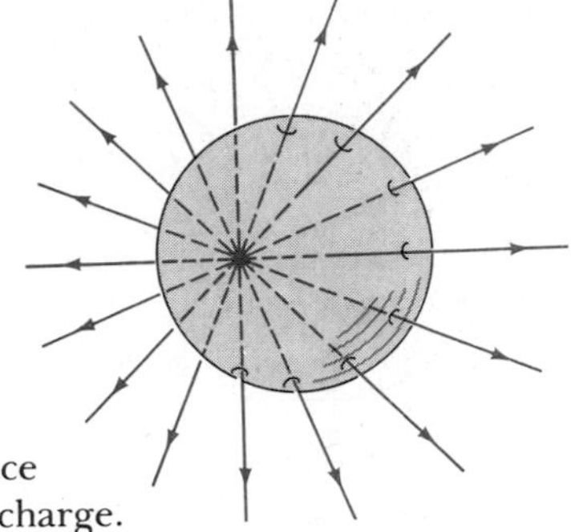

Fig. 24.6 A spherical surface surrounds a positive point charge.

24.2 Gauss' Law

We are now ready to prove an important theorem called **Gauss' Law:**

Gauss' Law

If the volume within an arbitrary closed surface holds a net charge Q, then the electric flux through the surface is Q/ε_0, that is,

$$\oint E \cos\theta \, dS = Q/\varepsilon_0 \qquad (4)$$

Here the small circle on the integral sign indicates that the integration is to be performed over a *closed* surface.

The proof of Gauss' Law is easy. The electric field appearing in Eq. (4) is a sum of the individual electric fields of some number of point charges. Some of the point charges are outside the closed surface and some are inside the closed surface. Let us consider the contribution to the flux from each individual electric field of each individual point charge. The individual electric field of a charge outside the closed surface generates no net flux through the closed surface — any field line of this field either does not touch the surface or else enters it at one point and leaves at another; neither case makes any contribution to the flux. However, the individual electric field of a positive or negative charge enclosed by the surface does contribute to the flux. For example, a positive charge q has q/ε_0 outward field lines, all of which will pierce the closed surface — such a charge therefore contributes a flux q/ε_0 (Figure 24.7). Taking into account that positive charges generate positive flux and negative charges negative flux, we see that the net flux through the surface, or the net number of lines piercing the surface, is equal to the net charge divided by ε_0. This completes the proof of Eq. (4).

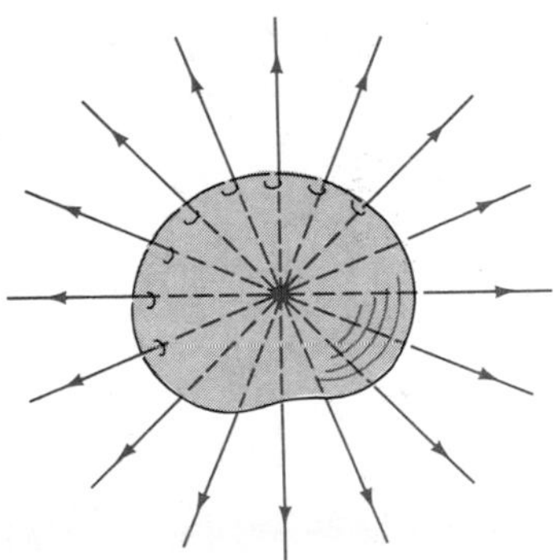

Fig. 24.7 A closed surface surrounds a positive point charge.

With the notation E_n for the component of **E** normal to the surface, we can put Gauss' Law in the somewhat more compact form

$$\oint E_n \, dS = Q/\varepsilon_0 \qquad (5)$$

An alternative notation frequently used associates a vector $d\mathbf{S}$ with the area dS; this vector has the magnitude of dS and the direction of the outward perpendicular to the surface (Figure 24.8). Then $E \cos\theta \, dS = \mathbf{E} \cdot d\mathbf{S}$ and Eq. (4) takes the form

Gauss' Law in vector notation

$$\oint \mathbf{E} \cdot d\mathbf{S} = Q/\varepsilon_0 \tag{6}$$

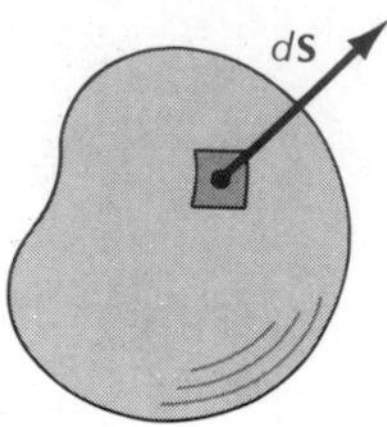

Fig. 24.8 A small area dS with its perpendicular vector $d\mathbf{S}$.

The closed mathematical surface over which the integrations in Eq. (4), (5), and (6) must be carried out is usually called a **Gaussian surface.**

Gauss' Law can be used to calculate the electric field provided the distribution of charge has a high degree of symmetry. Essentially, Gauss' Law can be regarded as a mathematical restriction imposed on the electric field. Symmetry conditions impose further restrictions on the field. By clever combination of all these restrictions one can often evaluate the electric field without the laborious process of integrating Coulomb's Law.

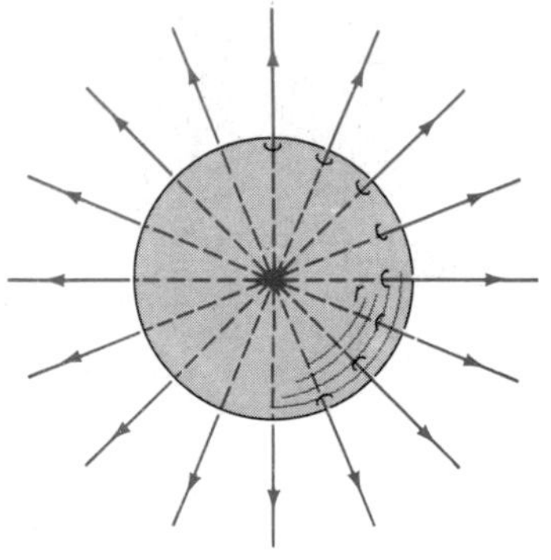

Fig. 24.9 A Gaussian spherical surface surrounds a positive point charge concentrically.

EXAMPLE 2. Find the electric field of a point charge q using Gauss' Law.

SOLUTION: For the purpose of this example we will pretend that we do not know the electric field of a point charge. We begin with the observation that the electric field must be spherically symmetric because the point charge is spherically symmetric; hence, at all points on the surface of a mathematical sphere of radius r centered on q, the field is in the radial direction and has a constant magnitude (Figure 24.9).

If the Gaussian surface is taken to coincide with the sphere of radius r, then $\cos\theta = 1$, since $\mathbf{E}$ is perpendicular to this surface. Furthermore, $\mathbf{E}$ has a constant magnitude over this surface. Therefore

$$\oint E \cos\theta \, dS = \oint E \, dS = E \oint dS$$

But $\oint dS$ is simply the area of the Gaussian surface, i.e., $4\pi r^2$. Hence Eq. (4) becomes

$$E(4\pi r^2) = q/\varepsilon_0$$

or

$$E = \frac{1}{4\pi\varepsilon_0} \frac{q}{r^2} \tag{7}$$

The proof of Eq. (4) shows that Gauss' Law is a consequence of the field-line concept, i.e., Gauss' Law is a consequence of Coulomb's Law. Conversely, Example 2 shows that Coulomb's Law is a consequence of Gauss' Law. These two laws are therefore equivalent, at least in regard to the electric fields of charges at rest. As we will see in a later chapter, it turns out that besides such static electric fields, there also exist time-dependent electric fields whose field lines are not attached to charges (the field lines form closed loops). Coulomb's Law does not apply to such time-dependent electric fields . . . but Gauss' Law does! The latter law is more general than the former. However, this subtle distinction need not concern us as long as all charges are static — as they are in the examples of the next section.

24.3 Some Examples

The following examples illustrate how Gauss' Law can be combined with symmetry requirements and used to evaluate the electric field of simple charge distributions. The solutions of these examples always involve two steps: first determine the *direction* of the electric field by appealing to the symmetry requirements imposed by the charge distribution, and then calculate the *magnitude* of the electric field from Gauss' Law. In all these examples the crucial trick lies in the choice of Gaussian surface — a good choice makes the calculation of the integral $\oint E \cos\theta \, dS$ easy and also permits one to "solve" Gauss' Law for the unknown value of E.

EXAMPLE 3. Charge is distributed uniformly along a very long thin line. If the amount of charge is λ coulomb per meter of line, what is the electric field?

SOLUTION: This example has already been worked out via Coulomb's Law (see Section 23.1); the following quick and elegant method is based on Gauss' Law. First we need to determine the direction of the electric field. The only direction consistent with the symmetry of the charge distribution is the radial direction (horizontal in Figure 24.10) — if the electric field had any other direction (up or down in Figure 24.10), the field would make an unacceptable distinction between the upper and lower ends of the line of charge. Furthermore, the rotational symmetry of the line of charge tells us that the electric field has constant magnitude over the lateral curved surface of any mathematical cylinder concentric with the line of charge.

Now take a Gaussian surface that coincides with such a cylinder, of radius x and height h (Figure 24.10). On the lateral, curved surface of the cylinder, $\cos\theta = 1$ and

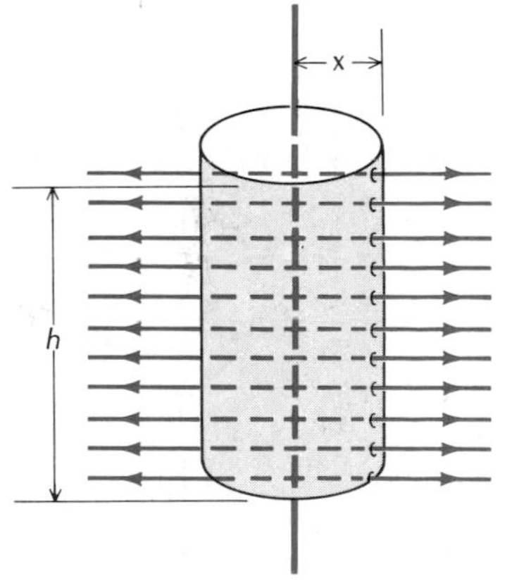

Fig. 24.10 A cylindrical Gaussian surface surrounds a line of charge.

$$\int_{\substack{\text{curved}\\\text{surface}}} E \cos\theta \, dS = \int E \, dS = E \int dS = E \times (2\pi x h) \qquad (8)$$

On each of the two circular bases of the cylinder, the electric field is tangent to the surface; hence $\cos\theta = 0$ and

$$\int_{\text{base}} E \cos\theta \, dS = 0 \qquad (9)$$

The integral over the entire surface of the cylinder (curved surface plus bases) is then $2\pi xhE$. By Gauss' Law this must equal the charge contained in the cylinder divided by ε_0, i.e.,

$$2\pi xhE = Q/\varepsilon_0 = \lambda h/\varepsilon_0$$

and

$$E = \frac{1}{2\pi\varepsilon_0} \frac{\lambda}{x} \qquad (10)$$

EXAMPLE 4. What is the electric field of a large uniform sheet of charge with a density of σ coulomb per square meter?

SOLUTION: This example also has been worked out earlier by a laborious integration procedure; this labor can be bypassed by means of Gauss' Law. Symmetry tells us that the electric field is everywhere perpendicular to the sheet of charge and it has a constant magnitude over any surface parallel to this sheet.

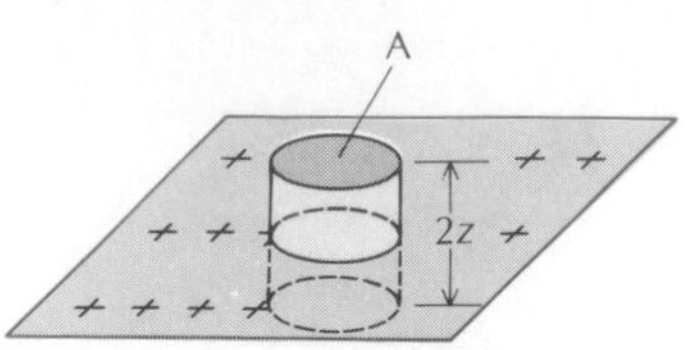

Fig. 24.11 A cylindrical Gaussian surface intersects a very large sheet of charge.

We take a Gaussian surface in the shape of a cylinder of height $2z$ and base area A (Figure 24.11). Note that since the bases are at the same distance from the sheet, the magnitude of E will be the same on both of them. On each base $\cos\theta = 1$ and, integrating over both, we find

$$\int_{\text{bases}} E\cos\theta\, dS = \int E\, dS = E \times (2A) \tag{11}$$

On the curved lateral surface of the cylinder, $\cos\theta = 0$ and therefore

$$\int_{\substack{\text{curved}\\ \text{surface}}} E\cos\theta\, dS = 0 \tag{12}$$

By Gauss' Law, the integral of $E\cos\theta$ over the complete surface must equal the charge within the cylinder divided by ε_0. Since the charge is σA, we obtain

$$2AE = Q/\varepsilon_0 = \sigma A/\varepsilon_0$$

and

$$E = \sigma/2\varepsilon_0 \tag{13}$$

EXAMPLE 5. A spherical region of radius R has a total charge q which is uniformly distributed over the volume of this region. (a) What is the electric field at points inside the sphere? (b) What is the electric field at points outside the sphere?

SOLUTION: (a) The charge density within the sphere is

$$\rho = [\text{charge}]/[\text{volume}] = q/(4\pi R^3/3) = 3q/4\pi R^3 \tag{14}$$

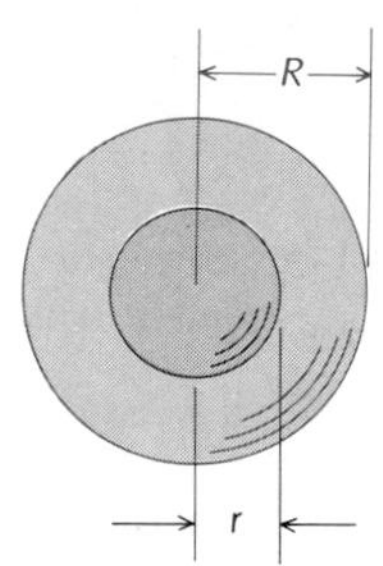

Fig. 24.12 A spherical volume with a uniform distribution of charge. The Gaussian surface has a radius $r < R$.

Since the charge distribution is spherically symmetric, the electric field must be radial and of constant magnitude over any spherical mathematical surface of given radius. To find the magnitude of the electric field inside the charge distribution, take a spherical Gaussian surface of radius r, where $r \leq R$ (Figure 24.12). On this surface $\cos\theta = 1$, so that

$$\oint E\cos\theta\, dS = \oint E\, dS = E\oint dS = 4\pi r^2 E \tag{15}$$

The charge inside this Gaussian surface is

$$Q = [\text{charge density}] \times [\text{volume}] = \rho \times \frac{4\pi}{3} r^3$$

$$= q\frac{r^3}{R^3} \tag{16}$$

Then Gauss' Law gives

$$4\pi r^2 E = \frac{q}{\varepsilon_0}\frac{r^3}{R^3}$$

i.e.,

$$E = \frac{1}{4\pi\varepsilon_0}\frac{qr}{R^3} \qquad \text{for } r \leq R \tag{17}$$

Note that this electric field increases linearly with the distance from the center and reaches a maximum value of $E = (1/4\pi\varepsilon_0)q/R^2$ when $r = R$.

(b) To find the electric field outside the sphere, take a spherical Gaussian surface of radius r (where $r \geq R$; Figure 24.13). With the usual symmetry arguments, the flux integral again has the form $4\pi r^2 E$ and Gauss' Law gives

$$4\pi r^2 E = q/\varepsilon_0$$

i.e.,

$$E = \frac{1}{4\pi\varepsilon_0}\frac{q}{r^2} \qquad \text{for } r \geq R \tag{18}$$

This means that outside the region containing charge, the electric field is exactly the same as it would be if all the charges were located at the center. Figure 24.14 is a plot of the electric field vs. distance.

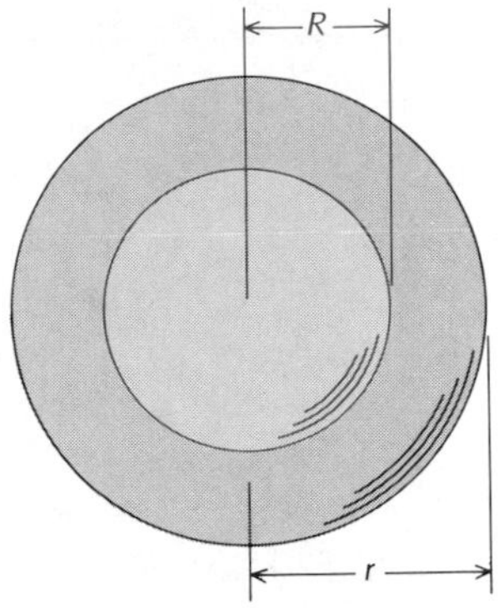

Fig. 24.13 The Gaussian surface has a radius $r > R$.

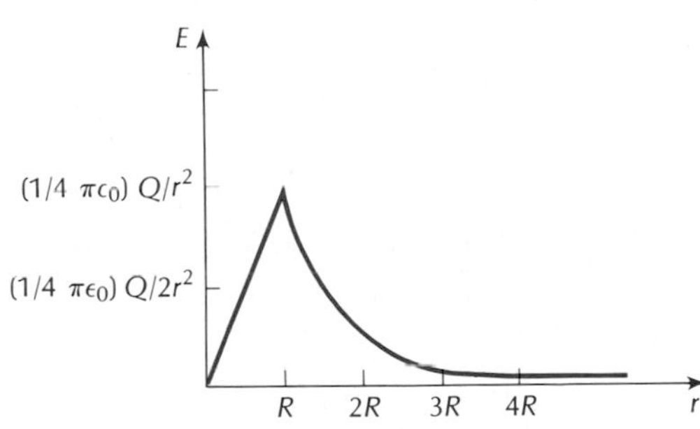

Fig. 24.14 Electric field as a function of radius for a uniformly charged sphere.

It is obvious that the argument of part (b) of the preceding example does not depend on the uniformity of the charge distribution — it only depends on its spherical symmetry. Hence outside any charge distribution that consists of a sequence of concentric shells of charge — so that the charge density is a function of the radius but not of the angular direction — the electric field will mimic that of a point charge. This result is similar to Newton's famous theorem concerning the gravitational forces exerted by a planet: the forces exerted by a spherically symmetric planet can be calculated as if all the mass were concentrated in a point at the planetary center. This similarity between electricity and gravitation reflects the similarity of the laws of force. Because of this similarity, we can use the same action–reaction argument as in Section 13.6 to show that a spherical charge distribution mimics a point charge not only in regard to the electric field that it produces, but also in regard to the force that it experiences when placed in some arbitrary external electric field. Thus, the electric force that an arbitrary electric field exerts on a spherical charge distribution can be calculated as if all of the charge were concentrated in a point at the center.

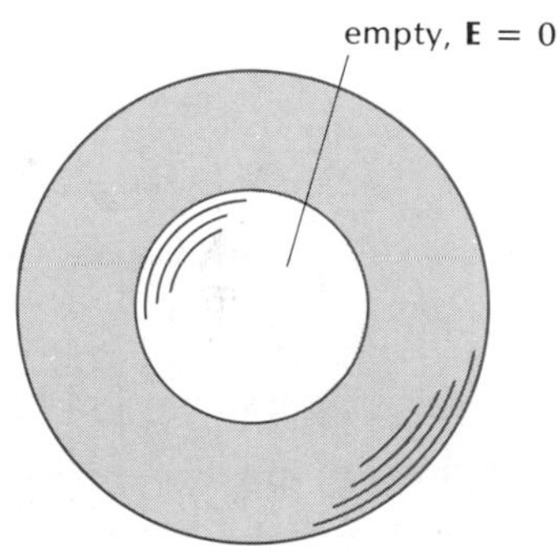

Fig. 24.15 A spherical shell (a sphere with a spherical cavity).

Note that Eq. (17) can be put in the form

Electric field of spherical charge distribution

$$E = \frac{1}{4\pi\varepsilon_0}\frac{Q(r)}{r^2} \tag{19}$$

where $Q(r)$ is the amount of charge inside the sphere of radius r. If the charge distribution is uniform, then $Q(r)$ is as given by Eq. (16); however, the expression (19) does not depend on the radial uniformity of the charge distribution — it only depends on spherical symmetry. According to Eq. (19), only the charge *inside* a radius less than r contributes to the electric field *at* the radius r. As a consequence, the electric field within the empty region inside a spherically symmetric shell of charge is exactly zero (Figure 24.15).

EXAMPLE 6. A proton is (approximately) a spherically symmetric ball of charge of radius 1.0×10^{-15} m. What is the electric field at the surface of the proton? If a second proton is brought within touching distance of the first, what is the repulsive electric force?

SOLUTION: For $q = e = 1.6 \times 10^{-19}$ C and $r = 1.0 \times 10^{-15}$ m, the electric field is

$$E = \frac{1}{4\pi\varepsilon_0}\frac{q}{r^2}$$

$$= \frac{9.0 \times 10^9 \text{ N} \cdot \text{m}^2 \cdot \text{C}^{-2} \times 1.6 \times 10^{-19} \text{ C}}{(1.0 \times 10^{-15} \text{ m})^2}$$

$$= 1.4 \times 10^{21} \text{ N/C} \tag{20}$$

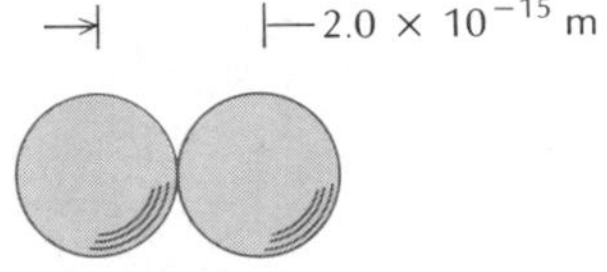

Fig. 24.16 Two protons in contact. Each proton may be regarded as a spherical ball of positive charge.

If the protons are touching (Figure 24.16), the center-to-center distance is $2 \times 1.0 \times 10^{-15}$ m. The electric field generated by one proton at this distance is one-quarter of the above value, i.e., $E = 3.6 \times 10^{20}$ N/C. The force exerted by one proton on the other is

$$F = eE = 1.6 \times 10^{-19} \text{ C} \times 3.6 \times 10^{20} \text{ N/C}$$

$$= 58 \text{ N} \tag{21}$$

This is about 13 lb of force; acting on a particle of a mass of only 10^{-27} kg, this represents a gigantic force! In the nucleus of an atom, the large repulsive electric force between the tightly packed protons is more than canceled by an even larger binding force (the "strong" force) that holds the nucleus together.

24.4 Conductors in an Electric Field

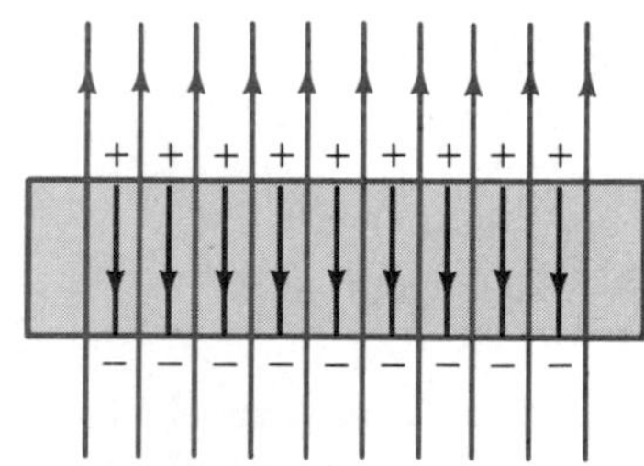

Fig. 24.17 Conductor immersed in an electric field. The charges that have accumulated on the surfaces generate an electric field (black) opposite to the original electric field (color).

As we pointed out in Section 22.4, in metallic conductors — such as copper, silver, aluminum — some of the electrons are free, that is, they can move without restraint within the volume of the metal. If such a conductor is immersed in an electric field, the free electrons move in response to the electric force. The electrons move in a direction opposite to the direction of the electric field and they continue moving until they reach the surface of the metal. As an excess of electrons accumulates on one part of the surface, a deficit of electrons will appear on another part of the surface: negative and positive charges are induced on the conductor. Within the volume of the conductor, the electric field of the induced charges tends to cancel the original electric field in which the conductor was immersed (Figure 24.17). The accumulation of negative and positive charge on the surface of the conductor continues until the electric field generated by these charges exactly cancels the original electric field. Consequently, *when the charge distribution on a conductor reaches static equilibrium, the net electric field within the material of the conductor is exactly zero.* The proof of this statement is by contradiction: if the electric field were different from zero, the free electrons would continue to move and the charge distribution would *not* (yet) be in equilibrium. For a good conductor

Electrostatic equilibrium

(copper, aluminum, etc.), the equilibrium is reached in a fairly short time, a small fraction of a second.[1]

In a conductor in static equilibrium, all the (extra) electric charge resides on the surface of the conductor. One can prove this by means of Gauss' Law: Consider a small closed surface inside the conducting material (Figure 24.18). Since $\mathbf{E} = 0$ everywhere in this material, the left side of Eq. (4) vanishes and therefore the right side must also vanish — which means that the charge enclosed by *any* arbitrary small surface is zero, that is, the charge in *any* small volume of the conductor is zero. Obviously, if the charges are not in the volume of the conductor, they must be on the surface.

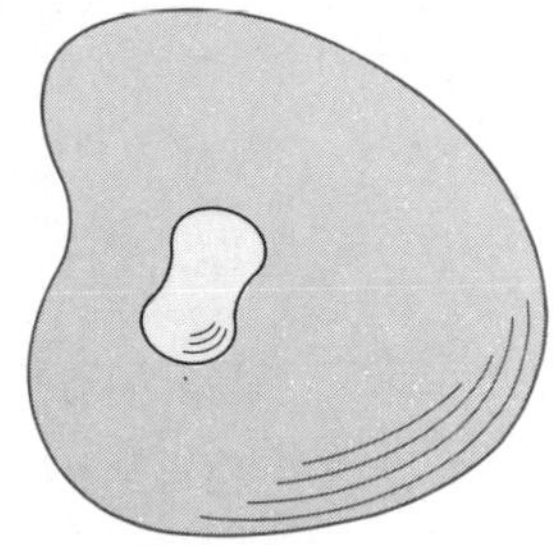

Fig. 24.18 Closed Gaussian surface inside a volume of conducting material.

Finally, we can say something about the electric field just outside a conductor: *the electric field at the surface of a conductor in static equilibrium is normal to the surface.* The proof is again by contradiction: if the electric field had a component tangential to the surface of the conductor, the free electrons would move along the surface and the charge distribution would *not* be in equilibrium.

Note that this argument does not exclude an electric field perpendicular to the surface of the conductor; such an electric field merely pushes the free electrons against the surface, where they are held in equilibrium by the combination of the force exerted by the electric field and the restraining force exerted by the surface of the conductor. Figure 24.19 displays an experimental demonstration of electric fields perpendicular to the surfaces of conductors. The flat plate and the cylinder shown in this figure are conductors, and we see that the field lines meet the surfaces of these conductors at right angles.

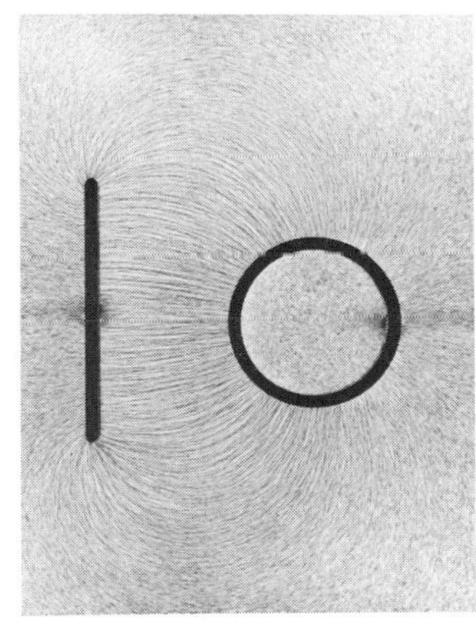

Fig. 24.19 Field lines in the space surrounding a charged flat plate and cylinder, made visible by small bits of thread suspended in oil. (Courtesy H. Waage, Princeton University.)

In the preceding paragraphs we have implicitly assumed that the material of the conductor is homogeneous, i.e., the material has uniform density and uniform chemical composition. If this is not the case, then our conclusions are not quite valid. For instance, suppose that the body of the conductor consists of two dissimilar metals joined together. Figure 24.20 shows a conductor made of a block of lead in contact with a block of silver. In this conductor there will exist some electric field at and near the interface of the metals and, what is more, such an electric field will exist even if the conductor carries no net charge. This electric field is created by the contact of the two dissimilar metals. To understand how the electric field comes about, imagine that at first the blocks of lead and of silver are separated. Each metal is neutral and contains within it a gas of free electrons. If the blocks are now brought into contact, some of the free electrons from the lead block will flow into the silver block. This is so because silver exerts a slightly stronger hold on free electrons than lead (the energy required to remove a free electron from silver is larger than that from lead). As the extra free electrons accumulate in the silver, their electric repulsions gradually bring the flow to a halt. At equilibrium, the silver will have an excess of electrons (negative charge) and the lead a deficiency of electrons (positive charge). These charges will then create an electric field at and near the interface of the joined blocks.

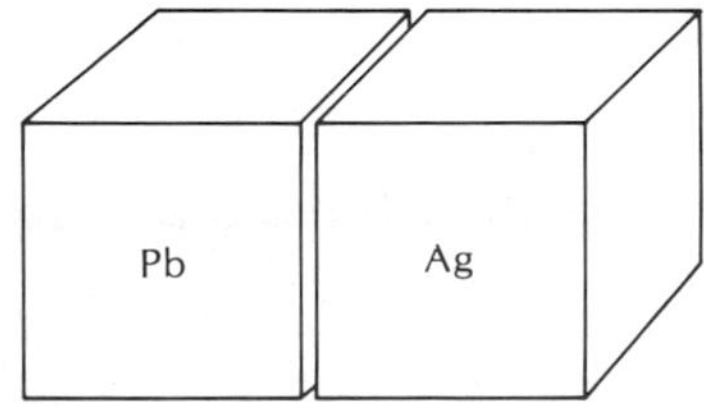

Fig. 24.20 A conductor made of two blocks of lead and silver joined together.

[1] The time for reaching equilibrium depends on details such as the shape and the size of the conductor and the characteristics of the conducting material. Paradoxically, a good conductor takes somewhat longer to reach equilibrium than a poor conductor. In the good conductor, the charges move very easily and they tend to overshoot their equilibrium positions — the charges slosh around on the conductor and they take a while to settle in their equilibrium positions.

The charges and the electric fields that are created by the contact of dissimilar metals are usually quite small. In the following chapters we will ignore these "contact" charges and fields. However, it is well to keep in mind that these charges and fields play a crucial role in the operation of transistors, solar cells, and other solid-state devices that consist of pieces of dissimilar conductors or semiconductors joined together.

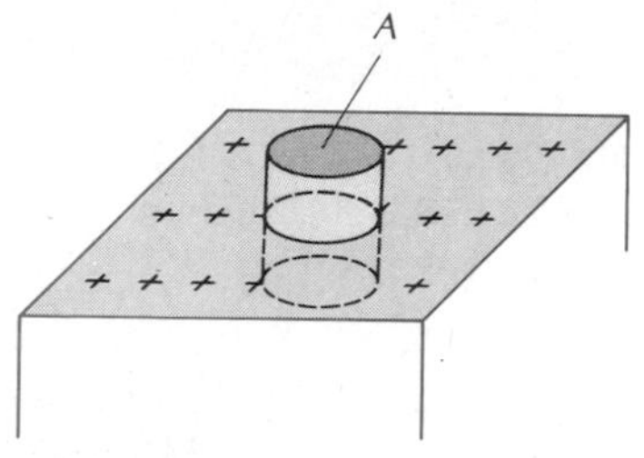

Fig. 24.21 A very large slab of conductor, with a uniform distribution of charge on its surface.

EXAMPLE 7. Find the electric field outside a very large flat conducting surface on which there is a surface charge density of σ coulomb per square meter.

SOLUTION: In view of the symmetry of the charge configuration, the electric field will be perpendicular to the conducting plane and it will have constant magnitude over any plane parallel to the conducting plane. As Gaussian surface, take the cylinder shown in Figure 24.21; the base area of the cylinder is A. The upper base contributes an amount

$$\int E \cos\theta \, dS = \int_{\substack{\text{upper}\\\text{base}}} E \, dS = E \int dS = EA \tag{22}$$

to the flux integral. The lower base, in the conductor, does not contribute to the flux integral, since $E = 0$ in this region. Finally, the curved lateral surface does not contribute, since $\cos\theta = 0$. The charge within the Gaussian surface is σA. Hence,

$$EA = \sigma A/\varepsilon_0$$

and

$$E = \sigma/\varepsilon_0 \tag{23}$$

This is a constant electric field, independent of distance from the conducting plane. Of course, if the plane is of finite size, then the electric field will have the constant value (23) only in a region very near the plane.

Note that, for a given surface-charge density, the electric field generated by a conducting surface is twice as strong as the electric field generated by a sheet of charge [see Eq. (13)]. The reason is obvious: in the former case all field lines that originate on the charges go to the same side of the surface, in the latter case half go to each side.

Over a small region, a curved conducting surface can be approximated by a flat surface. Hence the expression (23) can be used to find the electric field in a region very near any smooth curved conducting surface [the expression (23) is *not* a good approximation near sharp edges]; it is not even necessary that σ be constant — it can be some smooth function of position.

EXAMPLE 8. At the ground directly below a thundercloud, the electric field is 2×10^4 N/C and points upward. What is the surface charge density on the ground?

SOLUTION: For the purpose of this problem, we treat the ground as a good conductor. Equation (23) gives the relation between electric field and surface-charge density for points at or near the charged surface,

$$\sigma = \varepsilon_0 E = 8.85 \times 10^{-12} \frac{\text{C}^2}{\text{N} \cdot \text{m}^2} \times 2 \times 10^4 \text{ N/C}$$

$$= +1.8 \times 10^{-7} \text{ C/m}^2$$

SUMMARY

Electric flux through surface: $\Phi = \int E_n \, dS$

$$= \int \mathbf{E} \cdot d\mathbf{S}$$

Gauss' Law (for closed surface): $\oint \mathbf{E} \cdot d\mathbf{S} = Q/\varepsilon_0$

Conductor in electrostatic equilibrium:

The electric field within the conductor is zero.
The charge resides on the surface.
The electric field at the surface is perpendicular and of magnitude $E = \sigma/\varepsilon_0$.

QUESTIONS

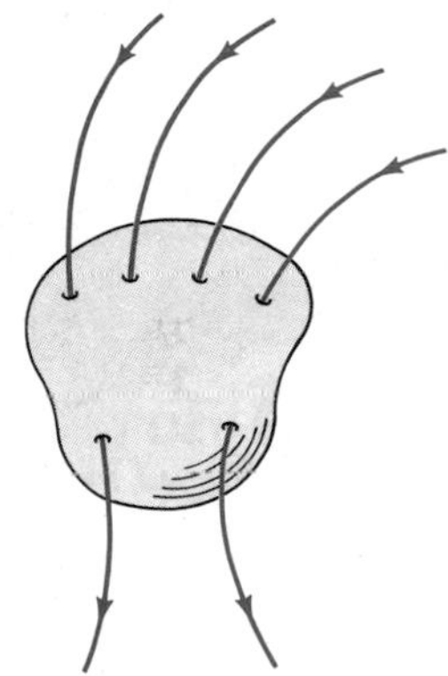

Fig. 24.22

1. A point charge Q is inside a spherical Gaussian surface, which is enclosed in a larger cubical Gaussian surface. Compare the fluxes through these two surfaces.

2. Figure 24.22 shows a Gaussian surface and lines of electric field entering and leaving this surface. Assuming that the number of lines has been normalized according to the convention of Section 23.3, what can you say about the magnitude and the sign of the electric charge within the surface?

3. Suppose that instead of $1/\varepsilon_0$ lines of force per unit charge, we had adopted some other normalization, say, k/ε_0 lines per unit charge. Would this change Gauss' Law?

4. A Gaussian surface contains an electric dipole, and no other charge. What is the electric flux through this surface?

5. Defining the gravitational field as the gravitational force per unit mass, formulate a Gauss' Law for gravity. Check that your law implies Newton's Law of universal gravitation.

6. Suppose that the electric field of a point charge were not exactly proportional to $1/r^2$, but rather proportional to $1/r^{2+a}$ where a is a small number, $a \ll 1$. Would Gauss' Law still be valid? (Hint: Consider Gauss' Law for the case of a point charge.)

7. Prove that if the electric field is uniform in some region, then the charge density must be zero in that region.

8. Problems soluble by Gauss' Law fall into three categories, according to their symmetry: spherical, cylindrical, and planar. Give some examples in each category.

9. A hemisphere of radius R has a charge Q uniformly distributed over its volume. Can we use Gauss' Law to find the electric field?

10. A spherical rubber balloon of radius R has a charge Q uniformly distributed over its surface. The balloon is placed in a uniform electric field of 120 N/C. What is the net electric field inside the rubber balloon?

11. The electric field at the surface of a conductor in static equilibrium is normal to the surface. Is the gravitational field at the surface of a mass in static equilibrium necessarily normal to the surface? What if the surface is that of a fluid, such as water?

12. Suppose we drop a charged ping-pong ball into a cookie tin and quickly close the lid. What happens to the portions of the electric field lines that are outside of the cookie tin when we close the lid?

13. When an electric current is flowing through a wire connected between a source and a sink of electric charge — such as the poles of a battery — there is an electric field inside the wire, even though the wire is a conductor. Why does our conclusion about zero electric field inside a conductor not apply to this case?

14. The free electrons belonging to a metal are uniformly distributed over the entire volume of the metal. Does this contradict the result we derived in Section 24.4, according to which the charges are supposed to reside on the surface of a conductor?

PROBLEMS

Sections 24.1 and 24.2

1. Consider the thundercloud described in Problem 23.14 and Figure 23.28. What is the total electric flux coming out of the surface of the cloud?

2. A point charge of 1.0×10^{-8} C is placed inside an uncharged metallic can (say, a closed beer can) insulated from the ground. How many flux lines will emerge from the surface of the can when the point charge is inside?

3. On a clear day, the Earth's atmospheric electric field near the ground has a magnitude of 100 N/C and points vertically down. Inside the ground, the electric field is zero, since the ground is a conductor. Consider a mathematical box of 1 m × 1 m × 1 m, half below the ground, and half above. What is the electric flux through the sides of this box? What is the charge enclosed by the box?

4. Suppose we suspend a small ball carrying a charge of 1.0×10^{-6} C in the middle of a safe and lock the door. The safe is made of solid steel; it has inside dimensions 0.3 m × 0.3 m × 0.3 m and outside dimensions 0.4 m × 0.4 m × 0.4 m. What is the electric flux through a cubical surface measuring 0.2 m × 0.2 m × 0.2 m centered on the ball? A cubical surface measuring 0.35 m × 0.35 m × 0.35 m? A cubical surface measuring 0.5 m × 0.5 m × 0.5 m?

5. Consider a mathematical surface having the shape of a cube of edge 5 cm. You do not know the electric charge or the electric field inside the cube, but you do know the electric field at the surface: at the top of the cube the electric field has a magnitude of 5×10^5 N/C and points perpendicularly out of the cube; at the bottom of the cube, the electric field has a magnitude of 2×10^5 N/C and points perpendicularly into the cube; on all other faces, the electric field is tangential to the surface of the cube.
   (a) How much electric charge is inside the cube?
   (b) Can you guess what charge distribution inside (and outside) the cube would generate this kind of electric field?

6. An electric field has the following form as a function of x, y, z:

$$E_x = 5.0x \qquad E_y = 0 \qquad E_z = 0$$

where E is in newtons per coulomb and x in meters. This represents an electric field in the x direction with a magnitude that increases in direct proportion to x. Show that such an electric field can only exist if space is filled with some electric charge density. Find the value of the required charge density as a function of x, y, z.

7. A point charge of 6.0×10^{-8} C sits at a distance of 0.30 m above the x–y plane. What is the electric flux that this charge generates through the (infinite) x–y plane?

Section 24.3

8. Charge is placed on a small metallic sphere which is surrounded by air. If the radius of the sphere is 0.5 cm, how much charge can be placed on the sphere before the air near the sphere suffers electric breakdown? The critical electric field strength that leads to breakdown in air is 3×10^6 N/C.

9. A uranium nucleus is a spherical ball of radius 7.4×10^{-15} m with a charge of $92e$ uniformly distributed over its volume. Plot the electric field produced by this charge distribution as a function of radius for $0 < r < 15 \times 10^{-15}$ m.

10. In symmetric fission, a uranium nucleus splits into two equal pieces each of which is a palladium nucleus. The palladium nucleus is spherical with a radius of 5.9×10^{-15} m and a charge of $46e$ uniformly distributed over its volume. Suppose that immediately after fission, the two palladium nuclei are barely touching (Figure 24.23). What is the value of the total electric field at the center of each? What is the repulsive force between them? What is the acceleration of each? The mass of a palladium nucleus is 1.99×10^{-25} kg.

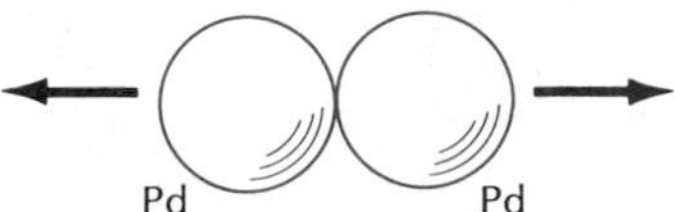

Fig. 24.23 Two palladium nuclei in contact. Each nucleus is a sphere.

11. Charge is distributed uniformly over the volume of a very long cylindrical plastic[2] rod of radius R. The amount of charge per meter of length of the rod is λ. Find a formula for the electric field at a distance r from the axis of the rod. Assume $r < R$.

12. What is the maximum amount of electric charge per unit length that one can place on a long, straight, human hair of diameter 8×10^{-3} cm if the surrounding air is not to suffer electrical breakdown? The air will suffer breakdown if the electric field exceeds 3×10^6 N/C.

13. A long plastic[2] pipe has inner radius a and outer radius b. Electric charge is uniformly distributed over the region $a < r < b$. The amount of charge is λ coulomb per meter of length of the pipe. Find the electric field in the regions $r < a$, $a < r < b$, and $r > b$.

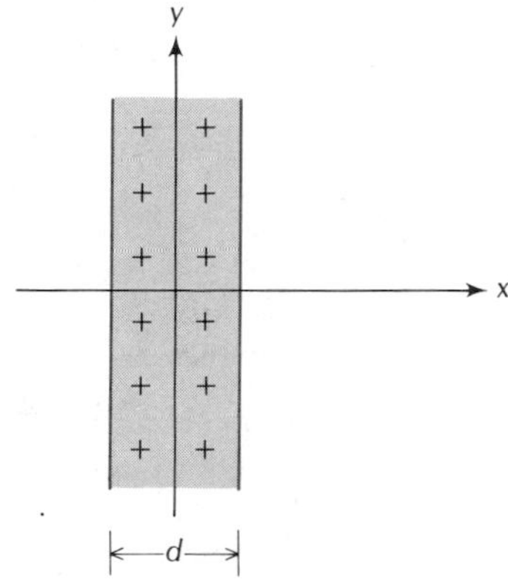

Fig. 24.24 A slab of plastic with electric charge uniformly distributed over its volume.

14. Charge is uniformly distributed over the volume of a large, plane slab of plastic[2] of thickness d. The charge density is ρ coulomb per cubic meter. The midplane of the slab is the y–z plane (Figure 24.24). What is the electric field at a distance x from the midplane? Consider both the cases $|x| < d/2$ and $|x| > d/2$.

15. The tube of a Geiger counter consists of a thin conducting wire of radius 1.3×10^{-3} cm stretched along the axis of a conducting cylindrical shell of radius 1.3 cm (Figure 24.25). The wire and the cylinder have equal and opposite charges of 7.2×10^{-10} C distributed along their length of 9.0 cm. Find a formula for the electric field in the space between the wire and the cylinder; pretend that the electric field is that of an infinitely long wire and cylinder. What is the magnitude of the electric field at the surface of the wire?

Fig. 24.25

16. A thick spherical shell of inner radius a, outer radius b has a charge Q uniformly distributed over its volume. Find the electric field in the regions $r < a$, $a < r < b$, and $r > b$.

17. According to a (crude) model, the neutron consists of an inner core of positive charge surrounded by an outer shell of negative charge. Suppose that the positive charge has a magnitude $+e$ and is uniformly distributed over a sphere of radius 0.50×10^{-15} m; suppose that the negative charge has a magnitude $-e$ and is uniformly distributed over a concentric shell of inner radius 0.50×10^{-15} m, outer radius 1.0×10^{-15} m (Figure 24.26). Find the magnitude and direction of the electric field at 1.0×10^{-15}, 0.75×10^{-15}, 0.50×10^{-15}, and 0.25×10^{-15} m from the center.

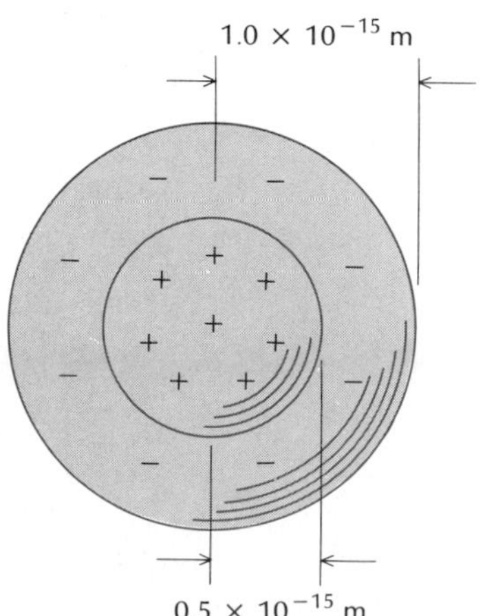

Fig. 24.26 Charge distribution inside a neutron.

[2] Assume that the plastic has no effect on the electric field.

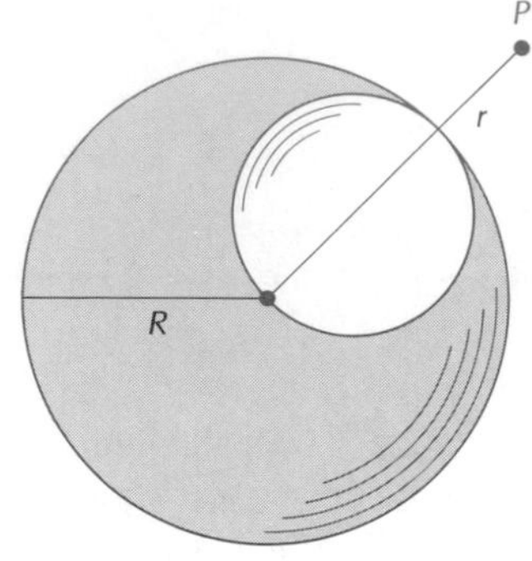

Fig. 24.27 Sphere with spherical cavity.

18. Positive charge Q is uniformly distributed over the volume of a solid sphere of radius R. Suppose that a spherical cavity of radius $R/2$ is cut out of the solid sphere, the center of the cavity being at a distance of $R/2$ from the center of the original solid sphere (Figure 24.27); the cut-out material and its charge are discarded. What new electric field does the sphere with the cavity produce at the point P at a distance r from the original center? Assume $r > R$.

19. According to an old (and erroneous) model due to J. J. Thomson, an atom consists of a cloud of positive charge within which electrons sit like plums in a pudding. The electrons are supposed to emit light when they vibrate about their equilibrium positions in this cloud. Assume that in the case of the hydrogen atom the positive cloud is a sphere of radius $R = 0.5$ Å with a charge of e uniformly distributed over the volume of this sphere. The (pointlike) electron is held at the center of this charge distribution by the electrostatic attraction.

(a) Show that the restoring force on the electron is $e^2r/(4\pi\varepsilon_0R^3)$ when the electron is at a distance r from the center ($r \leqslant R$).

(b) What is the frequency of small oscillations of the electron moving back and forth along a diameter? Give a *numerical* answer.

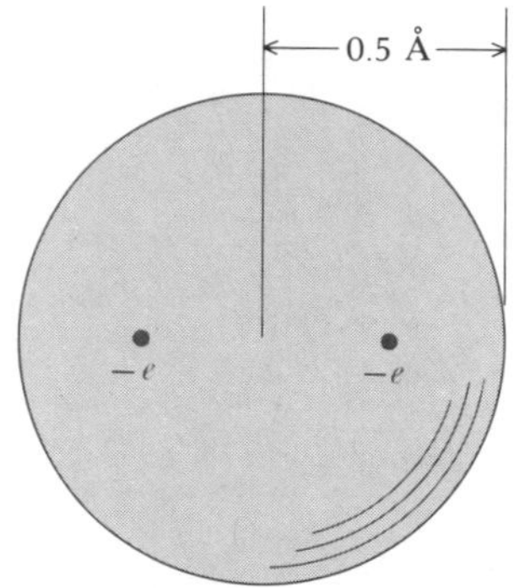

Fig. 24.28 Model of the helium atom.

20. According to the Thomson model (see also Problem 19), the atom of helium consists of a uniform spherical cloud of positive charge within which sit two electrons. Assume that the positive cloud is a sphere of radius 0.5 Å with a charge $2e$ uniformly distributed over the volume. The two electrons are symmetrically placed with respect to the center (Figure 24.28). What is the equilibrium separation of the electrons?

21. A charge distribution with spherical symmetry has a charge density ρ coulomb per cubic meter described by the formula $\rho = kr^n$, where k is a constant and $n > -3$.

(a) What is the amount of charge $Q(r)$ inside a sphere of radius r?

(b) What is the magnitude of the electric field as a function of r?

(c) For what value of n is the magnitude of the electric field constant?

(d) Why is it necessary to assume that $n > -3$?

22. The tau particle is a negatively charged particle similar to the electron, but of much larger mass — its mass is 3.18×10^{-27} kg, about 3490 times the mass of an electron. Nuclear material is transparent to the tau; thus, the tau can orbit around inside a nucleus, under the influence of the electric attraction of the nuclear charge. Suppose that a tau is in a circular orbit of radius 2.9×10^{-15} m inside a uranium nucleus. Treat the nucleus as a sphere of radius 7.4×10^{-15} m with a charge $92e$ uniformly distributed over its volume. Find the speed, the kinetic energy, the angular momentum, and the frequency of the orbital motion of the tau.

23. A very long cylinder of radius R has positive charge uniformly distributed over its volume. The amount of charge is λ coulomb per meter of length of the cylinder. A spherical cavity of radius $R' \leqslant R$, centered on the axis of the cylinder, has been cut out of this cylinder, and the charge in this cavity has been discarded.

(a) Find the electric field as a function of distance from the center of the sphere along the axis of the cylinder.

(b) Find the electric field as a function of distance along a line passing through the center of the sphere perpendicular to the axis of the cylinder. Consider both distances smaller and larger than R'.

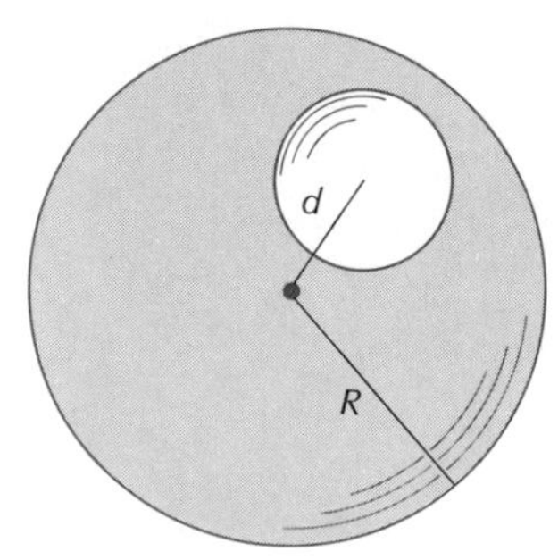

Fig. 24.29 Sphere with spherical cavity.

*24. A charge Q is uniformly distributed over the volume of a solid sphere of radius R. A spherical cavity is cut out of this solid sphere (Figure 24.29) and the material and its charge are discarded. Show that the electric field in the cavity will then be uniform, of magnitude $(1/4\pi\varepsilon_0)Qd/R^3$, where d is the distance between the centers of the spheres (Figure 24.29). Make a drawing of the lines of electric field in the cavity.

*25. When a point charge q moving at high speed passes by another stationary point charge q', the main effect of the electric forces is to give each charge a transverse impulse. Figure 24.30 shows the charge q moving at (almost) constant velocity v along the x axis in an almost straight line and shows the charge q' sitting at a distance R below the origin. The transverse impulse on q is

$$\int_{-\infty}^{\infty} F_y\, dt = \frac{q}{v} \int_{-\infty}^{\infty} E_y\, dx$$

Evaluate the integral $\int E_y\, dx$ by means of Gauss' Law and prove that

$$\int_{-\infty}^{\infty} F_y\, dt = \frac{q}{v} \frac{q'}{2\pi\varepsilon_0 R}$$

(Hint: Consider $2\pi R \int E_y\, dx$; show that this is the flux that q' produces through the infinite cylindrical surface indicated in Figure 24.30.)

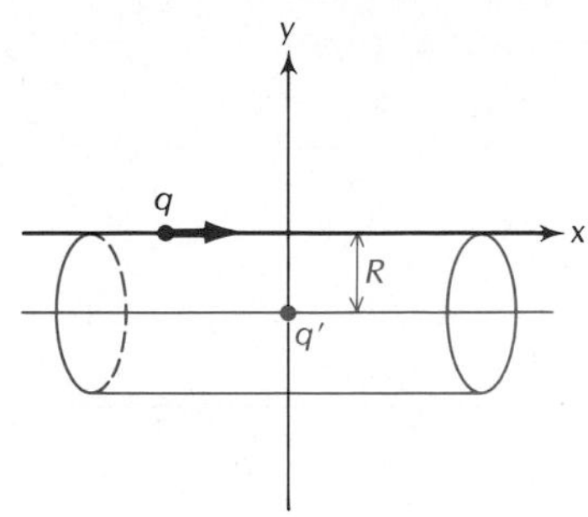

Fig. 24.30 The charge q moves along the x axis. The charge q' sits at a distance R below the x axis.

26. The formula derived in Problem 25 gives the transverse momentum that a high-speed charged particle acquires as it passes by a stationary charged particle.

(a) Calculate the transverse momentum that an electron of speed 4.0×10^7 m/s acquires as it passes by a stationary electron at a distance of 0.60×10^{-10} m.

(b) What transverse velocity corresponds to this transverse momentum?

(c) What will be the recoil velocity of the stationary electron (if it is free to move)?

Section 24.4

27. The surface of a long, cylindrical copper pipe has a charge of λ coulomb per meter (Figure 24.31). What is the electric field outside the pipe? Inside the pipe?

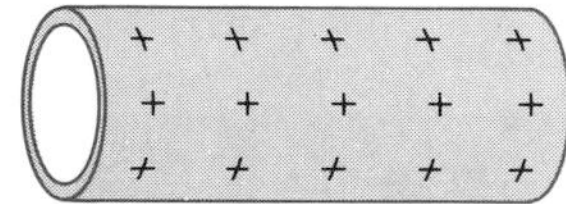

Fig. 24.31 A long pipe with positive charge on its surface.

28. A solid copper sphere of radius 3 cm carries a charge of 10^{-6} C. This sphere is placed concentrically within a spherical, thin copper shell of radius 15 cm carrying a charge of 3×10^{-6} C. Find a formula for the electric field in the space between the sphere and the shell. Find a formula for the electric field outside the shell. Plot these electric fields as a function of radius.

29. A thick spherical shell made of metal has an inner radius a, an outer radius b, and is initially uncharged. A point charge of 1.0×10^{-7} C is placed at the center of the shell. Find the electric field in the regions $r < a$, $a < r < b$, and $r > b$. Find the induced surface charge densities at $r = a$ and $r = b$.

30. On days of fair weather, the atmospheric electric field of the Earth is about 100 N/C; this field points vertically downward (compare Problem 23.2). What is the surface charge density on the ground? Treat the ground as a flat conductor.

31. You wish to generate a uniform electric field of 2.0×10^5 N/C in the space between two flat, parallel plates of metal placed face to face. The plates measure 0.30 cm $\times$ 0.30 cm. How much electric charge must you put on each plate? Assume that the gap between the plates is small so that the charge distribution and the electric field are approximately uniform, as for infinite plates.

CHAPTER 25

The Electrostatic Potential

In Chapter 8 we saw that to formulate a law of conservation of energy for a particle moving under the influence of a conservative force, we had to construct a potential energy. In this chapter we will construct the electrostatic potential energy for a charged particle moving under the influence of the electric force generated by a static charge distribution. This potential energy will help us in the calculation of the motion of the particle. Furthermore, this potential energy will give us yet another method for the calculation of the electric field of a given charge distribution. The method involves two steps: first, find the potential energy of a unit point charge placed somewhere near the charge distribution; second, find the electric field by differentiating this potential energy. Thus, we will have available three alternative methods for the calculation of the electric field: via Coulomb's Law, via Gauss' Law, and via the potential energy of a unit charge. If a problem does not yield to the simple and elegant method based on Gauss' Law, it is usually best to to use the method based on the potential energy because the mathematics are likely to be less cumbersome than with the method based on Coulomb's Law.

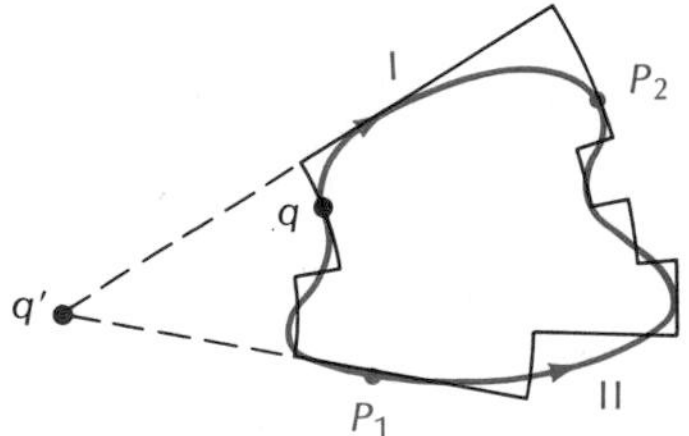

Fig. 25.1 A charge q moves from the point P_1 to the point P_2 while under the influence of the electric field of a charge q'. The path can be approximated by radial segments and circular arcs.

25.1 The Electrostatic Potential

According to the definition given in Chapter 8, a force is conservative if the work done by this force on a particle depends on the initial and final positions of the particle, but not on the shape of the path connecting these positions. The electric force exerted by a fixed point charge q' on a moving point charge q is a conservative force. The proof is the same as in the case of the gravitational force. Figure 25.1 shows a fixed point charge q' and a second point charge q that moves from P_1 to P_2; the figure also shows two alternative paths I and II from P_1 to P_2. The work done by the electric force along a path is $\int \mathbf{F} \cdot d\mathbf{l}$.[1] To evaluate this integral, we approximate each path by a sequence of infinitesimal radial segments and circular arcs (Figure 25.1). Along the circular arcs the work is zero because the force is perpendicular to the displacement. Along the radial segments the work is not zero. Since for each radial segment belonging to the first path, there is a corresponding radial segment (at the same radius) belonging to the second path, and since the magnitude of the force is the same at both these radial segments, the contributions to the work along the first and second paths are equal. This establishes that the work is independent of the shape of the path and that the force is conservative.

More generally, the net electric force exerted by any number of fixed point charges on another given point charge is conservative, since the net force is a sum of individual conservative forces.

For any conservative force we can construct a corresponding potential energy by means of the recipe given in Section 8.2: Take a reference point P_0 and at this point assign to the potential energy the value $U(P_0)$. At any other point, assign to the potential energy the value [see Eq. (8.4)]

$$U(P) = -\int_{P_0}^{P} \mathbf{F} \cdot d\mathbf{l} + U(P_0) \tag{1}$$

The **electrostatic potential** is defined as the potential energy of a charge q divided by q, i.e., it is defined as the potential energy per unit charge,

$$V(P) = \frac{U(P)}{q} \tag{2}$$

$$= -\int_{P_0}^{P} \frac{\mathbf{F}}{q} \cdot d\mathbf{l} + \frac{U(P_0)}{q} \tag{3}$$

Since the force per unit charge is the electric field ($\mathbf{E} = \mathbf{F}/q$), we can also write this as

$$\boxed{V(P) = -\int_{P_0}^{P} \mathbf{E} \cdot d\mathbf{l} + V(P_0)} \tag{4}$$

Electrostatic potential

In this equation $V(P_0)$ plays the role of an additive constant in the potential. In the calculations of potential differences between points (see below), the additive constant in the potential cancels. Hence, this constant is of no physical significance. We can make any convenient choice for this constant, but we must remain faithful to our choice throughout any subsequent calculation. We will see some examples of convenient choices of $V(P_0)$ in what follows.

---

[1] Note that we are now using the symbol $d\mathbf{l}$ for an infinitesimal displacement while in Chapters 7 and 8 we used the symbol $d\mathbf{r}$; we now want to reserve the letter $\mathbf{r}$ for the radial distance in Coulomb's Law.

Potential difference

The **potential difference** between two arbitrary points P_1 and P_2 is [compare Eq. (8.5)]:

$$V(P_2) - V(P_1) = -\int_{P_0}^{P_1} \mathbf{E} \cdot d\mathbf{l} + V(P_0) + \int_{P_0}^{P_2} \mathbf{E} \cdot d\mathbf{l} - V(P_0)$$

or

$$V(P_2) - V(P_1) = -\int_{P_1}^{P_2} \mathbf{E} \cdot d\mathbf{l} \tag{5}$$

Hence the potential difference is the work that *you* must do (against the electric force) in order to push one coulomb of charge from the point P_1 to the P_2.

Alessandro, Conte Volta, *1745–1827, Italian physicist, professor at Pavia. Volta established that the "animal electricity," observed by Liugi Galvani, 1737–1798, in experiments with frog muscle tissue placed in contact with dissimilar metals, was not due to any exceptional property of animal tissues, but was also generated whenever any wet body was sandwiched between dissimilar metals. This led him to develop the first "voltaic pile," or battery, consisting of a large stack of moist discs of cardboard (electrolyte) sandwiched between discs of metal (electrodes).*

The unit of electrostatic potential is the **volt** (V),[2]

$$1 \text{ volt} = 1 \text{ V} = 1 \text{ joule/coulomb} = 1 \text{ J/C} \tag{6}$$

The unit of electric field is 1 N/C. This can be expressed in terms of volts as follows:

$$1\,\frac{\text{N}}{\text{C}} = 1\,\frac{\text{N}\cdot\text{m}}{\text{C}\cdot\text{m}} = 1\,\frac{\text{J}}{\text{C}}\,\frac{1}{\text{m}} = 1\,\frac{\text{V}}{\text{m}} \tag{7}$$

Thus N/C and V/m are equal units; in practice, volt per meter is the preferred unit for the electric field.

EXAMPLE 1. A very large, flat conducting surface carries a uniform surface charge density which generates a constant electric field **E** (see Example 24.7). What is the electrostatic potential at some distance from this surface? Assume that the potential is zero at the surface.

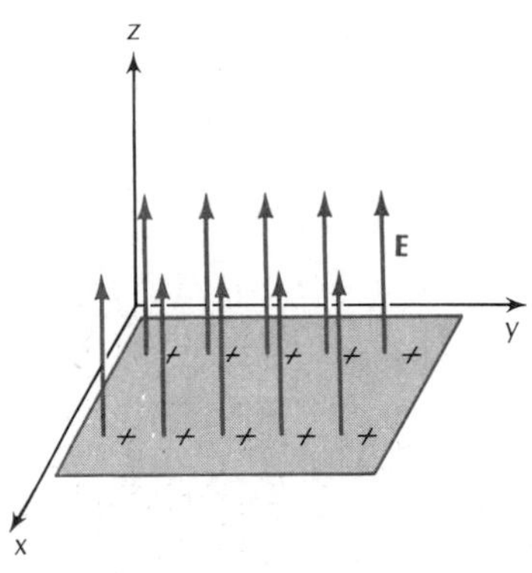

Fig. 25.2 Electric field of a very large charged surface lying in the x–y plane.

SOLUTION: In Figure 25.2, the charged surface coincides with the x–y plane. According to Example 24.7, the electric field generated by this surface is vertical and is independent of distance:

$$E_z = E_0 = [\text{constant}] \tag{8}$$

If P_0 is a point on the surface, then $V(P_0) = 0$ and

$$V(P) = -\int \mathbf{E} \cdot d\mathbf{l} = -\int E_x\, dx - \int E_y\, dy - \int E_z\, dz$$

$$= -\int_0^z E_0\, dz = -E_0 z \tag{9}$$

Hence the potential is directly proportional to the distance from the surface.

EXAMPLE 2. Near the ground directly below a thundercloud the electric field is 2×10^4 V/m and points upward. What is the potential difference be-

[2]Note that the same letter V is used in physics both as a symbol for potential and as an abbreviation for *volt.* This leads to confusing equations such as $V = 3.0$ V (which means $V = 3.0$ volt). If there is a possibility of confusion, it is best not to abbreviate *volt.*

tween the ground and a point in the air, 50 m above ground? Assume that the electric field is constant.

SOLUTION: From Eq. (9)

$$V(P) = -E_0 z = -2 \times 10^4 \text{ volt/m} \times 50 \text{ m} = -1 \times 10^6 \text{ volt}$$

Since the electric field in a conducting body in electrostatic equilibrium is zero, Eq. (5) tells us that the potential difference between any two points within a conducting body is zero. Thus, all points within a conducting body are at the same potential. For instance, since the Earth is a conductor, all points in the Earth or on the surface of the Earth are at the same electric potential. In experiments with electric circuits, it is usually convenient to adopt the convention that the potential of the Earth is zero, $V = 0$. The Earth is said to be the **electric ground,** and any conductors connected to it are said to be grounded.

Electric ground

25.2 The Electrostatic Potential of a Point Charge

To find the electrostatic potential of a fixed point charge q', we must substitute the radial electric field of such a point charge into the general formula for the potential,

$$V(P) = -\int_{P_0}^{P} \mathbf{E} \cdot d\mathbf{l} + V(P_0) \tag{10}$$

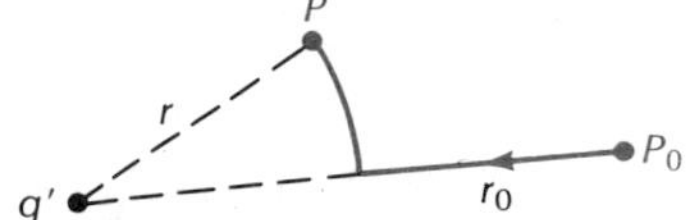

Fig. 25.3 A path from P_0 to P consisting of a radial segment and a circular arc.

Since the integral does not depend on the path between P_0 and P, we may take any path that is convenient. Figure 25.3 shows a path consisting of one radial segment and one circular arc. Along the circular arc, the integral receives no contribution. Along the radial segment, the electric field $\mathbf{E} = (1/4\pi\varepsilon_0)q'\hat{\mathbf{r}}/r^2$ is in the same direction as the displacement, since $d\mathbf{l} = \hat{\mathbf{r}}\,dr$; hence $\mathbf{E} \cdot d\mathbf{l} = (1/4\pi\varepsilon_0)(q'/r^2)\,dr$ and[3]

$$V(P) = -\int_{r_0}^{r} \frac{1}{4\pi\varepsilon_0}\frac{q'}{r'^2}\,dr' + V(P_0) \tag{11}$$

$$= \frac{q'}{4\pi\varepsilon_0}\left(\frac{1}{r} - \frac{1}{r_0}\right) + V(P_0) \tag{12}$$

For the reference point P_0 we choose a point at infinite distance from the fixed charge; for the value of the potential at this point we take $V(P_0) = 0$. The potential at the point P can then be written

$$\boxed{V(r) = \frac{1}{4\pi\varepsilon_0}\frac{q'}{r}} \tag{13}$$

Potential of point charge

Thus, the potential in the space surrounding a point charge is inversely proportional to the distance.

[3] The variable of integration in this integral has been written r' to distinguish it from the limit of integration r.

The potential energy of a charge q under the influence of the electric field of the charge q' is then

Potential energy of two point charges

$$U(r) = qV(r) = \frac{1}{4\pi\varepsilon_0}\frac{qq'}{r} \tag{14}$$

EXAMPLE 3. The electron in a hydrogen atom is at a distance of 0.53×10^{-10} m from the nucleus. What is the electrostatic potential generated by the nucleus at this position? What is the potential energy of the electron?

SOLUTION: The nucleus is (approximately) a point charge with $q' = e = 1.6 \times 10^{-19}$ C. The electrostatic potential generated by the nucleus is

$$V = \frac{1}{4\pi\varepsilon_0}\frac{q'}{r} = \frac{1}{4\pi\varepsilon_0} \times \frac{1.6 \times 10^{-19}\text{ C}}{0.53 \times 10^{-10}\text{ m}}$$

$$= 27 \text{ volt} \tag{15}$$

The charge of the electron is $q = -e = -1.6 \times 10^{-19}$ C. The potential energy of the electron is

$$U = qV = -e \times 27 \text{ volt} \tag{16}$$

$$= 1.6 \times 10^{-19}\text{ C} \times 27 \text{ volt} = -4.3 \times 10^{-18}\text{ J} \tag{17}$$

For the purposes of atomic physics, the joule is a rather large unit of energy and it is more convenient to leave the answer as in Eq. (16),

$$U = -27e \cdot \text{volt}$$

The product of the fundamental unit of atomic charge and the unit of potential, $e \cdot$ volt or eV, is a unit of energy. This unit of energy is called an **electron-volt.** It can be converted to joule by substituting the numerical value for e,

electron-volt, eV

$$1 \text{ eV} = 1 \times 1.60 \times 10^{-19}\text{ C} \times 1\text{V} = 1.60 \times 10^{-19}\text{ J} \tag{18}$$

In chemical reactions between atoms or molecules, the energy released or absorbed by each atom or molecule is typically 1 or 2 eV. Such reactions involve a change in the arrangement of the exterior electrons of the atoms, and the energy of 1 or 2 eV represents the typical amount of energy needed for this rearrangement.

EXAMPLE 4. An electron is initially at rest at a very large distance from a proton. Under the influence of the electric attraction, the electron falls toward the proton, which remains (approximately) at rest. What is the speed of the electron when it has fallen to within 0.53×10^{-10} m of the proton?

SOLUTION: The potential energy of the electron is $qV(r) = -eV(r)$. The total energy is the sum of the potential and kinetic energies,

$$E = \tfrac{1}{2}m_e v^2 - eV(r) \tag{19}$$

This total energy is conserved. The initial value of the energy is zero; hence the final value of the energy must also be zero:

$$\tfrac{1}{2}m_e v^2 - eV(r) = 0$$

and

$$v = \sqrt{\frac{2eV(r)}{m_e}} \tag{20}$$

According to Example 3, $V(r) = 27$ volt for $r = 0.53 \times 10^{-10}$ m, so that

$$v = \sqrt{\frac{2 \times 1.6 \times 10^{-19}\ \text{C} \times 27\ \text{V}}{9.1 \times 10^{-31}\ \text{kg}}} = 3.1 \times 10^{6}\ \text{m/s}$$

EXAMPLE 5. A sphere of radius R carries a total charge Q uniformly distributed over its volume. Find the electrostatic potential inside and outside the sphere.

SOLUTION: Outside the sphere ($r \geq R$), the electrostatic potential is the same as for a point charge,

$$V(r) = \frac{1}{4\pi\varepsilon_0}\frac{Q}{r} \qquad r \geq R \tag{21}$$

Inside the sphere ($r \leq R$), the electric field is [see Eq. (24.17)]

$$E(r) = \frac{1}{4\pi\varepsilon_0}\frac{Qr}{R^3} \tag{22}$$

The potential difference between R and r can be calculated from this field [see Eq. (5)]:

$$\begin{aligned} V(r) - V(R) &= -\int_R^r \frac{1}{4\pi\varepsilon_0}\frac{Qr'}{R^3}\,dr' \\ &= -\frac{Q}{4\pi\varepsilon_0}\left(\frac{r^2}{2R^3} - \frac{1}{2R}\right) \end{aligned} \tag{23}$$

Consequently,

$$V(r) = -\frac{Q}{4\pi\varepsilon_0}\left(\frac{r^2}{2R^3} - \frac{1}{2R}\right) + V(R)$$

Since $V(R) = (1/4\pi\varepsilon_0)\,(Q/R)$, this gives

$$V(r) = -\frac{1}{4\pi\varepsilon_0}\frac{Qr^2}{2R^3} + \frac{1}{4\pi\varepsilon_0}\frac{3Q}{2R} \qquad r \leq R \tag{24}$$

Figure 25.4 is a plot of the potential as a function of radius. The potential has a maximum at the center. The value of this maximum is

$$V(0) = \frac{1}{4\pi\varepsilon_0}\frac{3Q}{2R} \tag{25}$$

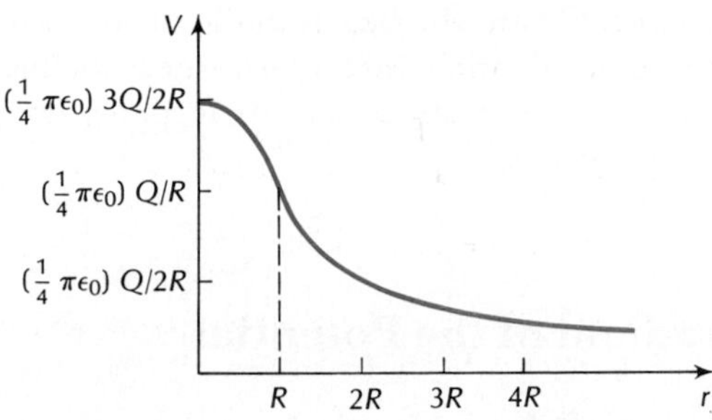

Fig. 25.4 Electrostatic potential of a uniformly charged sphere.

25.3 The Electrostatic Field as a Conservative Field

Conservative field

The electric field produced by a static charge distribution is said to be a **conservative field** because the integral $\int \mathbf{E} \cdot d\mathbf{l}$ between any two points P_1 and P_2 is independent of the path between these points. As we have seen in Section 8.1, this statement about the path independence of the integral is equivalent to the statement that the integral vanishes for any closed path, i.e.,

$$\oint \mathbf{E} \cdot d\mathbf{l} = 0 \tag{26}$$

where the circle on the integral sign indicates that the integration is around a closed path. Note that this equation implies that a field line can never form a closed loop — if it did, then the integral $\oint \mathbf{E} \cdot d\mathbf{l}$ around such a closed loop would not be zero. Consequently, any field line must start or end somewhere (on an electric charge). Since we derived the conservative property of the electrostatic field by examining the electric field produced by a point charge, it is not surprising that Eq. (26) should imply the existence of starting points and ending points for field lines. These sources and sinks of field lines are, of course, the positive and negative electric charges, and Gauss' Law tells us just how many field lines emerge from each electric charge. It can be shown that Eq. (26) together with Gauss' Law is exactly equivalent to Coulomb's Law, i.e., any electric field that satisfies both Eq. (26) and Gauss' Law is necessarily the electric field of a static distribution of point charges.

With Eq. (26) we can easily prove an interesting theorem about static electric fields: *within a closed, empty cavity inside a homogeneous conductor, the electric field is exactly zero.* Figure 25.5 shows such a cavity. If there were an electric field inside this cavity, then there would have to be field lines. Consider one of these field lines. Since the cavity is empty (contains no charge), the field line cannot end or begin within the cavity — it must therefore begin and end on the surface of the cavity (Figure 25.5). The field line cannot penetrate the conducting material, since the electric field is zero in this material. Now take a closed path consisting of one portion that follows the field line and a second portion that lies entirely in the conducting material. Along the first portion the integral $\int \mathbf{E} \cdot d\mathbf{l}$ is positive and along the second portion it is zero. Hence $\oint \mathbf{E} \cdot d\mathbf{l} \neq 0$ for this closed path. This is in contradiction to Eq. (26) and establishes the impossibility of such an electric field.

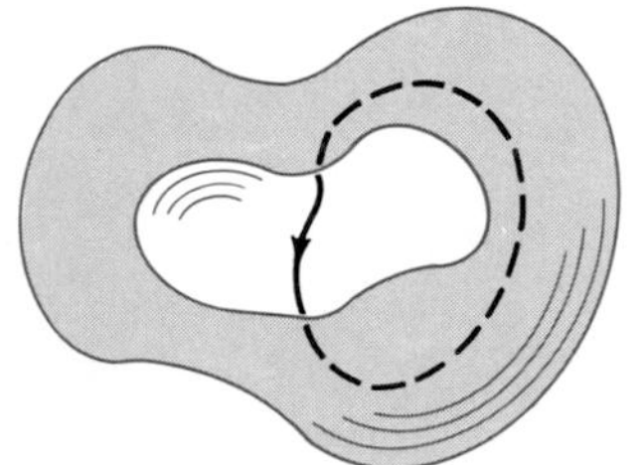

Fig. 25.5 Empty cavity in a volume of conducting material.

The absence of electrostatic fields in closed conducting cavities has important practical applications. Delicate electric instruments can be shielded from atmospheric electric fields, and other stray electric fields, by placing them in a box made of sheet metal. Such a box is called a **Faraday cage.** Often, the box is made of fine wire mesh rather than sheet metal; although such wire mesh does not have the perfect shielding properties of solid sheet metal, it provides good enough shielding for most purposes.

Faraday cage

25.4 The Gradient of the Potential

The electrostatic potential can be calculated from the electric field by integration,

$$V(P) = -\int_{P_0}^{P} \mathbf{E} \cdot d\mathbf{l} + V(P_0) \qquad (27)$$

Conversely, the electric field can be calculated from the potential by differentiation. Consider two nearby points separated by an infinitesimal displacement $d\mathbf{l}$. The change in potential between these points is then

$$dV = -\mathbf{E} \cdot d\mathbf{l} = -E \cos\theta \, dl \qquad (28)$$

where θ is the angle between the displacement $d\mathbf{l}$ and the electric field. Hence

$$\frac{dV}{dl} = -E \cos\theta \qquad (29)$$

This shows that the derivative dV/dl equals the negative of the component of E in the direction of dl.

If the displacement dl is in the x direction, $dl = dx$ and Eq. (29) becomes

Derivatives of the potential

$$\boxed{\frac{\partial V}{\partial x} = -E_x} \qquad (30)$$

Likewise

$$\boxed{\frac{\partial V}{\partial y} = -E_y} \qquad (31)$$

and

$$\boxed{\frac{\partial V}{\partial z} = -E_z} \qquad (32)$$

Thus, if the potential is a known function of position, the components of the electric field can be calculated by taking derivatives (compare Section 8.3).

In vector notation, we can express Eqs. (30)–(32) as

$$\frac{\partial V}{\partial x}\hat{\mathbf{x}} + \frac{\partial V}{\partial y}\hat{\mathbf{y}} + \frac{\partial V}{\partial z}\hat{\mathbf{z}} = -\mathbf{E} \qquad (33)$$

The quantity on the left side of this equation is called the **gradient** of the potential. The gradient is a vector with components $\partial V/\partial x$, $\partial V/\partial y$, and $\partial V/\partial z$. Thus, Eq. (33) asserts that the electric field is the negative of the gradient of the potential.

EXAMPLE 6. The electric potential generated by a uniformly charged flat conducting plate is

$$V = -E_0 z \tag{34}$$

where E_0 is a constant [see Eq. (9)]. Calculate the electric field from this potential.

SOLUTION: Equations (30)–(32) immediately give

$$E_x = -\frac{\partial V}{\partial x} = 0 \tag{35}$$

$$E_y = -\frac{\partial V}{\partial y} = 0 \tag{36}$$

$$E_z = -\frac{\partial V}{\partial z} = +E_0 \tag{37}$$

EXAMPLE 7. The electrostatic potential due to a point charge q' is

$$V = \frac{1}{4\pi\varepsilon_0}\frac{q'}{r} \tag{38}$$

Calculate the electric field from this potential.

SOLUTION: In order to use Eqs. (30)–(32) we would first have to express r in terms of x, y, z (i.e., $r = \sqrt{x^2 + y^2 + z^2}$). It is simpler to return to Eq. (29), which, with $dl = dr$, directly gives the component of **E** in the *radial direction,*

$$E_r = -\frac{dV}{dr} = -\frac{d}{dr}\left(\frac{1}{4\pi\varepsilon_0}\frac{q'}{r}\right) = \frac{1}{4\pi\varepsilon_0}\frac{q'}{r^2} \tag{39}$$

According to Eq. (29), the electric field only has components in those directions in which the potential changes. Since the potential (38) does not change in the tangential direction, the radial electric field given by Eq. (39) is the total field.

The results obtained in the two preceding examples do not tell us anything new; these calculations merely verify the consistency of our formulas.

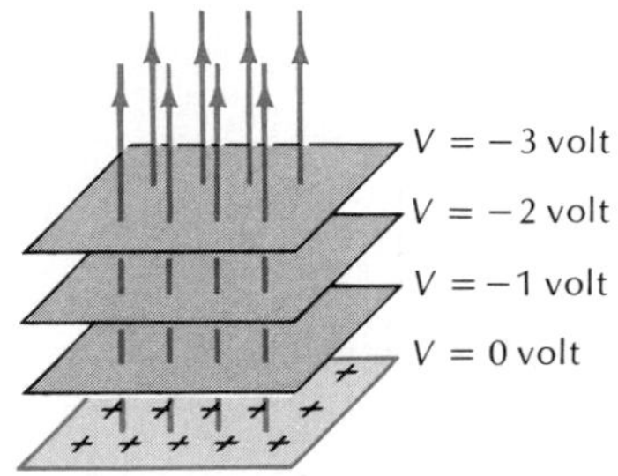

Fig. 25.6 Equipotential surfaces for a very large sheet with a uniform charge distribution.

A set of points at which the potential function has a fixed, constant value is called an **equipotential surface.** Figure 25.6 shows the equipotential surfaces belonging to the potential of a uniformly charged flat sheet — the equipotential surfaces are parallel planes. Figure 25.7

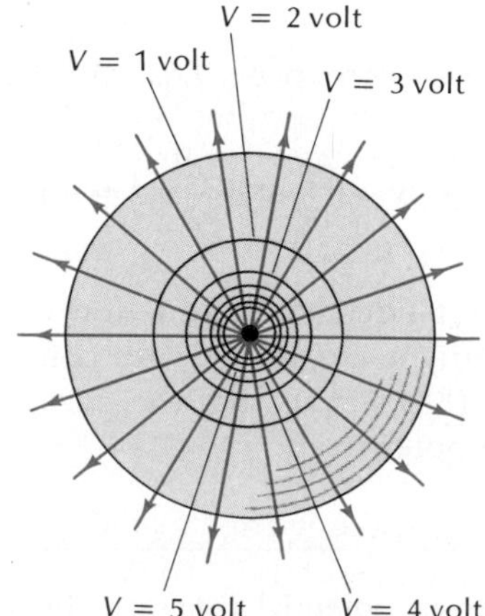

Fig. 25.7 Equipotential surfaces for a positive point charge. The equipotential surfaces are concentric spheres.

shows the equipotential surfaces belonging to the potential of a point charge — the equipotential surfaces are concentric spheres.

Note that the electric field is everywhere perpendicular to the equipotentials. This is a direct consequence of Eq. (29): along any equipotential surface the potential is constant, i.e., $dV = 0$; consequently the component of E along the surface must be zero and **E** must be perpendicular to the surface.

Conversely, if the electric field is everywhere perpendicular to a given surface, then this surface must be an equipotential. Since we already know (see Section 24.4) that along the surface of any conductor in electrostatic equilibrium **E** is perpendicular to the surface, we conclude immediately that any conducting surface is an equipotential surface. (This conclusion agrees with the general statement made at the end of Section 25.1: the potential is constant throughout any conductor.)

The calculation of the electric field of a static charge distribution can often be considerably simplified by first finding the electrostatic potential and then deriving the electric field from this. Equation (13) permits us to find directly the potential of any given charge distribution; we need only regard the charge distribution as a collection of point charges and sum, or integrate, the contributions the individual charges make to the potential. We can then calculate the components of the electric field by taking partial derivatives according to Eqs. (30)–(32). The advantage of this procedure is that it is much easier to sum the potentials of a distribution of point charges than it is to sum their electric fields; in the latter sum it is necessary to take into account the directions of the electric fields and to take into account the three electric field components — a rather messy procedure. The following is an example of a calculation of the electric field via the potential.

EXAMPLE 8. A charge Q is uniformly distributed along the circumference of a ring of radius R. Find the electric field on the axis of the ring.

SOLUTION: We have already solved this problem in Chapter 23 (see Example 23.3). Now we will solve it again by beginning with the electrostatic potential. At a height z above the plane of the ring (Figure 25.8) the potential contributed by a small charge element dQ is simply

$$dV = \frac{1}{4\pi\varepsilon_0}\frac{dQ}{r} = \frac{1}{4\pi\varepsilon_0}\frac{dQ}{(z^2 + R^2)^{1/2}} \tag{40}$$

Since all charge elements around the ring are at the same distance from the point z, they all contribute equally and the total potential is simply

$$V = \frac{1}{4\pi\varepsilon_0}\frac{Q}{(z^2 + R^2)^{1/2}} \tag{41}$$

Consequently, the z component of the electric field is

$$E_z = -\frac{\partial V}{\partial z} = \frac{1}{4\pi\varepsilon_0}\frac{Qz}{(z^2 + R^2)^{3/2}} \tag{42}$$

and the x and y components are zero. This result agrees with Eq. (23.20), but we now obtained it a bit more quickly.

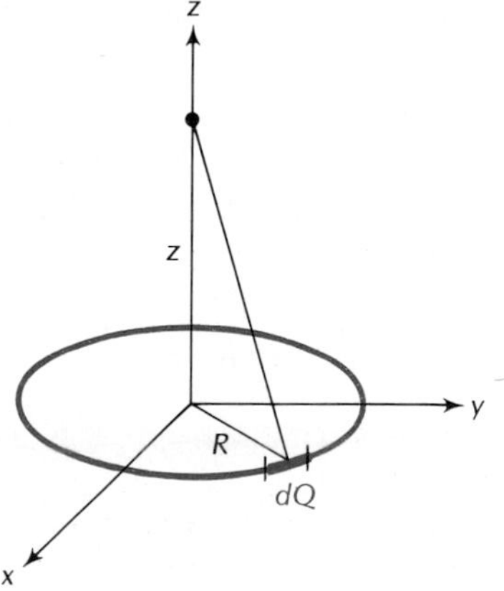

Fig. 25.8 A uniformly charged ring.

25.5 The Potential and Field of a Dipole

We will now calculate the electric field generated by an electric dipole. As in the above example, we will do this calculation via the potential.

Figure 25.9 shows an electric dipole consisting of charges $\pm Q$ on the z axis at $\pm l/2$. The net potential generated by this pair of charges is just the sum of the individual potentials,

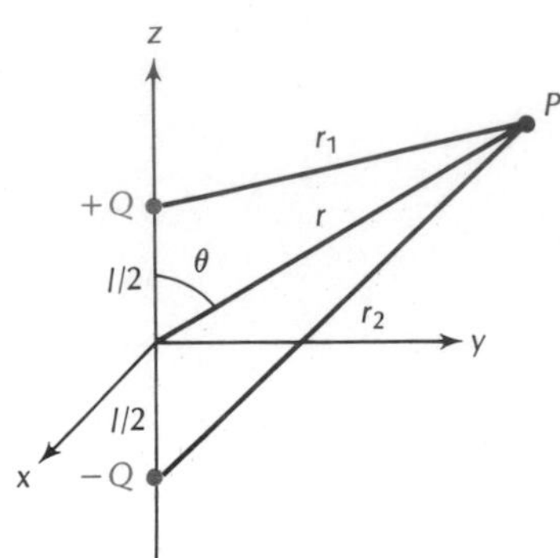

Fig. 25.9 Charges Q and $-Q$ on the z axis at $z = l/2$ and $z = -l/2$.

$$V = \frac{1}{4\pi\varepsilon_0}\frac{Q}{r_1} - \frac{1}{4\pi\varepsilon_0}\frac{Q}{r_2} \tag{43}$$

$$= \frac{Q}{4\pi\varepsilon_0}\frac{r_2 - r_1}{r_1 r_2} \tag{44}$$

Where r_1 and r_2 are the lengths of the lines QP and $-QP$. From this potential function we can calculate the electric field by taking derivatives. Before we do this, we will make the simplifying assumption that r_1 and r_2 are much larger than l, i.e., we will make the assumption that the field point P is at a large distance from the electric dipole. Figure 25.10 shows that under these conditions r_1 and r_2 are approximately equal,

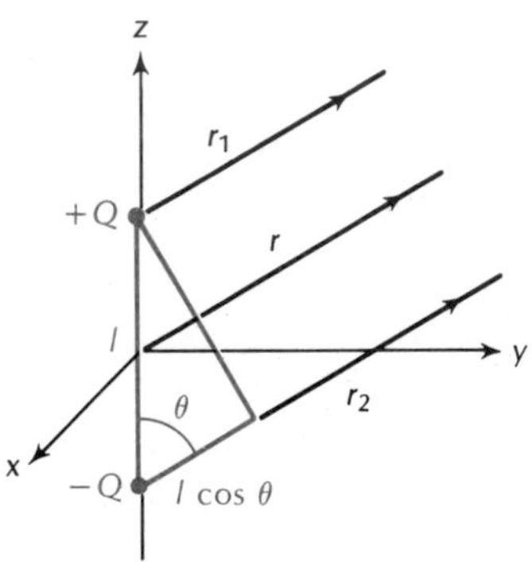

Fig. 25.10 If the point P is at a large distance (beyond the diagram), then the lines QP and $-QP$ are nearly parallel and the difference between their lengths r_1 and r_2 is $l \cos\theta$.

$$r_1 \cong r_2 \cong r \tag{45}$$

and their difference is a small quantity,

$$r_2 - r_1 \cong l\cos\theta \tag{46}$$

This permits us to write the following approximation for V:

$$V \cong \frac{Q}{4\pi\varepsilon_0}\frac{l\cos\theta}{r^2} \tag{47}$$

The product of Q and l is the dipole moment

$$p = lQ$$

and hence our approximation has the form

Potential of dipole

$$V = \frac{p}{4\pi\varepsilon_0}\frac{\cos\theta}{r^2} \tag{48}$$

To calculate the components of **E**, it is convenient to express everything in rectangular coordinates,

$$r = \sqrt{x^2 + y^2 + z^2} \tag{49}$$

$$\cos\theta = \frac{z}{\sqrt{x^2 + y^2 + z^2}} \tag{50}$$

so that

$$V = \frac{p}{4\pi\varepsilon_0}\frac{z}{(x^2 + y^2 + z^2)^{3/2}} \tag{51}$$

The components of the electric field are then

$$E_x = -\frac{\partial V}{\partial x} = \frac{p}{4\pi\varepsilon_0} \frac{3zx}{(x^2 + y^2 + z^2)^{5/2}} \tag{52}$$

$$E_y = -\frac{\partial V}{\partial y} = \frac{p}{4\pi\varepsilon_0} \frac{3zy}{(x^2 + y^2 + z^2)^{5/2}} \tag{53}$$

$$E_z = -\frac{\partial V}{\partial z}$$

$$= -\frac{p}{4\pi\varepsilon_0}\left(\frac{1}{(x^2 + y^2 + z^2)^{3/2}} - \frac{3z^2}{(x^2 + y^2 + z^2)^{5/2}}\right) \tag{54}$$

These expressions for the electric field are only approximations, but they are very good at large distances from the dipole. For instance, if the dipole is within a molecule, then l is very small — 10^{-10} m or less — and our approximation is valid whenever the distance is large compared to 10^{-10} m. Figure 25.11 shows the field lines for this electric field.

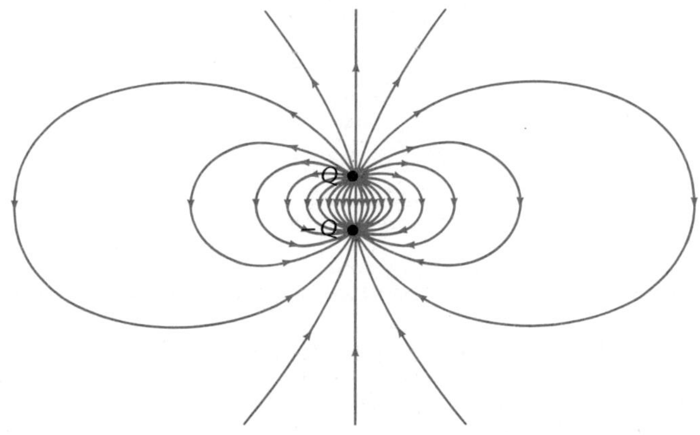

Fig. 25.11 Field lines of an electric dipole.

Equations (52)–(54) describe the electric field surrounding a water molecule, with a dipole moment $p = 6.1 \times 10^{-30}$ C · m (see Section 23.4). By means of this electric field the water molecule can act on other molecules — it can exert forces on the electric charges in other molecules. This is what makes water such a good solvent.

25.6 The Mean-Value Theorem[4]

There exists a very beautiful theorem concerning the electrostatic potential in a region free of electric charge. This theorem, called the **mean-value theorem,** states the following: *If S is the surface of a (mathematical) sphere whose interior is free of charge, then the potential at the center of the sphere equals the mean value of the potential over the surface.*

Mean-value theorem

The proof of the theorem is simple. Suppose the sphere has a radius R. Then the mean value of the potential over the surface of the sphere

[4] This section is optional.

is obtained by integrating the potential over the surface and dividing by the area of the surface,

$$\overline{V} = \frac{1}{4\pi R^2} \int V \, dS \tag{55}$$

The essential step of the proof is demonstrating that $\overline{V}$ is independent of R, i.e., $\overline{V}$ is independent of the size of the sphere. Consider the derivative of the integral (55),

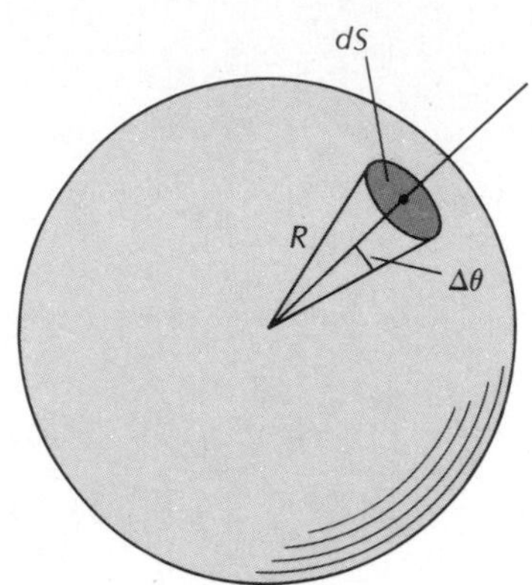

Fig. 25.12 Small circular area on surface of a sphere. The radius of this circular area is approximately $R\ \Delta\theta$.

$$\frac{d\overline{V}}{dR} = \frac{d}{dR}\left(\frac{1}{4\pi R^2} \int V \, dS\right) \tag{56}$$

Figure 25.12 shows an area element dS. This area is approximately the base of a cone of half-angle $\Delta\theta$; the base has an area $\pi(R\ \Delta\theta)^2$. The contribution to the integral (56) from this cone is

$$\frac{d}{dR}\left(\frac{1}{4\pi R^2} V \, dS\right) = \frac{d}{dR}\left[\frac{1}{4\pi R^2} V\pi(R\ \Delta\theta)^2\right]$$

$$= \frac{d}{dR}\left[\frac{(\Delta\theta)^2}{4} V\right] = \frac{(\Delta\theta)^2}{4}\frac{dV}{dR}$$

$$= \frac{1}{4\pi R^2}\frac{dV}{dR}\pi(R\ \Delta\theta)^2$$

$$= \frac{1}{4\pi R^2}\frac{dV}{dR} dS \tag{57}$$

The entire sphere can be regarded as a collection of such cones. Hence

$$\frac{d\overline{V}}{dR} = \frac{1}{4\pi R^2} \int \frac{dV}{dR} dS \tag{58}$$

But, according to Eq. (29), dV/dR is the negative of the component of $\mathbf{E}$ in the radial direction, i.e., it is the component of $\mathbf{E}$ in a direction perpendicular to the surface of integration. By Gauss' Law, the surface integral of this component is zero since the sphere S is free of charge, i.e.,

$$\frac{d\overline{V}}{dR} = \frac{1}{4\pi R^2} \int \frac{dV}{dR} dS = \frac{-1}{4\pi R^2} \int E_n \, dS = 0 \tag{59}$$

This establishes that $\overline{V}$ *is independent of* R. But if so, then the sphere of radius R and any smaller sphere must have the same value for the corresponding mean potential. In particular, since the potential at the center of the sphere can be regarded as the mean potential for a sphere of infinitesimal radius, it follows that the potential at the center must have the same value as the mean potential over the sphere of radius R. This concludes the proof.

The mean-value theorem has an important corollary: in a region free of charge, the electrostatic potential cannot have a minimum or a maximum. For suppose it had a minimum at a point P. This would mean that at *all* points in the immediate vicinity the potential is higher.

Hence, if we draw a small sphere centered on P, the mean value of the potential over the surface would have to be larger than the potential at the center. This would contradict the mean-value theorem and is therefore impossible. A similar argument shows that a maximum is also impossible. The absence of minima and maxima means that if the potential increases in some directions, it must decrease in others in such a way that the average of the changes in all directions is zero.

Finally, we will describe a very useful numerical method for the approximate calculation of the electrostatic potential. It will be easiest to describe this method in the context of a special example. Figure 25.13 shows a pair of large conducting plates with a kink. Figure 25.14 shows the plates edge on and gives the relevant dimensions. Suppose that the potential of the lower plate is 0 volt and that of the upper plate 3.0 volt. What is the potential in the space between the plates?

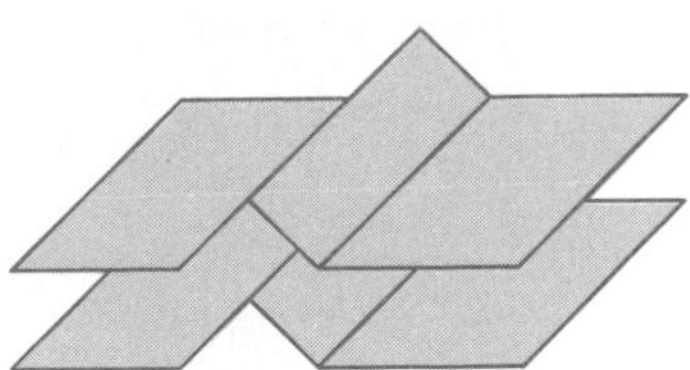

Fig. 25.13 Very large parallel plates with a kink.

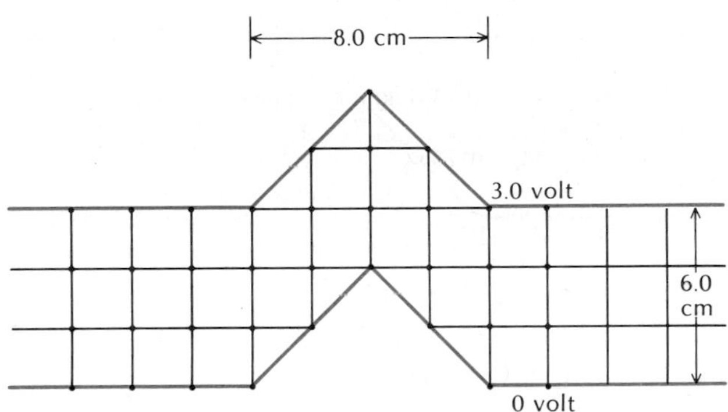

Fig. 25.14 Plates viewed edge on. A coordinate grid has been superimposed on the diagram; the squares of this grid are 2 cm × 2 cm.

In order to apply numerical methods to this problem, we must represent the space between the plates by a discrete, and finite, set of points. Figure 25.14 shows a coordinate grid superimposed on the picture of the plates. We will try to find the potential at the intersection points of this grid. This will give us only a rather rough description of the potential; for a more precise description we would have to take a finer grid. Note that the grid is two dimensional. The third dimension (out of the plane of the page in Figure 25.14) can be ignored since the potential at all points above or below the plane of the page is the same as that at the points shown in Figure 25.14.

To obtain the potential at the grid points, we proceed by the following method of successive approximations: We begin by making some reasonable first guess for the potential. Since for flat plates the potential in the space between would increase regularly from 0 volt to 3.0 volt, the values given in Figure 25.15a are a reasonable first approximation. To find the second approximation to the potential, we rely on the mean-value theorem. According to this theorem, the potential at any point should equal the average of the potentials of all the points that are at a distance of, say, 2 cm from the given point. In our grid, this average is approximated by an average over the four nearest neighbor points. We therefore obtain a second approximation by replacing the potential at each point by this average over the potentials of the four nearest neighbor points. This averaging procedure yields the values in Figure 25.15b. Next we obtain a third approximation by again replacing each of the potentials of Figure 25.15b with the average over the potentials of the four nearest neighbor points. This yields the values in Figure 25.15c, etc.

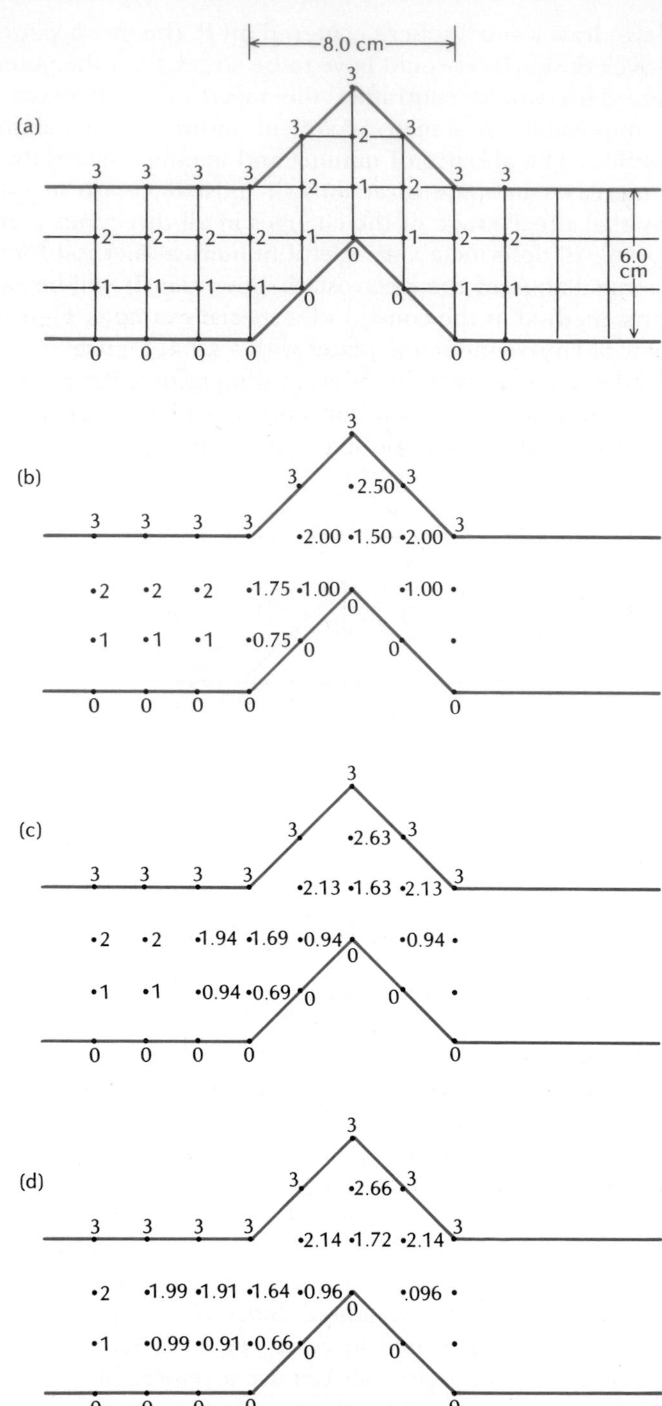

Fig. 25.15 The numbers at the grid points give the potential in volts.

After a few successive steps, this iteration procedure yields self-consistent values for the potential, i.e., it yields values that do not change appreciably from one step to the next. These values are an approximate answer to the problem. The components of the electric field can then be obtained from the values of the potential by numerical calculation of the derivatives of the potential in the horizontal and vertical directions.

As a check on the approximation, it is a good idea to repeat the calculation with the finer grid size. This tends to get tedious, but the method lends itself very well to programming on a digital computer.

SUMMARY

Definition of electrostatic potential:

$$V(P) = -\int_{P_0}^{P} \mathbf{E} \cdot d\mathbf{l} + V(P_0)$$

Potential of point charge: $V = \dfrac{1}{4\pi\varepsilon_0}\dfrac{q'}{r}$

Potential energy of two point charges: $U = \dfrac{1}{4\pi\varepsilon_0}\dfrac{qq'}{r}$

Derivatives of the potential: $\dfrac{dV}{dl} = -E\cos\theta$

$$\frac{\partial V}{\partial x} = -E_x, \quad \frac{\partial V}{\partial y} = -E_y, \quad \frac{\partial V}{\partial z} = -E_z$$

Mean-value theorem (for charge-free region): potential at center of any sphere equals mean potential on surface.

QUESTIONS

1. The potential difference between the poles of an automobile battery is 12 volt. Explain what this means in terms of the definition of potential as work per unit charge.

2. An old-fashioned word for electrostatic potential is electrostatic *tension.* Is it reasonable to think of the potential as analogous to mechanical tension?

3. If the electric field is zero in some region, must the potential also be zero? Give an example.

4. A bird sits on a high-voltage power line which is at a potential of 345,000 volt. Does this harm the bird?

5. How would you define the gravitational potential? Are the units of gravitational potential the same as the units of electric potential? According to your definition, what is the gravitational potential difference between the ground and a point 50 m above the ground?

6. Consider an electron moving in the vicinity of a proton. Where is the electrostatic potential produced by the proton highest? Where is the potential energy of the electron highest?

7. Suppose that the electrostatic potential has a minimum at some point. Is this an equilibrium point for a positive charge? For a negative charge? Is the equilibrium stable?

8. Consider a sphere of radius R with a charge Q uniformly distributed over its volume. Where does the potential have a maximum? Where does the magnitude of the electric field have a maximum?

9. Give an example of a conductor that is not an equipotential. Is this conductor in electrostatic equilibrium?

10. If the potential in a three-dimensional region of space is known to be constant, what can you conclude about the electric field in this region? If the potential on a two-dimensional surface is known to be constant, what can you conclude about the electric field on this surface?

11. In many calculations it is convenient to assign a potential of 0 volt to the ground. If so, what is the potential at the top of the Eiffel Tower? What is the potential at the top of your head? (Hint: Your body is a conductor.)

12. Is it true that the surface of a mass in static mechanical equilibrium is a gravitational equipotential surface? What if the surface is that of a fluid, such as water?

13. If a high-voltage power cable falls on top of your automobile, you will probably be safest if you remain inside the automobile. Why?

14. Suppose that several separate solid metallic bodies have been placed near a charge distribution. Is it necessarily true that all of these bodies will have the same potential?

15. If we surround some region with a conducting surface, we shield it from external electric fields. Why can we not shield a region from gravitational fields by a similar method?

16. The interior of an automobile provides good protection against lightning. Explain.

17. A cavity is completely surrounded by conducting material. Can you create an electric field in this cavity?

18. Show that different equipotential surfaces cannot intersect.

19. Consider the patterns of field lines shown in Figures 23.14 and 23.15. Roughly, sketch some of the equipotential surfaces for each case.

20. Sketch the equipotential surfaces for the potential described numerically in Figure 25.15d.

PROBLEMS

Section 25.1

1. In order to charge a typical 12-volt automobile battery fully, the charging device must force $+2.0 \times 10^5$ coulombs from the negative terminal of the battery to the positive terminal. How much work must the charging device do during this process?

2. A proton sits at the origin of coordinates. How much work (in electron-volts) must you do to push an electron from the point $x = 1.0$ Å, $y = 0$, $z = 0$ to the point $x = 0.5$ Å, $y = 0.5$ Å, $z = 0$?

3. On days of fair weather, the atmospheric electric field of the Earth is about 100 V/m; this field points vertically downward (compare Problem 23.2). What is the electric potential difference between the ground and an airplane flying at 600 m (2000 ft)? What is the potential difference between the ground and the tip of the Eiffel Tower? Treat the ground as a flat conductor.

4. Consider the arrangement of parallel sheets of charge described in Problem 23.22. Find the potential difference between the upper sheet and the lower sheet.

5. Suppose that, as a function of x, y, z, an electric field has components

$$E_x = 6x^2y \qquad E_y = 2x^3 + 2y \qquad E_z = 0$$

where E is measured in volts per meter and the distances are measured in meters.

(a) Find the potential difference between the origin and the point $x = 3$, $y = 0$, $z = 0$.

(b) Find the potential difference between the origin and the point $x = 0$, $y = 2$, $z = 0$.

6. At the Stanford Linear Accelerator (SLAC), electrons are accelerated from an energy of 0 eV to 20×10^9 eV as they travel in a straight evacuated tube 1600 m in length. The acceleration is due to a strong electric field pushing the electrons along. Assume that the electric field is uniform. What must be its strength?

7. The potential difference between the two poles of an automobile battery is 12.0 V. Suppose that you place such a battery in empty space and you release an electron at a point next to the negative pole of the battery. The electron will then be pushed away by the electric force and move off in some direction.

(a) If the electron strikes the positive pole of the battery, what will be its impact speed?

(b) If instead the electron moves away toward infinity, what will be its ultimate speed?

8. The gap between the electrodes of a spark plug in an automobile is 0.025 in. In order to produce an electric field of 3×10^6 V/m (required to initiate an electric spark), what minimum potential difference must you apply to the spark plug?

9. Prove that the plane midway between a positive and a negative point charge of equal magnitudes is an equipotential surface. Is this also true if both charges are positive?

Section 25.2

10. The nucleus of lead has a charge of $82e$ uniformly distributed over a spherical region of radius 7.1×10^{-15} m. What is the electrostatic potential at the nuclear surface? At the center?

11. An alpha particle of kinetic energy 1.7×10^{-12} J is shot directly toward a platinum nucleus. What will be the distance of closest approach? The electric charge of the alpha particle is $2e$ and that of the platinum nucleus is $78e$. Treat the alpha particle and the nucleus as spherical charge distributions and disregard the motion of the nucleus.

12. What is the minimum kinetic energy with which an alpha particle must be launched toward a plutonium nucleus if it is to make contact with the nuclear surface? The plutonium nucleus is a sphere of radius 7.5×10^{-15} m with a charge of $94e$ uniformly distributed over the volume. For the purpose of this problem, the alpha particle may be regarded as a particle (of negligible radius) with a charge of $2e$.

13. A thorium nucleus emits an alpha particle according to the reaction

$$\text{thorium} \rightarrow \text{radium} + \text{alpha}$$

Assume that the alpha particle is pointlike and that the residual radium nucleus is spherical with a radius of 7.4×10^{-15} m. The charge on the alpha particle is $2e$ and that on the radium nucleus is $88e$.

(a) At the instant the alpha particle emerges from the nuclear surface, what is its electrostatic potential energy?

(b) If the alpha particle has no initial kinetic energy, what will be its final kinetic energy and speed when far away from the nucleus? Assume that the radium nucleus does not move. The mass of the alpha particle is 6.7×10^{-27} kg.

14. Consider again the arrangement of charges within the thundercloud of Figure 23.28. Find the electric potential due to these charges at a point which is at a height of 8 km and on the vertical line passing through the charges. Find the electric potential at a second point which is at the same height and has a horizontal distance of 5 km from the first point.

15. In a helium atom, at some instant one of the electrons is at a distance of 0.3×10^{-10} m from the nucleus and the other electron is at a distance of

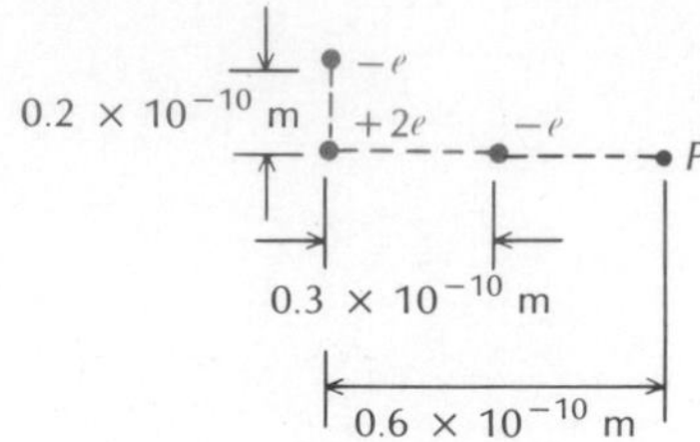

Fig. 25.16 Nucleus (charge $+2e$) and electrons (charge $-e$) of helium atom at one instant of time.

0.2×10^{-10} m, 90° away from the first (Figure 25.16). Find the electric potential produced jointly by the two electrons and the nucleus at a point P beyond the first electron and at a distance of 0.6×10^{-10} m from the nucleus.

*16. According to Bohr's theory of the atom (see also Problem 22.12), the electron in a hydrogen atom orbits around the nucleus in a circular orbit. The force that holds the electron in this orbit is the Coulomb force. The size of the orbit depends on the angular momentum — the smallest possible orbit has an angular momentum $\hbar = 1.05 \times 10^{-34}$ J · s; the next possible orbit has angular momentum $2\hbar$; the next $3\hbar$, etc.

(a) Show that if a circular orbit has angular momentum $n\hbar$ (where $n = 1, 2, 3, \ldots$), then its radius is

$$r = \frac{4\pi\varepsilon_0}{m_e e^2} n^2 \hbar^2$$

(b) Show that the orbital energy (kinetic and potential) of the electron in such an orbit is

$$E = -\frac{m_e e^4}{2(4\pi\varepsilon_0)^2 \hbar^2} \frac{1}{n^2}$$

(c) Evaluate this energy for $n = 1$; express your answer in electron-volts.

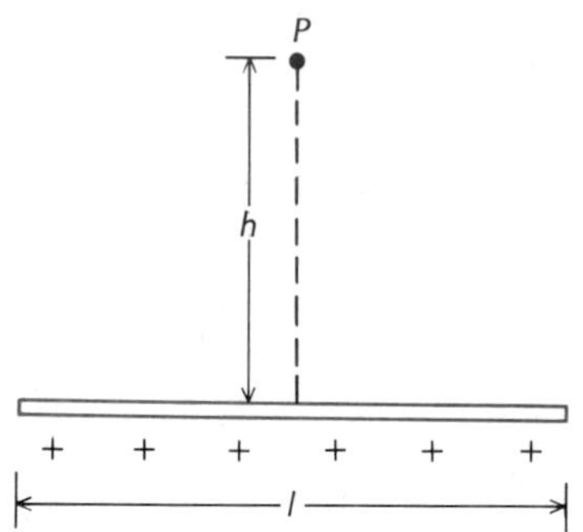

Fig. 25.17 A charged rod of length l.

Section 25.3

17. A total charge Q is distributed uniformly along a straight rod of length l. Find the potential at a point P at a distance h from the midpoint of the rod (Figure 25.17).

18. Three thin rods of glass of length l carry charges uniformly distributed along their lengths. The charges on the three rods are $+Q$, $+Q$, and $-Q$, respectively. The rods are arranged along the sides of an equilateral triangle. What is the electrostatic potential at the midpoint of this triangle?

19. A uniformly charged sphere of radius a is surrounded by a uniformly charged concentric spherical shell of inner radius b and outer radius c. The total charge on the sphere is Q and that on the shell $-Q$. Find the potential at $r = b$, at $r = a$, and at $r = 0$.

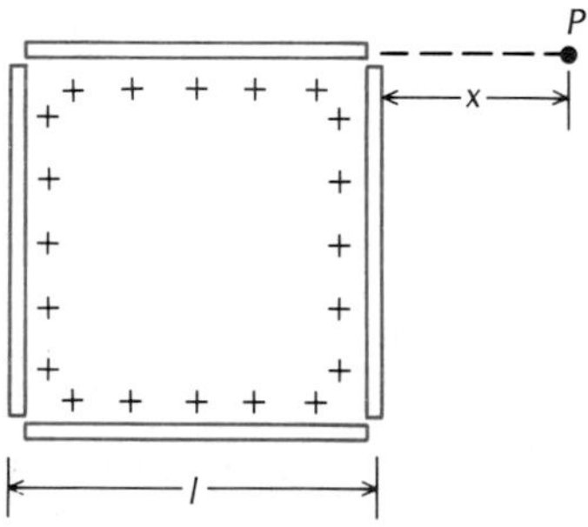

Fig. 25.18

20. Four rods of length l are arranged along the edges of a square. The rods carry charges $+Q$ uniformly distributed along their lengths (Figure 25.18). Find the potential at the point P at a distance x from one corner of the square.

21. Two semicircular rods and two short straight rods are joined in the configuration shown in Figure 25.19. The rods carry a charge of λ coulomb per meter. Calculate the potential at the center of this configuration.

22. A long straight wire of radius 0.80 mm is surrounded by an evacuated concentric conducting shell of radius 1.2 cm. The wire carries a charge of -5.5×10^{-8} coulomb per meter of length. Suppose that you release an electron at the surface of the wire. With what speed will this electron hit the conducting shell?

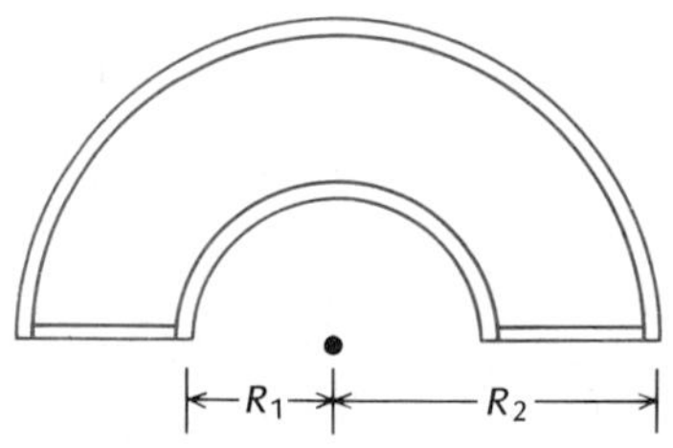

Fig. 25.19

23. A long plastic pipe has an inner radius a and an outer radius b. Charge is uniformly distributed over the volume $a < r < b$. The amount of charge is λ coulomb per meter of length of the tube. Find the potential difference between $r = b$ and $r = 0$. Assume that the plastic has no effect on the electric field.

24. A flat disk of radius R has charge Q uniformly distributed over its surface. Find a formula for the potential along the axis of the disk.

25. The tube of a Geiger counter consists of a thin straight wire surrounded by a coaxial conducting shell. The diameter of the wire is 0.001 in. and that of

the shell is 1.0 in. The length of the tube is 4.0 in; however in your calculation use the formula for the electric field of an infinitely long line of charge. If the potential difference between the wire and the shell is 1.0×10^3 volt, what is the electric field at the surface of the wire? At the cylinder?

*26. An infinite charge distribution with spherical symmetry has a charge density ρ coulomb per cubic meter given by the formula $\rho = kr^{-5/2}$ where k is a constant. Find the potential as a function of the radius. Assume $V = 0$ at $r = \infty$.

*27. A point charge Q is on the positive z axis at the point $z = h$. A point charge $-Q \times R/h$ (where R is a positive length, $0 < R < h$) is on the z axis at the point $z = R^2/h$. Show that the surface of the sphere of radius R about the origin is an equipotential surface.

Sections 25.4 and 25.5

28. In some region of space the electrostatic potential is the following function of x, y, and z:

$$V = x^2 + 2xy$$

where the potential is measured in volts and the distances in meters. Find the electric field at the point $x = 2$, $y = 2$.

29. In terms of x, y, and z, the potential of a point charge is

$$V(x, y, z) = \frac{1}{4\pi\varepsilon_0} \frac{q'}{\sqrt{x^2 + y^2 + z^2}}$$

(a) By differentiating this potential function, calculate the components E_x, E_y, and E_z of the electric field.
(b) Show that the magnitude $\sqrt{E_x^2 + E_y^2 + E_z^2}$ agrees with the usual expression for the electric field of a point charge.

30. A rod of length l has a charge Q uniformly distributed along its length (Figure 25.20). Find the potential at the point P at a distance x from one end of the rod. Find the electric field at this point.

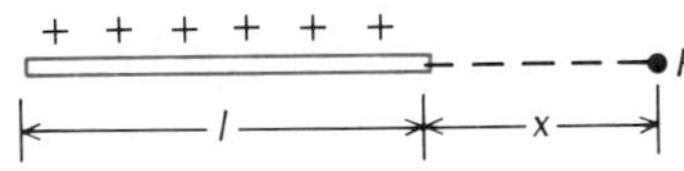

Fig. 25.20

31. Two rods of equal length l form a symmetric cross. The rods carry charges $\pm Q$ uniformly distributed along their lengths. Calculate the potential at the point P at a distance x from one end of the cross (Figure 25.21). Calculate the electric field at this point.

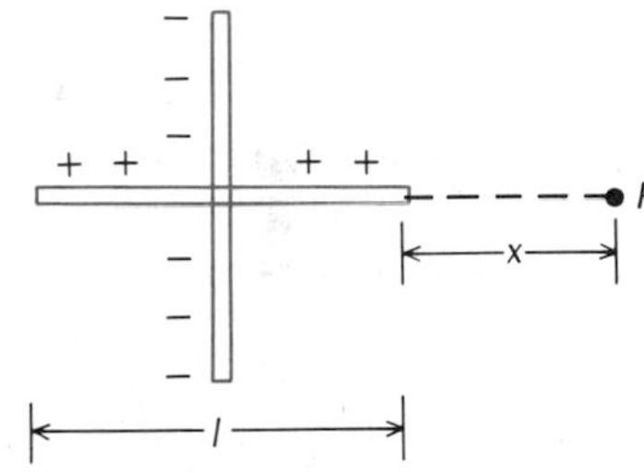

Fig. 25.21

32. An annulus (a disk with a hole) made of paper has an outer radius R and an inner radius $R/2$ (Figure 25.22). An amount Q of electric charge is uniformly distributed over the paper.
(a) Find the potential as a function of distance on the axis of the annulus.
(b) Find the electric field on the axis of the annulus.

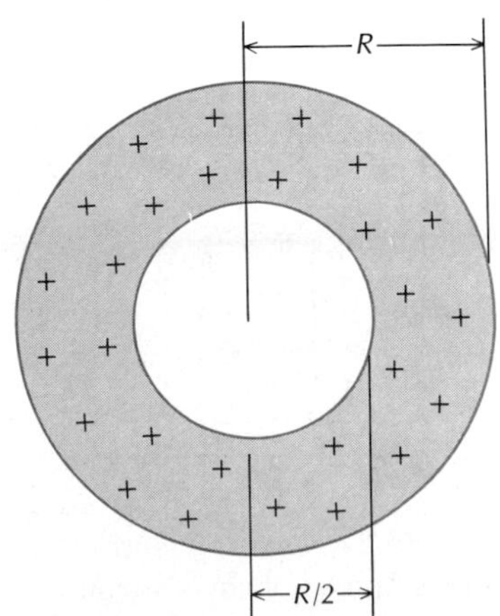

Fig. 25.22

33. A nucleus of carbon (charge $6e$) and one of helium (charge $2e$) are separated by a distance of 1.2×10^{-13} m and instantaneously at rest. The center of mass of this system is at a distance of 0.4×10^{-13} m from the carbon nucleus. Take this point as origin and take the x axis along the line joining the nuclei, with the carbon nucleus on the negative x axis.
(a) Find the potential V as a function of x, y, and z.
(b) Find E_x and E_y as a function of x, y, and z.

34. The water molecule has a dipole moment of 6.1×10^{-30} C · m.
(a) Find the magnitude and direction of the electric field at a point on the axis of the dipole at a distance of 12.0 Å from the molecule.
(b) Find the magnitude and direction of the electric field at a point on a line transverse to the axis of the dipole at the same distance.

*35. A thin cylindrical cardboard tube has a charge Q uniformly distributed over its surface. The radius of the tube is R and the length l.

(a) Find the potential at a point on the axis of the tube at a distance x from the midpoint. Assume $x > l$.
(b) Find the electric field at this point.

*36. A point charge $-2Q$ is at the origin of coordinates; two point charges $+Q$ are on the z axis, at $z = \pm l$, respectively (Figure 25.23).
(a) Show that, for $r \gg l$, the net potential of these charges is approximately

$$V = \frac{2Ql^2}{4\pi\varepsilon_0 r^3}\frac{(3\cos^2\theta - 1)}{2}$$

(b) Calculate E_x, E_y, and E_z, expressing these in terms of the coordinates x, y, and z.

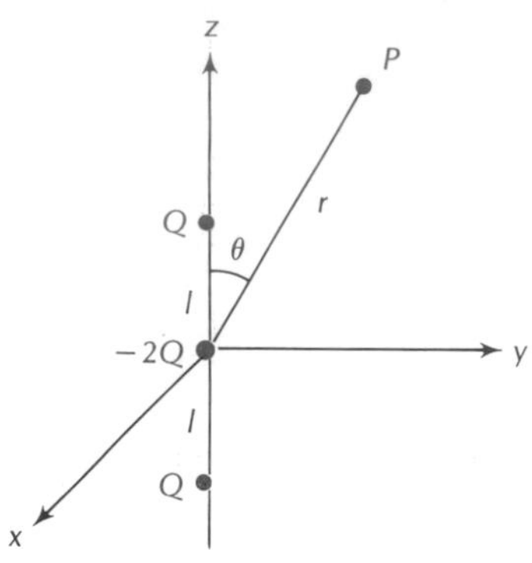

Fig. 25.23 Charges $-2Q$, Q, and Q on the z axis.

Section 25.6

37. Figure 25.24 shows two large, parallel, conducting plates seen in cross section. One of the plates has a kink. The potential difference between them is 2.0 V. Use the mean-value theorem to find the potential at the points of the grid in Figure 25.24 to at least two significant figures.

38. Figure 25.25 shows two large, parallel conducting plates seen in cross section. Both plates have rectangular kinks. The potential difference between the plates is 6.0 V. Use the mean-value theorem to find the potential at the grid points to within at least two significant figures.

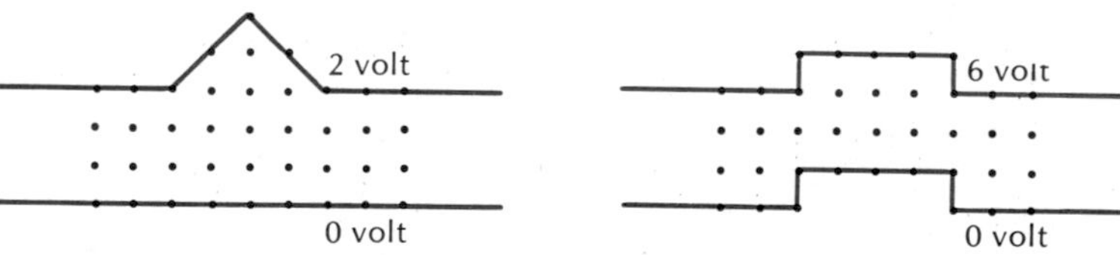

Fig. 25.24 Very large parallel plates, viewed edge on. One of the plates has a kink.

Fig. 25.25 Very large parallel plates, viewed edge on. Both plates have rectangular ridges.

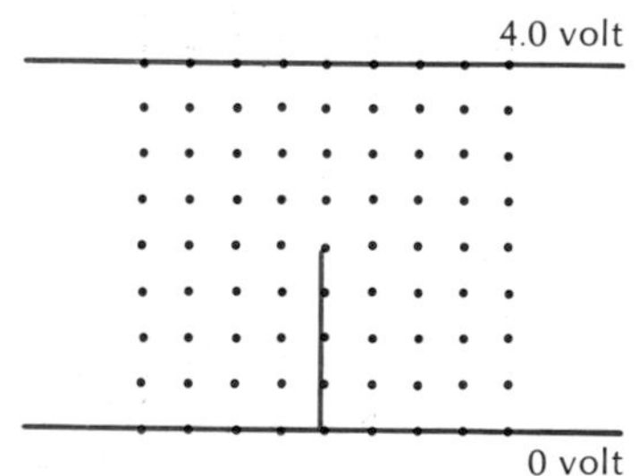

Fig. 25.26 Two very large parallel plates, seen edge on. The lower plate has a protruding thin ridge; the potential is $V = 0$ on this plate and on the ridge.

39. Two large, parallel conducting plates have a potential difference of 4.0 V between them. One of the plates carries a thin vertical ridge (Figure 25.26). Use the mean-value theorem to find the potential at the grid points to within two significant figures or better.

40. Given that each box of the grid of Figure 25.15d is 2 cm × 2 cm, evaluate numerically the derivative $\Delta V/\Delta z$ at each of the grid points and thereby find E_z. The z direction is the vertical direction (perpendicular to the flat portion of the plate).

41. Two long concentric tubes of sheet metal have a square cross section; Figure 25.27 shows their cross section. The outer tube is at a potential of 10 V; the inner tube is at a potential of 0 V. Use the mean-value theorem to find the potential at all the grid points shown to within two significant figures.

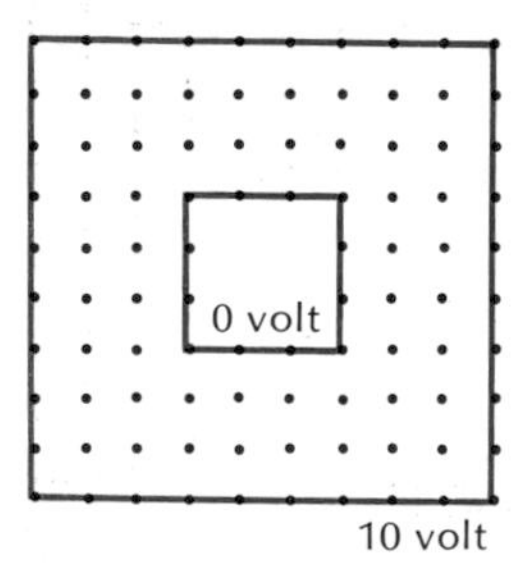

Fig. 25.27 Two concentric tubes (ducts) of sheet metal, seen in cross section.

42. When calculating the average potential in the method of successive approximations described in Section 25.6, we only took into account the nearest four points surrounding the given point in a *two-dimensional* grid. Since the mean-value theorem applies to a sphere surrounding the given point, we should actually take into account the six nearest points surrounding the given point in a *three-dimensional* grid, i.e., we should take into account an extra point in front of and an extra point behind the plane shown in Figure 25.14. Prove that for the problem discussed in Section 25.6, in which the potential in front of the given point and the potential behind the given point are the same as the potential at the given point, the average over the four nearest points in the two-dimensional grid coincides with the average over the six nearest points in the three-dimensional grid.

CHAPTER 26

Electric Energy

In the preceding chapter we calculated the electric potential energy of a charge — a test charge — placed in the electric field of a given charge distribution. Now we will calculate the potential energy of the charge distribution by itself, without the test charge. The charge distribution can be regarded as a collection of point charges; since all of these point charges exert forces on one another, it requires a certain amount of work to bring them together into their final configuration if they are initially separated by large distances. This amount of work is the potential energy of the charge distribution.

We will see that this potential energy is stored in the electric field. The energy is concentrated in those regions of space where the electric field is strong. We will see that the distribution of energy in the electric field can be described by an energy density. Since the electric field is endowed with energy, we must regard the field as a material object, a fifth state of matter.

26.1 Energy of a System of Point Charges

The electric potential energy of two point charges q_1, q_2 separated by a distance r is [see Eq. (25.14)]

$$U = \frac{1}{4\pi\varepsilon_0} \frac{q_1 q_2}{r} \tag{1}$$

This potential energy can be regarded as the work required to move q_1 from infinity to within a distance r of q_2 or, alternatively, the work required to move q_2 to within a distance r of q_1. It is a *mutual* potential

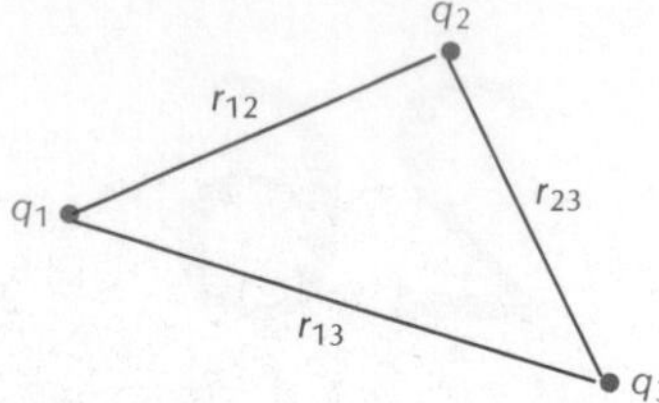

Fig. 26.1 Three point charges.

energy which belongs to both q_1 and q_2, i.e., it is an energy associated with the pair q_1, q_2.

For configurations consisting of more than two charges, the net potential energy can be calculated by writing down a term similar to that in Eq. (1) for *each pair* of charges. For instance, if we are dealing with three charges (Figure 26.1), we have three possible pairs, (q_1, q_2), (q_2, q_3), and (q_1, q_3), so that the net potential energy is

$$U = \frac{1}{4\pi\varepsilon_0}\frac{q_1 q_2}{r_{12}} + \frac{1}{4\pi\varepsilon_0}\frac{q_2 q_3}{r_{23}} + \frac{1}{4\pi\varepsilon_0}\frac{q_1 q_3}{r_{13}} \tag{2}$$

This is the work required to assemble the charges in the final configuration shown in Figure 26.1 starting from an initial condition of infinite separation.

Note that Eq. (2) is identically equal to

$$U = \frac{1}{2}\left(\frac{1}{4\pi\varepsilon_0}\frac{q_2}{r_{12}} + \frac{1}{4\pi\varepsilon_0}\frac{q_3}{r_{13}}\right) q_1 + \frac{1}{2}\left(\frac{1}{4\pi\varepsilon_0}\frac{q_1}{r_{12}} + \frac{1}{4\pi\varepsilon_0}\frac{q_3}{r_{23}}\right) q_2$$
$$+ \frac{1}{2}\left(\frac{1}{4\pi\varepsilon_0}\frac{q_1}{r_{13}} + \frac{1}{4\pi\varepsilon_0}\frac{q_2}{r_{23}}\right) q_3 \tag{3}$$

Here the potential energy of each pair of charges appears twice, each time with a factor of $\frac{1}{2}$. Equation (3) therefore leads to the following expression for the energy in terms of potentials:

$$U = \tfrac{1}{2}V_{\text{other}}(1)q_1 + \tfrac{1}{2}V_{\text{other}}(2)q_2 + \tfrac{1}{2}V_{\text{other}}(3)q_3 \tag{4}$$

where $V_{\text{other}}(1)$ is the electric potential produced at the position of charge 1 by the *other* charges (charges 2 and 3), i.e.,

$$V_{\text{other}}(1) = \frac{1}{4\pi\varepsilon_0}\frac{q_2}{r_{12}} + \frac{1}{4\pi\varepsilon_0}\frac{q_3}{r_{13}} \tag{5}$$

and similarly for $V_{\text{other}}(2)$ and $V_{\text{other}}(3)$.

By means of a generalization of this argument we can easily show that for a configuration consisting of any number of point charges, the electric potential energy is the sum

Energy of system of point charges

$$U = \tfrac{1}{2}V_{\text{other}}(1)q_1 + \tfrac{1}{2}V_{\text{other}}(2)q_2 + \tfrac{1}{2}V_{\text{other}}(3)q_3 + \tfrac{1}{2}V_{\text{other}}(4)q_4 + \cdots \tag{6}$$

This expression gives the work that must be done to bring the point charges to their final positions starting from initial positions at very large distances from each other. However, this expression is not the total potential energy because it does not take into account the energy that a point charge has when it is by itself, at a large distance from all other point charges. Such an isolated point charge has potential energy because it takes work to assemble the point charge out of infinitesimal pieces of charge. The energy needed to assemble the point charge is called the **self-energy** of the point charge.

The calculation of the self-energy of point charges, such as electrons, is one of the unsolved problems of physics. A straightforward calculation of the self-energy of an electron yields the absurd result that this energy is infinite [essentially, the calculated energy is infinite because the potential at the position of the point charge is infinite: $(1/4\pi\varepsilon_0)(q/r) \to \infty$ as $r \to 0$]. Although up to now physicists have found no satisfactory way to calculate the self-energy, they have invented several rules for bypassing this problem. These rules, called renormalization rules, give prescriptions on how to extract experimentally meaningful numbers from the theoretical calculations. In essence, the rules assert that the self-energy is a constant quantity that never has any effect on energy conservation and can therefore be ignored. The theoretical calculations based on this scheme have been extremely successful. For instance, by combining electromagnetic theory, relativity theory, and quantum theory, the electric energy of an electron in a hydrogen atom has been calculated to nine significant figures. But from the mathematical point of view this scheme has some shady aspects.

In the following we will always ignore the electric self-energy of point charges and pretend that Eq. (6) is the total electric energy.

26.2 Energy of a System of Conductors

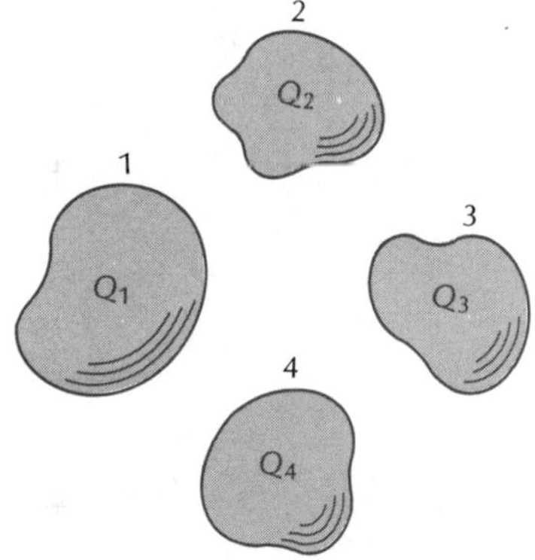

Fig. 26.2 Four conducting bodies carrying electric charges.

Equation (6) permits us to evaluate the electric energy of a system of charged conductors. Figure 26.2 shows several conductors of arbitrary shapes. Suppose that the charges on these conductors are Q_1, Q_2, Q_3, . . . , and that their potentials are V_1, V_2, V_3, . . . , respectively. The charge Q_1 consists of many point charges distributed over conductor 1. Each of these point charges is at potential V_1. This potential acting on a given point charge dQ_1 can be regarded as due to the *other* point charges (on conductor 1 and on other conductors) because the given point charge makes only an insignificant contribution to V_1. Hence, according to Eq. (6), the electric energy associated with a point charge dQ_1 on conductor 1 is $\frac{1}{2}(dQ_1)V_1$ and the potential energy associated with all the point charges on conductor 1 is simply $\frac{1}{2}Q_1V_1$. Since similar arguments apply to the other conductors, we conclude that net electric energy is

$$U = \tfrac{1}{2}Q_1V_1 + \tfrac{1}{2}Q_2V_2 + \tfrac{1}{2}Q_3V_3 + \cdots \qquad (7)$$

Energy of system of conductors

EXAMPLE 1. A metallic sphere of radius R carries a charge Q uniformly distributed over its surface. How much electric energy is stored in this charge distribution?

SOLUTION: The potential outside a spherically symmetric charge distribution is given by Eq. (25.21). At $r = R$, this potential is

$$V = \frac{1}{4\pi\varepsilon_0}\frac{Q}{R}$$

According to Eq. (7), the electric energy is then

$$U = \tfrac{1}{2}QV = \tfrac{1}{2}Q\frac{1}{4\pi\varepsilon_0}\frac{Q}{R} = \frac{1}{8\pi\varepsilon_0}\frac{Q^2}{R} \qquad (8)$$

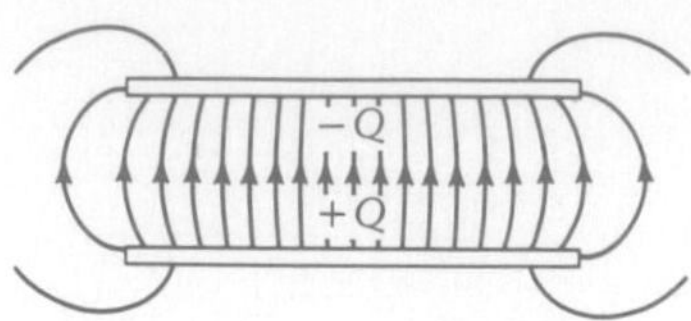

Fig. 26.3 Two very large parallel conducting plates with opposite electric charges.

EXAMPLE 2. Two large, parallel metallic plates of area A are separated by a distance d. Charges $+Q$ and $-Q$ are placed on the plates, respectively (Figure 26.3). What is the electric energy?

SOLUTION: The electric field between the plates is approximately the field of an infinite plate, i.e., the electric field is constant,

$$E = \frac{\sigma}{\varepsilon_0} = \frac{Q}{\varepsilon_0 A} \tag{9}$$

This expression fails near the edges of the plates, where there is an electric fringing field that is not constant (Figure 26.3). We will investigate the energy in such a fringing field in the next section. If the plates are large, then the edge region is only a very small fraction of the total region between the plates, and we can ignore this edge region without introducing excessive errors in our calculation.

The potential difference between the plates is

$$V_2 - V_1 = -Ed = -\frac{Qd}{\varepsilon_0 A} \tag{10}$$

and the electric potential energy is

$$U = \tfrac{1}{2}Q_1V_1 + \tfrac{1}{2}Q_2V_2 = \tfrac{1}{2}QV_1 - \tfrac{1}{2}QV_2$$

$$= \tfrac{1}{2}Q(V_1 - V_2) = \tfrac{1}{2}\frac{Q^2d}{\varepsilon_0 A} \tag{11}$$

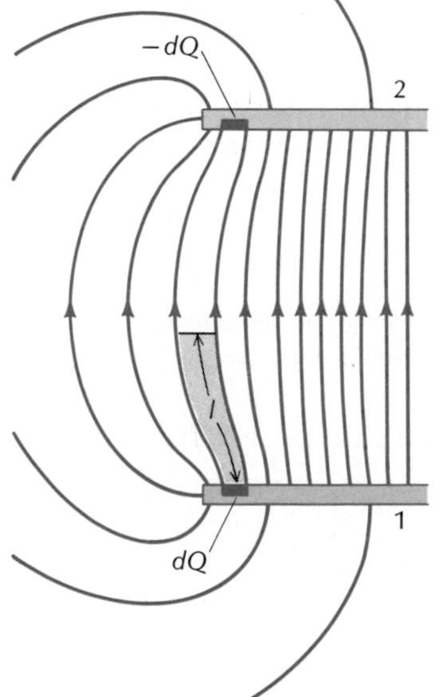

Fig. 26.4 Field lines start on the charge dQ on the lower plate and end on the charge $-dQ$ on the upper plate.

Note that Eq. (11) can be rewritten in the following interesting way:

$$U = \tfrac{1}{2}\varepsilon_0 \left(\frac{Q}{\varepsilon_0 A}\right)^2 Ad$$

$$= \tfrac{1}{2}\varepsilon_0 E^2 \times [\text{volume}] \tag{12}$$

where the "volume" is the volume between the plates, i.e., the volume of the region in which there is an electric field. The energy per unit volume of electric field is therefore $\tfrac{1}{2}\varepsilon_0 E^2$. This suggests that the electric energy is distributed over space, being concentrated in those regions where the electric field is strong. In the next section we will confirm that this is indeed the case.

26.3 The Energy Density

For the special case of the constant electric field in the region between parallel conducting plates, we found that the energy per unit volume of field is $\tfrac{1}{2}\varepsilon_0 E^2$. We will now prove that this result is also true for an electric field that is not constant, such as the fringing electric field at the edges of the conducting plates. Figure 26.4 shows this field at the edges. Consider a narrow bundle of field lines that start on one conductor and end on the other conductor. The bundle of field lines intercepts two small areas on the surfaces of the two conductors. The electric charges on these small areas are dQ and $-dQ$, respectively. The contribution to the net electrostatic energy from these charges $\pm dQ$ is

$$dU = \tfrac{1}{2}dQ\,V_1 - \tfrac{1}{2}dQ\,V_2 = \tfrac{1}{2}dQ(V_1 - V_2)$$

$$= \tfrac{1}{2}\,dQ \int_1^2 E(l)\,dl \qquad (13)$$

where l is the length measured along the bundle of field lines starting at conductor 1, and $E(l)$ is the strength of the electric field at the distance l. [Note that $E(l)\,dl = \mathbf{E} \cdot d\mathbf{l}$ because the path of integration coincides with a field line whose direction is always the same as that of the electric field.] Since dQ is constant, it can be placed inside the integral sign,

$$dU = \tfrac{1}{2} \int dQ\,E(l)\,dl \qquad (14)$$

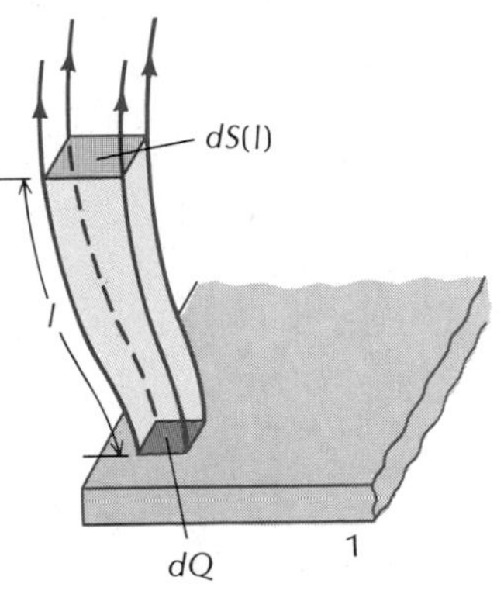

Fig. 26.5 The field lines that start on dQ form a tube of rectangular cross section.

Next we apply Gauss' Law to the tubular volume shown in Figure 26.5. The sides of the tube run along field lines; one cap of the tube is just inside the conductor, the other cap is perpendicular to field lines. The charge in the tube is dQ and the flux through the surface of the tube is entirely due to the field lines that emerge through the perpendicular cap $dS(l)$ at a distance l. Gauss' Law then tells us

$$\frac{dQ}{\varepsilon_0} = E(l)\,dS(l) \qquad (15)$$

Substituting this expression for dQ into Eq. (14), we obtain

$$dU = \tfrac{1}{2} \int \varepsilon_0 E(l) E(l)\,dS(l)\,dl \qquad (16)$$

But $dS(l)\,dl$ is the amount of volume in our tube between l and $l + dl$. Hence Eq. (16) can be written

$$dU = \tfrac{1}{2} \int \varepsilon_0 E^2\,dv \qquad (17)$$

where the integration extends over the volume of the tube from conductor 1 to conductor 2. This shows that the potential energy of the charges $\pm dQ$ on the conductors equals the integral of $\tfrac{1}{2}\varepsilon_0 E^2$ over the volume of the tube that lies within the bundle of field lines originating on these charges. Therefore the potential energy of all the charges on the conductors will equal the integral of $\tfrac{1}{2}\varepsilon_0 E^2$ over the volume of all the regions in which there is electric field.

Clearly, the above argument in no way depends on the shape of the conductors. Hence the result applies not only to parallel conducting plates, but to any number of conductors of arbitrary shape. Furthermore, although we have obtained this result for the case of field lines that start on one conductor and end on another, it is easy to see that we can obtain the same result for field lines that start at one conductor and go on to infinity. This establishes that for any arbitrary system of charged conductors the electric energy can be expressed as a volume integral of $\tfrac{1}{2}\varepsilon_0 E^2$, i.e.,

$$\boxed{U = \int \tfrac{1}{2}\varepsilon_0 E^2\,dv} \qquad (18)$$

Energy in electric field

where the integration extends over all the regions where there is electric field. Incidentally, this equation for the electric energy is also valid for charges placed on nonconductors; we will not deal with the proof of this assertion but we will take for granted that Eq. (18) is a general result for the energy associated with an electric field in vacuum.

Note that Eq. (7) expresses the energy as a sum over the electric charges, whereas Eq. (18) expresses the energy as an integral over the electric field. Thus, the former equation suggests that the energy is in the charges while the latter suggests it is in the field. To decide which of these alternatives is correct, we need some extra information. The clue is the existence of electric fields that are independent of electric charges. As we will see in Chapter 36, radio waves and light waves consist of electric and magnetic fields traveling through space. Although these fields are originally generated by electric charges, they persist even when the charges disappear. For example, a radio wave or a light beam continues to travel through space long after the radio transmitter has been shut down or the candle has been snuffed out. Obviously, the energy of a radio wave resides in the radio wave itself, in its electric and magnetic fields, and not in the electric charges in the antenna of the radio transmitter. But if energy is associated with the traveling electric fields of a radio wave, then energy should also be associated with the electric fields of a static charge distribution. We will therefore suppose that the energy (18) is in the electric field. The energy density, or energy per unit volume is

Energy density in electric field

$$\boxed{u = \tfrac{1}{2}\varepsilon_0 E^2} \tag{19}$$

EXAMPLE 3. Since energy has mass, the electric field should have not only an energy density but also a mass density. What is the corresponding mass density in a thundercloud where $E = 2 \times 10^6$ V/m?

SOLUTION:

$$u = \tfrac{1}{2}\varepsilon_0 E^2 = \tfrac{1}{2}\varepsilon_0 \times (2 \times 10^6 \text{ V/m})^2 = 18 \text{ J/m}^3$$

The mass density is the energy density divided by c^2, the square of the speed of light,

$$\frac{u}{c^2} = \tfrac{1}{2}\varepsilon_0 E^2/c^2 = (18 \text{ J/m}^3)/(3 \times 10^8 \text{ m/s})^2$$

$$= 2.0 \times 10^{-16} \text{ kg/m}^3$$

This is obviously much too small to be detectable.

EXAMPLE 4. To a good approximation, a uranium nucleus can be regarded as a sphere with charge uniformly distributed over its volume. The radius of the nucleus is 7.4×10^{-15} m and the electric charge is $92e$. What is the electric energy of the nucleus?

SOLUTION: According to Example 24.5, the electric field outside of the nucleus and inside the nucleus is, respectively,

$$E = \frac{1}{4\pi\varepsilon_0}\frac{q}{r^2} \qquad r \geq R \tag{20}$$

and

$$E = \frac{1}{4\pi\varepsilon_0}\frac{qr}{R^3} \qquad r \le R \tag{21}$$

The energy in the volume outside the nucleus is then

$$U_{\text{ext}} = \tfrac{1}{2}\varepsilon_0 \int E^2\, dv = \tfrac{1}{2}\varepsilon_0 \int \left(\frac{1}{4\pi\varepsilon_0}\frac{q}{r^2}\right)^2 dv \tag{22}$$

The amount of volume in a radial interval dr is $dv = 4\pi r^2\, dr$ and hence

$$U_{\text{ext}} = \tfrac{1}{2}\varepsilon_0 \int_R^\infty \left(\frac{1}{4\pi\varepsilon_0}\frac{q}{r^2}\right)^2 4\pi r^2\, dr$$

$$= \frac{q^2}{8\pi\varepsilon_0}\int_R^\infty \frac{1}{r^2}\, dr = \frac{q^2}{8\pi\varepsilon_0}\left[-\frac{1}{r}\right]_R^\infty = \frac{1}{8\pi\varepsilon_0}\frac{q^2}{R} \tag{23}$$

The energy in the volume inside the nucleus is

$$U_{\text{int}} = \tfrac{1}{2}\varepsilon_0 \int E^2\, dv = \tfrac{1}{2}\varepsilon_0 \int_0^R \left(\frac{1}{4\pi\varepsilon_0}\frac{qr}{R^3}\right)^2 4\pi r^2\, dr$$

$$= \frac{q^2}{8\pi\varepsilon_0}\frac{1}{R^6}\int_0^R r^4\, dr = \frac{q^2}{8\pi\varepsilon_0}\frac{1}{R^6}\left[\frac{r^5}{5}\right]_0^R$$

$$= \frac{1}{8\pi\varepsilon_0}\frac{q^2}{5R} \tag{24}$$

The total electrostatic energy is the sum of Eqs. (23) and (24),

$$U = U_{\text{ext}} + U_{\text{int}} = \frac{1}{8\pi\varepsilon_0}\frac{q^2}{R} + \frac{1}{8\pi\varepsilon_0}\frac{q^2}{5R} = \frac{1}{4\pi\varepsilon_0}\frac{3q^2}{5R} \tag{25}$$

Inserting numerical values, we obtain

$$U = \frac{1}{4\pi\varepsilon_0}\frac{3 \times (92 \times 1.6 \times 10^{-19}\ \text{C})^2}{5 \times 7.4 \times^{-15}\ \text{m}}$$

$$= 1.6 \times 10^{-10}\ \text{J} \tag{26}$$

Expressed in electron-volts, this energy amounts to about 9.8×10^8 eV. Compared to the typical electric energy of an electron in an atom (about 27 eV for a hydrogen atom, see Example 25.3), this is a very large amount of energy. The energy released in nuclear fission arises from this large electric energy of the nucleus (as described in Interlude H).

SUMMARY

Energy of a system of point charges:

$$U = \tfrac{1}{2}q_1V_{\text{other}}(1) + \tfrac{1}{2}q_2V_{\text{other}}(2) + \tfrac{1}{2}q_3V_{\text{other}}(3) + \cdots$$

Energy of a system of conductors:

$$U = \tfrac{1}{2}Q_1V_1 + \tfrac{1}{2}Q_2V_2 + \tfrac{1}{2}Q_3V_3 + \cdots$$

Energy density in electric field:

$$u = \tfrac{1}{2}\varepsilon_0 E^2$$

QUESTIONS

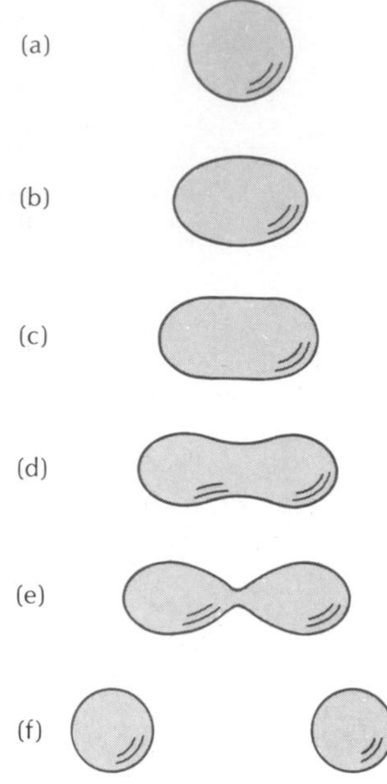

Fig. 26.6

1. Suppose we have a system of electric point charges with the electrical potential energy given by Eq. (6). By what factor will this energy change if we increase the values of all the electric charges by a factor of 2?

2. Consider a metallic sphere carrying a given amount of charge. Explain why the electric energy is large if the radius of the sphere is small. Would you expect a similar inverse proportion between the electric energy and the size of a conductor of arbitrary shape?

3. Suppose that we increase the separation between the metallic plates described in Example 2. Does this change the electric energy density? The net electric energy?

4. Consider the electric fields shown in Figures 23.14 and 23.15. In what regions of the latter figure is the electric energy density larger than in the former? Can you guess which of these electric fields has a larger energy density on the average?

5. Equation (6) suggests that the electric energy is located at the charges, whereas Eq. (18) suggests it is located in the field. How could we perform an experiment to test where the energy is located? (Hint: Energy gravitates.)

6. Figure 26.6 shows a sequence of deformations of a nucleus about to undergo fission. The volume of the nucleus and the electric charge remain constant during these deformations. Which configuration has the highest electric energy? The lowest?

7. Consider a sphere with a uniform distribution of charge over its volume. Where is the energy density within this sphere highest? Lowest?

8. What fraction of the electric energy of a sphere with a uniform distribution of charge is inside the sphere? Outside the sphere?

9. Suppose that a nucleus of charge Q, radius R, and electric energy $(1/4\pi\varepsilon_0)\,(3q^2/5R)$ fissions into two equal parts of charge $q/2$ each. The nuclear material in the original nucleus and in the final two nuclei has the same density. What is the radius of each of the two final nuclei? How does the sum of the individual electric energies of the two final nuclei compare with the initial electric energy?

10. Since the electric energy density is never negative, how can the mutual electric potential energy of a pair of opposite charges be negative?

PROBLEMS

Section 26.1

1. Consider once more the distribution of charges within the thundercloud shown in Figure 23.28. What is the electric potential energy of this charge distribution?

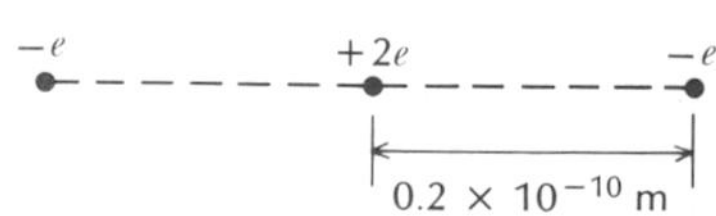

Fig. 26.7 The nucleus ($2e$) and the electrons ($-e$) of an atom of helium at an instant of time.

2. Problem 23.13 describes the arrangement of nuclear charges (positive charges) in a water molecule. Treating the nuclei as point charges, calculate the electric potential energy of this arrangement of three charges.

3. Suppose that at one instant the electrons and the nucleus of a helium atom occupy the positions shown in Figure 26.7; at this instant, the electrons are at a distance of 0.20×10^{-10} m from the nucleus. What is the electric potential energy of this arrangement? Treat the electrons and the nucleus as point charges.

4. According to the alpha-particle model of the nucleus, some nuclei consist of a regular geometric arrangement of alpha particles. For instance, the nucleus of ^{12}C consists of three alpha particles arranged on an equilateral triangle (Figure 26.8). Assuming that the distance between pairs of alpha particles is 3.0×10^{-15} m, what is the electric energy (in eV) of this arrangement of alpha particles? Treat the alpha particles as pointlike.

5. According to the alpha-particle model (see also the preceding problem), the nucleus of ^{16}O consists of four alpha particles arranged on the vertices of a tetrahedron (Figure 26.9). If the distance between pairs of alpha particles is 3.0×10^{-15} m, what is the electric energy (in eV) of this configuration of alpha particles? Treat the alpha particles as pointlike.

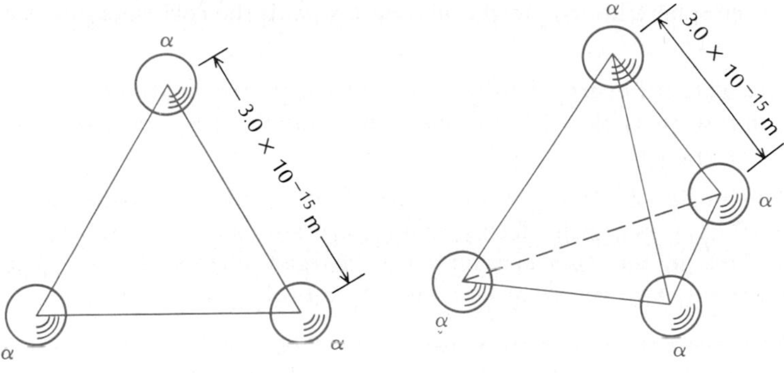

Fig. 26.8

Fig. 26.9

6. Problem 24.20 describes the Thomson model of the helium atom. The equilibrium separation of the electrons is 0.50 Å. Calculate the electric energy of this configuration. Take into account both the electric energy between the electrons and the positive charge, and the electric energy between the electrons; ignore the energy of the cloud and of the electrons by themselves.

7. Four equal particles of positive charges q and masses m are initially held at the four corners of a square of side L. If these particles are released simultaneously, what will be their speeds when they have separated by a very large distance?

*8. Two thin rods of length l carry equal charges Q uniformly distributed over their lengths. The rods are aligned, and their nearest ends are separated by a distance x (Figure 26.10). Calculate the mutual electric potential energy. Ignore the self-energy of each rod.

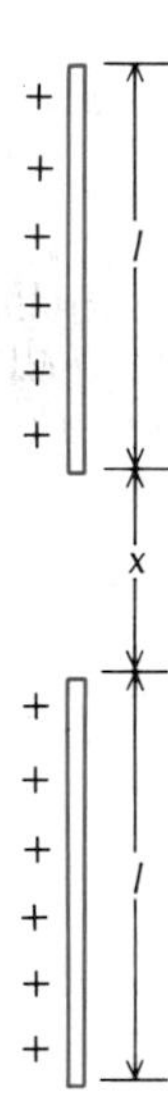

Fig. 26.10

Section 26.2

9. A pair of parallel plates, each measuring 30 cm × 30 cm, are separated by a gap of 1.0 mm. How much work must you do against the electric forces to charge these plates with $+1.0 \times 10^{-6}$ C and -1.0×10^{-6} C, respectively?

10. A charge of 7.5×10^{-6} C can be placed on a metallic sphere of radius 15 cm before the surrounding air suffers electrical breakdown. What is the electric energy of the sphere with this charge?

11. A sphere of radius R has a charge Q uniformly distributed over its volume. A thin conducting shell of radius $2R$ surrounds the sphere concentrically. The shell carries a charge $-Q$ on its interior surface. What is the electric energy of this system?

12. Pretend that an electron is a conducting sphere of radius R with a charge e distributed uniformly over its surface. In terms of e and the mass m_e of the electron, what must be the radius R if the electric energy is to equal the rest-mass energy m_ec^2 of the electron? Numerically, what is the value of R?

*13. Consider the Geiger-counter tube described in Problem 25.25. If the tube is initially uncharged, how much work must be done to bring the tube to its operating voltage of 1.0×10^3 V?

Section 26.3

14. Near the surface of the nucleus of a lead atom, the electric field has a strength of 3.4×10^{21} V/m. What is the energy density in this field?

15. The atmospheric electric field near the surface of the Earth has a strength of 100 V/m.
   (a) What is its energy density?
   (b) Assuming that the field has the same magnitude everywhere in the atmosphere up to a height of 10 km, what is the corresponding total energy?

16. Calculate the energy density in each of the electric fields, listed in the first four entries of Table 23.1. Calculate the mass densities that correspond to these energy densities.

17. The nuclei of ^{235}Pu, ^{235}Np, ^{235}U, and ^{235}Pa all have the same radii, about 7.4×10^{-15} m, but their electric charges are $94e$, $93e$, $92e$, and $91e$, respectively. Treating these nuclei as uniformly charged spheres, calculate their electric energies; express your answers in electron-volts.

18. One method for the determination of the radii of nuclei makes use of the known difference of electric energy between two nuclei of the same size but different electric charges. For instance, the nuclei ^{15}O and ^{15}N have the same size but their charges are $8e$ and $7e$, respectively. Given that the difference in electric energy is 3.7×10^6 eV, what is the nuclear radius?

19. A solid sphere of copper of radius 10 cm with a charge of 1.0×10^{-6} C is placed at the center of a thin, spherical copper shell of radius 20 cm with a charge of -1.0×10^{-6} C. Find a formula for the energy density in the space between the solid sphere and the shell. Find the total electric energy.

20. A proton can be crudely described as a uniformly charged sphere of charge e and radius 1.0×10^{-15} m. Find the electric self-energy of the proton. Express your answer in eV.

21. A spherical shell of inner radius a, outer radius b carries a charge Q uniformly distributed over its volume. What is the electric energy of this charge distribution?

22. In analogy to the electric field **E** (elecric force per unit mass) we can define a gravitational field **g** (gravitational force per unit mass).
   (a) The energy density in the electric field is $\varepsilon_0 E^2/2$. By analogy, show that the energy density in the gravitational field is $g^2/(8\pi G)$.
   (b) Calculate the gravitational field energy of the Moon due to its own gravity; treat this body as a sphere of uniform density. What is the ratio of the gravitational field energy to the rest-mass energy of the Moon?[1]

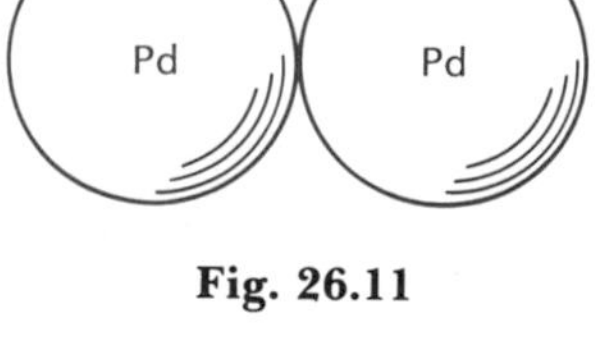

23. In symmetric fission, the nucleus of uranium (^{238}U) splits into two nuclei of palladium (^{119}Pd). The uranium nucleus is spherical with a radius of 7.4×10^{-15} m. Assume that the two palladium nuclei adopt a spherical shape immediately after fission; at this instant, the configuration is as shown in Figure 26.11. The size of the nuclei in Figure 26.11 can be calculated from the size of the uranium nucleus because the nuclear material maintains a constant density (the initial nuclear volume equals the final nuclear volume).

Fig. 26.11

[1] Warning: This problem does not take the gravitational *interaction* energy into account. The gravitational interaction energy density is [mass density] × [gravitational potential]. The total gravitational energy is the sum of the field energy and the interaction energy; this total gravitational energy is always negative.

(a) Calculate the electric energy of the uranium nucleus before fission.
(b) Calculate the total electric energy of the palladium nuclei in the configuration shown in Figure 26.11, immediately after fission. Take into account the mutual electric potential energy of the two nuclei and also the individual electric energies of the two palladium nuclei by themselves.
(c) Calculate the total electric energy a long time after fission when the two palladium nuclei have moved apart by a very large distance.
(d) Ultimately, how much electric energy is released into other forms of energy in the complete fission process (a) through (c)?
(e) If 1 kg of uranium undergoes fission, how much electric energy is released?

*24. Using the model described in Problem 24.17 for the charge distribution of a neutron, calculate the electric self-energy of a neutron. Express your answer in eV.

*25. A long rod of plastic of radius a has a charge of λ coulomb per unit length uniformly distributed over its volume. The rod is surrounded by a concentric cylinder of sheet metal of radius b with a charge of $-\lambda$ coulomb per unit length on its interior surface.
(a) What is the energy density (as a function of radius) in the space between the rod and the cylinder?
(b) What is the energy density in the volume of the rod?
(c) What is the total electric energy per unit length?

NUCLEAR FISSION[1]

The nucleus of the atom is very small and very dense. It consists of protons and neutrons closely packed together. Since the protons in the nucleus carry a positive electric charge, they exert large repulsive electric forces on one another; correspondingly, the electric energy stored within the nucleus is large. In a heavy nucleus, such as the nucleus of uranium, the disruptive electric forces that push the protons apart will sometimes overcome the cohesive "strong" forces that pull the protons and neutrons together — the nucleus splits or fissions into two or more fragments. When this happens, the large electric energy stored within the nucleus will be (partially) converted into kinetic energy of the fission fragments. Nuclear bombs and nuclear reactors derive almost all their power from this conversion of electric energy into kinetic energy. Thus, the "nuclear" energy released in fission devices is primarily electric energy.

H.1 THE NUCLEUS

Figure H.1 shows a nucleus of uranium (the isotope ^{238}U) consisting of 92 protons and 146 neutrons packed together in a spherical region. This is one of the largest, heaviest nuclei and yet its diameter is only 15×10^{-13} cm. Figures H.2–H.4 show some of the smallest, lightest nuclei: ordinary hydrogen (the isotope ^{1}H), heavy hydrogen or deuterium (the isotope ^{2}H, or D), and helium (the isotope ^{4}He); all of these are less than 4×10^{-13} cm across.

Fig. H.1 Uranium nucleus consisting of 92 protons (color) and 146 neutrons (white).

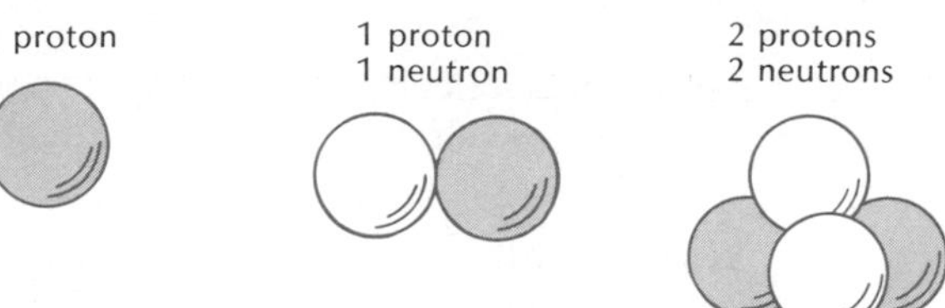

Fig. H.2 Hydrogen nucleus: one proton.

Fig. H.3 Deuterium nucleus: one proton and one neutron.

Fig. H.4 Helium nucleus: two protons and two neutrons.

The constituents of a nucleus — protons and neutrons — are usually called **nucleons.** Thus, the uranium nucleus of Figure H.1 has 238 nucleons. In general, the size of a nucleus depends on the number of nucleons that it contains — the more nucleons, the larger the size. Scattering experiments (such as the Rutherford experiment, described in Section C.1) as well as other experiments indicate that the radius of any given nucleus is proportional to the cube root of the number of its nucleons:

$$R = 1.2 \times 10^{-13} A^{1/3} \text{ cm} \tag{1}$$

where R is the radius and A the number of nucleons in the nucleus, often called the **mass number.**

The proportionality between R and $A^{1/3}$ implies that the number of nucleons per unit volume is the same for all nuclei,

$$\frac{A}{(4\pi/3)R^3} = \frac{A}{(4\pi/3)(1.2 \times 10^{-13} A^{1/3})^3 \text{ cm}^3}$$

$$= 1.38 \times 10^{38} \text{ nucleons/cm}^3 \tag{2}$$

Since the mass of each nucleon is about 1.7×10^{-24} g, the mass density of the nuclear material is $1.7 \times 10^{-24} \times 1.38\ 10^{38}$ g/cm^3 = 2.3×10^{14} g/cm^3 — one cubic centimeter of pure nuclear material would weigh 230 million metric tons! Note that the volume per nucleon is $1/(1.38 \times 10^{38})$ cm^3 and hence the average distance

[1] This chapter is optional.

from one nucleon to its nearest neighbor is $1/(1.38 \times 10^{38})^{1/3}$ cm $\cong 2 \times 10^{-13}$ cm. By comparing this with the radius of a proton or neutron, about 1×10^{-13} cm, we can see that inside the nucleus the nucleons are so tightly packed together that they almost touch.

Since the protons within a nucleus are at such short distances from one another, they exert very large repulsive electric forces on one another. Two neighboring protons, separated by a center-to-center distance of $\sim 2 \times 10^{-13}$ cm, experience a repulsive force of

$$F = \frac{1}{4\pi\varepsilon_0}\frac{e^2}{r^2} = 9.0 \times 10^9 \times \frac{(1.6 \times 10^{-19})^2}{(2 \times 10^{-15})^2}\ \text{N}$$

$$= 58\ \text{N} = 13\ \text{lb} \qquad (3)$$

Acting on a mass of (only) 10^{-24} g, this represents a colossal force.

Obviously, some extra force must be present in the nucleus to prevent it from instantaneously bursting apart under the influence of the mutual Coulomb repulsions of the protons. This extra force is the **strong force,** already mentioned in Section 6.1. This force acts equally between any two nucleons, regardless of whether they are protons or neutrons (the force is "charge independent"). Figure H.5 is a rough plot of the potential energy associated with the strong nucleon–nucleon force. For internucleon distances between $\sim 1 \times 10^{-13}$ cm and $\sim 2 \times 10^{-13}$ cm, the force is attractive and much larger than the Coulomb force — it can be more than 1000 lb. For internucleon distances smaller than $\sim 1 \times 10^{-13}$ cm, the force becomes repulsive, i.e., the nucleons have a hard core which resists interpenetration. For distances larger than $\sim 2 \times 10^{-13}$ cm, the force vanishes. Thus, in contrast to the Coulomb force, which fades only gradually and reaches out to large distances, the strong force cuts off sharply and has only a short range. In order to feel the strong force the nucleons must be touching or almost touching, i.e., the force acts only between nearest neighbors.

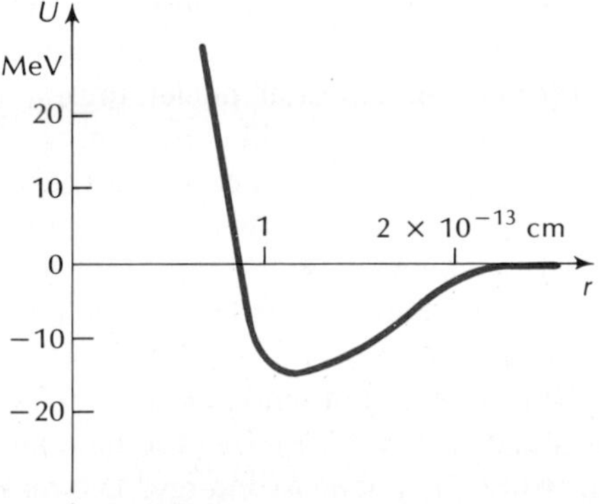

Fig. H.5 Potential energy as a function of the distance between two nucleons (for nucleons of parallel spin).

As a consequence of the short-range character of the strong force, a nucleon deep inside the nucleus does not experience any net force; the nucleon only interacts with its nearest neighbors, and since these pull it with equal force in almost all directions, the net force on the nucleon is zero or nearly zero (Figure H.6). However, a nucleon at the nuclear surface only has neighbors on the side that is toward the interior and hence these will exert a net force pulling the nucleon inward (Figure H.6). Altogether this means that nucleons are more or less free to wander about the interior of the nucleus, but whenever they approach the nuclear surface, the strong forces pull them back and prevent their escape.

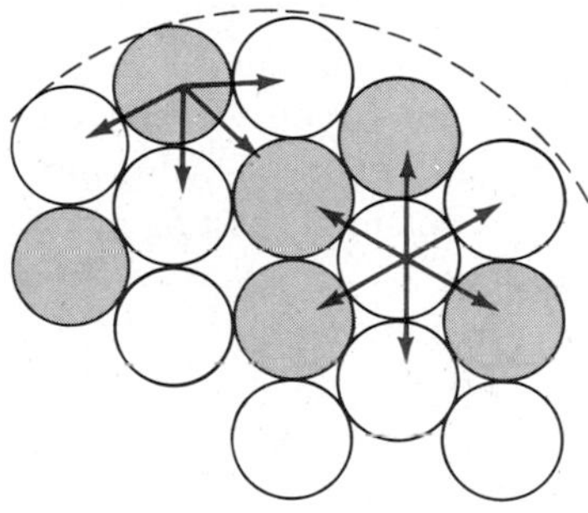

Fig. H.6 Forces on a nucleon at the nuclear surface, and forces on a nucleon in the nuclear interior.

This suggests that the nucleons in the nucleus behave somewhat like the water molecules in a drop of water; such molecules are free to wander through the volume of the drop, but when they approach the water surface, intermolecular forces hold them back. This similarity between nuclei and drops of water rests on a similarity of the laws of force. The intermolecular force has general features rather similar to those displayed in Figure H.5: it is attractive over a short range and then becomes strongly repulsive when the molecules begin to interpenetrate. The hard repulsive core of the potential makes the water nearly incompressible, while the short-range attraction provides a cohesive force that prevents water droplets from falling apart. The balance of attraction and repulsion encourages water molecules to stay at a particular distance from one another and this gives water a particular, uniform density.

Because of the similarities between a liquid and nuclear material, the nucleus can be crudely regarded as a droplet of incompressible "nuclear fluid" of uniform density. The fluid is, of course, made of nucleons, but for some purposes we can ignore the individual nucleons and calculate the properties of nuclei in terms of the gross properties of a liquid. For example, the spherical shape adopted by most nuclei can be easily understood as follows: any nucleon located on the surface of a globule of nuclear fluid experiences an inward force pulling it back into the volume and consequently the fluid tends to shrink its exposed surface to the smallest value compatible with its (fixed) volume. Since a sphere has the least surface area for a given volume, the globule of fluid will take the shape of a spherical droplet.

In a stable nucleus, the repulsive Coulomb forces between the protons are held in check by the attractive strong forces. To achieve this balance of forces, the presence of neutrons is an advantage: a nucleus with more neutrons will have a larger size and therefore a larger average distance between pairs of protons — the neutrons in the nucleus dilute the repulsive effect of the Coulomb forces. Consequently, all stable nuclei, with the exception of hydrogen and one isotope of helium, contain at least as many neutrons as protons; heavy nuclei, such as uranium, contain substantially more neutrons than protons.

Figure H.7 is a plot of the number of protons vs. the number of neutrons for all known nuclei. The number of protons is represented by Z and the number of neutrons by N (so that $A = Z + N$). On this plot, the black circles indicate the stable nuclei; the colored circles indicate the unstable nuclei, i.e., the radioactive isotopes. Note that there is no stable nucleus beyond bismuth ($Z = 83$). However, several elements beyond bismuth have some isotopes with very long half-lives; these are therefore almost stable and they occur naturally.

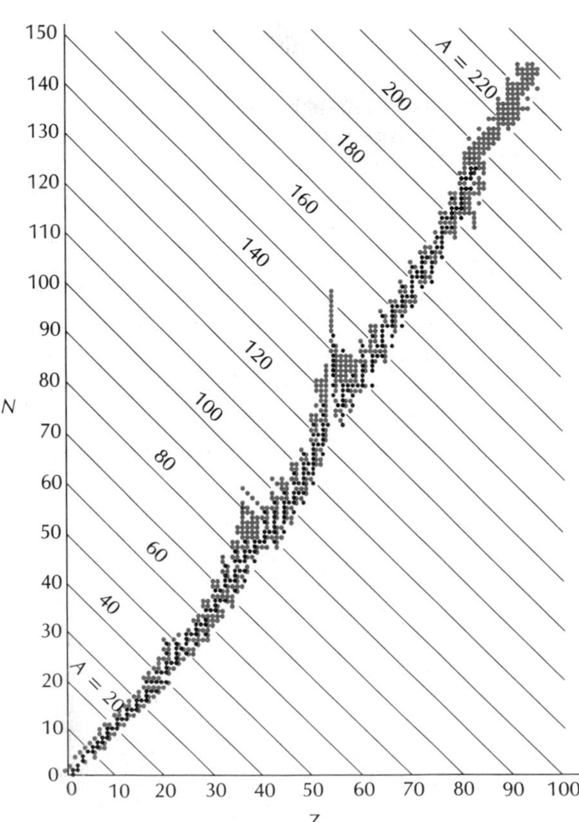

Fig. H.7 Number of protons (Z) vs. number of neutrons (N) for stable nuclei (black dots) and unstable nuclei (red dots).

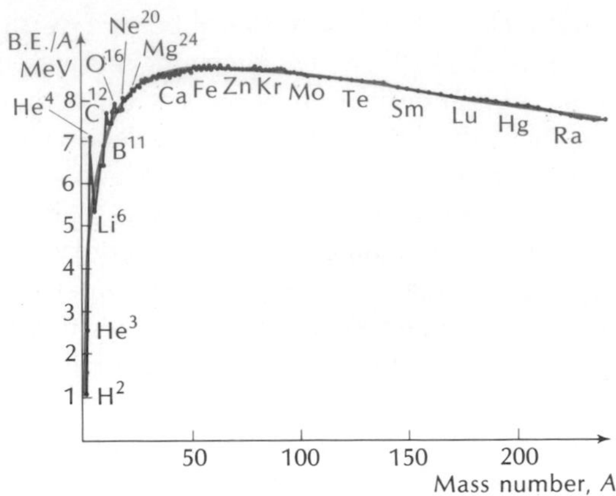

Fig. H.8 Average binding energy per nucleon vs. mass number.

The energy stored in a nucleus is a sum of the potential energies contributed by electric and strong forces and the kinetic energy of the nucleons. The negative of this energy is called the **binding energy** (B.E.); this is the amount of energy released when the nucleus is assembled out of its constituent nucleons. Figure H.8 is a plot of B.E./A, the binding energy divided by the number of nucleons, or the average binding energy per nucleon. The curve plotted in Figure H.8 is called the **curve of binding energy.** Incidentally, in this figure the energy unit is the MeV,

$$1 \text{ MeV} = 10^6 \text{ eV} = 1.6 \times 10^{-13} \text{ J} \tag{4}$$

This unit is widely used in nuclear physics.

The binding energy of a typical nucleus is a rather large amount of energy. As may be seen from Figure H.8, the average binding energy per nucleon is in the vicinity of 8 MeV for almost all nuclei; thus, a nucleus with a mass number A has a binding energy of about $A \times 8$ MeV. To put this number in perspective, we may compare it with the rest-mass energy of the nucleons. Each nucleon (neutron or proton) has a rest-mass energy m_nc^2; hence A nucleons have a rest-mass energy $A \times m_nc^2 = A \times 1.5 \times 10^{-10}$ J $= A \times 9.4 \times 10^2$ MeV. Thus the ratio of binding energy to rest-mass energy is about $8/(9.4 \times 10^2)$, i.e., the binding energy is almost 1% of the rest-mass energy! The mass associated with the binding energy is B.E./c^2; this mass is carried away by the energy released during the assembly of the nucleus from its constituent protons and neutrons. The mass of a typical nucleus is therefore about 1% less than the sum of the masses of these protons and neutrons. The mass difference is called the **mass defect,**

$$\begin{aligned}[\text{mass defect}] &= [\text{mass of } N \text{ neutrons and } Z \text{ protons}] \\ &\quad - [\text{mass of nucleus}] \\ &= \text{B.E.}/c^2 \end{aligned} \tag{5}$$

Experimental values of the binding energy of a nucleus usually are obtained via the mass defect. A precise measurement of the mass of the nucleus is compared with the sum of the masses of the constituent protons and neutrons; the difference, or mass defect, then gives the binding energy according to Eq. (5). This is how the experimental points in Figure H.8 were obtained.

H.2 FISSION

Although for most nuclei the binding energy per nucleon is in the vicinity of 8 MeV (see Figure H.8), for heavy nuclei (say, $A > 200$) the binding energy per nucleon is somewhat smaller. The reduction of the binding energy is due to the increasing importance of the electrostatic repulsion. This suggests that it may be energetically favorable for a heavy nucleus to split up into two fragments, forming two lighter nuclei. For example, consider the splitting of a ^{238}U nucleus into two equal fragments:

$$^{238}\text{U} \rightarrow {}^{119}\text{Pd} + {}^{119}\text{Pd} \tag{6}$$

This is a **symmetric fission** reaction: a nucleus of mass number 238 and charge 92e splits into two nuclei, each of mass number 119 and charge 46e.

The binding energy of the ^{238}U nucleus is approximately 2100 MeV and that of each ^{119}Pd nucleus is 1140 MeV. The energy released in the reaction (6) is therefore

$$-2100 \text{ MeV} + 2 \times 1140 \text{ MeV} = 180 \text{ MeV} \tag{7}$$

Thus, this fission reaction is *permitted* by energy conservation — but what is permitted is not compulsory. The process of fission begins with a slight elongation of the nucleus (Figure H.9b); the elongation then increases, developing two bulges joined by a waist (Figures H.9c and d); finally, the waist pinches off and the two fragments separate (Figures H.9e and f). Obviously, the attractive strong forces resist the elongation of the nucleus and it is only when the nucleus splits (Figures H.9e and f) that these strong forces irrevocably lose out to the repulsive Coulomb forces.

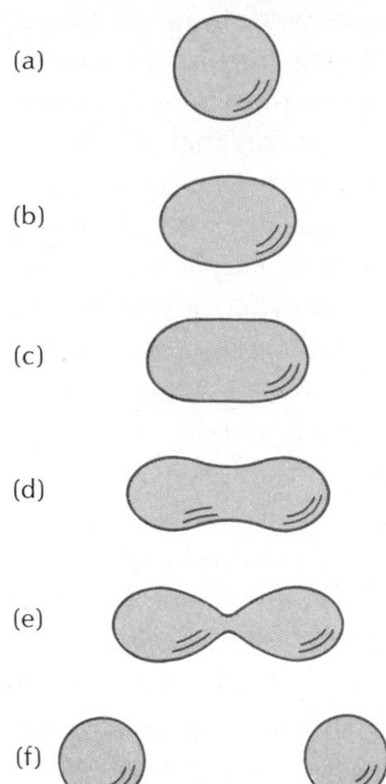

Fig. H.9 Stages of deformation in the fission of a nucleus.

Figure H.10 shows a plot of the potential energy of the drop of nuclear fluid as a function of its deformation; zero deformation corresponds to Figure H.9a; extreme deformation corresponds to Figure H.9f, in which the drop has split into two well-separated drops. Although the energy of the undeformed drop is higher than the energy of the separated drops, the intermediate stages of deformation have the highest energy; thus, there is a potential barrier which tends to prevent fission.

In the case of ^{238}U, spontaneous fission is a rare occurrence; only 1 nucleus in 2×10^6 will decay by spontaneous fission, the others will decay by alpha emission. Furthermore, when spontaneous fission does occur, it is rarely symmetric [as in Eq. (10)]; more often than not the fission is asymmetric, the nucleus splitting into two unequal fragments. However, the mass difference between the fragments is usually not all that large (typically 30%) and the energy released is not all that far from the value calculated in Eq. (7).

The energy of 180 MeV released by the reaction (6)

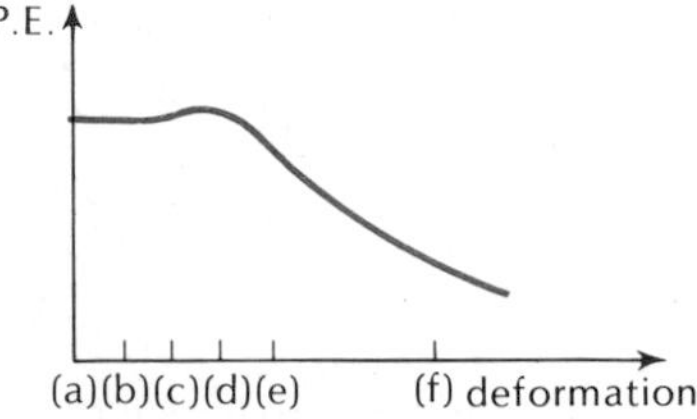

Fig. H.10 Potential energy as a function of deformation. The marks on the horizontal axis correspond to the stages of deformation shown in Figure H.9. The potential energy is maximum at stage (d).

represents the net work done by the electric and the strong forces during the fission process. This energy will emerge in the form of kinetic energy of the fission fragments, kinetic energy that is impressed on the fragments by the repulsive Coulomb force. The Coulomb force does positive work (it pushes the fragments apart, increasing their kinetic energy) while the strong force does negative work (it attempts to pull the fragments together, decreasing their kinetic energy). Thus, the initial electric energy of the nucleus must not only supply all the kinetic energy of the fragments, but it must also supply the energy absorbed by the strong force.

The total energy released as a consequence of the fission of a uranium nucleus is actually somewhat larger than 180 MeV. The fission fragments are very unstable — they contain both an unacceptable excess of internal (vibrational) energy and an unacceptable excess of neutrons. To eliminate these excesses, the fission fragments undergo a series of radioactive decays involving the ejection of neutrons, beta rays, gamma rays, and neutrinos. Some of these particles are ejected almost instantaneously, while others suffer a slight delay. These secondary emissions release roughly an extra 20 MeV and bring the total energy yield per fission to about 200 MeV. Table H.1 shows how the total energy is distributed, on the average, among fission fragments and other ejecta.

Table H.1 DISTRIBUTION OF ENERGY IN FISSION

Fission fragments (kinetic energy)	165 ± 5 MeV
Neutrons (kinetic energy)	5 ± 0.5
Gamma rays, instantaneous	7 ± 1
Beta rays, delayed	7 ± 1
Gamma rays, delayed	6 ± 1
Neutrinos, delayed	10
Total energy per fission	200 ± 6 MeV

Note that, according to this table, the average kinetic energy of the fission fragments is somewhat lower than was calculated above [see Eq. (7)]; this is so because asymmetric fission, which predominates, yields somewhat lower kinetic energies than symmetric fission.

We can gain a better appreciation of the large amount of energy released in fission by looking at the following numbers. The total energy released by the complete fission of 1 kg of uranium (a lump slightly larger than a golf ball) is 200 MeV times the number of atoms in 1 kg of uranium, i.e.,

$$200\,\frac{\text{MeV}}{\text{atom}} \times 6.02 \times 10^{23}\,\frac{\text{atoms}}{\text{mole}} \times \frac{1000}{238}\,\text{moles}$$

$$= 5.1 \times 10^{26}\ \text{MeV}$$

$$= 8.1 \times 10^{13}\ \text{J} \qquad (8)$$

This is equivalent to the energy released in the burning of 550,000 gal. of gasoline. It is also equivalent to the energy released in the explosion of about 20,000 tons of TNT.[2]

H.3 CHAIN REACTIONS

The rate of spontaneous fission in uranium is very low. Consequently the rate of release of energy is also very low and quite inadequate for applications such as nuclear reactors or nuclear bombs. However, the rate of fission increases drastically if the uranium is exposed to a flux of neutrons. The impact and absorption of a neutron by a uranium nucleus will deform the nucleus and set it in violent vibration which may split it apart (Figure H.11). This results in neutron-induced fission reactions such as

$$\text{neutron} + {}^{238}\text{U} \rightarrow {}^{145}\text{Ba} + {}^{94}\text{Kr} \qquad (9)$$

The rate of neutron-induced fission depends on the supply of neutrons and on their kinetic energy. What

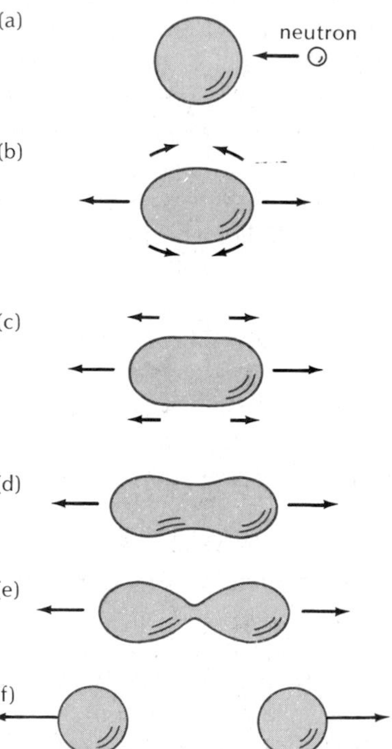

Fig. H.11 Impact of a neutron on a uranium nucleus causes vibrations that lead to fission.

makes fission reactions such as in Eq. (9) interesting is that the reaction itself supplies neutrons — both fission fragments in the reaction (9) are very neutron rich and they almost instantaneously eject two or three neutrons. Hence the net reaction is

[2] One ton of TNT releases 10^9 cal $= 4.2 \times 10^9$ J.

neutron + $^{238}U \rightarrow$ fission fragments

$$+ 2 \text{ or } 3 \text{ neutrons} \quad (10)$$

The neutrons released in this first fission can then strike other uranium nuclei and induce their fission, and the neutrons released there can induce further fissions, etc. If no neutrons, or only a few neutrons, are lost from this avalanche, then the result is a self-sustaining **chain reaction** (Figure H.12). The rate of fission and the rate of release of energy grow drastically with time. For example, if on the average two neutrons released in each fission reaction succeed in generating further fission reactions and further neutrons, then the numbers of fission reactions in successive steps of the chain will be 2, 4, 8, 16, 64, . . . If this geometric growth continues unchecked, the rate of release of energy will become explosive.

In the case of ^{238}U, conditions are not favorable for sustaining a chain reaction. Uranium-238 is a fairly stable nucleus; it will fission only if struck a hard blow by an energetic neutron — the kinetic energy of the incident neutron must be at least 1.2 MeV. The neutrons released by the fission of a uranium nucleus are energetic enough (see Table H.1), and when they collide with other nuclei they will occasionally induce a fission; but by far the most likely outcome of a collision is inelastic scattering, i.e., the neutron bounces off the nucleus with some loss of kinetic energy. Successive collisions of this kind gradually reduce the kinetic energy of the neutrons and remove their ability to induce fission — the neutrons are effectively lost from the fission chain.

In the case of ^{235}U, conditions for sustaining a chain reaction are much more favorable. This nucleus is less stable than ^{238}U; it will fission even if the kinetic energy of the incident neutron is very low. In fact, at low kinetic energy the fission reaction is more likely than at high kinetic energy; the reason for this is that the binding energy released when a neutron is captured by a ^{235}U nucleus is by itself quite sufficient to trigger the fission and, the kinetic energy of the neutron not being needed, it is advantageous for the neutron to move at low speed, since this lengthens the time it spends in the vicinity of the nucleus and therefore enhances the probability that a reaction will take place.

Besides ^{235}U, two other isotopes exist in which chain reactions are practicable. Both resemble ^{235}U in that the spontaneous decay rate is low — so that they can be held in storage without serious loss — and in that the rate of induced fission favors chain reactions: one is ^{233}U and the other is ^{239}Pu, an isotope of plutonium. The latter is very fissionable, but it is not found in ores on the Earth; it can only be obtained by artificial means, by nuclear alchemy.

In a given mass of ^{235}U or ^{239}Pu, neutrons produced by spontaneous fission or stray neutrons coming from elsewhere can initiate the first step of the chain reaction. Whether the reaction keeps going depends on the **multiplication factor,** i.e., the factor by which the number of neutrons increases between one step and the next along the fission chain. If no neutrons are lost from the fission chain, then the multiplication factor is simply equal to the average number of neutrons released per fission; but if some neutrons are absorbed by impurities or escape without inducing a fission, then the multiplication factor will be smaller. The mass of fissionable material is said to be in a **critical** condition if the multiplication factor is unity; in this case the chain reaction merely proceeds at a constant rate — as in a nuclear reactor. The mass is **supercritical** if the multiplication factor exceeds unity; in that case the chain reaction proceeds at an ever-increasing runaway rate leading to an explosion — as in a nuclear bomb.

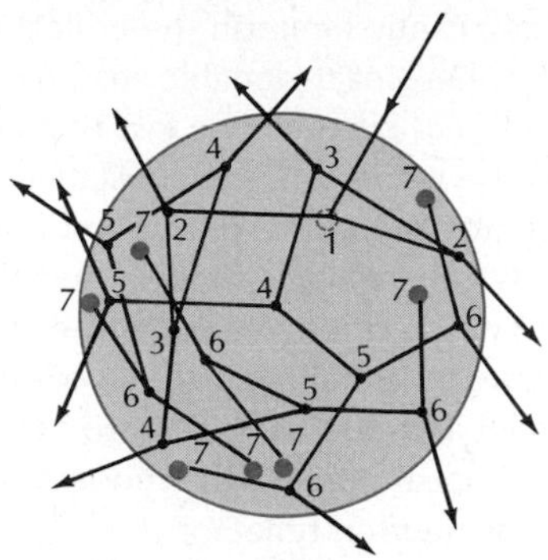

Fig. H.12 A chain reaction in a supercritical mass of fissionable material. The reaction starts at (1) with the arrival of a neutron from the outside. The diagram shows the first seven steps of the chain reaction. Each fission absorbs one neutron and releases two; occasionally a neutron escapes from the surface of the mass and is lost.

Neutrons can be lost from the fission chain by several mechanisms. For example, if the ^{235}U is not pure but contains an admixture of ^{238}U, then neutrons will be lost when they are absorbed by ^{238}U nuclei in transmutation reactions without fission. Even if the ^{235}U is pure, neutrons will still be lost when they escape from the surface of the piece of ^{235}U material. The rate of escape of the neutrons depends on the size and shape of the piece of ^{235}U material. The best shape is a sphere since in this case the surface is as distant as possible from the bulk of the material. Furthermore, neutrons are less likely to escape from a large sphere than from a small sphere — in the former a neutron released at some average point within the bulk of the material has a longer distance to travel to the surface and is therefore more likely to trigger a fission reaction while on the way. This indicates that a sphere of fissionable material must have a minimum size if it is to maintain fission. For ^{235}U, the sphere of minimum size has a diameter of 18 cm; the corresponding minimum mass — called the **critical mass** — is 53 kg.

We can significantly diminish the critical mass if we surround the sphere of fissionable material by a neutron "reflector" that prevents the loss of neutrons. The reflector must be made of some substance whose nuclei strongly scatter neutrons, but do not absorb them. In a nuclear bomb, a thick shell of beryllium metal makes a good neutron reflector; neutrons that escape from the sphere of fissionable material collide with the beryllium nuclei and bounce back into the fissionable material. In a nuclear reactor, the moderator (see Section H.6) acts as neutron reflector.

We can further diminish the critical mass by compressing the fissionable material to higher than normal densities. For example, compression of the material to twice its normal density diminishes the critical mass by a factor of 4. In a denser material, a neutron seeking to escape encounters more nuclei along its path to the surface and therefore is more likely to trigger a fission on the way.

H.4 THE BOMB

The simplest fission bomb, or **A-bomb,** consists of two pieces of ^{235}U such that separately their masses are less than the critical mass, but jointly their masses add up to more than the critical mass. To detonate such a bomb, the two pieces of ^{235}U, initially at a safe distance from one another, are suddenly brought closely together. The assembly of the two subcritical masses into a single supercritical mass must be carried out very quickly; if the two masses are brought together slowly, a partial explosion (predetonation) will push them apart prematurely, before the chain reaction can release its full energy — the explosion fizzles. The device commonly used for the assembly of the two pieces of uranium consists of a gun which propels one piece of uranium toward the other at high speed (Figure H.13); the propellant is an ordinary chemical high explosive.

A more sophisticated fission bomb consists of a (barely) subcritical mass of ^{239}Pu; if this is suddenly compressed to a higher than normal density, it will become supercritical. The sudden compression is

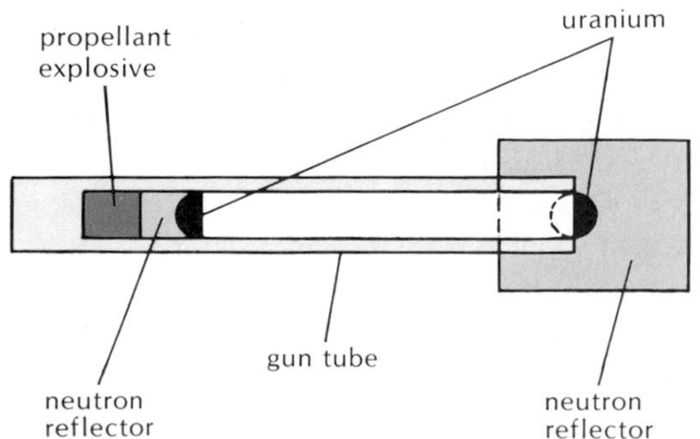

Fig. H.13 A fission bomb using a gun device.

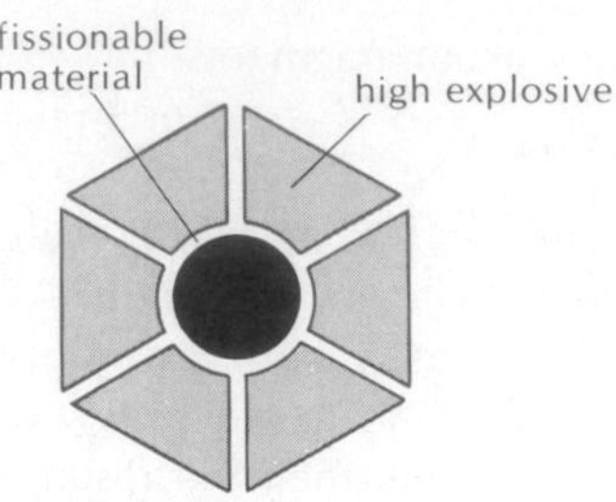

Fig. H.14 An implosion device.

achieved by the preliminary explosion of a chemical high explosive such as TNT. If this explosive has been carefully arranged in a shell around a sphere of ^{239}Pu (Figure H.14), then its detonation will crush the sphere of ^{239}Pu into itself; this implosion of the plutonium very suddenly brings its density to the supercritical value and triggers the chain reaction. The implosion technique is used with ^{239}Pu because this isotope has a strong tendency to predetonate; if one were to use the gun technique to bring together two subcritical masses of ^{239}Pu, the chain reaction would start while the masses were still moving toward one another; the consequent premature explosion would push the masses apart and prevent a full development of the chain reaction. The implosion technique assembles the supercritical mass much faster and therefore avoids the problem of a premature explosion.

During World War II, a scientific–military–industrial complex known as the Manhattan District produced three A-bombs: one plutonium bomb exploded at Alamogordo, New Mexico, on July 16, 1945, another plutonium bomb exploded at Nagasaki, Japan, on August 8, 1945, and one uranium bomb exploded at Hiroshima, Japan, on August 5, 1945. All of these bombs had yields of about 20 kilotons, i.e., an explosive energy equivalent to that of 20,000 tons of TNT. This is the energy released by the fission of 1 kg of uranium (or plutonium). Hence these devices were quite inefficient — only a small fraction of the total mass of fissionable material actually underwent fission; the rest was merely scattered in all directions, blown apart before it had a chance to react.

The uranium used for bombs has to be highly purified, "weapons-grade" ^{235}U. Uranium ores contain a mixture of 99.3% of the undesirable isotope ^{238}U and only 0.7% of the isotope ^{235}U. Since these isotopes are chemically identical, their separation is very laborious. The separation process depends on the small difference in the masses: in a gaseous compound of uranium, such as UF_6, at a given temperature, the molecules containing ^{235}U have a slightly higher average speed than the molecules containing ^{238}U, and they will diffuse slightly faster through a porous membrane; hence such a membrane acts as a (partial) filter that separates ^{235}U from ^{238}U. This is the basis of the

Fig. H.15 The bomb dropped on Hiroshima. This device was 120 in. long and weighed 7000 lb.

Fig. H.16 The bomb dropped on Nagasaki. This device was 128 in. long and weighed 10,000 lb.

gaseous-diffusion process which is still the main source of highly enriched ^{235}U.

Plutonium is not found in nature, except in insignificant trace amounts. It is made artificially, by transmutation of uranium in a nuclear reactor.

As can be seen from Figures H.15 and H.16, the bombs used in World War II were rather cumbersome. Advances in the technology of shaping chemical high explosives used for the implosion of plutonium have made it possible to construct bombs in which the lump of plutonium is only about as large as a golf ball; the overall diameter of such a bomb is only about a foot, and yet it releases an amount of energy in the kiloton range.

Much higher yields can be achieved by taking advantage of nuclear **fusion.** As Figure H.8 shows, the binding energy of light nuclei is relatively low — when two such light nuclei are made to fuse together to form a heavier nucleus, energy will be released. Fusion is the opposite of fission: in the former two nuclei merge into one, in the latter one nucleus splits into two. Furthermore, the energy released in fusion is strong energy while the energy released in fission is Coulomb energy. The strong force favors fusion since, by merging, the nuclei reduce their surface area. The Coulomb repulsion between the two nuclei opposes fusion, but in the case of light nuclei the strong force overcomes this opposition.

The heat given off by the Sun is due to a fusion reaction called hydrogen burning: hydrogen nuclei fuse together to make helium nuclei. This reaction involves several intermediate steps, which were discovered by theoretical calculations by H. Bethe.[3] This reaction cannot be duplicated on Earth because it will only proceed at extremely high temperatures and pressures, such as are found near the center of the Sun. However, some fusion reactions involving deuterium and tritium (the isotopes ^2H and ^3H) can be made to work on Earth:

$$^2H + {}^2H \rightarrow {}^3He + n \tag{11}$$

$$^2H + {}^2H \rightarrow {}^3H + p \tag{12}$$

$$\underline{^2H + {}^3H \rightarrow {}^4He + n} \tag{13}$$

$$5\,{}^2H \rightarrow {}^3He + {}^4He + p + 2n + 24.3\ \text{MeV}$$

The net result of the three reactions taken together is the disappearance of five ^2H nuclei and the formation of ^3He, ^4He, one free proton, and two neutrons, with the release of 24.3 MeV. The amount of energy released per nucleon of reactant is 24.3 MeV per 10 nucleons = 2.43 MeV per nucleon, whereas for the fission of uranium the energy released per nucleon of reactant is 200 MeV per 235 nucleons = 0.85 MeV per nucleon. Thus the fusion of a given mass of ^2H will yield about three times as much energy as the fisson of an equal mass of ^{235}U.

The reactions (11)–(13) are called **thermonuclear** because they will proceed only at very high temperature and pressure. The requisite temperatures and pressures are attained at the place of explosion of an A-bomb. Hence, the fusion reactions can be initiated by exploding a fission bomb next to a mass of heavy hydrogen; this results in self-sustained explosive "burning" of the hydrogen nuclei. This so-called **H-bomb** is really a fission–fusion device in which fission triggers fusion. What is more, the fusion reactions release a large number of energetic neutrons [see Eqs. (11) and (13)] which can be used to further enhance the vio-

[3] Hans A. Bethe, 1906–, German, later American, physicist. He formulated the theory of nuclear fusion in stars in 1937 and received the Nobel Prize for this in 1967; during the war he was director of the theoretical-physics division at Los Alamos.

lence of the explosion: the trick is to surround the fusion bomb by a blanket of cheap, natural uranium, consisting of mainly ^{238}U; although this isotope will not maintain a chain reaction, it will fission when exposed to the large flux of neutrons from the fusion. This kind of H-bomb is a fisson–fusion–fission device; typically, one-half of the total energy yield is due to fusion, one-half is due to fission. The fission of a large amount of uranium leaves behind a residue of highly radioactive fission products; hence fission–fusion–fission bombs are **dirty,** i.e., they generate a large volume of radioactive fallout.

The total energy yield of an H-bomb is of the order of one or several megatons[4] — roughly a thousand times the yield of an A-bomb. Explosions of up to 60 megatons have been tried with great success, and there seems to be no limit to the suicidal madness that nature will let us get away with.

H.5 THE EFFECTS OF NUCLEAR WEAPONS

The energy released by a nuclear bomb takes several forms as it emerges from the place of explosion: blast and shock in air, thermal radiation (including light), immediate nuclear radiation (neutrons and gamma rays), and delayed nuclear radiation (residual radioactivity, fallout). The exact distribution of the energy among these several forms depends on the type of bomb and on the altitude at which it explodes. For an air burst — an explosion in air at low or medium altitude — the energy distribution is typically as shown in Figure H.17.

The chronological sequence of phenomena in the air burst of a 1-megaton bomb is as follows (Figures H.18–H.23).

Within a few microseconds after the initiation of the explosion, the nuclear reactions will have released their energy. These reactions also release a large number of neutrons and gamma rays which immediately escape from the scene of the explosion (Figure H.18). Although these neutrons and gamma rays are gradually absorbed by the air, their energy is high and they penetrate the air for well over a mile with lethal intensity.

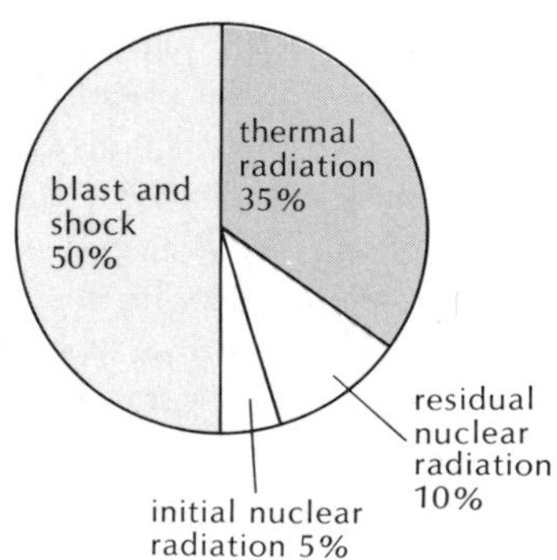

Fig. H.17 Distribution of energy released in a typical air burst at an altitude below 30,000 m. (From S. Glasstone, ed., *The Effects of Nuclear Weapons.*)

[4] One megaton = 1000 kilotons.

The energy released in the explosion raises the fragmented residues of the bomb to an enormous temperature — tens of millions of degrees — and vaporizes them instantly. The incandescent vapor radiates X rays which are quickly absorbed by the surrounding air, heating it. The heated air and the vaporized bomb residues form a spherical, hot, luminous mass of gas — the **fireball.** Figure H.24 is a high-speed photograph of such a fireball. For a 1-megaton explosion, at 1.8 s after time zero, the fireball is 1.2 mi across (Figure H.19); it continues to grow to a maximum diameter of 1.4 mi. The fireball attains a temperature of 6000 to 7000 K — hotter than the surface of the Sun. It emits intense thermal radiation, including intense light. Meanwhile a **blast wave,** caused by the sudden increase of pressure, travels outward from the center of explosion.

At 5 s after time zero, the blast wave has spread to a radius of almost 2 mi and begins to reflect from the ground (Figure H.20). The direct and the reflected wave merge and reinforce each other near the ground — they form a single, strong wave front called the **Mach front.** The overpressure at the wave front is 16 lb/in.^{2} at this time. The emission of thermal radiation gradually decreases as the fireball cools, but significant amounts continue to be emitted until 10 s after time zero.

At 11 s after time zero, the blast wave has spread out to a radius of 3 mi and the overpressure at the wave front has dropped to 6 lb/in.^{2} (Figure H.21). Behind the front, surface winds of 180 mi/h blow radially outward for a few seconds. Most of the thermal radiation from the fireball (about 80%) will already have been emitted by this time.

At 37 s after time zero, the blast wave reaches out to 10 mi, with an overpressure of about 1 lb/in.^{2} (Figure H.22). Meanwhile, the hot gas of the fireball begins to rise like a hot-air balloon, at about 250 mi/h. The rising gas generates a vertical updraft as well as horizontal afterwinds that rush radially inward along the surface. These winds push dirt and dust into a column or stem below the fireball. As the fireball expands and cools, the vaporized bomb residues condense and form a large radioactive cloud. This cloud and the stem below take on the characteristic mushroom shape (Figure H.25).

The mushroom grows, reaching a height of 7 mi 110 s after the explosion (Figure H.23). At this time, the mushroom cloud is still shooting upward at about 150 mi/h while along the surface the afterwinds reach 200 mi/h or more. Ultimately, the mushroom cloud at-

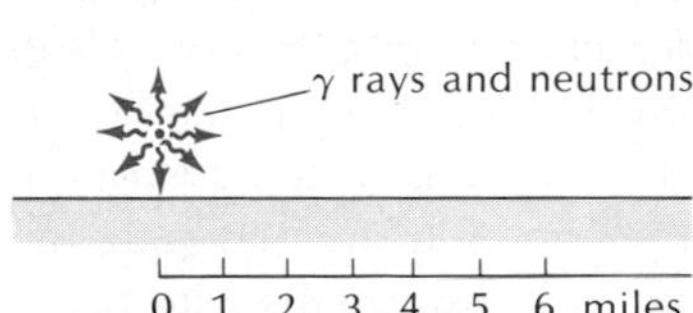

Fig. H.18 Chronological development of an air burst of 1 megaton; time 5×10^{-6} s.

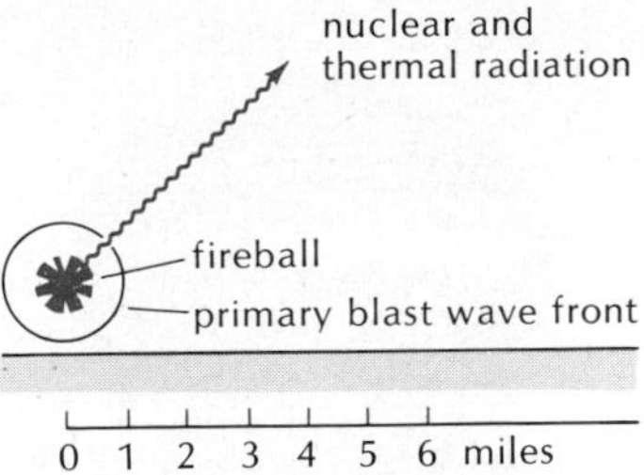

Fig. H.19 Chronological development of an air burst of 1 megaton; time 1.8 s. (Figures H.19–H.23, based on S. Glasstone, ed., *The Effects of Nuclear Weapons.*)

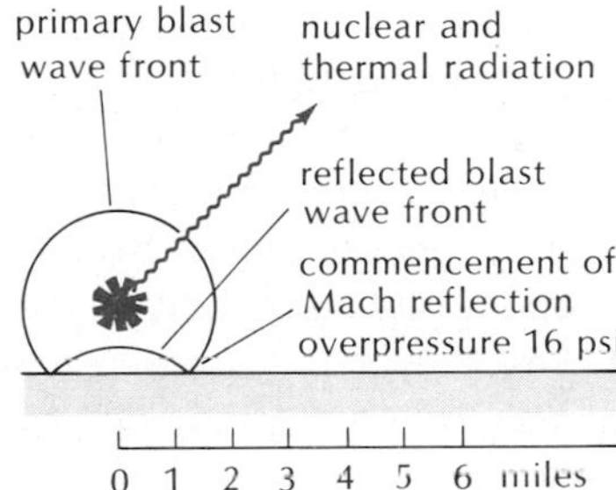

Fig. H.20 Chronological development of an air burst of 1 megaton; time 4.6 s.

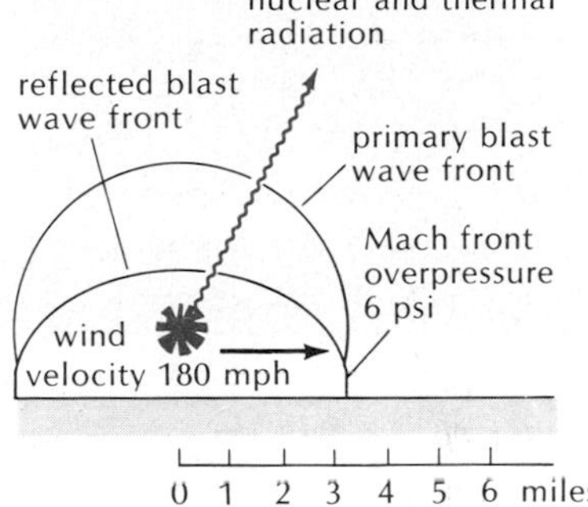

Fig. H.21 Chronological development of an air burst of 1 megaton; time 11 s.

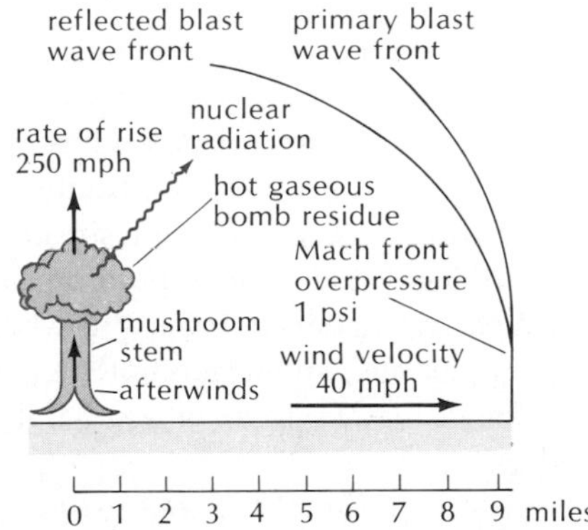

Fig. H.22 Chronological development of an air burst of 1 megaton; time 37 s.

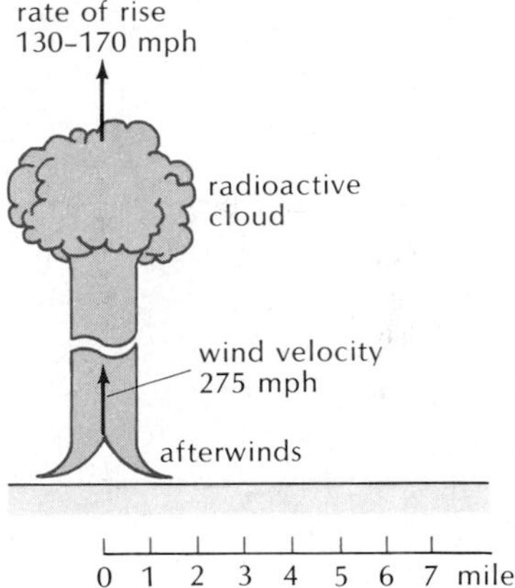

Fig. H.23 Chronological development of an air burst of 1 megaton; time 110 s.

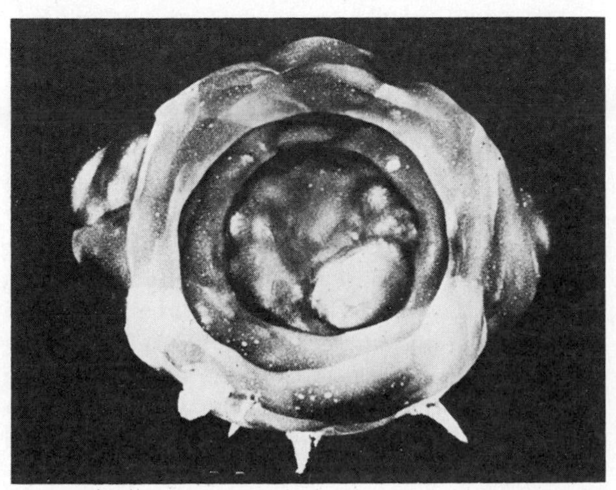

Fig. H.24 Fireball of a nuclear explosion of low yield. This high-speed photograph was taken from a large distance by means of a telescope. (Courtesy H. E. Edgerton, MIT.)

Fig. H.25 The mushroom cloud of a nuclear explosion of low yield.

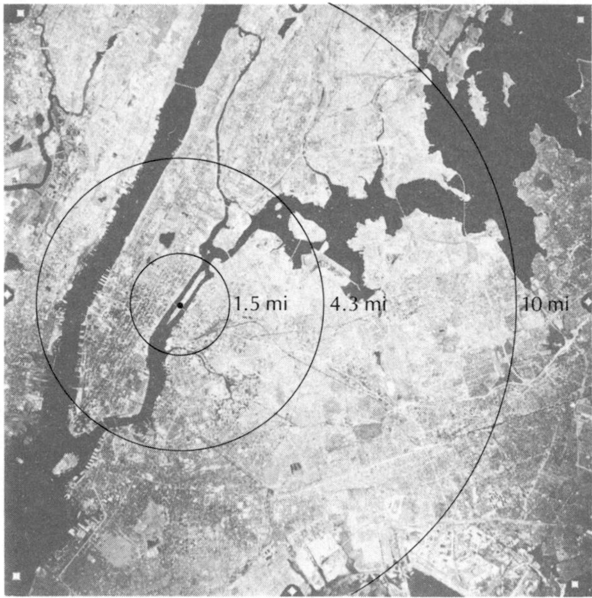

Fig. H.26 Critical radii for damage caused by 1-megaton explosion over New York City.

tains a height of 14 mi. Then it gradually begins to dissipate.

For a 1-megaton bomb bursting at an altitude of, say, 6500 ft, the violent effects of thermal radiation, nuclear radiation, and blast cover an area of radius 10 mi around ground zero. Figure H.26 shows a circle of radius 10 mi superimposed on an aerial photo of New York City; circles of radii 1.5 and 4.3 mi are also shown in this photo. Within a radius of 1.5 mi, the flux of the neutrons emitted by the initial nuclear reactions is so intense that any exposed victim absorbs a lethal dose of ionizing radiation. However, the physiological effects of overexposure to ionizing radiation are delayed; and the victims within this distance are more likely to suffer almost instantaneous death from the blast wave, which reaches them in a few seconds. Out to a radius of 4.3 mi the blast wave has an overpressure of more than 5 lb/in.^{2}, enough to level most buildings and crush the victims with flying and falling debris. Beyond 4.3 mi, the blast wave damages buildings more or less severely, but usually does not destroy them completely.

The flash of thermal radiation emitted by the fireball poses the most frightful danger in the zone from 4.3 to 10 mi from ground zero. The radiant heat is so intense that even at a distance of 10 mi, exposed combustible material will be ignited. At 10 mi, exposed human skin suffers second-degree burns (blisters); at 6 mi human skin is charred throughout its entire thickness. A victim caught in the open will be burnt fatally if at a distance of less than 8 mi; beyond this distance survival depends on the area of unprotected skin exposed to the flash of heat and light.

Since the lens of the eye focuses light on the retina, any victim looking at the fireball will suffer severe retinal damage. At distances of up to 20 mi, the result is total blindness; even at distances of 200 or 300 mi, retinal burns and partial loss of vision are likely.

Besides these primary effects of a nuclear explosion, there are secondary effects. A 1-megaton air burst over a city will set off numerous flash fires within a circle of a 10-mi radius. Such a large number of fires ignited over a wide area may cause a **firestorm:** hot air from the fires rises and creates both an updraft and strong surface winds (50 to 100 mi/h) which fan the flames. Consequently, annihilation of life and destruction of buildings may be total, or nearly total, within a 10-mi radius.

Another delayed effect is **fallout,** which consists of the radioactive residues in the mushroom cloud mixed with the dirt and dust of the mushroom stem. Wherever this contaminated dust falls on the surface of the Earth, the ground will become radioactive. Winds may carry the contaminated dust hundreds of miles as it gradually settles; the result is a fallout plume which covers a large area with dangerous radioactive contamination. The details of the distribution of fallout depend on the wind speed and direction; they also depend on the type of bomb and on the altitude of explosion — fallout is more intense for a surface burst (at ground level) than for an air burst. For a 1-megaton surface burst, whose yield is one-half from fission and one-half from fusion, fallout will be lethal over a downwind area about 150 mi long and 20 mi wide (Figure H.27). This assumes that the wind is steady at about 15 mi/h and that the victims are not sheltered from the ionizing radiation.

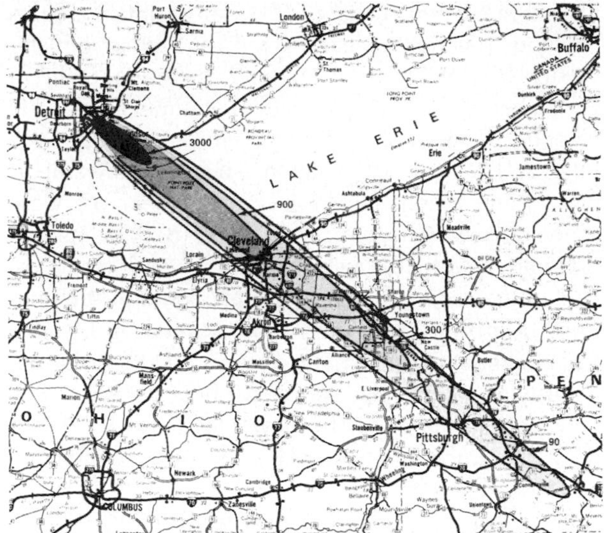

Fig. H.27 Fallout pattern for 1-megaton surface burst in Detroit, with 15-mi/h northwesterly wind. The radiation dose accumulated in a 7-day period is lethal within the second largest gray area. (From *The Effects of Nuclear War,* Office of Technology Assessment, Congress of the United States.)

The basement of a house offers some shelter from the radiation emitted by the radioactive dust blanketing the ground and the roof of the house — the radiation penetrating an underground basement is $\frac{1}{25}$ to $\frac{1}{100}$ of the outside radiation. In general, a massive layer of any material provides shielding against the radiation. Crude **fallout shelters** can be improvised with a layer of any dense material; 4 in. of concrete, 7 in. of earth, or 14 in. of books all give about the same protection (Figure H.28). The radioactivity of the fallout decreases with time (as a rough rule, any sevenfold increase in time reduces the intensity of the radiation by a factor of 10); but, depending on details of the fallout pattern, it may not be safe to leave the shelter for several weeks.

In view of the awesome immediate effects of a nuclear explosion, it is easy to underrate the effects of fallout. It should be kept in mind that even in the case of a limited nuclear attack on the United States — in which bombs are only directed against military installations rather than cities — there might be more than 12 million civilian casualties from the effects of fallout alone (Figure H.29). Besides these casualties from early, local fallout, there would be some further casualties from long-term, worldwide fallout involving isotopes of strontium (^{90}Sr) and cesium (^{137}Cs). These isotopes are produced in abundance by the fission of uranium and they have very long half-lives; therefore, they remain a hazard for a long time.

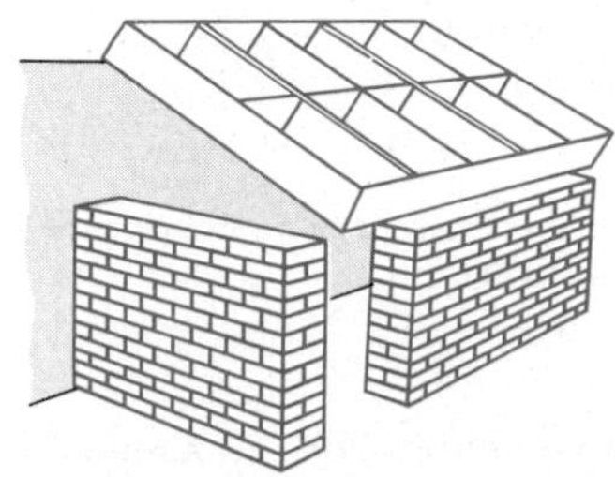

Fig. H.28 A simple fallout shelter in a basement. The bins are to be filled up with bricks or sand. (From *Radiological Defense Textbook,* Defense Civil Preparedness Agency.)

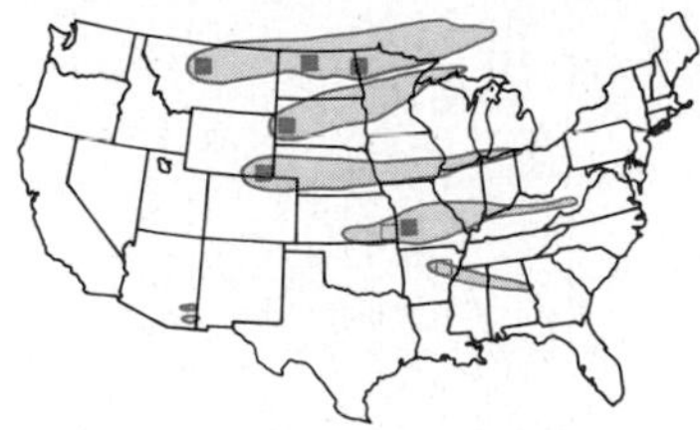

Fig. H.29 Fallout pattern for a massive nuclear attack on military installations in the United States, assuming typical westerly winds. More than one-half of the inhabitants of the colored areas would die. (Based on *Sci. Am.,* November 1976.)

H.6 NUCLEAR REACTORS

In a nuclear reactor, a fission chain reaction takes place under controlled conditions. The condition of the fission material in the reactor is *critical* rather than supercritical, i.e., the number neutrons and the number of fissions along each step of the chain reaction remains constant instead of increasing geometrically.

The most common type of reactor operates with "enriched" uranium consisting of a few percent ^{235}U mixed with 90-odd percent ^{238}U. Such a uranium mixture cannot by itself maintain a chain reaction — the ^{238}U soaks up too many of the fission neutrons. However, if the uranium is surrounded by a substance capable of slowing down the (fast) fission neutrons to a very low speed, then a chain reaction becomes viable. Slow neutrons are much more efficient at maintaining a chain reaction than fast neutrons. There are two reasons for this: slow neutrons are less likely to be absorbed by ^{238}U (the reaction rate for the capture of neutrons by ^{238}U is low at low neutron energies) and slow neutrons trigger the fission of ^{235}U more readily than fast neutrons (the reaction rate for induced fission of ^{235}U is high at low neutron energies).

The substance that slows down the neutrons is called the **moderator**. Inside the reactor the uranium is usually placed in long fuel rods and these are immersed in the bulk of the moderator (Figure H.30). Fast neutrons released by fissions travel from the fuel rods into the moderator; there they lose their kinetic energy by collisions with the moderator's nuclei; and then they wander back into one or another of the fuel

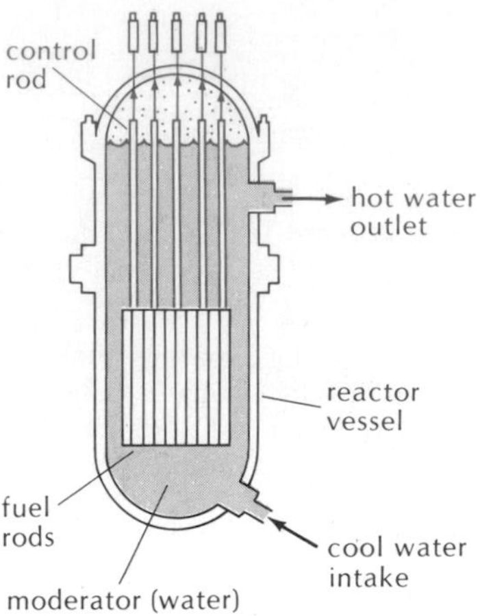

Fig. H.30 Nuclear reactor (schematic).

rods and induce further fissions. An efficient moderator must absorb the kinetic energy of the neutrons, but not absorb the neutrons themselves. The three best moderators are ordinary water (H_2O), heavy water (D_2O), and graphite (pure carbon). These materials are very efficient at slowing down neutrons because they contain nuclei of low mass, i.e., nuclei of mass equal to, or not much larger than, the mass of the neutron. As we saw in Section 10.2, the collision between two particles of equal mass transfers all of the kinetic energy to the initially stationary particle, whereas the collision between particles of unequal mass transfers only a fraction of the kinetic energy. After a few collisions with the moderator nuclei, the neutrons will have lost almost all their kinetic energy; they will only retain a kinetic energy equal to the kinetic energy of the random thermal motion of the moderator nuclei, i.e., a kinetic energy of $\frac{3}{2}kT$ [where T is the temperature of the moderator; see Eq. (19.23)]. Such neutrons are called **thermal neutrons.** Although some of these neutrons will escape from the reactor, in a critical reactor enough return to the uranium to keep the chain reaction going.

The configuration of the reactor — the size, number, and location of uranium rods, and the amount and shape of moderator — must be designed so that the reactor is critical or just barely supercritical. The number of neutrons can be finely adjusted to the point of criticality by control rods made of boron or cadmium; these substances readily soak up neutrons, and, by pushing the control rods in or pulling them out of the reactor, more or fewer neutrons can be removed from the fission chain, bringing the reactor from subcritical, to critical, or even to supercritical (brief instants of supercriticality are called **excursions**).

Both heavy water and graphite are such good moderators that in their presence even natural uranium, with its small percentage of ^{235}U, can maintain a chain reaction. The first nuclear reactor was built at the University of Chicago in 1942 under the direction of Enrico Fermi[5]; it contained natural uranium as fuel and graphite as moderator. Several similar reactors were built at Hanford, Washington, shortly afterward, as part of the Manhattan Project. These reactors were used as **converters,** i.e., they were used to convert ^{238}U into ^{239}Pu by the following sequence of reactions: Within the rods of natural uranium, some of the neutrons from the fission of ^{235}U are soaked up by ^{238}U, transmuting it into ^{239}U. The isotope ^{239}U then spontaneously undergoes two successive beta decays, which transmute it into ^{239}Pu. The few kilograms of ^{239}Pu for the Alamogordo and Nagasaki A-bombs were obtained by these means.

Nowadays reactors are extensively used to provide intense beams of neutrons for research; to produce radioisotopes for scientific, industrial, and medical applications (see Interlude B); and to generate mechanical or electric power. The latter reactors are called **power reactors.** In them, the heat energy released by the fission reactions is removed from the reactor core by a circulating coolant; this heat is then transferred to steam, which drives a steam turbine. Thus, the nuclear reactor serves the same purpose as the furnace of a conventional steam engine — and uranium replaces coal or oil.

Most power reactors in operation or under construction in the United States have water-filled cores; the water acts simultaneously as coolant and as moderator. Figure H.31 shows an outline of a **pressurized-water reactor** (PWR), the most common type of power reactor. The reactor core with its fuel rods is enclosed

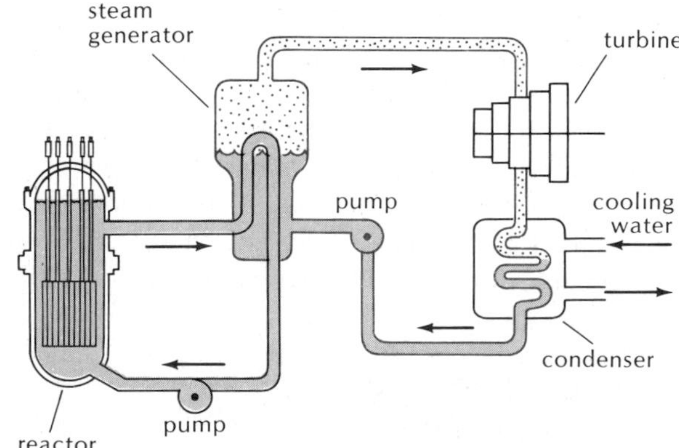

Fig. H.31 Schematic diagram of a nuclear power plant with a pressurized-water reactor.

[5] Enrico Fermi, 1901–1954, Italian, later American physicist. He was awarded the Nobel Prize in 1938 for his discovery of nuclear reactions initiated by neutron bombardment. The element *fermium* was named in his memory.

Fig. H.32 Reactor vessel.

in a massive reactor vessel. Water, driven by powerful pumps, circulates through the reactor vessel; this water is kept at a high pressure (2200 lb/in.^{2}) which prevents it from boiling even though the temperature within the reactor vessel is 600°F. The hot water circulates from the reactor vessel to a steam generator where it passes through a system of pipes immersed in water at a lower pressure. The low-pressure water boils off as steam, which drives a turbine; after the steam has done its work, it condenses and returns to the steam generator. Meanwhile the high-pressure water circulates back to the reactor vessel in a closed loop.

In a **boiling-water reactor** (BWR) the coolant is allowed to boil directly within the reactor vessel. Steam emerges from the top of the reactor and is fed into a turbine; it then condenses and circulates back to the reactor.

Figure H.32 shows the reactor vessel for a large power plant capable of generating 1200 MW of electricity; the height of the reactor vessel is 60 ft, the thickness of the steel walls is 10 in. at its thinnest, and the weight of the vessel is 1500 tons when empty. Smaller reactors are used for the propulsion of submarines, aircraft carriers, and icebreakers. For a submarine, a reactor power plant has the obvious advantage that it does not require oxygen to "burn" its fuel; besides, the reactor only needs to be refueled at very long intervals.

Both pressurized-water reactors and boiling-water reactors operate with ordinary water; they are **light-water reactors.** By contrast, **heavy-water reactors** operate with heavy water (D_2O) as moderator. Although heavy water is very expensive — the 750 tons of heavy water required for a single large reactor cost $80 million — it is such a good moderator that cheap, natural uranium can be used as fuel (see above). A type of heavy-water reactor developed in Canada (CANDU), contains heavy water as moderator and uses ordinary water, in a separate system of pipes, as coolant.

Another type of reactor, first developed in the United Kingdom and in France, is the **gas-cooled reactor** (GR). It contains graphite as moderator and uses gas as coolant. Gases that suit this purpose are carbon dioxide (low-temperature operation, 710°F) and helium (high-temperature operation, 1400°F). Since graphite is a good moderator, such reactors can be fueled by natural uranium; however, structural problems arising from the large size of these reactors (70-ft diameter) make it awkward to rely on natural uranium, and modern high-temperature gas-cooled reactors are fueled with highly enriched uranium, which permits a more compact design.

Power reactors now in operation obtain their energy from the fission of the isotope ^{235}U. Since the world's supply of this isotope is limited (see Interlude F) it would be desirable to build reactors that consume some other fissionable nucleus. One obvious choice is the very fissionable isotope ^{239}Pu. Although this isotope does not occur naturally, it can be readily manufactured by transmutation of the very abundant isotope ^{238}U. In fact, since the fuel of all power reactors is a mixture of ^{238}U and ^{235}U, the manufacture of ^{239}Pu is an automatic side effect of the operation of power reactors; the ^{239}Pu can subsequently be extracted by chemical reprocessing of the spent uranium fuel. A reactor fueled with ^{239}Pu not only makes good use of a material that would otherwise go to waste, but, if the reactor is supplied with a quantity of ^{238}U, it can also manufacture its own ^{239}Pu. What is more, the number of neutrons released in the fusion of ^{239}Pu is sufficiently large so that in an efficiently designed reactor slightly more than one of the neutrons released in an average fission reaction can be diverted from the fission chain to the transmutation of ^{238}U. This implies that the reactor produces *more* ^{239}Pu (from ^{238}U) than it consumes (from its original supply).

A reactor that produces more fissionable material than it consumes is called a **breeder.** Once the fuel cycle of breeder reactors has been started with an initial load of ^{239}Pu, only the abundant and cheap ^{238}U needs to be supplied to keep the fuel cycle going. Essentially, breeders extract energy indirectly from ^{238}U;

breeders would therefore be able to generate power for as long as our (abundant) supply of ^{238}U lasts (see Interlude F). Much effort has been expended on the development of breeders and a few experimental reactors have already been built. However, these reactors are afflicted with design and safety problems that have not yet been satisfactorily resolved.[6]

Further Reading

Nuclear Science and Society by B. L. Cohen (Doubleday, New York, 1974) is an excellent introduction to nuclear energy and its applications.

A concise and clear survey of nuclear physics will be found in *Secrets of the Nucleus* by J. S. Levinger (McGraw-Hill, New York, 1967). Very brief discussions of nuclear physics can be found in chapters in the books *Energy, Ecology, and the Environment* by R. Wilson and W. J. Jones (Academic Press, New York, 1974) and *The Atom and Its Nucleus* by G. Gamow (Prentice-Hall, Englewood Cliffs, 1965). *The Atomic Energy Deskbook,* edited by J. F. Hogerton (Reinhold, New York, 1963), and *Sourcebook on Atomic Energy,* edited by S. Glasstone (Van-Nostrand Reinhold, Princeton, 1967), are encyclopedias containing technical information.

The standard reference on nuclear explosions is *The Effects of Nuclear Weapons,* edited by S. Glasstone (United States Atomic Energy Commission, 1962). This contains a wealth of detail on nuclear explosions, their thermal radiation, blast, and radioactivity, and the injuries and damage wrought on people and buildings. *The Effects of Nuclear War* by the Office of Technology Assessment (Congress of the United States, Washington, D.C., 1979) presents case studies of hypothetical nuclear attacks on Soviet and U.S. industrial and civilian targets. *Radiological Defense* by the Defense Preparedness Agency (Department of Defense, 1974) is a textbook on protective measures against fallout. *Arsenal: Understanding Weapons in the Nuclear Age* by K. Tsipis (Simon and Schuster, New York, 1983) is a lucid survey of the principles and effects of nuclear bombs, the technology of delivery systems, and strategic implications.

The Cold and the Dark: The World After Nuclear War by P. R. Ehrlich, C. Sagan, D. Kennedy, and W. O. Roberts (W. W. Norton, New York, 1984) discusses the long-term worldwide consequences of nuclear war, especially the severe climatic changes ("nuclear winter") caused by the reduction of sunlight by the dust and smoke accumulated in the atmosphere, and the catastrophic effects of this on plants and animals. *Last Aid,* edited by E. Chivian, S. Chivian, R. J. Lifton, and J. E. Mack (Freeman and Co., San Francisco, 1982), is a chilling examination of the medical implications of nuclear war. *The Fate of the Earth,* by J. Schell, gives a frightening description of the global disasters that would follow a nuclear war and makes an impassioned plea for total disarmament. *Nuclear Nightmares* by N. Calder (Viking Press, New York, 1979) is a journalist's investigation of the threat of nuclear war; it deals with weapons technology, proliferation, and scenarios that might lead to war. *Survival and the Bomb,* edited by E. P. Wigner (Indiana University Press, Bloomington, 1969), presents the arguments for the implementation of a civil-defense program.

Manhattan Project by S. Groueff (Little, Brown, and Co., Boston, 1967) is a historical account of the development of the first A-bombs. *Enrico Fermi* by E. Segré (University of Chicago Press, Chicago, 1970) is a biography of the eminent physicist who designed and built the first nuclear reactor and triggered the first chain reaction. *Hans Bethe: Prophet of Energy* by J. Bernstein (Basic Books, New York, 1979) is a brilliant biography of another eminent physicist who discovered the cycles of nuclear reactions that release energy in stars. *The Curve of Binding Energy* by J. McPhee (Ballantine Books, New York, 1973) is a splendidly written account of the design and manufacturing problems involved in the production of small nuclear weapons.

The following is a list of magazine articles dealing with nuclear reactors, nuclear weapons, and related questions:

"Energy from Breeder Reactors," F. L. Culler and W. O. Harms, *Physics Today,* May 1972

"Natural-Uranium Heavy-Water Reactors," H. C. McIntyre, *Scientific American,* October 1975

"Civil Defense in Limited War — A Debate," A. A. Broyles and E. P. Wigner vs. S. D. Drell, *Physics Today,* April 1976

"A Natural Fission Reactor," G. A. Cowan, *Scientific American,* July 1976

"Limited Nuclear War," S. D. Drell and F. von Hippel, *Scientific American,* November 1976

"Superphénix: A Full-Scale Breeder Reactor," G. A. Vendryes, *Scientific American,* March 1977

"Nuclear Power and Nuclear-Weapons Proliferation," E. J. Moniz and T. L. Neff, *Physics Today,* April 1978

"Enhanced-Radiation Weapons," F. M. Kaplan, *Scientific American,* May 1978

"The Prompt and Delayed Effects of Nuclear War," K. L. Lewis, *Scientific American,* July 1979

"Catastrophic Releases of Radioactivity," S. A. Fetter and K. Tsipis, *Scientific American,* April 1981

"Gas-Cooled Nuclear Power Reactors," H. M. Agnew, *Scientific American,* June 1981

"Freeze on Nuclear Weapons Development and Deployment: Pro–Con," H. Feiveson and F. von Hippel vs. H. W. Lewis, *Physics Today,* January 1983

"Arms Limitation Strategies," H. F. York, *Physics Today,* March 1983

"Effects of Nuclear Weapons," L. Sartori, *Physics Today,* March 1983

"The Nuclear Arsenals of the US and the USSR," B. G. Levi, *Physics Today,* March 1983

"The Uncertainties of a Preemptive Nuclear Attack," M. Bunn and K. Tsipis, *Scientific American,* November 1983

"Weapons and Hope," F. J. Dyson, *The New Yorker,* February 6, 13, 20, and 27, 1984

"The Climatic Effects of Nuclear War," R. P. Turco, O. B. Toon, T. P. Ackerman, J. B. Pollack, and C. Sagan, *Scientific American,* August 1984

[6] For a discussion of reactor safety, see Section F.6.

Questions

1. According to Figure H.8, which isotope has the largest binding energy per nucleon? Which has the least binding energy and is therefore capable of releasing the most energy in a nuclear reaction?

2. By what factor is the amount of energy released per kilogram of reactant in a typical fission reaction larger than in a typical chemical reaction? (Hint: Compare fission of uranium with explosion of TNT.)

3. In 1933, Ernest Rutherford, the discoverer of the nucleus, declared that "the energy produced by the breaking down of the atom is a very poor kind of thing. Anyone who expects a source of power from the transformation of these atoms is talking moonshine." Taking into consideration that fission was unknown at the time, do you think Rutherford was making a fair assessment? What about heat released by natural radioactivity?

4. Free neutrons are unstable; they decay with an average lifetime of about 15 min. What effect does this have on a fission chain reaction?

5. Why is a critical mass needed for fission but not for fusion?

6. George Gamow, author of many delightful books on physics, was fond of saying that natural uranium is just as useless for carrying out a nuclear chain reaction as soaking-wet logs are for building a campfire. Explain this analogy.

7. A widely used modern method for the separation of the isotopes ^{235}U and ^{238}U relies on high-speed centrifuges. How does centrifugation affect a test tube full of a gaseous compound with molecules containing these isotopes?

8. Since the critical mass of ^{235}U is 53 kg, the Hiroshima bomb must have contained about that much uranium. If all of this uranium had undergone fission, what would have been the energy released?

9. Compare the shapes of the bombs in Figures H.15 and H.16. Why does the Hiroshima bomb have an elongated shape, and the Nagasaki bomb a rounded shape?

10. Why are high temperatures required to initiate fusion, but not to initiate fission?

11. Per kilogram of reactant, fusion releases about three times as much energy as fission. But the energy yield of a typical H-bomb is about a hundred times larger than that of a typical A-bomb. How can this be?

12. If an H-bomb is not surrounded by a blanket of natural uranium, it is a cleaner bomb, producing a smaller amount of radioactive fallout and a smaller energy yield. However, such a bomb, called a neutron bomb, releases a large number of energetic neutrons. Compare the effects of a neutron bomb with those of an ordinary H-bomb.

13. If a blast wave of overpressure 16 lb/in.^{2} strikes you, what is the force on the front of your body?

14. If you are at a distance of 8 mi from the place of a nuclear explosion, you have available about 30 s between the flash of light and the arrival of the blast wave. Look around the room you are in. Where could you take shelter in 30 s?

15. Contamination of food with radioactive strontium poses a severe hazard because strontium is chemically similar to calcium. Explain.

16. One of the effects of a large-scale nuclear war would be the accumulation in the atmosphere of dust and smoke from explosions and fires. Such a blanket of dust and smoke, covering the entire Earth for several months, would reduce the amount of sunlight reaching the ground, perhaps by 50% or more, so that the climate would become much cooler. Recent calculations suggest that even in summer the temperature would remain below freezing, a phenomenon that has been called the **nuclear winter**. What agricultural and other problems would arise from such a drastic modification of the climate?

17. Helium should make a good moderator, since it does not absorb neutrons and has a low mass. Why do we not use helium as moderator in a nuclear reactor?

18. Why must the fuel rods in a nuclear reactor be thin?

19. The uranium in the fuel rods in a nuclear reactor is enclosed in a metal pipe (cladding). What are the requirements that the material of the pipe must meet?

20. In a nuclear power plant, the water that passes through the reactor core flows through one loop, and the water that passes through the turbine flows in a separate loop (Figure H.31). Why do we not use a single loop that directly connects the reactor core to the turbine, as in an ordinary coal-burning power plant?

21. If you wanted to build a nuclear reactor of very small size, say, to power an artificial satellite, what isotope would you use?

22. The following question was asked by a reader of the *New York Times* (February 22, 1983): "Why is it less dangerous if the nuclear reactor of a falling satellite (such as the Soviet Cosmos 1402) burns up in space than if it falls to earth in one piece? Aren't radioactive atoms just as present one way as another?" How would you answer?

23. Breeder reactors generate more fuel than they consume. Does this violate the law of conservation of energy?

24. One of the disadvantages of breeder reactors is that their fuel can be diverted to the manufacture of bombs. Why is this not a problem with ordinary reactors?

CHAPTER 27

Capacitors and Dielectrics

Capacitor

Any arrangement of conductors that is used to store electric charge is called a **capacitor**, or condenser. Since work must be done during the charging process, the capacitor will also store electric potential energy. In our electric technology, capacitors find widespread application — they are part of the circuitry of radios, electronic calculators, automobile ignition systems, etc.

Dielectric

The first part of this chapter deals with the properties of capacitors. The second part deals with the properties of electric fields in regions of space filled with an insulating material, or **dielectric.** Since many capacitors are filled with such a dielectric material, the study of the mutual effects between the electric field and the dielectric material is closely linked to the study of capacitors. But the effects of electric fields and dielectric materials upon one another are also interesting in their own right. For instance, air is a dielectric material and we ought to inquire how the electric field in air differs from that in vacuum.

27.1 Capacitance

As a first example of a capacitor, consider an isolated metallic sphere of radius R. Obviously, charge can be stored on this sphere. If the amount of charge placed on the sphere is Q, then the potential of the sphere will be

$$V = \frac{1}{4\pi\varepsilon_0}\frac{Q}{R} \tag{1}$$

Thus, the amount of charge stored on the sphere is directly proportional to the potential.

This proportionality holds in general for any conductor of arbitrary shape. The charge on the conductor produces an electric field whose strength is directly proportional to the charge (twice the charge gives twice the field) and the electric field yields a potential which is directly proportional to the field strength (twice the field strength yields twice the potential); hence charge and potential are proportional. We write this relationship as

$$\boxed{Q = CV} \qquad (2)$$

Capacitance of a single conductor

where C is the constant of proportionality. This constant is called the **capacitance** of the conductor. The capacitance is large if the conductor is capable of storing a large amount of charge at a low potential. For instance, the capacitance of a sphere is

$$C = \frac{Q}{V} = \frac{Q}{(1/4\pi\varepsilon_0)(Q/R)} = 4\pi\varepsilon_0 R \qquad (3)$$

Thus, the capacitance of a sphere increases with its radius.

The unit of capacitance is the **farad** (F),

$$1 \text{ farad} = 1 \text{ F} = 1 \text{ coulomb/volt} \qquad (4)$$

farad, F

This unit of capacitance is rather large; in practice, electrical engineers prefer the **microfarad** and the **picofarad**. A microfarad equals 10^{-6} farad ($1\ \mu\text{F} = 10^{-6}$ F) and a picofarad equals 10^{-12} farad (1 pF $= 10^{-12}$ F).

EXAMPLE 1. What is the capacitance of an isolated metallic sphere of radius 20 cm?

SOLUTION: According to Eq. (3),

$$C = 4\pi\varepsilon_0 R = 4\pi \times 8.85 \times 10^{-12} \frac{(\text{coulomb})^2}{\text{N} \cdot \text{m}^2} \times 0.20 \text{ m}$$

$$= 2.2 \times 10^{-11} \frac{\text{coulomb}}{\text{volt}} = 2.2 \times 10^{-11} \text{ F} = 22 \text{ pF}$$

Note that $1 \text{ F} = 1 \text{ C/V} = 1 \text{ C}^2/\text{N} \cdot \text{m}$ so that the constant ε_0 can be written

$$\varepsilon_0 = 8.85 \times 10^{-12} \frac{\text{C}^2}{\text{N} \cdot \text{m}^2} = 8.85 \times 10^{-12} \text{ F/m} \qquad (5)$$

The latter expression is the one usually listed in tables of physical constants.

EXAMPLE 2. What is the capacitance of the Earth, regarded as a conducting sphere?

SOLUTION: The radius of the Earth is 6.4×10^6 m so that

$$C = 4\pi\varepsilon_0 R = 4\pi \times 8.85 \times 10^{-12}\ \text{F/m} \times 6.4 \times 10^6\ \text{m}$$

$$= 7.1 \times 10^{-4}\ \text{F}$$

As capacitances go, this is a rather large capacitance. However, note that it only takes a charge of $\sim 10^{-3}$ coulomb to alter the potential of the Earth by 1 volt.

In electrostatic experiments, the Earth is often used as a dump for unwanted positive or negative charge, which alters the potential of the Earth relative to infinity. This alteration conflicts with the convention adopted in Section 25.1, where we treated the ground as a body at a fixed potential, $V = 0$. But this conflict need not trouble us: for most terrestrial electric experiments, only the potential difference between the apparatus and the ground is relevant, and the alteration of the potential difference between the apparatus and infinity has no immediate effect.

The most common variety of capacitor consists of *two* metallic conductors, insulated from one another and carrying opposite amounts of electric charge $\pm Q$. The capacitance of such a pair of conductors is defined in terms of the *difference* of potential between the two conductors:

Capacitance of a pair of conductors

$$\boxed{Q = C\,\Delta V} \qquad (6)$$

In this expression, both Q and ΔV are taken as positive quantities. Note that the quantity Q is not the net charge in the capacitor, but the magnitude of the charge on each plate; the net charge in the two-conductor capacitor is zero.

Figure 27.1 shows such a two-conductor capacitor consisting of two large, parallel metallic plates, each of area A, separated by a distance d. The plates carry charges $+Q$ and $-Q$, respectively. The electric field in the region between the plates is (neglecting edge effects)

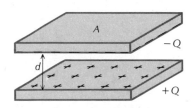

Fig. 27.1 Two parallel plates, with charges $+Q$ and $-Q$.

$$E = \frac{\sigma}{\varepsilon_0} = \frac{Q}{\varepsilon_0 A}$$

and the potential difference is

$$\Delta V = Ed = \frac{Qd}{\varepsilon_0 A} \qquad (7)$$

Hence the capacitance of this configuration is

Capacitance of parallel plates

$$\boxed{C = \frac{Q}{\Delta V} = \frac{Q}{Qd/\varepsilon_0 A} = \frac{\varepsilon_0 A}{d}} \qquad (8)$$

Thus, in order to store a large amount of charge at a low potential, we want a large plate area A, but a small plate separation d. Parallel-plate capacitors are usually manufactured out of two parallel sheets of aluminum foil, a few centimeters wide and several meters long. The sheets are placed very close together, but kept from contact by a thin sheet of plastic sandwiched between (Figure 27.2). For convenience, the entire sandwich is covered with another sheet of plastic and rolled up like a roll of toilet paper.

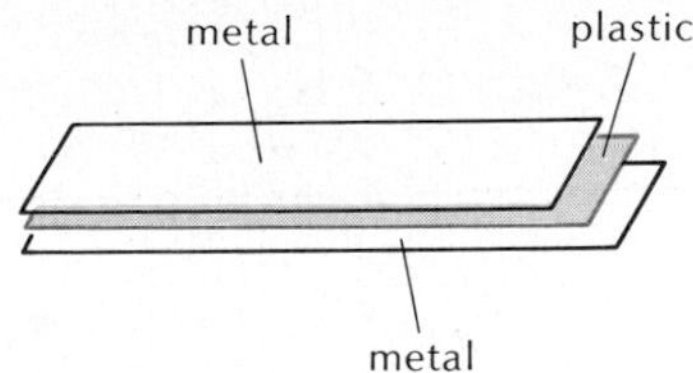

Fig. 27.2 Sheets of aluminum foil separated by a sheet of plastic.

EXAMPLE 3. A parallel-plate capacitor consists of two strips of aluminum foil, each with an area of 0.20 m², separated by a distance of 0.10 mm. The space between the foils is empty. A potential difference of 200 V is applied to this capacitor. What is the capacitance of this capacitor? What is the electric charge on each plate? What is the strength of the electric field between the plates?

SOLUTION: According to Eq. (8), the capacitance is

$$C = \frac{\varepsilon_0 A}{d} = \frac{8.85 \times 10^{-12}\ \text{F/m} \times 0.20\ \text{m}^2}{1.0 \times 10^{-4}\ \text{m}} = 1.8 \times 10^{-8}\ \text{F}$$

$$= 0.018\ \mu\text{F}$$

The charge on each plate is

$$Q = C\,\Delta V = 1.8 \times 10^{-8}\ \text{F} \times 200\ \text{volt} = 3.5 \times 10^{-6}\ \text{coulomb}$$

and the electric field between the plates is

$$E = \Delta V/d = 200\ \text{volt}/1.0 \times 10^{-4}\ \text{m} = 2.0 \times 10^{6}\ \text{volt/m}$$

27.2 Capacitors in Combination

Capacitors used in practical applications in electric circuitry commonly are of the two-conductor variety. Schematically, such capacitors are represented as two parallel plates with terminals emerging at their middles (Figure 27.3). In a circuit, several such capacitors are often wired together and it is then necessary to calculate the net capacitance of the combination. The simplest ways of wiring capacitors together are in **parallel** and in **series**.

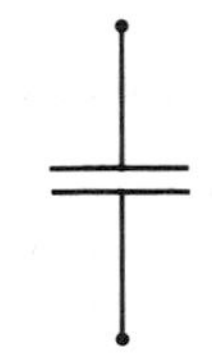

Fig. 27.3 Symbol for a capacitor in a circuit diagram.

Figure 27.4 shows two capacitors connected in *parallel*. If charge is fed into this combination via the two terminals, some of the charge will be stored on the first capacitor and some on the second. The net capacitance of the combination can be found as follows. Since the corresponding plates of the capacitors are joined by a conductor, the potential differences across both capacitors are the same,

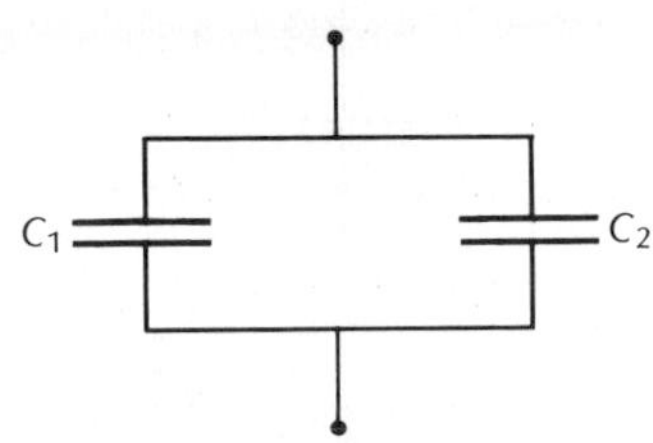

Fig. 27.4 Two capacitors connected in parallel.

$$\Delta V = \frac{Q_1}{C_1} \quad \text{and} \quad \Delta V = \frac{Q_2}{C_2} \tag{9}$$

Therefore the net charge can be expressed as

$$Q = Q_1 + Q_2 = C_1\,\Delta V + C_2\,\Delta V \tag{10}$$

i.e.,

$$Q = (C_1 + C_2)\,\Delta V \tag{11}$$

Comparing this with the definition for capacitance given in Eq. (6), we see that the combination is equivalent to a single capacitor of capacitance

$$C = C_1 + C_2 \tag{12}$$

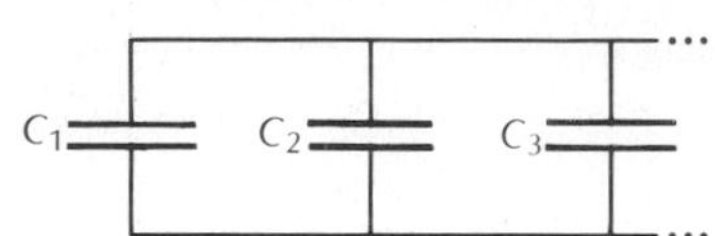

Fig. 27.5 Several capacitors connected in parallel.

Thus, the net capacitance of the parallel combination is simply the sum of the individual capacitances.

It is easy to obtain a similar result for any number of capacitors connected in parallel (Figure 27.5). The net capacitance is

Parallel combination of capacitors

$$\boxed{C = C_1 + C_2 + C_3 + \cdots} \tag{13}$$

Figure 27.6 shows two capacitors connected in *series.* Any charge fed into this combination via the two outside terminals will have to remain on the outside plates (the lower plate of the first capacitor (C_1) and the upper plate of the second (C_2), see Figure 27.6). Thus, the lowest plate will have a charge Q and the highest plate a charge $-Q$. But these charges on the outside plates will induce charges on the inside plates (the upper plate of the first capacitor and the lower plate of the second). The charge Q on the lowest plate will attract electrons to the facing plate and a charge $-Q$ will accumulate on this plate. Corresponding to the excess electrons on the upper plate of the first capacitor, there will be a deficit of electrons on the lower plate of the second capacitor and a charge $+Q$ will accumulate there. The capacitance of the combination can then be found as follows. The potential differences across the two capacitors are

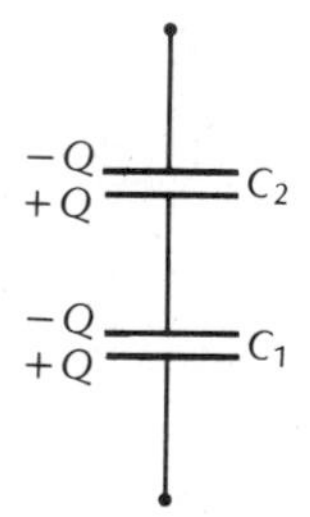

Fig. 27.6 Two capacitors connected in series.

$$\Delta V_1 = \frac{Q}{C_1} \quad \text{and} \quad \Delta V_2 = \frac{Q}{C_2} \tag{14}$$

The net potential difference between the external terminals is the sum of these,

$$\Delta V = \Delta V_1 + \Delta V_2 = \frac{Q}{C_1} + \frac{Q}{C_2} \tag{15}$$

i.e.,

$$\Delta V = Q\left(\frac{1}{C_1} + \frac{1}{C_2}\right) \tag{16}$$

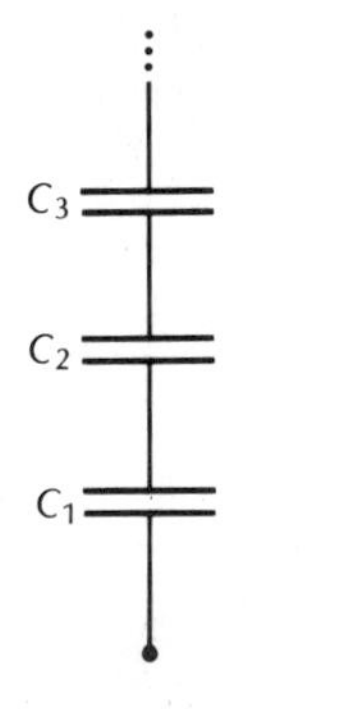

Fig. 27.7 Several capacitors connected in series.

From this it is clear that the combination has a net capacitance C given by

$$\frac{1}{C} = \frac{1}{C_1} + \frac{1}{C_2} \tag{17}$$

Thus, the net capacitance of the series combination is obtained by taking a sum of inverses. Note that the net capacitance is *less* than the individual capacitances. For example, if $C_1 = C_2$, then $C = \frac{1}{2}C_1 = \frac{1}{2}C_2$.

A similar result applies to any number of capacitors connected in series (Figure 27.7). The net capacitance is given by

$$\boxed{\frac{1}{C} = \frac{1}{C_1} + \frac{1}{C_2} + \frac{1}{C_3} + \cdots} \tag{18}$$

Series combination of capacitors

27.3 Dielectrics

So far, in dealing with problems of electrostatics we have assumed that the space surrounding the electric charge consisted of a vacuum, which has no effect on the electric field, or of air, which has only an insignificant effect on the electric field. However, in dealing with capacitors, we must take the effects of the medium into account. The space between the plates of a capacitor is usually filled with an insulator, or **dielectric**, which drastically changes the electric field from what it would be in a vacuum: the dielectric reduces the strength of the electric field.

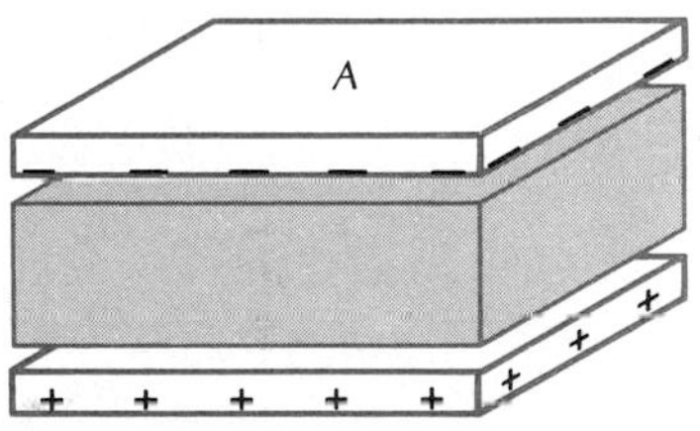

Fig. 27.8 A slab of dielectric between the plates of a capacitor.

To understand this, consider a parallel-plate capacitor whose plates carry some charge per unit area. Suppose that a slab of dielectric, such as glass or polyethylene, fills most of the space between the plates (Figure 27.8). This dielectric contains a large number of atomic nuclei and electrons but, of course, these positive and negative charges balance each other so that the material is electrically neutral. In an insulator, all the charges are **bound** — the electrons are confined within their atoms or molecules and they cannot wander about as in a conductor. Nevertheless, in response to the force exerted by the electric field, the charges will move very slightly without leaving their atoms. The electrons move in a direction opposite to that of the protons; consequently, atoms or molecules acquire a dipole moment in the direction of the original electric field. In most dielectrics, the magnitude of this dipole moment is directly proportional to the strength of the electric field; such dielectrics are said to be **linear**.

Bound charges

Linear dielectric

In some dielectrics the dipole moment results from a distortion of the molecules or atoms. By tugging on the electrons and protons in opposite directions, the electric field stretches the molecule and produces a small charge separation within it (Figure 27.9). The magnitude of the induced dipole moment is approximately proportional to the strength of the applied electric field.

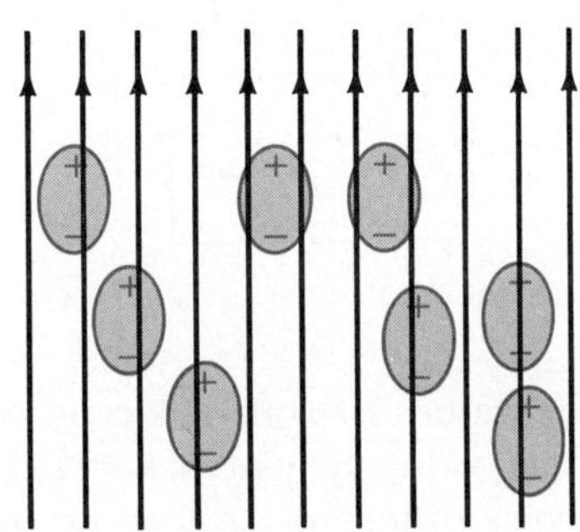

Fig. 27.9 The electric field produces a distortion of molecules.

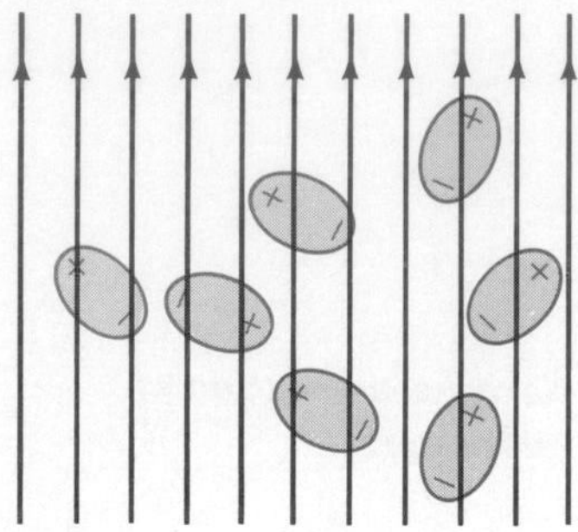

Fig. 27.10 The electric field produces a (partial) alignment of already distorted molecules.

In other dielectrics the dipole moment results (mainly) from a realignment of existing dipoles. In such dielectrics, usually gases or liquids, the molecules have permanent dipole moments that are randomly oriented when the dielectric is left by itself. But when the dielectric is placed in an electric field, these permanent dipoles experience a torque that tends to align them with the field (Figure 27.10). Random thermal motions oppose this alignment and the average amount of dipole moment in the direction of the applied electric field is again approximately proportional to the strength of the field.

Polarization

In any case, by the action of the electric field, the average positions of the positive and negative charges in the dielectric material become displaced relative to one another — the positive and negative charge distributions cease to overlap precisely (Figure 27.11). There will be an excess of positive charge on one surface of the slab of dielectric, and an excess of negative charge on the opposite surface, i.e., the slab of dielectric acquires layers of surface charge. The slab of dielectric is then said to be **polarized.** These surface charges act just like a pair of parallel planes of positive and negative charge; between the planes these charges generate an electric field that is *opposite* to the original applied electric field. The total electric field, consisting of the sum of the field of the free charges on the conducting plates plus the bound charges on the dielectric surfaces is therefore smaller than the field of the free charges alone (Figure 27.12).

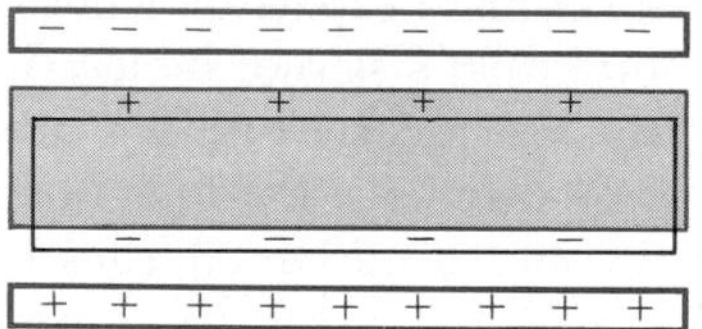

Fig. 27.11 The distributions of positive charge (color) and of negative charge (black) of the slab of dielectric do not overlap precisely.

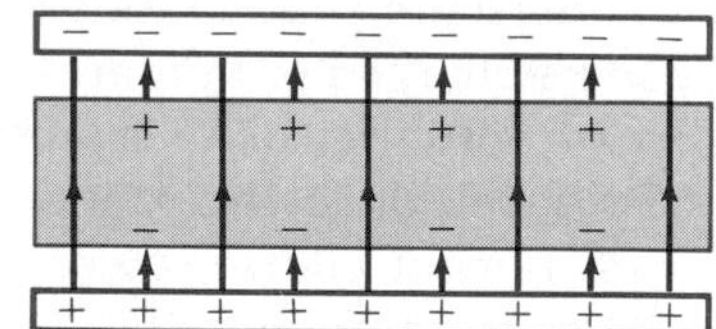

Fig. 27.12 Some electric field lines stop on the negative charges at the bottom of the slab of dielectric. The density of field lines is smaller in the dielectric than in the empty gaps adjacent to the plates.

Dielectric constant

In a linear dielectric, the amount by which the dielectric reduces the strength of the electric field can be characterized by the **dielectric constant** κ. This constant is merely the factor by which the electric field in the dielectric between the parallel plates is reduced, i.e., if E_{free} is the electric field that the free charges produce by themselves and E the electric field that the free charges and the bound charges produce together, then

Electric field in dielectric

$$\boxed{E = \frac{1}{\kappa} E_{\text{free}}} \tag{19}$$

where $\kappa \geq 1$.

Table 27.1 lists the values of the dielectric constant of some materials. Note that air has a value very near $\kappa = 1$, i.e., the dielectric properties of air are not very different from those of a vacuum.

Table 27.1. DIELECTRIC CONSTANTS OF SOME MATERIALS[a]

Material	κ
Vacuum	1
Helium	1.000068
Air	1.00054
Carbon dioxide	1.00098
Carbon tetrachloride	2.2
Paraffin	~2
Polyethylene	2.3
Rubber, hard	2.8
Transformer oil	~3
Plexiglas	3.4
Nylon	3.5
Epoxy resin	3.6
Paper	~4
Bakelite	~5
Pyrex glass	~5
Glass	~6
Porcelain	~7
Water, distilled	80

[a] At room temperature (20° C) and 1 atm.

Incidentally, a metal can be regarded as a dielectric with an infinite dielectric constant — if we substitute $\kappa = \infty$ into Eq. (19), we find that $E = 0$, as it should be inside the metal. This large value of the dielectric constant simply indicates that the material becomes very strongly polarized, i.e., the charges in the metal respond very strongly to an electric field.

If the slab of dielectric entirely fills the space between the plates, then the formula (19) for the reduction of the strength of the electric field applies throughout all of this space. Since the potential difference between the capacitor plates is directly proportional to the strength of the electric field, it follows that, for a given amount of free charge on the plates, the presence of the dielectric also reduces the potential difference by the factor κ,

$$\Delta V = \frac{1}{\kappa}\,\Delta V_0 \tag{20}$$

where ΔV_0 is the potential difference in the absence of the dielectric. Consequently, the presence of the dielectric increases the capacitance by a factor κ,

$$C = \frac{Q}{\Delta V} = \kappa\,\frac{Q}{\Delta V_0} = \kappa C_0 \tag{21}$$

where C_0 is the capacitance in the absence of dielectric. For example, the capacitance of a parallel-plate capacitor filled with dielectric is

$$C = \kappa C_0 = \kappa \varepsilon_0 A/d \tag{22}$$

By filling the space between the capacitor plates with dielectric, we can therefore obtain a substantial gain in capacitance. Furthermore, the dielectric can prevent electric breakdown in the space between the plates. If this space contains air, sparking will occur between the plates when the electric field reaches a value of about 3×10^6 V/m and the capacitor will discharge spontaneously. Some dielectrics are better insulators than air and they will tolerate an electric field that is appreciably larger than 3×10^6 V/m. For instance, Plexiglas will tolerate an electric field of up to 40×10^6 V/m before it suffers electric breakdown (Figure 27.13).

Fig. 27.13 Electric breakdown of a Plexiglas block in a very strong electric field caused minute perforations in the block and created this beautiful arboreal pattern.

The magnitude of the surface charge density on the slab of dielectric is given by the simple formula

Surface charge on dielectric

$$\boxed{\sigma_{\text{bound}} = -\frac{\kappa - 1}{\kappa}\,\sigma_{\text{free}}} \tag{23}$$

Here the negative sign on the right side of the formula indicates that the negatively charged surface of the dielectric adjoins the positively charged plate of the capacitor (see Figure 27.12). This formula is a consequence of Eq. (19). The total electric field E in the dielectric is the sum of the fields of the free charges and the bound charges,

$$E = E_{\text{free}} + E_{\text{bound}} \tag{24}$$

Hence

$$\frac{E_{\text{free}}}{\kappa} = E_{\text{free}} + E_{\text{bound}} \tag{25}$$

which, expressed in terms of charge densities, becomes

$$\frac{\sigma_{\text{free}}}{\varepsilon_0 \kappa} = \frac{\sigma_{\text{free}}}{\varepsilon_0} + \frac{\sigma_{\text{bound}}}{\varepsilon_0} \tag{26}$$

This is equivalent to Eq. (23).

Although we derived the results of this section for the special case of a parallel-plate capacitor, Eqs. (19), (20), (21), and (23) also hold for cylindrical or spherical conductors and concentric shells of dielectric filling the space between the conductors (Figure 27.14). If the dielectric does not entirely fill the space between the conductors, then our results must be modified.

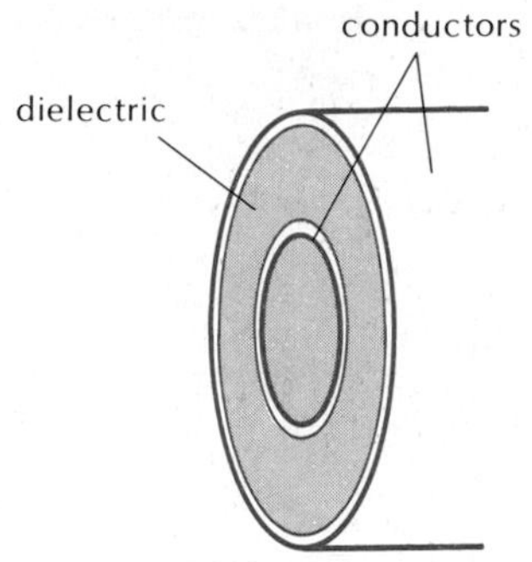

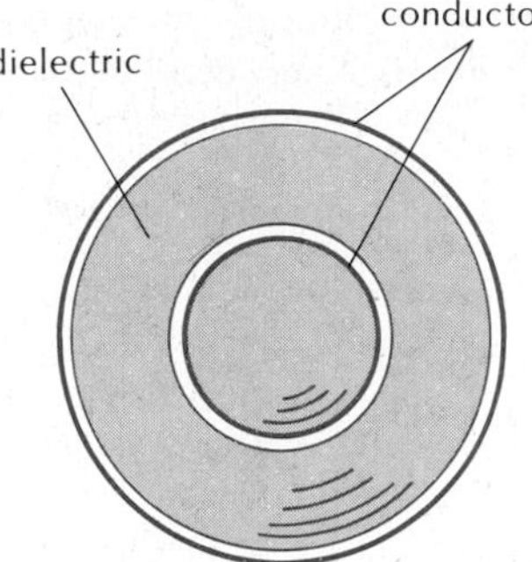

Fig. 27.14 A cylindrical capacitor and a spherical capacitor.

EXAMPLE 4. A parallel-plate capacitor, such as found in a radio, is made of two strips of aluminum foil with a plate area of 0.75 m². The plates are separated by a layer of polyethylene 2×10^{-5} m in thickness. Suppose that a potential difference of 30 V is applied to this capacitor. What is the magnitude of the free charge on each plate? What is the magnitude of the bound charge on the surface of the dielectric? What is the electric field in the dielectric?

SOLUTION: With $\kappa = 2.3$ (see Table 27.1), the capacitance is

$$C = \kappa\varepsilon_0 \frac{A}{d} = \frac{2.3 \times 8.85 \times 10^{-12}\ \text{F/m} \times 0.75\ \text{m}^2}{2 \times 10^{-5}\ \text{m}}$$

$$= 7.6 \times 10^{-7}\ \text{F}$$

The free charge on each plate is then

$$Q_{\text{free}} = C\,\Delta V = 7.6 \times 10^{-7}\ \text{F} \times 30\ \text{V} = 2.3 \times 10^{-5}\ \text{coulomb}$$

The bound charge on the surfaces of the dielectric can be found from Eq. (23):

$$Q_{\text{bound}} = A\sigma_{\text{bound}} = -\frac{\kappa - 1}{\kappa} A\sigma_{\text{free}} = -\frac{\kappa - 1}{\kappa} Q_{\text{free}}$$

$$= -\frac{2.3 - 1}{2.3} \times 2.3 \times 10^{-5}\ \text{coulomb}$$

$$= -1.3 \times 10^{-5}\ \text{coulomb}$$

Here, as in Eq. (23), the negative sign indicates that the bound charge on the dielectric is negative on the surface adjacent to the positive plate of the capacitor. The electric field in the dielectric is

$$E = \frac{1}{\kappa} E_{\text{free}} = \frac{1}{\kappa} \frac{\sigma_{\text{free}}}{\varepsilon_0} = \frac{1}{\kappa} \frac{Q_{\text{free}}}{\varepsilon_0 A}$$

$$= \frac{1}{2.3} \times \frac{2.3 \times 10^{-5}\ \text{coulomb}}{8.85 \times 10^{-12}\ \text{F/m} \times 0.75\ \text{m}^2} = 1.5 \times 10^6\ \text{volt/m}$$

Fig. 27.15 A slab of dielectric between the plates of a capacitor.

EXAMPLE 5. The space between two large parallel conducting plates is partially filled with a parallel slab of dielectric (Figure 27.15). The area of the plates is A, their separation is d, and the thickness of the slab is d'. The dielectric constant of the slab is κ. What is the capacitance of this arrangement?

SOLUTION: Consider a straight path from one plate to the other. A length $d - d'$ of this path is in a vacuum and a length d' is in the dielectric; the electric field has the value E_{free} in vacuum and E_{free}/κ in the dielectric. Hence the potential difference between the plates is

$$\Delta V = E_{\text{free}}(d - d') + \frac{1}{\kappa} E_{\text{free}} d'$$

$$= \left[1 - \frac{d'}{d}\left(1 - \frac{1}{\kappa}\right)\right] E_{\text{free}} d$$

$$= \left[1 - \frac{d'}{d}\left(1 - \frac{1}{\kappa}\right)\right] \Delta V_0 \tag{27}$$

The factor in brackets is less than one; hence ΔV is smaller than ΔV_0 and, correspondingly, the capacitance C is larger than C_0,

$$C = C_0 \Big/ \left[1 - \frac{d'}{d}\left(1 - \frac{1}{\kappa}\right)\right]$$

$$= \left(\frac{\varepsilon_0 A}{d}\right) \Big/ \left[1 - \frac{d'}{d}\left(1 - \frac{1}{\kappa}\right)\right] \tag{28}$$

27.4 Gauss' Law in Dielectrics

The electric field produced by the bound charges in a dielectric does, of course, obey Gauss' Law. Hence the *total* electric field **E** will obey Gauss' Law:

$$\varepsilon_0 \int \mathbf{E} \cdot d\mathbf{S} = Q_{\text{total}} = Q_{\text{free}} + Q_{\text{bound}} \tag{29}$$

Here the total charge is the sum of the free charge plus the bound charge.

Unfortunately, Q_{bound} is usually not known beforehand because the amount of polarization in the dielectric depends on the (unknown) strength of the electric field. Thus, the above form of Gauss' Law is not very helpful.

Since the free charge Q_{free} is usually known, it is better to devise a modified form of Gauss' Law that depends only on this free charge. For simplicity's sake, we will assume throughout the following discussion that the conductors, the distribution of free charge on them, and the dielectrics between them have enough symmetry so that Gauss' Law suffices for the calculation of the electric field. We can then proceed as follows.

Imagine that at first the dielectrics are absent, so that only the free charge produces an electric field. For this situation, Gauss' Law is

$$\varepsilon_0 \int \mathbf{E}_{\text{free}} \cdot d\mathbf{S} = Q_{\text{free}} \tag{30}$$

Next, imagine that the dielectrics are inserted into their proper places while the amount of free charge on the conductors is held constant. In general, we would expect that this will mess up the electric field because the bound charges on the dielectrics act back on the free charges and affect the distribution of these charges on the conductors. But if the arrangement of conductors has a high symmetry and the dielectrics have the *same* symmetry (parallel flat conductors with parallel flat slabs of dielectric; concentric cylindrical conductors with concentric cylindrical dielectrics; concentric spherical conductors with concentric spherical dielectrics), then the distribution of free charge is determined by the symmetry and the dielectric cannot disturb this distribution of free charge. Thus, the direction of the electric field at each point is unchanged. Only the strength of the electric field is altered: the new electric field will be smaller than the original one by a factor of κ,

$$\mathbf{E} = \mathbf{E}_{\text{free}}/\kappa \tag{31}$$

Substituting this into Eq. (30), we obtain

$$\boxed{\varepsilon_0 \int \kappa \mathbf{E} \cdot d\mathbf{S} = Q_{\text{free}}} \tag{32}$$

Gauss' Law in dielectrics

This is **Gauss' Law in dielectrics.** It relates the total electric field $\mathbf{E}$ to the *free* charge Q_{free}. The effect of the bound charge is implicitly contained in the factor κ appearing on the left side of the equation. Although the preceding discussion has focused on arrangements of conductors and dielectrics with high symmetry, the above modified version of Gauss' Law turns out to be valid for conductors and dielectrics of any shape whatsoever.

EXAMPLE 6. Two concentric spheres of sheet metal have radii r_1 and r_2 respectively. The space between these is filled with gas of dielectric constant κ (Figure 27.16). What is the capacitance of this contraption?

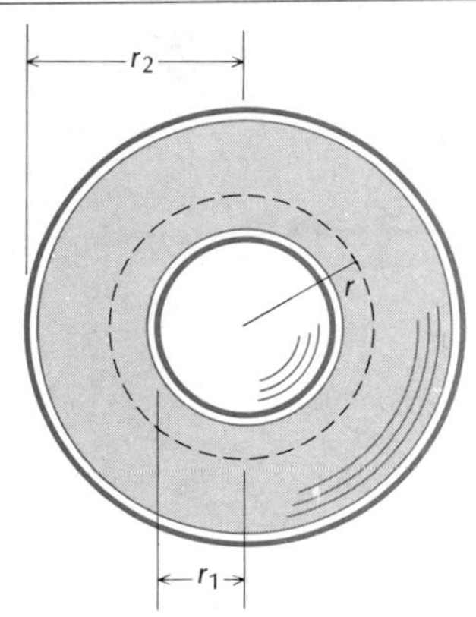

Fig. 27.16 Concentric conducting spheres.

SOLUTION: Suppose that the free charge on the inner sphere is Q_{free} and on the outer $-Q_{\text{free}}$. As Gaussian surface, take a sphere of radius r ($r_2 > r > r_1$, see Figure 27.16). Equation (32) then becomes

$$\varepsilon_0 \kappa E \times 4\pi r^2 = Q_{\text{free}}$$

or

$$E = \frac{1}{4\pi\kappa\varepsilon_0} \frac{Q_{\text{free}}}{r^2} \tag{33}$$

Note that this electric field differs from that of a point charge in vacuum by the factor $1/\kappa$.

The potential difference between r_1 and r_2 is then

$$\Delta V = \int_{r_1}^{r_2} \frac{1}{4\pi\kappa\varepsilon_0} \frac{q_{\text{free}}}{r^2}\, dr = \frac{Q_{\text{free}}}{4\pi\kappa\varepsilon_0}\left(\frac{1}{r_1} - \frac{1}{r_2}\right)$$

and the capacitance

$$C = \frac{Q_{\text{free}}}{\Delta V} = \frac{4\pi\kappa\varepsilon_0}{1/r_1 - 1/r_2} \tag{34}$$

In a more advanced study of dielectrics, it is useful to introduce a quantity **D** called the **electric displacement field,**

Electric displacement field

$$\mathbf{D} = \varepsilon_0 \kappa \mathbf{E}$$

In terms of this quantity, Gauss' Law becomes [see Eq. (32)]

$$\int \mathbf{D} \cdot d\mathbf{S} = Q_{\text{free}} \tag{35}$$

This version of Gauss' Law is of more general validity than Eq. (32); for instance, it remains valid even in nonlinear dielectrics, where Eq. (32) fails.

27.5 Energy in Capacitors

Capacitors not only store electric charge, but also electric energy. Consider a two-conductor capacitor with charges $\pm Q$ on its plates. If the capacitor contains no dielectric, then the electric potential energy can be calculated directly from Eq. (26.7):

$$U = \tfrac{1}{2}QV_2 + \tfrac{1}{2}(-Q)V_1 = \tfrac{1}{2}Q(V_2 - V_1)$$

where V_1 and V_2 are the potentials of the plates. Thus, the potential energy can be expressed in terms of charge and potential difference,

Energy in capacitor

$$\boxed{U = \tfrac{1}{2}Q\,\Delta V} \tag{36}$$

By means of the definition of capacitance, $Q = C\,\Delta V$, this can be put in the alternative forms

$$U = \tfrac{1}{2}C(\Delta V)^2 \tag{37}$$

or

$$U = \tfrac{1}{2}\,Q^2/C \tag{38}$$

If the capacitor contains a dielectric, then the calculation of the energy is a bit more involved. The trouble is that the dielectric, with its bound charges, contributes to the electric potential energy. However, in practice we are usually not interested in the total potential energy, but only in that part of the potential energy that changes as we charge (or discharge) the capacitor, i.e., we are interested only in the amount of work required to charge (or discharge) the capacitor. It turns out that this amount of work is correctly given by Eqs. (36)–(38), regardless of whether the capacitor contains a dielectric or not. The quantity Q in these equations is the charge on the plates, i.e., it is the *free* charge. To see this, let us derive Eq. (38) from a different starting point. Imagine that we charge the capacitor gradually, starting with an initial charge $q = 0$ and ending with a final charge $q = Q$. When the plates carry charges $\pm q$, the potential difference between them is q/C

and the work that we must perform to increase the charge on the plates by $\pm dq$ is

$$dU = \frac{q}{C}\,dq \tag{39}$$

The total work that we must perform to charge the capacitor is then

$$U = \int_0^Q \frac{q}{C}\,dq = \frac{1}{C}\int_0^Q q\,dq = \tfrac{1}{2}\,Q^2/C \tag{40}$$

This agrees with Eq. (38) and establishes its general validity. Note that as the free charges on the capacitor plate increase, the bound charges within the dielectric rearrange themselves, becoming more strongly polarized; the energy required for this rearrangement is already included in Eq. (40) (the properties of the dielectric enter into this formula via the capacitance).

In Section 26.3 we obtained a formula for the energy density in an electric field in a vacuum. That calculation was based on an examination of the electric field in the space between the plates of a parallel-plate capacitor. By repeating this calculation for a parallel-plate capacitor with dielectric, we readily find that the **energy density in the dielectric** is

$$\boxed{u = \tfrac{1}{2}\kappa\varepsilon_0 E^2} \tag{41}$$

Energy density in dielectric

Formally, this differs from Eq. (26.19) by an extra factor of κ. Note, however, that the electric field **E** in Eq. (41) is the actual electric field in the dielectric, which already contains an implicit dependence on κ.

EXAMPLE 7. Consider the parallel-plate capacitor of Example 4. What is the stored potential energy? What is the energy density in the dielectric?

SOLUTION: By Eq. (36),

$$U = \tfrac{1}{2}Q\,\Delta V = \tfrac{1}{2} \times 2.3 \times 10^{-5}\text{ coulomb} \times 30\text{ volt} = 3.4 \times 10^{-4}\text{ J}$$

The energy density can be calculated either from Eq. (41),

$$u = \tfrac{1}{2}\kappa\varepsilon_0 E^2 = \tfrac{1}{2} \times 2.3 \times 8.85 \times 10^{-12}\text{ F/m} \times (1.5 \times 10^6\text{ volt/m})^2$$

$$= 23\text{ J/m}^3$$

or else by taking the ratio of energy to volume,

$$u = \frac{U}{Ad} = \frac{3.4 \times 10^{-4}\text{ J}}{0.75\text{ m}^2 \times 2.0 \times 10^{-5}\text{ m}} = 23\text{ J/m}^3$$

SUMMARY

Capacitance of a pair of conductors: $C = Q/\Delta V$

Capacitance of parallel plates: $C = \varepsilon_0 A/d$

Parallel combination: $C = C_1 + C_2 + C_3 + \cdots$

Series combination: $1/C = 1/C_1 + 1/C_2 + 1/C_3 + \cdots$

Electric field in dielectric between parallel plates: $E = \frac{1}{\kappa} E_{\text{free}}$

Gauss' law in dielectric: $\int \kappa \mathbf{E} \cdot d\mathbf{S} = Q_{\text{free}}/\varepsilon_0$

Energy in capacitor: $U = \frac{1}{2} Q \, \Delta V$

Energy density in dielectric: $u = \frac{1}{2} \kappa \varepsilon_0 E^2$

QUESTIONS

1. Commercially available large capacitors have a capacitance of 1000 μF. How is it possible that the capacitance of such a device is larger than the capacitance of the Earth?

2. A single-conductor capacitor may be regarded as a two-conductor capacitor with the second plate consisting of a very large conducting shell of infinite radius. Show that for $r \to \infty$, Eq. (34), with $\kappa = 1$, reduces to Eq. (3).

3. Suppose we enclose the entire Earth in a conducting shell of a radius slightly larger than the Earth's radius. Explain why this would make the capacitance of the Earth much larger than the value calculated in Example 2.

4. Equation (8) shows that $C \to \infty$ as $d \to 0$. In practice, why can we not construct a capacitor of arbitrarily large C by making d sufficiently small? (Hint: What happens to E as $d \to 0$ while ΔV is held constant?)

5. If you put more charge on one plate of a parallel-plate capacitor than on the other, what happens to the extra charge?

6. Taking the fringing field into account, would you expect the capacitance of a parallel-plate capacitor to be larger or smaller than the value given by Eq. (8)? (Hint: How does the fringing affect the density of field lines between the plates?)

7. Explain why there must be a fringing field in the region near the edges of a pair of parallel plates. (Hint: Suppose there were no fringing field, so that the field lines look as in Figure 27.17. Is $\oint \mathbf{E} \cdot d\mathbf{l} = 0$ for the path shown in this figure?)

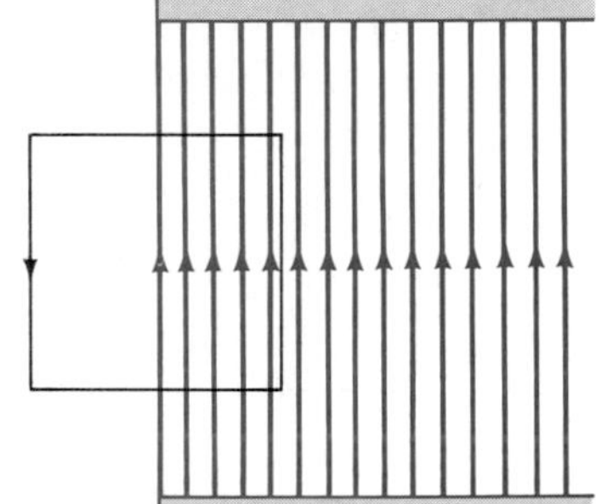

Fig. 27.17

8. Figure 27.18 shows a capacitor with a **guard rings.** These rings fit snugly around the edges of the capacitor plates, but they are not in electrical contact with the plates. In use, the potential on the rings is adjusted to the same value as the potential on the plates. Explain how the rings keep the field of the plates from fringing.

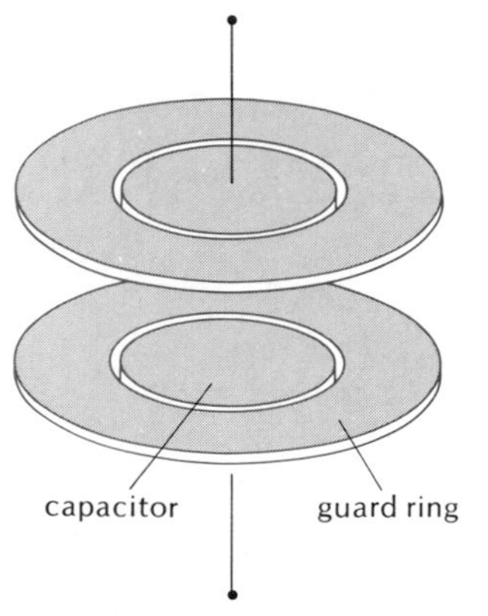

Fig. 27.18

9. Figure 27.19 shows the design of an adjustable capacitor used in the tuning circuit of a radio. This capacitor can be regarded as several connected capacitors. Are these several capacitors connected in series or in parallel? If we turn the tuning knob (and the attached colored plates) counterclockwise, does the capacitance increase or decrease?

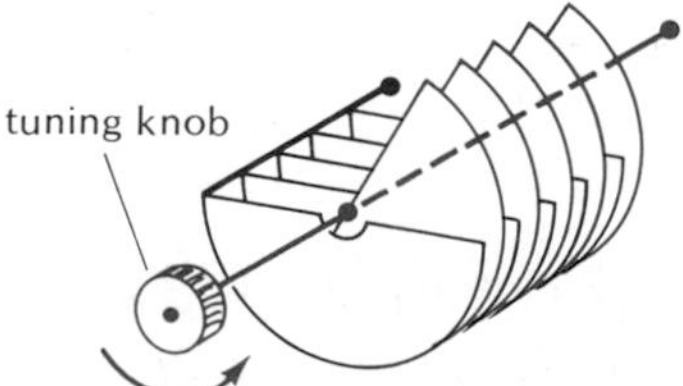

Fig. 27.19 The black plates are connected together, and the colored plates are connected together.

10. Consider a parallel-plate capacitor. Does the capacitance change if we insert a thin conducting sheet between the two plates, parallel to them?

11. Suppose we insert a thick slab of metal between the plates of a parallel-plate capacitor, parallel to the plates. Does the capacitance increase or decrease?

12. Consider a fluid dielectric that consists of molecules with permanent dipole moments. Will the dielectric constant increase or decrease as a function of temperature?

13. Figure 27.20 shows a dielectric slab partially inserted between the plates of a capacitor. Will the electric forces between the slab and the plates pull the slab into the region between the plates or push it out? (Hint: Consider the fringing field.)

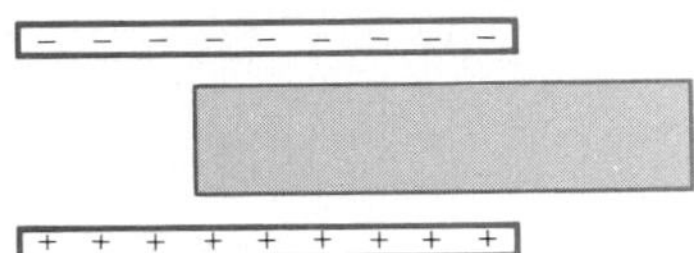

Fig. 27.20

14. If we increase the separation between the plates of a parallel-plate capacitor by a factor of 2, while holding the electric charge constant, by what factors will we change the electric field, the potential difference, the capacitance, and the electric energy?

15. Consider the parallel-plate capacitor with the slab of dielectric shown in Figure 27.15. How does the capacitance change if we move the slab up? If we move the slab to the right? If we tilt the slab?

16. Spell out the steps in the derivation of Eq. (41).

PROBLEMS

Section 27.1

1. Consider an isolated metallic sphere of radius R and another isolated metallic sphere of radius $3R$. If both spheres are at the same potential, what is the ratio of their charges? If both spheres carry the same charge, what is the ratio of their potentials?

2. The collector of an electrostatic machine is a metal sphere of radius 18 cm.
   (a) What is the capacitance of this sphere?
   (b) How many coulombs of charge must you place on this sphere to raise its potential to 2.0×10^5 V?

3. Your head is (approximately) a conducting sphere of radius 10 cm. What is the capacitance of your head? What will be the charge on your head if, by means of an electrostatic machine, you raise your head (and your body) to a potential of 100,000 V?

4. A capacitor consists of a metal sphere of radius 5 cm placed at the center of a thin metal shell of radius 12 cm. The space between is empty. What is the capacitance?

5. What is the capacitance of the Geiger-counter tube described in Problem 25.25? Pretend that the space between the conductors is empty.

Section 28.2

6. What is the combined capacitance if three capacitors of 3.0, 5.0, and 7.5 μF are connected in parallel? What is the combined capacitance if they are connected in series?

7. Three capacitors with capacitances $C_1 = 5.0\ \mu$F, $C_2 = 3.0\ \mu$F, and $C_3 = 8.0\ \mu$F are connected as shown in Figure 27.21. Find the combined capacitance.

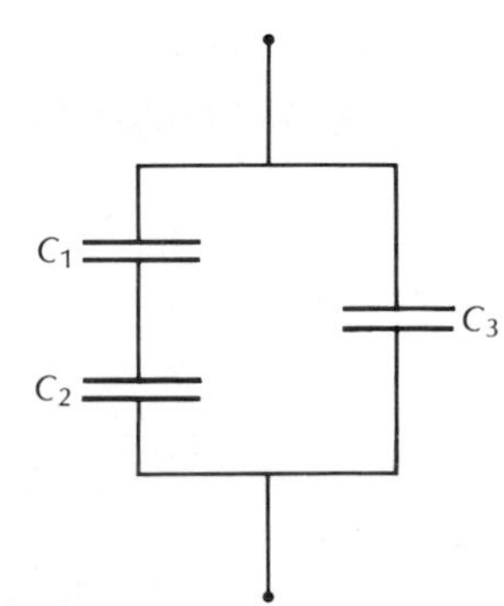

Fig. 27.21

8. Two capacitors, of 2.0 and 6.0 μF, respectively, are initially charged to 24 V by connecting each, for a few instants, to a 24-V battery. The battery is then removed and the charged capacitors are connected in a closed series circuit, the positive terminal of each capacitor being connected to the negative termi-

nal of the other (Figure 27.22). What will be the final charge on each capacitor?

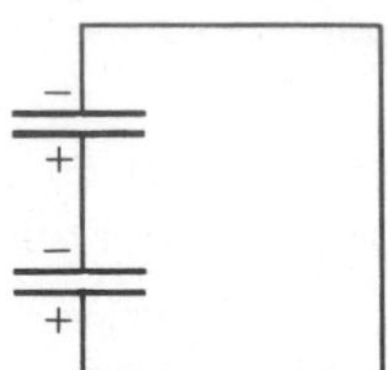

Fig. 27.22 Capacitors connected after they have been charged.

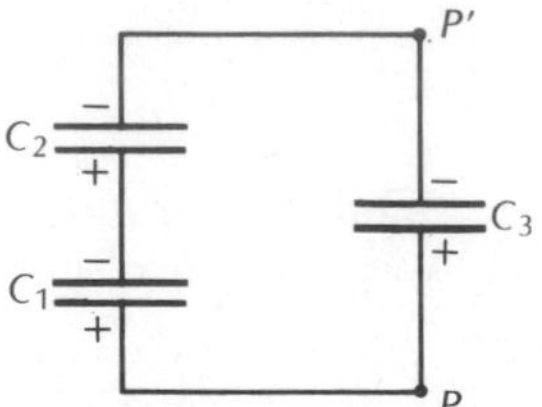

Fig. 27.23 Capacitors connected after they have been charged.

9. Three capacitors, of capacitances $C_1 = 2.0\ \mu$F, $C_2 = 5.0\ \mu$F, and $C_3 = 7.0\ \mu$F, are initially charged to 36 V by connecting each, for a few instants, to a 36-V battery. The battery is then removed and the charged capacitors are connected in a closed series circuit, with the positive and negative terminals joined as shown in Figure 27.23. What will be the final charge on each capacitor? What will be the voltage across the points *PP'* in Figure 27.23?

Section 27.3

10. You wish to construct a capacitor out of a sheet of polyethylene of thickness 5×10^{-2} mm and $\kappa = 2.3$ sandwiched between two aluminum sheets. If the capacitance is to be 3.0 μF, what must be the area of the sheets?

11. In order to measure the dielectric constant of a dielectric material, a slab of this material 1.5 cm thick is slowly inserted between a pair of parallel conducting plates separated by a distance of 2.0 cm. Before insertion of the dielectric, the potential difference across these capacitor plates is 3.0×10^5 V. During insertion, the charge on the plates remains constant. After insertion, the potential difference is 1.8×10^5 V. What is the value of the dielectric constant?

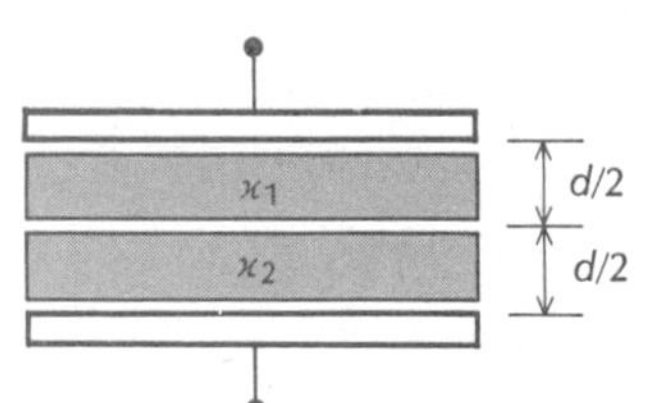

Fig. 27.24 Parallel-plate capacitor with two slabs of dielectric.

12. A parallel-plate capacitor of plate area *A* and spacing *d* is filled with two parallel slabs of dielectric of equal thickness with dielectric constants κ_1 and κ_2, respectively (Figure 27.24). What is the capacitance? (Hint: Check that the configuration of Figure 27.24 is equivalent to two capacitors in series.)

13. A capacitor with two large parallel plates of area *A* separated by a distance *d* is filled with two equal slabs of dielectric side by side (Figure 27.25). The dielectric constants are κ_1 and κ_2. What is the capacitance?

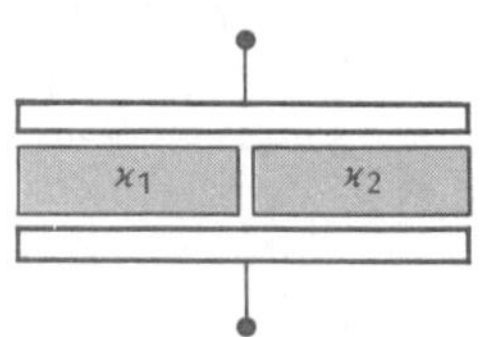

Fig. 27.25 Parallel-plate capacitor with two slabs of dielectric.

14. Show that the result of Example 5 can also be derived by regarding the capacitor partially filled with dielectric as two capacitors in series, one completely filled with dielectric, one empty.

15. A parallel-plate capacitor of plate area *A* and separation *d* contains a slab of dielectric of thickness $d/2$ (Figure 27.26) and dielectric constant κ. The potential difference between the plates is ΔV.
   (a) In terms of the given quantities, find the electric field in the empty region of space between the plates.
   (b) Find the electric field inside the dielectric.
   (c) Find the density of bound charge on the surface of the dielectric.

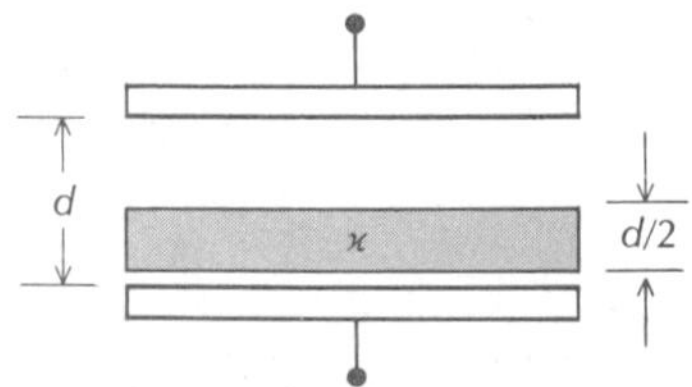

Fig. 27.26 A parallel-plate capacitor, partially filled with dielectric.

16. Within some limits, the difference between the dielectric constants of air and of vacuum is proportional to the pressure of the air, i.e., $\kappa - 1 \propto p$. Suppose that a parallel-plate capacitor is held at a constant potential difference by means of a battery. What will be the percentage change in the amount of charge on the plates as we increase the air pressure between the plates from 1.0 atm to 3.0 atm?

17. A parallel-plate capacitor is filled with carbon dioxide at 1 atm pressure. Under these conditions the capacitance is 0.5 μF. We charge the capacitor by means of a 48-V battery and then disconnect the battery so that the electric charge remains constant thereafter. What will be the change in the potential difference if we now pump the carbon dioxide out of the capacitor, leaving it empty?

*18. A parallel-plate capacitor is filled with a layer of distilled water 0.30 cm thick. The dipole moment of a water molecule is 6.1×10^{-30} C · m. Assume that the dipole moments of the water molecules are all perfectly aligned with the electric field. What is the surface charge density of bound charges on the surface of the layer of water?

Section 27.4

19. A spherical capacitor consists of a metallic sphere of radius R_1 surrounded by a concentric metallic shell of radius R_2. The space between R_1 and R_2 is filled with dielectric having a constant κ. Suppose that the free surface charge density on R_1 is $\sigma_{\text{free(1)}}$.

(a) What is the free surface charge density on the metallic sphere at R_2?
(b) What is the bound surface charge density on the dielectric at R_1?
(c) What is the bound surface charge density on the dielectric at R_2?

20. A long cylindrical copper wire of radius 0.20 cm is surrounded by a cylindrical sheath of rubber of inner radius 0.20 cm and outer radius 0.30 cm. The rubber has $\kappa = 2.8$. Suppose that the surface of the copper has a free charge density of 4.0×10^{-6} C/m^2.

(a) What will be the bound charge density on the inside surface of the rubber sheath? On the outside surface?
(b) What will be the electric field in the rubber near its inner surface? Near its outer surface?
(c) What will be the electric field just outside the rubber sheath?

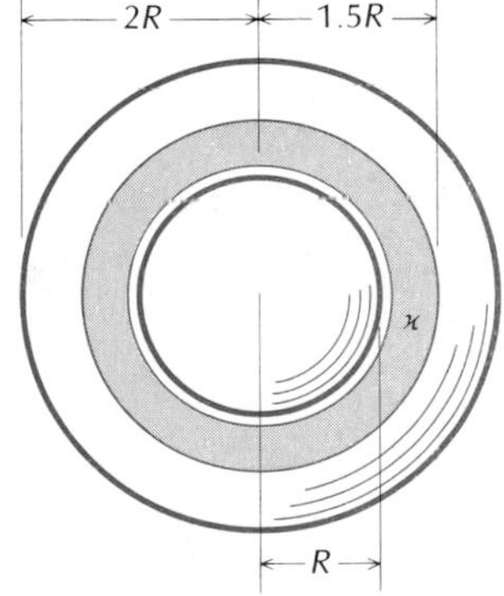

Fig. 27.27 A spherical capacitor, partially filled with dielectric.

21. A metallic sphere of radius R is surrounded by a concentric dielectric shell of inner radius R, outer radius $3R/2$. This is surrounded by a concentric, thin, metallic shell of radius $2R$ (Figure 27.27). The dielectric constant of the shell is κ. What is the capacitance of this contraption?

22. Two small metallic spheres are submerged in a large volume of transformer oil of dielectric constant $\kappa = 3.0$. The spheres carry electric charges of 2.0×10^{-6} C and 3.0×10^{-6} C, respectively, and the distance between them is 0.60 m. What is the force on each?

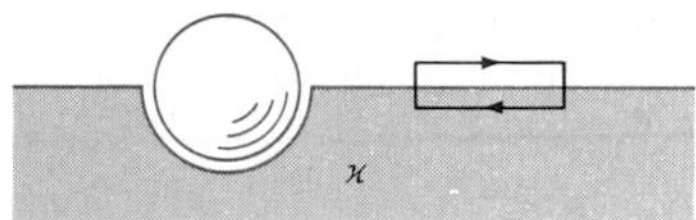

Fig. 27.28 A brass sphere afloat in a lake of oil.

*23. A sphere of brass floats in a large lake of oil of dielectric constant $\kappa = 3.0$. The sphere is exactly halfway immersed in the oil (Figure 27.28). The sphere has a net charge of 2.0×10^{-6} C. What fraction of this electric charge will be on the upper hemisphere? On the lower? (Hint: For the path shown in Figure 27.28, $\oint \mathbf{E} \cdot d\mathbf{l} = 0$; from this prove that the electric fields in the oil and in the air above the oil will be exactly the same.)

*24. A spherical capacitor consists of two concentric spheres of metal of radii R_1 and R_2. The space between these spheres is filled with two kinds of dielectric (Figure 27.29); the dielectric in the upper hemisphere has a constant κ_1 and the dielectric in the lower hemisphere has a constant κ_2. What is the capacitance of this device? (Hint: For the path shown in Figure 27.29 $\oint \mathbf{E} \cdot d\mathbf{l} = 0$; from this prove that the electric fields in both dielectrics are exactly the same.)

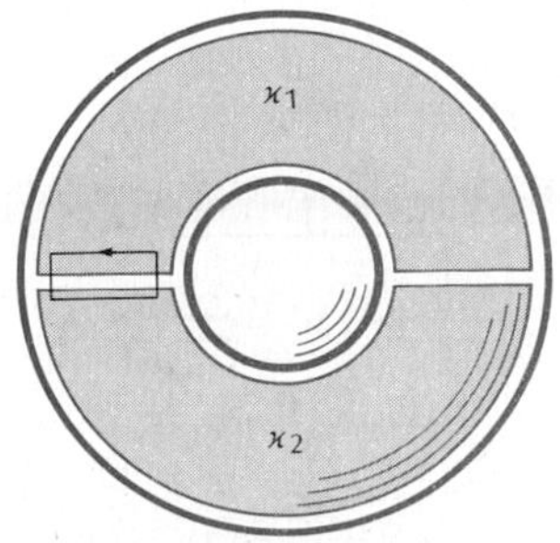

Fig. 27.29 A spherical capacitor with two kinds of dielectric.

*25. In a semiconductor with impurity ions, electrons will orbit around these ions. The sizes of the orbits are considerably larger than the spacings between the semiconductor atoms and hence an electron may be regarded as moving through a more or less uniform medium of a given dielectric constant.

(a) Show that the electric force of attraction between an electron and an ion of charge e immersed in a medium of dielectric constant κ is

$$F = \frac{1}{4\pi\varepsilon_0\kappa} \frac{e^2}{r^2}$$

(b) Calculate the orbital energy of an electron in a circular orbit according to Bohr's theory described in Problem 25.16.
(c) The dielectric constant of germanium is $\kappa = 15.8$. Evaluate the orbital energy of an electron moving around an ion embedded in germanium; assume that the electron is in the smallest Bohr orbit. By what factor does your result differ from what you would obtain for an electron moving around this ion in a vacuum?

Section 27.5

26. A parallel-plate capacitor has a plate area of 900 cm² and a plate separation of 0.50 cm. The space between the plates is empty.
(a) What is the capacitance?
(b) What is the potential difference if the charges on the plates are $\pm 6.0 \times 10^{-8}$ C?
(c) What is the electric field between the plates?
(d) The energy density?
(e) The total energy?

27. Repeat Problem 26 if the space between the plates is filled with Plexiglas.

28. A TV receiver contains a capacitor of 10 μF charged to a potential difference of 2×10^4 V. What is the amount of charge stored in this capacitor? The amount of energy?

29. Two parallel conducting plates of area 0.5 m² placed in a vacuum have a potential difference of 2.0×10^5 V when charges of $\pm 4.0 \times 10^{-3}$ C are placed on them, respectively.
(a) What is the capacitance of the pair of plates?
(b) What is the distance between them?
(c) What is the electric field between them?
(d) What is the electric energy?

30. Two capacitors of 5.0 μF and 8.0 μF are connected in series to a 24-V battery. What is the energy stored in the capacitors?

31. A parallel-plate capacitor without dielectric has an area A and a charge $\pm Q$ on each plate.
(a) What is the electric force F of attraction between the plates?
(b) How much work must you do against this force in order to increase the plate separation by an amount Δl?
(c) By means of Eq. (38), calculate the change of ΔU in potential energy during this change.
(d) By comparing (a) and (c) check that $F = -\Delta U/\Delta l$.
(Hint: One-half of the electric field between the plates is due to one plate and one-half is due to the other. Consequently, when calculating the electric force on a plate from the product of field times charge, only one-half of the field must be used; the other half gives the electric force of the plate on *itself* and is of no interest.)

32. Power companies are interested in the storage of surplus electric energy. Suppose we wanted to store 10^6 kW · h of electric energy (half a day's output for a large power plant) in a large parallel-plate capacitor filled with a plastic dielectric with $\kappa = 3.0$. If the dielectric can tolerate a maximum electric field of 5×10^7 V/m, what is the minimum total volume of dielectric needed to store this energy?

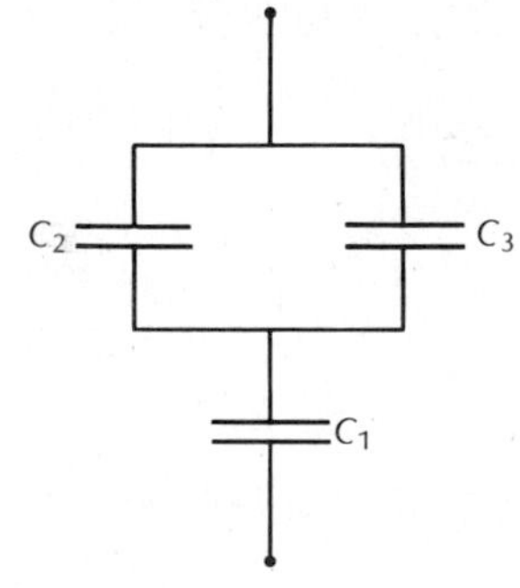

Fig. 27.30

33. Three capacitors are connected as shown in Figure 27.30. Their capacitances are $C_1 = 2.0$ μF, $C_2 = 6.0$ μF, and $C_3 = 8.0$ μF. If a voltage of 200 V is applied to the two free terminals, what will be the charge on each capacitor? What will be the energy in each?

CHAPTER 28

Currents and Ohm's Law

Under static conditions there can exist no electric field inside a conductor. But suppose that we suddenly deposit opposite amounts of electric charge on the opposite ends of a long metallic conductor, such as a wire. The conductor will then not be in electrostatic equilibrium and the charges at the ends will generate an electric field along and inside the conductor (Figure 28.1). This electric field propels the charges toward each other. When the charges meet, they cancel. The electric field then disappears — the conductor reaches equilibrium.

For a good conductor, such as copper, the approach to equilibrium is fairly rapid; typically, the time required to achieve equilibrium is a small fraction of a second. However, we can keep a conductor in a permanent state of disequilibrium if we continually supply more electric charge to its ends. For example, we can connect the two ends of a copper wire to the terminals of a battery or of an electric generator. The terminals of such a device act as source and sink of electric charge, just like the outlet and the intake of a pump act as source and sink of water. Under these conditions electric charge will continually flow from one terminal to the other, forming an electric current.[1]

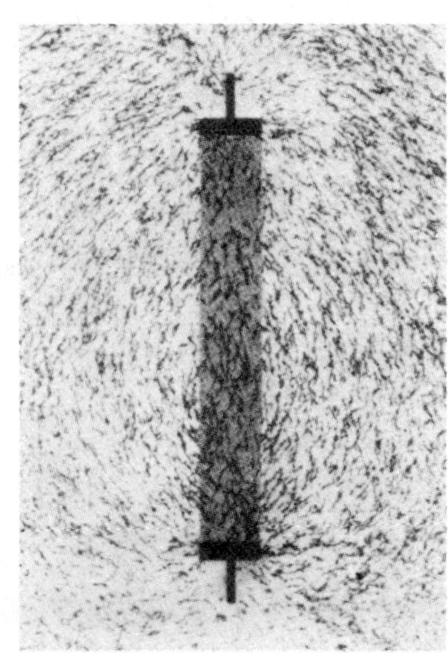

Fig. 28.1 Electric field lines in and near a straight conductor not in equilibrium. The field lines have been made visible by sprinkling grass seeds on the surface of the paper on which the conductor has been painted with conducting paint. (From O. Jefimenko, *Am. J. Phys.* **30,** 19, 1962.)

28.1 Electric Current

When a wire is connected between the two terminals of a battery or generator, the electric charges are propelled from one end of the wire to the other by the electric field that exists along and within the wire. Most of the field lines originate at the terminals of the battery or gen-

[1] We will discuss the inner workings of batteries, generators, and other "pumps" for electric charge in the next chapter.

erator, but some field lines originate at charges on the wire itself. As Figures 28.1 and 28.2 show, the field lines tend to concentrate within the conductor, and they tend to follow the conductor. If the conductor has no sharp kinks, the field lines are uniformly distributed over the cross-sectional area of the conductor. For instance, if the conductor is a more or less straight wire of constant thickness, then the electric field inside the wire will be of constant magnitude and of a direction parallel to the wire. If the length of the wire is l and if the battery or generator maintains a difference of potential ΔV across its ends, then the electric field in the wire is

Electric field in uniform wire

$$\boxed{E = \Delta V/l} \tag{1}$$

This electric field causes the flow of charge, or **electric current,** from one end of the wire to the other. Before we can explore the dependence of the current on the field, we need a precise definition of the current. Suppose that an amount of charge dq flows past some given point of the wire (e.g., the end of the wire) in a time dt; then the electric current is defined as charge divided by time,

Electric current

$$\boxed{I = \frac{dq}{dt}} \tag{2}$$

Note that if the sides of the wire do not leak (good insulation), then the conservation of electric charge requires that the current be the same everywhere along the wire, that is, the current is simply the rate at which charge enters the wire at one end or the rate at which charge leaves at the other end.

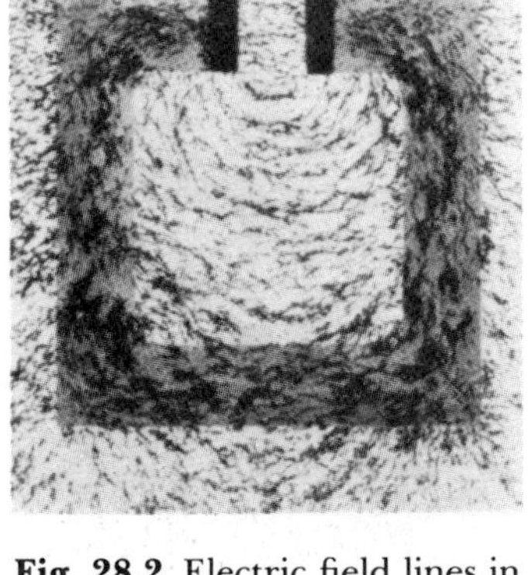

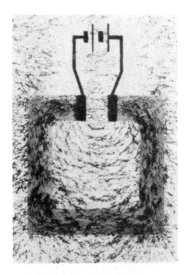

Fig. 28.2 Electric field lines in and near a rectangular conductor carrying an electric current. (From O. Jefimenko, *Am. J. Phys.* **30,** 19, 1962.)

The SI unit of current is the **ampere** (A); this is a flow of charge of one coulomb per second,

$$1 \text{ ampere} = 1 \text{ A} = 1 \text{ C/s} \tag{3}$$

In metallic conductors the charge carriers are electrons — a current in a metal is nothing but a flow of electrons. In electrolytes the charge carriers are positive ions, negative ions, or both — a current in such a conductor is a flow of ions. For the sake of uniformity, whenever we need to indicate the direction of the current along a conductor, we will follow the convention that the current has the direction of the positive flow of charge. This means that we pretend that the moving charges are always positive charges. Of course, in metals the moving charges are actually negative charges (electrons) and hence the above convention assigns to the current a direction opposite to that of the true motion of the charges. However, as regards the transfer of charge, the transport of negative charge in one direction is equivalent to the transport of positive charge in the opposite direction. Our convention for labeling the direction of the current takes advantage of this equivalence.

If we divide the current in a conductor by the cross-sectional area of the conductor (Figure 28.3) we obtain the **current density,**[2]

[2] Do not confuse the symbol A for area with the abbreviation A for ampere.

$$\boxed{j = I/A} \qquad (4)$$

Current density

This is really the *average* current density over the area A. We can also define a local current density in terms of the current dI that flows across an infinitesimal portion of dA of cross-sectional area (Figure 28.3),

$$j = \frac{dI}{dA} \qquad (5)$$

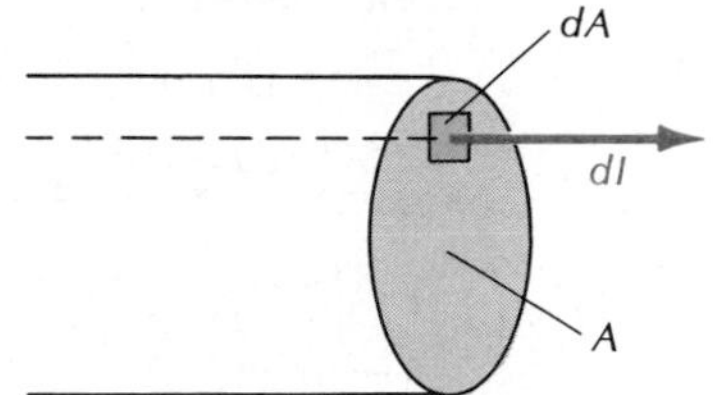

Fig. 28.3 A conductor carries a current from left to right. The conductor has been cut off on the right so as to give a clear view of its cross section.

Under normal conditions the current in a wire is uniformly distributed over the entire cross-sectional area of the wire. In terms of the local current density, this means that j is constant over the entire cross-sectional area.

28.2 Resistance and Ohm's Law

We will now examine in detail the behavior of a current in a metallic conductor. Such conductors contain a vast number of free electrons; for example, copper has about 8×10^{22} free electrons per cubic centimeter. These electrons form a gas which fills the entire volume of the metal. Of course, in a neutral conductor the negative charge of the free electrons is exactly balanced by the positive charge of the ions that make up the lattice of the metal. A current in a metallic conductor is simply a flow of the gas of electrons, while the ions remain at rest.

The flow of the gas of electrons along a metallic wire is analogous to the flow of water along a canal leading down a gentle slope. Under these conditions the force of gravity acting on the water has a component along the canal; this component pushes the water along. But the water does not accelerate because friction between the water and the walls of the canal opposes the motion — the water moves at a constant speed because the friction exactly matches the push of gravity.

Likewise, the electric field in the wire pushes the gas of electrons along. But the gas of electrons does not accelerate because friction between the gas and the body of the wire opposes the motion — the gas moves at a constant speed because the friction exactly matches the push of the electric field.

The analogy between the motion of water and the motion of the electron gas extends to the motion of individual water molecules and individual electrons. Although the water in a canal usually has a fairly low speed, perhaps a few meters per second, the individual molecules within the water have a rather high speed — the typical speed of the random thermal motion of water molecules is about 500 m/s at ordinary temperatures. But since this thermal motion consists of rapid zigzags which are just as likely to move the molecule backward as forward, this high speed does not contribute to the net downhill motion of the water. Figure 28.4 shows the motion of a water molecule in the canal; on a microscopic scale, this motion consists of rapid zigzags on which is superimposed a much slower "drift" along the canal.

Fig. 28.4 Path of a water molecule in canal. The molecule gradually drifts from left to right.

Likewise, the electron gas moves along the wire at a rather low speed, perhaps 10^{-2} m/s, but the individual electrons have a much higher speed — the typical speed of the random motion of electrons in a metal is about 10^6 m/s (this very high speed is due to quantum-mechanical effects, which we cannot discuss here). Thus, the net motion of an electron also consists of rapid zigzags on which is superimposed a much slower "drift" motion along the wire. Qualitatively, the motion resembles the path of a water molecule shown in Figure 28.4, but the amount of drift per zigzag is even less than shown in this figure.

The friction between the electron gas and the body of the wire is caused by collisions between the electrons and the ions of the crystal lattice of the wire. An electron moving through a piece of copper will suffer about 10^{14} collisions with ions per second. Each collision slows the electron down, brings it to a stop, or reverses its motion. Because of the disturbing effects of these collisions, the electron never gains much velocity from the electric field that is attempting to accelerate it. The collisions dissipate the kinetic energy that the electron receives from the electric field. The dissipated kinetic energy of the electrons remains in the crystal lattice in the form of random kinetic energy of the ions, i.e., it remains as heat.

The average velocity, or drift velocity, that an electron attains in the electric field is proportional to the strength of the electric field,

$$v_{\mathrm{d}} \propto E \tag{6}$$

This proportionality merely reflects the fact that if the electric field is strong, the electron gains more velocity between one collision and the next, and therefore attains a larger average velocity. The electric current carried by the wire is proportional to the average velocity of the electrons,

$$I \propto v_{\mathrm{d}} \propto E \tag{7}$$

The current is also proportional to the cross-sectional area of the wire, because a large cross-sectional area means that more electrons participate in the transport of charge. Hence

$$I \propto AE \tag{8}$$

With $E = \Delta V/l$, this proportionality becomes

$$I \propto \frac{A}{l}\,\Delta V \tag{9}$$

To transform this into an equality, we rewrite it as

$$I = \frac{1}{\rho}\frac{A}{l}\,\Delta V \tag{10}$$

where ρ is a constant of proportionality that characterizes the material of the wire. This constant is called the **resistivity** of the material.

Resistivity

It is customary to define the **resistance** of the wire as

Resistance

$$\boxed{R = \rho\frac{l}{A}} \tag{11}$$

Equation (10) can then be expressed in the convenient form

$$I = \frac{\Delta V}{R} \tag{12}$$

Ohm's Law

This equation is called **Ohm's Law.** It asserts that *the current is proportional to the potential difference* between the ends of the conductor. Note that in Eq. (12), the resistance plays the role of a constant of proportionality. For a wire of uniform cross section, the resistance can be calculated from the simple formula (11). But Ohm's Law is also valid for conductors of arbitrary shape — such as wires of nonuniform cross section — for which the resistance must be calculated from a more complicated formula tailored to the shape and the size of the conductor.

Ohm's Law is valid for metallic conductors and also for nonmetallic conductors (e.g., carbon) in which the current is carried by a flow of electrons.[3] It is even valid for plasmas and for electrolytes, in which the current is carried by a flow of both electrons and ions. However, we ought to keep in mind that in spite of its wide range of applicability, Ohm's Law is not a general law of nature — such as Gauss' Law — but only an assertion about the electrical properties of certain materials.

Georg Simon Ohm, *1787–1854, German physicist, professor at Munich. Ohm was led to his law by an analogy between the conduction of electricity and the conduction of heat: the electric field is analogous to the temperature gradient and the electric current is analogous to the heat flow. Equation (28.12) is then analogous to Eq. (20.5).*

28.3 The Flow of Free Electrons[4]

Before we deal with some applications of Ohm's Law, let us reexamine the derivation of this law and fill in the constants of proportionality that we left out in Eqs. (6)–(9). To do this, we need to calculate the average motion of the free electrons in a metal in some detail. The high-speed thermal motion of the electrons does not enter directly into the calculation of the average motion because it is random; however, it enters indirectly because it determines the collision rate. The large collision rate of a free electron in a metal — for instance, 10^{14} collisions per second for an electron in copper — results from the high speed of the random motion: the electron moves a large distance per second and therefore encounters many ions with which to collide. Since the extra speed that the electron gains from the electric field is very small compared with the random speed, the collision rate is nearly unaffected by the electric field. The average time per collision is therefore a constant τ that depends only on the characteristics of the metal. We can find the average motion of an electron by examining the losses and gains of momentum of this electron. If the average velocity, or drift velocity, of an electron is v_d, then the average momentum is $m_e v_d$. We expect that, on the average, a collision will absorb all of this momentum, i.e., a collision will destroy the forward drift velocity and leave the electron with only the random thermal motion. This means that, in a time interval τ, the electron loses a momentum $m_e v_d$; the

[3] The crucial difference between metals (good conductors) and nonmetals (poor conductors) is that the latter have very few free electrons.

[4] This section is optional.

average rate at which the electron loses momentum in collisions is therefore

$$\left(\frac{\Delta p}{\Delta t}\right)_{\text{loss}} = \frac{m_e v_d}{\tau} \tag{13}$$

On the other hand, the rate at which the electron gains momentum by the action of the electric force is

$$\left(\frac{\Delta p}{\Delta t}\right)_{\text{gain}} = -eE \tag{14}$$

Under steady-state conditions, the rate of loss of momentum must match the rate of gain. By setting the right sides of Eqs. (13) and (14) equal we immediately obtain

Drift velocity

$$\boxed{v_d = -eE\tau/m_e} \tag{15}$$

This is the average velocity with which the electron gas flows along the wire. As expected, this velocity is proportional to the strength of the electric field. The negative sign in Eq. (15) indicates that the direction of flow is opposite to the direction of the electric field.

EXAMPLE 1. A potential difference of 3.0 V is applied to the ends of a copper wire 0.5 m long. What is the drift velocity of the free electrons in the wire? In copper at room temperature, the average time interval between collisions is $\tau = 2.7 \times 10^{-14}$ s.

SOLUTION: The electric field in the wire is $E = 3.0$ V/0.5 m = 6.0 V/m. Hence Eq. (15) gives

$$v_d = -\frac{1.6 \times 10^{-19}\ \text{C} \times 6.0\ \text{V/m} \times 2.7 \times 10^{-14}}{9.1 \times 10^{-31}\ \text{kg}}$$

$$= -2.8 \times 10^{-2}\ \text{m/s}$$

Note that this speed is rather low — it takes an electron about a third of a minute to wander from one end of the wire to the other. Nevertheless, if a pulse of current (a signal) is suddenly injected into the wire at one end, a similar pulse of current will emerge from the far end *almost instantaneously.* What happens in this case is that, by means of their electric field, electrons push on neighboring electrons and a compressional wave travels through the electron gas; this ejects electrons out of the far end almost instantaneously. The phenomenon is analogous to the propagation of a wave on the water of a canal; the wave travels much faster than the flow of water.

To find the electric current that corresponds to the flow of the electron gas, we need to take into account the number of electrons. Suppose that the metal of the wire has n free electrons per unit volume. The quantity n depends on the metal; for copper $n = 8.5 \times 10^{28}/\text{m}^3$. If the wire has a cross-sectional area A and a length l, then its total number of free electrons is $n \times [\text{volume}] = nAl$ and the total charge associated with these electrons is

$$\Delta q = -enAl \tag{16}$$

It takes a time $\Delta t = l/|v_d|$ for all of these electrons to emerge at one end of the wire. Hence the current in the wire is

$$I = \frac{|\Delta q|}{\Delta t} = \frac{enAl}{l/|v_d|} = \frac{e^2 n\tau}{m_e} AE \qquad (17)$$

In this equation we have ignored the negative sign on the electric charge because the direction of the current is to be reckoned according to the convention described in section 28.1 and not according to the motion of the electrons. The current is in the same direction as the electric field.

According to Eq. (17), the current is directly proportional to the magnitude of the electric field. It is also directly proportional to the cross-sectional area of the wire. All of this agrees with our earlier, rough argument.

In terms of the potential difference $\Delta V = El$ across the ends of the wire, we can rewrite Eq. (17) as follows:

$$I = \frac{e^2 n\tau}{m_e} AE = \frac{e^2 n\tau}{m_e} \frac{A}{l} (El) \qquad (18)$$

$$= \left(\frac{e^2 n\tau}{m_e} \frac{A}{l}\right) \Delta V \qquad (19)$$

If we write the factor in parentheses as

$$R = \frac{m_e}{e^2 n\tau} \frac{l}{A} \qquad (20)$$

then Eq. (19) becomes

$$I = \frac{\Delta V}{R} \qquad (21)$$

which is Ohm's Law. Thus, Eq. (20) gives us a theoretical expression for the resistance in terms of the physical parameters associated with the free-electron gas. In practice, Eq. (20) is not very useful because there is no direct method for measuring the time per collision τ.

28.4 The Resistivity of Materials

As we saw in Section 28.2, the resistance of a wire of uniform cross section is related to the resistivity by the formula

$$R = \rho \frac{l}{A} \qquad (22)$$

We can use this formula to calculate the resistance if the resistivity of the material is known, and we can also use it to calculate the resistivity if the resistance is known. The latter calculation is important in the experimental determination of the resistivity of a material, which is done by measuring the potential difference and current in a wire of given length and cross section made of a sample of the material. This means that Ohm's Law is used both as a definition of resistance and as a law relating current, potential difference, and resistance. In the definitions of force (Section 5.2) and of temperature (Section 19.2) we have already encountered similar instances of such dual uses of laws.

As is obvious from Ohm's Law, the unit of resistance is 1 volt/ampere; this unit is called **ohm (Ω),**

ohm, Ω

$$1 \text{ ohm} = 1\ \Omega = 1 \text{ volt/ampere} \tag{23}$$

The unit of resistivity is the ohm-meter. Table 28.1 gives the resistivities of some conducting materials.

EXAMPLE 2. A wire commonly used for electrical installations in homes is No. 10 copper wire, which has a radius of 0.129 cm. What is the resistance of a piece of wire 30 m long? What is the potential drop along this wire if it carries a current of 10 A?

SOLUTION: The cross-sectional area of the wire is

$$A = \pi r^2 = \pi(0.129 \times 10^{-2} \text{ m})^2 = 5.2 \times 10^{-6} \text{ m}^2$$

By Eq. (22), the resistance will be

$$R = 1.7 \times 10^{-8}\ \Omega \cdot \text{m} \times 30 \text{ m}/5.2 \times 10^{-6} \text{ m}^2 = 0.098\ \Omega \tag{24}$$

For a current of 10 A, Ohm's Law then gives a potential drop

$$\Delta V = IR = 10 \text{ A} \times 0.098\ \Omega = 0.98 \text{ volt} \tag{25}$$

Note that by combining Eq. (12) and (22) we readily find

$$\frac{I}{A} = \frac{1}{\rho}\frac{\Delta V}{l} \tag{26}$$

Since I/A is the current density and $\Delta V/l$ is the electric field in the conductor, Eq. (26) can be written

$$j = \frac{1}{\rho} E \tag{27}$$

i.e., the current density is directly proportional to the electric field. This is an alternative expression for Ohm's Law. Although our derivation of Ohm's Law began with a wire of uniform cross section, Eq. (27) is valid for conductors of arbitrary shape. This equation relates two *local* quantities: the current density at one point within the conductor and the electric field at that point.

Dependence of resistivity on temperature

The resistivity depends somewhat on temperature. In ordinary metals, the resistivity increases slightly with temperature. This is due to an increase in the rate of collision between electrons and atoms of the lattice — at high temperature the atoms jump violently around their positions in the lattice and they are then more likely to disturb the motion of the electrons. The numbers in the first column of Table 28.1 give the resistivity at room temperature (20°C). The numbers in the second column give the percentage increase in the resistivity per degree Celsius.

Table 28.1 RESISTIVITIES OF METALS[a]

Material	ρ	*Increase in ρ per °C*
Silver	$1.6 \times 10^{-8}\ \Omega \cdot m$	0.38%
Copper	1.7×10^{-8}	0.39%
Aluminum	2.8×10^{-8}	0.39%
Brass	$\sim 7 \times 10^{-8}$	0.2%
Nickel	7.8×10^{-8}	0.6%
Iron	10×10^{-8}	0.5%
Steel	$\sim 11 \times 10^{-8}$	0.4%
Constantan	49×10^{-8}	0.001%
Nichrome	100×10^{-8}	0.04%

[a] At a temperature of 20°C.

EXAMPLE 3. Suppose that because of a current overload, the temperature of the copper wire of Example 2 increases from 20°C to 90°C. How much does the resistance increase?

SOLUTION: The temperature increase is 70°C. According to Table 28.1, the resistance of copper increases by 0.39% for a temperature increase of 1°C. Hence the resistance increases by $0.39 \times 70 = 27\%$ for a temperature increase of 70°C. The change of resistance is therefore

$$\Delta R = 0.098\ \Omega \times 0.27 = 0.026\ \Omega$$

and the new resistance of the wire will be

$$0.098\ \Omega + 0.026\ \Omega = 0.124\ \Omega$$

Superconductivity

At very low temperatures the resistivity of a metal will be substantially less than at room temperature. Some metals, such as lead, tin, zinc, and niobium, exhibit the phenomenon of **superconductivity**: their resistance vanishes completely as the temperature approaches absolute zero. For example, Figure 28.5 shows a plot of resistivity vs. temperature for tin; at a temperature of 3.72 K, the resistivity abruptly vanishes. The resistance of such a superconductor is *exactly* zero. In one experiment, a current of several hundred amperes was started in a superconducting ring; the current continued on its own with undiminished strength for over a year, without any battery or generator to maintain it. In Interlude K we will present a detailed discussion of the properties of superconductors.

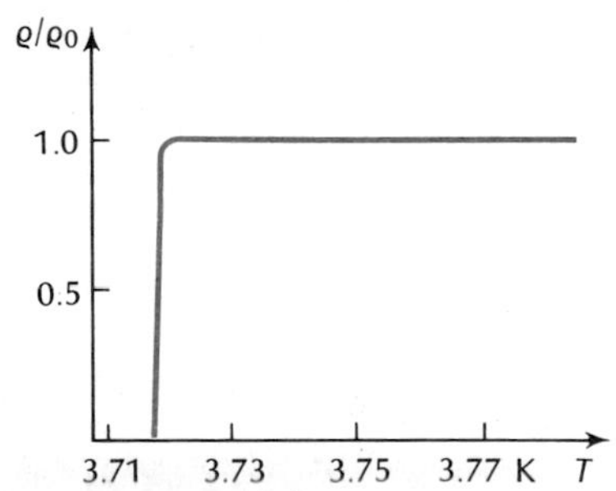

Fig. 28.5 Resistivity of tin as a function of temperature. Below 3.72 K, the resistivity is zero. The resistivity has been expressed as a fraction of the resistivity at 4.2 K, the temperature of liquefaction of helium.

According to the definition that we gave in Section 22.4, an ideal insulator is a material that does not permit the motion of electric charge. Real insulators, such as porcelain or glass, do permit some very slight motion of charge. What distinguishes them from conductors is their enormously large resistivity. Typically, the resistivity of insulators is more than 10^{20} times as large as that of conductors (see Table 28.2). This means that even when we apply a high voltage to a piece of glass, the flow of current will be insignificant (provided, of course, that the material does not suffer electrical breakdown). In fact, on a humid day it is likely that more current will flow along the microscopic film of water that tends to form on the surface of the insulator than through the insulator itself.

Table 28.2. RESISTIVITIES OF INSULATORS

Material	ρ
Polyethylene	$2 \times 10^{11}\ \Omega \cdot \text{m}$
Glass	$\sim 10^{12}$
Porcelain, unglazed	$\sim 10^{12}$
Rubber, hard	$\sim 10^{13}$
Epoxy	$\sim 10^{15}$

28.5 Semiconductors

A semiconductor is a material with a resistivity between that of conductors and insulators. The resistivities of semiconductors vary over a wide range; the resistivities may be 10^4 to 10^{15} times a large as the resistivities of conductors (see Table 28.3).

It is a characteristic feature of semiconductors that the addition of impurities to the material has a drastic effect on the resistivity. For instance, the silicon used in electronic devices is often "doped" with small amounts of arsenic or boron; the addition of just one part per million of arsenic will decrease the resistivity of silicon by a factor of more than 10^5. The manipulation of the resistivity of materials by intentional contamination with carefully selected impurities plays a crucial role in the manufacture of semiconductor devices such as diodes and transistors. Pure semiconductor materials are hardly ever used in practical applications. It is usually the presence of impurities that gives the semiconductor materials their interesting electric properties.

***n*-type and *p*-type semiconductors**

Semiconductors fall into two categories: n type and p type. In an n-type semiconductor, the carriers of current are free electrons, as in a metal. However, the resistance is higher because the semiconductor has fewer free electrons than a metal. Also, the semiconductor differs from a metal in that the resistivity *decreases* as the temperature increases. This curious behavior is due to an increase in the number of free electrons — as the temperature increases, more electrons shake loose from the atoms of the semiconductor and these extra free electrons more than compensate for the extra friction experienced by each at the higher temperature.

In a p-type semiconductor, the carriers of current are "holes" of positive charge. To understand what this means, consider Figure 28.6 showing an array of electrons and positive ions. In Figure 28.6a, these electrons and ions form neutral atoms. Suppose that the right end of this array is connected to the positive pole of a battery (not shown) and the left end to the negative pole. If the battery pulls an electron out of the right end, it will leave the array with a hole or missing electron at the position of the first atom (Figure 28.6b). The electrons will then play a game of musical chairs: the electron from the next atom will

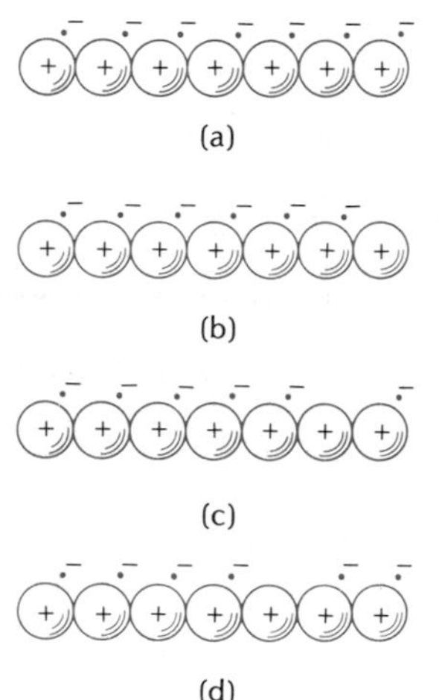

Fig. 28.6 A row of positive ions (balls marked with +) and electrons (color dots marked with −).

Table 28.3. RESISTIVITIES OF SEMICONDUCTORS

Material	ρ
Silicon	$2.6 \times 10^{3}\ \Omega \cdot \text{m}$
Germanium	4.2×10^{-1}
Carbon (graphite)	3.5×10^{-5}

jump into this hole, leaving a hole at the position of the second atom (Figure 28.6c); and then the electron from the next atom will jump, etc. The collective motion of the electrons from left to right can be conveniently described as the motion of a hole from right to left. The hole virtually carries positive charge from the right to the left. This is essentially the mechanism for conduction in a *p*-type semiconductor. Instead of a gas of free electrons, this type of semiconductor has a gas of free holes. A flow of current is then a flow of holes and the direction of the current is the same as the direction of motion of the holes.

Semiconductors usually contain both free electrons and free holes. Whether a semiconductor is *n* type or *p* type depends on which kind of charge carrier dominates. The concentration of free electrons and of free holes is largely determined by the impurities that are present in the material. **Donor** impurities consist of atoms that release their valence electrons when placed in the semiconductor and they thereby increase the number of free electrons. **Acceptor** impurities consist of atoms that trap electrons when placed in the semiconductor and they thereby generate holes. Hence, a semiconductor with donor impurities will be *n* type and one with acceptor impurities will be *p* type. For instance, silicon doped with arsenic is an *n*-type semiconductor and silicon doped with boron is a *p*-type semiconductor. Even though the added impurity atoms may only amount to a few parts per million, they completely change the conductivity because the semiconductor has so few current carriers to start with.

Donor and acceptor impurities

28.6 Resistances in Combination

The metallic wires of any electric circuit have some resistance. But in electronic devices — radios, televisions, amplifiers — the main contribution to the resistance is usually due to gadgets that are specifically designed to have a high resistance. These gadgets are **resistors**. They are commonly made out of a short piece of pure carbon (graphite) connected between two terminals (Figure 28.7a). Carbon has a high resistivity and hence a small piece of carbon can have a higher resistance than a long piece of metallic wire. Such resistors obey Ohm's Law (current proportional to potential difference) for a wide range of currents; of course, if the resistor is overloaded with current, it will heat up, possibly even burn, and Ohm's Law will fail.

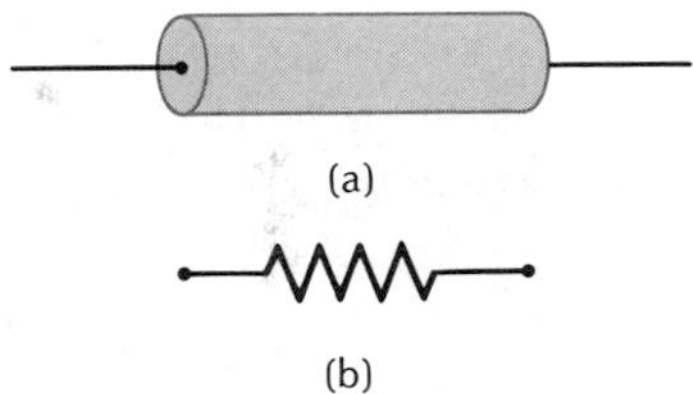

Fig. 28.7 (a) A resistor, consisting of a cylinder of carbon with two terminals attached. (b) Symbol for a resistor in a circuit diagram.

In circuit diagrams the symbol for a resistor is a zigzag line, reminiscent of the path of an electron inside a wire (Figure 28.7b).

Figure 28.8 shows two resistors connected in *series*. It is intuitively obvious that the net resistance of this combination is the sum of the individual resistances,

$$R = R_1 + R_2 \tag{28}$$

The formal derivation of this result begins with the observation that if the potential differences across the individual resistors are ΔV_1 and ΔV_2, then the net potential difference across the combination is

$$\Delta V = \Delta V_1 + \Delta V_2$$

Fig. 28.8 Two resistors connected in series.

Furthermore, the currents in both resistors must be exactly the same. Hence, by Ohm's Law

$$\Delta V = IR_1 + IR_2 = I(R_1 + R_2)$$

From this it is clear that the combination is equivalent to a single resistance given by Eq. (28).

The generalization of this result to any number of resistors in series (Figure 28.9) is obvious. The net resistance is

Series combination of resistors

$$\boxed{R = R_1 + R_2 + R_3 + \cdots} \tag{29}$$

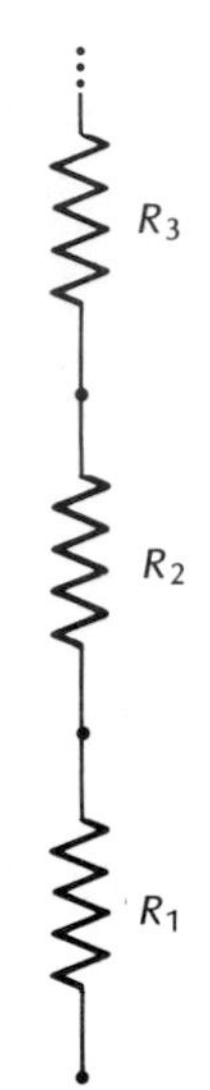

Fig. 28.9 Several resistors connected in series.

Figure 28.10 shows two resistors in *parallel.* The potential difference across each resistor is the same as the potential difference across the combination. Hence the currents are

$$I_1 = \frac{\Delta V}{R_1} \quad \text{and} \quad I_2 = \frac{\Delta V}{R_2} \tag{30}$$

The total current through the combination is

$$I = I_1 + I_2 = \frac{\Delta V}{R_1} + \frac{\Delta V}{R_2} = \left(\frac{1}{R_1} + \frac{1}{R_2}\right) \Delta V \tag{31}$$

The combination is therefore equivalent to a single resistance given by

$$\frac{1}{R} = \frac{1}{R_1} + \frac{1}{R_2} \tag{32}$$

Note that the resistance of the combination is less than each of the individual resistances. For example, if $R_1 = R_2 = 1.0\ \Omega$, then $R = 0.5\ \Omega$.

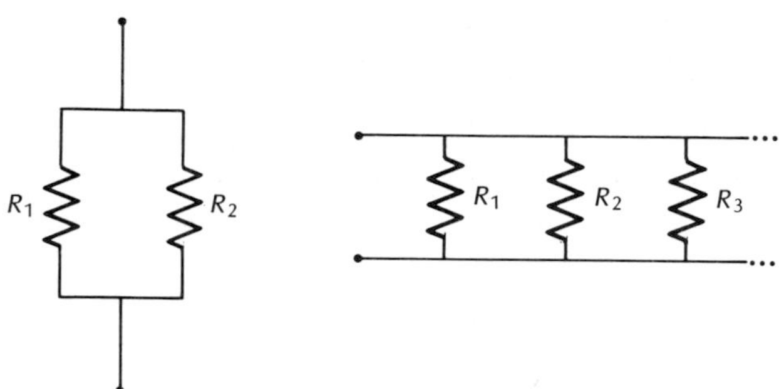

Fig. 28.10 Two resistors connected in parallel.

Fig. 28.11 Several resistors connected in parallel.

The generalization of this to any number of resistors in parallel (Figure 28.11) is again obvious. The net resistance is given by

Parallel combination of resistors

$$\boxed{\frac{1}{R} = \frac{1}{R_1} + \frac{1}{R_2} + \frac{1}{R_3} + \cdots} \tag{33}$$

EXAMPLE 4. Two resistors, with $R_1 = 10\ \Omega$ and $R_2 = 20\ \Omega$, are connected in parallel (see Figure 28.10). A total current of 1.8 A flows through this combination. What is the potential difference across the combination? What is the current in each resistor?

SOLUTION: According to Eq. (32), the net resistance is given by

$$\frac{1}{R}=\frac{1}{R_1}+\frac{1}{R_2}=\frac{1}{10\ \Omega}+\frac{1}{20\ \Omega}=\frac{30}{200\ \Omega}$$

i.e., $R = 6.67\ \Omega$. Hence, the potential difference is

$$\Delta V = IR = 1.8\ \text{A} \times 6.67\ \Omega = 12.0\ \text{volt}$$

and the individual currents are

$$I_1 = \Delta V/R_1 = 12.0\ \text{volt}/10\ \Omega = 1.20\ \text{A}$$

$$I_2 = \Delta V/R_2 = 12.0\ \text{volt}/20\ \Omega = 0.60\ \text{A}$$

SUMMARY

Electric field in uniform wire: $E = \Delta V/l$

Electric current: $I = dq/dt$

Current density: $j = I/A$

Resistance in terms of resistivity: $R = \rho\dfrac{l}{A}$

Ohm's Law: $I = \dfrac{\Delta V}{R}$

Resistivities: conductors: $\sim 10^{-8}\ \Omega\cdot\text{m}$
semiconductors: $\sim 10^{-4}$ to $10^{7}\ \Omega\cdot\text{m}$
insulators: $\sim 10^{11}$ to $10^{17}\ \Omega\cdot\text{m}$

Series combination of resistors: $R = R_1 + R_2 + R_3 + \cdots$

Parallel combination of resistors: $1/R = 1/R_1 + 1/R_2 + 1/R_3 + \cdots$

QUESTIONS

1. A wire is carrying a current of 15 A. Is this wire in electrostatic equilibrium?

2. Can a current flow in a conductor when there is no electric field? Can an electric field exist in a conductor when there is no current?

3. Figure 28.12 shows a putative mechanical analog of an electric conductor with resistance. A marble rolling down the inclined plane is stopped every so often by a collision with a pin and therefore maintains a constant average velocity v_d. Is this a good analog, i.e., is the average velocity v_d proportional to g?

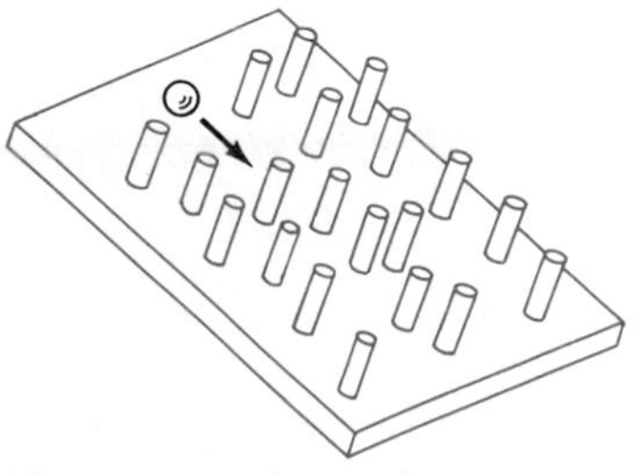

Fig. 28.12

4. By what factor must we increase the diameter of a wire to decrease its resistance by a factor of 2?

5. Show that for a wire of given length made of a given material, the resistance is inversely proportional to the mass of the wire.

6. What deviations from Ohm's Law do you expect if the current is very large?

7. Ohm's Law is an approximate statement about the electrical properties of a conducting body, just as Hooke's Law is an approximate statement about the mechanical properties of an elastic body. Is there an electric analog of the elastic limit?

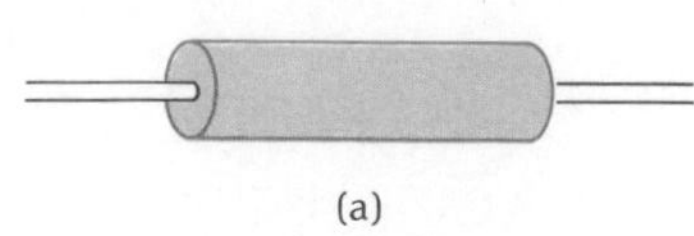

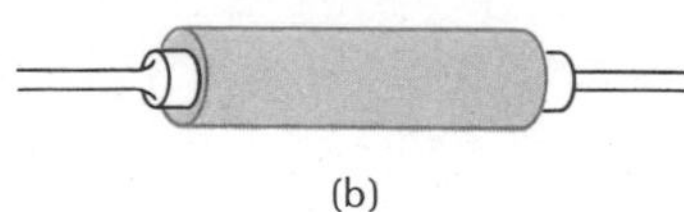

Fig. 28.13

8. An automobile battery has a potential difference of 12 V between its terminals even when there is no current flowing through the battery. Does this violate Ohm's Law?

9. Figure 28.13 shows two alternative experimental arrangements for determining the resistance of a carbon cylinder. A known potential difference is applied via the copper terminals, the current is measured, and the resistance is calculated from $R = \Delta V/I$. The arrangement in Figure 28.13a yields a higher resistance than that in Figure 28.13b. Why?

10. Why is it bad practice to operate a high-current appliance off an extension cord?

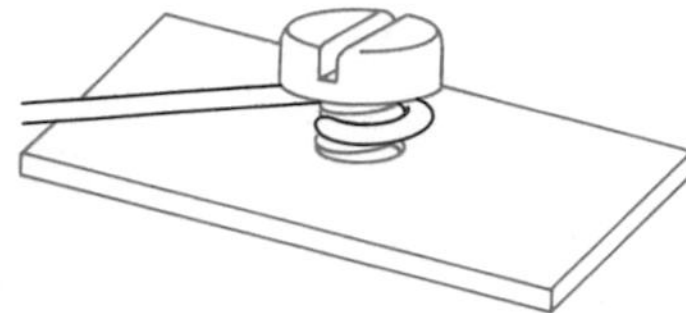

Fig. 28.14

11. The installation instructions for connecting an outlet to the wiring of a house recommend that the wire be wrapped at least three-quarters of the way around the terminal post (Figure 28.14). Explain.

12. The **temperature coefficient of resistivity** α is defined as the fractional increase of resistivity per degree Celsius, $\alpha = (1/\rho)\ d\rho/dT$. According to Table 28.1, what is the value of α for copper?

13. The resistivity of semiconductors and insulators *decreases* with temperature. Can you think of a likely explanation?

14. Figure 19.7 shows a platinum **resistance thermometer** consisting of a coil of fine platinum wire inside a glass tube that can be put in thermal contact with a body. How can this wire be used to determine the temperature of the body?

15. Aluminum wire should never be connected to terminals designed for copper wire. Why not? (Hint: Aluminum has a considerably higher coefficient of thermal expansion than copper.)

16. A wire of copper and a wire of silver are connected in parallel. Both wires have the same length and the same diameter. Which carries more current?

17. Two copper wires are connected in parallel. Both wires have the same length, but one has twice the diameter of the other. What fraction of the total current flows in each wire?

PROBLEMS

Sections 28.1–28.3

1. The effective resistance of a 150-W, 110-V light bulb is 0.73 Ω. What current passes through this light bulb when in operation? How many electrons per second does this amount to?

2. A circular loop of superconducting material has a radius of 2.0 cm. It carries a current of 4.0 A. What is the orbital angular momentum of the moving electrons in the wire? Take the center of the loop as origin.

3. A copper cable in a high-voltage transmission line has a diameter of 3.0 cm and carries a current of 750 A. What is the current density in the wire? What is the electric field in the wire?

4. When the starter motor of an automobile is in operation, the cable connecting it to the battery carries a current of 80 A. This cable is made of copper and is 0.50 cm in diameter. What is the current density in the cable? What is the electric field in the cable?

5. A table lamp is connected to an electric outlet by a copper wire of diameter 0.20 cm and length 2.0 m. Assume that the current through the lamp is 1.5 A, and that this current is steady. How long does it take an electron to travel from the outlet to the lamp?

6. A high-voltage transmission line consists of a copper cable of diameter 3.0 cm and length 250 km. Assume that the cable carries a steady current of 1500 A.
   (a) What is the electric field inside the cable?
   (b) What is the drift velocity of the free electrons?
   (c) How long does it take one electron to travel the full length of the cable?

Section 28.4

7. The electromagnet of a bell is constructed by winding copper wire around a cylindrical core, like thread on a spool. The diameter of the copper wire is 0.45 mm, the number of turns in the winding is 260, and the average radius of a turn is 5.0 mm. What is the resistance of the wire?

8. The following is a list of some types of copper wire manufactured in the United States:

Gauge No.	*Diameter*
8	0.3264 cm
9	0.2906
10	0.2588
11	0.2305
12	0.2053

For each type of wire, calculate the resistance for a 100-m segment.

9. To measure the resistivity of a metal, an experimenter takes a wire of this metal of diameter of 0.500 mm and length 1.10 m and applies a potential difference of 12.0 V to the ends. He finds that the resulting current is 3.75 A. What is the resistivity?

10. A high-voltage transmission line has an aluminum cable of diameter 3.0 cm, 200 km long. What is the resistance of this cable?

11. In silver, the number of free electrons is $n = 5.8 \times 10^{28}$ per cubic meter.
   (a) Using the value of the resistivity given in Table 28.1, calculate the average time interval between collisions of one of these electrons.
   (b) Assuming that the electric field in a current-carrying silver wire is 8.0 V/m, calculate the average drift velocity of an electron.

12. You want to make a resistor of 1.0 Ω out of a carbon rod of diameter 1.0 mm. How long a piece of carbon do you need?

13. The resistance of the wire in the windings of an electric starter motor for an automobile is 3.0×10^{-2} Ω. The motor is connected to a 12-V battery. What current will flow through the motor when it is stalled (does not turn)?

14. A lightning rod of iron has a diameter of 0.80 cm and a length 0.50 m. During a lightning stroke, it carries a current of 1.0×10^{4} A. What is the potential drop along the rod?

15. The air conditioner in a home draws a current of 12 A.
   (a) Suppose that the pair of wires connecting the air conditioner to the fuse box are No. 10 copper wire with a diameter of 0.259 cm and a length of 25 m each. What is the potential drop along each wire? Suppose that the voltage delivered to the home is exactly 110 V at the fuse box. What is the voltage delivered to the air conditioner?
   (b) Some older homes are wired with No. 12 copper wire with a diameter of 0.205 cm. Repeat the calculation of part (a) for this wire.

16. The electromagnet of a bell is wound with 8.2 m of copper wire of diameter 0.45 mm. What is the resistance of the wire? What is the current through the wire if the electromagnet is connected to a 12-V source?

17. The copper cable connecting the positive pole of a 12-V automobile battery to the starter motor is 0.60 m long and 0.50 cm in diameter.
   (a) What is the resistance of this cable?
   (b) When the starter motor is stalled, the current in the cable may be as much as 600 A. What is the potential drop along the cable under these conditions?

18. Although aluminum has a somewhat higher resitivity than copper, it has the advantage of having a considerably lower density. Find the weight of a 100-m segment of aluminum cable 3 cm in diameter. Compare this weight with that of a copper cable of the same length and the same resistance.

19. According to the National Electrical Code, the maximum permissible current in a No. 12 copper wire (diameter 0.21 cm) with rubber insulation is 20 A.
   (a) What is the current density in a wire carrying this current?
   (b) What is the potential drop along a 1-meter segment of the wire?

20. An aluminum wire of length 15 m is to carry a current of 25 A with a potential drop of no more than 5 V along its length. What is the minimum acceptable diameter of this cable?

21. According to safety standards set by the American Boat and Yacht Council, the potential drop along a copper wire connecting a 12-V battery to an item of electrical equipment should not exceed 10%, i.e., it should not exceed 1.2 V. Suppose that a 30-ft wire (length measured around the circuit) carries a current of 25 A; what gauge of wire is required for compliance with the above standard? Use the table of wire gauges given in Problem 8. Repeat the calculation for currents of 35 A and 45 A.

22. A parallel-plate capacitor with a plate area of 8.0×10^{-2} m^2 and a plate separation of 1.0×10^{-4} m is filled with polyethylene. If the potential difference between the plates is 2.0×10^4 V, what will be the current flowing through the polyethylene from one plate to the other?

23. Consider the aluminum cable described in Problem 10. If the temperature of this cable increases from 20°C to 50°C, how much will its resistance increase?

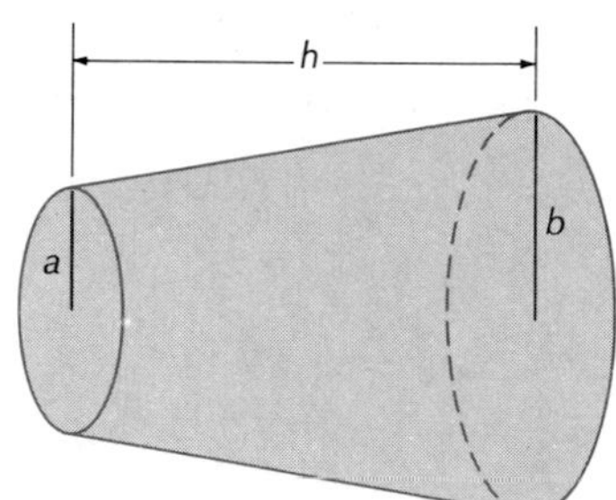

Fig. 28.15

24. What increase of temperature will increase the resistance of a nickel wire from 0.5 Ω to 0.6 Ω?

*25. A solid truncated cone is made of a material of resistivity ρ (Figure 28.15). The cone has a height h, a radius a at one end, and a radius b at the other end. Derive a formula for the resistance of this cone.

Section 28.6

26. A brass wire and iron wire of equal diameter and of equal length are connected in parallel. Together they carry a current of 6.0 A. What is the current in each?

27. Consider the brass and iron wires described in Problem 26. What is the current in each if they are at a temperature of 90°C instead of a temperature of 20°C?

28. An electric cable of length 12.0 m consists of a copper wire of diameter 0.30 cm surrounded by a cylindrical layer of rubber insulation of thickness 0.10 cm. A potential difference of 6.0 V is applied to the ends of the cable.
   (a) What will be the current in the copper?
   (b) Taking into account the conductivity of the rubber (see Table 28.2), what will be the current in the rubber?

29. A water pipe is made of iron with an outside diameter of 2.5 cm and an inside diameter of 2.0 cm. The pipe is used to ground an electric appliance. If a current of 20 A flows from the appliance into the water pipe, what fraction of this current will flow in the iron? What fraction in the water? Assume that water has a resistivity of $0.01\ \Omega \cdot \text{m}$.

30. A copper wire, of length 0.50 m and diameter 0.259 cm has been accidentally cut by a saw. The region of the cut is 0.4 cm long, and in this region the remaining wire has a cross-sectional area of only one-quarter of the original area. What is the percentage increase of the resistance of the wire caused by this cut?

31. The windings of high-current electromagnets are often made of copper pipe. The current flows in the walls of the pipe and cooling water flows in the interior of the pipe. Suppose the copper pipe has an outside diameter of 1.20 cm and an inside diameter of 0.80 cm. What is the resistance of 30 m of this copper pipe? What voltage must be applied to it if the current is to be 600 A?

32. An underground telephone cable, consisting of a pair of wires, has suffered a short somewhere along its length (Figure 28.16). The telephone cable is 5 km long and in order to discover where the short is, a technician first measures the resistance across the terminals *AB*; then he measures the resistance across the terminals *CD*. The first measurement yields $30\ \Omega$; the second $70\ \Omega$. Where is the short?

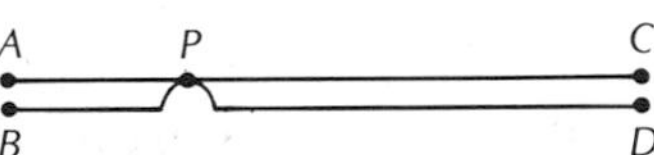

Fig. 28.16 A pair of wires with a short at the point *P* (the wires are joined at *P*).

33. The air of the atmosphere has a slight conductivity due to the presence of a few free electrons and positive ions.
(a) Near the surface of the Earth, the atmospheric electric field has a strength of about 100 V/m and the atmospheric current density is $4 \times 10^{-12}\ \text{A/m}^2$. What is the resistivity?
(b) The potential difference between the ionosphere (upper layer of atmosphere) and the surface of the Earth is 4×10^5 V. What is the total resistance of the atmosphere? (Hint: For the purposes of this problem you may assume that the Earth is flat.)

34. Three resistors with resistances of $3.0\ \Omega$, $5.0\ \Omega$, and $8.0\ \Omega$, are connected in parallel. If this combination is connected to a 12.0-V battery, what is the current through each resistor? What is the current through the combination?

35. A flexible wire for an extension cord for electric appliances is made of 24 strands of fine copper wire, each of diameter 0.053 cm, tightly twisted together. What is the resistance of a length of 1.0 m of this kind of wire?

36. Two copper wires of diameters 0.26 cm and 0.21 cm, respectively, are connected in parallel. What is the current in each if the combined current is 18 A?

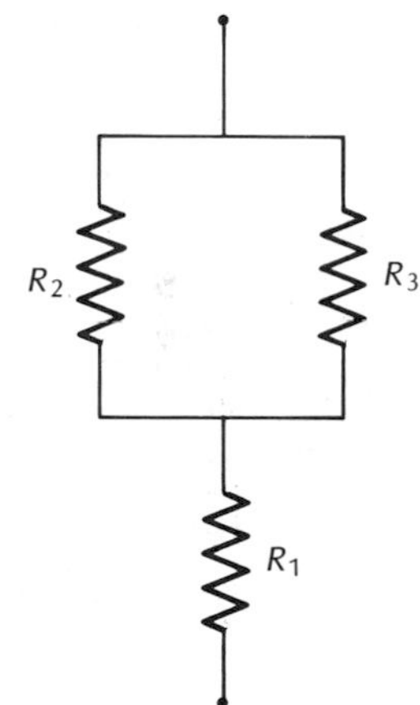

Fig. 28.17

37. Three resistors with $R_1 = 2.0\ \Omega$, $R_2 = 4.0\ \Omega$, and $R_3 = 6.0\ \Omega$ are connected as shown in Figure 28.17.
(a) Find the net resistance of the combination.
(b) Find the current that passes through the combination if a potential difference of 8.0 V is applied to the terminals.
(c) Find the potential difference and the current for each individual resistor.

38. Three resistors with $R_1 = 4.0\ \Omega$, $R_2 = 6.0\ \Omega$, and $R_3 = 8.0\ \Omega$ are connected as shown in Figure 28.18.
(a) Find the net resistance of the combination.
(b) Find the current that passes through the combination if a potential difference of 12.0 V is applied to the terminals.
(c) Find the potential difference and the current for each individual resistor.

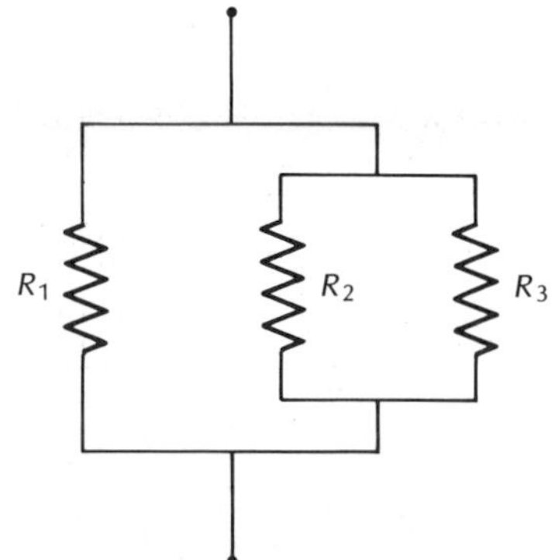

Fig. 28.18

39. Consider the combination of three resistors described in the preceding

problem. If we want a current of 6.0 A to flow through resistor R_2, what potential difference must we apply to the external terminals?

*40. What is the resistance of an infinite ladder of 1.0-Ω resistors connected as shown in Figure 28.19a? (Hint: The ladder can be regarded as made of two pieces connected in parallel; see Figure 28.19b.)

*41. Twelve resistors, each of resistance R, are connected along the edges of a cube (Figure 28.20). What is the resistance between diagonally opposite corners of this cube?

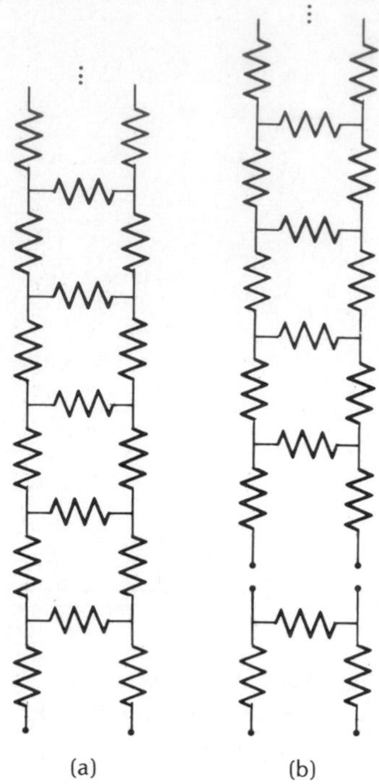

Fig. 28.19 (a) An infinite "ladder" of resistors. The terminals are marked by a pair of dots. (b) One rung has been cut off from the ladder.

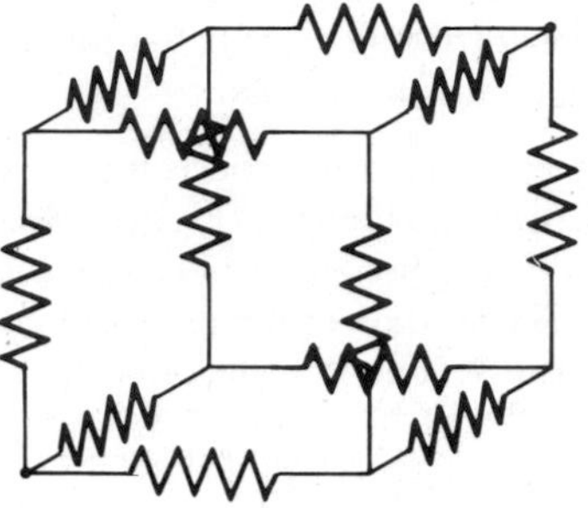

Fig. 28.20 A cube made of resistors.

CHAPTER 29

DC Circuits

If we want to keep a current flowing in a wire, we must connect its ends to a "pump of electricity," such as a battery or generator, that can continuously supply electric charges to one end of the wire and remove them from the other. Figure 29.1 shows a circuit consisting of a wire connected to the terminals of a battery. A current will flow around this circuit; in Figure 29.1 we have indicated the direction of the current according to our convention that it is in the direction of flow of (hypothetical) positive charges. As long as the "strength" of the battery and the resistance of the wire remain constant, the current will also remain constant. Such a steady, time-independent current is called a **direct current**, or DC.

Direct current, DC

29.1 Electromotive Force

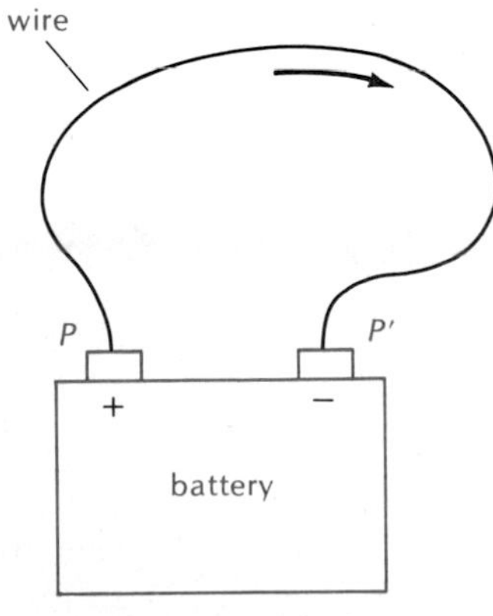

Fig. 29.1 A wire connected between the terminals of a battery.

The battery must do work on the charges in order to keep them flowing around the circuit shown in Figure 29.1. Suppose that a positive charge is originally at the point P, at one terminal of the battery. Pushed along by the electric field, the charge moves along the wire. On the average, the kinetic energy of the charge does not change — any kinetic energy that the charge gains from the electric field is dissipated by friction within the wire and the charge reaches the point P', at the other terminal of the battery, with its original kinetic energy.

Thus, only the potential energy changes. Since the electric field is directed along the wire, the electric potential steadily decreases with distance along the wire and the charge reaches the point P' with a potential energy lower than its original potential energy. In order to keep the current flowing, the battery must "pump" the charge from the

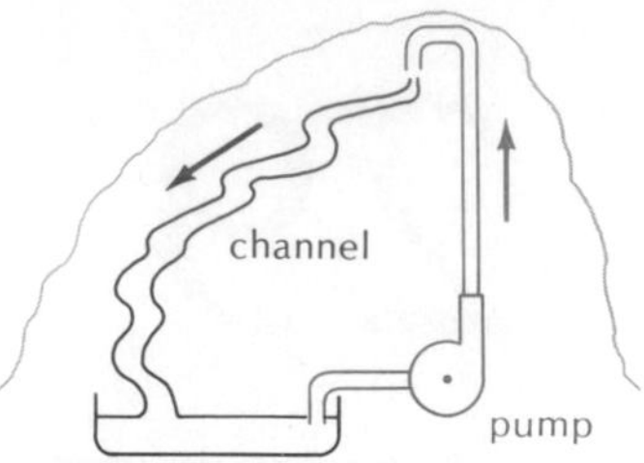

Fig. 29.2 Mechanical analog of the battery–wire circuit of Figure 29.1.

low-potential terminal to the high-potential terminal, i.e., the battery must supply electric potential energy to the charge. The role of the battery is analogous to that of a hydraulic pump system that lifts water from the bottom to the top of a hill (Figure 29.2). The wire is analogous to a channel by means of which the water runs down the hill returning to the pump. The water then flows in a closed hydraulic circuit, just as charge flows in a closed electric circuit. The hydraulic pump of Figure 29.2 can be regarded as a source of gravitational potential energy — it produces this energy from an external supply of chemical or mechanical energy. Likewise, the "pump of electricity" of Figure 29.1 can be regarded as a source of electric potential energy — it produces this energy from a supply of chemical energy.

Electromotive force, emf

To measure the "strength" of a source of electric potential energy, we introduce the concept of **electromotive force**, or **emf**. The emf of a source of electric potential energy is defined as *the amount of electric energy delivered by the source per couloumb of positive charge as this charge passes through the source from the low-potential terminal to the high-potential terminal.* Since the emf is energy per unit charge, its units are volts.

If a steady, time-independent current carries one coulomb of charge around the circuit of Figure 29.1 from P to P' along the wire and from P' to P through the source of electric potential energy, then the energy that this charge receives from the source must exactly match the energy it loses within the wire. If so, the charge returns to its starting point with exactly the same energy it had originally and it can repeat this round trip again and again in exactly the same manner. We can write this energy balance as

$$\mathscr{E} + \Delta V = 0 \tag{1}$$

where $\mathscr{E}$ represents the emf, or the increase of potential energy, due to the source and ΔV represents the decrease of potential energy along the wire ($\mathscr{E}$ is positive and ΔV is negative).

According to Eq. (1), the emf $\mathscr{E}$ has the same magnitude as the potential drop in the external circuit connected between the terminals of the source. For example, a battery with an emf of 1.5 V connected to an external circuit will do 1.5 J of work on a coulomb of positive charge that passes through the battery in a forward direction (from the − terminal to the + terminal) and the resistors and other devices in the external circuit will do −1.5 J of work on the charge as it flows around this circuit, from the + terminal to the − terminal. Because of the equality between the emf $\mathscr{E}$ and the potential drop ΔV, the emf is often simply called the **voltage** of the source.

Voltage

EXAMPLE 1. A fresh flashlight battery with a voltage of 1.5 V will deliver a current of 1 A for about 1 h before running down. How much work does the battery do in this time interval?

SOLUTION: The battery does 1.5 J of work on each coulomb that passes through. If the current is 1 A, the charge that passes through in 1 h is 1 A × 3600 s = 3600 C, and the total work is 1.5 J/C × 3600 C = 5400 J.

Note that if one coulomb of positive charge is forced through the battery in the reverse direction (from the + terminal to the − terminal), then the battery will take electric potential energy away from the charge. The charge will then emerge from the battery at a potential that is 1.5 volt lower than the potential with which it entered. The energy taken away from the charge will either be stored within the battery (if it is a reversible battery) or else it will merely be wasted as heat within the battery (if it is an irreversible battery).

29.2 Sources of Electromotive Force

The most important kinds of sources of emf are batteries, electric generators, fuel cells, and solar cells. We will now briefly discuss each of these.

BATTERIES These sources of emf convert chemical energy into electric energy. A very common type of battery is the **lead–acid battery** which finds widespread use in automobiles. In its simplest form, this battery consists of two plates of lead — the positive electrode and the negative electrode — immersed in a solution of sulfuric acid (Figure 29.3). The positive electrode is covered with a layer of lead dioxide, PbO_2. When the external circuit is closed, the following reactions, already mentioned in Section 22.3, take place at the immersed surfaces of the negative and positive electrodes, respectively:

Lead–acid battery

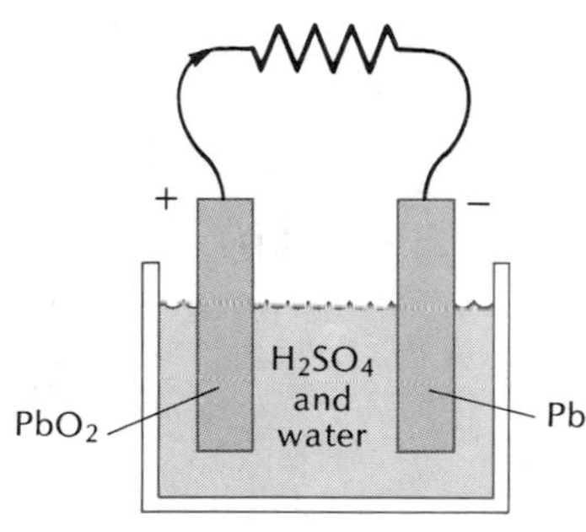

Fig. 29.3 A lead–acid battery.

$$Pb + SO_4^{--} \rightarrow PbSO_4 + 2e^- \tag{2}$$

$$PbO_2 + SO_4^{--} + 4H^+ + 2e^- \rightarrow PbSO_4 + 2H_2O \tag{3}$$

These reactions deposit electrons on the negative electrode and absorb electrons from the positive electrode. Thus, the battery acts as a pump for electrons — the negative electrode is the outlet, the positive electrode is the intake, and the electrons flow from one to the other via the external circuit.

The reactions (2) and (3) deplete the sulfuric acid in the solution and they deposit lead sulfate on the electrodes. The depletion of sulfuric acid finally halts the reaction — the battery is then "discharged."

The lead–acid battery can be "charged" by simply passing a current through it in the backward direction. This reverses the reactions (2) and (3) and restores the sulfuric acid solution. Note that what is stored in the battery during the "charging" process is not electric charge, but chemical energy. The number of positive and negative electric charges (protons and electrons) in the battery remains constant; what changes is the concentration of chemical compounds. A "charged" battery contains chemical compounds (lead, lead dioxide, sulfuric acid) of relatively high internal energy; a "discharged" battery contains chemical compounds (lead sulfate, water) of lower internal energy.

The single-cell battery shown in Figure 29.3 has an emf of 2.0 V. In an automobile battery, six such cells are stacked together and connected in series to give an emf of 12.0 V. The energy stored in such a battery is typically about 0.5 kW · h.[1] Large banks of batteries, weighing several hundred tons, for the propulsion of submarines store more than 5×10^3 kW · h.

Dry cell

Another familiar type of battery is the **dry cell.** The positive electrode consists of maganese dioxide and the negative electrode of zinc.

[1] Recall that $1\ kW \cdot h = 1000\ W \times 3600\ s = 3.6 \times 10^6\ J$.

The electrolyte in which these electrodes are "immersed" is a moist paste of ammonium chloride and zinc chloride. The chemical reactions at the electrodes convert chemical energy into electric energy and pump electrons from one electrode to the other via the external circuit. The emf of such a dry cell is 1.5 V. Since there is no liquid to slosh around, these batteries are particularly suitable for portable devices. The energy stored in an ordinary flashlight battery is typically of the order of 2×10^{-3} kW · h.

ELECTRIC GENERATORS Generators convert mechanical energy (kinetic energy) into electric energy. Their operation involves magnetic fields and the phenomenon of induction. We will leave the description of electric generators for Section 32.3.

Fuel cell

FUEL CELLS These resemble batteries in that they convert chemical energy into electric energy. However, in contrast to a battery, neither the high-energy chemicals nor the low-energy reaction products are kept inside the fuel cell. The former are supplied to the fuel cell from external tanks and the latter are ejected. Essentially, the fuel cell acts as a combustion chamber in which controlled combustion takes place. The fuel cell "burns" a high-energy fuel, but produces electric energy rather than heat energy.

Figure 29.4a shows a fuel cell that "burns" a hydrogen–oxygen fuel. The electrodes of the fuel cell are hollow cylinders of porous carbon; oxygen at high pressure is pumped into the positive electrode and hydrogen into the negative electrode. The electrodes are immersed in a potassium-hydroxide electrolyte. The reactions at the negative and positive electrode are, respectively,

$$2H_2 + 4OH^- \rightarrow 4H_2O + 4e^- \qquad (4)$$

$$O_2 + 2H_2O + 4e^- \rightarrow 4OH^- \qquad (5)$$

These reactions deposit electrons on the negative electrode and remove electrons from the positive electrode. This pumps electrons from one electrode to the other via the external circuit.

Note that the net result of the sequence of reactions (4) and (5) is the conversion of oxygen and hydrogen into water. This reaction is the reverse of the electrolysis of water (decomposition of water by an electric

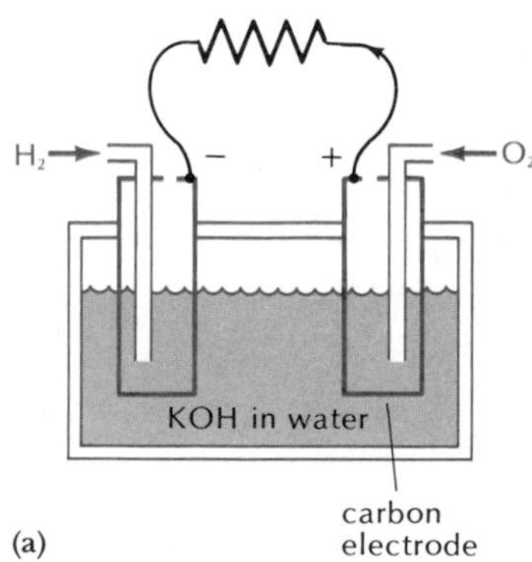

Fig. 29.4 (a) Schematic diagram of a fuel cell. (b) Fuel cell used on Skylab.

current). The excess water is removed from the cell in the form of water vapor.

All fuel cells produce a certain amount of waste heat. The best available fuel cells convert about 45% of the chemical energy of the fuel into electric energy and they waste the remainder. Fuel cells are still at an experimental stage; however, they have already been put to use as practical power sources aboard the Apollo spacecraft and on Skylab (Figure 29.4b). They are compact and clean; on Skylab the waste water eliminated from the fuel cell was used both for drinking and for showers.

Solar cell

SOLAR CELLS Solar cells convert the energy of sunlight directly into electric energy. They are made of thin wafers of a semiconductor, such as silicon. Figure 29.5a shows a cross section through a solar cell. It consists of a central core of n-type silicon surrounded by an outer layer of p-type silicon. The outer layer is very thin, only about 10^{-4} cm, and when sunlight penetrates this layer and reaches the boundary between the n-type silicon and the p-type silicon, it pumps electrons from the latter into the former.

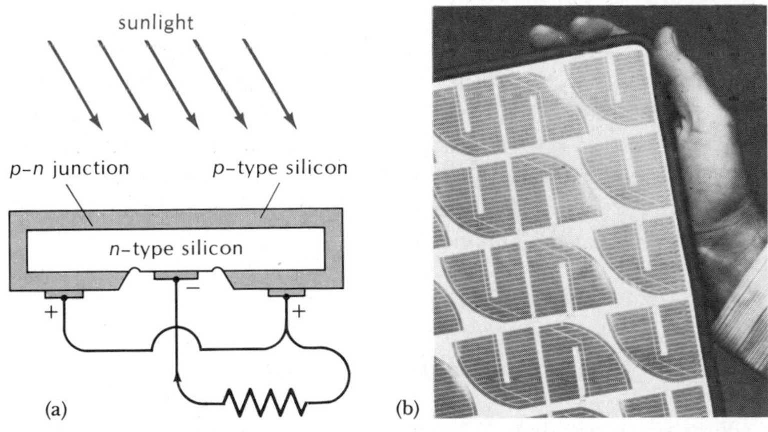

Fig. 29.5 (a) Schematic diagram of a solar cell. (b) An array of solar cells in a panel designed for charging automobile batteries.

p–n junction

The boundary between the two types of silicon is called a ***p–n* junction.** To explain why sunlight causes a current to flow across this junction, we must begin with a description of the behavior of electrons and holes. In n-type silicon the current carriers are electrons (negative charges), whereas in p-type silicon the current carriers are holes (positive charges). Figure 29.6a shows a piece of silicon of each type. When

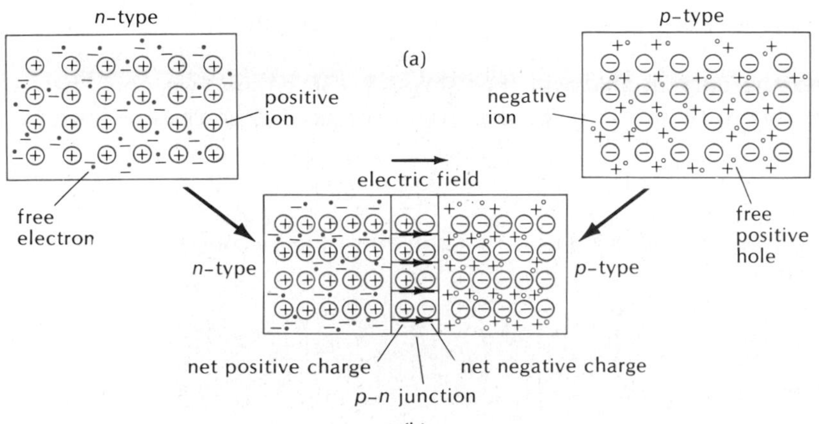

Fig. 29.6 (a) An n-type silicon and p-type silicon. (b) A $p–n$ junction.

these two pieces are joined, some electrons will diffuse across the boundary into the p-type material and some holes into the n-type material. This diffusion comes to a halt because the accumulated charges on the two sides of the boundary repel any other charges that attempt to cross. When equilibrium is established, the p-type material has a layer of negative charge (electrons) and the n-type material a layer of positive charge (holes). Between these layers there is a strong electric field (Figure 29.6b); this is an instance of a "contact" electric field, such as mentioned in Section 24.4.

The current of the solar cell is generated by the action of sunlight on the atoms within the region of this electric field of the p–n junction. When the sunlight strikes one of these atoms it will ionize it, i.e., it will kick an electron out of the atom. This process amounts to the creation of a free electron and of a free hole. Under the influence of the electric field at the p–n junction, the electron accelerates toward the n side of the junction and the hole toward the p side. The net effect is a positive current from the n side toward the p side. In the external circuit, the current will then flow from the p terminal to the n terminal (see Figure 29.5a), i.e., the former acts as the positive pole of a battery and the latter as the negative pole.

The emf of a silicon solar cell is about 0.6 V. However, the current that can be extracted is rather small. Even in full sunlight a single solar cell of surface area 5 cm^2 will only deliver 0.1 ampere. Solar cells are not very efficient; only about 11% of the energy of sunlight gets converted into electric energy. Large panels, containing very many solar cells, are needed for the generation of appreciable amounts of electric power (see Figures 29.5a and F.13).

29.3 Single-Loop Circuits

In schematic diagrams of electric circuits, a source with a time-independent emf is represented by a stack of parallel thick and thin lines suggesting the plates of a lead–acid battery. The high-potential terminal carries a plus sign and the low-potential terminal a minus sign. If the terminals of such a source are connected to a network of resistances, a steady current, also called a direct current or DC, will flow through the network.

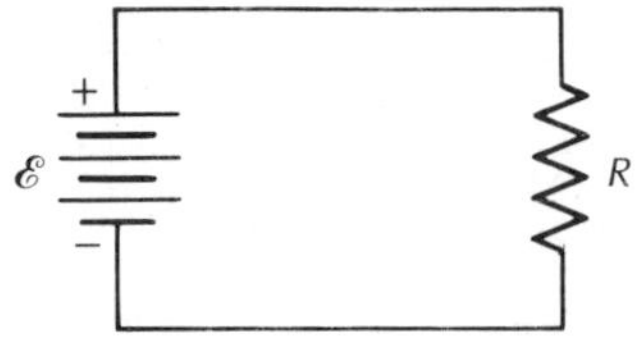

Fig. 29.7 A simple circuit with a source of emf and a resistor.

Figure 29.7 shows a very simple circuit consisting of a source of emf connected to a resistor. The wires from the resistor to the battery have negligible resistance (if higher accuracy is required, the resistance of the wires must be included in the total resistance of the circuit). To find the current that will flow through the circuit, we note that according to Ohm's Law the potential drop across the resistor must be

$$\Delta V = -IR \tag{6}$$

But, from Eq. (1), we know that the emf plus the potential drop must equal zero:

$$\mathscr{E} - IR = 0 \tag{7}$$

or

$$I = \frac{\mathscr{E}}{R} \tag{8}$$

Kirchhoff's (second) rule

Equation (7) is an instance of **Kirchhoff's rule** which states that *around any closed loop the sum of all the emfs and all the potential drops across resistors and other circuit elements must equal zero.*[2] In this sum, the emf of a source is reckoned as positive if the current flows through the source in the forward direction and negative if in the backward direction.

The proof of this general rule is similar to the proof of Eq. (1). If one coulomb of positive charge flows once around the closed loop, it will gain potential energy while passing through each emf and lose potential energy while passing through each resistor. Under steady conditions, the sum of gains and losses must equal zero, since the charge must return to its starting point with no change of energy. In this sum, the emf must be reckoned as negative (loss of potential energy) whenever the charge flows through the source in the backward direction (from the + terminal to the − terminal).

Gustav Robert Kirchhoff (keerkh-hoff), *1824–1887, German physicist, professor at Heidelberg and at Berlin. Mainly known for his development of spectroscopy, he also made many important contributions to mathematical physics, among them, his first and second rules for circuits.*

EXAMPLE 2. Figure 29.8 shows a circuit with two batteries and two resistors. The emfs of the batteries are $\mathscr{E}_1 = 12$ V and $\mathscr{E}_2 = 15$ V; the resistances are $R_1 = 4\ \Omega$ and $R_2 - 2\ \Omega$. What is the current in the circuit?

SOLUTION: To apply Kirchhoff's rule, we must decide in which direction the current flows around the loop. We will arbitrarily assume that the current flows in the clockwise direction. If this hypothesis is wrong, our calculation of the current will yield a negative value and this will indicate that the direction of the actual current is opposite to the hypothetical current.

The sum of all emfs and all potential drops across resistors is

$$\mathscr{E}_1 - IR_1 - \mathscr{E}_2 - IR_2 = 0 \tag{9}$$

Note that $\mathscr{E}_2$ enters with a negative sign into this equation since the hypothetical current passes through this source of emf in the backward direction. The solution of Eq. (9) leads to

$$I = \frac{\mathscr{E}_1 - \mathscr{E}_2}{R_1 + R_2} \tag{10}$$

or

$$I = \frac{12\text{ V} - 15\text{ V}}{4\ \Omega + 2\ \Omega} = -0.5\text{ A} \tag{11}$$

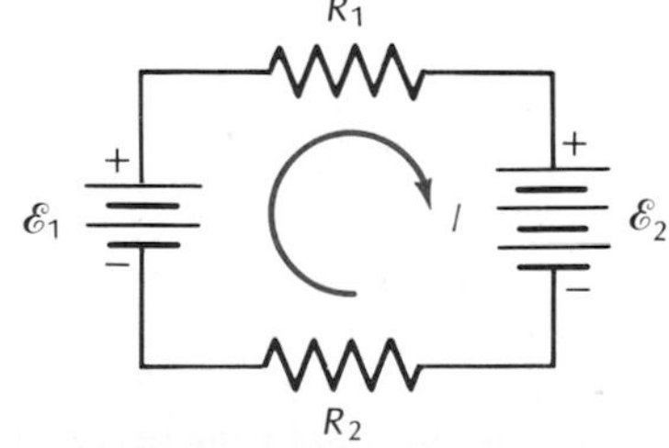

Fig. 29.8 Two sources of emf and two resistors.

The negative sign indicates that the current is *not* clockwise but counterclockwise. (We could of course have guessed that much. The stronger battery on the right will force the current backward through the weaker battery on the left.)

Incidentally, in Example 2 we have neglected the internal resistance of the batteries. The electrolyte in a battery always has some resistance and this causes the current to suffer a voltage drop even before it leaves the external terminal of the battery. The nominal emf $\mathscr{E}$ quoted

[2] This is often called Kirchhoff's second rule. We will become acquainted with his first rule in Section 29.4.

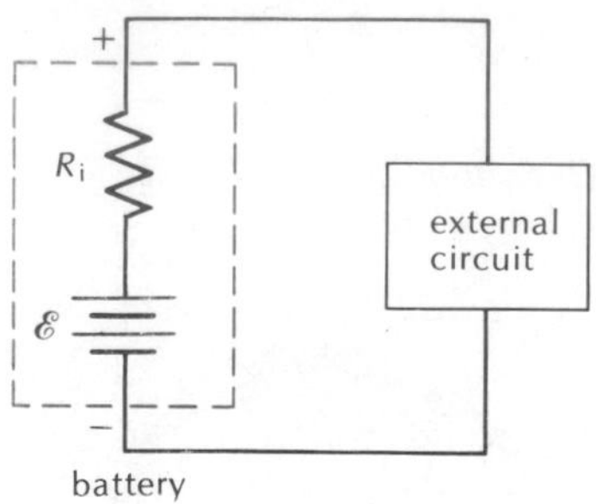

Fig. 29.9 The internal resistance is in series with the nominal emf.

on the labels of batteries refers to the potential difference between the terminals when no current, or only an infinitesimal current, is flowing. The internal resistance R_i of the battery may be regarded as connected in series with this nominal emf (Figure 29.9). When a current I is flowing, the voltage drops by $\Delta V = -IR_i$ across this internal resistance and hence the remaining voltage at the external terminals of the battery will be $\mathscr{E} - IR_i$. The internal resistance of a good battery is small and can often be neglected. But if we need to take it into account, we can do so by simply placing the appropriate internal resistance in series with each battery in the circuit diagram; then we can proceed with the usual calculation of the currents.

EXAMPLE 3. A flashlight battery of a nominal emf 1.5 V has an internal resistance of 0.05 Ω. What will be the potential difference across its terminals if the battery is delivering a current of 2 A? A current of 10 A?

SOLUTION: For a current of 2 A, the voltage across the terminals will be

$$\mathscr{E} - IR_i = 1.5 \text{ V} - 2 \text{ A} \times 0.05 \ \Omega = 1.4 \text{ V}$$

For a current of 10 A, the voltage across the terminals will be

$$\mathscr{E} - IR_i = 1.5 \text{ V} - 10 \text{ A} \times 0.05 \ \Omega = 1.0 \text{ V}$$

29.4 Multiloop Circuits

If several sources of emf and several resistors are connected in some complicated circuit with branches and loops, then the currents will flow along several alternative paths. To find these currents, we must solve a simultaneous set of several equations with the currents as unknowns.

The procedure for obtaining the necessary equations is as follows:

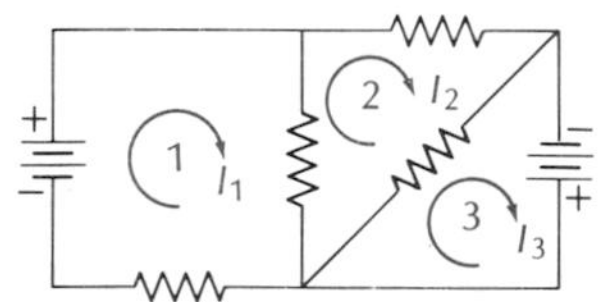

Fig. 29.10 A multiloop circuit. The currents I_1, I_2, I_3 are regarded as flowing in closed loops. Note that loops 1 and 2 share the middle resistor; and loops 2 and 3 share the diagonal resistor. The net current in the middle resistor is $I_1 - I_2$, and the net current in the diagonal resistor is $I_2 - I_3$.

a. Regard the given circuit as a collection of closed current loops. The loops may overlap, but each loop must have at least one portion that does not overlap with other loops (Figure 29.10).
b. Label the currents in the loops $I_1, I_2, I_3, \cdots$, and arbitrarily assign a direction to each of these currents.
c. Apply Kirchhoff's rule to each loop: the sum of all the emfs and of all the potential drops across resistors must add to zero around each loop. Note that when calculating the potential drop across a resistor, we must take the product of the resistance and the *net* current through the resistor; if the resistor belongs to two adjacent loops, then the *net* current is the algebraic sum of the two loop currents (Figure 29.10).

Loop method

This procedure, called the **loop method,** will result in the right number of equations for the unknown currents $I_1, I_2, I_3, \cdots$. We can then solve the equations for the unknowns by the standard mathematical methods for the solution of a system of equations with several unknowns. If a current turns out to be negative, its direction is opposite to the direction assigned in step (b).

EXAMPLE 4. Figure 29.11 shows a circuit with several batteries and resistors. Find the current in each of the resistors.

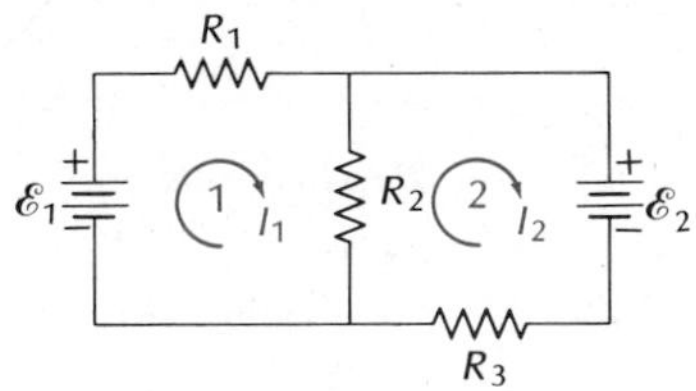

Fig. 29.11 The loops 1 and 2 share the resistor R_2. The currents I_1 and I_2 flow through R_2 in opposite directions; the net current through R_2 is therefore the difference between I_1 and I_2.

SOLUTION: Obviously, this circuit can be regarded as consisting of the two loops indicated by the arrows. We label the currents in the loops I_1 and I_2; these symbols have been written next to the arrows that indicate the directions. When calculating the potential drop in the resistor R_2 (which is included in both loops) we must take into account that both loop currents flow through R_2 simultaneously, in opposite directions. The net current through R_2 in the direction of the arrow of loop 1 is therefore $I_1 - I_2$. With this, Kirchhoff's rule for loop 1 yields

$$\mathscr{E}_1 - I_1R_1 - (I_1 - I_2)R_2 = 0 \tag{12}$$

Likewise, the net current through R_2 in the direction of the arrow of loop 2 is $I_2 - I_1$; and Kirchhoff's rule for loop 2 yields

$$-\mathscr{E}_2 - I_2R_3 - (I_2 - I_1)R_2 = 0 \tag{13}$$

This gives us two equations for the two unknowns I_1 and I_2.

Before proceeding to the solution of these equations it is convenient to substitute the numerical values for the known quantities $\mathscr{E}$ and R. For instance, if $\mathscr{E}_1 = 12.0$ V, $\mathscr{E}_2 = 8.0$ V, $R_1 = 4.0\ \Omega$, $R_2 = 4.0\ \Omega$, and $R_3 = 2.0\ \Omega$, then Eqs. (12) and (13) become

$$12 - 8I_1 + 4I_2 = 0 \tag{14}$$

$$-8 - 6I_2 + 4I_1 = 0 \tag{15}$$

with the solution

$$I_1 = 1.25\ \text{A} \qquad I_2 = -0.50\ \text{A} \tag{16}$$

The negative sign on I_2 indicates that the current in the second loop is opposite to the direction shown in Figure 29.11.

The current in the resistor R_1 is then $I_1 = 1.25$ A; the current in the resistor R_2 is $I_1 - I_2 = 1.75$ A; and the current in the resistor R_3 is $I_2 = -0.50$ A.

Kirchhoff's first rule

An alternative procedure for obtaining the equations for a multiloop circuit makes use of **Kirchhoff's first rule** which states that *the sum of all the currents entering any branch point of the circuit (where a wire merges with another wire) must equal the sum of currents leaving.* This rule expresses charge conservation — the amount of charge entering any branch point in some time interval equals the amount of charge leaving. Our previous procedure for circuits contained Kirchhoff's first rule implicitly, and hence we had no occasion to use this rule explicitly. The alternative procedure uses this Kirchhoff rule explicitly, as follows:

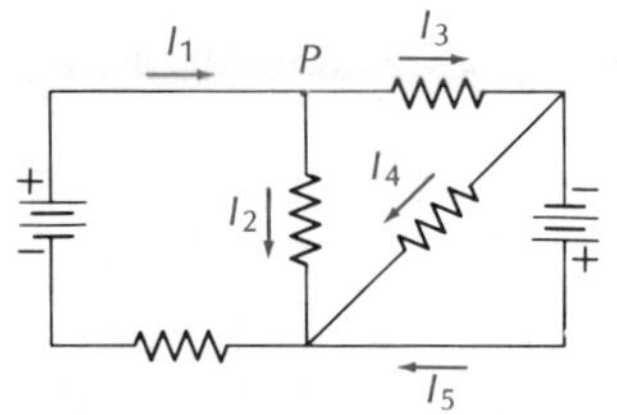

Fig. 29.12 A multiloop circuit. The currents in the distinct branches of the circuit are I_1, I_2, I_3, I_4, and I_5. The point P is a branch point, where wires merge.

a. Regard the given circuit as a collection of closed current loops (Figure 29.12).
b. Label the currents in the distinct branches I_1, I_2, I_3, . . . , and arbitrarily assign a direction to each of these currents.
c. Apply Kirchhoff's rule, now called Kirchhoff's second rule, to each loop: the sum of all the emfs and of all the potential drops across resistors must add to zero around each loop.

d. Apply Kirchhoff's first rule to the branch points: the sum of all the currents entering a branch point must equal the sum of the currents leaving.

Branch method

This procedure, called the **branch method**, will result in the right number of equations for the unknowns I_1, I_2, I_3, ...

EXAMPLE 5. Again consider the circuit of Figure 29.11. Repeat the calculation of the current in the resistors of this circuit using the branch method.

SOLUTION: The circuit has three distinct branches; we label the currents in these branches I_1, I_2, and I_3 (Figure 29.13). Kirchhoff's second rule for loop 1 yields

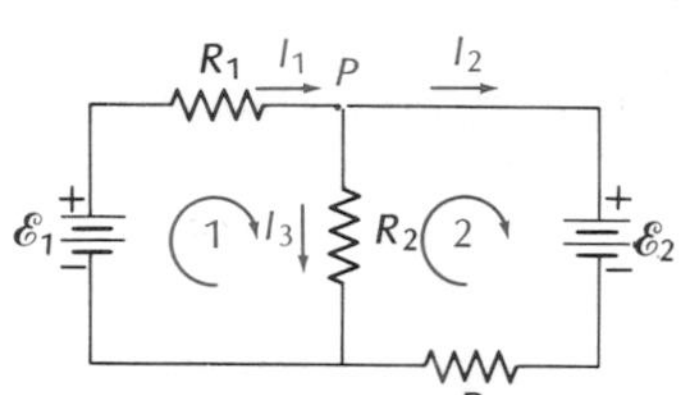

Fig. 29.13 The currents in the distinct branches of this circuit are I_1, I_2, and I_3.

$$\mathscr{E}_1 - I_1R_1 - I_3R_2 = 0 \tag{17}$$

Likewise, for loop 2

$$-\mathscr{E}_2 - I_2R_3 + I_3R_2 = 0 \tag{18}$$

The last term in Eq. (18) has been given a positive sign because the direction of the current I_3 is *opposite* to the direction of the arrow of loop 2.

Kirchhoff's first rule applied to the branch point P (Figure 29.13) yields

$$I_1 = I_2 + I_3 \tag{19}$$

Equations (17), (18), and (19) are three equations for the three unknowns I_1, I_2, and I_3. Note that in this example, the branch method gives us three equations with three unknowns, whereas the loop method gave us two equations with two unknowns. In general, the branch method always suffers from the disadvantage of giving us more unknowns and more equations. Solving Eq. (19) for I_3, we find

$$I_3 = I_1 - I_2 \tag{20}$$

When we substitute this into Eqs. (17) and (18), we obtain Eqs. (12) and (13) — the same equations for I_1 and I_2 as before. Hence the final answers obtained by the branch method are the same as those obtained by the loop method.

29.5 Energy in Circuits; Joule Heat

As we have seen in Section 29.1, to keep a current flowing in a circuit, the batteries or other sources of emf must do work. If an amount of charge dq passes through a source of an emf $\mathscr{E}$, the amount of work done will be

$$dW = \mathscr{E}\, dq \tag{21}$$

Hence the rate at which the source does work is

$$\frac{dW}{dt} = \mathscr{E}\,\frac{dq}{dt} \tag{22}$$

The rate of work is the power; the rate of flow of the charge is the current. Equation (22) therefore asserts that the electric power delivered by the source of emf to the current is

$$\boxed{P = \mathscr{E}I} \qquad (23)$$

Power delivered by source of emf

Note that in Eq. (23) we have not yet taken into account the algebraic sign of the power. Obviously, we will have to attach a positive sign to the power if the current passes through the source in the forward direction and a negative sign if the current passes through in the backward direction. In the former case the source delivers energy to the current and in the latter case the source receives energy from the current.

EXAMPLE 6. What power do the two batteries described in Example 4 deliver?

SOLUTION: The current through the first battery is 1.25 A. This current passes through the battery in the forward direction; hence the power delivered by the battery is positive,

$$P = 1.25 \text{ A} \times 12.0 \text{ V} = 15 \text{ W}$$

Likewise, the power delivered by the other battery is

$$P = 0.50 \text{ A} \times 8.0 \text{ V} = 4.0 \text{ W}$$

The net power delivered by both batteries is 19 W.

The electric energy acquired by the charges is continually dissipated in the resistors. If within a given resistor the charge dq suffers a potential drop ΔV (regarded as a positive quantity), then the loss of potential energy is $dU = \Delta V\, dq$ and the rate at which energy is dissipated is

$$\frac{dU}{dt} = \Delta V \frac{dq}{dt} \qquad (24)$$

i.e., the power dissipated in the resistor is

$$\boxed{P = (\Delta V)\, I} \qquad (25)$$

Power dissipated in resistor

By means of Ohm's Law, $\Delta V = IR$, we can also write this power as

$$P = I^2 R \qquad (26)$$

or as

$$P = (\Delta V)^2/R \qquad (27)$$

The energy lost by the charges during their passage through a resistor generates heat, i.e., it generates random microscopic motions of the atoms. This conversion of electric energy into thermal energy in a resistor is called **Joule heating.**

Joule heat

In any circuit consisting of several sources of emf and several resistors with steady currents, the total power delivered by the sources of emf must of course equal the total power dissipated in the resistors. This equality is a direct consequence of Kirchhoff's rule. The net result of the flow of current in such a circuit is therefore a conversion of energy of the sources of emf into an equal amount of energy of heat.

EXAMPLE 7. What is the rate at which Joule heat is produced in the resistors of Example 4?

SOLUTION: The resistances are $R_1 = 4.0\ \Omega$, $R_2 = 4.0\ \Omega$, and $R_3 = 2.0\ \Omega$; the corresponding currents are 1.25 A, 1.75 A, and 0.50 A. Equation (26) then gives the power dissipated in each resistor,

$$P_1 = (1.25\ \text{A})^2 \times 4.0\ \Omega = 6.25\ \text{W}$$

$$P_2 = (1.75\ \text{A})^2 \times 4.0\ \Omega = 12.25\ \text{W}$$

$$P_3 = (0.50\ \text{A})^2 \times 2.0\ \Omega = 0.50\ \text{W}$$

Note that the net power dissipated is 19.0 W, which agrees with the net power delivered by the batteries (see Example 6).

EXAMPLE 8. A high-voltage transmission line that connects a city to a power plant consists of a pair of copper wires, each with a resistance of 4 Ω. (a) The transmission line delivers to the city 1.7×10^5 kW of power at 2.3×10^5 V. What is the current in the transmission line? How much power is lost as Joule heat in the transmission line? (b) If the transmission line were to deliver the same 1.7×10^5 kW of power at 110 V, how much power would be lost as Joule heat? Is it more efficient to transmit power at high voltage or at low voltage?

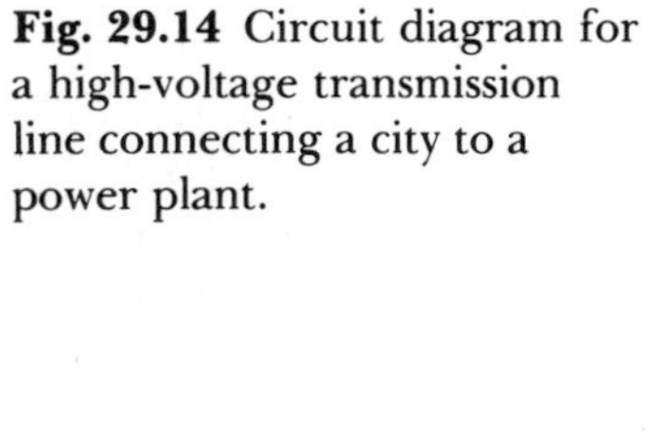

Fig. 29.14 Circuit diagram for a high-voltage transmission line connecting a city to a power plant.

SOLUTION: (a) Figure 29.14 shows the circuit consisting of power plant, transmission lines, and city. In terms of the power and the voltage delivered to the city, the current through the city is

$$I = \frac{P_{\text{delivered}}}{\Delta V_{\text{delivered}}} = \frac{1.7 \times 10^8\ \text{W}}{2.3 \times 10^5\ \text{V}} = 7.4 \times 10^2\ \text{A}$$

The current in both transmission lines must of course be the same. The combined resistance of both lines is $4\ \Omega + 4\ \Omega = 8\ \Omega$ and hence the power lost in the transmission lines is

$$P_{\text{lost}} = I^2R = (7.4 \times 10^2\text{A})^2 \times 8\ \Omega = 4.4 \times 10^6\ \text{W}$$

Thus, the power lost is 3% of the power delivered.

(b) For $\Delta V_{\text{delivered}} = 110$ V, the current is

$$I = \frac{1.7 \times 10^8\ \text{W}}{110\ \text{V}} = 1.6 \times 10^6\ \text{A}$$

and the power lost is

$$P_{\text{lost}} = I^2R = (1.6 \times 10^6\ \text{A})^2 \times 8\ \Omega = 1.9 \times 10^{13}\ \text{W}$$

Thus, the power lost is much larger than the power delivered! Obviously, transmission at high voltage is much more efficient than transmission at low voltage.

29.6 The Hazards of Electric Currents

As a side effect of the widespread use of electric machinery and devices in factories and homes, each year in the United States about 1000 people die by accidental electrocution. A much larger number suffer nonfatal electric shocks. Fortunately, the human skin is a fairly good insulator, which provides a protective barrier against injurious electric currents. The resistance of a square centimeter of dry human epidermis in contact with a conductor can be as much as $10^5\ \Omega$. However, the resistance varies in a rather sensitive way with the thickness, moisture, and temperature of the skin, and with the magnitude of the potential difference.[3]

The electric power supplied to factories and homes in the United States is usually in the form of alternating currents, or AC. These are oscillating currents, which periodically reverse direction (the standard period for the alternating current supplied by power companies is $\frac{1}{60}$ s). Since most accidental electric shocks involve alternating currents, the following discussion of the effects of currents on the human body will emphasize alternating currents. In the typical accidental electric shock, the current enters the body through the hands (in contact with one terminal of the source of emf) and exits through the feet (in contact with the ground, which constitutes the other terminal of the source of emf in most AC circuits). Thus the body plays the role of a resistor, closing an electric circuit.

The damage to the body depends on the magnitude of the current passing through it. An alternating current of about 0.001 A produces only a barely detectable tingling sensation. Higher currents produce pain and strong muscular contractions. If the victim has grasped an electric conductor with the hand, the muscular contraction may prevent the victim from releasing the hold on the conductor. The magnitude of the "let-go" current, at which the victim can barely release the hold on the conductor, is about 0.01 A. Higher currents lock the victim's hand to the conductor. Unless the circuit is broken within a few seconds, the skin in contact with the conductor will then suffer burns and blisters. Such damage to the skin drastically reduces its resistance, which can lead to a fatal increase of the current.

An alternating current of about 0.02 A flowing through the body from the hands to the feet produces a contraction of the chest muscles that halts breathing; this leads to death by asphyxiation if it lasts for a few minutes. A current of about 0.1 A lasting just a few seconds induces fibrillation of the heart. This is a rapid, uncoordinated flutter of the heart muscles, with cessation of the natural rhythm of the heartbeat and cessation of the pumping of blood. Fibrillation usually continues even when the victim is removed from the electric circuit; the

[3] The variation of resistance with potential difference implies that skin does not obey Ohm's Law.

consequences are fatal unless immediate medical assistance is available. The treatment for fibrillation involves the deliberate application of a severe electric shock to the heart by means of electrodes placed against the chest; this arrests the motion of the heart completely. When the shock ends, the heart usually resumes beating with its natural rhythm.

A current of a few amperes produces a block of the nervous system and paralysis of the respiratory muscles. Victims of such currents can sometimes be saved by prompt recourse to artificial respiration. At these high values of the current, the effects of AC and DC are not very different. But at lower values, a DC current poses less of a hazard than a comparable AC current, because the former does not trigger the strong muscular contractions triggered by the latter.

In the above we assumed that the path of the current through the body is from the hands to the feet. If the current enters and exits through the same arm or leg, or enters through one leg and exits through the other, no vital organs lie in its path, and the threat to life is lessened. However, an intense current through a limb tends to kill the tissue through which it passes, and may ultimately require the surgical excision of large amounts of dead tissue, and even the amputation of the limb.

Other things being equal, a higher voltage will result in a higher current. The hazard posed by contact with high-voltage sources is therefore obvious. But under exceptional circumstances, even sources of low voltage can be hazardous. Several cases of electrocution by contact with sources of a voltage as low as 12 V have been reported. It seems that in these cases death resulted from an unusually sensitive response of the nervous system; it is also conceivable that an unusually small skin resistance was a contributing factor. Thus, it is advisable to treat even sources of low voltage with respect!

First aid for electric shock

Aid to victims of electric shock should begin with switching off the current. When no switch, plug, or fuse for cutting off the current is accessible, the victim must be pushed or pulled away from the electric conductor by means of a piece of insulating material, such as a piece of *dry* wood or a rope. The rescuer must be careful to avoid electric contact. If the victim is not breathing, artificial respiration must be started at once. If there is no heartbeat, cardiac massage must be applied by trained personnel until the victim can be treated with a defibrillation apparatus.

SUMMARY

Kirchhoff's (second) rule: The sum of emfs and voltage drops around any closed loop in a circuit must equal zero.

Power delivered by source of emf: $P = I\mathscr{E}$

Power dissipated by resistor: $P = I\,\Delta V$

QUESTIONS

1. Can we use a capacitor as a pump of electricity in a circuit? In what way would such a pump differ from a battery?

2. Does a fully charged battery have the same emf as a partially charged battery?

3. The emf of a battery is sometimes called the "open-circuit voltage." Explain.

4. How would you measure the internal resistance of a battery?

5. Kirchhoff's second rule is equivalent to energy conservation in the electric circuit. Explain.

Ammeter

6. The **ammeter** is a device for measuring electric current. To measure the electric current at a point P in a circuit, the ammeter must be inserted as shown in Figure 29.15b. The ammeter has a very low internal resistance, so that it does not inhibit the flow of current. What would happen if we were to connect the ammeter incorrectly, as shown in Figure 29.15c?

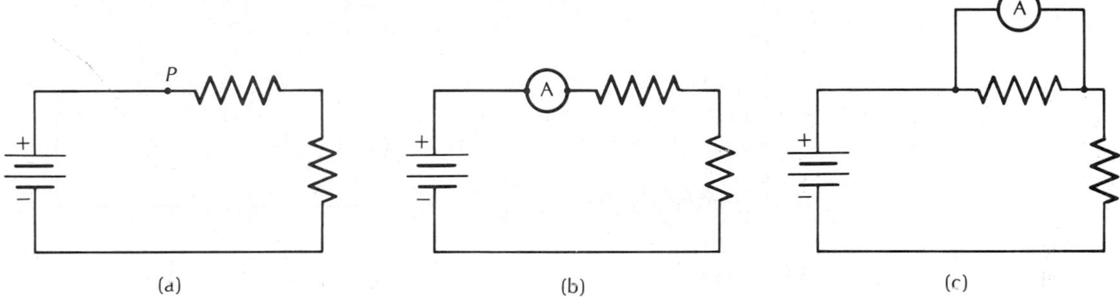

Fig. 29.15 (a) An electric circuit. (b) Correct connection of the ammeter. (c) Incorrect connection of the ammeter.

Voltmeter

7. The **voltmeter** is a device for measuring differences of potential. To measure the potential difference between points P and P' in a circuit, the voltmeter must be connected as shown in Figure 29.16a. The voltmeter has a very high internal resistance, so that it does not permit the redistribution of the current in the circuit. What would happen if we were to connect the voltmeter incorrectly, as shown in Figure 29.16b?

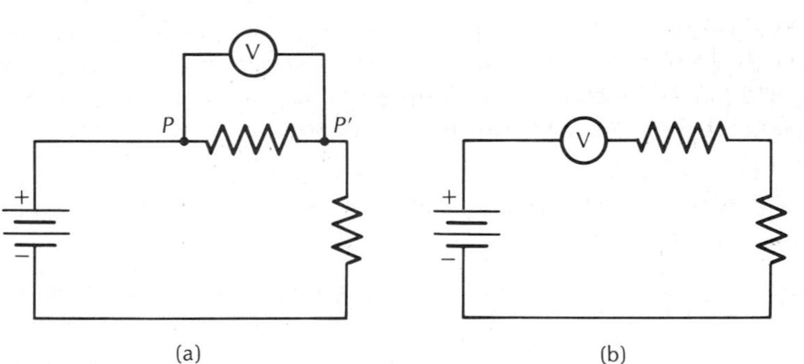

Fig. 29.16 (a) Correct connection of the voltmeter. (b) Incorrect connection of the voltmeter.

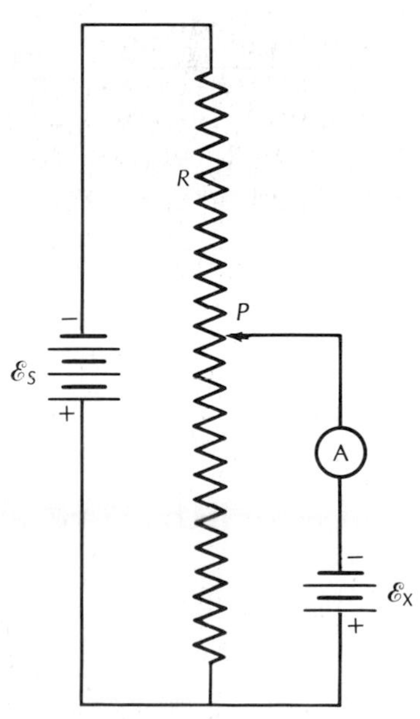

Fig. 29.17

8. A mechanic determines the internal resistance of an automobile battery by connecting a rugged, high-current ammeter (of nearly zero resistance) directly across the poles of the battery. The internal resistance is inversely proportional to the ammeter reading, $R_i \propto 1/I$. Explain.

Potentiometer

9. Precise comparisons of the voltages of two sources of emf can be performed with the circuit shown in Figure 29.17, called a **potentiometer.** In this circuit, $\mathscr{E}_S$ is a known, standard source of emf and $\mathscr{E}_X$ is the unknown source of emf. The resistance R is a long, uniform wire made of a metal of fairly high resistivity. The contact slider P is moved along the wire until the ammeter A

registers zero current (the potentiometer is then said to be balanced). Show that if the balance point is at the midpoint of the wire, then $\mathscr{E}_X = \frac{1}{2}\mathscr{E}_S$. What if the balance point is at three-quarters of the length of the wire?

Ohmmeter

10. An **ohmmeter** consists of a standard source of emf (a battery), connected in series with a standard resistance, and an ammeter. When the terminals of the ohmmeter are connected to an unknown resistor (Figure 29.18), the current registered by the ammeter permits the evaluation of the unknown resistance. Explain.

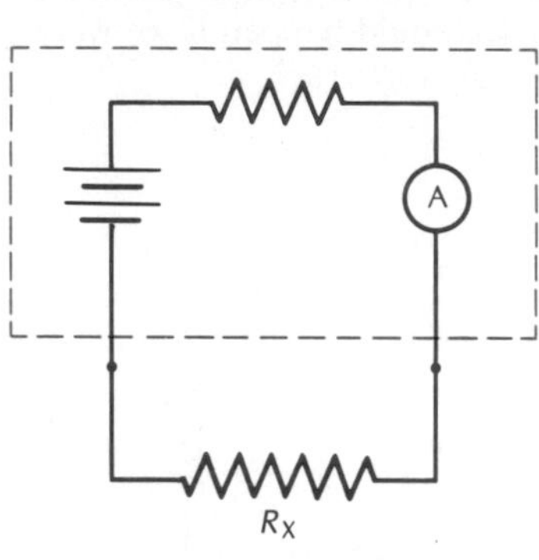

Fig. 29.18

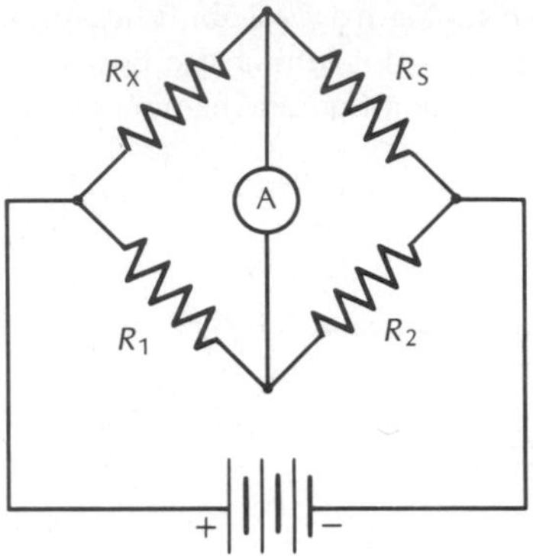

Fig. 29.19

Wheatstone bridge

11. The circuit shown in Figure 29.19 is called a **Wheatstone bridge.** It permits a very precise comparison of an unknown resistance with a known, standard resistance. In Figure 29.19 the unknown resistance is R_X, and the standard resistance is R_S. The resistances R_1 and R_2 are adjustable. In operation, the resistances R_1 and R_2 are adjusted until the ammeter A registers zero current. Show that under these conditions the ratios of the resistances are related as follows: $R_X/R_S = R_1/R_2$.

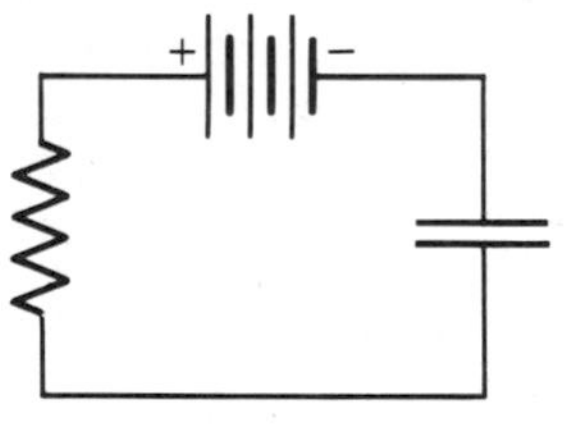

Fig. 29.20

12. At $t = 0$ you connect a battery and a resistor to a capacitor, as shown in Figure 29.20. The capacitor is initially uncharged. Qualitatively, describe the current as a function of time.

13. A homeowner argues that he should not pay his electric bill since he is not keeping any of the electrons that the power company delivers to his home — any electron that enters the wiring of his home sooner or later leaves and returns to the power station. How would you answer?

14. The spiral heating elements commonly used in electric ranges *appear* to be made of metal. Why do they not short circuit when you place an iron pot on them?

15. What are the advantages and what are the disadvantages of high-voltage power lines?

16. In many European countries, electric power is delivered to homes at 220 V, instead of the 110 V customary in the United States. What are the advantages and what are the disadvantages of 220 V?

17. How much does it cost you to operate a 100-watt light bulb for 24 hours? The price of electric energy is 8¢ per kilowatt-hour.

PROBLEMS

Section 29.1

1. The smallest batteries weigh 0.05 ounces and store an electric energy of about 5×10^{-6} kW · h. The largest batteries (used aboard submarines) weigh

300 tons and store an electric energy of 5×10^3 kw · h. What is the amount of energy stored per pound of battery in each case?

2. A size D flashlight battery will deliver 1.2 A · h at 1.5 V (i.e., it will deliver 1.2 A for 1 h or a larger or smaller current for a correspondingly shorter or longer time). An automobile battery will deliver 55 A · h at 12 V. The flashlight battery is cylindrical with a diameter of 1.3 in., a length of 2.2 in., and a weight of 0.19 lb. The automobile battery is rectangular with dimensions 12 in. × 6.5 in. × 9 in. and a weight of 50 lb.

(a) What is the available electric energy stored in each battery?
(b) What is the amount of energy stored per cubic inch of battery?
(c) What is the energy stored per pound of battery?

3. A heavy-duty 12-V battery for a truck is rated at 160 A · h, i.e., this battery will deliver 1 A for 160 h (or a larger current for a correspondingly shorter time). What is the amount of electric energy that this battery will deliver?

4. The electric starter motor in an automobile equipped with a 12-V battery draws a current of 80 A when in operation.

(a) Suppose it takes the starter motor 3.0 s to start the engine. What amount of electric energy has been withdrawn from the battery?
(b) The automobile is equipped with a generator that delivers 5.0 A to the battery when the engine is running. How long must the engine run so that the generator can restore the energy in the battery to its original level?

Section 29.3

5. A voltmeter of internal resistance $5.0 \times 10^4\ \Omega$ is connected across the poles of a 12-V battery of internal resistance 0.020 Ω.

(a) What is the current flowing through the battery?
(b) What is the voltage drop across the internal resistance of the battery?

6. A voltmeter reads 11.9 V when connected across the poles of a battery. The internal resistance of the battery is 0.020 Ω. What must be the minimum value of the internal resistance of the voltmeter if the reading of the instrument is to coincide with the emf of the battery to within better than 1%?

Section 29.4

7. Four resistors, with $R_1 = 25\ \Omega$, $R_2 = 15\ \Omega$, $R_3 = 40\ \Omega$, and $R_4 = 20\ \Omega$, are connected to a 12-V battery as shown in Figure 29.21.

(a) Find the combined resistance of the four resistors.
(b) Find the current in each resistor.

8. Consider the circuit shown in Figure 29.22. Given that $\mathscr{E}_1 = 6.0$ V, $\mathscr{E}_2 = 10$ V, and $R_1 = 2.0\ \Omega$, what must be the value of the resistance R_2 if the current through this resistance is to be 2.0 A?

9. Find the current in the two resistors shown in Figure 29.23. Find the power delivered by the 12-V battery. The resistances are $R_1 = 10\ \Omega$, $R_2 = 8\ \Omega$; the emfs are $\mathscr{E}_1 = 10$ V and $\mathscr{E}_2 = 12$ V.

10. Two batteries of emf $\mathscr{E}$ and of internal resistances R_i and R_i', respectively, are combined in parallel. The combination is connected to an external resistance R. Find the current through each battery.

11. Consider the circuit shown in Figure 29.11 with the given resistances and emfs. Suppose we replace the emf $\mathscr{E}_1 = 12.0$ V by a larger emf. How large must we make $\mathscr{E}_1$ if the current I_2 is to charge the battery $\mathscr{E}_2$?

12. Two batteries with internal resistances are connected as shown in Figure 29.24. Given that $R_1 = 0.50\ \Omega$, $R_2 = 0.20\ \Omega$, $\mathscr{E} = 12.0$ V, $\mathscr{E}' = 6.0$ V, $R_i = 0.025\ \Omega$, and $R_i' = 0.020\ \Omega$, find the currents in the resistances R_1 and R_2.

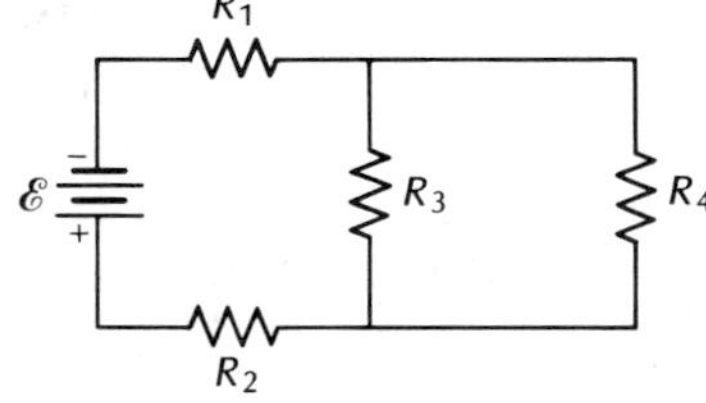

Fig. 29.21

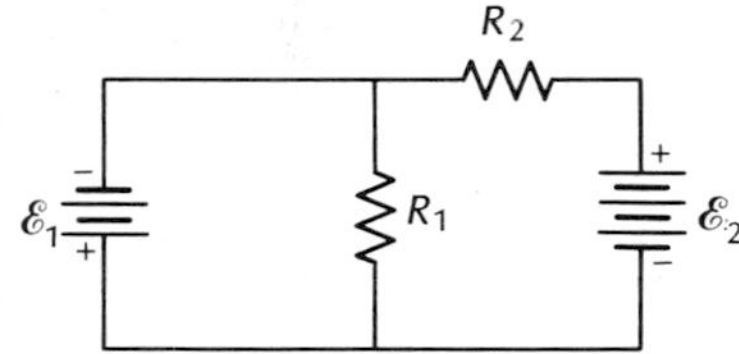

Fig. 29.22

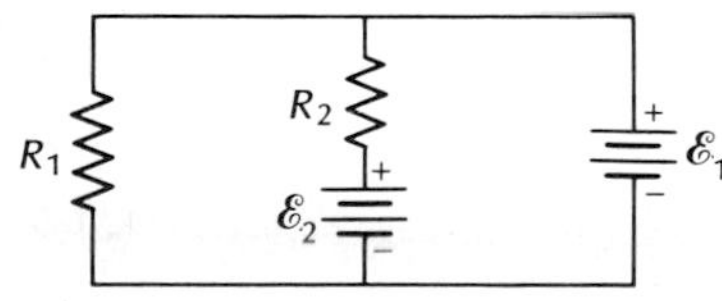

Fig. 29.23

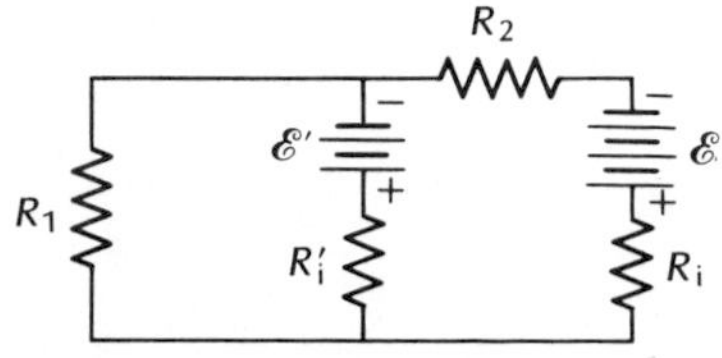

Fig. 29.24

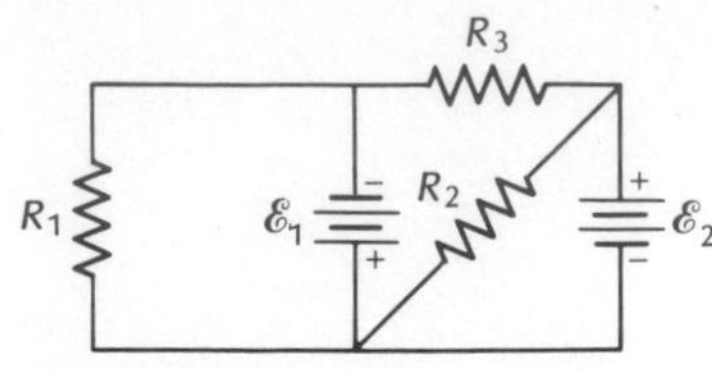

Fig. 29.25

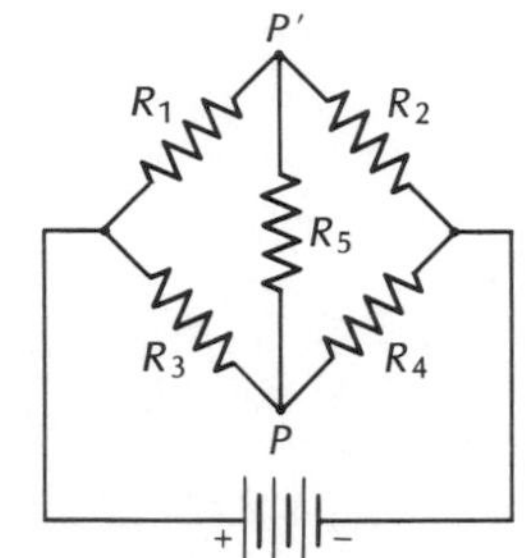

Fig. 29.26

*13. Two batteries, with $\mathscr{E}_1 = 6.0$ V and $\mathscr{E}_2 = 3.0$ V, are connected to three resistors, with $R_1 = 6.0\ \Omega$, $R_2 = 4.0\ \Omega$, and $R_3 = 2.0\ \Omega$, as shown in Figure 29.25. Find the current in each resistor and the current in each battery.

*14. Five resistors, of resistances $R_1 = 2.0\ \Omega$, $R_2 = 4.0\ \Omega$, $R_3 = 6.0\ \Omega$, $R_4 = 2.0\ \Omega$, and $R_5 = 3.0\ \Omega$, are connected to a 12-V battery as shown in Figure 29.26.

(a) What is the current in each resistor?

(b) What is the potential difference between the points P and P'?

Section 29.5

15. An electric toaster uses 1200 W at 110 V. What is the current through the toaster? What is the resistance of its heating coils?

16. A cyclotron accelerator produces a beam of protons of an energy of 700 million eV. The average current of this beam is 1.0×10^{-6} A. What is the number of protons per second delivered by the accelerator? What is the corresponding power delivered by the accelerator?

17. While cranking the engine, the starter motor of an automobile draws 80 A at 12.0 V for a time interval of 2.5 s. What is the electric power used by the starter motor? How many horsepower does this amount to? What is the electric energy used up in the given time interval?

18. An air conditioner operating on 110 V uses 1500 W of electric energy. What is the electric current through the air conditioner? What is the resistance of the air conditioner?

19. A small electric motor operating on 110 V delivers 0.75 hp of mechanical power. Ignoring friction losses within the motor, what current does this motor require?

20. A 12-V battery of internal resistance 0.20 Ω is being charged by an external source of emf delivering 6.0 A.

(a) What must be the minimum emf of the external source?

(b) What is the rate at which heat is developed in the internal resistance of the battery?

21. An electric car is equipped with an electric motor supplied by a bank of sixteen 12-V batteries. When fully charged, each battery stores an energy of 2.2×10^6 J.

(a) What current is required by the motor when it is delivering 12 hp? Ignore friction losses.

(b) With the motor delivering 12 hp, the car has a speed of 65 km/h (on a level road). How far can the car travel before its batteries run down?

22. The banks of batteries in a submarine store an electric energy of 5×10^3 kW · h. If the submarine has an electric motor developing 1000 hp, how long can it run on these batteries?

23. An electric toothbrush draws 7 watts. If you use it 4 minutes per day and if electric energy costs you 8¢/kW · h, what do you have to pay to use your toothbrush for one year?

24. In a small electrostatic generator, a rubber belt transports charge from the ground to a spherical collector at 2.0×10^5 V. The rate at which the belt transports charge is 2.5×10^{-6} C/s. What is the rate at which the belt does work against the electrostatic forces?

25. A solar panel (an assemblage of solar cells) measures 58 cm × 53 cm. When facing the sun, this panel generates 2.7 A at 14 V. Sunlight delivers an energy of 1.0×10^3 W/m^2 to an area facing it. What is the efficiency of this panel, i.e., what fraction of the energy in sunlight is converted into electric energy?

26. A 40-m cable connecting a lightning rod on a tower to the ground is made of copper with a diameter of 7 mm. Suppose that during a stroke of lightning the cable carries a current of 1×10^4 A.
   (a) What is the potential drop along the cable?
   (b) What is the rate at which Joule heat is produced?

27. The maximum recommended current for a No. 10 copper wire of diameter 0.259 cm is 25 A. For such a wire with this current, what is the rate of production of Joule heat per meter of wire? What is the potential drop per meter of wire?

28. The cable connecting the electric starter motor of an automobile with the 12.0-V battery is made of copper with a diameter of 0.50 cm and a length of 0.60 m. If the starter motor draws 500 A (while stalled), what is the rate at which Joule heat is produced in the cable? What fraction of the power delivered by the battery does this Joule heat represent?

29. Two heating coils have resistances of 12.0 Ω and 6.0 Ω, respectively.
   (a) What is the Joule heat generated in each if they are connected in parallel to a source of emf of 110 V?
   (b) What if they are connected in series?

30. A battery of emf $\mathscr{E}$ and internal resistance R_i is connected to an external circuit of resistance R. In terms of $\mathscr{E}$, R_i, and R, what is the power delivered by the battery to the external circuit? Show that this power is maximum if $R = R_i$.

31. A 3.0-V battery with an internal resistance of 2.5 Ω is connected to a light bulb of a resistance of 6.0 Ω. What is the voltage delivered to the light bulb? How much electric power is delivered to the light bulb? How much electric power is wasted in the internal resistance?

32. Suppose that a 12-V battery has an internal resistance of 0.40 Ω.
   (a) If this battery delivers a steady current of 1.0 A into an external circuit until it is completely discharged, what fraction of the initial stored energy is wasted in the internal resistance?
   (b) What if the battery delivers a steady current of 10.0 A? Is it more efficient to use the battery at low current or at high current?

33. An electric clothes dryer operates on a voltage of 220 V and draws a current of 20 A. How long does the dryer take to dry a full load of clothes? The clothes weigh 6.0 kg when wet and 3.7 kg when dry. Assume that all the electric energy going into the dryer is used to evaporate water (the heat of evaporation is 539 kcal/kg).

34. A large electromagnet draws a current of 200 A at 400 V. The coils of the electromagnet are cooled by a flow of water passing over them. The water enters the electromagnet at a temperature of 20°C, absorbs the Joule heat, and leaves at a higher temperature. If the water is to leave with a temperature no larger than 80°C, what must be the minimum rate of flow of water (in liters per minute) through the electromagnet?

ATMOSPHERIC ELECTRICITY[1]

The atmosphere is a great electric machine. Thunderstorms are the most spectacular manifestation of electric activity in the atmosphere, but even in fair weather the atmosphere is full of electric fields and electric currents. The thunderstorms act as giant electrostatic generators, delivering negative charge to the ground, and positive charge to the upper level of the atmosphere. This upper level of the atmosphere, or ionosphere, is a good conductor and the current reaching it quickly spreads laterally, over the entire globe. In fair-weather regions, this current gradually leaks down to the ground, completing the **atmospheric electric circuit** (Figure I.1).

There are roughly 2000 thunderstorms in action all over the Earth at any given time. The time-average current generated by a thunderstorm is about 1 A (the instantaneous current can of course be much larger — up to 20,000 A in a stroke of lightning). Thus, all the thunderstorms together contribute an average of 2000 A to the current in the atmospheric circuit. Bursting air bubbles at the ocean surface also contribute to the current, by spraying small positively charged droplets into the atmosphere; but this contribution is believed to be appreciably smaller than that of thunderstorms.

Figure I.2 is a schematic diagram of the atmospheric electric circuit. Both the ionosphere and the ground are good conductors, i.e., each is an equipotential surface; the potential difference between them is about 300,000 V. The resistance of the entire fair-weather atmosphere between ionosphere and ground is about 200 Ω. Most of this resistance is concentrated in the dense, low regions of the atmosphere; correspondingly, most of the 300,000-V potential drop occurs in the low regions of the atmosphere, within a few kilometers from the ground. On the average, the total electric power delivered to the global circuit by thunderstorm activity is about $2000 \text{ A} \times 300{,}000 \text{ V} = 6 \times 10^8$ W, i.e., nearly a million kW.

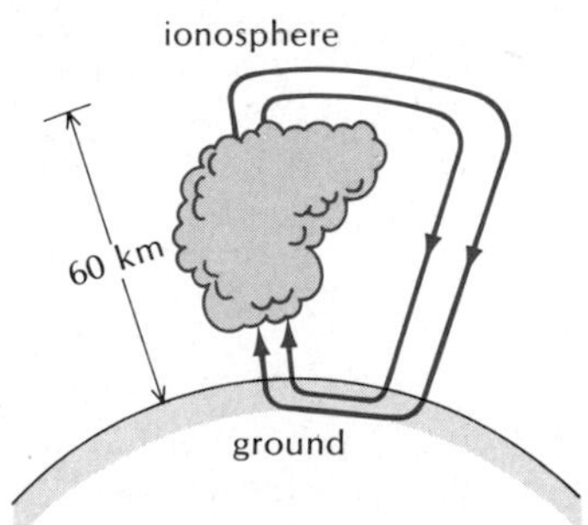

Fig. I.1 Flow of current in the Earth's atmosphere.

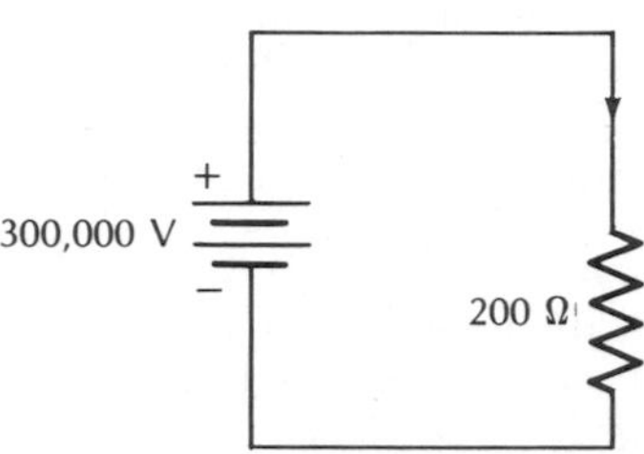

Fig. I.2 Schematic diagram of the atmospheric electric circuit.

I.1 THE FAIR-WEATHER ELECTRIC FIELD

The atmospheric current that, in the fair-weather regions of the Earth, flows from the ionosphere to the ground is carried by ions — atoms and molecules that have a positive or negative charge due to a deficit or excess in the number of their electrons. Such ions are always present in the atmosphere. They are continually produced by the impacts of cosmic rays on air molecules throughout the atmosphere, by the ultraviolet irradiation of the upper atmosphere, and by the natural and man-made radioactivity near the ground. Thus the current consists of a downward motion of positive ions and an upward motion of negative ions. The driving force that maintains this motion is an **atmospheric electric field** that points vertically downward; this electric field is analogous to the electric field that exists in a current-carrying wire and drives the electrons along the wire. Near the surface of the Earth, over open ground, the strength of this atmospheric electric field is between 100 and 200 V/m in fair weather. The exact value of the strength of the field depends on local conditions, such as dust in the

[1] This chapter is optional.

atmosphere, topography, time of day; the worldwide average value is 130 V/m.

The potential difference between the ground and a point 2 m above the ground (the height of a man) is therefore typically a few hundred volts. Does this mean that we can operate an appliance by plugging one terminal into the ground and the other terminal into air, 2 m above ground? Unfortunately, we cannot: the atmosphere will deliver only an infinitesimal current to the exposed terminal and therefore that terminal (and the entire appliance) will remain at the potential of the ground. The only consequence of sticking a wire, or any conductor, into the air is to deform the atmospheric equipotentials. The 0-V equipotential will follow the shape of the conductor; successively higher equipotentials will exhibit successively less deformation (Figure I.3). Buildings, trees, or human bodies are reasonably good conductors; they will deform the equipotentials, and the electric field, in a similar manner.

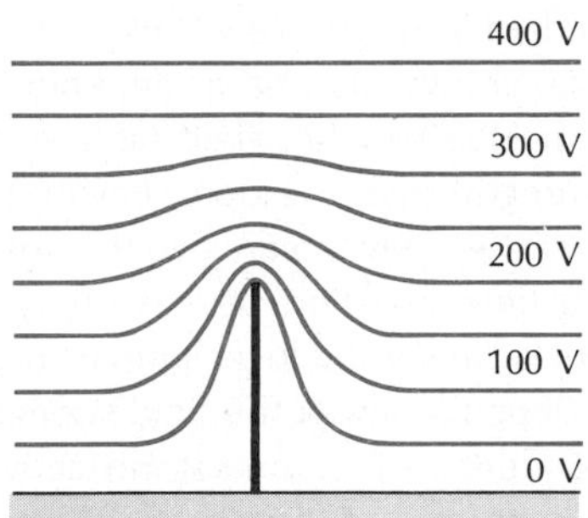

Fig. I.3 Equipotentials in air in the vicinity of a thin vertical conductor connected to the ground.

The lines of the atmospheric electric field end on the surface of the Earth; thus, there must be negative charge on the surface. The amount of charge per unit area is given by Eq. (24.23):

$$\sigma = \varepsilon_0 E = -8.85 \times 10^{-12}\ \text{F/m} \times 100\ \text{C/m} \cong -10^{-9}\ \text{C/m}^2$$

The total negative charge on the surface of the Earth (including the oceans) is about half a million coulombs.

The electric field near the ground can be measured with a **field meter.** A simple kind of field meter is constructed with a horizontal metallic plate placed near the ground and connected to the ground by means of a wire. The lines of electric field end on the surface of the metallic plate; this surface must therefore carry an electric charge. If a second metallic plate, also connected to the ground, is suddenly placed directly above the first plate, the lines of field will end on the second plate, i.e., the second plate shields the first plate from the electric field (Figure I.4). The charges on the first plate, released from the grip of the electric field, will then quickly flow through the wire into the ground. A sensitive ammeter connected to this wire will detect this flow charge; the total amount of charge that runs out of the plate is a measure of the strength of the electric field.

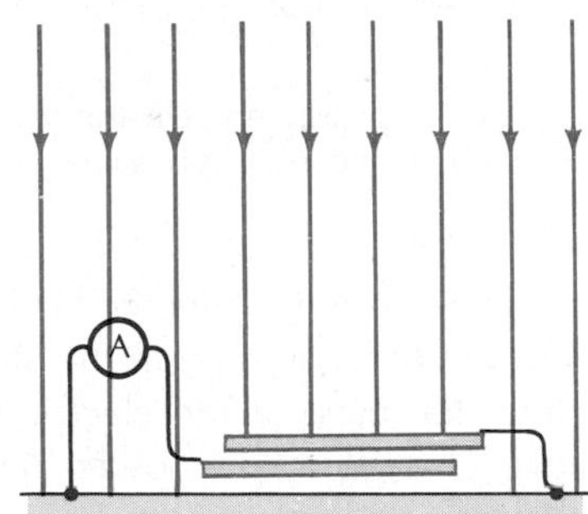

Fig. I.4 Two horizontal metallic plates, connected to the ground.

In practice, the two plates are often given the shape of a Maltese cross (Figure I.5). The upper plate is permanently kept above the lower plate and made to rotate in the horizontal plane. The arms of the upper plate then successively cover and uncover the arms of the lower plate; each covering and uncovering sends a pulse of current through the wire connecting the lower plate to the ground; the strength of this alternating current is a measure of the electric field. An instrument of this kind is called a **field mill.**

The strength of the atmospheric electric field decreases with altitude. This decrease is related to the decrease of the resistance of the atmosphere: at low altitude the resistance of the atmosphere is large, and at high altitude the resistance is small. Thus, at low altitude a large electric field is needed to maintain a given current, while at high altitude only a small electric field

Fig. I.5 A field mill. When in use, the cross is made to rotate at a high speed by a motor within the base.

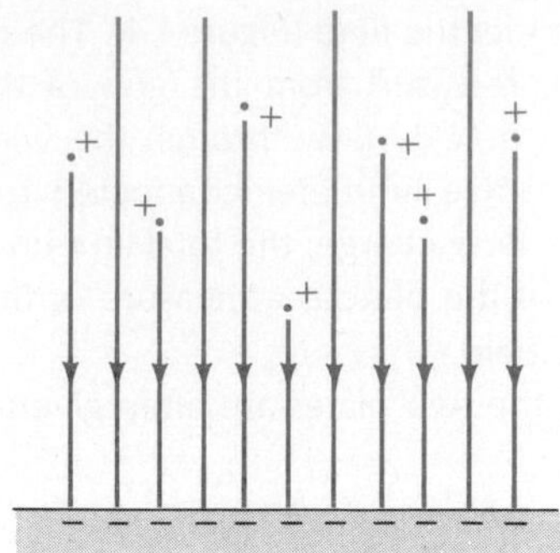

Fig. I.6 Field lines of the atmospheric electric field start on positive charges in the air and end on negative charges on the ground.

is needed. The decrease of the electric field is engendered by a positive charge density in the atmosphere (Figure I.6). Note that most of the electric field lines start on positive charges in the lower atmosphere, within a few kilometers from the ground; only very few field lines start on positive charges in the ionosphere. The total positive charge in the fair-weather atmosphere is of the same magnitude as the negative charge on the ground.

The strength of the fair-weather field depends to some extent on the time of day. The magnitude of the diurnal variation may be as large as 20%. Except for local effects due to comtamination of the atmosphere by smoke or dust, the variation is simultaneous at all places on the Earth; for example, the field strength is maximum at 18:00 local time (Greenwich Mean Time) in London and simultaneously at 13:00 local time (Eastern Standard Time) in New York.

Figure I.7a shows the time dependence of the strength of the fair-weather field (as a percentage of the mean strength) over the open ocean; these measurements over the open ocean give the best indication of the global time variation because they are free of spurious effects caused by local sources of pollution.

The time dependence of the field strength is due to a time dependence of thunderstorm activity. Worldwide, thunderstorm activity reaches a peak between 14:00 and 20:00 Greenwich Mean Time; this peak is mainly due to an abundance of mid-afternoon thunderstorms in the Amazon basin. Thus, the maximum in the electric field at 18:00 arises from the enhanced rate at which thunderstorms charge up the atmosphere at around this time (Figure I.7b).

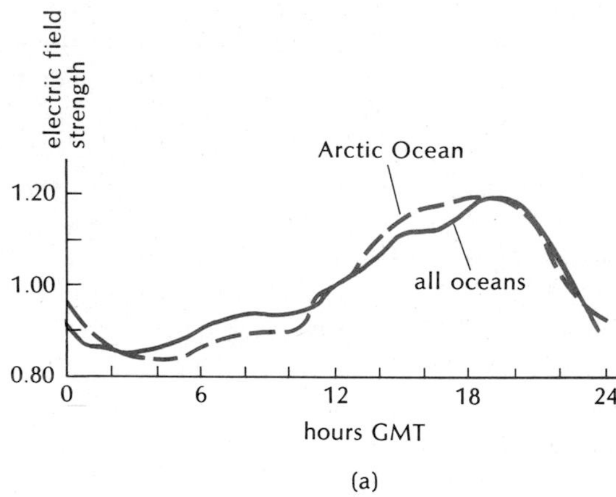

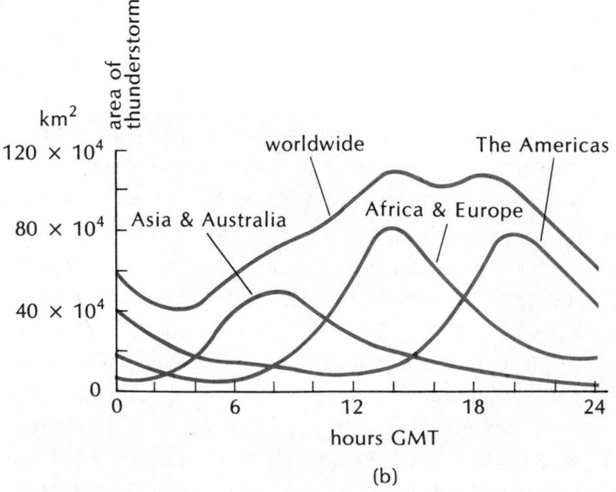

Fig. I.7 (a) Strength of electric field vs. Greenwich time. The strength of the electric field is expressed as a multiple of the average strength. (b) Thunderstorm activity vs. Greenwich time.

I.2 THUNDERSTORMS

Thunderstorms obtain the energy for their violent mechanical and electrical activity from humid air. Thunderstorms are heat engines; their heat reservoir is the heat of evaporation stored in water vapor. One cubic kilometer of air at 17°C, at atmospheric pressure, and at 100% relative humidity contains 1.6×10^7 kg of water vapor. Since the heat of evaporation for water is 586 kcal/kg (at 17°C), condensation of all the water vapor in 1 km^3 of air would release 9.2×10^9 kcal. This is equivalent to the energy released in the explosion of 9.2 kilotons of TNT. A typical thunderstorm involves many cubic kilometers of air and therefore the amount of stored energy (latent energy) is enormous. Although the efficiency for energy conversion is low, the release of just a small fraction of the energy stored in humid air suffices to account for the activity of a thunderstorm. Tornadoes and hurricanes also obtain their energy from humid air, and their destructive power reflects the large amount of available energy.

A thunderstorm is made of several **cells,** or thunderclouds, within each of which air moves upward or, in the later stages of the cell's development, downward. Each cell is some 8 km across; the base of the cell is at a height of about 1500 m and the top of the cell at a height that initially may be 7500 m but grows to 12,000 m or even 18,000 m. Within such a cell there

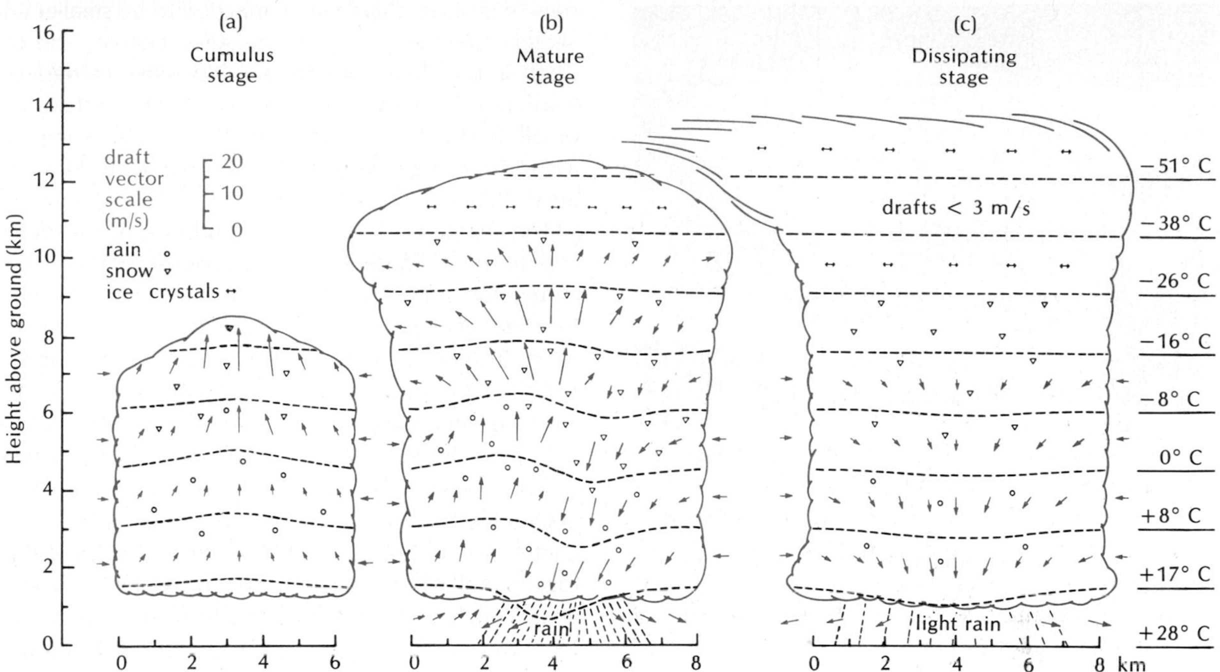

Fig. I.8 (a) A young thunderstorm cell. The vectors give the velocity of drafts within the cell. (b) A mature thunderstorm cell. (c) An old thunderstorm cell. (From A. J. Chisholm, *Meteorological Monographs,* November 1973, no. 36.)

are strong updrafts and, at later stages, strong downdrafts, commonly reaching vertical speeds of 12 m/s.

The rising motion of air in a storm cell is powered by the heat released in the condensation of water vapor. Dry air on the surface of the Earth will usually not rise; it is in stable (or neutral) equilibrium. If, because of some disturbance, a mass of dry air is pushed upward to a higher altitude, the reduced pressure of its surroundings leads to expansion of the air; the work done by the air during this expansion reduces its temperature; and this reduced temperature will match the normal reduced temperature of the environment at the higher altitude. Thus, dry air has no tendency to rise.

However, humid air is unstable with respect to vertical displacement. If a mass of humid air initially near the ground is pushed upward to a slightly higher altitude, the expansion and consequent reduction of temperature leads to condensation of a fraction of the water vapor. This supplies the air with extra heat. The air then will be warmer and less dense than the dry air that resides at the higher altitude — it continues to rise. Thus the air is unstable. Once the upward motion starts, it will continue faster and faster until all the water vapor has condensed, and even then the motion may continue, since the water can supply further energy by freezing into ice.

In a young thunderstorm cell the air rises as in a chimney, drawing in more and more humid air at the base while the top of the cell grows upward at speeds of 10 m/s or more (Figure I.8a).

A mature thunderstorm cell may extend to a height of 12,000 m or 18,000 m. Inside the cell small water droplets coalesce, forming raindrops; some of the rain freezes into ice and hail; snow may also form. Raindrops and ice particles that are too large and heavy to be supported by the updraft begin to fall. This downward motion drags some air along and starts a downdraft. Once the air begins to descend, it will continue descending — humid air is unstable with respect to downward displacement as well as upward displacement. If a mass of humid air (intermixed with water) is pushed downward to a slightly lower altitude and higher pressure, the compression tends to warm the air but the evaporation of water tends to cool it. The net result is that the air will be somewhat cooler and denser than the dry air surrounding the thunderstorm cell; therefore the mass of humid air continues to descend.

In the mature thunderstorm cell there are both strong updrafts and strong downdrafts. The top of the cell spreads out laterally, forming the characteristic anvil cloud, or **thunderhead** (Figure I.9). Observations made by looking down on thunderstorms from U-2 aircraft and satellites have revealed that above the anvil cloud, small turrets often sprout upward and reach into the stratosphere. These turrets consist of cloud masses thrown upward by violent updrafts with speeds as great as 100 m/s. During the mature stage, different regions of the thunderstorm cell acquire different electric charges, and lightning discharges ensue. Rain or hail falls out of the bottom of the cell, accom-

Fig. I.9 A thunderstorm with a well-developed anvil cloud.

panied by cool, descending air which generates a gusty wind near the ground (Figure I.8b).

In the late stage of a thunderstorm cell, the updraft ends and the downdraft takes over, covering the entire cell. Then the storm dissipates (Figure I.8c).

I.3 GENERATION OF ELECTRIC CHARGE

A mature thunderstorm cell typically has a charge distribution as shown in Figure I.10. There is a positive charge in the upper part of the cell and a negative charge in the lower part of the cell; furthermore, at the very bottom of the cell there is often an extra, small, positive charge.

Exactly how these changes are generated remains somewhat of a mystery. There are numerous theories invoking different mechanisms. Most of these theories blame the charge separation on a difference in the size of the charge carriers created in the cloud — the carriers of positive charge are supposed to be smaller and lighter than the carriers of negative charge. The updrafts in the cloud then blow the positive carriers upward, but the negative carriers either remain stationary or fall down. Thus the upper part of the cloud acquires a positive charge, while the lower part acquires a negative charge.

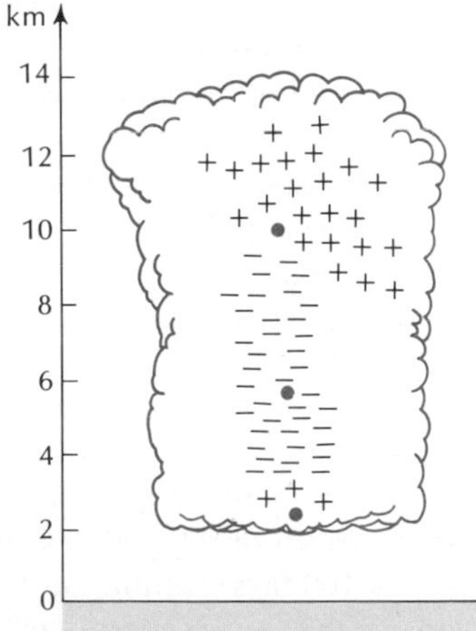

Fig. I.10 Charge distribution in a typical thundercloud.

The charge carriers may be raindrops, hail, ice crystals, or ions; different theories propose different scenarios for how any or all of these particles may become charged. The conjectural charging mechanisms involve some form of electrification by friction, collision, freezing, melting, or thermoelectricity. For instance, according to one theory, the charges are created by collisions between falling hailstones and small drops of water. The vertical downward atmospheric electric field polarizes the hailstones, inducing positive charges at their bottoms and negative charges at their tops. When such a falling hailstone strikes a drop of water, pushing it aside, the drop is likely to pick up a few positive charges during the contact with the hailstone (Figure I.11); this leaves the hailstone with a net negative charge. According to an alternative theory, the charges are created by collisions between falling hailstones and ions. These ions are present in the air, with equal average concentrations of negative and

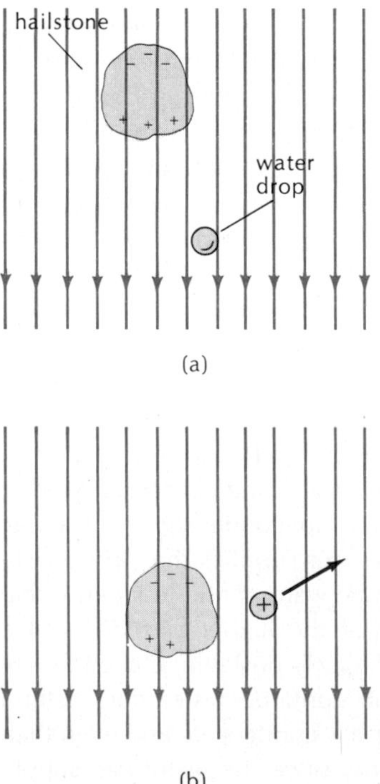

Fig. I.11 (a) A falling hailstone, polarized by the electric field, approaches a drop of water. (b) In the collision with the lower edge of the hailstone, the water drop picks up some positive charge.

positive ions. If the hailstone approaches a negative ion, it is likely to attract and capture the ion; if the hailstone approaches a positive ion, it is likely to merely repel the ion (Figure I.12). Thus the net result of such collisions is that the hailstone again tends to acquire a negative charge. The falling hailstone then carries this negative charge to the bottom of the cloud and updrafts carry the positive charge to the top of the cloud.

Another conjectural charging mechanism relies entirely on updrafts and downdrafts to accumulate charges in the cloud, after a small initial amount of charge has been generated in the cloud by other means. This mechanism can work in two ways. At the top of the cloud, the positive charge (see Figure I.10) attracts negative ions from the nearby air. If these ions become attached to droplets of water or particles of ice, downdrafts can carry them past the positive charge and deposit them in the lower portion of the cloud, adding to the negative charge already there. In a similar way, at the bottom of the cloud, the negative charge (see Figure I.10) attracts positive ions released by point discharge from sharp protuberances on the ground (see the next section). If these ions become attached to suitable water droplets or ice particles, updrafts can carry them to the upper portion of the cloud, adding to the positive charge already there.

Although all of these mechanisms, and others as well, presumably contribute to the charging of thunderclouds, it remains unclear which, if any, of the proposed mechanisms is dominant.

I.4 THE ELECTRIC FIELD OF A THUNDERCLOUD

Figure I.13a shows a crude model of the charge distributions in a thundercloud. For the purposes of this model, it has been assumed that the charge distributions are roughly spherical; the positive and negative charges can then be represented by point charges located at the centers of the charge distributions. The small positive charge at the bottom of the thundercloud has been ignored. The charges of ∓40 C are at heights of 5000 m and 10,000 m, respectively. At the

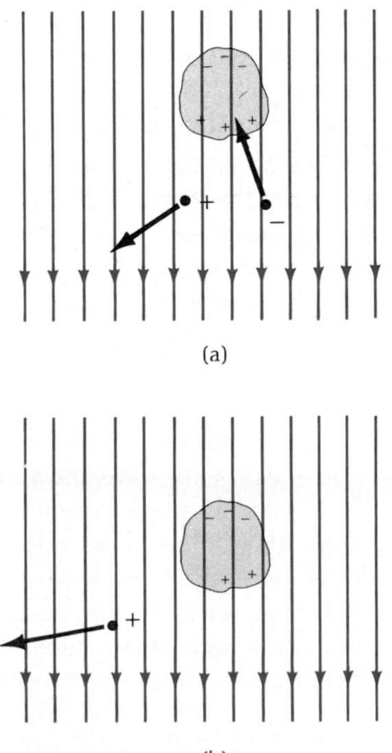

Fig. I.12 (a) A falling hailstone attracts negative ions, but repels positive ions. (b) Capture of a negative ion gives the hailstone a negative charge.

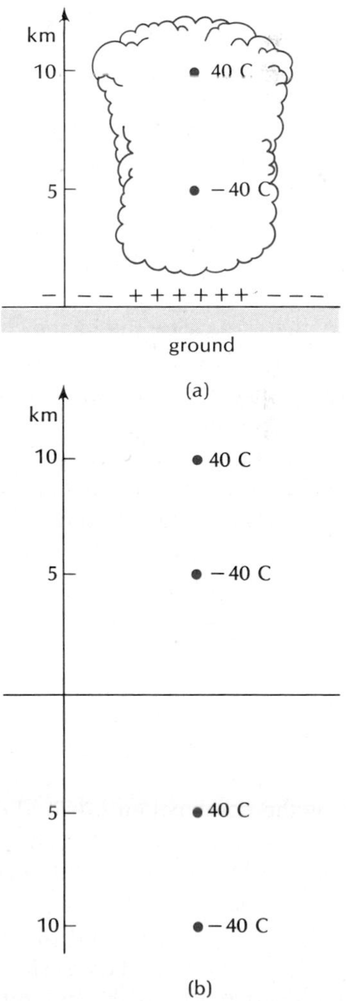

Fig. I.13 (a) In this crude model, the charge distribution has been approximated as consisting of two point charges. These point charges induce a charge distribution on the ground. (b) The effect of the induced charge distribution on the ground is equivalent to that of two virtual image charges.

ground, directly below these charges, they produce an upward electric field of magnitude

$$E = \frac{1}{4\pi\varepsilon_0}\left(\frac{Q_1}{r_1^2} + \frac{Q_2}{r_2^2}\right)$$

$$= 9.0 \times 10^9 \text{ F/m}\left(\frac{40\text{ C}}{(5\times 10^3\text{ m})^2} - \frac{40\text{ C}}{(10\times 10^3\text{ m})^2}\right)$$

$$= 11 \times 10^3 \text{ V/m} \tag{1}$$

But this is not the total electric field. The charges in the cloud induce charges on the ground (a conductor), and these produce an extra electric field. The induced charge distribution on the ground will cover a large area, perhaps some 100 km²; within this area, the charge density will be concentrated directly below the charges in the cloud, and it will gradually decrease with distance.

We can calculate the effect of this induced charge distribution by a clever trick involving three steps: First, replace the ground by a thin, flat conducting plate; this does not change the electric field in the space above the ground since the thickness of the conducting plate is irrelevant. Second, place two charges of ±40 C in the empty space below the conducting plate (Figure I.13b). These two new charges are at depths of 5000 m and 10,000 m below the conducting plate; they are called the **image** charges of the original charges in the cloud. The presence of these image charges below the plate does not change the electric field above the plate since the conductor shields these regions from each other. Third, slide the conducting plate away in a horizontal direction. This, again, does not change the electric field above the plate because, for the configuration of four charges shown in Figure I.13b, the midplane is an equipotential surface and the presence or absence of a coincident conducting surface makes no difference to the electric field.

Having taken these three steps, we recognize that the combination of cloud charges plus charges on the ground produces exactly the same electric field (above the ground) as the combination of cloud charges plus image charges; i.e., the induced charges on the ground are virtually equivalent to the image charges. Thus, we can calculate the total electric field at any given point on or above the ground by simply adding the electric fields of the four charges shown in Figure I.13b. At any given point on the ground, the image charges produce exactly the same electric field as the cloud charges — the total electric field is therefore *twice* the electric field of the cloud charges. Directly below the cloud charges, the electric field is then $2 \times 11 \times 10^3$ V/m $= 22 \times 10^3$ V/m. Incidentally, the above trick for the evaluation of the electric field of some given point charges in the presence of a conducting surface is called the **method of images;** the method only works under rather special conditions, but when it works, it permits a very quick and elegant evaluation.

We can also use this method to calculate the potential difference between the ground and the base of the thundercloud at a height of, say, 2 km (see Figure I.10). The potential of the ground is zero and the potential of the base of the thundercloud is a sum of four terms, corresponding to the four point charges of Figure I.13b:

$$V = \frac{1}{4\pi\varepsilon_0}\left(\frac{40\text{ C}}{8\times 10^3\text{ m}} - \frac{40\text{ C}}{3\times 10^3\text{ m}} + \frac{40\text{ C}}{7\times 10^3\text{ m}} - \frac{40\text{ C}}{12\times 10^3\text{ m}}\right)$$

$$= -5.4 \times 10^7 \text{ volt} \tag{2}$$

As these numbers show, the electric fields and the potential differences generated by thunderstorms are quite large.

In the above calculations we assumed a perfectly flat Earth. In reality there are protuberances on the surface of the Earth, and near these the electric field strength will be strongly enhanced.[2] It is a general rule that if a sharp conducting point protrudes from an equipotential surface, the electric field will be strong near the point. For example, Figure I.14 shows the electric field near a lightning rod; the high density of field lines near the tip of the rod indicates a strong field. It is easy to see that some enhancement of the field strength must occur for any kind of protruding point, regardless of the precise shape: the field lines must meet any conductor at right angles; hence they must bend toward the protruding conductor; hence their density increases.

The strong electric fields near any sharp spikes or edges on the ground produce **point discharge.** For example, experiments with trees have shown that a tree under a thunderstorm will carry a current of about 1 A out of the ground; the current flows into the atmosphere through the tips of the tree's leaves.

The point discharge is due to the electrical breakdown of the air — at a high electric field strength the air becomes a conductor. The sudden increase in the

[2] The field strength is also much greater in the vicinity of the charge concentrations in the cloud. On one occasion, a value of 340×10^3 V/m was measured at an airplane flying through a thunderstorm just before the airplane was struck by lightning.

conductivity of air is caused by ionization, i.e., the formation of positive ions and free electrons by the disruption of neutral atoms. There are always a few ions and free electrons present in the air, but not enough to conduct any large current. However, if the air is exposed to an intense electric field, then those few electrons are accelerated to high velocity, and when they smash into a neutral atom they can kick extra electrons out of the atom. A chain reaction sets in: each free electron generates an avalanche of free electrons. In dry air under atmospheric pressure, about 3×10^6 V/m are required to start such a discharge.

The discharges into the atmosphere can take a variety of forms: point discharge with no visible display, corona discharge (also called St. Elmo's fire) with its characteristic glow of visible light, sparks, or lightning. Although lightning is the most spectacular of these phenomena, in many thunderstorms point discharge actually plays the larger role in the transfer of charge from the thundercloud to the ground — it contributes a current to the atmospheric circuit which is several times larger than the (average) current contributed by lightning.

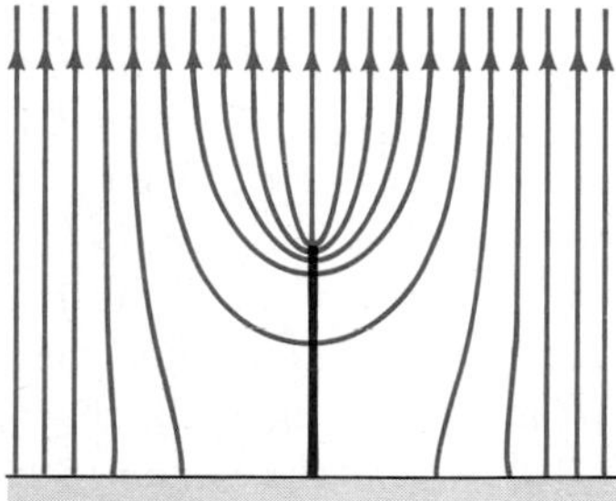

Fig. I.14 Electric field lines around a thin vertical conductor connected to the ground (lightning rod).

I.5 LIGHTNING

An electric field of 3×10^6 V/m is required to induce electric breakdown and produce small sparks in dry air at atmospheric pressure, but the presence of water drops and the lower-than-atmospheric pressure in a thundercloud favor breakdown and, therefore, sparking is likely to begin at somewhat lower electric fields. To initiate a flash of lightning, the electric field has to be very intense only in a small region near the cloud; once an avalanche of electrons starts, it can propagate into regions of less intense electric field. A lightning discharge can take place between the opposite charges inside the thundercloud, or between the thundercloud and clear air, or between the thundercloud and the ground. In the latter case the discharge usually takes place between the negative charge of the cloud

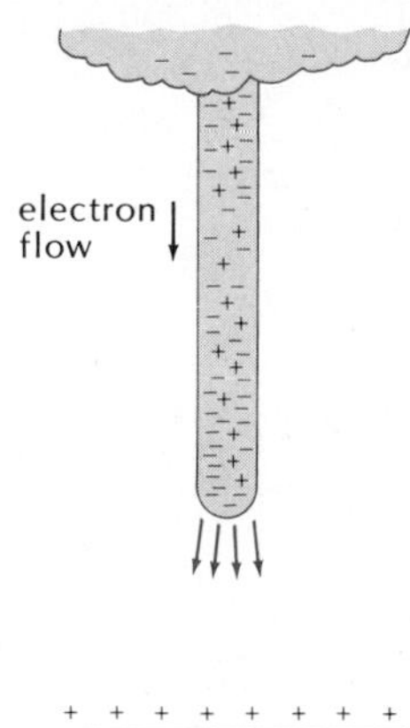

Fig. I.15 The leader. An electron avalanche from the cloud makes a channel of ionized air. Electrons fill this channel.

and the ground; discharges between the positive charge and the ground are rare.

The sequence of events in a flash of lightning occurs too fast to be resolved by the human eye; however, scientists have used techniques of high-speed photography to analyze the time development of flashes of lightning. A typical flash of lightning between a thundercloud and the ground begins with an electron avalanche in the intense electric field near the negative charge of the thundercloud. As the avalanche moves downward it leaves behind a channel of ionized air, or plasma; electrons from the cloud flow into this channel, giving it a negative charge. The concentration of negative charge near the tip of the channel generates strong electric fields which continue to push the avalanche along (Figure I.15). The channel follows a tortuous path, often with lateral branches whose twists and turns reflect random variations in the density of free electrons in the air ahead of the avalanche. The channel is called a **leader;** it has a radius of several meters (perhaps 5 m) but only its central region is (faintly) luminous. Figure I.16 shows a time sequence

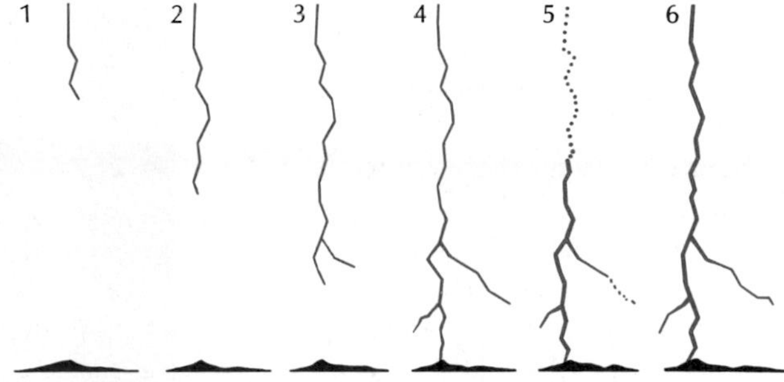

Fig. I.16 This sequence of diagrams, based on high-speed photographs, shows the downward motion of the leader (first four diagrams) and the upward motion of the return stroke (last two diagrams). The return stroke is much faster than the leader.

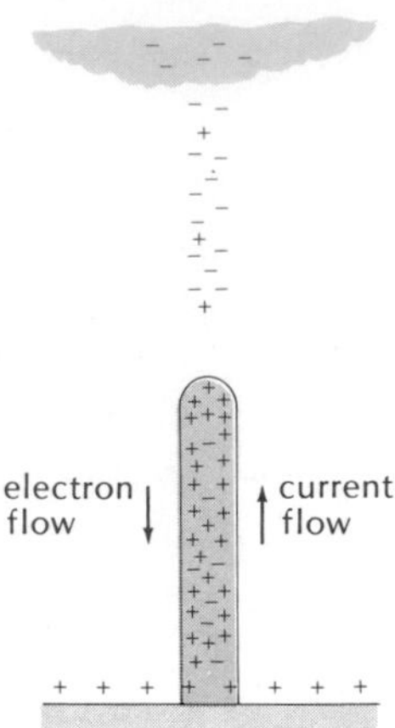

Fig. I.17 The return stroke. Electrons suddenly drain out of the bottom of the channel, leaving it with a positive charge.

of pictures of the leader. These pictures fail to illustrate one important feature of the leader: the leader does not move downward smoothly — it proceeds in steps of about 50 m with a pause of about 50 μs between steps. Because of this, it is sometimes called the **step leader.**

When the tip of the leader comes near the ground, its intense electric field initiates a discharge from the ground or from a pointed object on the ground. The upward-moving discharge meets the leader 20 m to 100 m above ground. At this instant the circuit between cloud and ground is complete — there is now a continuous conducting path and the negative charge can flow from cloud to ground with little resistance.

The electrons at the bottom of the channel are the first to move — they drain out into the ground, giving a large current. Then electrons from successively higher positions drain down. Thus, a "drainage front" moves upward toward the thundercloud. The drainage front is the head of a tube of intense current which snakes toward the thundercloud at as much as one-half the speed of light; this tube of current is the **return stroke** (Figure I.17). The intense current heats the air and produces the bright flash of light that we see as lightning (Figure I.18). The radius of the tube of current is quite small — it ranges from less than 1 cm to a few centimeters.

The peak current of the return stroke is about 10,000 A to 20,000 A. This strong current lasts less than 100 μs. When it ends, a weaker current (a few hundred amperes) continues to flow for several milliseconds. The strong current can cause severe damage to whatever body lies in its path (Figures I.19 and I.20).

The first return stroke is often succeeded by several more return strokes. A typical flash of lightning contains three to five return strokes with intervals of 40 ms between them. Each of the latter return strokes is preceded by its own **dart leader** which travels along the channel of the first return stroke. This channel retains its conductivity for some time and enables the dart leaders to travel downward very fast, without the pauses characteristic of the first step leader.

The total charge delivered to the ground by the complete flash of lightning with its several return strokes amounts to about −25 C (see Table I.1). Since the potential difference between the base of the cloud and the ground is about 5×10^7 V [see Eq. (2)], the energy dissipated in the lightning channel is about 10^9 J. Most of this energy goes into heat (Joule heating), although a small fraction goes into emission of light and of radio waves. Immediately after the current has passed, the plasma in the lightning channel will be at an extremely high temperature (about 30,000 K) and a correspondingly high pressure. The high-pressure plasma expands explosively, generating a shock wave

Fig. I.18 Lightning.

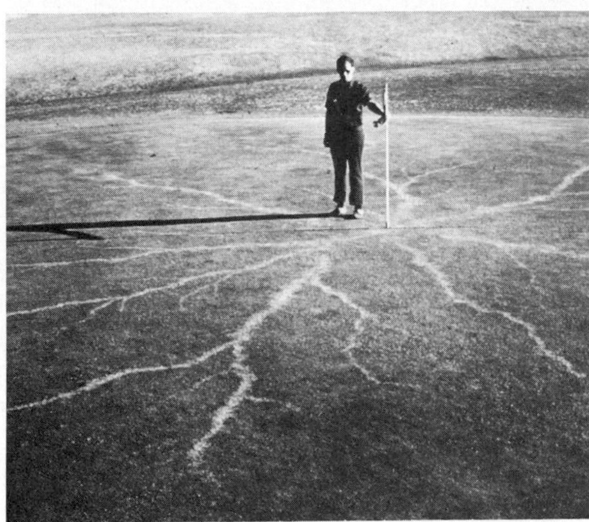

Fig. I.19 The grass of this golf course was burnt by the currents produced by a stroke of lightning. Note that the current branches out along the ground in all directions.

Table I.1. DATA ON LIGHTNING[a]

Step leader	
Length of step	50 m
Time between steps	50 μs
Speed	1.5×10^5 m/s
Charge in channel	−5 C
Dart leader	
Speed	2×10^6 m/s
Charge in channel	−1 C
Return stroke	
Speed	5×10^7 m/s
Peak current	10^4–2×10^4 A
Charge transferred to ground	−2.5 C
Complete lightning flash	
Number of strokes	3–5
Time between strokes	40 ms
Time duration of flash	0.2 s
Charge transferred to ground (including continuing current)	−25 C

[a] Based on M. A. Uman, *Lightning*. The numbers given are average values.

Fig. I.20 Tree exploded by lightning.

in the surrounding air. The shock wave dissipates within a few meters, gradually changing into a sound wave that propagates outward as a pulse. When this sound pulse reaches our ears, we hear it as thunder. The rumbling noises in thunder are due to the successive arrivals of sound pulses from different portions of the lightning bolt, which are at different distances from our ears.

At some distance from the lightning bolt, we will hear the sound after we see the flash of lightning — because of the high speed of light, the flash arrives almost instantaneously, while the sound takes an appreciable time to cover the intervening distance. Since the speed of sound is about $\frac{1}{3}$ km/s, we can use the following rule of thumb to reckon the distance (in kilometers) to the lightning bolt: count the seconds between flash and thunder and divide by 3.

The most spectacular displays of lightning occur in tornadoes. In many instances, eyewitnesses reported that the interior of the funnel of a tornado was brightly lit by constant flashes of lightning. Measurements of the radio waves emitted by the lightning in tornadoes[3] confirm these reports; from such measurements scientists have estimated that about 20 flashes of lightning occur each second. Since the energy per flash of lightning is about 10^9 J (see above), the electric power dissipated by a tornado is about $10^9 \text{ J} \times 20/\text{s} = 2 \times 10^{10}$ W — this is roughly $\frac{1}{20}$ of the combined output of all the electric power plants in the United States!

[3] These radio waves can be picked up on an ordinary TV set; this can give advance warning of the approach of a tornado.

Finally, some brief remarks on the weird phenomenon of ball lightning, or **Kugelblitz.** This consists of a glowing fireball that floats in midair. The sizes of these balls range from 10 cm to 100 cm, although airplane pilots have reported balls as large as 15 m to 30 m inside thunderclouds. The balls sometimes occur after a lightning stroke and sometimes they occur spontaneously, without apparent cause; they usually last only a few seconds. The balls sometimes fall down vertically from the sky, and sometimes drift horizontally near the ground. They have been known to enter buildings through doors, windows, or chimneys. Many of these balls disappear silently without a trace, but some disintegrate explosively with an ear-shattering bang.

It seems likely that the ball is created and maintained by electric effects, but we have no understanding of the mechanism involved in this phenomenon. Several theories have been proposed to explain the ball: according to one theory, it is a ball of plasma held together by magnetic fields; according to another, it is a miniature thundercloud of dust particles acting as a very efficient electrostatic generator. But owing to a lack of precise data and a lack of detailed calculations, the phenomenon remains a mystery.

Further Reading

The following books discuss thunderstorms and lightning at an introductory level:

The Lightning Book by P. E. Viemeister (Doubleday, Garden City, 1961)

The Flight of Thunderbolts by B. F. J. Schonland (Oxford, 1964)

The Nature of Violent Storms by L. J. Battan (Doubleday, Garden City, 1961)

Short but very informative articles at the same level are the following:

"Thundercloud Electricity," B. Vonnegut, *Discovery,* March 1965

"Thunder," A. A. Few, *Scientific American,* July 1975

"Thunder and Lightning," J. Latham, in *Forces of Nature,* edited by V. Fuchs (Thames and Hudson, London, 1977)

At a more technical level, the following books and articles provide a wealth of information:

Lightning by R. H. Golde (Academic Press, London, 1977)

Lightning by M. A. Uman (McGraw-Hill, New York, 1969)

Atmospheric Electricity by J. A. Chalmers (Pergamon Press, Oxford, 1967)

Physics of Lightning by D. J. Malan (English University Press, London, 1963)

"Atmospheric Electricity" by C. D. Stow, in *Reports on Progress in Physics,* Vol. 32, Part I, 1969

"Some Facts and Speculations Concerning the Origin and Role of Thunderstorm Electricity" by B. Vonnegut, in *Meteorological Monographs,* Vol. 5, No. 27, September 1963

Questions

1. It is possible to measure the potential differences between the atmosphere and the ground by shooting an arrow upward, with a fine wire trailing from its tail. A voltmeter connected between the lower end of the wire and the ground will then register the potential at the arrow. Why does this differ from sticking a stationary conductor into the air?

2. By comparing Figures I.7a and b, can you conclude that the bursting of bubbles on the ocean surface makes only a small contribution to the current in the atmospheric circuit?

3. In 1752, Benjamin Franklin flew a kite into a thundercloud and demonstrated that lightning is an electric phenomenon. A few months later, a scientist at St. Petersburg was killed by lightning while attempting to repeat this stunt. After that, scientists took the precaution of enclosing themselves in boxes of sheet metal while flying kites into thunderclouds. How does this help?

4. How could you construct a power plant that captures electric energy from a thundercloud?

5. What are the main dangers to an aircraft flying through a thundercloud?

6. A lightning rod serves not only to conduct lightning safely to the ground, but also to inhibit lightning by promoting point discharge. Explain.

7. What parts of an aircraft are most likely to be struck by lightning when flying in or near a thundercloud?

8. During a thunderstorm, would you be safer in an automobile with a sheet metal body or an automobile with a fiber glass body?

9. To protect the crew of a wood or fiber glass boat from the hazards of lightning, all metal parts on the boat should be connected to a thick conducting cable, terminating on a conducting plate on the outside of the hull, below the waterline. Explain.

10. What pattern of current flow do you expect below and on the water surface near the point of entry of lightning? Would you suffer damage if swimming nearby?

11. Is the conductivity of the ground due to a flow of electrons or a flow of ions?

12. The ancient Greeks noticed that when lightning strikes the ground, nearby cattle are killed, but nearby men often survive. The Greeks thought this indicated an affinity between men and gods. Can you think of a better explanation? (Hint: Chicken also survive.)

13. It is dangerous to take showers or baths when a thunderstorm is overhead. Why?

14. According to a familiar saying, lightning never strikes twice. According to Table I.1, how often does lightning strike, on the average?

15. Suppose you get caught in the open by a thunderstorm with severe lightning. Consider and discuss each of the following options: take shelter beneath a tree standing in isolation; lie down in the open, flat on the ground; sit in a ditch or depression, on your heels, with both feet close together.

16. When lightning strikes a tree, it often explodes the branches or the trunk. Explain. (Hint: The heating of water produces steam.)

17. Suppose that after a flash of lightning, you hear the first thunder in 3 s and the last in 9 s. What can you conclude about the distances of the nearest and the farthest portions of the lightning bolt?

CHAPTER 30

The Magnetic Force and Field

Hans Christian Oersted (örstad), *1777–1851, Danish physicist and chemist, professor at Copenhagen. He observed that a compass needle suffers a deflection when placed near a wire carrying an electric current. This discovery gave the first empirical evidence of a connection between electric and magnetic phenomena.*

The magnetic force that is most familiar in everyday experience is the force that the magnetic poles of the Earth exert on a compass needle. This magnetic force was known for many centuries, but only during the last century did experimenters discover that electric currents also exert magnetic forces on compass needles and that electric currents exert magnetic forces on one another. Finally, physicists came to understand that the magnetic force is nothing but an extra electric force acting between charges in motion. This means that between two charges in motion, there acts not only the Coulomb force, but also a force that is a function of the velocities of the charges.

In the next section we will write down an equation for the magnetic force between two moving charges. From this we will derive the equation for the magnetic force between two currents consisting of many moving charges. Our development goes counter to the historical development — the law of magnetic force between currents was discovered long before the law of magnetic force between individual charges. But our development follows the road taken in earlier chapters: we always begin with the fundamental laws that apply to *particles* and from these deduce the laws that apply to systems of particles (such as currents).

30.1 The Magnetic Force

According to Coulomb's Law, the electric force exerted by a point charge q' on a point charge q is

$$\mathbf{F} = \frac{1}{4\pi\varepsilon_0}\frac{qq'}{r^2}\,\hat{\mathbf{r}} \qquad (1)$$

where r is the distance between the charges and $\hat{\mathbf{r}}$ a unit vector pointing from q' toward q (Figure 30.1). The force on q' exerted by q is exactly the opposite of that in Eq. (1).

Equation (1) correctly gives the force acting on charges at rest. However, when the charges are *in motion*, there is an extra force acting on these charges. This extra force is called the **magnetic force.** This force depends on the relative positions of the charges and on their velocities. If the instantaneous velocities of the charges q and q' are $\mathbf{v}$ and $\mathbf{v}'$, respectively, then the magnetic force on charge q exerted by charge q' is

$$\mathbf{F} = [\text{constant}] \frac{qq'}{r^2} \mathbf{v} \times (\mathbf{v}' \times \hat{\mathbf{r}}) \qquad (2)$$

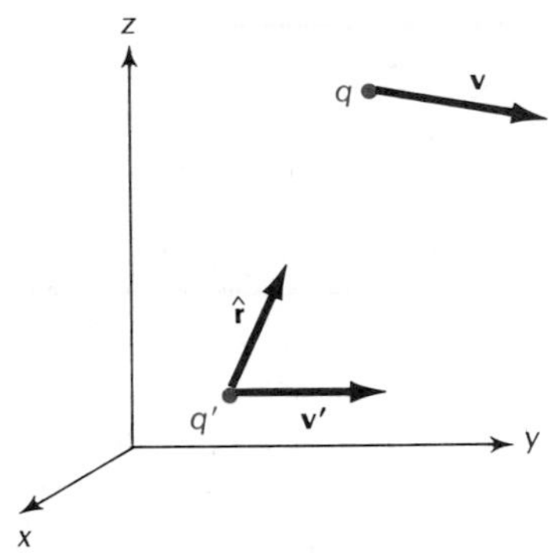

Fig. 30.1 The vectors $\mathbf{v}'$, $\hat{\mathbf{r}}$, and $\mathbf{v}$. In this example, all these vectors are in the z–y plane.

We must regard this equation for the magnetic force as a fundamental law of physics, which has the same status as Coulomb's Law or Newton's Law of Gravitation. Note that the magnetic force varies as the inverse square of the distance, like the electric force and the gravitational force. But the dependence on the velocity involves two cross products of vectors, and is rather complicated.

The direction of the force must be worked out by the right-hand rule for cross products: first we must cross-multiply $\hat{\mathbf{r}}$ by $\mathbf{v}'$ and then cross-multiply the result by $\mathbf{v}$ (see Figure 30.2). The magnetic force is always perpendicular to the velocity $\mathbf{v}$ of the charge q. Note that the magnetic force is zero unless both velocities are different from zero, i.e., the magnetic force acts only if *both* charges are in motion.

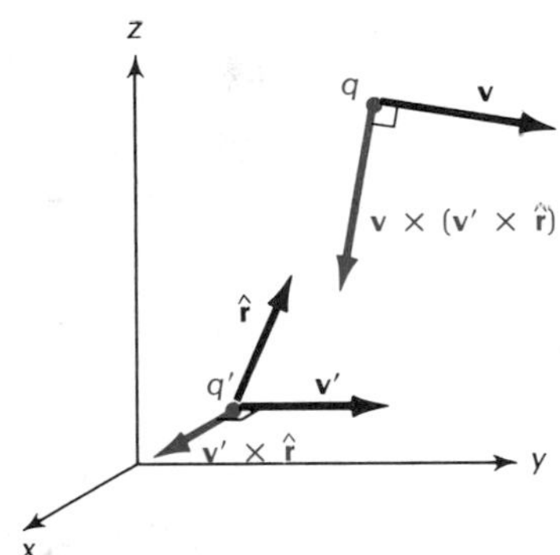

Fig. 30.2 With $\mathbf{v}'$, $\hat{\mathbf{r}}$, and $\mathbf{v}$ as in Figure 30.1, the cross product $\mathbf{v}' \times \hat{\mathbf{r}}$ is perpendicular to the z–y plane; and the cross product $\mathbf{v} \times (\mathbf{v}' \times \hat{\mathbf{r}})$ is in the z–y plane.

In the SI system of units, the numerical value of the constant in Eq. (2) is

$$[\text{constant}] = 1.00 \times 10^{-7} \frac{\text{N} \cdot \text{s}^2}{\text{C}^2} \qquad (3)$$

This constant is conventionally written in the form

$$[\text{constant}] = \frac{\mu_0}{4\pi} \qquad (4)$$

with

$$\mu_0 = 1.26 \times 10^{-6} \frac{\text{N} \cdot \text{s}^2}{\text{C}^2} \qquad (5)$$

Permeability constant

The quantity μ_0 is called the **permeability constant.**[1]

Our equation for the magnetic force on a point charge q exerted by q' then becomes

$$\boxed{\mathbf{F} = \frac{\mu_0}{4\pi} \frac{qq'}{r^2} \mathbf{v} \times (\mathbf{v}' \times \hat{\mathbf{r}})} \qquad (6)$$

Magnetic force on moving point charge

[1] The value of the constant in Eq. (3) is *exact*. Correspondingly, the value $\mu_0 = 4\pi \times 10^{-7}\ \text{N} \cdot \text{s}^2/\text{C}^2$ is also exact. (These values hinge on the definitions of current and of charge in the SI system.)

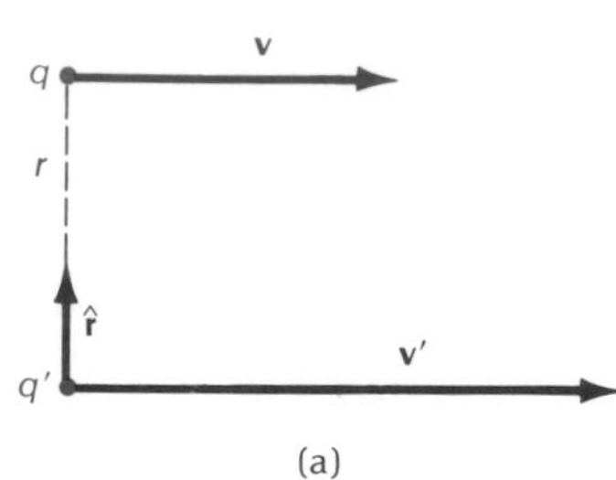

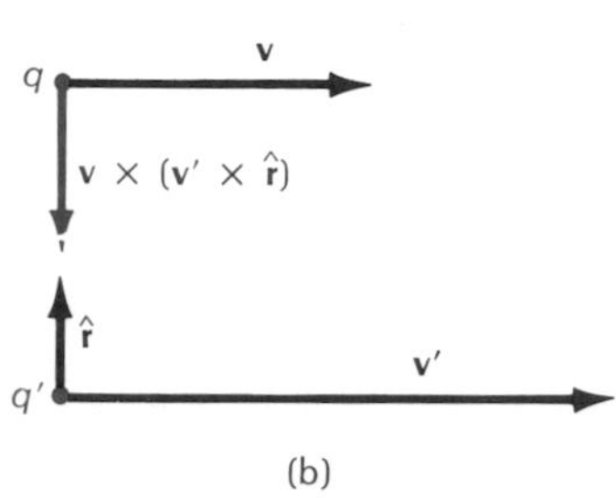

Fig. 30.3 (a) The velocity vectors **v** and **v′** are parallel. (b) The vector $\mathbf{v} \times (\mathbf{v}' \times \hat{\mathbf{r}})$ points from q toward q'.

EXAMPLE 1. Suppose that the (instantaneous) velocities of two positive point charges are parallel (Figure 30.3a). Compare the electric and magnetic forces.

SOLUTION: For the configuration shown in Figure 30.3a, the angles in both the cross products of Eq. (6) are 90° and hence the magnitude of the magnetic force is simply

$$F_{\text{mag}} = \frac{\mu_0}{4\pi} \frac{qq'}{r^2} vv' \tag{7}$$

By working out the directions of the cross products, we find that the force is attractive, i.e., the charge q is pulled toward q' (Figure 30.3b).

The magnitude of the electric force is

$$F_{\text{el}} = \frac{1}{4\pi\varepsilon_0} \frac{qq'}{r^2} \tag{8}$$

This force is, of course, repulsive. The net force is the sum of the magnetic and electric forces.

The ratio of the magnitudes of these two forces is

$$\frac{F_{\text{mag}}}{F_{\text{el}}} = \left(\frac{\mu_0}{4\pi} \frac{qq'}{r^2} vv'\right) \Big/ \left(\frac{1}{4\pi\varepsilon_0} \frac{qq'}{r^2}\right) = \mu_0\varepsilon_0 vv'$$

$$= 1.26 \times 10^{-6} \frac{\text{N} \cdot \text{s}^2}{\text{C}^2} \times 8.85 \times 10^{-12} \frac{\text{C}^2}{\text{N} \cdot \text{m}^2} vv'$$

$$= (1.12 \times 10^{-17}\ \text{s}^2/\text{m}^2) vv' \tag{9}$$

The numerical factor appearing in Eq. (9) is equal to $1/(3.0 \times 10^8\ \text{m/s})^2$. Hence Eq. (9) can be written

$$\frac{F_{\text{mag}}}{F_{\text{el}}} = \frac{v}{3.0 \times 10^8\ \text{m/s}} \frac{v'}{3.0 \times 10^8\ \text{m/s}} \tag{10}$$

This shows that F_{mag} will be small compared to F_{el} if v and v' are small compared to 3.0×10^8 m/s. Note that 3.0×10^8 m/s is the speed of light; why the speed of light should turn up in a calculation concerning electricity and magnetism is a puzzle which we will solve in Section 35.5.

According to Eq. (10), the magnetic force between point charges is small compared to the electric force unless the velocities of the charges approach the velocity of light. This suggests that the magnetic force only becomes significant when the velocities are so large that Newtonian physics fails. Once this happens, a variety of relativistic corrections will have to be included in the Newtonian equation of motion (see Section 17.7).[2] It would then seem that the magnetic force is no more than just another relativistic correction that has to be included in the Newtonian equation of motion of charged particles.

However, the magnetic force can become very important even when the velocities are low. This will happen whenever we are dealing with a charge distribution for which the electric forces cancel. For example, Figure 30.4 shows a moving point charge at some distance from a

Fig. 30.4 Point charge moving parallel to a current.

[2] If the velocities are large, some extra relativistic corrections will also be needed in Eqs. (1) and (2).

straight wire carrying a current. Under these conditions, there is no electric force on the point charge — the wire contains equal amounts of positive and negative charges and the electric attractions and repulsions on the point charge cancel. The only remaining force on the point charge is the magnetic force produced by the moving charges of the wire. Since the number of moving charges on the wire is very large, the magnetic force can be quite large.

Figure 30.5 shows two wires carrying parallel currents. The conditions here are similar to those of the preceding example. There is no electric force between these wires — the electric forces between the positive and negative charges on the two wires cancel. However, the magnetic forces do not cancel. The moving charges of one wire exert a magnetic force on the moving charges of the other wire. If the currents are parallel, as in Figure 30.5, this magnetic force is attractive (compare Example 1). If the currents are antiparallel, then the magnetic force is repulsive. The magnitude of the force between currents on wires can be calculated by integrating the forces of the individual point charges.

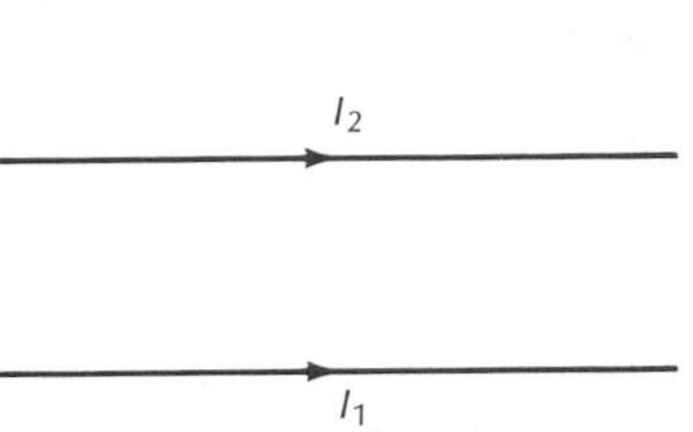

Fig. 30.5 Two parallel currents.

The magnetic force between the currents on two wires was discovered by Ampère in 1820, long before it was recognized that this force is really due to the magnetic force of Eq. (6) between individual moving charges on the wires. Experiments with individual moving charges (charged brass balls moving at high velocity) were carried out early in this century. It is very difficult to perform a precise measurement of the magnetic force between individual charges. However, it is fairly easy to perform precise measurements of the force between parallel wires carrying a current (such measurements are actually used for the *definition* of the ampere, which we will give in Section 31.4). This amounts to an indirect experimental verification of the magnetic-force law given by Eq. (6).

[As we will see in Section 30.4, the magnetic force can be regarded as generated by a transformation of reference frame applied to an electric force. Note that in a reference frame in which one of the two charges of Eq. (6) is at rest, there is no magnetic force. In this reference frame there is only an electric force. When we now transform to a different reference frame, the magnetic force suddenly makes its appearance — the transformation of reference frame generates an extra magnetic force from an electric force. In Section 30.4 we will use the principles of the theory of Special Relativity to derive the magnetic force from the electric force.]

Equation (6) tells us the force exerted by the point charge q' on the point charge q. To find the force exerted by q on q', we must exchange the velocities in Eq. (6) and we must replace $\hat{\mathbf{r}}$ by $\hat{\mathbf{r}}'$ (Figure 30.6; note that $\hat{\mathbf{r}} = -\hat{\mathbf{r}}'$). This gives the magnetic force on q,

$$\mathbf{F}' = \frac{\mu_0}{4\pi}\frac{qq'}{r^2}\mathbf{v}' \times (\mathbf{v} \times \hat{\mathbf{r}}') \tag{11}$$

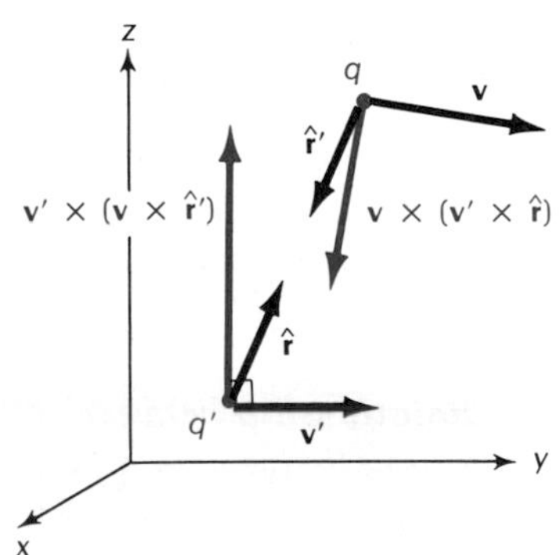

Fig. 30.6 With $\mathbf{v}'$, $\hat{\mathbf{r}}$, and $\mathbf{v}$ as in Figure 30.1, both $\mathbf{v} \times (\mathbf{v}' \times \hat{\mathbf{r}})$ and $\mathbf{v}' \times (\mathbf{v} \times \hat{\mathbf{r}}')$ are in the z–y plane, but they are not of equal magnitudes and opposite directions.

As Figure 30.6 shows, in general the magnetic force $\mathbf{F}'$ exerted by q on q' is *not* opposite and equal to the magnetic force $\mathbf{F}$ exerted on q' by q. For magnetic forces, *Newton's Third Law on the equality of action and reaction fails.* Consequently, the law of conservation of momentum will also fail — when magnetic forces act, the momentum of the system of particles is not constant.

This is a disaster for Newton, but is not a disaster for physics. If we take into account the electric field, we can see why a violation of momentum conservation is quite reasonable for electric charges in motion. We know that in the space between the electric charges there is an electric field, with an energy density. As the charges move, the field continuously changes and the energy of the field must continuously redistribute itself, i.e., the energy of the field must flow through space. We know that energy has mass; hence such a changing field involves a flow of field-mass from one part of space to another. Since any moving mass has momentum, we conclude that the moving, changing electric field has momentum. Once we recognize this, we can see why the momentum of the particles is not conserved. The particles not only exchange momentum with each other, but also with the electric field. What is conserved is not the momentum of the particles, but the total momentum of *particles and field.* Although Newton's Third Law fails, the law of conservation of momentum remains valid. The explicit calculation of the momentum contained in an electric field is fairly complicated and we will not attempt it here.

Incidentally, if the system of charged particles consists of wires with steady currents flowing in closed circuits, then we can prove that the net magnetic force of one circuit on another circuit does satisfy Newton's Third Law of the equality of action and reaction. Obviously, if steady currents are flowing in closed circuits, then the fields are constant in time and their momentum must also remain constant. There is then no possibility of a momentum transfer to the fields and the conservation of momentum would fail if Newton's Third Law did not hold.

Nikola Tesla, *1856–1943, American electrical engineer and inventor. He made many brilliant contributions to high-voltage technology, ranging from new motors and generators to transformers and a system for radio transmission. Tesla designed the power-generating station at Niagara Falls.*

30.2 The Magnetic Field

In Section 23.2 we presented some arguments in favor of the view that the electric force is communicated from one charge to another by action-by-contact, through an electric field. Likewise, the magnetic force is communicated from one charge to another through a **magnetic field**.

The definition of the magnetic field is as follows: To find the magnetic field at some point in the vicinity of moving charges or currents, place a test charge q at that point and give it some velocity $\mathbf{v}$. This charge will then experience a magnetic force depending on its velocity. The magnetic field $\mathbf{B}$ is implicitly defined by the equation for the magnetic force,

Relationship between magnetic force and field

$$\boxed{\mathbf{F} = q\mathbf{v} \times \mathbf{B}} \tag{12}$$

To find the magnitude and direction of the magnetic field at some point, we place a test charge q at this point and launch it with a velocity $\mathbf{v}$ in some direction. By repeating this procedure several times, we discover how the force depends on the direction of the velocity $\mathbf{v}$. In one direction, the force will be zero; this is the direction parallel or antiparallel to $\mathbf{B}$. In the direction perpendicular to this direction, the magnitude of the force will be $F = qvB$, from which we deduce that the magnitude of the magnetic field is $B = F/(qv)$. Hence, in the special case of velocity perpendicular to the magnetic field, the magnitude of the magnetic field is the force per unit charge and unit velocity.

The SI unit of magnetic field is N/(C · m/s); this unit is called the **tesla** (T),

$$1 \text{ tesla} = 1 \text{ T} = 1 \text{ N/(C} \cdot \text{m/s)} \tag{13}$$

tesla, T

An alternative name for this unit is weber/m^2. For weak magnetic fields, a smaller unit is often preferred; this is the **gauss,**

$$1 \text{ gauss} \Leftrightarrow 10^{-4} \text{ T} = 10^{-4} \text{ weber/m}^2 \tag{14}$$

We have written this relationship between gauss and tesla as an equivalence (⇔) rather than as an equality (=) because the gauss is a cgs unit of magnetic field rather than an SI unit; the definitions of magnetic field in the cgs and SI systems are somewhat different and the units do not carry the same dimensions and cannot be equated. Table 30.1 gives the values of some typical magnetic fields.

Karl Friedrich Gauss, *1777–1855, German mathematician, physicist, and astronomer. Gauss was the son of peasants, but his amazing mathematical abilities were recognized at an early age and he was granted a scholarship by the Duke of Göttingen. One of the greatest mathematicians of all time, Gauss' most celebrated work lay in number theory. Gauss was an indefatigable calculator, and he loved to perform enormously complicated computations, which today would be regarded as impossible without an electronic computer. He developed new methods for calculations in celestial mechanics, and successfully predicted the orbit of the asteroid Ceres, briefly seen and then lost by the astronomer G. Piazzi. Later, Gauss became interested in electric and magnetic phenomena, which he researched in collaboration with W. Weber. He also worked on geodetic surveys, and invented the electric telegraph.*

Table 30.1. SOME MAGNETIC FIELDS

At surface of nucleus	$\sim 10^{12}$ T
At surface of pulsar	$\sim 10^{8}$ T
Maximum achieved in laboratory:	
Explosive compression of field lines	1×10^{3} T
Steady	30 T
Large bubble-chamber magnet	2 T
In sunspot	~ 0.3 T
At surface of Sun	$\sim 10^{-2}$ T
Near small ceramic magnet	$\sim 2 \times 10^{-2}$ T
Near household wiring	$\sim 10^{-4}$ T
At surface of Earth	$\sim 5 \times 10^{-5}$ T
In sunlight (rms)	3×10^{-6} T
In Crab nebula	$\sim 10^{-8}$ T
In radio wave (rms)	$\sim 10^{-9}$ T
In interstellar galactic space	$\sim 10^{-10}$ T
Produced by human body	3×10^{-10} T
In shielded antimagnetic chamber	2×10^{-14} T

From Eq. (6) and (12), we see that the magnetic field generated by a point charge q' moving with velocity $\mathbf{v}'$ is

$$\boxed{\mathbf{B} = \frac{\mu_0}{4\pi} \frac{q'}{r^2} (\mathbf{v}' \times \hat{\mathbf{r}})} \tag{15}$$

Magnetic field of point charge

EXAMPLE 2. In a hydrogen atom, an electron moves in a circular orbit of radius 5.3×10^{-11} m. The velocity of the electron is 2.2×10^{6} m/s. What magnetic field does the electron produce at the center of the orbit?

SOLUTION: The velocity $\mathbf{v}'$ and the unit vector $\hat{\mathbf{r}}$ are perpendicular (Figure 30.7). Hence the magnitude of $\mathbf{B}$ is

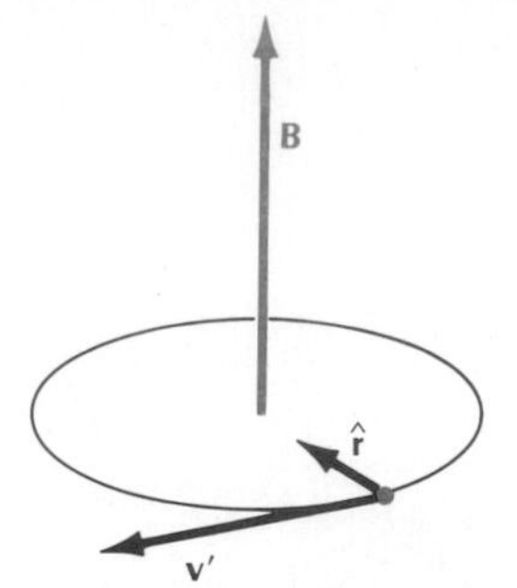

Fig. 30.7 Orbit of an electron around a nucleus, and magnetic field of the electron.

$$B = \frac{\mu_0}{4\pi}\frac{ev'}{r^2}$$

$$= \frac{1.00 \times 10^{-7}\ \text{N}\cdot\text{s/C}^2 \times 1.6 \times 10^{-19}\ \text{C} \times 2.2 \times 10^6\ \text{m/s}}{(5.3 \times 10^{-11}\ \text{m})^2}$$

$$= 13\ \text{T} \qquad (16)$$

The direction of **B** is perpendicular to the plane of the orbit.

The magnetic field can be represented graphically by field lines. As in the case of the electric field, the tangent to the field line indicates the direction of the field and the density of field lines indicates the strength of the field. Figure 30.8 shows the pattern of magnetic field lines around a moving positive charge. The velocity of the charge is directed perpendicularly out of the plane of the page. The pattern consists of concentric circles; the density of lines becomes infinite at the position of the charge. Ahead and behind the charge, the pattern of field lines also consists of concentric circles, but without the strong concentration near the center. Note that the direction of the field lines follows a simple right-hand rule: place the thumb along the velocity of the charge; the fingers will then curl around in the direction of the field lines. For a negative charge, the direction of the field lines in Figure 30.8 must of course be reversed.

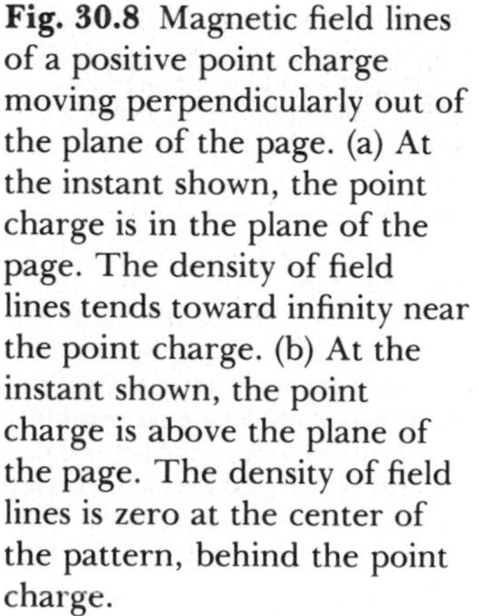

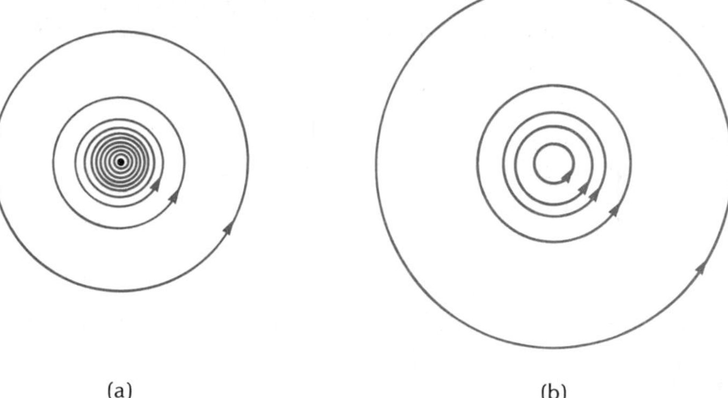

Fig. 30.8 Magnetic field lines of a positive point charge moving perpendicularly out of the plane of the page. (a) At the instant shown, the point charge is in the plane of the page. The density of field lines tends toward infinity near the point charge. (b) At the instant shown, the point charge is above the plane of the page. The density of field lines is zero at the center of the pattern, behind the point charge.

For computations with field lines we will adopt the convention that the number of lines per unit area equals the magnitude of the magnetic field. However, for the purpose of making drawings, this normalization is sometimes unwieldy and we will alter it in case of need.

The magnetic field lines of a moving charge form closed loops, i.e., the magnetic field lines do not begin or end anywhere in the way that the electric field lines begin and end on positive and negative electric charges. This indicates that there is no "magnetic charge" that acts as source or sink of magnetic field lines in the way that electric charge acts as source or sink of electric field lines.

Mathematically, we can express these features of the magnetic field lines in terms of a modified version of Gauss's Law. Consider a closed surface of arbitrary shape. The number of magnetic field lines that enter this surface is exactly equal to the number that leave, i.e., the magnetic flux through the closed surface is zero:

$$\boxed{\oint \mathbf{B} \cdot d\mathbf{S} = 0} \tag{17}$$

Gauss' Law for magnetic field

This equation is not only true for the magnetic field of a single moving charge, but also for any arbitrary magnetic field produced by any arbitrary number of moving charges. The net magnetic field produced by many charges acting together is the sum of their individual magnetic fields and since each of these individual fields satisfies Eq. (17), the net field will also.

Note that the argument of the preceding paragraph hinges on the **principle of linear superposition** for the magnetic field: if several moving charges $q_1, q_2, q_3, \ldots$, simultaneously generate magnetic fields $\mathbf{B}_1, \mathbf{B}_2, \mathbf{B}_3, \ldots$, then the net magnetic field generated by all these charges acting together is

Principle of linear superposition

$$\mathbf{B} = \mathbf{B}_1 + \mathbf{B}_2 + \mathbf{B}_3 + \cdots \tag{18}$$

We will exploit this superposition principle in the next section when we calculate the magnetic field of a current, that is, the magnetic field of a large number of point charges moving along a wire.

Jean Baptiste Biot (bio), *1774–1862, French physicist, professor at the Collège de France. His most important work dealt with the refraction and polarization of light, but he was also interested in a broad range of problems in the physical sciences. With Félix Savart, 1791–1841, he confirmed Oersted's discovery of magnetic fields generated by electric currents, and formulated the equation (30.25) for the strength of the magnetic field.*

30.3 The Biot–Savart Law

The magnetic field generated by a current is of much greater practical interest than the magnetic field generated by a single point charge. Figure 30.9 shows a thin wire of arbitrary shape carrying a steady current I. Following the usual convention, we will pretend that the current is due to a flow of positive charge. Consider a small segment dl of this wire. We can regard the moving charge dq' within dl as a point charge which produces a magnetic field

$$d\mathbf{B} = \frac{\mu_0}{4\pi} \frac{dq'}{r^2} (\mathbf{v}' \times \hat{\mathbf{r}}) \tag{19}$$

with a magnitude

$$dB = \frac{\mu_0}{4\pi} \frac{dq'}{r^2} v' \sin \theta \tag{20}$$

where θ is the angle between $\mathbf{v}'$ and $\hat{\mathbf{r}}$, i.e., it is the angle between the wire and $\hat{\mathbf{r}}$ (Figure 30.9). To relate dq' to the current, we begin with the definition of the current

$$dq' = I\, dt \tag{21}$$

Here dt is the time it takes the charge dq' to flow out of the small segment dl, i.e.,

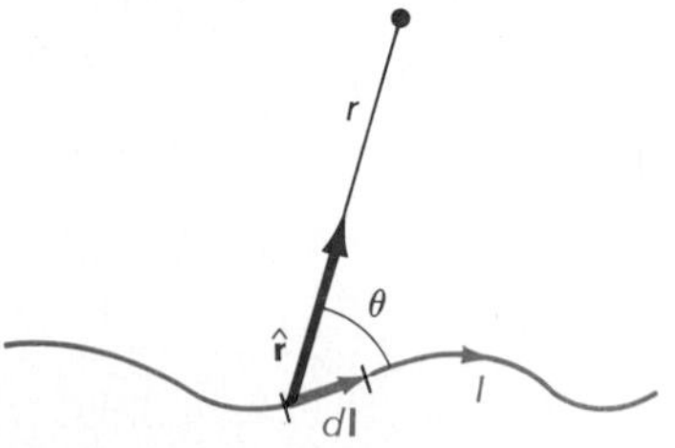

Fig. 30.9 Wire carrying a current I.

$$dt = dl/v' \tag{22}$$

Hence

$$dq' = I\, dl/v' \tag{23}$$

With this, Eq. (20) becomes

$$dB = \frac{\mu_0}{4\pi} \frac{I\, dl \sin\theta}{r^2} \tag{24}$$

To put this in vector form, we treat the segment of wire as a vector $d\mathbf{l}$ tangent to the wire and in the direction of the current (Figure 30.9). The magnetic field (19) can then be expressed as

Biot–Savart law

$$d\mathbf{B} = \frac{\mu_0}{4\pi} \frac{I\, d\mathbf{l} \times \hat{\mathbf{r}}}{r^2} \tag{25}$$

This equation holds true because the right side has the correct magnitude [compare with Eq. (24)] and it also has the correct direction [compare with Eq. (19) noting that $\mathbf{v}'$ and $d\mathbf{l}$ have the same direction].

Equation (25) gives us the magnetic field generated by a short segment of a wire. It is called the **law of Biot–Savart.** The magnetic field generated by a wire of any length and shape can be calculated by integrating this equation along the wire.

EXAMPLE 3. A very long and very thin straight wire carries a steady current I. What is the magnetic field at some distance from the wire?

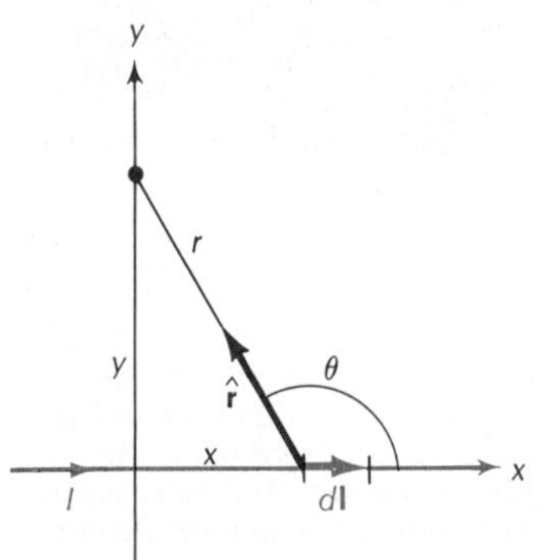

Fig. 30.10 Straight wire carrying a current I along the x axis.

SOLUTION: Figure 30.10 shows the wire lying along the x axis. The magnetic field lines are concentric circles around the wire; the lines come out of the plane of the page in the region above the wire and go into the plane of the page in the region below. We will calculate the magnetic field at a point which is at a distance y from the wire.

A small segment dx of the wire contributes a magnetic field of magnitude

$$dB = \frac{\mu_0}{4\pi} \frac{I\, dx \sin\theta}{r^2} \tag{26}$$

The directions of the contributions from all segments dx are parallel (at the point P the directions of all the contributions to the magnetic field are out of the plane of the page). Hence the magnitude of the net field is

$$B = \int_{-\infty}^{\infty} \frac{\mu_0}{4\pi} \frac{I \sin\theta}{r^2}\, dx \tag{27}$$

From Figure 30.10,

$$x = y \cot(\pi - \theta) = -y \cot\theta \tag{28}$$

so that

$$dx = y \csc^2\theta\, d\theta \tag{29}$$

Furthermore,

$$r = y/\sin(\pi - \theta) = y/\sin\theta \tag{30}$$

By means of Eqs. (29) and (30) we can change the variable of integration from x to θ; the new limits of integration are then $\theta = 0°$ and $\theta = 180°$, or, in radians, $\theta = 0$ and $\theta = \pi$,

$$B = \frac{\mu_0}{4\pi}\frac{I}{y}\int_0^{\pi} \sin\theta\, d\theta \tag{31}$$

$$= \frac{\mu_0}{4\pi}\frac{I}{y}\Big[-\cos\theta\Big]_0^{\pi} = \frac{\mu_0}{4\pi}\frac{I}{y} \times 2$$

which yields

$$\boxed{B = \frac{\mu_0}{2\pi}\frac{I}{y}} \tag{32}$$

Magnetic field of straight wire

Figure 30.11 shows the pattern of the magnetic field lines for this magnetic field. And Figure 30.12 shows iron filings sprinkled on a sheet of paper placed around a long straight wire carrying a strong current. The iron filings align in the direction of the magnetic field, and therefore make the pattern of field lines visible.

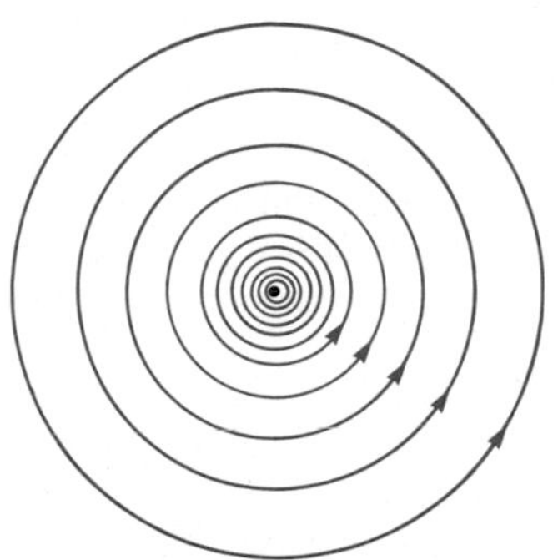

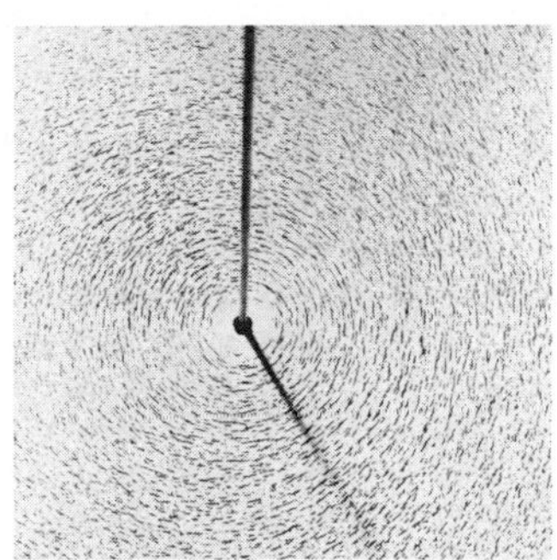

Fig. 30.11 (left) Magnetic field lines around a straight wire carrying a current. The wire is perpendicular to the plane of the page and the current emerges from this plane.

Fig. 30.12 (right) Magnetic field lines around a straight wire, made visible by iron filings sprinkled on a sheet of paper.

EXAMPLE 4. A square loop of wire of dimension $L \times L$ carries a current I. What is the magnetic field at the center of the loop?

SOLUTION: Figure 30.13a shows the loop. Each side of the loop makes the same contribution to the magnetic field, in magnitude and in direction. Hence the net magnetic field at the center is four times the magnetic field contributed by one side.

Figure 30.13b shows one side of the loop. The calculation of the magnetic field of this side is similar to the calculation of the magnetic field of the infi-

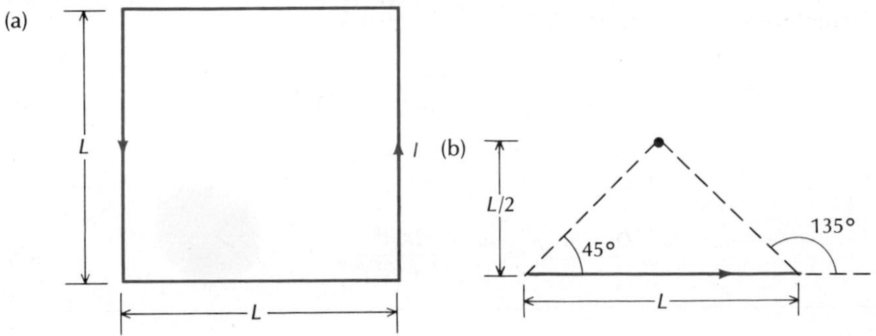

Fig. 30.13 A square loop carrying a current I.

nitely long wire, but the limits for the integration over x are now finite. In terms of the angle θ introduced in Eq. (28), the lower limit is $\theta = 45°$ (or $\theta = \pi/4$) and the upper limit is $\theta = 135°$ (or $\theta = 3\pi/4$). With these new limits, the integral in Eq. (31) becomes

$$B_1 = \frac{\mu_0}{4\pi}\frac{I}{L/2}\int_{\pi/4}^{3\pi/4} \sin\theta\, d\theta$$

$$= \frac{\mu_0}{4\pi}\frac{I}{L/2}\Big[-\cos\theta\Big]_{\pi/4}^{3\pi/4}$$

$$= \frac{\mu_0}{4\pi}\frac{I}{L/2}\sqrt{2} \tag{33}$$

Multiplying this by 4, we obtain the net magnetic field of the entire loop,

$$B = 4B_1 = \frac{\mu_0}{4\pi}\frac{8\sqrt{2}\,I}{L} \tag{34}$$

EXAMPLE 5. Find the magnetic field on the axis of a circular loop of wire of radius R carrying a current I.

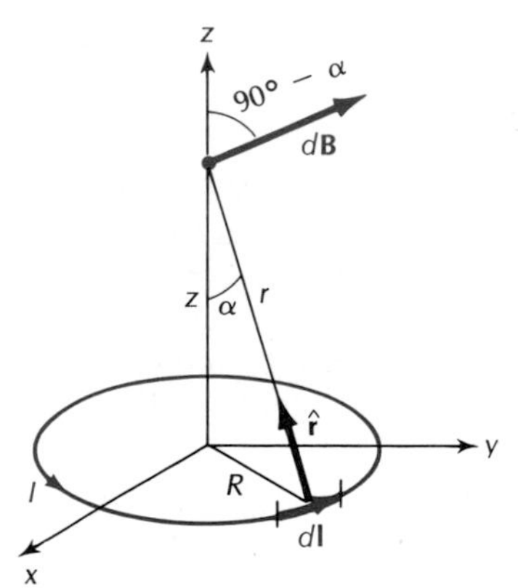

Fig. 30.14 Circular wire loop carrying a current I in the x–y plane.

SOLUTION: Figure 30.14 shows the loop and a point on its axis. A small segment dl of the ring produces a magnetic field of magnitude

$$dB = \frac{\mu_0}{4\pi}\frac{I\,dl}{z^2 + R^2} \tag{35}$$

The direction of this magnetic field is perpendicular to the line connecting dl and the point on the axis (Figure 30.14). Upon integration around the circle, the horizontal components of the magnetic field cancel because diametrically opposite segments dl contribute opposite horizontal components. Only the vertical component of **B** survives.

The vertical component of the magnetic field in Eq. (35) is

$$dB_z = dB\cos(90° - \alpha) = dB\sin\alpha = \frac{\mu_0}{4\pi}\frac{I\,dl}{z^2 + R^2}\sin\alpha \tag{36}$$

where, according to Figure 30.14,

$$\sin\alpha = R/\sqrt{z^2 + R^2} \tag{37}$$

Hence

$$B_z = \int \frac{\mu_0}{4\pi}\frac{I\,R\,dl}{(z^2 + R^2)^{3/2}} \tag{38}$$

Since all the terms in the integrand are constant, the integration amounts to multiplication of the integrand by the length of the path of integration (the circumference of the circle). Thus,

$$B_z = \frac{\mu_0}{4\pi}\frac{IR}{(z^2 + R^2)^{3/2}} \times 2\pi R \tag{39}$$

or

$$B_z = \frac{\mu_0}{2\pi}\frac{I\pi R^2}{(z^2 + R^2)^{3/2}} \tag{40}$$

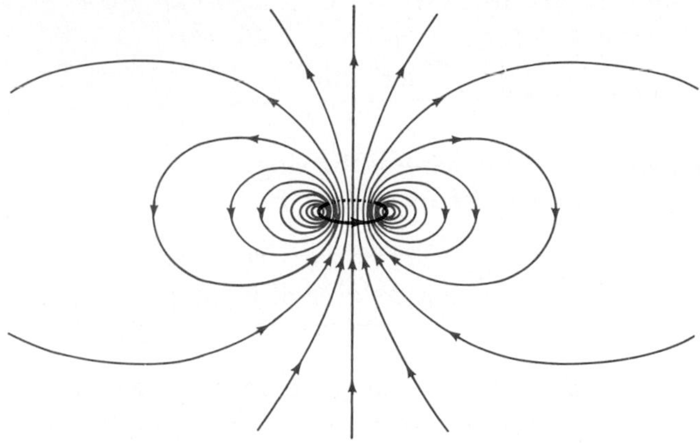

Fig. 30.15 Magnetic field lines of a circular loop of wire carrying a current.

The calculation of Example 5 only gives the magnetic field on the axis of the loop. The calculation of the magnetic field at other points is rather messy. Figure 30.15 shows the general pattern of magnetic field lines throughout space. Note that at large distance from the loop this pattern is quite similar to the pattern of electric field lines of an electric dipole (see Figure 25.11). Figure 30.16 shows the pattern of magnetic field lines, as made visible with iron filings.

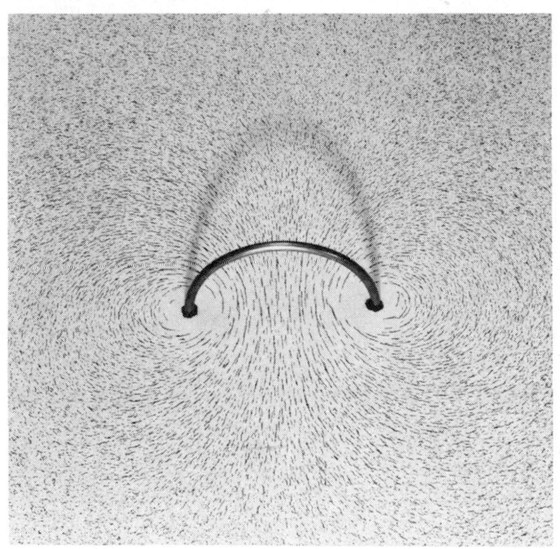

Fig. 30.16 Iron filings sprinkled on a sheet of a paper placed around a loop carrying a strong current.

Magnetic dipole

A small loop of current is called a **magnetic dipole**. In this context, *small* means that the size of the loop is very small compared to the distance at which we wish to calculate the magnetic field; in Eq. (40) it means $R \ll z$. We can then approximate the expression (40) by

$$B_z = \frac{\mu_0}{2\pi} \frac{I\pi R^2}{z^3} \tag{41}$$

This shows that the magnetic field of a loop of current decreases as the inverse cube of the distance. It is customary to write Eq. (41) as

$$B_z = \frac{\mu_0}{2\pi} \frac{\mu}{z^3} \tag{42}$$

where

$$\mu = I\pi R^2 \tag{43}$$

is the **magnetic dipole moment** of the ring.[3]

Note that Eq. (43) has the form

Magnetic dipole moment

$$\boxed{\mu = [\text{current}] \times [\text{area of loop}]} \tag{44}$$

This turns out to be a general expression for the magnetic dipole moment of a (plane) loop of arbitrary shape. For instance, the magnetic dipole moment of a square or rectangular loop of current can be calculated from Eq. (44) and the magnetic field at a large distance from the loop is then given by Eq. (42).

Electrons, protons, and many other elementary particles have magnetic dipole moments. These particles may be regarded crudely as small balls of electric charge spinning about an axis. Such a rotating charge distribution behaves like a collection of rings of current. This generates a magnetic field which, at a large distance from the particle, is essentially the dipole field of Figure 30.15. The dipole moment of an electron is $9.3 \times 10^{-24}\ \text{A} \cdot \text{m}^2$. Thus, the space surrounding an electron is filled not only with electric fields, but also with magnetic fields. Figure 30.17 shows the electric and magnetic fields of an electron. Of course, if the electron also has a translational motion, then there will be an extra magnetic field of the type given by Eq. (15) around the electron.

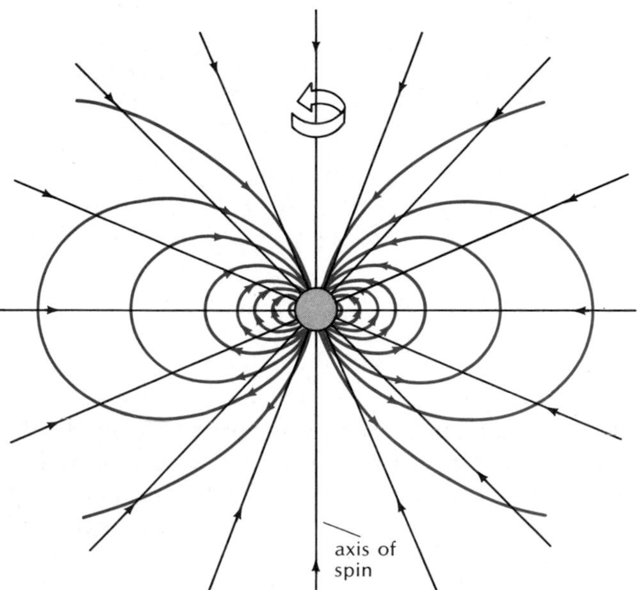

Fig. 30.17 Electric (black) and magnetic (colored) fields of an electron.

The Earth also has a magnetic dipole moment. This is presumably due to currents that flow around in loops deep inside the Earth, in the liquid iron of the core. Geophysicists do not yet know exactly what is

[3] The μ in Eq. (42) must not be confused with the μ_0 for permeability; they are not related.

the energy supply for the emf that drives these currents, but it is certain that the rotation of the Earth plays a crucial role in the generation of the currents. These currents create a magnetic field which, at large distances from the core, is nearly a dipole field. The magnetic moment of the Earth is $8.0 \times 10^{22}\ \mathrm{A \cdot m^2}$. Figure 30.18 shows the magnetic field lines. Note that the field lines emerge from the surface of the Earth near the geographic South Pole and they reenter the surface near the geographic North Pole. The force experienced by a compass needle is due to this magnetic field.

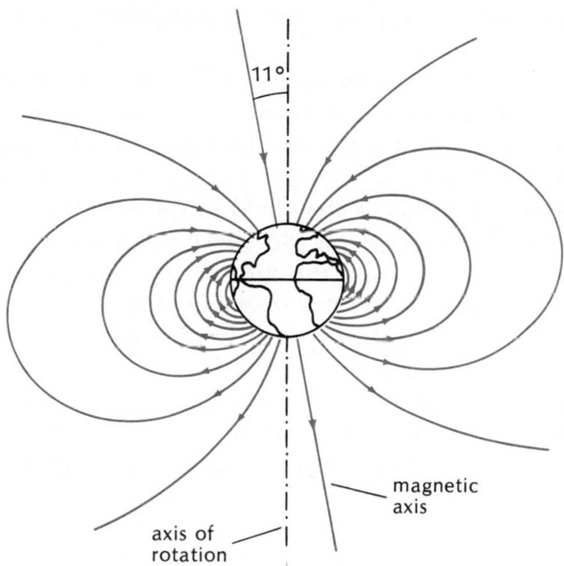

Fig. 30.18 Magnetic field of the Earth. The axis of the magnetic dipole makes an angle of 11° with the axis of rotation of the Earth.

30.4 Relativity and the Magnetic Field[4]

Since the magnetic force, $\mathbf{F} = q\mathbf{v} \times \mathbf{B}$, on a particle depends on the velocity, this force must necessarily depend on the reference frame with respect to which this velocity is reckoned. The following example illustrates this dependence in a drastic way.

Suppose that a positive charge q moves in the magnetic field generated by a current on a straight wire. We will assume that the (instantaneous) velocity $\mathbf{v}$ of the charge q is parallel to the wire. Figure 30.19 shows the situation in a reference frame in which the wire is at rest. The charge has a velocity $\mathbf{v}$ toward the right and the wire carries a current I toward the left. In this reference frame the current generates a magnetic field [see Eq. (32)]

$$B = \frac{\mu_0}{2\pi} \frac{I}{y} \tag{45}$$

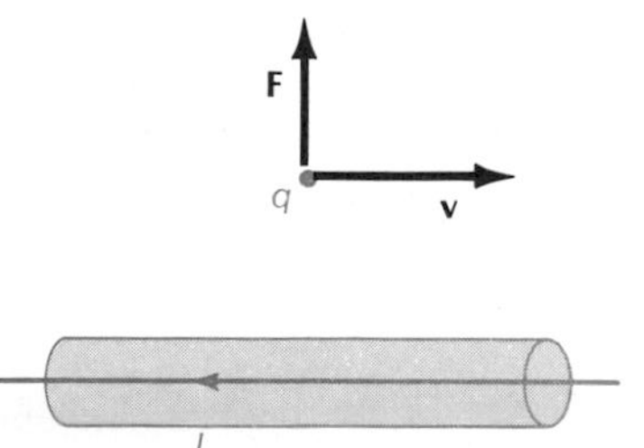

Fig. 30.19 In the rest frame of the wire, the charge q moves to the right with velocity $\mathbf{v}$.

and this magnetic field exerts a magnetic force

$$F = qvB = \frac{\mu_0}{2\pi} \frac{qvI}{y} \tag{46}$$

[4] This section is optional.

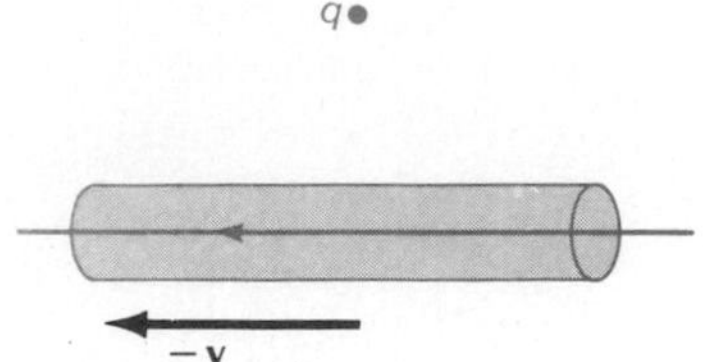

Fig. 30.20 In the rest frame of the charge q, the wire moves to the left with velocity $-\mathbf{v}$.

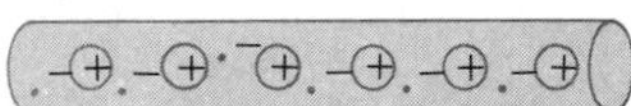

Fig. 30.21 In the reference frame of the wire, the positive and the negative charge densities on the wire are equal.

on the charge q. The force points radially away from the wire (Figure 30.19).

Next, let us examine the situation in a new reference frame in which the charge q is (instantaneously) at rest. Relative to the old reference frame, this new reference frame moves toward the right with velocity $\mathbf{v}$. Figure 30.20 shows the situation in this reference frame. The charge q is at rest and the wire moves toward the left with velocity $-\mathbf{v}$. In this new reference frame there can be no magnetic force on the charge q, since the velocity of the charge is zero. We are now faced with a paradox: in the old reference frame the charge experiences a magnetic force and hence an acceleration; in the new reference frame the charge experiences no magnetic force, hence no acceleration — and yet accelerations are supposed to be independent of the reference frame!

The resolution of this paradox hinges on a subtle relativistic effect. It turns out that in the new reference frame, the wire generates an *electric field* and the corresponding electric force on the charge gives it the required acceleration.

To understand where this electric field in the new reference frame comes from, let us begin by asking why there is no electric field in the old reference frame. Obviously the answer is that in the reference frame of the wire the positive and negative charge densities on the wire are of equal magnitude; hence their electric fields cancel exactly.[5] Figure 30.21 shows these charge distributions in the reference frame of the wire. The negative charge density is due to the free electrons carrying the current; the positive charge density is due to the positive ions fixed in the lattice of the wire. The charge density of the electrons is $-\lambda$ coulomb/m and that of the ions is $+\lambda$ coulomb/m.

In the reference frame of the wire, the negative charges are in motion and the positive charges are at rest. Since the current is flowing toward the left, the free electrons carrying this current must move toward the right. For the sake of simplicity, let us assume that the velocity of the free electrons coincides with the velocity of the charge q. This is a very special case — and not very likely to happen in reality. It is, of course, possible to solve the problem in general, but it greatly helps in the solution if we have to worry about only a single velocity rather than two different velocities.

Now consider the electric charge densities in the new reference frame. The crucial point is that in this reference frame the negative and positive densities will *not be equal.* The inequality arises from the length contraction effect of special relativity. We recall from Section 17.5 that if an object has a certain length in its own reference frame, then in any other reference frame the length is shorter by a factor

$$\sqrt{1 - v^2/c^2}$$

where c is the speed of light. Thus, if a given number of positive charges sitting on the wire occupy a length of 1 m in the old reference frame (rest frame of the wire), they will occupy a shorter length of

$$1 \text{ m} \times \sqrt{1 - v^2/c^2}$$

[5] For the present purposes, we ignore the small electric field needed to push the current along the wire.

in the new reference frame. Correspondingly, the density of these positive charges will be larger: if the density is λ coulomb/m in the old reference frame, it will be

$$\lambda/\sqrt{1-v^2/c^2} \text{ coulomb/m}$$

in the new reference frame. For the negative charge distribution, the length contraction has the opposite effect — the density of negative charges will be smaller: if the density of charge is $-\lambda$ coulomb/m in the old reference frame, it will be

$$-\lambda\sqrt{1-v^2/c^2} \text{ coulomb/m}$$

in the new reference frame. This is so because in the old reference frame (rest frame of the wire) the charge distribution of electrons is in motion; it therefore is a *contracted* charge distribution. In the new reference frame, the charge distribution is at rest and it is not contracted. The transformation from the old to the new reference frame therefore is a transformation from a reference frame in which the length of the negative charge distribution is already contracted to a reference frame in which it is not contracted. Correspondingly, the density of the negative charges will be smaller, as indicated above.

In the new reference frame, the net charge per unit length of the wire is then the sum of *unequal* positive and negative contributions,

$$\lambda_{\text{new}} = \frac{\lambda}{\sqrt{1-v^2/c^2}} - \lambda\sqrt{1-v^2/c^2} \tag{47}$$

If the speeds are small compared to the speed of light, we can use the approximations

$$1/\sqrt{1-v^2/c^2} \cong 1 + \tfrac{1}{2}v^2/c^2 \quad \text{and} \quad \sqrt{1-v^2/c^2} \cong 1 - \tfrac{1}{2}v^2/c^2$$

so that Eq. (47) becomes

$$\lambda_{\text{new}} = \frac{\lambda}{\sqrt{1-v^2/c^2}} - \lambda\sqrt{1-v^2/c^2} \cong \lambda v^2/c^2 \tag{48}$$

Such a charge density along the wire will generate a radial electric field [see Eq. (24.10)]

$$E = \frac{1}{2\pi\varepsilon_0}\frac{\lambda_{\text{new}}}{y} = \frac{1}{2\pi\varepsilon_0 c^2}\frac{\lambda v^2}{y} \tag{49}$$

This electric field exerts an electric force

$$F_{\text{new}} = \frac{1}{2\pi\varepsilon_0 c^2}\frac{q\lambda v^2}{y} \tag{50}$$

on the charge q. This force points away from the wire (Figure 30.22).

In order to compare this electric force in the new reference frame with the magnetic force in the old reference frame, we note that the product of the velocity v of the electrons and their charge density λ is the current I on the wire. Hence Eq. (50) can be written

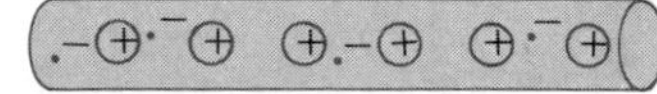

Fig. 30.22 In the reference frame of the charge q, the positive charge density on the wire exceeds the negative charge density.

$$F_{\text{new}} = \frac{1}{2\pi\varepsilon_0 c^2}\frac{qvI}{y} \tag{51}$$

Now examine the ratio of the forces in the old and the new reference frames,

$$\frac{F}{F_{\text{new}}} = \left(\frac{\mu_0}{2\pi}\frac{qvI}{y}\right) \Big/ \left(\frac{1}{2\pi\varepsilon_0 c^2}\frac{qvI}{y}\right) = \mu_0\varepsilon_0 c^2 \tag{52}$$

Inserting numerical values for the constants on the right side of Eq. (52), we find

$$\frac{F}{F_{\text{new}}} = 1.26 \times 10^{-6}\,\frac{\text{N}\cdot\text{s}^2}{\text{C}^2} \times 8.85 \times 10^{-12}\,\frac{\text{C}^2}{\text{N}\cdot\text{m}^2} \times (3.0 \times 10^8\ \text{m/s})^2$$

$$= 1.0 \tag{53}$$

that is, the forces are exactly equal. Note that this result hinges on the fact that the values of ε_0 and μ_0 are exactly right to cancel the factor of c^2 in Eq. (52); again, as in Section 30.1, we see that there is a deep connection between electricity, magnetism, and the speed of light.

What we conclude from the above calculation is then the following: the force that the charges on the wire exert on the positive point charge q is the same in both the old and the new reference frames. The transformation of reference frame does not change the magnitude or direction of the force (and of the acceleration) — it only changes the character of the force from purely magnetic to purely electric. Incidentally, in a reference frame moving toward the right with a speed of, say, $\frac{1}{2}v$, the force would still have the same magnitude and direction, but it would be partially magnetic and partially electric.

Although we obtained these results only for a very special and simple case, the main features have general validity. If a particle experiences a magnetic force in a given reference frame, a transformation to the rest frame of the particle will make this magnetic force disappear. But in the latter reference frame, charge distributions will appear at the locations of the currents, and the electric force of these charge distributions will replace the original magnetic force. The net force is the same in both reference frames.[6]

Electric and magnetic forces and fields transform into one another if we change the frame of reference. In the rest frame of a charged particle, only electric forces act on the particle. Hence we can regard the magnetic forces that act on the particle in any other reference frame as resulting from a transformation of the electric forces in the rest frame. In this sense, magnetic forces can be regarded as a consequence of electric forces and of relativity.

[6] This invariance of the force is true only if the relative speed of the reference frames is low ($v \ll c$). If the speed is high, then we must take into account the relativistic transformation law for force. It turns out that the transformation of electric and magnetic forces is consistent with that law.

SUMMARY

Magnetic force between moving point charges:

$$\mathbf{F} = \frac{\mu_0}{4\pi}\frac{qq'}{r^2}\,\mathbf{v}\times(\mathbf{v}'\times\hat{\mathbf{r}})$$

Magnetic field of point charge:

$$\mathbf{B} = \frac{\mu_0}{4\pi}\frac{q'}{r^2}(\mathbf{v}'\times\hat{\mathbf{r}})$$

Permeability constant: $\mu_0 = 1.26\times10^{-6}\ \mathrm{N\cdot s^2/C^2}$

$$\frac{\mu_0}{4\pi} = 1.00\times10^{-7}\ \mathrm{N\cdot s^2/C^2}$$

Definition of magnetic field: $\mathbf{F} = q\mathbf{v}\times\mathbf{B}$

Gauss' law for magnetism: $\oint \mathbf{B}\cdot d\mathbf{S} = 0$

Biot–Savart law: $d\mathbf{B} = \frac{\mu_0}{4\pi} I \frac{d\mathbf{l}\times\hat{\mathbf{r}}}{r^2}$

Magnetic dipole moment: $\mu = [\text{current}]\times[\text{area of loop}]$

QUESTIONS

1. List all the physical laws of force you can think of. Which of these laws are fundamental, and which can be derived from others?

2. Give an example of two moving charged particles with velocities such that the mutual magnetic forces do not obey Newton's Third Law. Give an example with velocities such that the mutual magnetic forces do obey Newton's Third Law.

3. How would the magnetic field lines shown in Figure 30.8 differ if the charge of the particle were negative instead of positive?

4. Theoretical physicists have proposed the existence of **magnetic monopoles**, which are sources and sinks of magnetic field lines, just as electric charges are sources and sinks of electric field lines. What would the pattern of magnetic field lines of a positive magnetic monople look like? Can you guess the pattern of *electric* field lines of a moving magnetic monopole? *Magnetic monopole*

5. The Earth's magnetic field at the equator is horizontal, in the northward direction. What is the direction of the magnetic force on an electron moving vertically up?

6. An electron with a vertical velocity passes through a magnetic field without suffering any deflection. What can you conclude about the magnetic field?

7. At an initial time, a charged particle is at some point P in a magnetic field and it has an initial velocity. Under the influence of the magnetic field, the particle moves to a point P'. If you now reverse the velocity of the particle, will it retrace its orbit and return to the point P?

8. An electron moving northward in a region of space is deflected toward the east by a magnetic field. What is the direction of the magnetic field?

9. A Faraday cage shields electric fields. Does it also shield magnetic fields?

10. Strong electric fields are hazardous — if you place some part of your body in a strong electric field you are likely to receive an electric shock. Are strong magnetic fields hazardous? Do they produce any effect on your body?

11. Figure 30.18 shows the magnetic field of the Earth. What must be the direction of the currents flowing in loops inside the Earth to give this magnetic field?

12. The needle of an ordinary magnetic compass indicates the direction of the horizontal component of the Earth's magnetic field. Explain why the magnetic compass is unreliable when used near the poles of the Earth.

Dip needle

13. A **dip needle** is a compass needle that swings about a horizontal axis. If the axis is oriented east–west, then the equilibrium direction of the dip needle is the direction of the Earth's magnetic field. The **dip angle** of the dip needle is the angle that it makes with the horizontal. How does the dip angle vary as you transport a dip needle along the surface of the Earth from the South Pole to the North Pole?

14. Suppose we replace the single loop shown in Figure 30.14 by a coil of N loops. How does this change the formula [Eq. (40)] for the magnetic field?

15. In order to eliminate or reduce the magnetic field generated by the pair of wires that connect a piece of electric equipment to an outlet, a physicist twists these wires tightly about each other. How does this help?

16. Consider a circular loop of wire carrying a current. Describe the direction of the magnetic field at different points in the plane of the loop, both inside and outside of the loop.

17. An infinite flat conducting sheet lies in the x–y plane. The sheet carries a current in the y direction; this current is uniformly distributed over the entire sheet. What is the direction of the magnetic field above the sheet? Below the sheet?

18. Suppose that an infinitely long straight wire lies along the axis of a circular loop. Both the wire and the loop carry currents I. Draw some field lines of the net magnetic field of the wire and the loop.

19. According to Eq. (42), the magnetic field at large distance from a square loop is approximately the same as the magnetic field of a circular loop of the same area. Why is the shape unimportant? [Hint: The circular loop can be regarded as made of many small (infinitesimal) square loops.]

20. The arguments of Section 30.4 show that the Eq. (51) for the magnetic force exerted on a charged particle by the current of a straight wire is a direct consequence of the relativistic length contraction. Can we therefore regard the experimental verification of this force law as an experimental verification of the length contraction?

PROBLEMS

Section 30.1

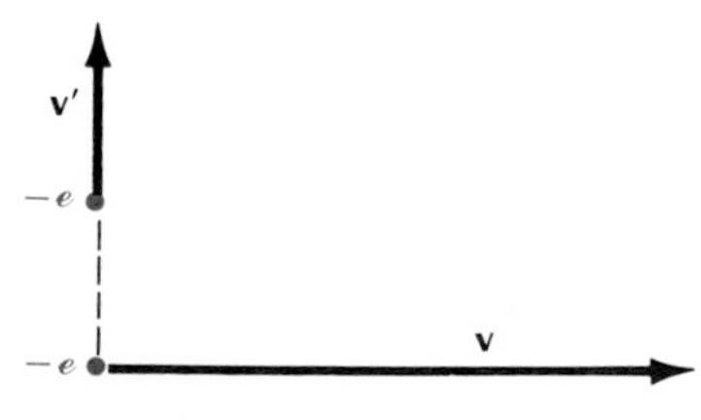

Fig. 30.23 Instantaneous positions and velocities of two electrons.

1. A positron (charge $+e$) and an electron (charge $-e$) are moving side by side with a uniform velocity of 2.0×10^6 m/s along parallel tracks. Is the magnetic force between them attractive or repulsive? By what percentage is the total force (electric and magnetic) acting between these moving charges larger or smaller than that acting between charges at rest at the same distance?

2. Two electrons are (instantaneously) moving at right angles with speeds $v = 3.0 \times 10^6$ m/s and $v' = 1.0 \times 10^6$ m/s in the same plane (Figure 30.23). Their distance is 0.80×10^{-10} m. Calculate the magnetic force of the first electron on the second and that of the second on the first. Show the directions of the forces on a diagram.

3. An electron (charge $-e$) and an antielectron (charge $+e$) are in a circular orbit about each other (Figure 30.24). The orbital radius is 0.53×10^{-10} m and the orbital speed is 1.1×10^6 m/s.
   (a) What is the magnitude of the electric force on each particle? Draw a diagram showing the direction of the electric force on each particle.
   (b) What is the magnitude of the magnetic force on each particle?

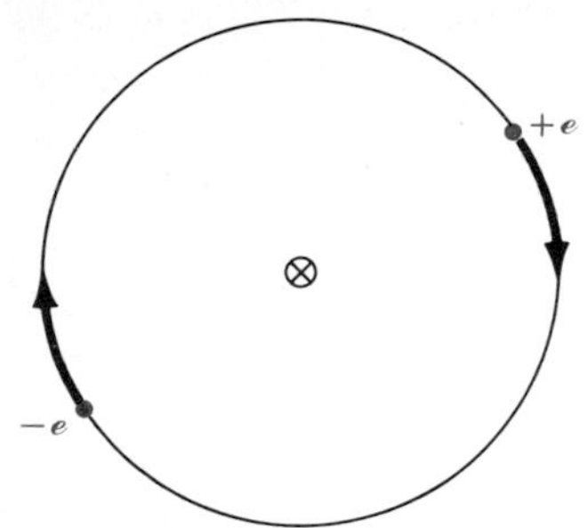

Fig. 30.24 Electron and antielectron in orbit about their common center of mass.

4. Two protons have equal speeds of 2.0×10^6 m/s. Their directions of motion are in the same plane and they make an angle of 60°. The distance between the protons is 2.4×10^{-10} m, and this distance is (instantaneously) perpendicular to the velocity $\mathbf{v}$ of one of the protons.
   (a) Calculate the magnetic force on each proton due to the other proton; show the directions of the forces on a diagram.
   (b) Calculate the instantaneous rate of change of the total momentum of the two-proton system.

Section 30.2

5. Show that the magnetic field of a point charge q' moving with a velocity $\mathbf{v}'$ [Eq. (15)] can be written in terms of the electric field of this point charge as $\mathbf{B} = \mu_0 \varepsilon_0 \mathbf{v}' \times \mathbf{E}$.

6. The magnetic field surrounding the Earth typically has a strength of 5×10^{-5} T. Suppose that a cosmic-ray electron of energy 3×10^4 eV is instantaneously moving in a direction perpendicular to the lines of this magnetic field. What is the force on this electron?

7. At a location where the strength of the Earth's magnetic field is 0.60×10^{-4} T, what must be the minimum speed of an electron if the magnetic force on it is to exceed its weight?

8. At the surface of a pulsar, or neutron star, the magnetic field may be as strong as 10^8 T. Consider the electron in a hydrogen atom on the surface of such a neutron star. The electron is at a distance of 0.53×10^{-10} m from the proton and has a speed of 2.2×10^6 m/s. Compare the electric force that the proton exerts on the electron with the magnetic force that the magnetic field of the neutron star exerts on the electron. Is it reasonable to expect that the hydrogen atom will be strongly deformed by the magnetic field?

9. In New York, the magnetic field of the Earth has a vertical (down) component of 0.60×10^{-4} T and a horizontal (north) component of 0.17×10^{-4} T. What are the magnitude and direction of the magnetic force on an electron of velocity 1.0×10^6 m/s moving (instantaneously) in an east to west direction in a television tube?

Section 30.3

10. The current in a lightning bolt may be as much as 2×10^4 A. What is the magnetic field at a distance of 1.0 m from a lightning bolt? The bolt can be regarded as a straight line of current.

11. The cable of a high-voltage power line is 25 m above the ground and carries a current of 1.8×10^3 A.
   (a) What magnetic field does this current produce at the ground?
   (b) The strength of the magnetic field of the Earth is 0.60×10^{-4} T at the location of the power line. By what factor do the fields of the power line and of the Earth differ?

12. In a motorboat, the compass is mounted at a distance of 0.80 m from a cable carrying a current of 20 A from an electric generator to a battery.
   (a) What magnetic field does this current produce at the location of the compass? Treat the cable as a long, straight wire.
   (b) The horizontal (north) component of the Earth's magnetic field is 0.18×10^{-4} T. Since the compass points in the direction of the net horizontal magnetic field, the current will cause a deviation of the compass.

The deviation will be maximum if the magnetic field of the current is horizontal and at right angles to the Earth's magnetic field. Under these worst circumstances, by how many degrees will the compass deviate from north?

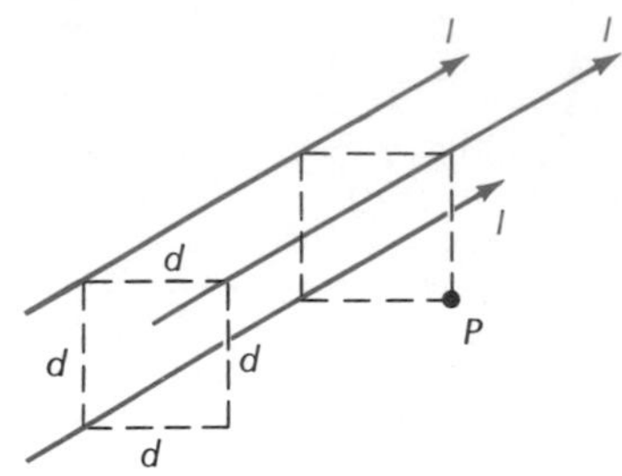

Fig. 30.25 Three long, parallel wires. The wires pass through three corners of a square.

13. A superconducting ring of diameter 3.0 cm carries a current of 12 A. What is the strength of the magnetic field at the center of the ring? Along the axis of the ring at a distance of 3.0 cm from the center?

14. A circular ring of wire of diameter 0.60 m carries a current of 35 A. What acceleration will the magnetic force generated by this ring give to an electron that is passing through the center of the ring with a velocity of 1.2×10^6 m/s in the plane of the ring?

15. Two very long, straight, parallel wires separated by distance d carry currents of magnitude I in opposite directions. Find the magnetic field at a point equidistant from the lines, with a distance $2d$ from each line. Draw a diagram showing the direction of the magnetic field.

16. Three parallel wires are spaced as shown in Figure 30.25. The wires carry equal currents in the same direction. What is the magnetic field at the point P? Draw a diagram giving the direction of the magnetic field.

17. Two very long parallel wires separated by a distance of 1.0 cm carry opposite currents of 8.0 A.

(a) Find the magnetic field at the midpoint between the wires.
(b) Find the magnetic field in the plane of the wires, at a distance of 2.0 cm from the midline.

18. A very long wire is bent at a right angle near its midpoint. One branch of it lies along the positive x axis and the other along the positive y axis (Figure 30.26). The wire carries a current I. What is the magnetic field at a point in the first quadrant of the x–y plane?

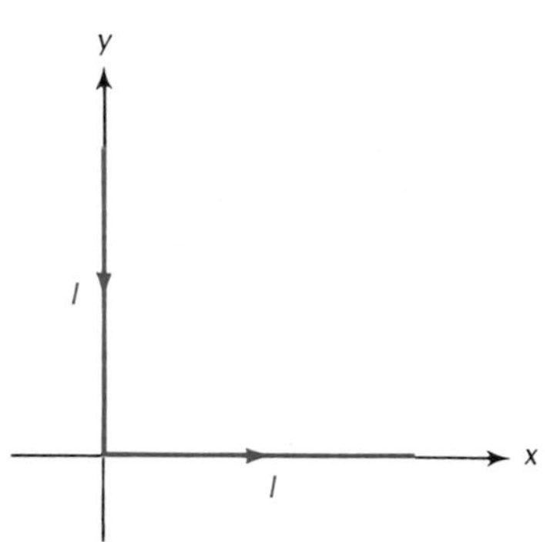

Fig. 30.26. A very long wire bent at a right angle.

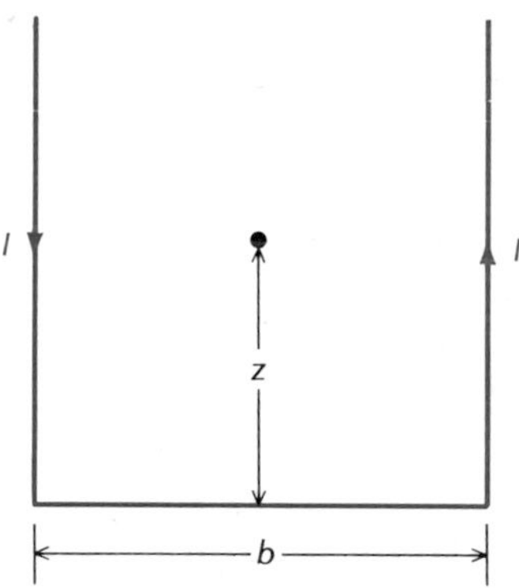

Fig. 30.27 A very long wire bent in the shape of a **U**.

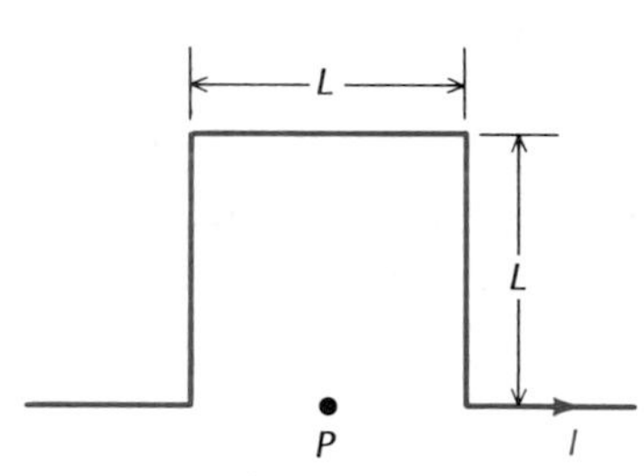

Fig. 30.28 A very long wire with square bends.

19. A very long wire carrying a current I is bent in the shape of a **U** (Figure 30.27). The two parallel segments are separated by a distance b. Find the magnetic field along the midline at a distance z from the bottom of the **U**. Consider both the case $z > 0$ and $z < 0$.

20. A very long, straight wire with a current I has a square bend near its midpoint (Figure 30.28). Each side of the square has a length L. What is the magnetic field at the point P, halfway between the two lower corners?

21. A loop of superconducting wire has the shape of a rectangle measuring $L \times 2L$. A current I flows around the wire. What is the strength of the magnetic field at the center of the rectangle?

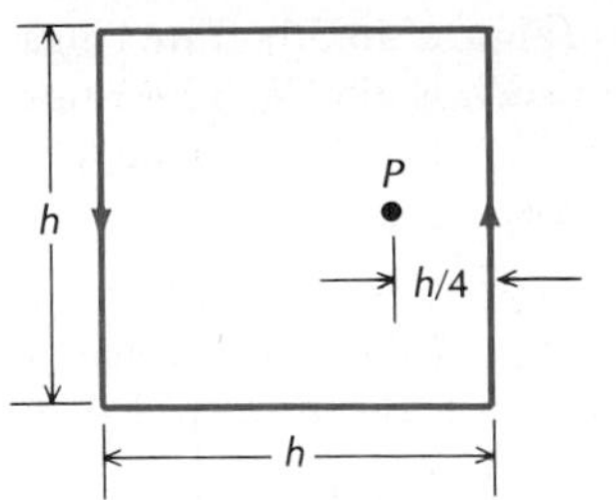

Fig. 30.29 A square loop.

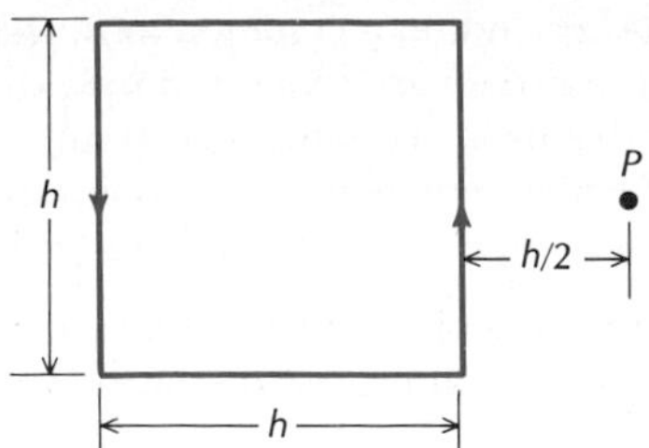

Fig. 30.30 A square loop.

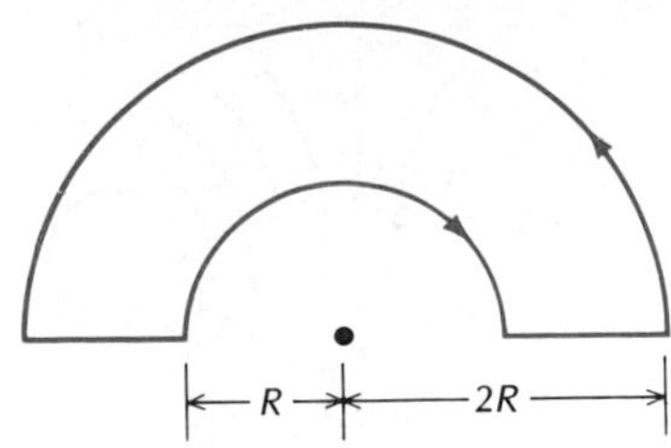

Fig. 30.31 A loop consisting of two semicircles and two straight segments.

22. A square loop of wire, measuring $h \times h$, carries a current I (Figure 30.29). Find the magnetic field at the point P at a distance $h/4$ from the center of the square.

23. A square loop of wire, measuring $h \times h$, carries a current I (Figure 30.30). Find the magnetic field at the point P at a distance h from the center of the square.

24. A loop of wire has the shape of two concentric semicircles connected by two radial segments (Figure 30.31). The loop carries a current I. Find the magnetic field at the center.

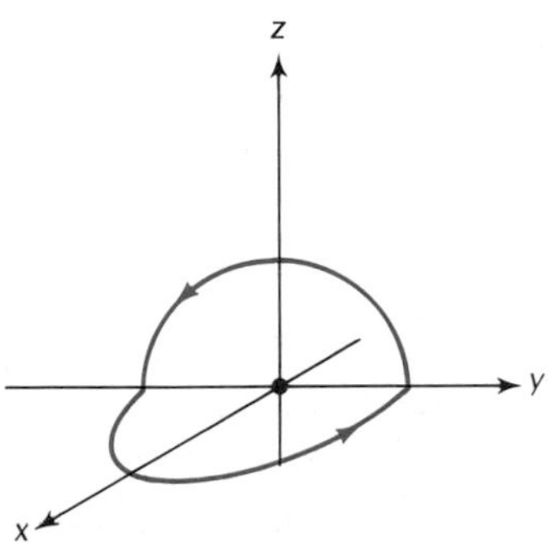

Fig. 30.32 A semicircle in the z–y plane and a semicircle in the x–y plane.

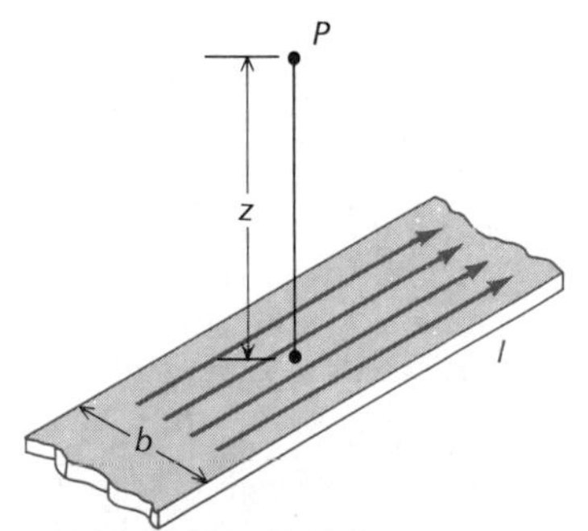

Fig. 30.33 A very long strip carrying a uniformly distributed current I.

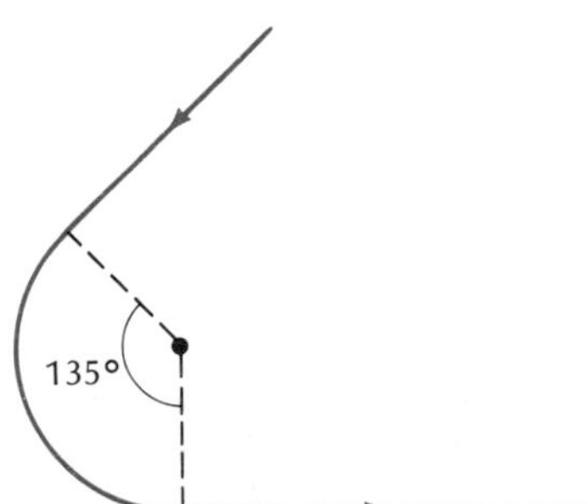

Fig. 30.34 Two very long wires joined by an arc of circle.

25. A circular loop of wire is folded along a diameter so as to form two semicircles of radius R intersecting at right angles (Figure 30.32). A current I flows around this loop. What is the magnetic field at the center? Draw a diagram showing the direction of the magnetic field.

26. A very long strip of copper of width b carries a current I uniformly distributed over the strip. What is the magnetic field at a distance z above the midline of this strip (Figure 30.33).

27. An infinite conducting sheet occupies the x–y plane. On this sheet a current flows in the y direction; the current is uniformly distributed over the sheet with σ ampere flowing across each 1-meter segment of the x axis. Find the magnetic field at a distance z from this sheet.

28. Two semi-infinite wires are in the same plane. The wires make an angle of 45° with each other and they are joined by an arc of circles of radius R (Figure 30.34). The wires carry a current I. Find the magnetic field at the center of the arc of circle.

29. Helmholtz coils are often used to make reasonably uniform magnetic fields in laboratories. These coils consist of two thin circular rings of wire par-

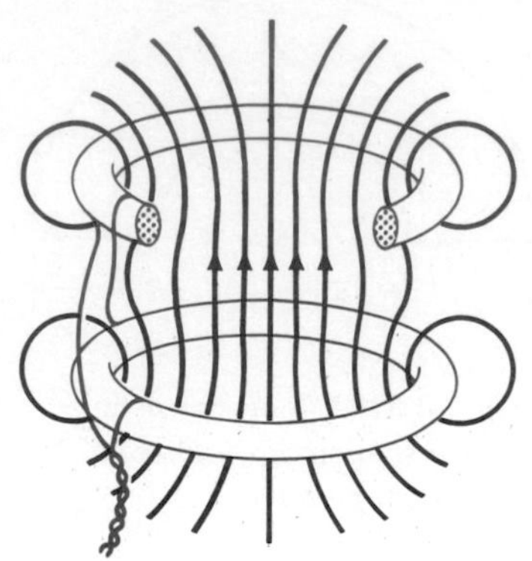

Fig. 30.35 Helmholtz coils.

allel to each other and on a common axis, the z axis (Figure 30.35). The rings have radius R and they are separated by a distance which is also R. The rings carry equal currents in the same direction.

(a) Find the magnetic field at any point of the z axis.

(b) Show that dB/dz and d^2B/dz^2 are both zero at $z = 0$.

30. The magnetic dipole moment of an electron is 9.3×10^{-24} A · m². What is the magnetic field on the axis of spin of the electron at a distance of 1.0 Å from its center?

31. The magnetic field of the Earth is approximately that of a magnetic dipole located at the center of the Earth. The strength of the magnetic field at the surface of the Earth at the magnetic north pole is 6.2×10^{-5} T. What is the strength of the magnetic field at an altitude of 1000 km above the pole? 2000 km? 3000 km? Make a plot of the strength of the magnetic field vs. altitude.

32. The magnetic field of the Earth is that of a magnetic moment of 8.0×10^{22} A · m² located deep within the Earth. Assume that the magnetic moment is due to a circular current flowing along the equator of the liquid core of the Earth; this core has a radius of 3500 km. What must be the current? What must be its direction (eastward or westward)?

33. It can be proved that the dependence on x, y, z of the magnetic field of a magnetic dipole is described by the same function as the dependence on x, y, z of the electric field of an electric dipole. Use this fact to convert the formulas of Eqs. (25.52)–(25.54) into formulas for a magnetic dipole.

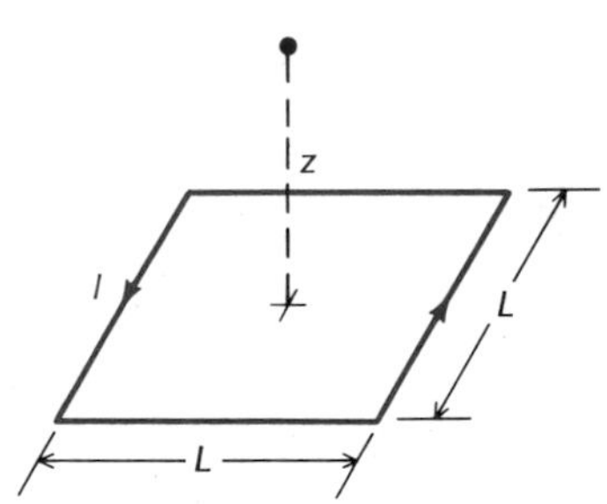

Fig. 30.36 A square loop.

34. A steady current I flows around a wire bent into a square of side L (Figure 30.36). Find the magnetic field at a distance z from the center of the square on a line perpendicular to the face of the square. Show that if $z \gg L$, your answer reduces to $B_z = (\mu_0/2\pi)\mu/z^3$, with a magnetic moment $\mu = IL^2$.

35. A pipe made of a superconducting material has a length of 0.30 m and a radius of 4.0 cm. A current of 4.0×10^3 A flows around the surface of the pipe; the current is uniformly distributed over the surface. What is the magnetic moment of this current distribution? (Hint: Treat the current distribution as a large number of rings stacked one on top of another.)

36. An amount of charge Q is uniformly distributed over a disk of paper of radius R. The disk spins about its axis with angular velocity ω. Find the magnetic dipole moment of the disk. Sketch the lines of magnetic field and of electric field in the vicinity of the disk.

*37. An amount of charge Q is uniformly distributed over a disk of paper of radius R. The disk spins about its axis with angular velocity ω. Find the magnetic field produced by this disk at a point on the axis of the disk, at a distance z from the center.

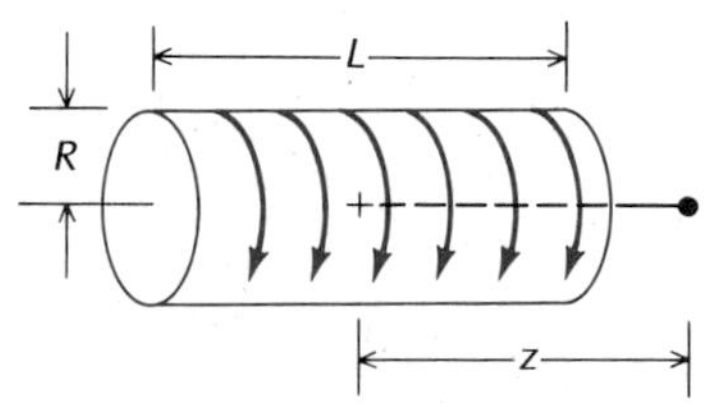

Fig. 30.37 A uniformly distributed current flows around the circumference of a copper pipe.

*38. A current I flows around the surface of a copper pipe of radius R and length L (Figure 30.37). Find a formula for the magnetic field on the axis of the pipe at a distance z from the center. Assume $z > L/2$.

*39. A semi-infinite wire lies along the positive x axis. Another semi-infinite wire lies along the y axis. The wires carry a current I from $y = \infty$ to $x = \infty$ (Figure 30.38).

(a) Find the magnetic field at a point in the first octant ($x > 0, y > 0, z > 0$).

(b) Find the magnetic field at a point in the second octant ($x < 0, y > 0, z > 0$).

*40. Two semi-infinite parallel wires lie in the x–y plane. The wires are joined by a semicircle of radius R whose center coincides with the origin of coordinates (Figure 30.39). The wires carry a current I in the direction shown in the figure. Find the x, y, and z components of the magnetic field at a point on the z axis.

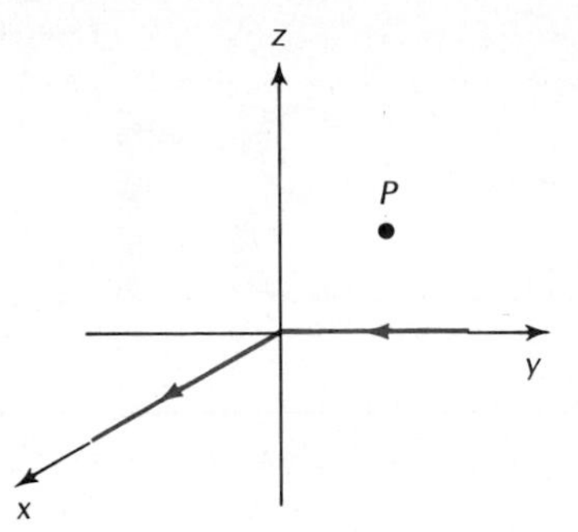

Fig. 30.38 Long straight wires along the x and the y axes.

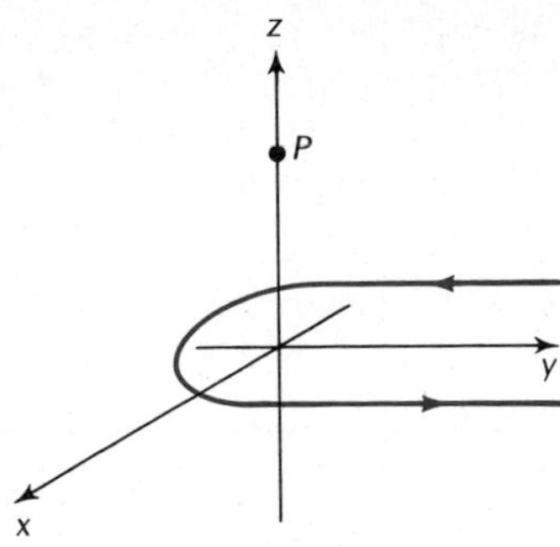

Fig. 30.39 Two long wires in the x–y plane joined by a semicircle.

Section 30.4

41. (a) In the reference frame of the laboratory, a proton moves with velocity 1.0×10^7 m/s parallel to a long straight wire at a distance of 0.10 m. The wire carries a current of 15 A. What is the magnetic force on the proton?
(b) In the reference frame of the proton, what is the magnetic force? What is the electric force? In this reference frame, what must be the charge density on the wire to give the correct value for the electric force?

CHAPTER 31

Ampère's Law

In Chapter 24 we saw that Gauss' Law gives us a shortcut for calculating the electric field of charge distributions that have a certain amount of symmetry. In this chapter we will see that there exists a corresponding method for calculating the magnetic field of current distributions. The method is based on Ampère's law, an important relation between the magnetic field and the current.

André Marie Ampère, *1775–1836, French physicist and mathematician, professor at the École Polytechnique and at the Collège de France. After Oersted's discovery of the generation of magnetic fields by electric currents, Ampère demonstrated experimentally that currents exert magnetic forces on each other. He carefully investigated the relationship between currents and magnetic fields, and he established that a magnet is equivalent to a distribution of currents. We are indebted to him not only for Ampère's law, but also for the equation (31.35) for the force exerted by a magnetic field on a current element.*

31.1 Ampère's Law

To obtain this relation, we begin by examining the magnetic field of a very long straight wire carrying a steady current I_0. According to Eq. (30.32), the magnetic field at a distance r from this wire has a magnitude

$$B = \frac{\mu_0}{2\pi} \frac{I_0}{r} \tag{1}$$

and the magnetic field lines are concentric circles around the wire. Consider now a closed mathematical path of arbitrary shape around this wire (Figure 31.1) and evaluate the integral $\oint \mathbf{B} \cdot d\mathbf{l}$ around this path. The path can be regarded as consisting of short radial segments and circular arcs. Since $\mathbf{B}$ is in the tangential direction, the radial segments do not contribute to the integral. Since the tangential segments have lengths $r\,d\theta$, we obtain

$$\oint \mathbf{B} \cdot d\mathbf{l} = \oint Br\,d\theta = \oint \frac{\mu_0}{2\pi} \frac{I_0}{r} r\,d\theta = \frac{\mu_0}{2\pi} I_0 \oint d\theta \tag{2}$$

The integral $\oint d\theta$ over the closed path is simply 2π. Hence

$$\oint \mathbf{B} \cdot d\mathbf{l} = \mu_0 I_0 \tag{3}$$

Although the above calculation was based on a planar path, the same result applies to nonplanar paths — any path segment parallel to the wire will be perpendicular to the magnetic field; hence, as in the above calculation, only the circular arcs of the path contribute to the integral.

It turns out that Eq. (3) is valid not only for the magnetic field produced by a current on a long wire, but also for the magnetic field produced by an arbitrary distribution of steady currents. This is **Ampère's Law:**

The integral of **B** *around any closed path equals* μ_0 *times the current intercepted by the area spanning the path,*

$$\oint \mathbf{B} \cdot d\mathbf{l} = \mu_0 I \tag{4}$$

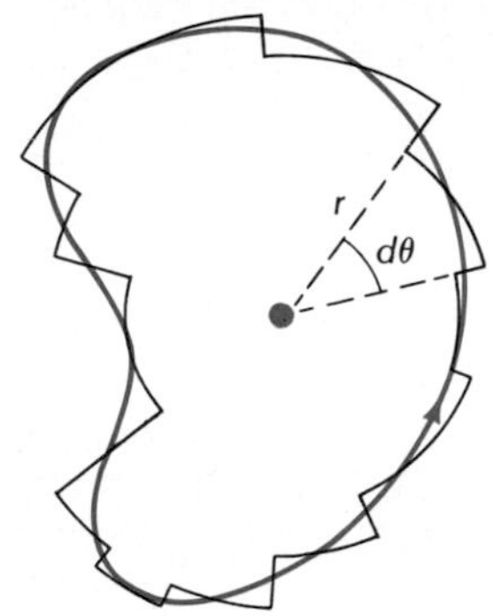

Fig. 31.1 Path of integration around a long, straight wire. The wire is perpendicular to the plane of the page. The path can be approximated by radial segments and circular arcs.

Ampère's Law

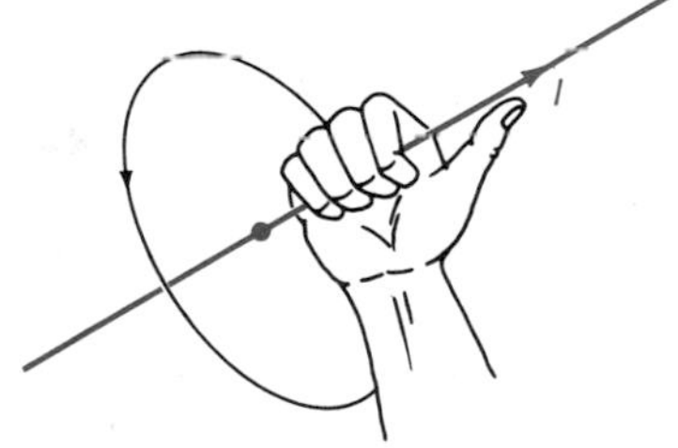

Fig. 31.2 The right-hand rule for Ampère's Law: if the fingers curl along the path of integration, the thumb points in the positive direction for the positive current.

Note that the current in Eq. (4) need not flow on thin wires; it may flow on a conductor of any shape, or it may even flow without any conductor (e.g., a beam of protons in vacuum). Furthermore, note that the current is to be reckoned as positive if it is related to the direction around the path by the right-hand rule (thumb points along current when fingers curl along path; Figure 31.2), and negative otherwise. Although Ampère's Law can be derived rigorously from our expression for the magnetic field of a small current element [see Eq. (30.25)], we will not attempt to give this general proof.

For the calculation of magnetic fields, Ampère's Law plays a role similar to that of Gauss' Law for the calculation of electric fields. Provided that a distribution of currents has sufficient symmetry, Ampère's Law completely determines the magnetic field.

EXAMPLE 1. Find the magnetic field of a very long, very thin, straight wire using Ampère's Law.

SOLUTION: By considerations of symmetry, the magnetic field lines of such a wire will have to be either concentric circles, or radial lines, or parallel lines in the same direction as the wire. The last two possibilities are inconsistent with the requirement that the magnetic field lines must form closed loops (see Section 30.2). Thus the field lines must be concentric circles, the magnetic field having a constant magnitude along each circle. If we integrate **B** along one of these circles we obtain

$$\oint \mathbf{B} \cdot d\mathbf{l} = \oint Br\, d\theta = 2\pi r B \tag{5}$$

and Ampère's Law then gives us

$$2\pi r B = \mu_0 I_0 \tag{6}$$

i.e.,

$$B = \frac{\mu_0}{2\pi} \frac{I_0}{r} \tag{7}$$

Thus, Ampère's Law gives back the formula with which we started this section.

EXAMPLE 2. A very long, straight wire has a circular cross section of radius R. The wire carries a current I_0 uniformly distributed over the cross-sectional area of the wire. What is the magnetic field inside the wire? Outside the wire?

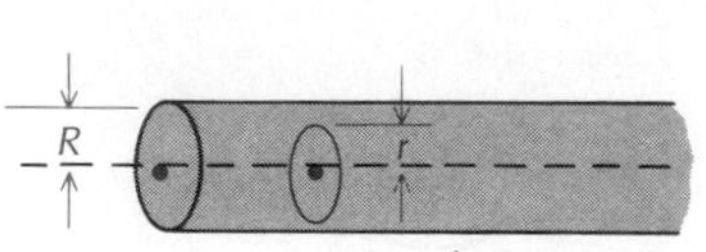

Fig. 31.3 Segment of a wire of radius R with a current I_0 uniformly distributed over the cross-sectional area.

SOLUTION: Consider a circular path of radius r inside the wire (Figure 31.3). Symmetry tells us that the magnetic field lines are circles. Hence

$$\oint \mathbf{B} \cdot d\mathbf{l} = 2\pi r B$$

Since the current is uniformly distributed over the volume of the wire, the amount of current I intercepted by the area within the circular path is in direct proportion to this area,

$$I = I_0 \frac{\pi r^2}{\pi R^2} = I_0 \frac{r^2}{R^2} \tag{8}$$

Ampère's Law then leads to

$$2\pi r B = \mu_0 I \tag{9}$$

or

$$2\pi r B = \mu_0 I_0 r^2/R^2 \tag{10}$$

and

$$B = \frac{\mu_0}{2\pi} \frac{I_0 r}{R^2} \tag{11}$$

Note that the magnetic field is zero at the center of the wire ($r = 0$) and reaches a maximum value at the surface of the wire ($r = R$).

Outside the wire, we can find the magnetic field by repeating the calculation of Example 1. This calculation only depends on cylindrical symmetry and not on the thickness of the wire. The magnetic field is therefore exactly the same as for a thin wire,

$$B = \frac{\mu_0}{2\pi} \frac{I_0}{r} \tag{12}$$

31.2 Solenoids

A solenoid is a conducting wire wound in a tight helical coil (Figure 31.4). A current in this wire will produce a strong magnetic field within the coil. Because of the similarity of the current distributions, a tight coil produces essentially the same magnetic field as a collection of loops stacked next to one another.[1] The magnetic field of such a collection of loops can be calculated by summing the magnetic fields of

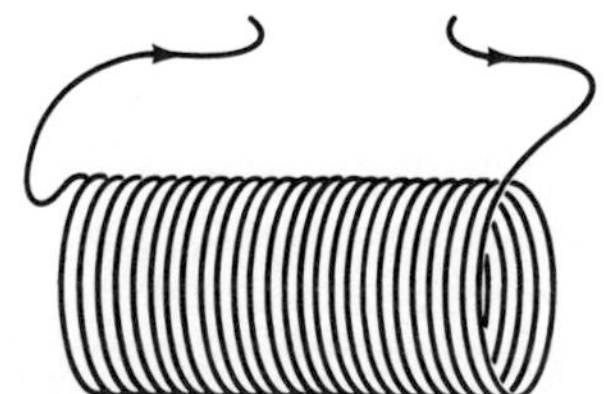

Fig. 31.4 A solenoid.

[1] There is, however, one small difference: in a cylindrical stack of loops the current flows in closed circles around the surface of the cylinder, whereas in a cylindrical solenoid the current has an extra component parallel to the axis of the cylinder (the current is helical). If the solenoid is tightly wound with thin wire, this axial component of the current is insignificant compared to the circular component and we can ignore it.

the individual loops. Figure 31.5 shows the pattern of field lines of the resultant magnetic field. Figure 31.6 shows the pattern of field lines of an actual solenoid with a rather loose coil, made visible with iron filings. The field is strong inside the solenoid but fairly weak outside. The detailed calculation of this magnetic field is rather messy because the individual magnetic fields of the loops of current must be added vectorially. We will not attempt this calculation here; instead, we will calculate the magnetic field of an **ideal solenoid**, that is, a very long (infinitely long) solenoid with very tightly wound coils so that the current distribution on the surface of the solenoid is nearly uniform.

Ideal solenoid

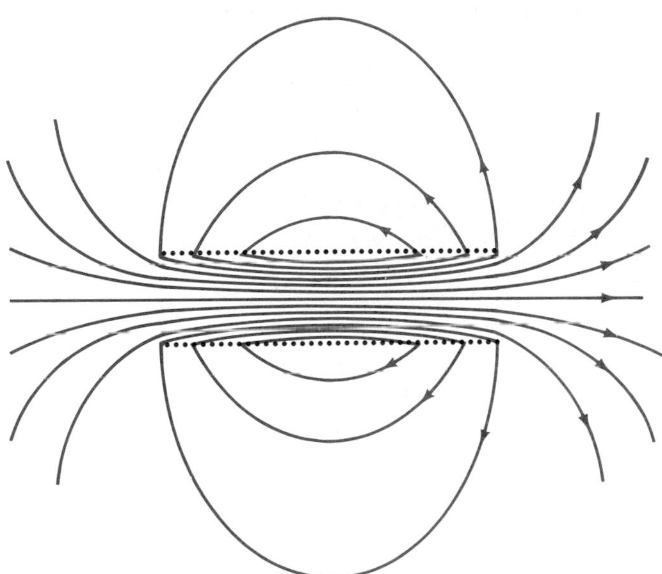

Fig. 31.5 Magnetic field lines of a solenoid.

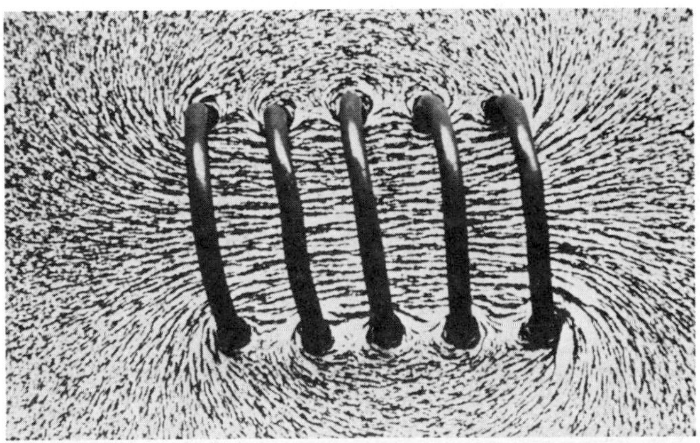

Fig. 31.6 Iron filings sprinkled on a sheet of paper inserted in a solenoid. Note that inside the solenoid, the distribution of field lines is nearly uniform.

To find this magnetic field, we begin with an appeal to symmetry, as in Example 1. The ideal solenoid has translational symmetry (along the axis of the solenoid) and rotational symmetry (around the axis). To be consistent with these symmetries, the magnetic field lines inside the solenoid will then have to be either concentric circles, or radial lines, or lines parallel to the axis. Concentric circles and radial lines are unacceptable; the former would require the presence of a current along the axis (compare Example 1) and the latter would require that the field lines start on the axis. Thus the field lines inside the solenoid must all be parallel to the axis. These field lines emerge from the end of the solenoid and return to the other end (Figure 31.5). For an ideal, very long solenoid, these external field lines will spread over a very large region of space; hence, the magnetic field outside of the solenoid is nearly zero.

We can now use Ampère's Law to find the magnitude of the magnetic field. Consider the rectangular path shown in Figure 31.7 and evaluate the integral of **B** along this path. The horizontal side external to the solenoid does not contribute to the integral (**B** is zero); the vertical sides do not contribute to the integral (**B** is perpendicular to the path). Hence only the horizontal side within the solenoid contributes. The magnetic field has some constant magnitude along this side and if the length of this side is l,

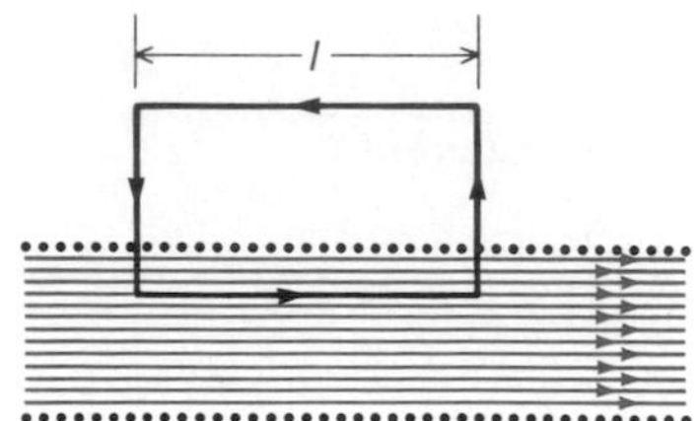

Fig. 31.7 Magnetic field lines of a very long solenoid.

$$\oint \mathbf{B} \cdot d\mathbf{l} = Bl \tag{13}$$

The total current intercepted by the area within the rectangle is the current I_0 in the wire multiplied by the number N of wires passing through this area. Hence Ampère's Law tells us

$$Bl = \mu_0 N I_0 \tag{14}$$

or

$$B = \mu_0 I_0 (N/l) \tag{15}$$

The quantity N/l is the number of turns of wire per unit length, commonly designated by n. Thus

Field in solenoid

$$\boxed{B = \mu_0 I_0 n} \tag{16}$$

Note that this result is independent of the length of the vertical sides of the path of integration; hence **B** has the same magnitude everywhere within the solenoid. This shows that the magnetic field within an ideal solenoid is perfectly uniform.

Electromagnet

An **electromagnet** is essentially a solenoid with a gap, or, what amounts to the same thing, a pair of solenoids with their ends placed close together (Figure 31.8). Magnetic field lines come out of one sole-

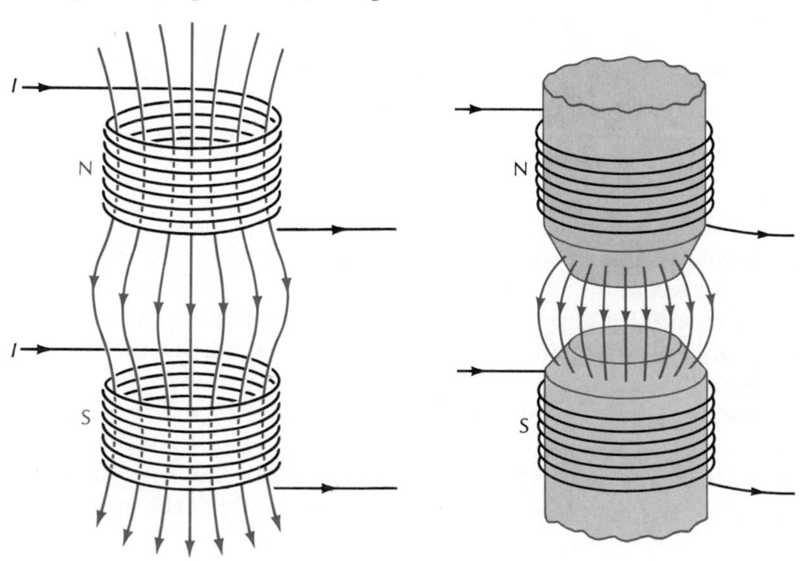

Fig. 31.8 (left) An electromagnet with two coils. The coil from which field lines emerge into the gap is called the north pole; the coil into which field lines enter is called the south pole.

Fig. 31.9 (right) An electromagnet with iron pole pieces.

noid and go into the other (of course, field lines will also have to come out of the solenoids at their other ends and close on themselves). The first solenoid is called the north pole of the electromagnet, and the second the south pole. If the gap is small, then the magnetic field in this region is almost the same as inside the solenoids. In most electromagnets the space inside the solenoids is filled with an iron core (Figure 31.9); as we will see in Section 33.3, the iron enhances the magnetic field, making it much stronger than the value given by Eq. (16).

EXAMPLE 3. A solenoid used for the investigation of the effect of magnetic fields on the propagation of light in a liquid consists of 180 turns of wire wound on a tube 19 cm long. The current in the wire is 5.0 A. What is the strength of the magnetic field within the tube?

SOLUTION: The number of turns per unit length is

$$n = 180/0.19 \text{ m} = 9.5 \times 10^2/\text{m} \tag{17}$$

Hence, according to Eq. (16),

$$\begin{aligned} B &= \mu_0 I_0 n \\ &= 1.26 \times 10^{-6} \text{ N} \cdot \text{m}^2/\text{C}^2 \times 5.0 \text{ A} \times 9.5 \times 10^2/\text{m} \\ &= 6.0 \times 10^{-3} \text{ T} \end{aligned}$$

EXAMPLE 4. A **toroid** is a conducting wire wound in a tight coil in the shape of a torus (a doughnut). It may be thought of as a solenoid that has been bent so that its ends meet (Figure 31.10a). What is the magnetic field inside the toroid?

SOLUTION: By symmetry, the magnetic field lines inside the toroid are closed circles (Figure 31.10b). If we integrate **B** along one of these circular field lines, we obtain

$$\oint \mathbf{B} \cdot d\mathbf{l} = 2\pi r B \tag{18}$$

The total current intercepted by the circular area within this path equals the total number N of wires sticking through the area multiplied by the current I_0 in each wire. Hence, from Ampère's Law,

$$2\pi r B = \mu_0 N I_0 \tag{19}$$

or

$$B = \frac{\mu_0}{2\pi} \frac{N I_0}{r} \tag{20}$$

Thus, the magnetic field depends on the radial distance from the axis of the torus.

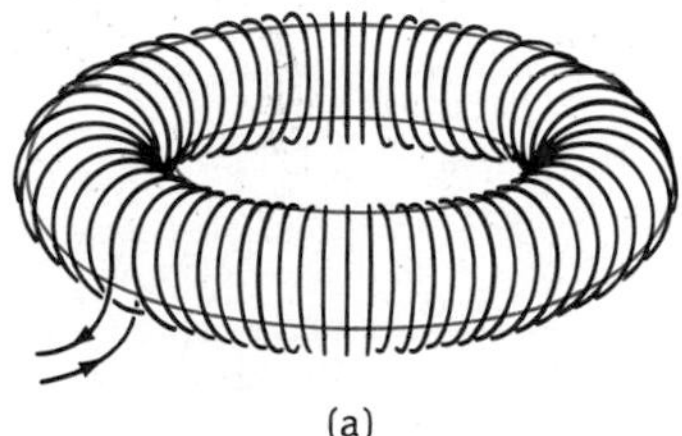

(a)

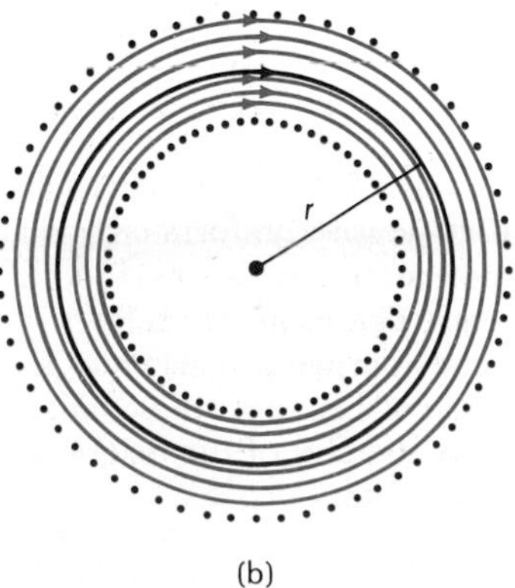

(b)

Fig. 31.10 (a) A toroid. (b) Magnetic field lines within a toroid.

31.3 Motion of Charges in Electric and Magnetic Fields

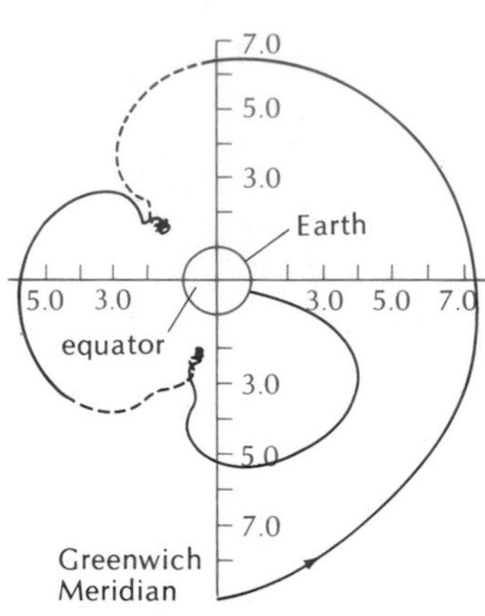

Fig. 31.11 Trajectory of a cosmic ray of momentum 5.66×10^{-19} kg · m/s in the magnetic field of the Earth. The trajectory is three dimensional; portions above the plane of the page are shown with a solid line, portions below the plane of the page are shown with a dashed line. The numbers along the axes give distances in Earth radii.

The force exerted by a magnetic field on a charged particle is

$$\mathbf{F} = q\mathbf{v} \times \mathbf{B} \tag{21}$$

This force is always at right angles to the velocity of the particle; consequently, the force changes the *direction* of the velocity of the particle, but not the *magnitude* of the velocity. The formal proof of this is easy:

$$\frac{d}{dt}v^2 = \frac{d}{dt}\mathbf{v}\cdot\mathbf{v} = 2\mathbf{v}\cdot\frac{d\mathbf{v}}{dt} = 2\mathbf{v}\cdot\mathbf{F}/m = 0 \tag{22}$$

where the last equality expresses the fact that force and velocity are perpendicular. Thus, the magnetic force only changes the momentum of the particle, but not the kinetic energy — the magnetic force *does no work* on the particle.

In general, the motion of a particle in a magnetic field is quite complex. For example, Figure 31.11 shows the trajectory of a cosmic-ray particle approaching the Earth and being deflected by the magnetic field (the effects of gravity on such a cosmic ray are insignificant). In what follows, we will only consider the relatively simple case of motion in a uniform magnetic field.

Figure 31.12 shows a uniform magnetic field, directed into the plane of the page. Suppose that a positively charged particle has an initial velocity in the plane of the page; this initial velocity is perpendicular to the magnetic field. The magnetic force $q\mathbf{v} \times \mathbf{B}$ is perpendicular to both $\mathbf{v}$ and $\mathbf{B}$; its direction is shown in Figure 31.12. The acceleration then has a constant magnitude

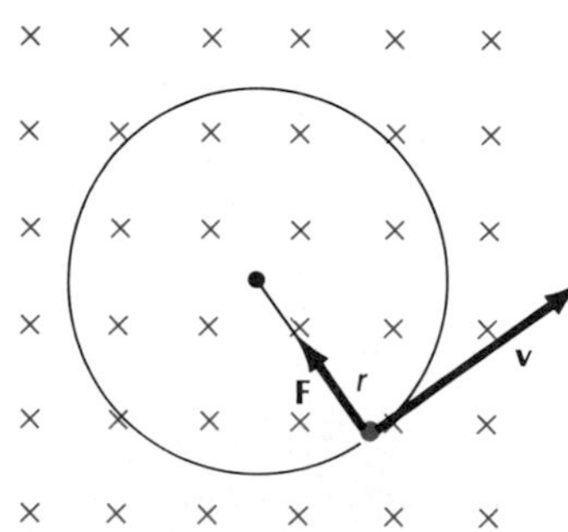

Fig. 31.12 Positively charged particle with uniform circular motion in a uniform magnetic field. The magnetic field points perpendicularly into the plane of the page; the crosses show the tails of the magnetic-field vectors.

$$a = F/m = qvB/m \tag{23}$$

and its direction is always perpendicular to the velocity. Such an acceleration corresponds to uniform circular motion. The particle will move in a circle of radius r and the acceleration given by Eq. (23) will play the role of centripetal acceleration, i.e.,

$$\frac{qvB}{m} = \frac{v^2}{r} \tag{24}$$

This leads to the following formula for the radius of the circular motion:

$$r = \frac{mv}{qB} \tag{25}$$

Figure 31.13 shows a beam of electrons executing such circular motion in a cathode-ray tube placed in a magnetic field.

The angular velocity of the circular motion is

$$\omega = \frac{v}{r} = \frac{qB}{m} \tag{26}$$

Fig. 31.13 Electrons moving in a circle in a cathode-ray tube in a magnetic field. The tube contains a gas at a very low pressure, and the atoms of the gas glow under the impact of the electrons; this makes the electron beam visible.

and the frequency

$$\nu = \frac{\omega}{2\pi} = \frac{qB}{2\pi m} \tag{27}$$

Cyclotron frequency

This is called the **cyclotron frequency** because the operation of cyclotrons (described in Example 5 below) involves particles moving with this frequency in a magnetic field. Note that the cyclotron frequency is independent of the speed of the circular motion — in a uniform magnetic field, slow particles and fast particles (of a given charge and mass) move around circles at the same frequency, but the slow particles move along smaller circles than the fast particles.

It is useful to write Eq. (25) in terms of the momentum $p = mv$; this leads to

$$r = \frac{p}{qB} \tag{28}$$

Circular orbit in magnetic field

The advantage of Eq. (28) is that it remains valid even when the particle moves with relativistic velocity. Thus Eq. (28) is more general than our derivation of it.

Cyclotron

EXAMPLE 5. A **cyclotron** is a device for acceleration of protons, deuterons, or other charged particles. It consists of an evacuated cavity placed between the poles of a large electromagnet; within the cavity there is a flat metallic can cut into two **D**-shaped pieces, or **dees** (Figures 31.14 and 31.15). An oscillating high-voltage generator is connected to the dees; this creates an oscillating electric field in the gap between the dees. The frequency of the voltage generator is adjusted so that it coincides with the cyclotron frequency of Eq. (27). An ion source at the center of the cyclotron releases protons. The electric field in the gap between the dees gives each of these protons a push and the uniform magnetic field in the cyclotron then makes the proton travel on a semicircle inside of the first dee. When the proton returns to the gap, the high-voltage generator will have reversed the electric field in the gap; the proton therefore re-

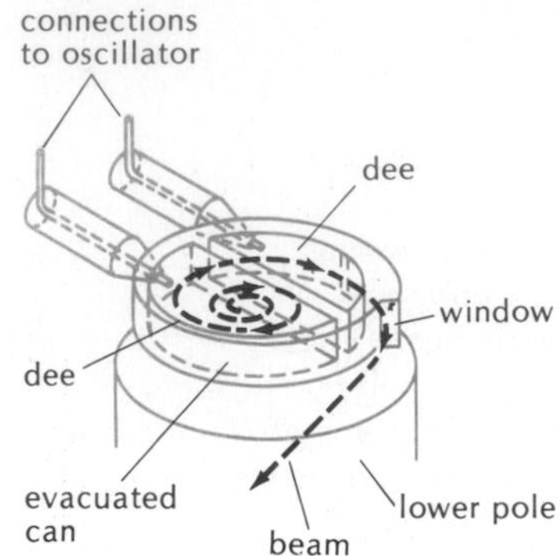

Fig. 31.14 Trajectory of a particle within the dees of a cyclotron. In this diagram, the upper pole of the electromagnet has been omitted for the sake of clarity.

Fig. 31.15 The dees of one of the early small cyclotrons built by E. O. Lawrence.

ceives an additional push which sends it into the second dee. There it travels on a semicircle of slightly larger radius corresponding to its slightly larger energy, etc. Each time the proton crosses the gap between the dees, it receives an extra push and extra energy. The proton travels along arcs of circles of stepwise increasing radius. When the protons reach the outer edge of the dees, they leave the cyclotron as a high-energy beam (Figure 31.14). One of the first cyclotrons, built by E. O. Lawrence at Berkeley in 1932, had dees with a diameter of 28 cm and its magnet was capable of producing a field of 1.4 T. What was the maximum energy of the protons accelerated by this cyclotron?

SOLUTION: When the proton reaches its maximum energy, its orbit has a radius of 14 cm. Since the magnetic field is 1.4 T, the momentum of such a proton is

$$p = mv = eBr$$

$$= 1.6 \times 10^{-19}\ \text{C} \times 1.4\ \text{T} \times 0.14\ \text{m}$$

$$= 3.1 \times 10^{-20}\ \text{kg} \cdot \text{m/s}$$

and the energy is

$$K = p^2/(2m)$$

$$= (3.1 \times 10^{-20}\ \text{kg} \cdot \text{m/s})^2 \,/\, (2 \times 1.67 \times 10^{-27}\ \text{kg})$$

$$= 2.9 \times 10^{-13}\ \text{J} = 1.8 \times 10^6\ \text{eV}$$

Nuclear physicists like to measure energies in MeV, where

$$1\ \text{MeV} = 10^6\ \text{eV} = 1.60 \times 10^{-13}\ \text{J}$$

In these units, the above kinetic energy is 1.8 MeV.

Synchrocyclotron

Cyclotrons of large size can be used to accelerate protons up to an energy of about 30 MeV. Above this energy, cyclotrons begin to fail because relativistic effects change the frequency of the orbital motion of the protons. Roughly, what happens is that at high energies the mass of the protons increases with their speed [as indicated by Eq. (17.54)] — the orbital frequency will decrease. In order to keep the pushes of the electric field in phase with the motion of the proton, we must gradually decrease the frequency of oscillation of the high-voltage generator. A modified cyclotron that automatically performs such an adjustment of frequency is called a **synchrocyclotron**. It accelerates a bunch of protons by matching its frequency of oscillation to the frequency of the orbital motion of the bunch; when it ejects the bunch, it begins to accelerate the next bunch, etc. Machines of this kind have been used to achieve energies up to several hundred MeV.

At even higher energies cyclotrons become impractical because they require excessively large magnets. It is then better to keep the protons moving around a circle of fixed radius in a large magnet of annular shape (Figure 31.16). As the protons gradually acquire more energy, the strength of the magnetic field must be gradually increased [see Eq. (28)]. Accelerator machines that automatically perform this increase of

Fig. 31.16 The Bevatron at the Lawrence Berkeley Laboratory.

magnetic field are called **synchrotrons**. The most powerful synchrotrons built to date are the machines at Fermilab (Batavia, Illinois) and at CERN (Centre Européenne de Recherche Nucléaire, on the Swiss–French border near Geneva). The Fermilab machine accelerates protons up to an energy of 1,000,000 MeV (see Interlude C).

Synchrotron

In our discussion of the motion of a charged particle in a uniform magnetic field, we assumed that the initial velocity was perpendicular to the magnetic field (see Figure 31.12). If this is not the case, then we can regard the velocity as consisting of two components: one component $\mathbf{v}_\perp$ perpendicular to **B** and one component $\mathbf{v}_\parallel$ parallel to **B** (Figure 31.17). Since the magnetic force is perpendicular to **B**, only the component $\mathbf{v}_\perp$ changes, while $\mathbf{v}_\parallel$ remains constant. The component $\mathbf{v}_\perp$ then gives rise to uniform circular motion with radius $r = mv_\perp/qB$ in a direction perpendicular to **B**, while the component $\mathbf{v}_\parallel$ gives rise to uniform translational motion in a direction parallel to **B**. The combination of these two motions is a helical motion with axis along the magnetic field — the particle spirals around the magnetic field lines (Figure 31.18). Such spiraling is a general feature of motion in a magnetic field; it will happen even if the magnetic field is a function of

Ernest Orlando Lawrence, *1901–1958, American experimental physicist, professor at Berkeley, and director of the Radiation Laboratory (now Lawrence Berkeley Laboratory). He was awarded the Nobel Prize in 1939 for the invention and development of the cyclotron.*

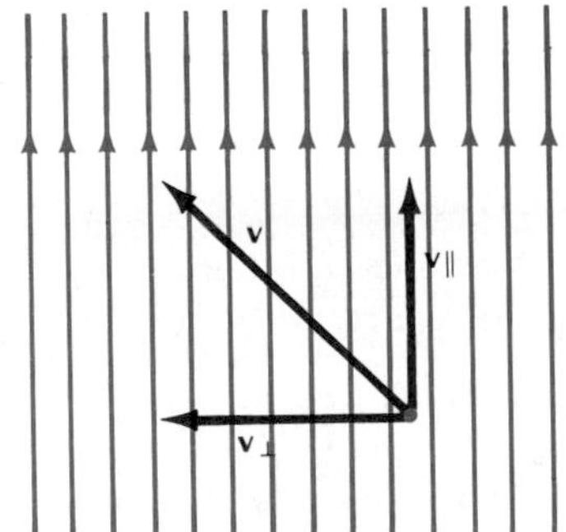

Fig. 31.17 The velocity of a particle has a component parallel to **B** and a component perpendicular to **B**.

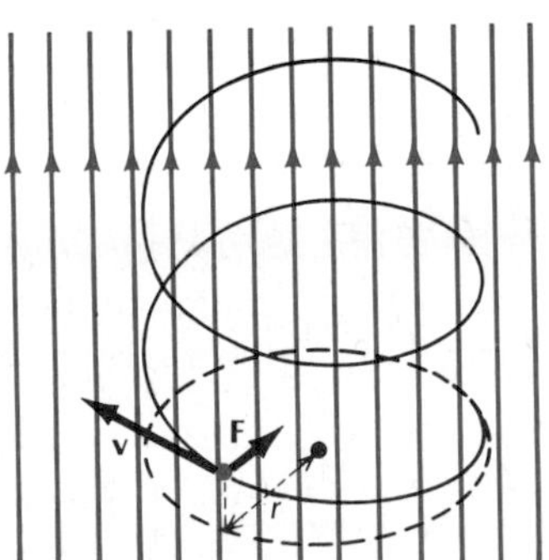

Fig. 31.18 Positively charged particle with helical motion in uniform magnetic field.

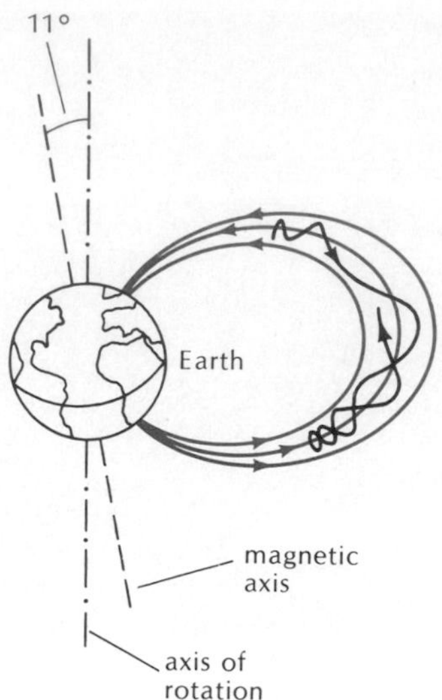

Fig. 31.19 Particle spiraling around the magnetic field lines of the Earth's field.

position. For example, Figure 31.19 shows a particle spiraling about the magnetic field lines of the field of the Earth. A detailed analysis shows that near the poles, where the field lines converge, the component $\mathbf{v}_{\parallel}$ of the velocity parallel to the field lines will decrease, then vanish, and then reverse (provided $\mathbf{v}_{\parallel}$ is not too large). Hence the particle spirals toward one pole, then halts its approach, and then spirals back toward the other pole, etc. The **Van Allen belts** surrounding the Earth (Figure 31.20) consist of a large number of electrons and protons spiraling back and forth along the field lines in this manner.

Van Allen belts

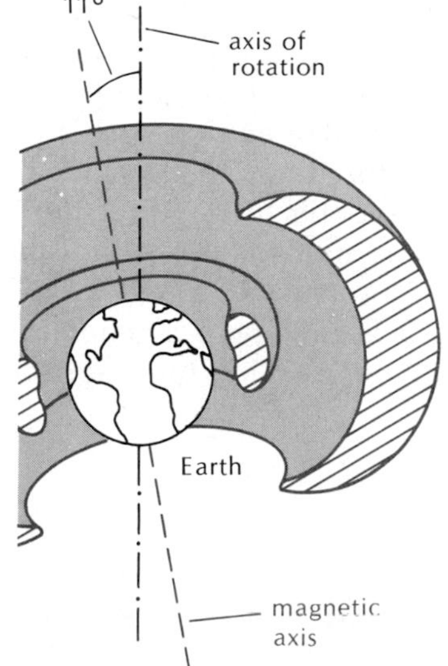

Fig. 31.20 The Van Allen radiation belts. Near the poles of the Earth, where the horns of the belts are close to the upper atmosphere, charged particles sometimes spill into the atmosphere, producing the luminous glow known as the Aurora Borealis.

If electric and magnetic fields act on a particle simultaneously, then the force has both an electric and a magnetic part,

Lorentz force

$$\boxed{\mathbf{F} = q\mathbf{E} + q\mathbf{v} \times \mathbf{B}} \tag{29}$$

This is called the **Lorentz force.**

Crossed fields

As an example of the simultaneous action of electric and magnetic fields, let us consider **crossed fields**, that is, uniform electric and magnetic fields at right angles to one another. Figure 31.21 shows electric field lines and magnetic field lines at right angles. The motion of a particle in such fields is usually fairly complicated, but under special circumstances the motion becomes very simple. Suppose that in Figure 31.21, a positively charged particle enters the field region from the left. The magnetic force $q\mathbf{v} \times \mathbf{B}$ is then opposite to the electric force $q\mathbf{E}$ and if the velocity has just the right magnitude, these forces will cancel and the particle will continue its original motion on a straight line. The condition for this is

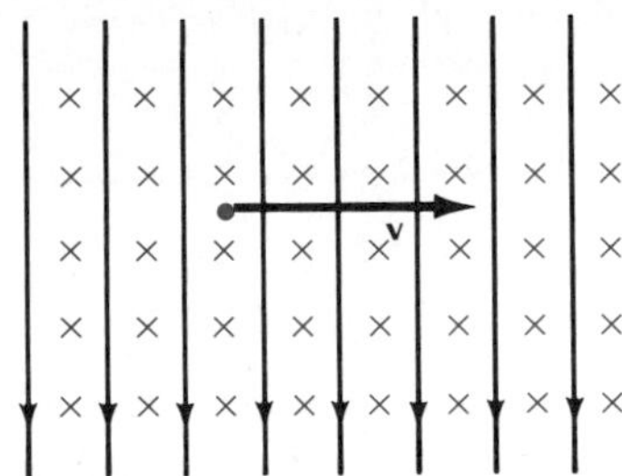

Fig. 31.21 Electric (black) and magnetic (color) fields at right angles. The crosses show the tails of the magnetic field vectors. A positively charged particle moves from left to right.

$$\mathbf{E} = -\mathbf{v} \times \mathbf{B} \tag{30}$$

or, considering magnitudes,

$$E = vB$$

The "right" velocity for cancellation of the forces is then

$$\boxed{v = E/B} \tag{31}$$

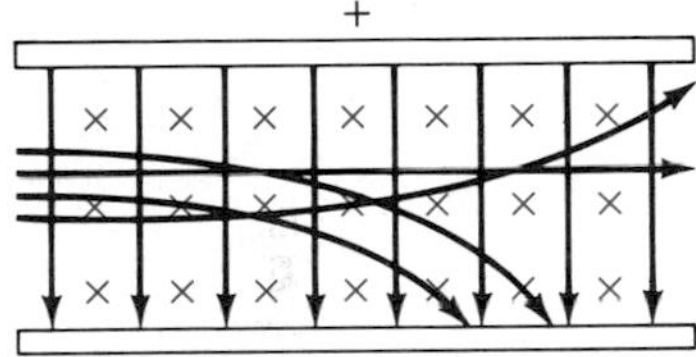

Fig. 31.22 A velocity selector with electric and magnetic fields at right angles. A beam of positively charged particles enters from the left. Particles of excessive velocity are deflected upward; particles of insufficient velocity are deflected downward.

This cancellation is the basic principle behind the **velocity selectors** (or velocity filters) often used in physics laboratories in order to select particles of some desirable velocity from a beam containing particles with a large variety of velocities. It is only necessary to shoot the beam into a region containing crossed **E** and **B** fields whose magnitudes are related to the desired velocity by Eq. (31). Particles with the right velocity will then proceed undeflected, all other particles will be deflected to one side or another and they are thereby eliminated from the beam (Figure 31.22).

Crossed **E** and **B** fields also play a role whenever a wire or some other conductor carries a current in a magnetic field. Figure 31.23 shows a metallic strip carrying a current; the strip is placed in a magnetic field, perpendicular to the surface of the strip. If the current is toward the right, the motion of the free electrons in the strip must be toward the left with some average velocity **v**. These free electrons experience a magnetic force $-e\mathbf{v} \times \mathbf{B}$, directed upward in Figure 31.23. This magnetic force will tend to push the electron toward the upper edge of the strip, and some electrons will accumulate there, leaving the lower edge with a deficit of electrons — the upper edge acquires a negative charge and the lower edge a positive charge. There will then exist an electric field in the strip, as shown in Figure 31.23. This electric field is perpendicular to the magnetic field, i.e., the **E** and **B** fields inside the strip are crossed fields. Under equilibrium conditions, the transverse magnetic force on an electron will be matched by the transverse electric force. The condition for this balance of forces is Eq. (30). Thus, inside the strip, there exists an electric field of magnitude $E = vB$; consequently, in a strip of width l, there is a potential difference

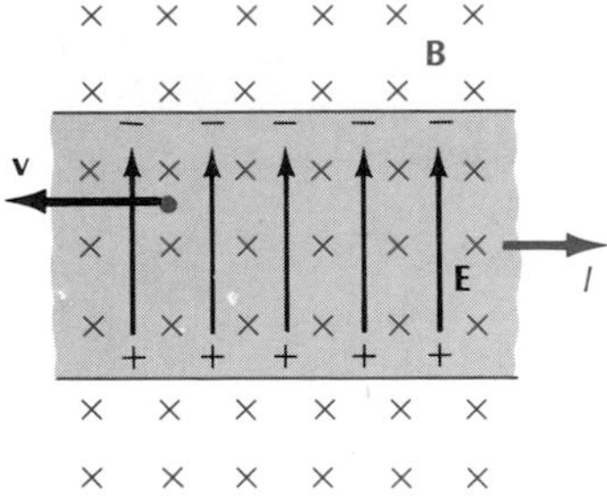

Fig. 31.23 A metallic strip carrying a current has been placed in a uniform magnetic field. The crosses show the tails of the magnetic field vectors. The current flows from left to right; the electrons making up this current move from right to left. The magnetic force on an electron is directed upward; the electric force is directed downward.

$$\Delta V = \int E\, dl = vBl \tag{32}$$

between the upper and the lower edge of the strip, the lower edge being at a higher potential. This generation of potential difference between opposite edges of a conductor carrying a current in a magnetic field is called the **Hall effect**.

Hall effect

We can express the Hall potential difference in terms of the current by means of Eq. (28.17):

$$v = \frac{I}{enA}$$

where A is the cross-sectional area of the conductor and n the number of free electrons per unit volume. This gives

$$\Delta V = \frac{IBl}{enA} \tag{33}$$

For example, a copper strip of cross-sectional area 5.0 cm × 0.02 cm carrying a current of 20 A in a magnetic field of 1.5 T develops a potential difference of

$$\Delta V = \frac{20 \text{ A} \times 1.5 \text{ T} \times 0.05 \text{ m}}{1.6 \times 10^{-19} \text{ C} \times 8.5 \times 10^{28}/\text{m}^3 \times 0.10 \times 10^{-4} \text{ m}^2}$$

$$= 1.1 \times 10^{-5} \text{ volt}$$

In practice, the Hall effect is often used to determine the density of free electrons n in a metal. Also, the Hall effect provides direct empirical evidence that the current carriers in metals are negative charges — if the current carriers were positive charges, then the direction of **v** in Figure 31.23 would be opposite to that shown and the positive charges would accumulate at the top of the strip rather than on the bottom.

31.4 Force on a Wire

If a wire carrying a current is placed in a magnetic field, the moving charges within the wire will experience a force; hence the wire will experience a force. According to Eq. (30.23), the amount of moving charge in a segment dl of the wire is

$$dq = I\frac{dl}{v}$$

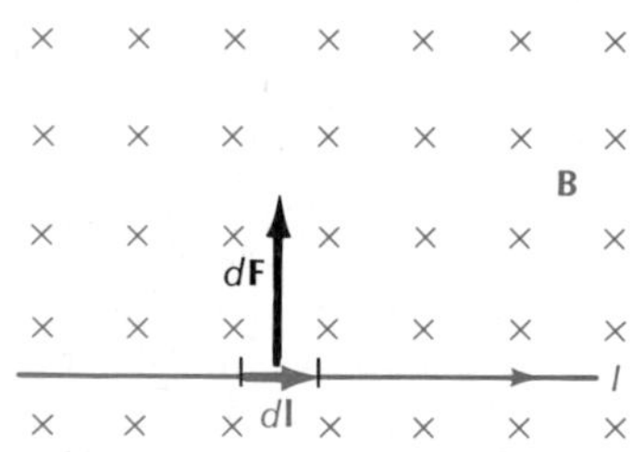

Fig. 31.24 A long straight wire in a uniform magnetic field. The force on a wire segment $d\mathbf{l}$ is perpendicular to $d\mathbf{l}$ and to **B**.

where, as always, we pretend that the moving charge is positive. If θ is the angle between the velocity and **B**, or, equivalently, the angle between the wire segment and **B** (Figure 31.24), then the force on the wire segment is

$$dF = dqvB \sin\theta = I\, dlB \sin\theta \tag{34}$$

This force can be expressed in vector form by an argument similar to that given in connection with Eq. (30.25):

Force on wire segment

$$\boxed{d\mathbf{F} = I\, d\mathbf{l} \times \mathbf{B}} \tag{35}$$

where the vector $d\mathbf{l}$ is along the wire in the direction of the current.

To find the net force on the entire wire we must integrate Eq. (35) along the wire.

EXAMPLE 6. Two very long, parallel wires separated by a distance r carry currents I and I', respectively. Find the magnetic force acting between them.

SOLUTION: Each wire generates a magnetic field, which exerts a force on the other wire. Figure 31.25 shows the magnetic field $\mathbf{B}'$ that the current I' produces in the space surrounding the current I. This magnetic field has a magnitude

$$B' = \frac{\mu_0}{2\pi}\frac{I'}{r} \tag{36}$$

and is perpendicular to the current I. By Eq. (35), the force on a segment dl of the wire carrying the current I is

$$dF = IB'\,dl = \frac{\mu_0}{2\pi}\frac{II'}{r}\,dl \tag{37}$$

The force acting between the wires is attractive if the currents are parallel, and repulsive if they are antiparallel. Of course, the net force between two infinitely long wires is infinite. It is therefore useful to consider the force per unit length; this is finite,

$$\frac{dF}{dl} = \frac{\mu_0}{2\pi}\frac{II'}{r} \tag{38}$$

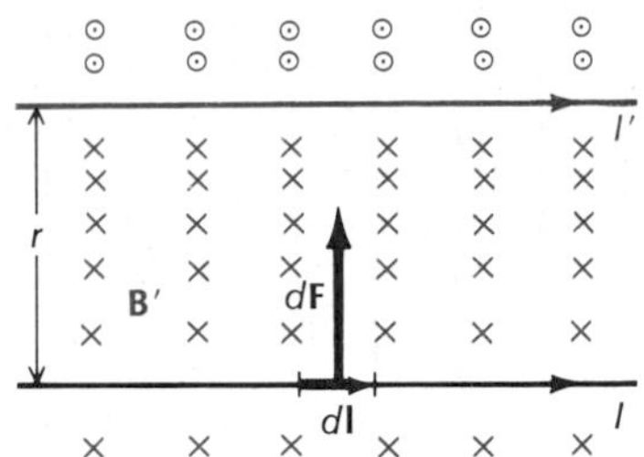

Fig. 31.25 A long straight wire carrying a current I in the magnetic field of a long straight wire carrying a current I'. The magnetic field lines of the current I' are circles perpendicular to the plane of the page. The crosses show the tails of the magnetic field vectors; the dots show the tips of the magnetic field vectors.

SI unit of current

The official definition of the **SI unit of current** is based on the force per unit length between two long parallel wires. This force can be measured very precisely by holding one wire stationary and suspending the other from a balance; the wires are connected in series so that the currents are exactly equal ($I = I'$). The force per unit length and the distance can be measured experimentally, and the value of I calculated from Eq. (38) is then the current in amperes. The constant μ_0 appearing in Eq. (38) is given the value $\mu_0 = 4\pi \times 10^{-7}\ \mathrm{N\cdot s^2/C^2}$ by *definition*.[2]

SI unit of electric charge

The official definition of the **SI unit of electric charge** is based on the unit of current. The coulomb is defined as the amount of charge that a current of one ampere will deliver in one second.

31.5 Torque on a Current Loop

The action of a magnetic field on a loop of wire carrying a current will in general result not only in a net force on the loop, but also in a torque. The forces and torques depend on the orientation and the shape of the loop. We will consider the special case of a rectangular loop placed in a uniform magnetic field.

Figure 31.26 shows the loop oriented perpendicular to the magnetic field. The forces on the four sides of the loop can be calculated from Eq. (35). The forces on opposite sides are opposite; hence the forces cancel in pairs and the loop will be in equilibrium.

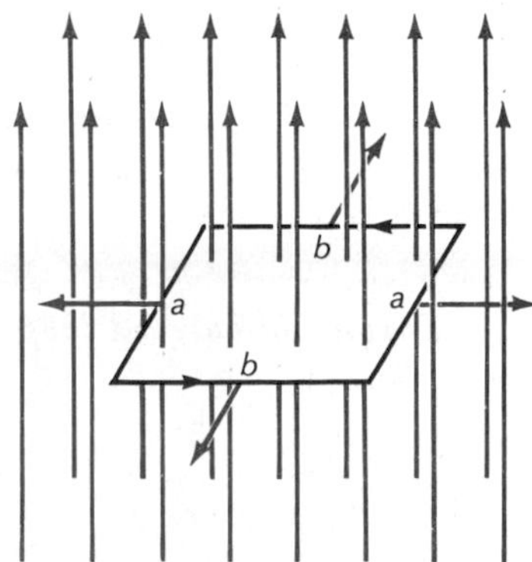

Fig. 31.26 Rectangular loop of current perpendicular to a uniform magnetic field.

[2] In contrast, the constant ε_0 appearing in Coulomb's Law must be determined by experiment.

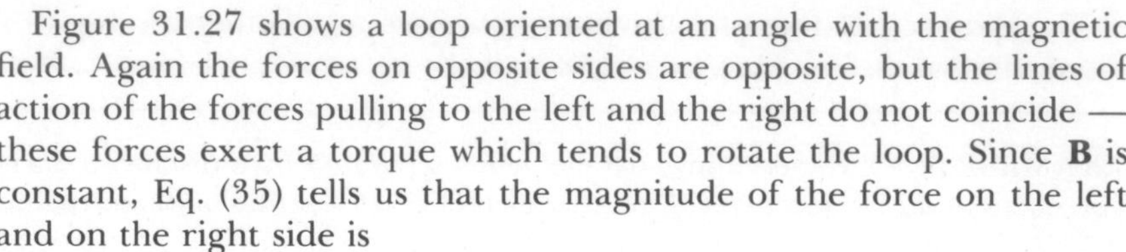

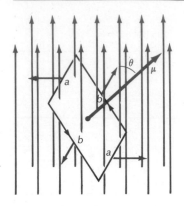

Fig. 31.27 Rectangular loop of current at an angle with a uniform magnetic field.

Figure 31.27 shows a loop oriented at an angle with the magnetic field. Again the forces on opposite sides are opposite, but the lines of action of the forces pulling to the left and the right do not coincide — these forces exert a torque which tends to rotate the loop. Since **B** is constant, Eq. (35) tells us that the magnitude of the force on the left and on the right side is

$$F = IaB \tag{39}$$

where a is the length of the side (Figure 31.27). The torque due to this pair of forces is

$$\tau = F(b/2)\sin\theta + F(b/2)\sin\theta$$

$$= IabB\sin\theta \tag{40}$$

where b is the length of the other side (Figure 31.27) and θ is the angle between the directions of **B** and of the normal to the loop.

According to the definition given in Section 30.3, the product of the area of the loop and the current around it is the **magnetic dipole moment**

$$\mu = Iab \tag{41}$$

Note that if the loop consists of several turns of wire, then Eq. (41) must be evaluated with the net current flowing around the rim of the loop. This net current equals the number N of turns of wire times the current I_0 in one turn so that

$$\mu = NI_0ab$$

In any case, the torque is

$$\tau = \mu B\sin\theta \tag{42}$$

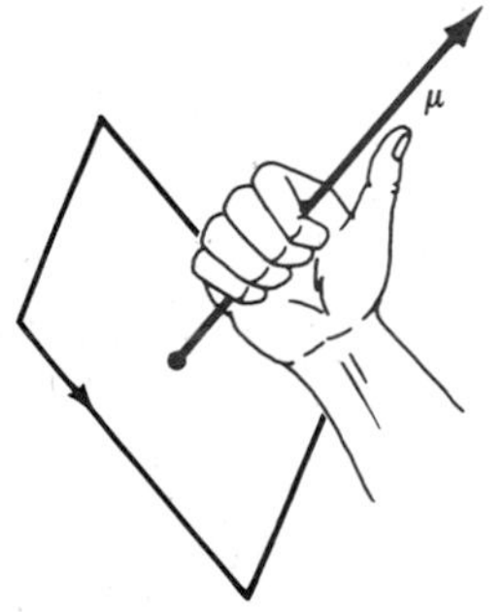

Fig. 31.28 Right-hand rule for the magnetic moment: if the fingers are curled around the loop in the direction of the current, the thumb points in the direction of μ.

This formula turns out to be valid not only for the rectangular loops, but also for loops of arbitrary shape. The direction of the torque is such that it tends to twist the loop into an orientation perpendicular to the magnetic field.

To express the torque in vector form, we must regard the magnetic dipole moment as a vector; this vector is perpendicular to the loop, in a direction given by the right-hand rule (fingers curled along current, thumb along μ; Figure 31.28). With this vector, we can write

Torque on current loop

$$\boxed{\tau = \mu \times \mathbf{B}} \tag{43}$$

This expression for the torque on a magnetic dipole in a magnetic field is mathematically similar to the expression for the torque on an electric dipole in an electric field [see Eq. (23.30)]. The torque tends to twist the dipole so as to bring μ into alignment with **B**.

We can also write a potential energy such that changes in this potential energy represent the work that you must do against the torque (43) when changing the orientation of the loop. The expression for the potential energy is

$$\boxed{U = -\boldsymbol{\mu} \cdot \mathbf{B}} \quad (44)$$

Potential energy of current loop

The derivation of this expression is entirely analogous to the derivation of the expression (23.32) for the potential energy of an electric dipole in an electric field. As expected, the potential energy (44) has a minimum for $\boldsymbol{\mu}$ parallel to **B**, and a maximum for $\boldsymbol{\mu}$ antiparallel to **B**.

A loop of current suitably pivoted on an axis acts as a compass needle; the normal of the loop seeks to align itself with the magnetic field. This similarity is no accident. As we will see in Section 33.3, a compass needle actually consists of a large number of current loops. The mechanism underlying the north-seeking behavior of a compass needle is essentially as described above.

EXAMPLE 7. A simple electric motor consists of a rectangular coil of wire that rotates on a longitudinal axle in a magnetic field of 0.50 T (Figure 31.29). The coil measures 10 cm × 20 cm; it has 40 turns of wire and the current in the wire is 8.0 A.

(a) As a function of the angle θ between the lines of **B** and the normal to the coil, what is the torque that the magnetic field exerts on the coil?

(b) In order to keep the sign of the torque constant, a switch (commutator) mounted on the axle reverses the current in the coil whenever θ passes through 0 and π radians. Plot this torque as a function of θ.

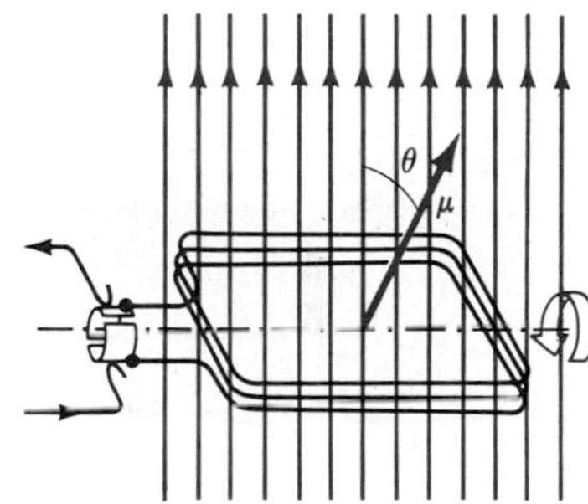

Fig. 31.29 An electric motor.

SOLUTION: (a) According to Eq. (42), the torque on the coil is

$$\tau = NI_0abB \sin\theta$$

$$= 40 \times 8.0 \text{ A} \times 0.10 \text{ m} \times 0.20 \text{ m} \times 0.50 \text{ T} \times \sin\theta$$

$$= 3.2 \text{ N} \cdot \text{m} \sin\theta$$

(b) If the torque always has the same sign (say, positive), we can write it as

$$\tau = 3.2 \text{ N} \cdot \text{m} \, |\sin\theta|$$

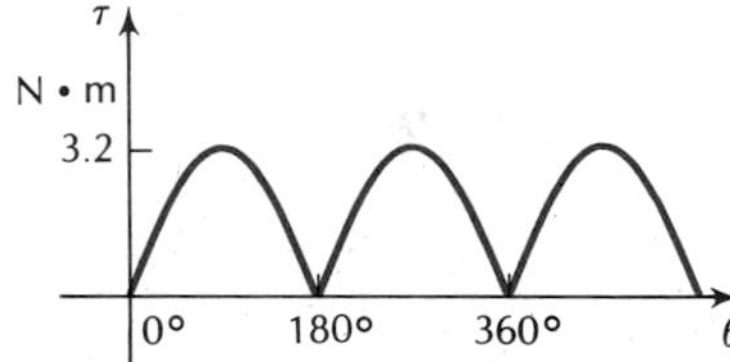

Fig. 31.30 Torque as a function of angle according to Example 7.

Figure 31.30 shows a plot of this torque.

Practical electric motors consist of many such coils, each with its commutator, arranged at regular intervals around the axle. The plot for the net torque of this arrangement is a sum of plots such as shown in Figure 31.30, but with different starting angles. This averages out the ups and downs of Figure 31.30 and yields a torque that is nearly constant as a function of angle.

SUMMARY

Ampère's Law: $\oint \mathbf{B} \cdot d\mathbf{l} = \mu_0 I$

Field in solenoid: $B = \mu_0 I_0 n$

Cyclotron frequency: $\nu = \dfrac{qB}{2\pi m}$

Circular orbit in magnetic field: $r = \dfrac{p}{qB}$

Lorentz force: $\mathbf{F} = q\mathbf{E} + q\mathbf{v} \times \mathbf{B}$

Force on wire segment: $d\mathbf{F} = I\, d\mathbf{l} \times \mathbf{B}$

Torque on current loop: $\boldsymbol{\tau} = \boldsymbol{\mu} \times \mathbf{B}$

Potential energy of current loop: $U = -\boldsymbol{\mu} \cdot \mathbf{B}$

QUESTIONS

1. Suppose you evaluate the integral $\oint \mathbf{B} \cdot d\mathbf{l}$ for the magnetic field of the Earth along a closed path that coincides with a meridian circle. What is the value of the integral?

2. Is the magnetic field a conservative field?

3. A long solenoid has been placed inside another long solenoid of larger radius. The solenoids are coaxial and both have the same number of turns per unit length and the same current. What is the formula for the magnetic field in the region within the smaller solenoid? Between the smaller and the larger solenoid?

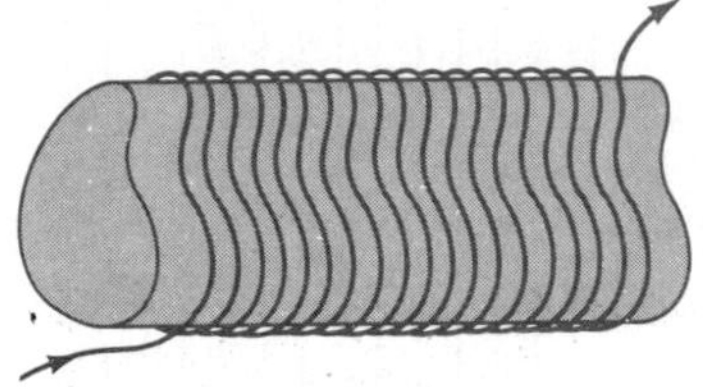

Fig. 31.31 Cylinder of arbitrary, noncircular cross section.

4. Figure 31.31 shows a solenoid of arbitrary cross section; this solenoid is a (noncircular) cylinder. Suppose that there are n turns of wire per unit length and that the current in the wire is I. Use Ampère's Law to show that the magnetic field in the solenoid is $\mu_0 In$, the same as for a circular cylinder.

5. Consider a long solenoid and a long straight wire along its axis, both carrying some current. Describe the field lines within the solenoid.

6. The drawing of Figure 31.8 shows the field lines near the gap of an electromagnet. Describe the pattern of field lines beyond the edges of the drawing.

7. Cosmic rays are high-speed charged particles — mostly protons — that crisscross interstellar space and strike the Earth from all directions. Why is it easier for the cosmic rays to penetrate through the magnetic field of the Earth near the poles than anywhere else?

8. Is it possible to define a magnetic potential energy for a charged particle moving in a magnetic field?

9. If we want a proton to orbit all the way around the equator in the Earth's magnetic field, must we send it eastward or westward?

10. Consider the strip of metal placed in a magnetic field, as shown in Figure 31.23. How does the Hall potential difference between the edges of the strip change if we reverse the current? If we reverse the magnetic field? If we reverse both?

11. If we replace the metallic strip in Figure 31.23 by a semiconducting strip of p-type semiconductor, will the lower edge remain at the higher potential?

12. The Earth's magnetic field at the equator is horizontal and in the northward direction. Assume that the atmospheric electric field is downward. The electric and magnetic fields are then crossed fields. In what direction must we launch an electron if it is to move without any deflection?

13. If a strong current flows through a thick wire, it tends to cause a compression of the wire. Explain.

14. If a wire carrying a current is placed in a magnetic field, the field exerts forces on the moving electrons in the wire. How does this cause a push on the wire?

15. A horizontal wire carries a current in the eastward direction. The wire is in a uniform magnetic field. What must be the direction of this field if it is to compensate for the weight of the wire?

16. In the SI system, we first define the unit of current (by means of the mag-

netic force between wires), and we then define the unit of charge in terms of the current. Could we proceed in the opposite manner: first define the unit of electric charge (by means of the electric force between charges), and then define the unit of current as a flow of charge? What would be the advantages and the disadvantages?

Ammeter

17. Figure 31.32 is a schematic diagram showing the mechanism of an **ammeter**. The mechanism consists of a coil suspended between the poles of a permanent magnet. The coil can rotate about an axis perpendicular to the magnetic field, and is held in its equilibrium position by a spiral spring. To operate the ammeter, the current of the external circuit is made to pass through the coil. Explain how this ammeter registers the current. Explain how a sensitive ammeter of this kind can be used as a voltmeter by connecting the coil in series with a large resistance.

Tangent galvanometer

18. The **tangent galvanometer** is an old form of ammeter, consisting of an ordinary magnetic compass mounted at the center of a coil whose axis is horizontal and oriented along the east–west line (Figure 31.33). If there is no current in the coil, the compass needle points north. Explain how the compass needle will deviate from north when there is a current in the coil.

19. A simple indicator of electric current, first used by H. C. Oersted in his early experiments, consists of a compass needle placed below a wire stretched in the northward direction. Explain how the angle of the compass needle indicates the electric current in the wire.

20. Positive charge is uniformly distributed over a sphere. If this sphere spins about a vertical axis in a counterclockwise direction as seen from above, what is the direction of the magnetic moment vector?

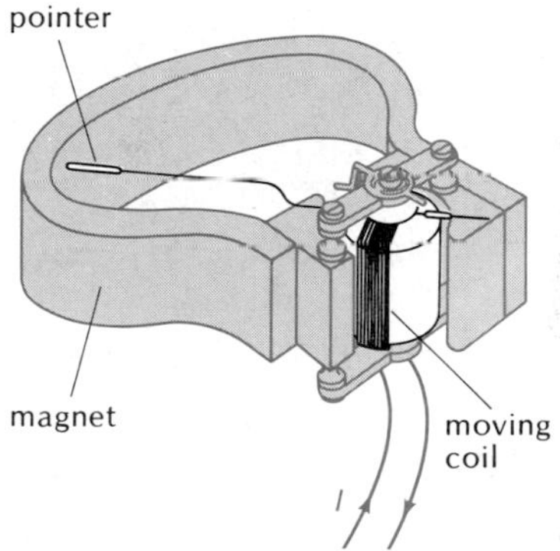

Fig. 31.32 Mechanism of the moving-coil ammeter.

PROBLEMS

Section 31.1

1. A wire of superconducting niobium, 0.20 cm in diameter, can carry a current of up to 1900 A. What is the strength of the magnetic field just outside of the wire when it carries this current?

2. A long, straight wire of copper with a radius of 1 mm carries a current of 20 A. What is the instantaneous magnetic force and the corresponding acceleration on one of the conduction electrons moving at 10^6 m/s along the surface of the wire in a direction opposite to that of the current? What is the direction of the acceleration?

3. A current I flows in a thin wire bent into a circle of radius R. The axis of the circle coincides with the z axis. What is the value of the integral $\int \mathbf{B} \cdot d\mathbf{l}$ along the z axis from $z = -\infty$ to $z = +\infty$?

4. In a proton accelerator, protons of velocity 3×10^8 m/s form a beam of current of 2×10^{-3} A. Assume that the beam has a circular cross section of radius 1 cm and that the current is uniformly distributed over the cross section. What is the magnetic field that the beam produces at its edge? What is the magnetic force on a proton at the edge of the beam?

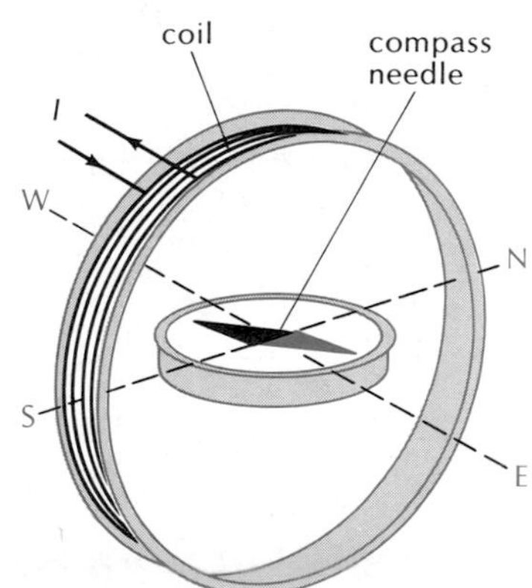

Fig. 31.33

5. A very large, thin conducting plate lies in the x–y plane. The plate carries a current in the y direction. The current is uniformly distributed over the plate with σ ampere flowing across each 1-m length along the x axis. Use Ampère's Law to find the magnetic field at some distance from the plate. (Hint: The magnetic field lines are parallel to the plate.)

6. One very large conducting plate coincides with the x–y plane. Another very large conducting plate coincides with the x–z plane. Each plate carries a uniformly distributed current with σ ampere flowing across each 1-m length per-

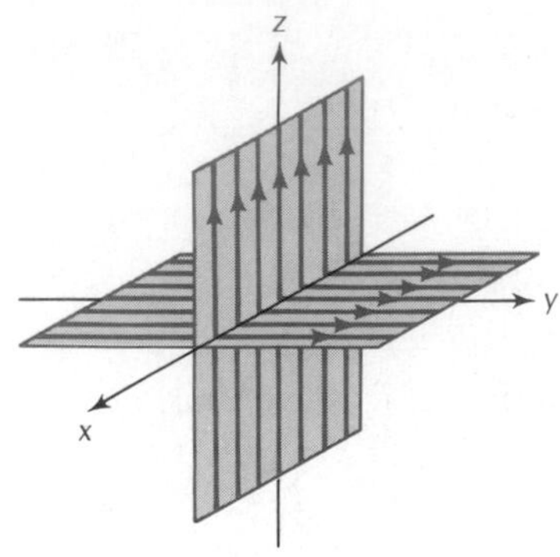

Fig. 31.34 Two large plates intersecting at right angles.

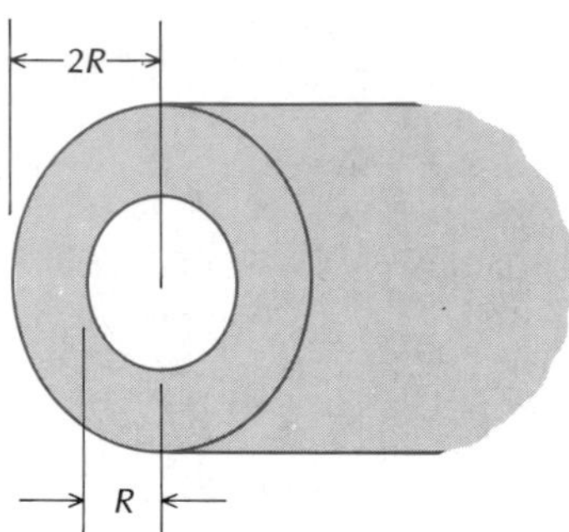

Fig. 31.35

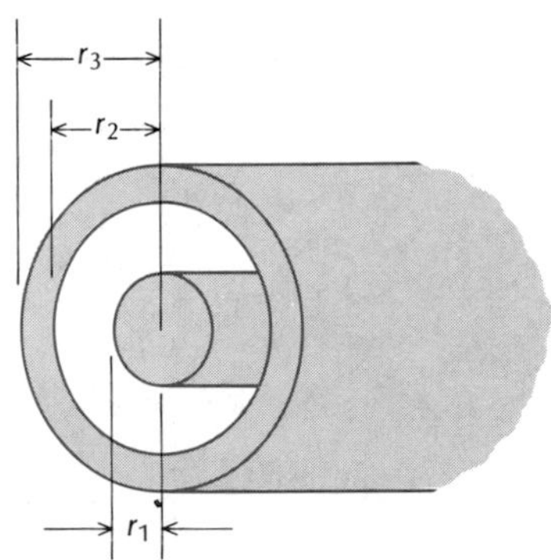

Fig. 31.36 Coaxial cylinder and cylindrical shell.

pendicular to the current (Figure 31.34). Find the magnetic field in each of the four quadrants. (Hint: See Problem 5.)

Section 31.2

7. A long solenoid has 15 turns per centimeter. What current must we put through its windings if we wish to achieve a magnetic field of 5.0×10^{-2} T in its interior?

8. The electromagnet of a small electric bell is a solenoid with 260 turns in a length of 2.0 cm. What magnetic field will this solenoid produce if the current is 8.0 A?

9. A long copper pipe with thick walls has an inner radius R and an outer radius $2R$ (Figure 31.35). A current I flows along this wall, uniformly distributed over the volume of copper. Find the magnetic fields at the radius $\frac{3}{2}R$ and at the radius $3R$.

10. A long solenoid of n turns per unit length carries a current I, and a long straight wire lying along the axis of this solenoid carries a current I'. Find the net magnetic field within the solenoid, at a distance r from the axis. Describe the shape of the magnetic field lines.

11. A coaxial cable consists of a long cylindrical wire of radius r_1 surrounded by a cylindrical shell of inner radius r_2, outer radius r_3 (Figure 31.36). The wire and the shell each carry equal and opposite currents I uniformly distributed over their volumes. Find formulas for the magnetic field in each of the regions $r < r_1$, $r_1 < r < r_2$, $r_2 < r < r_3$, $r > r_3$.

12. A toroid used in plasma research has 240 turns of wire carrying a current of 7.2×10^4 A. The inner radius of the toroid is 0.50 m and the outer radius is 1.50 m. What is the strength of the magnetic field at the inner radius? At the outer radius?

13. The coil of an electromagnet consists of a large number of layers of very thin wire wound on a cylindrical core. The inner radius of the windings is r_1 and the outer radius is r_2. The number of turns in each layer is n per unit length and the number of layers is n' per unit length in the radial direction. The current in the turns of wire is I. Find formulas for the magnetic field in the region $r < r_1$ and in the region $r_1 < r < r_2$.

Section 31.3

14. A bubble chamber, used to make the tracks of protons and other charged particles visible, is placed between the poles of a large electromagnet that produces a uniform magnetic field of 20 T. A high-energy proton passing through the bubble chamber makes a track that is an arc of a circle of radius 3.5 m. According to Eq. (28), what is the momentum of the proton?

15. A proton of energy 1.0×10^7 eV moves in a circular orbit in the magnetic field near the Earth. The strength of the field is 0.50×10^{-4} T. What is the radius of the orbit?

16. In principle, a proton of the right energy can orbit the Earth in an equatorial orbit under the influence of the Earth's magnetic field. If the orbital radius is to be 6.5×10^3 km and the magnetic field at this radius is 0.33×10^{-4} T, what must be the momentum of the proton?

17. In the Crab Nebula (the remnant of a supernova explosion), electrons of a momentum of up to about 10^{-16} kg · m/s orbit in a magnetic field of about 10^{-8} T. What is the orbital radius of such electrons? Note that it is necessary to use Eq. (28) for the calculation of the orbital radius.

18. At the Fermilab accelerator, protons of momentum 5.3×10^{-16} kg · m/s are held in a circular orbit of diameter 2.0 km by a vertical magnetic field. What is the strength of the magnetic field required for this?

19. Figure 31.37 shows the tracks of an electron (charge $-e$) and an antielectron (charge $+e$) created in a bubble chamber. When the particles made these tracks, they were under the influence of a magnetic field of a magnitude 1.0 T and of a direction perpendicular to and into the plane of the figure. What is the momentum of each particle? Assume that they are moving in the plane of the figure and that this figure is $\frac{1}{10}$ natural size. Which is the track of the electron and which is the track of the antielectron?

Fig. 31.37 Tracks of an electron and an antielectron in a bubble chamber. The tracks spiral because the particles suffer a loss of energy as they pass through the liquid. For the purposes of Problem 19, concentrate on the initial portions of the tracks.

20. You want to confine an electron of energy 3.0×10^4 eV by making it circle inside a solenoid of radius 10 cm under the influence of the force exerted by the magnetic field. The solenoid has 120 turns of wire per centimeter. What minimum current must you put through the wire if the electron is not to hit the wall of the solenoid?

21. In the Brookhaven AGS accelerator, protons are made to move around a circle of radius 128 m by a magnetic force exerted by a vertical magnetic field. The maximum field that the magnets of this accelerator can generate is 1.3 T.

(a) Calculate the maximum permissible momentum of the protons.

(b) Calculate the maximum kinetic energy (use the relativistic relation between energy and momentum).

(c) Calculate the orbital frequency of such protons.

22. Some astrophysicists believe that the radio waves of 10^9 Hz reaching us from Jupiter are emitted by electrons of fairly low (nonrelativistic) energies orbiting in Jupiter's magnetic field. What must be the strength of this field if the cyclotron frequency is to be 10^9 Hz?

23. A velocity selector with crossed **E** and **B** fields is to be used to select alpha particles of energy 2.0×10^5 eV from a beam containing particles of several energies. The electric field strength is 1.0×10^6 V/m. What must be the magnetic field strength?

24. The Earth's magnetic field at the equator has a magnitude of 5×10^{-5} T; its direction is horizontal toward the north. Suppose that the Earth's atmospheric electric field is 100 V/m; its direction is vertically down. With what velocity (magnitude and direction) must we launch an electron if the electric force is to cancel the magnetic force?

25. In a **mass spectrometer**, a beam of ions is first made to pass through a velocity selector with crossed **E** and **B** fields. The selected ions then are made to enter a uniform magnetic field **B′** where they move in arcs of circles (Figure 31.38). The radius of these circles depends on the masses of the ions. Assume that an ion has a single charge e.

(a) Show that in terms of E, B, B', and of the impact distance l marked in Figure 31.38, the mass of the ion is

$$m = \frac{eBB'l}{2E}$$

(b) Assume that in such a spectrometer, ions of the isotope ^{16}O impact at a distance of 29.20 cm and ions of a different isotope of oxygen impact at a distance of 32.86 cm. The mass of the ions of ^{16}O is 16.00 u. What is the mass of the other isotope?

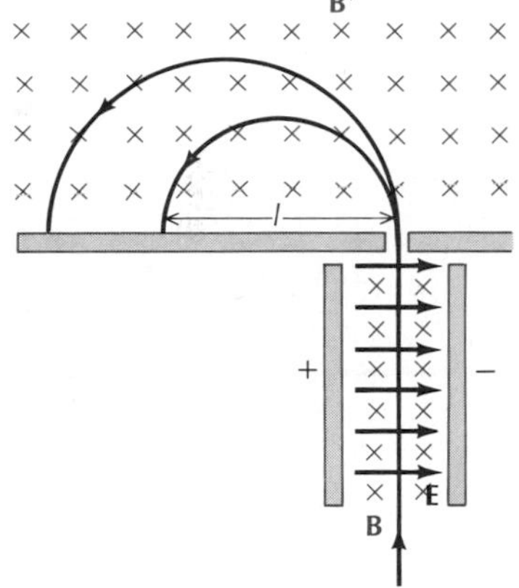

Fig. 31.38 Schematic diagram of a mass spectrometer. The magnetic fields **B** and **B′** are perpendicular to the plane of the page. The crosses show the tails of the magnetic field vectors.

Section 31.4

26. A straight wire is placed in a uniform magnetic field; the wire makes an angle of 30° with the magnetic field. The wire carries a current of 6.0 A and the magnetic field has a strength of 0.40 T. Calculate the force on a 10-cm segment of this wire. Show the direction of the force in a diagram.

27. Figure 31.39 shows a balance used for the measurement of a magnetic field. A loop of wire carrying a precisely known current is partially immersed in the magnetic field. The force that the magnetic field exerts on the loop can

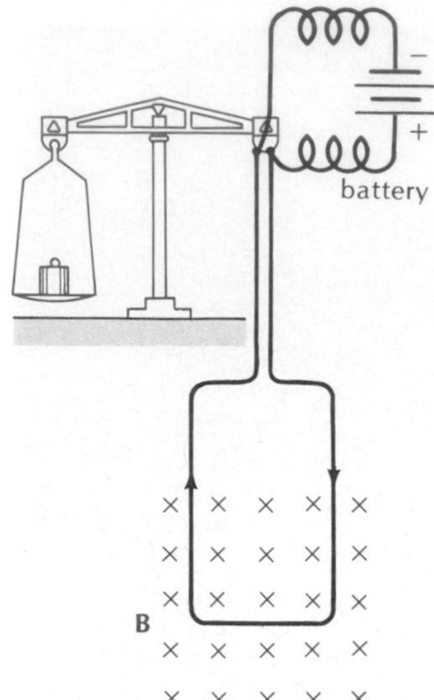

Fig. 31.39 A current balance. The magnetic field $\boldsymbol{B}$ is perpendicular to the plane of the page. The crosses show the tails of the magnetic field vectors.

be measured with the balance and this permits the calculation of the strength of the magnetic field. Suppose that the short side of the loop measures 10.0 cm, the current in the wire is 0.225 A, and the magnetic force is 5.35×10^{-2} N. What is the strength of the magnetic field?

28. The electric cable supplying an electric clothes dryer consists of two long straight wires separated by a distance of 1.2 cm. Opposite currents of 20 A flow on these wires. What is the magnetic force experienced by a 1.0-cm segment of wire due to the entire length of the other wire?

29. Two parallel cables of a high-voltage power line carry opposite currents of 1.8×10^3 A. The distance between the cables is 4.0 m. What is the magnetic force pushing on a 50-m segment of one of these cables? Treat both cables as very long, straight wires.

30. A rectangular loop of wire of dimensions 12 cm $\times$ 18 cm is near a long, straight wire. One of the short sides of the rectangle is parallel to the straight wire and at a distance of 6.0 cm; the long sides are perpendicular to the straight wire (Figure 31.40). A current of 40 A flows on the straight wire and a current of 60 A flows around the loop. What are the magnitude and direction of the net magnetic force that the straight wire exerts on the loop?

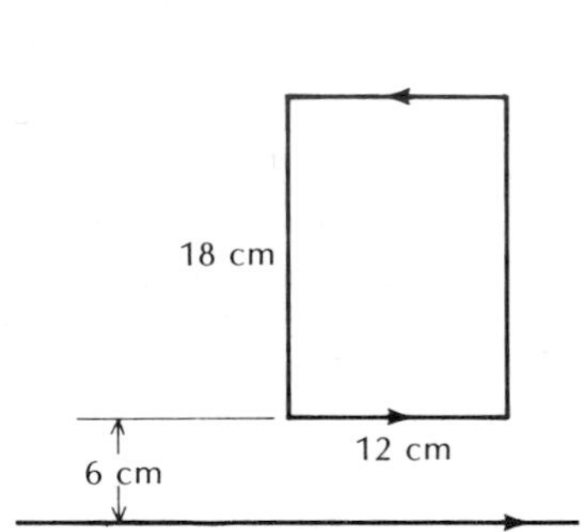

Fig. 31.40 Long, straight wire and rectangular loop.

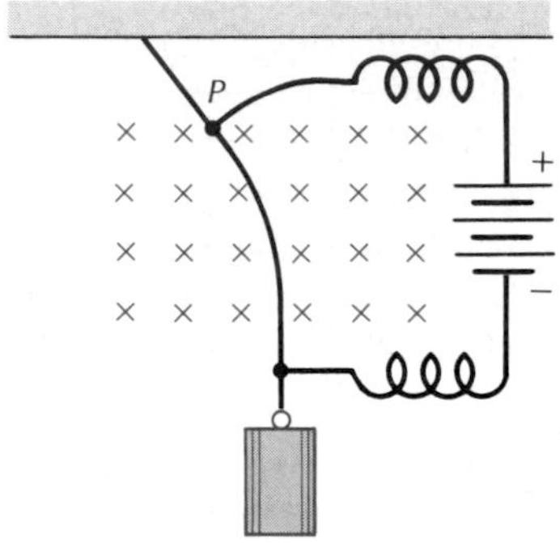

Fig. 31.41 A wire with a weight hanging in a magnetic field.

*31. A thin, flexible wire carrying a current I hangs in a uniform magnetic field $\mathbf{B}$ (Figure 31.41). A weight attached to one end of the wire provides a tension T. Within the magnetic field, the wire will adopt the shape of an arc of circle.

(a) Show that the radius of this arc of circle is $r = T/BI$.

(b) Show that if we remove the wire and launch a particle of charge $-q$ from the point P with a momentum $p = q\,T/I$ in the direction of the wire, it will move along the same arc of circle. (This means that the wire can be used to simulate the orbit of the particle. Experimental physicists sometimes use such wires to check the orbits of particles through systems of magnets.)

Section 31.5

32. A horizontal circular loop of wire of radius 20 cm carries a current of 25 A. At the location of the loop, the magnetic field of the Earth has a magnitude of 0.39×10^{-4} T and points down at an angle of 16° with the vertical. What is the magnitude of the torque that this magnetic field exerts on the loop?

33. The proton has a magnetic moment of 1.41×10^{-26} A · m².

(a) If this magnetic moment makes an angle of 45° with a uniform magnetic field of 0.80 T, what is the torque on the proton?

(b) If the magnetic moment is initially oriented antiparallel to the field, how much energy will be released when the proton flips into the parallel orientation?

34. Consider the electric motor of Example 7. What is the average value of the torque over a complete rotation? If this motor rotates at the rate of 50 rev/s what is the average horsepower that it delivers to its axle?

35. A molecule with a magnetic moment $\mu = 9.3 \times 10^{-24}$ A · m² is in a uniform magnetic field $B = 0.80$ T. What is the difference in the potential energy between the parallel and antiparallel orientations for μ and **B**? Express your answer in electron-volts.

*36. A rectangular loop of wire of dimension 12 cm × 25 cm faces a long, straight wire. The two long sides of the loop are parallel to the wire and the two short sides are perpendicular; the midpoint of the wire is 8.0 cm from the wire (Figure 31.42). Currents of 95 A and 70 A flow in the straight wire and the loop, respectively.

(a) What translational force does the straight wire exert on the loop?

(b) What torque about an axis parallel to the straight wire and through the center of the loop does the straight wire exert on the loop?

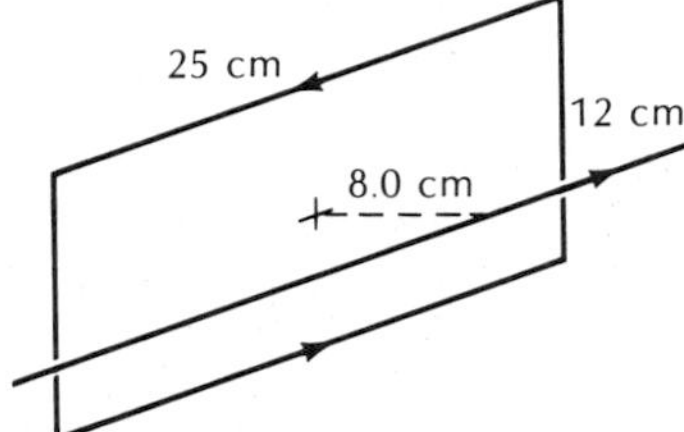

Fig. 31.42 Long, straight wire and rectangular loop.

*37. A compass needle is attached to an axle which permits it to turn freely in a horizontal plane so that only the horizontal component of the magnetic field affects its motion. The magnetic moment of the needle is 9.0×10^{-3} A · m², its moment of inertia is 2.0×10^{-8} kg · m², and the horizontal component of the magnetic field is 0.19×10^{-4} T.

(a) What is the torque on the needle as a function of the angle θ between the needle and the north direction?

(b) What is the frequency of small oscillations of the needle about the north direction? [Hint: Compare the equation for the rotational motion of the needle with Eq. (14.71) for the motion of a pendulum.]

CHAPTER 32

Electromagnetic Induction

In this chapter we will discover that electric fields can be generated not only by charges, but also by magnetic fields. Whenever the magnetic field lines move or change in any way, they will generate an electric field, an **induced electric field.** This kind of electric field exerts electric forces on charges — in this regard the induced electric field does not differ from an ordinary electrostatic electric field. However, the induced and electrostatic electric fields differ in that the latter is conservative while the former is not. This means that the path integral of **E** around a closed path is not zero if **E** is an induced electric field. The path integral of the induced electric field is called the **induced emf.** As we will see, a famous law formulated by Michael Faraday asserts that the magnitude of this induced emf is directly related to the rate at which the magnetic field lines cut across the path.

Induced emf

We begin with a study of the induced emf produced by the motion of a path in a constant magnetic field. Such a path cuts across field lines, and thereby generates an induced emf, called a motional emf.

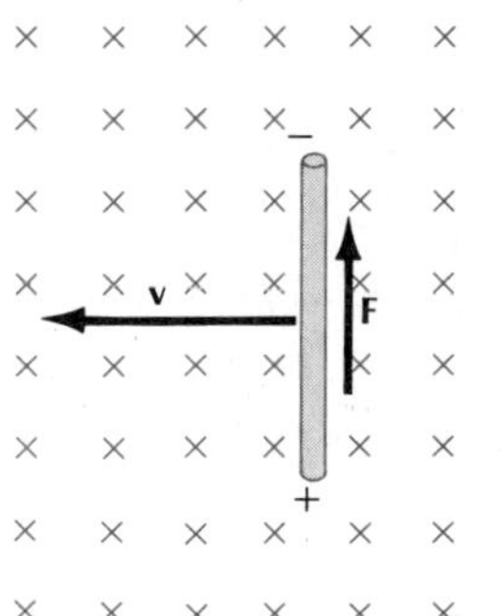

Fig. 32.1 Metallic rod moving with velocity **v** through a uniform magnetic field. A free electron within this rod experiences a force in the upward direction.

32.1 Motional Electromotive Force

Suppose that we push a rod of metal with some velocity through a uniform magnetic field, such as the magnetic field of a solenoid. If the rod and the velocity **v** of the rod are perpendicular to each other and to the magnetic field (Figure 32.1), then the free electrons in the metal will experience a magnetic force evB directed along the rod. The electrons will therefore flow along the rod, accumulating negative charge on the upper end and leaving positive charge on the lower end. This flow of charge will stop when the electric repulsion generated by the accumulated charges balances the magnetic force evB. However, if the ends of the rod are in sliding contact with a pair of long wires that provide a (stationary) return path (Figure 32.2), then the electrons will not accumulate on the ends of the rod — they will flow continuously around the circuit. Thus, the moving rod acts as a "pump of electricity." The rod is a source of emf. The upper end of the rod will act as the negative terminal of this source, and the lower end as the positive terminal.

The emf associated with the rod is the work done by the driving force on a unit positive charge that passes from the negative end of the rod to the positive. The driving force on a unit positive charge is vB and, if the length of the rod is l, the work done by this force is

$$\boxed{\mathscr{E} = vBl} \tag{1}$$

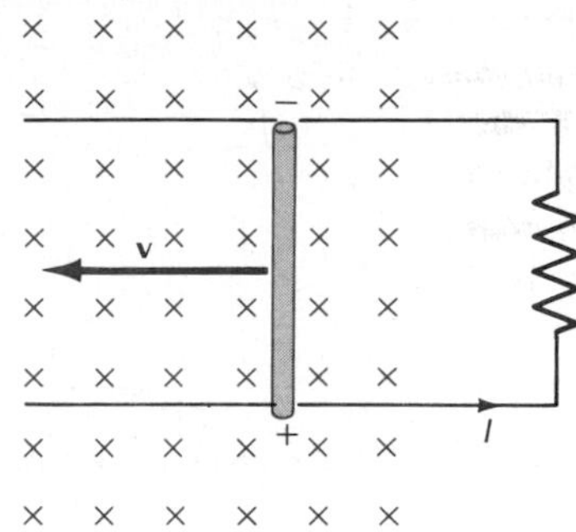

Fig. 32.2 If the ends of the rod are in sliding contact with a pair of wires, a current will flow around the circuit.

Motional emf

This is called a **motional emf** because it is generated by the motion of the rod through the magnetic field. Note that we have ignored the component of velocity associated with motion of the charge *along* the rod. This component leads to a magnetic force perpendicular to the length of the rod and that force is canceled by the constraining force that the sides of the rod exert on the charges, preventing their escape. The cancellation of one part of the magnetic force is crucial for the result stated in Eq. (1) — if it were not for this cancellation, the total magnetic force would do zero work [compare Eq. (31.22)]. Of course, the energy delivered to the current by the emf ultimately does not come from the magnetic field but from the mechanical device that pushes the rod. The magnetic field merely plays the role of mediator, transferring the mechanical energy into electric energy. If we did not apply an external push to the rod of Figure 32.2, it would slow down and stop as its kinetic energy is converted into electric energy and delivered to the current.

The quantity Blv appearing on the right side of Eq. (1) can be given the following interpretation: the product lv is the area that the rod sweeps across per unit time and B equals the number of lines of magnetic field per unit area;[1] hence Blv equals the number of magnetic field lines that the rod cuts across per unit time. We can then write Eq. (1) as

$$\mathscr{E} = [\text{rate of cutting of magnetic field lines}] \tag{2}$$

The advantage of Eq. (2) over Eq. (1) is that the former equation is of general validity — it holds for rods and other bodies of arbitrary shape moving through arbitrary magnetic fields (this can be proved by treating a rod of arbitrary shape as made of small straight segments chosen so that the magnetic field is approximately constant in the vicinity of each segment). Furthermore, as we will see in the next section, the relation between the induced emf and the rate of cutting of field lines holds even if the magnetic field is time dependent. Hence the concept of cutting of field lines, which is due to Faraday, not only provides us with a sharp mental picture of the mechanism of induction, but also helps us in the formulation of the general law of induction.

Michael Faraday, *1791–1867, English physicist and chemist. Faraday was apprentice to a bookbinder when he became interested in science. He attended lectures by Humphry Davy, the famous chemist, and prepared a set of lecture notes which so impressed Davy that he appointed Faraday his assistant at the laboratory of the Royal Institution. Ultimately, Faraday succeeded Davy as director of the laboratory. Faraday's earliest research lay in chemistry, but he soon turned to research in electricity and magnetism, making contributions of the greatest significance. His discovery of electromagnetic induction was no accident, but arose from a systematic experimental investigation of whether magnetic fields can generate electric currents. Although Faraday was essentially an experimenter, with no formal training in mathematics, he made an important theoretical contribution by introducing the concept of field lines and by recognizing that electric and magnetic fields are physical entities.*

[1] Note that this statement hinges on the normalization convention for magnetic field lines adopted in Section 30.2.

However, we must be careful not to interpret Faraday's vivid phrase *cutting of field lines* in a literal sense — the field lines are not physical objects, but merely mathematical constructs, and they suffer no damage when a rod passes through them.

Note that Eq. (2) only gives us the magnitude of the emf. To find the sign of the emf, we must remember that within the moving body that cuts across the field lines, a positive charge will be pushed in the direction of $\mathbf{v} \times \mathbf{B}$.

EXAMPLE 1. A straight metallic rod is rotating about its midpoint on an axis parallel to a uniform magnetic field (Figure 32.3a). The length of the rod is $2l$ and the angular velocity of rotation is ω. What is the induced emf between the midpoint of the rod and each end? What is the induced emf between the two ends?

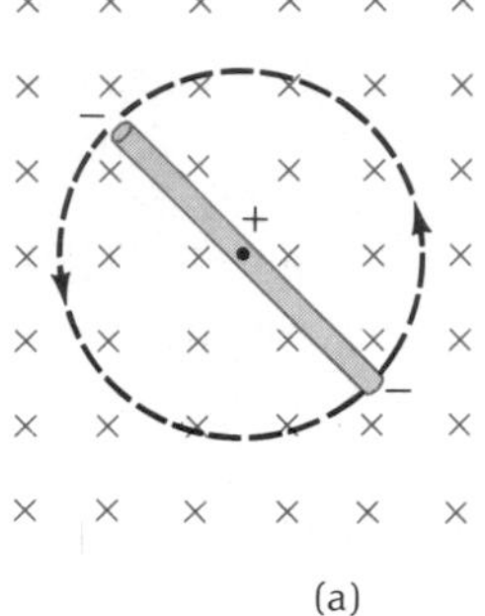

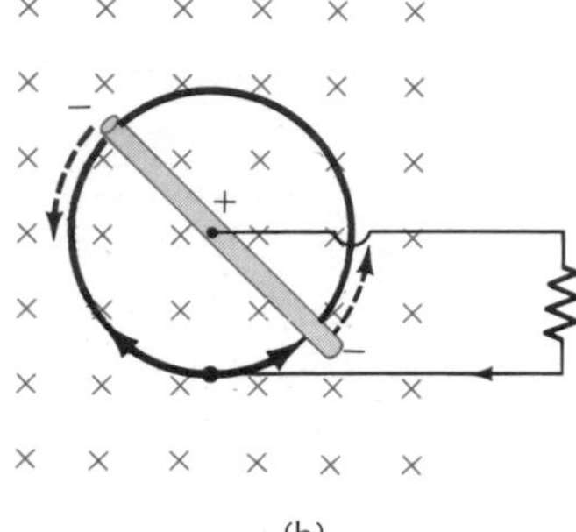

Fig. 32.3 (a) Rod rotating in a uniform magnetic field. (b) If the ends of the rod are in sliding contact with a circular track, a current will flow as shown.

SOLUTION: Consider one-half of the rod, from the midpoint to one end. This piece takes a time $2\pi/\omega$ to sweep out the circular area πl^2. Hence the rate at which this piece sweeps out area is $\pi l^2\omega/2\pi = l^2\omega/2$ and the rate at which it cuts across magnetic field lines is $Bl^2\omega/2$. Equation (2) then leads to an induced emf

$$\mathscr{E} = Bl^2\omega/2 \tag{3}$$

between the midpoint and each end. With the direction of motion as shown in Figure 32.3a, the midpoint will act as the positive terminal of a source of emf and each moving end as the negative terminal.

The emf between the two ends is zero because both ends act as negative terminals, with the same emf (the two halves of the rod act as two batteries connected in parallel). Note that a careless calculation with Eq. (2) would have led us to the wrong conclusion that the emf for the entire rod is twice that for each half, since the entire rod cuts twice as many field lines as each half — but we must take into account the signs of these emfs!

Homopolar generator

If the moving ends of the rod are in sliding contact with a circular track connected to the midpoint via an external circuit (Figure 32.3b), then the rod will generate a current in the circuit. The device shown in Figure 32.3b acts as a generator of DC voltage; this device is called a **homopolar generator.** For practical applications, homopolar generators are constructed with a rotating disk rather than a rotating rod. This does not affect the emf of the generator, but it helps to reduce its internal resistance. Homopolar generators are used in applications requiring a large current but only a fairly small voltage, such as in electroplating.

EXAMPLE 2. A **magnetohydrodynamic (MHD) generator** consists of a rectangular channel within which flows a hot, ionized gas, or plasma. The channel is placed in a strong magnetic field (Fig. 32.4). What emf will be induced between opposite sides of the stream of plasma? The width of the channel is 50 cm, the speed of the gas is 800 m/s, and the strength of the magnetic field is 6.0 T.

Magnetohydrodynamic generator

SOLUTION: Consider a straight path, fixed in the plasma, from one side of the stream of plasma to the other. This path cuts magnetic field lines, just as the straight rod discussed at the beginning of this section; hence Eq. (1) applies,

$$\mathscr{E} = vBl = 800 \text{ m/s} \times 6.0 \text{ T} \times 0.50 \text{ m} = 2.4 \times 10^3 \text{ V} \quad (4)$$

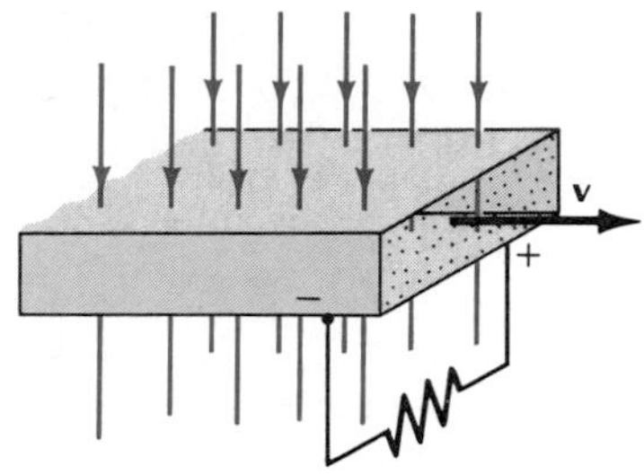

Fig. 32.4 A channel in which plasma flows toward the right. The channel is placed in a vertical magnetic field.

If, as shown in Figure 32.4, the magnetic field points downward, then the left side of the stream of plasma acts as the positive terminal of a source of emf and the right side as the negative terminal. This source of emf will drive a current if we connect an external circuit, as shown in Figure 32.4. Note that the two sides of the channel must be made of a conductor, so that they are in good electrical contact with the plasma, but the top and bottom of the channel must be made of an insulator.

Magnetohydrodynamic generators are expected to serve as auxiliary generators in power plants burning fossil fuel (Figure 32.5). The MHD generator uses the exhaust gas released by the burning of this fuel. For this purpose, the fuel must be burned in a special combustion chamber, similar to that of a jet engine, so as to produce a high-speed stream of very hot exhaust gas. After the hot gas emerges from the MHD generator, it can supply heat to a boiler providing steam for a conventional power generator.

Fig. 32.5 Experimental MHD generator at the Argonne National Laboratory. This generator has a circular channel of diameter ~ 1 m, which is surrounded by a powerful superconducting magnet.

Incidentally, note that the emf induced across a stream of plasma flowing in a channel is quite analogous to the voltage produced by the Hall effect across the stream of electrons flowing in a conducting strip (see Section 31.3). The Hall voltage can be regarded as an induced, motional emf.

32.2 Faraday's Law

As we saw in the preceding section, the induced emf between the ends of a rod moving through a magnetic field equals the rate at which the rod cuts across magnetic field lines. Of course, the rod will cut across

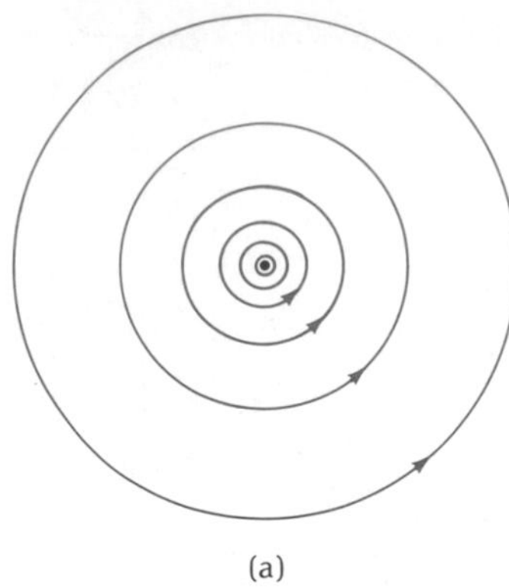

(a)

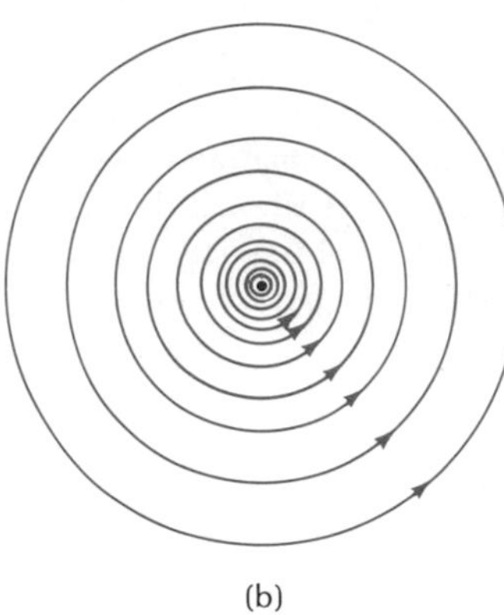

(b)

Fig. 32.6 (a) Magnetic field lines of a current on a very long wire. The wire is perpendicular to the plane of the page. (b) Magnetic field lines of a stronger current.

field lines whenever there is *relative motion* between rod and magnet. Relativity suggests that it should not make any difference whether we move the rod past a fixed magnet or move the magnet past a fixed rod — and experiments do indeed confirm that in both cases the induced emf is exactly the same.

But there is another way in which we can cut across field lines: we can hold the rod fixed and increase or decrease the strength of the magnetic field by increasing or decreasing the current. To understand why field lines will be cut under these conditions, we must first take a look at what happens to the field lines of a current when the current increases or decreases. Figure 32.6a shows the field lines produced by a current on a long, straight wire. If we increase the current, the magnetic field increases, i.e., the number of field lines increases. Figure 32.6b shows the field lines of a stronger current. Where do the extra field lines come from? Obviously, the current has to make them. When the current slowly increases, it makes new, small circles of field lines in its immediate vicinity; meanwhile the circles that already exist in Figure 32.6a expand and move outward, just as the ripples on the surface of a pond that are created when we drop a stone into the pond. Thus, the pattern shown in Figure 32.6a gradually grows into the new pattern shown in Figure 32.6b. Note that the pattern grows from the inside out.

If we decrease the current, field lines must disappear. But this is not quite the reverse of the creation of field lines — the circles of Figure 32.6b do not contract and disappear at the center. The pattern cannot change from outside in; it must first change *near* the current. When the current slowly decreases, it makes new small circles of field lines of *opposite* direction (negative field lines) in its immediate vicinity. These opposite circles expand and move outward, gradually canceling the original field lines.

In any case, a change in the strength of magnetic field involves moving field lines. If a rod is located in the vicinity of the current, the moving field lines will sweep across the rod. Hence an increase or decrease of the strength of magnetic field can cause field lines to cut across a *stationary* rod.

Does such a cutting of field lines induce an emf in the rod? This is a question that can only be resolved by experiment — and the answer is yes. The field lines cutting across the stationary rod induce an emf which is given by the same formula as before. The general statement is known as **Faraday's Law of induction:**

> *The induced emf along any moving or fixed path in a constant or changing magnetic field equals the rate at which magnetic field lines cut across the path,*

Faraday's Law

$$\boxed{\mathscr{E} = [\text{rate of cutting of magnetic field lines}]} \tag{5}$$

Note that we have expressed Faraday's Law as an assertion about a mathematical *path.* The emf will exist between the ends of this path regardless of whether we place a rod or wire along this path, i.e., whenever a unit positive charge moves along this path it will gain an amount $\mathscr{E}$ of energy from the induced electric field regardless of whether the charge moves on a rod or through empty space. The rod or wire merely serves as a convenient conduit for the charge. Of course, for the practical exploitation of the emf we must provide a rod, wire, or some other conductor along which the current can flow, and we will also have to provide a return path for the current.

For a closed path, Faraday's Law can be conveniently stated in terms of the **magnetic flux.** The magnetic flux through any given area is simply the number of magnetic field lines intercepted by the area. As in the case of electric flux, we can write the flux as an integral:

$$\boxed{\Phi_B = \int \mathbf{B} \cdot d\mathbf{S}} \qquad (6)$$

The unit of magnetic flux is the weber (Wb),

$$1 \text{ weber} = 1 \text{ Wb} = 1 \text{ T} \cdot \text{m}^2 \qquad (7)$$

When we reckon the net rate at which field lines cut across a closed path, we must take into account that a field line can make a positive or a negative contribution to the rate of cutting, depending on whether the field line enters the area bounded by the path or leaves it. The net rate of cutting of field lines is equal to the rate of change of the number of field lines intercepted by the area bounded by the path, i.e., it is equal to the rate of change of the magnetic flux through this area. Hence, Faraday's Law can be stated as follows:

Wilhelm Eduard Weber, *1804–1891, German physicist, professor at Göttingen. He worked on problems in magnetism, and devised a system of units for electric and magnetic quantities.*

Faraday's Law for closed loop

The induced emf around a closed path in a magnetic field is equal to the rate of change of the magnetic flux intercepted by the area within the path,

$$\boxed{\mathscr{E} = -\frac{d\Phi_B}{dt}} \qquad (8)$$

The minus sign that we have inserted in this equation indicates how the sign of the induced emf is related to the sign of the flux and of the rate of change of flux. The sign of the magnetic flux is fixed by the following **right-hand rule:** wrap the fingers around the path in the direction in which we are reckoning the emf; the magnetic flux is then positive if the magnetic field points in the direction of the thumb and negative otherwise. For example, in Figure 32.7, this right-hand rule indicates that the field lines shown make a positive contribution to the magnetic flux; consequently, if the strength of the magnetic field is increasing, $d\Phi_B/dt$ will be positive and, according to Eq. (8), the induced emf around the closed path will be negative. If the path in Figure 32.7 coincides with a conducting wire, the induced emf will therefore drive a current around this wire in a direction opposite to that indicated by the arrow.

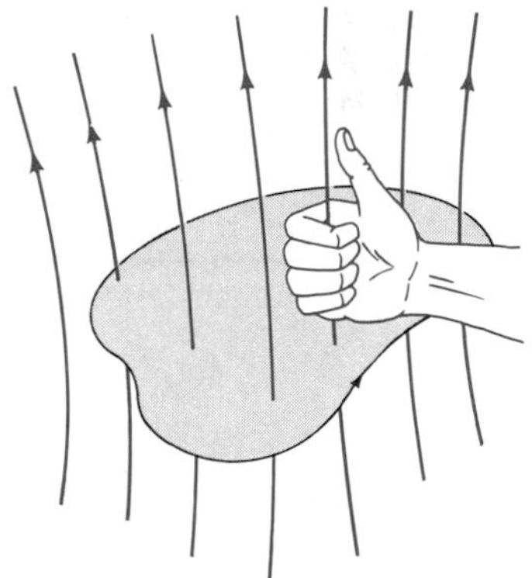

Fig. 32.7 A closed path in a changing magnetic field. The sign of the magnetic flux intercepted by the area within the path is given by a right-hand rule: if the fingers are curled around the path in the direction in which we are reckoning the emf (indicated by the black arrow), the thumb points in the direction that the magnetic field must have in order to give a positive flux.

32.3 Some Examples; Lenz' Law

In this section, we will look at some examples of the calculation of induced emfs. We will also lay down a simple rule for finding the direction of the emf.

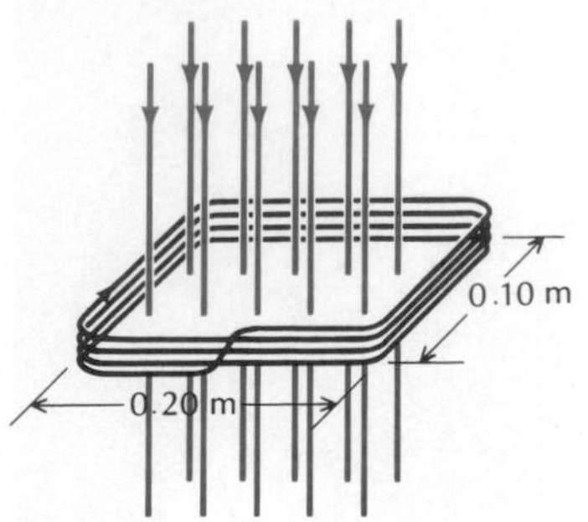

Fig. 32.8 A rectangular coil in a uniform, decreasing magnetic field.

EXAMPLE 3. A rectangular coil of 150 loops of wire forming a closed circuit measures 0.20 m × 0.10 m. The resistance of the coil is 5.0 Ω. The coil is placed in an electromagnet, face on to the magnetic field (Figure 32.8). Suppose that when the electromagnet is suddenly switched off, the strength of the magnetic field decreases at the rate of 20 T per second. What is the induced emf in the coil? The induced current? What is the direction of the induced current?

SOLUTION: The flux intercepted by the area within the coil is $\Phi_B = NAB$, where N is the number of turns, A the area, and B the magnetic field. Hence the induced emf is

$$\mathscr{E} = -\frac{d\Phi_B}{dt} = -NA\frac{dB}{dt} = 150 \times (0.20 \text{ m} \times 0.10 \text{ m}) \times 20 \text{ T/s}$$

$$= 60 \text{ V}$$

and the induced current is

$$I = \mathscr{E}/R = 60 \text{ V}/5.0\ \Omega = 12 \text{ amperes}$$

To determine the direction of the induced current, let us reckon the emf around the loop in the direction indicated by the arrow in Figure 32.8. By the right-hand rule, the field lines shown make a positive contribution to the magnetic flux. Since the magnetic field is *decreasing*, $d\Phi_B/dt$ will then be *negative* and, according to Eq. (8), the induced emf will be positive. This induced emf will therefore drive a current in the direction of the arrow.

EXAMPLE 4. A long solenoid has a circular cross section of radius R. The solenoid is connected to a source of alternating current so that the magnetic field inside the solenoid is

$$B = B_0 \sin \omega t \tag{9}$$

where B_0 and ω are constants. What is the induced emf around a concentric circular path of radius r inside the solenoid or outside the solenoid?

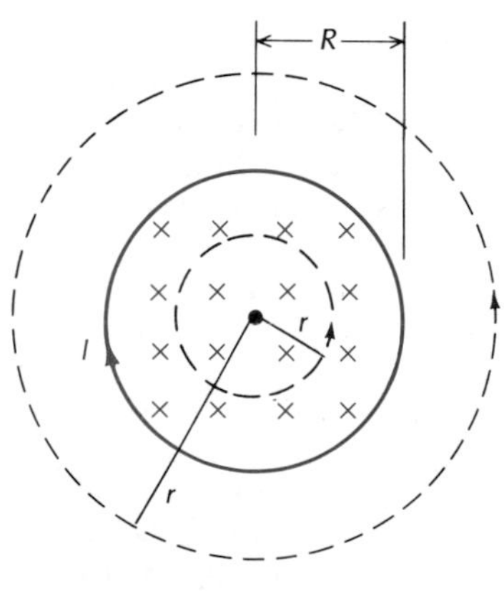

Fig. 32.9 Cross-sectional view of solenoid showing uniform magnetic field. The dashed circles are two alternative paths for reckoning the emf.

SOLUTION: Inside the solenoid ($r < R$), the magnetic field is uniform. Figure 32.9 shows a circle of radius r. The area of this circle intercepts a magnetic flux $-\pi r^2 B$. [If the positive direction for **B** is into the page and the direction in which we are reckoning the emf is counterclockwise (Figure 32.9), then, by the right-hand rule discussed in connection with Eq. (8), the flux will have a sign opposite to that of **B** — at the instant shown in Figure 32.9, the flux is negative.] The rate of change of the flux is $-\pi r^2 dB/dt$. Hence Faraday's Law gives us the induced emf

$$\mathscr{E} = -\frac{d\Phi_B}{dt} = \pi r^2 \frac{dB}{dt} = \pi r^2 \omega B_0 \cos \omega t \tag{10}$$

The emf increases in proportion to the area of the circle, reaching a maximum value if the radius of the circle coincides with the radius of the solenoid ($r = R$).

Note that as regards the time dependence, the emf is 90° out of phase with the magnetic field — when the magnetic field (9) has maximum magnitude ($\sin \omega t = \sin \pi/2 = 1$), the emf (10) has minimum magnitude ($\cos \omega t = \cos \pi/2 = 0$).

Outside the solenoid ($r > R$), there is no magnetic field. The magnetic flux intercepted by a circle of radius r (Figure 32.9) is therefore simply the flux $-\pi R^2 B$ of lines inside the solenoid and the rate of change in this flux is

$-\pi R^2\, dB/dt$.[2] Instead of Eq. (10) we now obtain

$$\mathscr{E} = -\frac{d\Phi_B}{dt} = \pi R^2 \frac{dB}{dt} = \pi R^2 \omega B_0 \cos \omega t \tag{11}$$

Thus the emf remains constant as we go away from the solenoid. Figure 32.10 shows a plot of the value of the induced emf as a function of the radius of the circular path.

Note that the solenoid and its current did not directly enter into our calculation — the induced emf is entirely determined by the magnetic field. This implies that we will get the same kind of induced emf if the magnetic field is that in the gap of an electromagnet (with round pole pieces) rather than that in a solenoid. Such induced emfs in electromagnets find an important application in the design of electron accelerators. In the **betatron** (Figure 32.11), electrons travel in a circular orbit between the poles of a large electromagnet. The electrons are held in their circular orbit by the magnetic field. The acceleration of the electron is accomplished by quickly increasing the strength of the magnetic field. This change of the magnetic field induces an emf around the circular orbits of the electrons, which accelerates the electrons, increasing their energy. This acceleration continues as long as the magnetic field is increasing; at the same time, the increasing magnetic field provides just the right amount of extra centripetal force to keep the electrons in a circular orbit of constant (or nearly constant) radius. In a typical betatron, the magnetic field increases over a time interval of a few milliseconds and in this time the electrons move around their circular orbit a few hundred thousand times, acquiring a final energy of 100 MeV or so. They are then suddenly guided out of the machine (by means of deflecting electric or magnetic fields) and the magnetic field is then permitted to decrease, readying the machine for the next cycle of acceleration.

Betatron

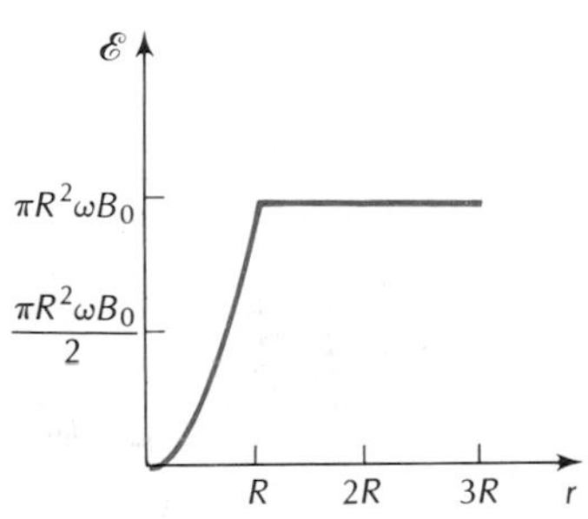

Fig. 32.10 Induced emf as a function of radius of the circular path (at time $t = 0$).

Fig. 32.11 The small machine on the cart in the center is the first betatron, built in 1940. The large machine in the background is the largest betatron. This machine, at the University of Illinois, attained an electron energy of 340 MeV.

[2] This leads to an intriguing question: the rate of change of flux ought to equal the rate of cutting of field lines by the circle, but if there is no magnetic field outside of the solenoid, how can the circle cut across field lines? The answer is that when we increase the current, a new field line will be created inside the solenoid and move toward the center; whenever this happens, the return portion of this field line must move from the current outward to infinity (the magnetic field lines form *closed* loops around the current). This outward-moving field line cuts across the outer circle of Figure 32.9.

The direction of the induced emf can always be figured out by an argument similar to that given in Examples 3 and 4. But there is a simple rule for finding the direction, a rule known as **Lenz' Law:**

Lenz' Law

The induced emfs are always in such a direction as to oppose the change that generated them.

Emil Khristianovich Lenz, *1804–1865, Russian physicist, rector of the University of St. Petersburg. He investigated induction and other electromagnetic phenomena. Independently of Joule, he established the law for the production of heat by a current in a resistor.*

The exact meaning of *change* depends on the problem at hand and is best made clear by some examples. For instance, the moving rod of Figure 32.12 generates a counterclockwise current. Consider now the magnetic force on this current in the rod; this force has the direction of $I\, d\mathbf{l} \times \mathbf{B}$, which points to the right in Figure 32.12. Thus, the force associated with the induced current *opposes* the motion of the rod, i.e., it opposes the change (motion) that generates the current. This is in agreement with Lenz' Law.

Likewise, if there were a circular wire loop within the solenoid of Figure 32.9, a counterclockwise current would flow in it [assuming that B of Eq. (9) is increasing, $0 < \omega t < \pi/2$]. Such a current flowing around the loop would produce a magnetic field within the loop opposite to the field B of Eq. (9). Thus, the induced current around the loop *opposes* the change of flux within the loop, i.e., it opposes the change that generates the current.

Note that Lenz' Law is consistent with energy conservation but that the contrary of Lenz' Law is not. If the contrary of Lenz' Law were true, then the force on the rod in Figure 32.12 would be toward the left so as to speed up the rod. This would develop into a runaway situation: as the rod acquires higher velocity, the force would increase and accelerate it more and more. There would then be no need for us to push the rod through the magnetic field . . . it would accelerate spontaneously! Clearly, this would be a violation of energy conservation.

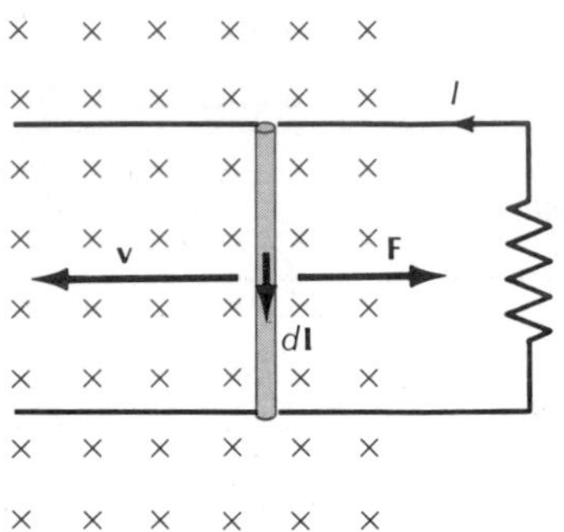

Fig. 32.12 The magnetic field exerts a force on the current in the rod.

EXAMPLE 5. An electromagnetic generator consists of a coil of N loops of wire that rotates about an axis perpendicular to a constant magnetic field. Sliding contacts connect the coil to an external circuit (Figure 32.13). What emf does the coil deliver to the external circuit? The coil has an area A and rotates with an angular frequency ω.

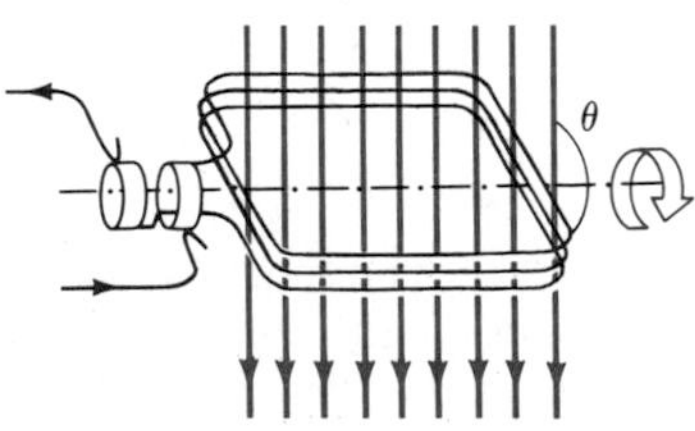

Fig. 32.13 Electromagnetic generator.

SOLUTION: The coil makes an angle $\theta = \omega t$ with the magnetic field. The magnetic flux through the coil is $\Phi = NAB \sin\theta = NAB \sin \omega t$. By Faraday's Law [see Eq. (8)] the induced emf is

$$\mathscr{E} = -\frac{d\Phi_B}{dt} = -\frac{d}{dt}(NAB \sin \omega t)$$

$$= -NAB\omega \cos \omega t \qquad (12)$$

At the instant shown in Figure 32.13, the magnetic flux through the coil is decreasing. According to Lenz' Law, the induced current must therefore flow around the coil clockwise so as to contribute an extra magnetic flux that tends to oppose the decrease of flux.

The function given by Eq. (12) is an **alternating emf,** or **AC voltage,** that oscillates between positive and negative values. A plot of this emf is shown in Figure 32.14. To obtain a constant emf, or DC voltage, from a generator, we can use a slotted sliding contact, or **commutator,** arranged so that the emf at the external terminals always has the same sign (Figure 32.15). A plot of the emf produced by such a generator is shown in Figure 32.16. This emf oscillates between zero and a maximum value but always remains positive. To elim-

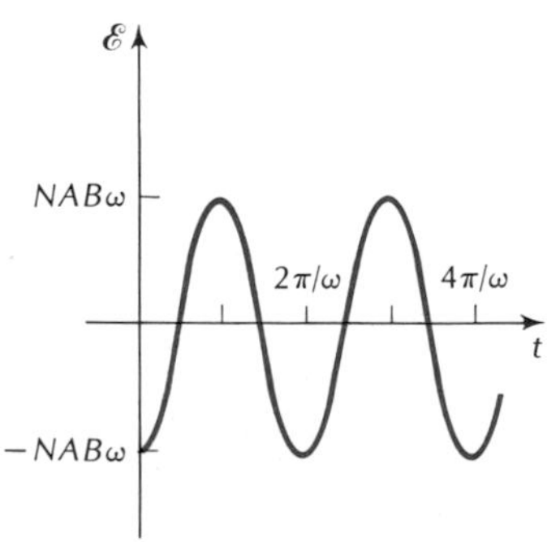

Fig. 32.14 Emf of the generator; this is an alternating voltage (AC).

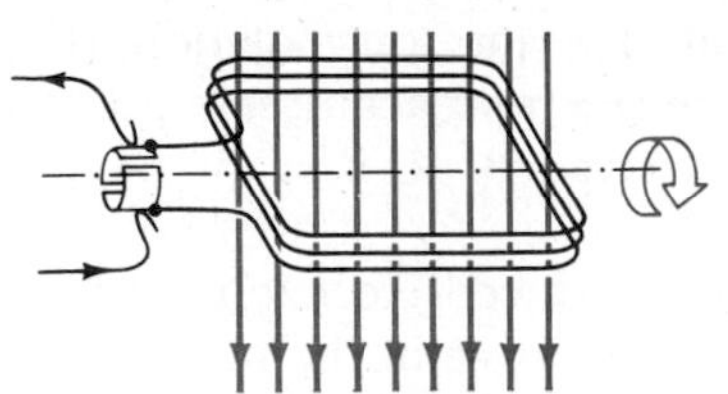
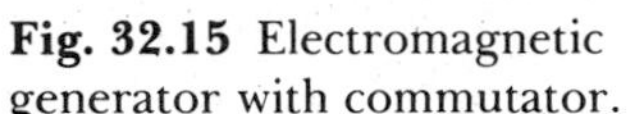

Fig. 32.15 Electromagnetic generator with commutator.

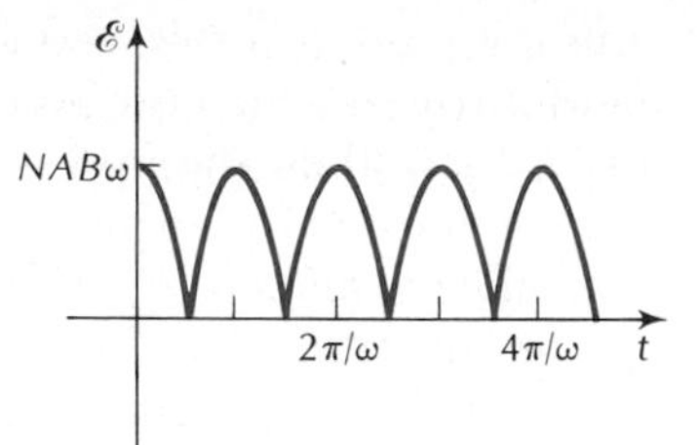

Fig. 32.16 Emf of generator with commutator.

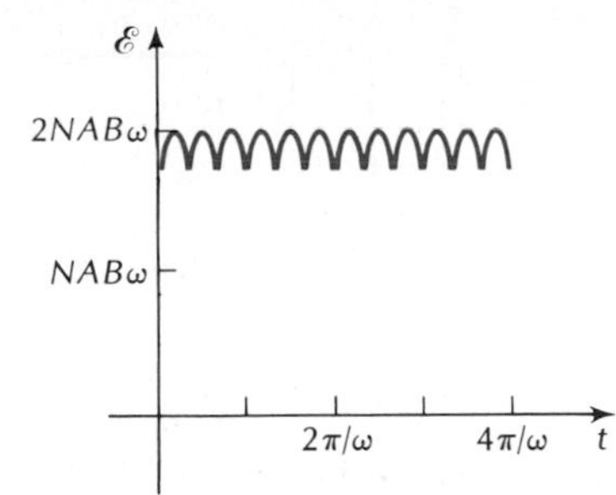

Fig. 32.17 Emf of generator with three coils 120° apart connected in series; each loop has a commutator. The ripple in the voltage is about 14% of the maximum voltage.

inate or reduce the oscillations of the emf, we can connect several generators in series, each with its coil at a slightly different initial angle to the magnetic field. In the sum of the emfs, the oscillations then tend to average out. For example, the plot of Figure 32.17 shows the combined emf of three coils whose positions are always 120° apart as they rotate. The emf delivered by this generator is fairly constant, with only a small ripple.

32.4 The Induced Electric Field

It is very instructive to reexamine the generation of the induced emf in a rod moving through a magnetic field from the point of view of a reference frame in which the rod is at rest. Consider again the rod of Figure 32.1 moving at constant velocity **v**. In a reference frame moving with the rod at the same velocity **v**, the free charges within the rod have no velocity and there is no *magnetic* force. However, the free charges must experience some other kind of force that pushes them along the rod. The question is then: What is this new kind of force in the moving reference frame that produces the same effect as the magnetic force in the stationary reference frame? Since the only kinds of force that act on electric charges are the magnetic force and the electric force, the "new" kind of force must be due to a "new" kind of electric field, an electric field that exists in the moving reference frame, but not in the stationary reference frame (Figure 32.18). This "new" electric field in the moving reference frame arises indirectly from the currents that are responsible for the magnetic field in the stationary reference frame — it turns out that although the conductors that carry these currents are electrically neutral relative to the stationary reference frame, they acquire an excess of positive or negative charge relative to the moving reference frame (for details of the mech-

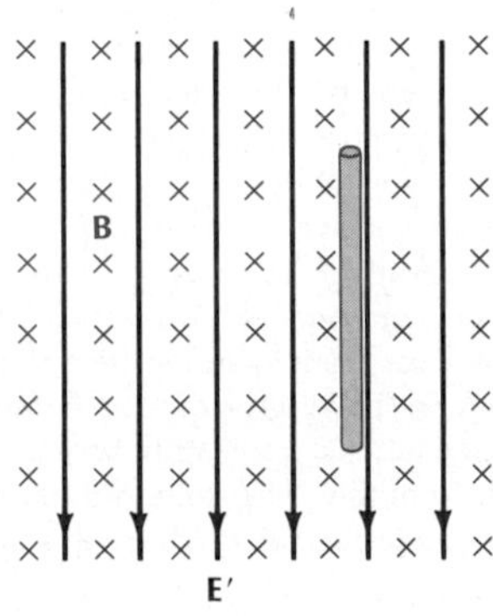

Fig. 32.18 In the moving reference frame of the rod, there is a magnetic field **B** (colored crosses) and also an electric field **E′** (black).

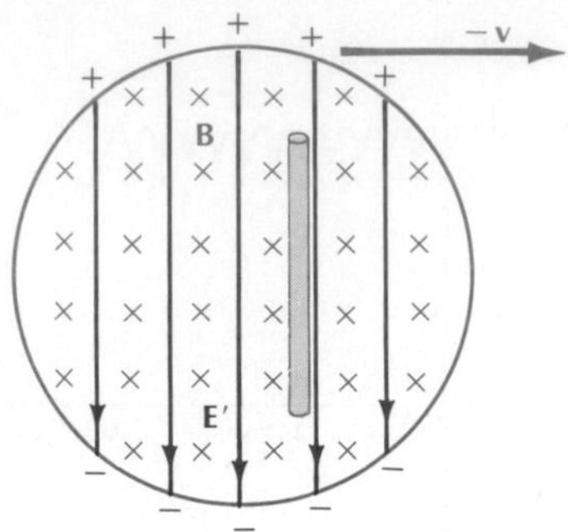

Fig. 32.19 In the moving reference frame, the solenoid has a magnetic field **B** (colored crosses) and an electric field **E′** (black). The edge of the solenoid has acquired an electric charge.

anism involved in this, see Section 30.4). For instance, if the magnetic field through which the rod moves is that of a very long solenoid (Figure 32.19), then the upper portion of the solenoid acquires a positive charge and the lower portion a negative charge, as seen in the moving reference frame.

We can determine the magnitude of the induced electric field from a consistency requirement: the "new" electric force in the moving reference frame must coincide with the magnetic force in the stationary reference frame. This tells us that the electric field in the moving reference frame must have a magnitude

$$E' = vB \tag{13}$$

This electric field does work on the free charges and therefore represents a source of emf.[3] The work done on a unit positive charge that passes through the rod is

$$\mathscr{E} = El \tag{14}$$

In view of Eq. (13), this value of the emf coincides with the value that we obtained Eq. (1). Thus, the motional emf can be calculated either in a stationary reference frame or else in a moving reference frame; in the former case it arises from a magnetic field and in the latter from an electric field.

In our simple example, the electric field **E′** is constant along the rod. More complicated problems may involve an electric field **E′** that depends on position. Equation (13) indicates that such a dependence can come about in two ways: if the magnetic field varies along the rod (a nonuniform magnetic field) or if the velocity varies along the rod (a rigid rod with rotational motion or perhaps a flexible rod undergoing a deformation). If **E′** depends on position, then we must of course replace Eq. (14) by an integral,

$$\mathscr{E} = \int \mathbf{E}' \cdot d\mathbf{l} \tag{15}$$

where the integration extends from one end of the rod to the other.

Induced electric field

The electric field **E′** that exists in the reference frame of the moving rod is called an **induced electric field.** One important feature of this kind of field is that it is *not conservative.* Figure 32.20 shows the electric field **E′** of a solenoid as seen in the reference frame of the moving rod. If we integrate **E′** around the closed path drawn in this figure, we will obviously get a result different from zero — only one side of the rectangle gives a contribution to the integral. This establishes that **E′** is not a conservative field since, for a conservative field, the integral has to be zero for *every* arbitrary closed path.[4]

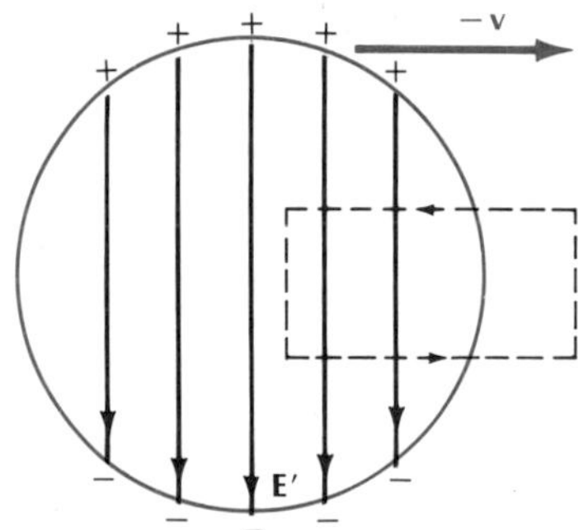

Fig. 32.20 Electric field of a long solenoid as seen in the moving reference frame. For the dashed path shown, which has one portion outside of the electric field **E′**, the integral $\oint \mathbf{E}' \cdot d\mathbf{l}$ is *not* zero.

[3] Note that this is an external electric field (generated by charges *not* on the rod). Besides there may be an internal electric field (generated by charges that accumulate on the ends of the rod); for now, we will not concern ourselves with this latter field.

[4] We might ask: Why can we not use a similar argument to establish that the electric field of a capacitor is nonconservative? The answer is that when we take a path that has one portion inside the capacitor and one portion outside, *both* portions contribute to the path integral because the capacitor has a fringing field that extends beyond the edges of the plates. In earlier chapters we have ignored this fringing field because it was unimportant, but for the evaluation of the path integral it is very important (see Question 27.7). In contrast, the ideal, infinitely long solenoid has no fringing field — the magnetic and the electric fields outside the solenoid are exactly zero.

In general, an induced electric field will exist in the reference frame of any path, or segment of a path, moving through a magnetic field. An induced electric field will also exist in the reference frame of a stationary path placed in a changing, time-dependent magnetic field. The induced electric field is always related to the induced emf by Eq. (15). Hence, in terms of the induced electric field, we can write Faraday's Law as

$$\int \mathbf{E}' \cdot d\mathbf{l} = [\text{rate of cutting of magnetic lines}] \tag{16}$$

for an arbitrary path or as

$$\boxed{\oint \mathbf{E}' \cdot d\mathbf{l} = -\frac{d\Phi_B}{dt}} \tag{17}$$

for a closed path. Here the induced electric field $\mathbf{E}'$ associated with some segment $d\mathbf{l}$ of the path must be reckoned in a reference frame in which the segment $d\mathbf{l}$ is (instantaneously) at rest.

EXAMPLE 6. Consider again the long solenoid with the time-dependent magnetic field described in Example 4. What is the induced electric field inside and outside the solenoid?

SOLUTION: We have already evaluated the induced emf in Example 4. Inside the solenoid ($r < R$), its value is given by Eq. (10); and outside the solenoid ($r > R$), its value is given by Eq. (11). The electric field is related to this emf by

$$\oint \mathbf{E} \cdot d\mathbf{l} = \mathscr{E} \tag{18}$$

where the path of integration is a closed concentric circle of radius r. Note that the electric field $\mathbf{E}$ in this equation is an electric field in the original reference frame (there is no moving reference frame involved in this problem). We therefore are entitled to denote the electric field by $\mathbf{E}$, rather than by $\mathbf{E}'$. To evaluate the left side of Eq. (18), we observe that the field lines of the electric field must be closed concentric circles because this is the only pattern of electric field lines consistent with the rotational symmetry of the solenoid. Hence the path of integration coincides with a field line and

$$\oint \mathbf{E} \cdot d\mathbf{l} = 2\pi r E \tag{19}$$

Combining Eqs. (10) and (19), we then obtain, inside the solenoid,

$$2\pi r E = \pi r^2 \omega B_0 \cos \omega t$$

or

$$E = \tfrac{1}{2} r \omega B_0 \cos \omega t \tag{20}$$

The electric field is zero on the axis of the solenoid ($r = 0$) and reaches a maximum value at the edge of the solenoid ($r = R$).

Combining Eqs. (11) and (19), we obtain, outside the solenoid,

$$2\pi r E = \pi R^2 \omega B_0 \cos \omega t$$

or

$$E = \frac{1}{2}\frac{R^2}{r}\,\omega B_0 \cos \omega t \tag{21}$$

The electric field decreases in strength ($\propto 1/r$) as we go away from the solenoid.

Figure 32.21 is a plot of the strength of the electric field as a function of r. Figure 32.22 shows a picture of the field lines. According to Eqs. (20) and (21), the electric field at the initial time ($t = 0$) is positive, i.e., it has the same direction as the direction of integration around the circular path in Figure 32.9 (counterclockwise).

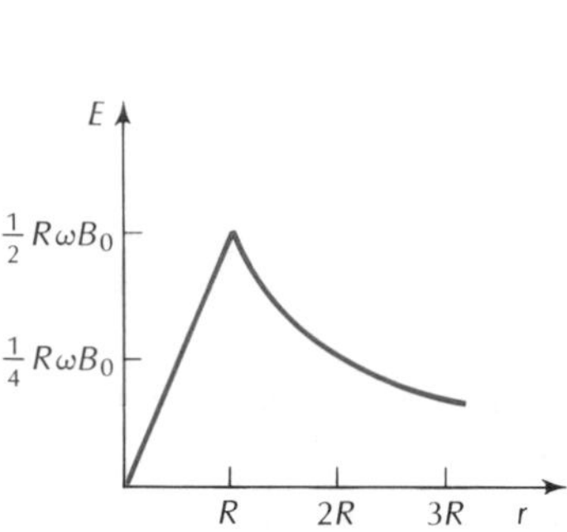

Fig. 32.21 Magnitude of the induced electric field as a function of radius (at $t = 0$).

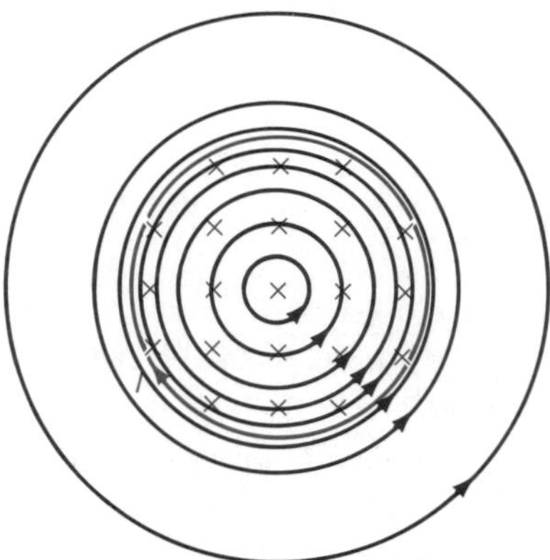

Fig. 32.22 Electric field lines (black) within solenoid and outside of solenoid.

The preceding example shows how a time-dependent magnetic field generates an electric field. This induced electric field differs from the electrostatic fields of the preceding chapters in that the field lines form closed loops — the field lines do not start on electric charges. Furthermore, the induced field is nonconservative; this becomes immediately obvious when we integrate the electric field along a closed field line [see Eq. (19)]. As we will see in Chapter 35, the electric fields of a light wave or radio wave also are instances of induced electric fields, with closed field lines.

32.5 Inductance

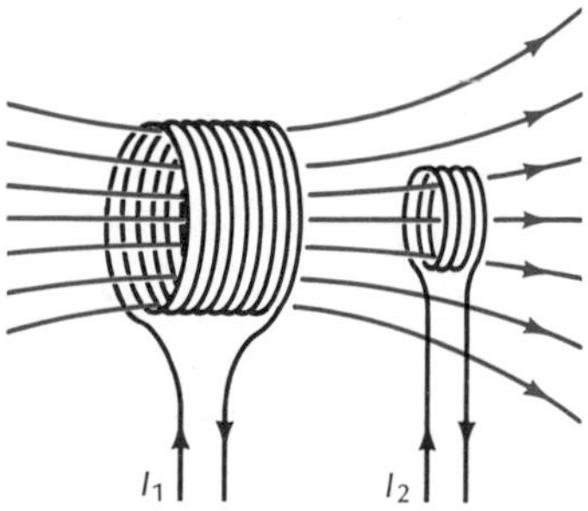

If a conductor carrying a time-dependent current is near some other conductor, then the changing magnetic field of the former can induce an emf in the latter. Thus, a time-dependent current in one conductor can induce a current in another nearby conductor. For instance, consider the two coils of Figure 32.23. One of the coils carries a time-dependent current, which generates a magnetic field. The changing flux Φ_{B_1} through the second coil induces an emf in this coil. The emf in the second coil is

Fig. 32.23 Coil 1 creates a magnetic field. Some of the field lines pass through coil 2.

$$\mathscr{E}_2 = -\frac{d\Phi_{B_1}}{dt} \tag{22}$$

The flux Φ_{B_1} depends on the strength of the magnetic field B_1 in the second coil produced by the current I_1 in the first coil. This field strength is directly proportional to I_1 [see Eq. (30.25)] for the dependence of magnetic field on current]; hence Φ_{B_1} is also proportional to I_1. We can write this relationship between Φ_{B_1} and I_1 as

$$\boxed{\Phi_{B_1} = L_{21} I_1} \tag{23}$$

and then Eq. (22) becomes

$$\boxed{\mathscr{E}_2 = -L_{21} \frac{dI_1}{dt}} \tag{24}$$

Here L_{21} is a constant of proportionality which depends on the size of the coils, their distance, and the number of turns in each, i.e., L_{21} depends on the geometry of Figure 32.23. This constant is called the **mutual inductance** of the coils.

Mutual inductance

Equation (24) states that the emf induced in coil 2 is proportional to the rate of change of current in coil 1. But the converse is also true: if coil 2 carries a current, then the emf induced in coil 1 is proportional to the rate of change of the current in coil 2,

$$\mathscr{E}_1 = -L_{12} \frac{dI_2}{dt} \tag{25}$$

The constants of proportionality L_{21} and L_{12} appearing in Eqs. (24) and (25) are exactly equal, i.e., $L_{21} = L_{12}$. Although we will accept this assertion without proof, we note that the result is quite reasonable: the mutual inductance reflects the geometry of the *relative* arrangement of the coils and that is of course the same in both cases.

The SI unit of inductance is called the **henry** (H),

$$1 \text{ henry} = 1 \text{ H} = 1 \text{ V} \cdot \text{s/A} \tag{26}$$

Incidentally, the permeability constant is commonly expressed in terms of this unit of inductance,

$$\mu_0 = 1.26 \times 10^{-6} \text{ H/m} \tag{27}$$

Joseph Henry, *1797–1878, American experimental physicist, professor at Princeton and first director of the Smithsonian Institution. He made important improvements in electromagnets by winding coils of insulated wire around iron pole pieces, and invented an electromagnetic motor and a new, efficient telegraph. He discovered self-induction, and investigated how currents in one circuit induce currents in another.*

EXAMPLE 7. A long solenoid has n turns per unit length. A ring of wire of radius r is placed within the solenoid, perpendicular to the axis (Figure 32.24). What is the mutual inductance of ring and solenoid?

SOLUTION: If the current in the solenoid windings is I_1, the magnetic field is $B_1 = \mu_0 n I_1$ [see Eq. (31.16)] and the flux through the ring is

$$\Phi_{B_1} = \pi r^2 \mu_0 n I_1$$

Comparing this with Eq. (23), we see that

$$L_{12} = \pi r^2 \mu_0 n \tag{28}$$

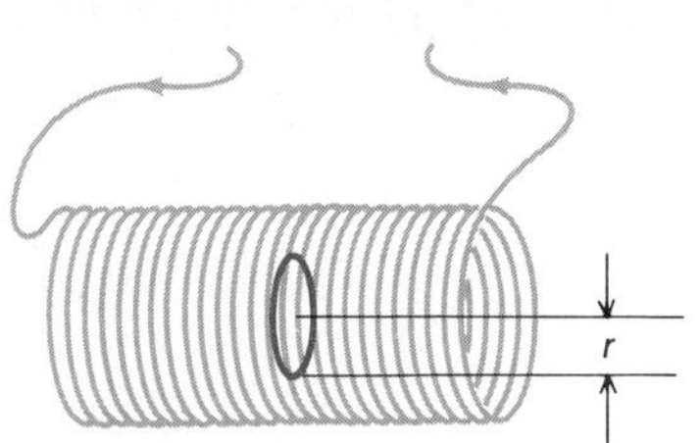

Fig. 32.24 A ring of wire inside a solenoid.

Self-inductance

A conductor by itself has a **self-inductance.** Consider a coil with a time-dependent current (Figure 32.25). The magnetic field of this coil will then produce a time-dependent magnetic flux and, by Faraday's Law, an induced emf. The net emf acting on the coil is then the sum of the external emf (applied to the terminals in Figure 32.25) and the self-induced emf. This means that whenever the current is time dependent, the coil will act back on the current and modify it (we will see how to calculate the net resulting current in Section 34.3). For this reason the self-induced emf is often called a **back emf.** From Lenz' Law we immediately recognize that the self-induced emf always acts in such a direction to *oppose* the change in the current, i.e., it attempts to maintain the current constant.

In terms of the flux through the circuit, the definition of the self-inductance is of the same form as Eq. (23),

$$\boxed{\Phi_B = LI} \tag{29}$$

and therefore the induced emf is

$$\boxed{\mathscr{E} = -L\frac{dI}{dt}} \tag{30}$$

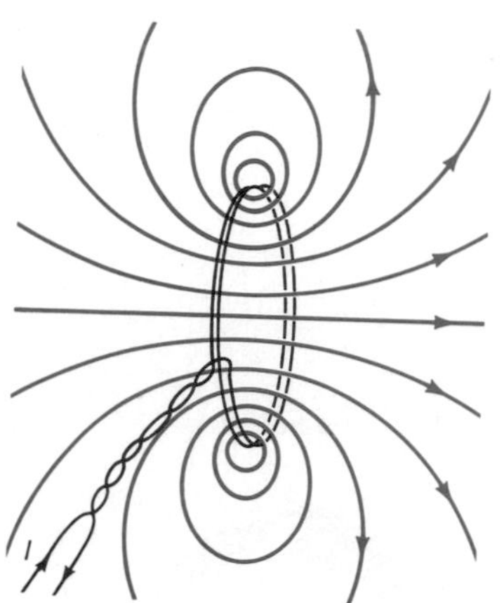

Fig. 32.25 A coil and its magnetic field.

EXAMPLE 8. A solenoid has n turns per unit length and a radius R. What is its self-inductance per unit length?

SOLUTION: The magnetic field inside the solenoid is $B = \mu_0 nI$. The number of loops in a length l is nl; each of these loops has a flux $\pi R^2 B$. Hence the flux through all the loops in a length l is

$$\Phi_B = \pi R^2 Bnl = \pi R^2 n^2 l \mu_0 I \tag{31}$$

and

$$L = \Phi_B / I = \mu_0 n^2 \pi R^2 l \tag{32}$$

The inductance per unit length is therefore $\mu_0 n^2 \pi R^2$.

32.6 Magnetic Energy

Inductors store magnetic energy, just as capacitors store electric energy. When we connect an external source of emf to an inductor and start a current through the inductor, the back emf will oppose the increase of the current and the external emf must do work in order to overcome this opposition and establish the flow of current. This work is stored in the inductor and it can be recovered by removing the external source of emf from the circuit. As the current begins to decrease, the inductor will supply an induced emf which tends to keep the current flowing for a while (at a decreasing rate). Thus, the inductor delivers energy to the current.

To calculate the amount of stored energy, we note that when the current increases at the rate of dI/dt, the back emf is

$$\mathscr{E} = -L\, dI/dt \tag{33}$$

The inductor does work on the current at the rate

$$I\mathscr{E} = -LI\, dI/dt \tag{34}$$

Here the negative sign implies that the energy is delivered by the current to the inductor rather than vice versa. In a time dt, the energy stored in the inductor is therefore

$$dU = -I\mathscr{E}\, dt = LI\, dI \tag{35}$$

If we integrate this from the initial value of the current ($I' = 0$) to the final value ($I' = I$), we obtain

$$U = \int_0^I LI'\, dI' = L \int_0^I I'\, dI' \tag{36}$$

or

$$\boxed{U = \tfrac{1}{2} LI^2} \tag{37}$$

Magnetic energy in inductor

This equation for the magnetic energy in an inductor is analogous to Eq. (27.40) for the electric energy in a capacitor.

EXAMPLE 9 A solenoid has a radius of 2.0 cm; its winding has one turn of wire per millimeter. A current of 10 A flows through the winding. What is the amount of energy stored per unit length of the solenoid?

SOLUTION: The inductance per unit length is [see Eq. (32)]

$$L/l = \mu_0 n^2 \pi R^2$$

$$= 1.26 \times 10^{-6}\ \text{H/m} \times (10^3/\text{m})^2 \times \pi \times (0.020\ \text{m})^2$$

$$= 1.6 \times 10^{-3}\ \text{H/m}$$

and the energy per unit length is

$$U/l = \tfrac{1}{2}(L/l)I^2$$

$$= \tfrac{1}{2} \times 1.6 \times 10^{-3}\ \text{H/m} \times (10\ \text{A})^2$$

$$= 7.9 \times 10^{-2}\ \text{J}$$

The energy stored in a solenoid can be expressed in terms of the magnetic field. Consider a portion of length l of the solenoid. The inductance of this portion is [see Eq. (32)]

$$L = \mu_0 n^2 \pi R^2 l \tag{38}$$

so that

$$U = \tfrac{1}{2}\mu_0 n^2 \pi R^2 l I^2 \tag{39}$$

Since for a solenoid $B = \mu_0 nI$, we can write this as

$$U = \frac{1}{2\mu_0} B^2 \pi R^2 l \tag{40}$$

or

$$U = \frac{1}{2\mu_0} B^2 \times [\text{volume}] \tag{41}$$

where the volume $\pi R^2 l$ is the volume of magnetic field in the portion of length l of the solenoid.

According to Eq. (41), the quantity

Energy density in magnetic field

$$\boxed{u = \frac{1}{2\mu_0} B^2} \tag{42}$$

can be regarded as the magnetic energy per unit volume. Although we have only derived this equation for the special case of a long solenoid, it turns out to be generally valid (in vacuum and in nonmagnetic materials). The magnetic field, just as the electric field, stores energy. The magnetic energy density is proportional to B^2 just as the electric density is proportional to E^2 [compare Eq. (26.19)].

EXAMPLE 10. Near the surface, the Earth's magnetic field typically has a strength of 0.3×10^{-4} T and the Earth's atmospheric electric field typically has a strength of 100 V/m. What is the energy density in each field?

SOLUTION: The magnetic energy density is

$$u_B = \frac{1}{2\mu_0} B^2 = \frac{(0.3 \times 10^{-4}\ \text{T})^2}{2 \times 1.26 \times 10^{-6}\ \text{H/m}} = 3.6 \times 10^{-4}\ \text{J/m}^3$$

and the electric energy density is

$$u_E = \frac{\varepsilon_0}{2} E^2$$

$$= \frac{(8.85 \times 10^{-12}\ \text{F/m}) \times (100\ \text{V/m})^2}{2}$$

$$= 4.4 \times 10^{-8}\ \text{J/m}^3$$

SUMMARY

Motional emf in a rod: $\mathscr{E} = vBl$

Faraday's Law: $\mathscr{E} =$ [rate of cutting of magnetic field lines]

Magnetic flux: $\Phi_B = \int \mathbf{B} \cdot d\mathbf{S}$

Faraday's Law for closed loop: $\mathscr{E} = -\dfrac{d\Phi_B}{dt}$

Lenz' Law: Induced emf opposes change.

Mutual inductance: $\Phi_{B_1} = L_{21} I_1$

$$\mathscr{E}_2 = -L_{21} \frac{dI_1}{dt}$$

Self-inductance: $\Phi_B = LI$

$$\mathscr{E} = -L \frac{dI}{dt}$$

Self-inductance of solenoid: $\mu_0 n^2 \pi R^2$ per unit length

Magnetic energy in inductor: $U = \frac{1}{2} LI^2$

Energy density in magnetic field: $u = \dfrac{1}{2\mu_0} B^2$

QUESTIONS

1. At the latitude of the United States, the magnetic field of the Earth has a downward component, larger than the northward component. Suppose that an airplane is flying due west in this magnetic field. Will there be an emf between its wingtips? Which wingtip will be positive? Will there be a flow of current?

2. What is the magnetic flux that the magnetic field of the Earth produces through the surface of the Earth?

3. Is Eq. (8) valid for an open path? Is Eq. (5) valid for a closed path?

4. A long straight wire carries a steady current. A square conducting loop is in the same plane as the wire. If we push the loop toward the wire, how is the direction of the current induced in the loop related to the direction of the current in the wire?

5. A **flip coil** serves to measure the strength of a magnetic field. It consists of a small coil of many turns connected to a sensitive ammeter. The coil is placed face on in the magnetic field and then suddenly flipped over. How does this indicate the presence of the magnetic field?

6. The **magneto** used in the ignition system of old automobile engines consists of a permanent magnet mounted on the flywheel of the engine. As the flywheel turns, the magnet passes by a stationary coil, which is connected to the spark plug. Explain how this device produces a spark.

7. A long straight wire carries a current that is increasing as a function of time. A rectangular loop is near the wire, in the same plane as the wire. How is the direction of the current induced in the loop related to the direction of the current in the wire?

8. Consider two adjacent rectangular circuits, in the same plane. If the current in one circuit is suddenly switched off, describe the direction of the current induced in the other circuit.

9. Figure 32.26 shows two coils of wire wound around a plastic cylinder. If the current in the left coil is made to increase what is the direction of the current induced in the coil on the right?

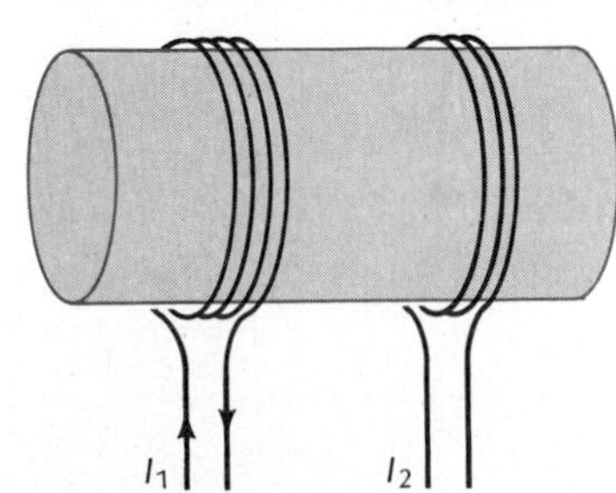

Fig. 32.26

10. A circular conducting ring is being pushed toward the north pole of a bar magnet. Describe the direction of the current induced in the ring.

11. Ganot's *Éléments de Physique*, a classic nineteenth-century textbook, states the following rules for the current induced in one loop face to face with another loop:

I. The distance remaining the same, a continuous and constant current does not induce any current in an adjacent conductor.
II. A current, at the moment of being closed, produces in an adjacent conductor an inverse current.
III. A current, at the moment it ceases, produces a direct current.
IV. A current which is removed, or whose strength diminishes, gives rise to a direct induced current.
V. A current which is approached, or whose strength increases, gives rise to an inverse induced current.

Explain these rules on the basis of Lenz' Law.

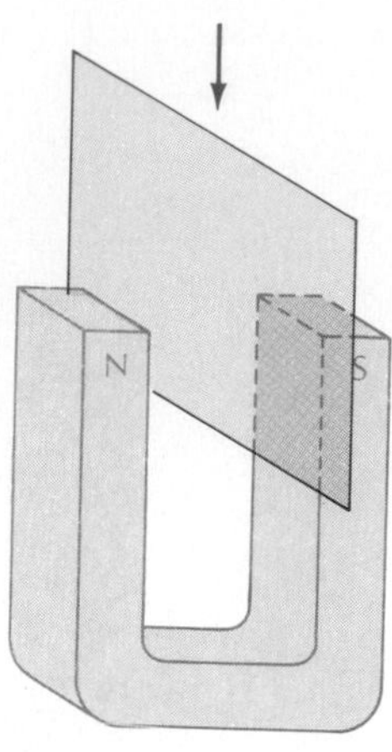

Fig. 32.27

12. A sheet of aluminum is being pushed between the poles of a horseshoe magnet (Figure 32.27). Describe the direction of flow of the induced currents, or **eddy currents**, in the sheet. Explain why there is a strong friction force that opposes the motion of the sheet.

13. Some beam balances use a magnetic damping mechanism to stop excessive swinging of the beam. This mechanism consists of a small conducting plate attached to the beam and a small magnet mounted on a fixed support near this plate. How does this damp the motion of the beam? (Hint: See the preceding question.)

14. A bar magnet, oriented vertically, is dropped toward a flat horizontal copper plate. Explain why there will be a repulsive force between the bar magnet and the copper plate. Is this an elastic force, i.e., will the bar magnet bounce if it is very strong?

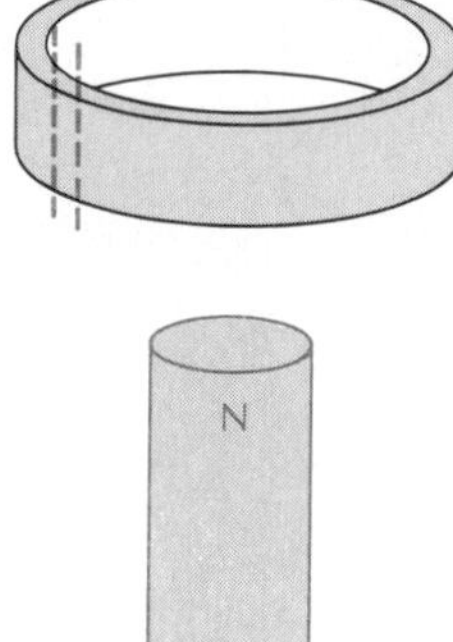

Fig. 32.28 A conducting ring falling toward a bar magnet. The dashed lines indicate where a slot will be cut through the ring.

15. A conducting ring is falling toward a bar magnet (Figure 32.28). Explain why there will be a repulsive force between the ring and the magnet. Explain why there will be no such force if the ring has a slot cut through it (see dashed lines in Figure 32.28).

16. You have two coils, one of slightly smaller radius than the other. To achieve maximum mutual inductance, should you place these coils face to face or one inside the other?

17. Two circular coils are separated by some distance. Qualitatively, describe how the mutual inductance varies as a function of the orientation of the coils.

18. If a strong current flows in a circuit and you suddenly break the circuit (by opening a switch), a large spark is likely to jump across the switch. Explain.

19. What arguments can you give in favor of the view that the magnetic energy of an inductor is stored in the magnetic field, rather than in the current?

PROBLEMS

Section 32.1

1. An automobile travels at 55 mi/h along a level road. The vertical downward component of the Earth's magnetic field is 0.58×10^{-4} T. What is the induced emf between the right and the left door handles separated by a distance of 7.0 ft? Which side is positive and which negative?

2. The DC-10 jet aircraft has a wingspan of 47 m. If such an aircraft is flying horizontally at 960 km/h at a place where the vertical component of the Earth's magnetic field is 0.60×10^{-4} T, what is the induced emf between its wingtips?

3. In order to detect the movement of water in the ocean, oceanographers sometimes rely on the motional emf generated by this movement of the water through the magnetic field of the Earth. Suppose that, at a place where the vertical magnetic field is 0.70×10^{-4} T, two electrodes are immersed in the water separated by a distance of 200 m measured perpendicularly to the movement of the water. If a sensitive voltmeter connected to the electrodes indicates a potential difference of 7.0×10^{-3} V, what is the speed of the water?

4. A homopolar generator consists of a metal disk rotating about a horizontal axis in a uniform horizontal magnetic field. The external circuit is connected to contact brushes touching the disk at the rim and at the axis. If the radius of the disk is 1.2 m and the strength of the magnetic field is 6.0×10^{-2} T, at what rate (rev/s) must you rotate the disk to obtain an emf of 6.0 V?

5. The rate of flow of a conducting liquid, such as detergent, tomato pulp, beer, liquid sodium, sewage, etc., can be measured with an electromagnetic **flowmeter** that detects the emf induced by the motion of the liquid in a magnetic field. Suppose that a plastic pipe of diameter 10 cm carries beer with a speed of 1.5 m/s. The pipe is in a transverse magnetic field of 1.5×10^{-2} T. What emf will be induced between the opposite sides of the column of liquid?

6. Pulsars or neutron stars rotate at a fairly high speed and they are surrounded by a strong magnetic field. The material in neutron stars is a good conductor and hence a motional emf is induced between the center of the neutron star and the rim (this is similar to the emf induced in a rotating metallic rod, see Example 1). Suppose that a neutron star of radius 10 km rotates at the rate of 30 rev/s and that the magnetic field has a strength of 10^8 T. What is the emf induced between the center of the star and a point on its equator?

7. A helicopter has blades of length 4.0 m rotating at 3 rev/s in a horizontal plane. If the vertical component of the Earth's magnetic field is 0.65×10^{-4} T, what is the induced emf between the tip of the blade and the hub?

*8. Sharks have delicate sensors on their bodies that permit them to sense small differences of potential. They can sense electrical disturbances created by other fish and they can also sense the Earth's magnetic field and use this for navigation. Suppose that a shark is swimming horizontally at 25 km/h at a place where the magnetic field has a strength of 4.7×10^{-5} T and points down at an angle of 40° with the vertical. Treat the shark as a cylinder of diameter 30 cm. What is the largest induced emf between diammetrically opposite points on the sides of the shark when heading east? North? Northeast? Make a rough plot of the induced emf vs. the direction of travel of the shark around all points of the compass. Indicate which side of the shark is positive, which negative. (Hint: The component of **B** perpendicular to the velocity **v** is proportional to $|\mathbf{v} \times \mathbf{B}|$.)

Sections 32.2 and 32.3

9. In Idaho, the magnetic field of the Earth points downward at an angle of 69° below the horizontal. The strength of the magnetic field is 0.59×10^{-4} T. What is the magnetic flux through 1 m² of ground in Idaho?

10. An electric generator consists of a rectangular loop of wire rotating about its longitudinal axis which is perpendicular to a magnetic field of 2.0×10^{-2} T. The loop measures 10.0 cm × 20.0 cm and it has 120 turns of wire. The ends of the wire are connected to an external circuit. At what speed (in rev/s) must you rotate this loop in order to induce an alternating emf of amplitude 12.0 V between the ends of the wire?

11. A very long train whose metal wheels are separated by a distance of 4 ft 9 in., is traveling at 80 mi/h on a level track. The vertical component of the magnetic field of the Earth is 0.62×10^{-4} T.

(a) What is the induced emf between the right wheels and the left wheels?
(b) The wheels are in contact with the rails and the rails are connected by

metal cross ties which close the circuit and permit a current to flow from one rail to the other. Since the number of cross ties is very large, their combined resistance is nearly zero; most of the resistance of the circuit is within the train. What is the current that flows in each axle of the train? The axles are cylindrical rods of iron of diameter 3 in. and length 4 ft 9 in.

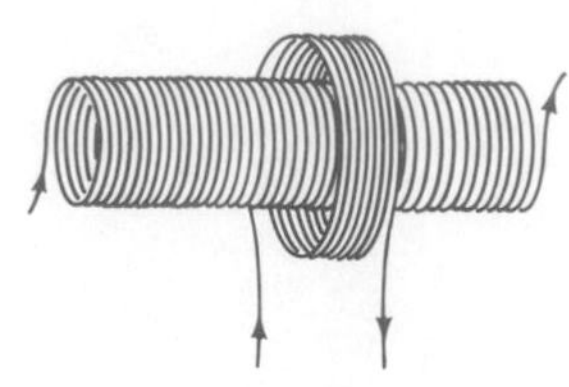

Fig. 32.29 A long solenoid surrounded by a circular coil.

12. (a) A long solenoid has 300 turns of wire per meter and it has a radius of 3.0 cm. If the current in the wire is increasing at the rate of 50 ampere per second, at what rate does the strength of the magnetic field in the solenoid increase?
(b) The solenoid is surrounded by a coil of wire with 120 turns (Figure 32.29). The radius of this coil is 6.0 cm. What induced emf will be generated in this coil while the current in the solenoid is increasing?

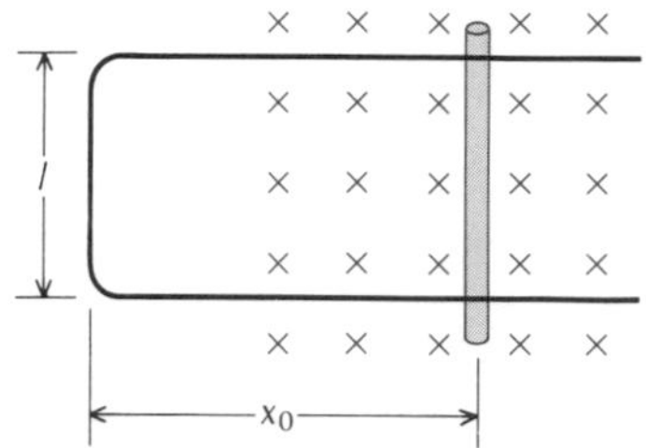

Fig. 32.30 Rod sliding on parallel tracks. The crosses show the tails of the magnetic field vectors.

13. A metal rod of length l and mass m is free to slide, without friction, on two parallel metal tracks. The tracks are connected at one end so that they and the rod form a closed circuit (Figure 32.30). The rod has a resistance R and the tracks have negligible resistance. A uniform magnetic field is perpendicular to the plane of this circuit. The magnetic field is increasing at a constant rate dB/dt. Initially the magnetic field has a strength B_0 and the rod is at rest at a distance x_0 from the connected end of the rails. Express the acceleration of the rod at this instant in terms of the given quantities.

14. A square loop of dimension 8.0 cm $\times$ 8.0 cm is made of copper wire of radius 1.0 mm. The loop is placed face on in a magnetic field which is increasing at the constant rate of 80 T/s. What induced current will flow around the loop? Draw a diagram showing the direction of the field and the induced current.

15. A very long solenoid with 20 turns per centimeter of radius 5.0 cm is surrounded by a rectangular loop of copper wire. The rectangular loop measures 10 cm $\times$ 30 cm and its wire has a radius of 0.05 cm. The resistivity of copper is $1.7 \times 10^{-8}\ \Omega \cdot \text{m}$. Find the induced current in the rectangular loop if the current in the solenoid is increasing at the rate of 5×10^4 A/s.

*16. A square loop of dimension $l \times l$ is moving at speed v toward a straight wire carrying a current I. The wire and the loop are in the same plane and two of the sides of the loop are parallel to the wire. What is the induced emf of the loop as a function of the distance d between the wire and the nearest side of the loop?

*17. A circular coil of insulated wire has a radius of 9.0 cm and contains 60 turns of wire. The ends of the wire are connected in series with a 15-Ω resistor closing the circuit. The normal to the loop is initially parallel to a constant magnetic field of 5.0×10^{-2} T. If the loop is flipped over, so that the direction of the normal is reversed, a pulse of current will flow through the resistor. What amount of charge will flow through the resistor? Assume that the resistance of the wire is negligible compared to that of the resistor.

*18. A flux meter used to measure magnetic fields consists of a small coil of radius 0.80 cm wound with 200 turns of fine wire. The coil is connected by means of a pair of tightly twisted trailing wires to a galvanometer that measures the electric charge flowing through the coil. The resistance of the coil–wire–galvanometer circuit is 6.0 Ω. Suppose that when the coil is quickly moved from a place outside the magnetic field to a place inside the magnetic field the galvanometer registers a flow of charge of 0.30 C. What is the magnetic flux through the coil? What is the component of **B** perpendicular to the face of the coil? Assume that **B** is uniform over the area of the coil.

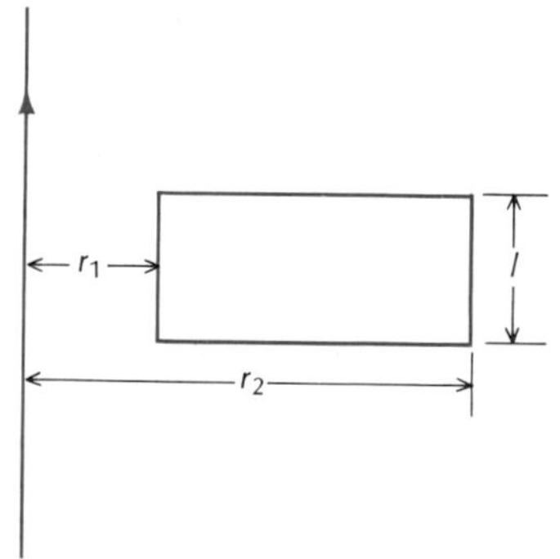

Fig. 32.31 Long straight wire and rectangular loop.

*19. A long, straight wire carries a current that increases at a steady rate dI/dt.
(a) What is the rate of increase of the magnetic field at a radial distance r?
(b) What is the induced emf around the loop shown in Figure 32.31?

Section 32.4

20. A long solenoid of radius 3.0 cm has 2×10^3 turns of wire per meter. The wire carries a current of 6.0 A. Suppose that this solenoid is moving relative to you at a velocity of 400 m/s in a direction perpendicular to its axis.
(a) What is the induced electric field inside the solenoid, as seen in your reference frame?
(b) What is the electric field in the reference frame of the solenoid?
(c) What is the induced emf between two points on the edge of the solenoid at opposite ends of a diameter perpendicular to the velocity?

21. The magnetic field of a betatron has an amplitude of oscillation of 0.90 T and a frequency of 60 Hz. What is the amplitude of oscillation of the induced electric field at a radius of 0.80 m? What is the amplitude of oscillation of the induced emf around a circular path of this radius?

22. A boxcar of a train is 2.5 m wide, 9.5 m long, and 3.5 m high; it is made of sheet metal and it is empty. The boxcar travels at 60 km/h on a level track at a place where the vertical component of the Earth's magnetic field is 0.62×10^{-4} T. Assume that the boxcar is not in electrical contact with the ground.
(a) What is the induced emf between the sides of the boxcar?
(b) Taking into account the electric field contributed by the charges that accumulate on the sides, what is the net electric field inside the boxcar (in the reference frame of the boxcar)?
(c) What is the surface charge density on each side? Treat the sides as two very large parallel plates.

23. A disk of metal of radius R rotates about its axis with a frequency ν. The disk is in a uniform magnetic field B, parallel to the axis of the disk.
(a) Find the induced emf between the center of the disk and a point on the disk at a radial distance r (where $r < R$).
(b) Find the strength of the induced electric field at this point.

Section 32.5

24. Two coils are arranged face to face, as in Figure 32.23. Their mutual inductance is 2.0×10^{-2} H. The current in coil 1 oscillates sinusoidally with a frequency of 60 Hz and an amplitude of 12 A,

$$I_1 = 12 \sin(120\pi t)$$

where the current is measured in amperes and the time in seconds.
(a) What is the magnetic flux that this current generates in coil 2 at time $t = 0$?
(b) What is the induced emf that this current induces in coil 2 at time $t = 0$?
(c) What is the direction of the induced current in coil 2 at time $t = 0$, according to Lenz's Law? Assume that the positive direction for the current I_1 is as shown by the arrows in the figure.

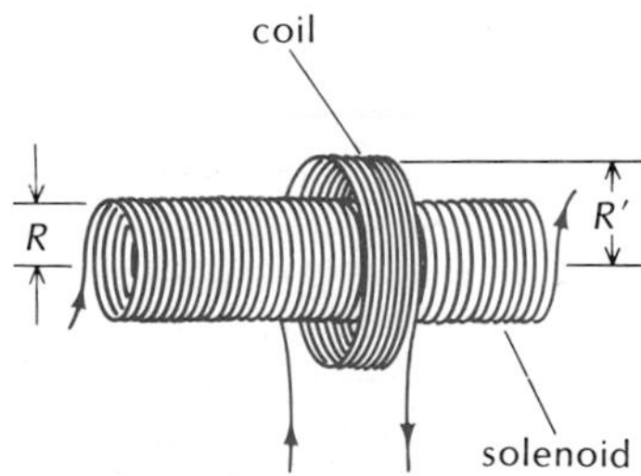

Fig. 32.32 A long solenoid and a circular coil.

25. A loop of wire carrying a current of 100 A generates a magnetic flux of 50 Wb through the area bounded by the loop.
(a) What is the self-inductance of the loop?
(b) If the current is decreased at the rate $dI/dt = 20$ A/s, what is the induced emf?

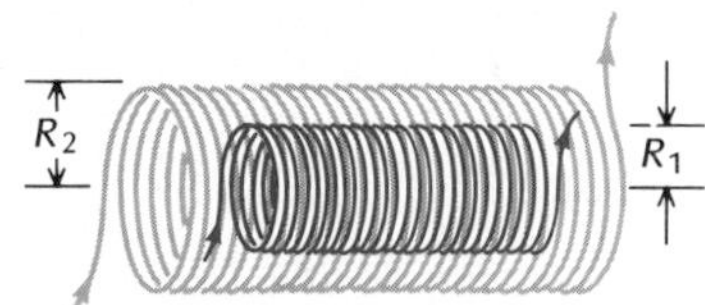

Fig. 32.33 Two long concentric solenoids.

26. A long solenoid of radius R has n turns per unit length. A circular coil of wire of radius R' with 200 turns surrounds the solenoid (Figure 32.32). What is the mutual inductance? Does the shape of the coil of wire matter?

27. Two long concentric solenoids of n_1 and n_2 turns per unit length have radii R_1 and R_2, respectively (Figure 32.33). What is the mutual inductance per unit length of the solenoids? Assume $R_1 < R_2$.

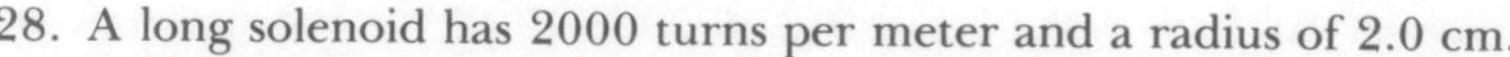

28. A long solenoid has 2000 turns per meter and a radius of 2.0 cm.
   (a) What is the self-inductance for a 1.0-m segment of this solenoid?
   (b) What back emf will this segment generate if the current in the solenoid is changing at the rate of 3.0×10^2 A/s?

29. A solenoid of inductance 2.2×10^{-3} H in which there is initially no current is suddenly connected in series with the poles of 24-V battery. What is the instantaneous rate of increase of the current in the solenoid?

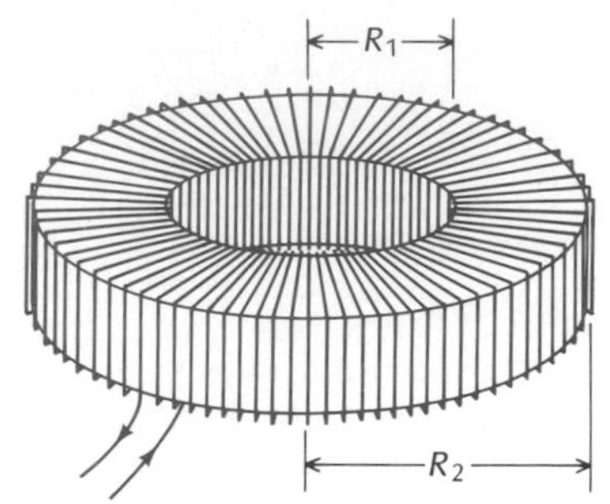

Fig. 32.34 A toroid with a square cross section.

*30. A toroid with a square cross section (Figure 32.34) has an inner radius R_1, an outer radius R_2. The toroid has N turns of wire carrying a current I; assume that N is very large, so that the current can be regarded as uniformly distributed over the surface of the toroid.
   (a) Find the magnetic field as a function of radius.
   (b) Find the flux passing through one turn, and find the self-inductance of the toroid.

*31. Two inductors of self-inductances L_1 and L_2 are connected in series. The inductors are magnetically shielded from one another so that neither produces flux in the other. Show that the self-inductance of the combination is $L = L_1 + L_2$.

*32. Two inductors of self-inductance L_1 and L_2 are connected in parallel. The inductors are magnetically shielded from one another so that neither produces flux in the other. Show that the self-inductance of the combination is given by $1/L = 1/L_1 + 1/L_2$.

Section 32.6

33. A ring of thick wire has a self-inductance of 4.0×10^{-8} H. How much work must you do to establish a current of 25 A in this ring?

34. The strongest magnetic field achieved in a laboratory is 10^3 T. This field can be produced only for a short instant by compressing the magnetic field lines with an explosive device. What is the energy density in this field?

35. For each of the first six entries in Table 30.1, calculate the energy density in the magnetic field.

36. For a crude estimate of the energy in the Earth's magnetic field, pretend that this field has a strength of 0.5×10^{-4} T from the ground up to an altitude of 6×10^6 m above ground. What is the total magnetic energy in this region?

37. A current of 5.0 A flows through a cylindrical solenoid of 1500 turns. The solenoid is 40 cm long and has a diameter of 3.0 cm.
   (a) Find the magnetic field in the solenoid. Treat the solenoid as very long.
   (b) Find the energy density in the magnetic field and find the magnetic energy stored in the space within the solenoid.

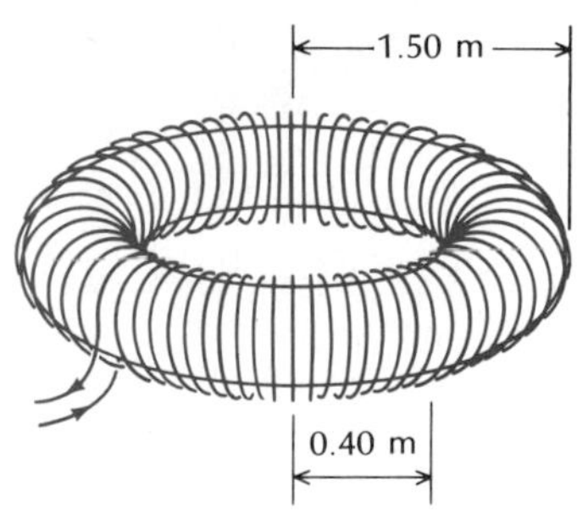

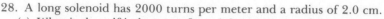

Fig. 32.35 A toroid.

38. A large toroid built for plasma research in the Soviet Union has a major radius of 1.50 m and a minor radius of 0.40 m (Figure 32.35). The average magnetic field within the toroid is 4.0 T. What is the magnetic energy?

39. Consider a point charge q moving at velocity $\mathbf{v}$ through empty space. What is the energy density in the magnetic field at a distance r ahead of the point charge? At a distance r to one side of the point charge?

40. According to one proposal, the surplus energy from a power plant could be temporarily stored in the magnetic field within a very large toroid. If the strength of the magnetic field is 10 T, what volume of the magnetic field would we need to store 1.0×10^5 kW·h of energy? If the toroid has roughly the proportions of a doughnut, roughly what size would it have to be?

*41. Two long, straight, concentric tubes made of sheet metal carry equal currents in opposite directions. The inner tube has a radius of 1.5 cm and the outer tube a radius of 3.0 cm. The current on the surface of each is 120 A. What is the magnetic energy in a 1.0-m segment of these tubes?

*42. A toroid of square cross section (Figure 32.34) has an inner radius R_1, and an outer radius R_2. The toroid has N turns of wire carrying a current I; assume that N is very large.

(a) Find the magnetic energy density as a function of radius.

(b) By integrating the energy density, find the total magnetic energy stored in the solenoid.

(c) Deduce the self-inductance from the formula $U = \frac{1}{2}LI^2$.

PLASMA[1]

Plasma is a very hot, ionized gas, a gas so hot that the violent thermal collisons dissociate all or many of its atoms into positive ions and electrons. If we reckon solid, liquid, and gas as the first three states of matter, then plasma is the fourth state of matter. We might call it pure fire since an ordinary flame is part plasma and part hot gas.

Most of the matter in the universe is plasma. The Sun and all the stars are giant balls of plasma — about 99% of the total mass of the universe is in these balls of plasma. Only in planets, in pulsars, and in some clouds of interstellar gas and dust do we find solids, liquids, and gases — these bodies make up only a small fraction of all the matter in the universe. In our immediate environment, naturally occurring plasma is quite rare, so much so that it was not recognized as a separate state of matter until late in the nineteenth century. Lightning bolts, St. Elmo's fire, the Aurora Borealis, and the ionosphere are plasmas — on the Earth, these are the only forms of naturally occurring plasmas. However, modern technology makes use of many forms of artificially produced plasmas. The gas in fluorescent tubes and in neon signs is plasma; the luminous arc of an electric welder is plasma; the exhaust fire of a rocket is plasma; and the fireball of a nuclear bomb is plasma.

J.1 THE FOURTH STATE OF MATTER

A plasma contains a mixture of positive ions, electrons, and neutral atoms. The extent of ionization depends on the temperature: if the temperature is low, the plasma will contain a substantial number of neutral atoms; if the temperature is high, almost all the atoms will be ionized.

An ordinary gas also contains some ions and electrons, but not enough to make it into a plasma. If we heat a gas to higher and higher temperatures, we will gradually transform it into a plasma; however, the transition from gas to plasma is not sharply defined — there is no sudden change of phase such as in the melting of a solid or the evaporation of a liquid. For example, the flame of a candle is on the borderline between hot gas and plasma; if it had more ions and electrons it would be a plasma, if it had less it would be an ordinary gas. The crucial distinction between an ordinary hot gas and a plasma lies in their electromagnetic properties. Plasma is an electric conductor and macroscopic electric and magnetic fields dominate its behavior, whereas an ordinary gas is an insulator and it does not respond in any drastic manner to electric and magnetic fields. Figure J.1 shows a prominence of plasma on the surface of the Sun. Obviously, this prominence has a complicated shape; it arcs up and down because its behavior is dominated by the magnetic field of the Sun rather than by gravity. In contrast, the flame of a candle has a simple shape; it streams upward because of the buoyancy of the hot gas. If we place the flame between the poles of a magnet, we find that its behavior is not changed in any noticeable way.

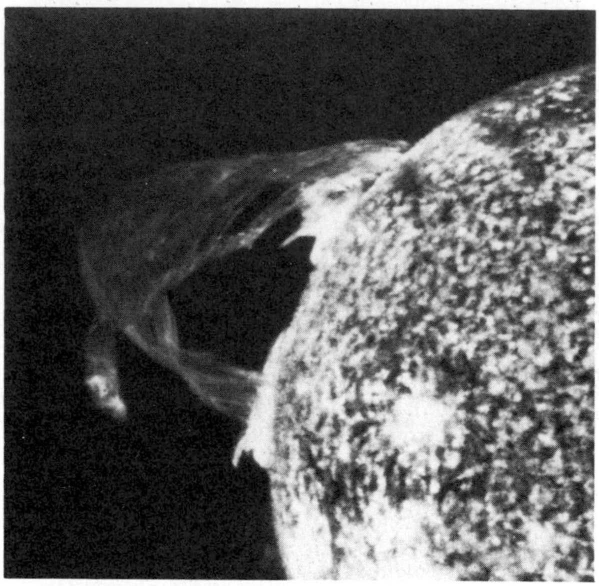

Fig. J.1 Giant prominence on the Sun photographed December 19, 1973, by an ultraviolet camera on Skylab. This prominence spans more than 588,000 km across the solar surface.

The criterion for a plasma is that the abundance of positive and negative charges must be so large that any local imbalance between the concentrations of

[1] This chapter is optional.

these charges is impossible. Any incipient accumulation of, say, positive charge rapidly attracts negative charges which restore the balance of charge. On a small scale, the positive and negative charges in a plasma move randomly, but on a large scale they move collectively — the positive and negative charges are strongly linked by their electric interaction and they move in unison so as to prevent any imbalance of charge. Thus, although the plasma contains a large number of free charges, it remains electrically neutral because, on the average, each unit volume contains equal amounts of positive and negative charge.

Matter in the plasma state has less order than matter in the solid, liquid, or gas state. However, the average electric neutrality of plasma represents some vestigial order. If we heat plasma to an extreme temperature, even this remaining order will be lost and the matter will turn into a chaotic mess of individual particles. This will happen when the thermal motion of the particles is so energetic that the electric forces cannot restrain them. Note that whether or not neutrality can be preserved depends on the scale of distance. On a small scale, neutrality always fails, while on a very large scale it hardly ever fails because it takes enormous amounts of energy to separate a large number of electric charges by a large distance. To gain some insight into what determines the critical scale of distance, let us examine the interaction between a plasma and a static electric field.

We know that, under static conditions, a good conductor will prevent an electric field from penetrating its interior — the conductor shields the electric field by accumulating some charge on its surface. This also happens in a plasma. If we insert a pair of electrodes into a plasma (Figure J.2), negative charges (electrons) will be drawn toward the positive electrode, and positive charges (ions) toward the negative electrode. Provided the electrodes are kept from direct contact with the plasma by a layer of dielectric insulator, the electrons and ions will accumulate, forming more or less static charge distributions around the electrodes and shielding their electric fields.

Because of thermal agitation, the charges cannot remain exactly static at the electrodes; instead, they

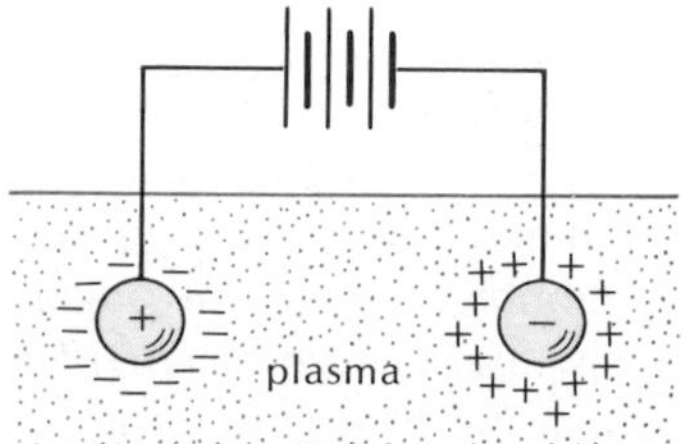

Fig. J.2 Two spherical electrodes immersed in a plasma.

form a restless cloud in the vicinity of each electrode. Only inside the cloud does the electrode disturb the plasma; outside of the cloud, the plasma hardly feels the presence of the electrode. The size of the cloud depends on the temperature — the cloud is large if the temperature is high because the random thermal back-and-forth motions of the charges will then tend to disperse them over larger distances.

The thickness of the cloud is called the shielding distance, or the **Debye distance.** It is given by the following formula:

$$D = \sqrt{\frac{\varepsilon_0 kT}{ne^2}} \tag{1}$$

where n is the number of electrons per unit volume, T the temperature, and k Boltzmann's constant (see Chapter 19). This distance determines how far an electric field penetrates within the plasma and it also determines the magnitude of the deviations from neutrality within a plasma. On a scale large compared to the Debye distance, the plasma remains electrically neutral because any concentrations of charge are immediately hidden in a cloud of opposite charge. Consequently, an ionized gas will exhibit plasma behavior on a scale of distance that is large compared to the Debye distance. For example, the ionized gas in an ordinary neon tube has an electron density of about $10^9/\text{cm}^3$ and these electrons have a temperature of about 20,000 K. Equation (1) then gives

$$D = \sqrt{\frac{8.85\times10^{-12}\ \text{F/m}\times1.38\times10^{-23}\ \text{J/K}\times20{,}000\ \text{K}}{10^{15}/\text{m}^3\times(1.6\times10^{-19}\ \text{C})^2}}$$

$$= 3\times10^{-4}\ \text{m} = 0.3\ \text{mm}$$

Hence a glob of this gas will behave as a plasma whenever it is larger than a few millimeters. Incidentally, the temperature of the electrons in a neon tube does not match the temperature of the ions — the electron temperature is about 20,000 K but the ion temperature is only about 2000 K. The reason for this discrepancy can be traced to the mechanism of energy transfer from electrons to ions. The electrons, being light and mobile, carry the electric current from one terminal of the tube to the other; the ions, being heavy and sluggish, do not contribute appreciably to this current. Thus the electrons receive all of the energy supplied to the tube by outside sources and they reach a high temperature. The ions can only acquire energy indirectly, by collisions with electrons. But in such a collision an electron will transfer only a very small fraction of its energy: when a light particle collides with a heavy particle, the light particle bounces off with next to no energy loss (see Section 10.2).

Table J.1 Some Plasmas

Plasma	*Electron temperature*	*Electron density*
Sun: center	2×10^7 K	$10^{26}/\text{cm}^3$
surface	5×10^3	10^6
corona	10^6	10^5
Fusion experiments (tokamak)	1×10^8	10^{14}
Fireball of atomic bomb	$\sim 10^7$	10^{20}
Solar wind	1×10^5	5
Lightning bolt	3×10^4	10^{18}
Glow discharge (neon tube)	2×10^4	10^9
Ionosphere	$\sim 2 \times 10^3$	10^5
Ordinary flame	2×10^3	10^8

Thus, the ions only receive small amounts of kinetic energy. The ions never reach thermal equilibrium with the electrons because the rate at which they receive energy barely matches the rate at which they lose energy in inelastic processes at the wall of the neon tube.

Table J.1 gives some examples of plasmas. Note that although most of the plasmas in Table J.1 are luminous, some are not. The reason is that some plasmas are so tenuous that they do not give off an appreciable amount of light even though their temperature is very high. The ionosphere of the Earth and the corona of the Sun are examples of such very faint plasmas. Only during a total eclipse does the faint light from the solar corona become visible (Figure J.3).

J.2 PLASMAS AND THE MAGNETIC FIELD

Many plasmas — both in nature and in our laboratories — are immersed in magnetic fields. The magnetic fields may originate from external sources or else from currents flowing within the plasma itself. In contrast to static electric fields which cannot exist inside a plasma, static magnetic fields can exist. However, such magnetic fields are subject to a severe restriction: the magnetic field lines are *frozen* in the plasma. If the plasma is stationary, then the field lines within the plasma are also stationary; if the plasma flows, then the field lines flow with it, each segment of field line following the motion of the small volume element of fluid in which it is embedded. For example, Figure J.4 shows the magnetic field lines passing through a ball of plasma initially placed between the poles of an electromagnet, and what happens to the field lines as we move the ball about or compress it.

Fig. J.3 Corona of Sun visible during the eclipse of June 8, 1918.

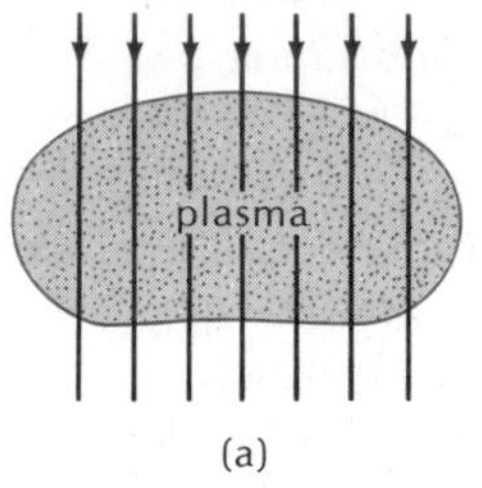

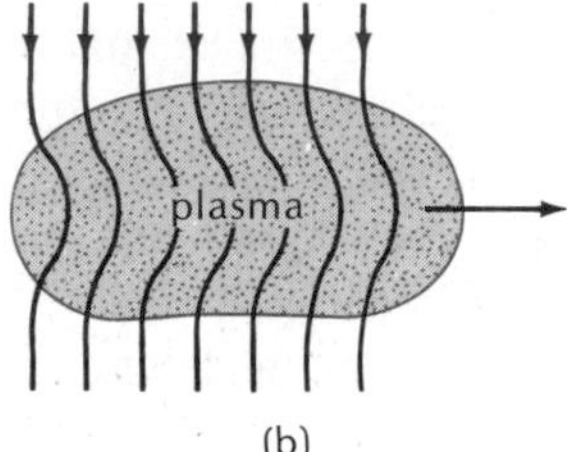

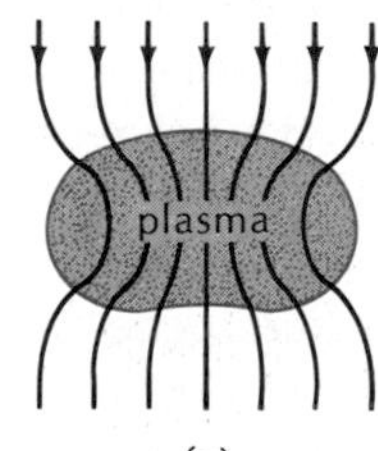

Fig. J.4 (a) Magnetic field lines in a ball of plasma. (b) If we move the ball of plasma, it drags the magnetic field lines along. (c) If we compress the ball of plasma, it compresses the field lines.

This initimate attachment of the field lines to the plasma is an immediate consequence of its high conductivity. In an ideal plasma, the conductivity is infinite and therefore the electric field in the rest frame of any given volume element of plasma must vanish. But we know, from Faraday's Law, that changes in the magnetic field induce electric fields. Since the electric field is forbidden, changes in the magnetic field are also forbidden. This means that in the rest frame of any given element of plasma, the magnetic field lines must remain fixed — they must move with the element of plasma. This freezing of the magnetic field lines within the plasma may be regarded as an extreme instance of Lenz' Law — any incipient change in the magnetic field immediately induces a current whose magnetic field combines with the original magnetic field in such a way that the net magnetic field remains constant. Note that the plasma not only keeps any interior field lines locked inside (as in Figure J.4), but it also keeps any exterior field lines outside. For exam-

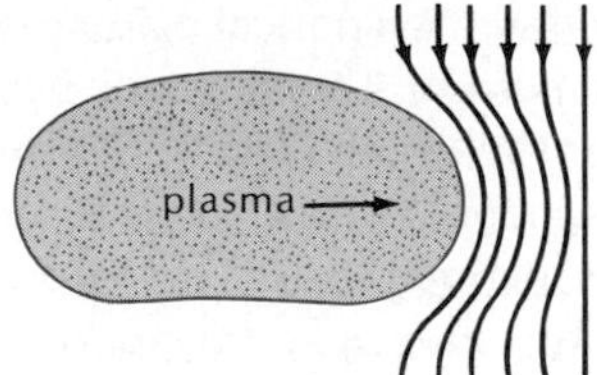

Fig. J.5 If we move a ball of plasma into a magnetic field, it pushes the field lines aside.

ple, if a ball of plasma with no field lines within it approaches a region with field lines, it will deform the latter and push them aside (Figure J.5).

This effect plays an important role in shaping the magnetic field of the Earth. If left to itself, the magnetic field of the Earth would be essentially a dipole field (Figure J.6). But this dipole field is affected by the solar wind, a stream of plasma blowing radially outward from the Sun. The solar wind consists of a neutral mixture of electrons and protons moving outward at a speed of about 400 km/s. The push of this wind against the magnetic field of the Earth causes a deformation — it compresses the field on the side facing the Sun and elongates the field on the side away from the Sun (Figure J.7). The region of space occupied by the magnetic field of the Earth is called the **magnetosphere.** Because of the action of the solar wind, the magnetosphere acquires the shape of an elongated raindrop with a tail extending to a distance of at least several hundred thousand kilometers.

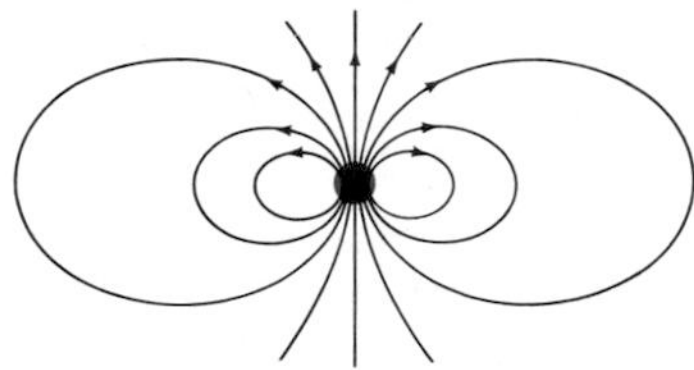

Fig. J.6 Magnetic field of an ideal dipole.

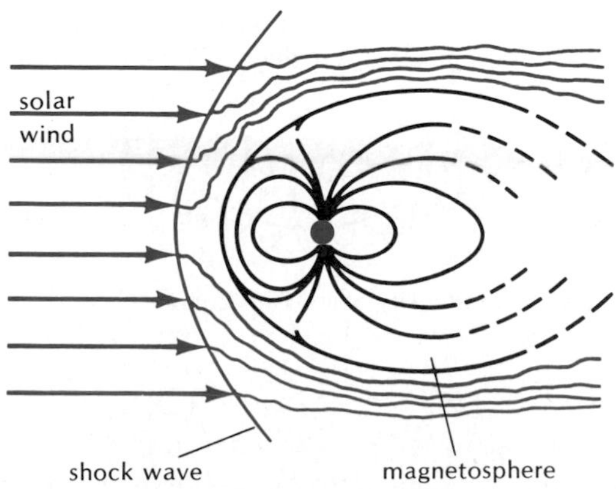

Fig. J.7 Magnetic field of Earth as distorted by the push of the solar wind. The speed of the wind is 400 m/s. At the leading edge of the magnetic field a shock wave forms.

Incidentally, the speed of the solar wind past the Earth is supersonic, i.e., the speed is larger than the speed of sound waves (compressional waves) in the plasma. Consequently, the supersonic motion of the Earth relative to the solar wind generates a shock wave in the plasma, just like the motion of a supersonic aircraft generates a shock wave in air (Figure J.7).

A deformation of the magnetic field lines, such as the deformations shown in Figures J.4 and J.5, involves an increase of the density of the magnetic field lines in some regions of space; thus it involves an increase of the energy stored in the magnetic field. In order to acquire this energy from the plasma, the magnetic field must exert a force on the plasma, a force that opposes the deformation. If the magnetic field is very strong, then the force with which it opposes the motion of the plasma may be so large that it prevents this motion altogether — the magnetic field lines then confine the plasma and hold it as though it were in a cage.

The force that the magnetic field exerts on the plasma can be described (in part) as a **magnetic pressure** acting within the plasma. It turns out that if the strength of the magnetic field is B, then this extra magnetic pressure within the plasma is $B^2/2\mu_0$. The net pressure within any region of the plasma is then the sum of the ordinary kinetic pressure of the plasma particles plus the extra magnetic pressure. Note that numerically the magnetic pressure equals the energy density of the magnetic field [see Eq. (32.42)]. This is no accident — the magnetic pressure arises precisely from the changes of magnetic energy, in the following way: If we compress some volume of plasma, we will also compress the magnetic field lines frozen in this volume; this increase of density of the field lines involves an increase of the energy stored in the magnetic field. By the argument given in the preceding paragraph, the magnetic field will then exert a force that opposes the compression, that is, a pressure.

Carefully designed arrangements of magnetic fields are used for the confinement of extremely hot plasmas in experiments on thermonuclear fusion. Such arrangements of magnetic fields are called **magnetic bottles;** they can hold a plasma that is too hot to be held by an ordinary vessel of metal or glass. For instance, Figure J.8 shows a magnetic bottle that confines a plasma by means of the magnetic field produced by a solenoid. In the central region, the coils of the solenoid are uniformly spaced, but at each end an extra-large coil intensifies the magnetic field and brings the field lines together. We can understand the confine-

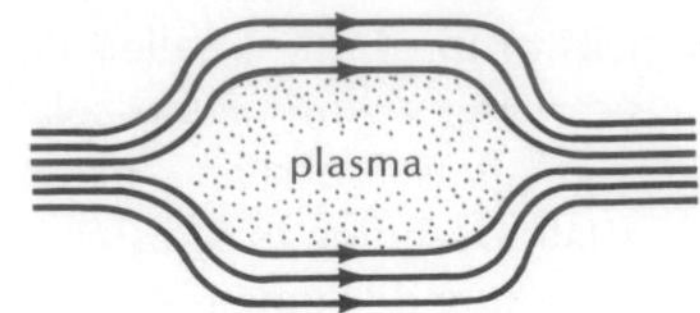

Fig. J.8 A magnetic bottle.

ment of the plasma by this bottle in terms of the magnetic pressure $B^2/2\mu_0$. This pressure is large in the strong magnetic field surrounding the plasma; hence this pressure tends to push the plasma inward, balancing the kinetic pressure that tends to push the plasma outward.

Alternatively, we can understand the confinement of the plasma in terms of the microscopic motions of the charged particles in the plasma. As we know from Section 31.3, a charged particle in a magnetic field moves in a helix around the magnetic field lines (Figure J.9). Thus, on the average, the particle gradually drifts along the direction of the field lines but does not wander off in a transverse direction. Near the ends of the bottle, where the field lines converge, the particle will be reflected. Figure J.10 shows how the converging field lines give rise to a component of force that halts the drift of the particle along the direction of the field lines and pushes it back (note that the circular motion of the particle does *not* stop or change direction; only the drift motion does). A magnetic field with convergent field lines is called a **magnetic mirror.**

A different method of confinement uses the magnetic field produced by a current flowing through the plasma, i.e., it uses the plasma's own magnetic field.

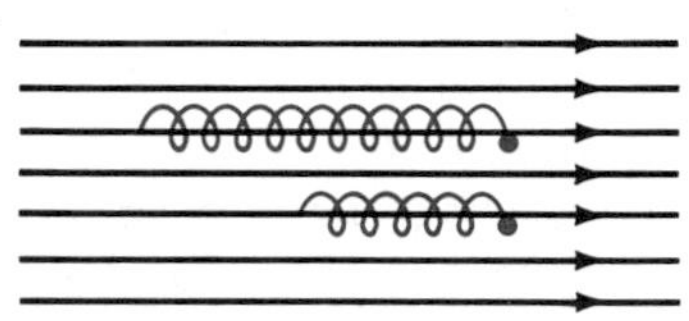

Fig. J.9 Charged particles with helical motion around the magnetic field lines.

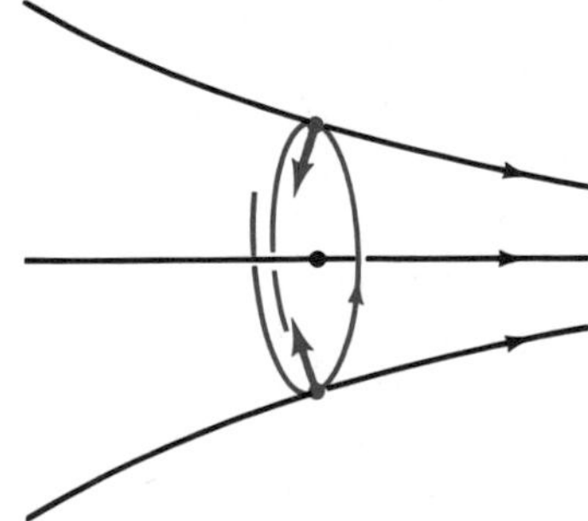

Fig. J.10 Charged particle and converging magnetic field lines. The magnetic force has a component along the axis of the helix.

Figure J.11 shows a cylindrical column of plasma, with a longitudinal current. The magnetic field produced by this current encircles the plasma and exerts a pressure on it. The magnetic field has its maximum value at the surface of the plasma; the magnetic pressure is therefore large at the surface and it pushes the plasma inward, balancing the kinetic pressure that pushes the plasma outward. The current used to hold together the very hot column of plasma in a thermonuclear fusion experiment may be as much as a million amperes.

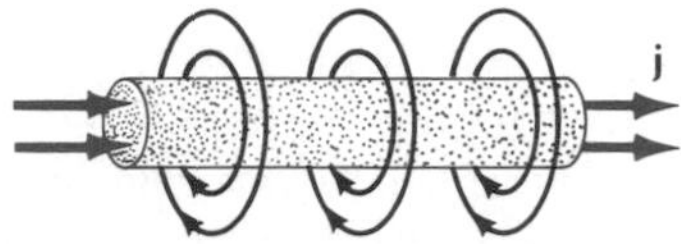

Fig. J.11 A cylindrical plasma column with an electric current passing through it.

We can also understand the confinement of such a column of plasma in terms of the attractive magnetic force between parallel current elements; obviously, this attractive force tends to compress the plasma and hold it together. The compression of a plasma by its own magnetic forces is called the **pinch effect.**

Unfortunately, the confinement provided by magnetic bottles is not perfect. The plasma tends to leak out at the ends and, what is worse, the plasma interacts with the magnetic field, giving rise to a variety of instabilities that destroy the delicate balance between the magnetic pressure and the kinetic pressure. For instance, Figures J.12 and J.13 show a kink instability and a sausage instability in a pinched plasma column. The cause of these instabilities can be qualitatively understood by examination of the field lines. In Figure J.12 the field lines are bunched together above the kink and spread apart below — hence the magnetic pressure pushes the plasma downward, further increasing the kinking of the column. In Figure J.13 the field lines are concentrated at the neck of the sausage; this squeezes the neck, further decreasing its diameter.

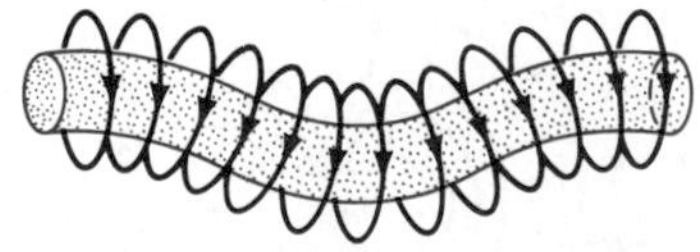

Fig. J.12 Kink instability of a plasma column.

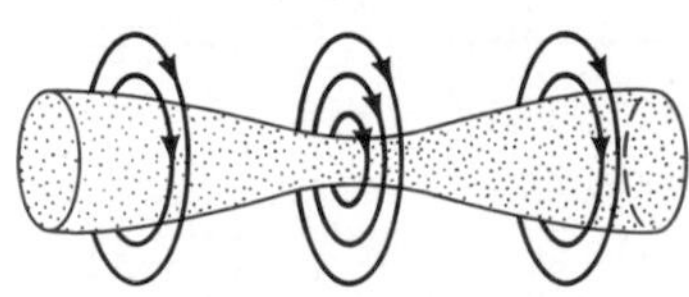

Fig. J.13 Sausage instability of a plasma column.

Both of these kinds of instability can be eliminated, or reduced, by means of an axial magnetic field in the plasma column (Figures J.14 and J.15). Such a field can be generated by placing the plasma in the core of a solenoid or by inducing an additional flow of current around the plasma column, effectively making it act as a solenoid. This magnetic field opposes the kink because deforming the axial magnetic field lines requires energy. Likewise, the magnetic field opposes the sausage constriction because compressing the magnetic field lines again requires energy.

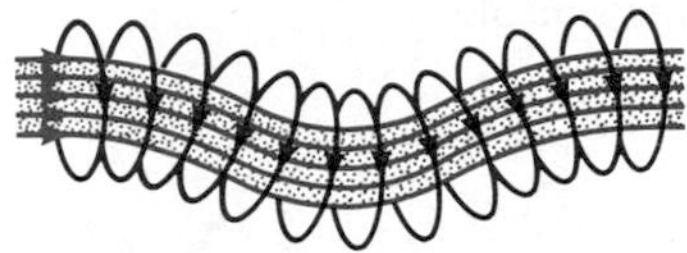

Fig. J.14 A plasma column with an extra axial magnetic field resists the kink instability.

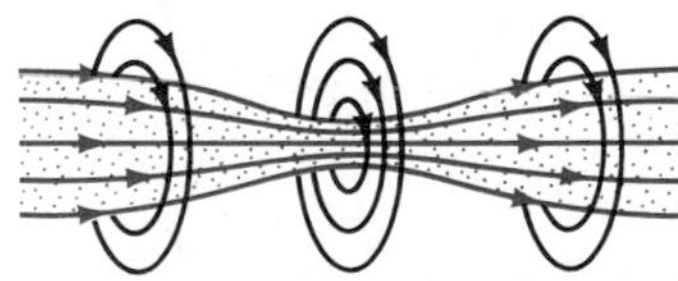

Fig. J.15 A plasma column with an extra axial magnetic field also resists the sausage instability.

Dozens of other, more complicated instabilities have been identified and studied by plasma physicists. The development of countermeasures to all these instabilities remains one of the central problems in plasma physics.

J.3 WAVES IN A PLASMA

The dynamical behavior of plasma is much more intricate than that of a gas. In the latter, the pressure completely determines the macroscopic motion, whereas in the former, electric and magnetic forces play a large role. For instance, consider the propagation of a sound wave in a plasma. This is a longitudinal wave with alternating zones of compression and rarefaction (Figure J. 16). In an ordinary gas, the restoring force in the sound wave is simply the excess pressure of the compressed zones. In a plasma, there is an additional restoring force: the concentrated positive electric charge of the ions of the plasma in the compressed zones gives rise to a repulsive electric force. Of course, the electrons of the plasma attempt to shield the ions and cancel their electric repulsion, but because the electrons have large random thermal motions (larger than those of the ions), the shielding is not quite perfect and a residual electric repulsive force remains.

Even more complicated dynamical effects can occur in a plasma immersed in a magnetic field. Suppose that the plasma is in an originally uniform magnetic field. A sound wave propagating in a direction perpendicular to this magnetic field will have alternating zones of compression and rarefaction and, since the magnetic field lines are frozen in the plasma, they must also have corresponding zones of compression and rarefaction (Figure J.17). The compression of the magnetic field lines generates an increase of the magnetic pressure. Thus, the restoring force that governs the propagation of this wave is due to a combination of kinetic pressure and magnetic pressure. A wave of this kind is called a **magnetosonic wave.**

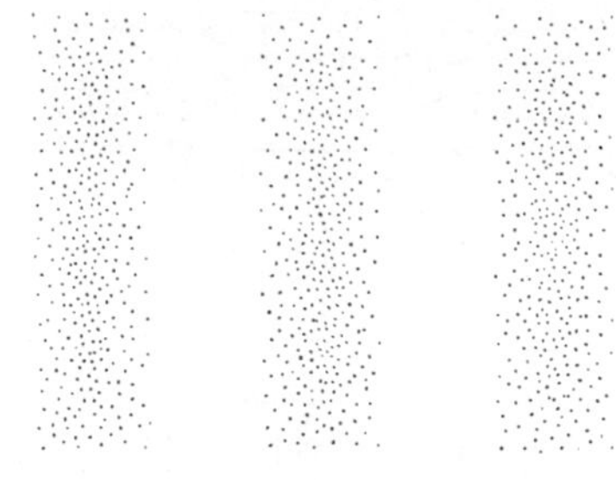

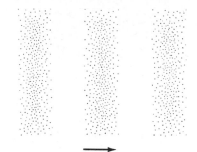

Fig. J.16 Sound wave in a gas or a plasma.

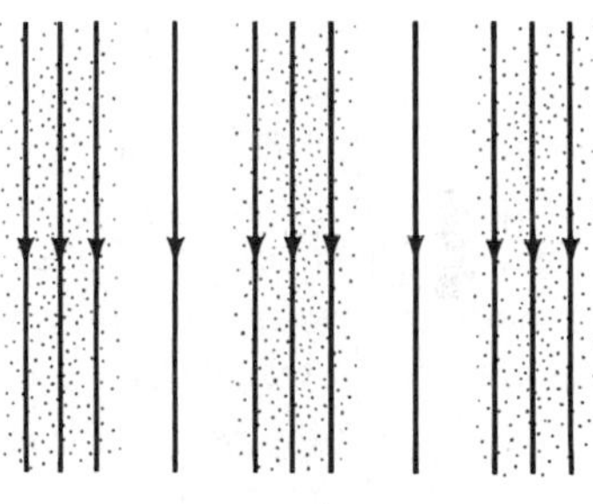

Fig. J.17 A sound wave in a plasma in the presence of a magnetic field. The compression of the plasma produces a compression of the magnetic field.

The magnetosonic wave is a longitudinal wave — it is merely a sound wave modified by magnetic effects. Surprisingly, it turns out that a plasma immersed in a magnetic field can also support a *transverse* wave. Consider a plasma in a uniform magnetic field. If one layer of this plasma is given a transverse displacement (Figure J.18), the magnetic field lines will suffer a transverse deformation. As is obvious from Figure J.18, this deformation leads to an increase of the density of field lines in the layer and hence to an increase of magnetic energy. Consequently, the embedded magnetic field exerts a restoring force on the plasma, opposing the deformation. Under the influence of the restoring force, the deformation propagates along the magnetic

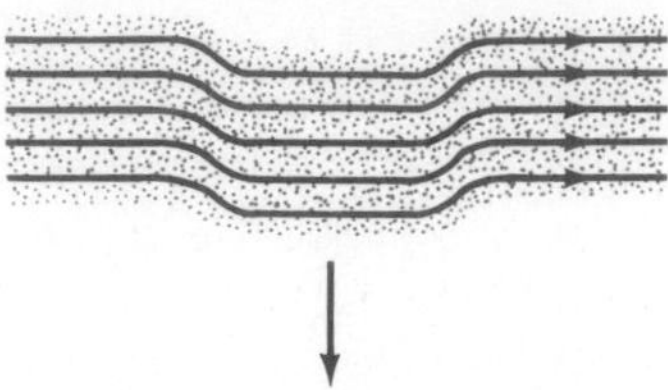

Fig. J.18 The displacement of a layer of plasma produces a deformation of the magnetic field.

field lines. This is a transverse wave, similar to a wave on a string (see Chapter 15); it propagates along the magnetic field lines just as though these lines were strings under tension. A wave of this kind is called an **Alfvén wave.**[2]

The propagation of radio waves in a plasma also has many intricate features. Since a plasma is a good conductor and shields electric fields, we expect it will shield the electric fields of a radio wave, i.e., it will reflect radio waves just as a metal does. This is true for radio waves of low frequency, but for radio waves of high frequency the shielding fails because the electrons do not have enough time to respond to the electric fields. Thus, high-frequency radio waves can penetrate into a plasma and propagate through it. The minimum frequency that can pass through a plasma is called the **cutoff frequency;** the value of this frequency depends on the electron density — it increases with the square root of the electron density. For instance, the ionosphere of the Earth will pass radio waves of frequency above about 30 MHz (radar and TV) but it will reflect waves of lower frequency (short waves, medium waves, and long waves).

The reflection of radio waves by the ionosphere is of great importance for radio communications. We can send radio waves from one end of the globe to the other by successively bouncing them back and forth between the ionosphere and the ground (Figure J.19).

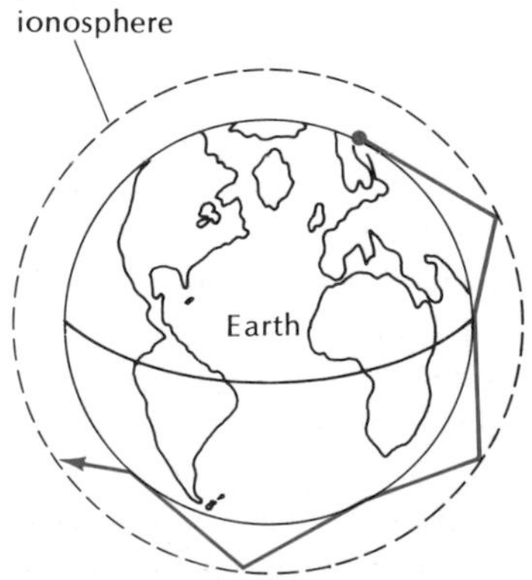

Fig. J.19 Path of a radio wave around the Earth.

[2] Hannes Alfvén, 1908–, Swedish physicist. He received the Nobel Prize in 1970 for his work in the theory of plasmas.

The ionosphere actually consists of several reflecting layers with characteristics that vary with day and night and also with the sunspot cycle (Figure J.20). Radio waves of different wavelength are reflected by different layers; for this reason, the choice of radio frequency band is critical in long-range radio communication. Incidentally, the radio blackout suffered by a space capsule during reentry of the atmosphere is also due to a plasma effect. The friction between the space capsule and the air gradually burns off the heat shield of the capsule, and the flames surround the capsule with a layer of plasma; this layer stops radio waves and prevents communication.

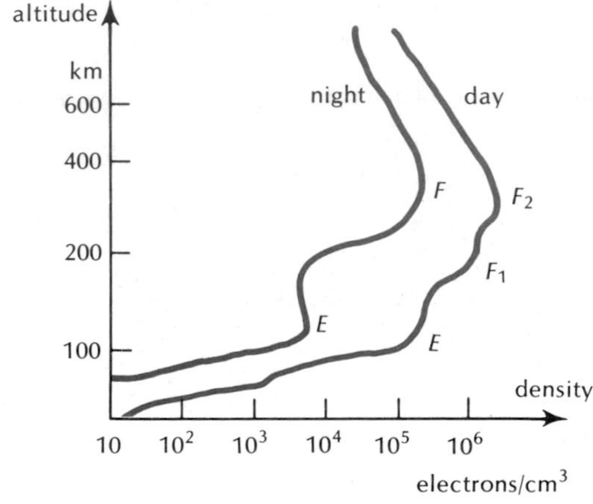

Fig. J.20 Density of free electrons in the atmosphere as a function of altitude (at a time of sunspot maximum). The peaks in the electron density at altitudes of ~120 km and ~300 km constitute the *E* and the *F* layers, respectively.

If the plasma is immersed in a magnetic field, then the behavior of radio waves depends in a complicated way on the frequency of the wave, the magnitude of the magnetic field, and the direction of propagation of the wave relative to the direction of the magnetic field. High-frequency radio waves can propagate in any direction (although their speed will depend on direction), but low-frequency radio waves can only propagate if their direction lies within a limited range of angles near the direction of the magnetic field. Hence magnetic field lines can act as guides for radio waves. This causes the weird phenomenon of whistlers that can be picked up with an audio amplifier at a few kilohertz; these are radio signals of low frequency emitted by flashes of lightning. The signals propagate in the tenuous plasma surrounding the Earth; they travel from one hemisphere to the other guided by the magnetic field lines of the Earth (Figure J.21). The high-frequency components of the signal travel faster than the low-frequency components. At the receiver, the result is a whistle consisting of a de-

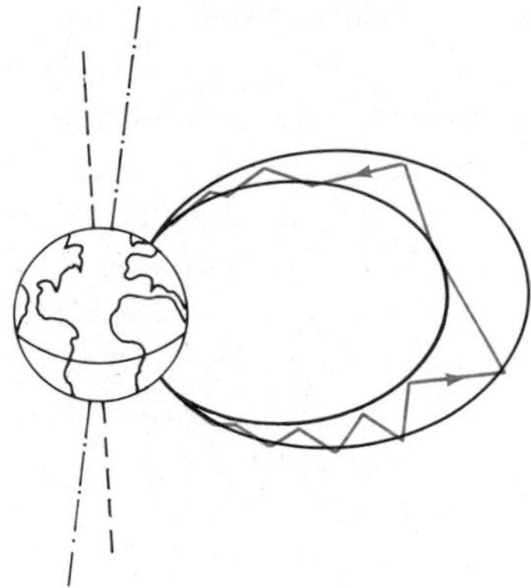

Fig. J.21 The magnetic field of the Earth acts as a waveguide for low-frequency radio waves.

scending glide tone (glissando) that begins at a high pitch and ends at low pitch.

J.4 THERMONUCLEAR FUSION

The Sun generates its heat by a nuclear fusion reaction burning hydrogen into helium. But for us on Earth, this reaction is not a viable energy source because it is a very slow reaction. A fusion reaction burning deuterium or tritium into helium is much more practical. In fact, the H-bomb derives its energy not from the fusion of hydrogen but from the fusion of deuterium and tritium (see Section H.4). Essentially, a fusion reactor brings the power of an H-bomb under control just like a fission reactor brings the power of an A-bomb under control.

Fusion is a very promising energy source because fuel for fusion is available in great abundance. Deuterium occurs naturally in molecules of heavy water (HDO) — about 0.03% of the water in the oceans is heavy water. Tritium is radioactive and does not occur naturally.[3] However, it can be easily produced by bombarding lithium with neutrons in a nuclear reactor. The available supply of deuterium is sufficient to satisfy all our energy requirements for a billion years; the supply of lithium is sufficient for a million years.

The most suitable reactions for the generation of fusion power are the **deuterium–deuterium reaction** and the **deuterium–tritium reaction.** The first of these involves the following sequence of four steps:

$$\begin{array}{l} \mathrm{D + D \rightarrow {}^3He + n} \\ \mathrm{D + D \rightarrow T + p} \\ \mathrm{D + T \rightarrow {}^4He + n} \\ \underline{\mathrm{D + {}^3He \rightarrow {}^4He + p}} \\ \mathrm{6D \rightarrow 2\,{}^4He + 2p + 2n + 43.1\ MeV} \end{array} \qquad (1)$$

The net result is the fusion of six deuterium nuclei into two helium nuclei with the release of two protons, two neutrons, and 43.1 MeV of energy. Note that the energy released for each fused deuterium nucleus is smaller than the energy released for each fissioned uranium nucleus — about 7 MeV per fused deuterium nucleus as compared to about 200 MeV per fissioned uranium nucleus. But weight for weight, the energy released in the fusion of deuterium is four times as large as in the fission of uranium. One further advantage of fusion over fission is that the reaction products from the former are harmless, whereas the reaction products from the latter are radioactive. In the reaction (1) the only potentially hazardous products are the neutrons; but these can be captured in a suitable absorber placed around the reactor vessel.

The deuterium–tritium reaction requires the presence of lithium and proceeds in two steps:

$$\begin{array}{l} \mathrm{D + T \rightarrow {}^4He + n} \\ \underline{\mathrm{n + {}^6Li \rightarrow {}^4He + T}} \\ \mathrm{D + {}^6Li \rightarrow 2\,{}^4He + 22.4\ MeV} \end{array} \qquad (2)$$

Thus, the net result is the fusion of deuterium and lithium nuclei into helium nuclei. The tritium only exists in an intermediate step in this reaction, being produced from the lithium. The technology required for the D–T reaction is much more complicated than that for the D–D reaction. The reactor chamber containing a mixture of deuterium and tritium would have to be surrounded by a blanket of lithium where neutrons can be absorbed and tritium can be generated; the tritium must afterward be extracted from the blanket and fed into the chamber. In spite of these complications, the D–T reaction is regarded as more promising than the D–D reaction because its ignition temperature is somewhat lower.

Both the D–D and D–T reactions will only proceed at extremely high temperature. The nuclei will only fuse if they are brought into contact by violent collisions — since the nuclei are positively charged, they experience a Coulomb repulsion, and to overcome this, the collision must be initiated with a large kinetic energy. The only feasible way to give the nuclei the necessary kinetic energy is by thermal motion. To attain a sufficiently large kinetic energy, a very high temperature is required — about 5×10^9 K for D–D fusion and about 1×10^8 K for D–T fusion. These temperatures are higher than at the center of the Sun. Nuclear reactions at such extreme temperatures are called **thermonuclear.** At these temperatures the deuterium or tritium atoms will be completely ionized and the nuclei and electrons will form a plasma.

To ignite the nuclear fire it is sufficient to mix the reactants and heat them to high temperature. But if

[3] Deuterium (D, or ^{2}H) and tritium (T, or ^{3}H) are two isotopes of hydrogen.

the nuclear fire is to yield a profitable amount of energy, it is also necessary to keep the reactants together for some minimum length of time. Obviously, to break even, we must gain an amount of energy equal to the amount that we invested to heat the plasma. A calculation shows that for the D–T reaction this requires a minimum time (in seconds)

$$\tau = \frac{2 \times 10^{14}}{n} \tag{3}$$

and for the D–D reaction a minimum time

$$\tau = \frac{5 \times 10^{15}}{n} \tag{4}$$

where n is the number of ions per cubic centimeter. For instance, according to Eq. (3), a D–T plasma of density $10^{14}/\text{cm}^3$ must be confined for at least 2 s if it is to yield a profitable amount of energy. And here is the great problem of plasma physics — the plasma tends to escape and disperse. The challenge is to invent some means of confining the plasma long enough to extract significant amounts of energy.

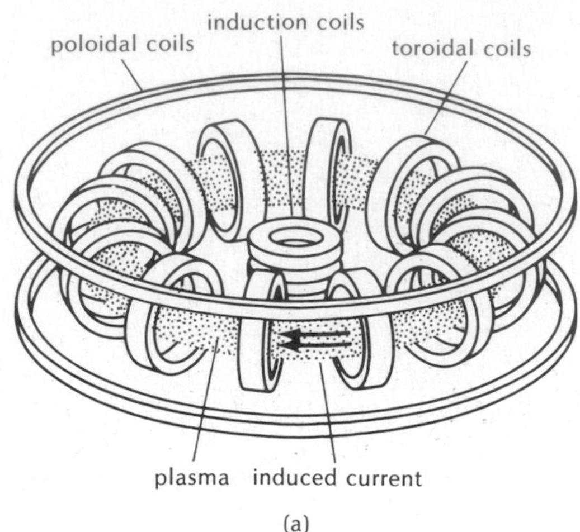

(a)

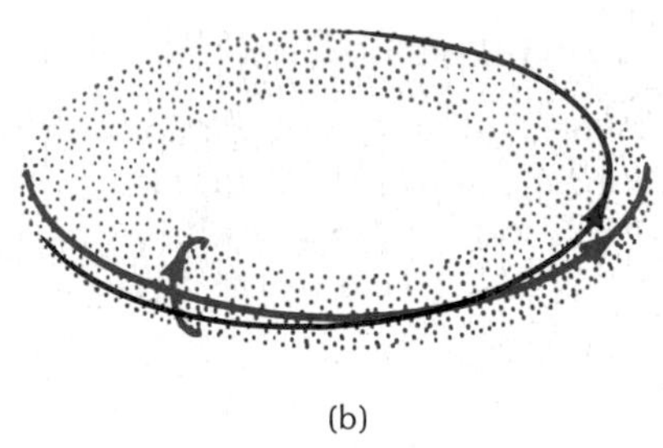

(b)

Fig. J.22 (a) Magnet coils in a tokamak. (b) The magnetic field of the toroidal coils lies along the plasma; the field lines are large horizontal circles (color). The magnetic field of the induced current flowing along the plasma column encircles the column; the field lines are small vertical circles (color). The net magnetic field consists of helical lines (black) wrapped around the plasma column. (The magnetic field of the poloidal coil has been neglected.)

J.5 PLASMA CONFINEMENT

Since the plasma in a thermonuclear fusion reactor must be kept very hot, we must prevent thermal contact with the walls of the reactor vessel. This can be achieved by suspending the plasma in the middle of the reactor vessel, in a magnetic field. But a magnetic field cannot confine the plasma completely. The minimum rate of leakage of the plasma is determined by diffusion caused by the random thermal motion. In practice, the leakage is much worse: the plasma tends to develop a variety of instabilities which make it squirt out of its magnetic confinement at a much faster rate than the minimum rate of leakage.

Hoping to conquer these instabilities, plasma physicists have constructed a variety of machines with fanciful names: Stellarator, Scylla, Scyllac, Tokamak, Alcator, etc. These machines are supposed to hold the plasma by means of intricate magnetic fields and they are supposed to heat the plasma to the point where fusion begins. Several of these machines have succeeded in triggering some fusion reactions, but none has generated any useful amount of power. Of these devices, the tokamak, originally developed in the Soviet Union and later copied in the United States, has so far proven the most successful.

In the **tokamak,** the plasma is confined to a toroidal region, i.e., a column bent into a ring. This avoids the loss of plasma from the ends of the column, since there are no ends. The magnetic fields confining the plasma are in part generated by external solenoidal windings and in part by a current induced in the plasma itself. Thus, both of the simple confinement mechanisms mentioned in Section J.2 come into play. Figure J.22 shows the arrangement of magnet coils in a typical tokamak. The current along the plasma column is generated by induction, by means of a changing magnetic field provided by central, poloidal coils. The current induced by this magnetic field is usually in excess of a million amperes.

A tokamak machine at the Princeton Plasma Physics Laboratory has achieved a temperature of 7×10^7 K, four times hotter than the temperature in the center of the Sun. The plasma in the torus had a density of about $2 \times 10^{13}/\text{cm}^3$ and remained confined for about $\frac{1}{20}$ of a second. Thus, the plasma was near its thermonuclear ignition temperature, but its density and confinement time did not reach the break-even point set

Fig. J.23 The torus of the Princeton Tokamak Fusion Test Reactor (TFTR) during construction.

Fig. J.24 Neutral-beam injectors (left) attached to the Princeton TFTR (right). The torus is almost completely hidden within the support frame.

by Eq. (3). Larger versions of this machine, now in operation, are expected to do better (Figure J.23). Incidentally, the Joule heat of the current induced along the toroidal plasma provides some of the heating needed to start the thermonuclear reaction. In the Princeton machine, extra heating is provided by injectors that shoot intense beams of high-speed neutral atoms into the torus (Figure J.24).

Another promising machine is the **tandem mirror** now under construction at the Lawrence Livermore National Laboratory. This machine is a straight solenoid closed off at each end by a magnetic mirror consisting of two magnets arranged in tandem (Figure J.25). One of these magnets has an ordinary circular coil, but the other magnet has several interlocking "ying-yang" coils, shaped somewhat like the seams on a baseball. In the magnetic field of such a tandem magnet, the electrons and the positive ions of the plasma tend to separate to some extent. This charge separation generates an electric field which aids in the confinement of the plasma. Hence the action of the tandem mirror is part magnetic and part electric. Figures J.26 and J.27 show the magnet coils for Livermore under construction.

Fusion reactors based on the Tokamak design or the mirror design aim to reach the break-even point [see Eqs. (3) and (4)] by confining a low-density plasma for a fairly long time. As the fuel burns, fresh fuel would be injected into the reactor chamber, either in the form of puffs of gas or in the form of frozen pellets of a D–T mixture. In contrast, fusion reactors based on the **Scylla** design aim to reach the break-even point with a much higher-density plasma and a much shorter confinement time. Such a reactor would burn its fuel in a burst lasting perhaps 10^{-3} s and then it would receive a fresh load of fuel for the next spurt of power. The pulsed release of heat in such a reactor is analogous to the pulsed release of heat in one of the cylinders of an internal combustion engine, which also burns its fuel in bursts and delivers spurts of power.

Instead of attempting to confine the plasma with magnetic fields, some experimenters are now exploring the possibility of a fusion reactor using the uncontrolled, explosive burning of small solid pellets of fuel. In such an **inertial confinement** reactor, one pellet at a time is dropped into a combustion chamber and suddenly vaporized into a very hot and very dense plasma by an intense pulse of light from a powerful laser or else by an intense pulse of beamed electrons or ions from an electron or ion gun. This sudden ignition will burn the deuterium and tritium before the plasma has a chance to expand and disperse. This is an extreme case of pulsed burning; it is explosive burning, lasting only about 10^{-9} s. During the explosion, the plasma is

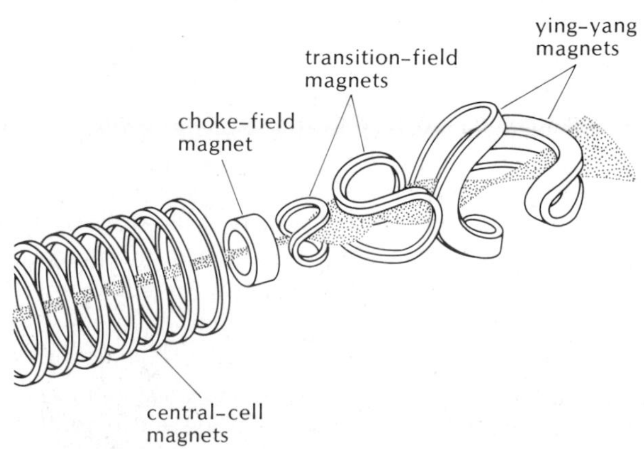

Fig. J.25 Magnet coils of the tandem mirror machine.

Fig. J.26 Yin-yang magnet coils for the tandem mirror machine under construction.

Fig. J.27 Circular magnet coils under construction.

confined to some extent by its own inertia — the outer layers of plasma do not directly participate in the fusion, but their inertia holds the core of the pellet together for a long enough time for the reactions to proceed. The success of such a fusion scheme depends on the development of very powerful and very efficient lasers (see Interlude L). Part of the output of the reactor will have to be used to energize the lasers.

Like the heat from an ordinary fission reactor, the heat from a fusion reactor can be removed from the reactor core and used to make steam to drive turbines generating mechanical and electric power. There is, however, an ingenious alternative method for the direct generation of electric power. The plasma in the combustion chamber of the fusion reactor is a mixture of electrons and positive ions. If this plasma is allowed to stream out of the chamber into a magnetic field, the negative and positive charges will separate. These opposite charges can be collected on two sets of plates; this results in a direct conversion of the plasma energy into electric energy. In principle, such a plasma generator is somewhat similar to a magnetohydrodynamic generator (see Example 32.2).

The fusion reactors of the future will probably be very large machines delivering more than 1000 MW of power (Figure J.28). But besides the confinement problem, there are several other problems that remain to be solved before we can exploit fusion power commercially. For instance, the design of the walls of the combustion chamber and the design of the magnets will require advances in technology. The walls are exposed to an abundant flux of neutrons emitted by the fusion reaction and also to the impact of charged particles escaping from the plasma. Metals exposed to such a flux of neutrons suffer from swelling and embrittlement. New, resistant alloys will have to be developed for the walls. Furthermore, the magnets surrounding the reactor vessel will probably have to be constructed out of a superconductor; otherwise their power consumption would be prohibitive. These magnets will have to be much larger than any magnets built to date and they will have to stand up to enormous magnetic forces between the currents in their windings.

As we already pointed out in Interlude F, a fusion reactor is much cleaner than a fission reactor in that it does not produce an abundance of radioactive residues. However, the flux of neutrons released by the fusion reaction induces radioactivity in the walls of the reactor chamber, in the magnets, and wherever else it is absorbed — the impact of neutrons transmutes ordi-

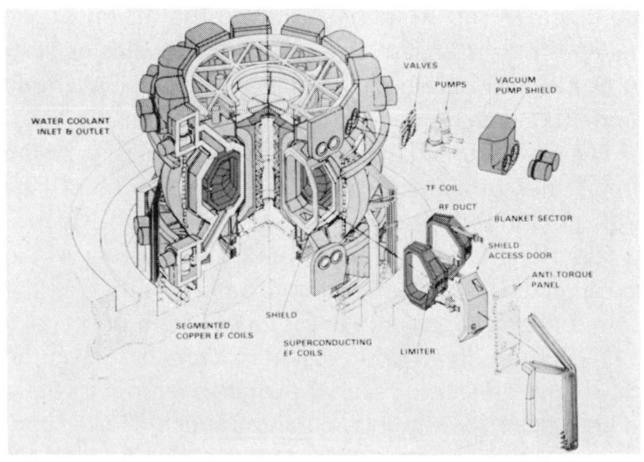

Fig. J.28 The design for the 1200-MW STARFIRE fusion reactor proposed by the Argonne National Laboratory.

nary stable nuclei into radioactive nuclei. The possible leakage of tritium from the reactor is an additional environmental hazard. Tritium is radioactive and, since it is chemically identical to hydrogen, it could easily contaminate our water supplies. If the confinement problem can be solved, these secondary problems will undoubtedly also be solved. We can then look forward to an abundant supply of relatively clean energy.

Further Reading

The Fourth State of Matter by B. Bova (New American Library, New York, 1974) is an elementary introduction to the properties and applications of plasmas. *A Physicist's ABC on Plasma* by L. A. Artsimovich (MIR Publishers, Moscow, 1978) is a very concise, slightly mathematical introduction to plasma physics, with emphasis on the magnetic confinement problem. The article "World Energy Reserves & Some Speculations on the Future of Nuclear Fusion Energy" by J. L. Tuck in *Cosmology, Fusion & Other Matters,* edited by F. Reines (Colorado Associated University Press, Boulder, 1972), gives a neat survey of available energy reserves and explains the advantages and principles of fusion reactors.

The following magazine articles deal with applications of plasma physics:

"VLF Emissions from the Magnetosphere," R. N. Sudan and J. Denavit, *Physics Today,* December 1973

"Fusion Reactors," B. B. Kadomtsev and T. K. Fowler, *Physics Today,* November 1975

"The Earth's Magnetosphere," S. Akasofu and L. J. Lanzerotti, *Physics Today,* December 1975

"Fusion Power with Particle Beams," G. Yonas, *Scientific American,* November 1978

"Recent Progress in Tokamak Experiments," M. Murakami and H. P. Eubank, *Physics Today,* May 1979

"Alternate Concepts in Magnetic Fusion," F. F. Chen, *Physics Today,* May 1979

"Progress toward a Tokamak Fusion Reactor," H. P. Furth, *Scientific American,* August 1979

"The Active Solar Corona," R. Wolfson, *Scientific American,* February 1983

"The Engineering of Magnetic Fusion Reactors," R. W. Conn, *Scientific American,* October 1983

Questions

1. Is there any plasma in the room you are in?

2. The ancient Greeks thought that there were five elements: earth, water, air, and fire. How does this list compare with the list of the four states of matter?

3. Can matter exist as a mixture of two or more of the four states? Can matter consisting of a single chemical element exist as a mixture of two or more states? Give examples.

4. According to the table of masses of the Sun and the planets printed on the endpapers of this book, roughly what fraction of the Solar System is plasma?

5. The chemical composition of the Sun is usually said to be 70% hydrogen and 30% helium. Are there actually any hydrogen or helium atoms on the Sun? What would be a more accurate description of the composition of the Sun?

6. How can the solar corona be hotter than the solar surface?

7. A neon tube contains plasma at very high temperature. Why does the tube not feel hot to the touch?

8. According to Table J.1, the temperature of the ionosphere is 2×10^3 K. Would a piece of paper burn if exposed to the ionosphere?

9. The free-electron gas in a metal has some of the features of a plasma. Does this mean we should regard a metal as plasma?

10. If you compress a gas (at fixed temperature), it becomes a liquid. If you compress a plasma, does it become a gas?

11. In the science fiction story *The Black Cloud* by the astronomer Fred Hoyle, a large cloud of plasma behaves as an intelligent being. How could the brain of such a creature store information in currents and magnetic fields? How could such a creature think?

12. Would you expect a kink or a sausage instability to develop in a plasma confined in a solenoidal magnetic field as shown in Figure J.8?

13. Explain why the magnetic mirrors shown in Figure J.8 permit some leakage along the center line.

14. Figure J.20 shows that the electron density in the ionosphere during the day is much larger than during the night (by up to a factor of 100). Can you guess why?

15. Consider a radio wave propagating around the Earth by bouncing back and forth between the ionosphere and the ground. Would you expect that a receiver at some given distance from the transmitter can be reached by two waves with different numbers of bounces?

16. What method could be used to separate the heavy water needed as fuel for a fusion reactor from ordinary water?

17. The Sun is a gigantic fusion reactor. What confines the plasma in this reactor? Why can we not use this method of confinement on Earth?

18. What would happen if the plasma in a fusion reactor were to come in contact with the wall of the reactor chamber?

19. Why is a runaway reaction possible in a fission reactor, but not in a fusion reactor?

20. Why does the reaction (1) require a higher temperature than the reaction (2)? (Hint: Which step in these reactions involves the most Coulomb repulsion?)

CHAPTER 33

Magnetic Materials

Within the atom and the nucleus, charged particles are continually moving about — electrons orbit around the nucleus and protons orbit around each other inside the nucleus. The orbital motions may be regarded as flows of electric currents within the atom, and these currents generate magnetic fields. Besides their orbital motions, the charged particles within atoms have rotational motions — electrons, protons, and neutrons all spin about their axes. The rotational motions may be regarded as flows of electric currents inside the particles, and these currents also generate magnetic fields.

The magnetic fields arising from currents flowing in loops inside atoms, nuclei, and particles can be described in terms of the corresponding magnetic dipole moments. If many of these small dipole moments within a sample of material are aligned, they will produce a strong magnetic field. Such an alignment can be achieved by placing the sample of material in the magnetic field of, say, an electromagnet. The magnetic field produced by the atomic and subatomic currents will then be strong, and it will modify the original magnetic field produced by the currents in the windings of the electromagnet. For instance, if a piece of iron is placed between the poles of an electromagnet, the magnetic field is strengthened drastically — it may become several thousand times stronger than the original magnetic field. Exactly how much the original magnetic field is increased or decreased depends on the response of the atomic and subatomic dipoles to the original magnetic field. According to the nature of their magnetic response, we can classify materials as paramagnetic, ferromagnetic, or diamagnetic. Before we discuss these types of magnetic materials, we will take a brief look at the magnitudes of the magnetic moments contributed by electrons, protons, and neutrons.

33.1 Atomic and Nuclear Magnetic Moments

An electron moving in an orbit around a nucleus produces an average current along its orbit. Strictly, the calculation of atomic orbits and currents requires quantum mechanics; but for the sake of simplicity, let us do this calculation with classical mechanics. If the electron has a circular orbit with radius r and speed v (Figure 33.1), then the time for one complete circular motion is $2\pi r/v$. The charge moved in this time is e and hence the average current along the orbit is

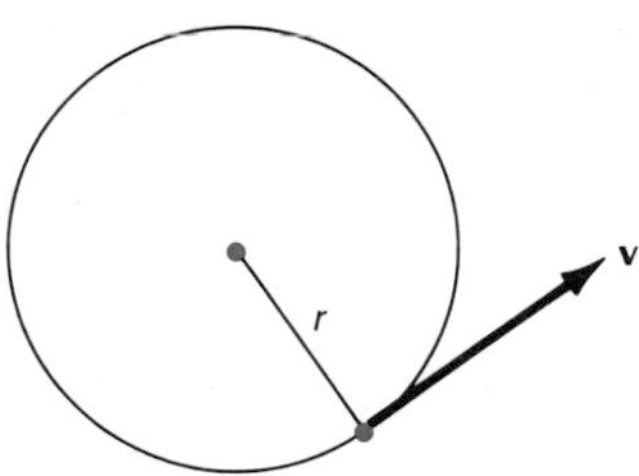

Fig. 33.1 Electron in a circular orbit around a nucleus.

$$I = \frac{e}{2\pi r/v} = \frac{ev}{2\pi r} \tag{1}$$

Such a circulating current will give rise to a magnetic moment [see Eq. (30.44)]

$$\mu = I \times [\text{area}] = \frac{ev}{2\pi r} \times \pi r^2 \tag{2}$$

$$= \frac{evr}{2} \tag{3}$$

In terms of the angular momentum $L = m_e vr$, the magnetic moment can be expressed as[1]

$$\boxed{\mu = \frac{e}{2m_e} L} \tag{4}$$

Orbital magnetic moment

This says that the magnetic moment of an orbiting electric charge is proportional to its angular momentum.

It turns out that the relation (4) is valid not only for circular orbits, but also for any other periodic orbit. Even more important: the relation (4) remains valid when we repeat the calculation using quantum mechanics. The net magnetic moment of the atom is the sum of the magnetic moments of all its electrons. Hence Eq. (4) can also be regarded as a relation between the net orbital angular momentum and the net magnetic moment of the entire atom.

It is a fundamental tenet of quantum mechanics that the magnitude of the orbital angular momentum is always some integral multiple of the constant value $\hbar = 1.06 \times 10^{-34}$ J · s.[2] Thus, the possible values of the orbital angular momentum are[3]

$$L = 0,\ \hbar,\ 2\hbar,\ 3\hbar,\ \ldots \tag{5}$$

Quantization of angular momentum

Because angular momentum only exists in discrete packets, it is said to

[1] In this equation, the magnetic moment μ must not be confused with the permeability μ_0.

[2] This is Planck's constant divided by 2π, i.e., $\hbar = h/2\pi = 6.63 \times 10^{-34}$ J · s/2π. See Appendix 8 for a more precise value of the constant $\hbar$.

[3] Strictly, these numbers are the possible values for the maximum component of the angular momentum in a chosen direction, say, the z direction.

be **quantized.** The constant $\hbar$ is the fundamental quantum of angular momentum, just as e is the quantum of electric charge. For instance, the oxygen atom has a net orbital angular momentum $L = 1\hbar$ and hence, according to Eq. (4), the orbital magnetic moment is

$$\mu = \frac{e}{2m_e}\hbar = 9.27 \times 10^{-24}\ \text{A}\cdot\text{m}^2 \qquad (6)$$

Besides the magnetic moment generated by the orbital motion of the electrons, we must also take into account that generated by the rotational motion of the electrons. Crudely, an electron may be thought of as a small ball of negative charge rotating about an axis at a fixed rate. The spin, or intrinsic angular momentum, of the electron has a value of $\hbar/2 = 0.53 \times 10^{-34}$ J · s. This kind of rotational motion again involves circulating charge and gives the electron a magnetic moment. This intrinsic magnetic moment has a fixed magnitude

Spin magnetic moment

$$\boxed{\mu_{\text{spin}} = \frac{e}{2m_e}\hbar = 9.27 \times 10^{-24}\ \text{A}\cdot\text{m}^2} \qquad (7)$$

Bohr magneton

which is called a **Bohr magneton.**[4] The direction of this magnetic moment is opposite to the direction of the spin angular momentum (Figure 33.2). Note that Eq. (4) is not valid for the spin magnetic moment. If we were to insert a spin angular momentum of $\hbar/2$ into Eq. (4), we would obtain a magnetic moment $e\hbar/(4m_e)$, one-half the value given by Eq. (7).

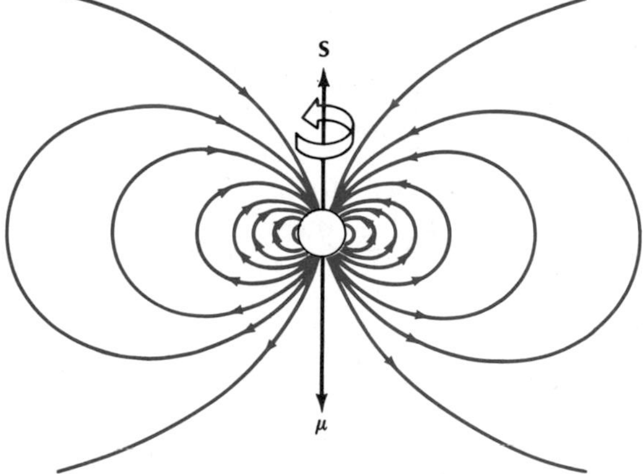

Fig. 33.2 Magnetic field lines of an electron. The magnetic moment μ is opposite to the spin angular momentum **S**.

The net magnetic moment of the atom is obtained by combining both the orbital and spin moments of all the electrons, taking into account the directions of these moments. In the oxygen atom, the net resultant magnetic moment is 13.9×10^{-24} A · m². In some atoms — such as helium and argon — the magnetic moments cancel. But in most atoms the net magnetic moment is different from zero. Thus, most atoms behave as small magnetic dipoles. Table 33.1 gives the magnetic moments of some atoms and ions.

[4] See Appendix 8 for more precise values of the magnetic moments of the elementary particles.

Table 33.1 MAGNETIC MOMENTS OF SOME ATOMS AND IONS

Atom	*Magnetic moment*
H	9.27×10^{-24} A · m^2
He	0
Li	9.27×10^{-24}
O	13.9×10^{-24}
Ne	0
Na	9.27×10^{-24}
Ce^{+++}	19.8×10^{-24}
Yb^{+++}	37.1×10^{-24}

The nucleus of the atom also has a magnetic moment. This is in part due to the orbital motion of the protons inside the nucleus, and in part due to the rotational motion of individual protons and neutrons. The spin of both these particles is $\hbar/2$, the same as that of an electron. The spin magnetic moment of a proton is 1.41×10^{-26} A · m^2 and that of a neutron is 0.97×10^{-26} A · m^2 (the former magnetic moment is parallel to the axis of spin and the latter is antiparallel).[5] The magnetic moment of a proton or a neutron is small compared to that of an electron, and in reckoning the total magnetic moment of an atom, the nucleus can usually be neglected.

33.2 Paramagnetism

The behavior of the magnetic dipoles of the atoms or ions determines whether the material will be paramagnetic, ferromagnetic, or diamagnetic.

Paramagnetic material

In a **paramagnetic material,** the atoms or ions have permanent magnetic dipole moments. When the material is left to itself, these dipoles are randomly oriented and their magnetic fields average to zero. However, if the material is immersed in the magnetic field of, say, an electromagnet, the torque on the dipoles tends to align them with the field (see Section 31.5). This alignment will not be perfect because of the disturbances caused by random thermal motions. But even a partial alignment of the dipoles will have an effect on the magnetic field. The material becomes **magnetized** and contributes an extra magnetic field that *enhances* the original magnetic field.

Magnetization

To see how such an increase of magnetic field comes about, consider a piece of paramagnetic material placed between the poles of an electromagnet. Figure 33.3 shows the alignment of magnetic dipoles in this material; for the sake of simplicity, Figure 33.3b shows a case of perfect alignment. The magnetic dipoles are due to small current loops within the atoms; in Figure 33.4a we see the aligned current loops. Now look at any point inside the magnetic material where two of these current loops (almost) touch. Since the currents at this point are opposite, they cancel. Thus, everywhere inside the material the current is effectively zero. However, at the surface of the material, the current does not cancel. The net result of the alignment of current

[5] See Appendix 8 for more precise values of the magnetic moments of the elementary particles.

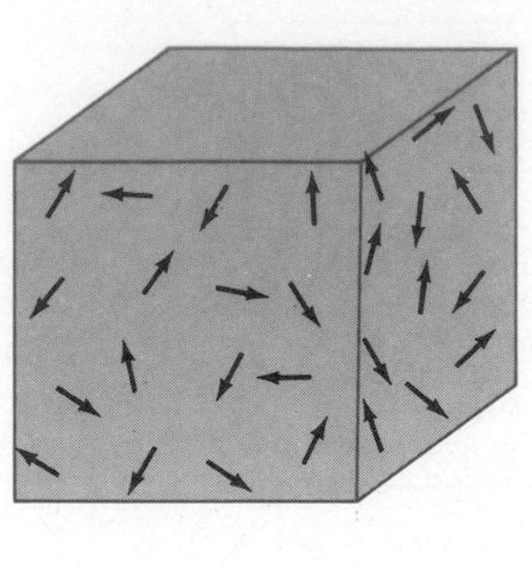
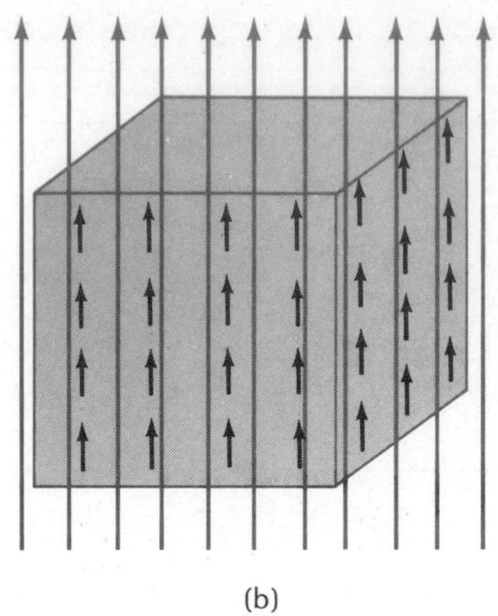

(a) (b)

Fig. 33.3 (a) A piece of paramagnetic material. In the absence of an external magnetic field, the magnetic dipoles are oriented at random. (b) In the presence of an external magnetic field, the magnetic dipoles align with the magnetic field. The figure shows an ideal case of perfect alignment; in practice, the alignment will only be partial.

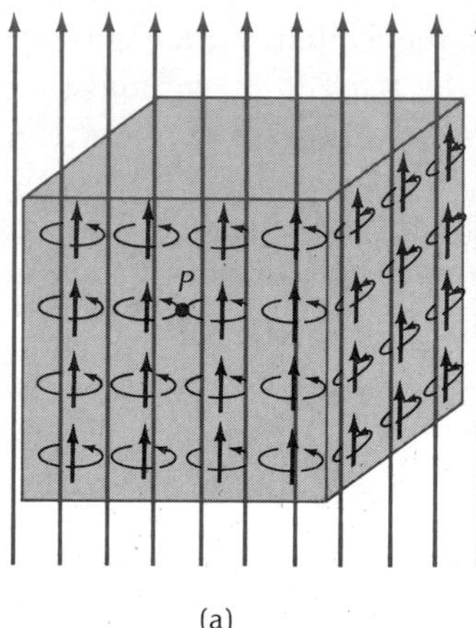

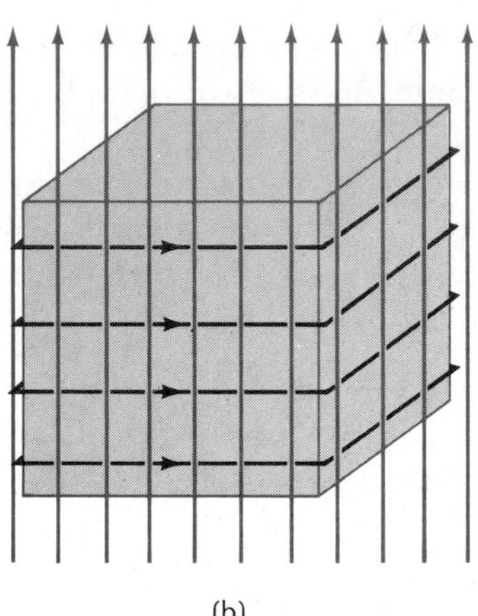

(a) (b)

Fig. 33.4 (a) The magnetic dipoles of Figure 33.3b can be regarded as small current loops. At the point P, the currents of adjacent loops are opposite, and they cancel. (b) The small aligned current loops of (a) are equivalent to a current along the surface of the piece of material.

loops is therefore a current running along the surface of the magnetized material (Figure 33.4b). The material consequently behaves like a solenoid — it produces an extra magnetic field in its interior. This extra magnetic field has the *same* direction as the original magnetic field (Figure 33.3b). Hence the total magnetic field in a paramagnetic material is larger than the magnetic field produced by the free currents of the electromagnet.

The alignment of the magnetic dipoles in a magnetized paramagnetic material is analogous to the alignment of the electric dipoles in a dielectric material. However, there is a crucial difference: the alignment of magnetic dipoles *increases* the original magnetic field while the alignment of electric dipoles *decreases* the original electric field.

The increase of the strength of the magnetic field by the paramagnetic material can be described by the **relative permeability constant** κ_m. This constant is simply the factor by which the magnetic field is increased. Thus, if the free currents of the electromagnet produce a field B_{free}, then these free currents and the currents in the paramagnetic material acting together produce a field

Relative permeability constant

$$\boxed{\mathbf{B} = \kappa_m \mathbf{B}_{free}} \qquad (8)$$

where $\kappa_m > 1$. Table 33.2 lists the values of κ_m for some paramagnetic materials.

Table 33.2 PERMEABILITIES OF SOME PARAMAGNETIC MATERIALS[a]

Material	κ_m
Air	1.000304
Oxygen	1.00133
Oxygen (−190°C, liquid)	1.00327
Manganese chloride	1.00134
Nickel monoxide	1.000675
Manganese	1.000124
Platinum	$1 + 13.8 \times 10^{-6}$
Aluminum	$1 + 8.17 \times 10^{-6}$

[a] At room temperature (20°C) and 1 atm unless otherwise noted.

Just as Gauss' Law in the presence of a dielectric material contains an extra factor κ [see Eq. (27.32)], Ampère's Law in the presence of a paramagnetic material contains an extra factor κ_m. A simple argument shows that the revised form of Ampère's Law is

$$\boxed{\oint \frac{1}{\kappa_m} \mathbf{B} \cdot d\mathbf{l} = \mu_0 I_{\text{free}}} \qquad (9)$$

Ampère's Law in magnetic materials

If the arrangement of currents and paramagnetic materials is sufficiently symmetric, then we can use Eq. (9) to calculate the magnetic field in the usual way.

EXAMPLE 1. A solenoid filled with air has a magnetic field of 1.20 T in its core. By how much will the magnetic field decrease if the air is pumped out of the core while the current is held constant?

SOLUTION: For air, $\kappa_m = 1.00030$. Hence the magnetic field with air is larger than that without by a factor 1.00030, i.e.,

$$B_{\text{free}} = B_{\text{air}}/1.00030 \cong B_{\text{air}}\,(1 - 0.00030)$$

where we have used the approximation $1/(1 + x) \cong 1 - x$ for small x. The decrease of the magnetic field is then

$$\Delta B = B_{\text{free}} - B_{\text{air}} = -0.00030 \times 1.20\ \text{T}$$

$$= -3.6 \times 10^{-4}\ \text{T}$$

33.3 Ferromagnetism

As is obvious from Table 33.2, the increase of magnetic field produced by a paramagnetic material is quite small. By contrast, the increase produced by a **ferromagnetic material** can be enormous. And what is more, such a material will remain magnetized even if it is not immersed in an external magnetic field. A material that retains magnetization will make a **permanent magnet.**

Ferromagnetic material

Permanent magnet

The intense magnetization in ferromagnetic materials is due to a strong alignment of the spin magnetic moments of electrons. In these materials, there exists a special force that couples the spins of the electrons in adjacent atoms in the crystal, a force created by some subtle quantum-mechanical effects (which we cannot discuss here). This spin–spin force tends to lock the spins of the electrons in a parallel configuration. This force acts in the crystals of only five chemical elements: iron, cobalt, nickel, dysprosium, and gadolinium; however, it also acts in the crystals of alloys and of oxides of a large number of other elements.

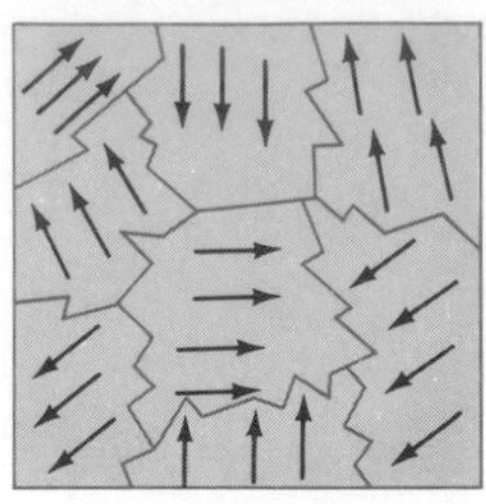

Fig. 33.5 Magnetic domains. Within each domain the magnetic dipoles have perfect alignment.

Since this special spin–spin force is fairly strong, we must ask why is it that ferromagnetic materials are ever found in a nonmagnetized state? Why is it that not every piece of iron is a permanent magnet? The answer is that on a microscopic scale ferromagnetic materials are *always* magnetized. A crystal of ordinary iron consists of a large number of small **domains** within which all the magnetic dipoles are perfectly aligned. But the direction of alignment varies from one domain to the next (Figure 33.5). Hence on a macroscopic scale, there is no discernible alignment because the domains are oriented at random. The sizes and shapes of the domains depend on the crystal. Typically, the sizes of domains range from a tenth of a millimeter to a few millimeters, although in a large, uniform crystal the length of a domain may be several centimeters. Figure 33.6 shows domains in a crystal of iron.

Domain

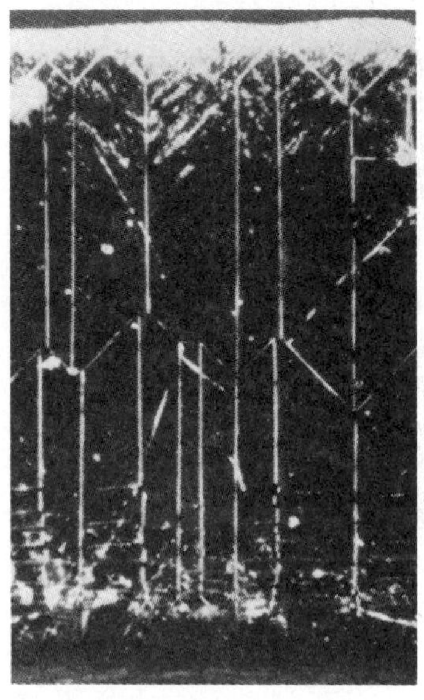

Fig. 33.6 Magnetic domains in a piece of iron. These domains are 0.1–0.3 mm across.

The formation of domains results from the tendency of the material to settle into the state of least energy (equilibrium state). The state of least energy for the spins would be a state of complete alignment. But such a complete alignment would generate a large magnetic field around the material, i.e., it would make the material into a (very strong) permanent magnet. Energetically, this is an unstable configuration, because there is a very large amount of energy in the magnetic field. The domain arrangement is a compromise. The spins align within the domains, but the domains do not align — the magnetic energy is then small because there is little magnetic field and the spin–spin energy is then also reasonably small because *most* adjacent spins are aligned.

However, if the material is immersed in an external magnetic field, all dipoles tend to align along this field. The domains will then change in two ways: those domains that already are more or less aligned with the field tend to grow in size at the expense of their neighbors and, furthermore, some domains will rotate their dipoles in the direction of the field.

If *all* the magnetic dipoles in a piece of ferromagnetic material align, their contribution to the magnetic field will be very large. For example, within a piece of magnetized iron, this contribution can be as much as 2.1 T, a rather strong magnetic field. Figure 33.7 is a plot of the actual magnetic field B in a piece of iron in a solenoid as a function of the magnetic field B_{free} contributed by the free current in the solenoid [$B_{\text{free}} = \mu_0 I_0 n$; see Eq. (31.16)]. If the value of B_{free} is 2.0×10^{-4} T, the value of B is about 1.0 T — the iron increases the magnetic field by a factor of about 5000!

The solid portion of the plot in Figure 33.7 was obtained by starting with a piece of unmagnetized iron (annealed iron), and subjecting it to an increasing magnetic field B_{free}. The dashed portion of the plot was obtained by gradually reducing the magnetic field B_{free} to zero, after it had reached the value indicated by the point P. The dashed plot shows that even when B_{free} is reduced to zero, the iron retains some magnetization — the iron becomes a permanent magnet. Thus, the iron retains some memory of the magnetic field to which it was exposed earlier. Such a dependence of the state of a system on its past history is called **hysteresis.**

Hysteresis

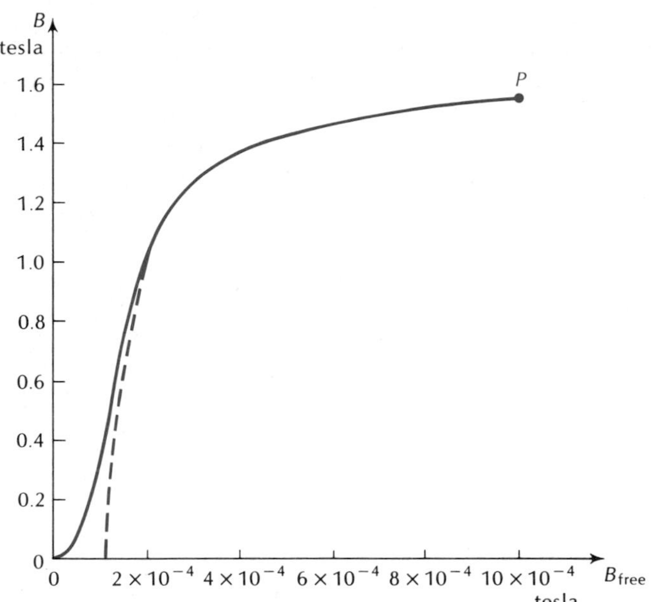

Fig. 33.7 Magnetic field B in annealed iron as a function of B_{free}. The solid curve shows B if the iron is immersed in an external magnetic field B_{free} that increases from an initial value of zero to a final value of 10×10^{-4} T. The dashed curve shows B if the external magnetic field is subsequently reduced to zero.

The hysteresis of a ferromagnetic material is due to a sluggishness in the rearrangement of the domains. Once the domains have become aligned in response to a strong external magnetic field, they tend to stay that way. If we remove the ferromagnetic material from the external magnetic field, the domains will suffer some rearrangement, but they will not lose their alignment completely. The remaining alignment gives the material a permanent distribution of magnetic dipole moments over its volume. The magnetic field of a permanent magnet is produced by these remaining aligned dipole moments.

As we saw in Section 33.2, the aligned dipoles effectively amount to a current running around the surface of the magnetized material (see Figure 33.4b). In a strong permanent magnet, this surface current may amount to several hundred amperes per centimeter of length of the magnet. The magnetic field produced by such a current distribution is obviously similar to that of a solenoid of finite length — the field lines emerge from the magnet at one end (the north pole) and reenter the magnet at the other end (the south pole); see Figures 33.8 and 33.9.

Incidentally, the maximum magnetization that a ferromagnetic material will retain after it has been removed from the external magnetic field depends on the temperature. The higher the temperature, the less the remaining magnetization. Above a certain critical temperature, called the **Curie temperature**, the magnetization disappears completely. For example, iron will not retain any magnetization if the temperature is in excess of 1043°C.

Curie temperature

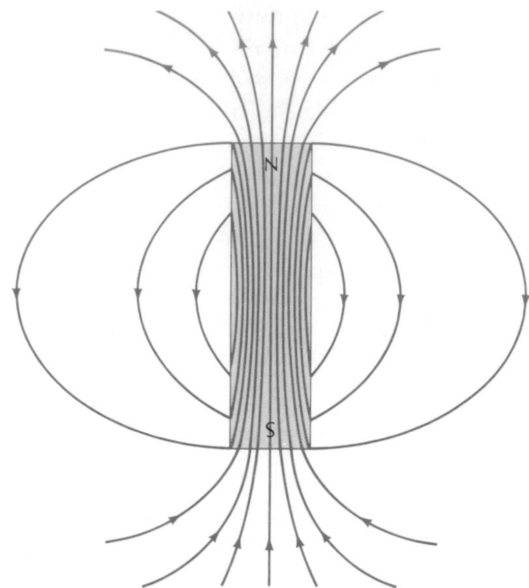

Fig. 33.8 Magnetic field lines of a permanent magnet.

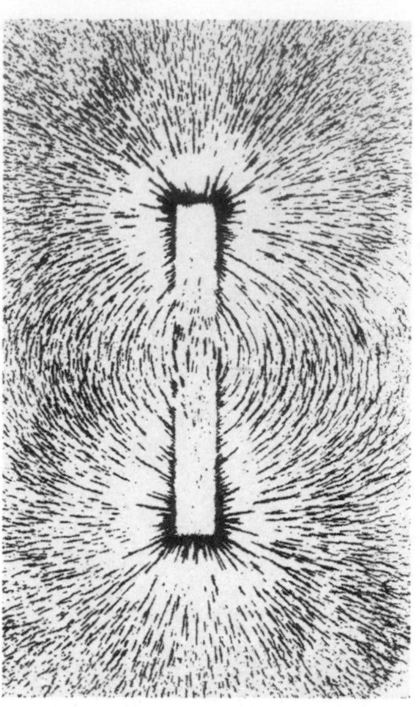

Fig. 33.9 Magnetic field lines of a permanent magnet, made visible by sprinkling iron filings on a sheet of paper.

33.4 Diamagnetism

In both paramagnetic and ferromagnetic materials the important effect is the alignment of *permanent* magnetic dipoles. This is quite analogous to the alignment of permanent electric dipoles in a dielectric. But we know that in some dielectrics the polarization is due to induced electric dipoles rather than permanent dipoles. Is it likewise possible for a material to acquire *induced* magnetic dipoles?

Diamagnetic material

In **diamagnetic materials** the magnetization arises from such induced magnetic dipoles. To see how the magnetic field can induce dipole moments in the atoms, imagine that a sample of some material is placed between the poles of an electromagnet which is initially switched off. If the electromagnet is now switched on, the external magnetic field must increase from its initial (zero) value to its final value. Thus, for a short while, the magnetic field will be time dependent and it will therefore induce an electric field within the sample. Let us consider the effect of this electric field on the motion of an electron. We will again pretend that classical mechanics applies to this problem.

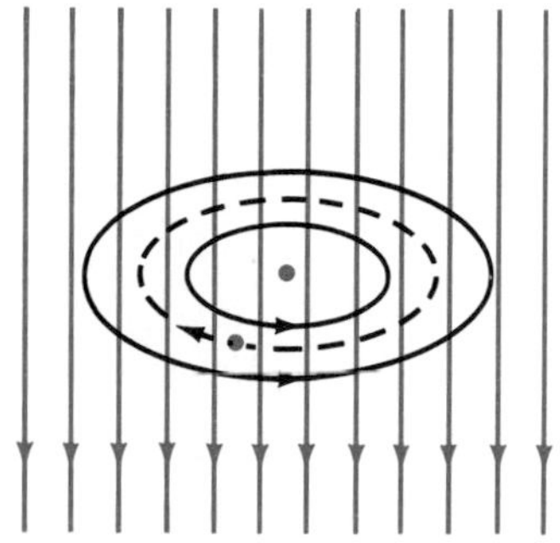

Fig. 33.10 Electron in a circular orbit around a nucleus immersed in an increasing magnetic field. The black circles indicate the induced electric field.

Figure 33.10 shows the orbit of an electron within an atom and also shows the increasing magnetic field. If the electron moves clockwise, as seen from above, the induced electric field (compare Example 32.4) will speed the electron up, giving it a larger orbital angular momentum and magnetic moment. The increment of angular momentum produces a magnetic field (within the orbit) that *opposes* the original magnetic field. This is obvious from Lenz' Law: the change in the motion of the electron amounts to an induced current, and the magnetic field associated with this current must be such as to oppose the original increasing magnetic field.

This argument establishes that induced magnetic moments reduce the strength of the magnetic field. The argument is quite general and applies to any kind of electron orbit and any kind of material. However, in paramagnetic and ferromagnetic materials the reduction of magnetic field due to induced magnetic moments is more than compensated by the increase of magnetic field due to alignment of permanent magnetic moments. If the material has no permanent magnetic moments, then the effect of the induced magnetic moments becomes noticeable and the material will be diamagnetic.

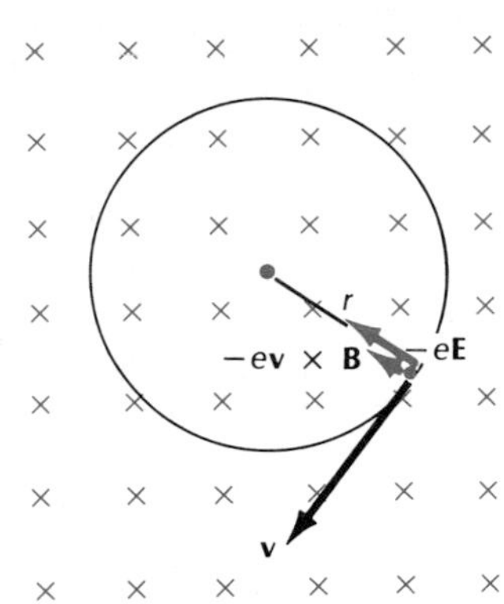

Fig. 33.11 Electron in a circular orbit around a nucleus immersed in a magnetic field. The centripetal force on the electron is the sum of the electric attraction of the nucleus and the magnetic force.

Diamagnetism is a very small effect. The following calculation gives some idea of the size of this effect. When the material is placed in a magnetic field $\mathbf{B}$, the electrons will experience a force $-e\mathbf{v} \times \mathbf{B}$ in addition to the usual electric force acting within the atom. Figure 33.11 shows the simple case of an electron in a circular orbit of radius r around a nucleus; the orbit is perpendicular to the magnetic field $\mathbf{B}$. If the nucleus produces an electric field $\mathbf{E}$, then the net force on the electron is $-e\mathbf{E} - e\mathbf{v} \times \mathbf{B}$. This force must match the product of mass and centripetal acceleration,

$$eE + evB = m_e v^2/r \tag{10}$$

In terms of the angular frequency of the motion, the equation of motion becomes

$$eE + e\omega rB = m_e \omega^2 r \tag{11}$$

Let us compare this with the equation of motion in the absence of the magnetic field (undisturbed atom, $B = 0$),

$$eE = m_e \omega_0^2 r \tag{12}$$

If we subtract Eq. (12) from Eq. (11), we obtain a relation between the frequencies ω and ω_0:

$$e\omega B = m_e\,(\omega^2 - \omega_0^2) \tag{13}$$

It is convenient to express this in terms of the increment of frequency,

$$\Delta\omega = \omega - \omega_0 \tag{14}$$

If the magnetic field is not excessively strong, $\Delta\omega$ will be small compared to ω_0 and hence

$$\begin{aligned} \omega^2 - \omega_0^2 &= (\omega_0 + \Delta\omega)^2 - \omega_0^2 \\ &= 2\omega_0\,\Delta\omega + (\Delta\omega)^2 \cong 2\omega_0\,\Delta\omega \end{aligned} \tag{15}$$

so that Eq. (13) becomes

$$e\omega B \cong 2m_e\omega_0\,\Delta\omega \tag{16}$$

Since ω and ω_0 are nearly equal, we can cancel them on both sides of this equation without introducing any additional errors. This leads to

$$\boxed{\Delta\omega = \frac{eB}{2m_e}} \tag{17}$$

Larmor frequency

This frequency is called the **Larmor frequency.** It tells us how much faster the electron will move around its orbit because of the presence of the magnetic field (of course, if the electron is initially moving in a direction opposite to that shown in Figure 33.11, then the electron will move *slower* by the same amount). Note that our calculation implicitly assumed that the orbital radius does not change as the magnetic field **B** is switched on. This assumption can be justified with an extra calculation that verifies that the work done on the electron by the induced emf changes the kinetic energy by just the amount required by the change of angular frequency at a fixed radius.

Corresponding to the change $\Delta\omega$ in the orbital frequency, there will be a change in the orbital magnetic moment. From Eq. (3),

$$\mu = \frac{evr}{2} = \frac{er^2\omega_0}{2} \tag{18}$$

Hence

$$\Delta\mu = \frac{er^2}{2}\,\Delta\omega \tag{19}$$

Dividing these equations into one another, we obtain an expression for the fractional change in the magnetic moment:

$$\frac{\Delta\mu}{\mu} = \frac{\Delta\omega}{\omega_0} \tag{20}$$

Let us insert some numbers. Typically the frequency of motion of an electron in an atom is $\omega_0 \cong 10^{15}$ s. If the magnetic field is fairly strong, say, $B = 1.0$ T, then

$$\Delta\omega = \frac{eB}{2m_e} = \frac{1.6\times10^{-19}\ \text{C}\times1.0\ \text{T}}{2\times9.1\times10^{-31}\ \text{kg}}$$

$$= 8.8\times10^{10}/\text{s} \tag{21}$$

Consequently,

$$\frac{\Delta\mu}{\mu} = \frac{\Delta\omega}{\omega_0} = \frac{8.8\times10^{10}/\text{s}}{10^{15}/\text{s}} \cong 10^{-4} \tag{22}$$

that is, the magnetic moment only changes by about 1 part in 10^4. This gives an indication of the small size of the diamagnetic effect.

The diamagnetic characteristics of a material can be described by a relative permeability κ_m that indicates by what factor the magnetic field is changed [compare Eq. (8)]. In the paramagnetic case $\kappa_m > 1$, but in the diamagnetic case $\kappa_m < 1$.

Table 33.3 lists some diamagnetic materials and the corresponding values of κ_m. In all cases, the value of κ_m is very near to 1.

Table 33.3 PERMEABILITIES OF SOME DIAMAGNETIC MATERIALS[a]

Material	κ_m
Bismuth	$1 - 1.9 \times 10^{-5}$
Beryllium	$1 - 1.3 \times 10^{-5}$
Methane	$1 - 3.1 \times 10^{-5}$
Ethylene	$1 - 2.0 \times 10^{-5}$
Ammonia	$1 - 1.4 \times 10^{-5}$
Carbon dioxide	$1 - 0.53 \times 10^{-5}$
Glass (heavy flint)	$1 - 1.5 \times 10^{-5}$

[a] At room temperature (20°C) and 1 atm.

SUMMARY

Magnetic moment of orbiting electron: $\mu = \dfrac{e}{2m_e} L$

Permeability constant: $B = \kappa_m B_{free}$

Magnetic materials: paramagnetic: $\kappa_m \gtrsim 1$
ferromagnetic: $\kappa_m \gg 1$
diamagnetic: $\kappa_m \lesssim 1$

Ampère's Law in paramagnetic and diamagnetic materials:

$$\oint \frac{1}{\kappa_m} \mathbf{B} \cdot d\mathbf{l} = \mu_0 I_{free}$$

Larmor frequency: $\Delta\omega = \dfrac{eB}{2m_e}$

QUESTIONS

1. Show that the SI unit of magnetic moment ($A \cdot m^2$) is the same as joule per tesla (J/T).

2. How would you measure the magnetic moment of a compass needle?

3. A bar magnet has a north pole and a south pole. If you break the bar magnet into two halves, do you obtain isolated north and south poles?

4. If we regard the Earth as a bar magnet, where is the magnetic north pole?

5. Consider a closed mathematical surface enclosing one of the poles of a bar magnet. What is the magnetic flux through this surface?

6. It is possible to magnetize an iron needle by pointing it north and giving it a few blows with a hammer. Explain.

7. If you drop a permanent magnet on a hard floor, it can become partially demagnetized. Explain.

8. Why does a magnet attract an (unmagnetized) piece of iron?

9. If you sprinkle iron filings on a sheet of paper placed in a magnetic field, the filings orient themselves along magnetic field lines (see Figure 33.9). Explain.

10. Other things being equal, a horseshoe magnet produces a stronger magnetic field than a bar magnet. Why?

11. You can make a chain of paper clips by touching one end of a clip to the pole of a magnet, then touching the free end to another paper clip, etc. Explain.

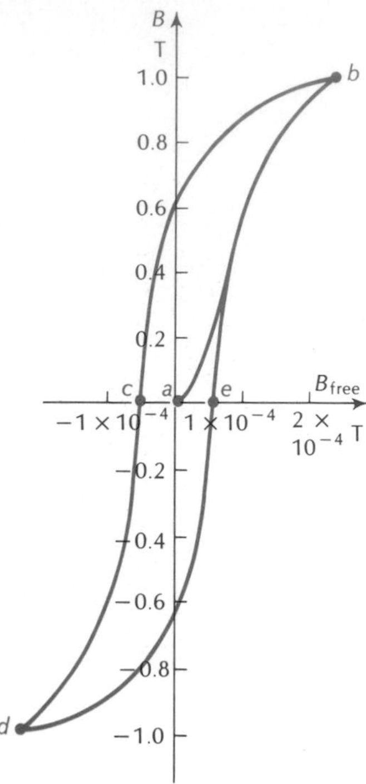

Fig. 33.12 Hysteresis loop for a piece of transformer steel. The configuration of the material initially corresponds to the point *a*. It is then successively brought to the points *b, c, d,* and *e* by suitable alterations of the external magnetic field B_{free}.

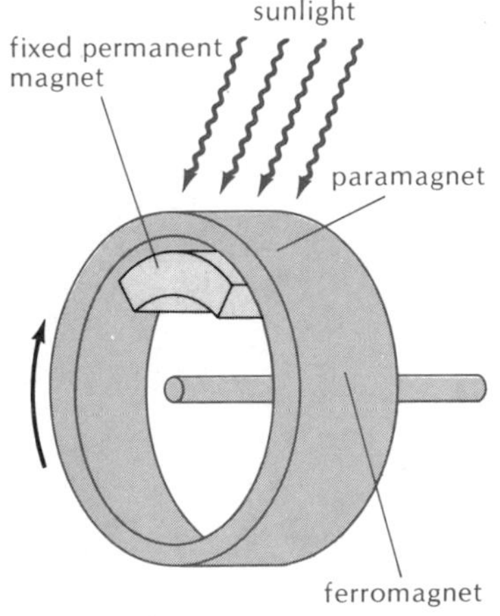

Fig. 33.13

12. When a mariner's compass is installed aboard an iron ship, it must be "adjusted" by placing several small permanent "correcting" magnets in its vicinity. What is the purpose of these magnets?

13. In the 1850s, Lord Kelvin redesigned the mariner's compass, partly by lengthy trials in his own yacht, and partly by theoretical analysis. He made the compass card (a circular disk, free to rotate, which carries the compass needles) lighter and concentrated most of its weight at the rim, and he used much smaller compass needles than had been customary before. What are the advantages of Kelvin's design?

14. The "keeper" for a horseshoe magnet is a small bar of iron that is placed across the poles of the magnet when not in use. What purpose does this serve?

15. Figure 33.12 shows a plot of B vs. B_{free} for a piece of transformer steel. This plot includes negative values of B and B_{free}. Qualitatively, explain the shape of this curve, called the **hysteresis loop.**

16. Why does the magnetization of a ferromagnet decrease with temperature?

17. Figure 33.13 shows a magnetic solar motor, called the Presnyakov wheel. The wheel is made of a ferromagnetic material. A fixed permanent magnet is installed near the top, where sunlight strikes the wheel. The heat of the sunlight converts the top portion of the ferromagnet temporarily into a paramagnet, with a much smaller value of the permeability. Explain why this wheel keeps turning after it has been given an initial push.

18. Does iron have diamagnetism?

19. The magnetic permeability of paramagnets and ferromagnets is strongly dependent on temperature, but the permeability of diamagnets is nearly independent of temperature. Why?

20. The practice of dowsing for water is still prevalent in some backward communities. The dowser holds a rod of wood, brass, or plastic in his hands in a peculiar way, and the rod supposedly dips wherever water is beneath the ground. The effect on the rod cannot be gravitational (a small body of water does not produce a perceptible disturbance of the Earth's average gravitational field). Hence, if the effect exists, it must be electric or magnetic. Discuss, keeping in mind that the ground is a conductor, and that the dowsing rod is nonmagnetic material.

PROBLEMS

Section 33.1

1. The electron in a hydrogen atom is 0.53×10^{-10} m away from the proton. What magnetic field does the magnetic moment of the proton produce at the position of the electron? Assume that the electron is instantaneously located on the axis of spin of the proton.

2. Two electrons are separated by a distance of 1.0×10^{-10} m. The first electron is on the axis of spin of the second.
   (a) What is the magnetic field that the magnetic moment of the second electron produces at the position of the first?
   (b) The potential energy of the magnetic moment of the first electron in this magnetic field depends on the orientation of the electrons. What is the potential energy (in electron-volts) if the spins of the two electrons are parallel? If antiparallel? Which orientation has the least energy?

3. Problem 22.12 gives the radii of the possible circular orbits of an electron in a hydrogen atom. For each such orbit, calculate the orbital magnetic moment.

4. The field of a fixed magnetic dipole located at the origin and oriented par-

allel to the z axis has the following components as a function of z and x in the plane $y = 0$:

$$B_x = \frac{\mu_0 \mu}{4\pi} \frac{3zx}{(x^2 + z^2)^{5/2}} \qquad B_y = 0,$$

$$B_z = -\frac{\mu_0 \mu}{4\pi} \left(\frac{1}{(x^2 + z^2)^{3/2}} - \frac{3z^2}{(x^2 + z^2)^{5/2}} \right)$$

Suppose that a second magnetic dipole of moment μ' is located at the point x, z (with $y = 0$); this dipole is free to rotate in the x–z plane.

(a) What is the orientation of least magnetic energy of this second dipole in the field of the first dipole, i.e., what is the angle that the second dipole will make with the z axis?

(b) What is the numerical value of the angle in the case $x = 0$, $z = r$? In the case $x = r$, $z = 0$? In the case $x = z = r/\sqrt{2}$?

5. Two magnetic dipoles are separated by a fixed distance r in the horizontal plane. The dipoles are free to rotate about a vertical axis (you may imagine that the dipoles are compass needles, but the magnetic field of the Earth is absent). If the dipoles settle into the configuration of least magnetic energy, what will be their orientation? Draw a diagram of the dipoles in this orientation. Prove your answer. (Hint: You may want to use the expression for the magnetic field given in Problem 4).

6. A compass needle has the shape of a thin rod of length 2.0 cm pivoted at its center so that it can swing freely in the horizontal plane. The compass needle has a mass of 0.12 g and a magnetic moment of 3.2×10^{-4} A · m². The compass needle is in Hawaii, where the horizontal, northerly component of the magnetic field is 2.9×10^{-5} T. What is the frequency of small rotational oscillations of the compass needle about the northerly direction?

7. Assume that the proton is a spherical ball within which the positive charge and the mass are uniformly distributed. Assume that the proton rotates rigidly with a spin angular momentum of $\hbar/2 = 0.53 \times 10^{-34}$ J · s. From this information, calculate the magnetic moment of a proton. [Hint: Use Eq. (4) with the mass of the proton. The result of this classical calculation differs by a factor of 5.6 from the actual value; this is due to a failure of classical physics when applied to subatomic particles.]

8. Assume that the charge distribution of a neutron is as indicated by the model described in Problem 24.17 and that the mass distribution is uniform. Assume that the neutron rotates rigidly with a spin angular momentum of $\hbar/2 = 0.53 \times 10^{-34}$ J · s. According to this crude model, what is the magnetic moment of a neutron? What is the direction of the magnetic moment relative to the direction of the spin angular momentum? [Hint: Apply Eq. (4) to the positive and negative charge distributions separately.]

*9. As described in Section 33.1, the proton has a magnetic moment of $\mu = 1.41 \times 10^{-26}$ A · m² parallel to the axis of its spin angular momentum. If the proton is in a magnetic field **B** and the magnetic moment makes an angle θ with this field, the torque exerted by this field on the magnetic moment will be $\tau = \mu B \sin\theta$ and the direction of this torque will be perpendicular to μ. Since the proton has a spin angular momentum **S** parallel to the magnetic moment, the torque will cause a precession of the spin about the direction of the magnetic field (see Section 12.6 for a discussion of the precession of a top under the influence of a torque).

(a) Show that the precession frequency of the proton is

$$\omega = \mu B/S$$

or, since $S = \hbar/2 = 0.53 \times 10^{-34}$ J · s,

$$\omega = 2\mu B/\hbar$$

(b) What is the precession frequency of a proton in a magnetic field of 0.20 T?

10. In a hydrogen atom, the electron orbits around the proton on a circular orbit of radius 0.53×10^{-10} m. As in Section 33.1, this orbiting electron can be regarded as a ring of current.

(a) Calculate the magnetic field that the ring of current produces at its center.

(b) Using the result of Problem 9, calculate the precession frequency of the proton in this magnetic field.

Section 33.2

11. A current of 25 A flows in a long solenoid of 1500 turns per meter.

(a) If the interior of this solenoid is empty (a vacuum), what is the strength of the magnetic field?

(b) If the interior is filled with liquid oxygen while the current stays constant, what will be the percentage change in the magnetic field?

12. Suppose that the dipole moments of all the atoms in a 20-g sample of lithium are perfectly aligned. What is the strength of the magnetic field on the axis of the dipoles at a distance of 1.0 m?

13. The space within a solenoid is to be filled with a mixture of air (paramagnetic) and methane (diamagnetic) so that the net permeability constant is exactly $\kappa_m = 1$. What percentage of air and methane should one use?

14. At a temperature of 20°C and a pressure of 1 atm the relative permeability of air is 1.000304. Calculate the relative permeability of air at the same temperature but a pressure of 3.0 atm. Assume that $\kappa_m - 1$ is proportional to the pressure.

15. Initially, the space within a long solenoid is empty. It is then filled with liquid oxygen, a paramagnetic material. What is the percentage change of the self-inductance of the solenoid? Does the self-inductance increase or decrease?

16. Show that the self-inductance per unit length of a very long solenoid filled with a paramagnetic material is $\kappa_m \mu_0 n^2 \pi R^2$, where n is the number of turns of wire per unit length and R is the radius of the solenoid.

17. Show that the energy density in the magnetic field in a very long solenoid filled with paramagnetic material is $u = B^2/(2\kappa_m \mu_0)$.

Section 33.3

18. A bar magnet of iron has a magnetic field of 0.03 T in its interior. The magnet is 15 cm long. What is the effective current running around its surface? Treat the magnet as though it were a very long cylindrical solenoid.

19. A long solenoid has 1200 turns per meter with a current of 6.0 A. The solenoid is filled with a ferromagnetic material. The value of the magnetic field B in this material is 2.0 T. What is the value of κ_m under these conditions?

20. In an iron crystal, two of the electrons of each atom participate in the alignment of spins; the magnetic field of a permanent magnet is caused by the magnetic moment of these electrons. Suppose that all of these electrons in a compass needle of mass 0.60 g are perfectly aligned. The compass needle is at a place where the horizontal, northerly component of the Earth's magnetic field is 2.4×10^{-5} T; the compass needle is free to swing in the horizontal plane. What is the torque on the compass needle when it is at an angle of 45° with the northerly component of the magnetic field? Does your answer depend on the shape of the compass needle?

21. Figure 33.7 is a plot of the magnetic field B in an iron-filled solenoid as a function of the magnetic field B_{free} that the solenoid would produce without

the iron. This plot has been prepared under the assumption that initially, when the current is zero, the iron is not magnetized. The value of the relative permeability depends on B_{free}. What is the value of κ_m when $B = 0.4$ T? When $B = 0.8$ T? When $B = 1.2$ T? Make a plot of κ_m vs. B_{free}. At what value of B_{free} is κ_m maximum?

22. Under conditions of maximum magnetization, the dipole moment per unit volume in cobalt is 1.5×10^5 A · m²/m³. Assuming that this magnetization is due to completely aligned electrons, how many such electrons are there per unit volume? How many aligned electrons per atom? The density of cobalt is 8.9×10^3 kg/m³ and the atomic mass is 58.9 g/mole.

23. In iron, two of the electrons of each atom participate in the alignment of spins that causes magnetization. Suppose that a cylindrical piece of iron, of radius 1.0 cm and length 8.0 cm, is completely magnetized along its axis, all the available electrons being in perfect alignment.

(a) What is the number of aligned electrons?

(b) The dipole moment of each electron is 9.27×10^{-24} A · m². What is the total dipole moment of all the aligned electrons?

(c) What surface current running around the surface of the cylinder will give the same total dipole moment?

(d) What magnetic field does this surface current produce in the interior of the iron?

24. The alignment of electron spins in a ferromagnetic material implies that the magnetized material has angular momentum. Suppose that a rod of iron 2.0 cm in diameter and 30 cm long is totally magnetized so that two of the electrons of each atom have their spins parallel to the axis of the rod. Suppose that the magnetization is suddenly reversed so that the spins become antiparallel to the axis. What is the change of angular momentum? The density of iron is 7.9 g/cm³ and the atomic mass is 55.8 g/mole.

*25. A long solenoid is filled with iron. The solenoid has 1800 turns per meter and the current in each turn is 50 A. Calculate the magnetic field inside the solenoid assuming that two of the electrons of each atom are completely aligned. (Hint: The magnetic field consists of two contributions: the field of the solenoid wire plus the field of the aligned electrons. The latter field can be calculated by replacing the electrons by a surface current as in Problem 23.)

Section 33.4

26. An electron in a hydrogen atom moves around a circular orbit of radius 0.53×10^{-10} m at a speed of 2.2×10^6 m/s. Suppose that the hydrogen atom is placed in a magnetic field of 0.50 T. The magnetic field is parallel to the orbital angular momentum.

(a) What is the change of the frequency of the motion of the electron? Does the frequency increase or decrease?

(b) What is the change of the speed of the electron? Assume the radius of the motion remains constant.

(c) What is the change of the energy of the electron?

(d) Check that the electric field induced by the change in magnetic flux through the orbit has the right direction to produce the change of speed and energy.

*27. A long, cylindrical bar magnet of diameter 1.0 cm has a magnetic field of 0.060 T in its interior. If you take a fine saw and cut the magnet in two pieces, the magnetic force will hold the pieces together. Estimate the magnitude of this magnetic force. [Hint: Suppose that you pull the pieces apart by a distance *dx*. The magnetic field in the gap between the pieces will then still be (nearly) 0.060 T. The magnetic energy in the gap equals the work that you must have done against the magnetic force while pulling the pieces apart.]

CHAPTER 34

AC Circuits

The current delivered by power companies to homes and factories is an oscillating function of time. This is called alternating current, or AC. Power companies prefer alternating currents to direct currents because of the ease with which alternating voltages can be stepped up or down by means of transformers. This makes it possible to step up the output of a power plant to several hundred thousand volts, transmit the power along a high-voltage line that minimizes the Joule losses, and finally step down the power to 220-volt AC or 110-volt AC just before delivery to the consumer.

All the appliances connected to ordinary outlets in homes therefore involve circuits with oscillating currents. Furthermore, electronic devices — such as radio transmitters and receivers — involve a variety of circuits with oscillating currents of high frequency. Many of these circuits have natural frequencies of oscillation. Such circuits exhibit the phenomenon of resonance when the natural frequency matches the frequency of a signal applied to the circuit. For instance, the tuning of a radio relies on an oscillating circuit whose frequency of oscillation is adjusted by means of a variable capacitor (attached to the tuning knob) so that it matches the frequency of the radio signal.

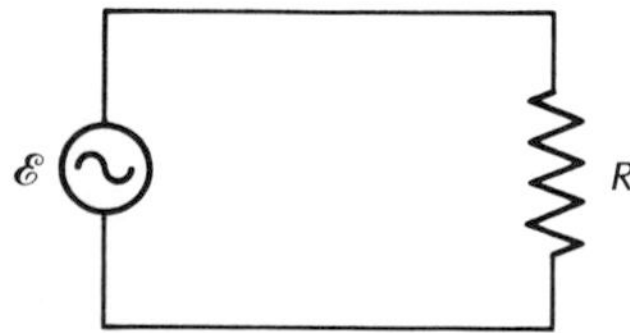

Fig. 34.1 Resistor connected to a source of alternating emf.

34.1 Simple AC Circuits with an External Electromotive Force

The simplest conceivable AC circuit consists of a pure resistor connected to an oscillating source of emf (Figure 34.1). In the circuit diagram, the source of emf is symbolized by a wavy line enclosed in a circle. This circuit might represent an electric heater or an incandescent lamp plugged into an ordinary wall outlet. In this case the emf is of the form

$$\mathscr{E} = \mathscr{E}_{\text{max}} \sin \omega t \tag{1}$$

where $\mathscr{E}_{\text{max}}$ is the amplitude of oscillation and ω the angular frequency. In the United States, the oscillating voltage available at the outlets of private homes has an amplitude $\mathscr{E}_{\text{max}} = 156$ V and a frequency of 60 Hz, that is, an angular frequency of $\omega = 60 \times 2\pi/\text{s}$. This kind of voltage is usually called "110-volt AC" for reasons that will become clear shortly.

Applied to the circuit of Figure 34.1, Kirchhoff's Law gives, at an instant of time,

$$\mathscr{E} - IR = 0 \tag{2}$$

or

$$\boxed{I = \frac{\mathscr{E}}{R} = \frac{\mathscr{E}_{\text{max}} \sin \omega t}{R}} \tag{3}$$

Current in resistor circuit

Thus the current oscillates in exactly the same way as the emf (Figure 34.2).

The instantaneous electric power dissipated in the resistor is

$$P = I\mathscr{E} = \frac{\mathscr{E}^2_{\text{max}} \sin^2 \omega t}{R} \tag{4}$$

The power oscillates between zero and a maximum value $\mathscr{E}^2_{\text{max}}/R$ (Figure 34.3). The time-average power can be obtained by averaging

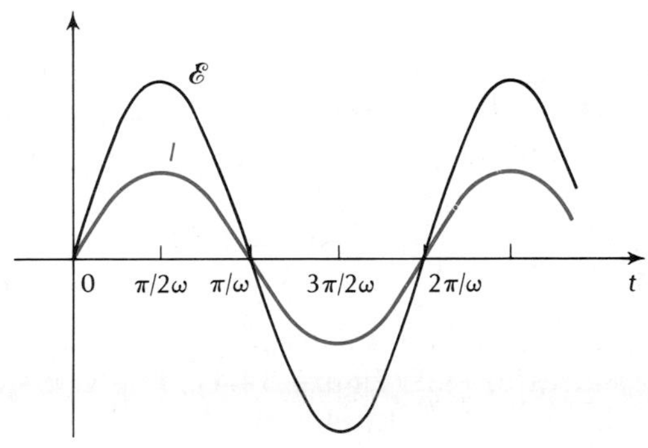

Fig. 34.2 The emf (black) and current (color) in the resistor circuit as a function of time.

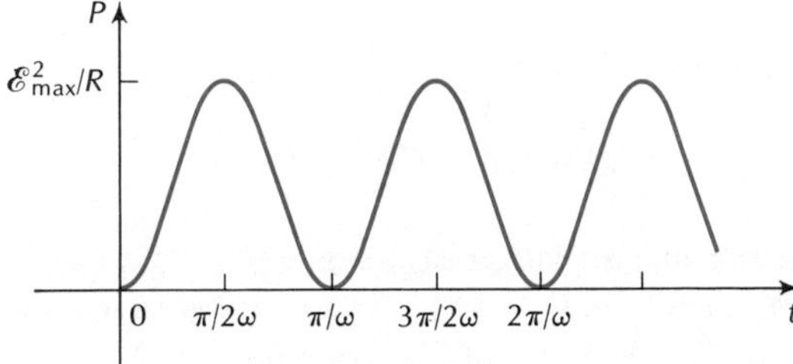

Fig. 34.3 Instantaneous power dissipated in the resistor as a function of time.

$\sin^2 \omega t$ over one cycle. We can easily verify that the average value of $\sin^2 \omega t$ is $\frac{1}{2}$,

$$\frac{1}{2\pi/\omega}\int_0^{2\pi/\omega} \sin^2 \omega t \, dt = \tfrac{1}{2} \tag{5}$$

Hence the average power is

Average power absorbed by resistor

$$\boxed{\overline{P} = \frac{\mathscr{E}^2_{max}}{2R}} \tag{6}$$

This is often written in the form

$$\boxed{\overline{P} = \frac{\mathscr{E}^2_{rms}}{R}} \tag{7}$$

where the quantity $\mathscr{E}_{rms}$, called the **root-mean-square voltage,** is the square root of the time average of the square of the voltage,

Root-mean-square voltage

$$\mathscr{E}_{rms} = \sqrt{\overline{\mathscr{E}^2}} = \sqrt{\frac{\mathscr{E}^2_{max}}{2}} = \frac{\mathscr{E}_{max}}{\sqrt{2}} \tag{8}$$

In engineering practice, an AC voltage is usually described in terms of $\mathscr{E}_{rms}$. For example, if $\mathscr{E}_{max} = 156$ V, then $\mathscr{E}_{rms} = 156/\sqrt{2}$ V $= 110$ V; an oscillating voltage with this value of $\mathscr{E}_{max}$ is described as "110-volt AC."

Comparison of Eqs. (7) and (29.23) shows that the average AC power delivered is equivalent to the DC power delivered by a steady voltage $\mathscr{E}_{rms}$. Thus, a 110-volt AC (with $\mathscr{E}_{max} = 156$ V) delivers the same average power as 110-volt DC.

EXAMPLE 1. A 110-V AC incandescent light bulb is rated at 150 W. What is the resistance of this light bulb (when at its operating temperature)?

SOLUTION: We have $\mathscr{E}_{rms} = 110$ V and $\overline{P} = 150$ W. Hence

$$R = \mathscr{E}^2_{rms}/\overline{P} = 80.7 \ \Omega \tag{9}$$

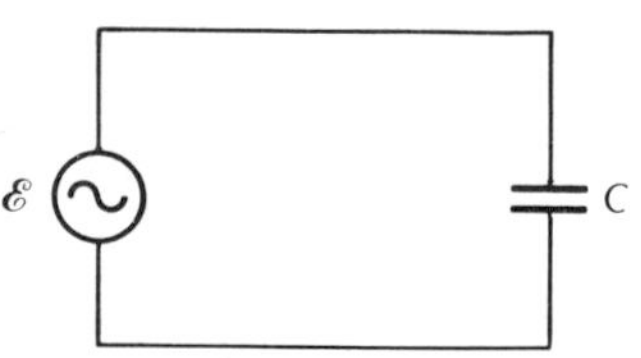

Fig. 34.4 Capacitor connected to a source of alternating emf.

Next, let us examine a circuit consisting of a capacitor connected to our oscillating source of emf (Figure 34.4). The voltage across the capacitor is Q/C and therefore Kirchhoff's Law gives

$$\mathscr{E} - Q/C = 0 \tag{10}$$

With Eq. (1) for $\mathscr{E}$, this yields

$$Q = C\mathscr{E} = C\mathscr{E}_{max} \sin \omega t \tag{11}$$

The current in the circuit is $I = dQ/dt$, or

$$I = \omega C\mathscr{E}_{max} \cos \omega t \tag{12}$$

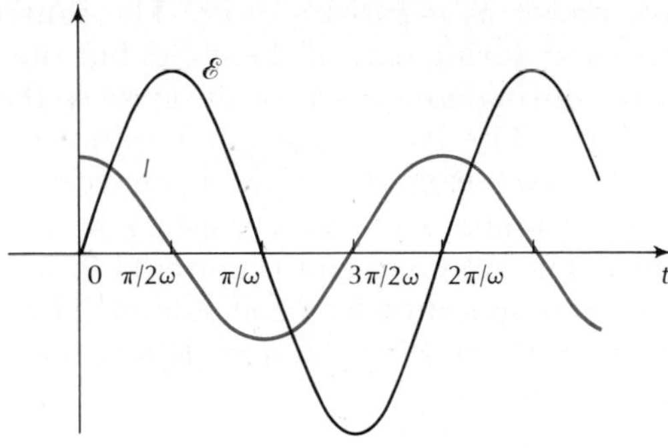

Fig. 34.5 The emf (black) and current (color) in the capacitor circuit as a function of time.

Comparison of Eqs. (12) and (1) shows that the current is a quarter cycle (90°) out of phase with the emf; for instance, at $t = 0$ the emf is minimum, and the current is maximum (Figure 34.5). Since the maxima in the current occur a quarter cycle *before* the maxima in the emf, we say that the current **leads** the emf.

It is customary to write Eq. (12) as

$$\boxed{I = \frac{\mathscr{E}_{max} \cos \omega t}{X_C}} \tag{13}$$

Current in capacitor circuit

where

$$X_C = \frac{1}{\omega C} \tag{14}$$

Capacitive reactance

is called the **capacitive reactance.** The quantity X_C plays roughly the same role for a capacitor in an AC circuit as does the resistance for a resistor [compare Eqs. (3) and (13)]. Note, however, that the reactance depends not only on the characteristics of the capacitor, but also on the frequency at which we are operating the circuit. The unit of reactance is the ohm, as it is for resistance.

The instantaneous power delivered to the capacitor is

$$P = I\mathscr{E} = \omega C \mathscr{E}_{max}^2 \cos \omega t \sin \omega t \tag{15}$$

The time dependence of this expression is contained in the factor $\cos \omega t \sin \omega t$. Since this equals $\frac{1}{2}\sin 2\omega t$, we recognize that the power oscillates at a frequency 2ω. But the important point is that the average power delivered is zero (Figure 34.6) — within one cycle, there is

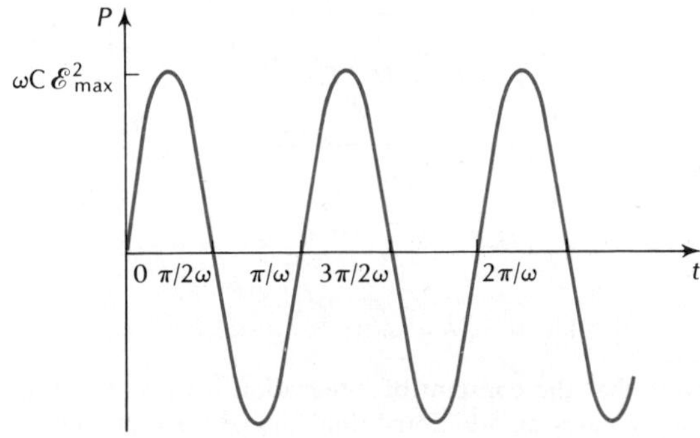

Fig. 34.6 Instantaneous power delivered to the capacitor as a function of time.

as much positive power as negative power. The source of emf does work on the capacitor during part of the cycle, but the capacitor does work on the source during other parts of the cycle so that, on the average, the *power is zero.* The ideal capacitor does not consume electric power because it has no means of dissipating electric energy.

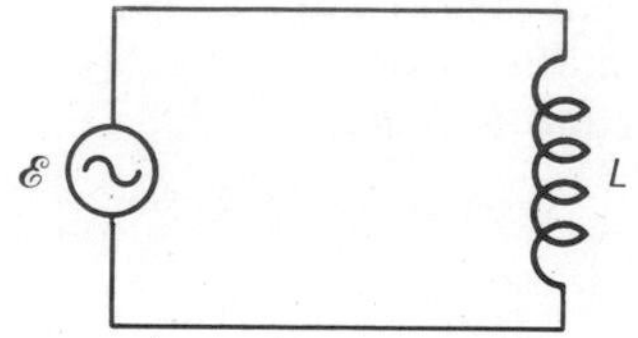

Fig. 34.7 Inductor connected to a source of alternating emf.

Finally, we will examine a circuit consisting of an inductor connected to an oscillating source of emf (Figure 34.7; in this circuit diagram the inductor is represented by a coiled line). The induced emf in the inductor (back emf) is $L\,dI/dt$, and by Kirchhoff's Law this must balance the applied emf,

$$\mathscr{E} - L\frac{dI}{dt} = 0 \tag{16}$$

which gives

$$\frac{dI}{dt} = \frac{\mathscr{E}}{L} = \frac{\mathscr{E}_{\text{max}} \sin \omega t}{L} \tag{17}$$

By integrating this we obtain[1]

$$I = -\frac{\mathscr{E}_{\text{max}} \cos \omega t}{\omega L} \tag{18}$$

Again, comparison of Eqs. (1) and (18) shows that the current is a quarter cycle (90°) out of phase with the emf (Figure 34.8). However, because of the minus sign in Eq. (18), the maxima in the current occur a quarter cycle *after* the maxima in the emf — the current **lags** the emf.

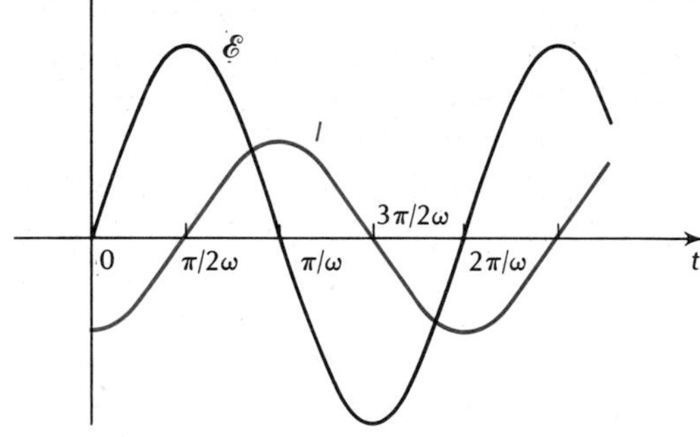

Fig. 34.8 The emf (black) and current (color) in the inductor circuit as a function of time.

We can write Eq. (18) as

Current in inductor circuit

$$I = -\frac{\mathscr{E}_{\text{max}} \cos \omega t}{X_L} \tag{19}$$

where

Inductive reactance

$$X_L = \omega L \tag{20}$$

[1] Here we assume that the constant of integration is zero. If this constant were not zero, the circuit would carry an additional time-independent current.

is the **inductive reactance.** The unit of this reactance is, again, the ohm.

The instantaneous power delivered to the inductor is

$$P = I\mathscr{E} = -\frac{1}{\omega L}\mathscr{E}_{\max}^2 \cos \omega t \sin \omega t \qquad (21)$$

As in the case of the capacitor, the average power is zero.

34.2 The Freely Oscillating LC Circuit

Figure 34.9 shows an LC circuit, which consists of an inductor and a capacitor connected in series. The circuit has no source of emf; nevertheless, a current will flow in this circuit provided that the capacitor is *initially charged.* The potential on one plate is then initially high and that on the other plate is low. A current will begin to flow around the circuit from the positive plate to the negative. If the circuit had no inductance, the current would merely neutralize the charge on the plates, i.e., the capacitor would discharge and that would be the end of the current. But the inductance makes a difference: the inductance initially opposes the buildup of the current, but once the current has become established, the inductance will keep it going for some extra time. Hence *more* charge flows from one capacitor plate to the other than required for neutrality and reversed charges accumulate on the capacitor plates. When the current finally does stop, the capacitor will again be fully charged, with reversed charges. And then a reversed current will begin to flow, etc. Thus the positive charge sloshes back and forth around the circuit.

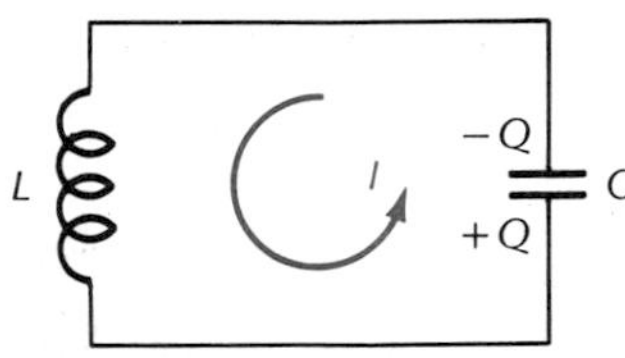

Fig. 34.9 Inductor and capacitor connected in series.

The LC system is analogous to a mass–spring system. The inductor is analogous to the mass — it tends to keep the current constant and provides "inertia." The charged capacitor is analogous to the stretched spring — it tends to accelerate the current and provides a "restoring force."

The equation of motion for the LC system follows from Kirchhoff's rule: the sum of emfs and voltage drops around the circuit must add to zero. Going around the circuit in the direction of the arrow shown in Figure 34.9, we find that the induced emf in the inductor (back emf) is

$$-L\frac{dI}{dt}$$

and the voltage across the capacitor is

$$-\frac{Q}{C}$$

Hence

$$-L\frac{dI}{dt} - \frac{Q}{C} = 0 \qquad (22)$$

Note that here Q is reckoned as positive when the charge on the lower plate is positive and I is reckoned as positive when the charge on the lower plate is increasing.

Since $I = dQ/dt$, we can also write Eq. (22) as

$$L\frac{d^2Q}{dt^2} + \frac{1}{C}Q = 0 \tag{23}$$

This equation has exactly the same mathematical form as the equation for the simple harmonic oscillator [Eq. (14.22)]:

$$m\frac{d^2x}{dt^2} + kx = 0$$

Obviously, Q plays the role of x, while L replaces m, and $1/C$ replaces k. Hence the solution of Eq. (23) can be immediately written down by recalling the solution for the simple harmonic oscillator [Eq. (14.25)]:

$$Q = Q_0 \cos\left(\frac{1}{\sqrt{LC}}t\right) \tag{24}$$

Here we have chosen the cosine solution because the initial condition of our problem specifies that at $t = 0$ the capacitor is fully charged ($Q = Q_0$) and the current is zero ($dQ/dt = 0$). Of course, other initial conditions can be accommodated by taking some suitable combination of cosine and sine solutions. From Eq. (24) we find that the current is

$$I = \frac{dQ}{dt} = -\frac{Q_0}{\sqrt{LC}}\sin\left(\frac{1}{\sqrt{LC}}t\right) \tag{25}$$

According to Eqs. (24) and (25), the charge and the current oscillate with a natural frequency

Natural frequency of LC circuit

$$\boxed{\omega_0 = \frac{1}{\sqrt{LC}}} \tag{26}$$

Figure 34.10 is a plot of the charge and the current in an LC circuit oscillating according to Eq. (25).

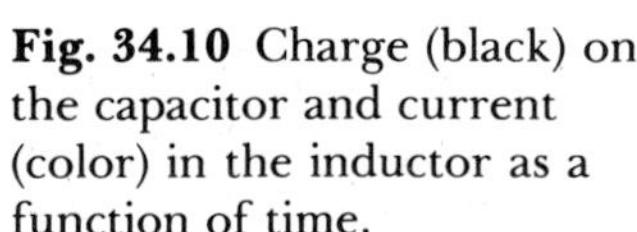

Fig. 34.10 Charge (black) on the capacitor and current (color) in the inductor as a function of time.

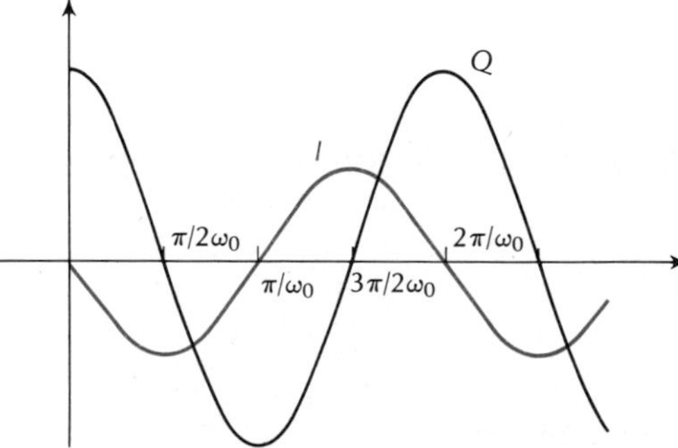

EXAMPLE 2. A primitive radio transmitter, such as those built in the early days of "wireless telegraphy," consists of an LC circuit oscillating at high fre-

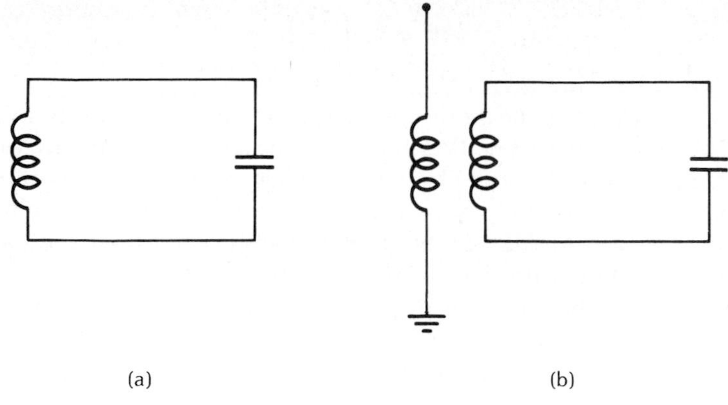

Fig. 34.11 (a) An LC circuit in a radio. (b) The LC circuit is coupled to the antenna by the mutual inductance of the two inductors.

quency (Figure 34.11a). The circuit is inductively coupled to an antenna (Figure 34.11b) so that the oscillating current in the circuit induces an oscillating current on the antenna; the latter current then radiates electromagnetic waves. Suppose that the inductance in the circuit of Figure 34.11a is 20 μH. What capacitance do we need if we want to produce oscillations of a frequency of 1500 kHz?

SOLUTION: The angular frequency is $2\pi \times 1500 \times 10^3$ radian/s. Hence, from Eq. (26)

$$C = \frac{1}{\omega_0^2 L}$$

$$= \frac{1}{(2\pi \times 1500 \times 10^3/\text{s})^2 \times 20 \times 10^{-6}\ \text{H}}$$

$$= 5.6 \times 10^{-10}\ \text{F} = 560\ \text{pF}$$

The energy of the LC system is the sum of the energies stored in the capacitor and in the inductor [see Eqs. (27.40) and (32.37)], that is,

$$U = \tfrac{1}{2}\frac{Q^2}{C} + \tfrac{1}{2}LI^2 \tag{27}$$

or

$$U = \tfrac{1}{2}\frac{Q^2}{C} + \tfrac{1}{2}L\left(\frac{dQ}{dt}\right)^2 \tag{28}$$

On physical grounds we expect that the energy remains constant during the oscillations. To prove this conservation theorem for the energy, we need only differentiate Eq. (28) with respect to time:

$$\frac{dU}{dt} = \frac{Q}{C}\frac{dQ}{dt} + L\frac{dQ}{dt}\frac{d^2Q}{dt^2} = \frac{dQ}{dt}\left(\frac{Q}{C} + L\frac{d^2Q}{dt^2}\right) \tag{29}$$

The expression on the right side is zero because of the relation between Q and d^2Q/dt^2 [see Eq. (23)].

The quantity $\frac{1}{2}Q^2/C$ is a potential energy (electrostatic potential energy). The quantity $\frac{1}{2}L(dQ/dt)^2$ may be regarded as a "kinetic" energy. Thus, the expression (28) is analogous to the expression

$$\tfrac{1}{2}kx^2 + \tfrac{1}{2}m\left(\frac{dx}{dt}\right)^2$$

for the energy of a harmonic oscillator. If the capacitor is initially charged but no current is flowing, then the energy is initially purely potential (it is stored in the electric fields in the capacitor). As the current begins to flow, the potential energy decreases and the "kinetic" energy increases. At the instant when the capacitor is completely discharged, the current reaches its maximum value. The potential energy is then zero and the energy is purely "kinetic" (it is stored in the magnetic fields in the inductor). Beyond this instant, the potential energy increases at the expense of the "kinetic" energy. When the capacitor is completely charged, with reversed charge, the current stops flowing. At this instant the energy is again purely potential. The process now repeats with the current flowing in the opposite direction. Figure 34.12 is a plot of potential energy and "kinetic" energy as a function of time for an LC circuit oscillating according to Eqs. (24) and (25).

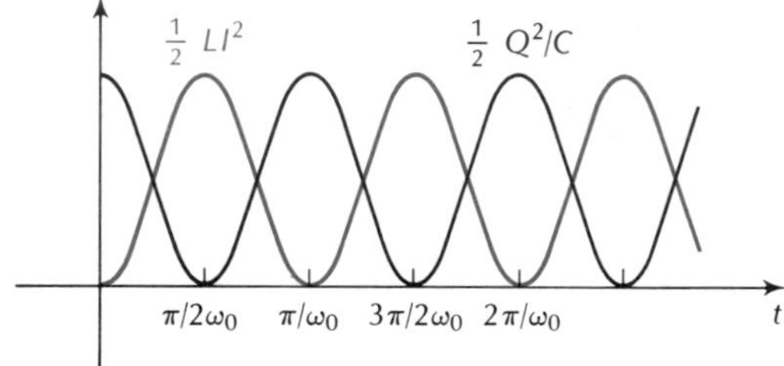

Fig. 34.12 Energy in the capacitor ($\frac{1}{2}Q^2/C$; black) and energy in the inductor ($\frac{1}{2}LI^2$; color) as a function of time.

So far we have assumed that our LC circuit contains no resistance. This is somewhat unrealistic since, at the very least, the wires connecting the circuit elements will have some resistance. The remainder of this section gives a description of the effects of resistance.

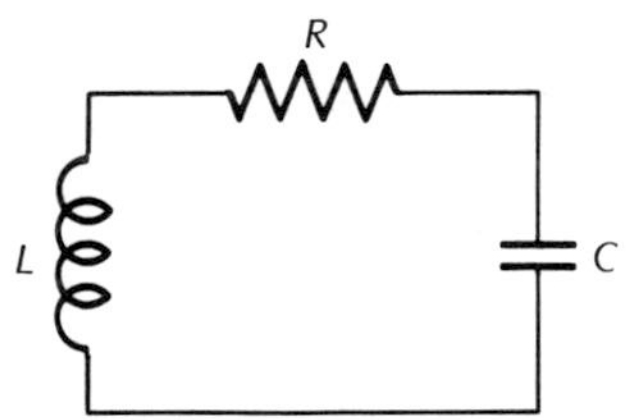

Fig. 34.13 Inductor, capacitor, and resistor connected in series.

Figure 34.13 shows an LCR circuit, i.e., an LC circuit with resistance. The resistance plays a role analogous to that of the friction force in the harmonic oscillator (see Section 14.6). The resistance gradually converts electric energy into heat; hence the electric energy decreases with time. This leads to damped oscillations of gradually decreasing amplitude.

Figure 34.14 shows the charge on the capacitor as a function of time for an LCR circuit. The charge can be described by the following equation:

$$Q = Q_0 e^{-\gamma t/2} \cos \omega_0 t \tag{30}$$

This, of course, is based on the assumption that the capacitor is fully charged at the initial time $t = 0$. The damping constant γ represents the frictional effects. It can be shown that γ is proportional to the resistance,

$$\gamma = \frac{R}{L} \tag{31}$$

As we saw in Section 14.6, the damping constant is directly related to the energy ΔU lost per period,

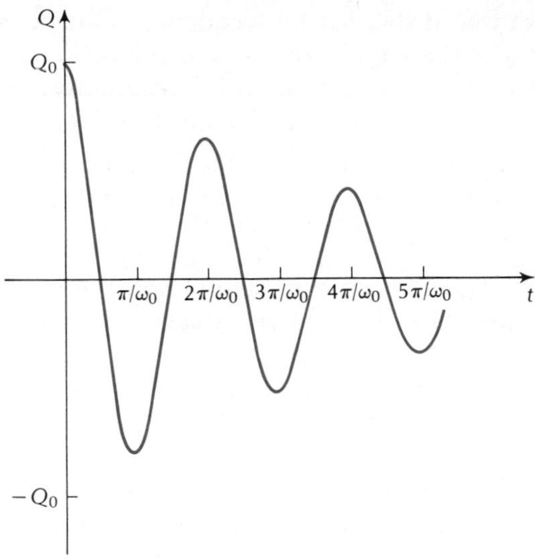

Fig. 34.14 Charge on the capacitor as a function of time.

$$\frac{\Delta U}{U} = -2\pi\frac{\gamma}{\omega_0} = -\frac{2\pi R}{\omega_0 L} \tag{32}$$

Electrical engineers usually express this energy loss in terms of the **"Q"** or the **quality factor** of the circuit,[2]

$$\boxed{\text{"}Q\text{"} = 2\pi|U/\Delta U| = \omega_0 L/R} \tag{33}$$

"Q" of freely oscillating LCR circuit

Typical circuits of low resistance in radio transmitters and receivers have a *"Q"* of up to 100.

Besides damping the amplitude of the oscillations, the resistance also produces another effect: it reduces the frequency of oscillation. Intuitively, we expect such a reduction of frequency because the friction in the resistance will slow the oscillations. Mathematically, one can show that the angular frequency for the natural oscillations of the LCR circuit is

$$\omega_0 = \frac{1}{\sqrt{LC}}\sqrt{1 - \frac{CR^2}{4L}} \tag{34}$$

Here $\sqrt{1 - CR^2/4L}$ represents the factor by which the frequency is decreased as compared to an LC circuit without resistance.

34.3 The LCR Circuit with an External Electromotive Force

Figure 34.15 shows a series LCR circuit with an oscillating source of emf. This emf acts as a driving force that pushes on the charge in the circuit. As in the case of the driven harmonic oscillator (see Section

Fig. 34.15 Inductor, capacitor, and resistor connected in series to a source of alternating emf.

[2] This *"Q"* must not be confused with electric charge.

14.6), we expect that if the driving frequency coincides with the natural frequency ω_0 of the circuit, then the oscillations will build up to a very large value. This is the condition for **resonance.**

Resonance

Suppose that the oscillating emf driving the circuit has an angular frequency ω:

$$\mathscr{E} = \mathscr{E}_{\text{max}} \sin \omega t \tag{35}$$

Under steady-state conditions the current will then oscillate with the same angular frequency. This implies that the current will be of the form

$$I = I_{\text{max}} \sin(\omega t + \phi) \tag{36}$$

We are now faced with the task of finding how the amplitude I_{max} and the phase ϕ of the current are related to the known parameters of the circuit.

We begin with Kirchhoff's rule: the external emf $\mathscr{E}$ must match the sum of the voltages across the resistor, the capacitor, and the inductor,

$$\mathscr{E} = \Delta V_{\text{R}} + \Delta V_{\text{C}} + \Delta V_{\text{L}} \tag{37}$$

To evaluate the three terms on the right side of this equation, we make use of the results of Section 34.1. The voltage across the resistor is in phase with the current:

$$\Delta V_{\text{R}} = IR = RI_{\text{max}} \sin(\omega t + \phi) \tag{38}$$

The voltage across the capacitor is one-quarter cycle behind the current,

$$\Delta V_{\text{C}} = -X_{\text{C}} I_{\text{max}} \cos(\omega t + \phi) \tag{39}$$

and the voltage across the inductor is one-quarter cycle ahead of the current,

$$\Delta V_{\text{L}} = X_{\text{L}} I_{\text{max}} \cos(\omega t + \phi) \tag{40}$$

Substitution of these into Eq. (37) yields

$$\mathscr{E}_{\text{max}} \sin \omega t = RI_{\text{max}} \sin(\omega t + \phi) - (X_{\text{C}} - X_{\text{L}}) I_{\text{max}} \cos(\omega t + \phi) \tag{41}$$

With the trigonometric identities for the sine and the cosine of the sum of two angles this becomes

$$\begin{aligned} \mathscr{E}_{\text{max}} \sin \omega t = {} & RI_{\text{max}}(\sin \omega t \cos \phi + \cos \omega t \sin \phi) \\ & - (X_{\text{C}} - X_{\text{L}}) I_{\text{max}} (\cos \omega t \cos \phi - \sin \omega t \sin \phi) \\ = {} & [R \cos \phi + (X_{\text{C}} - X_{\text{L}}) \sin \phi] I_{\text{max}} \sin \omega t \\ & + [R \sin \phi - (X_{\text{C}} - X_{\text{L}}) \cos \phi] I_{\text{max}} \cos \omega t \end{aligned} \tag{42}$$

If we examine this equation at time $t = 0$, we find that

$$0 = R \sin \phi - (X_C - X_L)\cos \phi \tag{43}$$

or

$$\boxed{\tan \phi = \frac{X_C - X_L}{R}} \tag{44}$$

Phase angle for series LCR circuit

From this we obtain

$$\sin \phi = \frac{1}{\sqrt{1 + 1/\tan^2 \phi}} = \frac{X_C - X_L}{\sqrt{R^2 + (X_C - X_L)^2}} \tag{45}$$

$$\cos \phi = \frac{1}{\sqrt{1 + \tan^2 \phi}} = \frac{R}{\sqrt{R^2 + (X_C - X_L)^2}} \tag{46}$$

With these expressions for $\sin \phi$ and $\cos \phi$, Eq. (42) reduces to

$$\mathscr{E}_{max} \sin \omega t = \frac{R^2 + (X_C - X_L)^2}{\sqrt{R^2 + (X_C - X_L)^2}} I_{max} \sin \omega t \tag{47}$$

or

$$I_{max} = \frac{\mathscr{E}_{max}}{\sqrt{R^2 + (X_C - X_L)^2}} \tag{48}$$

Equations (44) and (48) are the desired expressions for I_{max} and ϕ in terms of the known parameters of the circuit.

The quantity

$$Z = \sqrt{R^2 + (X_C - X_L)^2}$$

or

$$\boxed{Z = \sqrt{R^2 + (1/\omega C - \omega L)^2}} \tag{49}$$

Impedance for series LCR circuit

is called the **impedance** of the series LCR circuit. In terms of this quantity the current is

$$\boxed{I = \frac{\mathscr{E}_{max} \sin(\omega t + \phi)}{Z}} \tag{50}$$

Current in series LCR circuit

The relationships among the voltages and the currents in the circuit elements can be represented graphically by a **phasor diagram.** In such a diagram the amplitude of a sinusoidal function is represented by a line segment of length equal to the amplitude of oscillation, and the phase is represented by the angle between this line segment and the horizontal axis. The line segment is called a **phasor.** For instance, the function $I = I_{max} \sin(\omega t + \phi)$ is represented by the phasor shown in Figure 34.16. Obviously, the projection of this phasor on the vertical

Phasor

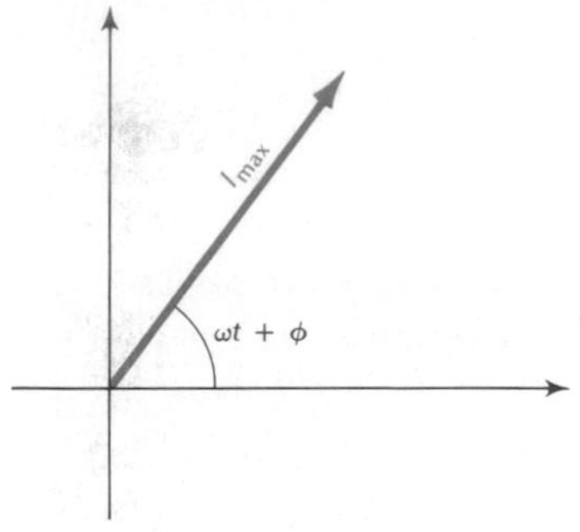

Fig. 34.16 Phasor representing the function $I = I_{max}\sin(\omega t + \phi)$. The phasor makes an angle of $\omega t + \phi$ with the horizontal axis. The projection of the phasor on the vertical axis is $I_{max}\sin(\omega t + \phi)$.

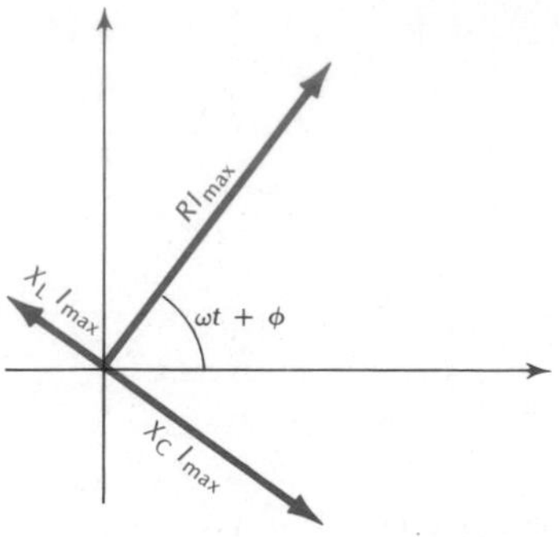

Fig. 34.17 The three phasors representing ΔV_R, ΔV_C, and ΔV_L. Their phase angles are $\omega t + \phi$, $\omega t + \phi - 90°$, and $\omega t + \phi + 90°$, respectively.

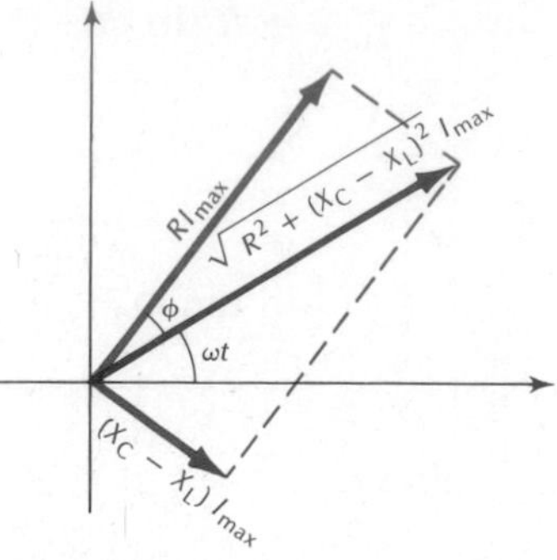

Fig. 34.18 In the phasor diagram, the sum of the voltages ΔV_R, ΔV_L, and ΔV_C is represented by the vector sum (black) of the three vectors (color) shown in Figure 34.17. From the diagram we see that this vector sum has a magnitude $\sqrt{R^2 + (X_C - X_L)^2}\, I_{max}$. The vector sum makes an angle ϕ with the phasor ΔV_R (or with the phasor representing I); from the diagram we see that $\tan\phi = (X_C - X_L)/R$.

axis is $I_{max}\sin(\omega t + \phi)$, i.e., the projection is exactly I. Likewise, we can represent each of the voltages on the right side of Eq. (37) by phasors (Figure 34.17). The vector sum of these phasors then represents the sum of these voltages. Kirchhoff's rule demands that this sum be equal to the external emf (Figure 34.18). We can then derive the expressions for the amplitude and for the phase angle of the current by direct inspection of Figure 34.18.

The amplitude of the oscillations of the current in the circuit depends critically on the frequency. With $X_C = 1/\omega C$ and $X_L = \omega L$, Eq. (48) exhibits the following dependence on frequency:

$$I_{max} = \frac{\mathscr{E}_{max}}{\sqrt{R^2 + (1/\omega C - \omega L)^2}} \tag{51}$$

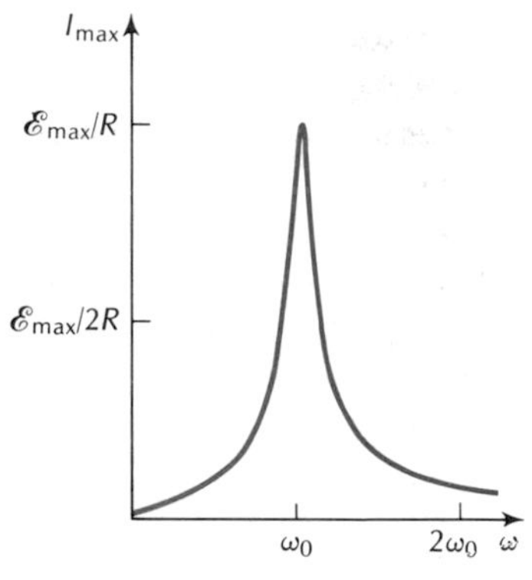

Fig. 34.19 Maximum current as a function of angular frequency.

Figure 34.19 is a plot of this maximum current as a function of frequency. The current is weak when ω is near zero, and it is also weak when ω is very large. However, when ω is near $\omega_0 = 1/\sqrt{LC}$, the current becomes very strong. This is a resonance phenomenon: the oscillations become large when the driving force pushes the circuit with the same frequency as that of the natural oscillations [see Eq. (26)]. When the frequency ω exactly matches the natural frequency ($\omega = \omega_0 = 1/\sqrt{LC}$), the amplitude of the oscillations of the current reaches the value

$$I_{max} = \frac{\mathscr{E}_{max}}{R} \tag{52}$$

Let us now examine the phase of the current given by Eq. (50). If the driving frequency is below resonance ($\omega < \omega_0$), then $X_C - X_L$ is positive, and ϕ is also positive [see Eq. (44)]. The current then leads the external emf, that is, the maxima in the current occur earlier than those in the emf (see Figure 34.20a). If the driving frequency is above resonance, ($\omega > \omega_0$), then $X_C - X_L$ is negative, and ϕ is also negative [see Eq. (44)]. The current then lags the external emf, that is, the max-

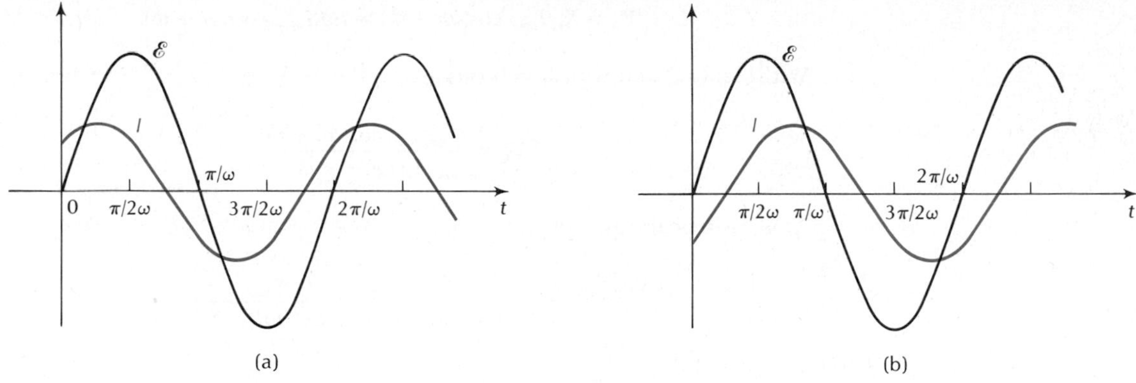

Fig. 34.20 The emf (black) and current (color) in the LCR circuit as a function of time: (a) $\omega < \omega_0$ and (b) $\omega > \omega_0$.

ima in the current occur later than those in the emf (Figure 34.20b).

Finally, what is the power delivered by the source of emf to the LCR circuit? The instantaneous power is

$$P = I\mathscr{E} = \frac{1}{Z}\mathscr{E}_{\max}^2 \sin(\omega t + \phi)\sin \omega t \tag{53}$$

or

$$P = \frac{1}{Z}\mathscr{E}_{\max}^2(\sin^2 \omega t \cos \phi + \cos \omega t \sin \omega t \cos \phi) \tag{54}$$

To find the time-average power, we note that the time-average of $\sin^2 \omega t$ is $\frac{1}{2}$ [see Eq. (5)] and the time-average of $\cos \omega t \sin \omega t$ is zero. Hence

$$\overline{P} = \frac{\mathscr{E}_{\max}^2}{2Z} \cos \phi \tag{55}$$

which we can also write as

$$\boxed{\overline{P} = \frac{\mathscr{E}_{\text{rms}}^2}{Z} \cos \phi} \tag{56}$$

Average power absorbed by LCR circuit

EXAMPLE 3. An LC circuit with $L = 3.0 \times 10^{-4}$ H and $C = 2.0 \times 10^{-6}$ F is being driven by an oscillating source delivering an AC emf of amplitude 0.40 V at a frequency of 5.0×10^4 radian/s. What is the peak instantaneous voltage across the inductor? Across the capacitor?

SOLUTION: The natural frequency of this circuit, which we will need later in our calculation, is

$$\begin{aligned} \omega_0 &= \frac{1}{\sqrt{LC}} \\ &= \frac{1}{\sqrt{3.0 \times 10^{-4}\ \text{H} \times 2.0 \times 10^{-6}\ \text{F}}} \\ &= 4.1 \times 10^4 \text{ radian/s} \end{aligned} \tag{57}$$

The instantaneous voltage across the inductor is

$$V_L = X_L\, I_{max} \cos(\omega t + \phi) = \omega L I_{max} \cos(\omega t + \phi)$$

With Eq. (51) and with $R = 0$ this gives

$$V_L = \frac{\omega L \mathscr{E}_{max} \cos(\omega t + \phi)}{1/\omega C - \omega L} \tag{58}$$

This has a peak value of

$$\frac{\omega L \mathscr{E}_{max}}{1/\omega C - \omega L} = \frac{\mathscr{E}_{max}}{1/(\omega^2 LC) - 1} = \frac{\mathscr{E}_{max}}{\omega_0^2/\omega^2 - 1}$$

$$= \frac{0.40\ \text{V}}{\dfrac{(4.1 \times 10^4/\text{s})^2}{(5.0 \times 10^4/\text{s})^2} - 1} = -1.20\ \text{V}$$

The instantaneous voltage across the capacitor is

$$V_C = -X_C I_{max} \cos(\omega t + \phi)$$

$$= -\frac{1}{\omega C} \frac{\mathscr{E}_{max} \cos(\omega t + \phi)}{1/\omega C - \omega L} \tag{59}$$

This has a peak value of

$$-\frac{1}{\omega C}\frac{\mathscr{E}_{max}}{1/\omega C - \omega L} = -\frac{\mathscr{E}_{max}}{1 - \omega^2/\omega_0^2}$$

$$= -\frac{0.40\ \text{V}}{1 - \dfrac{(5.0 \times 10^4/\text{s})^2}{(4.1 \times 10^4/\text{s})^2}} = 0.80\ \text{V}$$

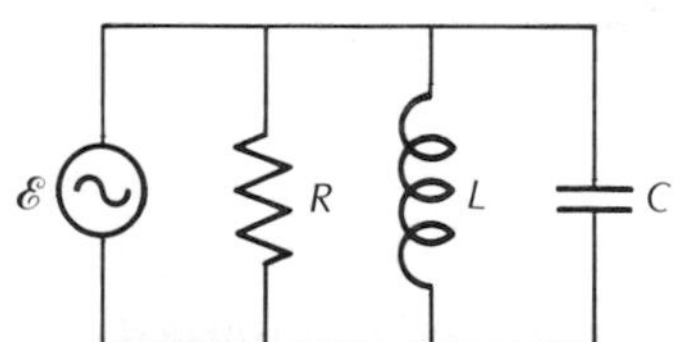

Fig. 34.21 Inductor, capacitor, and resistor connected in parallel to a source of alternating emf.

EXAMPLE 4. Figure 34.21 shows a parallel LCR circuit with an oscillating source of emf. What is the net current delivered by the source of emf?

SOLUTION: If the emf of the source is $\mathscr{E} = \mathscr{E}_{max} \sin \omega t$, then this is also the emf acting on the inductor, the capacitor, and the resistor individually. The currents in the inductor, capacitor, and resistor are then, respectively [see Eqs. (19), (13), and (3)]:

$$I_L = -\frac{\mathscr{E}_{max} \cos \omega t}{X_L}$$

$$I_C = \frac{\mathscr{E}_{max} \cos \omega t}{X_C}$$

$$I_R = \frac{\mathscr{E}_{max} \sin \omega t}{R}$$

The net current delivered by the source of emf is the sum of these currents. To evaluate this sum, we can use a phasor diagram. Figure 34.22 shows the phasors representing the currents I_L, I_C, and I_R. The net current is represented by the vector sum of these phasors. By inspection of Figure 34.22 we see that the vector sum has a magnitude

$$I_{max} = \mathscr{E}_{max}\sqrt{\frac{1}{R^2} + \left(\frac{1}{X_C} - \frac{1}{X_L}\right)^2} \tag{60}$$

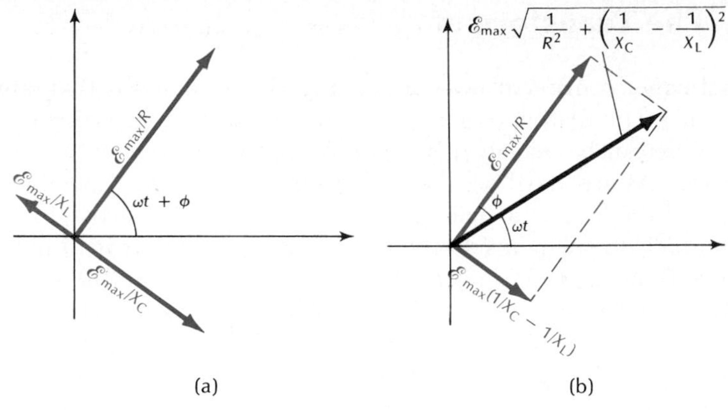

Fig. 34.22 (a) The three phasors (color) representing I_L, I_C, and I_R. (b) The vector sum (black) of these phasors represents the net current.

or

$$I_{max} = \mathscr{E}_{max}\sqrt{\frac{1}{R^2} + \left(\omega C - \frac{1}{\omega L}\right)^2} \tag{61}$$

The angle between the phasor representing the net current and the phasor representing the current in the resistor (or the phasor representing the emf) is given by

$$\boxed{\tan\phi = \frac{1/X_C - 1/X_L}{1/R}} \tag{62}$$

Phase angle for parallel LCR circuit

We can then write the current as

$$I = I_{max}\sin(\omega t + \phi) = \mathscr{E}_{max}\sqrt{\frac{1}{R^2} + \left(\omega C - \frac{1}{\omega L}\right)^2}\,\sin(\omega t + \phi)$$

or as

$$\boxed{I = \frac{\mathscr{E}_{max}\sin(\omega t + \phi)}{Z}} \tag{63}$$

Current in parallel LCR circuit

where

$$\boxed{Z = 1\Big/\sqrt{\frac{1}{R^2} + \left(\omega C - \frac{1}{\omega L}\right)^2}} \tag{64}$$

Impedance for parallel LCR circuit

is the impedance of the parallel LCR circuit.

Note that if the frequency coincides with the resonant frequency $\omega = \omega_0 = 1/\sqrt{LC}$, the current I specified by Eq. (63) is *minimum*. This is so because at this frequency, the currents in the inductor and the capacitor are of equal magnitudes; since these currents are always in opposite directions, the equality of their magnitudes implies their cancellation.

34.4 The Transformer

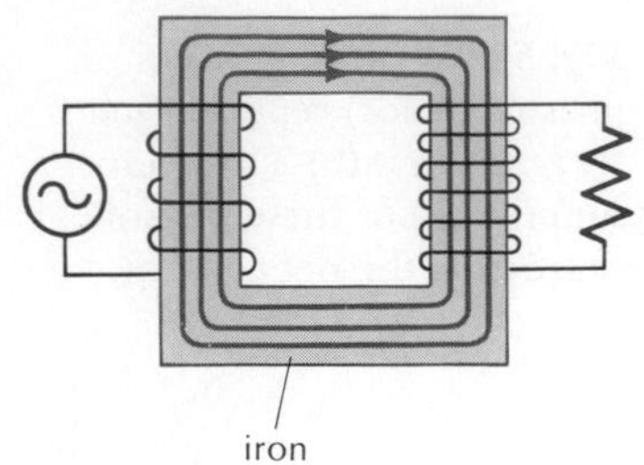

Fig. 34.23 A transformer.

A transformer consists of two coils arranged in such a way that (almost) all the magnetic field lines generated by one of them pass through the other. This can be achieved by winding both the coils on a common iron core. As we have seen in Chapter 33, the iron increases the strength of the magnetic field in its interior by a large factor. Since the field is much stronger inside the iron than outside, the field lines will concentrate inside the iron; thus the iron tends to keep the field lines together and acts as a conduit for the field lines (Figure 34.23).

Each coil is part of a separate electric circuit (Figure 34.24). The **primary** circuit has a source of alternating emf and the **secondary** circuit has a resistance or some other load that consumes electric power. The alternating current in the primary circuit induces an alternating emf in the secondary circuit. We will show that the induced emf $\mathscr{E}_2$ in the secondary circuit is related as follows to the emf $\mathscr{E}_1$ in the primary circuit:

emf in primary and secondary circuits of transformer

$$\boxed{\mathscr{E}_2 = \mathscr{E}_1 \frac{N_2}{N_1}} \tag{65}$$

where N_1 and N_2 are, respectively, the numbers of turns in the primary and secondary coils.

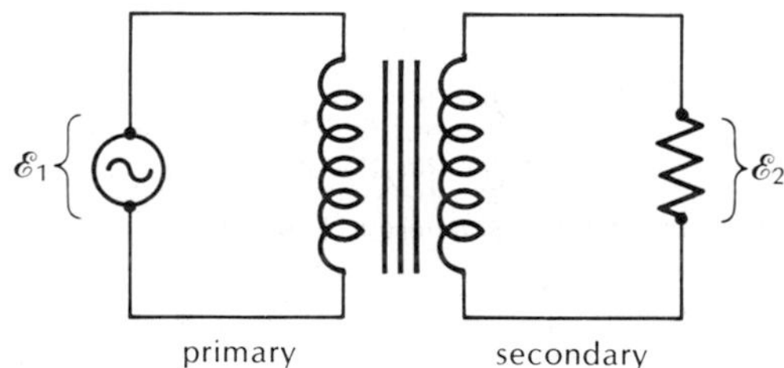

Fig. 34.24 Circuit diagram for the transformer. The parallel lines represent the mutual inductance.

To prove Eq. (65), we begin with Kirchhoff's rule as it applies to the primary circuit: the emf $\mathscr{E}_1$ of the source must equal the induced emf $\mathscr{E}_{1,\text{ind}}$ across the primary coil. But by Faraday's Law, the induced emf equals the rate of change of flux,

$$\mathscr{E}_1 = \mathscr{E}_{1,\text{ind}} = -\frac{d\Phi_1}{dt} \tag{66}$$

Likewise, the emf $\mathscr{E}_2$ delivered to the load must equal the induced emf $\mathscr{E}_{2,\text{ind}}$ in the secondary coil which, in turn, equals the rate of change of flux in that coil,

$$\mathscr{E}_2 = \mathscr{E}_{2,\text{ind}} = -\frac{d\Phi_2}{dt} \tag{67}$$

Since the same numbers of magnetic field lines pass through both coils, the fluxes and their rates of change are necessarily in the ratio N_2/N_1,

$$\frac{d\Phi_2}{dt} = \frac{N_2}{N_1}\frac{d\Phi_1}{dt} \tag{68}$$

From Eqs. (66), (67), and (68) we obtain

$$\mathscr{E}_2 = -\frac{N_2}{N_1}\frac{d\Phi_1}{dt} = \frac{N_2}{N_1}\mathscr{E}_1 \tag{69}$$

which we wanted to prove.

If $N_2 > N_1$, we have a step-up transformer and if $N_2 < N_1$, a step-down transformer.

EXAMPLE 5. Door bells and buzzers usually are designed for 12-volt AC and they are powered by small transformers which step down 110-volt AC to 12-volt AC. Suppose that such a transformer has a primary winding with 1500 turns. How many turns are there on the secondary winding?

SOLUTION: Equation (65) applies to the instantaneous voltages. It is therefore also valid for the rms voltages. With our numerical values

$$N_2 = N_1\frac{\mathscr{E}_2}{\mathscr{E}_1} = 1500 \times \frac{12\text{ V}}{110\text{ V}} = 164\text{ turns}$$

As long as the secondary circuit is open and carries no current ($I_2 = 0$), an ideal transformer does not consume electric power. Under these conditions, the primary circuit consists of nothing but the source of emf and an inductance, i.e., it is a pure L circuit. In such a circuit the power delivered by the source of emf averages to zero (see Section 34.1).

If the secondary circuit is closed, a current will flow ($I_2 \neq 0$). This current contributes to the magnetic flux in the transformer and induces a current in the primary circuit. The current in the latter is then different from that in a pure L circuit and the power will *not* average to zero over a cycle. In an ideal transformer, the electric power that the primary circuit takes from the source of emf exactly matches the power that the secondary circuit delivers to the load. Good transformers approach this ideal condition fairly closely: about 99% of the power supplied to the input terminals emerges at the output terminals; the difference is lost as heat in the iron core and in the windings.

Transformers play a large role in our electric technology. As we saw in Section 29.5, transmission lines for electric power operate much more efficiently at high voltage since this reduces the Joule losses. To take advantage of this high efficiency, power lines are made to operate at several hundred kilovolt. The voltage must be stepped up to this value at the power plant and, for safety's sake, it must be stepped down just before it reaches the consumer. For these operations, large banks of transformers are needed at both ends.

SUMMARY

Average AC power absorbed by resistor: $\overline{P} = \dfrac{\mathscr{E}^2_{max}}{2R}$

$$= \frac{\mathscr{E}^2_{rms}}{R}$$

Natural frequency of LC circuit: $\omega_0 = \dfrac{1}{\sqrt{LC}}$

Energy loss in freely oscillating series LCR circuit: $Q = 2\pi\,|U/\Delta U| = \omega_0 L/R$

Impedance of series LCR circuit: $Z = \sqrt{R^2 + (1/\omega C - \omega L)^2}$

Impedance of parallel LCR circuit: $Z = 1\Big/\sqrt{\dfrac{1}{R^2} + \left(\omega C - \dfrac{1}{\omega L}\right)^2}$

Transformer: $\mathscr{E}_2 = \mathscr{E}_1 \dfrac{N_2}{N_1}$

QUESTIONS

1. You can perceive the 120-Hz flicker (two peaks of intensity per AC cycle) in a fluorescent light tube (by sweeping your eye quickly across the tube), but you cannot perceive any such flicker in an incandescent light bulb. Explain.

2. Do the electrons from the power station ever reach the wiring of your house?

3. Some electric motors operate only on DC, others only on AC. What is the difference between these motors?

4. If you connect a capacitor across a 110-V outlet, does any current flow through the connecting wires? Through the space between the capacitor plates? Does the outlet deliver instantaneous electric power? Average electric power?

5. It is sometimes said that a capacitor becomes a short circuit at high frequencies, and an inductor becomes an open circuit at high frequencies. Explain.

6. Can you blow a fuse by connecting a very large capacitor across an ordinary 110-V outlet?

7. How could you use an LC circuit to measure the capacitance of a capacitor?

8. In Section 34.3 we asserted that under steady-state conditions the current in an LCR circuit has the same frequency as the driving emf. Is this also true if the circuit is not in steady state? Give an example.

9. Consider a series LCR circuit. Can the voltage across the capacitor ever be larger than $\mathscr{E}_{max}$? Across the inductor? Across the resistor? (Hint: Inspect the phasor diagram.)

10. Roughly plot the phase angle ϕ given by Eq. (44) as a function of ω. What is the phase angle at resonance?

11. If you use an AC voltmeter to measure the driving emf and the voltages across the inductor, the capacitor, and the resistor in a series LCR circuit, you will find that the emf is larger than the voltage across the resistor, but smaller than the sum of the voltages across the inductor, capacitor, and resistor. Explain.

12. What is the impedance of a series LCR circuit at resonance?

13. If we substitute $R = 0$ in Eq. (50), we obtain the equation for the current in a driven LC circuit. What must we substitute to obtain the equation for a driven CR circuit? A driven LR circuit?

14. Show that the equation for the current in a series LCR circuit [Eq. (50)]

includes Eqs. (3), (13), and (19) as special cases. What are the individual impedances of a resistor, a capacitor, and an inductor?

15. Show that the equation for the current in a parallel LCR circuit [Eq. (63)] includes Eqs. (3), (13), and (19) as special cases.

16. Consider the parallel LCR circuit described in Example 4. If the driving frequency is greater than $1/\sqrt{LC}$, does the net current lead or lag the emf? If the driving frequency is smaller than $1/\sqrt{LC}$?

17. Show that the time-average power absorbed by a series LCR circuit [Eq. (56)] is maximum at resonance.

18. In a parallel LCR circuit, the time-average power absorbed is $\mathscr{E}^2_{max}/2R$. Explain.

19. Why can we not use a transformer to step up the voltage of a battery?

20. Does an electric motor absorb more electric power when pulling a mechanical load than when running freely?

PROBLEMS

Section 34.1

1. An electric heater plugged into a 110-V AC outlet uses an average electric power of 1200 W.
   (a) What is the rms current and the maximum instantaneous current through the heater?
   (b) What is the maximum instantaneous power and the minimum instantaneous power?

2. An immersible heating element used to boil water consumes an (average) electric power of 400 W when connected to a source of 110 volts AC. Suppose that you connect this heating element to a source of 110 volts DC. What power will it consume?

3. A high-voltage power line operates on an rms voltage of 230,000 volts AC and delivers an rms current of 740 A.
   (a) What are the maximum instantaneous voltage and current?
   (b) What are the maximum instantaneous power and the average power delivered?

4. An AC current of 20 A flows in a copper wire of diameter 0.30 cm connected to an electric outlet. The drift velocity of the free electrons in the copper will then oscillate at 60 Hz. What is the maximum value of the instantaneous drift velocity? What is the maximum value of the acceleration of the drift velocity?

5. An electric heater operating with a 115-V AC power supply delivers 1200 W of heat.
   (a) What is the rms current through this heater?
   (b) What is the maximum instantaneous current?
   (c) What is the resistance of this heater?

6. The GG-1 electric locomotive develops 4600 hp; it runs on an AC voltage of 1100 V.
   (a) What rms current does this locomotive draw?
   (b) Why is it advantageous to supply the electric power for locomotives at high voltage (and fairly low current)?

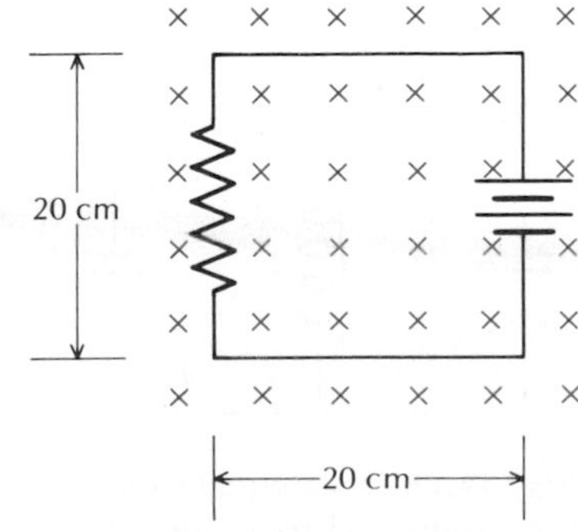

Fig. 34.25

7. A circuit consists of a resistor connected in series to a battery; the resistance is 5 Ω and the emf of the battery is 12 V. The wires (of negligible resistance) connecting these circuit elements are laid out along a square of 20 cm × 20 cm (Figure 34.25). The entire circuit is placed face on in an oscillating mag-

netic field. The instantaneous value of the magnetic field is

$$B = B_0 \sin \omega t$$

with $B_0 = 0.15$ T and $\omega = 360$ radian/s.

(a) Find the instantaneous current in the resistor.
(b) Find the average power dissipated in the resistor.

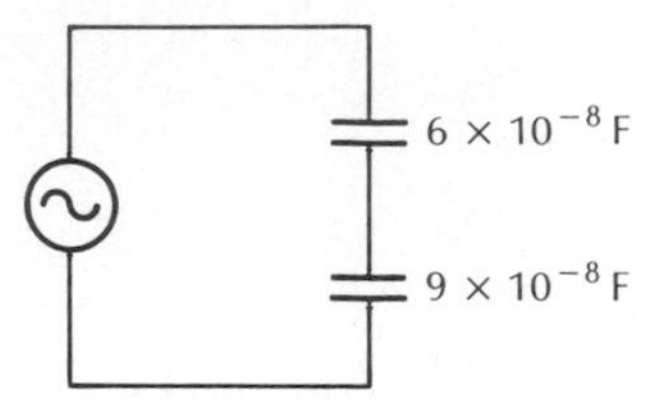

Fig. 34.26

8. A circuit consists of two capacitors of 6.0×10^{-8} F and 9.0×10^{-8} F connected in series to an oscillating source of emf (Figure 34.26). This source delivers a sinusoidal emf $\mathscr{E} = 1.8 \sin(120\pi t)$, where $\mathscr{E}$ is in volts and t in seconds.

(a) Find the charge on each capacitor as a function of time.
(b) At what time is the charge on the capacitors maximum? At what time minimum?
(c) What is the maximum energy in the capacitors? What is the time-average energy?

9. An inductor of 1.6×10^{-3} H is connected to a source of alternating emf. The current in the inductor is $I = I_0 \cos \omega t$, with $I_0 = 180$ A and $\omega = 120\pi$ radian/s.

(a) What is the potential difference across the inductor at time $t = 0$? At time $t = 1/240$ s?
(b) What is the energy in the inductor at time $t = 0$? At time $t = 1/240$ s?
(c) What is the instantaneous power delivered by the source of emf to the inductor at time $t = 0$? At time $t = 1/240$ s?

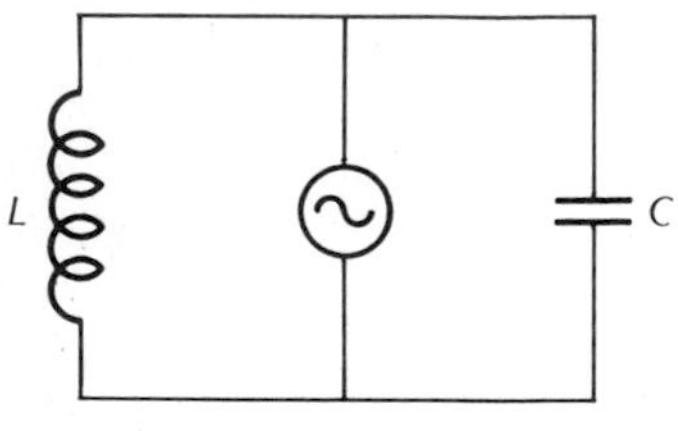

Fig. 34.27 Inductor and capacitor connected to a source of alternating emf.

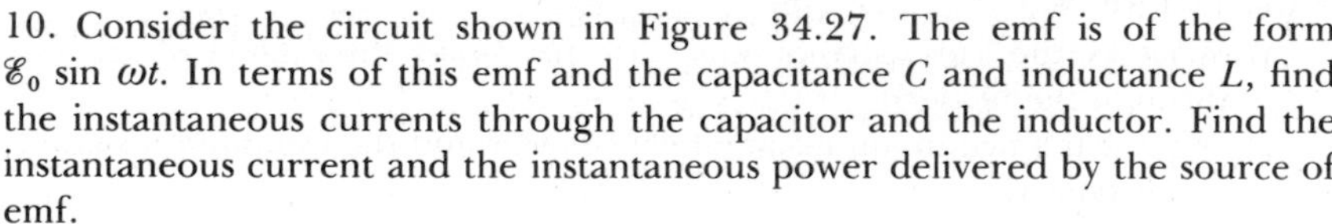

10. Consider the circuit shown in Figure 34.27. The emf is of the form $\mathscr{E}_0 \sin \omega t$. In terms of this emf and the capacitance C and inductance L, find the instantaneous currents through the capacitor and the inductor. Find the instantaneous current and the instantaneous power delivered by the source of emf.

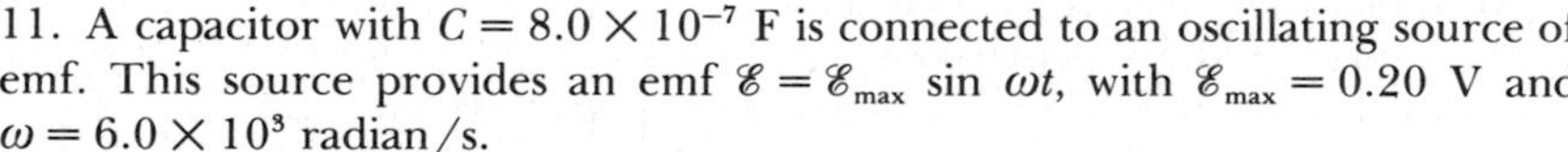

11. A capacitor with $C = 8.0 \times 10^{-7}$ F is connected to an oscillating source of emf. This source provides an emf $\mathscr{E} = \mathscr{E}_{max} \sin \omega t$, with $\mathscr{E}_{max} = 0.20$ V and $\omega = 6.0 \times 10^3$ radian/s.

(a) What is the reactance of a capacitor?
(b) What is the maximum current in the circuit?
(c) What is the current at time $t = 0$? At time $t = \pi/4\omega$?

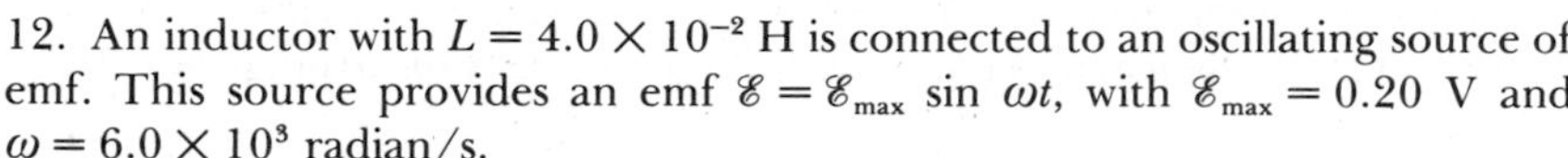

12. An inductor with $L = 4.0 \times 10^{-2}$ H is connected to an oscillating source of emf. This source provides an emf $\mathscr{E} = \mathscr{E}_{max} \sin \omega t$, with $\mathscr{E}_{max} = 0.20$ V and $\omega = 6.0 \times 10^3$ radian/s.

(a) What is the reactance of the inductor?
(b) What is the maximum current in the circuit?
(c) What is the current at time $t = 0$? At time $t = \pi/4\omega$?

Section 34.2

13. What is the natural frequency for an LC circuit consisting of a 2.2×10^{-6} F capacitor and a 8.0×10^{-2} H inductor?

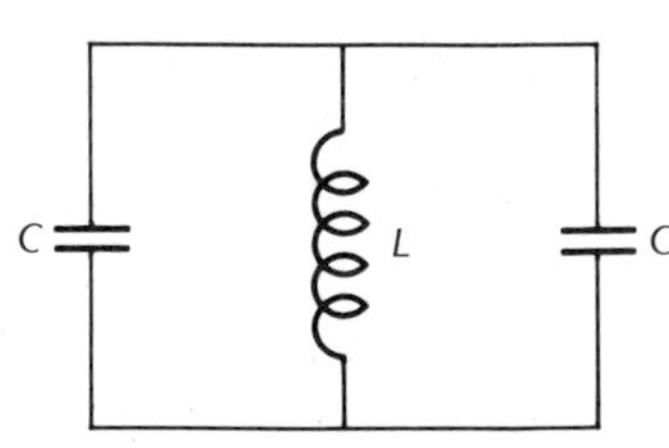

Fig. 34.28 Two equal capacitors connected to an inductor.

14. A radio receiver contains an LC circuit whose natural frequency of oscillation can be adjusted, or tuned, to match the frequency of incoming radio waves. The adjustment is made by means of a variable capacitor. Suppose that the inductance of the circuit is 15 μH. Over what range of capacitances must the capacitor be adjustable if the frequencies of oscillation of the circuit are to span the range from 530 kHz to 1600 kHz?

15. What is the natural frequency of oscillation of the circuit shown in Figure 34.28? The capacitances are 2.4×10^{-5} F each and the inductance is 1.2×10^{-3} H.

16. An LC circuit has an inductance of 5.0×10^{-2} H and a capacitance of 5.0×10^{-6} F. At $t = 0$, the capacitor is fully charged so that $Q_0 = 1.2 \times 10^{-4}$ C. What is the energy in this circuit? At what time, after $t = 0$, will the energy be purely magnetic? At what *later* time will it be purely electric?

17. Consider the LC circuit described in Problem 16. If we add a resistance of 120 Ω to this circuit (in series), what will be the frequency of the natural oscillations? By what factor is this smaller than the frequency of the circuit without resistance?

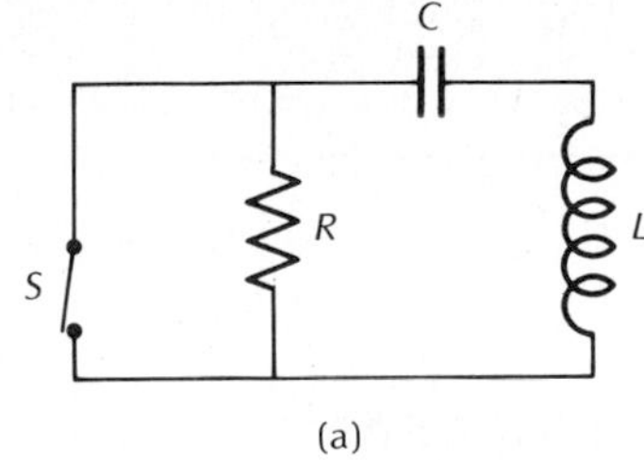

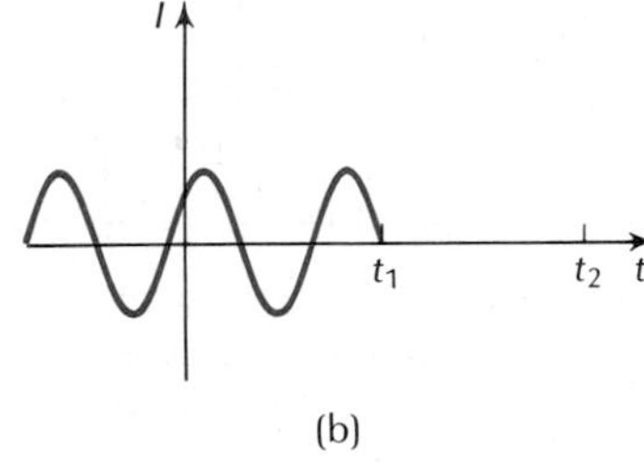

Fig. 34.29 (a) Inductor, capacitor, and resistor connected in a circuit. The switch S is initially closed. (b) Current in the circuit as a function of time. At $t = t_1$, the switch S is opened.

18. The circuit of Figure 34.29a is oscillating with the switch S closed. The graph of current vs. time is shown in Figure 34.29b.
   (a) At time t_1 the switch S is suddenly opened. Is the frequency of oscillation increased, decreased, or unchanged? In the space on the left in Figure 34.29b sketch the graph of current for times after t_1.
   (b) At time t_2 the switch S is closed. Sketch the graph of current after this time.

19. Consider an RC circuit consisting of a capacitor and a resistor in series (Figure 34.30). The capacitor is initially charged.
   (a) Show that Kirchhoff's rule leads to the following equation for this system:

$$R\frac{dQ}{dt} + \frac{1}{C}Q = 0$$

   (b) Verify that the solution of this equation is

$$Q = Q_0 e^{-t/RC}$$

   where Q_0 is the initial charge at time $t = 0$.
   (c) Show that the current is

$$I = -\frac{Q_0}{RC}e^{-t/RC}$$

   (d) Suppose that the resistance is 3.0 Ω. Suppose that the capacitance of the capacitor is 1.0×10^{-4} F and that the initial voltage across its terminals is 3.0 V. For this special case, plot the current as a function of time. What is the current at the initial instant? After how many seconds will the current have dropped to one-half of its initial value?

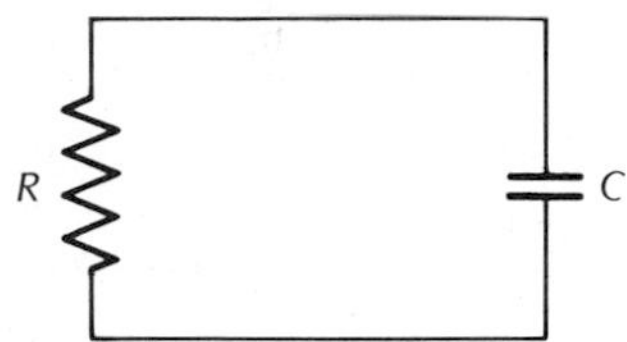

Fig. 34.30 Capacitor and resistor connected in series.

20. An LCR series circuit has $L = 0.5$ H, $C = 2.0 \times 10^{-5}$ F, and $R = 10$ Ω. At time $t = 0$, the capacitor is fully charged with a voltage of 24 V across its plates.
   (a) What is the initial energy in the circuit?
   (b) What is the percentage loss of energy per period?
   (c) At what time after $t = 0$ will the energy in the circuit have fallen to one-half of its initial value? One-tenth of its initial value?
   (d) Plot the energy as a function of time for the time interval $0 \text{ s} \leq t \leq 0.1$ s.

Section 34.3

21. A series LC circuit with $L = 3.0 \times 10^{-4}$ H and $C = 2.0 \times 10^{-5}$ F is being driven by an oscillating source of emf delivering an AC voltage of amplitude 0.40 V and frequency of 1.6×10^4 radian/s. What is the maximum instantaneous value of the current? What is the maximum instantaneous voltage across the inductor? Across the capacitor?

22. A series LC circuit is being driven by an audio generator that delivers an emf of amplitude 0.50 V. When the generator delivers the emf at a frequency of 2.0×10^3 radian/s, the maximum current is 1.0×10^{-1} A; when the generator delivers the same emf at a frequency of 1.5×10^3 radian/s, the maximum current is 2.7×10^{-2} A. From this information deduce the values of the inductance and the capacitance.

23. A capacitor of $C = 24.0\ \mu$F and an inductor of $L = 0.180$ H are connected

in series with an oscillating source that delivers an emf $\mathscr{E} = 12.0 \sin 377t$, where t is measured in seconds and $\mathscr{E}$ in volts.

(a) What is the maximum instantaneous current?
(b) What is the maximum instantaneous voltage across the capacitor?
(c) Across the inductor?

24. A series LC circuit with $C = 1.5 \times 10^{-7}$ F and $L = 2.5 \times 10^{-4}$ H is being driven by an audio generator that delivers a sinusoidal emf with an amplitude of 0.80 V and a frequency 2.2×10^4 Hz.

(a) Plot the emf as a function of time.
(b) Plot the current as a function of time.

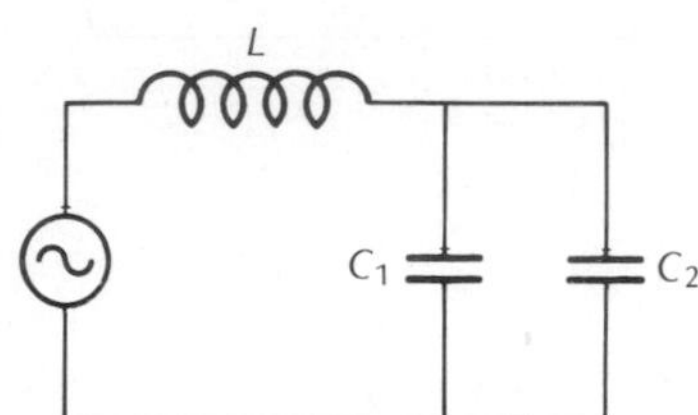

Fig. 34.31 Two capacitors and an inductor connected to a source of alternating emf.

25. A driven LC circuit consists of an inductor of 3.0×10^{-3} H, a capacitor of 2.0×10^{-8} F, and an oscillating source of emf operating at a frequency of 1.0×10^5 radian/s, all connected in series. If the amplitude of the current in the circuit is to be 6.0×10^{-3} A, what must be the amplitude of the alternating emf?

26. Consider the circuit shown in Figure 34.31. The oscillating source delivers a sinusoidal emf of amplitude 0.80 V and frequency of 400 Hz. The inductance is 5.0×10^{-2} H and the capacitances are 8.0×10^{-7} F and 16.0×10^{-7} F. Find the maximum instantaneous current in each capacitor.

27. Consider the circuit described in Example 3. If we add a resistor with $R = 10\ \Omega$ in series with the other elements of this circuit, what will be the peak instantaneous voltage across the inductor? Across the capacitor? Across the resistor? What will be the average power dissipated in the resistor?

28. (a) Consider the driven LCR circuit described by Eqs. (35)–(51). If $L = 6.0 \times 10^{-2}$ H, $C = 3.0 \times 10^{-6}$ F, $R = 1.2 \times 10^2\ \Omega$, $\mathscr{E}_{max} = 24.0$ V, and $\omega = 2.5 \times 10^3$ radian/s, calculate the maximum value of the current in this circuit under steady-state conditions.
(b) At what time after $t = 0$ does the alternating emf of the source reach its maximum value? At what time does the current reach its maximum value?
(c) Make a plot of $\mathscr{E}$ as a function of time. On top of this plot, make a plot of I as a function of time.

29. An LCR circuit consists of an inductor of 1.5×10^{-2} H, a capacitor of 2.8×10^{-6} F, and a resistor of 5.0 Ω connected in series to a source of alternating emf. The source delivers a voltage of $\mathscr{E} = \mathscr{E}_{max} \sin \omega t$, with $\mathscr{E}_{max} = 0.60$ V and $\omega = 6.0 \times 10^4$ radian/s.

(a) What is the impedance of this circuit?
(b) What is the maximum current in this circuit?
(c) What is the phase angle of the current?

30. An LCR circuit consists of an inductor of 1.2×10^{-2} H, a capacitor of 2.4×10^{-6} F, and a resistor of 2.0 Ω connected in series to a source of alternating emf with an amplitude of 0.80 V.

(a) At what frequency will this circuit be in resonance with the driving voltage?
(b) What is the maximum current in the circuit at resonance?
(c) What is the average dissipation of power in the circuit at resonance?

31. An LR circuit consists of an inductor with $L = 2.0 \times 10^{-4}$ H and a resistor with $R = 1.2\ \Omega$ connected in series to an oscillating source of emf. This source generates a voltage $\mathscr{E} = \mathscr{E}_{max} \sin \omega t$, with $\mathscr{E}_{max} = 0.50$ V and $\omega = 3 \times 10^3$ radian/s. Find the maximum current in the circuit. Find the phase angle of the current. Find the average dissipation of power in the resistor.

32. An RC circuit consists of a resistor with $R = 0.80\ \Omega$ and a capacitor with $C = 1.5 \times 10^{-4}$ F connected in series to an oscillating source of emf. The source generates a voltage $\mathscr{E} = \mathscr{E}_{max} \sin \omega t$, with $\mathscr{E}_{max} = 0.40$ V and $\omega = 9 \times 10^3$ radian/s. Find the maximum current in the circuit. Find the phase angle of the current. Find the average dissipation of power in the resistor.

33. Consider the circuit of Example 3. What is the maximum energy in this circuit at one instant of time? The minimum energy?

34. Derive an expression for the instantaneous power absorbed by the parallel LCR circuit of Example 4. Show that the time-average power is $\mathscr{E}_{max}^2/(2R)$.

*35. Consider an LC circuit (Figure 34.32) with a driving emf $\mathscr{E} = \mathscr{E}_{max} \sin \omega t$.
(a) Show that if $\omega \gg 1/\sqrt{LC}$, the amplitude of the potential difference across the terminals in Figure 34.32 is $\Delta V \cong (X_C/X_L)\,\mathscr{E}_{max}$.
(b) Show that ΔV is much smaller than $\mathscr{E}_{max}$. Thus, this circuit can be used as a **filter** that strongly attenuates high-frequency components in the driving emf.

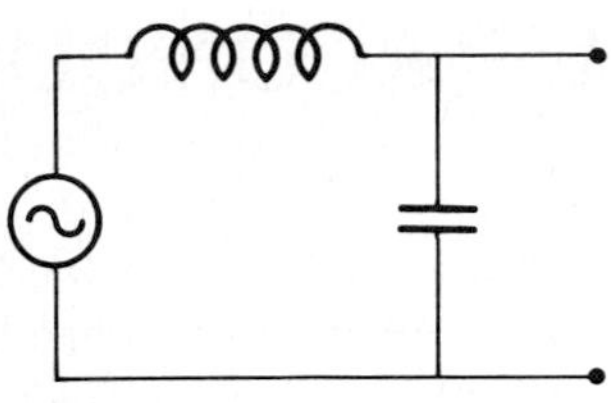

Fig. 34.32

Section 34.4

36. A transformer used to step up 110 V to 5000 V has a primary coil of 100 turns. What must be the number of turns in the secondary coil?

37. A transformer operating on a primary voltage of 110 volts AC delivers a secondary voltage of 6.0 volts AC to a small electric buzzer. If the current in the secondary circuit is 3.0 A, what is the rms current in the primary circuit? Assume that no electric power is lost in the transformer.

38. The generators of a large power plant deliver an electric power of 2000 MW at 22 kilovolts AC. For transmission, this voltage is stepped up to 400 kV by a transformer. What is the rms current delivered by the generators? What is the rms current in the transmission line? Assume that the transformer does not waste any power.

39. A power station feeds 1.0×10^8 W of electric power at 760 kV into a transmission line. Suppose that 10% of this power is lost in Joule heat in the transmission line. What percentage of the power would be lost if the power station were to feed 340 kV into the transmission line instead of 760 kV, other things being equal?

40. The largest transformer ever built handles a power of 1.50×10^9 W. This transformer is used to step down 765 kV to 345 kV. What is the rms current in the primary? What is the current in the secondary? Assume that no electric power is lost by the transformer.

SUPERCONDUCTIVITY[1]

Of all gases, helium has the lowest liquefaction temperature. All other gases had been liquified during the nineteenth century, but helium resisted the best efforts of low-temperature physicists until 1908 when it was finally liquified by H. Kammerlingh Onnes[2] at the University of Leiden. The temperature of boiling helium is 4.2 K (at atmospheric pressure).

The availability of liquid helium laid the realm of low-temperature physics open to exploration. Any material can be cooled to 4.2 K merely by immersing it in liquid helium. Furthermore, lower temperatures can be attained by pumping on the helium; this means that with a vacuum pump connected to a closed vessel containing the liquid helium, the space above the liquid is partially evacuated to a low pressure. The liquid then cools by evaporation until it reaches a lower temperature corresponding to the lower pressure (recall that the boiling point of a liquid is lowered by a reduction of pressure; see Section 20.4). By this method, temperatures slightly below 1 K can be attained.

At such extremely low temperatures, materials develop very unusual properties. Many metals and alloys become **superconductors,** i.e., their resistance to electric currents vanishes completely. Liquid helium, at a temperature of 2.2 K, becomes a **superfluid,** i.e., its internal friction disappears and it can flow without drag through very fine capillary holes. These strange properties are a manifestation of a high degree of order within the material. As we know from the Third Law of Thermodynamics, the entropy vanishes at absolute zero, i.e., the disorder caused by random thermal disturbances disappears. The material then settles into a definite microscopic state. This microscopic state is a quantum state that cannot be adequately described by classical mechanics. Thus, superconductivity and superfluidity are macroscopic manifestations of the underlying quantum behavior of matter.

[1] This chapter is optional.

[2] Heike Kammerlingh Onnes, 1853–1926, Dutch physicist and professor at Leiden. He was awarded the Nobel Prize in 1913 for his investigations of the properties of matter at low temperatures.

K.1 ZERO RESISTANCE

Superconductivity was discovered by Onnes in 1911 during electrical experiments with a sample of frozen mercury. Figure K.1 shows a plot of the measured values of the resistivity of mercury as a function of temperature. At a temperature of 4.15 K, the resistivity drops sharply, and below this critical temperature, the resistivity is zero.

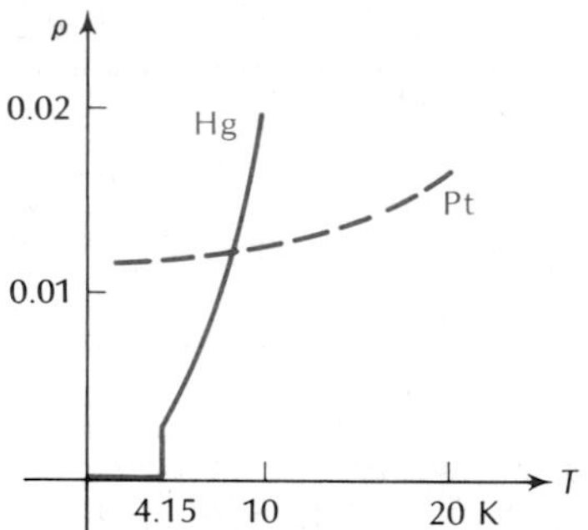

Fig. K.1 Resistivity of a sample of mercury as a function of temperature. The resistivity has been expressed as a fraction of the resistivity at 273 K. Below 4.15 K, mercury is a superconductor. For comparison, the dashed curve shows the resistivity of platinum, which is a normal conductor.

The sudden change of resistivity indicates that the material has suffered some drastic alteration of state. Experimentation reveals that in superconducting mercury, as in any metal, the carriers of electric current are free electrons. Hence, it must be that these free electrons suffer some alteration in their state. We will discuss what happens to the electrons in a later section.

Besides mercury, several other metals exhibit superconductivity at low temperatures. Table K.1 lists some superconducting metals and their transition temperatures. Furthermore, a large variety of compounds and alloys exhibit superconductivity. The highest known transition temperature is found in an alloy containing niobium and germanium (Nb_3Ge); this transition temperature is 23 K.

Table K.1 SOME SUPERCONDUCTORS

Element	T_c
Aluminum	1.20 K
Indium	3.40
Lead	7.19
Mercury	4.15
Niobium	9.26
Osmium	0.66
Tin	3.72
Tungsten	0.012
Vanadium	5.30
Zinc	0.87

A superconductor is a perfect conductor — its resistance is truly zero. Even the most precise experiments have not been able to detect any residual resistance in a superconductor. If a closed loop of superconducting wire initially has a current, then this current will keep flowing around the loop on its own accord as long as the wire is kept cold. Such a steady current that flows without any resistive loss is called a **persistent current.** In one case, a current that was started in a superconducting loop kept on flowing for two and a half years with undiminished strength, and it would probably still be flowing today if the experimenters had not run out of liquid helium for cooling their apparatus. Of course, in view of measurement errors, we cannot prove that the resistivity in superconductors is exactly zero; but if it is not zero, it is certainly extremely small — no more than 10^{-15} times the resistivity of the best normal conductors.

Persistent currents induced by changing magnetic fields bring about some spectacular levitation effects. If a small bar magnet is dropped toward a superconducting lead dish (Figure K.2), the magnetic field induces persistent currents along the surface of the dish; by Lenz' Law, the direction of these currents is such that their magnetic force on the magnet is repulsive. When the magnet is close to the dish, this magnetic force is large enough to support the weight of the magnet — the magnet floats forever above the lead dish.

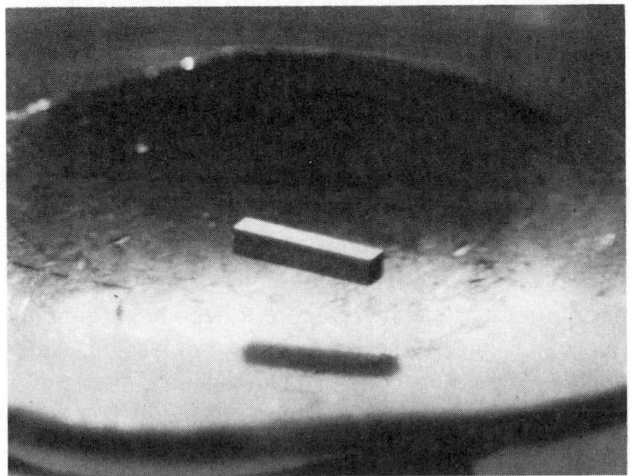

Fig. K.2 A small bar magnet levitated above a superconducting dish. (Courtesy A. Leitner, Rensselaer Polytechnic Institute.)

A similar effect can be demonstrated with a superconducting lead ring and sphere (Figure K.3). The ring is stationary and it carries an initial current that has been induced by means of an external magnetic field. If now a lead sphere is dropped toward the ring, it will remain floating at some height above the ring. This is analogous to the levitation effect described in the preceding paragraph; the ring with its current and associated magnetic field plays the role of the magnet and the lead sphere plays the role of the lead dish.

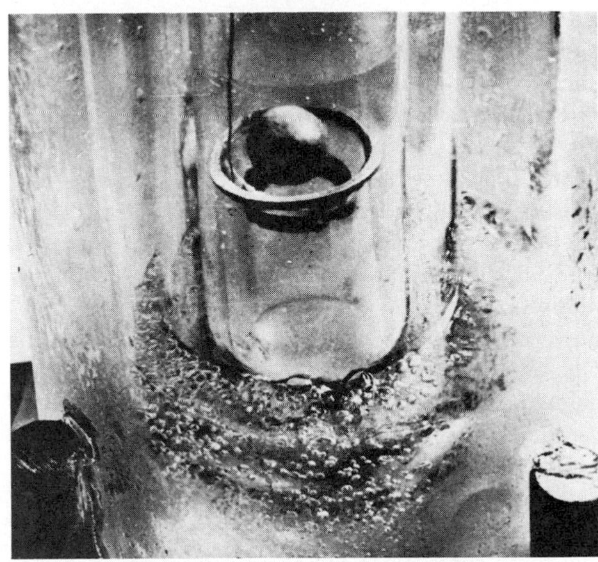

Fig. K.3 A small superconducting sphere levitated above a superconducting ring carrying a current.

K.2 THE CRITICAL MAGNETIC FIELD

Superconducting loops with persistent currents make magnetic fields — they are magnets. Such superconducting magnets are similar to permanent magnets in that they do not require any electric power supply to maintain their current and their magnetic field. Only an initial energy input is needed to get the persistent current started. This suggests that superconductors should permit us to produce extremely intense magnetic fields with little expenditure of energy.

Unfortunately, intense magnetic fields have an adverse effect on superconductors: intense magnetic fields destroy superconductivity. For instance, at a

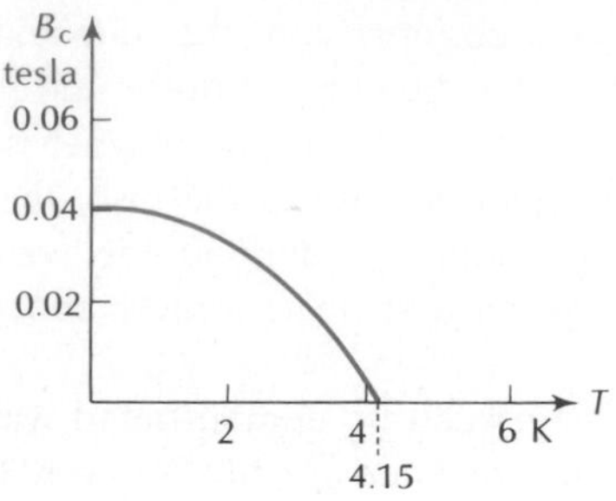

Fig. K.4 Critical magnetic field as a function of temperature for mercury.

temperature near absolute zero, a magnetic field of 0.041 T will destroy the superconductivity of mercury. At a temperature near the critical temperature (4.15 K), an even smaller magnetic field suffices to destroy the superconductivity. The minimum magnetic field that will quench the superconductivity of a material is called the **critical magnetic field, B_c**. Its strength depends on the temperature; Figure K.4 shows a plot of critical field strength for mercury as a function of temperature.

This breakdown of superconductivity imposes serious restrictions on the maximum current that can be carried by a superconductor. The current in, say, a wire will itself generate a magnetic field and, if this magnetic field is intense enough, it will cause a breakdown of the superconductivity of the wire. For example, a superconducting wire of mercury, 0.2 cm in diameter at a temperature near absolute zero, can carry a current of at most 200 A; a larger current will lead to a breakdown of the superconductivity. Such restrictions must be kept in mind in the design of superconducting magnets. We will see in Section K.4 that some compounds and alloys can tolerate substantially larger magnetic fields.

K.3 THE MEISSNER EFFECT

As we know from our study of ordinary conductors (see Chapter 28), a steady current in such a conductor requires an electric field to overcome the resistance. The electric field within a conductor carrying a given current is directly proportional to the resistance — a conductor of large resistance has a large electric field and a conductor of small resistance has a small electric field (see Ohm's Law in Section 28.2). A superconductor, with zero resistance, always has zero electric field in its interior. Furthermore, it follows from this that the rate of change of magnetic field in a superconductor must always be zero; if it were not, then the changing magnetic flux would induce an electric field, in contradiction with the requirement that the electric field remain zero. For example, if we transport a superconducting cylinder into a magnetic field (Figure K.5), it will push the magnetic field lines aside so that none of these penetrate the cylinder. What happens here is that as the cylinder touches the magnetic field, currents are induced on the surface of the cylinder and the magnetic field of these currents produces just the right deformation of the magnetic field lines to prevent their penetration into the cylinder. This behavior is characteristic of a perfect conductor; for instance, a ball of plasma exhibits the same behavior and for the same reason — a ball of plasma is an (almost) perfect conductor.

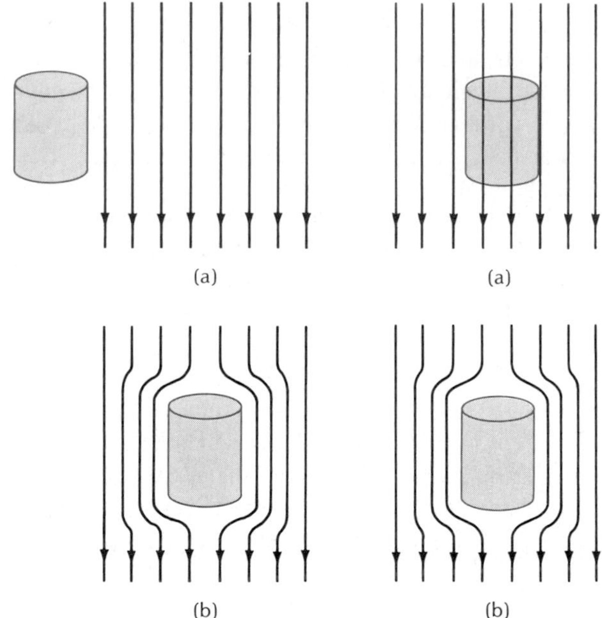

Fig. K.5 (left) Superconducting cylinder transported into a magnetic field.

Fig. K.6 (right) Conducting cylinder in a magnetic field: (a) normal metal, before T_c is reached, and (b) superconductor, after T_c is reached.

But it turns out that a superconductor is more than just a perfect conductor. A superconductor not only prevents the penetration of magnetic field lines that are initially outside of the superconducting material, but it also *expels* any magnetic field lines that are initially inside the material (before it becomes superconducting). Figure K.6 shows a cylinder of lead, above its critical temperature, placed in a magnetic field. This is an ordinary conductor and the magnetic field lines penetrate it without hindrance. However, if we cool the lead to its critical temperature, it will expel these field lines and, when the lead reaches its superconducting state, the magnetic field within it will be zero. This behavior is to be contrasted with that of a mere perfect conductor. Figure K.7 shows a ball of gas, placed in a magnetic field. If we convert this gas into a plasma (by ionizing it), the magnetic field lines are trapped (or frozen) in the plasma — when we switch the magnet off, the magnetic field lines inside the plasma remain unchanged, the induced currents in the plasma generating just enough magnetic field to keep the flux constant.

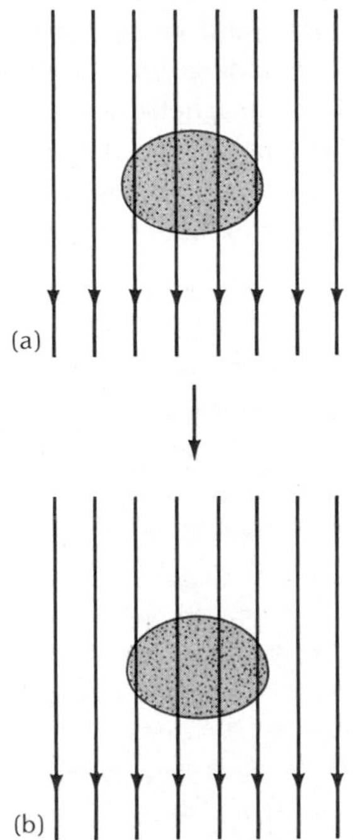

Fig. K.7 Ball of gas in a magnetic field: (a) normal gas, before ionization, and (b) plasma, after ionization.

The expulsion of magnetic flux from a metal during the transition from the normal to the superconducting state is called the **Meissner effect.** It means that a superconductor is not only a perfect conductor, but also a perfect diamagnet. (Recall that a diamagnetic material immersed in a magnetic field tends to reduce the strength of the magnetic field in its interior; see Section 33.4).

The elimination of the magnetic field from the interior of the superconductor is brought about by currents that flow along the surface of the superconductor; the magnetic field generated by these currents cancels the magnetic field generated by the external sources.

To see that the currents carried by a superconductor must always be surface currents, we note that if any currents were to flow within the volume of the superconductor, then Ampère's Law would demand a nonzero value of **B** within the volume (Figure K.8); such a nonzero value of **B** would be in contradiction to the Meissner effect. From this we can draw the general conclusion that *any* current carried by a superconductor must flow on the surface; this applies to induced currents as well as to currents originating from other sources of emf. Although from a macroscopic point of view these superconductor currents are surface currents, from a microscopic point of view they do not flow exactly on the surface but rather in a thin layer or skin. The thickness of this layer carrying the current is typically about 10^{-5} cm. Within the surface layer the magnetic field is not quite zero; the Meissner effect is incomplete and some magnetic flux penetrates.

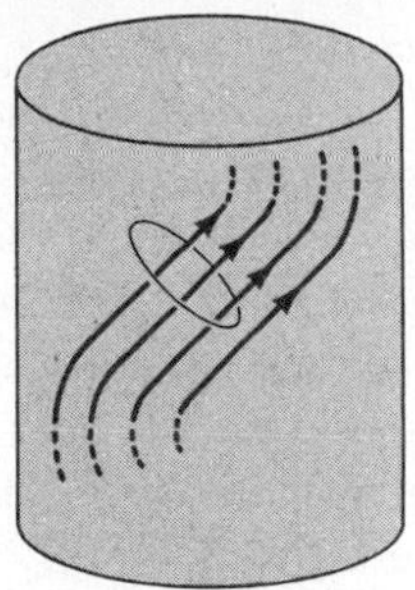

Fig. K.8 The lines with arrows show the flow of a hypothetical current within the volume of a superconductor. The closed loop is the path of integration for Ampère's Law; since some current passes through this loop, $\int \mathbf{B} \cdot d\mathbf{l}$ cannot be zero.

In Figure K.6 we illustrated the Meissner effect for a superconducting cylinder. Strictly, the ideal Meissner effect, with complete expulsion of the magnetic flux from the entire volume of the metal, occurs only if the metal has the shape of a very long cylinder (a wire) aligned with the magnetic field. For other shapes, the extent of the expulsion of the magnetic flux depends on the geometry. In general, the volume of the metal splits into domains of superconducting material and normal material. If we increase the strength of the magnetic field, the size of the normal domains increases at the expense of the superconducting domains and, when the field reaches the critical strength, the entire volume of metal becomes normal.

K.4 SUPERCONDUCTORS OF THE SECOND KIND

In most pure superconducting metals, the expulsion of magnetic flux from each of the superconducting domains in the metal is an all or nothing affair: if the metal is held at a fixed temperature and immersed in a magnetic field, it will prevent the penetration of the magnetic flux as long as the magnetic field is weaker than the critical value; but the metal will suddenly cease to be a superconductor when the magnetic field becomes stronger than the critical value and it will then freely permit the penetration of the magnetic flux.

But in niobium, in vanadium, and in alloys of other metals, the expulsion of magnetic flux from a superconducting domain is a much more complicated affair. The alloy will permit a partial penetration of flux if the magnetic field is of intermediate strength; and it will finally cease to be a superconductor and permit complete penetration of flux when the magnetic field becomes stronger. Thus, an alloy has two critical values of magnetic field strength: at a value B_{c_1} the flux begins to penetrate and at a value B_{c_2} the flux penetrates completely and superconductivity breaks down. For example, at a temperature of 4.2 K, the niobium–tin alloy Nb_3Sn has critical values $B_{c_1} = 0.019$ T and $B_{c_2} = 22$ T. The high value of B_{c_2} is of great practical importance — the alloy retains its superconductivity even in a very strong magnetic field where any pure metal would lose its superconductivity.

Materials such as Nb_3Sn that permit a partial penetration of the magnetic flux when immersed in a magnetic field of intermediate strength are called **superconductors of the second kind,** or of type II. When immersed in a magnetic field of intermediate strength, such a superconductor is in a mixed state: the bulk of the material is superconducting, but it is threaded by very thin filaments of normal material; these filaments are oriented parallel to the external magnetic field and they serve as conduits for the penetrating lines of this external magnetic field (Figure K.9). A current circulates around the perimeter of each filament; this current shields the bulk of the superconductor from the magnetic field in the filament. The flow of this current has the character of a vortex; because of this, the filaments are usually called **vortex lines.**

It turns out that the amount of flux associated with each vortex line has a fixed value related to Planck's constant and the electric charge of the electron,[3]

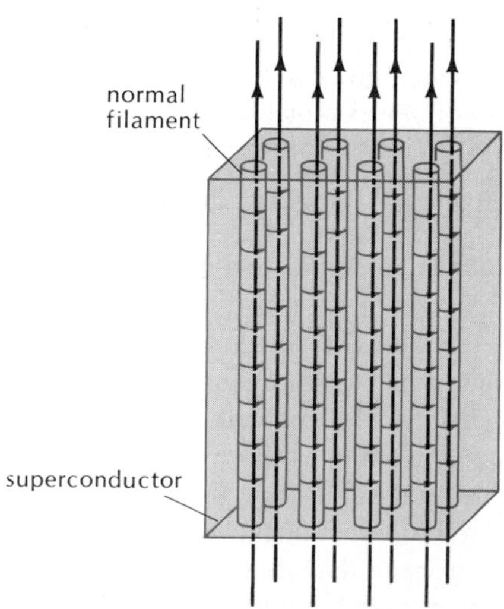

Fig. K.9 Filaments of normal material in a superconductor. The filaments serve as conduits for the magnetic field lines.

[3] See Appendix 8 for a more precise value of the magnetic flux quantum.

$$\Phi_0 = \frac{h}{2e} = 2.07 \times 10^{-15}\ \text{T} \cdot \text{m}^2 \qquad (1)$$

This means that the flux is *quantized* — Φ_0 represents a **quantum of flux**[3] just as the electric charge e of an electron represents a quantum of electric charge. In a superconductor of the second kind, an increase of the strength of the external magnetic field will not cause an increase of the flux associated with each vortex line; instead it will cause an increase in the number of vortex lines threading the superconductor. The stronger the external magnetic field, the more densely will the vortex lines be packed. Figure K.10 shows the vortex lines (viewed end on) in a sample of lead–indium alloy; the vortex lines are packed in a regular triangular pattern.

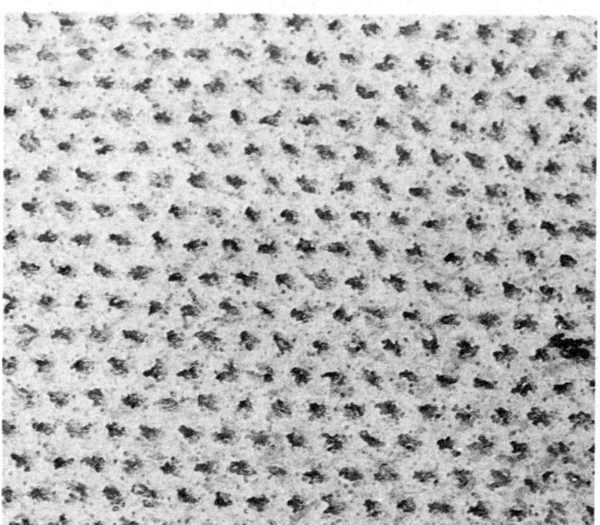

Fig. K.10 Ends of vortex lines at the surface of a sample of superconducting lead–indium. The vortex lines have been made visible by dusting with powdered iron. The separation between the vortex lines is about 0.005 cm.

K.5 THE BCS THEORY

We know from Chapter 28 that the resistivity of a metal is caused by collisions between the free electrons and the ions of the crystal lattice of the metal. The resistivity depends on temperature because the random thermal motion of the ions of the lattice increases the likelihood that an electron will suffer collisions. In fact, in a perfect lattice with no thermal vibrations an electron can travel without ever suffering any collisions. If we adopt the naive classical picture of an electron as a pointlike particle, we can see that travel without collisions is possible if the electron moves along a straight line between the rows of atoms. This classical picture of the motion of electrons has been shown to be invalid by modern quantum mechanics, according to which electrons have wave properties (see Section 41.6); however it turns out that the quantum-mechanical picture of the motion of electron waves leads to a similar conclusion. The electron wave can travel through a perfect crystal lattice without suffering any scattering (deviation) because whatever effect is caused by one atom is canceled by the effects of other atoms. The quantum-mechanical picture indicates that in a perfect crystal lattice the electron wave can travel unhindered in *any* direction. Imperfections in the regularity of the position of the atoms of the lattice will hinder the propagation of the electron wave and give the metal a finite resistivity. Thus, the random thermal vibrations of the ions contribute to the resistivity. At low temperatures, the reduction of these thermal vibrations brings about a reduction of resistivity; at zero temperature, when the thermal vibrations disappear, the resistivity should also disappear (except for a residual contribution due to impurities and dislocations of the lattice).

It therefore came as no surprise to physicists that the experimental values of resistivity were small at low temperatures — what came as a surprise was that the resistivity of some metals vanished completely at a few degrees above absolute zero.

The details of the mechanism underlying superconductivity were finally spelled out in the **Bardeen–Cooper–Schrieffer** (BCS)[4] **theory** of superconductivity, some 50 years after the discovery of this phenomenon. The key to this mechanism is the formation of electron pairs (Cooper pairs). After many false starts, theoretical physicists recognized that the free electrons in a metal are not quite free, but they interact with one another via the lattice. The negative charge of each free electron exerts an attractive force on the positive charges of the ions of the lattice; consequently, the nearby ions contract slightly toward the electron. This slight concentration of positive charge, in turn, attracts other electrons. The net effect is that a free electron exerts a small attractive force on another free electron. Although this attractive force is too small to be of any consequence at room temperature, it is strong enough to permanently bind two electrons into a pair when the temperature is within a few degrees of absolute zero, where the thermal disturbances nearly disappear. In a superconducting metal in electrostatic equilibrium (no current), each **Cooper pair** consists of two electrons of exactly opposite momenta. Obviously such a configuration makes no sense from a classical point of view — if two particles have opposite and constant momenta, they will travel away from each other in opposite directions; they will then cease to interact and they cannot remain bound. However, the configuration makes sense from a quantum-mechanical point of view, where each particle is described by a wave — if two waves have opposite directions of motion they can continue to overlap for a long time and they can continue to interact. Thus, the BCS theory is intrinsically quantum mechanical and the language of classical physics can only convey a crude outline of the theory.

[4] John Bardeen, 1908–, Leon N. Cooper, 1930–, and J. Robert Schrieffer, 1931–, American physicists. They shared the Nobel Prize in 1972 for their theory of superconductivity.

In a superconducting metal that is carrying a current, the electron pairs have a net momentum in the direction opposite to that of the current, and they transport electric charge. The electron pairs move through the lattice without resistance because whenever the lattice scatters one of the electrons and changes its momentum, it will also scatter the other electron of the pair and change its momentum by an opposite amount. Consequently, the lattice cannot change the net momentum of a pair — it can neither slow down nor speed up the average motion of the paired electrons.

The quantum character of superconductivity shows up quite explicitly in the quantization of the current and of the magnetic flux in closed superconducting loops. Delicate experiments with small superconducting loops have shown that the current is restricted to a discrete set of values defined by the condition that the flux intercepted by the area within the loop is always a multiple of the basic quantum of flux, $\Phi_0 = 2.07 \times 10^{-15}\ \mathrm{T \cdot m^2}$. Even more spectacular quantum effects occur in a **Josephson junction,**[5] consisting of a thin layer of insulator placed between two adjacent pieces of superconductor (Figure K.11). Classically, the layer of insulator constitutes an inpenetrable barrier for electrons; but quantum-mechanically the electron waves can tunnel through this barrier. Consequently, the thin layer of insulator permits the passage of a sizable fraction of the superconducting current. The interplay of the electron waves on the two sides of the Josephson junction gives rise to some remarkable in-

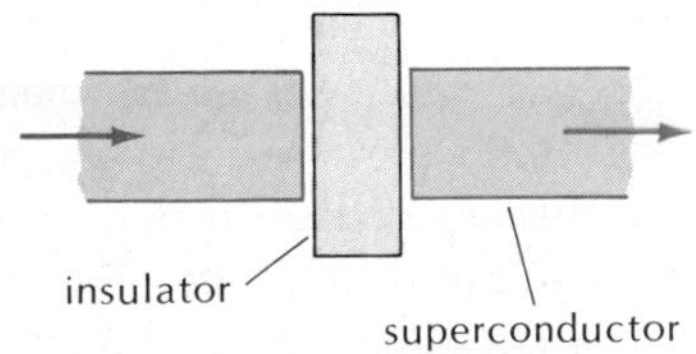

Fig. K.11 A Josephson junction. The thickness of the layer of insulator, an oxide, is usually less than 100 Å.

[5] Brian D. Josephson, 1940–, English physicist, and Ivar Giaever, 1929–, American physicist, shared the Nobel Prize in 1973 for the prediction of the properties of the Josephson junction and for the discovery of the tunneling of superconducting currents through insulators, respectively.

terference phenomena. For instance, if we apply a DC potential difference ΔV across the junction, the result is an AC current of frequency $2e\,\Delta V/h$. And if we apply an AC potential difference of frequency $2e\,\Delta V/h$ across the junction, the result is a combination of AC and DC currents. The measurement of the frequency of the oscillating current produced by a steady potential difference has been used for a new determination of the ratio of the fundamental constants e and h, with unprecedented precision. Such measurements also provide the most precise method for the determination of potential differences.

Other remarkable interference phenomena arise when two Josephson junctions are connected in parallel (Figure K.12), an arrangement called a **SQUID** (*su*perconductive *qu*antum *i*nterference *d*evice). The net current passing through this device depends on the magnetic flux intercepted by the area spanned by the loop: the current is zero whenever the flux is a half-integer multiple of Φ_0, and the current is maximum whenever the flux is an integer multiple of Φ_0. This sensitive dependence of the current on the magnetic flux can be exploited for extremely precise measurements of magnetic fields.

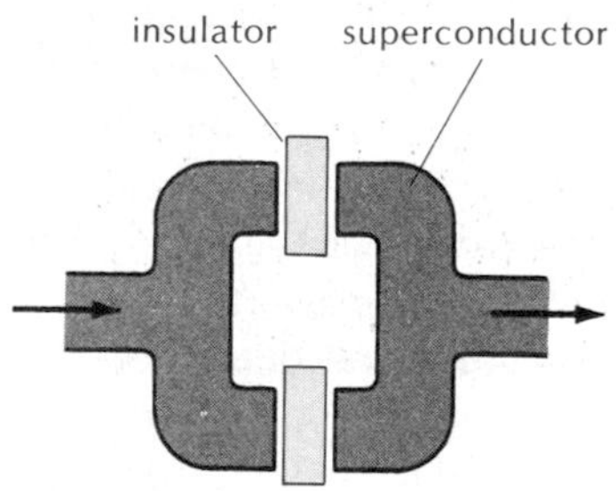

Fig. K.12 Two Josephson junctions connected in parallel form a SQUID.

K.6 TECHNOLOGICAL APPLICATIONS

Practical applications of superconductivity have focused on the development of electromagnets with superconductive coils capable of generating intense magnetic fields. Conventional electromagnets, with copper windings and iron cores, are among the least energy-efficient devices produced by our technology. They consume enormous amounts of electric power merely to keep a constant current running around in circles. Except for the magnetic energy initially stored in the electromagnet at startup, all of the electric power is wasted as heat. By contrast, a magnet with superconducting coils requires no electric power at all to keep running. Of course, the superconducting magnet must be cooled to a temperature near absolute zero and maintained at this temperature. The initial cooling and the continual removal of any heat that leaks into the magnet from the environment require refrigeration and a supply of liquid helium.

Fig. K.13 A high-field superconducting magnet. This magnet produces a field of 15 T in a bore of 7 cm.

Figure K.13 shows a superconducting magnet capable of generating a magnetic field of 15 T; a magnetic field of this strength is very difficult to attain with a conventional magnet. In Figure K.14 we see a dramatic

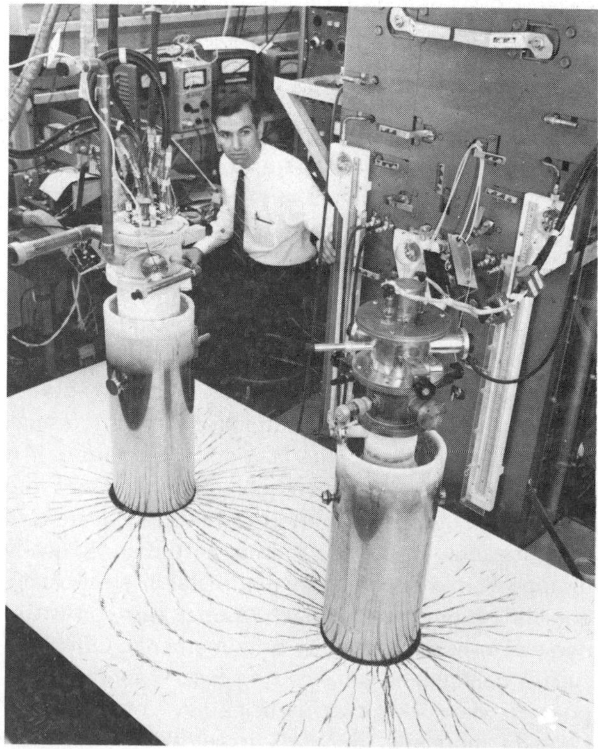

Fig. K.14 Nails scattered on a board align with the field lines of the intense magnetic field generated by two superconducting magnets. The coils of these magnets are immersed in stainless-steel dewars full of liquid helium.

Fig. K.15 The coils of this large superconducting magnet for a bubble chamber at Argonne National Laboratory are made of niobium–titanium alloy embedded in copper.

display of the strength of the magnetic fields of superconducting magnets: thousands of iron nails align along the field lines connecting two such magnets.

Figure K.15 shows the coils of a large superconducting magnet for a bubble chamber at the Argonne National Laboratory. The coils are 4.8 m across and they produce a magnetic field of 1.8 T. After the initial current has been established in this magnet, the only power required is 190 kW for a liquid-helium refrigerator; in contrast, a conventional magnet of similar size would require about 10,000 kW of electric power. The construction costs for this superconducting magnet are about the same as for a conventional magnet of similar size, but the yearly running costs are lower by a factor of 10.

Superconducting magnets have recently been installed at the Fermilab accelerator. This machine was originally built with conventional magnets, but in order to make the machine capable of handling particles of higher energy, new superconducting magnets producing more intense magnetic fields were required. Fermilab developed 7-m-long supermagnets producing a magnetic field of up to 4.5 T. Figure K.16b shows one of these magnets. The complete accelerator ring, 6.2 km in circumference, has 774 of these magnets plus 240 extra magnets used for focusing the beam of high-energy particles. The ring of supermagnets is placed just below the ring of ordinary magnets (Figure K.16a). Particles are first accelerated in the ordinary ring and then fed into the super ring, where they reach a final energy of 10^6 MeV.

The construction of superconducting coils must overcome some tricky manufacturing problems. The superconducting material commonly employed in these coils is either an alloy of niobium with titanium (NbTi) or else the compound Nb_3Sn. The latter material can withstand the highest current densities and magnetic fields (up to 2×10^5 A/cm^2 and up to 15 T), but it is extremely brittle and cannot be drawn into wires to be wound around a magnet core. Instead, it is necessary to first wind a coil with separate strands of niobium wrapped in tin and then heat the entire coil to fuse these materials into the Nb_3Sn compound.

Figure K.17 shows a cross section through one strand of a superconducting cable of NbTi alloy. The alloy is in the form of very thin filaments embedded in a copper matrix. A bundle of thin filaments has more surface area than a single thick wire of the same cross section as the bundle; since the supercurrents always run on the surface of the superconductor, the large area of the filaments permits the flow of a large current. The complete cable consists of several strands of NbTi alloy and several wires of copper wound around

(a)

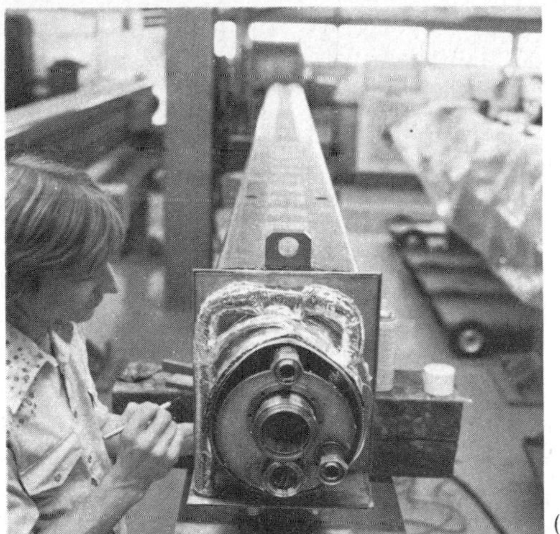

(b)

Fig. K.16 (a) The tunnel of the main accelerator at Fermilab. The upper ring consists of ordinary magnets; the lower ring consists of superconducting magnets. (b) One of the superconducting magnets at Fermilab.

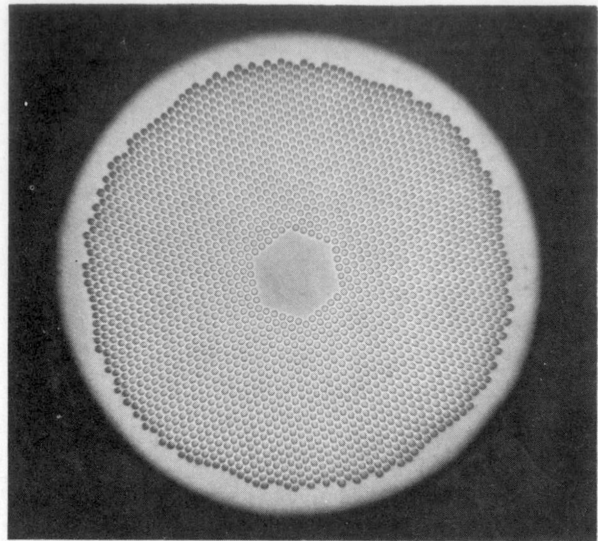

Fig. K.17 This strand of a superconducting cable consists of 2100 thin filaments of NbTi alloy embedded in a copper matrix. The diameter of this strand is about 0.7 mm.

each other. As long as the NbTi alloy is in its superconducting state, the copper plays no role; in fact, relative to the superconductor, the copper behaves as an insulator and all the current flows in the superconductor. But if the superconductor should ever fail because of inadequate cooling or excessive magnetic fields, the current can continue to flow in the parallel wires of copper. This protects the coils of the magnet from the explosive conversion of magnetic energy into heat that would occur if a large current (10^5 A or more) were suddenly halted by a large resistance.

Whereas electromagnets are among the least efficient of our technological devices, electric generators and electric motors are among the most efficient. Typically, an ordinary electric generator will convert close to 99% of the supplied mechanical power into electric power. Electric generators with superconducting coils could probably be even more efficient, but their main advantage is their smaller size and smaller weight — for a given power output, the weight of a superconducting generator can be 10 times less than that of a conventional generator. Several prototypes of superconducting generators and motors have been built (Figure K.18). Superconducting generators may find application in nuclear power plants because conventional generators capable of handling the very large power output of these plants must be of huge size and they therefore face serious structural problems because of the weight of the moving parts. Superconducting motors may also find applications in marine propulsion where, again, weight is a problem.

The transmission and storage of electric power are another important application of superconductors. Resistive losses in conventional transmission lines are of the order of 10% of the total power transmitted. A superconducting cable, with zero resistance, would of course eliminate the loss. However, from a financial point of view, the elimination of the resistive loss makes only a minor contribution to the thriftiness of superconducting cables; what makes a major contribution is their compact size — large amounts of power

Fig. K.18 Rotor (center) of an experimental superconducting electric generator (right), capable of delivering 20,000 kW.

can be carried by a single cable of fairly small dimensions that can be installed in an underground tunnel not requiring the expensive strips of real estate on which conventional transmission lines are erected. Figure K.19 shows an experimental version of a superconducting transmission line; the outside diameter of the pipe enclosing the cable is 40 cm and it can carry a power of 1000 MW, roughly the total power of a nuclear generating station.

Incidentally, an AC current in a superconductor of the second kind is not entirely free of resistive losses. The trouble is that the periodic oscillation of the current produces a similar oscillation of the strength of the magnetic field and consequently lines of magnetic flux (vortex lines) move periodically in and out of the superconductor. This motion of the vortex lines involves some "friction" and it generates some heat. To avoid this loss, the AC current must be carried by a superconductor of the first kind, without magnetic flux in its interior.

On conventional transmission lines, the electric power is carried at very high voltage to minimize the resistive loss. In a superconducting cable, high voltage is not necessary — one can dispense with the awkward step-up and step-down transformers at the generating plant and at the consumer station. Furthermore, if transformation is not required, it becomes advantageous to transmit DC power rather than AC power. A superconducting cable for DC power is cheaper than a cable for AC power; it can be made of a superconductor of the second kind which will tolerate higher current densities and, furthermore, the design can be kept simpler.

Another possible application of superconductivity is the storage of electric energy. Because of daily fluctuations in the demand for electric power by a city, it is desirable to hold large amounts of electric energy temporarily in some kind of storage system. The power can be fed into the storage system when the demand is low and released when the demand is high. Table K.2 compares the energy density in several systems. It is obvious from this table that magnetic fields constitute one of the most concentrated forms of stored energy.

Table K.2 ENERGY STORAGE

System	*Energy density*
Magnetic fields, 10 T	11 kW · h/m^3
Electric field, 10^5 kV/m	0.01
Water in reservoir, height 100 m	0.27
Compressed air, 50 atm	5
Hot water, 100°C	18[a]

[a] This number takes into account that the thermodynamic conversion efficiency for this heat into mechanical energy is at most 20%.

Fig. K.19 Superconducting power transmission line at Brookhaven National Laboratory.

In order to store 100 MW · h, the volume of the magnetic field (at 10 T) would have to be about 10^4 m^3. A large toroid about 5 m thick and 100 m across would provide this volume. The windings of the toroid would of course have to be made of superconductor. The toroid would have to withstand enormous forces due to the magnetic interaction of the currents. And it would be exposed to a severe danger: if, by some failure, one part of the winding became resistive, the sudden generation of heat would probably cause the entire toroid to become resistive and the sudden conversion of all of the magnetic energy into heat would then melt the toroid and blow it apart with great violence.

As a final example of an application of superconductivity, we briefly consider magnetic suspension of trains. In Section K.1 we saw that a permanent magnet will remain suspended over a superconducting sheet — the former induces currents in the latter and the magnetic force on the currents is repulsive. If we replace the permanent magnet by a superconducting coil with a current (a superconducting magnet), we will get the same repulsion effect. Taking advantage of this, we could attach superconducting coils to the underside of a train and levitate it on superconducting rails. The suspension is frictionless — the train sits on a cushion of magnetic fields and slides without friction. In practice, this ideal arrangement is not feasible because it would be much too expensive to lay down miles and miles of superconducting rails. However, it is also possible to obtain levitation with rails made of an ordinary conductor. In this case, there is no repulsive force between the coils and the rails when they are at rest, but there is a force when they are in motion relative to one another because the moving magnetic field induces currents in the rails. These currents suffer some resistive loss, but the effective friction would be reasonably small. Engineers estimate that at speeds in excess of 200 km/h, a train riding on superconductive coils would be much safer than a train on iron wheels.

Further Reading

The Quest for Absolute Zero by K. Mendelssohn (Halsted Press, New York, 1977) is a splendid account of the exciting discoveries in low-temperature physics; it includes a chapter on superconductivity. *Near Zero* by D. K. C. MacDonald (Doubleday, Garden City, 1961) is a much briefer and very elementary introduction. *Cryophysics* by K. Mendelssohn (Interscience Publishers, New York, 1960) is an advanced textbook that gives a concise survey of low-temperature physics; although the book is aimed a senior-level students, many sections are accessible to lower-level students.

Superconductivity by A. W. B. Taylor (Wykeham, London, 1970) is a concise introductory survey with a minimum of mathematics. *Superconductivity* by D. Schoenberg (Cambridge University Press, Cambridge, 1965) is a somewhat more advanced survey; written long before the days of the BCS theory, it is somewhat outdated, but its classic description of the phenomenological properties of superconductors remains useful.

The following magazine articles deal with applications of superconductivity:

"Large-scale Applications of Superconductivity," B. B. Schwartz and S. Foner, *Physics Today*, July 1977

"Superconductors in Electric-Power Technology," T. H. Geballe and J. K. Hulm, *Scientific American*, November 1980

"Superconducting Electronics," D. G. McDonald, *Physics Today*, February, 1981

"The Road to Superconducting Materials," J. K. Hulm, J. E. Kunzler, and B. T. Matthias, *Physics Today*, January 1981

"The Superconducting Computer," J. Matisoo, *Scientific American*, May, 1980

Questions

1. Why would you expect helium to be more difficult to liquify than any other gas?

2. The graph in Figure 28.5 gives the impression that the resistance of tin is constant for $T > 3.72$ K. Is this true?

3. Is Ohm's Law valid for a superconductor?

4. Is it reasonable to regard superconductors as another state of matter? If so, how many states of matter are there?

5. Is a ring of superconductor with a persistent current a perpetual motion machine, i.e., does it violate the law of conservation of energy?

6. Suppose that a closed ring of superconducting wire is initially outside of a magnetic field. If we push the ring into the magnetic field, the magnetic flux through the ring remains zero. Explain.

7. Figure K.6b shows a superconducting cylinder that has expelled the magnetic field by the Meissner effect. Describe the currents that must be flowing along the surface of this cylinder.

8. Suppose that a piece of metal immersed in a magnetic field is initially a normal conductor, and is then cooled so that it becomes a superconductor. To expel the magnetic field by the Meissner effect, the metal must acquire electric currents. Where does the energy for these currents come from?

9. If the resistance of a superconductor were small but not zero, could the Meissner effect be sustained for a long time?

10. The levitated magnet shown in Figure K.2 was placed near the superconducting dish *after* the dish became superconducting. Consequently, does Figure K.2 demonstrate infinite conductivity, or the Meissner effect, or both?

11. (a) If you push a ring of superconductor into a magnetic field, a current will be induced along the ring; if you then remove the ring from the magnetic field, the current disappears. Explain. (b) If you place a ring of normal conductor in a magnetic field, then cool it so it becomes a superconductor, and then remove it from the magnetic field, the ring will be left with a persistent current. Explain.

12. Suppose that a very large flat metallic plate is placed between the poles of an electromagnet. If this plate is cooled so that it becomes superconducting, it will not expel the magnetic field lines. Why not?

13. The equation describing the graph in Figure K.4 is $B_c(T) = (1 - T^2/T_c^2)B_c(0)$. At what temperature is $B_c(T) = \frac{1}{2}B_c(0)$?

14. The existence of a critical magnetic field implies the existence of a critical current. For a cylindrical wire of radius r, the critical current is given by the equation $I_c = 2\pi r B_c/\mu_0$, known as the Silsbee rule. Explain why the critical current is proportional to the radius of the wire.

15. Would you expect that a noncrystalline material, such as glass, could exhibit superconductivity?

16. Consider a pair of electrons moving around a circular superconducting ring carrying a current. According to the discussion in Section 41.4, the orbital angular momentum of the electrons is quantized. Explain how this leads to the quantization of the electric current and the magnetic flux associated with this current.

17. A superconducting cable carrying a large current and a copper cable are connected in parallel. Is it really true that the copper cable carries no current at all? (Hint: What is the potential difference from one end of the superconducting cable to the other?)

CHAPTER 35

The Displacement Current and Maxwell's Equations

James Clerk Maxwell, *1831–1879, Scottish physicist. He was professor at King's College, London, and later professor at Cambridge, where he supervised the construction of the Cavendish Laboratory. Maxwell at first attempted to explain the behavior of electric and magnetic fields in terms of a complicated mechanical model, according to which all of space was filled with an elastic medium, or ether, consisting of a multitude of small rotating vortex cells. But ultimately Maxwell discarded the mechanical model and treated the electric and magnetic fields as physical entities that exist in their own right, without any need for an underlying medium. He published his electromagnetic theory in 1873 in his celebrated* Treatise on Electricity and Magnetism. *In this book he laid down a complete and consistent set of laws for all electromagnetic phenomena, and thereby accomplished for electricity and magnetism what Newton had accomplished for mechanics.*

We already know that a changing magnetic field will induce an electric field. In this chapter we will discover that the converse is also true: a changing electric field will induce a magnetic field. The law describing this induction effect of electric fields was formulated by James Clerk Maxwell, who thereby achieved a wide-ranging unification of the laws of electricity and magnetism. The next four chapters of this book are nothing but applications of these basic laws.

The mutual induction of electric and magnetic fields gives rise to the phenomenon of self-supporting electromagnetic oscillation in empty space. If, initially, there exists an oscillating electric field, it will induce a magnetic field, and this will induce a new electric field, and this will induce a new magnetic field, etc. Thus, these fields can perpetuate each other. Of course, an oscillating charge or current is needed to get the fields started, but after this initiation the fields continue on their own. These self-supporting oscillations are **electromagnetic waves,** either traveling waves or standing waves. We will examine such electromagnetic waves in the space within a conducting cavity and in the space surrounding an accelerated electric charge.

35.1 The Displacement Current

Ampère's Law asserts that the integral of **B** around any closed path is related to the current intercepted by an arbitrary surface that spans this path:

$$\oint \mathbf{B} \cdot d\mathbf{l} = \mu_0 I \tag{1}$$

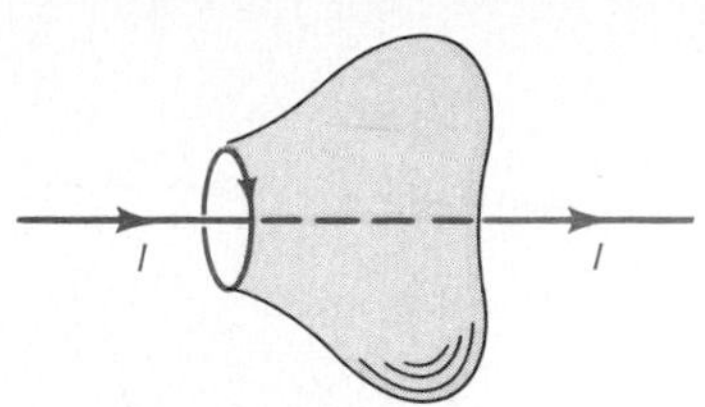

Fig. 35.1 A circular path (color) is spanned by a baglike mathematical surface. The wire carrying the current enters the "mouth" of the bag and is intercepted by the bottom of the bag. Alternatively, the circular path can be spanned by a flat circular surface. The same current will then be intercepted by this flat circular surface.

If all the currents flow along closed, continuous circuits, then the shape of the surface used to intercept the current is irrelevant. For example, Figure 35.1 shows a circular path of integration spanned by a baglike curved surface with a circular mouth. This baglike surface intercepts the same current as a flat circular surface spanning the mouth — both surfaces give the same result in Eq. (1).

However, if the wire carrying the current is interrupted by a capacitor, then Ampère's Law develops a flaw. Suppose that a small capacitor is connected to long, straight wires as in Figure 35.2. Suppose further that the wires carry a constant current I which steadily charges up the capacitor. (This of course requires that the voltage of the source of emf driving the current steadily increases so as to allow for the steadily increasing voltage on the capacitor.) The distribution of current on the interrupted wire of Figure 35.2 differs from that on the continuous wire of Figure 35.1 only in that a short piece of current is missing (between the capacitor plates). Except in the immediate vicinity of the capacitor, the magnetic field in Figure 35.2 will therefore be about the same as in Figure 35.1. Let us now apply Ampère's Law to the baglike surface shown in Figure 35.2. This surface passes between the capacitor plates and it therefore intercepts *no current*. Hence the right side of Eq. (1) is zero, but the left side is obviously not zero — Ampère's Law fails.

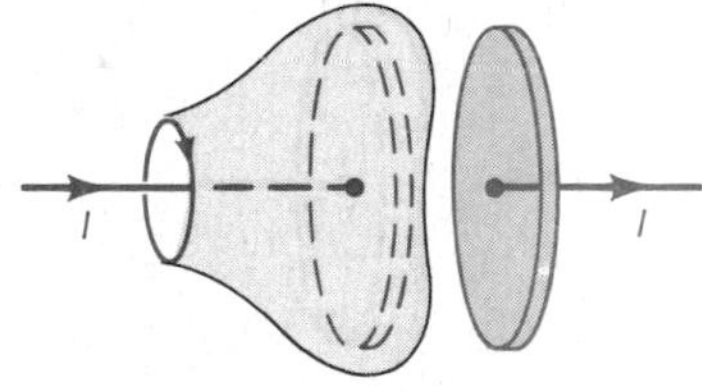

Fig. 35.2 A similar baglike surface that passes between the plates of a capacitor. The baglike surface intercepts no current.

The question is then, can we modify Ampère's Law so that it gives the right answer even when the surface passes between capacitor plates? To discover the required modification, we note that although the surface does not intercept current in the region between the capacitor plates, it does intercept *electric flux* (Figure 35.3). The positive plate of the capacitor is a source of electric field lines and the negative plate is a sink; field lines going from one plate to the other cross the baglike surface. This suggests that the required correction to Ampère's Law involves the electric flux Φ.

It is not hard to see that to achieve this correction for our baglike surface, the right side of Eq. (1) should be replaced by $\mu_0 \varepsilon_0 \, d\Phi/dt$,

$$\oint \mathbf{B} \cdot d\mathbf{l} = \mu_0 \varepsilon_0 \frac{d\Phi}{dt} \tag{2}$$

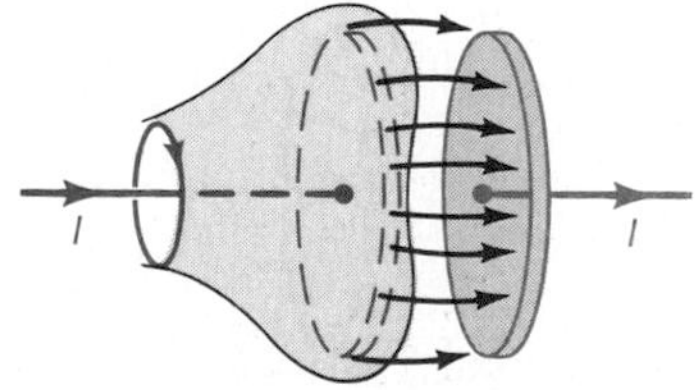

Fig. 35.3 The baglike surface intercepts electric flux.

To check that this works, let us calculate the amount of flux. If the electric charge on the positive plate is Q, then Q/ε_0 lines of field start on this plate. All these lines cross the baglike surface, i.e., the flux is

$$\Phi = \frac{Q}{\varepsilon_0} \tag{3}$$

The rate of change of the flux is then

$$\frac{d\Phi}{dt} = \frac{1}{\varepsilon_0} \frac{dQ}{dt} = \frac{1}{\varepsilon_0} I \tag{4}$$

Hence the term $\mu_0 \varepsilon_0 \, d\Phi/dt$ on the left side of Eq. (2) equals

$$\mu_0 \varepsilon_0 \frac{d\Phi}{dt} = \mu_0 \varepsilon_0 \frac{1}{\varepsilon_0} I = \mu_0 I \tag{5}$$

This shows that the calculation with the baglike surface passing between the plates and intercepting the electric flux [Eq. (2)] gives the same result as a calculation with some other surface cutting across the wire and intercepting the current [Eq. (1)].

It can of course also happen that electric flux and current are present at the same place; for example, if the space between the capacitor plates is filled with some slightly conducting material (a leaky capacitor), then a surface passing between the plates will intercept both some electric flux and some current. In such a case, this electric flux and this current must be combined,

Maxwell's modification of Ampère's Law

$$\oint \mathbf{B} \cdot d\mathbf{l} = \mu_0 I + \mu_0 \varepsilon_0 \frac{d\Phi}{dt} \tag{6}$$

where it is understood that I is the current intercepted by whatever surface is used in the calculation [thus, the current I in Eq. (6) is not necessarily the same as the current on the wires]. Equation (6) is Maxwell's modification of Ampère's Law. The great importance of this equation lies in its general validity — Maxwell boldly proposed that this equation is valid not only for a capacitor, but also for any arbitrary system of electric fields, currents, and magnetic fields. Incidentally, the sign of the electric flux in Eq. (6) is fixed by the following right-hand rule: wrap the fingers around the path of integration, in the direction of integration; the flux is to be reckoned as positive if the electric field is in the direction of the thumb.

The quantity of $\varepsilon_0\, d\Phi/dt$ is called the **displacement current,**

Displacement current

$$I_d = \varepsilon_0 \frac{d\Phi}{dt} \tag{7}$$

This quantity is of course not a true current, but in Eq. (6) it has the same effect as though it were:

$$\oint \mathbf{B} \cdot d\mathbf{l} = \mu_0(I + I_d) \tag{8}$$

Note that the displacement current between the capacitor plates has the same direction as the true current arriving at the plates; the displacement current is the continuation of the real current "by other means."

Apart from the term $\mu_0 I$, the Maxwell–Ampère Law of Eq. (6) is analogous to the Faraday Law of Eq. (32.12). The latter relates the path integral of the electric field to the rate of change of magnetic flux. The former relates the path integral of the magnetic field to the rate of change of the electric flux. Hence these laws indicate a certain symmetry in the effects of electric and magnetic fields on one another.

Just as a time-dependent magnetic field can induce an electric field, a time-dependent electric field can induce a magnetic field. The following example is quite analogous to Example 32.3.

EXAMPLE 1. A parallel-plate capacitor has circular plates of radius R. The capacitor is connected to a source of alternating emf so that the electric field between the plates oscillates according to

$$E = E_0 \sin \omega t \tag{9}$$

where E_0 and ω are constants. What is the induced magnetic field inside and outside the capacitor? Assume that the electric field is uniform inside the capacitor; neglect the fringing of the field at the edges.

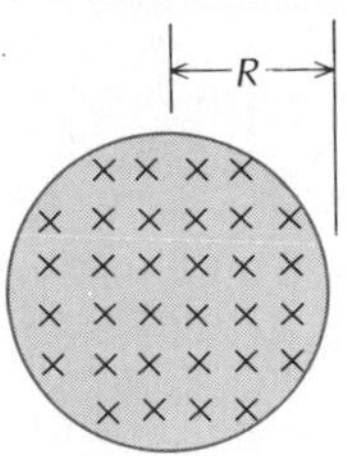

Fig. 35.4 The electric field between the capacitor plates. The crosses show the tails of the electric field vectors.

SOLUTION: Inside the capacitor ($r < R$), the electric field is uniform. Figure 35.4 shows this electric field. To find the induced magnetic field, take a circle of radius r. The flux through this circle is $\pi r^2 E$ and the rate of change of this flux is $\pi r^2\, dE/dt$. According to Eq. (6), we then have

$$\oint \mathbf{B} \cdot d\mathbf{l} = \mu_0 \varepsilon_0 \pi r^2\, dE/dt \tag{10}$$

The field lines of **B** must be closed concentric circles, because this is the only pattern of field lines consistent with the rotational symmetry of the electric field shown in Figure 35.4. Since the path of integration coincides with a field line, we can immediately evaluate the left side of Eq. (10),

$$\oint \mathbf{B} \cdot d\mathbf{l} = 2\pi r B \tag{11}$$

Combining Eq. (10) and (11), we then obtain

$$2\pi r B = \mu_0 \varepsilon_0 \pi r^2 \frac{dE}{dt} \tag{12}$$

or

$$B = \frac{\mu_0 \varepsilon_0}{2} r \frac{dE}{dt} \tag{13}$$

$$= \frac{\mu_0 \varepsilon_0}{2} r \omega E_0 \cos \omega t \tag{14}$$

Outside the capacitor ($r > R$), the electric field is zero. Hence the electric flux through a circle of radius r is $\pi R^2 E$. We can then obtain the magnetic field by going through the same steps as above, with the result

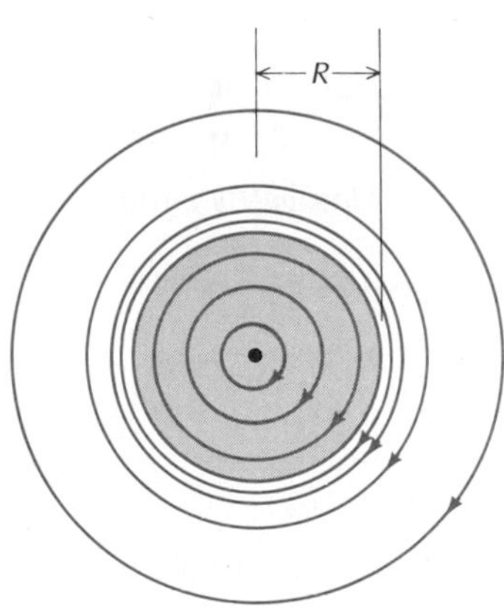

Fig. 35.5 The magnetic field between the capacitor plates.

$$B = \frac{\mu_0 \varepsilon_0}{2} \frac{R^2}{r} \frac{dE}{dt} \tag{15}$$

$$= \frac{\mu_0 \varepsilon_0}{2} \frac{R^2}{r} \omega E_0 \cos \omega t \tag{16}$$

The formulas (14) and (16) are obviously analogous to (32.20) and (32.21).

Figure 35.5 shows the magnetic field lines. The direction of these lines is related to that of the current flowing toward the capacitor plates by the usual right-hand rule. Note that the magnetic field given by Eq. (16) outside the capacitor is exactly that of a long, straight wire ($B \propto 1/r$). As far as this magnetic field is concerned, the removal of a short piece of wire and its replacement by a capacitor makes no difference at all.[1]

[1] However, a more precise calculation, taking into account the fringing of the electric field at the edges of the capacitor plates, shows that near the plates the magnetic field differs slightly from that of a straight wire.

35.2 Maxwell's Equations

Equation (6) is the last of the fundamental laws that we need to completely describe the behavior of electric and magnetic fields. There are four fundamental laws, as follows[2]:

Gauss' Law for electricity [Eq. (24.6)]

$$\oint \mathbf{E} \cdot d\mathbf{S} = Q/\varepsilon_0 \tag{17}$$

Gauss' Law for magnetism [Eq. (30.17)]

$$\oint \mathbf{B} \cdot d\mathbf{S} = 0 \tag{18}$$

Faraday's Law [Eq. (32.12)]

$$\oint \mathbf{E} \cdot d\mathbf{l} = -\frac{d\Phi_B}{dt} \tag{19}$$

Maxwell–Ampère's Law [Eq. (6)]

$$\oint \mathbf{B} \cdot d\mathbf{l} = \mu_0 I + \mu_0 \varepsilon_0 \frac{d\Phi}{dt} \tag{20}$$

Maxwell's equations

Taken as a whole, these laws are known as **Maxwell's equations,** because Maxwell added the missing link between the magnetic and the electric fields [Eq. (20)], and thereby brought electromagnetic theory to completion. Maxwell recognized that these equations imply a dynamic interplay between electric and magnetic fields, an interplay that couples and unifies electric and magnetic phenomena.

The physical basis for each of these four equations may be briefly summarized as follows. Gauss' Law for electricity is based on Coulomb's Law describing the forces of attraction and repulsion between stationary charges. Gauss' Law for magnetism asserts that there exist no magnetic monopoles. Faraday's Law describes the induction of an electric field by motion or by a changing magnetic field. And finally, Maxwell–Ampère's Law is based on the law of magnetic force between moving charges and it also contains the induction of a magnetic field by a changing electric field.

The above equations must be supplemented by the equation for the **Lorentz force** on a point charge:

[2] These equations are valid for vacuum. The equations in the presence of dielectric or magnetic materials must be modified either by implicitly including the bound charges of the dielectric in Q and the currents of the magnetic material in I, or else by inserting the dielectric constant and the relative permeability into the appropriate terms of the equations.

$$\boxed{\mathbf{F} = q\mathbf{E} + q\mathbf{v} \times \mathbf{B}} \tag{21}$$

Lorentz force

This expression is based on the definition of **E** and **B** [see Eqs. (23.15) and (30.12)].

Maxwell's equations give a complete description of the interactions among charges, currents, electric fields, and magnetic fields. All the properties of the fields can be deduced by mathematical manipulation of these equations. If the distribution of charges and currents is known, then these equations uniquely determine the corresponding fields. Even more important, Maxwell's equations uniquely determine the time evolution of the fields, starting from a given initial condition of these fields. Thus, these equations accomplish for the dynamics of electromagnetic fields what Newton's equations of motion accomplish for the dynamics of particles.

Calculations with Maxwell's equations are often best done by converting the integral equations (17)–(20) into *differential equations;* however, the latter involve partial derivatives and for this reason we will not attempt to use them here.

Although the empirical foundation on which we based the development of Maxwell's equations was restricted to charges at rest or charges in uniform motion, these equations also govern the fields of accelerated charges and the fields of light and radio waves. In the remaining sections of this chapter and in the next chapters we will calculate these fields from our equations and we will see that the results are in agreement with the observed properties of light and radio waves.

35.3 Cavity Oscillations

As we saw in Example 1, the time-dependent electric field in a parallel-plate capacitor will induce a magnetic field. This implies that the capacitor has some self-inductance, since the changing, time-dependent magnetic field will, in turn, induce an electric field, i.e., it will induce an emf. We therefore expect that a parallel-plate capacitor placed in a circuit with a source of alternating emf will behave very much like the driven LC circuit of Section 34.3. The following is a calculation of the oscillations of a parallel-plate capacitor.[3] However, instead of treating this system as an LC circuit with some effective inductance, we will directly examine the electric and magnetic fields between the plates and their dependence on position and time.

Figure 35.6 shows a parallel-plate capacitor with circular plates connected to a source of alternating emf of frequency ω. The electric field between the plates will then oscillate at the same frequency. As a first approximation, we ignore the induction effects within the capacitor. The electric field between the plates is then uniform[4] and parallel to the axis. We use the symbol $E_{(1)}$ for this field and we write

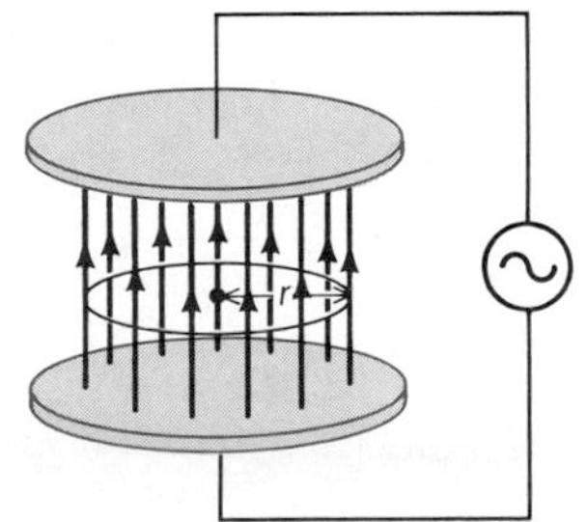

Fig. 35.6 Electric field lines in a parallel-plate capacitor connected to a source of alternating emf. The fringing of the electric field lines has been ignored.

$$E_{(1)} = E_0 \sin \omega t \tag{22}$$

where E_0 is a constant.

[3] Based on *The Feynman Lectures on Physics* by R. P. Feynman, R. B. Leighton, and M. Sands.

[4] As always, we disregard the fringing effects at the edges of the capacitor.

According to Example 1, such a time-dependent electric field induces a magnetic field [see Eq. (14)]

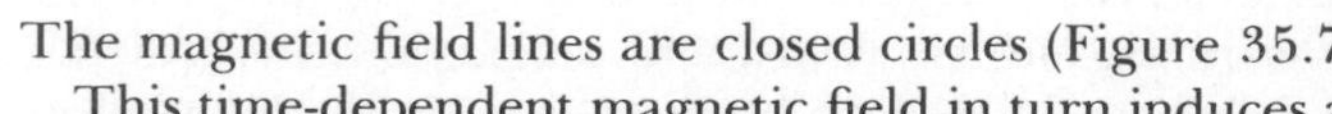

$$B_{(1)} = \frac{\mu_0 \varepsilon_0}{2} r\omega E_0 \cos \omega t \tag{23}$$

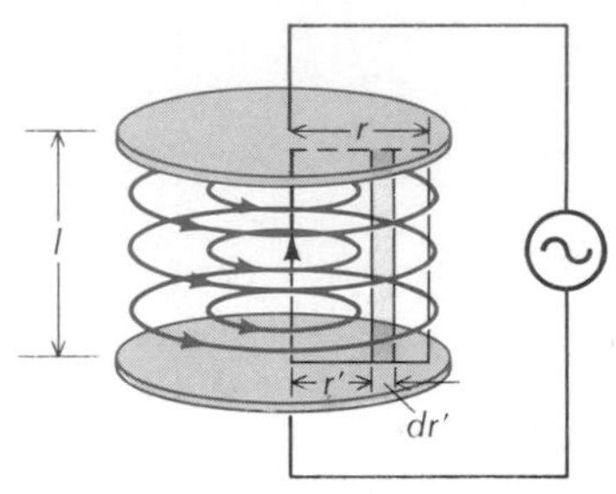

Fig. 35.7 Magnetic field lines. The path of integration is a rectangle of height l and width r. The direction of integration is clockwise.

The magnetic field lines are closed circles (Figure 35.7).

This time-dependent magnetic field in turn induces an extra electric field, which will have to be added to the field $E_{(1)}$ of Eq. (22). The direction of the extra field is again parallel to the axis. To find an expression for this field, let us apply Faraday's Law to the path shown in Figure 35.7. If we use the symbol $\mathbf{E}_{(2)}$ for the induced electric field, then

$$\int \mathbf{E}_{(2)} \cdot d\mathbf{l} = -\frac{d\Phi_{B_{(1)}}}{dt} \tag{24}$$

The integral on the left side does not receive any contribution from the horizontal segments (Figure 35.7), since there the field is perpendicular to the path. Furthermore, we will suppose that $E_{(2)} = 0$ along the axis of the capacitor,[5] so that the integral receives no contribution from the vertical segment along the axis (Figure 35.7). Only the other vertical segment remains. If the latter is at a distance r from the axis, the left side of Eq. (24) becomes

$$-E_{(2)}(r)\, l \tag{25}$$

where l is the distance between the plates (the minus sign is needed because we reckon the electric field as positive if it is directed upward, whereas the segment of path is directed downward).

To evaluate the right side of Eq. (24), we begin by calculating the flux $\Phi_{B_{(1)}}$. Consider a strip of height l and width dr'; the flux through this strip is $B_{(1)} l\, dr'$ and the flux through all such strips is the integral

$$\Phi_{B_{(1)}} = \int_0^r B_{(1)}\, l\, dr' \tag{26}$$

$$= \int_0^r \left(\frac{\mu_0 \varepsilon_0}{2} r' \omega E_0 \cos \omega t \right) l\, dr'$$

$$= \frac{\mu_0 \varepsilon_0}{2} l \omega E_0 \cos \omega t \int_0^r r'\, dr'$$

$$= \frac{\mu_0 \varepsilon_0}{4} l r^2 \omega E_0 \cos \omega t \tag{27}$$

Hence

$$\frac{d\Phi_{B_{(1)}}}{dt} = -\frac{\mu_0 \varepsilon_0}{4} l r^2 \omega^2 E_0 \sin \omega t \tag{28}$$

---

[5] The choice $E_{(2)} = 0$ along the axis is a matter of definition. It means that we regard $E_{(1)}$ of Eq. (22) as the *exact* electric field along the axis of the capacitor, while we regard the deviations from $E_{(1)}$ occurring at all other points as induced corrections which remain to be calculated.

Upon inserting Eqs. (25) and (28) into Eq. (24), we immediately find the induced electric field

$$E_{(2)} = -\frac{\mu_0\varepsilon_0}{4} r^2\omega^2 E_0 \sin \omega t \tag{29}$$

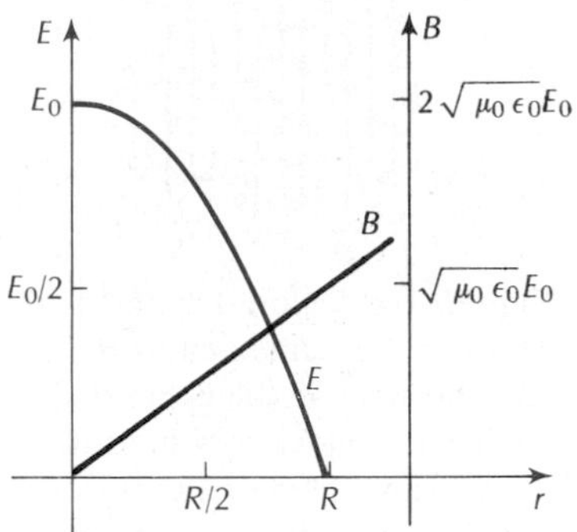

Fig. 35.8 Electric field (black) and magnetic field (color) in a parallel-plate capacitor as a function of radius.

Thus, the electric field $E_{(1)}$ must be corrected by adding to it the electric field $E_{(2)}$:

$$E = E_0 \sin \omega t - \frac{\mu_0\varepsilon_0}{4} r^2\omega^2 E_0 \sin \omega t + \cdots$$

$$= \left(1 - \frac{\mu_0\varepsilon_0}{4} \omega^2 r^2\right) E_0 \sin \omega t + \cdots \tag{30}$$

But the correction to the electric field implies a corresponding correction to the induced magnetic field, and this implies a further correction to the electric field, etc. This calculation therefore leads to an infinite sequence of successive approximations for the fields. The result takes the form of a power series in ascending powers of the radial variable r. The two terms in parentheses in Eq. (30) are merely the first two terms of the power series; the other terms have been indicated by the three dots in Eq. (30). For a start let us ignore all the extra corrections and use Eq. (30) as it stands. One interesting feature of this solution is that the electric field vanishes at a radius R such that

$$1 - \frac{\mu_0\varepsilon_0}{4} \omega^2 R^2 = 0 \tag{31}$$

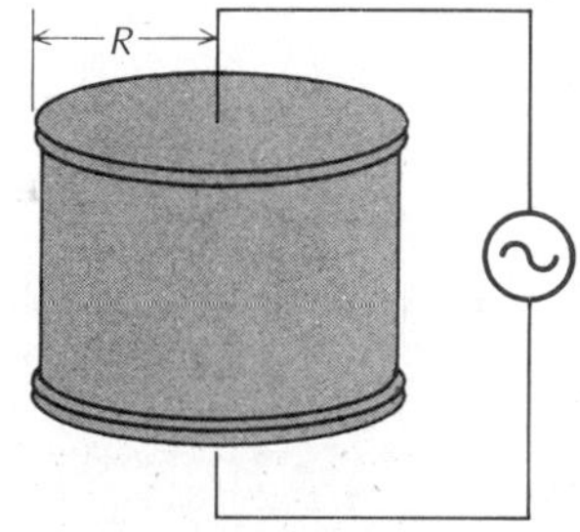

Fig. 35.9 A conducting cylindrical wall of radius R closes off the sides of the parallel-plate capacitor.

i.e.,

$$R = \frac{2}{\sqrt{\mu_0\varepsilon_0}} \frac{1}{\omega} \tag{32}$$

A more exact calculation taking into account the extra terms in Eq. (30) shows that the precise radius at which the electric field vanishes is somewhat larger,

$$R = \frac{2.405}{\sqrt{\mu_0\varepsilon_0}} \frac{1}{\omega} \tag{33}$$

Figure 35.8 is a plot of the radial variation of the electric and magnetic fields.

Cavity oscillations

The above solution for the fields between capacitor plates also gives us the solution for another problem, one of much greater practical importance: the problem of the electromagnetic oscillations of a closed conducting cavity. We can convert our pair of capacitor plates into a closed cavity by inserting a conducting cylindrical wall of radius R between the plates (Figure 35.9). The cylindrical wall does not interfere with the electric field of Eq. (30) because this electric field is zero anyhow at the radius R. Thus our solution for the fields between capacitor plates is also the solution for the fields in such a cavity. Note that here the choice of the correct radius is crucial: if the cylindrical wall had a radius different from R, then the electric field would not "fit" this cylinder.

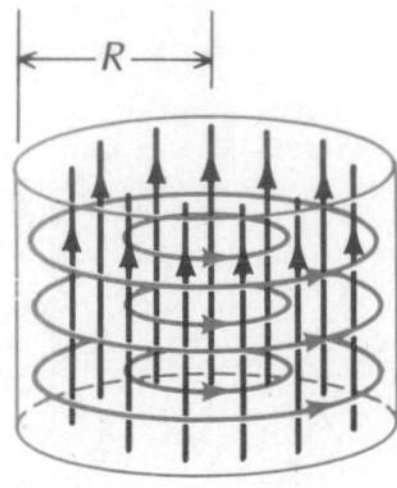

Fig. 35.10 A closed cylindrical can with the same radius R, and the electric (black) and magnetic (color) field lines in its interior. This diagram does not show the time dependence of the fields (both fields reverse direction every half cycle; when the electric field is maximum, the magnetic field is zero and conversely).

Figure 35.10 shows a closed cylindrical can and the lines of electric and magnetic field within it. The external circuit with its driving emf has been omitted in this figure. The currents can flow from one horizontal end face of the cylinder to the other via the vertical wall; the external wire of Figure 35.6 is therefore not needed. And the external source of emf is not needed either — once the oscillations of the cavity have been started, they are self-sustaining. If charge is initially placed on the end faces,[6] then the cavity begins to oscillate spontaneously just as the LC circuit of Section 34.2 begins to oscillate when charge is initially placed on the capacitor. And, just as in the LC circuit, the energy sloshes back and forth between potential energy (in electric fields) and "kinetic" energy (in magnetic fields).

The electromagnetic oscillations in a conducting cavity are analogous to the acoustic vibrations in a closed organ pipe. Equation (33) determines the resonant frequency of a cavity of given radius. This is analogous to Eq. (16.6), which determines the resonant frequency of an organ pipe. The electromagnetic oscillations in a cavity can be regarded as a standing wave (we will discuss electromagnetic waves in the next chapter) and Eq. (33) can then be regarded as the condition for a proper fit of the wave in the cavity. We know from the study of sound that an organ pipe has several possible harmonic standing waves of different resonant frequencies. It turns out that this is also true for an oscillating elecromagnetic cavity, but in order to find the higher harmonics and their frequencies, we would have to calculate the higher-order terms in Eq. (30).

Electromagnetic cavities have many practical applications in the technology of microwaves, that is, high-frequency radio and radar waves. For instance, one of the devices commonly used to generate high-frequency waves is the **klystron**, consisting of a cavity in which electromagnetic oscillations are excited by an incident beam of electrons (Figures 35.11 and 35.12). The excitation of electromagnetic oscillations in a cavity by a beam of electrons is analogous to the excitation of sound oscillations in an organ pipe by a stream of air. The

Fig. 35.11 One of the 245 large klystrons in the gallery at the Stanford Linear Accelerator Center. These klystrons generate the electromagnetic waves used in the accelerator, which is buried 25 ft below the floor.

[6] The charge must be placed on the *interior* of the end faces. If the charge were placed on the exterior, it would not produce any field inside the cavity.

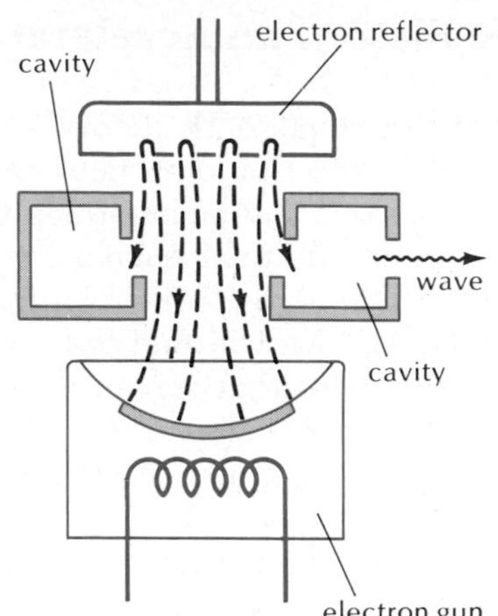

Fig. 35.12 Schematic diagram of a reflex klystron. An electron beam (dashed lines) emerges from an electron gun and moves upward, passing by the open gaps in the walls of the cavity; it is then reflected by an electric field and moves downward, again passing by the gaps. The electric fields of this beam excite oscillations in the cavity.

electrons give up their kinetic energy to the electric and magnetic fields and they thereby increase the strength of the oscillations. As we mentioned above, the oscillations can be regarded as standing waves. A small opening on the side of the klystron (Figure 35.12) permits some of these waves to spill out into space in the form of a traveling radio wave.

Incidentally, in practice the radio waves are not released into space directly from the generator, but are first guided into a dishlike antenna (a radar antenna) mounted at some suitable height. The waves are carried from the generator to the antenna by a hollow conducting tube or **waveguide** (Figure 35.13). Such a tube is nothing but a cavity with two open ends. The mathematical analysis of the electromagnetic oscillations in a waveguide is quite similar to that in a closed cavity — the electric and magnetic fields induce each other just as in the above calculation. However, these oscillations take the form of traveling waves rather than that of standing waves. This can be readily understood in terms of the analogy with sound waves. We know that a standing sound wave in, say, a closed organ pipe consists of two traveling waves of opposite directions (see Section 15.6). If we suddenly open the two ends of such an organ pipe, the standing wave will separate into its traveling parts, each spilling out of one end of the organ pipe.

Waveguides for microwaves are usually copper pipes of rectangular cross section. Figure 35.14 shows the pattern of field lines of a traveling wave in such a waveguide.

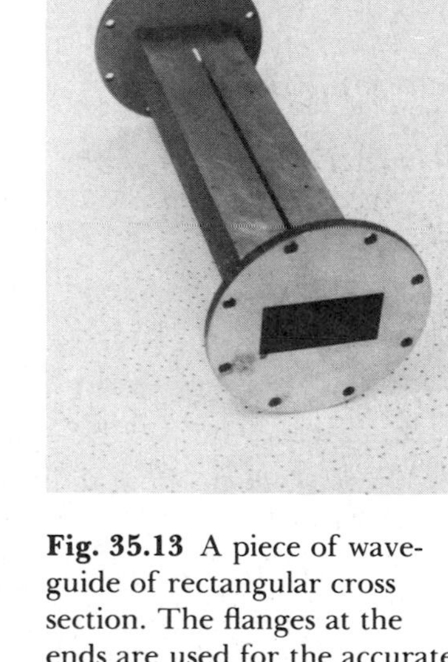

Fig. 35.13 A piece of waveguide of rectangular cross section. The flanges at the ends are used for the accurate joining of adjacent pieces.

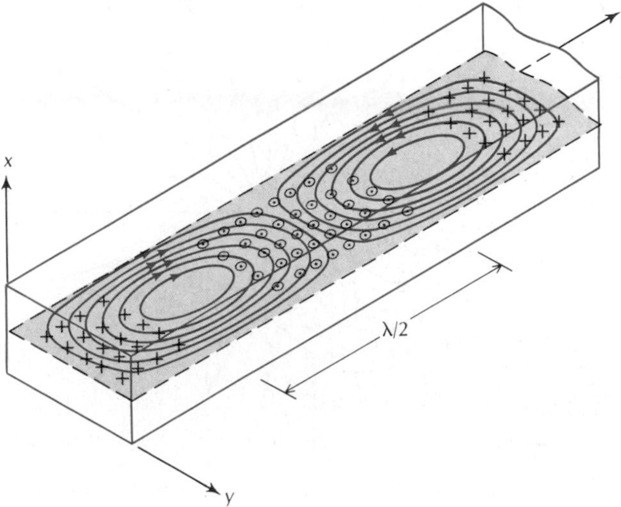

Fig. 35.14 Electric (black) and magnetic (color) field lines in a waveguide. The wave is traveling to the right. The dots and the crosses are the tips and the tails of the electric field lines, respectively. The closed loops are the magnetic field lines.

35.4 The Electric Field of an Accelerated Charge

We started our study of electromagnetic theory with the electric field of a charge at rest — this is the Coulomb field given by Eq. (23.16). Later, we set down the fields of a charge in motion with uniform velocity — in addition to the Coulomb field, such a charge has a magnetic field given by Eq. (30.15). Now we will investigate the fields of a charge with *accelerated motion.* We will find that in this case there are extra electric and magnetic fields that spread outward from the position of the charge, like ripples on a pond in which a stone has been dropped, and carry away energy and momentum. These fields are called **radiation fields.**

We can derive a formula for the electric field surrounding an accelerated charge by the following argument. Suppose that the charge is initially at rest, then is quickly accelerated for some short time interval τ, and then continues to move with a constant final velocity. We can summarize the motion thus:

$t < 0$	charge is at rest	
$0 \leq t < \tau$	charge has acceleration a	
$t \geq \tau$	charge moves at constant velocity $v = a\tau$	(34)

The initial electric field lines originate at the initial position of the charge. But the field lines at some later time ($t > \tau$) must originate on the new position of the charge. The field lines cannot change from their initial configuration to their final configuration instantaneously; rather, a disturbance must travel outward from the position of the charge and gradually change the field lines. The disturbance takes the form of a kink connecting the old and the new field lines.

The disturbance travels at some speed c (we will prove below that the speed of the disturbance is actually the speed of light; our notation anticipates this result). Figure 35.15 shows the situation at some later time. The disturbance begins at the time 0 when the acceleration begins. The leading edge of the disturbance (outer edge of kink)

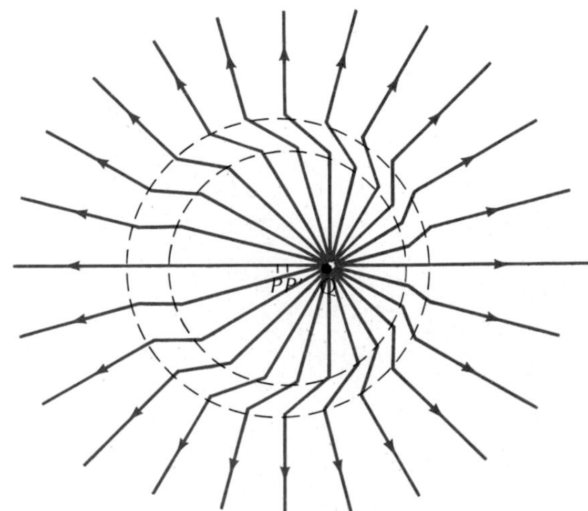

Fig. 35.15 Electric field lines of a charge that has suffered an acceleration. Here Q is the present position of the charge; P is the initial position of the charge. Between P and P' the charge suffered a constant acceleration. Between P' and Q the charge moved at constant velocity. The outer dashed sphere (outer edge of kink) has radius ct and is centered on P; the inner dashed sphere (inner edge of kink) has radius $c(t - \tau)$ and is centered on P'.

travels outward from the initial position of the charge and in a time t it reaches out to a distance ct. Beyond the sphere of radius ct, the electric field is still the old field with field lines centered on the initial postion of the charge.

The disturbance ceases as soon as the acceleration ceases. The field in the vicinity of the uniformly moving charge then settles into the new radial configuration centered on the new position of the charge. The trailing edge of the disturbance (inner edge of kink) marking the cessation of the acceleration travels outward from the position that the charge has at the time τ, and by the time t it reaches out to a distance $c(t - \tau)$. Within the sphere of radius $c(t - \tau)$ the electric field is the new radial field of the uniformly moving charge.

The disturbance produced by the accelerated charge is confined to the space between the larger and smaller spheres in Figure 35.15. The field lines in this zone must connect the lines of the new field of the uniformly moving charge with the lines of the old field of the stationary charge. We have drawn the connecting line segments in Figure 35.15 as straight segments. Strictly, this is only an approximation based on the rationale that a curve can be approximated by a straight line. (A careful argument shows that the segments are actually slightly curved, but the curvature is unimportant if the speed of the charge is much smaller than the speed of light, $v \ll c$).

In the zone of the kink, the electric field has both a radial component and a tangential, or **transverse,** component. This transverse component is characteristic of the field of an accelerated charge.

Figure 35.16 shows one electric field line in detail. This figure relies on the assumption that a long time has elapsed since the acceleration interval, that is, $t \gg \tau$. Consequently, on the scale of this diagram, the distance that the charge has covered while under acceleration is very small compared to the distance that the charge has covered at uniform velocity. The inner and outer circles of Figure 35.16 are then nearly concentric and the distance from the initial position of the charge to the position at time t is nearly vt.[7] Figure 35.16 also contains the implicit assumption that the final speed of the charge is much less than the speed of light, that is, $v \ll c$. (As we remarked above, if this assumption does not hold, then the field line in the zone of the kink is appreciably curved and the straight-line approximation of Figure 35.16 is not satisfactory.)

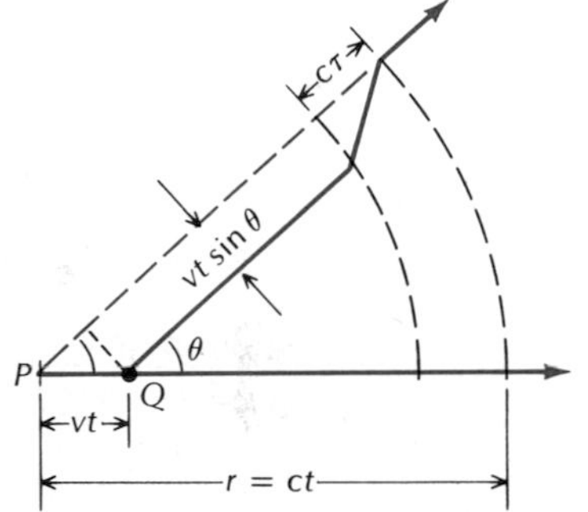

Fig. 35.16 One of the electric field lines (color). This line makes an angle θ with the direction of motion of the charge. Here we have neglected the small distance PP' (see Figure 35.15) that the charge covers while under acceleration; with this approximation, the distance PQ is $\sim vt$.

By inspection of Figure 35.16, we see that the ratio of the transverse and radial components of the electric field in the kink is

$$\frac{E_\theta}{E_r} = \frac{vt \sin\theta}{c\tau} = \frac{at \sin\theta}{c} \tag{35}$$

where, of course, $a = v/\tau$. The radial electric field equals the number of lines passing through a unit area on a sphere; this number is not affected by the angle that the lines make with the radial direction. Thus, E_r is simply the usual Coulomb field:

$$E_r = \frac{1}{4\pi\varepsilon_0} \frac{q}{r^2} \tag{36}$$

Inserting this in Eq. (35), we obtain

[7] The exact distance is $\frac{1}{2}v\tau + v(t - \tau)$ which, for $t \gg \tau$, indeed reduces to $\sim vt$.

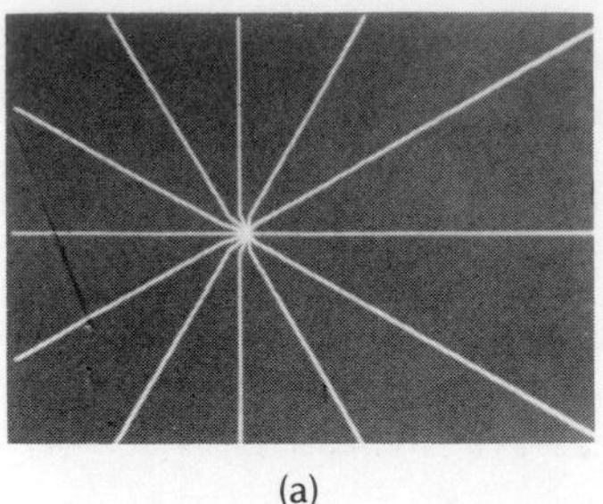

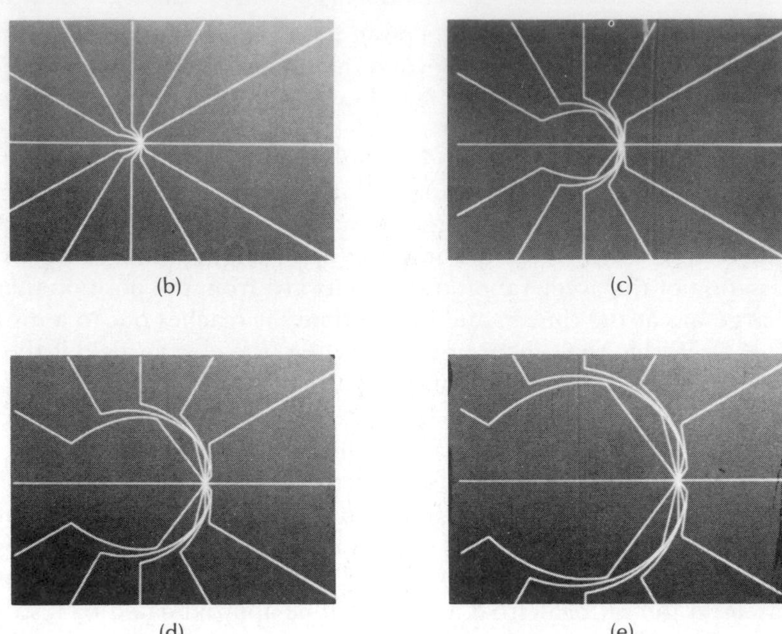

(a) (b) (c) (d) (e)

Fig. 35.17 Time-dependent electric field lines for a charge that suddenly accelerates and then moves with uniform velocity. (a) The charge is at rest. (b) The charge has begun to accelerate. (c)–(e) The charge moves with uniform velocity. Note that the field lines in the region of the kink are curved; this is due to the very high speed of the motion.

$$E_\theta = \frac{1}{4\pi\varepsilon_0}\frac{q}{r^2}\frac{at\sin\theta}{c} \tag{37}$$

Since $t \gg \tau$, the radii of the inner and outer circles in Figure 35.16 are nearly the same, that is, the difference $\Delta r = c\tau$ between them is very small compared to the radius $r = ct$. We can therefore substitute this value of the radius in Eq. (37) so that

Radiation field of accelerated charge

$$\boxed{E_\theta = \frac{1}{4\pi\varepsilon_0}\frac{qa\sin\theta}{c^2 r}} \tag{38}$$

This transverse electric field is the **radiation field** of the accelerated charge. Note that it is directly proportional to the acceleration a. Also note that it varies as the inverse distance, not the inverse square of the distance. The radiation field therefore decreases less sharply with distance than the Coulomb field — it remains significant at large distances where the Coulomb field practically disappears.

The radiation field produced by an acceleration a acting at time 0 reaches the distance r at a time r/c after time 0, i.e., the field needs time to propagate from the position of the charge to a remote point. Mathematically, we can express this retardation by writing the acceleration as a function of time:

$$E_\theta(t, r) = \frac{1}{4\pi\varepsilon_0}\frac{q\sin\theta}{c^2 r}\,a\left(t - \frac{r}{c}\right) \tag{39}$$

This formula indicates that the value of the function E_θ at time t is related to the value of the function a at the earlier time $t - r/c$. For example, if $r = 3.8 \times 10^8$ m (the distance from the Earth to the Moon), then $r/c = 3.8 \times 10^8$ m/(3.0×10^8 m/s) = 1.3 s. Thus the radiation field that an accelerated charge on the Earth produces at the Moon at time t depends on the acceleration at time $t - 1.3$ s, i.e., it depends on

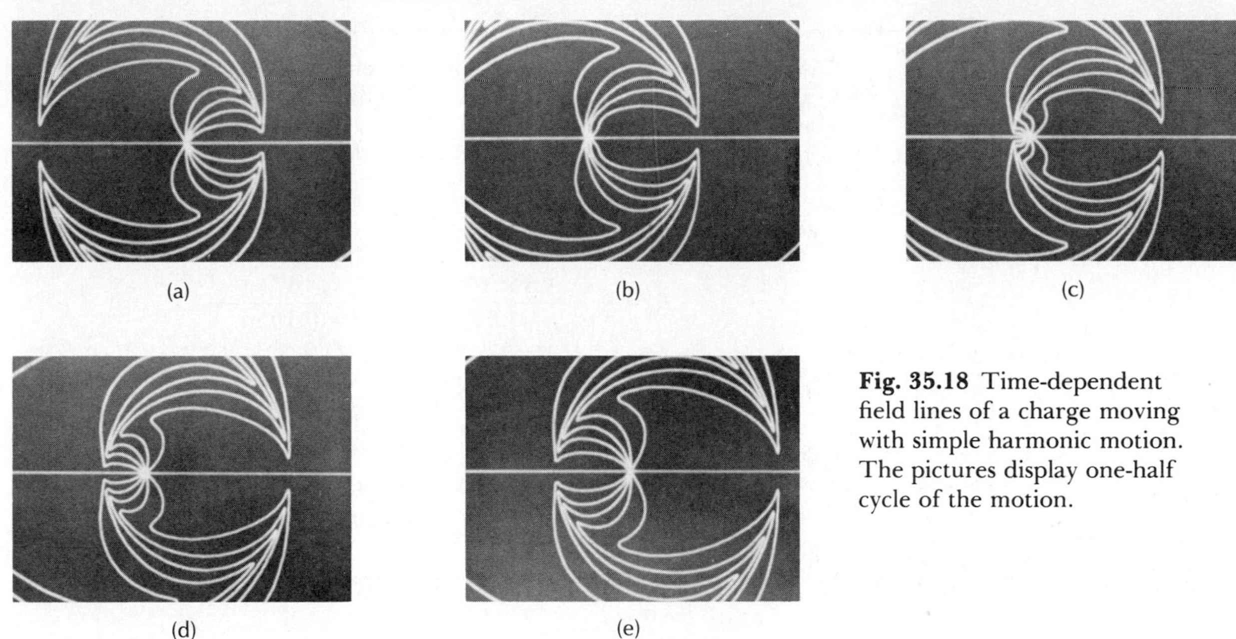

Fig. 35.18 Time-dependent field lines of a charge moving with simple harmonic motion. The pictures display one-half cycle of the motion.

the acceleration at a time 1.3 s earlier, the radiation field needing this much time to propagate from the charge to the Moon.

If the acceleration lasts only a short time, as in the special case shown in Figure 35.15, then the electric radiation field at any point also lasts only a short time. The radiation field described by Eq. (39) then behaves like a single wave pulse that propagates outward from the place at which the acceleration occurred. Figure 35.17 shows a sequence of snapshots from a filmed computer simulation of the electric field lines of an accelerated charge, whose acceleration only lasted a short time. These snapshots give a clear impression of the outward propagation of the kinks in the field lines.

The formula (39) for the radiation field is perfectly general and applies even if the acceleration is not constant and lasts for any length of time. For instance, if the charge oscillates back and forth with simple harmonic motion of frequency ω so that

$$a = a_0 \sin \omega t \tag{40}$$

then

$$\boxed{E_\theta(t,r) = \frac{1}{4\pi\varepsilon_0}\frac{q \sin\theta}{c^2 r} a_0 \sin \omega\left(t - \frac{r}{c}\right)} \tag{41}$$

Radiation field of oscillating charge

This represents a (spherical) harmonic wave of frequency ω traveling in the radial direction with a speed c. The amplitude of the wave gradually decreases ($\propto 1/r$) as the wave spreads out. As we will see in the next chapter, the function (41) describes a radio wave emitted from an antenna or a light wave emitted from an atom.

Figure 35.18 shows a sequence of snapshots from a computer simulation of the electric field lines of a charge oscillating back and forth with simple harmonic motion, as indicated by Eq. (40). The folds in the field lines can be seen to propagate outward, forming a regular succession of wave peaks and wave troughs.

EXAMPLE 2. In a head-on collision with an atom, a fast-moving electron suffers a deceleration of 4.0×10^{23} m/s². What is the electric radiation field generated by the electron at a distance of 10 cm at right angles to the direction of motion? Pretend that classical mechanics can be applied to this problem.

SOLUTION: With $\theta = 90°$, Eq. (38) gives

$$E_\theta = \frac{1}{4\pi\varepsilon_0} \frac{1.6 \times 10^{-19}\ \text{C} \times 4.0 \times 10^{23}\ \text{m/s}^2}{(3.0 \times 10^8\ \text{m/s})^2 \times 0.10\ \text{m}}$$

$$= 6.4 \times 10^{-2}\ \text{V/m}$$

Bremsstrahlung

Radiation emitted by the collision of a beam of fast-moving electrons with the atoms in a target (block of metal) is called **Bremsstrahlung.**[8] This is essentially the mechanism by means of which X rays are produced in an X-ray tube (see Figure C.1).

35.5 The Magnetic Field of an Accelerated Charge[9]

When a charge accelerates, the disturbance traveling outward consists not only of electric fields, but also of magnetic fields. The initial magnetic field of a charge at rest is zero while the final magnetic field of a moving charge is that given by Eq. (30.15). Hence a disturbance must move outward and gradually change the magnetic field from its initial to its final value.

The magnetic field can be regarded as an induced field obeying the Maxwell–Ampère Law

$$\oint \mathbf{B} \cdot d\mathbf{l} = \mu_0 \varepsilon_0 \frac{d\Phi}{dt} \tag{42}$$

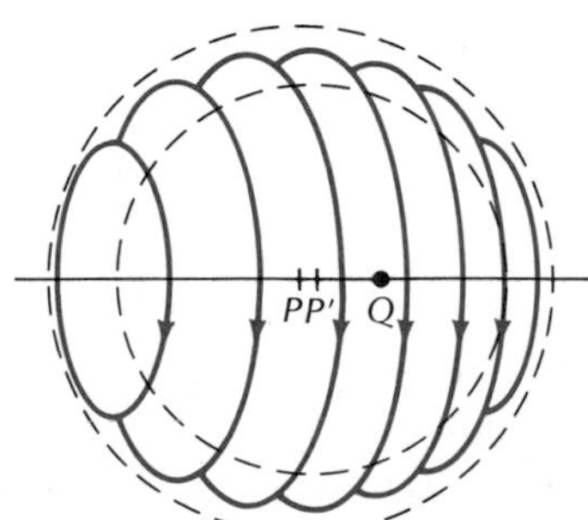

Fig. 35.19 Magnetic field lines in the zone of the kink. These lines are circles around the line of motion. At any point, the magnetic field is perpendicular to the electric field shown in Figure 35.15.

By symmetry, the magnetic field lines are circles about the direction of motion of the charge. Figure 35.19 shows the magnetic field lines in the zone of the kink. Note that the magnetic field is everywhere perpendicular to the electric field. Let us evaluate Eq. (42) for a circular path that follows one of the magnetic field lines (Figure 35.20). The radius of the circle is $r \sin \theta$ and hence the left side of Eq. (42) is

$$\oint \mathbf{B} \cdot d\mathbf{l} = 2\pi r (\sin \theta) B \tag{43}$$

In the evaluation of the right side of Eq. (42) we need to take into account two contributions to the flux: from the radial electric field E_r and from the transverse electric field E_θ. It turns out that the flux produced by the radial electric field induces the familiar magnetic field dependent on velocity, i.e., the magnetic field given by Eq. (30.15). We will not present the detailed calculation establishing this result; instead we leave this calculation as a problem (see Problem 29). The flux produced by the transverse electric field induces a magnetic field dependent on the acceleration. Since in the context of this section we are

[8] German for "braking radiation."

[9] This section is optional.

only interested in the magnetic field characteristic of accelerated motion, we will only take into account the contribution of the flux from this transverse electric field E_θ.

Taking a flat circular area for the evaluation of the flux, we see that this area intercepts the transverse electric field E_θ only in a narrow angular region of radius $r \sin \theta$ and width $c\tau/\sin \theta$ (see Figure 35.20). The area of this region is

$$[\text{area}] \cong [\text{circumference}] \times [\text{width}] = 2\pi r \sin \theta \times \frac{c\tau}{\sin \theta} = 2\pi r c\tau \quad (44)$$

The transverse field E_θ makes an angle of θ with this area and hence the flux is

$$\Delta\Phi = E_\theta \sin \theta \times [\text{area}] = 2\pi r c\tau E_\theta \sin \theta \quad (45)$$

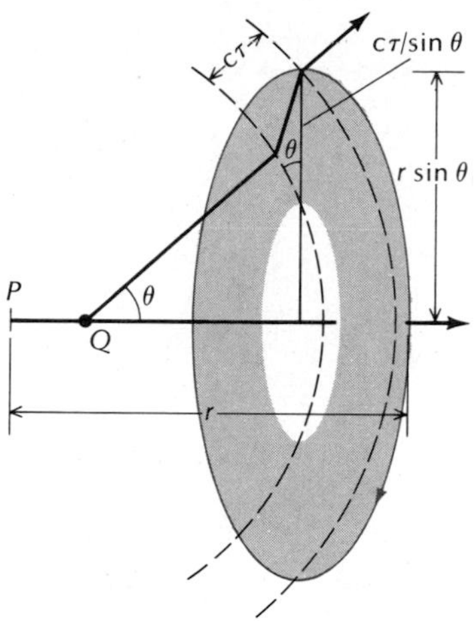

Fig. 35.20 The magnetic field line (color) is a circle of radius $r \sin \theta$ perpendicular to the plane of the page. The area bounded by this field line is a circle of the same radius perpendicular to the plane of the page. This circle intercepts the transverse electric field in the annular region marked with color.

In a time τ, all of this flux disappears since the kink moves beyond the annular area. Hence

$$\frac{d\Phi}{dt} = \frac{\Delta\Phi}{\tau} = 2\pi r c E_\theta \sin \theta \quad (46)$$

Using Eqs. (43) and (46) in Eq. (42), we obtain

$$B = c\mu_0\varepsilon_0 E_\theta \quad (47)$$

This equation expresses the magnetic field of the pulse of radiation in terms of the transverse electric field, the speed of propagation c, and the familiar constants ε_0 and μ_0. As we have pointed out above, the speed of propagation of the pulse coincides with the speed of light,

$$c = 3.00 \times 10^8 \text{ m/s} \quad (48)$$

However, we can do better than this. In what follows we will derive a theoretical expression for the speed of propagation and will confirm that the theoretical prediction agrees with the experimental value given by Eq. (48).

The preceding calculation regarded the magnetic field **B** as induced by $\mathbf{E}_\theta$. But the converse is also true: the electric field $\mathbf{E}_\theta$ is induced by the magnetic field **B** according to Faraday's Law

$$\oint \mathbf{E} \cdot d\mathbf{l} = -\frac{d\Phi_B}{dt} \quad (49)$$

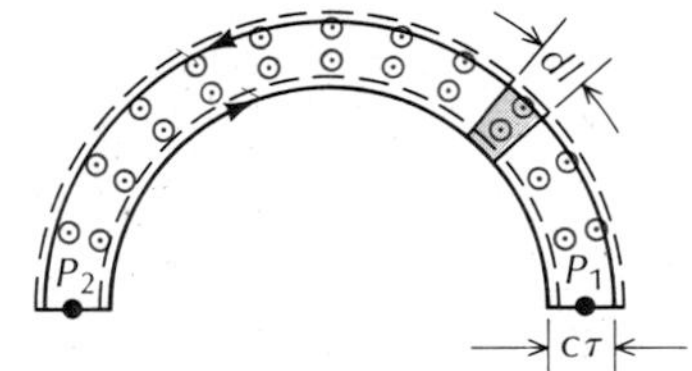

Fig. 35.21 The path of integration for Eq. (49) is marked with a solid line. As in the other figures, the boundaries of the zone of the kink are marked with dashed lines. The larger semicircle of the path is in the zone of the kink, but the smaller semicircle is not. The magnetic field is perpendicular to the plane of the page; the colored dots show the tips of the magnetic field vectors.

To extract some useful information from this law, it is convenient to evaluate the integral for the path shown in Figure 35.21; this path consists of two semicircles joined by short radial segments. The transverse electric field is only different from zero along the larger semicircle. Hence the left side of Eq. (49) is

$$\oint \mathbf{E} \cdot d\mathbf{l} = \int_{P_1}^{P_2} E_\theta \, dl \quad (50)$$

where the points P_1 and P_2 are shown in Figure 35.21. The magnetic flux on the right side can be calculated by taking small area elements of the shape shown in Figure 35.21 with dimensions $c\tau \times dl$:

$$\Delta\Phi_B = \int \mathbf{B}\cdot d\mathbf{S} = \int Bc\tau\, dl = c\tau \int_{P_1}^{P_2} B\, dl \tag{51}$$

In a time τ, the two dashed circles of Figure 35.21, with the magnetic field between them, sweep beyond the fixed path shown in this figure. Hence

$$\frac{d\Phi_B}{dt} = \frac{-\Delta\Phi_B}{\tau} = -c \int_{P_1}^{P_2} B\, dl \tag{52}$$

Faraday's Law now becomes

$$\int_{P_1}^{P_2} E_\theta\, dl = c\int_{P_1}^{P_2} B\, dl$$

In view of Eq. (47), this can be written

$$\int_{P_1}^{P_2} E_\theta\, dl = c^2\mu_0\varepsilon_0 \int_{P_1}^{P_2} E_\theta\, dl \tag{53}$$

Obviously, the consistency of this equation requires that

$$1 = c^2\mu_0\varepsilon_0 \tag{54}$$

or

Speed of electromagnetic wave

$$\boxed{c = \frac{1}{\sqrt{\mu_0\varepsilon_0}}} \tag{55}$$

We have therefore derived a theoretical expression for the speed of propagation of an electromagnetic disturbance in vacuum. The numerical value of the right side of Eq. (55) is

$$\begin{aligned} c &= 1/(1.26 \times 10^{-6}\ \text{H/m} \times 8.85 \times 10^{-12}\ \text{F/m})^{1/2} \\ &= 3.00 \times 10^{8}\ \text{m/s} \end{aligned} \tag{56}$$

We see that this agrees with the experimental value (48) for the speed of light (in section 36.1 we will make a more precise comparison of the theoretical and experimental values of the speed of light). The derivation of Eq. (55) was one of the great and early triumphs of the electromagnetic theory formulated by Maxwell in his Eqs. (17)–(20).

By means of Eq. (55), we can somewhat simplify the relation between B and E_θ. We have

$$B = c\mu_0\varepsilon_0 E_\theta = \frac{c}{c^2} E_\theta \tag{57}$$

or

$$\boxed{B = \frac{E_\theta}{c}} \qquad (58)$$

Relation between electric and magnetic radiation fields

EXAMPLE 3. In Example 2 we found that the sudden deceleration of an electron during a collision produces a transverse electric field of 6.4×10^{-2} V/m at a certain distance. What is the magnetic field produced by the deceleration?

SOLUTION: According to Eq. (58),

$$\begin{aligned} B &= E_\theta/c \\ &= (6.4 \times 10^{-2}\ \text{V/m})/(3.0 \times 10^{8}\ \text{m/s}) \\ &= 2.1 \times 10^{-10}\ \text{T} \end{aligned}$$

SUMMARY

Displacement current: $I_d = \varepsilon_0 \dfrac{d\Phi}{dt}$

Maxwell's equations:

$$\oint \mathbf{E} \cdot d\mathbf{S} = Q/\varepsilon_0$$

$$\oint \mathbf{B} \cdot d\mathbf{S} = 0$$

$$\oint \mathbf{E} \cdot d\mathbf{l} = -\frac{d\Phi_B}{dt}$$

$$\oint \mathbf{B} \cdot d\mathbf{l} = \mu_0 I + \mu_0 \varepsilon_0 \frac{d\Phi}{dt}$$

Transverse electric field of accelerated charge:

$$E_\theta = \frac{1}{4\pi\varepsilon_0} \frac{qa \sin\theta}{c^2 r}$$

Speed of electromagnetic wave: $c = 1/\sqrt{\mu_0 \varepsilon_0}$

Transverse magnetic field: $B = \dfrac{E_\theta}{c}$

QUESTIONS

1. The displacement current between the plates of a capacitor has the same magnitude as the conduction current in the wires connected to the capacitor, and yet the magnetic field produced by the former current near and within the capacitor is much weaker than that produced by the latter current near and within the wire. Explain.

2. Consider the electric field of a single positive electric charge moving at constant velocity. What is the direction of the displacement current intercepted by a circular area perpendicular to the velocity in front of the charge? Behind the charge?

3. A capacitor is connected to an alternating source of emf. Does the displacement current between the capacitor plates lead or lag the emf?

4. Which of Maxwell's equations permits us to deduce the electric Coulomb field of a static charge? Which of Maxwell's equations permits us to deduce the magnetic field of a charge moving with uniform velocity?

5. Suppose that there exist magnetic monopoles, i.e., postive and negative magnetic charges that act as sources and sinks of magnetic field lines, analogous to positive and negative electric charges. Which of Maxwell's equations would have to be modified to take into account such monopoles? Qualitatively, what are the required modifications?[10]

6. Given Eq. (30) for the electric field between the capacitor plates described in Section 35.3, is there a well-defined potential difference between the plates?

7. Some microwave ovens have rotating turntables that continually turn the meat while cooking. What is the purpose of this arrangement? (Hint: Microwave ovens are cavities with standing electromagnetic waves.)

8. Cavities are capable of oscillating at much higher frequencies than ordinary LC circuits. What limits the maximum frequency attainable with LC circuits?

9. In practice, the electromagnetic oscillations in a cavity are damped by "frictional" losses. How does this "friction" arise?

10. Efficient waveguides are manufactured out of a very good conductor, such as copper, sometimes with a silver lining. Why is high conductivity essential for high efficiency?

11. As stated in Section 35.5, the magnetic field of a charge moving at uniform velocity can be regarded as an induced magnetic field produced by the changing electric flux in the space surrounding the charge. Check that the direction of the induced magnetic field agrees with the direction given by Eq. (30.15).

12. Since the magnetic field of a charge moving at uniform velocity can be regarded as an induced magnetic field produced by the changing electric flux, we might expect that the electric field (Coulomb field) can be regarded as an induced electric field produced by a changing magnetic flux. Is this the case?

13. Is the transverse electric radiation field E_θ a conservative field?

14. In Figure 35.16 the electric field lines have a sharp discontinuity in slope. Is this discontinuity unphysical? On what is this discontinuity to be blamed?

15. Roughly sketch the electric field lines for a positive charge that is initially moving at uniform velocity and suddenly stops.

16. In a collision with an atom, an electron suddenly stops. Describe the directions of the electric and magnetic radiation fields at some distance from the electron at right angles to the acceleration.

17. Given that the magnitudes of the magnetic and electric radiation fields of an accelerated charge are related by $B = E_\theta/c$, compare the energy densities in these fields.

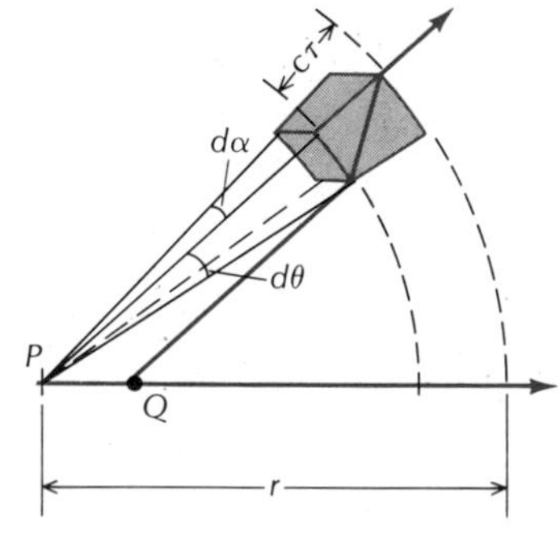

Fig. 35.22

18. Consider a small angular patch (Figure 35.22) in the radiation field of the accelerated charge described in Section 35.4. The volume of this small patch is [area] × [thickness] = $r^2\ d\theta\ d\alpha \times c\tau$ and the energy in this volume is $\frac{1}{2}\varepsilon_0 E_\theta^2 \times (r^2\ d\theta\ d\alpha \times c\tau)$. Show that this energy remains constant as the pulse of radiation propagates outward.

19. A charged particle moving around a circular orbit at uniform speed has a centripetal acceleration and therefore produces a radiation field. However, a uniform current flowing around a circular loop does *not* produce a radiation field. Is this a contradiction? Explain.

[10] For a quantitative discussion of these modifications, see Problem 10.

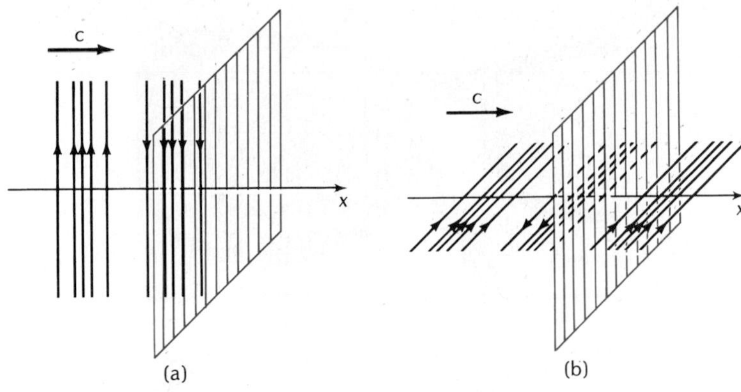

Fig. 36.7 An array of vertical wires (a) blocks the passage of a microwave of vertical polarization, but (b) permits the passage of a microwave of horizontal polarization.

other (Figure 36.7). The preferential direction of polarization that permits the passage of the electric field of a wave is then *perpendicular* to the direction of the wires because the wires have very little effect on a perpendicular electric field; on the other hand, an electric field parallel to the wires causes strong currents to flow in the wires, which both reflect the wave and dissipate its energy.

EXAMPLE 1. The wave reaching a point at some distance from a radio transmitter has an electric field with an amplitude of 2.0×10^{-3} V/m. What is the amplitude of the magnetic field?

SOLUTION: By Eq. (18) the amplitude of the magnetic field is E_0/c,

$$B_0 = \frac{E_0}{c} = \frac{2.0 \times 10^{-3}\ \text{V/m}}{3.0 \times 10^8\ \text{m/s}} = 6.7 \times 10^{-12}\ \text{T}$$

EXAMPLE 2. Suppose that a light wave polarized in the y direction is incident on a sheet of Polaroid whose preferential direction (direction of passage) makes an angle θ with the y axis (Figure 36.8). By what factor is the intensity of the light reduced? Pretend that the Polaroid sheet acts as an ideal polarizing filter, with 100% transmission for the component of the wave parallel to the preferential direction, and total absorption for the component perpendicular to this direction.

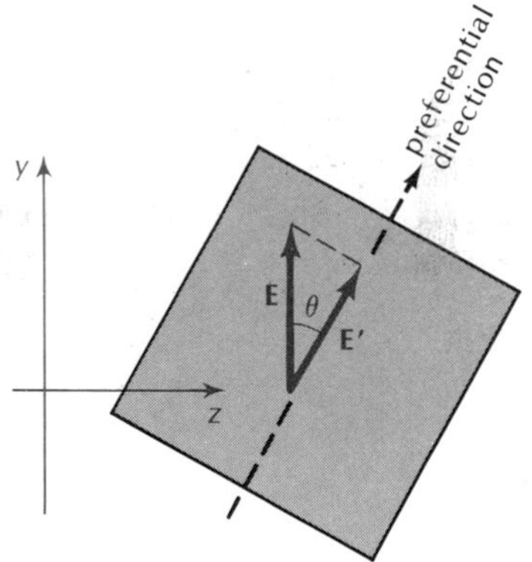

Fig. 36.8 A light wave polarized in the y direction is incident on a sheet of Polaroid whose preferential direction makes an angle θ with the y direction. The transmitted wave has a reduced amplitude and is polarized in the preferential direction of the sheet. The vectors **E** and **E′** indicate the directions of polarization of the incident and the transmitted waves.

SOLUTION: If the magnitude of the electric field of the incident wave is E_0, then the component of the electric field parallel to the preferential direction is (Figure 36.8)

$$E' = E_0 \cos\theta \tag{21}$$

This is the only component that will pass through the sheet. Thus, the transmitted wave will be polarized at an angle θ with respect to the incident wave, and it will have the amplitude given by Eq. (21), i.e., it will have an amplitude smaller than that of the incident wave by a factor of $\cos\theta$. Since the intensity of a wave is proportional to the square of the amplitude, the intensity of the transmitted wave is smaller than that of the incident wave by a factor $\cos^2\theta$,

$$\boxed{[\text{transmitted intensity}] = \cos^2\theta \times [\text{incident intensity}]} \tag{22}$$

This relation between the incident and the transmitted intensities of a wave passing through a polarizing filter is called the **law of Malus.** Note that for

Law of Malus

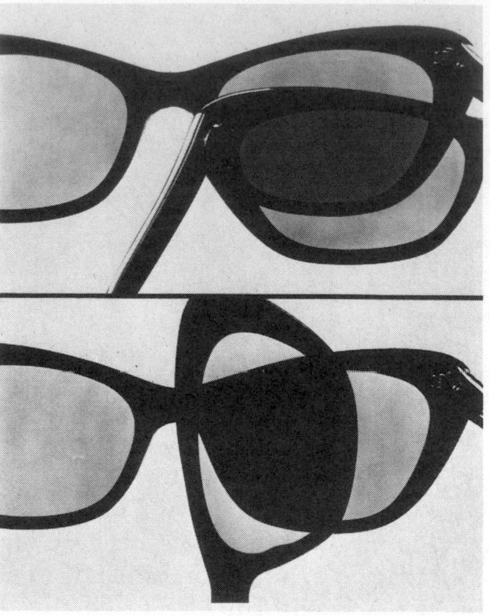

Fig. 36.9 The light that passes through the distant pair of sunglasses becomes polarized in the vertical direction, because this is the preferential direction for the Polaroid in the sunglasses. If the near pair of sunglasses is oriented parallel to the distant pair ($\theta = 0°$), it permits the passage of the polarized light. But if the near pair of sunglasses is oriented perpendicular to the distant pair ($\theta = 90°$), it stops the polarized light completely.

$\theta = 90°$, the transmitted intensity is zero, i.e., the Polaroid sheet stops the light wave completely (Figure 36.9).

36.3 The Generation of Electromagnetic Waves

Since the speed of the electromagnetic wave described by Eqs. (17)–(20) is c, the speed of light, the wavelength is

$$\lambda = \frac{2\pi}{\omega} c \tag{23}$$

or, if we replace the angular frequency ω by the ordinary frequency $\nu = \omega/2\pi$,

$$\boxed{\lambda = \frac{c}{\nu}} \tag{24}$$

This shows how the wavelength depends on the frequency of the source. For instance, the charges on the antenna of an FM radio station typically oscillate back and forth with a frequency of 10^8 Hz (or 100 MHz); correspondingly, the wavelength of the radiation emitted by these accelerated charges has a wavelength of

$$\lambda = \frac{c}{\nu} = \frac{3.0 \times 10^8 \text{ m/s}}{10^8\text{/s}} = 3.0 \text{ m} \tag{25}$$

The oscillations of the charges on the antenna are produced by means of a resonating LC circuit coupled to the antenna by a mutual inductance (see Figure 34.11b). This is essentially the method used to generate **long waves, medium waves** (AM), and **short waves** (including FM), as well as **TV waves**; such radio waves span a wavelength range from 10^5 m to a few centimeters.

Kinds of electromagnetic radiation

Waves of shorter wavelength, or **microwaves,** are best generated by a resonating cavity such as a klystron (see Section 35.3); the antenna in this case is merely a horn, connected to the cavity by a waveguide, that permits the waves to spill out into space. This method can be used to generate waves of wavelength as short as about a millimeter. Shorter wavelengths cannot be generated with currents oscillating on macroscopic laboratory equipment; however, short wavelengths are easily generated by electrons vibrating within molecules and atoms subjected to stimulation by heat or by an electric current. Depending on the details of the motion, the electrons in molecules and in atoms will emit **infrared** radiation, **visible** light, **ultraviolet** radiation, or **X rays;** the corresponding wavelengths range from 10^{-3} m to 10^{-11} m. X Rays can also be generated by the acceleration that high-speed electrons suffer during impact on a target; this is **Bremsstrahlung** (see Example 35.2). Radiations of even shorter wavelengths are emitted by protons and neutrons moving within a nucleus; these are **gamma rays** with wavelengths as short as 10^{-13} m. Of course, the motion of subatomic particles and their emission of radiation cannot be calculated by classical mechanics or classical electromagnetic theory; such calculations require quantum mechanics and quantum electrodynamics.

Bremsstrahlung

In an ordinary light source, such as a neon tube, the individual atoms or molecules radiate independently. The emerging light consists of a superposition of many individual light waves with random phase differences, random directions of polarization, and diverging directions of propagation (light waves with random, unpredictable phase differences are said to be **incoherent**). In a **laser,** the atoms or molecules radiate in unison, by a quantum-mechanical phenomenon called **stimulated emission.** The emerging light is a superposition of light waves with exactly the same phases, the same directions of polarization, and the same directions of propagation (light waves with no phase differences, or with predictable phase differences, are said to be **coherent**). Since the individual light waves in this kind of light combine constructively, the light beam emerging from the laser is very intense, and it is also very sharply collimated.

Stimulated emission

Another important mechanism for the generation of electromagnetic waves is **cyclotron emission.** This involves high-speed electrons undergoing centripetal acceleration while spiraling in a magnetic field (see Section 31.3). Depending on the speed of the electron and the strength of the magnetic field, the radiation may consist of radio waves, X rays, or anything in between. Most of the radio waves reaching us from stars, pulsars, and radio galaxies are generated by this process.

Figure 36.10 displays the wavelength and the frequency bands of electromagnetic radiation. The bands overlap to some extent because the names assigned to the different ranges of wavelength depend not only on the value of the wavelength, but also on the method used to generate and/or detect the radiation. For example, radiation of a wavelength of a tenth of a millimeter will be called a radio wave (microwave) if detected by a radio receiver, but it will be called infrared radiation if detected by a heat sensor.

Visible light is electromagnetic radiation of a wavelength between

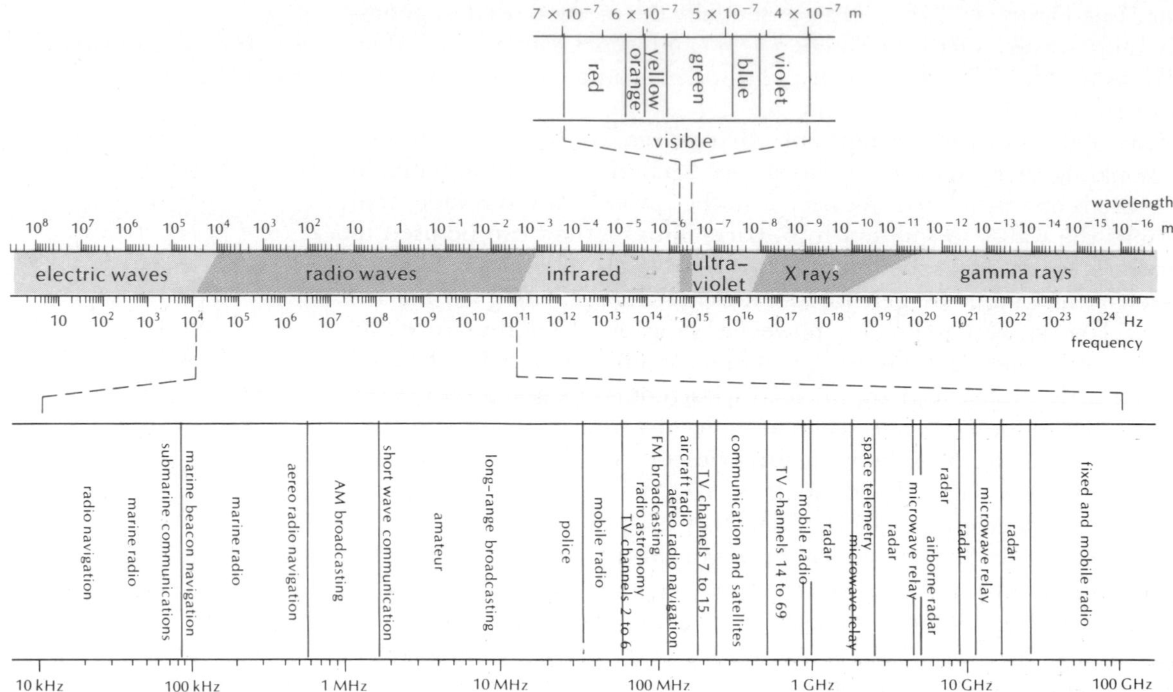

Fig. 36.10 Wavelength and frequency bands of electromagnetic radiation.

about 7×10^{-7} m and 4×10^{-7} m. This is the range of wavelengths to which our eyes are sensitive. We perceive different wavelengths within the visible region as having different colors. Figure 36.11 shows how colors are correlated with wavelength.

Incidentally, our eyes are almost completely insensitive to the polarization of light waves. We can only detect the polarization with special equipment such as Polaroid sunglasses. Radio and TV antennas are of course very sensitive to the direction of polarization of radio waves and they must have the proper orientation to pick up a strong signal.

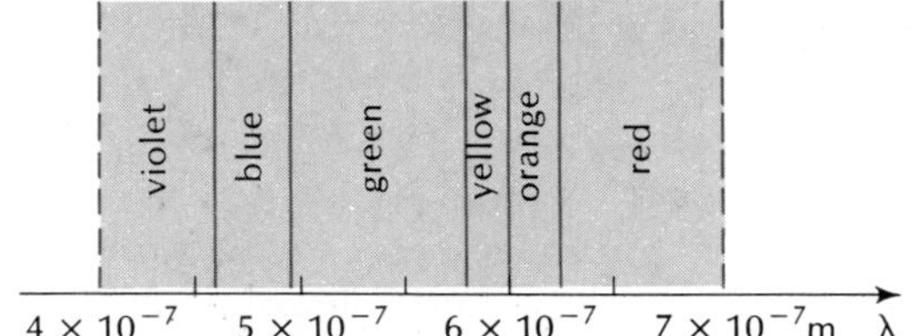

Fig. 36.11 Colors of visible light.

36.4 Energy of a Wave

The electric and magnetic fields of a wave contain energy. As the wave moves along, so does this energy — the wave transports energy.

Let us calculate the flow of energy in a plane wave. Suppose the wave is of the kind described by Eqs. (17) and (18), i.e., it is a wave propagating in the positive x direction. The densities of electric and magnetic energy are, respectively [see Eqs. (26.19) and (32.42)],

$$u_E = \frac{\varepsilon_0}{2} E^2 \quad \text{and} \quad u_B = \frac{1}{2\mu_0} B^2 \tag{26}$$

Within a thin rectangular slab of this wave, of thickness dx and frontal area A (Figure 36.12), the fields are nearly constant. The total amount of energy in this slab is

$$U = (u_E + u_B) \times [\text{volume}] \tag{27}$$

$$= \left(\frac{\varepsilon_0}{2} E^2 + \frac{1}{2\mu_0} B^2\right) \times A\, dx \tag{28}$$

Since $E = cB$ and $\varepsilon_0 = 1/\mu_0 c^2$ we can write this as

$$dU = \left[\frac{1}{2\mu_0 c^2} E(cB) + \frac{1}{2\mu_0} B\left(\frac{E}{c}\right)\right] A\, dx \tag{29}$$

$$= \frac{1}{\mu_0 c} EBA\, dx \tag{30}$$

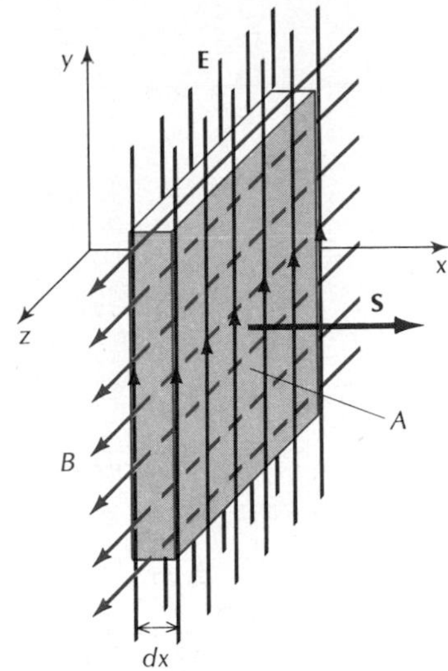

Fig. 36.12 A slab of electric (black) and magnetic (color) fields in a plane wave propagating toward the right. The slab of thickness dx and frontal area A moves with the wave.

Note that the two terms on the right side of Eq. (29) are equal, i.e., the electric and magnetic energy densities in an electromagnetic wave are equal.

The amount of energy dU moves out of the volume $A\, dx$ in a time $dt = dx/c$. Hence the rate of flow of energy is

$$\frac{dU}{dt} = \frac{1}{\mu_0} EBA \tag{31}$$

and the rate of flow of energy per unit frontal area, or the **energy flux,** is

Energy flux

$$\frac{1}{A}\frac{dU}{dt} = \frac{1}{\mu_0} EB \tag{32}$$

In Figure 36.12 the energy flows toward the right, along the direction of propagation of the wave. We can describe the energy flux in terms of a vector that has a magnitude equal to the right side of Eq. (32) and a direction along the flow. The vector satisfying these conditions is

Poynting vector

$$\boxed{\mathbf{S} = \frac{1}{\mu_0} \mathbf{E} \times \mathbf{B}} \tag{33}$$

Since **E** and **B** are perpendicular, the magnitude of the vector **S** given by Eq. (33) coincides with the magnitude given by Eq. (32). Furthermore the direction of **E** × **B** in Figure 36.12 is toward the right, as desired. The vector **S** is called the **Poynting vector.** It turns out that this vector gives the energy flux not only in a plane wave, but in any arbitrary electromagnetic field. The units of energy flux are J/(m$^2 \cdot$ s), or W/m^2.

John Henry Poynting, *1852–1914, British physicist, professor at the University of Birmingham. He introduced the Poynting vector while presenting a proof of a general theorem on energy transfers in the electromagnetic fields.*

Since both the electric and magnetic fields of a wave oscillate in time, so does the energy flux. If the value of x in Eqs. (17) and (18) is fixed, say, $x = 0$, we have

$$E = E_0 \sin \omega t \quad \text{and} \quad B = \frac{E_0}{c} \sin \omega t \tag{34}$$

so that

$$S = \frac{1}{\mu_0 c} E_0^2 \sin^2 \omega t \tag{35}$$

Thus, the energy flux oscillates between zero and a maximum value $E_0^2/\mu_0 c$. The time-average energy flux is

Time-average magnitude of Poynting vector

$$\boxed{\overline{S} = \frac{1}{2\mu_0 c} E_0^2} \tag{36}$$

EXAMPLE 3. At a distance of 6.0 km from a radio transmitter, the amplitude of the oscillating electric field of the radio wave is $E_0 = 0.13$ V/m. What is the time-average energy flux? What is the total power radiated by the radio transmitter? Pretend that the transmitter radiates uniformly in all directions.

SOLUTION: From Eq. (36) we find

$$\overline{S} = \frac{1}{2\mu_0 c} E_0^2$$

$$= \frac{(0.13 \text{ V/m})^2}{2 \times 1.26 \times 10^{-6} \text{ H/m} \times 3 \times 10^8 \text{ m/s}}$$

$$= 2.2 \times 10^{-5} \text{ W/m}^2$$

To obtain the total power, we must multiply the power per unit area by the area of a sphere of radius 6.0 km:

$$\overline{P} = 4\pi r^2 \overline{S} \tag{37}$$

$$= 4\pi \times (6.0 \times 10^3 \text{ m})^2 \times 2.2 \times 10^{-5} \text{ W/m}^2$$

$$= 1.0 \times 10^4 \text{ W} = 10 \text{ kW}$$

Note that, according to Eq. (37), the energy flux $\overline{S}$ for a spherical wave spreading out from a source is inversely proportional to the square of the distance,

$$\overline{S} \propto \frac{1}{r^2} \tag{38}$$

This decrease of the flux is consistent with Eq. (1), according to which the amplitude of the electric field is inversely proportional to the distance, $E_0 \propto 1/r$. We must take this dependence of flux and amplitude on distance into account whenever we want to investigate the spreading of the wave over a large range of distances (but we can ignore this dependence over a small range of distances).

36.5 Momentum of a Wave

An electromagnetic wave carries not only energy but also momentum. This can be readily understood in terms of Einstein's mass–energy relation, which tells us that the energy in the wave has mass. As the energy moves so does the mass, and such moving mass has momentum.

The following simple calculation permits us to derive a quantitative formula for the amount of momentum in an electromagnetic wave. Imagine that a plane electromagnetic wave polarized in the y direction and traveling in the x direction strikes a charged particle (Figure 36.13). The electric and magnetic fields of the wave will then exert forces on the particle, i.e., they will change the momentum of the particle. Let us look at the x component of the rate of change of momentum. The electric field is entirely in the y direction — it exerts no force in the x direction. The magnetic field is entirely in the z direction — it exerts the following force in the x direction:

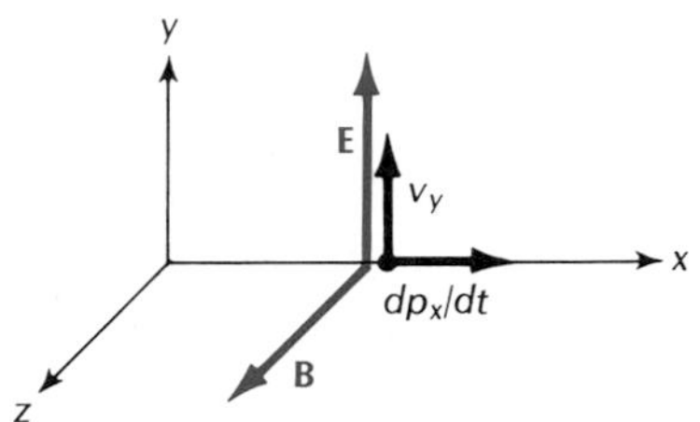

Fig. 36.13 A plane wave traveling toward the right strikes a particle. The electric and magnetic fields of the wave exert forces on the particle.

$$\frac{dp_x}{dt} = q(\mathbf{v} \times \mathbf{B})_x = q(v_y B_z - v_z B_y) = q v_y B_z$$

Since $B_z = E_y/c$ [see Eqs. (17) and (18)], we can also write this rate of change of momentum as

$$\frac{dp_x}{dt} = \frac{q}{c} v_y E_y \tag{39}$$

Let us compare this with the rate at which the particle acquires energy from the wave. Only the electric field does work on the particle. The rate of work, or the rate of increase of the energy of the particle, is

$$\frac{dW}{dt} = q\mathbf{E} \cdot \mathbf{v} = qE_y v_y \tag{40}$$

(here we have used the symbol W for the energy of the particle because the symbol E is reserved for the electric field). If we compare the right sides of Eqs. (39) and (40), we recognize that

$$\frac{dp_x}{dt} = \frac{1}{c}\frac{dW}{dt}$$

i.e., the rate at which the particle acquires x momentum from the wave matches the rate at which it acquires energy. Whenever the particle absorbs energy from the wave, it will also absorb x momentum,

$$dp_x = \frac{1}{c}\,dW \tag{41}$$

Any gain of energy and momentum by the particle implies a corresponding loss of energy and momentum by the wave. If we imagine that the wave loses all of its energy and momentum to the particle (and disappears),[2] we recognize that the total amount of energy and the

[2] Actually, a single charged particle cannot absorb all of a wave (the particle will scatter the wave); but a cloud of many charged particles can completely absorb a wave.

total amount of x momentum stored in the wave must be related by a formula similar to Eq. (41):

Momentum of wave

$$\boxed{P_x = \frac{U}{c}} \tag{42}$$

This is the general relation between the momentum and the energy of a plane wave. This relation is also valid for the amounts of momentum and energy in some portion of the wave.

Since we already know that the energy flows with the wave, we can conclude that the momentum also flows with the wave. By means of Eq. (31), we can then express the momentum flow within a frontal area A of the wave as

$$\frac{dP_x}{dt} = \frac{1}{c}\frac{dU}{dt} = \frac{1}{c\mu_0} EBA \tag{43}$$

This shows that the momentum flow oscillates in the same way as the energy flow. The time-average momentum flow is

$$\overline{\frac{dP_x}{dt}} = \frac{1}{2\mu_0 c^2} E_0^2 A \tag{44}$$

Whenever a wave strikes a body and is absorbed by it, the wave will exert a force on the body and transfer momentum to it. If the body has a frontal area A facing the wave, the rate of change of momentum, or the force, will simply be given by Eq. (43),

$$F_x = \frac{dP_x}{dt} = \frac{1}{c\mu_0} EBA$$

or

$$F_x = \frac{1}{c} SA$$

where S is the magnitude of the Poynting vector. The force per unit frontal area is then

Pressure of radiation

$$\boxed{\frac{F_x}{A} = \frac{S}{c}} \tag{45}$$

This is called the **pressure of radiation** or the pressure of light. Note that the formula (45) hinges on the assumption that the wave is completely absorbed. If the wave is partially or totally reflected, then the force is *larger* than given by Eq. (45). For instance, if the wave is totally reflected, then the body must not only absorb all of the initial momentum, but also supply the momentum for the reversed motion of the wave. The force is then twice as large as that given by Eq. (45).

The magnitude of the radiation-pressure force is usually insignificant compared with other ordinary forces that act on a body.

EXAMPLE 4. The average energy flux in the sunlight incident on the Earth is $\overline{S} = 1.4 \times 10^3$ W/m². What force does the pressure of light exert on the Earth? How does this compare with the gravitational force that the Sun exerts on the Earth? Pretend that all the light striking the Earth is absorbed.

SOLUTION: The cross-sectional area that the Earth offers to the stream of sunlight is πR_E^2. The total power absorbed by the Earth is then

$$\frac{dU}{dt} = \overline{S}A$$

$$= 1.4 \times 10^3 \text{ W/m}^2 \times \pi \times (6.4 \times 10^6 \text{ m})^2$$

$$= 1.8 \times 10^{17} \text{ W} \qquad (46)$$

Hence

$$F = \frac{1}{c}\frac{dU}{dt} = \frac{1.8 \times 10^{17} \text{ W}}{3.0 \times 10^8 \text{ m/s}} = 6.0 \times 10^8 \text{ N}$$

The gravitational force that the Sun exerts on the Earth is much greater:

$$F = \frac{GM_S M_E}{r^2}$$

$$= \frac{6.7 \times 10^{-11} \text{ N} \cdot \text{m}^2/\text{kg}^2 \times 2.0 \times 10^{30} \text{ kg} \times 6.0 \times 10^{24} \text{ kg}}{(1.5 \times 10^{11} \text{ m})^2}$$

$$= 3.6 \times 10^{22} \text{ N}$$

The force of radiation pressure on the Earth is insignificant compared to the gravitational pull of the Sun. However, on an object of very small mass, such as a grain of dust in interplanetary space, the force of radiation pressure can be as large as, or even larger than, the gravitational pull of the Sun. A simple calculation shows that if a grain of dust is smaller than about 10^{-6} m across, then radiation pressure overcomes gravitation and blows the grain away from the Sun.[3] The tails of comets are a spectacular demonstration of this effect. These tails consist of a plume of dust that is blown away from the comet's head by the pressure of sunlight (Figure 36.14). Regardless of the direction of motion of the comet, the tail always points away from the Sun, much as a pennant fluttering from the mast of a ship always points away from the wind.

Fig. 36.14 Comet Mrkos and its tails, photographed in August 1957. This comet has both a dust tail (haze, upper right) and an ion tail (streaks, upper center). The dust tail is produced by the pressure of sunlight; the ion tail is produced by the pressure of the solar wind.

36.6 The Doppler Shift of Light

The frequency of light waves, like that of waves of any kind, suffers a Doppler shift whenever the emitter of waves is in motion relative to the receiver. However, in contrast to sound waves in air or surface

[3] For *very* small grains of dust (less than 10^{-7} m across) and for molecules this will not happen because such very small objects scatter light instead of absorbing it. This cuts down the force exerted by light.

waves in water, light waves in a vacuum are not propagating in any material medium that can serve as a preferred reference frame with respect to which we can reckon their velocity. Light waves have the same speed c with respect to *all* inertial reference frames (this well-established observational fact lies at the root of the theory of Special Relativity, as discussed in Chapter 17). Hence, for the calculation of the Doppler shift we must first decide which reference frame is the most suitable. Since we want to find the frequency as seen by the *receiver,* it is clear that the relevant reference frame is the rest frame of the receiver.

In this reference frame, the calculation of the Doppler shift can be done exactly as in Section 16.3. Since the receiver is at rest in the "medium" (reference frame) in which light has the speed c, the formula for the Doppler shift of light must be [see Equation (16.13)]

$$\nu = \frac{1}{1 \pm v/c}\,\nu_0 \qquad (47)$$

where, as before, ν_0 is the frequency radiated by the emitter, ν is the frequency detected by the receiver, and v is the speed of the emitter (the positive sign is to be used if the emitter is receding, the negative sign if approaching). Since the calculations of Section 16.3 relied on Newtonian physics, which is valid only if the speed is low compared to the speed of light($v \ll c$), Eq. (47) can be replaced by the approximation

$$\nu \cong (1 \mp v/c)\nu_0 \qquad (48)$$

without any significant loss of accuracy. From this we see that the frequency shift $\Delta\nu = \nu - \nu_0$ is given by

Approximate Doppler shift

$$\boxed{\frac{\Delta\nu}{\nu_0} \cong \mp\frac{v}{c}} \qquad \begin{array}{l} +\text{ for approaching emitter} \\ -\text{ for receding emitter} \end{array} \qquad (49)$$

Measurements of the Doppler shift of light play a crucial role in astronomy. The velocities of remote stars relative to us can be determined via the Doppler shift of the starlight reaching us. The rotational velocities of stars can likewise be determined by a careful measurement of the difference between the Doppler shifts of the light from one edge of the star and the opposite edge.

Although Eq. (47) is adequate whenever the speed of the source is small compared to the speed of light, it must be modified for relativistic effects if the speed of the source is large. According to Section 17.4, the emitter is subject to a relativistic time-dilation effect. This, in itself, would *reduce* the frequency of the emitter by a factor $\sqrt{1 - v^2/c^2}$. We must therefore insert this extra factor in Eq. (47),

$$\nu = \sqrt{\frac{1 - v^2/c^2}{1 \pm v/c}}\,\nu_0 \qquad (50)$$

This is the exact formula for the Doppler shift of light. Since $1 - v^2/c^2 = (1 - v/c)(1 + v/c)$, this formula can be simplified by appropriate cancellations:

$$\nu = \sqrt{\frac{1 - v/c}{1 + v/c}}\,\nu_0 \quad \text{for receding emitter} \tag{51}$$

$$\nu = \sqrt{\frac{1 + v/c}{1 - v/c}}\,\nu_0 \quad \text{for approaching emitter} \tag{52}$$

Exact Doppler shift

For low speeds there is of course no appreciable difference between these exact formulas and the approximate formula (47).

The velocities of recession of galaxies and quasars partaking of the expansion of the universe (see Interlude D) give the light and radio waves from these objects very large Doppler shifts. For instance, the light from the quasar 3C 147 (Figure 36.15) has a frequency shift amounting to a factor of 1.55. If we insert this factor in Eq. (51),

$$\frac{\nu}{\nu_0} = \frac{1}{1.55} = \sqrt{\frac{1 - v/c}{1 + v/c}} \tag{53}$$

we find a velocity of recession $v = 0.41c$, that is, 41% of the speed of light! The velocities of recession of some other quasars are even higher.

Fig. 36.15 The quasar 3C 147.

SUMMARY

Electric and magnetic fields of plane wave:

$$E_y = E_0 \sin\left(\omega t - \frac{\omega x}{c}\right)$$

$$B_z = \frac{E_0}{c} \sin\left(\omega t - \frac{\omega x}{c}\right)$$

Poynting vector: $\mathbf{S} = \dfrac{1}{\mu_0}\,\mathbf{E} \times \mathbf{B}$

Time-average energy flux of plane wave: $\overline{S} = \dfrac{1}{2\mu_0 c} E_0^2$

Momentum of wave: $P_x = \dfrac{U}{c}$

Radiation pressure: $\dfrac{F_x}{A} = \dfrac{S}{c}$

Doppler shift: $\dfrac{\Delta\nu}{\nu_0} \cong \mp \dfrac{v}{c}$ at low speed

$$\nu = \sqrt{\frac{1 - v/c}{1 + v/c}}\,\nu_0 \quad \text{(receding)}$$

$$\nu = \sqrt{\frac{1 + v/c}{1 - v/c}}\,\nu_0 \quad \text{(approaching)}$$

QUESTIONS

1. Since the speed of light has now been adopted as the standard of speed and has been assigned the value 2.99792458×10^8 m/s *by definition,* why is it still meaningful to compare this number with Maxwell's theoretical prediction for the speed of light?

2. The round-trip travel time for a radio signal to the Moon is 2.6 s. Does this have a noticeable effect on radio communications with astronauts on the Moon?

3. A severe limitation on the speed of computation of large electronic computers is imposed by the speed of light because the electric signals on the connecting wires within the computer are electromagnetic waves ("guided waves"), which travel at a speed roughly equal to the speed of light. If the computer measures about 1 m across, what is the minimum travel time required for a typical signal sent from one part of the computer to another? What is the maximum number of signals that can be sent back and forth (sequentially) per second? Is there any way to avoid the limitation imposed by the travel time of signals?

4. Describe the direction of polarization of the radiation field of an accelerated charge at a few typical points in the space surrounding the charge. Show that the direction of polarization is never perpendicular to the direction of acceleration.

5. Consider transverse waves on a string. How many independent directions of polarization do such waves have? How could you construct a mechanical polarization filter that only permits the passage of a wave polarized in a preferential direction?

6. If the preferential directions of two adjacent sheets of Polaroid are at right angles, no light will pass through. However, if you now slip a third sheet of Polaroid between the other two and orient its preferential direction so that it lies between the directions of the other two, then some light will pass through the three sheets. Explain.

7. It has been proposed that we could eliminate the glare of the headlights of approaching automobiles by covering the windshields and the headlights with sheets of Polaroid. What orientation should we pick for the sheets of Polaroid installed on windshields and on headlights so that the light of every approaching automobile is blocked out, but our own light is not?

8. Malus discovered that when unpolarized light is incident on any surface, the reflected light is partially polarized in a direction parallel to the surface. This phenomenon, called **polarization by reflection,** arises from a dependence of the strength of reflection on the direction of polarization. Taking this phenomenon into account, explain how Polaroid sunglasses can eliminate a large fraction of the reflected glare (Figure 36.16), and explain why it is advantageous to arrange the preferential direction in Polaroid sunglasses vertically.

9. Some small-boat sailors like to wear Polaroid sunglasses because these make disturbances of the water surface stand out with exceptional contrast, and thereby make it easier to spot approaching gusts of wind. Why are Polaroid sunglasses better for this purpose than ordinary sunglasses? (Hint: Consider polarization by reflection; see preceding question.)

10. Suppose you are given a sheet of Polaroid that has no markings identifying its preferential direction. You have available a beam of unpolarized light. How can you determine the preferential direction of the sheet? (Hint: Cut the sheet in two and place one sheet behind the other rotated by 90° around the axis of the beam, so that the light is completely blocked. What will happen if

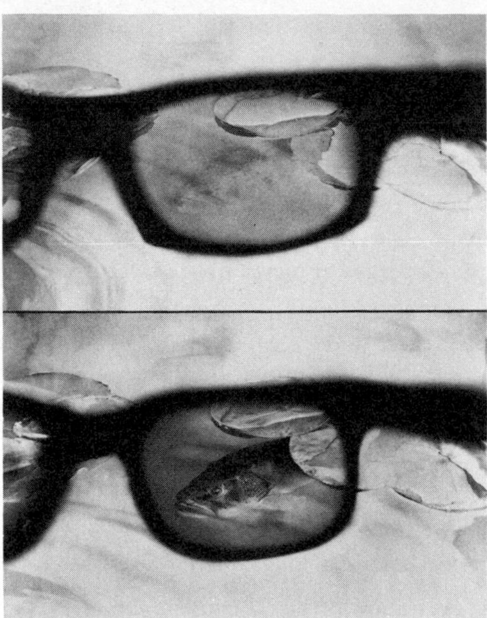

Fig. 36.16 When viewed through ordinary sunglasses (top), the fish is almost completely hidden by the glare reflected by the water. When viewed through Polaroid sunglasses (bottom), the fish becomes visible because the glare is reduced.

you now rotate one sheet about a *transverse axis,* i.e., an axis perpendicular to the beam?

11. The scattered light reaching you from the blue sky is (partially) polarized: if you look straight up, the direction of polarization is perpendicular to the direction of the Sun. How does this polarization arise? [Hint: Consider a beam of (unpolarized) sunlight passing overhead. The electric field of the light waves in this beam accelerates electrons in the molecules of air, and the radiation emitted by these electrons constitutes the scattered light of the sky. Since the direction of acceleration is perpendicular to the incident beam, what can you say about the polarization of the radiation emitted downward, toward you?]

12. Figure 36.17 shows the sensitivity of the human eye as a function of the wavelength of light. The sensitivity is maximum at about 5.5×10^{-7} m and drops to about 1% at 6.9×10^{-7} m and at 4.3×10^{-7} m. Suppose that the sensitivity of your eye were constant over the entire interval of wavelengths shown in Figure 36.17. How would this alter your visual perception of some of the things you see in your everyday life?

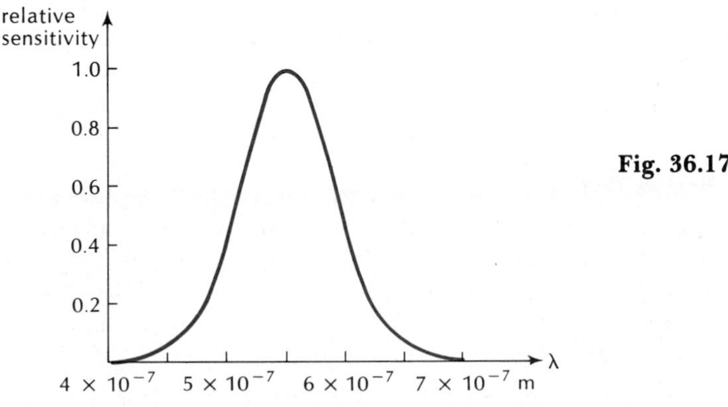

Fig. 36.17

13. An atom radiates visible light of a wavelength several thousand times longer than the size of the atom. How is this possible?

14. Why does the radio reception fade in the receiver of your automobile when you enter a tunnel?

15. Short-wave radio waves are reflected by the ionosphere of the Earth; this makes them very useful for long-range communication. Explain.

16. Why is the effect of the pressure of sunlight on a grain of dust floating in interplanetary space more significant if the grain is very small?

17. Could we propel a spaceship by shining a very intense light out of its back? What would be the advantages of such a propulsion scheme?

18. Astronauts of the future could travel all over the Solar System in a spaceship suspended from a large "sail" coated with a reflecting material. The pressure of sunlight on the sail could propel the spaceship in any direction. How would the astronauts have to orient their (flat) sail to move radially away from the Sun? To move at right angles to the radial direction? To move radially toward the Sun?

19. Suppose that a light wave is totally reflected by a moving mirror. Would you expect the energy of the light wave to change during this reflection? (Hint: Consider the work done by the radiation pressure.)

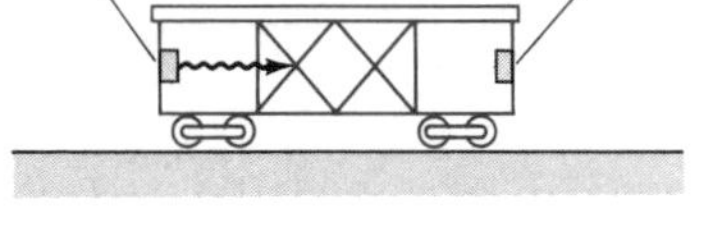

Fig. 36.18

20. Einstein's first derivation of his famous equation $E = mc^2$ relied on the equation for the momentum of light. Einstein considered a closed boxcar, free to roll on frictionless rails, with a light source at one end and an absorber at the other end (Figure 36.18). If the source at one end emits a light pulse of energy E, the recoil sets the boxcar in motion; this motion stops when the light pulse reaches the other end and is absorbed. Given that the center of mass of the closed system should remain stationary during this process, how could Einstein conclude that the transfer of the energy E from one end to the other involves a transfer of mass?

21. In the seventeenth century, the Danish astronomer Ole Roemer noticed that the orbital periods of the moons of Jupiter, as observed from the Earth, exhibit some systematic irregularities: the periods are slightly longer when the Earth is moving away from Jupiter, and slightly shorter when the Earth is moving toward Jupiter. Roemer attributed this apparent irregularity to the finite speed of propagation of light, and used it to make the first determination of the speed of light. Explain how one can use the lengthening or the shortening of the observed period to deduce the speed of light. (This shift of period may be regarded as the earliest discovery of a Doppler shift.)

22. Equations (51) and (52) give the Doppler shift for an emitter that is approaching or receding along the line of sight. Taking into account relativistic effects, would you expect a Doppler shift for an emitter moving at right angles to the line of sight? (This is called the transverse Doppler shift.)

PROBLEMS

Section 36.1

1. At a distance of 6.0 km from a radio transmitter, the amplitude of the electric radiation field of the emitted radio wave is $E_0 = 0.13$ V/m. Taking into account the decrease of the amplitude of the wave with distance, what will be the amplitude of the radio wave when it reaches a distance of 12.0 km? A distance of 18.0 km?

2. Consider the plane wave pulse shown in Figures 36.1 and 36.2. If the magnitude of the electric field in this pulse is 4×10^{-3} V/m, what is the magnitude of the displacement current flowing along the front surface of the pulse per meter of length measured perpendicularly to the current?

3. Two plane-wave pulses of the kind described in Section 36.1 are traveling

in opposite directions. Their polarizations are parallel and the magnitudes of their electric fields are 2×10^{-3} V/m.

(a) What are the electric energy density and the magnetic energy density in each pulse?
(b) Suppose that at one instant the two pulses overlap. What are the magnitudes of the electric field and the magnetic field in this superposition?
(c) What are the electric energy density and the magnetic energy density?

4. (a) One type of antenna for a radio receiver consists of a short piece of straight wire; when the electric field of a radio wave strikes this wire it makes currents flow along it, which are detected and amplified by the receiver. Suppose that the electric field of a radio wave is vertical. What must be the orientation of the wire for maximum sensitivity?
(b) Another type of antenna consists of a circular loop; when the magnetic field of a radio wave strikes this loop it induces currents around it. Suppose that the magnetic field of a radio wave is horizontal. What must be the orientation of the loop for maximum sensitivity?

Section 36.2

5. At a distance of several kilometers from a radio transmitter, the electric field of the emitted radio wave has a magnitude of 0.12 V/m at one instant of time. What is the energy density in this electric field? What is the energy density in the magnetic field of the radio wave?

6. A plane electromagnetic wave travels in the eastward direction. At one instant the electric field at a given point has a magnitude of 0.60 V/m and points down. What are the magnitude and direction of the magnetic field at this instant? Draw a diagram showing the electric field, the magnetic field, and the direction of propagation.

7. An electromagnetic wave traveling along the x axis consists of the following superposition of two waves polarized along the y and z directions, respectively:

$$\mathbf{E} = \hat{\mathbf{y}} E_0 \sin(\omega t - \omega x/c) + \hat{\mathbf{z}} E_0 \cos(\omega t - \omega x/c)$$

This electromagnetic wave is said to be **circularly polarized.**

(a) Show that the magnitude of the electric field is E_0 at all points of space at all times.
(b) Consider the point $x = y = z = 0$. What is the angle between **E** and the z axis at time $t = 0$? $t = \pi/2\omega$? $t = \pi/\omega$? $t = 3\pi/2\omega$? Draw a diagram showing the y and z axes and the direction of **E** at these times. In a few words, describe the behavior of **E** as a function of time.

8. An electromagnetic wave has the form

$$\mathbf{E} = \hat{\mathbf{x}} E_0 \sin(\omega t + \omega z/c) + 2\hat{\mathbf{y}} E_0 \sin(\omega t + \omega z/c)$$

(a) What is the direction of propagation of this wave?
(b) What is the direction of polarization, i.e., what angle does the direction of polarization make with the x, y, and z axes?
(c) Write down a formula for the magnetic field of this wave as a function of space and time.

9. The preferential directions of two adjacent sheets of Polaroid make an angle of 45°. A beam of polarized light, whose direction of polarization coincides with the preferential direction of the *second* sheet, is incident on the *first* sheet. By what factor is the intensity of the transmitted beam emerging from the second sheet reduced compared to the intensity of the incident beam? Assume that the sheets act as ideal polarizing filters.

10. Suppose that the preferential directions of two adjacent sheets of Polaroid make an angle θ with each other. If unpolarized light is incident on these

sheets, what is the transmitted intensity as a function of the angle θ? Assume that the sheets act as ideal polarizing filters.

11. If the preferential directions of two adjacent sheets of Polaroid are at right angles, they will completely block a light beam. However, if you insert a third sheet of Polaroid between the other two, then some light will pass through. Derive a formula for the dependence of the intensity of the transmitted light as a function of the angle that the preferential direction of the inserted sheet makes with that of the first sheet. Assume that the incident light is unpolarized, and that the sheets act as ideal polarizing filters. For what orientation of the inserted sheet is the transmitted intensity maximum?

*12. Example 2 shows that when a light wave passes through a sheet of Polaroid, its direction of polarization rotates, albeit with some loss of intensity. Consider an infinite number of adjacent sheets of Polaroid, each tilted by an infinitesimal angle with respect to the preceding sheet (Figure 36.19). Show that such a stack rotates the plane of polarization of the light wave through a finite angle without loss of intensity. Assume that each sheet of Polaroid acts as an ideal polarizing filter.

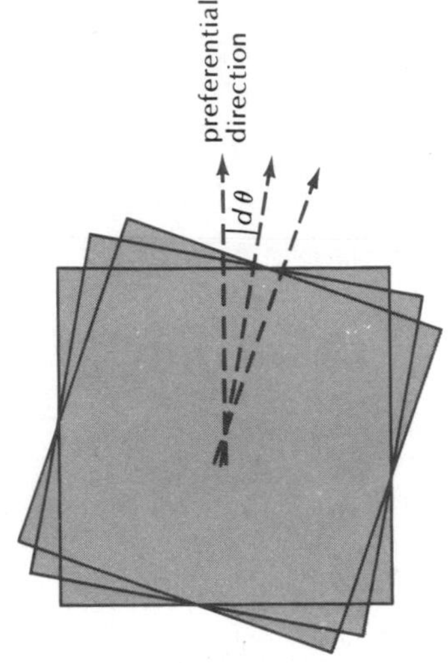

Fig. 36.19

Section 36.3

13. An ordinary radio receiver, such as found in homes across the country, has an AM dial and an FM dial. The AM dial covers a range from 530 to 1600 kHz and the FM dial a range from 88 to 108 MHz. What is the range of wavelengths for AM? For FM?

14. At many coastal locations, radio stations of the National Weather Service transmit continuous weather reports at a frequency of 162.5 MHz. What is the wavelength of these transmissions?

Section 36.4

15. A plane electromagnetic wave travels in the northward direction. At one instant, the electric field at a given point has a magnitude of 0.50 V/m and is in the eastward direction. What are the magnitude and direction of the magnetic field at the given point? What are the magnitude and direction of the Poynting vector?

16. The average energy flux of sunlight incident on the top of the Earth's atmosphere is 1.4×10^3 W/m^2. What are the corresponding amplitudes of oscillation of the electric and magnetic fields?

17. In the United States, the accepted standard for the safe maximum level of continuous whole-body exposure to microwave radiation is 10 milliwatts/cm^2.[4]

(a) For this energy flux, what are the corresponding amplitudes of oscillation of the electric and magnetic fields?

(b) Suppose that a man of frontal area 1.0 m^2 completely absorbs microwaves with an intensity of 10 milliwatts/cm^2 incident on this area and that the microwave energy is converted to heat within his body. What is the rate (in calories per second) at which his body develops heat?

18. A silicon solar cell of frontal area 13 cm^2 delivers 0.20 A at 0.45 V when exposed to full sunlight of energy flux 1.0×10^3 W/m^2. What is the efficiency for conversion of light energy into electric energy?

19. A magnifying glass of diameter 10 cm focuses sunlight into a spot of diameter 0.50 cm. The energy flux in the sunlight incident on the lens is 0.10 W/cm^2.

(a) What is the energy flux in the focal spot? Assume that all points in the spot receive the same flux.

[4] It is of interest that in the Soviet Union, where many experiments on the effects of microwaves on the human body have been performed, the accepted standard is much lower, 10 microwatts/cm^2.

(b) Will newspaper ignite when placed at the focal spot? Assume that the flux required for ignition is 2 W/cm^2.

20. Binoculars are usually marked with their magnification and lens size. For instance, 7 × 50 binoculars magnify angles by a factor of 7 and their collecting lenses have an aperture of diameter 50 mm. Your pupil, when dark adapted, has an aperture of diameter 7.0 mm. When observing a distant pointlike light source at night, by what factor do these binoculars increase the energy flux penetrating your eye?

21. At night, the naked, dark-adapted eye can see a star provided the energy flux reaching the eye is 8.8×10^{-11} W/m^2.

(a) Under these conditions, how many watts of power enter the eye? The diameter of the dark-adapted pupil is 7.0 mm.

(b) Assume that in our neighborhood there are, on the average, 3.5×10^{-3} stars per cubic light-year and that each of these emits the same amount of light as the Sun (3.9×10^{26} W). If so, how many stars could we see in the sky with the naked eye? How far would the faintest visible star be?

22. The beam of a powerful laser has a diameter of 0.2 cm and carries a power of 6 kW. What is the time-average Poynting vector in this beam? What are the amplitudes of the electric and the magnetic fields?

23. Calculate the instantaneous Poynting vector for the electromagnetic wave described in Problem 7.

24. A radio transmitter emits a time-average power of 5 kW in the form of a radio wave with uniform intensity in all directions. What are the amplitudes of the electric and magnetic fields of this radio wave at a distance of 10 km from the transmitter?

25. A radio receiver has a sensitivity of 2×10^{-4} V/m. At what maximum distance from a radio transmitter emitting a time-average power of 10 kW will this radio receiver still be able to detect a signal? Assume that the transmitter radiates uniformly in all directions.

26. A TV transmitter emits a spherical wave, i.e., a wave of the form given by Eq. (1). At a distance of 5 km from the transmitter, the amplitude of the wave is 0.22 V/m. What is the time-average power emitted by the transmitter?

27. A steady current of 12 A flows in a copper wire of radius 0.13 cm.

(a) What is the longitudinal electric field[5] in the wire?

(b) What is the magnetic field at the surface of the wire?

(c) What is the magnitude of the radial Poynting vector at the surface of the wire?

(d) Consider a 1.0-m segment of this wire. According to the Poynting vector, what amount of power flows into this piece of wire from the surrounding space?

(e) Show that the power calculated in part (d) coincides with the power of the Joule heat developed in the 1.0-m segment of wire.

28. A stationary electric point charge is located in a uniform magnetic field. Draw a diagram showing the lines of electric and magnetic field. Sketch the direction of the Poynting vector at a few places. Roughly sketch the flow lines for the energy according to the Poynting vector.

*29. A coaxial cable transmits DC power from a source of emf to a load (Figure 36.20). The cable consists of a long conducting straight wire of radius a surrounded by a conducting shell of radius b; assume that these conductors have zero resistance. The source has an emf $\mathscr{E}$ and delivers a power P.

(a) Show that the electric field within the coaxial cable is

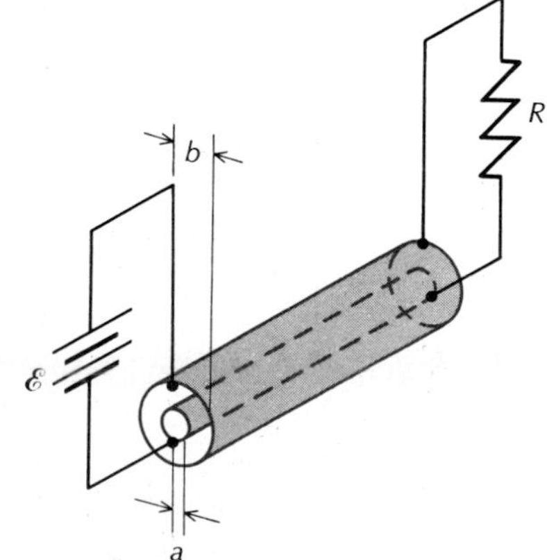

Fig. 36.20 A coaxial cable consisting of a central wire surrounded by a cylindrical shell. The radius of the wire is a; the radius of the shell is b.

[5] At the surface and outside the wire, the electric field also has a radial component. This contributes a *longitudinal* component to the Poynting vector, corresponding to the transport of energy *along* the wire.

$$E = \frac{\mathscr{E}}{\ln(b/a)} \frac{1}{r} \qquad \text{for } a < r < b$$

(b) What is the current in each conductor of the cable? Show that the magnetic field within the coaxial cable is

$$B = \frac{\mu_0}{2\pi} \frac{P}{\mathscr{E}} \frac{1}{r} \qquad \text{for } a < r < b$$

(c) What is the Poynting vector within the coaxial cable?
(d) Integrate the Poynting vector over the annular surface $a < r < b$ and show that the result equals P.

Section 36.5

30. According to a proposed scheme, solar energy is to be collected by a large power station on a satellite orbiting the Earth. The energy is then to be transmitted down to the surface of the Earth as a beam of microwaves. At the surface of the Earth, the beam is to have a width of about 10 km $\times$ 10 km and it is to carry a power of 5×10^9 W.
 (a) What would be the time-average Poynting vector in this beam?
 (b) What would be the amplitudes of the electric and magnetic fields in the beam?
 (c) What pressure would the beam exert on the surface of the Earth?

31. According to another proposal for the exploitation of solar energy, a large mirror is to be placed in orbit around the Earth. The mirror reflects sunlight and focuses it on a collector on the surface of the Earth.
 (a) One version of this proposal calls for a mirror 370 km in diameter. What force does the pressure of sunlight exert on such a mirror when face on to the Sun?
 (b) To prevent the mirror from drifting away under the influence of this force, it will have to be controlled with rocket thrusters. The mass of the mirror and its control system has been estimated as 5.7×10^9 kg. If the rocket thrusters are *not* switched on, how far would the mirror drift in one hour, starting from rest? Ignore the orbital motion of the mirror.

32. A grain of dust is spherical, with a diameter of 1.0×10^{-6} m. It consists of material with a density 2.0×10^3 kg/m^3. The grain is floating in interplanetary space, its distance from the Sun equal to the Earth–Sun distance.
 (a) What is the attractive force of the Sun's gravity on the grain?
 (b) What is the repulsive force of the Sun's radiation pressure? Which force is larger? Assume that the grain completely absorbs the incident sunlight.
 (c) Repeat the above calculations if the diameter of the grain is 0.50×10^{-6} m.
 (d) Show that the ratio of the forces of gravity and radiation pressure is independent of the distance of the grain from the Sun.

33. Astronauts of the future could travel all over the Solar System in a spaceship equipped with a large "sail" coated with a reflecting material. Such a "sail" would act as a mirror; the pressure of sunlight on this mirror could support and propel the spaceship. How large a "sail" do we need to support a spaceship of 70 metric tons (equal to the mass of Skylab) against the gravitational pull of the Sun? Ignore the mass of the "sail."

34. A beam of light is incident on a mirror at an angle of 30° with the normal and is entirely reflected. The energy incident in some time interval is 10^5 J. How much momentum is transferred to the mirror in this time interval?

35. A sinusoidal electromagnetic wave travels to the right in a coaxial waveguide consisting of an inner wire surrounded by an outer shell. The radius of the inner conductor is a and the radius of the outer conductor is b. Figure

36.21 shows the electric field lines and the magnetic field lines at one instant of time.

(a) Sketch the direction of the Poynting vector.
(b) The electric and magnetic fields in the region between the conductors have the mathematical forms

$$E = \frac{A}{r}\sin\left(\frac{\omega}{c}x - \omega t\right)$$

$$B = \frac{A}{cr}\sin\left(\frac{\omega}{c}x - \omega t\right)$$

where A is a constant, r is the radial distance from the axis of the wire ($b > r > a$), x is the distance along the wire (direction of propagation), and the other symbols have their usual meaning. Find the instantaneous power that comes out of the coaxial waveguide at its far end at $x = L$. Find the average power.
(c) Suppose that at the far end of the waveguide, the wave is absorbed by a perfect absorber. What is the average force exerted by the wave on the absorber?

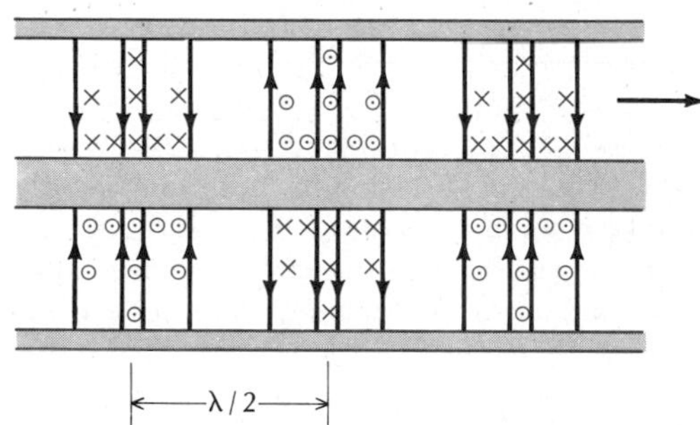

Fig. 36.21 Electric and magnetic field lines of an electromagnetic wave traveling toward the right in a coaxial waveguide. The drawing shows a longitudinal cross section of the waveguide. The electric field lines are radial lines. The magnetic field lines are circles around the central conductor; the dots and the crosses mark the heads and the tails of the magnetic field vectors.

36. Suppose that a magnetic monopole (source of magnetic field lines) and an electric point charge (source of electric field lines) are at some distance from one another.

(a) Draw a diagram showing the overlapping electric and magnetic field lines. Sketch the direction of the Poynting vector at a few places. Describe in words or pictures along what paths the energy flows in the space around the monopole and the charge.
(b) The Poynting vector gives not only the flow of energy, but also that of momentum. Accordingly, do you expect that the fields sketched in part (a) possess angular momentum?

*37. Question 20 describes how Einstein derived his equation $E = mc^2$, which expresses the equivalence of energy and mass. Formulating the arguments presented in this question mathematically, calculate how much mass the light pulse of energy E transports from one end of the boxcar to the other, and thereby derive Einstein's equation.

Section 36.6

38. In the spectrum of the light reaching us from the quasar PKS 0106+01 astronomers find some distinctive light emitted by hydrogen atoms. The wavelength of this light as received on Earth is 3776 Å. The wavelength as emitted by the atoms on the quasar is 1216 Å. Find the velocity of recession that will produce this shift of wavelength.

39. The light reaching us from the nearby galaxy NGC 221 has a Doppler shift, mainly due to the orbital motion of the Sun around the center of our Galaxy. A wavelength $\lambda_0 = 3968.5$ Å generated by calcium atoms in NGC 221 has shifted to $\lambda = 3965.8$ Å when detected by us on Earth. What is the radial speed of NGC 221 relative to us? Are we moving toward or away from this galaxy?

40. A spent Scout rocket, falling toward the surface of the Earth at 7.0 km/s, emits a radio signal at a frequency of 2203.08 MHz. What is the frequency of the radio signal received on the surface of the Earth?

41. The light reaching us from all the distant galaxies exhibits a red shift caused by the motion of recession of these galaxies. This is called the **cosmological red shift.** Astronomers usually measure this red shift by means of a wavelength generated by calcium atoms; in the rest frame of the atoms the wavelength is 3968.5 Å.[6] The following table lists several distant galaxies and the corresponding Doppler-shifted wavelength detected on the Earth. The table also lists the distances of these galaxies.

Galaxy	*Wavelength*	*Distance*
In Virgo Cluster	3984 Å	78×10^6 light-years
In Ursa Major Cluster	4167	980
In Corona Borealis Cluster	4254	1400
In Bootes Cluster	4485	2500
In Hydra Cluster	4776	4000

(a) Calculate the velocities of recession of these galaxies. Make a plot of the distances vs. the velocities.
(b) Assuming that the galaxies had the same velocities at earlier times, calculate at what time all of the galaxies were at zero distance from our Galaxy.

42. Doppler radar units, employed by police to measure the speed of automobiles, consist of a transmitter of microwaves and a receiver. The transmitter sends a wave of fixed frequency toward the target and the receiver detects the reflected wave sent back by the target. During this process the wave suffers two Doppler shifts: first when reaching the target and second when returning to the receiver.
(a) Suppose that an automobile approaches a Doppler radar unit at a speed of 60 mi/h. Calculate the fractional change of frequency between the transmitted and the received waves.
(b) In the radar unit, the transmitted and received frequencies are allowed to beat against one another. Suppose that the frequency of the transmitted microwaves is 8.0×10^9 Hz. What is the beat frequency?

43. Astronomers measure the velocity of rotation of stars by observing the Doppler shifts of light emitted by the approaching and the receding edge of the star. Suppose that the surface of a star emits a spectral line of a wavelength of 4101.74 Å. On the Earth, the wavelength received from the opposite edges of the surface are 4101.77 Å and 4101.71 Å.[7] What are the velocities of ap-

[6] This is a dark spectral line, i.e., an absorption line.

[7] The light received from other points of the surface will have a Doppler shift somewhere between the Doppler shifts of the light from the edges. Thus, the light received on Earth contains a certain continuous range of wavelengths — the sharp spectral line is *broadened* by the Doppler shift. Thermal motions of the atoms on the surface of the star contribute a further broadening.

proach and recession? What is the angular velocity of rotation of the star if its radius is 7.0×10^8 m and its axis of rotation is perpendicular to the line of sight?

44. Suppose that an astronomer attempting to measure the velocity of rotation of a star by the method described in Problem 43 finds that the wavelengths he receives from opposite edges of the star are 4103.82 Å and 4103.76 Å, respectively. As in Problem 43, the wavelength of this light is 4101.74 Å when emitted by the surface of the star. What can the astronomers conclude from this regarding the rotational and translational velocities of the star? Assume that the axis of rotation is perpendicular to the line of sight.

45. What is the percentage difference between the Doppler-shifted frequencies predicted by the formulas (48) and (51) for a receding emitter with $v/c = 0.10$? For a receding emitter with $v/c = 0.90$?

CHAPTER 37

Reflection and Refraction

So far we have examined the propagation of electromagnetic waves only in a vacuum. There, a plane wave will simply propagate in a fixed direction at the constant speed c. But if the wave encounters a region filled with matter — a sheet of metal, a pane of glass, or a layer of water — then the wave will interact with the matter and suffer changes in speed, direction, intensity, and polarization. These changes can of course be calculated from Maxwell's equations, taking into account the electric charges and currents induced in matter. Sometimes the calculations get extremely complicated and, furthermore, Maxwell's equations often tell us more than we want to know. For instance, if a wave is incident on a water surface we may wish to compute the angle at which it penetrates, but we often do not need to know the changes in intensity and polarization. We will see that a simple rule, called Huygens' Construction, permits us to discover, without complicated calculations, how the change of direction of propagation is related to the change of speed of the wave. Huygens' Construction is a geometric construction, bypassing Maxwell's equations. This construction serves as the basis for **geometric optics,** which studies the propagation of light under the assumption that light propagates in a fixed direction (rectilinearly) while in a uniform medium, and suffers changes of direction only when it encounters an interface between two different media.

Geometric optics

37.1 Huygens' Construction

The propagation of a wave can be conveniently described by means of the *wave fronts,* or wave crests, that is, the points at which the wave has maximum amplitude[1] at some instant of time. For example, Figure 37.1 shows the instantaneous wave fronts of the radio wave emitted by an antenna. With the passing of time, each of these wave fronts travels in the outward direction.

Huygens' Construction

The rule governing the propagation of wave fronts is **Huygens' Construction:**

> *To find the change of position of a wave front in a small time interval Δt, draw many small spheres of radius [wave speed] $\times$ Δt with centers on the old wave front. The new wave front is the surface of tangency to these spheres.*

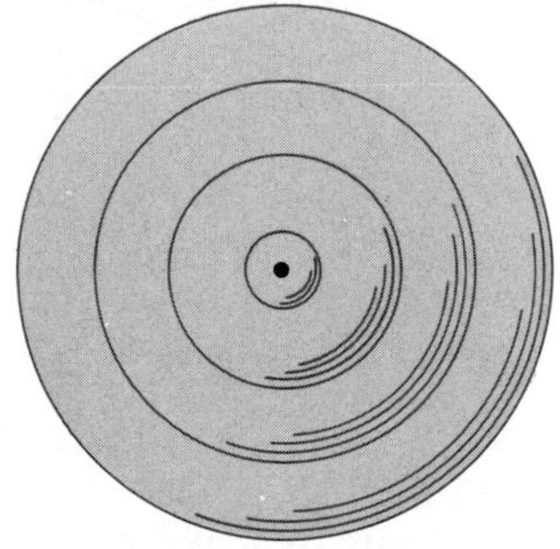

Fig. 37.1 Spherical wave fronts at one instant of time. At a later time, each of these wave fronts will have moved outward by some distance.

The small spheres employed in this construction are called **wavelets.** Figure 37.2 shows how Huygens' Construction applies to the propagation of one of the spherical wave fronts of Figure 37.1. The wave speed in this example is simply c and hence the radius of the wavelets is $c\ \Delta t$. Note that the wavelets of Figure 37.2 have two tangent surfaces, one in the outward direction and one in the inward direction. Obviously, only the former surface is appropriate to our problem (the latter surface would be appropriate if we were dealing with a convergent wave sent toward, say, a radio transmitter and precisely focused on it; this wave would be the time reverse of the wave sent out by the transmitter).

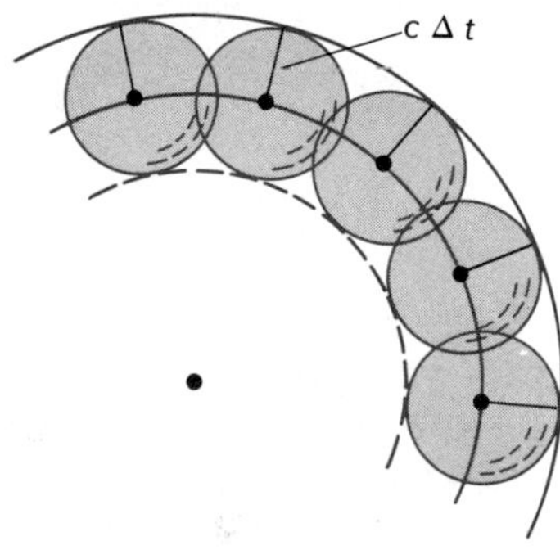

Fig. 37.2 Huygens' Construction for the propagation of a wave front. The inner solid arc shows the wave front at a time t; the outer solid arc shows the propagated wave front at time $t + \Delta t$. The dashed arc shows the propagated wave front for a time-reversed (convergent) wave.

Huygens' Construction applies not only to propagation of electromagnetic waves in a vacuum, but also to propagation in any transparent medium. As we will see in the following sections, this construction allows us to derive the laws of reflection and refraction. Although our emphasis will be on the propagation of light, Huygens' Construction is a general feature of wave propagation — it applies just as well to sound waves and water waves. The laws of reflection and refraction for all such waves are essentially the same and some of our examples will take advantage of this.

37.2 Reflection

When a light wave encounters a material surface — such as the surface of a pane of glass, or the surface of a pond — part of the wave penetrates the surface and part is reflected. If the surface consists of very smoothly polished metal — such as the silvered surface of a mirror — almost all of the wave is reflected. In this section we will deal with the reflected part of the wave; in the next section we will deal with the part of the wave that penetrates from one medium into the other.

Law of reflection

The **law of reflection** for a wave incident on a flat surface is well known: *the angle of incidence equals the angle of reflection.* (The same law of reflection is valid for a wave incident on a small portion of a curved surface, which can be approximated by a flat, tangent surface.) To derive this law from Huygen's Construction, we begin with Figure 37.3a, which shows wave fronts approaching a reflecting surface; one edge of the leading wave front is barely touching the surface at the point P. Figure 37.3b shows some Huygens' wavelets a short time later; the portions of these wavelets below the reflecting surface have been omitted as irrelevant. The new wave front touches the reflecting surface at

[1] By amplitude of an electromagnetic wave we will hereafter always mean the amplitude of the electric field.

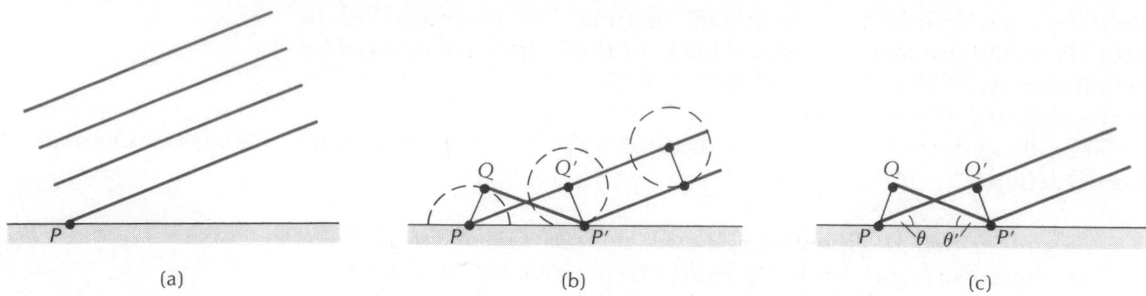

Fig. 37.3 (a) Wave fronts approaching a reflecting surface. The leading wave front barely touches the reflecting surface. (b) Huygens wavelets erected on the leading wave front of part (a). (c) The incident wave front PQ' makes an angle θ with the reflecting surface; the reflected wave front QP' makes an angle θ' with this surface.

the point P'. Obviously, to the right of the point P', the new wave front is simply parallel to the old wave front, i.e., this part of the wave has not yet been reflected. To find the new wave front to the left of the point P', we draw a straight line that starts at P' and is tangent to the wavelet centered on P. This straight line represents the part of the wave that has already been reflected. To see that the incident wave front and the reflected wave front make the same angle with the reflecting surface, we appeal to Figure 37.3c. The right triangles $PQ'P'$ and $P'QP$ are congruent since they have a common side (PP') and their short sides (PQ and $P'Q'$) are equal. Hence the angles θ and θ' are equal.

Rays

The direction of propagation of a wave is often described by the **rays** of the wave. These are lines perpendicular to the wave fronts. For example, Figure 37.4 shows the rays associated with the incident and reflected wave fronts.

Angle of incidence and of reflection

The angle θ (or θ') between the wave front and a reflecting surface is obviously equal to the angle between the ray and the normal to the surface. The angles θ and θ' are called the **angles of incidence and of reflection** (Figure 37.5). Thus, from the Huygens' Construction we have deduced that the angle of incidence equals the angle of reflection, which is the law of reflection. Note that in all of the above, we have implicitly taken it for granted that the incident and the reflected rays are coplanar. This is evident from considerations of symmetry, and it can also be deduced from Huygens' Construction by taking into account that, in three dimensions, the wave fronts in Figures 37.3a and b are planes and the wavelets are spheres.

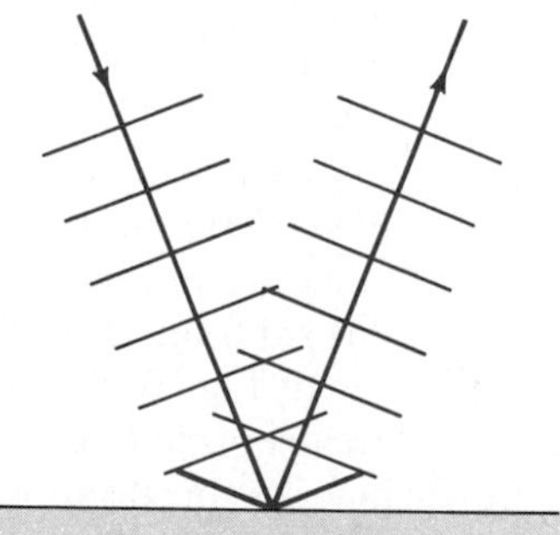

Fig. 37.4 The rays of the wave are perpendicular to the wave fronts.

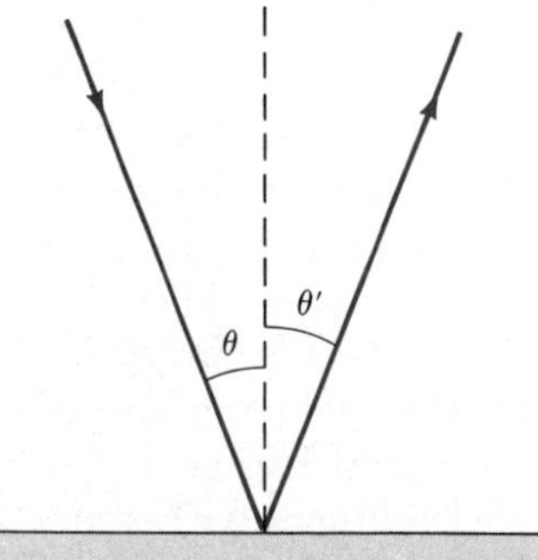

Fig. 37.5 The angle of incidence θ and the angle of reflection θ'. These angles are the same as in Figure 37.3c.

When light from some source strikes a flat mirror, the reflection of the light leads to formation of an image of the source. Figure 37.6 shows a point source of light and the rays emerging from it; it also shows the reflected rays. If we extrapolate the reflected rays to the far side of the mirror, we find that they all appear to come from a point source of light. This apparent point source is the **image.** To an eye looking into the mirror, the image looks like the original object — the eye perceives the mirror image as existing in the space beyond the mirror. But the mirror image is an illusion; the light does not come from beyond the mirror. This kind of illusory image that gives the impression that light rays emerge from where they do not is called a **virtual image.**

Virtual image

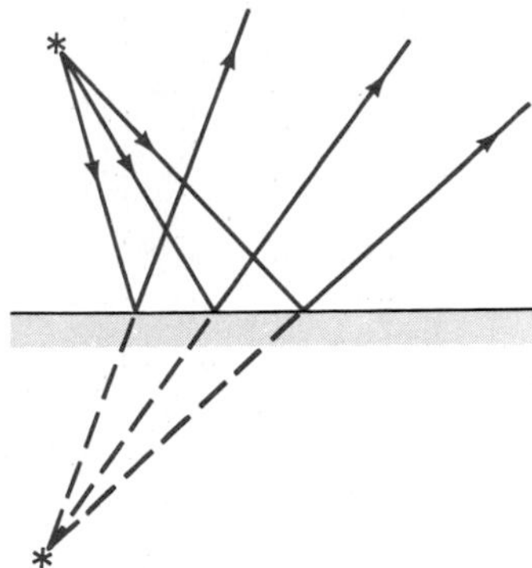

Fig. 37.6 Rays emerging from a point source are reflected by a mirror. The extrapolated rays (dashed) appear to come from a point source beyond the mirror.

If, instead of a single luminous point, our light source consists of an extended object made of many luminous points, then the mirror image will also be an extended object. Note that the mirror image of an object is a mirror-reversed object. For instance, Figure 37.7 shows some written letters and their mirror images. This reversal is commonly referred to as a reversal of left and right. However, it is more accurately described as a reversal of front to back — mirror writing is ordinary writing seen from behind. And the mirror image of, say, a hand facing north is a hand facing south (the reversal is not an ordinary "about face," but involves passing the front of the hand through its back, thereby converting a right hand into a left hand, and vice versa; Figure 37.8).

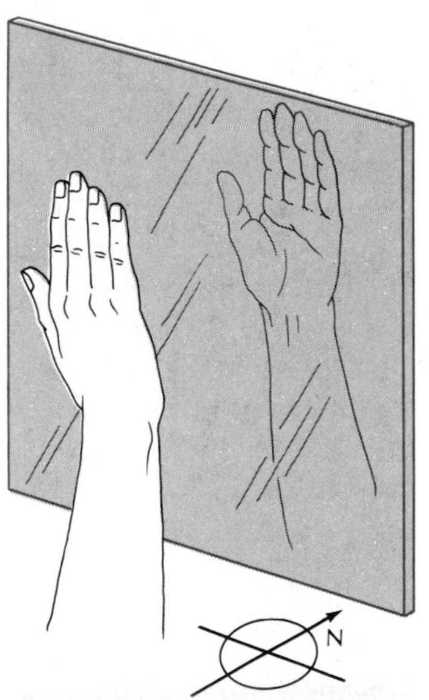

Fig. 37.7 (left) Some letters and their images in a mirror.

Fig. 37.8 (right) A hand facing north and its image in a mirror.

Two mirrors at right angles form a **corner reflector.** A ray of light incident on this reflector is sent back in exactly the same direction it came from. Figure 37.9 shows how this works in two dimensions. But the same principle also applies to three dimensions; here it involves three mirrors at right angles to each other (a corner) and the light ray is sent back upon three successive reflections. Reflectors on automobiles and bicycles make use of such reflectors arranged in an array of a large number of small corners. Radar reflectors on boats apply the same principle; Figure 37.10 shows such a reflector to be hung on the

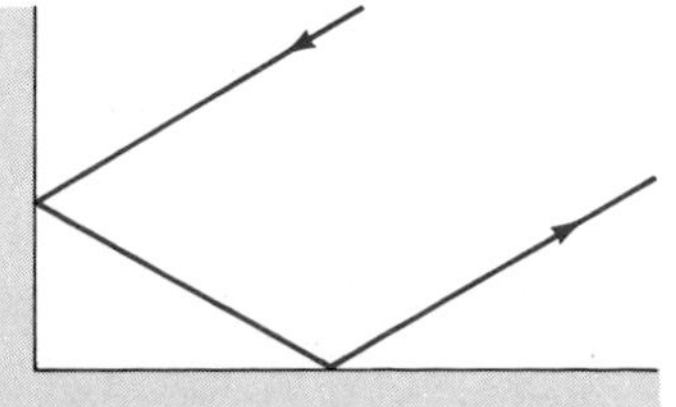

Fig. 37.9 A corner reflector.

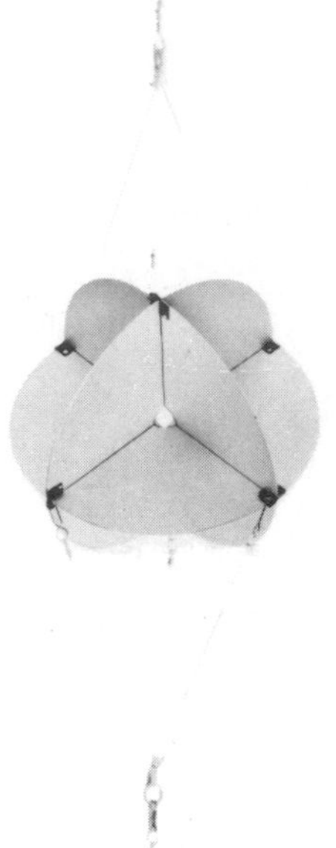

Fig. 37.10 Radar reflector for a sailboat.

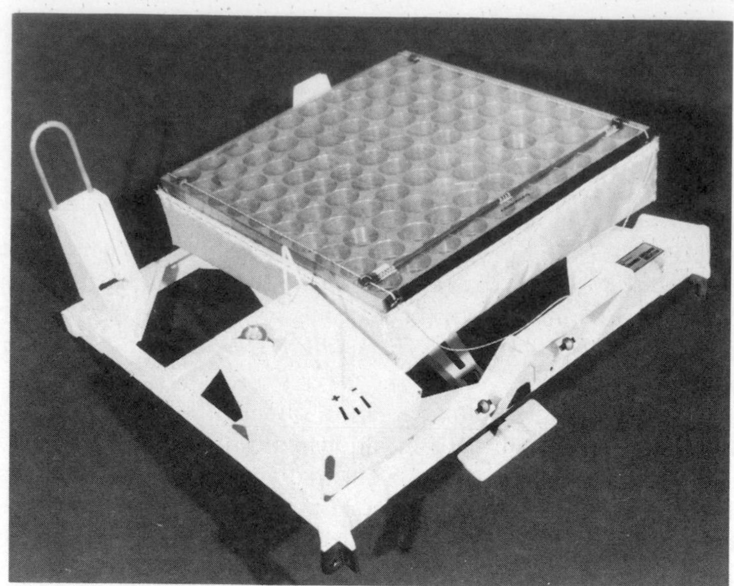

Fig. 37.11 Corner reflectors for installation on the surface of the Moon.

mast of a sailboat. Finally, Figure 37.11 shows an array of corner reflectors deposited on the Moon by the Apollo 11 astronauts. This was used as a target for a laser beam sent from the Earth to the Moon, in an experiment that determined the Earth–Moon distance with very high precision (±15 cm) by measuring the travel time of a pulse of laser light sent from the Earth to the Moon and reflected back to the Earth (see also Interlude L.3).

37.3 Refraction

The speed of light in a material medium — such as air, water, or glass — differs from the speed of light in a vacuum. We can recognize this immediately by recalling the theoretical formula for the speed of light derived from Maxwell's equations,

$$c = \frac{1}{\sqrt{\varepsilon_0 \mu_0}} \tag{1}$$

We know from Chapters 27 and 33 that in a material medium with given dielectric and magnetic characteristics, the quantities ε_0 and μ_0 in Maxwell's equations get replaced by $\kappa\varepsilon_0$ and $\kappa_m\mu_0$ [compare Eqs. (27.32) and (33.9)]. Hence, the formula (1) for the speed of light likewise gets replaced by

$$v = \frac{1}{\sqrt{\kappa\kappa_m \varepsilon_0 \mu_0}} \tag{2}$$

This is usually written as

$$\boxed{v = \frac{c}{n}} \tag{3}$$

Speed of light in a material medium

where

$$n = \sqrt{\kappa \kappa_m} \tag{4}$$

The quantity n is called the **index of refraction** of the material. In most materials the value of κ_m is very near 1 except, of course, in ferromagnetic materials, where light does not propagate anyhow (see Tables 33.1 and 33.3). Therefore the factor κ_m in Eq. (4) can often be omitted. This permits us to write

$$\boxed{n = \sqrt{\kappa}} \tag{5}$$

Index of refraction

One important warning: the value of the dielectric κ depends on the frequency of the electric field. Hence the values of the dielectric constants from Table 27.1 cannot be inserted in Eq. (5), because the former values only apply to static fields whereas we are now concerned with the high-frequency fields of a light wave.

Table 37.1 gives the values of the index of refraction for a few materials. The values in this table apply to light waves of medium frequency (yellow-green light). The index of refraction is slightly larger for blue light and slightly smaller for red light; we will deal with this complication later in this section.

With the wave speed $v = c/n$, the relation between frequency and wavelength becomes

$$\lambda \nu = v = \frac{c}{n}$$

or

$$\boxed{\lambda = \frac{c}{n}\frac{1}{\nu}} \tag{6}$$

Table 37.1 INDICES OF REFRACTION OF SOME MATERIALS[a]

Material	*n*
Air, 1 atm, 0°C	1.00029
1 atm, 15°C	1.00028
1 atm, 30°C	1.00026
Water	1.33
Ethyl alcohol	1.36
Castor oil	1.48
Quartz, fused	1.46
Glass, crown	1.52
light flint	1.58
heavy flint	1.66

[a] For light of wavelength ~5500 Å.

For example, if a wave penetrates from air into water, where $n = 1.33$, its speed is reduced by a factor of 1.33, but its frequency remains constant. Consequently, Eq. (6) shows that its wavelength will be reduced by a factor of 1.33. The fact that the frequency of the wave remains constant can be understood in terms of the atomic mechanism underlying the change of wave velocity. When the wave strikes the water surface, it shakes the electrons of the water molecules; this acceleration of electric charges produces extra waves, which combine with the original wave. The net electromagnetic wave within the water is then a superposition of the original wave plus the extra waves, and it is this superposition that makes up the refracted wave with its changed direction. The superposition has the same frequency as the original wave because the shaking of electrons proceeds at the original frequency and therefore the extra waves generated also will have exactly this frequency.

Incidentally, Eq. (6) does not imply that a light source changes color when immersed in water. The color we perceive depends on the *frequency* of the light reaching our eyes; and this frequency is independent of whether the light source, our eyes, or both are immersed in water or in air.

When a wave strikes the surface of a dielectric, part of it is reflected and part of it penetrates the dielectric. In the preceding section we have investigated the direction of propagation of the reflected wave; now let us investigate the penetrating wave. Again, we will use Huygens' Construction to find out what the wave does when it strikes the dielectric surface. In vacuum the speed of light is c; in the dielectric it is c/n. Figure 37.12a shows a wave front at one instant of time; the left edge of this wave front barely touches the surface. Figure 37.12a also shows the Huygens wavelets that determine the position of the wave front at some later time; only the forward portions of these wavelets are relevant. Above the dielectric surface, the wavelet has a radius $c\,\Delta t$; below the dielectric surface, the wavelet has a smaller radius $(c/n)\,\Delta t$. The reduction of the speed of propagation of one side of the wave front causes the wave front to swing around, changing its direction of advance. The right triangles $PP'Q$ and $PP'Q'$ have the side PP' in common (Figure 37.12b). In terms of the length l of this common side, the sines of the angles between the wave fronts and the dielectric surface are

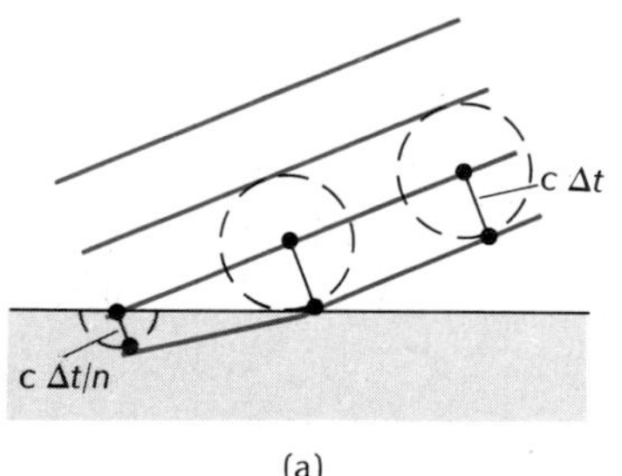

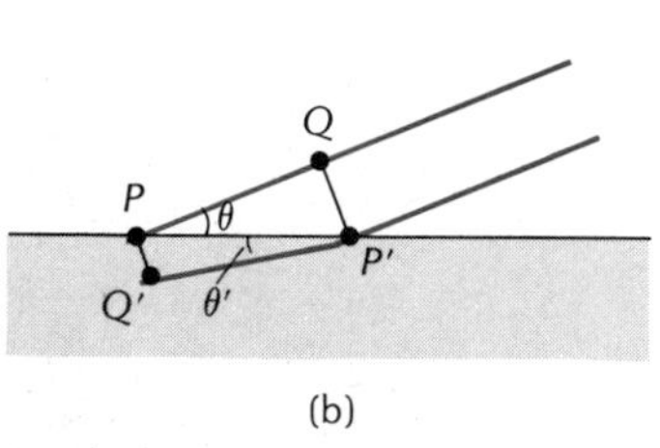

Fig. 37.12 (a) Huygens wavelets erected on a wave front whose edge barely touches the dielectric surface. (b) The incident wave front makes an angle θ with the dielectric surface; the refracted wave front makes an angle θ'.

$$\sin\theta = \frac{c\,\Delta t}{l} \tag{7}$$

and

$$\sin\theta' = \frac{(c/n)\,\Delta t}{l} \tag{8}$$

The ratio of this pair of equations is

$$\frac{\sin\theta}{\sin\theta'} = \frac{c}{c/n} \tag{9}$$

or

$$\boxed{\sin\theta = n\sin\theta'} \qquad (10)$$

Law of refraction

This equation describes the change of direction of a wave upon penetration into a dielectric. This change of direction is called **refraction.** Equation (10) is called the **law of refraction,** or **Snell's Law.** The angle θ is the angle of incidence, and θ' is the angle of refraction. It is usually convenient to measure these angles between the rays and the normal to the dielectric surface; Figure 37.13 shows the incident and refracted rays, and the angles of incidence and refraction. Note that the ray in the dielectric is bent toward the normal ($\theta' < \theta$). Also note that, as in the case of reflection, the incident and the refracted rays are coplanar.

Our formula (10) describing refraction at the interface between vacuum and dielectric is a special case of a general formula describing refraction at the interface of two dielectrics. If the indices of refraction are n_1, n_2 and the angles between the rays and the normals are θ_1, θ_2, then

$$n_1 \sin\theta_1 = n_2 \sin\theta_2 \qquad (11)$$

This equation can be derived by some simple changes in the argument that led to Eq. (10).

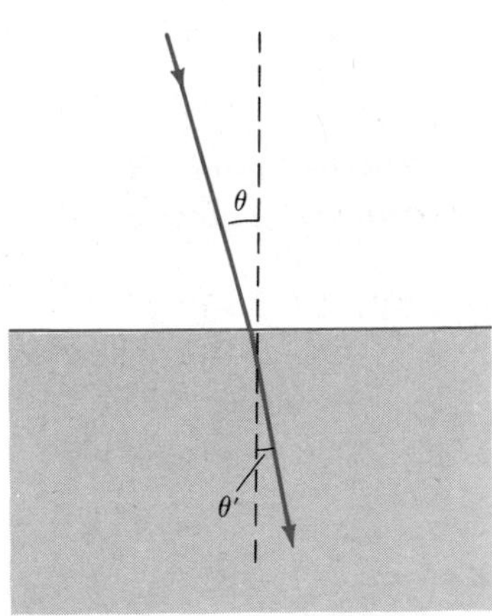

Fig. 37.13 The angle of incidence θ and the angle of refraction θ'.

EXAMPLE 1. A small, shiny fish is in the water 1.0 m below the surface. Where does a fisherman looking down into the water see the fish; that is, where is the image of the fish?

SOLUTION: Figure 37.14 shows two light rays from the fish to the eye of the fisherman. The first light ray is perpendicular to the surface and is not bent. The second light ray is bent away from the normal. Since the angles are small, Eq. (10) gives

$$\theta = n\theta' = 1.33\theta' \qquad (12)$$

Extrapolation of the refracted ray into the water shows that it intersects the vertical ray at the point P', above the point P. Hence the image is above the object. The image distance OP' and the object distance OP are related as follows (Figure 37.14):

$$OP \tan\theta' = OP' \tan\theta \qquad (13)$$

For small angles this yields the approximation

$$\frac{OP'}{OP} = \frac{\theta'}{\theta} = \frac{1}{1.33} \qquad (14)$$

Hence the image distance is smaller than the object distance by a factor of 1.33. If $OP = 1$ m, then $OP' = 0.75$ m. The fish seems to be nearer to the surface than it is.

Note that the above calculation hinges on the assumption that the angles are small (so that $\sin\theta \cong \theta$, $\tan\theta \cong \theta$).

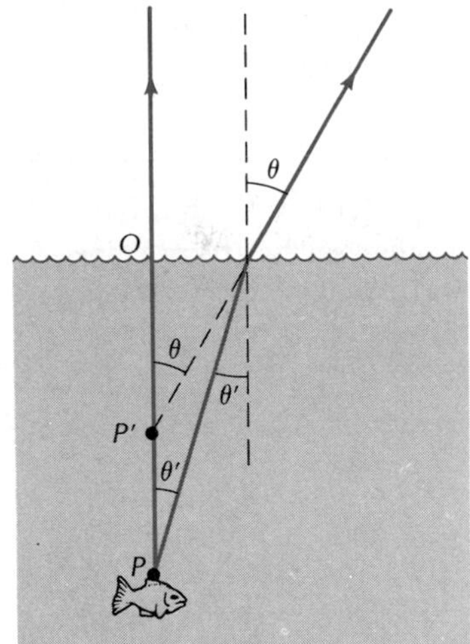

Fig. 37.14 A small shiny fish acts as a source of light. Note that the direction of propagation of the ray is opposite to that shown in Figure 37.13; but this does not affect the validity of Eq. (10). The extrapolated ray (dashed) appears to come from the point P'.

For a ray of light attempting to leave water, there is a critical angle beyond which refraction is impossible. As the light ray emerges into air it is bent away from the normal; in the extreme case it is bent so much that it lies almost along the water surface (Figure 37.15). This extreme case corresponds to $\theta = 90°$ in Eq. (10),

$$n \sin \theta' = \sin 90° = 1 \tag{15}$$

The critical angle for this extreme form of refraction is therefore

Critical angle for total internal reflection

$$\boxed{\theta'_c = \sin^{-1}\left(\frac{1}{n}\right)} \tag{16}$$

which for $n = 1.33$ gives

$$\theta'_c = \sin^{-1}\left(\frac{1}{1.33}\right) = 48.75° \tag{17}$$

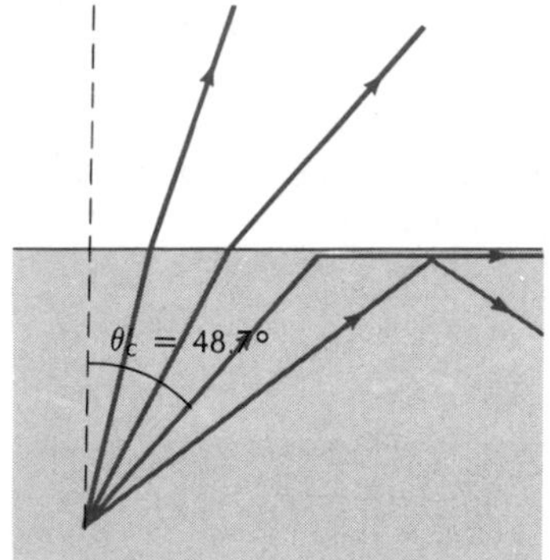

Fig. 37.15 A ray approaching a water surface from below with an angle of incidence $\theta'_c = 48.7°$ is refracted along the water surface.

If a light ray strikes a water surface from below at an angle larger than this, refraction is impossible. The only alternative is reflection — the water surface behaves as a perfect mirror. This phenomenon is called **total internal reflection.** It will occur whenever the index of refraction of the medium containing the light ray is larger than the index of refraction of the adjacent medium.

Total internal reflection has many important applications in optics. For instance, in a periscope the light is reflected down the tube by internal reflection in a prism (Figure 37.16); this gives a much better image than reflection in a mirror. In an optical fiber, light moves along a thin rod made of a transparent material; the light zigzags back and forth between the walls of the rod, undergoing a sequence of total internal reflections (Figure 37.17). Such optical fibers are being used to replace telephone cables. The electrical impulses normally carried on a cable are converted into pulses of light which can be transmitted via an optical fiber. The efficiency of optical fibers is very high because they can carry many conversations simultaneously; in modern telephone systems, single optical fibers are being used to carry 670 telephone conversations simultaneously.

In most materials the index of refraction depends somewhat on the wavelength of light. Usually, the index of refraction increases as the wavelength decreases. For instance, Figures 37.18 and 37.19 are plots of the index of refraction of light in water and in flint glass as a function of wavelength (the wavelengths plotted along the horizontal axis in these figures are measured in air, before the light penetrates the

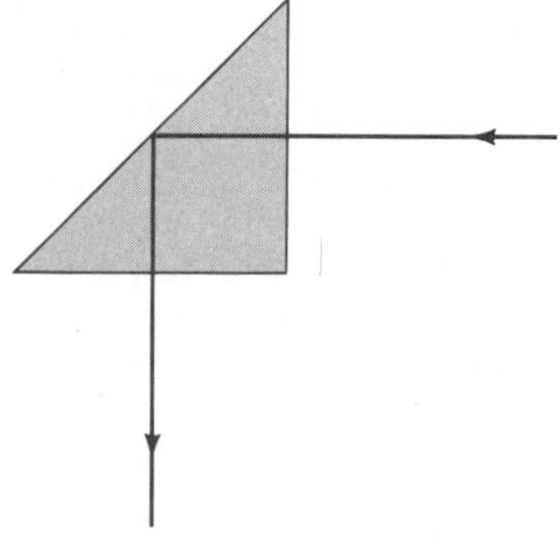

Fig. 37.16 Internal reflection in a prism.

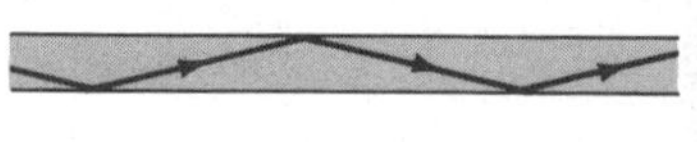

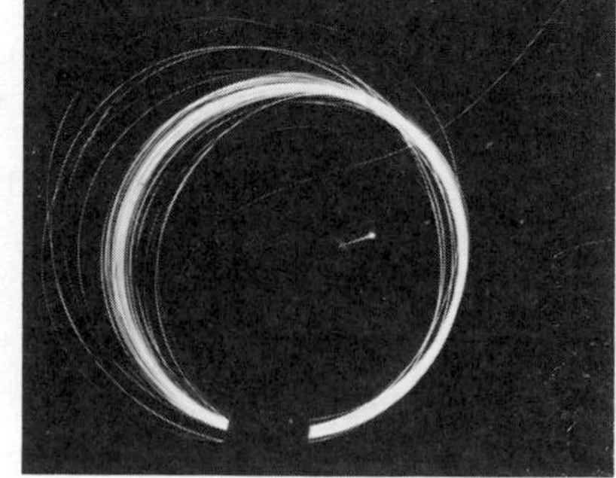

Fig. 37.17 (above) Internal reflection in an optical fiber. (right) Light enters the optical fiber at the top right and emerges at the center.

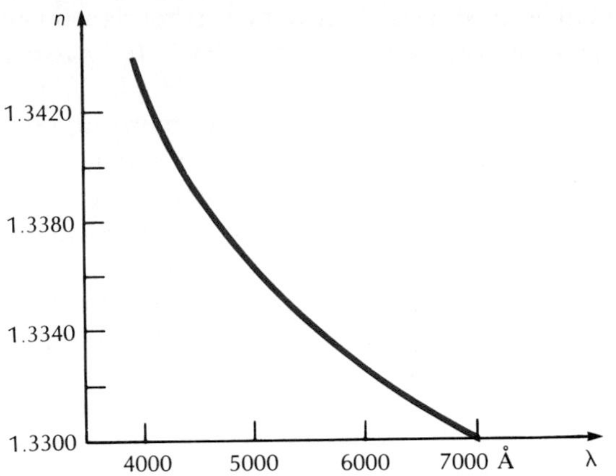

Fig. 37.18 Index of refraction of light in water as a function of wavelength. The index of refraction varies by about 1% over the range of visible wavelengths.

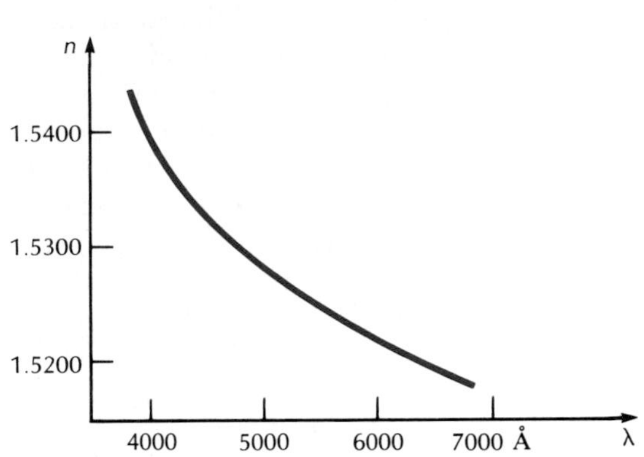

Fig. 37.19 Index of refraction of light in glass (Schott telescope flint glass) as a function of wavelength.

material). When a ray of light containing several wavelengths, or colors, is refracted by a medium with an index of refraction that depends on wavelength, the refracted rays of different colors will emerge at somewhat different angles. The separation of a ray by refraction into distinct rays of different colors is called **dispersion.**

Dispersion

EXAMPLE 2. The index of refraction for red light in water is 1.330 and for violet light it is 1.342. Suppose that a ray of light approaches a water surface with an angle of incidence of 80°. What are the angles of refraction for red light and for violet light?

SOLUTION: For $n = 1.330$, Eq. (10) yields

$$\sin \theta' = \frac{\sin 80°}{1.330} = 0.740 \tag{18}$$

and

$$\theta' = 47.8°$$

For $n = 1.342$, Eq. (10) yields

$$\sin \theta' = \frac{\sin 80°}{1.342} = 0.734 \tag{19}$$

and

$$\theta' = 47.2°$$

Thus the violet light is bent more toward the normal than the red light (Figure 37.20). For $\theta = 80°$, the difference in the angles of refraction is about 0.6°. Thus, refraction in water slightly separates a light ray according to colors. A beautiful demonstration of this effect is found in rainbows, which are produced by the refraction of sunlight in water droplets.

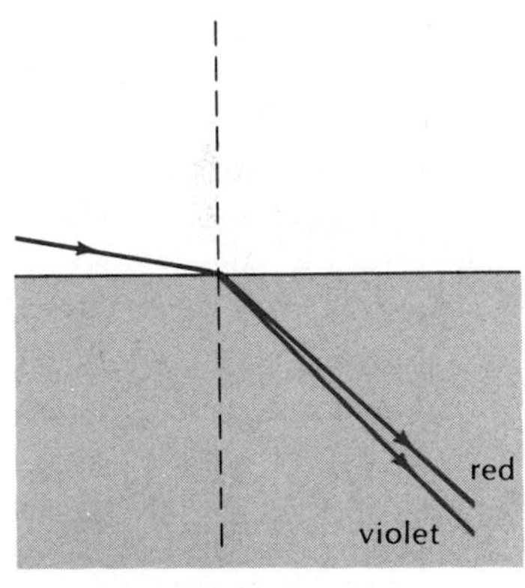

Fig. 37.20 Refraction of red and of violet light in water. The difference between the angles of the refracted rays has been exaggerated for the sake of clarity.

Prism

A **prism** is the traditional device employed to separate light rays into their constituent colors. The basic mechanism is the same as that discussed in Example 2: the glass in the prism has different indices of refraction for light of different wavelengths and hence it bends rays of different colors by different amounts (Figure 37.21). In passing through a prism, the light is refracted twice: first at the air–glass interface and then at the glass–air interface. Under normal operating conditions, a good prism will introduce a difference of several degrees between the angular directions of the emerging red and violet light rays. The pattern of colors produced by the analysis of a light ray by means of a prism is called the **spectrum** of the light. The white light emitted by the Sun has a continuous spectrum consisting of a mixture of all the colors. The colored light emitted by the atoms of a chemical element in an electric discharge tube, such as a neon tube, has a discrete spectrum consisting of just a few discrete colors. Each of these discrete colors is absolutely pure, i.e., it is light of a single wavelength. For example, hydrogen atoms emit the following discrete colors: red (6563 Å), blue-green (4861 Å), blue-violet (4340 Å), and violet (4102 Å). These discrete colors are called **spectral lines.**[2] Figure 37.22 shows the spectral lines of hydrogen as displayed by means of a prism illuminated with light from a fine slit. Each of the lines in this figure is a separate image of the slit made by a separate color after refraction by the prism (Figure 37.23).

Spectrum

Spectral lines

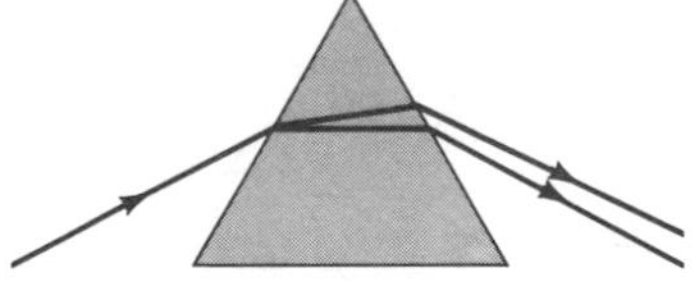

Fig. 37.21 Refraction of red and of violet light by a prism.

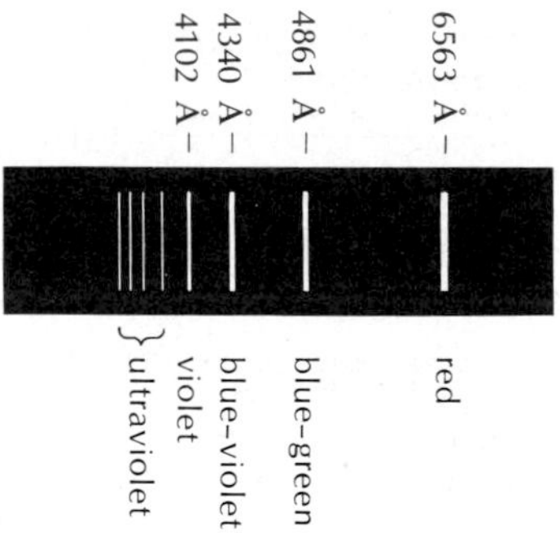

Fig. 37.22 The spectral lines of hydrogen. (A color print of these spectral lines appears between pages 850 and 851.)

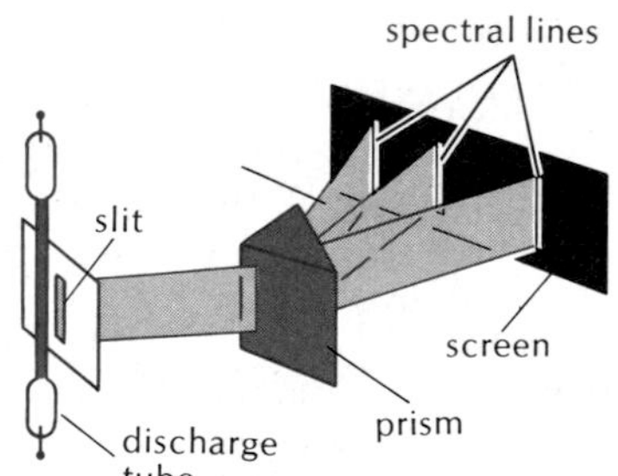

Fig. 37.23 Analysis of light by means of a prism.

37.4 Spherical Mirrors

Focal point

A mirror with a surface curved like the surface of a sphere can focus a beam of light to a point. Figure 37.24 shows a *concave* spherical mirror and wave fronts of light incident and reflected on this mirror; the reflected waves converge to a point, the **focal point** of the mirror. We can describe the direction of propagation of the waves by rays. Figure 37.25 shows the incident and the reflected rays. The reflected rays converge at the focal point.

The focal point of the spherical mirror is halfway between the mirror and the center of the spherical surface. To prove this, we use Figure 37.26, which shows the path of a single ray of light. The focal point is the intersection of this ray with the axial line *CA*. To find the

[2] Hydrogen also emits spectral lines in the ultraviolet and in the infrared; see Chapter 41.

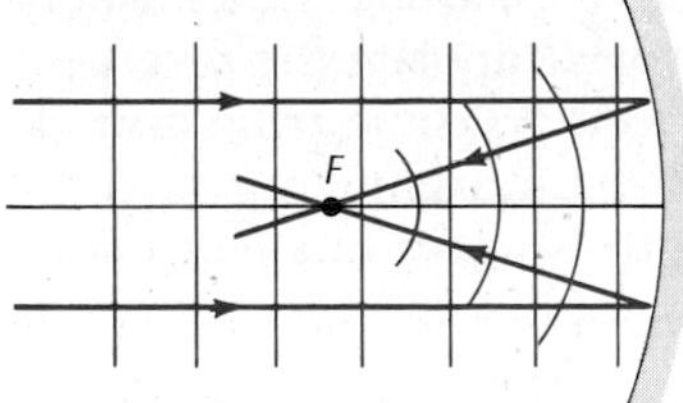

Fig. 37.24 A concave spherical mirror focuses an incident plane wave on a point.

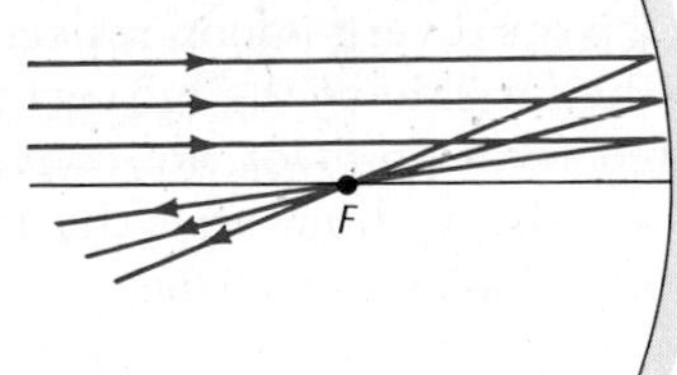

Fig. 37.25 Reflection of parallel rays by a concave spherical mirror.

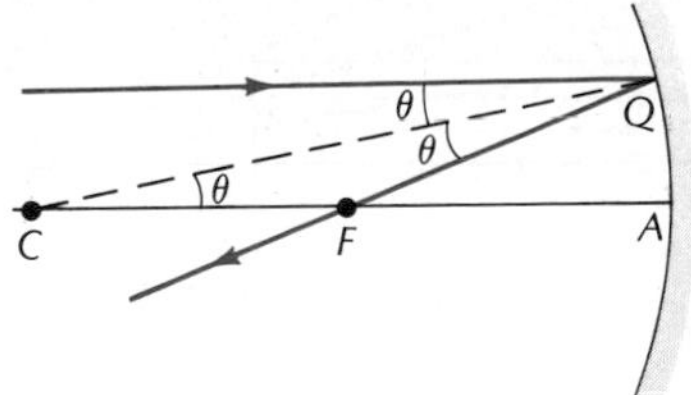

Fig. 37.26 Reflection of a single ray.

distance FA, called the **focal length,** we begin with the observation that in the isosceles triangle CFQ the length CF equals FQ. Under the assumption that the angle θ is small (equivalently, that the incident ray is near the axial line), the length FQ is approximately equal to FA. Hence

$$CF = FA \tag{20}$$

that is, the point F is halfway between the mirror (A) and the center (C) of the spherical surface. The focal length FA is therefore one-half of the radius of the spherical surface. Designating the former by f and the latter by R, we can write

Focal length of spherical mirror

$$\boxed{f = \tfrac{1}{2}R} \tag{21}$$

Note that this focusing of a beam of light rests on the approximation that the beam is narrow so that all the rays are near the axial line. If a ray strikes the mirror at some appreciable distance from the axial line, then the reflected ray will miss the focal point (Figure 37.27). This defect in the focusing properties of a mirror is called **spherical aberration.** For better focusing, it is necessary to replace the spherical mirror by a paraboloidal mirror (Figure 37.28).

Spherical aberration

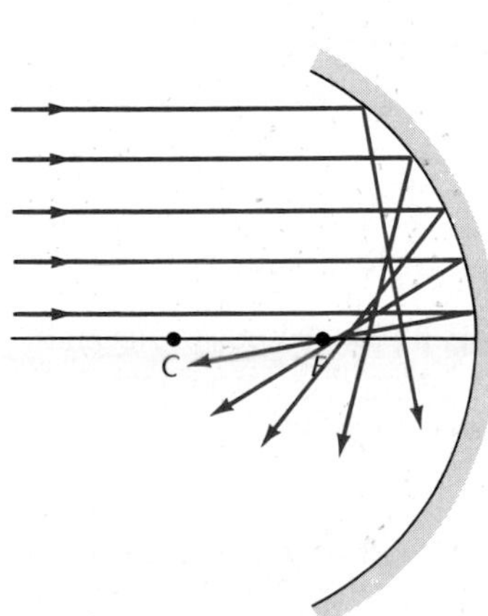

Fig. 37.27 Reflection of rays far from the axis of the mirror. Such rays miss the focal point.

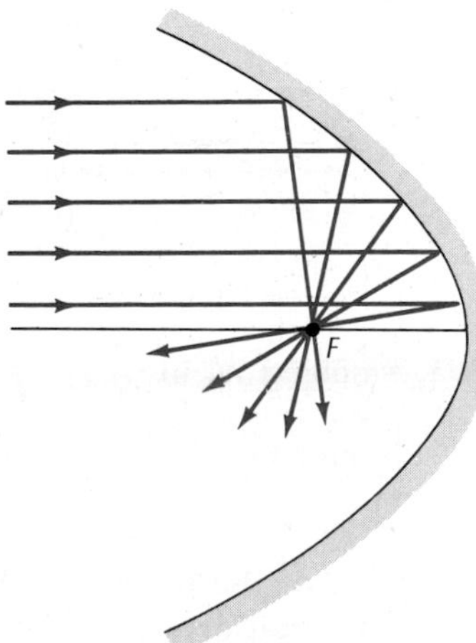

Fig. 37.28 Reflection of parallel rays by a paraboloidal mirror. All rays converge at the focal point.

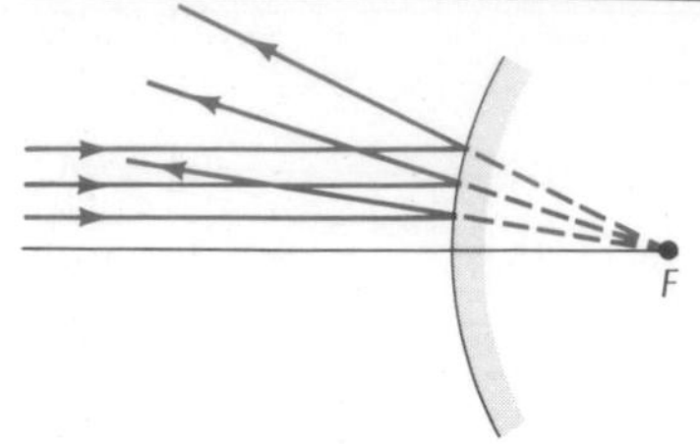

Fig. 37.29 Reflection of parallel rays by a convex spherical mirror.

Figure 37.29 shows a *convex* spherical mirror. Parallel rays incident on this mirror diverge upon reflection. If we extrapolate the divergent rays to the far side of the mirror, they all seem to come from a single point, the focal point of the convex mirror. An argument similar to that given above shows that the focal length is again one-half of the radius of the spherical surface,

$$\boxed{f = -\tfrac{1}{2}R} \tag{22}$$

A negative sign has been inserted in Eq. (22) to indicate that the focal point is on the far side of the mirror.

Both concave and convex mirrors will form images of objects placed in front of them. Figure 37.30 shows a point source of light in front of a concave mirror. To find the position of the image, we must trace some of the rays of light. The three rays that are easiest to trace are (i) the ray PC through the center, (ii) a ray PR parallel to the axis, and (iii) a ray PF through the focal point (Figure 37.30). The first of these rays strikes the mirror perpendicularly and is therefore reflected on itself; the second ray passes through the focus after being reflected; and the third ray emerges parallel to the axis after being reflected. All these rays, and any other rays originating at P, come together at P'. This point is the image of the point source. Note that to locate the image, two out of the three rays mentioned above are already sufficient — the third is redundant but serves as a useful check.

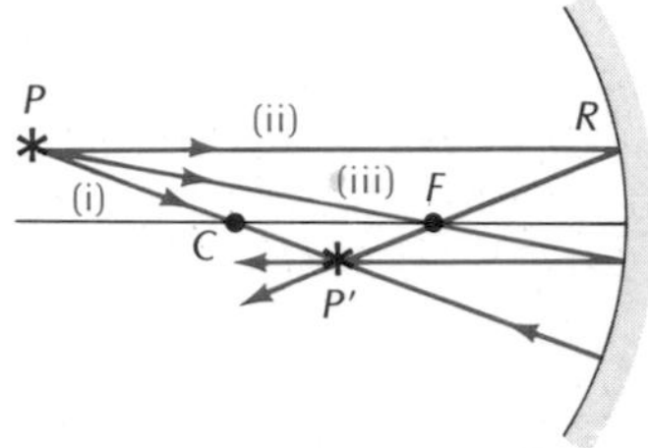

Fig. 37.30 A point source of light P in front of a concave mirror. The rays (i), (ii), and (iii) intersect at the image P'.

If the source of light is an extended object, then we must find the image of each of its points. For instance, a luminous object in the shape of an arrow has an image as shown in Figure 37.31. We can easily verify this by drawing the rays that emerge from, say, the midpoint of the arrow, the tail of the arrow, etc.

The ray-tracing technique summarized in Figure 37.30 is a graphical method for finding the image of a source. This method also applies to convex mirrors; an example of this is shown in Figure 37.32.

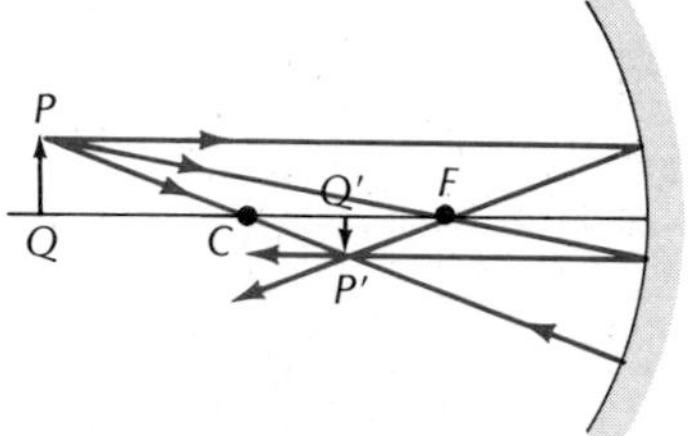

Fig. 37.31 An object PQ in the shape of an arrow and its image $P'Q'$ formed by a concave mirror.

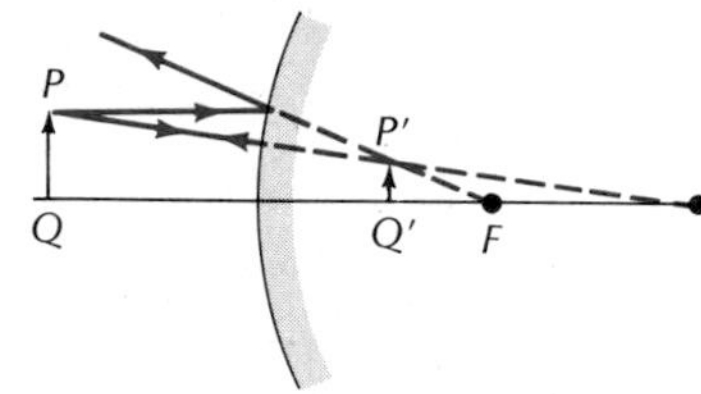

Fig. 37.32 An object PQ and its image $P'Q'$ formed by a convex mirror.

The position of the image can be calculated algebraically by means of the **mirror equation**

Mirror equation

$$\boxed{\frac{1}{s} + \frac{1}{s'} = \frac{1}{f}} \tag{23}$$

Here s is the distance from the object to the mirror and s' is the distance from the image to the mirror. The distance s or s' is positive if the object or image is in front of the mirror; the distance s or s' is negative if the object or image is behind the mirror.[3] As already indicated above, f is positive for a concave mirror, negative for a convex mirror.

To derive Eq. (23), we make use of Figure 37.33, which shows an object PQ, its image $P'Q'$, and two rays. The ray PCP' passes through the center of the spherical surface and is reflected on itself; the ray PAP' strikes the center of the mirror and is reflected symmetrically with respect to the axial line so that the angles θ and θ' are equal. The triangles PQA and $P'Q'A$ are similar; hence

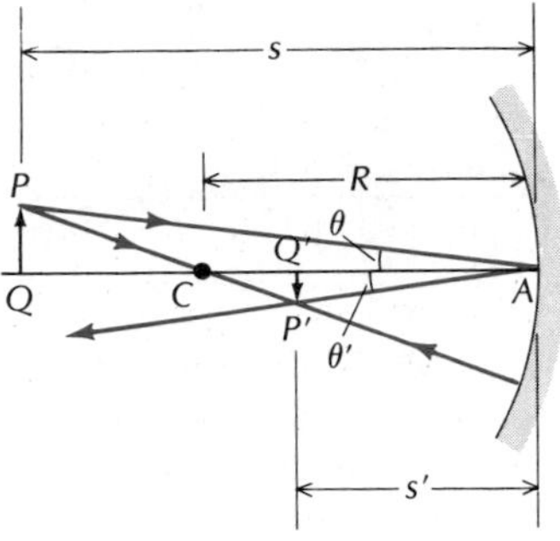

Fig. 37.33 The angles θ and θ' are equal, hence the right triangles PQA and $P'Q'A$ are similar.

$$\frac{PQ}{P'Q'} = \frac{s}{s'} \tag{24}$$

The triangles PQC and $P'Q'C$ are also similar; hence

$$\frac{PQ}{P'Q'} = \frac{QC}{Q'C} \tag{25}$$

or, since $QC = s - R$ and $Q'C = R - s'$,

$$\frac{PQ}{P'Q'} = \frac{s-R}{R-s'} \tag{26}$$

Combining Eqs. (24) and (26), we find

$$\frac{s}{s'} = \frac{s-R}{R-s'} \tag{27}$$

By a bit of algebraic manipulation, we can rearrange this equation to read

$$\frac{1}{s} + \frac{1}{s'} = \frac{2}{R} \tag{28}$$

Obviously, Eq. (28) is equivalent to Eq. (23).

EXAMPLE 3. A candle is placed 41 cm in front of a convex spherical mirror of radius 60 cm. Where is the image?

SOLUTION: With $s = 41$ cm and $f = 30$ cm, Eq. (23) gives

$$\frac{1}{41\text{ cm}} + \frac{1}{s'} = \frac{1}{30\text{ cm}} \tag{29}$$

or $s' = 112$ cm. The positive sign indicates that the image is on the same side of the mirror as the object (Figure 37.34).

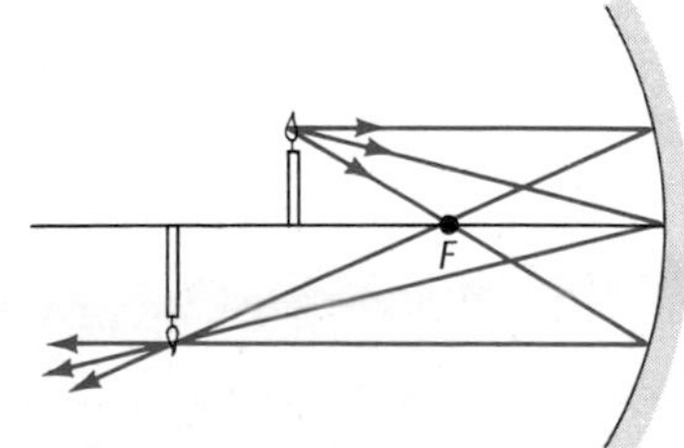

Fig. 37.34 The real image lies in the space in front of the mirror.

Real image

The image in the preceding example is a **real image.** The light rays not only seem to come from this image, but they actually do. As Figure

[3] The object can be behind the mirror if what serves as object for the mirror is actually an image produced by another mirror or by a lens.

37.34 shows, the light rays pass through the image and diverge from it, just as they diverge from the object. The real image is in front of the mirror. Visually, it gives the impression of a ghostly replica of the object floating in midair.

37.5 Thin Lenses

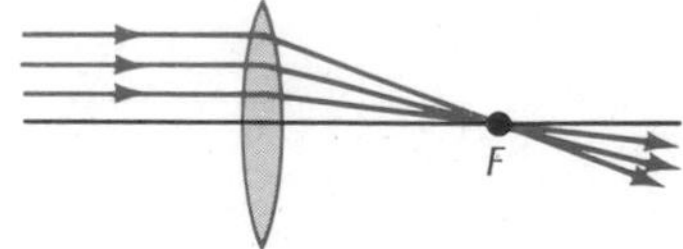

Fig. 37.35 Refraction of rays by a convex lens. The focal length is positive.

A lens made of a refracting material with two spherical surfaces will focus a beam of light to a point (Figure 37.35). For a thin lens, the focal length is given by the **lens-maker's formula**

Lens-maker's formula

$$\frac{1}{f} = (n-1)\left(\frac{1}{R_1} + \frac{1}{R_2}\right) \tag{30}$$

where n is the index of refraction of the material of the lens and R_1, R_2 are the radii of the two spherical surfaces making up the lens. This equation is based on the assumption that the lens is thin (its thickness is small compared to R_1, R_2) and that the incident rays are near the axial line. A ray that enters the lens at some appreciable distance from the axis will miss the focus; this is a form of aberration analogous to spherical aberration of a mirror.

Equation (30) can be derived by tracing rays through the lens, taking into account their refraction at the two curved surfaces. We will not perform this tedius calculation and only note that the focusing depends on the fact that rays far from the axial line strike the surface of the lens with a larger angle of incidence than rays near the axial line. Thus, the far rays are refracted through a larger angle, i.e., they are bent more sharply toward the axis. This is of course exactly what is required to make these far rays cross the axis at the same point (the focus) as the near rays.

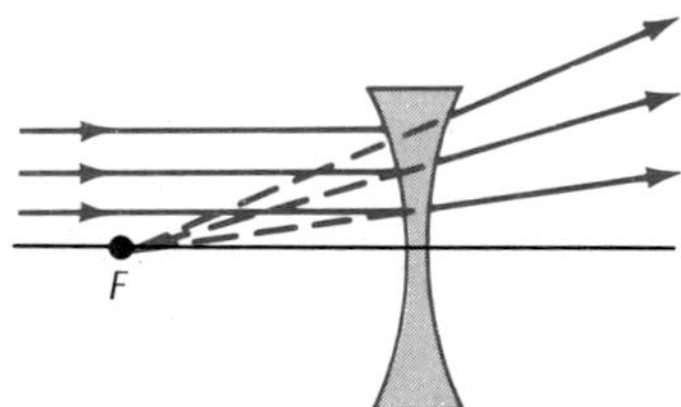

Fig. 37.36 Refraction of rays by a concave lens. The focal length is negative.

Equation (30) may also be applied to a concave lens (Figure 37.36). In this case, the radii R_1, R_2 must be reckoned as *negative* and the focal distance f is then also negative. The meaning of a negative value of f is the same as in the case of mirrors: parallel rays incident on the lens diverge when they emerge from the lens, and the focal point is the point at which the extrapolated rays appear to intersect (Figure 37.36). Furthermore, Eq. (30) can be applied to a concave–convex lens (Figure 37.37). Whether such a lens produces net convergence or divergence depends on whether the positive radius (convex) or the negative radius (concave) is smaller; for instance, the lens of Figure 37.37 will produce convergence.

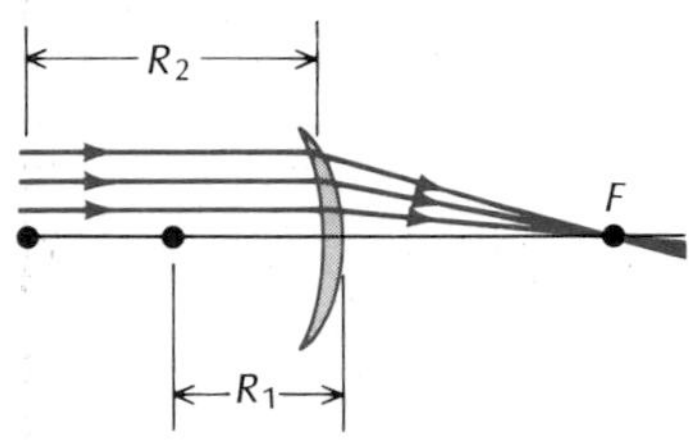

Fig. 37.37 Refraction of rays by a concave–convex lens. The convex surface (right surface) has a smaller radius than the concave surface (left surface). In Eq. (30), the radius of the former surface is reckoned positive ($R_1 > 0$) and the radius of the latter surface is negative ($R_2 < 0$). The sum $1/R_1 + 1/R_2$ is positive.

Note that a lens has *two* focal points at equal distances right and left of the lens. The point on the right of a converging lens is the focus for a parallel beam coming from the left and, conversely, the point on the left is the focus for a parallel beam coming from the right.

To find the image of an object placed near the lens, we can use a ray-tracing technique similar to that used for mirrors. Figure 37.38 shows three rays that are easy to trace: (i) the ray PQP' that starts parallel to the axis and ultimately passes through the focus F, (ii) the ray PCP' that passes undeflected through the center of the lens, and (iii) the ray $PQ'P'$ that passes through the focus F' and emerges parallel to the axis. All these rays intersect at the point P', the image point.

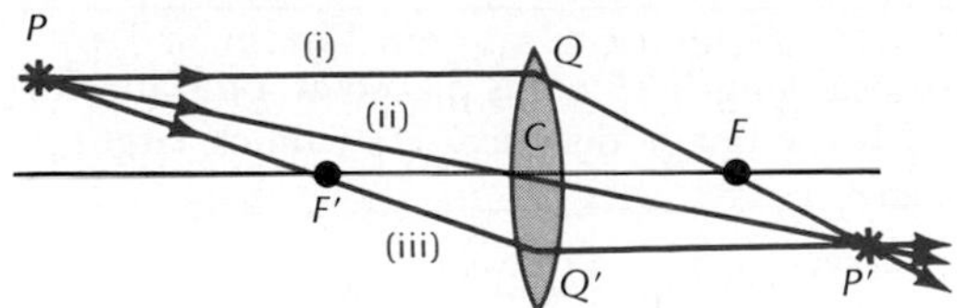

Fig. 37.38 A point source of light P in front of a convex lens. The rays (i), (ii), and (iii) intersect at the image P'.

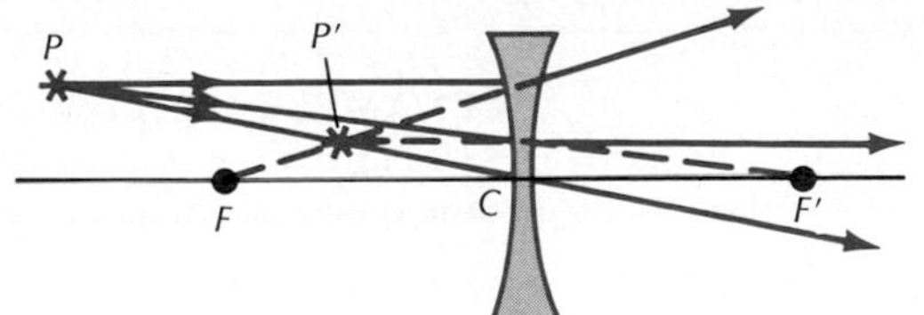

Fig. 37.39 A point source of light P in front of a concave lens, and the image P'.

As in the case of mirrors, two of the above three rays are already sufficient to locate the image. And, of course, the same ray-tracing technique can be applied to concave lenses (Figure 37.39).

The equation to be used for the algebraic calculation of the image distances is the same as Eq. (23),

$$\boxed{\frac{1}{s}+\frac{1}{s'}=\frac{1}{f}} \qquad (31)$$

Lens equation

but the sign conventions are slightly different. The object distance s is positive if the object is on the near side of the lens and negative if it is on the far side; the image distance s' is positive if the image is on the far side of the lens and negative if it is on the near side. In this context, the "near" side is the side from which the light rays are incident on the lens and the "far" side is the other side. The sign of f is positive for a convex lens, negative for a concave lens.

Although the derivation of the lens equation (31) can be based on a geometric argument similar to that used for the mirror formula (23), we can bypass this labor by a trick. We begin by noting that a concave mirror is equivalent to one-half of a convex lens placed directly in front of a flat mirror. If the concave mirror and the (entire) convex lens have the same value of f, then the two arrangements shown in Figures 37.40a and b have exactly the same optical properties — in both cases the image distances are the same. If we now remove the flat mirror in Figure 37.40a and replace the one-half lens by the entire lens, the image distance will remain the same, but the image will form on the opposite side of the lens, i.e., the sign of the image distance will be reversed. Consequently, the same equation (23) must apply to the concave mirror and the convex lens; the only difference is that the sign of the image distance is reversed. This reversal of sign has already been taken into account in our description of the sign conventions associated with Eqs. (23) and (31) — for the mirror, s' is taken as *positive* if on the near side of the mirror, whereas for the lens, s' is taken as *negative* if on the near side of the lens. There is of course a similar correspondence between a convex mirror and a concave lens.

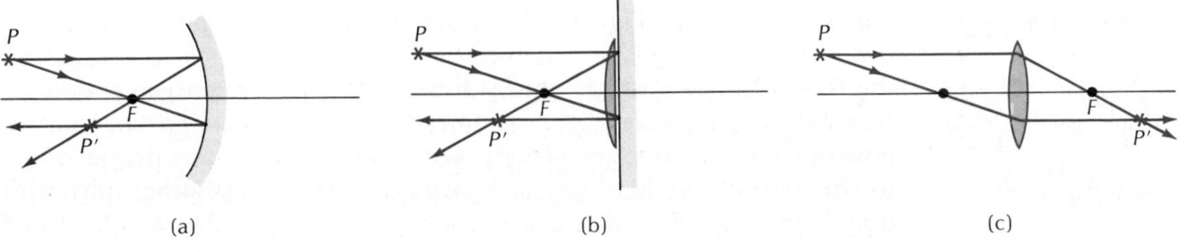

Fig. 37.40 (a) A point source of light and its image formed by a concave mirror. (b) A similar image is formed by one-half of a convex lens placed in front of a flat mirror. Note that each ray has to pass through the one-half lens twice: once before reflection by the mirror, once after. The deflection suffered by a ray in two passages through one-half of a lens is the same as that in a single passage through the entire lens. (c) A similar image is also formed by the entire lens, but the image is now on the other side of the lens.

EXAMPLE 4. A convex lens of focal length 25 cm is placed at a distance of 10 cm from a printed page. What is the image distance? How much larger is the image of the page than the page?

SOLUTION: With $s = 10$ cm and $f = 25$ cm, Eq. (31) gives

$$\frac{1}{10\text{ cm}} + \frac{1}{s'} = \frac{1}{25\text{ cm}}$$

which yields $s' = -16.7$ cm. The negative sign indicates that the image is on the near side of the lens (Figure 37.41). The image is virtual.

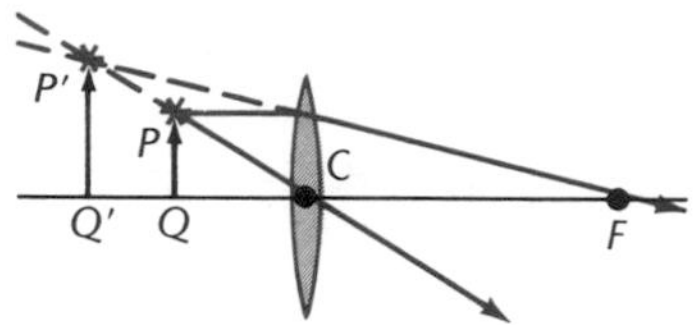

Fig. 37.41 An object PQ and its image $P'Q'$.

Since the triangles $P'Q'C$ and PQC are similar, the sizes of image and object are in the ratio

$$\frac{P'Q'}{PQ} = \frac{Q'C}{QC} = \frac{|s'|}{s} \tag{32}$$

$$= \frac{16.7\text{ cm}}{10\text{ cm}} = 1.67$$

i.e., the image is larger than the object by a factor of 1.67. This is the principle involved in the magnifying glass.

37.6 Optical Instruments

Spherical and chromatic aberrations

The simple lenses described in the preceding section suffer from diverse kinds of aberrations. Parallel rays that strike the lens at an appreciable distance from the axis will miss the focal point; this is **spherical aberration,** similar to the spherical aberration of a mirror. Rays of different colors will be refracted differently by the glass of the lens, and they will therefore be focused differently; this is **chromatic aberration.** Furthermore, the images of objects of large size will display complicated three-dimensional distortions. All these troublesome aberrations can be partially eliminated by combining several simple lenses of different shapes made of different kinds of glass into a compound lens. For instance, Figure 37.42 shows the Zeiss "Tessar" lens, one of the most best-known photographic lenses. In such a compound lens, the individual lenses are carefully designed so that their aberrations tend to cancel mutually.

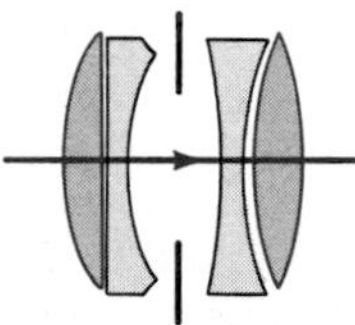

Fig. 37.42 Zeiss "Tessar" lens. The outer lenses (color) are made of crown glass, and the inner lenses (gray) of flint glass.

Most optical instruments — cameras, magnifiers, microscopes, and telescopes — employ compound lenses. However, in the following discussion of optical instruments, we will schematically represent the compound lenses by single lenses of appropriate focal lengths.

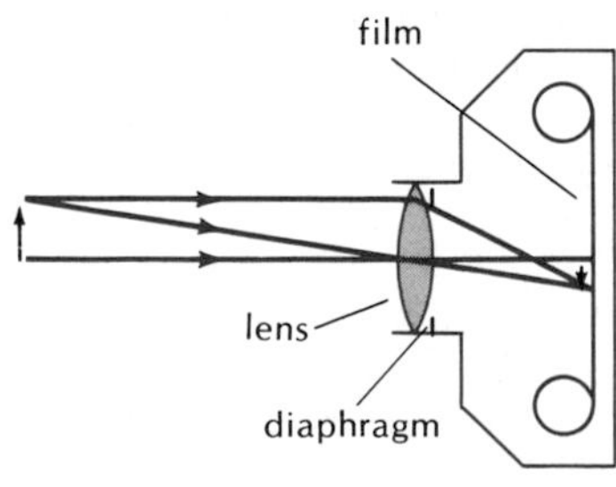

Fig. 37.43 A photographic camera.

THE PHOTOGRAPHIC CAMERA The lens of the camera forms a real image of the object on the photographic film and thereby imprints this image on the film (Figure 37.43). The distance between the lens and the film is adjustable so that the image can always be made to fall on the film, regardless of the object distance. A shutter controls the exposure time during which light is admitted to the camera. For a given exposure time, the amount of light entering the camera is proportional to the area of the lens. Thus, a large lens permits photography with dim light. The size of a camera lens is commonly labeled by the **f**

f number

number, which is defined as the ratio of the focal length of the lens to its diameter. For instance, a lens of focal length 55 mm and diameter 32 mm has an f number of 55:32, or 1.7. A lens of small f number is said to be "fast" because it collects sufficient light for a photograph with a short exposure time. Good cameras have an adjustable iris diaphragm that can be used to block part of the area of the lens and thereby alter the effective f number. If the diaphragm is closed down so that only a small central portion of the lens remains unblocked, the camera will require a long exposure time, but it will have a large depth of field — it simultaneously forms sharp images for objects spanning a large range of object distances. This is so because the rays emitted from any point of an object enter the camera within a narrow cone, and they therefore intersect at the image within a narrow cone; thus these rays will be close together on the photographic film even if the position of the image does not fall exactly on the film.

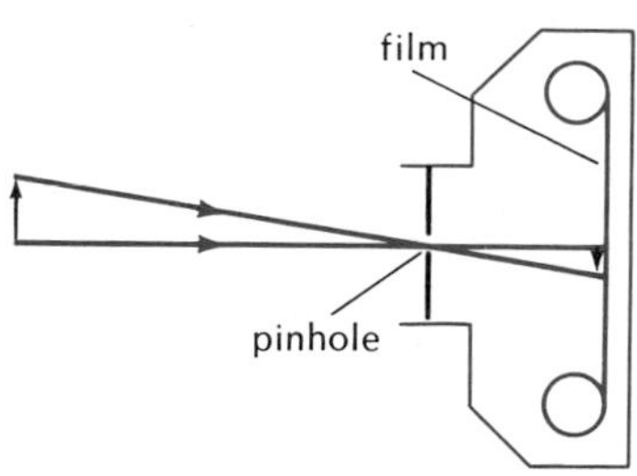

Fig. 37.44 A pinhole camera.

Pinhole camera

A **pinhole camera** can be regarded as an ordinary camera with the iris diaphragm closed down to a point. The lens can then be discarded since the rays pass through its center, where they suffer no deflection (Figure 37.44). To obtain a sharp image, we must use an extremely small pinhole; hence such a camera requires very bright light or a very long exposure time. Figure 37.45 is a picture taken with a pinhole camera.

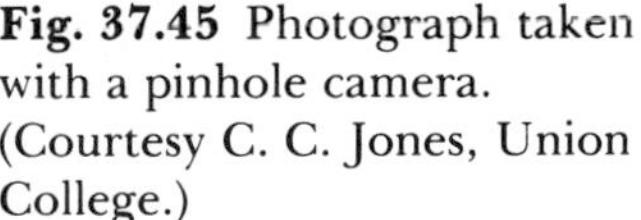

Fig. 37.45 Photograph taken with a pinhole camera. (Courtesy C. C. Jones, Union College.)

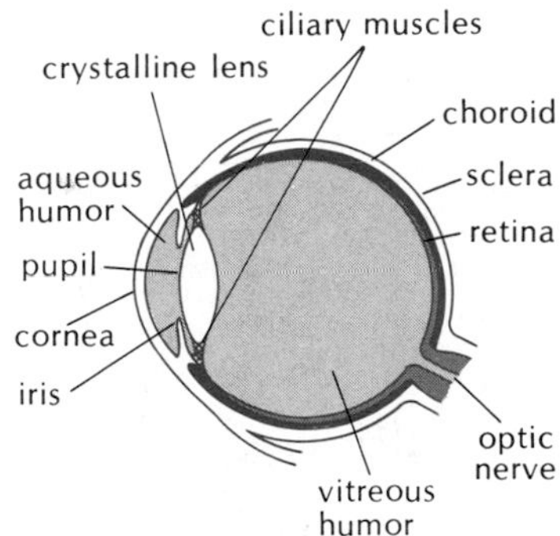

Fig. 37.46 The human eye (natural size). The space between the cornea and the crystalline lens is filled with a transparent jelly (the aqueous humor). The main body of the eye is also filled with a transparent jelly (the vitreous humor). The indices of refraction of the humors are 1.34, nearly the same as for water. The index of refraction of the crystalline lens is 1.44. The sclera is the thick white outer casing of the eye. The choroid is a pigmented black membrane that absorbs stray light, like the black paint in cameras.

THE EYE In principle, the eye is similar to a camera. The lens of the eye forms a real image on the retina, a delicate membrane packed with light-sensitive cells which send nerve impulses to the brain. Figure 37.46 shows a section through the human eye; the diameter of the eyeball is typically 2.3 cm. The cornea and the aqueous humor act as a lens; they provide most of the refraction for rays entering the eye. The crystalline lens merely provides the fine adjustment of focal length required to make the image of an object fall on the retina, regardless of the object distance. The crystalline lens is flexible; its focal length is adjusted by the ciliary muscles. If the eye is viewing a distant object, the muscles are relaxed and the lens is fairly flat, with a long focal length. If the eye is viewing a nearby object, the muscles are contracted and the lens is more rounded, with a shorter focal length. The shortest attainable focal length determines the shortest distance at which an object can be placed from the eye and still be seen sharply. This shortest distance is called the **near point.** For a normal young adult, the near point is typically 25 cm. With advancing age the lens loses its flexibility and the near point recedes; for instance, at an age of 60 years, the near point is typically around 200 cm.

Nearsightedness and farsightedness

The two most common optical defects of the eye are nearsightedness and farsightedness. In a **nearsighted** (myopic) eye, the focal length is excessively short, even when the ciliary muscles are completely relaxed. Thus, parallel rays from a distant object come to a focus in front of the retina and fail to form a sharp image on the retina — vision of distant objects is blurred. This condition can be corrected by eyeglasses with divergent lenses. In a **farsighted** (hyperopic) eye, the focal length is excessively long, even when the ciliary muscles are fully contracted (in other words, the near point of the eye is farther away than normal). Hence rays from a nearby object converge toward an image beyond the retina and fail to form a sharp image on the retina. This condition can be corrected by eyeglasses with convergent lenses. In old age, both of these conditions often occur simultaneously, through the loss of flexibility of the crystalline lens and the weakening of the ciliary muscles. The correction then requires bifocal lenses, with a lower convergent portion for near vision, and an upper divergent portion for far vision.

THE MAGNIFIER In order to see fine detail with the naked eye, we must bring the object very close to the eye, so that the angular size of the object is large and, correspondingly, the image on the retina is large (Figure 37.47). This means we want to bring the object to the near point, at a typical distance of 25 cm for the eye of a young adult. To see finer detail, we need a magnifier. This consists of a strongly convergent lens placed adjacent to the eye (Figure 37.48).[4] Such a lens permits us to bring the object closer to the eye, and thereby increase the size of the image on the retina.

The angular magnification of the magnifier is defined as the ratio of the angular size of the image at infinity produced by the magnifier (as in Figure 37.48) to the angular size of the object seen by the naked eye at the standard distance of 25 cm (as in Figure 37.47). Since the angles in Figures 37.47 and 37.48 are small, the angles are approximately equal to the tangents:

$$\theta \cong \tan\theta \cong h/25\ \text{cm} \tag{33}$$

and

$$\theta' \cong \tan\theta' = h/f \tag{34}$$

where h is the size of the object. Taking the ratio of these angles, we find

[4] Note that such a magnifier is *not* the same thing as a magnifying glass. In common use, the magnifying glass is placed at an appreciable distance from the eye, near the object to be magnified, because this maximizes the magnification. A magnifying glass can be regarded as a magnifier (in the technical sense of this word) only if it is placed next to the eye.

$$[\text{angular magnification}] = \theta'/\theta = \frac{25\text{ cm}}{f} \tag{35}$$

Angular magnification of magnifier

This tells us the magnification relative to the (typical) naked eye, i.e., it tells us how much better the magnifier is than the naked eye. For example, a magnifier with $f = 5$ cm has an angular magnification of 25 cm/5 cm = 5. Note that this result is valid only under the assumptions that the magnifier is placed adjacent to the eye, and that the object is placed near the focus of the magnifier so that the image is at infinity. The second of these assumptions is not crucial — if the object is placed closer than the focus, the magnification will be changed only slightly. But the first assumption is crucial — if the magnifier is placed at some appreciable distance from the eye, then the magnification will be very different!

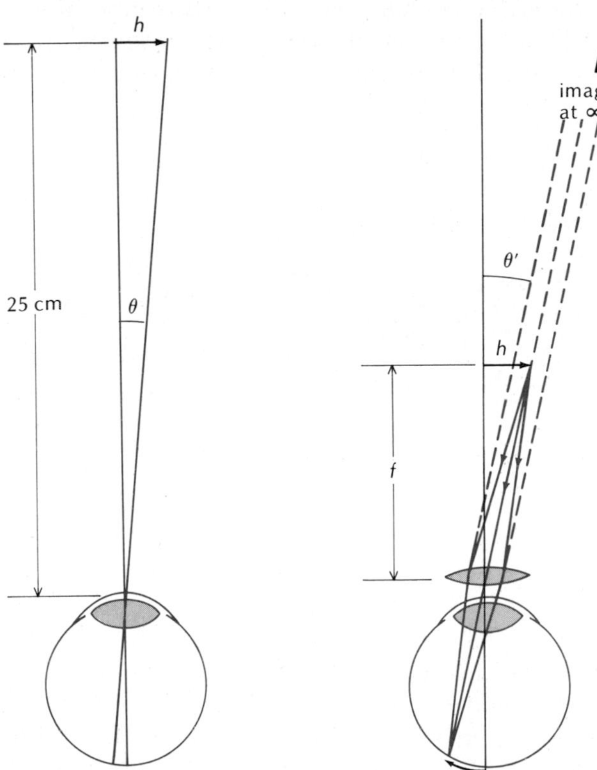

Fig. 37.47 (left) The angular size of the object determines the size of the image on the retina. Here the object has been placed at a distance of 25 cm from the eye.

Fig. 37.48 (right) The magnifier is adjacent to the eye. The object has been placed at a distance slightly shorter than the focal distance, so that the eye sees the image at infinity.

THE MICROSCOPE The microscope consists of two lenses: the objective and the ocular, or eyepiece. Both of these lenses have very short focal lengths. The objective is placed near the object, and it forms a real, magnified image of the object. This image serves as object for the ocular, which acts as a magnifier and forms a virtual image at infinity (Figure 37.49). Thus, both the objective and the ocular contribute to the magnification of the microscope. The net angular magnification of the microscope is the angular magnification of the ocular multiplied by the magnification of the objective. The angular magnification of the ocular is given by Eq. (35); and the magnification of the objective is given by Eq. (32), where s and s' are, respectively, the object and image distances for the objective (these distances are shown in Figure 37.49). Hence the net angular magnification of the microscope is

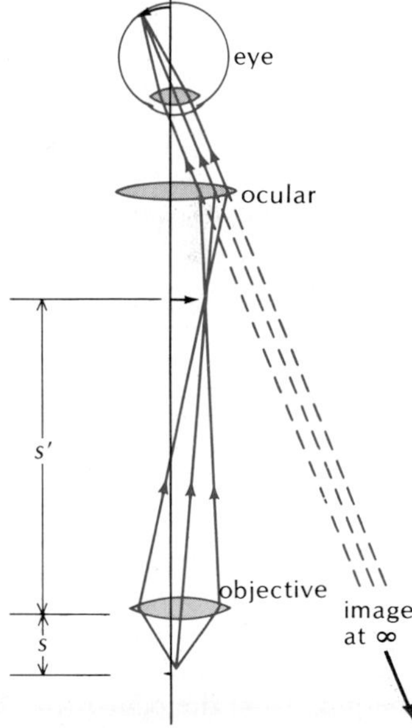

Fig. 37.49 Arrangement of lenses in a microscope. The object to be magnified is below the objective and the eye is just above the ocular. The ocular forms a virtual image at infinity, and the lens of the eye focuses on the retina the parallel rays emerging from the ocular.

Angular magnification of microscope

$$[\text{angular magnification}] = \frac{25 \text{ cm}}{f_{oc}} \times \frac{s'}{s} \tag{36}$$

Note that, as in the case of the magnifier, this tells us the magnification relative to the (typical) naked eye. Good microscopes operate at magnifications of up to 1400. Although higher magnifications can be achieved, this serves little purpose because the diffraction of the light waves at the objective limits the detail that can be resolved. To overcome this limitation we need to use waves of shorter wavelength, such as the electron waves used in electron microscopes.

THE TELESCOPE A simple astronomical telescope consists of an objective of very long focal length and an ocular of short focal length. These two lenses are separated by a distance (nearly) equal to the sum of their individual focal lengths, so that their focal points coincide. The objective forms a real image of a distant object. This image serves as object for the ocular, which forms a magnified virtual image at infinity (Figure 37.50). To find the angular magnification produced by this telescope, we begin by noting that the lens equation with $s = \infty$ applied to the objective gives

$$\frac{1}{\infty} + \frac{1}{s'} = \frac{1}{f_{ob}}$$

i.e.,

$$s' = f_{ob} \tag{37}$$

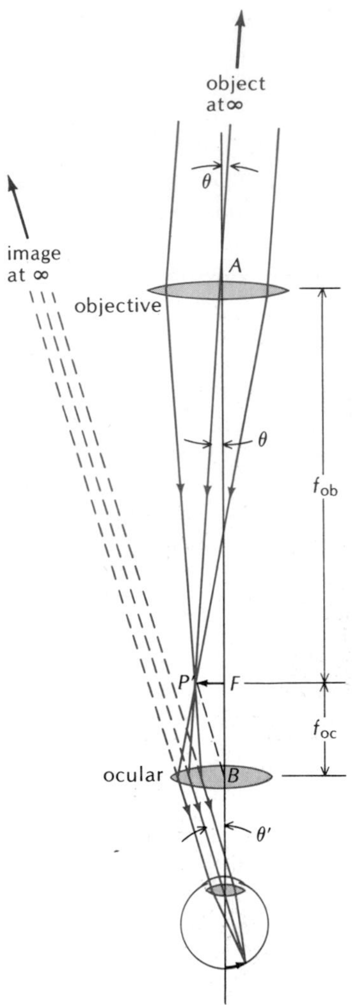

Fig. 37.50 An astronomical telescope. The object is at a large distance above. The observer's eye is below the ocular. The ocular forms a virtual image at infinity, and the lens of the eye focuses on the retina the parallel rays emerging from the ocular.

This means that the image is at the focal distance, and verifies that the location of the image (FP') is correctly shown in Figure 37.50. We can therefore use the geometric relationships contained in this figure. The angular magnification is the ratio of the angles θ' and θ that represent, respectively, the angular sizes of the object and the final image viewed by the eye. Since both angles are small,

$$[\text{angular magnification}] = \frac{\theta'}{\theta}$$

$$\cong \frac{\tan\theta'}{\tan\theta} = \frac{FP'/FB}{FP'/FA} = \frac{FA}{FB} \tag{38}$$

i.e., the angular magnification is the ratio of the focal length of the objective to that of the ocular,

Angular magnification of telescope

$$[\text{angular magnification}] = \frac{f_{ob}}{f_{oc}} \tag{39}$$

For example, an astronomical telescope with $f_{ob} = 120$ cm and $f_{oc} = 2.5$ cm has an angular magnification of 120 cm/2.5 cm = 48.

Many astronomical telescopes are reflecting telescopes in which a concave mirror plays the role of the objective. The mirror forms a real image that serves as object for the ocular. Of course, the ocular must

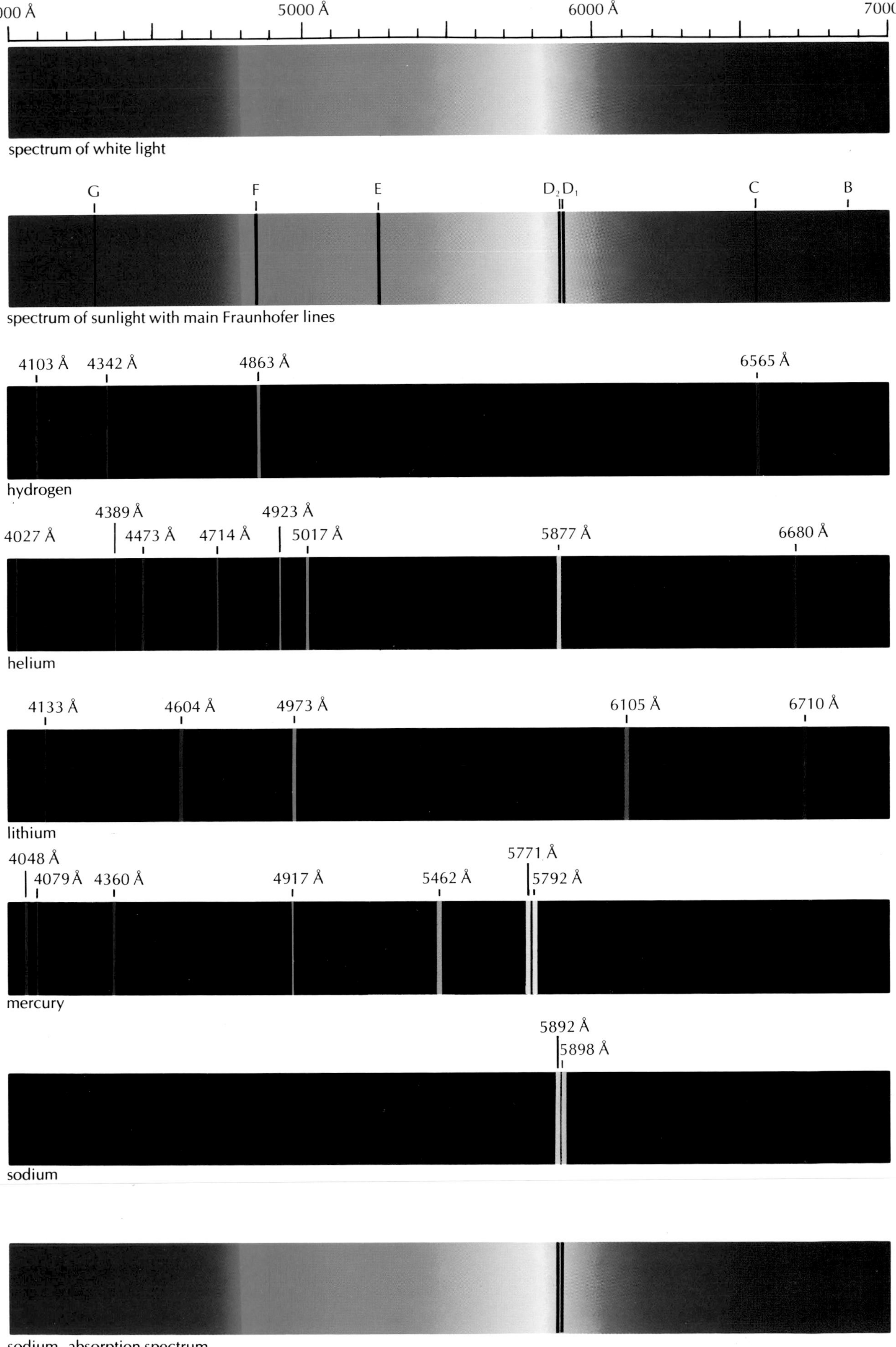

(All wavelengths are measured in vacuum.)

Courtesy Eastman Kodak Company

be placed in front of the mirror (and blocks out some of the light). Figure 37.51 shows this arrangement. Note that Figure 37.51 is essentially the same as Figure 37.50 folded over at the position of the objective lens; hence the geometry of the light rays in reflecting and refracting astronomical telescopes is essentially the same. Because large mirrors of good quality, free of aberrations, are easier to manufacture than large lenses of good quality, the largest astronomical telescopes all use mirrors. For instance, the telescope on Mt. Palomar uses a mirror of diameter 510 cm (200 in.) and of a focal length 1680 cm; the mirror is aspherical for the sake of better imaging (Figure 37.52).

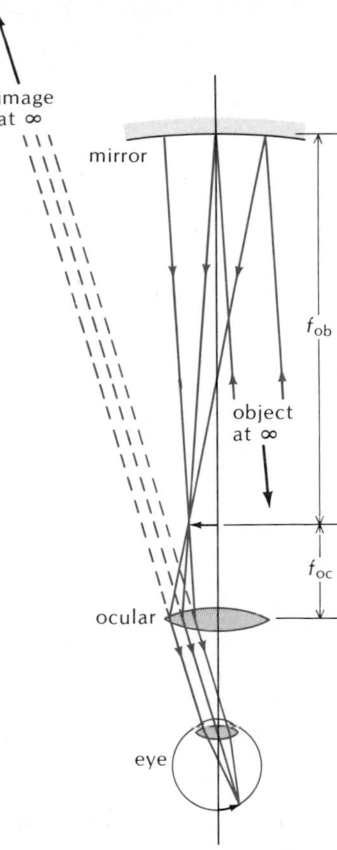

Fig. 37.51 A reflecting telescope. The object has been placed below, to facilitate comparison with Figure 37.50.

Fig. 37.52 The 200-in. telescope on Mt. Palomar. The mirror can be seen under the observation cage.

SUMMARY

Huygens' Construction: The new wave front is the surface of tangency of the wavelets erected on the old wave front.

Law of reflection: The angle of incidence equals the angle of reflection.

Index of refraction: $v = c/n$

$$n = \sqrt{\kappa}$$

Law of refraction: $\sin\theta = n \sin\theta'$

Critical angle for total internal reflection:

$$\theta'_c = \sin^{-1}\left(\frac{1}{n}\right)$$

Focal length of spherical mirror: $f = \pm\frac{1}{2}R$ (f is positive for concave mirror, negative for convex)

Mirror equation: $\frac{1}{s} + \frac{1}{s'} = \frac{1}{f}$ (s or s' is positive if object or image is in front of mirror, negative if behind)

Lens-maker's formula: $\frac{1}{f} = (n-1)\left(\frac{1}{R_1} + \frac{1}{R_2}\right)$ (f is positive for convex lens, negative for concave)

Lens equation: $\frac{1}{s} + \frac{1}{s'} = \frac{1}{f}$ (s is positive if on near side of lens, negative if on far side; s' is positive if on far side, negative if on near side)

Angular magnification of magnifier: 25 cm$/f$

Angular magnification of microscope: $(25 \text{ cm}/f_{oc}) \times (s'/s)$

Angular magnification of telescope: f_{ob}/f_{oc}

QUESTIONS

1. When light is incident on a smooth surface — a glass surface, a painted surface, a water surface — the reflection is strongest if the angle of incidence is near 90° (grazing incidence). Can Huygen's Construction explain this?

2. In celestial navigation, the navigator measures the angle between the Sun, or some other celestial body, and the horizon with a sextant. If the navigator is on dry land, where the horizon is not visible, he can measure instead the angle between the Sun and its reflection in a pan full of water, and divide this angle by two. Explain.

3. Suppose we release a short flash of light in the space between two parallel mirrors placed face to face. Why does this light flash not travel back and forth between the two mirrors forever?

4. What is the minimum size of a mirror hanging on a wall such that you can see your entire body when standing in front of it?

5. Artists are notorious for making mistakes when drawing or painting mirror images. What is wrong with the position and orientation of the mirror images shown in the cartoon in Figure 37.53?

Fig. 37.53

6. Two parallel mirrors are face to face. Describe what you see if you stand between these mirrors.

7. Figure 37.54 shows spots of sunlight on a wall in the shade of a tree. The spots were made by sunlight that has passed through very small gaps between the leaves of the tree. Explain why all the spots are round and of nearly the same size, even though the gaps are of irregular shape and size. (Hint: The Sun is round.)

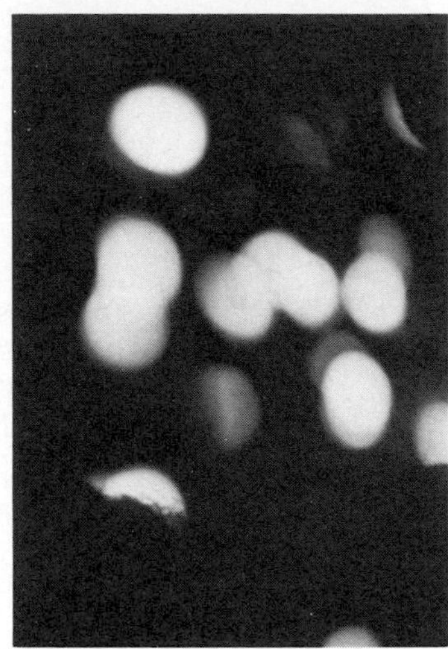

Fig. 37.54

8. If you immerse one-half of a stick in the water, it will appear bent. Explain.

9. At sunset, the image of the Sun remains visible for some time after the actual position of the Sun has sunk below the horizon. Explain.

10. After a navigator measures the angle between the Sun and the horizon with a sextant, he must make a correction for the refraction of sunlight by the atmosphere of the Earth. Does this refraction increase or decrease the apparent angle between the Sun and the horizon?

11. Figure 37.55 shows water waves refracted in the shallows near a headland. This refraction deflects the waves toward the headland. Explain this, keeping in mind that the speed of water waves in shallow water is proportional to the square root of the depth ($v = \sqrt{gh}$).

Fig. 37.55

12. At amusement parks you find mirrors that make you look very short and fat or very tall and thin. What kinds of mirrors achieve these effects?

13. Storeowners often install convex mirrors at strategic locations in their stores to supervise the customers. What is the advantage of a convex mirror over a flat mirror?

14. You look toward a lens or a mirror and you see the image of an object. How can you tell whether this image is real or virtual?

15. Hand mirrors are sometimes concave, but never convex. Why?

16. If you place a book in front of a concave mirror, will it show you mirror writing? Does the answer depend on the distance of the book from the mirror?

17. You place an object in front of a concave mirror, at a distance smaller than the focal length. Is the image real or virtual? Erect or inverted? Magnified or reduced? What if the distance is greater than the focal length? What if the mirror is convex?

18. How could you make a lens that focuses sound waves?

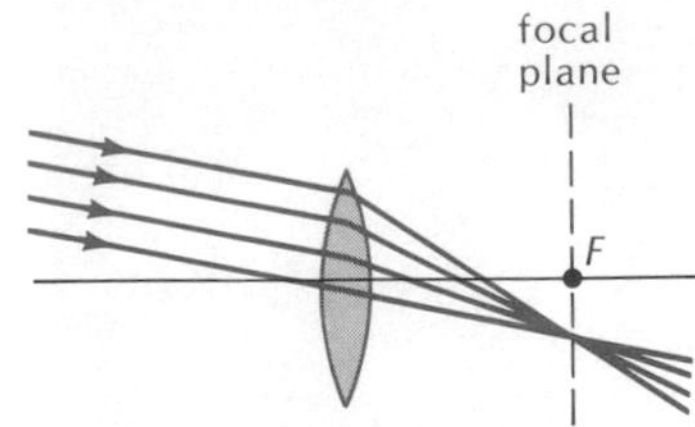

Fig. 37.56 The dashed line indicates the focal plane, seen edge on.

19. A convex lens is made of glass of index of refraction 1.2. If you immerse this lens in water, will it produce convergence or divergence of incident parallel rays?

20. If you place a small light bulb at the focus of a convex lens and look toward the lens from the other side, what will you see?

21. Consider Figure 37.38. How do we know that the ray $PQ'P'$ emerges parallel to the axis of the lens?

22. Are the distances to the two focal points of a thick lens necessarily the same?

23. Suppose you place a magnifying glass against a flat mirror and look into the glass. What do you see if your face is very near the glass? If it is not very near?

24. Consider a beam of parallel rays incident on a convex lens at a small angle with the axis of the lens (Figure 37.56). Show that these parallel rays will be focused at a point in a plane that is perpendicular to the axis and passes through the focal point. This plane is called the **focal plane.**

25. You place an object in front of a convex lens at a distance smaller than the focal length. Is the image real or virtual? Erect or inverted? Magnified or reduced? What if the distance is greater than the focal length? What if the lens is concave?

26. Lenses, like mirrors, suffer from aberration. Consider rays incident on a convex lens, parallel to the axis. If the point of incidence is far from the axis, would you expect the ray to pass in front of the focus or behind?

27. Figure 37.57 shows a large **Fresnel lens** used in the lantern of a lighthouse. The lens consists of annular segments, each with a curved surface similar to the curved surface of an ordinary lens. Why is this arrangement better than a single curved surface?

Fig. 37.57 Fresnel lens of the lighthouse at Point Reyes, California.

28. The telescopes built by Galileo Galilei consisted of a convex objective lens and a concave ocular lens. The lenses are arranged so that the focal point of the objective coincides with the focal point of the ocular (Figure 37.58). Explain how this telescope produces an angular magnification. Show that the eye sees an erect image. (Hint: The concave lens placed *before* the point F in Figure 37.58 has the same effect as the convex lens placed *beyond* the point F in Figure 37.50 — the lens produces an image at infinity. Is this image erect or inverted?)

29. Binoculars use prisms to reflect the light back and forth (Figure 37.59). What is the purpose of these prisms, and what is their advantage over mirrors?

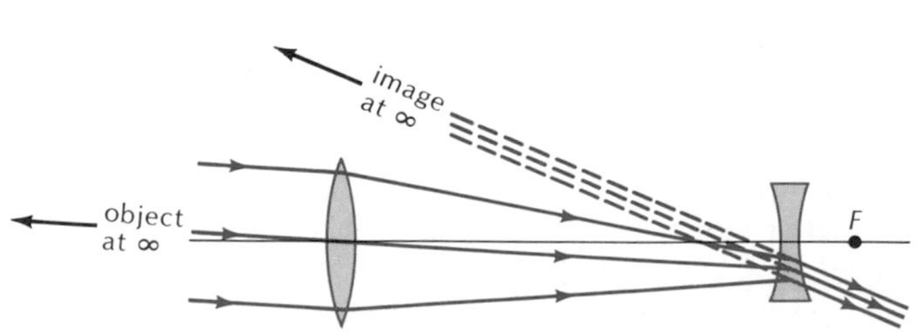

Fig. 37.58 Galilean telescope.

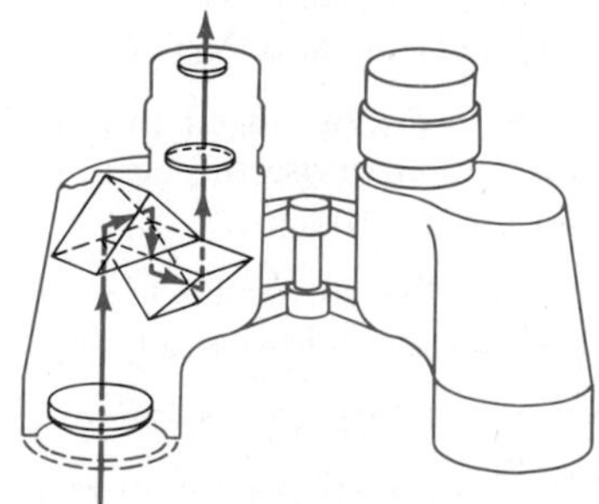

Fig. 37.59 Arrangement of prisms in a binocular.

PROBLEMS

Section 37.2

1. According to a (questionable) story, Archimedes set fire to the Roman ships besieging Syracuse by focusing the light of the Sun on them with mirrors. Suppose that Archimedes used flat mirrors. How many flat mirrors must simultaneously reflect sunlight at a piece of canvas if it is to catch fire? The energy flux of the sunlight at the surface of the Earth is 0.1 W/cm^2 and the energy flux required for ignition of canvas is 4 W/cm^2. Assume that the mirrors reflect the sunlight without loss.

2. A vertical mirror, oriented toward the Sun, throws a rectangular patch of sunlight on the floor in front of the mirror. The size of the mirror is 0.5 m $\times$ 0.5 m and its bottom rests on the floor. If the Sun is 50° above the horizon, what is the size of the patch of sunlight on the floor?

3. Two mirrors meeting at a right angle make a corner reflector (see Figure 37.9). Prove that a ray of light reflected successively by both mirrors will emerge on a path antiparallel to its original path.

Section 37.3

4. The speed of sound in air is 340 m/s and in water it is 1500 m/s. If a sound wave in air approaches a water surface with an angle of incidence of 10°, what will be the angle of refraction?

5. Make a plot of the angle of incidence vs. the angle of refraction for light rays incident on a water surface. What is the maximum angle of refraction?

6. A ship's navigator observes the position of the Sun with his sextant and measures that the Sun is exactly 39° away from the vertical. Taking into account the refraction of the Sun's light by air, what is the true angular position of the Sun? For the purpose of this problem assume that the Earth is flat and that the atmosphere can be regarded as a flat, transparent plate of uniform density and an index of refraction 1.0003.

7. A ray of light strikes a plate of window glass of index of refraction 1.5 and thickness 2.0 mm with an angle of incidence of 50°.
   (a) Show that the transmitted ray leaving the glass on the other side is parallel to the incident ray.
   (b) Find the lateral displacement between the transmitted ray and the extrapolation of the incident ray.[5]

8. A point source of light is placed above a thick plate of glass of index of refraction n (Figure 37.60). The distance from the source to the upper surface of the plate is l and the thickness of the plate is d. A ray of light from the source may suffer either a single reflection at the upper surface, or a single reflection at the lower surface, or multiple alternating reflections at the lower and upper surfaces. Thus, each ray splits into several rays, giving rise to multiple images. In terms of l and d, find the distance of the first, second, and third images below the upper surface of the plate. Assume that the angle of incidence of the ray is small.[5]

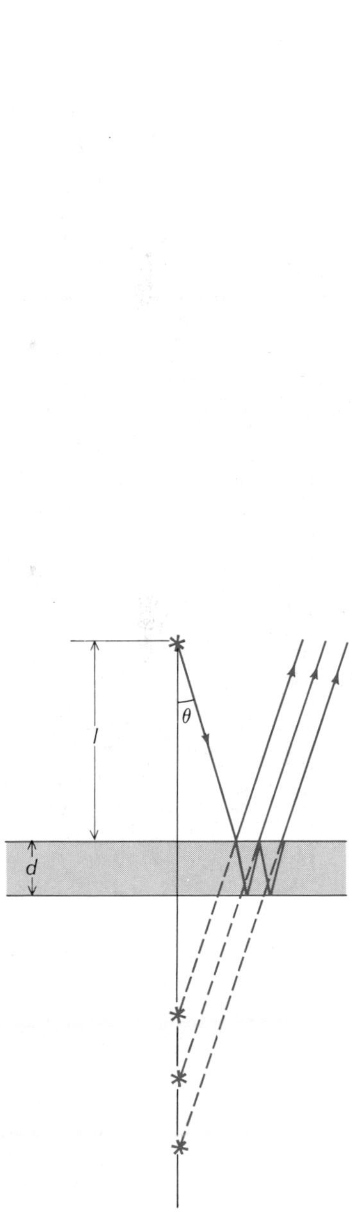

Fig. 37.60 Ray of light reflected and refracted by a plate of glass.

9. The bottom half of a beaker of depth 20 cm is filled with water ($n = 1.33$) and the top half is filled with oil ($n = 1.48$). If you look into this beaker from above, how far below the upper surface of the oil does the bottom of the beaker seem to be?

10. Because of refraction in the plate of glass, an object viewed through an ordinary window will seem somewhat nearer than its actual distance.
   (a) Consider the rays that strike the glass at a small angle of incidence

[5] In this problem assume that the index of refraction of air is 1.

(nearly normal). Show that if an object is at a distance l from a window, the image is at a distance that is shorter by an amount $\Delta l = (1 - 1/n)d$ where n is the index of refraction and d the thickness of the glass.

(b) What change in distance does this formula give if the windowpane is ordinary glass with $d = 2.0$ mm and $n = 1.5$?

(c) What change in distance does this formula give if the window pane is heavy plate glass with $d = 8.0$ mm and $n = 1.5$?

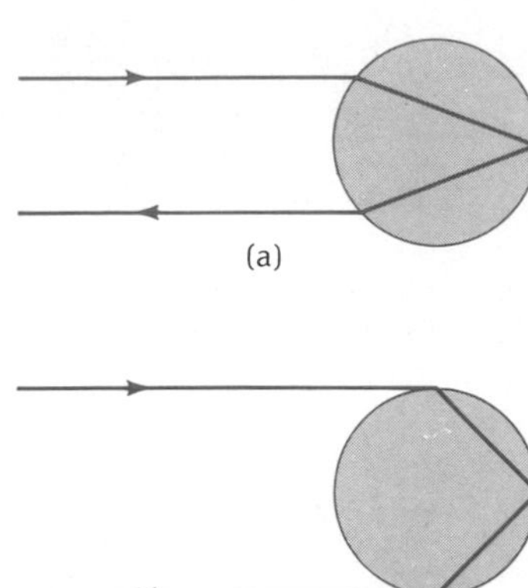

Fig. 37.61 (a) Path of a light ray that reverses direction in a transparent sphere. (b) Path of a light ray that penetrates at maximum distance from the axis of the sphere.

11. The highly reflective paint used on highway signs contains small glass beads which reverse the direction of propagation of a light ray, throwing it back toward the source of light (see Figure 37.61a). The reversal of direction will occur only for those light rays incident on the sphere of glass at a selected distance from the axis, a distance that depends on the index of refraction of the glass. Show that if the reversal of direction is to occur at all, the index of refraction of the glass must be in the range $1 < n \leq \sqrt{2}$. (Hint: Consider the light ray shown in Figure 37.61b; what index of refraction does the reversal of this light ray require?)

12. A signal rocket explodes at a height of 200 m above a ship on the surface of a smooth lake. The explosion sends out sound waves in all directions. Since the speed of sound in water (1500 m/s) is larger than the speed of sound in air (340 m/s), a sound wave can suffer total reflection at a water surface if it strikes at a sufficiently large angle of incidence. At what minimum distance from the ship will a sound wave from the explosion suffer total reflection?

13. An optical fiber is made of a thin strand of glass of index of refraction 1.5. If a ray of light is to remain trapped within this fiber, what is the largest angle it may make with the surface of the fiber?

14. (a) A transparent medium of index of refraction n adjoins a transparent medium of index of refraction n'. Assuming $n > n'$, show that the critical angle for total internal reflection of a ray attempting to leave the first medium is

$$\theta_c = \sin^{-1}\left(\frac{n'}{n}\right)$$

(b) A layer of kerosene ($n' = 1.2$) floats on a water surface. In this case, what is the critical angle for total internal reflection within the water?

15. A layer of oil, of index of refraction n', floats on a surface of water. A ray of light coming from below attempts to pass from the water to the oil and from there to the air above. What is the maximum angle of incidence of the ray on the water–oil surface that will permit the ultimate escape of the ray into the air? Does you answer depend on n'? (Hint: Use the formula derived in Problem 14.)

16. A seagull sits on the (smooth) surface of the sea. A shark swims toward the seagull at a constant depth of 5 m. How close (measured horizontally) can the shark approach before the seagull can see it?

17. To discover the percentage of sucrose (cane sugar) in an aqueous solution, a chemist determines the index of refraction of the solution very precisely and then finds the percentage in a table giving the dependence of index of refraction on sucrose concentration. He determines the index of refraction by immersing a glass prism in the sucrose solution and measuring the critical angle for total internal reflection of a light ray inside the glass prism.

(a) Suppose that with a prism of index of refraction 1.6640 the critical angle is 57.295°. Use the result of Problem 14 to find the index of refraction of the sucrose solution.

(b) Use the following table, interpolating if necessary, to find the concentration of sucrose to four significant figures.

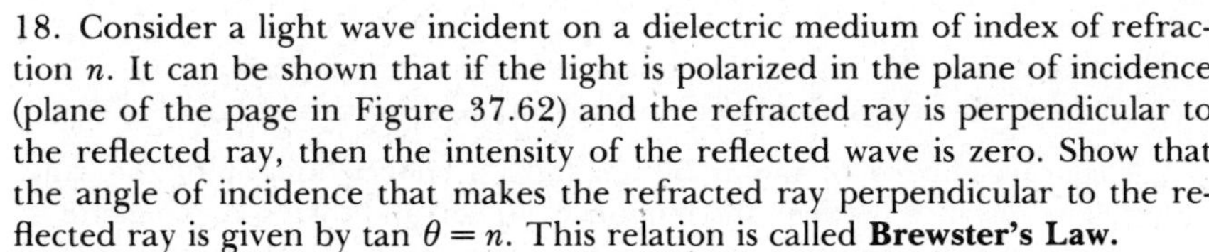

Concentration	n
40.00%	1.3997
40.10	1.3999
40.20	1.4001
40.30	1.4003

18. Consider a light wave incident on a dielectric medium of index of refraction n. It can be shown that if the light is polarized in the plane of incidence (plane of the page in Figure 37.62) and the refracted ray is perpendicular to the reflected ray, then the intensity of the reflected wave is zero. Show that the angle of incidence that makes the refracted ray perpendicular to the reflected ray is given by $\tan\theta = n$. This relation is called **Brewster's Law.**

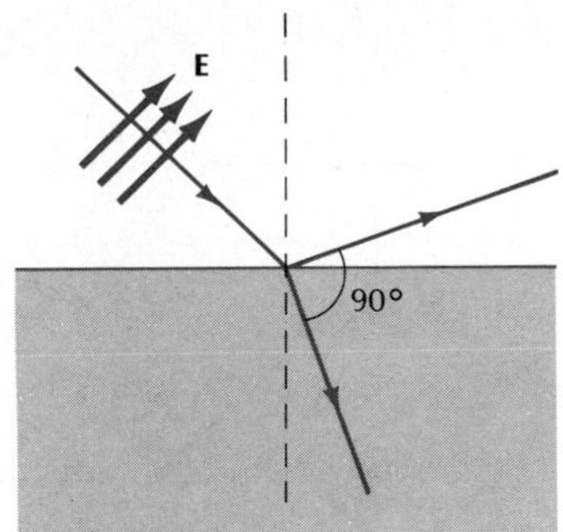

Fig. 37.62 The light wave is polarized in the plane of the page. The refracted ray is perpendicular to the reflected ray.

*19. Rainbows are produced by the refraction of sunlight by drops of water. Figure 37.63 shows a ray of light entering a spherical drop of water. The ray is refracted at A, reflected at B, and refracted at C. The angle of incidence at A (between the ray and the normal to the surface) is θ and the angle of refraction is θ'.

(a) By geometry, show that the angles of incidence and reflection at B coincide with θ' and that the angles of incidence and refraction at C coincide with θ' and θ, respectively.

(b) Show that the angular deflection of the ray from its path is $\theta - \theta'$ at A, $\pi - 2\theta'$ at B, and $\theta - \theta'$ at C. These angles are all in radians and they are measured clockwise from the incident path at each point.

(c) The total angular deflection of the ray by the raindrop is $\Delta = 2(\theta - \theta') + \pi - 2\theta'$. A rainbow will form when all the rays within an infinitesimal range $d\theta$ of angles of incidence (on different drops) suffer the same angular deflection, i.e., when the derivative $d\Delta/d\theta = 2 - 4d\theta'/d\theta$ is zero. If this condition is satisfied, the rays sent back by the raindrops are concentrated, producing a bright zone in the sky. Show that the critical angle θ_c, at which $d\Delta/d\theta = 0$, is given by

$$\cos^2\theta_c = \tfrac{1}{3}(n^2 - 1)$$

where n is the index of the refraction of water.

(d) The index of refraction for red light in water is 1.330. Find θ_c and find Δ in degrees and minutes of arc. Draw a diagram showing a red ray coming from the Sun, hitting the drop, and reaching the eye of a rainbow watcher.

(e) The index of refraction for violet light in water is 1.342. Find θ_c and find Δ. On top of the preceding diagram, draw a violet ray coming from the Sun, hitting the drop (at a different point), and reaching the eye of the rainbow watcher. Will the watcher see the red color above or below the violet?

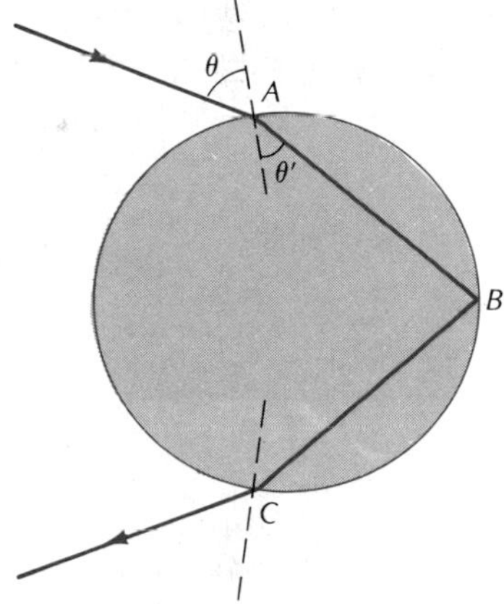

Fig. 37.63 Path of light ray in a drop of water.

Section 37.4

20. At what distance from a concave mirror of radius R must you place an object if the image is to be at the same position as the object?

21. A concave mirror has a radius of curvature R. If you want to form a real image, within what range of distances from the mirror must you place the object? If you want to form a virtual image, within what range of distances must you place the object?

22. A woman's hand mirror is to show a (virtual) image of her face magnified 1.5 times when held at a distance of 20 cm from the face. What must be the radius of curvature of a spherical mirror that will serve the purpose? Must it be concave or convex?

23. The surface of a highly polished doorknob of brass has a radius of curva-

ture of 4.5 cm. If you hold this doorknob 15 cm away from your face, where is the image that you see? By what factor does the size of the image differ from the size of your face?

24. A concave mirror of radius 30 cm faces a second concave mirror of radius 24 cm. The distance between the mirrors is 80 cm and their axes coincide. A light bulb is suspended between the mirrors, at a distance of 20 cm from the first mirror.
(a) Where does the first mirror form an image of the light bulb?
(b) Where does the second mirror form an image of this image?

25. A concave mirror of radius 60 cm faces a convex mirror of the same radius. The distance between the mirrors is 50 cm, and their axes coincide. A candle is held between the mirrors, at a distance of 10 cm from the convex mirror. Consider rays of light that first reflect off the concave mirror and then off the convex mirror. Where do these rays form an image?

Section 37.5

26. The crystalline lens of a human eye has two convex surfaces with radii of curvature of 10 mm and 6.0 mm. The index of refraction of its material is 1.45. Treating it as a thin lens, what is its focal length when removed from the eye and placed in air?

27. A thin lens of flint glass with $n = 1.58$ has one concave surface of radius 15 cm and one flat surface.
(a) What is the focal length of this lens?
(b) If you place this lens at a distance of 40 cm from a candle, where will you find the image of the candle?

28. A thin, symmetric, convex lens of crown glass with index of refraction $n = 1.52$ is to have a focal length of 20 cm. What are the correct radii of the spherical surfaces of the lens?

29. A slide projector has a lens of focal length 13 cm. The slide is at a distance of 2.0 m from the screen. What must be the distance from the slide to the lens if a sharp image of the slide is to be seen on the screen?

30. The convex lens of a magnifying glass has a focal length of 20 cm. At what distance from a postage stamp must you hold this lens if the image of the stamp is to be twice as large as the stamp?

31. If you place a convex lens of focal length 18 cm at a distance of 30 cm from a small light bulb, where will you find the image of the light bulb? Is this a real or virtual image? Is it upright or inverted?

32. Show that if a thin lens of index of refraction n is placed in a medium (e.g., water) of index of refraction n', then the lens-maker's formula (30) must be modified as follows:

$$\frac{1}{f} = \left(\frac{n}{n'} - 1\right)\left(\frac{1}{R_1} + \frac{1}{R_2}\right)$$

[Hint: In the law of refraction — Eq. (11) — only the ratio of the indices of refraction is relevant.]

33. Show that if two thin lenses of focal lengths f_1 and f_2 are placed next to one another (in contact), the net focal length f is given by

$$\frac{1}{f} = \frac{1}{f_1} + \frac{1}{f_2}$$

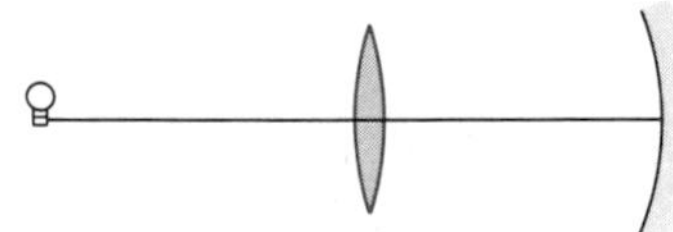

Fig. 37.64 Light bulb, convex lens, and concave mirror.

34. A convex lens of focal length 25 cm is at a distance of 60 cm from a concave mirror of focal length 20 cm. A light bulb is 80 cm from the lens (Figure 37.64).

(a) Where does the lens form an image of the light bulb?
(b) Where does the mirror form an image of this image?

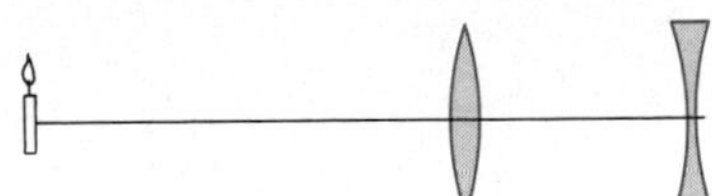

Fig. 37.65 Candle, convex lens, and concave lens.

35. Two lenses, one concave and one convex, have equal focal lengths of 30 cm. The lenses are separated by a distance of 10 cm. A candle is 20 cm from the convex lens (Figure 37.65).
(a) Where does the convex lens form an image?
(b) Where does the concave lens form an image of this image?

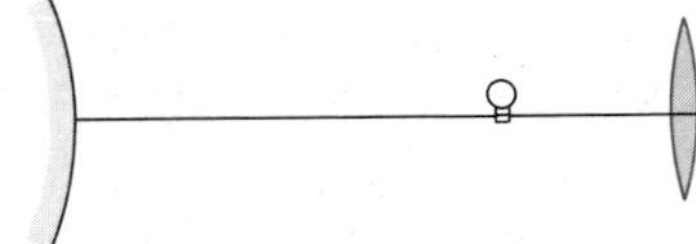

Fig. 37.66 Light bulb, convex mirror, and convex lens.

36. A light bulb is 15 cm in front of a convex mirror of radius 10 cm. A convex lens of focal length 25 cm is 5 cm beyond the light bulb (Figure 37.66). Where do you see the light bulb if you look through the convex lens at the mirror?

*37. Figure 37.67a shows a Fresnel lens consisting of a large number of annular segments each of which is an annular portion of an ordinary lens. Fresnel lenses of diameters more than 1 m are commonly used in the lamps of lighthouses where ordinary lenses without segments would be much too thick and too heavy. Consider one of these annular segments (Figure 37.67b); for the sake of simplicity, assume that the width of the segment is infinitesimal. Derive a formula for the angle of inclination θ of the surface of this segment in terms of the index of refraction n, the distance r of the segment from the axis of the lens, and the focal length of the lens.

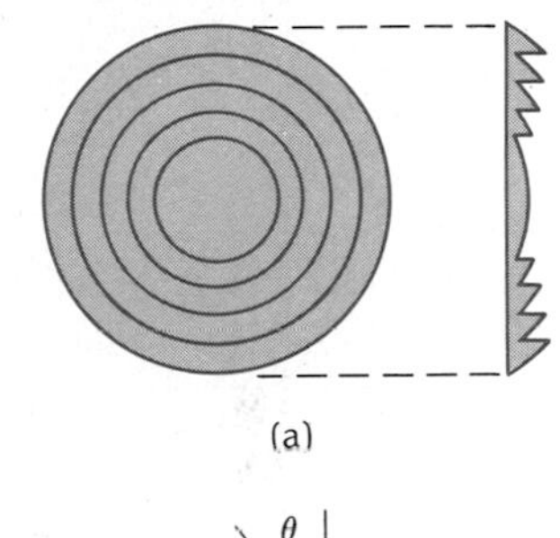

(a)

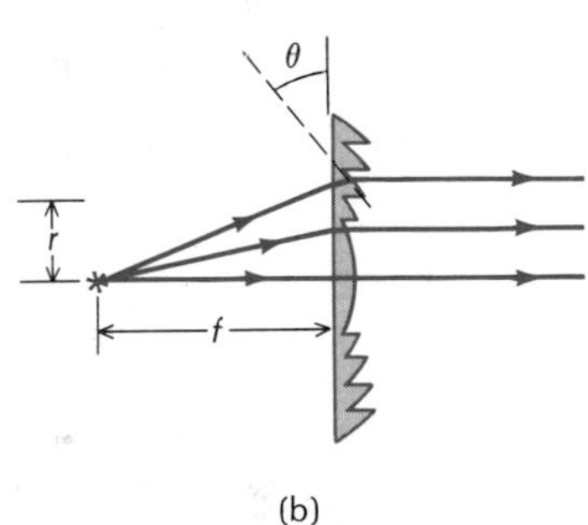

(b)

Fig. 37.67 (a) Fresnel lens, frontal view and cross section. (b) A ray passing through one of the annular segments.

Section 37.6

38. The lens of a 35-mm camera has a focal length of 55 mm. The distance of the lens from the film is adjustable over a range from 55 mm to 62 mm. Over what range of object distances (measured from the lens) is this camera capable of producing sharp pictures?

39. A miniature Minox camera has a lens of focal length 15 mm. This camera can be focused on an object as close as 20 cm, or as far away as infinity. What must be the distance from the lens to the film if the camera is set for 20 cm? What if the camera is set for infinity?

40. The light meter of a 35-mm camera with a lens of f number 1.7 indicates that the correct exposure time for a photograph is $\frac{1}{250}$ s. If the iris diaphragm is closed down so that the f number becomes 4, what will be the correct exposure time?

41. Pretend that the cornea and the crystalline lens of the human eye act together as a single thin lens placed at a distance of 2.2 cm from the retina (Figure 37.68). This lens is deformable; it can change its focal length by changing its shape.
(a) What must be the focal length if the eye is viewing an object at a very large distance?
(b) What must be the focal length if the eye is viewing an object at a distance of 25 cm?

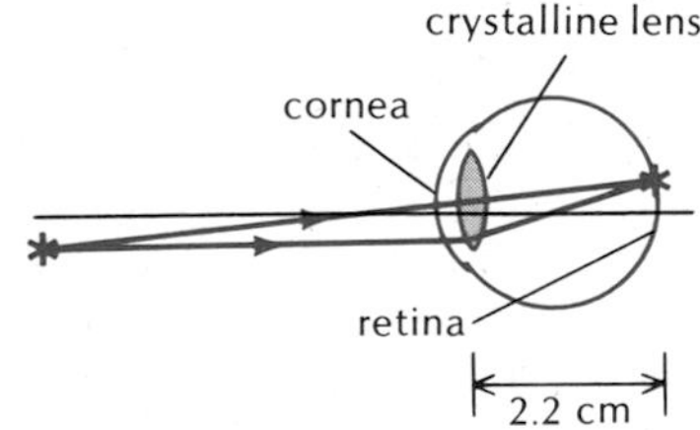

Fig. 37.68 Lens of human eye forms image on retina.

42. In a nearsighted eye the (relaxed) lens has an abnormally short focal length and consequently the eye fails to form an image of a distant object on the retina. This defect can be corrected with a contact lens. Pretend that both the lens of the eye and the contact lens are thin lenses so that the formula given in Problem 33 applies.
(a) Suppose that the focal length of the eye is 2.0 cm. What must be the focal length of the contact lens if it is to increase the net focal length to 2.2 cm? Should it be convergent or divergent?
(b) The radius of curvature of the side of the contact lens next to the eye should be −0.80 cm so as to fit tightly on the cornea. What must be the radius of curvature of the other side? The contact lens is made of plastic with an index of refraction 1.33.

43. Equation (35) gives the angular magnification for a magnifier if the object is so placed that the image is at infinity. Show that if the object is so placed

that the image is at a distance of 25 cm, then the angular magnification is $1 + 25\ \text{cm}/f$.

44. A microscope has an objective of focal length 4.0 mm. This lens forms an image at a distance of 224 mm from the lens. If we want to attain a net angular magnification of 550, what choice must we make for the angular magnification of the ocular?

45. A microscope has an objective of focal length 1.9 mm and an ocular of focal length 25 mm. The distance between these lenses is 180 mm.
 (a) At what distance must the object be placed from the objective so that the ocular forms an image at infinity, as shown in Figure 37.49?
 (b) What is the net angular magnification of this microscope?

46. A telescope has an objective of focal length 160 cm and an ocular of focal length 2.5 cm. If you look into the *objective* (i.e., into the wrong end) of this telescope, you will see distant objects *reduced* in size. By what factor will the angular size of objects be reduced?

47. An amateur astronomer uses a telescope with an objective of focal length 90.0 cm and an ocular of focal length 1.25 cm. What is the angular magnification of this telescope?

48. The large reflecting telescope on Mt. Palomar has a mirror of focal length 1680 cm. If this telescope is operated with an ocular of focal length 1.25 cm, what is the angular magnification?

CHAPTER 38

Interference

Geometric optics relies on the assumption that the sizes of the mirrors or lenses and the separations between them are much larger than the wavelength of light. Under these conditions, we can adequately describe the propagation of light by rays that are rectilinear, except when refracted by the surfaces of dielectric media. Thus, in geometric optics, the wave properties of light do not show up explicitly (although these wave properties enter into the derivation of the laws of reflection and refraction).

Wave optics, or physical optics, deals with the propagation of light in the general case, without any restrictive assumptions as to the sizes of the bodies through or around which the light is propagating. If we let light waves interact with a body or an obstacle of a size comparable to a wavelength, the light waves will display their wave properties explicitly through the phenomena of interference and diffraction. As we saw in Chapter 16, interference is the constructive or destructive combination of two or more waves meeting at one place; diffraction is the bending and spreading of waves around obstacles.

Thomas Young, *1773–1829, English physicist, physician, and Egyptologist. Young worked on a wide variety of scientific problems, ranging from the structure of the eye and the mechanism of vision, to the decipherment of the Rosetta stone. He revived the wave theory of light, and recognized that interference phenomena provide proof of the wave properties of light.*

In this chapter we will examine the interference of light waves and of other electromagnetic waves. Like all electric and magnetic fields, the fields of electromagnetic waves obey the principle of linear superposition: if two waves meet at some point, the resultant electric or magnetic field is simply the vector sum of the individual fields. If two waves of equal amplitude meet crest to crest, they combine and produce a wave of doubled amplitude; if they meet crest to trough, they cancel and give a wave of zero amplitude. The former case is called **constructive interference** and the latter **destructive interference.** We will encounter both cases of interference in the following sections. The first section involves the interference between two waves propagating in opposite directions; the other sections involve waves propagating in the same or nearly the same direction.

Constructive and destructive interference

38.1 The Standing Electromagnetic Wave

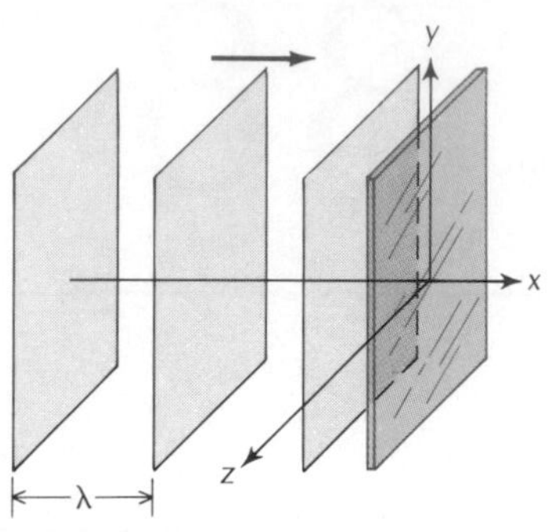

Fig. 38.1 Wave fronts of a plane electromagnetic wave striking a mirror.

Suppose that an electromagnetic wave strikes a mirror perpendicularly and is totally reflected. The space in front of the mirror will then contain two overlapping waves: the incident wave and the reflected wave. Figure 38.1 shows the mirror in the y–z plane. We will assume that the incident wave is a plane harmonic wave, polarized in the y direction. The electric fields of the incident and the reflected waves are then

$$E_{\text{in}} = E_0 \cos(\omega t - \omega x/c) \tag{1}$$

$$E_{\text{ref}} = -E_0 \cos(\omega t + \omega x/c) \tag{2}$$

Both these electric fields are in the y direction. The first equation represents a wave of frequency ω traveling toward the right, and the second a wave traveling toward the left. A negative sign has been inserted in front of the right side of Eq. (2) so as to satisfy the boundary condition at the mirror; since the mirror at $x = 0$ is a conducting surface, the net electric field must vanish at this point at all times:

$$E_{\text{tot}} = E_{\text{in}} + E_{\text{ref}} = E_0 \cos \omega t - E_0 \cos \omega t = 0 \tag{3}$$

The negative sign in Eq. (2) means that the reflecting surface not only reverses the direction of propagation of the incident wave, but also reverses the electric field. It can be shown that this reversal of the electric field is a general feature of reflection whenever the index of refraction of the medium in which the wave is propagating is smaller than the index of refraction of the medium off which the wave reflects.[1]

In the region $x < 0$, the superposition of the incident and reflected waves gives

$$\begin{aligned} E_{\text{tot}} &= E_{\text{in}} + E_{\text{ref}} \\ &= E_0[\cos(\omega t - \omega x/c) - \cos(\omega t + \omega x/c)] \end{aligned} \tag{4}$$

With the trigonometric identity $\cos(\alpha - \beta) - \cos(\alpha + \beta) = 2 \sin \alpha \sin \beta$, this simplifies to

$$E_{\text{tot}} = 2E_0 \sin \omega t \sin \omega x/c \tag{5}$$

Standing wave

This is a **standing wave,** i.e., a wave whose peaks do not travel right or left but remain fixed in space while the entire wave increases and decreases in unison — the wave pulsates (compare Section 15.6). The maxima of the wave are at the points where

$$\frac{\omega x}{c} = \frac{\pi}{2}, \frac{3\pi}{2}, \frac{5\pi}{2}, \ldots \tag{6}$$

Since the wavelength of the wave is $\lambda = 2\pi c/\omega$, we can also write this as

[1] As was pointed out in Section 27.3, a conducting medium (such as the metal of a mirror) can be regarded as having an infinite value of κ and hence, according to Eq. (37.5), an infinite value of n.

$$x = \tfrac{1}{4}\lambda,\ \tfrac{3}{4}\lambda,\ \tfrac{5}{4}\lambda,\ \ldots \tag{7}$$

At these points the incident and reflected waves interfere constructively during one part of the cycle and destructively during another part of the cycle so that the standing wave oscillates with an amplitude $2E_0$. The minima of the wave are at the points where

$$x = 0,\ \tfrac{1}{2}\lambda,\ \lambda,\ \tfrac{3}{2}\lambda,\ \ldots \tag{8}$$

Here the waves interfere destructively at all times so that the standing wave has zero amplitude.

The magnetic fields associated with the waves described by Eqs. (1) and (2) are in the z direction:

$$B_{\text{in}} = \frac{E_0}{c}\cos\left(\omega t - \frac{\omega x}{c}\right) \tag{9}$$

$$B_{\text{ref}} = \frac{E_0}{c}\cos\left(\omega t + \frac{\omega x}{c}\right) \tag{10}$$

Note that there is no negative sign in Eq. (10); this is so because the right-hand rule tells us that if the electric field is in the negative y direction [see Eq. (2)] and the wave is traveling in the negative x direction, the magnetic field must be in the positive z direction.

The superposition of the magnetic fields gives

$$B_{\text{tot}} = B_{\text{in}} + B_{\text{ref}} = 2\frac{E_0}{c}\cos\omega t \cos\omega x/c \tag{11}$$

This of course is also a standing wave. Its maxima are at the points where

$$\frac{\omega x}{c} = 0,\ \pi,\ 2\pi,\ 3\pi,\ \ldots \tag{12}$$

$$x = 0,\ \tfrac{1}{2}\lambda,\ \lambda,\ \tfrac{3}{2}\lambda,\ \ldots \tag{13}$$

and its minima at

$$x = \tfrac{1}{4}\lambda,\ \tfrac{3}{4}\lambda,\ \tfrac{5}{4}\lambda,\ \ldots \tag{14}$$

Thus the maxima of the magnetic field are displaced a distance of $\frac{1}{4}$ wavelength from the maxima of the electric field. Furthermore, note that the time dependence of the magnetic field ($\cos\omega t$) is a quarter of a cycle out of phase with the time dependence of the electric field ($\sin\omega t$). Figure 38.2 shows a plot of the electric and magnetic fields of the standing wave.

The interference between the incident and the reflected waves can be readily observed in a standing microwave. Such a wave can be set up by aiming a microwave generator at a metallic plate, which acts as a mirror. The positions of the maxima and minima of the electric and

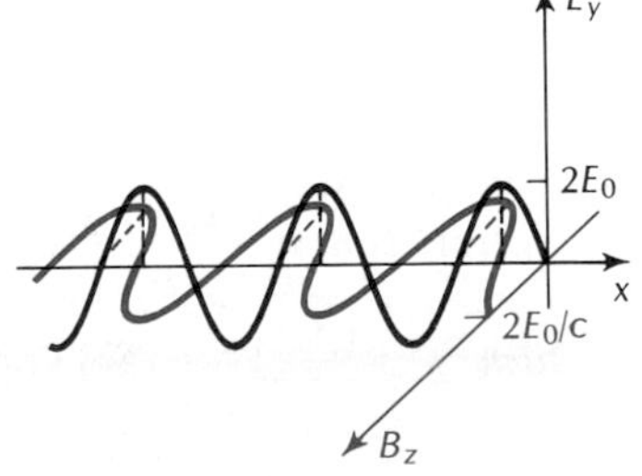

Fig. 38.2 Electric (black) and magnetic (color) fields of the standing wave plotted as a function of position (at a fixed time). The electric field is in the y direction and the magnetic field in the z direction.

magnetic fields can be investigated by probing the fields with small antennas connected to a microwave receiver.

38.2 Thin Films

The interference between a light wave incident on a mirror and the light wave reflected by the mirror is very difficult to observe. As we found in the preceding section, the two waves traveling in opposite directions make up a standing wave with destructive interference at points one-half wavelength apart [see Eq. (8)]. Since the wavelength of visible light is quite small, our eyes cannot perceive the individual minima (or maxima) of the standing wave — we only see the average intensity without noticeable interference effects.

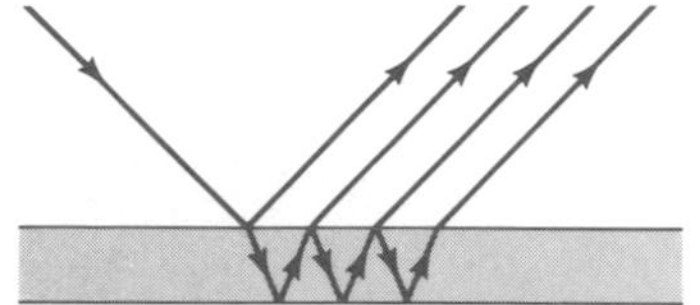

Fig. 38.3 Incident ray and multiple reflected and refracted rays produced by a thin film.

Very spectacular interference effects may become visible when a light wave is reflected by a thin film, such as a thin film of oil floating on water. When the wave strikes the upper surface of the film (Figure 38.3), it will set up a multitude of reflected waves due to reflection on the upper surface, reflection on the lower surface, and multiple zigzags between the surfaces. These reflected waves all travel in the same direction and they can interfere destructively or constructively over a large region of space. Note that we are now interested only in the interference between the several reflected waves — the incident wave plays no direct role in this.

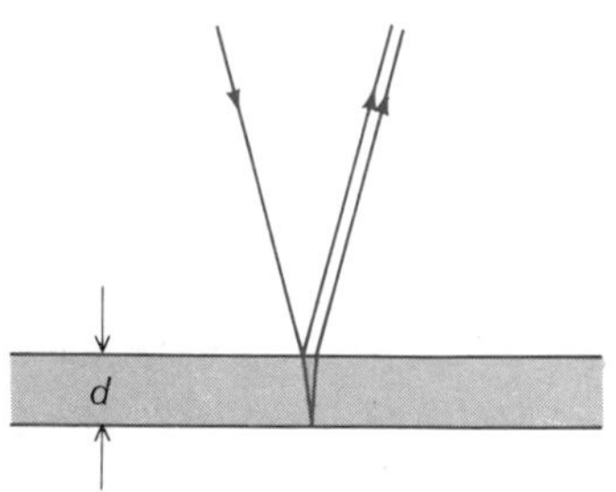

Fig. 38.4 Incident ray and reflected rays for nearly perpendicular incidence and reflection.

To find the conditions for constructive and destructive interference between the waves reflected by a thin film, let us make the simplifying assumption that the direction of propagation of the wave is nearly perpendicular to the surface of the film. The two most intense waves are those that suffer only one reflection: the wave that reflects only on the upper surface and the wave that reflects only on the lower surface. Figure 38.4 shows the rays corresponding to these two waves. Under what conditions will these waves interfere constructively in the space above the film? Obviously, the wave that is reflected at the lower surface has to travel an extra distance to emerge from the film. If the thickness of the film is d, and if the direction of propagation is nearly perpendicular to the film, then the extra distance the wave has to travel is $2d$. Provided that this extra distance is equal to one, two, three, etc., wavelengths, the wave reflected at the lower surface will meet crest to crest with the wave reflected from the upper surface. The condition for constructive interference is therefore

Constructive and destructive interference for wave reflected by thin film

$$2d = \lambda,\ 2\lambda,\ 3\lambda,\ \ldots \tag{15}$$

Likewise, the condition for destructive interference is

$$2d = \tfrac{1}{2}\lambda,\ \tfrac{3}{2}\lambda,\ \tfrac{5}{2}\lambda,\ \ldots \tag{16}$$

Thus, depending on the thickness of the film and on the wavelength, the reflection can be either very strong or very weak. Note that the wavelength λ in Eqs. (15) and (16) is the wavelength of the light within the film; the wavelength outside the film will differ from this by a factor depending on the index of refraction.

EXAMPLE 1. A film of kerosene 4500 Å thick floats on water. White light, a mixture of all visible colors, is vertically incident on this film. Which of the wavelengths contained in the white light will give maximum intensity upon reflection? Which will give minimum intensity? The index of refraction of kerosene is 1.2.

SOLUTION: For maximum intensity we need

$$\lambda = 2d, \tfrac{2}{2}d, \tfrac{2}{3}d, \ldots = 9000\ \text{Å}, 4500\ \text{Å}, 3000\ \text{Å}, \ldots \tag{17}$$

These are the wavelengths in kerosene; to obtain the wavelengths in air, we must multiply by the index of refraction of kerosene, $n = 1.2$. Of the resulting wavelengths, the only one in the visible region is 1.2×4500 Å = 5400 Å.

For minimum intensity we need

$$\lambda = 4d, \tfrac{4}{3}d, \tfrac{4}{5}d, \ldots$$

$$= 18{,}000\ \text{Å}, 6000\ \text{Å}, 3600\ \text{Å}, \ldots \tag{18}$$

Upon multiplication by 1.2 we find that the only wavelength in air in the visible region is 1.2×3600 Å = 4320 Å.

A wavelength of 5400 Å corresponds to a yellow-green color. The film will therefore be seen to have this color in reflected light.

The color displays seen on oil slicks and on soap bubbles arise from such interference effects. Different portions of an oil or soap film usually have different thicknesses, and they therefore give constructive interference for different wavelengths. This results in a pattern of bright colored bands, or colored **fringes.**

Fringes

Similar interference effects can also arise in a narrow gap between two adjacent glass surfaces; such a gap can be regarded as a thin film of air. For instance, Figure 38.5 shows a photograph of the interference fringes produced by the thin film of air between a flat glass plate and a lens of large radius of curvature. The convex surface of the lens is in contact with the plate at the center, but leaves a gradually widening gap for increasing distances from the center. The photograph was taken with monochromatic light. At the bright rings, the width of the gap is such as to give constructive interference of the reflected light.

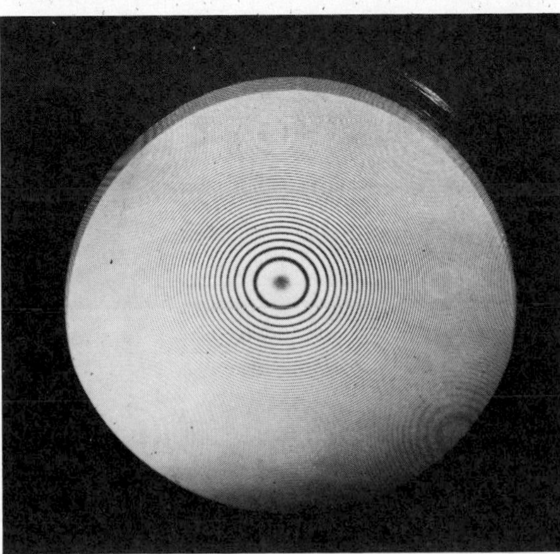

Fig. 38.5 Fringes of constructive and destructive interference seen in the light reflected in the gap between a flat glass plate and a spherical lens in contact.

Newton's rings

At the dark rings, the width is such as to give destructive interference. The rings in Figure 38.5 are called **Newton's rings.**

Note that in the derivation of Eqs. (15) and (16) we have not taken into account the reversal, or change of phase, of the electric field upon reflection. As stated in the preceding section, this change of phase will occur if (and only if) the index of refraction of the medium in which the wave is propagating is smaller than the index of refraction of the medium off which the wave reflects. In the case of a kerosene film resting on water, both the wave reflected at the upper surface and the one reflected at the lower surface suffer this change of phase; consequently, the *relative* phase between the waves is unaffected. However, in the case of an oil film or soap film suspended in air, the wave reflected at the lower surface does not suffer a change of phase; this introduces an extra phase difference of 180° between the two waves. Such a phase difference has the same effect as an extra path difference of one-half wavelength. Consequently, Eqs. (15) and (16) must be interchanged: the former equation now applies to destructive interference and the latter to constructive interference.

Thin films are of great practical importance in the manufacture of optical instruments. The lenses of high-quality instruments are often coated with a thin film of a transparent dielectric so that undesirable reflection of light is eliminated. Of course, with one thin film we can achieve destructive interference only at one wavelength; but with several layers of thin films, we can achieve destructive interference at several wavelengths.

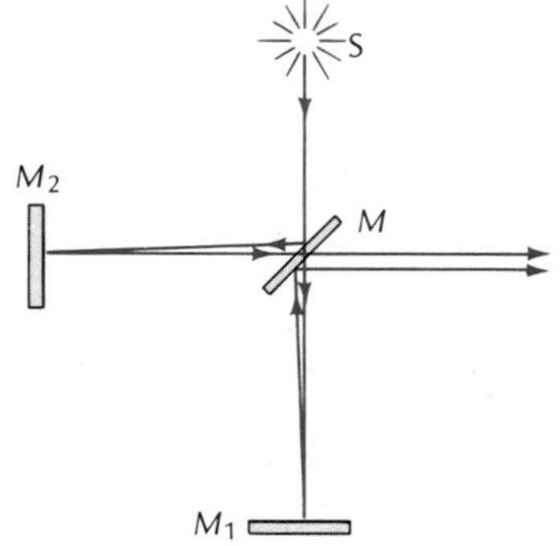

Fig. 38.6 Paths of rays in a Michelson interferometer. For the sake of clarity, rays are shown reaching the mirrors M_1 and M_2 with a small angle between them; they are actually parallel and they overlap.

38.3 The Michelson Interferometer

The Michelson interferometer takes advantage of the interference between two light waves to achieve an extremely precise comparison of two lengths. Figure 38.6 is a schematic diagram of the essential parts of such an interferometer. The apparatus consists of two arms at the ends of which are mounted mirrors M_1 and M_2. Light waves from a monochromatic source S fall on a semitransparent mirror M (a half-silvered mirror). This mirror splits the light wave into two parts: one part continues straight ahead and reaches mirror M_1, the other part is reflected and reaches mirror M_2. These mirrors reflect the waves back toward the central mirror M and, upon reflection or transmission by this mirror, the waves emerge from the interferometer. When they emerge, they interfere constructively or destructively. Suppose that the lengths MM_1 and MM_2 differ by d. Then one of the waves must travel an extra distance $2d$ and the condition for the constructive interference is

Constructive and destructive interference in Michelson interferometer

$$\boxed{2d = 0,\ \lambda,\ 2\lambda,\ \ldots} \qquad (19)$$

and for destructive interference it is

$$\boxed{2d = \tfrac{1}{2}\lambda,\ \tfrac{3}{2}\lambda,\ \tfrac{5}{2}\lambda,\ \ldots} \qquad (20)$$

To achieve this interference, the mirrors M_1 and M_2 must be very precisely aligned so that the image of M_1 seen in M is exactly parallel to M_2. The alignment can be achieved by means of adjusting screws on the backs of the mirrors.

The mirror M_1 is usually mounted on a carriage that can be moved along a track by means of a carefully machined screw. If the mirror is slowly moved inward or outward, the interference of the emerging waves will change back and forth between constructive and destructive whenever the mirror is displaced by $\frac{1}{4}$ wavelength — and the intensity of the emerging light will change back and forth between maxima and minima. Thus, the displacement of the mirror can be measured very precisely by counting fringes and fractions of fringes. This measurement expresses the displacement in terms of the wavelength of the light. Modern interferometers, such as that illustrated in Figure 1.10, are designed to count the fringes automatically with a photoelectric device. Such an interferometer is capable of counting 19,000 fringes per second and its mirror is capable of a displacement of up to 1 m.

Albert Abraham Michelson, *1852–1931, American experimental physicist, professor at the Case Institute of Technology and at the University of Chicago. He made precise measurements of the speed of light, and used interferometric methods to determine the length of spectral lines in terms of the standard meter. He first performed the "Michelson–Morley" experiment on his own in 1881, and then repeated it several times in collaboration with E. W. Morley, with increasing accuracy. Michelson received the Nobel Prize in 1907.*

As was pointed out in Section 17.1, interferometers have played an important role in the test of the dependence of the speed of light on the motion of the Earth. This test is the famous **Michelson–Morley experiment,** first performed in 1881. The principle behind this experiment is as follows: If light were to propagate in a manner analogous to sound, then we would expect that the motion of the Earth toward or away from a light wave would affect the speed of light relative to the Earth, just as the motion of a train toward or away from a source of sound affects the speed of sound relative to the train. Such an alteration of the speed of light could be detected with an interferometer by orienting one of the arms parallel to the direction of motion of the Earth and the other arm perpendicular. A difference in the speed of light c along the arms would entail a difference in the corresponding wavelengths ($\lambda = 2\pi c/\omega$) and alter the conditions (19) and (20) for bright and dark fringes. The easiest way to detect the speed difference is by rotating the interferometer so that the arm that had been parallel to the motion becomes perpendicular and vice versa. If the speed were different in the two directions, this rotation could shift the fringes from bright to dark or vice versa.

Michelson and Morley found that there was no observable fringe shift to within the accuracy of their experiment. Taking into account possible experimental errors, they established that the effect of the motion of the Earth on the speed of light was at most ± 5 km/s. This value is substantially less than the speed of the Earth around the Sun (~ 30 km/s) and proved beyond all reasonable doubt that the propagation of light through space is *not* analogous to the propagation of sound through air. Figure 38.7 shows Michelson and Morley's interferometer.

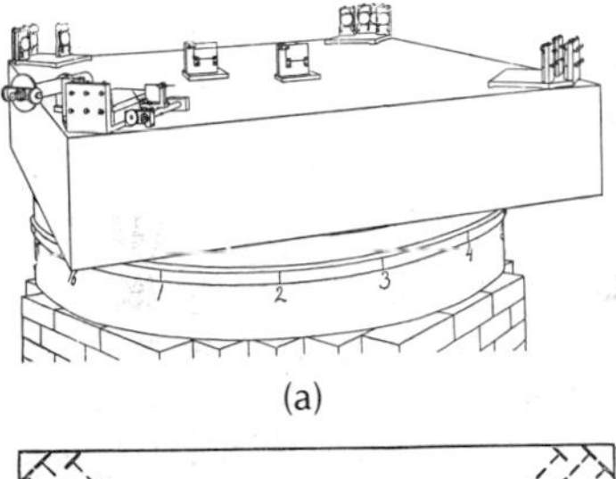

(a)

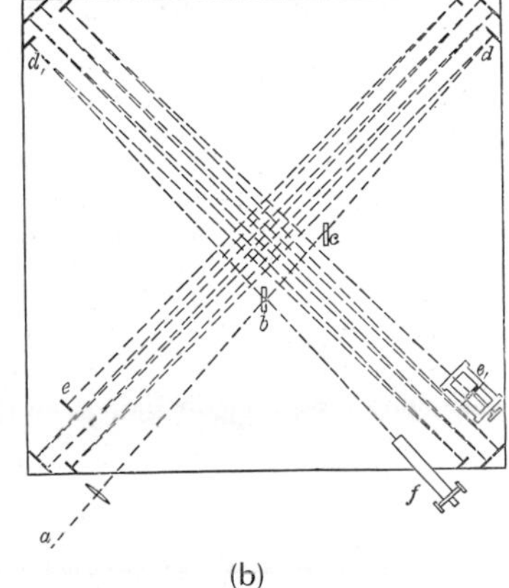

(b)

Fig. 38.7 (a) The interferometer used in the experiment of Michelson and Morley. (b) The many mirrors reflect the light beams back and forth several times, increasing the path length of the light.

38.4 Interference from Two Slits

A very clear experimental demonstration of interference effects in light can be performed with two small light sources. The light waves spread out from the sources, run into one another, and interfere constructively or destructively, giving rise to a pattern of bright and dark zones. These interference effects were discovered by Thomas Young around 1800 and, in conjunction with diffraction effects (see next chapter), they convinced physicists of the wave nature of light.

Coherence

If the interference pattern is to remain stationary in space, the two light sources must be **coherent,** that is, they must emit waves of the same frequency and the same phase (a *constant* phase difference is also acceptable; the crucial thing is that the relative phase must not fluctuate in time). Such coherent sources are easily manufactured by aiming a monochromatic wave with plane or spherical wave fronts at an opaque plate with two small slits or holes; the waves diverging from the two slits are then coherent because they arise from a single original wave (Figure 38.8). An extended object, such as an ordinary light bulb, can be used to illuminate the slits provided it is placed at a *very large* distance. The light bulb will then effectively act as a point source, illuminating the slits with light waves that consist of a succession of nearly plane wave fronts. If the light bulb is placed too close to the slits, then light waves arrive at the slit from several directions at once and this tends to wash out the interference pattern. For convenience the light bulb is sometimes placed fairly near the slits; but it must then be covered with a shield perforated with a single pinhole so that the emerging light comes from just one point on the surface of the light bulb; this again makes the light source into a point source. In modern practice, a laser is often used to illuminate the slits because it provides a very intense plane wave, so that even very faint interference and diffraction effects become visible (lasers and the light emitted by them will be discussed in Interlude L).

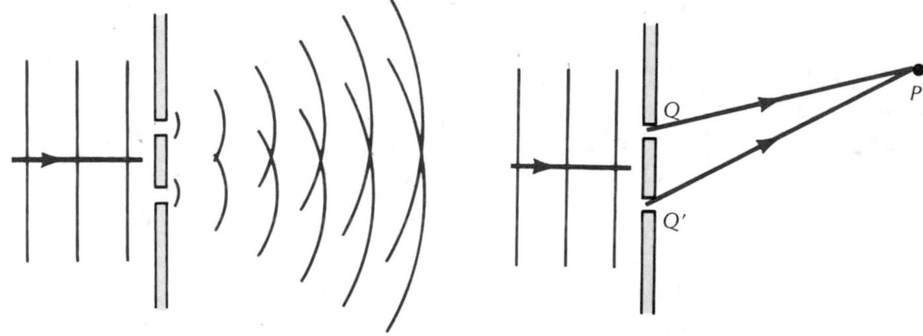

Fig. 38.8 (left) A plane light wave strikes a plate with two very narrow slits. The slits act as two coherent light sources. Light waves diverge from the two slits.

Fig. 38.9 (right) The waves reaching the point P have different path lengths QP and $Q'P$.

If the slits in the plate are very narrow, less than one wavelength in width, they will act as pointlike sources. As expected for a pointlike source, the wave diverges radially and spreads out over a wide angle beyond each slit (see Figure 38.8). Incidentally, this spreading of the wave is a diffraction effect; we will study this in detail in the next chapter.

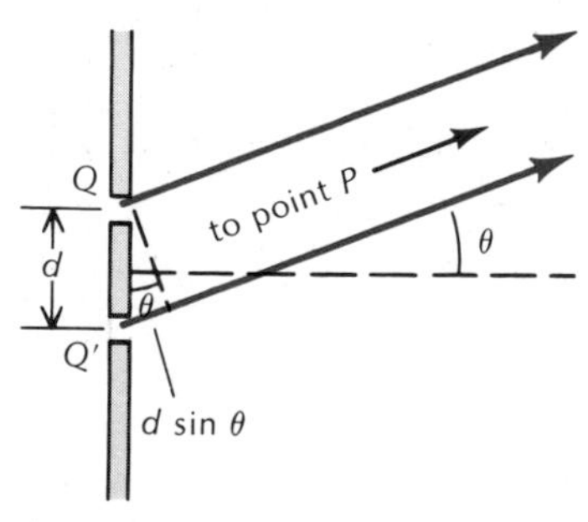

Fig. 38.10 If P is far away, QP and $Q'P$ are nearly parallel. The lengths QP and $Q'P$ then differ by $d \sin \theta$.

To find the interference maxima and minima in the space beyond the slits, we need to reckon the path difference between the rays from each of the slits. Figure 38.9 shows a light wave incident on the plate from the left and the rays QP and $Q'P$ leading from the slits to a point P on the right. We will assume that the light source is either a laser or else some other light source placed very far from the plate so that the incident wave is a plane wave (this assumption is implicit in all the calculations of this chapter). The waves emerging from the slits and reaching P interfere constructively if the difference between the lengths QP and $Q'P$ is zero, or one wavelength, or two wavelengths, etc.; and they interfere destructively if this difference is one-half wavelength, or three-halves wavelength, etc.

We can obtain a simple formula for the angular position of the maxima and minima if we make the additional assumption that the point P is at a very large distance from the plate (to be precise, QP is very large compared to QQ'). If so, then the rays QP and $Q'P$ are nearly parallel and, as Figure 38.10 shows, the difference between the lengths of these rays is approximately $d \sin \theta$, where d is the distance between the slits and θ the angle between P and the perpendicular midline.[2] Our condition for maximum intensity is then

Maxima and minima for two-slit interference pattern

$$d \sin \theta = 0, \lambda, 2\lambda, \ldots \tag{21}$$

Likewise, the condition for minimum intensity is

$$d \sin \theta = \tfrac{1}{2}\lambda, \tfrac{3}{2}\lambda, \tfrac{5}{2}\lambda, \ldots \tag{22}$$

These equations give the angular positions of the interference maxima and minima.

The regions of high intensity have the shape of beams fanning out in the space beyond the slits. The beams of high intensity are separated by lines of zero intensity; these are called nodal lines, analogous to the nodal points of the wave on a string (see Chapter 15). Figure 38.11 is a photograph of such a pattern of beams produced by the interference of water waves spreading out from two pointlike sources [the formulas (21) and (22) apply to water waves and to any other kinds of waves]. With light waves, we cannot photograph the entire pattern of beams at once; instead, we must be content with Figure 38.12a, which shows the pattern of bright and dark fringes recorded on a photographic film that intercepts the beams at some fixed distance beyond the slits.

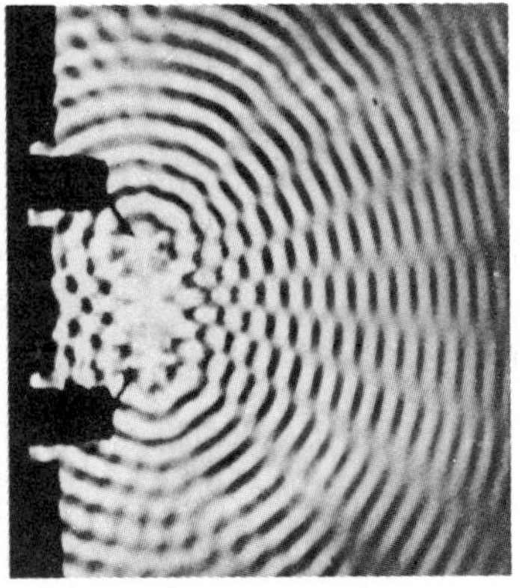

Fig. 38.11 Interference between water waves spreading out from two coherent pointlike sources in a ripple tank.

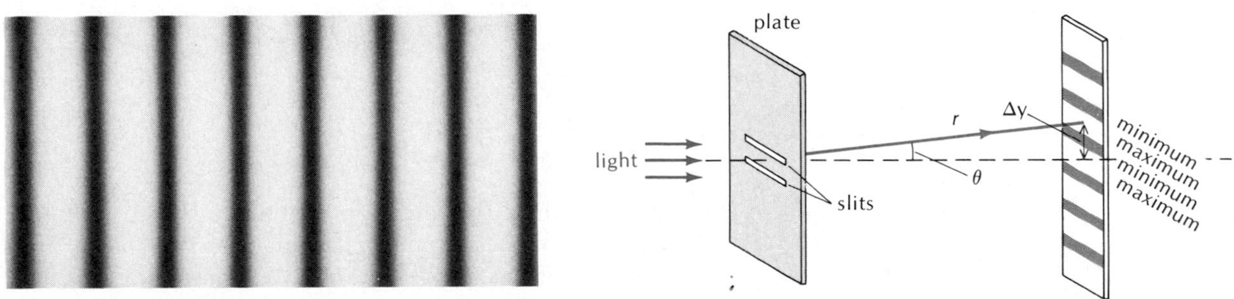

Fig. 38.12 (a) A photographic film placed beyond two illuminated narrow slits records a regular pattern of bright and dark fringes. (Courtesy C. C. Jones, Union College.) (b) Placement of the photographic film beyond the slits.

[2] This is called the **Fraunhofer approximation** for the difference between the lengths of the rays. Sometimes a lens is placed beyond the slits so as to focus their light on the point P. The approximation $d \sin \theta$ for the difference between the lengths is then justified even if the point P is not at a very large distance, because the lens brings together rays that are parallel when leaving the slits. For the sake of simpicity, we will assume throughout the following discussion that P is at a large distance and that no lens is required.

EXAMPLE 2. Two narrow slits separated by a distance of 0.12 mm are illuminated with light of wavelength 5890 Å from a sodium lamp. What is the angular position of the first lateral maximum? If the light is intercepted by a photographic film placed 2.00 m beyond the slits, what is the distance along the film between this maximum and the central maximum?

SOLUTION: With $d = 1.2 \times 10^{-4}$ m and $\lambda = 5.89 \times 10^{-7}$ m, Eq. (21) gives

$$\sin\theta = \frac{\lambda}{d} = \frac{5.89 \times 10^{-7}\ \text{m}}{1.2 \times 10^{-4}\ \text{m}} = 4.9 \times 10^{-3} \tag{23}$$

The corresponding angle is $\theta = 4.9 \times 10^{-3}$ radian.

The distance between the points with $\theta = 0$ and $\theta = 4.9 \times 10^{-3}$ radian on the photographic film is (see Figure 38.12b)

$$\Delta y \cong r\theta = 2.00\ \text{m} \times 4.9 \times 10^{-3} = 9.8 \times 10^{-3}\ \text{m}$$

Thus the maxima are separated by nearly 1 cm.

Let us now calculate the intensity distribution as a function of angle. The electric fields of the spherical waves spreading out from points in the two slits[3] and reaching the point P are [compare Eq. (36.1)]

$$E_1 = \frac{A}{r_1}\cos\left(\omega t - \frac{\omega r_1}{c}\right) \tag{24}$$

and

$$E_2 = \frac{A}{r_2}\cos\left(\omega t - \frac{\omega r_2}{c}\right) \tag{25}$$

In these equations, A is a constant that depends on the strength of the wave incident on the slits; r_1 is the distance QP and r_2 is the distance $Q'P$. With the same large-distance approximation as above, we have

$$r_1 = r_0 - \frac{d}{2}\sin\theta \tag{26}$$

$$r_2 = r_0 + \frac{d}{2}\sin\theta \tag{27}$$

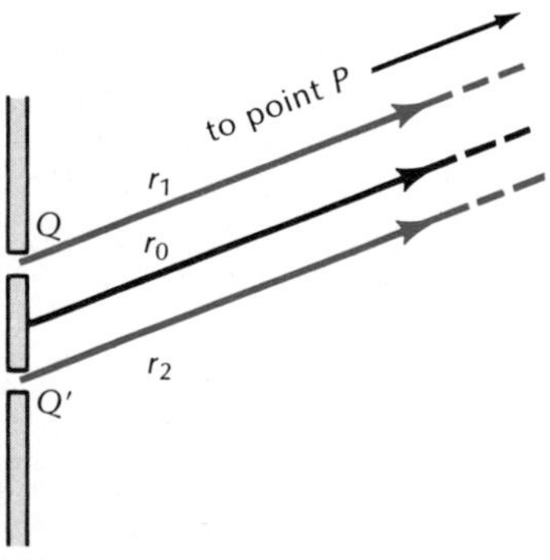

where r_0 is the distance from the midpoint of the two slits to P (Figure 38.13). We can substitute these approximations into the cosine functions in the numerators of Eqs. (24) and (25). We should also substitute these approximations into the denominators of these equations, but these denominators are not all that sensitive to small changes in the distances and the crude approximation $r_1 \cong r_2 \cong r_0$ suffices here:

$$E_1 = \frac{A}{r_0}\cos\left(\omega t - \frac{\omega r_0}{c} + \frac{\omega d}{2c}\sin\theta\right) \tag{28}$$

$$E_2 = \frac{A}{r_0}\cos\left(\omega t - \frac{\omega r_0}{c} - \frac{\omega d}{2c}\sin\theta\right) \tag{29}$$

Fig. 38.13 Rays from the slits to the point P.

[3] In the following calculation we will only consider the waves emanating from one point in each slit (the points in the plane of Figure 38.13); other points in the slits contribute similar waves.

The superposition of the two electric fields E_1 and E_2 then gives

$$E = E_1 + E_2$$

$$= \frac{A}{r_0}\left[\cos\left(\omega t - \frac{\omega r_0}{c} + \frac{\omega d}{2c}\sin\theta\right) + \cos\left(\omega t - \frac{\omega r_0}{c} - \frac{\omega d}{2c}\sin\theta\right)\right] \quad (30)$$

With the trigonometric identity $\cos(\alpha + \beta) + \cos(\alpha - \beta) = 2\cos\alpha\cos\beta$ this becomes

$$E = \frac{2A}{r_0}\cos\left(\omega t - \frac{\omega r_0}{c}\right)\cos\left(\frac{\omega d}{2c}\sin\theta\right) \quad (31)$$

The factor $\cos(\omega t - \omega r_0/c)$ is the usual oscillating function of space and time, characteristic of a wave propagating outward. The factor $\cos[(\omega d/2c)\sin\theta]$ indicates how the amplitude of this wave depends on the position angle θ.

The intensity of the wave is proportional to E^2. Since we are only interested in the time-average intensity, we can replace the factor $\cos^2(\omega t - \omega r_0/c)$ appearing in E^2 by its time-average value $\frac{1}{2}$. Hence

$$[\text{intensity}] \propto \frac{1}{r_0^2}\cos^2\left(\frac{\omega d}{2c}\sin\theta\right) \quad (32)$$

With $\lambda = 2\pi c/\omega$, we can also write this as

$$\boxed{[\text{intensity}] \propto \frac{1}{r_0^2}\cos^2\left(\frac{\pi d}{\lambda}\sin\theta\right)} \quad (33)$$

Intensity for two-slit interference pattern

This formula describes the intensity pattern produced by the double slits. The factor $\cos^2[(\pi d/\lambda)\sin\theta]$ gives the distribution of intensity as a function of angle. Figure 38.14 is a plot of this factor. This plot represents the intensity of light that reaches different angles at some constant distance r_0 from the midpoint of the slits. If we want to measure this amount of light by intercepting it with a screen or a photographic film, we have to bend the screen along a circular arc of radius r_0 (Figure 38.15); however, the observations are usually restricted to small angles θ and then a flat screen will serve well enough.

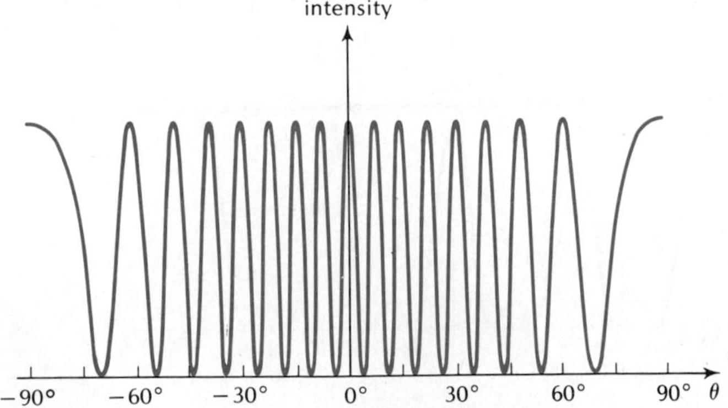

Fig. 38.14 (below, left) Intensity as a function of the angle θ for the light arriving at a screen placed beyond two narrow slits. The distance between the slits is $d = 8\lambda$.

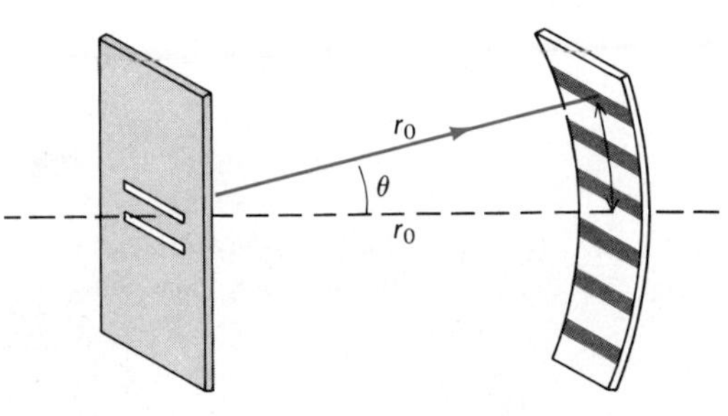

Fig. 38.15 (below, right) Zones of maximum intensity and minimum intensity on a curved screen. All points of this screen are at the same distance r_0.

38.5 Interference from Multiple Slits

It is easy to see that Eq. (21) also applies to the case of multiple slits. Figure 38.16 shows an opaque plate with three evenly spaced slits. If waves from *adjacent* slits interfere constructively, then *all* the waves from all three slits interfere constructively. Hence the condition for maximum intensity is the same as Eq. (21).

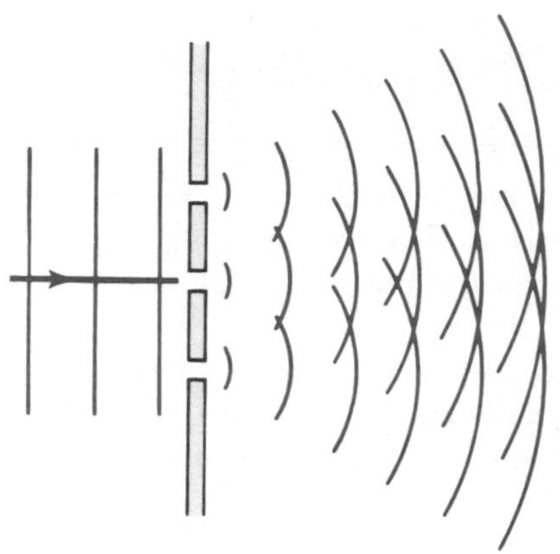

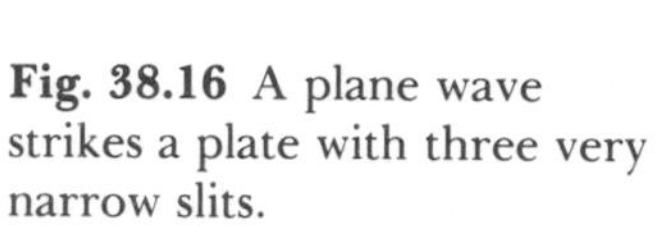

Fig. 38.16 A plane wave strikes a plate with three very narrow slits.

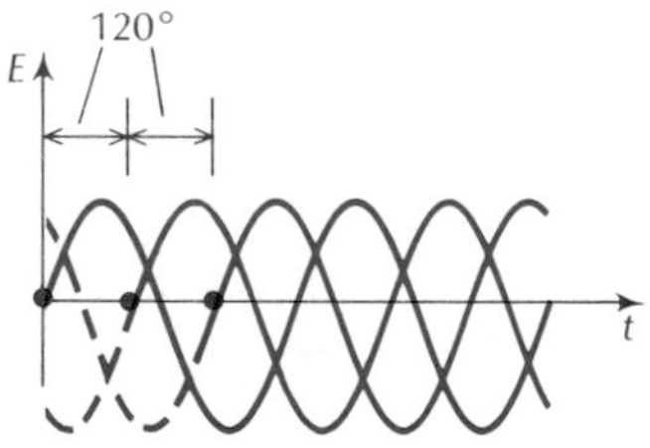

Fig. 38.17 Electric field as a function of time at the point P. The three waves from the three slits differ in phase by 120°; their sum is zero.

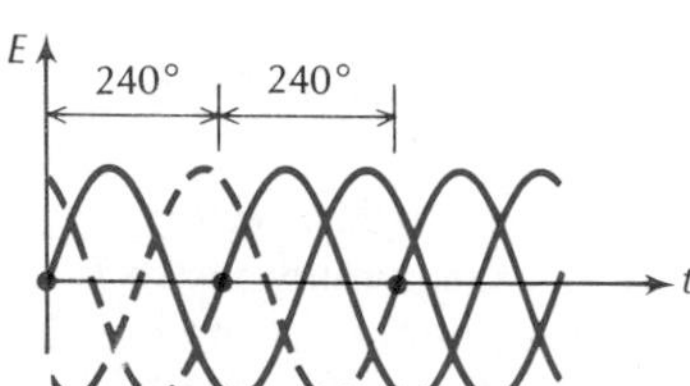

Fig. 38.18 The three waves differ in phase by 240°; their sum is zero.

However, the condition for minimum intensity is *not* Eq. (22). For instance, in the case of three slits, destructive interference of waves from adjacent slits will lead to cancellation of the waves originating from one pair of slits, but the wave from the remaining slit will not suffer cancellation. The condition for destructive interference between three waves (of equal amplitude) is that the phases differ by 120° from one wave to the next; the sum of two of the waves then always cancels against the third wave (Figure 38.17). Hence a triple slit will yield a minimum if $d \sin\theta = \lambda/3$, where, of course, d is the separation between adjacent slits. The triple slit will also yield a minimum if $d \sin\theta = 2\lambda/3$; this increased value of the path difference merely amounts to an extra shift of each wave crest by $\frac{1}{3}$ of a period and does not alter the relative distribution of these crests (compare Figures 38.17 and 38.18). Pursuing this argument, we find further minima at $d \sin\theta = 4\lambda/3$, $5\lambda/3$, etc. Figure 38.19 is a plot of the intensity as a function of θ for the case of three slits (for small values of θ). Between any two principal maxima, there are two minima. Between the two minima there is a weak maximum, a *secondary* maximum. Figure 38.20 is a photograph of this interference pattern.

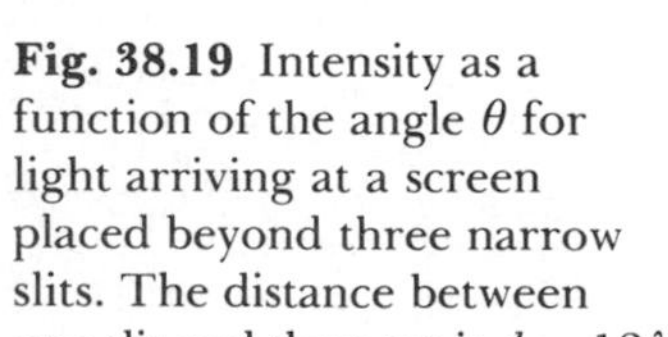

Fig. 38.19 Intensity as a function of the angle θ for light arriving at a screen placed beyond three narrow slits. The distance between one slit and the next is $d = 12\lambda$.

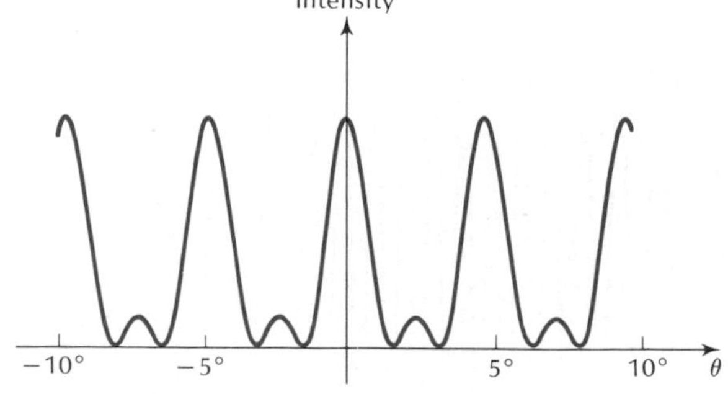

Fig. 38.20 A photographic film placed beyond the slits shows the strong principal maxima and the weaker secondary maxima. (Courtesy C. C. Jones, Union College.)

For the case of N slits, we can summarize the important features of the interference pattern as follows. The principal maxima are given by

Maximа and minima for multiple-slit interference pattern

$$d \sin \theta = n\lambda \qquad n = 0, 1, 2, \ldots \tag{34}$$

The minima are given by

$$d \sin \theta = m\lambda / N \qquad \begin{array}{l} m = 1, 2, 3, \ldots \\ \text{(with } m = N, 2N, \ldots, \text{ excluded)} \end{array} \tag{35}$$

The secondary maxima are (approximately) halfway between the minima.

Arrangements of multiple slits are commonly used to analyze light into its colors. If a light beam containing several wavelengths passes through such multiple slits, the maxima for these different wavelengths will form beams at different angles — the beams of long-wavelength light will be found at larger angles than the beams of short-wavelength light [see Eq. (34)]. Thus, the system of slits separates the light according to color and produces a spectrum in much the same way that a prism does. There is, however, one important difference between the spectra formed by a prism and by a system of slits: in the prism the long-wavelength light (red) suffers the least deflection; in the system of slits the long-wavelength light suffers the most deflection.

The system of slits will produce one complete spectrum for each value of n in Eq. (34), i.e., each principal maximum, except the central maximum, gets spread out by color. These spectra are called the first-order spectrum ($n = 1$), second-order spectrum ($n = 2$), etc. Sometimes these spectra overlap, e.g., the red end of the second-order spectrum may show up at the same angle as the blue end of the third-order spectrum (Figure 38.21).

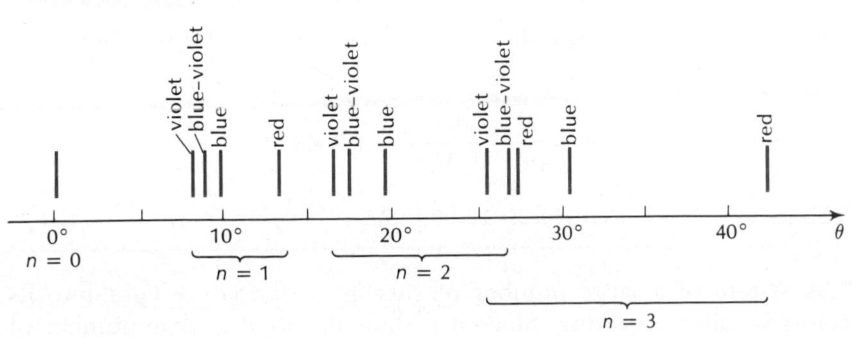

Fig. 38.21 First-, second-, and third-order spectra of hydrogen light produced by a system of N slits. The lines correspond to the principal maxima; the secondary maxima are weak and can be ignored. Each spectrum consists of a violet, blue-violet, blue, and red spectral line (compare Figure 37.22). The pattern of spectral lines for negative values of θ (negative n) is similar.

In order to achieve a very sharp separation of colors within each spectrum, we must use a very large number of slits. This makes the principal maxima very narrow and therefore reduces the overlap between maxima formed by two colors differing but little in wavelength. To understand how this helps, note that the distance from the center of a principal maximum to the next minimum is, according to Eq. (35),

Width of principal maximum

$$\Delta\theta = \frac{1}{N}\frac{\lambda}{d} \tag{36}$$

where we assume that θ is small so that $\sin\theta \cong \theta$. For a change $\Delta\lambda$ in wavelength, the change in the angular position of the principal maximum is, according to Eq. (34),

$$\Delta\theta = n\frac{\Delta\lambda}{d} \tag{37}$$

If the angular shift (37) is equal to the angular width (36), then the changed wavelength will produce a maximum at the minimum of the original wavelength. Under these conditions the two maxima are obviously well separated and can be told apart very clearly (Figure 38.22). Setting Eqs. (37) and (36) equal, we therefore find that the wavelength difference that can be clearly resolved by our arrangements of slits is given by

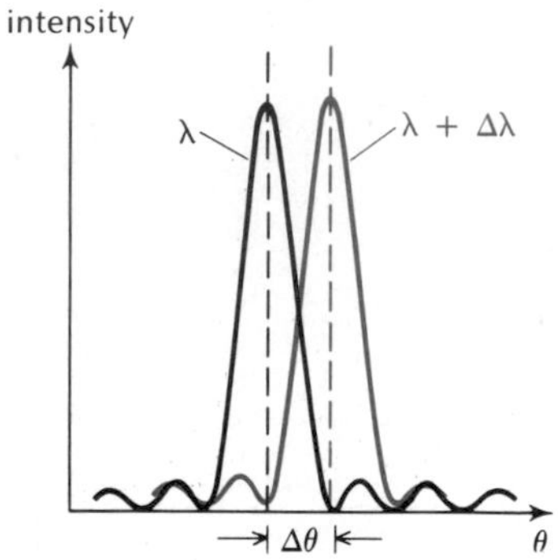

Fig. 38.22 Two distinct principal maxima produced by two different spectral lines of wavelengths λ and $\lambda + \Delta\lambda$. The maximum produced by one of these wavelengths coincides with the minimum produced by the other.

$$\frac{1}{N}\frac{\lambda}{d} = n\frac{\Delta\lambda}{d} \tag{38}$$

or

Resolving power

$$\frac{\lambda}{\Delta\lambda} = Nn \tag{39}$$

The ratio $\lambda/\Delta\lambda$ is called the **resolving power** of the system of slits. The formula (39) shows that if we want to detect a small wavelength difference, we need a large value of N.

EXAMPLE 3. Among the wavelengths emitted by atoms of iron are $\lambda = 5005.72$ Å and $\lambda = 5006.13$ Å. If we want to separate these two wavelengths clearly in the first-order spectrum of a system of slits, what is the minimum number of slits required?

SOLUTION: The wavelength difference is $\Delta\lambda = 0.41$ Å and $n = 1$. Hence Eq. (39) gives

$$N = \frac{\lambda}{\Delta\lambda} = \frac{5006 \text{ Å}}{0.41 \text{ Å}} = 1.2 \times 10^4$$

This means that we need at least 12,000 slits.

A system of a large number of slits used to analyze light into its colors is called a **grating.** Since it is difficult to cut a large number of slits in an opaque plate, gratings are usually manufactured by cutting fine parallel grooves in a glass or metal surface with a diamond stylus guided by a special ruling machine. When illuminated by a light wave, the edges of the grooves act as light sources in much the same way as slits. High-quality gratings used in spectroscopy have 10^5 or more lines with distances of about 10^{-6} m between them.

Radiotelescope

Radiotelescopes, consisting of a regular array of evenly spaced antennas, act as "gratings" for radio waves; the same formulas apply to these as to ordinary gratings used with light. Of course, there is a difference between the operation of a radiotelescope and that of an ordinary grating; in the former the radio waves from a distant point *enter* the antennas and interfere constructively or destructively within the radio receiver, while in the latter the light waves *emerge* from the slits and travel to a distant point where they interfere. Nevertheless, the condition for, say, maximum intensity can still be expressed in terms of the direction of the wave by Eq. (34) because this condition only hinges on the phase relationships among the waves, and these relationships are independent of the direction of propagation (for instance, in Figure 38.16 it makes no difference whether the waves are traveling toward or away from the slits).

EXAMPLE 4. One branch of the VLA (Very Large Array) radiotelescope at Socorro, New Mexico, has nine antennas arranged on a straight line with a distance of 2.0 km between one antenna and the next (Figure 38.23). These antennas are all connected to a single radio receiver by waveguides of equal length; the receiver then registers a maximum intensity if the radio waves incident on all antennas are in phase. Radio waves of wavelength 21 cm from a pointlike source in the sky strike this radiotelescope. When a source is at the zenith, the intensity in the radio receiver is maximum. What must be the angular displacement of a source from the zenith to make the intensity minimum?

SOLUTION: According to Eq. (36), the angular distance from the center of a principal maximum to the adjacent minimum is

$$\Delta\theta = \frac{1}{N}\frac{\lambda}{d} = \frac{1}{9}\frac{0.21 \text{ m}}{2.0 \times 10^3 \text{ m}} = 1.2 \times 10^{-5} \text{ radian}$$

This angle is about 3 seconds of arc.

Fig. 38.23 The VLA radiotelescope. The branch in the center consists of nine antennas. The distance between the antennas can be adjusted by rolling them along the track.

SUMMARY

Constructive interference: Waves meet crest to crest.

Destructive interference: Waves meet crest to trough.

Change of phase by reflection: 180° if medium off which wave reflects has higher index of refraction

Two-slit interference pattern:

maxima: $d \sin \theta = 0, \lambda, 2\lambda, \ldots$

minima: $d \sin \theta = \frac{1}{2}\lambda, \frac{3}{2}\lambda, \frac{5}{2}\lambda, \ldots$

Multiple-slit interference pattern:

principal maxima: $d \sin \theta = 0, \lambda, 2\lambda, \ldots$

minima: $d \sin \theta = \lambda/N, 2\lambda/N, \ldots$

Width of principal maximum of grating: $\Delta\theta = \dfrac{1}{N}\dfrac{\lambda}{d}$

Resolving power of grating: $\dfrac{\lambda}{\Delta\lambda} = Nn$

QUESTIONS

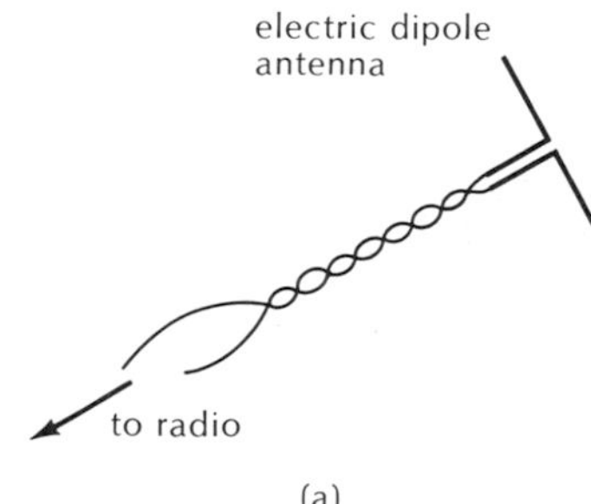

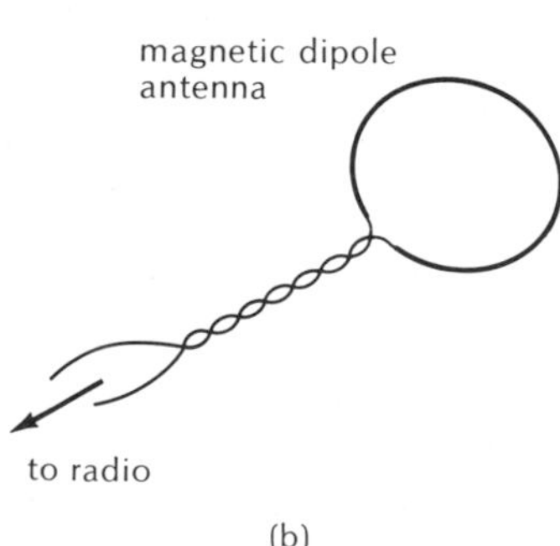

Fig. 38.24 (a) Electric dipole antenna. (b) Magnetic dipole antenna.

1. According to Eq. (11), the magnetic field of the standing wave is maximum at the surface of the mirror. The magnetic field is zero inside the mirror. What is the direction of the current that must flow along the surface of the mirror to produce this sudden drop of the magnetic field?

2. The light wave inside the tube of a laser is a standing wave, reflected at both ends of the tube. If the distance between the ends is L, what condition must the wavelength of the standing wave satisfy?

3. Would you expect sound waves reflected by a wall to produce a standing wave? How could you test this experimentally?

4. The radar wave emitted by a stationary police radar unit is reflected by an approaching automobile. Does this set up a standing wave?

5. In a traveling electromagnetic wave, the maxima of the electric and magnetic fields occur at the same place. A standing wave is a superposition of two traveling waves. How, then, can the maxima of the electric and magnetic fields in this wave occur at different places?

6. The maxima and the minima of the electric and magnetic fields in a standing wave produced by a microwave generator can be detected with an electric dipole antenna or a magnetic dipole antenna connected to a radio receiver (Figure 38.24). How must you orient each of these antennas relative to the directions of the electric and magnetic fields for maximum sensitivity?

7. When two waves interfere destructively at one place, what happens to their energy?

8. When two coherent waves of equal intensity interfere constructively at one place, the energy density at that place becomes four times as large as the energy density of each individual wave. Is this a violation of the law of conservation of energy?

9. Suppose that N waves of equal intensity meet at one place. If the waves are

coherent, the net intensity is N^2 times that of each individual wave. If the waves are incoherent, the (average) intensity is N times that of each individual wave. Explain, and give an example of each case.

10. Why do we not see colored fringes in thick oil slicks?

11. When light strikes a window pane, some rays will be reflected back and forth between the two glass surfaces. Why do these reflected rays not produce visible colored interference fringes?

12. Suppose that a lens is covered with an antireflective coating that eliminates the reflection of perpendicularly incident light of some given color. Will this coating also eliminate the reflection of light incident at an angle?

13. Is it possible to cover the surface of an aircraft with an antireflective coating so that it does not reflect radar waves of wavelength, say, 5 cm?

14. Consider the light *transmitted* by a thin film. For a light wave with a direction of propagation perpendicular to the surface of the film, what is the condition for constructive interference between the direct wave and the wave that is reflected once by the lower surface and once by the upper surface? Would you expect that this condition for constructive interference in transmission coincides with the condition for destructive interference in reflection?

15. Why is the central spot in Newtons' rings (Figure 38.5) dark?

16. Explain how Newtons's rings may be used as a sensitive test of the rotational symmetry of a lens.

17. Two flat plates of glass are in contact at one edge and separated by a thin spacer at the other edge (Figure 38.25a). Explain why we see parallel interference fringes in the reflected light if we illuminate these plates from above (Figure 38.25b).

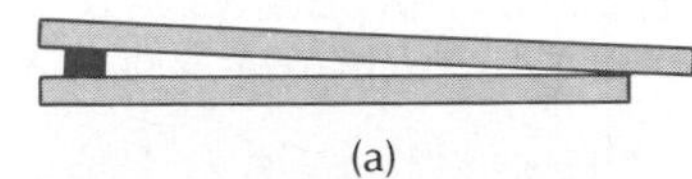

(a)

(b)

Fig. 38.25 (a) Two flat plates of glass, separated by a thin wedge of air. (b) Interference fringes.

18. In the experiment that sought to test the dependence of the speed of light on the motion of the Earth, Michelson and Morley used an interferometer with very long arms (about 11 m, obtained by multiple reflections back and forth between sets of mirrors). Why does this make the instrument more sensitive?

19. If you stand next to your TV receiver, your body will sometimes affect the reception. Why?

20. Consider (1) sunlight, (2) sunlight passed through a monochromatic filter selecting one wavelength, (3) light from a neon tube, (4) light from a laser, (5) starlight passed through a monochromatic filter, and (6) radio waves emitted by a radio station. Which of these kinds of light or electromagnetic radiation is sufficiently coherent so that when it is used to illuminate two slits such as shown in Figure 38.8, it will give rise to an interference pattern?

21. When installing a pair of stereo loudspeakers, the terminals of the loudspeakers should be connected to the amplifier in the same way, so that the loudspeakers are in phase. What would happen to sound waves of long wavelength if the loudspeakers were out of phase?

22. What happens to the interference pattern plotted in Figure 38.14 if $d < \lambda$?

23. Suppose we cover the left side of one of the slits in Figure 38.10 with a glass plate. Since the wavelength in glass is shorter than that in air, the path in the glass includes more wavelengths than the path in air, and the phases of the two waves emerging from the slits will be different. If the phase difference between these waves is exactly 180°, how does this change the location of maxima and minima?

24. Consider the interference pattern shown in Figure 38.14. How would this pattern change if the entire interference apparatus were immersed in water?

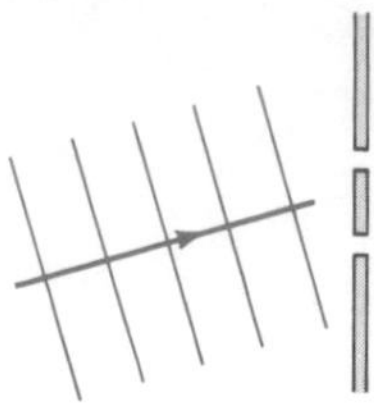

Fig. 38.26 The plane wave is incident on the plate at a slant.

25. Suppose that a plane wave is incident at an angle on a plate with two narrow slits (Figure 38.26). In what direction will we then find the central maximum?

26. Several radio antennas are arranged at regular intervals along a straight line; the antennas are connected to the same radio transmitter, so that they radiate coherently. How can this array be used to concentrate the radio emission in a selected direction?

27. When a crystal is illuminated with X rays, each atom acts as a pointlike source of scattered X rays. The typical separation between adjacent atoms in a crystal is 1 Å. Roughly what must be the wavelength of the X rays if they are to exhibit distinct interference effects?

PROBLEMS

Section 38.1

1. A light wave of wavelength 6943 Å from a laser strikes a mirror at normal incidence. This will set up a standing wave in front of the mirror.
   (a) At what distance from the mirror will the electric field have nodes? Antinodes?
   (b) At what distance from the mirror will the magnetic field have nodes? Antinodes?

2. In 1887, Hertz gave the first direct experimental demonstration of the existence of the electromagnetic waves that had been predicted by Maxwell's theory. For this experiment, Hertz placed a high-frequency spark generator at some distance in front of a large vertical zinc plate. The wave emitted by the spark reflected off the zinc plate, setting up a standing wave. Hertz found that when the generator was operating at a frequency of about 4×10^7 cycles per second, the distance between a node and an antinode in the standing wave pattern was about 2 m. What value of the speed of propagation of the electromagnetic disturbances could he deduce from this?

3. With a microwave generator, you can set up a standing electromagnetic wave between two large, flat, parallel plates of metal. The standing wave must have zero electric field at the surfaces of the plates at all times. If your microwave generator produces waves of frequency 2×10^9 Hz, what minimum separation between the plates do you need for a standing wave? Where does this standing wave have a maximum amplitude of the electric field? Of the magnetic field?

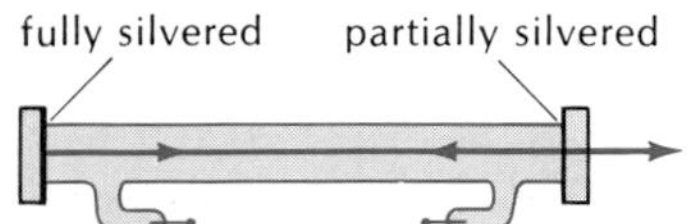

Fig. 38.27 Tube of a laser with mirrors at its ends.

4. The tube of a laser has a mirror at each end (Figure 38.27). A standing electromagnetic wave fills the space between these mirrors. This wave must satisfy the boundary condition that its electric field is zero at all times at the position of the mirrors.
   (a) Show that the distance between the mirrors and the wavelength of the wave must be related by

$$d = \left(\frac{n+1}{2}\right)\lambda \qquad n = 0, 1, 2, \ldots$$

   (b) A He–Ne laser with $d = 30.00$ cm operates at $\lambda = 6328$ Å. What is the value of n? How many minima and maxima does the standing wave have between the two mirrors?
   (c) At what distance from each of the mirrors is the nearest maximum of the electric field? The nearest maximum of the magnetic field?

Section 38.2

5. The wall of a soap bubble floating in air has a thickness of 4000 Å. If sunlight strikes the wall perpendicularly, what colors in the reflected light will be

strongly enhanced as seen in air? The index of refraction of the soap film is 1.35.

6. A lens made of flint glass with an index of refraction of 1.61 is to be coated with a thin layer of magnesium fluoride with an index of refraction of 1.38.
   (a) How thick should the layer be so as to give destructive interference for the perpendicular reflection of light of wavelength 5500 Å seen in air?
   (b) Does your choice of thickness permit constructive interference for the reflection of light of some other wavelength in the visible spectrum? (If it does, you ought to make a better choice.)

7. A thin oil slick, of index of refraction 1.3, floats on water. When a beam of white light strikes this film vertically, the only colors enhanced in the reflected beam seen in air are orange-red (about 6500 Å) and violet (about 4300 Å). From this information, deduce the thickness of the oil slick.

8. Two flat, parallel plates of glass are separated by thin spacers so as to leave a gap of width d (Figure 38.28). If light of wavelength λ is normally incident on these plates, what is the condition for constructive interference between the rays reflected by the lower surface of the top plate and the upper surface of the bottom plate?

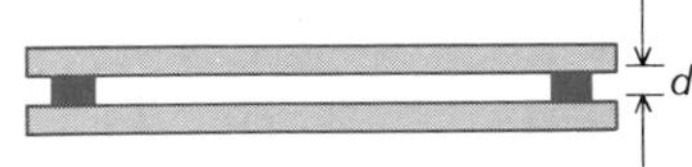

Fig. 38.28

9. A layer of oil of thickness 2000 Å floats on top of a layer of water of thickness 4000 Å resting on a flat, metallic mirror. The index of refraction of the oil is 1.24 and that of the water 1.33. A beam of light is normally incident on these layers. What must be the wavelength of the beam if the light reflected by the top surface of the oil is to interfere destructively with the light reflected by the mirror?

*10. A lens with one flat surface and one convex surface rests on a flat plate of glass (Figure 38.29). A light ray, normally incident on the lens, will be partially reflected by the curved surface of the lens and partially reflected by the flat plate of glass. The interference between these reflected rays will be constructive or destructive, depending on the height of the air gap between the lens and the plate. The interference gives rise to the pattern of Newton's rings, shown in Figure 38.5.
   (a) Why is the center of the pattern dark?
   (b) Show that the radius of the nth bright ring is

$$r = \sqrt{n\lambda R - n^2\lambda^2/4}$$

   where λ is the wavelength of the light and R is the radius of the convex surface of the lens.
   (c) What is the radius of the first dark ring if $\lambda = 5000$ Å, $R = 3.0$ m?

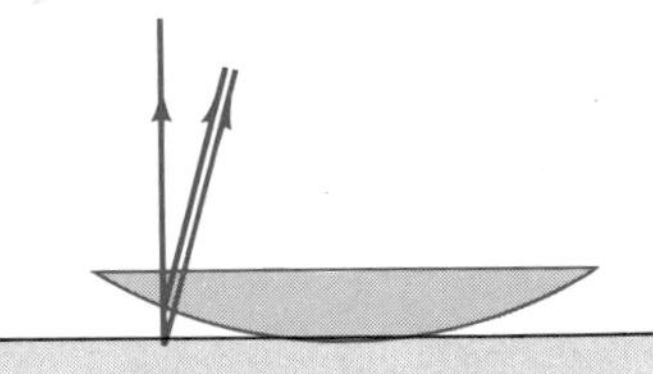
Fig. 38.29 Lens resting on flat glass plate.

Section 38.3

11. The **Fabry–Perot interferometer** consists of two parallel half-silvered mirrors. A ray of light entering the space between the mirrors may pass straight through, or be reflected once or several times by each mirror (Figure 38.30). Show that the condition for constructive interference of the emerging light (at large distance from the mirrors) is

$$2d \cos \theta = 0, \lambda, 2\lambda, \ldots$$

where d is the distance between the mirrors, λ the wavelength of light, and θ the angle of incidence of the light.

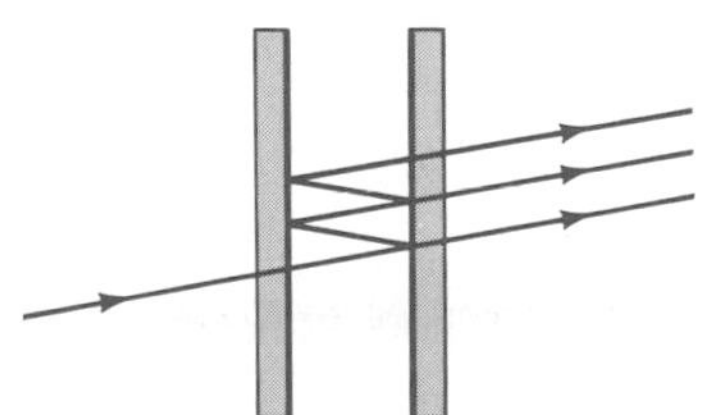
Fig. 38.30 Fabry–Perot interferometer.

12. The interferometer of the Bureau International de Poids et Mesures can count 19,000 bright fringes (maxima) per second. To achieve this count rate, what must be the speed of motion of the moving mirror (mirror M_1 in Figure 38.6)? Assume that the interferometer operates with krypton light of wavelength $\lambda = 6058$ Å.

13. An **etalon** consists of two mirrors held a fixed distance apart by means of a

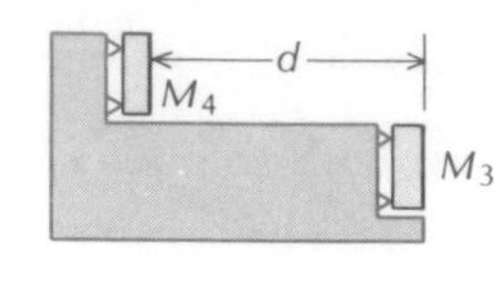

(a)

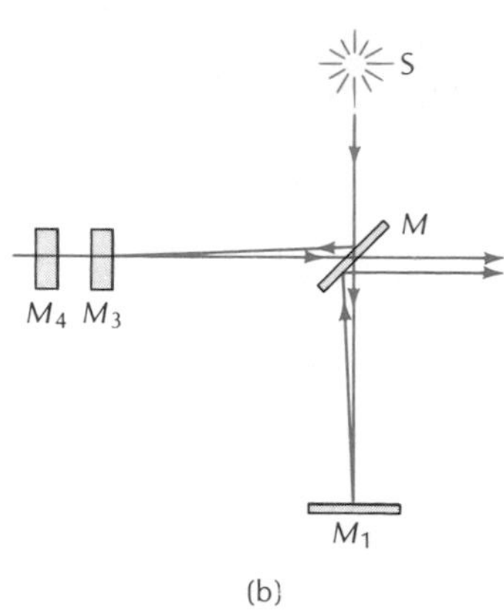

(b)

Fig. 38.31 (a) An etalon. (b) Etalon (M_3, M_4) installed in Michelson interferometer.

rigid support (Figure 38.31a). The distance between the two mirrors can serve as a standard of length. To measure this distance in terms of wavelengths of light, the etalon is installed in a Michelson interferometer, replacing the fixed mirror M_2 (Figure 38.31b). For a start, the distances MM_1 and MM_3 are made exactly equal; then the mirror M_1 is slowly moved outward, producing a sequence of interference maxima and minima until the distances MM_1 and MM_4 are exactly equal. [The exact equality of distances can be checked by replacing the usual monochromatic light source of the interferometer with a white light source. White light will give an interference maximum if and only if the distance d in Eq. (19) is exactly zero because only then will Eq. (19) be satisfied for *all* wavelengths.] Suppose that you operate a Michelson interferometer with krypton light of wavelength 6057.802 Å and you count 36,484.8 interference maxima while moving the mirror M_1 from its initial to its final position. To within six significant figures, what is the length of the etalon?

Section 38.4

14. Suppose that the two slits, the film, and the light source described in Example 2 are immersed in water. What will now be the distance between the central maximum and the first lateral maximum?

15. A piece of aluminum foil with two narrow slits is being illuminated with red light of wavelength 6943 Å from a laser. This yields a row of evenly spaced bright bands on a screen placed 3.00 m beyond the slits. The interval between the bright bands is 1.4 cm. What is the distance between the two slits?

16. Consider the water waves shown in Figure 38.11. With a ruler, measure the wavelength of the waves and the distance between the sources; with a protractor, measure the angular positions of the nodal lines (minima). Do these measured quantities satisfy Eq. (22)?

17. Two radio beacons are located on an east–west line and separated by a distance of 6.0 km. The radio beacons emit synchronous (in phase) sinusoidal waves of a frequency of 1.0×10^5 Hz. The navigator of a ship wants to determine his position relative to the radio beacons. His radio receiver indicates zero signal strength at the position of the ship. What is the angular bearing of the ship relative to the radio beacons? Assume that the distance between the ship and the radio beacons is much larger than 6 km. Note: This problem has *several* answers.

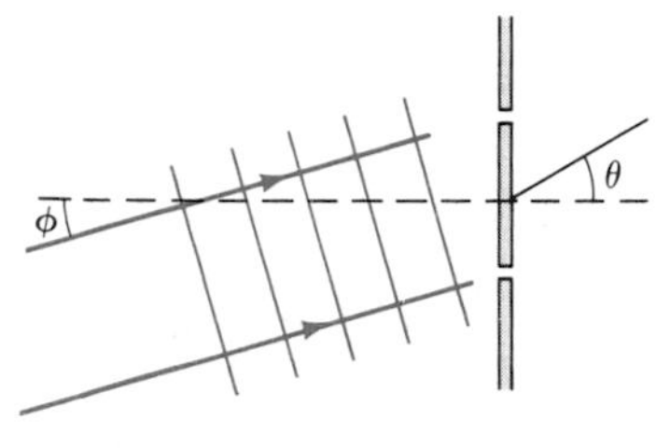

Fig. 38.32

18. Light of wavelength λ is obliquely incident on a pair of narrow slits separated by a distance d. The angle of incidence of the light on the slits is ϕ (Figure 38.32).

(a) Show that the diffracted light emerging at an angle θ interferes constructively if

$$d \sin \theta - d \sin \phi = 0, \lambda, 2\lambda, \ldots$$

and destructively if

$$d \sin \theta - d \sin \phi = \tfrac{1}{2}\lambda, \tfrac{3}{2}\lambda, \tfrac{5}{2}\lambda, \ldots$$

(b) Show that if θ is small, the angular separation between the interference maxima and minima is independent of the angle ϕ.

19. When a beam of monochromatic light is incident on two narrow slits separated by a distance 0.15 mm, the angle between the central beam and the third lateral maximum in the interference pattern is 0.52°. What is the wavelength of the light?

20. An interferometric radiotelescope consists of two antennas separated by a distance of 1.0 km. The two antennas feed their signals into a common receiver tuned to a frequency of 2300 MHz. The receiver will detect a maximum (constructive interference) if the wave sent out by a radio source in the

sky arrives at the two antennas with the same phase. What possible angular positions of the radio source will result in such a maximum? Measure the angular position from the vertical line erected at the midpoint of the antennas. Treat the antennas as pointlike.

21. The radio wave from a transmitter to a receiver may follow either a direct path or else an indirect path involving a reflection on the ground (Figure 38.33). This can lead to destructive interference of the two waves and a consequent fading of the radio signal at certain locations. Suppose that a transmitter and a receiver operating at 98 MHz are both at a height of 60 m on tall buildings with bare ground between. What is the maximum distance between the buildings that will lead to destructive interference? Be careful to take into account the phase change upon reflection.

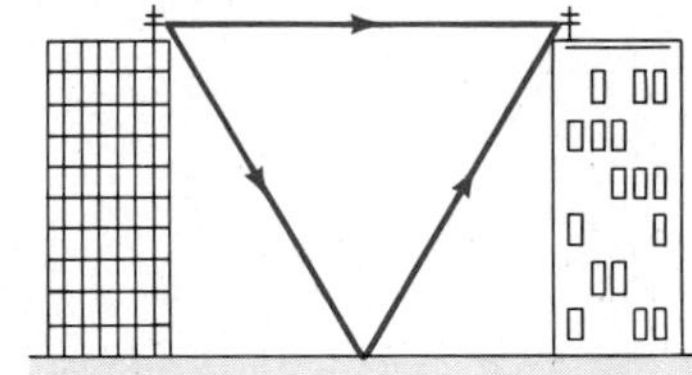

Fig. 38.33 Antennas on two buildings.

22. Two radio beacons emit waves of frequency 2.0×10^5 Hz. The beacons are on a north–south line, separated by a distance of 3.0 km. The southern beacon emits waves $\frac{1}{4}$ of a cycle later than the northern beacon. Find the angular directions for constructive interference. Measure angles relative to the east–west line and assume that the distance between the beacons and the point of observation is large.

23. A device called **Lloyd's mirror** produces interference between a ray reaching a vertical screen directly and a ray reaching the screen after reflection by a horizontal mirror. Show that in terms of the distances z, z_0, and l defined in Figure 38.34, the condition for constructive interference is

$$\sqrt{l^2 + (z + z_0)^2} - \sqrt{l^2 + (z - z_0)^2} = \tfrac{1}{2}\lambda,\ \tfrac{3}{2}\lambda,\ \tfrac{5}{2}\lambda,\ \ldots$$

and that for destructive interference is

$$\sqrt{l^2 + (z + z_0)^2} - \sqrt{l^2 + (z - z_0)^2} = 0,\ \lambda,\ 2\lambda,\ \ldots$$

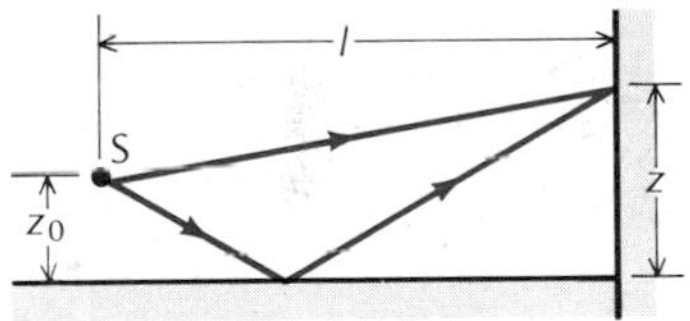

Fig. 38.34 Lloyd's mirror.

Note that in this calculation you must take into account the reversal, or change of phase, of the wave during reflection.

*24. In the first application of interferometric methods in radio astronomy, Australian astronomers observed the interference between a radio wave arriving at their antenna on a direct path from the Sun and on a path involving one reflection on the surface of the sea (Figure 38.35). Assuming that the radio waves have a frequency of 6.0×10^7 Hz and that the radio receiver is at a height of 25 m above the level of the sea, what is the least angle of the source above the horizon that will give destructive interference of the waves at the receiver?

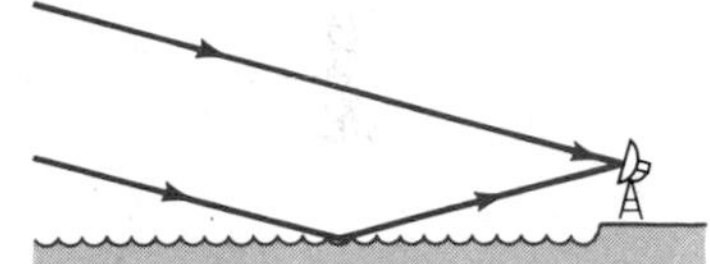

Fig. 38.35

25. Two radio beacons transmit waves of the same phase and frequency. The transmitters are on the x axis, at $\pm x_0$. Show that the interference is constructive at those points of the x–y plane satisfying the condition

$$\sqrt{(x + x_0)^2 + y^2} - \sqrt{(x - x_0)^2 + y^2} = n\lambda$$

where $n = 0, 1, 2, \ldots$. Show that for a given nonzero value of n, this is the equation of a hyperbola. (To show this, either use graphical methods to plot the curve, or else use your knowledge of analytic geometry.)

26. Light of wavelength 6943 Å from a ruby laser is incident on two narrow parallel slits cut in a thin sheet of metal. The slits are separated by a distance of 0.11 mm. A screen is placed 1.5 m beyond the slits. Find the intensity, relative to the central maximum, at a point on the screen 1.2 cm to one side of the central maximum.

Section 38.5

27. A grating has 5000 lines per centimeter. What are the angular positions of the principal maxima produced by this grating when illuminated with light of wavelength 6500 Å?

28. Sodium light with wavelengths 5889.9 Å and 5895.9 Å is incident on a grating with 5500 lines per centimeter. A screen is placed 3.0 m beyond the grating. What is the distance between the two spectral lines in the first-order spectrum on the screen? In the second-order spectrum?

29. The red line in the spectrum of hydrogen has a wavelength of 6563 Å; the blue line in this spectrum has a wavelength of 4340 Å. If hydrogen light falls on a grating of 6000 slits per centimeter, what will be the angular separation (in degrees) of these two spectral lines as seen in the first-order spectrum?

30. One of the gratings made by H. Rowland — a celebrated maker of gratings, some of which remain unsurpassed to this day for their uniformity — had 14,438 lines per inch and was 6.0 in. wide. What was the resolving power of this grating in the first order?

31. A grating has 40,000 lines uniformly spaced over a width of 8.0 cm. What is the width of the principal maxima formed by this grating with light of wavelength 5500 Å?

32. A thin curtain of fine batiste consists of vertical and horizontal threads of cotton forming a net that has a regular array of square holes. While looking through this curtain at the red (6700 Å) taillight of an automobile, a physicist notices that the taillight appears as a multiple array of images (an array of principal maxima). The angular separation between adjacent images is 2×10^{-3} radian. From this information deduce the spacing between the threads in the batiste curtain.

33. A good grating cut in speculum metal has 5900 lines per centimeter. If this grating is illuminated with white light ranging over wavelengths from 4000 Å to 7000 Å, it will produce a spectrum ranging over some interval of angles. From what angle to what angle does the first-order spectrum extend? The second-order spectrum? The third-order spectrum? Do these angular intervals overlap? Is the third-order spectrum complete?

34. Consider the Rowland grating described in Problem 30. In the interference pattern generated by this grating, how many secondary maxima are there between two adjacent principal maxima?

35. A carbon arc emits, among others, two spectral lines of wavelengths 4267.02 Å and 4267.27 Å. You wish to resolve these spectral lines in the second-order spectrum of a grating of total width 5.0 cm. How many slits per centimeter do you need in your grating?

CHAPTER 39

Diffraction

We saw in Chapter 16 that waves can bend and spread around obstacles. For instance, a water wave will spread out beyond a gap in a breakwater (see Figure 16.29) and it will spread around an island (see Figure 16.32). Such deviations from rectilinear propagation constitute **diffraction.** These deviations become quite large if the size of the gap, island, or other obstruction is of the same order of magnitude as the wavelength.

Diffraction

Since light is a wave, it displays diffraction effects. But the wavelength of light is very short, and hence these effects are difficult to detect (Figure 39.1). The diffraction effects become pronounced only when light passes through extremely narrow slits or when it strikes extremely small opaque obstacles. Diffraction effects with light had been noticed in the seventeenth century. Nevertheless, Newton held to the belief that light is of a corpuscular nature, and consists of a stream of particles. The importance of diffraction effects in establishing the wave nature of light was not fully appreciated until 1818, when Augustin Fresnel mathematically formulated the theory of diffraction. The brilliant success of this theory finally convinced physicists that light is a wave. In this chapter we will calculate the intensity pattern produced by a light wave diffracted by a narrow slit.

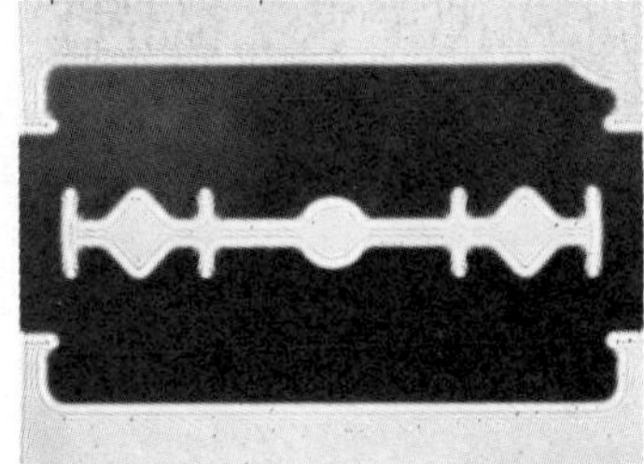

Fig. 39.1 Diffraction of light around a razor blade. The diffracted light generates a complex pattern of fine fringes at the edges of the shadow. This photograph was prepared by illuminating the razor blade with a distant light source, and throwing its shadow on a distant screen.

39.1 Diffraction by a Single Slit

Figure 39.2 shows a plane light wave approaching a slit in an opaque plate. To find the distribution of light in the space beyond the slit, we will use the following prescription, called the **Huygens–Fresnel Principle:**

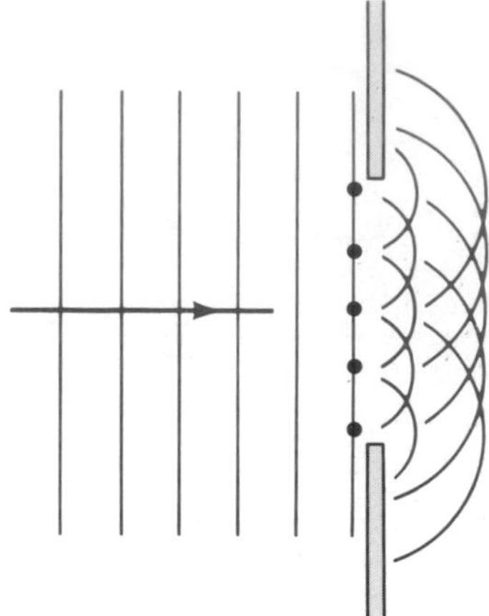

Fig. 39.2 A plane wave approaches a slit in an opaque plate. Each point on the wave front at the slit gives rise to a spherical wave that spreads out beyond the slit.

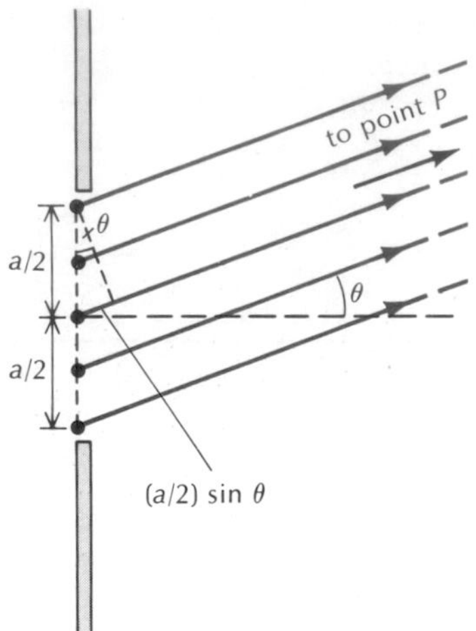

Fig. 39.3 Rays from points in the slit to the point P. The path difference between the uppermost ray and the middle ray is $(a/2)\sin\theta$.

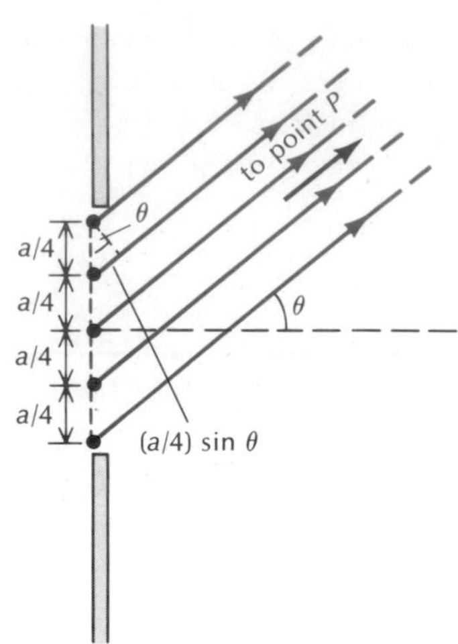

Fig. 39.4 The path difference between the uppermost ray and the ray starting at a distance $a/4$ below the uppermost ray is $(a/4)\sin\theta$.

Huygens–Fresnel Principle

Pretend that each point of the wave front reaching the slit can be regarded as a point source of light emitting a spherical wave; the net wave in the region beyond the slit is simply the superposition of all these waves.

Augustin Fresnel (frenel), *1788–1827, French physicist and engineer. His brilliant experimental and mathematical investigations firmly established the wave theory of light. Fresnel was comissioner of lighthouses, for which he designed large compound lenses (Fresnel lenses) as a replacement for systems of mirrors.*

This prescription has some obvious similarities to Huygens' Construction. However, the latter is merely a geometric construction for finding the successive positions of the wave fronts and does not yield any information about the distribution of intensity, whereas the aim of our new prescription is precisely the calculation of this distribution of intensity. Although the new prescription has a strong intuitive appeal, it turns out that it is not all that easy to justify it rigorously. The trouble is that light waves are not really sources of light waves — only accelerated charges are sources of light waves. The intensity pattern on the far side of the slit arises from the superposition of the original wave on all the waves radiated by the electric charges sitting in the opaque plate when accelerated by the original wave. Thus, our prescription lacks a simple physical basis. Nevertheless, it gives the right answer, or almost the right answer. This can be shown by a somewhat sophisticated mathematical argument, to be presented in Section 39.3.

We begin our calculation by finding the positions of the maxima and minima in the diffraction pattern of the slit. Figure 39.3 shows some rays associated with the waves that spread out from the points of a wave front at the slit. All the rays lead to a point P in the space beyond the slit. As in our calculation of the two-slit pattern, we will assume that P is very far away so that all the rays are nearly parallel. These rays can be divided into two equal groups: those that come from the upper part of the slit and those that come from the lower. Rays from the first group have a shorter distance to travel to the point P than rays from the second group. Consider the uppermost ray from the first group and the uppermost ray from the second. The path difference between these is $(a/2)\sin\theta$, where a is the width of the slit. If this path difference is $\frac{1}{2}\lambda$, the two rays will interfere destructively. Furthermore, pairs of rays that originate at an equal distance below each of these two uppermost rays in the two groups will also interfere destructively. This shows that all the waves cancel in pairs when

$$\frac{a}{2}\sin\theta = \tfrac{1}{2}\lambda \tag{1}$$

or

$$a\sin\theta = \lambda \tag{2}$$

This is the condition for the first minimum.

To find the next minimum, we divide the rays into four equal groups and consider the uppermost ray from the first and the second groups (Figure 39.4). These rays, and other pairs of rays, will interfere destructively if their path difference is $\frac{1}{2}\lambda$:

$$\frac{a}{4}\sin\theta = \tfrac{1}{2}\lambda \tag{3}$$

or

$$a\sin\theta = 2\lambda \tag{4}$$

By continuing this argument, we find a general condition for minima:

$$\boxed{a\sin\theta = \lambda,\ 2\lambda,\ 3\lambda,\ \ldots} \tag{5}$$

Minima for a single slit

As regards the maxima, there is of course a strong central maximum ($\theta = 0$). The secondary maxima cannot be found by any simple argument; roughly, their position is halfway between the successive minima. On a photographic film placed at some distance, the maxima and minima show up as a pattern of light and dark fringes (Figure 39.5).

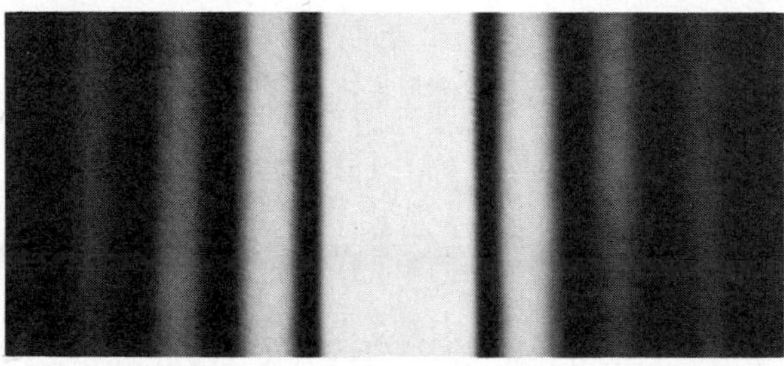

Fig. 39.5 A photographic film placed beyond an illuminated slit records a strong central maximum and successively weaker secondary maxima. (Courtesy C. C. Jones, Union College.)

EXAMPLE 1. Equation (5) applies not only to light waves, but also to radio waves and other waves. Suppose that radio waves from a TV transmitter with a wavelength of 0.80 m strike the wall of a large building. In this wall there is a very wide window with a height of 1.4 m (a horizontal slit). The wall is opaque to radio waves and the window is transparent. What is the angular width of the central maximum of the diffraction pattern formed by the waves inside the building?

SOLUTION: According to Eq. (5), the first minimum is at

$$\theta = \sin^{-1}(\lambda/a) = \sin^{-1}(0.80 \text{ m}/1.4 \text{ m}) = 34.8°$$

The central maximum extends from −34.8° to +34.8°, i.e., it has a width of 69.6°.

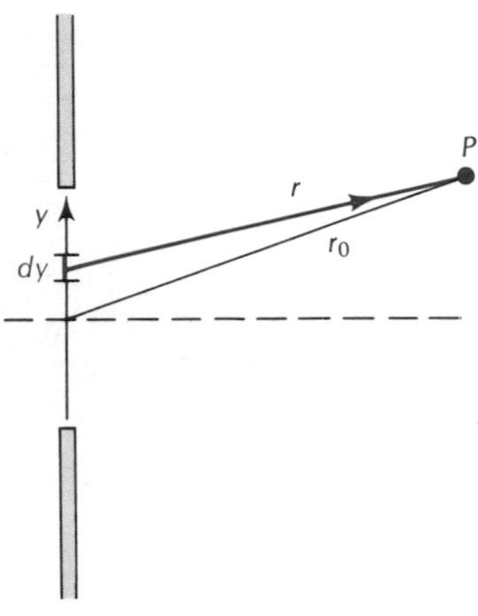

Fig. 39.6 A short segment *dy* of the slit.

To calculate the complete intensity distribution as a function of angle we must sum, or rather integrate, all the spherical waves originating from all the points of the wave front at the slit.[1] A short segment dy of the slit (Figure 39.6) contributes an amplitude

$$dE = \frac{A}{r}\sin\left(\omega t - \frac{\omega r}{c}\right) dy \tag{6}$$

at the point P. This expression is a spherical wave starting at the point y [compare Eq. (38.24)]. The constant A is not determined by our prescription, but this is of no importance if we concentrate our attention on the intensity *pattern,* that is, the distribution of relative intensity.

With the usual assumption that the distance r is large, the rays are nearly parallel and we can approximate r as in Eq. (38.26),

$$r = r_0 - y \sin\theta \tag{7}$$

where r_0 is the distance measured from the midpoint of the slit. The distance r appears in both the denominator and the numerator of Eq. (6). The denominator is not very sensitive to the value of r and the crude approximation $r \cong r_0$ suffices there. The numerator is more sensitive and we must use the better approximation (7). Thus

$$dE = \frac{A}{r_0}\cos\left(\omega t - \frac{\omega r_0}{c} + \frac{\omega y}{c}\sin\theta\right) dy \tag{8}$$

We can easily integrate this over y; the limits of integration are $y = \pm a/2$:

$$E = \int_{-a/2}^{a/2} \frac{A}{r_0}\cos\left(\omega t - \frac{\omega r_0}{c} + \frac{\omega y}{c}\sin\theta\right) dy \tag{9}$$

$$= \frac{A}{r_0}\frac{1}{(\omega/c)\sin\theta}\left[\sin\left(\omega t - \frac{\omega r_0}{c} + \frac{\omega a}{2c}\sin\theta\right) - \sin\left(\omega t - \frac{\omega r_0}{c} - \frac{\omega a}{2c}\sin\theta\right)\right] \tag{10}$$

We can combine the two terms in the bracket by means of the trigonometric identity

[1] As in the calculation of Section 38.4, we only consider the waves emanating from points in the slit that are in the plane of Figure 39.6; other points in the slit contribute similar waves, but do not affect our final results.

$$\sin\alpha - \sin\beta = 2\cos\tfrac{1}{2}(\alpha+\beta)\sin\tfrac{1}{2}(\alpha-\beta) \tag{11}$$

This yields

$$E = \frac{2A}{r_0(\omega/c)\sin\theta}\cos\left(\omega t - \frac{\omega r_0}{c}\right)\sin\left(\frac{\omega a}{2c}\sin\theta\right) \tag{12}$$

To find the intensity of light, we must evaluate E^2. A calculation similar to what we did for the double slit [see Eqs. (38.31)–(38.33)] gives us

$$\boxed{[\text{intensity}] \propto \frac{1}{r_0^2}\frac{\sin^2[(\pi a/\lambda)\sin\theta]}{\sin^2\theta}} \tag{13}$$

Intensity for single-slit diffraction pattern

This is the formula for the intensity pattern produced by a single slit. Figure 39.7 is a plot of the intensity as a function of θ for the special case $a = 3\lambda$. As expected, there is a maximum at $\theta = 0$ and there are minima at the positions given by Eq. (5).

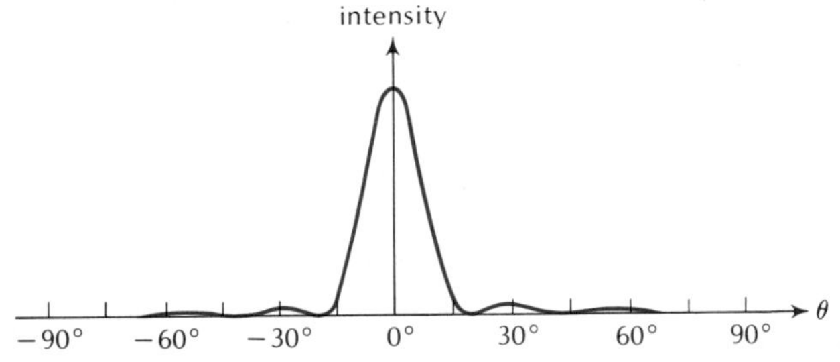

Fig. 39.7 Intensity as a function of θ for light arriving at a screen placed behind a narrow slit of width $a = 3\lambda$.

Incidentally, note that the width of the central peak *increases* as the width of the slit *decreases*. The central peak extends from the first minimum on the left to the first minimum on the right (Figure 39.7). The angular half-width of the peak is therefore given by the formula for the first minimum,

$$\sin\theta = \frac{\lambda}{a} \tag{14}$$

This angle θ increases as a decreases. If $a < \lambda$, then there is no minimum, i.e., the central peak is so wide that it covers all angles.

Our two-slit calculation of the preceding chapter was based on the implicit assumption that each slit is very narrow ($a \ll \lambda$) so that the diffraction pattern is very wide. Each slit by itself would then give essentially uniform illumination over all angles. If so, the light distribution is completely determined by the interference of the waves that spread out from the two slits.

However, if the slits are not very narrow, then the light distribution is determined by an interplay of diffraction at each slit and interference between slits. It turns out that the resulting pattern is the product of an interference curve [Eq. (38.33)] and a diffraction curve [Eq. (13)]. For example, Figure 39.8 shows the intensity pattern produced by two slits of width $a = 8\lambda$ separated by a center-to-center distance $d = 26\lambda$; Figure 39.9 is a photograph of this intensity pattern.

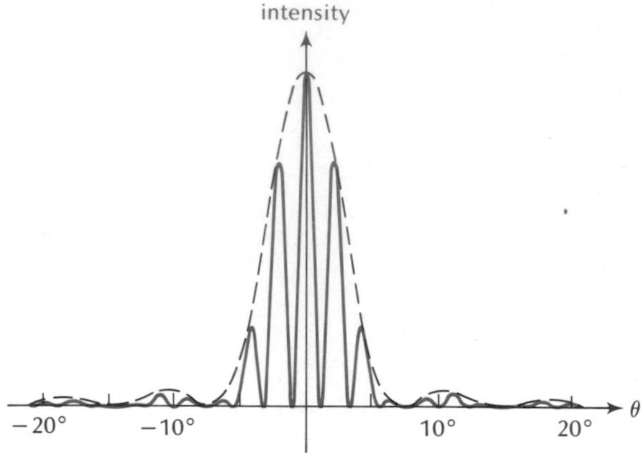

Fig. 39.8 Intensity as a function of θ for light arriving at a screen placed behind two slits of width $a = 8\lambda$ separated by a center-to-center distance of $d = 26\lambda$. The dashed curve (black) is the diffraction curve for a single slit.

Fig. 39.9 A photographic film placed beyond the two slits records a complex pattern of bands. (Courtesy C. C. Jones, Union College.)

39.2 Diffraction by a Circular Aperture; Rayleigh's Criterion

The diffraction of light by a circular aperture is in principle no different from the diffraction by a slit (a very long rectangular aperture). To calculate the light distribution we must sum the waves originating at all points of a wave front at the circular aperture. The evaluation of the resulting integral is a bit messy and we will not attempt it. Figure 39.10 shows a plot of the intensity distribution as a function of the

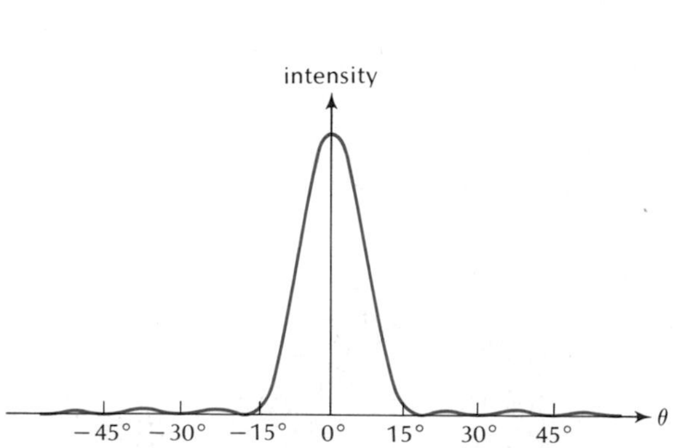

Fig. 39.10 Intensity as a function of θ for light arriving at a screen placed behind a circular aperture of diameter $a = 4\lambda$. Note that this is somewhat similar to Figure 39.7.

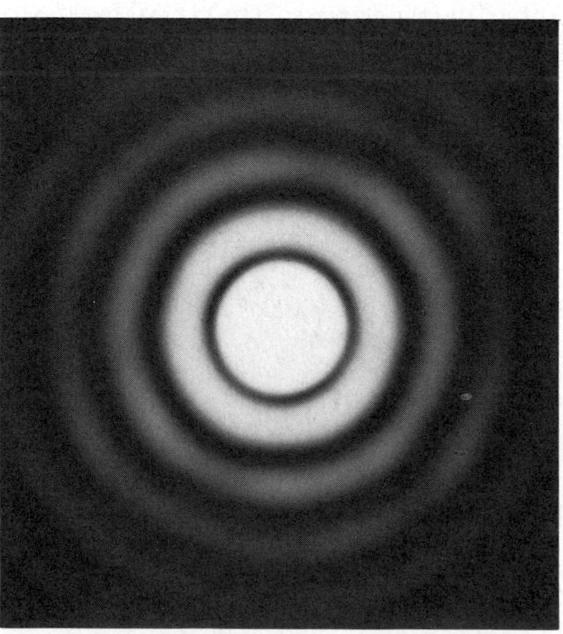

Fig. 39.11 A photographic film placed beyond a circular aperture records a strong central maximum and annular secondary maxima. (Courtesy C. C. Jones, Union College.)

usual angle θ; the distribution is of course axially symmetric. The overall shape of the curve plotted in Figure 39.10 is quite similar to that of Figure 39.7; there is a strong central maximum and a sequence of minima and secondary maxima. The angular position of the first minimum is given by

First minimum for circular aperture

$$\boxed{\sin\theta = 1.22\lambda/a} \tag{15}$$

where a is the diameter of the circular aperture.

Figure 39.11 is a photograph of the diffraction pattern produced by a circular aperture.

Many optical instruments — telescopes, microscopes, cameras, etc. — have circular apertures and these will diffract light. For instance, the objective lens of an astronomical telescope will act like a circular opening in a plate; the parallel wave fronts arriving from some distant star will suffer diffraction effects and produce an intensity distribution such as that plotted in Figure 39.10. The image of the star as seen through this telescope will then not be a bright point, but a disk surrounded by concentric rings as in Figure 39.11. For example, seen through a telescope with an objective lens 6 cm in diameter, stars look like small disks, about 2×10^{-5} radian (or 4 seconds of arc) in diameter.

This spreading out of the image puts a limit on the detail that can be perceived through this telescope. If two stars are very close together, their images tend to merge and it may be impossible to tell them apart. Figures 39.12a–d show the images produced by a pair of pointlike light sources upon diffraction by a circular aperture. In the first of these figures, the angular separation of the sources is so small that the two images look like one image. In the second figure, the angular separation is large enough to give a clear indication of the existence of two separate images.

The angular separation of the sources in Figure 39.12b is such that the central maximum of the diffraction pattern of one source coincides

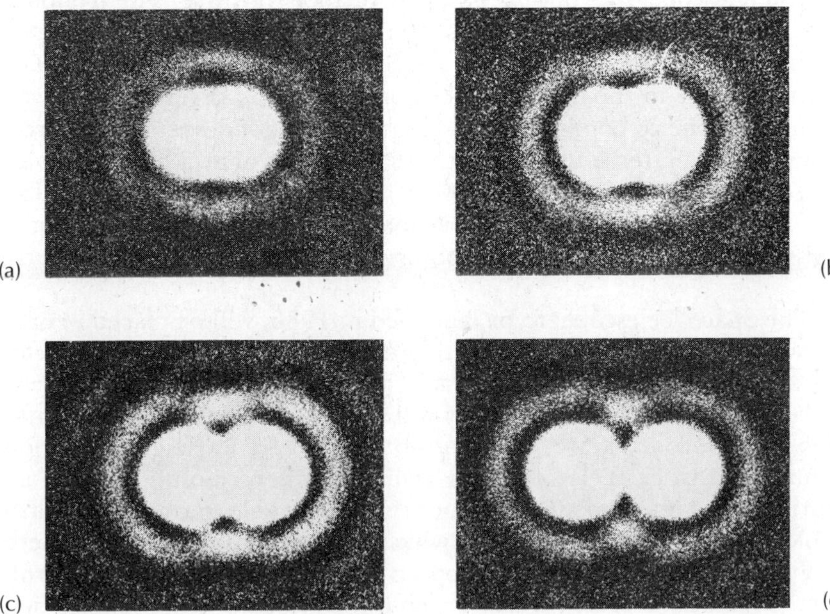

Fig. 39.12 When the light waves from two pointlike sources arrive at a circular aperture simultaneously, each set of light waves will produce its own diffraction pattern. If the angular separation between the two sources is small, the diffraction patterns overlap. In photograph (a), the angular separation is very small. In photographs (b), (c), and (d), the angular separation is progressively larger. In the first photograph, the two light sources are not resolved (not distinct). In the second photograph, they are just barely resolved, according to Rayleigh's criterion. (Courtesy C. C. Jones, Union College.)

with the minimum of the diffraction pattern of the other source. Since we are now dealing with small angles, Eq. (15) tells us that this angular separation must be

Rayleigh's criterion

$$\boxed{\Delta\theta = 1.22\,\frac{\lambda}{a}} \qquad (16)$$

We will regard this as the critical angle that decides whether the two sources are clearly distinguishable: the telescope (or other optical device) can resolve the sources if their angular separation is larger than that in Eq. (16), and it cannot resolve them if the separation is smaller. This is **Rayleigh's criterion.**

EXAMPLE 2. The star ζ Orionis is a binary star, that is, it consists of two stars very close together. The angular separation of the stars is 2.8 seconds of arc. Can the stars be resolved with a telescope having an objective lens 6 cm in diameter? Assume that the wavelength of the starlight is 5500 Å.

SOLUTION: According to Rayleigh's criterion, a telescope of this aperture can resolve stars as close as

$$\Delta\theta = 1.22\,\frac{\lambda}{a} = \frac{1.22 \times 5.5 \times 10^{-7}\ \text{m}}{0.06\ \text{m}}$$

$$= 1.1 \times 10^{-5}\ \text{radian} \qquad (17)$$

This is equal to 2.3 seconds of arc. Hence the telescope can resolve the double stars.

John William Strutt, 3rd Baron Rayleigh, *1842–1919, English physicist, professor at Cambridge and at the Royal Institution. He is best known for his extensive mathematical investigations of sound and of light. He also investigated the behavior of gases at high densities, and discovered argon; for this he was awarded the Nobel Prize in 1904.*

The ability of a telescope to resolve stars or other objects of small angular separation improves with the size of the telescope — a telescope of 30-cm diameter can resolve angular separations as small as 0.5 seconds of arc. However, beyond 30 cm, the resolution of an Earth-bound telescope does not increase any further with size. The trouble is that fluctuations in the density of the atmosphere produce slight perturbations in the path of light rays coming down from the sky. This causes a jitter in the apparent position of the star. Even under favorable atmospheric conditions — designated as **good seeing** by astronomers — the jitter is usually at least 0.5 seconds of arc. Thus, the very large aperture of the 5.1-m (200-in.) telescope on Mt. Palomar does not improve the resolution, but it does collect a very large amount of light and therefore permits the observation of very faint objects in the sky.

The Space Telescope, to be launched in 1986, will be placed in orbit above the atmosphere of the Earth and it will therefore not be affected by atmospheric seeing conditions. This telescope (Figure 39.13) will have an aperture of 2.4 m and it will achieve an angular resolution of about 0.1 second of arc, close to the limit set by Rayleigh's criterion. Thus, the Space Telescope will achieve higher resolution than any Earth-bound telescope. Furthermore, it will be able to observe at ultraviolet and infrared wavelengths, which are blocked by the atmosphere.

For Earth-bound radiotelescopes, the atmosphere poses no problem — radio waves do not suffer from the effects of atmospheric den-

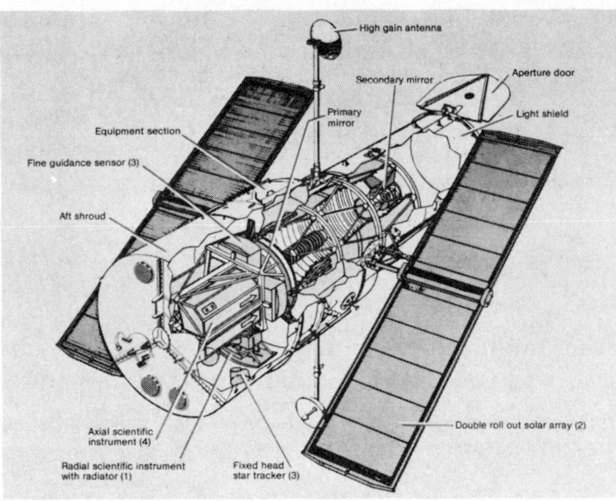

Fig. 39.13 The Space Telescope.

sity fluctuations.[2] Hence, with increasing size, the resolution of a radiotelescope improves indefinitely. Figure 39.14 shows the large radiotelescope at Arecibo, Puerto Rico. The concave "mirror" of this telescope has an aperture of 1000 ft and a radius of curvature which is also 1000 ft. The shortest wavelength at which this telescope has been operated is 4 cm. For this wavelength, Rayleigh's criterion gives a limiting angular resolution of

$$\theta = 1.22\,\frac{\lambda}{a} = 1.22 \times \frac{0.04\ \text{m}}{1000\ \text{ft} \times 0.30\ \text{m/ft}}$$

$$= 1.6 \times 10^{-4}\ \text{radian}$$

which is about 30 seconds of arc.

Fig. 39.14 Arecibo radiotelescope.

[2] The density fluctuations occur over such small distances that radio waves do not notice them.

A very substantial improvement in resolution is attainable by employing two radiotelescopes separated by a very large distance. If the signals detected by these two telescopes are brought into superposition,[3] then the system will act essentially like the two-slit system of Section 38.4. The limiting angular resolution is roughly the angle between a maximum and the adjacent minimum,

$$\Delta\theta \cong \frac{\lambda}{d} \tag{18}$$

where d is the distance between the telescopes. In one application of this method astronomers employed radiotelescopes separated by a distance of more than 10,000 km: they were in Greenbank, West Virginia, and in the Crimea.

39.3 Babinet's Principle[4]

In this last section we will present the theoretical justification for the Huygens–Fresnel Principle. According to this principle, the amplitude of the light in the region beyond an opaque plate with an aperture of a given shape (single slit, double slit, circular hole, etc.) can be calculated by pretending that the points of the aperture act as point sources of light. To gain some insight into the actual origins of the amplitude of the light, consider first an opaque plate whose aperture has been closed with a plug that exactly fits the opening (Figure 39.15). Obviously, when we place such a plugged plate in the path of a light wave, the amplitude of the light beyond the plate will be zero. It is very instructive to inquire: How does the plate stop the light wave? The absence of light beyond the plate is a cancellation effect. When the wave strikes the plate, it shakes the electrons within the atoms and causes them to radiate light of the same wavelength as that of the incident light. The original wave passes through the plate and reaches the region beyond the plate. The radiated waves from all the electrons also reach this region. The net amplitude of the light in this region is then the sum of the contribution of the original wave plus the contribution from all the electrons — in order that the plugged plate be truly opaque, these two contributions must cancel exactly.

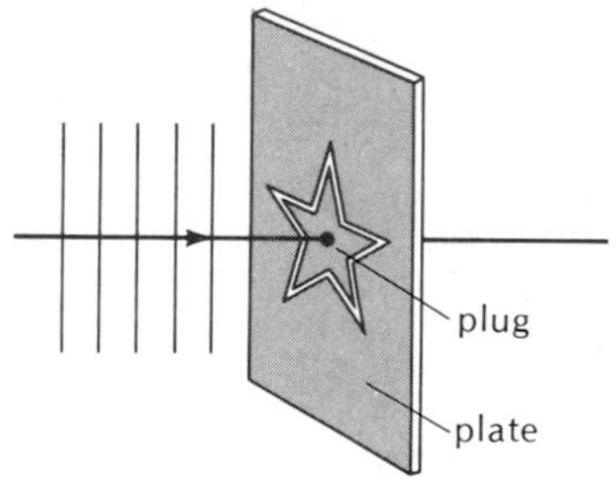

Fig. 39.15 An opaque plate with an aperture. The aperture has been closed with a plug.

Mathematically we can express this as follows: At some point beyond the plate, the electric field contributed by the radiating electrons has an amplitude $\mathbf{E}_{rad}$ and the electric field contributed by the original wave has an amplitude $\mathbf{E}_0$. These must be opposite,

$$\mathbf{E}_{rad} = -\mathbf{E}_0 \tag{19}$$

The field contributed by the electrons can be split into two parts: that due to the electrons in the perforated plate (without the plug) and that in the plug. Thus, Eq. (19) becomes[5]

[3] For radiotelescopes separated by an intercontinental distance, it is not practical to connect both antennas to the same, single-radio receiver. Instead, each antenna is connected to its own radio receiver, recordings of the received signals are made on magnetic tapes, and the signals are later superposed electronically.

[4] This section is optional.

[5] This equation is not exact. An unperforated plate, made of a single piece of material, is not completely equivalent to a perforated plate plus a separate plug. The motion

$$\boxed{\mathbf{E}_{\text{rad, plate}} + \mathbf{E}_{\text{rad, plug}} = -\mathbf{E}_0} \tag{20}$$

This relationship between the fields radiated from the perforated plate and the plug is called **Babinet's Principle.**[6] The plug can be regarded as a plate of a shape *complementary* to that of the perforated plate. With this terminology, Babinet's Principle simply states that the sum of the fields radiated from a plate and from its complement equals the negative of the field of the original wave.

With these preliminaries, we are now ready to examine the amplitude beyond the plate when the plug is *absent.* This amplitude has a contribution from the radiating electrons in the perforated plate and a contribution from the original wave:

$$\mathbf{E}_{\text{beyond plate}} = \mathbf{E}_{\text{rad, plate}} + \mathbf{E}_0 \tag{21}$$

But, in view of the identity (20) the right side can be replaced by $-\mathbf{E}_{\text{rad,plug}}$. We then obtain

$$\mathbf{E}_{\text{beyond plate}} = -\mathbf{E}_{\text{rad, plug}} \tag{22}$$

Thus, apart from a minus sign, the amplitude beyond the plate happens to be just what we would calculate if we were to pretend that the points in the aperture radiate as point sources (or electrons). Of course, this is exactly what the Huygens–Fresnel Principle tells us to do — we have therefore obtained its justification. The minus sign on the right side of Eq. (22) is left out of account by the Huygens–Fresnel Principle, but this is of no importance because ultimately we are only interested in the *intensity* and this depends only on the square of the electric field.

Finally, let us rephrase the Babinet Principle in terms of the amplitude of the diffracted light. Suppose we remove the perforated plate from the light beam and instead place the plug there, i.e., we replace the plate by the complementary plate. Then an argument similar to that which led to Eq. (22) yields

$$\mathbf{E}_{\text{beyond plug}} = -\mathbf{E}_{\text{rad, plate}} \tag{23}$$

By substituting Eqs. (22) and (23) into Eq. (20) we obtain

$$\boxed{\mathbf{E}_{\text{beyond plate}} + \mathbf{E}_{\text{beyond plug}} = \mathbf{E}_0} \tag{24}$$

Babinet's Principle

This identity is another form of Babinet's Principle. It states that the sum of the amplitudes of light diffracted by a plate and by its complement equals the amplitude of the original wave.

---

of the electrons on the perforated plate and on the plug is subject to the constraint that the electrons cannot cross the boundary separating the perforated plate and the plug; the motion on the unperforated plate is not subject to this constraint. This difference leads to small deviations from Eq. (20); but these deviations are only important in regions fairly near the plate–plug boundary. As in the preceding sections, we will only deal with points at a large distance from the plate and there Eq. (20) is a good approximation.

[6] For another version of Babinet's Principle, see below.

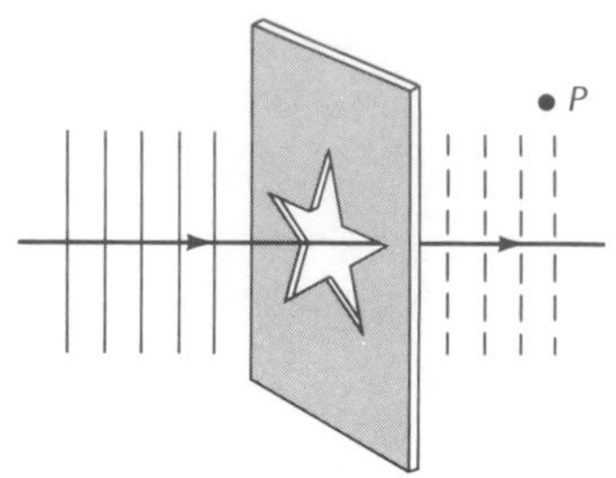

Fig. 39.16 The point P is not in the path of the incident light wave.

Equation (24) has an interesting application. Suppose that we want to calculate the intensity of light in some region outside of the path of the original light wave (Figure 39.16). In this region $\mathbf{E}_0 = 0$. Hence Eq. (24) gives

$$\mathbf{E}_{\text{beyond plate}} = -\mathbf{E}_{\text{beyond plug}} \tag{25}$$

Since the intensity of light depends on the square of the electric field, Eq. (25) shows that a plate and its complement produce the same diffraction pattern! For instance, an arrangement of star-shaped holes in a plate and the complementary arrangement of opaque stars suspended in midair (Figure 39.17) produce the same diffraction pattern. Figure 39.18 shows photographs of these diffraction patterns. Of course they differ in the central region which is in the path of the original wave; but elsewhere they are identical.

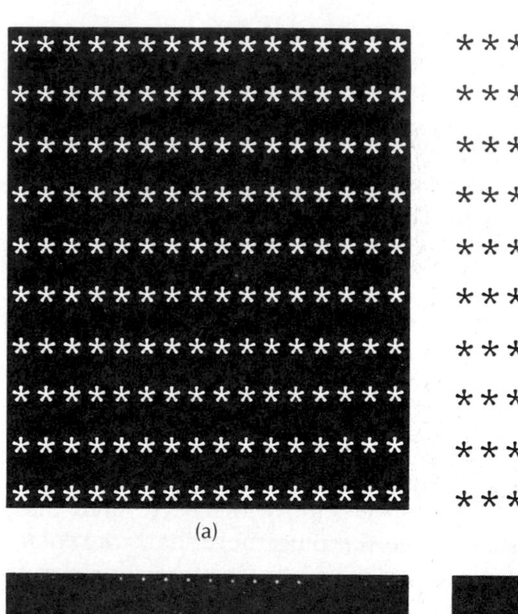

(a)

(b)

Fig. 39.17 (a) An opaque plate with star-shaped holes. (b) The complementary plate consists of opaque stars. (Courtesy C. C. Jones, Union College.)

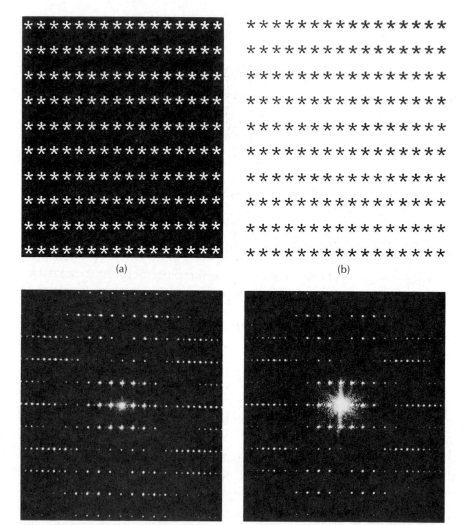

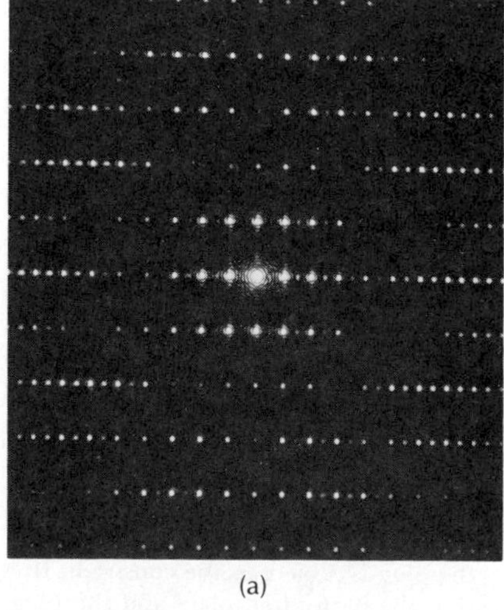

(a)

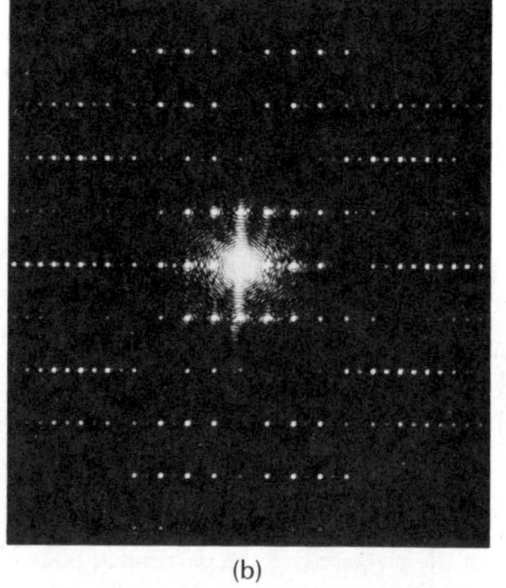

(b)

Fig. 39.18 (a) Photograph of the diffraction pattern of the plate shown in Figure 39.17a. (b) Photograph of the diffraction pattern of the complementary plate shown in Figure 39.17b. The two diffraction patterns are identical, except in the central region, where the undiffracted light beam passes through the complementary plate with high intensity. (Courtesy C. C. Jones, Union College.)

SUMMARY

Minima for single slit: $a \sin\theta = \lambda, 2\lambda, 3\lambda, \ldots$

First minimum for circular aperture: $a \sin\theta = 1.22\lambda$

Rayleigh's criterion: $\Delta\theta = 1.22\lambda/a$

Babinet's Principle: $\mathbf{E}_{\text{beyond plate}} + \mathbf{E}_{\text{beyond plug}} = \mathbf{E}_0$

QUESTIONS

1. Gratings used for analyzing light are often called diffraction gratings, but it would be more accurate to call them interference gratings. Why?

2. Consider the diffraction pattern shown in Figure 39.7. How would this pattern change if the entire apparatus were immersed in water?

3. Figure 16.28 shows the diffraction of water waves at the entrance of a harbor. Could this diffraction be eliminated by making the entrance smaller?

4. What would you expect the diffraction pattern of a rectangular aperture to look like?

5. In the center of the shadow of a disk or a sphere there is a small bright spot, called the **Poisson spot** (Figure 39.19). This spot is very faint near the disk, but becomes more noticeable at large distances. Qualitatively, explain why the diffraction of the light waves around the edges of the disk gives rise to this spot.

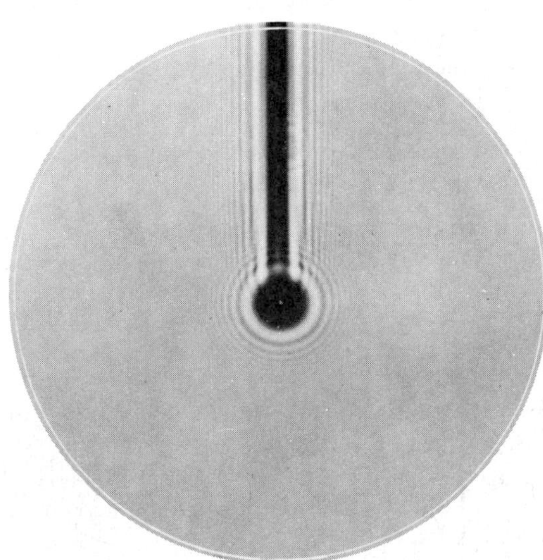

Fig. 39.19 The Poisson spot.

6. Besides good angular resolution, what other advantage does a telescope of large aperture have over a telescope of small aperture?

7. Spy satellites use cameras with lenses of very large diameter, 30 cm or more. Why are such large diameters necessary?

8. What might be the advantages of building a radiotelescope in space, in orbit around the Earth?

9. In order to beam a sound wave sharply in one direction with a loud-hailer, should the horn of the loud-hailer have a small aperture or a large aperture?

10. The manufacturers of the Questar telescope claim that this telescope can distinguish two stars even if their separation is somewhat smaller than that specified by Rayleigh's criterion. Why is this possible?

11. The maximum useful magnification of an optical microscope is determined by diffraction effects at the objective lens. Explain.

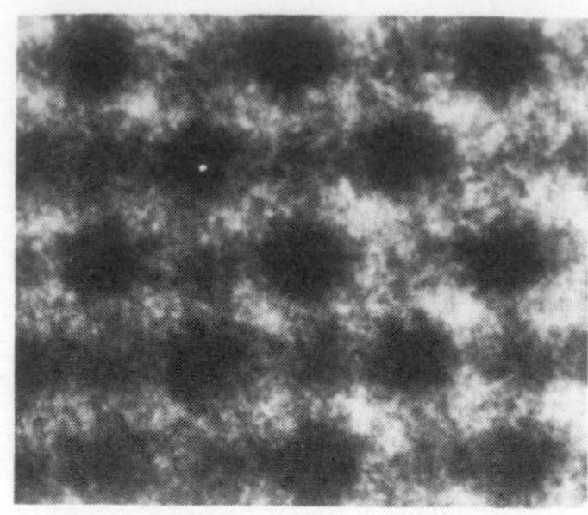

Fig. 39.20 Atoms in a barium titanate crystal.

12. Other things being equal, how much resolution can you gain by operating an optical microscope with blue light instead of red light? Why can you not operate the microscope with ultraviolet light?

13. The picture of atoms in Figure 39.20 was taken with an electron microscope using electron waves of extremely short wavelength. Roughly, how short must the wavelength be to make individual atoms visible?

14. The compound eye of insects consists of a large number of small eyes, or ommatidia (Figure 39.21). Each ommatidium is typically 0.03 mm across; it does not form an image, but merely acts as a sensor of the intensity of light arriving from a narrow cone of directions. What are some of the advantages and disadvantages of such a compound eye as compared to the camera eye of vertebrates?

15. The corona formed by diffraction of sunlight by small droplets of water in clouds consists of a bright ring surrounding the Sun. What is the color of the outer edge of the corona? The inner edge?

16. Antennas used for the transmisson of microwaves in communication links consist of metal dishes. What factors determine the size of these dishes?

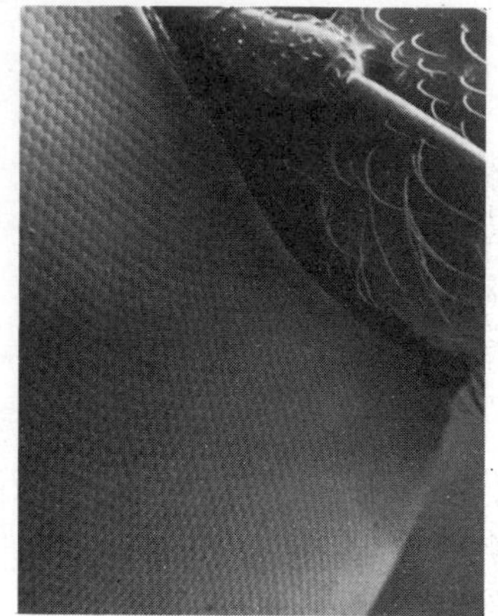

Fig. 39.21 A portion of the compound eye of a dragonfly. This eye consists of about 28,000 ommatidia. (Courtesy R. Olberg, Union College.)

17. In the joint operation of two radiotelescopes separated by an intercontinental distance, it is not practical to connect both of these antennas to the same, single radio receiver. Instead, each antenna is connected to its own radio receiver, and recordings of the received signals are made on magnetic tapes. How can we extract the interference patterns from the information stored on these tapes?

18. What would the diffraction pattern of a narrow opaque strip look like? Of a small opaque disk?

19. How would the diffraction patterns of Figures 39.18a and b change if all the stars in Figure 39.17 were made smaller?

20. What feature of the plate shown in Figure 39.17a determines the spacing between the dots in Figure 39.18a? What feature determines the gradual modulation of the intensity of these dots?

PROBLEMS

Section 39.1

1. Light of wavelength 6328 Å from a He–Ne laser illuminates a single slit of width 0.10 mm. What is the width of the central maximum formed on a screen placed 2.0 m beyond the slit?

2. Consider the water waves shown in Figure 39.22. With a ruler, measure the wavelength of the waves and the length of the gap; with a protractor, measure the angular positions of the two nodal lines (minima). Check whether these quantities satisfy Eq. (5).

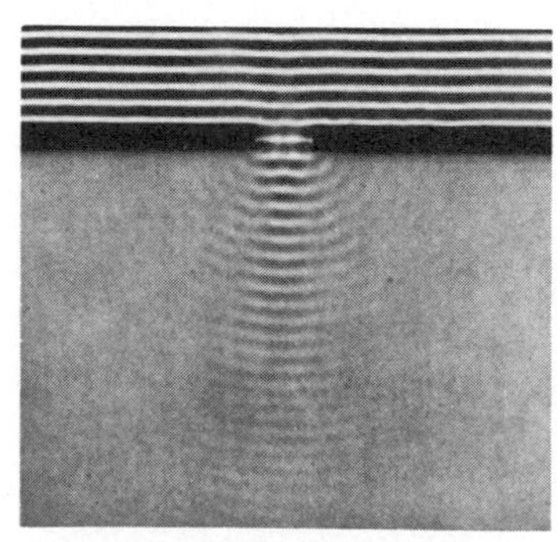

Fig. 39.22

3. A sound wave of frequency 820 Hz passes through a doorway of width 1.0 m. What are the angular directions of the minima of the diffraction pattern?

4. Water waves of wavelength 20 m approach a harbor entrance 50 m across at right angles to their path. What is the angular width of the central beam of diffracted waves beyond the entrance?

5. Light of wavelength λ is obliquely incident on a slit of width a. The angle of incidence of the light on the slit is ϕ (Figure 39.23).

(a) Show that the diffracted light emerging at an angle θ interferes destructively if

$$a \sin \theta - a \sin \phi = \lambda, 2\lambda, 3\lambda, \ldots$$

(b) Show that, for small values of θ, the angular separation between the directions of destructive interference is independent of ϕ.

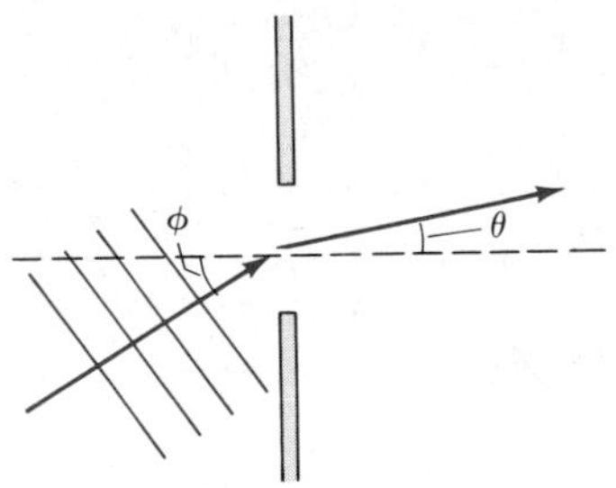

Fig. 39.23

6. A slit of width 0.11 mm cut in a sheet of metal is illuminated with light of wavelength 5770 Å from a mercury lamp. A screen is placed 4.0 m beyond the slit.
   (a) Find the width of the central maximum in the diffraction pattern on the screen, i.e., find the distance between the first minimum on the left and on the right.
   (b) Find the width of the secondary maximum, i.e., find the distance between the first minimum and the second minimum on the same side.

7. Plot intensity vs. angle for the light diffracted by a pair of slits of width $a = 4\lambda$ separated by a center-to-center distance $d = 8\lambda$.

8. Two narrow slits of width a are separated by a center-to-center distance d. Suppose that the ratio of d to a is an integer, $d/a = n$.
   (a) Show that in the diffraction pattern produced by this arrangement of slits, the nth interference maximum (corresponding to $d \sin \theta = n\lambda$) is suppressed because of coincidence with a diffraction minimum. Show that this is also true for the $2n$th, $3n$th, etc., interference maxima.
   (b) How many interference maxima are there between one diffraction minimum on one side and the next on the same side?

9. Figure 38.19 shows the intensity curve for the interference pattern produced by light emerging from three very thin slits (three pointlike sources) separated by distances of 12λ. Suppose the three very thin slits are replaced by three new slits of width 6λ. This will change the intensity curve.
   (a) On top of Figure 38.19 (or on a copy of Figure 38.19), plot the intensity curve for the diffraction pattern produced by *one* of these new slits.
   (b) By suitably combining these two curves, obtain the complete intensity curve for the system of the three new slits.

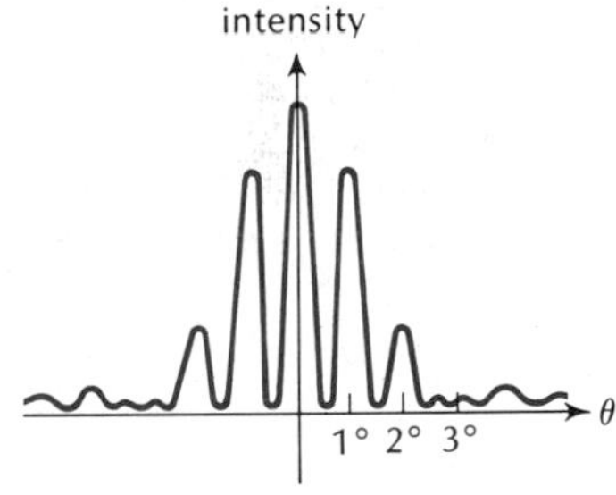

Fig. 39.24 Intensity as a function of θ in the diffraction pattern of two slits.

10. Figure 39.24 is a plot of the intensity of the light on a screen placed beyond an opaque plate with two parallel narrow slits. The distance between the slits is 0.12 mm. Deduce the width of each of the slits.

Section 39.2

11. According to a recent proposal (see Section F.4), solar energy is to be collected by a large power station on an artificial satellite orbiting the Earth at an altitude of 35,000 km. The power is to be beamed down to the surface of the Earth in the form of microwaves. If the microwaves have a wavelength of 10 cm and if the antenna emitting the microwaves is 1.5 km in diameter, what is the angular width of the central beam emerging from this antenna? What will be the transverse dimension of the beam when it reaches the surface of the Earth?

12. When the eye looks at a star (a point of light), diffraction at the pupil spreads the image of the star on the retina into a small disk.
   (a) When opened to maximum size, the diameter of the pupil of a human eye is 7.0 mm. Assuming the starlight has a wavelength of 5500 Å, what is the angular size of the image on the retina?
   (b) The distance from pupil to retina is 23 mm. What is the linear size of the image of the star?
   (c) At the midpoint on the retina (fovea), there are 150,000 light-sensitive cells (rods) per mm^2. How many of these cells are illuminated when the eye looks at a star?

13. A sailor uses a speaking trumpet to concentrate his voice into a beam. The opening at the front end of the speaking trumpet has a diameter of 25 cm. If the sailor emits a sound of wavelength 15 cm (this is very roughly the wavelength a man emits when yelling "eeeee . . ."), what is the angular width of the central maximum of the beam of sound?

14. The antenna of a small radar transmitter operating at 1.5×10^{10} Hz consists of a circular dish of diameter 1.0 m. What is the angular width of the central maximum of the radar beam? What is the linear width at a distance of 5.0 km from the transmitter?

15. A microwave antenna used to relay communication signals has the shape of a circular dish of diameter 1.5 m. The antenna emits waves with $\lambda = 4.0$ cm.
   (a) What is the width of the central maximum of the beam of this antenna at a distance of 30 km?
   (b) The power emitted by the antenna is 1.5×10^3 W. What is the energy flux directly in front of the antenna? What is the energy flux at a distance of 30 km? Assume that the power is evenly distributed over the width of the central beam.

16. For an optically perfect lens, the size of the focal spot is limited only by diffraction effects. Suppose that a lens of diameter 10 cm and focal length 18 cm is illuminated with parallel light of wavelength 5500 Å. What is the angular width of the central maximum in the diffraction pattern? What is the corresponding linear width at the focal distance?

17. The Space Telescope that will be placed into an orbit above the atmosphere of the Earth has an aperature of 2.4 m. According to Rayleigh's criterion, what angular resolution can this telescope achieve with visible light of wavelength 5500 Å? With ultraviolet light of wavelength 1200 Å? How much better is this than the angular resolution of 0.5 seconds of arc achieved by telescopes on the surface of the Earth?

18. The radiotelescope at Jodrell Bank (England) is a dish with a circular aperture of diameter 76 m. What angular resolution can this radiotelescope achieve when operating at a wavelength of 21 cm?

19. According to newspaper reports, a photographic camera on a "Blackbird" reconnaissance jet flying at an altitude of 17 mi can distinguish detail on the ground as small as the size of a man.
   (a) Roughly, what angular resolution does this require?
   (b) According to Rayleigh's criterion, what minimum diameter must the lens of the camera have?

20. Rumor has it that a photographic camera on a spy satellite can read the license plate of an automobile on the ground.
   (a) If the altitude of the satellite is 100 mi, roughly what angular resolution does the camera need to read a license plate? Assume that the reading requires a linear resolution of about 2 in.
   (b) To attain this angular resolution, what must be the diameter of the aperture of the camera?

21. (a) According to Rayleigh's criterion, what is the angular resolution that the human eye can achieve for light of wavelength 5500 Å? The fully distended pupil of the human eye has a diameter of 7.0 mm.
   (b) Even during steady fixation, the eye has a spontaneous tremor that swings it through angles of 20 or 30 seconds of arc. Compare this angular tremor with the angular resolution that you found in part (a). Would the elimination of the tremor greatly improve the acuity of the eye?

22. At night, on a long stretch of straight road in Nevada, a truckdriver sees the distant headlights of another truck. How close must he be to the other truck in order for his eyes to resolve two headlights? Assume that the pupils of the truckdriver have a diameter of 5.0 mm, that the headlights are separated by 1.8 m, and that the light has a wavelength of 5500 Å.

23. Some spy satellites carry cameras with lenses 30 cm in diameter and with a focal length of 2.4 m.

(a) What is the angular resolution of the camera according to Rayleigh's criterion? Assume that the wavelength of light is 5500 Å.
(b) If such a satellite looks down on the Earth from a height of 150 km, what is the distance between two points on the ground that the camera can barely resolve?
(c) The lens projects images of the two points on a film at the focal plane of the lens. What is the distance between the two images projected on the film?

24. 7 × 50 binoculars magnify angles by a factor of 7 and their objective lenses have an aperture of 50-mm diameter.
(a) According to Rayleigh's criterion, what is the intrinsic angular resolution of these binoculars? Assume that the light has a wavelength of 5500 Å.
(b) At best, the pupil of your eye has an aperture of 7.0-mm diameter. Compare the angular resolution of your eye divided by a factor of 7 with the intrinsic angular resolution of the binoculars. Which of the two numbers determines the actual angular resolution you can achieve while looking through the binoculars?

25. When exposed to strong light, the pupil of the eye of a cat narrows to a fine slit, about 0.3 mm across. Suppose that the cat is looking at two white mice 20 m away and separated by a distance of 5 cm. Can the cat distinguish one mouse from the other?

26. Show that the optimum diameter for the pinhole of a pinhole camera (see Section 37.6) is approximately $\sqrt{2.44\lambda L}$, where λ is the wavelength of the light, and L is the distance from the pinhole to the film. (Hint: Consider light arriving from a distant point source. If the pinhole is excessively small, diffraction at the pinhole spreads the light over a spot diameter $2.44\lambda L/a$ on the film. If the pinhole is excessively large, then the light from a distant point source illuminates a spot diameter a. What choice of a makes the smallest spot on the film?

Section 39.3

27. A narrow, opaque strip, 0.060 mm across, is placed in the light beam of wavelength 4579 Å from a laser. The diffraction pattern formed by this strip on a screen 4.8 m beyond shows bright and dark fringes. What is the distance between the second and the third dark fringe on one side of the central beam?

28. A circular opaque disk of diameter 0.12 mm is placed in the beam of a He–Ne laser emitting a wavelength of 6328 Å. The disk makes a diffraction pattern of bright and dark rings on a screen placed 5.0 m beyond. What is the diameter of the first dark circular ring in this diffraction pattern?

29. A thin strip of opaque material 1.5 mm wide is illuminated by light of wavelength 6940 Å. At what distance beyond the strip is the width of the central maximum of the diffracted light as large as the width of the geometric shadow (1.5 mm), so that the shadow disappears?

CHAPTER 40

Quanta of Light

In the two preceding chapters we examined the wave properties of light. We saw that light exhibits interference and diffraction, in agreement with Maxwell's theory, according to which light is a wave consisting of electric and magnetic fields with a smooth distribution of energy. In this chapter we will discover that light has particle properties. We will describe experimental evidence that shows that a light beam consists of a stream of particle-like energy packets. These energy packets are called **quanta of light** or **photons.**

The fact that light has the dual attributes of wave and of particle indicates that neither the wave concept of classical physics nor the particle concept of classical physics gives an adequate description of light. We have to think of light as a wave–particle object, sometimes called a **wavicle.** This object sometimes behaves pretty much as a classical wave, sometimes pretty much as a classical particle, and sometimes as a bit of both. Furthermore, we will see in the next chapter that electrons, protons, neutrons, and all the other known "particles" also exhibit such dual attributes of wave and of particle — we must regard all of them as wavicles.

40.1 Blackbody Radiation

The first hint of a failure of classical physics emerged around 1900 from the study of thermal radiation. When we heat a body to high temperature, it glows; for instance, when we heat a bar of iron to 1300 K, it glows in a deep red. This glow is thermal radiation ("radiant heat"). The spectrum of thermal radiation emitted by a hot body is continuous — if we analyze the light with a prism, we find that the energy is smoothly distributed over all wavelengths.

Spectral emittance

A quantitative description of the distribution of energy over the different wavelengths is provided by the **spectral emittance** S_λ. This is the energy flux (or power per unit area) emitted by the surface of the glowing body per unit wavelength interval; thus, $S_\lambda\, d\lambda$ is the flux emitted in a small wavelength interval $d\lambda$. The spectral emittance is a function of wavelength. Measurements of the thermal radiation emitted by a glowing body show that the energy flux at very long and at very short wavelengths is quite small, and that the energy flux has a maximum at some intermediate wavelength. The position of this maximum depends on the temperature. For example, an iron bar at 1300 K has a maximum emittance at 22,000 Å (infrared), whereas the Sun, with a surface temperature of 5800 K, has a maximum emittance at 5000 Å (blue-green). Figure 40.1 is a plot of the emittance of the Sun as a function of the wavelength; this figure may be regarded as a plot of the intensity distribution in the smear of light that an (ideal) prism produces when exposed to sunlight.

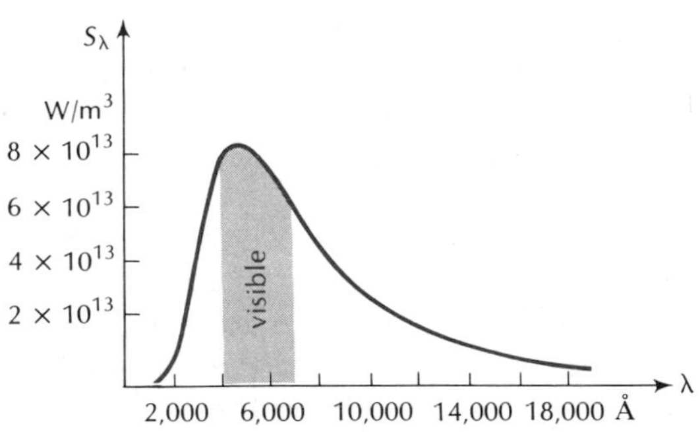

Fig. 40.1 Spectral emittance of the Sun. In this plot the discrete dark spectral lines (Fraunhofer lines), which result from the blocking out of some of the thermal radiation by gas in the solar atmosphere, have been neglected.

The thermal radiation emerging from the surface of a glowing body is generated within the volume of the body by the random thermal motions of atoms and electrons. Before the radiation reaches the surface and escapes, it is absorbed and re-emitted many times and it attains thermal equilibrium with the atoms and electrons. This equilibration process shapes the continuous spectrum of the radiation, completely washing out all of the original spectral features of the radiation.

The flux of thermal radiation emerging from the surface of a glowing body depends to some extent on the characteristics of the surface. The surface usually permits the escape of only a fraction of the flux reaching it from the inside of the body. Correspondingly, if the body is irradiated with an equal flux of thermal radiation from the outside, the surface permits the ingress of only an equal fraction of this flux, reflecting the rest. This equality of the emissive and absorptive characteristics of the surface can be deduced by an argument based on thermodynamics. Thus we are led to a general rule: *a good absorber is a good emitter; and a poor absorber is a poor emitter.* With this rule we can understand how thermos bottles or dewars provide such excellent thermal insulation. These bottles are constructed with a double glass wall, and the space between these walls is evacuated (Figure 40.2). Heat cannot flow across the evacuated space by conduction or convection; it can only flow by radiation. To inhibit radiation, the glass walls

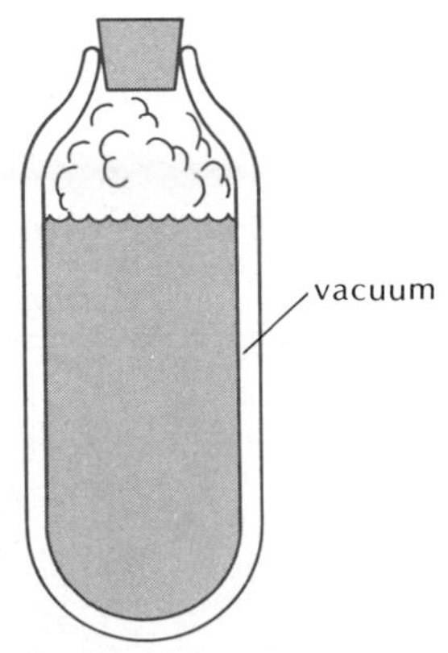

Fig. 40.2 A thermos bottle.

are silvered and thereby made into mirrorlike reflecting surfaces; these highly reflective surfaces are very poor absorbers and emitters of radiation. This keeps the heat transfer between the walls very small.

Blackbody

A body with a perfectly absorbing (and emitting) surface is called a **blackbody;** such a body would look black under illumination from the outside. When a blackbody is hot, its surface emits more thermal radiation than that of any other hot body at the same temperature. In practice the characteristics of an ideal blackbody are most easily achieved by a trick: take a body with a cavity, such as a hollow cube, and drill a small hole in one side of the cube (Figure 40.3). The hole then acts as a blackbody — any radiation incident on the hole from outside will be completely absorbed. Because of this equivalence between a blackbody and a hole in a cavity, the terms *blackbody radiation* and *cavity radiation* are used interchangeably.

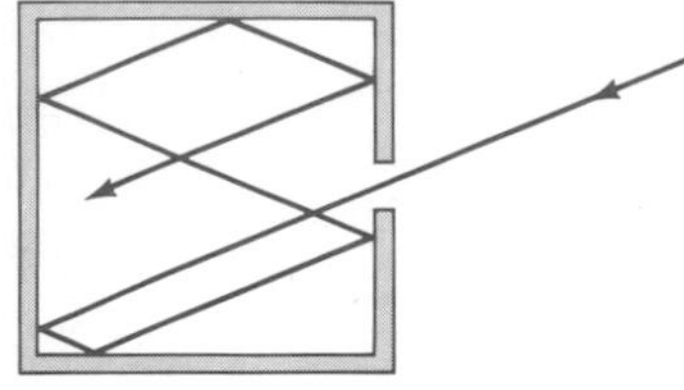

Fig. 40.3 A cavity with a small hole. Any radiation entering the hole is trapped; it will suffer multiple reflections and it will ultimately be absorbed.

The blackbody plays a special role in the study of thermal radiation because its spectral emittance does not depend on the material of which it is made or on any other characteristics of the body — the spectral emittance depends only on the temperature of the body. We can prove this by appealing to the Second Law of Thermodynamics, according to which heat cannot flow spontaneously from a colder body to a warmer body. Consider two cavities at equal temperature with holes of equal size (Figure 40.4). The cavity on the left radiates into the cavity on the right and vice versa. If the flux emitted by the cavity on the left were larger than that emitted by the cavity on the right, the radiative heat transfer would lead to an increase of temperature on the right and a decrease on the left. Heat would then be flowing from a colder to a warmer body, in contradiction to the Second Law. This argument establishes that the fluxes emitted by both cavities are the same. Furthermore, a slight refinement of this argument establishes that the fluxes in any given small wavelength interval $d\lambda$ are also the same. We need only make a slight alteration in the arrangement shown in Figure 40.4 by inserting a filter for light between the two cavities, a filter that only permits the passage of radiation of wavelengths in the interval λ to $\lambda + d\lambda$. Our thermodynamic argument then leads to the conclusion that the fluxes in this chosen wavelength interval must be the same.

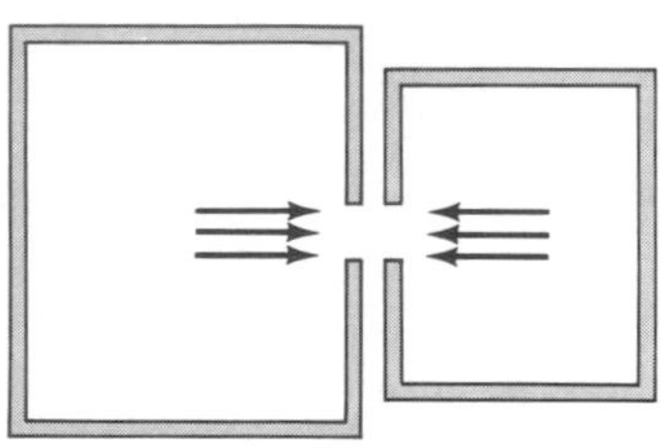

Fig. 40.4 Two cavities with holes of equal size.

Thus for a blackbody, the spectral emittance S_λ is a universal function of the wavelength λ and of the temperature T, and of nothing else. In the last years of the nineteenth century, physicists engaged in an intensive experimental and theoretical effort to find the function S_λ.

40.2 Energy Quanta

One of the first attempts at a theoretical derivation of the blackbody spectrum was made by Lord Rayleigh. Since the energy flux of the radiation emerging from the hole in a cavity is directly proportional to the energy density of the radiation inside the cavity,[1] Rayleigh decided to calculate the latter quantity. He began by noting that the radiation

[1] According to Eqs. (36.30) and (36.32), for a plane wave the constant of proportionality between the energy flux (Poynting vector) and the energy density is c, the speed of light. For thermal radiation, with many plane waves traveling in random directions, the constant of proportionality is $c/4$.

in a cavity is made up of a large number of standing waves; Figure 40.5 shows some of these standing waves. Each of these standing waves can be regarded as a mode of vibration of the cavity. Rayleigh then appealed to the equipartition theorem, according to which, at thermal equilibrium, each mode of vibration has an average thermal energy of kT (in Section 19.4 we stated a special case of the equipartition theorem for free translational or rotational motion). Thus, each of the standing waves of Figure 40.5 ought to have an energy kT, and from this one can calculate the spectral emissivity. Although this calculation gave reasonable results at the long-wavelength end of the blackbody spectrum, it gave disastrous results at the short-wavelength end: the number of possible standing-wave modes of very short wavelength is infinitely large and, if each of these modes had an energy kT, the total energy in the cavity would be infinite! This disastrous failure of classical theory has been called the **ultraviolet catastrophe.**

Ultraviolet catastrophe

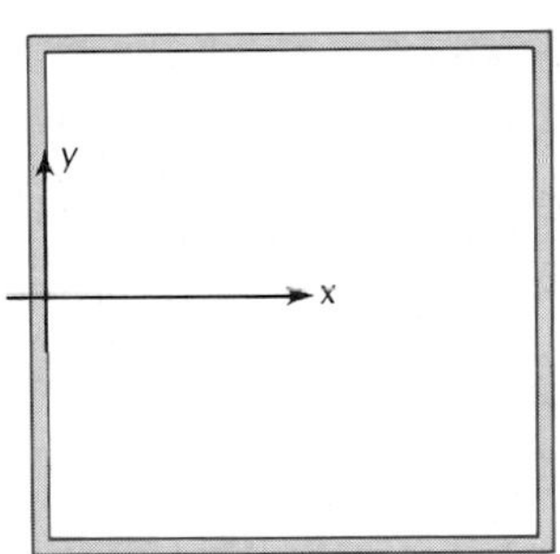

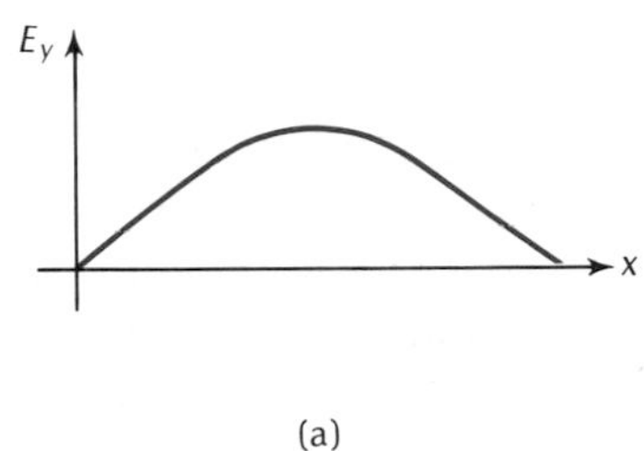

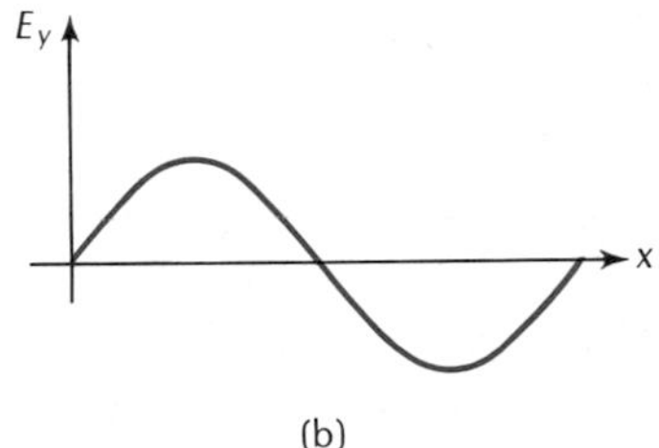

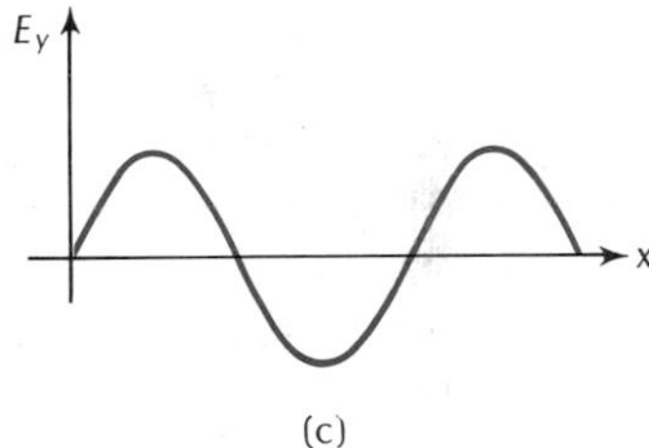

Fig. 40.5 Some of the possible standing electromagnetic waves in a closed cavity. For the sake of simplicity, only waves with a horizontal direction of propagation are shown. The plots give the electric field as a function of x at one instant of time.

The correct formula for the spectral emittance of a blackbody was finally obtained by Max Planck in 1900. By some inspired guesswork, based on thermodynamics, Planck hit upon the following formula:

$$S_\lambda = \frac{2\pi c^2 h}{\lambda^5} \frac{1}{e^{hc/kT\lambda} - 1} \tag{1}$$

Spectral emittance of blackbody

where h is **Planck's constant,**[2]

$$h = 6.63 \times 10^{-34}\ \text{J} \cdot \text{s} \tag{2}$$

Planck's constant, h

and k is Boltzmann's constant. Experimental investigation quickly established that the values of the spectral emittance calculated from Eq.

[2] See Appendix 8 for a more precise value of Planck's constant.

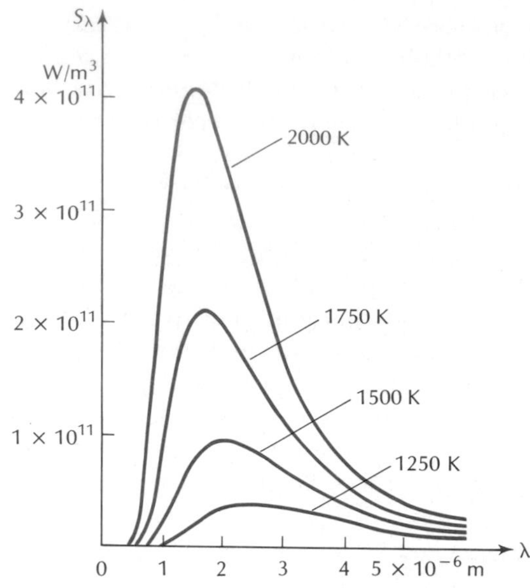

Fig. 40.6 Spectral emittance of a blackbody according to Planck's Law. Each curve is labeled with the appropriate temperature.

Max Planck, *1858–1947, German physicist, professor at Berlin and president of the Kaiser Wilhelm Society for the Advancement of Science (now the Max Planck Institute). Planck made significant contributions to thermodynamics before he became involved with the problem of blackbody radiation. He was a very conservative scientist, and he regarded his quantization postulate as "an act of desperation" into which he was forced because "a theoretical explanation had to be found at any cost, whatever the price." Planck received the Nobel Prize in 1918.*

(1) agreed very precisely with the measured values. Figure 40.6 shows the spectral emittance for several values of the temperature.

At first, Planck had proposed his formula merely as an empirical law that happens to give a good fit to the experimentally measured points. But then he searched for a theoretical justification of this law. The search led Planck to a revolutionary discovery: the quantization of energy. This discovery was to bring about the overthrow of classical Newtonian physics and the rise of quantum physics. For Planck, the postulate of the quantization of energy was "an act of desperation" which he committed because "a theoretical explanation had to be found at any cost, whatever the price."

Planck's derivation of the blackbody radiation formula involves some sophisticated statistical mechanics and we cannot reproduce the details of his derivation here. We will merely give a sketchy outline of this derivation. Planck began by making a theoretical model of the walls of the cavity; he regarded the atoms in the walls as small harmonic oscillators with electric charges. Although this is a rather crude model of an atom, it was adequate for his purposes since, by the thermodynamic argument of the preceding section, the radiation in a cavity is completely independent of the characteristics of the wall. The random thermal motions of the oscillators result in the emission of electromagnetic radiation. This radiation fills the cavity and acts back on the oscillators. When thermal equilibrium is attained, the average rate of emission of radiation energy by the oscillators matches the rate of absorption of radiation energy. Thus, the oscillators share their energy with the radiation in the cavity, and Planck was able to show that, under equilibrium conditions, the average radiation energy at some frequency ν (or at a wavelength $\lambda = c/\nu$) is directly proportional to the average energy of an oscillator of frequency ν. These steps of Planck's calculation involved nothing but classical mechanics. But in the next step of the calculation, Planck departed radically from classical physics. He postulated that the energy of the oscillators is quantized according to the following rule: *in an oscillator of frequency* ν, *the only permitted values of the energy are*

$$E = 0,\ h\nu,\ 2h\nu,\ 3h\nu,\ \ldots \tag{3}$$

All other values of the energy are forbidden. The constant h in Eq. (3) is Planck's constant, the same as appears in Eqs. (1) and (2). The energy $h\nu$ is called an **energy quantum;** according to the quantization rule, the energy of an oscillator is always some multiple of the basic energy quantum,

Energy quantum

$$E = nh\nu, \qquad n = 0, 1, 2, 3, \ldots \tag{4}$$

Energy quantization of oscillator

The integer n is called the **quantum number** of the oscillator.

Quantum number

With this quantization condition, Planck calculated the average energy of the oscillators; and from that he derived his radiation formula. Although we cannot go into the details of this derivation, we can achieve a rough understanding of how Planck's calculation avoids the ultraviolet catastrophe. The thermal energy of the walls of the cavity is shared at random among all the oscillators in these walls. Some of these oscillators have high frequencies, some have low frequencies. For an oscillator of very high frequency, the energy quantum $h\nu$ is very large. If this oscillator is initially quiescent ($n = 0$), it cannot begin to move unless it acquires one energy quantum; but since this energy quantum $h\nu$ is very large, the random thermal disturbances will be insufficient to provide it — the oscillator will remain quiescent. Thus, the quantization of energy tends to inhibit the thermal excitation of the high-frequency oscillators. If the high-frequency oscillators remain quiescent, then they will not supply energy to the corresponding high-frequency standing waves in the cavity, and there will be no ultraviolet catastrophe.

Note that for an oscillator with a frequency of $\nu \simeq 10^{15}$/s, which is typical for atomic vibrations, the energy quantum is $h\nu = 6.6 \times 10^{-34}\,\text{J}\cdot\text{s} \times 10^{15}/\text{s} = 6.6 \times 10^{-19}\,\text{J}$. Since this is a very small amount of energy, quantization does not make itself felt at a macroscopic level. But quantization plays a pervasive role at the atomic level.

Unfortunately, Planck could not offer any basic justification for his postulate of quantization of energy. His postulate took care of the radiation law, but raised serious questions about classical physics. Obviously, quantization of energy makes no sense in classical physics — there is nothing in the laws of Newton that would prevent an oscillator from acquiring energy in any amount whatsoever. Planck could only justify his postulate by its consequences; but (in physics) the end does not justify the means. A deeper explanation of the quantization of energy only emerged much later, with the development of quantum mechanics (see Chapter 41).

From Planck's formula one can show that the spectral emittance of a blackbody has a maximum at a wavelength given by

$$\lambda_{\text{max}} = (2.90 \times 10^{-3}\,\text{m}\cdot\text{K}) \times \frac{1}{T} \tag{5}$$

Wien's Law

This inverse proportionality of the wavelength of maximum emittance and the temperature is called **Wien's Law.**

By integrating Planck's formula one obtains the total flux, or power per unit area, emitted by the blackbody at all wavelengths:

$$S = \int S_\lambda\, d\lambda = \int \frac{2\pi c^2 h}{\lambda^5} \frac{1}{e^{hc/kT\lambda} - 1}\, d\lambda \tag{6}$$

One can show that this total flux is proportional to the fourth power of the temperature,

Stefan–Boltzmann Law

$$\boxed{S = \sigma T^4} \tag{7}$$

where $\sigma = 5.67 \times 10^{-8}\ \mathrm{W/(m^2 \cdot K^4)}$.[3] This is called the **Stefan–Boltzmann Law.** The laws of Wien and of Stefan and Boltzmann had both been discovered empirically many years before Planck wrote down his formula.

EXAMPLE 1. On a clear night, the surface of the Earth loses heat by radiation. Suppose that the temperature of the ground is 10°C and that the ground radiates like a blackbody. What is the rate of loss of heat per square meter?

SOLUTION: The absolute temperature of the ground is 283 K. Hence, the Stefan–Boltzmann Law tells us that the radiated flux, or power per unit area, is

$$S = \sigma T^4 = 5.67 \times 10^{-8}\ \mathrm{W/(m^2 \cdot K^4)} \times (283\ \mathrm{K})^4$$

$$= 364\ \mathrm{W/m^2}$$

This amounts to 87 cal/m² per second.

EXAMPLE 2. Astronomers sometimes determine the size of a star by a method that relies on the Stefan–Boltzmann Law. Determine the radius of the star Capella from the following data: the flux of the starlight reaching the Earth is $1.2 \times 10^{-8}\ \mathrm{W/m^2}$, the distance of the star is 4.3×10^{17} m, and its surface temperature is 5200 K. Assume the star radiates like a blackbody.

SOLUTION: According to the Stefan–Boltzmann Law, the flux, or power per unit area, emerging from the surface of the star is σT^4. The surface area of the star is $4\pi R^2$ and hence the total emitted power is

$$4\pi R^2 \times \sigma T^4 \tag{8}$$

This must match the total power crossing a sphere of radius 4.3×10^{17} m centered on the star. Since at each point of this sphere the power per unit area is $1.2 \times 10^{-8}\ \mathrm{W/m^2}$, the total power is

$$4\pi (4.3 \times 10^{17}\ \mathrm{m})^2 \times 1.2 \times 10^{-8}\ \mathrm{W/m^2} = 2.8 \times 10^{28}\ \mathrm{W} \tag{9}$$

Comparing Eqs. (8) and (9), we obtain

$$4\pi R^2 \sigma T^4 = 2.8 \times 10^{28}\ \mathrm{W} \tag{10}$$

or

$$R = \left(\frac{2.8 \times 10^{28}\ \mathrm{W}}{4\pi \times 5.67 \times 10^{-8}\ \mathrm{W/(m^2 \cdot K^4)} \times (5200\ \mathrm{K})^4} \right)^{1/2}$$

$$= 7.3 \times 10^9\ \mathrm{m}$$

This is about 10 times the radius of the Sun. Capella is a giant star.

---

[3] See Appendix 8 for a more precise value of the Stefan–Boltzmann constant.

40.3 Photons and the Photoelectric Effect

In 1905, Einstein showed that Planck's formula could be understood much more simply in terms of a direct quantization of the energy of the radiation. Planck had postulated that the oscillators in the wall of the cavity have discrete quantized energies, but he had treated the electromagnetic radiation as a smooth, continuous distribution of energy, exactly as it is supposed to be according to classical electromagnetic theory. In contrast, Einstein proposed that electromagnetic radiation consists of particle-like packets of energy. He regarded a wave of some given frequency ν as a stream of more or less localized energy packets, each with one quantum of energy $h\nu$. The wave then has an energy $h\nu$ if it contains only one such quantum, $2h\nu$ if it contains two, etc. The particle-like energy packets in light are called **photons.** The thermal radiation in a cavity, with waves traveling randomly in all directions, can then be regarded as a gas of photons. Einstein applied statistical mechanics to calculate the energy spectrum of this gas and he thereby obtained the energy spectrum of the cavity radiation.

Photon

EXAMPLE 3. The energy flux of sunlight reaching the surface of the Earth is 1.0×10^3 W/m^2. How many photons reach the surface of the Earth per square meter per second? For the purposes of this calculation assume that all the photons in sunlight have an average wavelength 5000 Å.

SOLUTION: The energy of a photon of wavelength 5000 Å is

$$E = h\nu = \frac{hc}{\lambda}$$

$$= \frac{6.63 \times 10^{-34}\ \text{J} \cdot \text{s} \times 3.00 \times 10^{8}\ \text{m/s}}{5.0 \times 10^{-7}\ \text{m}} = 4.0 \times 10^{-19}\ \text{J}$$

The energy incident per square meter per second is 1.0×10^3 J. To obtain the number of photons, we must divide this by the energy per photon, 4.0×10^{-19} J. This gives $1.0 \times 10^3\ \text{J}/4.0 \times 10^{-19}\ \text{J} = 2.5 \times 10^{21}$ photons per square meter per second.

Because the number of photons in sunlight and other common light sources is so large, we do not perceive the grainy character of the energy distribution in light at the macroscopic level.

With this concept of light as a stream of photons, Einstein was also able to offer an explanation of the **photoelectric effect.** In the early experiments on the production of radio waves by sparks, Hertz had noticed that light shining on an electrode tended to promote the formation of sparks. Subsequent careful experimental investigations demonstrated that the impact of light on an electrode can eject electrons. The electrons emerge with a kinetic energy that increases directly with the frequency of the light.

Photoelectric effect

Figure 40.7 is a schematic diagram of the apparatus used in the investigation of the photoelectric effect. Light from a lamp illuminates an electrode of metal *(C)* enclosed in an evacuated tube. Electrons ejected from this electrode travel to the collecting electrode *(A)*, and then flow around the external circuit. A galvanometer *(G)* detects this

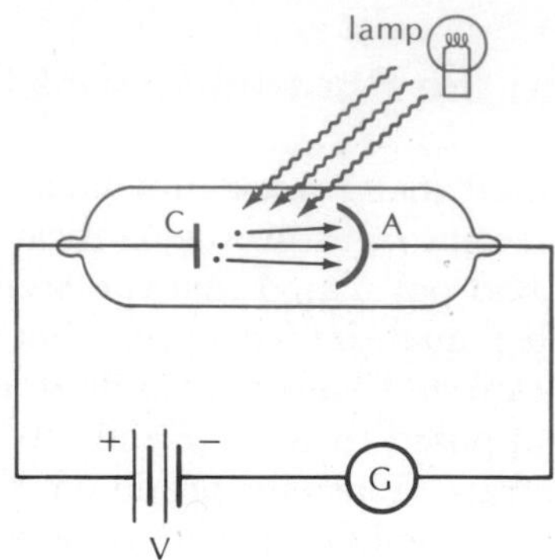

Fig. 40.7 Schematic diagram of the apparatus for the investigation of the photoelectric effect. The light from the lamp ejects electrons from the electrode *C* (cathode) and they travel to the collecting electrode *A* (anode).

flow of electrons. The kinetic energy of the ejected photoelectrons can be determined by applying a potential difference between the emitting and the collecting electrodes by means of an adjustable source of emf (*V*). With the polarity shown in the figure, the collector has a negative potential relative to the emitter, i.e., the collector exerts a repulsive force on the photoelectrons. If the potential energy matches or exceeds the initial kinetic energy of the photoelectrons, then the flow of these electrons will stop. The corresponding potential is called the **stopping potential;** the measured value of this stopping potential gives us the kinetic energy of the electrons:

Stopping potential

$$K = eV_{\text{stop}} \tag{11}$$

Experimentally, one finds that the kinetic energy determined in this way increases linearly with the frequency of the incident light. For example, Figure 40.8 is a plot of the kinetic energy vs. the frequency of the light for photoelectrons ejected from sodium. Note that if the frequency is below 4.4×10^{14} Hz, then the light is incapable of ejecting electrons. This frequency is called the **threshold frequency.**

Threshold frequency

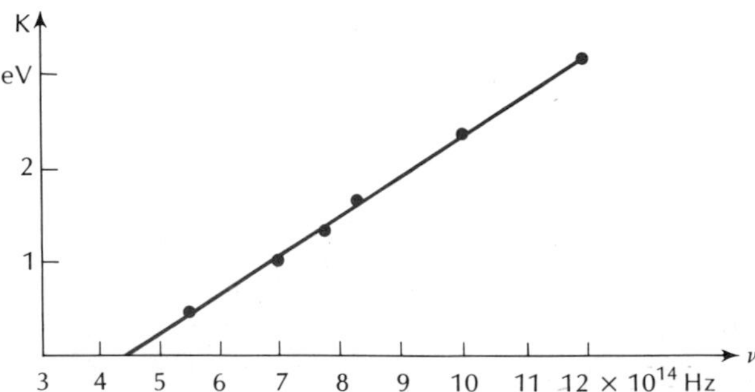

Fig. 40.8 Kinetic energies (in electron-volts) of photoelectrons ejected from sodium by light of different frequencies.

Einstein's quantum theory of light accounts for these experimental observations as follows. The electrons in the illuminated electrode absorb quanta of light, one at a time. When an electron absorbs a quantum, it acquires an energy $h\nu$. But before this electron can emerge from the electrode, it must overcome the restraining forces that bind it to the metal of the electrode. The energy required for this is called the **work function** of the metal, designated by ϕ. The energy left with

Work function

the electron is then $h\nu - \phi$ and this must be the kinetic energy of the emerging electron,[4]

Einstein's photoelectric equation

$$\boxed{K = h\nu - \phi} \quad (12)$$

This is Einstein's photoelectric equation. It shows that the kinetic energy is indeed a linearly increasing function of the frequency, in agreement with the data of Figure 40.8. According to Eq. (12), the slope of the straight line in Figure 40.8 should equal Planck's constant.

Einstein's photoelectric equation was verified in detail by a long series of meticulous experiments by R. A. Millikan (the data in Figure 40.8 are due to him). In order to obtain reliable results, Millikan found it necessary to take extreme precautions to avoid contamination of the surface of the photosensitive electrode. Since the surfaces of metals exposed to air quickly accumulate a layer of oxide, he developed a technique for shaving the surfaces of his metals in a vacuum by means of a magnetically operated knife.

The results of these experiments gave strong support to the quantum theory of light. This success of Einstein's theory was all the more striking in view of the failure of the classical wave theory of light to account for the features of the photoelectric effect. According to the wave theory, the crucial parameter that determines the ejection of a photoelectron should be the intensity of light. If an intense electromagnetic wave strikes an electron, it should be able to jolt it loose from the metal, regardless of the frequency of the wave. Furthermore, the kinetic energy of the ejected electron should be a function of the intensity of the wave. The observational evidence contradicts these predictions of the wave theory: A wave with a frequency below the threshold frequency never ejects an electron, regardless of its intensity. And, furthermore, the kinetic energy depends on the frequency, and not on the intensity. High-intensity light ejects more photoelectrons, but does not give the individual electrons more kinetic energy.

Today, the photoelectric effect finds many practical applications in sensitive electronic devices for the detection of light. For instance, in a photomultiplier tube, an incident photon ejects an electron from an electrode; this electron is accelerated toward a second electrode (called a dynode; see Figure 40.9) where its impact ejects several secondary electrons; these, in turn, are accelerated toward a third electrode where their impact ejects tertiary electrons, etc. Thus, one electron from the first electrode generates an avalanche of electrons. In a high-gain photomultiplier tube, a pulse of 10^9 electrons emerges from the last electrode, delivering a measurable pulse of current to an external circuit. In this way, the photomultiplier tube can detect the arrival of individual photons. Some sensitive television cameras, such as the image orthicon, rely on the same multiplier principle to convert the arrival of a photon at a photosensitive faceplate into a measurable pulse of current.

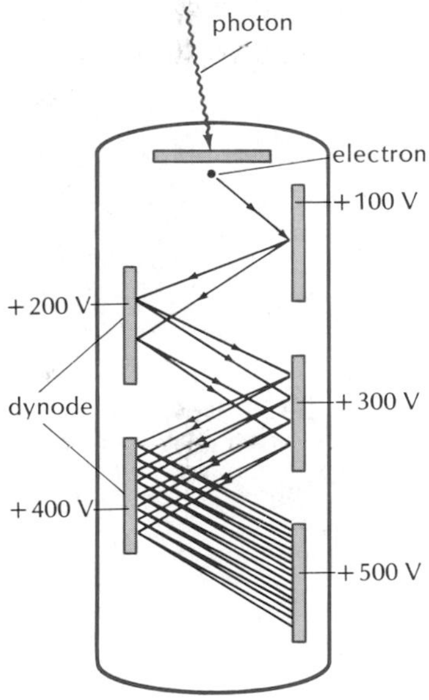

Fig. 40.9 Schematic diagram of a photomultiplier tube. The secondary electrodes are called dynodes. For the purpose of this diagram it has been assumed that each electron impact on a dynode releases two electrons. The arrows show an avalanche of electrons.

EXAMPLE 4. The work function for platinum is 9.9×10^{-19} J. What is the threshold frequency for the ejection of photoelectrons from platinum?

[4] Some electrons suffer extra energy losses in collisions within the metal before they emerge. Thus, the expression actually gives the *maximum* kinetic energy with which electrons can emerge.

SOLUTION: If an electron absorbs a photon at the threshold frequency, it will just barely have enough energy to overcome the binding forces holding it in the metal, and it will emerge with zero kinetic energy. By Eq. (12) this corresponds to

$$0 = h\nu - \phi$$

or

$$\nu = \frac{\phi}{h} = \frac{9.9 \times 10^{-19}\ \text{J}}{6.63 \times 10^{-34}\ \text{J}\cdot\text{s}} = 1.5 \times 10^{15}\ \text{Hz}$$

40.4 The Compton Effect

Arthur Holly Compton, *1892–1962, American experimental physicist, professor at the University of Chicago. For his discovery of the Compton effect, he received the Nobel Prize in 1927.*

Very clear experimental evidence for the particle-like behavior of photons was uncovered by A. H. Compton in 1922. Compton had been investigating the scattering of X rays by a target of graphite. When he bombarded the graphite with monochromatic X rays, he found that the scattered (deflected) X rays had a wavelength somewhat larger than that of the original X rays. Compton soon recognized that this effect could be understood in terms of collisions of photons with electrons, collisions in which the photons behave like particles.

In this collision the electron of a carbon atom can be regarded as free because the force binding the electron to the atom is insignificant compared to the force exerted by the photon. When the photon bounces off the electron, the electron recoils and thereby picks up some of the photon's energy — the deflected photon is left with reduced energy. Since the energy of the photon is $E = h\nu = hc/\lambda$, a reduction of energy implies an increase of wavelength. Qualitatively, we expect that those photons deflected through the largest angles should lose the most energy and therefore emerge with the longest wavelength. And this is just what Compton found in his experiments.

For a quantitative discussion of the photon–electron collision, we need an expression for the momentum of a photon. From Eq. (36.42) we know that the momentum and the energy of light are related by $p = E/c$. A photon of energy $E = h\nu$ therefore has a momentum

Momentum of photon

$$\boxed{p = \frac{h\nu}{c}} \qquad (13)$$

[Alternatively, we can derive this equation from the relativistic formula for the momentum of a particle. Since the speed of the photon is c, it must be regarded as an ultrarelativistic particle. For such a particle, Eq. (17.53) tells us that $p = E/c$, which leads, again, to $p = h\nu/c$. Incidentally, note that according to Eq. (17.51) the energy of a particle with a speed equal to the speed of light can be finite only if $m = 0$; thus photons must be regarded as particles of zero mass.]

Before the collision, the electron is at rest and the photon has a frequency ν and momentum $h\nu/c$. After the collision, the electron has a recoil momentum $m_e v$ and the photon has a reduced frequency $\nu - \Delta\nu$ and momentum $h(\nu - \Delta\nu)/c$. Figure 40.10a shows the initial momentum vector of the photon and the final momentum vectors of the electron and the photon. Conservation of momentum demands that the sum of the two final momentum vectors equal the initial momentum vector; the three momentum vectors therefore form a triangle (Figure 40.10b). Applying the law of cosines to this triangle, we find

$$(m_e v)^2 = \left(\frac{h\nu}{c}\right)^2 + \left(\frac{h(\nu - \Delta\nu)}{c}\right)^2 - \frac{2h\nu}{c}\frac{h(\nu - \Delta\nu)}{c}\cos\theta \qquad (14)$$

or

$$(m_e v)^2 = \left(\frac{h\nu}{c}\right)^2\left[1 + \left(1 - \frac{\Delta\nu}{\nu}\right)^2 - 2\left(1 - \frac{\Delta\nu}{\nu}\right)\cos\theta\right] \qquad (15)$$

$$= \left(\frac{h\nu}{c}\right)^2\left[2 - 2\frac{\Delta\nu}{\nu} + \left(\frac{\Delta\nu}{\nu}\right)^2 - 2\left(1 - \frac{\Delta\nu}{\nu}\right)\cos\theta\right] \qquad (16)$$

If we neglect the term $(\Delta\nu/\nu)^2$, we can rewrite this as

$$\tfrac{1}{2}m_e v^2 = \frac{1}{m_e}\left(\frac{h\nu}{c}\right)^2\left(1 - \frac{\Delta\nu}{\nu}\right)(1 - \cos\theta) \qquad (17)$$

Conservation of energy demands that the kinetic energy of the electron equals the energy lost by the photon,

$$\tfrac{1}{2}m_e v_e^2 = h\,\Delta\nu \qquad (18)$$

Comparing Eqs. (17) and (18), we see that

$$\Delta\nu = \frac{h\nu^2}{m_e c^2}\left(1 - \frac{\Delta\nu}{\nu}\right)(1 - \cos\theta) \qquad (19)$$

or

$$\frac{c\,\Delta\nu}{\nu(\nu - \Delta\nu)} = \frac{h}{m_e c}(1 - \cos\theta) \qquad (20)$$

The complicated expression on the left side of this equation is simply the change of wavelength:

$$\frac{c\,\Delta\nu}{\nu(\nu - \Delta\nu)} = \frac{c}{\nu - \Delta\nu} - \frac{c}{\nu} = = (\lambda + \Delta\lambda) - \lambda = \Delta\lambda \qquad (21)$$

where we have taken into account that a decreased frequency $\nu - \Delta\nu$ implies an increased wavelength $\lambda + \Delta\lambda$. Our equation for the change of wavelength of the photon as a function of the angle of deflection is then

$$\boxed{\Delta\lambda = \frac{h}{m_e c}(1 - \cos\theta)} \qquad (22)$$

Wavelength shift of photon in Compton effect

This result, first obtained by Compton, remains valid even if the speed

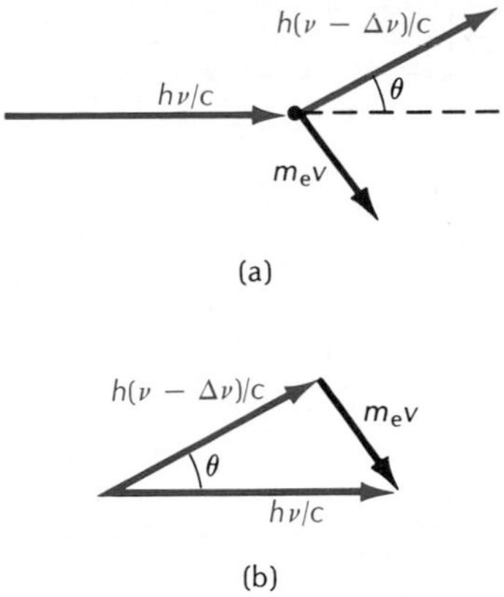

Fig. 40.10 (a) Momentum vector of photon before the collision and momentum vectors of photon and electron after the collision. Before the collision the photon has a momentum $h\nu/c$, and the electron is at rest; after the collision the photon has a momentum $h(\nu - \Delta\nu)/c$, and the electron has a momentum $m_e v$. (b) The triangle of the momentum vectors.

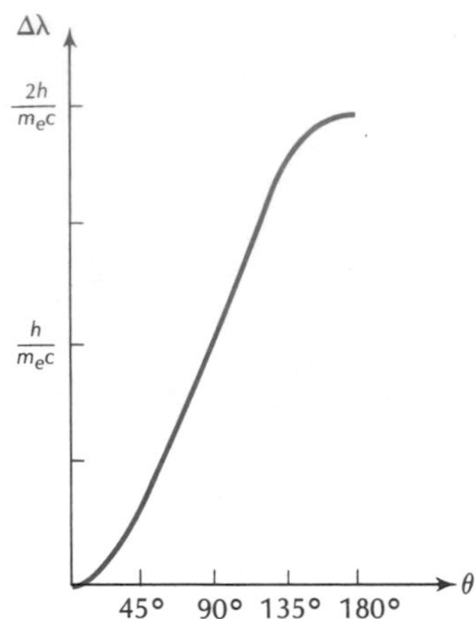

Fig. 40.11 Wavelength shift of the photon as a function of the deflection angle.

of recoil of the electron is relativistic [in fact, the relativistic calculation shows that Eq. (22) is *exact*].

Figure 40.11 is a plot of $\Delta\lambda$ as a function of θ. For $\theta = 180°$ (a head-on collision), the wavelength shift is maximum, $\Delta\lambda = 2h/(m_e c) = 0.0485$ Å; but even this maximum is quite small. Such a small wavelength shift is difficult to detect unless the wavelength of the X rays is itself quite small, so that $\Delta\lambda$ is an appreciable fraction of λ. In his experiments Compton used X rays of a wavelength $\lambda = 0.71$ Å; thus the maximum wavelength shift amounted to about 7% of the wavelength.

40.5 Wave vs. Particle

From the photoelectric effect and the Compton effect we learned that photons have particle properties. On the other hand, we know that light displays interference and diffraction phenomena which prove that the photons also have wave properties. Thus photons are neither classical particles nor classical waves. They are some new kind of object, unknown to classical physics, with a subtle combination of both wave and particle properties. B. Hoffman has coined the name **wavicle** for this new kind of object. It is difficult to achieve a clear understanding of the character of a wavicle because these objects are very remote from our everyday experience. We all have an intuitive grasp of the concept of a classical particle or of a classical wave from our experience with, say, billiard balls and water waves, but we have no such experience with wavicles.

Wavicle

We can gain some insight into the interplay between the particle behavior and the wave behavior of a photon by a new, detailed examination of a simple diffraction experiment. Figure 40.12 shows a light beam striking a plate with a narrow slit. In the usual diffraction experiments described in Chapter 39, we installed a screen or a photographic film at the far right, and with this we recorded the intensity of the light in the diffraction pattern. Instead, in our new arrangement we will install the faceplate of a very sensitive TV camera in place of the customary screen. With this, we can detect the individual photons of the light in the diffraction pattern.

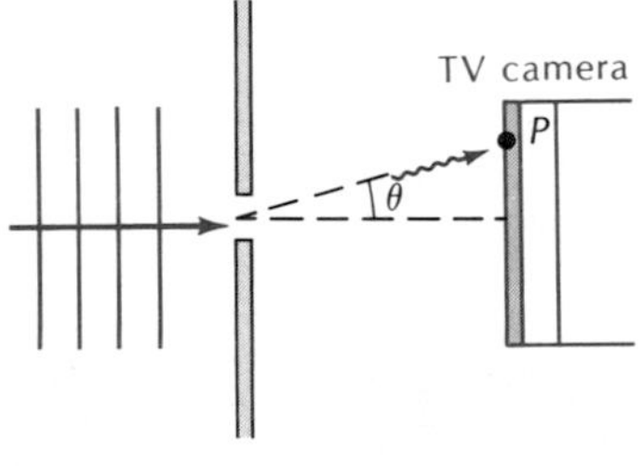

Fig. 40.12 An interference experiment. A light beam is incident from the left on a plate with two narrow slits. A sensitive TV camera detects photons on the right.

If the incident light beam has a very low intensity, so that there is only one photon passing through the slit at a time, we can watch the photons arriving one by one at the faceplate of our TV camera. Figure 40.13a shows a typical pattern of impacts of 28 photons. The pattern seems quite random. If the photons behaved like classical particles, they would travel along a straight line and they would reach only those points on the faceplate that are within the geometric image of the slit. The widely scattered impacts prove that the photons are certainly not traveling along such straight lines. Figure 40.13b shows the pattern of accumulated impacts for 1000 photons; and Figure 40.13c shows it for 10,000 photons. In these figures we can recognize a tendency of the photons to cluster in bandlike zones. These zones correspond to the maxima of the diffraction pattern predicted by the wave theory of light. Finally, Figure 40.13d shows the pattern of accumulated impacts for a very large number of photons; this is simply the familiar intensity pattern of light diffracted by a slit (see also Figure 39.5).

From this diffraction experiment we learn that the behavior of the photons is governed by a probabilistic law. The point of impact of an

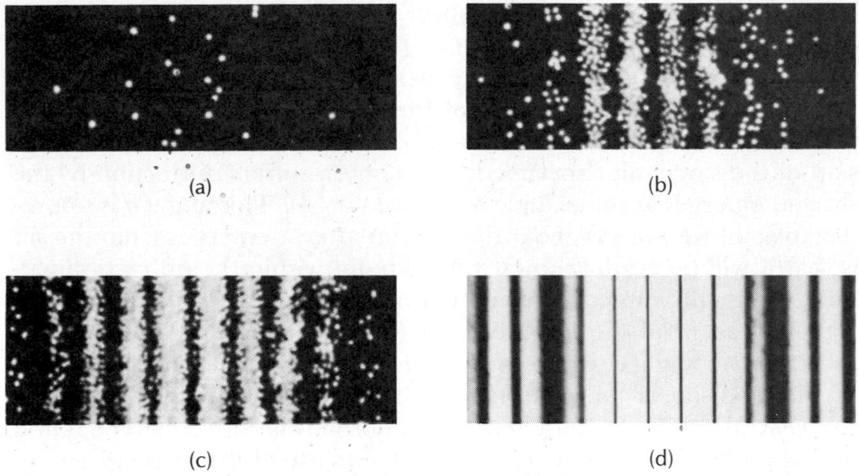

Fig. 40.13 (a) Pattern of impacts of 28 photons (simulation). (b) Pattern of impacts of 1000 photons (simulation). (c) Pattern of impacts of 10,000 photons (simulation). (d) Intensity pattern of light recorded on a photographic plate by a very large number of photons. The intensity of light at each point is proportional to the number of photons arriving at that point.

individual photon on the faceplate is unpredictable. Only the average distribution of impacts of a large number of photons is predictable: the distribution of photons matches the intensity distribution calculated from the wave theory of light. Thus, the probability that a photon arrives at a given point on the faceplate is proportional to the calculated intensity of the wave at that point, i.e.,

Probability interpretation of wave

$$\boxed{[\text{probability for photon at point } P] \propto E^2(P)} \qquad (23)$$

Here we have a connection between the wave and the particle aspects of the photon: *the intensity of the photon wave at some point determines the probability that there is a photon particle at that point.* This probability interpretation of the intensity of the wave was discovered by Max Born.

Our single-slit experiment can also teach us something about the limitations that quantum theory imposes on the ultimate precision of measurement of the position of a wavicle. Suppose we have a light wave consisting of one photon and we want to measure the position of this photon. Of course, the position has x, y, and z components; we will concentrate on the y component, perpendicular to the direction of propagation. Figure 40.14 shows the wave propagating in the horizontal direction; the y direction is vertical. To detemine this vertical position of the photon, we use a narrow slit placed in the path of the wave. If the photon succeeds in passing through this slit, then we will have achieved a determination of the vertical position to within an uncertainty

$$\Delta y = a \qquad (24)$$

where a is the width of the slit. If the photon fails to pass through this slit, then our measurement is inconclusive and will have to be repeated.[5]

By making the slit very narrow, we can make the uncertainty of our

Max Born, *1882–1970, German, and later British, theoretical physicist, professor at Göttingen and at Edinburgh. He was awarded the Nobel Prize somewhat tardily in 1954 for his discovery of the probabilistic interpretation of quantum waves in 1926.*

[5] An alternative method for measuring the position of a photon is to observe its impact on the faceplate of a sensitive TV camera. But this measurement is destructive (the photon is absorbed by the photoelectron and disappears). Besides, the uncertainties in this measurement are not easily analyzed.

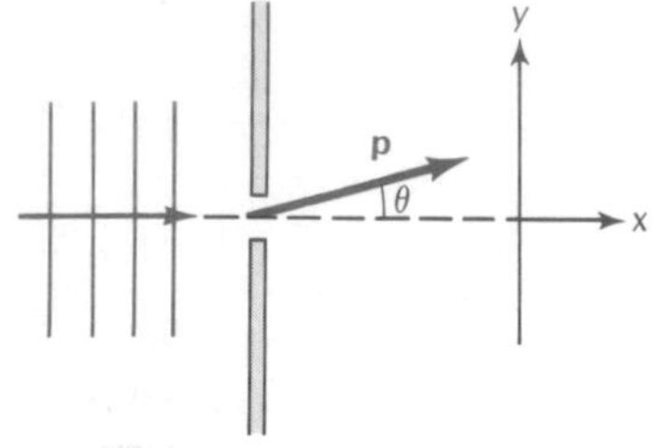

Fig. 40.14 Photon passes through a narrow slit and emerges at an angle θ.

determination of the y coordinate very small. But this has a surprising consequence for the y component of the momentum of the photon: if we make the uncertainty in the y coordinate small, we will make the uncertainty in the y component of the momentum large. To see how this comes about, let us recall that according to our preceding discussion of the single-slit experiment, the photon suffers diffraction by the slit and emerges at some angle θ (Figure 40.14). This angle θ is unpredictable; all we can say about the photon after it emerges from the slit is that it will be heading toward some point within the diffraction pattern. Thus, the direction of motion of the photon is uncertain. As a rough measure of the magnitude of this uncertainty in direction, we can take the angular width of the central diffraction maximum (most of the intensity of the photon wave is gathered within the region of this central maximum, and hence the photon is most likely to be found in this region). This estimate of the uncertainty of the angle gives us

$$\Delta\theta \gtrsim \lambda/a \tag{25}$$

The y component of the momentum is $p_y = p \sin\theta$; since we are concerned with a small angle, we can approximate this by $p_y = p\theta$. The uncertainty in p_y is then

$$\Delta p_y = p\,\Delta\theta \gtrsim p\lambda/a \tag{26}$$

But, from Eq. (13),

$$p = \frac{h\nu}{c} = \frac{h}{\lambda} \tag{27}$$

which gives us

$$\Delta p_y \gtrsim h/a \tag{28}$$

Werner Heisenberg, *1901–1976, German theoretical physicist, professor at Leipzig, and later director of the Max Planck Institute for Physics in Munich. He was one of the founders of the new quantum mechanics, and received the Nobel Prize in 1932.*

Thus, if the slit is very narrow, the uncertainty in the y component of the momentum will be very large! Comparing Eqs. (28) and (24), we find that

$$\boxed{\Delta y\,\Delta p_y \gtrsim h} \tag{29}$$

This equation states that Δy and Δp_y cannot both be small; if one is small then the other must be large, so that their product equals or exceeds Planck's constant.

Heisenberg's uncertainty relation for y and p_y

Equation (29) is one of **Heisenberg's uncertainty relations.** There are corresponding relations for the other components of position and momentum. Although we have obtained the uncertainty relation (29) by examining the special case of a position measurement by means of a slit, it turns out that this relation is actually of general validity for any kind of position measurement. The Heisenberg uncertainty relations tell us that there exist ultimate, insuperable limitations in the precision of our measurements. At the macroscopic level, the quantum uncertanties in our measurements can be neglected. But at the atomic level, these quantum uncertainties are often so large that it is completely meaningless to speak of the position or momentum of a wavicle.

SUMMARY

Spectral emittance of blackbody:

$$S_\lambda = \frac{2\pi c^2 h}{\lambda^5} \frac{1}{e^{hc/kT\lambda} - 1}$$

$$h = 6.63 \times 10^{-34}\ \mathrm{J \cdot s}$$

Energy quantization of an oscillator: $E = nh\nu$

Wien's Law: $\lambda_{\max} \propto \dfrac{1}{T}$

Stefan–Boltzmann Law: $S \propto T^4$

Energy and momentum of a photon: $E = h\nu, \qquad p = h\nu/c$

Kinetic energy of photoelectron: $K = h\nu - \phi$

Wavelength shift of photon (Compton effect):

$$\Delta\lambda = \frac{h}{m_e c}(1 - \cos\theta)$$

Probability interpretation of wave:

[probability for presence of photon] $\propto$ [intensity of wave]

Heisenberg's uncertainty relation for y and p_y:

$$\Delta y\, \Delta p_y \gtrsim h$$

QUESTIONS

1. Is the light emitted by a neon tube thermal radiation? The light emitted by an ordinary incandescent light bulb?

2. Does your body emit thermal radiation?

3. Other things being equal, on a clear night, the ground is likely to become much colder than on a cloudy night. Explain.

4. The insulation used in the walls of homes consists of a thick blanket of fiber glass covered on one side by a thin aluminum foil. What is the purpose of these two layers?

5. Black velvet looks much blacker than black paint. Why?

6. For protection against the heat of sunlight, parts of the Lunar Lander (and some other spacecraft) were wrapped in shiny aluminum foil. Why is shiny foil useful for this purpose?

7. If you look into a kiln containing pottery heated to a temperature equal to that of the walls of the kiln, you can scarcely see the pottery. Explain.

8. According to Figure 40.6, at what wavelength is the spectral emittance maximum for a body at 2000 K? At 1750 K? At 1500 K? At 1250 K? Do these wavelengths satisfy Wien's Law?

9. The quantization of electric charge is consistent with classical physics, but the quantization of energy is not. Does this make sense?

10. Suppose that Planck's constant were much larger than it is, say, 10^{34} times larger. What strange behavior would you notice in a simple harmonic oscillator consisting of a mass hanging on a spring?

11. Consider a seconds pendulum, i.e., a pendulum that has a period of 2 seconds. What is the magnitude of one energy quantum for such a pendulum? Would you expect that quantum effects are noticeable in such a pendulum?

12. Suppose that two stars have the same size, but the temperature of one is twice that of the other. By what factor will the thermal power radiated by the hotter star be larger than that radiated by the cooler star?

13. Why do we not notice the discrete quanta of light when we look at a light bulb?

14. Day-glo paints achieve their exceptionally bright orange or red color by converting short-wavelength photons into long-wavelength (red) photons. Why can we not make such a paint in a blue or violet color?

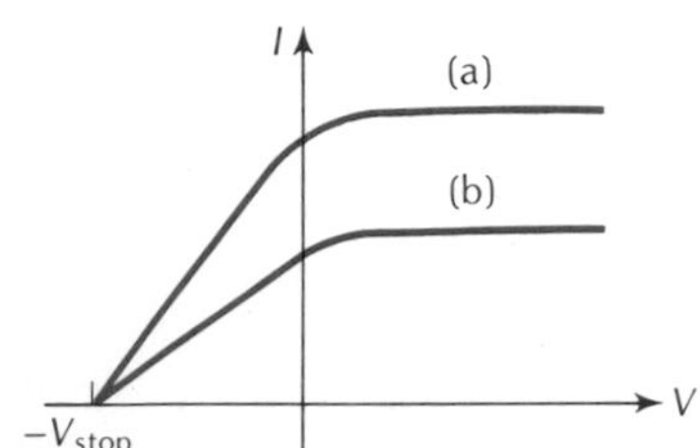

Fig. 40.15 I vs. V for two different values of the intensity of light: (a) high intensity and (b) low intensity.

15. Figure 40.15 shows a plot of current vs. applied potential for the photoelectric current emitted by the surface of a metal illuminated with light of a given wavelength. Qualitatively, explain why the current is zero if $V < -V_{stop}$, explain why the current levels off for a large positive V, and explain why the curves differ for different intensities of the light.

16. When light of a given wavelength ejects photoelectrons from the surface of a metal, why is it that not all of these photoelectrons emerge with the same kinetic energy?

17. Can a particle of mass zero ever be at rest?

18. According to Eq. (22), a photon suffers a maximum change of wavelength in a collision with an electron if it emerges at an angle $\theta = 180°$, and a minimum change of wavelength (no change) if it emerges at an angle $\theta = 0°$. Is this reasonable?

19. Suppose that a photon and an electron have the same momentum. Which has the larger energy, taking into account both the rest-mass energy and the kinetic energy?

20. Can the Compton effect occur with visible light? Would it be observable?

21. Photons of short wavelength are more particle-like than photons of long wavelength. Why?

22. Give an example of an experiment in which photons behave like waves. Give an example of an experiment in which they behave like particles.

23. What happens to the y momentum that a photon gains or loses in the diffraction experiment described in Figure 40.14?

24. If photons were classical particles, what pattern of impact points would we find in the diffraction experiment that led to the results described in Figures 40.13a–d?

25. According to Eq. (23) we can only predict the ***probability*** that a photon will be found at some given point. Does this mean that quantum physics is not deterministic?

PROBLEMS

Sections 40.1 and 40.2

1. Incandescent light bulbs have a tungsten filament whose temperature is typically 3200 K. At what wavelength does such a filament radiate a maximum flux? Assume that the filament acts as a blackbody.

2. Interplanetary and interstellar space is filled with thermal radiation of a temperature of 3 K left over from the Big Bang.

(a) At what wavelength is the flux of this radiation maximum?

(b) What is the power incident on the surface of the Earth due to this radiation?

3. The tungsten filament of a light bulb is a wire of diameter 0.080 mm and length 5.0 cm. The filament is at a temperature of 3200 K. Calculate the power radiated by the filament. Assume the filament acts as a blackbody.

4. The spectral emittance of the Sun is maximum at 5000 Å. By what factor is the emittance smaller at 7000 Å? At 4000 Å?

5. Prove that the spectral emissivity given by Eq. (1) does have a maximum at $\lambda_{max} = (2.9 \times 10^{-3}\ \text{m} \cdot \text{K})/T$. (Hint: Substitute this wavelength into the mathematical condition for a maximum.)

6. Derive the Stefan–Boltzmann Law

$$S \propto T^4$$

from Planck's Law. [Hint: Consider the integral given in Eq. (6); change the variable of integration to $x = hc/\lambda kT$ and show that the result has the form $S = [\text{constant}] \times T^4$; you do not have to evaluate the constant.]

7. Show that the flux radiated by a blackbody in a frequency interval $d\nu$ is

$$S_\nu\, d\nu = \frac{2\pi h}{c^2} \frac{\nu^3}{e^{h\nu/kT} - 1}\, d\nu$$

Does the maximum of S_ν coincide with the maximum of S_λ?

8. The star Procyon B is at a distance of 11 light-years from Earth. The flux of its starlight reaching us is $1.7 \times 10^{-12}\ \text{W/m}^2$ and the surface temperature of the star is 6600 K. Calculate the size of the star.

9. At the Earth, the flux of sunlight per unit area facing the Sun is $1.34 \times 10^3\ \text{W/m}^2$. The Earth absorbs heat from the sunlight and reradiates heat as thermal infrared radiation. For equilibrium, the power arriving from the Sun must equal the average power radiated by the surface of the Earth. What average surface temperature for the Earth can you deduce from this? Assume that the Earth radiates like a blackbody.

10. Deduce the average surface temperature of Pluto by the method described in Problem 9. You will find data on Pluto in the table printed on the endpapers.

11. If you stand naked in a room, your skin and the walls of the room will exchange heat by radiation. Suppose the temperature of your skin is 33°C; the total area of your skin is 1.5 m². The temperature of the walls is 15°C. Assume your skin and the walls behave like blackbodies.
(a) What is the rate at which your skin radiates heat?
(b) What is the rate at which your skin absorbs heat? What is your net rate of loss of heat?

12. An oscillator of frequency 2×10^{15} Hz consists of a mass of 9.1×10^{-31} kg attached to a spring. What is the amplitude of oscillation of this oscillator if its energy of oscillation is one energy quantum? Two energy quanta?

13. We know from Chapter 14 that at small amplitudes a pendulum behaves like an oscillator. Suppose that a pendulum consists of a mass of 0.10 kg attached to a (massless) string of length 1.0 m.
(a) Taking into account the quantization of energy, what is the least (nonzero) amount of energy that this pendulum can have?
(b) What is the amplitude of oscillation of the pendulum with this least amount of energy?

*14. In Problem 24.19 you will find a description of the Thomson model of the hydrogen atom.
(a) What is the frequency of oscillation of the electron in this atom?

(b) The electron can be regarded as an oscillator. Show that if the energy of the electron is one quantum, the amplitude of oscillation exceeds the radius (0.5 Å) of the atom.

Section 40.3

15. Photons of green light have a wavelength of 5500 Å. What is the energy and what is the momentum of one of these photons?

16. Show that if we express the energy of a photon in keV and the wavelength in angstroms, then

$$E = 12.4/\lambda$$

17. A radio transmitter radiates 10 kW at a frequency of 8.0×10^5 Hz. How many photons does the transmitter radiate per second?

18. The energy flux in the starlight reaching us from the bright star Capella is 1.2×10^{-8} W/m^2. If you are looking at this star, how many photons per second enter your eyes? The diameter of your pupil is 0.70 cm. Assume that the average wavelength of the light is 5000 Å.

19. If you want to make a very faint light beam that delivers only 1 photon per square meter per second, what must be the amplitude of the electric field in this light beam? The wavelength of the light is 5000 Å.

20. The energy density of starlight in intergalactic space is 10^{-15} J/m^3. What is the corresponding density of photons? Assume the average wavelength of the photons is 5000 Å.

21. An incandescent light bulb radiates 40 W of thermal radiation from a filament of temperature 3200 K. Estimate the number of photons radiated per second; assume that the photons have an average wavelength equal to the λ_{max} given by Wien's Law.

22. Show that the flux of photons, or the number of photons per unit area and unit time, emitted by a blackbody in the frequency interval $d\nu$ is

$$\frac{2\pi}{c^2} \frac{\nu^2}{e^{h\nu/kT} - 1} d\nu$$

23. A photon has an energy of 5 eV in the reference frame of the laboratory. What is the energy of this photon in the reference frame of a proton moving through the laboratory at a speed of $\frac{1}{2}c$ in the same direction as the photon?

24. According to Figure 40.8, what is the work function of sodium? Express your answer in electron-volts.

25. The work function of potassium is 2.26 eV. What is the threshold frequency for the photoelectric effect in potassium?

26. The work functions of K, Cr, Zn, and W are 2.26, 4.37, 4.24, and 4.49 eV, respectively. Which of these metals will emit photoelectrons when illuminated with red light ($\lambda = 7000$ Å)? Blue light ($\lambda = 4000$ Å)? Ultraviolet light ($\lambda = 2800$ Å)?

27. By inspection of Figure 40.8, find the slope of the line in eV/Hz. Convert these units into J · s and verify that the slope is the same as Planck's constant.

28. The binding energy of an electron in a hydrogen atom is 13.6 eV. Suppose that a photon of wavelength 400 Å strikes the atom and gives up all of its energy to the electron. With what kinetic energy will the electron be ejected from the atom?

Section 40.4

29. X Rays emitted by molybdenum have a wavelength of 0.72 Å. What are the energy and the momentum of one of the photons in these X rays?

30. In a collision with an initially stationary electron, a photon suffers a wavelength increase of 0.022 Å. What must have been the deflection angle of this photon?

31. What is the maximum energy that a free electron (initially stationary) can acquire in a collision with a photon of energy 4.0×10^3 eV?

32. X rays of wavelength 0.30 Å collide with free electrons at rest. Calculate the wavelength of the X rays that emerge from this collision with a deflection of 60°. Calculate the wavelength of the X rays that emerge from this collision with a deflection of 120°.

33. In a collision with a free electron, a photon of energy 2.0×10^3 eV is deflected by 90°. What energy does the electron acquire in this collision?

34. A photon of energy 1.6×10^8 eV collides with a *proton* initially at rest. The photon is deflected by 45°. What is its new energy?

35. A photon of initial wavelength 0.40 Å suffers two successive collisions with two electrons. The deflection in the first collision is 90° and in the second collision it is 60°. What is the final wavelength of the photon?

*36. A photon of energy 5.0×10^3 eV collides head on with a free electron of energy 2.0×10^3 eV. After the collision the photon moves in a direction opposite to its initial direction. Find the energy of the photon. Find the energy of the electron.

Section 40.5

37. A photon passes through a horizontal slit of width 5×10^{-6} m. What uncertainty in the vertical position will this photon have as it emerges from the slit? What uncertainty in vertical momentum?

38. Consider a radio wave in the form of a pulse lasting 0.001 s. This pulse then has a length of $0.001 \text{ s} \times c = 3 \times 10^5$ m. Since an individual photon of this radio wave can be anywhere within this pulse, the uncertainty in the position of the photon is $\Delta x = 3 \times 10^5$ m along the direction of propagation.

(a) According to Heisenberg's relation, what is the corresponding uncertainty in the momentum of the photon?

(b) What is the uncertainty in the frequency of the photon?

CHAPTER 41

Atomic Structure and Spectral Lines

The photographs of the Prelude reproduced in Figure 41.1 give convincing visual evidence that solids, liquids, and gases are made of atoms, small grains of matter with a diameter of about 10^{-10} m. These photographs were prepared with powerful microscopes of a special design. Unfortunately, none of these microscopes is sufficiently powerful to reveal the inside of the atom. Hence, for the exploration of the internal structure of the atom, we still have to rely on the technique developed by Ernest Rutherford and his associates around 1910: bombard the atom with a beam of particles and use this beam as a probe to "feel" the interior of the atom.

By 1910, most physicists had come to believe that atoms are made of some combination of positive and negative electric charges, and that the attractions and repulsions between these electric charges are the basis for all the chemical and physical phenomena observed in solids,

(a)

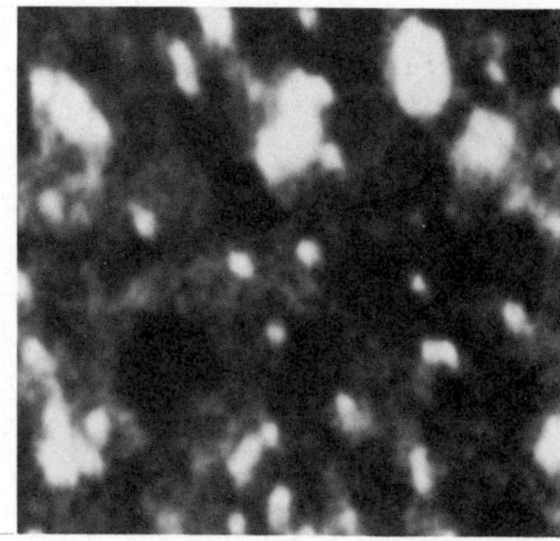

(b)

Fig. 41.1 (a) Platinum atoms in the tip of a fine needle as seen with an ion microscope (magnification $\sim 5 \times 10^5 \times$). (b) Uranium atom seen with an electron microscope (magnification $\sim 5 \times 10^6 \times$).

liquids, and gases. Since electrons were known to be present in all of these forms of matter, it seemed reasonable to suppose that each atom consists of a combination of electrons and positive charge. The vibrational motions of the electrons within such an atom would then result in the radiation of electromagnetic waves; this was supposed to account for the emission of light by the atom. However, both the arrangement of the electric charges within the atom and the mechanism that accounts for the characteristic colors of the emitted light remained mysteries until Rutherford's discovery of the nucleus and Niels Bohr's discovery of the quantization of atomic states. In this chapter we will look at these two crucial discoveries.

The study of the internal structure of the atom led to the inescapable conclusion that in the atomic realm Newton's laws of motion are not valid. Electrons and other subatomic particles obey new equations of motion that are drastically different from the old equations of motion obeyed by planets, billiard balls, or shotgun pellets. The new theory of motion is called **quantum mechanics;** it rules the realm of the atom. In contrast, the old theory of Newton is called **classical mechanics.** The discovery of an entirely new set of laws of motion was without doubt the greatest scientific revolution of this century.

Quantum mechanics

41.1 Spectral Lines

The earliest attempts at a theory of atomic structure ended in failure — they were not able to explain the characteristic colors of the light emitted by atoms. These colors show up very distinctly when a small sample of gas is made to emit light by the application of heat or of an electric current. For instance, if we put a few grains of ordinary salt into a flame, the sodium vapor released by the salt will glow with a characteristic yellow color. If we put neon gas into an evacuated glass tube and connect the ends of the tube to a high-voltage generator (Figure 41.2), the gas will glow with the familiar orange-red color of neon signs.

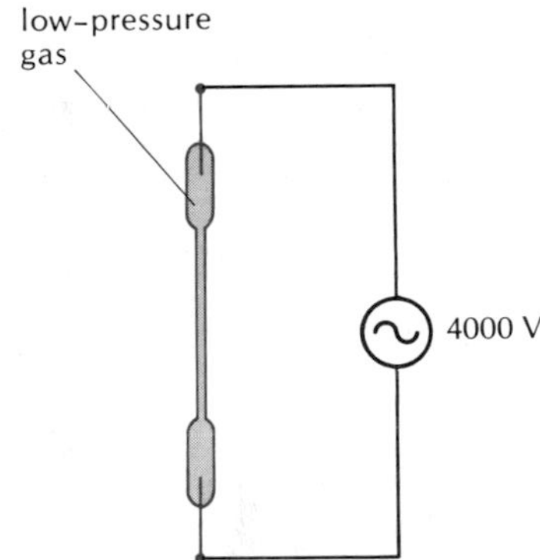

Fig. 41.2 An electric discharge tube. The tube contains gas at a very low pressure. When the terminals are connected to a high-voltage generator, an electric current flows through the gas and makes it glow.

The light emitted by an atom can be precisely analyzed with a prism (Figure 41.3); this breaks the light up into its component colors. In the arrangement shown in Figure 41.3, each separate color generates a bright line. These are the **spectral lines.** Each kind of atom has its own discrete spectral lines. The color plate between pages 850–851 shows the spectral lines of hydrogen; the numbers next to the spectral lines give the wavelengths in angstroms. Hydrogen has four spectral lines in the visible region (already mentioned in Chapter 37) and many ultraviolet and infrared lines not visible to the human eye. The color plate

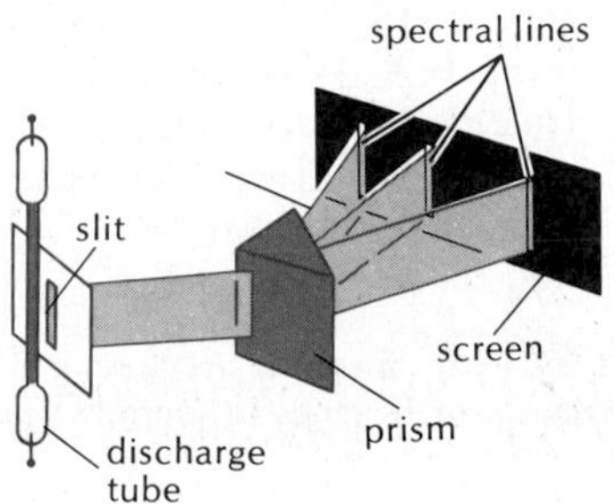

Fig. 41.3 Analysis of light by means of a prism. In this arrangement, each separate color of the light emerging from the slit gives rise to a separate spectral line on the screen.

also shows the spectral lines of sodium, helium, and mercury. Obviously, the set of spectral lines, or **spectrum,** belonging to hydrogen is unmistakably different from those of sodium, helium, and mercury — the spectrum of an atom can serve as a fingerprint in its identification.

Spectrum

In spectroscopy laboratories, scientists often perform the quantitative analysis of a sample of atoms with absorption lines rather than emission lines. It so happens that an atom capable of emitting light of a given wavelength is also capable of absorbing light of that wavelength. When we illuminate a sample of atoms with white light (a mixture containing all colors or wavelengths), the atoms will absorb light of their characteristic wavelengths and, upon analyzing the remaining light with a prism, we find dark absorption lines in the uniform background generated by the white light. The last picture in the color plate shows such an absorption spectrum for sodium vapor. The dark lines of this absorption spectrum coincide with the bright lines in the emission spectrum (see the second picture in the color plate).[1]

Joseph von Fraunhofer, *1787–1826, German optician and physicist. Starting as an apprentice to a glazier, he became a famous maker of optical instruments and a member of the Bavarian Academy of Sciences. He rediscovered the dark lines in the solar spectrum, observed previously by W. H. Wollaston, and introduced many improvements in the design of the spectroscope. He was one of the first makers of diffraction gratings.*

One advantage of spectroscopy over chemistry is that the analysis can be performed even on minuscule amounts of material. What is more, atoms can be identified at a distance. For example, we can identify the atoms on the surface of the Sun by careful analysis of the distribution of colors in sunlight — we do not need to pluck a sample of atoms from the Sun. The power of this technique is best illustrated by the story of the discovery of helium (the "Sun element"). In 1868, the gas was yet unknown to chemists when an astronomer discovered it on the Sun by means of its light; 30 years later chemists finally found traces of helium in minerals on the Earth. By spectroscopic techniques astronomers can identify atoms in remote stars, clouds of interstellar gas, galaxies, and quasars. Figure 41.4 shows the spectrum of the star Caph in the constellation Cassiopeia; the spectral lines indicate the presence of hydrogen, calcium, iron, manganese, chromium, etc.

Note that only a small part of the light from a star is in the form of discrete spectral lines from individual atoms on the stellar surface. Most of the starlight is white light, a more or less uniform mixture of all colors. This is thermal radiation produced in the stellar interior. As we know from the preceding chapter, this kind of light does not retain the fingerprint of the atoms that produced it. Light originating in the stellar interior cannot escape directly, but is first tossed back and forth (scattered) many times by the restless atoms of the hot stellar gas. The random motion of these atoms communicates random changes of wavelength to the light and what finally emerges from the stellar interior is a continuous mixture of a wide range of wavelengths.

Figure 41.5 displays a portion of the spectrum of the white light from our Sun. White light is a nearly uniform mixture of all colors. However, there are many dark lines in this spectrum, caused by absorption in the gas surrounding the Sun. These are called the **Fraunhofer lines.**

Fraunhofer lines

Careful examination of the sets of spectral lines in the color plate between pages 850–851 reveals certain systematic regularities in the spacing of the lines. For instance, in the hydrogen spectrum we find that the spacing of the lines and their intensity decrease systematically

[1] The latter spectrum sometimes has extra lines that are absent in the former spectrum. This is so because the absorption process tends to suppress some lines; they become so faint as to be unnoticeable.

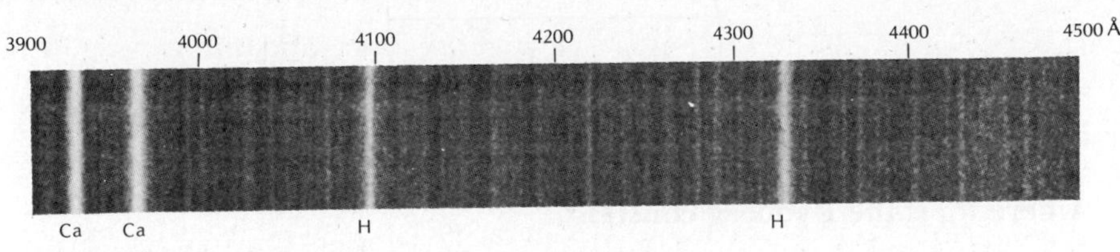

Fig. 41.4 A portion of the spectrum of the star Caph (β Cassiopeiae), from 3900 Å to 4500 Å. The two strong absorption lines on the left are due to ionized calcium. The other two strong lines (middle and right) are due to hydrogen.

Fig. 41.5 A portion of the spectrum of the Sun, from 3900 Å to 4500 Å. Besides the strong absorption lines of calcium and of hydrogen, similar to those in Figure 41.4, there are also strong lines of iron (close pair, right of center).

as we look at shorter and shorter wavelengths; we will see in the next section that the spacing can be described by a simple mathematical formula. These hydrogen lines are said to form a **spectral series.** In the spectra of other elements, we find similar series; however, the spectra usually contain several overlapping series, and this makes it a bit harder to perceive the regularities in the spacing.

Spectral series

41.2 The Balmer Series and Other Spectral Series

The systematic pattern in the spacing of the spectral lines of hydrogen suggests that the wavelengths of these lines should be described by some simple mathematical formula. Table 41.1 lists the wavelengths of the first few of these spectral lines; there actually is an infinite number of spectral lines, the spacing between them becoming smaller and smaller at shorter wavelengths. In 1885, Johann Balmer toyed with the numbers in such a table and discovered that the wavelengths accurately fit the formula

$$\lambda = 911.76\ \text{Å} \times \frac{4n^2}{n^2 - 4} \qquad (1)$$

with $n = 3, 4, 5, 6$, etc. This infinite series of spectral lines is called the **Balmer series.** Note that for $n \to \infty$, the wavelength approaches the asymptotic value $\lambda = 3647.0$ Å; this is called the **series limit.**

Balmer's formula was purely descriptive, or phenomenological; it did not explain the atomic mechanism responsible for the production of the spectral lines. Nevertheless, it proved very fruitful because it led to more general formulas describing other series of spectral lines. It is best to rewrite the formula in terms of the frequency,

$$\nu = \frac{c}{\lambda} = \frac{c}{911.76\ \text{Å}}\left(\frac{1}{4} - \frac{1}{n^2}\right) \qquad (2)$$

or

Table 41.1 THE BALMER SERIES IN THE HYDROGEN SPECTRUM

Wavelength λ[a]
6564.7 Å
4862.7
4341.7
4102.9
3971.2
3890.2
3836.5
3799.0
etc.

[a] Wavelengths are measured in vacuum.

Balmer series of hydrogen

$$\nu = cR_H\left(\frac{1}{2^2} - \frac{1}{n^2}\right) \tag{3}$$

where R_H is the **Rydberg constant,**

Rydberg constant

$$R_H = \frac{1}{911.76\ \text{Å}} = 109{,}678\ \text{cm}^{-1}$$

Balmer proposed that there might be other series in the hydrogen spectrum, with the 2 in Eq. (3) replaced by 1, or 3, or 4, etc. This yields the frequencies

$$\nu = cR_H\left(\frac{1}{1^2} - \frac{1}{n^2}\right) \qquad n = 2, 3, 4, \ldots \tag{4}$$

$$\nu = cR_H\left(\frac{1}{3^2} - \frac{1}{n^2}\right) \qquad n = 4, 5, 6, \ldots \tag{5}$$

$$\nu = cR_H\left(\frac{1}{4^2} - \frac{1}{n^2}\right) \qquad n = 5, 6, 7, \ldots \tag{6}$$

$$\nu = cR_H\left(\frac{1}{5^2} - \frac{1}{n^2}\right) \qquad n = 6, 7, 8, \ldots \tag{7}$$

These four series of spectral lines were actually discovered many years after Balmer proposed them; they are called, respectively, the **Lyman,** the **Paschen,** the **Brackett,** and the **Pfund series** (Figure 41.6). We can combine all these formulas into the single general formula

Spectral series of hydrogen

$$\nu = cR_H\left(\frac{1}{n_2^2} - \frac{1}{n_1^2}\right) \tag{8}$$

where n_1 and n_2 are positive integers and $n_1 > n_2$.

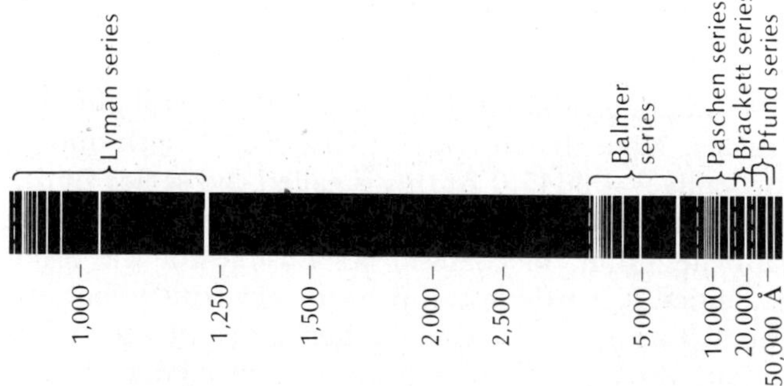

Fig. 41.6 The series of lines in the spectrum of hydrogen.

EXAMPLE 1. What is the shortest wavelength that a hydrogen atom will emit or absorb?

SOLUTION: To find the shortest wavelength, we must choose n_1 and n_2 in Eq. (8) so as to obtain the highest frequency. Obviously, this demands $n_1 = \infty$ and $n_2 = 1$, which gives

$$\nu = cR_H\left(\frac{1}{1} - \frac{1}{\infty}\right) = cR_H \tag{9}$$

or

$$\lambda = c/\nu = 1/R_H = 911.76 \text{ Å}$$

Note that according to Eq. (8) the frequencies of hydrogen are written as differences between two terms, cR_H/n_2^2 and cR_H/n_1^2. Therein lies a crucial clue to the atomic mechanism responsible for the production of the spectral lines — as we will see in Section 41.4, these term differences correspond to energy differences. Careful examination of the frequencies of the spectral lines of other atoms shows that in all cases we can write the frequencies as differences between two terms, although the terms for these other atoms do not have as simple a mathematical form as those for the hydrogen atom. Furthermore, whenever we take any difference between two terms, there actually exists a spectral line that corresponds to this difference; this rule is called the **Rydberg–Ritz combination principle.**

Sir Joseph John Thomson, *1856–1940, English experimental physicist, director of the Cavendish Laboratory at Cambridge, and president of the Royal Society. His discovery of the electron marks the beginning of modern experimental physics. Thomson received the Nobel Prize in 1906 for his investigations on electric discharges in gases. His experiments with beams of positive ions led to the first separation of the isotopes of a chemical element.*

41.3 The Nuclear Atom

The regularity in the series of spectral lines of the atom must be due to an underlying regularity in the structure of the atom. We may think of an atom as analogous to a musical instrument, such as a flute. The atom can only emit a discrete set of spectral lines, just as the flute can only emit a discrete set of tones which make up a musical scale. The regularity in the spacing of tones in this musical scale is due to an underlying regularity in the structure of the flute — the tube of the instrument has regularly spaced blowholes that determine what kind of standing waves can build up within the tube and what kind of waves will be radiated.

J. J. Thomson, the discoverer of the electron, made one of the first attempts at explaining the emission of light in terms of the structure of the atom. Having established that electrons are a ubiquitous component of matter, Thomson proposed the following picture: An atom consists of a number of electrons, say, Z electrons, embedded in a cloud of positive charge. The cloud is heavy, carrying almost all of the mass of the atom. The positive charge in the cloud is $+Ze$, so that it exactly neutralizes the negative charge $-Ze$ of the electrons. In an undisturbed atom the electrons will sit at their equilibrium positions, where the attraction of the cloud on the electrons balances their mutual repulsion (Figure 41.7). But if the electrons are disturbed by, say, a collision, then they will vibrate around their equilibrium positions and emit light. This model of the atom, called the "plum-pudding model," does yield frequencies of vibration of the same order of magnitude as the frequency of light, but it does not yield the observed spectral series; for instance, on the basis of this model, hydrogen should have only one single spectral line, in the far ultraviolet. And in 1910, experiments by

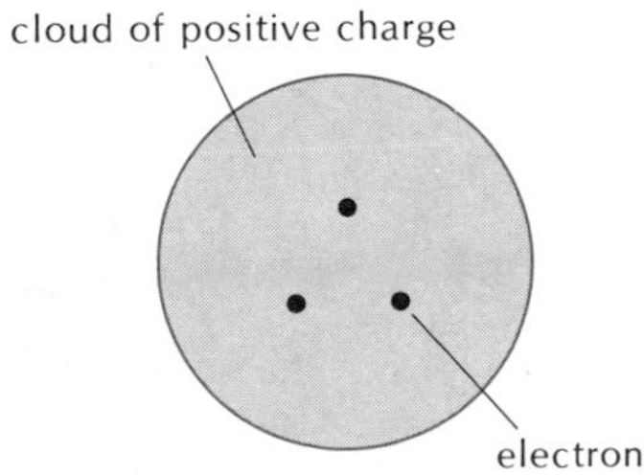

Fig. 41.7 The lithium atom according to the "plum-pudding" model. The three electrons sit at their equilibrium positions.

Rutherford and his collaborators established conclusively that most of the mass of the atom is not spread out over a cloud — instead, the mass is concentrated in a small kernel, or **nucleus,** at the center of the atom.

Rutherford had been studying the emission of alpha particles from radioactive substances. These alpha particles carry a positive charge $2e$ and they have a mass of 6.64×10^{-27} kg, about four times the mass of a proton (alpha particles have the same structure as nuclei of helium atoms; see Section B.1). Some radioactive substances, such as radioactive polonium and radioactive bismuth, spontaneously emit alpha particles with energies of several million electron-volts. These energetic alpha particles readily pass through thin foils of metal, or thin sheets of glass, or other materials. Rutherford was much impressed by the penetrating power of these alpha particles and it occurred to him that a beam of these particles can serve as a probe to "feel" the interior of the atom. When a beam of alpha particles strikes a foil of metal, the alpha particles penetrate the atoms and they are deflected by collisions with the subatomic structures; the magnitude of these deflections gives a clue about the subatomic structures. For example, if the interior of the atom had the "plum-pudding" structure proposed by J. J. Thomson, then the alpha particles would suffer only very small deflections since neither the electrons, with their small masses, nor the diffuse cloud of positive charge would be able to disturb the motion of a massive and energetic alpha particle.

Sir Ernest Rutherford, *1871–1937, British experimental physicist, professor at McGill and at Manchester, and director of the Cavendish Laboratory at Cambridge, where he succeeded J. J. Thomson. Rutherford identified alpha and beta rays. He founded nuclear physics with his discoveries of the nucleus and of transmutation of elements by radioactive decay; he also produced the first artificial nuclear reaction. He was awarded the Nobel Prize for Chemistry in 1908.*

The crucial experiments were performed by H. Geiger and E. Marsden working under Rutherford's direction. They used thin foils of gold and of silver as targets and bombarded these with a beam of alpha particles from a radioactive source. After the alpha particles passed through the foil, they were detected on a zinc sulfide screen which registers the impact of each particle by a faint scintillation (Figure 41.8). To Rutherford's amazement, some of the alpha particles were deflected by such a large angle that they came out backward. In Rutherford's own words: "It was quite the most incredible event that has ever happened to me in my life. It was almost as incredible as if you fired a 15-inch shell at a piece of tissue paper and it came back and hit you." Rutherford immediately recognized that the large deflection must be produced by a close encounter between the alpha particle and a very small but very massive kernel inside the atom. He therefore proposed the following picture: An atom consists of a small nucleus of charge $+Ze$ containing almost all of the mass of the atom; this nucleus is surrounded by a swarm of Z electrons. Thus, the atom is like a solar sys-

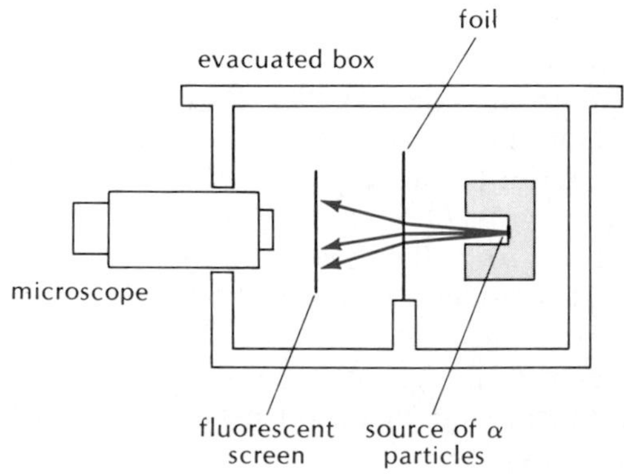

Fig. 41.8 Rutherford's apparatus.

tem — the nucleus plays the role of Sun and the electrons play the role of planets.

On the basis of this nuclear model of the atom, Rutherford calculated what fraction of the beam of alpha particles should be deflected through what angle. If an alpha particle passes close to the nucleus it will experience a large electric repulsion and it will be deflected by a large angle; if it passes far from the nucleus it will only be deflected by a small angle. Figure 41.9 shows the trajectories of several alpha particles approaching a nucleus; these trajectories are hyperbolas. The perpendicular distance between the nucleus and the original (undeflected) line of motion is called the **impact parameter.** In order to suffer a large deflection, the alpha particle must hit an atom with a very small impact parameter, 10^{-13} m or less; since the alpha particles in the beam strike the foil of metal at random, only very few of them will score such a close hit.

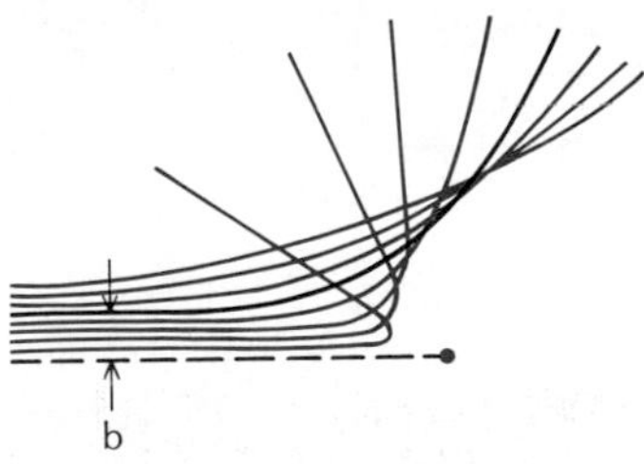

Fig. 41.9 Hyperbolic trajectories of several alpha particles with different impact parameters passing by a nucleus.

EXAMPLE 2. An alpha particle of energy E approaches a nucleus of charge Ze. The impact parameter of the alpha particle is b. What is the distance of the closest approach between the alpha particle and the nucleus?

SOLUTION: Figure 41.10 shows the hyperbolic orbit of the alpha particle and the point of closest approach. When the alpha particle is far away, its velocity is $\mathbf{v}$ and the perpendicular distance between the origin and the line of motion is b; hence the initial angular momentum is mvb. When the alpha particle is at the point of closest approach, its velocity is $\mathbf{v}'$ and its radius vector is $\mathbf{r}'$; since these vectors are perpendicular, the angular momentum is $mv'r'$. Conservation of angular momentum tells us

$$mv'r' = mvb \tag{10}$$

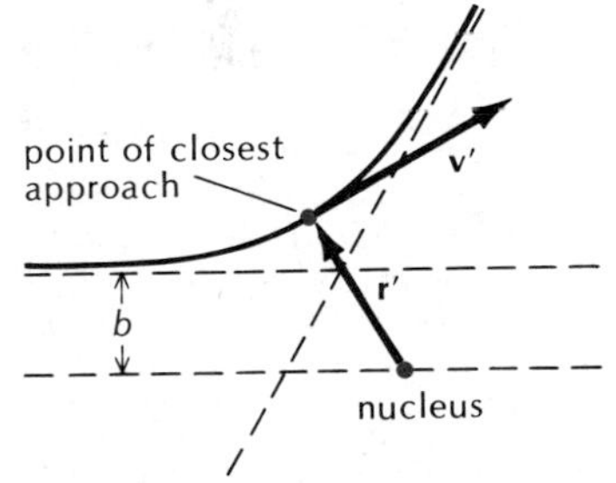

Fig. 41.10 At the point of closest approach, the alpha particle has a velocity $\mathbf{v}'$ perpendicular to its position vector $\mathbf{r}'$.

When the alpha particle is far away its energy is $\frac{1}{2}mv^2$. When the alpha particle is at the point of closest approach its energy is a sum of kinetic and potential energies,

$$\tfrac{1}{2}mv'^2 + \frac{(2e)(Ze)}{4\pi\varepsilon_0}\frac{1}{r'}$$

Conservation of energy then tells us

$$\tfrac{1}{2}mv'^2 + \frac{2Ze^2}{4\pi\varepsilon_0}\frac{1}{r'} = \tfrac{1}{2}mv^2 \tag{11}$$

Using Eq. (10), we can eliminate v' from Eq. (11),

$$\tfrac{1}{2}mv^2\frac{b^2}{r'^2} + \frac{2Ze^2}{4\pi\varepsilon_0}\frac{1}{r'} = \tfrac{1}{2}mv^2$$

We can regard this as a quadratic equation for the unknown $1/r'$. The standard formula for the solution of such a quadratic equation then gives

$$\frac{1}{r'} = \frac{-2Ze^2/4\pi\varepsilon_0 \pm \sqrt{(2Ze^2/4\pi{\varepsilon_0}^2) + m^2v^4b^2}}{mv^2b^2}$$

Only the positive square-root sign makes sense (the negative sign gives a negative value of $1/r'$). In terms of the energy $E = \frac{1}{2}mv^2$ of the alpha particle, we can express the distance of the closest approach as

$$r' = \frac{Eb^2}{-(Ze^2/4\pi\varepsilon_0) + \sqrt{(Ze^2/4\pi\varepsilon_0)^2 + E^2b^2}} \tag{12}$$

For example, if an alpha particle of energy $E = 1.2 \times 10^{-12}$ J approaches a gold nucleus ($Z = 79$) with an impact parameter of 5.0×10^{-14} m, then

$$\frac{Ze^2}{4\pi\varepsilon_0} = \frac{79 \times (1.6 \times 10^{-19}\ \text{C})^2}{4\pi\varepsilon_0} = 1.8 \times 10^{-26}\ \text{J}\cdot\text{m}$$

and the distance of closest approach will be

$$r' = \frac{1.2 \times 10^{-12}\ \text{J} \times (5.0 \times 10^{-14}\ \text{m})^2}{-1.8\times10^{-26}\ \text{J}\cdot\text{m} + \sqrt{(1.8\times10^{-26}\ \text{J}\cdot\text{m})^2 + (1.2\times10^{-12}\ \text{J}\times5.0\times10^{-14}\ \text{m})^2}}$$

$$= 6.7 \times 10^{-14}\ \text{m}$$

41.4 Bohr's Theory

Niels Bohr, *1885–1962, Danish theoretical physicist. He worked under J. J. Thomson and Rutherford in England and then became professor at Copenhagen and director of the Institute of Theoretical Physics, for the foundation of which he was largely responsible. After formulating the quantum theory of the atom, he played a leading role in the further development of the new quantum mechanics. He received the Nobel Prize in 1922.*

Rutherford's experiments did reveal the gross arrangement of the electrons in the atom, but not the details of their motion. Since the electrons make up the outer layers of an atom, their arrangement and motion should determine the chemical properties of the atom and the emission of light. But when physicists tried to calculate the electron motion according to the laws of classical mechanics and electromagnetism, they immediately ran into trouble.

To gain some insight into the source of this trouble, let us examine the case of the hydrogen atom. Suppose that the single electron of this atom is moving, according to the laws of classical mechanics, in a circular orbit of radius $\sim 10^{-10}$ m. The electron would then have a centripetal acceleration which is very large, about 10^{23} m/s^2. Because of this acceleration, the electron would emit high-frequency electromagnetic radiation, i.e., it would emit light. The energy carried away by the light must be supplied by the electron. Hence the emission process has the same effect on the electron as a friction force — it removes energy from the electron. This kind of friction would cause the electron to leave its circular orbit and gradually spiral in toward the nucleus, just like the residual atmospheric friction on an artificial satellite in a low-altitude orbit around the Earth causes it to spiral down toward the ground. A calculation using the laws of classical mechanics and electricity shows that the rate of emission of light by the orbiting electron in a hydrogen atom would be quite large. Correspondingly, the rate of energy loss of the electron would be large — the electron would spiral inward and collide with the nucleus within a time as short as 10^{-10} s!

Thus, our classical calculation leads us to the troublesome conclusion that hydrogen atoms, and other atoms, ought to be unstable — all the electrons ought to collapse into the nucleus almost instantaneously. Furthermore, the light that the electron emits during the spiraling motion ought to be a wave of continually increasing amplitude and increasing frequency (in musical terminology, crescendo and glissando); this is so because the closer the electron comes to the nucleus, the larger its acceleration and the higher its frequency of orbital motion. Hydrogen atoms do not behave as this calculation predicts. Hydrogen atoms are stable and when they do emit light, they emit discrete frequencies (spectral lines) instead of a continuum of frequencies.

These irreconcilable differences between the observed properties of atoms and the calculated properties gave evidence of a serious breakdown of the classical mechanics of Newton and the classical theory of electromagnetism. Although these theories had proved very successful on a macroscopic scale, they were in need of some drastic modification on an atomic scale.

In 1913, Niels Bohr took a bold step toward resolving these difficulties. He made the radical proposal that, at the atomic level, the laws of classical mechanics and of classical electromagnetism must be replaced or supplemented by other laws. Bohr expressed these new laws of atomic mechanics in the form of several postulates:

Bohr's postulates

1. The orbits and the energies of the electrons in an atom are quantized, i.e., only certain discrete orbits and energies are permitted. When an electron is in one of the quantized orbits, it does not emit any electromagnetic radiation; thus, the electron is said to be in a **stationary state.** The electron can make a discontinuous transition, or **quantum jump,** from one stationary state to another. During this transition it does emit radiation.
2. The laws of classical mechanics apply to the orbital motion of the electrons in a stationary state, but these laws do not apply during the transition from one state to another.
3. When an electron makes a transition from one stationary state to another, the excess energy ΔE is released as a single photon of frequency $\nu = \Delta E/h$.
4. The permitted orbits are characterized by quantized values of the orbital angular momentum. This angular momentum is always an integral multiple of $h/2\pi$:

Quantization of angular momentum

$$\boxed{L = nh/2\pi \qquad n = 1, 2, 3, \ldots} \tag{13}$$

Let us now see how to calculate the stationary states and the spectrum of the hydrogen atom on the basis of these postulates. For the sake of simplicity, we will assume that the electron moves in a circular orbit around the proton which remains at rest (Figure 41.11). Since the electric force of attraction between the electron and the proton is $e^2/4\pi\varepsilon_0 r^2$, the equation of motion for the electron in a circular orbit is

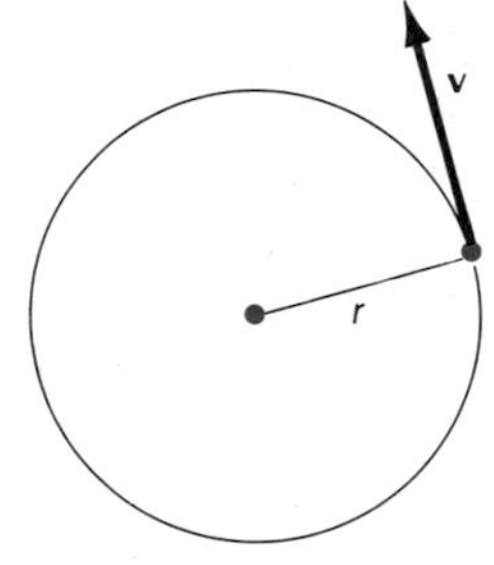

Fig. 41.11 Electron in circular orbit around proton.

$$\frac{m_e v^2}{r} = \frac{1}{4\pi\varepsilon_0}\frac{e^2}{r^2} \tag{14}$$

According to Bohr's postulate, the orbital angular momentum must be $h/2\pi$ multiplied by an integer,

$$L = m_e vr = nh/2\pi \qquad n = 1, 2, 3, \ldots \tag{15}$$

or

$$m_e vr = n\hbar \tag{16}$$

where $\hbar$ (pronounced "h bar") is Planck's constant divided by 2π:

$$\hbar = h/2\pi = 1.06 \times 10^{-34}\ \text{J} \cdot \text{s}$$

Angular-momentum quantum number

The number n is called the **angular-momentum quantum number.** From Eq. (16)

$$v^2 = \frac{n^2\hbar^2}{m_e^2 r^2} \tag{17}$$

which, when substituted into Eq. (14), yields an expression for the radius of the orbit,

$$r = \frac{4\pi\varepsilon_0 n^2\hbar^2}{m_e e^2} \tag{18}$$

Thus, the radius of the smallest permitted orbit ($n = 1$) is

$$r_1 = \frac{4\pi\varepsilon_0\hbar^2}{m_e e^2} = 0.529 \times 10^{-10}\ \text{m} = 0.529\ \text{Å}$$

Bohr radius

This is called the **Bohr radius,**[2] usually designated by a_0,

$$\boxed{a_0 = \frac{4\pi\varepsilon_0\hbar^2}{m_e e^2}} \tag{19}$$

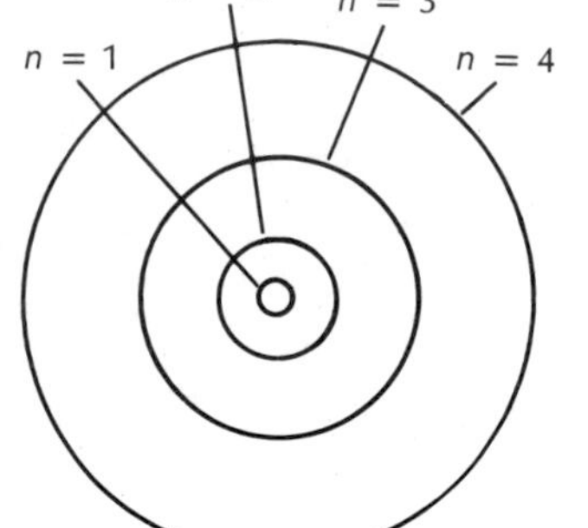

Fig. 41.12 The possible Bohr orbits of an electron in the hydrogen atom.

Figure 41.12 shows the permitted circular orbits, drawn to scale.

The energy of the electron in one of these orbits is a sum of kinetic and potential energies:

$$E = \tfrac{1}{2}m_e v^2 - \frac{e^2}{4\pi\varepsilon_0 r}$$

$$= \tfrac{1}{2}m_e\left(\frac{n^2\hbar^2}{m_e^2}\right)\left(\frac{m_e e^2}{4\pi\varepsilon_0 n^2\hbar^2}\right)^2 - \frac{e^2}{4\pi\varepsilon_0}\left(\frac{m_e e^2}{4\pi\varepsilon_0 n^2\hbar^2}\right) \tag{20}$$

or

Energy of stationary states of hydrogen

$$\boxed{E = -\frac{m_e e^4}{2(4\pi\varepsilon_0)^2\hbar^2}\frac{1}{n^2}} \tag{21}$$

Thus, the energy of the stationary state of least energy ($n = 1$) is

$$E_1 = -\frac{m_e e^4}{2(4\pi\varepsilon_0)^2\hbar^2}$$

$$= -2.18 \times 10^{-18}\ \text{J} = -13.6\ \text{eV} \tag{22}$$

The energies of the other stationary states are

[2] See Appendix 8 for a more precise value of the Bohr radius.

$$\boxed{E_n = -\frac{13.6 \text{ eV}}{n^2}} \qquad (23)$$

Energy-level diagram

Figure 41.13 displays these quantized energies in an **energy-level diagram.** Each horizontal line represents one of the energies given by Eq. (23). According to Bohr's assumptions, the electron radiates when it makes a quantum jump from one stationary state to a lower stationary state. Such quantum jumps have been indicated by arrows in Figure 41.13. The stationary state of lowest energy is called the **ground state;** the next one is called the **first excited state,** etc. Ordinarily, the electron of the hydrogen atom is in the ground state, i.e., the circular orbit of radius a_0. This is the configuration of least energy into which the atom tends to settle when it is left undisturbed. As long as the atom remains in the ground state it does not emit light. To bring about the emission of light, we must first kick the electron into one of the excited states, i.e., a circular orbit of larger radius. We can do this by heating a sample of atoms or by passing an electric current through the sample. Collisions between the atoms will then disturb the electronic motions and occasionally kick an electron into a larger orbit. From there, the electron will spontaneously jump into a smaller orbit, giving off a quantum of light. Note that the quantum jumps shown in Figure 41.13 form several series: one series consists of all those jumps that end in the ground state, another series consists of all those jumps that end in the first excited state, etc. These series of jumps give rise to the series of spectral lines: the Lyman series, the Balmer series, etc.

Ground state and excited states

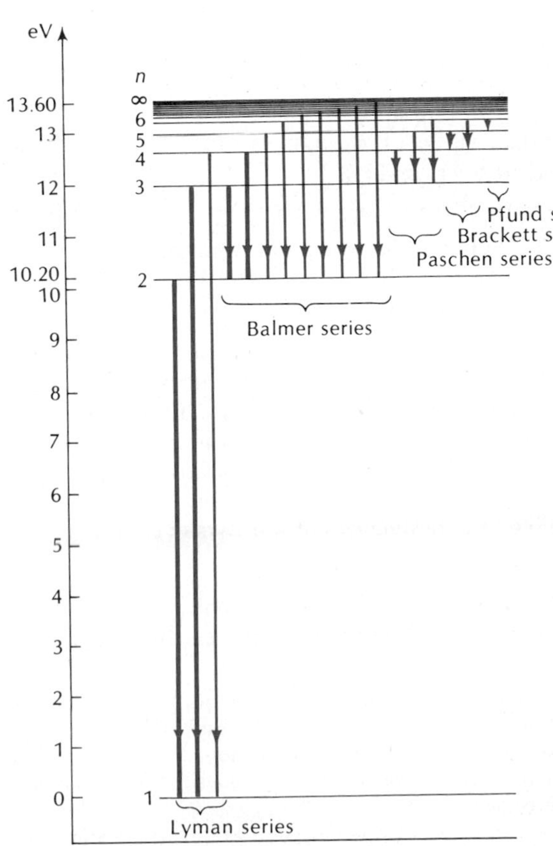

Fig. 41.13 Energy-level diagram for the hydrogen atom. The arrows show the possible quantum jumps for the electron. Note that in this diagram the energies are given relative to the ground state, which is assigned an energy of zero.

Let us now calculate the frequency of the light emitted in a quantum jump from some initial state i to a final state f. In this jump the electron releases an energy

$$\Delta E = E_i - E_f$$

$$= \frac{m_e e^4}{2(4\pi\varepsilon_0)^2\hbar^2}\left(\frac{1}{n_f^2} - \frac{1}{n_i^2}\right) \tag{24}$$

According to Bohr's postulate, this energy is radiated as a single photon of frequency $\nu = \Delta E/h$, i.e.,

Frequency of photon emitted in transition

$$\boxed{\nu = \frac{E_i - E_f}{h} = \frac{m_e e^4}{4\pi(4\pi\varepsilon_0)^2\hbar^3}\left(\frac{1}{n_f^2} - \frac{1}{n_i^2}\right)} \tag{25}$$

This equation looks just like the general formula (8) for the frequencies of the spectral series. Comparison of Eqs. (25) and (8) yields the following theoretical formula for the Rydberg constant:

$$R_H = \frac{m_e e^4}{4\pi(4\pi\varepsilon_0)^2\hbar^3 c} \tag{26}$$

Upon insertion of the accurate values of the fundamental constants given in Appendix 8, we obtain

$$R_H = \frac{9.10953 \times 10^{-31}\ \text{kg} \times (1.602189 \times 10^{-19}\ \text{C})^4}{4\pi(4\pi \times 8.854178 \times 10^{-12}\ \text{F/m})^2 \times (1.054589 \times 10^{-34}\ \text{J}\cdot\text{s})^3 \times 2.997925 \times 10^8\ \text{m/s}}$$

$$= 109{,}737\ \text{cm}^{-1}$$

This theoretical value of R_H agrees quite well with the experimental value quoted in Section 41.2.[3]

EXAMPLE 3. Suppose that the atoms in a sample of hydrogen gas are initially in the ground state. If we illuminate these atoms with light (from some kind of lamp), what frequencies will the atoms absorb?

SOLUTION: Absorption of light is the reverse of emission. When an electron in an atom absorbs a photon (supplied by the lamp), it jumps from the initial state to a state of higher energy. The energy of the photon must match the energy difference between the states. Thus, the frequencies of the photons that the electrons can absorb when jumping upward from the ground state are exactly those frequencies that they emit when jumping downward into the ground state, i.e., the frequencies of the Lyman series.

With his theory Bohr attained the goal of explaining the regularities in the spectrum of hydrogen in terms of the regularities of the struc-

[3] The small disagreement between the theoretical value of R_H given in Eq. (26) and the experimental value given in Section 41.2 is due to the motion of the nucleus of the atom, which we have neglected in our calculation. A careful calculation that takes into account the motions of electron and proton about their common center of mass eliminates the disagreement.

ture of the atom. By showing that this structure is based on a simple numerical sequence, he fulfilled the ancient dream of Pythagoras of a universe based on simple numerical ratios, a dream that arose from an analogy with musical instruments. Bohr's theory tells us how the atom plays its tune.

41.5 The Correspondence Principle

We saw in Chapter 40 that in an ordinary light wave we usually do not notice the individual quanta of energy — the number of quanta is so large that the energy distribution seems continuous. Likewise, when the electron in a hydrogen atom has a large angular momentum, an increase or decrease of this angular momentum by one quantum $\hbar$ represents such a small fraction of the whole that the change in angular momentum seems continuous. For instance, if an electron in a Bohr orbit with quantum number $n = 5000$ jumps down, step by step, to the next orbit ($n = 4999$), and to the next orbit ($n = 4998$), and etc., the changes of angular momentum, energy, and radius will seem quite continuous because each step is small compared to the remaining angular momentum, energy, and radius. Under these conditions the electron will behave pretty much like a classical particle — the quantized character of the angular momentum and energy, and the discrete character of the orbits will not be very noticeable.

Correspondence Principle

This is an instance of Bohr's **Correspondence Principle.** This principle states that *in the limiting case of large quantum numbers, the results obtained from quantum theory must agree with those obtained from classical theory.*

We can check that the behavior of an electron in a hydrogen atom does obey this principle both qualitatively and quantitatively, by calculating the frequency of the emitted light; we will see that in the limiting case of large n, the results of quantum theory and of classical theory agree. According to Eq. (25), the frequency emitted in a transition from the state n to the state $n - 1$ is

$$\nu = \frac{m_e e^4}{4\pi(4\pi\varepsilon_0)^2\hbar^3}\left[\frac{1}{(n-1)^2} - \frac{1}{n^2}\right] \tag{27}$$

$$= \frac{m_e e^4}{4\pi(4\pi\varepsilon_0)^2\hbar^3}\,\frac{2n-1}{n^2(n-1)^2} \tag{28}$$

If n is very large, $(2n-1)/[n^2(n-1)^2] \cong 2n/n^4 = 2/n^3$ so that

$$\nu \cong \frac{m_e e^4}{4\pi(4\pi\varepsilon_0)^2\hbar^3}\,\frac{2}{n^3} \tag{29}$$

This is the frequency according to quantum theory. To find the frequency according to classical theory, we note that for a charge in accelerated motion, classical electromagnetism predicts that the frequency of the emitted light coincides with the frequency of the motion. For our electron in a circular orbit, the frequency of the motion is $v/2\pi r$. From Eqs. (16) and (18) we then find that the frequency of the emitted light is

$$\nu_{\text{class}} = \frac{v}{2\pi r} = \frac{n\hbar/m_e r}{2\pi r} = \frac{n\hbar}{2\pi m_e}\frac{1}{r^2}$$

$$= \frac{n\hbar}{2\pi m_e}\left(\frac{m_e e^2}{4\pi\varepsilon_0 n^2\hbar^2}\right)^2 \tag{30}$$

If we simplify the right side of this equation, we recognize that it coincides with the right side of Eq. (29), i.e., the result of the classical calculation agrees with the result of the quantum-mechanical calculation. Note, however, that this agreement holds only for large values of n; if n is *not* large, then the classical frequency [Eq. (30)] is smaller than the quantum-mechanical frequency [Eq. (28)].

Louis Victor, prince de Broglie (de broy), *1892– , French theoretical physicist, professor at the University of Paris. He discovered Eq. (41.31) by reasoning that if waves have particle properties, then maybe particles have wave properties. For his discovery of the wave properties of matter he was awarded the Nobel Prize in 1929, after the existence of these wave properties was confirmed experimentally.*

41.6 Quantum Mechanics

Bohr's theory is a hybrid. It relies on some basic classical features (orbits) and grafts onto these some quantum features (quantum jumps, quanta of light). In the 1920s the cooperative efforts of several brilliant physicists — L. de Broglie, E. Schrödinger, W. Heisenberg, M. Born, P. Jordan, P. A. M. Dirac — established that the remaining classical features had to be eradicated from the theory of the atom. Bohr's semiclassical theory had to be replaced by a new quantum mechanics with an entirely different equation of motion.

The basis of the new quantum mechanics was laid by the discovery that electrons — as well as protons, neutrons, and all the other "particles" found in nature — have not only particle properties but also wave properties. When a beam of electrons is made to pass through an extremely narrow slit, the electrons exhibit diffraction. This means that electrons are neither classical particles nor classical waves. Electrons, just like photons, are a new kind of object with a subtle combination of particle and wave properties. Electrons are *wavicles.* The wavelength associated with an electron or some other wavicle is inversely proportional to its momentum:

De Broglie wavelength

$$\boxed{\lambda = h/p} \tag{31}$$

This is called the **de Broglie wavelength.**

Schrödinger equation

In quantum mechanics, the motion of an electron is described by a wave equation, the **Schrödinger equation.** This Schrödinger equation plays the same role for electrons as the Maxwell equations play for photons. The intensity of the electron wave at some point determines the probability that there is an electron particle at that point [compare Eq. (40.23)]. Furthermore, as a consequence of their wave properties, electrons obey the Heisenberg uncertainty relations for position and momentum [see Eq. (40.29)]. These quantum uncertainties are of crucial importance for the behavior of an electron inside an atom. For such an electron, the uncertainty in the position is very large — about as large as the size of the atom. This implies that the electron follows no definite orbit. It is therefore not surprising that the Bohr theory should have failed in its attempt at calculating the electron motion in the helium atom and in other atoms with several electrons; what is surprising is that this theory should have succeeded as well as it did in the case of the hydrogen atom.

EXAMPLE 4. Consider an electron in the ground state of hydrogen. Show that a well-defined orbit is inconsistent with the Heisenberg uncertainty relations.

SOLUTION: If the electron is to follow a well-defined orbit, the uncertainty in its momentum (in any direction) must be much smaller than the magnitude of the momentum. According to Eq. (16), the speed of the electron in the smallest circular orbit ($n = 1$) is $v = \hbar/m_e r$ and the magnitude of the momentum is $p = m_e v = \hbar/r$. For a well-defined orbit we therefore require

$$\Delta p_y \ll \hbar/r \qquad (32)$$

Furthermore, we require that the uncertainty in the position must be much smaller thanthe size of the orbit,

$$\Delta y \ll r \qquad (33)$$

Taking the product of Eqs. (32) and (33), we find

$$\Delta y \, \Delta p_y \ll \hbar \qquad (34)$$

This is inconsistent with the Heisenberg uncertainty relation [Eq. (40.29)].

Although electrons are wavicles, they will sometimes behave pretty much like classical particles. Roughly, we can say that classical mechanics will be a good approximation whenever the quantum uncertainties are small compared to the relevant magnitudes of positions and momenta. For instance, for the electrons in the beam of a TV tube, the quantum uncertainty in the momentum is negligible compared with the magnitude of the momentum. Under these conditions, classical mechanics gives an adequate description of the motion of the electrons.

Erwin Schrödinger, *1887–1961, Austrian theoretical physicist, professor at Berlin and at Vienna. Another of the founders of the new quantum mechanics, he received the Nobel Prize in 1933.*

What we have said about the quantum mechanics of electrons also applies to other "particles" found in nature — they all have wave properties and they all have quantum uncertainties in their position and momentum. Strictly speaking, even large macroscopic bodies have wave properties. For example, an automobile is a wavicle and it has some quantum uncertainty in its position. However, it turns out that the quantum uncertainties are very small whenever the mass of the body is large compared to atomic masses — the quantum uncertainty in the position of the body of an automobile is typically no more than about 10^{-18} m, a number that can be ignored for all purposes. Hence, for automobiles and other macroscopic bodies, quantum effects are completely insignificant and classical mechanics gives an excellent description of the motion of these bodies.

SUMMARY

Spectral series of hydrogen:

$$\nu = cR_{\text{H}}\left(\frac{1}{n_2^2} - \frac{1}{n_1^2}\right); \qquad R_{\text{H}} = 109{,}678 \text{ cm}^{-1}$$

Quantization of angular momentum: $L = n\hbar$

Bohr radius: $a_0 = \dfrac{4\pi\varepsilon_0\hbar^2}{m_e e^2}$

Energy of stationary states of hydrogen:

$$E = -\frac{m_e e^4}{2(4\pi\varepsilon_0)^2\hbar^2}\frac{1}{n^2}$$

$$= -\frac{13.6 \text{ eV}}{n^2}$$

Frequency of photon emitted in transition: $\nu = \dfrac{E_i - E_f}{h}$

Correspondence Principle: In the limiting case of large quantum numbers, the results obtained from quantum theory must agree with those obtained from classical theory.

De Broglie wavelength: $\lambda = h/p$

QUESTIONS

1. Do the spectral lines seen in a stellar spectrum (e.g., Figure 41.4) tell us anything about the chemical composition of the stellar interior?

2. The spectrum of hydrogen shown in the color plate between pages 850–851 displays all of the spectral lines simultaneously. Since a hydrogen atom emits only one spectral line at a time, how can all the lines be visible simultaneously?

3. What is the longest wavelength that a hydrogen atom will emit or absorb?

4. The target used in Rutherford's scattering experiment was a very thin foil of metal. What is the advantage of a thin foil over a thick foil in this experiment?

5. How can Rutherford's experiment tell us something about the size of the nucleus?

6. Why is Bohr's postulate of stationary states in direct contradiction with classical mechanics and electromagnetism?

7. If an electron in a hydrogen atom makes a transition from some state to a lower state, does its kinetic energy increase or decrease? Its potential energy? Its orbital angular momentum?

8. At low temperatures, the absorption spectrum of hydrogen displays only the spectral lines of the Lyman series. At higher temperatures, it also displays other series. Explain.

9. The planets move around the Sun in circular orbits. Is their orbital angular momentum quantized?

10. In a muonic atom, a muon orbits around the nucleus. The mass of the muon is 207 times the mass of the electron. What is the Bohr radius for a muonic atom with a hydrogen nucleus?

11. Given that the orbital angular momentum of an atom is quantized, can we conclude that the orbital magnetic moment is also quantized?

Fine-structure constant

12. The quantity $\hbar/(m_e c)$ is called the **Compton wavelength.** The quantity $e^2/(4\pi\varepsilon_0 m_e c^2)$ is called the **"classical electron radius."** Show that the Bohr radius, the Compton wavelength, and the classical electron radius are in the ratio $1{:}\alpha{:}\alpha^2$, where $\alpha = e^2/(4\pi\varepsilon_0\hbar c)$. The quantity α is called the **fine-structure constant.** What is the numerical value of this constant?

13. Would you expect that Bohr's theory of the hydrogen atom can be adapted to the singly ionized helium atom, i.e., the helium atom with one missing electron? To what other ionized atoms can Bohr's theory be adapted?

14. According to the **Complementarity Principle,** formulated by Bohr, a wavicle has both wave properties and particle properties, but these properties are never exhibited simultaneously: if the wavicle exhibits wave properties in an experiment, then it will not exhibit particle properties, and conversely. Give some examples of experiments in which wave or particle properties (but not both simultaneously) are exhibited.

Complementarity Principle

15. Show that photons obey the de Broglie relation.

16. If the de Broglie wavelengths of two electrons differ by a factor of 2, by what factor must their energies differ?

17. According to the de Broglie relation, the wavelength of an electron of very small momentum is very large. Could we take advantage of this to design an experiment that makes the wave properties of the electron obvious?

18. An electron and a proton have the same energy. Which has the longer de Broglie wavelength?

19. Describe the interference pattern expected for an electron wave incident on a plate with two very narrow parallel slits separated by a small distance.

20. Electron microscopes achieve high resolution because they use electron waves of very short wavelength, usually less than 0.1 Å. Why can we not build a microscope that uses *photons* of equally short wavelength?

PROBLEMS

Section 41.2

1. Use Eq. (4) to calculate the wavelengths of the first four lines of the Lyman series.

2. Which of the spectral lines of the Brackett series is closest in wavelength to the first spectral line ($n = 6$) of the Pfund series? By how much do the wavelengths differ?

3. Show that the spectral lines of the Balmer series all have a higher frequency than the spectral lines of the Paschen series. Do the spectral lines of the Paschen series all have a higher frequency than those of the Brackett series?

4. When astronomers examine the light of a distant galaxy, they find that all the wavelengths of the spectral lines of the atoms are longer than those of the atoms here on Earth by a common multiplicative factor. This is the *red shift* of light; it is a Doppler shift caused by the motion of recession of the galaxy, away from the Earth. In the light of a galaxy beyond the constellation Virgo, astronomers find spectral lines of wavelengths 4117 Å and 4357 Å.
   (a) Assume that these are two spectral lines of hydrogen, with the wavelengths multiplied by some factor. Identify these lines. What is the factor by which these wavelengths are longer than the normal wavelengths of the two spectral lines?
   (b) What is the speed of recession of the galaxy?

5. One of the spectral series of the lithium atom is the **principal series,** with the following wavelengths: 6707.9 Å, 3232.6 Å, 2741.3 Å, 2562.5 Å, 2475.3 Å. Show that these wavelengths approximately fit the formula

$$\frac{1}{\lambda} = R\left[\frac{1}{(1+s)^2} - \frac{1}{(n+p)^2}\right] \qquad n = 2, 3, 4, \ldots$$

where $R = 109{,}728\ \text{cm}^{-1}$ is the Rydberg constant for lithium, and s and p are constants characteristic of the series. Given that $p = -0.041$, what value of s must you use to make the wavelengths fit the formula?

6. Another of the spectral series of the lithium atom is the **diffuse series,** with the following wavelengths: 6103.5 Å, 4603.0 Å, 4132.3 Å, 3915.0 Å, 3794.7 Å. These wavelengths approximately fit the formula

$$\frac{1}{\lambda} = R\left[\frac{1}{(2+p)^2} - \frac{1}{(n+d)^2}\right] \qquad n = 3, 4, 5, \ldots$$

where, as in the preceding problem, $R = 109{,}728\ \text{cm}^{-1}$ and p and d are constants.

(a) Given that $d = -0.0015$, what value of p must you use to make the wavelengths fit the formula?
(b) The principal series (see Problem 5) and the diffuse series of lithium are analogous to two spectral series of hydrogen. Which two series?

Section 41.3

7. What is the distance of closest approach for a 5.5-MeV alpha particle in a head-on collision with a gold nucleus? With an aluminum nucleus?

8. The nucleus of platinum has a radius 6.96×10^{-15} m and an electric charge of $78e$. What must be the minimum energy of an alpha particle in a head-on collision if it is to just barely reach the nuclear surface? Assume the alpha particle is pointlike.

9. An alpha particle of energy 5.5 MeV is incident on a silver nucleus with an impact parameter 8.0×10^{-15} m. Find the distance of closest approach of the particle. Find the speed at the point of closest approach.

10. Prove that the distance of closest approach given by Eq. (13) can be expressed as

$$r' = \frac{b^2}{-r^*/2 + \sqrt{(r^*/2)^2 + b^2}}$$

where r^* is the distance of closest approach for a head-on collision (with $b = 0$).

*11. A foil of gold, 2.1×10^{-5} cm thick, is being bombarded by alpha particles of energy 7.7 MeV. The particles impact at random over an area of 1 cm^2 of the foil of gold.

(a) How many atoms are there within the volume $1\ \text{cm}^2 \times 2.1 \times 10^{-5}$ cm under bombardment? The density of gold is 19.3 g/cm^2 and the mass of one atom is 3.27×10^{-25} kg.
(b) It can be shown that to suffer a deflection of more than 30°, an alpha particle must strike within 5.5×10^{-14} m of the center of a gold nucleus. What is the probability for this to happen?
(c) If 10^{10} alpha particles impact on the foil, how many will suffer deflections of more than 30°?

Section 41.4

12. What is the speed of an electron in the smallest ($n = 1$) Bohr orbit? Express your answer as a fraction of the speed of light.

13. What is the frequency of the orbital motion for an electron in the smallest ($n = 1$) Bohr orbit? In the next ($n = 2$) Bohr orbit? Do either of these frequencies coincide with the frequency of the light emitted during the transition $n = 2$ to $n = 1$?

14. If a hydrogen atom is in the ground state, what is the *longest* wavelength it will absorb?

15. What is the ionization energy of hydrogen (i.e., what energy must you supply to remove the electron from the atom when it is in the ground state)? Express the answer in electron-volts.

16. A hydrogen atom emits a photon of wavelength 1026 Å. From what stationary state to what lower stationary state did the electron jump?

17. Suppose that the electron in a hydrogen atom is initially in the second excited state ($n = 3$). What wavelength will the atom emit if the electron jumps directly to the ground state? What two wavelengths will the atom emit if the electron jumps to the first excited state and then to the ground state?

18. Find the orbital radius, the speed, the angular momentum, and the centripetal acceleration for an electron in the $n = 2$ orbit of hydrogen.

19. If you bombard hydrogen atoms in their ground state with a beam of particles, the collisions will (sometimes) kick atoms into one of their excited states. What must be the minimum kinetic energy of the bombarding particles if they are to achieve such an excitation?

20. A hydrogen atom is initially in the ground state. In a collision with an argon atom, the electron of the hydrogen atom absorbs an energy of 15.0 eV. With what speed will the electron be ejected from the hydrogen atom?

21. The singly ionized helium atom (usually designated HeII) has one electron in orbit around a nucleus of charge $2e$.

(a) Apply Bohr's theory to this atom and find the energies of the stationary states. What is the value of the ionization energy, i.e, the energy that you must supply to remove the electron from the atom when it is in the ground state? Express the answer in electron-volts.

(b) Show that for every spectral line of the hydrogen atom, the ionized helium atom has a spectral line of identical wavelength.

22. Doubly ionized lithium (usually designated LiIII) has one electron in orbit around a nucleus of charge $Z = 3e$. What is the radius of the smallest Bohr orbit in doubly ionized lithium? What is the energy of this orbit?

23. The muon (or mu meson) is a particle somewhat similar to an electron; it has a charge $-e$ and a mass 206.8 times as large as the mass of the electron. When such a muon orbits around a proton, they form a **muonic hydrogen atom,** similar to an ordinary hydrogen atom, but with the muon playing the role of the electron. Calculate the Bohr radius of this muonic atom and calculate the energies of the stationary states. What is the energy of the photon emitted when the muon makes a transition from $n = 2$ to $n = 1$?

24. Assume that, as proposed by J. J. Thomson, the hydrogen atom consists of a cloud of positive charge e, uniformly distributed over a sphere of radius R. However, instead of placing the electron in static equilibrium at the center of the sphere, assume that the electron orbits around the center with uniform circular motion under the influence of the electric centripetal force $(e^2/4\pi\varepsilon_0)\,(r/R^3)$. If the angular momentum of this orbiting electron is quantized according to Bohr's theory (so that $L = n\hbar$), what are the radii and the energies of the quantized orbits? What are the frequencies of the photons emitted in transitions from one quantized orbit to another? What must be the value of R if at least two orbits are to fit inside this atom?

*25. In our calculation of the energies of the stationary states of hydrogen we pretended that the proton remains at rest. Actually, both the electron and the proton orbit about their common center of mass. Show that the energies of the stationary states, taking into account this motion of the proton, are given by

$$E_n = -\frac{\mu e^4}{2(4\pi\varepsilon_0)^2\hbar^2}\frac{1}{n^2}$$

where

$$\mu = \frac{m_e m_p}{m_e + m_p}$$

(Hint: the electron and the proton move in circles of radii

$$r_e = r\frac{m_p}{m_p + m_e} \quad \text{and} \quad r_p = r\frac{m_e}{m_p + m_e}$$

where r is the distance between the electron and the proton. According to Bohr's theory, the net angular momentum of this system of two particles is quantized, $L = n\hbar$.)

*26. The atom of **positronium** consists of an electron and a positron (or anti-electron) orbiting about their common center of mass. According to Bohr's theory, the net angular momentum of this system is quantized, $L = n\hbar$. What is the radius of the smallest possible circular orbit of this system? What is the wavelength of the photon released in the transition $n = 2$ to $n = 1$?

Section 41.5

27. In principle, Bohr's theory also applies to the motion of the Earth around the Sun. The Earth plays the role of the electron, the Sun that of the nucleus, and the gravitational force that of the electric force.
   (a) Find a formula analogous to Eq. (18) for the radii of the permitted circular orbits of the Earth around the Sun.
   (b) The actual radius of the Earth's orbit is 1.50×10^{11} m. What value of the quantum number n does this correspond to?
   (c) What is the radial distance between the Earth's actual orbit and the next larger orbit?

28. According to classical electrodynamics, an electron in an elliptical orbit emits not only radiation at the orbital frequency, but also radiation at multiples of the orbital frequency; thus, if the orbital frequency is ν, the radiation contains the fundamental frequency ν and also the harmonic frequencies 2ν, 3ν, 4ν, etc. Does this contradict the Correspondence Principle? (Hint: Calculate what frequencies the electron will emit in the transition n to $n - 2$, n to $n - 3$, n to $n - 4$, etc.)

29. What must be the energy of an electron if its wavelength is to equal the wavelength of visible light, about 5500 Å?

30. A photon and an electron each have an energy of 6.0×10^3 eV. What are their wavelengths?

31. An electron microscope operates with electrons of energy 40,000 eV. What is the wavelength of such electrons? By what factor is this wavelength smaller than that of visible light?

32. What is the de Broglie wavelength of a tennis ball of mass 0.060 kg moving at a speed of 1.0 m/s?

33. What is the de Broglie wavelength, in angstroms, of an electron in the ground state of hydrogen? In the first excited state?

34. Use Eq. (16) to derive an expression for the de Broglie wavelength of an electron in the nth state of hydrogen. Show that the circumference of the orbit equals n times the de Broglie wavelength.

Section 41.6

35. Interferometric methods permit us to measure the position of a macroscopic body to within $\pm 10^{-12}$ m (see Section 1.3). Suppose we perform a position measurement of such a precision on a body of mass 0.050 kg. What

uncertainty in momentum is implied by the Heisenberg relation? What uncertainty in velocity?

36. Suppose that the velocity of an electron has been measured to within an uncertainty of ± 1 cm/s. What minimum uncertainty in the position of the electron does this imply?

37. If the position of a parked automobile of mass 2×10^3 kg is uncertain by $\pm 10^{-18}$ m, what is the corresponding uncertainty in its velocity?

38. The nucleus of aluminum has a diameter of 7.2×10^{-15} m. Consider one of the protons in this nucleus. The uncertainty in the position of this proton is necessarily less than 7.2×10^{-15} m. What is the minimum uncertainty in its momentum and velocity?

39. Consider an electron in a circular orbit of quantum number n in a hydrogen atom. The orbit is well defined provided that $\Delta p_y \ll p$ *and* $\Delta y \ll r$. Show that if $n \gg 1$, these requirements are *not* in conflict with the Heisenberg uncertainty relations. Thus, *large* orbits in the hydrogen atom are well defined (this is in accord with the Correspondence Principle).

LASER LIGHT[1]

Man is a visual animal. Man's intake of visual information is about nine times larger than his combined intake of all other kinds of information. It is therefore no wonder that, up to a few years ago, we regarded light as illumination — our main direct use of light was to make our environment visible. The most powerful light sources were searchlights and lighthouses with an output of up to 1 kW.

The invention of the laser in 1958 changed all that. We now have available light beams of fearsome intensity — up to 10^{10} kW in short pulses. Light beams from lasers are now used in industry to cut and weld metal, cloth, and plastic; in medicine to cut and cauterize human tissues; and in plasma research to vaporize and ignite the fuel for fusion reactors. Light has become a tool and, in military applications, it has become a weapon.

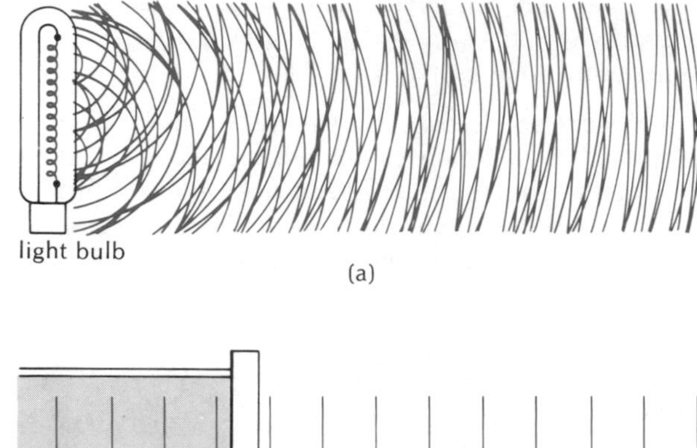

Fig. L.2 (a) Light waves emitted by an ordinary light bulb. (b) Light wave emitted by a laser.

L.1 STIMULATED EMISSION

Light is emitted by the electrons in the atoms of the light source. As we saw in Chapter 41, the electrons in an atom can only move in certain selected, quantized orbits. As long as the electron stays in one of these orbits, it does not emit light. But when the electron jumps from one orbit to a smaller orbit, it emits a packet of light, or photon, with an energy equal to the energy difference between the two orbits (Figure L.1).

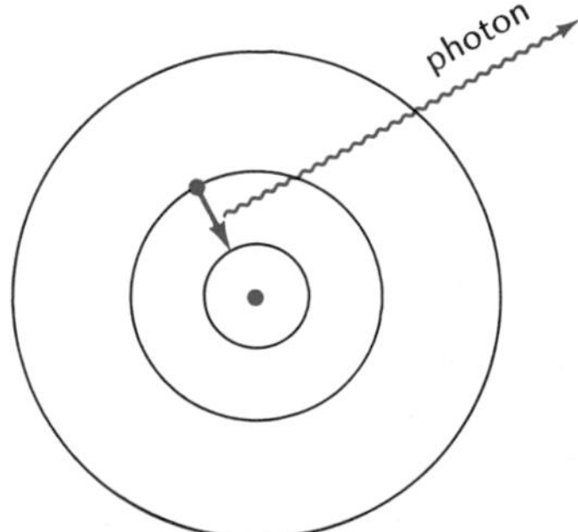

Fig. L.1 The electron in an atom emits a photon during a quantum jump.

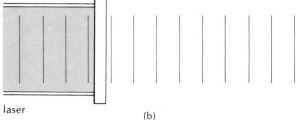

[1] This chapter is optional.

In an ordinary light source, such as a light bulb, the atoms emit their light independently. The atom suffers concussions from thermal collisions and this excites one or more electrons from lower orbits to higher orbits. At some later, unpredictable time the electron spontaneously jumps back down, emitting a pulse of light. Because the atoms emit at random, the light emerging from an ordinary light bulb is a confused combination of many different waves with no special directions of propagation and no special phase relationships (Figure L.2a). On the average, the total intensity of the light is simply the sum of the intensities contributed by the individual atoms; therefore, the total intensity is simply proportional to the number of atoms.

In a laser, the atoms emit their light in unison. The electrons in different atoms either jump down at the same time or else they jump with a time difference of one or several periods of oscillation of the light wave. Furthermore, the electrons emit their light waves in the same direction. The result is that the light emerging from a laser is a **coherent** combination of waves (Figure L.2b). All the light waves from different atoms are in phase — the light wave contributed by each atom combines crest to crest with the light waves contributed by the other atoms. The total amplitude is then proportional to the number of atoms, and therefore the total intensity is proportional to the *square* of the number of atoms. Since the number of atoms in even a fairly small light source is more than 10^{16}, the coherent emission from these atoms can be enormously stronger than the incoherent emission.

What keeps the electrons in different atoms in a laser in step is the phenomenon of **stimulated emission.** Imagine that an excited electron in one of the atoms jumps to its lower orbit, releasing a wave of light. As this wave passes by some other atom with an excited electron, it will jiggle the electron; this causes the electron to resonate and to jump down in unison with the wave instead of waiting to jump spontaneously. The passage of the light wave triggers, or stimulates, the emission of an additional coherent light wave.

The mechanism behind this process involves quantum theory and we cannot present the details here, but it turns out that the additional wave will have the same direction of propagation and the same phase as the original light wave. The two waves therefore combine constructively and they proceed to stimulate the emission of more and more waves from other excited atoms. The process has some features of a chain reaction: whenever an excited atom joins in the emission, it strengthens the wave and thereby increases the probability that the remaining excited atoms will also join in the emission.

The word *laser* is an acronym for *l*ight *a*mplification by *s*timulated *e*mission of *r*adiation. From the above discussion it is obvious that the key to the operation of a laser is the initial presence of a large number of atoms in an excited configuration. This is called a **population inversion** because under normal conditions atoms tend to settle into an unexcited configuration. Indeed, there must be more atoms in the excited configuration than in the unexcited configuration — otherwise the light wave will lose energy by stimulated absorption rather than gain energy by stimulated emission. One method for achieving the required population inversion uses an intense flash of (ordinary) light to lift the electrons into excited orbits; this is called **optical pumping.** Unfortunately, the flash of light that pumps the electrons upward into an excited orbit can also pump them downward; hence direct pumping into an excited orbit does not produce a population inversion. This problem can be circumvented by making the electron jump upward by an indirect route: it first jumps to some high excited orbit and then spontaneously jumps down into a somewhat lower excited orbit, where it remains for some time awaiting stimulated emission (Figure L.3). The lasing action begins when a sufficiently large number of electrons have accumulated in these lower excited orbits.

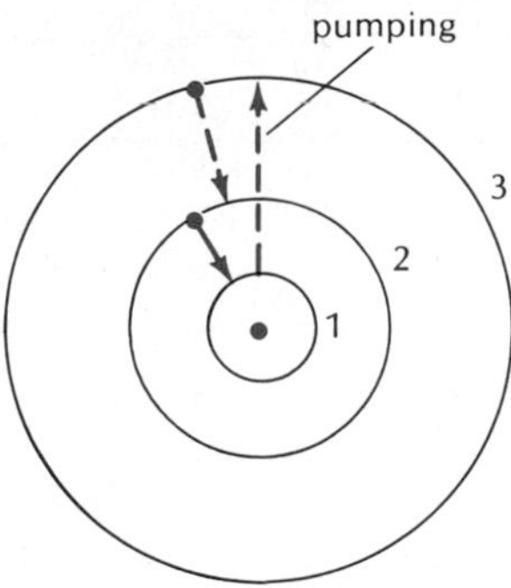

Fig. L.3 Upward and downward jumps of an electron in an atom subjected to optical pumping. A flash of light lifts the electron from the lowest orbit (1) to a high excited orbit (3). From there the electron spontaneously jumps to a somewhat lower excited orbit (2). The lasing action occurs between this orbit and the lowest orbit.

The fundamental theoretical principles involved in the operation of a laser were first published by C. H. Townes and A. L. Schawlow.[2] Their speculations on the generation of coherent light were motivated by an earlier discovery of the generation of coherent microwaves in a maser, a device similar to a laser but operating with radiowaves rather than light waves.

L.2 LASERS

The first laser designed according to the above principles was a **ruby laser** built by T. H. Maiman in 1960 (Figure L.4). This kind of laser, still in common use today, generates light in a long cylindrical crystal of synthetic ruby. The crystal contains corundum, an oxide of aluminum, with a few chromium impurities; the red color of ruby is due to these impurities. In the ruby laser, only the chromium atoms lase. To pump the electrons in these atoms into their excited orbits, the crystal is surrounded by a flash lamp (Figure L.4); the high-intensity light from the flash lamp supplies the energy for the initial upward jump of the electrons. The first electron that engages in the spontaneous emission of light will trigger the stimulated emission of light by other electrons.

In order to ensure that all, or almost all, the excited atoms participate in stimulated emission, it is best to make the light wave traverse the ruby rod several times so that any excited atoms not triggered on the first pass are triggered on a later pass. For this purpose,

[2] Charles H. Townes, 1915–, and Arthur L. Schawlow, 1921–, American physicists. Townes shared the Nobel Prize in 1964 with the Soviet physicists Nicolai G. Basov and Alexander M. Prochorov for the discovery of the maser. Schawlow was awarded the 1981 Nobel Prize for his work on lasers.

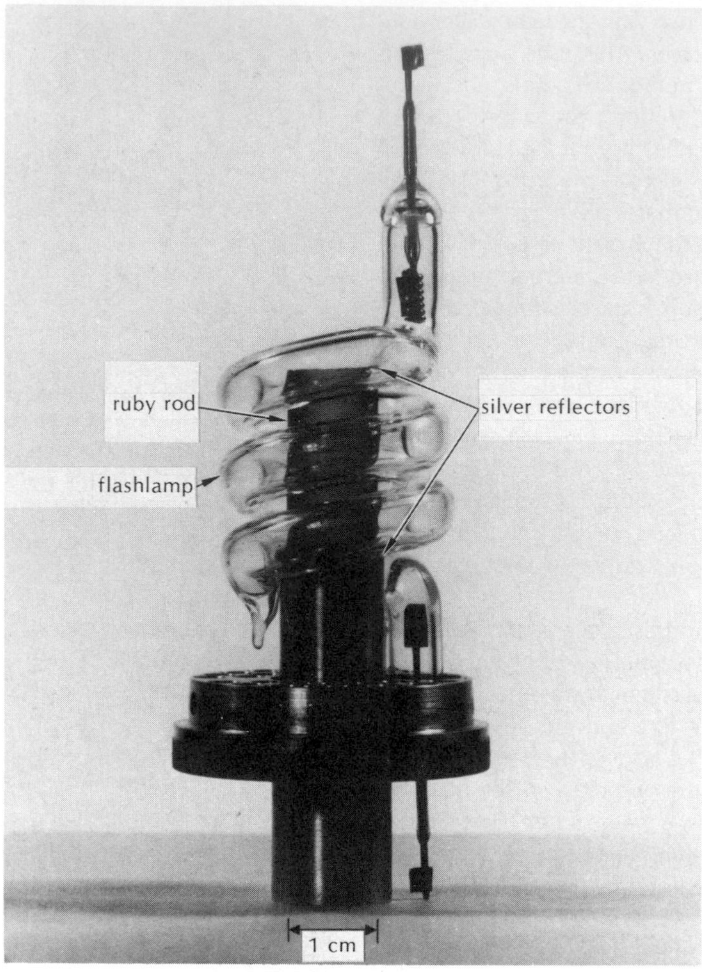

Fig. L.4 The ruby laser of Maiman.

the ends of the rod are polished flat and silvered, making them into mirrors. One end is only partially silvered so that the light can ultimately escape. The back-and-forth reflections drastically enhance the lasing action for those light waves that are emitted exactly parallel to the axis of the rod and that have a wavelength fitting into the length of the rod exactly an integral number of times. For such light waves, the rod acts as a resonant cavity oscillating in one of its normal modes. In this cavity, the light wave is a standing wave and the stimulated emission by the excited atoms feeds more and more energy into this standing wave. Some of the wave leaks out of the partially silvered end of the rod, forming the useful external beam of the laser. The chromium atoms in ruby will lase at one or another of several wavelengths between 6930 Å and 7000 Å, in the red part of the spectrum. The emerging beam is essentially unidirectional since it originates from light waves traveling exactly parallel to the axis of the rod. Typically, the beam of a laser has an angular spread of a minute of arc or less.

The beam from a ruby laser emerges as a pulse lasting as long as excited atoms remain in the rod, about 10^{-3} s. When the atoms become exhausted, they must be pumped again with a new flash from the flash lamp. For some investigations of the effect of intense light on matter, the concentration of power into a succession of short pulses of laser light is very useful. Neodymium glass lasers designed for thermonuclear fusion experiments generate pulses as short as 10^{-12} s with a peak power of 10^9 kW.

However, for many other applications it is desirable to build lasers with a continuous output of light. Obviously, this requires that, on the average, the electrons in the atoms be pumped up into the excited orbits at the same rate as they jump down by stimulated emission. The most commonly employed laser with a continuous output of light is the **helium–neon laser.** It consists of a glass tube with a mixture of helium and neon at low pressure. The tube has a silvered and a partially silvered mirror at the two ends, and also two electrodes (Figure L.5). The lasing action is due to the neon; the helium merely serves to pump the neon. When the electrodes are connected to a high-voltage

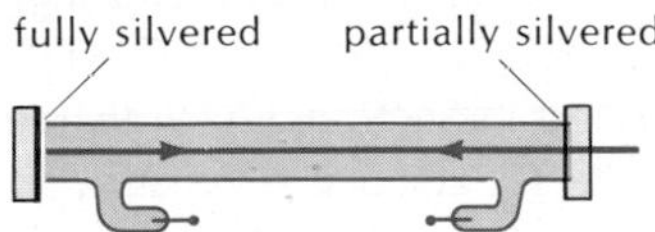

Fig. L.5 Schematic diagram of a helium–neon laser. The terminals are connected to a high-voltage supply.

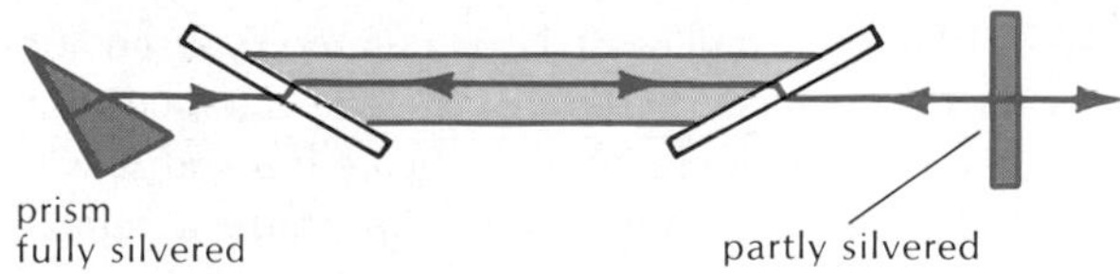

Fig. L.6 Laser with a prism for the selection of one color.

power supply, a current of electrons flows through the tube. These electrons collide with the helium atoms and kick their atomic electrons into excited orbits. The excited helium atoms in turn collide with the neon atoms. In such a collision it is very likely that the helium atom will transfer its excess internal energy to the neon atom. This pumps the electron in the neon into an excited orbit suitable for stimulated emission. The neon atoms will lase at a wavelength of 6328 Å (red light) and also at several wavelengths in the infrared region of the spectrum (there exist several techniques to restrict the lasing action to just a single wavelength; for instance, see below).

A laser capable of very high power output is the **carbon dioxide–nitrogen laser.** This laser operates in much the same way as the helium–neon laser — the carbon dioxide molecules lase and the nitrogen molecules transfer energy to them. Lasers of this kind can easily deliver 10 kW of continuous power. But they can also be operated in a pulsed fashion with a much higher power concentrated in very short pulses. The carbon dioxide laser emits infrared light at several wavelengths, but no visible light. The high power and relatively high efficiency (~30%) attained with this laser is largely due to the long wavelength of the emitted infrared light. The downward energy jump during emission by a molecule is small and, correspondingly, the upward energy jump during pumping is also small. The pumping mechanism can supply these small energy quanta much more easily than the larger energy quanta required for lasers operating with shorter-wavelength visible light.

The resonant cavity of a ruby, helium–neon, or carbon dioxide laser has very many different normal modes and, unless special precautions are taken, the atoms will simultaneously lase in all the normal modes whose frequencies coincide with the frequencies of light generated by jumps from the available excited states. This means that the light emerging from the laser is a mixture of several colors. To obtain truly monochromatic light, a special selective device must be attached to the laser. Figure L.6 shows one of the devices that can be used for this purpose. The rear end of the laser tube is left transparent and a prism with one silvered face is placed beyond it. The prism refracts rays of different colors at different angles. When the multicolored light from the laser penetrates the prism, only one of the rays, of a selected color, strikes the silvered face at a right angle and is reflected back into the laser. Hence only standing waves of one selected color are possible, and lasing will occur only at this one color. The emerging laser beam will then be perfectly monochromatic. Note that by turning the prism through some angle, we can select a different color for lasing. This is the principle of tunable lasers that produce light of adjustable color.

If laser light illuminates a slightly irregular surface — a wall or a sheet of paper — the patch of light on the surface will look grainy to the eye, appearing to have bright specks and dark specks. This is called the **speckle effect** (Figure L.7). The apparent irregularity of the illumination of the wall is an illusion. The laser beam illuminates the entire patch of surface uniformly, but small irregular points of the surface reflect light with somewhat different phases; thus, the incident light is a uniform plane wave, but the reflected light is a superposition of many spherical waves with somewhat different phases. When these waves reach the retina of the eye or a photographic plate, they interfere constructively or destructively, making bright spots and dark spots.

Incidentally, looking directly into a laser beam can damage the eye. This danger is of course obvious in the case of powerful lasers that are capable of burning holes in metals, but the danger subsists even in the

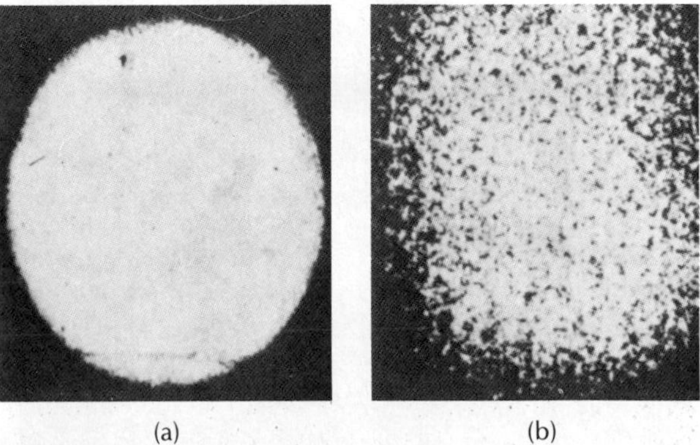

Fig. L.7 Bright spot made by a beam of light on a cement surface: (a) ordinary light from a mercury arc and (b) laser light from a helium–neon laser.

case of the small milliwatt laser commonly used for optical experiments. The lens of the eye can focus the laser beam on a single spot of the retina and it will then not take very much power to cause a retinal burn.

L.3 SOME APPLICATIONS

Laser light is useful because of its directionality, intensity, pure color, and coherence. The applications of laser light exploit one or several of these characteristics.

Surveys and Distance Measurements

Because of its sharp directionality, the beam from a laser is a very convenient tool for laying out straight lines over large distances. The traditional and tedious method for doing this involves setting up a row of markers along the line of sight of a survey telescope. If we replace the telescope by a laser, we can dispense with the cumbersome markers since the laser beam itself can serve as a marker. For example, to dig a straight horizontal trench, we can aim a laser beam parallel to the ground and proceed to dig in its direction; we can check the depth of the trench by intercepting the beam with a vertical meter stick resting against the bottom of the trench. Laser beams can similarly be used in the construction of large aircraft and ships to check the alignment of ribs and frames.

Laser pulses have found application in rangefinders, especially for military purposes. The rangefinder consists of a laser and a light detector, both linked to a timing device. The laser sends a short pulse of light out to the target and the target reflects part of this pulse back. The travel time for this round trip indicates the distance.

As was mentioned in Section 37.2, this method has also been used with spectacular success in an accurate determination of the Earth–Moon distance. Both the Apollo 11 and the Apollo 14 astronauts installed corner reflectors on the surface of the Moon during their missions (Figure L.8). By means of a telescope at MacDonald Observatory, Texas, a pulse from a powerful laser was aimed at one of these reflectors. By the time the pulse reached the Moon, it had spread out to a diameter of about 3 km, but enough light was reflected back to the Earth to be picked up by a sensitive phototube. The precise measurement of the elapsed time for the round trip gave the distance to the Moon to within about 15 cm, i.e., about nine significant figures!

Light-Wave Communications

As we saw in Section 37.3, light pulses can be piped through long, thin glass fibers. If the sound of the human voice is encoded as a series of light pulses, then these optical fibers can be used to transmit telephone conversations. In commercial telephone installations of this kind, the light pulses are generated either by small lasers or by light-emitting diodes. The laser light is much more monochromatic than the diode light, and this is an advantage because it helps to preserve the shape of the light pulses. If a light pulse containing a mixture of colors is sent along an optical fiber, the dispersion of the medium will cause the short-wavelength colors to fall behind and the long-wavelength colors to get ahead; this tends to

Fig. L.8 Corner reflector placed on the surface of the Moon by the Apollo 14 astronauts.

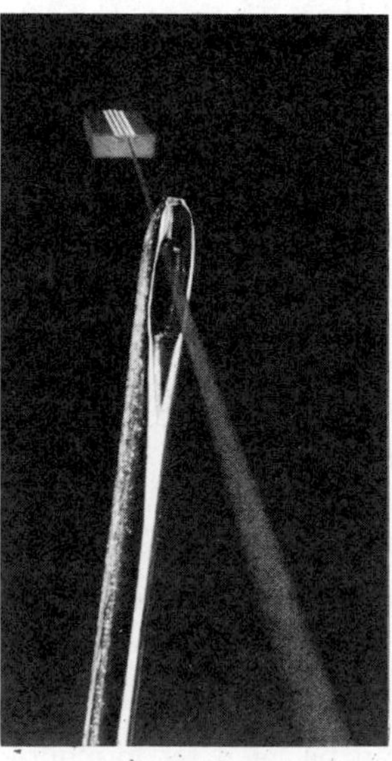

Fig. L.9 Solid-state laser.

Fig. L.10 Robot laser welding machine at an automobile assembly line.

Fig. L.11 Laser beam cutting a thick steel plate.

spread the pulse out, making it indistinct. For laser light the spreading effect is about 10 times smaller than for diode light.

The lasers that generate the pulses for such telephone installations are solid-state devices made of a sandwich of semiconductor, roughly the size of a grain of salt (Figure L.9); their power is about 0.5 mW. In an optical telephone line, each laser feeds 5×10^7 pulses per second into its attached optical fiber; this is enough to encode 672 one-way telephone speeches.

Laser as Torch

Many applications take advantage of the high concentration of energy in laser light to melt, weld, cut, or vaporize materials. For industrial applications, a carbon dioxide laser with a power of several kilowatts makes an excellent welding torch; it has a high welding speed, requires little or no filler metal, and produces joints of excellent quality. Figure L.10 shows a laser welding system in service at an automobile factory. High-power lasers are also used to cut holes in very hard materials, e.g., steel (Figure L.11). In the clothing industry, laser beams are now used to cut out patterns in stacks of several hundreds of layers of cloth all at once.

In medicine, lasers have been used for many years for surgery of the eye. A laser beam aimed through the (transparent) lens of the eye and focused on the retina can "weld" a detached patch of retina into its proper position. In some other surgical operations, a laser beam can serve as a scalpel; this is particularly suitable for operations on blood-rich tissues, such as the liver, where the immediate cauterization of the tissues by the laser beam prevents excessive bleeding.

Laser Inteferometry

Lasers make excellent light sources for interference experiments. As we saw in Section 38.4, when an ordinary light source is used to illuminate the slits or holes, a shield with a pinhole must be placed around the light source so as to select waves of a single direction from the confused combination of waves produced by the light source; this, of course, entails a drastic reduction of the light intensity. The light from a laser already has a sharply defined direction and no pinhole is required.

In an interferometer, such as that of Michelson, the great advantage of laser light is its large-scale coherence. An ordinary light beam, even when selected in one direction, is only coherent over a length of a few meters. This is so because the individual wave trains of light emitted by individual atoms are only about that long. Hence portions of the light beam separated by more than a few meters consist of waves produced by different atoms; these waves are not coherent — they have no regular phase relationships and they will not produce interference patterns (Figure L.12a). When the light beam is split into two by a half-silvered mirror, it will display interference patterns only if each portion travels roughly the same distance to the place of recombination, because only then will the wave pulse of a given atom be able to interfere with itself. On the other hand, in the light beam of a laser all the waves are coherent and the entire beam is a single coherent wave (Figure L.12b). Therefore any segment of the wave train will display interference with any other segment when they are brought together.

The light from ordinary lasers is nearly monochromatic, but not perfectly monochromatic because the thermal motion of the emitting atoms gives the light a fluctuating Doppler shift. To eliminate or reduce these Doppler shifts, scientists have developed **stabilized lasers** in which a feedback device monitors the wave-

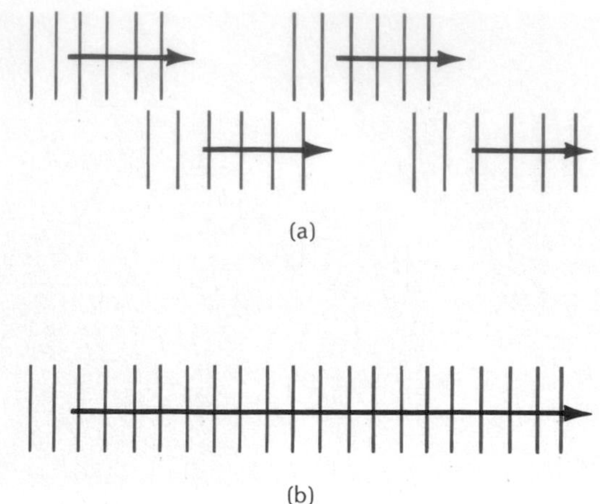

Fig. L.12 (a) The waves of light produced by different atoms are not coherent. (b) The waves produced by a laser are coherent.

Fig. L.13 Several of the lasers used at the National Bureau of Standards for a precise comparison of frequencies. (Courtesy K. M. Evenson, National Bureau of Standards.)

length. A laser of this kind can serve as an excellent standard of length and also of frequency.

As mentioned in Chapter 1, stabilized lasers are used to implement the new definition of the meter. For this purpose, the frequency of the laser light must be measured in terms of the frequency of the cesium atomic clock that serves as our standard of time. The wavelength of the laser light can then be calculated by multiplying the inverse of the frequency by the standard value of the speed of light specified in the definition of the meter, $c = 2.99792458 \times 10^8$ m/s. However, it is difficult to measure the frequency of a laser with precision: even for a long-wavelength laser (infrared laser) the frequency is several orders of magnitude higher than that of the cesium clock, and direct comparison is therefore impossible. Scientists at the National Bureau of Standards overcame this obstacle by using several pairs of lasers of different frequencies and cleverly exploiting beat frequencies and harmonic frequencies generated when the light waves struck suitable detectors of light (diodes). With a chain of four pairs of lasers (Figure L.13), they were able to span the range of frequencies from the cesium atomic clock to a methane-stabilized laser, and thereby determine the frequency of the laser light to nine significant figures.

Laser Spectroscopy

The development of tunable lasers has led to spectacular advances in spectroscopy. **Tunable lasers** operate with molecules of organic dyes, which emit a large number of closely spaced spectral lines; each of the spectral lines contains a spread of frequencies[3] and therefore overlaps to some extent with the adjacent spectral lines. Thus the frequencies available for lasing in an organic dye span a continuous range, and the selective device attached to the laser — such as the prism described in Section L.2 — can be used to tune the laser to any chosen frequency in this range. When the beam from the laser strikes a sample of atoms, it will trigger resonant quantum jumps of the electrons, provided that the frequency of the laser light coincides with the frequency of a spectral line of these atoms. Thus, the laser light will be strongly absorbed if, and only if, its frequency coincides with that of an atomic spectral line; this provides a sensitive and convenient method for the measurement of the frequency of the spectral line. One of the great advantages of this method is that, with a clever arrangement of two laser beams passing through the sample in opposite directions, it is possible to selectively trigger absorption in only those atoms of the sample that have zero velocity. This eliminates the Doppler shift usually present in spectral lines, and therefore permits an extremely precise measurement of their frequencies. Recent measurements of the frequencies of the spectral lines of hydrogen atoms by this technique have led to the best available determination of the Rydberg constant.

[3] All spectral lines contain some small spread of frequencies; in organic dyes, this spread is exceptionally large.

Laser Fusion

The most powerful lasers have been developed in an attempt to extract energy from the thermonuclear fusion of heavy hydrogen (deuterium and tritium). At a temperature of 10^8 K, the violent thermal collisions between the nuclei of heavy hydrogen merge them together into helium. This reaction releases a large amount of heat (see Interlude J for details of this reaction).

In the laser-fusion reactors now under develop-

Fig. L.14 Neodymium lasers at Lawrence Livermore Laboratory.

Fig. L.15 Target chamber at Lawrence Livermore Laboratory. Laser beams, within tubes, converge toward the center of this chamber.

ment, a small pellet of deuterium–tritium mixture is dropped into a combustion chamber where the intense beam of a very powerful laser is suddenly focused on it. The sudden influx of energy not only heats the pellet to 10^8 K, but also compresses its inner core because the sudden vaporization of the outer layers generates high pressure, which causes these layers to explode outward and implode inward. This increases the density of the core by a factor of about 1000. The high temperature and high density ignite thermonuclear reactions within the pellet. These reactions go on for a short time; then the pellet flies apart. Thus, the fusion energy is released in a miniature explosion — the pellet acts as a miniature H-bomb. The essential difference between the H-bomb and the pellet is the method of ignition; fusion in the H-bomb is ignited by an A-bomb, whereas fusion in the pellet is ignited by the laser beam.

An experimental laser fusion device, called **Shiva,** at the Lawrence Livermore Laboratory employs 20 neodymium lasers focused into a target chamber (Figures L.14 and L.15). The lasers deliver a total power of more than 2×10^{10} kW in a short pulse. The radiant energy strikes a deuterium–tritium pellet about 0.1 mm in diameter. Figure L.16 shows some of these pellets; they are small glass balloons filled with deuterium and tritium under high pressure; each holds an energy equivalent to that of a barrel of oil. Figure L.17 shows

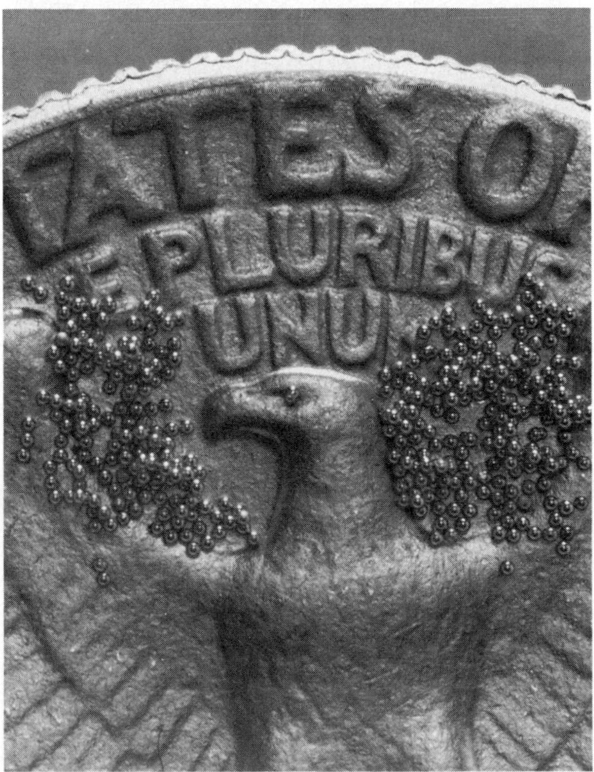

Fig. L.16 Small glass balloons, or microballoons, filled with a mixture of deuterium and tritium.

the compression of one of these pellets. The Shiva device is able to trigger miniature thermonuclear explosions, but the energy output is below the energy input required by the lasers. Livermore will build a new device, called **Nova,** with lasers 10 times more powerful. This device is expected to reach the break-even point — the energy output of the explosions is expected to equal the energy input of the lasers. Of course, it will be necessary to go beyond this point and it will also be necessary to build accessories that convert the heat from the fusion into electric energy.

L.4 HOLOGRAPHY

The large-scale coherence and high intensity of laser light has made it possible to take three-dimensional photographic pictures of objects. The photographic plate, or **hologram,** is prepared by illuminating the object with laser light. Figure L.18 shows the arrangement of light source, object, and photographic plate. The laser beam is split into a reference beam and an object beam, and these are allowed to interfere at the photographic plate. The plate records the pattern of interference fringes (Figure L.19). After developing the plate, we shine laser light on it. If we then view the plate from the far side, we will see an exact replica of the original object. The image is three dimensional; by moving our eye up, down, or sideways we can bring the top, bottom, or sides of the object into view (Figure L.20). But, of course, the image is virtual — if we try to touch the ghostly thing behind the hologram with a finger, we find that nothing is there.

The hologram is produced by the interference between the complicated wave fronts emerging from the

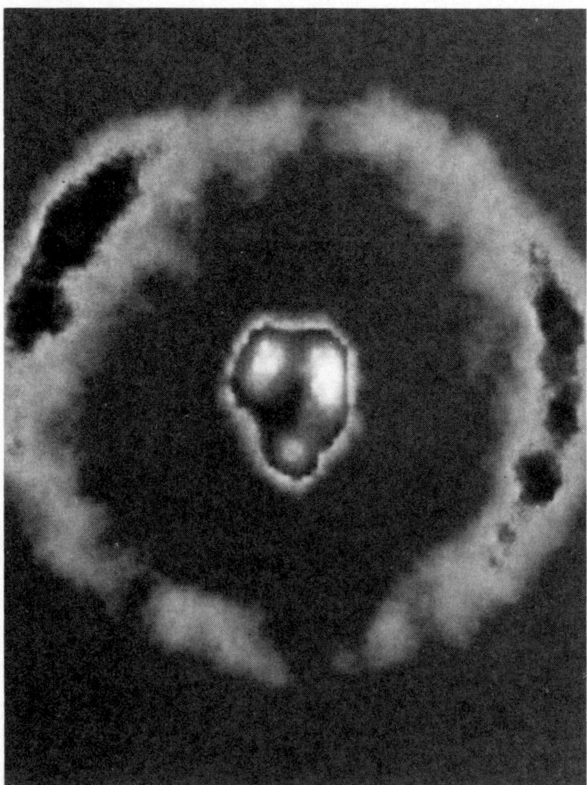

Fig. L.17 X-ray picture of a deuterium–tritium fuel pellet that has been compressed by an intense pulse of light from the Shiva laser at Livermore. The pellet has reached a temperature of 10^7 K and it shines in its own X-ray light. Fusion reactions are occurring within the pellet.

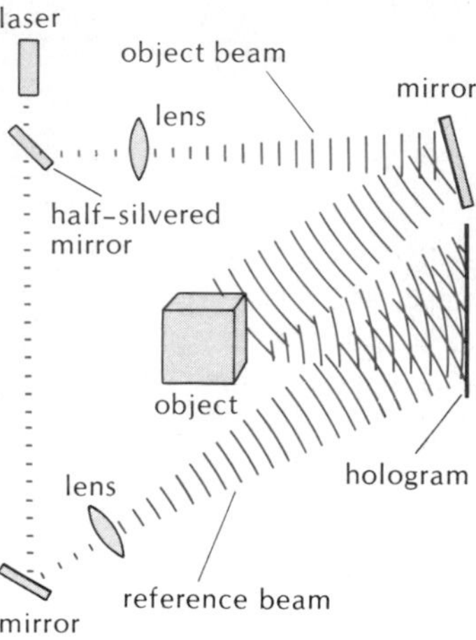

Fig. L.18 Arrangement for making a hologram. All points of the illuminated object act as sources of spherical waves. For the sake of simplicity, only one spherical wave originating at one point of the object has been shown.

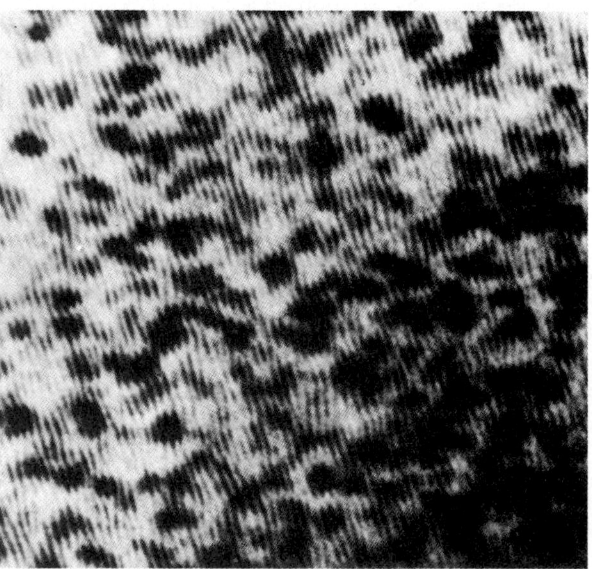

Fig. L.19 A hologram. The dark and the light fringes on this photographic plate record the constructive and the destructive interference of reference and object beams.

illuminated object and the plane wave fronts of the reference beam. To understand how this hologram generates a three-dimensional image, suppose that instead of a complicated object we have a very simple object: a plane mirror (Figure L.21). The wave fronts emerging from the illuminated mirror are then simply plane waves and, at the photographic plate, these plane waves interfere with the plane waves of the reference beam. Obviously, this interference will give a fringe pattern consisting of parallel bright and dark bands. If the angle of incidence of the object beam is α, then the distance d between one dark fringe and the next is related to the wavelength λ by (Figure L.22)

$$d \sin \alpha = \lambda \qquad (1)$$

The developed photographic plate will then look like a grating with a distance d between the slits. If we illuminate this grating with laser light, constructive interference from these slits will yield several diffracted beams of maximum intensity (Figure L.23). The angle of the first-order beam will satisfy the condition [see Eq. (38.21)]

$$d \sin \theta = \lambda \qquad (2)$$

Comparing Eqs. (1) and (2), we recognize that the angle of emergence of the diffracted beam (produced by the illuminated hologram) coincides with the angle of incidence of the original beam (produced by the illuminated mirror). Thus, the hologram *reconstructs* the wave fronts — it generates a light wave that has exactly the same characteristics as the light wave emitted by the object that was photographed. If we place our eyes beyond the hologram (Figure L.20), we will see exactly what we would see if instead of the illuminated hologram we had the illuminated object in front of our eyes.

It can be demonstrated that the same reconstruction of wave fronts obtains if the illuminated object has some complicated shape, so that the emitted wave fronts also have some complicated shape. The illuminated hologram will then generate a wave with exactly

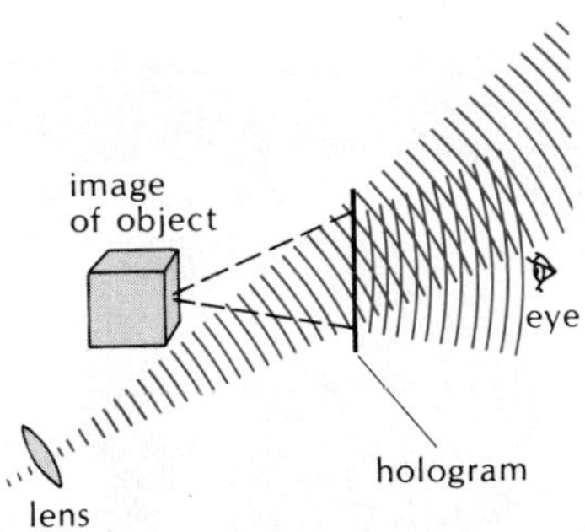

Fig. L.20 Arrangement for viewing a hologram.

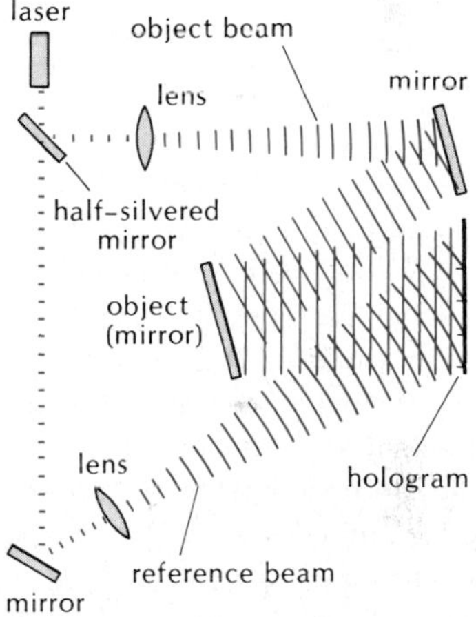

Fig. L.21 Arrangement for making a hologram. The object is a mirror. The marks on the hologram show the points of constructive interference.

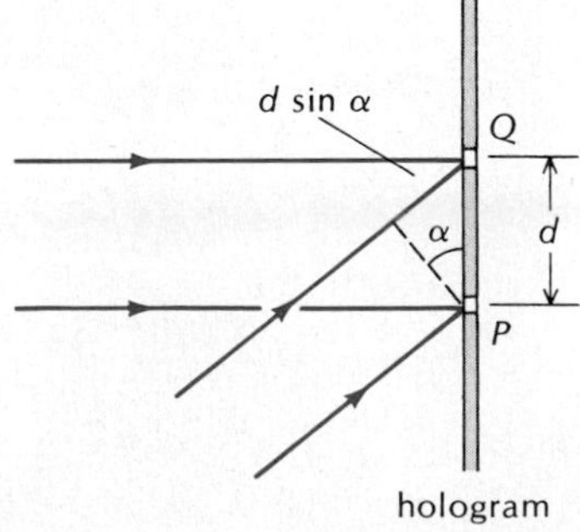

Fig. L.22 The object beam (horizontal) and the reference beam (at an angle α) reaching the photographic plate. If there is constructive interference at P, then there is also constructive interference at Q, provided the extra distance $d \sin \alpha$ equals one wavelength.

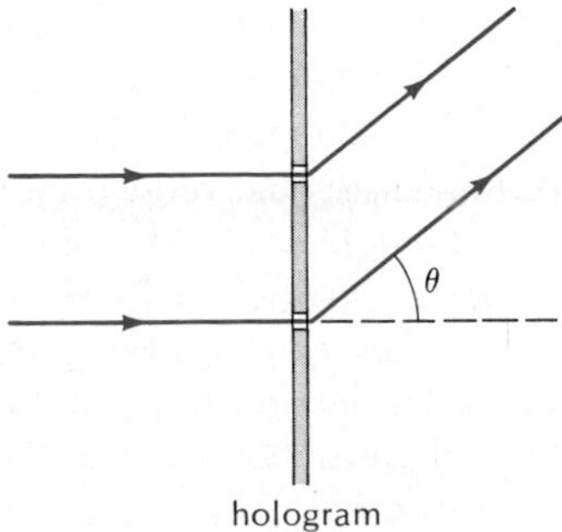

Fig. L.23 Diffracted beam (at an angle θ) emerging from the illuminated hologram.

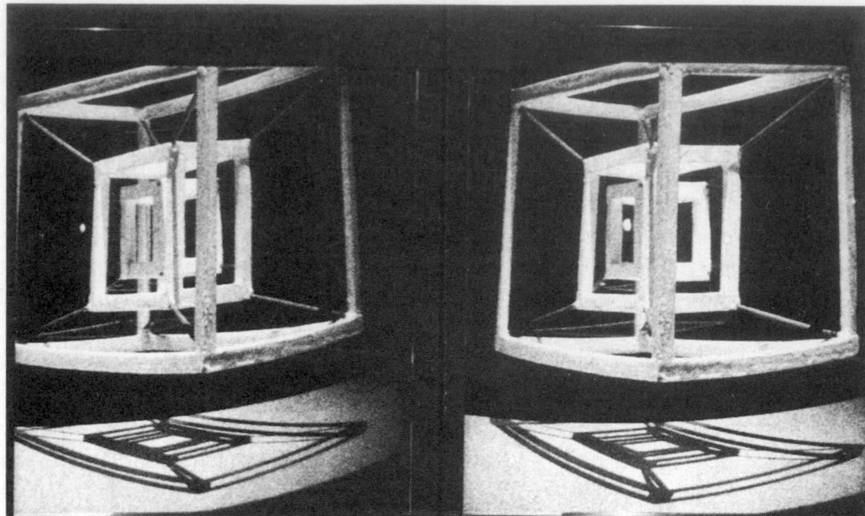

Fig. L.24 The holographic image of a piece of modern sculpture has been photographed from two different directions, giving different points of view. (If you can de-couple your eyes, and view the left image with the left eye and the right image with the right eye, you will be able to perceive these photos as three dimensional.)

the same complicated shape, and our eyes cannot tell the difference between this reconstructed wave and the wave emitted by the object itself — when we view the hologram, we believe we see the object with all its three-dimensional features (Figure L.24).

A remarkable aspect of this imaging process is that each piece of a hologram contains enough information for the reconstruction. We can cut a hologram in two or more pieces and each piece will retain the capability of giving an entire view of the three-dimensional image. However, small pieces provide a view from only a narrow range of directions, like a view through a small window.

Note that, in principle, we could make a hologram with light from an ordinary monochromatic light source (e.g., a mercury lamp) instead of light from a laser. As in the investigation of interference and diffraction by thin slits, such a light source must be placed either at a very large distance from the object or else covered with a shield with a single pinhole (see Section 38.4). But this would reduce the intensity of the light and would require very long and very impractical exposure times. Even with light from a laser, the exposure times for holography are much longer than for photography. Because of this, it is usually necessary to take special precautions to hold the object stationary during exposure.

Holographic images make spectacular displays for demonstrations, but they also have many practical applications. They can be used for sensitive measurements of small deformations and displacements. For

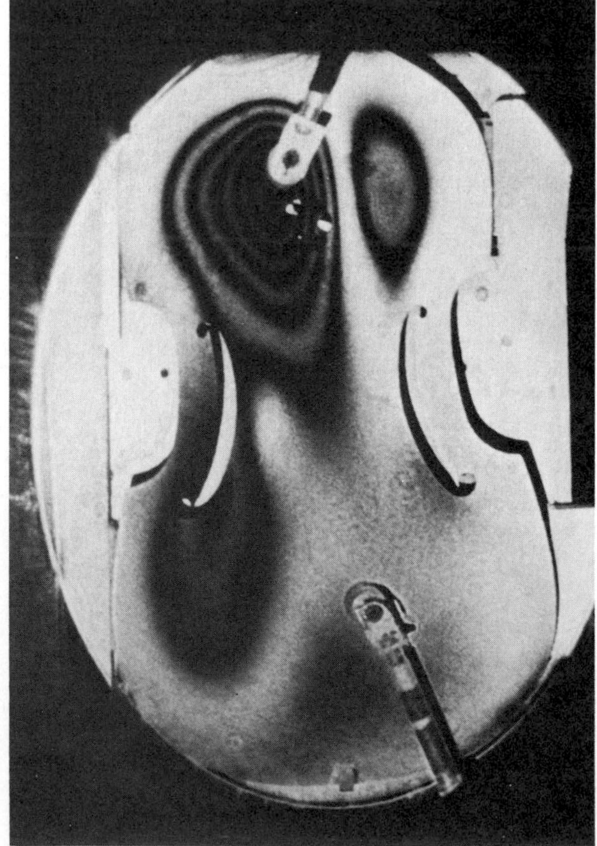

Fig. L.25 Image of the vibrating top plate of a viola generated by a time-exposure hologram.

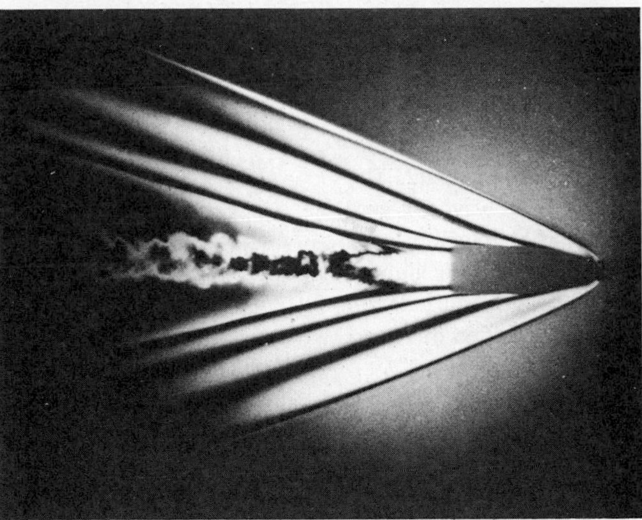

Fig. L.26 Image of shock waves around a bullet in flight, generated by a doubly exposed hologram.

example, Figure L.25 is a holographic image of the vibrating body of a viola. The interference fringes indicate contours of equal height and therefore reveal the pattern of vibration of the viola. The exposure time in Figure L.25 was much longer than the period of vibration of the viola; since a vibrating body spends most of its time at the turning points of the motion (where the velocity is zero), a time exposure is then essentially equivalent to a double exposure — one exposure at each turning point. When we illuminate such a doubly exposed hologram, it will reconstruct simultaneously the wave fronts emitted by the body at the two separate instants of exposure. These two wave fronts will interfere constructively or destructively, depending on how far the surface of the body has moved between one exposure and the next; thus, in the holographic image, the displacement of the surface becomes visible by interference fringes. Scientists hope that the investigation of the pattern of vibration of the viola will lead to improvements in its loudness, dynamic range, and playing ease.

A double-exposure hologram can also make visible small alterations of the density of air. Figure L.26 shows the shock waves surrounding a bullet speeding through air. The hologram was first exposed before the bullet entered the field of view, and then again when the bullet arrived. The compression and rarefaction of the air near the shock fronts alters the index of refraction and hence changes the phase of light between one exposure and the next. When the reconstructed wave fronts from this hologram interfere, the variations of index of refraction show up as bright and dark zones. Incidentally, the exposures were made with high-intensity laser pulses of very short duration so as to "stop" the action at one instant.

Another very promising application of holography is information storage. Obviously, we can store the information of, say, a printed page by taking a holographic photograph of the page. The hologram can store the information at high density because it can be made as small or smaller than an ordinary microfilm photograph, and it has the further advantage that specks of dust or small defects in the film will not block out the words on the page (remember that each part of the hologram can reconstruct the entire image). In principle, a three-dimensional hologram would permit the most efficient storage of information. In a three-dimensional medium, such as a thick photographic emulsion or a light-sensitive crystal, many different holograms can be stored successively by means of multiple exposures with light waves incident at different angles. To recover any one of the holographic images, it is necessary only to illuminate the medium with a light wave incident at the same angle as the original wave. About 1000 different holograms could be stored in a layer about 1 mm thick. With the right technology, it should be possible to store an entire 3D movie in a cube of transparent material as small as a cube of sugar.

Holography was originally invented by D. Gabor[4] as a means for increasing the power of microscopes. Gabor intended to photograph holographically the image produced by a microscope with light of short

[4] Dennis Gabor, 1900–1979, British physicist. He received the Nobel Prize in 1971 for the discovery of the principles of holography.

wavelength and afterward illuminate the hologram with light of long wavelength. The reconstructed image would then be larger than the original image in the ratio of the wavelengths. This technique of holographic microscopy was employed to make the photograph of the neon atom shown in Figure 22.2. The atom was first illuminated with electron waves of very short wavelength,[5] which generated a hologram on a photographic plate. The hologram was then illuminated with visible light, of a wavelength much larger than that of the electron waves. This yielded the enormous magnification in Figure 22.2.

Further Reading

Lasers and Holography by W. E. Kock (Doubleday, Garden City, 1969) is an elementary introduction to coherent light, its generation by lasers, and its application in holography. *The Amazing Lazer* by B. Bova (Westminster, Philadelphia, 1971) is another elementary, and very readable, introduction.

Engineering Applications of Lasers and Holography by W. E. Kock (Plenum Press, New York, 1975) and *Laser* by J. Hecht and D. Teresi (Ticknor & Fields, New Haven, 1982) give broad surveys of many practical applications of lasers.

The following magazine articles describe lasers and their applications:

"Laser Light," A. L. Schawlow, *Scientific American*, September 1968

"Applications of Laser Light," D. R. Herriott, *Scientific American*, September 1968

"The Lunar Laser Reflector," J. E. Faller and E. J. Wampler, *Scientific American*, March 1970

"Progress in Holography," E. N. Leith and J. Upatnieks, *Physics Today*, March 1972

"Metal-Vapor Lasers," W. T. Silfvast, *Scientific American*, February 1973

"Ultrafast Phenomena in Liquids and Solids," R. R. Alfano and S. L. Shapiro, *Scientific American*, June 1973

"Laser-induced Thermonuclear Fusion," J. Nuckolls, J. Emmet, and L. Wood, *Physics Today*, August 1973

"Fusion Power by Laser Implosion," J. L. Emmet, J. Nuckolls, and L. Wood, *Scientific American*, June 1974

"X-Ray Lasers," G. Chapline and L. Wood, *Physics Today*, June 1975

"White-Light Holograms," E. N. Leigh, *Scientific American*, October 1980

"Processing Materials with Lasers," E. M. Breinan, B. H. Kear, and C. M. Banas, *Physics Today*, November 1976

"Laser Separation of Isotopes," R. N. Zare, *Scientific American*, February 1977

"Light-Wave Communications," W. S. Boyle, *Scientific American*, August 1977

[5] As was pointed out in Chapter 41, on the atomic scale electrons behave as waves; their wavelength is usually much shorter than that of light. The operation of electron microscopes takes advantage of these electron waves.

"Laser Fusion," C. M. Stickley, *Physics Today*, May 1978

"Cosmic Masers," D. F. Dickinson, *Scientific American*, June 1978

"Guided-Wave Optics," A. Yariv, *Scientific American*, January 1979

"Laser Chemistry," A. M. Ronn, *Scientific American*, May 1979

"Counting the Atoms," G. S. Hurst, M. G. Pane, S. D. Kramer, and C. H. Chen, *Physics Today*, September 1980

"Laser Weapons," K. Tsipis, *Scientific American*, December 1981

"Laser Applications in Manufacturing," A. V. La Rocca, *Scientific American*, March 1982

"Laser Annealing of Silicon," J. M. Poate and W. L. Brown, *Physics Today*, June 1978

"The Feasibility of Inertial-Confinement Fusion," J. N. Nuckolls, *Physics Today*, September 1982.

"Lasers and Physics: A Pretty Good Hint," A. L. Schawlow, *Physics Today*, December 1982

The Resource Letter L-1 in the *American Journal of Physics*, October 1981, gives a comprehensive list of sources of information on lasers.

Questions

1. Consider two overlapping waves with equal amplitudes and with a random phase difference. Explain why the superposition of these waves has an average intensity equal to twice the average intensity of each wave.

2. Consider two overlapping waves with equal amplitudes and with equal phases. Explain why the superposition of these waves has an intensity four times as large as the intensity of each individual wave. Does this violate energy conservation?

3. Suppose that a light source consists of 10^{16} atoms. If these atoms radiate coherently, by what factor is the average intensity larger than if they radiate incoherently?

4. Can you guess what the acronym *maser* stands for?

5. Why can the flash of light that pumps electrons upward into excited orbits also pump them downward?

6. It is easiest to attain a population inversion by optical pumping if the lower excited orbit (see orbit 2, Figure L.3) is a **metastable** orbit, i.e., an orbit in which the electron tends to remain for a long time without performing a spontaneous downward jump. Why does this help?

7. Explain how a grating can be used as a color-selective device in a laser, instead of the prism shown in Figure L.6.

8. The beam of a laser can be made visible by blowing chalk dust or smoke in its path. Can this also be done with a beam of ordinary, incoherent light?

9. For which of the applications of laser light described in Section L.3 is the coherence essential? For which is the pure color essential?

10. Because of the pull of gravity, a tightly stretched string

sags downward slightly, whereas a laser beam "sags" upward (however, on the Earth the amount of this "sag" is too small to be measurable). How does this difference arise?

11. The U.S. Department of Defense is trying to develop a laser gun. What might be the advantages and disadvantages of such a gun over a conventional gun? Explain how a cloud would stop the beam of a laser gun, even though it cannot stop the bullet from a conventional gun.

12. One obstacle to shooting intense laser beams through the atmosphere at a distant target is that the beam heats the air, changing its index of refraction. This causes the beam to spread out laterally, an effect called **thermal blooming.** Explain.

13. Could an intense laser beam be used to propel a rocket? For this purpose would a laser beam be any better than a light beam?

14. If the individual wave trains of light emitted by individual atoms have a length of a few meters in space, how long are they in time?

15. The wavelength of the light emitted by a laser can be determined to nine significant figures. If you use the light from this laser in conjunction with an interferometer to measure a displacement of roughly 1 m, what precision can you expect? Assume that the interferometer introduces no additional uncertainties.

16. The frequency of a methane-stabilized laser is 8.84×10^{13} Hz. By what factor is this larger than the frequency of the cesium atomic clock?

17. Suppose that a hologram has been prepared with green light. If you illuminate this hologram simultaneously with green light and with red light, what do you see?

18. Why is the image produced by a very small piece of a hologram somewhat fuzzy? (Hint: A hologram can be regarded as a grating. According to Section 38.5, the angular resolution of a grating increases in direct proportion with the number of slits, that is, with the size of the grating.)

19. The automatic checkout system found in large supermarkets sweeps many beams of laser light over the bottom and sides of a package, so that at least one of the beams is reflected off the product-code stripes with sufficient intensity to register in the light detector. To generate these many beams of light, the system requires a complicated array of mirrors; but instead of actual mirrors, the system uses a hologram. Explain how you could make a hologram that simulates a complicated array of mirrors.

APPENDIX 1: INDEX TO TABLES

Note: In the two-volume edition, all pages after 530 are in Vol. 2 except those in parentheses, which are in Vol. 1. Interlude F is the last in Vol. 1.

Accelerators, some large C-6
Accidents, reactor F-13
Acute radiation doses, effects of B-9
Angular momenta, some 260
Areas 946 (534)
Artificial satellites of the Earth, the first 313
Atmospheric pollutants from fossil fuels F-10
Atoms 972 (560)
Balmer series in the hydrogen spectrum 923
Baryons C-10
Charges, electric, of some particles 537
Charges, electric, of electrons, protons, and neutrons 534
Chemical elements and periodic table 972–73 (560–61)
Chromatic musical scale 393
Comets, some 318
Conductivities, some thermal 491
Conserved quantities, forces and G-4
Constants, fundamental, best values of 970 (558)
Conversion efficiencies, energy F-2
Conversion factors 965–69 (553–57)
Densities of some fluids 447
Derivatives, some 954 (542)
Derived units, names of 964 (552)
Diamagnetic materials, permeabilities of some 749
Dielectric constants of some materials 619
Earth endpaper
Electric charges of some particles 537
Electric charges of protons, electrons, and neutrons 534
Electric fields, some 550
Electric forces (qualitative) 533
Elements, periodic table of 972 (560)
Energies, some 185
Energy and entropy F-2
Energy conversion efficiencies F-2
Energy in fission, distribution of H-5
Energy storage K-10
Entropy, energy and F-2
Expansion, coefficients of 488
Fields and their quanta G-5
Fission, distribution of energy in H-5
Fluids, densities of some 447
Forces and conserved quantities G-4
Forces, some 96
Forces, the four fundamental G-2
Fossil fuels, atmospheric pollutants from F-10
Fossil fuels, worldwide supply of F-4
Friction coefficients 131
Fuels, supply of nuclear F-5
Fuels, worldwide supply of fossil F-4
Fundamental constants, best values of 970 (558)
Fundamental forces, the four 121
Fusion, heats of 493
g, variation with latitude 34
Gases, specific heats of some 496
Greek alphabet endpaper
Heats of fusion and vaporization 493
Hydrogen spectrum, Balmer series 923
Indices of refraction, of some materials 835
Inertia, some moments of 259
Insulators, resistivities of 640
Integrals, some 546
Isotopes, chart of Prelude-17
Jupiter, main moons of 313
Kinetic energies, some 163
Leptons C-9
Lightning data I-10
Magnetic fields, some 673
Magnetic moments of some atoms and ions 741
Mathematical symbols and formulas 944–46 (532–34)
Mesons C-10
Metals, resistivities of 639
Moments, magnetic 741
Moments of inertia, some 259
Moon endpaper
Moons of Jupiter, main 313
Moons of Saturn, some 331
Musical scale, chromatic 393
Nuclear fuels, supply of F-5
Paramagnetic materials, permeabilities of some 743
Particles, subatomic C-4
Penetrating radiations B-3
Perimeters, areas, and volumes 946 (534)
Period of simple pendulum, increase of 349
Periodic table of the elements 972 (560)
Permeabilities of some diamagnetic materials 749
Permeabilities of some paramagnetic materials 743
Planets 313
Plasmas, some J-3
Pollutants, atmospheric, from fossil fuels F-10
Powers, some 190
Prefixes for units 964 (552)
Pressures, some 452
Quanta, fields and their G-5
Quarks C-14
Radiation doses, effects of acute B-9
Radiation doses from diagnostic X rays B-10
Radiation doses per inhabitant of the United States, average annual B-9
Radiations, penetrating B-3
Radioactive sources, strengths of B-5
RBE values B-8
Reactor accidents F-13
Refraction, indices of, of some materials 835
Resistivities of insulators 640
Resistivities of metals 639
Resistivities of semiconductors 640
Satellites of the Earth, the first artificial 313
Saturn, some moons of 331
Semiconductors, resistivities of 640
Simple pendulum, increase of period of 349
Sound intensities, some 392
Sound, speed of, in some materials 395
Specific heats of some gases 496
Specific heats, some 485
Speed of sound in some materials 395
Storage, energy K-10
Subatomic particles C-4
Sun endpaper
Superconductors, some K-2
Temperatures, some 475
Thermal conductivities, some 491
Trigonometric identities 951 (539)
Trigonometric functions, values of 950 (538)
Vaporization, heats of 493
Volumes 946 (534)

APPENDIX 2: MATHEMATICAL SYMBOLS AND FORMULAS

A2.1 Symbols

$a = b$ means a equals b
$a \neq b$ means a is not equal to b
$a > b$ means a is greater than b
$a < b$ means a is less than b
$a \geq b$ means a is not less than b
$a \leq b$ means a is not greater than b
$a \propto b$ means a is proportional to b
$a \cong b$ means a is approximately equal to b
$a \sim b$ means a is of the order of magnitude of b, i.e., a is within a factor of 10 or so of b
$a \gg b$ means a is much larger than b
$a \ll b$ means a is much less than b
$\Sigma_i a_i$ stands for the sum $a_1 + a_2 + a_3 + a_4 + \cdots$
$n!$ (or "n factorial") stands for the product $1 \cdot 2 \cdot 3 \cdots n$
$\pi = 3.14159\ldots$
$\mathrm{e} = 2.71828\ldots$

A2.2 The Quadratic Equation

The quadratic equation $ax^2 + bx + c = 0$ has two solutions:

$$x = \frac{-b \pm \sqrt{b^2 - 4ac}}{2a} \tag{1}$$

A2.3 Some Approximations

The following approximations are valid for small values of x, that is, $x \ll 1$:

$$(1 + x)^n \cong 1 + nx \tag{2}$$

$$(1+x)^{\frac{1}{2}} \cong 1+\frac{x}{2} \tag{3}$$

$$\frac{1}{(1+x)} \cong 1-x \tag{4}$$

$$\frac{1}{(1+x)^{\frac{1}{2}}} \cong 1-\frac{x}{2} \tag{5}$$

These approximations can be derived from the binomial expansion

$$(1+x)^n = 1+nx+\frac{n(n-1)}{2!}x^2+\frac{n(n-1)(n-2)}{3!}x^3+\cdots \tag{6}$$

by neglecting all powers of x, except the first power.

A2.4 The Exponential and the Logarithmic Functions

The **exponential function** $\exp(x)$ is defined by the following infinite series:

$$\exp(x) = 1+x+\frac{x^2}{2!}+\frac{x^3}{3!}+\frac{x^4}{4!}+\cdots \tag{7}$$

This function is equivalent to raising the constant $e = 2.71828\ldots$ to the power x,

$$\exp(x) = e^x \tag{8}$$

The **natural logarithmic function** is the inverse of the exponential function, so that

$$x = e^{\ln x} \tag{9}$$

or

$$x = \ln(e^x) \tag{10}$$

Natural logarithms obey the usual rules for logarithms,

$$\ln(x \cdot y) = \ln x + \ln y \tag{11}$$

$$\ln\left(\frac{x}{y}\right) = \ln x - \ln y \tag{12}$$

$$\ln(x^a) = a \ln x \tag{13}$$

Note that

$$\ln e = 1 \tag{14}$$

and

$$\ln 10 = 2.3026\ldots \tag{15}$$

If we designate the **common,** or base-10, **logarithm** by $\log x$, then the relationship between the two kinds of logarithm is as follows:

$$\ln x = \ln(10^{\log x}) = (\log x)(\ln 10) = 2.3026 \log x \tag{16}$$

APPENDIX 3: PERIMETERS, AREAS, AND VOLUMES

[perimeter of a circle of radius r] = $2\pi r$
[area of a circle of radius r] = πr^2
[area of a triangle of base b, altitude h] = $hb/2$
[surface of a sphere of radius r] = $4\pi r^2$
[volume of a sphere of radius r] = $4\pi r^3/3$
[curved surface of a cylinder of radius r, height h] = $2\pi rh$
[volume of a cylinder of radius r, height h] = $\pi r^2 h$

APPENDIX 4: BRIEF REVIEW OF TRIGONOMETRY

A4.1 Angles

The angle between two intersecting straight lines is defined as the fraction of a complete circle included between these lines (Figure A4.1). To express the angle in **degrees,** we assign an angular magnitude of 360° to the complete circle; any arbitrary angle is then an appropriate fraction of 360°. To express the angle in **radians,** we assign an angular magnitude of 2π radian to the complete circle; any arbitrary angle is then an appropriate fraction of 2π. For example, the angle shown in Figure A4.1 is $\frac{1}{8}$ of a complete circle, that is, 45°, or $\pi/4$ radian. In view of the definition of angle, the length of arc included between the two intersecting straight lines is proportional to the angle θ between these lines; if the angle is expressed in radians, then the constant of proportionality is simply the radius:

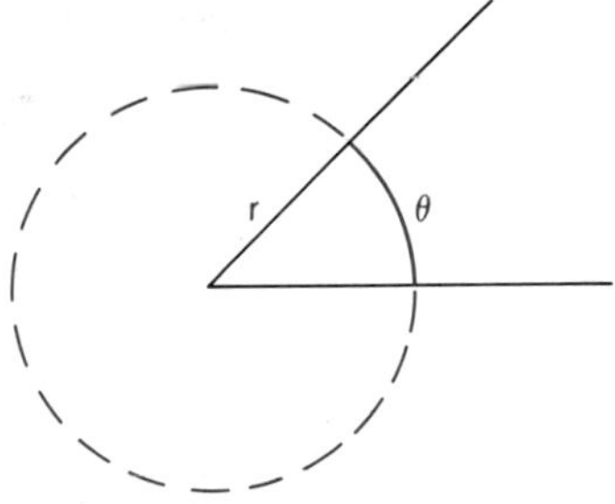

Fig. A4.1

$$s = r\theta \tag{1}$$

Since 2π radian = 360°, it follows that

$$1 \text{ radian} = \frac{360°}{2\pi} = \frac{360°}{2 \times 3.14159} = 57.2958° \tag{2}$$

Each degree is divided into 60 minutes of arc (arcminutes), and each of these into 60 seconds of arc (arcseconds). In degrees, minutes of arc, and seconds of arc, the radian is

$$1 \text{ radian} = 57°\ 17'\ 44.8'' \tag{3}$$

A4.2 The Trigonometric Functions

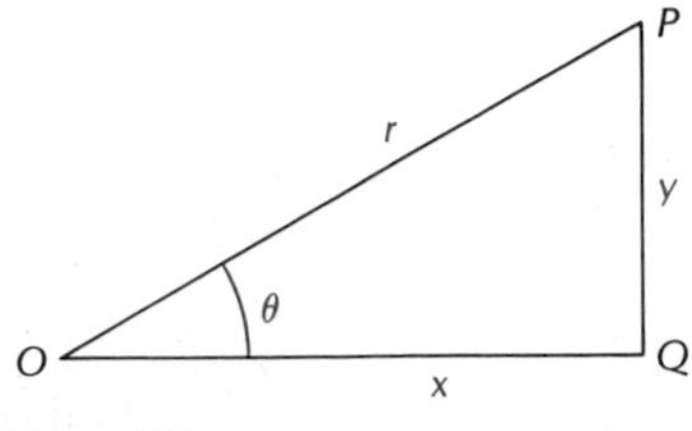

Fig. A4.2

The trigonometric functions of an angle are defined as ratios of the lengths of the sides of a right triangle erected on this angle. Figure A4.2 shows an acute angle θ and a right triangle, one of whose angles coincides with θ. The adjacent side OQ has a length x, the opposite side

QP a length y, and the hypothenuse OP a length r. The **sine, cosine, tangent, cotangent, secant,** and **cosecant** of the angle θ are then defined as follows:

$$\sin \theta = y/r \tag{4}$$

$$\cos \theta = x/r \tag{5}$$

$$\tan \theta = y/x \tag{6}$$

$$\cot \theta = x/y \tag{7}$$

$$\sec \theta = r/x \tag{8}$$

$$\csc \theta = r/y \tag{9}$$

EXAMPLE 1. Find the sine, cosine, and tangent for angles of 0°, 90°, and 45°.

SOLUTION: For an angle of 0°, the opposite side is zero ($y = 0$), and the adjacent side coincides with the hypothenuse ($x = r$). Hence

$$\sin 0° = 0 \qquad \cos 0° = 1 \qquad \tan 0° = 0 \tag{10}$$

For an angle of 90°, the adjacent side is zero ($x = 0$), and the opposite side coincides with the hypothenuse ($y = r$). Hence

$$\sin 90° = 1 \qquad \cos 90° = 0 \qquad \tan 90° = \infty \tag{11}$$

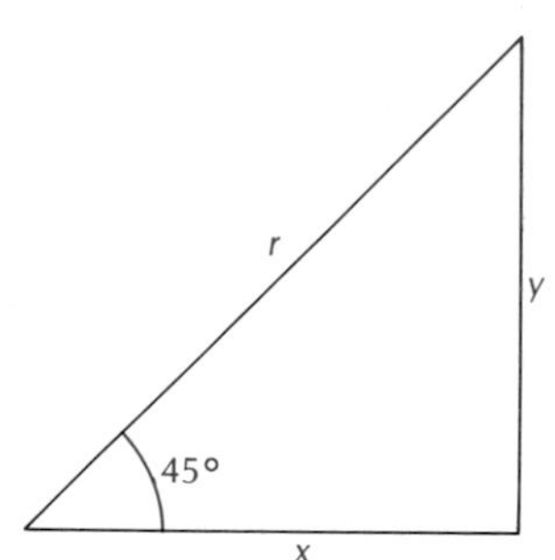

Fig. A4.3

Finally, for an angle of 45° (Figure A4.3), the adjacent and the opposite sides have the same length ($x = y$) and the hypothenuse has a length of $\sqrt{2}$ times the length of either side ($r = \sqrt{2}x = \sqrt{2}y$). Hence

$$\sin 45° = \frac{1}{\sqrt{2}} \qquad \cos 45° = \frac{1}{\sqrt{2}} \qquad \tan 45° = 1 \tag{12}$$

Table A4.1 lists the values of the trigonometric functions sine, cosine, and tangent for angles ranging from 0° to 90°.

The definitions (4)–(9) are also valid for angles greater than 90°, such as the angle shown in Figure A4.4. In the general case, the quantities x and y must be interpreted as the rectangular coordinates of the point P. For any angle larger than 90°, one or both of the coordinates x and y are negative. Hence some of the trigonometric functions will also be negative. For instance,

$$\sin 135° = \frac{1}{\sqrt{2}} \qquad \cos 135° = -\frac{1}{\sqrt{2}} \qquad \tan 135° = -1 \tag{13}$$

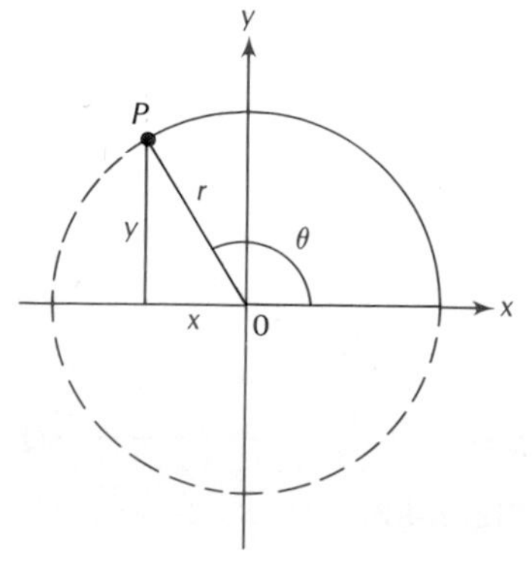

Fig. A4.4

Figure A4.5 shows plots of the sine, cosine, and tangent vs. θ for the range 0°–360°.

If the angle θ is small, say, $\theta < 0.2$ radian, or $\theta < 10°$, then the length of the opposite side is approximately equal to the length of the circular arc (Figure A4.6). If we express the angle in radians, this length is [see Eq. (1)]

$$y \cong s = r\theta \tag{14}$$

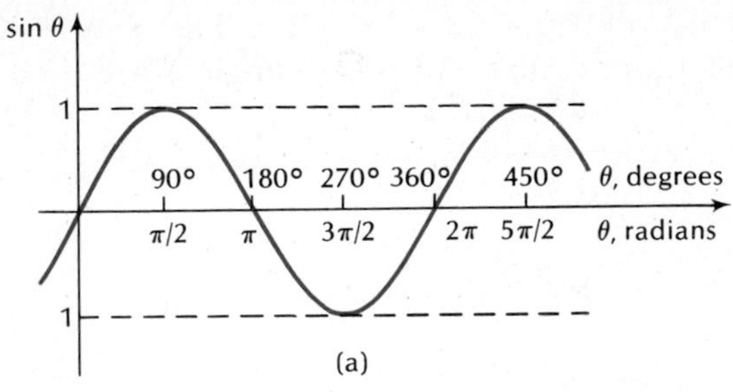

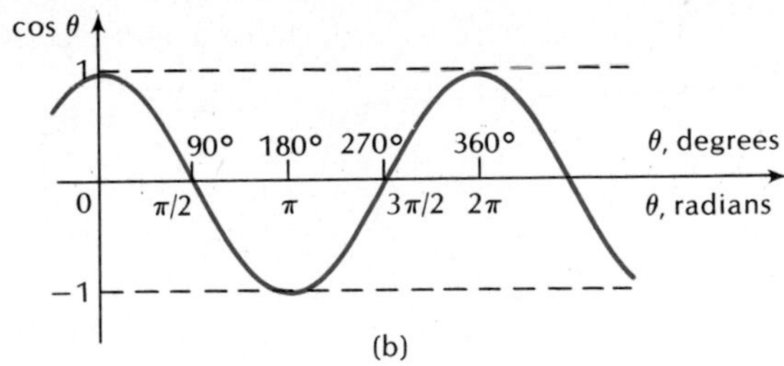

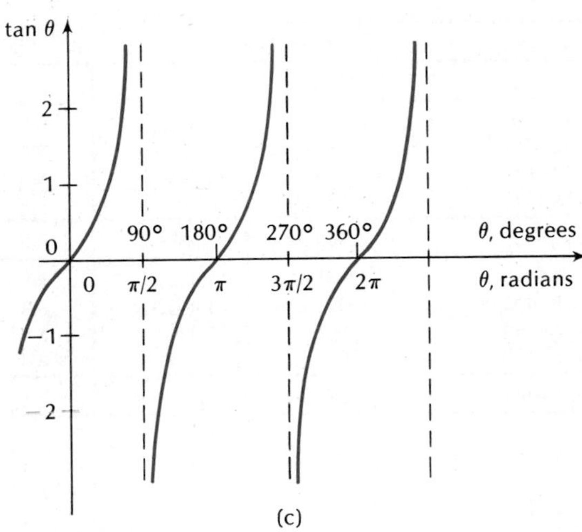

Fig. A4.5

and therefore

$$\sin\theta \cong \theta \tag{15}$$

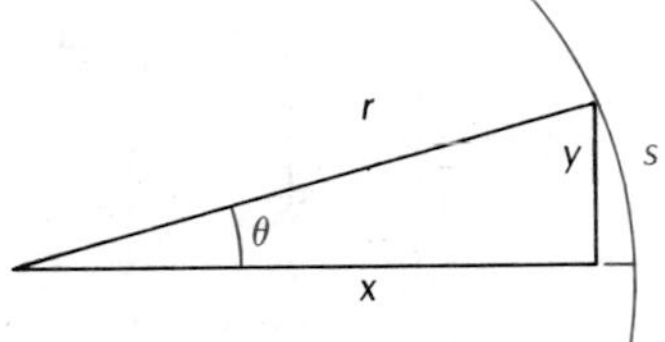

Fig. A4.6

To obtain a corresponding approximation for $\cos\theta$, we use the Pythagorean theorem to express the adjacent side in terms of the hypotenuse r and the opposite side $y \cong r\theta$,

$$x = \sqrt{r^2 - y^2} \cong \sqrt{r^2 - r^2\theta^2} \tag{16}$$

$$\cong r\sqrt{1-\theta^2} \tag{17}$$

For small θ, Eq. (A2.3) gives us the approximation

$$\sqrt{1-\theta^2} \cong 1 - \theta^2/2 \tag{18}$$

so that

$$\cos\theta = x/r \cong 1 - \theta^2/2 \tag{19}$$

When using the approximate formulas (15) and (19), we must always remember that the angle θ is expressed in radians!

The inverse trigonometric functions $\mathbf{sin^{-1}}$, $\mathbf{cos^{-1}}$, and $\mathbf{tan^{-1}}$ give the angle that corresponds to a specified value of the sine, cosine, or tangent, respectively. Thus, if

$$\sin\theta = u$$

then

$$\theta = \sin^{-1} u$$

Table A4.1 VALUES OF TRIGONOMETRIC FUNCTIONS

Degrees	Radians	Sine	Tangent	Cotangent	Cosine		
0	0	0	0	∞	1.0000	1.5708	90
1	.0175	.0175	.0175	57.290	.9998	1.5533	89
2	.0349	.0349	.0349	28.636	.9994	1.5359	88
3	.0524	.0523	.0524	19.081	.9986	1.5184	87
4	.0698	.0698	.0699	14.301	.9976	1.5010	86
5	.0873	.0872	.0875	11.430	.9962	1.4835	85
6	.1047	.1045	.1051	9.5144	.9945	1.4661	84
7	.1222	.1219	.1228	8.1443	.9925	1.4486	83
8	.1396	.1392	.1405	7.1154	.9903	1.4312	82
9	.1571	.1564	.1584	6.3138	.9877	1.4137	81
10	.1745	.1736	.1763	5.6713	.9848	1.3963	80
11	.1920	.1908	.1944	5.1446	.9816	1.3788	79
12	.2094	.2079	.2126	4.7046	.9781	1.3614	78
13	.2269	.2250	.2309	4.3315	.9744	1.3439	77
14	.2443	.2419	.2493	4.0108	.9703	1.3265	76
15	.2618	.2588	.2679	3.7321	.9659	1.3090	75
16	.2793	.2756	.2867	3.4874	.9613	1.2915	74
17	.2967	.2924	.3057	3.2709	.9563	1.2741	73
18	.3142	.3090	.3249	3.0777	.9511	1.2566	72
19	.3316	.3256	.3443	2.9042	.9455	1.2392	71
20	.3491	.3420	.3640	2.7475	.9397	1.2217	70
21	.3665	.3584	.3839	2.6051	.9336	1.2043	69
22	.3840	.3746	.4040	2.4751	.9272	1.1868	68
23	.4014	.3907	.4245	2.3559	.9205	1.1694	67
24	.4189	.4067	.4452	2.2460	.9135	1.1519	66
25	.4363	.4226	.4663	2.1445	.9063	1.1345	65
26	.4538	.4384	.4877	2.0503	.8988	1.1170	64
27	.4712	.4540	.5095	1.9626	.8910	1.0996	63
28	.4887	.4695	.5317	1.8807	.8829	1.0821	62
29	.5061	.4848	.5543	1.8040	.8746	1.0647	61
30	.5263	.5000	.5774	1.7321	.8660	1.0472	60
31	.5411	.5150	.6009	1.6643	.8572	1.0297	59
32	.5585	.5299	.6249	1.6003	.8480	1.0123	58
33	.5760	.5446	.6494	1.5399	.8387	.9948	57
34	.5934	.5592	.6745	1.4826	.8290	.9774	56
35	.6109	.5736	.7002	1.4281	.8192	.9599	55
36	.6283	.5878	.7265	1.3764	.8090	.9425	54
37	.6458	.6018	.7536	1.3270	.7986	.9250	53
38	.6632	.6157	.7813	1.2799	.7880	.9076	52
39	.6807	.6293	.8098	1.2349	.7771	.8901	51
40	.6981	.6428	.8391	1.1918	.7660	.8727	50
41	.7156	.6561	.8693	1.1504	.7547	.8552	49
42	.7330	.6691	.9004	1.1106	.7431	.8378	48
43	.7505	.6820	.9325	1.0724	.7314	.8203	47
44	.7679	.6947	.9657	1.0355	.7193	.8029	46
45	.7854	.7071	1.0000	1.0000	.7071	.7854	45
		Cosine	Cotangent	Tangent	Sine	Radians	Degrees

A4.3 Trigonometric Identities

From the definitions (4)–(9) we immediately find the following identities:

$$\tan\theta = \sin\theta/\cos\theta \tag{20}$$

$$\cot\theta = 1/\tan\theta \tag{21}$$

$$\sec\theta = 1/\cos\theta \tag{22}$$

$$\csc\theta = 1/\sin\theta \tag{23}$$

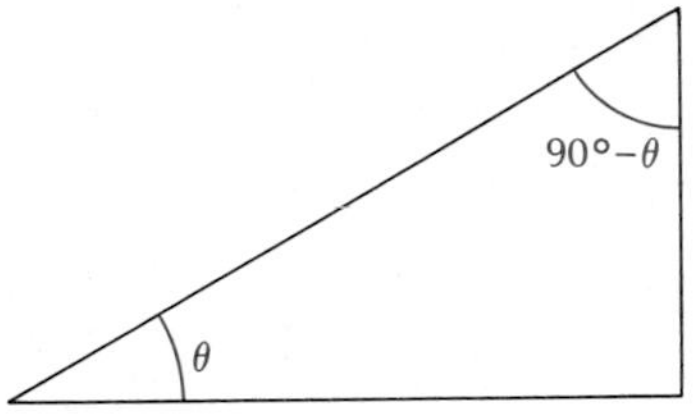

Fig. A4.7

Figure A4.7 shows a right triangle with angles θ and $90° - \theta$. Since the adjacent side for the angle θ is the opposite side for the angle $90° - \theta$ and vice versa, we see that the trigonometric functions also obey the following identities:

$$\sin(90° - \theta) = \cos\theta \tag{24}$$

$$\cos(90° - \theta) = \sin\theta \tag{25}$$

$$\tan(90° - \theta) = \cot\theta = 1/\tan\theta \tag{26}$$

According to the Pythagorean theorem, $x^2 + y^2 = r^2$. With $x = r\cos\theta$ and $y = r\sin\theta$ this becomes $r^2\cos^2\theta + r^2\sin^2\theta = r^2$, or

$$\cos^2\theta + \sin^2\theta = 1 \tag{27}$$

The following are a few other trigonometric identities, which we state without proof:

$$\sec^2\theta = 1 + \tan^2\theta$$

$$\csc^2\theta = 1 + \cot^2\theta$$

$$\sin 2\theta = 2\sin\theta\cos\theta$$

$$\cos 2\theta = 2\cos^2\theta - 1$$

$$\sin\tfrac{1}{2}\theta = \sqrt{(1 - \cos\theta)/2}$$

$$\cos\tfrac{1}{2}\theta = \sqrt{(1 + \cos\theta)/2}$$

$$\sin(\alpha + \beta) = \sin\alpha\cos\beta + \cos\alpha\sin\beta$$

$$\cos(\alpha + \beta) = \cos\alpha\cos\beta - \sin\alpha\sin\beta$$

$$\tan(\alpha + \beta) = \frac{\tan\alpha + \tan\beta}{1 - \tan\alpha\tan\beta}$$

$$\sin\alpha + \sin\beta = 2\sin\tfrac{1}{2}(\alpha + \beta)\cos\tfrac{1}{2}(\alpha - \beta)$$

$$\cos\alpha + \cos\beta = 2\cos\tfrac{1}{2}(\alpha + \beta)\cos\tfrac{1}{2}(\alpha - \beta)$$

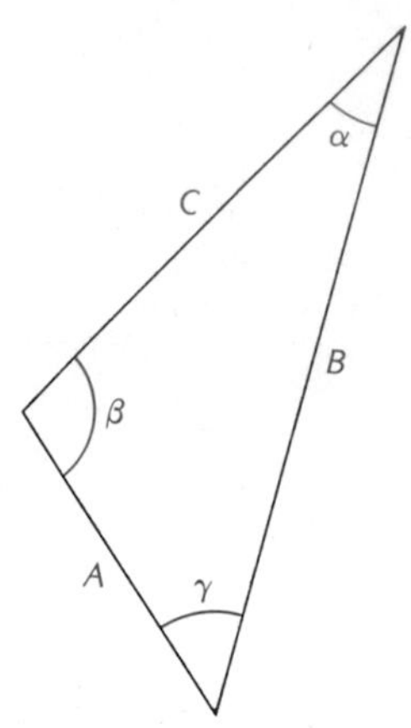

Fig. A4.8

A4.4 The Laws of Cosines and of Sines

In an arbitrary triangle the lengths of the sides and the angles obey the laws of cosines and of sines. The law of cosines states that if the lengths of two sides are A and B and the angle between them is γ (Figure A4.8), then the length of the third side is given by

$$C^2 = A^2 + B^2 - 2AB\cos\gamma \tag{28}$$

The law of sines states that the sines of the angles of the triangle are in the same ratio as the lengths of the opposite sides (Figure A4.8):

$$\frac{\sin\alpha}{A} = \frac{\sin\beta}{B} = \frac{\sin\gamma}{C} \tag{29}$$

Both of these laws are very useful in the calculation of unknown lengths or angles of a triangle.

A5.1 Derivatives

We saw in Section 2.3 that if the position of a particle is some function of time, say $x = x(t)$, then the instantaneous velocity of the particle is the derivative of x with respect to t,

$$v = \frac{dx}{dt} \tag{1}$$

This derivative is defined by first looking at a small increment Δx that results from a small increment Δt, and then evaluating the ratio $\Delta x/\Delta t$, in the limit when both Δx and Δt tend toward zero. Thus

$$\frac{dx}{dt} = \lim_{\Delta t \to 0} \frac{\Delta x}{\Delta t} \tag{2}$$

Graphically, in a plot of the worldline, the derivative dx/dt is the slope of the straight line tangent to the (curved) worldline at the time t (see Figure 2.6).

In general, if $f = f(u)$ is some given function of a variable u, the **derivative** of f with respect to u is defined by

$$\frac{df}{du} = \lim_{\Delta u \to 0} \frac{\Delta f}{\Delta u} \tag{3}$$

In a plot of f vs. u, this derivative is the slope of the straight line tangent to the curve representing $f(u)$ (see Figure A5.1).

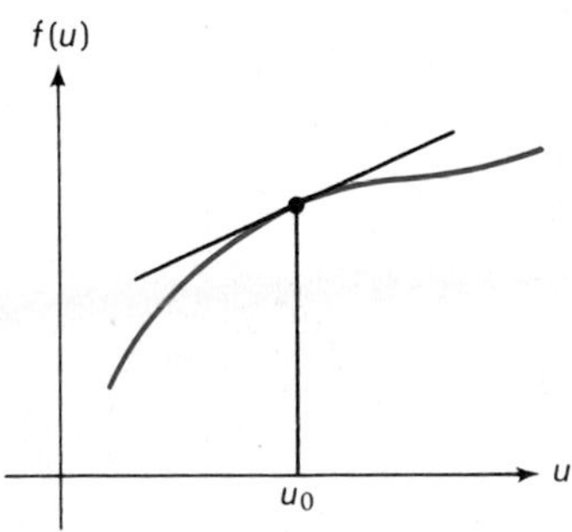

Fig. A5.1 The derivative of $f(u)$ at u_0 is the slope of the straight line tangent to the curve at u_0.

Starting with the definition (3) we can find the derivative of any function (provided the function is sufficiently smooth so that the derivative exists!). For example, consider the function $f(u) = u^2$. If we increase u to $u + \Delta u$, the function $f(u)$ increases to

$$f + \Delta f = (u + \Delta u)^2 \tag{4}$$

so that

$$\begin{aligned}\Delta f &= (u + \Delta u)^2 - f = (u + \Delta u)^2 - u^2 \\ &= 2u\,\Delta u + (\Delta u)^2\end{aligned} \tag{5}$$

The derivative df/du is then

$$\frac{df}{du} = \lim_{\Delta u \to 0} \frac{\Delta f}{\Delta u} = \lim_{\Delta u \to 0} \frac{2u\,\Delta u + (\Delta u)^2}{\Delta u} \tag{6}$$

$$= \lim_{\Delta u \to 0} (2u) + \lim_{\Delta u \to 0} (\Delta u) \tag{7}$$

The second term on the right side vanishes in the limit $\Delta u \to 0$; the first term is simply $2u$. Hence

$$\frac{df}{du} = 2u \tag{8}$$

or

$$\frac{d}{du}(u^2) = 2u \tag{9}$$

This is one instance of the general rule for the differentiation of u^n,

$$\frac{d}{du}(u^n) = nu^{n-1} \tag{10}$$

This general rule is valid for any positive or negative number n, including zero. The proof of this rule can be constructed by an argument similar to that above. Table A5.1 lists the derivatives of the most common functions.

Table A5.1 SOME DERIVATIVES

$\frac{d}{du} u^n = nu^{n-1}$

$\frac{d}{du} \ln u = \frac{1}{u}$

$\frac{d}{du} e^u = e^u$

$\frac{d}{du} \sin u = \cos u$ (where u is in *radians*)

$\frac{d}{du} \cos u = -\sin u$ (where u is in *radians*)

$\frac{d}{du} \tan u = \sec^2 u$ (where u is in *radians*)

$\frac{d}{du} \cot u = -\csc^2 u$ (where u is in *radians*)

$\frac{d}{du} \sec u = \tan u \sec u$ (where u is in *radians*)

$\frac{d}{du} \csc u = -\cot u \csc u$ (where u is in *radians*)

$\frac{d}{du} \sin^{-1} u = 1/\sqrt{1-u^2}$ (where u is in *radians*)

$\frac{d}{du} \cos^{-1} u = -1/\sqrt{1-u^2}$ (where u is in *radians*)

$\frac{d}{du} \tan^{-1} u = \frac{1}{1+u^2}$ (where u is in *radians*)

A5.2 Other Important Rules for Differentiation

i. Derivative of a constant times a function:

$$\frac{d}{du}(cf) = c\,\frac{df}{du} \tag{11}$$

For instance,

$$\frac{d}{du}(6u^2) = 6\,\frac{d}{du}(u^2) = 6 \times 2u = 12u$$

ii. Derivative of the sum of two functions:

$$\frac{d}{du}(f+g) = \frac{df}{du} + \frac{dg}{du} \tag{12}$$

For instance,

$$\frac{d}{du}(6u^2 + u) = \frac{d}{du}(6u^2) + \frac{d}{du}(u) = 12u + 1$$

iii. Derivative of the product of two functions:

$$\frac{d}{du}(f \times g) = g\,\frac{df}{du} + f\,\frac{dg}{du} \tag{13}$$

For instance,

$$\frac{d}{du}(u^2 \sin u) = \sin u\,\frac{d}{du}\,u^2 + u^2\,\frac{d}{du}\sin u$$

$$= \sin u \times 2u + u^2 \times \cos u$$

iv. Chain rule for derivatives: If f is a function of g and g is a function of u, then

$$\frac{d}{du}f(g) = \frac{df}{dg}\,\frac{dg}{du} \tag{14}$$

For instance, if $g = 2u$ and $f(g) = \sin g$, then

$$\frac{d}{du}\sin(2u) = \frac{d\sin(2u)}{d(2u)}\,\frac{d(2u)}{du}$$

$$= \cos(2u) \times 2$$

A5.3 Integrals

We have learned that if the position is known as a function of time, then we can find the instantaneous velocity by differentiation. What about the converse problem: if the instantaneous velocity is known as a function of time, how can we find the position? In Section 2.5 we learned how to deal with this problem in the special case of motion with constant acceleration. The velocity is then a fairly simple function of time [see Eq. (2.17)]

$$v = v_0 + at \tag{15}$$

and the position deduced from this velocity is [see Eq. (2.22)]

$$x = x_0 + v_0 t + \tfrac{1}{2}at^2 \tag{16}$$

where x_0 and v_0 are the initial position and velocity at the initial time $t_0 = 0$. Now we want to deal with the general case of a velocity that is an arbitrary function of time,

$$v = v(t) \tag{17}$$

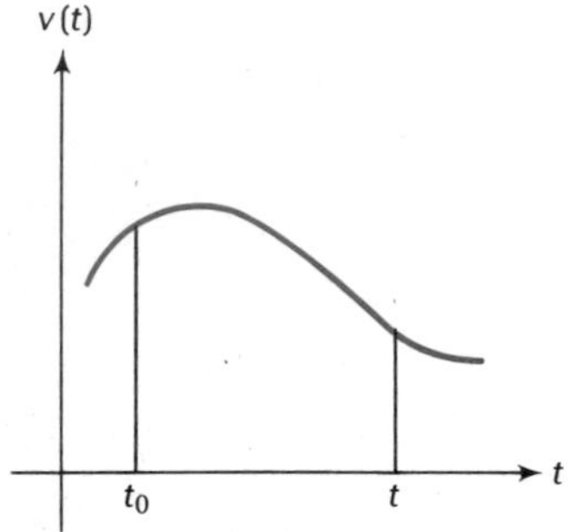

Fig. A5.2 Plot of a function $v(t)$.

Figure A5.2 shows what a plot of v vs. t might look like. At the initial time t_0, the particle has an initial position x_0 (for the sake of generality we now assume that $t_0 \neq 0$). We want to find the position at some later time t. For this purpose, let us divide the time interval $t - t_0$ into a large number of small time intervals, each of the duration Δt. The total number of intervals is N, so that $t - t_0 = N\,\Delta t$. The first of these intervals lasts from t_0 to $t_0 + \Delta t$; the second from $t_0 + \Delta t$ to $t_0 + 2\,\Delta t$; etc.

In Figure A5.3 the beginnings and the ends of these intervals have been marked t_0, t_1, t_2, etc., with $t_1 = t_0 + \Delta t$, $t_2 = t_0 + 2\Delta t$, etc. If Δt is sufficiently small, then during the first time interval the velocity is approximately $v(t_0)$; during the second, $v(t_1)$; etc. This amounts to replacing the smooth function $v(t)$ by a series of steps (see Figure A5.3). Thus, during the first time interval, the displacement of the particle is approximately $v(t_0)\,\Delta t$; during the second interval, $v(t_1)\,\Delta t$; etc. The net

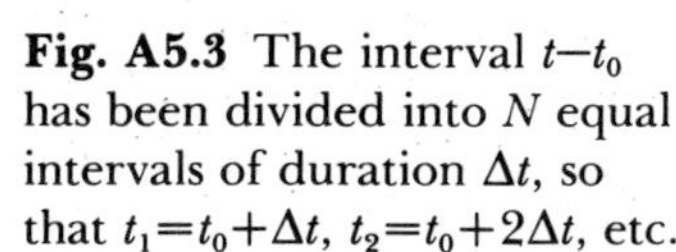

Fig. A5.3 The interval t–t_0 has been divided into N equal intervals of duration Δt, so that $t_1 = t_0 + \Delta t$, $t_2 = t_0 + 2\Delta t$, etc.

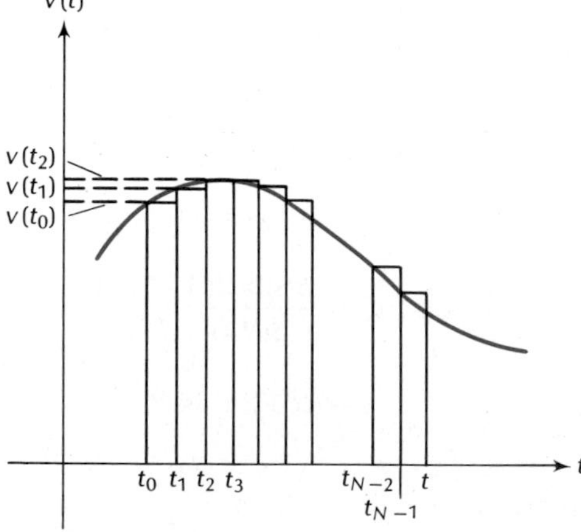

displacement of the particle during the entire interval $t - t_0$ is the sum of all these small displacements,

$$x(t) - x_0 \cong v(t_0)\,\Delta t + v(t_1)\,\Delta t + v(t_2)\,\Delta t + \cdots \tag{18}$$

Using the standard mathematical notation for summation, we can write this as

$$x(t) - x_0 \cong \sum_{i=0}^{N-1} v(t_i)\,\Delta t \tag{19}$$

We can give this sum the following graphical interpretation: since $v(t_i)\,\Delta t$ is the area of the rectangle of height $v(x_i)$ and width Δt, the sum is the net area of all the rectangles shown in Figure A5.3, i.e., it is approximately the area under the velocity curve. Note that if the velocity is negative, the area must be reckoned as negative!

Of course, Eq. (19) is only an approximation. To find the exact displacement of the particle we must let the step size Δt tend to zero (while the number of steps N tends to infinity). In this limit, the steplike horizontal and vertical line segments in Figure A5.3 approach the smooth curve. Thus,

$$x(t) - x_0 = \lim_{\substack{\Delta t \to 0 \\ N \to \infty}} \sum_{i=0}^{N-1} v(t_i)\,\Delta t \tag{20}$$

In the notation of calculus, the right side of Eq. (20) is usually written in the following fashion:

$$x(t) - x_0 = \int_{t_0}^{t} v(t')\,dt' \tag{21}$$

The right side is called the **integral** of the function $v(t)$. The subscript and the superscript on the integration symbol $\int$ are called, respectively, the lower and the upper limit of integration; and t' is called the variable of integration (the prime on the variable of integration t' merely serves to distinguish that variable from the limit of integration t). Graphically, the integral is the exact area under the velocity curve between the limits t_0 and t in a plot of v vs. t (see Figure A5.4). Areas below the t axis must be reckoned as negative.

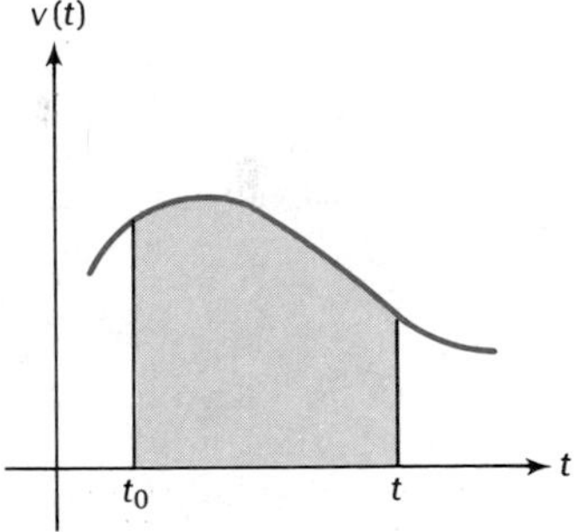

Fig. A5.4 The area under the velocity curve.

In general, if $f(u)$ is a function of u, then the integral of this function is defined by a limiting procedure similar to that described above for special case of the function $v(t)$. The integral over an interval from $u = a$ to $u = b$ is

$$\int_{a}^{b} f(u)\,du = \lim_{\substack{\Delta u \to 0 \\ N \to \infty}} \sum_{i=0}^{N-1} f(u_i)\,\Delta u \tag{22}$$

where $u_i = a + N\,\Delta u$. As in the case of the integral of $v(t)$, this integral can again be interpreted as an area: it is the area under the curve between the limits a and b in a plot of f vs. u.

For the explicit evaluation of integrals we can take advantage of the connection between integrals and antiderivatives. An **antiderivative** of a function $f(u)$ is simply a function $F(u)$ such that $dF/du = f$. For example, if $f(u) = u^n$ and $n \neq -1$, then an antiderivative of $f(u)$ is $F(u) = u^{n+1}/(n+1)$. The fundamental theorem of calculus states that the integral of any function $f(u)$ can be expressed in terms of antiderivatives:

$$\int_a^b f(u)\,du = F(b) - F(a) \tag{23}$$

In essence, this means that integration is the inverse of differentiation. We will not prove this theorem here, but we remark that such an inverse relationship between integration and differentiation should not come as a surprise. We have already run across an obvious instance of such a relationship: we know that velocity is the derivative of the position, and we have seen above that the position is the integral of the velocity.

We will sometimes write Eq. (23) as

$$\int_a^b f(u)\,du = [F(u)]_a^b \tag{24}$$

where the notation $[\quad]_a^b$ means that the function enclosed in the brackets is to be evaluated at a and at b, and these values are to be subtracted. For example, if $n \neq -1$,

$$\int_a^b u^n\,du = \left[\frac{u^{n+1}}{n+1}\right]_a^b = \frac{b^{n+1}}{n+1} - \frac{a^{n+1}}{n+1} \tag{25}$$

Table A5.2 lists some frequently used integrals. In this table, the limits of integration belonging with Eq. (24) have been omitted for the sake of brevity.

Table A5.2 SOME INTEGRALS

$\int u^n\,du = \dfrac{u^{n+1}}{n+1}$ for $n \neq -1$

$\int \dfrac{1}{u}\,du = \ln u$ for $u > 0$

$\int e^{ku}\,du = \dfrac{e^{ku}}{k}$

$\int \ln u\,du = u \ln u - u$

$\int \sin(ku)\,du = -\dfrac{1}{k}\cos(ku)$ (where ku is in *radians*)

$\int \cos(ku)\,du = \dfrac{1}{k}\sin(ku)$ (where ku is in *radians*)

$\int \dfrac{du}{1+ku} = \dfrac{1}{k}\ln(1+ku)$

$\int \dfrac{du}{\sqrt{k^2-u^2}} = \sin^{-1}\left(\dfrac{u}{k}\right)$

$\int \dfrac{du}{\sqrt{u^2 \pm k^2}} = \ln(u + \sqrt{u^2 \pm k^2})$

$\int \dfrac{du}{k^2+u^2} = \dfrac{1}{k}\tan^{-1}\left(\dfrac{u}{k}\right)$

$\int \dfrac{du}{u\sqrt{k^2 \pm u^2}} = -\dfrac{1}{k}\ln\left(\dfrac{k+\sqrt{k \pm u^2}}{u}\right)$

A5.4 The Most Important Rules Regarding Integration

i. Integral of a constant times a function:

$$\int_a^b cf(u)\, du = c \int_a^b f(u)\, du$$

For instance,

$$\int_a^b 5u^2\, du = 5 \int_a^b u^2\, du = 5\left(\frac{b^3}{3} - \frac{a^3}{3}\right)$$

ii. Integral of a sum of two functions:

$$\int_a^b [f(u) + g(u)]\, du = \int_a^b f(u)\, du + \int_a^b g(u)\, du$$

For instance,

$$\int_a^b (5u^2 + u)\, du = \int_a^b 5u^2\, du + \int_a^b u\, du = 5\left(\frac{b^3}{3} - \frac{a^3}{3}\right) + \left(\frac{b^2}{2} - \frac{a^2}{2}\right)$$

iii. Change of limits of integration:

$$\int_a^b f(u)\, du = \int_a^c f(u)\, du + \int_c^b f(u)\, du$$

$$\int_a^b f(u)\, du = -\int_b^a f(u)\, du$$

iv. Change of variable of integration:
If u is a function of v, then

$$\int_a^b f(u)\, du = \int_{v(a)}^{v(b)} f(u) \frac{du}{dv}\, dv$$

For instance, with $u = v^2$,

$$\int_a^b u^3\, du = \int_a^b v^6\, du = \int_{\sqrt{a}}^{\sqrt{b}} v^6(2v)\, dv$$

Finally, let us apply these general results to some specific examples of integration of the velocity.

EXAMPLE 1. A particle with constant acceleration has the following velocity as a function of time [compare Eq. (15)]:

$$v(t) = v_0 + at \tag{26}$$

where v_0 is the velocity at $t = 0$.

By integration, find the position as a function of time.

SOLUTION: According to Eq. (21), with $t_0 = 0$,

$$x(t) - x_0 = \int_0^t v(t')\, dt' = \int_0^t (v_0 + at')\, dt'$$

Using rule ii and rule i, we find that this equals

$$x(t) - x_0 = \int_0^t v_0 dt' + \int_0^t at'\, dt' = v_0 \int_0^t dt' + a \int_0^t t'\, dt'$$

The first entry listed in Table A5.2 gives $\int dt' = t'$ (for $n = 0$) and $\int t'\, dt' = t'^2/2$ (for $n = 1$). Thus

$$x(t) - x_0 = v_0[t']_0^t + s[t'^2/2]_0^t$$

$$= v_0 t + at^2/2 \tag{27}$$

This, of course, agrees with Eq. (16).

EXAMPLE 2. The instantaneous velocity of a projectile traveling through air is the following function of time:

$$v(t) = 2152 - 200.6t + 10.7t^2$$

where $v(t)$ is measured in feet per second and t is measured in seconds. Assuming that $x = 0$ at $t = 0$, what is the position as a function of time? What is the position at $t = 3.0$ s?

SOLUTION: With $x_0 = 0$ and $t_0 = 0$, Eq. (21) becomes

$$x(t) = \int_0^t (2152 - 200.6t' + 10.7t'^2)\, dt'$$

$$= 2152 \int_0^t dt' - 200.6 \int_0^t t'\, dt' + 10.7 \int_0^t t'^2\, dt'$$

$$= 2152\, [t']_0^t - 200.6[t'^2/2]_0^t + 10.7\, [t'^3/3]_0^t$$

$$= 2152t - 200.6t^2/2 + 10.7t^3/3$$

When evaluated at $t = 3.0$ s, this yields

$$x(1.0) = 2152 \times 3.0 - 200.6 \times (3.0)^2/2 + 10.7 \times (3.0)^3/3$$

$$= 5.6 \times 10^3 \text{ ft}$$

EXAMPLE 3. The acceleration of a mass pushed back and forth by an elastic spring is

$$a(t) = B \cos \omega t \tag{28}$$

where B and ω are constants. Find the position as a function of time. Assume $v = 0$ and $x = 0$ at $t = 0$.

SOLUTION: The calculation involves two steps: first we must integrate the acceleration to find the velocity, then we must integrate the velocity to find the position. For the first step we use an equation analogous to Eq. (21),

$$v(t) - v_0 = \int_{t_0}^{t} a(t')\, dt' \tag{29}$$

This equation becomes obvious if we remember that the relationship between acceleration and velocity is analogous to that between velocity and position. With $v_0 = 0$ and $t_0 = 0$, we obtain from Eq. (29)

$$v(t) = \int_0^t B \cos \omega t' dt' = B\left[\frac{1}{\omega} \sin \omega t'\right]_0^t$$

$$= \frac{B}{\omega} \sin \omega t \tag{30}$$

Next,

$$x(t) = \int_0^t v(t')\, dt' = \int_0^t \frac{B}{\omega} \sin \omega t'\, dt' = \frac{B}{\omega}\left[-\frac{1}{\omega} \cos \omega t'\right]_0^t$$

$$= -\frac{B}{\omega^2} \cos \omega t + \frac{B}{\omega^2} \tag{31}$$

A5.4 The Taylor Series

Suppose that $f(u)$ is a smooth function of u in some neighborhood of a given point $u = a$, so that the function has continuous derivatives of all orders. Then the value of the function at an arbitrary point near a can be expressed in terms of the following infinite series, where all the derivatives are evaluated at the point a:

$$f(u) = f(a) + \frac{df}{du}(x - a) + \frac{1}{2!}\frac{d^2f}{du^2}(x - a)^2 + \frac{1}{3!}\frac{d^3f}{du^3}(x - a)^3 + \cdots \tag{32}$$

This is called the **Taylor series** for the function $f(u)$ about the point a. The series converges, and is valid, provided x is sufficiently close to a. How close is "sufficiently close" depends on the function f and on the point a. Some functions, such as sin u, cos u, and e^u, are extremely well behaved, and their Taylor series converge for any choice of x and of a. The Taylor series gives us a convenient method for the approximate evaluation of a function.

EXAMPLE 4. Find the Taylor series for sin u about the point $u = 0$.

SOLUTION: The derivatives of sin u at $u = 0$ are

$$\frac{d}{du} \sin u = \cos u = 1$$

$$\frac{d^2}{du^2} \sin u = \frac{d}{du} \cos u = -\sin u = 0$$

$$\frac{d^3}{du^3}\sin u = \frac{d}{du}(-\sin u) = -\cos u = -1$$

$$\frac{d^4}{du^4}\sin u = \frac{d}{du}(-\cos u) = \sin u = 0, \qquad \text{etc.}$$

Hence Eq. (32) gives

$$\begin{aligned}\sin u &= 0 + 1 \times (u-0) + \frac{1}{2!}\times 0 \times (u-0)^2 + \frac{1}{3!}\times(-1)\times(u-0)^3 \\ &\quad + \frac{1}{4!}\times 0 \times (u-0)^4 + \cdots \\ &= u - \tfrac{1}{6}u^3 + \cdots\end{aligned}$$

Note that for very small values of u, we can neglect all higher powers of u, so that $\sin u \cong u$, which agrees with the approximation given in Appendix 4.

APPENDIX 6: THE INTERNATIONAL SYSTEM OF UNITS (SI)

A6.1 Base Units

The SI system of units is the modern version of the metric system. The SI system recognizes seven fundamental, or base, units for length, mass, time, electric current, thermodynamic temperature, amount of substance, and luminous intensity.[a] The following definitions of the base units were adopted by the Conférence Générale des Poids et Mesures in the years indicated:

Meter (m) "The metre is the length of the path travelled by light in vacuum during a time interval of 1/299 792 458 of a second." (Adopted in 1983.)

Kilogram (kg) "The kilogram is . . . the mass of the international prototype of the kilogram." (Adopted in 1889 and in 1901.)

Second (s) "The second is the duration of 9 192 631 770 periods of the radiation corresponding to the transition between the two hyperfine levels of the ground state of the cesium-133 atom." (Adopted in 1967.)

Ampere (A) "The ampere is that constant current which, if maintained in two straight parallel conductors of infinite length, of negligible circular cross section, and placed one meter apart in vacuum, would produce between these conductors a force equal to 2×10^{-7} newton per meter of length." (Adopted in 1948.)

Kelvin (K) "The kelvin . . . is the fraction 1/273.16 of the thermodynamic temperature of the triple point of water." (Adopted in 1967.)

Mole "The mole is the amount of substance of a system which contains as many elementary entities as there are atoms in 0.012 kilogram of carbon-12." (Adopted in 1967.)

Candela (cd) "The candela is the luminous intensity, in a given direction, of a source that emits monochromatic radiation of frequency 540×10^{12} Hz and that has a radiant intensity in that direction of $\frac{1}{683}$ watt per steradian." (Adopted in 1979.)

Besides these seven base units, the SI system also recognizes two supplementary units of angle and solid angle:

Radian (rad) "The radian is the plane angle between two radii of a circle which cut off on the circumference an arc equal in length to the radius."

Steradian (sr) "The steradian is the solid angle which, having its vertex in the center of a sphere, cuts off an area equal to that of a [flat] square with sides of length equal to the radius of the sphere."

[a] At least two of the seven base units of the SI system are redundant. The mole is merely a certain number of atoms or molecules, in the same sense that a dozen is a number; there is no need to designate this number as a unit. The candela is equivalent to $\frac{1}{683}$ watt per steradian; it serves no purpose that is not served equally well by watt per steradian. Two other base units could be made redundant by adopting new definitions of the unit of temperature and of the unit of electric charge. Temperature could be measured in energy units because, according to the equipartition theorem, temperature is proportional to the energy per degree of freedom. Hence the kelvin could be defined as a derived unit, with 1 K $= \frac{1}{2} \times 1.38 \times 10^{-23}$ joule per degree of freedom. Electric charge could also be defined as a derived unit, to be measured with a suitable combination of the units of force and distance, as is done in the cgs system.

Furthermore, the definitions of the supplementary units — radian and steradian — are gratuitous. These definitions properly belong in the province of mathematics and there is no need to include them in a system of physical units.

A6.2 Derived Units

The derived units are formed out of products and ratios of the base units. Table A6.1 lists those derived units that have been glorified with special names. (Other derived units are listed in the tables of conversion factors in Appendix A.7.)

Table A6.1 NAMES OF DERIVED UNITS

Quantity	*Derived Unit*	*Name*	*Symbol*
frequency	1/s	hertz	Hz
force	kg · m/s²	newton	N
pressure	N/m²	pascal	Pa
energy	N · m	joule	J
power	J/s	watt	W
electric charge	A · s	coulomb	C
electric potential	J/C	volt	V
electric capacitance	C/V	farad	F
electric resistance	V/A	ohm	Ω
conductance	A/V	siemens	S
magnetic flux	V · s	weber	Wb
magnetic field	V · s/m²	tesla	T
inductance	V · s/A	henry	H
temperature	K	degree Celsius	°C
luminous flux	cd · sr	lumen	lm
illuminance	cd · sr/m²	lux	lx
radioactivity	1/s	becquerel	Bq
absorbed dose	J/kg	gray	Gy
dose equivalent	J/kg	sievert	Sv

A6.3 Prefixes

Multiples and submultiples of SI units are indicated by prefixes, such as the familiar ***kilo, centi,*** and ***milli*** used in ***kilometer, centimeter,*** and ***millimeter,*** etc. Table A6.2 lists all the accepted prefixes. Some enjoy more popularity than others; it is best to avoid the use of uncommon prefixes, such as ***atto*** and ***exa,*** since hardly anybody will recognize those.

Table A6.2 PREFIXES FOR UNITS

Multiplication factor	*Prefix*	*Symbol*
10^{18}	exa	E
10^{15}	peta	P
10^{12}	tera	T
10^{9}	giga	G
10^{6}	mega	M
10^{3}	kilo	k
10^{2}	hecto	h
10	deka	da
10^{-1}	deci	d
10^{-2}	centi	c
10^{-3}	milli	m
10^{-6}	micro	μ
10^{-9}	nano	n
10^{-12}	pico	p
10^{-15}	femto	f
10^{-18}	atto	a

APPENDIX 7: CONVERSION FACTORS

The units for each quantity are listed alphabetically, except that the SI unit is always listed first. The numbers are based on "American National Standard; Metric Practice" published by the Institute of Electrical and Electronics Engineers, 1982.

ANGLE

1 radian = 57.30° = $3.438 \times 10^{3}\,'$ = $\frac{1}{2\pi}$ rev = $2.063 \times 10^{5}\,''$

1 degree (°) = 1.745×10^{-2} **radian** = 60′ = 3600″ = $\frac{1}{360}$ rev

1 minute of arc (′) = 2.909×10^{-4} **radian**= $\frac{1}{60}$° = 4.630×10^{-5} rev = 60″

1 revolution (rev) = 2π **radian** = 360° = $2.160 \times 10^{4}\,'$ = $1.296 \times 10^{6}\,''$

1 second of arc (″) = 4.848×10^{-6} **radian** = $\frac{1}{3600}$° = $\frac{1}{60}$′ = 7.716×10^{-7} rev

LENGTH

1 meter (m) = 1×10^{10} Å = 6.685×10^{-12} AU = 100 cm = 1×10^{15} F = 3.281 ft = 39.37 in. = 1×10^{-3} km = 1.057×10^{-16} light-year = 1×10^{6} μm = 5.400×10^{-4} nmi = 6.214×10^{-4} mi = 3.241×10^{-17} pc = 1.094 yd

1 angstrom (Å) = 1×10^{-10} m = 1×10^{-8} cm = 1×10^{5} F = 3.281×10^{-10} ft = 1×10^{-4} μm

1 astronomical unit (AU) = 1.496×10^{11} m = 1.496×10^{13} cm = 1.496×10^{8} km = 1.581×10^{-5} light-year = 4.848×10^{-6} pc

1 centimeter (cm) = 0.01 m = 1×10^{8} Å = 1×10^{13} F = 3.281×10^{-2} ft = 0.3937 in. = 1×10^{-5} km = 1.057×10^{-18} light-year = 1×10^{4} μm

1 fermi (F) = 1×10^{-15} m = 1×10^{-13} cm = 1×10^{5} Å

1 foot (ft) = 0.3048 m = 30.48 cm = 12 in. = 3.048×10^{5} μm = 1.894×10^{-4} mi = $\frac{1}{3}$ yd

1 inch (in.) = 2.540×10^{-2} m = 2.54 cm = $\frac{1}{12}$ ft = 2.54×10^{4} μm = $\frac{1}{36}$ yd

1 kilometer (km) = 1×10^{3} m = 1×10^{5} cm = 3.281×10^{3} ft = 0.5400 nmi = 0.6214 mi = 1.094×10^{3} yd

1 light-year = 9.461×10^{15} m = 6.324×10^{4} AU = 9.461×10^{17} cm = 9.461×10^{12} km = 5.879×10^{12} mi = 0.3066 pc

1 micron, or micrometer (μm) = 1×10^{-6} m = 1×10^{4} Å = 1×10^{-4} cm = 3.281×10^{-6} ft = 3.937×10^{-5} in.

1 nautical mile (nmi) = 1.852×10^{3} m = 1.852×10^{5} cm = 6.076×10^{3} ft = 1.852 km = 1.151 mi

1 parsec (pc) = 3.086×10^{16} m = 2.063×10^{5} AU = 3.086×10^{18} cm = 3.086×10^{13} km = 3.262 light-years

1 statute mile (mi) = 1.609×10^{3} m = 1.609×10^{5} cm = 5280 ft = 1.609 km = 0.8690 nmi = 1760 yd

1 yard (yd) = 0.9144 m = 91.44 cm = 3 ft = 36 in. = $\frac{1}{1760}$ mi

TIME

1 second (s) = 1.157×10^{-5} day = $\frac{1}{3600}$ h = $\frac{1}{60}$ min = 1.161×10^{-5} sidereal day = 3.169×10^{-8} yr

1 day = 8.640×10^4 s = 24 h = 1440 min = 1.003 sidereal days = 2.738×10^{-3} yr

1 hour (h) = 3600 s = $\frac{1}{24}$ day = 60 min = 1.141×10^{-4} yr

1 minute (min) = 60 s = 6.944×10^{-4} day = $\frac{1}{60}$ h = 1.901×10^{-6} year

1 sidereal day = 8.616×10^4 s = 0.9973 day = 23.93 h = 1.436×10^3 min = 2.730×10^{-3} yr

1 year (yr) = 3.156×10^7 s = 365.24 days = 8.766×10^3 h = 5.259×10^5 min = 366.24 sidereal days

MASS

1 kilogram (kg) = 6.024×10^{26} u = 5000 carats = 1.543×10^4 grains = 1000 g = 1×10^{-3} t = 35.27 oz.-mass = 2.205 lb-mass = 1.102×10^{-3} short ton-mass = 6.852×10^{-2} slug

1 atomic mass unit (u) = 1.6605×10^{-27} kg = 1.6605×10^{-24} g

1 carat = 2×10^{-4} kg = 0.2 g = 7.055×10^{-3} oz.-mass = 4.409×10^{-4} lb-mass.

1 grain = 6.480×10^{-5} kg = 6.480×10^{-2} g = 2.286×10^{-3} oz.-mass = $\frac{1}{7000}$ lb-mass

1 gram (g) = 1×10^{-3} kg = 6.024×10^{23} u = 5 carats = 15.43 grains = 1×10^{-6} t = 3.527×10^{-2} oz.-mass = 2.205×10^{-3} lb-mass = 1.102×10^{-6} short ton-mass = 6.852×10^{-5} slug

1 metric ton (t) = 1×10^3 kg = 1×10^6 g = 2.205×10^3 lb-mass = 1.102 short ton-mass = 68.52 slugs

1 ounce-mass (oz.-mass) = 2.835×10^{-2} kg = 141.7 carats = 437.5 grains = 28.35 g = $\frac{1}{16}$ lb-mass

1 pound-mass (lb-mass)[b] = 0.4536 kg = 453.6 g = 4.536×10^{-4} t = 16 oz.-mass = $\frac{1}{2000}$ short ton-mass = 3.108×10^{-2} slug

1 short ton-mass = 907.2 kg = 9.07×10^5 g = 0.9072 t = 2000 lb-mass

1 slug = 14.59 kg = 1.459×10^4 g = 32.17 lb-mass.

AREA

1 square meter (m^2) = 1×10^4 cm^2 = 10.76 ft^2 = 1.550×10^3 $in.^2$ = 1×10^{-6} km^2 = 3.861×10^{-7} mi^2 = 1.196 yd^2

1 barn = 1×10^{-28} m^2 = 1×10^{-24} cm^2

1 square centimeter (cm^2) = 1×10^{-4} m^2 = 1.076×10^{-3} ft^2 = 0.1550 $in.^2$ = 1×10^{-10} km^2 = 3.861×10^{-11} mi^2

1 square foot (ft^2) = 9.290×10^{-2} m^2 = 929.0 cm^2 = 144 $in.^2$ = 3.587×10^{-8} mi^2 = $\frac{1}{9}$ yd^2

1 square inch ($in.^2$) = 6.452×10^{-4} m^2 = 6.452 cm^2 = $\frac{1}{144}$ ft^2

1 square kilometer (km^2) = 1×10^6 m^2 = 1×10^{10} cm^2 = 1.076×10^7 ft^2 = 0.3861 mi^2

1 square statute mile (mi^2) = 2.590×10^6 m^2 = 2.590×10^{10} cm^2 = 2.788×10^7 ft^2 = 2.590 km^2

1 square yard (yd^2) = 0.8361 m^2 = 8.361×10^3 cm^2 = 9 ft^2 = 1296 $in.^2$

VOLUME

1 cubic meter (m^3) = 1×10^6 cm^3 = 35.31 ft^3 = 264.2 gal. = 6.102×10^4 $in.^3$ = 1×10^3 liters = 1.308 yd^3

1 cubic centimeter (cm^3) = 1×10^{-6} m^3 = 3.531×10^{-5} ft^3 = 2.642×10^{-4} gal. = 6.102×10^{-2} $in.^3$ = 1×10^{-3} liter

1 cubic foot (ft^3) = 2.832×10^{-2} m^3 = 2.832×10^4 cm^3 = 7.481 gal. = 1728 $in.^3$ = 28.32 liters = $\frac{1}{27}$ yd^3

1 gallon (gal.)[c] = 3.785×10^{-3} m^3 = 0.1337 ft^3

1 cubic inch ($in.^3$) = 1.639×10^{-5} m^3 = 16.39 cm^3 = 5.787×10^{-4} ft^3

---

[b] This is the "avoirdupois" pound. The "troy" or "apothecary" pound is 0.3732 kg, or 0.8229 lb avoirdupois.

[c] This is the U.S. gallon; the U.K. and the Canadian gallon are 4.546×10^{-3} m^3, or 1.201 U.S. gallons.

1 liter (l) = 1×10^{-3} m^3 = 1000 cm^3 = 3.531×10^{-2} ft^3
1 cubic yard (yd^3) = 0.7646 m^3 = 7.646×10^5 cm^3 = 27 ft^3 = 202.0 gal.

DENSITY

1 kilogram per cubic meter (kg/m^3) = 1×10^{-3} g/cm^3 = 6.243×10^{-2} lb-mass/ft^3 = 8.345×10^{-3} lb-mass/gal. = 3.613×10^{-5} lb-mass/in.^{3} = 8.428×10^{-4} short ton-mass/yd^3 = 1.940×10^{-3} slug/ft^3
1 gram per cubic centimeter (g/cm^3) = 1×10^3 kg/m^3 = 62.43 lb-mass/ft^3 = 8.345 lb-mass/gal. = 3.613×10^{-2} lb-mass/in.^{3} = 0.8428 short ton-mass/yd^3 = 1.940 slug/ft^3
1 lb-mass per cubic foot (lb-mass/ft^3) = 16.02 kg/m^3 = 1.602×10^{-2} g/cm^3 = 0.1337 lb-mass/gal. = 1.350×10^{-2} short ton-mass/yd^3 = 3.108×10^{-2} slug/ft^3
1 pound-mass per gallon (1 lb-mass/gal.) = 119.8 kg/m^3 = 7.481 lb-mass/ft^3 = 0.2325 slug/ft^3
1 short ton-mass per cubic yard (short ton-mass/yd^3) = 1.187×10^3 kg/m^3 = 74.07 lb-mass/ft^3
1 slug per cubic foot (slug/ft^3) = 515.4 kg/m^3 = 0.5154 g/cm^3 = 32.17 lb-mass/ft^3 = 4.301 lb-mass/gal.

SPEED

1 meter per second (m/s) = 100 cm/s = 3.281 ft/s = 3.600 km/h = 1.944 knot = 2.237 mi/h
1 centimeter per second (cm/s) = 0.01 m/s = 3.281×10^{-2} ft/s = 3.600×10^{-2} km/h = 1.944×10^{-2} knot = 2.237×10^{-2} mi/h
1 foot per second (ft/s) = 0.3048 m/s = 30.48 cm/s = 1.097 km/h = 0.5925 knot = 0.6818 mi/h
1 kilometer per hour (km/h) = 0.2778 m/s = 27.78 cm/s = 0.9113 ft/s = 0.5400 knot = 0.6214 mi/h
1 knot, or nautical mile per hour = 0.5144 m/s = 51.44 cm/s = 1.688 ft/s = 1.852 km/h = 1.151 mi/h
1 mile per hour (mi/h) = 0.4470 m/s = 44.70 cm/s = 1.467 ft/s = 1.609 km/h = 0.8690 knot

ACCELERATION

1 meter per second squared (m/s^2) = 100 cm/s^2 = 3.281 ft/s^2 = 0.1020 gee
1 centimeter per second squared (cm/s^2) = 0.01 m/s^2 = 3.281×10^{-2} ft/s^2 = 1.020×10^{-3} gee
1 foot per second squared (ft/s^2) = 0.3048 m/s^2 = 30.48 cm/s^2 = 3.108×10^{-2} gee
1 gee = 9.807 m/s^2 = 980.7 cm/s^2 = 32.17 ft/s^2

FORCE

1 newton (N) = 1×10^5 dynes = 0.1020 kp = 0.2248 lb = 1.124×10^{-4} short ton
1 dyne = 1×10^{-5} N = 1.020×10^{-6} kp = 2.248×10^{-6} lb = 1.124×10^{-9} short ton
1 kilopond, or kilogram force (kp) = 9.807 N = 9.807×10^5 dynes = 2.205 lb = 1.102×10^{-3} short ton
1 pound (lb) = 4.448 N = 4.448×10^5 dynes = 0.4536 kp = $\frac{1}{2000}$ short ton
1 short ton = 8.896×10^3 N = 8.896×10^8 dynes = 907.2 kp = 2000 lb

ENERGY

1 joule (J) = 9.478×10^{-4} Btu = 0.2388 cal = 1×10^7 ergs = 6.242×10^{18} eV = 0.7376 ft · lb = 2.778×10^{-7} kW · h
1 British thermal unit (Btu)[d] = 1.055×10^3 J = 252.0 cal = 1.055×10^{10} ergs = 778.2 ft · lb = 2.931×10^{-4} kW · h

[d] This is the "International Table" Btu; there are several other Btus.

1 calorie (cal)[e] = 4.187 J = 3.968×10^{-3} Btu = 4.187×10^{7} ergs = 3.088 ft · lb = 1×10^{-3} kcal = 1.163×10^{-6} kW · h
1 erg = 1×10^{-7} J = 9.478×10^{-7} Btu = 2.388×10^{-8} cal = 6.242×10^{11} eV = 7.376×10^{-8} ft · lb = 2.778×10^{-14} kW · h
1 electron-volt (eV) = 1.602×10^{-19} J = 1.602×10^{-12} erg = 1.182×10^{-19} ft · lb
1 foot-pound (ft · lb) = 1.356 J = 1.285×10^{-3} Btu = 0.3239 cal = 1.356×10^{7} ergs = 8.464×10^{18} eV = 3.766×10^{-7} kW · h
1 kilocalorie (kcal), or **large calorie** (Cal) = 4.187×10^{3} J = 1×10^{3} cal
1 kilowatt-hour (kW · h) = 3.600×10^{6} J = 3412 Btu = 8.598×10^{5} cal = 3.6×10^{13} ergs = 2.655×10^{6} ft · lb

POWER

1 watt (W) = 3.412 Btu/h = 0.2388 cal/s = 1×10^{7} ergs/s = 0.7376 ft · lb/s = 1.341×10^{-3} hp
1 British thermal unit per hour (Btu/h) = 0.2931 W = 7.000×10^{-2} cal/s = 0.2162 ft · lb/s = 3.930×10^{-4} hp
1 calorie per second (cal/s) = 4.187 W = 14.29 Btu/h = 4.187×10^{7} erg/s = 3.088 ft · lb/s = 5.615×10^{-3} hp
1 erg per second (erg/s) = 1×10^{-7} W = 2.388×10^{-8} cal/s = 7.376×10^{-8} ft · lb/s = 1.341×10^{-10} hp
1 foot-pound per second (ft · lb/s) = 1.356 W = 0.3238 cal/s = 4.626 Btu/h = 1.356×10^{7} ergs/s = 1.818×10^{-3} hp
1 horsepower (hp)[f] = 745.7 W = 2.544×10^{3} Btu/h = 178.1 cal/s = 550 ft · lb/s
1 kilowatt (kW) = 1×10^{3} W = 3.412×10^{3} Btu/h = 238.8 cal/s = 737.6 ft · lb/s = 1.341 hp

PRESSURE

1 newton per square meter (N/m²), or **pascal** (Pa) = 9.869×10^{-6} atm = 1×10^{-5} bar = 7.501×10^{-4} cmHg = 10 dynes/cm² = 2.089×10^{-2} lb/ft² = 1.450×10^{-4} lb/in.² = 7.501×10^{-3} torr
1 atmosphere (atm) = 1.013×10^{5} N/m² = 76.00 cmHg = 1.013×10^{6} dynes/cm² = 2.116×10^{3} lb/ft² = 14.70 lb/in.²
1 bar = 1×10^{5} N/m² = 0.9869 atm = 75.01 cmHg
1 centimeter of mercury (cmHg) = 1.333×10^{3} N/m² = 1.316×10^{-2} atm = 1.333×10^{-2} bar = 1.333×10^{4} dynes/cm² = 27.85 lb/ft² = 0.1934 lb/in.² = 10 torr
1 dyne per square centimeter (dyne/cm²) = 0.1 N/m² = 9.869×10^{-7} atm = 7.501×10^{-5} cmHg = 2.089×10^{-3} lb/ft² = 1.450×10^{-5} lb/in.²
1 kilopond per square centimeter (kp/cm²) = 9.807×10^{4} N/m² = 0.9678 atm = 9.807×10^{5} dynes/cm² = 14.22 lb/in.²
1 pound per square inch (lb/in.², or psi) = 6.895×10^{3} N/m² = 6.805×10^{-2} atm = 6.895×10^{4} dynes/cm² = 7.031×10^{-2} kp/cm²
1 torr, or **millimeter of mercury** (mmHg) = 1.333×10^{2} N/m² = 0.1 cmHg

ELECTRIC CHARGE[g]

1 coulomb (C) ⇔ 2.998×10^{9} statcoulombs, or esu of charge ⇔ 0.1 abcoulomb, or emu of charge

---

[e] This is the "International Table" calorie, which equals exactly 4.1868 J. There are several other calories; for instance, the thermochemical calorie, which equals 4.184 J.

[f] There are several other horsepowers; for instance, the metric horsepower, which equals 735.5 J.

[g] The dimensions of the electric quantities in SI units, electrostatic units (esu), and electromagnetic units (emu) are different; hence the relationships among these units are correspondences (⇔) rather than equalities (=).

ELECTRIC CURRENT

1 ampere (A) ⇔ 2.998×10^{9} statamperes, or esu of current ⇔ 0.1 abampere, or emu of current

ELECTRIC POTENTIAL

1 volt (V) ⇔ 3.336×10^{-3} statvolt, or esu of potential ⇔ 1×10^{8} abvolts, or emu of potential

ELECTRIC FIELD

1 volt per meter (V/m) ⇔ 3.336×10^{-5} statvolt/cm ⇔ 1×10^{6} abvolts/cm

MAGNETIC FIELD

1 tesla (T), or **weber per square meter** (Wb/m^2) ⇔ 1×10^{4} gauss

ELECTRIC RESISTANCE

1 ohm (Ω) ⇔ 1.113×10^{-12} statohm, or esu of resistance ⇔ 1×10^{9} abohms, or emu of resistance

ELECTRIC RESISTIVITY

1 ohm-meter (Ω · m) ⇔ 1.113×10^{-10} statohm-cm ⇔ 1×10^{11} abohm-cm

CAPACITANCE

1 farad (F) ⇔ 8.988×10^{11} statfarads, or esu of capacitance ⇔ 1×10^{-9} abfarad, or emu of capacitance

INDUCTANCE

1 henry (H) ⇔ 1.113×10^{-12} stathenry, or esu of inductance ⇔ 1×10^{9} abhenrys, or emu of inductance

FUNDAMENTAL CONSTANTS

Compiled by E. R. Cohen and B. N. Taylor under the auspices of the CODATA Task Group on Fundamental Constants. This set has been officially adopted by CODATA and is taken from J. Phys. Chem. Ref. Data, Vol. 2, No. 4, p. 663 (1973) and CODATA Bulletin No. 11 (December 1973).

Quantity	Symbol	Numerical Value *	Uncert. (ppm)	SI † ← Units →	cgs ‡
Speed of light in vacuum	c	299792458(1.2)	0.004	$m \cdot s^{-1}$	10^2 $cm \cdot s^{-1}$
Permeability of vacuum	μ_0	4π		10^{-7} $H \cdot m^{-1}$	
		$=12.5663706144$		10^{-7} $H \cdot m^{-1}$	
Permittivity of vacuum, $1/\mu_0 c^2$	ϵ_0	8.854187818(71)	0.008	10^{-12} $F \cdot m^{-1}$	
Fine-structure constant, $[\mu_0 c^2/4\pi](e^2/\hbar c)$	α	7.2973506(60)	0.82	10^{-3}	10^{-3}
	α^{-1}	137.03604(11)	0.82		
Elementary charge	e	1.6021892(46)	2.9	10^{-19} C	10^{-20} emu
		4.803242(14)	2.9		10^{-10} esu
Planck constant	h	6.626176(36)	5.4	10^{-34} $J \cdot s$	10^{-27} $erg \cdot s$
	$\hbar = h/2\pi$	1.0545887(57)	5.4	10^{-34} $J \cdot s$	10^{-27} $erg \cdot s$
Avogadro constant	N_A	6.022045(31)	5.1	10^{23} mol^{-1}	10^{23} mol^{-1}
Atomic mass unit, $10^{-3} kg \cdot mol^{-1} N_A^{-1}$	u	1.6605655(86)	5.1	10^{-27} kg	10^{-24} g
Electron rest mass	m_e	9.109534(47)	5.1	10^{-31} kg	10^{-28} g
		5.4858026(21)	0.38	10^{-4} u	10^{-4} u
Proton rest mass	m_p	1.6726485(86)	5.1	10^{-27} kg	10^{-24} g
		1.007276470(11)		u	u
Ratio of proton mass to electron mass	m_p/m_e	1836.15152(70)	0.38		
Neutron rest mass	m_n	1.6749543(86)	5.1	10^{-27} kg	10^{-24} g
		1.008665012(37)	0.037	u	u
Electron charge to mass ratio	e/m_e	1.7588047(49)	2.8	10^{11} $C \cdot kg^{-1}$	10^{7} $emu \cdot g^{-1}$
		5.272764(15)	2.8		10^{17} $esu \cdot g^{-1}$
Magnetic flux quantum, $[c]^{-1}(hc/2e)$	Φ_0	2.0678506(54)	2.6	10^{-15} Wb	10^{-7} $G \cdot cm^2$
	h/e	4.135701(11)	2.6	10^{-15} $J \cdot s \cdot C^{-1}$	10^{-7} $erg \cdot s \cdot emu^{-1}$
		1.3795215(36)	2.6		10^{-17} $erg \cdot s \cdot esu^{-1}$
Josephson frequency-voltage ratio	$2e/h$	4.835939(13)	2.6	10^{14} $Hz \cdot V^{-1}$	
Quantum of circulation	$h/2m_e$	3.6369455(60)	1.6	10^{-4} $J \cdot s \cdot kg^{-1}$	$erg \cdot s \cdot g^{-1}$
	h/m_e	7.273891(12)		10^{-4} $J \cdot s \cdot kg^{-1}$	$erg \cdot s \cdot g^{-1}$
Faraday constant, $N_A e$	F	9.648456(27)	2.8	10^{4} $C \cdot mol^{-1}$	10^{3} $emu \cdot mol^{-1}$
		2.8925342(82)	2.8		10^{14} $esu \cdot mol^{-1}$
Rydberg constant, $[\mu_0 c^2/4\pi]^2(m_e e^4/4\pi\hbar^3 c)$	R_∞	1.097373177(83)	0.075	10^{7} m^{-1}	10^{5} cm^{-1}
Bohr radius, $[\mu_0 c^2/4\pi]^{-1}(\hbar^2/m_e e^2) = \alpha/4\pi R_\infty$	a_0	5.2917706(44)	0.82	10^{-11} m	10^{-9} cm
Classical electron radius, $[\mu_0 c^2/4\pi](e^2/m_e c^2) = \alpha^3/4\pi R_\infty$	$r_e = \alpha \lambda\!\!\!^{-}_C$	2.8179380(70)	2.5	10^{-15} m	10^{-13} cm
Thomson cross section, $(8/3)\pi r_e^2$	σ_e	0.6652448(33)	4.9	10^{-28} m^2	10^{-24} cm^2
Free electron g-factor, or electron magnetic moment in Bohr magnetons	$g_e/2 = \mu_e/\mu_B$	1.0011596567(35)	0.0035		
Free muon g-factor, or muon magnetic moment in units of $[c](e\hbar/2m_\mu c)$	$g_\mu/2$	1.00116616(31)	0.31		
Bohr magneton, $[c](e\hbar/2m_e c)$	μ_B	9.274078(36)	3.9	10^{-24} $J \cdot T^{-1}$	10^{-21} $erg \cdot G^{-1}$
Electron magnetic moment	μ_e	9.284832(36)	3.9	10^{-24} $J \cdot T^{-1}$	10^{-21} $erg \cdot G^{-1}$
Gyromagnetic ratio of protons in H_2O	γ'_p	2.6751301(75)	2.8	10^{8} $s^{-1} \cdot T^{-1}$	10^{4} $s^{-1} \cdot G^{-1}$
	$\gamma'_p/2\pi$	4.257602(12)	2.8	10^{7} $Hz \cdot T^{-1}$	10^{3} $Hz \cdot G^{-1}$
γ'_p corrected for diamagnetism of H_2O	γ_p	2.6751987(75)	2.8	10^{8} $s^{-1} \cdot T^{-1}$	10^{4} $s^{-1} \cdot G^{-1}$
	$\gamma_p/2\pi$	4.257711(12)	2.8	10^{7} $Hz \cdot T^{-1}$	10^{3} $Hz \cdot G^{-1}$
Magnetic moment of protons in H_2O in Bohr magnetons	μ'_p/μ_B	1.52099322(10)	0.066	10^{-3}	10^{-3}
Proton magnetic moment in Bohr magnetons	μ_p/μ_B	1.521032209(16)	0.011	10^{-3}	10^{-3}
Ratio of electron and proton magnetic moments	μ_e/μ_p	658.2106880(66)	0.010		
Proton magnetic moment	μ_p	1.4106171(55)	3.9	10^{-26} $J \cdot T^{-1}$	10^{-23} $erg \cdot G^{-1}$
Magnetic moment of protons in H_2O in nuclear magnetons	μ'_p/μ_N	2.7927740(11)	0.38		
μ'_p/μ_N corrected for diamagnetism of H_2O	μ_p/μ_N	2.7928456(11)	0.38		
Nuclear magneton, $[c](e\hbar/2m_p c)$	μ_N	5.050824(20)	3.9	10^{-27} $J \cdot T^{-1}$	10^{-24} $erg \cdot G^{-1}$
Ratio of muon and proton magnetic moments	μ_μ/μ_p	3.1833402(72)	2.3		
Muon magnetic moment	μ_μ	4.490474(18)	3.9	10^{-26} $J \cdot T^{-1}$	10^{-23} $erg \cdot G^{-1}$
Ratio of muon mass to electron mass	m_μ/m_e	206.76865(47)	2.3		

Quantity	Symbol	Numerical Value *	Uncert. (ppm)	SI † (Units)	cgs ‡ (Units)
Muon rest mass	m_μ	1.883566(11)	5.6	10^{-28} kg	10^{-25} g
		0.11342920(26)	2.3	u	u
Compton wavelength of the	λ_C	2.4263089(40)	1.6	10^{-12} m	10^{-10} cm
electron, $h/m_e c=\alpha^2/2R_\infty$	$\bar{\lambda}_C=\lambda_C/2\pi=\alpha a_0$	3.8615905(64)	1.6	10^{-13} m	10^{-11} cm
Compton wavelength of the	$\lambda_{C,p}$	1.3214099(22)	1.7	10^{-15} m	10^{-13} cm
proton, $h/m_p c$	$\bar{\lambda}_{C,p}=\lambda_{C,p}/2\pi$	2.1030892(36)	1.7	10^{-16} m	10^{-14} cm
Compton wavelength of the	$\lambda_{C,n}$	1.3195909(22)	1.7	10^{-15} m	10^{-13} cm
neutron, $h/m_n c$	$\bar{\lambda}_{C,n}=\lambda_{C,n}/2\pi$	2.1001941(35)	1.7	10^{-16} m	10^{-14} cm
Molar volume of ideal gas at s.t.p.	V_m	22.41383(70)	31	10^{-3} $m^3 \cdot mol^{-1}$	10^3 $cm^3 \cdot mol^{-1}$
Molar gas constant, $V_m p_0/T_0$	R	8.31441(26)	31	$J \cdot mol^{-1} \cdot K^{-1}$	10^7 $erg \cdot mol^{-1} \cdot K^{-1}$
($T_0\equiv 273.15$ K; $p_0\equiv 101325$ Pa$\equiv$1atm)		8.20568(26)	31	10^{-5} $m^3 \cdot atm \cdot mol^{-1} \cdot K^{-1}$	10 $cm^3 \cdot atm \cdot mol^{-1} \cdot K^{-1}$
Boltzmann constant, R/N_A	k	1.380662(44)	32	10^{-23} $J \cdot K^{-1}$	10^{-16} $erg \cdot K^{-1}$
Stefan-Boltzmann constant, $\pi^2 k^4/60\hbar^3 c^2$	σ	5.67032(71)	125	10^{-8} $W \cdot m^{-2} \cdot K^{-4}$	10^{-5} $erg \cdot s^{-1} \cdot cm^{-2} \cdot K^{-4}$
First radiation constant, $2\pi hc^2$	c_1	3.741832(20)	5.4	10^{-16} $W \cdot m^2$	10^{-5} $erg \cdot cm^2 \cdot s^{-1}$
Second radiation constant, hc/k	c_2	1.438786(45)	31	10^{-2} $m \cdot K$	$cm \cdot K$
Gravitational constant	G	6.6720(41)[h]	615	10^{-11} $m^3 \cdot s^{-2} \cdot kg^{-1}$	10^{-8} $cm^3 \cdot s^{-2} \cdot g^{-1}$
Ratio, kx-unit to ångström, $\Lambda=\lambda(\text{Å})/\lambda(\text{kxu})$; $\lambda(\text{Cu}K\alpha_1)\equiv 1.537400$ kxu	Λ	1.0020772(54)	5.3		
Ratio, Å* to ångström, $\Lambda^*=\lambda(\text{Å})/\lambda(\text{Å}^*)$; $\lambda(\text{W}K\alpha_1)\equiv 0.2090100$ Å*	Λ^*	1.0000205(56)	5.6		

ENERGY CONVERSION FACTORS AND EQUIVALENTS

Quantity	Symbol	Numerical Value *	Units	Uncert. (ppm)
1 kilogram ($kg \cdot c^2$)		8.987551786(72)	10^{16} J	0.008
		5.609545(16)	10^{29} MeV	2.9
1 Atomic mass unit ($u \cdot c^2$)		1.4924418(77)	10^{-10} J	5.1
		931.5016(26)	MeV	2.8
1 Electron mass $m_e \cdot c^2$)		8.187241(42)	10^{-14} J	5.1
		0.5110034(14)	MeV	2.8
1 Muon mass ($m_\mu \cdot c^2$)		1.6928648(96)	10^{-11} J	5.6
		105.65948(35)	MeV	3.3
1 Proton mass ($m_p \cdot c^2$)		1.5033015(77)	10^{-10} J	5.1
		938.2796(27)	MeV	2.8
1 Neutron mass ($m_n \cdot c^2$)		1.5053738(78)	10^{-10} J	5.1
		939.5731(27)	MeV	2.8
1 Electron volt		1.6021892(46)	10^{-19} J	2.9
			10^{-12} erg	2.9
	1 eV/h	2.4179696(63)	10^{14} Hz	2.6
	1 eV/hc	8.065479(21)	10^5 m^{-1}	2.6
			10^3 cm^{-1}	2.6
	1 eV/k	1.160450(36)	10^4 K	31
Voltage-wavelength conversion, hc		1.986478(11)	10^{-25} $J \cdot m$	5.4
		1.2398520(32)	10^{-6} $eV \cdot m$	2.6
			10^{-4} $eV \cdot cm$	2.6
Rydberg constant	$R_\infty hc$	2.179907(12)	10^{-18} J	5.4
			10^{-11} erg	5.4
		13.605804(36)	eV	2.6
	$R_\infty c$	3.28984200(25)	10^{15} Hz	0.075
	$R_\infty hc/k$	1.578885(49)	10^5 K	31
Bohr magneton	μ_B	9.274078(36)	10^{-24} $J \cdot T^{-1}$	3.9
		5.7883785(95)	10^{-5} $eV \cdot T^{-1}$	1.6
	μ_B/h	1.3996123(39)	10^{10} $Hz \cdot T^{-1}$	2.8
	μ_B/hc	46.68604(13)	$m^{-1} \cdot T^{-1}$	2.8
			10^{-2} $cm^{-1} \cdot T^{-1}$	2.8
	μ_B/k	0.671712(21)	$K \cdot T^{-1}$	31
Nuclear magneton	μ_N	5.505824(20)	10^{-27} $J \cdot T^{-1}$	3.9
		3.1524515(53)	10^{-8} $eV \cdot T^{-1}$	1.7
	μ_N/h	7.622532(22)	10^6 $Hz \cdot T^{-1}$	2.8
	μ_N/hc	2.5426030(72)	10^{-2} $m^{-1} \cdot T^{-1}$	2.8
			10^{-4} $cm^{-1} \cdot T^{-1}$	2.8
	μ_N/k	3.65826(12)	10^{-4} $K \cdot T^{-1}$	31

* Note that the numbers in parentheses are the one standard-deviation uncertainties in the last digits of the quoted value computed on the basis of internal consistency, that the unified atomic mass scale $^{12}C \triangleq 12$ has been used throughout, that u=atomic mass unit, C=coulomb, F=farad, G=gauss, H=henry, Hz=hertz=cycle/s, J=joule, K=kelvin (degree Kelvin), Pa=pascal=$N \cdot m^{-2}$, T=tesla (10^4 G), V=volt, Wb=weber=$T \cdot m^2$, and W=watt. In cases where formulas for constants are given (e.g., R_∞), the relations are written as the product of two factors. The second factor, in parentheses, is the expression to be used when all quantities are expressed in cgs units, with the electron charge in electrostatic units. The first factor, in brackets, is to be included only if all quantities are expressed in SI units. We remind the reader that with the exception of the auxiliary constants which have been taken to be exact, the uncertainties of these constants are correlated, and therefore the general law of error propagation must be used in calculating additional quantities requiring two or more of these constants.

† Quantities given in u and atm are for the convenience of the reader; these units are not part of the International System of Units (SI).

‡ In order to avoid separate columns for "electromagnetic" and "electrostatic" units, both are given under the single heading "cgs Units." When using these units, the elementary charge e in the second column should be understood to be replaced by e_m or e_e, respectively.

[h] This value has been superseded by a new, more accurate measurement giving $G = 6.6726(5) \times 10^{-11}\ m^3 \cdot s^{-2} \cdot kg^{-1}$. The uncertainty is 65 ppm.

APPENDIX 9: THE CHEMICAL ELEMENTS AND THE PERIODIC TABLE

Table A9.1 THE PERIODIC TABLE OF THE CHEMICAL ELEMENTS[i]

Table P.1 THE PERIODIC TABLE OF CHEMICAL ELEMENTS[a]

IA	IIA	IIIB	IVB	VB	VIB	VIIB	VIII	VIII	VIII	IB	IIB	IIIA	IVA	VA	VIA	VIIA	0
1 H 1.0079																	2 He 4.00260
3 Li 6.941	4 Be 9.01218											5 B 10.81	6 C 12.011	7 N 14.0067	8 O 15.9994	9 F 18.99840	10 Ne 20.179
11 Na 22.98977	12 Mg 24.305											13 Al 26.98154	14 Si 28.0855	15 P 30.97376	16 S 32.06	17 Cl 35.453	18 Ar 39.948
19 K 39.098	20 Ca 40.08	21 Sc 44.9559	22 Ti 47.90	23 V 50.9414	24 Cr 51.996	25 Mn 54.9380	26 Fe 55.847	27 Co 58.9332	28 Ni 58.71	29 Cu 63.546	30 Zn 65.38	31 Ga 69.72	32 Ge 72.59	33 As 74.9216	34 Se 78.96	35 Br 79.904	36 Kr 83.80
37 Rb 85.4678	38 Sr 87.62	39 Y 88.9059	40 Zr 91.22	41 Nb 92.9064	42 Mo 95.94	43 Tc 98.9062	44 Ru 101.07	45 Rh 102.9055	46 Pd 106.4	47 Ag 107.868	48 Cd 112.40	49 In 114.82	50 Sn 118.69	51 Sb 121.75	52 Te 127.60	53 I 126.9045	54 Xe 131.30
55 Cs 132.9054	56 Ba 137.34	57–71 Rare Earths	72 Hf 178.49	73 Ta 180.947	74 W 183.85	75 Re 186.2	76 Os 190.2	77 Ir 192.22	78 Pt 195.09	79 Au 196.9665	80 Hg 200.59	81 Tl 204.37	82 Pb 207.2	83 Bi 208.9804	84 Po (210)	85 At (210)	86 Rn (222)
87 Fr (223)	88 Ra 226.0254	89-103 Acti- nides	104 Rf (257)	105 Ha (260)	106 (263)	107 (262)		109 (266)									

57 La 138.9055	58 Ce 140.12	59 Pr 140.9077	60 Nd 144.24	61 Pm (145)	62 Sm 150.4	63 Eu 151.96	64 Gd 157.25	65 Tb 158.9254	66 Dy 162.50	67 Ho 164.9304	68 Er 167.26	69 Tm 168.9342	70 Yb 173.04	71 Lu 174.97	Rare Earths (Lanthanides)
89 Ac (227)	90 Th 232.0381	91 Pa 231.0359	92 U 238.029	93 Np 237.0482	94 Pu (244)	95 Am (243)	96 Cm (247)	97 Bk (247)	98 Cf (251)	99 Es (254)	100 Fm (257)	101 Md (258)	102 No (259)	103 Lr (256)	Actinides

[i] In each box, the upper number is the *atomic number.* The lower number is the *atomic mass,* i.e., the mass (in grams) of one mole or, alternatively, the mass (in atomic mass units) of one atom. Numbers in parentheses denote the atomic masses of the most stable or best-known isotope of the element; all other numbers represent the average masses of a mixture of several isotopes as found in naturally occurring samples of the element.

Table A9.2 THE CHEMICAL ELEMENTS

Element	*Chemical symbol*	*Atomic number*	*Atomic mass*[j]
Hydrogen	H	1	1.0079 u
Helium	He	2	4.00260
Lithium	Li	3	6.941
Beryllium	Be	4	9.01218
Boron	B	5	10.81
Carbon	C	6	12.011
Nitrogen	N	7	14.0067
Oxygen	O	8	15.9994
Fluorine	F	9	18.99840
Neon	Ne	10	20.179
Sodium	Na	11	22.98977
Magnesium	Mg	12	24.305
Aluminum	Al	13	26.98154
Silicon	Si	14	28.0855
Phosphorus	P	15	30.97376
Sulfur	S	16	32.06
Chlorine	Cl	17	35.453
Argon	Ar	18	39.948
Potassium	K	19	39.098
Calcium	Ca	20	40.08
Scandium	Sc	21	44.9559
Titanium	Ti	22	47.90
Vanadium	V	23	50.9414
Chromium	Cr	24	51.996
Manganese	Mn	25	54.9380
Iron	Fe	26	55.847
Cobalt	Co	27	58.9332
Nickel	Ni	28	58.71
Copper	Cu	29	63.546
Zinc	Zn	30	65.38
Gallium	Ga	31	69.72
Germanium	Ge	32	72.59
Arsenic	As	33	74.9216
Selenium	Se	34	78.96
Bromine	Br	35	79.904
Krypton	Kr	36	83.80
Rubidium	Rb	37	85.4678
Strontium	Sr	38	87.62
Yttrium	Y	39	88.9059
Zirconium	Zr	40	91.22
Niobium	Nb	41	92.9064
Molybdenum	Mo	42	95.94
Technetium	Tc	43	98.9062
Ruthenium	Ru	44	101.07
Rhodium	Rh	45	102.9055
Palladium	Pd	46	106.4
Silver	Ag	47	107.868
Cadmium	Cd	48	112.40
Indium	In	49	114.82
Tin	Sn	50	118.69
Antimony	Sb	51	121.75
Tellurium	Te	52	127.60
Iodine	I	53	126.9045

[j] Numbers in parentheses denote the atomic masses of the most stable or best-known isotope of the element; all other numbers represent the average masses of a mixture of several isotopes as found in naturally occurring samples of the element.

Table A9.2 THE CHEMICAL ELEMENTS

Element	*Chemical symbol*	*Atomic number*	*Atomic mass*
Xenon	Xe	54	131.30
Cesium	Cs	55	132.9054
Barium	Ba	56	137.34
Lanthanum	La	57	138.9055
Cerium	Ce	58	140.12
Praseodymium	Pr	59	140.9077
Neodymium	Nd	60	144.24
Promethium	Pm	61	(145)
Samarium	Sm	62	150.4
Europium	Eu	63	151.96
Gadolinium	Gd	64	157.25
Terbium	Tb	65	158.9254
Dysprosium	Dy	66	162.50
Holmium	Ho	67	164.9304
Erbium	Er	68	167.26
Thulium	Tm	69	168.9342
Ytterbium	Yb	70	173.04
Lutetium	Lu	71	174.97
Hafnium	Hf	72	178.49
Tantalum	Ta	73	180.947
Tungsten	W	74	183.85
Rhenium	Re	75	186.2
Osmium	Os	76	190.2
Iridium	Ir	77	192.22
Platinum	Pt	78	195.09
Gold	Au	79	196.9665
Mercury	Hg	80	200.59
Thallium	Tl	81	204.37
Lead	Pb	82	207.2
Bismuth	Bi	83	208.9804
Polonium	Po	84	(210)
Astatine	At	85	(210)
Radon	Rn	86	(222)
Francium	Fr	87	(223)
Radium	Ra	88	226.0254
Actinium	Ac	89	(227)
Thorium	Th	90	232.0381
Protactinium	Pa	91	231.0359
Uranium	U	92	238.029
Neptunium	Np	93	237.0482
Plutonium	Pu	94	(244)
Americium	Am	95	(243)
Curium	Cm	96	(247)
Berkelium	Bk	97	(247)
Californium	Cf	98	(251)
Einsteinium	Es	99	(254)
Fermium	Fm	100	(257)
Mendelevium	Md	101	(258)
Nobelium	No	102	(259)
Lawrencium	Lr	103	(256)
Rutherfordium	Rf	104	(257)
Hahnium	Ha	105	(260)
?		106	(263)
?		107	(262)
?		109	(266)

Chapters 1–21:

$v = dx/dt$

$a = dv/dt = d^2x/dt^2$

$x = x_0 + v_0 t + \frac{1}{2}at^2$

$a(x - x_0) = \frac{1}{2}(v^2 - v_0^2)$

$A_x = A \cos \theta_x$

$A = \sqrt{A_x^2 + A_y^2 + A_z^2}$

$\mathbf{A} \cdot \mathbf{B} = AB \cos \phi = A_x B_x + A_y B_y + A_z B_z$

$|\mathbf{A} \times \mathbf{B}| = AB \sin \phi$

$a = v^2/r$

$\mathbf{v}' = \mathbf{v} - \mathbf{V}$

$m\mathbf{a} = \mathbf{F}$

$\mathbf{p} = m\mathbf{v}$

$\mathbf{L} = \mathbf{r} \times \mathbf{p}$

$\frac{d\mathbf{L}}{dt} = \mathbf{r} \times \mathbf{F}$

$w = mg$

$f_k = \mu_k N$

$f_s \leq \mu_s N$

$F = -kx$

$W = \mathbf{F} \cdot \Delta\mathbf{r}$

$W = \int \mathbf{F} \cdot d\mathbf{r}$

$K = \frac{1}{2}mv^2$

$U = mgz$

$E = K + U = [\text{constant}]$

$U(P) = -\int_{P_0}^{P} \mathbf{F} \cdot d\mathbf{r} + U(P_0)$

$U = \frac{1}{2}kx^2$

$E = mc^2$

$P = dW/dt$

$P = \mathbf{F} \cdot \mathbf{v}$

$\mathbf{r}_{CM} = \frac{1}{M}\int \mathbf{r}\rho \, dv$

$\mathbf{I} = \int_0^{\Delta t} \mathbf{F} \, dt$

$v_1' = \frac{m_1 - m_2}{m_1 + m_2} v_1; \quad v_2' = \frac{2m_1}{m_1 + m_2} v_1$

$\omega = d\phi/dt$

$\alpha = d\omega/dt = d^2\phi/dt^2$

$v = R\omega$

$K = \frac{1}{2}I\omega^2$

$I = \int \rho R^2 \, dV$

$I_{CM} = MR^2$ (hoop); $\frac{1}{2}MR^2$ (disk); $\frac{2}{5}MR^2$ (sphere); $\frac{1}{12}ML^2$ (rod)

$I = I_{CM} + Md^2$

$L_z = I\omega$

$I\alpha = \tau_z$

$P = \tau_z \omega$

$F = GMm/r^2$

$v^2 = GM/r$

$g = GM_E/R_E^2$

$U = -GMm/r$

$x = A \cos(\omega t + \delta)$

$T = 2\pi/\omega; \quad \nu = 1/T = \omega/2\pi$

$m \, d^2x/dt^2 = -kx$

$\omega = \sqrt{k/m}$

$\omega = \sqrt{g/l}; \; T = 2\pi\sqrt{l/g}$

$\omega = \sqrt{mgl/I}$

$\omega = \sqrt{\kappa/I}$

$y = A \cos k(x - vt) = A \cos(kx - \omega t)$

$\lambda = 2\pi/k; \quad \nu = v/\lambda; \quad \omega = 2\pi\nu$

$v = \sqrt{F/\mu}$

$P \propto v\omega^2 A^2$

$\nu_{beat} = \nu_1 - \nu_2$

$\nu' = \nu(1 \pm V_R/v)$

$\nu' = \nu/(1 \mp V_E/v)$

$\sin \theta = v/V_E$

$x' = \frac{x - Vt}{\sqrt{1 - V^2}}; \quad t' = \frac{t - Vx}{\sqrt{1 - V^2}}$

$\Delta t = \frac{\Delta t'}{\sqrt{1 - V^2}}$

$\Delta x = \sqrt{1 - V^2} \, \Delta x'$

$v_x' = \frac{v_x - V}{1 - v_x V}$

$p = \frac{mv}{\sqrt{1 - v^2/c^2}}; \quad E = \frac{mc^2}{\sqrt{1 - v^2/c^2}}$

$p - p_0 = -\rho g z$

$\frac{1}{2}\rho v^2 + \rho g z + p = [\text{constant}]$

$pV = NkT$

$T_C = T - 273.15$

$v_{rms} = \sqrt{3kT/m}$

$pV^\gamma = [\text{constant}]; \quad \gamma = C_p/C_V$

$\Delta E = \Delta Q - \Delta W$

$e = 1 - T_2/T_1$

$S(A) = \int_{A_0 \text{ rev.}}^{A} \frac{dQ}{T} + S(A_0)$

$S(B) - S(A) \geq \int_{A \text{ irrev.}}^{B} \frac{dQ}{T}$

$g = 9.81$ m/s^2
$G = 6.67 \times 10^{-11}$ N · m^2/kg^2
$M_E = 5.98 \times 10^{24}$ kg
$R_E = 6.37 \times 10^6$ m

$m_e = 9.11 \times 10^{-31}$ kg
$m_p = 1.67 \times 10^{-27}$ kg
$c = 3.00 \times 10^8$ m/s

$N_A = 6.02 \times 10^{23}$/mole
$k = 1.38 \times 10^{-23}$ J/K
1 cal = 4.19 J

Chapters 22–41:

$$F=\frac{1}{4\pi\varepsilon_0}\frac{qq'}{r^2}$$

$$E=\frac{1}{4\pi\varepsilon_0}\frac{q'}{r^2}$$

$$E=\sigma/2\varepsilon_0$$

$$p=lQ$$

$$\tau=\mathbf{p}\times\mathbf{E}$$

$$U=-\mathbf{p}\cdot\mathbf{E}$$

$$\oint E_n\,dS=\frac{Q}{\varepsilon_0}$$

$$V(P_2)-V(P_1)=-\int_{P_1}^{P_2}\mathbf{E}\cdot d\mathbf{l}$$

$$V=\frac{1}{4\pi\varepsilon_0}\frac{q'}{r}$$

$$\frac{\partial V}{\partial x}=-E_x,\quad \frac{\partial V}{\partial y}=-E_y,\quad \frac{\partial V}{\partial z}=-E_z$$

$$U=\tfrac{1}{2}Q_1V_1+\tfrac{1}{2}Q_2V_2+\tfrac{1}{2}Q_3V_3+\cdots$$

$$u=\tfrac{1}{2}\varepsilon_0E^2$$

$$C=Q/\Delta V$$

$$C=\varepsilon_0A/d$$

$$E=E_{\text{free}}/\kappa$$

$$\int\kappa E_n\,dS=\frac{Q_{\text{free}}}{\varepsilon_0}$$

$$u=\tfrac{1}{2}\kappa\varepsilon_0E^2$$

$$I=\Delta V/R$$

$$R=\rho l/A$$

$$P=I\mathscr{E}$$

$$P=I\,\Delta V$$

$$\mathbf{F}=\frac{\mu_0}{4\pi}\frac{qq'}{r^2}\mathbf{v}\times(\mathbf{v}'\times\hat{\mathbf{r}})$$

$$\mathbf{B}=\frac{\mu_0}{4\pi}\frac{q'}{r^2}(\mathbf{v}'\times\hat{\mathbf{r}})$$

$$\mathbf{F}=q\mathbf{v}\times\mathbf{B}$$

$$d\mathbf{B}=\frac{\mu_0}{4\pi}I\frac{d\mathbf{l}\times\hat{\mathbf{r}}}{r^2}$$

$$\oint\mathbf{B}\cdot d\mathbf{l}=\mu_0I$$

$$B=\mu_0I_0n$$

$$r=\frac{p}{qB}$$

$$d\mathbf{F}=I\,d\mathbf{l}\times\mathbf{B}$$

$$\mu=I\cdot[\text{area of loop}]$$

$$\tau=\mu\times\mathbf{B}$$

$$U=-\mu\cdot\mathbf{B}$$

$$\mathscr{E}=vBl$$

$$\mathscr{E}=-\frac{d\Phi_B}{dt}$$

$$\Phi_B=\int\mathbf{B}\cdot d\mathbf{S}$$

$$\Phi_B=LI$$

$$\mathscr{E}=-L\frac{dI}{dt}$$

$$U=\tfrac{1}{2}LI^2$$

$$u=\frac{1}{2\mu_0}B^2$$

$$\mu=\frac{e}{2m_e}L$$

$$B=\kappa_mB_{\text{free}}$$

$$\omega_0=1/\sqrt{LC}$$

$$Z=\sqrt{R^2+\left(\frac{1}{\omega C}-\omega L\right)^2}$$

$$Z=1\Big/\sqrt{\frac{1}{R^2}+\left(\omega C-\frac{1}{\omega L}\right)^2}$$

$$\mathscr{E}_2=\mathscr{E}_1\frac{N_2}{N_1}$$

$$\oint\mathbf{B}\cdot d\mathbf{l}=\mu_0I+\mu_0\varepsilon_0\frac{d\Phi}{dt}$$

$$E_\theta=\frac{1}{4\pi\varepsilon_0}\frac{qa\sin\theta}{c^2r}$$

$$B=E_\theta/c$$

$$\mathbf{S}=\frac{1}{\mu_0}\mathbf{E}\times\mathbf{B}$$

$$P_x=U/c$$

$$\nu=\sqrt{\frac{1-v/c}{1+v/c}}\,\nu_0$$

$$v=c/n$$

$$\sin\theta=n\sin\theta'$$

$$f=\pm\tfrac{1}{2}R$$

$$\frac{1}{s}+\frac{1}{s'}=\frac{1}{f}$$

Minima (interference):

$$d\sin\theta=\tfrac{1}{2}\lambda,\ \tfrac{3}{2}\lambda,\ \tfrac{5}{2}\lambda,\ \ldots$$

Maxima (interference):

$$d\sin\theta=0,\ \lambda,\ 2\lambda,\ \ldots$$

Minima (diffraction):

$$a\sin\theta=\lambda,\ 2\lambda,\ 3\lambda,\ \ldots$$

$$a\sin\theta=1.22\lambda$$

$$E=h\nu$$

$$p=h\nu/c$$

$$\Delta y\,\Delta p_y\gtrsim h$$

$$L=n\hbar$$

$$E=-\frac{m_ee^4}{2(4\pi\varepsilon_0)^2\hbar^2}\frac{1}{n^2}=-\frac{13.6\text{ eV}}{n^2}$$

$$\lambda=h/p$$

$e=1.60\times10^{-19}$ C
$\varepsilon_0=8.85\times10^{-12}$ F/m

$\mu_0=1.26\times10^{-6}$ H/m
$c=1/\sqrt{\mu_0\varepsilon_0}=3.00\times10^8$ m/s

$m_e=9.11\times10^{-31}$ kg
$m_p=1.67\times10^{-27}$ kg
$h=2\pi\hbar=6.63\times10^{-34}$ J·s

APPENDIX II: ANSWERS TO PROBLEMS

Chapter 1

2. 6×10^{-5} m
4. 8×10^{-7}, 4×10^{-9}, 3×10^{-13}, and 8×10^{-14} in.
6. 6.9×10^{8} m
8. 1.00×10^{7} m; 9.01×10^{6} m
10. 6.3×10^{6} m
12. 1.4×10^{17} s
14. 0.25 minute of arc, 0.29 mi
16. 0.134%, 99.866%
18. 0.021%, 99.979%
20. (a) 8.4×10^{24}; (b) 4.3×10^{46}; (c) 1.6×10^{3}
22. 9.22×10^{56}
24. 6.7×10^{27}
26. 8.3 light-minutes; 1.3 light-seconds
28. 8.9×10^{3} kg/m^3, 5.6×10^{2} lb-mass/ft^3, 0.32 lb-mass/in.^{3}, 17 slug/ft^3
30. 73×10^{-3} m^3
32. 0.038 m^3/s; 38 kg/s
34. 2.3×10^{8} tons/cm^3
36. 6.0×10^{7} tons/cm^3
38. 5.4×10^{3}, 5.2×10^{3}, 5.5×10^{3}, 3.9×10^{3}, 1.2×10^{3}, 0.63×10^{3}, 1.3×10^{3}, 0.17×10^{3}, and 10×10^{3} kg/m^3, respectively, for Mercury, Venus, Earth, Mars, Jupiter, Saturn, Uranus, Neptune, and Pluto

Chapter 2

2. 23 mi/h
4. 1.9×10^{10} years
6. 13 m/s, 0.77 s
8. (a) 13.8 s, 388 m; (b) 72 m
10. 6.4 m/s; 0 m/s
12. 32.4 m/s
14. (a) 4.25 m, 3.0 m/s, −6.0 m/s^2; (b) 2.0 m, −6.0 m/s, −6.0 m/s^2; (c) −1.5 m/s, −6.0 m/s^2
16. (b) 5650 ft; (c) approximately 5650 ft
18. (b) 1.6 s and also at times earlier or later than this by a multiple of 3.16 s, ±2.0 m/s, 0 m/s^2; (c) 0 s and also at times earlier or later than this by a multiple of 3.16 s, 0 m/s, ±2.0 m/s^2
20. 2.4 m/s^2
22. 4.0 years; 6.2×10^{8} m/s
24. 350 m/s^2
26. (a) 65.2 ft/s^2; (b) at constant acceleration the distance would have been 345 yd; (c) 319 mi/h
28. 7.1 m/s^2
32. 1.60 ft/s^2; 13.7 s
34. 76 km/h
36. 66 m
38. 44 m
40. 34 m; 25 liters
42. 1.6×10^{4} m/s^2
44. 802 m/s; 1.89 s
46. (a) $n\sqrt{2h/g}$; (b) $\frac{3}{4}$ h; (c) $\frac{2}{3}$ h

Chapter 3

2. 11.2 km at 2.3° north of east
6. (b) 2.12×10^{11} m
10. 5550 km
12. 9.2 km; 7.7 km
14. (b) $C_x = 2$ cm, $C_y = 5$ cm
16. $A_x = 4.2$, $A_y = 0.5$, $A_z = \pm 4.2$
18. 3.74
20. 5.9; 32°, −60°, 80°
22. $17\hat{\mathbf{x}} + 3\hat{\mathbf{y}} - 16\hat{\mathbf{z}}$
24. 2.24×10^{6} m^2
26. $0.44\hat{\mathbf{x}} - 0.22\hat{\mathbf{y}} - 0.88\hat{\mathbf{z}}$
28. 2.4; 1.8
30. 47.3 m down
32. $-12\hat{\mathbf{x}} - 14\hat{\mathbf{y}} - 9\hat{\mathbf{z}}$
34. $0.45\hat{\mathbf{x}} - 0.59\hat{\mathbf{y}} - 0.67\hat{\mathbf{z}}$
38. (a) 2.80 mi, 4.15 mi; (b) 3.59 mi, 3.48 mi
42. rotate coordinate system by an angle $\theta = -26.6°$
44. (a) $x'' = x'$, $y'' = y' \cos\phi + z' \sin\phi$, $z'' = -y' \sin\phi + z' \cos\phi$; (b) $x'' = x\cos\theta + y\sin\theta$, $y'' = -x\sin\theta\cos\phi + y\cos\theta\cos\phi + z\sin\phi$, $z'' = x\sin\theta\sin\phi - y\cos\theta\sin\phi + z\cos\phi$

Chapter 4

2. 29.9 km/s; 19.0 km/s
4. (a) $8\hat{\mathbf{x}} + 10\hat{\mathbf{y}}$, 12.8 m/s; (b) $4\hat{\mathbf{x}} + 6\hat{\mathbf{y}}$, 7.2 m/s^2
6. 2.65 ft
8. 89 mi/h
10. (a) 2.5×10^4 m; 5.0×10^4 m; (c) no
12. 216 ft/s; 306 ft/s
14. (a) 7.25°; (b) 13 m
16. 25 ft/s; 63°
18. (a) 22 m/s; (b) 3.5 rev/s
20. 230 ft; 9.4 gal.
22. (a) 6.0 m; (b) 1.5 m; (c) 1.1 s
24. 0.11 m
26. (a) 230 ft/s; (b) 14 ft; (c) 26 ft
28. 43°; 35 ft/s
30. (a) 17.5°; (b) 6 m
32. (a) 270 m; (b) air resistance
34. no; 12.2 m
36. 9.95°; 205 m, or 42 minutes of arc
38. 2.4×10^2 ft/s; 1.8×10^5 ft/s^2
40. 8.9 m/s^2
42. 7.4×10^3 ft/s^2
44. 13 m/s at 40° from vertical
46. 27 m/s at 68° from vertical
48. 329 m/s
50. 27 km/h at 43° east of north; 33°
52. 85°
54. 15 km/h at 15° east of north
56. $v = 4v_0^2t/(4v_0^2t^2 + h^2)^{1/2}$
58. 528 km/h

Chapter 5

2. 54 ft/s^2
4. 6.6×10^2 kp
6. 1.2×10^4 N
8. 3.31×10^5 lb
10. 6.4×10^2 lb; 4.1×10^2 lb
12. $F = -623 + 66.5t$ measured in lb
14. no, since with 1150 children on each side the tension would be 34,500 lb
16. 4.7×10^{20} N at 25° with the Sun–Moon line
18. 175 lb toward dock
20. 5.9×10^2 N; 7.8×10^2 N
22. 6.9×10^5 N; 2.8×10^3 N
24. 40 lb
26. 8.2×10^2 kp
28. 37 lb at 20° downward from horizontal
30. 150 N
32. 3.6×10^3 slug · ft/s; 5.0 mi/h
34. 1.28 slug · ft/s; 1.96 slug · ft/s
36. 7.4×10^2 N
38. 5.3×10^{-13} kg · m^2/s
40. 2.1×10^{12} kg · m^2/s in north direction
42. (a) 1.1×10^{14} kg · m^2/s; (b) 4.3×10^{36} kg · m^2/s
44. (a) 0 kg · m^2/s; (b) 2.7×10^{40} kg · m^2/s, 5.4×10^{40} kg · m^2/s, 2.7×10^{40} kg · m^2/s, no
46. 5.26×10^{12} m
48. $2\hbar/[(1 - 2\sqrt{2}/3)m_e a_0] = 7.6 \times 10^7$ m/s; $2\hbar/[(1 + 2\sqrt{2}/3)m_e a_0)] = 2.2 \times 10^6$ m/s

Chapter 6

2. (a) 0.018 oz; (b) no, since price is based on oz-mass
4. 9.2×10^2 N; 1.4×10^2 N
6. (a) [weight] = 7.4×10^2 N, [normal force] = 6.0×10^2 N, [resultant] = 4.2×10^2 N; (b) 5.6 m/s^2
8. 64 m; 5.1 s
10. 9.8 m/s^2
12. 7.0°
14. $a_1 = g(4m_2m_3 - m_1m_2 - m_1m_3)/(4m_2m_3 + m_1m_2 + m_1m_3)$, $a_2 = g(3m_1m_3 - 4m_2m_3 - m_1m_2)/(4m_2m_3 + m_1m_2 + m_1m_3)$, $a_3 = g(3m_1m_2 - 4m_2m_3 - m_1m_3)/(4m_2m_3 + m_1m_2 + m_1m_3)$, $T_1 = 2T_2 = 8m_1m_2m_3g/(4m_2m_3 + m_1m_2 + m_1m_3)$
16. 9.6×10^2 lb
18. 3.9 m/s^2
20. 1.3×10^2 ft
22. 53 m
24. 0.54 ft/s^2
26. 3.8×10^2 ft; 6.4 s
28. 27°
30. 4.4 m/s^2, 1.3 N
32. $T = \mu_k mg/\sqrt{1 + \mu_k^2}$
34. $a_1 = F/m_1 - \mu_1 g$, $a_2 = \mu_1 m_1 g/m_2 - \mu_2 g(m_1 + m_2)/m_2$
36. 6.8×10^2 N
38. 74 N
40. 0.15 m
42. 180 N/m, 360 N/m
46. 3.4 s
48. 4.4×10^2 N
50. 7.6×10^2 N; 8.1×10^2 N
52. 3.0 m/s
54. 67°
56. 47 mi/h
58. 6.4°
60. $v = \sqrt{gl \tan\theta \sin\theta}$
62. $T = mv^2/(2\pi r)$
64. $\phi = \tan^{-1}[\tan\theta/(1 - v_E^2/gR_E)] - \theta$ or approximately $(\sin\theta\cos\theta)v_E^2/gR_E$, where R_E and v_E are the equatorial radius and speed of the Earth and θ is the latitude angle; 0.099°

Chapter 7

2. 3.6×10^3 ft · lb
4. 8.8×10^3 J
6. 2.6×10^3 J
8. (a) 1.2×10^4 J; (b) 290 N, 1.2×10^4 J
10. 24 J
12. (a) 7.1×10^3 N; (b) 2.2×10^5 J, 8.1×10^3 N
14. curved ramp: $W = mgR[(1 - \sqrt{2}/2) + \mu_k\sqrt{2}/2]$; straight ramp: $W = mgR(1 - \sqrt{2}/2)[1 + \mu_k \cos(45°/2)]$, which is less
16. 2.2×10^{-18} J
18. (a) 4.0×10^5 J; (b) 2.5×10^4 J; (c) 1.2×10^6 J
20. 7.2×10^6 ft · lb, 4.2×10^6 ft · lb; 3.0×10^6 ft · lb
22. 8.6×10^4 ft · lb, 4.2×10^3 ft · lb; 20 times
24. (a) 1.2×10^4 N; (b) 39 m; (c) 4.7×10^5 J; (d) 4.7×10^5 J

26. 1.4×10^5 ft · lb
28. 1.6×10^9 ft · lb, 4.8×10^9 ft · lb
30. 8.2×10^6 m^3
32. 5.1 m; yes, since the jumper does some extra work with his arms
34. 99 m/s; 9.8×10^{10} J; 23 tons
36. (a) 2.5×10^4 N; (b) 1.2×10^7 J; (c) 15 m/s
38. 50 J, 17 J
40. 9.9 m/s; 26 m/s
42. 53° from top

Chapter 8

4. $U = K/(3x^3)$
6. 217 J
8. (a) 47 J; (b) 0.89 m
10. 0.26 m
12. $t = v/\mu_k g$; $x = v^2/2\mu_k g$; $mv^2/2$
14. $\mathbf{F} = (2K/x^3)\hat{\mathbf{x}}$
16. 0.19 Å, 0.80 Å
18. (a) none and 0.2 m for E_1, 3.1 m and 0.3 m for E_2, 1.3 m and 0.5 m for E_3; (b) absolute maximum at 0.9 m, absolute minimum at turning points, local maximum at 2.2 m, local minimum at 1.7 m; (c) unbound for E_1, bound for E_2 and E_3
20. 11 eV
22. 2.2×10^6, 1.1×10^7, 2.7×10^6, 5.8×10^5, 3.6×10^6, and 9.8×10^6 J; bus; snowmobile
24. 1.0×10^6 eV
26. 9.4×10^8 eV
28. 540 kcal
30. 18 kW · h
32. 1.1×10^3 W; 0 W
34. 0.61 hp
36. 550 ft · lb
38. 9.1 gal./h
40. 4.2×10^5 W
42. (a) 1.2×10^{10} ft · lb; (b) 17 min; (c) 17 min, 2 or 3 mi
44. 1.2×10^{-4} W
46. 20 mi/h
48. 53 hp
50. 37%
52. (a) 3.2×10^4 W; (b) 7.8×10^2 W; (c) 3.1×10^4 W
54. 2.5×10^3 km^2
56. 2.4×10^5 ft · lb/s
58. 4.3×10^9 kg/s; 1.4×10^{17} kg

Chapter 9

2. 4.3 ft/s
4. 14 mi/h; 26 mi/h and 54 mi/h
6. 150 bullets
8. $\sqrt{2}v$, at 135° with respect to the direction of the other fragments
10. (a) 42 km/h at 56° east of north; (b) 3.6×10^5 J
12. 74 lb
14. (a) 0.10 kg/s; (b) 2.3 kg m/s, 2.3 N
16. 0.62 ft from center of seesaw
18. 2.7×10^{-4} m
20. 2.28 Å from H
22. (950, 180, 820)
24. on diagonal of cube, at $\sqrt{3}/3L$ from vertex
26. $L/(64\pi - 4)$ from center of cube, above hole
30. 950 m
32. 6.5×10^7 ft · lb
34. 2120 lb
36. 1.0 m
38. 7.1×10^5 m from first fragment
40. (a) 6.8×10^6 ft · lb, 0 ft · lb; (b) 6.8×10^8 ft · lb, 1.3×10^6 ft · lb, extra energy comes from explosive chemical reactions
42. (a) 3.4×10^5 J, 3.6×10^5 J; (b) 3.4×10^5 J, 0 J
44. 2.81×10^{34} kg · m^2/s; 3.46×10^{32} kg · m^2/s
46. $1 - 1/e$

Chapter 10

2. -7.8×10^8 N; 1.1 m/s^2
4. 3.4×10^3 ft/s^2; 1.5×10^4 lb
6. 14 m/s
8. 2.6 km/h, 12.6 km/h
10. (a) 18 m/s; (b) 30 m/s
12. 39 m/s
14. 0.57 J
16. 1.8×10^4 m/s, at 139°
18. 4.0×10^{-13} J
20. (a) $h/9$, $4h/9$; (b) h, 0
22. $v/3$ to left, $2v/9$ to right, $8v/9$ to right
24. 45 ft/s
26. (a) 3.0 m/s; (b) 2.1×10^2 m/s^2
28. 21 m/s
30. $v/5$ to left, $2\sqrt{3}v/5$ up 30° to right, and $2\sqrt{3}v/5$ down 30° to right
32. $v_2' = 7.1 \times 10^6$ m/s, $\theta_2' = 52°$
34. 45° each, 4.0×10^{-13} J each
36. 1.9×10^5 eV, 70°
38. 1.25×10^{-13} J
40. 4.3×10^{-11} J

Chapter 11

2. 7.3×10^{-5} radian/s; 460 m/s, 350 m/s
4. 0.13 in./min; 1.2×10^3 in./min; 5.2×10^2 in./min
6. 81 radian/s; 13 rev/s
8. (b) 5.4×10^{-3} m
10. 12 radian/s^2
12. 9.6×10^{-22} radian/s^2
14. 1.21×10^{-10} m
16. 6.50×10^{-46} kg · m^2
18. $\frac{1}{2}M(R_1^2 + R_2^2)$
20. -1.6×10^{22} kg · m^2
22. $0.38M_E^2R_E^2$
24. $\frac{1}{12}ML^2 \sin^2 \theta$
28. $\frac{1}{8}MR^2$
30. $\frac{1}{2}MR^2 - (r^2M/R^2)(\frac{1}{2}r^2 + d^2)$
32. $\frac{1}{2}MR^2 - (2r^2M/R^2)(r^2 + \frac{1}{2}R^2)$
34. $0.338M_E^2R_E^2$
36. (a) 2.2×10^2 kg · m^2; (b) 4.4×10^3 J
38. 2.61×10^5 ft · lb; 3.83%

42. 2.74×10^3 J; 13.1 J; 4.74×10^{-3}
44. $\frac{1}{6}Ml^2$
48. 3.8×10^{-2} slug · ft^2/s^2
50. 5.6×10^{41} J · s; 3.1×10^{43} J · s; 1.8%
52. 1.8×10^{22} kg · m^2/s^2

Chapter 12

2. 400 N
4. 1.2×10^4 N; 7.2×10^3 N · m; depress
6. 2.7×10^4 N · m
8. 180 lb
10. 9.7 m/s^2
12. (a) 110 J · s; (b) 34 kg · m^2/s^2; (c) 34 N · m
14. (a) 76°; (b) 290 N, 250 N
16. 1.0×10^{18} J · s; 3.0×10^{12} J · s/s; 3.0×10^{12} N · m
18. (a) $\omega_1 R_1/R_2$; (b) $(T - T')R_1$, $(T - T')R_2$; (c) $(T - T')R_1\omega_1$, $(T - T')R_2\omega_2$, yes
20. 27 N · m
22. 4.6 radian/s
24. (a) 8.4×10^2 ft · lb; (b) 6.9 ft/s^2
26. (a) 2.4×10^{19} J · s; (b) 3.0×10^{-19} radian/s
28. 7.2°
30. (a) 5.0 m/s; (b) 9.0×10^{-3} radian/s; (c) 19 times
32. [height of end] $= m^2v^2/[2g(M/3 + m)(M/2 + m)]$
34. $v_{CM} = mv/(m + M)$, $\omega = v_{CM}/[\frac{1}{3}l + \frac{1}{4}l/(1 + m/M)^2 + \frac{1}{4}lm/M]$
38. 9.6 m/s^2
40. 100 N in *forward* direction; 300 N in forward direction
42. $2a/7$
44. $a = g(\sin\theta - \mu_k \cos\theta)$; $\alpha = (2/R)g\mu_k \cos\theta$
46. $a = (g \sin\theta)(m + 4M)/(m + 6M)$
48. 0.85 radian/s
50. 22 lb at 23° right of vertical at the top, and 23° left of vertical at the bottom
52. $2mg$, $mg/\sqrt{3}$, $mg/\sqrt{3}$
54. 1.7×10^3 N; 1.5×10^3 N
56. $0.11Mg$
58. (a) $U = 5.3 \times 10^2 \text{ J} \times \{\sin(\theta + \tan^{-1} 2) +$ [constant]$\}$; (b) 27°; (c) 56 J
60. 38% front, 62% rear
62. 0.31 m; 590 N

Chapter 13

2. 4.7×10^{-35} N
4. 8.88 m/s^2, 3.71 m/s^2, 3.71 m/s^2
6. 4.0×10^{-4} radian/s, or 0.23 rev/h
8. 2×10^8 years, 3×10^5 m/s
10. Lincoln, Nebraska, 22.6° west of New York City
12. $m_1/m_2 = 1.6$
14. 3.0×10^{10} m
16. (a) 7.5×10^3 m/s, 8.32×10^3 m/s; (b) 3.94×10^8 J, 4.85×10^8 J
18. 3.5 days
20. 2.66×10^{33} J, -5.31×10^{33} J; -2.66×10^{33} J
22. (a) 6.0×10^{-14} m/s^2, 9.1×10^{-14} m/s^2; (b) 1.1×10^5 m/s, 1.6×10^5 m/s; (c) 3.4×10^{51} J, 5.1×10^{51} J, 7.7×10^{51} J, yes
24. (a) 1.1×10^4 m/s; (b) 1.2×10^{11} J, 29 tons; (c) 1.2×10^5 m/s^2
26. (a) -1.1×10^{11} J, -2.2×10^{11} J, -1.1×10^{11} J; (b) yes, yes
28. (a) 1.1×10^4 m/s; (b) 2.4×10^3 m/s
30. (a) I and III elliptical, II circular
32. 2.5×10^6 m; 8.2×10^2 m/s
34. (a) 4.21×10^4 m/s; (b) 7.19×10^4 m/s, 1.23×10^4 m/s
36. 7.78×10^7 m
40. (a) 2.6×10^3 m/s; (b) 2.7×10^3 m/s; (c) 0.40 year; (d) at launch, Venus must be 90° behind Earth as seen from Sun
42. 4.92, 7.67, 10.8, 9.8, 9.9, and 9.8 m/s^2
44. $U = GMm(r^2/R^2 - 3)/2R$

Chapter 14

2. (a) 1.7×10^4 m/s^2, 27 m/s; 2.0×10^4 N
4. (a) $A = 3.0$ m, $\nu = 0.32$ Hz, $\omega = 2.0$ radian/s, $T = 3.1$ s; (b) 0.26 s, 1.05 s (and also at times larger and smaller by a multiple of one-half period)
6. (a) $0.20 \cos 6\pi t + (4.0/6\pi) \sin 6\pi t$; (b) 0.043 s, -1.0×10^2 m/s^2
8. 1.9×10^4 N
10. 1.13×10^{14} Hz
12. (a) 4.5 Hz; (b) 8.5 m/s, strong vibrations
14. (a) $x_1 = -x_2 = (-v_1/\omega) \cos\omega t$ with $\omega = \sqrt{3k/m}$; (b) $x_1 = x_2 = (-v_1/\omega)\cos\omega t$ with $\omega = \sqrt{k/m}$
16. (a) 6.6 J; (b) $t = 0.30$ s, $t = 0$; (c) $t = 0.15$ s
18. 18 eV; no
22. (a) 0.73 min; (b) 1.0 mm
24. 0.19 Hz
26. (a) 1.3×10^{-3} J; (b) 0.14 m/s
28. (a) 9.4 J; (b) 1.6×10^{-5} W; (c) 6.0×10^{-3} N · m/radian
32. 1.988 s
34. 1.5 s
36. (b) $l = R/\sqrt{2}$
38. 3.6 radian/s; 11 ft/s
40. $2\pi\sqrt{4l/5g}$
42. (a) $T = mg[1 - \frac{1}{2}A^2 + \frac{3}{2}A^2 \sin^2(\sqrt{g/l}\, t)$; (b) $t = \frac{1}{2}\pi/\sqrt{g/l}$ gives $T = mg(1 + A^2)$
44. $\gamma = 2.5 \times 10^{-3}$/s

Chapter 15

2. 0.114 Hz, 0.716 radian/s, 0.0524 /m, 13.7 m/s
4. (a) 6.0×10^{-3} m, 0.31 m, 20 /m, 0.64 Hz, 4.0 radian/s; (b) -0.26 s
6. 7.8 m/s
8. 5.5×10^{-7} m; 4.2×10^{-7} m
10. 0.69 s
12. 1.6 s
16. 1:2.4
18. $A = 5.0$; $\delta = -\tan^{-1} \frac{4}{3}$
20. 0.007 Hz
22. 1.00/day, 0.92/day; 1.00 day, 1.09 day; first is due to Sun, second due to Moon
24. 392 Hz, 588 Hz, 784 Hz, 980 Hz

26. 1.6 Hz, 3.2 Hz
28. 71 N
30. 3.7 m/s; 6.8×10^3 m/s^2
32. (a) 9.6×10^3 N; (b) 12 Hz
34. 1.25×10^{-7} m; 2.5×10^{-7} m
36. 0.018 J/m $\times$ ($\sin^2 \omega t \cos^2 kx + \cos^2 \omega t \sin^2 kx$), where $\omega = 3.7 \times 10^3$ radian/s and $k = 28$/m; location of maxima depends on time: maxima are at 0.057 m, 0.17 m, 0.28 m when $\sin^2 \omega t < \cos^2 \omega t$, and maxima are at 0 m, 0.11 m, 0.23 m, 0.34 m when $\sin^2 \omega t > \cos^2 \omega t$; 0.018 J/m $\times \sin^2 \omega t$ or 0.018 J/m $\times \cos^2 \omega t$

Chapter 16

2. 55 Hz, 4187 Hz
4. 1.9, 1.8, 1.7, 1.6, 1.5, 1.5, 1.3, 1.3, 1.2, 1.1, 1.1, and 1.0 cm
6. 3.1 m
8. 130 times; 21 dB
10. 9.8 s; use visual signal
12. 150 m
14. (a) 1.2×10^{-4} s; (b) on the glass
18. (a) 0.63 m; (b) 3.5, 3.3, 3.2, 3.0, 2.8, 2.7, 2.5, 2.4, 2.2, 2.1, 2.0, and 1.9 cm
22. 71 ft/s; 0.22 Hz
24. 594 Hz, 596 Hz
26. 476 Hz
28. 481 Hz
30. (a) 33.4°; (b) 30.4 s
32. 16.1 ft/s; 6.76 ft/s^2
34. 1.7×10^2 m
36. (a) 2.3 /h; (b) 320 km, 740 km/h
38. (a) 31 m/s; (b) 0.11 /h, 9.0 h; (c) probably
40. (b) 1.6 m, 3.2 m
42. 4.0×10^3 km
44. 7.2
46. (a) 6.3 m/s; (b) 1.0 m

Chapter 17

2. (b) 2.4, $v' = 0.42$
4. (a) $x = 2.3 \times 10^8$ light-seconds, $y = 4.0 \times 10^8$ light-seconds, $z = 0$, $t = -1.7 \times 10^8$ s; (b) before
8. $(1 + 3.4 \times 10^{-10})$; 3.0×10^{-5} s
10. 3.7×10^2 m/s; $(1 + 7.5 \times 10^{-13})$
12. $V = (1 - 1.0 \times 10^{-11})$; 2.2×10^6 years
14. (a) 8.2×10^8 s; (b) 7.6×10^8 s
16. 4×10^{-13}%
18. $0.999c$
20. $0.77c$; 67°
22. $0.87c$
26. 4.0×10^{-3}%
28. 130 m/s; 5.3×10^{-16} kg · m/s
30. 1.4×10^{-15} m/s
32. (a) 9.0×10^{13} m/s^2; (b) 1.8×10^{-24} kg; (c) 1.6×10^{-10} N
34. $0.83c$
38. (a) $0.99962c$; (b) $0.973c$; (c) 6.5×10^{-10} J

Chapter 18

2. (a) 11 m; (b) 8.1 cm, 11.7 cm
4. 1.4×10^3 lb; layer of air under the paper exerts an opposite pressure force
6. 20 in.^{2}
8. 34.1 lb/in.^{2}
10. 5.1×10^4 N
12. 51 atm
14. (a) 360 N; (b) 330 N
16. 0.86 m
18. 10.3 m
20. (b) 5.0×10^{11} N/m^2
22. (a) 4.7×10^7 m^3; (b) 4.8×10^{10} kg
24. yes
26. (a) 9.5 m; (b) 28.4 m
28. 0.115%; 0.011%; 0.002%
30. $\rho = M/[\frac{4}{3}\pi R^3 + R'^2 (l - h)]$
32. 19 cm
34. 29 N/m^2
36. (a) 51 ft/s; (b) 16 lb/in.^{2} (overpressure)
38. (a) $v = \sqrt{2g(h_2 - h_1)}$; (b) $p_{atm} - \rho g h_2$; (c) $p_{atm}/\rho g$
40. 2.3 hp
44. $(A/A')\sqrt{2l/g}$

Chapter 19

2. 58.0°C = 331.2 K; −88.3°C = 184.9 K
4. 2.69×10^{19} molecules
6. 3.4 atm
8. 1.4×10^{-9} N/m^2
10. 11.6 kg/m^3
12. 3.6×10^6 N/m^2
14. 0.094 kg
16. 6.4×10^3 lb/in.^{2}
18. 96.3 g
20. 2.1×10^3 kg
22. (a) 1.2 m; (b) 2.5 kg at 2.5 atm
24. 29.0 g
28. 5.7×10^{-21} J
30. 6×10^6; 30 cm
32. 0.43%
34. 1.9×10^5 J; $\frac{3}{5}$; $\frac{2}{5}$
36. 284 K

Chapter 20

2. 97 kW
4. 25°C
6. 97 W
8. 4.3 cal/s
10. 3.5 h
12. 1.4 liters/min
14. 0.21 m
16. ±0.08°C
18. (a) 0.43 in.; (b) 1.6 ft
22. (a) 709 kg/m^3; (b) 43¢/kg, 45¢/kg
26. 100.32°C
28. 0.085 kg/h
30. 1.8 cm

32. 0.45 cm/h
36. (a) 1.0×10^{11} kcal; (b) 5 A-bombs
38. 22°C
40. 0.65 kg/h
42. 2.5×10^3 J; 50 m/s
44. 1.19×10^2 kcal
46. 1.5×10^4 kcal
48. (a) 2.62 min; (b) 2.64 min, no
50. 30.8 atm; 507°C
52. −225°C
54. 8.1×10^3 ft · lb

Chapter 21

2. $\Delta Q = 1.96$ J, $\Delta W = 1.96$ J, $\Delta E = 0$
4. 518 kcal/kg
6. 0.19 kg/s
10. (a) 2.0×10^4 J; (b) 3.9×10^4 J, 1.9×10^4 J
12. $1 - 3 \times 10^{-9}$, or 99.9999997%
14. 3.9×10^2 W
16. (a) 3.1×10^2 MW; (b) 1.0×10^3 MW
18. 7.1 kW
20. 1.7×10^2 J
22. 8.2×10^3 cal/K · s
24. 3.7×10^2 cal/K · day
26. -2.9×10^2 cal/K
28. 39 cal/K
30. (a) 6.9×10^3 cal/K; (b) 4.6×10^2 cal/K

Chapter 22

2. 2.89×10^{-9} N
4. 17 N; 2.6×10^{27} m/s²
6. $-6.9 \times 10^{-4}\,\hat{\mathbf{x}} + 1.7 \times 10^{-3}\,\hat{\mathbf{y}}$; $6.9 \times 10^{-4}\,\hat{\mathbf{x}} - 1.7 \times 10^{-3}\,\hat{\mathbf{y}}$
8. 10^{-9}; 10^{-5}
10. 2.6×10^{-39} C; 1:1.8×10^5; attractive
14. 8.4×10^{22}
16. 2.4×10^{28} of each
18. 6.8×10^{32} electrons on Earth and 1.9×10^{32} electrons on Moon
20. 9.4×10^{18}/s
22. 4 electrons

Chapter 23

2. 6.2 m/s² upward
4. 1.3×10^{-13} N; 1.4×10^{17} m/s²
6. 9.0×10^3 N/C
8. 6.2×10^{-7} m
10. 5.1×10^{11} N/C
12. 5.1×10^{12} N/C in positive x direction
14. 2.9×10^4 N/C at 71° below horizontal
16. 2.3×10^4, 1.7×10^4, 0.88×10^4, 0.35×10^4, 0.11×10^4, and 0.017×10^4 N/C
18. $\sqrt{15}\lambda/(8\pi\varepsilon_0 d)$
20. $E_x = -\dfrac{1}{4\pi\varepsilon_0}\dfrac{\lambda}{\sqrt{x^2+y^2}}$, $E_y = \dfrac{1}{4\pi\varepsilon_0}\dfrac{\lambda}{y}\left(1 + \dfrac{x}{\sqrt{x^2+y^2}}\right)$
22. 1.1×10^5 N/C up, 1.1×10^5 N/C down, 1.1×10^5 N/C down, 3.4×10^5 N/C down
24. 2.4×10^5 N/C
26. (a) $E = \dfrac{Q}{4\pi\varepsilon_0}\dfrac{1}{x(x+l)}$ in positive x direction

    (b) $E = \dfrac{Q}{2\pi\varepsilon_0}\dfrac{1}{y\sqrt{l^2+4y^2}}$ in positive y direction
28. $E = 2Q/(l^2\pi\varepsilon_0)$ toward lower right corner
30. $E = \dfrac{Q}{2\pi\varepsilon_0}\dfrac{1}{R^2}\left(1 - \dfrac{z}{\sqrt{x^2+R^2}}\right)$ along axis
32. $E_x = \dfrac{\lambda}{4\pi\varepsilon_0}\left[\dfrac{1}{x}\left(1 + \dfrac{y}{\sqrt{x^2+y^2}}\right) - \dfrac{1}{\sqrt{x^2+y^2}}\right]$

    $E_y = \dfrac{\lambda}{4\pi\varepsilon_0}\left[\dfrac{1}{y}\left(1 + \dfrac{x}{\sqrt{x^2+y^2}}\right) - \dfrac{1}{\sqrt{x^2+y^2}}\right]$
34. $\dfrac{Q^2}{4\pi\varepsilon_0 L^2}\ln\left[\dfrac{(x+L)^2}{x(x+2L)}\right]$
36. (a) 8.5×10^{-30} C · m; (b) 0 C · m
38. 4.8×10^{-24} N · m

Chapter 24

2. 1.1×10^3 N · m²/C
4. 1.1×10^5 N · m²/C; 0 N · m²/C; 1.1×10^5 N · m²/C
6. 4.4×10^{-11} C/m³
8. 8.3×10^{-9} C
10. 4.8×10^{20} N/C; 3.5×10^3 N; 1.8×10^{28} m/s²
12. 7×10^{-9} C/m
14. $\mathbf{E} = (\rho x/\varepsilon_0)\hat{\mathbf{x}}$ for $x < d/2$; $\mathbf{E} = \pm(\rho d/2\varepsilon_0)\hat{\mathbf{x}}$ for $\pm x > d/2$
16. $E = 0$ for $r < a$, $E = \dfrac{Q}{4\pi\varepsilon_0}\dfrac{r^3 - a^3}{b^3 - a^3}\dfrac{1}{r^2}$ for $a < r < b$, $E = \dfrac{Q}{4\pi\varepsilon_0}\dfrac{1}{r^2}$ for $r > b$
18. $E = \dfrac{Q}{4\pi\varepsilon_0}\left[\dfrac{1}{r^2} - \dfrac{1}{8(r - R/2)^2}\right]$ in radial direction
20. 0.5 Å
22. 1.2×10^7 m/s, 2.2×10^{-13} J, 1.1×10^{-34} J · s, 6.5×10^{20} Hz
26. (a) 1.9×10^{-25} kg · m/s; (b) 2.1×10^5 m/s; (c) 2.1×10^5 m/s
28. 10^{-6} C/$(4\pi\varepsilon_0 r^2)$; 4×10^{-6} C/$(4\pi\varepsilon_0 r^2)$
30. 8.8×10^{-10} C/m²

Chapter 25

2. −6.0 eV
4. 4.5×10^3 V
6. 1.3×10^7 V/m
8. 1.9×10^3 V/m
10. 1.7×10^7 V; 2.5×10^7 V/m
12. 5.8×10^{-12} J
14. 7.5×10^7 V; 1.7×10^7 V
16. (c) −13.6 eV
18. $V = \dfrac{1}{4\pi\varepsilon_0}\dfrac{Q}{l}\ln\left(\dfrac{2+\sqrt{3}}{2-\sqrt{3}}\right)$
20. $V = \dfrac{1}{4\pi\varepsilon_0}\dfrac{Q}{l}\ln\left[\dfrac{x+l}{x}\,\dfrac{l+\sqrt{x^2+l^2}}{x}\,\dfrac{l+\sqrt{(x+l)^2+l^2}}{x+l}\,\dfrac{l+x+\sqrt{(x+l)^2+l^2}}{x+\sqrt{x^2+l^2}}\right]$

22. 3.1×10^7 m/s

24. $V = \dfrac{Q}{2\pi\varepsilon_0}\dfrac{1}{R^2}(\sqrt{R^2+z^2}-z)$

26. $V = 4k/(\varepsilon_0 r^{1/2})$

28. $E = -8\hat{\mathbf{x}} - 4\hat{\mathbf{y}}$ in volt/meter

30. $V = \dfrac{1}{4\pi\varepsilon_0}\dfrac{Q}{l}\ln\left(\dfrac{x+l}{x}\right); \quad E = \dfrac{1}{4\pi\varepsilon_0}\dfrac{Q}{x(x+l)}$

 in positive x direction

32. (a) $V = \dfrac{2Q}{3\pi\varepsilon_0 R^2}(\sqrt{z^2+R^2} - \sqrt{z^2+R^2/4})$

 (b) $E = \dfrac{2Qz}{3\pi\varepsilon_0 R^2}\left(-\dfrac{1}{\sqrt{z^2+R^2}} + \dfrac{1}{\sqrt{z^2+R^2/4}}\right)$

 along axis

34. (a) 6.4×10^7 V/m parallel to **p**;
 (b) 3.2×10^7 V/m antiparallel to **p**

36. (b) $E_x = \dfrac{Ql^2}{4\pi\varepsilon_0}\dfrac{x}{r^5}\left[2 + \dfrac{5(2z^2-x^2-y^2)}{r^2}\right]$

 $E_y = \dfrac{Ql^2}{4\pi\varepsilon_0}\dfrac{y}{r^5}\left[2 + \dfrac{5(2z^2-x^2-y^2)}{r^2}\right]$

 $E_z = \dfrac{Ql^2}{4\pi\varepsilon_0}\dfrac{z}{r^5}\left[-4 + \dfrac{5(2z^2-x^2-y^2)}{r^2}\right]$

 where $r^2 = x^2 + y^2 + z^2$

Chapter 26

2. 4.0×10^{-17} J
4. 5.8×10^6 eV
6. -50 eV
8. $U = \dfrac{1}{4\pi\varepsilon_0}\dfrac{Q^2}{l^2}\left[x \ln\dfrac{(x+2l)x}{(x+l)^2} + 2l \ln\dfrac{x+2l}{x+l}\right]$
10. 1.7 J
12. 1.41×10^{-15} m
14. 5.1×10^{31} J/m^3
16. 1.8×10^{31}, 4×10^{16}, 1.6×10^{12}, and 1.1×10^2 J/m^3; 2.0×10^{14}, 0.5, 1.8×10^{-5}, and 1.2×10^{-15} kg/m^3
18. 3.5×10^{-15} m
20. 1.1×10^6 eV
22. (b) 1.2×10^{29} J; 1:5.3×10^{10}
24. $0.980\ e^2/8\pi\varepsilon_0 R) = 7.1 \times 10^5$ eV

Chapter 27

2. (a) 2.0×10^{-11} F; (b) 4.0×10^{-6} C
4. 9.5 pF
6. 15.5 μF; 1.5 μF
8. 2.4×10^{-5} C, 7.2×10^{-5} C
10. 7.4 m^2
12. $C = \dfrac{\varepsilon_0 A}{d}\dfrac{2\kappa_1\kappa_2}{\kappa_1+\kappa_2}$
16. 0.11%
18. 0.20 C/m^2
20. (a) -2.6×10^{-6} C/m^2, 1.7×10^{-6} C/m^2;
 (b) 1.6×10^5 V/m, 1.1×10^5 V/m;
 (c) 3.0×10^5 V/m
22. 5.0×10^{-2} N
24. $C = \dfrac{2\pi\varepsilon_0(\kappa_1+\kappa_2)}{1/R_1 - 1/R_2}$
26. (a) 1.6×10^{-10} F; (b) 3.8×10^2 V;
 (c) 7.5×10^4 V/m; (d) 2.5×10^{-2} J/m^3;
 (e) 1.1×10^{-5} J
28. 0.2 C, 2×10^3 J
30. 8.9×10^{-4} J
32. 1.1×10^8 m^3

Chapter 28

2. 5.7×10^{-14} J·s
4. 4.1×10^6 A/m^2; 6.9×10^{-2} V/m
6. (a) 3.6×10^{-2} V/m; (b) 1.7×10^{-4} m/s; 46 years
8. 0.201, 0.256, 0.323, 0.407, and 0.514 Ω
10. 7.9 Ω
12. 2.2 cm
14. 1.0×10^{-7} V
16. 0.88 Ω; 14 A
18. 191 kg vs. 384 kg
20. 0.164 cm
22. 8×10^{-5} A
24. 33° C
26. 3.8 A, 2.2 A
28. (a) 2.1×10^2 A; (b) 2.7×10^{-19} A
30. 2.4%
32. 1.5 km from AB
34. 4.0 A, 2.4 A, 1.5 A; 7.9 A
36. 10.9 A, 7.1 A
38. (a) 1.85 Ω; (b) 6.5 A;
 (c) $\Delta V_1 = \Delta V_2 = \Delta V_3 = 12$ V, $I_1 = 3.0$ A, $I_2 = 2.0$ A, $I_3 = 1.5$ A
40. $(1+\sqrt{3})$ Ω

Chapter 29

2. (a) 6.5×10^3 J, 2.4×10^6 J;
 (b) 2.2×10^3 J/in.^{3}, 3.4×10^3 J/in.^{3};
 (c) 3.4×10^4 J/lb, 4.8×10^4 J/lb
4. 2.9×10^3 J; (b) 48 s
6. 2.0 Ω
8. 8.0 Ω
10. $R_i'\mathscr{E}/(RR_i + RR_i' + R_iR_i')$, $R_i\mathscr{E}/(RR_i + RR_i' + R_iR_i')$
12. 12.5 A, 25.5 A
14. (a) 2.6, 1.7, 1.3, 2.1, and 0.86 A; (b) 2.6 V
16. 6.2×10^{12} protons per second; 7.0×10^2 W
18. 14 A, 8.1 Ω
20. (a) 13.2 V; (b) 7.2 W
22. 6.7 h
24. 0.50 W
26. (a) 1.8×10^2 V; (b) 1.8×10^6 W
28. 1.3×10^2 W; 0.022
32. (a) 1/30; (b) 1/3
34. 19 liters per second

Chapter 30

2. 1.2×10^{-12} N, 0 N
4. (a) $|\mathbf{F}| = 0.89 \times 10^{-13}$ N, $|\mathbf{F}'| = 1.78 \times 10^{-13}$ N;
 (b) 0.89×10^{-13} N in direction of **v**
6. 8×10^{-16} N in direction of $-\mathbf{v} \times \mathbf{B}$
8. 8.2×10^{-8} N and 3.5×10^{-5} N; yes
10. 4.0×10^{-3} T

12. (a) 5.0×10^{-6} T; (b) 16°
14. 1.5×10^{13} m/s^2
16. $B = \dfrac{\mu_0}{2\pi} \dfrac{3\sqrt{2}}{2} \dfrac{I}{d}$
18. $B = \dfrac{\mu_0 I}{4\pi} \dfrac{\sqrt{x^2+y^2}+x+y}{xy}$ in positive z direction
20. $B = \dfrac{\sqrt{5}\,\mu_0 I}{2\pi L}$ into plane
22. $B = (\sqrt{5} + \sqrt{13}/3) \dfrac{\mu_0 I}{\pi h}$ out of plane
24. $B = \mu_0 I/8R$ into plane
26. $B = \dfrac{\mu_0}{\pi} \dfrac{I}{b} \tan^{-1} \dfrac{b}{2z}$ perpendicular to z and to current
28. $B = \dfrac{\mu_0 I}{4\pi R}(3\pi/4 + 2)$ out of plane
30. 1.9 T
32. 2.1×10^9 A; westward
34. $B = \dfrac{4\mu_0}{\pi} \dfrac{IL^2}{(4z^2+L^2)\sqrt{4z^2+2L^2}}$ in positive z direction
36. $Q\omega R^2/4$
38. $B = \dfrac{\mu_0 I}{2L}\left[\dfrac{z+L/2}{\sqrt{(z+L/2)^2+R^2}} - \dfrac{z-L/2}{\sqrt{(z-L/2)^2+R^2}}\right]$ in positive z direction
40. $B_x = 0$, $B_y = -\dfrac{\mu_0 I}{2\pi} \dfrac{Rz}{(z^2+R^2)^{3/2}}$
$B_z = \dfrac{\mu_0 I}{2\pi}\left[\dfrac{\pi}{2} \dfrac{R^2}{(z^2+R^2)^{3/2}} + \dfrac{R}{R^2+z^2}\right]$

Chapter 31

2. 6.4×10^{-16} N, 7.0×10^{14} m/s^2; into wire
4. 4.0×10^{-8}; 1.9×10^{-18} N
6. 0, $\mu_0\sigma\hat{\mathbf{x}}$, 0, $-\mu_0\sigma\hat{\mathbf{x}}$
8. 0.13 T
10. $\mu_0\sqrt{n^2I^2 + (I'/2\pi r)^2}$; helical
12. 6.9 T; 2.3 T
14. 1.1×10^{-17} kg · m/s
16. 3.4×10^{-17} kg · m/s
18. 3.3 T
20. 0.39 A
22. 3.6×10^{-2} T
24. 2×10^6 m/s eastward
26. 0.12 N
28. 6.7×10^{-5} N
30. 7.2×10^{-4} N toward the straight wire
32. 3.4×10^{-5} N · m
34. 2.0 N · m; 0.86 hp
36. (a) 4.0×10^{-3} N; (b) 3.2×10^{-4} N · m

Chapter 32

2. 0.75 V
4. 22 rev/s
6. 9×10^{17} V
8. 9.8×10^{-5} V, 7.5×10^{-5} V, 8.7×10^{-5} V
10. 40 rev/s
12. (a) 1.9×10^{-2} T/s; (b) 6.4×10^{-3} V
14. 3.0×10^2 A
16. $\mathscr{E} = \dfrac{\mu_0 I v}{2\pi} \dfrac{l}{(l+d)d}$
18. 45 T
20. (a) 6.0 V/m; (b) 0 V/m; (c) 0.36 V
22. (a) 2.6×10^{-3} V/m; (b) 0 V/m; (c) 9.2×10^{-15} C/m^2
24. (a) 0 Wb; (b) 90 V; (c) negative
26. $200\,\mu_0 n\pi R^2$; no
28. (a) 6.3×10^{-3} H; (b) −1.9 V
30. (a) $B = \dfrac{\mu_0}{2\pi} \dfrac{NI}{r}$; (b) $\Phi_B(1) = \dfrac{\mu_0 NI}{2\pi}(R_2 - R_1) \times \ln(R_2/R_1)$, $L = \dfrac{\mu_0 N^2}{2\pi}(R_2 - R_1)\ln(R_2/R_1)$
34. 4×10^{11} J/m^3
36. 7×10^{18} J
38. 3.6×10^7 J
40. 9.1×10^3 m^3; about 40 m across
42. (a) $u = \dfrac{\mu_0}{8\pi^2}\left(\dfrac{NI}{r}\right)^2$; (b) $U = \dfrac{\mu_0}{4\pi} N^2I^2(R_2 - R_1) \times \ln(R_2/R_1)$; (c) $L = \dfrac{\mu_0}{2\pi} N^2(R_2 - R_1)\ln(R_2/R_1)$

Chapter 33

2. (a) 1.9 T; (b) -1.1×10^{-4} eV, 1.1×10^{-4} eV, parallel
4. (a) $\tan^{-1}[3zx/(2z^2 - x^2)]$; (b) 0°, 0°, 72°
6. 0.24 Hz
8. $(3/14)\, e\hbar/m_n = 2.2 \times 10^{-27}$ A · m^2; opposite
10. (a) 13 T; (b) 3.3×10^9 radian/s
12. 3.2×10^{-6} T
14. 1.000912
18. 3.6×10^3 A
20. 2.0×10^{-6} N · m; no
22. 1.6×10^{28} electrons/m^3; 0.18 electron/atom
24. 3.4×10^{-9} J · s
26. (a) 4.4×10^{10} radian/s; (b) 2.3 m/s; (c) 4.7×10^{-24} J

Chapter 34

2. 400 W
4. 3.2×10^{-4} m/s; 0.12 m/s^2
6. (a) 3.1×10^3 A; (b) less loss in transmission line
8. (a) $Q = 6.5 \times 10^{-8} \sin(120\pi t)$; (b) 1/240 s, 0 s; (c) 5.8×10^{-8} J, 2.9×10^{-8} J
10. $I = \omega C\mathscr{E}_0 \cos\omega t$, $I = -\mathscr{E}_0/(\omega L)\cos\omega t$; $I = (\omega C - 1/\omega L)\,\mathscr{E}_0 \cos\omega t$, $P = (\omega C - 1/\omega L)\,\mathscr{E}_0^2 \cos\omega t \sin\omega t$
12. (a) 240 Ω; (b) 8.3×10^{-4} A; (c) -8.3×10^{-4} A, -5.9×10^{-4} A
14. 6.0×10^{-9} F to 6.6×10^{-10} F
16. 1.4×10^{-3} J; 7.9×10^{-4} s; 15.7×10^{-4} s
18. (a) circuit oscillates at decreased frequency with gradually diminishing amplitude; (b) circuit oscillates at original frequency, but with much diminished amplitude
20. (a) 5.8×10^{-3} J; (b) 40%; (d) 0.035 s, 0.12 s
22. 1.0×10^{-2} H, 2.0×10^{-5} F
24. (a) $0.80 \sin(2\pi \times 2.2 \times 10^4\, t)$; (b) $5.9 \times 10^{-2} \cos(2\pi \times 2.2 \times 10^4\, t)$

26. 13.4×10^{-3} A, 6.6×10^{-3} A
28. (a) 0.20 A; (b) 6.3×10^{-4} s, 6.8×10^{-4} s
30. (a) 5.9×10^3 Hz; (b) 0.40 A; (c) 0.16 W
32. 0.37 A; 43°; 5.4×10^{-2} W
34. $\frac{1}{2}\mathscr{E}_{max}^2(\omega C - 1/\omega L)\sin 2\omega t + (\mathscr{E}_{max}^2/R)\sin^2 \omega t$
36. 4550 turns
38. 9.1×10^4 A; 5.0×10^3 A
40. 1.96×10^3 A; 4.35×10^3 A

Chapter 35

2. (a) 1.0×10^6 V/m s; (b) 8.3×10^{-13} T, 1.7×10^{-12} T
4. (a) 2.0×10^{-3} A; (b) 1.0 s; (c) 0.89×10^{-9} T, 1.8×10^{-9} T, 2.7×10^{-9} T
6. (a) $(V_0/R)\sin \omega t$; (b) $(\varepsilon_0 A\omega V_0/d)\cos \omega t$; (c) $(V_0/R)\sin \omega t + (\varepsilon_0 A\omega V_0/d)\cos \omega t$; (d) $(\mu_0/2\pi)[(V_0/rR)\sin \omega t + (\varepsilon_0 \pi r\omega V_0/d)\cos \omega t]$
10. $\oint \mathbf{E}\cdot d\mathbf{S} = Q/\varepsilon_0$, $\oint \mathbf{B}\cdot d\mathbf{S} = q_m$, $\oint \mathbf{E}\cdot d\mathbf{l} = -I_m - d\Phi_B/dt$, $\oint \mathbf{B}\cdot d\mathbf{l} = \mu_0 I + \mu_0\varepsilon_0\, d\Phi/dt$
12. (a) $B_{(2)} = -\frac{1}{8}\mu_0^2\varepsilon_0^2 r^3\omega^3 E_0 \cos \omega t$; (b) $B = \frac{1}{2}\mu_0\varepsilon_0 r\omega(1 - \frac{1}{4}\mu_0\varepsilon_0\omega^2 r^2)E_0 \cos \omega t$; (c) $B = \sqrt{\mu_0\varepsilon_0}\, E_0 \cos \omega t$
14. (a) $(2\pi/\mu_0\omega^2)E_0 \sin \omega t$; (b) $(2\pi/\mu_0\omega)E_0 \cos \omega t$
16. 0.11 V/m; 6.7×10^{-10} s
18. (a) 7.4×10^{17} m/s²; (b) 7.9×10^{-9} V/m
20. (a) 4.8×10^{-15} m, 6.0×10^{27} m/s²; (b) 3.2×10^{11} V/m; (c) 0 V/m
22. 4.7×10^4 m/s²
24. (a) 2.5×10^{-12} V/m, 8.3×10^{-21} T
26. (a) 1.1×10^{29} m/s²; (b) 7.1×10^{13} V/m, 2.4×10^5 T
28. (a) 6.8×10^9 m/s²; (b) 2.7×10^{-17} V/m, 9.1×10^{-26} T; (c) 6.1×10^5 V/m, 2.1×10^{-3} T; (d) 3.8×10^{-9} T
30. (a) $q^2va/(12\pi\varepsilon_0 c^3)$; (b) $q^2va/(12\pi\varepsilon_0 c^3)$

Chapter 36

2. 5.3×10^{-6} A/m
4. (a) vertical; (b) vertical and perpendicular to **B**
6. 2.0×10^{-9} T, northward
8. (a) negative z; (b) 63°, 27°, 90°; (c) $2\hat{\mathbf{x}}(E_0/c)\sin(\omega t + \omega z/c) - \hat{\mathbf{y}}(E_0/c)\sin(\omega t + \omega z/c)$
10. $S = \frac{1}{2}S_0 \cos^2\theta$
14. 1.85 m
16. 1.0×10^3 V/m, 3.4×10^{-6} T
18. 6.9%
20. 51
22. 1.9×10^9 W/m²; 1.2×10^6 V/m, 4.0×10^{-3} T
24. 5.5×10^{-2} V/m, 1.8×10^{-10} T
26. 2.0×10^4 W
30. (a) 50 W/m²; (b) 1.9×10^2 V/m, 6.5×10^{-7} T; (c) 1.7×10^{-7} N/m²
32. (a) 6.2×10^{-18} N; (b) 3.7×10^{-18} N, gravity is larger; (c) 7.7×10^{-19} N, 9.2×10^{-19} N, radiation pressure is larger
34. 5.8×10^{-4} kg · m/s
38. $0.81c$
40. 2203.13 MHz
42. (a) 1.8×10^{-7}; (b) 1.4×10^3 Hz
44. 2.2×10^3 m/s, 1.50×10^5 m/s

Chapter 37

2. 0.50 m × 0.42 m
4. 50°
6. 39° 0′ 50″
8. l, $l + 2d/n$, $l + 4d/n$
10. (a) 0.67 mm; (b) 2.67 mm
12. 47 m
14. (b) 64°
16. 5.7 m
20. R
22. 120 cm; concave
24. (a) 60 cm in front of first mirror; (b) 30 cm in front of second mirror
26. 8.3 mm
28. 21 cm
30. 10 cm
34. (a) 36 cm to right of lens; (b) 130 cm to left of mirror
36. 475 cm to left of lens
38. 49 cm to infinity
40. 0.022 s
42. (a) −22 cm, divergent; (b) 0.90 cm
44. 10
46. 64
48. 1344

Chapter 38

2. 3×10^8 m/s
4. (b) 9.48×10^5, 9.48×10^5; (c) 1582 Å, 3164 Å
6. (a) 996 Å; (b) no
8. $2d = \frac{1}{2}\lambda, \frac{3}{2}\lambda, \frac{5}{2}\lambda, \ldots$
10. (c) 1.2 mm
12. 5.8 mm/s
14. 7.4 mm
16. $\lambda/d \cong 0.20$ implies nodal lines at 12°, 24°, 37°, and 53°
20. $\pm 1.3 \times 10^{-4}$ radian, $\pm 2.6 \times 10^{-4}$ radian, $\pm 3.9 \times 10^{-4}$ radian, etc.
22. 61°, 22°, −7°, −39°
24. 5.7°
26. 0.45
28. 1.0 mm; 2.6 mm
30. 8.7×10^4
32. 0.3 mm
34. $N - 2 = 8.7 \times 10^4$

Chapter 39

2. $\lambda/a \cong 0.44$ implies nodal lines at 26° and 63°
4. 47°
6. (a) 4.2 cm; (b) 2.1 cm
8. (b) $n - 1$
10. 0.040 mm
12. (a) 1.9×10^{-4} radian; (b) 4.4×10^{-3} mm; (c) 2.3
14. 2.8°; 2.4×10^2 m
16. 1.3×10^{-5} radian; 2.4×10^{-4} cm

18. 3.4×10^{-3} radian, or 12′
20. (a) 3×10^{-7} radian; (b) 2 m
22. 13 km
24. (a) 1.3×10^{-5} radian; (b) 1.4×10^{-5} radian, eye
28. 6.4 cm

Chapter 40

2. (a) 1 mm; (b) 2×10^{9} W
4. 0.78; 0.87
8. 1.3×10^{7} m
10. 44 K
12. 1.4×10^{-10} m; 1.9×10^{-10} m
14. (a) 7.2×10^{15} Hz
18. 1.2×10^{6}/s
20. $2.5 \times 10^{3}/\text{m}^{3}$
24. 1.8 eV
26. none; K; K, Cr, Zn
28. 17.5 eV
30. 85°
32. 0.31 Å; 0.34 Å
34. 1.5×10^{8} eV
36. 5.8×10^{3} eV; 1.2×10^{3} eV
38. (a) 2×10^{-39} kg · m/s; (b) 1×10^{3} Hz

Chapter 41

2. first ($n = 5$); 34,076 Å
4. (a) 4102.9 Å and 4341.7 Å in Balmer series; (b) 1.1×10^{6} m/s
6. (a) $p = -0.0409$; (b) Lyman and Balmer
8. 3.22×10^{7} eV
12. $7.30 \times 10^{-3}\, c$
14. 1215.7 Å
16. $n = 3$ to $n = 1$
18. 2.12×10^{-10} m, 1.09×10^{6} m/s, 2.10×10^{-34} J · s, 5.66×10^{21} m/s^{2}
20. 7.0×10^{5} m/s
22. 0.176 Å; −122 eV
24. $r = (a_0 R^3 n^2)^{1/4}$, $E = \dfrac{n\hbar e}{\sqrt{4\pi\varepsilon_0 m_e R^3}} - \dfrac{e^2}{4\pi\varepsilon_0}\dfrac{3}{2R}$, $\nu = \dfrac{n_2 - n_1}{2\pi}\dfrac{e}{\sqrt{4\pi\varepsilon_0 m_e R^3}}$; $4a_0$
26. 1.06 Å, 2431 Å
30. 2.1 Å, 0.16 Å
32. 1.1×10^{-32} m
34. $\lambda = 2\pi r/n$
36. ±7 cm
38. 9.2×10^{-20} kg · m/s, 5.5×10^{7} m/s

PHOTOGRAPH CREDITS

PRELUDE

page 1 James L. Mairs
page 2 (top) James L. Mairs; (bottom) James L. Mairs
page 3 (top) James L. Mairs; (center) Lockwood, Kessler, and Bartlett, Inc.; (bottom) NASA
page 4 (top) U.S. Geological Survey; (center) General Electric Space Systems; (bottom) NASA
page 5 NASA
page 9 (top) Hale Observatories; (bottom) Hale Observatories
page 10 (top) Hale Observatories; (bottom) Hale Observatories
page 11 (top) J. R. Kuhn, Dept. of Physics, Princeton Univ.; (bottom) James L. Mairs
page 12 (top) Zeiss; (center) Dr. Ronald Radius, The Eye Institute, Milwaukee, Wis.; (bottom) From *Tissues and Organs: A Test Atlas of Scanning Electron Microscopy* by Richard G. Kessel and Randy H. Kardon, W. H. Freeman & Co., San Francisco, © 1979
page 13 (top) From *Tissue and Organs: A Test Atlas of Scanning Electron Microscopy* by Richard G. Kessel and Randy H. Kardon, W. H. Freeman & Co., San Francisco, © 1979; (bottom) Prof. T. T. Tsong, Pennsylvania State Univ.
page 14 A. V. Crewe and M. Utlaut, Univ. of Chicago
page 15 L. S. Bartell, Univ. of Michigan

CHAPTER 1

Fig. 1.7 BIPM–Picture
Fig. 1.9 U.S. National Bureau of Standards
Fig. 1.10 BIPM–Picture
Fig. 1.11 U.S. National Bureau of Standards
Fig. 1.12 U.S. National Bureau of Standards
Fig. 1.13 Hewlett-Packard Company
Fig. 1.14 U.S. National Bureau of Standards
Fig. 1.15 U.S. National Bureau of Standards

CHAPTER 2

Fig. 2.11 *PSSC Physics,* 2nd ed., 1965, D. C. Heath & Co. with the Education Development Center, Newton, Mass.
Fig. 2.12 BIPM–Picture
Fig. 2.15 NASA
page 33 (Galileo Galilei) AIP Niels Bohr Library

INTERLUDE A

Fig. A.1 Reproduced by gracious permission of Her Majesty Queen Elizabeth II
Fig. A.2 Duvall Corporation
Fig. A.3 R. Gronsky, Lawrence Berkeley Laboratory
Fig. A.4 Martin J. Buerger, Institute Professor, MIT
Fig. A.5 Prof. T. T. Tsong, Pennsylvania State Univ.
Fig. A.6 R. P. Goehner, General Electric
Fig. A.7 R. P. Goehner, General Electric
Fig. A.8 David Scharf, 1977, through Schocken Books, Inc.
Fig. A.9 *Earth Scenes* © Breck P. Kent
Fig. A.10 W. A. Bentley; reproduced from *Snow Crystals* by W. A. Bentley and W. J. Humphreys, Dover, New York, 1962
Fig. A.11 Hans C. Ohanian
Fig. A.12 Courtesy H. B. Huntington, RPI
Fig. A.13 © Beeldrecht, Amsterdam/VAGA, New York, Collection Haags Gemeentemuseum, The Hague
Fig. A.14 © Beeldrecht, Amsterdam/VAGA, New York, Collection Haags Gemeentemuseum, The Hague
Fig. A.15 © Beeldrecht, Amsterdam/VAGA, New York, Collection Haags Gemeentemuseum, The Hague
Fig. A.17 © Beeldrecht, Amsterdam/VAGA, New York, Collection Haags Gemeentemuseum, The Hague
Fig. A.27 C. S. Smith, MIT
Fig. A.33 Otsuka Kogeisha & Co., Ltd.
Fig. A.34 C. S. Smith, MIT
Fig. A.35 Metropolitan Museum of Art

CHAPTER 4

Fig. 4.5 *PSSC Physics,* 2nd ed., 1965, D. C. Heath & Co. with the Education Development Center, Newton, Mass.
Fig. 4.7 U.S. Naval Institute
Fig. 4.8 Dr. Harold E. Edgerton, MIT
Fig. 4.12 Schenectady (N.Y.) *Gazette*
Fig. 4.16 NASA

CHAPTER 5

Fig. 5.2 The Smithsonian Institution
Fig. 5.4 *Road & Track*
Fig. 5.7 NASA
Fig. 5.25 The Science Museum, London
page 92 (Sir Isaac Newton) Burndy Library, courtesy AIP Niels Bohr Library
page 94 (Ernst Mach) AIP Niels Bohr Library

CHAPTER 6

Fig. 6.2 Hans C. Ohanian
Fig. 6.3 NASA
Fig. 6.4 NASA
Fig. 6.11 S. J. Calabrese, RPI
Fig. 6.12 S. J. Calabrese, RPI
Fig. 6.31 Leo De Wys, Inc. © Larry Miller
Fig. 6.34 U.S. National Bureau of Standards
page 119 (Pierre Simon, Marquis de LaPlace) AIP Niels Bohr Library
page 131 (Leonardo da Vinci) NMAH Archives Center, Smithsonian Institution

CHAPTER 7

page 154 (James P. Joule) AIP Niels Bohr Library
page 165 (Christiaan Huygens) AIP Niels Bohr Library

CHAPTER 8

Fig. 8.9 U.S. Air Force Photo
page 176 (Joseph Louis LaGrange) The Granger Collection
page 185 (Hermann von Helmholtz) AIP Niels Bohr Library
page 187 (James Watt) NBS Archives, courtesy AIP Niels Bohr Library

CHAPTER 9

Fig. 9.5 *PSSC Physics,* 2nd ed., 1965, D. C. Heath & Co. with the Education Development Center, Newton, Mass.
Fig. 9.11 The Bettmann Archive, Inc.
Fig. 9.13 NASA
Fig. 9.17 NASA
Fig. 9.18 Duomo Photography, Inc.

CHAPTER 10

Fig. 10.1 Mercedes-Benz of North America, Inc.
Fig. 10.9 Reprinted from *Philosophical Magazine,* R. H. Brown et al., vol. 40, 1949
Fig. 10.10 Hans C. Ohanian
Fig. 10.11 W. B. Hamilton, U.S. Geological Survey
Fig. 10.18 C. F. Powell and G. P. S. Occhialini, *Nuclear Physics in Photographs,* Oxford Univ. Press, New York

INTERLUDE B

Fig. B.3 Reprinted from the *Proceedings of the Royal Society,* P. M. S. Blackett and D. S. Lees, 1932
Fig. B.8 Reprinted from the *Proceedings of the Royal Society,* P. M. S. Blackett and D. S. Lees, 1932
Fig. B.13 Nuclear Medicine Section, Dept. of Radiology, Hospital of the Univ. of Pennsylvania
Fig. B.14 Michael G. Velchik, M.D., Nuclear Medicine Section, Dept. of Radiology, Hospital of the Univ. of Pennsylvania
Fig. B.15 Michael G. Velchik, M.D., Nuclear Medicine Section, Dept. of Radiology, Hospital of the Univ. of Pennsylvania
Fig. B.16 Michael G. Velchik, M.D., Nuclear Medicine Section, Dept. of Radiology, Hospital of the Univ. of Pennsylvania
Fig. B.17 Michael G. Velchik, M.D., Nuclear Medicine Section, Dept. of Radiology, Hospital of the Univ. of Pennsylvania
Fig. B.18 M. D. Anderson Hospital and Tumor Institute, Univ. of Texas, Houston

CHAPTER 11

Fig. 11.1 Dr. Harold E. Edgerton, MIT

CHAPTER 12

Fig. 12.12 Sperry Flight Systems
Fig. 12.15 NASA
Fig. 12.19 *PSSC Physics,* 2nd ed., 1965, D. C. Heath & Co. with the Education Development Center, Newton, Mass.
Fig. 12.32 Wide World Photos
Fig. 12.46 NASA

INTERLUDE C

Fig. C.2 Leybold-Heraeus GMBH & Co.
Fig. C.3 "Lent to Science Museum, London," by the late J. J. Thomson, Trinity College, Cambridge, in the case of Cathode Ray Tube of J. J. Thomson
Fig. C.7 (a) The MIT Museum; (b) Cavendish Laboratory, Cambridge, England
Fig. C.8 Lawrence Berkeley Laboratory, Univ. of California
Fig. C.9 Fermilab Photo
Fig. C.10 Fermilab Photo
Fig. C.11 Fermilab Photo
Fig. C.12 Stanford Linear Accelerator Center
Fig. C.14 CERN
Fig. C.15 Stanford Linear Accelerator Center
Fig. C.16 CERN
Fig. C.17 CERN
Fig. C.18 MIT Bubble Chamber Group
Fig. C.19 CERN
Fig. C.20 CERN
Fig. C.21 Fermilab Photo
Fig. C.24 Brookhaven National Laboratory

CHAPTER 13

Fig. 13.5 From Cavendish, *Philosophical Transactions of the Royal Society 18* (1798): 388.
Fig. 13.14 (a) Sovfoto Agency
Fig. 13.15 NASA
Fig. 13.35 NASA
Fig. 13.38 (a) National Film Board of Canada; (b) National Film Board of Canada
Fig. 13.41 NASA
Fig. 13.45 The Bettmann Archive, Inc.
page 308 (Henry Cavendish) NBS Archives, courtesy AIP Niels Bohr Library
page 309 (Nicolas Copernicus) National Portrait Gallery, Smithsonian Institution
page 312 (Johannes Kepler) AIP Niels Bohr Library

INTERLUDE D

Fig. D.1 (a) Palomar Observatory Photograph; (b) Palomar Observatory Photograph
Fig. D.2 Palomar Observatory Photograph
Fig. D.3 Palomar Observatory Photograph
Fig. D.4 Palomar Observatory Photograph
Fig. D.5 Palomar Observatory Photograph
Fig. D.6 Palomar Observatory Photograph
Fig. D.7 Palomar Observatory Photograph
Fig. D.8 Palomar Observatory Photograph
Fig. D.9 Palomar Observatory Photograph
Fig. D.11 Palomar Observatory Photograph
Fig. D.12 AT&T Bell Laboratories

CHAPTER 14

Fig. 14.6 NASA
Fig. 14.8 From *Energy, A Sequel to IPS* by Uri Haber-Schaim, Prentice-Hall, Englewood Cliffs, N.J., 1983
Fig. 14.14 Hans C. Ohanian
Fig. 14.17 U.S National Bureau of Standards
Fig. 14.20 Dr. J. E. Faller, Univ. of Colorado
Fig. 14.21 U.S. National Bureau of Standards
Fig. 14.25 Photo Researchers, Inc.

CHAPTER 15

Fig. 15.19 *PSSC Physics,* 2nd ed., 1965, D. C. Heath & Co. with the Education Development Center, Newton, Mass.
Fig. 15.20 *PSSC Physics,* 2nd ed., 1965, D. C. Heath & Co. with the Education Development Center, Newton, Mass.
Fig. 15.21 *PSSC Physics,* 2nd ed., 1965, D. C. Heath & Co. with the Education Development Center, Newton, Mass.
Fig. 15.22 UPI © Bashford Thompson, Tacoma, Wash.

CHAPTER 16

Fig. 16.1 Fundamental Photographs, New York
Fig. 16.2 *Engineering Applications of Lasers and Holography* by Winston E. Kock, Plenum Publishing Co., New York, 1975
Fig. 16.8 S. Gilmore, General Electric Research and Development Center
Fig. 16.9 C. F. Quate and L. Lam, Hansen Laboratories, Stanford Univ.
Fig. 16.18 Dr. Harold E. Edgerton, MIT
Fig. 16.23 Mrs. Harry A. Simms, Sr.
Fig. 16.25 Californian Historical Society, Arnold Genthe, New York
Fig. 16.26 National Oceanic and Atmospheric Administration
Fig. 16.28 John S. Shelton
Fig. 16.29 *PSSC Physics,* 2nd ed., 1965, D. C. Heath & Co. with the Education Development Center, Newton, Mass.
Fig. 16.30 *PSSC Physics,* 2nd ed., 1965, D. C. Heath & Co. with the Education Development Center, Newton, Mass.
Fig. 16.31 Education Development Center, Newton, Mass.
Fig. 16.32 Education Development Center, Newton, Mass.

CHAPTER 17

Fig. 17.2 AIP Niels Bohr Library
page 419 (Albert Einstein) The Bettmann Archive, Inc.
page 428 (Hendrik Lorentz) AIP Niels Bohr Library

INTERLUDE E

Fig. E.3 NASA
Fig. E.10 From Arthur Eddington, *Space, Time, and Gravitation,* courtesy Cambridge Univ. Press
Fig. E.23 Hewlett-Packard Company
Fig. E.25 R. F. Vassot and M. L. Levine, Smithsonian Astrophysical Observatory
Fig. E.30 Riccardo Giacconi, Harvard Smithsonian Center for Astrophysics
Fig. E.31 Hale Observatories
Fig. E.32 After Eardley and Press, *Annual Review of Astronomy and Astrophysics,* 1975

CHAPTER 18

Fig. 18.1 U.S. Geological Survey
Fig. 18.4 J. M. LeBlanc and J. R. Wilson, Lawrence Livermore Laboratory
Fig. 18.8 A. D. Moore, Univ. of Michigan, from *Introduction to Electric Fields* by W. E. Rogers, McGraw-Hill Book Co., New York, 1954
Fig. 18.9 D. C. Hazen and R. F. Lehnert, Subsonic Aerodynamics Laboratory, Princeton
Fig. 18.10 D. C. Hazen and R. F. Lehnert, Subsonic Aerodynamics Laboratory, Princeton
Fig. 18.11 MCO Properties, Inc.
Fig. 18.13 Dept. of the Navy, Naval Historical Center
Fig. 18.30 Hans C. Ohanian
Fig. 18.32 NASA
Fig. 18.34 U.S. Navy Photo
Fig. 18.39 Robert Azzi, Woodfin Camp & Assocs.
page 451 (Blaise Pascal) National Portrait Gallery, Smithsonian Institution
page 458 (Daniel Bernoulli) National Portrait Gallery, Smithsonian Institution

CHAPTER 19

Fig. 19.7 U.S. National Bureau of Standards
Fig. 19.8 U.S. National Bureau of Standards
page 470 (Lord Kelvin) Wellcome Institute Library, London
page 471 (Robert Boyle) AIP Niels Bohr Library
page 479 (Ludwig Boltzmann) AIP Niels Bohr Library

CHAPTER 20

Fig. 20.9 Robert Azzi, Woodfin Camp & Assocs.
Fig. 20.10 VANSCAN® thermogram by Daedalus Enterprises, Inc.
page 485 (Benjamin Rumford) NBS Archives, courtesy AIP Niels Bohr Library

CHAPTER 21

Fig. 21.18 Vorpal Galleries: New York, San Francisco, Laguna Beach, Calif.
page 512 (Sadi Carnot) AIP Niels Bohr Library
page 516 (Rudolph Clausius) National Portrait Gallery, Smithsonian Institution
page 524 (Walther Nernst) NMAH Archives Center, Smithsonian Institution

INTERLUDE F

Fig. F.6 Bill Pierce/Contact-Camp
Fig. F.7 Los Alamos National Laboratory
Fig. F.8 PG & E Photographs
Fig. F.9 NASA
Fig. F.10 French Embassy Press and Information Division
Fig. F.11 SYGMA/Zimberoff
Fig. F.12 (b) Advanced Energy Technologies
Fig. F.13 NASA
Fig. F.14 Free Energy Systems, Inc.
Fig. F.15 PG & E Photographs
Fig. F.17 PG & E Photographs
Fig. F.18 French Embassy Press and Information Division
Fig. F.19 New York Power Authority
Fig. F.20 PG & E Photographs
Fig. F.22 Henry O. Navratil, Sewickley, Penna.

CHAPTER 22

Fig. 22.2 L. S. Bartell, Univ. of Michigan
Fig. 22.6 National Center for Atmospheric Research / National Science Foundation
Fig. 22.9 Leybold-Heraeus
page 533 (Benjamin Franklin) Library of Congress
page 534 (Charles de Coulomb) AIP Neils Bohr Library

CHAPTER 23

Fig. 23.22 Robert Matthews, Princeton Univ.
Fig. 23.23 Robert Matthews, Princeton Univ.

INTERLUDE G

Fig. G.1 *Philosophical Magazine* Ser. 7, Vol. 40, P1. X, 1949
Fig. G.2 CERN
Fig. G.3a CERN
Fig. G.3b CERN
Fig. G.4 Lawrence Berkeley Laboratory
Fig. G.8 CERN
Fig. G.13 CERN

CHAPTER 24

Fig. 24.19 Robert Matthews, Princeton Univ.

INTERLUDE H

Fig. H.15 Los Alamos National Laboratory
Fig. H.16 Los Alamos National Laboratory
Fig. H.24 Dr. Harold E. Edgerton, MIT
Fig. H.25 Los Alamos National Laboratory
Fig. H.26 NASA
Fig. H.27 From *The Effects of Nuclear War,* Office of Technology Assessment, Congress of the U.S.
Fig. H.32 General Electric

CHAPTER 25

page 582 (Alessandro Volta) Lande Collection, AIP Niels Bohr Library

CHAPTER 27

Fig. 26.13 Earl Williams and Chatham Cook, High Voltage Laboratory, MIT

CHAPTER 28

Fig. 28.1 O. Jelimenko, *American Journal of Physics 30,* no. 19, 1962

Fig. 28.2 O. Jelimenko, *American Journal of Physics 30,* no. 19, 1962

page 635 (Georg Ohm) E. Scott Barr Collection, AIP Neils Bohr Library

CHAPTER 29

Fig. 29.4b NASA

Fig. 29.5b ARCO-Solar, Inc.

page 655 (Gustav Kirchoff) NMAH Archives Center, Smithsonian Institution

INTERLUDE I

Fig. I.5 Hans C. Ohanian

Fig. I.9 National Severe Storms Lab/NOAA, W. David Rust

Fig. I.18 National Center for Atmospheric Research, Dave Baumhefner

Fig. I.19 E. P. Krider, The Univ. of Arizona

Fig. I.20 U.S. Department of Agriculture, Forest Division

CHAPTER 30

Fig. 30.12 Education Development Center, Newton, Mass.

Fig. 30.16 Education Development Center, Newton, Mass.

page 668 (Hans Oersted) NMAH Archives Center, Smithsonian Institution

page 673 (Karl Gauss) AIP Niels Bohr Library

page 672 (Nikola Tesla) NMAH Archives Center, Smithsonian Institution

page 675 (Jean Baptiste Biot) NMAH Archives Center, Smithsonian Institution

CHAPTER 31

Fig. 31.6 Manfred Kage / Peter Arnold, Inc.

Fig. 31.13 Leybold-Heraeus

Fig. 31.15 Lawrence Berkeley Laboratory

Fig. 31.16 Lawrence Berkeley Laboratory

Fig. 31.36 Lawrence Berkeley Laboratory

page 692 (André Ampère) AIP Niels Bohr Library

page 701 (Ernest Lawrence) Lawrence Berkeley Laboratory

CHAPTER 32

page 715 (Michael Faraday) AIP Niels Bohr Library

page 719 (Wilhelm Weber) NMAH Archives Center, Smithsonian Institution

page 722 (Emil Lenz) E. Scott Barr Collection, AIP Niels Bohr Library

page 727 (Joseph Henry) The Smithsonian Institution, courtesy AIP Niels Bohr Library

INTERLUDE J

Fig. J.1 NASA

Fig. J.3 Hale Observatories

Fig. J.23 Princeton Plasma Physics Laboratory

Fig. J.24 Princeton Plasma Physics Laboratory

Fig. J.26 Lawrence Livermore Laboratory

Fig. J.26 Lawrence Livermore Laboratory

Fig. J.38 Argonne National Laboratory

CHAPTER 33

Fig. 33.6 H. J. Williams, R. M. Bozorth, and W. Shockley, *Physical Review* 75 (1949): 155.

Fig. 33.9 Education Development Center, Newton, Mass.

INTERLUDE K

Fig. K.2 A. Leitner, RPI

Fig. K.3 Cryogenic Technology, Inc.

Fig. K.10 A. D. Little, Inc., Cambridge, Mass.

Fig. K.13 IGC Intermagnetic General

Fig. K.14 General Electric

Fig. K.15 Argonne National Laboratory

Fig. K.16a Fermilab Photo

Fig. K.16b Fermilab Photo

Fig. K.17 Intermagnetics General Corporation

Fig. K.18 General Electric

Fig. K.19 Brookhaven National Laboratories

CHAPTER 35

Fig. 35.11 Stanford Linear Accelerator Center

Fig. 35.13 Hans C. Ohanian

Fig. 35.17 Education Development Center, Newton, Mass.

page 778 (James Maxwell) AIP Niels Bohr Library

CHAPTER 36

Fig. 36.9 Polaroid Corporation

Fig. 36.14 Hale Observatories

Fig. 36.15 Hale Observatories

Fig. 36.16 Polaroid Corporation

page 802 (Heinrich Hertz) Berlin Univ., Westphal Collection, courtesy AIP Niels Bohr Library

CHAPTER 37

Fig. 37.10 Davis Instruments Corporation

Fig. 37.11 NASA

Fig. 37.17 AT&T Bell Laboratories

Fig. 37.45 C. C. Jones, Union College

Fig. 37.52 Hale Observatories

Fig. 37.53 Drawing by Chas. Addams; © 1983 The New Yorker Magazine, Inc.

Fig. 37.54 Hans C. Ohanian

Fig. 37.55 John Shelton

Fig. 37.57 Point Reyes National Seashore, Point Reyes, Calif.

CHAPTER 38

Fig. 38.5 Bausch & Lomb

Fig. 38.7 (a) AIP Niels Bohr Library

Fig. 38.11 Education Development Center, Newton, Mass.

Fig. 38.12 (a) C. C. Jones, Union College

Fig. 38.20 C. C. Jones, Union College

Fig. 38.23 National Radio Astronomy Observatory, operated by Associated Universities, Inc., under contract with the National Science Foundation

Fig. 38.25 (b) Bausch & Lomb

page 867 (Albert Michelson) University of Chicago

page. 861 (Thomas Young) NMAH Archives Center, Smithsonian Institution

CHAPTER 39

Fig. 39.1 F. W. Sears, *Optics* © 1949, Addison-Wesley, Reading, Mass., Fig. 9.8. Reprinted with permission.

Fig. 39.5 C. C. Jones, Union College

Fig. 39.9 C. C. Jones, Union College

Fig. 39.11 C. C. Jones, Union College

Fig. 39.12 C. C. Jones, Union College

Fig. 39.13 Space Telescope Science Institute, Baltimore, Md.

Fig. 39.14 Cornell Univ.

Fig. 39.17 C. C. Jones, Union College

Fig. 39.18 C. C. Jones, Union College

Fig. 39.19 Sears, Zemansky, & Young, *University Physics,* © 1982, Addison-Wesley, Reading, Mass. Reprinted with permission. Page 789, Fig. 41–18.

Fig. 39.20 R. Gronsky, Lawrence Berkeley Laboratory

Fig. 39.21 R. Olberg, Union College

Fig. 39.22 Education Development Center, Newton, Mass.

page 884 (Augustin Fresnel) AIP Neils Bohr Library
page 890 (Rayleigh) The Royal Society, London, courtesy AIP Niels Bohr Library

CHAPTER 40

Fig. 40.13 (a, b, c) E. R. Huggins, New York, 1968; (d) C. C. Jones, Union College
page 913 (Max Born) AIP Niels Bohr Library
page 910 (Arthur Compton) AIP Niels Bohr Library
page 914 (Werner Heisenberg) Bainbridge Collection, AIP Niels Bohr Library
page 904 (Max Planck) AIP Niels Bohr Library

CHAPTER 41

Fig. 41.9 Hale Observatories Photograph
page 928 (Niels Bohr) Princeton University, courtesy AIP Niels Bohr Library
page 934 (Louis Victor, Prince de Broglie) AIP Meggers Gallery of Nobel Laureates
page 922 (Joseph von Fraunhofer) NMAH Archives Center, Smithsonian Institution
page 926 (Ernest Rutherford) AIP Niels Bohr Library
page 935 (Erwin Schrödinger) Ullstein, courtesy AIP Niels Bohr Library
page 925 (Joseph John Thomson) Bell Telephone Laboratories, courtesy AIP Niels Bohr Library

INTERLUDE L

Fig. L.4 Hughes Aircraft
Fig. L.7 The Warder Collection
Fig. L.8 NASA
Fig. L.9 Bell Laboratories
Fig. L.10 Arnold Zann © 1980 Black Star
Fig. L.11 United Technologies
Fig. L.13 U.S. National Bureau of Standards
Fig. L.14 Lawrence Livermore National Laboratory
Fig. L.15 Lawrence Livermore National Laboratory
Fig. L.16 Lawrence Livermore National Laboratory
Fig. L.17 Lawrence Livermore National Laboratory
Fig. L.19 From H. M. Smith, *Principles of Holography,* 2nd ed., © John Wiley & Sons, Inc., 1975
Fig. L.24 Museum of Holography, New York City, © Ronald R. Erickson, 1983
Fig. L.25 Karl Stetson
Fig. L.26 TRW

INDEX

aberration:
  chromatic, 846
  spherical, 841
A-bomb, H7
absolute acceleration, 93
absolute motion, 416
absolute temperature scale, 470, 473
absolute time, 3–5, 80
absolute zero, entropy at, 524
absorbed dose, B7
accelerated charge:
  electric field of, 788–92
  induced magnetic field of, 795
  magnetic field of, 792–95
  radiation field of, 788, 790
acceleration, 29–37
  absolute, 93
  angular, *see* angular acceleration
  average, 29–30
  average, in three dimensions, 69
  of center of mass, 208
  centrifugal, 79
  centripetal, 78, 93–94, 141–44
  conversion factors for, 967 (555)
  g as, 34, 36–37
  instantaneous, 30–31
  instantaneous, in three dimensions, 69
  motion with constant, 31–33, 70–71
  transformation equations for, 82
acceleration of gravity, 33–37, 307
  measurement of, 37, 352
  universality of, 33–34
  variation of, with altitude, 307
  variation of, with latitude, 94
accelerators:
  linear, C5
  for particles, 700–701, C3–C6, K8
accelerometers, 110
acceptor impurities, 641
accidents:
  automobile, 239
  nuclear, F12–F15
  reactor, F12–F15
AC circuits, 752–77
  simple, with an external electromagnetic force, 754–59
accretion, by black hole, E13
AC current, 754
  hazards of, 661–62
acoustic micrograph, 394–95
action and reaction, 101–4
action-at-a-distance, 121, 306, 546, 549
action-by-contact, 550
activity, radioactive, B4
AC voltage, 722–23
Adams, J. C., 304
addition law for velocity, Galilean, 417
addition of vectors, 48–51
  associative law of, 50–51
  commutative law of, 50
  by components, 54–55
addition of velocities, relativistic, 433–34
adiabatic equation, for gas, 497–99
adiabatic process, 496
age:
  of chemical elements, D6
  of globular clusters, D5–D6
  of the universe, D5–D6
air, dry and humid, stability of, I4
air burst, of bomb, H9
airfoil, flow around, 449, 458
air friction, 33, 35
air pollution, F10
air resistance, in projectile motion, 76
Alamagordo, New Mexico, H7
Alcator, J9
Alfvén wave, J7
Alpha Centauri, Prelude-8
alpha particles, scattering of, 926, C2–C3
alpha rays, B1
  radiation damage by, B5
alternating current, 754
  hazards of, 661–62
alternating emf, 722–23
alternative energy sources, F4–F10
ammeter, 663, 709
ampere (A), 15, 632, 963 (551)
Ampère, André Marie, 671, 692
Ampère's Law, K4, 692–94
  displacement current and, 778–80
  electric flux and, 779
  in magnetic materials, 743
  modified by Maxwell, 780, 782
  right-hand rule for, 693
amplitude:
  of motion, 338
  of wave, 370
Angers, France, bridge collapse at, 358–59
angle:
  bearing, 2
  conversion factors for, 965 (553)
  definition of, 947 (535)
  elevation, 74–75
  of incidence and of reflection, 832
  latitude, 2
  longitude, 2
angular acceleration:
  average, 250
  constant, equations for, 250
  instantaneous, 250
  rotational motion with constant, 251–53
angular frequency, 338
  of simple harmonic oscillator, 342
  of wave, 371–72
angular magnification:
  of magnifier, 849
  of microscope, 850
  of telescope, 850
angular momentum, 106–9
  component of, along axis of rotation, 262
  cross product and, 106, 259
  in elliptical orbit, 318
  quantization of, 739–40, 929
  rate of change of, 108, 212
  rate of change of, relative to center of mass, 285
  of a rigid body, 259–63
  of a system of particles, 212–13
angular momentum, conservation of, 107–9, 213, G4
  central force and, 107–9
  law of, 108, 213
  in planetary motion, 312
  in rotational motion, 278–82
angular-momentum quantum number, 930
angular resolution, of telescope, 889–92
angular velocity:
  average, 249
  instantaneous, 249
antibaryons, C10
antiderivative, 957 (545)
antimatter-matter annihilation, 538, G5
antimesons, C10
antinodes and nodes, 382
antiparticles, C10–C11
antiquarks, C14
anvil cloud, in thunderstorm, I5
aphelion, 311
  of comets, 318
  of planets, 313
apogee, Prelude-5
  of artificial satellites, 313
apparent weight, 143–44

In the two-volume edition, all pages after 530 are in Volume Two, except those in parentheses, which are in Volume One. Interlude F is the last one in Volume One.

Archimedes' Principle, 456
area, 946 (534)
  conversion factors for, 966 (554)
Arecibo radiotelescope, 891
artificial satellites, 310, 313
  apogee of, 313
  perigee of, 313
  period of, 313
associative law of vector addition, 50–51
astrology, 330
atmosphere, 451
  as electric machine, I1
atmospheric electric circuit, I1
atmospheric electric current, I1
atmospheric electric field, I1
  time dependence of, I3
atmospheric electricity, I1–I12
atmospheric pressure, 455
atom, C1–C5
  electron distribution in, 531
  nuclear model of, 927
  nucleus of, *see* nucleus
  photograph of, Prelude-14, Prelude-15, L13
  stationary states of, 930–31
  structure of, 531
  table of, Prelude-14
atomic clock, Cesium, 10
atomic clocks, portable, 11
atomic magnetic moments, 739–41
atomic mass unit, 13
atomic number, C3
atomic standard of mass, 13
atomic standard of time, 10
atomic structure, 920–27
atom smashers, C1
Atwood's machine, 147, 275
Aurora Borealis, 702
automobile accidents, 239
automobile battery, 538–39
automobile collision, 224, 239
automobiles, energy conversion in, F1–F2
average acceleration, 29–30
  in three dimensions, 69
average angular acceleration, 250
average angular velocity, 249
average power, 187
average speed, 24–25
average velocity, 25–26
  in three dimensions, 67–68
Avogadro's number, 12–13, 472
axis, instantaneously fixed, 282–83
axis of symmetry, A5

Babinet's Principle, 892–94
back emf, 728
background radiation, cosmic, D6
balance, 94, 97–98, 123
  beam, 97, 123
  Cavendish, 308, 354, E1
  spring, 97, 123
balance wheel, 354
ballistic curve, 76
ballistic missile, 138
ballistic pendulum, 231–32
ball lighting, I11
Balmer series, 923–24
banked curve, 143
Bardeen-Cooper-Schrieffer (BCS) theory, K5–K6
barometer, mercury, 455
Barton, O., 452
baryon, C10, G1
  multiplets of, C12–C13
baryon decuplet, C12–C13
baryon number, C11–C12, G4
baryon octet, C12
base units, 14
  of SI system, 963 (551)
bassoon, 397
bathyscaphe, 463
bathysphere, 452
battery:
  automobile, 538–39
  dry cell, 651–52
  internal resistance of, 655–56
  lead-acid, 538–39, 651
BCS (Bardeen-Cooper-Schrieffer) theory, K5–K6
beam balance, 97, 123
bearing angle, 2
beat frequency, 379–80
beats, of a wave, 379–80
Becquerel, A. H., B4*n*
Beebe, C. W., 452
bending of light rays, 4, E3–E4
Bernoulli, Daniel, 458
Bernoulli's equation, 458
best values, of fundamental constants, 970 (558)
beta decay, C11
beta rays, B1
  radiation damage by, B5–B6
betatrons, 721
Bethe, Hans A., H8*n*
Big Bang, D1, D5–D6, G8
  element formation during, D6
  radiant heat of, D6
bimetallic strip thermometers, 474, 489
binary star system, 315
  resolution of telescope and, 890
binding energy of nucleus, H3
  curve of, H3
binoculars, 854
Biot, Jean Baptiste, 675
Biot-Savart Law, 675–76
blackbody, spectral emittance of, 903
blackbody radiation, 900–902
  cosmic, D6
black hole, D8, E10–E13
  accretion by, E13
  singularity in, E12
blackout, 146
blast waves, nuclear, H9
blowhole, 397
blue, C15
body-centered cube, A10
body-mass measurement device, 98
Bohr, Niels, 929
Bohr magnetron, 740
Bohr radius, 930
Bohr's Complementarity Principle, 937
Bohr's Correspondence Principle, 933
Bohr's postulates, 929
boiling-water reactor, H14
Boltzmann's constant, 472
bomb:
  A-, H7
  air burst of, H9
  fission, H7
  H-, H8
  nuclear, H6–H8
  plutonium, H7
  uranium, H7
boom, sonic, 401
Born, Max, 912
bottom quark, C15
boundary conditions, 382
bound charges, 617
bound orbit, 183
Boyle, Robert, 471
Boyle's law, 471
Brackett series, 924
Braginsky, V. B., E1–E2
brake, hydraulic, 453
brake horsepower, 278
branch method, for circuits, 658
breeder reactors, F5, H14
Bremsstrahlung, 811
bridge collapse:
  at Angers, France, 358–59
  at Tacoma Narrows, 384
British system of units, 9, 14, 24
British thermal unit (Btu), 485
broken symmetry, G8
Brown's Ferry, Alabama, reactor accident at, F13
bubble chamber, C7–C8
buoyant force, 456
Bureau International de l'Heure, 11
Bureau International des Poids et Mesures (BIPM), 7, 12, 37

cable, superconducting, K8–K10
calculus, 953–62 (541–50)
Callysto, 313
caloric, 484
calorie, 484, 486
camera:
  photographic, 846–47
  pinhole, 847, B10
candela, 15, 963 (551)
capacitance, 612–17
  of the Earth, 614
  of a single conductor, 613
capacitative reactance, 757
capacitors, 612–17
  circuit with, 757
  cylindrical, 621
  energy in, 624–25
  guard rings for, 626
  in parallel, 615–16
  parallel-plate, 614–15
  in series, 615–16
  spherical, 621
Capella, 906
Caph, spectrum of, 923
capillary wave, 401
carbon dioxide-nitrogen laser, L4
Carnot, Sadi, 512
Carnot cycle, 512–13, 516
Carnot engine, 512–15
  efficiency of, 514–15
  Second Law of Thermodynamics and, 521
Carnot's theorem, 520
cat, landing on feet by, 282
cathode rays, C1–C2
cathode-ray tube, C2

In the two-volume edition, all pages after 530 are in Volume Two, except those in parentheses, which are in Volume One. Interlude F is the last one in Volume One.

Cavendish, Henry, 308
Cavendish balance, 308, 354, E1
cavity oscillation, 783–87
  electromagnetic, 785–86
cavity radiation, 902
celestial mechanics, 304
  accuracy of, 304
cell:
  thunderstorm, I3–I4
  triple-point, 473
Celsius temperature scale, 474–75
center of force, 138
center of mass, 202–9
  acceleration of, 208
  of continuous mass distribution, 204
  gravitational force acting on, 288
  motion of, 207–9
  rate of change of angular momentum relative to, 285
  rotation about, 285
  velocity of, 207, 230
central force, conservation of angular momentum and, 107–9
Centre Européen de la Recherche Nucléaire (CERN), 430, 433–34, C5–C8
  bubble chamber at, C7–C8
centrifugal acceleration, 79
centrifuge, 78
centripetal acceleration, 78, 93–94, 141–44
  Newton's Second Law and, 142
centroid, 204
Čerenkov counters, C7
Cerro Prieto (Mexico), F9
Cesium atomic clock, 10
Cesium standard of time, 10
cgs system of units, 542
Chadwick J., C3
chain reactions, H5–H7
Chalk River, Canada, reactor accident in, F13
changes of state, 492–93
charge, electric, 533–39
  associated with current, 682–83
  bound, 617
  bound vs. free, in dielectrics, 617
  conservation of, 538–39, G4
  conversion factors for, 969 (557)
  of electron, 534
  of electron, measurement of, 560
  elementary, 534
  of elementary particles, C10
  generation of, in thunderstorms, I5–I6
  image, I7
  motion of, in magnetic field, 698
  of particles, 533–34, 537
  point, 535
  of proton, 534
  quantization of, 537, G8*n*
  SI unit of, 705
  static equilibrium of, 572–73
  surface, on dielectric, 620
charged surface, flat, potential of, 582
Charles' Law, 471
charm, of quarks, C15
chemical composition of universe, D6
chemical elements, 973–74 (561–62)
  age of, D6
  during Big Bang, D6
  Periodic Table of, Prelude-14, 972 (560)
  transmutation of, B4
chemical reactions, conservation of charge in, 539
China syndrome, F14
chromatic aberration, 846
chromatic musical scale, 393
circuit, electric:
  AC, 754–77
  atmospheric, I1
  branch method for, 658
  with capacitor, 757
  with inductor, 758–59
  LC, 759–69
  LCR, *see* LCR circuit
  loop method for, 656
  multiloop, 656–58
  with resistor, 755
  single-loop, 654–56
circular aperture:
  diffraction by, 888–92
  minimum in diffraction pattern of, 889
circular orbits, 308–10
  energy for, 318
  in magnetic field, 699
classical electron radius, 936
classical mechanics, quantum mechanics vs., 921
classical particle, 1
classical physics, 1, 4
Clausius, Rudolph, 516
Clausius statement of Second Law of Thermodynamics, 521
Clausius' theorem, 516, 521–22
clipper ship, power of, F1
clock:
  Cesium atomic, 10
  pendulum, 349, 352
  portable atomic, 11
  synchronization of, 6, 11, 419–21
close packing, A8–A11
coal:
  production of, F4
  reserves of, F4
Cockcroft, J. D., C3–C4
coefficient of cubical thermal expansion, 487–88
coefficient of kinetic friction, 130–32
coefficient of linear thermal expansion, 487
coefficient of restitution, 243
coefficient of static friction, 131–34
coefficients of friction, 129–34
coherence of light, 868, L1, L6
coherent radiation, L1
colliding beams, C7
collimator, B10
collisions, 223–45
  automobile, 224, 239
  impulsive forces and, 223–27
  of nuclei, 236–39
  of particles, 236–39
collisions, elastic, 227–30
  conservation of energy in, 227–28, 233
  conservation of momentum in, 227–28, 233
  in one dimension, 227–30
  one-dimensional, speeds after, 228–30
  in two dimensions, 232–34
collisions, inelastic, 230
  internal kinetic energy in, 238
  rest-mass energy in, 237
  in two dimensions, 234–36
color:
  of quarks, C15, G9–G11
  of visible light, 812
combination of velocities, relativistic, 433–34
combination principle, Rydberg-Ritz, 925
comets, 318
  Halley's 332
  orbits of, 319
  perihelion of, 318
  period of, 318
  tails of, 817
common logarithm, 946 (534)
communication satellites, 310
commutative law of vector addition, 50
compass needle, 668
Complementarity Principle, 937
complimentary plate, Babinet's Principle and, 893
component of angular momentum along axis of rotation, 262
components, 53–57, 59
Compton, A. H., 910
Compton effect, 910–12
Compton wavelength, 936
concave spherical mirror, 840
Concorde SST, sonic boom of, 401
concrete, thermal expansion of, 489
conduction of heat, 490–92
conductivity, thermal, 490–91
conductors, 539
  in an electric field, 572–74
  energy of a system of, 603–4
  insulators vs., 639
conical pendulum, 152
conservation laws, 153–54
  elementary particles and, C11–C12
  symmetry and, 154, G4
conservation of angular momentum, 107–9, 213, G4
  central force and, 107–9
  law, of, 108, 213
  in planetary motion, 312
  in rotational motion, 278–82
conservation of electric charge, 538–39, G4
  in chemical reactions, 539
conservation of energy, 154, 174–97, 318, G4
  general law of, 185
  in Lenz' Law, 722
  in one-dimensional elastic collision, 227–28
  in simple harmonic motion, 345, 348–49
  in two-dimensional elastic collision, 233
conservation of mass, 153
conservation of mechanical energy, 165–66, 177–78
  law of, 165–66, 177–78

In the two-volume edition, all pages after 530 are in Volume Two, except those in parentheses, which are in Volume One. Interlude F is the last one in Volume One.

conservation of mechanical energy (*continued*)
in rotational motion, 276
conservation of momentum, 105, 199, 227, 230, G4
in elastic collisions, 227–28, 233
in fields, 549
law of, 199
conservative electric field, 586
conservative force, 174–80
gravity as, 316
path independence and, 175
potential energy of, 176–80
conserved quantities, 105, C11–C12
forces and, G4
constant:
fine-structure, 936
fundamental, best values of, 970 (558)
constant-volume gas thermometer, 473
constructive and destructive interference, 378, 861
in standing electromagnetic wave, 863
for wave reflected by thin film, 864–66
"contact" electric field, 654
"contact" forces, 533
contact potential, 574
containment shell, F13
contamination, radioactive, F14
continuity equation, 449
contraction, of universe, D7
control rod, in nuclear reactor, F12, H13
convection, 492
conversion factors, 16, 965–69 (533–58)
conversion of energy, *see* energy conversion
converter reactor, H13
convex spherical mirror, 842
Conway granite, E5
cooling towers, F11–F12
Cooper pairs, K6
coordinate grid, 2, 5–7, 80–81
coordinates:
cylindrical, 2–3
Galilean transformation of, 417, 424
Lorentz transformation of, 429
origin of, 2, 5–6
rotation of, 60–61
transformation of, 60–61
Copernicus, Nicolas, 309
core:
of Earth, 405
of nuclear reactor, F12
corner reflector, 833–34
on Moon, L5
corona discharge, I8
Correspondence Principle, 933
$\cos^{-1}$, 949 (537)
cosecant, 948 (536)
cosine, 948 (536)
cosines, law of, 952 (540)
cosmic background radiation, D6
cosmic ray, 236, B2
primary, B2
secondary, B2
trajectory of, 698
Cosmological Principle, D4
cosmological red shift, 828
cosmology, D1
cotangent, 948 (536)
coulomb (C), 534
Coulomb's Law, 534–36
accuracy of, 536–37
in vector notation, 535
critical angle, for total internal reflection, 838
critical magnetic field, K2–K3
critical mass, H6
critical radii, for nuclear explosion, H11
crossed electric and magnetic fields, 702–3
cross product, 57–59
angular momentum and, 106, 259
by components, 590
in definition of torque, 270
of unit vectors, 58
crystal, A1–A14
close packing and, A8–A11
defects in, A11–A13
of virus particles, A2
crystal lattice, A1–A2
crystallites, A4
crystallography, A1
crystal structure, A1–A5
cube:
body-centered, A10
face-centered, A9
symmetries of, A9
curie (Ci), B4
Curie, Marie Sklowdowska, B4*n*
Curie, Pierre, B4*n*
Curie temperature, 745
current, electric, 631–33
AC, 754
AC, hazards of, 661–62
associated electric charge and, 682
atmospheric, I1
in capacitor circuit, 757
conversion factors for, 969 (557)
DC, 649–67
DC, hazards of, 661–62
displacement, 778, 780
generating magnetic field, 675–76
in inductor circuit, 758–59
"let-go," 661
magnetic force between, 671
in parallel LCR circuit, 769
persistent, K2
in series LCR circuit, 765
SI unit of, 705
current density, 632–33
current loop:
potential energy of, 707
torque on, 705–6
current resistor circuit, 755
curvature of space, 4, E3–E6
perihelion precession and, E7
curve:
banked, 143
of binding energy, H3
curved spacetime, E7
cutoff frequency, in plasma, J7
cycloid, 283
cyclotron, 699–700, C4
cyclotron emission, 811
cyclotron frequency, 699
Cygnus X-1, 331, E12–E13
cylindrical capacitor, 621
cylindrical coordinates, 2–3

damped harmonic motion, 356
damped oscillations, 356–59
damped oscillator:
driving force on, 358
energy loss of, 357
Q of, 357–58
resonance or sympathetic oscillation of, 358–59
ringing time of, 357–58
dart leader, in lightning, I9
daughter material, B4
da Vinci, Leonardo, 130–31
day:
mean solar, 10
sidereal, 93*n*
solar, 10, 93*n*
DC current, 649–67
hazards of, 661–62
DC-10 airliner crash near Orly, 465
de Broglie wavelength, 934
Debye distance, J2
decay of particles, G3–G4
deceleration, 31
decibel, 391–92
decuplet, baryon, C12–C13
dees, 699
defects, in crystals, A11–A13
deflection of light, by Sun, E3–E4
degree absolute, 470
degree-days, F16
degree of arc, definition of, 947 (535)
degrees, 947 (535)
density, 14, 204
conversion factors for, 967 (555)
critical, for universe, D7–D8
of fluid, 446–47
of nucleus, H1–H2
deoxyribonucleic acid, Prelude-13, Prelude-15
depth finder, 408
derivative:
definition of, 953 (541)
partial, 181
of the potential, 587
rules for, 955 (543)
derived unit, 14
of SI system, 964 (552)
destructive and constructive interference, 378, 861
in standing electromagnetic wave, 863
for wave reflected by thin film, 864–66
determinant, 59
determinism, 119
deuterium:
abundance of, in universe, D8
as nuclear fuel, F6
deuterium-deuterium reaction, J8
deuterium-tritium reaction, J8
Dialogues (Galileo), 39
diamagnetic materials, 746–49
permeabilities of, 748–49
diamagnetism, 746–49
diamond, A10
diatomic gas, energy of, 478
diatomic molecule, 181
vibrations of, 254–55
Dicke, R. H., experiments on universality of free fall by, E1–E2
dielectric, 612, 617–25
electric field in, 618
energy density in, 625
Gauss' Law in, 622–24
linear, 617

In the two-volume edition, all pages after 530 are in Volume Two, except those in parentheses, which are in Volume One. Interlude F is the last one in Volume One.

surface charge on, 620
dielectric constant, 618–19
differential equation, numerical
solution of, 137–41
differentiation, 953 (541)
rules for, 955 (543)
diffraction, 406–7, 883–99
at a breakwater, 406
by circular aperture, 888–92
by a single slit, 883–87
of sound waves, 407
of water waves, 406
X-ray, A3
diffraction pattern, 406
for separation of isotopes, H7
of single slit, 885, 887
diodes, A13
Dione, 331
dip angle, 686
dip needle, 686
dipole:
electric field of, 590–91
magnetic, 679
potential of, 590
potential energy of, 557
torque on, 557
dipole moment, 556–58
magnetic, 680, 706
permanent and induced, 557
direct current, 649–67
hazards of, 661–62
discharge, electric, C1
dislocation, A11–A12
dispersion, of light, 839
dispersive medium, 374
displacement, 8
displacement current, 778, 780
displacement field, electric, 624
displacement vector, 46–48
dissipation of energy:
in U.S., F3–F4
in U.S., per capita, F3
distributive law, 56*n*
DNA, Prelude-13, Prelude-15
domains, magnetic, 744
donor impurities, 641
doping, A13
Doppler shift, 397–401, E10
of light, 817–19
doses, radiation, B9
effects of acute, B9
dot product, 56
by components, 56–57
in definition of work, 158
of unit vectors, 56
double-exposure hologram, L12
"doubling the angle on the bow," 63
down quark, C14
drift velocity:
of free-electron gas, 634
of free electrons, 636
driving force, on damped oscillator, 358
dry cell battery, 651–52
dynamics, 91–152
fluid, 457–61
of rigid body, 270–88
dynodes, 909

early universe, G8
Earth, 313
capacitance of, 614
core of, 405
escape velocity from, 320
magnetic field of, 680–81
magnetic moment of, 681
mantle of, 405
moment of inertia of, 262–63, 266, 268
perihelion of, 329
reference frame of, 93–94
rotation of, 10, 93
spin of, 262–63
earthquake:
in Alaska, 404
in San Francisco, 405
eddy currents, 732
effects of acute radiation doses, B9
efficiency:
of Carnot engine, 514–15
of energy conversion, F2–F3
of engines, 512
eigenfrequencies, 383
Eightfold Way, C12–C13
Einstein, Albert, 5, 185, 416, 419, 907, E1, E3
mass-energy equation of, 822
photoelectric equation of, 908
elastic body, 135
elastic collision, 227–30
conservation of energy in, 227–28, 233
conservation of momentum in, 227–28, 233
in one dimension, 227–30
speeds after one-dimensional, 228–30
in two dimensions, 232–34
elastic force, 135–36
elastic limit, 136
electric charge, 533–39
associated with current, 682–83
bound, 617
bound vs. free, in dielectrics, 617
conservation of, 538–39, G4
conversion factors for, 969 (557)
of electron, 534
of electron, measurement of, 560
elementary, 534
of elementary particles, C10
generation of, in thunderstorms, I5–I6
image, I7
motion of, in magnetic field, 698
of particles, 533–34, 537
point, 535
of proton, 534
quantization of, 537, G8*n*
SI unit of, 705
static equilibrium of, 572–73
surface, on dielectric, 620
electric circuit, *see* circuit, electric
electric dipole, 556–58
electric discharge, C1
electric displacement field, 624
electric energy, 601–11
in capacitors, 624–25
in nucleus, H1–H2
of spherical charge distribution, 607
in uranium nucleus, 607
electric energy density in dielectric, 625
electric field, 546, 549–64
of accelerated charge, 788–92
atmospheric, I1, I3
conductors in, 572–74
conservative, 586
"contact," 654
conversion factors for, 969 (557)
definition of, 550
in dielectric, 618
of dipole, 590–91
electric dipole in, 556–58
electric force and, 550
energy density of, 604–7
fair-weather, I1–I3
of flat sheet, 552
induced, 714, 723–26, 784
induced, in solenoid, 725–26
magnetic field and, related by relativity, 684
of plane wave, 806–7
of point charge, 551
of spherical charge distribution, 571–72
of thundercloud, I6–I8
in uniform wire, 632
electric field lines, 553–56
made visible, 558
made visible, with grass seeds, 631
of point charge, 553–55
sources and sinks of, 555
electric flux, 565–67
Ampère's Law and, 779
electric force, 531–34
compared to gravitational force, 532, 535
Coulomb's law for, 534–36
electric field and, 550
in nucleus, H1–H2
qualitative summary of, 533
superposition of, 546–49
electric fringing field, 604, 626
electric generators, 652
electric ground, 583
electricity:
atmospheric, I1–I12
frictional, 532, 540–41
Gauss' Law for, 782
electric motor, 707
electric shock, 661–62
electrolytes, 540
electromagnet, 696
electromagnetic flowmeter, 733
electromagnetic force, 120–21, 532, G1
electromagnetic generator, 722–23
electromagnetic induction, 714–37
electromagnetic interactions, G2
electromagnetic oscillation of cavity, 785–86
electromagnetic radiation:
kinds of, 811–12
wavelength and frequency bands of, 812
electromagnetic wave, 778
energy flux in, 813–14
generation of, 810–12
momentum of, 815–17
speed of, 794, 805
standing, 862–64
electromagnetic wave pulse, 802–6
electromotive force, *see* emf (electromotive force)
electron, C2–C4
charge of, 534
distribution of, in atoms, 531

In the two-volume edition, all pages after 530 are in Volume Two, except those in parentheses, which are in Volume One. Interlude F is the last one in Volume One.

electron (*continued*)
electric and magnetic fields of, 680
free, 539–40, 635–36
free of gas, 540, 634, A9
magnetic moment of, 680
mass of, 99
measurement of charge of, 560
spin of, 680
electron avalanche, I8
electron field, G6
electron-holography microscope, Prelude-15, 531, L13
electron microscope, Prelude-12, A2
electron-volts (eV), 184, 584
electrostatic equilibrium, 572
approach to, 573
electrostatic force, 532–33
electrostatic induction, 541–42
electrostatic potential, 581
electroweak force, G7
elementary charge, 534
elementary particles, C1–C16
conservation laws and, C11–C12
electric charges of, C10
family relationships of, C12–C13
masses of, C10
spins of, C10
elements, chemical, 973–74 (561–62)
age of, D6
during Big Bang, D6
Periodic Table of, Prelude-14, 972 (560)
transmutation of, B4
elevation angle, 74–75
elevator with counterweight, 129
ellipse, semimajor axis of, 311
elliptical galaxies, D2
elliptical orbits, 311–16
angular momentum in, 318
energy in, 318
vs. parabolic orbit for projectile, 315–16
emergency cooling systems, F14
emf (electromotive force), 649–54
alternating, 722–23
induced, 714, 718–23, 726–28
LCR circuit with, 763
motional, 714–17
power delivered by, 659
in primary and secondary circuits of transformer, 770
sources of, 651–54
emission, stimulated, 811, L2
empirical evidence, 3
emulsion, photographic, 237, G2
energy, 153–97
alternative sources of, F4–F10
alternative units for, 184–85
in capacitor, 624–25
for circular orbit, 318
conservation of mechanical, in rotational motion, 276
conversion factors for, 967–68 (555–56)
of diatomic gas, 478
dissipation of, in U.S., F3–F4
electric, *see* electric energy
in elliptical orbit, 318
entropy and, F2
geothermal, F9
gravitational potential, of a body, 207
heat, 211
of ideal gas, 477–79
internal kinetic, 210–11, 230
internal kinetic, in inelastic collision, 238
kinetic, *see* kinetic energy
in LC circuits, 761–62
Lorentz transformations for, 437–38
magnetic, 728–30
magnetic, in inductor, 729
mass and, 185–87
mechanical, *see* mechanical energy
momentum and, relativistic transformation for, 437–38
nuclear, H1
nuclear, from fission, F5
nuclear, from fusion, F6
of photon, 907
potential, *see* potential energy
released in fission, H5
resources of, F4
rest-mass, 186, 437
rest-mass, in inelastic collision, 237
of rotational motion, 276
of rotational motion of gas, 478–79
in simple harmonic motion, 344–45
of simple pendulum, 348
solar, F7–F9
of stationary states of hydrogen, 930–31
of system of conductors, 603–4
of system of particles, 209–11
of system of point charges, 601–3
thermal, 484
threshold, 239
tidal, F9–F10
total relativistic, 437
in wave, 374–77, 812–14
energy, conservation of, 154, 174–97, 318, G4
general law of, 185
in Lenz' Law, 722
in one-dimensional elastic collision, 227–28
in simple harmonic motion, 345, 348–49
in two-dimensional elastic collision, 233
energy, conservation of mechanical, 165–66, 177–78
in rotational motion, 276
energy conversion, F1–F3
in automobiles, F1–F2
entropy and, F2
in thunderstorm, I3
energy conversion efficiencies, F2–F3
energy density:
in dielectric, 625
in electric field, 604–7
in magnetic field, 730
of wave, 375–76
energy dissipation, in U.S., per capita, F3
energy distribution:
in fission, H5
in nuclear explosion, H9
energy flux, in electromagnetic wave, 813
energy level, 183
of hydrogen, 931
energy-level diagram, 932
energy loss, of damped oscillator, 357
energy quantization, of oscillator, 905
energy quantum, 902–6
energy storage, K10
energy use, in U.S., F3
energy-work theorem, 162–63
engine, efficiency of, 512
enriched uranium, H12
entropy, 516–20
at absolute zero, 524
change, in free expansion of gas, 518
change, in isothermal expansion of gas, 518
of different kinds of energy, F2
disorder and, 524
energy conversion and, F2
of ideal gas, 517
irreversible process, 523
negative, 527
Eötvös, Lorand von, 326, E1
Eötvös experiments, 326, E1
epicenter, 404–5
equation of motion, 119–20, 137–41
for rotation, 273–74
of simple harmonic oscillator, 342
of simple pendulum, 347
see also Newton's Second Law
equilibrium:
electrostatic, 572, 573
of fluid, 452
stable and unstable, 351
thermodynamic, 511
equilibrium point, 183
equinoxes, precession of, 279*n*
equipartition theorem, 479
equipotential surface, 588–89
Equivalence, Principle of, 327, E1–E3
equivalence of inertial and gravitational mass, 326
equivalent absorbed dose, B8
escape velocity:
from Earth, 320
from Sun, 320
for universe, D7
Escher, M. C., A5–A8
etalon, 879–80
ether, 417–18
ether wind, 418–19
Euclidean geometry, E5
deviations from, E6
Euclidean space, 3–5
Euler's theorem, 246
Europa, 313
Evenson, K. M., 802
events, 1
excited states, 932
expansion, free, of a gas, 510, 518
expansion, thermal:
of concrete, 489
linear, coefficient of, 487
of solids and liquids, 487–89
of water, 488–89
expansion of universe, D1, D4–D5
Explorer I, 313
Explorer III, 313
explosion, nuclear, *see* nuclear explosion
exponential function, 945 (533)
external field, 556
external forces, 201

In the two-volume edition, all pages after 530 are in Volume Two, except those in parentheses, which are in Volume One. Interlude F is the last one in Volume One.

eye:
myopic and hyperopic, 848
nearsighted and farsighted, 848
as optical instrument, 847–48

Fabry-Perot interferometer, 879
face-centered cube, A9
Fahrenheit temperature scale, 474–75
fair-weather electric field, I1–I3
fallout, from nuclear explosion, H11
fallout pattern, for nuclear explosion, H12
fallout shelters, H12
farad (F), 613
Faraday, Michael, 714
Faraday cage, 586
Faraday's constant, 544
Faraday's Law, 782, J3
Faraday's Law of induction, 718–19, 725
farsightedness, 848
Fermi, Enrico, H13*n*
Fermi National Accelerator Laboratory (Fermilab), C5–C7
accelerator at, K8
Tevatron, C6
ferromagnetic materials, 743–46
ferromagnetism, 743–46
Feynman diagram, G6
fibers, optical, L5
fibrillation, 661–62
field, 121, 546
electric, *see* electric field
electron, G6
forces and, G5
magnetic, *see* magnetic field
momentum in, 672
quanta and, G4–G7
radiation, of accelerated charge, 788, 790
radiation, of oscillating charge, 791
field lines, 553–56
made visible, 558
of point charge, 553–55, 674
sources and sinks of, 555
field meter, I2
field mill, I2
filter, polarizing, 808–10
fine-structure constant, 936
fireball:
of nuclear explosion, H9–H10
primordial, D6
fireball radiation, cosmic, D6
firestorm, of nuclear explosion, H11
first harmonic mode, 382
First Law of Thermodynamics, 509–11
fission, H1, H4–H5
deformation of nucleus during, H4
distribution of energy in, H5
ejection of particles in, H5
energy released in, H5
multiplication factor for, H6
neutron-induced, H5
neutrons released in, H5–H6
potential energy during, H4
reflector in, H7
as source of nuclear energy, F5
spontaneous, H4
symmetric, H4
fission bomb, H7
fission-fusion-fission device, H9
flash burns, from nuclear explosion, H11
flat sheet, electric field of, 552
flow:
around an airfoil, 449, 458
of free electrons, 635
of heat, 490
incompressible, 447–48
methods for visualizing, 449
in a nozzle, 460
plastic, 445
of rock, 445
of solids, 445
from source to sink, 449
steady, 447–48
velocity of, 446
around a wing, 449–50, 458
flowmeter:
electromagnetic, 733
Venturi, 459, 468
fluid, 445
density of, 446–47
equilibrium of, 452
incompressible, 448
nuclear, H2
perfect, 446
static, 452
fluid dynamics, 457–61
fluid mechanics, 445–68
fluorescent tube, flicker in, 772
flute, 397
flux:
electric, 565–67
magnetic, 719
flux quantum, K5
f number, 847
focal length, of spherical mirror, 841
focal plane, 854
focal point:
of lenses, 844
of a mirror, 840–41
foot, 9
foot-pound, 155
force, 95–104
buoyant, 456
calculated from potential energy, 180–82
center of, 138
central, angular momentum and, 107–9
color, C15, G9–G11
conservative, 174–80, 316
conserved quantities and, G4
contact, 533
conversion factors for, 967 (555)
as derivative of potential energy, 180–82
elastic, 135–36
electric, *see* electric force
electromagnetic, 120–21, 532, G1
electromotive, *see* emf (electromotive force)
electrostatic, 532–33
electroweak, G7
equation of motion and, 119–20
exchange of virtual particles and, G6
failure of Newton's Third Law for, 671
fields and, G5
frictional, on rope, 291–92
fundamental, 120–21, G1–G2
gravitational, *see* gravitational force
impulsive, 223–27
internal and external, 201
inverse-square, 137–38
Lorentz, 702, 783
motion with constant, 125–29
motion with variable, 137–41
N as, 125
net, 99–100
power delivered by, 188
between quarks, C15
range of, 121
restoring, 135–36
resultant, 99–100
of a spring, 135–36
"strong," 121, G1, G9, H1–H2
tube of, 559
units of, 96–97
"weak," 121, G1
force, magnetic, 532, 668–72
between currents, 671
due to magnetic field, 672
failure of Newton's Third Law for, 671
on moving point charge, 669
on relativistic correction, 670
on wire, 704
forced oscillations, 356, 358
fossil fuel:
pollutants from, F10
reserves of, F4
supply of, F4
Foucault effect, 346*n*
Foucault pendulum, 93
Fourier series, 380
Fourier's theorem, 380, 392
frame of reference, 6, 79–81, 419
in calculation of work, 156
of Earth, 93–94
freely falling, 123–24
inertial, 92–94
local inertial, E2
power and, 189
for rotational motion, 247, 282
Franklin, Benjamin, I11
Frauenhofer approximation, 869*n*
Frauenhofer Lines, 922
free body, 92
"free-body" diagram, 126
free charges, 617, 624
free-electron gas, 540, A9
drift velocity of, 634
friction in, 634
free electrons, 539–40
drift velocity of, 636
flow of, 635
free expansion of a gas, 510
entropy change in, 518
free fall, 326
universality of, 326, 349, E1–E3
weightlessness in, 123–24
free-fall motion, 34–35, 123–24
frequency, 249–50
normal, 383
proper, 383
of simple harmonic motion, 338
threshold, 908
of wave, 371
frequency bands, of electromagnetic radiation, 812
Fresnel, Augustin, 407, 883
Fresnel lens, 854, 859

In the two-volume edition, all pages after 530 are in Volume Two, except those in parentheses, which are in Volume One. Interlude F is the last one in Volume One.

friction, 129–34
  air, 33, 35
  coefficient of kinetic, 130–32
  coefficients of, 129–34
  in flow of free-electron gas, 634
  on ice and snow, 132
  loss of mechanical energy by, 180
  microscopic and macroscopic area of contact and, 131
  sliding (kinetic), 129–33
  static, 133–34
  static, coefficient of, 131–34
frictional electricity, 532, 540–41
frictional force, on rope, 291–92
fringes, 865
fringing field, 604, 626
fuel, fossil:
  pollutants from, F10
  reserves of, F4
  supply of, F4
fuel cells, 652–53
  on Skylab, 653
fuel pellets, for laser fusion, L8–L9
fuel rods, of nuclear reactor, F12, H13
functions, inverse trigonometric, 949 (537)
fundamental constants, best values for, 970 (558)
fundamental forces, 120–21, G1–G2
  range of, 121
  strength of, 121
fundamental mode, 382
Fundy, Bay of, 327
fusion, 492–93
  heat of, 492–93
  inertial confinement in, F6
  laser, L7–L9
  nuclear, F6, H8, J9–J12
  as source of nuclear energy, F6
  thermonuclear, J8–J12, L7
  triggered by laser, F6
fusion reactions, H8
fusion reactor, F6, J9–J12
future of universe, D6–D8

g, 34, 36–37
  measurement of, 37, 307–8
  variation of, with latitude, 94
Gabor, Dennis, L12–L13
galaxies, D2–D5
  elliptical, D2
  spiral, D2
Galaxy of the Milky Way, Prelude-9, D1
Galilean addition law for velocity, 417
Galilean coordinate transformations, 417, 424
Galilean Relativity, 109–11
Galilean telescope, 854
Galilean transformations, 81
  for momentum, 434–35
  for velocity, 417
Galileo Galilei, 37, 39, 91
  experiments on universality of free fall by, 326, 349, 360, E2
  isochronism of pendulum and, 364
  pendulum experiments by, 360
  tide theory of, 84
gamma rays, B1
  radiation damage by, B6
Ganymede, 313

gas:
  adiabatic equation for, 497–99
  Boyle's law for, 471
  energy of ideal, 477–79
  entropy change in free expansion of, 518
  entropy change in isothermal expansion of, 518
  of free electrons, 540, 634, A9
  free expansion of, 510
  Gay-Lussac's Law for, 471
  ideal, *see* ideal gas
  interstellar, Prelude-7
  natural, F4
  root-mean-square speed of, 476–77
  specific heat of, 493–96
gas constant, universal, 470
gas-cooled reactor, H14
  kinetic theory of, 469
gas state, Prelude-2
gas thermometer, constant-volume, 473
gauge blocks, 7
gauge pressure, 456
gauss (G), 673
Gauss, Karl Friedrich, 673
Gauss' Law, 565, 567–72
  in dielectrics, 622–24
  for electricity, 782
  for magnetic field, 675
  for magnetism, 782
Gaussian surface, 568
Gay-Lussac's Law, 471
gee forces, 36–37, 144
Geiger, H., 926
Gell-Mann, M., C12, C14
general law of conservation of energy, 185
General Relativity theory, 5, 314, D1, E1, E6–E8
generation of electric charge in thunderstorm, I5–I6
generator:
  electric, 652
  electromagnetic, 722–23
  homopolar, 716
  magnetohydrodynamic, 717
  superconducting, K9
geometric optics, 830
geometrodynamics, E6
geometry, deviations from Euclidean, E6
geostationary orbit, 310
geostationary satellite, 310
geothermal energy, F9
Geysers, the (California), F9
G force, 144
gimbals, 278
glide line, A7
globular clusters, age of, D5–D6
gluons, G8, G10–G11
gradient, of the potential, 587
graphite, A10
grating, 874–75
  principal maximum of, 874
  resolving power of, 874
gravimeter, pendulum, 352
gravitation, 304–36
  law of universal, 305–7
gravitational constant, 305
  measurement of, 307–8
gravitational force, 120–21, 305–7, G1
  acting on center of mass, 288
  compared to electric force, 532, 535
  exerted by spherical mass, 322–25
  motion of recession of galaxies and, D7
gravitational interactions, G2
gravitational mass, 325–26
gravitational potential energy, 164–65, 316–22
  of a body, 207
  of spherical mass, 323
gravitational red shift, E8–E10
gravitational singularity, E12
gravitational time dilation, E8–E10
  experiments on, E8–E9
graviton, G5
gravity:
  action and reaction and, 102
  as conservative force, 316
  of Earth, *see* weight
  work done by, 160–61
gravity, acceleration of, 33–37, 307
  measurement of, 37, 352
  universality of, 33–34
  variation of, with altitude, 307
  variation of, with latitude, 94
gravity wave, 401–2
  speed of, 402
green , C15
greenhouse effect, F11
  Earth's temperature increase and, F11
Greenwich Mean Time, 10–11
ground, electric, 583
ground state, 932
group velocity, of a wave, 374
G suit, 146–47
guard rings, for capacitors, 626
Guericke, Otto von, 112–13
guidance system, inertial, 110, 279
guitar, 397
Gulf Stream, 82
gun device, H7
gyroscope, 278–79
  directional, 279
  precession of, 286–88

hadron, C9, G1
hairsprings, 354
Hale telescope, 851
half-life, B4
Hall effect, 703–4
Halley's comet, 332
Hanford, Washington, H13
harmonic function, 338
harmonic mode, 382
harmonic motion:
  damped, 356
  see also simple harmonic motion
harmonic wave, 370–71
H-bomb, H8
headlight method, D4
hearing, threshold of, 391
heat, 184, 484–507
  of evaporation, in water vapor, I3
  of fusion, 492–93
  mechanical equivalent of, 486
  specific, *see* specific heat
  temperature changes and, 486
  transfer, by convection, 492
  transfer, by radiation, 492
  of vaporization, 492–93
  waste, F11
heat capacity, specific, 485

In the two-volume edition, all pages after 530 are in Volume Two, except those in parentheses, which are in Volume One. Interlude F is the last one in Volume One.

heat conduction, 490–92
  equation of, 491
heat current, 490
heat energy, 211
heat flow, 490
heat pump, 515
heat reservoir, 511
heavy water, D8
heavy-water reactor, H14
height, maximum, of projectiles, 73–74
Heisenberg's uncertainty relation, 914
helical motion, in magnetic field, 701
helicentric system, 309
helium and hydrogen abundance of universe, D6
helium-neon laser, L3–L4
Helmholtz, Hermann von, 185
henry, 727
hertz (Hz), 339
Hertz, Heinrich, 802, 907
high-voltage transmission line, power dissipated in, 660–61
Hiroshima, Japan, H7–H8
Hoffman, B., 912
"holes," in semiconductors, 640
hologram, L9–L13
  double-exposure, L12
holographic images, L10
holographic three-dimensional images, L10–L11
holography, L9–L13
homopolar generator, 716
Hooke, Robert, 135
Hooke's law, 135–36, 341
horizon, E11
horsepower (hp), 187–88
hot rod, 146
Hubble, Edwin, D1
Hubble's Law, D4
Hubble time, D5
Huygens, Christiaan, 165
Huygens' Construction, 830–31
Huygens-Fresnel Principle, 883, 892
Huygens' tilted pendulum, 360
Huygens' wavelets, 831
hydraulic brake, 453
hydraulic press, 453
hydroelectric power, F7
hydrogen:
  atom, stationary states of, 930–31
  burning, in Sun, H8
  energy levels of, 931
  and helium abundance of universe, D6
  isotopes of, H1
  molecule, oscillations of, 346
  spectral series of, 924–25, 931
hydrogen spectrum, 923
  produced by grating, 873
hydrometer, 467
hydrostatic pressure, 454
hyperbola, 319
hyperbolic orbit, 319
hyperopic eye, 848
hysteresis, 745
hysteresis loop, 750

Iapetus, 331
Idaho Falls, Idaho, reactor accident at, F13
ideal gas, 470–71
  energy of, 477–79
  entropy of, 517
ideal-gas law, 470–73
ideal-gas temperature scale, 473
ideal particle, 1
ideal solenoid, 695
image, 833
  method of, I7
  real, 843–44
  three-dimensional, from hologram, L10–L11
image charges, I7
image orthicon, 909
impact parameter, 927
impedance:
  for LCR circuit, 765
  for parallel LCR circuit, 769
implosion device, H7
implosion of universe, D7
impulse, 224–25
impulsive force, 223–27
impurities, acceptor and donor, 641
incidence, angle of, 832
incompressible flow, 447–48
incompressible fluid, 448
indeterminacy relation, 914
index of refraction, 835
induced dipole moments, 557
induced electric and magnetic fields between capacitor plates, 784
induced electric field, 714, 723–26
  in solenoid, 725–26
induced emf, 714, 718–23
  in adjacent coils, 726–28
  in solenoids, 720–21
induced magnetic field, 780–81
  of accelerated charge, 795
  in capacitor, 780–81
inductance, 726–28
  conversion factors for, 969 (557)
  mutual, 726
  self-, 728
induction:
  electromagnetic, 714–37
  electrostatic, 541–42
  Faraday's Law of, 718–19, 725
inductive reactance, 758–59
inductor:
  circuit with, 758–59
  current in, 758–59
inelastic collision, 230
  internal kinetic energy in, 238
  rest-mass energy in, 237
  in two dimensions, 234–36
inertia, law of, 92
inertia, moment of, *see* moment of inertia
inertial confinement, J10–J11
  in fusion, F6
inertial guidance system, 110, 279
inertial mass, 325–26
inertial reference frames, 92–94
  local, E2
infrared radiation, 811
infrared slavery, G9
inner product, of vectors, 56–57
instantaneous acceleration, 30–31
  in three dimensions, 69
instantaneous angular acceleration, 250
instantaneous angular velocity, 249
instantaneously fixed axis, 282–83
instantaneous power, 187
instantaneous velocity, 26–29
  analytic method for, 28–29
  as derivative, 29
  graphical method for, 28
  in three dimensions, 68–69
instantaneous velocity vector, 68
insulating material, *see* dielectric
insulators, 539
  conductors vs., 639
  resistivities of, 640
integral:
  definition of, 957 (545)
  for work, 157
integration, 157, 955–61 (543–49)
  rules for, 959 (547)
intensity:
  of sound waves, 391
  for two-slit interference pattern, 871
intensity level of sound, 391–92
interaction:
  electromagnetic, G2
  gravitational, G2
  "strong," G2
  "weak," G2
interaction region, 233
interference, 861–82
  constructive and destructive, 378, 861, 863–66
  maxima and minima for, 869
  maxima and minima, for multiple slits, 873
  from multiple slits, 872–75
  in standing electromagnetic wave, 863
  in thin films, 864–66
  two-slit, pattern for, 869, 871
  from two slits, 867–71
interferometer, 7
  Fabry-Perot, 879
  Michelson, 866–67, L6
interferometry, laser, L6
intergalactic space, mass in, D8
internal forces, 201
internal kinetic energy, 210–11, 230
  in inelastic collision, 238
internal resistance, of batteries, 655–56
international standard meter bar, 7
International System of Units (SI), 15, 963 (551)
  see also SI units
interstellar gas, Prelude-7
interstitial defects, A11
interstitial space, A10
invariance of speed of light, 419
inverse-square force, 137–38
inverse trigonometric functions, 949 (537)
inversion symmetry, A5–A6
Io, 313
ionization, B5
ionizing radiation, B5
ion microscope, Prelude-13, A2–A3
ionosphere, I1, J7
ions, in electrolytes, 540
irreversible process, 522
  entropy change in, 523
isochronism, 343
  of simple pendulum, limitations of, 349
isospin, C11, G4
isothermal expansion of gas, entropy change in, 518

In the two-volume edition, all pages after 530 are in Volume Two, except those in parentheses, which are in Volume One. Interlude F is the last one in Volume One.

isotopes, Prelude-17, B3
chart of, Prelude-17
of hydrogen, H1
radioactive, B3
of uranium, H7, H13

J/ψ meson, C15
Jodrell Bank radiotelescope, 898
Josephson junction, K6–K7
joule (J), 155
Joule, James Prescott, 154
Joule heat, 660
Joule's experiment, 486
Jupiter, 313
main moons of, 313

"keeper," of magnet, 750
kelvin, 15, 470, 963 (551)
Kelvin-Planck statement of Second Law of Thermodynamics, 520–21
Kepler, Johannes, 312
Kepler's Laws, 311–16
of areas, 312
limitations of, 316
for motion of moons and satellites, 313–14
for motion of stars, 315–16
Kepler's First Law, 311
Kepler's Second Law, 312
Kepler's Third Law, 312
killer smog, F10
kilocalorie, 185
kilogram, 7, 12, 15, 963 (551)
multiples and submultiples of, 13
standard, 94
kilopond (kp), 96
kilowatt-hours, 185
kinematics, 23
of rigid body, 246–69
in three dimensions, 67–90
kinetic energy, 161–64
as function of momentum, 164
of ideal monatomic gas, 477–78
internal, 210–30, 238
relativistic, 436
of rotation, 253–59
in simple harmonic motion, 344–45
of simple pendulum, 348
of a system of particles, 210–211
kinetic friction, 129–33
coefficient of, 130–32
kinetic pressure, 475–77
kinetic theory, of gases, 469
Kirchoff's first rule, 657
Kirchoff's second rule, 654–55
klystrom, 786–87
Krüger 60, 315
krypton wavelength standard, 7–8
Kugelblitz, I11

Lagoona Beach, Michigan, reactor accident at, F13
Lagrange, Joseph Louis, Comte, 176
Laplace, Pierre Simon, Marquis de, 119
black holes and, E10
deflection of light and, E4
La Rance river, F9–F10
Larmor frequency, 748
laser, 811, L2–L5
carbon dioxide-nitrogen, L4
definition of speed of light and, L7
fusion triggered by, F6
helium-neon, L3–L4
medical use of, L6
neodymium glass, L3
resonant cavity in, L3–L4
ruby, L2–L3
solid-state, L5
stabilized, 7–8, L6–L7
as torch, L6
tunable, L7
laser fusion, L7–L9
laser-fusion reactors, L7
laser interferometry, L6
laser light, 868, L1–L14
laser spectroscopy, L7
latitude angle, 2
lattice, crystal, A1–A2
lattice points, A6–A7
Laue spots, A3–A4
law of conservation of angular momentum, 108, 213
law of conservation of mechanical energy, 165–66, 177–78
law of conservation of momentum, 199
law of cosines, 952 (540)
law of inertia, 92
law of sines, 952 (540)
Lawrence, E. O., 700–701, C4
laws of planetary motion, 311–16
LC circuit, 759–69
energy in, 761–62
freely oscillating, 759–69
natural frequency of, 760
LCR circuit, 762–63
current in, 765, 769
with emf, 763
impedance for, 765, 769
phase angle for, 765, 769
power absorbed in, 767
Q of, 763
resonance in, 764
lead-acid battery, 538–39, 651
leader, in lightning, I8–I9
dart, I9
leap second, 11
length:
conversion factors for, 965 (553)
precision of measurement of, 8
standard of, 7–8
length contraction, 422, 431–32
visual appearance and, 432
lens:
focal point of, 844
objective, 849–51
ocular, 849–51
"Tessar," 846
thin, 844–46
lens equation, 845
lens-maker's formula, 844
Lenz' Law, 722, 746, J3
lepton, C9, G1
lepton number, C11–C12, G4
"let-go" current, 661
Leverrier, U. J. J., 304
Lichtenstein, R., 328
light:
coherent, 868, L1, L6
deflection of, by Sun, 4, E3–E4
dispersion of, 839
Doppler shift of, 817–19
from laser, 868, L1–L14
polarized, 808–10
quanta of, 900–919
reflection of, 831–34
refraction of, 834–40
retardation of, by Sun, E5
spectral lines of, 840
spectrum of, 840, 873, 923
as standard of speed, L7
ultraviolet, 811
unpolarized, 808
visible, 811–12
lightning, I8–I11
ball, I11
dart leader in, I9
data on, I10
leader in, I8–I9
radio waves from, I10
return stroke in, I8–I9
step leader in, I8–I9
thunder and, I10
in tornado, I10
lightning bolt, I10
light rays, bending of, by Sun, 4, E3–E4
light-second, 424–25
light, speed of, 8, 417–19, 789, 793–94
invariance of, 419
laser standard for, L7
in material medium, 835–36
measurement of, 802
universality of, 419
light-water reactor, H14
light-wave communications, L5–L6
light waves, 801–29
incoherent, 811
linear accelerator, C5
linear dielectric, 617
linear superposition, principle of, 547, 675
lines of electric field, 553–56
made visible, 558, 631
liquid drop model of nucleus, H2
liquids, thermal expansion of, 487–89
liquid state, Prelude-2
lithium, spectral series of, 937–38
Lloyd's mirror, 881
Local Group, Prelude-10
local inertial reference frame, E2
Local Supercluster, Prelude-10
logarithmic function, 945 (533)
longitude angle, 2
longitudinal wave, 369
long wave, 811
looping the loop, 143–44
loop method, for circuits, 656
Lorentz, Hendrik Antoon, 428
Lorentz force, 702, 783
Lorentz transformations, 423, 427–29
for coordinates, 429
for momentum and energy, 437–38
for velocity, 433–34
Los Alamos National Laboratories, fusion experiments at, F6
lunar tides, 327-
Lyman series, 924

MacDonald Observatory, L5
Mach, Ernst, 94
Mach cone, 400–401

In the two-volume edition, all pages after 530 are in Volume Two, except those in parentheses, which are in Volume One. Interlude F is the last one in Volume One.

Mach front, of nuclear explosion, H9
macroscopic and microscopic parameters, 469
magnet:
  "keeper" of, 750
  permanent, 743
  superconducting, K7–K8
magnetic bottles, J4–J5
magnetic dipole, 679
  field of, 679
magnetic dipole moment, 680, 706
  of Earth, 681
  of electron, 680
magnetic domains, 744
magnetic energy, 728–30
  in inductor, 729
magnetic field, 672–84
  of accelerated charge, 792–95
  circular orbit in, 699
  conversion factors for, 969 (557)
  critical, K2–K3
  derived from relativity, 681–84
  of Earth, 680–81
  electric field and, related by relativity, 684
  of electron, 680
  energy density in, 730
  Gauss' Law for, 675
  generated by current, 675–76
  helical motion in, 701
  induced, of accelerated charge, 795
  induced, in capacitor, 780–81
  made visible with iron filings, 677
  of magnetic dipole, 679
  magnetic force and, 672
  motion of charge in, 698
  of plane wave, 806–7
  plasma and, J3–J6
  of point charge, 673
  of point charge, represented by field lines, 674
  principle of linear superposition for, 675
  of solenoid, 696
  of straight wire, 677
magnetic field lines:
  of circular loop, 679
  made visible with iron filings, 679, 695
magnetic flux, 719
magnetic flux quantum, K5
magnetic force, 532, 668–72
  between currents, 671
  failure of Newton's Third Law for, 671
  magnetic field and, 672
  on moving point charge, 669
  on relativistic correction, 670
  on wire, 704
magnetic materials, 738–53
magnetic mirrors, J5
magnetic moment:
  of Earth, 681
  of electron, 680
  nuclear, 739–41
  orbital, 739
  spin, 740
magnetic monopole, 685
magnetic pressure, J4
magnetic suspension, K10–K11
magnetism, Gauss' Law for, 782
magnetization, 741
magnetohydrodynamic (MHD) generator, 717
magnetosonic waves, J6
magnetosphere, J4
magnifier, 848–49
Maiman, T. H., L2
Malus, Law of, 809–10
mandolin, 397
Manhattan District, H7
Manned Spacecraft Center, 78
manometer, 456
mantle of Earth, 405
many-body problem, 316
Mariner 6, E5
Mariner 7, E5
Marly, France, waterworks at, F1
Mars, 313
Marsden, E., 926
maser, L2
mass, 12–13
  atomic standard of, 13
  center of, *see* center of mass
  conservation of, 153
  conversion factors for, 966 (554)
  critical, H6
  definition of, 94–95
  of electron, 99
  of elementary particles, C10
  energy and, 185–87
  equivalence of inertial and gravitational, 326
  gravitational, 325–26
  inertial, 325–26
  in intergalactic space, D8
  moment of inertia of continuous distribution of, 255
  of neutron, 99
  of proton, 99
  rest, 186
  standard of, 12, 94
  supercritical, H6
  weight vs., 122–23
mass defect, H4
mass density, in universe, D8
mass-energy relation, 436–37, 822
mass number, B3, H1
mass spectrometer, 98–99, 711
matter-antimatter annihilation, 538, G5
maximum height, of projectiles, 73–74
Maxwell, James Clerk, 801–2
  formula for speed of light by, 805
Maxwell-Ampère's Law, 780, 782
Maxwell's demon, 527
Maxwell's equations, 782–83
mean solar day, 10
mean-value theorem, 591–93
mechanical energy, 165–67, 177–78
  conservation of, in rotational motion, 276
  law of conservation of, 165–66, 177–78
  loss of, by friction, 180
mechanical equivalent of heat, 486
mechanics:
  celestial, 304
  classical vs. quantum, 921
  fluid, 445–68
medium, dispersive, 374
medium wave, 811
Meissner effect, K3–K4
Mercury, 313
  perihelion precession of, 304, E1, E7
mercury barometer, 455
mercury-bulb thermometers, 474, 489
meson, C10, G1
  multiplets of, C12–C13
metal:
  resistivities of, 639
  temperature dependence of resistivity of, 638
meter, 7–9, 15, 963 (551)
  cubic, 14
method of images, 17
metric system, 7, 24, 963 (551)
MHD (magnetohydrodynamic generator), 717
Michelson, A. A., 418, 802
Michelson interferometer, 866–67, L6
Michelson-Morley experiment, 418, 867
microballoon, L8
microcrystal, A4
microfarad, 613
micrograph, acoustic, 394–95
microscope, 849–50
  angular magnification of, 850
  electron, Prelude-12, A2
  electron-holographic, Prelude-15, 531, L13
  holographic, Prelude-15, 531, L13
  ion, Prelude-13, A2–A3
  two-wavelength, A2
  ultrasonic, 394
microscopic and macroscopic parameters, 469
middle C, 393
Milky Way Galaxy, Prelude-9, D1
millibar, 451
Millikan, R. A., 909
Millikan's experiment, 560
Minas-Cobequid Basin, Canada, F10
Minkowski diagram, 423–24
mirror equation, 842–43
mirror:
  focal point of, 840–41
  image formed by, 833
  Lloyd's, 881
  magnetic, J5
  paraboloidal, 841
  plane, 833
  reflection by, 833–34
  spherical, 840–44
  in telescope, 850–51
missile, ballistic, 138
mode:
  fundamental, 382
  harmonic, 382
moderator, in nuclear reactor, H12
modulation, of wave, 379
mole, 12–13, 963 (551)
moment, magnetic dipole, 680, 706
moment arm, 271
moment of force, 270
moment of inertia:
  of continuous mass distribution, 255
  of Earth, 262–63, 266, 268
  of nitric acid molecule, 266
  of oxygen molecule, 266
  of system of particles, 253–59
  of water molecule, 266
momentum, 104–9

In the two-volume edition, all pages after 530 are in Volume Two, except those in parentheses, which are in Volume One. Interlude F is the last one in Volume One.

momentum (*continued*)
angular, *see* angular momentum
of an electromagnetic wave, 815–17
energy and, relativistic transformation for, 437–38
in fields, 672
Galilean transformations for, 434–35
kinetic energy as function of, 164
Lorentz transformations for, 437–38
Newton's Laws and, 104–6
of a photon, 910
rate of change of total, 201–2
relativistic, 435
of a system of particles, 198–202, 208
momentum, conservation of, 105, 199, 227, 230, G4
in elastic collisions, 227–28, 233
in fields, 549
law of, 199
monatomic gas, kinetic energy of ideal, 477–78
Moon-Earth distance measurement, L5
Moon simulator, 124–25
moons of Saturn, 331
Morley, E. W., 418
motion, 25–45
absolute, 416
amplitude of, 338
of center of mass, 207–9
of charges in magnetic field, 698
with constant acceleration, 31–33
with constant acceleration, in three dimensions, 70–71
with constant angular acceleration, 251–53
with constant force, 125–29
equation of, *see* equation of motion
free-fall, 34–35, 123–24
helical, in magnetic field, 701
harmonic, 356
in one dimension, 25
parabolic, 73
periodic, 337
planetary, 311–16
of projectiles, 71–76
as relative, 25, 79–82, 416
of rigid bodies, 246–48
of rockets, 213–15
rolling, 282–86
rotational, *see* rotational motion
simple harmonic, *see* simple harmonic motion
three-dimensional, 67
translational, 23, 246, 274*n*, 282
uniform circular, 76–79, 141–44
with variable force, 137–41
wave, 368
motional emf, 714–17
motion of recession, D4
of galaxies, gravitational force, and, D7
Mt. Palomar telescope, 851
resolution of, 890
multiloop circuits, 656–58
multiplets, of baryons or mesons, C12–C13
multiplication factor, for fission, H6
muon, B2
mushroom, of nuclear explosion, H9
mutual inductance, 726
muzzle velocity, 68, 74
myopic eye, 848

N (normal force), 125
Nagasaki, Japan, H7–H8
National Bureau of Standards, 10–12
natural gas, F4
natural logarithmic function, 945 (533)
Naval Observatory, 11
near point, 848
nearsightedness, 848
Ne'eman, Y., C12
negentropy, 527
neodymium glass laser, L3
neon tube, C1
Neptune, 313
discovery of, 304
Nernst, Walther Hermann, 524
net force, 99–100
neutral-beam injectors, J10
neutral equilibrium, 351
neutron, Prelude-17, C3–C4
fission induced by, H5
mass of, 99
in nuclear reactor, H12
quark structure of, C14
radiation damage by, B6–B7
released in fission, H5–H6
thermal, H13
neutron sources, B3
newton (N), 96
Newton, Isaac, 4, 91–92, 315–16
on action-at-a-distance, 546
bucket experiment of, 461
experiments on universality of free fall by, 326, 349, 360, E1
pendulum experiments by, 360
Newtonian physics, 4
Newtonian relativity, 109–11
newton-meter, 271–72
Newton's Laws, 91–119
in rotational motion, 270
Newton's First Law, 91–94
momentum and, 104–5
Newton's Second Law, 94–100, 153
centripetal acceleration and, 142
momentum and, 105
see also equation of motion
Newton's Third Law, 101–4
failure of, for magnetic force, 671
momentum and, 105–6
Newton's Law of Universal Gravitation, 305–7
Newton's rings, 865–66
nitric acid molecule, moment of inertia of, 266
nodes and antinodes, 382
noise, white, 392
normal force (N), 125
normal frequencies, 383
Northern Lights, 702
Nova, L9
n-type semiconductor, 640
nuclear accidents, F12–F15
nuclear bomb, H6–H8
air burst of, H9
nuclear density, H1–H2
nuclear energy, H1
from fission, F5
from fusion, F6
nuclear explosion, H9–H12
blast wave of, H9
chronological development of, H10
critical radii of, H11
distribution of energy released in, H9
fallout from, H11
fallout pattern for, H12
fireball of, H9–H10
firestorm of, H11
flash burns from, H11
Mach front of, H9
mushroom of, H9
underground, 405
nuclear fission, *see* fission
nuclear fluid, H2
nuclear fuel:
deuterium as, F6
reprocessing of, H14
supply of, F5
uranium as, F5
nuclear fusion, F6, H8, J9–J12
nuclear magnetic moments, 739–41
nuclear model of atom, 927
nuclear power plant, H13
safety of, F12–F15
nuclear radius, H1
nuclear reactor, F5–F6, F12, H12–H15
accidents in, F12–F15
accidents in, scenarios for, F14
boiling-water, H14
breeder, F5, H14
control rods in, F12, H13
converter, H13
core of, F12
fuel rods in, F12, H13
fusion, F6, J9–J12
gas-cooled, H14
heavy-water, H14
laser-fusion, L7
light-water, H14
moderator in, H12
neutrons in, H12
plutonium (^{239}Pu) produced in, F5, H13–H14
power, H13
pressurized-water, H13
rods in, H13
in submarines, H14
nuclear reactor vessel, H14
nuclear weapons, H7, H8
effects of, H9–H12
nuclear winter, H15–H16
nucleons, H1
nucleus, Prelude-16, C3
binding energy of, H3
collisions of, 236–39
deformation of, during fission, H4
density of, H1–H2
electric energy in, H1–H2
electric force and, H1–H2
liquid-drop model of, H2
stable and unstable, H3
table of, Prelude-17
numerical solution, of differential equation, 137–41

objective lens, 849–51
oboe, sound wave emitted by, 392
observable universe, D5
ocean waves:
diffraction of, 406
speed of, 402
octave, 393
octet, baryon, C12
ocular lens, 849–51

In the two-volume edition, all pages after 530 are in Volume Two, except those in parentheses, which are in Volume One. Interlude F is the last one in Volume One.

Odeillo, France, solar furnace at, F7
Oersted, Hans Christian, 668
ohm (Ω), 638
ohmmeter, 664
Ohm's Law, 635
oil, production of, F4
oil reserves, F4
oil shales, F4
Ω^-(omega particle), discovery of, C13
one-way membrane, E11
optical fibers, L5
optical instruments, 846
optical pumping, L2
optics:
  geometric, 830
  geometric vs. wave, 861
  physical, 861
  wave, 861
orbit:
  bound, 183
  circular, *see* circular orbits
  of comets, 319
  elliptical, 311–16, 318
  geostationary, 310
  hyperbolic, 319
  parabolic, 319
  period of, 309
  planetary, 309, E7
  planetary, data on, 313
  synchronous, 310
  unbound, 183
orbital magnetic moment, 739
organ, 397
organ pipe, standing waves in, 396–97
origin of coordinates, 2, 5–6
Orion Spiral Arm, Prelude-9
orthicon, 909
oscillating charge, radiation field of, 791
oscillating systems, 350–55
oscillations, 337–67
  electromagnetic, of a cavity, 785–86
  forced, 356, 358
oscillator:
  damped, *see* damped oscillator
  energy quantization of, 905
  simple harmonic, 341–44
overpressure, 456
oxygen molecule, moment of inertia of, 266

parabola, 319
parabolic motion, 73
parabolic orbit, 319
parabolic vs. elliptical orbit for projectile, 315–16
paraboloidal mirror, 841
parallel-axis theorem, 256–57
parallel-plate capacitor, 614–15
paramagnetic material, 741–43
  permeabilities of, 742–43
paramagnetism, 741–43
parapsychology, 145
parent material, B4
parity, G4
partial derivative, 181
particle:
  accelerators for, C3–C6
  alpha, 926, C2–C3
  angular momentum of system of, 212–13
  center of mass of system of, 202–9
  classical, 1
  collisions of, 236–39
  decays of, G3–G4
  ejection of, in fission, H5
  electric charges of, 533–34, 537
  elementary, *see* elementary particles
  ideal, 1
  kinetic energy of system of, 210–11
  momentum of system of, 208
  moment of inertia of system of, 253–59
  Ω^- (omega), C13
  in quantum mechanics, 935
  scattering of, C2–C3
  stable, C11
  strongly interacting, C9
  subatomic, 263, C4
  system of, *see* system of particles
  W, G5, G8
  wave vs., 912–14
  Z, G5, G8
pascal, 451
Pascal, Blaise, 451
Pascal's Principle, 453
Paschen series, 924
Passamaquody Bay, F10
pendulum, 166–67, 360
  ballistic, 231–32
  conical, 152
  Foucault, 93
  Huygens' tilted, 360
  isochronism of, 364
  physical, 351–53
  reversible, 352
  simple, *see* simple pendulum
  torsional, 353–54
pendulum clocks, 349, 352
pendulum gravimeter, 352
penetrating radiation, B1–B3
  sources of, B2
Penzias, A. A., D6
perfect fluid, 446
perigee, Prelude-5
  of artificial satellites, 313
perihelion, 311
  of comets, 318
  of Earth, 329
  of planets, 313
perihelion precession, 316, E7
  curvature of space and, E7
  of Mercury, 304, E1, E7
perimeters, 946 (534)
period, 249
  of artificial satellites, 313
  of comets, 318
  increase of, for simple pendulum, 349
  of orbit, 309
  of planets, 313
  of simple harmonic motion, 338
  of wave, 371
periodic motion, 337
Periodic Table of Chemical Elements, Prelude-14, 972 (560)
permanent dipole moments, 557
permanent magnet, 743
permeability constant, 669
  relative, 742
permittivity constant, 535
perpendicular-axis theorem, 257–58
perpetual motion machine:
  of the first kind, 508–9
  of the second kind, 509
persistent current, K2
Pfund series, 924
phase, 339
phase angle:
  for parallel LCR circuit, 769
  for series LCR circuit, 765
phase constant, 339
phase velocity, of a wave, 374
phasor, 765
phasor diagram, 765–66
photoelectric effect, 907–10
photoelectric equation, 909
photographic camera, 846–47
photographic emulsion, 237, G2
photomultiplier tube, 909
photons, 900, 907–10, G5
  in Compton effect, 911
  emitted in atomic transition, 932
  energy of, 907
  momentum of, 910
physical optics, 861
physical pendulum, 351–53
picofarad, 613
pinch effect, J5
pinhole camera, 847, B10
pion, G5–G6
pion field, G6
pitch, 247
Pitot tube, 468
Planck, Max, 520, 904–6
Planck's constant, 263, 739–40, 903, K5
Planck's Law, 903
plane mirror, 833
planetary motion:
  conservation of angular momentum in, 312
  Kepler's laws of, 311–16
planetary orbit, 309
  data on, 313
  perihelion of, 313
  perihelion precession and, E7
  period of, 313
plane wave, 391
  electric and magnetic fields of, 806–7
plane wave pulse, electromagnetic, 802–6
plasma, 540, J1–J12
  cutoff frequency in, J7
  in fusion reactor, E6
  magnetic field and, J3–J6
  in universe, J1
  waves in, J6–J8
plasma confinement, J9–J12
plasma state, Prelude-6
plastic flow, 445
Pluto, 313
  mass of, 329
plutonium (^{239}Pu):
  in chain reaction, H6
  produced in nuclear reactor, F5, H13–14
plutonium bomb, H7
point charge, 535
  electric field of, 551
  electric field lines of, 553–55, 674
  energy of a system of, 601–3
  magnetic field of, 673–74
  moving, magnetic force and, 669
  potential of, 583–84
  self-energy of, 602–3

In the two-volume edition, all pages after 530 are in Volume Two, except those in parentheses, which are in Volume One. Interlude F is the last one in Volume One.

point discharge, I7
point groups, A8*n*
Poisson spot, 895
polarization, 618, 808
  by reflection, 820
polarized light, 808–10
polarized plane waves, electric and magnetic fields of, 806–7
polarizing filter, 808–10
Polaroid, 808–10
pollution, F10–F11
  thermal, F11
polycrystalline solids, A4
polymer, A2
population inversion, L2
portable atomic clocks, 11
position vector, 52
potential:
  conversion factors for, 969 (557)
  derivatives of, 587
  of dipole, 590
  electrostatic, 581
  of flat charged surface, 582
  gradient of, 587
  maxima and minima of, 592–93
  numerical method for calculation of, 593–94
  of point charge, 583–84
  of spherical charge distribution, 585
potential difference, 582
potential energy, 164
  of conservative force, 176–80
  of current loop, 707
  curve of, 182–84
  of dipole, 557
  during fission, H4
  force calculated from, 180–82
  gravitational, 164–65, 316–22
  gravitational, of a body, 207
  gravitational, of spherical mass, 323
  mutual, 601–2
  in simple harmonic motion, 345
  of simple pendulum, 348
  of a spring, 178
  of strong force, H2
  turning points and, 182–84
potentiometer, 663–64
Potsdam, determination of acceleration of gravity at, 352
pound (lb), 14, 96–97
pound-foot, 272
power, 187–90
  absorbed by resistor, 756
  absorbed in LCR circuit, 767
  average, 187
  of clipper ship, F1
  conversion factors for, 968 (556)
  delivered by force, 188
  delivered by source of emf, 659
  dissipated in high-voltage transmission line, 660–61
  dissipated in resistor, 659–60
  dissipation of, in U.S., F3
  hydroelectric, F7
  instantaneous, 187
  from nuclear fission, F5
  from nuclear fusion, F6
  reference frame and, 189
  in rotational motion, 276–77
  from solar cells, F8–F9
  transported by a wave, 376–77
  from windmills and waterwheels, F1, F7
power plant, nuclear, H13
  safety of, F12–F15
power reactor, H13
power transmission line, superconducting, K10
Poynting vector, 813–14
precession, 287
  of the equinoxes, 279*n*
  of a gyroscope, 286–88
  perihelion, 316, E7
  perihelion, of Mercury, 304, E1, E7
precession frequency, 287
prefixes for units, 964 (552)
pressure, 446, 451–52
  atmospheric, 455
  conversion factors for, 968 (556)
  gauge, 456
  hydrostatic, 454
  kinetic, 475–77
  magnetic, J4
  of radiation, 816
  standard temperature and, 472
  in a static fluid, 452–57
pressurized-water reactor, H13
primary, of a transformer, 770
primary cosmic rays, B2
primitive cell, A6
primitive vectors, A6
primordial fireball, D6
primordial universe, D6
Princeton Tokamak Fusion Test Reactor (TFTR), J9–J10
principal maximum, of grating, 874
Principia Mathematica (Newton), 4
Principle of Equivalence, 327, E1–E3
principle of linear superposition, 547
  for magnetic field, 675
principle of relativity, 419–23
principle of superposition, 99–100
  for waves, 377
prism, 839–40
probability interpolation of wave, 913
production:
  of coal, F4
  of oil, F4
projectile:
  maximum height of, 73–74
  parabolic vs. elliptical orbit for, 315–16
  range of, 73–74
  time of flight of, 73–74
  trajectory of, as function of elevation angle, 75
projectile motion, 71–76
  with air resistance, 76
prominence, on the Sun, J1
Prony dynamometer, 278
proper frequencies, 383
proton, Prelude-17, C3–C4
  charge of, 534
  mass of, 99
  quark structure of, C14
proton field, G6
Proxima Centauri, Prelude-7
pseudo-gravity, 328
psychokinesis, 145
Ptolemaic system, 309
p-type semiconductor, 640
p-V diagram, 513
P wave, 405

Q (ringing time):
  of damped oscillator, 357–58
  of freely oscillating LCR circuit, 763
quadratic equation, 944 (532)
quantization:
  of angular momentum, 739–40, 929
  of electric charge, 537, G8*n*
  of energy, 905
quantum, G5
  creation and destruction of, G6
  of energy, 905
  exchange of, G6
  fields and, G4–G7
  of flux, K5
  of light, 900–19
quantum chromodynamics, G8–G11
quantum electrodynamics (QED), G6
quantum jump, 929
quantum mechanics, 921, 934–35
quantum number, 905, 930
quark, Prelude-18, C13–C15
  bottom, C15
  charm of, C15
  charmed, C15
  color of, C15, G9–G11
  down, C14
  electric charges of, 537
  "flavors" of, G9–G11
  force between, C15
  strange, C14
  top, C15
  up, C14
quark fields, G6
quark structure of proton and neutron, C14
quark triplet, C14
quasar, deflection of radio waves and, E4
quasar QSO (OH471), D2

rad, B7
radar echo, from Venus, E5
radian, 947 (535), 963 (551)
  definition of, 947 (535)
radiant heat, of Big Bang, D6
radiation, 492, B1–B14
  blackbody, 900–902
  cavity, 902
  coherent, L1
  cosmic background, D6
  cosmic fireball, D6
  electromagnetic, 811–12
  electromagnetic, wavelength and frequency bands of, 812
  heat transfer by, 492
  infrared, 811
  ionizing, B5
  penetrating, B1–B3
  physiological effects of, B8–B10
  pressure of, 816
  thermal, 901, 922
radiation damage:
  by alpha rays, B5
  by beta rays, B5–B6
  by gamma rays, B6
  in molecules and living cells, B7–B8
  by neutrons, B6–B7

In the two-volume edition, all pages after 530 are in Volume Two, except those in parentheses, which are in Volume One. Interlude F is the last one in Volume One.

radiation doses, B9
acute, effects of, B9
radiation field:
of accelerated charge, 788, 790
of oscillating charge, 791
radiation pressure, 816
radioactive contamination, F14
radioactive decay, B3–B4
radioactive isotopes, B3
radioactive materials, B1
radioactivity, B4
radioisotopes, B10
radio station CHU, 11
radio station WWV, 10–11
radiotelescope, 875
at Arecibo, 891
at Jodrell Bank, 898
Very Large Array, 875
radio waves, 801–29
from lightning, I10
quasars and, E4
range:
of a force, 121
of fundamental forces, 121
of projectiles, 73–74
Rasmussen Report, F14
rate of change:
of angular momentum, 108, 212
of angular momentum, relative to center of mass, 285
of momentum, 201–2
Rayleigh, Lord, 902–3
Rayleigh's criterion, 890
rays, 832
alpha, B1, B5
beta, B1, B5–B6
cathode, C1–C2
cosmic, 236, 698, B2
gamma, B1, B6
X, 811, A3, B1, E12–E13
RBE (Relative Biological Effectiveness), B8
reactance:
capacitative, 757
inductive, 758–59
reaction and action, 101–4
reactor, nuclear, F5–F6, F12, H12–H15
boiling-water, H14
breeder, F5, H14
control rods in, F12, H13
converter, H13
core of, F12
fuel rods in, F12, H13
fusion, F6, J9–J12
gas-cooled, H14
heavy-water, H14
laser-fusion, L7
light-water, H14
moderator in, H12
neutrons in, H12
plutonium (^{239}Pu) produced in, F5, H13–H14
power, H13
pressurized-water, H13
rods in, H13
in submarines, H14
reactor accidents, F12–F15
scenarios for, F14
real image, 843–44
recession, motion of, D4
recoil, 200
rectangular coordinates, 2
red, C15
redout, 146
red shift:
cosmological, 828
gravitational, E8–E10
red-shift experiments, E9–E10
red-shift method, D2
reference frame, 6, 79–81, 419
in calculation of work, 156
of Earth, 93–94
freely falling, 123–24
inertial, 92–94
local inertial, E2
power and, 189
for rotational motion, 247, 282
reflection, 831–34
angle of, 832
critical angle for total internal, 838
law of, 831
by mirror, 833–34
polarization by, 820
total internal, 838
reflection symmetry, A5
reflector:
corner, 833–34, L5
in fission, H7
on Moon, L5
refraction, 834–40
index of, 835
law of, 837
Relative Biological Effectiveness (RBE), B8
relative coordinates, 5*n*
relative measurement, 5
relative permeability constant, 742
relativistic kinetic energy, 436
relativistic momentum, 435
relativistic total energy, 437
relativistic transformation:
of electric and magnetic fields, 684
for momentum and energy, 437–38
relativity:
general theory of, 5, 314, D1, E1, E6–E8
magnetic field and, 681–84
of motion, 25, 79–82, 416
Newtonian, 109–11
principle of, 419–23
of simultaneity, 420–21
special theory of, 5, 416–44
of synchronization of clocks, 421
renormalization rules, 603
reserves:
of coal, F4
of fossil fuel, F4
of oil, F4
resistance, 634
in combination, 641–43
conversion factors for, 969 (557)
of human skin, 661
internal, of batteries, 655–56
resistance thermometers, 474, 644
resistivity, 634
conversion factors for, 969 (557)
of insulators, 640
of materials, 637–40
of materials, temperature dependence of, 638
of metals, 639
of semiconductors, 640
of semiconductors, temperature dependence of, 640
temperature coefficient of, 644
resistors, 641–43
circuit with, 755
in parallel, 642
power absorbed by, 756
power dissipated in, 659–60
in series, 642
resolution, angular, of telescope, 889–92
resolving power, of grating, 874
resonance, C11
of damped oscillator, 358–59
in LCR circuit, 764
in musical instruments, 397
resonant cavity, in laser, L3–L4
resources:
of energy, F4
limits of, F4
of nuclear fuels, F5
rest frame, 6
restitution, coefficient of, 243
rest mass, 186
rest-mass energy, 186, 437
in inelastic collision, 237
restoring force, 135–36
resultant, of vectors, 49
resultant force, 99–100
retardation of light, by Sun, E5
return stroke, in lightning, I8–I9
reversible pendulum, 352
reversible process, 512
entropy for, 517
Rhea, 331
Richter, B., C15
Richter magnitude scale, 405
right-hand rule, 58
for induced emf, 719
for Ampère's Law, 693
rigid body, 246–303
angular momentum of, 259–63
component of angular momentum along axis of rotation of, 262
dynamics of, 270–88
kinematics of, 246–69
kinetic energy of rotation of, 253–59
moment of inertia of, 253–59
motion of, 246–48
parallel-axis theorem for, 256–57
perpendicular-axis theorem for, 257–58
rotational motion of, 246
some moments of inertia for, 259
spin of, 262
statics of, 288–92
translational motion of, 246
ringing time (Q):
of damped oscillator, 357–58
of freely oscillating LCR circuit, 763
rock, flow of, 445
rocket:
motion of, 213–15
thrust of, 214
rocket equation, 215
Roemer, Ole, measurement of speed of light and, 822
roll, 247
rolling motion, 282–86

In the two-volume edition, all pages after 530 are in Volume Two, except those in parentheses, which are in Volume One. Interlude F is the last one in Volume One.

Röntgen, Wilhelm Konrad, B1*n*
root-mean-square speed, of gas, 476–77
root-mean-square voltage, 756
rope, frictional force on, 291–92
rotation:
  about center of mass, 285
  of coordinates, 60–61
  of the Earth, 10, 93
  frequency of, 249–50
  kinetic energy of, 253–59
  noncommutative, 248
  period of, 249
rotational motion, 246–53
  conservation of angular momentum in, 278–82
  conservation of mechanical energy in, 276
  with constant angular acceleration, 251–53
  energy in, 276
  equation of, 273–74
  about a fixed axis, 248–51
  of gas, 478–79
  Newton's laws in, 270
  power in, 276–77
  reference frame for, 247, 282
  of rigid body, 246
  translational motion as analogous to, 274*n*
  translation and, in rolling motion, 282
  work in, 276
rotation symmetry, A5–A7
ruby laser, L2–L3
Rumford, Count, 484
Rutherford, Ernest, 920, 925–27, C2–C3, H16
Rutherford scattering, 926
Rydberg constant, 924
  measurement of, L7
Rydberg-Ritz combination principle, 925

St. Elmo's fire, I8
satellites:
  artificial, 310, 313
  communication, 310
  geostationary, 310
Saturn, 313
  moons of, 331
scalar, 48
scalar product, 56
scale:
  chromatic musical, 393
  Richter magnitude, 405
scans:
  brain, B11
  thyroid, B11
scattering, 236
  of alpha particles, 926, C2–C3
  of particles, C2–C3
Schawlow, A. L., L2
Schick stabilizer, 294
Schrödinger equation, 934
Schwarzschild radius, E11
scintillation counters, C7
screw axis, A8
Scylla, J9
Scyllac, J9
secant, 948 (536)
second, 7, 10–12, 15, 963 (551)
  multiples and submultiples of, 12
secondary, of a transformer, 770
secondary cosmic rays, B2
second harmonic mode, 382
Second Law of Thermodynamics, 520–24
seiche, 409
seismic waves, 404–5
seismometer, 409
self-energy, of point charge, 602–3
self-field, 556
self-inductance, 728
semiconductors, 640–41, A13
  "holes" in, 640
  n-type, 640
  p-type, 640
  resistivities of, 640
  temperature dependence of resistivity of, 640
semimajor axis of ellipse, 311
  related to energy, 318
series limit, 923
Sèvres, France, 7
Shiva, L8
shock waves, L12
Shortt pendulum clock, 349, 352
short wave, 811
sidereal day, 93*n*
silicon, *p*-type and *n*-type, 653–54
silicon solar cells, 532–54
simple harmonic motion, 337–40
  conservation of energy in, 345, 348–49
  frequency of, 338
  kinetic energy in, 344–45
  period of, 338
  phase of, 339
  potential energy in, 345
simple harmonic oscillator, 341–44
  angular frequency of, 342
  equation of motion of, 342
simple pendulum, 346–49
  energy of, 348
  equation of motion for, 347
  increase of period of, 349
  isochronism of, 349
  kinetic energy of, 348
  potential energy of, 348
simultaneity, relative, 420–21
$\sin^{-1}$, 949 (537)
sine, 948 (536)
  law of, 952 (540)
single-loop circuits, 654–56
single slit:
  diffraction by, 883–87
  diffraction pattern of, 885, 887
  minima and maxima in diffraction pattern of, 883
singularity, gravitational, E12
sinks, of field lines, 555
siphon, 468
SI units, 15, 963 (551)
  of current, 705
  derived, 964 (552)
  of electric charge, 705
skin, human, resistance of, 661
sky diver, 37
Skylab mission, 94, 97–98, 281–82, 326, E2–E3
  body-mass measurement device on, 341
  fuel cell on, 653
  solar panel on, F8
sky walks, collapse of, 294
sliding friction, 129–33
slope, 18
  of worldline, 27–28
slowing of light, by Sun, E5
slug, 14, 97
smog, F10–F11
Snell's Law, 837
Sobral, Brazil, eclipse at, E4
solar cells, 653–54
  as power source, F8–F9
  silicon, 653–54
solar collectors, F7–F8
solar day, 10, 93*n*
solar energy, F7–F9
  indirect, F6–F7
solar heating system, F8
Solar One power station, F8
solar panels, F8
solar power station, F7–F8
Solar System, 305, 311
  data on, 313
solar tides, 327
solar wind, J4
solenoid, 694–97
  induced electric field, 725–26
  induced emf in, 720–21
  magnetic field of, 696
solid:
  flow of, 445
  thermal expansion of, 487–89
solid solutions, A10
solid state, Prelude-2
solid-state electronics, A13
solid-state laser, L5
sonic boom, 401
sound:
  intensity level of, 391–92
  speed of, 395–97
sound waves:
  in air, 391–95
  diffraction of, 407
  intensity of, 391
sources:
  alternative, for energy, F4–F10
  of electromotive force, 651–54
  of field lines, 555
  of penetrating radiation, B2
space, curvature of, 4, E3–E6
  perihelion precession and, E7
space, interstitial, A10
space groups, A8*n*
Space Shuttle, 124
space station, 328
Space Telescope, 890, 898
spacetime, 423
  curved, E7
spacetime diagram, 423–24
spark chambers, C8
Special Relativity theory, 5, 416–44
specific heat:
  at constant pressure, 494–95
  at constant volume, 494
  of a gas, 493–96
specific heat capacity, 485
speckle effect, L4

In the two-volume edition, all pages after 530 are in Volume Two, except those in parentheses, which are in Volume One. Interlude F is the last one in Volume One.

spectra, color plate, 922
spectral emittance, 901
 of blackbody, 903
spectral lines, 840, 921–23
spectral series, 923–25
 of hydrogen, 924–25, 931
 of lithium, 937–38
spectroscopy, laser, L7
spectrum, 922
 of Caph, 923
 produced by grating, 873
 produced by prism, 840
 of Sun, 923
speed:
 average, 24–25
 conversion factors for, 967 (555)
 of electromagnetic wave, 794, 805
 after one-dimensional elastic collision, 228–30
 as relative, 25
 of sound, 395–97
 standard of, 8, L7
 terminal, 37
 unit of, 24
 velocity vs., 26, 69
 of water waves, 402–3
 of waves, on a string, 372–74
speed of light, 8, 417–19, 789, 793–94
 invariance of, 419
 laser standard for, L7
 in material medium, 835–36
 measurement of, 802
 universality of, 419
Sperry C-12 directional gyroscope, 279
spherical aberration, 841
spherical capacitor, 621
spherical charge distribution:
 electrical field of, 571–72
 electric energy of, 607
 potential of, 585
spherical coordinates, 2–3
spherical mass:
 gravitational force exerted by, 322–25
 gravitational potential energy of, 323
spherical mirrors, 840–44
 concave, 840
 convex, 842
 focal length of, 841
spin, 262
 of Earth, 262–63
 of electron, 680
 of elementary particles, C10
 of rigid body, 262
 of subatomic particles, 263, C4
spin magnetic moment, 740
spin-spin force, 744
spiral galaxies, D2
spontaneous breakdown of symmetry, G8
spring balance, 97, 123
spring constant, 135–37
springs, 135–37, 341
 force of, 135–36
 potential energy of, 178
spring tides, 330
Sputnik I, 313–14, 334
Sputnik II, 313
Sputnik III, 313
SQUID (superconductive quantum interference device), K7
stabilized laser, 7–8, L6–L7
stable equilibrium, 351
stable particle, C11
standard kilogram, 94
standard meter bar, international, 7
standard of length, 7–8
standard of mass, 12, 94
standard of speed, 8
 established with laser, L7
standard of time, 10
standard temperature and pressure (STP), 472
Standards, National Bureau of, 10–12
standing electromagnetic wave, 862–64
standing wave, 381–84, 862
 in a tube, 396–97
Stanford Linear Accelerator (SLAC), C5, C13
states of atoms, stationary, 930–31
states of matter, Prelude-2, Prelude-6
 fourth, J1–J3
static equilibrium, 288–92
 of electric charge, 572–73
 examples of, 289–92
static fluid, 452
static friction, 133–34
 coefficient of, 131–34
statics, of rigid body, 288–92
stationary state, 929
 of hydrogen, 930–31
steady flow, 447–48
steel:
 crystal structure of, A12
 hard and soft, A12–A13
Stefan-Boltzmann Law, 906
Stellarator, J9
step leader, in lightning, I8–I9
steradian, 963 (551)
stimulated emission, 811, L2
stopping potential, 908
storage, of energy, K10
strangeness, C11–C12, G4
strange quark, C14
streamlines, 447–49
stream tube, 449
stringed instruments, 397
"strong" force, 121, G1, G9
 in nucleus, H1–H2
 potential energy of, H2
strong interactions, G2
strongly interacting particles, C9
subatomic particles, spin of, 263, C4
sublimation, 493
substitutional impurities, A11
suction pump, 466
Sun:
 eclipse of, E4
 escape velocity from, 320
 generation of energy in, H8
 hydrogen burning in, H8
 light rays and, 4, E3–E5
 prominence on, J1
 spectrum of, 923
sunglasses, Polaroid, 810
Supercluster, Local, Prelude-10
superconducting cable, K8–K10
superconducting generators, K9
superconducting magnet, K7–K8
superconducting power transmission line, K10
superconductivity, 639, K1–K12
superconductors, K1–K3
 of the second kind, K4–K5
supercritical mass, H6
superfluid, K1
supernovas, D5–D6
superposition:
 of electric forces, 546–49
 of waves, 377
superposition principle, 99–100
 for waves, 377
supersonic aircraft, sonic boom and, 401
supply of fossil fuels, F4
surface charge, on dielectric, 620
surface wave, 405
suspension, magnetic, K10–K11
S wave, 405
sword blades, Japanese, A12
symmetric fission, H4
symmetry, in physics, A1, A5–A8
 broken, G8
 conservation laws and, 154, G4
 of a cube, A9
 inversion, A5–A6
 reflection, A5
 rotation, A5–A7
 spontaneous breakdown of, G8
 translation, A6–A7
symmetry axis, A5
sympathetic oscillation, of damped oscillator, 358–59
synchrocyclotron, 700
synchronization of clocks, 6, 11, 419–20
 relativity of, 421
synchronous (geostationary) orbit, 310
synchrotron, 700–701
system of particles, 198–222
 angular momentum of, 212–13
 center of mass in, 202–9
 energy of, 209–11
 kinetic energy of, 210–11
 moment of inertia of, 253–59
 momentum of, 198–202, 208
system of units (SI), 15, 963 (551)
 see also SI units

Tacoma Narrows, 383–84
tails of comets, 817
$\tan^{-1}$, 949 (537)
tangent, 948 (536)
tangent galvanometer, 709
Taylor series, 961 (549)
telescope, 850–51
 angular magnification of, 850
 angular resolution of, 889–92
 Arecibo radio-, 891
 Galilean, 854
 Hale, 851
 Jodrell Bank radio-, 898
 mirror, 850–51
 Mt. Palomar, 851, 890
 radio-, 875, 891, 898
 Space, 890, 898
 Very Large Array radio-, 875
teletherapy unit, B12
television cameras, 909
temperature:
 coefficient of resistivity, 644
 lowest attained, 525
 standard pressure and, 472

In the two-volume edition, all pages after 530 are in Volume Two, except those in parentheses, which are in Volume One. Interlude F is the last one in Volume One.

temperature scales, 473–75
absolute, 470, 473
Celsius, 474–75
comparison of, 475
Fahrenheit, 474–75
ideal-gas, 473
thermodynamic, 516
tension, 102–4
tensor analysis, 60
terminal speed, 37
tesla (T), 673
"Tessar" lens, 846
Tethys, 331
Tevatron, Fermilab, C6
thermal conductivity, 490–91
thermal energy, 484
thermal engine, efficiency of, 512
thermal expansion:
of concrete, 489
linear, coefficient of, 487
of solids and liquids, 487–89
of water, 488–89
thermal neutrons, H13
thermal pollution, F11
thermal radiation, 901, 922
thermal units, 185
thermocouples, 474
thermodynamic equilibrium, 511
thermodynamics, 508–30
First Law of, 509–11
Second Law of, 520–24
Third Law of, 524–25
Zeroth Law of, 525–26
thermodynamic temperature scale, 516
thermometer:
bimetallic strip, 474, 489
constant-volume gas, 473
mercury-bulb, 474, 489
resistance, 474, 644
thermocouple, 474
thermonuclear fusion, J8, J12, L7
thermonuclear fusion reactor, J9–J12
thermonuclear reactions, F6, H8, J8–J9, L8
thermos bottle, 901
thin films, interference in, 864–66
thin lenses, 844–46
Third Law of Thermodynamics, 524–25
Thomson, J. J., 925, C1–C2
thorium, as nuclear fuel, F5
three-dimensional space, 2–3
Three Mile Island, Pennsylvania, reactor accident at, F13
threshold, of hearing, 391
threshold energy, 239
threshold frequency, 908
thrust, 214
thunder, I10
thundercloud, electric field of, I6–I8
thunderhead, I4
thunderstorm, I1, I3–I11
anvil cloud in, I5
energy conversion in, I3
generation of electric charge in, I5–I6
turrets in, I4
thunderstorm cells, I3–I4
tidal energy, F9–F10
tidal wave, 403–4
tides, 84, 327
lunar, 327
solar, 327
spring, 330
time:
absolute, 3–5, 80
atomic standard of, 10
Cesium standard of, 10
conversion factors for, 966 (554)
standard of, 10
unit of, 10–12
Universal Coordinated, 10–11
time dilation, 4–5, 421, 429–31
gravitational, E8–E10
gravitational, experiments on, E8–E9
time of flight, of projectiles, 73–74
time signals, 10–11
Ting, S., C15
Titan, 331
Tokamak, J9–J10
top, 286
top quark, C15
torch, laser as, L6
tornado:
lightning in, I10
pressure in, 464
toroid, 697
torque, 109, 270–72
cross product in definition of, 270
on a current loop, 705–6
on dipole, 557
torr, 451
torsional constant, 353
torsional pendulum, 353–54
torsional suspension fiber, 353
torsion balance, 308
total internal reflection, 838
Townes, C. H., L2
trajectory of projectiles, as function of elevation angle, 75
transfer of heat, 490, 492
transformation equation, 60, 82
Galilean, 81, 417, 434–35
Lorentz, 423, 427–29, 433–34, 437–38
transformation of coordinates, 60–61
transformer, 770–71
transistors, A13
translational motion, 23, 246
rotational motion as analogous to, 274*n*
rotation and, in rolling motion, 282
translation symmetry, A6–A7
transmission line:
power dissipated in, 660–61
superconducting, K10
transmutation of elements, B4
transverse radiation field of accelerated charge, 789
transverse wave, 369
traveling wave, 369–72
Trieste, 463
trigonometric functions, 947 (535)
trigonometric identities, 951 (539)
trigonometry, 947 (535)
triple-point cell, 473
triple point of water, 473
triplet, quark, C14
tritium, as nuclear fuel, F6
trumpet, 397
tsunamis, 403–4
tube of force, 559
tumors, B11
tunable laser, L7
turbulence, 448*n*
turning points, 182
turrets, in thunderstorm, I4
TV waves, 811
twin paradox, 431
two-slit interference, 867–71
pattern for, 869, 871
two-wavelength microscope, A2

UFO, 409
ultrasonic microscope, 394
ultrasound, 391
ultraviolet catastrophe, 903
ultraviolet freedom, G9
ultraviolet light, 811
unbound orbit, 183
uncertainty relation, 914
underground nuclear explosions, 405
unified field theory, G1
unified theory of weak and electromagnetic forces, G7–G8
uniform circular motion, 76–79, 141–44
United States:
energy dissipation in, F3–F4
power dissipation in, F3
unit of length, 6–7
Units, International System of, (SI), 15
base, 963 (551)
of current, 705
derived, 964 (552)
of electric charge, 705
units of force, 96–97
unit vector, 52–53
cross product of, 58
dot product of, 56
Universal Coordinated Time (UTC), 10–11
universal gas constant, 470
universal gravitation, law of, 305–7
universality of acceleration of gravity, 33–34
universality of free fall, 326, 349, E1–E3
universality of speed of light, 419
universe:
age of, D5–D6
chemical composition of, D6
contraction of, D7
critical density for, D7–D8
deuterium abundance in, D8
early, G8
escape velocity for, D7
expansion of, D1, D4–D5
future evolution of, D6–D8
hydrogen and helium abundance of, D6
implosion of, D7
mass density in, D8
observable, D5
plasma in, J1
primordial, D6
unpolarized light, 808
unstable equilibrium, 351
up quark, C14
uranium:
in chain reaction, H6

In the two-volume edition, all pages after 530 are in Volume Two, except those in parentheses, which are in Volume One. Interlude F is the last one in Volume One.

enriched, H12
isotopes of, H7, H13
natural, F5
as nuclear fuel, F5
nucleus of, electrical energy in, 607
"weapons grade," H7
uranium bomb, H7
Uranus, 313

vacancies, A11
Van Allen belts, 702
Van de Graaff, R. J., C3–C4
Vanguard I, 313–14
vaporization, 492–93
vector addition, 48–51
associative law of, 50–51
commutative law of, 50
by components, 54–55
vector product, 57–59
vectors, 46–66
components of, 53–57, 59
cross product of, 57–59
definition of, 48
displacement, 46–48
dot (scalar, inner) product of, 56–57
instantaneous velocity, 68
multiplication of, 52, 55–59
position, 52
Poynting, 813–14
primitive, A6
resultant of, 49
rigorous definition of, 61
subtraction of, 51
unit, *see* unit vector
vector triangle, 49
velocity:
angular, average and instantaneous, 249
average, 25–26
average, in three dimensions, 67–68
of center of mass, 207, 230
of flow, 446
Galilean addition law for, 417
instantaneous, *see* instantaneous velocity
Lorentz transformations for, 433–34
muzzle, 68, 74
relativistic combination of, 433–34
speed vs., 26, 69
transformation equations for, 81
velocity field, 448
velocity filters, 703
velocity selectors, 703
velocity transformation, Galilean, 417
Venturi flowmeter, 459, 468
Venus, 313
radar echo from, E5
Verne, Jules, 332–33
Very Large Array (VLA) radiotelescope, 875
vibrations, of diatomic molecule, 354–55
violin, 397
sound wave emitted by, 392
Virgo Cluster, Prelude-10
virtual image, 833
virtual particles, G6
virus particles, arranged in crystal, A2
viscosity, 448*n*
visible light, 811
colors and wavelengths of, 812
voltage, 650
AC, 722–23
voltmeter, 663
volume, 14, 946 (534)
conversion factors for, 966–67 (554–55)
vortex, 397
vortex lines, K5

Wairakei (New Zealand), F9
Walton, E. T. S., C3–C4
W and Z particles, discovery of, G8
WASH-740 Report, F14
waste heat, F11
water:
heavy, D8
molecule, moment of inertia for, 266
thermal expansion of, 488–89
triple point of, 473
water waves, 401–4
diffraction of, 406
speed of, 402–3
waterwheel, 163, 280–81
power from, E1, F7
watt (W), 187
Watt, James, 187
wave, 368–415
Alfvén, J7
amplitude of, 370
angular frequency of, 371–72
beats of, 379–80
capillary, 401
constructive and destructive interference for, 378
on Earth's surface, 405
electromagnetic, *see* electromagnetic wave
energy in, 374–77, 812–14
energy density of, 375–76
frequency of, 371
gravity, 401–2
group velocity of, 374
harmonic, 370–71
light and radio, 801–29
long, 811
longitudinal, 369
magnetosonic, J6
measurement of momentum and position of, 914
medium, 811
modulation of, 379
ocean, 402, 406
P, 405
particle vs., 912–14
period of, 371
phase velocity of, 374
plane, 391, 806–7
in a plasma, J6–J8
power transported by, 376–77
probability interpolation and, 913
radio, 801–29
radio, from lightning, I10
radio, quasars and, E4
S, 405
seismic, 404–5
short, 811
sound, *see* sound waves
speed of, on a string, 372–74
standing, 381–84, 862
standing, in a tube, 396–97
superposition of, 377
surface, 405
tidal, 403–4
transverse, 369
traveling, 369–72
TV, 811
water, *see* water waves
wave crests, 370
wave fronts, 390–91, 393
circular, 390–91
plane, 391
spherical, 390–91
waveguide, 787
wavelength, 7, 371
of visible light, 812
wavelength bands, of electromagnetic radiation, 812
wavelength shift, of photon, in Compton effect, 911
wavelets, 831
wave motion, 368
wave number, 370
wave optics, 861
wave pulse, 368–69
electromagnetic, 802–6
wave trough, 371
wavicle, 900, 912
measurement of, 914
"weak" force, 121, G1
weak interactions, G2
"weapons grade" uranium, H7
weber, 719
weight, 12, 14, 122–25
apparent, 143–44
mass vs., 122–23
weightlessness, 123–24
simulated, 462
welding, by laser, L6
Wheatstone bridge, 664
Wheeler, John, E6
whistlers, J7–J8
white noise, 392
Wien's Law, 905
Wilson, R. W., D6
wind instruments, 397
windmills, power from, F1, F7
Windscale, England, reactor accident at, F13
wing, flow around, 449–50, 458
wire:
magnetic force on, 704
straight, magnetic field of, 677
uniform, electric field in, 632
work, 153–97
definition of, 154
done by constant force, 154, 158
done by gravity, 160–61
done by variable force, 157, 159
dot product in definition of, 158
frame of reference in calculation of, 156
as integral, 157
internal, in muscles, 156
in one dimension, 154–58
in rotational motion, 276
in three dimensions, 158–61

In the two-volume edition, all pages after 530 are in Volume Two, except those in parentheses, which are in Volume One. Interlude F is the last one in Volume One.

work energy theorem, 162–63
work function, 908
worldline, 25
  slope of, 27–28
W particles, G5, G8

X ray, 811, B1
  from Cygnus X-1, E12–E13
X-ray diffraction, A3

yaw, 247
ying-yang coil, J10
Young, Thomas, 407, 867–68

Z and W particles, discovery of, G8
Zeiss "Tessar" lens, 846
zero gee, 123–24
zero-point motion, 524*n*
zero resistance, K1–K2
Zeroth Law of Thermodynamics, 525–26
Z particles, G5, G8
Zweig, G., C14

In the two-volume edition, all pages after 530 are in Volume Two, except those in parentheses, which are in Volume One. Interlude F is the last one in Volume One.

Fundamental Constants

Speed of light	$c = 3.00 \times 10^8$ m/s
Planck's constant	$h = 6.63 \times 10^{-34}$ J · s
	$\hbar = h/2\pi = 1.05 \times 10^{-34}$ J · s
Gravitational constant	$G = 6.67 \times 10^{-11}$ N · m²/kg²
Permeability constant	$\mu_0 = 1.26 \times 10^{-6}$ H/m
Permittivity constant	$\varepsilon_0 = 8.85 \times 10^{-12}$ F/m
Boltzmann constant	$k = 1.38 \times 10^{-23}$ J/K
Electron charge	$-e = -1.60 \times 10^{-19}$ C
Electron mass	$m_e = 9.11 \times 10^{-31}$ kg
Proton mass	$m_p = 1.673 \times 10^{-27}$ kg
Neutron mass	$m_n = 1.675 \times 10^{-27}$ kg
Rydberg constant	$R_H = 1.10 \times 10^7$/m
Bohr radius	$4\pi\varepsilon_0\hbar^2/m_e e^2 = 5.29 \times 10^{-11}$ m
Compton wavelength	$\hbar/m_e c = 3.86 \times 10^{-13}$ m

Miscellaneous Physical Constants

Acceleration of gravity	1 gee = 9.81 m/s² = 32.2 ft/s²
Atomic mass unit	1 u = 1.66×10^{-27} kg
Avogadro's number	$N_A = 6.02 \times 10^{23}$/mole
Density of dry air	1.29 kg/m³ (0°C, 1 atm)
Molecular mass of air	28.98 g/mole
Speed of sound in air	331 m/s (0°C, 1 atm)
Density of water	1000 kg/m³
Heat of vaporization of water	539 kcal/kg
Heat of fusion of ice	79.7 kcal/kg
Mechanical equivalent of heat	1 cal = 4.19 J
Solar constant	1.4 kW/m²
Index of refraction of water	1.33

The Planets[a]

Planet	*Mean distance from Sun*	*Period of revolution*	*Mass*	*Equatorial radius*	*Surface gravity*	*Period of rotation*
Mercury	57.9×10^6 km	0.241 year	3.30×10^{23} kg	2,439 km	0.38 gee	58.6 days
Venus	108	0.615	4.87×10^{24}	6,052	0.91	243
Earth	150	1.00	5.98×10^{24}	6,378	1.00	0.997
Mars	228	1.88	6.42×10^{23}	3,397	0.38	1.026
Jupiter	778	11.9	1.90×10^{27}	71,398	2.53	0.41
Saturn	1,430	29.5	5.67×10^{26}	60,000	1.07	0.43
Uranus	2,870	84.0	8.70×10^{25}	25,400	0.92	0.65
Neptune	4,500	165	1.03×10^{26}	24,300	1.19	0.77
Pluto	5,910	248	6.6×10^{23}	2,500	0.72	6.39

[a] Based on *The Astronomical Almanac*, 1984.